HELPING TEACHERS AND STUDENTS SUCCEED TOGETHER

WILEY

# Anatomy and Physiology

## from Science to Life

# Anatomy and Physiology

## from Science to Life

### Third Edition

Gail W. Jenkins
Montgomery College

Gerard J. Tortora
Bergen Community College

John Wiley & Sons, Inc.

| | |
|---|---|
| Vice President & Publisher | Kaye Pace |
| Executive Editor | Bonnie Roesch |
| Project Editor | Lorraina Raccuia |
| Editorial Assistant | Christina Picciano |
| Development Editor | Karen Trost |
| Executive Marketing Manager | Clay Stone |
| Production Manager | Juanita Thompson/Dorothy Sinclair |
| Production Editor | Sandra Dumas |
| Senior Illustration Editor | Anna Melhorn |
| Illustration Coordinator | Claudia Volano |
| Text Designer | Maureen Eide |
| Cover Designer | Wendy Lai |
| Senior Media Editor | Linda Muriello |
| Media Specialist | Svetlana Barskaya |
| Photo Department Manager | Hilary Newman |
| Production Management Services | Ingrao Associates |

Front cover photo: Scott Tysick/Getty Images, Inc.
Back cover photos: (nurse holding touchscreen pad): Mutlu Kurtbas/iStockphoto; (body scan on touchscreen pad) Simon Fraser/Photo Researchers, Inc.

This book was typeset in 10.5/12.5 Times at Aptara and printed and bound by Quad Graphics/DuBuque. The cover was printed by Quad Graphics/DuBuque.

Founded in 1807, John Wiley & Sons, Inc. has been a valued source of knowledge and understanding for more than 200 years, helping people around the world meet their needs and fulfill their aspirations. Our company is built on a foundation of principles that include responsibility to the communities we serve and where we live and work. In 2008, we launched a Corporate Citizenship Initiative, a global effort to address the environmental, social, economic, and ethical challenges we face in our business. Among the issues we are addressing are carbon impact, paper specifications and procurement, ethical conduct within our business and among our vendors, and community and charitable support. For more information, please visit our website: *www.wiley.com/go/citizenship*.

The paper in this book was manufactured by a mill whose forest management programs include sustained yield -harvesting of its timberlands. Sustained yield harvesting principles ensure that the number of trees cut each year does not exceed the amount of new growth.

This book is printed on acid-free paper.



ISBN 13 978-0470-59891-7
ISBN 13 978-1118-12920-3

Printed in the United States of America.

10 9 8 7 6 5 4 3 2 1

## FROM THE CLASSROOM TO THE CLINIC

An anatomy and physiology course can be the gateway to a gratifying career in a whole host of health-related professions. It can also be an incredible challenge. Through years of teaching and collaboration with instructors and students, we know that students begin an anatomy and physiology course with great expectations, fully motivated to succeed and move toward a bright future in nursing, physical therapy, or any number of other allied health career choices. We also realize that it does not take long for students to begin to feel completely overwhelmed by the amount of content to learn and the time investment needed for study. Our goal with ***Anatomy and Physiology: From Science to Life*** is to help students maintain their enthusiasm for—and interest in—the science of anatomy and physiology by focusing them on the core concepts that provide the knowledge and skills needed for understanding and making meaningful connections to the life endeavors they seek and will encounter.

The third edition of ***Anatomy and Physiology: From Science to Life*** integrated with ***WileyPLUS*** is full of enhancements and revisions that—we believe—strengthen our presentation of content, offering a unique solution for teaching and learning. From the carefully revised manageable modules of content that put a conceptual focus on chapter material, new or significantly revised Case Stories, the addition of short Clinical Connections, numerous new visuals, and refined questions that promote critical thinking—to new animations, dynamic interactive activities, and the grounding of all content and assessment in learning outcomes matched to the guidelines developed by the Human Anatomy and Physiology Society, everything is designed with the goal of helping instructors like you teach in a way that inspires confidence and resilience in your students.

On the following pages students will discover the tips and tools needed to make the most of their time studying using the integrated text and media. For instructors, an overview of the changes to this edition and resources available to create dynamic classroom experiences as well as build meaningful assessment opportunities are all highlighted. Both will be interested in the outstanding resources available to seamlessly link laboratory activity with lecture presentation and study time.

# NOTES TO STUDENTS

The challenges of learning anatomy and physiology can be complex and time consuming. This textbook and *WileyPLUS* for Anatomy & Physiology have been carefully designed to maximize your time studying by simplifying the choices you make in deciding what to study, how to study it, and in assessing your understanding of the content.

## Anatomy and Physiology Is a Visual Science

Studying the figures in this book is as important as reading the narrative. The tools described here will help you understand the concepts being presented in any figure and assure you get the most out of the visuals.

1. **LEGEND** Read this first. It explains what the figure is about.
2. **KEY CONCEPT STATEMENT** Indicated by a "key" icon, this reveals a basic idea portrayed in the figure.
3. **ORIENTATION DIAGRAM** Added to many figures, this small diagram helps you understand the perspective from which you are viewing a particular piece of anatomical art.
4. **FUNCTION BOXES** Included with selected figures, these provide brief summaries of the functions of the anatomical structure or system depicted.

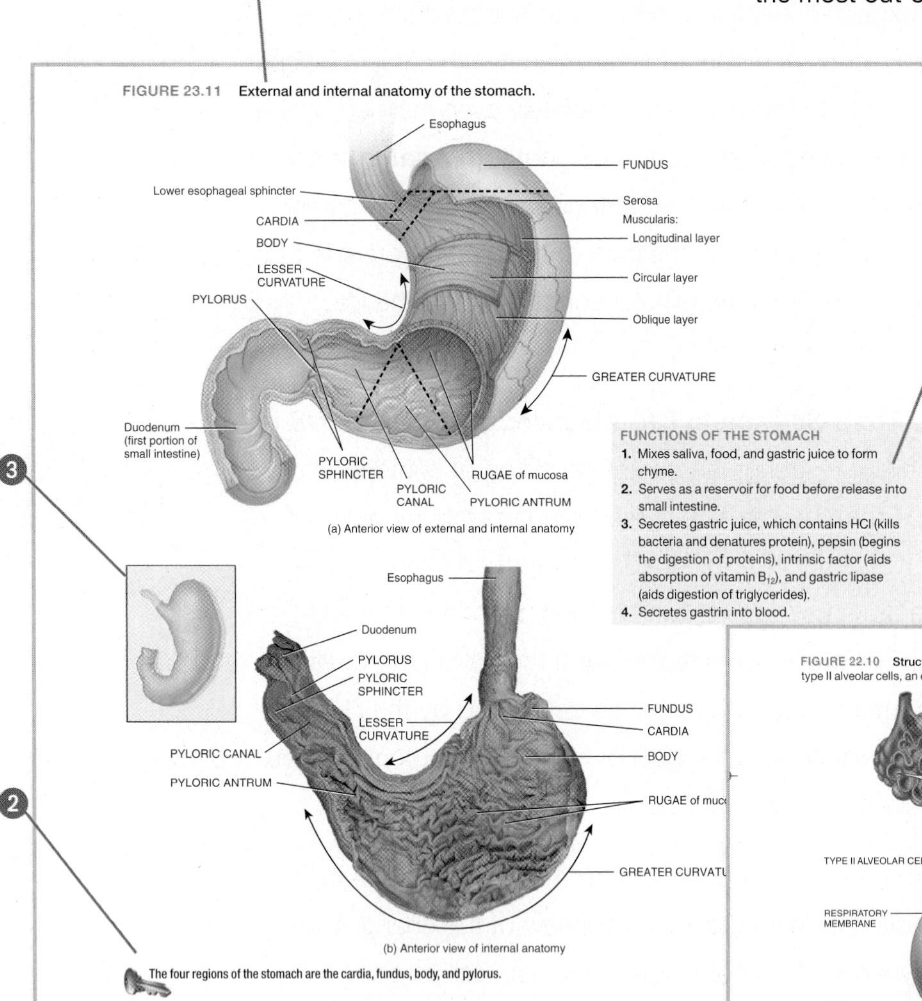


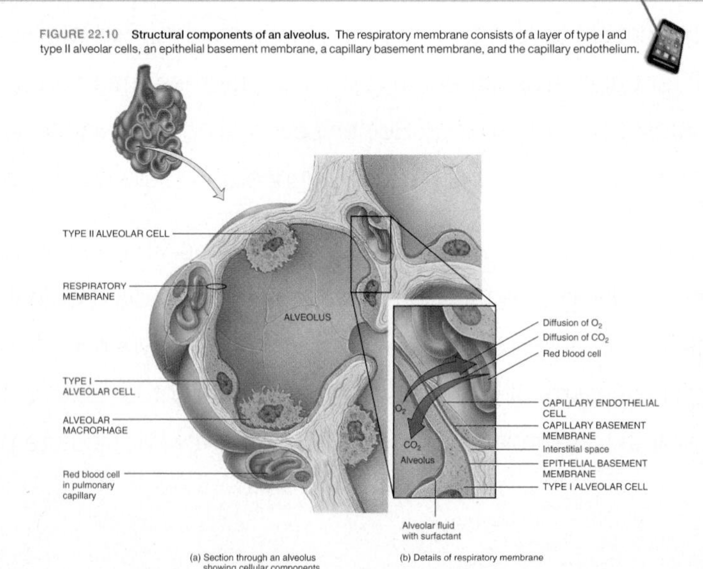

5. **MP3 DOWNLOADS** In each chapter you will find that several illustrations are marked with this icon. This indicates that an audio file which narrates and discusses the important elements of that particular illustration is available. You can access these downloads on the student companion website or within *WileyPLUS*.

# NOTES TO STUDENTS

Studying physiology requires an understanding of the sequence of processes. Correlation of sequential processes in text and art is achieved through the use of special numbered lists in the narrative that correspond to numbered segments in the accompanying figure. This approach is used extensively throughout the book to lend clarity to the flow of complex processes.

## Physiology of Hearing

The following events are involved in hearing (Figure 16.22):

1. The auricle directs sound waves into the external auditory canal.
2. When sound waves strike the tympanic membrane, the alternating high and low pressure of the air causes the tympanic membrane to vibrate back and forth. The tympanic membrane vibrates slowly in response to low-frequency (low-pitched) sounds and rapidly in response to high-frequency (high-pitched) sounds. It vibrates more forcefully in response to higher intensity (louder) sounds, more gently in response to lower intensity (quieter) sounds.
3. The central area of the tympanic membrane connects to the malleus, which also starts to vibrate. The vibration is transmitted from the malleus to the incus, and then to the stapes.
4. As the stapes moves back and forth, its oval-shaped footplate vibrates in the oval window. The vibrations at the oval window are about 20 times more vigorous than the

FIGURE 16.22 **Events in the stimulation of auditory receptors in the right ear.** The cochlea has been uncoiled to more easily visualize the transmission of sound waves and their distortion of the vestibular and basilar membranes of the cochlear duct.

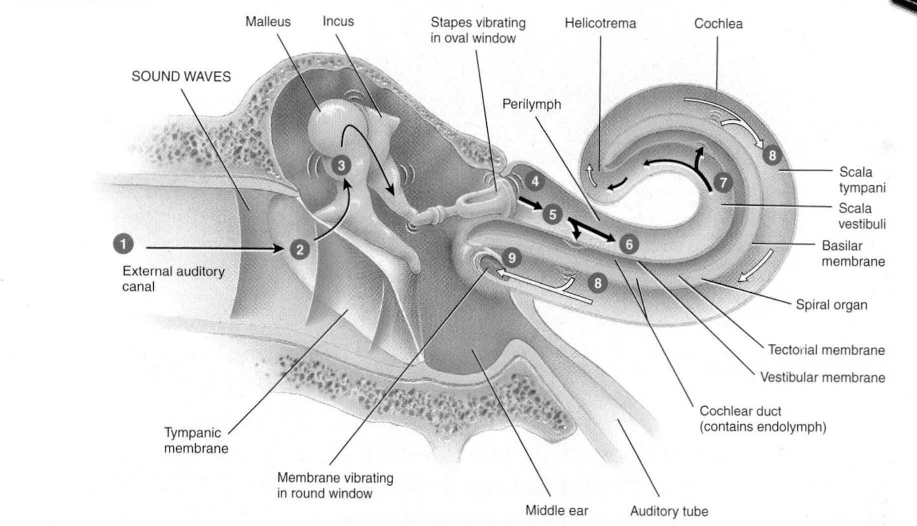


**Hair cells of the spiral organ convert a mechanical vibration (stimulus) into an electrical signal (receptor potential).**

There are many visual resources within *WileyPLUS*, in addition to the art from your text. These can help you master the topic you are studying. Examples closely integrated with the reading material include *animations* and *cadaver video clips. Anatomy Drill and Practice* lets you test your knowledge of structures with simple-to-use drag and drop labeling exercises, or fill-in-the-blank labeling. You can drill and practice on these activities using illustrations from the text, cadaver photographs, histology micrographs, or lab models.

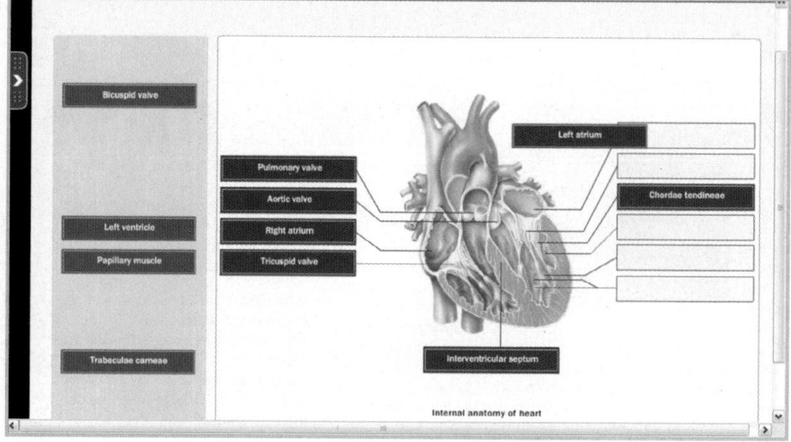

# NOTES TO STUDENTS

## Case Stories in Each Chapter Bring the Science to Life

The Case Stories found in each chapter of this text are a unique opportunity for you to readily answer for yourself why you need to learn many of the facts, terminology, and complex processes while studying anatomy and physiology.

CHAPTER 7

Fernando's Story

Elena loves her mother's father, Fernando, and always spends Saturdays with him. Elena's mother, Isabel, works on Saturdays, so Fernando always comes to their house on Friday night to spend the night and be there to care for 9-year-old Elena the next day. This had been the arrangement for as long as Elena could remember. Grandpa Fernando would let Elena sleep late, prepare French toast, her

Today they plan to go to a fair where they will ride the Ferris wheel and the merry-go-round, visit the new lambs and piglets, and eat some fried dough and cotton candy. Every year when the fair comes to town they go, even though every year Grandpa says it will be his last, complaining that it is too much walking for an old man. After breakfast, Elena dresses quickly and helps her grandfather clean up the kitchen so they can get going.

"I'm moving as fast as I can," grumbles Fernando. Neither one of them sees the bicyclist hurrying around the corner to make the turn before the light turns red. The bicycle goes down with a crash. Its rider is unhurt, but Fernando is knocked toward the curb, where he hits his head face-first on the lamppost. Elena cries out, "Grandpa, are you OK?" but gets no answer. Elena is shaking but

Each chapter opens with a case story that is closely related to the content you are about to study. This case will develop as you progress through the chapter. As you read modules within the chapter, you will become armed with information required to understand the story and to answer the questions posed.

Each case returns at appropriate points within the chapter and adds more detail and interest to the story now that you have developed some content knowledge. Questions included with these returns help you synthesize your knowledge of anatomy and physiology with the situation presented in the case story.

RETURN TO **Fernando's Story**

Isabel arrives at the emergency room shortly after the ambulance carrying Fernando and Elena. After Isabel gives her daughter a big hug and praises her for her quick response to the accident, Dr. Mueller calls them into his office to update them on Fernando's condition. "Fernando is still unconscious but his vital signs are stable. He is currently undergoing a CAT scan to determine if there is any brain damage, because his x-rays revealed a skull fracture over his left ear. I will be admitting him to the Intensive Care Unit for observation after the CAT scan. And one more thing–he has also sustained many facial cuts and abrasions, and his face is quite swollen. You can go see him after the test, but he won't look like his normal self."

When Isabel and Elena reach the ICU, the nurse informs them that Fernando settled, they can see him. After a short time, the nurse returns to the waiting area and escorts them to Fernando's room. Isabel and Elena are glad that the doctor warned them—they would not have recognized him because his face is so swollen and red. Some of the hair on his head has been shaved and there are stitches in his left eyebrow and over his left ear. Elena takes Fernando's hand and and smiles as she tells know he is going to be we are here. He just hand." This surprises t picks up his other hand to squeeze it. When he tells them that it's a very he is now conscious a to commands.

A. *Fernando has a fracture ear. Which bone or bone fractured?*
B. *Fernando has stitches in eyebrow. Which bone is eyebrow?*
C. *The left side of Fernando scraped from his forehe chin. Which bones are b scrapes?*

ory

EPILOGUE AND DISCUSSION

r-old by a way his . He cious frac- oken , and uries to his face and zed until he regains his injuries begin to ospital he develops s from a combination bed, the two broken ribs, and from a kyphotic thoracic curve. He is lucky that he did not sustain injury to his spine when he fell, because a fracture in an already deformed vertebra will not heal in normal alignment and the back curvature can become worse, with pain and possibly paralysis.

Elena and Isabel take Fernando home after he is discharged from the hospital. All of the discoloration and bruising gradually clear up, and the swelling and pain disappear. However, Fernando's kyphosis remains, as there is no corrective treatment. If he does postural exercises he can prevent it from getting worse, but for the rest of his life Fernando will appear "crooked."

G. *What effect does Fernando's kyphosis have on the thoracic cage?*
H. *What are the components of the thoracic cage?*
I. *Fernando has fractures of the left seventh and eighth ribs. To what structures do these ribs attach?*

At the end of each chapter, there is an Epilogue and Discussion for the case story. These discussion questions will help you make the transition from memorizing facts to applying what you have learned, an essential tool for you as a foundation for succeeding in your chosen career path.

# NOTES TO STUDENTS

## Clinical Discussions Make your Study Relevant

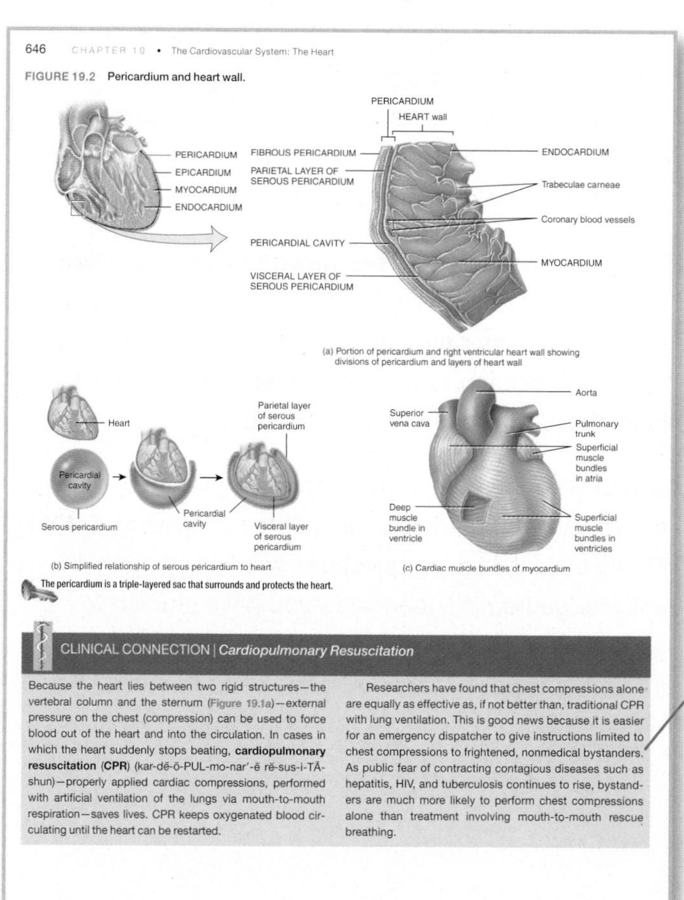

646 CHAPTER 19 • The Cardiovascular System: The Heart

FIGURE 19.2 Pericardium and heart wall.



(a) Portion of pericardium and right ventricular heart wall showing divisions of pericardium and layers of heart wall

(b) Simplified relationship of serous pericardium to heart

(c) Cardiac muscle bundles of myocardium

The pericardium is a triple-layered sac that surrounds and protects the heart.

CLINICAL CONNECTION | *Cardiopulmonary Resuscitation*

Because the heart lies between two rigid structures—the vertebral column and the sternum (Figure 19.1a)—external pressure on the chest (compression) can be used to force blood out of the heart and into the circulation. In cases in which the heart suddenly stops beating, **cardiopulmonary resuscitation (CPR)** (kar-dē-ō-PUL-mo-nar′-ē rē-sus-i-TĀ-shun)—properly applied cardiac compressions, performed with artificial ventilation of the lungs via mouth-to-mouth respiration—saves lives. CPR keeps oxygenated blood circulating until the heart can be restarted.

Researchers have found that chest compressions alone are equally as effective as, if not better than, traditional CPR with lung ventilation. This is good news because it is easier for an emergency dispatcher to give instructions limited to chest compressions to frightened, nonmedical bystanders. As public fear of contracting contagious diseases such as hepatitis, HIV, and tuberculosis continues to rise, bystanders are much more likely to perform chest compressions alone than treatment involving mouth-to-mouth rescue breathing.

The relevance of the anatomy and physiology that you are studying is best understood when you make the connection between normal structure and function and what happens when the body doesn't work the way it should. Throughout the chapters of the text you will find **Clinical Connections** that introduce you to interesting clinical perspectives related to the text discussion.

*WileyPLUS* offers you opportunities to explore even more interesting clinical applications. An additional folder with extra Clinical Connections is provided with your chapter resources. And for an interesting and engaging break from traditional study routines, check out the animated and interactive case studies called Homeostatic Imbalances also included with each chapter's resources.

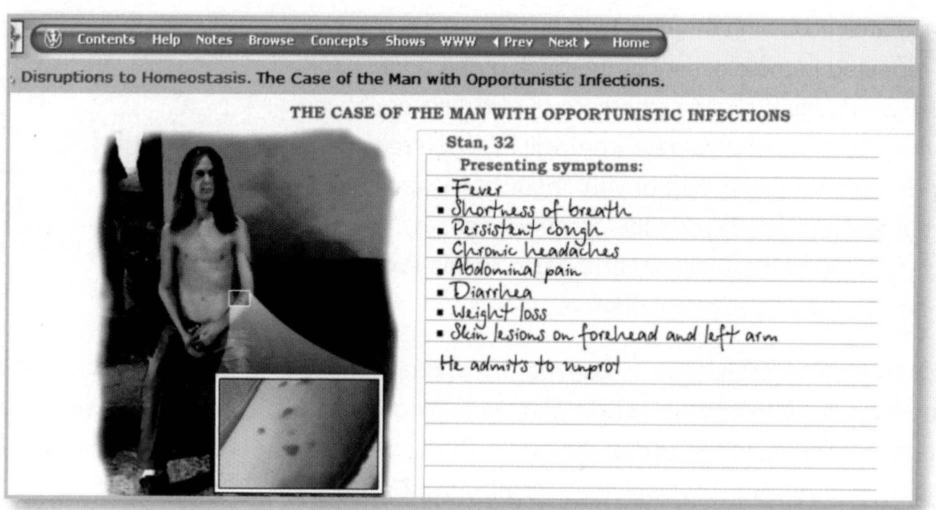

# NOTES TO STUDENTS

## Chapter Resources Help You Focus and Review

Your book has a variety of special features that will make your time studying anatomy and physiology a more rewarding experience. These have been developed based on feedback from students like you who have used previous editions of the text. Their effectiveness is even further enhanced within *WileyPLUS* for Anatomy and Physiology.

- **Key Concepts** and a short **Introduction** are listed at the start of each chapter. Before you begin, read and review these. This will set the stage for you so that you can anticipate what you will be focused on while reading and studying.
- **Checkpoint Questions** at the end of each concept help you assess if you have absorbed what you have read. Take time to review and answer these before progressing to the next concept module.
- **Mnemonics** are a memory aid that can be particularly helpful when learning specific anatomical features. Mnemonics are included throughout the text, some displayed in figures, or tables, and some included within the text discussion. We encourage you to not only use the mnemonics provided, but also to create your own to help you learn the multitude of terms involved in your study of human anatomy.
- **Chapter Review and Resources** is a helpful table at the end of chapters that offers you a concise summary of the important concepts from the chapter and links each section to the media resources available in *WileyPLUS* for Anatomy & Physiology.
- **Understanding the Concepts** gives you an opportunity to evaluate your understanding of the chapter as a whole. These critical thinking questions ask you to apply the knowledge from the concepts you have studied to specific situations. There is also a **Practice Quiz** for you to assess your knowledge of each chapter in *WileyPLUS*.

## Mastering the Language of Anatomy and Physiology

Throughout the text we have included phonetic **Pronunciations** and, sometimes, **Word Roots**, for many terms that may be new to you. These appear in parentheses immediately following the new words. The pronunciations are repeated in the glossary at the back of the book. Look at the words carefully and say them out loud several times. Learning to pronounce a new word will help you remember it and make it a useful part of your medical vocabulary. Take a few minutes to read the pronunciation key, found at the beginning of the Glossary at the end of this text, so it will be familiar as you encounter new words.

To provide more assistance in learning the language of anatomy, a full **Glossary** of terms with phonetic pronunciations also appears at the end of the book. The basic building blocks of medical terminology—**Combining Forms, Word Roots, Prefixes, and Suffixes**—are listed inside the back cover, as is a listing of **Eponyms,** traditional terms that include reference to a person's name, along with the current terminology.

*WileyPLUS* houses help for you in building your new language skills as well. The **Audio Glossary**, which is always available to you, lets you hear all these new, unfamiliar terms pronounced. Throughout the e-text, these terms can be clicked on and heard pronounced as you read. In addition, you can use the helpful **Mastering Vocabulary** program which creates electronic flashcards for you of the key terms within each chapter for practice, as well as the ability to take a self-quiz specifically on the terms introduced in each chapter.

# NOTES TO INSTRUCTORS

As active teachers of the course, we recognize both the rewards and challenges in providing a strong foundation for understanding the complexities of the human body. We believe that teaching goes beyond just sharing information. *How* we share information makes all the difference–especially, if like us, you have an increasingly diverse population of students with varying degrees of learning abilities. As we revised this text we focused on those areas that we knew we could enhance to provide greater impact in terms of better learning outcomes. Feedback from many of you, as well as the students we interact with in our own classrooms, guided us in assuring that the revisions to the text along with the powerful enhancements in *WileyPLUS* for Anatomy and Physiology support the needs and challenges you face day-to-day in your own classrooms.

We focused on several key areas for revision: the all important visuals, both drawings and photographs; providing new and revised case stories based on user feedback and interest; adding some new, and revising many, tables to increase their effectiveness; updating and adding clinical material that helps students relate what they are learning to their desired career goals and the world around them; and narrative changes aimed at increasing student engagement with—and comprehension of—the material. For a detailed list of revisions for each chapter, please visit our website at www.wiley.com/college/sc/aandp and click on the text cover.

## The Art of Anatomy and Physiology

**Illustrations** throughout the text have been refined. The color palette for the skulls in Chapter 7, and for the brain and spinal cord throughout the text, has been adjusted for greater impact. Illustrations in each chapter have been revised and updated, providing greater clarity and more saturated colors. Particular emphasis was placed on revised drawings of joints, muscles, and blood vessels.

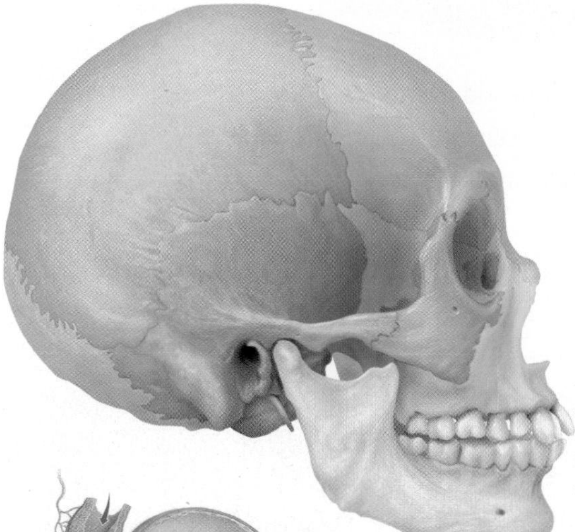

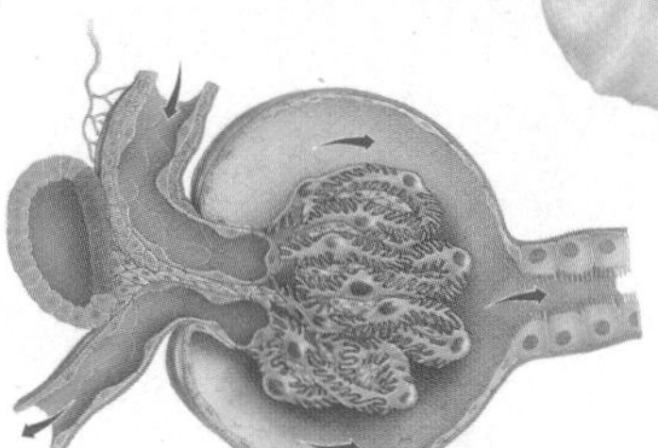

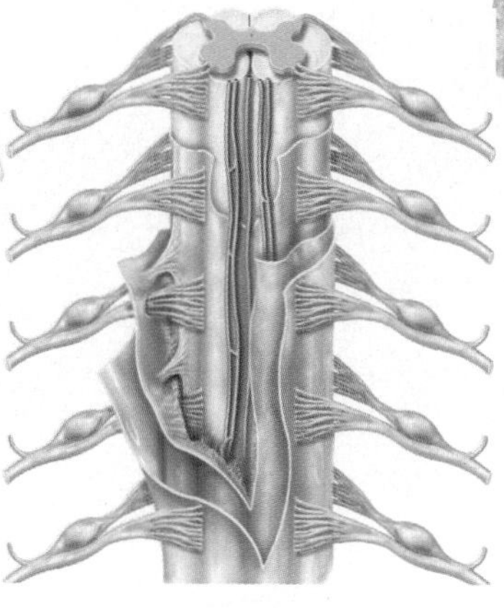

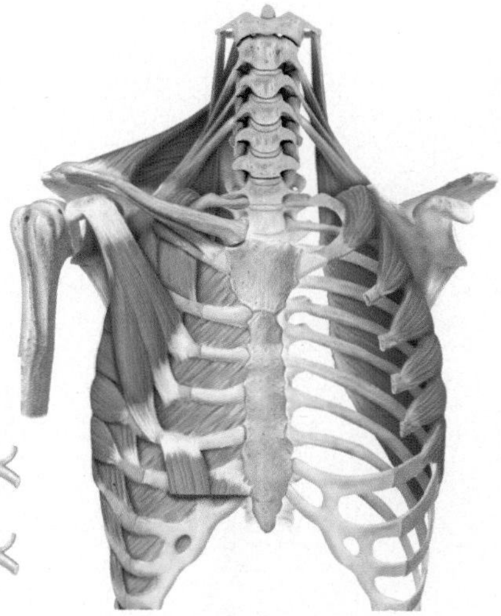

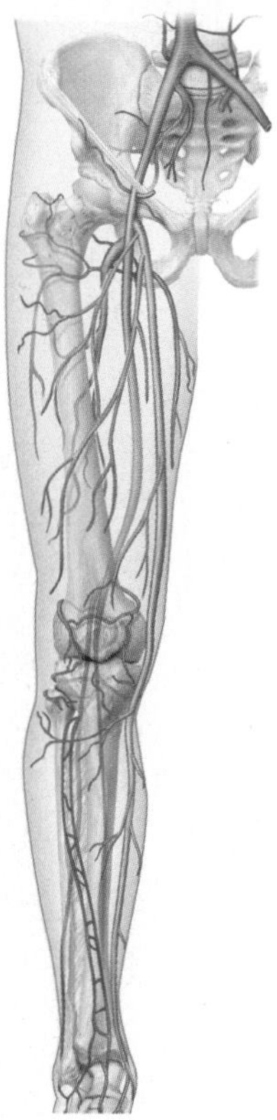

# NOTES TO INSTRUCTORS

**Cadaver Photographs** are included throughout the text to help students relate the content to real life images. These are often paired with diagrams to help make structural connections. Most of the meticulous dissections and outstanding photography come from Mark Nielsen's lab at the University of Utah.

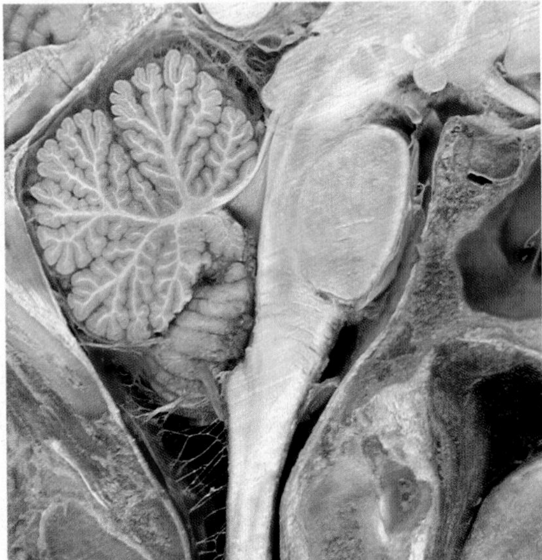

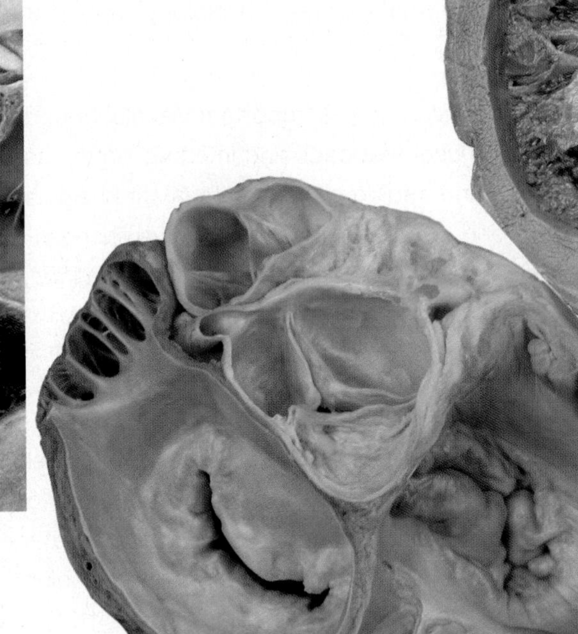

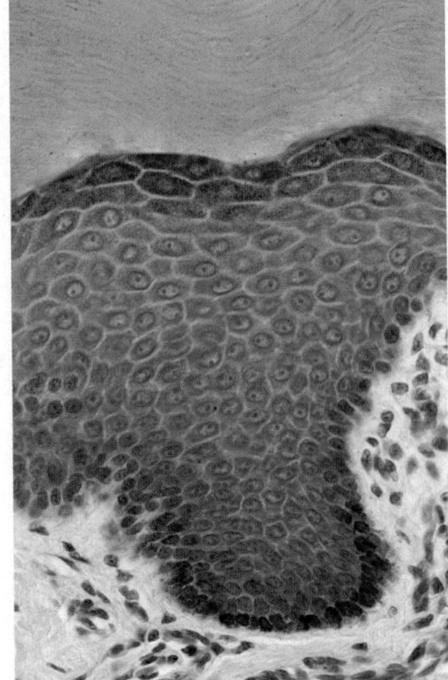

**Photomicrographs**—Most tissue photomicrographs have been replaced with exceptionally clear photomicrographs with high magnification blowouts.

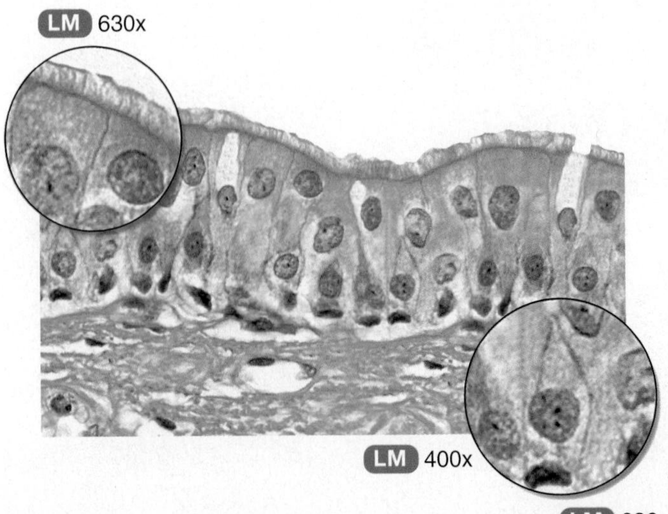

# NOTES TO INSTRUCTORS

## Case Stories and Critical Thinking

Unique to our text is the use of Case Stories in each chapter. These not only help to engage your students and guide them in making the connections between the science of anatomy and physiology and real life situations, but also are the perfect platform for building the all important critical thinking skills that students need to succeed. For this edition we have included twelve new case stories, with topics such as multiple sclerosis, Parkinson's disease, heart attack, Crohn's disease, and sexually transmitted disease. The remaining case stories have been revised and streamlined based on feedback from students and professors.

As students progress through the chapter and begin to grasp the major concepts being presented in each module of content, the case story expands in detail to match their growing knowledge. With each 'Return' to the case, questions are posed that link the case directly to the material in the preceding pages of the chapter. These questions help students develop their skills, allowing them to apply their knowledge to the situations presented. At the end of the chapter, the Epilogue and Discussion of the case brings everything together and continues to guide the students in their abilities to move beyond memorization of material to application.

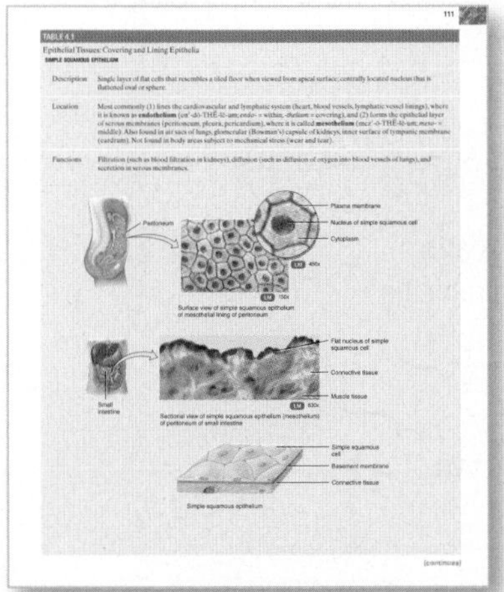

Also new for this edition is the development of a special set of PowerPoint slides matched with clicker questions that you can use to include discussion of the case and its implications into your lectures either as a way to launch into the topics at hand, or as a capstone for completion of the lectures on the material. These are available to you in the Prepare and Present section of *WileyPLUS*, as well as on the Instructor's Companion Website.

## Tables

New **Tables** including Skin Glands, Summary of the Levels of Organization within a Skeletal Muscle, and Summary of the Respiratory System have been added in addition to refinement of many of the existing tables with either new illustrations or rewritten text.

## Clinical Connections

Your students are fascinated by the clinical connections to the normal anatomy and physiology that they are learning. At the request of many users of the text, we have added this feature back into the body of the text for each chapter. The engaging discussions cover a variety of clinical scenarios from diseases to tests and procedures and are all fully up to date and reflect current information. Additional Clinical Connections are available in a folder for each chapter within *WileyPLUS*. A complete reference list of the Clinical Connections for each chapter follows the Table of Contents.

# NOTES TO INSTRUCTORS

## *WileyPLUS* and You

***WileyPLUS*** is designed to simplify both the teaching and learning experience with an effective and efficient environment for accessing content and resources in dynamic and interactive ways. For students, a wealth of resources, including a complete eBook, allow them to have everything they need at their fingertips. Understanding that students will choose different resources to support their study depending on how they learn best, everything is simply organized into manageable groups of **See, Hear, and Do**. Instructors have a wide choice of resources as well, whether it's to help build effective lecture presentations, provide students with a directed course of study, make homework assignments, or administer quizzes and tests. **QuickStart** is a tool that allows you to quickly implement a course plan for each chapter without having to search and review through multiple choices. A full gradebook is also available for managing your course. These resources and tools allow you to streamline your own course or share your course with other instructors and adjuncts.

### FOR STUDENTS

**eBook**
- Integrated animations
- Integrated audio pronunciations
- Integrated audio discussions of key figures

**See**
- Anatomy Overviews, an interactive review of structures for each body system
- Animations of Key Physiological Processes
- Muscles in Motion 3D animations
- Additional Clinical Connections
- Cadaver Videos

**Hear**
- Audio downloads keyed to significant illustrations within the text
- Mastering Vocabulary with audio glossary and flashcards

**Do**
- Anatomy Drill and Practice, interactive drag-and-drop and labelling exercises using illustrations, cadavers, histology, and laboratory models
- Practice Quizzes
- Interactive and Animated Exercises
- Interactive Concept Maps
- Homeostatic Imbalances, animated and interactive clinical case studies
- Worksheets linked to Animations

### FOR INSTRUCTORS

**Visual Resources**
- Anatomy and Physiology Visual Library, a searchable database of illustrations and photographs in labelled, unlabelled, and leader-line only formats
- Lecture PowerPoints
- Case Story PowerPoints
- Editable Art PowerPoints
- Animation PowerPoints
- Table PowerPoints

**Teaching Guides and Answer Keys**
- Case Story Resources include answers to the questions and teaching summaries
- Chapter Overview
- What's New
- Suggested Lecture Outline and Objectives
- Teaching Tips
- A Guide to Using *QuickStart*
- A Guide to the Real Anatomy DVD
- Answer Keys
  - o Interaction Worksheets
  - o Real Anatomy Worksheets

**Assessment**
- Test Bank
- Pre-lecture questions
- Post-lecture questions
- Labelling questions, based on Anatomy Drill and Practice
- Case Story Questions
- Clicker Questions

# RESOURCES FOR INTEGRATING LABORATORY EXPERIENCES

## Laboratory Manual for Anatomy and Physiology, 4e

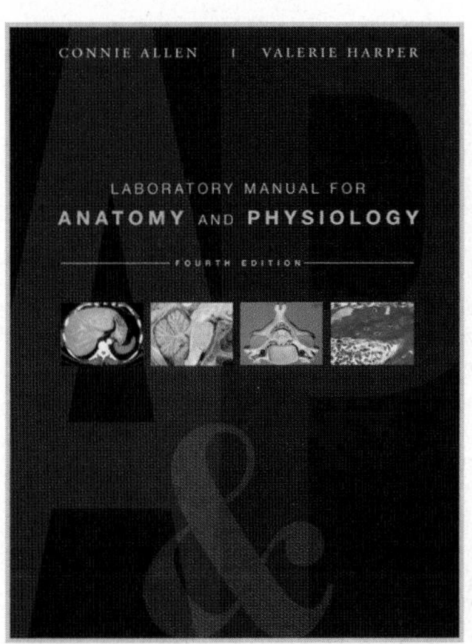


**Connie Allen and Valerie Harper**

Newly revised, The ***Laboratory Manual for Anatomy and Physiology*** with ***WileyPLUS*** engages your students in active learning and focuses on the most important concepts in A&P. Exercises reflect the multiple ways in which students learn and provide guidance for anatomical exploration and application of critical thinking to analyzing physiological processes. A concise narrative, self-contained exercises that include a wide variety of activities and question types, and two types of lab reports for each exercise, keep students focused on the task at hand. Depending on your needs, a Cat Dissection Manual or Fetal Pig Dissection Manual accompanies the main text. Rich media within *WileyPLUS* further enhances student experience and includes dissection videos, animations, and illustrated drill and practice exercises with illustrations, micrographs, cadaver photos, and popular lab models. Each lab text comes with access to *PowerPhys 2.0.*

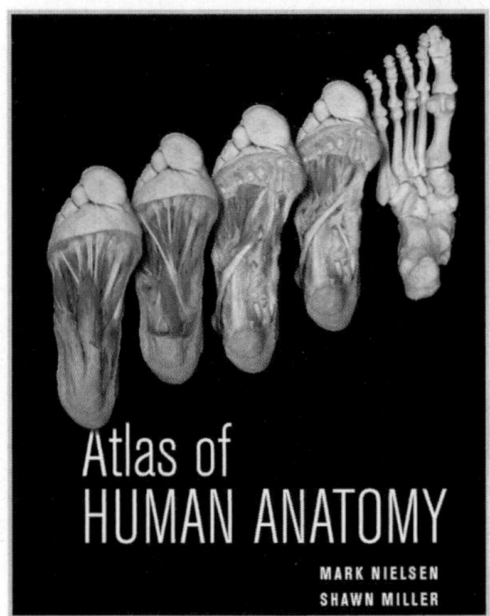


## Atlas of Human Anatomy, 1e

**Mark Nielsen and Shawn Miller**

This new atlas filled with outstanding photographs of meticulously executed dissections of the human body has been developed to be a strong teaching and learning solution, not just a catalog of photographs. Organized around body systems, each chapter includes a narrative overview of the body system and is then followed with detailed photographs that accurately and realistically represent the anatomical structures. Histology is included. *Atlas of Human Anatomy* will work well in your laboratories, as a study companion to your textbook, and a print companion to using the Real Anatomy DVD.

## Photographic Atlas of the Human Body, 2e

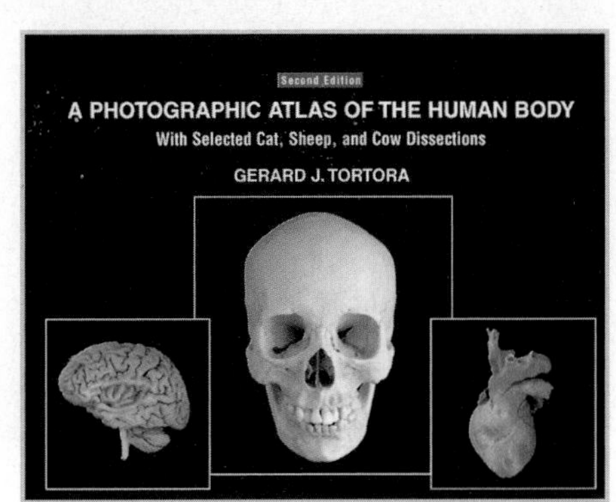


**Gerard J. Tortora**

Like the new atlas from Nielsen and Miller, this popular atlas is also systemic in its approach to the photographic review of the human body. In addition to the excellent cadaver photographs and micrographs, this atlas also contains selected cat and sheep heart dissections. The high quality imagery can be used in the classroom, laboratory, or for study and review.

# RESOURCES FOR INTEGRATING LABORATORY EXPERIENCES

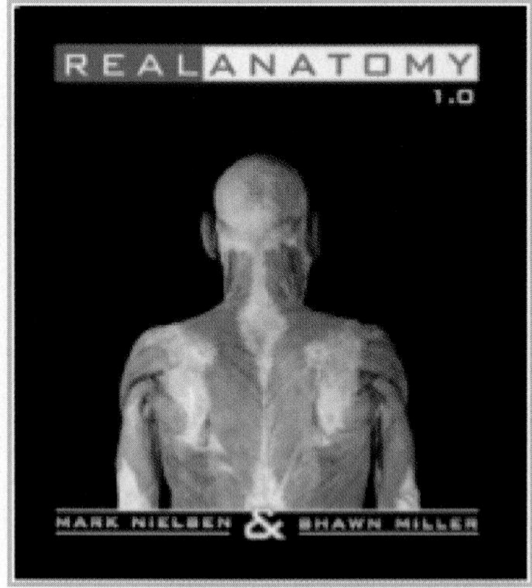


## Real Anatomy

**Mark Nielsen and Shawn Miller**

Real Anatomy is 3-D imaging software that allows you to dissect through multiple layers of a three-dimensional real human body to study and learn the anatomical structures of all body systems.

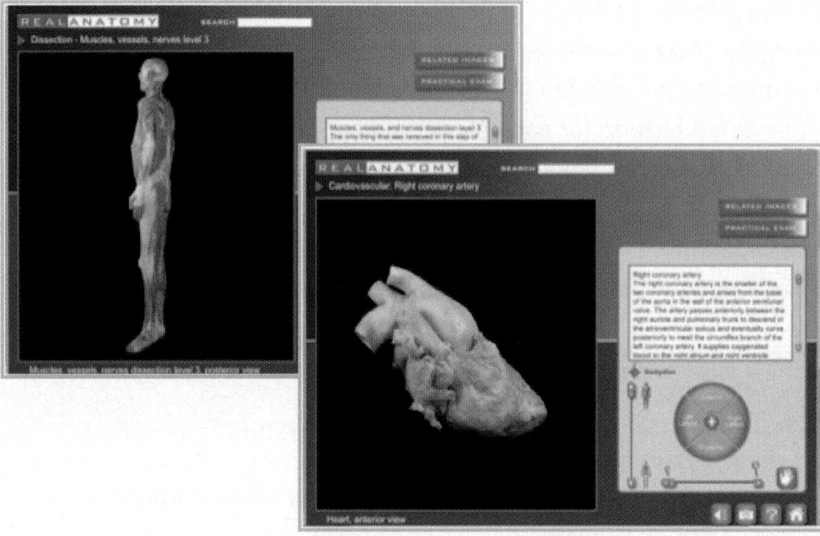


- Dissect through up to 40 layers of the body and discover the relationships of the structures to the whole
- Rotate the body, as well as major organs to view the image from multiple perspectives

- Use a built in "zoom" feature to get a closer look at detail
- A unique approach to highlighting and labeling structures does not obscure the real anatomy in view

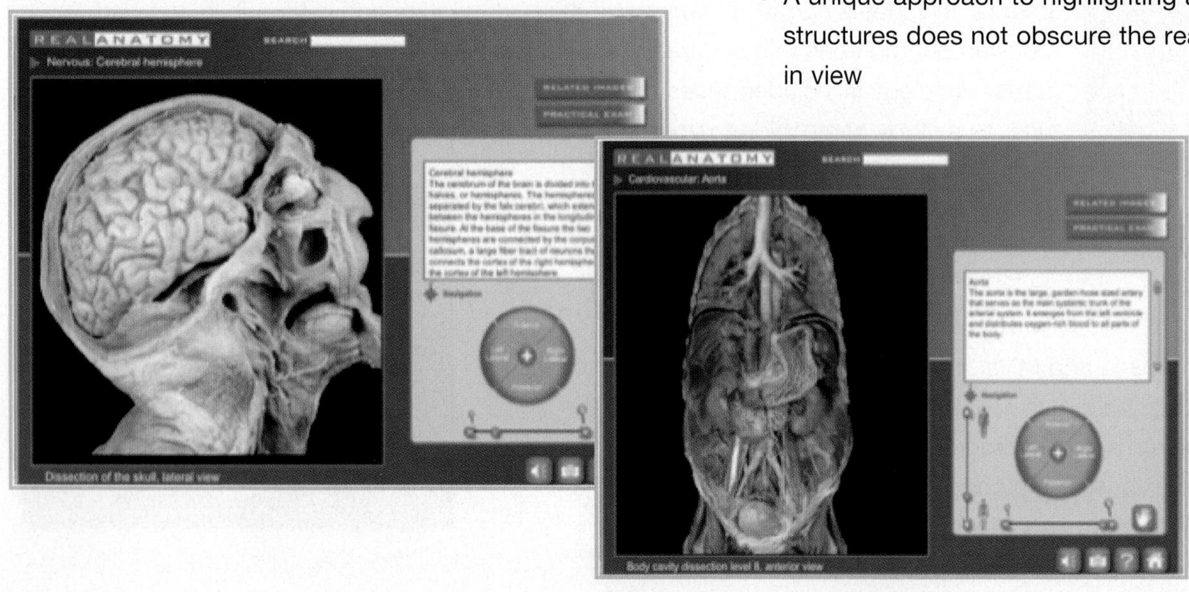

# RESOURCES FOR INTEGRATING LABORATORY EXPERIENCES

- Related Images provide multiple views of structures being studied

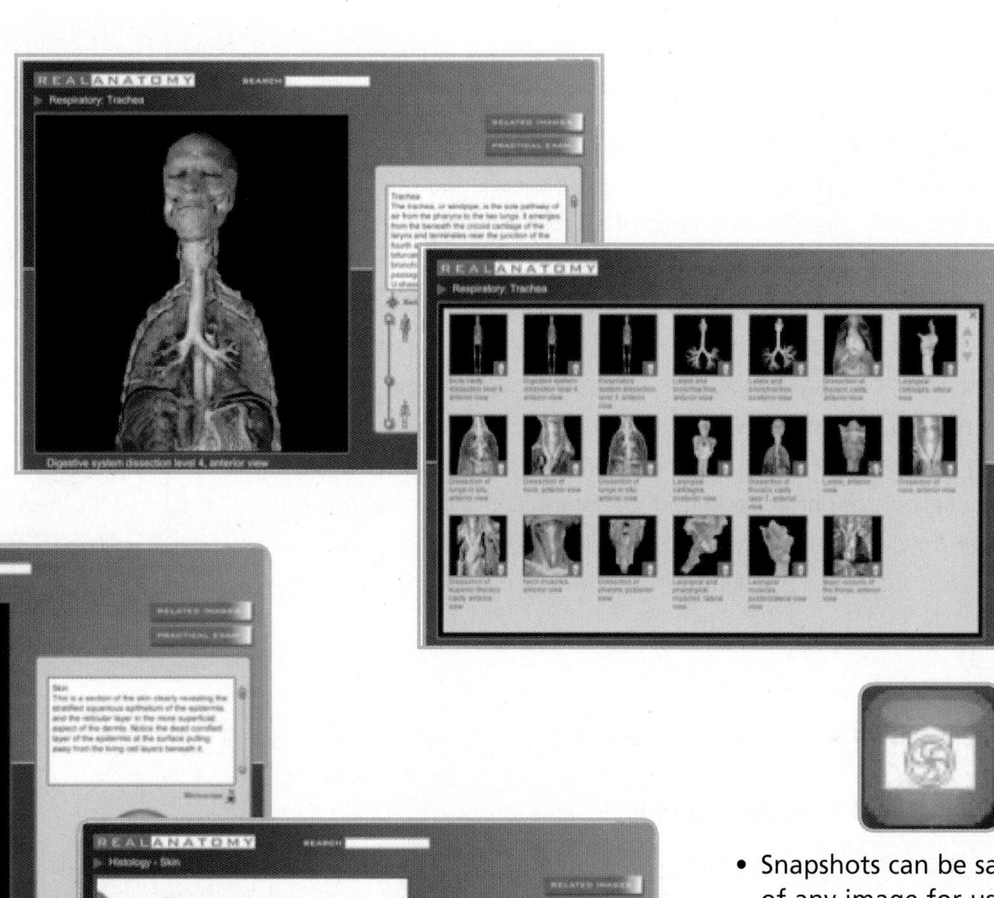


- View histology micrographs at varied levels of magnification with the virtual microscope

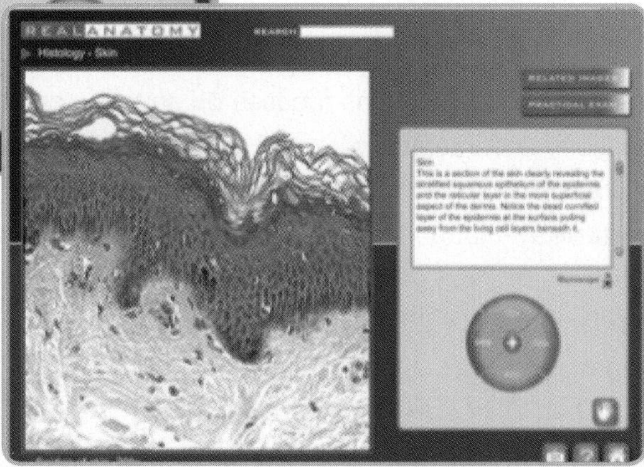


- Snapshots can be saved of any image for use in PowerPoints, quizzes, or handouts

- Audio pronunciation of all labeled structures is readily available

***Virtual Dissection – 100% Real***

REAL ANATOMY

# RESOURCES FOR INTEGRATING LABORATORY EXPERIENCES

## PowerPhys 2.0

**Connie Allen, Valerie Harper, Thomas Lancraft, Yuri Ivlev**

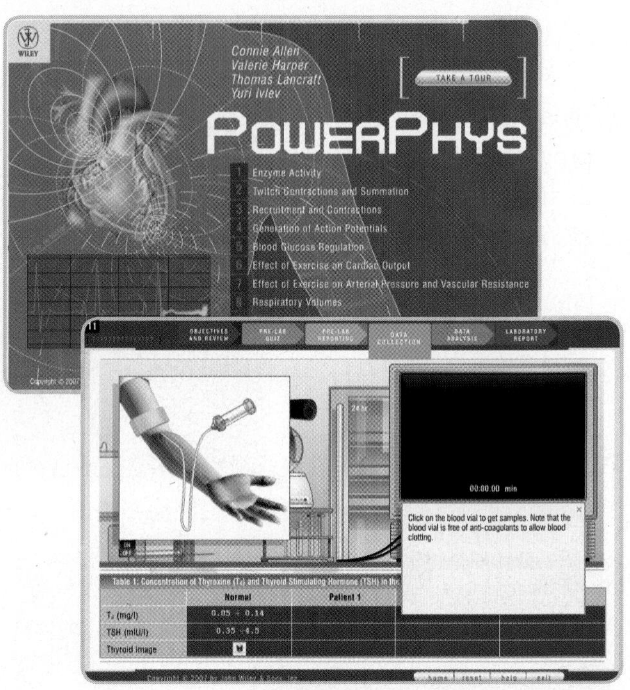


***PowerPhys 2.0*** provides a simulated laboratory experience for students, giving them the opportunity to review their knowledge of core physiological concepts, predict outcomes of an experiment, collect data, analyze it, and report on their findings. This revised edition features a new activity on Homeostatic Imbalance of Thyroid Function and revised lab report questions throughout. An easy-to-use and intuitive interface guides students through the experiments from basic review to laboratory reports. All experiments contain randomly generated data, allowing students to experiment multiple times but still arrive at the same conclusions. A perfect addition to distant learning or hybrid courses, *PowerPhys 2.0* is a stand-alone web-based program as well as fully integrated with Allen and Harper's laboratory manual.

## Interactions: Exploring the Functions of the Human Body 3.0

**Thomas Lancraft and Frances Frierson**

***Interactions 3.0*** is the most complete program of interactive animations and activities available for anatomy and physiology. A series of modules encompassing all body systems focuses on a review of anatomy, the examination of physiological processes using animations and interactive exercises, and clinical correlations to enhance student understanding. At the heart of ***Interactions*** is a focus on core principles—**homeostasis; communication; energy flow; fluid flow; and boundaries**—that underscore the key relationships between structure and function as well as interrelationships between systems. It is the reinforcement of these fundamental organizing principles that sets this series apart from others. ***Interactions*** is available on DVD, web-based, or fully integrated within *WileyPLUS.*

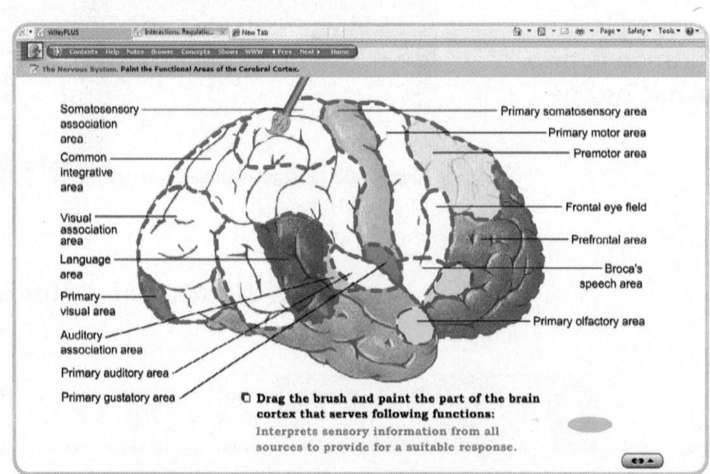

# ACKNOWLEDGMENTS

We wish to especially thank several academic colleagues for their indispensable contributions to this edition.

A talented group of educators has contributed to the high quality of the diverse materials that accompany this text. We wish to acknowledge each and thank them for their work. Thanks to Celina Bellanceau, University of South Florida; Jody E. Johnson, Arapahoe Community College, Raymond Larsen, Bowling Green State University; Daniel Moore, Southern Maine Community College; Thomas Sarro, Mount Saint Mary College; Scott Raschulte, Ivy Tech Community College; Terry Thompson, Wor-Wic Community College; and Diane L. Wood, Southeast Missouri State University. Special thanks to Molly Cochran Clay, MS, CRNP, Montgomery College; Thomas Lehman, Coconino Community College; and Jennifer Gima for their contributions to the case stories.

We would like to express our heartfelt thanks to our Executive Editor, Bonnie Roesch, who believed in the vision of the book and in us as its authors. Bonnie rode shotgun on the entire process and her expertise was invaluable. It is impossible to know how to thank Bonnie for all her guidance, creativity, and dedication in making our vision for this book become a reality. Bonnie has inspired and influenced every vital component of the book from front cover to end of book content. Bonnie, we can only say, "You're the best!"

We would also like to thank other vital members of our winning John Wiley & Sons, Inc. project team who often worked diligently days, evenings, and weekends to ensure the success of this book. Their commitment to and passion for this textbook is so obvious and appreciated. In particular, Lorraina Raccuia, Project Editor, masterfully orchestrated this endeavor by helping us implement the revision plan, and managed the endless day-to-day details to ensure that the book's content, look, and feel reflect the needs of current anatomy and physiology curricula. Lorraina is uniquely able to balance warm compassion and considerate flexibility with efficient professionalism. Lorraina deserves personal thanks for the many pep talks and virtual pats on the back that kept us going. Karen Trost, our Freelance Developmental Editor, carefully edited and mapped the text and art, and effectively managed the numerous revisions and reviewer feedback. Karen misses nothing, and with accuracy and details, reigns supreme! Thanks also to Suzanne Ingrao, Freelance Production Manager, whose determination, grit, and superb coordination organized all stages of the production of our book. We really value the talents of Harry Nolan, Director of Creative Services, and Maureen Eide, Senior Designer, who laid out the pages to be visually attractive, pedagogically effective, and flow in a student-friendly style that makes the book come alive. Clay Stone, Executive Marketing Manager, inspired us by his tireless enthusiasm for this particular publication. Hilary Newman, our diligent photo researcher, brought faces to the case stories and transitioned our mental images into real world photographs. We're still amazed that Hilary always managed to quickly find exactly the right photo that our mind's eye envisioned. Special thanks also to Linda Muriello, our Senior Media Editor, who really gets all the high tech minutiae, and to Christina Picciano, our Editorial Assistant, who never failed to find the right answer to every request. We are truly appreciative!

Special thanks from Gail Jenkins to those at Montgomery College: Dr. Brad Stewart, Vice President and Provost, for his ongoing support of and counsel in this authoring project; Nelson Bennett and Dr. Stephen Cain for the many collegial discussions about teaching methodologies; Donna Schena for her understanding and humor while being rudely ignored during this endeavor; and to the many campus faculty and students whose input strongly influenced this edition.

Finally, we particularly want to express our gratitude to the following reviewers who took the time to read and evaluate the manuscript prior to production. Our reviewers generously provided their expertise to help us to continue to ensure the book's accuracy, clarity, and focus on the needs of today's anatomy and physiology instruction. Their insightful suggestions were appreciated and were well utilized in this edition of the book. Thank you for your invaluable contributions.

Teresa Alvarez, *St. Louis Community College*

Ali O. Azghani, *University of Texas—Tyler*

John Boucher, *University of Texas—Tyler*

Stephen Burnett, *Clayton State University*

Gina L. Cano-Monreal, *Texas State Technical College—Harlingen*

Beth Campbell, *Itawamba Community College*

Misty G. Carriger, *Northeast State Technical Community College*

Jim Collier, *Truckee Meadows Community College*

Robert E. Dudock, *Mott Community College*

Joseph D. Gar, *West Kentucky Community and Technical College*

J. Mark Danley, *Central New Mexico Community College*

Sarah Darhower, *Southern Maine Community College*

Angela Edwards, *Trident Technical College*

John F. Harms, *Messiah College*

Clare Hays, *Metropolitan State College of Denver*

D. J. Hennager, *Kirkwood Community College*

Dale R. Horeth, *Tidewater Community College-Virginia Beach*

William F. Huber, *St. Louis Community College*

Melinda Hutton, *McNeese State University*

Walter Johnson, *Merritt College*

Brittany Joseph, *Owens Community College*

Patricia C. Lager, *Lincoln Land Community College*

Anthony Lamanna, *Paradise Valley Community College*

Brenda Leady, *Lourdes College*

Jeffrey Lee, *Essex County College*

David G. Little, *Tri-County Technical College*

Mary Katherine Lockwood, *University of New Hampshire*

John Kell, *Radford University*

Karen L. Keller, *Frostburg State University*

Thomas Kober, *Cincinnati State Technical Community College*

Peter E. Malo, *Richard J. Daley College*

Amanda Mitchell, *Hinds Community College*

Sandra Perez, *El Paso Community College*

Sidney L. Palmer, *Brigham Young University*

Penny S. Perkins-Johnston, *California State University—San Marcos*

John Placyk, *University of Texas—Tyler*

Wendy M. Rappazzo, *Harford Community College*

James Rayburn, *Jacksonville State University*

Caroline Rivera, *Tidewater Community College-Norfolk*

Dean Scherer, *Oklahoma State University*

Heiko Schoenfuss, *St. Cloud State University*

James Schwartz, *Front Range Community College*

Marilyn Shopper, *Johnson County Community College*

William Stewart, *Middle Tennessee State University*

Lisa Strong, *Northwest Mississippi Community College*

Yong Tang, *Front Range Community College*

Teresa Trendler, *Pasadena City College*

Marisa Twiner, *Baker College*

Andrzej Wierasko, *The College of Staten Island/CUNY*

Roger Young, *Southern Maine Community College*

# ABOUT THE AUTHORS

**Gail W. Jenkins** is Professor of Biology at Montgomery College in Maryland, where she teaches human anatomy and physiology as well as general biology and microbiology. She received her bachelor's degree in botany from the University of California in Davis, with a minor in medical technology and completed a graduate biological sciences instructor credential program. Her master's degree in biological sciences was from California State University in Sacramento, where she focused on anatomy with research in neuroembryology conducted at the University of Kentucky College of Medicine in Lexington.

Gail is passionately devoted to assisting students in the learning process and in their preparation for health science vocations. She was the recipient of the Montgomery College Outstanding Faculty Award in 1999 and the 1997 National Institute for Staff and Organizational Development Excellence Award for Outstanding Contribution to Teaching and Learning from the University of Texas. At Montgomery College, Gail is Course Coordinator of the human anatomy and physiology curriculum, course curriculum liaison to college Health Science programs, and has served as the Chair of the Department of Biology, Physical Education, and the Health Science; Chair of the Faculty Council (faculty governance organization); Phi Theta Kappa Advisor, Administrative Associate to the Vice President and Provost, co-developer of the Physical Therapist Assistant Program, and mentor for adjunct science faculty through the college Center for Teaching and Learning.

Gail has been active in the Human Anatomy and Physiology Society (HAPS) for 15 years, serving 8 years as a member of the Executive Committee, and has been an active participant in annual and regional conference planning. Gail founded and directed a program at Stanford University's Department of Anatomy to integrate human cadaveric materials and medical imaging into pre-health science curricula, co-founded the Northern California Society of Anatomists, developed a hospital laboratory work/learn internship program for health science students, was a Federal Liaison Officer in Washington D.C. working with Congress for higher education funding, and served as an educational consultant to Stanford University's Advanced Media Research Group, and to publishing and software companies.

*To my students whose dedication, fascination, and passion for learning human body design inspire my teaching. – G. W. J.*

**Gerard J. Tortora** is Professor of Biology and former Biology Coordinator at Bergen Community College in Paramus, New Jersey, where he teaches human anatomy and physiology as well as microbiology. He received his bachelor's degree in biology from Fairleigh Dickinson University and his master's degree in science education from Montclair State College. He is a member of many professional organizations, including the Human Anatomy and Physiology Society (HAPS), the American Society of Microbiology (ASM), American Association for the Advancement of Science (AAAS), National Education Association (NEA), and the Metropolitan Association of College and University Biologists (MACUB).

Above all, Jerry is devoted to his students and their aspirations. In recognition of this commitment, Jerry was the recipient of MACUB's 1992 President's Memorial Award. In 1996, he received a National Institute for Staff and Organizational Development (NISOD) excellent award from the University of Texas and was selected to represent Bergen Community College in a campaign to increase awareness of the contributions of community colleges to higher education.

Jerry is the author of several best-selling science textbooks and laboratory manuals, a calling that often requires an additional 40 hours per week beyond his teaching responsibilities. Nevertheless, he still makes time for four or five weekly aerobic workouts that include biking and running. He also enjoys attending college basketball and professional hockey games and performances at the Metropolitan Opera House.

*To my mother, Angelina M. Tortora. (August 20, 1913–August 14, 2010). Her love, guidance, faith, support, and example continue to be the cornerstone of my personal and professional life. – G. J. T.*

# BRIEF CONTENTS

# CONTENTS

## 17 THE ENDOCRINE SYSTEM 576



## 18 THE CARDIOVASCULAR SYSTEM: THE BLOOD 614

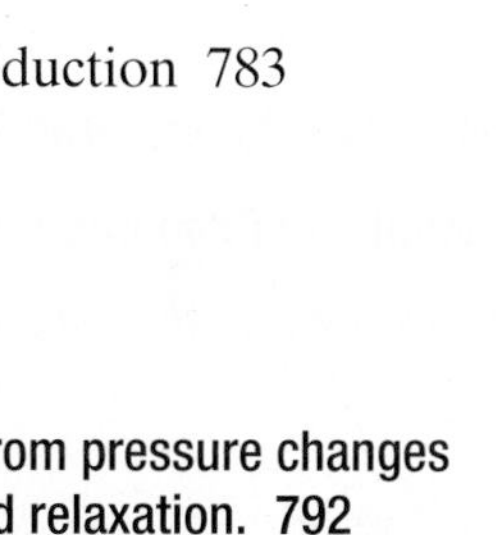

# CLINICAL CONNECTIONS

# Anatomy and Physiology

## from Science to Life

# CHAPTER 1

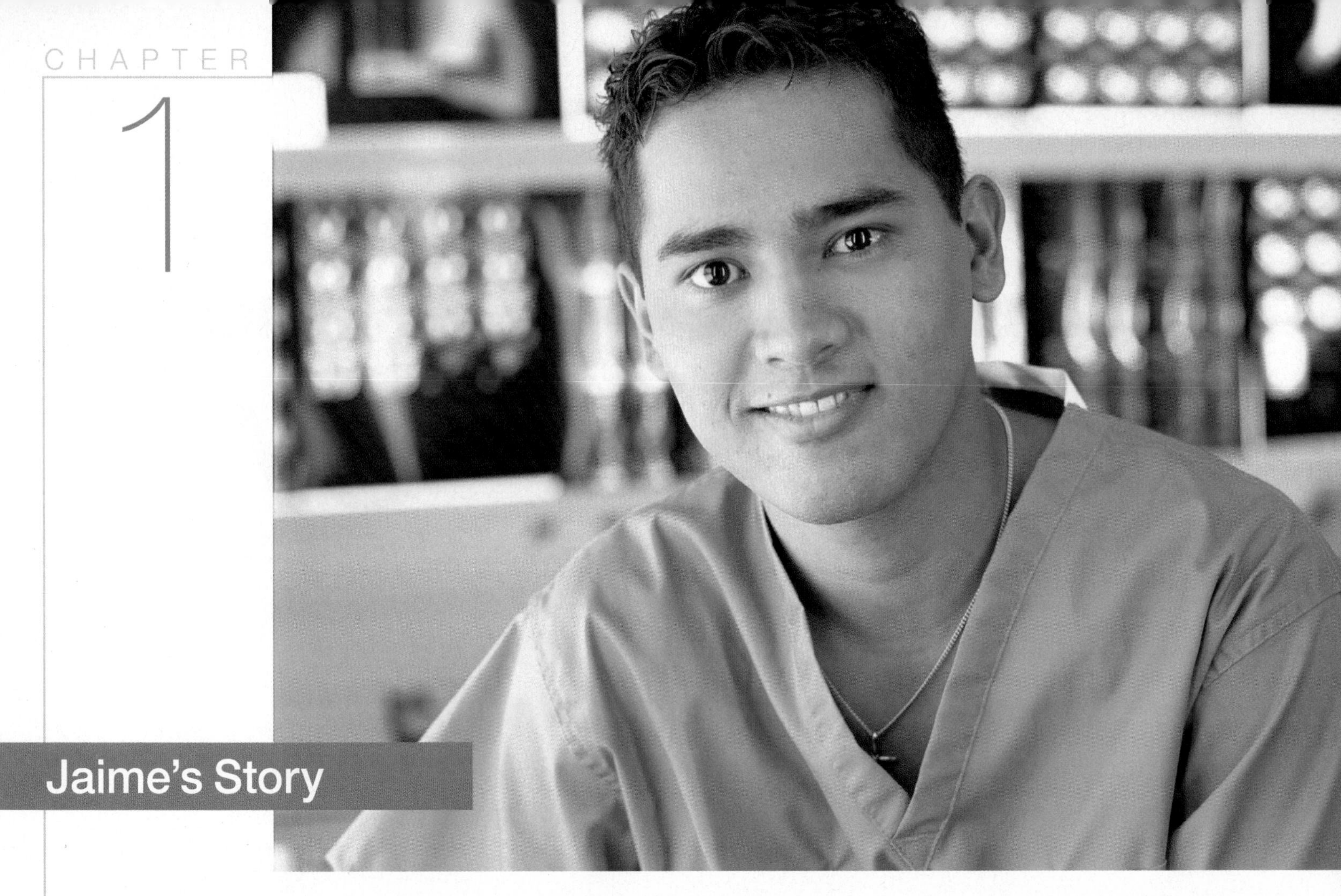

## Jaime's Story

Jaime is nervous as he walks up to the front desk of the radiology department. He clears his throat and says, "Hi. My name's Jaime. I'm the new transporter."

"Hi. I'm Ellen. Nice to meet you – just a minute," replies Ellen without looking up from what she is doing. She finishes typing something on the computer and dials the phone. "Jan, we've got a 9 o'clock C-spine and a 10 o'clock cranial series coming your way. Thanks." After she hangs up the phone Ellen turns back toward Jaime. "Sorry about that. We have a couple of CT scans coming in and I didn't want to forget."

"Not a problem. C-spine. . . . That's the cervical spine, right? I just started an A&P class and we were just learning about that." Ellen nods as she stands up and hands him some folders. She waves for him to follow her as she heads down the hall.

"That's right. You're going to get a lot of chances to put your learning into practice here. You won't just be transporting patients. You'll get a chance to help the x-ray techs, develop X-ray films, and do scheduling, too. You can follow me around until you're comfortable with everything. Right now, we're going to take an x-ray of a little girl's broken arm. Here's the chart. Write down the name, time, and procedure here. All of this is inputted into the computer, but we always keep a backup on paper. Let's go."

After taking the x-rays, Jaime helps Ellen at the desk, scheduling a lower-right ultrasound for a gentleman with back pain. "I didn't know ultrasound was used for anything other than pregnancy." Ellen laughs and explains that ultrasonography can be used for most hollow spaces in the body. She explains that the doctor wants an ultrasound image of the patient's kidneys to determine if a CT scan is necessary.

A basic understanding of the human body is necessary not only for the health-care provider, as we will find following Jaime's story, but also for anyone concerned about their own health and well-being.

# An Introduction to the Human Body

## INTRODUCTION

In recent years the explosion of information in science and medicine has led to dramatic advances in the improvement of human health. To understand how your body maintains a healthy balance in all of its systems, you must learn about its structures and how they function. Our exploration of the human body begins with an introduction to the meanings of anatomy (structure) and physiology (function). We will then consider how living things are organized and explore the properties that we share with animals, plants, and even bacteria. The study of human anatomy and physiology will provide you with basic information about the human body and a common vocabulary that will help you speak about the body in a way that is understood by scientists and health-care professionals alike.

The sciences of anatomy and physiology are the foundations for understanding the structure and function of the human body. **Anatomy** (a-NAT-ō-mē; *ana-* = up; *-tomy* = process of cutting) is the science of body structures and the relationships among structures. **Physiology** (fiz′-ē-OL-ō-jē; *physio-* = nature; *-logy* = study of) is the science of body functions—how the body parts work. Because function can never be separated completely from structure, we can understand the human body best by studying anatomy and physiology together. In this book we will look at how each part of the body is designed to carry out particular functions and how the structure of a part often reflects its functions. For example, the bones of the skull join tightly to form a rigid case that protects the brain. The bones of the fingers, by contrast, are more loosely joined to allow a variety of movements.

## CONCEPTS

**1.1** The human body is composed of six levels of structural organization and contains eleven systems.

**1.2** The human body carries on basic life processes that distinguish it from nonliving objects.

**1.3** Homeostasis is controlled through feedback systems.

**1.4** The human body is described using the anatomical position and specific terms.

**1.5** Body cavities are spaces within the body that help protect, separate, and support internal organs.

**1.6** Serous membranes line the walls of body cavities and cover the organs within them.

**1.7** The abdominopelvic cavity is divided into regions or quadrants.

## 1.1 The human body is composed of six levels of structural organization and contains eleven systems.

The structures of the human body are organized on several levels. Our exploration of the human body will extend from some of the smallest body structures and their functions to the largest structure—an entire person. From the smallest to the largest, six levels of organization will help you to understand anatomy and physiology: the chemical, cellular, tissue, organ, system, and organismal levels (**Figure 1.1**).

1. **Chemical level**. This very basic level includes **atoms**, the smallest components of a chemical element that retain the properties of the element, and **molecules**, two or more atoms joined together. Certain atoms, such as carbon (C), hydrogen (H), oxygen (O), nitrogen (N), phosphorus (P), calcium (Ca), and sulfur (S), are essential for life. Two familiar examples of molecules found in the body are

FIGURE 1.1 Levels of structural organization in the human body.

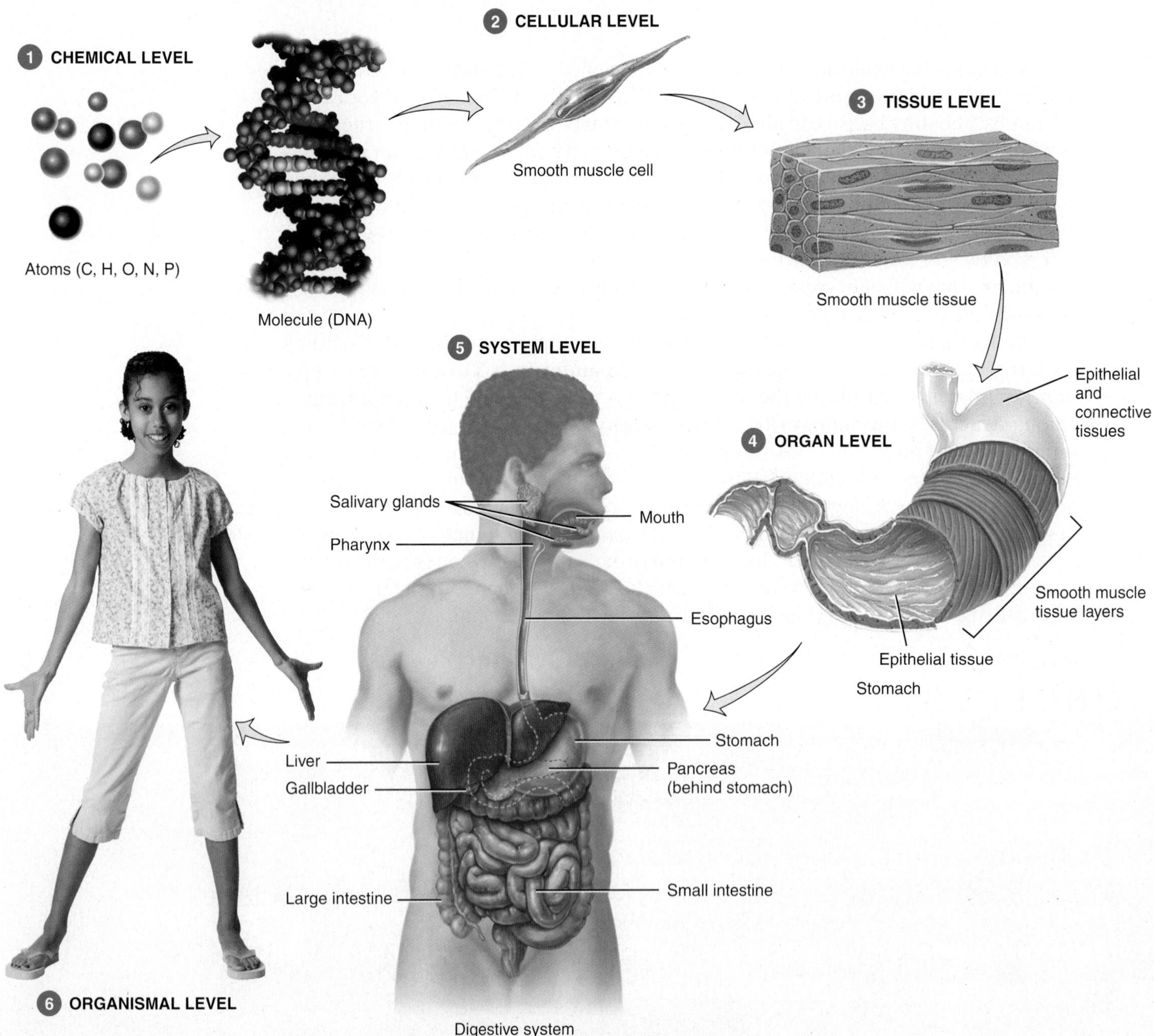


The levels of structural organization are chemical, cellular, tissue, organ, system, and organismal.

deoxyribonucleic acid (DNA), the genetic material passed from one generation to the next, and glucose, a sugar readily used by the body. Chapter 2 focuses on the chemical level of organization.

2. **Cellular level**. Molecules combine to form **cells**, the basic structural and functional units of an organism that are composed of chemicals. Cells are the smallest living units in the human body. Among the many types of cells in your body are muscle cells, nerve cells, and blood cells. The cellular level of organization is the focus of Chapter 3.
3. **Tissue level**. **Tissues** are groups of cells and the materials surrounding them that work together to perform a particular function. There are just four basic types of tissue in your body: *epithelial tissue, connective tissue, muscle tissue,* and *nervous tissue*. Note in Figure 1.1 that smooth muscle tissue consists of tightly packed smooth muscle cells. Chapter 4 describes the tissue level of organization in greater detail.
4. **Organ level**. At the organ level different types of tissues join together to form body structures. **Organs** usually have a recognizable shape, are composed of two or more different types of tissues, and have specific functions. Examples of organs include the skin, bones, stomach, heart, liver, lungs, and brain. Figure 1.1 shows how several tissues make up the stomach. The stomach's outer covering is a layer of epithelial tissue and connective tissue that reduces friction when the stomach moves and rubs against other organs. Underneath are three muscle tissue layers, which contract to churn and mix food and push it into the next digestive organ, the small intestine. The innermost lining of the stomach is an epithelial tissue layer that produces fluid and chemicals responsible for digestion in the stomach.
5. **System level**. A **system** consists of related organs that have a common function. One example of the system level, also called the organ system level, is the digestive system, which breaks down and absorbs food. Its organs include the mouth, salivary glands, pharynx (throat), esophagus, stomach, small intestine, large intestine, liver, gallbladder, and pancreas. Sometimes an organ is part of more than one system. For example, the pancreas is part of both the digestive system and the hormone-producing endocrine system.
6. **Organismal level**. An **organism** (OR-ga-nizm) is any living individual. All parts of the human body functioning together constitute the total organism—a single living person.

In the chapters that follow, you will study the anatomy and physiology of the major body systems. Table 1.1 introduces the components and functions of these systems in the order they are discussed in this book. As you study the body systems, you will discover how they work together to maintain health, protect you from disease, and allow for reproduction of the human species. You will also discover that all body systems influence one another.

As an example, consider how just two body systems—the integumentary system and skeletal system—cooperate to function at the organismal level. The integumentary system, which includes the skin, hair, and nails, protects all other systems, including the skeletal system, which includes all of the bones and joints of the body. The skin serves as a barrier between the outside environment and internal tissues and organs, such as those that make up the skeletal system. The skin also participates in the production of vitamin D, which the body needs to absorb the calcium in foods such as milk. Calcium is used to build bones and teeth. The skeletal system, in turn, provides support for the integumentary system, serves as a reservoir for calcium by storing it in times of plenty and releasing it in times of need, and produces white blood cells that help the skin resist invasion by disease-causing microbes such as bacteria and viruses.

## CLINICAL CONNECTION | *Noninvasive Diagnostic Techniques*

Health-care professionals and students of anatomy and physiology commonly use several noninvasive diagnostic techniques to assess certain aspects of body structure and function. A **noninvasive diagnostic technique** is one that does not involve insertion of an instrument or device through the skin or a body opening. In **inspection**, the examiner observes the body for any changes that deviate from normal. For example, a physician may examine the mouth cavity for evidence of disease. Following inspection, one or more additional techniques may be employed. In **palpation** (pal-PĀ-shun; *palp-* = gently touching) the examiner feels body surfaces with the hands. An example is palpating the abdomen to detect enlarged or tender internal organs or abnormal masses. In **auscultation** (aws-kul-TĀ-shun; *auscult-* = listening) the examiner listens to body sounds to evaluate the functioning of certain organs, often using a stethoscope to amplify the sounds. An example is auscultation of the lungs during breathing to check for crackling sounds associated with abnormal fluid accumulation. In **percussion** (pur-KUSH-un; *percus-* = beat through) the examiner taps on the body surface with the fingertips and listens to the resulting echo. For example, percussion may reveal the abnormal presence of fluid in the lungs or air in the intestines. It may also provide information about the size, consistency, and position of an underlying structure. An understanding of anatomy is important for the effective application of most of these diagnostic techniques.

## TABLE 1.1

The Eleven Systems of the Human Body

**INTEGUMENTARY SYSTEM (CHAPTER 5)**

***Components:*** **Skin** and associated structures, such as **hair**, **fingernails** and **toenails**, **sweat glands**, and **oil glands**.

***Functions:*** Protects body; helps regulate body temperature; eliminates some wastes; helps make vitamin D; detects sensations such as touch, pain, warmth, and cold.

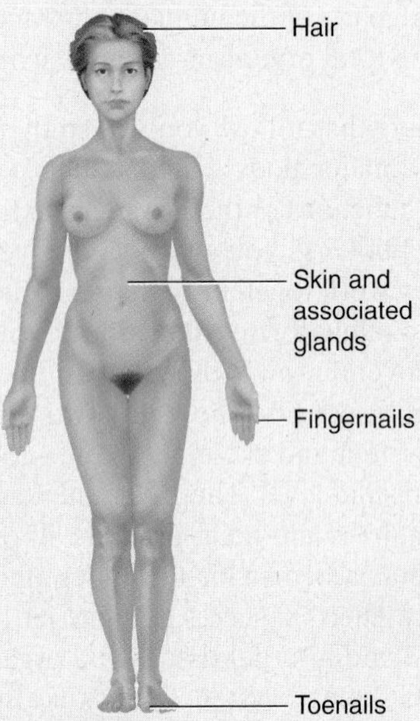


**SKELETAL SYSTEM (CHAPTERS 6–9)**

***Components:*** **Bones** and **joints** of the body and their associated **cartilages**.

***Functions:*** Supports and protects body; provides surface area for muscle attachments; aids body movements; houses cells that produce blood cells; stores minerals and lipids (fats).

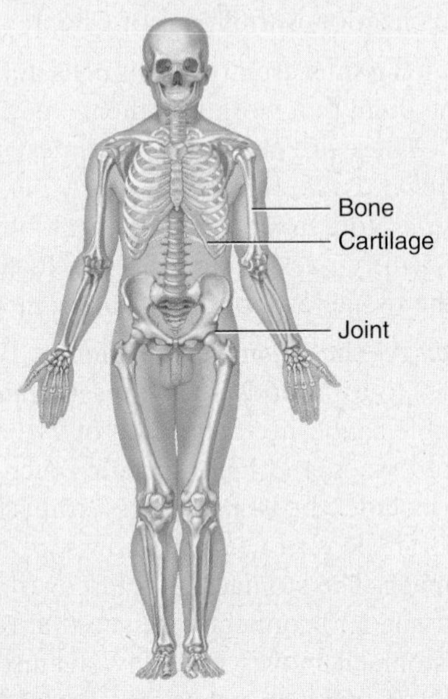


**MUSCULAR SYSTEM (CHAPTERS 10, 11)**

***Components:*** **Skeletal muscles**—voluntarily controlled muscles usually attached to bones (smooth and cardiac muscles are components of other body systems).

***Functions:*** Participates in body movements, such as walking; maintains posture; produces heat.

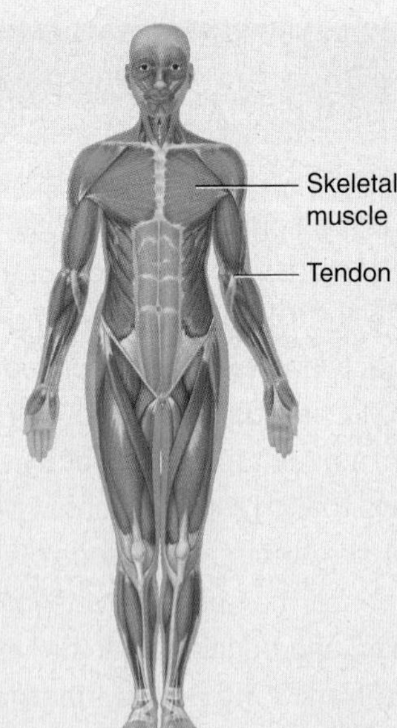


**NERVOUS SYSTEM (CHAPTERS 12–16)**

***Components:*** **Brain**, **spinal cord**, **nerves**, and special sense organs, such as **eyes** and **ears**.

***Functions:*** Generates action potentials (nerve impulses) to regulate body activities; detects changes in body's internal and external environments, interprets changes, and responds by causing muscular contractions or glandular secretions.

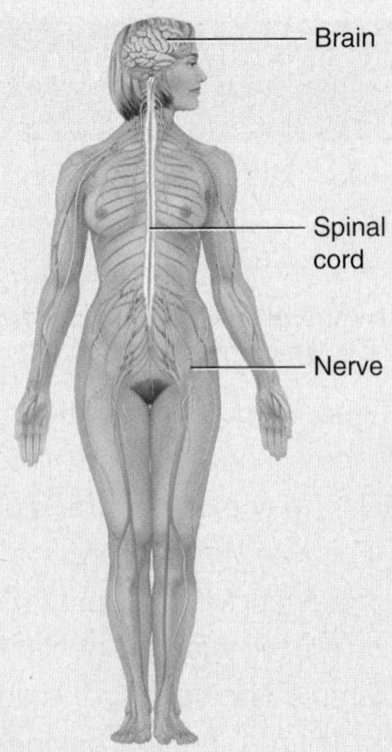

ENDOCRINE SYSTEM (CHAPTER 17)

***Components:*** Hormone-producing glands (**pineal gland**, **hypothalamus**, **pituitary gland**, **thymus**, **thyroid gland**, **parathyroid glands**, **adrenal glands**, **pancreas**, **ovaries**, and **testes**) and hormone-producing cells in several other organs.

***Functions:*** Regulates body activities by releasing hormones (chemical messengers transported in blood from endocrine gland or tissue to target organ).

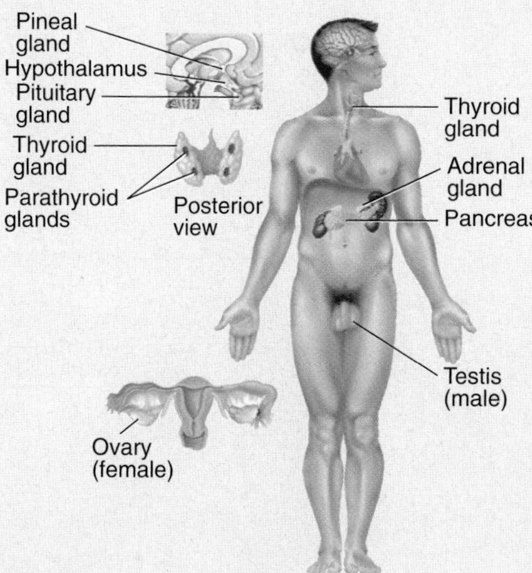


CARDIOVASCULAR SYSTEM (CHAPTERS 18–20)

***Components:*** **Blood**, **heart**, and **blood vessels**.

***Functions:*** Heart pumps blood through blood vessels; blood carries oxygen and nutrients to cells and carbon dioxide and wastes away from cells and helps regulate acid–base balance, temperature, and water content of body fluids; blood components help defend against disease and repair damaged blood vessels.

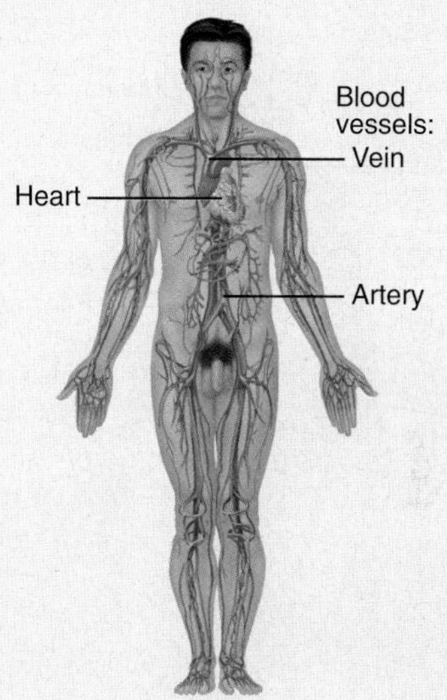


LYMPHATIC SYSTEM AND IMMUNITY (CHAPTER 21)

***Components:*** **Lymphatic fluid** and **lymphatic vessels**; **spleen**, **thymus**, **lymph nodes**, and **tonsils**; cells that carry out immune responses (**B cells**, **T cells**, and others).

***Functions:*** Returns proteins and fluid to blood; carries lipids from gastrointestinal tract to blood; contains sites of maturation and proliferation of B cells and T cells that protect against disease-causing microbes.

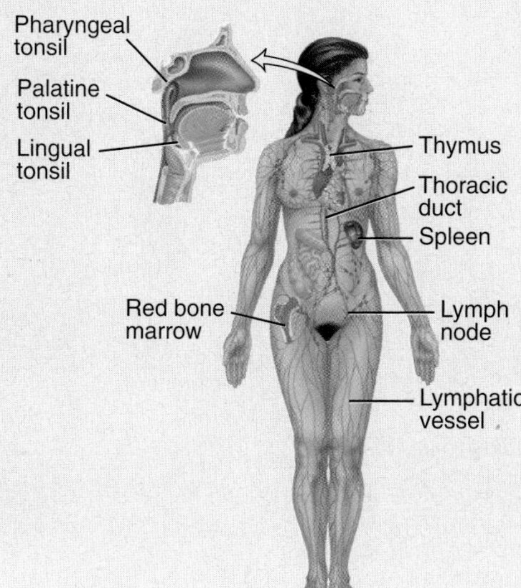


RESPIRATORY SYSTEM (CHAPTER 22)

***Components:*** **Lungs** and air passageways such as the **pharynx** (throat), **larynx** (voice box), **trachea** (windpipe), and **bronchial tubes** leading into and out of lungs.

***Functions:*** Transfers oxygen from inhaled air to blood and carbon dioxide from blood to exhaled air; helps regulate acid–base balance of body fluids; air flowing out of lungs through vocal cords produces sounds.

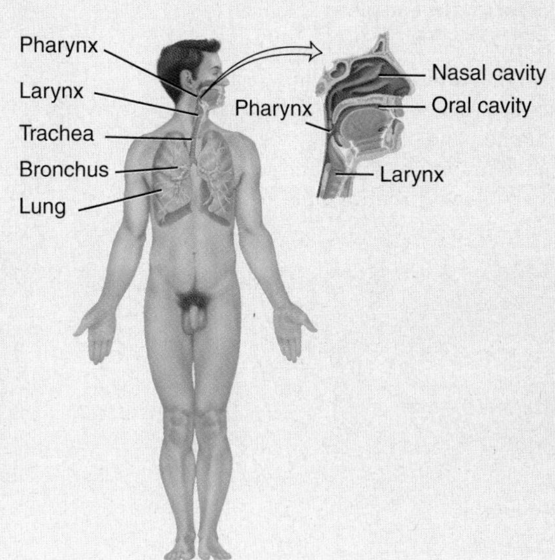


(continues)

**TABLE 1.1 CONTINUED**

The Eleven Systems of the Human Body

**DIGESTIVE SYSTEM (CHAPTER 23)**

***Components:*** Organs of gastrointestinal tract, a long tube that includes the **mouth**, **pharynx** (throat), **esophagus** (food tube), **stomach**, **small** and **large intestines**, and **anus**; also includes accessory organs that assist in digestive processes, such as **salivary glands**, **liver**, **gallbladder**, and **pancreas**.

***Functions:*** Achieves physical and chemical breakdown of food; absorbs nutrients; eliminates solid wastes.

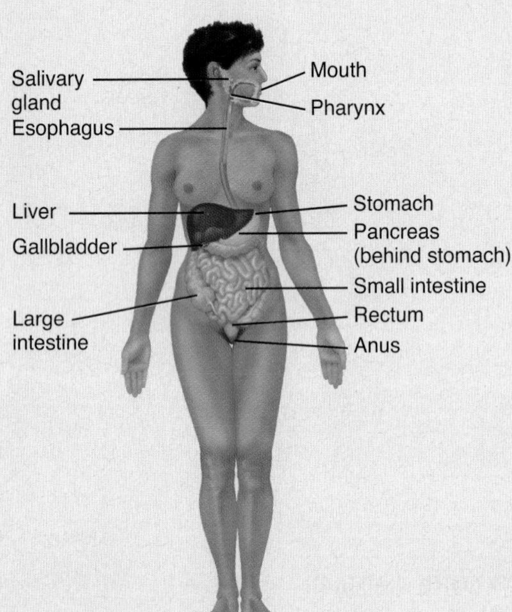


**URINARY SYSTEM (CHAPTER 24)**

***Components:*** **Kidneys**, **ureters**, **urinary bladder**, and **urethra**.

***Functions:*** Produces, stores, and eliminates urine; eliminates wastes and regulates volume and chemical composition of blood; helps maintain the acid–base balance of body fluids; maintains body's mineral balance; helps regulate production of red blood cells.

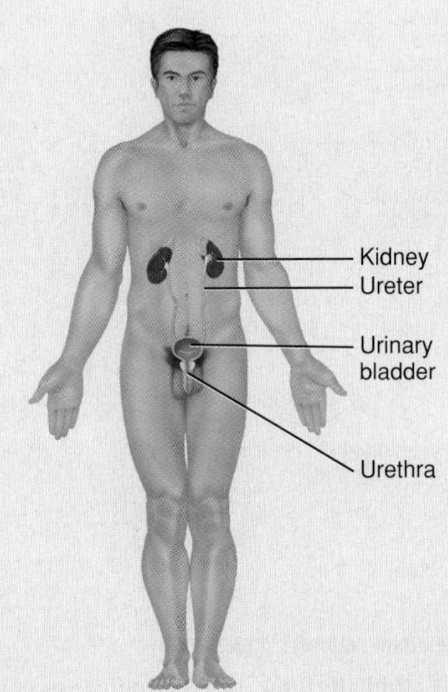


**REPRODUCTIVE SYSTEMS (CHAPTER 25)**

***Components:*** **Gonads** (**testes** in males and **ovaries** in females) and associated organs (**uterine tubes**, **uterus**, **vagina**, and **mammary glands** in females and **epididymides**, **ductus deferens**, **seminal vesicles**, **prostate**, and **penis** in males).

***Functions:*** Gonads produce gametes (sperm or oocytes) that unite to form a new organism; gonads also release hormones that regulate reproduction and other body processes; associated organs transport and store gametes; mammary glands produce milk.

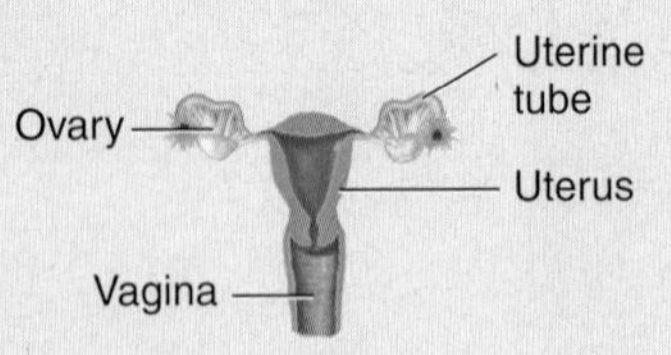


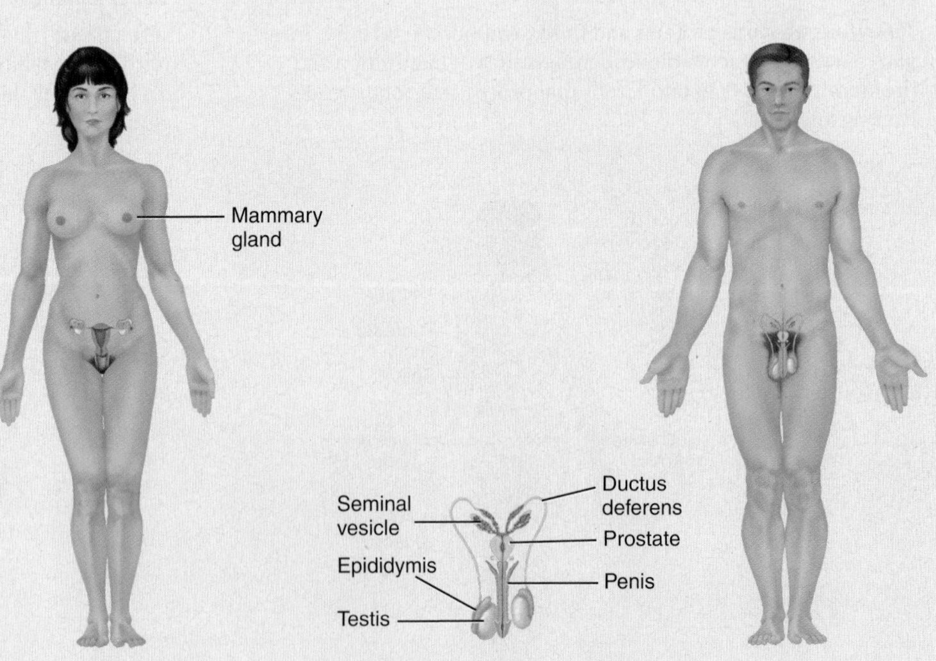

✓ CHECKPOINT

1. Two or more atoms join together to form what? To which level of organization does this belong?
2. Which levels of structural organization are you unable to see with your naked eye?
3. Which level of organization is composed of two or more different types of tissues that work together to perform a specific function?
4. How are organs and tissues different?

## 1.2 The human body carries on basic life processes that distinguish it from nonliving objects.

### Basic Life Processes

All living organisms have certain characteristics that distinguish them from nonliving things. The following are the six most important life processes of the human body:

**Metabolism** (me-TAB-ō-lizm) is the sum of all chemical processes that occur in the body. One phase of metabolism is **catabolism** (ka-TAB-ō-lizm; *catabol-* = throwing down; *-ism* = a condition), the breaking down of complex chemical substances into simpler ones. The other phase of metabolism is **anabolism** (a-NAB-ō-lizm; *anabol-* = a raising up), the building up of complex chemical substances from smaller, simpler ones. For example, catabolism splits proteins in food into amino acids, which can be used during anabolism as the building blocks for new proteins that make up muscles and bones. In metabolic processes, oxygen taken in by the respiratory system and nutrients broken down in the digestive system provide the chemical energy to power cellular activities.

**Responsiveness** is the body's ability to detect and respond to changes in its internal or external environment. An increase in body temperature during a fever would be an example of a change in the internal environment (inside your body). Turning your head toward the sound of squealing brakes is an example of responsiveness to a change in the external environment (outside your body) to prepare for a potential threat. Different cells in the body respond to environmental changes in characteristic ways. Nerve cells respond by generating electrical signals known as nerve impulses. Muscle cells respond to nerve impulses by contracting, which generates force to move body parts.

**Movement** includes motion of the whole body, individual organs, single cells, and even tiny organelles inside cells. For example, the coordinated action of several muscles enables you to move your body from one place to another by walking or running. After you eat a meal that contains fats, your gallbladder contracts and releases bile into the gastrointestinal tract to help in the digestion of fats. When a body tissue is damaged or infected, certain white blood cells move from the bloodstream into the affected tissue to help clean up and repair the area. Even inside individual cells, various cellular parts move from one position to another to carry out their functions.

**Growth** is an increase in body size. It may be due to an increase in (1) the size of existing cells, (2) the number of cells, or (3) the amount of material surrounding cells. In a growing bone, for example, mineral deposits accumulate between bone cells, causing the bone to grow in length and width.

**Differentiation** (dif′-er-en-shē-Ā-shun) is the process in which unspecialized cells become specialized cells. Such unspecialized cells, which can divide and give rise to cells that undergo differentiation, are known as **stem cells**. Each type of cell in the body has a specialized structure and function. Specialized cells differ in structure and function from the unspecialized cells that gave rise to them. For example, red blood cells and several types of white blood cells differentiate from the same unspecialized cells in red bone marrow. Similarly, a single fertilized human egg (ovum) undergoes tremendous differentiation to develop into an embryo, and then into a fetus, an infant, a child, and finally an adult.

**Reproduction** refers to either (1) the formation of new cells for tissue growth, repair, or replacement or (2) the production of a new individual. Cell formation occurs continuously throughout life; life continues from one generation to the next through the production of a new individual through the fertilization of an ovum by a sperm cell.

Although not all of these processes occur all of the time in every cell of the body, when they cease to occur properly, cell death may occur. When cell death is extensive and leads to organ failure, without intervention the result is death of the organism. Clinically, loss of the heartbeat, absence of spontaneous breathing, and loss of brain functions indicate death in the human body.

### Homeostasis

The French physiologist Claude Bernard (1813–1878) first proposed that the cells of many-celled organisms flourish because they live in the relative constancy of *le milieu interieur*—the environment of the interior of the body—despite continual changes in the organisms' external environment. The American physiologist Walter B. Cannon (1871–1945) coined the term *homeostasis* to describe this dynamic constancy. **Homeostasis** (hō′-mē-ō-STĀ-sis; *homeo-* = sameness; *-stasis* = standing still) is the maintenance of relatively stable conditions in the body's internal environment. Through the constant interaction of the body's many regulatory processes, homeostasis ensures that the body's internal environment remains stable despite changes inside and outside the body. Each body structure, from the chemical level to the system level, contributes in some way to keeping the internal environment of the body within normal limits.

Homeostasis is a dynamic process. As conditions change, the body makes adjustments within a narrow range that is compatible with maintaining life. For example, the pancreas normally helps maintain the level of glucose in blood between 70 and 110 milligrams of glucose per 100 milliliters of blood. When blood glucose levels begin to fall, the pancreas secretes the hormone glucagon, which stimulates cells to release glucose into the blood. As blood glucose start to rise, the pancreas releases another hormone, insulin, which stimulates cells to remove glucose from the blood.

## Body Fluids

An important aspect of homeostasis is maintaining the volume and composition of **body fluids**, dilute, watery solutions containing dissolved chemicals that are found inside cells and surrounding them. The fluid within cells is **intracellular fluid** (*intra-* = inside). The fluid outside body cells is **extracellular fluid** (*extra-* = outside). Dissolved in the water of intracellular and extracellular fluids are oxygen, nutrients, proteins, and a variety of *ions* (electrically charged chemical particles). All these substances are needed to maintain life. The extracellular fluid that fills the narrow spaces between cells of tissues is known as **interstitial fluid** (in′-ter-STISH-al; *inter-* = between). As you progress with your studies, you will learn that the extracellular fluid within blood vessels is termed **blood plasma**, that within lymphatic vessels is called **lymph**, that in and around the brain and spinal cord is called **cerebrospinal fluid**, and that in joints is called **synovial fluid**.

As Bernard predicted, the proper functioning of body cells depends on precise regulation of the composition of the interstitial fluid surrounding them. Because interstitial fluid surrounds all body cells, it is often called the body's *internal environment*. The composition of interstitial fluid changes as substances move back and forth between it and blood plasma. Such exchange of materials occurs across the thin walls of the smallest blood vessels in the body, the *blood capillaries*. This movement in both directions across capillary walls provides needed materials, such as glucose, oxygen, ions, and so on, to tissue cells. It also removes wastes, such as carbon dioxide, from interstitial fluid.

### ✓ CHECKPOINT

5. What are some examples of movements that occur inside the human body?
6. Shivering when a cold wind blows over you is an example of what basic life process?
7. Where would you find intracellular fluid, extracellular fluid, and interstitial fluid?
8. A spinal tap is a procedure in which fluid is drawn from the space between the membranes that surround the spinal cord. What type of body fluid is being extracted?

### RETURN TO Jaime's Story

Jaime can't believe how quickly the day has flown by. He thinks about his first day at work as he hurries to his evening anatomy and physiology class. Jaime had helped with several x-rays of different parts of the body, including the chest of a suspected tuberculosis case, a sprained ankle of an elderly woman, and the abdomen of a child who had swallowed a coin. Ellen had shown him how to develop the x-ray film cassettes, and he had tagged along as she transported patients. She had explained most of the terms that she was using and quizzed him to see which he knew and which he didn't.

Jaime finds Amy and Lindsey waiting for him outside the lecture hall. "So, save any lives today?" Amy asks as they settle into their seats in the classroom.

"No, not exactly. But, it's amazing. I think I'm really going to like working there. The x-ray techs are all nice, and Dr. Sebbo, the radiologist, bought everyone lunch today. He was happy to answer a bunch of questions for me while he took some x-rays and pointed out a lot of amazing stuff. He even showed me some films from last week, including a gallbladder packed with stones and an inflamed kidney full of urine."

Lindsey makes a face and says, "Ugh, nasty. Did they figure out what was wrong with the patient?"

"He had an enlarged prostate gland that was blocking release of the urine in his urinary bladder. It was backing up into the kidney and causing a lot of inflammation and swelling. Dr. Sebbo showed me the healthy kidney versus the inflamed one and explained how the increased pressure and concentration of the urine could damage the kidney if they didn't do something about it. They used a catheter to relieve pressure from the urinary bladder. It was incredible. I can't wait to go back tomorrow."

A. *Jaime is helping move patients and develop x-rays. What levels of structural organization is he managing?*

B. *The gallbladder and kidney described by Jaime are found in which organ system(s)?*

C. *Would the urine in the patient's inflamed kidney be considered an intracellular or extracellular fluid?*

D. *The prostate patient suffered from decreased fluid movement through the kidney. Which basic life processes would be affected by this condition?*

# 1.3 Homeostasis is controlled through feedback systems.

Homeostasis in the human body is continually being disturbed. Some disruptions come from the external environment in the form of physical insults such as the intense heat of a Texas summer or a lack of enough oxygen for that two-mile run. Other disruptions originate in the internal environment, such as a blood glucose level that falls too low when you skip breakfast. Homeostatic imbalances may also occur due to psychological stresses in your social environment—the demands of work and school, for example. In most cases the disruption of homeostasis is mild and temporary, and the body responds quickly to restore balance in the internal environment. In some cases the disruption of homeostasis may be intense and prolonged, as in poisoning, overexposure to temperature extremes, severe infection, or major surgery. Under these circumstances, regulation of homeostasis may fail.

Fortunately, the body has many regulating systems that can usually bring the internal environment back into balance. Most often, the nervous system and the endocrine system, working together or independently, provide the needed corrective measures. The nervous system regulates homeostasis by sending electrical signals known as *nerve impulses* to organs that can counteract changes from the balanced state. The endocrine system includes many glands that secrete messenger molecules called *hormones* into the blood. Nerve impulses typically cause rapid changes, but hormones usually work more slowly. However, both means of regulation work toward the maintenance of homeostasis using processes you will learn about next.

## Feedback Systems

The body can regulate its internal environment through many feedback systems. A **feedback system** is a cycle of events in which the status of a body condition is monitored, evaluated, changed, remonitored, reevaluated, and so on. Each monitored variable, such as body temperature, blood pressure, or blood glucose level, is termed a *controlled condition*. Any disruption that changes a controlled condition is called a *stimulus*. Three basic components make up a feedback system—a receptor, a control center, and an effector (Figure 1.2):

1. A **receptor** is a body structure that monitors changes in a controlled condition and sends input to a control center. Typically, the input is in the form of nerve impulses or chemical signals. Nerve endings in the skin that sense temperature are one of hundreds of different kinds of receptors in the body.
2. A **control center** in the body sets the range of values within which a controlled condition should be maintained, evaluates the input it receives from receptors, and generates output commands when they are needed. Output typically occurs as nerve impulses or chemical signals. In our skin temperature example, the brain acts as the control center, receiving nerve impulses from the skin receptors and generating nerve impulses as output.
3. An **effector** is a body structure that receives output from the control center and produces a *response* or effect that changes the controlled condition. Nearly every organ or tissue in the body can behave as an effector. When your body temperature drops sharply, your brain (control center) sends nerve impulses to your skeletal muscles (effectors) that cause you to shiver, which generates heat and raises your body temperature.

FIGURE 1.2 **Operation of a feedback system.** The dashed return arrow symbolizes negative feedback.

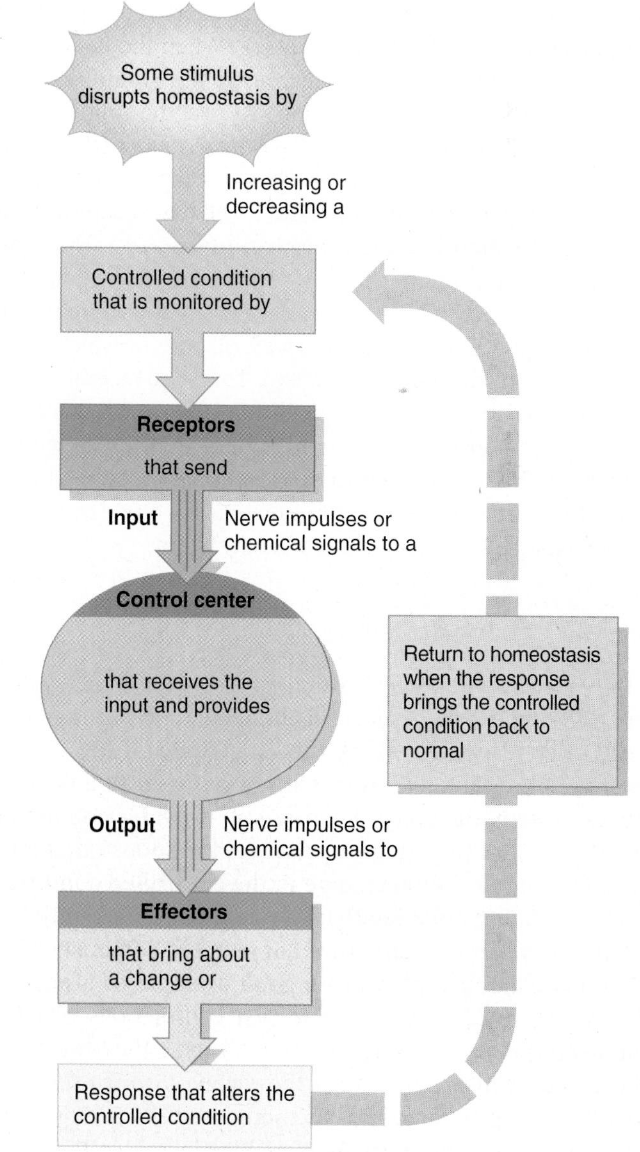


The three basic components of a feedback system are the receptors, a control center, and effectors.

A group of receptors and effectors communicating with their control center forms a feedback system that can regulate a controlled condition in the body's internal environment. In a feedback system, the response of the system "feeds back" information to change the controlled condition in some way, either inhibiting it (negative feedback) or enhancing it (positive feedback).

### Negative Feedback Systems

A **negative feedback system** *reverses* a change in a controlled condition (negates the change). First, a stimulus disrupts homeostasis by altering the controlled condition. The receptors that are part of the feedback system detect the change and send input to a control center. The control center evaluates the input and, if necessary, issues output commands to an effector. The effector produces a physiological response that is able to return the controlled condition to its normal state.

Consider one negative feedback system that helps regulate blood pressure. *Blood pressure* is the force exerted by *blood* as it presses against the walls of blood vessels. When the heart beats faster or harder, blood pressure increases. If a stimulus causes blood pressure (controlled condition) to rise, the following sequence of events occurs (Figure 1.3). *Baroreceptors* (the receptors), pressure-sensitive nerve cells located in the walls of certain blood vessels, detect the higher pressure. The baroreceptors send nerve impulses (input) to the brain (control center). The brain interprets the impulses and responds by sending nerve impulses (output) to the heart and blood vessels (the effectors). Heart rate decreases and blood vessels dilate (widen), both of which cause blood pressure to decrease (response). This sequence of events quickly returns the controlled condition—blood pressure—to normal, and homeostasis is restored. Notice that the activity of the effector produces a result (a drop in blood pressure) that is opposite to the original stimulus (an increase in blood pressure). This is why it is called a negative feedback system.

**FIGURE 1.3 Homeostatic regulation of blood pressure by a negative feedback system.** Note that the response is fed back into the system, and the system continues to lower blood pressure until there is a return to normal blood pressure (homeostasis).

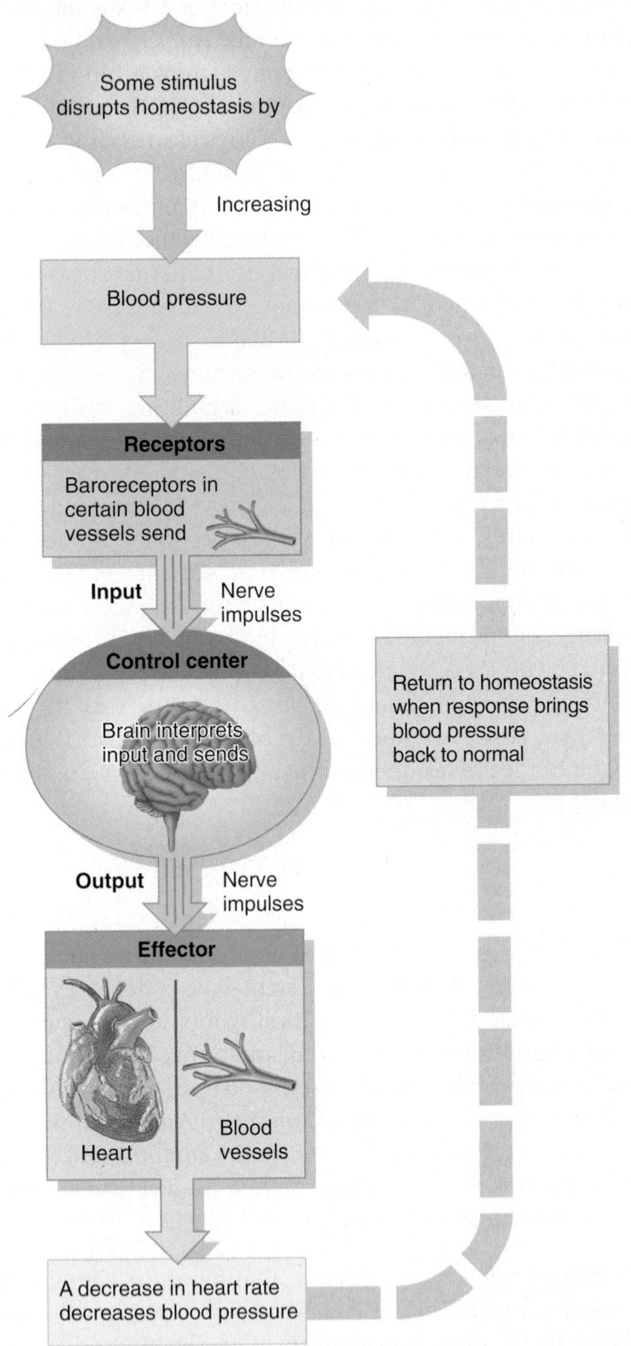
Some stimulus disrupts homeostasis by
Increasing
Blood pressure
Receptors
Baroreceptors in certain blood vessels send
Input
Nerve impulses
Control center
Brain interprets input and sends
Output
Nerve impulses
Effector
Heart
Blood vessels
A decrease in heart rate decreases blood pressure
Return to homeostasis when response brings blood pressure back to normal

**If the response reverses the stimulus, a system is operating by negative feedback.**

### Positive Feedback Systems

Unlike a negative feedback system, a **positive feedback system** tends to *strengthen* or *reinforce* a change in one of the body's controlled conditions, making the magnitude of the change larger or, in mathematical terms, more positive. In a positive feedback system, the control center still provides commands to an effector, but the effector produces a response that adds to or *reinforces* the initial change in the controlled condition. The action of a positive feedback system continues until it is interrupted by some mechanism that restores homeostasis.

Normal childbirth provides a good example of a positive feedback system (Figure 1.4). The first contractions of labor (stimulus) push part of the fetus into the cervix, the lowest part of the uterus, which opens into the vagina. Stretch-sensitive nerve cells (receptors) monitor the amount of stretching of the cervix (controlled condition). As stretching increases, they send more nerve impulses (input) to the brain (control center), which in turn releases the hormone oxytocin (output) into the blood. Oxytocin causes muscles in the wall of the uterus (effector) to contract even more forcefully. The contractions push the fetus farther down the uterus, which stretches the cervix even more. The cycle of stretching, hormone release, and ever-stronger

FIGURE 1.4 **Positive feedback control of labor contractions during birth of a baby.** The solid return arrow symbolizes positive feedback.

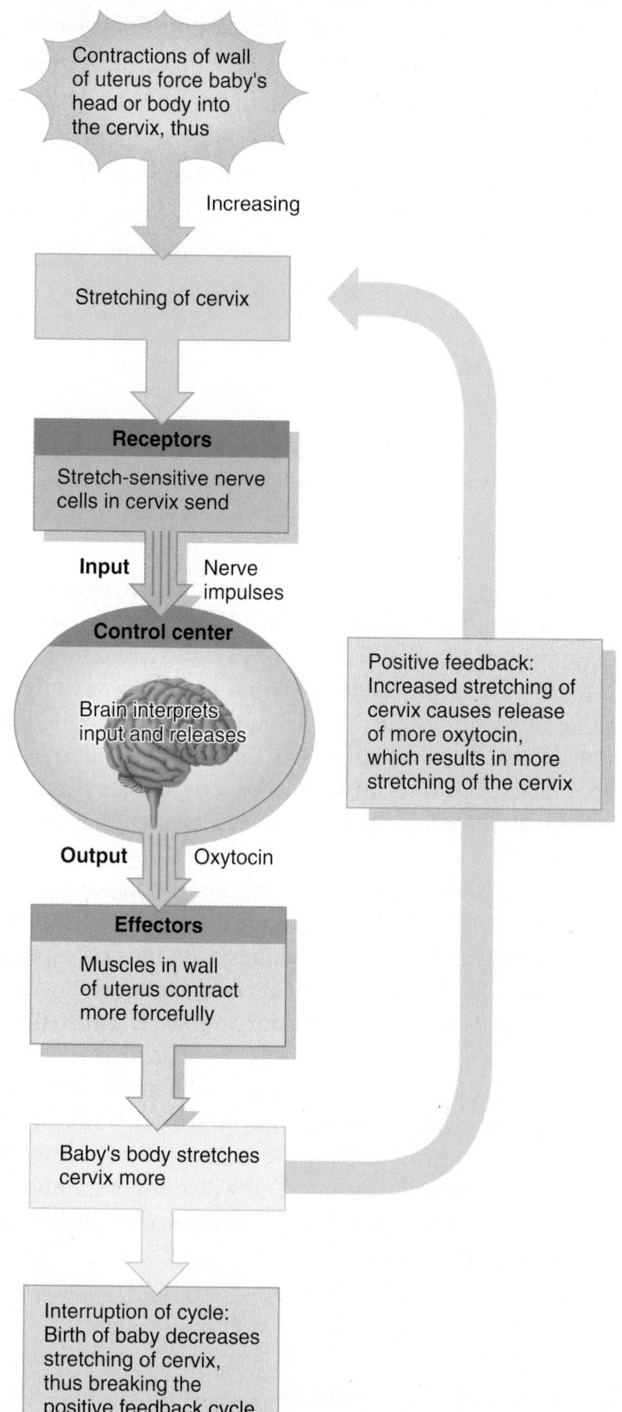


If the response enhances or intensifies the stimulus, a system is operating by positive feedback.

contractions is interrupted only by the birth of the baby. Then stretching of the cervix ceases and oxytocin is no longer released.

Because a positive feedback system continually reinforces a change in a controlled condition, some event outside the system must shut it off. If the action of a positive feedback system is not stopped, it can "run away," and may even produce life-threatening conditions in the body. The action of a negative feedback system, by contrast, slows and then stops as the controlled condition returns to its normal state. Usually, positive feedback systems reinforce conditions that do not happen very often, and negative feedback systems regulate conditions in the body that remain fairly stable over long periods.

## Homeostatic Imbalances

As long as all of the body's controlled conditions remain within certain narrow limits, body cells function efficiently, negative feedback systems maintain homeostasis, and the body stays healthy. Should one or more components of the body lose their ability to contribute to homeostasis, however, the normal balance among all of the body's processes may be disturbed. If the homeostatic imbalance is moderate, a disorder or disease may occur; if it is severe, death may result.

A **disorder** is any abnormality of structure or function. **Disease** is a more specific term for an illness characterized by a recognizable set of signs and symptoms. A *local disease* affects one part or a limited region of the body (for example, a sinus infection); a *systemic disease* affects either the entire body or several of its systems (for example, influenza). Diseases alter body structures and functions in characteristic ways, usually producing a recognizable cluster of signs and symptoms. **Signs** are objective changes that a clinician can observe and measure. Signs of disease can be either anatomical, such as swelling or a rash, or physiological, such as fever, high blood pressure, or paralysis. **Symptoms** are subjective changes in body functions that are not apparent to an observer, such as headache, nausea, and anxiety.

### ✓ CHECKPOINT

9. Which two body systems are largely responsible for maintaining homeostasis?
10. What is the name for the body structure that responds to the control center signal in a feedback system?
11. The response to a stimulus in either a negative or positive feedback system is initiated by what body systems?
12. How are negative and positive feedback systems similar? How are they different? Which one bears primary responsibility for maintaining homeostasis?
13. Why do positive feedback systems that are part of a normal physiological response include a mechanism to stop them?

## CLINICAL CONNECTION | *Diagnosis of Disease*

**Diagnosis** (dī′-ag-NŌ-sis; *dia-* = *through;* -gnosis = knowledge) is the science and skill of distinguishing one disorder or disease from another. The patient's symptoms and signs, his or her medical history, a physical exam, and laboratory tests provide the basis for making a diagnosis. Taking a *medical history* consists of collecting information about events that might be related to a patient's illness. These include the chief complaint (primary reason for seeking medical attention), history of present illness, past medical problems, family medical problems, social history, and review of symptoms. A *physical examination* is an orderly evaluation of the body and its functions. This process includes the noninvasive techniques of inspection, palpation, auscultation, and percussion that you learned about earlier in the chapter, along with measurement of vital signs (temperature, pulse, respiratory rate, and blood pressure), and sometimes laboratory tests.

# 1.4 The human body is described using the anatomical position and specific terms.

Scientists and health-care professionals use a common language of special terms when referring to body structures and their functions. The language of anatomy they use has precisely defined terms that allow us to communicate clearly. For example, is it correct to say, "The wrist is above the fingers"? This might be true if your upper limbs (described shortly) are at your sides. But if you hold your hands up above your head, your fingers would be above your wrists. To prevent this kind of confusion, anatomists use a standard anatomical position and a specific vocabulary for relating body parts to one another.

## Body Positions

Descriptions of any part of the human body assume that the body is in a standard position of reference called the **anatomical position** (an′-a-TOM-i-kul). In the anatomical position, the subject stands erect facing the observer, with the head level and the eyes facing forward. The feet are flat on the floor and directed forward, and the upper limbs are at the sides with the palms turned forward (Figure 1.5). In the anatomical position, the body is upright. Two terms describe a reclining body: face down, it is in the **prone** position; face up, it is in the **supine** position.

## Regional Names

The human body is divided into several major regions that can be identified externally. These are the head, neck, trunk, upper limbs, and lower limbs (Figure 1.5). The **head** consists of the skull and face. The **skull** encloses and protects the brain, and the **face** is the front portion of the head that includes the eyes, nose, mouth, forehead, cheeks, and chin. The **neck** supports the head and attaches it to the trunk. The **trunk** consists of the chest, abdomen, and pelvis. Each **upper limb** is attached to the trunk and consists of the shoulder, armpit, arm (portion of the limb from the shoulder to the elbow), forearm (portion of the limb from the elbow to the wrist), wrist, and hand. Each **lower limb** is also attached to the trunk and consists of the buttock, thigh (portion of the limb from the buttock to the knee), leg (portion of the limb from the knee to the ankle), ankle, and foot. The *groin* is the area on the front surface of the body marked by a crease on each side, where the trunk attaches to the thighs.

Figure 1.5 shows the anatomical and common names of major parts of the body. The anatomical term for each part appears first and is followed in parentheses by the corresponding common name. For example, if you receive a tetanus shot in your *gluteal region*, it is an injection in your *buttock*. Because the anatomical term for a body part usually is based on a Greek or Latin word, it may look different from the common name for the same part or area. For example, the Latin word for *axilla* (ak-SIL-a) is the common name armpit. Thus, the axillary nerve is one of the nerves passing within the armpit. You will learn more about the Greek and Latin word roots of anatomical and physiological terms as you read this book.

## Directional Terms

To locate various body structures, anatomists use specific **directional terms**, words that describe the position of one body part relative to another. Several directional terms are grouped in pairs that have opposite meanings, such as anterior (front) and posterior (back). It is important to understand that directional terms have relative meanings; they only make sense when used to describe the position of one structure in relation to another. For example, your knee is superior to your ankle, even though both are located in the inferior half of the body.

FIGURE 1.5 **The anatomical position.** The anatomical names and corresponding common terms (in parentheses) are indicated for specific body regions. For example, the cephalic region is the head.

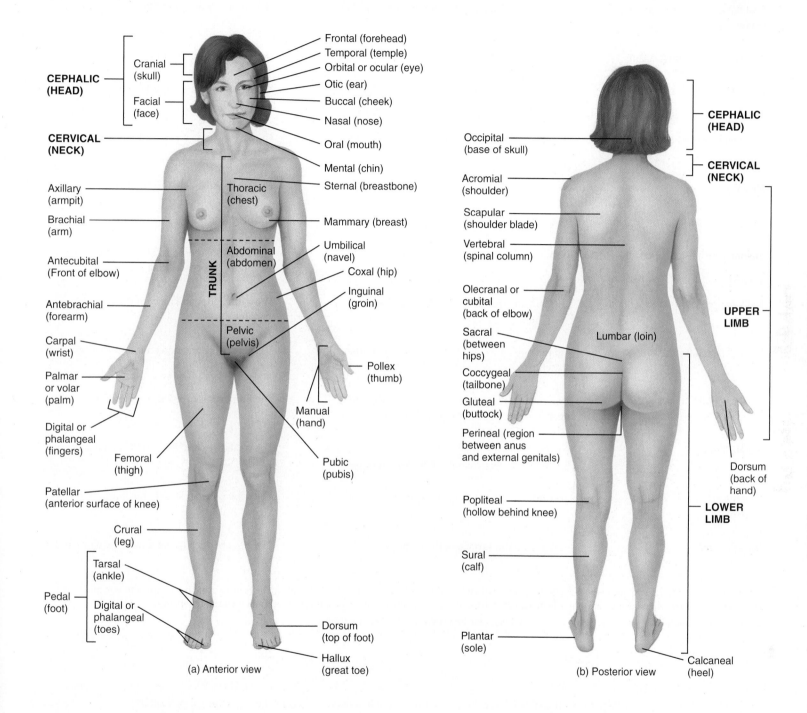



In the anatomical position, the subject stands erect facing the observer with the head level and the eyes facing forward. The feet are flat on the floor and directed forward, and the upper limbs are at the sides with the palms facing forward.

**TABLE 1.2**

Directional Terms

| DIRECTIONAL TERM | DEFINITION | EXAMPLE OF USE |
|---|---|---|
| **Superior** (soo′-PĒR-ē-or) **(cephalic or cranial)** | Toward the head, or the upper part of a structure | The heart is superior to the liver. |
| **Inferior** (in′-FĒR-ē-or) **(caudal)** | Away from the head, or the lower part of a structure | The stomach is inferior to the lungs. |
| **Anterior** (an′-TĒR-ē-or) **(ventral)*** | Nearer to or at the front of the body | The sternum (breastbone) is anterior to the heart. |
| **Posterior** (pos′-TĒR-ē-or) **(dorsal)** | Nearer to or at the back of the body | The esophagus is posterior to the trachea (windpipe). |
| **Medial** (MĒ-dē-al) | Nearer to the midline† | The ulna is medial to the radius. |
| **Lateral** (LAT-er-al) | Farther from the midline | The lungs are lateral to the heart. |
| **Intermediate** (in′-ter-MĒ-dē-at) | Between two structures | The transverse colon is intermediate to the ascending and descending colons. |
| **Ipsilateral** (ip-si-LAT-er-al) | On the same side of the body as another structure | The gallbladder and ascending colon are ipsilateral. |
| **Contralateral** (CON-tra-lat-er-al) | On the opposite side of the body from another structure | The ascending and descending colons are contralateral. |
| **Proximal** (PROK-si-mal) | Nearer to the attachment of a limb to the trunk; nearer to the origination of a structure | The humerus is proximal to the radius. |
| **Distal** (DIS-tal) | Farther from the attachment of a limb to the trunk; farther from the origination of a structure | The phalanges are distal to the carpals. |
| **Superficial** (soo′-per-FISH-al) **(external)** | Toward or on the surface of the body | The ribs are superficial to the lungs. |
| **Deep (internal)** | Away from the surface of the body | The ribs are deep to the skin of the chest and back. |

*The terms *anterior* and *ventral* mean the same thing in humans. However, in four-legged animals *ventral* refers to the belly side and is therefore *inferior*. Similarly, the terms *posterior* and *dorsal* mean the same thing in humans, but in four-legged animals *dorsal* refers to the back side and is therefore *superior*.

†Recall that the **midline** is an imaginary vertical line that divides the body into equal right and left sides.

Table 1.2 and Figure 1.6 present the main directional terms. Study the directional terms in Table 1.2 and the example of how each is used. As you read the examples, look at Figure 1.6 to see the location of each structure.

FIGURE 1.6 Directional terms.

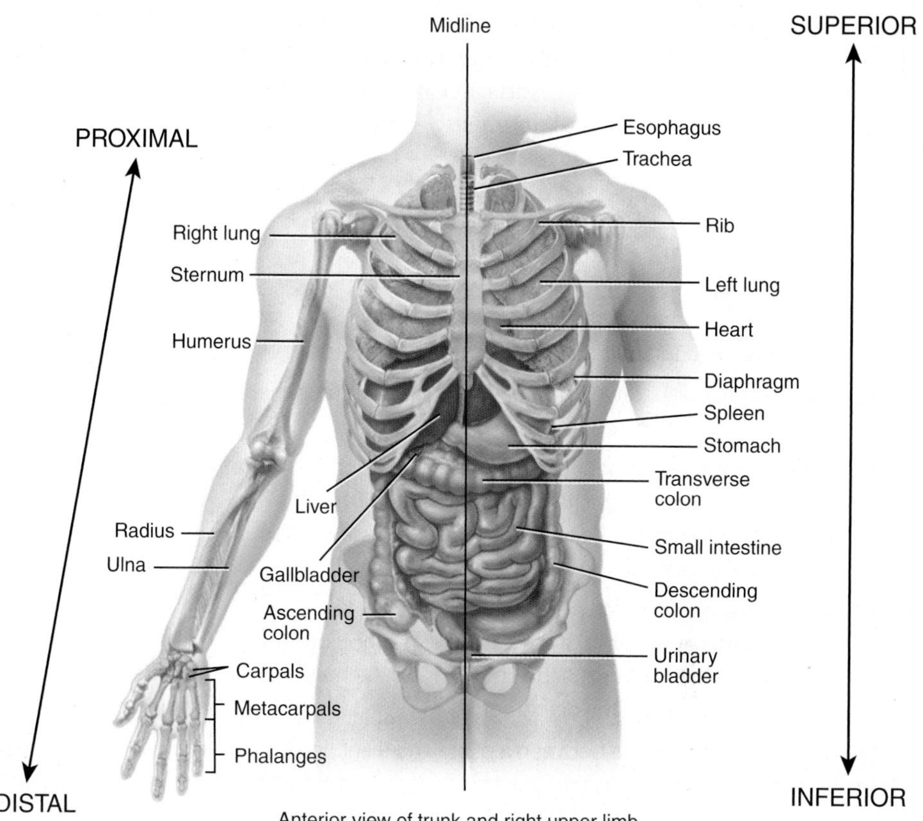


Anterior view of trunk and right upper limb

**Directional terms precisely locate various parts of the body relative to one another.**

## Planes and Sections

You will also study parts of the body relative to **planes**—imaginary flat surfaces that pass through body parts (Figure 1.7). A **sagittal plane** (SAJ-i-tal; *sagitt-* = arrow) is a vertical plane that divides the body or organ into right and left sides. More specifically, when such a plane passes through the midline of the body or an organ and divides it into *equal* right and left sides, it is called a **midsagittal (median) plane**. The *midline* is an imaginary vertical line that divides the body into equal left and right sides. If the sagittal plane does not pass through the midline but instead divides the body or an organ into *unequal* right and left sides, it is called a **parasagittal plane** (*para-* = near). A **frontal (coronal) plane** (kō-RŌ-nal; *corona* = crown) divides the body or organ into anterior and posterior portions. A **transverse (horizontal) plane** divides the body or organ into superior and inferior portions. Sagittal, frontal, and transverse planes are all at right angles to one another. An **oblique plane**, by contrast, passes through the body or an organ at an oblique angle (any angle other than a 90° angle).

FIGURE 1.7 Planes through the human body.

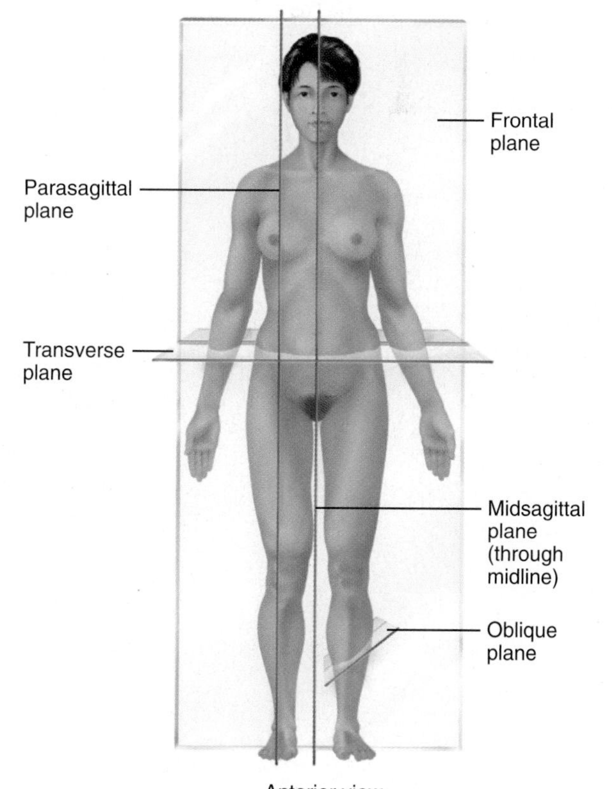


Anterior view

**Frontal, transverse, sagittal, and oblique planes divide the body in specific ways.**

When you study a body region, you often view it in section. A **section** is a cut of the body or one of its organs made along one of the planes just described. It is important to know the plane of the section so you can understand the anatomical relationship of one part to another. Figure 1.8 indicates how three different sections—*transverse, frontal*, and *midsagittal*—provide different views of the brain.

**✓ CHECKPOINT**

**14.** Describe the anatomical position in your own words.

**15.** Blood is typically drawn from the antecubital space. Where is this?

**16.** Is the esophagus anterior or posterior to the trachea? Is the urinary bladder medial or lateral to the ascending colon? Is the radius proximal or distal to the humerus? Are the ribs superficial or deep to the lungs?

**17.** Which directional terms can be used to specify the relationships between (1) the elbow and the shoulder, (2) the left and right shoulders, (3) the sternum and the humerus, and (4) the heart and the diaphragm?

**18.** What is the difference between a plane and a section?

**FIGURE 1.8 Planes and sections through different parts of the brain.** The diagrams (left) show the planes, and the photographs (right) show the resulting sections. Note: The arrows in the diagrams indicate the direction from which each section is viewed. This aid is used throughout the book to indicate viewing perspectives.

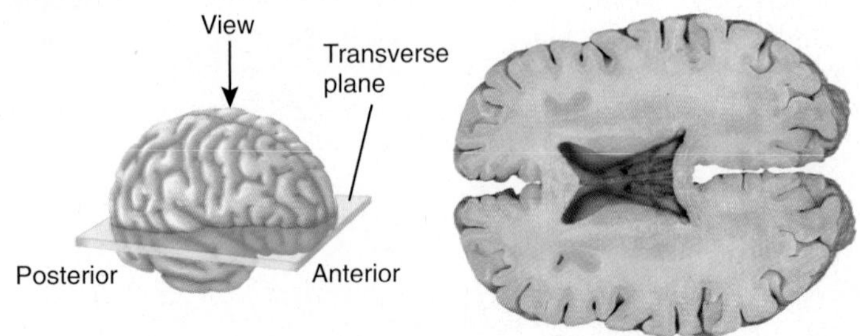


(a)

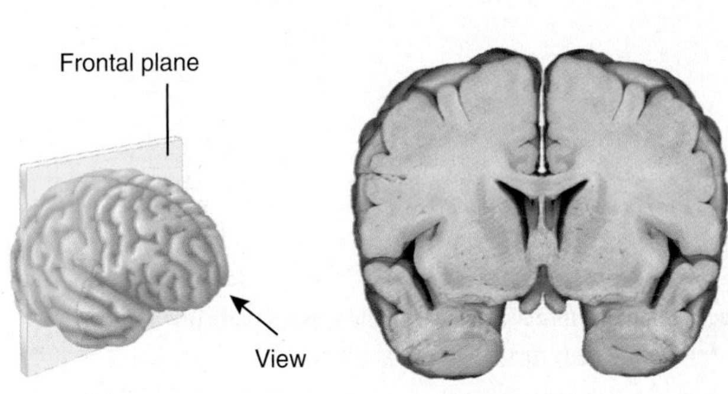


(b)

Midsagittal plane

View

Midsagittal section

(c)

Planes divide the body in various ways to produce sections.

# 1.5 Body cavities are spaces within the body that help protect, separate, and support internal organs.

You have spaces within your body called **body cavities** that protect and support internal organs and isolate them from one another. Bones, muscles, and ligaments separate the various body cavities. Here we discuss several of the larger body cavities (Figure 1.9).

The **cranial cavity** (KRĀ-nē-al) is formed by the cranial bones and contains the brain. The **vertebral canal** (VERT-te-bral) is formed by the bones of the vertebral column (backbone), which contains the spinal cord. The cranial cavity and vertebral canal are continuous with one another.

The major body cavities of the trunk are the thoracic and abdominopelvic cavities. The **thoracic cavity** (thor-AS-ik; *thorac-* = chest), or chest cavity, is encircled by the ribs, the muscles of the chest, the sternum (breastbone), and the thoracic portion of the vertebral column (Figure 1.10). Within the thoracic cavity are the **pericardial cavity** (per′-i-KAR-dē-al; *peri-* = around; *-cardial* = heart) containing the heart and two **pleural cavities** (PLOOR-al; *pleur-* = rib or side), each containing one lung. The central portion of the thoracic cavity is an anatomical region called the **mediastinum** (mē′-dē-as-TĪ-num; *media-* = middle; *-stinum* = partition). It is located between the lungs and extends from the sternum to the vertebral column, and from the first rib to the diaphragm. The mediastinum contains all thoracic organs except the lungs themselves. Among the structures in the mediastinum are the heart, esophagus, trachea, thymus, and several large blood vessels that enter and exit the heart. The **diaphragm** (DĪ-a-fram = partition or wall) is a dome-shaped muscle that separates the thoracic cavity from the abdominopelvic cavity.

The **abdominopelvic cavity** (ab-dom′-i-nō-PEL-vik) extends from the diaphragm to the pelvic floor and is enclosed by the abdominal wall and the bones and muscles of the pelvis

FIGURE 1.9 **Body cavities.** The dashed black lines indicate the border between the abdominal and pelvic cavities.

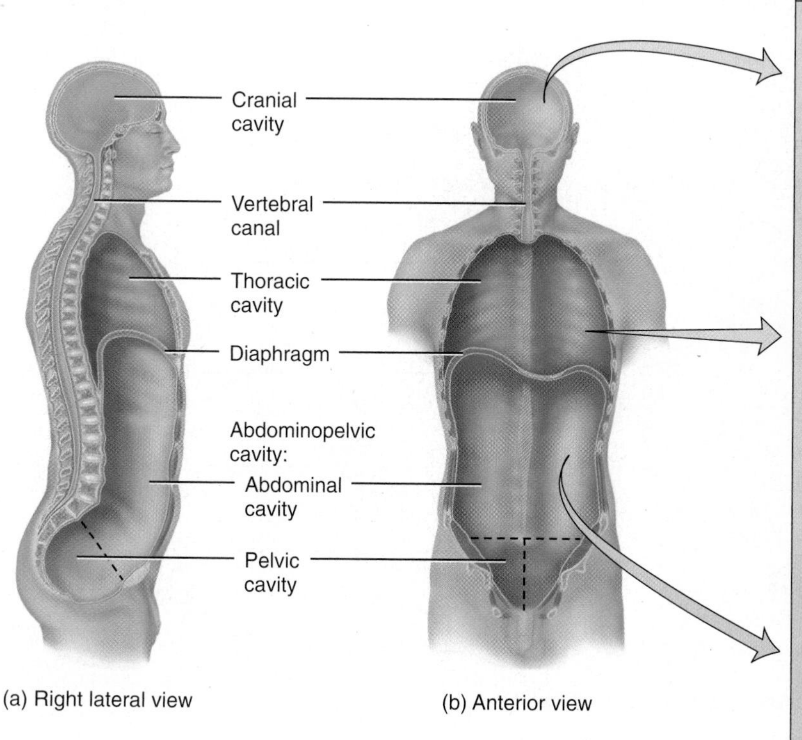


| CAVITY | COMMENTS |
|---|---|
| **Cranial cavity** | Formed by cranial bones and contains brain. |
| **Vertebral canal** | Formed by vertebral column and contains spinal cord and the beginnings of spinal nerves. |
| **Thoracic cavity*** | Chest cavity; contains pleural and pericardial cavities and mediastinum. |
| *Pleural cavity* | Each surrounds a lung; the serous membrane of each pleural cavity is the pleura. |
| *Pericardial cavity* | Surrounds the heart; the serous membrane of the pericardial cavity is the pericardium. |
| *Mediastinum* | Central portion of thoracic cavity between the lungs; extends from sternum to vertebral column and from first rib to diaphragm; contains heart, thymus, esophagus, trachea, and several large blood vessels. |
| **Abdominopelvic cavity** | Subdivided into abdominal and pelvic cavities. |
| *Abdominal cavity* | Contains stomach, spleen, liver, gallbladder, small intestine, and most of large intestine; the serous membrane of the abdominal cavity is the peritoneum. |
| *Pelvic cavity* | Contains urinary bladder, portions of large intestine, and internal organs of reproduction. |

* See Figure 1.10 for details of the thoracic cavity.

**The major cavities of the trunk are the thoracic and abdominopelvic cavities.**

**FIGURE 1.10 The thoracic cavity.** The dashed lines indicate the borders of the mediastinum. Note: When transverse sections are viewed inferiorly (from below), the anterior aspect of the body appears on top and the left side of the body appears on the right side of the illustration.

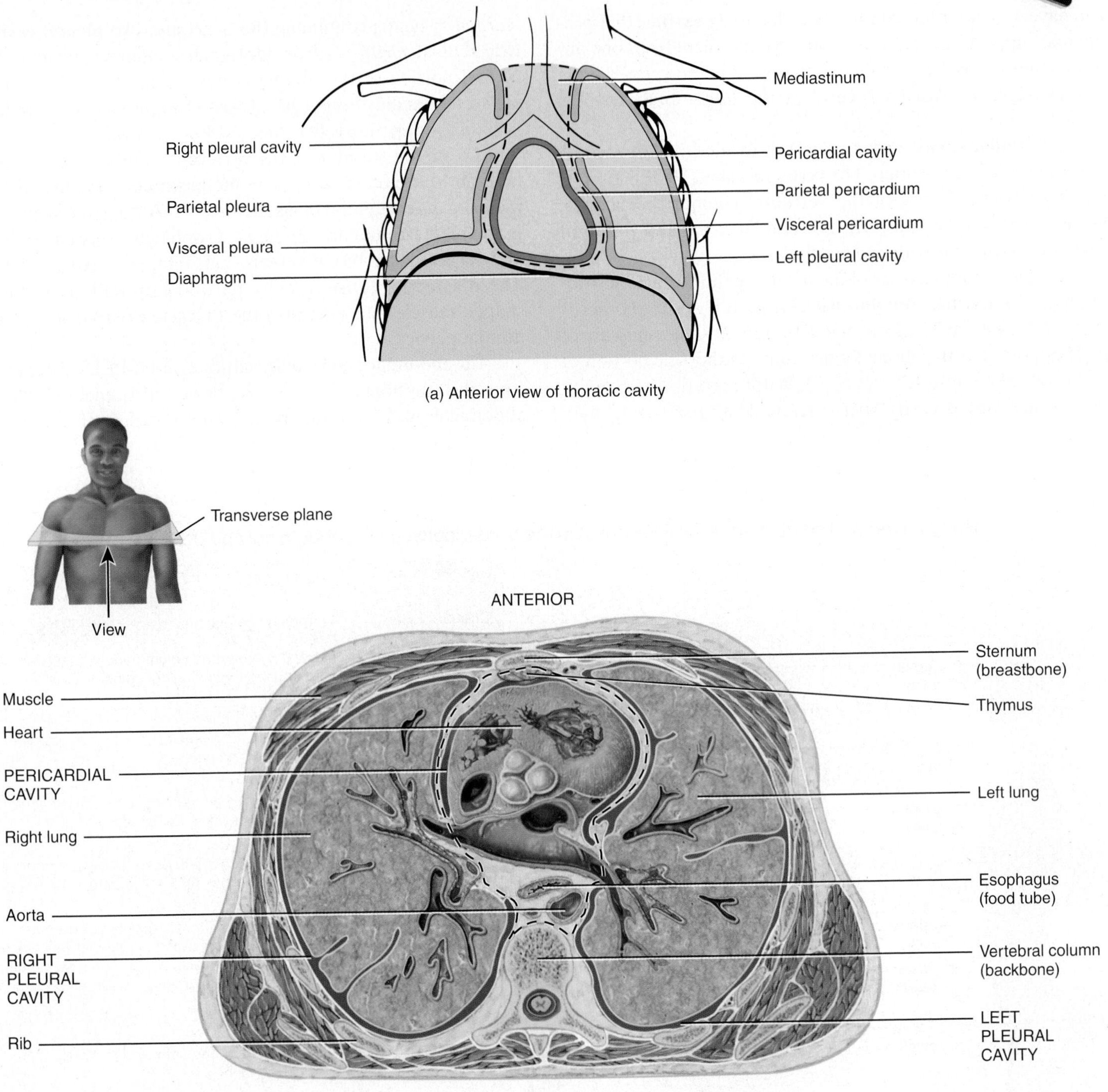

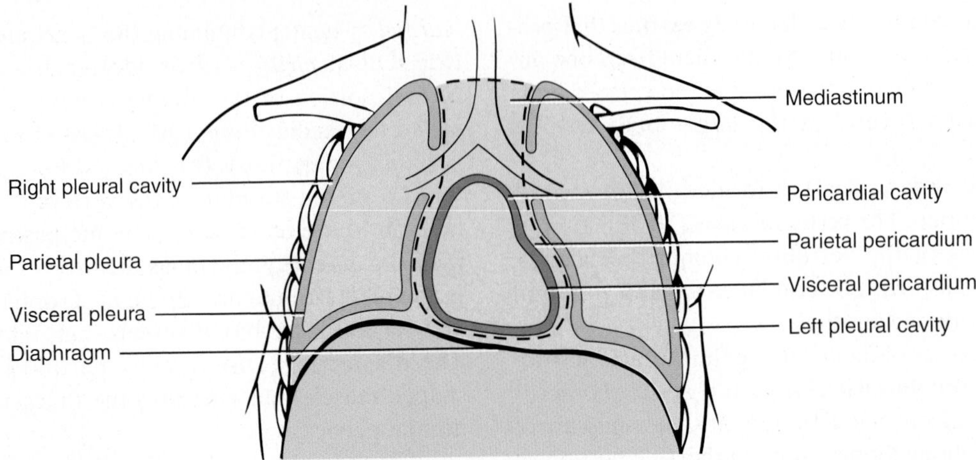


(a) Anterior view of thoracic cavity

Transverse plane
View

ANTERIOR
Sternum (breastbone)
Muscle
Thymus
Heart
PERICARDIAL CAVITY
Left lung
Right lung
Esophagus (food tube)
Aorta
Vertebral column (backbone)
RIGHT PLEURAL CAVITY
LEFT PLEURAL CAVITY
Rib
POSTERIOR

(b) Inferior view of transverse section of thoracic cavity

**The thoracic cavity contains three smaller cavities and the mediastinum.**

(Figure 1.11). As the name suggests, the abdominopelvic cavity is divided into two portions, even though no wall separates them. The superior portion, the **abdominal cavity** (*abdomin-* = belly), contains the stomach, spleen, liver, gallbladder, small intestine, and most of the large intestine. The inferior portion, the **pelvic cavity** (PEL-vik; *pelv-* = basin), contains the urinary bladder, portions of the large intestine, and internal organs of the reproductive system. Organs inside the thoracic and abdominopelvic cavities are called **viscera** (VIS-er-a).

## ✓ CHECKPOINT

19. The diaphragm separates what two cavities?
20. In which cavities are the following organs located: urinary bladder, stomach, heart, small intestine, lungs, thymus, and liver? Use the following symbols for your response: T = thoracic cavity, A = abdominal cavity, or P = pelvic cavity.
21. Which of the following structures are contained in the mediastinum: right lung, heart, esophagus, spinal cord, trachea, ribs, thymus, left pleural cavity?
22. What is the name of the cavity that surrounds the heart? Which cavities surround the lungs?
23. What organ lies distinctly in both the abdominal and pelvic cavities?

FIGURE 1.11 **The abdominopelvic cavity.** The lower dashed line shows the approximate boundary between the abdominal and pelvic cavities.

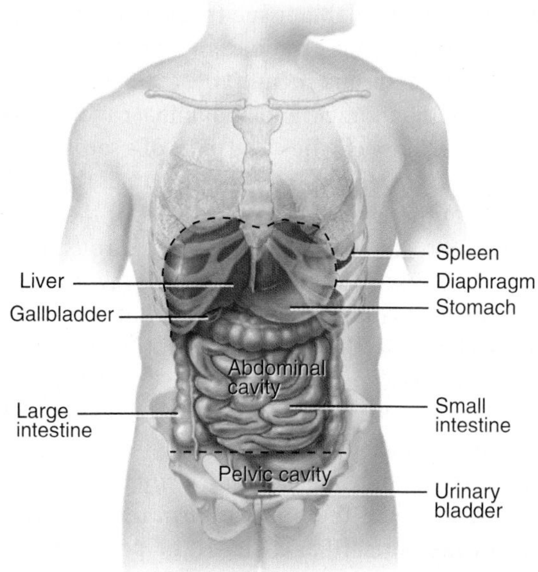


Anterior view

**The abdominopelvic cavity extends from the diaphragm to the groin.**

## RETURN TO Jaime's Story

The next day, Jaime shows up at work and finds Terry, one of the x-ray technicians he had met the previous day, sitting at the desk. "Morning, Terry. What have you got for me?"

"Good morning, Jaime. I'm going to need your help with a 9 o'clock patient. Doc wants a bunch of new films on him. Motorcycle versus truck accident!" Jaime glances at the x-ray films lying on the desk. "Those his films?" Jaime asks.

"Yeah, check this out. He broke his neck in several places. Doc wants a lateral view and an anterior view. He also broke his collarbone and humerus, and shattered his elbow. We'll need an acromial series, an olecranal series, and an oblique view. You'll need to help position him. It's going to hurt . . . a lot. Why don't you see if this guy is ready now." Jaime heads off to check on the patient.

After lunch, Jaime helps take a series of x-rays of an elderly man with suspected pneumonia. The patient coughs repeatedly whenever he moves, which appears to induce more coughing. When Jaime is finally able to get a few clear x-rays taken and developed, he returns the patient to his room. The patient complains of back pain and the chills, so Jaime checks with a nurse and finds a blanket and an extra pillow for him. On his way back to the desk, Jaime is called to bring an ultrasound device to admitting for a victim in a car accident who may have internal bleeding in her spleen. As he watches the doctor quickly scan the patient's abdomen with the ultrasound, Jaime tries to keep up as she rattles off the organs found medial and inferior to the spleen, commenting on which appear intact and which look suspicious. After that, Jaime returns to the radiology department with the equipment and helps Terry with more patients.

That evening, Jaime describes his experiences to classmates in his study group at the library. "It was unbelievable. Her spleen appeared intact, but her stomach might have been punctured by a broken rib. I tell you, it's incredible how much of this stuff you see on the job."

E. *If Terry and Jaime need to obtain a lateral view of the motorcycle rider's neck, from what direction would they take the x-ray picture?*
F. *Where are the acromial and olecranal structures found on the motorcyclist's body?*
G. *The coughing patient's movements appeared to prompt more coughing. Is this a type of feedback mechanism? Explain your answer.*
H. *What are the elderly patient's signs and symptoms?*
I. *Using anatomical directions, describe the relationship between the car accident victim's stomach and her spleen.*

## 1.6 Serous membranes line the walls of body cavities and cover the organs within them.

A **membrane** is a thin, pliable tissue that covers, lines, partitions, or connects structures. One example is a slippery, double-layered membrane called a **serous membrane**, a thin epithelium that lines the walls of the thoracic and abdominal cavities and covers the viscera within those cavities. The parts of a serous membrane are (1) the *parietal layer* (pa-RĪ-e-tal) that lines the walls of the cavities, and (2) the *visceral layer* (VIS-er-al) that covers and adheres to the organs within the cavities. Serous membranes secrete a small amount of lubricating fluid (serous fluid) into the space between the two layers. Serous fluid reduces friction, allowing the viscera to slide somewhat during movements—as when the lungs inflate and deflate during breathing.

The serous membrane of the pleural cavities is called the **pleura** (PLOO-ra) (see Figure 1.10a). The *visceral pleura* clings to the surface of the lungs, the *parietal pleura* lines the chest wall, and the space between them is the *pleural cavity*. The serous membrane of the pericardial cavity is the **pericardium** (per′-i-KAR-dē-um) (see Figure 1.10a). The *visceral pericardium* covers the surface of the heart, the *parietal pericardium* lines the chest wall, and the *pericardial cavity* is in between them. The **peritoneum** (per-i-tō-NĒ-um) is the serous membrane of the abdominal cavity. The *visceral peritoneum* covers the abdominal viscera, and the *parietal peritoneum* lines the abdominal wall. Between them is the *peritoneal cavity*. Most abdominal organs are located in the peritoneal cavity. Some are located between the parietal peritoneum and the posterior abdominal wall. Such organs are said to be *retroperitoneal* (re′-trō-per-i-tō-NĒ-al; *retro* = behind). Examples of retroperitoneal organs include the kidneys, pancreas, and duodenum of the small intestine.

A summary of body cavities and their membranes is presented in the table included in Figure 1.9.

✓ **CHECKPOINT**

**24.** What is the function of serous membranes?

**25.** What is the meaning of the term "retroperitoneal"?

## 1.7 The abdominopelvic cavity is divided into regions or quadrants.

To describe the location of the many abdominal and pelvic organs more easily, anatomists and clinicians use two methods of dividing the abdominopelvic cavity into smaller areas (Figure 1.12). In the first method, two horizontal lines and two vertical lines, aligned like a tic-tac-toe grid, partition this cavity into nine **abdominopelvic regions** (Figure 1.12a, b). The top horizontal line is drawn just inferior to the rib cage, across the stomach; the bottom horizontal line is drawn just inferior to the tops of the hip bones. Two vertical lines are drawn through the midpoints of the clavicles (collar bones). The four lines divide the abdominopelvic cavity into nine regions: the **right hypochondriac**, **epigastric**, **left hypochondriac**, **right lumbar**, **umbilical**, **left lumbar**, **right inguinal**, **hypogastric**, and **left inguinal**.

In the second, simpler method a midsagittal line and a transverse line are passed through the **umbilicus** (um-bi-LĪ-kus; *umbilic-* = navel) or *belly button*, to divide the abdominopelvic cavity into **quadrants** (KWOD-rantz; *quad-* = one-fourth), as shown in Figure 1.12c. The abdominopelvic quadrants are the **right upper quadrant**, **left upper quadrant**, **right lower quadrant**, and **left lower quadrant**. The nine-region division is more widely used for anatomical studies, and quadrants are more commonly used by clinicians for describing the site of abdominopelvic pain, a tumor, or other abnormality.

✓ **CHECKPOINT**

**26.** In which abdominopelvic region is each of the following found: most of the liver, transverse colon, urinary bladder, spleen?

**27.** When anatomical terms refer to right and left, do we mean your right and left or those of the person you are facing? (*Hint: See Figure 1.12.*)

FIGURE 1.12 **Regions and quadrants of the abdominopelvic cavity.**

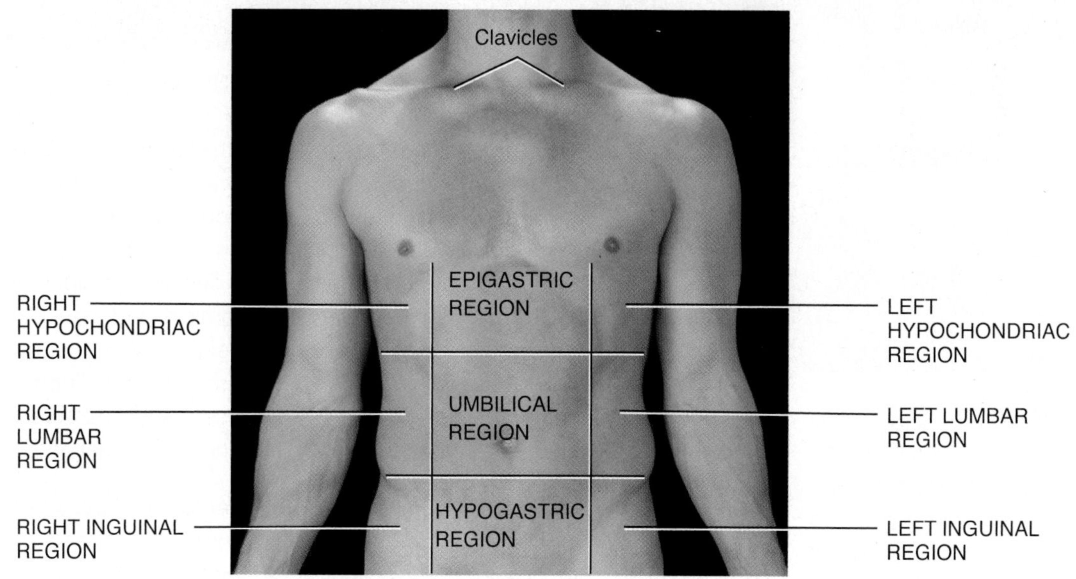


(a) Anterior view showing abdominopelvic regions

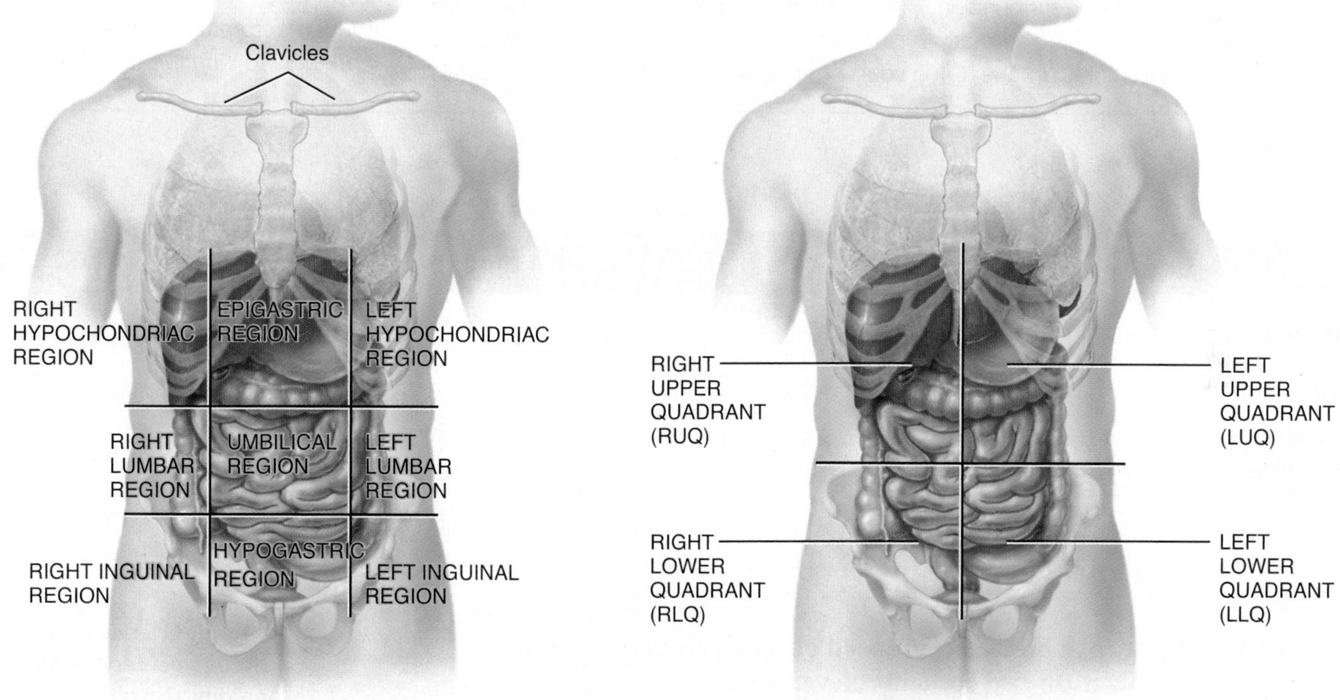


(b) Anterior view showing location of abdominopelvic regions

(c) Anterior view showing location of abdominopelvic quadrants

**The nine-region designation is used for anatomical studies; the quadrant designation is used by clinicians to locate the site of pain, a tumor, or some other abnormality.**

## Jaime's Story

## EPILOGUE AND DISCUSSION

At the end of the week, Jaime spends an hour with Dr. Sebbo, reviewing the structures of the mediastinum in a series of x-rays to better read the x-rays and understand the instructions for positioning the x-ray equipment. They use a recent CT scan of a patient with appendicitis to review the quadrants and regions of the abdominopelvic cavities. "As you can see," Dr. Sebbo say, "the appendix is located here inferior to the junction between the small and large intestines. In a normal individual, it would be hidden posterior to the junction, but in this poor fellow, it was so swollen that it was starting to tear through the surrounding membrane." When Jaime is able to correctly identify the various structures in the CT scan, Dr. Sebbo compliments him on picking up on the terminology and techniques so quickly. Jaime grins and says, "I admit that it's all coming faster than I expected, but everyone's been really helpful with this."

Jaime has found that his study of anatomy and physiology has direct application to his new job. Gaining knowledge about anatomy and physiology allows us to understand the relationship of body structures to their functions. As you read through this book keep in mind that, like Jaime's experience, what you are learning will have meaning to you in both your professional life and your personal life. So far, you have learned how the body is organized, what characteristics determine life, and some basic anatomical terminology that names body parts and describes their locations.

In each chapter, we will explore a new story designed to illustrate the application of anatomical and physiological principles in your everyday life. Although the case stories themselves are fictional, the situations, illnesses, and characters are based on the experiences of real people just like you.

J. *Name the structures found within the mediastinum seen on the x-ray.*

K. *Describe the quadrant and regions in which you would expect to find the CT patient's and a normal person's appendix.*

L. *What membrane was the inflamed appendix tearing through?*

# Concept and Resource Summary

| Concept | Resources  |
|---|---|
| **Introduction**<br>1. **Anatomy** is the science of body structures and their relationships.<br>2. **Physiology** is the science of body functions. | |
| **Concept 1.1 The human body is composed of six levels of structural organization and contains eleven systems.**<br>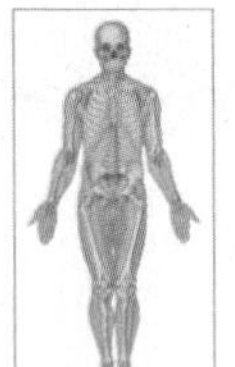<br>1. The six structural levels of body organization, from smallest to largest, are chemical, cellular, tissue, organ, system, and organismal.<br>2. The chemical level includes **atoms** and **molecules**.<br>3. The cellular level occurs when molecules build **cells**, the basic units of structure and function in living organisms.<br>4. Group of cells unite to form the tissue level composed of four basic types of **tissues**: epithelial, connective, muscle, and nervous. | Anatomy Overview—The Integumentary System<br>Anatomy Overview—The Skeletal System<br>Anatomy Overview—The Muscular System<br>Anatomy Overview—The Nervous System<br>Anatomy Overview—The Endocrine System<br>Anatomy Overview—The Cardiovascular System<br>Anatomy Overview—The Lymphatic System and Disease Resistance |

## Concept

5. Different tissues join together to form the organ level.
6. Different **organs** join together to form the system level.
7. All the **systems** form the largest level, the organismal level. An **organism** is a living individual.
8. There are eleven body systems in the human body (see Table 1.1).

## Resources WILEY PLUS

Anatomy Overview—The Respiratory System
Anatomy Overview—The Digestive System
Anatomy Overview—The Urinary System
Anatomy Overview—The Reproductive Systems
Exercise—Concentrate on Systemic Functions
Exercise—Find the System Outsiders

## Concept 1.2 The human body carries on basic life processes that distinguish it from nonliving objects.

1. All living organisms undergo chemical processes referred to as **metabolism**, including **catabolism**, the breaking down of large chemical compounds to smaller ones, and **anabolism**, the building of large compounds from smaller ones.
2. Living organisms respond to changes in their environment (a characteristic called **responsiveness**), and they exhibit **movement** of the whole body or structures within the body.
3. **Growth**, an increase in body size, and **differentiation**, such as that of **stem cells** into specialized cells, are two additional basic life processes.
4. Living organisms undergo **reproduction** to form new cells for tissue growth or for production of a new individual.
5. **Homeostasis** is the maintenance of relatively stable internal body conditions despite changes that occur inside and outside the body.
6. Homeostasis maintains the volume and composition of **body fluids** inside and outside of cells.
7. The fluid inside cells is **intracellular fluid**, and the fluid outside of cells is **extracellular fluid**. **Interstitial fluid**, **blood plasma**, **lymph**, **cerebrospinal fluid**, and **synovial fluid** are extracellular fluids.

Animation—Homeostatic Relationships
Clinical Connection—Autopsy

## Concept 1.3 Homeostasis is controlled through feedback systems.

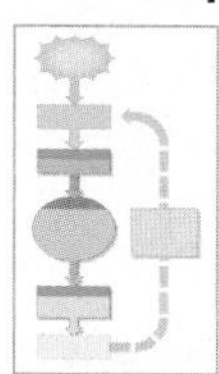

1. The body is continually confronted with internal and external stresses. The principal regulatory systems, the nervous system and endocrine system, work to provide corrective measures to achieve homeostasis and restore balance to the body's internal environment.
2. A **feedback system** is a cycle of events that monitor, evaluate, and change disruptions to a controlled condition in the body.
3. Three components comprise a feedback system: receptor, control center, and effector.
4. A **receptor** monitors changes and sends input to a **control center**, where input is evaluated and output commands are sent to **effectors** that produce a response.
5. Responses that reverse the original stimulus involve **negative feedback**.
6. Responses that intensify the original stimulus involve **positive feedback**.
7. The inability of one or more body components to contribute to homeostasis may result in disease or death.
8. A **disorder** is an abnormality of structure and/or function.
9. A **disease** refers to an illness accompanied by a set of clinically observable, measurable changes called **signs**, and subjective changes experienced by the patient referred to as **symptoms**.

Animation—Communication, Regulation, and Homeostasis
Animation—Negative Feedback Control of Blood Pressure
Animation—Negative Feedback Control of Temperature
Animation—Positive Feedback Control of Labor
Concepts and Connections—Regulation and Communication
Concepts and Connections—Negative Feedback Loop
Exercise—Waterways
Concepts and Connections—Homeostasis

| Concept | Resources  |
|---|---|

## Concept 1.4 The human body is described using the anatomical position and specific terms.

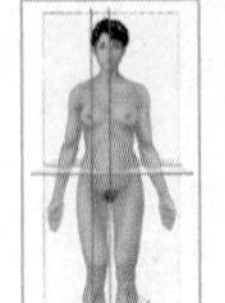

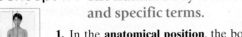

1. In the **anatomical position**, the body is erect, facing forward, head level, eyes forward, feet flat and forward, and arms at the sides with palms forward.
2. The body is in the **prone** position when lying face down, and in the **supine** position when lying face up.
3. The human body is divided into several major regions identified externally. These are the **head**, **skull**, **face**, **neck**, **trunk**, **upper limbs**, and **lower limbs**.
4. In order to describe the position of one body part relative to another body part, specific **directional terms** must be used (see Table 1.2 and Figure 1.6).
5. The body and body parts can be cut along imaginary flat surfaces referred to as **planes**.
6. A **sagittal plane** divides the body or organ into right and left sections. When the plane passes directly through the midline resulting in equal right and left sides, it is a **midsagittal (median) plane**. A **parasagittal plane** results in unequal right and left sides.
7. A **frontal (coronal) plane** divides the body or organ into front and back sections.
8. A **transverse (horizontal) plane** results in upper and lower sections.
9. An **oblique plane** passes through the body or organ at an angle.
10. Body regions are often studied in sections. A **section** is one flat surface of the three-dimensional structure or cut along a plane for which is it named.

Resources: Figure 1.5—The Anatomical Position; Figure 1.6—Directional Terms

## Concept 1.5 Body cavities are spaces within the body that help protect, separate, and support internal organs.

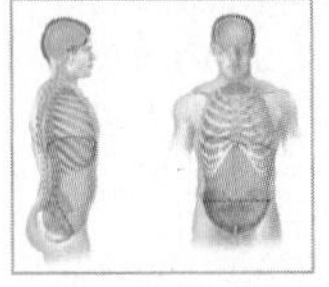

1. **Body cavities** are separated from one another by bones, muscles, and ligaments.
2. The **cranial cavity** houses the brain, and the **vertebral canal** contains the spinal cord.
3. The **diaphragm** separates the superior thoracic cavity and the inferior abdominopelvic cavity.
4. The **thoracic cavity** houses three smaller cavities: the **pericardial cavity**, around the heart; and two **pleural cavities**, each surrounding a lung. Located between the lungs, extending from the neck to the diaphragm, is the **mediastinum**, which contains all thoracic viscera except the lungs.
5. The **abdominopelvic cavity** has a superior region, the **abdominal cavity**, containing the stomach, spleen, liver, gallbladder, small intestine, and most of the large intestine. The inferior region, the **pelvic cavity**, contains the urinary bladder, portions of the large intestine, and reproductive organs.

Resources: Figure 1.10—The Thoracic Cavity

## Concept 1.6 Serous membranes line the walls of body cavities and cover the organs within them.

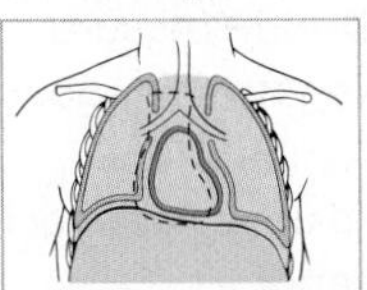

1. The thoracic and abdominal cavity walls and viscera are covered with a thin, slippery **serous membrane**. The part of the serous membrane that lines cavity walls is the parietal layer, and the visceral layer covers the organs within the cavity.
2. The pleura, pericardium, and peritoneum are serous membranes.
3. The **pleura** is the serous membrane of the pleural cavities around the lungs; the **pericardium** is the serous membrane of the pericardial cavity around the heart; and the **peritoneum** is the serous membrane of the abdominal cavity around abdominal viscera.
4. Organs such as the kidneys that are located posterior to the parietal peritoneum are referred to as retroperitoneal.

Resources: Anatomy Overview—Serous Membrane

| Concept | Resources |
|---|---|
| **Concept 1.7** The abdominopelvic cavity is divided into regions or quadrants. | |

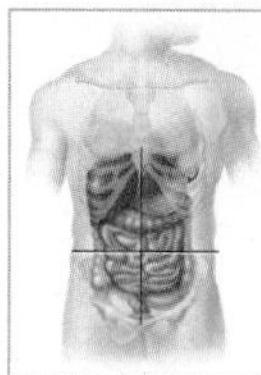

1. Anatomists and clinicians use two methods of dividing the abdominopelvic cavity into smaller areas in order to describe the location of organs.
2. One method divides the cavity into nine **abdominopelvic regions: right hypochondriac, epigastric, left hypochondriac, right lumbar, umbilical, left lumbar, right inguinal, hypogastric**, and **left inguinal**.
3. The other method divides the cavity into **quadrants: right upper quadrant, left upper quadrant, right lower quadrant**, and **left lower quadrant**.

Clinical Connection–Medical Imaging

## Understanding the Concepts

1. Referring to Table 1.1, which body systems help eliminate wastes? Which regulate body activities? Which protect the body? Which maintain acid–base balance in the body?
2. Which life process in the human body is utilized when any of the others occur?
3. Why would a weight-loss supplement that stimulates "metabolism" not necessarily help you lose weight? (Hint: Think about the definition of metabolism.)
4. Referring to Figure 1.3, what would happen to your heart rate if some stimulus caused your blood pressure to drop? Would this occur by positive or negative feedback?
5. Sometimes a synthetic form of the hormone oxytocin, called Pitocin, is given to pregnant women to induce labor. Referring to Figure 1.4, which part of the feedback system is bypassed when this drug is given? Explain your answer.
6. A television commercial promoting a new drug advises parents to look for certain symptoms. Is the advertiser using the term "symptoms" correctly? Explain your answer.
7. Would a sore throat be a symptom or a sign of a potential disease? Explain your answer.
8. On which part of your body would you find a plantar wart? The occipital bone of the skull? An inguinal hernia? Carpal tunnel syndrome?
9. Why are the pedals on a bicycle most likely called "pedals"?
10. Which plane divides the heart into anterior and posterior portions? The brain into unequal right and left portions? What plane would be used to visualize the appendix?
11. Why is meningitis, an inflammation of membranes in the cranial cavity and vertebral canal, particularly dangerous?
12. Which part of the peritoneum covers the stomach? Which part lines the abdominal wall?
13. In which abdominopelvic quadrant would pain from appendicitis (inflammation of the appendix) be felt?

CHAPTER

# 2

## Eugene's Story

The doors to the emergency department swung open with a thud as the paramedics wheeled in an unconscious man on a gurney. "Is Dr. Kim working tonight?" asked one of the paramedics.

"Yes she is. Who do we have here?" responded the nurse.

"Well, Linda, this is Eugene. He was found passed out in the alley behind the Kozy Korner bar. Apparently he's had a bit too much to drink. We brought him in several months ago and Dr. Kim saw him at that time. He's a chronic alcoholic, diabetic too, I think," said the paramedic.

Linda touched the man's forehead, then gently lifted his eyelids. "Eugene, can you hear me? You're at the hospital. We need you to wake up!" Eugene mumbled and then coughed. "I don't think he's waking up anytime soon. Put him in bed two and I'll see if Dr. Kim is working tonight. You're in luck, Eugene, Dr. Kim is on duty tonight." Just then Dr. Kim happened in.

She leaned over, looking at Eugene. "Hmm. Linda, get a CBC and blood chemistries started right away."

Our blood is a key indicator of the body's homeostasis. Blood chemistry analysis is used to assess various chemical products, both organic and inorganic, that are related to the function of the organs and organ systems.

Imbalances in blood chemistry can help in the diagnosis of specific disease states or other homeostatic imbalances; as you will learn in this chapter, they can also provide other important insights into the functions of the human body.

# The Chemical Level of Organization

## INTRODUCTION

The chemical level of organization, the lowest level of structural organization, consists of atoms and molecules. These small particles ultimately combine to form body organs and systems of astonishing size and complexity. In this chapter, we will consider how atoms bond together to form molecules and how atoms and molecules release or store energy in processes known as chemical reactions. We will also learn about the vital importance of water, which accounts for nearly two-thirds of body weight, in chemical reactions and the maintenance of homeostasis. Finally, we present several groups of molecules whose unique properties contribute to assembly of the body's structures or to powering the processes that enable you to live.

Because your body is composed of chemicals and all body activities are chemical in nature, it is important to become familiar with the language and basic ideas of chemistry to understand human anatomy and physiology. **Chemistry** (KEM-is-trē) is the science of the structure and interactions of matter. All living and nonliving things consist of **matter**, which is anything that occupies space and has **mass**. Mass is the amount of matter in any object, which does not change. *Weight*, the force of gravity acting on matter, does change. When objects are farther from Earth, the pull of gravity is weaker; this is why the weight of an astronaut is close to zero in outer space.

## CONCEPTS

**2.1** Chemical elements are composed of small units called atoms.

**2.2** Atoms are held together by chemical bonds.

**2.3** Chemical reactions occur when atoms combine with or separate from other atoms.

**2.4** Inorganic compounds include water, salts, acids, and bases.

**2.5** Organic molecules are large carbon-based molecules that carry out complex functions in living systems.

**2.6** Carbohydrates function as building blocks and sources of energy.

**2.7** Lipids are important for cell membrane structure, energy storage, and hormone production.

**2.8** Proteins are amino acid complexes serving many diverse roles.

**2.9** Nucleic acids contain genetic material and function in protein synthesis.

**2.10** Adenosine triphosphate (ATP) is the principal energy-transferring molecule in living systems.

## 2.1 Chemical elements are composed of small units called atoms.

Matter exists in three states: solid, liquid, and gas. *Solids*, such as bones and teeth, are compact and have a definite shape and volume. *Liquids*, such as blood plasma, have a definite volume and assume the shape of their container. *Gases*, like oxygen and carbon dioxide, have neither a definite shape nor volume. All forms of matter—both living and nonliving—are made up of a limited number of building blocks called **chemical elements**. Each element is a substance that cannot be split into a simpler substance by ordinary chemical means. Scientists now recognize 117 elements. Of these, 92 occur naturally on Earth. The rest have been produced from the natural elements using particle accelerators or nuclear reactors. Each element is designated by a **chemical symbol**, one or two letters of the element's name in English, Latin, or another language. Examples of chemical symbols are H for hydrogen, C for carbon, O for oxygen, N for nitrogen, Ca for calcium, and Na for sodium (*natrium* = sodium).*

* The periodic table of elements, which lists all of the known chemical elements, can be found in Appendix B.

Twenty-six different chemical elements normally are present in your body. Just four elements—oxygen, carbon, hydrogen, and nitrogen—make up about 96 percent of the body's mass. Eight others—calcium, phosphorus (P), potassium (K), sulfur (S), sodium (Na), chlorine (Cl), magnesium (Mg), and iron (Fe)—contribute 3.8 percent of the body's mass. An additional 14 elements—the **trace elements**—are present in tiny amounts. Together, trace elements account for the remaining 0.2 percent of the body's mass. Several trace elements have important functions in the body. For example, iodine is needed to make thyroid hormones. The functions of some trace elements are unknown. Table 2.1 lists the main chemical elements of the human body.

### Structure of Atoms

Each element is made up of **atoms**, the smallest units of matter that retain the properties and characteristics of the element (Figure 2.1). Atoms are extremely small. Two hundred thousand

**TABLE 2.1**

Main Chemical Elements in the Body

| CHEMICAL ELEMENT (SYMBOL) | % OF TOTAL BODY MASS | SIGNIFICANCE |
|---|---|---|
| **MAJOR ELEMENTS** | (About 96) | |
| **Oxygen (O)** | 65.0 | Part of water and many organic (carbon-containing) molecules; used to generate ATP, a molecule used by cells to temporarily store chemical energy |
| **Carbon (C)** | 18.5 | Forms backbone chains and rings of all organic molecules: carbohydrates, lipids (fats), proteins, and nucleic acids (DNA and RNA) |
| **Hydrogen (H)** | 9.5 | Constituent of water and most organic molecules; ionized form ($H^+$) makes body fluids more acidic |
| **Nitrogen (N)** | 3.2 | Component of all proteins and nucleic acids |
| **LESSER ELEMENTS** | (About 3.6) | |
| **Calcium (Ca)** | 1.5 | Contributes to hardness of bones and teeth; ionized form ($Ca^{2+}$) needed for blood clotting, release of some hormones, contraction of muscle, and many other processes |
| **Phosphorus (P)** | 1.0 | Component of nucleic acids and ATP; required for normal bone and tooth structure |
| **Potassium (K)** | 0.35 | Ionized form ($K^+$) is most plentiful cation (positively charged particle) in intracellular fluid; needed to generate action potentials |
| **Sulfur (S)** | 0.25 | Component of some vitamins and many proteins |
| **Sodium (Na)** | 0.2 | Ionized form ($Na^+$) is most plentiful cation in extracellular fluid; essential for maintaining water balance; needed to generate action potentials |
| **Chlorine (Cl)** | 0.2 | Ionized form ($Cl^-$) is most plentiful anion (negatively charged particle) in extracellular fluid; essential for maintaining water balance |
| **Magnesium (Mg)** | 0.1 | Ionized form ($Mg^{2+}$) needed for action of many enzymes, molecules that increase the rate of chemical reactions in organisms |
| **Iron (Fe)** | 0.005 | Ionized forms ($Fe^{2+}$ and $Fe^{3+}$) are part of hemoglobin (oxygen-carrying protein in red blood cells) and some enzymes |
| **TRACE ELEMENTS** | (About 0.4) | Aluminum (Al), boron (B), chromium (Cr), cobalt (Co), copper (Cu), fluorine (F), iodine (I), manganese (Mn), molybdenum (Mo), selenium (Se), silicon (Si), tin (Sn), vanadium (V), and zinc (Zn) |

of the largest atoms would fit on the period at the end of this sentence. Hydrogen atoms, the smallest atoms, have a diameter less than 0.1 nanometer ($0.1 \times 10^{-9}$ m = 0.0000000001 m); the largest atoms are only five times larger.

Dozens of different **subatomic particles** compose individual atoms. However, only three types of subatomic particles are important for understanding the chemical reactions in the human body: protons, neutrons, and electrons (Figure 2.1). The dense central core of an atom is its **nucleus**. Within the nucleus are positively charged **protons ($p^+$)** and uncharged (neutral) **neutrons ($n^0$)**. The tiny negatively charged **electrons ($e^-$)** move about in a large space surrounding the nucleus. They do not follow a fixed path or orbit but instead form a negatively charged "cloud" that envelops the nucleus (Figure 2.1a).

Even though their exact positions cannot be predicted, specific groups of electrons are most likely to move about within certain regions around the nucleus. These regions, called **electron shells**, are depicted as simple circles around the nucleus. Each electron shell can hold a specific number of electrons (Figure 2.1b). The first electron shell (nearest the nucleus) never holds more than 2 electrons. The second shell holds a maximum of 8 electrons, and the third can hold up to 18 electrons. The electron shells fill with electrons in a specific order, beginning with the first shell. For example, notice in Figure 2.2 that sodium (Na), which has 11 electrons total, contains 2 electrons in the first shell, 8 electrons in the

FIGURE 2.1 **Two representations of the structure of an atom.** Electrons move about the nucleus, which contains neutrons and protons. (a) In the electron cloud model of an atom, the shading represents the chance of finding an electron in regions outside the nucleus. (b) In the electron shell model, filled circles represent individual electrons, which are grouped into concentric circles according to the shells they occupy.

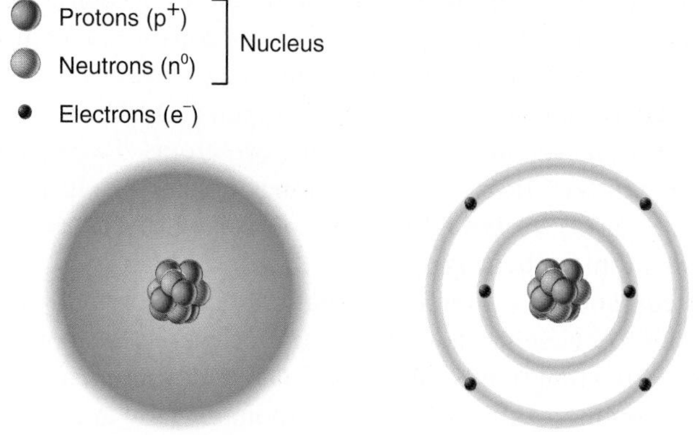


(a) Electron cloud model of carbon (b) Electron shell model of carbon

**An atom is the smallest unit of matter that retains the properties and characteristics of its element.**

FIGURE 2.2 **Atomic structures of several atoms that have important roles in the human body.**

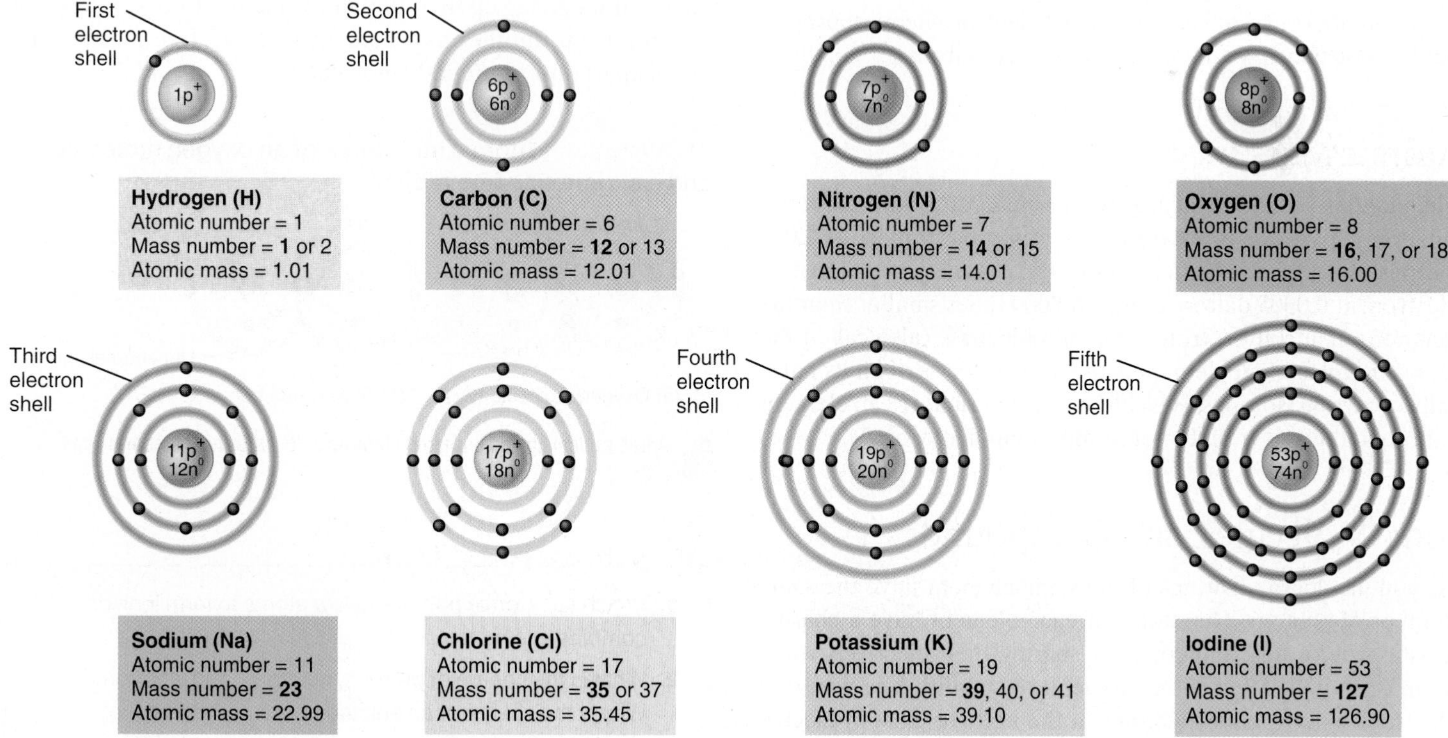


Atomic number = number of protons in an atom
Mass number = number of protons and neutrons in an atom (boldface indicates most common isotope)
Atomic mass = average mass of all stable atoms of a given element in daltons

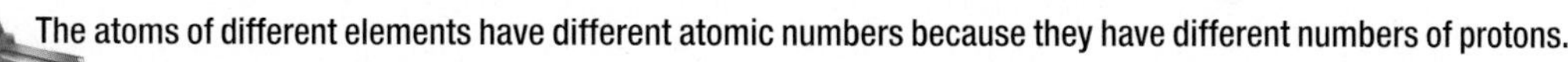

**The atoms of different elements have different atomic numbers because they have different numbers of protons.**

second shell, and 1 electron in the third shell. The most massive element present in the human body is iodine, which has a total of 53 electrons: 2 in the first shell, 8 in the second shell, 18 in the third shell, 18 in the fourth shell, and 7 in the fifth shell.

The number of electrons in an atom of an element always equals the number of protons. Because each electron and proton carries one charge, the negatively charged electrons and the positively charged protons balance each other. As a result, each atom is electrically neutral, meaning its total charge is zero.

## Atomic Number and Mass Number

The *number of protons* in the nucleus of an atom is that atom's **atomic number**. Figure 2.2 shows that atoms of different elements have different atomic numbers because they have different numbers of protons. For example, oxygen has an atomic number of 8 because its nucleus has 8 protons, and sodium has an atomic number of 11 because its nucleus has 11 protons.

The **mass number** of an atom is the sum of its protons and neutrons. Because sodium has 11 protons and 12 neutrons, its mass number is 23 (Figure 2.2). Although all atoms of one element have the same number of protons, they may have different numbers of neutrons. **Isotopes** are atoms of an element that have different numbers of neutrons and therefore different mass numbers. In a sample of oxygen, for example, most atoms have 8 neutrons, and a few have 9 or 10, but all have 8 protons and 8 electrons. As you will discover shortly, the number of electrons of an atom determine its chemical properties. Although the isotopes of an element have different numbers of neutrons, they have identical chemical properties because they have the same number of electrons.

## Atomic Mass

The standard unit for measuring the mass of atoms and their subatomic particles is a **dalton**. A neutron has a mass of 1.008 daltons, and a proton has a mass of 1.007 daltons. The mass of an electron, at 0.0005 dalton, is almost 2000 times smaller than the mass of a neutron or proton. The **atomic mass** (also called the *atomic weight*) of an element is the average mass of all its naturally occurring isotopes. Typically, the atomic mass of an element is close to the mass number of its most abundant isotope.

## Ions, Molecules, and Compounds

As you just learned, atoms of the same element have the same number of protons. The atoms of each element have a characteristic way of losing, gaining, or sharing their electrons when interacting with other atoms to achieve stability. The way that electrons behave enables atoms in the body to exist in electrically charged forms called ions, or to join together into complex combinations as molecules or compounds.

If an atom either *gives up* or *gains electrons*, it becomes an **ion** (Ī-on). An ion is an atom that has a positive or negative charge because it has unequal numbers of protons and electrons. An ion of an atom is symbolized by writing its chemical symbol followed by the number of its positive (+) or negative (−) charges. Thus, $Ca^{2+}$ stands for a calcium ion that has two positive charges because it has lost two electrons.

When two or more atoms *share* electrons, the resulting combination is called a **molecule** (MOL-e-kūl). A *molecular formula* indicates the elements and the number of atoms of each element that make up a molecule. A molecule may consist of two atoms of the same kind, such as an oxygen molecule (Figure 2.3a). The molecular formula for a molecule of oxygen is $O_2$. The subscript 2 indicates that the molecule contains two atoms. Two or more different kinds of atoms may also form a molecule, as in a water molecule ($H_2O$). In $H_2O$, one atom of oxygen shares electrons with two atoms of hydrogen.

A **compound** is a molecule that contains atoms of two or more different elements. Most of the atoms in the body are joined into compounds. Water ($H_2O$) and sodium chloride (NaCl), common table salt, are compounds. However, a molecule of oxygen ($O_2$) is not a compound because it consists of atoms of only one element. So all compounds are molecules, but not all molecules are compounds.

A **free radical** is an electrically charged atom or group of atoms with an unpaired electron in the outermost shell. A common example is superoxide, which is formed by the addition of an electron to an oxygen molecule (Figure 2.3b). Having an unpaired electron makes a free radical unstable, highly reactive, and destructive to nearby molecules. Free radicals become stable by either giving up their unpaired electron to, or taking on an electron from, another molecule. In so doing, free radicals may break apart important body molecules.

**FIGURE 2.3 Atomic structures of an oxygen molecule and a superoxide free radical.**

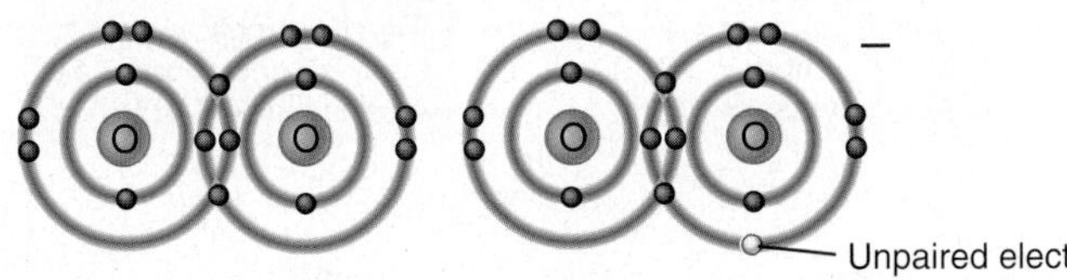


(a) Oxygen molecule ($O_2$) (b) Superoxide free radical ($O_2^-$)

**A free radical has an unpaired electron in its outermost electron shell.**

### ✓ CHECKPOINT

1. Which subatomic particles allow atoms to form ions or combine into molecules?
2. What is the charge of protons, neutrons, and electrons? Why is the charge of an electrically neutral atom zero?
3. What are the atomic number, mass number, and atomic mass of carbon? How are they related?
4. What are isotopes and free radicals?

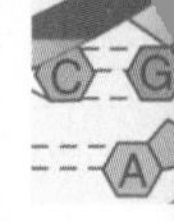

CLINICAL CONNECTION | *Free Radicals and Antioxidants*

In our bodies, several processes can generate free radicals, including exposure to ultraviolet radiation in sunlight, exposure to x-rays, and some reactions that occur during normal metabolic processes. Certain harmful substances, such as carbon tetrachloride (a solvent used in dry cleaning), also give rise to free radicals when they participate in metabolic reactions in the body. Among the many disorders, diseases, and conditions linked to oxygen-derived free radicals are cancer, atherosclerosis, Alzheimer's disease, emphysema, diabetes mellitus, cataracts, macular degeneration, rheumatoid arthritis, and deterioration associated with aging. Consuming more **antioxidants**—substances that inactivate oxygen-derived free radicals—is thought to slow the pace of damage caused by free radicals. Important dietary antioxidants include selenium, zinc, beta-carotene, and vitamins C and E. Red, blue, or purple fruits and vegetables contain high levels of antioxidants.

## 2.2 Atoms are held together by chemical bonds.

The forces that hold together the atoms of a molecule or compound are **chemical bonds**. The likelihood that an atom will form a chemical bond with another atom depends on the number of electrons in its outermost shell, also called the **valence shell**. An atom with a valence shell holding eight electrons is *chemically stable*, which means it is unlikely to form chemical bonds with other atoms. Neon, for example, has eight electrons in its valence shell, and for this reason it does not bond easily with other atoms. The valence shell of hydrogen and helium is the first electron shell, which holds a maximum of two electrons. Because helium has two valence electrons, it too is stable and seldom bonds with other atoms. Hydrogen, on the other hand, has only one valence electron (see Figure 2.2), so it binds readily with other atoms.

The atoms of most biologically important elements do not have eight electrons in their valence shells. Under the right conditions, two or more atoms can interact in ways that produce a chemically stable arrangement of eight electrons in the valence shell of each atom. This chemical principle, called the *octet rule* (*octet* = set of eight), helps explain why atoms interact in predictable ways. One atom is more likely to interact with another atom if doing so will leave both with eight valence electrons. For this to happen, an atom either empties its partially filled valence shell, fills it with donated electrons, or shares electrons with other atoms. The way that valence electrons are distributed determines what kind of chemical bond results. We will consider three types of chemical bonds: ionic bonds, covalent bonds, and hydrogen bonds.

### Ionic Bonds

When atoms lose or gain one or more valence electrons, ions are formed. Positively and negatively charged ions are attracted to one another—opposites attract. The force of attraction that holds together ions with opposite charges is an **ionic bond**. Consider sodium and chlorine atoms, the components of common table salt. Sodium has one valence electron (Figure 2.4a). If sodium *loses* this electron, it is left with the eight electrons in its second shell, which becomes the valence shell. As a result, the total number of protons (11) now exceeds the number of electrons (10), and the sodium atom becomes a **cation** (KAT-ī-on), or positively charged ion. A sodium ion has a charge of +1 and is written $Na^+$. By contrast, chlorine has seven valence electrons (Figure 2.4b). If chlorine *gains* an electron from a neighboring atom, it will have 8 electrons in its valence shell. When this happens, the total number of electrons (18) exceeds the number of protons (17), and the chlorine atom becomes an **anion** (AN-ī-on), a negatively charged ion. The ionic form of chlorine is called a *chloride* ion. It has a charge of −1 and is written $Cl^-$. When an atom of sodium donates its sole valence electron to an atom of chlorine, the resulting positive and negative charges pull both ions tightly together into an ionic bond (Figure 2.4c). The resulting ionic compound is sodium chloride, written NaCl.

In general, ionic compounds exist as solids, with an orderly, repeating arrangement of the ions, as in a crystal of NaCl (Figure 2.4d). A crystal of NaCl may be large or small—the total number of ions can vary—but the ratio of $Na^+$ to $Cl^-$ is always 1:1. In the body, ionic bonds are found mainly in teeth and bones, where they give great strength to these important structural tissues. An ionic compound that breaks apart into positive and negative ions in solution is called an **electrolyte** (e-LEK-trō-līt), so named because in solution it can conduct an electric current. Most ions in the body are dissolved in body fluids as electrolytes. As you will see in later chapters, electrolytes have many important functions. For example, they are critical for controlling water movement within the body, maintaining acid–base balance, and producing nerve and muscle impulses. Table 2.2 lists the names and symbols of the most common ions in the body.

**FIGURE 2.4 Ions and ionic bond formation.** (a) A sodium atom can attain the stability of eight electrons in its outermost shell by losing one valence electron; it then becomes a sodium ion, $Na^+$. (b) A chlorine atom can attain the stability of eight electrons in its outermost shell by accepting one electron; it then becomes a chloride ion, $Cl^-$. (c) An ionic bond may form between oppositely charged ions. (d) In a crystal of NaCl, each $Na^+$ is surrounded by six $Cl^-$. In (a), (b), and (c), the electron that is lost or accepted is colored red.

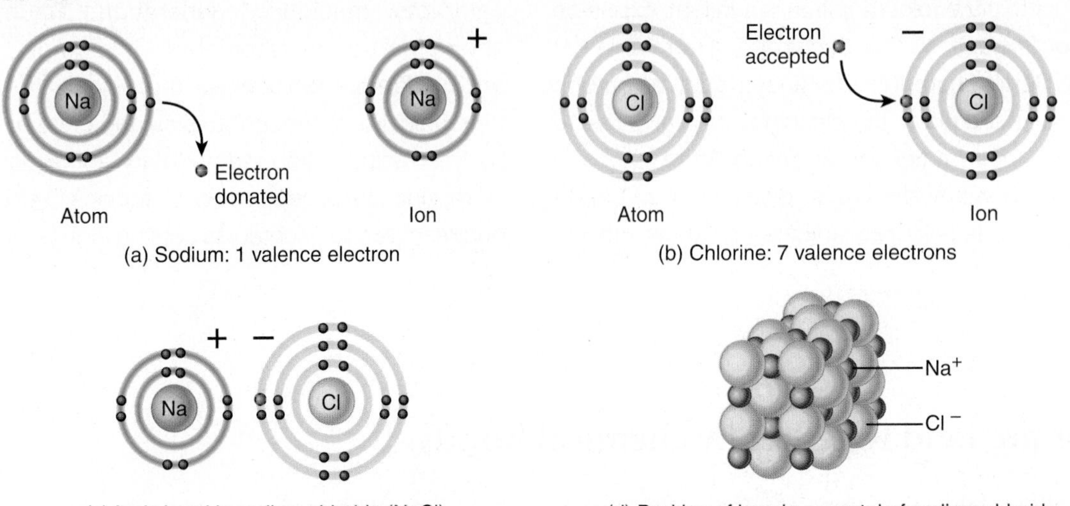


**An ionic bond is the force of attraction that holds together oppositely charged ions.**

**TABLE 2.2**

Common Ions in the Body

| CATIONS | | ANIONS | |
|---|---|---|---|
| NAME | SYMBOL | NAME | SYMBOL |
| **Hydrogen ion** | $H^+$ | Fluoride ion | $F^-$ |
| **Sodium ion** | $Na^+$ | Chloride ion | $Cl^-$ |
| **Potassium ion** | $K^+$ | Iodide ion | $I^-$ |
| **Ammonium ion** | $NH_4^+$ | Hydroxide ion | $OH^-$ |
| **Magnesium ion** | $Mg^{2+}$ | Bicarbonate ion | $HCO_3^-$ |
| **Calcium ion** | $Ca^{2+}$ | Oxide ion | $O^{2-}$ |
| **Iron(II) ion** | $Fe^{2+}$ | Sulfate ion | $SO_4^{2-}$ |
| **Iron(III) ion** | $Fe^{3+}$ | Phosphate ion | $PO_4^{3-}$ |

## Covalent Bonds

When a **covalent bond** forms, two or more atoms *share* electrons rather than gaining or losing them. The larger the number of electrons shared between two atoms, the stronger is the covalent bond. Covalent bonds may form between atoms of the same element or between atoms of different elements. They are the most common chemical bonds in the body, and the compounds that result from them form most of the body's structures.

A **single covalent bond** results when two atoms share one electron pair. For example, a molecule of hydrogen forms when two hydrogen atoms share their single valence electrons (Figure 2.5a), which allows both atoms to have a full valence shell. A **double covalent bond** results when two atoms share two pairs of electrons, as happens in an oxygen molecule (Figure 2.5b). A **triple covalent bond** occurs when two atoms share three pairs of electrons, as in a molecule of nitrogen (Figure 2.5c). Notice in the *structural formulas* for covalently bonded molecules in Figure 2.5 that the number of lines between the chemical symbols for two atoms indicates whether the bond is a single (—), double (=), or triple (≡) covalent bond.

The same principles of covalent bonding that apply to atoms of the same element also apply to covalent bonds between atoms of different elements. The gas methane ($CH_4$) contains covalent bonds formed between the atoms of two different elements, one carbon and four hydrogens (Figure 2.5d). The valence shell of the carbon atom can hold eight electrons but has only four of its own. The single electron shell of a hydrogen atom can hold two electrons, but each hydrogen atom has only one of its own. A methane molecule contains four separate single covalent bonds. Each hydrogen atom shares one pair of electrons with the carbon atom.

In some covalent bonds, two atoms share the electrons equally—one atom does not attract the shared electrons more strongly than the other atom. This type of bond is a **nonpolar covalent bond**. The bonds between two identical atoms are always nonpolar covalent bonds (Figure 2.5a–c). The bonds between carbon and hydrogen atoms are also nonpolar, such as the four C—H bonds in a methane molecule (Figure 2.5d).

In a **polar covalent bond**, the sharing of electrons between two atoms is unequal—the nucleus of one atom attracts the shared electrons more strongly than the nucleus of the other atom. When polar covalent bonds form, the resulting molecule has a partial negative charge near the atom that attracts electrons more strongly. At

FIGURE 2.5 **Covalent bond formation.** The red electrons are shared equally in (a)–(d) and unequally in (e). In writing the structural formula of a covalently bonded molecule, each straight line between the chemical symbols for two atoms denotes a pair of shared electrons. In molecular formulas, the number of atoms in each molecule is noted by subscripts.

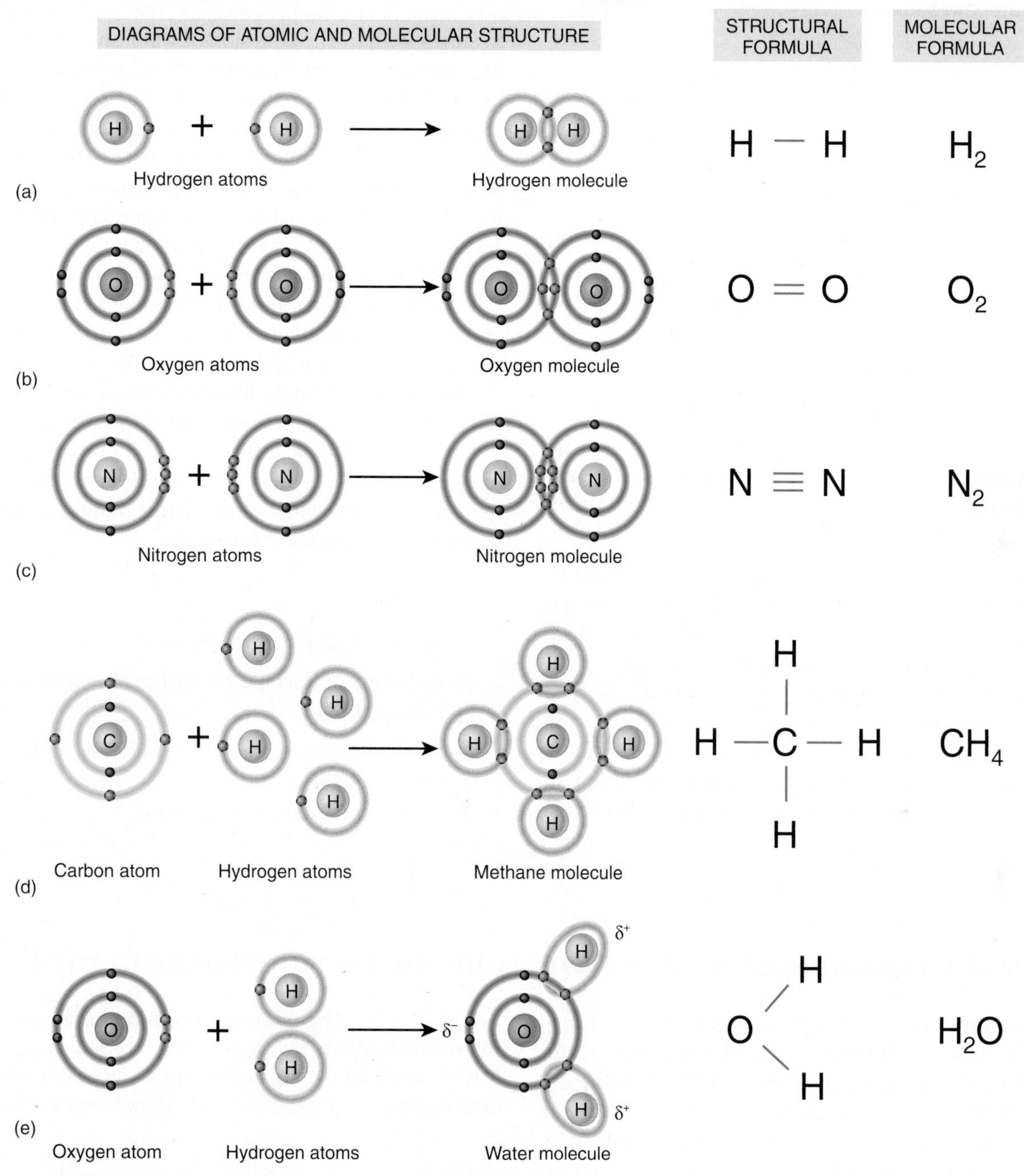


**In a covalent bond, two atoms share one, two, or three pairs of valence electrons.**

least one other atom in the molecule then will have a partial positive charge. The partial charges are indicated by a lowercase Greek delta with a minus or plus sign: $\delta^-$ or $\delta^+$. An important example of a polar covalent bond in living systems is the bond between oxygen and hydrogen in a molecule of water (Figure 2.5e); in this molecule, the nucleus of the oxygen atom attracts the electrons more strongly than the nuclei of the hydrogen atoms, so the oxygen atom has a partial negative charge. Later in the chapter, we will see how polar covalent bonds allow water to dissolve many molecules that are important to life. Bonds between nitrogen and hydrogen and those between oxygen and carbon are also polar bonds.

## Hydrogen Bonds

The polar covalent bonds that form between hydrogen atoms and other atoms can give rise to a third type of chemical bond,

**FIGURE 2.6 Hydrogen bonding among water molecules.** Each water molecule forms hydrogen bonds, indicated by dotted lines, with three to four neighboring water molecules.

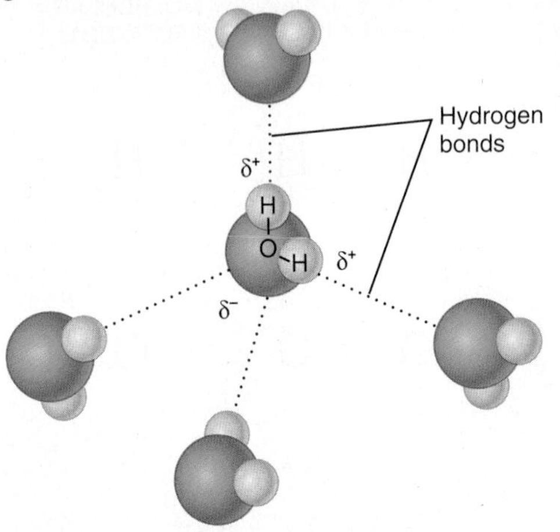


**Hydrogen bonds occur because hydrogen atoms in one water molecule are attracted to the partial negative charge of the oxygen atom in another water molecule.**

a hydrogen bond (Figure 2.6). A **hydrogen bond** forms when a hydrogen atom with a partial positive charge ($\delta^+$) attracts the partial negative charge ($\delta^-$) of neighboring electronegative atoms, most often larger oxygen or nitrogen atoms. Thus, hydrogen bonds result from attraction of oppositely charged parts of molecules, rather than from sharing of electrons as in covalent bonds, or the loss or gain of electrons as in ionic bonds. Hydrogen bonds are weak compared to ionic and covalent bonds. Thus, they cannot bind atoms into molecules. However, hydrogen bonds do establish important links between molecules or between different parts of a large molecule, such as a protein or nucleic acid (both discussed later in this chapter).

The hydrogen bonds that link neighboring water molecules give water considerable *cohesion*, the tendency of like particles to stay together. The cohesion of water molecules creates a very high **surface tension**, a measure of the difficulty of stretching or breaking the surface of a liquid. At the boundary between water and air, water's surface tension is high because the water molecules are much more attracted to one another than they are attracted to molecules in the air. This is readily seen when a spider walks on water or a leaf floats on water. The influence of water's surface tension on the body can be seen in the way it increases the work required for breathing. Because a thin film of watery fluid coats the air sacs of the lungs, each inhalation must have enough force to overcome the opposing effect of surface tension as the air sacs stretch and enlarge when taking in air.

Even though single hydrogen bonds are weak, very large molecules may contain thousands of these bonds. Acting collectively, hydrogen bonds provide considerable strength and stability and help determine the three-dimensional shape of large molecules. As you will see later in this chapter, a large molecule's shape determines how it functions.

**✓ CHECKPOINT**

5. Which electron shell is the valence shell of an atom, and what is its significance?
6. What is the principal difference between an ionic bond and a covalent bond?
7. How does a hydrogen bond form?

## 2.3 Chemical reactions occur when atoms combine with or separate from other atoms.

A **chemical reaction** occurs when new bonds form or existing bonds break between atoms. Chemical reactions are the foundation of all life processes, and as we have seen, the interactions of valence electrons are the basis of all chemical reactions. Consider how hydrogen and oxygen molecules react to form water molecules (Figure 2.7). The starting substances—two $H_2$ and one $O_2$—are known as the **reactants**. The ending substances—two molecules of $H_2O$—are the **products**. The arrow in the figure indicates the direction in which the reaction proceeds. In a chemical reaction, the total number of atoms of each element is the same before and after the reaction. However, because the atoms are rearranged, the reactants and products have different chemical properties. Through thousands of different chemical reactions, body structures are built and body functions are carried out. The term **metabolism** refers to all of the chemical reactions occurring in the body.

**FIGURE 2.7 The chemical reaction between two hydrogen molecules ($H_2$) and one oxygen molecule ($O_2$) to form two molecules of water ($H_2O$).** Note that the reaction occurs by breaking old bonds and making new bonds.

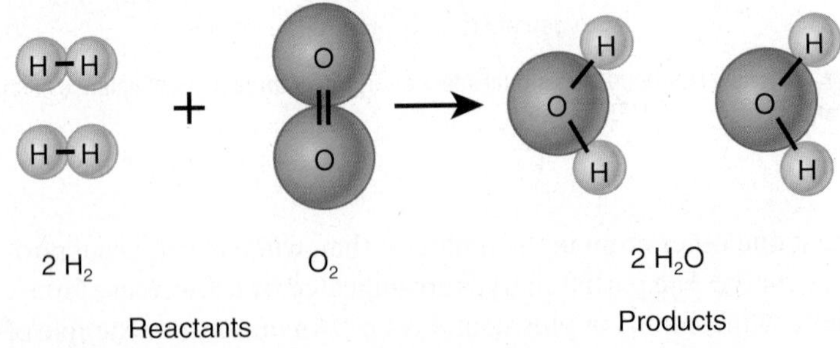


**The number of atoms of each element is the same before and after a chemical reaction.**

## Forms of Energy and Chemical Reactions

Each chemical reaction involves energy changes. **Energy** (*en-* = in; *-ergy* = work) is the capacity to do work. Two principal forms of energy are **potential energy**, energy stored by matter due to its position, and **kinetic energy**, the energy associated with matter in motion. For example, the energy stored in water behind a dam or in a person poised to jump down some steps is potential energy. When the gates of the dam are opened or the person jumps, potential energy is converted into kinetic energy. **Chemical energy** is a form of potential energy that is stored in the bonds of compounds and molecules. The total amount of energy present at the beginning and end of a chemical reaction is the same. Although energy can be neither created nor destroyed, it may be converted from one form to another. For example, some of the chemical energy in the foods we eat is eventually converted into various forms of kinetic energy, such as mechanical energy used to walk and talk. Conversion of energy from one form to another generally releases heat, some of which is used to maintain normal body temperature.

FIGURE 2.8 Activation energy.

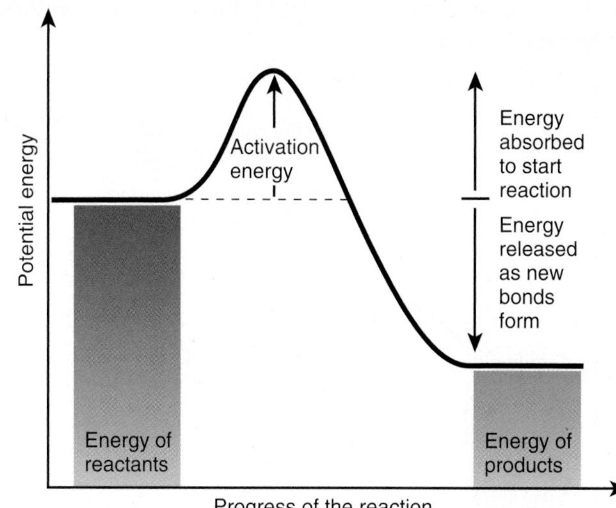


**Activation energy is the energy needed to break chemical bonds in the reactant molecules so a reaction can begin.**

## Energy Transfer in Chemical Reactions

Chemical bonds represent stored chemical energy and chemical reactions occur when new bonds are formed or old bonds are broken between atoms. The *overall reaction* may either release energy or absorb energy. **Exergonic reactions** (*ex-* = out) release more energy than they absorb. By contrast, **endergonic reactions** (*end-* = within) absorb more energy than they release.

A key feature of the body's metabolism is the coupling of exergonic reactions and endergonic reactions. Energy released from an exergonic reaction often is used to drive an endergonic one. In general, exergonic reactions occur as nutrients, such as glucose, are broken down. Some of the energy released may be trapped in the covalent bonds of adenosine triphosphate (ATP), which we describe more fully later in this chapter. The energy transferred to the ATP molecules is then used to drive endergonic reactions needed to build body structures, such as muscles and bones. The energy in ATP is also used to do the mechanical work involved in the contraction of muscle or the movement of substances into or out of cells.

### Activation Energy

Because particles of matter such as atoms, ions, and molecules have kinetic energy, they are continuously moving and colliding with one another. A sufficiently forceful collision can disrupt the movement of valence electrons, causing an existing chemical bond to break or a new one to form. The collision energy needed to break the chemical bonds of the reactants is called the **activation energy** (Figure 2.8). This initial energy "investment" is needed to start a reaction. The reactants must absorb enough energy for their chemical bonds to become unstable and for their valence electrons to form new combinations. Then, as new bonds form, energy is released to the surroundings.

Both the concentration of particles and the temperature influence the chance that a collision will occur and cause a chemical reaction.

- ***Concentration***. The more particles of matter present in a confined space, the greater is the chance that they will collide (think of people crowding into a subway car at rush hour). The concentration of particles increases when more are added to a given space or when the pressure on the space increases, which forces the particles closer together so that they collide more often.
- ***Temperature***. As temperature rises, particles of matter move about more rapidly. Thus, the higher the temperature of matter, the more forcefully particles will collide, and the greater is the chance that a collision will produce a reaction.

### Catalysts

As we have seen, chemical reactions occur when chemical bonds break or form after atoms, ions, or molecules collide with one another. However, body temperature and the concentrations of molecules in body fluids are far too low for most chemical reactions to occur rapidly enough to maintain life. Raising the temperature and the number of reacting particles of matter in the body could increase the frequency of collisions and thus increase the rate of chemical reactions, but doing so could also damage or kill the body's cells.

Substances called catalysts solve this problem. **Catalysts** are chemical compounds that speed up chemical reactions by lowering

**FIGURE 2.9 Comparison of energy needed for a chemical reaction to proceed with a catalyst (blue curve) and without a catalyst (red curve).**

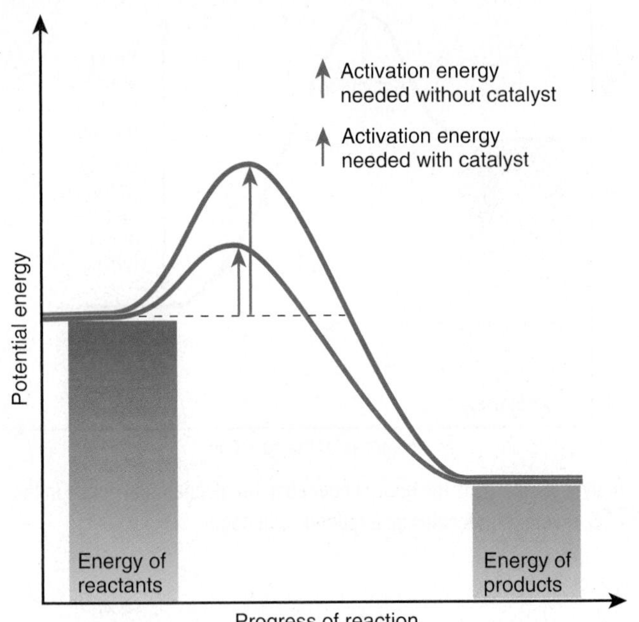


**Catalysts speed up chemical reactions by lowering the activation energy required to initiate them.**

the activation energy needed for a reaction to occur (Figure 2.9). The most important catalysts in the body are enzymes, which we will discuss later in this chapter.

A catalyst does not alter the difference in potential energy between the reactants and the products. Rather, it lowers the amount of energy needed to start the reaction. For chemical reactions to occur, some particles of matter—especially large molecules—must not only collide with sufficient force, but they must "hit" one another at precise spots. A catalyst helps to properly orient the colliding particles. Thus, they interact at the spots that make the reaction happen. Although the action of a catalyst helps to speed up a chemical reaction, the catalyst itself is unchanged at the end of the reaction. A single catalyst molecule can assist one chemical reaction after another.

# Types of Chemical Reactions

After a chemical reaction takes place, the atoms of the reactants are rearranged to yield products with new chemical properties. In this section we will look at the types of chemical reactions so important to the operation of the human body that are discussed throughout the book.

## Synthesis Reactions—Anabolism

**Synthesis reactions** combine two or more atoms, ions, or molecules to form new and larger molecules. The word *synthesis* means "to put together." A synthesis reaction can be expressed as follows:

$$\underset{\text{Atom, ion, or molecule A}}{A} + \underset{\text{Atom, ion, or molecule B}}{B} \xrightarrow{\text{Combine to form}} \underset{\text{New molecule AB}}{AB}$$

One example of a synthesis reaction is the reaction between two hydrogen molecules and one oxygen molecule to form two molecules of water (see Figure 2.7). Another example of a synthesis reaction is the formation of ammonia from nitrogen and hydrogen:

$$\underset{\text{One nitrogen, molecule}}{N_2} + \underset{\text{Three hydrogen molecules}}{3H_2} \xrightarrow{\text{Combine to form}} \underset{\text{Two ammonia molecules}}{2NH_3}$$

All of the synthesis reactions that occur in your body are collectively referred to as **anabolism** (a-NAB-ō-lizm) or anabolic reactions. Overall, anabolic reactions are usually endergonic because they absorb more energy than they release. Combining simple molecules like amino acids (discussed shortly) to form large molecules such as proteins is an example of anabolism.

## Decomposition Reactions—Catabolism

**Decomposition reactions** split up large molecules into smaller atoms, ions, or molecules. A decomposition reaction is expressed as follows:

$$\underset{\text{Molecule AB}}{AB} \xrightarrow{\text{Breaks down into}} \underset{\text{Atom, ion, or molecule A}}{A} + \underset{\text{Atom, ion, or molecule B}}{B}$$

The decomposition reactions that occur in your body are collectively referred to as **catabolism** (ka-TAB-ō-lizm) or catabolic reactions. Overall, catabolic reactions are usually exergonic because they release more energy than they absorb. For instance, during digestion large starch molecules are broken down into many small glucose molecules by catabolic reactions.

## Exchange Reactions

Many reactions in the body are **exchange reactions**; they consist of both synthesis and decomposition reactions. One type of exchange reaction works like this:

$$AB + CD \longrightarrow AD + BC$$

The bonds between A and B and between C and D break (decomposition), and new bonds then form (synthesis) between A and D and between B and C. An example of an exchange reaction is

$$\underset{\text{Hydrochloric acid}}{HCl} + \underset{\text{Sodium bicarbonate}}{NaHCO_3} \longrightarrow \underset{\text{Carbonic acid}}{H_2CO_3} + \underset{\text{Sodium chloride}}{NaCl}$$

Notice that the ions in both compounds have "switched partners": The hydrogen ion ($H^+$) from HCl has combined with the bicarbonate ion ($HCO_3^-$) from $NaHCO_3$, and the sodium ion ($Na^+$) from $NaHCO_3$ has combined with the chloride ion ($Cl^-$) from HCl.

### Reversible Reactions

Some chemical reactions proceed in only one direction, from reactants to products, as previously indicated by the single arrow. Other chemical reactions may be reversible. In a **reversible reaction**, the products can revert to the original reactants. A reversible reaction is indicated by two half-arrows pointing in opposite directions:

$$AB \underset{\text{Combines to form}}{\overset{\text{Breaks down into}}{\rightleftharpoons}} A + B$$

Some reactions are reversible only under special conditions:

$$AB \underset{\text{Heat}}{\overset{\text{Water}}{\rightleftharpoons}} A + B$$

In that case, whatever is written above the arrow indicates the condition needed for the forward reaction to occur, and whatever is written below the arrow designates the condition needed for the reverse reaction to occur. In the example shown above, AB breaks down into A and B only when water is added, and A and B react to produce AB only when heat is applied. Often, different enzymes guide the reactions in opposite directions.

✓ **CHECKPOINT**

8. What is the relationship between reactants and products in a chemical reaction? What is the collective term for chemical reactions occurring in the body?
9. Why is the reaction illustrated in Figure 2.8 exergonic?
10. How do catalysts affect activation energy?
11. Define anabolism and catabolism. Which of the two usually involve endergonic reactions?
12. What are reversible reactions?

## 2.4 Inorganic compounds include water, salts, acids, and bases.

Most of the chemicals in your body exist in the form of compounds. Biologists and chemists divide these compounds into two principal classes: inorganic compounds and organic compounds. **Inorganic compounds** usually lack carbon and are structurally simple. They include water and many salts, acids, and bases. Inorganic compounds may have either ionic or covalent bonds. They include water; many salts, acids, and bases; and two carbon-containing compounds, carbon dioxide ($CO_2$) and bicarbonate ion ($HCO_3$). Water makes up 55–60 percent of a lean adult's total body mass; all other inorganic compounds add 1–2 percent. **Organic compounds**, by contrast, always contain carbon, usually contain hydrogen, and always have covalent bonds. Most are large molecules and many are made up of long chains of carbon atoms. Organic compounds make up the remaining 38–43 percent of the human body.

## Water

**Water** is the most important and abundant inorganic compound in all living systems. Although you might be able to survive for weeks without food, without water you would die in a matter of days. Nearly all the body's chemical reactions occur in a watery medium. Water has many properties that make it an indispensable compound for life. The most important property of water is its polarity—the uneven sharing of valence electrons that confers a partial negative charge near the one oxygen atom and two partial positive charges near the two hydrogen atoms in a water molecule (see Figure 2.5e). This property alone makes water an excellent solvent for other ionic or polar substances, gives water molecules cohesion (the tendency to stick together), and allows water to resist temperature changes.

### Water as a Solvent

In medieval times people searched in vain for a "universal solvent," a substance that would dissolve all other materials. They found nothing that worked as well as water. Although it is the most versatile solvent known, water is not the universal solvent sought by medieval alchemists. If it were, no container could hold it because it would dissolve all potential containers! What exactly is a solvent? A **solvent** is a substance in which some other substance, called a **solute**, can be dissolved. The combination of solvent plus solute is called a **solution**. Usually there is more solvent than solute in a solution. For example, your sweat is a dilute solution of water (the solvent) plus small amounts of salts (the solutes).

The versatility of water as a solvent for ionized or polar substances is due to its polar covalent bonds and its bent shape, which allows each water molecule to interact with several neighboring ions or molecules. Solutes that are charged or contain polar covalent bonds are **hydrophilic** (*hydro-* = water; *-philic* = loving), which means they dissolve easily in water. Common examples of hydrophilic solutes are sugar and salt. Molecules that contain mainly nonpolar covalent bonds, by contrast, are **hydrophobic** (*-phobic* = fearing). They are not very water soluble. Examples of hydrophobic compounds include animal fats and vegetable oils.

To understand the dissolving power of water, consider what happens when a crystal of a salt such as sodium chloride (NaCl)

**FIGURE 2.10 How water molecules dissolve salts and polar substances.** When a crystal of sodium chloride is placed in water, the slightly negative oxygen end (red) of water molecules is attracted to the positive sodium ions ($Na^+$), and the slightly positive hydrogen portions (gray) of water molecules are attracted to the negative chloride ions ($Cl^-$).

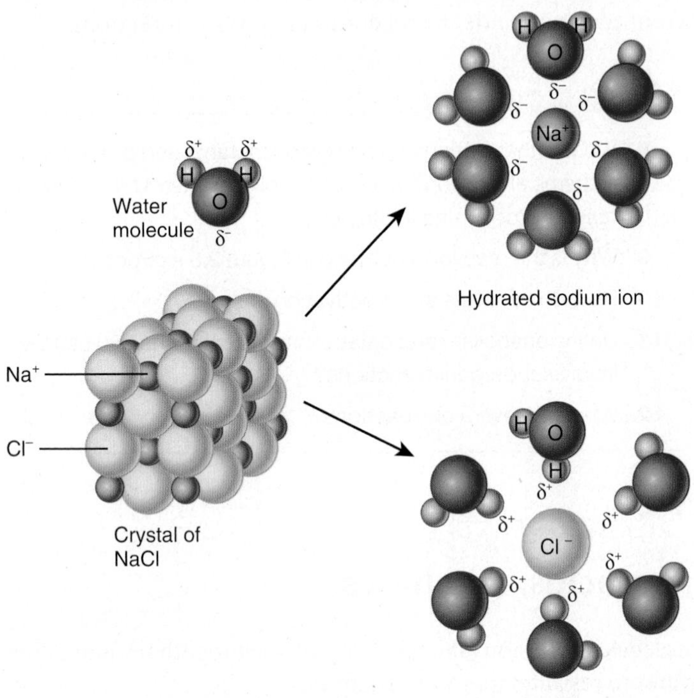


Water is a versatile solvent because its polar covalent bonds, in which electrons are shared unequally, create positive and negative regions.

is placed in water (Figure 2.10). The slightly negative oxygen atom in water molecules attracts the sodium ions ($Na^+$), and the slightly positive hydrogen atoms in water molecules attract the chloride ions ($Cl^-$). Soon, water molecules surround and separate some $Na^+$ and $Cl^-$ ions from each other at the surface of the crystal, breaking the ionic bonds that held NaCl together. The water molecules surrounding the ions also lessen the chance that $Na^+$ and $Cl^-$ ions will come together and reform an ionic bond.

The ability of water to form solutions is essential to health and survival. Because water can dissolve so many different substances, it is an ideal medium for metabolic reactions. Water enables dissolved reactants to collide and form products. Water also dissolves waste products, which allows them to be flushed out of the body in the urine.

### Water in Chemical Reactions

Water serves as the medium for most chemical reactions in the body and participates as a reactant or product in certain reactions. During digestion, for example, decomposition reactions break down large nutrient molecules into smaller molecules by the addition of water molecules. This type of reaction is called **hydrolysis** (hī-DROL-i-sis; *hydro-* = *water*; *-lysis* = to break apart). Hydrolysis reactions allow dietary nutrients to be absorbed into the body. By contrast, when two smaller molecules join to form a larger molecule in a **dehydration synthesis reaction** (*de-* = from, down, or out; *-hydra-* = water), a water molecule is one of the products formed. See Figure 2.20 for an example of such reactions occurring during the synthesis of proteins.

### Thermal Properties of Water

In comparison to most substances, water can absorb or release a relatively large amount of heat with only a slight change in its own temperature. The reason for this property is the large number of hydrogen bonds in water. As water absorbs heat energy, some of the energy is used to break hydrogen bonds. Less energy is then left over to increase the motion of water molecules, which would otherwise increase the water's temperature. The large amount of water in the body lessens the impact of environmental temperature changes, thereby helping maintain the homeostasis of body temperature.

Water also requires a large amount of heat to change from a liquid to a gas. As water evaporates from the surface of the skin, it removes a large quantity of heat, providing an important cooling mechanism.

### Water as a Lubricant

Water is a major component of saliva, mucus, and other lubricating body fluids. Lubrication is especially necessary in the thoracic and abdominal cavities where internal organs touch and slide over one another. It is also needed at joints, where bones, ligaments, and tendons rub against one another. Inside the gastrointestinal tract, mucus and other watery secretions moisten foods, which aids their smooth passage through the digestive system.

## Solutions, Colloids, and Suspensions

A **mixture** is a combination of elements or compounds that are physically blended together but not bound by chemical bonds. For example, the air you are breathing is a mixture of gases that includes nitrogen, oxygen, argon, and carbon dioxide. Three common liquid mixtures are solutions, colloids, and suspensions.

Once mixed together, solutes in a **solution** remain evenly dispersed among the solvent molecules. Because the solute particles in a solution are very small, a solution looks transparent.

A **colloid** differs from a solution mainly because of the size of its particles. The solute particles in a colloid are large enough to scatter light, just as water droplets in fog scatter light from the beams of a car's headlights. For this reason, colloids usually appear translucent or opaque. Milk is an example of a liquid that is both a colloid and a solution: The large milk proteins make milk a colloid, and the dissolved calcium salts, milk sugar (lactose), ions, and other small particles make it a solution.

The solutes in both solutions and colloids do not settle out and accumulate at the bottom of the container. In contrast, the suspended material in a **suspension** may mix with the liquid or suspending medium for some time, but will eventually settle out.

Blood is an example of a suspension. When freshly drawn from the body, blood has an even, reddish color due to the suspension of red blood cells in the liquid portion of blood. After blood sits for a while in a test tube, red blood cells settle out of the suspension and drift to the bottom of the tube (see Figure 18.1a). The upper layer, the liquid portion of blood, appears pale yellow and is called blood plasma. Blood plasma is both a solution of small solutes and a colloid due to the presence of larger plasma proteins.

## Inorganic Acids, Bases, and Salts

Many inorganic compounds can be classified as acids, bases, or salts. When inorganic acids, bases, or salts dissolve in water, they **dissociate** (dis-SŌ-sē-āt), or separate, into ions and become surrounded by water molecules. An **acid** is a substance that dissociates into one or more **hydrogen ions ($H^+$)** and one or more anions (Figure 2.11a). A **base** dissociates into one or more **hydroxide ions ($OH^-$)** and one or more cations (Figure 2.11b).

A **salt**, when dissolved in water, dissociates into cations and anions, neither of which is $H^+$ or $OH^-$ (Figure 2.11c). In the body, salts such as potassium chloride are electrolytes that are important for carrying electrical currents (ions flowing from one place to another), especially in nerve and muscle tissues. The ions of salts also provide many essential chemical elements in body fluids such as blood, lymph, and the interstitial fluid of tissues.

Acids and bases react with one another to form salts. For example, the reaction of hydrochloric acid (HCl) and potassium hydroxide (KOH), a base, produces the salt potassium chloride (KCl) and water ($H_2O$). This exchange reaction can be written as follows:

$$\underset{\text{Acid}}{HCl} + \underset{\text{Base}}{KOH} \longrightarrow \underbrace{H^+ + Cl^- + K^+ + OH^-}_{\text{Dissociated ions}} \longrightarrow \underset{\text{Salt}}{KCl} + \underset{\text{Water}}{H_2O}$$

**FIGURE 2.11 Dissociation of inorganic acids, bases, and salts.**

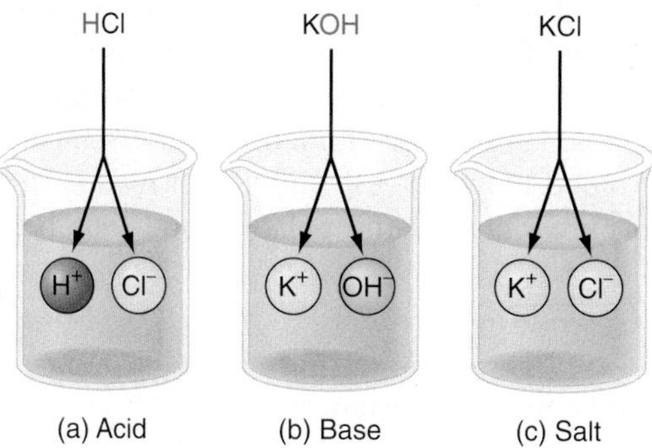


**Dissociation is the separation of inorganic acids, bases, and salts into ions in a solution.**

## Acid–Base Balance: The Concept of pH

To ensure homeostasis, intracellular and extracellular fluids must contain balanced quantities of acids and bases. The more hydrogen ions ($H^+$) dissolved in a solution, the more acidic the solution; the more hydroxide ions ($OH^-$) dissolved in a solution, the more basic (alkaline) the solution. The chemical reactions that take place in the body are very sensitive to even small changes in the acidity or alkalinity of the body fluids in which they occur. Any departure from the narrow limits of normal $H^+$ and $OH^-$ concentrations greatly disrupts body functions.

A solution's acidity or alkalinity is expressed on the **pH scale** (Figure 2.12). This scale is based on the concentration of hydrogen ions in a solution. The pH scale extends from 0 to 14 and is logarithmic, meaning that a change of one whole number on the pH scale represents a *tenfold* change in the hydrogen ion concentration. A pH of 6 denotes 10 times more $H^+$ than a pH of 7, and a pH of 8 indicates 10 times fewer $H^+$ than a pH of 7 and 100 times fewer $H^+$ than a pH of 6.

The midpoint of the pH scale is 7, where the concentrations of $H^+$ and $OH^-$ are equal. A solution with a pH of 7, such as pure water, is neutral—neither acidic nor alkaline. A solution that has more $H^+$ than $OH^-$ is an **acidic solution** and has a pH below 7. A solution that has more $OH^-$ than $H^+$ is a **basic (alkaline) solution** and has a pH above 7.

**FIGURE 2.12 The pH scale.** A pH below 7 indicates an acidic solution—more $H^+$ than $OH^-$. $[H^+]$ = hydrogen ion concentration; $[OH^-]$ = hydroxide ion concentration.

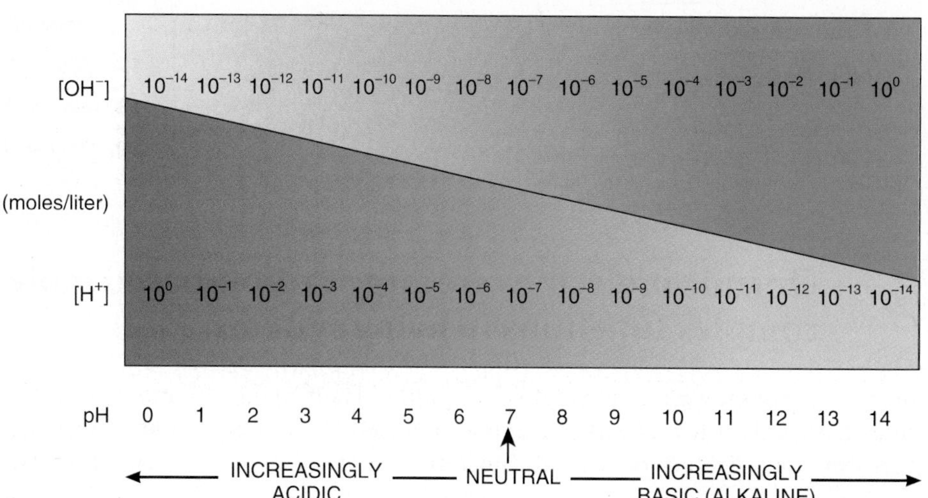


**The lower the numerical value of the pH, the more acidic is the solution because the $H^+$ concentration becomes progressively greater. A pH above 7 indicates a basic (alkaline) solution; that is, there are more $OH^-$ than $H^+$. The higher the pH, the more basic is the solution.**

## Maintaining pH: Buffer Systems

Although the pH of body fluids may differ, as we have discussed, the normal limits for each fluid are quite narrow. Table 2.3 shows the pH values for certain body fluids along with those of common household substances. Homeostatic mechanisms maintain the pH of blood between 7.35 and 7.45, which is slightly more basic than pure water. If the pH of blood falls below 7.35, a condition called acidosis occurs, and if the pH rises above 7.45, it results in a condition called alkalosis; both conditions can seriously compromise homeostasis. Saliva is slightly acidic, and semen is slightly basic. Because the kidneys help remove excess acid from the body, urine can be quite acidic.

**TABLE 2.3**

pH Values of Selected Substances

| SUBSTANCE* | pH VALUE |
|---|---|
| • Gastric juice (found in the stomach) | 1.2–3.0 |
| Lemon juice | 2.3 |
| Vinegar | 3.0 |
| Carbonated soft drink | 3.0–3.5 |
| Orange juice | 3.5 |
| • Vaginal fluid | 3.5–4.5 |
| Tomato juice | 4.2 |
| Coffee | 5.0 |
| • Urine | 4.6–8.0 |
| • Saliva | 6.35–6.85 |
| Milk | 6.8 |
| Distilled (pure) water | 7.0 |
| • Blood | 7.35–7.45 |
| • Semen (fluid containing sperm) | 7.20–7.60 |
| • Cerebrospinal fluid (fluid associated with nervous system) | 7.4 |
| • Pancreatic juice (digestive juice of the pancreas) | 7.1–8.2 |
| • Bile (liver secretion that aids fat digestion) | 7.6–8.6 |
| Milk of magnesia | 10.5 |
| Lye (sodium hydroxide) | 14.0 |

*Bullets (•) denote substances in the human body.

Even though strong acids and bases are continually taken into and formed by the body, the pH of fluids inside and outside cells remains almost constant. One important reason is the presence of **buffer systems**, which function to convert strong acids or bases into weak acids or bases. Strong acids (or bases) ionize easily and contribute many $H^+$ (or $OH^-$) to a solution. Therefore, they can change pH drastically, which can disrupt the body's metabolism. Weak acids (or bases) do not ionize as much and contribute fewer $H^+$ (or $OH^-$). Hence, they have less effect on the pH. **Buffers** are chemical compounds that convert strong acids or bases into weak acids or bases by removing or adding protons ($H^+$).

One important buffer system in the body is the **carbonic acid–bicarbonate buffer system**. Carbonic acid ($H_2CO_3$) can act as a weak acid, and the bicarbonate ion ($HCO_3^-$) can act as a weak base. Hence, this buffer system can compensate for either an excess or a shortage of $H^+$. For example, if there is an excess of $H^+$ (an acidic condition), $HCO_3^-$ can function as a weak base and remove the excess $H^+$, as follows:

$$\underset{\text{Hydrogen ion}}{H^+} + \underset{\text{Bicarbonate ion (weak base)}}{HCO_3^-} \longrightarrow \underset{\text{Carbonic acid}}{H_2CO_3}$$

By contrast, if there is a shortage of $H^+$ (an alkaline condition), $H_2CO_3$ can function as a weak acid and provide needed $H^+$ as follows:

$$\underset{\text{Carbonic acid (weak acid)}}{H_2CO_3} \longrightarrow \underset{\text{Hydrogen ion}}{H^+} + \underset{\text{Bicarbonate ion}}{HCO_3^-}$$

### ✓ CHECKPOINT

13. How do inorganic compounds differ from organic compounds?
14. What functions does water perform in the body?
15. What is the difference between a solvent and a solute?
16. How do hydrophilic molecules differ from hydrophobic molecules?
17. What is hydrolysis?
18. Which pH is more acidic, 6.82 or 6.91?
19. Why are buffers important in maintaining homeostasis?

# 2.5 Organic molecules are large carbon-based molecules that carry out complex functions in living systems.

Inorganic compounds are relatively simple. Their molecules have only a few atoms and cannot be used by cells to perform complicated biological functions. Many organic molecules, by contrast, are relatively large and have unique characteristics that allow them to carry out complex functions. Important categories of organic compounds include carbohydrates, lipids, proteins, nucleic acids, and adenosine triphosphate (ATP).

Organic compounds always contain carbon. Carbon has several properties that make it particularly useful to living organisms. It can form bonds with one to thousands of other carbon atoms to produce large molecules that can have many different shapes. Due to this property of carbon, the body can build many different organic compounds, each of which has a unique structure and function. Moreover, the large size of most carbon-containing molecules and the fact that some do not dissolve easily in water make them useful materials for building body structures.

Organic compounds are usually held together by covalent bonds. Carbon has four electrons in its outermost (valence) shell. It can bond covalently with a variety of atoms, including other carbon atoms, to form rings and straight or branched chains. Other elements that most often bond with carbon in organic compounds are hydrogen, oxygen, nitrogen, sulfur, and phosphorus.

Because organic molecules often are big, there are shorthand methods for representing their structural formulas. Figure 2.13 shows two ways to indicate the structure of the sugar glucose, a molecule with a ring-shaped chain of carbon atoms that has several hydroxyl groups attached.

Small organic molecules can combine into very large molecules called **macromolecules** (*macro-* = large). Macromolecules are usually **polymers** (*poly-* = many; *-mers* = parts). A polymer is a large molecule formed by the covalent bonding of many identical or similar small building-block molecules called **monomers** (*mono-* = one). Usually, the reaction that joins two monomers is a dehydration synthesis. In this type of reaction, a hydrogen atom is removed from one monomer and a hydroxyl group is removed from the other to form a molecule of water (see Figure 2.14). Macromolecules such as carbohydrates, lipids, proteins, and nucleic acids are assembled in cells via dehydration synthesis reactions.

FIGURE 2.13 **Alternative ways to write the structural formula for glucose.**

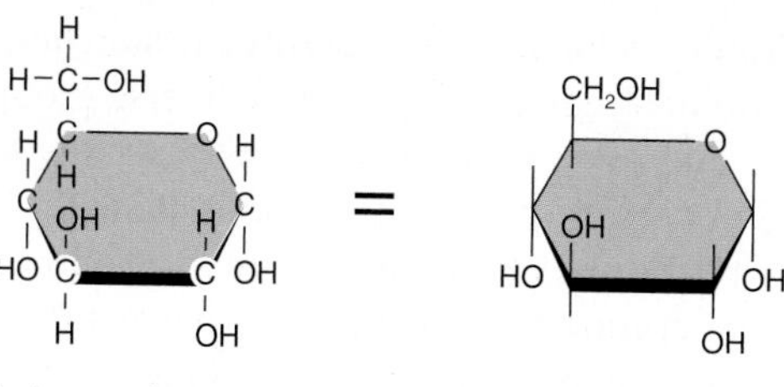


**In standard shorthand, carbon atoms are understood to be at locations where two bond lines intersect, and single hydrogen atoms are not indicated.**

✓ **CHECKPOINT**

**20.** How do polymers and monomers differ?

## RETURN TO Eugene's Story

As the paramedics followed Linda to bed two, Dr. Kim gave instructions. "Linda, get an IV started right away. I think I'd like to get some glucose and B-complex vitamins into Eugene right away. We'll need ABGs (arterial blood gases), too. With his history, Eugene's acid–base balance may be off, especially with the alcohol consumption. Let's get a head and neck CT scan. He may have fallen and hit his head."

Linda helped the paramedics transfer Eugene to the bed. She then began checking Eugene's vital signs: pulse rate 110/min, blood pressure 135/80, respirations 25/min, and blood oxygen saturation 97 percent.

It didn't take long for the lab results to return. Dr. Kim quickly scanned the results.

"His electrolytes are all off. Sodium, potassium, magnesium, and phosphorus levels are all below normal. Linda, keep an eye on his heart monitor and be alert for any sign of seizure activity. Electrolyte disturbance is common in chronic alcoholics, especially when there is damage to the liver. Eugene was in with a bout of pancreatitis a couple of months ago. Might be another flare up. His arterial blood gases look okay, but he is a bit acidotic."

"Linda, let's get some bicarbonate into Eugene; his blood pH is 7.28. It may be from the acute alcohol intoxication, but there may be something else going on.

A. *What are electrolytes and why is the doctor concerned about them with respect to Eugene's heart and brain?*
B. *Why would administering bicarbonate to Eugene be advisable if his blood pH was too low? What is the bicarbonate going to do to adjust Eugene's blood pH?*
C. *Compile the information you currently know about Eugene. What are his vitals? What history do you have? What exam results do you have so far? What further information would help you determine if Eugene's loss of consciousness was from a chemical imbalance or something else?*

## 2.6 Carbohydrates function as building blocks and sources of energy.

**Carbohydrates** include sugars, glycogen, starches, and cellulose. Even though they are a large and diverse group of organic compounds and have several functions, carbohydrates represent only 2–3 percent of your total body mass. In humans and animals, carbohydrates function mainly as a source of chemical energy for generating ATP needed to drive metabolic reactions. Only a few carbohydrates are used for building structural units. One example is deoxyribose, a type of sugar that is a building block of deoxyribonucleic acid (DNA), the molecule that carries inherited genetic information.

Carbon, hydrogen, and oxygen are elements found in carbohydrates. Although there are exceptions, carbohydrates generally contain one water molecule for each carbon atom. This is the reason they are called carbohydrates, which means "watered carbon." The three major groups of carbohydrates, based on their size, are monosaccharides, disaccharides, and polysaccharides (Table 2.4).

**TABLE 2.4**

Major Carbohydrate Groups

| TYPE OF CARBOHYDRATE | EXAMPLES |
|---|---|
| **Monosaccharides** (simple sugars that contain from 3 to 7 carbon atoms) | Glucose (the main blood sugar)<br>Fructose (found in fruits)<br>Galactose (in milk sugar)<br>Deoxyribose (in DNA)<br>Ribose (in RNA) |
| **Disaccharides** (simple sugars formed from the combination of two monosaccharides by dehydration synthesis) | Sucrose (table sugar) = glucose + fructose lactose (milk sugar) = glucose + galactose Maltose = glucose + glucose |
| **Polysaccharides** (from tens to hundreds of monosaccharides joined by dehydration synthesis) | Glycogen (stored form of carbohydrates in animals)<br>Starch (stored form of carbohydrates in plants and main carbohydrates in food)<br>Cellulose (part of cell walls in plants that cannot be digested by humans but aids movement of food through intestines) |

### Monosaccharides and Disaccharides: The Simple Sugars

Monosaccharides and disaccharides are known as **simple sugars**. The monomers of carbohydrates, **monosaccharides** (mon′-ō-SAK-a-rīds; *sacchar-* = sugar), contain from three to seven carbon atoms. They are designated by names ending in "–ose".

A **disaccharide** (dī-SAK-a-rīd; *di-* = two) is a molecule formed from the combination of two monosaccharides by dehydration synthesis. For example, molecules of the monosaccharides glucose and fructose combine to form a molecule of the disaccharide sucrose (table sugar), as shown in Figure 2.14. Notice that the formula for sucrose is $C_{12}H_{22}O_{11}$, not $C_{12}H_{24}O_{12}$, because a molecule of water is removed as the two monosaccharides are joined. Disaccharides can also be split into smaller, simpler molecules by hydrolysis. For example, a molecule of sucrose may be hydrolyzed into its components, glucose and fructose, by the addition of water. Figure 2.14, which is a reversible reaction, also illustrates the hydrolysis of sucrose into glucose and fructose.

### Polysaccharides

The third major group of carbohydrates is the **polysaccharides** (pol′-ē-SAK-a-rīds). Each polysaccharide molecule contains tens or hundreds of monosaccharides joined through dehydration synthesis reactions. Unlike simple sugars, polysaccharides usually are insoluble in water and do not taste sweet. The main polysaccharide in the human body is **glycogen**, which is made entirely of glucose monomers linked to one another in branching chains

FIGURE 2.14 **Structural and molecular formulas for the monosaccharides glucose and fructose, and the disaccharide sucrose.** In dehydration synthesis (read from left to right), two smaller molecules, glucose and fructose, are joined to form a larger molecule of sucrose. Note the loss of a water molecule. In hydrolysis (read from right to left), the addition of a water molecule to the larger sucrose molecule breaks the disaccharide into two smaller molecules, glucose and fructose.

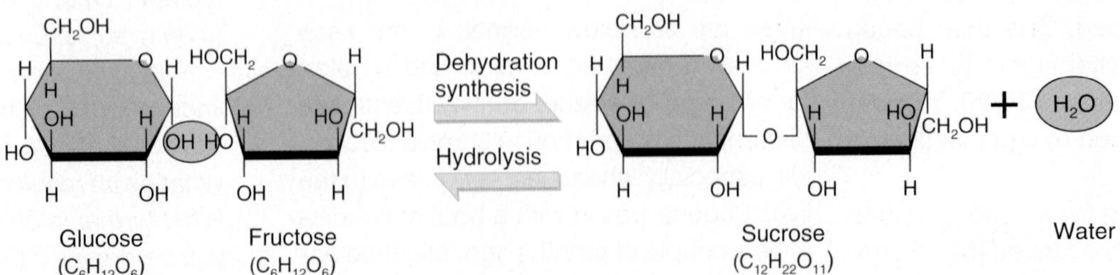


Monosaccharides are the monomers used to build carbohydrates.

## CLINICAL CONNECTION | *Artificial Sweeteners*

Some individuals use **artificial sweeteners** to limit their sugar consumption for medical reasons, while others do so to avoid calories that might result in weight gain. Examples of artificial sweeteners include aspartame (Nutrasweet® and Equal®), saccharin (Sweet 'N Low®), and sucralose (Splenda®). Aspartame is 200 times sweeter than sucrose and it adds essentially no calories to the diet because only small amounts of it are used to produce a sweet taste. Saccharin is about 400 times sweeter than sucrose, and sucralose is 600 times sweeter than sucrose. Both saccharin and sucralose have zero calories because they pass through the body without being metabolized. Artificial sweeteners are also used as sugar substitutes because they do not cause tooth decay. In fact, studies have shown that using artificial sweeteners in the diet helps reduce the incidence of dental cavities.

(Figure 2.15). A limited amount of carbohydrates is stored as glycogen in the liver and skeletal muscles. **Starches** are polysaccharides formed from glucose by plants. They are found in foods such as pasta and potatoes and are the major carbohydrates in the diet. Like disaccharides, polysaccharides such as glycogen and starches can be broken down into monosaccharides through hydrolysis reactions. For example, when the blood glucose level falls, liver cells break down glycogen into glucose and release it into the blood, making glucose available to body cells, which break it down to synthesize ATP. **Cellulose** is a polysaccharide formed from glucose by plants that cannot be digested by humans, but it does provide bulk to help eliminate feces.

**✓ CHECKPOINT**

**21.** What is the primary function of carbohydrates?

**22.** Which body cells store glycogen?

FIGURE 2.15 **Part of a glycogen molecule, the main polysaccharide in the human body.**

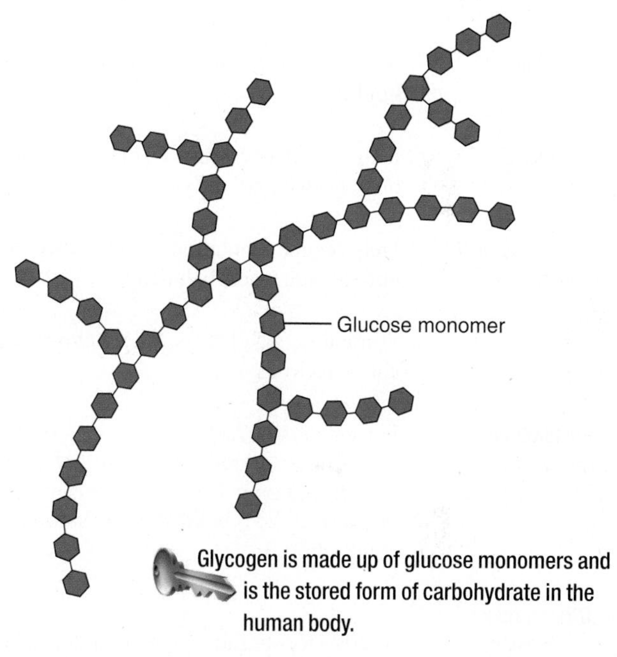


**Glycogen is made up of glucose monomers and is the stored form of carbohydrate in the human body.**

## 2.7 Lipids are important for cell membrane structure, energy storage, and hormone production.

A second important group of organic compounds is **lipids** (*lip-* = fat). Lipids make up 18–25 percent of body mass in lean adults. Like carbohydrates, lipids contain carbon, hydrogen, and oxygen. The proportion of electronegative oxygen atoms in lipids is usually smaller than in carbohydrates, so there are fewer polar covalent bonds. As a result, most lipids are insoluble in polar solvents such as water; they are *hydrophobic*. Because they are hydrophobic, only the smallest lipids (some fatty acids) can dissolve in watery blood plasma. To become more soluble in blood plasma, larger lipid molecules join with hydrophilic protein molecules. The resulting lipid/protein complexes are termed **lipoproteins**. Lipoproteins are soluble because the proteins are on the outside and the lipids are on the inside.

The diverse lipid family includes fatty acids, triglycerides (fats and oils), phospholipids (lipids that contain phosphorus), steroids (lipids that contain rings of carbon atoms), fat-soluble vitamins (vitamins A, D, E, and K), and lipoproteins. Table 2.5 introduces the various types of lipids and highlights their roles in the human body.

### Fatty Acids

Among the simplest lipids are the **fatty acids**, which are used to synthesize triglycerides and phospholipids. Fatty acids can be either saturated or unsaturated. A **saturated fatty acid** contains only *single covalent bonds* between carbon atoms. Because

**TABLE 2.5**

Types of Lipids in the Body

| TYPE OF LIPID | FUNCTIONS |
|---|---|
| **Fatty acids** | Used to synthesize triglycerides and phospholipids or catabolized to generate adenosine triphosphate (ATP). |
| **Triglycerides** *(fats and oils)* | Protection, insulation, energy storage. |
| **Phospholipids** | Major lipid component of cell membranes. |
| **Steroids** | |
| *Cholesterol* | Minor component of all animal cell membranes; precursor of bile salts, vitamin D, and steroid hormones. |
| *Bile salts* | Needed for digestion and absorption of dietary lipids. |
| *Vitamin D* | Helps regulate calcium level in body; needed for bone growth and repair. |
| *Adrenocortical hormones* | Help regulate metabolism, resistance to stress, and salt and water balance. |
| *Sex hormones* | Stimulate reproductive functions and sexual characteristics. |
| **Eicosanoids** *(prostaglandins and leukotrienes)* | Have diverse effects on modifying responses to hormones, blood clotting, inflammation, immunity, stomach acid secretion, airway diameter, lipid breakdown, and smooth muscle contraction. |
| **Other lipids** | |
| *Carotenes* | Needed for synthesis of vitamin A (used to make visual pigments in eye); function as antioxidants. |
| *Vitamin E* | Promotes wound healing, prevents tissue scarring, contributes to normal structure and function of nervous system, and functions as antioxidant. |
| *Vitamin K* | Required for synthesis of blood-clotting proteins. |
| *Lipoproteins* | Transport lipids in blood, carry triglycerides and cholesterol to tissues, and remove excess cholesterol from blood. |

they lack double bonds, each carbon atom is *saturated with hydrogen atoms* (see, for example, palmitic acid in Figure 2.16a). An **unsaturated fatty acid** contains one or more *double covalent bonds* between carbon atoms. Thus, the fatty acid is not completely saturated with hydrogen atoms (see, for example, oleic acid in Figure 2.16a). The unsaturated fatty acid has a kink (bend) at the site of the double bond. If the fatty acid has just one double bond in the hydrocarbon chain, it is *monounsaturated* and it has just one kink. If a fatty acid has more than one double bond in the hydrocarbon chain, it is *polyunsaturated* and it contains more than one kink.

## Triglycerides

The most plentiful lipids in your body and in your diet are the **triglycerides** (trī-GLI-cer-īds; *tri-* = three). A triglyceride consists of a single glycerol molecule and three fatty acid molecules. The three-carbon **glycerol** molecule forms the "backbone" of a triglyceride (Figure 2.16b, c). Three fatty acids are attached by dehydration synthesis reactions, one to each carbon of the glycerol backbone. The reverse reaction, hydrolysis, breaks down a single molecule of a triglyceride into three fatty acids and glycerol.

Triglycerides can be either solids or liquids at room temperature. A **fat** is a triglyceride that is a solid at room temperature. The fatty acids of a fat are mostly saturated. Because these saturated fatty acids lack double bonds in their hydrocarbon chains, they can closely pack together and solidify at room temperature. A fat that mainly consists of saturated fatty acids is called a **saturated fat**. Although saturated fats occur mostly in meats (especially red meats) and nonskim dairy products (whole milk, cheese, and butter), they are also found in a few plant products, such as cocoa butter, palm oil, and coconut oil. Diets that contain large amounts of saturated fats are associated with disorders such as heart disease and colorectal cancer.

An **oil** is a triglyceride that is a liquid at room temperature. The fatty acids of an oil are mostly unsaturated. Recall that unsaturated fatty acids contain one or more double bonds in their hydrocarbon chains. The kinks at the sites of the double bonds prevent the unsaturated fatty acids of an oil from closely packing together and solidifying. The fatty acids of an oil can be either monounsaturated or polyunsaturated. **Monounsaturated fats** contain triglycerides that mostly consist of monounsaturated fatty acids. Olive oil, peanut oil, canola oil, most nuts, and avocados are rich in triglycerides with monounsaturated fatty acids.

**Polyunsaturated fats** contain triglycerides that mostly consist of polyunsaturated fatty acids. Corn oil, safflower oil, sunflower oil, soybean oil, and fatty fish (salmon, tuna, and mackerel) contain a high percentage of polyunsaturated fatty acids. Both monounsaturated and polyunsaturated fats are believed to decrease the risk of heart disease.

Triglycerides are the body's most highly concentrated form of chemical energy. Triglycerides provide more than twice as much energy per gram as do carbohydrates and proteins. Our capacity to store triglycerides in adipose (fat) tissue is unlimited for all practical purposes. Excess dietary carbohydrates, proteins, fats, and oils all have the same fate: They are deposited in adipose tissue as triglycerides.

**FIGURE 2.16 Fatty acid structure and triglyceride synthesis.** Shown in (a) are the structures of a saturated fatty acid and an unsaturated fatty acid. Each time a glycerol and a fatty acid are joined in dehydration synthesis (b), a molecule of water is removed. The fatty acids vary in length and in the number and location of double bonds between carbon atoms (C=C). Shown here (c) is a triglyceride molecule that contains two saturated fatty acids and a monounsaturated fatty acid. The kink (bend) in the oleic acid occurs at the double bond.

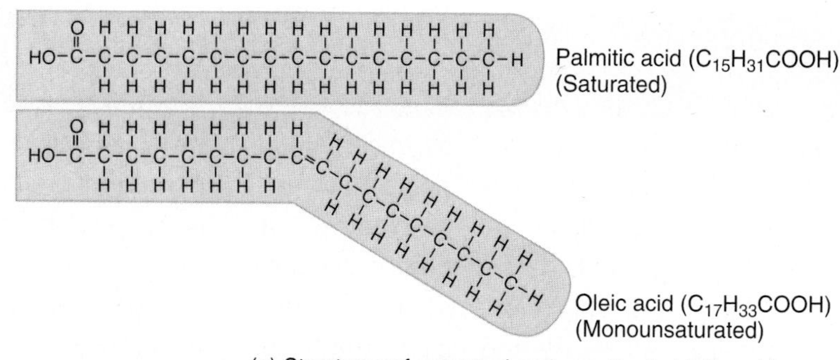


(a) Structures of saturated and unsaturated fatty acids

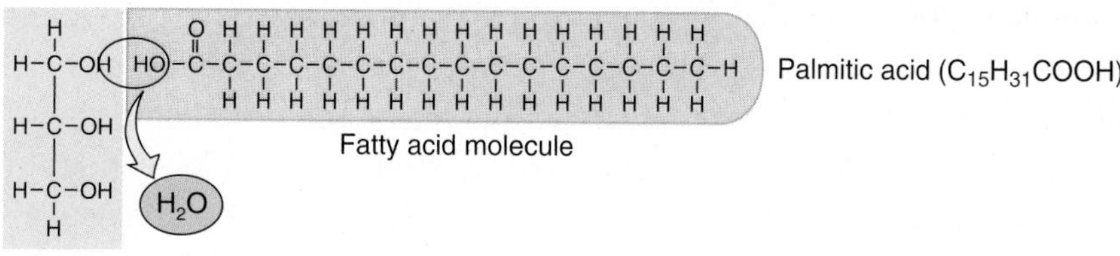


(b) Dehydration synthesis involving glycerol and a fatty acid

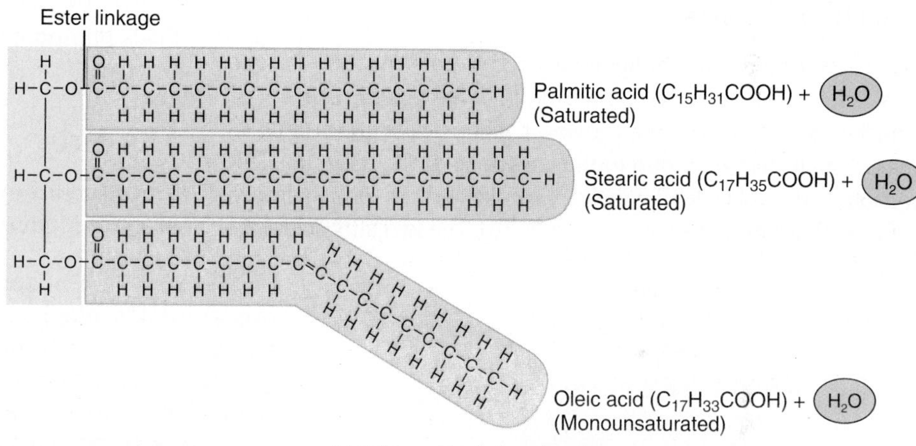


(c) Triglyceride (fat) molecule

**One glycerol and three fatty acids are the monomers of triglycerides.**

## CLINICAL CONNECTION | *Fatty Acids in Health and Disease*

As its name implies, a group of fatty acids called **essential fatty acids (EFAs)** is essential to human health. However, they cannot be made by the human body and must be obtained from foods or supplements. Among the more important EFAs are *omega-3 fatty acids, omega-6 fatty acids,* and *cis-fatty acids.*

Omega-3 and omega-6 fatty acids are polyunsaturated fatty acids that are believed to work together to promote health. They may have a protective effect against heart disease and stroke by lowering total cholesterol, raising HDL (high-density lipoproteins or "good cholesterol"), and lowering LDL (low-density lipoproteins or "bad cholesterol"). In addition, omega-3 and omega-6 fatty acids decrease bone loss by increasing calcium utilization by the body; reduce symptoms of arthritis due to inflammation; promote wound healing; improve certain skin disorders (psoriasis, eczema, and acne); and improve mental functions. Primary sources of omega-3 fatty acids include flaxseed, fatty fish, oils that have large amounts of polyunsaturated fatty acids, fish oils, and walnuts. Primary sources of omega-6 fatty acids include most processed foods (cereals, breads, white rice), eggs, baked goods, oils with large amounts of polyunsaturated fatty acids, and meats (especially organ meats, such as liver).

(continues)

## CLINICAL CONNECTION | *Fatty Acids in Health and Disease (Continued)*

Note in Figure 2.16a that the hydrogen atoms on either side of the double bond in oleic acid are on the same side of the unsaturated fatty acid. Such an unsaturated fatty acid is called a *cis*-fatty acid. *Cis*-fatty acids are nutritionally beneficial unsaturated fatty acids that are used by the body to produce hormone-like regulators and cell membranes. However, when *cis*-fatty acids are heated, pressurized, and combined with a catalyst (usually nickel) in a process called *hydrogenation,* they are changed to unhealthy *trans*-fatty acids. In *trans*-fatty acids the hydrogen atoms are on opposite sides of the double bond of an unsaturated fatty acid. Hydrogenation is used by manufacturers to make vegetable oils solid at room temperature and less likely to turn rancid. Hydrogenated or *trans*-fatty acids are common in commercially baked goods (crackers, cakes, and cookies), salty snack foods, some margarines, and fried foods (donuts and french fries). When oil is used for frying and if the oil is reused (like in fast food french fry machines), *cis*-fatty acids are converted to *trans*-fatty acids. If a product label contains the words hydrogenated or partially hydrogenated, then the product contains *trans*-fatty acids. Among the adverse effects of *trans*-fatty acids are an increase in total cholesterol, a decrease in HDL, an increase in LDL, and an increase in triglycerides. These effects, which can increase the risk of heart disease and other cardiovascular diseases, are similar to those caused by saturated fats.

## Phospholipids

Like triglycerides, **phospholipids** have a glycerol backbone and two fatty acid chains attached to the first two carbons (Figure 2.17a). Attached to the third carbon, however, a phosphate group ($PO_4^{3-}$) links a small charged group to the backbone of the molecule. This portion of the molecule (the "head") is polar and can form hydrogen bonds with water molecules (Figure 2.17b). The two fatty acids (the "tails"), by contrast, are nonpolar and can interact only with other lipids. Phospholipids line up tails to tails in a double row to make up much of the membrane that surrounds each cell (Figure 2.17c).

## Steroids

The structure of **steroids** differs considerably from that of triglycerides. Steroids have four rings of carbon atoms (colored

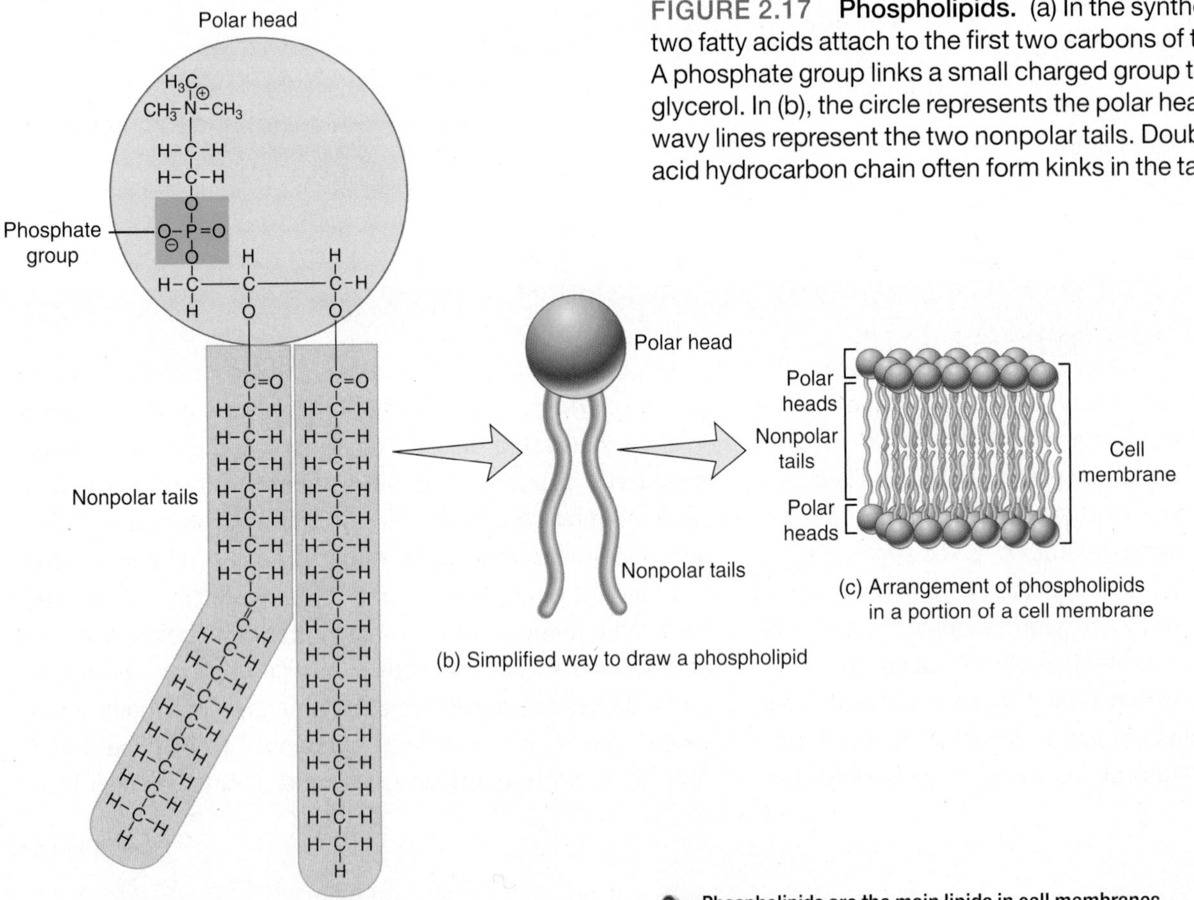


(a) Chemical structure of a phospholipid

(b) Simplified way to draw a phospholipid

(c) Arrangement of phospholipids in a portion of a cell membrane

FIGURE 2.17 **Phospholipids.** (a) In the synthesis of phospholipids, two fatty acids attach to the first two carbons of the glycerol backbone. A phosphate group links a small charged group to the third carbon in glycerol. In (b), the circle represents the polar head region, and the two wavy lines represent the two nonpolar tails. Double bonds in the fatty acid hydrocarbon chain often form kinks in the tail.

**Phospholipids are the main lipids in cell membranes.**

gold in Figure 2.18). Body cells synthesize other steroids from cholesterol (Figure 2.18a). In the body, commonly encountered steroids are cholesterol, estrogens, testosterone, cortisol, bile salts, and vitamin D. Cholesterol is needed for cell membrane structure; estrogens and testosterone are required for regulating sexual functions; cortisol is necessary for maintaining normal blood sugar levels; bile salts are needed for lipid digestion and absorption; and vitamin D is related to bone growth.

FIGURE 2.18 **Steroids.** All steroids have four rings of carbon atoms. The individual rings are designated by the letters A, B, C, and D.

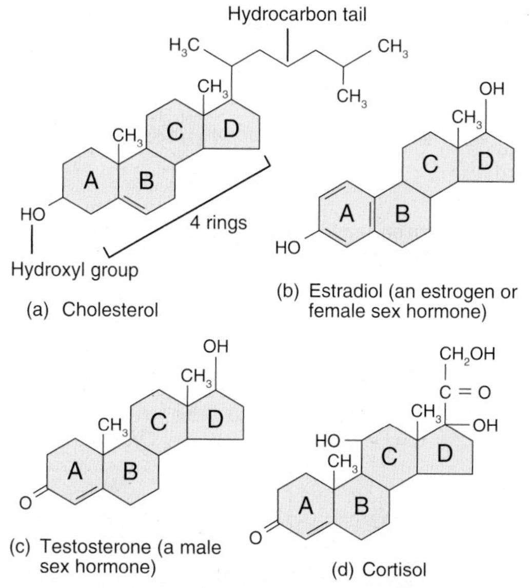

**Cholesterol is the starting material for synthesis of other steroids in the body.**

## Other Lipids

**Prostaglandins** (pros′-ta-GLAN-dins) are lipids with a wide variety of functions. They modify responses to hormones, contribute to the inflammatory response (Chapter 21), prevent stomach ulcers, dilate (enlarge) airways to the lungs, regulate body temperature, and influence formation of blood clots, to name just a few. Other lipids include fat-soluble vitamins such as beta-carotenes (the yellow-orange pigments in egg yolk, carrots, and tomatoes that are converted to vitamin A); vitamins D, E, and K; and lipoproteins.

**✓ CHECKPOINT**

**23.** How do saturated, monounsaturated, and polyunsaturated fats differ in structure? What is one dietary source of each type of triglyceride?

**24.** How does a phospholipid differ structurally from a triglyceride? Which portion of a phospholipid is hydrophilic, and which portion is hydrophobic?

**25.** What is the importance to the body of triglycerides, phospholipids, and steroids?

# 2.8 Proteins are amino acid complexes serving many diverse roles.

**Proteins** are large molecules that contain carbon, hydrogen, oxygen, and nitrogen. Some proteins also contain sulfur. A normal, lean adult body is 12–18 percent protein. Much more complex in structure than carbohydrates or lipids, proteins have many roles

**TABLE 2.6**

Functions of Proteins

| TYPE OF PROTEIN | FUNCTIONS |
|---|---|
| **Structural** | Form structural framework of various parts of body<br>*Examples:* collagen in bone and other connective tissues; keratin in skin, hair, and fingernails |
| **Regulatory** | Function as hormones that regulate various physiological processes; control growth and development; as neurotransmitters, mediate responses of nervous system<br>*Examples:* the hormone insulin (regulates blood glucose level); the neurotransmitter known as substance P (mediates sensation of pain in nervous system) |
| **Contractile** | Allow shortening of muscle cells, which produces movement<br>*Examples:* myosin; actin |
| **Immunological** | Aid responses that protect body against foreign substances and invading pathogens<br>*Examples:* antibodies; interleukins |
| **Transport** | Carry vital substances throughout body<br>*Example:* hemoglobin (transports most oxygen and some carbon dioxide in blood) |
| **Catalytic** | Act as enzymes that regulate biochemical reactions<br>*Examples:* salivary amylase; sucrase; ATPase |

in the body and are largely responsible for the structure of body tissues. Enzymes are proteins that speed up most biochemical reactions. Other proteins work as "motors" to drive muscle contraction. Antibodies are proteins that defend against invading microbes. Some hormones that regulate homeostasis also are proteins. Table 2.6 describes several important functions of proteins.

## Amino Acids and Polypeptides

The monomers of proteins are **amino acids** (a-MĒ-nō). Each of the 20 different amino acids has three important functional groups attached to a central carbon atom (Figure 2.19a): (1) an amino group ($-NH_2$), (2) an acidic carboxyl group ($-COOH$), and (3) a side chain (R group). At the normal pH of body fluids, both the amino group and the carboxyl group are ionized (Figure 2.19b). The different side chains give each amino acid its distinctive chemical identity (Figure 2.19c).

A protein is synthesized in stepwise fashion—one amino acid is joined to a second, a third is then added to the first two, and so on. The covalent bond joining each pair of amino acids is a **peptide bond**. It always forms between the carbon of the carboxyl group ($-COOH$) of one amino acid and the nitrogen of the amino group ($-NH_2$) of another. As the peptide bond is formed, a molecule of water is removed (Figure 2.20), making this a dehydration synthesis reaction. Breaking a peptide bond, as occurs during digestion of dietary proteins, is a hydrolysis reaction (Figure 2.20).

When two amino acids combine, a **dipeptide** results. Adding another amino acid to a dipeptide produces a **tripeptide**. Further additions of amino acids result in the formation of a chainlike **peptide** (4–9 amino acids) or **polypeptide** (10–2000 or more amino acids). Proteins are polypeptide chains that contain as few as 50 amino acids to as many as thousands of amino acids.

Because each variation in the number or sequence of amino acids can produce a different protein, a great variety of proteins is possible. The situation is similar to using an alphabet of 20 letters to form words. Each different amino acid is like a letter, and their various combinations give rise to a seemingly endless diversity of words (peptides, polypeptides, and proteins).

FIGURE 2.19 **Amino acids.** (a) In keeping with their name, amino acids have an amino group (shaded blue) and a carboxyl (acid) group (shaded red). The side chain (R group) is shaded gold and is different in each type of amino acid. (b) At pH close to 7, both the amino group and the carboxyl group are ionized. (c) Glycine is the simplest amino acid; the side chain is a single H atom. Cysteine is one of two amino acids that contain sulfur (S). The side chain in tyrosine contains a six-carbon ring. Lysine has a second amino group at the end of its side chain.

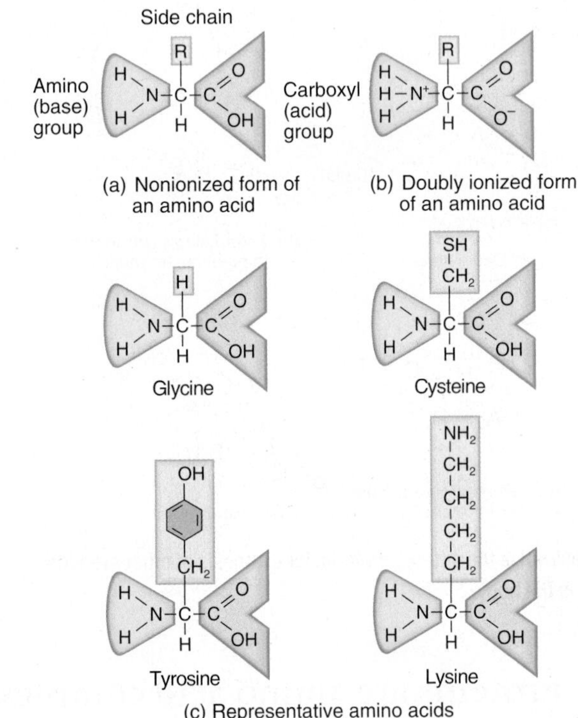


**Body proteins contain 20 different amino acids, each of which has a unique side chain.**

## Levels of Structural Organization in Proteins

Proteins exhibit four levels of structural organization. The **primary structure** is the unique sequence of amino acids that are linked by covalent peptide bonds to form a polypeptide chain (Figure 2.21a).

FIGURE 2.20 **Formation of a peptide bond between two amino acids during dehydration synthesis.** When two amino acids are chemically united by dehydration synthesis (read from left to right), the resulting covalent bond between them is called a peptide bond. The peptide bond is formed at the point where water is lost. Here, the amino acids glycine and alanine are joined to form the dipeptide glycylalanine. Breaking a peptide bond occurs by hydrolysis (read from right to left).

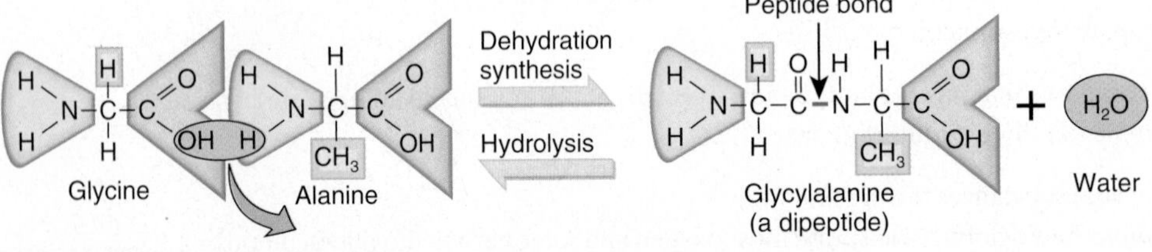


**Amino acids are the monomers used to build proteins.**

FIGURE 2.21 **Levels of structural organization in proteins.** (a) The primary structure is the sequence of amino acids in the polypeptide. (b) Common secondary structures include alpha helixes and beta pleated sheets. For simplicity, the amino acid side groups are not shown here. (c) The tertiary structure is the overall folding pattern that produces a distinctive, three-dimensional shape. (d) The quaternary structure in a protein is the arrangement of two or more polypeptide chains relative to one another.

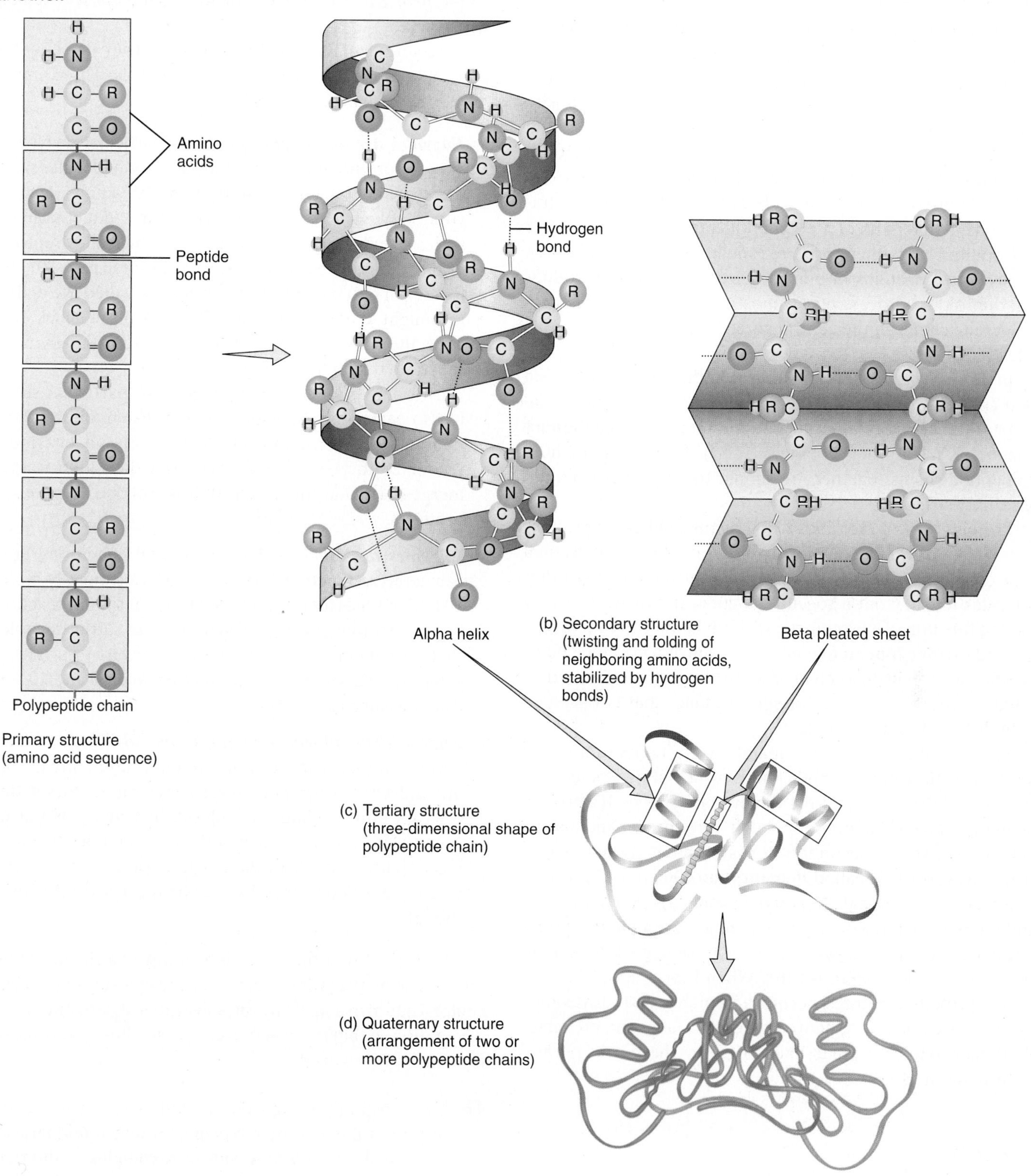


**The unique shape of each protein permits it to carry out specific functions.**

The **secondary structure** of a protein is the repeated twisting or folding of neighboring amino acids in the polypeptide chain (Figure 2.21b). The secondary structure of a protein is stabilized by hydrogen bonds, which form at regular intervals along the polypeptide backbone.

The **tertiary structure** (TUR-shē- er′-ē) refers to the three-dimensional shape of a polypeptide chain. Each protein has a unique tertiary structure that determines how it will function. The tertiary folding pattern may allow amino acids at opposite ends of the chain to be close neighbors (Figure 2.21c). Several types of bonds can contribute to a protein's tertiary structure, including covalent bonds between sulfur atoms called *disulfide bridges*, hydrogen bonds, ionic bonds, and hydrophobic interactions. Some parts of a polypeptide are attracted to water (hydrophilic), and other parts are repelled by it (hydrophobic). Because most proteins in our body exist in watery surroundings, the folding process places most amino acids with hydrophobic side chains in the central core, away from the protein's surface.

In those proteins that contain more than one polypeptide chain (not all of them do), the arrangement of the individual polypeptide chains relative to one another is the **quaternary structure** (KWA-ter-ner′-ē; Figure 2.21d). The bonds that hold polypeptide chains together are similar to those that maintain the tertiary structure.

Proteins vary tremendously in structure. Different proteins have different architectures and different three-dimensional shapes. A protein's unique shape permits it to interact with other molecules to carry out a specific function. In practically every case, the function of a protein depends on its ability to recognize and bind to some other molecule. For example, a hormone binds to a specific protein on a cell in order to alter its function, and an antibody protein binds to a foreign substance that has invaded the body to isolate and disable it.

Homeostatic mechanisms maintain the temperature and chemical composition of body fluids, which allows body proteins to keep their proper three-dimensional shapes. If a protein encounters an altered environment, it may unravel and lose its characteristic shape (secondary, tertiary, and quaternary structure). This process is called **denaturation**. Denatured proteins are no longer functional. Although in some cases denaturation can be reversed, a frying egg is a common example of permanent denaturation. In a raw egg the soluble egg-white protein (albumin) is a clear, viscous fluid. When heat is applied to the egg, the protein denatures, becomes insoluble, and turns white. Just as we are not able to "unfry" an egg, proteins can no longer revert back to their original shapes after they have been permanently denatured.

## Enzymes

In living cells, most catalysts are protein molecules called **enzymes** (EN-zīms). The names of enzymes usually end in the suffix *-ase*. All enzymes can be grouped according to the types of chemical reactions they catalyze. For example, *oxidases* add oxygen, *kinases* add phosphate, *dehydrogenases* remove hydrogen, *ATPases* split ATP, *anhydrases* remove water, *proteases* break down proteins, and *lipases* break down triglycerides.

Enzymes catalyze specific reactions. They do so with great efficiency and with many built-in controls. Three important properties of enzymes are as follows:

- ***Enzymes are highly specific.*** Each particular enzyme binds only to specific **substrates**—the reactant molecules on which the enzyme acts. Of the more than 1000 known enzymes in your body, each has a characteristic three-dimensional shape with a specific surface configuration, which allows it to recognize and bind to certain substrates. In some cases, the part of the enzyme that catalyzes the reaction, called the **active site**, is thought to fit the substrate like a key fits in a lock. In other cases the active site changes its shape to fit snugly around a substrate that enters the site.

  Not only is an enzyme matched to a particular substrate, it also catalyzes a specific reaction. From among the large number of diverse molecules in a cell, an enzyme must recognize the correct substrate and then take it apart or merge it with another substrate to form one or more specific products.
- ***Enzymes are efficient.*** Under optimal conditions, enzymes can catalyze reactions at rates that are from 100 million to 10 billion times more rapid than those of similar reactions occurring without enzymes. The number of substrate molecules that a single enzyme molecule can convert to product molecules in one second is generally between 1 and 10,000 and can be as high as 600,000.
- ***Enzymes are subject to cellular controls.*** Their rate of synthesis and their concentration of enzymes at any given time are under the control of a cell's genes. Substances within the cell may either enhance or inhibit the activity of a given enzyme. Many enzymes have both active and inactive forms in cells. The rate at which the inactive form becomes active or vice versa is determined by the chemical environment inside the cell.

Enzymes lower the activation energy of a chemical reaction by decreasing the "randomness" of the collisions between molecules. They also help bring the substrates together in the proper orientation so that the reaction can occur. Figure 2.22 illustrates how an enzyme works:

1. The substrates make contact with the active site on the surface of the enzyme molecule, forming a temporary compound called the **enzyme–substrate complex**. In this reaction, the two substrate molecules are the sucrose (a disaccharide) and water.

❷ The substrate molecules are transformed by the rearrangement of existing atoms, the breakdown of the substrate molecule, or the combination of several substrate molecules into the products of the reaction. Here the products are two monosaccharides: glucose and fructose.

❸ After the reaction is completed and the reaction products move away from the enzyme, the unchanged enzyme is free to attach to other substrate molecules.

**FIGURE 2.22** How an enzyme works.

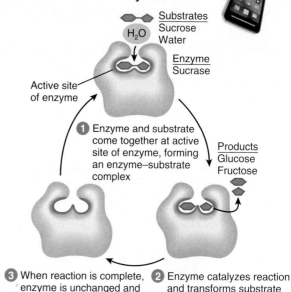

Substrates
Sucrose
Water
$H_2O$
Enzyme
Sucrase
Active site of enzyme
❶ Enzyme and substrate come together at active site of enzyme, forming an enzyme–substrate complex
Products
Glucose
Fructose
❸ When reaction is complete, enzyme is unchanged and free to catalyze same reaction again on new substrates
❷ Enzyme catalyzes reaction and transforms substrate into products

**An enzyme speeds up a chemical reaction without being altered or consumed.**

## ✓ CHECKPOINT

**26.** How are the components of proteins different from those in carbohydrates or lipids?

**27.** What is a peptide bond? How many peptide bonds would there be in a tripeptide?

**28.** What is the difference between a peptide, a polypeptide, and a protein?

**29.** List the important properties of enzymes.

**30.** Describe the mechanism by which enzymes are able to speed up the rate of chemical reactions.

## RETURN TO Eugene's Story

"Were you able to get the CT scan arranged?" asked Dr. Kim.

Linda responded, "Yes, they're sending up someone to transport him to radiology right now."

"Good," she replied. "Let's try to wake him up again. Give me a hand, Linda. Hey, Eugene, can you hear me? Wake up!" Dr. Kim gently nudged Eugene's arm. Linda shook his other arm.

"Huh? What do you want?" asked Eugene lethargically.

"Eugene, I'm Dr. Kim. Someone found you passed out. How do you feel? Are you in any pain?"

Eugene blinked slowly and replied, "I don't know, my stomach was hurting before, but I don't feel it much now."

Dr. Kim looked at Linda. "Might be his pancreas again. Call radiology and have them also get an ultrasound of the pancreas."

The radiology transport aide came in shortly after Eugene woke up and took him for his tests. Several hours passed before he was returned to the emergency department from radiology. Linda made sure Eugene was comfortable before she checked in with Dr. Kim. "So how did Eugene's tests come back, Dr. Kim?"

"Well, he has an enlarged fatty liver and an inflamed pancreas, but no head trauma. Let's draw some more blood and check his electrolytes again. Linda, when you get a chance, please call someone from Social Services. We should try to get him into a rehabilitation program. Eugene's going to be in a lot of pain when the alcohol wears off. Plus he'll have the withdrawal to deal with. He's done some serious damage to his body."

Linda sighed as she glanced at Eugene's chart. "Wow, his liver enzymes are really elevated, lipase too, and his blood sugar was really low when he came in. Why is his cholesterol level so high, Dr. Kim? He seems so malnourished."

D. *Linda mentioned Eugene's elevated cholesterol. Which kind of lipid is cholesterol? What other molecules related to cholesterol may be affected by an imbalance of cholesterol in Eugene's body? How might this be related to Eugene's low blood sugar?*

E. *Dr. Kim finds that Eugene's blood lipase levels are elevated; lipase is an enzyme released by the pancreas that assists in breaking down triglycerides. Which type of reaction would lipase normally be catalyzing?*

F. *Investigate the function of the liver and pancreas. What types of enzymes or organic molecules do these two organs produce? What are the functions of these molecules in the body?*

## 2.9 Nucleic acids contain genetic material and function in protein synthesis.

**Nucleic acids** (nū-KLĒ-ic), so named because they were first discovered in the nuclei of cells, are huge organic molecules that contain carbon, hydrogen, oxygen, nitrogen, and phosphorus. Nucleic acids are of two varieties. The first, **deoxyribonucleic acid (DNA)** (dē-ok′-sē-rī-bō-nū-KLĒ-ik), forms the inherited genetic material inside each human cell (Figure 2.23). In humans, each **gene** is a segment of a DNA molecule. Our genes determine the traits we inherit, and by controlling protein synthesis, they regulate most of the activities that take place in body cells throughout our lives. When a cell divides, its hereditary information passes on to the next generation of cells. **Ribonucleic acid (RNA)**, the second type of nucleic acid, relays instructions

FIGURE 2.23 **DNA molecule.** (a) A nucleotide consists of a base, a five-carbon sugar, and a phosphate group. (b) The paired bases project toward the center of the double helix. The structure is stabilized by hydrogen bonds (dotted lines) between each base pair. There are two hydrogen bonds between adenine and thymine and three between cytosine and guanine.

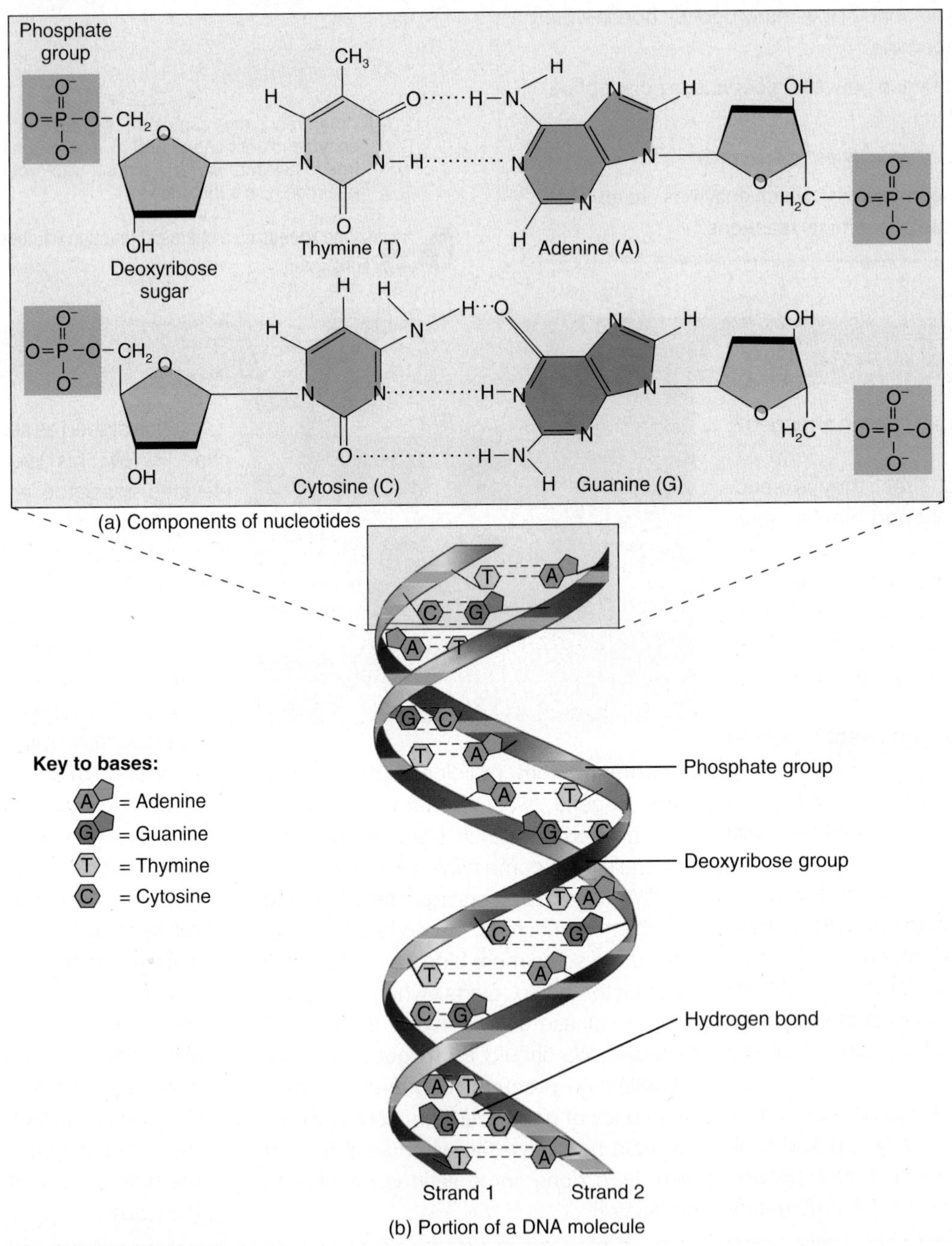


Nucleotides are the monomers of nucleic acids.

from the genes to guide each cell's synthesis of proteins from amino acids.

A nucleic acid is a chain of repeating monomers called **nucleotides**. Each nucleotide of DNA consists of three parts (Figure 2.23a):

1. ***Nitrogenous base.*** DNA contains four different nitrogenous bases, which contain atoms of C, H, O, and N. In DNA the four **nitrogenous bases** are adenine (A), thymine (T), cytosine (C), and guanine (G). Adenine and guanine are larger, double-ring bases, and thymine and cytosine are smaller, single-ring bases.
2. ***Five-carbon sugar.*** A five-carbon sugar called **deoxyribose** attaches to each base in DNA.
3. ***Phosphate group.*** **Phosphate groups** ($PO_4^{3-}$) alternate with deoxyribose to form the "backbone" of each DNA strand; the bases project inward from the backbone chain (Figure 2.23b).

In 1953, F. H. C. Crick of Great Britain and J. D. Watson, a young American scientist, published a brief paper describing how these three components might be arranged in DNA. Their insights into data gathered by others led them to construct a model so elegant and simple that the scientific world immediately knew it was correct! In the Watson–Crick **double helix** model, DNA resembles a spiral ladder (Figure 2.23b). Two strands of alternating phosphate groups and deoxyribose sugars form the uprights of the ladder. Paired bases, held together by hydrogen bonds, form the rungs. Adenine always pairs with thymine, and cytosine always pairs with guanine. Each time DNA is copied, as when living cells divide to increase their number, the two strands unwind. Each strand serves as the template or mold on which to construct a new second strand. Any change that occurs in the base sequence of a DNA strand is called a *mutation*. Some mutations can result in the death of a cell, cause cancer, or produce genetic defects in future generations.

RNA, the second variety of nucleic acid, differs from DNA in several respects. In humans, RNA is single-stranded. The sugar in the RNA nucleotide is **ribose**, and RNA contains the base uracil (U) instead of the thymine found in DNA. Cells contain three different kinds of RNA: messenger RNA, ribosomal RNA, and transfer RNA. Each has a specific role to perform in carrying out the instructions coded in DNA (described in Chapter 3).

**✓ CHECKPOINT**

**31.** What are the functions of DNA and RNA?

**32.** What are the components of a nucleotide?

## 2.10 Adenosine triphosphate (ATP) is the principal energy-transferring molecule in living systems.

**Adenosine triphosphate (ATP)** (a-DEN-ō-sēn trī-FOS-fāt) is the "energy currency" of living systems. ATP transfers the energy liberated in exergonic catabolic reactions to power cellular activities that require energy (endergonic reactions). Among these cellular activities are muscular contractions, movement of chromosomes during cell division, movement of structures within cells, transport of substances across cell membranes, and synthesis of larger molecules from smaller ones. As its name implies, ATP consists of three phosphate groups attached to adenosine, a unit composed of adenine and the five-carbon sugar ribose (Figure 2.24).

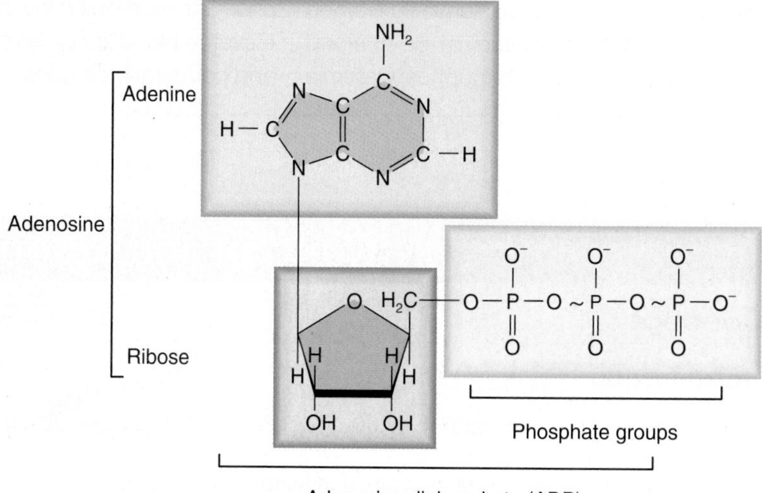


FIGURE 2.24 **Structures of ATP and ADP.** "Squiggles" (~) indicate the two phosphate bonds in the phosphate group that can be used to transfer energy. Energy transfer typically involves hydrolysis of the last phosphate bond of ATP.

**ATP transfers chemical energy to power cellular activities.**

When a water molecule is added to ATP, the third phosphate group ($PO_4^{3-}$), symbolized by Ⓟ in the following discussion, is removed, and the overall reaction liberates energy. The enzyme that catalyzes the hydrolysis of ATP is called *ATPase*. Removal of the third phosphate group produces a molecule called **adenosine diphosphate (ADP)** in the following reaction:

$$\underset{\text{Adenosine triphosphate}}{\text{ATP}} + \underset{\text{Water}}{H_2O} \xrightarrow{\textit{ATPase}} \underset{\text{Adenosine diphosphate}}{\text{ADP}} + \underset{\text{Phosphate group}}{\text{P}} + \underset{\text{Energy}}{\text{E}}$$

The energy supplied by the breakdown of ATP into ADP is constantly being used by the cell. As the supply of ATP at any given time is limited, a mechanism exists to replenish it: The enzyme *ATP synthase* catalyzes the addition of a phosphate group to ADP in the following reaction:

$$\underset{\text{Adenosine diphosphate}}{\text{ADP}} + \underset{\text{Phosphate group}}{\text{P}} + \underset{\text{Energy}}{\text{E}} \xrightarrow{\textit{ATP synthase}} \underset{\text{Adenosine triphosphate}}{\text{ADP}} + \underset{\text{Water}}{H_2O}$$

As you can see from the above reaction, energy is required to produce ATP. The energy needed to attach a phosphate group to ADP is supplied mainly by the catabolism of glucose in cellular respiration, a process that will be discussed in Chapters 10 and 23.

### ✓ CHECKPOINT

**36.** What are some cellular activities that depend on energy supplied by ATP?

**37.** What is the composition of ATP?

## Eugene's Story

## EPILOGUE AND DISCUSSION

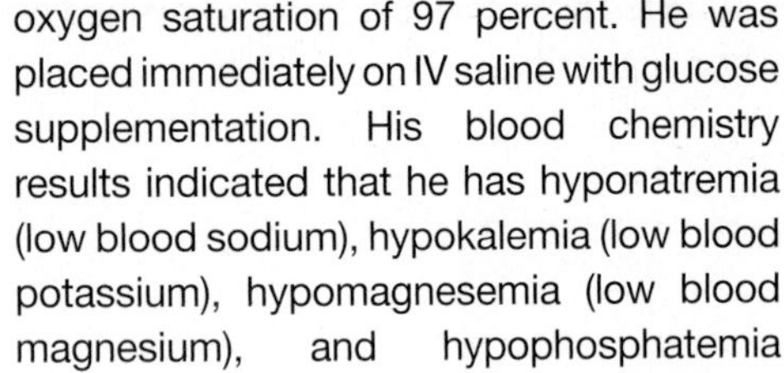

Eugene, a man with a history of chronic alcohol abuse, was brought to the emergency room in an inebriated and unresponsive state. He was found unconscious. Vital signs indicated a pulse rate of 110/min, blood pressure of 135/80, respirations of 25/min, and blood oxygen saturation of 97 percent. He was placed immediately on IV saline with glucose supplementation. His blood chemistry results indicated that he has hyponatremia (low blood sodium), hypokalemia (low blood potassium), hypomagnesemia (low blood magnesium), and hypophosphatemia (low blood phosphorus levels). His blood pH was found to be acidotic (too acidic), and bicarbonate was administered to bring his blood pH back to normal limits. No head trauma was found during Dr. Kim's examination. However, Eugene's liver was enlarged (hepatomegaly) and his pancreas was inflamed (pancreatitis). Further investigation of Eugene's blood chemistries revealed low blood albumin and elevated blood lipase levels, indicating liver and pancreatic damage.

As this story illustrates, damage to the body and its organs and body systems leads to imbalances in homeostasis at the chemical level of organization. This is an important concept to keep in mind as you progress through your studies and your career. An understanding of the chemical level of organization of the human body is needed to diagnose and treat disease appropriately.

G. *Understanding chemistry is important for truly understanding homeostasis. Provide examples from Eugene's story that illustrate the relationship between chemical homeostasis and homeostasis of body systems and the human organism.*

# Concept and Resource Summary

| Concept | Resources  |
|---|---|
| **Introduction**<br>1. The lowest level of structural organization of the body is the chemical level, which consists of atoms that build molecules.<br>2. All living and nonliving things consist of **matter**, anything that occupies space and has **mass**.<br>3. **Chemistry** is the science of the structure and interactions of matter. | |

## Concept | Resources

### Concept 2.1 Chemical elements are composed of small units called atoms.

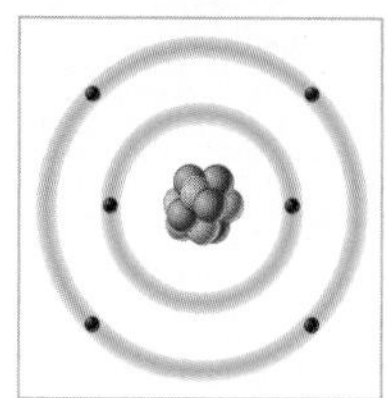

1. Matter exists as solids, liquids, and gases. All matter is made up of building blocks called **chemical elements**, which are substances that cannot be split into simpler substances by ordinary chemical means. Of the 117 elements, only four elements—oxygen, carbon, hydrogen, and nitrogen—make up about 96 percent of body mass.
2. **Atoms** are the building blocks of elements. Many **subatomic particles** make up an atom. The three most important types include positively charged **protons**, uncharged **neutrons**, and negatively charged **electrons**.
3. The protons and neutrons are located in the atomic core, or **nucleus**; the electrons form a moving cloud around the nucleus in regions referred to as **electron shells**.
4. The atoms of each element have a unique **atomic number**, the number of protons in an atom's nucleus. The **mass number** of an atom is the sum of its protons and neutrons. All atoms of a particular element have the same atomic number but can have different mass numbers due to differences in numbers of neutrons; such atoms are called **isotopes**.
5. The mass of an atom is measured in **daltons**. The **atomic mass** (atomic weight) of an element is close to the mass number of its most abundant isotope.
6. Atoms can lose, gain, or share electrons by interacting with other atoms. An **ion** is an atom that has a positive or negative charge due to the loss or gain of electrons.
7. Two or more atoms that share electrons combine to form a **molecule**. A **compound** is a molecule that contains two or more different elements.
8. A **free radical** is an electrically charged atom or group of atoms with an unpaired electron in its outermost shell.

Resources: Animation—Atomic Structure and the Basis of Bonds

### Concept 2.2 Atoms are held together by chemical bonds.

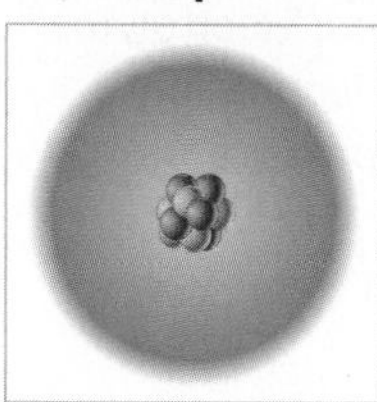

1. **Chemical bonds** are forces that hold together atoms of molecules and compounds. The number of electrons in the **valence** (outermost) **shell** influences formation of chemical bonds.
2. Atoms become charged ions when they gain electrons, resulting in negatively charged **anions**, or when they lose electrons, resulting in positively charged **cations**. Oppositely charged atoms can form an **ionic bond**.
3. An ionic compound that dissociates into positive and negative ions in solution is called an **electrolyte**.
4. A **covalent bond** forms between atoms that share one, two, or three pairs of valence electrons, resulting in a **single covalent bond**, a **double covalent bond**, or a **triple covalent bond**, respectively.
5. Atoms that share electrons equally form a **nonpolar covalent bond**; atoms that share electrons unequally form a **polar covalent bond**.
6. A **hydrogen bond** forms between hydrogen atoms with a partial positive charge and, typically, oxygen or nitrogen atoms with a partial negative charge. Hydrogen bonds between water molecules create the high **surface tension** of water.

Resources: Animation—Chemical Bonding
- Figure 2.6—Hydrogen Bonding among Water Molecules
- Exercise—Bond Boulevard
- Concepts and Connections—Chemical Bonds

### Concept 2.3 Chemical reactions occur when atoms combine with or separate from other atoms.

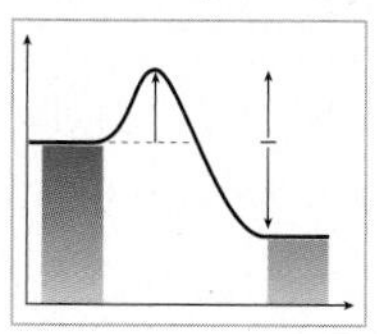

1. A **chemical reaction** involves formation of new bonds between atoms or breaking of existing bonds. In chemical reactions, the starting substances are the **reactants** and the ending substances are the **products**.
2. All of the chemical reactions of the body are collectively referred to as **metabolism**.
3. There are two principal forms of **energy**, the capacity to do work: **potential energy** is stored energy, and **kinetic energy** is the energy of motion. **Chemical energy** is the potential energy stored in chemical bonds.

Resources: Animation—Types of Reactions and Equilibrium
- Exercise—Reaction Race

| Concept | Resources |
|---|---|

4. A chemical reaction that releases energy is **exergonic**; one that absorbs energy is **endergonic**. These two types of reactions are often coupled. Released energy is transferred to the formation of ATP molecules, which the body can use to drive endergonic reactions and to perform mechanical work.
5. Atoms, ions, and molecules have kinetic energy and continuously move and collide with one another. **Activation energy** is the collision energy needed to break chemical bonds. Temperature and the concentration of the particles influence the chance that a collision will produce a reaction.
6. **Catalysts** are chemical compounds, such as enzymes, that speed up chemical reactions by lowering the activation energy needed for a reaction to occur.
7. Important chemical reactions in the body include **synthesis reactions (anabolism)**, **decomposition reactions (catabolism)**, **exchange reactions** (exchange of the atoms of different molecules among compounds), and **reversible reactions** (in which the products can revert to reactants under the right conditions).

## Concept 2.4 Inorganic compounds include water, salts, acids, and bases.

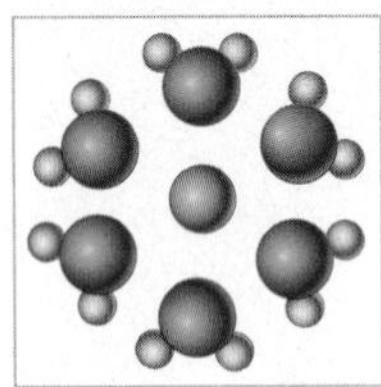

1. Compounds can be divided into two principal classes. **Inorganic compounds** are structurally simple and lack carbon; **organic compounds** always contain carbon and are typically large molecules.
2. **Water** is an important and abundant inorganic compound in the body. Its polarity makes it an excellent solvent in body fluids. A **solvent** is a liquid or gas in which the **solutes** of a **solution** dissolve. Solutes that are charged or contain polar covalent bonds are **hydrophilic** and dissolve in water; molecules with nonpolar covalent bonds are **hydrophobic** and are not very water-soluble.
3. **Hydrolysis** reactions in the body involve the breakdown of large nutrient molecules into smaller molecules by the addition of water; **dehydration synthesis reactions** involve the removal of a water molecule as two smaller molecules form a larger molecule.
4. Water can absorb and release a large amount of heat with only a slight change in its own temperature. It also allows evaporative cooling because it requires a large amount of heat to vaporize water from a liquid (in sweat) to a gas.
5. Water is a lubricant in saliva, mucus, and other body fluids.
6. A **mixture** is a combination of elements or compounds. The three common liquid mixtures are solutions, colloids, and suspensions. Solute particles in a **solution** are small and evenly dispersed and do not settle out. A **colloid** has larger solute particles that do not settle out. A **suspension** has large solute particles that do settle out.
7. In water, inorganic acids, bases, and salts separate into ions (**dissociate**). **Acids** dissociate into **hydrogen ions ($H^+$)** and anions. **Bases** dissociate into **hydroxide ions ($OH^-$)** and cations. **Salts** dissociate into cations and anions that are not $H^+$ or $OH^-$. Salts in the body are electrolytes that carry electrical currents.
8. The acidity or alkalinity of a solution is expressed on a **pH scale** from 0 to 14 that is based on the number of hydrogen ions in a solution. At pH 7, $H^+$ and $OH^-$ are equal and the solution is neutral. **Acidic solutions** have more $H^+$ than $OH^-$, and **basic (alkaline) solutions** have more $OH^-$ than $H^+$. A change of one whole number on the pH scale represents a 10-fold change in the concentration of $H^+$ ions.
9. Body fluids have normal pH ranges that must be maintained by homeostatic mechanisms. **Buffers** are chemical compounds that help stabilize the pH of a solution by adding or removing protons ($H^+$). **Buffer systems** in body fluids convert strong acids or bases into weak acids and bases. The **carbonic acid–bicarbonate buffer system** functions in the homeostasis of blood pH, which has a normal range of 7.35 to 7.45.

**Resources (Concept 2.4):**

Anatomy Overview—Common Biomolecules: Water
- Animation—Polarity and Solubility of Molecules
- Animation—Water and Fluid Flow
- Animation—Acids and Bases
- Anatomy Overview—Common Biomolecules: Blood Gases
- Anatomy Overview—Common Biomolecules: Electrolytes
- Exercise—Destination: Acid/Base Balance

**Concept** | **Resources** WILEY PLUS

**Concept 2.5** Organic molecules are large carbon-based molecules that carry out complex functions in living systems.

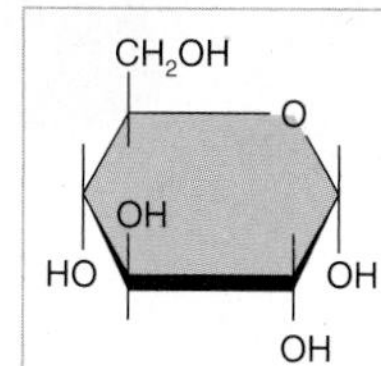


1. Organic compounds always contain carbon atoms, are often covalently bonded to form long carbon chains, and are bonded to other atoms, particularly hydrogen, oxygen, and nitrogen.
2. Small organic molecules can combine into large **macromolecules** such as **polymers**, which are formed by covalent bonding of identical or similar **monomers**.

Concepts and Connections—Organic Molecules

**Concept 2.6** Carbohydrates function as building blocks and sources of energy.

1. **Carbohydrates** include sugars, starches, glycogen, and cellulose. Carbon, oxygen, and hydrogen are the elements in carbohydrates; the ratio of hydrogen to oxygen atoms is usually 2:1. The three major groups of carbohydrates are monosaccharides, disaccharides, and polysaccharides.
2. **Simple sugars—monosaccharides** and **disaccharides**—contain from three to seven carbon atoms per monomer and are the building blocks of carbohydrates.
3. The monosaccharide glucose is a major source of chemical energy for generating ATP. Other monosaccharides include ribose and deoxyribose.
4. Disaccharides such as sucrose are formed by dehydration synthesis from two monosaccharide molecules.
5. **Polysaccharides** (complex carbohydrates) contain many monosaccharides joined by dehydration synthesis reactions. They typically differ from monosaccharides and disaccharides because they don't taste sweet and are insoluble in water. **Glycogen**, which is made of glucose monomers, is stored in liver and muscle cells. Other polysaccharides used by the body include **starches** and **cellulose** from plants in the diet.

Anatomy Overview—Carbohydrates

**Concept 2.7** Lipids are important for cell membrane structure, energy storage, and hormone production.

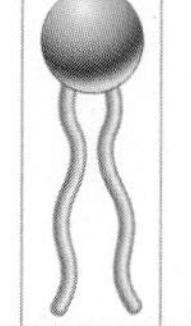

1. **Lipids** have carbon, hydrogen, and some oxygen. They are insoluble in water. Major lipids include triglycerides, phospholipids, steroids, fatty acids, and fat-soluble vitamins (A, D, E, and K).
2. **Triglycerides** include **fats** and **oils**. They serve as a highly concentrated form of chemical energy in the body. **Glycerol** and three **fatty acids** are the building blocks of triglycerides.
3. **Saturated fats** are usually solid at room temperature and contain only single covalent bonds between fatty acid carbon atoms. **Monounsaturated fats** contain fatty acids with one double covalent bond between two fatty acid carbon atoms. **Polyunsaturated fats** contain more than one double covalent bond between fatty acid carbon atoms.
4. **Phospholipids** are found in all plasma membranes arranged in a double row. They have glycerol, two fatty acid chains, and a polar phosphate group.
5. **Steroids** have four rings of carbon atoms. Body steroids include cholesterol, estrogens, testosterone, cortisol, bile salts, and vitamin D.

Anatomy Overview—Lipids

**Concept 2.8** Proteins are amino acid complexes serving many diverse roles.

1. **Proteins** are large, complex molecules that contain carbon, hydrogen, oxygen, and nitrogen, and many contain sulfur. Proteins have diverse roles in the body including hormones, catalysts, and antibodies.
2. **Amino acids** are monomers that serve as the building blocks of proteins; there are 20 different amino acids. Each amino acid has a central carbon atom to which the following are attached: (1) an amino group (—$NH_2$); (2) a carboxyl group (—COOH), and (3) a side chain (R).

Anatomy Overview—Proteins
- Animation—Enzyme Functions and ATP
- Figure 2.20—Formation of a Peptide Bond between Two Amino Acids during Dehydration Synthesis

**Concept** | **Resources**

3. Amino acids are bonded by **peptide bonds** during dehydration synthesis, resulting in **dipeptides**, **tripeptides**, and **polypeptides**.
4. The **primary structure** of a protein is its unique sequence of amino acids forming a polypeptide chain. The **secondary structure** is the twisting or folding of amino acids in the polypeptide chain into alpha helixes and beta pleated sheets. The **tertiary structure** involves further folding that determines the function of the protein. The **quaternary structure** occurs in proteins having two or more polypeptide chains and involves bonds joining the polypeptide chains similar to those that create tertiary structure.
5. When proteins are in certain unfavorable environments, they may unravel and lose their unique shapes and functions in a process called **denaturation**.
6. Most catalysts in body cells are protein molecules called **enzymes**.
7. Enzymes have specificity; they can only bind to specific **substrates**. The **active site** of an enzyme is the part that catalyzes the reaction. Another property of enzymes is efficiency; they can catalyze reactions at extremely rapid rates. Cellular controls, such as genes, affect the rate of synthesis and the concentration of enzymes.
8. Enzymes lower the activation energy of a chemical reaction and work in a three-step process: (1) the **enzyme–substrate complex** forms, (2) substrate molecules are transformed into products, and (3) unchanged enzymes release the reaction products and become free to catalyze the next reaction.

Resources:
- Figure 2.22—How an Enzyme Works
- Exercise—Enzyme Anticipation

## Concept 2.9 Nucleic acids contain genetic material and function in protein synthesis.

1. **Nucleic acids** include **deoxyribonucleic acid (DNA)**, which forms the genetic material of cells, and **ribonucleic acid (RNA)**, which functions in protein synthesis.
2. A nucleic acid is a chain of monomers called **nucleotides**.
3. Each nucleotide in DNA consists of three parts: a **nitrogenous base** (adenine, thymine, cytosine, or guanine); the five-carbon sugar **deoxyribose**; and a **phosphate group**. DNA has two strands of nucleotides in a **double helix**, so it is referred to as double-stranded.
4. RNA is single-stranded. Each of its nucleotides consist of three parts: a nitrogenous base (adenine, cytosine, guanine, uracil); the five-carbon sugar **ribose**; and a phosphate group.
5. There are three types of RNA: messenger RNA, ribosomal RNA, and transfer RNA.

Resources: Anatomy Overview—Nucleic Acids

## Concept 2.10 Adenosine triphosphate (ATP) is the principal energy-transferring molecule in living systems.

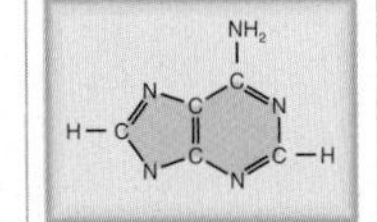

1. **Adenosine triphosphate (ATP)** is the "energy currency" of cells. ATP transfers the energy liberated in exergonic catabolic reactions to energy-requiring cellular activities, including endergonic reactions and mechanical work.
2. ATP is composed of adenine, ribose, and three phosphate groups.
3. ATPase catalyzes the hydrolysis of ATP to ADP and phosphate; ATP synthase catalyzes the addition of a phosphate group to **adenosine diphosphate (ADP)** to form ATP.

Resources: Animation—Enzyme Functions and ATP

## Understanding the Concepts

1. Carbon dioxide ($CO_2$) and nitrogen ($N_2$) are gases involved in respiratory physiology. Would each of these gases best be described as a molecule or a compound? Explain your answer.
2. Which atom in a water molecule has a greater ability to attract electrons? Why?
3. Will the element potassium (K) be more likely to form an anion or a cation? Why? (Hint: See Figure 2.2 for the atomic structure of K.)
4. What form of energy is found in an apple you just ate? A vase sitting on a high shelf? A vase falling off a high shelf? How is energy conserved in each of these scenarios?
5. Are you creating a solution, colloid, or suspension in each of the following situations? Explain each of your answers.
   - As you stir a teaspoon of table sugar into a glass of iced tea, you can see sugar crystals spinning around undissolved.
   - After stirring a teaspoon of table sugar into a cup of hot tea, you notice that the liquid appears clear, indicating the table sugar crystals have dissolved in the tea.
   - While preparing a salad dressing, the vinegar and oil appear to stay mixed together as long as you keep vigorously stirring, but if you quit, the vinegar separates from the oil and settles to the bottom of the container.
6. The compound $CaCO_3$ (calcium carbonate) dissociates into a calcium ion ($Ca^{2+}$) and a carbonate ion ($CO_3^{2-}$). Is it an acid, a base, or a salt? What about $H_2SO_4$, which dissociates into two $H^+$ and one $SO_4^{2-}$? Explain your answers.
7. Why is carbon an essential component of all organic molecules?
8. What are the structural differences among monosaccharides, disaccharides, and polysaccharides?
9. Why are most lipids insoluble in water?
10. How do DNA and RNA differ functionally?
11. Do all proteins have a tertiary structure? Do they all have a quaternary structure?
12. Which levels of structural organization are lost when a protein is denatured?
13. Referring to Figure 2.22, why can't sucrase catalyze the formation of sucrose from glucose and fructose?
14. In the reaction catalyzed by ATP synthase shown in Concept 2.10, identify the substrates and products. Is this an exergonic reaction or an endergonic reaction? Explain your answer.

CHAPTER

# 3

## Joseph's Story

The bright summer sun shone down on Joseph and his son Marcus as they began setting up camp for a weekend father–son getaway. Joseph had been looking forward to it for weeks. He'd been busy working long hours and the stress of his job was wearing on his body. He had felt tightness in his chest, he couldn't sleep, his mind raced, and he had begun smoking again after three smoke-free years. He needed this time off. "Dad, let's take a break and play catch for a while," said Marcus.

"Okay, Marc, but take it easy on your old dad," said Joseph. Marcus was only 12 years old and wasn't much of a match for his 38-year-old dad, but liked to think he was. "Pitch that ball in here, slugger," said Joseph as he squatted down to catch the ball. After he threw the ball as hard as he could, Marcus thought his father was acting odd. Joseph grunted; as he reached for the ball, his face contorted and he collapsed on the ground in a heap.

"Dad, come on, knock it off. Quit fooling around and throw me the ball. You didn't even try to catch that one!" Marcus jogged toward Joseph. "Dad, get up." Marcus ran the last few steps to Joseph's side, and saw that his father's face was ashen gray. He wasn't breathing. Marcus panicked and screamed. "Dad! Someone help! Help me! Something's wrong with my dad!"

Most of us tend to think of humans at the organismal level. Who we are is more than simply a collection of chemicals, cells, and organ systems. However, the cellular level is essential to the homeostasis of the entire organism. In Joseph's case, you will see that each cell of his body is fighting a battle for survival.

# The Cellular Level of Organization

## INTRODUCTION

In Chapter 1, you learned that the human body has various levels of organization, and Chapter 2 helped you understand the chemistry of the human body. In this chapter you will see how atoms and molecules are organized on the cellular level to form the structures comprising cells and to perform cellular activities.

**Cells** are living structural and functional units enclosed by a membrane. About 200 different types of specialized cells carry out a multitude of unique biochemical or structural roles that support homeostasis and contribute to the many coordinated functional capabilities of the human body. **Cell biology** is the study of cellular structure and function. As you study the various parts of a cell and their relationships to each other, you will learn that cell structure and function are intimately related. In this chapter, you will also learn that cells carry out a dazzling array of chemical reactions to create and maintain life processes—in part, by isolating specific types of chemical reactions within specialized cellular structures.

## CONCEPTS

**3.1** The principal parts of a cell are the plasma membrane, the cytoplasm, and the nucleus.

**3.2** The plasma membrane contains the cytoplasm and regulates exchanges with the extracellular environment.

**3.3** Transport of a substance across the plasma membrane occurs by both passive and active processes.

**3.4** Cytoplasm consists of the cytosol and organelles.

**3.5** The nucleus contains nucleoli and genes.

**3.6** Cells make proteins by transcribing and translating the genetic information contained in DNA.

**3.7** Cell division allows the replacement of cells and the production of new cells.

# 3.1 The principal parts of a cell are the plasma membrane, the cytoplasm, and the nucleus.

Figure 3.1 provides an overview of the typical structures found in human cells. Most cells have many of the structures shown in this diagram, but no one cell has all of them. For ease of study, we divide the cell into three main parts: plasma membrane, cytoplasm, and nucleus.

- The **plasma membrane** forms the cell's flexible outer surface, separating the cell's internal environment (everything inside the cell) from the external environment (everything outside the cell). This selective barrier regulates the flow of materials into and out of a cell to help establish and maintain the appropriate environment for normal cellular activities. The plasma membrane also plays a key role in communication among cells and between cells and their external environment.
- The **cytoplasm** (SĪ-tō-plasm; *-plasm* = formed or molded) consists of all the cellular contents between the plasma membrane and the nucleus. The cytoplasm consists of cytosol and organelles. **Cytosol** (SĪ-tō-sol), the fluid portion of cytoplasm, contains water, dissolved solutes, and suspended particles. It is also called **intracellular fluid**. Within the cytosol are several different types of **organelles** (or-ga-NELZ = little organs). Each type of organelle has a characteristic shape and specific functions. Examples include the endoplasmic reticulum, Golgi complex, lysosomes, peroxisomes, and mitochondria.
- The **nucleus** (NOO-klē-us = nut kernel) is a large organelle that houses most of a cell's DNA. Within the nucleus, each **chromosome** (KRŌ-mō-sōm; *chromo-* = colored), a single molecule of DNA associated with several proteins, contains thousands of hereditary units called **genes** that control most aspects of cellular structure and function.

## ✓ CHECKPOINT

1. What is the difference between cytoplasm and cytosol?

FIGURE 3.1 Typical structures found in many cells.

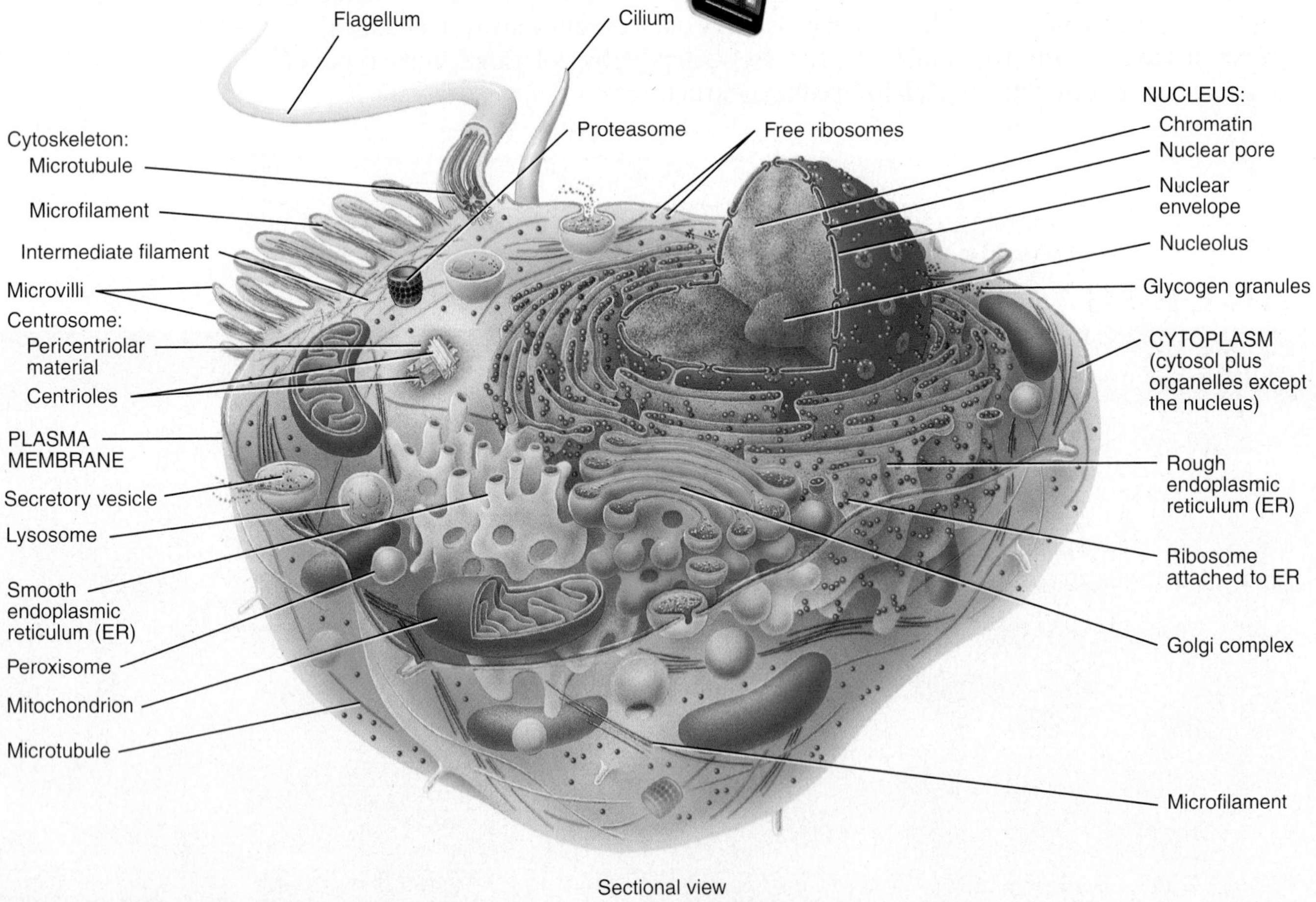

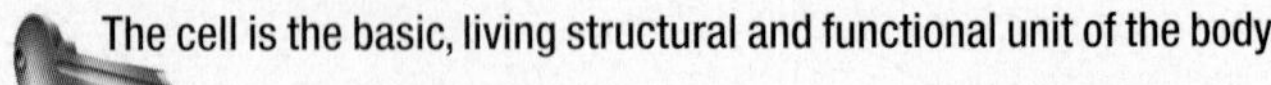
The cell is the basic, living structural and functional unit of the body.

# 3.2 The plasma membrane contains the cytoplasm and regulates exchanges with the extracellular environment.

The **plasma membrane**, a flexible yet sturdy barrier that surrounds and contains the cytoplasm of a cell, is best described by using a structural model called the *fluid mosaic model*. According to this model, the molecular arrangement of the plasma membrane resembles a continually moving sea of lipids that contains a mosaic of many different proteins (Figure 3.2). Some proteins float freely like icebergs in the lipid sea, while others are anchored at specific locations like islands. The membrane lipids allow passage of several types of lipid-soluble molecules but act as a barrier to the entry or exit of charged or polar substances. Some of the proteins in the plasma membrane allow movement of polar molecules and ions into and out of the cell. Other proteins may act as receptors or link neighboring cells together.

## The Lipid Bilayer

The basic structural framework of the plasma membrane is the **lipid bilayer**, two back-to-back layers made up of three types of lipid molecules—primarily phospholipids with smaller amounts of cholesterol and glycolipids (Figure 3.2). The bilayer arrangement occurs because the lipids have both polar and nonpolar parts. The "head" of each **phospholipid**, a lipid that contains phosphorus (see Figure 2.17), is polar and *hydrophilic* (*hydro-* = water; *-philic* = loving). The two fatty acid "tails" of the phospholipid are nonpolar and *hydrophobic* (*-phobic* = fearing). Because "like seeks like," the phospholipid molecules orient themselves into a bilayer, with their hydrophilic heads facing outward. In this way, the heads face a watery fluid on either side of the bilayer—cytosol on one side and extracellular fluid on the other side. The hydrophobic fatty acid tails in each half of the bilayer point toward one another and away from water, forming a nonpolar, hydrophobic core in the plasma membrane's interior.

**Cholesterol** molecules are found in both layers of the plasma membrane. Cholesterol molecules are steroids with an attached —OH (hydroxyl) group. The tiny —OH group (see Figure 2.18a), which is the only polar region of cholesterol, forms hydrogen bonds with the polar heads of phospholipids and glycolipids. Most of the cholesterol molecule is nonpolar and fits among the fatty acid tails of the phospholipids and glycolipids.

**Glycolipids** are lipids with attached carbohydrate groups that form a polar "head"; their fatty acid "tails" are nonpolar.

FIGURE 3.2 **The fluid mosaic arrangement of lipids and proteins in the plasma membrane.**

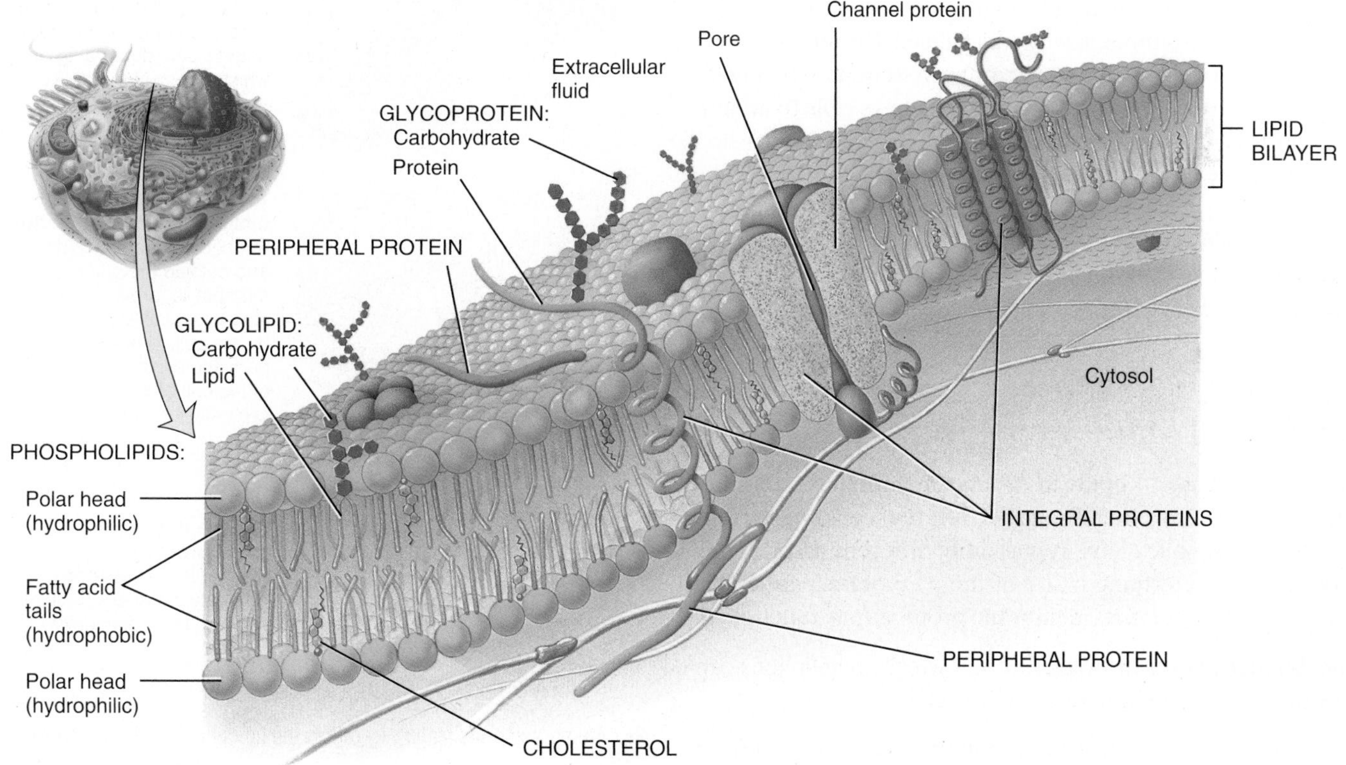


**Membranes are fluid structures because the lipids and many of the proteins are free to rotate and move sideways in the bilayer.**

Glycolipids appear only in the layer of the plasma membrane that faces the extracellular fluid, which is one reason the two sides of the bilayer are asymmetric, or different.

## Arrangement of Membrane Proteins

Membrane proteins are classified as integral or peripheral according to whether or not they are firmly embedded in the plasma membrane (Figure 3.2). **Integral proteins** extend into or through the lipid bilayer among the fatty acid tails and are firmly embedded in it. Most integral proteins are *transmembrane proteins* (*trans-* = across), which means that they span the entire lipid bilayer and protrude into both the cytosol and extracellular fluid. Like membrane lipids, integral membrane proteins have hydrophilic regions that protrude into either the watery extracellular fluid or the cytosol, and hydrophobic regions that extend among the fatty acid tails.

As their name implies, **peripheral proteins** (pe-RIF-er-al) are not as firmly embedded in the membrane. They are attached to the polar heads of membrane lipids or to integral proteins at the inner or outer surface of the plasma membrane.

Many integral proteins are **glycoproteins**, proteins with carbohydrate groups attached to the ends that protrude into the extracellular fluid. The carbohydrate portions of glycolipids and glycoproteins form an extensive sugary coat called the **glycocalyx** (glī-kō-KĀL-iks). Because the pattern of carbohydrates in the glycocalyx varies from one cell to another, the glycocalyx acts like a molecular "signature" that allows cells to recognize one another. For example, a white blood cell's ability to detect a "foreign" glycocalyx helps our immune system destroy invading organisms. In addition, the glycocalyx allows cells to adhere to one another in some tissues and protects cells from being digested by enzymes in the extracellular fluid. The hydrophilic properties of the glycocalyx attract a film of fluid to the surface of many cells. This action makes red blood cells slippery as they flow through narrow blood vessels and protects cells that line the airways and the gastrointestinal tract from drying out.

## Functions of Membrane Proteins

Generally, the types of lipids in cellular membranes vary only slightly. In contrast, the membranes of different cells and various intracellular organelles have remarkably different assortments of proteins that determine many of the membrane's functions (Figure 3.3). To summarize, membrane proteins may function as:

- **Ion channels**, *pores* or holes through which specific ions can flow thorough to get into or out of the cell.
- **Carriers** that selectively move a polar substance or ion from one side of the membrane to the other by changing shape. For example, amino acids, needed to synthesize new proteins, enter body cells via carriers.

FIGURE 3.3 **Functions of membrane proteins.**

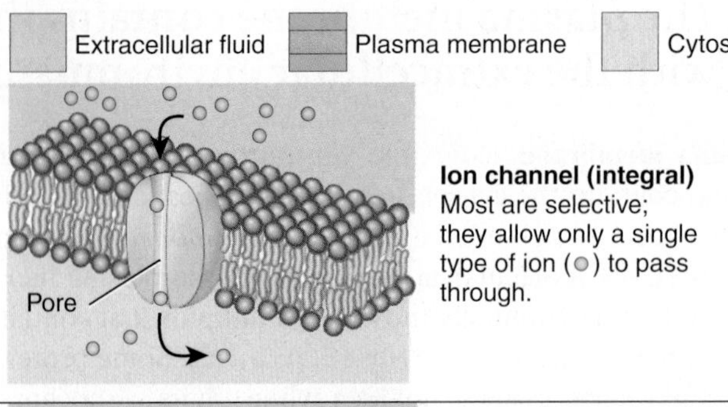


**Ion channel (integral)**
Most are selective; they allow only a single type of ion (○) to pass through.

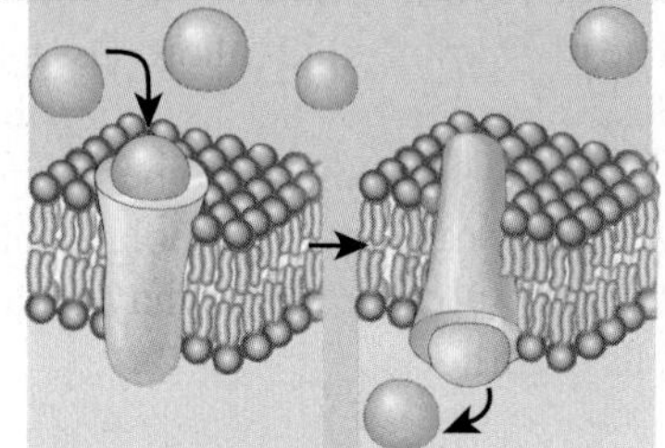

**Carrier (integral)**
Transports specific substances (○) across membrane by changing shape. Carrier proteins are also known as *transporters*.

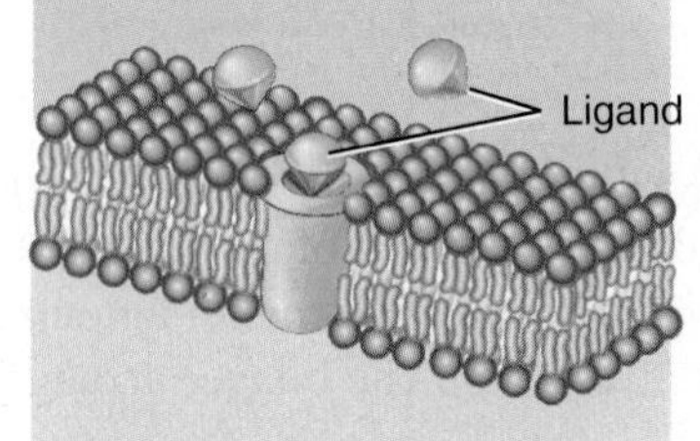


**Receptor (integral)**
Recognizes specific ligand (▽) and alters cell's function in some way.

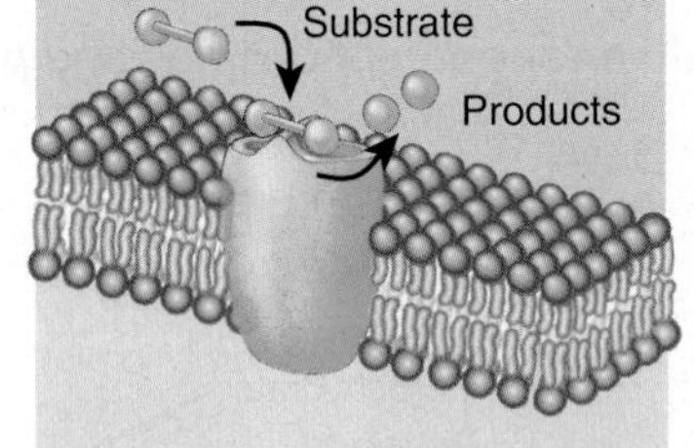


**Enzyme (integral and peripheral)**
Catalyzes reaction inside or outside cell (depending on which direction the active site faces).

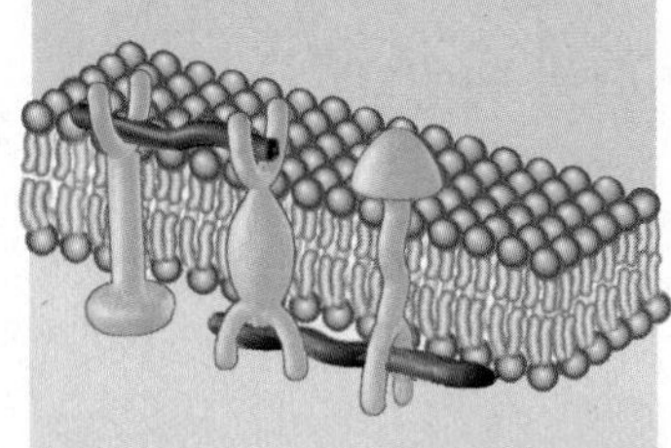

**Linker (integral and peripheral)**
Anchors filaments inside and outside the plasma membrane, providing structural stability and shape for the cell. May also participate in movement of the cell or link two cells together.

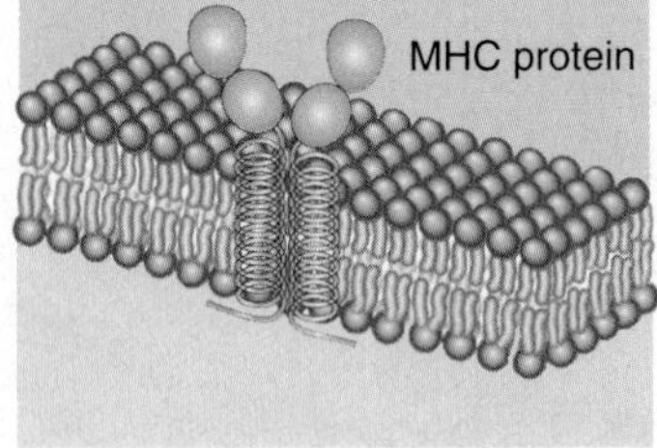


**Cell identity marker (glycoprotein)**
Distinguishes your cells from anyone else's (unless you are an identical twin).

**Membrane proteins largely reflect the functions a cell can perform.**

- **Receptors** that serve as cellular recognition sites. For example, antidiuretic hormone binds to receptors in the kidneys and changes the water permeability of certain plasma membranes. A specific molecule that binds to a receptor is called a *ligand* (LĪ-gand; *liga* = tied) of that receptor.
- **Enzymes** that catalyze specific chemical reactions at the inside or outside surface of the cell. For example, lactase protruding from epithelial cells lining your small intestine splits the disaccharide lactose in the milk you drink.
- **Linkers** that anchor membrane proteins of neighboring cells to one another or to protein filaments inside and outside the cell. Linkers help support the plasma membrane, provide temporary binding sites that assist movement of materials and organelles within the cell, assist movement of the cell itself, and allow changes in cell shape in dividing and contracting cells.
- **Cell-identity markers** that allow a cell to recognize other cells of the same kind during tissue formation or to recognize and respond to potentially dangerous foreign cells. The ABO blood type markers are one example of cell-identity markers. When you receive a blood transfusion, the blood type must be compatible with your own.

## Membrane Fluidity

Membranes are fluid structures; that is, most of the membrane lipids and many of the membrane proteins easily rotate and move sideways in their own half of the bilayer. Neighboring lipid molecules exchange places about 10 million times per second and may wander completely around a cell in only a few minutes! Membrane fluidity depends on the number of double bonds in the fatty acid tails of the lipids that make up the bilayer. Each double bond puts a "kink" in the fatty acid tail (see Figure 2.17), which increases membrane fluidity by preventing lipid molecules from packing tightly in the membrane.

Membrane fluidity is an excellent compromise for the cell; a rigid membrane would lack mobility, and a completely fluid membrane would lack the structural organization and mechanical support required by the cell. Membrane fluidity allows interactions to occur within the plasma membrane, such as the assembly of membrane proteins. It also allows the movement of the membrane components responsible for cellular processes such as cell movement, growth, division, and secretion, and the formation of cellular junctions. Fluidity allows the lipid bilayer to self-seal if torn or punctured. When a needle is pushed through a plasma membrane and pulled out, the puncture site seals spontaneously, and the cell does not burst. This property of the lipid bilayer allows fertilization of an oocyte (egg cell) by injecting a sperm cell through a tiny syringe to help infertile couples conceive a child. It also permits removal and replacement of a cell's nucleus in cloning experiments, such as the one that created Dolly, the famous cloned sheep.

Despite the great mobility of membrane lipids and proteins in their own half of the bilayer, they seldom flip-flop from one half of the bilayer to the other because it is difficult for the hydrophilic parts of membrane molecules to pass through the hydrophobic core of the membrane. This difficulty contributes to the asymmetry of the membrane bilayer.

## Membrane Permeability

When a membrane is referred to as *permeable*, it means that the membrane allows substances to pass through it, while an *impermeable* membrane does *not* allow substances through it (*im-* = not). The permeability of the plasma membrane to different substances varies. Plasma membranes permit some substances to pass more readily than others, a property called **selective permeability** (per′-mē-a-BIL-i-tē).

The lipid bilayer portion of the membrane is permeable to nonpolar, uncharged molecules such as oxygen, carbon dioxide, and steroids; impermeable to ions and large, uncharged polar molecules such as glucose; and *slightly* permeable to small, uncharged polar molecules such as water and *urea*, a waste product from the breakdown of amino acids. Water and urea are thought to pass through the lipid bilayer in the following way: As the fatty acid tails of membrane phospholipids and glycolipids randomly move about, small gaps briefly appear in the hydrophobic environment of the membrane's interior; water and urea molecules are small enough to move from one gap to another until they succeed in crossing the membrane.

Transmembrane proteins that act as channels and carriers increase the plasma membrane's permeability to a variety of ions and uncharged polar molecules that, unlike water and urea molecules, cannot cross the lipid bilayer unassisted. Channels and carriers are very selective. Each one helps a specific molecule or ion cross the membrane. Macromolecules, such as proteins, are so large that they are unable to pass across the plasma membrane except by endocytosis and exocytosis (discussed later in this chapter).

## Gradients across the Plasma Membrane

The selective permeability of the plasma membrane allows a living cell to maintain different concentrations of certain substances on either side of the plasma membrane. A **concentration gradient** is a difference in the concentration of a chemical from one place to another, such as from the cytosol side to the extracellular fluid side of the plasma membrane. For example, oxygen molecules and sodium ions are more concentrated in the extracellular fluid than in the cytosol; the opposite is true of carbon dioxide molecules and potassium ions.

The plasma membrane also creates a difference in the distribution of positively and negatively charged ions between the two sides of the plasma membrane. Typically, the inner surface of the plasma membrane is more negatively charged and the outer surface is more positively charged. A difference in electrical charges between two regions is called an **electrical gradient**.

As you will see shortly, the concentration gradient and the electrical gradient are both important because they help move substances across the plasma membrane. In many cases a substance will move across a plasma membrane *down its concentration gradient*. That is to say, a substance will move "downhill" to reach equilibrium, from where it is more concentrated to where it is less concentrated. Similarly, a positively charged substance will tend to move toward a negatively charged area, and a negatively charged substance will tend to move toward a positively charged area. The combined influence of the concentration gradient and the electrical gradient on movement of a particular ion is referred to as its **electrochemical gradient**.

**✓ CHECKPOINT**

2. How do hydrophobic and hydrophilic regions govern the arrangement of membrane lipids in a bilayer?
3. What is the glycocalyx?
4. Which substances can diffuse through the lipid bilayer? Which cannot?
5. Explain in your own words the concept of selective permeability of the plasma membrane.
6. What factors contribute to an electrochemical gradient?

## 3.3 Transport of a substance across the plasma membrane occurs by both passive and active processes.

Transport of materials across the plasma membrane is essential to the life of a cell. Certain substances must move into the cell to support metabolic reactions. Other substances that have been produced by the cell for export or as cellular waste products must move out of the cell. Substances generally move across cellular membranes via transport processes that can be classified as passive or active, depending on whether they require cellular energy. In **passive processes**, a substance moves down its concentration or electrical gradient to cross the membrane using only its own *kinetic energy* (energy of motion). Kinetic energy is intrinsic to the particles that are moving; there is no input of energy from the cell. Simple diffusion is an example of a passive process. In **active processes**, cellular energy is used to drive the substance "uphill" against its concentration or electrical gradient; the cellular energy is usually in the form of adenosine triphosphate (ATP). Active transport is one example of an active process.

### Passive Processes

#### The Principle of Diffusion

Learning why materials diffuse across membranes requires an understanding of how diffusion occurs in a solution. **Diffusion** (di- FŪ-zhun; *diffus-* = spreading) is a passive process in which the random mixing of particles in a solution occurs because of the particles' kinetic energy. Both the *solutes*, the dissolved substances, and the *solvent*, the liquid that does the dissolving, undergo diffusion. If a particular solute is present in high concentration in one area of a solution and in low concentration in another area, solute molecules will diffuse toward the area of lower concentration—they move *down their concentration gradients*. After some time, the particles become evenly distributed throughout the solution and the solution is said to be *at equilibrium* (ē-kwi-LIB-rē-um). The particles continue to move about randomly due to their kinetic energy, but their concentrations do not change.

For example, when you place a crystal of dye in a water-filled container (Figure 3.4), the color is most intense in the area closest to the dye because its concentration is higher there. At increasing distances, the color is lighter and lighter because the dye concentration is lower. Some time later, the solution of water and dye will have a uniform color, because the dye molecules and water molecules have diffused down their concentration gradients until they are evenly mixed in solution—they are at equilibrium.

FIGURE 3.4 **The principle of diffusion.** At the beginning of our experiment, a crystal of dye is placed in a cylinder of water and dissolves (a) and then diffuses from the region of higher dye concentration to regions of lower dye concentration (b). At equilibrium (c), the dye concentration is uniform throughout, although random movement continues.

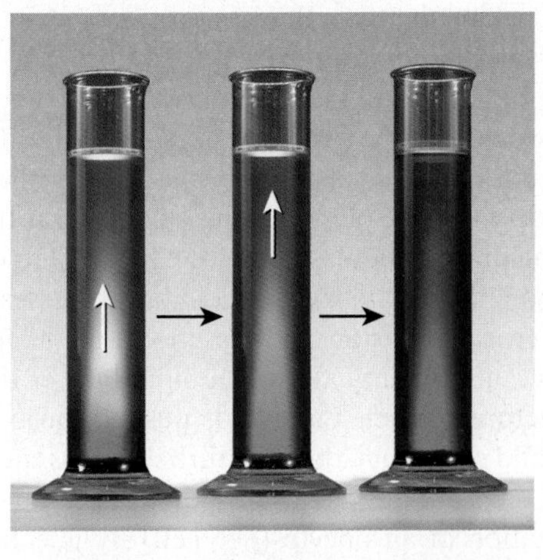


**In diffusion, a substance moves down its concentration gradient.**

In this simple example, no membrane was involved. Substances may also diffuse through a membrane if the membrane is permeable to them. Several factors influence the diffusion rate of substances across plasma membranes:

- ***Steepness of the concentration gradient.*** The greater the difference in concentration between the two sides of the membrane, the higher the rate of diffusion. When charged particles are diffusing, the steepness of the electrochemical gradient determines the diffusion rate across the membrane.
- ***Temperature.*** The higher the temperature, the faster the rate of diffusion. All of the body's diffusion processes occur more rapidly in a person with a fever.
- ***Mass of the diffusing substance.*** The larger the mass of the diffusing particle, the slower its diffusion rate. Smaller molecules diffuse more rapidly than larger ones.
- ***Surface area.*** The larger the membrane surface area available for diffusion, the faster the diffusion rate. For example, the air sacs of the lungs have a large surface area available for diffusion of oxygen from the air into the blood. Some lung diseases, such as emphysema, reduce the surface area. This slows the rate of oxygen diffusion and makes breathing more difficult.
- ***Diffusion distance.*** The greater the distance over which diffusion must occur, the longer it takes. Diffusion across a plasma membrane takes only a fraction of a second because the membrane is so thin. In pneumonia, fluid collects in the lungs; the additional fluid increases the diffusion distance because oxygen must move through both the built-up fluid and the membrane to reach the bloodstream.

Now that you have a basic understanding of the nature of diffusion, we will consider three types of diffusion: simple diffusion, facilitated diffusion, and osmosis.

## Simple Diffusion

**Simple diffusion** is a passive process in which substances move freely through the lipid bilayer of the plasma membranes of cells without the help of membrane transport proteins (Figure 3.5). Nonpolar molecules, hydrophobic molecules, and uncharged polar molecules move across the lipid bilayer by simple diffusion. Simple diffusion through the lipid bilayer is important in the movement of oxygen and carbon dioxide between blood and cells, and between blood and air within the lungs during breathing. It is also the route for absorption of some nutrients and excretion of some wastes by cells.

## Facilitated Diffusion

Solutes that are too polar or highly charged to move through the lipid bilayer by simple diffusion can cross the plasma membrane by a passive process called **facilitated diffusion**. In this process, an integral membrane protein assists a specific substance across the plasma membrane. The integral membrane protein can be either a membrane channel or a carrier.

**Channel-mediated Facilitated Diffusion.** In **channel-mediated facilitated diffusion**, a solute moves down its concentration gradient across the lipid bilayer through a membrane channel (Figure 3.5). Most membrane channels are ion channels, integral transmembrane proteins that allow passage of small ions that are too hydrophilic to penetrate the nonpolar interior of the lipid bilayer. Diffusion of ions through channels is generally slower than simple diffusion because channels occupy a smaller fraction of the plasma membrane's total surface area than lipids. Still, channel-mediated facilitated diffusion is a very fast process: More than a million potassium ions can flow through a $K^+$ channel in one second!

FIGURE 3.5 **Types of diffusion.**

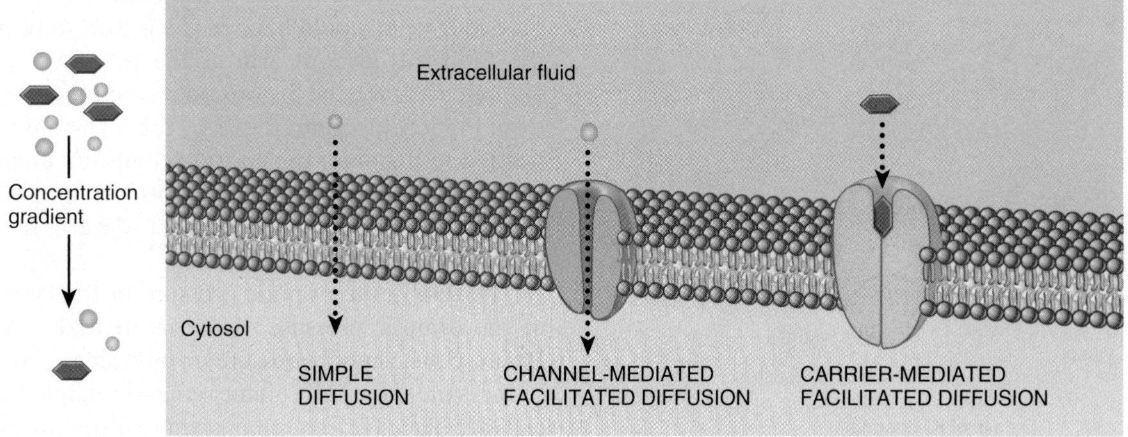


**In simple diffusion, a substance moves across the lipid bilayer of the plasma membrane without the help of membrane transport proteins. In facilitated diffusion, a substance moves across the lipid bilayer aided by a channel protein or a carrier protein.**

A channel is said to be *gated* when part of the channel protein acts as a "gate," changing shape in one way to open the pore and in another way to close it (Figure 3.6). Some gated channels randomly alternate between the open and closed positions; others are regulated by chemical or electrical changes inside or outside the cell. When the gates of a channel are open, ions diffuse into or out of cells down their electrochemical gradients. The plasma membranes of different types of cells may have different numbers of ion channels and thus display different permeabilities to various ions.

Carrier-mediated Facilitated Diffusion. In **carrier-mediated facilitated diffusion**, a carrier moves a solute down its concentration gradient across the plasma membrane (see Figure 3.5). Since this is a passive process, no cellular energy is required. The solute binds to a specific carrier on one side of the membrane and is released on the other side after the carrier undergoes a change in shape. The solute binds more often to the carrier on the side of the membrane with a higher concentration of solute. Once the solute reaches equilibrium, arriving at the same concentration on both sides of the membrane, solute molecules bind to the carrier on the cytosol side and move out to the extracellular fluid as rapidly as they bind to the carrier on the extracellular fluid side and move into the cytosol. The rate of carrier-mediated facilitated diffusion (how quickly it occurs) is determined by the steepness of the concentration gradient across the membrane.

## Osmosis

**Osmosis** (oz-MŌ-sis) is a type of diffusion in which a solvent, such as water, moves through a selectively permeable membrane.

FIGURE 3.6 **Channel-mediated facilitated diffusion of potassium ions ($K^+$) through a gated $K^+$ channel.** A gated channel is one in which a portion of the channel protein acts as a gate to open or close the channel's pore to the passage of ions.

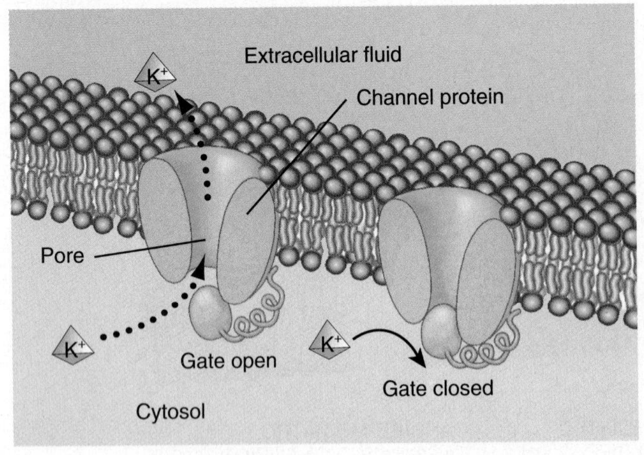


Details of $K^+$ channel

**Channels are integral membrane proteins that allow specific, small ions to pass across the membrane by facilitated diffusion.**

Like the other types of diffusion, osmosis is a passive process. Water moves by osmosis across plasma membranes from an area of *higher water concentration* to an area of *lower water concentration*. Another way to understand this idea is to consider the solute concentration: In osmosis, water moves through a selectively permeable membrane from an area of *lower solute concentration* to an area of *higher solute concentration*. During osmosis through a plasma membrane, water molecules can pass directly through the lipid bilayer or move through integral membrane proteins that function as water channels.

Osmosis occurs only when a membrane is permeable to water but is not permeable to certain solutes. A simple experiment can demonstrate osmosis. Consider a U-shaped tube in which a selectively permeable membrane separates the left and right arms of the tube (Figure 3.7). A volume of pure water is poured into the left arm, and the same volume of a solution containing a solute that cannot pass through the membrane is poured into the right arm (Figure 3.7a). Because the water concentration is higher on the left and lower on the right, net movement of water molecules—osmosis—occurs from left to right, down water's concentration gradient. At the same time, the membrane prevents diffusion of the solute from the right arm into the left arm. As a result, the volume of water in the left arm decreases, and the volume of solution in the right arm increases (Figure 3.7b).

You might think that osmosis would continue until no water remained on the left side, but this is not what happens. In this experiment, the higher the column of solution in the right arm becomes, the more pressure it exerts on its side of the membrane. Pressure exerted in this way by a liquid, known as **hydrostatic pressure**, forces water molecules to move back into the left arm. Equilibrium is reached when just as many water molecules move from right to left due to the hydrostatic pressure as move from left to right due to osmosis (Figure 3.7b).

The solution with the impermeable solute also exerts a force, called the **osmotic pressure**. The osmotic pressure of a solution is proportional to the concentration of the solute particles that cannot cross the membrane—the higher the solute concentration, the higher is the solution's osmotic pressure. Consider what would happen if a piston were pressed against the fluid in the right arm of the tube in Figure 3.7c. With enough pressure, the piston could stop the movement of water from the left tube into the right tube. The amount of pressure needed to maintain the starting condition equals the osmotic pressure. So, in our experiment osmotic pressure is the pressure needed to stop the movement of water from the left tube into the right tube.

Normally, the osmotic pressure of the cytosol is the same as the osmotic pressure of the interstitial fluid outside cells. Because the osmotic pressure on both sides of the plasma membrane is the same, cell volume remains relatively constant. When cells are placed in a solution having a different osmotic pressure than cytosol, however, the shape and volume of the cells change. As water moves by osmosis into or out of the cells, their volume increases or decreases. The solution's ability to change the

FIGURE 3.7 **Principle of osmosis.** Water molecules move through the selectively permeable membrane; the solute molecules in the right arm cannot pass through the membrane. (a) As the experiment starts, water molecules move from the left arm into the right arm, down the water concentration gradient. (b) After some time, the volume of water in the left arm has decreased and the volume of solution in the right arm has increased. At equilibrium, there is no net osmosis: Hydrostatic pressure forces just as many water molecules to move from right to left as osmosis forces water molecules to move from left to right. (c) If pressure is applied to the solution in the right arm, the starting conditions can be restored. This pressure, which stops osmosis, is equal to the osmotic pressure.

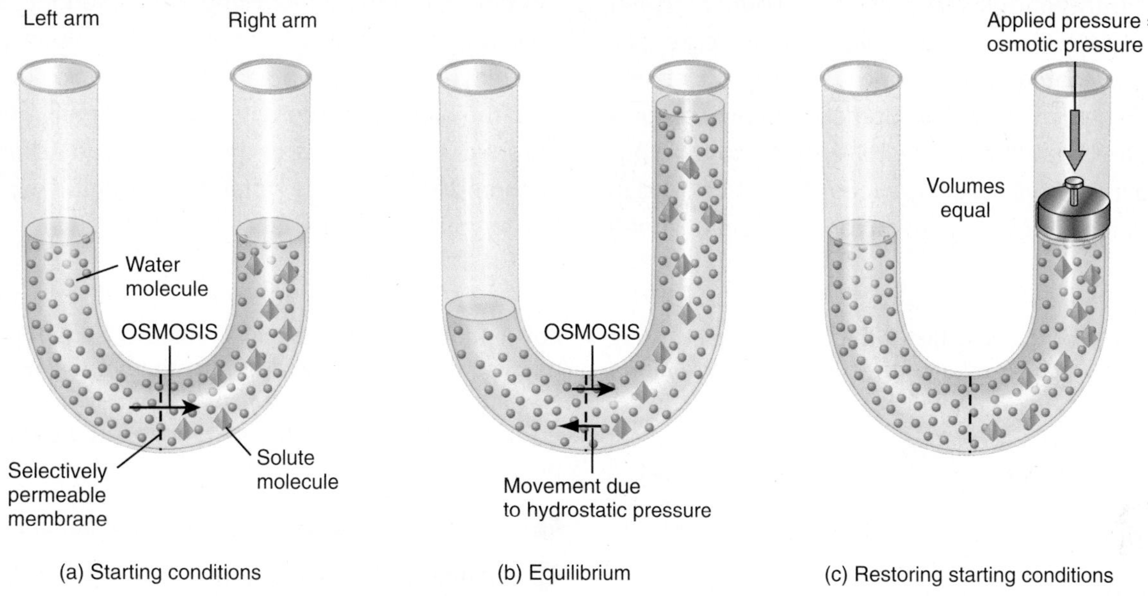


**Osmosis is the movement of water molecules through a selectively permeable membrane.**

volume of cells by altering their water content is called **tonicity** (tō-NIS-i-tē; *tonic* = tension).

Any solution in which a cell—for example, a red blood cell (RBC)—maintains its normal shape and volume is an **isotonic solution** (ī-sō-TON-ik; *iso-* = same) (Figure 3.8). In isotonic solutions the concentrations of solutes that cannot cross the plasma membrane are the same on both sides of the membrane. For instance, a 0.9 percent NaCl solution, called a *normal saline solution*, is isotonic for RBCs. The RBC plasma membrane permits the water to move back and forth, but prevents a net movement of the solutes $Na^+$ and $Cl^-$. (Any $Na^+$ or $Cl^-$ ions that enter the cell are immediately moved back out by active transport or other means.) When RBCs are bathed in 0.9 percent NaCl, water molecules enter and exit at the same rate, allowing the RBCs to keep their normal shape and volume.

A different situation results if RBCs are placed in a **hypotonic solution** (hī-pō-TON-ik; *hypo-* = less than), a solution that has a *lower* concentration of solutes than the cytosol inside the RBCs (Figure 3.8). In this case, water molecules enter the cells faster than they leave, causing the RBCs to swell and eventually to burst. The rupture of RBCs in this manner is called **hemolysis** (hē-MOL-i-sis; *hemo-* = blood; *-lysis* = to loosen or split apart); the rupture of other types of cells is referred to simply as **lysis**. Pure water is very hypotonic and causes rapid hemolysis.

A **hypertonic solution** (hī-per-TON-ik; *hyper-* = greater than) has a *higher* concentration of solutes than does the cytosol inside RBCs (Figure 3.8). One example of a hypertonic solution is a 2 percent NaCl solution. In such a solution, water

FIGURE 3.8 **Tonicity and its effects on red blood cells (RBCs).** The arrows indicate the direction and degree of water movement into and out of the cells. One example of an isotonic solution for RBCs is 0.9 percent NaCl.

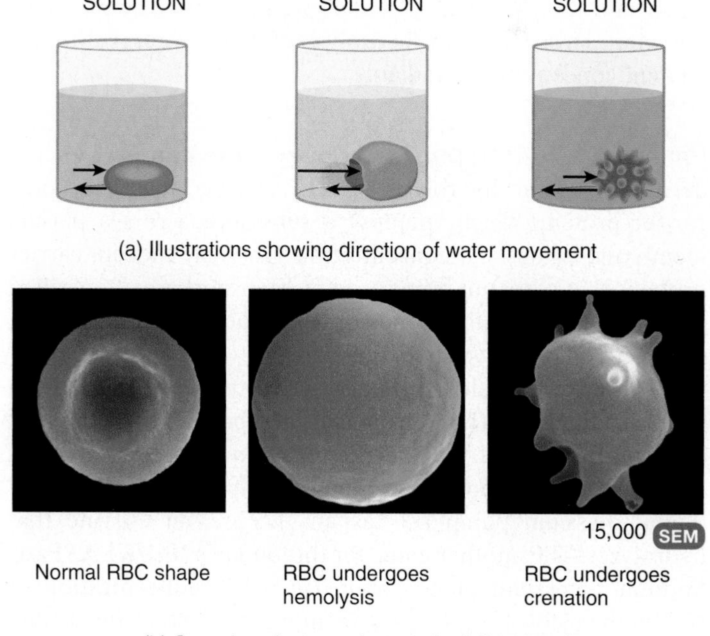


**Cells placed in an isotonic solution maintain their shape because there is no net water movement into or out of the cell.**

## CLINICAL CONNECTION | *Medical Uses of Isotonic, Hypertonic, and Hypotonic Solutions*

RBCs and other body cells may be damaged or destroyed if exposed to hypertonic or hypotonic solutions. For this reason, most **intravenous (IV) solutions,** liquids infused into the blood of a vein, are isotonic. Examples are isotonic saline (0.9 percent NaCl) and D5W, which stands for dextrose 5 percent in water. Sometimes infusion of a hypertonic solution such as mannitol is useful to treat patients who have *cerebral edema*, excess interstitial fluid in the brain. Infusion of such a solution relieves fluid overload by causing osmosis of water from interstitial fluid into the blood. The kidneys then excrete the excess water from the blood into the urine. Hypotonic solutions, given either orally or through an IV, can be used to treat people who are dehydrated. The water in the hypotonic solution moves from the blood into interstitial fluid and then into body cells to rehydrate them. Water and most sports drinks that you consume to "rehydrate" after a workout are hypotonic relative to your body cells.

molecules move out of the cells faster than they enter, causing the cells to shrink. Such shrinkage of cells is called **crenation** (kre-NĀ-shun).

## Active Processes

### Active Transport

Some polar or charged solutes that must enter or leave cells cannot cross the plasma membrane through any form of passive transport because they would need to move "uphill," against their concentration gradients. Such solutes may be able to cross the membrane by a process called **active transport**. Active transport is considered an active process because energy is required for carrier proteins to move solutes across the membrane against a concentration gradient. Cellular energy to drive active transport comes from either the hydrolysis of ATP or energy stored in an ionic concentration gradient.

**Primary Active Transport.** In **primary active transport**, energy derived from the hydrolysis of ATP changes the shape of a carrier protein, which "pumps" a substance across a plasma membrane against its concentration gradient. Indeed, carrier proteins that carry out primary active transport are often called **pumps**. A typical cell expends about 40 percent of the ATP it generates on primary active transport.

The most prevalent primary active transport mechanism expels sodium ions ($Na^+$) from cells and brings potassium ions ($K^+$) in. Because of the specific ions it moves, this carrier is called the **sodium–potassium pump**. Because a part of the sodium–potassium pump acts as an *ATPase*, an enzyme that hydrolyzes ATP, another name for this pump is **$Na^+/K^+$ ATPase**. Sodium–potassium pumps maintain a low concentration of $Na^+$ in the cytosol by pumping them into the extracellular fluid against the $Na^+$ concentration gradient. At the same time, the pumps move $K^+$ into cells against the $K^+$ concentration gradient. Because $K^+$ and $Na^+$ slowly leak back across the plasma membrane down their electrochemical gradients, the sodium–potassium pumps must work nonstop to maintain a low concentration of $Na^+$ and a high concentration of $K^+$ in the cytosol.

Figure 3.9 depicts the operation of the sodium–potassium pump.

1. Three $Na^+$ in the cytosol bind to the pump protein.
2. Binding of $Na^+$ triggers the hydrolysis of ATP into ADP, a reaction that also attaches a phosphate group Ⓟ to the pump protein. This chemical reaction changes the shape of the pump protein, expelling the three $Na^+$ into the extracellular fluid. Now the shape of the pump protein favors binding of two $K^+$ in the extracellular fluid.
3. The binding of $K^+$ triggers release of the phosphate group from the pump protein. This reaction again causes the shape of the pump protein to change.
4. As the pump protein reverts to its original shape, it releases $K^+$ into the cytosol. At this point, the pump is again ready to bind three $Na^+$, and the cycle repeats.

The different concentrations of $Na^+$ and $K^+$ in cytosol and extracellular fluid are crucial for maintaining normal cell volume and for the ability of some cells to generate electrical signals such as action potentials. Because sodium ions that diffuse into a cell or enter through secondary active transport are immediately pumped out, it is as if they never entered. In effect, sodium ions behave as if they cannot penetrate the membrane. Thus, sodium ions are an important contributor to the tonicity of the extracellular fluid. A similar condition holds for $K^+$ in the cytosol. By helping to maintain normal tonicity on each side of the plasma membrane, the sodium–potassium pumps ensure that cells neither shrink nor swell due to the movement of water by osmosis out of or into cells.

**Secondary Active Transport.** In **secondary active transport**, the energy stored in a $Na^+$ or $H^+$ concentration gradient is used to drive other substances across the membrane against their own concentration gradients. Because a $Na^+$ or $H^+$ gradient is

FIGURE 3.9 **The sodium–potassium pump.** The sodium-potassium pump ($Na^+/K^+$ ATPase) expels sodium ions ($Na^+$) and brings potassium ions ($K^+$) into the cell.

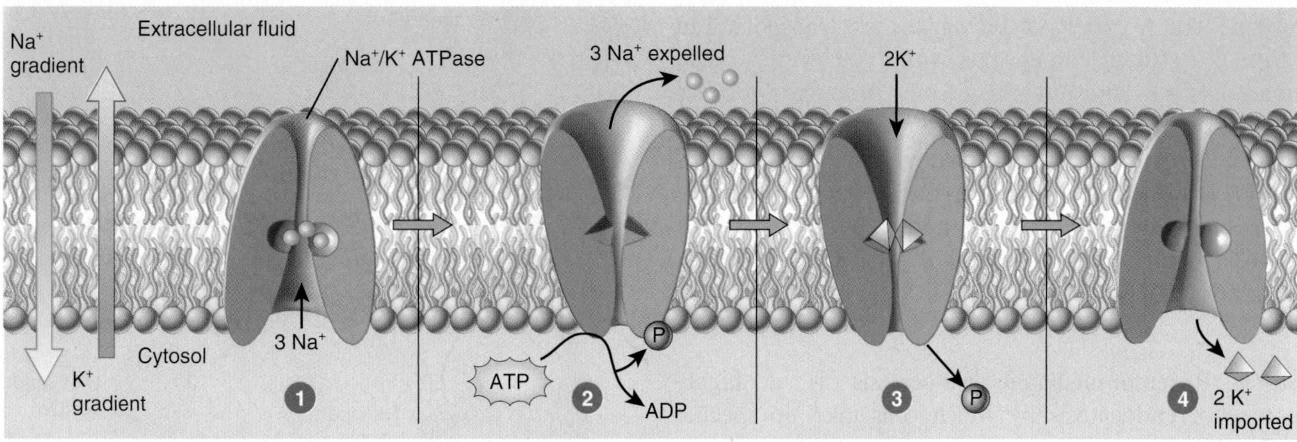


**Sodium–potassium pumps maintain a low intracellular concentration of sodium ions.**

established by primary active transport, secondary active transport indirectly uses energy obtained from the hydrolysis of ATP.

The sodium–potassium pump maintains a steep concentration gradient of $Na^+$ across the plasma membrane. As a result, the sodium ions have stored or potential energy, just like water behind a dam. Accordingly, if there is a route for $Na^+$ to leak back in, some of the stored energy can be used to transport other substances *against their concentration gradients*. In essence, secondary active transport proteins harness the energy in the $Na^+$ concentration gradient by providing routes for $Na^+$ to leak into cells. In secondary active transport, a carrier protein simultaneously binds to both $Na^+$ and another substance and then changes its shape so that both substances cross the plasma membrane at the same time. If these transporters move two substances in the same direction they are called **symporters** (sim-PORT-ers; *sym-* = same); **antiporters** (an′-tē-PORT-ers), in contrast, move two substances in opposite directions across the membrane (*anti-* = against).

Plasma membranes contain several antiporters and symporters that are powered by the $Na^+$ gradient. For example, the concentration of calcium ions ($Ca^{2+}$) is low in the cytosol because antiporters eject calcium ions (Figure 3.10a). Likewise, $Na^+/H^+$ antiporters help regulate the cytosol's pH ($H^+$ concentration) by expelling excess $H^+$. By contrast, glucose and amino acids are absorbed by symporters into cells that line the small intestine (Figure 3.10b). In each case, sodium ions are moving down their concentration gradient while the other solutes move "uphill," against their concentration gradients. Keep in mind that all of these symporters and antiporters can do their jobs because the sodium–potassium pumps maintain a low concentration of $Na^+$ in the cytosol.

FIGURE 3.10 **Secondary active transport mechanisms.** (a) Antiporters carry two substances across the membrane in opposite directions. (b) Symporters carry two substances across the membrane in the same direction.

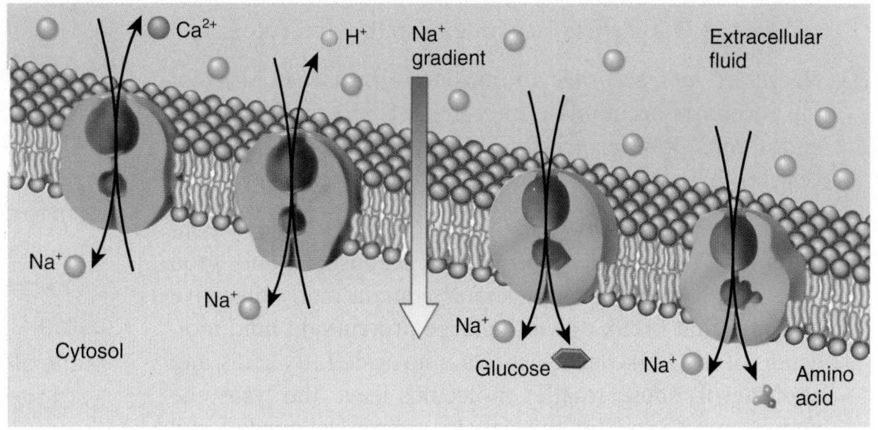


(a) Antiporters (b) Symporters

**Secondary active transport mechanisms use the energy stored in an ionic concentration gradient (here, for $Na^+$). Because primary active transport pumps that hydrolyze ATP maintain the gradient, secondary active transport mechanisms consume ATP indirectly.**

## Transport in Vesicles

A **vesicle** (VES-i-kul = little blister or bladder) is a small, spherical membrane sac. A variety of substances are transported in vesicles from one structure to another within cells. Vesicles also import materials into and out of the cell. During **endocytosis** (en′-dō-sī-TŌ-sis; *endo-* = within), materials move *into* a cell within a vesicle formed from the plasma membrane. In **exocytosis** (ek-sō-sī-TŌ-sis; *exo-* = out), materials move *out of* a cell by the fusion of vesicles with the plasma membrane. Both endocytosis and exocytosis require energy supplied by ATP. Thus, transport in vesicles is an active process.

Endocytosis. **Receptor-mediated endocytosis** is a highly selective type of endocytosis by which cells take up specific ligands. (Recall that *ligands* are molecules that bind to specific receptors.) A vesicle forms after a receptor protein in the plasma membrane recognizes and binds to a particular substance in the extracellular fluid. Receptor-mediated endocytosis of ligands such as cholesterol-containing low-density lipoproteins (LDLs) occurs as follows (Figure 3.11):

1. ***Binding.*** On the extracellular side of the plasma membrane, an LDL particle binds to a specific receptor in the plasma membrane to form a receptor–LDL complex. The receptors are concentrated in recessed regions of the plasma membrane called *clathrin-coated pits.* While binding to LDL, the clathrin molecules come together around the receptor–LDL complexes, causing the plasma membrane to invaginate (fold inward).
2. ***Vesicle formation.*** The invaginated edges of the plasma membrane around the clathrin-coated pit fuse, pinching off a small piece of the plasma membrane. The resulting clathrin-coated vesicle contains the receptor–LDL complexes.
3. ***Uncoating.*** Almost immediately after the vesicle forms, it loses its clathrin coating as clathrin molecules return to the plasma membrane.
4. ***Fusion with endosome.*** The uncoated vesicle quickly fuses with a vesicle known as an *endosome.* Once within an endosome, the LDL particles separate from their receptors.
5. ***Recycling of receptors to plasma membrane.*** Most of the receptors accumulate in elongated protrusions of the endosome. These protrusions pinch off, forming transport vesicles that return the receptors to the plasma membrane.
6. ***Degradation in lysosomes.*** Other transport vesicles containing LDL particles bud off the endosome and soon fuse with a lysosome. Lysosomes contain many digestive enzymes that break down the large protein and lipid molecules of the LDL particle into amino acids, fatty acids, and cholesterol. These smaller molecules leave the lysosome and become used by the cell to synthesize needed cell components.

**FIGURE 3.11 Receptor-mediated endocytosis of a low-density lipoprotein (LDL) particle.**

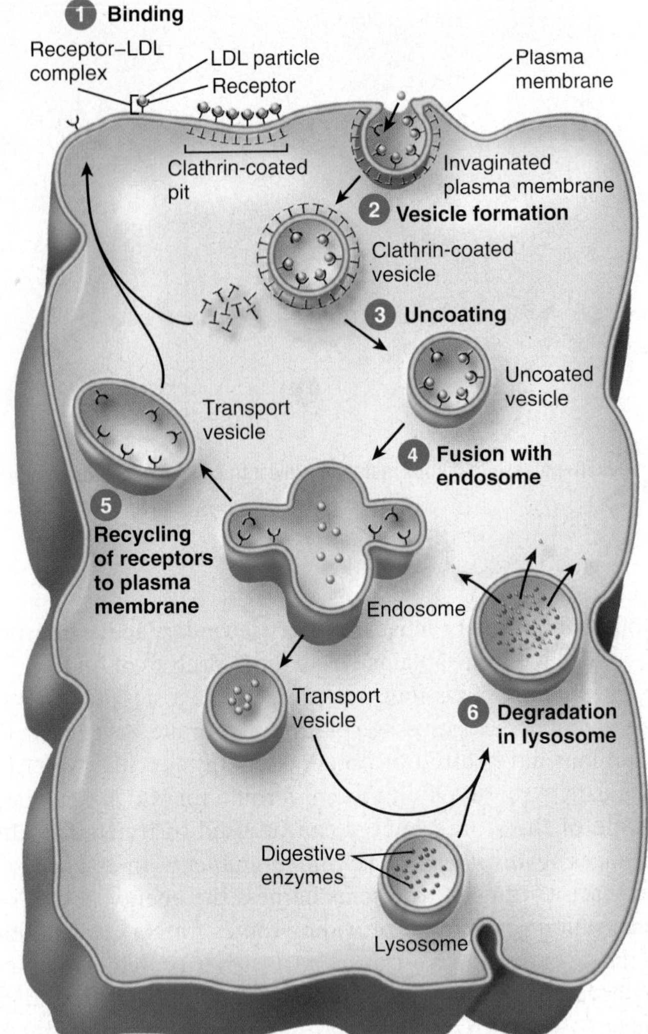

**Receptor-mediated endocytosis imports materials that are needed by cells.**

### CLINICAL CONNECTION | *Viruses and Receptor-mediated Endocytosis*

Although receptor-mediated endocytosis normally imports needed materials, some viruses are able to use this mechanism to enter and infect body cells. For example, the human immunodeficiency virus (HIV), which causes acquired immunodeficiency syndrome (AIDS), can attach to a receptor called CD4. This receptor is present in the plasma membrane of white blood cells called helper T cells. After binding to CD4, HIV enters the helper T cell via receptor-mediated endocytosis.

**Phagocytosis** (fag′-ō-sī-TŌ-sis; *phago-* = to eat) is a form of endocytosis in which the cell engulfs large solid particles, such as worn-out cells, whole bacteria, or viruses (Figure 3.12). Only a few cells, termed **phagocytes** (FAG-ō-sīts), are able to carry out phagocytosis. Phagocytosis begins when the particle binds to a receptor on the plasma membrane of the phagocyte. The phagocyte extends a **pseudopod** (SOO-dō-pod; *pseudo-* = false; *-pod* = foot), a projection of its plasma membrane and cytoplasm, which surrounds the particle, fuses into a vesicle called a *phagosome*, and then enters the cytoplasm. The phagosome fuses with one or more lysosomes, and lysosomal enzymes break down the ingested material. The remaining phagosome and any undigested materials are stored indefinitely as a *residual body* or secreted via exocytosis.

FIGURE 3.12 **Phagocytosis.** Pseudopods surround a particle, and the membranes fuse to form a phagosome.

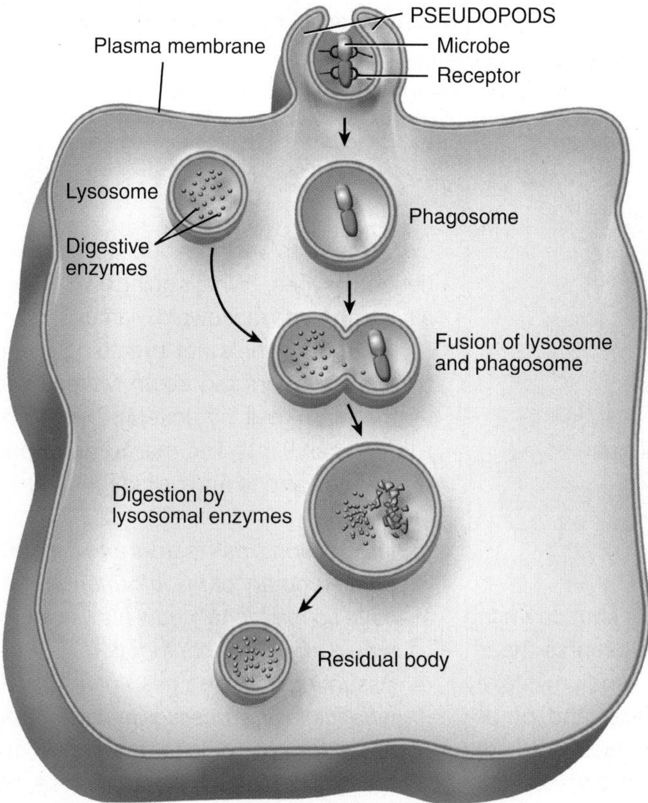


(a) Diagram of phagocytosis

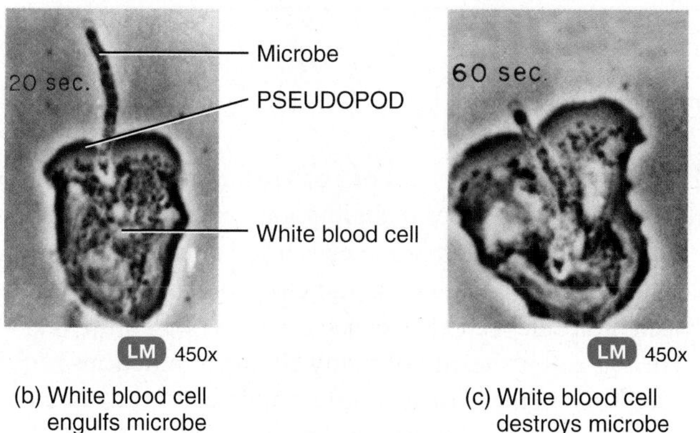


(b) White blood cell engulfs microbe

(c) White blood cell destroys microbe

**Phagocytosis is a vital defense mechanism that helps protect the body from disease.**

Most cells carry out **bulk-phase endocytosis**, also called pinocytosis (pi-nō-sī-TŌ-sis; *pino-* = to drink), a form of endocytosis in which tiny droplets of extracellular fluid are taken into the cell (Figure 3.13). No receptor proteins are involved; all solutes dissolved in the extracellular fluid are brought into the cell. During bulk-phase endocytosis, the plasma membrane folds inward and forms a vesicle containing a droplet of extracellular fluid. The vesicle pinches off from the plasma membrane and enters the cytosol. Within the cell, the vesicle fuses with a lysosome, where enzymes degrade the engulfed solutes. The resulting smaller molecules leave the lysosome to be used elsewhere in the cell.

**Exocytosis.** In contrast with endocytosis, which brings materials into a cell, **exocytosis** releases materials from a cell. Cells carry out exocytosis to release secretions such as digestive enzymes, hormones, mucus, or cellular wastes. During exocytosis, membrane-enclosed vesicles called secretory vesicles form inside the cell, fuse with the plasma membrane, and release their contents into the extracellular fluid.

FIGURE 3.13 **Bulk-phase endocytosis.** The plasma membrane folds inward, forming a vesicle.

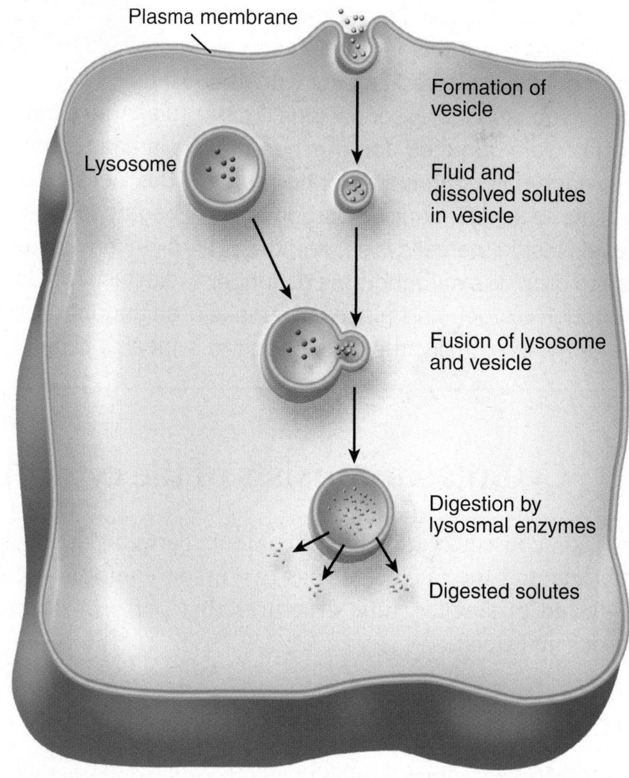


**Most cells carry out bulk-phase endocytosis, the nonselective uptake of tiny droplets of extracellular fluid.**

Segments of the plasma membrane lost through endocytosis are recovered or recycled by exocytosis. The balance between endocytosis and exocytosis keeps the surface area of a cell's plasma membrane relatively constant. Membrane exchange is quite extensive in certain cells. In your pancreas, for example, the cells that secrete digestive enzymes can recycle an amount of plasma membrane equal to the cell's entire surface area in 90 minutes.

**Transcytosis.** Transport in vesicles may also be used to successively move a substance into, across, and out of a cell. In this active process, called **transcytosis** (tranz′-sī-TŌ-sis), vesicles undergo endocytosis on one side of a cell, move across the cell, and then undergo exocytosis on the opposite side. As vesicles fuse with the plasma membrane during exocytosis, their contents are released into the extracellular fluid. Transcytosis occurs most often across the endothelial cells that line blood vessels and is a means for materials to move between blood plasma and interstitial fluid.

Table 3.1 summarizes the processes by which materials move into and out of cells.

✓ **CHECKPOINT**

7. What is the key difference between passive and active transport?
8. Which factors can increase the rate of diffusion?
9. What is osmotic pressure?
10. How do primary active transport and secondary active transport differ?
11. How do symporters and antiporters carry out their functions?
12. What is transcytosis?

## RETURN TO Joseph's Story

As he lay unconscious in the park, inside his heart Joseph's cardiac muscle cells struggled to survive the lack of blood flow caused by his atherosclerosed coronary arteries. Joseph's family had a history of vascular disease, but he hadn't thought about it much. His weight had climbed; he always found an excuse to justify eating the fatty foods he loved. Now, the cells in his heart struggled to work. As the flow of life-giving blood slowed, along with the oxygen, glucose, and essential ions it carried that were required for the heart to pump, each cell continued working, rapidly using up its ATP supply. Carbon dioxide levels inside the cells rose, and the pH began to drop; the mitochondria no longer had enough oxygen and glucose to make ATP energy. Eventually, the cells began to die. Active transport pumps shut down the plasma membranes of Joseph's heart. The cells became leaky; sodium slowly began to leak into the cells, and potassium leaked out. As the chemical gradients diminished, Joseph's heart convulsed in an uncontrolled spasm of arrhythmic electrical activity.

Time was running out. Joseph's cells needed oxygen. "What's going on?!" shouted a tall man as he ran toward Marcus.

"My dad—he's not breathing. He just fell down!" Tears ran down Marcus's face as the man knelt by Joseph, then turned. "Jenny, call 911!" The man's friend pulled out her cell phone and called for help.

A. *List Joseph's risk factors and create a brief summary of the information you have so far. Identify how his risk factors would affect cellular function.*

B. *Assuming Joseph's heart has stopped, what cellular processes and membrane functions are going to be affected by the loss of oxygen, blood glucose, and waste removal?*

# 3.4 Cytoplasm consists of the cytosol and organelles.

Cytoplasm consists of all cellular contents between the plasma membrane and the nucleus. It has two major components: the cytosol and organelles, tiny structures that perform different functions in the cell.

## Cytosol

The **cytosol (intracellular fluid)** is the fluid portion of the cytoplasm that surrounds organelles (see Figure 3.1) and constitutes about 55 percent of total cell volume. Although it varies in composition and consistency from one part of a cell to another, cytosol is mostly water plus various solutes including ions, glucose, proteins, lipids, ATP, and waste products. Some cells also contain various organic molecules that aggregate into masses for storage such as clusters of glycogen molecules called *glycogen granules* (see Figure 3.1).

The cytosol is the site of many chemical reactions required for a cell's existence. For example, enzymes in cytosol catalyze *glycolysis*, a series of chemical reactions that produce ATP from glucose. Other types of cytosolic reactions provide the building blocks for maintenance of cell structures and for cell growth.

**TABLE 3.1**

Transport of Materials Into and Out of Cells

| TRANSPORT PROCESS | DESCRIPTION | SUBSTANCES TRANSPORTED |
|---|---|---|
| **PASSIVE PROCESSES** | Movement of substances down a concentration gradient until equilibrium is reached; do not require cellular energy in the form of ATP | |
| **Diffusion** | Movement of molecules or ions down a concentration gradient due to their kinetic energy until they reach equilibrium | |
| **Simple diffusion** | Passive movement of a substance down its concentration gradient through the lipid bilayer of the plasma membrane without the help of membra ne transport proteins | Nonpolar, hydrophobic solutes: oxygen, carbon dioxide, and nitrogen gases; fatty acids; steroids; and fat-soluble vitamins<br>Polar molecules such as water, urea, and small alcohols |
| **Facilitated diffusion** | Passive movement of a substance down its concentration gradient through the lipid bilayer by transmembrane proteins that function as channels or carriers | Polar or charged solutes: glucose; fructose; galactose; some vitamins; and ions such as $K^+$, $Cl^-$, $Na^+$, and $Ca^{2+}$ |
| **Osmosis** | Passive movement of water molecules across a selectively permeable membrane from an area of higher to lower water concentration until equilibrium is reached | Solvent: water in living systems |
| **ACTIVE PROCESSES** | Movement of substances against a concentration gradient; requires cellular energy in the form of ATP | |
| **Active Transport** | Active process in which a cell expends energy to move a substance across the membrane against its concentration gradient by transmembrane proteins that function as carriers | Polar or charged solutes |
| **Primary active transport** | Active process in which a substance moves across the membrane against its concentration gradient by pumps (carriers) that use energy supplied by hydrolysis of ATP | $Na^+$, $K^+$, $Ca^{2+}$, $H^+$, $I^-$, $Cl^-$, and other ions |
| **Secondary active transport** | Coupled active transport of two substances across the membrane using energy supplied by a $Na^+$ or $H^+$ concentration gradient maintained by primary active transport pumps<br>Antiporters move $Na^+$ (or $H^+$) and another substance in opposite directions across the membrane; symporters move $Na^+$ (or $H^+$) and another substance in the same direction across the membrane | Antiport: $Ca^{2+}$, $H^+$ out of cells<br>Symport: glucose, amino acids into cells |
| **Transport in Vesicles** | Active process in which substances move into or out of cells in vesicles that bud from plasma membrane; requires energy supplied by ATP | |
| **Endocytosis** | Movement of substances into a cell in vesicles | |
| **Receptor-mediated endocytosis** | Ligand–receptor complexes trigger infolding of a clathrin-coated pit that forms a vesicle containing ligands | Ligands: transferrin, low-density lipoproteins (LDLs), some vitamins, certain hormones, and antibodies |
| **Phagocytosis** | "Cell eating"; movement of a solid particle into a cell after pseudopods engulf it to form a phagosome | Bacteria, viruses, and aged or dead cells |
| **Bulk-phase endocytosis** | "Cell drinking"; movement of extracellular fluid into a cell by infolding of plasma membrane to form a vesicle | Solutes in extracellular fluid |
| **Exocytosis** | Movement of substances out of a cell in secretory vesicles that fuse with the plasma membrane and release their contents into the extracellular fluid | Neurotransmitters, hormones, and digestive enzymes |
| **Transcytosis** | Movement of a substance through a cell as a result of endocytosis on one side and exocytosis on the opposite side | Substances, such as antibodies, across endothelial cells; common route for substances to pass between blood plasma and interstitial fluid |

## Organelles

**Organelles** are specialized structures within the cell that have characteristic shapes and perform specific functions in cellular growth, maintenance, and reproduction. Each type of organelle has its own set of enzymes that carry out specific reactions and serves as a functional compartment for specific biochemical processes. Despite the many chemical reactions going on in a cell at any given time, there is little interference among reactions because they are confined to different organelles. The numbers and types of organelles vary in different cells, depending on the cell's function. Although they have different functions, organelles often cooperate to maintain homeostasis. Even though the nucleus is a large organelle, it is discussed in a separate section because of its special importance in directing the life of a cell.

### The Cytoskeleton

The **cytoskeleton** is a network of protein filaments that extends throughout the cytosol (see Figure 3.1). Three types of filamentous proteins contribute to the cytoskeleton's structure. In order of their increasing diameter, these structures are microfilaments, intermediate filaments, and microtubules.

**Microfilaments** (mī-krō-FIL-a-ments), the thinnest elements of the cytoskeleton, are composed of the proteins *actin* and *myosin* and are most prevalent at the periphery of a cell (Figure 3.14a). Microfilaments help generate cellular movements.

FIGURE 3.14 **Cytoskeleton.**

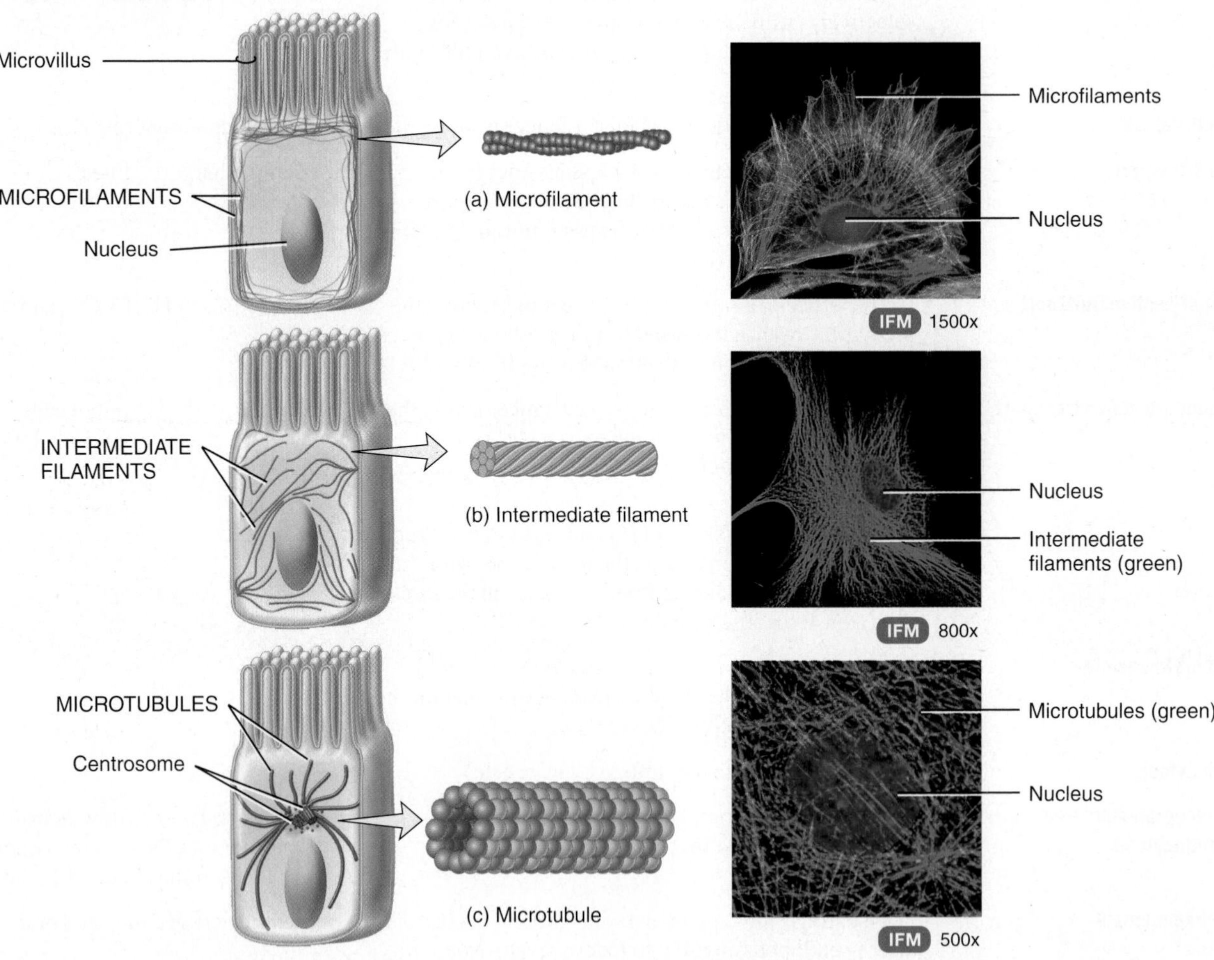


FUNCTIONS
1. Serves as a scaffold that helps to determine a cell's shape and organize the cellular contents.
2. Aids movement of organelles within the cell, of chromosomes during cell division, and of whole cells such as phagocytes.

**The cytoskeleton is a network of three types of protein filaments that extend throughout the cytoplasm: microfilaments, intermediate filaments, and microtubules.**

Microfilaments are involved in muscle contraction, cell division, and cell locomotion, such as occurs in the migration of embryonic cells during development, the invasion of tissues by white blood cells to fight infection, or the migration of skin cells during wound healing.

Microfilaments also provide much of the mechanical support that is responsible for the basic strength and shapes of cells. They anchor the cytoskeleton to integral proteins in the plasma membrane. Microfilaments also provide mechanical support for cell extensions called **microvilli** (mī-krō-VIL-ī; *micro-* = small; *-villi* = tufts of hair; singular is *microvillus*), nonmotile, microscopic fingerlike projections of the plasma membrane. Within each microvillus is a core of parallel microfilaments that supports it. Because they greatly increase the surface area of the cell, microvilli are abundant on cells involved in absorption, such as the epithelial cells that line the small intestine.

As their name suggests, **intermediate filaments** are thicker than microfilaments but thinner than microtubules (Figure 3.14b). Several different proteins can compose intermediate filaments, which are exceptionally strong. They resist mechanical stress, help stabilize the position of organelles such as the nucleus, and help attach cells to one another.

The largest of the cytoskeletal components, **microtubules** (mī-krō-TOO-būls) are long, unbranched hollow tubes composed mainly of the protein *tubulin* (Figure 3.14c). The assembly of microtubules begins in an organelle called the centrosome (discussed shortly). Microtubules grow outward from the centrosome toward the periphery of the cell. Microtubules help determine cell shape. They also function in the movement of organelles (such as secretory vesicles), chromosomes (during cell division), and specialized cell projections (such as cilia and flagella).

## Centrosome

The **centrosome** (SEN-trō-sōm) is located near the nucleus and consists of two components: a pair of centrioles and pericentriolar material (Figure 3.15a). The two **centrioles** (SEN-trē-ōls) are cylindrical structures, each composed of nine clusters of three microtubules (triplets) arranged in a circular pattern (Figure 3.15b). The long axis of one centriole is at a right angle to the long axis of the other (Figure 3.15c). Surrounding the centrioles is **pericentriolar material** (per′-ē-sen′-trē-Ō-lar), composed of the protein *tubulin*. Pericentriolar material is used as organizing centers for growth of the mitotic spindle, which plays a critical role in cell division, and for microtubule formation in nondividing cells. During cell division, centrosomes replicate so that succeeding generations of cells have the capacity to divide.

FIGURE 3.15 **Centrosome.**

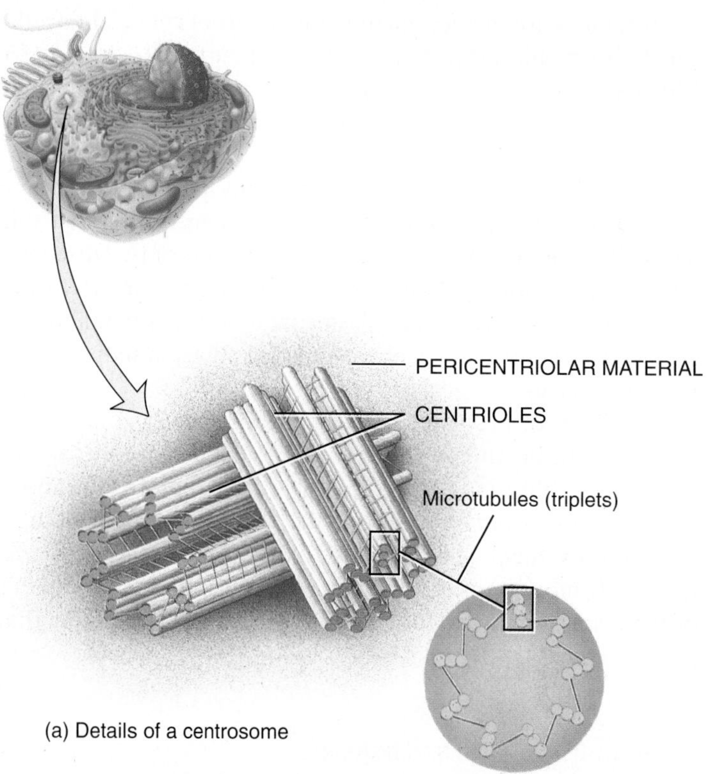


(a) Details of a centrosome

(b) Arrangement of microtubules in centriole

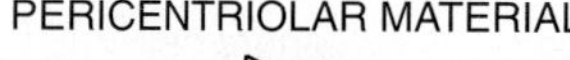


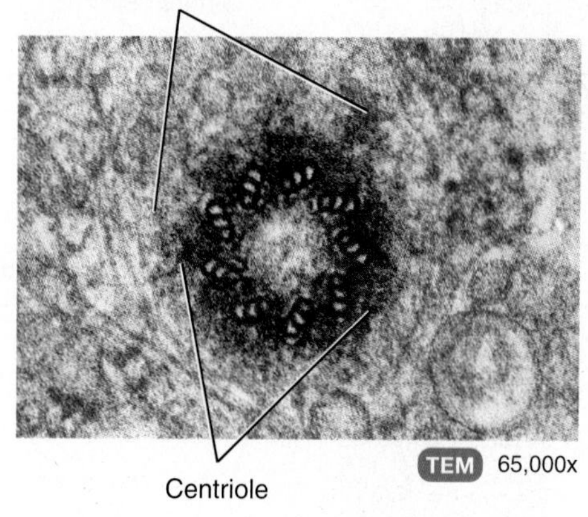


(c) CENTRIOLE

**FUNCTIONS**

1. Builds microtubules in nondividing cells.
2. Forms the mitotic spindle during cell division.

**Located near the nucleus, the centrosome consists of a pair of centrioles and pericentriolar material.**

## Cilia and Flagella

Microtubules are the dominant components of cilia and flagella, which are motile projections of the cell surface (Figure 3.16). **Cilia** (SIL-ē-a = eyelashes; singular is *cilium*) are numerous, short, hairlike projections that extend from the surface of the cell (see Figure 3.1 and Figure 3.16b). Each cilium contains a core of 20 microtubules surrounded by plasma membrane (Figure 3.16a). The microtubules are arranged such that one pair in the center is surrounded by nine clusters of two fused microtubules (doublets). Each cilium is anchored to a *basal body* just below the surface of the plasma membrane. A basal body is similar in structure to a centriole and functions in initiating the assembly of cilia and flagella.

A cilium displays an oarlike pattern of beating; it is relatively stiff during the power stroke (oar digging into the water), but more flexible during the recovery stroke (oar moving above the water preparing for a new stroke) (Figure 3.16d). The coordinated movement of many cilia on the surface of a cell causes the steady movement of fluid along the cell's surface. Many cells of the respiratory tract, for example, have hundreds of cilia that help sweep foreign particles trapped in mucus away from the lungs (Figure 3.16b). The movement of cilia is paralyzed by nicotine in cigarette smoke. For this reason, smokers cough often to remove foreign particles from their airways. Cells that line the uterine (fallopian) tubes also have cilia that sweep oocytes toward the uterus, and females who smoke have an increased risk of ectopic (outside the uterus) pregnancy.

**Flagella** (fla-JEL-a = whip; singular is *flagellum*) are similar in structure to cilia but are typically much longer. Flagella usually move an entire cell. A flagellum generates forward motion along its axis by rapidly wiggling in a wavelike pattern (Figure 3.16e). The only example of a flagellum in the human body is a sperm cell's tail, which propels the sperm toward the oocyte in the uterine tube (Figure 3.16c).

## Ribosomes

**Ribosomes** (RĪ-bō-sōms; *-somes* = bodies) are the sites of protein synthesis. The name of these tiny structures reflects their high content of one type of ribonucleic acid, **ribosomal RNA (rRNA)**. Structurally, a ribosome consists of two subunits, one

FIGURE 3.16 **Cilia and flagella.**

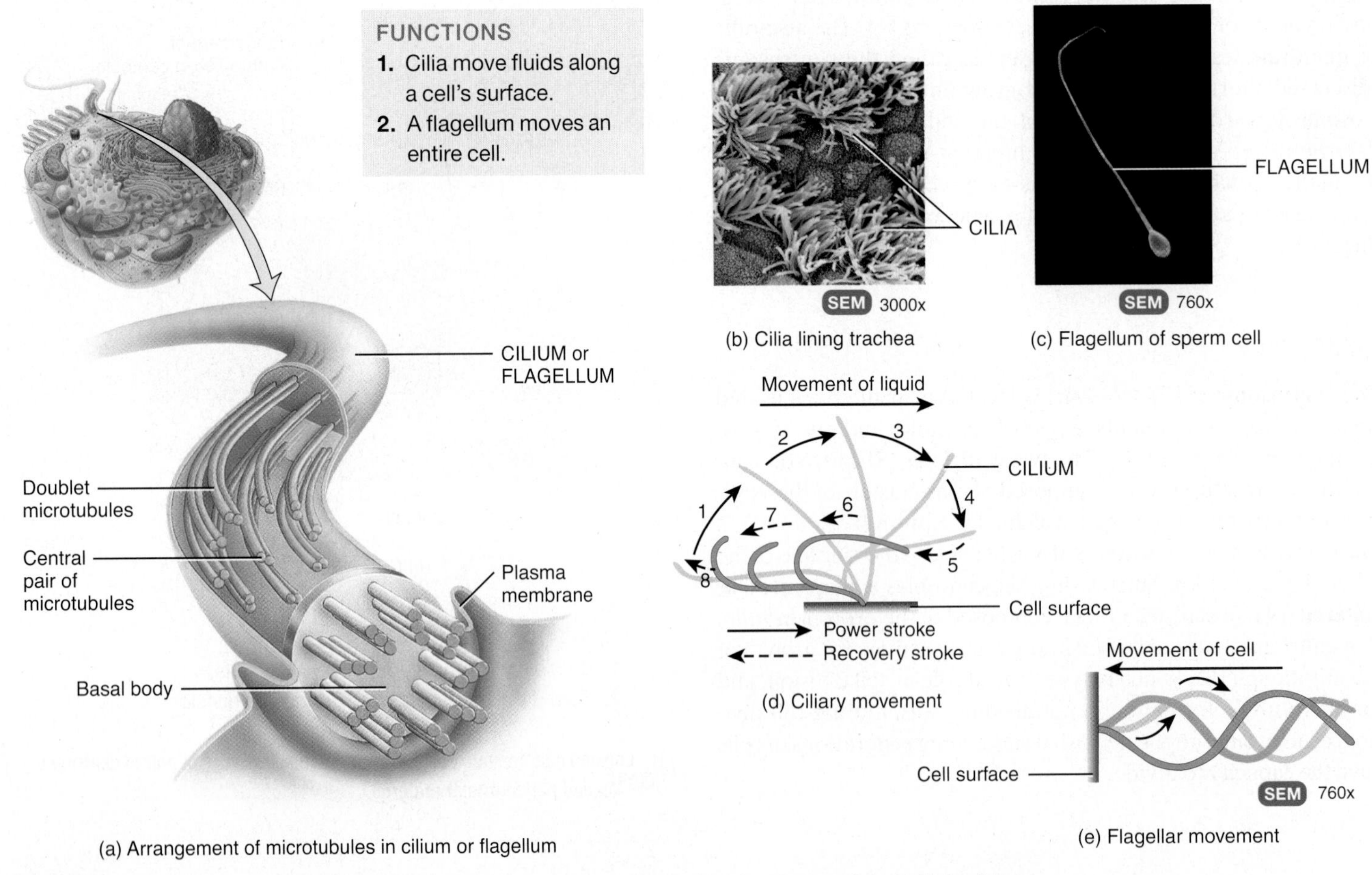

(a) Arrangement of microtubules in cilium or flagellum

(b) Cilia lining trachea

(c) Flagellum of sperm cell

(d) Ciliary movement

(e) Flagellar movement

**A cilium contains a core of microtubules with one pair in the center surrounded by nine clusters of doublet microtubules.**

FIGURE 3.17 **Ribosomes.**

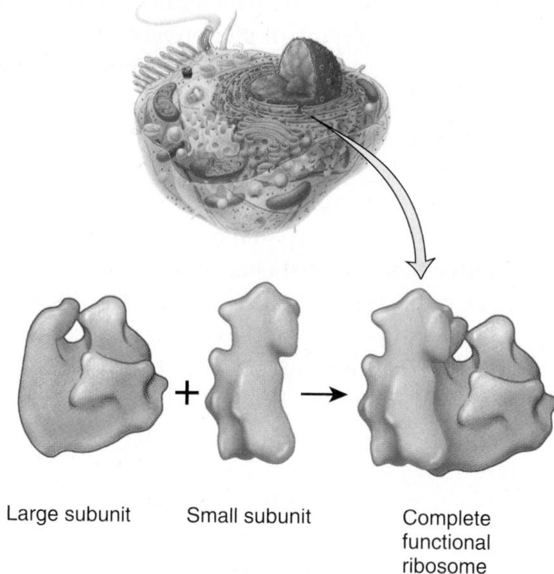


Details of ribosomal subunits

**FUNCTIONS**

1. Attached ribosomes synthesize proteins destined for insertion in the plasma membrane or secretion from the cell.
2. Free ribosomes synthesize proteins used in the cytosol.

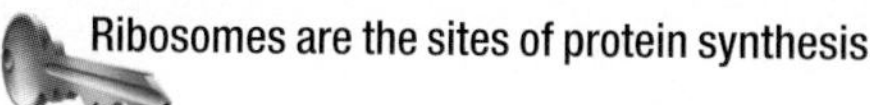

**Ribosomes are the sites of protein synthesis.**

FIGURE 3.18 **Endoplasmic reticulum.**

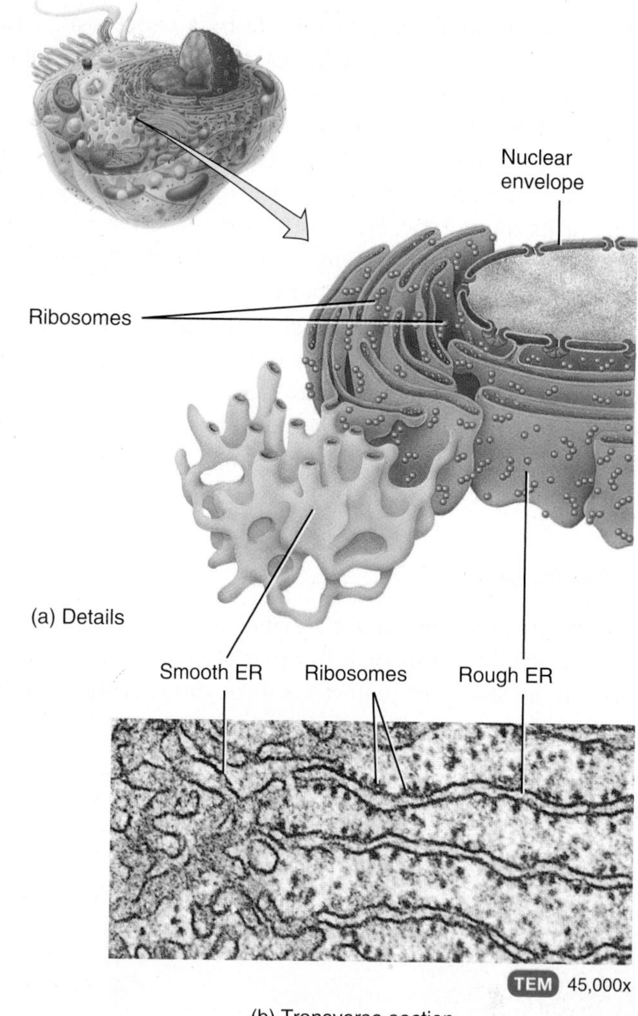


(b) Transverse section

**FUNCTIONS**

1. Rough ER synthesizes glycoproteins and phospholipids that are transferred into cellular organelles, inserted into the plasma membrane, or secreted during exocytosis.
2. Smooth ER synthesizes fatty acids and steroids, releases glucose into the bloodstream, inactivates or detoxifies drugs and other potentially harmful substances, and stores and releases calcium ions that trigger muscle contraction.

**The endoplasmic reticulum is a network of membrane-enclosed sacs or tubules that extend throughout the cytoplasm and connect to the nuclear envelope.**

about half the size of the other (Figure 3.17). The large and small subunits are made separately in the nucleolus, a spherical body inside the nucleus. Once produced, the large and small subunits exit the nucleus separately, then come together in the cytoplasm.

Some ribosomes are attached to the outer surface of the nuclear membrane and to an extensively folded membrane called the endoplasmic reticulum. These ribosomes synthesize proteins destined for specific organelles, for insertion in the plasma membrane, or for export from the cell. Other ribosomes are "free" or unattached to other cytoplasmic structures. Free ribosomes synthesize proteins used in the cytosol. Ribosomes are also located within mitochondria, where they synthesize mitochondrial proteins.

## Endoplasmic Reticulum

The **endoplasmic reticulum** (en′-dō-PLAS-mik re-TIK-ū-lum; *-plasmic* = cytoplasm; *reticulum* = network) or **ER** is a network of membranes in the form of flattened sacs or tubules (Figure 3.18). The ER extends from the nuclear envelope (membrane around the nucleus), to which it is connected, throughout the cytoplasm. The ER is so extensive that it constitutes more than half of the membranous surfaces within the cytoplasm of most cells.

Cells contain two distinct forms of ER that differ in structure and function. **Rough ER** is continuous with the nuclear envelope and usually is folded into a series of flattened sacs. The outer surface of rough ER is studded with ribosomes, the sites of protein synthesis. Proteins synthesized by ribosomes attached to rough ER enter spaces within the ER for processing and sorting. In some cases, enzymes attach the proteins to carbohydrates to form glycoproteins. In other cases, enzymes attach the proteins to phospholipids, also synthesized by rough ER. The resulting glycoproteins and phospholipids may be incorporated into the membranes of organelles, inserted into the plasma membrane, or secreted via exocytosis. Thus, rough ER produces many organellar proteins, plasma membrane proteins, and secretory proteins.

**Smooth ER** extends from the rough ER to form a network of membrane tubules (Figure 3.18). Unlike rough ER, smooth ER appears "smooth" because it lacks ribosomes on the outer surfaces of its membrane. Because it lacks ribosomes, smooth ER does not synthesize proteins, but it does synthesize fatty acids and steroids, such as estrogens and testosterone. In liver cells, the smooth ER helps release glucose into the bloodstream and inactivate or detoxify a variety of potentially harmful substances, such as alcohol, pesticides, and *carcinogens* (cancer-causing agents). In muscle cells, the calcium ions ($Ca^{2+}$) that trigger contraction are released from the sarcoplasmic reticulum, a form of smooth ER.

## Golgi Complex

Most of the proteins synthesized by ribosomes attached to rough ER are ultimately transported to other regions of the cell. The first step in the transport pathway is through the **Golgi complex** (GOL-jē). It consists of 3 to 20 **cisternae** (sis-TER-nē = cavities; singular is *cisterna*), small, flattened membranous sacs with bulging edges that resemble a stack of pita bread (Figure 3.19). The cisternae are often curved, giving the Golgi complex a bowl-like shape. Most cells have several Golgi complexes. Golgi complexes are more extensive in cells that secrete proteins, a clue to the organelle's role in the cell.

The cisternae at the opposite ends of a Golgi complex differ from each other in size, shape, and enzymatic activity. The convex **entry face** is a cisterna that faces the rough ER. The concave **exit face** is a cisterna that faces the plasma membrane. Sacs between the entry and exit faces are called **medial cisternae**. Transport vesicles (described shortly) from the ER merge to form the entry face. From the entry face, the cisternae are thought to mature, in turn becoming medial and then exit cisternae.

Different enzymes in the cisternae allow the Golgi complex to modify, sort, and package proteins for transport to different destinations. The entry face receives and modifies proteins produced by the rough ER. The medial cisternae further modify the proteins by adding carbohydrates or lipids. The exit face modifies the molecules further and then sorts and packages them for transport to their destinations.

**FIGURE 3.19** **Golgi complex.**

**FUNCTIONS**

1. Modifies, sorts, packages, and transports proteins received from the rough ER.
2. Generates secretory vesicles that secrete processed proteins during exocytosis.
3. Forms membrane vesicles that ferry new molecules to the plasma membrane.
4. Produces transport vesicles that carry molecules to other organelles.

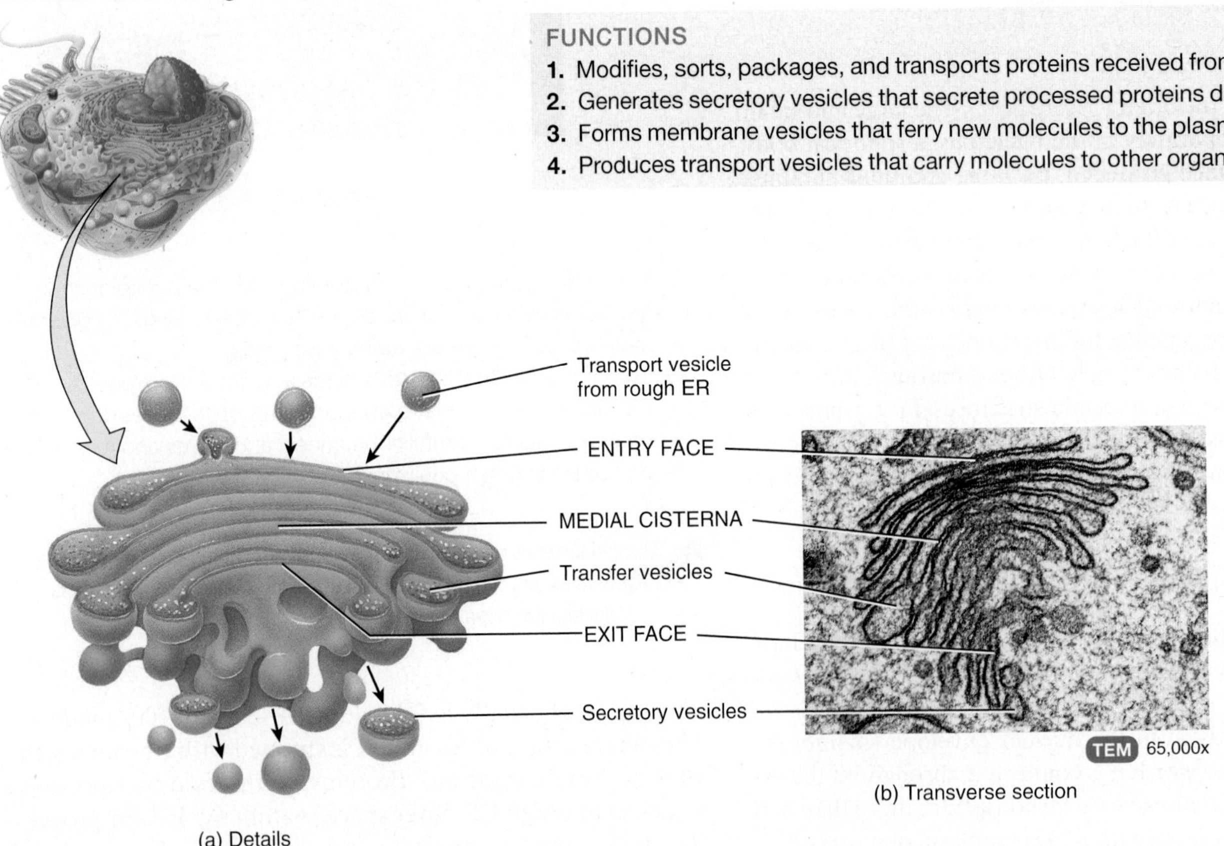


(a) Details

(b) Transverse section

Most proteins synthesized by ribosomes attached to rough ER pass through the Golgi complex for processing.

Proteins arriving at, passing through, and exiting the Golgi complex do so through maturation of the cisternae and exchanges that occur via transfer vesicles (Figure 3.20):

1. Proteins synthesized by ribosomes on the rough ER are surrounded by a piece of the ER membrane, which eventually buds from the membrane surface to form **transport vesicles**.
2. Transport vesicles move toward the entry face of the Golgi complex.
3. Fusion of several transport vesicles creates an additional entry face of the Golgi complex while releasing proteins into its lumen (space).
4. The proteins move from the entry face into one or more medial cisternae. Enzymes in the medial cisternae add carbohydrates or lipids to the proteins. **Transfer vesicles** that bud from the edges of the cisternae move specific enzymes back toward the entry face and move partially modified proteins toward the exit face.
5. The partially modified proteins enter the lumen of the exit face.
6. Within the exit face cisterna, the proteins are further modified and are sorted and packaged.
7. Some of the processed proteins leave the exit face in **secretory vesicles**. These vesicles deliver the proteins to the plasma membrane, where they fuse with the plasma membrane and discharge the proteins by exocytosis into the extracellular fluid.
8. Other processed proteins leave the exit face in **membrane vesicles** that deliver their contents to the plasma membrane for the purpose of adding new segments of plasma membrane as existing segments are lost and to modify the number and distribution of membrane molecules.
9. Finally, some processed proteins leave the exit face in transport vesicles that will carry the proteins to another cellular destination. For instance, **transport vesicles** carry digestive enzymes to lysosomes; the structure and functions of these important organelles are discussed next.

## Lysosomes

**Lysosomes** (LĪ-sō-sōms; *lyso-* = dissolving; *-somes* = bodies) are membrane-enclosed vesicles that form from the Golgi complex

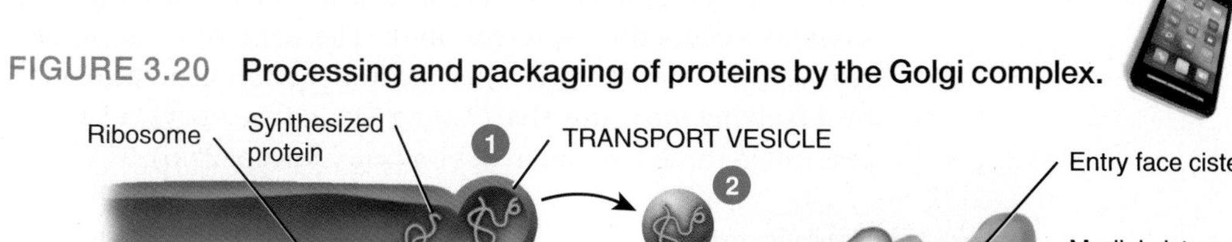

FIGURE 3.20 **Processing and packaging of proteins by the Golgi complex.**

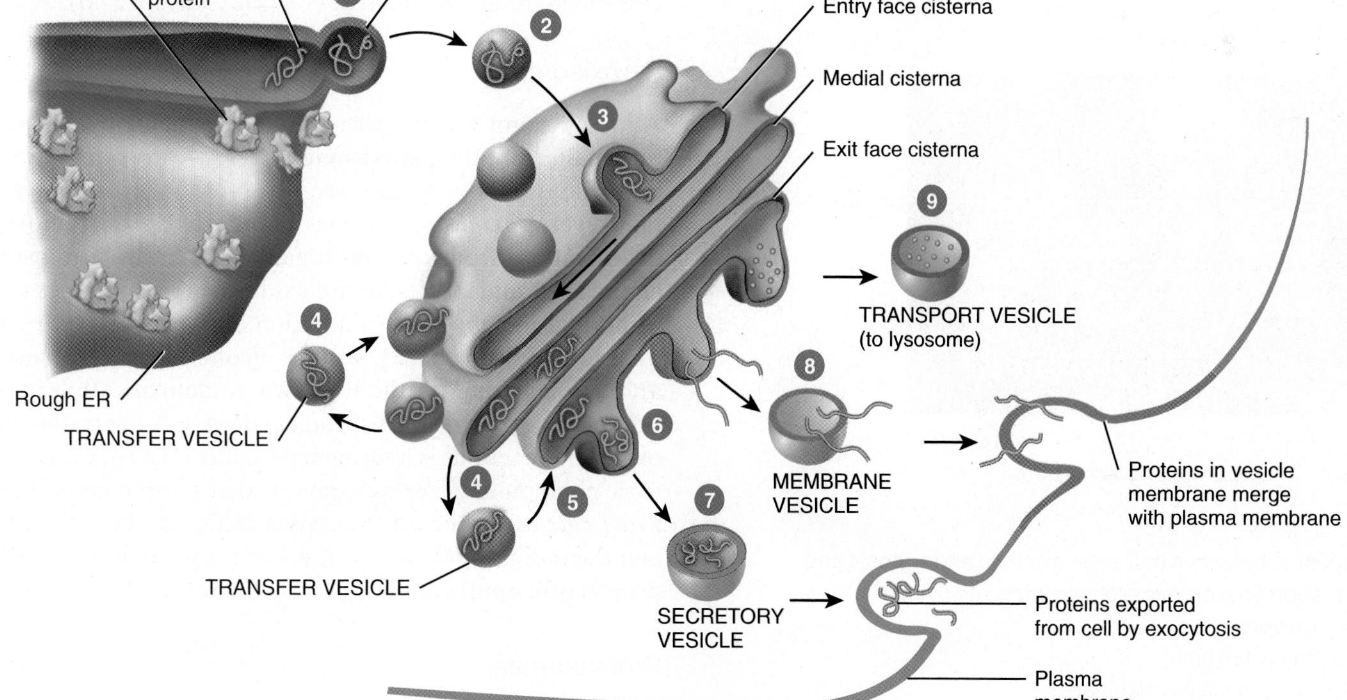

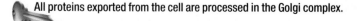

**All proteins exported from the cell are processed in the Golgi complex.**

(Figure 3.21). Inside, powerful enzymes can break down a wide variety of molecules once lysosomes fuse with vesicles formed during endocytosis. Because lysosomal enzymes work best at an acidic pH, the lysosomal membrane includes active transport pumps that import hydrogen ions ($H^+$). Thus, the lysosomal interior has a pH of 5, which is 100 times more acidic than the pH of the cytosol (pH 7). The lysosomal membrane also includes carriers that move the digested molecules into the cytosol.

Lysosomal enzymes also help recycle worn-out cell structures. A lysosome can engulf another organelle, digest it, and return the digested components to the cytosol for reuse. In this way, old organelles are continually replaced. Lysosomal enzymes may also destroy the entire cell, a process known as **autolysis** (aw-TOL-i-sis). Autolysis occurs in some pathological conditions and is also responsible for the tissue deterioration that occurs immediately after death.

As we just discussed, most lysosomal enzymes act within a cell. However, some operate in extracellular digestion. One example occurs during fertilization. The head of a sperm cell releases lysosomal enzymes that aid its penetration of the oocyte by dissolving the protective coating around the oocyte in a process called the acrosomal reaction (see Concept 25.6).

**FIGURE 3.21 Lysosomes.**

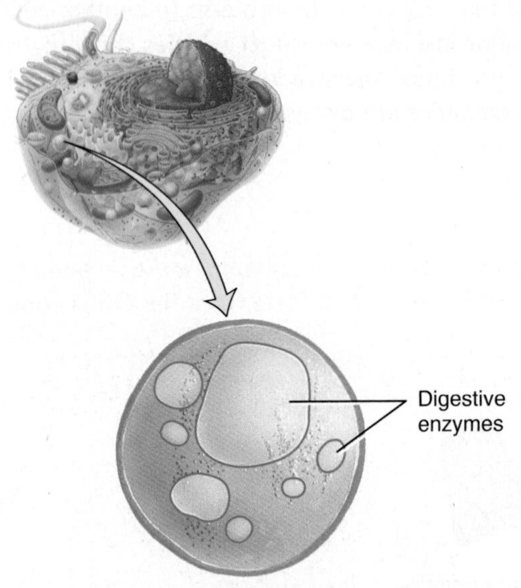


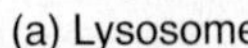
(a) Lysosome

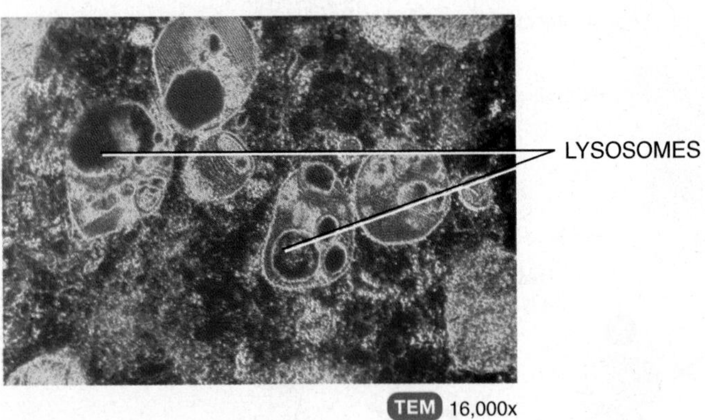


(b) Several lysosomes

**FUNCTIONS**

1. Digest substances that enter a cell via endocytosis and transport final products of digestion into cytosol.
2. Digest worn-out organelles.
3. Destroy entire cell.
4. Digest extracellular materials.

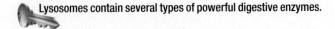

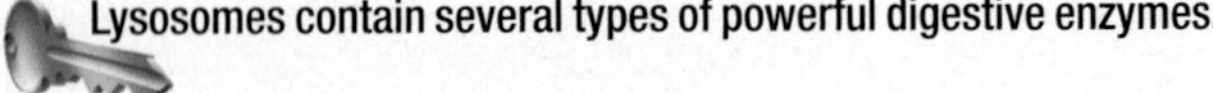
**Lysosomes contain several types of powerful digestive enzymes.**

### CLINICAL CONNECTION | *Tay-Sachs Disease*

Some disorders are caused by faulty or absent lysosomal enzymes. For instance, **Tay-Sachs disease** (TĀ-SAKS), which most often affects children of Ashkenazi (eastern European Jewish) descent, is an inherited condition characterized by the absence of a single lysosomal enzyme called Hex A. This enzyme normally breaks down a membrane glycolipid called ganglioside $G_{M2}$ that is especially prevalent in nerve cells. As the excess ganglioside $G_{M2}$ accumulates, the nerve cells function less efficiently. Children with Tay-Sachs disease typically experience seizures and muscle rigidity. They gradually become blind, demented, and uncoordinated and usually die before the age of 5. Tests can now reveal whether an adult is a carrier of the defective gene.

## Peroxisomes

Another group of organelles similar in structure to lysosomes, but smaller, are the **peroxisomes** (pe-ROKS-i-sōms; *peroxi-* = peroxide; *somes* = bodies; see Figure 3.1). Peroxisomes contain several *oxidases*, enzymes that can oxidize (remove hydrogen atoms from) various organic substances. For instance, amino acids and fatty acids are oxidized in peroxisomes as part of normal metabolism. In addition, enzymes in peroxisomes oxidize toxic substances, such as alcohol. Thus, peroxisomes are very abundant in the liver, where detoxification of alcohol and other damaging substances occurs. A byproduct of the oxidation reactions is hydrogen peroxide ($H_2O_2$), a potentially toxic compound. However, peroxisomes also contain the enzyme *catalase*, which decomposes $H_2O_2$. By both producing and degrading $H_2O_2$ within the same organelle, peroxisomes protect other parts of the cell from the toxic effects of $H_2O_2$.

## Proteasomes

Although lysosomes degrade proteins delivered to them in vesicles, proteins within the cytosol also require disposal at certain times in the life of a cell. Continuous destruction of unneeded, damaged, or faulty proteins is the function of tiny structures called **proteasomes** (PRŌ-tē-a-sōms = protein bodies). For example, proteins that are part of metabolic pathways need to be degraded after they have accomplished their function. Such protein destruction plays a part in negative feedback by halting a pathway once the appropriate response has been achieved. Proteasomes contain *proteases*, enzymes that cut proteins into small peptides. The peptide products are broken apart by other cellular enzymes into amino acids, which can be recycled into new proteins.

## Mitochondria

Because they are the site of most ATP production, **mitochondria** (mī-tō-KON-drē-a; *mito-* = thread; *-chondria* = granules; singular is *mitochondrion*) are referred to as the "powerhouses" of the cell. A cell may have as few as a hundred or as many as several thousand mitochondria, depending on the activity of the cell. Active cells, such as those found in muscles, use ATP at a high rate and have a large number of mitochondria.

A mitochondrion consists of an **outer mitochondrial membrane** and an **inner mitochondrial membrane**, with a small fluid-filled space between them (Figure 3.22). Both membranes are similar in structure to the plasma membrane. The inner mitochondrial membrane is arranged in a series of folds called **cristae** (KRIS-tē = ridges). The central fluid-filled cavity of a mitochondrion, enclosed by the inner mitochondrial membrane, is the **matrix**. The elaborate folds of the cristae provide an enormous surface area for the chemical reactions that are part of the aerobic phase of *cellular respiration*, the reactions that produce most of a cell's ATP. The enzymes that catalyze these reactions are located on the cristae and in the matrix of the mitochondria.

Mitochondria self-replicate, a process that occurs during times of increased cellular energy demand or before cell division. Synthesis of some proteins needed for mitochondrial functions occurs on the ribosomes that are present in the mitochondrial matrix. Mitochondria even contain their own DNA with a small number of genes; these genes control synthesis of the proteins that build mitochondrial components.

FIGURE 3.22 **Mitochondria.**

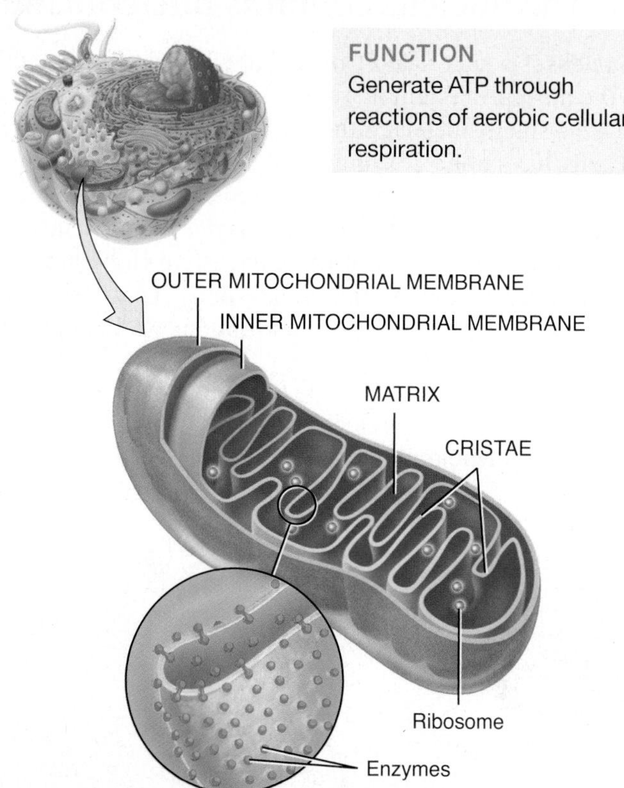


(a) Details

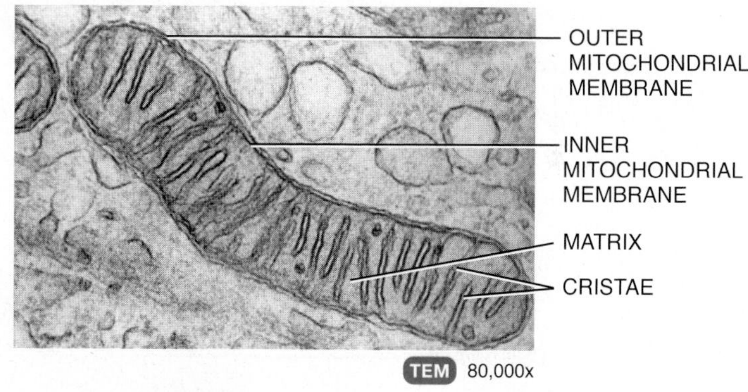


(b) Transverse section

Within mitochondria, chemical reactions of aerobic cellular respiration generate ATP.

## ✓ CHECKPOINT

13. Which cytoskeletal components help form the structure of microvilli? Which help form centrioles, cilia, and flagella?
14. What is the functional difference between cilia and flagella?
15. Where are subunits of ribosomes synthesized? Where are they assembled?
16. What are the structural and functional differences between rough and smooth ER?
17. What are the three general destinations for proteins that leave the Golgi complex?
18. What is autolysis?
19. What happens on the cristae and in the matrix of mitochondria?

# 3.5 The nucleus contains nucleoli and genes.

The **nucleus** is a spherical or oval-shaped structure that usually is the most prominent feature of a cell (Figure 3.23). Most cells have a single nucleus, although some, such as mature red blood cells, have none. In contrast, skeletal muscle cells and a few other types of cells have multiple nuclei. A double membrane called the **nuclear envelope** separates the nucleus from the cytoplasm. Both layers of the nuclear envelope are lipid bilayers similar to the plasma membrane. The outer membrane of the nuclear envelope is continuous with rough ER. Many openings called **nuclear pores** extend through the nuclear envelope. Each nuclear pore consists of a circular arrangement of proteins surrounding a large central opening that is about 10 times wider than the pore of a channel protein in the plasma membrane.

Nuclear pores control the movement of substances between the nucleus and the cytoplasm. Small molecules and ions move through the pores passively by diffusion. Most large molecules, such as RNAs and proteins, pass through the nuclear

FIGURE 3.23 **Nucleus.**

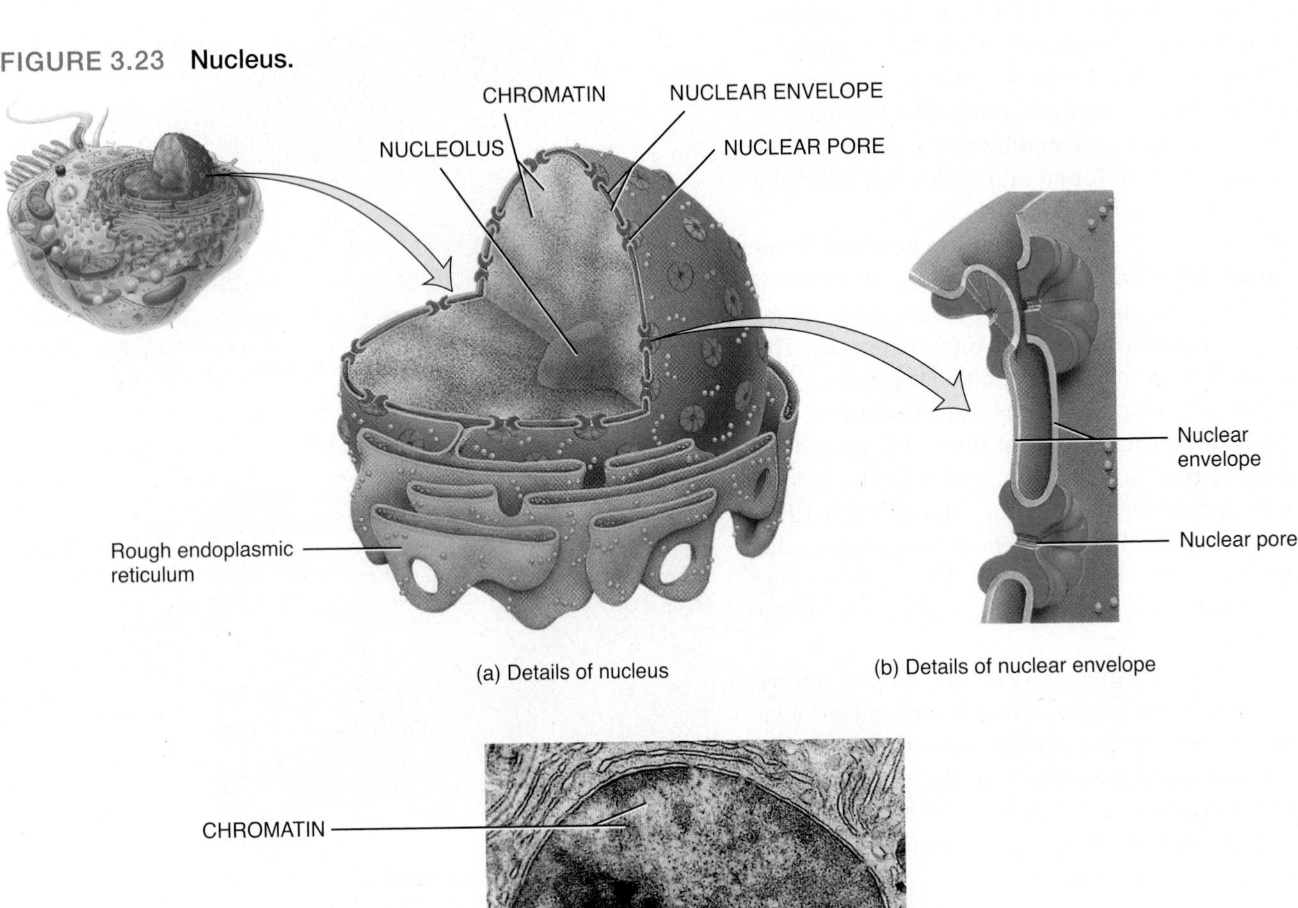


FUNCTIONS
1. Controls cellular structure.
2. Directs cellular activities.
3. Produces ribosomal subunits in nucleoli.

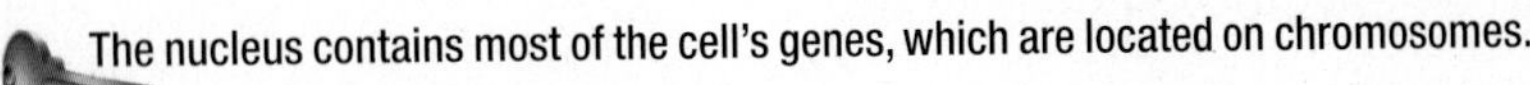
The nucleus contains most of the cell's genes, which are located on chromosomes.

pores by an active transport process in which the molecules are recognized and selectively transported through the nuclear pore into or out of the nucleus. For example, proteins needed for nuclear functions move into the nucleus from the cytosol; newly formed RNA molecules move out of the nucleus into the cytosol.

Inside the nucleus are one or more spherical bodies called **nucleoli** (noo-KLĒ-ō-lī; singular is *nucleolus*) that function in producing ribosomes. Each nucleolus is simply a cluster of protein, DNA, and RNA; it is not enclosed by a membrane. Nucleoli are the sites of synthesis of rRNA and its assembly into ribosomal subunits. Nucleoli are quite prominent in cells that synthesize large amounts of protein, such as muscle and liver cells.

Within the nucleus are most of the cell's hereditary units, called **genes**, which control cellular structure and direct cellular activities. Genes are arranged along **chromosomes** (KRŌ-mō-sōms; *chromo-* = colored). Human somatic (body) cells have 46 chromosomes, 23 inherited from each parent. Each chromosome is a long molecule of DNA that is coiled together with several proteins (Figure 3.24). This complex of DNA, proteins, and some RNA is called **chromatin** (KRŌ-ma-tin). The total genetic information carried in a cell or an organism is its **genome** (JĒ-nōm).

In cells that are not dividing, the chromatin appears as a diffuse, granular mass. Electron micrographs reveal that chromatin has a beads-on-a-string structure. Each bead is a **nucleosome** that consists of double-stranded DNA wrapped twice around a core of eight proteins called **histones**, which help organize the coiling and folding of DNA. The string between the beads is **linker DNA**, which holds adjacent nucleosomes together. In cells that are not dividing, another histone promotes coiling of nucleosomes into a larger diameter **chromatin fiber**, which then folds into large loops. Just before cell division takes place, however, the DNA replicates (duplicates) and the loops condense even more, forming a pair of **chromatids**. As you will see shortly, during cell division a pair of chromatids constitutes a chromosome.

The main parts of a cell, their descriptions, and their functions are summarized in Table 3.2.

### ✓ CHECKPOINT

**20.** How do large particles enter and exit the nucleus?

**21.** How is DNA packed inside the nucleus?

**22.** What are the components of a nucleosome?

**FIGURE 3.24 Packing of DNA into a chromosome in a dividing cell.** When packing is complete, two identical DNA molecules and their histones form a pair of chromatids, which are held together by a centromere.

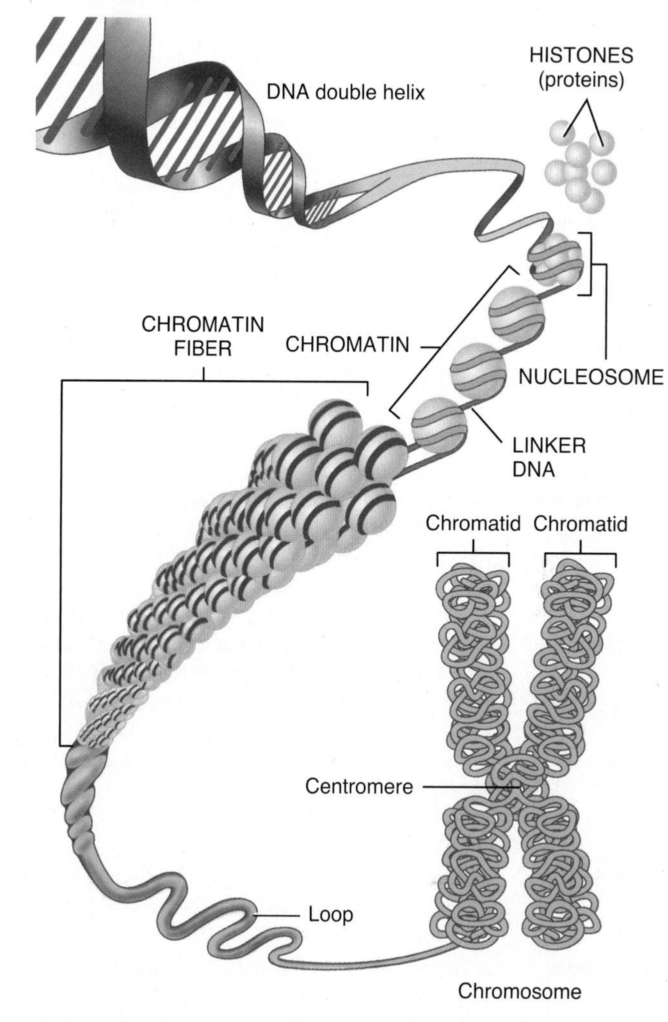


(a) Illustration

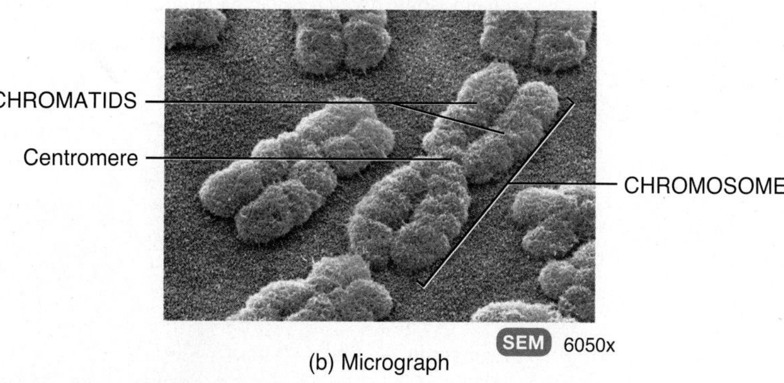


(b) Micrograph

A chromosome is a highly coiled and folded DNA molecule that is combined with protein molecules.

## TABLE 3.2

### Cell Parts and Their Functions

| PART | DESCRIPTION | FUNCTIONS |
|---|---|---|
| **PLASMA MEMBRANE** | Fluid-mosaic lipid bilayer (phospholipids, cholesterol, and glycolipids) studded with proteins; surrounds cytoplasm | Protects cellular contents; makes contact with other cells; contains channels, carriers, receptors, enzymes, cell-identity markers, and linker proteins; mediates the entry and exit of substances |
| **CYTOPLASM** | Cellular contents between the plasma membrane and nucleus—cytosol and organelles | Site of all intracellular activities except those occurring in the nucleus |
| **Cytosol** | Composed of water, solutes, suspended particles, lipid droplets, and glycogen granules | Medium in which many of cell's metabolic reactions occur |
| **Organelles** | Specialized structures with characteristic shapes | Each organelle has specific functions |
| **Cytoskeleton** | Network of three types of protein filaments: microfilaments, intermediate filaments, and microtubules | Maintains shape and general organization of cellular contents; responsible for cellular movements |
| **Centrosome** | A pair of centrioles plus pericentriolar material | The pericentriolar material contains tubulins, which are used for growth of the mitotic spindle and microtubule formation |
| **Cilia and flagella** | Motile cell surface projections that contain 20 microtubules and a basal body | Cilia move fluids over a cell's surface; flagella move an entire cell |
| **Ribosome** | Composed of two subunits containing ribosomal RNA and proteins; may be free in cytosol or attached to rough ER | Protein synthesis |
| **Endoplasmic reticulum (ER)** | Membranous network of flattened sacs or tubules<br>Rough ER is covered by ribosomes and is attached to the nuclear envelope<br>Smooth ER lacks ribosomes | Rough ER synthesizes glycoproteins and phospholipids that are transferred to cellular organelles, inserted into the plasma membrane, or secreted during exocytosis<br>Smooth ER synthesizes fatty acids and steroids; inactivates or detoxifies drugs; removes the phosphate group from glucose 6-phosphate; and stores and releases calcium ions in muscle cells |
| **Golgi complex** | Consists of 3–20 flattened membranous sacs called cisternae; structurally and functionally divided into entry face, medial cisternae, and exit face | Entry face accepts proteins from rough ER; medial cisternae form glycoproteins, glycolipids, and lipoproteins; exit face modifies the molecules further, then sorts and packages them for transport to their destinations |
| **Lysosome** | Vesicle formed from Golgi complex; contains digestive enzymes | Fuses with and digests contents of endosomes, pinocytic vesicles, and phagosomes and transports final products of digestion into cytosol; digests worn-out organelles (autophagy), entire cells (autolysis), and extracellular materials. |
| **Peroxisome** | Vesicle containing oxidases (oxidative enzymes) and catalase (decomposes hydrogen peroxide); new peroxisomes bud from preexisting ones | Oxidizes amino acids and fatty acids; detoxifies harmful substances, such as alcohol |
| **Proteasome** | Tiny barrel-shaped structure that contains proteases (proteolytic enzymes) | Degrades unneeded, damaged, or faulty proteins by cutting them into small peptides |
| **Mitochondrion** | Consists of an outer and an inner mitochondrial membrane, cristae, and matrix; new mitochondria form from preexisting ones | Site of aerobic cellular respiration reactions that produce most of a cell's ATP |
| **NUCLEUS** | Consists of a nuclear envelope with pores, nucleoli, and chromosomes, which exist as a tangled mass of chromatin in interphase cells | Nuclear pores control the movement of substances between the nucleus and cytoplasm, nucleoli produce ribosomes, and chromosomes consist of genes that control cellular structure and direct cellular functions |

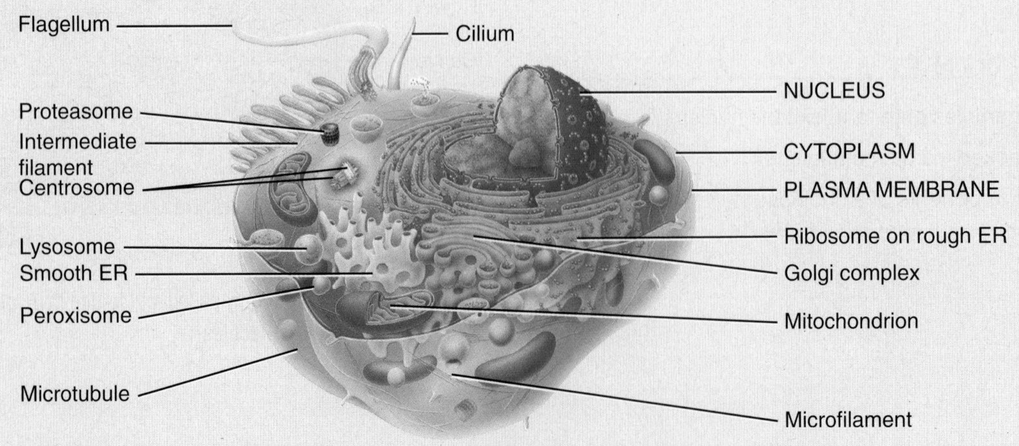

## CLINICAL CONNECTION | *Genomics*

In the last decade of the twentieth century, the genomes of humans, mice, fruit flies, and more than 50 microbes were sequenced. As a result, research in the field of **genomics,** the study of the relationships between the genome and the biological functions of an organism, has flourished. The Human Genome Project began in June 1990 as an effort to sequence all of the nearly 3.2 billion nucleotides of our genome, and was completed in April 2003. More than 99.9 percent of the nucleotide bases are identical in everyone. Less than 0.1 percent of our DNA (1 in each 1000 bases) accounts for inherited differences among humans. Surprisingly, at least half of the human genome consists of repeated sequences that do not code for proteins, so-called "junk" DNA. The average gene consists of 3000 nucleotides, but sizes vary greatly. The largest known human gene, with 2.4 million nucleotides, codes for the protein dystrophin. Scientists now know that the total number of genes in the human genome is about 30,000, far fewer than the 100,000 previously predicted to exist. Information regarding the human genome and how it is affected by the environment seeks to identify and discover the functions of the specific genes that play a role in genetic diseases. Genomic medicine also aims to design new drugs and to provide screening tests to enable physicians to provide more effective counseling and treatment for disorders with significant genetic components such as hypertension (high blood pressure), obesity, diabetes, and cancer.

## 3.6 Cells make proteins by transcribing and translating the genetic information contained in DNA.

Although cells synthesize many chemicals to maintain homeostasis, much of the cellular machinery is devoted to synthesizing large numbers of diverse proteins. The proteins in turn determine the physical and chemical characteristics of cells and, therefore, the organisms formed from them. Some proteins help assemble cellular structures such as plasma membrane proteins and the cytoskeleton. Others serve as hormones, antibodies, or contractile elements in muscle tissue. Still others act as enzymes regulating the rates of the numerous chemical reactions that occur in cells, or transporters carrying various materials in the blood.

In **gene expression**, a gene's DNA is used as a template for synthesis of a specific protein (Figure 3.25). First, in a process aptly named transcription, the information encoded in a specific region of DNA is *transcribed* (copied) to produce a specific molecule of RNA (ribonucleic acid). In a second process, referred to as *translation*, the RNA attaches to a ribosome, where the information contained in RNA is *translated* into a corresponding sequence of amino acids to form a new protein molecule.

DNA and RNA store genetic information as sets of three-nucleotide sequences. A sequence of three such nucleotides in DNA is called a **base triplet**. Each DNA base triplet is transcribed as a complementary sequence of three RNA nucleotides, called a **codon**. A given codon specifies (or *codes for*) a particular amino acid.

FIGURE 3.25 **Overview of gene expression.** Synthesis of a specific protein requires transcription of a gene's DNA into RNA and translation of RNA into a corresponding sequence of amino acids.

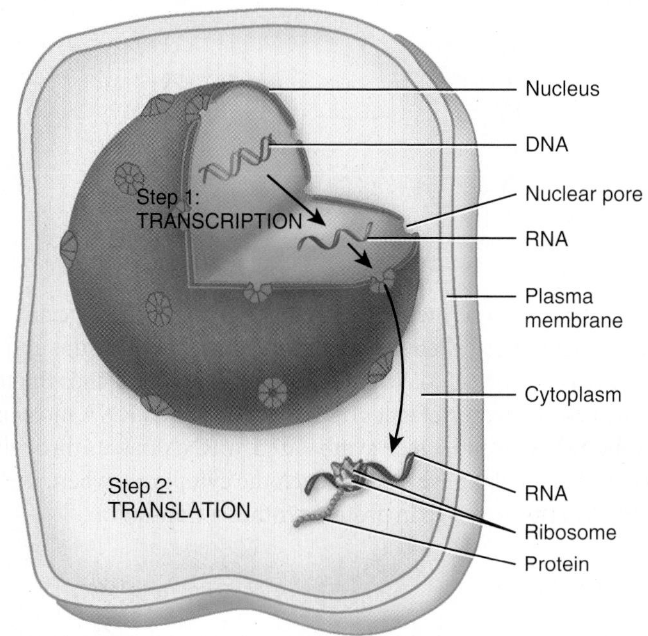


**Transcription occurs in the nucleus; translation occurs in the cytoplasm.**

## Transcription

**Transcription** occurs in the nucleus. During transcription, genetic information in a sequence of DNA base triplets is used as a template to copy a complementary sequence of RNA codons. Three types of RNA are made from the DNA template:

- **Messenger RNA (mRNA)**, the traveling copy of a gene, enters the cytoplasm to direct the synthesis of a specific protein.
- **Ribosomal RNA (rRNA)** joins with ribosomal proteins to make ribosomes.
- **Transfer RNA (tRNA)** binds to an amino acid and holds it in place on a ribosome until it is incorporated into a protein during translation, a process you will learn about next. One end of the tRNA carries a specific amino acid, and the opposite end consists of a triplet of nucleotides called an **anticodon** (see Figure 3.28). By pairing between complementary bases, the tRNA anticodon attaches to the mRNA codon.

The enzyme **RNA polymerase** (po-LIM-er-ās) catalyzes the transcription of DNA. However, the enzyme must be instructed where to start the transcription process and where to end it. Near the beginning of a gene is a special nucleotide sequence called a **promoter** (Figure 3.26a). The promoter is where RNA polymerase attaches to the DNA to start transcription.

During transcription, bases pair in a complementary manner: The bases cytosine (C), guanine (G), and thymine (T) in the DNA template pair with guanine, cytosine, and adenine (A), respectively, in the RNA strand (Figure 3.26b). However, adenine in the DNA template pairs with uracil (U), not thymine, in RNA:

| Template DNA base sequence | | Complementary RNA base sequence |
|---|---|---|
| A | | U |
| T | | A |
| G | ⟶ | C |
| C | | G |
| A | | U |
| T | | A |

Transcription of the DNA strand ends at another special nucleotide sequence called a **terminator**, which specifies the end of the gene (Figure 3.26a). When RNA polymerase reaches the terminator, the enzyme detaches from the transcribed RNA molecule and the DNA strand. Once synthesized, mRNA passes through a pore in the nuclear envelope to reach the cytoplasm, where it participates in the next step in protein synthesis, translation.

FIGURE 3.26 **Transcription.** DNA transcription begins at a promoter and ends at a terminator.

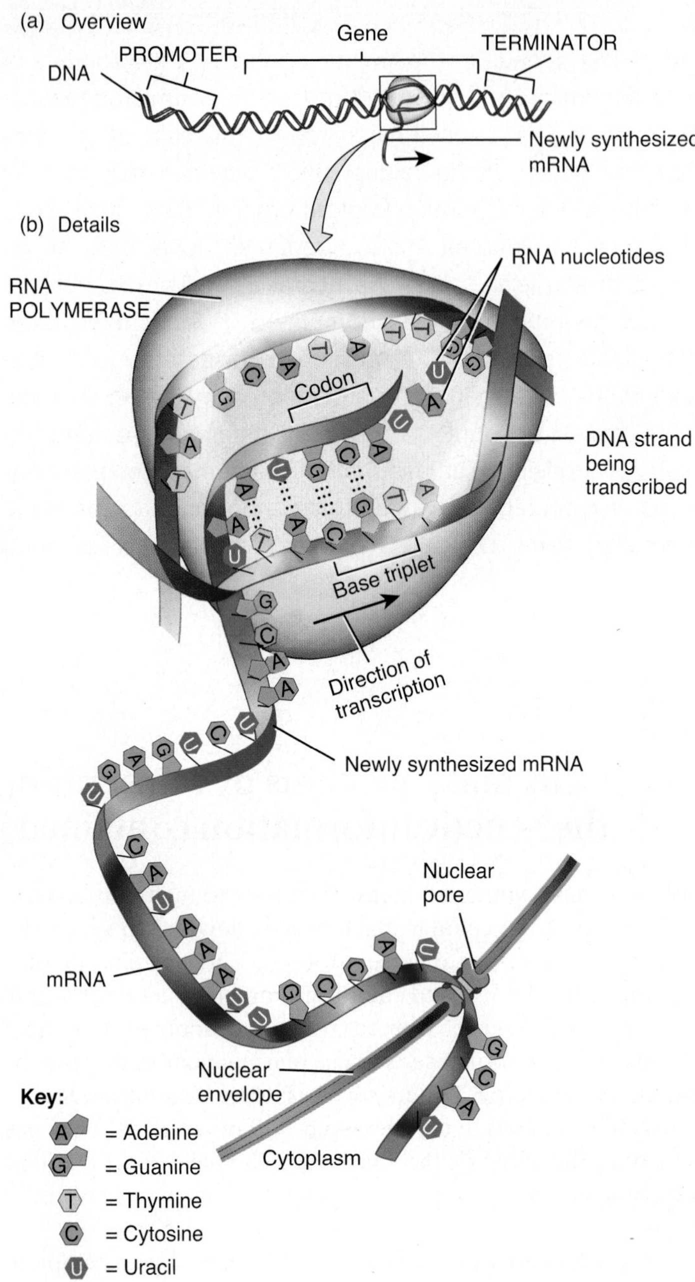


During transcription, the genetic information in DNA is copied to RNA.

## Translation

In the process of **translation**, the nucleotide sequence in an mRNA molecule specifies the amino acid sequence of a particular protein. Ribosomes in the cytoplasm carry out translation. The small subunit of a ribosome has a *binding site* for mRNA; the large subunit has two binding sites for tRNA molecules, a *P site* and an *A site* (Figure 3.27). The first tRNA molecule bearing its specific amino acid attaches to mRNA at the P site. The A site holds the

FIGURE 3.27 **Translation.** During translation, an mRNA molecule binds to a ribosome. Then the mRNA nucleotide sequence specifies the amino acid sequence of a protein.

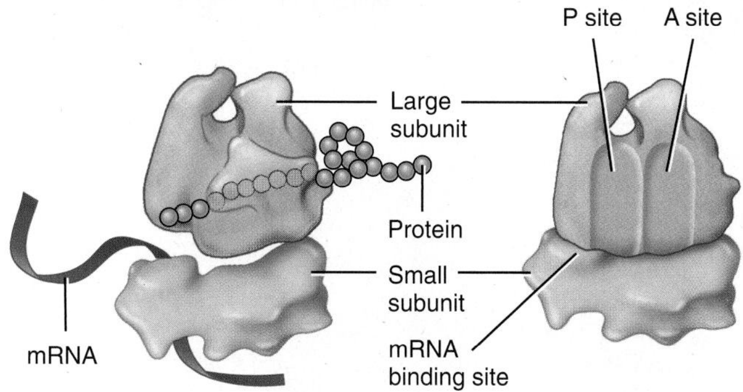


(a) Components of a ribosome and their relationship to mRNA and protein during translation

(b) Interior view of tRNA binding sites

**Ribosomes have a binding site for mRNA and a P site and an A site for attachment of tRNA.**

next tRNA molecule bearing its amino acid. Translation occurs in the following way (Figure 3.28):

1. An mRNA molecule binds to the small ribosomal subunit at the mRNA binding site. A tRNA carrying the first amino acid binds to the start codon on mRNA, where translation begins. The tRNA anticodon attaches to the mRNA codon by pairing between the complementary bases.
2. Next, the large ribosomal subunit attaches to the small ribosomal subunit–mRNA complex, creating a functional ribosome. The first tRNA, with its amino acid, fits into the P site of the ribosome.
3. The anticodon of another tRNA with its attached amino acid pairs with the second mRNA codon at the A site of the ribosome.
4. A peptide bond is formed between the amino acid carried by the first tRNA and second tRNA.
5. After peptide bond formation, the tRNA at the P site detaches from the ribosome, and the ribosome shifts the mRNA strand by one codon. The tRNA in the A site, now bearing the two-peptide protein, shifts into the P site, allowing another tRNA with its amino acid to bind to a newly exposed codon at the A site. Steps 3 through 5 occur repeatedly, and the protein lengthens progressively.
6. Protein synthesis ends when the ribosome reaches a stop codon at the A site, which causes the completed protein to detach from the final tRNA. When the tRNA vacates the A site, the ribosome splits into its large and small subunits.

As the ribosome moves along the mRNA and before it completes synthesis of the whole protein, another ribosome may attach behind it and begin translation of the same mRNA strand. The simultaneous movement of several ribosomes along the same mRNA molecule allows translation of one mRNA into several identical proteins at the same time.

## CLINICAL CONNECTION | *Recombinant DNA*

Scientists have developed techniques for inserting genes from other organisms into a variety of host cells. Manipulating the cell in this way can cause the host organism to produce proteins it normally does not synthesize. Organisms so altered are called **recombinants** (rē-KOM-bi-nants), and their DNA—a combination of DNA from different sources—is called **recombinant DNA.** When recombinant DNA functions properly, the host will synthesize the protein specified by the new gene it has acquired. The technology that has arisen from the manipulation of genetic material is referred to as **genetic engineering**.

The practical applications of recombinant DNA technology are enormous. Strains of recombinant bacteria now produce large quantities of many important therapeutic substances, including *human growth hormone (hGH)*, required for normal growth and metabolism; *insulin*, a hormone that helps regulate blood glucose level and is used by diabetics; *interferon (IFN)*, an antiviral (and possibly anticancer) substance; and *erythropoietin (EPO)*, a hormone that stimulates production of red blood cells.

## ✓ CHECKPOINT

**23.** What is the difference between transcription and translation?

**24.** What are the functions of mRNA, rRNA, and tRNA?

**25.** Which roles do the P and A sites serve?

FIGURE 3.28 **Protein elongation and termination of protein synthesis during translation.**

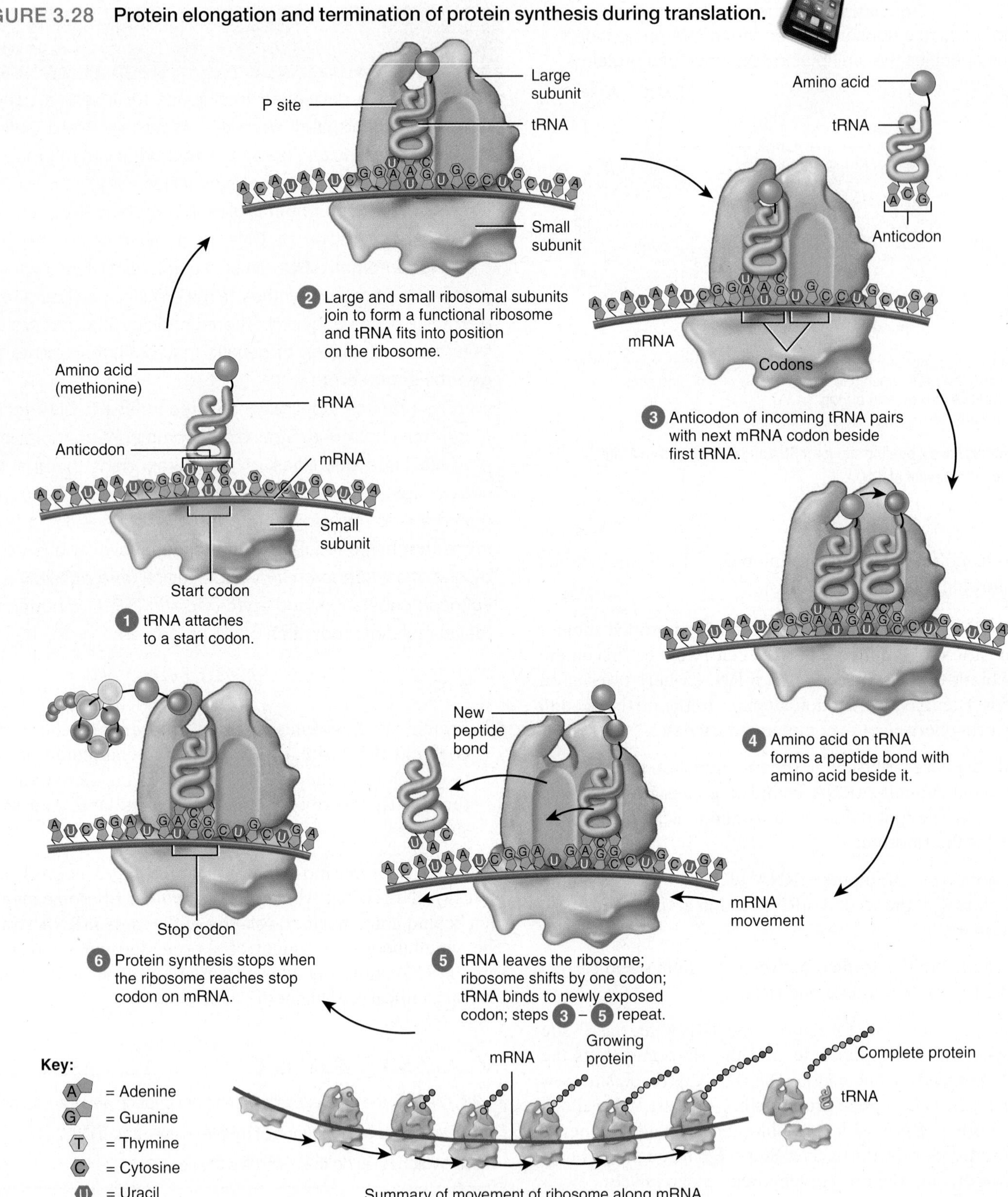

Large subunit
P site
tRNA
Small subunit
2 Large and small ribosomal subunits join to form a functional ribosome and tRNA fits into position on the ribosome.
Amino acid
tRNA
Anticodon
mRNA
Codons
3 Anticodon of incoming tRNA pairs with next mRNA codon beside first tRNA.
Amino acid (methionine)
tRNA
Anticodon
mRNA
Small subunit
Start codon
1 tRNA attaches to a start codon.
4 Amino acid on tRNA forms a peptide bond with amino acid beside it.
New peptide bond
mRNA movement
5 tRNA leaves the ribosome; ribosome shifts by one codon; tRNA binds to newly exposed codon; steps 3 – 5 repeat.
Stop codon
6 Protein synthesis stops when the ribosome reaches stop codon on mRNA.
Key:
A = Adenine
G = Guanine
T = Thymine
C = Cytosine
U = Uracil
mRNA
Growing protein
Complete protein
tRNA
Summary of movement of ribosome along mRNA

**During protein synthesis the small and large ribosomal subunits join to form a functional ribosome.**

## RETURN TO Joseph's Story

Joseph lay motionless except for the chest compressions being applied by the stranger who had come to help. Minutes had passed, but with each cycle of CPR, some fresh oxygen diffused into Joseph's lungs, and some carbon dioxide was forced out. Even with these efforts, many cells in Joseph's body had begun to die. The lack of ATP had not only affected the plasma membrane pumps, it also meant that special calcium ATPases had stopped moving calcium from the cytosol into the endoplasmic reticulum of his cardiac muscle cells. As the intracellular calcium levels rose, they caused proteases to spill into the interior of the cell, attacking the cytoskeleton. Lysosomal enzymes normally bound safely inside vesicles began to digest the plasma membranes and the membranes of the organelles.

After what seemed an eternity to Marcus, the paramedics finally arrived. His dad's face had regained some color, but he was still unconscious. The paramedics quickly did their work, restarting Joseph's heart with an electrical jolt from their portable defibrillator. Stabilizing Joseph as best they could, they all headed off to the emergency department.

Joseph's wife Sally had been contacted by the Sheriff's Department and was waiting at the hospital when they brought him in. Marcus joined her in the waiting room, where they waited for over half an hour before the doctor finally came to talk with them. "Your husband has suffered a massive heart attack. Right now he's stable and responsive, but he's very fatigued. From what we can tell, he has a blockage of one of the coronary arteries. We'll need to unblock the vessel to reestablish blood flow to the heart."

Sally put her arm around Marcus. "Is he going to be alright?" she asked.

"Well, the body has an amazing capacity to heal itself," said the doctor. "There are probably parts of the heart muscle that have been permanently damaged, but what we need to do right now is to provide oxygen and blood to the cells of the heart that are still alive."

C. *Which intracellular organelles have membranes as part of their structure? How would the breakdown of the membranes of these structures affect the function of Joseph's heart cells?*

D. *Two important pieces of information—the instructions Joseph's body needs to repair itself and his predisposition for vascular disease—are both contained within the cell on which structures?*

E. *Joseph's heart attack has caused the function of his cells to change. What types of proteins in the cell membrane were involved in the homeostatic imbalances of his heart cells?*

# 3.7 Cell division allows the replacement of cells and the production of new cells.

Most cells of the human body undergo **cell division**, the process by which cells reproduce themselves. The two types of cell division—somatic cell division and reproductive cell division—accomplish different goals for the organism.

A **somatic cell** (sō-MAT-ik; *soma* = body) is any cell of the body other than a gamete (sperm or oocyte) or any precursor cell destined to become a gamete. In **somatic cell division**, a cell undergoes a nuclear division called **mitosis** (mī-TŌ-sis; *mitos* = thread) followed by a cytoplasmic division called **cytokinesis** (sī′-tō-ki-NĒ-sis; *cyto* = cell; *kinesis* = movement) to produce two genetically identical cells, each with the same number and kind of chromosomes as the original cell. Somatic cell division replaces dead or injured cells and adds new cells during tissue growth.

**Reproductive cell division** is the mechanism that produces gametes, the cells needed to form the next generation of sexually reproducing organisms. This process consists of a special two-step division called **meiosis**, in which the number of chromosomes in the nucleus is reduced by half.

## Somatic Cell Division

Human cells, such as those in the brain, stomach, and kidneys, contain 23 pairs of chromosomes, for a total of 46. Because somatic cells contain two sets of chromosomes, they are called **diploid cells** (DIP-loyd; *dipl-* = double; *-oid* = form), symbolized **2n**. One member of each chromosome pair is inherited from each parent. The two chromosomes that make up each pair are called **homologous chromosomes** (hō-MOL-ō-gus; *homo-* = same); they contain similar genes arranged in the same (or almost the same) order. When examined under a light microscope, homologous chromosomes generally look very similar. The exception to this rule is one pair of chromosomes called the **sex**

**chromosomes**, designated X and Y. In females the homologous pair of sex chromosomes consists of two large X chromosomes; in males the pair consists of an X chromosome and a smaller Y chromosome. When a cell reproduces, it must duplicate all its chromosomes to pass its genes to the next generation of cells.

The **cell cycle** is an orderly sequence of events by which a somatic cell duplicates its contents and divides into two cells (Figure 3.29). The cell cycle consists of two major periods: interphase, when a cell is not dividing, and the mitotic (M) phase, when a cell is dividing.

## Interphase

During **interphase** (IN-ter-fāz) the cell replicates its DNA. It also produces additional organelles and cytosolic components in anticipation of cell division. Interphase is a state of high metabolic activity; it is during this time that the cell does most of its growing. Interphase consists of three phases: $G_1$, S, and $G_2$ (Figure 3.29). Because the G phases are periods when there is no activity related to DNA duplication, they are thought of as *gaps* or interruptions in DNA duplication. The S stands for *synthesis* of DNA.

The **$G_1$ phase** is the interval between the mitotic phase and the S phase. During $G_1$, the cell is metabolically active; it replicates most of its organelles and cytosolic components but not its DNA. Replication of centrosomes also begins in the $G_1$ phase. Virtually all of the cellular activities described in this chapter happen during $G_1$. For a cell with a total cell cycle time of 24 hours, $G_1$ lasts 8 to 10 hours. However, the duration of this phase is quite variable. It is very short in many embryonic cells or cancer cells. Once a cell enters the S phase, however, it is committed to go through the rest of the cell cycle.

The **S phase**, the interval between $G_1$ and $G_2$, lasts about 8 hours in a 24-hour cell cycle. During the S phase, DNA replication occurs. As a result, the two identical cells formed during cell division later in the cell cycle will have the same genetic material. When DNA replicates during the S phase, its helical structure partially uncoils, and the two strands separate at the points where hydrogen bonds connect base pairs (Figure 3.30). Each exposed base of the old DNA strand then pairs with the

FIGURE 3.29 **The cell cycle.** Not illustrated is cytokinesis, division of the cytoplasm, which usually occurs during late anaphase of the mitotic phase.

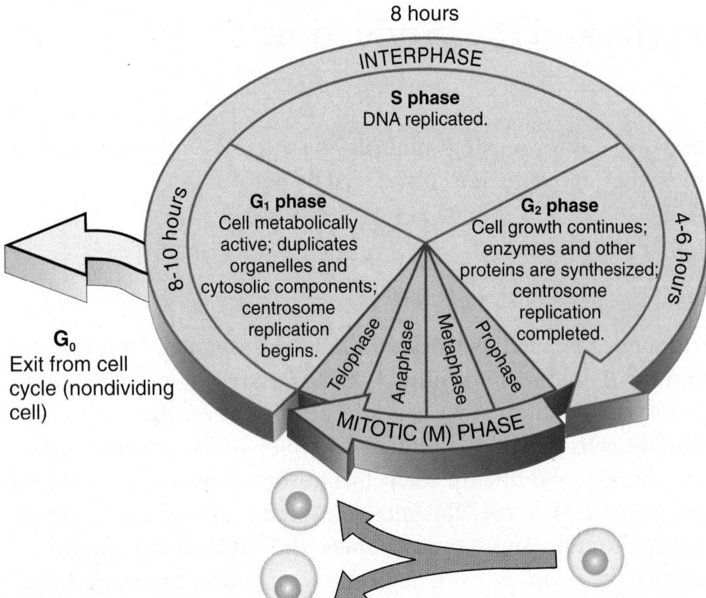


In a complete cell cycle, a cell duplicates its contents and divides into two identical cells.

FIGURE 3.30 **Replication of DNA.** The two strands of the double helix separate by breaking the hydrogen bonds (shown as dotted lines) between nucleotides. New, complementary nucleotides attach at the proper sites, and a new strand of DNA is synthesized alongside each of the original strands. Arrows indicate hydrogen bonds forming again between pairs of bases.

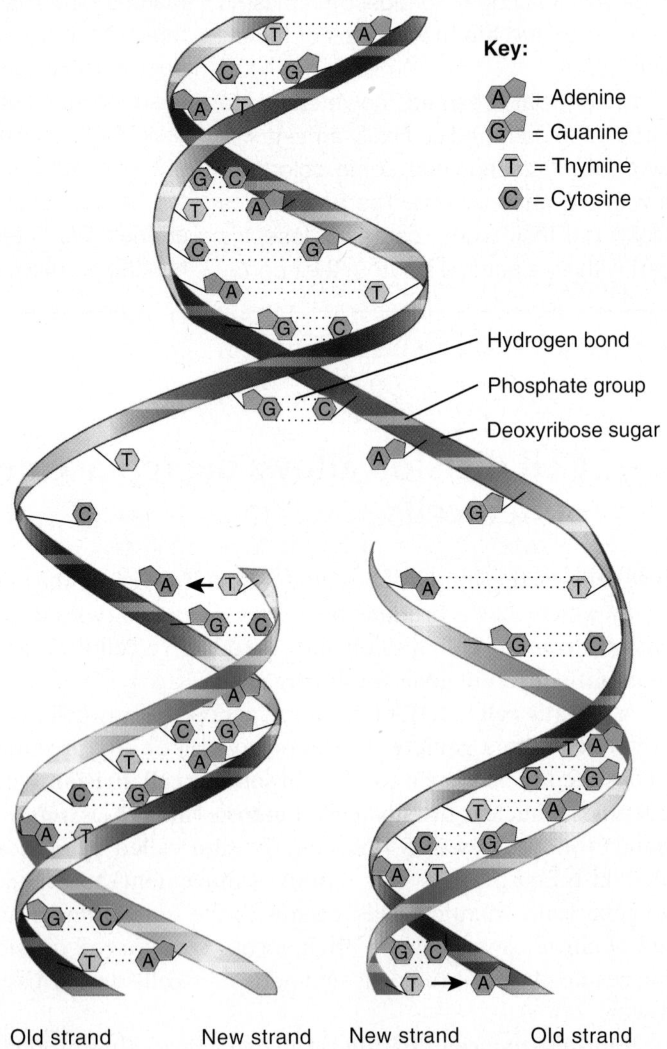


Replication doubles the amount of DNA.

complementary base of a newly synthesized nucleotide. A new DNA strand takes shape as chemical bonds form between neighboring nucleotides. The uncoiling and complementary base pairing continues until each of the two original DNA strands is joined with a newly formed complementary DNA strand. The original DNA molecule has become two identical DNA molecules.

The **$G_2$ phase**, the interval between the S phase and the mitotic phase, lasts 4 to 6 hours of a 24-hour cell cycle. During $G_2$, cell growth continues, enzymes and other proteins are synthesized in preparation for cell division, and replication of centrosomes is completed.

A microscopic view of a cell during interphase shows a clearly defined nuclear envelope, a nucleolus, and a tangled mass of chromatin (Figure 3.31a). Once a cell completes its activities during the $G_1$, S, and $G_2$ phases of interphase, the mitotic phase begins.

### Mitotic Phase

The **mitotic (M) phase** of the cell cycle consists of a nuclear division (mitosis) and a cytoplasmic division (cytokinesis) to form two identical cells. The events that occur during mitosis and cytokinesis are plainly visible under a microscope because chromatin condenses into discrete chromosomes.

Nuclear Division: Mitosis. **Mitosis**, as noted earlier, is the distribution of two sets of chromosomes into two separate nuclei. The process results in the exact partitioning of genetic information. For convenience, biologists divide the process into four stages: prophase, metaphase, anaphase, and telophase. However, mitosis is a continuous process; one stage merges directly into the next.

1. **Prophase** (PRŌ-fāz). During early prophase, the chromatin fibers condense and shorten into chromosomes that are visible under the light microscope (Figure 3.31b). The condensation process may prevent entangling of the long DNA strands as they move during mitosis. Because DNA replication took place during the S phase of interphase, each prophase chromosome consists of a pair of identical strands called *chromatids*. A constricted region called a **centromere** (SEN-trō-mēr) holds the chromatid pair together. Later in prophase, the centrosomes start to form the **mitotic spindle**, a football-shaped assembly of microtubules that attach to each centromere (Figure 3.31b). As the microtubules lengthen, they push the centrosomes to the poles (ends) of the cell so that the spindle extends from pole to pole. The mitotic spindle is responsible for the separation of chromatids to opposite poles of the cell. As prophase ends, the nucleolus disappears and the nuclear envelope breaks down.

2. **Metaphase** (MET-a-fāz). During metaphase, the microtubules of the mitotic spindle align the centromeres of the chromatid pairs at the exact center of the mitotic spindle (Figure 3.31c). This midpoint region is called the *metaphase plate*.

3. **Anaphase** (AN-a-fāz). During anaphase, the centromeres split, separating the two members of each chromatid pair (Figure 3.31d). Once separated, the chromatids are termed chromosomes. As the chromosomes are pulled by the microtubules of the mitotic spindle during anaphase, they appear V-shaped because the centromeres lead the way, dragging the trailing arms of the chromosomes toward the pole.

4. **Telophase** (TEL-ō-fāz). The final stage of mitosis, telophase, begins after chromosomal movement stops (Figure 3.31e). The identical sets of chromosomes, now at opposite poles of the cell, uncoil and revert to the threadlike chromatin form. A nuclear envelope forms around each chromatin mass, nucleoli reappear in the identical nuclei, and the mitotic spindle breaks up.

Cytoplasmic Division: Cytokinesis. As noted earlier, division of a cell's cytoplasm and organelles into two identical cells is called **cytokinesis**. This process usually begins in late anaphase with the formation of a **cleavage furrow**, a slight indentation of the plasma membrane, and is completed after telophase. The cleavage furrow usually appears midway between the centrosomes and extends around the periphery of the cell (❺ Figure 3.31d, e). Microfilaments that lie just inside the plasma membrane form a contractile ring that pulls the plasma membrane progressively inward. The ring constricts the center of the cell like a belt around the waist, and ultimately pinches the cell in two. Because the plane of the cleavage furrow is always perpendicular to the mitotic spindle, the two sets of chromosomes end up in separate cells. When cytokinesis is complete, each new cell enters interphase (❻ Figure 3.31f).

The sequence of events can be summarized as

$$G_1 \longrightarrow \text{S phase} \longrightarrow G_2 \text{ phase} \longrightarrow \text{mitosis} \longrightarrow \text{cytokinesis}$$

Table 3.3 summarizes the events of the cell cycle in somatic cells.

## Reproductive Cell Division

In sexual reproduction, a new organism results from the union of two different gametes (fertilization), one produced by each parent. If gametes had the same number of chromosomes as somatic cells, the number of chromosomes would double at fertilization. **Meiosis** (mī-Ō-sis; *mei-* = lessening; *-osis* = condition of), the reproductive cell division that occurs in the gonads (ovaries and testes), produces gametes in which the number of chromosomes is reduced by half. As a result, gametes contain a single set of 23 chromosomes and thus are **haploid cells** (HAP-loyd; *hapl-* = single symbolized ***n***). Fertilization restores the diploid number of chromosomes.

### Meiosis

Unlike mitosis, which is complete after a single round, meiosis occurs in two successive stages: meiosis I and meiosis II. During

**FIGURE 3.31 Cell division: mitosis and cytokinesis.** Begin the sequence at 1 at the top of the figure and read clockwise to complete the process.

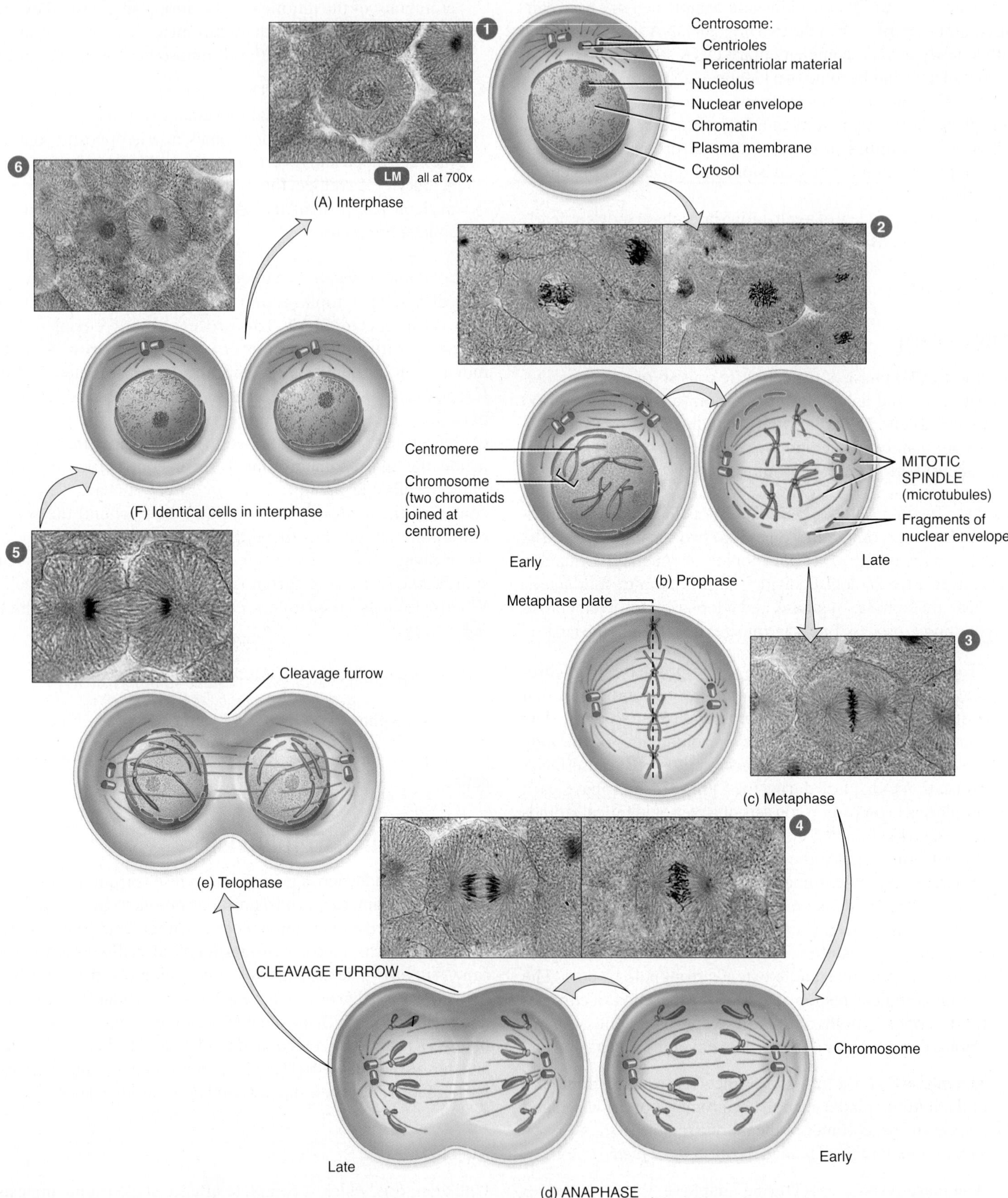


In somatic cell division, a single diploid cell divides to produce two identical diploid cells.

**TABLE 3.3**

Events of the Somatic Cell Cycle

| PHASE | ACTIVITY |
|---|---|
| **Interphase** | Period between cell divisions; chromosomes not visible under light microscope |
| $G_1$ phase | Metabolically active cell duplicates most of its organelles and cytosolic components; replication of chromosomes begins (Cells that remain in the $G_1$ phase for a very long time, and possibly never divide again, are said to be in the $G_0$ phase.) |
| S phase | Replication of DNA and centrosomes |
| $G_2$ phase | Cell growth, enzyme and protein synthesis continue; replication of centrosomes complete |
| **Mitotic Phase** | Parent cell produces identical cells with identical chromosomes; chromosomes visible under light microscope |
| **Mitosis** | Nuclear division; distribution of two sets of chromosomes into separate nuclei |
| Prophase | Chromatin fibers condense into paired chromatids; nucleolus and nuclear envelope disappear; each centrosome moves to an opposite pole of the cell |
| Metaphase | Centromeres of chromatid pairs line up at metaphase plate |
| Anaphase | Centromeres split; identical sets of chromosomes move to opposite poles of cell |
| Telophase | Nuclear envelopes and nucleoli reappear; chromosomes resume chromatin form; mitotic spindle disappears |
| **Cytokinesis** | Cytoplasmic division; contractile ring forms cleavage furrow around center of cell, dividing cytoplasm into separate and equal portions |

the interphase that precedes meiosis I, the chromosomes of the diploid cell replicate. As a result of replication, each chromosome consists of two sister (genetically identical) chromatids, which are attached at their centromeres. This replication of chromosomes is similar to the one that precedes mitosis in somatic cell division.

Meiosis I. **Meiosis I**, which begins once chromosomal replication is complete, consists of four phases: prophase I, metaphase I, anaphase I, and telophase I (Figure 3.32a). Prophase I is an extended phase in which the chromosomes shorten and thicken, the nuclear envelope and nucleoli disappear, and the mitotic spindle forms. Two events that are not seen in mitotic prophase occur during prophase I of meiosis. First, the two sister (genetically identical) chromatids of each pair of homologous chromosomes pair off. Second, parts of the chromatids of two paired homologous chromosomes may be exchanged with one another. Such an exchange between parts of nonsister (genetically different) chromatids is termed **crossing-over** (Figure 3.32b). Due to crossing-over, the resulting cells are genetically unlike each other and genetically unlike the starting cell that produced them. Crossing-over results in genetic *recombination*—that is, the formation of new combinations of genes—and accounts for part of the great genetic variation among humans and other organisms that form gametes via meiosis.

In metaphase I, the homologous pairs of chromosomes line up along the metaphase plate of the cell, with homologous chromosomes side by side (Figure 3.32a). During anaphase I, the members of each homologous pair of chromosomes separate as they are pulled to opposite poles of the cell by the microtubules attached to the centromeres. The paired chromatids, held by a centromere, remain together. (Recall that during mitotic anaphase, the centromeres split and the sister chromatids separate.) Telophase I and cytokinesis of meiosis are similar to telophase and cytokinesis of mitosis. The net effect of meiosis I is that each resulting cell contains the haploid number of chromosomes because it contains only one member of each pair of the homologous chromosomes present in the starting cell.

Meiosis II. The second stage of meiosis, **meiosis II**, also consists of four phases: prophase II, metaphase II, anaphase II, and telophase II (Figure 3.32a). These phases are similar to those that occur during mitosis; the centromeres split, the sister chromatids separate and move toward opposite poles of the cell, and a nuclear envelope forms around each chromatin mass.

In summary, meiosis I begins with a diploid starting cell and ends with two haploid cells, each with half the number of chromosomes. During meiosis II, each of the two haploid cells divides, resulting in four haploid gametes that are genetically different from the original diploid starting cell.

Figure 3.33 compares the events of meiosis and mitosis.

**FIGURE 3.32 Meiosis, reproductive cell division.** Details of events are discussed in the text.

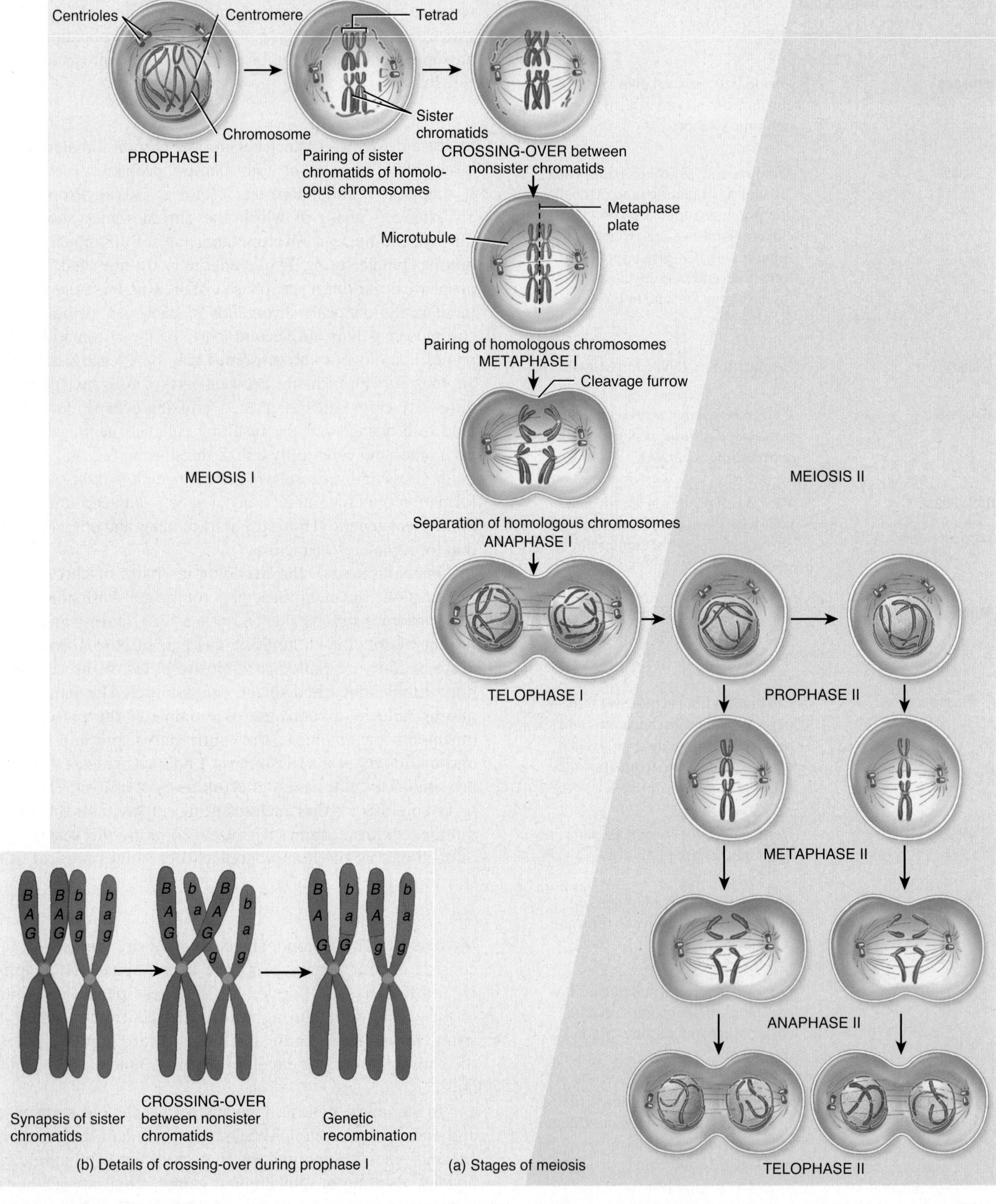


In reproductive cell division, a single diploid starting cell undergoes meiosis I and meiosis II to produce four haploid gametes that are genetically different from the starting cell that produced them.

FIGURE 3.33 **Comparison between mitosis (left) and meiosis (right) in which the starting cell has two pairs of homologous chromosomes.**

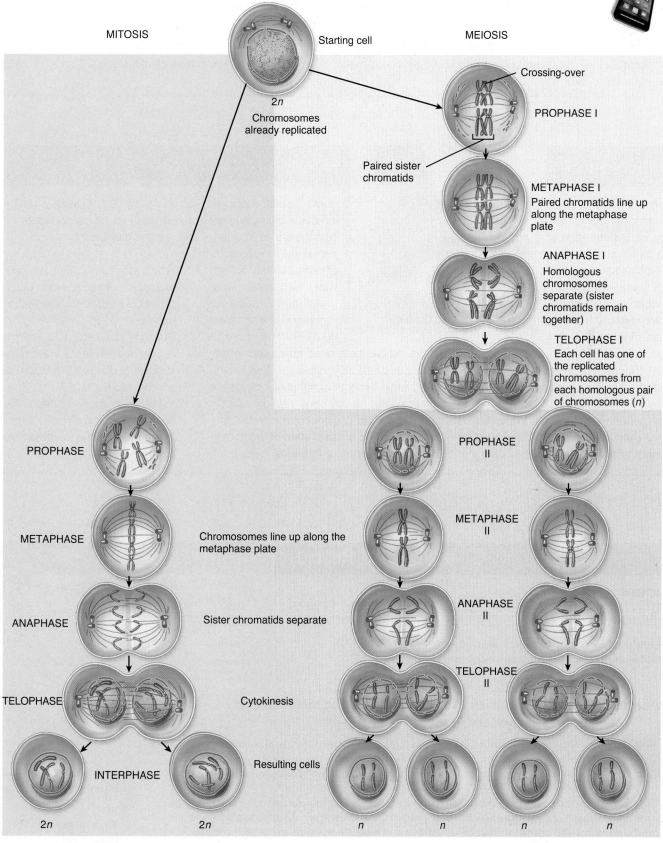

MITOSIS

Starting cell

2*n*

Chromosomes already replicated

MEIOSIS

Crossing-over

PROPHASE I

Paired sister chromatids

METAPHASE I

Paired chromatids line up along the metaphase plate

ANAPHASE I

Homologous chromosomes separate (sister chromatids remain together)

TELOPHASE I

Each cell has one of the replicated chromosomes from each homologous pair of chromosomes (*n*)

PROPHASE

PROPHASE II

METAPHASE

Chromosomes line up along the metaphase plate

METAPHASE II

ANAPHASE

Sister chromatids separate

ANAPHASE II

TELOPHASE

Cytokinesis

TELOPHASE II

INTERPHASE

Resulting cells

2*n* 2*n* *n* *n* *n* *n*

Somatic cells with diploid number of chromosomes (not replicated)

Gametes with haploid number of chromosomes (not replicated)

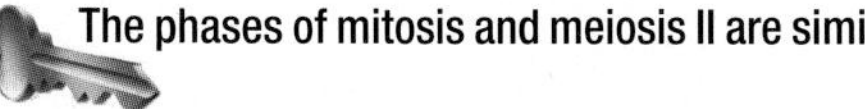

**The phases of mitosis and meiosis II are similar.**

✓ **CHECKPOINT**

**26.** What is the difference between somatic and reproductive cell division? What is the importance of each?

**27.** What are the major events of each phase of interphase?

**28.** What are the major events of each stage of the mitotic phase of the cell cycle?

**29.** When does cytokinesis begin during the cell cycle?

**30.** How does anaphase I of meiosis differ from anaphase of mitosis and anaphase II of meiosis?

**31.** What is the significance of meiosis I?

## Joseph's Story

## EPILOGUE AND DISCUSSION

**J**oseph, a 38-year-old male smoker with a history of hypertension, has suffered a heart attack. He has several risk factors associated with cardiovascular disease, including a family history of heart disease, hypertension, and poor diet. As we have learned in this chapter, his heart attack directly impacted cellular processes. Oxygen is required by cells to produce ATP, the energy currency of the cell. Without ATP, oxygen, and nutrients, multiple cellular processes affecting homeostasis can shut down; water balance, ionic equilibrium, and pH balance are all key factors in the homeostasis of individual cells. Joseph was fortunate that oxygen flow to the cells of his body was maintained by CPR. The damage to the cellular processes discussed in this story can cause irreversible cell death in the brain within a matter of minutes if oxygen is not available. As you continue your studies of anatomy and physiology, remember that each tissue and organ system of the human body is composed of individual cells, each of which functions to keep you alive. Health and disease are ultimately determined at the cellular and molecular levels.

F. *Why was reestablishing oxygen flow to Joseph's body so important? What processes would be affected by lack of oxygen?*

G. *Consider once again Joseph's health history and risk factors. What could you suggest to Joseph to reduce his risk for another heart attack?*

H. *Explain why Joseph's heart failed based on what you have learned so far about the function of cells in the human body.*

# Concept and Resource Summary

| Concept | Resources  |
|---|---|
| **Introduction**<br>1. **Cells** are the basic, living structural and functional units of the body.<br>2. **Cell biology** is the study of cellular structure and function. | |
| **Concept 3.1 The principal parts of a cell are the plasma membrane, the cytoplasm, and the nucleus.**<br>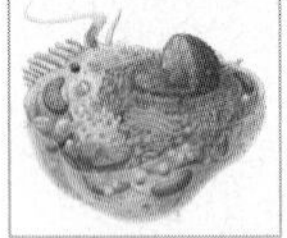<br>1. The cell is separated from the external environment by its **plasma membrane**, a flexible surface that serves as a selective barrier, regulating exchanges of material with the extracellular environment and facilitating communication with other cells.<br>2. The **cytoplasm** is the region between the plasma membrane and nucleus; its two components are the cytosol and organelles. The **cytosol** is the fluid portion containing water, dissolved solutes, and suspended particles. The cytosol surrounds several types of **organelles**, which have specific shapes and carry out specific cellular functions. | Anatomy Overview—Cell Structure and Function<br>Figure 3.1—Typical Structures Found in Many Cells<br> |

**Concept** | **Resources**

3. The cell's DNA is housed in a large organelle called the **nucleus**. Each **chromosome** in the nucleus is a DNA molecule associated with several proteins. **Genes** within the DNA molecule are hereditary units that control cellular structure and function.

---

## Concept 3.2 The plasma membrane contains the cytoplasm and regulates exchanges with the extracellular environment.

1. The **plasma membrane** is a flexible barrier that surrounds the cell's cytoplasm. It has a fluid mosaic model structure consisting of a sea of fluid lipids containing a mosaic of different proteins.
2. The **lipid bilayer** contains **phospholipids**, **cholesterol**, and **glycolipids**. The phospholipids have a hydrophilic, polar phosphate "head," and a hydrophobic, nonpolar part containing two long fatty acid "tails." The hydrophilic heads face outward and the hydrophobic tails point toward one another in the membrane's interior. Cholesterol molecules are scattered throughout the lipid layers. Glycolipids are associated with the lipid layer facing the extracellular fluid.
3. There are two types of membrane proteins. **Integral proteins** extend into the membrane; many of these are transmembrane proteins, extending all the way through the membrane. **Peripheral proteins** are associated with the polar heads of membrane lipids or with integral proteins at the inner or outer membrane surfaces. **Glycoproteins** protrude into the extracellular fluid; the carbohydrate portions of the glycoproteins and glycolipids form the **glycocalyx**. The glycocalyx functions in cell-to-cell recognition, cell adherence, and attraction of a film of surface fluid around the cell.
4. The specific types of membrane proteins vary considerably among cells. Proteins determine many membrane functions, including the following: forming **ion channels**, acting as **carriers**, serving as **receptors**, acting as **enzymes**, serving as **linkers**, and acting as **cell-identity markers**.
5. Most of the membrane lipids and many membrane proteins rotate and move laterally in the bilayer. Membrane fluidity is increased by the numbers of double bonds in the fatty acid tails of the lipids. Membrane fluidity allows interactions to occur within the membrane; enables cell movement, growth, division, secretion, and formation of cellular junctions; and helps the membrane to self-seal if punctured.
6. Plasma membranes have **selective permeability**; the lipid bilayer is permeable to nonpolar, uncharged molecules and slightly permeable to small, uncharged polar molecules, but it is impermeable to ions and large, uncharged polar molecules. Transmembrane proteins acting as channels and carriers increase permeability to ions and uncharged polar molecules that cannot cross the lipid bilayer.
7. Selective permeability of the plasma membrane allows the living cell to maintain **concentration gradients**, different concentrations of certain ions and molecules on either side of the plasma membrane. Typically, the inner surface of the plasma membrane is more negatively charged and the outer surface is more positively charged due to a difference in distribution of charged ions; this creates an **electrical gradient**. The combined influence of the concentration gradient and the electrical gradient on movement of a particular ion is referred to as its **electrochemical gradient**.

**Resources:**

Animation—Membrane Functions
Anatomy Overview—Plasma Membrane
Figure 3.2—The Fluid Mosaic Arrangement of Lipids and Proteins in the Plasma Membrane
Exercise—Paint a Cell Membrane
Concepts and Connections—Membrane Functions

---

## Concept 3.3 Transport of a substance across the plasma membrane occurs by both passive and active processes.

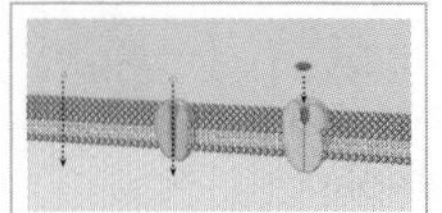

1. **Diffusion** is a passive transport process in which the solutes and solvents are driven by their own kinetic energy to move from an area of higher concentration toward an area of lower concentration down their concentration gradient until evenly distributed. They continue to move about due to kinetic energy but their concentrations do not change. Factors that affect diffusion rates include steepness of the concentration gradient, temperature, mass of the diffusing substance, surface area of the membrane, and diffusion distance.
2. In **simple diffusion**, nonpolar, hydrophobic molecules diffuse freely through the lipid bilayer without help of the membrane transport proteins. Water and urea are uncharged polar molecules that diffuse in small amounts through the lipid bilayer.

**Resources:**

Clinical Connection—Digitalis Increases $Ca^{2+}$ in Heart Muscle Cells

**Concept** | **Resources** WILEY PLUS

3. **Facilitated diffusion** is a passive transport process in which a polar or highly charged solute utilizes channels or carriers to cross the other side of the plasma membrane as it moves down its concentration gradient. In **channel-mediated facilitated diffusion**, integral transmembrane proteins form ion channels for passage of small, hydrophilic ions. Many ion channels are gated; they either open and close randomly or are regulated by chemical and electrical changes inside and outside the cells. Carriers can also ferry solutes across the plasma membrane by changing their shape in a process called **carrier-mediated facilitated diffusion**.
4. The net movement of a solvent through a selectively permeable membrane is called **osmosis**. Water molecules move by osmosis down their concentration gradient, across plasma membranes. Osmosis only occurs when a membrane is permeable to water but not permeable to certain solutes.
5. **Tonicity** is the measure of a solution's ability to change the volume of cells by altering their water content. **Isotonic solutions** allow a cell to maintain normal shape and volume. In **hypotonic solutions**, which have a lower concentration of solutes than the cell's cytosol, water molecules enter the cell faster than they leave it, causing the cell to swell and undergo **lysis**. Because **hypertonic solutions** have a higher concentration of solutes than the cell's cytosol, water molecules leave the cell faster than they enter, causing the cell to shrink or undergo **crenation**.
6. **Active transport** occurs when polar or charged solutes move against their concentration gradients using carrier proteins and cellular energy. In **primary active transport**, hydrolysis of ATP supplies energy to change the shape of the carrier protein, which "pumps" the substance against its concentration gradient. One such carrier, called the **sodium–potassium pump**, hydrolyzes ATP to expel $Na^+$ and bring $K^+$ into the cell. In **secondary active transport**, the energy stored in a $Na^+$ or $H^+$ concentration gradient is used to drive other substances across the membrane against their own gradients.
7. **Endocytosis** is the movement of materials into a cell within a **vesicle** formed from the plasma membrane. The three types of endocytosis are receptor-mediated endocytosis, phagocytosis, and bulk-phase endocytosis (pinocytosis). **Receptor-mediated endocytosis** is a highly selective process in which ligands must attach to specific membrane receptors for vesicular transport into the cell. **Phagocytosis** occurs when the **phagocyte** (cell) engulfs large, solid particles or whole cells. In **bulk-phase endocytosis**, the plasma membrane folds inward to form a vesicle around tiny droplets of dissolved substances that then enter the cytoplasm.
8. **Exocytosis** is the movement of vesicle-packaged materials out of the cell by fusion of the vesicle with the plasma membrane; the contents are then released into the extracellular fluid.
9. **Transcytosis** involves endocytosis of vesicles on one side of a cell, movement of the vesicles across the cell, and exocytosis on the other side.

## Concept 3.4 Cytoplasm consists of the cytosol and organelles.

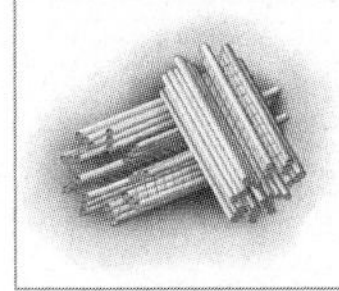

1. The cytoplasm consists of all the cellular contents between the plasma membrane and the nucleus; it is made up of the cytosol and organelles. The **cytosol (intracellular fluid)**, the fluid portion of the cytoplasm that surrounds the organelles, consists of water with dissolved and suspended components. The cytosol is the site of many chemical reactions. The **cytoskeleton** acts as a structural framework, helps determine cell shape, helps organize cellular contents, and aids movement within the cell and of whole cells. The cytoskeleton consists of microfilaments, intermediate tubules, and microtubules. **Microfilaments** supply mechanical support for cell strength and shape, and generate movements. **Intermediate filaments** help give stability to the position of organelles. **Microtubules** help determine cell shape, move organelles, and are components of cilia and flagella.
2. **Organelles** are structures with specific shapes that perform specific cell functions.
3. Near the nucleus is the **centrosome**, composed of two **centrioles** and **pericentriolar material**, which plays a role in growth of the mitotic spindle in cell division.
4. **Cilia** and **flagella** are motile extensions of the cell surface that contain microtubules surrounded by plasma membrane. Cilia are numerous, short, hairlike projections with an oarlike pattern of beating that moves fluids across the cell surface. Flagella are single, long, whiplike structures that move an entire cell.

Anatomy Overview—Cell Structure and Function
Clinical Connection—Smooth ER and Drug Tolerance
Clinical Connection—Mitochondrial Cytopathies
Figure 3.20—Processing and Packaging of Proteins by the Golgi Complex

Exercise—Concentrate on Cellular Functions
Exercise—Target Practice
Concepts and Connections—Human Cell
Concepts and Connections—Membranous Organelles

| Concept | Resources |
|---|---|

5. **Ribosomes**, composed of one small subunit and one larger subunit, are the sites of protein synthesis. Ribosomes attached to the surface of the endoplasmic reticulum synthesize proteins for insertion on the plasma membrane or for export from the cell; free, unattached ribosomes synthesize proteins used in the cytosol.
6. The **endoplasmic reticulum (ER)** is a membranous network of tubules throughout the cytoplasm with connections to the nuclear envelope. **Rough ER** is studded with ribosomes that synthesize proteins for cell secretion or use in the cell. **Smooth ER** synthesizes fatty acid and steroids, detoxifies drugs, and stores calcium in muscle cells.
7. The **Golgi complex** consists of small, flattened, stacked membranous sacs, or **cisternae**. Enzymes of the Golgi complex modify, sort, and package proteins in vesicles for transport out of the cell, to the plasma membrane, or to other locations within the cell.
8. The cytoplasm contains a number of different vesicular organelles. **Lysosomes** are membrane-enclosed vesicles, containing digestive enzymes used to break down worn-out organelles or cells. **Peroxisomes** contain oxidases for oxidizing fatty acids, amino acids, and toxic substances.
9. **Mitochondria** have a smooth outer membrane and an inner membrane arranged in folds called **cristae**. Within mitochondria most of the cell's ATP is produced during the aerobic phase of cellular respiration. Mitochondria can self-replicate and have their own DNA.

## Concept 3.5 The nucleus contains nucleoli and genes.

1. The most prominent organelle is the (usually single) **nucleus**. The **nuclear envelope** is a double-layered membrane, pierced with **nuclear pores**. One or more **nucleoli**, which function in producing ribosomes, are located in the nucleus. The nucleus is the cell's control center and houses the cell's hereditary units, called **genes**, arranged along **chromosomes**. All of the genetic information of a cell or organism is called its **genome**.
2. Chromosomes are long DNA molecules coiled with proteins. Each beadlike unit of DNA wrapped around a core of eight **histone** proteins is called a **nucleosome**. Between nucleosomes are strings of **linker DNA**. The entire complex of DNA and proteins is called **chromatin**. Prior to cell division, the DNA replicates and condenses into linear chromosomes in which there are two identical **chromatids** per chromosome.

Anatomy Overview—The Nucleus

## Concept 3.6 Cells make proteins by transcribing and translating the genetic information contained in DNA.

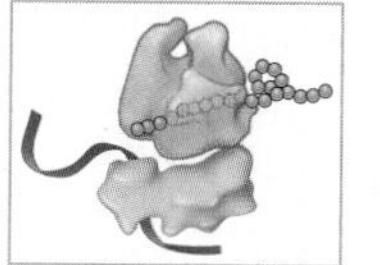

1. In **gene expression**, a gene's DNA is the template for synthesis of a specific protein.
2. Genetic information carried in DNA and RNA occurs in sets of three-nucleotide sequences called a **base triplet** in DNA and a **codon** in transcribed RNA. Each set of three nucleotides represents a specific amino acid in the protein undergoing synthesis.
3. Protein synthesis is facilitated by specific RNA molecules in two steps called transcription and translation. **Transcription** occurs in the nucleus, where RNA polymerase catalyzes transcription of DNA into **messenger RNA (mRNA)**. Bases are paired in a complementary manner. Newly synthesized mRNA passes through a nuclear pore to the cytoplasm, where it participates in translation.
4. **Translation** occurs on ribosomes out in the cytoplasm, where amino acids are joined to form a specific protein. The small ribosome subunit has a binding site for mRNA; the large subunit has two binding sites for **transfer RNA (tRNA)** that carry amino acids to the ribosomes.
5. Translation begins when mRNA binds to the small ribosomal subunits; the large ribosomal subunit attaches, forming a complete ribosome. As tRNAs shuttle in appropriate amino acids by following the codons on the mRNA, peptide bonds form a growing polypeptide. When the ribosome reaches a "stop" codon on the mRNA, protein synthesis stops and the protein detaches.

Animation—Protein Synthesis
Figure 3.28—Protein Elongation and Termination of Protein Synthesis During Translation
Exercise—Protein-Producing Processes
Exercise—Synthesize and Transport a Protein
Concepts and Connections—Protein Synthesis

## Concept

## Resources

### Concept 3.7 Cell division allows the replacement of cells and the production of new cells.

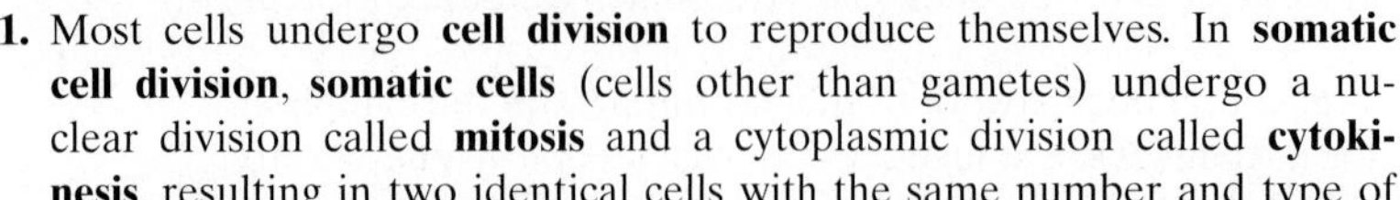

1. Most cells undergo **cell division** to reproduce themselves. In **somatic cell division**, **somatic cells** (cells other than gametes) undergo a nuclear division called **mitosis** and a cytoplasmic division called **cytokinesis**, resulting in two identical cells with the same number and type of chromosomes as the original cell. Gametes are produced by **reproductive cell division** or **meiosis**, a two-step process that reduces the chromosome number by half.
2. Somatic cells are called **diploid cells** because they contain two sets of chromosomes. The two chromosomes of each pair are called **homologous chromosomes**; one member of each pair is inherited from each parent. Human cells have 23 pairs of chromosomes, including the one pair called the **sex chromosomes**. During cell reproduction, the cell must replicate all its chromosomes to pass them on to the newly formed cells.
3. The **cell cycle** is the sequence of events from the time a somatic cell is formed until it divides. It consists of two parts: interphase, when the cell is not dividing; and the mitotic (M) phase, when the cell is dividing.
4. During **interphase**, the cell is metabolically active. During the **$G_1$ phase**, the cell replicates organelles and centrosomes. During the the **S phase**, DNA replication occurs. The DNA helix uncoils, and each exposed base of the old DNA strand pairs with a complementary base of a newly synthesized nucleotide. At the end of replication, each identical DNA molecule is composed of one old and one newly synthesized strand. The **$G_2$ phase** is the time when growth continues and final preparation for cell division occurs.
5. The **mitotic (M) phase** is a time of nuclear division and cytokinesis, and results in two identical cells. **Mitosis** is the distribution of two sets of chromosomes into two separate nuclei. It occurs in four stages: **prophase**, **metaphase**, **anaphase**, and **telophase**.
6. Late in mitosis the process of cytoplasmic division, called **cytokinesis**, takes place. After completion of cytokinesis there are two new and separate cells, with equal portions of cytoplasm and organelles and identical chromosomes.
7. Sexual reproduction produces an organism from the union of two gametes, one from each parent. Gametes must be produced with half the chromosome number; otherwise the chromosome number would double each time fertilization occurred. **Meiosis**, or reproductive cell division, consists of two consecutive nuclear divisions resulting in four daughter cells; each of the daughter cells is a **haploid cell** containing a single set of 23 chromosomes.
8. During the interphase preceding meiosis I, the chromosomes of the diploid cell replicate. **Meiosis I** consists of prophase I, metaphase I, anaphase I, and telophase I. During prophase I, the two sister chromatids of each pair of homologous chromosomes pair off. **Crossing-over** of parts of the chromatids may occur; these new combinations of genes allow for genetic variability among the offspring. Homologous pairs of chromosomes line up in metaphase, then separate in anaphase I. Cytokinesis occurs in telophase I. Meiosis I results in two cells, each containing a haploid number of chromosomes.
9. **Meiosis II** consists of prophase II, metaphase II, anaphase II, and telophase II. These phases result in the separation of members of each pair of chromatids produced in meiosis I, with a final product of four haploid gametes that are genetically different from the original diploid cell.

Animation—The Cell Cycle and Division Processes

Figure 3.33—Comparison Between Mitosis and Meiosis

Clinical Connection—Tumor-Suppressor Genes

Clinical Connection—Progeria and Werner Syndrome

Exercise—Mitosis Matchup

Exercise—Meiosis Matchup

Exercise—Cell Cycle Quiz

Clinical Connection—Cancer

## Understanding the Concepts

1. What are the three principal parts of a cell?
2. "The proteins present in a plasma membrane determine the functions that a membrane can perform." Is this statement true or false? Explain your answer.
3. When stimulating a cell, the hormone insulin first binds to a protein in the plasma membrane. This action best represents which membrane protein function?
4. In Figure 3.7, will the fluid level in the right arm rise until the water concentrations are the same in both arms? Why or why not?
5. Will a 2% solution of NaCl cause hemolysis or crenation of RBCs? Explain your answer.
6. In which ways are endocytosis and exocytosis similar? How are they different?
7. What does cytoplasm have that cytosol lacks?
8. What is the function of the nucleolus?
9. If a DNA template has the base sequence AGCT, what will be the mRNA base sequence?
10. If you observed that a cell did not have a centrosome, what could you predict about its capacity for cell division?
11. Why is it crucial that DNA replication occurs before cytokinesis in somatic cell division?
12. How does crossing-over affect the genetic content of the haploid gametes?

CHAPTER

# 4

## John Doe's Story

A light snow fell in the valley, dusting the ground. It was cold, with little wind. Tina and Eric climbed cautiously up the uneven mountain trail, stepping over large, jagged rocks. "Let's rest up on those big boulders. We'll get a good view of the valley from there," said Eric.

He bounded up the trail to where two large rocky outcrops emerged from the side of the hill, took off his pack, and began digging in the side pocket for his binoculars. "Wow, what a view!" exclaimed Tina, sitting down next to Eric as he scanned the horizon.

"We should see some mountain goats up here. They like this area. What in the world is that? Tina, take a look. Does that look like a plane to you?"

Tina reached for the glasses and adjusted the focus. "It is! I can see the registration numbers on the wings. But I don't remember anything being reported about a plane going down around here."

The plane rested against a large tree. Its nose was smashed in; one wing was broken off, the other bent at an unnatural angle. "I think there's someone in there," Tina whispered.

"We need to call the police," said Eric. "Come on, Tina. Let's go!"

Two hikers discover a wrecked plane, a body, and a mystery. In this chapter, you will study the tissue level of organization. In this story, we will follow the investigation of a mysterious event, and learn how an understanding of anatomy and physiology at the tissue level is used by medical investigators to uncover clues that aid in understanding crime scenes and unexplained deaths.

# The Tissue Level of Organization

## INTRODUCTION

As you learned in Chapter 3, a cell is a complex collection of compartments, each of which carries out a host of biochemical reactions that make life possible. Few cells function as isolated units in the human body. In this chapter we will learn how cells work together in groups called **tissues** that function together to carry out specialized activities. The structure and properties of a specific tissue are influenced by factors such as the nature of the extracellular material that surrounds the tissue cells and the connections between the cells that compose the tissue. Tissues may be hard, semi-solid, or even liquid in their consistency, a range exemplified by bone, fat, and blood. In addition, tissues vary tremendously with respect to the kinds of cells present, how the cells are arranged, and the types of fibers present, if any.

**Histology** (his′-TOL-ō-jē; *hist-* = tissue; *-ology* = study of) is the science that deals with the study of tissues. A **pathologist** (pa-THOL-ō-gist; *patho-* = disease) is a physician who examines cells and tissues to help other physicians make accurate diagnoses. One of the principal functions of a pathologist is to examine tissues for any changes that might indicate disease.

## CONCEPTS

**4.1** Human body tissues can be classified as epithelial, connective, muscle, or nervous.

**4.2** Cell junctions hold cells together to form tissues.

**4.3** Epithelial tissue covers body surfaces, lines organs and body cavities, or secretes substances.

**4.4** Connective tissue binds organs together, stores energy reserves as fat, and helps provide immunity.

**4.5** Epithelial and connective tissues have obvious structural differences.

**4.6** Membranes cover the surface of the body, line body cavities, and cover organs.

**4.7** Muscle tissue generates the physical force needed to make body structures move.

**4.8** Nervous tissue consists of neurons and neuroglia.

**4.9** The ability of an injured tissue to repair itself depends on the extent of damage and the regenerative ability of the injured tissue.

## 4.1 Human body tissues can be classified as epithelial, connective, muscle, or nervous.

Body tissues can be classified into four basic types according to their structure and function (Figure 4.1):

- **Epithelial tissue** covers body surfaces and lines hollow organs, body cavities, and ducts. It also forms glands. Epithelial tissue allows the body to interact with its internal and external environments.
- **Connective tissue** protects and supports the body and its organs. Various types of connective tissue bind organs together, store energy reserves as fat, and help provide immunity against disease-causing organisms.
- **Muscle tissue** is composed of cells that can contract to generate the physical force needed to move body structures and generate heat that warms the body.
- **Nervous tissue** detects changes in a variety of conditions inside and outside the body and responds by generating electrical signals called action potentials that activate muscular contractions and glandular secretions.

Epithelial tissue and most types of connective tissue have a wide distribution in the body and are discussed in some detail in this chapter. The general features of bone and blood are introduced here, but their detailed discussions are presented in Chapters 6 and 18, respectively. Similarly, the structure and function of muscular tissue and nervous tissue are introduced here and examined in detail in Chapters 10 and 12, respectively.

✓ **CHECKPOINT**

1. Define a tissue.
2. What are the four basic types of body tissues?

FIGURE 4.1 Types of tissues.

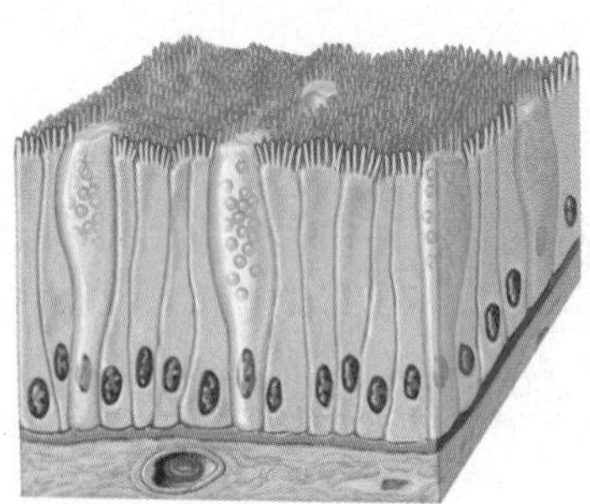

(a) Epithelial tissue

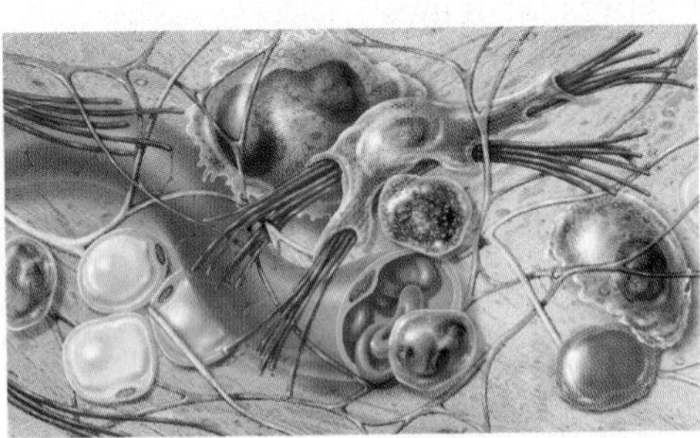

(b) Connective tissue

(c) Muscle tissue

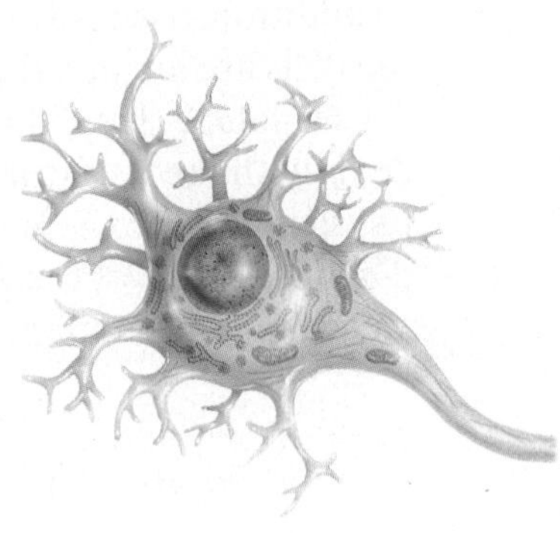

(d) Nervous tissue

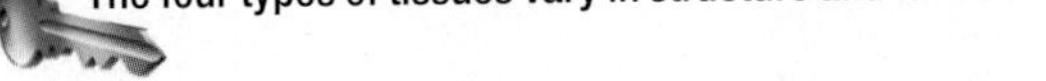

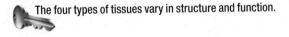

The four types of tissues vary in structure and function.

## 4.2 Cell junctions hold cells together to form tissues.

Most cells within a tissue remain anchored to other cells or structures. Only a few cells, such as phagocytes, move freely through the body, searching for invaders to destroy. Many cells are tightly joined into functional units by contact points between their plasma membranes called **cell junctions**.

**Tight junctions** consist of weblike strands of transmembrane proteins that fuse together the outer surfaces of adjacent plasma membranes to seal off passageways between adjacent cells (Figure 4.2a). Cells of epithelial tissues that line the stomach, intestines, and urinary bladder have many tight junctions. They form tight seals between adjacent cells, like the zip-lock closure at the top of a plastic storage bag, that prevent substances from passing between cells and leaking out of these organs.

**Desmosomes** (DEZ-mō-sōms; *desmo-* = band) have transmembrane glycoproteins that extend into the intercellular space between adjacent cell membranes, attaching cells to one another. *Plaque* (PLAK), a dense layer of proteins on the cytosol side of the plasma membrane, attaches to intermediate filaments of the cytoskeleton (Figure 4.2b). The intermediate filaments extend from the plaque on one side of the cell across the cytosol to the plaque on the opposite side of the cell. Desmosomes contribute to the stability of cells and anchor adjacent cells within a tissue. These spot-weld-like junctions are common among skin cells, preventing them from separating when skin is pulled, and muscle cells in the heart, keeping the tissue from pulling apart during contraction.

FIGURE 4.2 Cell junctions.

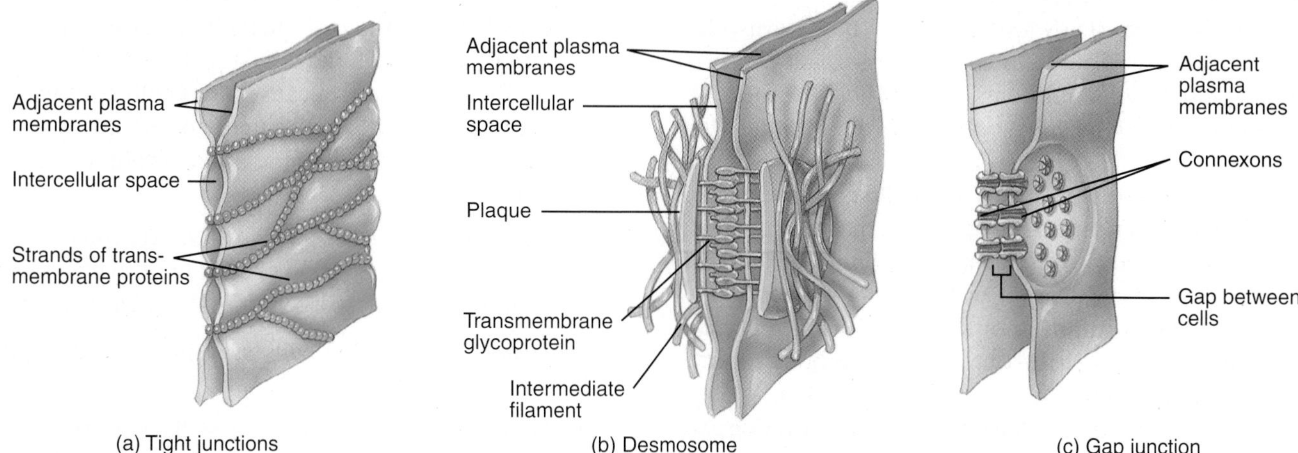


**Most epithelial cells and some muscle and nerve cells are connected by cell junctions.**

At **gap junctions**, transmembrane proteins form tiny fluid-filled tunnels called *connexons* that connect neighboring cells (Figure 4.2c). The plasma membranes of gap junctions are separated by a very narrow intercellular gap (space). Ions and small molecules can diffuse through the connexons from one cell to the next. Gap junctions allow the cells in a tissue to communicate with one another. Gap junctions also enable action potentials to spread rapidly among cells, a process that is crucial for muscle contraction and the normal operation of the nervous system.

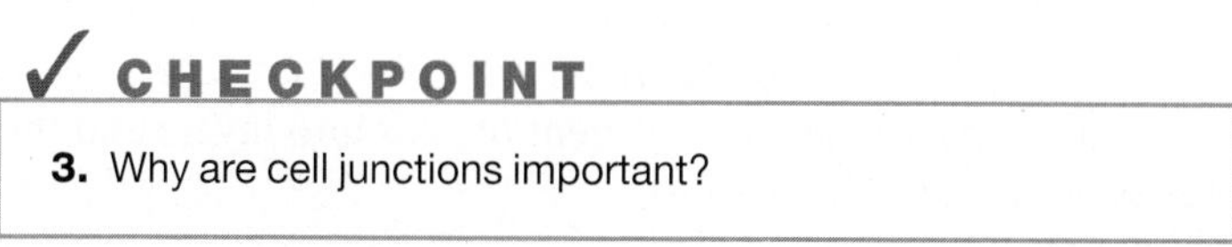

✓ CHECKPOINT

**3.** Why are cell junctions important?

## 4.3 Epithelial tissue covers body surfaces, lines organs and body cavities, or secretes substances.

An **epithelial tissue** (ep-i-THĒ-lē-al), or **epithelium** (ep-i-THĒ-lē-um; plural is *epithelia*), consists of cells arranged in continuous sheets, in either single or multiple layers. The cells are closely packed with little intercellular space between adjacent plasma membranes and are held tightly together by many cell junctions. Therefore, epithelial tissue is an excellent protective barrier for body surfaces, such as the skin.

Epithelial cells have an **apical surface** (Ā-pi-kul) that faces the body surface, a body cavity, or the lumen (interior space) of an organ (Figure 4.3). Apical surfaces may contain cilia or microvilli. The **lateral surfaces** face adjacent cells on either side and may contain cell junctions. The **basal surface** is opposite the apical surface and adheres to extracellular materials. In discussing epithelia with multiple layers, the term *apical layer* refers to the most superficial layer of cells, and the *basal layer* is the deepest layer of cells.

The **basement membrane** is a thin extracellular layer that commonly consists of the basal lamina and reticular lamina. The *basal lamina* (LAM-i-na = thin layer) contains mostly glycoproteins secreted by the epithelial cells. The *reticular lamina* is deep to the basal lamina and contains primarily collagen proteins produced by connective tissue cells called fibroblasts. The basement membrane attaches to and supports the overlying epithelial tissue and anchors it to the underlying connective tissue. In addition, it forms a surface along which epithelial cells migrate

FIGURE 4.3 Surfaces of epithelial cells and the structure and location of the basement membrane.

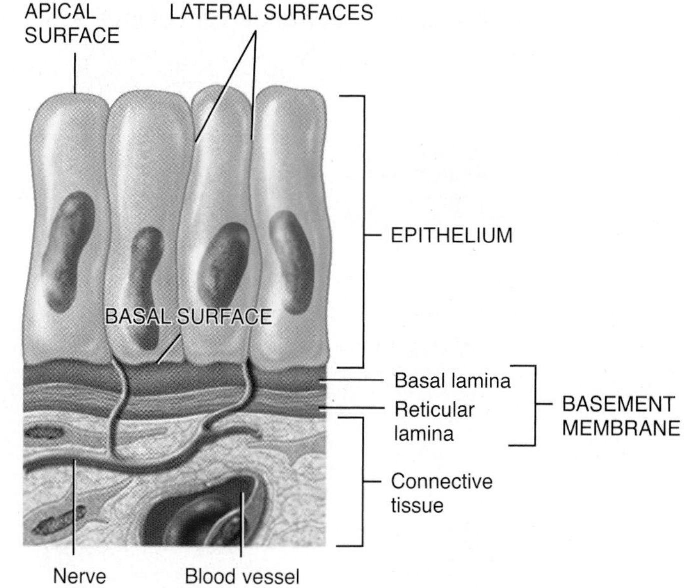


**The basement membrane is found between epithelium and connective tissue.**

during wound healing and restricts passage of molecules between the epithelium and connective tissue.

Epithelial tissue has its own nerve supply but is **avascular** (ā-VAS-kū-ler; *a-* = without; *-vascular* = vessel)—it lacks its own blood vessels. Needed nutrients and epithelial wastes diffuse between the epithelium and blood vessels of the underlying connective tissue.

Because epithelial tissue forms boundaries between the body's organs, or between the body and the external environment, it is repeatedly subjected to physical stress and injury. Epithelial tissue has a high rate of cell division, which allows it to constantly repair itself by sloughing off dead or injured cells and replacing them with new ones. Epithelial tissue roles in the body include protection, filtration, secretion, absorption, and excretion.

Epithelial tissue may be divided into two types. **Covering and lining epithelium** forms the outer covering of the skin and outer covering of some organs. It also lines body cavities and hollow organs. **Glandular epithelium** constitutes the secreting portion of glands such as the thyroid gland, adrenal glands, and sweat glands.

## Covering and Lining Epithelium

Covering and lining epithelial tissues are classified according to two characteristics: the arrangement of cells into layers and the shapes of the cells (Figure 4.4).

1. ***Arrangement of cells in layers.*** The cells are arranged in one or more layers depending on function:
   - *Simple epithelium* is a single layer of cells that functions in diffusion, osmosis, filtration, secretion, or absorption. **Secretion** is the production and release of substances such as mucus, sweat, or enzymes. **Absorption** is the intake of substances, such as digested food from the intestinal tract.
   - *Pseudostratified epithelium* (soo′-dō-STRA-te-fīd; *pseudo-* = false) appears to have multiple layers of cells because the cell nuclei lie at different levels and not all cells reach the apical surface, but it is actually a simple epithelium because all its cells rest on the basement membrane. Cells that do extend to the apical surface may contain cilia; others (goblet cells) secrete mucus.
   - *Stratified epithelium* (*stratum* = layer) consists of two or more layers of cells that protect underlying tissues in locations where there is considerable wear and tear.
2. ***Cell shapes.*** Epithelial cells vary in shape depending on their function.
   - *Squamous cells* (SKWĀ-mus = flat) are thin, which allows substances to pass rapidly through them.
   - *Cuboidal cells* are as tall as they are wide and are shaped like cubes. They may have microvilli at their apical surface and function in either secretion or absorption.
   - *Columnar cells* are much taller than they are wide and protect underlying tissues. Their apical surface may have cilia or microvilli, and they often are specialized for secretion and absorption.
   - *Transitional cells* change shape; they "transition" between squamous and cuboidal, allowing organs such as the urinary bladder to stretch to a larger size and then collapse to a smaller size.

As noted earlier, covering and lining epithelium forms outer coverings of the skin and some internal organs and inner linings of hollow organs and body cavities. Table 4.1 describes covering and lining epithelium in more detail. The description, location,

**FIGURE 4.4 Cell shapes and arrangement of layers for covering and lining epithelium.**

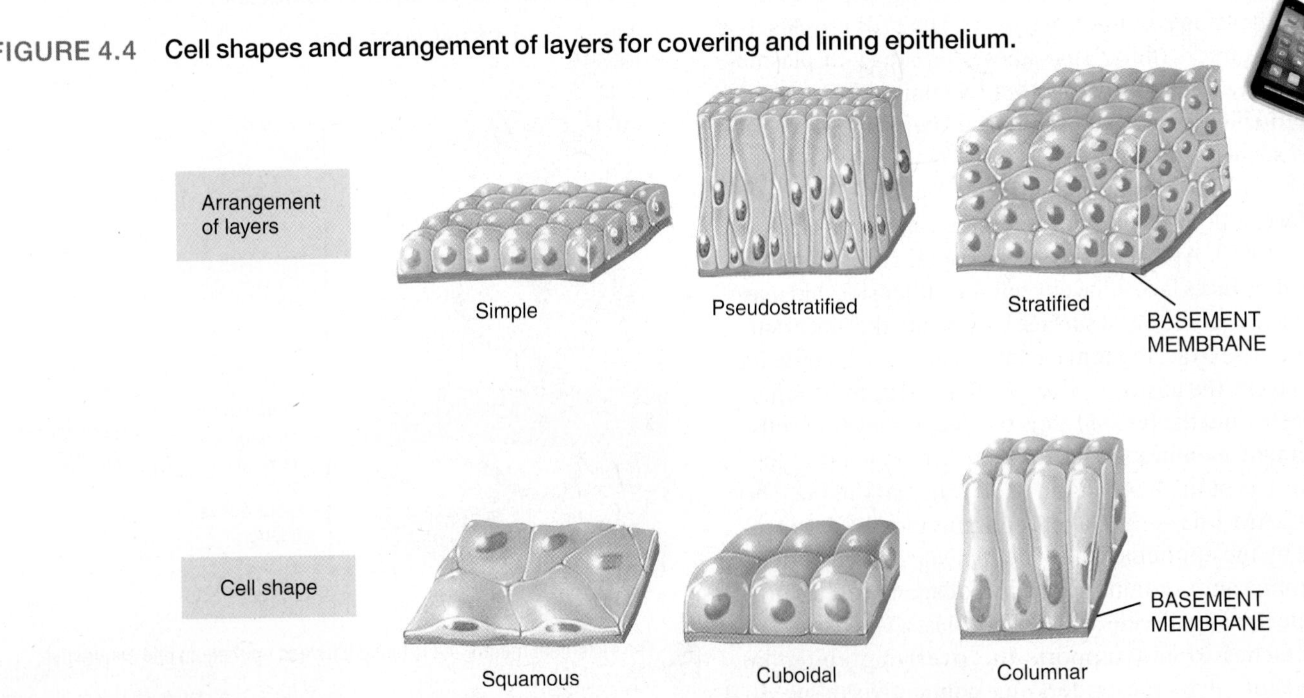


**Cell shape and arrangement of layers are the bases for classifying covering and lining epithelium.**

## TABLE 4.1

### Epithelial Tissues: Covering and Lining Epithelia

**SIMPLE SQUAMOUS EPITHELIUM**

| | |
|---|---|
| Description | Single layer of flat cells that resembles a tiled floor when viewed from apical surface; centrally located nucleus that is flattened oval or sphere. |
| Location | Most commonly (1) lines the cardiovascular and lymphatic system (heart, blood vessels, lymphatic vessel linings), where it is known as **endothelium** (en′-dō-THĒ-lē-um; *endo-* = within; *-thelium* = covering), and (2) forms the epithelial layer of serous membranes (peritoneum, pleura, pericardium), where it is called **mesothelium** (mez′-ō-THĒ-lē-um; *meso-* = middle). Also found in air sacs of lungs, glomerular (Bowman's) capsule of kidneys, inner surface of tympanic membrane (eardrum). Not found in body areas subject to mechanical stress (wear and tear). |
| Functions | Filtration (such as blood filtration in kidneys), diffusion (such as diffusion of oxygen into blood vessels of lungs), and secretion in serous membranes. |

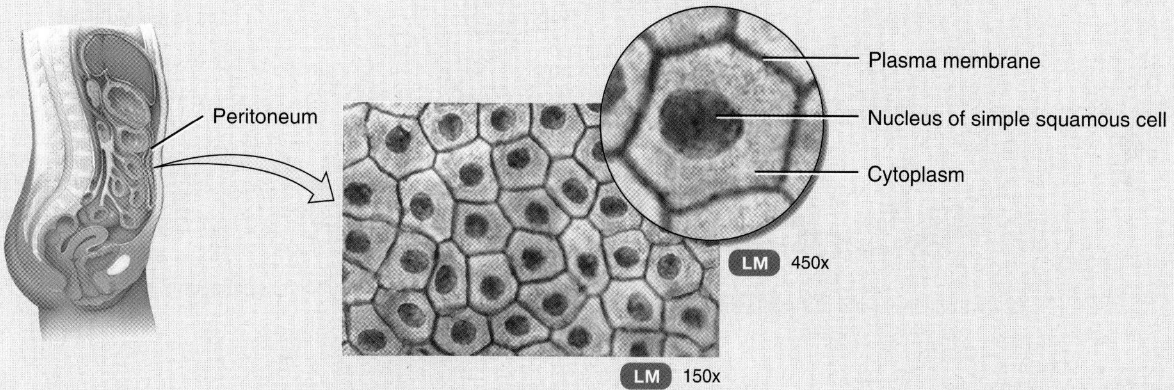


Surface view of simple squamous epithelium of mesothelial lining of peritoneum

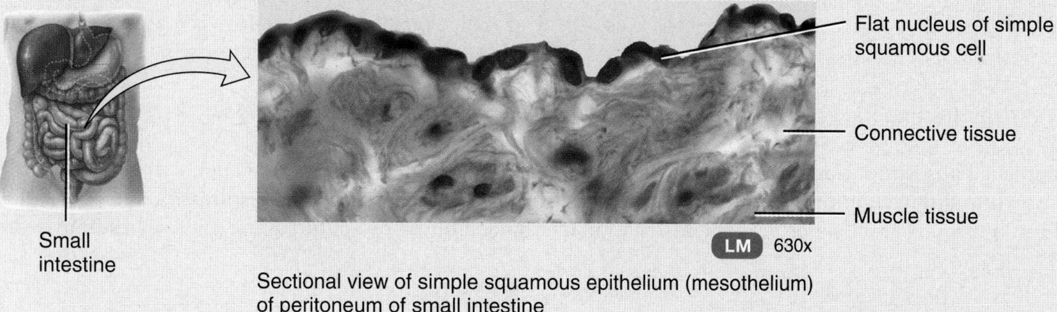


Sectional view of simple squamous epithelium (mesothelium) of peritoneum of small intestine

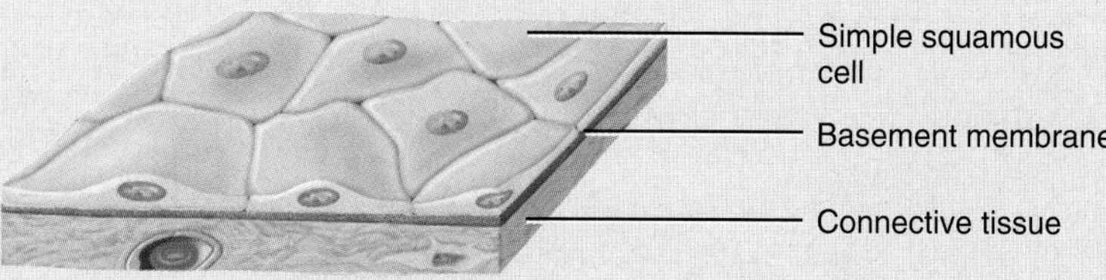


Simple squamous epithelium

(continues)

**TABLE 4.1 CONTINUED**

**SIMPLE CUBOIDAL EPITHELIUM**

| | |
|---|---|
| Description | Single layer of cube-shaped cells; round, centrally located nucleus. Cuboidal cell shape is obvious when tissue is sectioned and viewed from the side. (Note: Strictly cuboidal cells could not form small tubes; these cuboidal cells are more pie-shaped but still nearly as high as they are wide at the base.) |
| Location | Covers surface of ovary; lines anterior surface of capsule of lens of the eye; forms pigmented epithelium at posterior surface of retina of the eye; lines kidney tubules and smaller ducts of many glands; makes up secreting portion of some glands, such as thyroid gland and ducts of some glands such as pancreas. |
| Functions | Secretion and absorption. |

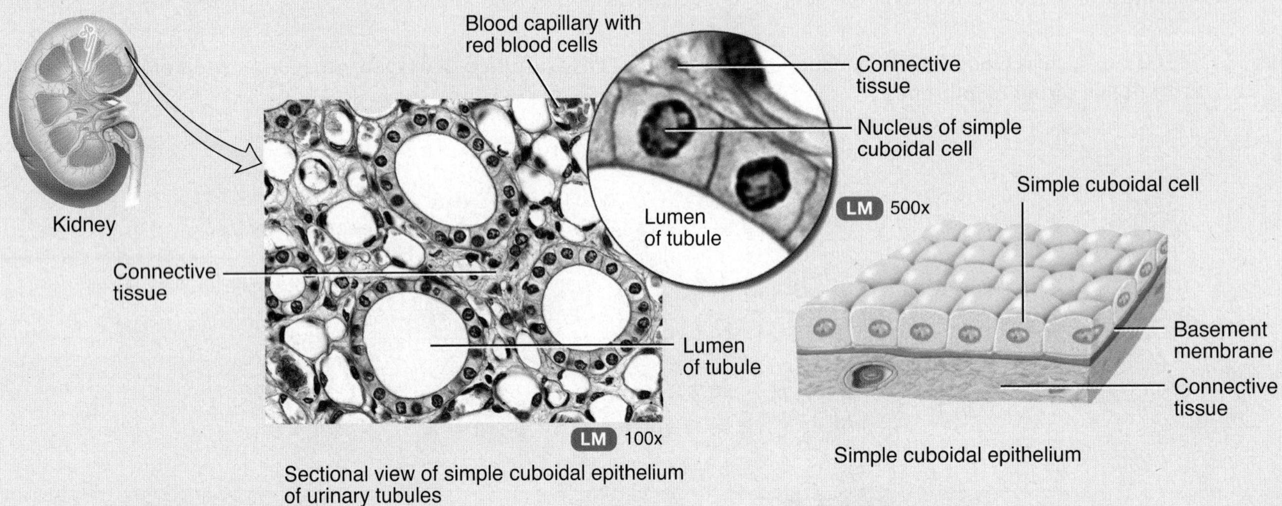


Sectional view of simple cuboidal epithelium of urinary tubules

Simple cuboidal epithelium

**NONCILIATED SIMPLE COLUMNAR EPITHELIUM**

| | |
|---|---|
| Description | Single layer of nonciliated column-like cells with oval nuclei near base of cells; contains (1) columnar epithelial cells with microvilli at apical surface and (2) goblet cells. **Microvilli**, fingerlike cytoplasmic projections, increase surface area of plasma membrane (see Figure 3.1), thus increasing cell's rate of absorption. **Goblet cells** are modified columnar epithelial cells that secrete *mucus*, a slightly sticky fluid, at their apical surfaces; before release, mucus accumulates in upper portion of cell, causing it to bulge and making the whole cell resemble a goblet or wine glass. |
| Location | Lines gastrointestinal tract (from stomach to anus), ducts of many glands, and gallbladder. |
| Functions | Secretion and absorption; larger columnar cells contain more organelles and thus are capable of higher levels of secretion and absorption than cuboidal cells. Secreted mucus lubricates linings of digestive, respiratory, and reproductive tracts, and most of urinary tract; helps prevent destruction of stomach lining by acidic gastric juice secreted by stomach. |

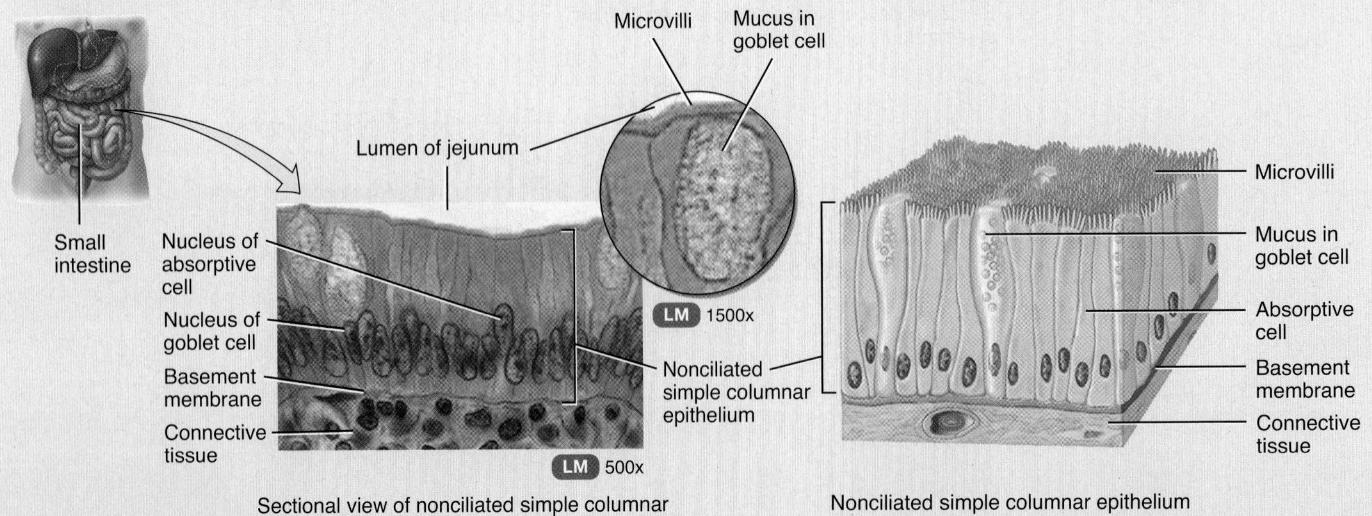


Sectional view of nonciliated simple columnar epithelium of lining of jejunum of small intestine

Noncilated simple columnar epithelium

**CILIATED SIMPLE COLUMNAR EPITHELIUM**

| | |
|---|---|
| Description | Single layer of ciliated column-like cells with nuclei near base of cells. Goblet cells are usually interspersed among ciliated columnar epithelia. |
| Location | Lines some bronchioles (small tubes) of respiratory tract, uterine (fallopian) tubes, uterus, some paranasal sinuses, central canal of spinal cord, and ventricles of brain. |
| Functions | Cilia beat in unison, moving mucus and foreign particles toward throat, where they can be coughed up and swallowed or spit out. Coughing and sneezing speed up movement of cilia and mucus. Cilia also help move oocytes expelled from ovaries through uterine (fallopian) tubes into uterus. |

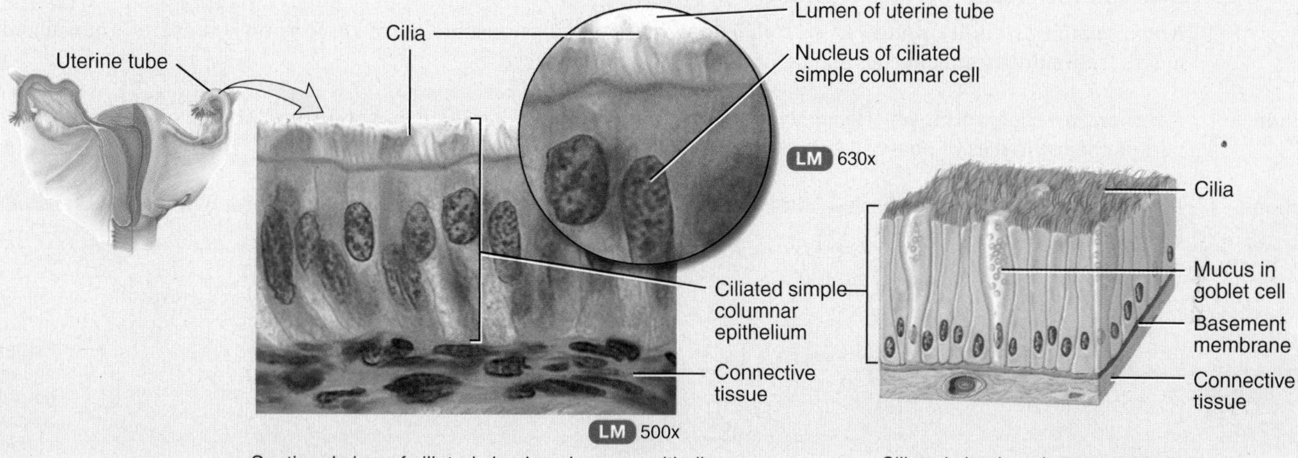


Sectional view of ciliated simple columnar epithelium of uterine tube

Ciliated simple columnar epithelium

**PSEUDOSTRATIFIED COLUMNAR EPITHELIUM**

| | |
|---|---|
| Description | Appears to have several layers because cell nuclei are at various levels. All cells are attached to basement membrane in a single layer, but some cells do not extend to apical surface. When viewed from side, these features give false impression of a multilayered tissue (thus the name pseudostratified; *pseudo* = false). *Pseudostratified ciliated columnar epithelium* contains cells that extend to surface and secrete mucus (goblet cells) or bear cilia. *Pseudostratified nonciliated columnar epithelium* contains cells without cilia and lacks goblet cells. |
| Location | Ciliated variety lines airways of most of upper respiratory tract; nonciliated variety lines larger ducts of many glands, epididymis, and part of male urethra. |
| Functions | Ciliated variety secretes mucus that traps foreign particles, and cilia sweep away mucus for elimination from body; nonciliated variety functions in absorption and protection. |

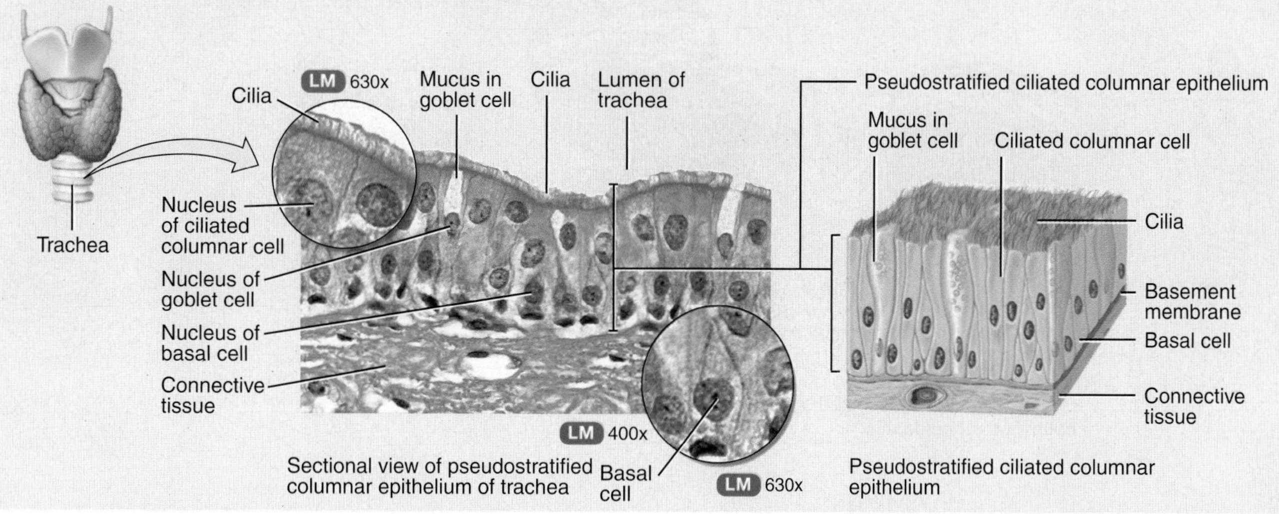


Sectional view of pseudostratified columnar epithelium of trachea

Pseudostratified ciliated columnar epithelium

(continues)

## TABLE 4.1 CONTINUED

### STRATIFIED SQUAMOUS EPITHELIUM

| | |
|---|---|
| Description | Two or more layers of cells; cells in apical layer and several layers deep to it are squamous; cells in deeper layers vary from cuboidal to columnar. As basal cells divide, daughter cells arising from cell divisions push upward toward apical layer. As they move toward surface and away from blood supply in underlying connective tissue, they become dehydrated and less metabolically active. Tough proteins predominate as cytoplasm is reduced, and cells become tough, hard structures that eventually die. At apical layer, after dead cells lose cell junctions they are sloughed off, but they are replaced continuously as new cells emerge from basal cells.<br>*Keratinized stratified squamous epithelium* (KER-a-ti-nīzd) develops tough layer of keratin in apical layer of cells and several layers deep to it. (**Keratin** [KER-a-tin] is a tough, fibrous intracellular protein that helps protect skin and underlying tissues from heat, microbes, and chemicals.) Relative amount of keratin increases in cells as they move away from nutritive blood supply and organelles die.<br>*Nonkeratinized stratified squamous epithelium* does not contain large amounts of keratin and is constantly moistened by mucus from salivary and mucous glands; organelles are not replaced. |
| Location | Keratinized variety forms superficial layer of skin; nonkeratinized variety lines wet surfaces (lining of mouth, esophagus, part of epiglottis, part of pharynx, and vagina) and covers tongue. |
| Functions | Protects against abrasion, water loss, ultraviolet radiation, and foreign invasion. Both types form first line of defense against microbes. |

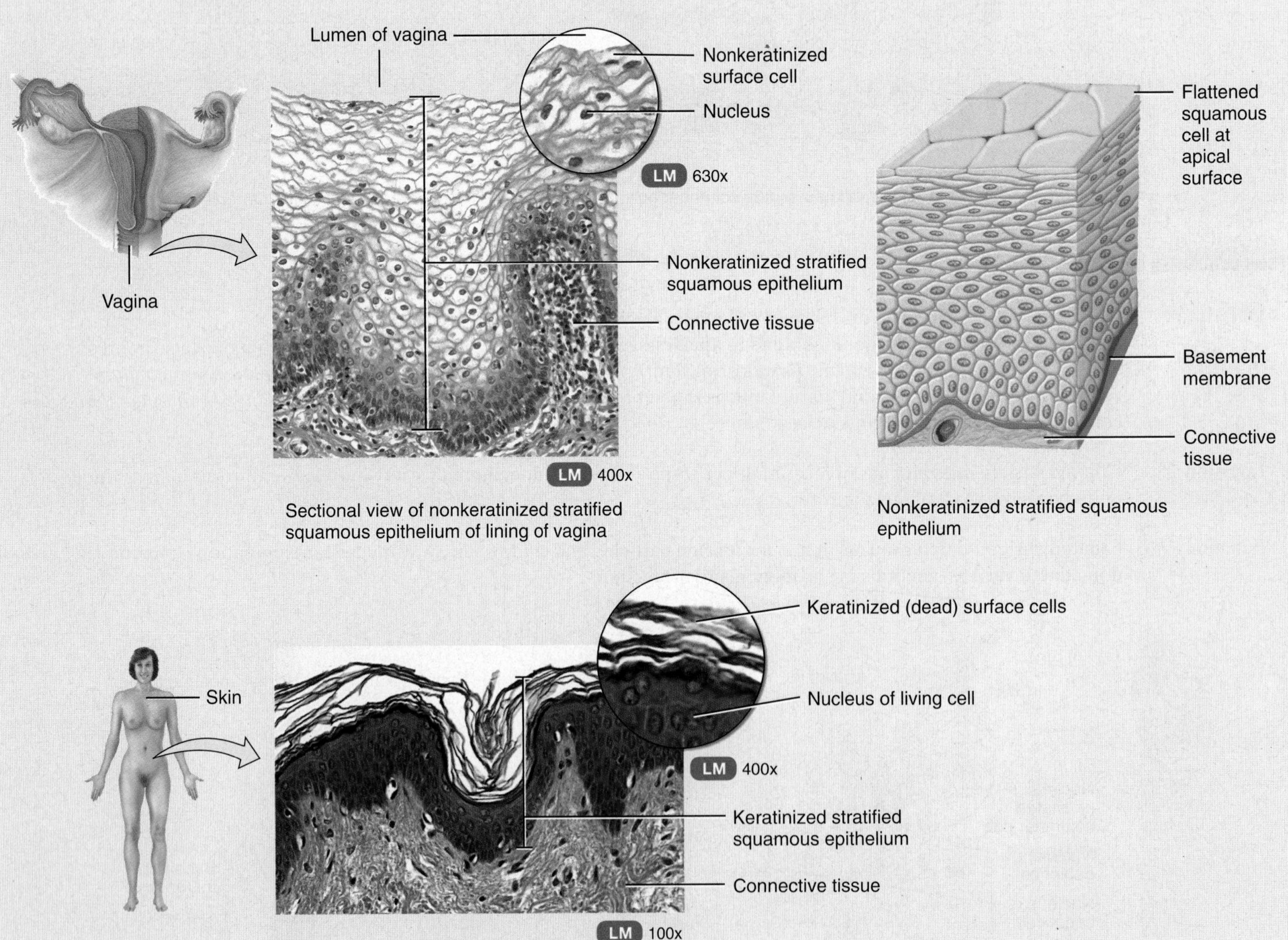


Sectional view of nonkeratinized stratified squamous epithelium of lining of vagina

Nonkeratinized stratified squamous epithelium

Sectional view of keratinized stratified squamous epithelium of epidermis

**TRANSITIONAL EPITHELIUM**

| | |
|---|---|
| Description | Variable appearance (transitional). In relaxed or unstretched state, looks like stratified cuboidal epithelium, except apical layer cells tend to be large and rounded. As tissue is stretched, cells become flatter, giving the appearance of stratified squamous epithelium. Multiple layers and elasticity make it ideal for lining hollow structures (urinary bladder) subject to expansion from within. |
| Location | Lines urinary bladder and portions of ureters and urethra. |
| Functions | Allows urinary organs to stretch and maintain protective lining while holding variable amounts of fluid without rupturing. |

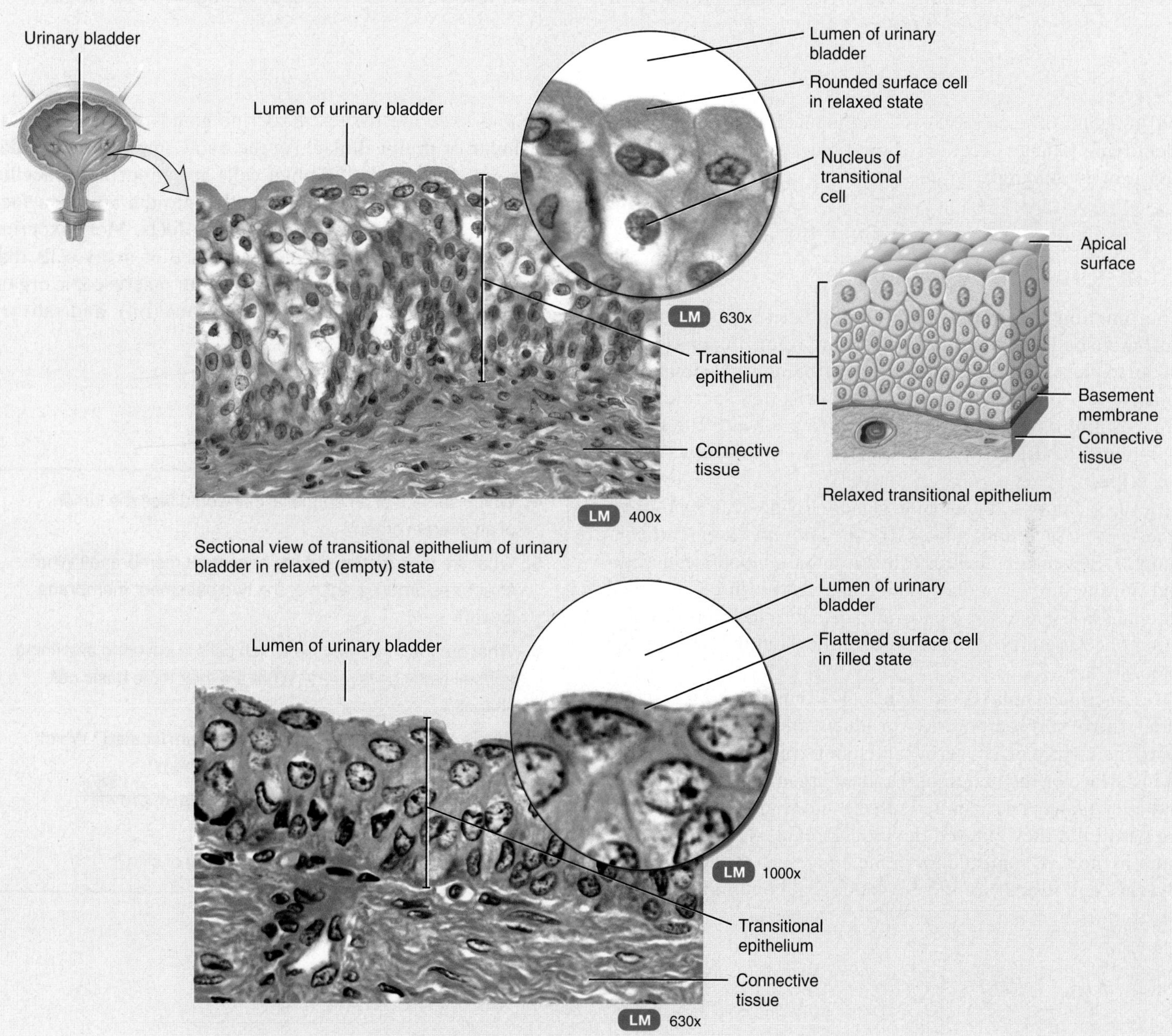


Sectional view of transitional epithelium of urinary bladder in relaxed (empty) state

Relaxed transitional epithelium

Sectional view of transitional epithelium of urinary bladder in filled state

### CLINICAL CONNECTION | *Papanicolaou Test*

A **Papanicolaou test** (pa-pa-NI-kō-lō), also called a **Pap test** or **Pap smear,** involves collection and microscopic examination of epithelial cells that have been scraped off the apical layer of a tissue. A very common type of Pap test involves examining the cells from the nonkeratinized stratified squamous epithelium of the vagina and cervix (inferior portion) of the uterus. This type of Pap test is performed mainly to detect early changes in the cells of the female reproductive system that may indicate a precancerous condition or cancer. In performing a Pap smear, a physician collects cells, which are then smeared on a microscope slide. The slides are then sent to a laboratory for analysis. Pap tests should be started within three years of the onset of sexual activity, or age 21, whichever comes first. Annual screening is recommended for females ages 21–30 and every 2–3 years for females age 30 or older following three consecutive negative Pap tests.

and function of each tissue is accompanied by an illustration identifying a major location of the tissue, a photomicrograph, and a corresponding diagram.

## Glandular Epithelium

The function of glandular epithelium is secretion, which is accomplished by glandular cells that often lie in clusters deep to the covering and lining epithelium. A **gland** may consist of a single cell or a group of cells that secretes substances into *ducts* (tubes), onto a surface, or into the blood. All glands of the body are classified as either endocrine or exocrine.

The secretions of **endocrine glands** (EN-dō-krin; *endo-* = *inside*, *-crine* = secretion; Table 4.2), called *hormones*, enter the interstitial fluid and then diffuse directly into the bloodstream without flowing through a duct. Endocrine glands will be described in detail in Chapter 17. Endocrine secretions have far-reaching effects because they are distributed throughout the body by the bloodstream.

**Exocrine glands** (EX-ō-krin; *exo-* = outside; *-crine* = secretion; Table 4.2) secrete their products into ducts that empty onto the surface of a covering and lining epithelium such as the skin surface or the lumen of a hollow organ. The secretions of exocrine glands have limited effects and some of them would be harmful if they entered the bloodstream. As you will learn later in the text, some glands of the body, such as the pancreas, ovaries, and testes, are mixed glands that contain both endocrine and exocrine tissues. Exocrine glands are classified as unicellular or multicellular. As the name implies, **unicellular glands** are single-celled. Goblet cells are important unicellular glands that secrete mucus directly onto the apical surface of a lining epithelium rather than into ducts. Most exocrine glands are **multicellular glands**, composed of many cells that form a distinctive microscopic structure or macroscopic organ. Examples include sudoriferous, sebaceous (oil), and salivary glands.

### ✓ CHECKPOINT

**4.** Which surface of an epithelial cell would face the lumen of an internal organ?

**5.** What are the functions of the basement membrane? What structures produce each of the two basement membrane layers?

**6.** What are the three ways in which cells in covering and lining epithelia may be layered? What are their three basic cell shapes?

**7.** Where are endothelium and mesothelium located? Which type of epithelium composes each of them?

**8.** What are the functions of goblet cells, microvilli, and cilia?

**9.** Why is keratin needed in the epithelium of skin?

## TABLE 4.2

### Epithelial Tissues: Glandular Epithelia

**ENDOCRINE GLANDS**

| | |
|---|---|
| Description | Secretions (*hormones*) enter interstitial fluid and diffuse directly into bloodstream without flowing through a duct. Endocrine glands will be described in detail in Chapter 17. |
| Location | Examples: pituitary gland at base of brain, pineal gland in brain, thyroid and parathyroid glands near larynx (voice box), adrenal glands superior to kidneys, pancreas near stomach, ovaries in pelvic cavity, testes in scrotum, thymus in thoracic cavity. |
| Functions | Hormones secreted by endocrine glands regulate many metabolic and physiological activities to maintain homeostasis. |

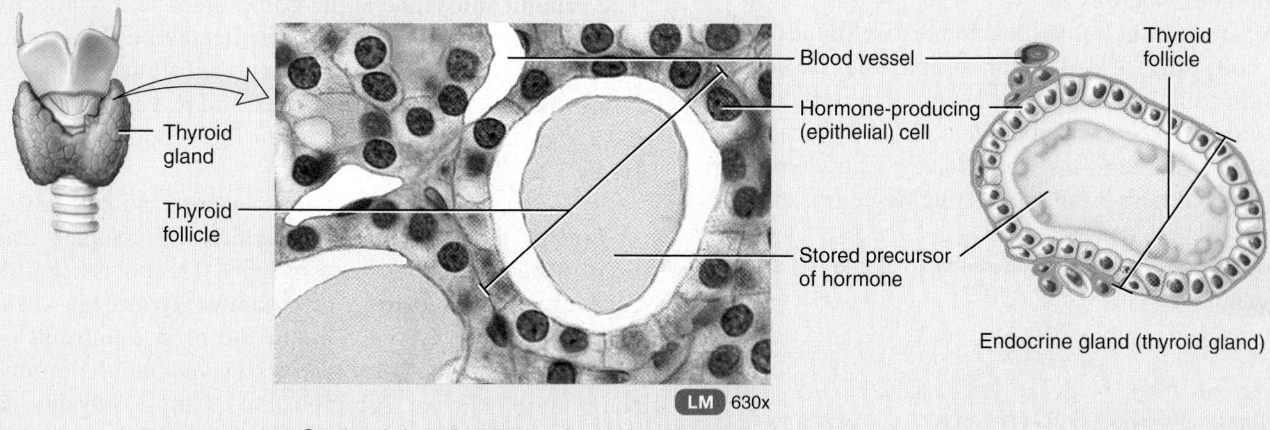


Sectional view of endocrine gland (thyroid gland)

**EXOCRINE GLANDS**

| | |
|---|---|
| Description | Secrete substances into ducts that empty onto surface of a covering and lining epithelium, such as skin surface or lumen of hollow organ. |
| Location | Sweat, oil, and earwax glands of skin; digestive glands such as salivary glands (secrete into mouth cavity) and pancreas (secretes into small intestine). |
| Functions | Produce substances such as sweat (to help lower body temperature), oil and earwax (for lubrication and protection), saliva and digestive enzymes (to assist in lubrication and digestion of nutrients). |

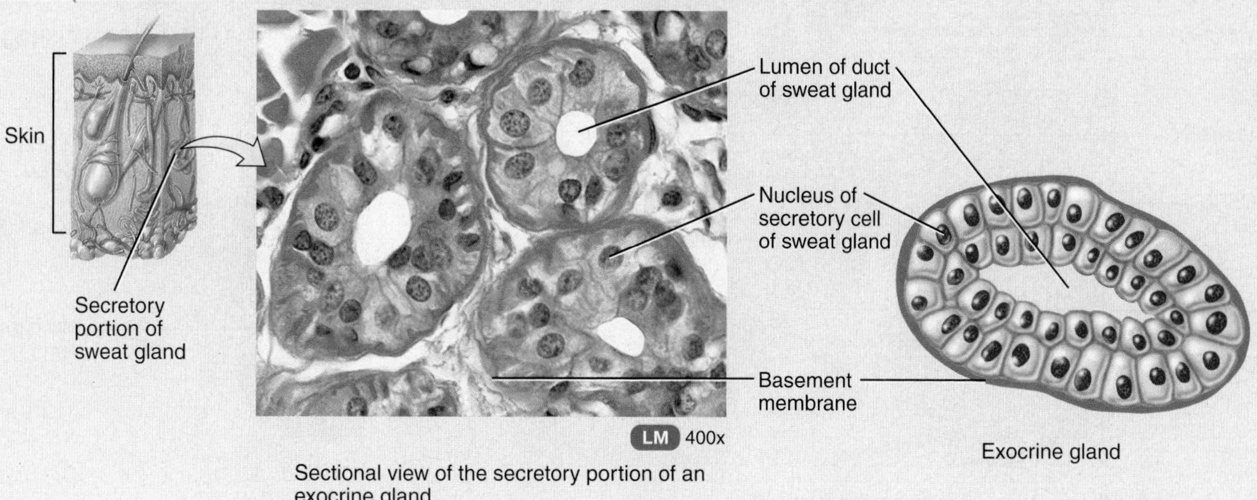


Sectional view of the secretory portion of an exocrine gland

# 4.4 Connective tissue binds organs together, stores energy reserves as fat, and helps provide immunity.

**Connective tissue** is one of the most abundant and widely distributed tissues in the body (Figure 4.5). In its various forms, connective tissue has a variety of functions. It binds together, supports, and strengthens other body tissues; protects and insulates internal organs; compartmentalizes structures such as skeletal muscles; serves as the major transport system within the body (blood, a fluid connective tissue); is the primary location of stored energy reserves (adipose, or fat, tissue); and is the main source of immune responses.

In contrast to epithelial tissues, connective tissues are not located on body surfaces. Also unlike epithelial tissues, connective tissues usually are highly vascular; that is, they have a rich blood supply. Exceptions include cartilage, which is completely avascular, and tendons, which have a scanty blood supply. Like epithelial tissues, all connective tissues except cartilage are supplied with nerves.

Connective tissue consists of two basic elements: extracellular matrix and cells.

## Connective Tissue Extracellular Matrix

As its name implies, a connective tissue's **extracellular matrix** (MĀ-triks) is the material located between its widely spaced cells. The extracellular matrix is secreted by the connective tissue cells and accounts for many of the functional properties of the tissue. For example, the extracellular matrix of cartilage is firm but pliable. By contrast, the extracellular matrix of bone is hard and inflexible. The extracellular matrix consists of ground substance and fibers.

### Ground Substance

The **ground substance** is the component of a connective tissue between the cells and fibers. It may be fluid, semifluid, gelatinous, or calcified. The ground substance supports cells, binds them together, stores water, and provides a medium for exchange of substances between the blood and cells of the tissue.

Ground substance contains water and an assortment of organic molecules, many of which are complex combinations of polysaccharides and proteins. The polysaccharides are collectively referred to as **glycosaminoglycans** (glī-kos-a-mē′-nō-GLĪ-kans) or **GAGs**. One of the most significant properties of GAGs is that they trap water, making the ground substance more jellylike. An important example is **hyaluronic acid** (hī′-a-loo-RON-ik), a viscous, slippery substance that binds cells together, lubricates joints, and helps maintain the shape of the eyeballs. White blood cells, sperm cells, and some bacteria produce *hyaluronidase*, an enzyme that breaks apart hyaluronic

**FIGURE 4.5** **Representative cells and fibers present in connective tissue.**

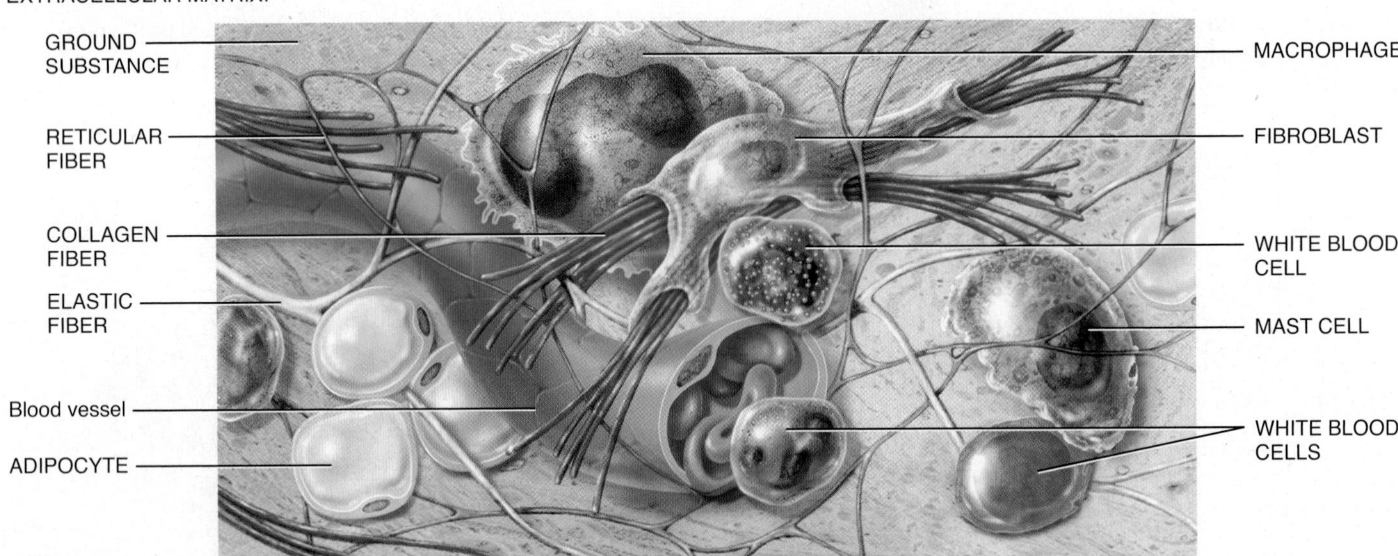


**Fibroblasts are usually the most numerous connective tissue cells.**

acid, thus causing the ground substance of connective tissue to become more liquid. The ability to produce hyaluronidase helps white blood cells move more easily through connective tissues to reach sites of infection and aids penetration of an oocyte by a sperm cell during fertilization. Hyaluronidase also accounts for the rapid spread of bacteria through connective tissues.

### Fibers

Three types of **fibers** are embedded in the extracellular matrix between the cells: collagen fibers, elastic fibers, and reticular fibers (Figure 4.5). They function to strengthen and support connective tissues.

**Collagen fibers** (KOL-a-jen; *colla* = glue) are very strong and resist pulling forces, but they are not stiff, which allows tissue flexibility. The properties of different types of collagen fibers vary from tissue to tissue. For example, the collagen fibers found in cartilage attract more water molecules than those in bone, which gives cartilage a more cushioning consistency. Collagen fibers often occur in parallel bundles (see Table 4.4, dense regular connective tissue). The bundle arrangement gives the tissue great *tensile strength*, the ability to resist stretching. Chemically, collagen fibers consist of the protein *collagen,* which is the most abundant protein in your body, representing about 25 percent of the total protein. Collagen fibers are found in most types of connective tissues, especially bone, cartilage, tendons (which attach muscle to bone), and ligaments (which attach bone to bone).

**Elastic fibers**, which are smaller in diameter than collagen fibers, branch and join together to form a fibrous network within a tissue. Elastic fibers contain the protein *elastin*, which allows them to stretch up to 150 percent of their relaxed length without breaking. Equally important, elastic fibers have the ability to return to their original shape after being stretched, a property called **elasticity**. Elastic fibers are plentiful in skin, blood vessel walls, and lung tissue.

**Reticular fibers** (*reticul-* = net) consist of collagen arranged as fine, branching, interwoven fibers that provide support in the walls of blood vessels and form a network around the cells in some tissues, such as areolar connective tissue, adipose tissue, and smooth muscle tissue. Reticular fibers are much thinner than collagen fibers and form branching networks. Like collagen fibers, reticular fibers provide support and strength. Reticular fibers are plentiful in reticular connective tissue, which forms the **stroma** (= bed or covering) or supporting framework of many soft organs, such as the spleen and lymph nodes.

## Connective Tissue Cells

Each major type of connective tissue contains an immature class of cells with a name ending in *blast*, which means "to bud or sprout." These immature cells are called *fibroblasts* in loose and dense connective tissue, *chondroblasts* in cartilage, and *osteoblasts* in bone. Blast cells retain the capacity for cell division and secrete the extracellular matrix that is characteristic of the tissue. In cartilage and bone, once the extracellular matrix is produced, the blast cells differentiate into mature cells with names ending in *-cyte*, namely chondrocytes and osteocytes. Mature cells have reduced capacities for cell division and extracellular matrix formation and are mostly involved in maintaining the extracellular matrix.

The types of connective tissue cells vary according to the type of tissue and include the following (Figure 4.5):

- **Fibroblasts** (FĪ-brō-blasts; *fibro-* = fibers) are large, flat cells with branching processes. They are present in several connective tissues, and usually are the most numerous. Fibroblasts migrate through the connective tissue, secreting the fibers and ground substance of the extracellular matrix.
- **Macrophages** (MAK-rō-fā-jez; *macro-* = large; *-phages* = eaters) have an irregular shape with short branching projections and are capable of engulfing bacteria and cellular debris by phagocytosis.
- **Mast cells** produce histamine, a chemical that dilates small blood vessels as part of the inflammatory response, the body's reaction to injury or infection. In addition, researchers have recently discovered that mast cells can ingest and kill bacteria.
- **Adipocytes** (A-di-pō-sīts), also called fat cells, are connective tissue cells that store triglycerides (fats).
- **White blood cells** are not found in significant numbers in normal connective tissue. However, in response to certain conditions they migrate from blood into connective tissues to mediate immune system responses. The various types of white blood cells will be described in detail in Chapter 18.

## Types of Connective Tissue

**Mature connective tissue** is present in the newborn. Its cells develop primarily from *mesenchyme* (MEZ-en-kīm), an embryonic connective tissue that arises during the early weeks of pregnancy. The five types of mature connective tissue are loose connective tissue, dense connective tissue, cartilage, osseous tissue, and liquid connective tissue (blood and lymph).

### Loose Connective Tissue

The fibers of **loose connective tissue** are *loosely* arranged between cells. The types of loose connective tissue include areolar connective tissue, adipose tissue, and reticular connective tissue (Table 4.3).

**TABLE 4.3**

## Loose Connective Tissues

### AREOLAR CONNECTIVE TISSUE (a-RĒ-ō-lar; *areol-* = small space)

| | |
|---|---|
| Description | One of the most widely distributed connective tissues; consists of fibers (collagen, elastic, reticular) arranged randomly and several kinds of cells (fibroblasts, macrophages, adipocytes, mast cells, and a few white blood cells) embedded in semifluid ground substance. |
| Location | In and around nearly every body structure (thus, called "packing material" of the body). Hypodermis deep to skin; lamina propria of mucous membranes; around blood vessels, nerves, and body organs. |
| Functions | Strength, elasticity, support. |

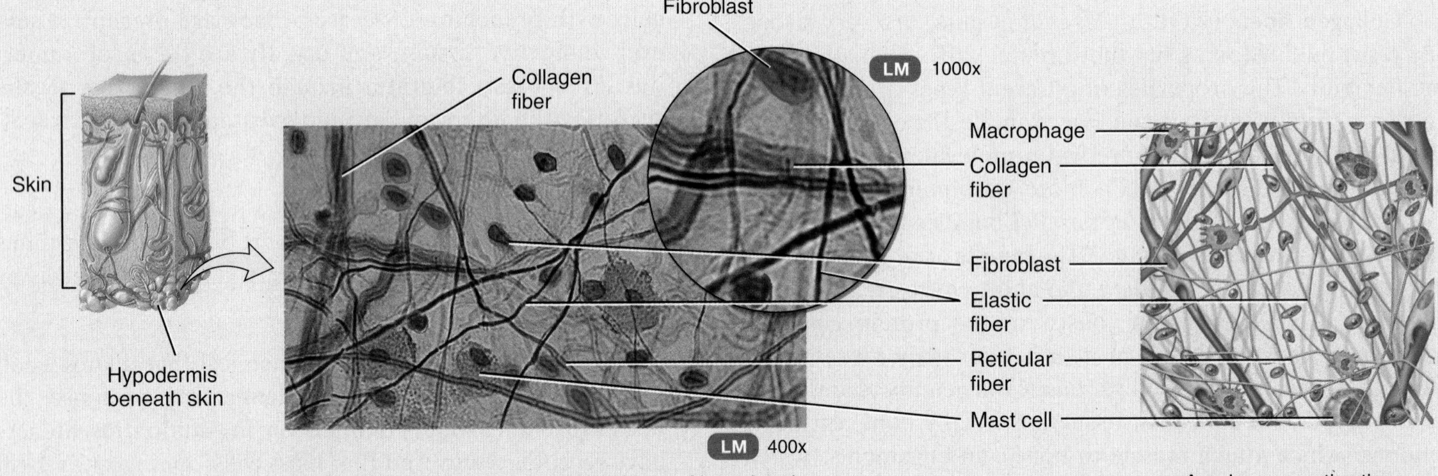


Sectional view of areolar connective tissue of hypodermis

Areolar connective tissue

### ADIPOSE TISSUE

| | |
|---|---|
| Description | Contains **adipocytes**, cells specialized to store triglycerides (fats) as a large centrally located droplet. Cell fills up with a single, large triglyceride droplet, and cytoplasm and nucleus are pushed to periphery of cell. |
| Location | Subcutaneous layer deep to skin; around heart and kidneys; yellow bone marrow; padding around joints and behind eyeball in eye socket. |
| Functions | Reduces heat loss through skin; serves as an energy reserve; supports and protects organs. |

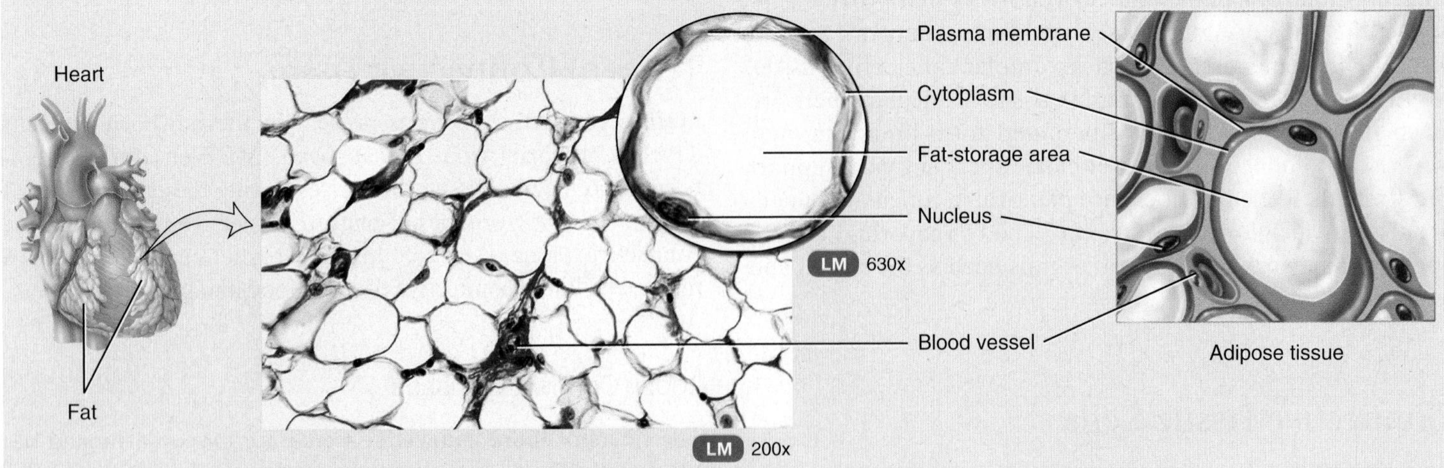


Sectional view of adipose tissue showing adipocytes and details of an adipocyte

Adipose tissue

**RETICULAR CONNECTIVE TISSUE**

| | |
|---|---|
| Description | Fine interlacing network of reticular fibers and reticular cells. |
| Location | *Stroma* (supporting framework) of liver, spleen, lymph nodes. |
| Functions | Forms stroma of organs; filters and removes worn-out blood cells in spleen and microbes in lymph nodes. |



Sectional view of reticular connective tissue of a lymph node

Reticular connective tissue

## Dense Connective Tissue

**Dense connective tissue** contains more fibers, which are thicker and more *densely* packed, but has considerably fewer cells than loose connective tissue. There are three types: dense regular connective tissue, dense irregular connective tissue, and elastic connective tissue (Table 4.4).

## CLINICAL CONNECTION | *Liposuction*

A surgical procedure called **liposuction** (*lip-* = fat) or **suction lipectomy** (*-ectomy* = to cut out) involves suctioning out small amounts of adipose tissue from various areas of the body. After an incision is made in the skin, the fat is removed through a stainless steel tube, called a cannula, with the assistance of a powerful vacuum pressure unit that suctions out the fat. The technique can be used as a body-contouring procedure in regions such as the thighs, buttocks, arms, breasts, and abdomen, and to transfer fat to another area of the body. Postsurgical complications that may develop include fat that may enter blood vessels broken during the procedure and obstruct blood flow, infection, loss of feeling in the area, fluid depletion, injury to internal structures, and severe postoperative pain.

**TABLE 4.4**

## Dense Connective Tissues

### DENSE REGULAR CONNECTIVE TISSUE

| | |
|---|---|
| Description | Consists mainly of collagen fibers *regularly* arranged in parallel bundles with fibroblasts in rows between bundles. |
| Location | Forms tendons (attach muscle to bone); most ligaments (attach bone to bone); aponeuroses (sheetlike tendons that attach muscle to muscle or muscle to bone). |
| Function | Provides strong attachment between various structures. |

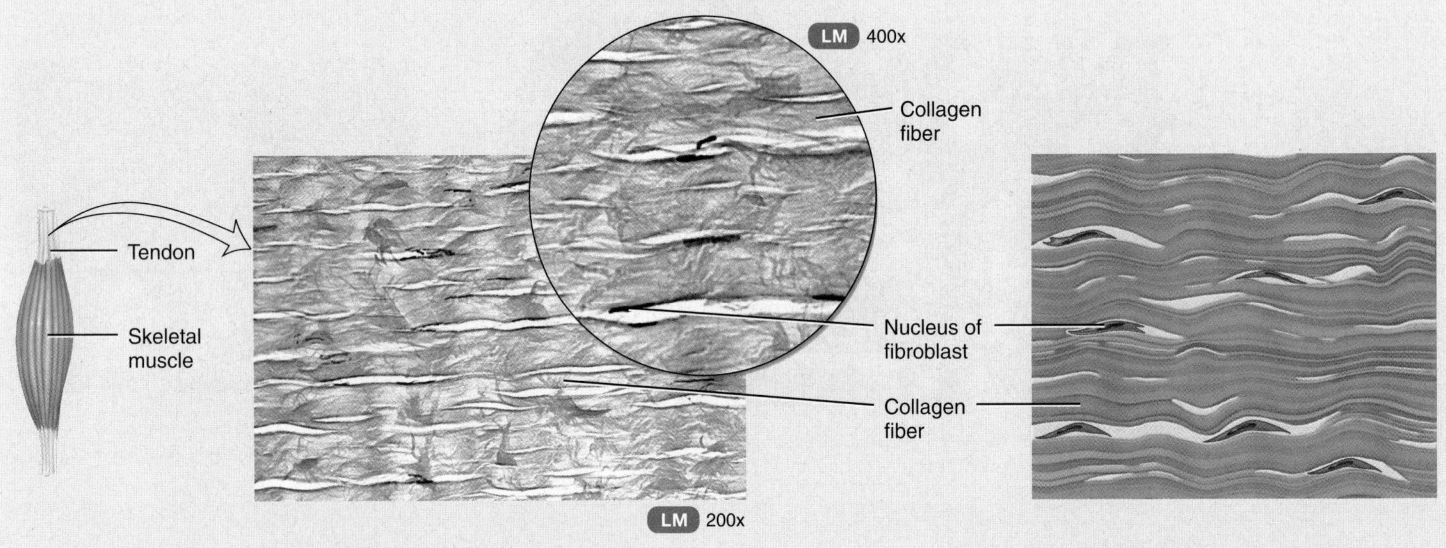


Sectional view of dense regular connective tissue of a tendon

Dense regular connective tissue

### DENSE IRREGULAR CONNECTIVE TISSUE

| | |
|---|---|
| Description | Collagen fibers; usually *irregularly* arranged with a few fibroblasts. |
| Location | *Fasciae* (FASH-ē-ē; tissues around muscles and other organs); dermis of skin; fibrous capsules of organs and joints; heart valves. |
| Function | Provides tensile (pulling) strength in many directions. |

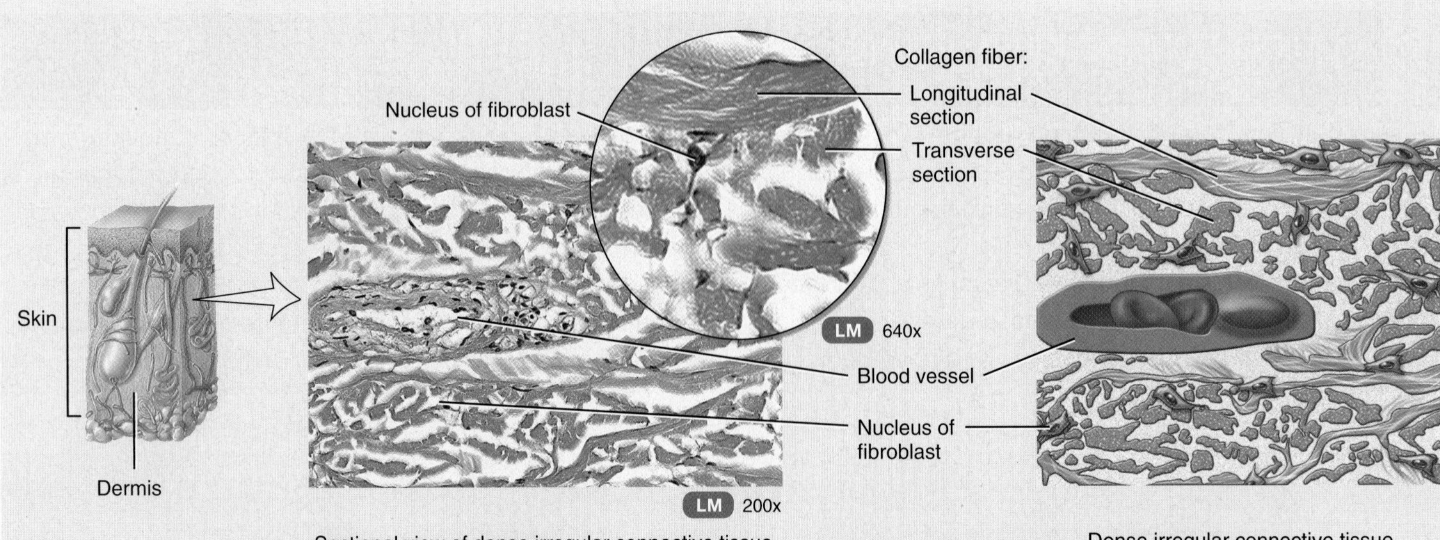


Sectional view of dense irregular connective tissue of reticular region of dermis

Dense irregular connective tissue

**ELASTIC CONNECTIVE TISSUE**

| | |
|---|---|
| Description | Predominantly elastic fibers with fibroblasts between fibers. |
| Location | Walls of elastic arteries; lung tissue; trachea; bronchial tubes; ligaments between vertebrae. |
| Function | Allows stretching of various organs; is strong and can recoil to original shape after being stretched. Elasticity is important to normal functioning of lung tissue (recoils in exhaling) and elastic arteries (recoil between heartbeats to help maintain blood flow). |

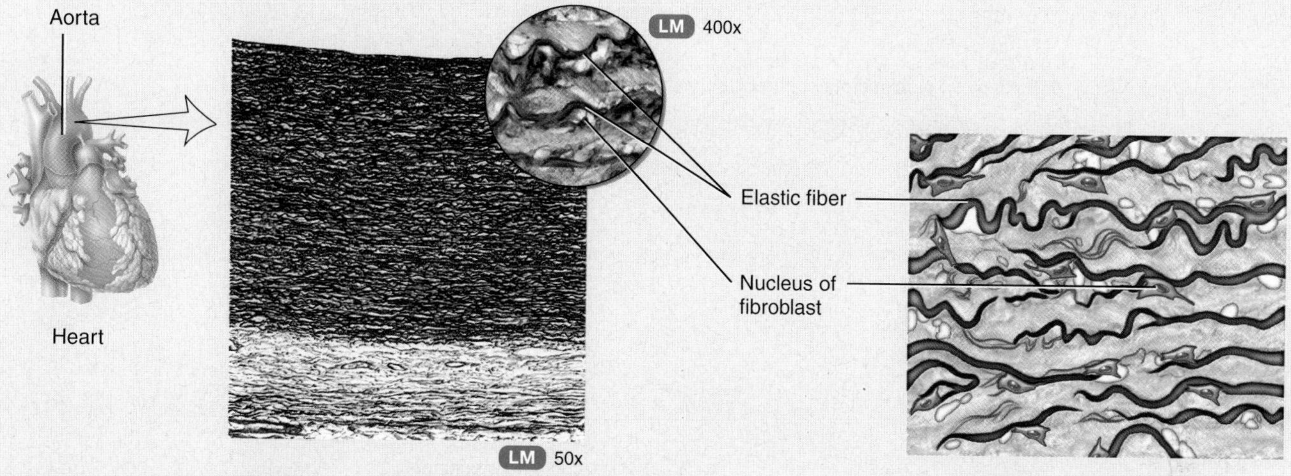

Aorta

Heart

LM 400x

Elastic fiber

Nucleus of fibroblast

LM 50x

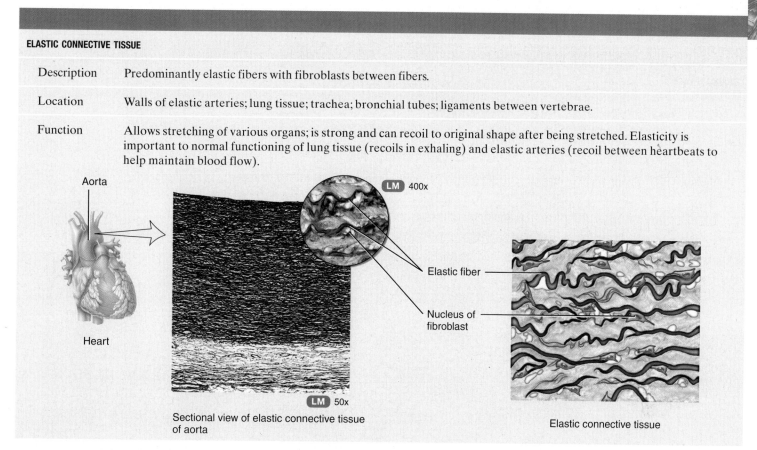

Sectional view of elastic connective tissue of aorta

Elastic connective tissue

## Cartilage

**Cartilage** (KAR-ti-lij) consists of a dense network of collagen fibers and elastic fibers firmly embedded in *chondroitin sulfate*, a gel-like component of the ground substance (Table 4.5). Cartilage can endure considerably more stress than loose and dense connective

**TABLE 4.5**

Cartilage

**HYALINE CARTILAGE**

| | |
|---|---|
| Description | Hyaline (*hyalinos* = glassy) cartilage contains a resilient gel as ground substance and appears in body as bluish-white, shiny substance (can stain pink or purple when prepared for microscopic examination). Fine collagen fibers are not visible with ordinary staining techniques; chondrocytes are found in lacunae. |
| Location | Ends of long bones; anterior ends of ribs; nose; trachea; larynx; bronchi; embryonic and fetal skeleton. |
| Functions | Reduces friction and absorbs shock at joints, provides flexibility and support; weakest type of cartilage. |

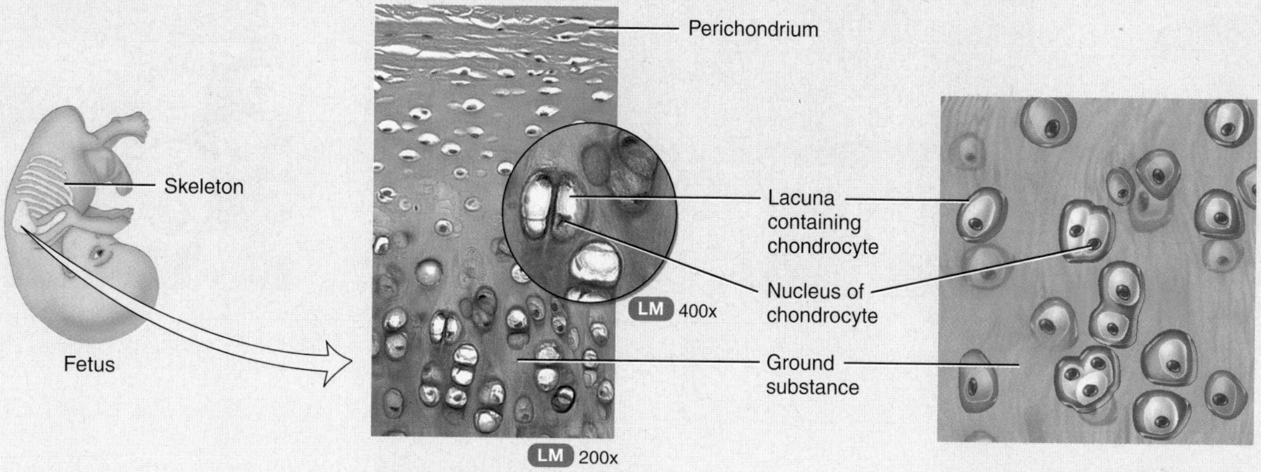

Perichondrium

Skeleton

Fetus

Lacuna containing chondrocyte

Nucleus of chondrocyte

LM 400x

Ground substance

LM 200x

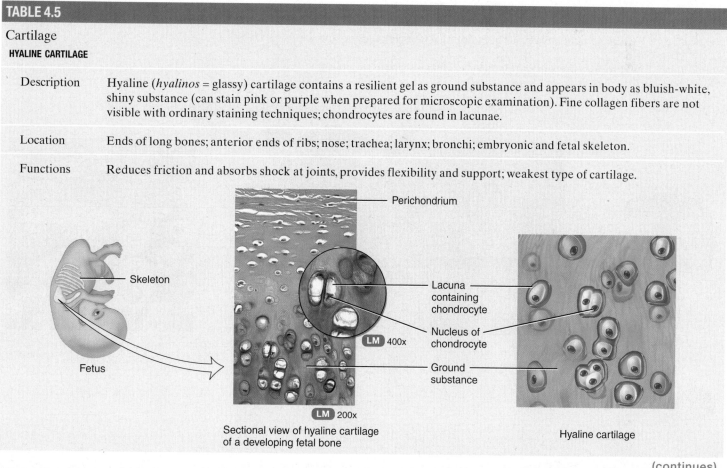

Sectional view of hyaline cartilage of a developing fetal bone

Hyaline cartilage

(continues)

**TABLE 4.5 CONTINUED**

**FIBROCARTILAGE**

| | |
|---|---|
| Description | Chondrocytes scattered among clearly visible thick bundles of collagen fibers within extracellular matrix. |
| Location | Pubic symphysis (where hip bones join anteriorly); intervertebral discs, menisci (cartilage pads) of knee; portions of tendons. |
| Functions | Support and joining structures together. Strength and rigidity make it the strongest type of cartilage. |

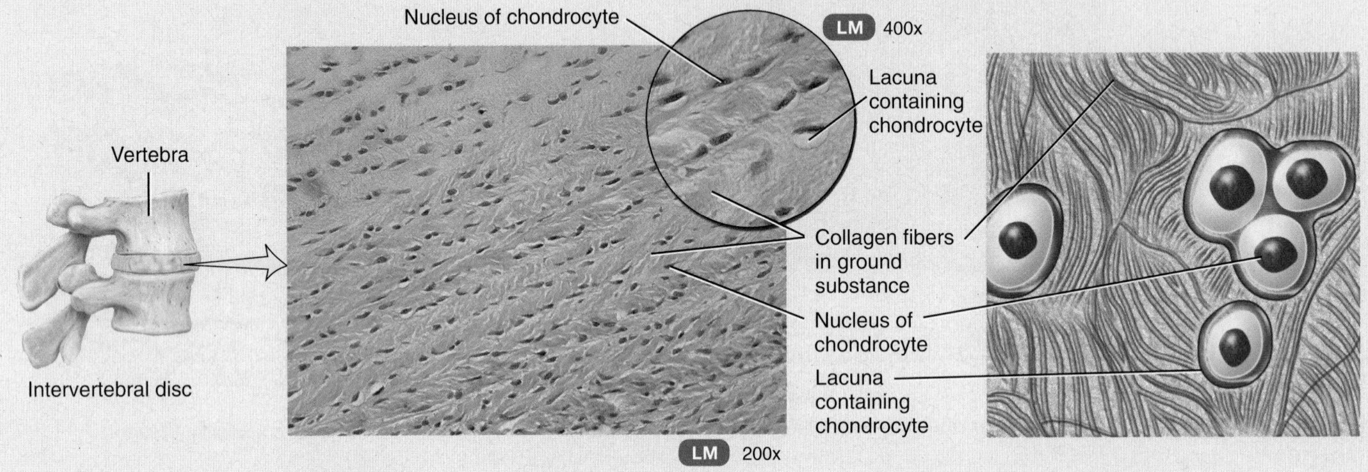


Sectional view of fibrocartilage of intervertebral disc

Fibrocartilage

**ELASTIC CARTILAGE**

| | |
|---|---|
| Description | Chondrocytes in threadlike network of elastic fibers within extracellular matrix. |
| Location | Epiglottis of larynx; part of external ear (auricle); auditory tubes. |
| Functions | Provides strength and elasticity; maintains shape of structures. |

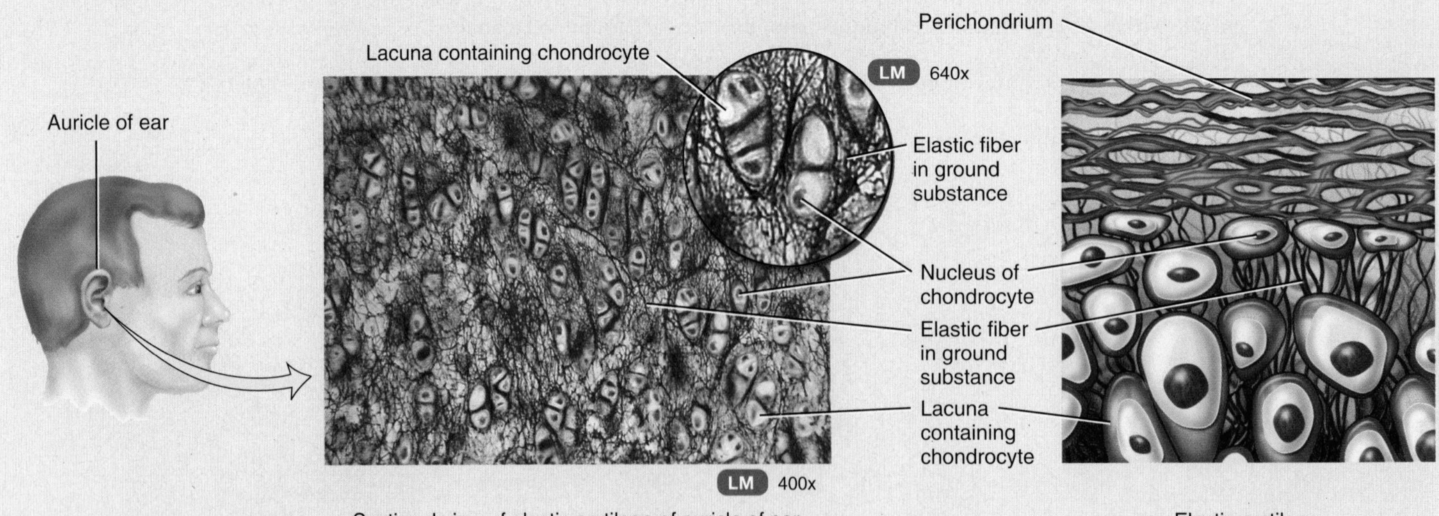


Sectional view of elastic cartilage of auricle of ear

Elastic cartilage

tissues. The extracellular matrix of cartilage forms a strong, firm material that resists tension (stretching), compression (squeezing), and shear (pushing in opposite directions). The chondroitin sulfate in the extracellular matrix is largely responsible for cartilage's resilience (ability to assume its original shape after deformation). Because of these properties, cartilage plays an important role as a support tissue in the body. It is also a precursor to bone, forming almost the entire embryonic skeleton. Though bone gradually replaces cartilage during further development, cartilage persists after birth as the growth plates within bone that allow bones to increase in length during the growing years. Cartilage also persists throughout life as the lubricated articular surfaces of most joints.

Like other connective tissues, cartilage has few cells and large quantities of extracellular matrix. It differs from other connective tissues, however, in not having nerves or blood vessels in its extracellular matrix. Because cartilage is avascular, it heals slowly following an injury. The cells of mature cartilage, called **chondrocytes** (KON-drō-sīts; *chondro-* = cartilage), occur singly or in groups within spaces called **lacunae** (la-KOO-nē = little lakes; singular is *lacuna,* pronounced la-KOO-na) in the extracellular matrix.

There are three types of cartilage: hyaline cartilage, fibrocartilage, and elastic cartilage (Table 4.5). The locations of cartilage are summarized in Figure 4.6.

FIGURE 4.6 **Cartilage locations in the adult.**

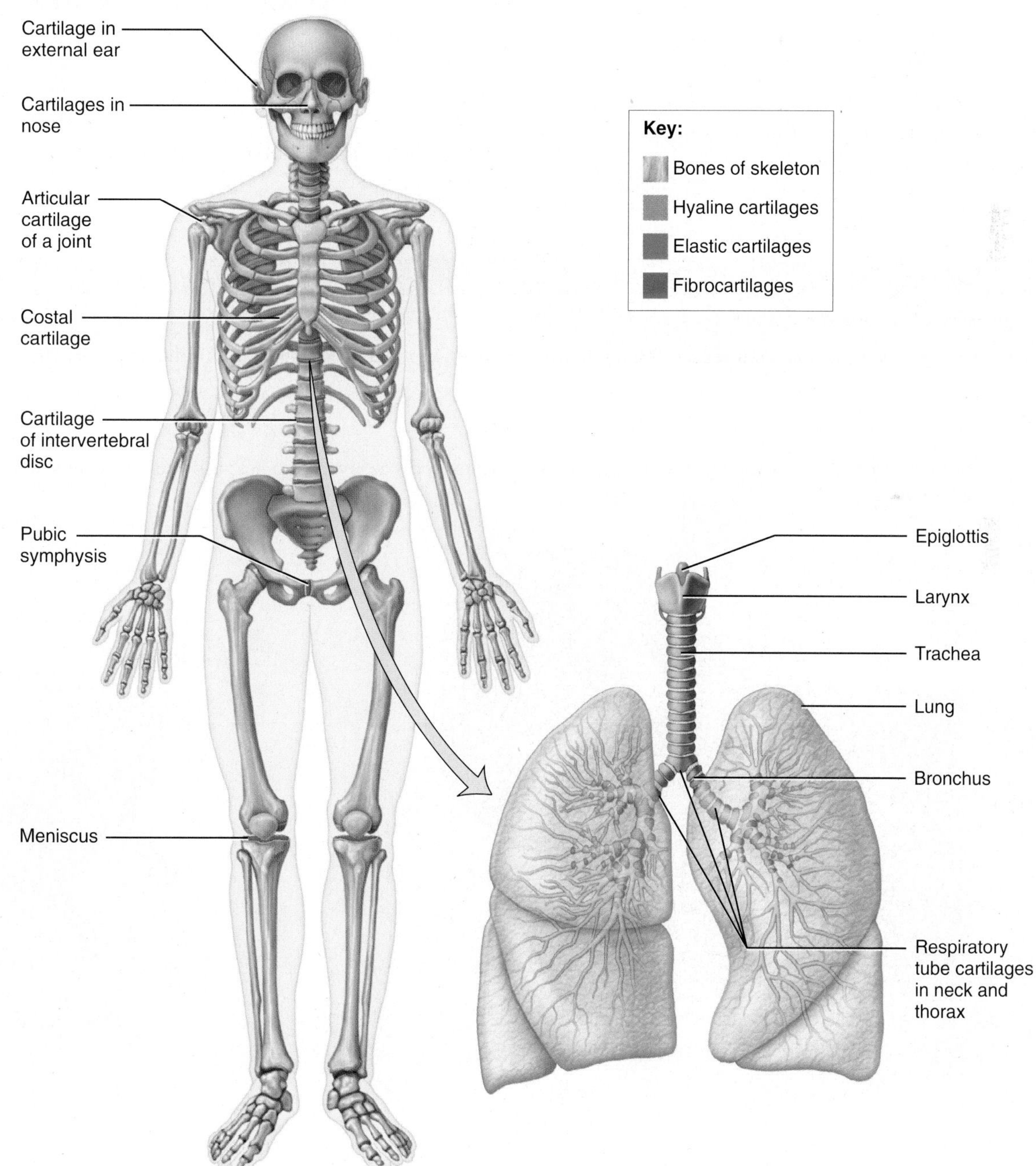


**Cartilage provides flexible, resilient support.**

### Osseous Tissue (Bone Tissue)

Cartilage, joints, and bones make up the skeletal system. The skeletal system supports soft tissues, protects delicate structures, and works with skeletal muscles to generate movement. Bones store calcium and phosphorus, produce blood cells, and store triglycerides. Bones are organs composed of primarily **osseous tissue** (OS-ē-us). The extracellular matrix of osseous tissue consists of mineral salts (mostly calcium and phosphates), which give bone its hardness and compressive strength, and collagen fibers, which give bone its tensile strength. The extracellular matrix is arranged in concentric rings or lattice-like thin columns (Table 4.6). Mature bone cells, called **osteocytes**, are located within lacunae in the extracellular matrix. Chapter 6 presents the histology of osseous tissue in more detail.

### Liquid Connective Tissue

In a liquid connective tissue, the cells are suspended in a liquid extracellular matrix. The two types of liquid connective tissues are blood and lymph.

Blood. Blood is a connective tissue with a liquid extracellular matrix and formed elements. The extracellular matrix is called **blood plasma**, and consists mostly of water with a wide variety of dissolved substances—nutrients, wastes, enzymes, plasma proteins, hormones, respiratory gases, and ions. Suspended in the blood plasma are **formed elements**—red blood cells, white blood cells, and platelets (Table 4.7). **Red blood cells** transport oxygen to body cells and remove carbon dioxide from them. **White blood cells** are involved in phagocytosis, immunity, and allergic reactions. **Platelets** (PLĀT-lets) participate in blood clotting. The details of blood are considered in Chapter 18.

Lymph. **Lymph** is a connective tissue that flows in lymphatic vessels. It consists of several types of cells in a clear liquid extracellular matrix that is similar to blood plasma, but with much less protein. The composition of lymph varies from one part of the body to another. For example, lymph leaving lymph nodes includes many lymphocytes, a type of white blood cell, but lymph from the small intestine has a high content of newly absorbed dietary lipids. The details of lymph are considered in Chapter 21.

**TABLE 4.6**

Mature Connective Tissues: Osseous Tissue

| | |
|---|---|
| Description | Osteocytes in lacunae and an extracellular matrix arranged in concentric rings or thin columns. |
| Location | Bones. |
| Functions | Support, protection, storage; houses blood-forming tissue; serves as levers that act with muscle tissue to enable movement. |

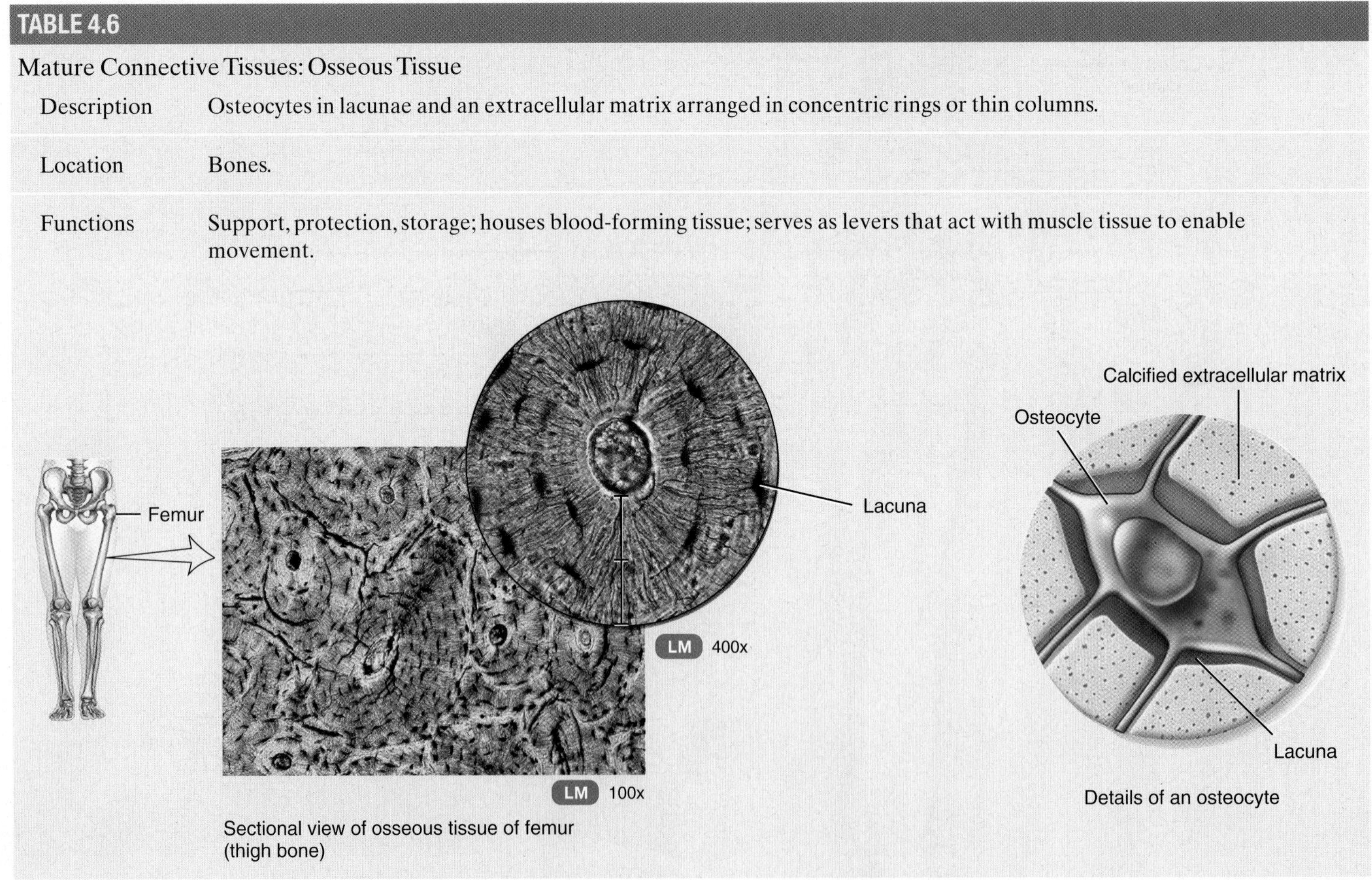


Sectional view of osseous tissue of femur (thigh bone)

Details of an osteocyte

**TABLE 4.7**

Mature Connective Tissues: Blood

| | |
|---|---|
| Description | Blood plasma and formed elements: red blood cells, white blood cells, platelets. |
| Location | Within blood vessels; within chambers of heart. |
| Functions | Red blood cells transport oxygen and some carbon dioxide; white blood cells carry on phagocytosis and are involved in allergic reactions and immune system responses; platelets are essential for blood clotting. |

White blood cell
Blood plasma
Platelet
Red blood cell
LM 630x
Red blood cells
White blood cells
Platelets

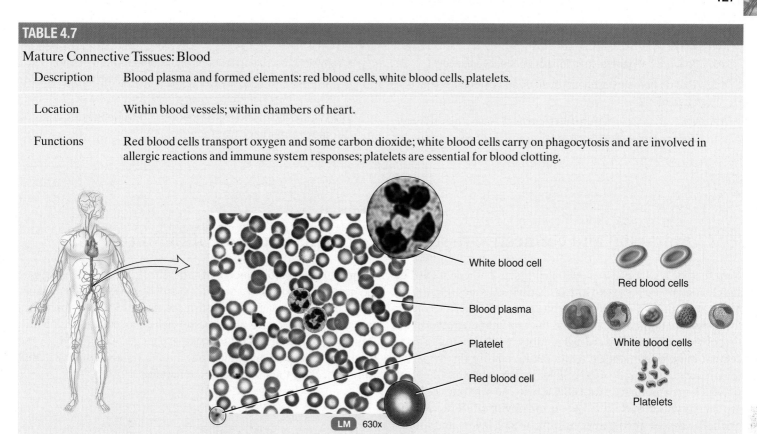

Blood in blood vessels

Blood smear (all enlargements are 1500x)

## CLINICAL CONNECTION | *Tissue Engineering*

**Tissue engineering** is a technology that combines synthetic material with cells and has allowed scientists to grow new tissues in the laboratory to replace damaged tissues in the body. Tissue engineers have already developed laboratory-grown versions of skin and cartilage using scaffolding beds of biodegradable synthetic materials or collagen as substrates that permit body cells to be cultured. As the cells divide and assemble, the scaffolding degrades; the new, permanent tissue is then implanted in the patient. Other structures currently under development include bones, tendons, heart valves, bone marrow, and intestines. Work is also under way to develop insulin-producing cells for diabetics, dopamine-producing cells for Parkinson's disease patients, and even entire livers and kidneys.

**✓ CHECKPOINT**

**10.** What are the general functions of connective tissue?

**11.** What do fibroblasts, chondroblasts, and osteoblasts have in common?

**12.** Which connective tissue cells store fats? Have a role in fighting infection (immunity)? Secrete fibers and ground substance?

**13.** Which fibers keep tendons and ligaments from pulling apart? Form a netlike internal framework for filtering organs such as the spleen?

**14.** Which kind of cartilage is found in intervertebral discs? In the external ear? In the nose?

## 4.5 Epithelial and connective tissues have obvious structural differences.

Now that we have reviewed epithelial tissues and connective tissues, let's compare these two widely distributed tissues (Figure 4.7). Major structural differences between epithelial tissue and connective tissue are immediately obvious under a light microscope. The first obvious difference is the cell number in relation to the extracellular matrix. In epithelial tissue, many cells are tightly packed together with little or no extracellular matrix, while in connective tissue a large amount of extracellular material separates cells that are usually widely scattered. The second obvious difference is that epithelial tissues almost always form surface layers and are not covered by another tissue. An exception is the epithelial lining of blood vessels where blood constantly passes over the epithelium. Another key difference is that epithelial tissue has no blood vessels, while most connective tissues have significant networks of blood vessels. While these structural distinctions account for some of the major functional differences between these tissue types, they also lead to a common bond. Because epithelial tissues lack blood vessels and form surfaces, they are always found immediately adjacent to blood-vessel-rich connective tissues, which enables epithelial tissues to make exchanges with blood for the delivery of oxygen and nutrients and the removal of wastes that are critical for their survival and function.

**✓ CHECKPOINT**

**15.** What relationship between epithelial tissue and connective tissue is important for the survival and function of epithelial tissues?

FIGURE 4.7 **Comparison between epithelial tissue and connective tissue.**

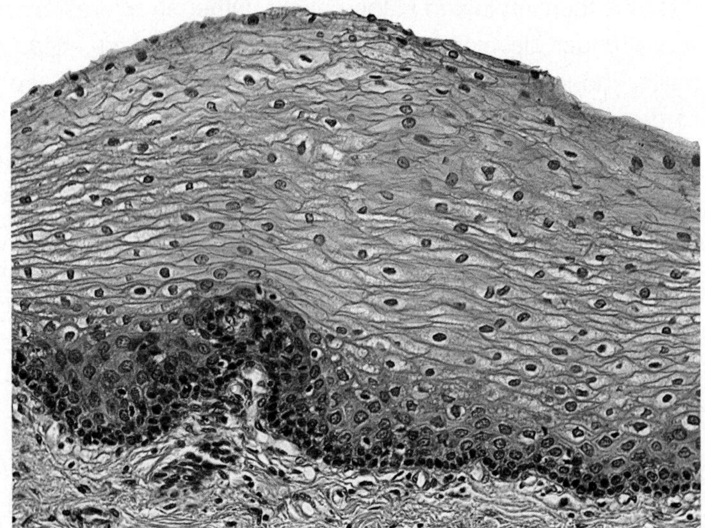

(a) Epithelial tissue with many cells tightly packed together and little to no extracellular matrix

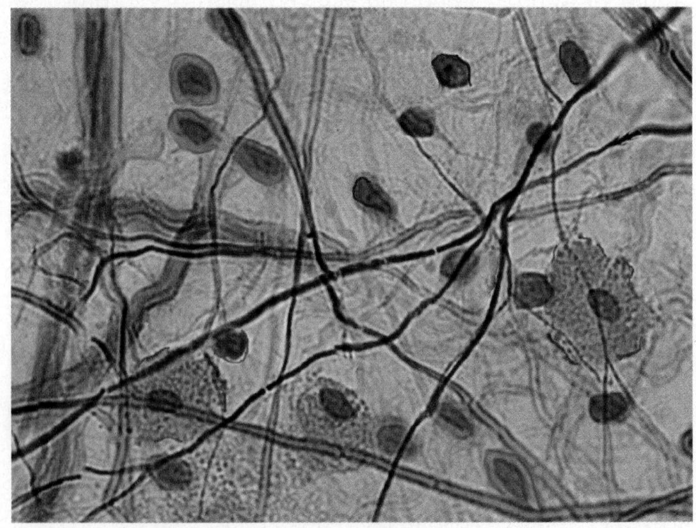

(b) Connective tissue with a few scattered cells surrounded by large amounts of extracellular matrix

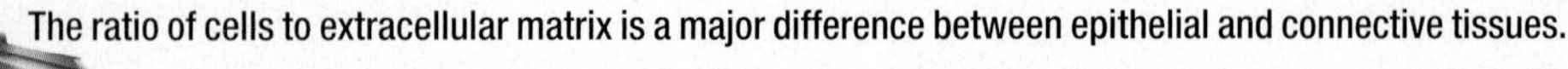

**The ratio of cells to extracellular matrix is a major difference between epithelial and connective tissues.**

## RETURN TO John Doe's Story

The sheriff and several deputies had cordoned off the area of the crash. The medical examiner, FAA inspectors, and EMS personnel had all arrived, forming a small, busy crowd. The medical examiner, Dr. Greg Diego, didn't waste any time. He photographed the crash scene and body from various angles, performed a quick examination of the body, and collected samples from the body and the area around it.

"Well, Doc, any idea how long he's been here?" the sheriff asked, holding his nose with one hand.

"I'll have to perform an autopsy to determine exact time of death. Been dead a while, several weeks at least. He was probably alive when he crashed, though. Bruises don't form on dead people. Any idea who he is?"

The sheriff sighed. "Not yet. The FAA inspectors made some calls; they're checking the registration numbers."

Dr. Diego nodded his head toward the plane. "Can you give me a hand getting the body out?"

As Dr. Diego grasped the body by the wrists, he noticed that the skin slipped and pulled away from the underlying connective tissue. The mountain weather and decay of the corpse had taken their toll on the tissues.

Back at the morgue, Dr. Diego examined insect larvae he had taken from the body at the crash site. Their developmental stage told Diego that the John Doe had been dead for approximately 14 to 16 days.

Dr. Diego's assistant helped position the dead man on the autopsy table. The man's skin was discolored to a gray-bluish tinge, indicating early decay. Diego noted no external wounds and some bruising around the face and neck, indicating that the victim sustained injuries in the crash.

He made a few notes on his clipboard, then reached for what looked like a long, very sharp bread knife. "Let's see what we can find out on the inside."

A. *What types of tissues form the surface of the dead man's skin? How does the structure of the skin normally protect the body?*

B. *As the body was being moved, skin slid away from the underlying connective tissue. What is the normal function of connective tissue? Would the slippage of the dead man's skin be due to decomposition of cell junctions or underlying connective tissues?*

C. *Would the bruises on the dead man have formed in the epithelia or in the underlying connective tissue? Explain your answer.*

## 4.6 Membranes cover the surface of the body, line body cavities, and cover organs.

**Membranes** are flat sheets of pliable tissue that cover or line a part of the body. The majority of membranes are **epithelial membranes** consisting of an epithelial layer and an underlying connective tissue layer. The principal epithelial membranes of the body are mucous membranes, serous membranes, and the cutaneous membrane, or skin. Another type of membrane, the **synovial membrane**, lines joints and contains connective tissue but no epithelium.

### Epithelial Membranes

#### Mucous Membranes

A **mucous membrane** or **mucosa** (mū-KŌ-sa) lines a body cavity that opens directly to the exterior. Mucous membranes line the gastrointestinal tract, respiratory tract, reproductive tract, and much of the urinary tract. They consist of both a lining layer of epithelium and an underlying layer of connective tissue (Figure 4.8a).

The epithelial layer of a mucous membrane is an important feature of the body's defense mechanisms because it is a barrier that microbes and other pathogens have difficulty penetrating. Usually, tight junctions connect the cells, so materials cannot leak in between them. Goblet cells and other cells of the epithelial layer of a mucous membrane secrete mucus, and this slippery fluid prevents the cavities from drying out. Mucus also traps particles in the respiratory passageways and lubricates food as it moves through the gastrointestinal tract. In addition, the epithelial layer secretes some of the enzymes needed for digestion and is the site of food and fluid absorption in the gastrointestinal tract. The epithelia of mucous membranes vary greatly in different parts of the body. For example, the mucous membrane of the small intestine is nonciliated

**FIGURE 4.8 Membranes.**

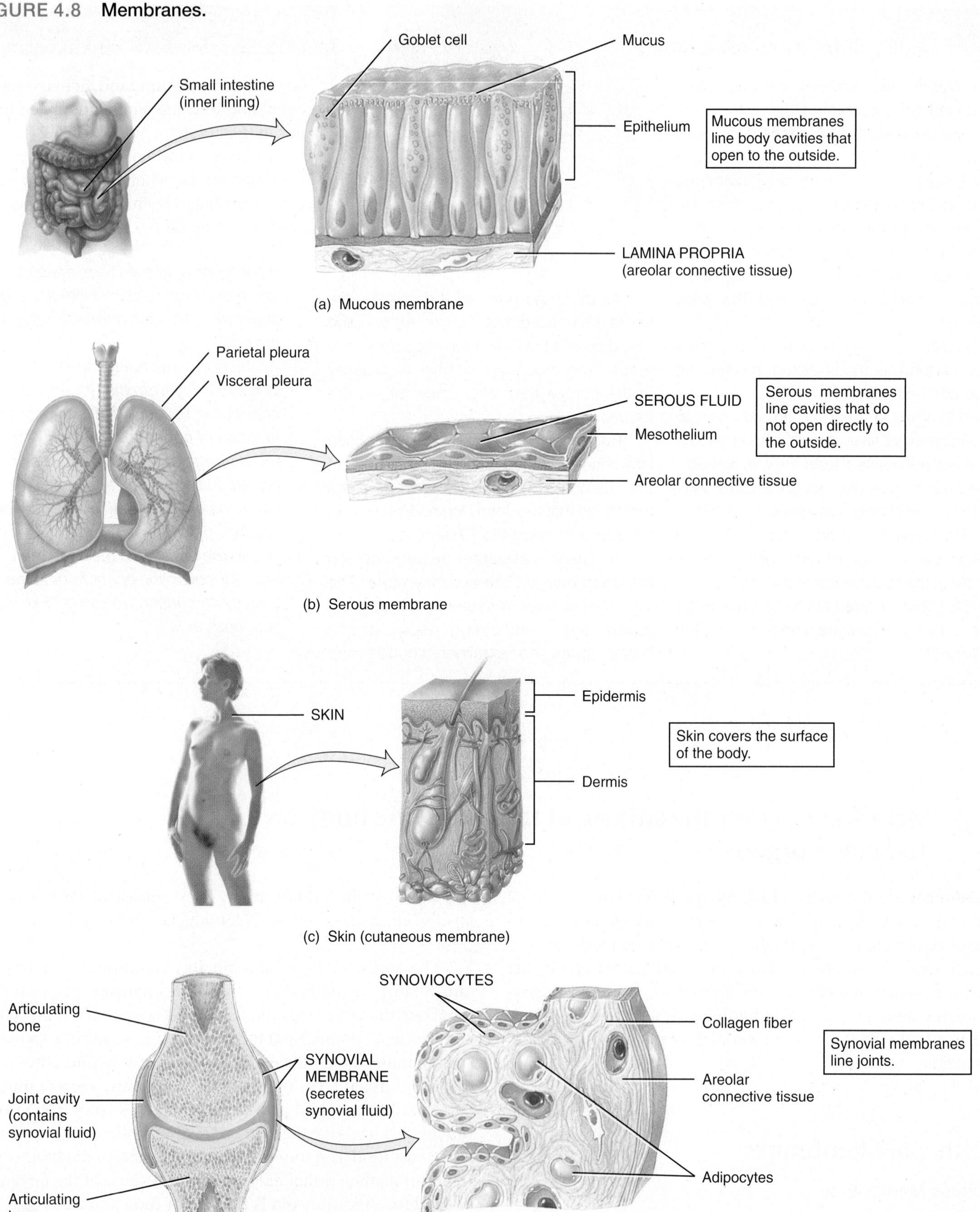


**A membrane is a flat sheet of pliable tissue that covers or lines a part of the body.**

simple columnar epithelium, and the large airways to the lungs are pseudostratified ciliated columnar epithelium (see Table 4.1).

The connective tissue layer of a mucous membrane is areolar connective tissue and is called the **lamina propria** (LAM-i-na PRŌ-prē-a; *propria* = one's own, referring here to the mucous membrane's own layer). The lamina propria supports the epithelium, binds it to the underlying structures, and affords some protection for underlying structures. It also holds blood vessels in place and is the vascular source for the overlying epithelium.

### Serous Membranes

A **serous membrane** (SĒR-us; *serous* = watery) or **serosa** lines a body cavity that does not open directly to the exterior (thoracic or abdominal cavities), and it covers the organs that are within the cavity. Serous membranes consist of areolar connective tissue covered by mesothelium (simple squamous epithelium) (Figure 4.8b). Serous membranes have two layers: The layer lining the cavity wall is called the **parietal layer** (pa-RĪ-e-tal; *pariet-* = wall); the layer that covers the organs inside the cavity is the **visceral layer** (*viscer-* = body organ). The mesothelium of a serous membrane secretes **serous fluid**, a watery lubricant that allows organs to glide easily over one another or to slide against the walls of cavities.

The serous membrane lining the thoracic cavity and covering the lungs is the **pleura** (see its layers, the visceral pleura and parietal pleura, in Figure 1.10a). The serous membrane lining the heart cavity and covering the heart is the **pericardium** (see its layers, the visceral pericardium and parietal pericardium, in Figure 1.10a). The serous membrane lining the abdominal cavity and covering the abdominal organs is the **peritoneum**.

### Cutaneous Membrane

The **cutaneous membrane** or **skin** covers the surface of the body and consists of a superficial portion called the *epidermis* and a deeper portion called the *dermis* (Figure 4.8c). The skin is described in detail in Chapter 5.

## Synovial Membranes

**Synovial membranes** (sin-Ō-vē-al; *syn-* = together, referring here to a place where bones come together) line the cavities of freely movable joints (joint cavities). Like serous membranes, synovial membranes line structures that do not open to the exterior. Unlike mucous, serous, and cutaneous membranes, they lack an epithelium and are therefore not epithelial membranes. Synovial membranes are composed of a layer of cells called **synoviocytes** (si-NŌ-vē-ō-sīts), which are closer to the synovial cavity (space between the bones) and a deeper layer of areolar connective tissue rich with adipocytes (Figure 4.8d). Synoviocytes secrete synovial fluid. **Synovial fluid** lubricates and nourishes the cartilage covering the bones at movable joints and contains macrophages that remove microbes and debris from the joint cavity.

**✓ CHECKPOINT**

**16.** What is an epithelial membrane?

**17.** Describe the location in the body and the function of each type of membrane.

**18.** What functions are served by mucus? Which membranes secrete mucus? Which cells produce mucus?

**19.** Which part of the serosa covers the stomach and intestines? What is its function?

**20.** Where are synovial membranes found? What are the functions of the synovial fluid they secrete?

## 4.7 Muscle tissue generates the physical force needed to make body structures move.

**Muscle tissue** consists of elongated cells called *muscle fibers* that produces body movements by contracting. Based on its location and certain structural and functional features, muscle tissue is classified into three types: skeletal, cardiac, and smooth (Table 4.8). Chapter 10 provides a detailed discussion of muscle tissue.

**TABLE 4.8**

## Muscle Tissues

### SKELETAL MUSCLE TISSUE

| | |
|---|---|
| Description | Long, cylindrical, striated fibers (*striations* are alternating light and dark bands within fibers that are visible under a light microscope). Skeletal muscle fibers vary greatly in length, from a few centimeters in short muscles to 30–40 cm (about 12–16 in.) in the longest muscles. A muscle fiber is roughly cylindrical and has many nuclei at the periphery of the cell. Skeletal muscle is *voluntary* because it can be made to contract or relax by conscious control. |
| Location | Skeletal muscles. |
| Function | Motion, posture, heat production, protection. |

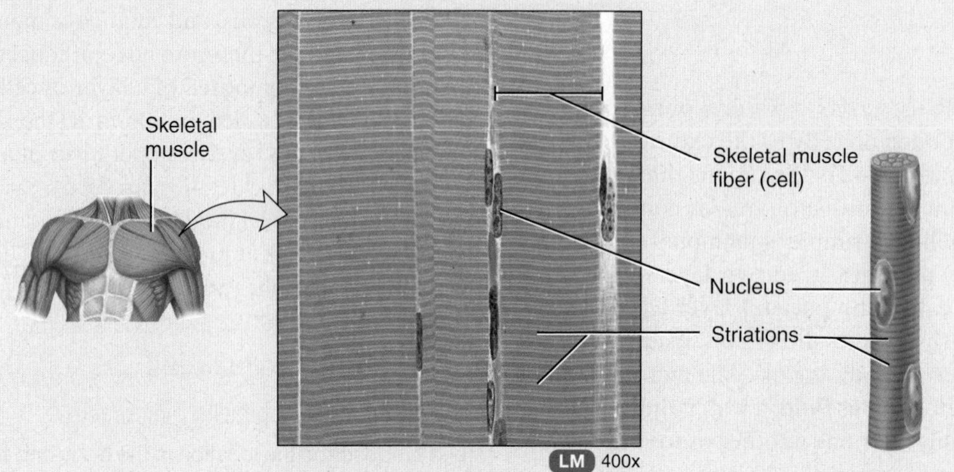


Longitudinal section of skeletal muscle tissue

Skeletal muscle fiber

### CARDIAC MUSCLE TISSUE

| | |
|---|---|
| Description | Branched, striated fibers with usually only one centrally located nucleus (occasionally two). Attached end to end by transverse thickenings of plasma membrane called *intercalated discs* (in-TER-ka-lāt-ed; *intercalate* = to insert between), which contain desmosomes and gap junctions. Desmosomes strengthen tissue and hold fibers together during vigorous contractions. Gap junctions provide route for quick conduction of electrical signals (muscle action potentials) throughout heart. Cardiac muscle is *involuntary*; its contractions are not consciously controlled. |
| Location | Heart wall. |
| Function | Pumps blood to all parts of body. |

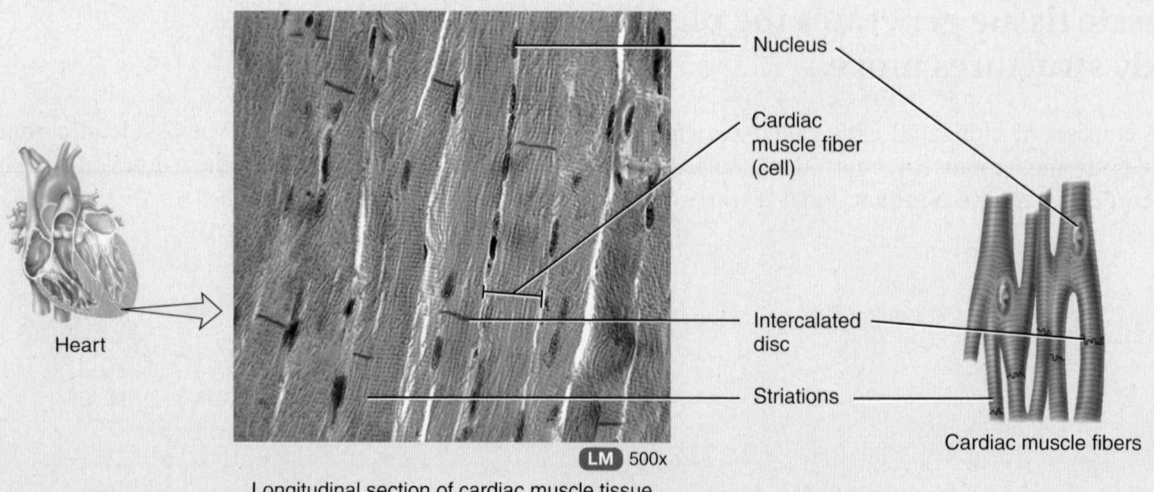


Longitudinal section of cardiac muscle tissue

Cardiac muscle fibers

**SMOOTH MUSCLE TISSUE**

| | |
|---|---|
| Description | Fibers usually *involuntary,* nonstriated (lack striations, hence the term *smooth)*. Smooth muscle fiber is a small spindle-shaped cell thickest in the middle, tapering at each end, and containing a single, centrally located nucleus. Gap junctions connect many individual fibers in some smooth muscle tissues (for example, in intestinal wall) to allow powerful contractions as many muscle fibers contract in unison. Where gap junctions are absent, such as iris of eye, smooth muscle fibers contract individually, like skeletal muscle fibers. |
| Location | Iris of eyes; walls of hollow organs such as blood vessels, stomach, intestines, urinary bladder, and uterus. |
| Function | Motion (propulsion of foods through gastrointestinal tract, contraction of stomach and urinary bladder). |

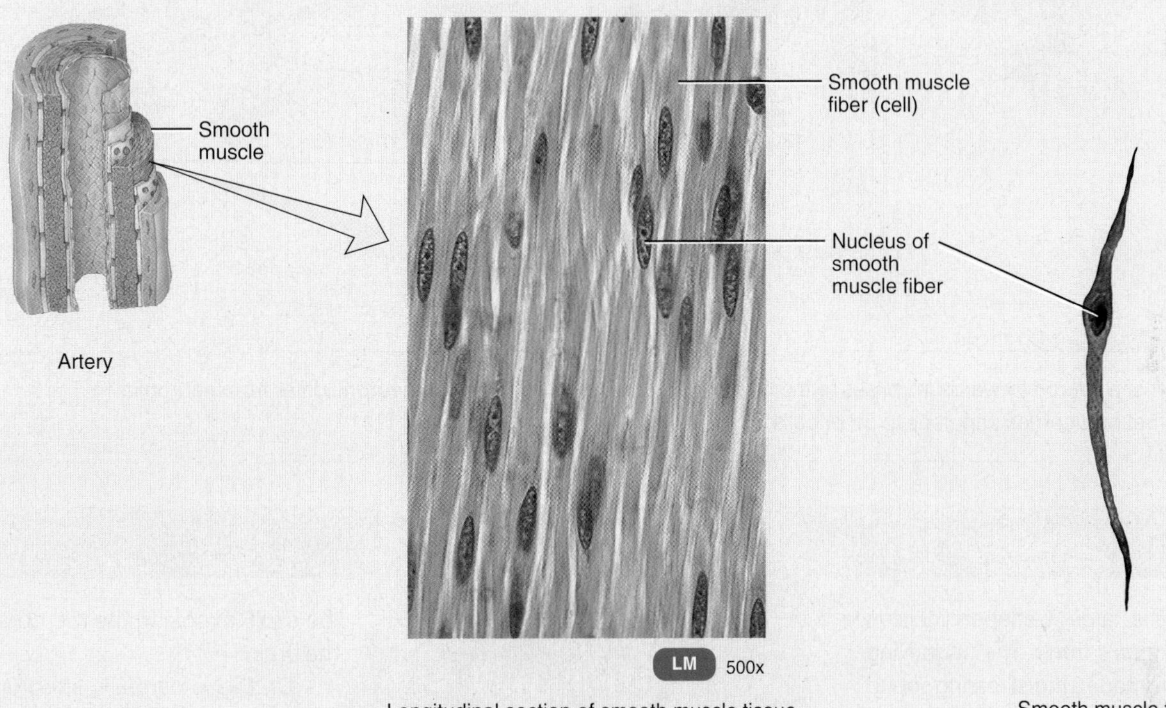


Longitudinal section of smooth muscle tissue

Smooth muscle fiber

## ✓ CHECKPOINT

**21.** List the functions of the three types of muscle tissue.

**22.** If you were presented with prepared slides, what characteristics could you use to identify skeletal muscle tissue? Cardiac muscle tissue? Smooth muscle tissue?

**23.** Which muscle tissue moves food through the intestines? Produces heart contractions? Moves bones?

**24.** What are the functions of intercalated discs?

# 4.8 Nervous tissue consists of neurons and neuroglia.

Despite the awesome complexity of the nervous system, it consists of only two principal types of cells: neurons and neuroglia. **Neurons** (NOO-rons; *neuro-* = nerve) are sensitive to various stimuli. They convert stimuli into electrical signals called **action potentials (impulses)** and conduct these impulses to other neurons, to muscle tissue, or to glands. Most neurons consist of three basic parts: a cell body, dendrites, and axons (Table 4.9). The **cell body** contains the nucleus and other organelles. **Dendrites** (*dendr-* = tree) are tapering, highly branched, and usually short cell processes (extensions). They are the major receiving or input portion of a neuron. The **axon** (*axo-* = axis) of a neuron is a single, thin, cylindrical process that may be very long. It is the output portion of a neuron, conducting impulses toward another neuron or to some other tissue.

Even though **neuroglia** (noo-RŌG-lē-a; *-glia* = glue) do not generate or conduct impulses, these cells do have many important supportive functions. The detailed structure and function of neurons and neuroglia are discussed in Chapter 12.

**TABLE 4.9**

Nervous Tissue

| | |
|---|---|
| Description | Neurons (nerve cells), which consist of cell body and processes extending from cell body (one to multiple dendrites and a single axon); and neuroglia, which do not generate or conduct impulses but have other important supporting functions. |
| Location | Nervous system. |
| Function | Exhibits sensitivity to various types of stimuli; converts stimuli into impulses (action potentials); conducts impulses to other neurons, muscle fibers, or glands. |

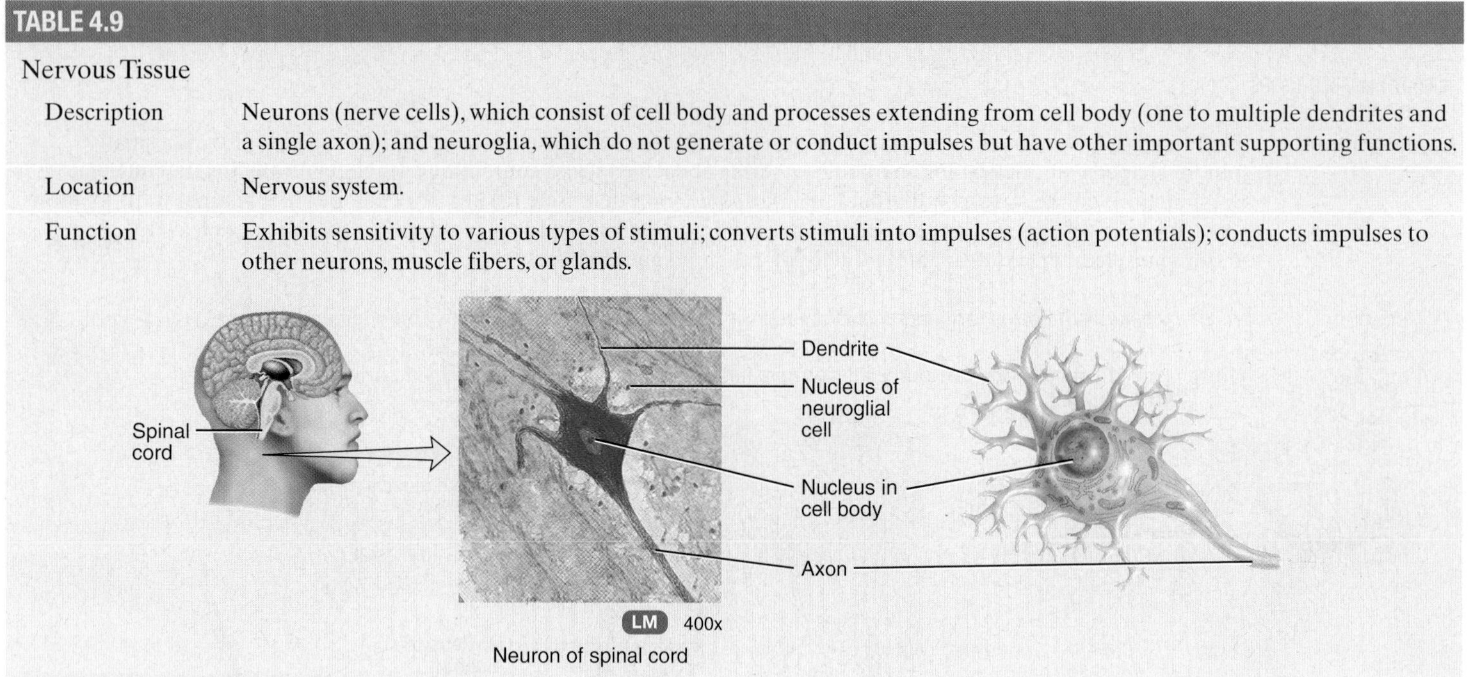

Neuron of spinal cord

## ✓ CHECKPOINT

25. What part of a neuron forwards impulses to the neuron cell body? What part carries impulses to other cells?

26. How do neurons differ from neuroglia?

## RETURN TO John Doe's Story

Dr. Diego made a large Y-shaped incision along the dead man's trunk. The underlying connective tissue made a moist tearing sound as he peeled back the skin and underlying muscles of the chest and abdominal wall.

Using an electric saw, Dr. Diego made a cut from the bottom of the rib cage up each side of the body, severing the ribs, muscles, and connective tissue. Lifting off the chest plate, he exposed the membranes covering the heart and lungs.

Dr. Diego grasped the membrane surrounding the heart, tearing it to expose the organ. He cut the pulmonary trunk vessels where blood had once traveled to the lungs. "The thoracic cavity is full of blood. Tentatively, I'd say he died as a result of internal bleeding sustained from trauma in the crash." Dr. Diego's assistant leaned over the body to look. "Plane versus mountain, mountain wins. But what caused him to lose control of the plane and crash?"

Dr. Diego grasped the windpipe with one hand and cut above the larynx, severing the esophagus and trachea in one smooth motion. The hyaline cartilage rings of the trachea provided a relatively stable handle to grip as he pulled the organs down and out of the chest.

Dr. Diego cut the dome-shaped diaphragm next and pulled the abdominal organs out, handing them to his assistant. Clotted blood dripped off the organs as he carried the tissues to a dissecting table near the sink.

Dr. Diego's assistant began dissecting the mass of organs, while Diego began the process of removing John Doe's brain. With a quick swipe of his scalpel, Diego cut through the skin of the head and pulled the skin forward over the man's face. A reciprocating bone saw quickly sawed through the skull, exposing the membranes around the brain.

Dr. Diego carefully lifted out the brain, severing the spinal cord just below the brain stem, and suspended it in a large jar filled with preservative.

D. *Why do the cartilaginous structures in the airway make a good "handle" to grab onto in removing the lungs and heart?*

E. *As the medical examiner pulls out the thoracic and abdominal organs, the membranes within the body cavity provide little resistance. Why? What are these membranes called and what are they made of?*

F. *The dead man's body cavities contain clotted blood released when organs ruptured during the plane crash. What cells might you expect to find in blood? Which component of blood participates in blood clot formation?*

G. *Dr. Diego was careful to preserve the brain tissue, and didn't dissect and analyze it immediately during his preliminary autopsy. Why is nervous tissue soft and fragile?*

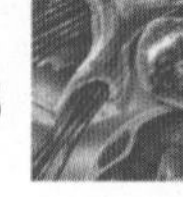

## 4.9 The ability of an injured tissue to repair itself depends on the extent of damage and the regenerative ability of the injured tissue.

Tissue repair is the replacement of worn-out, damaged, or dead cells. In adults, each of the four basic tissue types (epithelial, connective, muscle, and nervous) has a different capacity for replenishing cells lost by injury, disease, or other processes.

Epithelial cells, which endure considerable wear and tear (and even injury) in some locations, have a continuous capacity for renewal. In some cases, immature, undifferentiated cells called **stem cells** divide to replace lost or damaged cells. For example, stem cells residing in the epithelia of the skin replenish cells cast off from the apical layer, and stem cells in some areas of bone marrow continually provide new blood cells and platelets.

Some connective tissues also have a continuous capacity for renewal. One example is osseous tissue, which has an ample blood supply that allows broken bones to heal. Other connective tissues, such as cartilage, replenish cells more slowly, in part because of a smaller blood supply.

Muscle tissue has a relatively poor capacity for renewal of lost cells. Even though skeletal muscle tissue contains stem cells, they do not divide rapidly enough to replace extensively damaged muscle fibers. Cardiac muscle tissue lacks stem cells, and existing cardiac muscle fibers do not undergo mitosis to form new cells. Recent evidence suggests that stem cells do migrate into the heart from the blood. There, they can replace a limited number of cardiac muscle fibers and endothelial cells lining heart blood vessels. Smooth muscle fibers can proliferate to some extent, but they do so much more slowly than the cells of epithelial or connective tissues.

Nervous tissue has the poorest capacity for renewal. Although experiments have revealed the presence of some stem cells in the brain, they normally do not replace damaged neurons. Discovering why this is so is a major goal of researchers who seek ways to repair nervous tissue damaged by injury or disease.

During tissue repair, new cells originate by cell division from the **stroma**, the supporting connective tissue, or from the **parenchyma** (pa-REN-ki-ma), cells that constitute the functioning part of the tissue or organ. The restoration of an injured tissue or organ to normal structure and function depends entirely on whether parenchymal cells are active in the repair process. If parenchymal cells accomplish the repair, **tissue regeneration** is possible, and a near-perfect reconstruction of the injured tissue may occur. However, if fibroblasts of the stroma are active in the repair, the fibroblasts will synthesize extracellular matrix materials that aggregate to form scar tissue, a process known as **fibrosis**. Because scar tissue is not specialized to perform the functions of the parenchymal tissue, the original function of the tissue or organ is impaired.

When tissue damage is extensive, as in large, open wounds, both the parenchyma and stroma are active in repair. During this process, fibroblasts manufacture new collagen fibers to provide structural strength. Blood capillaries sprout new buds to supply needed materials to the healing tissue. The actively growing connective tissue, called **granulation tissue** (gran-ū-LĀ-shun), forms across a wound to provide a framework (stroma) that supports the epithelial cells that migrate into the open area and fill it. The newly formed granulation tissue also secretes a fluid that kills bacteria.

Tissue repair is affected by nutrition and blood circulation. Nutrition is vital because the healing process places a great demand on the body's store of nutrients. Adequate protein in the diet is essential because most of the structural components of a tissue are proteins. Several vitamins also play a direct role in wound healing and tissue repair. For example, vitamin C directly affects the normal production and maintenance of extracellular matrix materials, especially collagen, and promotes the formation of new blood vessels. In a person with vitamin C deficiency, even superficial wounds fail to heal, and the walls of the blood vessels become fragile and are easily ruptured. Proper blood circulation is essential to transport oxygen, nutrients, antibodies, and many defensive cells to the injured site during tissue repair. Blood also plays an important role in the removal of bacteria, foreign bodies, and debris, elements that would otherwise interfere with healing.

### ✓ CHECKPOINT

27. Rank epithelial, connective, muscle, and nervous tissues in order from most to least able to fully repair.
28. Why does osseous tissue heal faster than cartilage?
29. What is the difference between the parenchyma and the stroma?
30. How do stromal and parenchymal tissue repair differ?
31. What is the importance of granulation tissue?

## John Doe's Story — EPILOGUE AND DISCUSSION

Two weeks have passed since the autopsy, enough time for John Doe's brain to harden in the fixative. Dr. Diego can now finish his autopsy. He hadn't found any evidence of foul play, and no specific cause of death in the John Doe case. The sheriff had called a few days earlier and said the plane had not failed mechanically. So they were leaning toward pilot error as a cause of the crash.

Dr. Diego put on rubber gloves and lifted the jar containing John Doe's brain from a shelf. Opening the jar, he carried the brain over to the metal dissecting table and sectioned it into 1-cm slices. As the pieces of brain tissue began to form stacks, he looked at one of them closely and said to himself, "Well! Now there's something you don't see every day." Circular chunks of dead fibrous tissue of various sizes were scattered throughout the man's brain like holes in a sponge. Some of the circular cysts were hard nodules. One nodule had been cut through, showing a small white mass inside a hollow shell.

Dr. Diego recognized the pathology. This was neurocysticerosis, the result of an intracranial infection of a parasitic tapeworm, *Taenia solium*, within the brain. It wasn't common in the United States; on the other hand, it wasn't unheard of. Certain forms of the parasite can be ingested in undercooked pork or poor sanitary practices can result in ingestion of feces from an infected human. Because humans are not the normal host for the parasite, we seem to confuse the beast. In unfamiliar territory, the tapeworm can escape through the intestinal wall and end up in the brain, forming cysts. The body responds to the invasion by sealing the tapeworm within a shell of scar tissue that eventually hardens into scarlike nodules.

John Doe most likely had lost consciousness and crashed. "Well, it looks like the mystery is solved," said Dr. Diego.

H. *What types of cells would be found in the dead man's brain? Why didn't the man's brain regenerate new tissue to replace the tissue damaged by the parasite?*

I. *How is an understanding of tissues essential to the job performance of a medical professional like Dr. Diego?*

# Concept and Resource Summary

| Concept | Resources  |
|---|---|
| **Introduction**<br>1. A **tissue** is a group of cells that have the same specialized function and embryonic origin.<br>2. The study of tissues is called **histology**.<br>3. A physician who studies cells and tissues is a **pathologist**. | |
| **Concept 4.1** Human body tissues can be classified as epithelial, connective, muscle, or nervous.<br>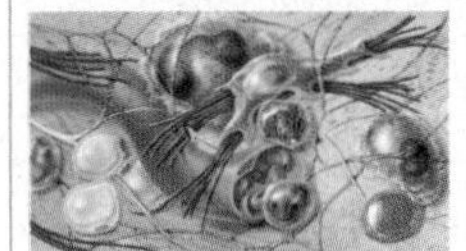<br>1. **Epithelial tissue** covers body surfaces, lines cavities, and forms glands.<br>2. **Connective tissue** binds organs together, stores energy, and participates in body defenses.<br>3. **Muscle tissue** produces the physical force for body movements.<br>4. **Nervous tissue** detects and responds to changes in internal and external body conditions. | Anatomy Overview—Tissues<br>Clinical Connection—Biopsy |

## Concept

## Resources

### Concept 4.2 Cell junctions hold cells together to form tissues.

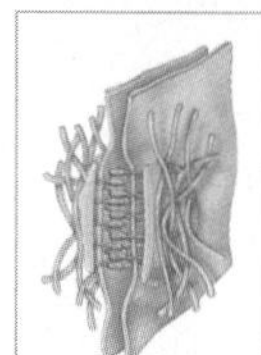

1. **Cell junctions** are points of contact between adjacent plasma membranes.
2. **Tight junctions** form fluid-tight seals between cells; **desmosomes** anchor cells to one another or to the basement membrane, and **gap junctions** permit electrical and chemical signals to pass between cells.

Animation—Intercellular Junctions

Exercise—Getting It Together

### Concept 4.3 Epithelial tissue covers body surfaces, lines organs and body cavities, or secretes substances.

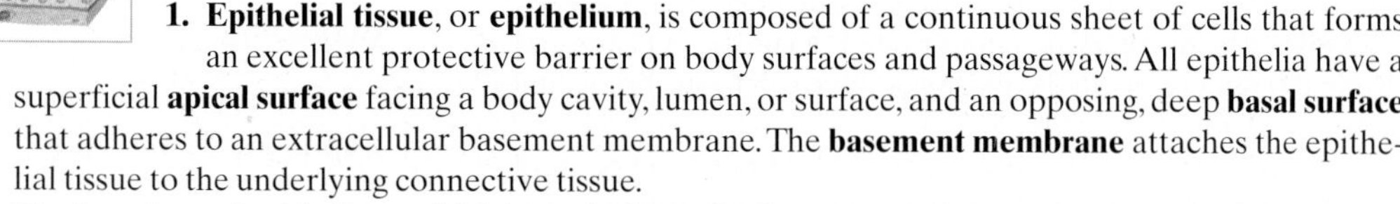

1. **Epithelial tissue**, or **epithelium**, is composed of a continuous sheet of cells that forms an excellent protective barrier on body surfaces and passageways. All epithelia have a superficial **apical surface** facing a body cavity, lumen, or surface, and an opposing, deep **basal surface** that adheres to an extracellular basement membrane. The **basement membrane** attaches the epithelial tissue to the underlying connective tissue.
2. The functions of epithelium, which is **avascular** and relies on underlying connective tissue blood vessels for obtaining nutrients and removing wastes, include protection, filtration, secretion, absorption, and excretion. There are two types of epithelia: **covering and lining epithelium** and **glandular epithelium**.
3. Two characteristics are used to classify covering and lining epithelium: arrangement of cells in layers and cell shapes. The cells of covering and lining epithelium can be arranged as a single layer called simple epithelium; two or more layers called stratified epithelium; or a single layer that appears to have multiple layers called pseudostratified epithelium. The cell shapes of covering and lining epithelia include squamous cells, cuboidal cells, columnar cells, and transitional cells. The latter change shape from cuboidal to flat in response to pressure.
4. **Simple squamous epithelium** is a single layer of broad, flat cells, located where filtration and diffusion occur. The simple squamous epithelium lining blood vessels, the heart, and lymphatic vessels is called **endothelium**; the simple squamous epithelium of serous membranes is called **mesothelium**.
5. **Simple cuboidal epithelium** is a single layer of cube-shaped cells, located where secretion and absorption occur.
6. **Simple columnar epithelium** is a single layer of tall, column-shaped cells that can be ciliated or nonciliated. Nonciliated simple columnar epithelium contains cells that have apical **microvilli** for increased absorption, and mucus-secreting **goblet cells** for lubrication. **Ciliated simple columnar epithelium** contains cells with apical cilia that sweep material across their surfaces and goblet cells.
7. **Pseudostratified columnar epithelium** is composed of a single layer of cells attached to the basement membrane but appears falsely stratified because the nuclei occur at various depths. Cells that reach up to the apical surface have cilia or are goblet cells.
8. **Stratified squamous epithelium** has flat apical layer cells and basal layer cells that continually divide, get pushed upward, die, slough off, and get replaced by new cells. It can be divided into two types: keratinized stratified squamous epithelium, which contains **keratin**, a tough, fibrous protein; and nonkeratinized stratified squamous epithelium which remains moist.
9. **Transitional epithelium** stretches in response to pressure, causing the apical cells to become flat and then return to a larger, rounded appearance when pressure subsides.
10. **Glandular epithelia** form **glands** that produce secretions onto a surface, into a duct, or into the blood. Body glands are either endocrine glands or exocrine glands. **Endocrine glands** secrete hormones that enter interstitial fluid and then the blood. The secretions of **exocrine glands** enter ducts that empty secretions onto the skin surface or into the lumen of an organ.

Anatomy Overview—Epithelial Tissues

Figure 4.4—Cell Shapes and Arrangement of Layers for Covering and Lining Epithelium

Clinical Connection—Basement Membranes and Disease

Clinical Connection—Sjögren's Syndrome

## Concept

### Concept 4.4 Connective tissue binds organs together, stores energy reserves as fat, and helps provide immunity.

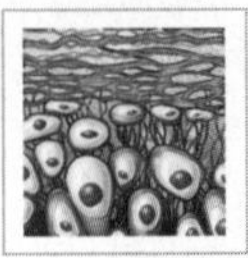

1. **Connective tissue** is widely distributed and abundant in the body, and is usually vascular. Its functions include binding, supporting, and strengthening other body tissues, and protecting, insulating, and compartmentalizing structures. The two structural components of all connective tissues are cells and an extracellular matrix. The **extracellular matrix** is composed of protein fibers embedded in a ground substance.
2. The **ground substance** supports and binds together cells embedded in it; and allows for material exchange between blood and cells. The composition of the ground substance includes water, proteins, and polysaccharides called **glycosaminoglycans**.
3. Three types of **fibers** can be embedded in the extracellular matrix: collagen fibers, elastic fibers, and reticular fibers. **Collagen fibers** are very strong, flexible, and composed of the protein collagen. They often occur in tightly packed, parallel bundles. **Elastic fibers** are composed of the protein elastin, and can stretch and recoil back to shape. **Reticular fibers** consist of collagen and form delicate, branching networks that can form a supporting framework for soft organs.
4. **Fibroblasts** are the most numerous connective tissue cells; they secrete the fibers and ground substance of the matrix. Other connective tissue cells include phagocytic **macrophages**, inflammatory **mast cells**, fat-storing **adipocytes**, and defensive **white blood cells**.
5. There are five types of **mature connective tissue:** (1) loose connective tissue, (2) dense connective tissue, (3) cartilage, (4) osseous (bone) tissue, and (5) liquid connective tissue (blood and lymph).
6. **Loose connective tissue** has loosely intertwined fibers and many cells. It includes areolar connective tissue, adipose tissue, and reticular connective tissue. **Areolar connective tissue** is widely distributed in the body. It contains various cells including fibroblasts, macrophages, mast cells, and adipocytes. Collagen, elastic, and reticular fibers are scattered throughout the tissue. **Adipose tissue** is composed of **adipocytes** that store triglycerides (fat). It is an insulator and the body's major energy reserve. **Reticular connective tissue** is made of fine interlacing reticular fibers and cells. It forms the filtering stroma of the liver, spleen, and lymph nodes.
7. **Dense connective tissue** has more fibers and fewer cells than loose connective tissue. It includes dense regular connective tissue, dense irregular connective tissue, and elastic connective tissue. **Dense regular connective tissue**, which contains regularly arranged parallel bundles of collagen fibers, forms the body's tendons and ligaments. **Dense irregular connective tissue** contains irregularly arranged collagen fibers and is found in the dermis, fibrous capsules of organs and joints, and heart valves. **Elastic connective tissue** contains elastic fibers. It can stretch and recoil to its original shape and is important to the functioning of the lungs and elastic arteries.
8. **Cartilage** has a dense network of collagen and elastic fibers embedded in a gel-like ground substance. Cartilage contains **chondrocytes** found in **lacunae** and is avascular. There are three types of cartilage: hyaline cartilage, fibrocartilage, and elastic cartilage. **Hyaline cartilage** contains prominent chondrocytes and not visible collagen fibers, and provides flexible support and shock absorption. **Fibrocartilage**, the strongest type, is composed of chondrocytes scattered among visible bundles of collagen fibers. **Elastic cartilage**, which is elastic and resilient, contains chondrocytes scattered among elastic fibers.
9. **Osseous tissue** or bone tissue contains **osteocytes** within lacunae of an extracellular matrix containing mineral salts and collagen fibers. Osseous tissue is a primary component of bones. Bones store minerals, form the skeleton, and work with skeletal muscles to produce movement.
10. **Liquid connective tissue** includes blood and lymph, and has cells suspended in a liquid extracellular matrix. Blood has an extracellular matrix called **plasma** in which the formed elements (**red blood cells**, **white blood cells**, and **platelets**) are suspended. **Lymph** has an extracellular matrix similar to plasma along with several types of cells including lymphocytes.

## Resources

Anatomy Overview—Connective Tissues

Figure 4.5—Representative Cells and Fibers Present in Connective Tissue

Clinical Connection—Marfan Syndrome

Clinical Connection—Systemic Lupus Erythematosus

| Concept | Resources |
|---|---|

## Concept 4.5 Epithelial and connective tissue have obvious structural differences.

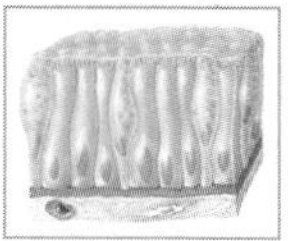

1. Epithelial tissues have many cells tightly packed together and are avascular.
2. Connective tissues have relatively few cells with lots of extracellular material.

Anatomy Overview—Tissues

## Concept 4.6 Membranes cover the surface of the body, line body cavities, and cover organs.

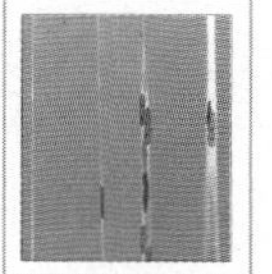

1. **Membranes** are flat sheets of flexible tissue that cover or line a body structure.
2. **Epithelial membranes** are composed of an epithelial layer and an underlying connective tissue layer. Mucous membranes, serous membranes, and the cutaneous membrane are epithelial membranes.
3. Body cavities that open to the exterior are lined with a **mucous membrane** or **mucosa**. The epithelial layer of the mucosa functions as a protective barrier where goblet cells secrete mucus to prevent dehydration, trap particles, and allow lubrication. The underlying connective tissue layer of the mucosa is called the **lamina propria** and is composed of a supporting areolar connective tissue.
4. Body cavities not opening to the exterior are lined with a **serous membrane** or **serosa** that also covers cavity organs. Serous membranes have two layers: the **parietal layer** attached to the cavity wall and a **visceral layer** attached to cavity organs. **Serous fluid**, secreted by the mesothelium, is a lubricant that prevents organ friction. Serous membranes include the **pleura** lining the thoracic cavity and covering the lungs; the **pericardium** lining the heart cavity and covering the heart; and the **peritoneum** lining the abdominal cavity and covering abdominal organs.
5. The **skin**, or **cutaneous membrane,** is composed of an epithelial layer, the epidermis, and a connective tissue layer, the dermis.
6. **Synovial membranes** line joint cavities and consist of areolar connective tissue; they do not have an epithelial layer. Synovial membranes secrete a lubricant called **synovial fluid**.

Anatomy Overview—Epithelial Membranes

## Concept 4.7 Muscle tissue generates the physical force needed to make body structures move.

1. **Muscle tissue** is composed of elongated cells called muscle fibers.
2. Muscle fibers generate the force to produce body movements, maintain posture, and generate heat.
3. There are three types of muscle tissue: skeletal, cardiac, and smooth.
4. **Skeletal muscle tissue** is attached to bones and is striated, multinucleated, and voluntary.
5. **Cardiac muscle tissue** forms most of the heart wall and is involuntary. Cardiac muscle cells are striated, have one nucleus and have intercalated discs.
6. **Smooth muscle tissue** is nonstriated, with small tapered muscle fibers containing one central nucleus per cell; it is involuntary and is located in the walls of hollow organs.

Anatomy Overview—Muscle Tissue

| Concept | Resources WILEY PLUS |
|---|---|

## Concept 4.8 Nervous tissue consists of neurons and neuroglia.

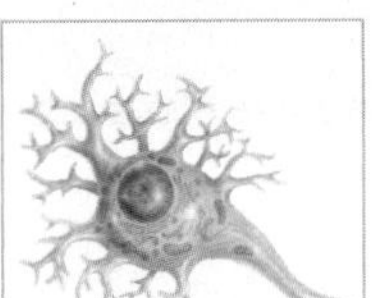

1. The cells that make up nervous tissue are neurons and neuroglia.
2. **Neurons** convert stimuli into electrical signals that they conduct to other neurons, muscle tissue, or glands.
3. The three parts of most neurons are the **cell body**; short cell processes called **dendrites**; and a single, elongated **axon.**
4. Neuroglia do not generate or conduct impulses but do support neurons.

Anatomy Overview—Nervous Tissue

## Concept 4.9 The ability of an injured tissue to repair itself depends on the extent of damage and the regenerative ability of the injured tissue.

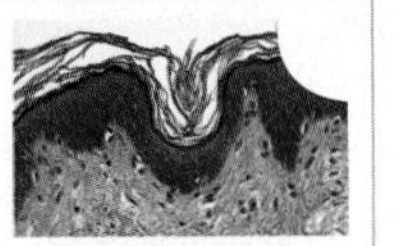

1. Tissue repair is the process of replacing worn-out, damaged, or dead cells with healthy ones.
2. Epithelial cells have the capacity for continuous renewal.
3. **Stem cells**, found in some tissues, are immature, undifferentiated cells that divide as needed for tissue repair.
4. Connective tissues have different capacities for cell repair, from complete renewal to minimal repair. Muscle tissue has poor capacity for cell renewal. Nervous tissue has the poorest capacity for cell renewal.
5. When the functioning part of a tissue or organ, called the **parenchyma**, participates in cell repair, fully functioning **tissue regeneration** is possible.
6. When the supporting connective tissue, or **stroma,** of a tissue or organ is involved in repair, **fibrosis** occurs as fibroblasts produce scar tissue, impairing the function of the tissue or organ.
7. Extensive tissue damage involves repair by both the parenchyma and the stroma, resulting in new blood capillaries and a **granulation tissue** that spreads across the open wound.
8. Good nutrition and blood circulation are vital to tissue repair.

Clinical Connection—Adhesions

## Understanding the Concepts

1. Which type of tissue connects the stomach to the underlying pancreas? Responds to a sudden sound? Moves an arm?
2. Which type of cell junction functions in communication between adjacent cells?
3. If epithelial tissue is avascular, how do nutrients and oxygen get to the tissue and waste products leave?
4. Why is it important for epithelial tissue to repair itself quickly?
5. During an anatomy and physiology exam, you are asked to identify epithelial tissue from prepared slides. Based on the observations below, give the name and functions of each.
   - A single layer of cells that are about as high as they are wide.
   - A tissue that appears to contain multiple layers of cells because nuclei are at different levels. Cilia and goblet cells are visible.
   - A single layer of tall cells with nuclei near the basal surface of the cells. Goblet cells are occasionally visible.
   - Multiple layers of cells with flat cells in the apical layer.
   - A single layer of flat cells with centrally located oval or round nuclei.
   - Multiple layers of cells with bulging, rounded cells in the apical layer.
6. How do stratified and pseudostratified epithelia differ? (Hint: Pseudo *means "false."*)
7. How do the prefixes *exo-* and *endo-* give you a clue to the function of exocrine and endocrine glands?
8. A friend tells you that he has been diagnosed with *hyperhydrosis,* a disorder characterized by excessive sweating. He asks you if the glands secreting the excess sweat also secrete hormones. Explain your answer.
9. During an anatomy and physiology exam, you are asked to view and identify connective tissues from prepared slides. What is the name and function of each of the following connective tissues?
   - Cells appear empty in the center with nuclei visible in an outer ring of cytoplasm.
   - Cells inside lacunae; extracellular matrix contains bundles of thick fibers.
   - Cells inside lacunae; concentric rings of extracellular matrix.
   - Flattened cells located between densely packed, parallel bundles of thick fibers.
   - Random arrangement of several fiber and cell types.

10. What type of connective tissue connects skin to underlying structures? (Hint: *It is found in the subcutaneous layer.*)
11. What type of connective tissue can be found in the lungs, allowing them to expand during inhalation and snap back to a smaller size during exhalation?
12. What characteristics would you look for when viewing a tissue under a light microscope to determine if the tissue is epithelial or connective tissue?
13. How do the extracellular matrixes of cartilage, bone, and blood differ?
14. Why does mucosa stay attached to underlying structures?
15. How do synovial membranes and epithelial membranes differ in structure?
16. Although movement of food through the gut is involuntary, initiation of swallowing is voluntary. Which type of muscle tissue lets you swallow a bite of pizza?
17. Why would a glioma (a cancer of neuroglial cells) not necessarily disrupt the transmission of electrical impulses within the nervous system?
18. Why are stem cells of such intense interest to many scientists?
19. Why might replacement of damaged organ tissue with scar tissue compromise the function of an organ?

CHAPTER

# 5

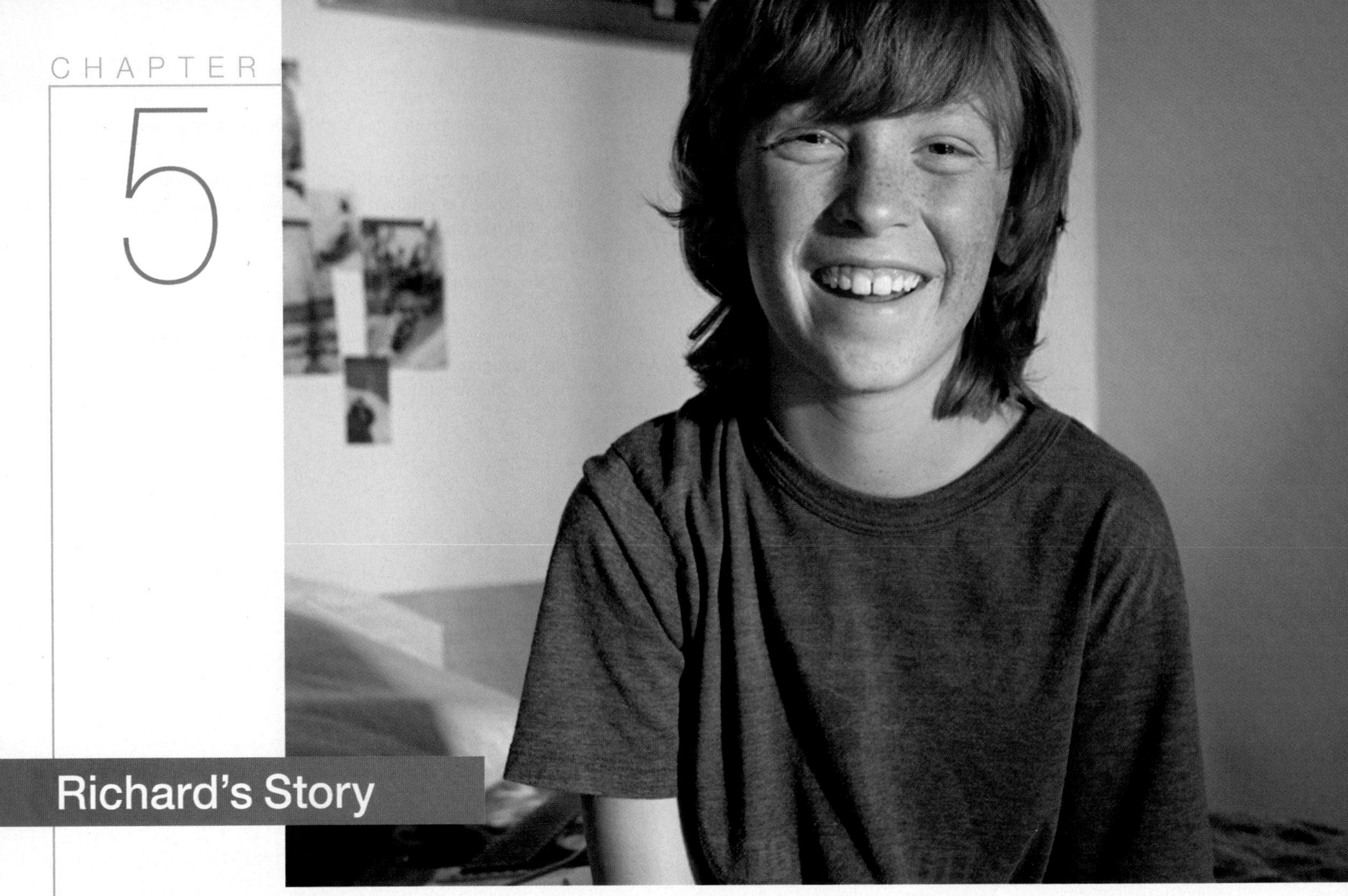

## Richard's Story

The dream had been so pleasant, until that car began beeping nonstop. Richard had been on the beach with the warm sun beating down on him. As he began to regain awareness, he realized he'd been dreaming and recognized the car beeping for what it really was—the piercing squeal of a smoke detector. The attic bedroom of the old Minnesota farmhouse where he slept was unusually warm, and he smelled smoke.

"Mom, Dad, wake up! I smell smoke!" he screamed as he dashed down the narrow staircase to the second floor. He could feel the heat of the fire as he looked into his parents' bedroom. No one was there. "They must have gotten out," he said to himself. He dashed down the stairs to the first floor and saw his father battling in vain to put out the flames that crept along the old wooden floors and were consuming the curtains of the living room. "Dad, where's Mom and the girls?" Richard shouted.

"They're outside already! Get out of here! The place is burning like a pile of matches!"

"I'll get the fire extinguisher—it's in the kitchen!" called Richard.

As he dashed toward the kitchen door, his father yelled out, "No! Richard—stop!"

As Richard pushed on the thin wooden door to the kitchen, he felt the air suck past him into the kitchen; it was followed by a silence that seemed to last forever. Then the back draft hit him. He saw a wall of flame, and he felt searing heat and instant pain like nothing he had ever imagined.

In this chapter you will study the integumentary system. The skin, or integument, is an important system that we often take for granted until it is compromised. As you will see in Richard's story, the skin has many important functions that are crucial to maintaining the homeostasis of the human body.

# The Integumentary System

## INTRODUCTION

Recall from Chapter 1 that a system consists of a group of organs working together to perform specific activities. The **integumentary system** (in-teg-ū-MEN-tar-ē; *in* = inward; *tegere* = to cover) is composed of the skin, hair, skin glands, and nails. The integumentary system protects the body, helps maintain a constant body temperature, and provides sensory information about the surrounding environment. Of all of the body's organs, none is more easily inspected or more exposed to infection, disease, and injury than the skin. Although its location makes it vulnerable to damage from trauma, sunlight, microbes, and pollutants in the environment, the skin's protective features ward off such damage. Because of its visibility, skin reflects our emotions (frowning, blushing) and some aspects of normal physiology (sweating). Changes in skin color may also indicate homeostatic imbalances in the body. For example, the bluish skin color associated with hypoxia (oxygen deficiency at the tissue level) is one sign of heart failure as well as other disorders. Abnormal skin eruptions or rashes such as chickenpox, cold sores, or measles may reveal systemic infections or diseases of internal organs, while other conditions, such as warts, age spots, or pimples, may involve the skin alone. So important is the skin to self-image that many people spend a great deal of time and money to restore it to a more normal or youthful appearance.

## CONCEPTS

**5.1** Skin is composed of a superficial epidermis and a deeper dermis.

**5.2** The layers of the epidermis include the stratum basale, stratum spinosum, stratum granulosum, stratum lucidum, and stratum corneum.

**5.3** The dermis contains blood vessels, nerves, sensory receptors, hair follicles, and glands.

**5.4** Skin color is a result of the pigments melanin, carotene, and hemoglobin.

**5.5** The functions of hair, skin glands, and nails include protection and body temperature regulation.

**5.6** The two major types of skin are thin skin and thick skin.

**5.7** Skin regulates body temperature, protects underlying tissues, provides cutaneous sensations, excretes body wastes, and synthesizes vitamin D.

**5.8** Skin damage sets in motion a sequence of events that repairs the skin.

# 5.1 Skin is composed of a superficial epidermis and a deeper dermis.

The **skin**, or **cutaneous membrane** (kū-TĀ-nē-us), covers the external surface of the body. It is the largest organ of the body in surface area and weight. In adults, the skin covers an area of about 2 square meters (22 square feet) and weighs 4.5–5 kg (10–11 lb), about 16 percent of total body weight. It ranges in thickness from 0.5 mm (0.02 in.) on the eyelids to 4.0 mm (0.16 in.) on the heels. However, over most of the body it is 1–2 mm (0.04–0.08 in.) thick.

Structurally, the skin consists of two principal parts (Figure 5.1). The superficial, thinner portion, which is composed of *epithelial tissue*, is the **epidermis** (ep′-i-DERM-is; *epi-* = above). The deeper, thicker *connective tissue* portion is the **dermis**. While the epidermis is avascular, the dermis is vascular. For this reason, if you cut the epidermis there is no bleeding, but if you cut the dermis there is bleeding.

Deep to the dermis, but not part of the skin, is the **hypodermis** (*hypo-* = below), or **subcutaneous layer**. This layer consists of areolar and adipose tissues. Fibers that extend from the dermis anchor the skin to the hypodermis, which in turn attaches to underlying *fascia*, the connective tissue around muscles and

FIGURE 5.1 **Components of the integumentary system.**

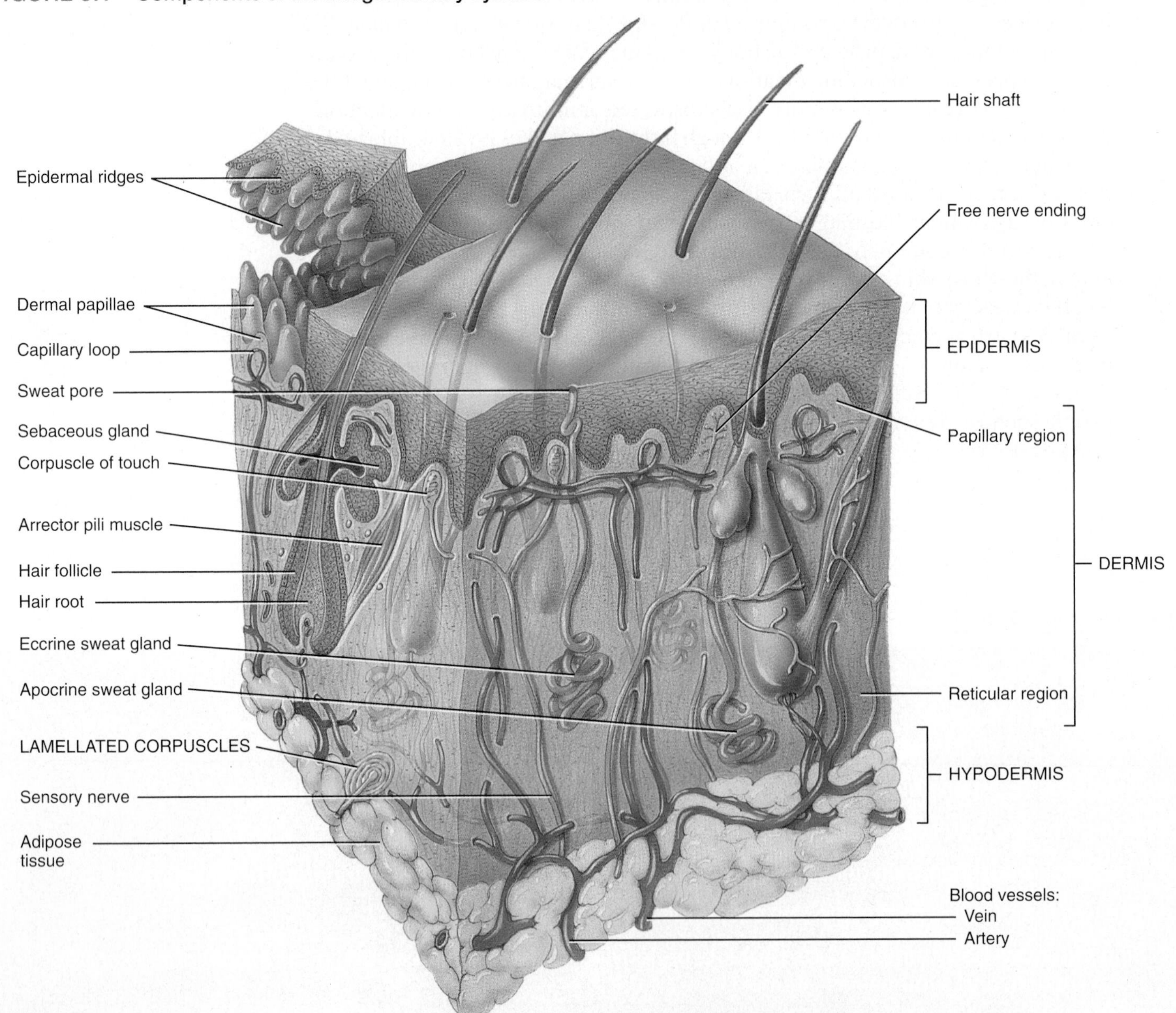


(a) Sectional view of skin and subcutaneous layer

bones. The hypodermis serves as a storage depot for fat and contains large blood vessels that supply the skin. This region (and sometimes the dermis) also contains nerve endings called **lamellated (pacinian) corpuscles** (pa-SIN-ē-an) that are sensitive to pressure.

✓ **CHECKPOINT**

1. List the components of the integumentary system.
2. Which tissues compose the hypodermis?

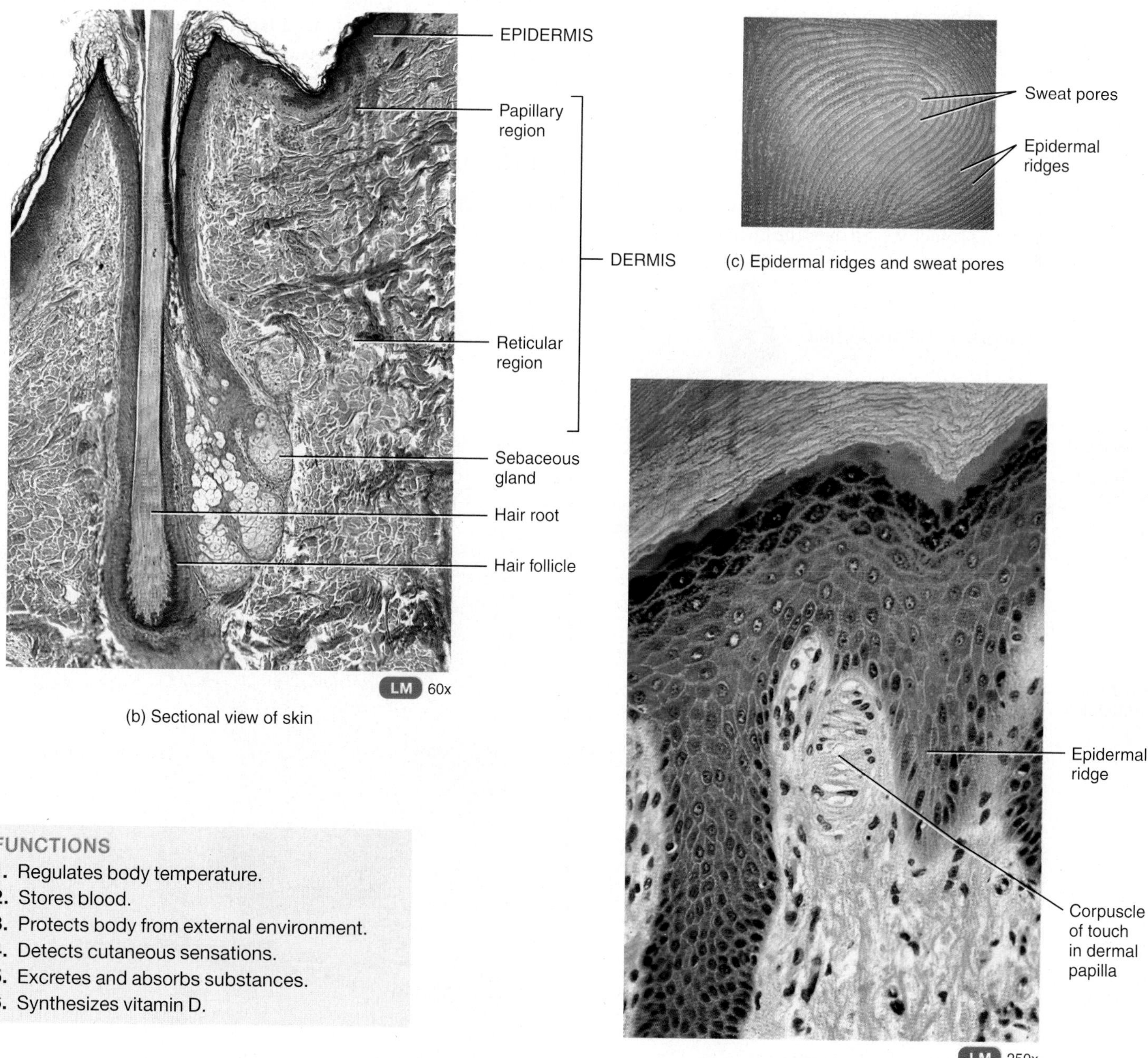


(b) Sectional view of skin

(c) Epidermal ridges and sweat pores

(d) Sectional view of dermal papillae and epidermal ridges

**FUNCTIONS**

1. Regulates body temperature.
2. Stores blood.
3. Protects body from external environment.
4. Detects cutaneous sensations.
5. Excretes and absorbs substances.
6. Synthesizes vitamin D.

**The integumentary system includes the skin, hair, skin glands, and nails.**

# 5.2 The layers of the epidermis include the stratum basale, stratum spinosum, stratum granulosum, stratum lucidum, and stratum corneum.

## Cells of the Epidermis

The **epidermis** is composed of keratinized stratified squamous epithelium. It contains four principal types of cells: keratinocytes, melanocytes, Langerhans cells, and tactile cells (Figure 5.2). **Keratinocytes** (ker-a-TIN-ō-sīts; *keratino-* = hornlike; *-cytes* = cells), the most numerous epidermal cells, are arranged in four or five layers and produce **keratin** (KER-a-tin), a tough, fibrous protein that helps protect the skin and underlying tissues from abrasions, heat, microbes, and chemicals. Keratinocytes also produce lamellar granules, which release a waterproofing sealant that decreases water entry and loss.

**Melanocytes** (MEL-a-nō-sīts; *melano-* = black) have long, slender projections that extend between the keratinocytes and transfer melanin to them. **Melanin** (MEL-a-nin) is a yellow-red or brown-black pigment that contributes to skin color and absorbs damaging ultraviolet (UV) light. Once inside keratinocytes, melanin forms a protective veil over the nucleus on the side facing the skin surface. In this way melanin shields the nuclear DNA from UV light. Although keratinocytes gain some protection from melanin, melanocytes themselves are particularly susceptible to damage by UV light.

**Langerhans cells** (LANG-er-hans) participate in immune responses against microbes that invade the skin by helping other immune system cells recognize an invading microbe and destroy it. Langerhans cells are easily damaged by UV light.

**Tactile (Merkel) cells** (MER-kel) are the least numerous of the epidermal cells. They are located in the deepest layer of the epidermis, where they contact a **tactile (Merkel) disc**, a flattened process of a sensory neuron (nerve cell). Tactile cells and their associated tactile discs function together to detect touch sensations.

## Strata of the Epidermis

Several distinct layers of keratinocytes in various stages of development form four or five strata, or layers of the epidermis (Figure 5.2).

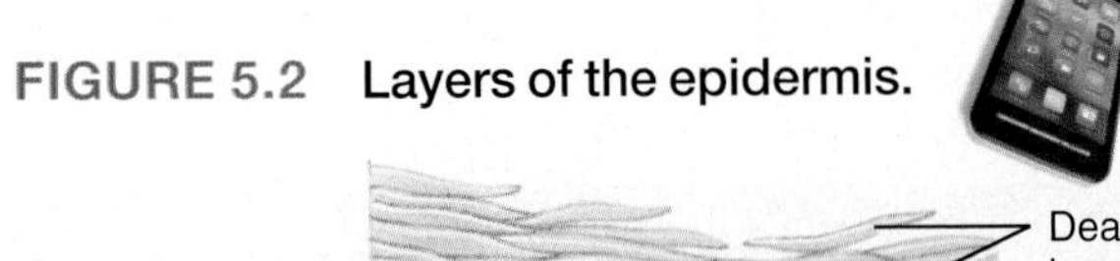

**FIGURE 5.2 Layers of the epidermis.**

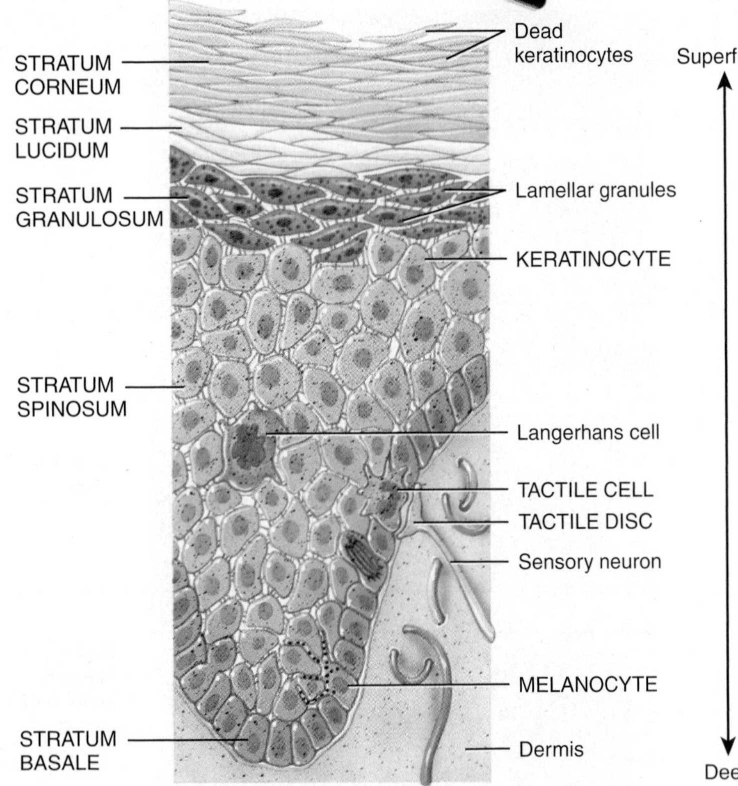


(a) Four principal cell types in epidermis

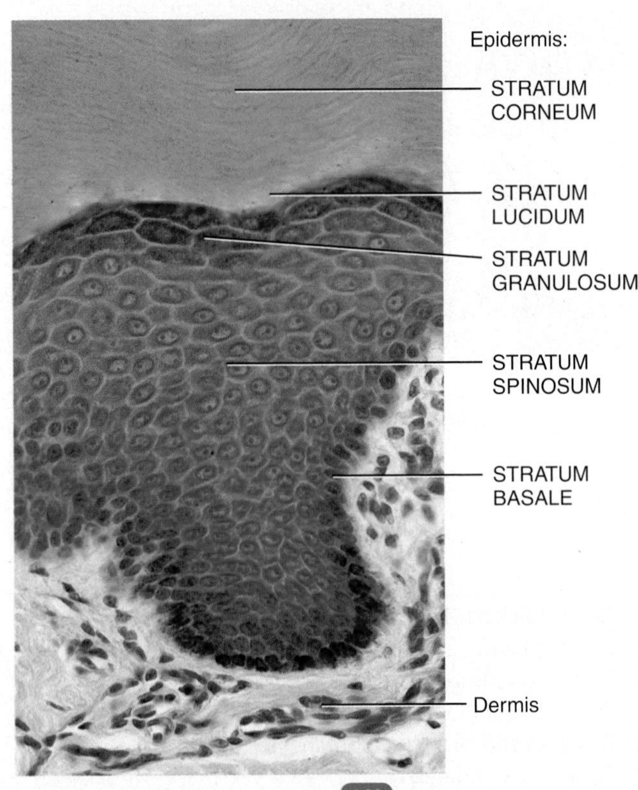


(b) Photomicrograph of portion of skin

**Most of the epidermis consists of keratinocytes, which produce the protein keratin (protects underlying tissues) and lamellar granules (contain a waterproof sealant).**

The deepest layer of the epidermis, the **stratum basale** (ba-SA-lē; *basal-* = base), is composed of a single row of cuboidal or columnar keratinocytes. Some cells in this layer are *stem cells* that undergo cell division to continually produce new keratinocytes. Melanocytes and tactile cells (with their associated tactile discs) are scattered among the keratinocytes of the stratum basale layer.

Superficial to the stratum basale is the **stratum spinosum** (spi-NŌ-sum; *spinos-* = thornlike), arranged in 8 to 10 layers of many-sided keratinocytes fitting closely together. Cells in the more superficial layers become somewhat flattened. Keratinocytes of the straum spinosum shrink and pull apart when prepared for microscopic examination, so that they appear to be covered with thornlike spines (hence the name "spinosum"). Langerhans cells and projections of melanocytes are present in this layer.

At about the middle of the epidermis, the **stratum granulosum** (gran-ū-LŌ-sum; *granulos-* = little grains) consists of three to five layers of flattened keratinocytes. As the keratinocytes of the stratum granulosum move farther from their source of nutrition (the dermal blood vessels), they can no longer carry on vital metabolic reactions, and die. Thus, the stratum granulosum marks the transition between the deeper, metabolically active strata and the dead cells of the more superficial strata. A distinctive feature of this layer is the presence of membrane-enclosed **lamellar granules** within the keratinocytes. Lamellar granules release a lipid-rich secretion that is deposited in the spaces between cells of the stratum granulosum, stratum lucidum, and stratum corneum. The lipid-rich secretion acts as a water-repellent sealant, retarding loss of water and entry of foreign materials.

The **stratum lucidum** (LOO-si-dum; *lucid-* = clear) is present only in the thick skin of the fingertips, palms, and soles. It consists of four to six layers of flattened clear, dead keratinocytes that contain large amounts of keratin and thickened plasma membranes.

The **stratum corneum** (KOR-nē-um; *corne-* = horn or horny) consists of 25 to 30 layers of flattened dead keratinocytes. These cells are continuously shed and replaced by cells from the deeper strata. The cells are extremely thin, flat, plasma-membrane enclosed packages of keratin that no longer contain a nucleus or any internal organelles. The cells within each layer overlap one another like the scales on the skin of a snake. In this outer stratum of the epidermis, cells are continuously shed and replaced by cells from the deeper strata. Its multiple layers of dead cells help the stratum corneum to protect deeper layers from injury and microbial invasion.

## Growth of the Epidermis

The stratum basale cells are closest to blood vessels in the dermis, and therefore receive the most nutrients and oxygen. These metabolically active cells continuously undergo cell division to produce new keratinocytes. Newly formed cells are slowly pushed from one epidermal strata to the next by the continuing cell division of the stratum basale. As keratinocytes are pushed toward the surface, they receive fewer nutrients from the blood supply and eventually die. They also accumulate more and more keratin through a process called **keratinization** (ker-a-tin-i-ZĀ-shun). Eventually the dead, keratinized cells slough off the surface of the stratum corneum and are replaced by underlying cells. The whole process by which cells form in the stratum basale, rise to the surface, become keratinized, and slough off takes about four weeks.

Table 5.1 summarizes the distinctive features of the epidermal strata.

### ✓ CHECKPOINT

3. Which tissue type makes up the epidermis?
4. What is the function of lamellar granules? Of melanin?
5. Which epidermal layer includes stem cells that continually undergo cell division?
6. Which visual characteristics allow you to distinguish each epidermal layer?

**TABLE 5.1**

Epidermal Strata (see Figure 5.2)

| STRATUM | DESCRIPTION |
|---|---|
| **Basale** | Deepest layer, composed of single row of cuboidal or columnar keratinocytes; stem cells undergo cell division to produce new keratinocytes; melanocytes and tactile cells associated with tactile discs are scattered among keratinocytes |
| **Spinosum** | Multiple layers of many-sided keratinocytes; contains armlike projections of melanocytes and Langerhans cells |
| **Granulosum** | Multiple layers of flattened keratinocytes that are beginning to degenerate; cells contain lamellar granules, release a water-repellent secretion, and contain keratin |
| **Lucidum** | Present only in skin of fingertips, palms, and soles; consists of multiple rows of clear, flat, dead keratinocytes with large amounts of keratin |
| **Corneum** | Multiple rows of dead, flat keratinocytes that contain mostly keratin |

## CLINICAL CONNECTION | *Psoriasis*

**Psoriasis** (sō-RĪ-a-sis) is a common and chronic skin disorder in which keratinocytes divide and move more quickly than normal from the stratum basale to the stratum corneum. They are shed prematurely in as little as 7 to 10 days. The immature keratinocytes make an abnormal keratin, which forms flaky, silvery scales at the skin surface, most often on the knees, elbows, and scalp (dandruff). Effective treatments—various topical ointments and ultraviolet phototherapy—suppress cell division, decrease the rate of cell growth, or inhibit keratinization.

## 5.3 The dermis contains blood vessels, nerves, sensory receptors, hair follicles, and glands.

The second, deeper part of the skin, the **dermis**, is composed of dense irregular connective tissue containing a woven network of collagen and elastic fibers that provide great *tensile strength* (resistance to pulling or stretching forces). The dermis is much thicker than the epidermis. Because the dermis is typically thinner in women than in men, many women have the appearance of dimples in the skin, referred to as *cellulite.* Leather, which we use for belts, shoes, and basketballs, is the treated dermis of other animals. Blood vessels, nerves, glands, and hair follicles are embedded in the dermis. The dermis can be divided into a thin superficial papillary region and a thick deeper reticular region.

The **papillary region** makes up about one-fifth of the thickness of the total dermis (see Figure 5.1). It contains thin collagen and fine elastic fibers. Its surface area is greatly increased by **dermal papillae** (pa-PIL-ē = nipples), small nipple-shaped projections into the undersurface of the epidermis. Dermal papillae contain **capillary loops** (blood capillaries) and can contain sensory receptors: **Corpuscles of touch (Meissner corpuscles)** that are sensitive to touch and/or **free nerve endings**, which initiate signals that produce sensations of warmth, coolness, pain, tickling, and itching.

The **reticular region** (*reticul-* = netlike) is attached to the hypodermis and contains bundles of thick collagen fibers, some elastic fibers, scattered fibroblasts and adipose cells, and various wandering cells (such as macrophages). Blood vessels, nerves, hair follicles, sebaceous (oil) glands, and sudoriferous (sweat) glands occupy the spaces between fibers. The combination of collagen and elastic fibers in the reticular region provides the skin with strength, **extensibility** (ek-sten′-si-BIL-i-tē) (the ability to stretch), and **elasticity** (e-las-TISS-i-tē) (the ability to return to original shape after stretching). The extensibility of skin can be readily seen around joints and in pregnancy and obesity.

The surfaces of the palms, fingers, soles, and toes have a series of ridges and grooves. They appear either as straight lines or as a pattern of loops and whorls, as on the tips of the fingers. These **epidermal ridges** project downward from the epidermis into the dermis between the dermal papillae (see Figure 5.1c, d). Epidermal ridges and dermal papillae fit together like complementary teeth of a zipper forming an extremely strong bond to resist lateral forces that attempt to separate the epidermis from the dermis. The epidermal ridges also increase the surface area of the epidermis and thus increase the grip of the hand or foot by increasing friction. Because the ducts of sweat glands open on the tops of the epidermal ridges as sweat pores, the sweat and ridges form *fingerprints* (or *footprints*) upon touching a smooth object. The epidermal ridge pattern is genetically determined and is unique for each individual, and thus can serve as the basis for identification.

Table 5.2 summarizes the structural features of the regions of the dermis.

### ✓ CHECKPOINT

7. How do the papillary and reticular regions of the dermis differ structurally?
8. What are the functions of fibers in the reticular region of the dermis?
9. How could you explain that a slight paper cut to the skin does not bleed?

**TABLE 5.2**

Papillary and Reticular Regions of the Dermis (see Figure 5.1)

| REGION | DESCRIPTION |
|---|---|
| **Papillary** | Superficial portion of dermis (about one-fifth); consists of dense irregular connective tissue with thin collagen and fine elastic fibers; contains dermal papillae that house capillaries, corpuscles of touch, and free nerve endings |
| **Reticular** | Deeper portion of dermis (about four-fifths); consists of dense irregular connective tissue with bundles of thick collagen and some coarse elastic fibers; spaces between fibers contain some adipose cells, hair follicles, nerves, sebaceous glands, and sudoriferous glands |

## RETURN TO Richard's Story

Richard was unconscious for almost four days after the fire. His mother, father, and two sisters came to visit him in the burn unit every day, and his mother stayed overnight. Richard's burns were serious, and the doctor was very cautious about Richard's prognosis. "Mrs. Anderson, your son has severe burns over one-third of his body. His hands and face have second-degree partial thickness burns. This type of injury damages the dermis of the skin but doesn't completely destroy it; it's very painful but should heal fairly well given enough time. I am more concerned about the burns on his chest and abdomen. Those burns are third-degree full thickness burns, involving all the layers of the skin down to the underlying subcutaneous layers. In these types of injuries the tissue is damaged down into the underlying muscle. They often require skin grafting and a long recovery time."

Mrs. Anderson thought about the fleece top that Richard always wore to bed. "The attic gets cold, Mom," Richard had said. The flammable material had caught fire after the explosion and melted into Richard's skin, burning his chest severely.

A. *The doctor describes Richard as having two types of burns, "partial thickness" and "full thickness." Based on what you have learned about the skin, explain why a partial thickness burn is extremely painful and why it would heal faster than a full thickness burn.*

## 5.4 Skin color is a result of the pigments melanin, carotene, and hemoglobin.

Melanin, carotene, and hemoglobin are three pigments that give skin a wide variety of colors. The amount of **melanin**, which is located mostly in the epidermis, causes the skin's color to vary from pale yellow to reddish-brown to black. Because the *number* of melanocytes, the melanin-producing cells, is about the same in all people, differences in skin color are due mainly to the *amount of melanin* the melanocytes produce and transfer to keratinocytes. In some people, melanin tends to accumulate in patches called *freckles*. Freckles typically are reddish or brown and tend to be more visible in the summer than the winter. As a person ages, *age (liver) spots* may develop. These flat blemishes have nothing to do with the liver. These flat blemishes look like freckles and range in color from light brown to black. Like freckles, age spots are accumulations of melanin. Age spots are darker than freckles and build up over time due to exposure to sunlight. Age spots do not fade away during the winter months and are more common in adults over forty. A *nevus* (NĒ-vus), or a *mole*, is a round, flat, or raised area that represents a benign localized overgrowth of melanocytes.

Exposure to ultraviolet (UV) light stimulates melanocytes to produce melanin, which gives the skin a tanned appearance and helps protect the body against further UV radiation. A tan is lost when the melanin-containing keratinocytes are shed from the stratum corneum. Melanin absorbs UV radiation and prevents damage to DNA in epidermal cells. Thus, within limits, melanin serves a protective function. As you will see later, exposing the skin to a *small* amount of UV light is actually necessary for the skin to begin the process of vitamin D synthesis. However, repeatedly exposing the skin to a *large* amount of UV light may cause skin cancer.

**Carotene** (KAR-ō-tēn; *carot-* = carrot) is a yellow-orange pigment that gives egg yolks and carrots their color. Carotene is stored in the stratum corneum and dermis after eating carotene-rich foods.

Dark-skinned individuals have large amounts of melanin in the epidermis, so their skin color ranges from yellow to reddish-brown to black. Light-skinned individuals have little melanin in the epidermis, making their epidermis appear translucent with a skin color ranging from pink to red depending on the level of oxygen in the blood moving through capillaries in the dermis. The red color is due to **hemoglobin**, the oxygen-carrying pigment in red blood cells.

### ✓ CHECKPOINT

**10.** What do age spots and freckles have in common?

**11.** What are the three pigments in the skin and how do they contribute to skin color?

## CLINICAL CONNECTION | *Skin Color as a Diagnostic Clue*

The color of skin and mucous membranes can provide clues for diagnosing certain conditions. When blood is not picking up an adequate amount of oxygen from the lungs, as in someone who has stopped breathing, the mucous membranes, nail beds, and skin appear bluish or **cyanotic** (sī-a-NOT-ik; *cyan-* = blue). **Jaundice** (JON-dis; *jaund-* = yellow) is due to a buildup of the yellow pigment bilirubin in the skin. This condition gives a yellowish appearance to the skin and the whites of the eyes, and usually indicates liver disease. **Erythema** (er-e-THĒ-ma; *eryth-* = red), redness of the skin, is caused by engorgement with blood of capillaries in the dermis due to skin injury, exposure to heat, infection, inflammation, or allergic reactions. **Pallor** (PAL-or), or paleness of the skin, may occur in conditions such as shock and anemia. All skin color changes are observed most readily in people with lighter-colored skin and may be more difficult to discern in people with darker skin. However, examination of the nail beds and gums can provide some information about circulation in individuals with darker skin.

# 5.5 The functions of hair, skin glands, and nails include protection and body temperature regulation.

The accessory structures of the skin—hair, skin glands, and nails—have a host of important functions. For example, hair and nails protect the body, and sweat glands help regulate body temperature. Together, the skin and accessory structures compose the integumentary system.

## Hair

**Hairs**, or **pili** (PĪ-lī), are present on most skin surfaces but are absent from such areas as the lips, palms, soles, and parts of the external genitalia. In adults, hair usually is most heavily distributed across the scalp, in the eyebrows, in the axillae (armpits), and around the external genitalia. Genetic and hormonal influences largely determine the thickness and pattern of hair distribution.

Although the protection it offers is limited, hair on the head guards the scalp from injury and the sun's rays. It also decreases heat loss from the scalp. Eyebrows and eyelashes protect the eyes from foreign particles, similar to the way hair in the nostrils and in the external ear canal defend those structures. Touch receptors (hair root plexuses) associated with hair follicles are activated whenever a hair is moved even slightly. Thus, hairs also function in sensing light touch.

### Anatomy of a Hair

Each hair is composed of columns of dead, keratinized epidermal cells bonded together by extracellular proteins. The **shaft** is the superficial portion of the hair that projects from the surface of the skin (Figure 5.3a). The **root** is the portion of the hair deep to the shaft that penetrates into the dermis, and sometimes into the hypodermis. The shaft and root both consist of three concentric layers of hair cells: medulla, cortex, and cuticle (Figure 5.3c, d). The inner *medulla* is composed of cells that contain pigment in dark hair but mostly air spaces between the cells in gray hair and the absence of pigment in white hair. The middle *cortex* forms the major part of the shaft and consists of elongated cells. The *cuticle* of the hair, the outermost layer, consists of a single layer of thin, flat cells that are arranged like shingles on the roof of a house, with their free edges pointing toward the distal end of the hair (Figure 5.3b).

Surrounding the root of the hair is the **hair follicle** (FOL-i-kul), which is made up of an external root sheath and an internal root sheath (Figure 5.3c, d). The *external root sheath* is a downward continuation of the epidermis. The *internal root sheath* is produced by the hair matrix (described shortly) and forms a tubular sheath of epithelium between the external root sheath and the hair. The dermis surrounding the hair follicle forms the **dermal root sheath**.

The base of each hair follicle is enlarged into an onion-shaped structure, the **bulb** (Figure 5.3c). The bulb houses a nipple-shaped indentation, the **papilla** of the hair, which contains many blood vessels that nourish the growing hair follicle. The bulb also contains a germinal layer of cells called the **hair matrix**, the site of hair cell division. Hence, hair matrix cells are responsible for the growth of existing hairs, and they produce new hairs when old hairs are shed.

Sebaceous (oil) glands (discussed shortly), smooth muscle cells, and sensory receptors are also associated with hairs (Figure 5.3a). The smooth muscle cell bundle, called the **arrector pili** (a-REK-tor PĪ-lī; *arrect-* = to raise), extends from the superficial dermis of the skin to the dermal root sheath around the hair follicle. In its normal position, hair emerges at an angle to the surface of the skin. Under physiological or emotional stress, such as cold or fright, autonomic nerve endings stimulate the arrector pili to contract, which pulls the hair shafts perpendicular to the skin surface. This action causes "goose bumps" or "gooseflesh" as the skin around the shaft forms slight elevations. Surrounding each hair follicle is a **hair root plexus**, a sensory nerve ending that is sensitive to touch (Figure 5.3a). The hair root plexuses generate impulses if the hair shaft is moved, which happens, for example, when an insect bumps into a hair as it crawls across your arm.

FIGURE 5.3 **Hair.**

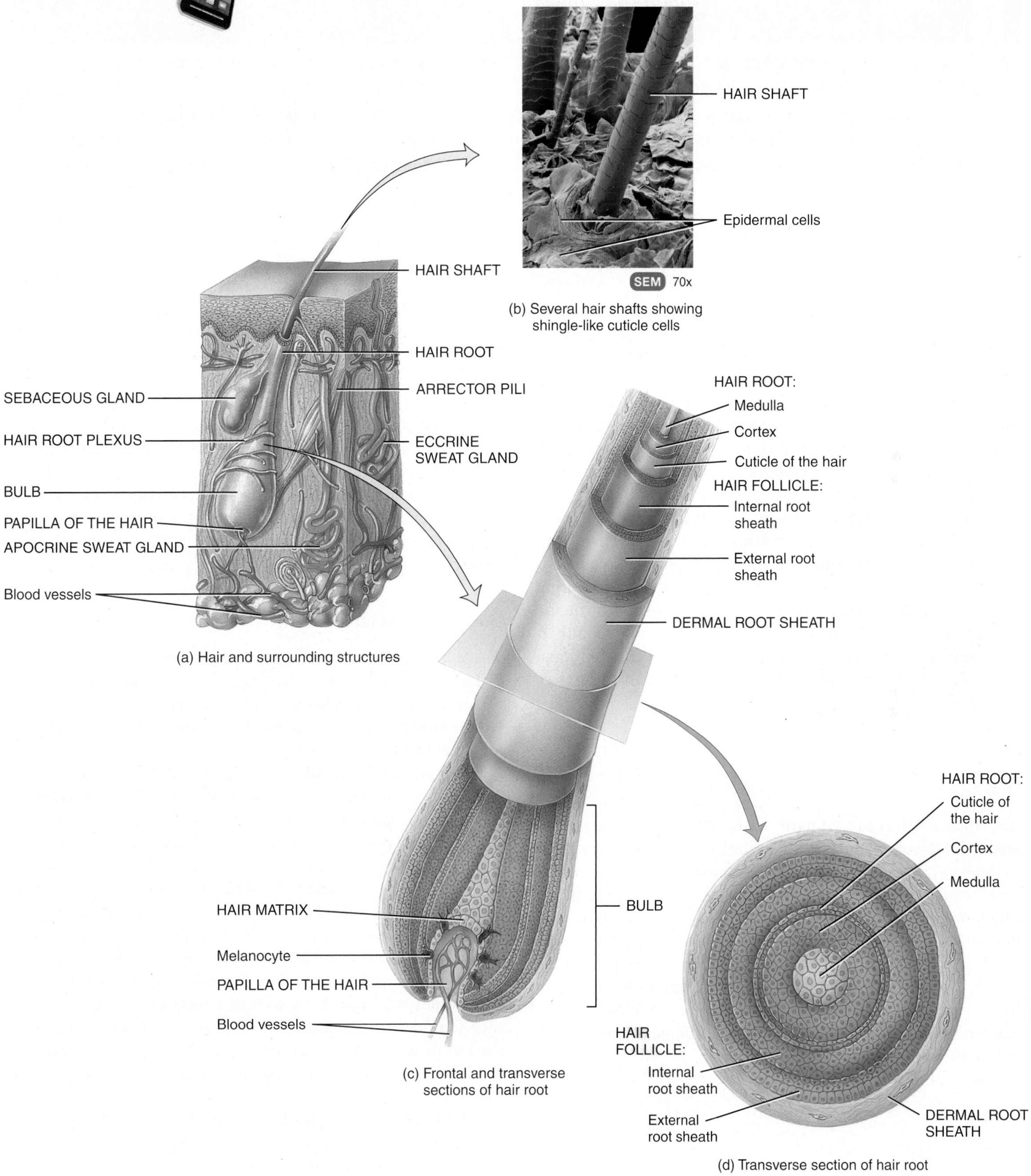

HAIR SHAFT

Epidermal cells

SEM 70x

(b) Several hair shafts showing shingle-like cuticle cells

HAIR SHAFT

HAIR ROOT

SEBACEOUS GLAND

ARRECTOR PILI

HAIR ROOT PLEXUS

ECCRINE SWEAT GLAND

BULB

PAPILLA OF THE HAIR

APOCRINE SWEAT GLAND

Blood vessels

(a) Hair and surrounding structures

HAIR ROOT:
Medulla
Cortex
Cuticle of the hair

HAIR FOLLICLE:
Internal root sheath
External root sheath

DERMAL ROOT SHEATH

HAIR MATRIX

Melanocyte

PAPILLA OF THE HAIR

Blood vessels

BULB

(c) Frontal and transverse sections of hair root

HAIR ROOT:
Cuticle of the hair
Cortex
Medulla

HAIR FOLLICLE:
Internal root sheath
External root sheath

DERMAL ROOT SHEATH

(d) Transverse section of hair root

**Hairs are growths of epidermis composed of dead, keratinized cells.**

## CLINICAL CONNECTION | *Chemotherapy and Hair Loss*

**Chemotherapy** is the treatment of disease, usually cancer, by means of chemical substances or drugs. Chemotherapeutic agents interrupt the life cycle of rapidly dividing cancer cells. Unfortunately, the drugs also affect other rapidly dividing cells in the body, such as the hair matrix cells of a hair. It is for this reason that individuals undergoing chemotherapy experience hair loss. Since about 15 percent of the hair matrix cells of scalp hairs are in the resting stage, these cells are not affected by chemotherapy. Once chemotherapy is stopped, the hair matrix cells replace lost hair follicles and hair growth resumes.

### Hair Growth

Each hair follicle goes through a growth cycle. At the start of each growth cycle, cells of the hair matrix divide, adding new hair cells to the base of the hair root, which extends the hair as existing cells of the hair root are pushed upward. As hair cells are being pushed upward, they become keratinized and die. In time, the cells of the hair matrix stop dividing, the hair follicle atrophies (shrinks), and the hair stops growing. As a new growth cycle begins, the old hair root falls out or is pushed out of the hair follicle as a new hair begins to grow in its place. Scalp hair grows for 2 to 6 years and rests for about 3 months. At any time, about 85 percent of scalp hairs are growing. Visible hair is dead, but until the hair is pushed out of its follicle by a new hair, portions of its root within the scalp are alive. Normal hair loss in the adult scalp is about 70–100 hairs per day. Both the rate of growth and the replacement cycle may be altered by illness, radiation therapy, chemotherapy, age, genetics, gender, and severe emotional stress. Rapid weight-loss diets that severely restrict calories or protein increase hair loss.

### Hair Color

The color of hair is due primarily to the amount and type of melanin in its cells. Melanin is synthesized by melanocytes scattered in the hair matrix of the bulb and passes into cells of the cortex and medulla (Figure 5.3c). Various forms of melanin produce a range of hair colors: black, brown, red, or blond. Dark-colored hair contains mostly true melanin (brown to black). Blond and red hair contains variants of melanin (yellow to red) in which there is iron and more sulfur. Gray hair contains less melanin because of a progressive decline in its production. White hair results from the lack of melanin and the accumulation of air bubbles in the hair shaft.

## Glands of the Skin

Recall from Chapter 4 that glands are epithelial cells that secrete a substance. Several kinds of exocrine glands are associated with the skin: sebaceous (oil) glands, sudoriferous (sweat) glands, and ceruminous glands.

### Sebaceous Glands

**Sebaceous glands** (se-BĀ-shus; *sebace-* = greasy), or **oil glands**, with few exceptions, are connected to hair follicles (see Figures 5.1 and 5.3a). The secreting portion of the gland lies in the dermis and usually opens into the hair follicle. Sebaceous glands are found in the skin over all regions of the body except the palms and soles. Sebaceous glands secrete an oily substance called **sebum** (SĒ-bum), a mixture of triglycerides, cholesterol, proteins, and inorganic salts. Sebum coats the surface of hairs and helps keep them from drying and becoming brittle. Sebum also prevents excessive evaporation of water from the skin, keeps the skin soft and pliable, and inhibits the growth of some bacteria.

### Sudoriferous Glands

There are 3 to 4 million **sudoriferous glands** (soo′-dor-IF-er-us; *sudori-* = sweat; *-ferous* = bearing), or **sweat glands**, in the body. Sudoriferous glands release sweat, or *perspiration*, into hair follicles or onto the skin surface through pores. Sudoriferous glands are divided into two main types: eccrine and apocrine.

**Eccrine sweat glands** (EK-rin; *eccrine* = secreting outwardly) are distributed throughout the skin of most regions of the body, especially in the skin of the forehead, palms, and soles. The secretory portion is located in the dermis, with its duct projecting upward to open as a *pore* at the surface of the epidermis (see Figures 5.1a and 5.3a). The sweat produced by eccrine sweat glands (about 600 mL per day) consists of water, ions (mostly $Na^+$ and $Cl^-$), urea, uric acid, ammonia, amino acids, glucose, and lactic acid. The main function of eccrine sweat is to help regulate body temperature through evaporation. As sweat evaporates, large quantities of heat energy leave the body surface. Eccrine sweat also plays a small role in eliminating wastes such as urea, uric acid, and ammonia, although the kidneys are primarily responsible for excreting these waste products from the body.

**Apocrine sweat glands** (AP-ō-krin; *apo* = separated from) are found mainly in the skin of the axilla (armpit), groin, areolae (pigmented areas around the nipples) of the breasts, and bearded regions of the face in adult males. The secretory portion of these sweat glands is located mostly in the hypodermis, and their excretory ducts open into hair follicles (see Figures 5.1a and 5.3a). Their secretion is slightly viscous

## CLINICAL CONNECTION | *Acne*

During childhood, sebaceous glands are relatively small and inactive. At puberty, androgens from the testes, ovaries, and adrenal glands stimulate sebaceous glands to grow in size and increase their production of sebum. **Acne** is an inflammation of sebaceous glands that usually begins at puberty, when the sebaceous glands are stimulated by androgens. Acne occurs predominantly in sebaceous follicles that have been colonized by bacteria, some of which thrive in the lipid-rich sebum. The infection may cause a cyst or sac of connective tissue cells to form, which can destroy and displace epidermal cells. This condition, called **cystic acne**, can permanently scar the epidermis. Treatment consists of gently washing the affected areas once or twice daily with a mild soap, topical antibiotics (such as clindamycin and erythromycin), topical drugs such as benzoyl peroxide or tretinoin, and oral antibiotics (such as tetracycline, minocycline, erythromycin, and isotretinoin). Contrary to popular belief, foods such as chocolate or fried foods do not cause or worsen acne.

compared to eccrine secretions and contains the same components as eccrine sweat plus lipids and proteins. Sweat secreted from apocrine sweat glands is odorless. However, when bacteria on the surface of the skin metabolize its components, apocrine sweat develops a musky and often unpleasant odor that is referred to as body odor. Eccrine sweat glands start to function soon after birth, but apocrine sweat glands do not begin to function until puberty.

Both eccrine sweat glands and apocrine sweat glands are active during emotional sweating. In addition, apocrine sweat glands secrete sweat during sexual activities. In contrast to eccrine sweat, apocrine sweat does not play a significant role in regulating body temperature.

### Ceruminous Glands

Modified sweat glands in the external ear, called **ceruminous glands** (se-RŪ-mi-nus; *cer-* = wax), produce a waxy secretion. Their excretory ducts open onto the surface of the external auditory canal (ear canal). The combined secretion of the ceruminous and sebaceous glands is a yellowish material called **cerumen** (se-ROO-men), or earwax. Cerumen, together with hairs in the external auditory canal, provides a sticky barrier that impedes the entrance of foreign bodies. Cerumen also waterproofs the canal and prevents bacteria and fungi from entering cells.

Table 5.3 presents a summary of skin glands.

**TABLE 5.3**

Skin Glands (See Figures 5.1 and 5.3)

| FEATURE | SEBACEOUS GLANDS | ECCRINE SWEAT GLANDS | APOCRINE SWEAT GLANDS | CERUMINOUS GLANDS |
|---|---|---|---|---|
| **Distribution** | Throughout skin of most regions of body; absent in palms and soles | Throughout skin of most regions of body, especially skin of forehead, palms, and soles | Skin of axilla, groin, areolae, bearded regions of face, clitoris, and labia minora | External auditory canal |
| **Location of secretory portion** | Dermis | Dermis | Mostly in hypodermis | Hypodermis |
| **Termination of excretory duct** | Hair follicle | Surface of epidermis | Hair follicle | Surface of external auditory canal |
| **Secretion** | Sebum (mixture of triglycerides, cholesterol, proteins, and inorganic salts) | Less viscous; consists of water, ions ($Na^+$, $Cl^-$), urea, uric acid, ammonia, amino acids, glucose, and lactic acid | More viscous; consists of same components as eccrine sweat glands plus lipids and proteins | Cerumen, a waxy material |
| **Functions** | Prevent hairs from drying out, prevent water loss from skin, keep skin soft, inhibit growth of some bacteria | Regulation of body temperature, waste removal, stimulated during emotional stress | Stimulated during emotional stress and sexual excitement | Impede entrance of foreign bodies and insects into external ear canal, waterproof canal, and prevent microbes from entering cells |
| **Onset of function** | Relatively inactive during childhood; activated during puberty | Soon after birth | Puberty | Soon after birth |

FIGURE 5.4 **Nails.** Shown is a fingernail.

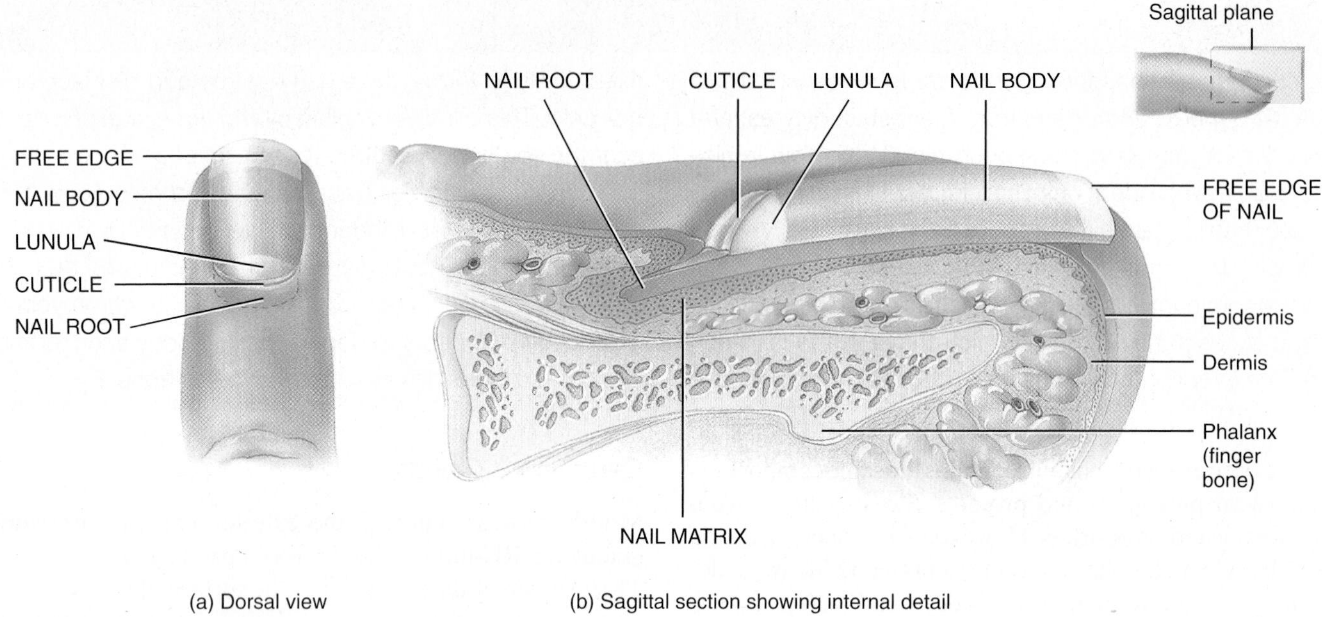


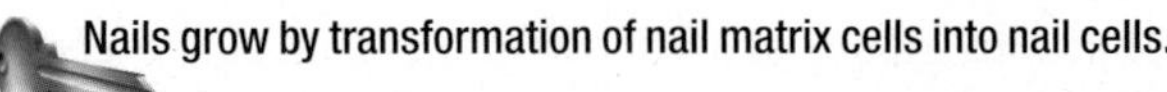

Nails grow by transformation of nail matrix cells into nail cells.


## Nails

**Nails** (Figure 5.4) are plates of tightly packed, hard, keratinized epidermal cells that form a clear, solid covering over the dorsal surfaces of the distal portions of the fingers and toes. Nails help us grasp and manipulate small objects, provide protection against trauma to the ends of the fingers and toes, and allow us to scratch various parts of the body.

Each nail consists of a nail body, a free edge, and a nail root (Figure 5.4). The **nail body** is the visible portion of the nail. Most of the nail body appears pink because of blood flowing through underlying capillaries. The **free edge** is the part of the nail body that may extend past the distal end of the finger or toe. The free edge is white because there are no underlying capillaries. The **nail root** is the proximal portion of the nail that is buried in a fold of skin. The whitish, crescent-shaped proximal end of the nail body is the **lunula** (LOO-noo-la = little moon). It appears whitish because the thickened nail matrix (described shortly) obscures the vascular tissue underneath. The **cuticle**, a narrow band of epidermis, extends over the margin (border) of the nail body. The proximal portion of the epithelium deep to the nail root is the **nail matrix**. Nail matrix cells divide to produce new nail cells. In the process, the harder outer layer is pushed distally toward the free edge.

Nails protect the distal end of the fingers and toes, they allow us to grasp and manipulate small objects, and they can be used to scratch and groom the body in various ways.

### ✓ CHECKPOINT

**12.** What are the functions of hair?

**13.** List the layers of a hair from the innermost component to the outermost component.

**14.** What purpose does the papilla of the hair serve?

**15.** How are goose bumps produced?

**16.** What are the functions of sebum, eccrine sweat, and cerumen?

**17.** What are the functions of nails?

## 5.6 The two major types of skin are thin skin and thick skin.

Although the skin over the entire body is similar in structure, there are quite a few local variations related to thickness of the epidermis, strength, flexibility, degree of keratinization, distribution and type of hair, density and types of glands, pigmentation, vascularity (blood supply), and innervation (nerve supply). Two major types of skin are recognized on the basis of certain structural and functional properties: thin (hairy) skin and thick (hairless) skin. The greatest contributor to epidermal thickness is the increased

**TABLE 5.4**

Comparison of Thin and Thick Skin

| FEATURE | THIN SKIN | THICK SKIN |
|---|---|---|
| **Distribution** | All parts of body except areas such as palms, palmar surface of fingers and toes, and soles | Areas such as palms, palmar surface of fingers and toes, and soles |
| **Epidermal thickness** | 0.10–0.15 mm (0.004–0.006 in.) | 0.6–4.5 mm (0.024–0.18 in.), due mostly to a thicker stratum corneum |
| **Epidermal strata** | Stratum lucidum lacking; thinner strata spinosum and corneum | Stratum lucidum present; thicker strata spinosum and corneum |
| **Epidermal ridges** | Lacking due to poorly developed and fewer dermal papillae | Present due to well-developed and more numerous dermal papillae organized in parallel rows |
| **Hair follicles and arrector pili muscles** | Present | Absent |
| **Sebaceous glands** | Present | Absent |
| **Sudoriferous glands** | Fewer | More numerous |
| **Sensory receptors** | Sparser | Denser |

number of layers in the stratum corneum. This arises in response to the greater mechanical stress in regions of thick skin.

Table 5.4 compares the features of thin and thick skin.

**18.** What feature of skin contributes the most to its thickness?

# 5.7 Skin regulates body temperature, protects underlying tissues, provides cutaneous sensations, excretes body wastes, and synthesizes vitamin D.

Now that you have a basic understanding of the structure of the skin, you can better appreciate its functions. The numerous functions of the integumentary system (mainly the skin) include thermoregulation, storage of blood, protection, cutaneous sensations, excretion and absorption, and synthesis of vitamin D.

## Regulation of Body Temperature

The skin helps regulate body temperature in two ways: by liberating sweat at its surface and by adjusting the flow of blood in the dermis. In response to *high* environmental temperature or heat produced by exercise, sweat secretion from eccrine sweat glands increases; the resulting evaporation of sweat from the skin surface helps lower body temperature. In addition, blood vessels in the dermis of the skin dilate (become wider); consequently, more blood flows through the dermis, which increases the amount of heat radiated from the body. In response to *low* environmental temperature, body heat is conserved through decreased sweat secretion from eccrine sweat glands. Also, blood vessels in the dermis constrict (become narrow), which decreases blood flow through the skin and reduces heat loss from the body.

## Blood Reservoir

The dermis houses an extensive network of blood vessels that carry 8 to 10 percent of the total blood flow in a resting adult. For this reason, the skin acts as a *blood reservoir*.

## Protection

The skin provides protection to the body in various ways. Keratin protects underlying tissues from microbes, abrasion, heat, and chemicals, and the tightly interlocked keratinocytes resist invasion by microbes. Lipids released by lamellar granules inhibit evaporation of water from the skin surface, thus guarding against dehydration; they also retard entry of water across the skin surface when you bathe or swim. The oily sebum from the sebaceous glands keeps skin and hairs from drying out and contains bactericidal chemicals that kill surface bacteria. The acidic pH of perspiration retards the growth of some microbes. Melanin provides some protection against the damaging effects of ultraviolet light. Two types of immune system cells carry out protective functions. Langerhans cells (see Figure 5.2a) in the epidermis alert the immune system to the presence of potentially harmful microbial invaders. Macrophages in the dermis phagocytize bacteria and viruses that manage to bypass the Langerhans cells of the epidermis.

## Cutaneous Sensations

*Cutaneous sensations* are sensations that arise in the skin. These include tactile sensations—touch, pressure, vibration, and tickling—as well as thermal sensations such as warmth and coolness. Another cutaneous sensation, pain, usually is an indication of impending or actual tissue damage. There are a wide variety of sensory receptors distributed throughout the skin, including tactile discs in the epidermis, the corpuscles of touch in the dermis, and hair root plexuses around each hair follicle (see Figures 5.1, 5.2, and 5.3). Chapter 15 provides more details about cutaneous sensations.

## Excretion and Absorption

The skin has a small role in *excretion*, the elimination of substances from the body, and *absorption*, the passage of materials from the external environment into body cells. Despite the almost waterproof nature of the stratum corneum, about 400 mL of water evaporates through it daily. Besides removing water and heat from the body, sweat is the vehicle for excretion of small amounts of salts, carbon dioxide, ammonia, and urea.

The absorption of water-soluble substances through the skin is negligible, but certain lipid-soluble materials do penetrate the skin. These include fat-soluble vitamins (A, D, E, and K), certain drugs, and the gases oxygen and carbon dioxide. Toxic materials that can be absorbed through the skin include organic solvents such as acetone (in some nail polish removers) and carbon tetrachloride (dry-cleaning fluid); salts of heavy metals such as lead, mercury, and arsenic; and substances in poison ivy and poison oak. Since topical (applied to the skin) steroids, such as cortisone, are lipid-soluble, they move easily into the dermis where they exert their anti-inflammatory properties.

## Synthesis of Vitamin D

Synthesis of vitamin D requires activation of a precursor molecule in the skin by ultraviolet (UV) rays in sunlight. Enzymes in the liver and kidneys then modify the activated molecule, finally producing *calcitriol*, the most active form of vitamin D. Calcitriol is a hormone that aids in calcium absorption from foods in the gastrointestinal tract. Only a small amount of exposure to UV light (about 10 to 15 minutes at least twice a week) is required for vitamin D synthesis.

**✓ CHECKPOINT**

**19.** How does the skin help regulate body temperature?

**20.** How does the skin serve as a protective barrier?

**21.** Which sensations arise from stimulation of neurons in the skin?

### RETURN TO **Richard's Story**

The nurses cleaned Richard's burns and applied antibiotic ointments daily, and kept IV fluids and pain medication flowing into his body. His legs and arms were propped up on pillows to keep them higher than his torso. Blisters, collections of fluid that separate skin layers, formed on the outer edges of the burns as the wounds began to form scabs.

The healing process had been painful and seemed very slow, at least to Richard. Every day became a monotonous, excruciating routine: the cleansing of the burns, the *debridement* (scraping and removal) of the dead tissue, and the administration of the IV antibiotics and nausea-inducing pain killers. Richard sometimes felt like dying in the fire would have been less of an ordeal. His fingers were red and raw because he had picked at his cuticles out of boredom. To add insult to injury, the nurses scolded him for it, saying that he was just creating another place where bacteria could enter his body.

His compression bandages tightly hugged the burns and gave him some comfort, so he couldn't really feel as much on his chest and stomach as he did on his face and arms. The pain eventually began to subside.

B. *Hospital staff members continually administer intravenous fluids, clean his wounds, administer antibiotics, and give him pain medication. Based on these observations, what functions of Richard's skin have been compromised? What other functions of skin would be of concern to the medical staff?*

C. *Which component of skin is being damaged as Richard picks at his cuticles?*

## 5.8 Skin damage sets in motion a sequence of events that repairs the skin.

Skin damage sets in motion a sequence of events that repairs the skin to its normal (or near-normal) structure and function. Epidermal wound healing occurs following injuries that affect only the epidermis; deep wound healing occurs following wounds that penetrate the dermis.

### Epidermal Wound Healing

Common types of epidermal wounds include abrasions, in which a portion of skin has been scraped away, and minor burns.

Even though an epidermal wound may extend to the dermis, there is usually only slight damage to epidermal cells at the edges of the wound. In response to an epidermal injury, stratum basale cells of the epidermis surrounding the wound break contact with the basement membrane (Figure 5.5a). The stratum basale cells migrate across the wound until advancing cells from opposite sides of the wound meet. Migration of the stratum basale cells stops completely when each is finally in contact with other stratum basale cells on all sides.

As the stratum basale cells migrate, a hormone called *epidermal growth factor* stimulates basal stem cells to divide and replace the ones that have moved into the wound. The relocated basal cells divide to build new epidermal strata, thus thickening the new epidermis (Figure 5.5b).

FIGURE 5.5 **Skin wound healing.**

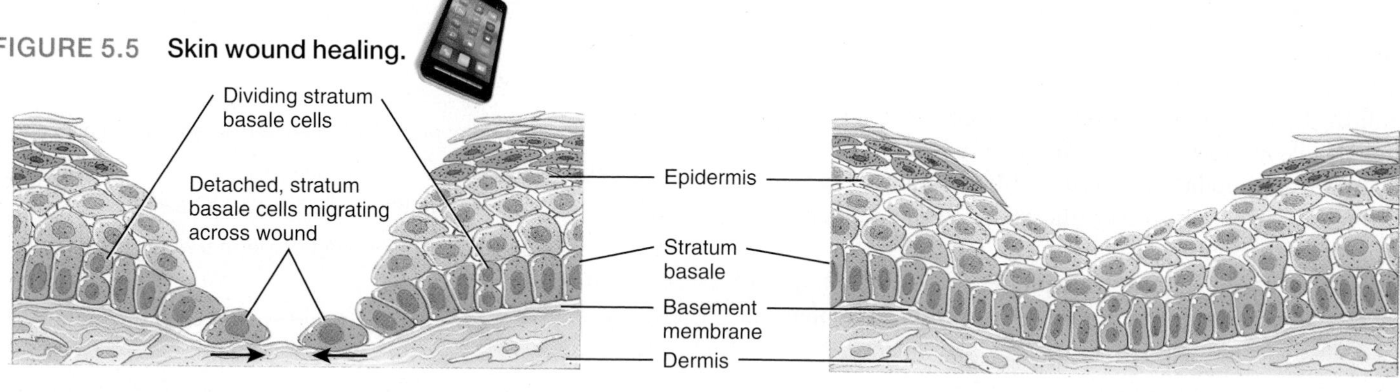


(a) Division of stratum basale cells and migration across wound

(b) Thickening of epidermis

Epidermal wound healing

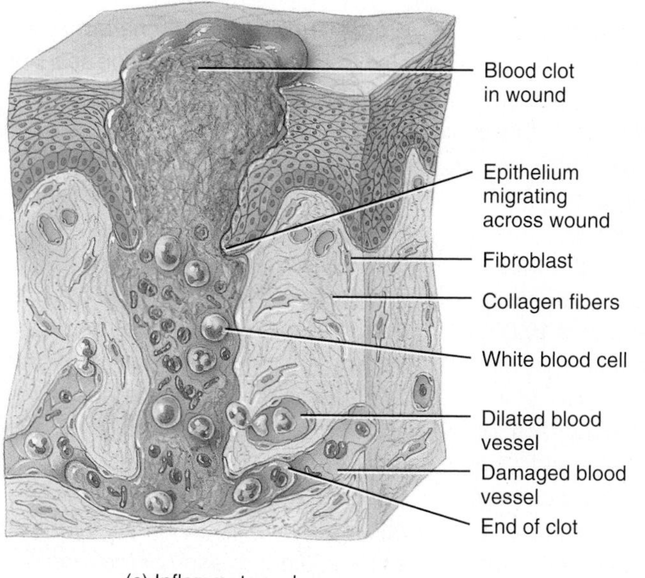


(c) Inflammatory phase

Scab
Resurfaced epithelium
Collagen fibers
Scar tissue
Fibroblast
Restored blood vessel

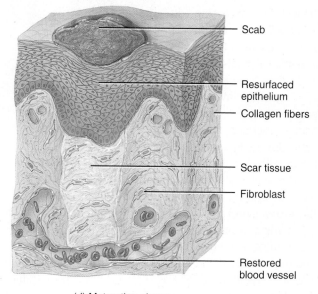

(d) Maturation phase

Deep wound healing

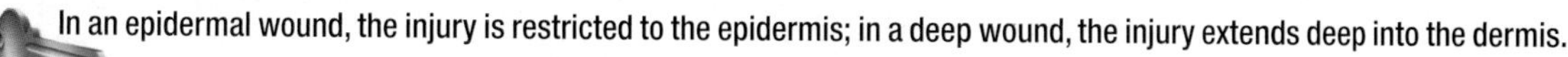
In an epidermal wound, the injury is restricted to the epidermis; in a deep wound, the injury extends deep into the dermis.

## Deep Wound Healing

Deep wound healing occurs when an injury extends to the dermis and hypodermis. Because multiple tissue layers must be repaired, the healing process is more complex than in epidermal wound healing. In addition, because scar tissue forms, the healed tissue loses some of its normal function. Deep wound healing occurs in four phases: an inflammatory phase, a migratory phase, a proliferative phase, and a maturation phase.

During the **inflammatory phase**, bleeding produces a blood clot in the wound, which loosely unites the wound edges (Figure 5.5c). As its name implies, this phase of deep wound healing involves **inflammation**, a response that helps eliminate microbes, foreign material, and dying tissue in preparation for repair. Inflammation increases the permeability and diameter of local blood vessels (vasodilation), enhancing delivery of helpful cells. These include white blood cells, which phagocytize microbes in the wound area, and mesenchymal cells, which develop into fibroblasts.

The three phases that follow do the work of repairing the wound. In the **migratory phase**, the clot becomes a scab, and epithelial cells migrate beneath the scab to bridge the wound. Deep to the epithelial cell bridge, fibroblasts migrate into the wound area and begin synthesizing collagen fibers and glycoproteins to become scar tissue, and damaged blood vessels begin to regrow. During this phase, the tissue filling the wound is called **granulation tissue**. The **proliferative phase** is characterized by extensive growth of epithelial cells beneath the scab, deposition of collagen fibers by fibroblasts, and continued growth of blood vessels. Finally, during the **maturation phase**, the scab sloughs off as the epidermis is restored to normal thickness. In the dermis, collagen fibers become more organized, fibroblasts decrease in number, and blood vessels are restored to normal (Figure 5.5d).

The process of scar tissue formation is called **fibrosis**. Sometimes, so much scar tissue is formed during deep wound healing that a *raised scar*—one that is elevated above the normal epidermal surface—results. Scar tissue differs from normal skin in that its collagen fibers are more densely arranged; it also has less elasticity, fewer blood vessels, and may not contain the same number of hairs, skin glands, or sensory structures as undamaged skin. Because of the arrangement of collagen fibers and the scarcity of blood vessels, scars usually are lighter in color than normal skin.

### ✓ CHECKPOINT

**22.** During wound healing, which epidermal strata can divide to replace lost strata?

**23.** How does inflammation assist in wound healing?

**24.** What is fibrosis? Why doesn't epidermal wound healing result in fibrosis?

## Richard's Story — EPILOGUE AND DISCUSSION

"How bad is this going to look? You know, when the scabs fall off and it finally heals?" he had asked the nurse.

"It's different with each person," she said. "Richard, they can do really great things with skin grafts these days. You have good granulation tissue at the edges of the wounds, and with the skin grafts, well, we'll have to wait and see."

It wasn't what Richard wanted to hear. "Well, maybe I'll have a really hairy chest, and that will cover up the scars." The nurse shook her head.

Richard's healing will be slow, and painful. Not just physically, but emotionally as well. As you have learned, the skin is an organ system with multiple functions. Burns, whether from fires, chemicals, radiation, or electricity, compromise those functions. More than 2 million people each year suffer burn injuries; of these, around 20,000 cases are as severe as Richard's, and over 10,000 people die each year as a result of severe burns.

The skin is not only the surface of the body we present to the world; the integumentary system provides complex functions including protection, sensation, temperature regulation, water and salt balance, and even the synthesis of vitamin D. Being aware of and protecting this important organ system is crucial to maintaining the overall homeostasis of the human body.

D. *Why is it unlikely that Richard will be able to grow hair to cover the scars on his chest?*

E. *What takes place during the first phase of deep wound healing? Relate this process to what Richard has experienced during the initial stages of his burn healing.*

F. *The nurse notes that Richard has granulation tissue forming at the edges of his wounds. Will the formation of granulation tissue lead to normal appearing and normal functioning skin as Richard heals? Explain your answer.*

G. *Describe the possible long-term consequences of Richard's injuries as they relate to each of the functions of the integument described in Concept 5.7.*

# Concept and Resource Summary

**Concept**

**Resources** 

## Introduction

1. The **integumentary system** is composed of the skin, hair, oil and sweat glands, and nails.

## Concept 5.1 Skin is composed of a superficial epidermis and a deeper dermis.

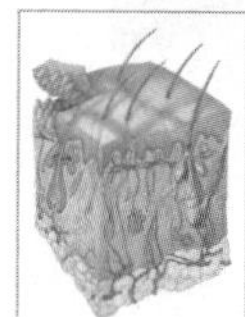

1. The **skin**, the largest organ of the body, is a protective barrier that has sensory receptors, excretes excess salts, synthesizes vitamin D, and regulates body temperature.
2. The two principal parts of the skin are the upper, thin **epidermis** and the deeper, thicker **dermis**. Skin rests on the **hypodermis** that stores fat and has large blood vessels.

Anatomy Overview—The Integumentary System

## Concept 5.2 The layers of the epidermis include the stratum basale, stratum spinosum, stratum granulosum, stratum lucidum, and stratum corneum.

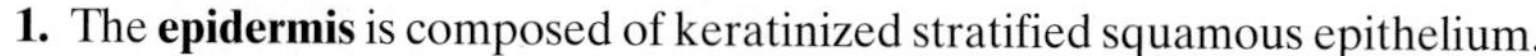

1. The **epidermis** is composed of keratinized stratified squamous epithelium.
2. The most numerous epithelial cells are **keratinocytes** that produce the protective protein **keratin**. **Melanocytes** produce the skin pigment **melanin**. Melanin enters keratinocytes and protects their DNA from UV light damage. **Langerhans cells** defend against microbes, and **tactile cells** function in the sensation of touch.
3. The **stratum basale** is the deepest epidermal layer. It contains keratinocytes, melanocytes, and tactile cells. Some keratinocytes are stem cells that divide to make new keratinocytes.
4. The **stratum spinosum** is superficial to the stratum basale; the **stratum granulosum** is the middle layer, followed by the **stratum lucidum**, which is present only in thick skin. The most superficial layer is the **stratum corneum**. The protective, water-repellent stratum corneum has layers of continually shed dead keratinocytes that get replaced by cells pushing up from deeper strata.
5. **Keratinization** is a process in which keratinocytes gain more keratin as they move toward the surface.

Figure 5.2—Layers of the Epidermis

Clinical Connection—Skin Cancer

## Concept 5.3 The dermis contains blood vessels, nerves, sensory receptors, hair follicles, and glands.

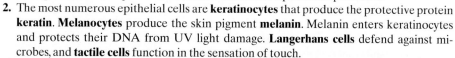

1. The **dermis** is composed of dense irregular connective tissue containing collagen and elastic fibers.
2. The upper **papillary region** contains thin collagen and fine elastic fibers, **corpuscles of touch**, with fingerlike projections called **dermal papillae** that project upward into the epidermis.
3. Dermal papillae contain **capillary loops**, **corpuscles of touch**, and **free nerve endings** that sense temperature, pain, and itching.
4. The deeper **reticular region** is attached to the hypodermis and contains collagen and elastic fibers, fibroblasts, adipose cells, macrophages, hair follicles, blood vessels, nerves, and glands.
5. **Epidermal ridges** project down between dermal papillae on skin of the palms, fingers, soles, and toes. These are the source of fingerprints and footprints.

Clinical Connection—Tension Lines and Surgery

Clinical Connection—Burns

## Concept 5.4 Skin color is a result of the pigments melanin, carotene, and hemoglobin.

1. **Melanin** gives the skin a pale yellow to reddish-brown to black color. The number of melanocytes is the same among people, but the amount of melanin produced differs and leads to different skin colors.
2. **Carotene** is a yellow-orange pigment that can give skin an orange color during dietary excess.
3. **Hemoglobin**, an oxygen-carrying red pigment inside red blood cells in dermal blood vessels, gives skin a reddish hue.

Clinical Connection—Tattooing and Body Piercing

## Concept

## Resources

### Concept 5.5 The functions of hair, skin glands, and nails include protection and body temperature regulation.

1. **Hairs**, which are composed of dead, keratinized epidermal cells held together by extracellular proteins, are present on most skin surfaces except lips, palms, soles, and parts of the external genitalia. The thickness and pattern of hair distribution are determined by hormones and genetics. Hairs offer a limited amount of protection from sunlight, heat loss, and foreign particles. Hairs also sense touch.
2. The **shaft** projects from the skin surface; the deeper **root** penetrates down into the dermis.
3. The hair root is surrounded by the **hair follicle**. The expanded base of the hair follicle is the **bulb**, which houses a cluster of blood vessels within the **papilla**, and the **hair matrix**, which produces new hair cells. Sebaceous (oil) glands, **arrector pili**, and **hair root plexuses** are associated with hair follicles. New hairs develop from division of hair matrix cells in the bulb; hair replacement and growth occur in a cyclical pattern of follicle activity followed by inactivity. Melanocytes in the hair matrix produce the melanin for the various colors of hair.
4. **Sebaceous glands** occur all over the skin except the palms and soles and secrete an oily substance called **sebum** into hair follicles or onto the skin surface. Sebum lubricates the hair and skin, inhibits some bacteria, and prevents water loss.
5. The two types of **sudoriferous glands** are eccrine and apocrine sweat glands. **Eccrine sweat glands** have an extensive distribution; their ducts terminate at pores at the surface of the epidermis. Eccrine sweat helps regulate body temperature through evaporative cooling, is involved in waste removal, and is secreted during emotional stress.
6. **Apocrine sweat glands** are limited to the skin of the axillae, groin, and areolae; their ducts open into hair follicles. Apocrine sweat glands are stimulated during emotional stress and sexual excitement.
7. **Ceruminous glands** are modified sudoriferous glands that secrete **cerumen**. They are found in the external auditory canal (ear canal).
8. **Nails** are hard, tightly packed keratinized epidermal cells that cover the dorsal, distal surfaces of fingers and toes. Nails consist of a **nail body**, **free edge**, and **nail root**. A band of epidermis overlapping the edge of the nail body is the **cuticle**, and a nearby whitish semilunar area is the **lunula**. Deep to the nail root is an area of actively dividing cells called the **nail matrix** from which the nail grows.

Anatomy Overview—Hair
Anatomy Overview—Nails
Figure 5.3—Hair

Clinical Connection—Hair and Hormones
Clinical Connection—Impacted Cerumen
Clinical Connection—Hair Removal

### Concept 5.6 The two major types of skin are thin skin and thick skin.

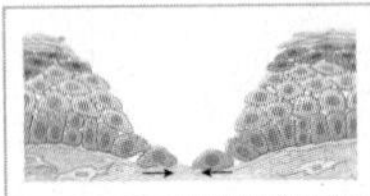

1. Thin skin covers all parts of the body except for the palms, palmar surfaces of the fingers and toes, and the soles.
2. Thick skin covers the palms, palmar surfaces of the fingers and toes, and the soles. Unlike thin skin, thick skin lacks hair and sebaceous glands, and it has more sudoriferous glands.

### Concept 5.7 Skin regulates body temperature, protects underlying tissues, provides cutaneous sensations, excretes body wastes, and synthesizes vitamin D.

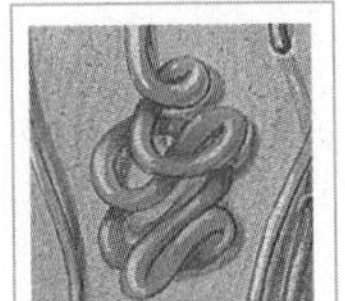

1. The skin helps regulate body temperature by liberating sweat at its surface and by adjusting the flow of blood in the dermis.
2. The skin stores blood.
3. Skin provides barriers that help protect the body. Keratin is a barrier to microbes, chemicals, heat, and abrasions; lipid secretions prevent dehydration; sebum lubricates skin and hair, and is antibacterial; melanin protects against UV light damage; and Langerhans cells and macrophages defend against microbes.
4. The skin has many sensory receptors for tactile, temperature, and pain sensations.
5. Sweat allows for excretion of water, salts, urea, and carbon dioxide. Absorption of fat-soluble vitamins, drugs, oxygen, carbon dioxide, and toxic substances occurs through the skin.
6. UV light from sunlight activates a precursor to vitamin D needed for calcium absorption in the GI tract.

Anatomy Overview—The Skin and Disease Resistance
Animation—Nonspecific Disease Resistance
Clinical Connection—Sun Damage, Sunscreens, and Sunblocks

**Concept** | **Resources**

## Concept 5.8 Skin damage sets in motion a sequence of events that repairs the skin.

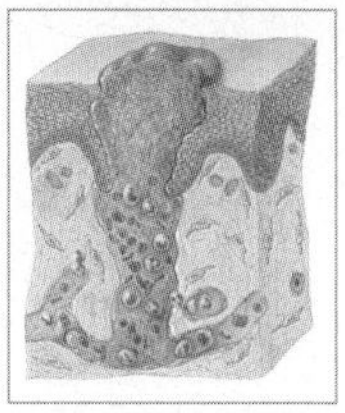

1. Two types of wound healing can occur when skin is damaged: epidermal wound healing and deep wound healing.
2. Healing of epidermal wounds begins as cells from the stratum basale migrate to cover the wound until stopped by contact with other migrating stratum basale cells. Epidermal growth factor stimulates stratum basale cells to divide and replace migrating cells.
3. Deep wounds extend into the dermis or hypodermis, resulting in scar tissue; healing occurs in four phases: the inflammatory phase, migratory phase, proliferative phase, and maturation phase.
4. During the **inflammatory phase**, a blood clot unites the wound edges, epithelial cells migrate across the wound, vasodilation and increased permeability of blood vessels enhance delivery of white blood cells to phagocytize microbes, and mesenchymal cells develop into fibroblasts.
5. During the **migratory phase**, the clot becomes a scab, epithelial cells migrate to form a repair bridge, and fibroblasts migrate into the wound area and produce scar tissue.
6. During the **proliferative phase**, epithelial cells, collagen fibers, and blood vessels grow.
7. The end of repair occurs during the **maturation phase**, when the scab falls off as the epidermis is restored to normal thickness. In the dermis, fibroblasts begin to disappear, and blood vessels are restored to normal.
8. **Fibrosis** produces scar tissue. Scar tissue has a different structure than normal tissue.

Figure 5.5—Skin Wound Healing

Clinical Connection—Pressure Ulcers

## Understanding the Concepts

1. Cells at the surface of skin are keratinized and dead. Why is this a survival advantage?
2. Which epithelial cells can sense that someone is touching your skin? Help protect you from the damaging rays of the sun? Guard against surface bacteria? Help waterproof your skin?
3. Which component of skin lets you know your hand is touching a glass of water? Helps you grip the wet glass? Informs you that the glass is cold?
4. If the number of melanocytes is the same in peoples of all races, what causes their skin color to differ so significantly?
5. Why does it hurt when you pluck a hair but not when you have a haircut? What makes your hair grow longer?
6. Distinguish between eccrine glands and apocrine glands. Which ones are associated with hair follicles?
7. Which part of a nail grows new nail? Which part shows underlying blood capillaries?
8. How does skin on the bottom of the feet differ from that on the face? Why is that difference a survival advantage?
9. How is the skin involved in excretion? In nutrition?
10. Which type of tissue initially fills a deep wound? What does this tissue eventually become?

CHAPTER

# 6

## Cathy's Story

"Good morning Cathy!" calls Mel, as Cathy enters his butcher shop. Cathy smiles, and Mel notices that she has her cane with her today. It must be one of her bad days. She orders a beautiful brisket and examines the ham, but dismisses it with a wave of her hand because she can no longer handle carrying the weight. She can barely make it up the one flight of stairs to her apartment without pain. Mel wraps up the brisket, mock hurt all over his face.

"Cathy, how long have you been coming to me? Over twenty years! My son will be by with your meat within the hour." Cathy pretends to make a fuss, but this is their normal routine. She will grudgingly consent to the purchase and make her way slowly and carefully back to her empty apartment, where she will have a nice tip waiting for Mel's son. As the door closes, Mel's smile fades. He is worried about his friend. When they met, they were both already getting old. He saw her through her children leaving home, the loss of her husband, and, more recently, a noticeable pattern of broken bones. In her late fifties, she broke only a single bone—her wrist—but in her sixties and seventies, she has broken the other wrist, her arm, and several vertebrae. She has taken it all as well as anyone could, always saying, "this is just the price you pay to watch your grandchildren grow up, and otherwise I am still healthy." It is obvious that Cathy is one of millions of women living with primary, age-related osteoporosis. Her accelerated bone loss due to age, exacerbated by early menopause, has caused deterioration in her bone structure and may be affecting all of the bones in her body. In this chapter we will focus on the structure and function of the skeletal system and learn about the homeostatic mechanisms that govern bone growth and remodeling.

# Introduction to the Skeletal System

## INTRODUCTION

Despite their simple appearance, bones are complex and dynamic living tissues that are continuously growing, remodeling, and repairing themselves—as you read these pages, new bone is being built in your body, and old bone is being broken down. A bone is composed of several different tissues working together: osseous tissue, cartilage, dense connective tissues, epithelium, adipose tissue, and nervous tissue. For this reason, each individual bone is considered an organ. Osseous tissue, a complex and dynamic living tissue, continually engages in a process called *remodeling*—the construction of new bone tissue and breaking down of old bone tissue. The entire framework of bones and their cartilages, along with ligaments and tendons, constitute the **skeletal system**. In this chapter we will explore the various components of bones, how bones perform their many functions, and how they develop and remodel.

## CONCEPTS

**6.1** Skeletal system functions include support, protection, movement, mineral homeostasis, blood cell production, and energy storage.

**6.2** Bones are classified as long, short, flat, irregular, or sesamoid.

**6.3** Long bones have a diaphysis, a medullary cavity, epiphyses, metaphyses, and periosteum.

**6.4** Osseous tissue can be arranged as compact bone tissue or spongy bone tissue.

**6.5** Bones are richly supplied with blood vessels and nerves.

**6.6** The two types of bone formation are intramembranous ossification and endochondral ossification.

**6.7** Bones grow longer at the epiphyseal plate and increase in diameter by the addition of new osseous tissue around the outer surface.

**6.8** Bone remodeling renews osseous tissue, redistributes bone extracellular matrix, and repairs bone injuries.

**6.9** Dietary and hormonal factors influence bone growth and remodeling.

## 6.1 Skeletal system functions include support, protection, movement, mineral homeostasis, blood cell production, and energy storage.

The skeletal system performs several basic functions:

- ***Support.*** The skeleton serves as the structural framework for the body by supporting soft tissues and providing attachment points for the tendons of most skeletal muscles.
- ***Protection.*** The skeleton protects the most important internal organs from injury. For example, cranial bones protect the brain, vertebrae (backbones) protect the spinal cord, and the rib cage protects the heart and lungs.
- ***Assistance in movement.*** Most skeletal muscles attach to bones; when they contract, they pull on bones to produce movement. Skeletal muscle movement is discussed in detail in Chapter 10.
- ***Mineral homeostasis.*** Osseous tissue stores several minerals, especially calcium and phosphorus, which contribute to the strength of bone. Osseous tissue stores about 99 percent of the body's calcium. Bones can release minerals on demand into the bloodstream to maintain critical mineral balances (homeostasis) and to distribute minerals to other parts of the body.
- ***Blood cell production.*** Within certain bones, a connective tissue called **red bone marrow** produces red blood cells, white blood cells, and platelets by a process called **hemopoiesis** (hēm-ō-poy-Ē-sis; *hemo-* = blood; *-poiesis* = making). Red bone marrow is present in certain bones, such as the hip (pelvic) bones, ribs, sternum (breastbone), vertebrae, skull, and ends of the bones of the humerus (arm bone) and femur (thigh bone).
- ***Triglyceride storage.*** **Yellow bone marrow** consists mainly of adipocytes, which store triglycerides. The stored triglycerides are a potential chemical energy reserve. In the newborn, all bone marrow is red and is involved in hemopoiesis. With increasing age, much of the bone marrow changes from red to yellow.

✓ **CHECKPOINT**

**1.** Which minerals are stored in bones?

## 6.2 Bones are classified as long, short, flat, irregular, or sesamoid.

Bones can be classified into five principal types: long, short, flat, irregular, and sesamoid (Figure 6.1).

**Long bones** have greater length than width and consist of a shaft and two extremities (ends). They are slightly curved for strength. A curved bone absorbs the stress of the body's weight at several different points so that the stress is evenly distributed. If long bones were straight, the weight of the body would be unevenly distributed and the bone would fracture more easily. Long bones include the femur, tibia and fibula (leg bones), humerus, ulna and radius (forearm bones), and phalanges (finger and toe bones).

**Short bones** are somewhat cube-shaped and are nearly equal in length and width. Examples of short bones are most carpal (wrist) bones and most tarsal (ankle) bones.

**Flat bones** are generally thin, afford considerable protection, and provide extensive surfaces for muscle attachment. Flat bones include the cranial bones, which protect the brain; the sternum (breastbone) and ribs, which protect organs in the thorax; and the scapulae (shoulder blades).

**Irregular bones**, which have complex shapes and cannot be grouped into any of the above categories, include the vertebrae, hip bones, certain facial bones, and the calcaneus (heel bone).

**Sesamoid bones** (SES-a-moyd = shaped like a sesame seed) develop in certain tendons where there is considerable friction, tension, and physical stress, such as those of the palms

FIGURE 6.1 **Types of bones based on shape.** The bones are not drawn to scale.

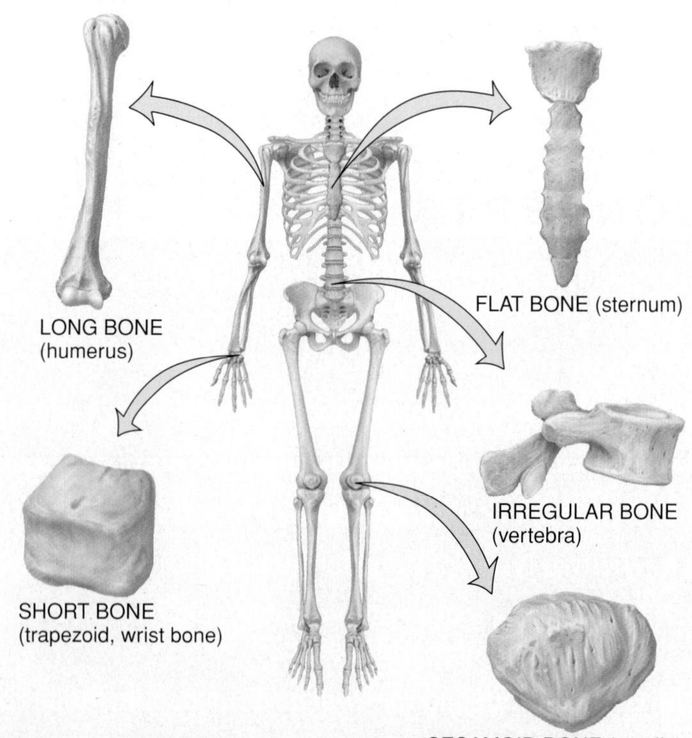


**The shapes of bones largely determine their functions.**

and soles. They typically measure only a few millimeters in diameter except for the two patellae (kneecaps), the largest of the sesamoid bones. Sesamoid bones vary in number from person to person except for the patellae, which are normally present in all individuals. Sesamoid bones protect tendons from excessive wear and tear, and they often change the direction of pull of a tendon, which improves the mechanical advantage at a joint.

✓ **CHECKPOINT**

2. Which type of bone has a boxy shape? Which has an elongated shape?
3. Which type of bone provides a large, flat surface area for muscle attachment?
4. Which type of bone is small and located in tendons?

## 6.3 Long bones have a diaphysis, a medullary cavity, epiphyses, metaphyses, and a periosteum.

The structure of a bone may be examined by considering the components of a long bone such as the humerus (the arm bone) (Figure 6.2). A typical long bone consists of the following parts:

- The **diaphysis** (dī-AF-i-sis = growing between) is the bone's shaft, or body—the long, cylindrical, main portion of the bone. Blood is supplied to a bone through numerous blood vessels such as the nutrient artery (described shortly).

FIGURE 6.2 **Parts of a long bone.**

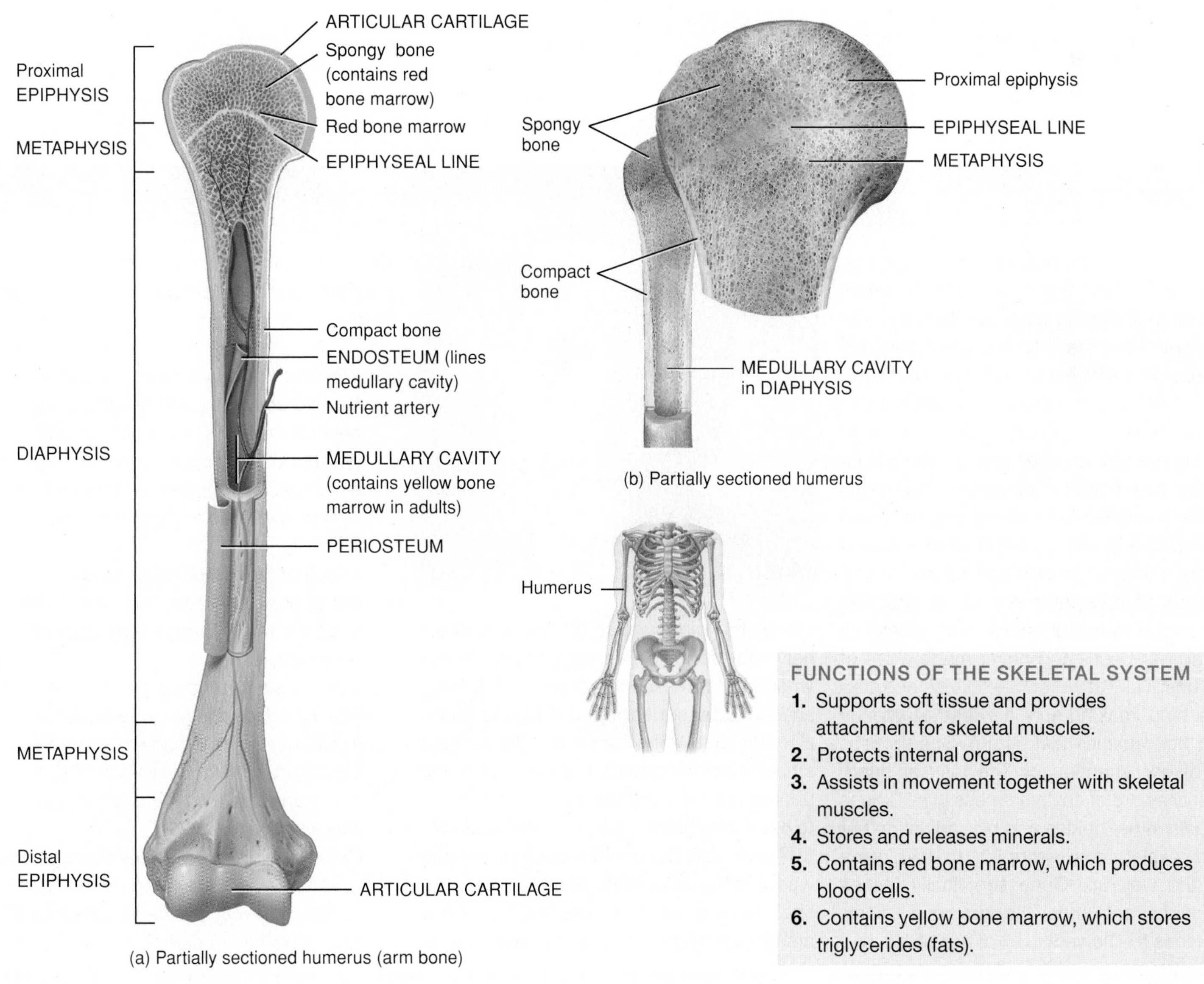


**A long bone is covered by articular cartilage at its proximal and distal epiphyses and by periosteum around the diaphysis.**

- The **epiphyses** (e-PIF-i-sēz = growing over; singular is *epiphysis*) are the proximal and distal ends of the bone.
- The **metaphyses** (me-TAF-i-sēz; *meta-* = between; singular is *metaphysis*) are the regions between the diaphysis and the epiphyses. In a growing bone, each metaphysis contains an **epiphyseal plate** (ep′-i-FIZ-ē-al), a layer of hyaline cartilage that allows the diaphysis of the bone to grow in length (described later in the chapter). When a bone ceases to grow in length at about ages 18–21, the cartilage in the epiphyseal plate is replaced by osseous tissue and the resulting bony structure is known as the **epiphyseal line**.
- The **articular cartilage** is a thin layer of hyaline cartilage covering the part of the epiphysis where the bone forms an articulation (joint) with another bone. Articular cartilage reduces friction and absorbs shock at freely movable joints.
- The **periosteum** (per′-ē-OS-tē-um; *peri-* = around) is a tough connective tissue sheath that surrounds the bone surface wherever it is not covered by articular cartilage. The periosteum contains bone-forming cells that enable bone to grow in thickness. The periosteum also protects the bone, assists in fracture repair, helps nourish osseous tissue, and serves as an attachment point for ligaments and tendons.
- The **medullary cavity** (MED-ū-lar′-ē; *medulla-* = marrow, pith), or **marrow cavity**, is a hollow, cylindrical space within the diaphysis that contains yellow bone marrow in adults.
- The **endosteum** (en-DOS-tē-um; *endo-* = within), a thin membrane that lines the internal bone surface facing the medullary cavity, contains bone-forming cells.

### ✓ CHECKPOINT

**5.** Where are the diaphysis, epiphyses, and metaphyses each located on a long bone?

**6.** Describe the location, composition, and function of the epiphyseal plate, articular cartilage, and periosteum.

## RETURN TO Cathy's Story

Cathy stops at the bottom of the steps that lead up to her apartment. She transfers her three shopping bags—all that she can manage now—to her left shoulder and grasps the railing firmly with her right hand.

"This is my Everest," she says aloud to no one in particular, willing herself to undertake yet another trip up the stairs. Cathy has become extremely conscious of her movements in an attempt to avoid falling. She is very aware of what happens to the elderly after a broken hip and femur. Friends of hers have ended up enduring prolonged hospital stays and extensive rehab—at best—and never made it out of the hospital—at worst. She climbs one step at a time, making sure her feet are steady and her muscles feel strong. She focuses all of her attention on the task at hand, keenly aware of any weakness in her knees or pain in her ankles and hip, adjusting her body weight and stopping to rest twice on the way up. Once she finally makes it inside, she puts the groceries away, replaces the flowers in all of her vases, and sits by the window with a book to wait for her meat delivery. Feeling a twinge in her wrist, she assesses the bones in her wrist and hand by squeezing and feeling them. Her doctor told her once that there were small holes or dents on the surfaces of her bones, but they still feel hard and smooth to her. Cathy has shrunk from 5'3" to barely 5'1" and can't bend and move in ways she was able to just less than five years ago, but she's determined to believe that her bones are still as strong as they feel. With that in mind she remembers to take her calcium supplements and opens the curtains wider to let in more sunlight.

A. *Why would Cathy's osteoporosis affect her ability to carry her groceries, walk up the stairs, and hold a book? What are the functions of bones, and what specific function is Cathy helping by taking vitamin and mineral supplements?*

B. *Which of the five principal types of bones has Cathy already broken? What does this tell you about the nature of osteoporosis?*

C. *Cathy is a shorter than average person. Which part of her long bones is most likely shorter than most people's? Explain how the composition of Cathy's long bones have changed since she was a young child.*

D. *Cathy's doctor has told her that she has "pits" or "dents" on the surface of her bones. Why may this be a concern with respect to bone health?*

## 6.4 Osseous tissue can be arranged as compact bone tissue or spongy bone tissue.

Like other connective tissues, **osseous tissue** (OS-ē-us) contains an abundant extracellular matrix that surrounds widely separated cells. The extracellular matrix is about 15 percent water, 30 percent collagen fibers, and 55 percent crystallized mineral salts. The most abundant mineral salt is calcium phosphate [$Ca_3(PO_4)_2$]. It combines with another mineral salt, calcium hydroxide [$Ca(OH)_2$], to form crystals of **hydroxyapatite** [$Ca_{10}(PO_4)_6(OH)_2$] (hī-drok-sē-AP-a-tīt). As these mineral salts are deposited in the framework formed by the collagen fibers of the extracellular matrix, they crystallize and the tissue hardens. This process of **calcification** (kal′-si-fi-KĀ-shun) is initiated by bone-building cells called osteoblasts (described shortly).

The combination of crystallized salts and collagen fibers is responsible for the characteristics of bone. Although a bone's *hardness* depends on the crystallized inorganic mineral salts, a bone's *flexibility* depends on its collagen fibers. Like reinforcing metal rods in concrete, collagen fibers and other organic molecules provide *tensile strength*, resistance to being stretched or torn apart.

### Cells of Osseous Tissue

Four types of cells are present in osseous tissue: osteogenic cells, osteoblasts, osteocytes, and osteoclasts (Figure 6.3).

- **Osteogenic cells** (os′-tē-ō-JEN-ik; *-genic* = producing) are unspecialized stem cells (undifferentiated cells that can divide to produce new cells). They are the only bone cells to undergo cell division; the resulting cells develop into osteoblasts. Osteogenic cells are found in the periosteum, in the endosteum, and in the canals within bones that contain blood vessels.
- **Osteoblasts** (OS-tē-ō-blasts′; *-blasts* = buds or sprouts) are bone-building cells. They synthesize and secrete collagen

FIGURE 6.3 Types of cells in osseous tissue.

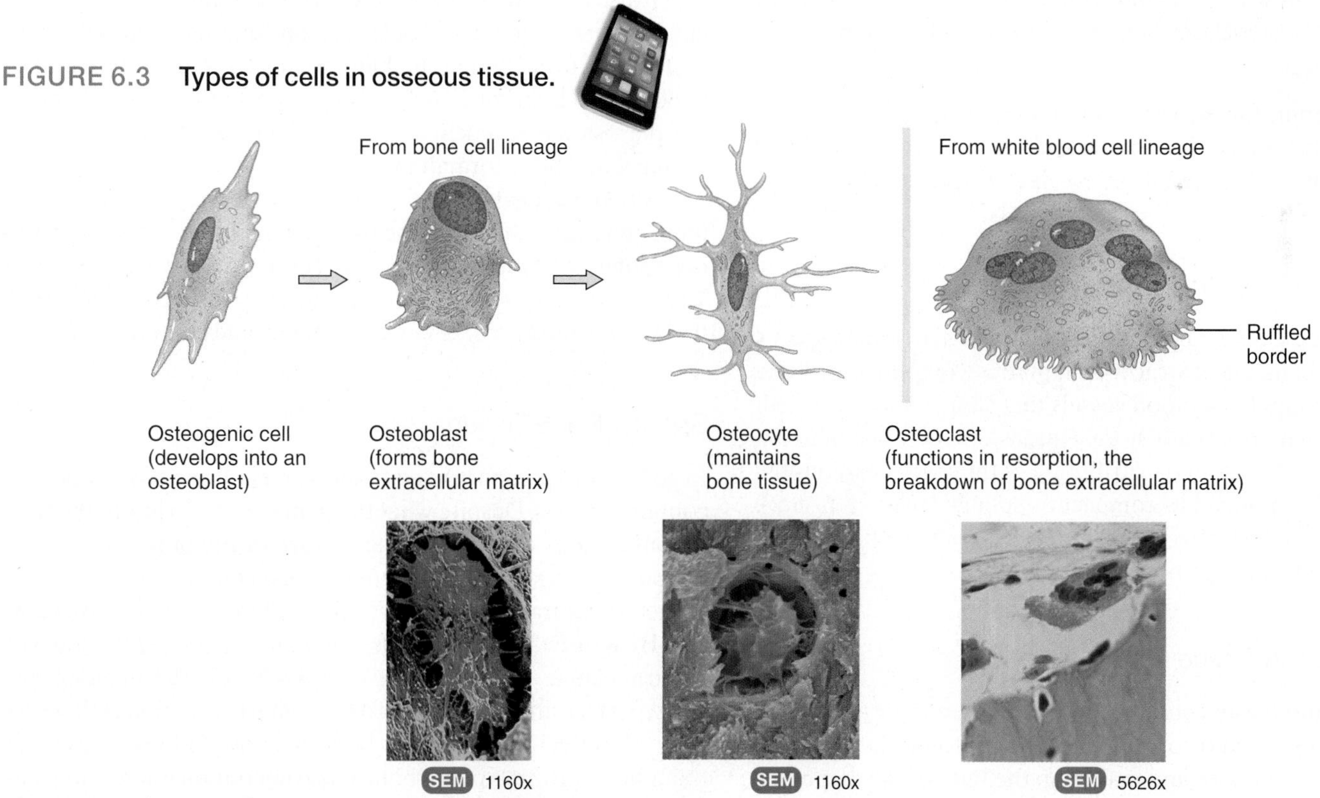


Osteogenic cells develop into osteoblasts, which secrete the extracellular matrix of osseous tissue.

fibers and other organic components needed to build the extracellular matrix of osseous tissue, and they initiate calcification (described shortly). As osteoblasts surround themselves with extracellular matrix, they become trapped in their secretions and become osteocytes. (Note: Cells with the ending *blasts* in their name means that the cells secrete extracellular matrix.)

- **Osteocytes** (OS-tē-ō- sīts′; *-cytes* = cells), mature bone cells, are the most numerous cells in osseous tissue and maintain its daily metabolism, such as the exchange of nutrients and wastes with the blood. Like osteoblasts, osteocytes do not undergo cell division. (Note: Cells with the ending *cytes* in the name means that the cells maintain the tissue.)
- **Osteoclasts** (OS-tē-ō-clasts′; *-clast* = break) are huge cells derived from the fusion of as many as 50 monocytes (a type of white blood cell) and are concentrated in the endosteum. On the side of the cell that faces the bone surface, the osteoclast's plasma membrane is deeply folded into a *ruffled border*. Here the osteoclast releases powerful lysosomal enzymes and acids that digest the underlying extracellular matrix of osseous tissue. This breakdown of the extracellular matrix, termed *resorption* (rē-SORP-shun), is part of the normal development, maintenance, and repair of bones. (Note: Cells with the ending *clast* in the name means that the cell breaks down extracellular matrix.) In response to certain hormones, osteoclasts help regulate blood calcium levels (see Concept 6.9).

A mnemonic that will help you remember the difference between the function of osteoblasts and osteoclasts is as follows: osteo**B**lasts **B**uild bone, while osteo**C**lasts **C**arve out bone.

## Types of Osseous Tissue

Osseous tissue is not completely solid but has many small spaces between its cells and extracellular matrix components. Some spaces are channels for blood vessels that supply bone cells with nutrients. Other spaces are storage areas for red bone marrow. Depending on the size and distribution of the spaces, osseous tissue may be categorized as compact or spongy (see Figure 6.2). Overall, about 80 percent of the skeleton is compact bone tissue and 20 percent is spongy bone tissue.

### Compact Bone Tissue

**Compact bone tissue** contains few spaces (see Figure 6.2) and is the strongest form of osseous tissue. It is found beneath the periosteum of all bones and makes up the bulk of the diaphyses of long bones. Compact bone tissue provides protection and support and resists the stresses produced by weight and movement.

Compact bone tissue is composed of repeating structural units called **osteons** or **haversian systems** (ha-VER-shun) (Figure 6.4a). Each osteon consists of **concentric lamellae** (la-MEL-ē)—rings of calcified extracellular matrix much like the growth rings of a sliced tree trunk that are arranged around a **central** or **haversian canal** (ha-VER-shun). Osteons generally form a series of cylinders that tend to run parallel to the long axis of the bone. Between lamellae are small spaces called **lacunae** (la-KOO-nē = little lakes; singular is *lacuna*), which contain osteocytes.

Radiating in all directions from the lacunae are tiny **canaliculi** (kan′-a-LIK-ūlī = small channels) filled with extracellular fluid. Inside the canaliculi are slender fingerlike processes of osteocytes (see inset at the right of Figure 6.4a). Neighboring osteocytes communicate via gap junctions. The canaliculi connect lacunae with one another and with the central canals, forming an intricate, miniature system of interconnected canals throughout the bone. This system provides many routes for nutrients and oxygen to reach the osteocytes and for the removal of wastes.

Osteons are aligned in the same direction and are parallel to the length of the diaphysis. As a result, the shaft of a long bone resists bending or fracturing even when considerable force is applied from either end. The lines of stress in a bone are not static. They change as a person learns to walk and in response to repeated strenuous physical activity, such as weight training. The lines of stress in a bone can also change in response to fractures or physical deformity. Thus, the organization of osteons changes over time in response to the physical demands placed on the skeleton.

The areas between osteons contain **interstitial lamellae**, which also have lacunae with osteocytes and canaliculi. Interstitial lamellae are fragments of older osteons that have been partially destroyed during bone rebuilding or growth. **Circumferential lamellae** (ser′-kum-fer-EN-shē-al) encircle the bone just beneath the periosteum or encircle the medullary cavity. They develop during initial bone formation.

Blood vessels, lymphatic vessels, and nerves from the periosteum penetrate compact bone tissue through transverse **perforating canals**, or *Volkmann's canals* (FŌLK-mans) (Figure 6.4a). The vessels and nerves of the perforating canals connect with those of the medullary cavity, periosteum, and central canals.

### Spongy Bone Tissue

In contrast to compact bone tissue, **spongy bone tissue** does not contain osteons. Despite what the name seems to imply, the term "spongy" does not refer to the texture of the bone, only its appearance (Figure 6.4b, c). Spongy bone consists of lamellae arranged in an irregular lattice of thin columns called **trabeculae** (tra-BEK-ū-lē = little beams; singular is *trabecula*). Between the trabeculae are spaces that are visible to the unaided eye. These spaces are filled with red bone marrow in bones that produce blood cells, and yellow bone marrow (adipose tissue) in other bones. Both types of bone marrow contain numerous small blood vessels that provide nourishment to the osteocytes. Each trabecula consists of concentric lamellae, osteocytes that lie in lacunae, and canaliculi that radiate outward from the lacunae.

Spongy bone tissue is always located in the interior of a bone where it is protected by a covering of compact bone tissue. It makes

**FIGURE 6.4 Histology of compact and spongy bone tissue.** (a) Sections through the diaphysis of a long bone, from surrounding periosteum on the right, to compact bone tissue in the middle, to spongy bone tissue and the medullary cavity on the left. The inset at upper right shows an osteocyte in a lacuna. (b and c) Details of spongy bone tissue.

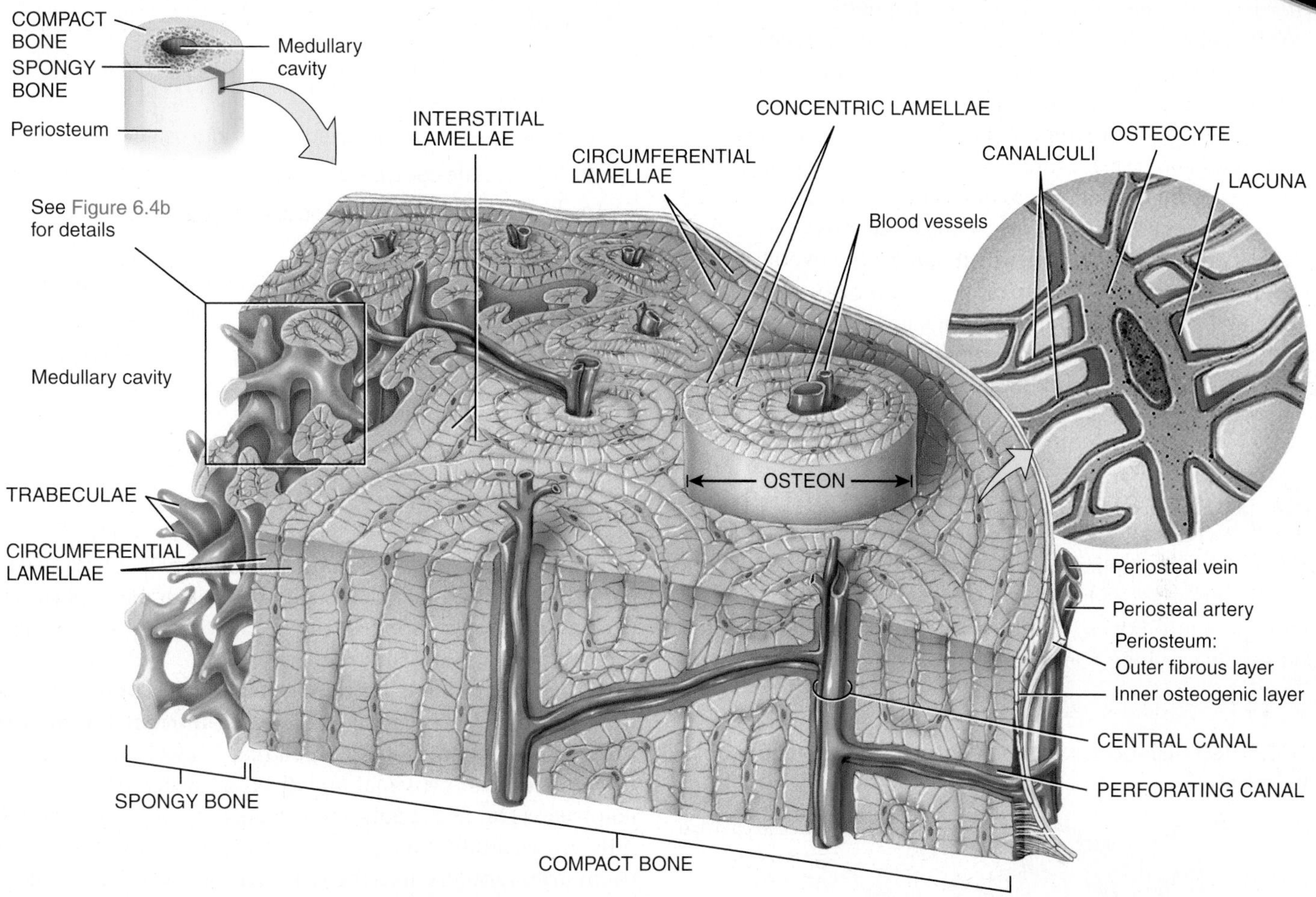

(a) Osteons (haversian systems) in compact bone and trabeculae in spongy bone

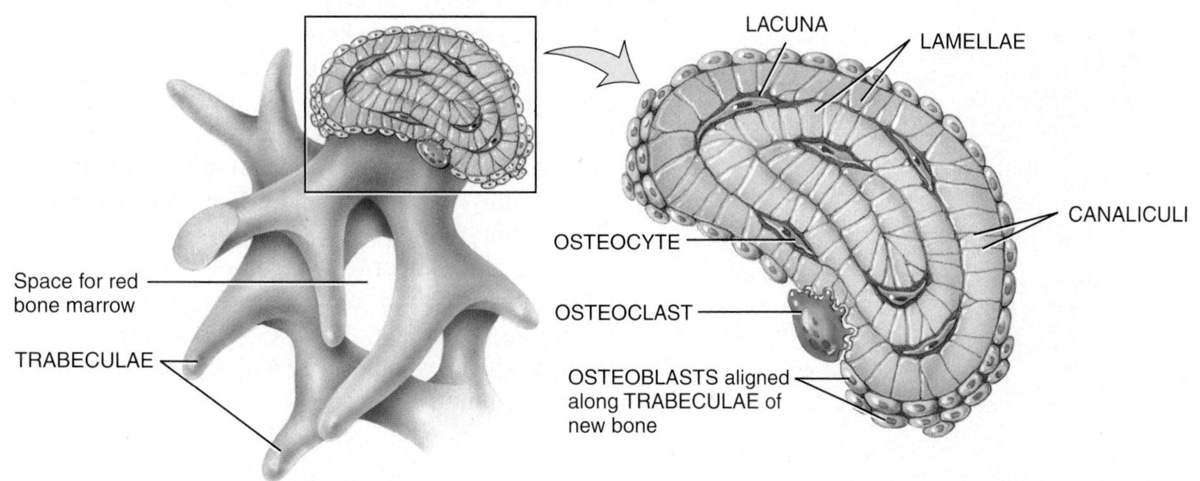


(b) Enlarged aspect of spongy bone trabeculae

(c) Details of a section of a trabecula

**Compact bone tissue is arranged in concentric circles around a central canal; spongy bone tissue is arranged irregularly in trabeculae.**

up most of the interior of short, flat, irregular, and sesamoid bones. In long bones, spongy bone tissue forms the core of the epiphyses beneath a paper-thin layer of compact bone, and forms a narrow rim bordering the medullary cavity of the diaphysis.

At first glance, the trabeculae of spongy bone tissue may appear to be less organized than the osteons of compact bone tissue. However, the trabeculae of spongy bone tissue are precisely oriented along lines of stress, a characteristic that helps bones resist stresses and transfer force without breaking. Spongy bone tissue tends to be located where bones are not heavily stressed or where stresses are applied from many directions.

Spongy bone tissue is light, which reduces the overall weight of a bone and allows the bone to move more readily when pulled by a skeletal muscle. The trabeculae of spongy bone tissue support and protect red bone marrow. Spongy bone tissue in the hip bones, ribs, sternum (breastbone), vertebrae, and the proximal ends of long bones contain red bone marrow, the site of hemopoiesis (blood cell production) in adults.

✓ **CHECKPOINT**

7. What is the composition of the extracellular matrix of osseous tissue?
8. Which cells turn into osteoblasts? Into osteocytes?
9. Which bone cell type is derived from white blood cells?
10. List the functions of osteoblasts, osteocytes, and osteoclasts.
11. How are compact and spongy bone tissues different in microscopic appearance, location, and function?
12. Which kind of bone marrow is in spongy bone tissue?

## 6.5 Bones are richly supplied with blood vessels and nerves.

Bones are richly supplied with blood. Blood vessels, which are especially abundant in portions of bones containing red bone marrow, pass into bones from the periosteum. We consider the blood supply of a long bone such as the tibia (shin bone) shown in Figure 6.5.

FIGURE 6.5 **Blood supply of a mature long bone, the tibia (shin bone).**

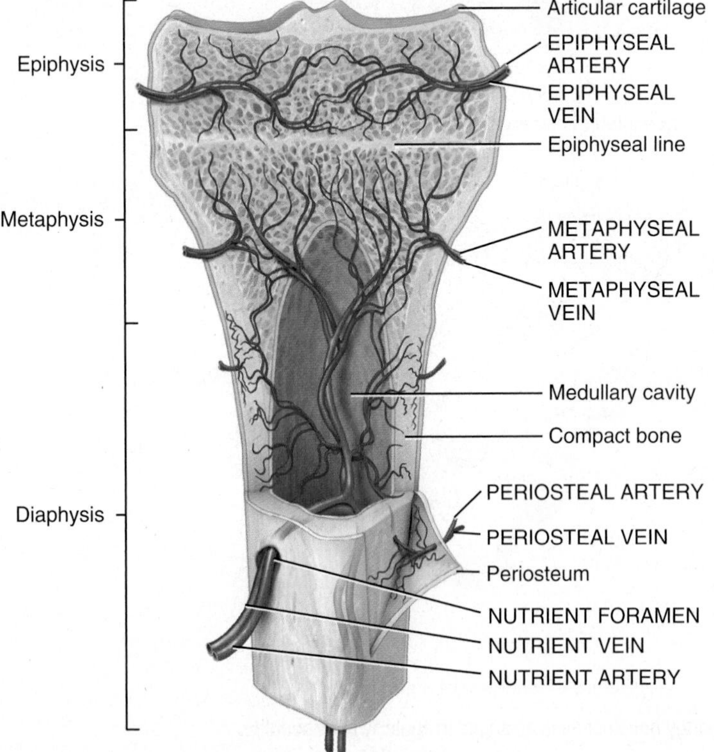


Partially sectioned tibia (shin bone)

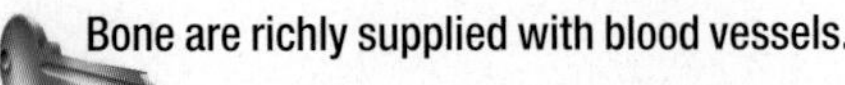
**Bone are richly supplied with blood vessels.**

**Periosteal arteries** (per′-ē-OS-tē-al), small arteries accompanied by nerves, enter the diaphysis through many perforating canals and supply blood to the periosteum and outer part of the compact bone tissue (see Figure 6.4a). Near the center of the diaphysis, a large **nutrient artery** passes through a hole in the compact bone tissue called the **nutrient foramen** (see Figure 6.2). On entering the medullary cavity, the nutrient artery divides into proximal and distal branches that supply the diaphysis as far as the epiphyseal plates (or epiphyseal lines in a fully-grown adult). Some bones, like the tibia, have only one nutrient artery; others, like the femur, have several. **Metaphyseal arteries** (met-a-FIZ-ē-al) enter the metaphyses of a long bone and, together with the nutrient artery, supply blood to the metaphyses. **Epiphyseal arteries** (ep′-i-FIZ-ē-al) supply blood to the epiphyses of a long bone.

Veins accompany their respective arteries to carry blood away from bones. One or two **nutrient veins** exit through the diaphysis. Numerous **epiphyseal veins** exit long bones through the epiphyses, while **metaphyseal veins** exit through the metaphysis. Many small **periosteal veins** exit through the periosteum.

Nerves accompany the blood vessels that supply bones. The periosteum is rich in sensory nerves, some of which carry pain sensations. These nerves are especially sensitive to tearing or tension, which explains the severe pain resulting from a fracture or a bone tumor.

✓ **CHECKPOINT**

13. Where do periosteal arteries enter a long bone?
14. Where does the nutrient artery enter a long bone?

## CLINICAL CONNECTION | *Bone Scan*

A **bone scan** is a diagnostic procedure that takes advantage of the fact that bone is living tissue. A small amount of a radioactive tracer compound that is readily absorbed by bone is injected intravenously. The degree of uptake of the tracer is related to the amount of blood flow to the bone. A scanning device (gamma camera) measures the radiation emitted from the bones, and the information is translated into a photograph that can be read like an x-ray on a monitor. Normal osseous tissue is identified by a consistent gray color throughout because of its uniform uptake of the radioactive tracer. Darker or lighter areas may indicate bone abnormalities. Darker areas called "hot spots" are areas of increased metabolism that absorb more of the radioactive tracer due to increased blood flow. Hot spots may indicate bone cancer, abnormal healing of fractures, or abnormal bone growth. Lighter areas called "cold spots" are areas of decreased metabolism that absorb less of the radioactive tracer due to decreased blood flow. Cold spots may indicate problems such as degenerative bone disease, decalcified bone, fractures, bone infections, Paget's disease, and rheumatoid arthritis. A bone scan detects abnormalities three to six months sooner than standard x-ray procedures and exposes the patient to less radiation. A bone scan is the standard test for bone density screening, and is particularly important in screening for osteoporosis in females.

## RETURN TO **Cathy's Story**

"Cathy, we've received the results of your bone mineral density (BMD) tests and x-rays. I'm afraid that you do indeed have osteoporosis." Staring out the window of her apartment, Cathy remembers hearing Dr. Andrews' words and trying to figure out what they meant. The test had been performed after she experienced a number of broken bones over the course of several years, starting with her wrist. The doctor had explained that bones are always changing, and for multiple reasons, now Cathy's bones were breaking down much faster than they could be built back up. Dr. Andrews had also told her that osteoporosis has the greatest effects on spongy bone tissue, the most fragile part of bones, but that losses in all types of bone tissues are to be expected.

The doorbell rings, pulling Cathy back into the present, and she rises slowly and carefully from her chair. On her way to the door, she passes through the hallway, which is lined by ornate picture frames holding her wedding photos and portraits of her four children and six grandchildren. An empty spot waits for a photo of her next grandchild, due any day now. She looks down at her wrists and flexes her fingers, straightens her back, and stretches cautiously. Everything broken so far has healed and she has much to be thankful for. She knows she shouldn't be living alone anymore, but she can't bring herself to give up, letting the disease win. As Dr. Andrews said, her "bones are always changing," and, well, this is just another change. She opens the door, accepts her package, and hands the grinning delivery boy some money. She smiles as he bounds down the same stairs it took her forever to climb just a little while ago. As she closes the door and shuffles to the kitchen to preheat the oven, she smiles to herself because she knows that she is still winning.

E. *Why did Dr. Andrews perform a "bone mineral density" test? What would you expect her doctor to have seen in her x-rays that led her to the diagnosis of osteoporosis? What types of molecules are most likely "missing" from Cathy's bones?*

F. *Dr. Andrews told Cathy that her bones are breaking down much faster than they can be built back up. Which cells are involved in this process and how are they involved in Cathy's osteoporosis?*

G. *In what type of osseous tissue will Cathy experience the greatest structural loss? Why would this type of osseous tissue be more vulnerable to damage than the other type?*

# 6.6 The two types of bone formation are intramembranous ossification and endochondral ossification.

The process by which bones form is called **ossification** (os′-i-fi-KĀ-shun; *ossi-* = bone; *-fication* = making). Bone formation occurs in three principal situations: (1) the initial formation of bones in an embryo and fetus, (2) the growth of bones until their adult sizes are reached, and (3) the remodeling and repair of bones. (Remodeling is the replacement of old osseous tissue with new osseous tissue.)

We will first consider the initial formation of bones in an embryo and fetus. The embryonic "skeleton" is originally composed of mesenchyme in the general shape of bones. **Mesenchyme**

(MEZ-en-kīm) is the tissue from which other connective tissues arise. These bone precursors are the site of ossification during the sixth week of embryonic development.

Bone formation follows one of two patterns. During **intramembranous ossification** (in′-tra-MEM-bra-nus; *intra-* = within; *-membran* = membrane), a bone develops directly within mesenchyme, which is arranged in sheetlike layers that resemble membranes. In the second method, **endochondral ossification** (en′-dō-KON-dral; *endo-* = within; *-chondral* = cartilage), a bone forms within hyaline cartilage that develops from mesenchyme.

## Intramembranous Ossification

Intramembranous ossification is the simpler of the two methods of bone formation. The flat bones of the skull, the mandible (lower jawbone), and clavicle (collar bone) are formed in this way. Also, the "soft spots" that help the fetal skull pass through the birth canal later harden as they undergo intramembranous ossification. Intramembranous ossification occurs as follows (Figure 6.6):

1. ***Development of the ossification center.*** At the site where the bone will develop, specific chemical messages cause the cells in mesenchyme to cluster together and differentiate, first into osteogenic cells and then into osteoblasts. The site of such a cluster is called an **ossification center**. Osteoblasts secrete bone extracellular matrix until they are surrounded by it.
2. ***Calcification.*** After secretion of extracellular matrix stops, the cells, now called osteocytes, lie in lacunae and extend their narrow cytoplasmic processes into canaliculi that radiate in all directions. Within a few days, calcium and other mineral salts are deposited and the extracellular matrix hardens as it calcifies (calcification). The mesenchyme continue to produce osteogenic cells, which form a peripheral layer of osteoblasts.

**FIGURE 6.6 Intramembranous ossification.** Illustrations 1 and 2 show a smaller field of vision at higher magnification than illustrations 3 and 4.

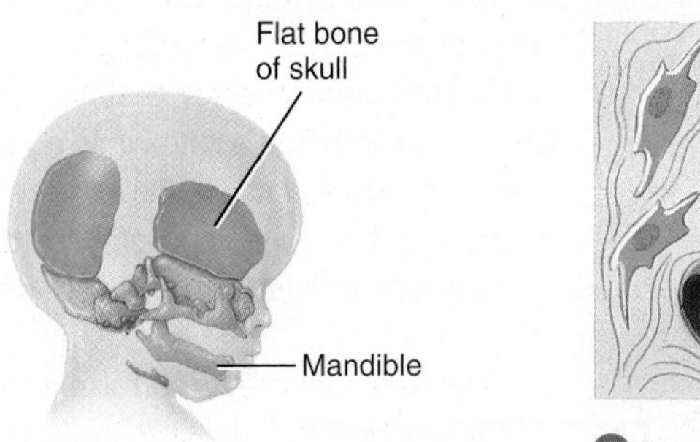


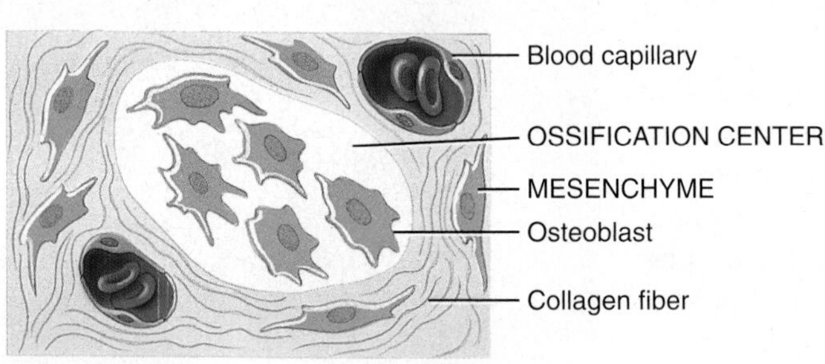


1 Development of ossification center

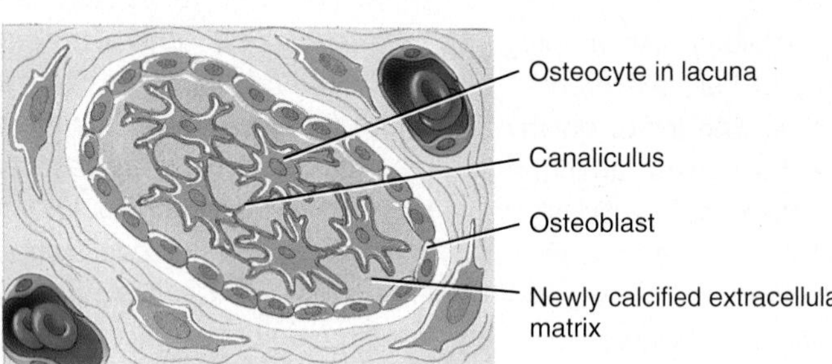


2 Calcification

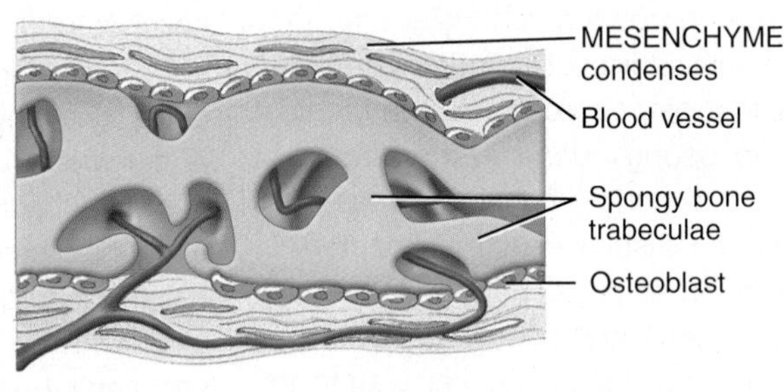


3 Formation of trabeculae

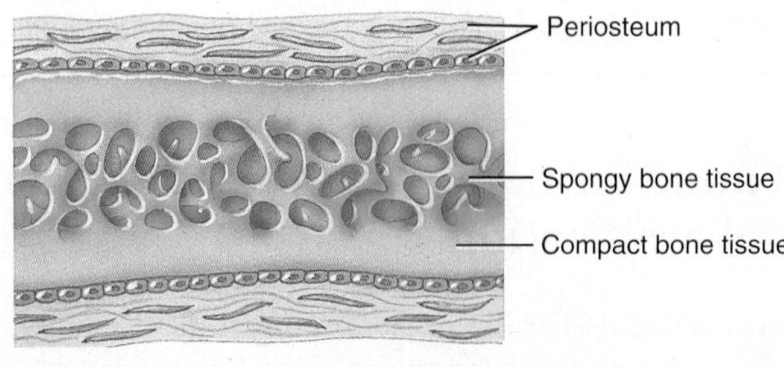


4 Development of periosteum

**Intramembranous ossification involves the formation of a bone within mesenchyme arranged in sheetlike layers that resemble membranes.**

3. ***Formation of trabeculae.*** As the bone extracellular matrix forms, it develops into trabeculae that fuse with one another to form spongy bone tissue. Blood vessels grow into the spaces between the trabeculae and deposit connective tissue cells that differentiate into red bone marrow.
4. ***Development of the periosteum.*** At the periphery, the mesenchyme condenses and develops into the periosteum. Within the periosteum, osteoblasts produce a thin layer of compact bone tissue, but spongy bone tissue remains in the center. Much of the newly formed bone is remodeled as the bone is transformed into its adult size and shape.

## Endochondral Ossification

The replacement of cartilage by osseous tissue is called endochondral ossification. Although most bones of the body are formed in this way, the process is best observed in a long bone. It proceeds as follows (Figure 6.7):

1. ***Development of the cartilage model.*** At the site where the bone is going to form, specific chemical messages cause the cells in mesenchyme to crowd together in the general shape of the future bone, and then develop into **chondroblasts**, immature cells that secrete the extracellular matrix

FIGURE 6.7 Endochondral ossification.

1 Development of cartilage model

2 Growth of cartilage model

3 Development of primary ossification center

4 Development of the medullary cavity

5 Development of secondary ossification centers

6 Formation of articular cartilage and epiphyseal plate

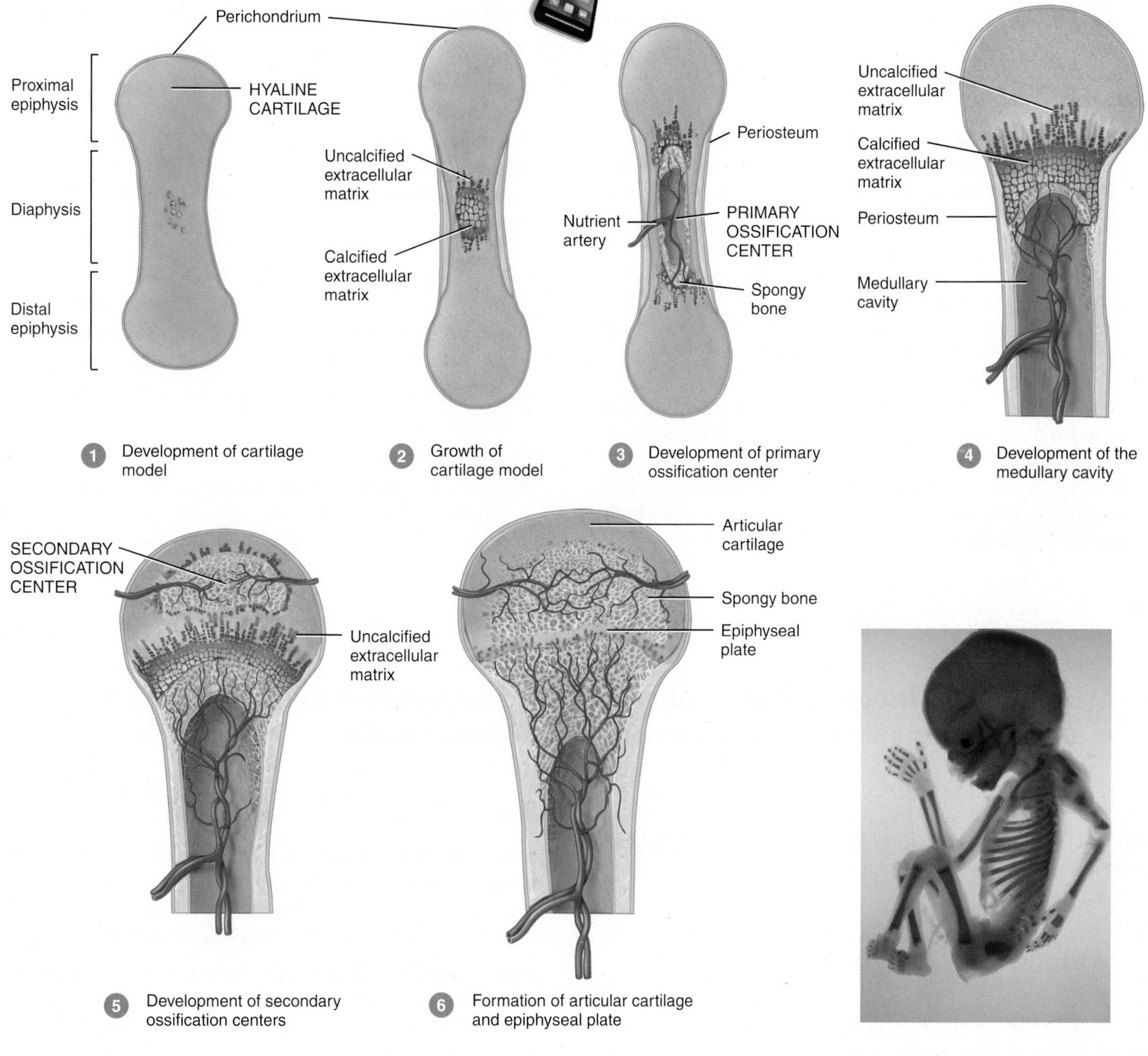

(a) Sequence of events

(b) Twelve-week fetus. The red areas represent bones that are forming (calcified). Yellow areas represent cartilage (uncalcified).

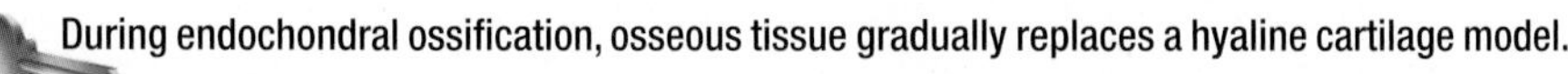

During endochondral ossification, osseous tissue gradually replaces a hyaline cartilage model.

of cartilage. The chondroblasts secrete cartilage extracellular matrix, producing a **hyaline cartilage model**. A covering called the **perichondrium** (per′-i-KON-drē-um) develops around the hyaline cartilage model.

2 *Growth of the cartilage model*. Once chondroblasts become deeply buried in the cartilage extracellular matrix, they are called chondrocytes. The cartilage model grows by continual cell division of chondrocytes accompanied by further secretion of the cartilage extracellular matrix. As the cartilage model continues to grow, chondrocytes in its mid-region *hypertrophy* (increase in size) and stimulate the surrounding extracellular matrix to calcify. Other chondrocytes within the calcifying cartilage die because nutrients can no longer diffuse quickly enough through the extracellular matrix. As these chondrocytes die, the spaces left behind by dead chondrocytes merge into small cavities called lacunae.

3 ***Development of the primary ossification center***. A nutrient artery penetrates the perichondrium and the calcifying cartilage model through a nutrient foramen in the mid-region of the cartilage model, stimulating cells within the perichondrium to differentiate into osteoblasts instead of chondroblasts. The osteoblasts secrete a thin shell of compact bone tissue deep to the perichondrium. Once osseous tissue forms beneath the perichondrium, the covering is known as the *periosteum*. Near the middle of the model, periosteal capillaries grow into the disintegrating calcified cartilage, inducing growth of a **primary ossification center**, a region where osseous tissue will replace cartilage. Osteoblasts then begin to deposit bone extracellular matrix over the remnants of calcified cartilage, forming spongy bone trabeculae. Primary ossification spreads toward both ends of the cartilage model.

4 ***Development of the medullary cavity***. As the primary ossification center grows toward the ends of the bone, osteoclasts break down some of the newly formed spongy bone trabeculae. This activity leaves a cavity, the medullary or marrow cavity, in the diaphysis. Invading blood vessels fill the medullary cavity with red bone marrow, which is progressively replaced by yellow bone marrow. Eventually, most of the wall of the diaphysis is replaced by compact bone.

5 ***Development of the secondary ossification centers***. When branches of the epiphyseal artery enter the epiphyses, **secondary ossification centers** develop. Bone formation in the secondary ossification centers is similar to that in primary ossification centers. One difference, however, is that spongy bone tissue remains in the interior of the epiphyses (no medullary cavities are formed).

6 ***Formation of articular cartilage and the epiphyseal plate***. The hyaline cartilage that covers the epiphyses becomes the articular cartilage. Prior to adulthood, hyaline cartilage remains between the diaphysis and epiphysis as the epiphyseal plate, which is responsible for the lengthwise growth of long bones and is discussed next.

✓ **CHECKPOINT**

**15.** What are the major events of intramembranous ossification? Of endochondral ossification?

**16.** How do the primary and secondary ossification centers differ?

**17.** How do the medullary cavities in a long bone form?

**18.** Which ossification method would form a femur? Which would form most skull bones?

## 6.7 Bones grow longer at the epiphyseal plate and increase in diameter by the addition of new osseous tissue around the outer surface.

During infancy, childhood and adolescence, long bones lengthen by the addition of osseous tissue on the diaphyseal side of the epiphyseal plate and bones throughout the body grow in thickness at the outer surface.

### Growth in Length

The growth in length of long bones involves two major events: (1) growth of cartilage on the epiphyseal side of the epiphyseal plate and (2) replacement of cartilage on the diaphyseal side of the epiphyseal plate with osseous tissue. The *epiphyseal plate*, a layer of hyaline cartilage in the metaphysis of a growing bone, consists of four zones (Figure 6.8):

- ***Zone of resting cartilage.*** This layer is nearest the epiphysis and consists of small, scattered chondrocytes. The cells do not function in bone growth (thus the term *resting*). Instead, they anchor the epiphyseal plate to the epiphysis of the bone.
- ***Zone of proliferating cartilage.*** Slightly larger chondrocytes in this zone are arranged like stacks of coins. These chondrocytes secrete cartilage extracellular matrix and divide to replace chondrocytes that die at the diaphyseal side of the epiphyseal plate.

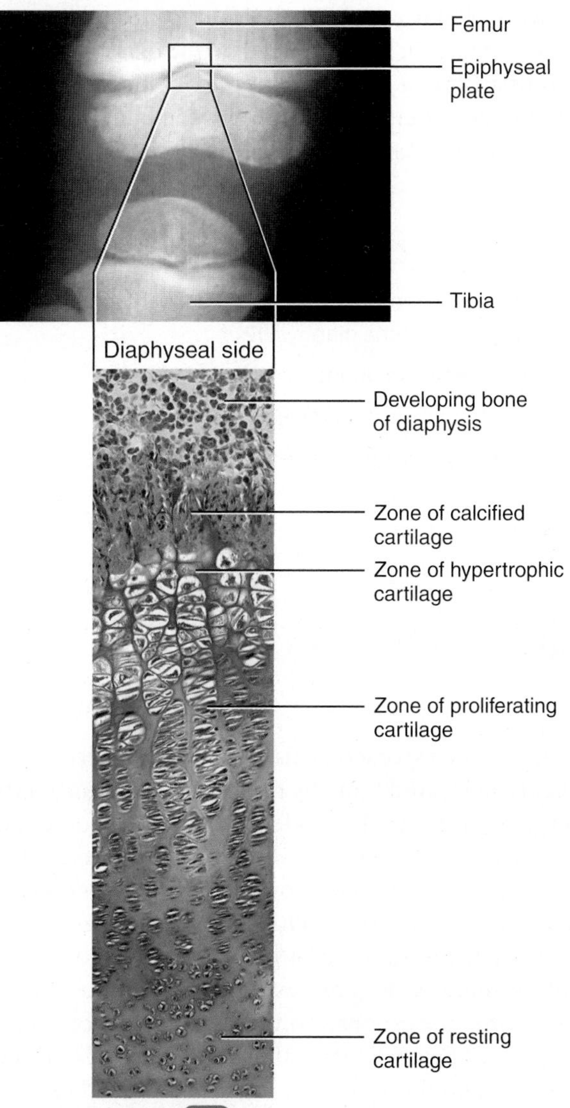

FIGURE 6.8 **Epiphyseal plate.** The epiphyseal plate, a layer of hyaline cartilage in the metaphysis of a growing bone, appears as a dark band between whiter calcified areas in the radiograph (x-ray) shown in (a).

- ***Zone of hypertrophic cartilage*** (hī-per-TRŌ-fik). This layer consists of large, maturing chondrocytes arranged in columns.
- ***Zone of calcified cartilage.*** The final zone of the epiphyseal plate is only a few cells thick and consists mostly of dead chondrocytes because the extracellular matrix around them has calcified. Osteoclasts dissolve the calcified cartilage, and osteoblasts and capillaries from the diaphysis invade the area. Osteoblasts lay down bone extracellular matrix, replacing the calcified cartilage by endochondral ossification. As a result, the zone of calcified cartilage becomes "new diaphysis" that is firmly cemented to the rest of the diaphysis of the bone.

The activity of the epiphyseal plate is the only way that the diaphysis can increase in length. As a bone grows, new chondrocytes are formed on the epiphyseal side of the plate, while old chondrocytes on the diaphyseal side of the plate are replaced by osseous tissue. In this way the thickness of the epiphyseal plate cartilage remains relatively constant, but the bone on the diaphyseal side increases in length (Figure 6.8c).

At about age 18 in females and 21 in males, the epiphyseal plates close. The epiphyseal cartilage cells stop dividing, and bone replaces all of the cartilage. The epiphyseal plate fades, leaving a bony structure called the *epiphyseal line*. The appearance of the epiphyseal line signifies that the bone has stopped growing in length. The clavicle is the last bone to stop growing. Closure of the epiphyseal plate is a gradual process that is useful in determining bone age to establish age at death from skeletal remains, especially those of infants, children, and adolescents.

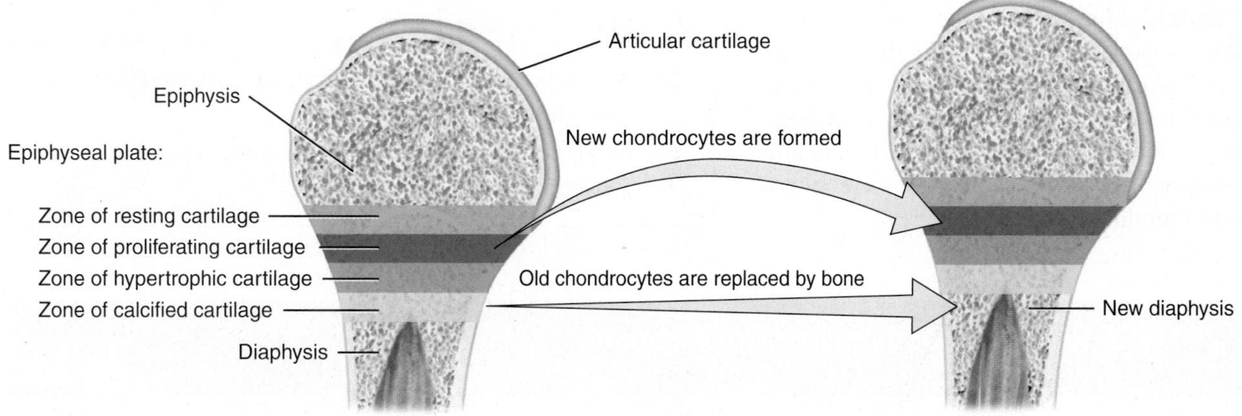


(c) Lengthwise growth of bone at epiphyseal plate

**The epiphyseal plate allows the diaphysis of a bone to increase in length.**

For example, an open epiphyseal plate indicates a younger person than an individual with a partially closed epiphyseal plate or one that is completely closed.

If a bone fracture damages the epiphyseal plate, the fractured bone may be shorter than normal once adult stature is reached. Because damaged cartilage is avascular, it repairs much more slowly than damaged osseous tissue, which is highly vascular. Epiphyseal plate damage accelerates its ossification, thus inhibiting lengthwise growth of the bone, which is one reason fractures involving the epiphyseal plate are of particular concern in children and adolescents.

## Growth in Thickness

Bones grow in thickness (diameter) by the deposition of osseous tissue on the outer surface of the bone. At the bone surface, cells in the periosteum differentiate into osteoblasts. These osteoblasts secrete the extracellular matrix of osseous tissue, adding lamellae to the surface of the bone, and new osteons of compact bone tissue are formed. As the osteoblasts become surrounded by extracellular matrix, they develop into osteocytes. As new osseous tissue is being deposited on the outer surface of bone, osseous tissue lining the medullary cavity is destroyed by osteoclasts in the endosteum. In this way, the medullary cavity enlarges as the bone increases in thickness.

✓ **CHECKPOINT**

**19.** Which activities of the epiphyseal plate account for the lengthwise growth of the diaphysis?

**20.** How does the epiphyseal line develop?

**21.** How do bones grow in diameter?

**22.** How does the medullary cavity enlarge during bone growth?

# 6.8 Bone remodeling renews osseous tissue, redistributes bone extracellular matrix, and repairs bone injuries.

Even after bones have reached their adult shapes and sizes, they continue to be renewed. **Bone remodeling** is the ongoing replacement of old osseous tissue by new osseous tissue. During remodeling, osteoclasts destroy the extracellular matrix and then osteoblasts deposit new extracellular matrix. At any given time, about 5 percent of the total bone mass in the body is being remodeled. The renewal rate for compact bone tissue is about 4 percent per year and for spongy bone tissue is about 20 percent per year. Remodeling also takes place at different rates in various body regions. The distal portion of the femur is replaced about every four months. By contrast, bone in certain areas of the diaphysis of the femur will not be completely replaced during an individual's life.

During remodeling, an osteoclast attaches its ruffled border to the bone surface at the endosteum or periosteum and releases enzymes and several acids (see Figure 6.3). The enzymes digest collagen fibers and other organic substances, while the acids dissolve the bone minerals. Working together, several osteoclasts carve out a small tunnel in the old bone. The degraded bone proteins and extracellular matrix minerals, mainly calcium and phosphorus, diffuse into nearby blood capillaries. Osteoblasts follow the osteoclasts to rebuild the bone in that area.

Remodeling renews osseous tissue before deterioration sets in and heals an injured bone by replacing the damaged tissue with new osseous tissue. Remodeling has several other advantages. New osseous tissue is less brittle and stronger than old osseous tissue. Remodeling also redistributes bone extracellular matrix along lines of mechanical stress (muscle pull and gravity) so that the shape of a bone is altered to keep it strongest where there is the greatest need for support. Since a bone will remodel in response to the demands placed on it, subjecting a bone to heavy loads will produce new osseous tissue that is thicker and therefore stronger than the old osseous tissue.

✓ **CHECKPOINT**

**23.** What is bone remodeling? What roles do osteoblasts and osteoclasts play in the process?

**24.** Why is bone remodeling necessary when a person reaches adulthood?

## CLINICAL CONNECTION | *Remodeling and Orthodontics*

**Orthodontics** (or-thō-DON-tiks) is the branch of dentistry concerned with the prevention and correction of poorly aligned teeth. The movement of teeth by braces places a stress on the bone that forms the sockets that anchor the teeth. In response to this artificial stress, osteoclasts and osteoblasts remodel the sockets so that the teeth align properly.

### CLINICAL CONNECTION | *Paget's Disease*

A delicate balance exists between the actions of osteoclasts and osteoblasts. Should too much new tissue be formed, the bones become abnormally thick and heavy. If too much mineral material is deposited in the bone, the surplus may form thick bumps, called *spurs*, on the bone that interfere with movement at joints. Excessive loss of calcium or tissue weakens the bones, and they may break, as occurs in osteoporosis, or they may become too flexible, as in rickets and osteomalacia. (For more on these disorders, see your WileyPlus resources.) Abnormal acceleration of the remodeling process results in a condition called *Paget's disease*, in which the newly formed bone, especially that of the pelvis, limbs, lower vertebrae, and skull, becomes enlarged, hard, and brittle and fractures easily. In Paget's disease, there is an excessive proliferation of osteoclasts, so bone resorption occurs faster than bone deposition. In response, osteoblasts attempt to compensate, but the new bone is weaker because it has a higher proportion of spongy to compact bone, mineralization is decreased, and the newly synthesized extracellular matrix contains abnormal proteins.

## 6.9 Dietary and hormonal factors influence bone growth and remodeling.

Normal bone growth in the young and bone remodeling in the adult depend on adequate dietary intake of minerals and vitamins, as well as sufficient levels of several hormones.

- ***Minerals***. Large amounts of calcium and phosphorus and smaller amounts of magnesium, fluoride, and manganese are needed while bones are growing or remodeling.
- ***Vitamins***. Vitamin A stimulates activity of osteoblasts. Vitamins C, K, and $B_{12}$ are needed for synthesis of osseous tissue proteins. Vitamin D helps build osseous tissue by increasing the absorption of calcium from foods in the gastrointestinal tract.
- ***Hormones***. During childhood, insulinlike growth factors stimulate osteoblasts, promote cell division at the epiphyseal plate and in the periosteum, and enhance synthesis of proteins needed to build new osseous tissue. Insulinlike growth factors are produced by the liver and osseous tissue in response to the secretion of human growth hormone from the pituitary gland. Thyroid hormones from the thyroid gland promote bone growth by stimulating osteoblasts. In addition, insulin from the pancreas promotes bone growth by increasing the synthesis of bone proteins.

A moderate level of weight-bearing exercise also maintains sufficient strain on bones to increase and maintain their density.

At puberty, the secretion of **sex hormones** has a dramatic impact on bone growth. The sex hormones include estrogens (produced by the ovaries) and testosterone (produced by the testes). These hormones increase osteoblast activity and synthesis of bone extracellular matrix and are responsible for the sudden "growth spurt" that occurs during the teenage years. Females produce much higher levels of estrogens than androgens; males secrete higher levels of androgens than estrogens. Differences in sex hormone levels promote the typical femininization or masculinization of our skeletons. For example, estrogens account for the widening of the pelvis in females. Ultimately, sex hormones induce conversion of the epiphyseal plates into epiphyseal lines, stopping bone elongation. Adult women are typically shorter than adult men because their higher levels of sex hormones stop lengthwise growth of bones at an earlier age.

Calcium homeostasis also influences bone growth and remodeling. Calcium is essential to many physiological processes. For example, nerve and muscle cells need stable levels of calcium ions in the surrounding extracellular fluid to function properly. Blood clotting also requires calcium. For these reasons, blood calcium level is very closely regulated between 9 and 11mg/100mL. Even small changes in calcium concentration outside this range may prove fatal—the heart may stop (cardiac arrest) if the concentration goes too high, or breathing may cease (respiratory arrest) if the level falls too low.

Bones are the body's major calcium reservoir, storing 99 percent of total body calcium. One way to maintain the level of calcium in the blood is to control the movement of calcium between bones and blood. Osteoclasts release calcium into blood plasma when blood calcium level decreases, and osteoblasts absorb calcium when blood calcium level rises.

In response to decreasing blood calcium, the parathyroid glands release parathyroid hormone into the blood. **Parathyroid hormone** raises blood calcium level to normal by (1) increasing osteoclast activity to release calcium from bone, (2) stimulating the kidneys to decrease calcium loss in the urine, and (3) activating vitamin D. Vitamin D promotes absorption of calcium from foods in the gastrointestinal tract. The actions of parathyroid hormone and activated vitamin D elevate blood calcium level.

When blood calcium rises above normal, the thyroid gland secretes calcitonin. **Calcitonin** (kal-si-TŌ-nin) inhibits the activity of osteoclasts and accelerates blood calcium uptake

## CLINICAL CONNECTION | *Hormonal Abnormalities That Affect Height*

Excessive or deficient secretion of hormones that normally control bone growth can cause a person to be abnormally tall or short. Oversecretion of hGH during childhood produces **gigantism**, in which a person becomes much taller and heavier than normal. Undersecretion of hGH produces **pituitary dwarfism**, in which a person has short stature. (The usual adult height of a *dwarf* is under 4 feet 10 inches.) Although the head, trunk, and limbs of a pituitary dwarf are smaller than normal, they are proportionate. The condition can be treated medically with hGH until epiphyseal plate closure. Oversecretion of hGH during adulthood is called **acromegaly** (ak′-rō-MEG-a-lē). Although hGH cannot produce further lengthening of the long bones because the epiphyseal (growth) plates are already closed, the bones of the hands, feet, and jaws thicken and other tissues enlarge. In addition, the eyelids, lips, tongue, and nose enlarge, and the skin thickens and develops furrows, especially on the forehead and soles. **Achondroplasia** (a-kon-drō-PLĀ-zē-a; *a* = without; *chondro* = cartilage; *-plasia* = to mold) is an inherited condition in which the conversion of cartilage to bone is abnormal. It results in the most common type of dwarfism, called **achondroplastic dwarfism**. These individuals are typically about four feet tall as adults. They have an average-size trunk, short limbs, and a slightly enlarged head with a prominent forehead and flattened nose at the bridge. The condition is essentially untreatable, although some individuals opt for limb-lengthening surgery.

into bones. The net result is that calcitonin promotes bone formation and decreases blood calcium level. Despite these effects, the role of calcitonin in normal calcium homeostasis is uncertain because it can be completely absent without causing symptoms.

Chapter 17 provides a more detailed discussion of hormonal regulation of blood calcium level.

### ✓ CHECKPOINT

**25.** Which nutrients are important for bone growth and remodeling?

**26.** How does human growth hormone regulate bone growth?

**27.** What body functions depend on proper levels of calcium?

## Cathy's Story — EPILOGUE AND DISCUSSION

Women, on average, live longer and have a lower bone density than men, and they also face a time period of rapid bone mineral loss during menopause. Cathy's situation as an older woman, living alone, with lower than average bone mineral density, is unfortunately all too common. Dr. Andrews did not catch Cathy's osteoporosis immediately after the first break because broken wrists are not unusual in any age range, and it took several other breaks over a 10-year period before a BMD test was performed. By this time, Cathy's osteoclasts had degraded a significant amount of the extracellular matrix of her bones, and as we have seen, the overall function of Cathy's skeletal system has been compromised. If Cathy's greatest fears come true and she breaks her hip, it will be a long and painful road to recovery, and it is possible that Cathy may not recover from a particularly bad break. Considering her situation, Cathy has the best possible mind-set. She is fully aware of her disease and takes every precaution to prevent injury. She has made appropriate lifestyle changes, and pays attention to the signals her body is sending her. Cathy has many things yet to look forward to in her life, and she is doing exactly that: looking forward.

H. *Cathy has many grandchildren and even one on the way. She is very interested in bones, bone growth, and regeneration. Please answer these questions for her:*
- *My youngest daughter is seven months pregnant. Is the same "building up and breaking down" process Dr. Andrews described happening in her baby to produce bones?*
- *Some of my grandchildren have already broken bones. Will broken bones affect the overall growth of the child?*

I. *What did Dr. Andrews mean when he told Cathy that "bones are always changing"? What advice would you give to her daughters and granddaughters to help them prevent osteoporosis?*

# Concept and Resource Summary

**Concept** | **Resources**

## Introduction

1. Bones are composed of several different tissues working together: osseous tissue, cartilage, dense connective tissues, epithelium, adipose tissue, and nervous tissue.
2. All of the bones and their cartilages, along with ligaments and tendons, constitute the **skeletal system**.

## Concept 6.1 Skeletal system functions include support, protection, movement, mineral homeostasis, blood cell production, and energy storage.

1. The skeleton supports soft tissues and provides attachment points for tendons of skeletal muscles.
2. The skeleton protects the most important internal organs.
3. Bones assist in movement by acting as levers for skeletal muscles.
4. Osseous tissue stores and releases minerals, particularly calcium and phosphorus.
5. Bones can contain **red bone marrow**, which produces blood cells, or **yellow bone marrow**, which stores triglycerides, a source of potential energy.

*Resources:* Anatomy Overview—Bone Structure and Tissues

## Concept 6.2 Bones are classified as long, short, flat, irregular, or sesamoid.

1. Bones are classified by shape into five principal types: **long**, **short**, **flat**, **irregular**, and **sesamoid**.

*Resources:* Real Anatomy Viewpoint—Bones by Type

## Concept 6.3 Long bones have a diaphysis, a medullary cavity, epiphyses, metaphyses, and periosteum.

1. The cylindrical shaft of a long bone is called the **diaphysis**. The space within the diaphysis that contains yellow marrow is called the **medullary cavity**.
2. The distal and proximal ends of a long bone are called the **epiphyses**. They are covered with **articular cartilage** composed of hyaline cartilage.
3. The regions where the diaphysis joins the epiphyses are called the **metaphyses**. In growing bone, each metaphysis has an **epiphyseal plate** essential for bone growth. Once growth stops, the epiphyseal plate becomes the **epiphyseal line**.
4. The **periosteum** is a connective tissue sheath surrounding the bone, and the **endosteum** is a thin membrane lining the medullary cavity. The periosteum and endosteum contain bone-forming cells.

*Resources:* Exercise—Growing Long Bone

## Concept 6.4 Osseous tissue can be arranged as compact bone tissue or spongy bone tissue.

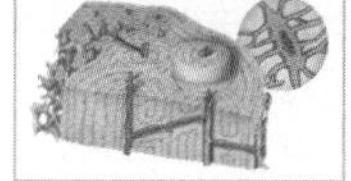

1. **Osseous tissue** is a connective tissue containing an extracellular matrix of water, collagen fibers, and crystallized mineral salts called **hydroxyapatite**.
2. The four principal types of cells in osseous tissue are **osteogenic cells** (undifferentiated cells that give rise to osteoblasts), osteoblasts (bone-building cells), osteocytes (maintain daily activity of osseous tissue), and osteoclasts (bone-destroying cells).
3. **Osteoblasts** initiate **calcification**, the process in which mineral salts are deposited around collagen fibers in the extracellular matrix, then crystallize into a hard tissue.
4. **Compact bone tissue** has few spaces, forms the external layer of all bones, and comprises most of the diaphysis of long bones. It is composed of **osteons**. Each osteon has a **central canal** where blood vessels, lymphatic vessels, and nerves are found. The central canal is surrounded by **concentric lamellae** with **lacunae** containing osteocytes that are connected to one another and to the **central canal** by radiating **canaliculi**.
5. **Spongy bone tissue** does not contain osteons. It consists of **trabeculae** consisting of irregular lamellae and osteocytes in lacunae. Spaces between trabeculae are filled with red bone marrow, a tissue that produces blood cells. Spongy bone predominates in short, flat, and irregularly shaped bones; most of the epiphyses of long bones; and a narrow rim around the medullary cavity of long bones.

*Resources:*
- Anatomy Overview—Compact Bone
- Anatomy Overview—Spongy Bone
- Animation—Bone Dynamics and Tissues
- Figure 6.3—Types of Cells in Osseous Tissue
- Figure 6.4—Histology of Compact and Spongy Bone Tissue

| Concept | Resources WILEY PLUS |
|---|---|

## Concept 6.5 Bones are richly supplied with blood vessels and nerves.

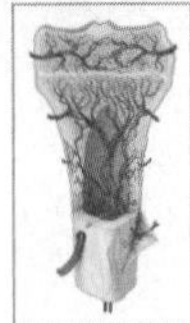

1. Bones are highly vascular, with blood vessels passing into them from the periosteum.
2. In a long bone, **periosteal arteries** and nerves enter the diaphysis through **perforating canals** to supply the periosteum and outer compact bone tissue. At least one large **nutrient artery** enters the diaphysis and branches in the medullary cavity. Bone also contains **metaphyseal** and **epiphyseal arteries**, **nutrient veins**, **periosteal veins**, and **epiphyseal veins**.

Resources:

Clinical Connection—Bone Marrow Examination

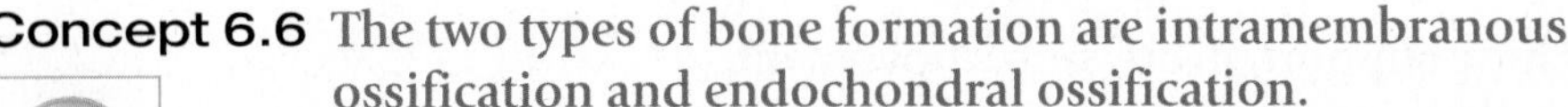

## Concept 6.6 The two types of bone formation are intramembranous ossification and endochondral ossification.

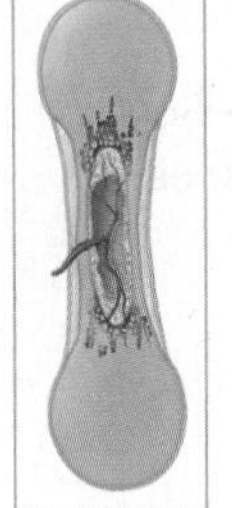

1. The formation of bone, called **ossification**, occurs in three principal situations: (1) the initial formation of bones before birth; (2) the growth of bones until their adult sizes are reached; (3) the remodeling and repair of bones.
2. The two types of ossification, intramembranous and endochondral, involve the replacement of a preexisting connective tissue with bone.
3. **Intramembranous ossification** begins when cells in the mesenchymal cluster to form an **ossification center**, where they differentiate into osteoblasts that secrete bone extracellular matrix as trabeculae of spongy bone tissue. Mesenchyme at the periphery develop into periosteum. A thin layer of compact bone tissue develops over the deeper spongy bone tissue.
4. **Endochondral ossification** refers to bone formation within hyaline cartilage that develops from mesenchyme. The **primary ossification center** begins in the mid-region of the cartilage model as cartilage degenerates. Osteoblasts begin to secrete extracellular matrix of spongy bone trabeculae. Osteoclasts break down some of the newly formed trabeculae, forming the medullary cavity. Next, **secondary ossification centers** develop in the epiphyses, where bone replaces cartilage. Hyaline cartilage remains as the articular cartilage and epiphyseal plate.

Resources:

Animation—Bone Formation

Figure 6.6—Intramembranous Ossification

Figure 6.7—Endochondral Ossification

Exercise—Observe the Ossification

## Concept 6.7 Bones grow longer at the epiphyseal plate and increase in diameter by the addition of new osseous tissue around the outer surface.

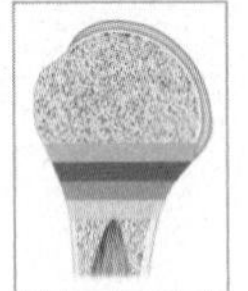

1. Throughout childhood and adolescence, long bones grow in length when cartilage is replaced by osseous tissue on the diaphyseal side of the epiphyseal plate. An epiphyseal line results as growth in length stops.
2. Bone grows in diameter as osteoblasts in the periosteum secrete extracellular matrix on the bone surface while the medullary cavity is enlarged by osteoclasts of the endosteum.

Resources:

Animation—Bone Elongation and Bone Widening

Animation—Regulation of Bone Growth

Figure 6.8—The Epiphyseal Plate

Exercise—Bone Growth Sequencing

Exercise—Concentrate on Bone Tissue

## Concept 6.8 Bone remodeling renews osseous tissue, redistributes bone extracellular matrix, and repairs bone injuries.

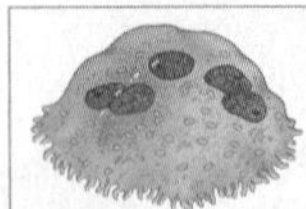

1. **Bone remodeling** continues throughout life as osteoclasts destroy old osseous tissue and then osteoblasts rebuild it.
2. Remodeling renews osseous tissue before deterioration occurs, heals injured bones, and redistributes bone extracellular matrix along lines of mechanical stress.

Resources:

Animation—Bone Remodeling

Animation—Regulation of Blood Calcium

Animation—Bone Dynamics and Tissues

Clinical Connection—Fracture and Repair of Bone

Clinical Connection—Treatments for Fractures

| Concept | Resources |
| --- | --- |
| **Concept 6.9** Dietary and hormonal factors influence bone growth and remodeling.<br>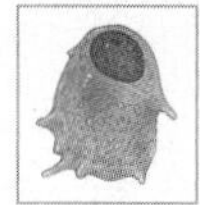<br>1. Dietary minerals (especially calcium and phosphorus) and vitamins (A, C, D, K, and $B_{12}$) are needed for bone growth and maintenance.<br>2. Insulinlike growth factors (IGFs), human growth hormone, thyroid hormones, and insulin stimulate bone growth during childhood.<br>3. The **sex hormones**, estrogens and testosterone, stimulate the growth spurt in adolescence and lead to ossification that transforms the epiphyseal plate into an epiphyseal line.<br>4. Bone is the major reservoir for calcium in the body. **Parathyroid hormone** secreted by the parathyroid gland increases blood calcium level. Vitamin D enhances absorption of calcium from food and thus raises the blood calcium level. **Calcitonin** from the thyroid gland has the potential to decrease blood calcium level. | Exercise—Regulation Sequences<br><br>Clinical Connection—Osteoporosis<br>Clinical Connection—Rickets and Osteomalacia<br>Clinical Connection—Exercise and Osseous Tissue<br>Clinical Connection—Summary of Factors That Affect Bone Growth |

## Understanding the Concepts

1. How do red and yellow bone marrow differ in function?
2. What purpose do sesamoid bones serve?
3. Why is osseous tissue considered a connective tissue?
4. What is the path a nutrient would travel through compact bone tissue from its diffusion out of a blood vessel in the periosteum to an osteocyte located within the second osteon in from the surface of the bone?
5. What effect would the gravity-free environment of space have on an astronaut's bones?
6. Compare and contrast the microscopic appearance, location, and function of spongy and compact bone.
7. How does blood leave a bone?
8. Explain why the two ossification processes are called "intramembranous ossification" and "endochondral ossification."
9. As we approach adulthood, which hormone is primarily responsible for ending the elongation of bones? Explain your answer.
10. If radiographs of an 18-year-old basketball star show clear epiphyseal plates but no epiphyseal lines, is she likely to grow taller? Why or why not?
11. What would be the effect on your skeleton if your osteoclasts were more active than your osteoblasts? If your osteoblasts were more active than your osteoclasts?

CHAPTER

# 7

## Fernando's Story

Elena loves her mother's father, Fernando, and always spends Saturdays with him. Elena's mother, Isabel, works on Saturdays, so Fernando always comes to their house on Friday night to spend the night and be there to care for 9-year-old Elena the next day. This had been the arrangement for as long as Elena could remember. Grandpa Fernando would let Elena sleep late, prepare French toast, her favorite breakfast, and then wake her up to eat. After breakfast they would do something special, like go to the zoo or to a movie. As far as Elena was concerned, Saturday was the best day of the week.

Today they plan to go to a fair where they will ride the Ferris wheel and the merry-go-round, visit the new lambs and piglets, and eat some fried dough and cotton candy. Every year when the fair comes to town they go, even though every year Grandpa says it will be his last, complaining that it is too much walking for an old man. After breakfast, Elena dresses quickly and helps her grandfather clean up the kitchen so they can get going.

They walk down the sidewalk and start to cross the street toward the bus stop.

"Hurry Grandpa, the light is changing!" Elena shouts as she skips up onto the curb.

"I'm moving as fast as I can," grumbles Fernando. Neither one of them sees the bicyclist hurrying around the corner to make the turn before the light turns red. The bicycle goes down with a crash. Its rider is unhurt, but Fernando is knocked toward the curb, where he hits his head face-first on the lamppost. Elena cries out, "Grandpa, are you OK?" but gets no answer. Elena is shaking but manages to use her cell phone to dial 911 and then to call her mother.

In this chapter you will learn the names of the bones of the axial skeleton and discover how trauma and aging can affect the skeletal system.

# The Axial Skeleton

## INTRODUCTION

Without bones, you could not survive. You would be unable to perform movements such as walking or grasping, and the slightest blow to your head or chest could damage your brain or heart. Because the skeletal system forms the framework of the body, a familiarity with the names, shapes, and positions of individual bones will help you locate and name many other anatomical features. For example, the radial artery, the site where the pulse is usually taken, is named for its closeness to the radius, the lateral bone of the forearm. The frontal lobe of the brain lies deep to the frontal (forehead) bone. Parts of certain bones also help locate structures within the skull and outline the lungs, heart, and abdominal and pelvic organs.

Movements such as throwing a ball, biking, and walking require interactions between bones and muscles. To understand how muscles produce different movements, you will learn where the muscles attach on individual bones as well as the types of joints acted on by the contracting muscles. The bones, muscles, and joints together form an integrated system called the **musculoskeletal system**.

## CONCEPTS

**7.1** Bones of the axial skeleton and appendicular skeleton have characteristic surface markings.

**7.2** The skull provides attachment sites for muscles and membranes, and protects and supports the brain and sense organs.

**7.3** The cranial bones include the frontal, parietal, temporal, occipital, sphenoid, and ethmoid bones.

**7.4** Facial bones include the nasal bones, maxillae, zygomatic bones, lacrimal bones, palatine bones, inferior nasal conchae, vomer, and mandible.

**7.5** Unique features of the skull include the nasal septum, orbits, sutures, paranasal sinuses, and fontanels.

**7.6** The hyoid bone supports the tongue and attaches to muscles of the tongue, pharynx, and larynx.

**7.7** The vertebral column protects the spinal cord, supports the head, and is a point of attachment for bones and muscles.

**7.8** A vertebra usually consists of a body, a vertebral arch, and several processes.

**7.9** Vertebrae in the different regions of the vertebral column vary in size, shape, and detail.

**7.10** The thoracic cage protects vital organs in the thorax and upper abdomen and provides support for the bones of the upper limbs.

# 7.1 Bones of the axial skeleton and appendicular skeleton have characteristic surface markings.

The adult human skeleton consists of 206 named bones. The skeletons of infants and children have more than 206 bones because some of their bones, such as the hip bones, fuse later in life.

There are two principal divisions in the adult skeleton: This chapter focuses on the **axial skeleton** and Chapter 8 will explore the **appendicular skeleton**. Figure 7.1 shows how the two divisions join to form the complete skeleton (the bones of the axial skeleton are shown in blue). The axial skeleton lies around the longitudinal *axis* of the human body, an imaginary vertical line that runs through the body's center of gravity from the head to the space between the feet. The axial skeleton consists of the skull bones, auditory ossicles, hyoid bone (see Figure 7.5), ribs, sternum (breastbone), and bones of the vertebral column. Functionally, the auditory ossicles, bones in the middle ear that vibrate in response to sound waves that strike the eardrum, are not part of either the axial or appendicular skeleton, but they are grouped with the axial skeleton for convenience. The appendicular skeleton contains the bones of the **upper** and **lower limbs**, plus the bones forming the **girdles** that connect the limbs to the axial skeleton. Table 7.1 presents the 80 bones of the axial skeleton and the 126 bones of the appendicular skeleton.

Bones have characteristic **surface markings**, structural features adapted for specific functions. Most are not present at birth but develop in response to certain forces and are most prominent in the adult skeleton. In response to tension on a bone surface from tendons, ligaments, aponeuroses, and fasciae, new osseous tissue is deposited as raised or roughened areas. Conversely, compression on a bone surface results in a depression. There are two major types of surface markings: (1) *depressions and openings* usually allow passage of blood vessels and nerves or help form joints, and (2) *processes*, which are projections or outgrowths that either help form joints or serve as attachment points for connective tissue (such as ligaments and tendons). Table 7.2 describes the various surface markings and provides examples of each.

### ✓ CHECKPOINT

1. Classify the following structures as part of the axial skeleton or the appendicular skeleton: skull, clavicle, vertebral column, shoulder girdle, humerus, pelvic girdle, and femur.
2. What is the function of depressions and openings?
3. What is the function of processes?

**TABLE 7.1**

The Bones of the Adult Skeletal System

| DIVISION OF THE SKELETON | STRUCTURE | NUMBER OF BONES |
|---|---|---|
| **Axial skeleton** | **Skull** | |
| | Cranium | 8 |
| | Face | 14 |
| | **Hyoid** | 1 |
| | **Auditory ossicles** | 6 |
| | **Vertebral column** | 26 |
| | **Thorax** | |
| | Sternum | 1 |
| | Ribs | 24 |
| | **Number of bones = 80** | |

| DIVISION OF THE SKELETON | STRUCTURE | NUMBER OF BONES |
|---|---|---|
| **Appendicular skeleton** | **Pectoral (shoulder) girdles** | |
| | Clavicle | 2 |
| | Scapula | 2 |
| | **Upper limbs** | |
| | Humerus | 2 |
| | Ulna | 2 |
| | Radius | 2 |
| | Carpals | 16 |
| | Metacarpals | 10 |
| | Phalanges | 28 |
| | **Pelvic (hip) girdle** | |
| | Hip, pelvic, or coxal bone | 2 |
| | **Lower limbs** | |
| | Femur | 2 |
| | Patella | 2 |
| | Fibula | 2 |
| | Tibia | 2 |
| | Tarsals | 14 |
| | Metatarsals | 10 |
| | Phalanges | 28 |
| | **Number of bones = 126** | |

**Total bones in an adult skeleton = 206**

**FIGURE 7.1 Divisions of the skeletal system.** The axial skeleton is indicated in blue. Note the position of the hyoid bone in Figure 7.5.

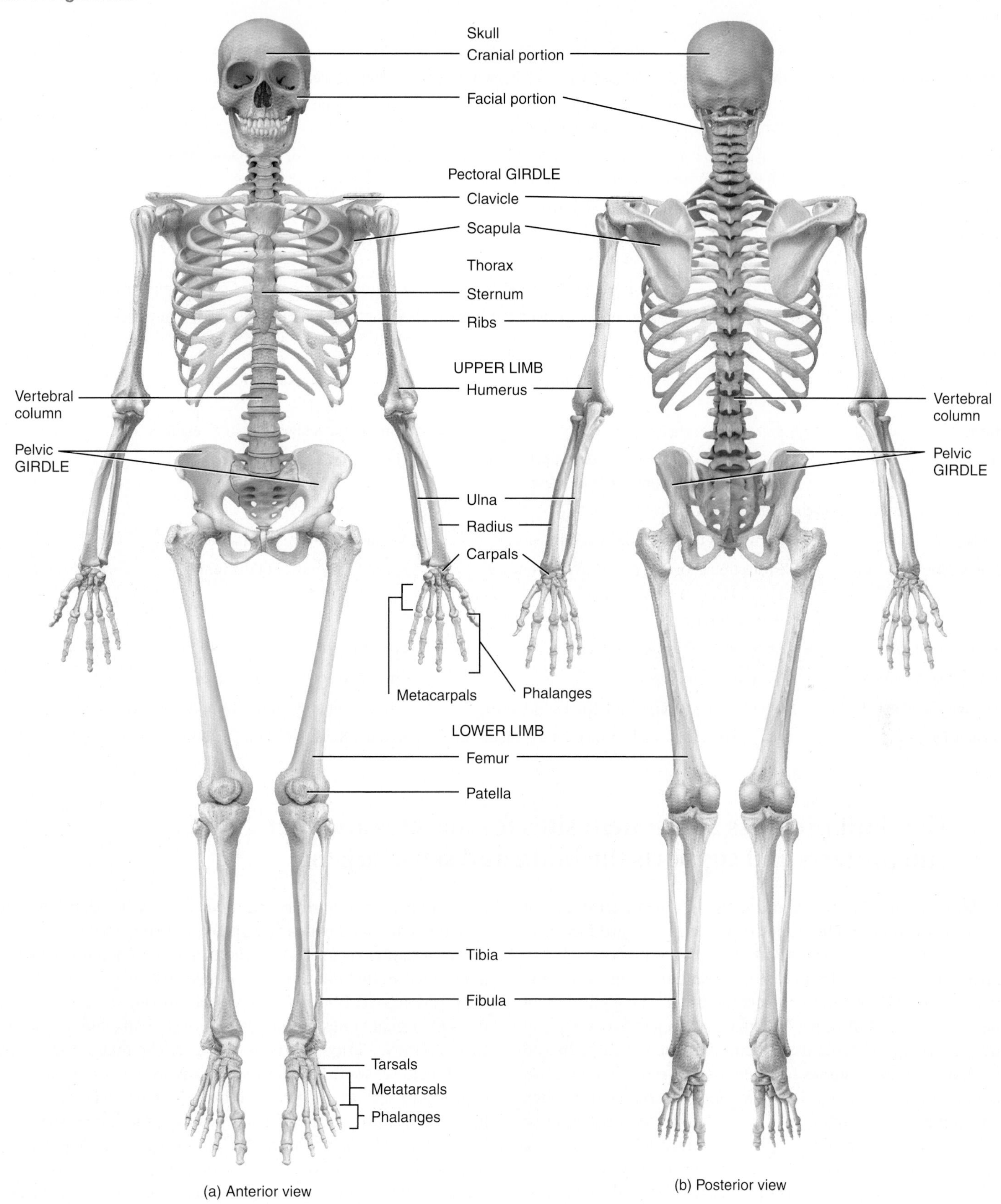


(a) Anterior view

(b) Posterior view

**The adult human skeleton consists of 206 bones grouped into axial and appendicular divisions.**

**TABLE 7.2**

Bone Surface Markings

| MARKING | DESCRIPTION | EXAMPLE |
|---|---|---|
| **DEPRESSIONS AND OPENINGS: SITES ALLOWING THE PASSAGE OF SOFT TISSUE (NERVES, BLOOD VESSELS, LIGAMENTS, TENDONS) OR FORMATION OF JOINTS** | | |
| **Fissure** (FISH-ur) | Narrow slit between adjacent parts of bones through which blood vessels or nerves pass | Superior orbital fissure of the sphenoid bone (Figure 7.11) |
| **Foramen** (fō-RĀ-men; plural is **foramina)** | Opening (*foramen* = hole) through which blood vessels, nerves, or ligaments pass | Optic foramen of the sphenoid bone (Figure 7.11) |
| **Fossa** (FOS-a) | Shallow depression (*fossa* = trench) | Coronoid fossa of the humerus (Figure 8.5a) |
| **Sulcus** | Furrow along a bone surface that accommodates a blood vessel, nerve, or tendon | Intertubercular sulcus of the humerus (Figure 8.5a) |
| **Meatus** (mē-Ā-tus) | Tubelike opening (*meatus* = passageway) | External auditory meatus of the temporal bone (Figure 7.3) |
| **PROCESSES: PROJECTIONS OR OUTGROWTHS ON BONE THAT FORM JOINTS OR ATTACHMENT POINTS FOR CONNECTIVE TISSUE, SUCH AS LIGAMENTS AND TENDONS** | | |
| Processes that form joints: | | |
| **Condyle** (KON-dil) | Large, round protuberance (*condylus* = knuckle) at the end of a bone | Lateral condyle of the femur (Figure 8.13b) |
| **Facet** | Smooth flat articular surface | Superior articular facet of a vertebra (Figure 7.17b) |
| **Head** | Rounded articular projection supported on the neck (constricted portion) of a bone | Head of the femur (Figure 8.13c) |
| Processes that form attachment points for connective tissue: | | |
| **Crest** | Prominent ridge or elongated projection | Median sacral crest of the sacrum (Figure 7.20b) |
| **Epicondyle** | Projection above (*epi-* = above) a condyle | Medial epicondyle of the femur (Figure 8.13a) |
| **Line** (linea) | Long, narrow ridge or border (less prominent than a crest) | Linea aspera of the femur (Figure 8.13b) |
| **Spinous process** | Sharp, slender projection | Spinous process of a vertebra (Figure 7.16) |
| **Trochanter** (trō-KAN-ter) | Very large projection | Greater trochanter of the femur (Figure 8.13b) |
| **Tubercle** (TOO-ber-kul) | Small, rounded projection (*tuber-* = knob) | Greater tubercle of the humerus (Figure 8.5a) |
| **Tuberosity** | Large, rounded, usually roughened projection | Ischial tuberosity of the hip bone (Figure 8.10b) |

## 7.2 The skull provides attachment sites for muscles and membranes, and protects and supports the brain and sense organs.

The **skull** contains 22 bones and rests on the superior end of the vertebral column. The bones of the skull are grouped into two categories: cranial bones and facial bones (Table 7.3). The **cranial bones** (*crani-* = brain case), or **cranium**, form the **cranial cavity**, which encloses and protects the brain. The eight cranial bones are the frontal bone, two parietal bones, two temporal bones, the occipital bone, the sphenoid bone, and the ethmoid bone. Fourteen **facial bones** form the face: two nasal bones, two maxillae (or maxillas; singular is *maxilla*), two zygomatic bones, the mandible, two lacrimal bones, two palatine bones, two inferior nasal conchae, and the vomer. Figures 7.2 through 7.11 illustrate the bones of the skull from different viewing directions.

Besides forming the large cranial cavity, the skull also forms several smaller cavities, including the nasal cavity and orbits (eye sockets), which both open to the exterior. Certain skull bones also contain cavities called paranasal sinuses that are lined with mucous membranes and open into the nasal cavity. Also within the skull are small middle and inner ear cavities that house the structures involved in hearing and equilibrium (balance).

In addition to protecting the brain, the cranial bones have inner surfaces that attach to membranes (meninges) that stabilize the positions of the brain, blood vessels, and nerves. The outer surfaces of cranial bones provide large areas of attachment for muscles that move various parts of the head. The bones also provide attachment for some muscles that are involved in producing facial expressions such as the puckered brow look of concentration you have when studying. The facial bones form the framework of the face and provide support for the entrances to the digestive and respiratory systems. Together, the cranial and facial bones protect and support the delicate special sense organs for vision, taste, smell, hearing, and equilibrium.

Other than the auditory ossicles (tiny bones involved in hearing, which are located within the temporal bones and described further in Concept 16.7), the mandible is the only

**TABLE 7.3**

Bones of the Adult Skull

| CRANIAL BONES | FACIAL BONES |
|---|---|
| Frontal (1)* | Nasal (2) |
| Parietal (2) | Maxillae (2) |
| Temporal (2) | Zygomatic (2) |
| Occipital (1) | Mandible (1) |
| Sphenoid (1) | Lacrimal (2) |
| Ethmoid (1) | Palatine (2) |
| | Inferior nasal conchae (2) |
| | Vomer (1) |

*The numbers in parentheses indicate how many of each bone are present.

movable bone of the skull. Most of the skull bones are held together by immovable joints, called **sutures**, which are especially noticeable on the outer surface of the skull.

The skull also has many surface markings such as **foramina** (rounded passageways; singular is foramen) and **fissures** (slit-like openings) through which blood vessels, nerves, and ligaments pass. We will mention most of the foramina of the skull in the descriptions of the bones that they penetrate. Structures that pass through the foramina will be described in *WileyPLUS.*

✓ **CHECKPOINT**

4. What are the functions of the cranial bones? The facial bones?
5. What function is served by foramina?

# 7.3 The cranial bones include the frontal, parietal, temporal, occipital, sphenoid, and ethmoid bones.

## Frontal Bone

The **frontal bone** forms the forehead (the anterior part of the cranium), the roofs of the *orbits* (eye sockets), and most of the anterior part of the cranial floor (Figure 7.2). Soon after birth the left and right sides of the frontal bone are united by the *metopic suture*, which usually disappears between the ages of six and eight. The frontal bone gradually slopes inferiorly from the coronal suture on top of the skull, then angles abruptly and

**FIGURE 7.2** Anterior view of the skull.

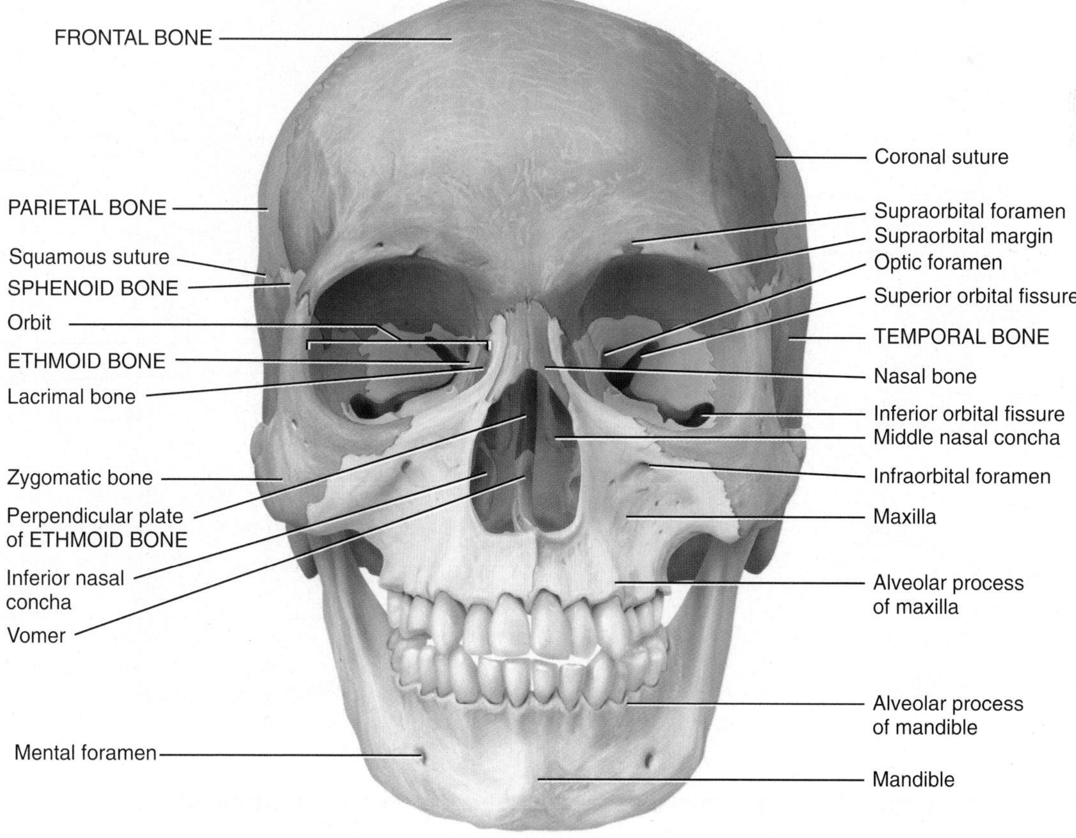


The skull consists of cranial bones and facial bones.

becomes almost vertical. Superior to the orbits the frontal bone thickens, forming the *supraorbital margin* (*supra-* = above; *-orbital* = wheel rut). From this margin the frontal bone extends posteriorly to form the roof of the orbit and part of the floor of the cranial cavity. Within the supraorbital margin is the *supraorbital foramen*, a passageway for nerves of the eyebrow and eyelid. The *frontal sinuses* lie deep to the frontal bone.

## Parietal Bones

The two **parietal bones** (pa-RĪ-e-tal; *pariet-* = wall) form the greater portion of the sides and roof of the cranial cavity (Figure 7.3). The internal surfaces of the parietal bones contain many protrusions and depressions that accommodate the blood vessels supplying the dura mater, the superficial membrane covering the brain.

## Temporal Bones

The paired **temporal bones** (*tempor-* = temple) form the inferior lateral sides of the cranium and part of the cranial floor (Figure 7.3). Projecting from the inferior portion of the temporal bone is the *zygomatic process*, which articulates (forms a joint) with the temporal process of the zygomatic (cheek) bone. Together, the zygomatic process of the temporal bone and the temporal process of the zygomatic bone form the *zygomatic arch*. On the inferior posterior surface of the zygomatic process of the temporal bone is a socket called the *mandibular fossa*. The mandibular fossa articulates with the mandible (lower jawbone) to form the *temporomandibular joint (TMJ)*.

Inferior to the zygomatic process is an external opening in the temporal bone, the *external auditory meatus* (*meatus* = passageway), or ear canal, which directs sound waves into the ear (Figure 7.3). The *mastoid process* is a rounded projection of

FIGURE 7.3 **Right lateral view of the skull.**

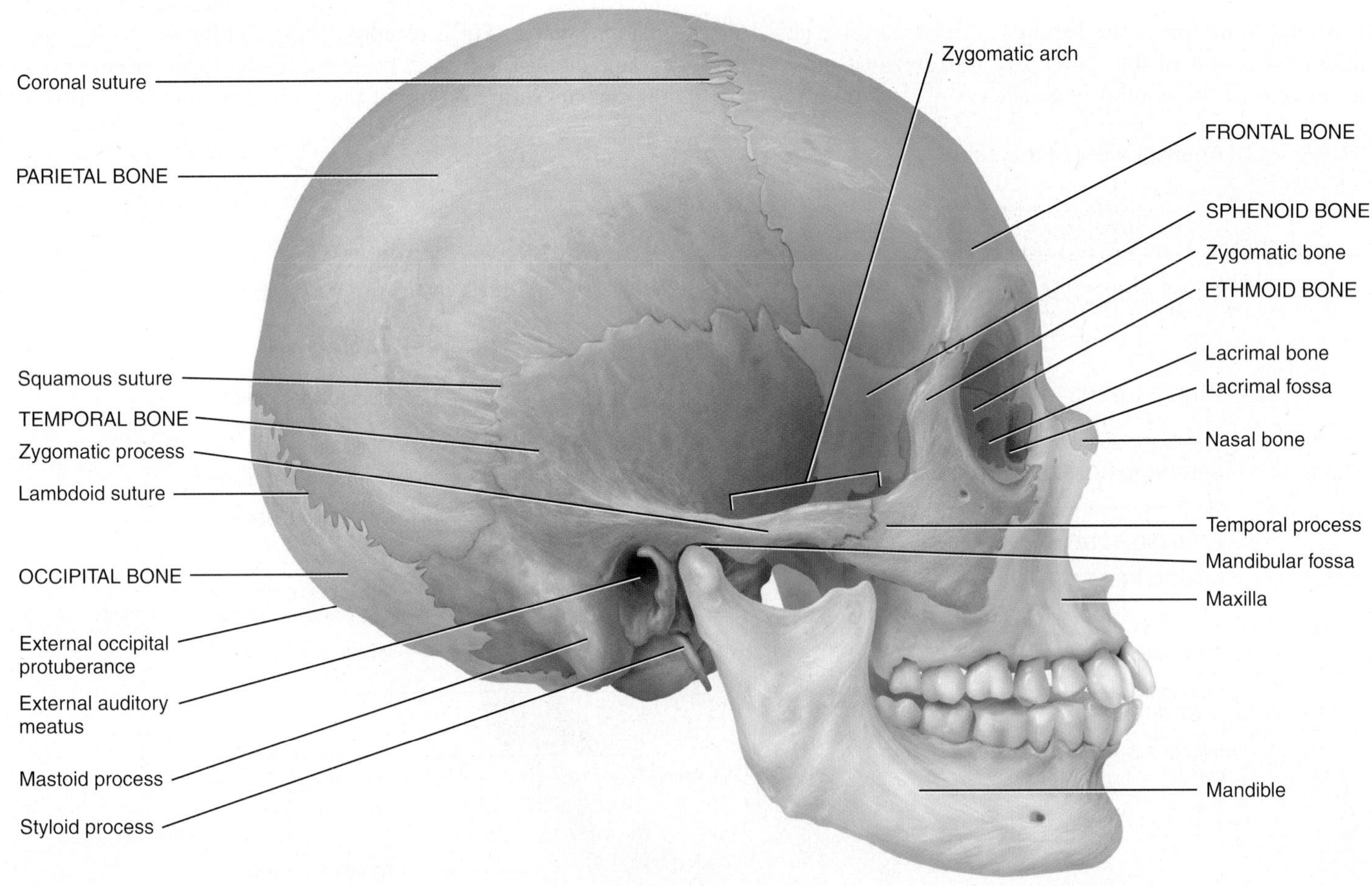


**The zygomatic arch is formed by the zygomatic process of the temporal bone and the temporal process of the zygomatic bone.**

the temporal bone posterior to the external auditory meatus. It serves as a point of attachment for several neck muscles. The *styloid process* (*styl-* = stake or pole) projects from the inferior surface of the temporal bone and serves as a point of attachment for muscles and ligaments of the tongue and neck (Figures 7.3 and 7.4). Between the styloid process and the mastoid process is the *stylomastoid foramen* (Figure 7.4) through which a nerve passes that controls movement of facial muscles.

At the floor of the cranial cavity (see Figure 7.7a) is the *petrous portion* (*petrous* = rock) of the temporal bone. The petrous portion is triangular and located at the base of the skull between the sphenoid and occipital bones. The petrous portion houses the internal ear and the middle ear, structures involved in hearing and equilibrium (balance). The *internal auditory meatus* (see Figures 7.5 and 7.7) is the opening in the petrous portion through which impulses for hearing and equilibrium are carried to the brain. The petrous portion also contains the *carotid foramen*, through which the internal carotid artery carries blood to the brain (Figure 7.4). Located in the suture between the petrous portion and the occipital bone is the *jugular foramen*, a passageway for the internal jugular vein carrying blood away from the brain.

## Occipital Bone

The **occipital bone** (ok-SIP-i-tal; *occipit-* = back of head) forms the posterior part and most of the base of the skull (see Figures 7.3, 7.4, and 7.6). The *foramen magnum* (= large hole)

FIGURE 7.4 **Inferior view of the skull.** The mandible (lower jawbone) has been removed.

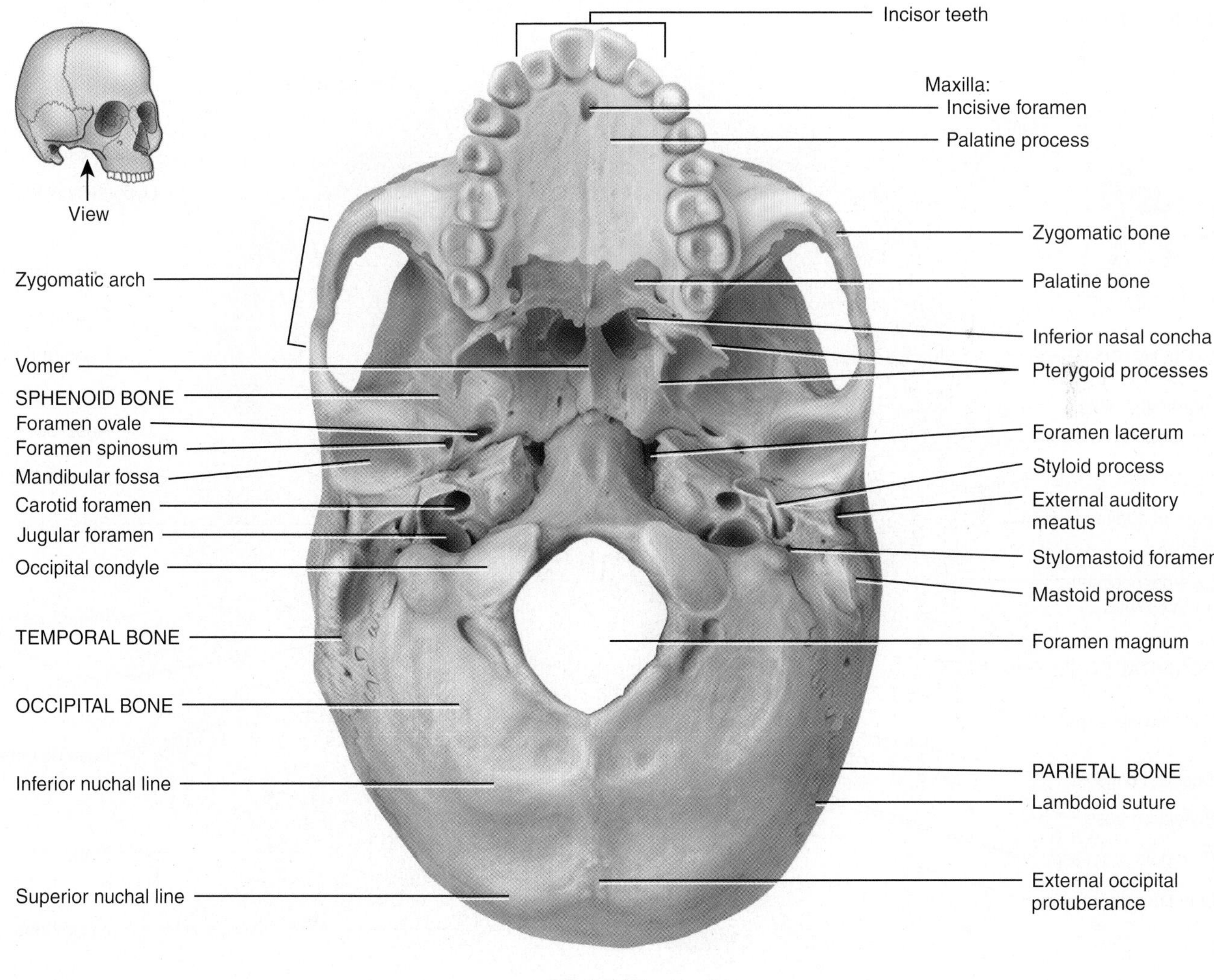


**The occipital condyles of the occipital bone articulate with the first cervical vertebra to allow you to nod your head.**

is in the inferior part of the bone (Figure 7.4). Within this foramen, the medulla oblongata (inferior part of the brain) connects with the spinal cord. The vertebral and spinal arteries also pass through this foramen. The *occipital condyles* are oval processes with convex surfaces, one on either side of the foramen magnum (Figure 7.4). They articulate with the first cervical vertebra (atlas), which allows you to nod your head "yes." Superior to each occipital condyle on the inferior surface of the skull is the *hypoglossal canal* (*hypo-* = under; *-glossal* = tongue), a passageway for impulses that control movement of the tongue (Figure 7.5).

The *external occipital protuberance* is a prominent midline projection on the posterior surface of the occipital bone just superior to the foramen magnum. You may be able to feel this structure as a definite bump on the back of your head, just above your neck (Figure 7.3). Extending laterally from the protuberance are two curved lines, the *superior nuchal lines*, and below these are two *inferior nuchal lines*, which are areas of muscle attachment (Figure 7.6).

## Sphenoid Bone

The **sphenoid bone** (SFĒ-noyd = wedge-shaped) lies at the middle of the base of the skull (Figures 7.4 and 7.7). This bone is the keystone of the cranial floor because it articulates with all the other cranial bones, holding them together. Viewing the cranial floor from above (see Figure 7.7a), note the sphenoid bone joins anteriorly with the frontal bone, laterally with the parietal and temporal bones, and posteriorly with the occipital

FIGURE 7.5 **Medial view of sagittal section of the skull.** Although the hyoid bone is not part of the skull, it is included in the illustration for reference.

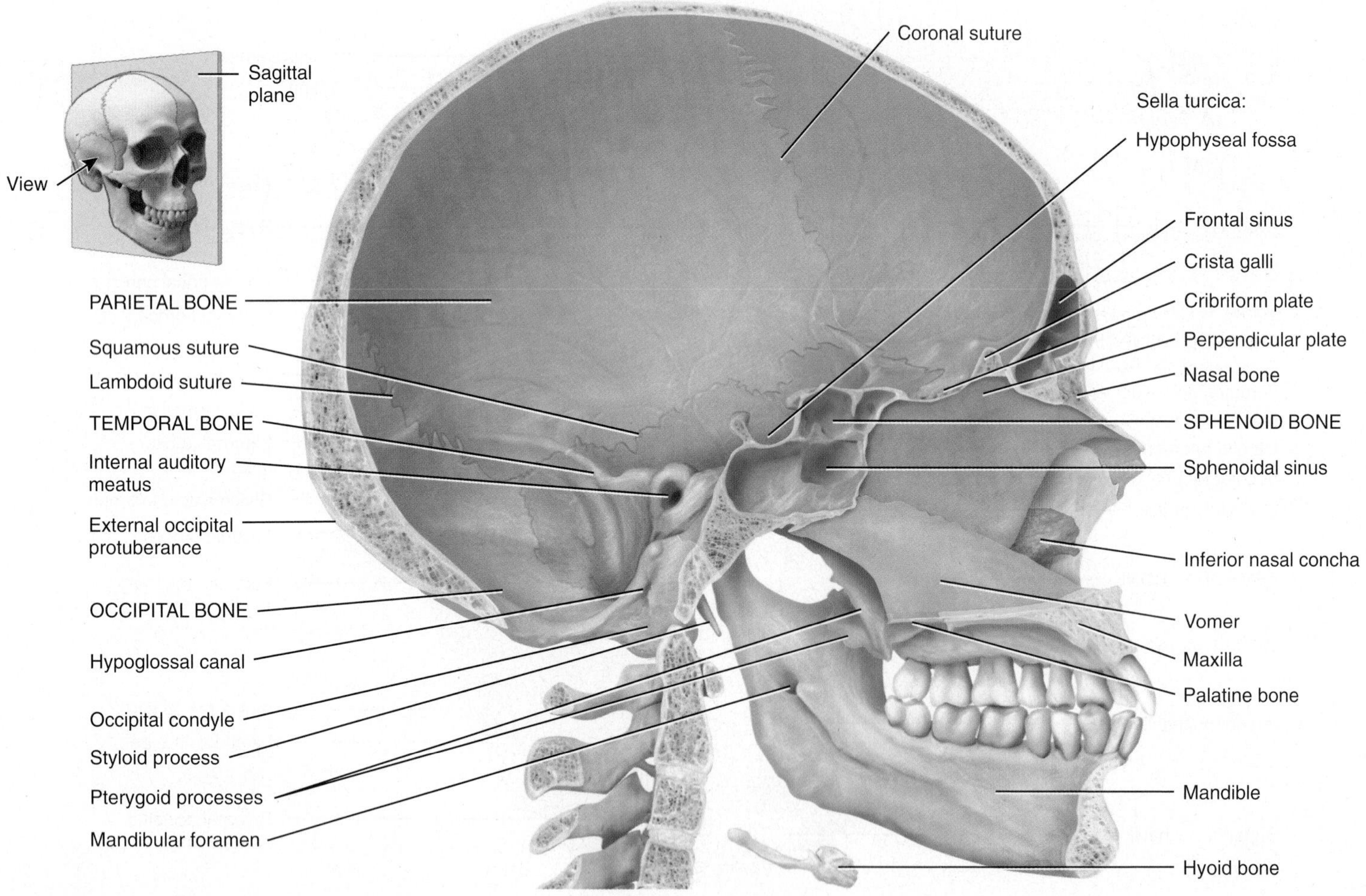


Medial view of sagittal section

**The cranial bones are the frontal, parietal, temporal, occipital, sphenoid, and ethmoid bones. The facial bones are the nasal bone, maxillae, zygomatic bones, lacrimal bones, palatine bones, mandible, and vomer.**

FIGURE 7.6 **Posterior view of the skull.** The sutures are exaggerated for emphasis.

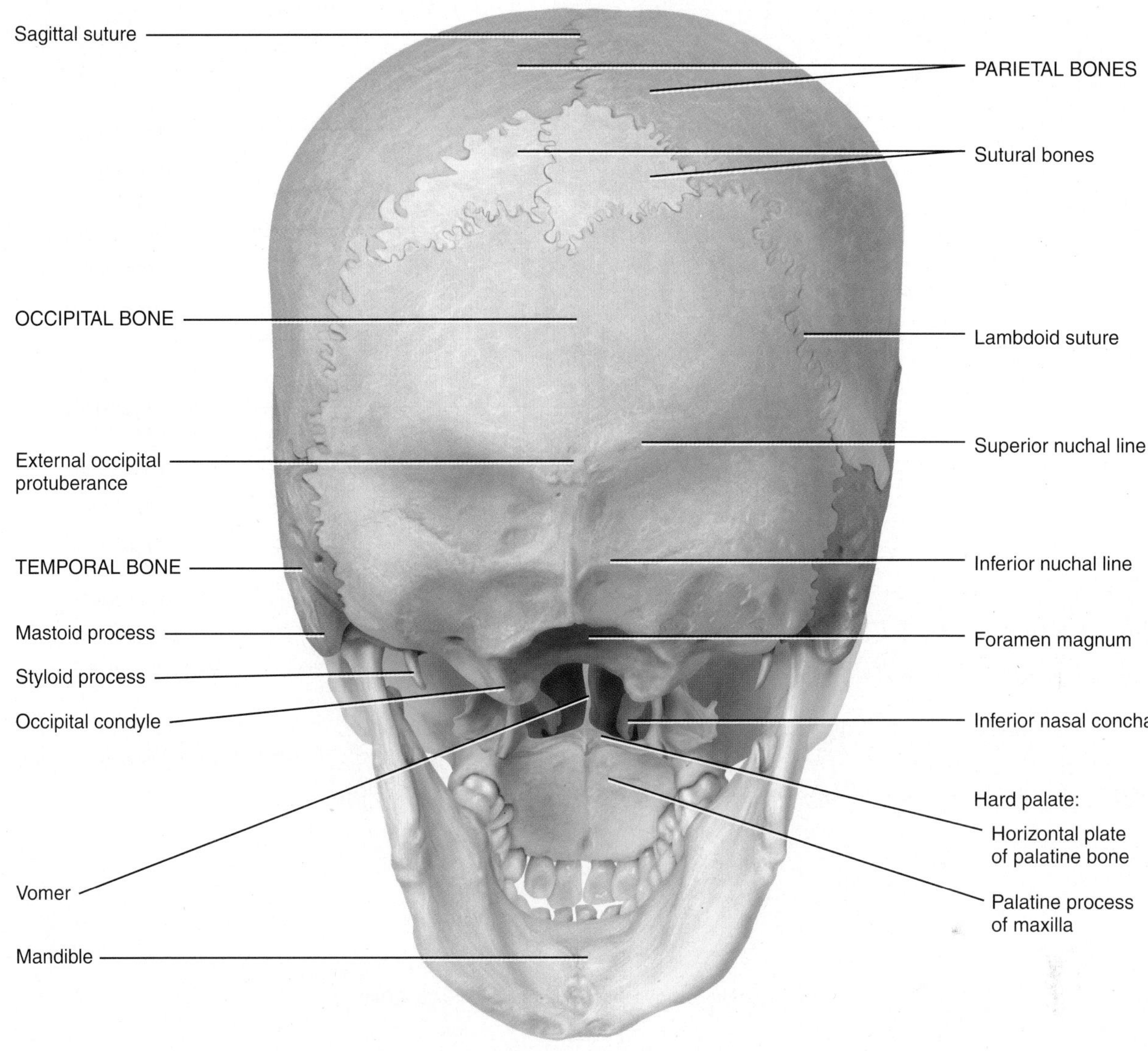


**The occipital bone forms most of the posterior and inferior portions of the skull.**

bone. It lies posterior and slightly superior to the nasal cavity and forms part of the floor, side walls, and rear wall of the orbit (see Figure 7.11).

The shape of the sphenoid bone resembles a butterfly with outstretched wings (Figure 7.7b). The *body* of the sphenoid bone is the hollowed cubelike medial portion between the ethmoid and occipital bones. The space inside the body is the *sphenoidal sinus*, which drains into the nasal cavity (see Figure 7.12). The *sella turcica* (SEL-a TUR-si-ka; *sella* = saddle; *turcica* = Turkish) is a bony saddle-shaped structure on the superior surface of the body of the sphenoid bone (Figure 7.7a). The seat of the saddle is a depression, the *hypophyseal fossa* (hī-pō-FIZ-ē-al), which contains the pituitary gland.

The *greater wings* of the sphenoid bone project laterally from the body, forming the anterolateral cranial floor (Figure 7.7a) and part of the lateral wall of the skull just anterior to the temporal bone (see Figure 7.3). The *lesser wings*, which are smaller than the greater wings, form a ridge of bone anterior and superior to the greater wings. They form part of the cranial floor (Figure 7.7a) and the posterior part of the orbit of the eye (see Figure 7.11). In Figures 7.4 and 7.7b, you can see the *pterygoid processes* (TER-i-goyd = winglike) extending inferiorly from the sphenoid bone. Some of the muscles that move the mandible attach to the pterygoid processes.

Between the body and lesser wing just anterior to the sella turcica is the *optic foramen* (*optic* = eye), through which impulses

**FIGURE 7.7 Sphenoid bone.**

View

Transverse plane

Coronal suture

Superior orbital fissure

Foramen rotundum

Foramen ovale

Foramen spinosum

Internal auditory meatus

Hypoglossal canal

Foramen magnum

OCCIPITAL BONE

FRONTAL BONE

ETHMOID BONE:

Crista galli

Olfactory foramina

Cribriform plate

SPHENOID BONE:

Lesser wing

Greater wing

Hypophyseal fossa of sella turcica

Foramen lacerum

Squamous suture

TEMPORAL BONE:

Petrous portion

Jugular foramen

PARIETAL BONE

Lambdoid suture

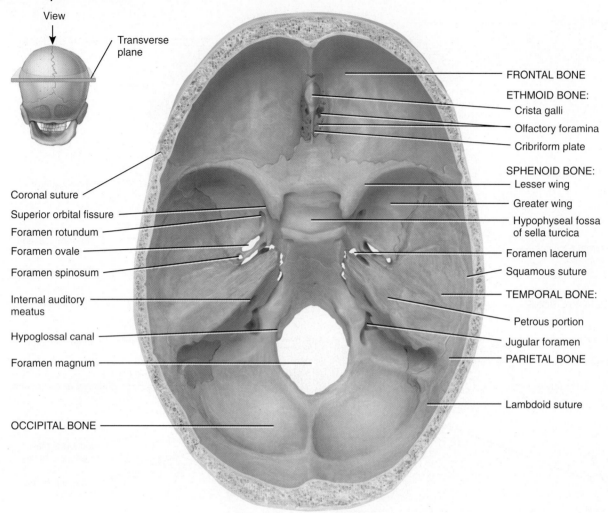

(a) Superior view of sphenoid bone in floor of cranium

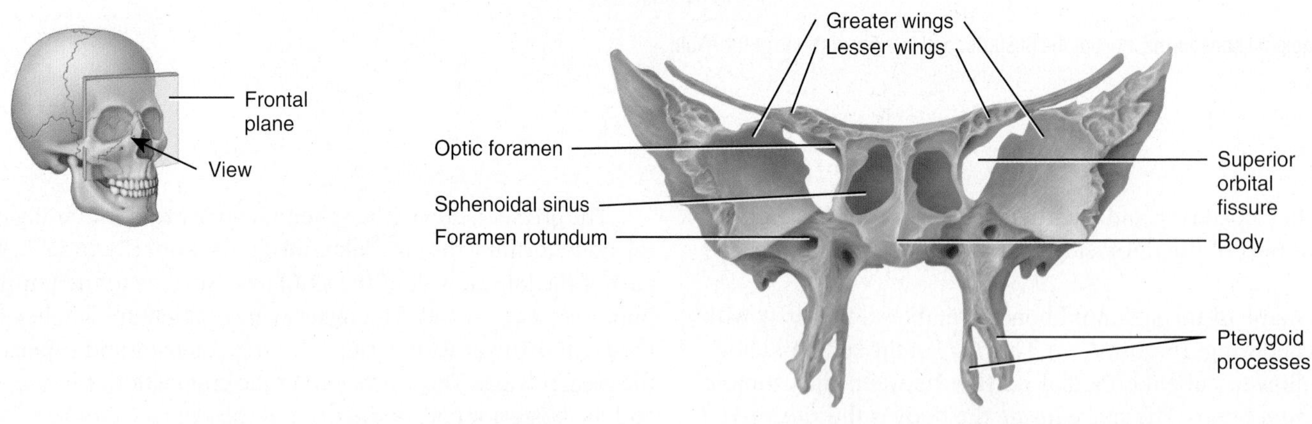


(b) Anterior view of sphenoid bone

**The sphenoid bone is called the keystone of the cranial floor because it articulates with all other cranial bones, holding them together.**

of vision are carried to the brain (Figure 7.7b and see Figure 7.11). Lateral to the body between the greater and lesser wings is a triangular slit called the *superior orbital fissure*, through which impulses are transmitted that control eyeball movement (Figure 7.7b and see Figure 7.11). In the greater wing, three foramina, from anterior to posterior, are the *foramen rotundum* (= round), through which passes a nerve responsible for facial sensations, the *foramen ovale* (= oval hole), through which passes a nerve that controls movement of the mandible, and the *foramen spinosum* (= resembling a spine), through which a blood vessel carries blood to the membranes covering the brain (Figure 7.7a). The *foramen lacerum* (= lacerated) is a jagged passageway between the sphenoid and occipital bones for a blood vessel supplying the brain (Figures 7.4 and 7.7a).

## Ethmoid Bone

The **ethmoid bone** (ETH-moyd = like a sieve) is spongelike in appearance and is located in the anterior part of the cranial floor medial to the orbits (Figure 7.8). It is anterior to the sphenoid and posterior to the nasal bones. The ethmoid bone forms (1) part of the anterior cranial floor; (2) the medial wall of the orbits; (3) the superior portions of the nasal septum, which divides the nasal cavity into right and left sides; and (4) most of the superior sidewalls of the nasal cavity.

The *cribriform plate* (*cribri-* = sieve) of the ethmoid bone lies in the anterior cranial floor and forms the roof of the nasal cavity (Figures 7.5 and 7.7a). The cribriform plate contains the *olfactory foramina* (*olfact-* = to smell) through which smell sensations are transmitted to the brain. Projecting superiorly from the cribriform plate is a triangular process called the *crista galli* (*crista* = crest; *galli* = cock) (Figures 7.5 and 7.8). This structure serves as a point of attachment for the membranes that separate the two sides of the brain. Projecting inferiorly from the cribriform plate is the *perpendicular plate*, which forms the superior portion of the nasal septum (see Figure 7.9).

The ethmoid bone contains 3 to 18 air spaces, or "cells." The ethmoidal cells together form the *ethmoidal sinuses* (see Figure 7.12). Also part of the ethmoid bone are two thin, scroll-shaped projections on either side of the nasal septum (Figure 7.8). These are the *superior nasal concha* (KONG-ka = shell) and the *middle nasal concha*. The plural term is *conchae* (KONG-kē). A third pair of conchae, the inferior nasal

FIGURE 7.8 Ethmoid bone.

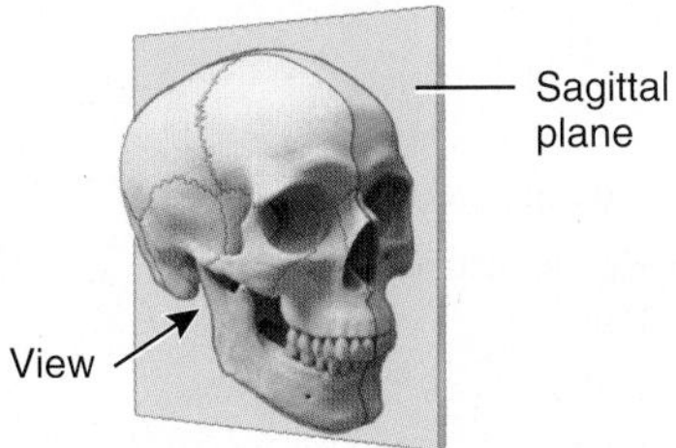


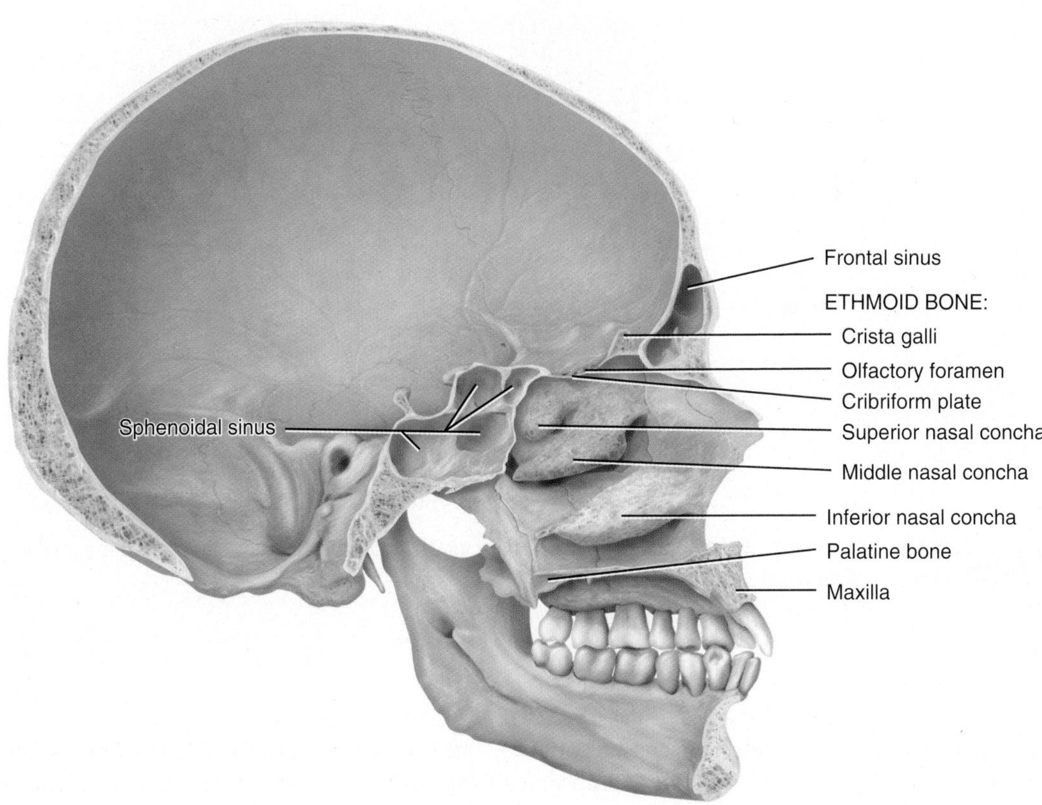


(a) Medial view of sagittal section

(continues)

**FIGURE 7.8 (continued)**

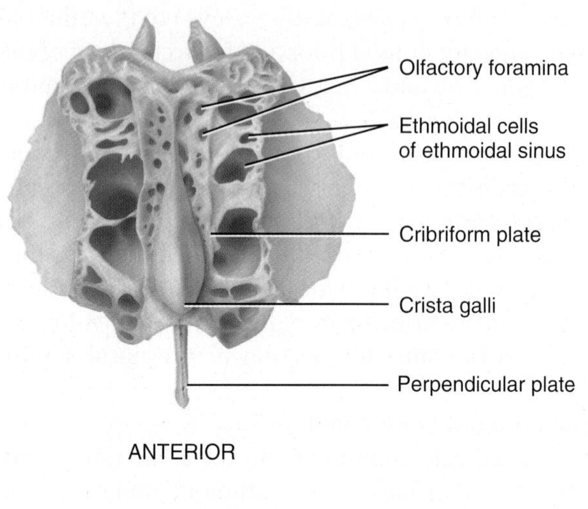


(b) Superior view

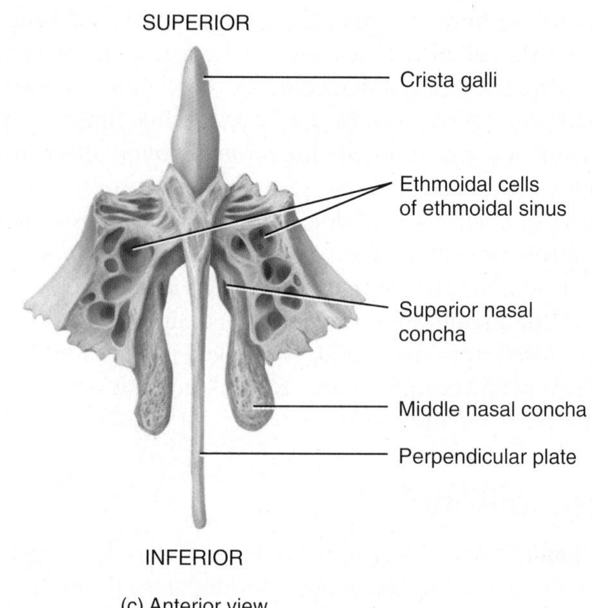


(c) Anterior view

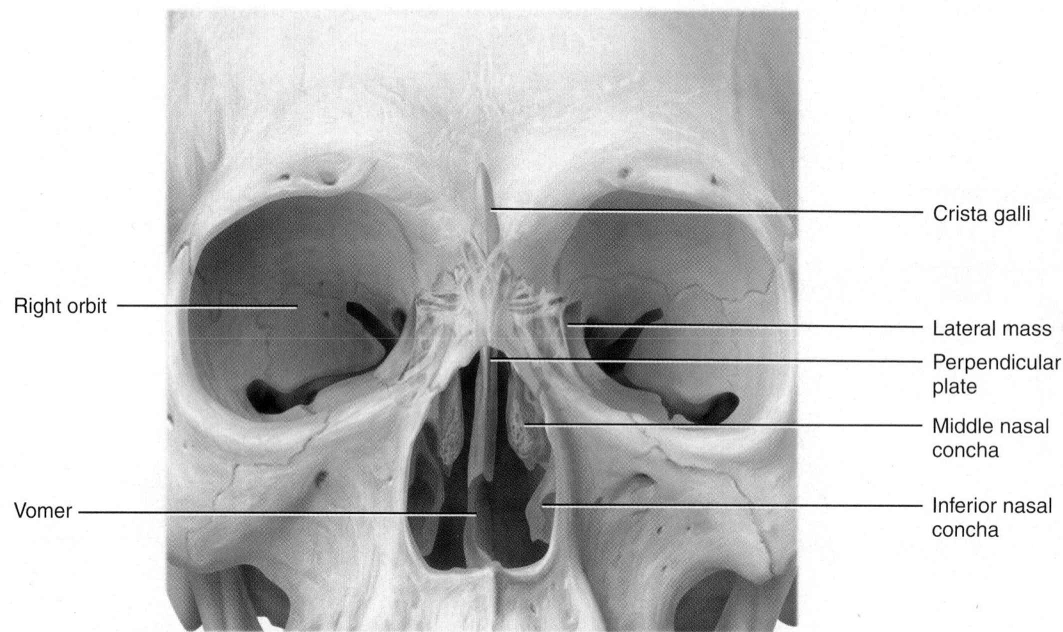


(d) Anterior view of position of ethmoid bone in skull (projected to surface)

**The ethmoid bone forms part of the anterior portion of the cranial floor, the medial wall of the orbits, the superior portions of the nasal septum, and most of the sidewalls of the nasal cavity.**

conchae, are separate bones (discussed shortly). The conchae increase the vascular and mucous membrane surface area in the nasal cavity, which warms and moistens inhaled air before it passes into the lungs. The conchae also cause inhaled air to swirl, and the result is that many inhaled particles become trapped in the mucus that lines the nasal cavity. This action of the conchae helps cleanse inhaled air before it passes into the rest of the respiratory passageways. The superior nasal conchae aid the sense of smell because the mucous membrane covering them contains sensory receptors for olfaction (smell).

✓ CHECKPOINT

6. List the cranial bones, and indicate which ones are single and which ones are paired.
7. With which bones does the parietal bone articulate? With which bones does the temporal bone articulate?
8. What are the components of the zygomatic arch? The temporomandibular joint?
9. Which marking of the skull articulates with the first cervical vertebra? Which one forms a bump on the back of your head?
10. Starting at the ethmoid bone and going in a clockwise direction, list the bones that articulate with the sphenoid bone.
11. What functions are served by the internal auditory meatus and external auditory meatus? On which cranial bone are these markings located?
12. What functions are served by nasal conchae? Which nasal conchae are part of the ethmoid bone?

## 7.4 Facial bones include the nasal bones, maxillae, zygomatic bones, lacrimal bones, palatine bones, inferior nasal conchae, vomer, and mandible.

The shape of the face changes dramatically during the first two years after birth. The brain and cranial bones expand, the first set of teeth form and erupt (emerge), and the paranasal sinuses increase in size. Growth of the face ceases at about 16 years of age. The 14 facial bones include two nasal bones, two maxillae, two zygomatic bones, the mandible, two lacrimal bones, two palatine bones, two inferior nasal conchae, and the vomer.

### Nasal Bones

The paired **nasal bones** are small, flattened, rectangular-shaped bones that form the bridge of the nose (see Figure 7.2). These small bones protect the upper entry to the nasal cavity and provide attachment for a couple of thin muscles of facial expression. For those who wear glasses, the nasal bones form the resting place for the bridge of the glasses. The major structural portion of the nose consists of cartilage (Figure 7.9).

### Maxillae

The paired **maxillae** (mak-SIL-ē = jawbones; singular is *maxilla*) unite to form the upper jawbone. They articulate with every bone of the face except the mandible, or lower jawbone (see Figures 7.2, 7.3, and 7.4). The maxillae form part of the floors of the orbits, part of the lateral walls and floor of the nasal cavity, and most of the hard palate. The *hard palate* is the bony roof of the mouth and is formed by the palatine processes of the maxillae and horizontal plates of the palatine bones. The hard palate separates the nasal cavity from the oral cavity.

FIGURE 7.9 Nasal cavity.

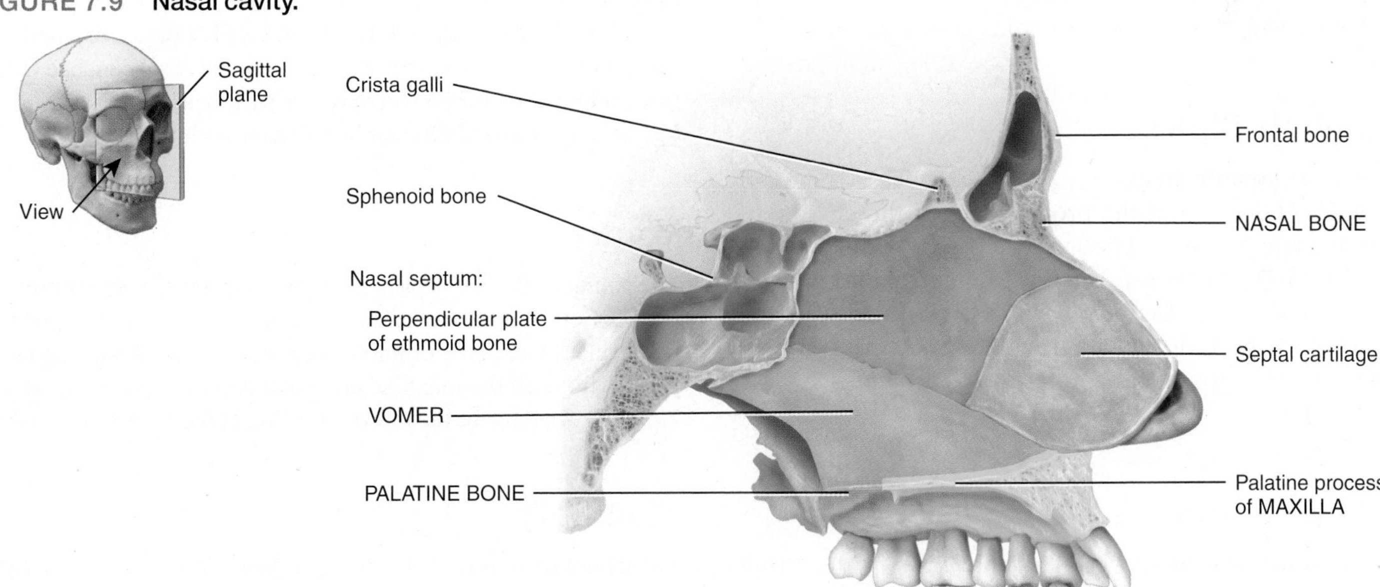


The structures that form the nasal septum are the perpendicular plate of the ethmoid bone, the vomer, and the septal cartilage.

## CLINICAL CONNECTION | *Cleft Palate and Cleft Lip*

Usually the palatine processes of the maxillae unite during weeks 10 to 12 of embryonic development. Failure to do so can result in one type of **cleft palate**. The condition may also involve incomplete fusion of the horizontal plates of the palatine bones (see Figure 7.4). Another form of this condition, called **cleft lip**, involves a split in the upper lip. Cleft lip and cleft palate often occur together. Depending on the extent and position of the cleft, speech and swallowing may be affected. In addition, children with cleft palate tend to have many ear infections, which can lead to hearing loss. Facial and oral surgeons recommend closure of cleft lip during the first few weeks following birth, and surgical results are excellent. Repair of cleft palate typically is completed between 12 and 18 months of age, ideally before the child begins to talk. Because the palate is important for pronouncing consonants, speech therapy may be required, and orthodontic therapy may be needed to align the teeth. Recent research strongly suggests that supplementation with folic acid (one of the B vitamins) during early pregnancy decreases the incidence of cleft palate and cleft lip. The mechanism behind this birth defect is not yet understood.

Each maxilla contains a large *maxillary sinus* that empties into the nasal cavity (see Figure 7.12). The *alveolar process* (al-VĒ-ō-lar; *alveol-* = small cavity) of the maxilla is a ridgelike arch that contains the alveoli (sockets) for the maxillary (upper) teeth (see Figure 7.2). The *palatine process* is a horizontal projection of the maxilla that forms the anterior portion of the hard palate (see Figures 7.4 and 7.6). The union and fusion of the maxillary bones normally is completed before birth.

The *infraorbital foramen* (*infra-* = below; *orbital* = orbit), which can be seen in the anterior view of the skull in Figure 7.2, is a passageway in the maxilla below the orbit for a nerve to the face. Another prominent foramen, the *incisive foramen* (= incisor teeth), seen in Figure 7.4, is a passageway in the hard palate just posterior to the incisor teeth for blood vessel and nerve supply to the hard palate. Nerves and blood vessels to the face pass through the *inferior orbital fissure*, located within the orbit between the maxilla and sphenoid bone (see Figure 7.11).

## Zygomatic Bones

The two **zygomatic bones** (*zygo-* = yokelike), commonly called *cheekbones*, form the prominences of the cheeks and part of the lateral wall and floor of each orbit (see Figures 7.2, 7.3, and 7.11). They articulate with the frontal, maxilla, sphenoid, and temporal bones. The *temporal process* of the zygomatic bone. projects posteriorly and articulates with the zygomatic process of the temporal bone to form the *zygomatic arch* (see Figure 7.3).

## Lacrimal Bones

The paired **lacrimal bones** (LAK-ri-mal; *lacrim-* = teardrops) roughly resemble a fingernail in size and shape (see Figures 7.2, 7.3, and 7.11). These bones, the smallest bones of the face, are posterior and lateral to the nasal bones and form a part of the medial wall of each orbit. The lacrimal bones each contain a *lacrimal fossa*, a vertical tunnel formed with the maxilla, through which tears pass into the nasal cavity (see Figures 7.3 and 7.11).

## Palatine Bones

The two L-shaped **palatine bones** (PAL-a-tīn) form the posterior portion of the hard palate, part of the floor and lateral wall of the nasal cavity, and a small portion of the floors of the orbits (see Figures 7.4 and 7.9).

## Inferior Nasal Conchae

The two **inferior nasal conchae** are inferior to the middle nasal conchae of the ethmoid bone and are separate bones that are not part of the ethmoid bone (see Figures 7.2 and 7.8a). These scroll-like bones form a part of the inferior lateral wall of the nasal cavity and project into the nasal cavity. All three pairs of nasal conchae (superior, middle, and inferior) increase the surface area of the nasal cavity and help swirl and filter air before it passes into the lungs.

## Vomer

The **vomer** (VŌ-mer = plowshare) is a roughly triangular bone on the floor of the nasal cavity that articulates superiorly with the perpendicular plate of the ethmoid bone and sphenoid bone and inferiorly with both the maxillae and palatine bones (see Figures 7.2, 7.4, and 7.9). It forms the inferior portion of the bony nasal septum.

## Mandible

The **mandible** (*mand-* = to chew), or lower jawbone, is the largest, strongest facial bone (see Figures 7.2, 7.3, and 7.10). It is the only freely movable skull bone (other than the auditory ossicles, the small bones of the ear). In the lateral view shown in Figure 7.10, you can see that the mandible consists of a curved, horizontal

FIGURE 7.10 **Mandible.**

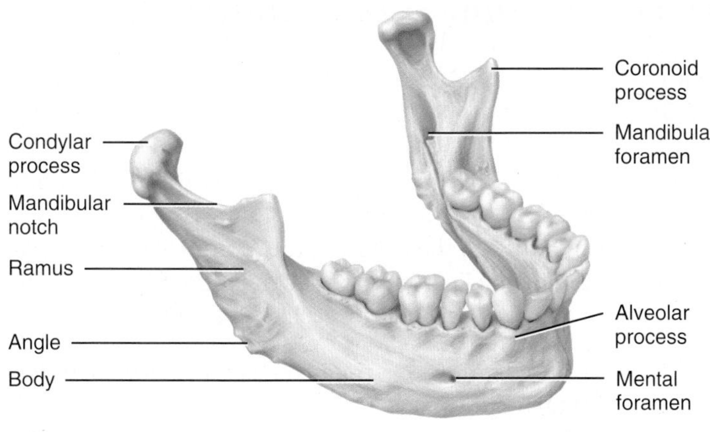


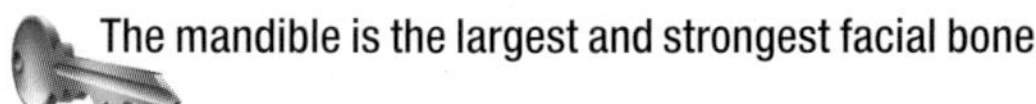
**The mandible is the largest and strongest facial bone.**

portion, the *body*, and two perpendicular portions, the *rami* (RĀ-mī = branches; singular is *ramus*). The *angle* of the mandible is the area where each *ramus* meets the body. Each ramus has a posterior *condylar process* (KON-di-lar) that articulates with the mandibular fossa of the temporal bone (see Figure 7.3) to form the *temporomandibular joint (TMJ)*. It also has an anterior *coronoid process* (KOR-ō-noyd) to which the temporalis muscle attaches (Figure 7.10). The depression between the coronoid and condylar processes is called the *mandibular notch*. The *alveolar process* is an arch containing the *alveoli* (sockets) for the mandibular (lower) teeth.

The *mental foramen* (*ment-* = chin) is a hole in the mandible that dentists use when injecting anesthetics. Another foramen associated with the mandible is the *mandibular foramen* on the medial surface of each ramus, another site often used by dentists to inject anesthetics. Through the mandibular foramen pass nerves and blood vessels that supply the mandibular teeth.

## ✓ CHECKPOINT

**13.** On which bones do eyeglasses normally rest?

**14.** What bones form the roof of the mouth? What is the anatomical name for the bony roof of the mouth?

**15.** Through which bones do tears travel from the eyes to the nasal cavity?

**16.** What anatomical structure do the maxillae and mandible have in common?

**17.** Which bone forms the inferior part of the nasal septum?

**18.** Which bone attaches to other skull bones by freely movable joints?

**19.** What are the components of the temporomandibular joint?

## RETURN TO **Fernando's Story**

Isabel arrives at the emergency room shortly after the ambulance carrying Fernando and Elena. After Isabel gives her daughter a big hug and praises her for her quick response to the accident, Dr. Mueller calls them into his office to update them on Fernando's condition. "Fernando is still unconscious but his vital signs are stable. He is currently undergoing a CAT scan to determine if there is any brain damage, because his x-rays revealed a skull fracture over his left ear. I will be admitting him to the Intensive Care Unit for observation after the CAT scan. And one more thing–he has also sustained many facial cuts and abrasions, and his face is quite swollen. You can go see him after the test, but he won't look like his normal self."

When Isabel and Elena reach the ICU, the nurse informs them that Fernando has just arrived and as soon as he is settled, they can see him. After a short time, the nurse returns to the waiting area and escorts them to Fernando's room. Isabel and Elena are glad that the doctor warned them—they would not have recognized him because his face is so swollen and red. Some of the hair on his head has been shaved and there are stitches in his left eyebrow and over his left ear. Elena takes Fernando's hand and speaks to him. She feels a slight squeeze and smiles as she tells her mother, "I know he is going to be OK. He knows we are here. He just squeezed my hand." This surprises the nurse, who picks up his other hand and asks him to squeeze it. When he does, the nurse tells them that it's a very good sign that he is now conscious and responding to commands.

A. *Fernando has a fracture above his left ear. Which bone or bones could be fractured?*

B. *Fernando has stitches in his left eyebrow. Which bone is beneath the eyebrow?*

C. *The left side of Fernando's face is scraped from his forehead to his chin. Which bones are beneath those scrapes?*

# 7.5 Unique features of the skull include the nasal septum, orbits, sutures, paranasal sinuses, and fontanels.

The skull exhibits several unique features and a structural organization not seen in other bones of the body. These include the nasal septum, orbits, paranasal sinuses, sutures, and fontanels.

## Nasal Septum

The inside of the nose, called the nasal cavity, is divided into right and left sides by a vertical partition called the **nasal septum**. The three components of the nasal septum are the vomer, septal cartilage, and perpendicular plate of the ethmoid bone (see Figure 7.9). The anterior border of the vomer articulates with the septal cartilage, which is hyaline cartilage, to form the anterior portion of the nasal septum. The superior border of the vomer articulates with the perpendicular plate of the ethmoid bone to form the remainder of the nasal septum. The term *broken nose*, in most cases, refers to damage to the septal cartilage rather than the nasal bones themselves.

## Orbits

Seven bones of the skull join to form each **orbit** (eye socket), which contains the eyeball and associated structures (Figure 7.11). The three cranial bones of the orbit are the frontal, sphenoid, and ethmoid; the four facial bones are the palatine, zygomatic, lacrimal, and maxilla. Associated with each orbit are five openings:

- The *supraorbital foramen* is found on the supraorbital margin of the frontal bone.
- The *optic foramen* is in the sphenoid bone.
- The *superior orbital fissure* is in the sphenoid bone.
- The *inferior orbital fissure* is located between the sphenoid bone, zygomatic bone, and maxilla.
- The *lacrimal fossa* is in the lacrimal bone.

FIGURE 7.11 Details of the orbit.

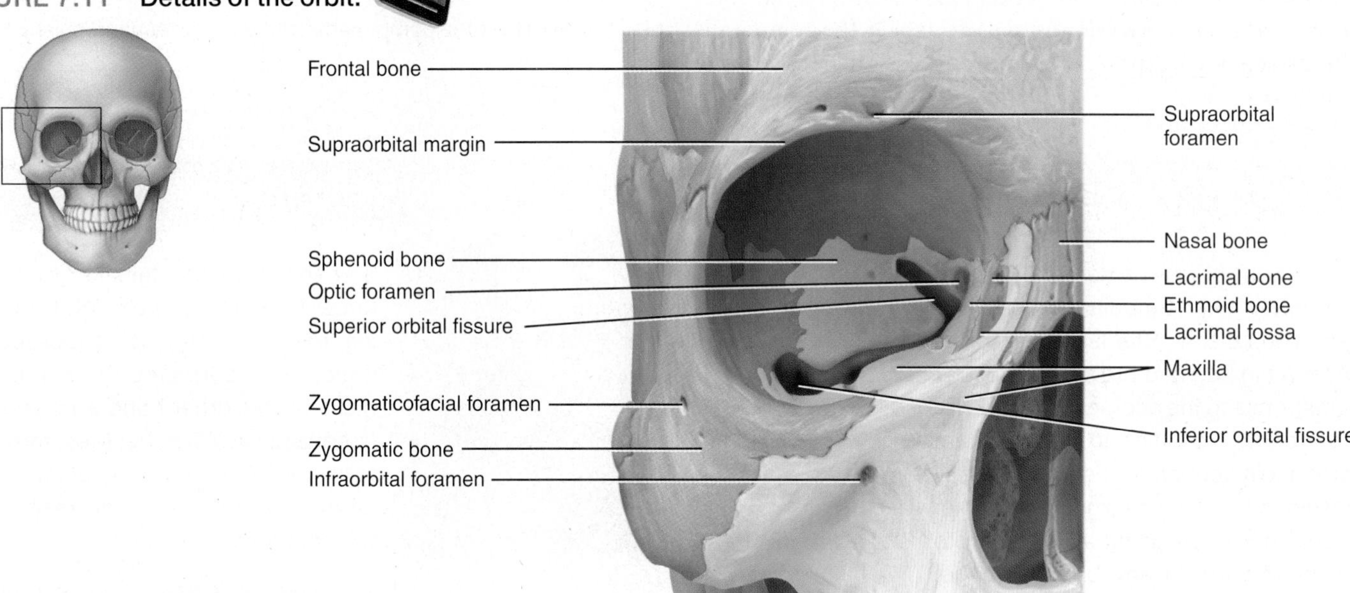


The orbit is a cavity that contains the eyeball and associated structures.

## CLINICAL CONNECTION | *Deviated Nasal Septum*

A **deviated nasal septum** is one that does not run along the midline of the nasal cavity. It deviates (bends) to one side. A blow to the nose can easily damage, or break, this delicate septum of bone and displace and damage the cartilage. Often, when a broken nasal septum heals, the bones and cartilage deviate to one side or the other. This deviated septum can block airflow into the constricted side of the nose, making it difficult to breathe through that half of the nasal cavity. The deviation usually occurs at the junction of the vomer bone with the septal cartilage. Septal deviations may also occur due to developmental abnormality. If the deviation is severe, it may block the nasal passageway entirely. Even a partial blockage may lead to infection. If inflammation occurs, it may cause nasal congestion, blockage of the paranasal sinus openings, chronic sinusitis, headache, and nosebleeds. The condition usually can be corrected or improved surgically.

## Paranasal Sinuses

The **paranasal sinuses** (par′-a-NĀ-zal SĪ-nus-ez; *para-* = beside), mucous membrane-lined cavities near the nasal cavity, are found within the frontal, sphenoid, ethmoid, and maxillary bones (Figure 7.12). The paranasal sinuses are lined with mucous membranes that are continuous with the lining of the nasal cavity. Secretions produced by the mucous membranes of the paranasal sinuses drain into the nasal cavity. The paranasal sinuses lighten the mass of the skull and increase the surface area of the nasal mucosa to help moisten and cleanse inhaled air. In addition, the paranasal sinuses serve as resonating (echo) chambers that intensify and prolong sounds, thereby enhancing the quality of the voice. The influence of the paranasal sinuses on your voice becomes obvious when you have a cold; the passageways through which sound travels into and out of the paranasal sinuses become blocked by excess mucus production, changing the quality of your voice.

## Sutures

**Sutures** (SOO-churs = seams), found only between skull bones, hold most skull bones together. Sutures in the skulls of infants and children often are movable, but those in an adult usually form immovable joints. Of the many sutures found in the skull, we will identify only four prominent ones:

- The **coronal suture** (kō-RŌ-nal; *coron-* = crown) unites the frontal bone and both parietal bones (see Figure 7.3).

FIGURE 7.12 **Paranasal sinuses.**

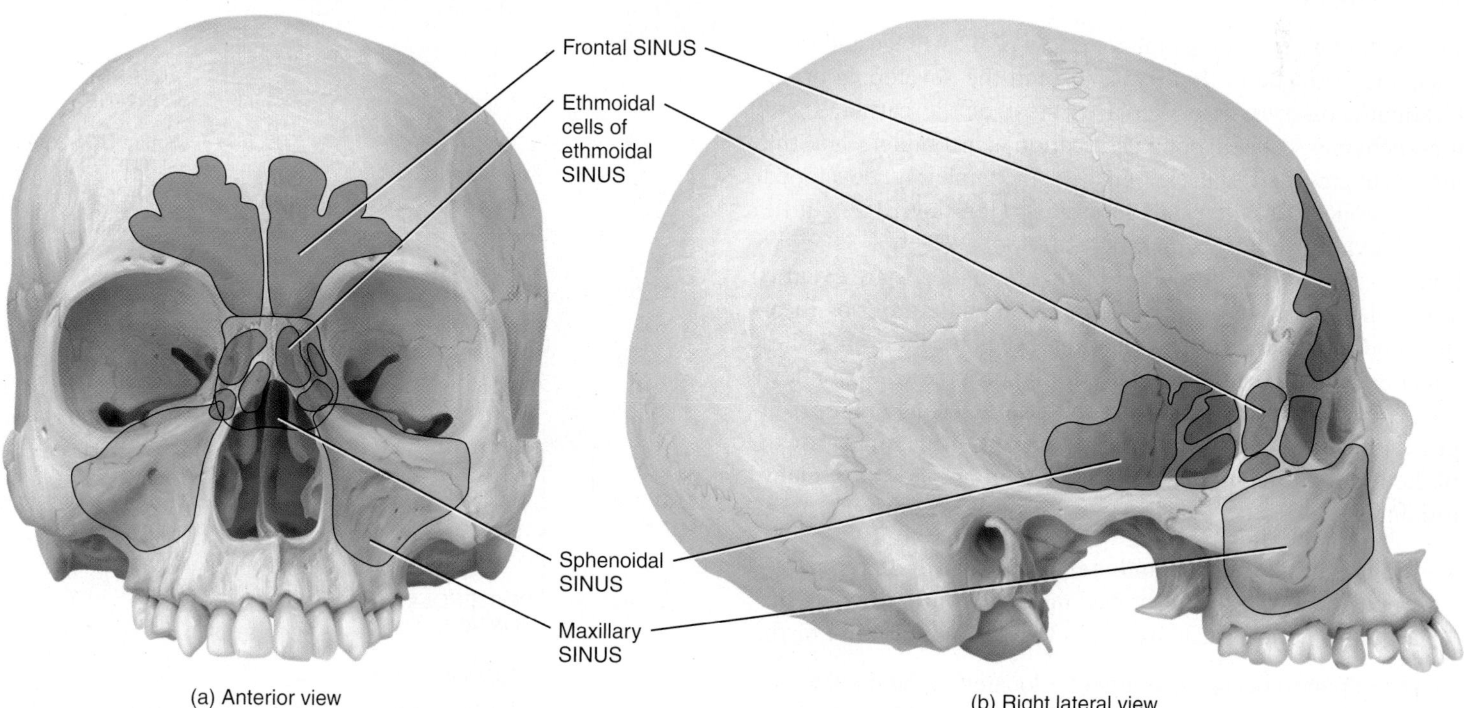


Paranasal sinuses are mucous membrane-lined spaces in the frontal, sphenoid, ethmoid, and maxillary bones that connect to the nasal cavity.

## CLINICAL CONNECTION | *Sinusitis*

**Sinusitis** (sīn-ū-SĪ-tis) is an inflammation of the mucous membrane of one or more paranasal sinuses. It may be caused by a microbial infection (virus, bacterium, or fungus), allergic reactions, nasal polyps, or a severely deviated nasal septum. If the inflammation or an obstruction blocks the drainage of mucus into the nasal cavity, fluid pressure builds up in the paranasal sinuses, and a sinus headache may develop. Other symptoms may include nasal congestion, inability to smell, fever, and cough. Treatment options include decongestant sprays or drops, oral decongestants, nasal corticosteroids, antibiotics, analgesics to relieve pain, warm compresses, and surgery.

- The **sagittal suture** (SAJ-i-tal; *sagitt-* = arrow) unites the two parietal bones on the superior midline of the skull (see Figure 7.6). The sagittal suture is so named because in the infant, before the bones of the skull are firmly united, the suture and the fontanels (soft spots) associated with it resemble an arrow.
- The **lambdoid suture** (LAM-doyd) unites the two parietal bones to the occipital bone. This suture is so named because of its resemblance to the Greek letter lambda (λ), as can be seen in Figure 7.6 (with the help of a little imagination). The sagittal and lambdoid sutures may contain small bones called sutural bones (SOO-chur-al; *sutur-* = seam).
- The two **squamous sutures** (SKWĀ-mus; *squam-* = flat, like the flat overlapping scales of a snake) unite the parietal and temporal bones on the lateral sides of the skull (see Figure 7.3).

## Fontanels

The skull of a developing embryo consists of cartilage and mesenchyme arranged in thin plates around the developing brain. Gradually, osseous tissue replaces most of the cartilage and mesenchyme. At birth, bone ossification is incomplete, and the mesenchyme-filled spaces between incompletely developed cranial bones are called called **fontanels** (fon-ta-NELZ = little fountains), or, more commonly, "soft spots" (Figure 7.13). As bone formation continues after birth, the fontanels are eventually replaced with osseous tissue by intramembranous ossification, and the junctions that remain between neighboring bones become the sutures. Functionally, fontanels provide some flexibility to the fetal skull, allowing the skull to change shape as it passes through the birth canal and later permitting rapid growth of the brain during infancy. Although an infant may have many fontanels at birth, the form and location of six are fairly constant:

- The unpaired **anterior fontanel**, located at the midline between the two parietal bones and the frontal bone, is the largest fontanel. It usually closes 18 to 24 months after birth.
- The unpaired **posterior fontanel** is located at the midline between the two parietal bones and the occipital bone. It generally closes about 2 months after birth.
- The paired **anterolateral fontanels**, located laterally between the frontal, parietal, temporal, and sphenoid bones, are small. Normally, they close about 3 months after birth.
- The paired **posterolateral fontanels** are located laterally between the parietal, occipital, and temporal bones. They begin to close 1 to 2 months after birth, but closure is generally not complete until 12 months.

The amount of closure in fontanels helps a physician gauge the degree of brain development. In addition, the anterior fontanel serves as a landmark for withdrawal of blood for analysis from the superior sagittal sinus (a large vein on the midline surface of the brain).

FIGURE 7.13 **Fontanels at birth.**

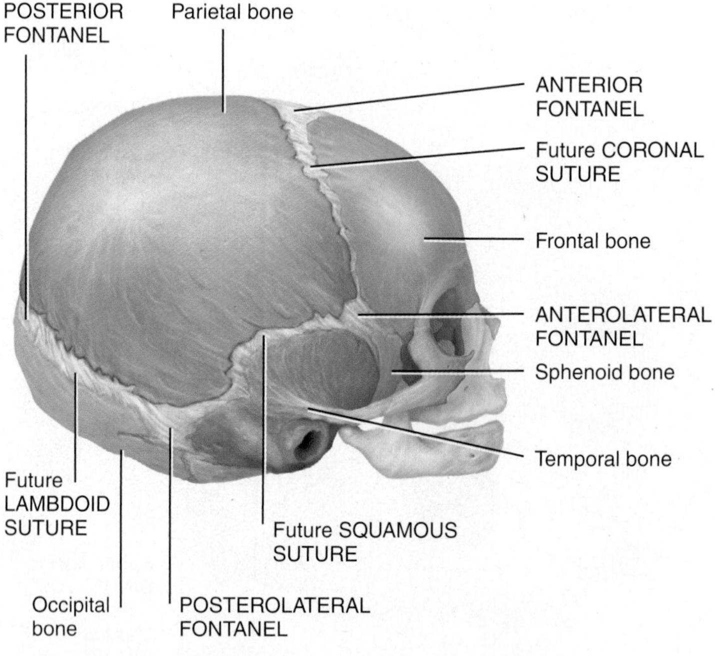


**Fontanels are mesenchyme-filled spaces between cranial bones that are present at birth.**

## ✓ CHECKPOINT

20. What structures make up the nasal septum? What kind of tissue composes the anterior portion of the nasal septum?
21. What are the locations and functions of the paranasal sinuses?
22. What bones are joined by the squamous suture? The lambdoid suture? The coronal suture? The sagittal suture?
23. Which fontanel is bordered by four different skull bones?
24. Following the birth of a baby, how many months are needed before all fontanels are ossified into sutures?

# 7.6 The hyoid bone supports the tongue and attaches to muscles of the tongue, pharynx, and larynx.

The single **hyoid bone** (= U-shaped) is a unique component of the axial skeleton because it does not articulate with any other bone (see Figure 7.5). Rather, it is suspended from the styloid processes of the temporal bones by ligaments and muscles. Located in the anterior neck between the mandible and larynx (Figure 7.14a), the hyoid bone supports the tongue and provides attachment sites for some tongue muscles and for muscles of the pharynx and larynx. The hyoid bone consists of a *body* and paired projections called the *lesser horns* and the *greater horns* (Figure 7.14b, c). Muscles and ligaments attach to these paired projections.

The hyoid bone and cartilages of the larynx and trachea are often fractured during strangulation. Thus, they are carefully examined at autopsy when strangulation is a suspected cause of death.

## ✓ CHECKPOINT

25. What are the functions of the hyoid bone?

FIGURE 7.14 Hyoid bone.

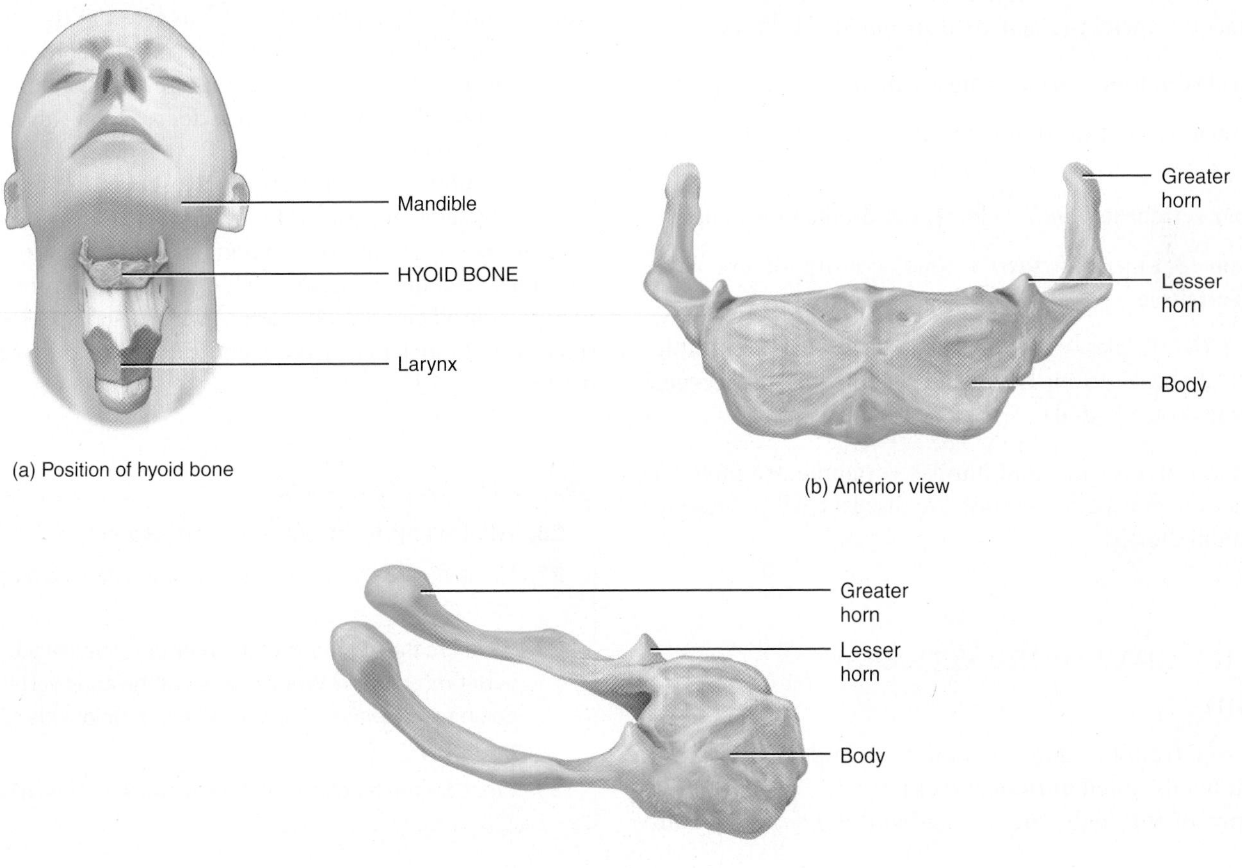


**The hyoid bone supports the tongue and provides attachment sites for muscles of the tongue, neck, and pharynx.**

# 7.7 The vertebral column protects the spinal cord, supports the head, and is a point of attachment for bones and muscles.

The **vertebral column**, also called the *spine*, makes up about two-fifths of your total height and is composed of a series of bones called **vertebrae** (VER-te-brē; singular is *vertebra*) and connective tissue. The vertebral column, sternum, and ribs form the skeleton of the trunk of the body. The length of the column is about 71 cm (28 in.) in an average adult male and about 61 cm (24 in.) in an average adult female. The vertebral column functions as a strong, flexible rod with elements that can rotate and move forward, backward, sideways, and rotate. In addition to enclosing and protecting the spinal cord, the vertebral column supports the head, and serves as a point of attachment for the ribs, pelvic girdle, and muscles of the back and upper limbs.

The vertebral column is divided into five regions (Figure 7.15a). Beginning superiorly and moving inferiorly, the regions are cervical, thoracic, lumbar, sacral, and coccygeal. Note that vertebrae in each region are numbered in sequence, from superior to inferior.

The total number of vertebrae during early development is 33. As a child grows, several vertebrae in the sacral and coccygeal regions fuse. As a result, the adult vertebral column typically contains 26 vertebrae that are distributed as follows:

- 7 **cervical vertebrae** (*cervic-* = neck) are in the neck region.
- 12 **thoracic vertebrae** (*thorax* = chest) are posterior to the thoracic cavity.
- 5 **lumbar vertebrae** (*lumb-* = loin) support the lower back.
- 1 **sacrum** (SĀ-krum = sacred bone) consists of five fused **sacral vertebrae**.
- 1 **coccyx** (KOK-siks = cuckoo, because the shape resembles the bill of a cuckoo bird), consists of four fused **coccygeal vertebrae** (kok-SIJ-ē-al).

The cervical, thoracic, and lumbar vertebrae are movable, but the sacrum and coccyx are not. We discuss each of these regions in detail shortly.

## Normal Curves of the Vertebral Column

When viewed from the side, the adult vertebral column shows four slight bends called **normal curves** (Figure 7.15b). Relative to the front of the body, the *cervical* and *lumbar curves* are convex (bulging anteriorly), and the *thoracic* and *sacral curves* are concave (curving posteriorly). The curves of the vertebral column increase its strength, help maintain balance in the upright position, absorb shocks during walking, and help protect the vertebrae from fracture.

## Intervertebral Discs

**Intervertebral discs** (in′-ter-VER-te-bral; *inter* = between) are found between the bodies of adjacent vertebrae from the second cervical vertebra to the sacrum (Figure 7.15c). Each disc has an outer fibrous ring consisting of fibrocartilage called the *annulus fibrosus* (*annulus* = ring or ringlike) and an inner soft, pulpy, highly elastic substance called the *nucleus pulposus* (*pulposus* = pulplike). The discs form strong joints, permit various movements of the vertebral column, and absorb vertical shock. Under compression, they flatten and broaden. During the course of a day the discs compress so that we are a bit shorter at night. While we are sleeping there is less compression so that we are taller when we awaken in the morning. As we age, the nucleus pulposus hardens and becomes less elastic. Decrease in vertebral height with age results from bone loss in the vertebral bodies and not a decrease in thickness of the intervertebral discs.

Since intervertebral discs are avascular, the annulus fibrosus and nucleus pulposus rely on blood vessels from the bodies of vertebrae to obtain oxygen and nutrients and remove wastes. Certain stretching exercises, such as yoga, decompress discs and increase blood circulation, both of which increase the uptake of oxygen and nutrients by discs and the removal of wastes from discs.

### ✓ CHECKPOINT

**26.** What are the functions of the vertebral column?

**27.** Name the five types of vertebrae and state the number of each type.

**28.** What are the functions of the normal curves of the vertebral column? Which curves of the adult vertebral column are concave (relative to the anterior side of the body)?

**29.** What are the functions of the intervertebral discs?

**FIGURE 7.15 Vertebral column.** The numbers in parentheses in (a) indicate the number of vertebrae in each region. In (c), the relative size of the disc has been enlarged for emphasis.

CERVICAL VERTEBRAE (7)
THORACIC VERTEBRAE (12)
LUMBAR VERTEBRAE (5)
INTERVERTEBRAL DISC
SACRUM (1)
COCCYX (1)

(a) Anterior view showing regions of vertebral column

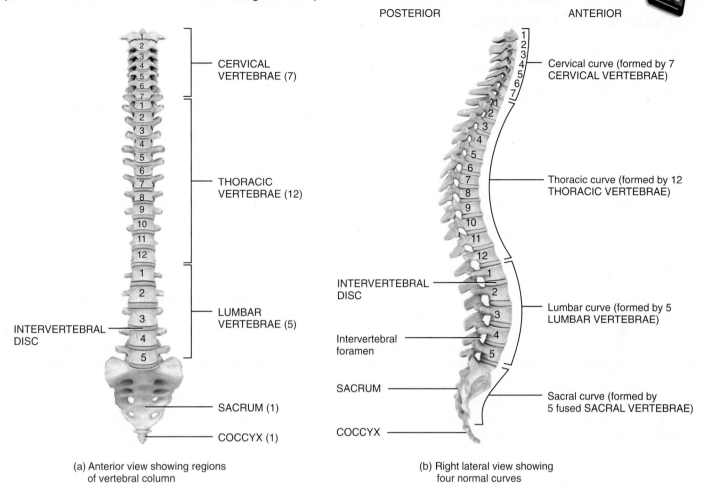

POSTERIOR
ANTERIOR
Cervical curve (formed by 7 CERVICAL VERTEBRAE)
Thoracic curve (formed by 12 THORACIC VERTEBRAE)
INTERVERTEBRAL DISC
Lumbar curve (formed by 5 LUMBAR VERTEBRAE)
Intervertebral foramen
SACRUM
Sacral curve (formed by 5 fused SACRAL VERTEBRAE)
COCCYX

(b) Right lateral view showing four normal curves

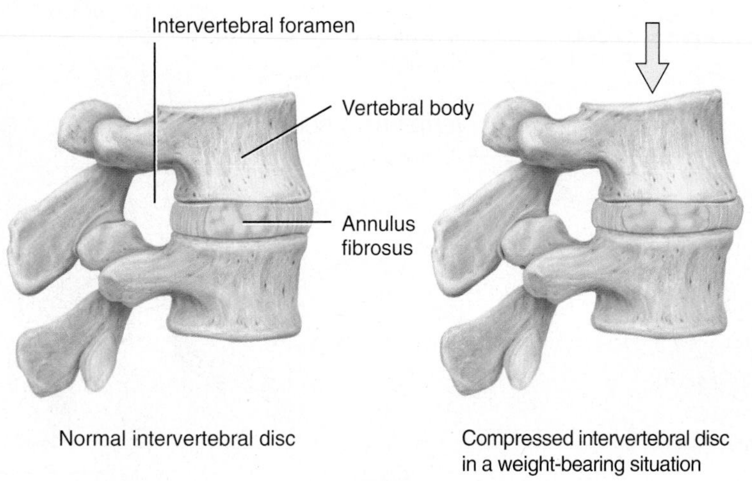


Normal intervertebral disc

Compressed intervertebral disc in a weight-bearing situation

(c) Intervertebral disc

**The adult vertebral column typically contains 26 vertebrae.**

# 7.8 A vertebra usually consists of a body, a vertebral arch, and several processes.

Vertebrae in different regions of the vertebral column vary in size, shape, and detail, but they are similar enough that we can discuss the structures and functions of a typical vertebra (Figure 7.16a). Vertebrae typically consist of a vertebral body, a vertebral arch, and several processes.

## Body

The **vertebral body** is the thick, disc-shaped anterior portion that is the weight-bearing part of a vertebra. Its superior and inferior surfaces are roughened for the attachment of intervertebral discs. The anterior and lateral surfaces contain nutrient foramina through which blood vessels supply the osseous tissue.

## Vertebral Arch

Two short, thick processes, the *pedicles* (PED-i-kuls = little feet), project posteriorly from the vertebral body and then unite with the flat *laminae* (LAM-i-nē = thin layers) to form the **vertebral arch**. Together, the vertebral body and the vertebral arch surround the spinal cord by forming the *vertebral foramen.* The vertebral foramen contains the spinal cord, adipose tissue, areolar connective tissue, and blood vessels. Collectively, the vertebral foramina of all vertebrae form the **vertebral canal**. When the vertebrae are stacked on top of one another, they form an opening between adjoining vertebrae on both sides of the column. Each opening, called an *intervertebral foramen*, permits the passage of a single spinal nerve carrying information to and from the spinal cord (see Figure 7.15c).

## Processes

Seven **processes** arise from the vertebral arch (Figure 7.16). At the point where a lamina and pedicle join, a *transverse process* extends laterally on each side. A single *spinous process* projects posteriorly from the junction of the laminae. These three processes serve as points of attachment for muscles. The remaining

FIGURE 7.16 **Structure of a typical vertebra, as illustrated by a thoracic vertebra.** In (b), only one spinal nerve has been included, and it has been extended beyond the intervertebral foramen for clarity.

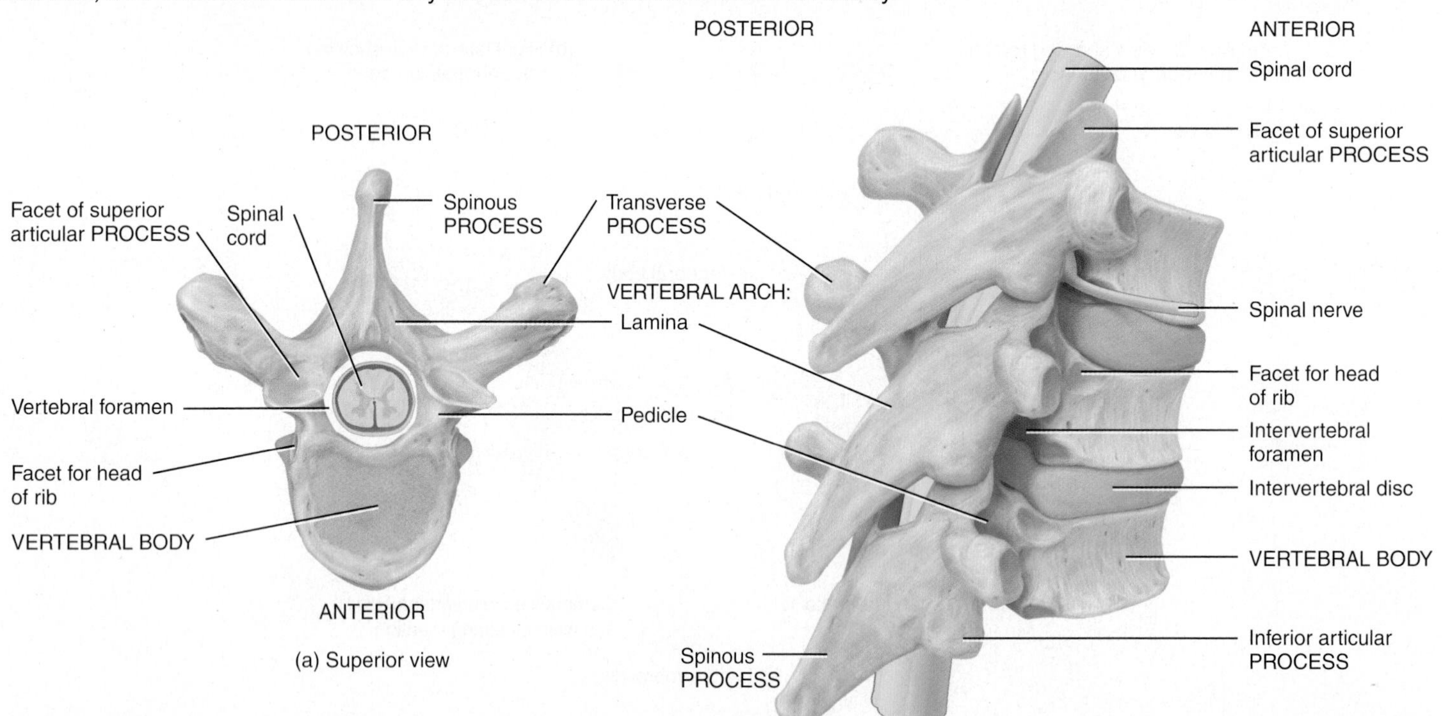


(a) Superior view

(b) Right posterolateral view of articulated vertebrae

A vertebra consists of a body, a vertebral arch, and several processes.

four processes form joints with other vertebrae above or below. The two *superior articular processes* of a vertebra articulate (form joints) with the two inferior articular processes of the vertebra immediately above them. In turn, the two *inferior articular processes* of that vertebra articulate with the two superior articular processes of the vertebra immediately below them. The articulating surfaces of the articular processes are referred to as *facets* (FAS-ets or fa-SETS = little faces) and are covered with hyaline cartilage.

**✓ CHECKPOINT**

**30.** What part of a vertebra is most responsible for supporting a person's weight?

**31.** What are the components of the vertebral arch?

**32.** Through what foramen does the spinal cord pass?

**33.** To which processes do muscles attach? Which processes attach to other vertebrae? Which parts of processes physically articulate with other bones?

## RETURN TO **Fernando's Story**

Elena and Isabel visit Fernando every day. Over several days, the swelling in Fernando's face gradually declines and each day he looks more like himself except for the purple and blue bruising.

"You should see the other guy," he jokes. He can only open his eyes partially for the neurological examination and to see. As his level of consciousness improves, he begins to complain of severe pain in his nose and his left side. Dr. Mueller requests a magnetic resonance imaging (MRI) test of Fernando's nose and x-rays of his ribs. These tests show that, on top of everything else, Fernando has fractures of his seventh and eighth ribs, along with a broken nose. Only the septal cartilage in his nose was fractured; the nasal bones, vomer, and ethmoid are intact. Dr. Mueller prescribes some pain medication and reassures Fernando that his ribs and nose will heal.

Five days after the accident, Fernando is moved out of the intensive care unit onto a regular trauma floor. The first thing he does is to ask for a double portion of dinner because he is so hungry. "They starve you in the ICU," he tells his nurse. The first night out of the intensive care unit he feels hot and rings for the nurse to ask for a cold drink. The nurse takes his temperature and tells him that he has a fever of 101°F. The fever does not diminish with administration of acetaminophen, so the next morning Dr. Mueller orders a chest x-ray, which reveals some pockets of atelectasis (unexpanded air sacs in the lungs) that could develop into pneumonia. Dr. Mueller explains to Fernando that he must get out of bed and walk and take regular deep breaths to expand his lungs. He attributes the atelectasis to the length of time that Fernando had been in bed, and two things that are keeping his lungs from fully inflating: the rib fractures that inhibited deep breathing, and the fact that over the years Fernando had developed an abnormal exaggerated curvature of the thoracic spine called kyphosis.

"You are very lucky that you did not injure your spine when you fell," said Dr. Mueller.

"Kyphosis, huh? So that's why I look crooked," replied Fernando.

D. *What are the bones that make up the nasal septum?*

E. *Fernando's eyes are swollen shut, but when he was examined it was determined that he did not have an orbital fracture. Which bones comprise the eye socket?*

F. *Fernando has kyphosis or humpback. What could be an explanation for this abnormality?*

## 7.9 Vertebrae in the different regions of the vertebral column vary in size, shape, and detail.

We turn now to the five regions of the vertebral column, beginning superiorly and moving inferiorly. The regions are the cervical, thoracic, lumbar, sacral, and coccygeal. Note that vertebrae in each region are numbered in sequence, from superior to inferior. When you actually view the bones of the vertebral column, you will notice that the transition from one region to the next is not abrupt but gradual, a feature that helps vertebrae fit together.

## Cervical Region

The bodies of *cervical vertebrae* (C1–C7) are smaller than all other vertebrae except those that form the coccyx (Figure 7.17). All cervical vertebrae have three foramina: one vertebral foramen and two transverse foramina (Figure 7.17b, c). The vertebral foramina of cervical vertebrae are the largest in the spinal column because they house the cervical enlargement of the spinal cord. The *transverse foramina* serve as a passageway for the vertebral artery that supplies blood to the brain. The spinous processes of C2 through C6 are often *bifid*—that is, they branch into two small projections at the tips (Figure 7.17a, c).

The first two cervical vertebrae differ considerably from the others. The **atlas** (C1), named after the mythological Atlas who supported the world on his shoulders, is the first cervical vertebra and supports the head (Figure 7.17a, b). The atlas lacks a body and a spinous process. Instead, it consists of a ring of bone with *anterior* and *posterior arches*. The superior, lateral surfaces contain *superior articular facets* that articulate with the occipital condyles of the occipital bone. These articulations permit you to move your head to signify "yes." The inferior lateral surfaces contain the *inferior articular facets* that articulate with the second cervical vertebra.

The second cervical vertebra (C2), called the **axis** (Figure 7.17a, d, e), does have a body. A peglike process on the body, the *dens* (= tooth), projects superiorly through the anterior portion of the vertebral foramen of the atlas. The dens makes a pivot on which the atlas and head rotate. This arrangement permits side-to-side movement of the head, as when you move your head to signify "no." In some instances of trauma, the dens of the axis may be driven into the medulla oblongata of the brain. This type of injury is the usual cause of death from whiplash injuries.

The third through sixth cervical vertebrae (C3–C6), represented by the vertebra in Figure 7.17c, correspond to the structural pattern of the typical cervical vertebra previously described. The seventh cervical vertebra (C7), called the *vertebra prominens*, is somewhat different. It has a single large spinous process that may be seen and felt at the base of the posterior neck (Figure 7.17a).

**FIGURE 7.17** **Cervical vertebrae.**

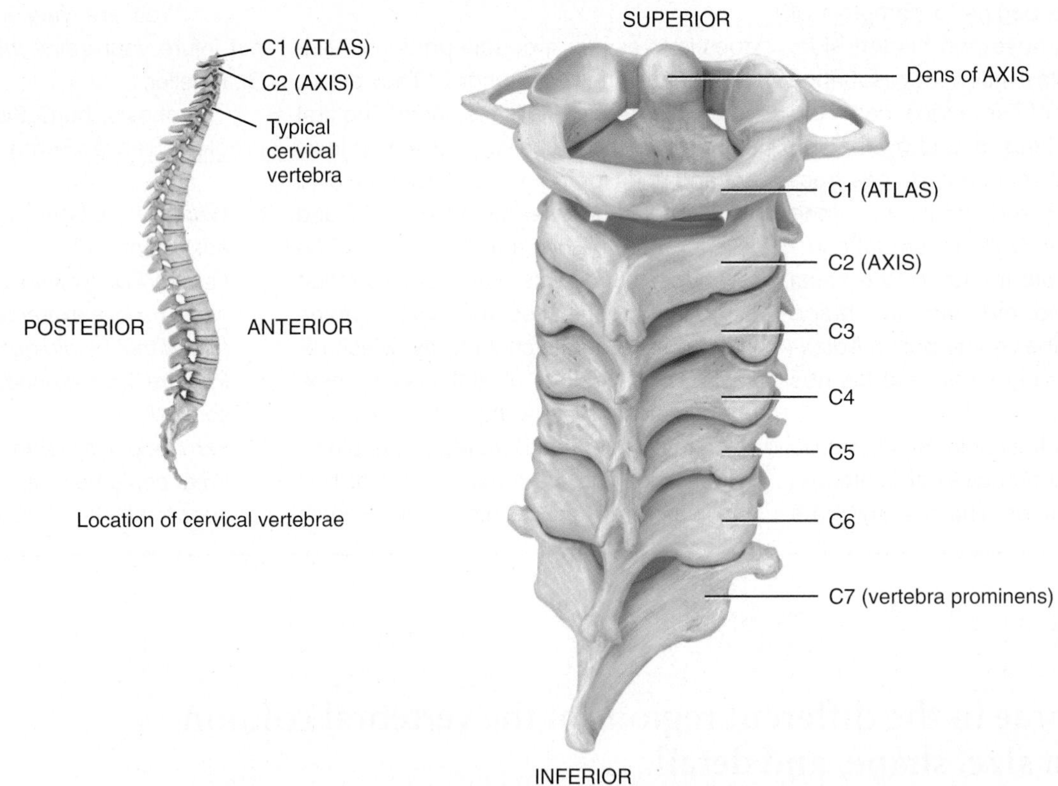


(a) Posterior view of articulated cervical vertebrae

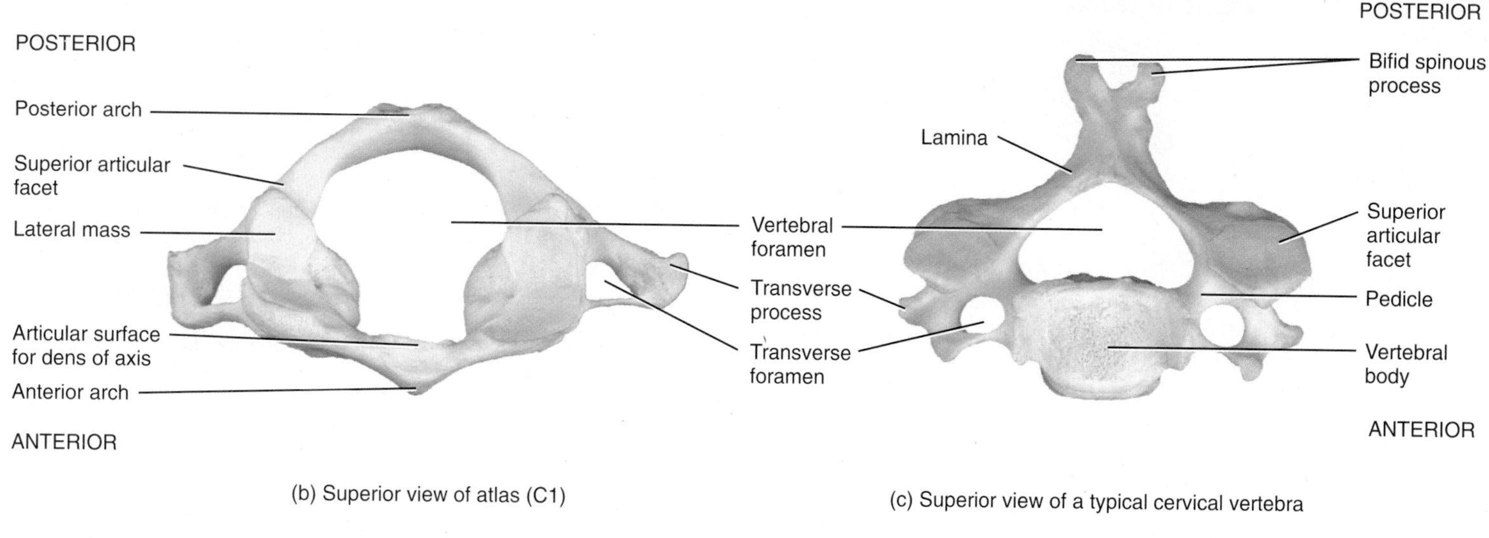


(b) Superior view of atlas (C1)

(c) Superior view of a typical cervical vertebra

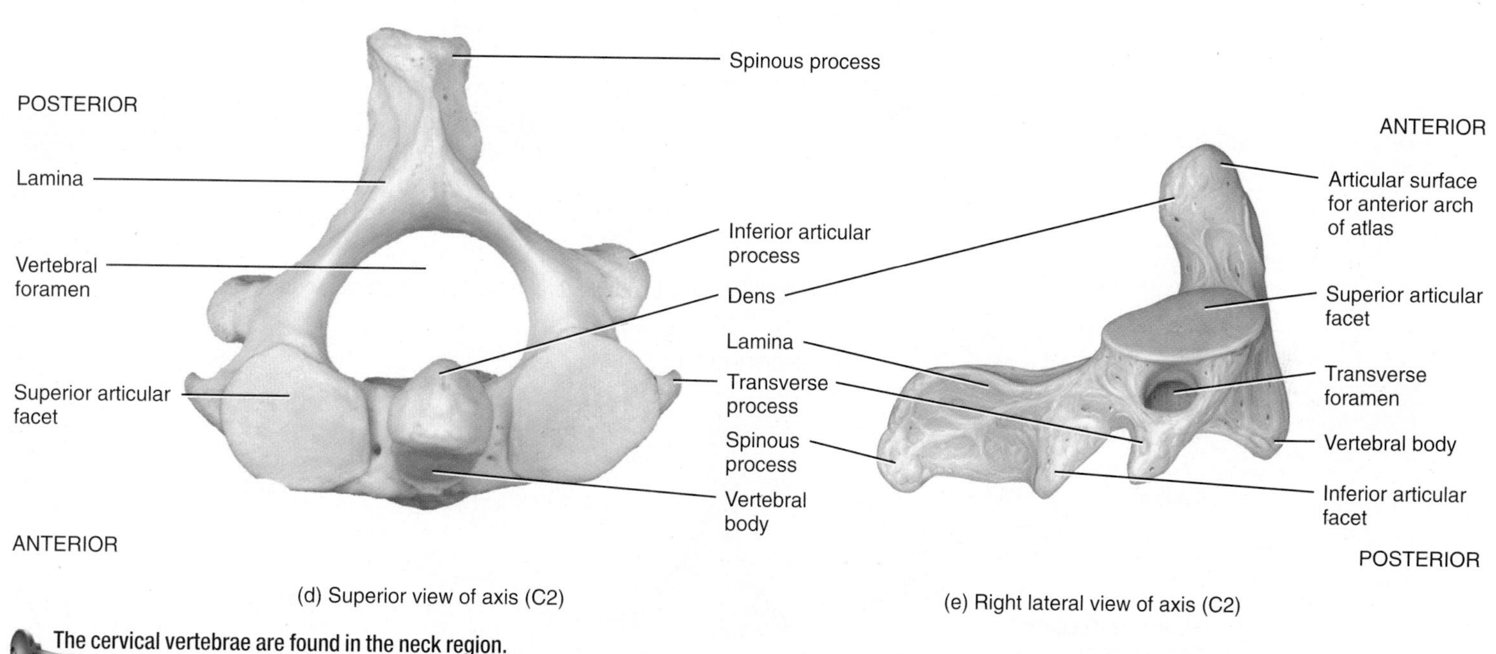


(d) Superior view of axis (C2)

(e) Right lateral view of axis (C2)

**The cervical vertebrae are found in the neck region.**

## Thoracic Region

*Thoracic vertebrae* (T1–T12; Figures 7.16 and 7.18) are considerably larger and stronger than cervical vertebrae. In addition, the spinous processes on T1 through T10 are long, laterally flattened, and directed inferiorly. In contrast, the spinous processes on T11 and T12 are shorter, broader, and directed more posteriorly. Compared to cervical vertebrae, thoracic vertebrae have longer transverse processes (see Figure 7.15a).

The most distinguishing feature of thoracic vertebrae is that they articulate with the ribs. As you can see in Figure 7.18, the bodies of thoracic vertebrae have either *facets* or *demifacets*

**FIGURE 7.18 Thoracic vertebrae.**

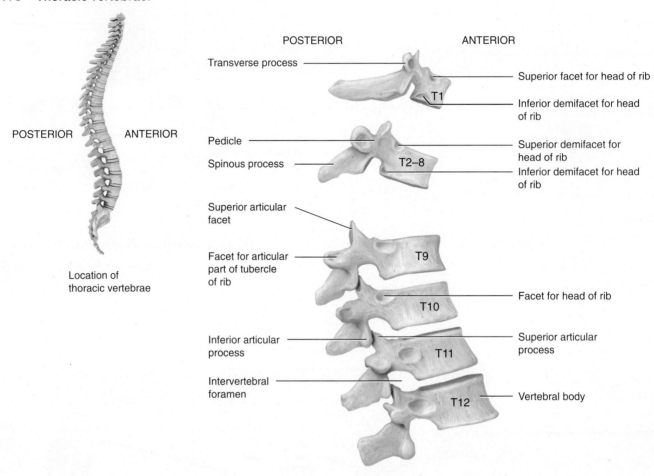

POSTERIOR
ANTERIOR
Location of thoracic vertebrae

POSTERIOR
ANTERIOR
Transverse process
Superior facet for head of rib
T1
Inferior demifacet for head of rib
Pedicle
Spinous process
T2–8
Superior demifacet for head of rib
Inferior demifacet for head of rib
Superior articular facet
Facet for articular part of tubercle of rib
T9
Facet for head of rib
T10
Inferior articular process
Superior articular process
T11
Intervertebral foramen
T12
Vertebral body

(a) Right lateral view of several articulated thoracic vertebrae

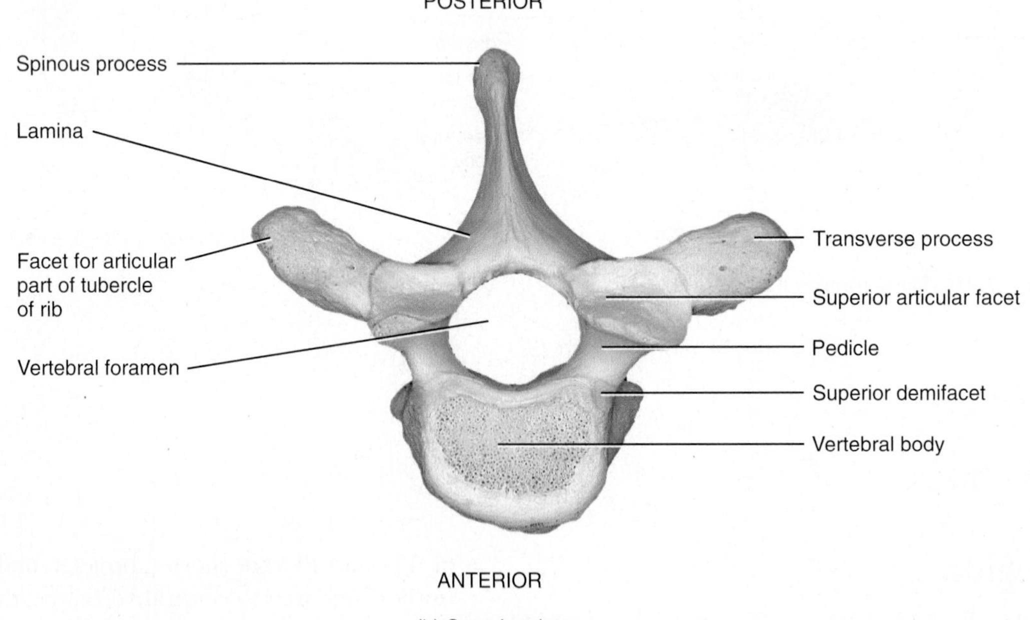


(b) Superior view

(half facets) that articulate with the *heads* of the ribs, and their transverse processes have facets that articulate with the *tubercles* of the ribs (see Figure 7.22c). Movements of the thoracic region are limited by the attachment of the ribs to the sternum.

## Lumbar Region

The *lumbar vertebrae* (L1–L5) are the largest and strongest of the unfused bones in the vertebral column (Figure 7.19a-c) because the amount of body weight supported by the vertebrae increases toward the inferior end of the spine. Their various projections are short and thick. The superior articular processes are directed medially instead of superiorly, and the inferior articular processes are directed laterally instead of inferiorly. The spinous processes are thick and broad, and project nearly straight posteriorly. The spinous processes are well adapted for the attachment of the large back muscles.

A summary of the major structural differences among cervical, thoracic, and lumbar vertebrae is presented in Table 7.4.

## Sacrum

The *sacrum* is a triangular bone formed by the union of the five sacral vertebrae (Figure 7.20). The sacral vertebrae begin to fuse between ages 16 and 18; this process is usually completed by age 30. Positioned at the posterior portion of the pelvic cavity medial to the two hip bones, the sacrum serves as a strong foundation for the pelvic girdle. To accommodate pregnancy and childbirth, the female sacrum is shorter, wider, and more curved than the male sacrum.

The concave, anterior side of the sacrum faces the pelvic cavity. It is smooth and contains four *transverse lines* that mark the joining of the sacral vertebral bodies (Figure 7.20a). Lateral to these lines are four pairs of *sacral foramina* that open to both the anterior and posterior surfaces and through which nerves and blood vessels pass. Extending laterally on either side from the superior surface of the sacrum is a winglike *sacral ala* (ĀL-a = wing).

The convex, posterior surface of the sacrum contains a *median sacral crest*, which is composed of the fused spinous processes of sacral vertebrae, and a *lateral sacral crest*, which results from the fusion of the transverse processes of sacral vertebrae (Figure 7.20b). The *sacral canal* is continuous with the vertebral canal. The laminae of the fifth sacral vertebra, and sometimes the fourth, fail to meet. This leaves an inferior entrance to the vertebral canal called the *sacral hiatus* (hī-Ā-tus = opening).

The anteriorly projecting border of the superior sacrum, called the *sacral promontory* (PROM-on-tō-rē), is a landmark used for measuring the pelvis. On both lateral surfaces, the sacrum has a large ear-shaped *auricular surface* (*auricular* = ear) that articulates with the ilium of each hip bone to form the sacroiliac joint, which contains depressions for the attachment of ligaments. The auricular surface articulates with the hip bones. The *superior articular processes* of the sacrum articulate with the inferior articular processes of the fifth lumbar vertebra.

## Coccyx

The *coccyx*, like the sacrum, is triangular in shape. It is formed by the fusion of four coccygeal vertebrae, indicated in Figure 7.20 as Co1–Co4. The coccygeal vertebrae fuse between the ages of 20 and 30. The *superior articular processes* of the first coccygeal vertebra articulates superiorly with the sacrum. In females, the coccyx points inferiorly to allow the passage of a baby during birth; in males, it points anteriorly.

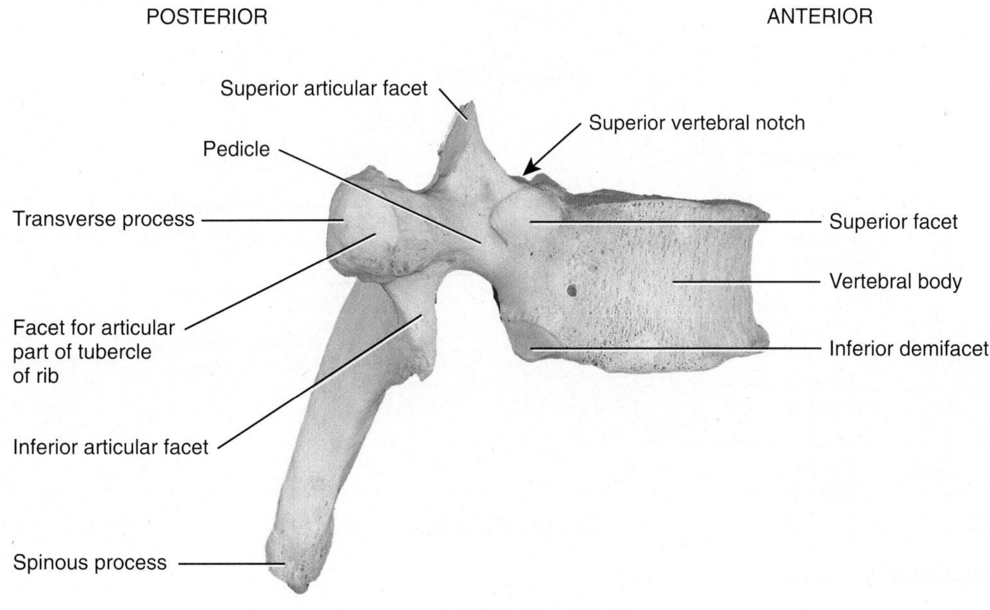


(c) Right lateral view

**The thoracic vertebrae are found in the chest region and articulate with the ribs.**

**FIGURE 7.19 Lumbar vertebrae.**

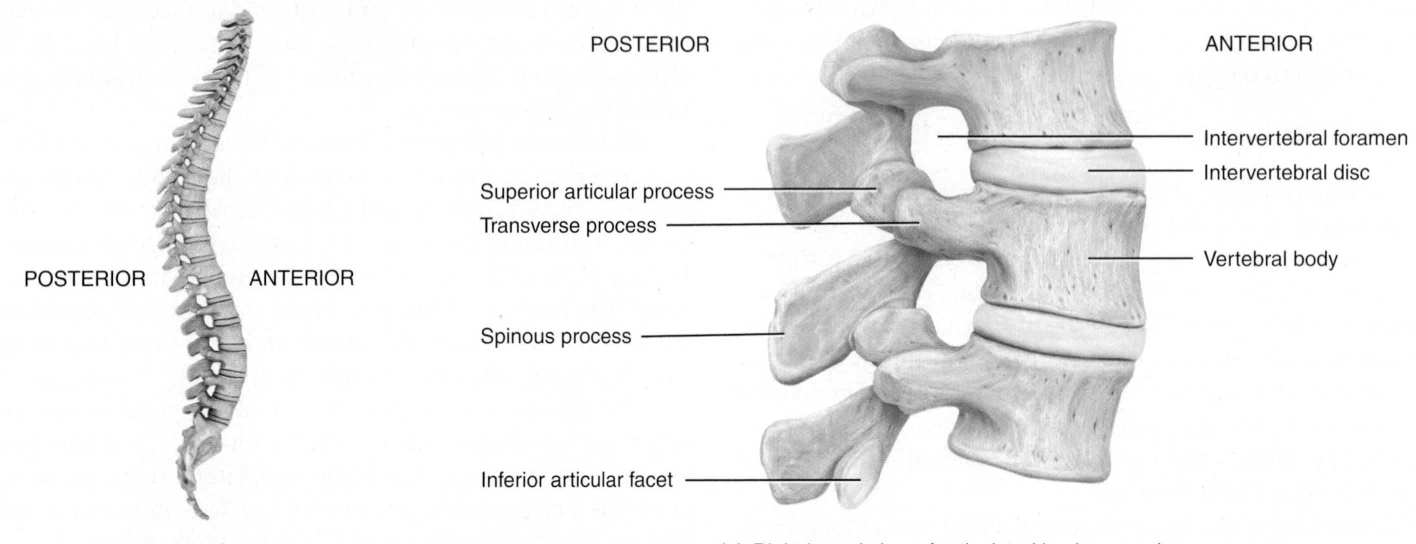


(a) Right lateral view of articulated lumbar vertebrae

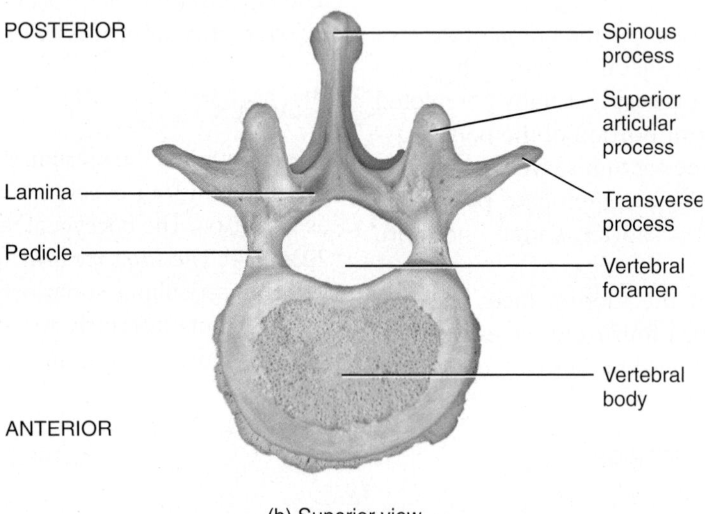


(b) Superior view

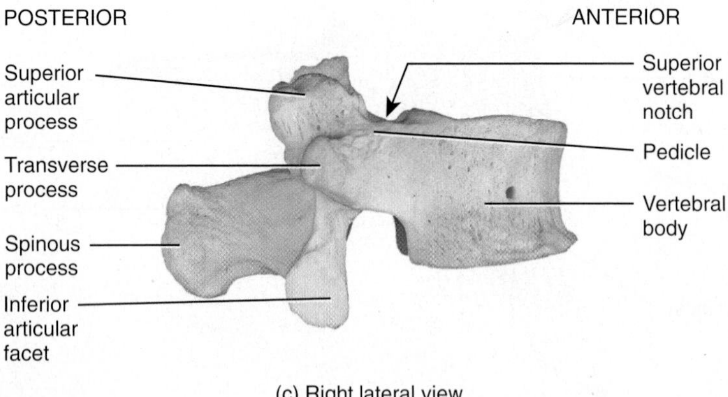


(c) Right lateral view

**Lumbar vertebrae are found in the lower back.**

## TABLE 7.4

Comparison of Major Structural Features of Cervical, Thoracic, and Lumbar Vertebrae

| CHARACTERISTIC | CERVICAL | THORACIC | LUMBAR |
|---|---|---|---|
| **Overall structure** | | | |
| **Size** | Small | Larger | Largest |
| **Foramina** | One vertebral and two transverse | One vertebral | One vertebral |
| **Spinous processes** | Slender, often bifid (C2–C6) | Long, fairly thick (most project inferiorly) | Short, blunt (project posteriorly rather than inferiorly) |
| **Transverse processes** | Small | Fairly large | Large and blunt |
| **Articular facets for ribs** | Absent | Present | Absent |
| **Direction of articular facets** | | | |
| **Superior** | Posterosuperior | Posterolateral | Medial |
| **Inferior** | Anteroinferior | Anteromedial | Lateral |
| **Size of intervertebral discs** | Thick relative to size of vertebral bodies | Thin relative to size of vertebral bodies | Thickest |

FIGURE 7.20 Sacrum and coccyx.

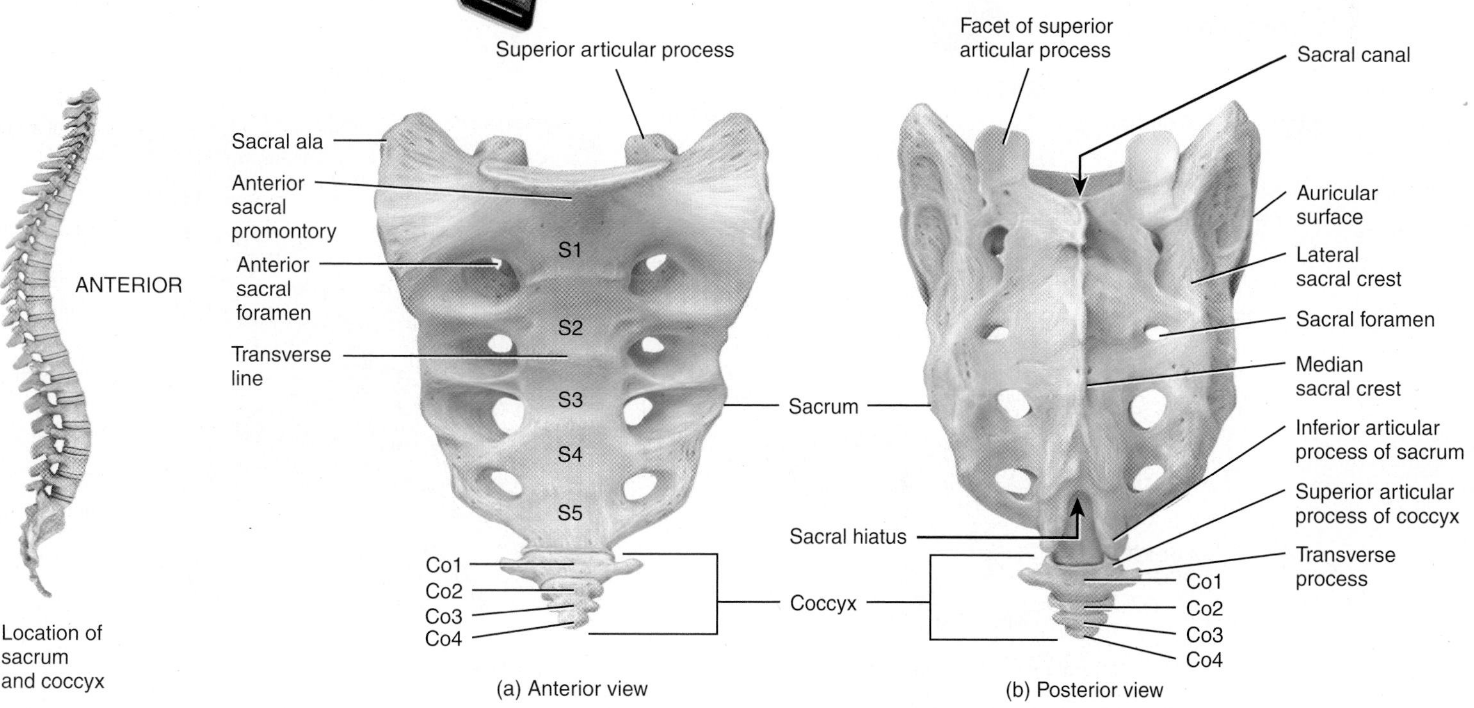


**The sacrum is formed by the union of five sacral vertebrae, and the coccyx is formed by the union of four coccygeal vertebrae.**

## ✓ CHECKPOINT

**34.** Which vertebrae contain transverse foramina? What function is served by transverse foramina?

**35.** How are the spinous processes different among cervical, thoracic, and lumbar vertebrae? Which cervical vertebra has the most prominent spinous process?

**36.** What are the names of the first and second cervical vertebrae?

**37.** Which bones permit nodding of the head to indicate "yes"? Which permit the rotational movement of the head to indicate "no"?

**38.** Which vertebrae articulate with ribs? Which parts of those vertebrae articulate with the ribs?

**39.** Name the superior and inferior openings of the sacrum that extend the vertebral column. Which sacral structures unite with the hip bones?

# 7.10 The thoracic cage protects vital organs in the thorax and upper abdomen and provides support for the bones of the upper limbs.

The term **thorax** refers to the entire chest region. The skeletal part of the thorax, the **thoracic cage**, is a bony enclosure formed by the sternum, ribs and their costal cartilages, and the bodies of the thoracic vertebrae (Figure 7.21). The thoracic cage is narrower at its superior end and broader at its inferior end. The thoracic cage encloses and protects the organs in the thoracic and superior abdominal cavities, provides support for the bones of the upper limbs, and, as you will see in Chapter 22, plays a role in breathing.

## Sternum

The **sternum**, or breastbone, is a flat, narrow bone located in the center of the anterior thoracic wall that measures about 15 cm (6 in.) in length and consists of three parts (Figure 7.21). The superior part is the **manubrium** (ma-NOO-brē-um = handlelike); the middle and largest part is the **body**; and the inferior, smallest part is the **xiphoid process** (ZĪ-foyd = sword-shaped). The parts of the sternum typically fuse by age 25, and the points of fusion are marked by transverse ridges.

FIGURE 7.21 Skeleton of the thorax.

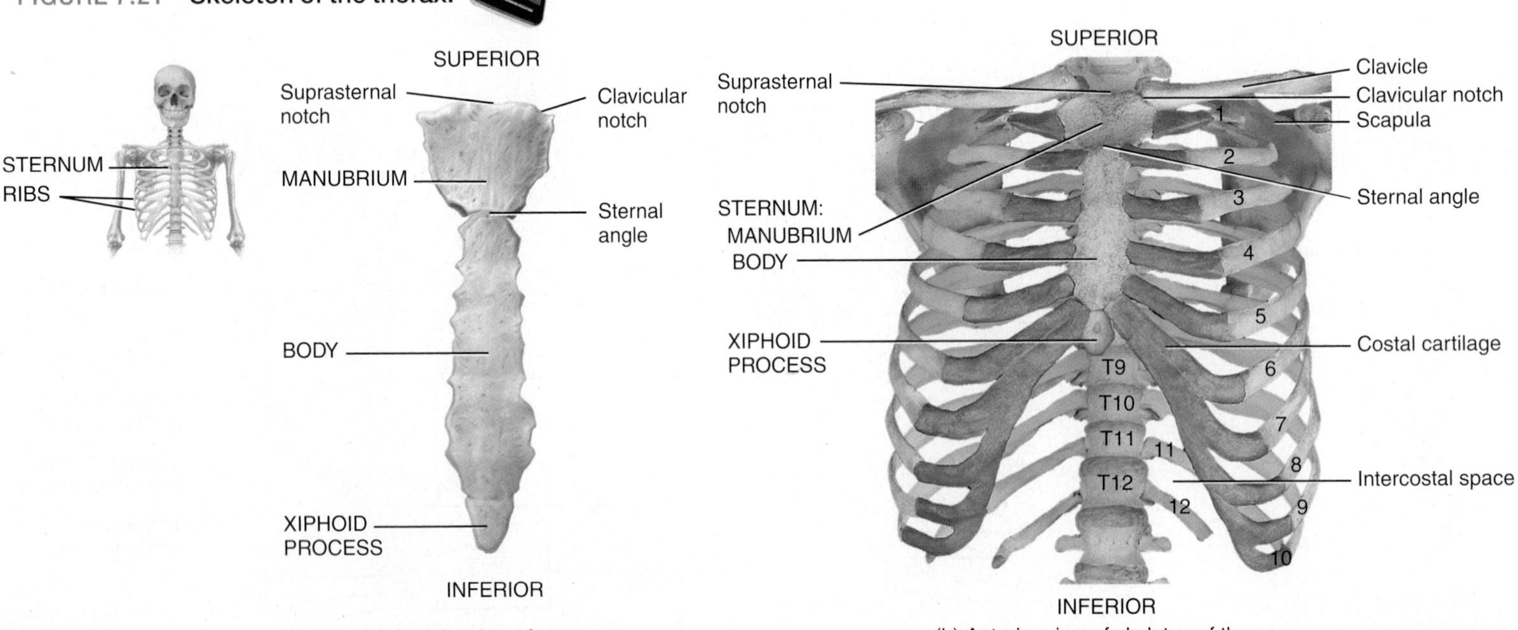


(a) Anterior view of sternum

(b) Anterior view of skeleton of thorax

The bones of the thorax enclose and protect organs in the thoracic cavity and upper abdominal cavity.

The junction of the manubrium and body forms the *sternal angle*. The sternal angle is a common clinical landmark for identifying the second rib, a starting point for counting ribs to assist in locating underlying structures such as heart valves. The manubrium has a depression on its superior surface, the *suprasternal notch*. Lateral to the suprasternal notch are *clavicular notches* that articulate with the medial ends of the clavicles. The manubrium also articulates with the costal cartilages of the first and second ribs.

The body of the sternum articulates directly or indirectly with the costal cartilages of the second through tenth ribs. The xiphoid process consists of hyaline cartilage during infancy and childhood and does not ossify completely until about age 40. No ribs are attached to it, but the xiphoid process provides attachment for some abdominal muscles. Incorrect positioning of the hands of a rescuer during cardiopulmonary resuscitation (CPR) may fracture the xiphoid process, driving it into internal organs.

## Ribs

Twelve pairs of **ribs**, numbered 1–12 from superior to inferior, give structural support to the sides of the thoracic cavity (Figure 7.21b). The ribs increase in length from the first through seventh, and then decrease in length to the twelfth rib. Each rib articulates posteriorly with its corresponding thoracic vertebra.

The first through seventh pairs of ribs have a direct anterior attachment to the sternum by a strip of hyaline cartilage called *costal cartilage* (*cost-* = rib). The costal cartilages contribute to the elasticity of the thoracic cage to allow breathing and to prevent various blows to the chest from fracturing the sternum and/or ribs. The ribs that have costal cartilages and attach directly to the sternum are called *true ribs*. The remaining five pairs of ribs are termed *false ribs* because their costal cartilages either attach indirectly to the sternum or do not attach to the sternum at all. The cartilages of the eighth, ninth, and tenth pairs of false ribs attach to one another and then to the cartilages of the seventh pair of ribs. The eleventh and twelfth false ribs are also known as *floating ribs* because the costal cartilage at their anterior ends does not attach to the sternum at all. Floating ribs attach only posteriorly to the thoracic vertebrae.

Figure 7.22a shows the parts of a typical (third through ninth) rib. The *head* is a projection at the posterior end of the rib that contains *superior* and *inferior facets*. The *neck* is a constricted portion just lateral to the head. A knoblike structure on the posterior surface, where the neck joins the body, is called the *tubercle* (TOO-ber-kul). The *body* is the main shaft of the rib. A short distance beyond the tubercle, an abrupt change in the curvature of the body occurs. This point is called the *costal angle*. The inner surface of the rib has a *costal groove* that protects blood vessels and a small nerve.

The posterior portion of the rib connects to a thoracic vertebra by its head and the tubercle. The facet of the head fits into a facet on the body of a single vertebra or into the demifacets of two adjoining vertebrae. The facet of the tubercle articulates with the facet of a transverse process of the vertebra (Figure 7.22c).

Spaces between ribs, called *intercostal spaces*, are occupied by intercostal muscles, blood vessels, and nerves. Surgical access to the lungs or other structures in the thoracic cavity is commonly obtained through an intercostal space. Special rib retractors are used to create a wide separation between ribs. The costal cartilages are sufficiently elastic in younger individuals to permit considerable bending without breaking.

FIGURE 7.22 **The structure of ribs.**

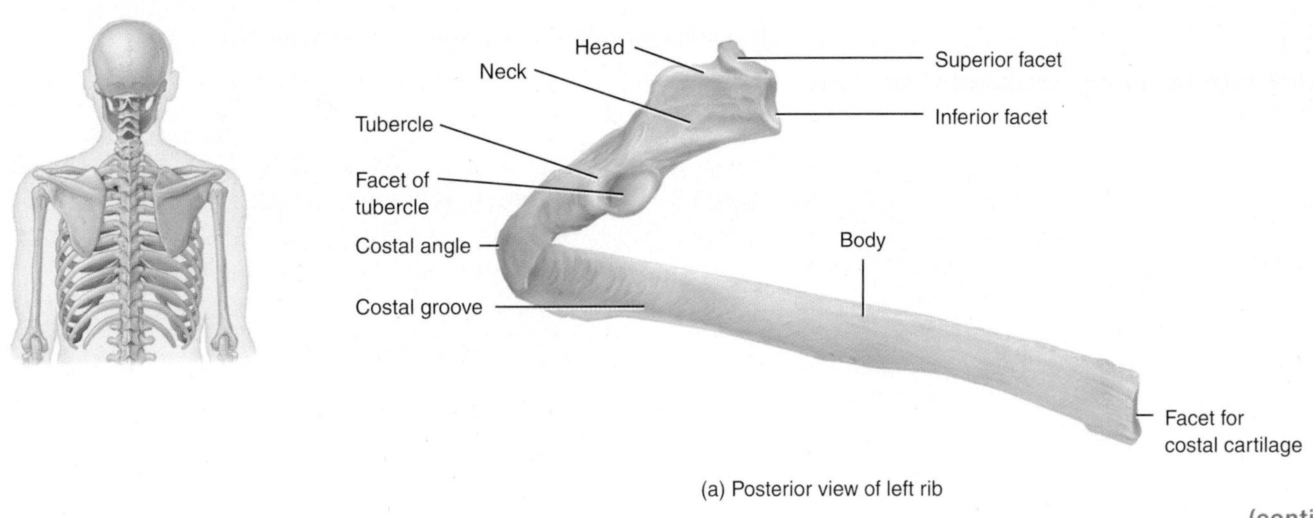


(a) Posterior view of left rib

(continues)

**FIGURE 7.22 (continued)**

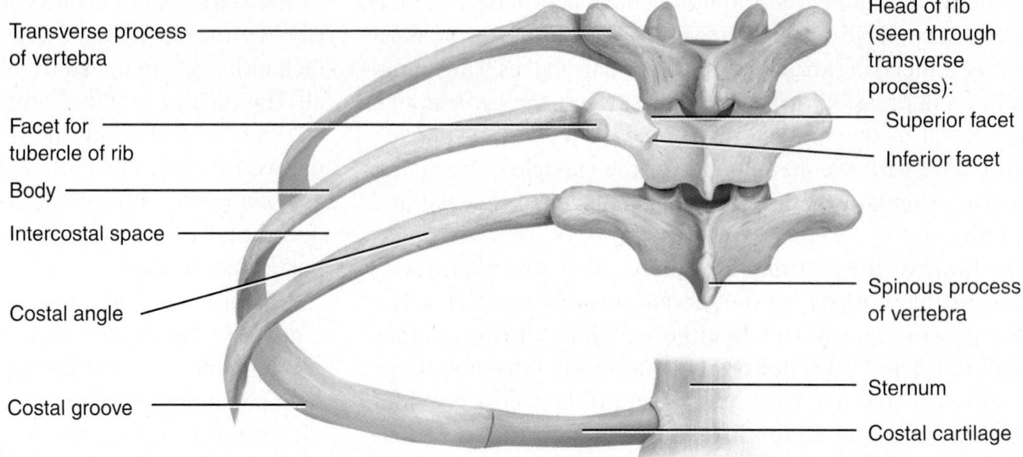


(b) Posterior view of left ribs articulated with thoracic vertebrae and sternum

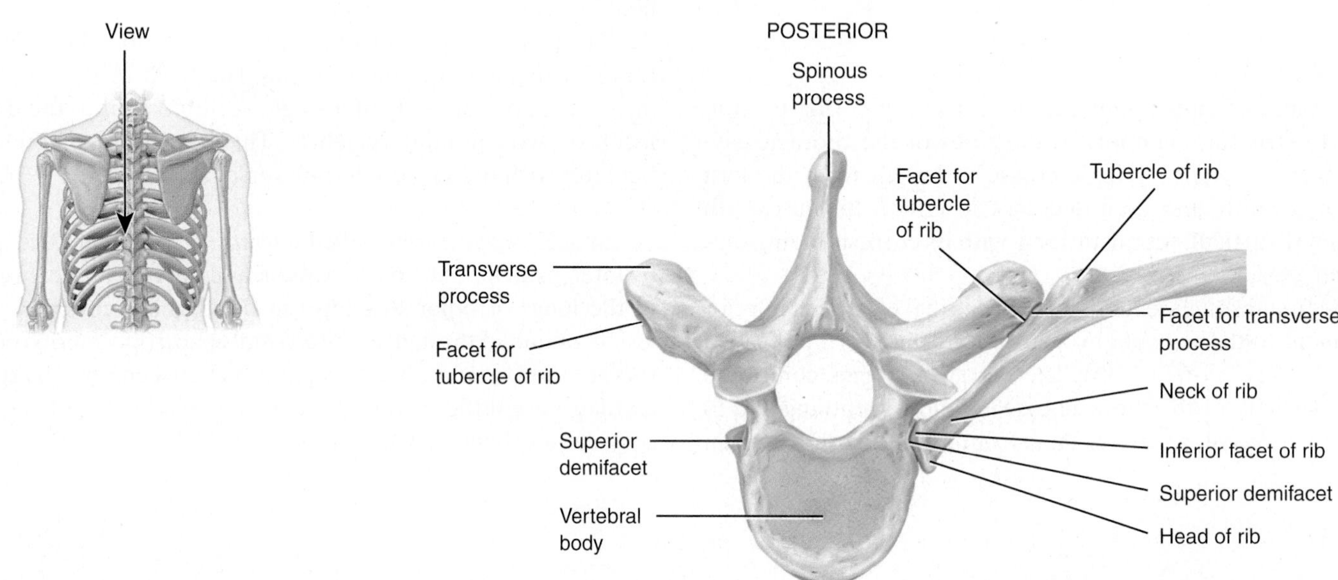


(c) Superior view of left rib articulated with thoracic vertebra

**Each rib articulates posteriorly with its corresponding thoracic vertebra.**

## CLINICAL CONNECTION | *Rib Fractures, Dislocations, and Separations*

**Rib fractures** are the most common chest injuries. They usually result from direct blows, most often from impact with a steering wheel, falls, or crushing injuries to the chest. Ribs tend to break at the point where the greatest force is applied, but they may also break at their weakest point—the site of greatest curvature, just anterior to the costal angle. The middle ribs are the most commonly fractured. In some cases, fractured ribs may puncture the heart, great vessels of the heart, lungs, trachea, bronchi, esophagus, spleen, liver, and kidneys. Rib fractures are usually quite painful. Rib fractures are no longer bound with bandages because of the pneumonia that would result from lack of proper lung ventilation.

**Dislocated ribs**, which are common in body contact sports, involve displacement of a costal cartilage from the sternum, with resulting pain, especially during deep inhalations.

**Separated ribs** involve displacement of a rib and its costal cartilage; as a result, a rib may move superiorly, overriding the rib above and causing severe pain.

## ✓ CHECKPOINT

**40.** From superior to inferior, what are the three components of the sternum?

**41.** What bones articulate with the manubrium? What structures articulate with the body?

**42.** What is the function of the xiphoid process?

**43.** What component of the thoracic cage allows the thorax to expand and contract during breathing?

**44.** How does a rib articulate with a thoracic vertebra? What is the space between two ribs called, and what is its clinical importance?

## Fernando's Story

## EPILOGUE AND DISCUSSION

Fernando is a 70-year-old man who is struck by a bicycle while on his way to a festival with his granddaughter, Elena. He is knocked unconscious and sustains a skull fracture as well as a broken nose, two broken ribs, and multiple soft tissue injuries to his face and head. He is hospitalized until he regains consciousness and his injuries begin to heal. While in the hospital he develops some collapsed air sacs from a combination of too much time in bed, the two broken ribs, and from a kyphotic thoracic curve. He is lucky that he did not sustain injury to his spine when he fell, because a fracture in an already deformed vertebra will not heal in normal alignment and the back curvature can become worse, with pain and possibly paralysis.

Elena and Isabel take Fernando home after he is discharged from the hospital. All of the discoloration and bruising gradually clear up, and the swelling and pain disappear. However, Fernando's kyphosis remains, as there is no corrective treatment. If he does postural exercises he can prevent it from getting worse, but for the rest of his life Fernando will appear "crooked."

G. *What effect does Fernando's kyphosis have on the thoracic cage?*

H. *What are the components of the thoracic cage?*

I. *Fernando has fractures of the left seventh and eighth ribs. To what structures do these ribs attach?*

# Concept and Resource Summary

| Concept | Resources |
|---|---|
| **Introduction**<br>1. The bones, muscles, and joints form an integrated system called the musculoskeletal system. | |
| **Concept 7.1 Bones of the axial skeleton and appendicular skeleton have characteristic surface markings.**<br>1. The 206 bones of the adult human skeleton can be categorized into two divisions: the **axial skeleton** and the **appendicular skeleton**.<br>2. The axial skeleton consists of bones around the body's longitudinal axis including skull bones, ribs, sternum, vertebral column, auditory ossicles, and hyoid bone.<br>3. The appendicular skeleton is composed of the bones of the **upper** and **lower limbs**, and the shoulder and hip **girdles**.<br>4. Bone **surface markings** are important anatomical features that indicate sites of muscle attachment, ligaments, joints, and passage of blood vessels and nerves. | Anatomy Overview—The Axial Skeleton |
| **Concept 7.2 The skull provides attachment sites for muscles and membranes, and protects and supports the brain and sense organs.**<br>1. The **skull** contains 22 bones, 8 **cranial bones** and 14 **facial bones**.<br>2. The cranial bones protect the brain and sense organs, and allow attachment of membranes and muscles.<br>3. The facial bones form the framework of the face, allow muscle attachment, and protect and support the entrances to the digestive and respiratory tracts.<br>4. The immovable joints in the skull are called **sutures**; the openings for blood vessels and nerves are called **foramina** and **fissures**. | Anatomy Overview—The Skull<br>Anatomy Overview—Fibrous Joints |
| **Concept 7.3 The cranial bones include the frontal, parietal, temporal, occipital, sphenoid, and ethmoid bones.**<br>1. The **frontal bone** forms the forehead and anterior cranial floor, helps form the orbits, and houses the frontal sinuses.<br>2. The two **parietal bones** make up a large portion of the sides and roof of the cranial cavity.<br>3. The two **temporal bones** form the inferior lateral sides of the skull and cranial floor. The mandibular fossa is a socket that articulates with the mandible, forming the temporomandibular joint (TMJ).<br>4. The **occipital bone** forms the posterior part and most of the base of the skull. Inferiorly, the foramen magnum allows the brain to connect with the spinal cord.<br>5. The **sphenoid bone** lies in the middle of the base of the skull and articulates with all other cranial bones. It contains sphenoidal sinuses.<br>6. The **ethmoid bone** is in the anterior part of the cranial floor, between the orbits; the ethmoid bone forms the anterior cranial floor, part of the orbits, part of the nasal septum, and most of the superior side walls of the nasal cavity. | Clinical Connection—Black Eye<br>Clinical Connection—Temporomandibular Joint Syndrome |
| **Concept 7.4 Facial bones include the nasal bones, maxillae, zygomatic bones, lacrimal bones, palatine bones, inferior nasal conchae, vomer, and mandible.**<br>1. The two **nasal bones** unite medially to form the bridge of the nose.<br>2. Paired **maxillae** form the upper jawbone, articulate with every facial bone except the mandible, and contribute to the orbits, nasal cavity, and hard palate.<br>3. The two **zygomatic bones** form the cheeks and part of the orbits. | Anatomy Overview—Facial Bones |

4. The smallest facial bones, the paired **lacrimal bones**, contribute to the orbits; each contains the lacrimal fossa through which tears pass to the nasal cavity.
5. Paired **palatine bones** help form the hard palate, nasal cavity, and orbits.
6. Scroll-like **inferior nasal conchae** are inferior to the conchae of the ethmoid bone.
7. The **vomer** helps to form the nasal septum.
8. The strongest facial bone, and only freely movable skull bone, is the **mandible**, or lower jawbone.

## Concept 7.5 Unique features of the skull include the nasal septum, orbits, sutures, paranasal sinuses, and fontanels.

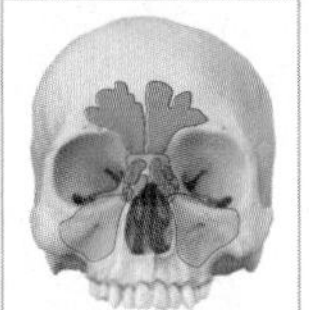

1. The nasal cavity has a vertical partition called the **nasal septum** formed by the vomer, the perpendicular plate of the ethmoid, and septal cartilage.
2. Each **orbit** houses the eyeball and comprises parts of seven bones: frontal, sphenoid, ethmoid, palatine, zygomatic, lacrimal, and maxilla.
3. **Paranasal sinuses** are cavities lined by mucous membranes found in the frontal, sphenoid, ethmoid, and maxillary bones.
4. The immovable joints between most skull bones are called **sutures**. Four principal sutures are the **coronal suture**, the **sagittal suture**, the **lambdoid suture**, and the **squamous sutures**.
5. **Fontanels** are mesenchyme-filled spaces between skull bones that are not yet fully ossified in the fetus and infant. They allow for skull flexibility during delivery and accommodate brain growth.

Resources:

- Anatomy Overview—Fibrous Joints and Sutures
- Figure 7.11—Details of the Orbit
- Clinical Connection—Foramina of the Skull

## Concept 7.6 The hyoid bone supports the tongue and attaches to muscles of the tongue, pharynx, and larynx.

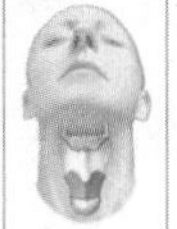

1. The **hyoid bone** in the anterior neck does not articulate with any other bones.
2. It supports the tongue and provides attachment for some tongue muscles and for some muscles of the pharynx and neck.

Resources:

- Anatomy Overview—Muscles for Speech, Swallowing, and Chewing

## Concept 7.7 The vertebral column protects the spinal cord, supports the head, and is a point of attachment for bones and muscles.

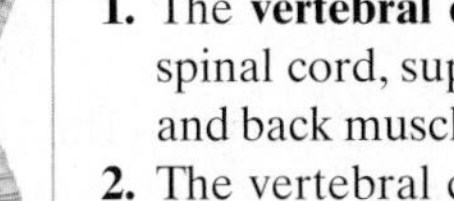

1. The **vertebral column** is composed of a series of **vertebrae** that surround and protect the spinal cord, support the head, and serve as a point of attachment for the ribs, pelvic girdle, and back muscles.
2. The vertebral column includes 7 **cervical vertebrae**, 12 **thoracic vertebrae**, 5 **lumbar vertebrae**, a **sacrum**, and a **coccyx**.
3. Except for the sacrum and coccyx, the vertebrae are freely movable.
4. The adult vertebral column has four **normal curves** that increase strength, absorb shock, help maintain balance, and protect against vertebral fracture.
5. **Intervertebral discs** occur between bodies of adjacent vertebrae; they allow vertebral column movement and absorb shock.

Resources:

- Anatomy Overview—The Vertebral Column
- Figure 7.15—Vertebral Column
- Clinical Connection—Abnormal Curves of the Vertebral Column
- Clinical Connection—Herniated (Slipped) Disc
- Clinical Connection—Osteoporosis
- Clinical Connection—Spina Bifida

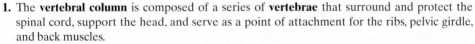

## Concept

## Resources

### Concept 7.8 A vertebra usually consists of a body, a vertebral arch, and several processes.

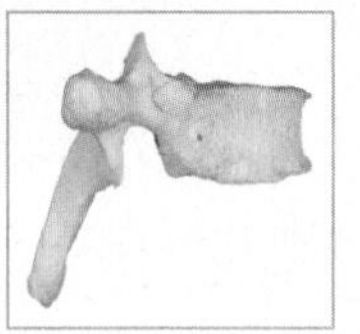

1. The **vertebral body** is the weight-bearing portion of the vertebra.
2. The **vertebral arch** has two pedicles and two laminae.
3. The body and vertebral arch enclose an opening called the vertebral foramen. The foramina from all the vertebrae form the **vertebral canal** for the spinal cord.
4. Lateral openings, called intervertebral foramina, occur between articulating vertebrae on both sides of the column to allow passage of spinal nerves.
5. The vertebral **processes** include two transverse processes, a spinous process, two superior articular processes, and two inferior articular processes.

Anatomy Overview—The Vertebral Column

### Concept 7.9 Vertebrae in the different regions of the vertebral column vary in size, shape, and detail.

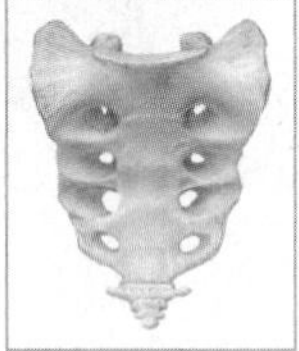

1. Each transverse process of most cervical vertebrae has a transverse foramen through which the vertebral arteries pass to supply blood to the brain.
2. The **atlas** (C1) articulates with the occipital condyles of the skull; lacks a body and spinous process; and has modifications that allow head movement to signify "yes."
3. The **axis** (C2) has the dens, which projects superiorly to act as a pivot for rotation of the atlas to signify "no."
4. The unique feature of the thoracic vertebrae is their articulation with ribs. Their bodies and transverse processes have facets and demifacets for articulation with ribs.
5. The lumbar vertebrae have short, thick processes and are the largest, strongest vertebrae.
6. The sacrum is formed by the fusion of the five sacral vertebrae. It articulates laterally with the two hip bones and sits posteriorly in the pelvis.
7. The coccyx, formed by the fusion of four coccygeal vertebrae, points inferiorly in females and anteriorly in males.

Anatomy Overview—The Vertebral Column

Clinical Connection—Caudal Anesthesia

Table 7.4—Comparison of Major Structural Features of Cervical, Thoracic, and Lumbar Vertebrae

Figure 7.20—Sacrum and Coccyx

Clinical Connection—Fractures of the Vertebral Column

### Concept 7.10 The thoracic cage protects vital organs in the thorax and upper abdomen and provides support for the bones of the upper limbs.

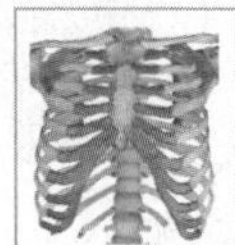

1. The **thoracic cage** is composed of the sternum, ribs and their costal cartilages, and bodies of thoracic vertebrae.
2. The **sternum** is in the midline of the anterior thoracic wall. It includes three parts: **manubrium**, **body**, and **xiphoid process**.
3. Twelve pairs of **ribs** articulate posteriorly with the thoracic vertebrae: There are seven pairs of true ribs and five pairs of false ribs, with the last two pairs being called floating ribs.

Anatomy Overview—The Thorax

Figure 7.21—Skeleton of the Thorax

Exercise—Bones, Bones, Bones

## Understanding the Concepts

1. Which type of opening is shaped like a tube? A narrow slit? A shallow depression?
2. Which type of process projects from a narrowed part of a bone, or "neck"? Forms a pointed, narrow projection? Is large with a roughened surface?
3. What do the auditory ossicles and mandible have in common?
4. What components of the nervous system join together within the largest foramen in the skull? Where is it located?
5. Which opening in the skull is a passageway for impulses related to the sense of vision? For impulses related to eyeball movement? In which cranial bone is each of these openings located?
6. What part of the ethmoid bone forms the superior part of the nasal septum? Forms the roof of the nasal cavity? Is a passageway for impulses related to the sense of smell? Is an attachment site for membranes separating the two sides of the brain?
7. Which foramina would you want your dentist to know the exact location of prior to initiating potentially painful dental work? Why? In which facial bone can these foramina be found?
8. How does the tissue composition of fontanels justify their common name, "soft spots"? What is the functional significance of fontanels?
9. In what way is the hyoid bone different from all the other bones of the axial skeleton?
10. How are the vertebrae of the sacrum and coccyx functionally different from those of the rest of the vertebral column?
11. If the body of a vertebra were to degenerate and collapse, as occurs in some musculoskeletal diseases, what foramen would compress the spinal nerve?
12. Why must the lumbar vertebrae be the largest and strongest in the vertebral column?
13. What are the components of the thoracic cage? What functions does it serve?
14. Why are "kidney punches" (blows to the lowermost ribs in boxing) considered dangerous and illegal?

CHAPTER

# 8

## Hassan's Story

The sun rose quickly. Hassan had been up early, excited by the finds of the previous day: several chunks of wood, evidence of an ancient campfire, animal bones, and the remnants of a leather sandal. To the untrained eye, it would have meant nothing, but Hassan and his team knew they were on the right track. No one had believed in him when he had decided to pursue this dig site, but Hassan had an uncanny knack for putting together clues.

The flap of his tent flew open. "Hassan! We have found the tomb. You must see!" exclaimed his colleague Baru. The two men reached the excavation site quickly. Peering down into the hole, Hassan saw what looked like stacks of sticks. Human skeletons projected out of the ground; ribs, the rounded domes of skulls, and limbs were scattered in a jumbled heap.

As he gazed into the pit, a profound sadness tempered Hassan's excitement. These loyal servants had trusted their king and queen, and then were brutally sacrificed to be the sentinels of this tomb. But why? Now that the tomb had been found, the ancient people would answer through their bones.

Baru and Hassan mobilized as many students as they could. The process of uncovering the tomb entrance was slow and meticulous. A sarcophagus was discovered immediately inside the entrance, with the image of a young woman holding a baby painted on its surface, along with a royal seal and a series of hieroglyphs.

"These scenes depict some kind of tragedy. The land had been plagued by drought, famine, and disease, and the death of the princess is repeated through the scenes. To appease the gods, the King offered up his firstborn child. Those were truly desperate times."

Hassan motioned for his men to place their steel pry bars beneath the lid. He stepped back as the men struggled to lift the lid slowly and carefully. Inside were two skeletons, a small, delicately framed woman with arms folded around what appeared to be a child.

Because skeletal remains may persist for thousands of years, it is possible to trace patterns of disease and nutrition, evaluate the effects of social and economic changes, and deduce patterns of reproduction and mortality. Skeletal remains also may reveal an individual's sex, age, height, and race. In this chapter we will explore the anatomy of the appendicular skeleton and learn how understanding its structure can serve as a tool for medical science and other disciplines.

# The Appendicular Skeleton

## INTRODUCTION

As noted in Chapter 7, the two main divisions of the skeletal system are the axial skeleton and the appendicular skeleton. The general function of the axial skeleton is the protection of internal organs; the primary function of the appendicular skeleton, the focus of this chapter, is movement. As you progress through this chapter, you will see how the bones of the appendicular skeleton are connected with one another and with skeletal muscles, making possible a wide array of movements. This arrangement permits you to do things such as walk, write, use a computer, dance, swim, and play a musical instrument.

The appendicular skeleton includes the bones that make up the upper and lower limbs as well as the bones, arranged in formations called girdles that attach the limbs to the axial skeleton.

## CONCEPTS

**8.1** Each pectoral girdle, which consists of a clavicle and scapula, attaches an upper limb to the axial skeleton.

**8.2** The bones of each upper limb include the humerus, ulna, radius, carpals, metacarpals, and phalanges.

**8.3** The pelvic girdle supports the vertebral column and pelvic viscera and attaches the lower limbs to the axial skeleton.

**8.4** Male pelves are generally larger, heavier, and have more prominent markings; female pelves are generally wider and shallower.

**8.5** The bones of each lower limb include the femur, patella, tibia, fibula, tarsals, metatarsals, and phalanges.

# 8.1 Each pectoral girdle, which consists of a clavicle and scapula, attaches an upper limb to the axial skeleton.

The **pectoral** (PEK-tō-ral) or **shoulder girdles** attach the bones of the upper limbs to the axial skeleton (Figure 8.1). Each of the two pectoral girdles consists of a clavicle and a scapula. The clavicle, the anterior bone, articulates with the manubrium of the sternum; the scapula, the posterior bone, articulates with the clavicle and with the humerus. The pectoral girdles do not articulate (form joints) with the vertebral column and are held in position and stabilized by a group of large muscles that extend from the vertebral column and ribs to the scapula.

## Clavicle

Each slender, S-shaped **clavicle** (KLAV-i-kul = key), or *collarbone*, is a long, slender bone that lies horizontally across the anterior part of the thorax superior to the first rib (Figure 8.2). The bone is S-shaped because the medial half of the clavicle is convex anteriorly (curves toward you when viewed in the anatomical position), and the lateral half is concave anteriorly (curves away from you). The medial end of the clavicle, the *sternal end*, is rounded and articulates with the manubrium of the sternum; the

FIGURE 8.1 Right pectoral girdle.

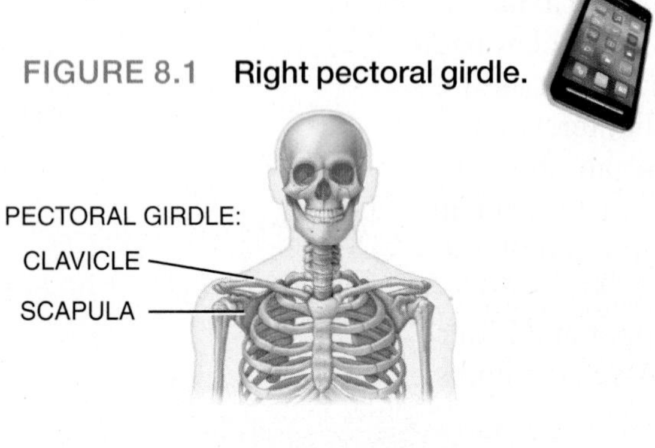


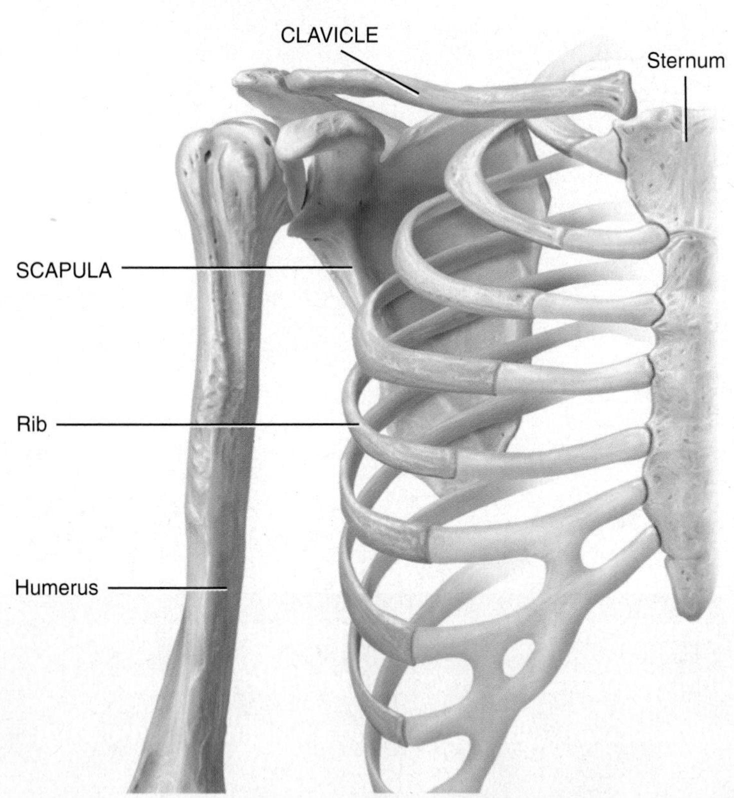


(a) Anterior view of pectoral girdle

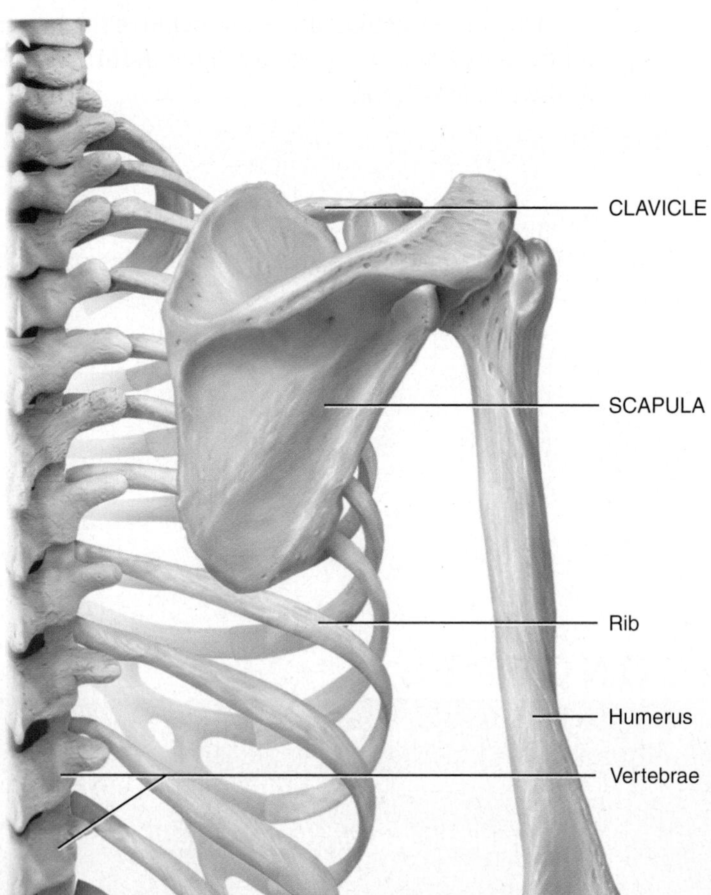


(b) Posterior view of pectoral girdle

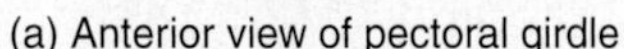

The clavicle is the anterior bone of the pectoral girdle and the scapula is the posterior bone.

FIGURE 8.2 **Right clavicle.**

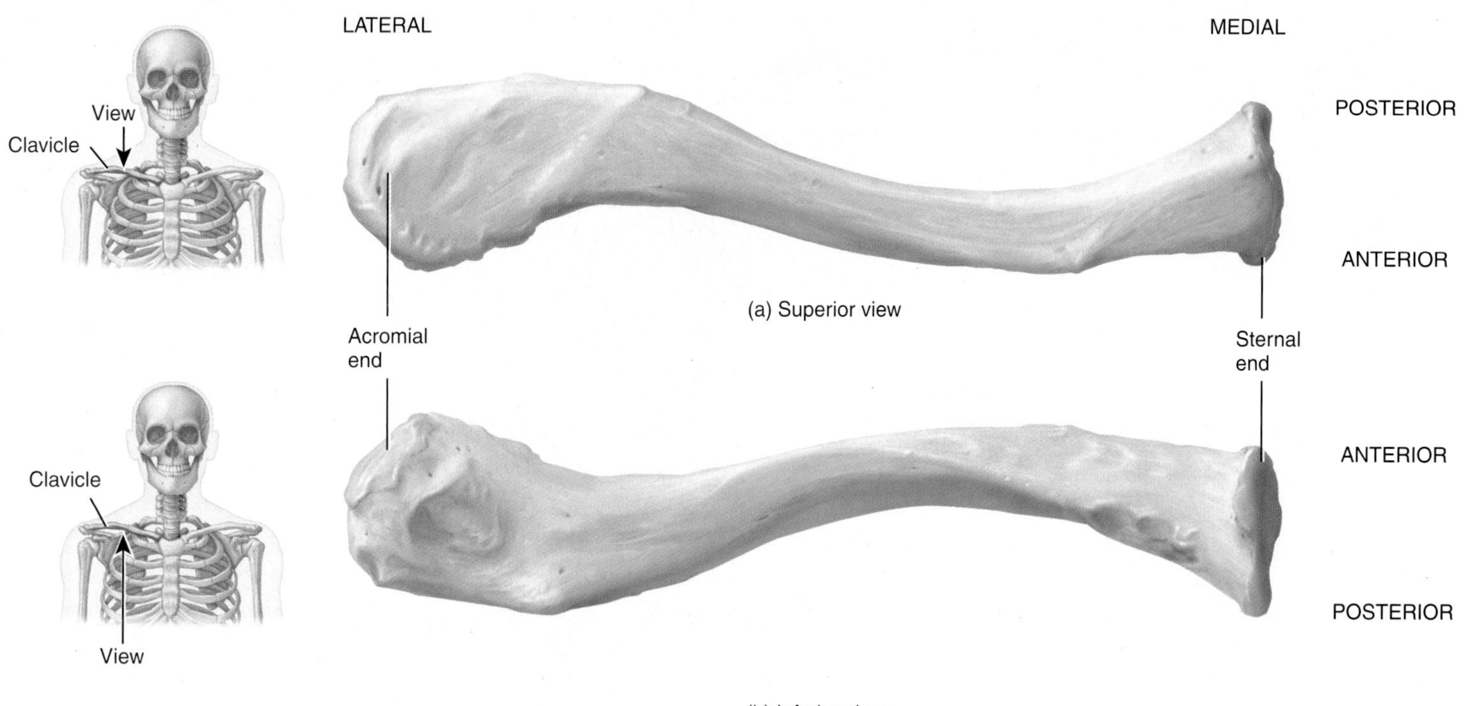


**The clavicle articulates medially with the manubrium of the sternum and laterally with the acromion of the scapula.**

broad, flat, lateral end, the *acromial end* (a-KRŌ-mē-al), articulates with the acromion of the scapula (Figure 8.1).

## Scapula

Each **scapula** (SCAP-ū-la; plural is *scapulae*), or *shoulder blade*, is a large, triangular, flat bone situated in the superior part of the posterior thorax (Figure 8.3). A prominent ridge called the *spine* runs diagonally across the posterior surface of the scapula (Figure 8.3b). The lateral end of the spine projects as a flattened, expanded process called the *acromion* (a-KRŌ-mē-on; *acrom-* = topmost; *omos* = shoulder), easily felt as the high point of the shoulder. Tailors measure the length of the upper limb from the acromion. The acromion articulates with the acromial end of the clavicle. Inferior to the acromion is a shallow depression, the *glenoid cavity*, that articulates with the head of the humerus (arm bone).

The thin edge of the scapula closer to the vertebral column is called the *medial border* (Figure 8.3). The thick edge closer to the arm is called the *lateral border*. The medial and lateral borders join at the *inferior angle*. The superior edge of the scapula, called the *superior border*, joins the medial border at the *superior angle*.

At the lateral end of the superior border of the scapula is a projection of the anterior surface, called the *coracoid process*

### CLINICAL CONNECTION | *Fractured Clavicle*

The clavicle transmits mechanical force from the upper limb to the trunk. If the force transmitted to the clavicle is excessive, as when you fall on your outstretched arm, a **fractured clavicle** may result. A fractured clavicle may also result from a blow to the superior part of the anterior thorax, for example, as a result of an impact following an automobile accident. The clavicle is one of the most frequently broken bones in the body. Because the junction of the two curves of the clavicle is its weakest point, the clavicular midregion is the most frequent fracture site. Even in the absence of fracture, compression of the clavicle as a result of automobile accidents involving the use of shoulder harness seatbelts often causes damage to the brachial plexus (the network of nerves that enter the upper limb), which lies between the clavicle and the second rib. A fractured clavicle is usually treated with a figure-of-eight sling to keep the arm from moving outward.

**FIGURE 8.3 Right scapula.**

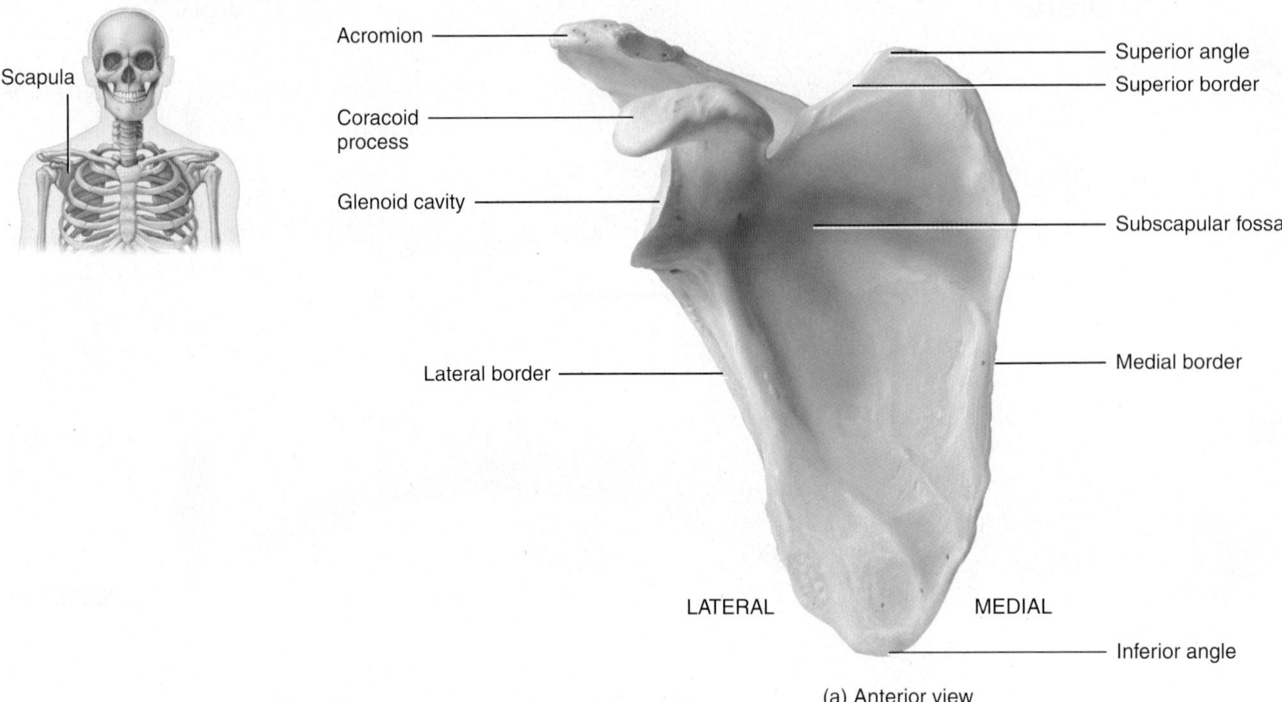


(a) Anterior view

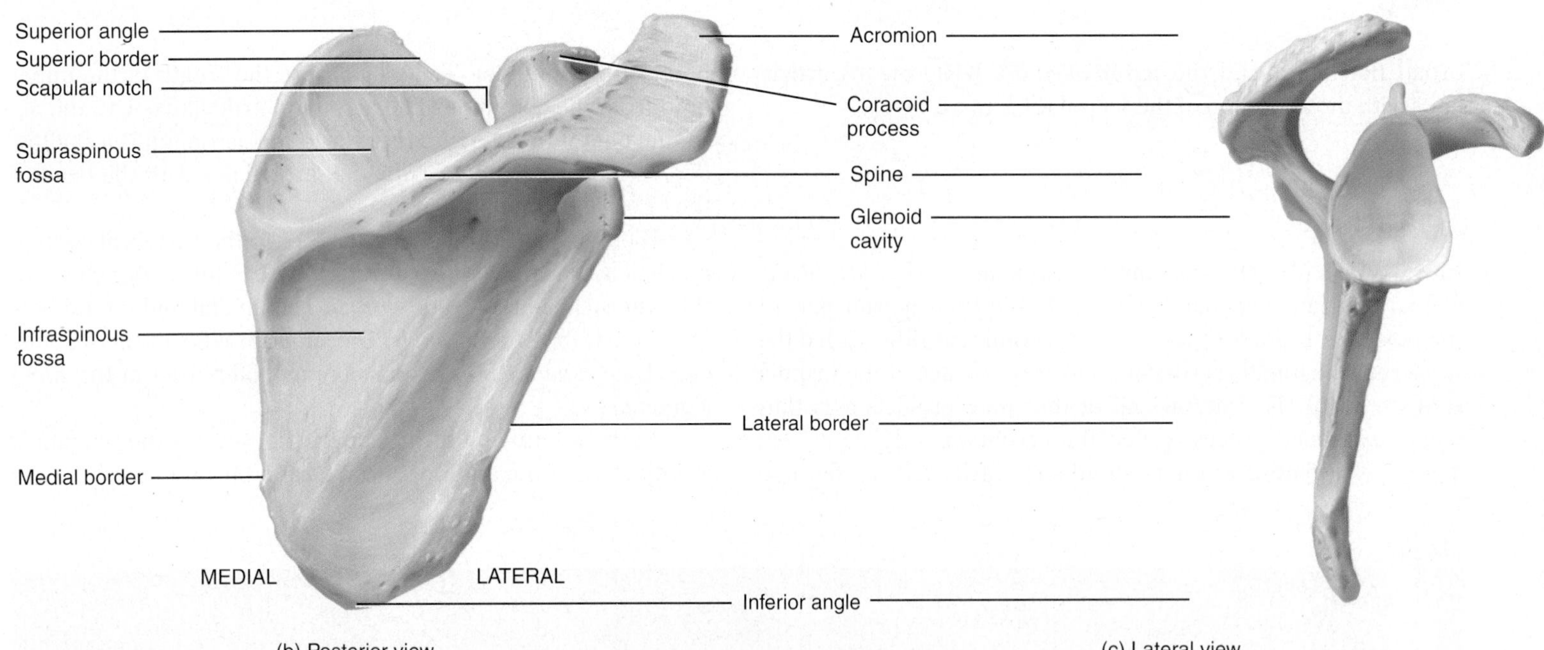


(b) Posterior view

(c) Lateral view

**The glenoid cavity of the scapula articulates with the head of the humerus to form the shoulder joint.**

(KOR-a-koyd = like a crow's beak), to which the tendons of muscles and ligaments attach. Superior and inferior to the spine are two fossae: the *supraspinous fossa* (sū-pra-SPĪ-nus) and the *infraspinous fossa* (in-fra-SPĪ-nus), respectively. Each serves as a surface of attachment for a shoulder muscle. On the anterior surface is a slightly hollowed-out area called the *subscapular fossa*, also a surface of attachment for a shoulder muscle.

**✓ CHECKPOINT**

1. Which parts of the clavicle and sternum articulate? Which parts of the clavicle and scapula articulate?
2. Which parts of the scapula and humerus articulate?
3. Which two fossae are separated by the spine of the scapula?

# 8.2 The bones of each upper limb include the humerus, ulna, radius, carpals, metacarpals, and phalanges.

Each **upper limb** consists of 30 bones: a humerus in each arm, ulna and radius in each forearm, 8 carpals in each wrist, 5 metacarpals in each palm, and 14 phalanges in the fingers of each hand (Figure 8.4).

FIGURE 8.4 **Right upper limb.**

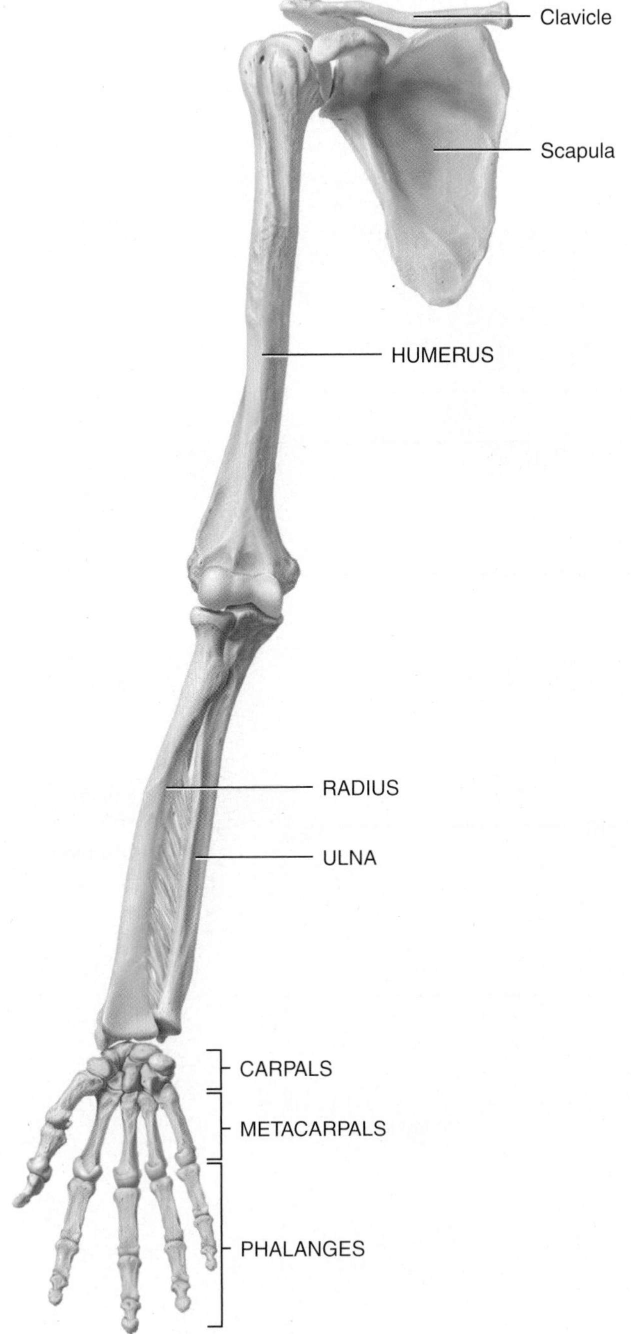


Anterior view of upper limb

There are a total of 30 bones in the arm, forearm, wrist, and hand.

## Humerus

The **humerus** (HŪ-mer-us), or arm bone, is the longest and largest bone of the upper limb (Figure 8.5). It articulates proximally with the scapula and distally at the elbow joint with both the ulna and the radius.

The proximal end of the humerus features a rounded *head* that articulates with the glenoid cavity of the scapula to form the shoulder joint (Figure 8.4). Distal to the head is the *anatomical neck*, which is visible as an oblique groove. The *greater tubercle* is a lateral projection distal to the anatomical neck. It is the most laterally palpable bony landmark of the shoulder region. The *lesser tubercle* projects anteriorly. Between both tubercles runs an *intertubercular sulcus*. The *surgical neck* is a constriction in the humerus just distal to the tubercles, where the head tapers to the shaft; it is so named because fractures often occur here.

The *body* (diaphysis) of the humerus contains a roughened, V-shaped area called the *deltoid tuberosity*. This area serves as a point of attachment for the tendons of the deltoid muscle.

Several prominent features are evident at the distal end of the humerus. The *capitulum* (ka-PIT-ū-lum; *capit-* = head) is a rounded knob on the lateral aspect of the bone that articulates with the head of the radius. The *radial fossa* is an anterior depression above the capitulum that articulates with the head of the radius when the elbow joint is flexed (bent). The *trochlea* (TRŌK-lē-a = pulley), located medial to the capitulum, is a spool-shaped surface that articulates with the trochlear notch of the ulna. The *coronoid fossa* (KOR-ō-noyd = crown-shaped) is an anterior depression that receives the coronoid process of the ulna when the elbow joint is flexed. The *olecranon fossa* (ō-LEK-ra-non = elbow) is a large posterior depression that receives the olecranon of the ulna when the elbow joint is extended (straightened). The *medial epicondyle* and *lateral epicondyle* are rough projections on either side of the distal end of the humerus to which most muscles of the forearm are attached. The ulnar nerve may be palpated by rolling a finger over the skin surface above the posterior surface of the medial epicondyle. This superficial nerve makes you feel a severe pain when you hit your elbow, which for some reason is commonly referred to as the funny bone even though this event is anything but funny.

**FIGURE 8.5** **Right humerus in relation to the scapula, ulna, and radius.**

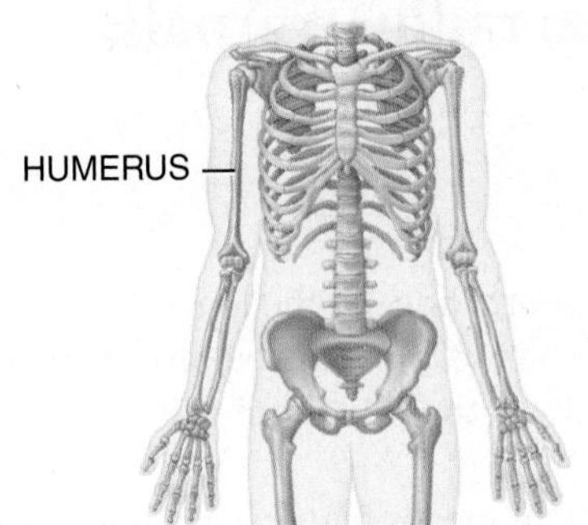


Anatomical neck
Greater tubercle
Lesser tubercle
Intertubercular sulcus
Head
Surgical neck
Scapula
HUMERUS
Deltoid tuberosity
Body
Radial fossa
Lateral epicondyle
Capitulum
Head
RADIUS
Coronoid fossa
Medial epicondyle
Trochlea
Coronoid process
ULNA

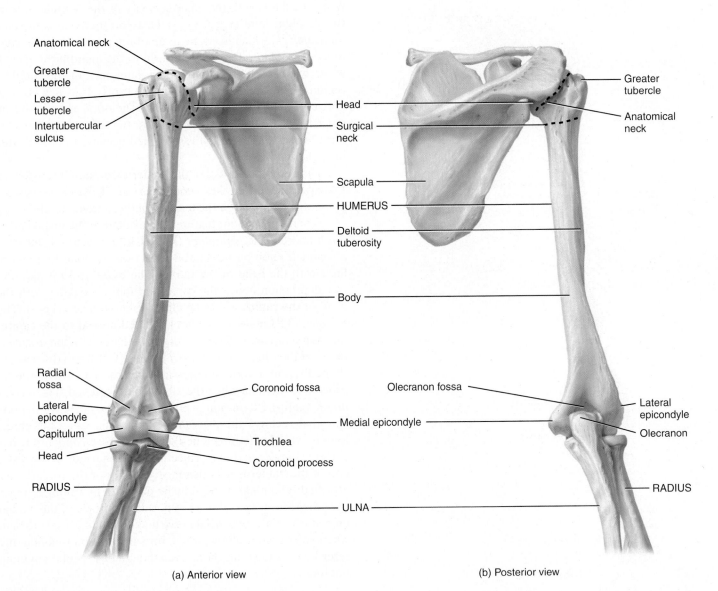

(a) Anterior view

Greater tubercle
Anatomical neck
Olecranon fossa
Lateral epicondyle
Olecranon
RADIUS

(b) Posterior view

**The humerus is the longest and largest bone of the upper limb.**

## Ulna and Radius

The **ulna** is located on the medial aspect (the little-finger side) of the forearm and is longer than the radius (Figure 8.6). You may find it convenient to use an aid to learn new or unfamiliar information. Such an aid is called a *mnemonic* (nē-MON-ik = memory). One such mnemonic to help you remember the location of the ulna in relation to the hand is "p.u." (the **p**inky is on the **u**lna side). At the proximal end of the ulna (Figures 8.6b and 8.7b) is the *olecranon*, which forms the prominence of the

FIGURE 8.6 **Right ulna and radius in relation to the humerus and carpals.**

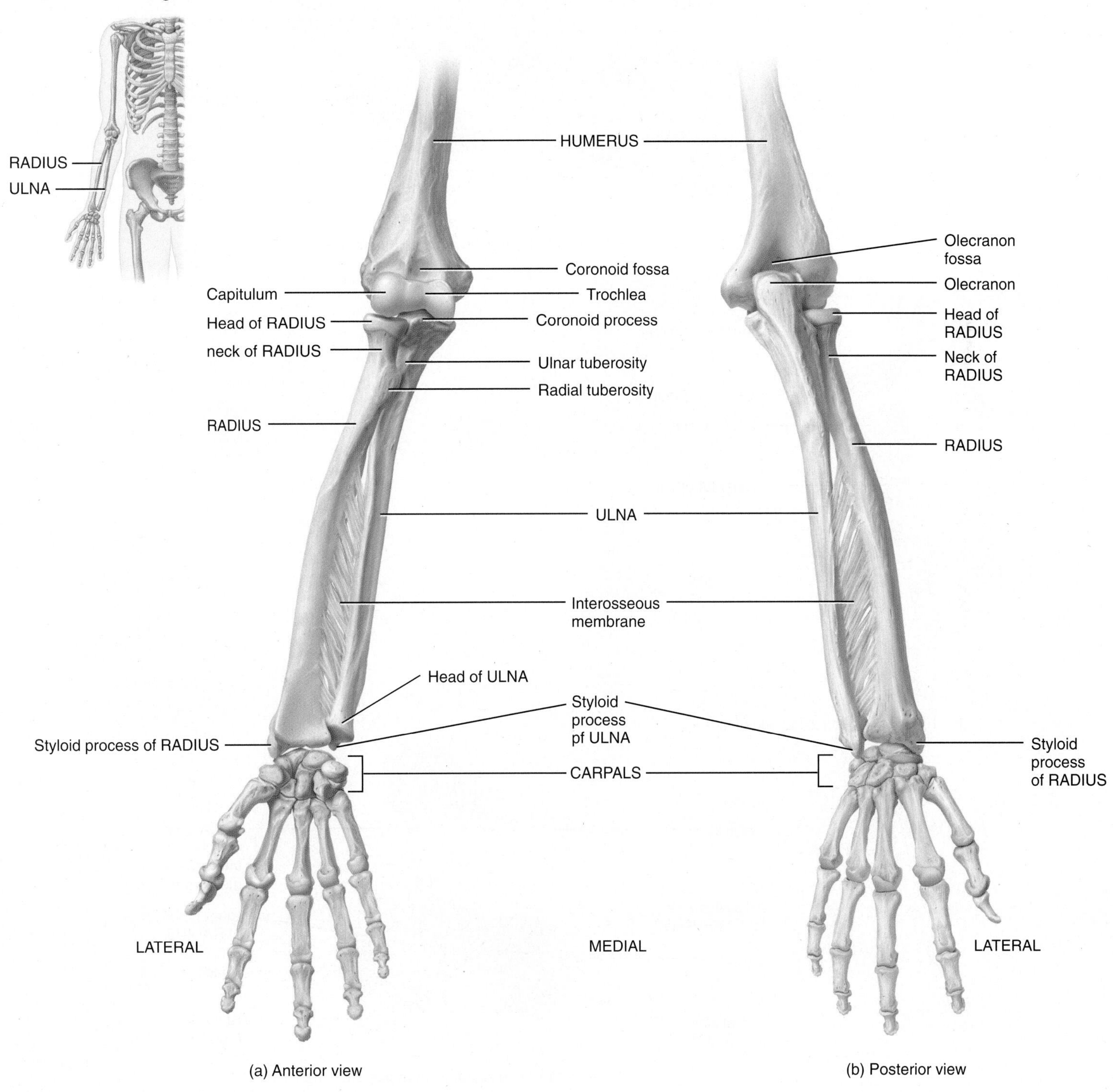


(a) Anterior view

(b) Posterior view

In the forearm, the ulna is on the medial side, and the radius is on the lateral side.

**FIGURE 8.7 Articulations formed by the right ulna and radius.** (a) Right elbow joint. (b) Joint surfaces at proximal end of the ulna. (c) Joint surfaces at distal ends of radius and ulna.

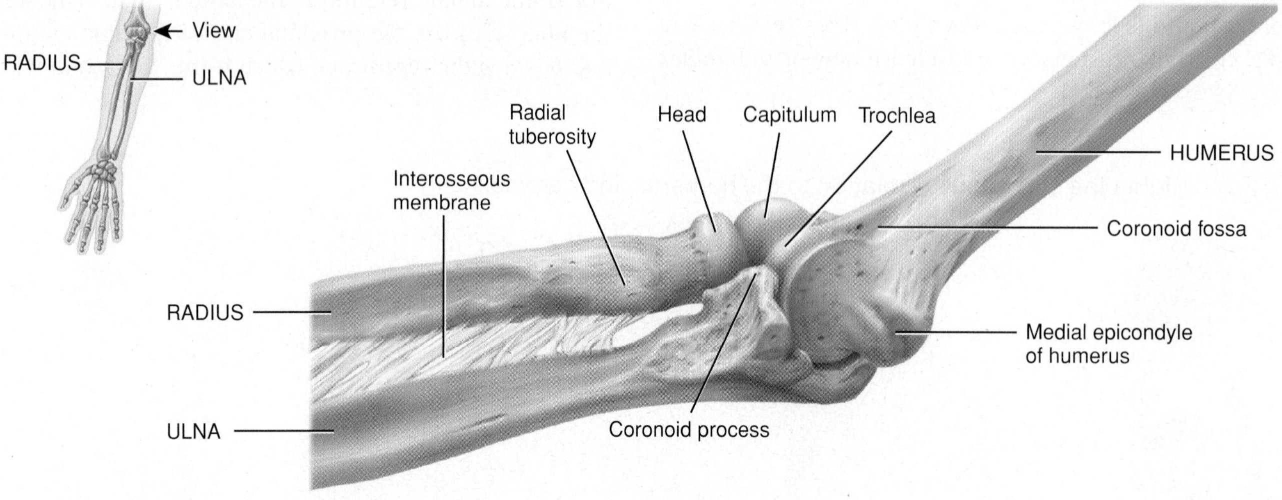

(a) Medial view in relation to humerus

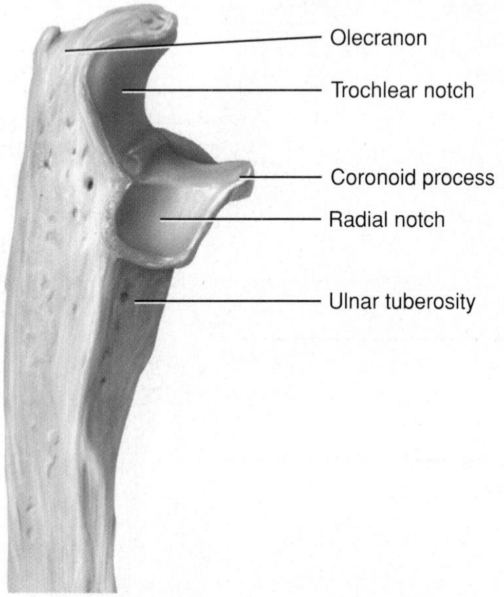

(b) Lateral view of proximal end of ulna

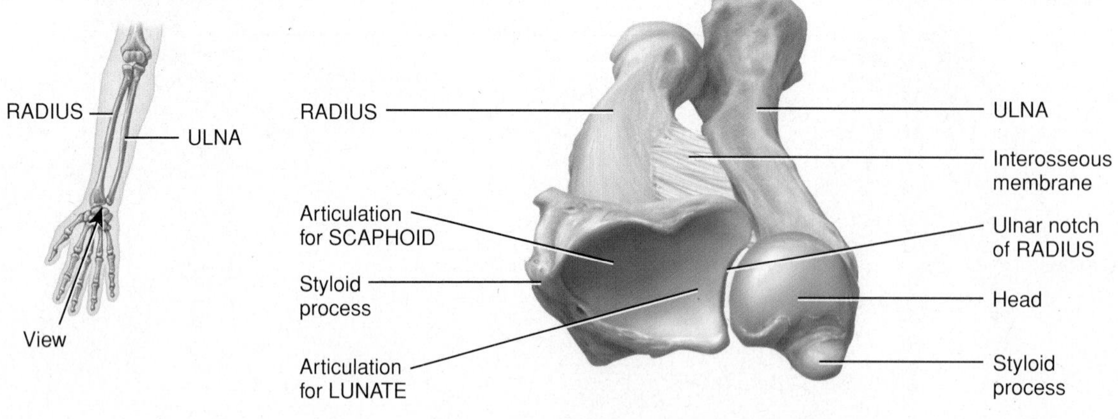



(c) Inferior view of distal ends of radius and ulna

**The elbow joint is formed by two articulations: (1) the trochlear notch of the ulna with the trochlea of the humerus and (2) the head of the radius with the capitulum of the humerus.**

elbow. The *coronoid process* (Figures 8.6a and 8.7b) is an anterior projection distal to the olecranon. The *trochlear notch* (Figure 8.7b) is a large curved area between the olecranon and coronoid process that articulates with the trochlea of the humerus at the elbow joint. The *radial notch* (Figure 8.7b) is a depression that is lateral and inferior to the trochlear notch that articulates with the head of the radius. The distal end of the ulna consists of a *head* on the anterior side and a *styloid process* on the posterior side (Figure 8.6). The styloid process provides attachment for the ulnar collateral ligament to the wrist.

The **radius** is located on the lateral aspect (thumb side) of the forearm (Figure 8.6). The proximal end of the radius has a disc-shaped *head* that articulates with the capitulum of the humerus (Figure 8.6a) and the radial notch of the ulna. Inferior to the head is the constricted *neck*. A roughened area inferior to the neck on the medial side, called the *radial tuberosity*, is a point of attachment for the biceps brachii muscle. The shaft of the radius widens distally where it contains a depression, the *ulnar notch*, on the medial side and a *styloid process* on the lateral side, which can be felt proximal to the thumb. The styloid process provides attachment for the brachioradialis muscle and wrist bones. Fracture of the distal end of the radius is the most common fracture in adults older than 50 years.

The ulna and radius articulate with the humerus at the elbow joint. The articulation occurs in two places: where the head of the radius articulates with the capitulum of the humerus, and where the trochlear notch of the ulna receives the trochlea of the humerus (Figure 8.7).

The ulna and the radius connect with one another at three sites. First, a broad, flat, fibrous connective tissue called the *interosseous membrane* (in′-ter-OS-ē-us; *inter-* = between, *-osseous* = bone) joins the shafts of the two bones. The ulna and radius also articulate at their proximal and distal ends. Proximally, the head of the radius articulates with the ulna's *radial notch* (Figure 8.7b). Distally, the head of the ulna articulates with the *ulnar notch* of the radius (Figure 8.7c). Finally, the distal end of the radius articulates with three bones of the wrist—the lunate, the scaphoid, and the triquetrum—to form the *wrist joint*.

## Carpals, Metacarpals, and Phalanges

The **carpus** (wrist) is the proximal region of the hand and consists of eight small bones, the **carpals**, joined to one another by ligaments (Figure 8.8). The carpals are arranged in two transverse rows of four bones each. Their names reflect their shapes. The carpals in the proximal row, from lateral to medial, are the **scaphoid** (SKAF-oyd = boatlike), **lunate** (LOO-nāt = moon-shaped), **triquetrum** (trī-KWĒ-trum = three-cornered), and **pisiform** (PIS-i-form = pea-shaped). The proximal row of carpals articulates with the distal ends of the ulna and radius to form the *wrist joint*. The carpals in the distal row, from lateral to medial, are the **trapezium** (tra-PĒ-zē-um = four-sided figure with no two sides parallel), **trapezoid** (TRAP-e-zoyd = four-sided figure with two sides parallel), **capitate** (KAP-i-tāt = head-shaped), and **hamate** (HAM-āt = hooked). The capitate is the largest carpal bone; its rounded projection, the head, articulates with the lunate. The hamate is named for a large hook-shaped projection on its anterior surface. In about 70 percent of carpal fractures, only the scaphoid is broken. This is because the force of a fall on an outstretched hand is transmitted from the capitate through the scaphoid to the radius.

The anterior concave space formed by the pisiform and hamate (on the ulnar side), and the scaphoid and trapezium (on the radial side), with the roof-like covering of the *flexor retinaculum* (fibrous bands of fascia) is the **carpal tunnel**. Tendons for muscles to the digits and thumb and the median nerve pass through the carpal tunnel. Narrowing of the carpal tunnel, due to such factors as inflammation, may give rise to a condition called carpal tunnel syndrome (described in Concept 11.7).

A useful mnemonic for learning the names of the carpal bones is provided in Figure 8.8. The first letter of the carpal bones from lateral to medial (proximal row, then distal row) corresponds to the first letter of each word in the mnemonic.

The **metacarpus** (*meta-* = beyond), or palm, is the intermediate region of the hand and consists of five bones called **metacarpals**. The metacarpals are numbered I to V (or 1–5), starting with the thumb, from lateral to medial (Figure 8.8). The proximal ends articulate with the distal row of carpal bones; the distal ends articulate with the proximal phalanges. The distal ends of the metacarpals, commonly called *knuckles*, are readily visible in a clenched fist.

The **phalanges** (fa-LAN-jēz; *phalanx* = a battle line), or bones of the fingers, make up the distal region of the hand. There are 14 phalanges in the fingers of each hand and, like the metacarpals, the fingers are numbered I to V (or 1–5), beginning with the thumb, from lateral to medial. A single bone of a finger is referred to as a *phalanx* (Fā-lanks). There are two phalanges (proximal and distal) in the thumb and three phalanges (proximal, middle, and distal) in each of the other four fingers. In order from the thumb, these other four fingers are commonly referred to as the index finger, middle finger, ring finger, and little finger. The *proximal* phalanges articulate with the metacarpal bones and middle phalanges. The *middle* phalanges articulate with the proximal row and *distal* phalanges. The distal phalanges articulate with the middle phalanges.

**FIGURE 8.8 Right wrist and hand in relation to the ulna and radius.**

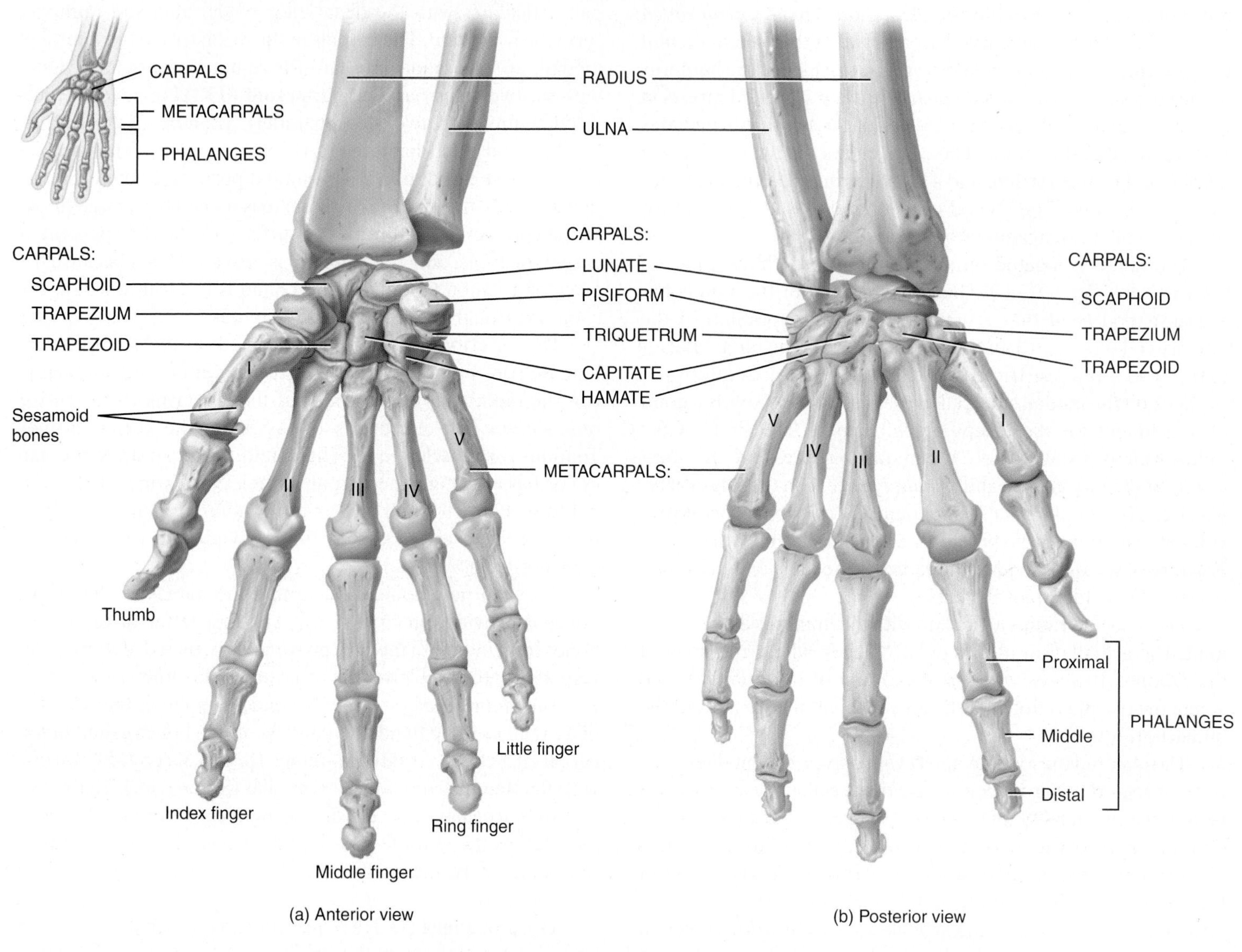


(a) Anterior view

(b) Posterior view

MNEMONIC for carpal bones*:

**S**top **L**etting **T**hose **P**eople **T**ouch **T**he **C**adaver's **H**and.

| **S**caphoid | **L**unate | **T**riquetrum | **P**isiform | **T**rapezium | **T**rapezoid | **C**apitate | **H**amate |
|---|---|---|---|---|---|---|---|
| Proximal row | | | | Distal row | | | |
| Lateral → Medial | | | | Lateral → Medial | | | |

* *Edward Tanner, University of Alabama, School of Medicine*

**The skeleton of the hand consists of the proximal carpals, the intermediate metacarpals, and the distal phalanges.**

## ✓ CHECKPOINT

4. How many bones make up the upper limb?
5. Which part of which bone forms the most lateral projection of the shoulder?
6. Which part of the ulna makes up the projecting point of the elbow?
7. Which parts of the humerus, ulna, and radius articulate at the elbow joint? Which depression of the humerus and projection of the ulna articulate when the elbow is flexed? When the elbow is straight?
8. Which carpal bones form a concave space that is part of the carpal tunnel?
9. Which bones form the palm of the hand? Are commonly called the knuckles?

## RETURN TO Hassan's Story

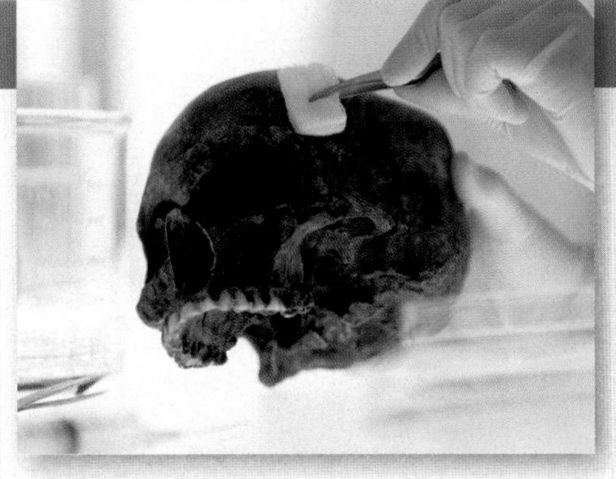

Stefan groaned as he hoisted the large wooden crate onto his workbench. The crate was dusty, marked with bold, black lettering indicating the university name and "Attention: Department of Archaeology." Hassan, director of the Ishtamira archaeological dig site, had been sending such boxes for months now. For Stefan, it had meant unpacking, categorizing, photographing, and sorting.

Hassan had been very excited about a particular box and here it was, the crate that held the remains of the princess. He picked up a large metal pry bar from his workbench and began removing the lid. Inside the box were bones carefully and neatly wrapped in soft cotton padding. Stefan removed the cotton and lifted out the bones one at a time: a humerus, a radius, an ulna, and several small cube-shaped bones. These were followed by shorter bones that looked like those left over after a chicken dinner. A flat, triangular-shaped bone was revealed. "Right scapula," he thought, and placed it above the now-articulated right upper limb. He aligned the socket of the shoulder blade with the head of the humerus. Immediately under the right scapula in the shipping crate was the left scapula. "Where are your collarbones, Princess?" he mused, as he searched carefully through the packing material in the crate.

Very neatly, Stefan began arranging the bones on the table. Stefan knew his anatomy, and, slowly, a skeletal upper limb began to appear. Stefan always enjoyed this part of his job. Rearticulating the skeleton was like piecing together an ancient puzzle. "Ah, here we go." Stefan pulled out two bundles, each containing a narrow, S-shaped bone. Turning them over in his hands, he quickly decided which was right and which left, then placed each clavicle by its neighboring scapula.

Arranging the right and left humerus, radius, and ulna was easy for Stefan using their obvious landmarks—the head, deltoid tuberosity, and olecranon fossa of the humerus, the olecranon and pointed styloid process of the ulna, and the circular head and wide styloid process of the radius. The deltoid tuberosity on the right humerus was somewhat larger than on the left. "Hmm, must have been right-handed," Stefan noted.

The wrist and hand bones were more difficult to put back in order. Stefan unwrapped each individual bone, identifying them first as phalanges, metacarpals, or carpals. Stefan focused on the wrist bones. "Pisiform goes here, scaphoid, and... let's see here, this is...."

A. *Which clue would tell Stefan which scapular surface was anterior and which was posterior? What is the name of the shallow, oval socket of the scapula that Stefan placed next to the humerus?*
B. *Which bone is Stefan referring to as the "collarbone"?*
C. *Which surface markings could Stefan use to distinguish the right humerus from the left?*
D. *Why would Stefan think that an enlarged right deltoid tuberosity might indicate right-handedness?*

## 8.3 The pelvic girdle supports the vertebral column and pelvic viscera and attaches the lower limbs to the axial skeleton.

The **pelvic (hip) girdle** consists of the two **hip bones**, also called **coxal bones** (KOK-sal; *cox-* = hip) (Figure 8.9). The hip bones unite anteriorly at a joint called the **pubic symphysis** (PŪ-bik SIM-fi-sis). They unite posteriorly with the sacrum at the *sacroiliac joints*. The complete ring composed of the hip bones, pubic symphysis, and sacrum forms a deep, basinlike structure called the **pelvis** (plural is *pelves* [PEL-vēz]; *pelv-* = basin). Functionally, the pelvis provides a strong and stable support for the vertebral column and protects pelvic and lower abdominal organs. The pelvic girdle of the pelvis also connects the bones of the lower limbs to the axial skeleton.

There are some significant differences between the pectoral girdle and pelvic girdle. The pectoral girdle does not directly articulate with the vertebral column, but the pelvic girdle does so via the sacroiliac joint. The pectoral girdle sockets (glenoid cavities) that articulate with the upper limbs are shallow and maximize movement, in contrast to the pelvic girdle sockets (acetabula) that articulate with the lower limbs, which are deep and allow less movement. Overall, the structure of the pectoral girdle offers more mobility than strength, and that of the pelvic girdle offers more strength than mobility.

Each of the two hip bones of a newborn consists of three bones separated by cartilage: a superior *ilium*, an inferior and anterior *pubis*, and an inferior and posterior *ischium*. By age 23, the three separate bones fuse together (see Figure 8.10a). Although the hip bones function as single bones, anatomists commonly discuss each hip bone as three separate bones.

FIGURE 8.9 **Pelvis.** Shown here is the female pelvis.

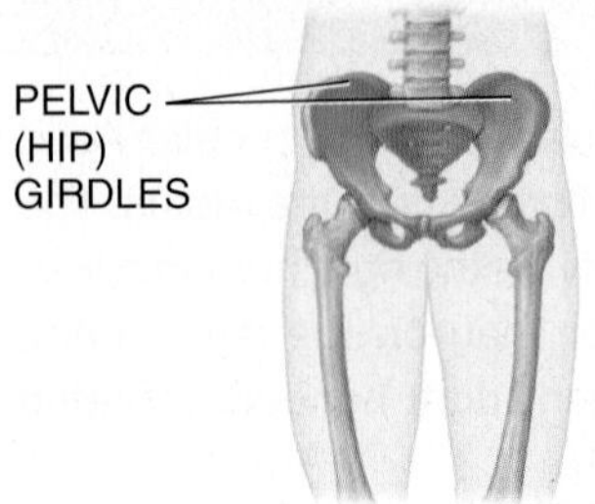


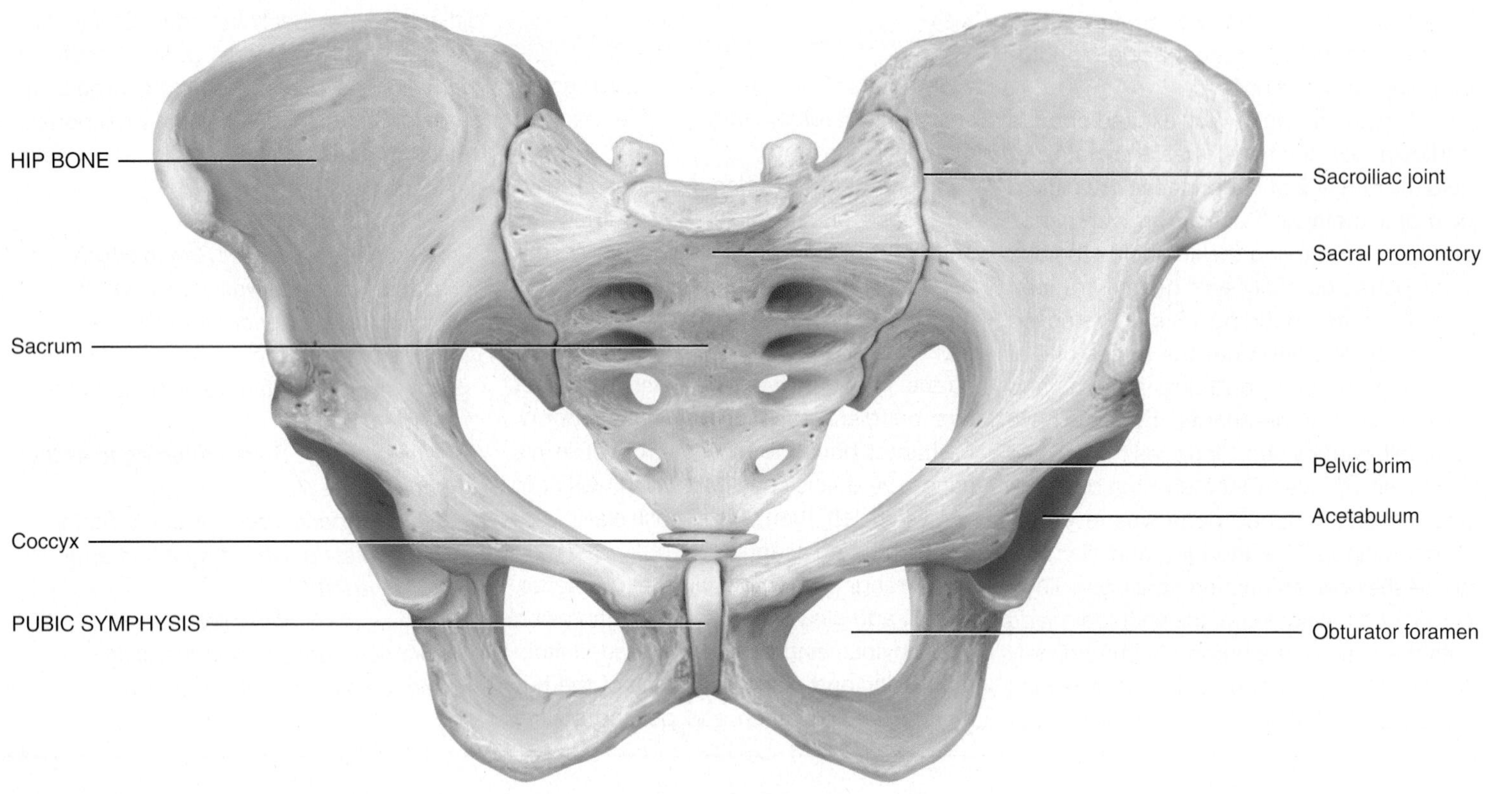


Anterosuperior view of pelvic girdle

**The hip bones are united anteriorly at the pubic symphysis and posteriorly at the sacrum to form the pelvis.**

## Ilium

The **ilium** (IL-ē-um = flank) is the largest of the three components of the hip bone (Figure 8.10). It is comprised of a superior *ala* (= wing) and an inferior *body*. The body is one of the components of the *acetabulum*, the socket for the head of the femur. The superior border of the ilium, the *iliac crest*, ends anteriorly in a blunt *anterior superior iliac spine* (Figure 8.10b, c). Below this spine is the *anterior inferior iliac spine*. Posteriorly, the iliac crest ends in a sharp *posterior superior iliac spine*. Below this spine is the *posterior inferior iliac spine*. The spines serve as points of attachment for the muscles of the trunk, hip, and thighs. Below the posterior inferior iliac spine is the *greater sciatic notch* (sī-AT-ik), through which the sciatic nerve, the longest nerve in the body, passes.

The medial surface of the ilium contains a concavity, the *iliac fossa*, a site of muscle attachment. Posterior to this fossa is the *auricular surface* (*auric-* = ear-shaped), which articulates with the sacrum (Figure 8.9). Projecting anteriorly and inferiorly from the auricular surface of the ilium is a ridge called the *arcuate line* (AR-kū-āt; *arc-* = bow) (Figure 8.10c).

The other conspicuous markings of the ilium are three arched lines on its lateral surface called the *posterior gluteal line* (*glut-* = buttock), the *anterior gluteal line*, and the *inferior gluteal line*. The gluteal muscles attach to the ilium between these three lines.

## Ischium

The **ischium** (IS-kē-um = hip), the inferior, posterior portion of the hip bone (Figure 8.10), comprises a superior *body* and

FIGURE 8.10 **Right hip bone.** The lines of fusion of the ilium, ischium, and pubis depicted in (a) are not always visible in an adult.

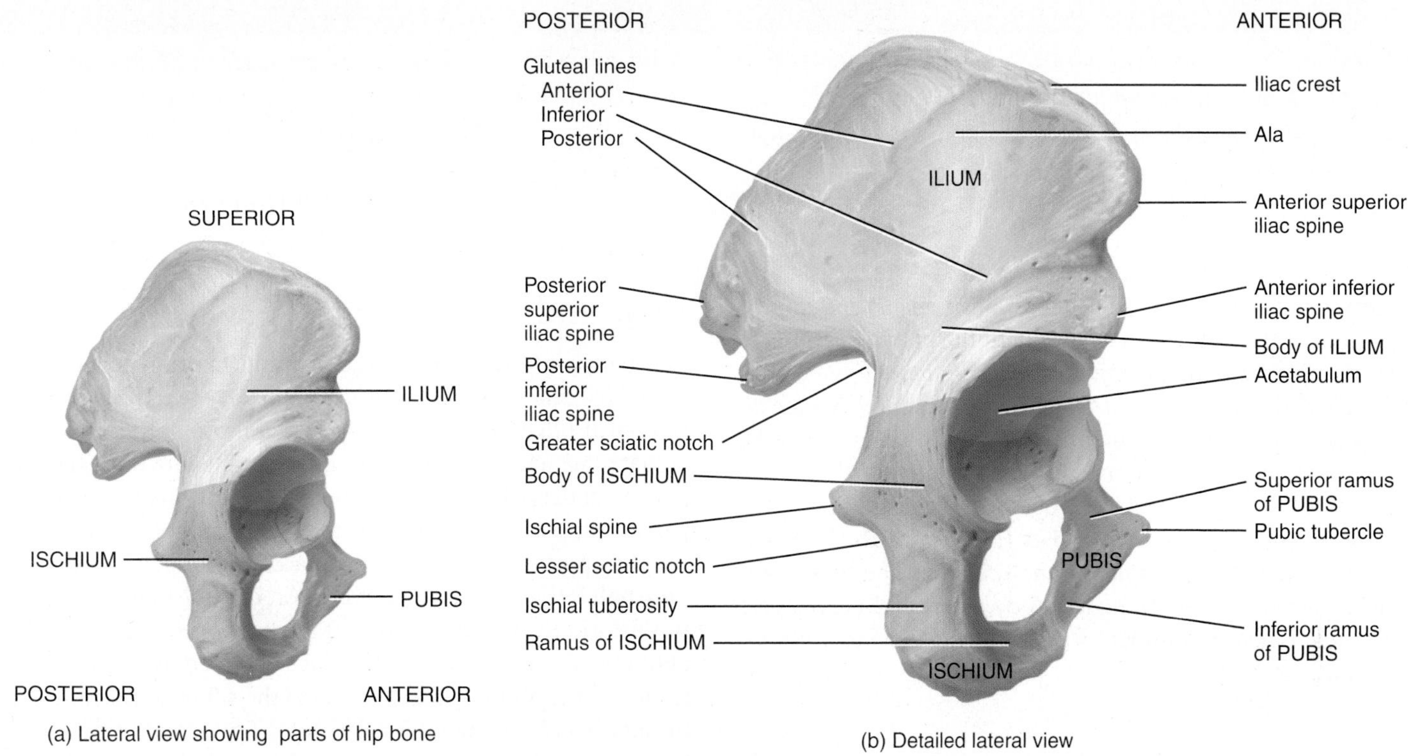


(a) Lateral view showing parts of hip bone

(b) Detailed lateral view

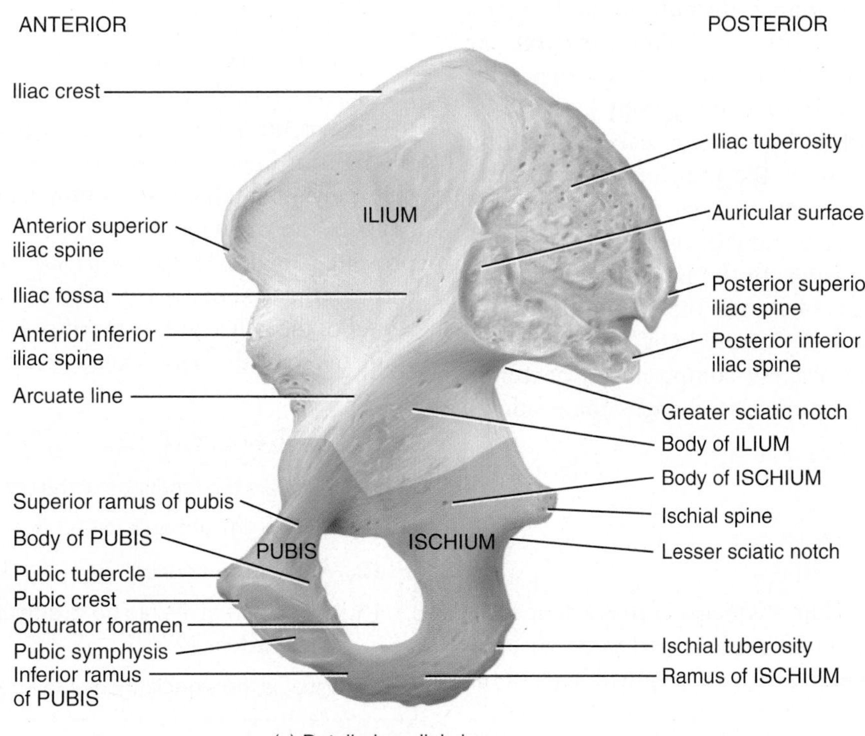


(c) Detailed medial view

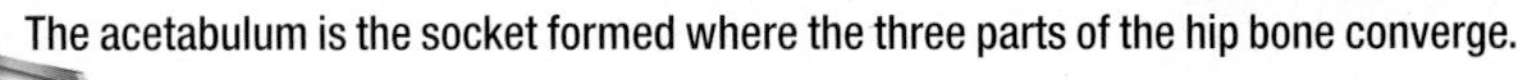
**The acetabulum is the socket formed where the three parts of the hip bone converge.**

## CLINICAL CONNECTION | *Pelvimetry*

**Pelvimetry** is the measurement of the size of the inlet and outlet of the birth canal, which may be done by ultrasonography or physical examination. Measurement of the pelvic cavity in pregnant females is important because the fetus must pass through the narrower opening of the pelvis at birth. A cesarean section is usually planned if it is determined that the pelvic cavity is too small to permit passage of the baby.

an inferior *ramus* (*ram-* = branch; plural is *rami*). The ramus is the portion of the ischium that fuses with the pubis. Features of the ischium include the prominent *ischial spine*, a *lesser sciatic notch* below the spine, and a rough and thickened *ischial tuberosity*. Because this prominent tuberosity is just deep to the skin, it commonly begins hurting after a relatively short time when you sit on a hard surface. Together, the ischium and the pubis surround the *obturator foramen* (OB-too-rā-tōr; *obtur-* = closed up), the largest foramen in the skeleton. The foramen is so named because, even though blood vessels and nerves pass through it, it is nearly completely closed by the fibrous *obturator membrane*.

### Pubis

The **pubis** (PŪ-bis), or **pubic bone**, is the anterior and inferior part of the hip bone (Figure 8.10b, c). A *superior ramus*, an *inferior ramus*, and a *body* between the rami comprise the pubis. The anterior, superior border of the body is the *pubic crest*, and at its lateral end is a projection called the *pubic tubercle*. The *pubic symphysis* is the joint between the two hip bones (Figure 8.9). It consists of a disc of fibrocartilage. Inferior to this joint, the inferior rami of the two pubic bones converge to form the *pubic arch*. In the later stages of pregnancy, the hormone relaxin increases the flexibility of the pubic symphysis to ease delivery of the baby. Weakening of the joint, together with an already compromised center of gravity due to an enlarged uterus, also changes the mother's gait during pregnancy.

### Acetabulum

The *acetabulum* (as-e-TAB-ū-lum = vinegar cup) is a deep fossa (depression) formed by the ilium, ischium, and pubis. It functions as the socket that articulates with the rounded head of the femur to form the *hip joint*.

### False and True Pelves

The pelvis is divided into superior and inferior portions by a boundary called the *pelvic brim* (Figure 8.11a). The pelvic brim is an oval ridge that runs from the sacral promontory, along the arcuate line and then the pubic crest, and finally to the superior portion of the pubic symphysis. Because of the tilt of the pelvis, the pelvic brim is higher in the back than in the front (Figure 8.11b).

The portion of the pelvis superior to the pelvic brim is the **false pelvis** (Figure 8.11b). It is bordered by the lumbar vertebrae posteriorly, the upper portions of the hip bones laterally, and the abdominal wall anteriorly. The space enclosed by the false pelvis is part of the abdomen and contains the superior portion of the urinary bladder (when it is full) and the lower intestines in both genders, and the uterus, ovaries, and uterine tubes of the female.

The portion of the pelvis inferior to the pelvic brim is the **true pelvis** (Figure 8.11b). It is bounded by the sacrum and coccyx posteriorly, inferior portions of the ilium and ischium laterally, and the pubic bones anteriorly. The true pelvis surrounds the pelvic cavity (see Figure 1.9). The superior opening of the true pelvis, bordered by the pelvic brim, is called the *pelvic inlet*; the inferior opening of the true pelvis is the *pelvic outlet*. The true pelvis contains the rectum and urinary bladder in both genders, the vagina and cervix of the uterus in females, and the prostate in males. The *pelvic axis* is an imaginary line that curves through the true pelvis from the center of the pelvic inlet to the center of the pelvic outlet. During childbirth the pelvic axis is the route taken by the baby's head as it descends through the pelvis.

#### ✓ CHECKPOINT

10. What are the functions of the pelvic girdle and the pelvis?
11. Distinguish among a hip bone, pelvic girdle, and pelvis.
12. What are the four spines of the ilium? What is their function?
13. Which part of the hip bone articulates with the sacrum? With the femur?
14. What is the significance of the pelvic axis?

**FIGURE 8.11 True and false pelves.** Shown here is the female pelvis. For simplicity, in (a) the landmarks of the pelvic brim are shown only on the left side of the body, and the outline of the pelvic brim is shown only on the right side. The entire pelvic brim is shown in Table 8.1.

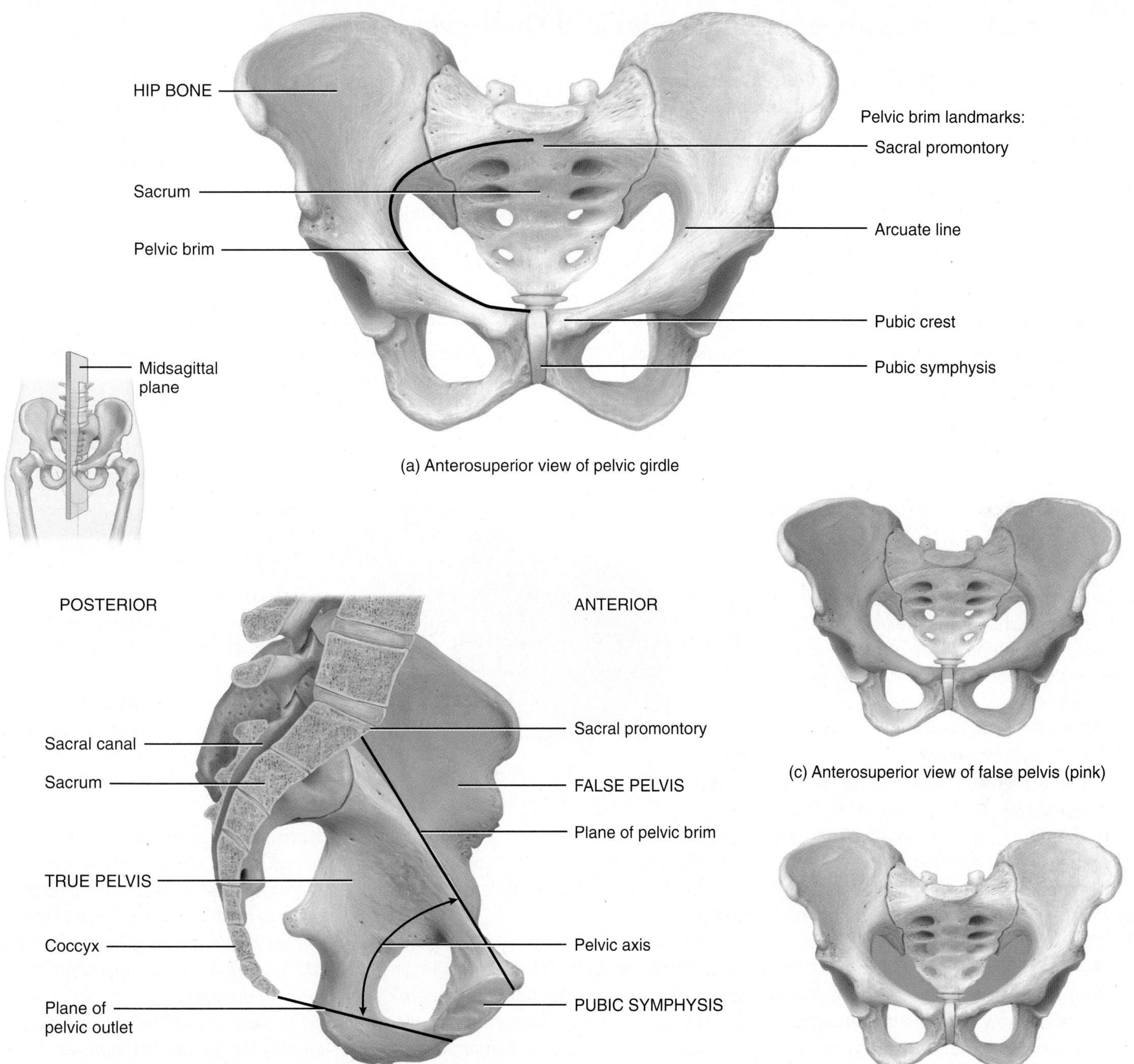


**The true and false pelves are separated by the pelvic brim.**

## 8.4 Male pelves are generally larger, heavier, and have more prominent markings; female pelves are generally wider and shallower.

Generally, the bones of males are larger and heavier and possess larger surface markings than those of females of comparable age and physical stature (Table 8.1). Sex-related differences in the features of bones are readily apparent when comparing the female and male pelves. Most of the structural differences in the pelves are adaptations to the requirements of pregnancy and childbirth. The female's pelvis is wider and shallower than the male's. Consequently, there is more space in the true pelvis of the female, especially in the pelvic inlet and pelvic outlet, to accommodate the passage of the infant's head at birth. Other significant structural differences between pelves of females and males are described in Table 8.1.

### ✓ CHECKPOINT

**15.** Why are there structural differences between female and male pelves?

**16.** What adaptations of the female pelvis assist the infant's head exiting at birth?

### RETURN TO Hassan's Story

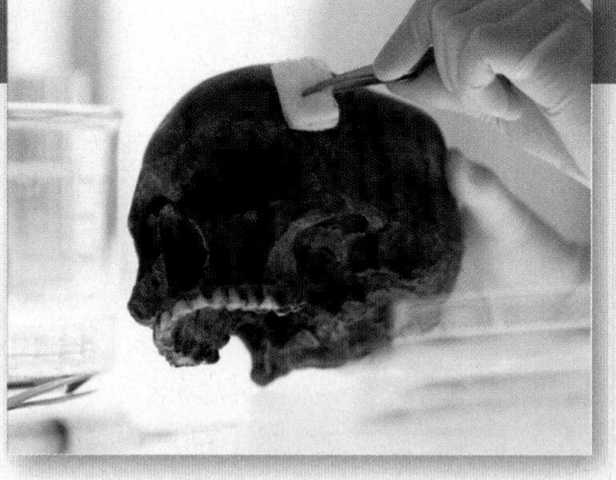

Stefan kept at his work diligently until he had the remains completely articulated. He had made some notes of his observations and decided it was time to telephone Professor Hassan.

"Hello, Professor Hassan?"

"Yes," came the reply.

"This is Stefan. I have the remains of the princess here and Octavio will be looking at them tonight or tomorrow. They are remarkably well-preserved."

"I agree. We are most fortunate. What do you think, Stefan? Are they mother and child?"

Stefan shifted in his chair; his back hurt from hours of leaning over the remains. "The lack of brow ridge development and pelvic appearance support the gender of both as female. The adult is possibly the mother; the pubic symphysis had defined, discrete lesions that could have occurred by pulling of the cartilage of the pubic symphysis during childbirth."

Hassan was silent for a moment. "Have Octavio prepare his report as soon as he can, then have him call me. Baru thought that the child was probably about two to three years old at the time of death. It could be her child."

"Professor, we should also be getting some of the results from the genetic testing soon. What do you think happened there?"

Hassan paused again, then replied. "We have found some evidence suggesting intentional homicide, but many of the remains do not give such an indication. There could have been a plague, but most diseases produce soft tissue injury and act quickly, leaving no evidence in bones. I have sent several samples directly to the genetics lab to test for some possible suspects. There may be spores left in the soil, or perhaps on the bones of some of the remains. Be careful, Stefan. No food or drink in the lab."

Stefan felt his stomach tighten and made a mental note to wash up thoroughly when his work was finished. "Of course, Professor."

E. *What is the location of the pubic symphysis Stefan refers to in the story?*

F. *Which adaptation would have taken place in the pubic symphysis of the female skeleton during the later stages of her pregnancy in preparation for the birthing process?*

G. *What clues could Stefan have used to identify the gender of the pelvis?*

H. *How would Stefan have distinguished between the right and the left hip bones?*

I. *Would the bones of each hip bone be fused in the female child's skeleton?*

**TABLE 8.1**

Comparison of Female and Male Pelves

| POINT OF COMPARISON | FEMALE | MALE |
|---|---|---|
| **General structure** | Light and thin | Heavy and thick |
| **False pelvis** | Shallow | Deep |
| **Pelvic brim (inlet)** | Larger and more oval | Smaller and heart-shaped |
| **Acetabulum** | Small and faces anteriorly | Large and faces laterally |
| **Obturator foramen** | Oval | Round |
| **Pubic arch** | Greater than 90° angle | Less than 90° angle |

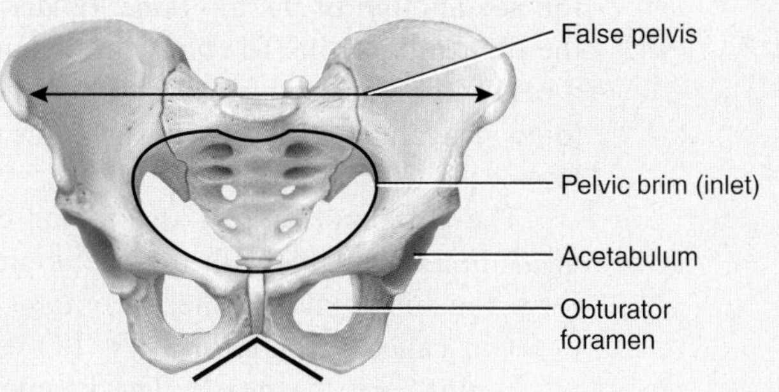




Anterior views

| POINT OF COMPARISON | FEMALE | MALE |
|---|---|---|
| **Iliac crest** | Less curved | More curved |
| **Ilium** | Less vertical | More vertical |
| **Greater sciatic notch** | Wide | Narrow |
| **Coccyx** | More movable and more curved anteriorly | Less movable and less curved anteriorly |
| **Sacrum** | Shorter, wider (see anterior views), and less curved anteriorly | Longer, narrower (see anterior views), and more curved anteriorly |

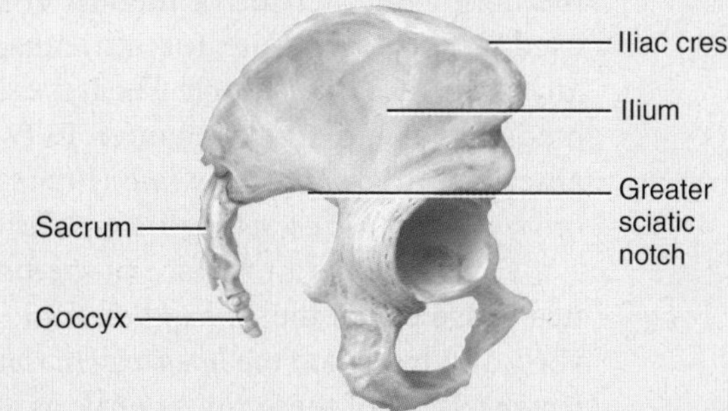


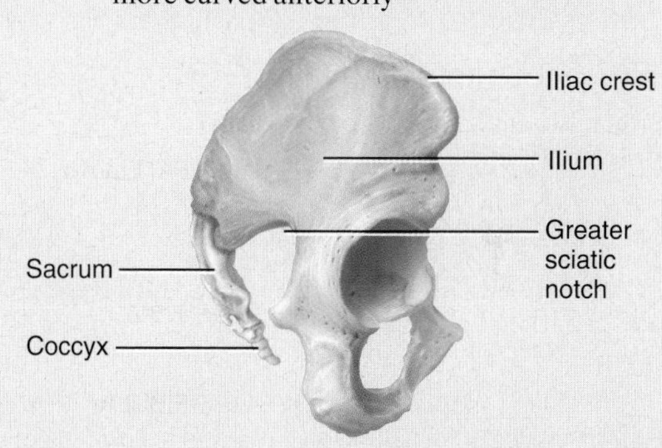


Right lateral views

| POINT OF COMPARISON | FEMALE | MALE |
|---|---|---|
| **Pelvic outlet** | Wider | Narrower |
| **Ischial tuberosity** | Shorter, farther apart, and more medially projecting | Longer, closer together, and more laterally projecting |



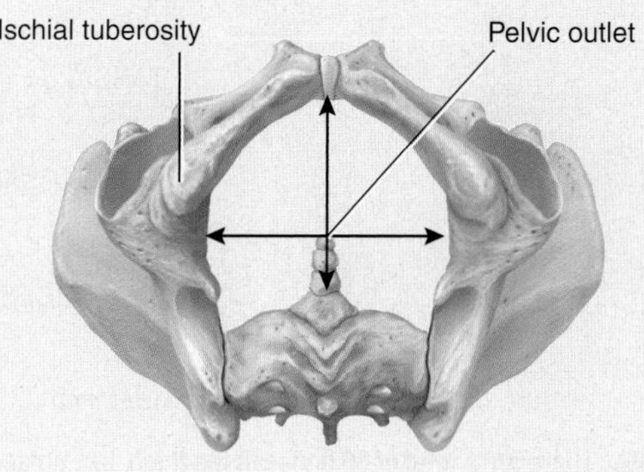


Inferior views

# 8.5 The bones of each lower limb include the femur, patella, tibia, fibula, tarsals, metatarsals, and phalanges.

Each **lower limb** is composed of 30 bones: the femur in the thigh, the patella (kneecap), the tibia and fibula in the leg, 7 tarsals in the ankle, 5 metatarsals in the foot, and 14 phalanges in the toes (Figure 8.12).

FIGURE 8.12 Right lower limb.

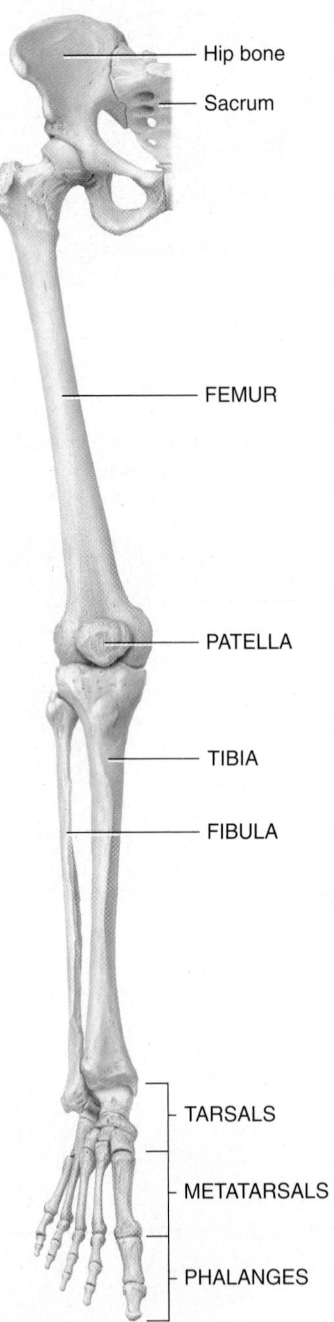


Anterior view of lower limb

There are a total of 30 bones in the thigh, leg, ankle, and foot.

## Femur

The **femur**, or thigh bone, is the longest, heaviest, and strongest bone in the body (Figure 8.13). Its proximal end articulates with the acetabulum of the hip bone. Its distal end articulates with the tibia and patella. The *body* of the femur angles medially; as a result, the knee joints are closer to the midline than the hip joints. The angle of the femur is greater in females because the female pelvis is broader.

The proximal end of the femur consists of a rounded *head* that articulates with the acetabulum of the hip bone to form the hip joint. The head contains a small centered depression called the *fovea capitis* (FŌ-vē-a CAP-i-tis; *fovea* = pit; *capitis* = of the head). The ligament of the head of the femur connects the fovea capitis of the femur to the acetabulum of the hip bone. The *neck* of the femur is a constricted region distal to the head. A "broken hip" is more often associated with a fracture in the neck of the femur than fractures of the hip bones. The *greater trochanter* (trō-KAN-ter) and *lesser trochanter* are projections of the femur that serve as points of attachment for thigh and buttock muscles. The greater trochanter is the prominence felt and seen anterior to the hollow on the side of the hip. It is a landmark commonly used to locate the site for intramuscular injections into the lateral surface of the thigh. The lesser trochanter is inferior and medial to the greater trochanter. Between the trochanters is an anterior *intertrochanteric line* (Figure 8.13a) and a posterior *intertrochanteric crest* (Figure 8.13b).

On the posterior surface of the body of the femur is a vertical ridge called the *gluteal tuberosity* that blends into another vertical ridge called the *linea aspera* (LIN-ē-a AS-per-a; *asper-* = rough). Both of these ridges serve as attachment points for several thigh muscles.

The distal end of the femur expands to include the *medial condyle* (= knuckle) and the *lateral condyle*. These articulate with the medial and lateral condyles of the tibia to form the *knee joint*. Superior to the condyles are the *medial epicondyle* and the *lateral epicondyle*. A depressed area between the condyles on the posterior surface is called the *intercondylar fossa* (in-ter-KON-di-lar). The *patellar surface*, located anteriorly between the condyles, is a smooth surface over which the patella glides.

FIGURE 8.13 **Right femur in relation to the hip bone, patella, tibia, and fibula.**

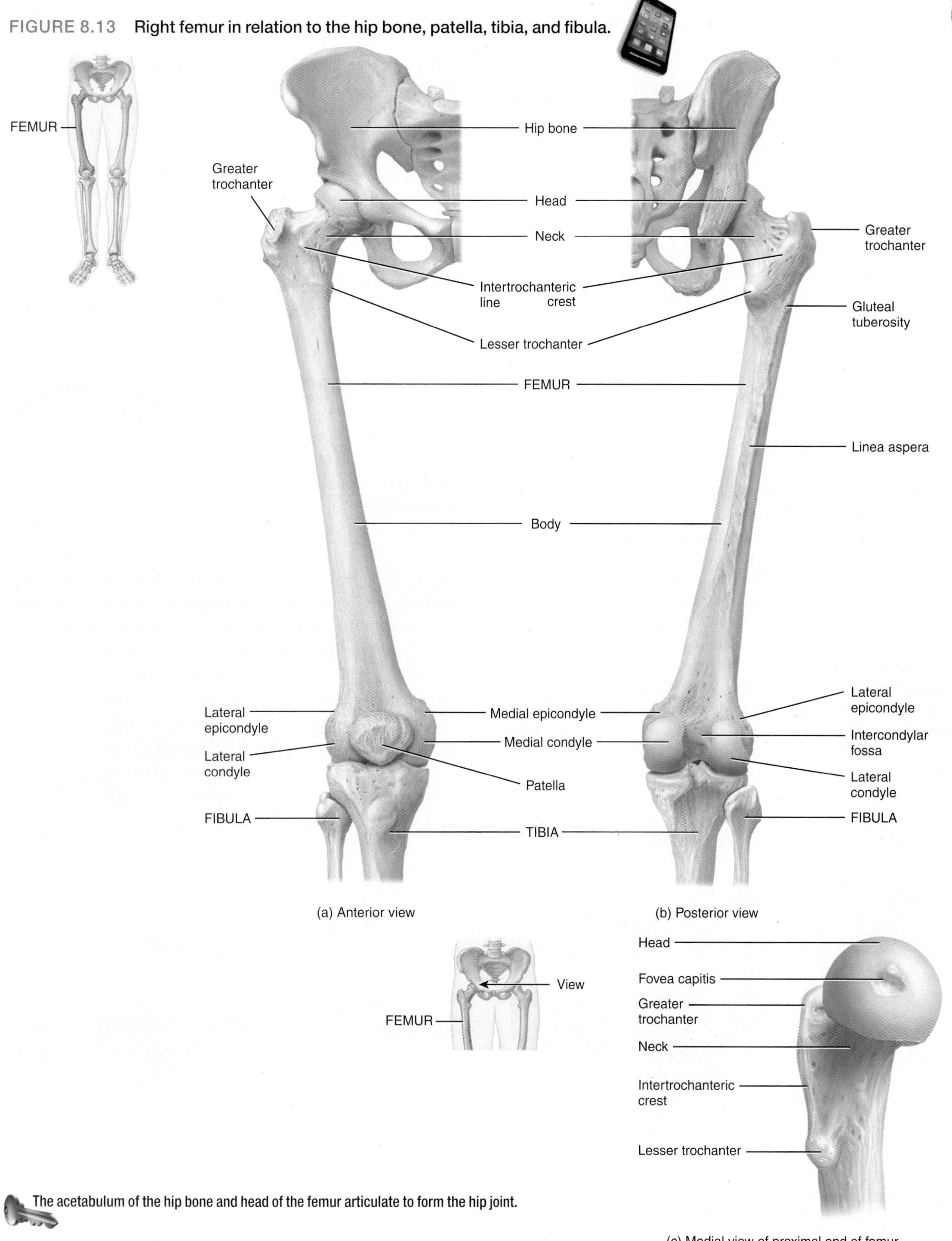


(a) Anterior view

(b) Posterior view

(c) Medial view of proximal end of femur

The acetabulum of the hip bone and head of the femur articulate to form the hip joint.

## Patella

The **patella** (= little dish), or kneecap, is a small, triangular sesamoid bone located anterior to the knee joint (Figure 8.14). The broad superior end of the patella is called the *base* and the pointed inferior end is the *apex*. The posterior surface contains two *articular facets*, one for each condyle of the femur. During normal flexion and extension of the knee, the patella glides up and down in the groove between the two femoral condyles. The patella is enclosed superiorly within the tendon of the quadriceps femoris (an anterior thigh muscle) and inferiorly is attached to the tibial tuberosity of the tibia by the patellar ligament. The patella increases the leverage of the quadriceps femoris, maintains the position of the tendon when the knee is flexed (bent), and protects the knee joint.

## Tibia and Fibula

The **tibia**, or shin bone, is the larger, medial, weight-bearing bone of the leg (Figure 8.15). The tibia articulates at its proximal end with the femur and fibula and at its distal end with the fibula and the talus of the ankle. The tibia and fibula, like the ulna and radius, are connected by an interosseous membrane.

The proximal end of the tibia is expanded into a *lateral condyle* and a *medial condyle*, which articulate with the condyles of the femur to form the knee joint. The inferior surface of the lateral condyle articulates with the head of the fibula. The slightly concave condyles are separated by an upward projection called the *intercondylar eminence* (Figure 8.15b). The *tibial tuberosity* on the anterior surface is a point of attachment for the patellar ligament. Inferior to and continuous with the tibial tuberosity is a sharp ridge that can be felt below the skin, known as the *anterior border* or shin.

The medial surface of the distal end of the tibia forms the *medial malleolus* (mal-LĒ-ō-lus = hammer). This structure articulates with the talus of the ankle and forms the prominence that can be felt on the medial surface of the ankle. The *fibular notch* (Figure 8.15c) articulates with the distal end of the fibula.

The **fibula** is parallel and lateral to the tibia, but is considerably smaller. Figure 8.15 includes a mnemonic describing the relative positions of the tibia and fibula. Unlike the tibia, the fibula does not articulate with the femur, but it does help stabilize the ankle joint. The *head* of the fibula, the proximal end, articulates with the inferior surface of the lateral condyle of the tibia below the level of the knee joint. The distal end has a projection, the *lateral malleolus*, that articulates with the talus of the ankle. The lateral malleolus is the prominence on the lateral surface of the ankle. As noted previously, the fibula also articulates with the tibia at the fibular notch.

### CLINICAL CONNECTION | *Patellofemoral Stress Syndrome*

**Patellofemoral stress syndrome** ("runner's knee") is one of the most common problems runners experience. During normal flexion and extension of the knee, the patella tracks (glides) superiorly and inferiorly in the groove between the femoral condyles. In patellofemoral stress syndrome, normal tracking does not occur; instead, the patella tracks laterally as well as superiorly and inferiorly, and the increased pressure on the joint causes aching or tenderness around or under the patella. The pain typically occurs after a person has been sitting for awhile, especially after exercise. It is worsened by squatting or walking down stairs. One cause of runner's knee is constantly walking, running, or jogging on the same side of the road. Other predisposing factors include running on hills, running long distances, and being knock-kneed (a deformity in which the knees are abnormally close together).

FIGURE 8.14 **Right patella.**

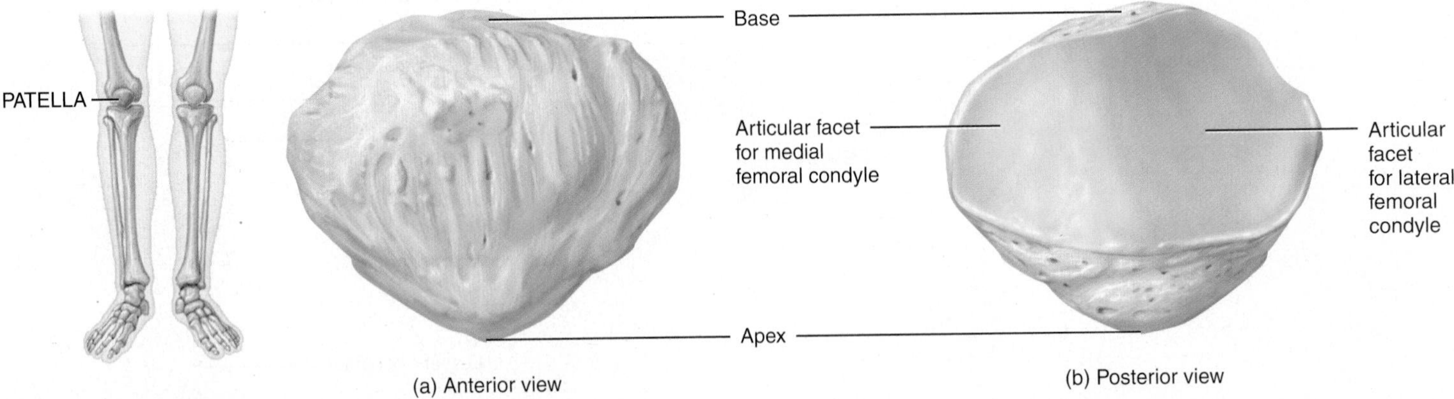


**The patella articulates with the lateral and medial condyles of the femur.**

FIGURE 8.15 Right tibia and fibula in relation to the femur, patella, and talus.

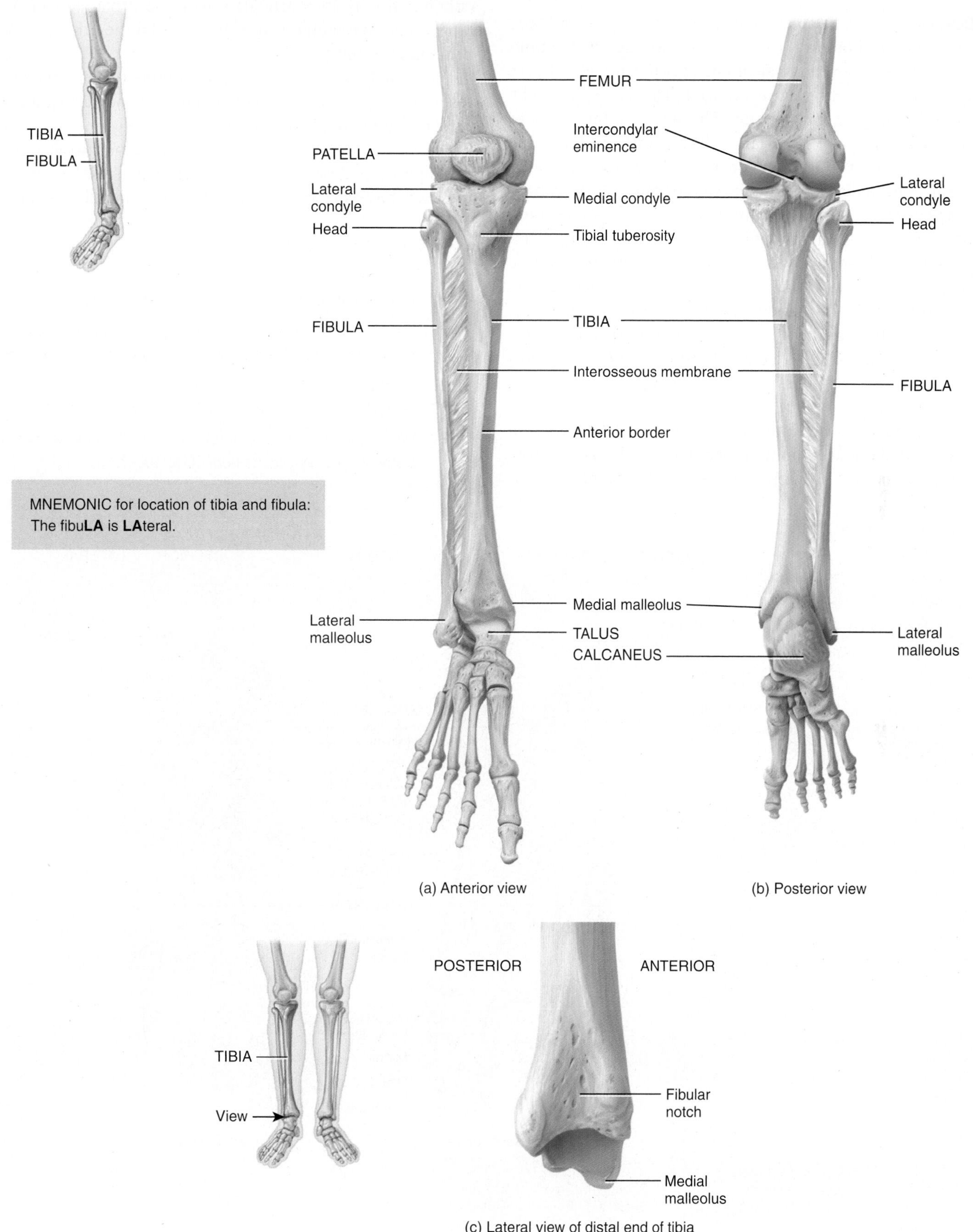


MNEMONIC for location of tibia and fibula: The fibu**LA** is **LA**teral.

(a) Anterior view

(b) Posterior view

(c) Lateral view of distal end of tibia

**The tibia articulates with the femur and fibula proximally, and with the fibula and talus distally.**

## Tarsals, Metatarsals, and Phalanges

The **tarsus** (ankle) is the proximal region of the foot and consists of seven **tarsal bones** (Figure 8.16). They include the **talus** (TĀ-lus = ankle bone) and **calcaneus** (kal-KĀ-nē-us = heel), located in the posterior part of the foot. The calcaneus is the largest and strongest tarsal bone. The anterior tarsal bones are the **navicular** (na-VIK-ū-lar = like a little boat), three **cuneiform bones** (KYOO-nē-i-form = wedge-shaped) called the **lateral, intermediate**, and **medial cuneiforms**, and the **cuboid** (KŪ-boyd = cube-shaped). A mnemonic to help you remember the names of the tarsal bones is included in Figure 8.16. The talus, the most superior tarsal bone, is the only bone of the foot that articulates with the fibula and tibia. It articulates on one side with the medial malleolus of the tibia and on the other side with the lateral malleolus of the fibula to form the *ankle joint*. As you walk, the talus transmits about half the weight of your body to the calcaneus. The remainder is transmitted to the other tarsal bones.

The **metatarsus** is the intermediate region of the foot and consists of five **metatarsal bones** numbered I to V (or 1–5) from the medial to lateral position (Figure 8.16). The metatarsals articulate proximally with the cuneiform bones and with the cuboid. Distally, they articulate with the proximal row of phalanges. The first metatarsal is thicker than the others because it bears more weight.

The **phalanges** comprise the distal component of the foot and resemble those of the hand both in number and arrangement. The toes are numbered I to V (or 1–5) beginning with the great toe, from medial to lateral. The great or big toe has two large, heavy phalanges called proximal and distal phalanges. The other four toes each have three phalanges—proximal, middle, and distal.

### ✓ CHECKPOINT

**17.** Which bone is the strongest bone in the body?

**18.** Which part or parts of the femur are attachment sites for buttock muscles? Is a lateral landmark for intramuscular injections?

**19.** The patella is classified as which type of bone? Why?

**20.** Which tarsal bone articulates with the tibia and fibula? Is the bony component of the heel of the foot?

**FIGURE 8.16 Right foot.**

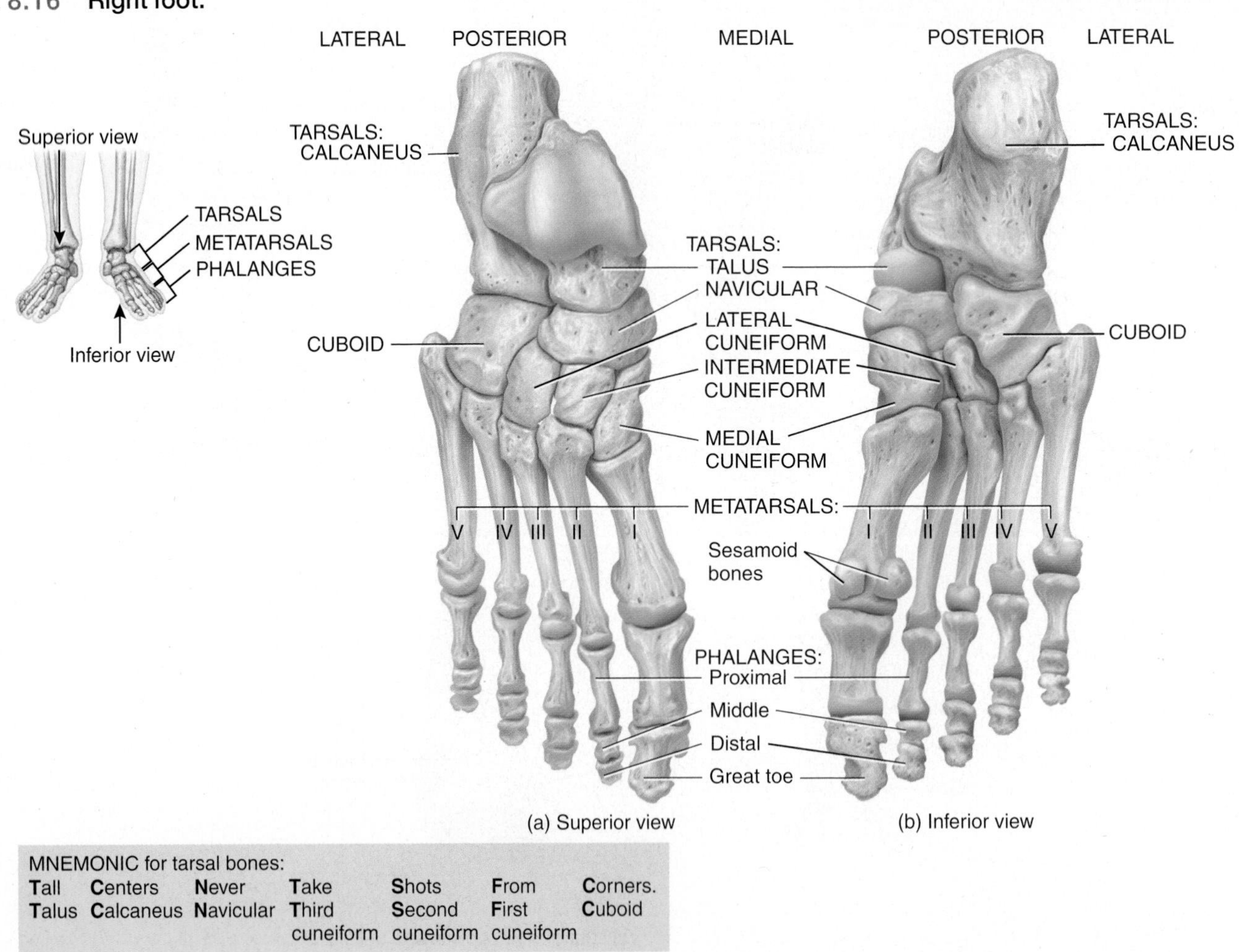


**The skeleton of the foot consists of the proximal tarsals, the intermediate metatarsals, and the distal phalanges.**

## Hassan's Story

## EPILOGUE AND DISCUSSION

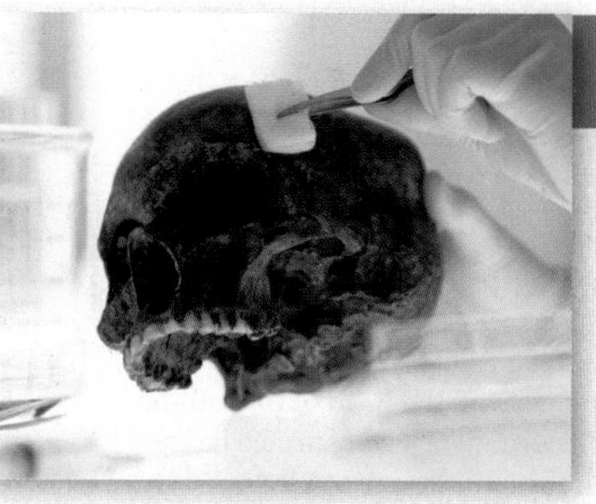

"Professor Hassan, this is Octavio."

"Hello, Octavio," Hassan replied. "What have you found?"

"Based on the teeth of the female adult, I estimate an age of 15 to 17 years. The epiphyseal plates of the phalanges of the foot had not yet closed. The lack of arthritic growths and generally youthful appearance of the calcaneus and talus would also seem to support this age. The child was probably about two to three years old at the time of death. The results from the DNA lab came back this morning. It is her child. The fractures in the child's cervical vertebrae appear to be postmortem. The bones likely were damaged after the soft tissue had decayed. I don't think her neck was intentionally broken."

Octavio continued. "There was evidence of periostitis in the princess. There are bilateral lesions on both tibia. It is likely that the tuberculosis you suspected had become a systemic infection. It's hard to say if tuberculosis killed her. I think it's more likely that these people succumbed to a virulent, infectious disease that killed them before a bony response could take place—something that spread quickly, and that affected people of all different ages. The lab results have confirmed the presence of *Mycobacterium tuberculosis* DNA. We have also found high concentrations of lead in many of the bone samples."

After a brief pause Hassan replied, "Interesting findings. They seem to fit with our translations of the hieroglyphs. The sarcophagus inscriptions indicate that this is the princess of Ishtamira and the child was her daughter. Something happened, a plague of some sort. The princess and her child were 'taken by the gods.' So many were dying that the tomb intended to be the final resting place of the king became a mass grave for his people."

"Something puzzles me, Professor. I understand that they would have needed to bury the bodies of the dead, but why the sacrifices? I didn't think that human sacrifice was endemic to this region, even that long ago."

"Difficult to say, Octavio. Many of the cooking vessels, cups, and utensils we recovered had lead in them, which is a potent neurotoxin. Lead ingestion may have caused a low level of lead poisoning. Perhaps the king went mad and ordered the sacrifices. We will never know for sure."

You have seen that the anatomy and physiology of bone reveal much about who we are. Nutritional status, activity level, exposure to certain diseases, gender, age, and many other pieces of information can be deduced from skeletal remains. In thinking about what you have learned about the skeletal system, see if you can answer the following questions:

J. *How do infants' and children's skeletons differ from those of adults? How do male and female skeletons differ overall?*

K. *Where in bone could DNA be located?*

L. *What information do the surface markings of bones reveal about an individual's activities during life?*

M. *How would the mineral content of bone reflect an individual's health status?*

N. *Octavio mentions that the phalangeal epiphyseal plates were unclosed, an observation he used to help him estimate the age of the remains. What are the epiphyseal plates, and why are they useful for determining age?*

# Concept and Resource Summary

| Concept | Resources |
|---|---|

### Introduction

1. The appendicular skeleton is composed of the upper and lower limbs, and the girdles that attach the limbs to the axial skeleton.

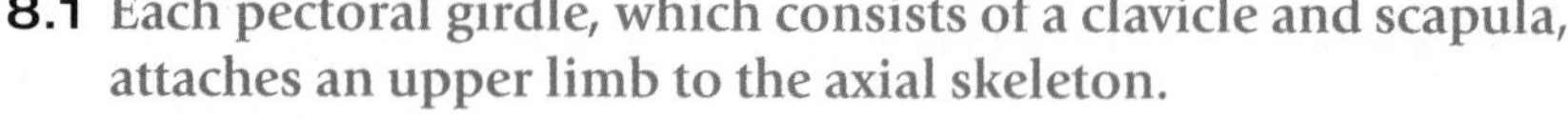

### Concept 8.1 Each pectoral girdle, which consists of a clavicle and scapula, attaches an upper limb to the axial skeleton.

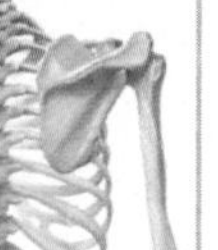

1. Each of the **pectoral girdles** is composed of a clavicle and scapula and attaches an upper limb to the axial skeleton.
2. Each **clavicle**, or collarbone, articulates medially with the sternum and laterally with the scapula.
3. Each **scapula** has an acromion that articulates with a clavicle. The scapula's glenoid cavity articulates with the head of the humerus, and the large surfaces of each scapula serve as attachment points for shoulder muscles.

**Resources:**

Anatomy Overview—The Appendicular Skeleton—The Pectoral Girdle

Figure 8.1—Right Pectoral Girdle

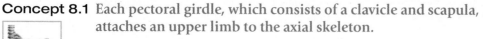

Clinical Connection—Fractured Clavicle

| Concept | Resources  |
|---|---|

## Concept 8.2 The bones of each upper limb include the humerus, ulna, radius, carpals, metacarpals, and phalanges.

1. The 30 bones of each **upper limb** include the following: a humerus in the arm; an ulna and radius in the forearm; 8 carpals in the wrist; 5 metacarpals in the palm; and 14 phalanges in the fingers.
2. The **humerus** articulates proximally by its head with the glenoid cavity of the scapula. Distally, at the elbow joint the capitulum of the humerus articulates with the head of the radius, and the trochlea of the humerus articulates with the ulna. The distal posterior surface has an olecranon fossa that receives the olecranon of the ulna when the elbow is extended.
3. The forearm bones, the medial **ulna** and the lateral **radius**, articulate proximally (at the elbow) with the humerus; distally, the radius articulates with three of the carpals. The ulna's olecranon forms the prominence of the elbow.
4. The **carpus** contains eight **carpals**. The concave space formed by several carpals plus the flexor retinaculum is the **carpal tunnel**. Tendons of muscles to the digits and thumb, and the median nerve, pass through this tunnel.
5. The **metacarpus**, or palm, consists of five **metacarpals** that articulate proximally with the carpals and distally with the phalanges.
6. The **phalanges** are the bones of the fingers. There are three phalanges in each finger, except the thumb, which has two phalanges.

Resources: Anatomy Overview—The Upper Limb; Figure 8.5—Right Humerus in Relation to the Scapula, Ulna, and Radius

## Concept 8.3 The pelvic girdle supports the vertebral column and pelvic viscera and attaches the lower limbs to the axial skeleton.

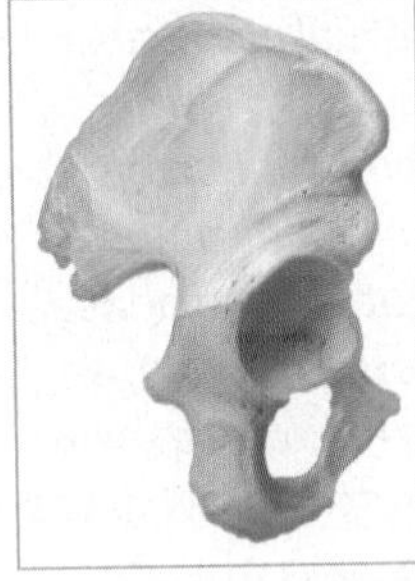

1. The **pelvic girdle** is composed of two **hip bones** that unite anteriorly at the **pubic symphysis** and articulate posteriorly with the sacrum. The hip bones, pubic symphysis, and sacrum form a basinlike structure called the **pelvis**, which gives strong support for the vertebral column, and protects lower abdominal and pelvic viscera. The pelvic girdle connects the lower limbs to the axial skeleton. Each hip bone is composed of three fused bones: the **ilium**, **ischium**, and **pubis**.
2. The ilium, the largest of the three components of the hip bone, comprises a superior ala and an inferior body.
3. The ischium, the inferior, posterior portion of the hip bone, comprises a superior body and an inferior ramus, which joins to the pubis.
4. The parts of the pubis include the superior ramus, inferior ramus, and body.
5. The point where these three bones fuse on the lateral surface of each hip bone is a socket, called the acetabulum, which accepts the head of the femur.
6. The pelvis is divided by the pelvic brim into a superior **false pelvis** and an inferior **true pelvis**. The pelvic brim is a ridge that runs from the sacral promontory to the arcuate lines to the superior region of the pubic symphysis.
7. The superior opening of the true pelvis, bordered by the pelvic brim, is called the pelvic inlet; its inferior opening is called the pelvic outlet.

Resources: Anatomy Overview—The Pelvic Girdle; Anatomy Overview—Cartilaginous Joints

## Concept 8.4 Male pelves are generally larger, heavier, and have more prominent markings; female pelves are generally wider and shallower.

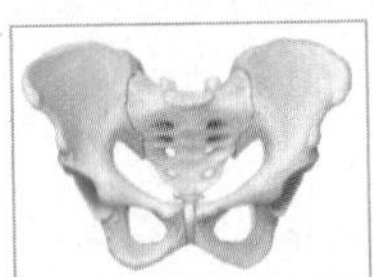

1. The male pelvis is larger, heavier, and narrow compared to the female pelvis.
2. Most structural differences in the female are to accommodate pregnancy and childbirth. The female pelvis is wide and shallow with more space in the true pelvis (see Table 8.1).

Resources: Clinical Connection—Pelvimetry

## Concept

## Resources

**Concept 8.5** The bones of each lower limb include the femur, patella, tibia, fibula, tarsals, metatarsals, and phalanges.

1. The 30 bones of each **lower limb** include the femur in the thigh, the patella, the tibia and fibula in the leg, 7 tarsals in the ankle, 5 metatarsals in the foot, and 14 phalanges in the toes.
2. The **femur** is the longest, strongest bone in the body. Its head articulates with the acetabulum and its distal end articulates with the tibia and patella.
3. The **patella** is a sesamoid bone, enclosed in the tendon of the quadriceps, and located anterior to the knee joint.
4. The **tibia** is the more medial and weight-bearing leg bone. It articulates proximally with the femur to form the knee joint, and articulates distally with the talus of the ankle.
5. The **fibula** is lateral and parallel to the tibia. Its proximal end articulates with the tibia below the level of the knee joint, and the fibula articulates distally with the talus of the ankle.
6. The **tarsus** consists of seven tarsal bones.
7. There are five **metatarsal bones** of the **metatarsus** of the foot.
8. There are fourteen **phalanges**, three in each toe except for the big toe, which has two phalanges.

Anatomy Overview—The Lower Limb

Figure 8.13—Right Femur in Relation to the Hip Bone, Patella, Tibia, and Fibula

Exercise—Bones, Bones, Bones

Clinical Connection—Hip Fracture

Clinical Connection—Fractures of the Metatarsals

Clinical Connection—Flatfoot and Clawfoot

Clinical Connection—Bone Grafting

Clinical Connection—Patellofemoral Stress Syndrome

Clinical Connection—Rickets and Osteomalacia

## Understanding the Concepts

1. Why is the acromion used as a landmark in obtaining accurate measurements of sleeve length?
2. The skeletal component projecting from the medial side of our elbows is commonly called the funny bone because, when we accidentally hit it, we feel a strange, tingling pain due to stimulation of the ulnar nerve. Which part of which bone is the funny bone? With what other bone or bones does it articulate?
3. When raising your arm to ask a question in class, your deltoid muscle lifts your arm by pulling which part of the humerus? When flexing your elbow to raise a cup to your mouth, your biceps brachii muscle lifts your forearm by pulling which part of the radius?
4. How many carpals do you have in each wrist? Name each one.
5. On the way to class you slip on ice, landing hard on your right buttock. Which part of which hip bone is likely to take the brunt of the fall?
6. How would you distinguish between the false and true pelves? Between the pelvic inlet and pelvic outlet?
7. Some bicycle seats are designed for gender comfort; seats for female riders have two depressions that are farther apart than seats for male riders. Why?
8. Which part of which leg bone is injured when you accidentally bang your shin on a low table? Which of the two leg bones bears the weight of the body?
9. The part of which bone forms the bony prominences found on the medial surface of the ankle? On the lateral surface of the ankle?

CHAPTER

# 9

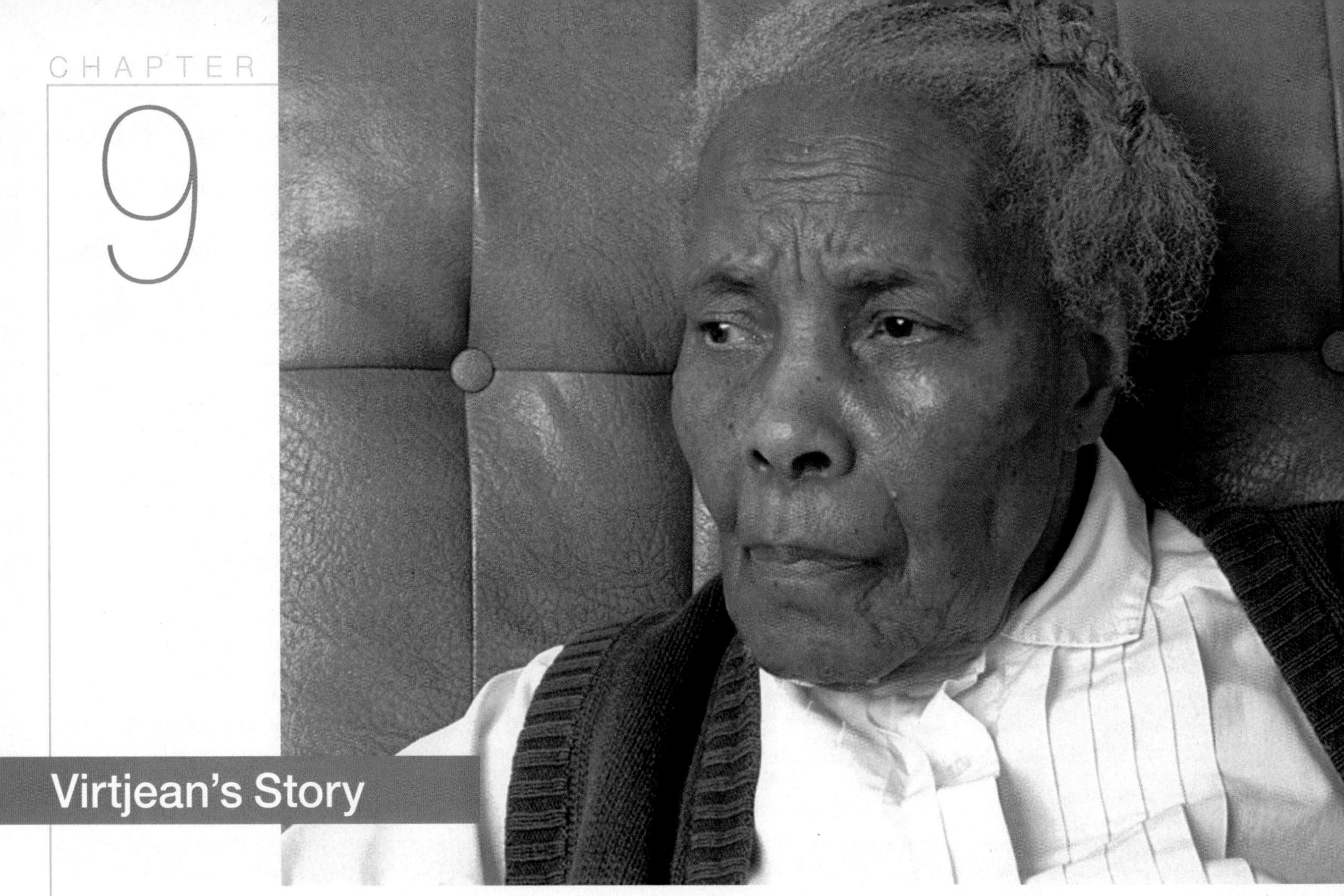

## Virtjean's Story

The alarm went off at 5 A.M. Virtjean didn't hear it; her hearing wasn't that great anymore. But the flashing of the red digital clock's large numbers blinked in her face insistently. As she reached for the alarm clock, her hand, gnarled from arthritis, glowed red in the light of the clock's display.

Normally, Virtjean didn't need an alarm clock to wake her up. She had always been an early riser. "Early to bed and early to rise. Isn't that how the saying goes?" she would joke with her granddaughter. But today she was having visitors and wanted extra time to get ready. The occupational therapist at her new "home" said Virtjean could use the kitchen facility to make special goodies for her great-grandchildren. Virtjean knew her way around the kitchen. Seven children and a lifetime on a farm had made it a requirement as the matriarch of her clan.

But first she would have to get up. Every morning it was hard. She had often remarked to her friends that it was now just a part of life. Virtjean could no longer remember a time when her joints did not ache. She had first started noticing intermittent signs of arthritis when she was in her fifties. Now, at age 84, her hands were almost unusable. But she had learned to adapt and she was not going to let anything stop her today. She didn't get visits very often and she wanted this time to be special.

Virtjean has rheumatoid arthritis, a painful, chronic joint disease. It is an autoimmune disorder in which the immune system attacks the cartilage and linings of freely movable joints (synovial joints). Disorders that affect joints can limit our mobility and reduce our quality of life. As we will see in this chapter, proper joint functioning is important for movement of the body and manipulation of our environment.

# Articulations

## INTRODUCTION

Bones are too rigid to bend without being damaged. Fortunately, flexible connective tissues form joints that hold bones together while still permitting, in most cases, some degree of movement. A **joint**, also called an **articulation** (ar-tik-ū-LĀ-shun), is a point of contact between two bones, between bone and cartilage, or between bone and teeth. When we say one bone *articulates* with another bone, we mean that the bones form a joint. Because most movements of the body occur at joints, you can appreciate their importance if you imagine how a cast over your knee joint makes walking difficult, or how a splint on a finger limits your ability to manipulate small objects. In this chapter we will explore the different types of joints, their structures, and the way in which joints allow the myriad movements that the human body can produce.

## CONCEPTS

**9.1** Joints are classified structurally and functionally.

**9.2** Fibrous joints lack a synovial cavity and are held together by dense connective tissue.

**9.3** Cartilaginous joints lack a synovial cavity and are held together by cartilage.

**9.4** Articulating surfaces of bones at a synovial joint are covered with articular cartilage and enclosed within a synovial cavity.

**9.5** Synovial joint movement terminology indicates the direction of movement or the relationships of body parts during movement.

**9.6** Synovial joints are described as plane, hinge, pivot, condyloid, saddle, or ball-and-socket.

**9.7** The shoulder, elbow, hip, and knee joints provide examples of synovial joint components, classifications, and movements.

## 9.1 Joints are classified structurally and functionally.

Joints are classified structurally based on their anatomical characteristics and functionally based on the type of movement they permit.

The structural classification of joints is based on (1) the presence or absence of a space between the articulating bones, called a synovial cavity, and (2) the type of connective tissue that holds the bones together. Structurally, joints are classified as one of the following types:

- **Fibrous joints** (FĪ-brus): The bones are held together by dense connective tissue that is rich in collagen fibers, and there is no synovial cavity.
- **Cartilaginous joints** (kar-ti-LAJ-i-nus): The bones are held together by cartilage, and there is no synovial cavity.
- **Synovial joints** (si-NŌ-vē-al; *syn-* = together): The bones are united by the dense connective tissue of an articular capsule and often by accessory ligaments, and there is a synovial cavity.

The functional classification of joints relates to the degree of movement they permit. Functionally, joints are classified as one of the following types:

- **Synarthrosis** (sin′-ar-THRŌ-sis): An immovable joint. The plural is *synarthroses*.
- **Amphiarthrosis** (am′-fē-ar-THRŌ-sis; *amphi-* = on both sides): A slightly movable joint. The plural is *amphiarthroses*.
- **Diarthrosis** (dī-ar-THRŌ-sis = movable joint): A freely movable joint. The plural is *diarthroses*. All diarthroses are synovial joints. They have a variety of shapes and permit several different types of movements.

The following sections present the joints of the body according to their structural classifications. As we examine the structure of each type of joint, we will also explore its functional attributes.

**✓ CHECKPOINT**

1. On what basis are joints classified?
2. Which kind of joint is united by cartilage? Which kind is immovable? Which kind is highly movable?

## 9.2 Fibrous joints lack a synovial cavity and are held together by dense connective tissue.

As previously noted, fibrous joints lack a synovial cavity, and the articulating bones are held together very closely together by dense connective tissue. Fibrous joints permit little or no movement. The three types of fibrous joints are sutures, syndesmoses, and interosseous membranes.

### Sutures

A **suture** (SOO-chur; *suturi-* = seam) is a fibrous joint composed of a thin layer of dense connective tissue. Sutures occur only between bones of the skull. An example is the coronal suture between the parietal and frontal bones (Figure 9.1a). The irregular, interlocking edges of sutures give them added strength and decrease their chance of fracturing. In older individuals, sutures are immovable (synarthroses), but in infants and children they are slightly movable (amphiarthroses) (Figure 9.1b). Sutures play important roles in shock absorption in the skull.

Sutures form as the numerous bones of the skull come in contact during development. Some sutures, although present during childhood, are completely replaced in the adult by osseous tissue across the suture line. Such a suture is called a **synostosis** (sin′-os-TŌ-sis; *os-* = bone) and is classified as a synarthrosis because it is immovable. For example, the frontal bone grows in halves that are joined together by a suture that usually completely fuses by age 6. If the suture persists beyond age 6, it is called a **frontal** or **metopic suture** (me-TŌ-pik; *metopon* = forehead).

### Syndesmoses

A **syndesmosis** (sin′-dez-MŌ-sis; *syndesmo-* = band or ligament; plural is *syndesmoses*) is a fibrous joint in which there is a greater distance between the articulating bones and dense connective tissue than in a suture. The dense connective tissue is typically arranged as a bundle (ligament), allowing the joint to permit limited movement. One example of a syndesmosis is the distal tibiofibular joint, where the anterior tibiofibular ligament connects the tibia and fibula (Figure 9.1c, left). It permits slight movement (amphiarthrosis).

Another example of a syndesmosis is a **gomphosis** (gom-FŌ-sis; *gompbo-* = bolt or nail) in which a cone-shaped peg fits into a socket. The only examples of gomphoses are the articulations between the roots of the teeth and their sockets (alveoli) in the maxillae and mandible (Figure 9.1c, right). The dense connective tissue between a tooth and its socket is the thin periodontal ligament. A gomphosis is classified functionally as a synarthrosis, an immovable joint.

FIGURE 9.1 Fibrous joints.

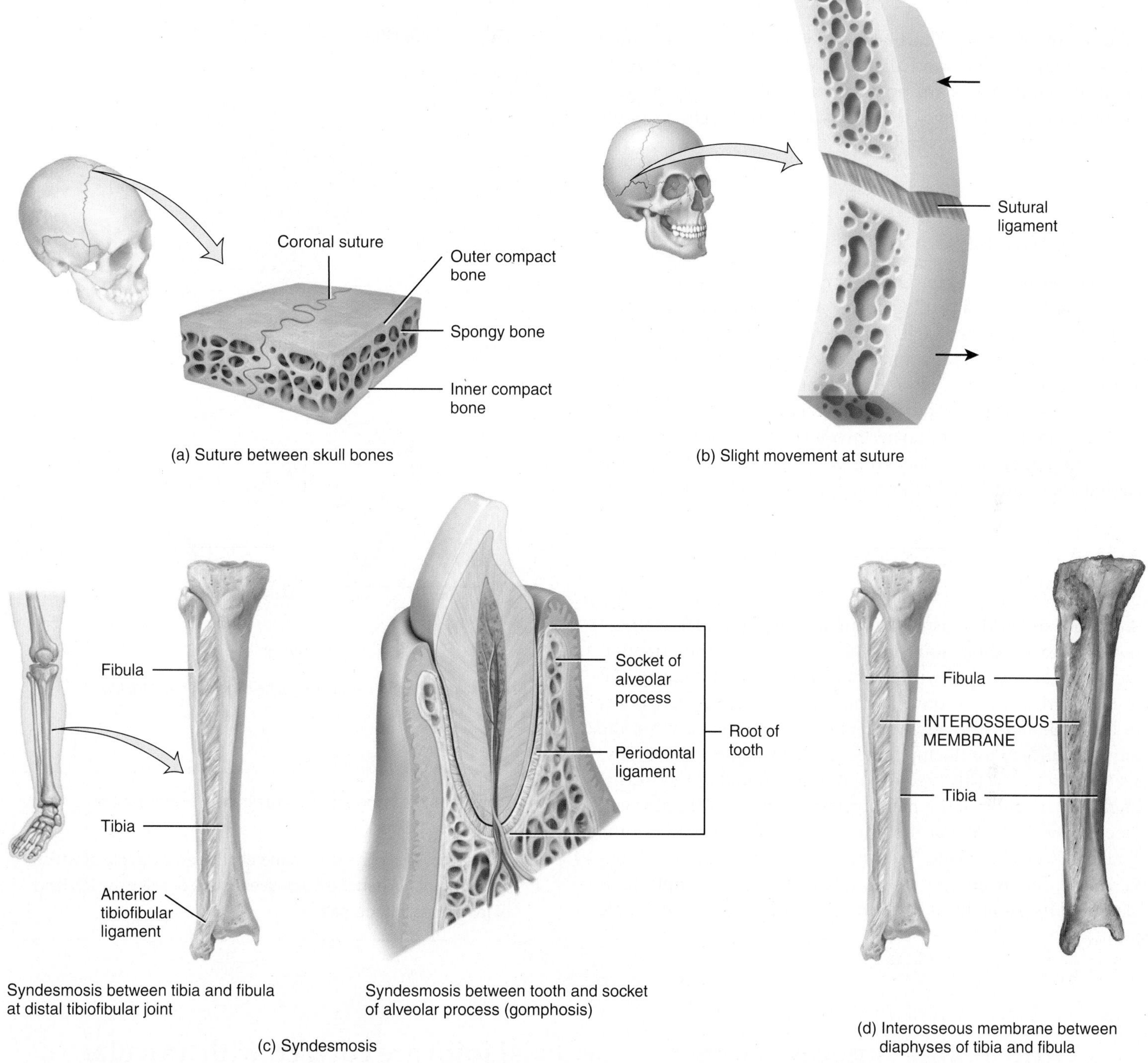


**At a fibrous joint the bones are held together by dense connective tissue.**

## Interosseous Membranes

The final category of fibrous joint is the **interosseous membrane** (in′-ter-OS-ē-us), which is a substantial sheet of dense irregular connective tissue that binds neighboring long bones and permits slight movement (amphiarthrosis). There are two principal interosseous membrane joints in the human body. One occurs between the radius and ulna in the forearm (see Figure 8.6) and the other occurs between the tibia and fibula in the leg (Figure 9.1d).

### ✓ CHECKPOINT

3. Where in the body can sutures be found? Where can gomphoses be found?

## 9.3 Cartilaginous joints lack a synovial cavity and are held together by cartilage.

Like a fibrous joint, a cartilaginous joint (kar′-ti-LAJ-i-nus) lacks a synovial cavity and allows little or no movement. Here the articulating bones are tightly connected, either by hyaline cartilage or fibrocartilage. The two types of cartilaginous joints are synchondroses and symphyses.

FIGURE 9.2 **Cartilaginous joints.**

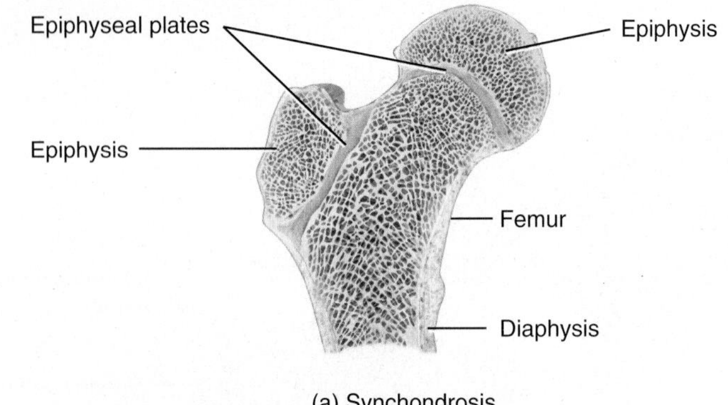


(a) Synchondrosis

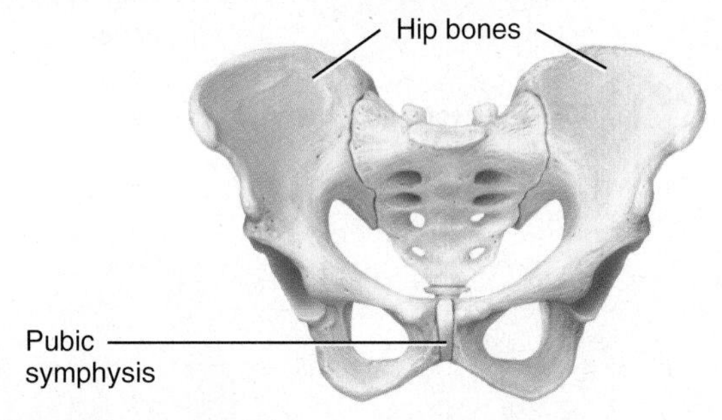


(b) Symphysis

**At a cartilaginous joint the bones are held together by cartilage.**

### Synchondroses

A **synchondrosis** (sin′-kon-DRŌ-sis; *chondro-* = cartilage; plural is *synchondroses*) is a cartilaginous joint in which the connecting material is hyaline cartilage. An example of a synchondrosis is the epiphyseal plate that connects the epiphysis and diaphysis of a growing bone (Figure 9.2a). A photomicrograph of the epiphyseal plate is shown in Figure 6.8a. Functionally, a synchondrosis is an immovable joint (synarthrosis). When bone elongation ceases, osseous tissue replaces the hyaline cartilage, and the synchondrosis becomes an immovable synostosis, a bony joint.

### Symphyses

A **symphysis** (SIM-fi-sis = growing together; plural is *symphyses*) is a cartilaginous joint in which the ends of the articulating bones are covered with hyaline cartilage, but a broad, flat disc of fibrocartilage connects the bones. All symphyses occur in the midline of the body. The pubic symphysis between the anterior surfaces of the hip bones is one example of a symphysis (Figure 9.2b). This type of joint is also found at the sternal angle, the junction of the manubrium and body of the sternum (see Figure 7.21), and at the intervertebral joints between the bodies of vertebrae (see Figure 7.19a). A portion of the intervertebral disc is composed of fibrocartilage. A symphysis is classified functionally as an amphiarthrosis, a slightly movable joint.

✓ **CHECKPOINT**

4. How does the connective tissue in synchondroses and symphyses differ?
5. The epiphyseal plate in a young child is an example of which type of joint? When the epiphyseal plate ossifies, which kind of joint does it become?

## 9.4 Articulating surfaces of bones at a synovial joint are covered with articular cartilage and enclosed within an articular (synovial) cavity.

Synovial joints (si-NŌ-vē-al) have certain characteristics that distinguish them from other joints. The unique characteristic of a synovial joint is the presence of a space called a **synovial (joint) cavity** between the articulating bones (Figure 9.3). Because the synovial cavity allows considerable movement at a joint, all synovial joints are classified functionally as freely movable (diarthroses). The bones at a synovial joint are covered by a layer of hyaline cartilage called **articular cartilage** that provides a smooth, slippery surface that reduces friction between bones in the joint during movement and helps to absorb shock.

### Articular Capsule

A sleevelike **articular (joint) capsule** surrounds the synovial joint, encloses the synovial cavity, and unites the articulating bones. The articular capsule is composed of two layers, an outer

FIGURE 9.3 **Structure of a typical synovial joint.** Note the two layers of the articular capsule—the fibrous membrane and the synovial membrane.

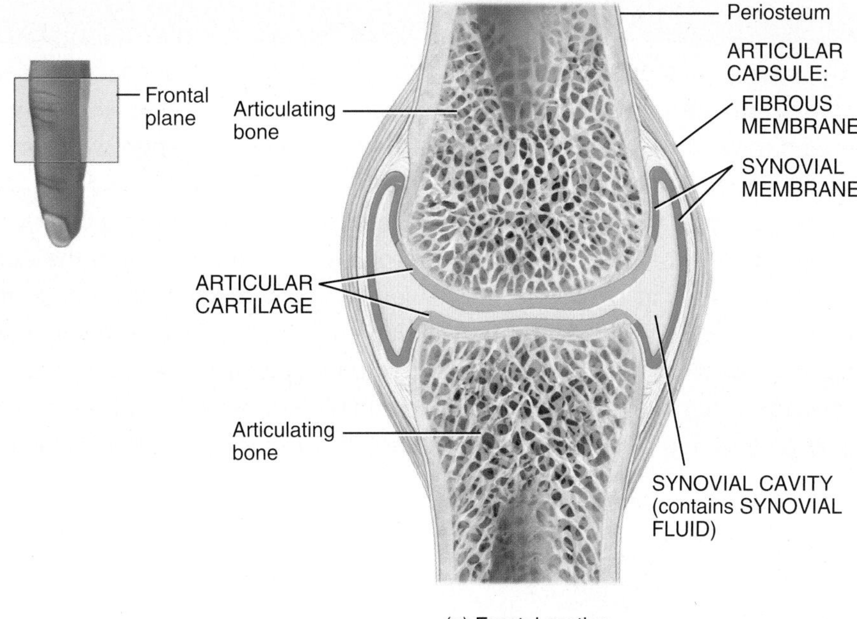


(a) Frontal section

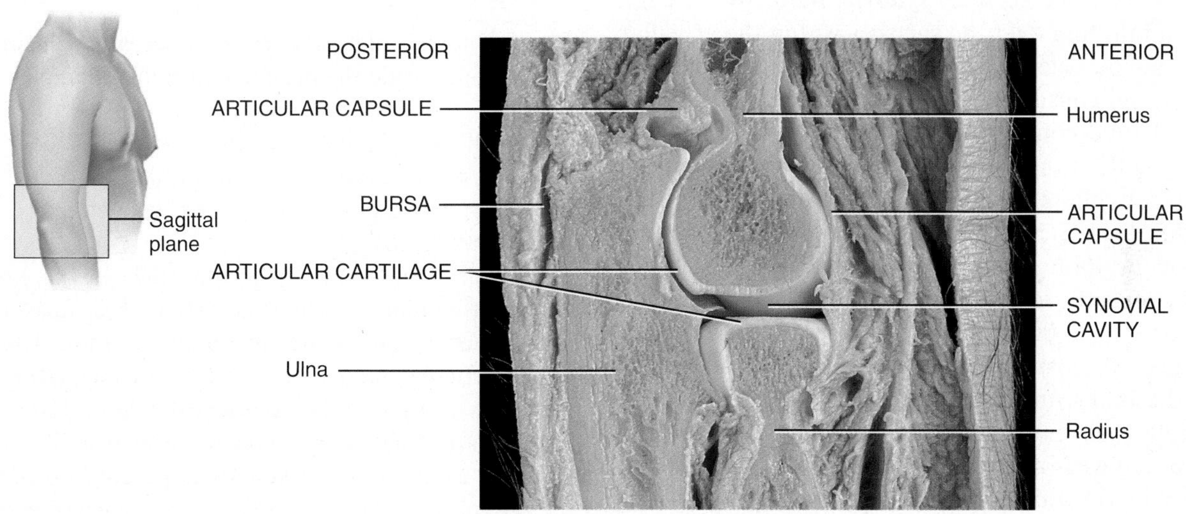


(b) Sagittal section of right elbow joint

**The distinguishing feature of a synovial joint is the synovial cavity between the articulating bones.**

fibrous membrane and an inner synovial membrane (Figure 9.3). The **fibrous membrane** usually consists of dense connective tissue (mostly collagen fibers) that attaches to the periosteum of the articulating bones. The flexibility of the fibrous membrane permits considerable movement at a joint while its great tensile strength (resistance to stretching) helps prevent the bones from dislocating. The fibers of some fibrous membranes are arranged in parallel bundles that are highly adapted for resisting strains. Such fiber bundles, called *ligaments* (*liga-* = bound or tied), are one of the principal mechanical factors that hold bones close together in a synovial joint. The inner layer of the articular capsule, the **synovial membrane**, is composed of areolar connective tissue. Many synovial joints include cushioning accumulations of adipose tissue called **articular fat pads**. An example is the *infrapatellar fat pad* in the knee (see Figure 9.14c).

## CLINICAL CONNECTION | *Torn Cartilage and Arthroscopy*

The tearing of menisci in the knee, commonly called **torn cartilage**, occurs often among athletes. Such damaged cartilage will begin to wear and may cause arthritis to develop unless the damaged cartilage is treated surgically. Years ago, if a patient had torn cartilage, the entire meniscus was removed by a procedure called a *meniscectomy* (men′-i-SEK-tō-mē). The problem was that over time the articular cartilage was worn away more quickly. Currently, surgeons perform a partial meniscectomy, in which only the torn segment of the meniscus is removed. Surgical repair of the torn cartilage may be assisted by **arthroscopy** (ar-THROS-kō-pē; *-scopy* = observation). This minimally invasive procedure involves examination of the interior of a joint, usually the knee, with an arthroscope, a lighted, pencil-thin instrument used for visualizing the nature and extent of damage. Arthroscopy is also used to monitor the progression of disease and the effects of therapy. The insertion of surgical instruments through other incisions also enables a physician to remove torn cartilage and repair damaged cruciate ligaments in the knee; obtain tissue samples for analysis; and perform surgery on other joints, such as the shoulder, elbow, ankle, and wrist.

## Synovial Fluid

The synovial membrane secretes **synovial fluid** (*ov-* = egg), which forms a thin film over the surfaces within the articular capsule. This viscous, clear fluid was named for its similarity in appearance and consistency to uncooked egg white (albumin). Synovial fluid consists of hyaluronic acid secreted by fibroblast-like cells in the synovial membrane and interstitial fluid filtered from blood plasma. It forms a thin film over the surfaces within the articular capsule that reduces friction by lubrication of the joint and absorbs shocks; it also supplies oxygen and nutrients to and removes carbon dioxide and metabolic wastes from the articular cartilage. (Recall that cartilage is an avascular tissue, so it does not have blood vessels to perform the latter function.) Synovial fluid also contains phagocytic cells that remove microbes and the debris that results from normal wear and tear in the joint. When a synovial joint is immobile for a time, the fluid becomes quite viscous (gel-like), but as joint movement increases, the fluid thins, becoming less viscous. One of the benefits of warming up before exercise is that it stimulates the production and secretion of synovial fluid; more fluid means less stress on the joints during exercise.

We are all familiar with the cracking sounds heard as certain joints move, or the popping sounds that arise when a person pulls on the fingers to "crack" the knuckles. According to one theory, when the synovial cavity expands, the pressure inside the synovial cavity decreases, creating a partial vacuum. This suction draws carbon dioxide and oxygen out of blood vessels in the synovial membrane, forming bubbles in the fluid. When the fingers are flexed, the bubbles burst, creating the cracking or popping sound as the gases are driven back into solution.

## Accessory Ligaments and Articular Menisci

Many synovial joints also contain **accessory ligaments** that lie outside and inside the articular capsule. Examples of accessory ligaments that lie outside the articular capsule are the fibular and tibial collateral ligaments of the knee joint (see Figure 9.14d). Examples of accessory ligaments inside the articular capsule are the anterior and posterior cruciate ligaments of the knee joint (see Figure 9.14d).

Inside some synovial joints, such as the knee, pads of fibrocartilage lie between the articular surfaces of the bones and are attached to the fibrous membrane. These pads are called **articular menisci** (me-NIS-sī or me-NIS-kī; singular is *meniscus*). Figure 9.14d depicts the lateral and medial menisci in the knee joint. The menisci usually subdivide the synovial cavity into two separate spaces. This separation can allow separate movements to occur in each space. By modifying the shape of the joint surfaces of the articulating bones, menisci allow two bones of different shapes to fit more tightly together. Menisci also help to maintain the stability of the joint and direct the flow of synovial fluid across the articular surfaces of the joint.

## Bursae and Tendon Sheaths

The various movements of the body create friction between moving parts. Saclike structures called **bursae** (BER-sē = purses; singular is *bursa*) are strategically situated to alleviate friction in some joints, such as the shoulder and knee joints (see Figures 9.11a and 9.14c). Bursae are not strictly parts of synovial

## CLINICAL CONNECTION | *Bursitis*

An acute or chronic inflammation of a bursa, called **bursitis** (bur-SĪ-tis), is usually caused by irritation from repeated, excessive exertion of a joint. The condition may also be caused by trauma, by an acute or chronic infection (including syphilis and tuberculosis), or by rheumatoid arthritis. Symptoms include pain, swelling, tenderness, and limited movement. Treatment may include oral anti-inflammatory agents and injections of cortisol-like steroids.

joints, but they do resemble joint capsules because their walls consist of connective tissue lined by a synovial membrane. They are also filled with a small amount of fluid that is similar to synovial fluid. Bursae can be located between the skin and bones, tendons and bones, muscles and bones, or ligaments and bones. The fluid-filled bursal sacs cushion the movement of these body parts against one another.

Structures called **tendon sheaths** also reduce friction at joints. These tubelike bursae wrap around tendons that experience considerable friction. Tendon sheaths are found where tendons pass through synovial cavities, such as the tendon of the biceps brachii muscle at the shoulder joint (see Figure 9.11c). Tendon sheaths are also found in the fingers and toes, where there is a great deal of movement (see Figure 11.17a).

### ✓ CHECKPOINT

6. Nutritional supplements that claim to aid joint movement by rebuilding knee cartilage are most likely referring to help with which component of the synovial joint?
7. What are the functions of synovial fluid?
8. What is the functional classification of synovial joints?
9. As you may have noticed in the chapters on the skeleton, the ends of long bones are often rounded. What structures add stability to synovial joints and compensate for this anatomical phenomenon?
10. How are bursae similar to and different from joint capsules? What is the function of bursae?

## RETURN TO Virtjean's Story

Virtjean got out of bed slowly, turning first onto her side, and then pushing herself up with her arms. As she bent forward to retrieve her slippers, she gasped in response to a sharp twinge of pain in her lower back. She paused for a moment, then reached for her slippers with a small gnarled hand. Her fingers were bent at odd angles, the knuckles enlarged and swollen.

Janey, the occupational therapist at Shady Acres, had spent quite a bit of time showing Virtjean how to use her remaining strength and mobility to manipulate her environment, such as using built-up handles on her toothbrush and eating utensils and getting dressed using a special "grabber" tool. Today Janey was going to help Virtjean in the kitchen so she could make cookies for her great-grandchildren.

Virtjean walked down the hall, passing several open doors. "Good morning, Gertrude," she called as she passed by.

"Good morning, Virj. Need any help in the kitchen?" Gertrude asked.

"I wouldn't mind the company."

Virtjean proceeded to get out the spoons and measuring cups, and Gertrude retrieved the flour and sugar from the cabinets. Virtjean was able to open the utensil drawer by hooking her middle and index fingers through the handle. She grasped a wooden spoon with a large-diameter foam handle, and struggled to pull apart a set of measuring cups, finally spilling them onto the counter with a clatter.

"I just can't get these old hands to cooperate," Virtjean sputtered.

Gertrude put an arm around her. "Honey, you're going to make cookies for your great-grandchildren today, arthritis or not."

A. *Would the sutures in Virtjean's skull be affected by her rheumatoid arthritis? Why or why not?*

B. *What functional classification of joints is found at Virtjean's metacarpophalangeal joints (knuckles)?*

C. *What properties of synovial joints allow healthy interphalangeal joints to move freely? What damage could have occurred inside Virtjean's interphalangeal joints to reduce their mobility?*

D. *The articular capsules of Virtjean's metacarpophalangeal joints are deformed and damaged. Describe the articular capsule, and explain its role in the synovial joint.*

## 9.5 Synovial joint movement terminology indicates the direction of movement or the relationships of body parts during movement.

Anatomists, physical and occupational therapists, and kinesiologists use specific terminology to designate movements that can occur at synovial joints. These precise terms may indicate the form of motion, the direction of movement, or the relationship of one body part to another during movement. The term **range of motion** (**ROM**) refers to the range, measured in degrees in a circle, through which the bones of a joint can be moved. Movements at synovial joints are grouped into four main categories: (1) gliding, (2) angular movements, (3) rotation, and (4) special movements that occur only at certain joints.

### Gliding

**Gliding** is a simple movement in which nearly flat bone surfaces move back and forth and from side to side with respect to one another (Figure 9.4). There is no significant alteration of the angle between the bones. Gliding movements are limited in range due to the structure of the articular capsule and associated ligaments and bones.

### Angular Movements

In **angular movements**, there is an increase or a decrease in the angle between articulating bones. The principal angular movements are flexion, extension, lateral flexion, hyperextension, abduction, adduction, and circumduction. These movements are always discussed with respect to the body in the anatomical position (see Figure 1.5).

FIGURE 9.4 **Gliding movements at synovial joints.**

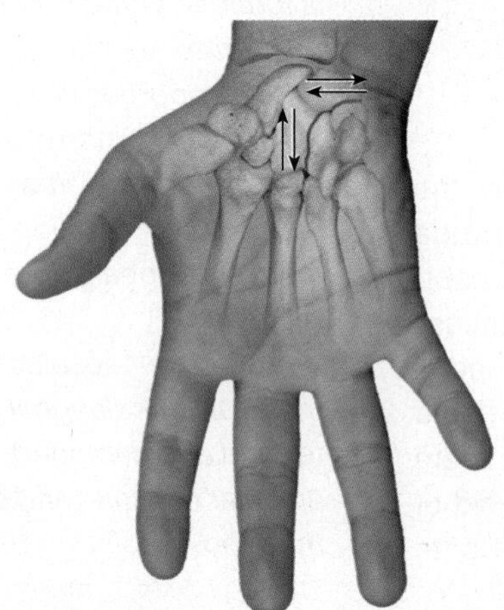

Gliding between intercarpals (arrows)

**Gliding motion consists of side to side and back and forth movements.**

#### Flexion, Extension, Lateral Flexion, and Hyperextension

Flexion and extension are opposite movements. In **flexion** (FLEK-shun; *flex-* = to bend) there is a decrease in the angle between articulating bones; in **extension** (eks-TEN-shun; *exten-* = to stretch out) there is an increase in the angle between articulating bones, often to restore a part of the body to the anatomical position after it has been flexed (Figure 9.5). All of the following are examples of flexion (as you have probably already guessed, extension is simply the reverse of these movements):

- Bending the head toward the chest at the atlanto-occipital joint between the atlas (the first vertebra) and the occipital bone of the skull, and at the intervertebral joints between the cervical vertebrae (Figure 9.5a).
- Bending the trunk forward at the intervertebral joints, as occurs when bending over to touch your toes.
- Moving the humerus forward at the shoulder joint, as in swinging the arms forward while walking (Figure 9.5b).
- Moving the forearm toward the arm at the elbow joint between the humerus, ulna, and radius (Figure 9.5c).
- Moving the palm toward the forearm at the wrist joint between the radius and carpals, as in preparing to shoot a basketball (Figure 9.5d).
- Bending at the interphalangeal joints between phalanges, as when making a fist of the fingers or when curling the toes.
- Moving the femur forward at the hip joint between the femur and hip bone, as in walking (Figure 9.5e).
- Moving the leg toward the thigh at the knee joint between the tibia, femur, and patella, as occurs when bending the knee (Figure 9.5f).

Flexion and extension usually occur along the sagittal plane, although there are a few exceptions. An example is movement of the trunk sideways to the right or left at the waist as in a side bend. This movement, which occurs along the frontal plane and involves the intervertebral joints, is called **lateral flexion** (Figure 9.5g).

The movement that returns body parts to the anatomical position is extension (Figure 9.5a–f). Continuation of extension beyond the anatomical position is called **hyperextension** (hī-per-ek-STEN-shun; *hyper-* = beyond or excessive). Examples of hyperextension include the following:

- Bending the head backward at the atlanto-occipital and cervical intervertebral joints (Figure 9.5a).
- Bending the trunk backward at the intervertebral joints, as in a back bend.

FIGURE 9.5 Angular movements at synovial joints—flexion, extension, hyperextension, and lateral flexion.

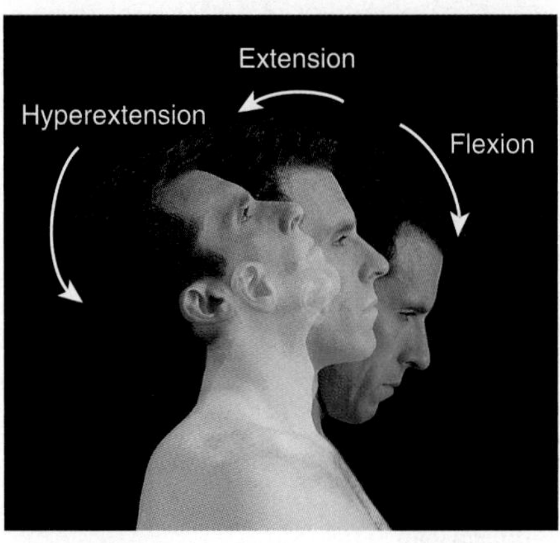


(a) Atlanto-occipital and cervical intervertebral joints

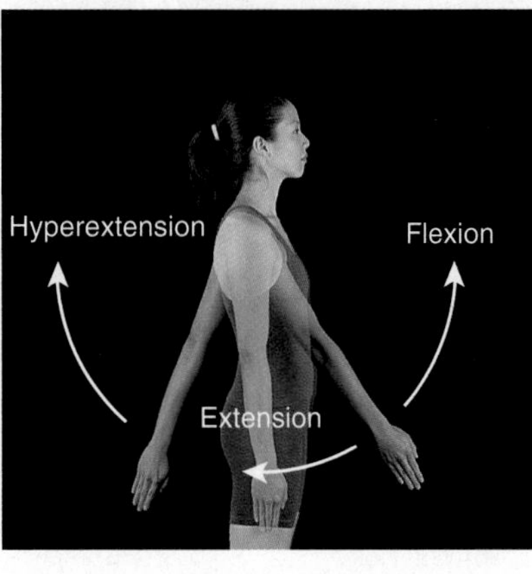


(b) Shoulder joint

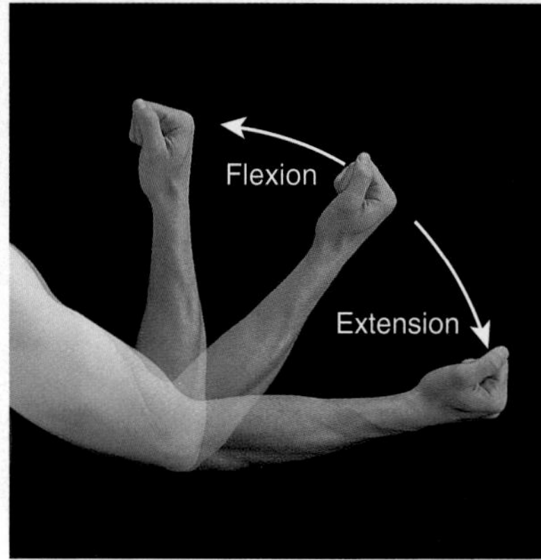


(c) Elbow joint

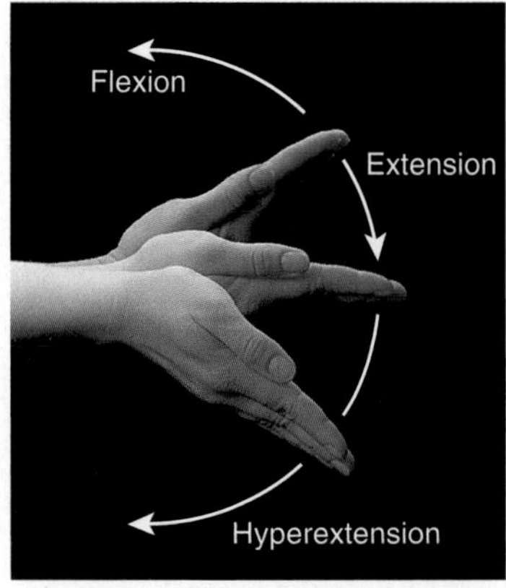


(d) Wrist joint

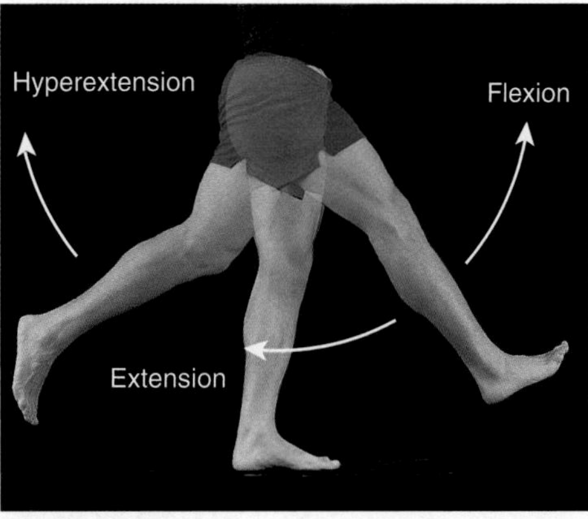


(e) Hip joint

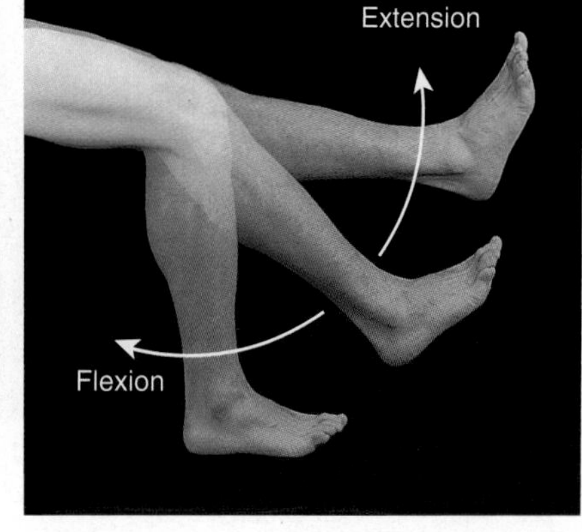


(f) Knee joint

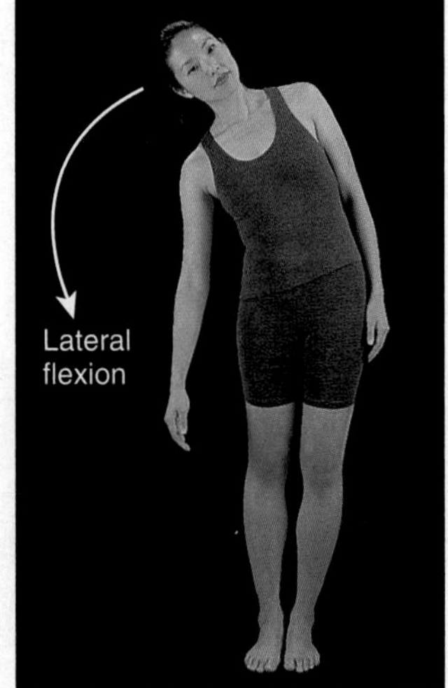


(g) Intervertebral joints

In angular movements, there is an increase or decrease in the angle between articulating bones.

- Moving the humerus backward at the shoulder joint, as in swinging the arms backward while walking (Figure 9.5b).
- Moving the palm backward at the wrist joint (Figure 9.5d).
- Moving the femur backward at the hip joint, as in walking (Figure 9.5e).

Hyperextension of the elbow, interphalangeal, and knee joints is usually prevented by the arrangement of ligaments and the anatomical alignment of the bones.

## Abduction, Adduction, and Circumduction

**Abduction** (ab-DUK-shun; *ab-* = away; *-duct-* = to lead) is the movement of a bone away from the midline; **adduction** (ad-DUK-shun; *ad-* = toward) is the movement of a bone toward the midline. Both movements usually occur along the frontal plane. Examples of abduction include moving the humerus laterally at the shoulder joint, moving the palm laterally (away from the body) at the wrist joint, and moving the femur laterally at the hip

FIGURE 9.6 **Angular movements at synovial joints—abduction and adduction.**

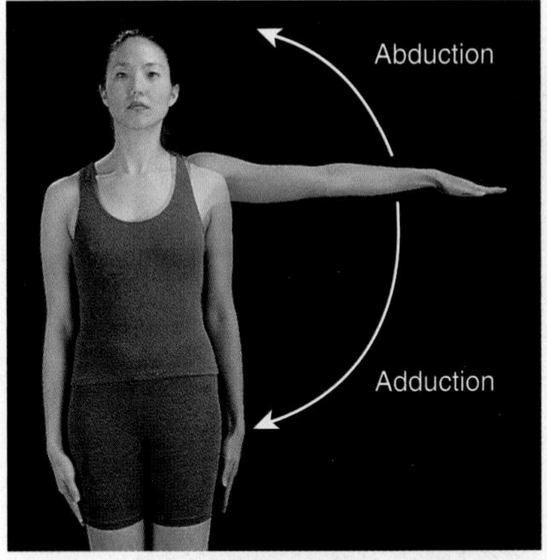


(a) Shoulder joint

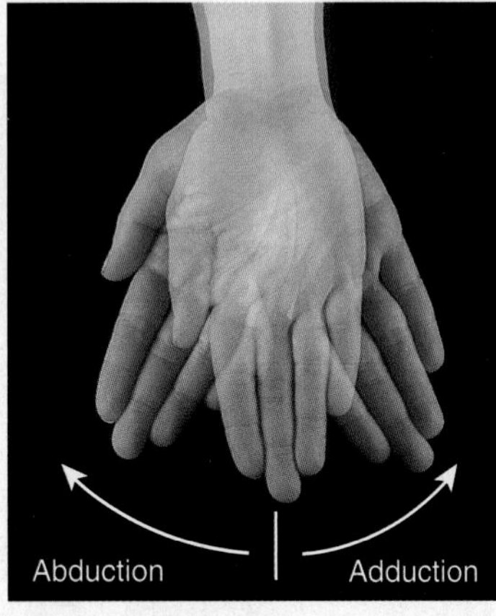


(b) Wrist joint

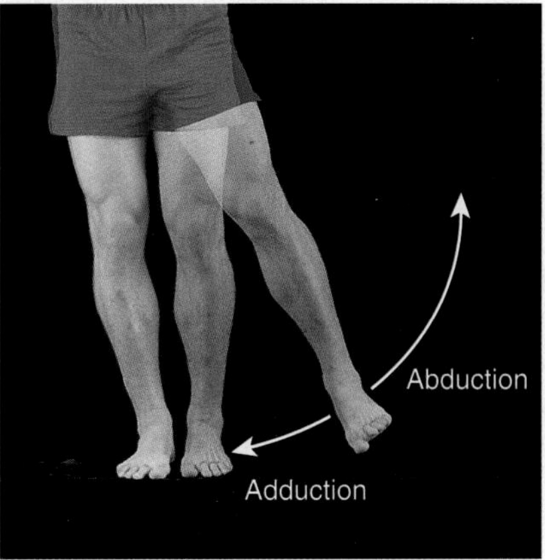


(c) Hip joint

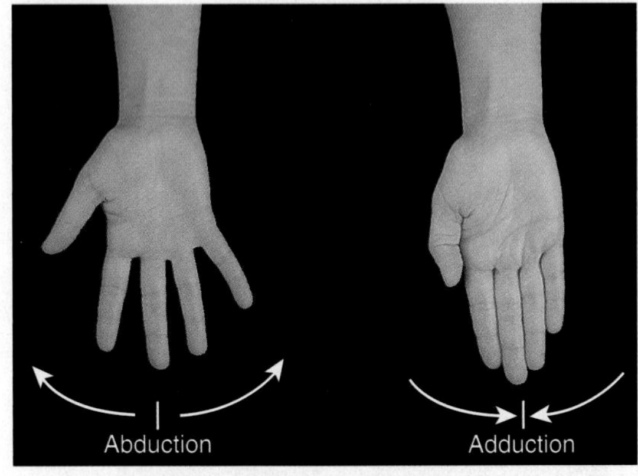


(d) Metacarpophalangeal joints of fingers (not thumb)

**Abduction and adduction usually occur along the frontal plane.**

joint (Figure 9.6a–c). The movement that returns each of these body parts to the anatomical position is adduction (Figure 9.6a–c).

The midline of the body is *not* used as a point of reference for abduction and adduction of the fingers. In abduction of the fingers, an imaginary line is drawn through the longitudinal axis of the middle (longest) finger, and the fingers move away (spread out) from the middle finger (Figure 9.6d). Abduction of the toes is relative to an imaginary line drawn through the second toe. Adduction of the fingers and toes involves returning them to the anatomical position.

**Circumduction** (ser-kum-DUK-shun; *circ-* = circle) is movement of the distal end of a body part in a circle (Figure 9.7). Circumduction is not an isolated movement by itself but rather a continuous sequence of flexion, abduction, extension, and adduction. Examples of circumduction are moving the humerus in a circle at the shoulder joint (Figure 9.7a), moving the hand in a circle at the wrist joint, moving the thumb in a circle at the carpometacarpal joint (between the carpal and metacarpal), moving a finger in a circle at the metacarpophalangeal joint (between the metacarpal and proximal phalanx), and moving the femur in a circle at the hip joint (Figure 9.7b). Both the shoulder and hip joints permit circumduction, but circumduction is more limited at the hip joint due to greater tension on the ligaments and muscles and the depth of the acetabulum in the hip joint (see Figures 9.11 and 9.13).

## Rotation

In **rotation** (rō-TĀ-shun; *rota-* = revolve), a bone revolves around its own longitudinal axis. One example is turning the head from side to side at the atlanto-axial joint (between the atlas and axis), as when you shake your head "no" (Figure 9.8a). Another is turning

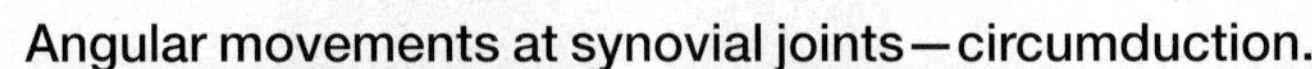

FIGURE 9.7 **Angular movements at synovial joints—circumduction.**

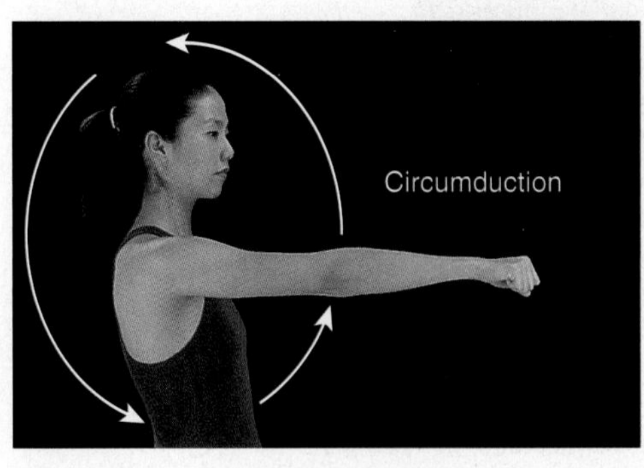


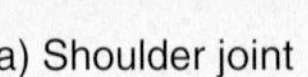

(a) Shoulder joint

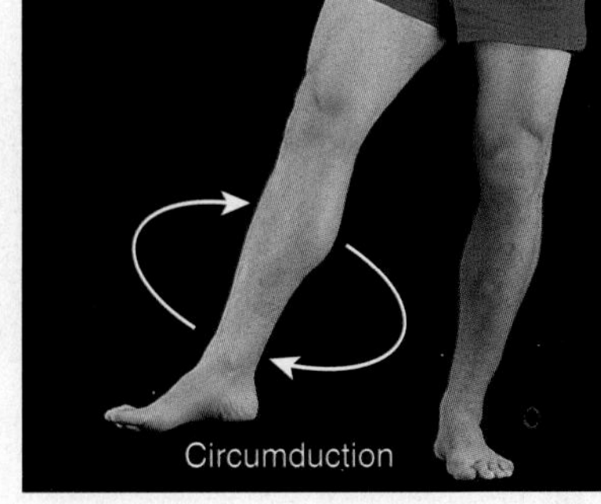


(b) Hip joint

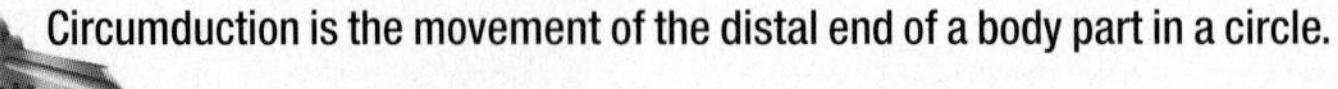

**Circumduction is the movement of the distal end of a body part in a circle.**

FIGURE 9.8 **Rotation at synovial joints.**

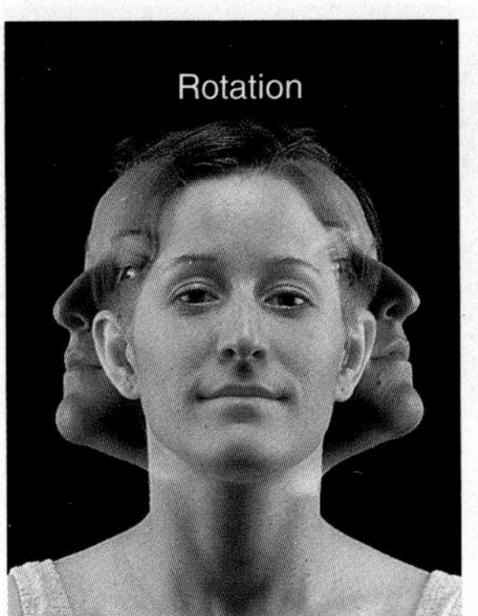


(a) Atlanto-axial joint

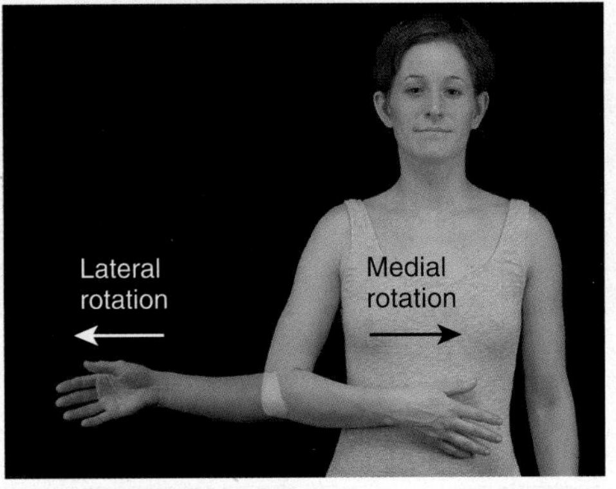


(b) Shoulder joint

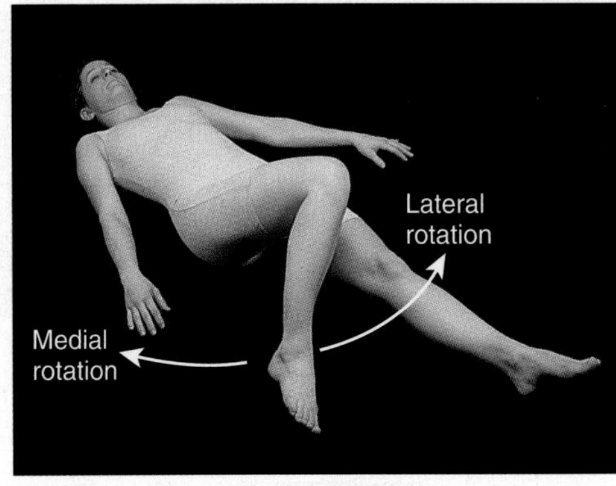


(c) Hip joint

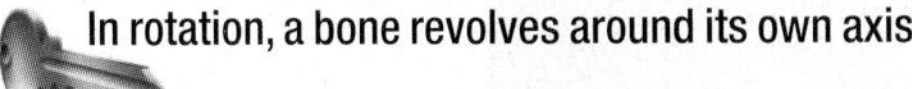
**In rotation, a bone revolves around its own axis.**

the trunk from side to side at the intervertebral joints while keeping the hips and lower limbs in the anatomical position.

In the limbs, rotation is defined relative to the midline, and specific qualifying terms are used. If the anterior surface of a bone of the limb is turned toward the midline, the movement is called *medial rotation*. You can medially rotate the humerus at the shoulder joint as follows: Starting in the anatomical position, flex your elbow and then move your palm across the chest (Figure 9.8b). You can medially rotate the femur at the hip joint as follows: Lie on your back, bend your knee, and then move your leg and foot laterally from the midline. Although you are moving your leg and foot laterally, the femur is rotating medially (Figure 9.8c). If the anterior surface of the bone of a limb is turned away from the midline, the movement is called *lateral rotation* (Figure 9.8b, c).

## Special Movements

**Special movements** occur only at certain joints. They include elevation, depression, protraction, retraction, inversion, eversion, dorsiflexion, plantar flexion, supination, pronation, and opposition (Figure 9.9):

- **Elevation** (el-e-VĀ-shun = to lift up) is an upward movement of a part of the body, such as closing the mouth at the temporomandibular joint (between the mandible and temporal bone) to elevate the mandible (Figure 9.9a) or shrugging the shoulders at the acromioclavicular joint (between the acromion of the scapula and the clavicle) to elevate the scapula. Its opposing movement is depression.
- **Depression** (de-PRESH-un = to press down) is a downward movement of a part of the body, such as opening the mouth to depress the mandible (Figure 9.9b) or returning shrugged shoulders to the anatomical position to depress the scapula.
- **Protraction** (prō-TRAK-shun = to draw forth) is a movement of a part of the body anteriorly in the transverse plane. You can protract your mandible at the temporomandibular joint by thrusting it outward (Figure 9.9c). Its opposing movement is retraction.
- **Retraction** (rē-TRAK-shun = to draw back) is a movement of a protracted part of the body back to the anatomical position (Figure 9.9d).
- **Inversion** (in-VER-zhun = to turn inward) is movement of the sole of the foot medially at the intertarsal joints (between the tarsals) so that it faces the other foot (Figure 9.9e). Its opposing movement is eversion. Physical therapists also refer to inversion of the feet as *supination*.
- **Eversion** (ē-VER-zhun = to turn outward) is a movement of the sole laterally at the intertarsal joints so that it faces away from the other foot (Figure 9.9f). Physical therapists also refer to eversion of the feet as *pronation*.
- **Dorsiflexion** (dor-si-FLEK-shun) refers to bending of the foot at the ankle in the direction of the dorsum (superior surface) (Figure 9.9g). Dorsiflexion occurs when you stand on your heels. Its opposing movement is plantar flexion.
- **Plantar flexion** (PLAN-tar) involves bending of the foot at the ankle joint in the direction of the plantar or inferior surface (Figure 9.9g), as when you elevate your body by standing on your toes.

**FIGURE 9.9 Special movements at synovial joints.**

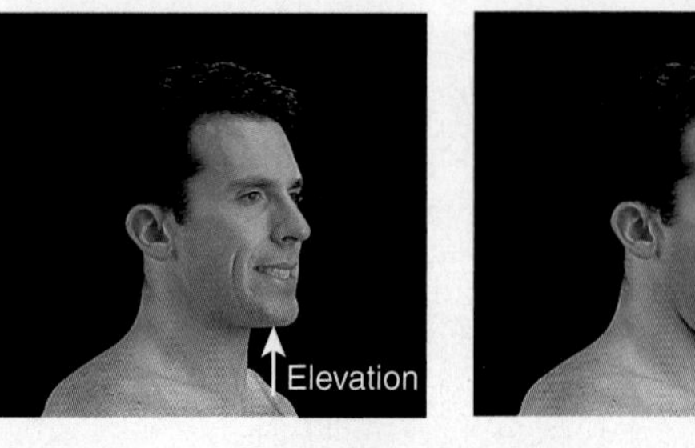


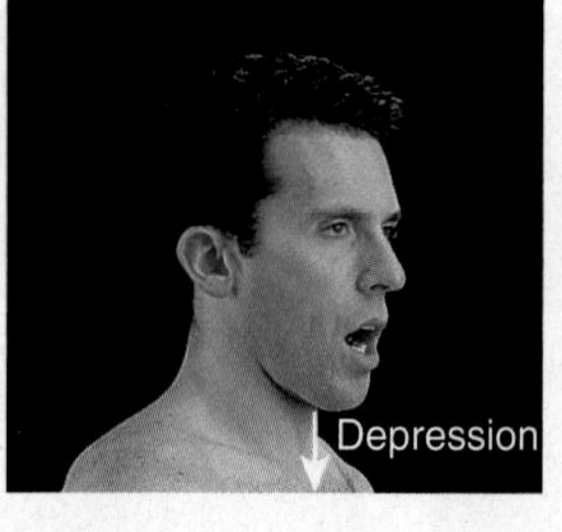


(a) Temporomandibular joint (b)

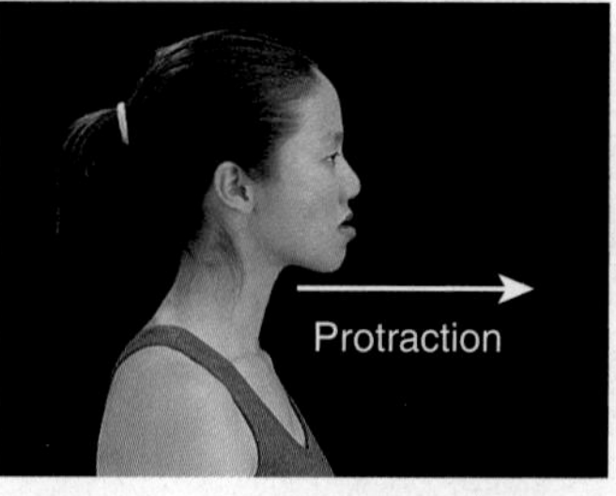


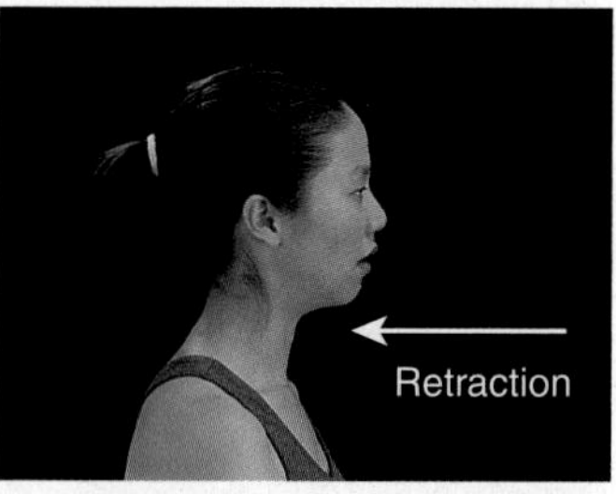

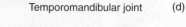

(c) Temporomandibular joint (d)

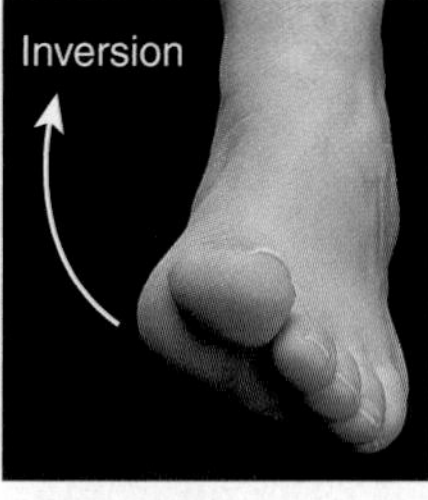


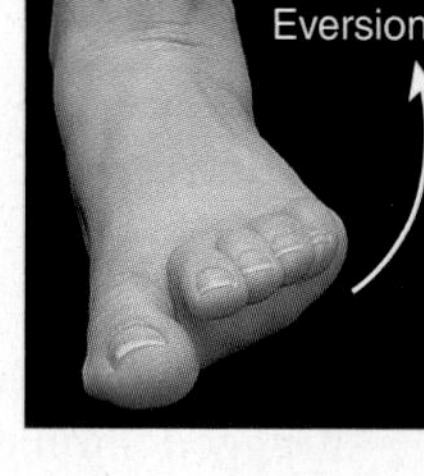


(e) Intertarsal joints (f)

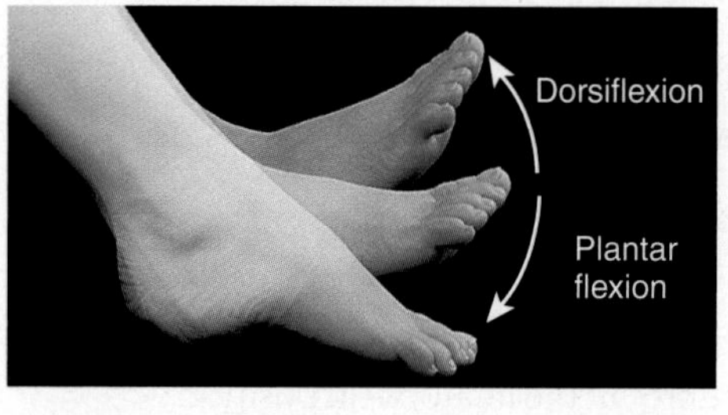


(g) Ankle joint

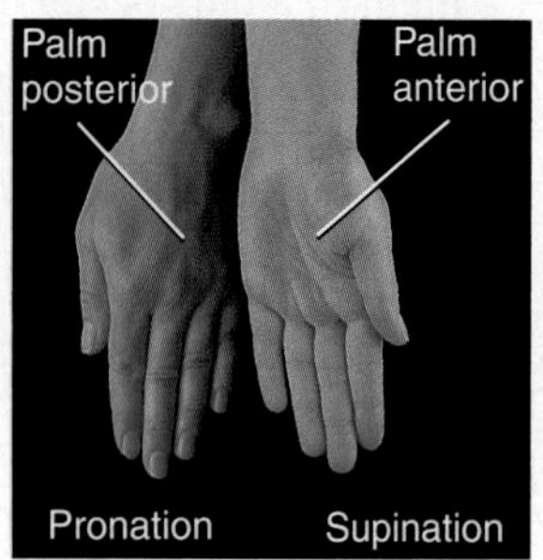


(h) Radioulnar joint

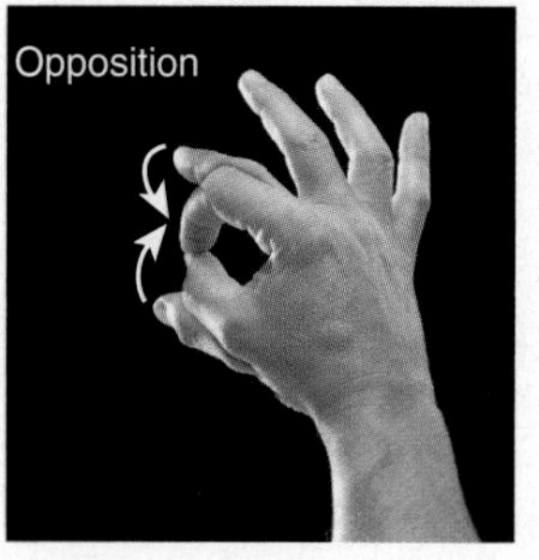


(i) Carpometacarpal joint

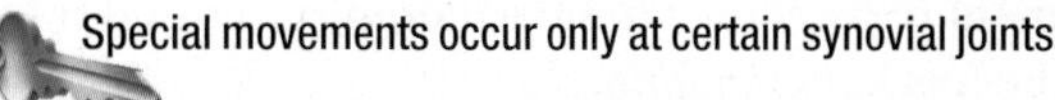

**Special movements occur only at certain synovial joints.**

- **Supination** (soo-pi-NĀ-shun) is a movement of the forearm at the proximal and distal radioulnar joints in which the palm is turned anteriorly, or forward (Figure 9.9h). This position of the palms is one of the defining features of the anatomical position. Its opposing movement is pronation.
- **Pronation** (prō-NĀ-shun) is a movement of the forearm so that the distal end of the radius crosses over the distal end of the ulna and the palm is turned posteriorly (Figure 9.9h).
- **Opposition** (op-ō-ZISH-un) is the movement of the thumb at the carpometacarpal joint (between the trapezium and metacarpal of the thumb) in which the thumb moves across the palm to touch the tips of the fingers on the same hand (Figure 9.9i). These "opposable thumbs" allow the distinctive hand movement that gives humans and other primates the ability to grasp and manipulate objects very precisely.

A summary of the movements that occur at synovial joints is presented in Table 9.1.

### ✓ CHECKPOINT

**11.** Which type of angular movement describes what is happening at a joint when someone:

- **a.** Straightens the knees to move from a squatting position to a standing position?
- **b.** Draws a big circle during art class?
- **c.** Uses the hip joint to move the lower limb so that the toes point laterally?
- **d.** Brings the other fingers toward the middle finger to slip his hand through the sleeve of a coat?
- **e.** Closes her mouth?

**12.** In what way is considering adduction as "adding your limb to your trunk" an effective learning device?

**13.** A young bird is trying to fly. Which two movements is it producing as it flaps its wings up and down?

**14.** A common test of dexterity is to have someone touch, as fast as he can, his fingertips, one at a time, to the thumb on the same hand. Which type of movement is this?

**TABLE 9.1**

Movements at Synovial Joints

| MOVEMENT | DESCRIPTION |
|---|---|
| **Gliding** | Movement of relatively flat bone surfaces back and forth and side to side over one another; little change in angle between bones |
| **Angular** | Increase or decrease in angle between bones |
| Flexion | Decrease in angle between articulating bones, usually in sagittal plane |
| Lateral flexion | Movement of trunk in frontal plane |
| Extension | Increase in angle between articulating bones, usually in sagittal plane |
| Hyperextension | Extension beyond anatomical position |
| Abduction | Movement of bone away from midline, usually in frontal plane |
| Adduction | Movement of bone toward midline, usually in frontal plane |
| Circumduction | Flexion, abduction, extension, adduction, and rotation in succession (or in the opposite order); distal end of body part moves in circle |
| **Rotation** | Movement of bone around longitudinal axis; in limbs, may be medial (toward midline) or lateral (away from midline) |
| **Special** | Occurs at specific joints |
| Elevation | Superior movement of body part |
| Depression | Inferior movement of body part |
| Protraction | Anterior movement of body part in transverse plane |
| Retraction | Posterior movement of body part in transverse plane |
| Inversion | Medial movement of sole |
| Eversion | Lateral movement of sole |
| Dorsiflexion | Bending foot in direction of dorsum (superior surface) |
| Plantar flexion | Bending foot in direction of plantar surface (sole) |
| Supination | Movement of forearm that turns palm anteriorly |
| Pronation | Movement of forearm that turns palm posteriorly |
| Opposition | Movement of thumb across palm to touch fingertips on same hand |

## 9.6 Synovial joints are described as plane, hinge, pivot, condyloid, saddle, or ball-and-socket.

Although all synovial joints are similar in structure, the shapes of the articulating surfaces vary, allowing a variety of movements. Synovial joints are divided into six categories based on type of movement: plane, hinge, pivot, condyloid, saddle, and ball-and-socket.

### Plane Joints

The articulating surfaces of bones in a **plane joint** (PLĀN) are flat or slightly curved (Figure 9.10a). Plane joints primarily permit side to side and back and forth gliding movements between the flat surfaces of bones, but they may also rotate against one another. Examples of plane joints are the intercarpal joints (between carpal bones at the wrist), intertarsal joints (between tarsal bones at the ankle), sternoclavicular joints (between the manubrium of the sternum and the clavicle), acromioclavicular joints (between the acromion of the scapula and the clavicle), and vertebrocostal joints (between the ribs and transverse processes of thoracic vertebrae).

### Hinge Joints

In a **hinge joint**, the convex surface of one bone fits into the concave surface of another bone (Figure 9.10b). As the name implies, hinge joints produce an angular, opening-and-closing motion like

**FIGURE 9.10 Types of synovial joints.** For each type, a drawing of the actual joint and a simplified diagram are shown.

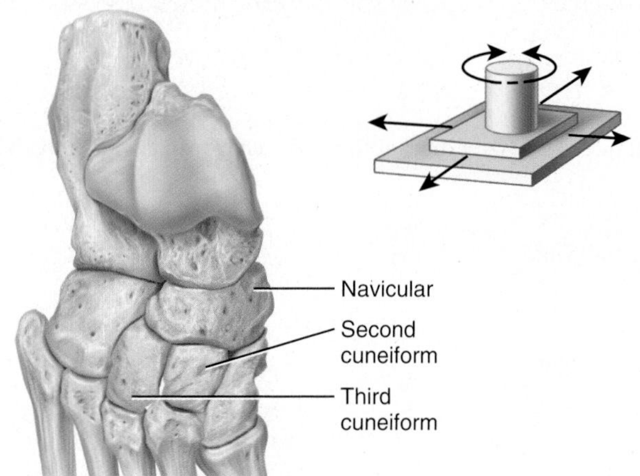


(a) Plane joint between navicular and second and third cuneiforms of tarsus in foot

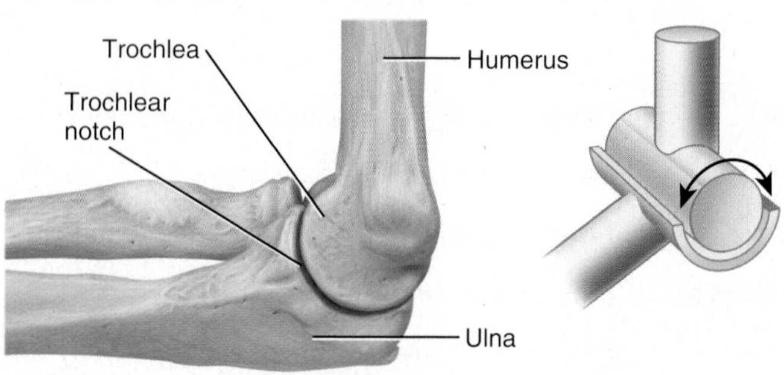


(b) Hinge joint between trochlea of humerus and trochlear notch of ulna at elbow

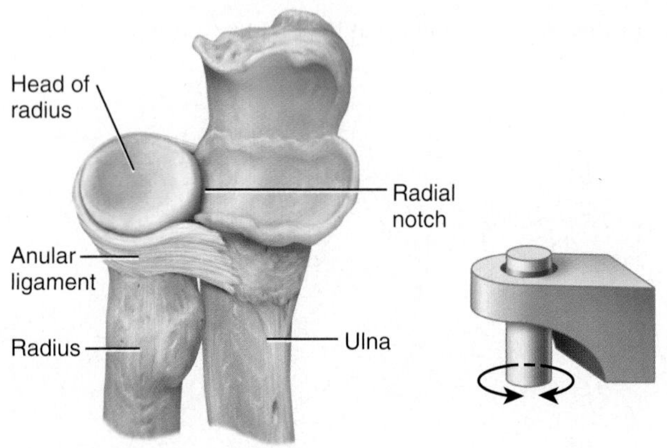


(c) Pivot joint between head of radius and radial notch of ulna

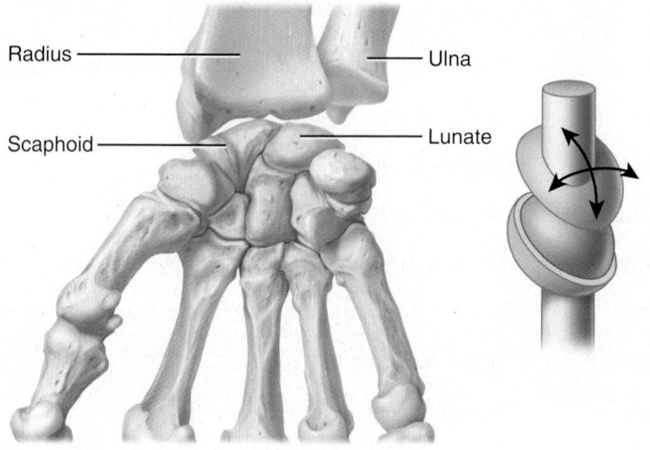


(d) Condyloid joint between radius and scaphoid and lunate bones of wrist

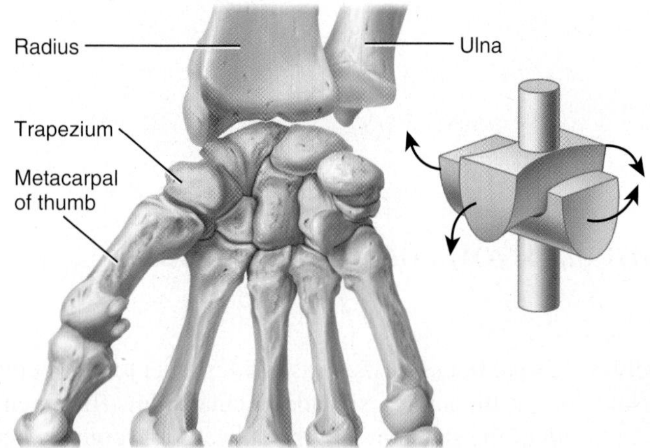


(e) Saddle joint between trapezium of wrist and metacarpal of thumb

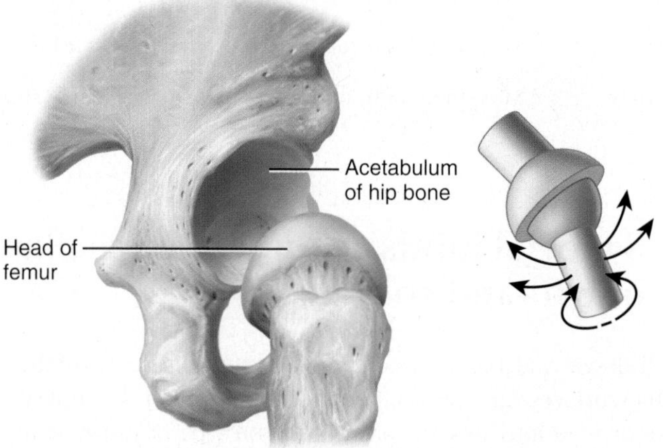


(f) Ball-and-socket joint between head of femur and acetabulum of hip bone

**Synovial joints are classified on the basis of the shapes of the articulating bone surfaces.**

that of a hinged door. In most joint movements, one bone remains in a fixed position while the other moves around an axis. An *axis* is a straight line around which a bone moves. Hinge joints allow motion along a single axis. Hinge joints permit only flexion and extension. Examples of hinge joints are the knee, elbow, ankle, and interphalangeal joints (between the phalanges of the fingers and toes).

## Pivot Joints

In a **pivot joint**, the rounded or pointed surface of one bone articulates with a ring formed partly by another bone and partly by a ligament (Figure 9.10c). A pivot joint allows rotation only around its own axis. Examples of pivot joints are the atlanto-axial joint, in which the atlas rotates around the axis and permits the head to turn from side to side as when you shake your head "no" (see Figure 9.8a), and the radioulnar joints that enable the palms to turn anteriorly and posteriorly as the head of the radius pivots around its axis in the radial notch of the ulna (see Figure 9.9h).

## Condyloid Joints

In a **condyloid joint** (KON-di-loyd; *condyl-* = knuckle), the convex oval-shaped projection of one bone fits into the oval-shaped depression of another bone (Figure 9.10d). A condyloid joint permits movement around two axes (flexion–extension and abduction–adduction), plus limited circumduction (remember that circumduction is not an isolated movement). Examples of condyloid joints are the wrist and the metacarpophalangeal joints (between the metacarpals and phalanges) of the second through fifth fingers.

## Saddle Joints

In a **saddle joint**, the articular surface of one bone is saddle-shaped (the concave surface); the articular surface of the other bone fits into the "saddle" like a rider sitting on a horse (Figure 9.10e). Saddle joints permit movement around three axes: side to side, up and down, and rotation. Like a condyloid joint, movements at a saddle joint are around two axes (flexion–extension and abduction–adduction) plus limited circumduction. An example of a saddle joint is the carpometacarpal joint between the trapezium of the wrist and metacarpal of the thumb.

## Ball-and-Socket Joints

A **ball-and-socket joint** consists of the ball-like surface of one bone fitting into a cuplike depression of another bone (Figure 9.10f). Such joints permit movement around three axes (flexion–extension, abduction–adduction, and rotation). Examples of ball-and-socket joints are the shoulder and hip joints. At the shoulder joint the head of the humerus fits into the glenoid cavity of the scapula. At the hip joint the head of the femur fits into the acetabulum of the hip bone.

Table 9.2 summarizes the structural and functional categories of joints.

### ✓ CHECKPOINT

15. Skip is asked by his instructor if he knows what a pivot joint is and he vigorously shakes his head side-to-side, indicating "no." His instructor responds "Good job. Nice example." Skip is confused—why is the instructor happy?
16. Why are condyloid and saddle joints important to many musicians?
17. Which types of synovial joints allow movement around one axis, two axes, three axes, no axes?

**TABLE 9.2**

Structural and Functional Classifications of Joints

| STRUCTURAL CLASSIFICATION | DESCRIPTION | FUNCTIONAL CLASSIFICATION | EXAMPLE |
|---|---|---|---|
| **FIBROUS** No Synovial Cavity; Articulating Bones Held Together by Fibrous Connective Tissue | | | |
| **Suture** | Articulating bones united by thin layer of dense irregular connective tissue, found between skull bones; with age, some sutures replaced by synostosis (separate cranial bones fuse into single bone) | Synarthrosis (immovable) and amphiarthrosis (slightly movable) | Coronal suture |
| **Syndesmosis** | Articulating bones united by more dense irregular connective tissue, usually a ligament | Amphiarthrosis (slightly movable) | Distal tibiofibular joint |
| **Interosseous membrane** | Articulating bones united by substantial sheet of dense irregular connective tissue | Amphiarthrosis (slightly movable) | Between tibia and fibula |
| **CARTILAGINOUS** No Synovial Cavity; Articulating Bones United by Hyaline Cartilage or Fibrocartilage | | | |
| **Synchondrosis** | Connecting material: hyaline cartilage; becomes synostosis when bone elongation ceases | Synarthrosis (immovable) | Epiphyseal plate between diaphysis and epiphysis of long bone |
| **Symphysis** | Connecting material: broad, flat disc of fibrocartilage | Amphiarthrosis (slightly movable) | Pubic symphysis and intervertebral joints |
| **SYNOVIAL** Characterized by Synovial Cavity, Articular Cartilage, and Articular Capsule; May Contain Accessory Ligaments, Articular Discs, and Bursae | | | |
| **Plane** | Articulated surfaces flat or slightly curved | Many biaxial diarthroses (freely movable): back-and-forth and side-to-side movements Some triaxial diarthroses: back-and-forth, side-to-side, rotation | Intercarpal, intertarsal, sternocostal (between sternum and second to seventh pairs of ribs), and vertebrocostal joints |
| **Hinge** | Convex surface fits into concave surface | Uniaxial diarthrosis: flexion–extension | Knee (modified hinge), elbow, ankle, and interphalangeal joints |
| **Pivot** | Rounded or pointed surface fits into ring formed partly by bone and partly by ligament | Uniaxial diarthrosis: rotation | Atlanto-axial and radioulnar joints |
| **Condyloid** | Oval-shaped projection fits into oval-shaped depression | Biaxial diarthrosis: flexion–extension, abduction–adduction | Radiocarpal and metacarpophalangeal joints |
| **Saddle** | Articular surface of one bone is saddle-shaped; articular surface of other bone "sits" in saddle | Biaxial diarthrosis: flexion–extension, abduction–adduction | Carpometacarpal joint between trapezium and metacarpal of thumb |
| **Ball-and-socket** | Ball-like surface fits into cuplike depression | Triaxial diarthrosis: flexion–extension, abduction–adduction, rotation | Shoulder and hip joints |

## RETURN TO Virtjean's Story

Janey walked into the kitchen. "I was hoping that you could show me some of the special kitchen appliances you told me about. I'm so clumsy with these old hands," Virtjean sighed.

Janey smiled. "From what I can see, it looks like you're off to a good start. There's an electric mixer for mixing your dough, and you and Gertrude can probably find most of what you need on your own. I would like to just watch how you're doing, Virtjean. That way, I can assess any difficulties, come up with a treatment plan, and think about other adaptive equipment."

Janey took some notes: *84-year-old female, history of rheumatoid arthritis, currently on pain medication (check with nursing staff on meds), has stage 3 R.A., deformity and limited mobility of fingers. Check arm strength, appears to have difficulty in extension of forearm at elbow, metacarpophalangeal joints of index and middle fingers may be subluxed (partially dislocated).*

Virtjean used a wooden spoon with a built-up handle to spoon cookie dough onto the sheet. Janey noticed that she remembered her instructions, hyperextending her hand at the wrist so that her fingers naturally flexed palmward. This simple trick of the arm's anatomy allowed someone with a weak grip like Virtjean to close their fingers to their palm.

Virtjean had trouble abducting her arms as well, so the women worked at a table that was adjusted to a lower than normal height. She stood with an awkward posture; her knees were flexed, her thighs somewhat medially rotated at the hips, and her feet everted. Janey wrote on her notepad: *New shoes to address postural concerns?*

E. *Which type of synovial joint is the metacarpophalangeal joint? Which type of movement does it normally permit that Virtjean would now have difficulty doing?*

F. *Which movements would Virtjean normally have to make in order to spoon dough from a bowl onto a cookie sheet?*

G. *When Virtjean's knees are flexed and her thighs are rotated inward at the hips, are her knees abducted or adducted?*

H. *Virtjean has trouble abducting her arms. Explain what this means in terms that she could understand.*

## 9.7 The shoulder, elbow, hip, and knee joints provide examples of synovial joint components, classifications, and movements.

In Chapters 7 and 8, we discussed the major bones and their markings. In this chapter we have examined how joints are classified according to their structure and function, and we have introduced the movements that occur at joints. Table 9.3 (selected joints of the axial skeleton) and Table 9.4 (selected joints of the appendicular skeleton) will help you integrate the information you learned about bones with the information about joint classifications and movements. These tables list some of the major joints of the body according to their articular components (the bones that enter into their formation), their structural and functional classifications, and the type(s) of movement that occur(s) at each joint.

Next we will examine four selected synovial joints in more detail. The joints described are the shoulder (glenohumeral) joint, elbow joint, hip (coxal) joint, and knee joint. Because these joints are described in detail in the text, they are not included in Tables 9.3 and 9.4. The temporomandibular joint is described in detail in *WileyPLUS* resources and is, therefore, also omitted from Table 9.3. Each discussion of a selected synovial joint is accompanied by figures that illustrate that joint. Structures that are labeled in all capital letters are specifically referred to in the joint discussion. Additional related structures are also labeled as a reference source.

**TABLE 9.3**

Selected Joints of the Axial Skeleton

| JOINT | ARTICULAR COMPONENTS | CLASSIFICATION | MOVEMENTS |
|---|---|---|---|
| **Suture** | Between skull bones | *Structural:* fibrous<br>*Functional:* amphiarthrosis and synarthrosis | None |
| **Atlanto-occipital** | Between superior articular facets of atlas and occipital condyles of occipital bone | *Structural:* synovial (condyloid)<br>*Functional:* diarthrosis | Flexion and extension of head; slight lateral flexion of head to either side |
| **Atlanto-axial** | (1) Between dens of axis and anterior arch of atlas;<br>(2) Between lateral masses of atlas and axis | *Structural:* synovial (pivot) between dens and anterior arch; synovial (planar) between lateral masses;<br>*Functional:* diarthrosis | Rotation of head |
| **Intervertebral** | (1) Between vertebral bodies<br>(2) Between vertebral arches | *Structural:* cartilaginous (symphysis) between vertebral bodies; synovial (planar) between vertebral arches<br>*Functional:* amphiarthrosis between vertebral bodies; diarthrosis between vertebral arches | Flexion, extension, lateral flexion, and rotation of vertebral column |
| **Vertebrocostal** | (1) Between facets of heads of ribs and facets of bodies of adjacent thoracic vertebrae and intervertebral discs between them<br>(2) Between articular part of tubercles of ribs and facets of transverse processes of thoracic vertebrae | *Structural:* synovial (planar)<br>*Functional:* diarthrosis | Slight gliding |
| **Sternocostal** | Between sternum and first seven pairs of ribs | *Structural:* cartilaginous (synchondrosis) between sternum and first pair of ribs; synovial (plane) between sternum and second through seventh pair of ribs<br>*Functional:* synarthrosis between sternum and first pair of ribs; diarthrosis between sternum and second through seventh pair of ribs | None between sternum and first pair of ribs; slight gliding between sternum and second through seventh pair of ribs |
| **Lumbosacral** | (1) Between body of fifth lumbar vertebra and base of sacrum<br>(2) Between inferior articular facets of fifth lumbar vertebra and superior articular facets of first vertebra of sacrum | *Structural:* cartilaginous (symphysis) between body and base; synovial (planar) between articular facets<br>*Functional:* amphiarthrosis between body and base; diarthrosis between articular facets | Flexion, extension, lateral flexion, and rotation of vertebral column |

**TABLE 9.4**

Selected Joints of the Appendicular Skeleton

| JOINT | ARTICULAR COMPONENTS | CLASSIFICATION | MOVEMENTS |
|---|---|---|---|
| **Sternoclavicular** | Between sternal end of clavicle, manubrium of sternum, and first costal cartilage | *Structural:* synovial (plane, pivot)<br>*Functional:* diarthrosis | Gliding, with limited movements in nearly every direction |
| **Acromioclavicular** | Between acromion of scapula and acromial end of clavicle | *Structural:* synovial (plane)<br>*Functional:* diarthrosis | Gliding and rotation of scapula on clavicle |
| **Radioulnar** | Proximal radioulnar joint between head of radius and radial notch of ulna; distal radioulnar joint between ulnar notch of radius and head of ulna | *Structural:* synovial (pivot)<br>*Functional:* diarthrosis | Rotation of forearm |
| **Wrist (radiocarpal)** | Between distal end of radius and scaphoid, lunate, and triquetrum of carpus | *Structural:* synovial (condyloid)<br>*Functional:* diarthrosis | Flexion, extension, abduction, adduction, circumduction, and slight hyperextension of wrist |
| **Intercarpal** | Between proximal row of carpal bones, distal row of carpal bones, and between both rows of carpal bones (midcarpal joints) | *Structural*: synovial (plane), except for hamate, scaphoid, and lunate (midcarpal) joint, which is synovial (saddle)<br>*Functional:* diarthrosis | Gliding plus flexion, extension, abduction, adduction, and slight rotation at midcarpal joints |
| **Carpometacarpal** | Carpometacarpal joint of thumb between trapezium of carpus and first metacarpal; carpometacarpal joints of remaining digits formed between carpus and second through fifth metacarpals | *Structural:* synovial (saddle) at thumb; synovial (plane) at remaining digits<br>*Functional:* diarthrosis | Flexion, extension, abduction, adduction, and circumduction at thumb; gliding at remaining digits |
| **Metacarpophalangeal and metatarsophalangeal** | Between heads of metacarpals (or metatarsals) and bases of proximal phalanges | *Structural:* synovial (condyloid)<br>*Functional:* diarthrosis | Flexion, extension, abduction, adduction, and circumduction of phalanges |
| **Interphalangeal** | Between heads of phalanges and bases of more distal phalanges | *Structural:* synovial (hinge)<br>*Functional:* diarthrosis | Flexion and extension of phalanges |
| **Sacroiliac** | Between auricular surfaces of sacrum and ilia of hip bones | *Structural:* synovial (plane)<br>*Functional:* diarthrosis | Slight gliding (even more so during pregnancy) |
| **Pubic symphysis** | Between anterior surfaces of hip bones | *Structural:* cartilaginous (symphysis)<br>*Functional:* amphiarthrosis | Slight movements (even more so during pregnancy) |
| **Tibiofibular** | Proximal tibiofibular joint between lateral condyle of tibia and head of fibula; distal tibiofibular joint between distal end of fibula and fibular notch of tibia | *Structural:* synovial (plane) at proximal joint; fibrous (syndesmosis) at distal joint<br>*Functional:* diarthrosis at proximal joint; amphiarthrosis at distal joint | Slight gliding at proximal joint; slight rotation of fibula during dorsiflexion of foot |
| **Ankle (talocrural)** | (1) Between distal end of tibia and its medial malleolus and talus (2) Between lateral malleolus of fibula and talus | *Structural:* synovial (hinge)<br>*Functional:* diarthrosis | Dorsiflexion and plantar flexion of foot |
| **Intertarsal** | Subtalar joint between talus and calcaneus of tarsus; talocalcaneonavicular joint between talus and calcaneus and navicular of tarsus; calcaneocuboid joint between calcaneus and cuboid of tarsus | *Structural:* synovial (plane) at subtalar and calcaneocuboid joints; synovial (saddle) at talocalcaneonavicular joint<br>*Functional:* diarthrosis | Inversion and eversion of foot |
| **Tarsometatarsal** | Between three cuneiforms of tarsus and bases of five metatarsal bones | *Structural:* synovial (plane)<br>*Functional:* diarthrosis | Slight gliding |

## The Shoulder Joint

The **shoulder joint** (Figure 9.11) is a ball-and-socket joint formed by the head of the humerus and the glenoid cavity of the scapula (Figure 9.11c). It also is referred to as the *glenohumeral joint.*

### Anatomical Components

The anatomical components of the shoulder joint include the following:

- *Articular capsule.* Thin, loose sac that completely envelops the joint and extends from the glenoid cavity to the anatomical neck of the humerus (Figure 9.11).
- ***Coracohumeral ligament*** (kor′-ā-kō-Ū-mer-al). Ligament that strengthens the superior part of the articular capsule and extends from the coracoid process of the scapula to the greater tubercle of the humerus (Figure 9.11a, b). The ligament strengthens the superior part of the articular capsule and reinforces the anterior aspect of the articular capsule.
- ***Glenohumeral ligaments*** (glē-nō-HŪ-mer-al). Three thickenings of the articular capsule over the anterior surface of the joint that extend from the glenoid cavity to the lesser tubercle and anatomical neck of the humerus. These ligaments are often indistinct or absent and provide only minimal strength (Figure 9.11a, b). They play a role in joint stabilization when the humerus approaches or exceeds its limits of motion.
- ***Glenoid labrum.*** Narrow rim of fibrocartilage around the edge of the glenoid cavity that slightly deepens and enlarges the socket (Figure 9.11b, c).
- ***Bursae.*** Four *bursae* help reduce friction where body parts move across the shoulder joint. They are the *subscapular bursa* (Figure 9.11a), *subdeltoid bursa, subacromial bursa* (Figure 9.11a–c), and *subcoracoid bursa.*

### Movements

The shoulder joint allows flexion, extension, abduction, adduction, medial rotation, lateral rotation, and circumduction of the arm (see Figures 9.5–9.8). It has more freedom of movement than any other joint of the body. This freedom results from the looseness of the articular capsule and shallowness of the glenoid cavity in relation to the large size of the head of the humerus. The loose fit of the shoulder joint certainly allows mobility, but at the sacrifice of stability; hence, shoulder dislocations are common.

Although the ligaments of the shoulder joint strengthen it to some extent, most of the strength results from the muscles that surround the joint, especially the *rotator cuff muscles.* These muscles (supraspinatus, infraspinatus, teres minor, and subscapularis) join the scapula to the humerus (see also Figure 11.14). The tendons of the rotator cuff muscles encircle the joint and fuse with the articular capsule. The rotator cuff muscles work as a group to hold the head of the humerus in the glenoid cavity.

**FIGURE 9.11 Right shoulder joint.**

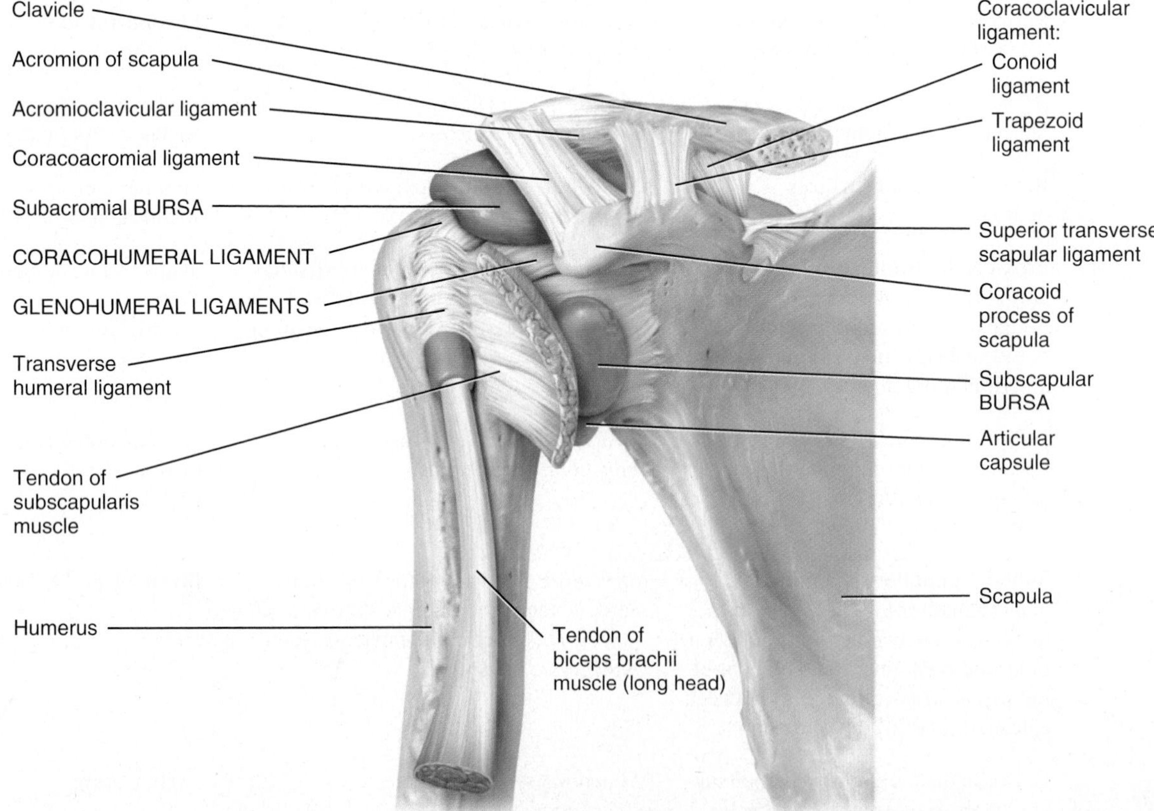


(a) Anterior view

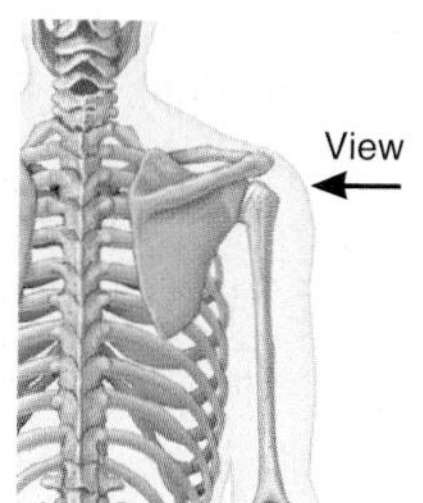


(b) Lateral view (opened)

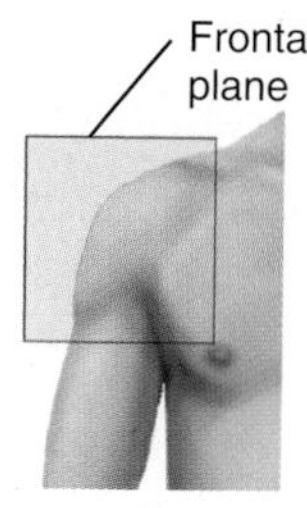


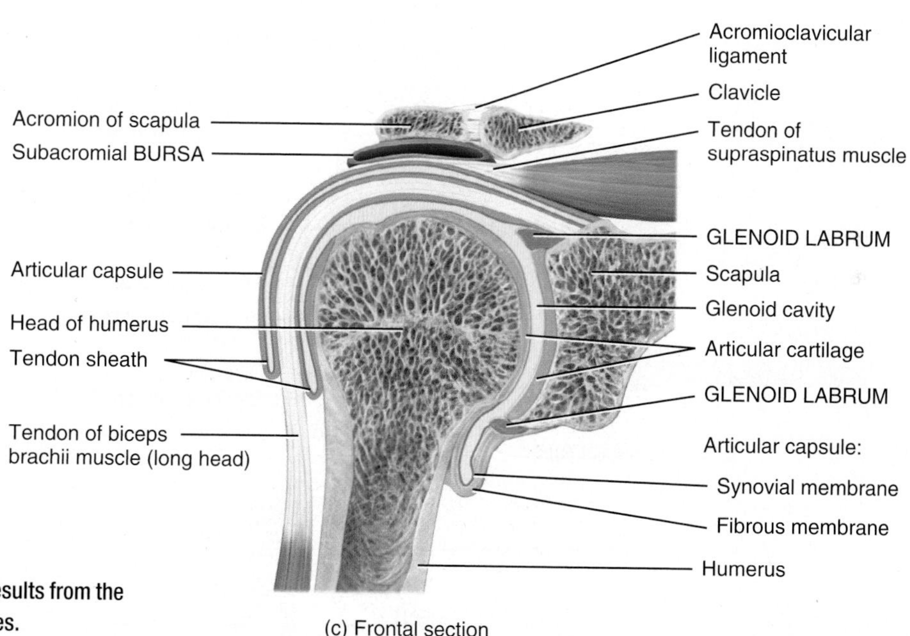


(c) Frontal section

**Most of the stability of the shoulder joints results from the arrangement of the rotator cuff muscles.**

## CLINICAL CONNECTION | *Rotator Cuff Injury, Dislocated and Separated Shoulder, and Torn Glenoid Labrum*

**Rotator cuff injury** is a strain or tear in the rotator cuff muscles and is a common injury among baseball pitchers, volleyball players, racket sports players, swimmers, and violinists, due to shoulder movements that involve vigorous circumduction. It also occurs as a result of wear and tear, aging, trauma, poor posture, improper lifting, and repetitive motions in certain jobs, such as placing items on a shelf above your head. Most often, there is tearing of the supraspinatus muscle tendon of the rotator cuff. This tendon is especially predisposed to wear and tear because of its location between the head of the humerus and acromion of the scapula, which compresses the tendon during shoulder movements. Poor posture and poor body mechanics also increase compression of the supraspinatus muscle tendon.

The joint most commonly dislocated in adults is the shoulder joint because its socket is quite shallow and the bones are held together by supporting muscles. Usually in a **dislocated shoulder**, the head of the humerus becomes displaced inferiorly, where the articular capsule is least protected. Dislocations of the mandible, elbow, fingers, knee, or hip are less common. Dislocations are treated with rest, ice, pain relievers, manual manipulation, or surgery followed by use of a sling and physical therapy.

A **separated shoulder** actually refers to an injury not to the shoulder joint but to the acromioclavicular joint, a joint formed by the acromion of the scapula and the acromial end of the clavicle. This condition is usually the result of forceful trauma to the joint, as when the shoulder strikes the ground in a fall. Treatment options are similar to those for treating a dislocated shoulder, although surgery is rarely needed.

In a **torn glenoid labrum**, the fibrocartilaginous labrum may tear away from the glenoid cavity. This causes the joint to catch or feel like it's slipping out of place. The shoulder may indeed become dislocated as a result. A torn labrum is reattached to the glenoid surgically with anchors and sutures. The repaired joint is more stable.

## The Elbow Joint

The **elbow joint** (Figure 9.12) is a hinge joint formed by the trochlea of the humerus, the trochlear notch of the ulna, and the head of the radius.

### Anatomical Components

Anatomical components of the elbow joint include the following:

- *Articular capsule*. The anterior part extends from the humerus to the ulna and the anular ligament that encircles

FIGURE 9.12 **Right elbow joint.** (See also Figure 9.3b.)

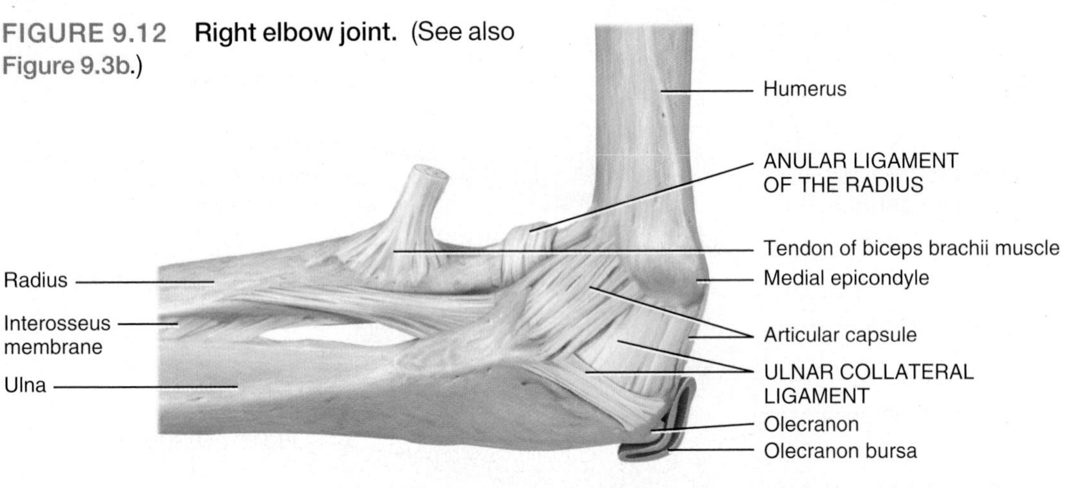


(a) Medial aspect

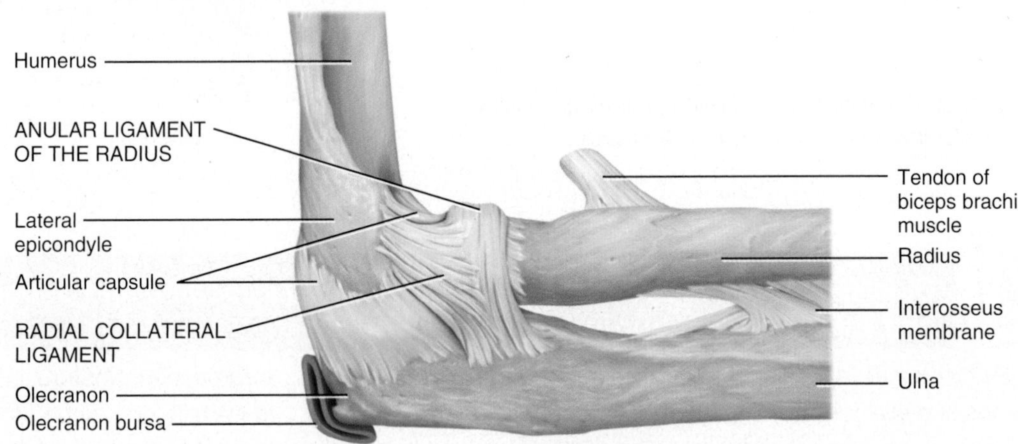


(b) Lateral aspect

The elbow joint is formed by parts of three bones: humerus, ulna, and radius.

## CLINICAL CONNECTION | *Tennis Elbow, Little-League Elbow, and Dislocation of the Radial Head*

**Tennis elbow** most commonly refers to pain at or near the lateral epicondyle of the humerus, usually caused by an improperly executed backhand. The extensor muscles strain, resulting in pain. **Little-league elbow** is an inflammation of the epiphyseal (growth) plate of the medial epicondyle as a result of overuse of the tendons of the forearm muscles and ligaments in the elbow used in throwing. It typically develops as a result of a heavy pitching schedule and/or a schedule that involves throwing curve balls, especially among youngsters. In this disorder, there is pain and the elbow may enlarge, fragment, or separate.

A **dislocation of the radial head (nursemaid's elbow)** is the most common upper limb dislocation in children. In this injury, the head of the radius slides past or ruptures the radial anular ligament, a ligament that forms a collar around the head of the radius at the proximal radioulnar joint. Dislocation is most apt to occur when a strong pull is applied to the forearm while it is extended and supinated, for instance while swinging a child around with outstretched arms.

the head of the radius. The posterior part extends from the capitulum, olecranon fossa, and lateral epicondyle of the humerus to the anular ligament of the radius, the olecranon of the ulna, and the ulna posterior to the radial notch (Figure 9.12a, b).

- ***Ulnar collateral ligament.*** Ligament that extends from the medial epicondyle of the humerus to the coronoid process and olecranon of the ulna (Figure 9.12a).
- ***Radial collateral ligament.*** Ligament that extends from the lateral epicondyle of the humerus to the anular ligament of the radius and the radial notch of the ulna (Figure 9.12b).
- ***Anular ligament of the radius.*** Strong band that encircles the head of the radius. It holds the head of the radius in the radial notch of the ulna (Figure 9.12a, b).

### Movements

The elbow joint allows flexion and extension of the forearm (see Figure 9.5c).

## The Hip Joint

The **hip joint** *(coxal joint)* (Figure 9.13) is a ball-and-socket joint formed by the head of the femur and the acetabulum of the hip bone.

FIGURE 9.13 **Right hip joint.**

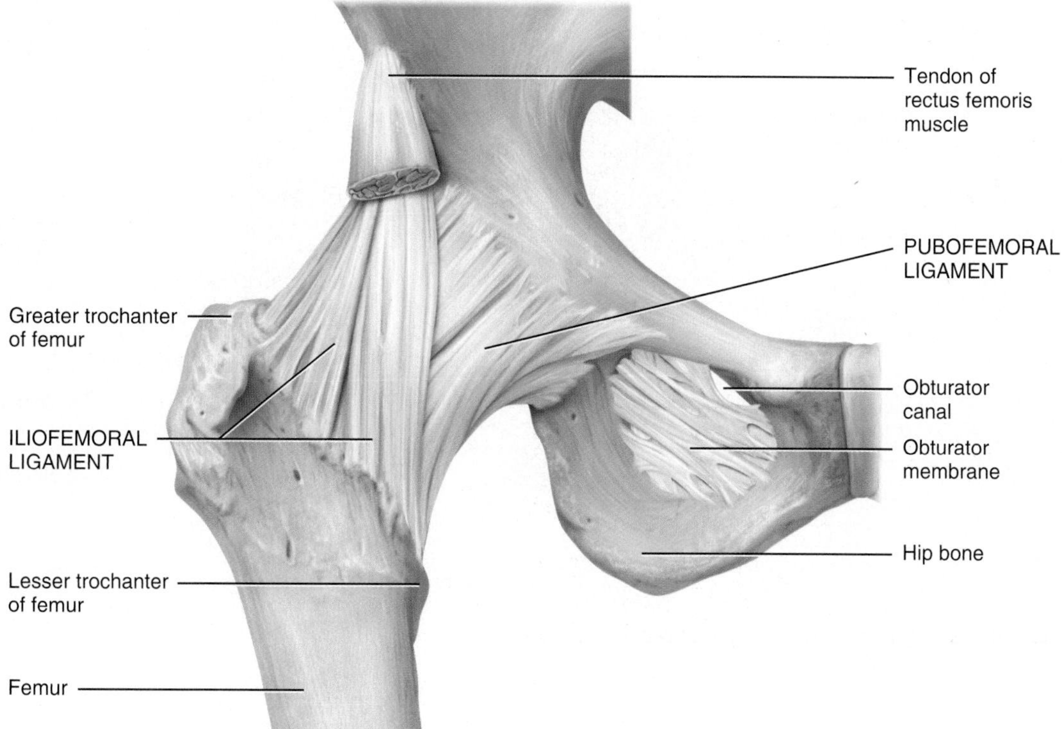


(a) Anterior view

(continues)

**FIGURE 9.13 (continued)**

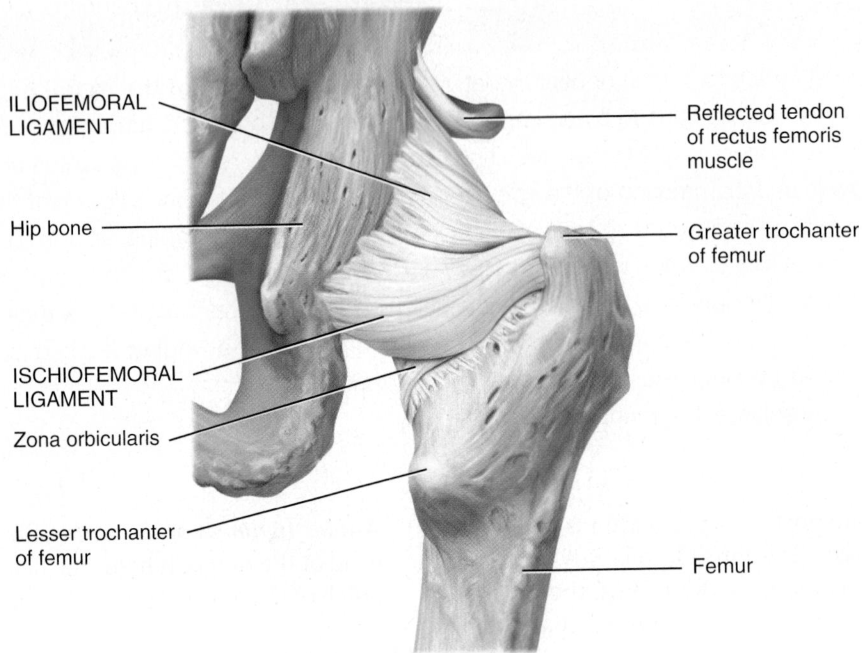


(b) Posterior view

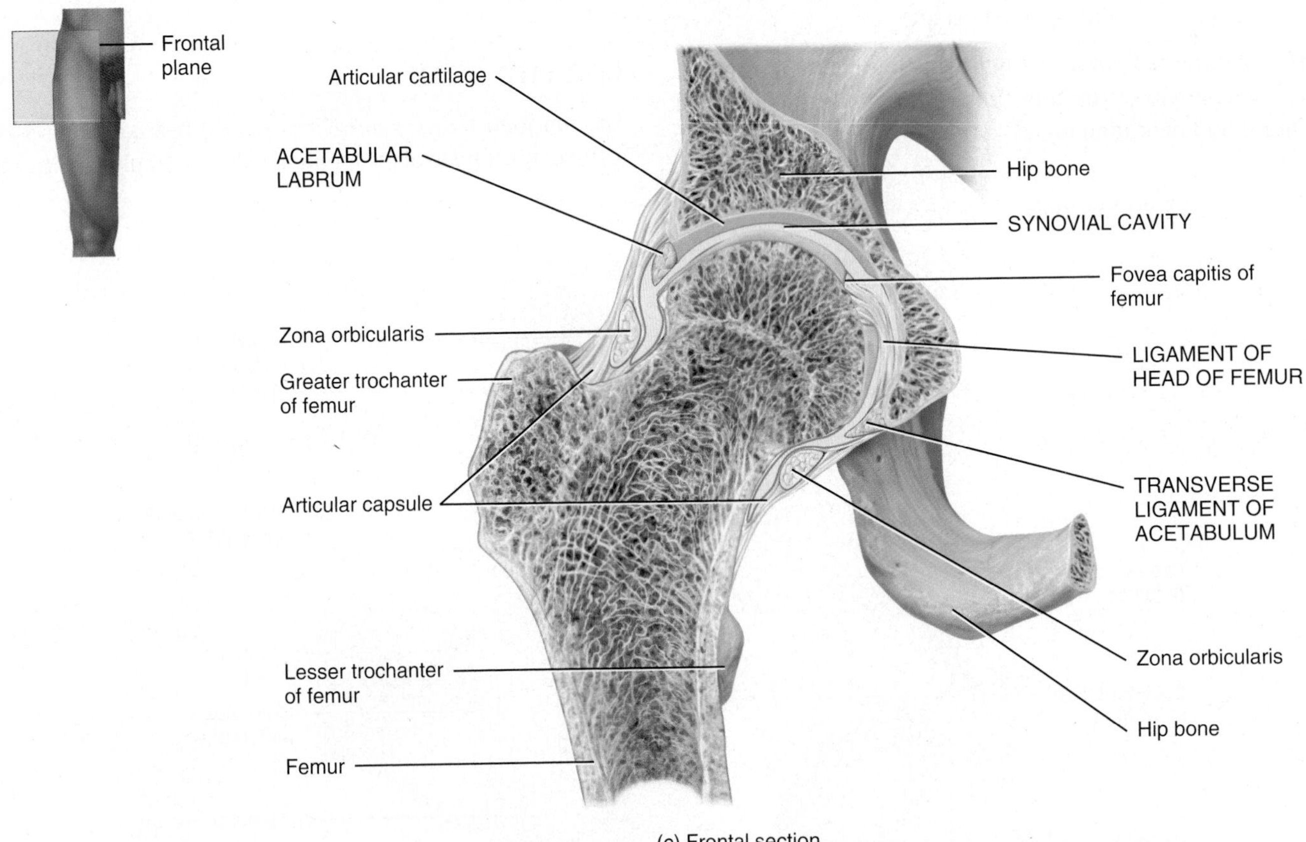


(c) Frontal section

The articular capsule of the hip joint is one of the strongest structures in the body.

### Anatomical Components

Anatomical components of the hip joint include the following:

- *Articular capsule*. Very dense and strong capsule that extends from the rim of the acetabulum to the neck of the femur (Figure 9.13c). This capsule, which is one of the strongest structures of the body, consists of overlapping circular and longitudinal fibers. The circular fibers, called the zona orbicularis, form a collar around the neck of the femur.
- ***Iliofemoral ligament*** (il′-ē-ō-FEM-o-ral). Portion of the articular capsule that extends from the ilium of the hip bone to the intertrochanteric line of the femur (Figure 9.13a, b). This ligament is said to be the body's strongest and prevents hyperextension of the femur at the hip joint during standing.
- ***Pubofemoral ligament*** (pū′-bō-FEM-o-ral). Portion of the articular capsule that extends from the pubic wall of the acetabulum to the neck of the femur (Figure 9.13a). This ligament prevents overabduction of the femur at the hip joint and strengthens the articular capsule.
- ***Ischiofemoral ligament*** (is′-kē-ō-FEM-o-ral). Portion of the articular capsule that extends from the ischial wall of the acetabulum to the neck of the femur (Figure 9.13b). This ligament slackens during adduction, tenses during abduction, and strengthens the articular capsule.
- ***Ligament of the head of the femur***. Flat band that extends from the fossa of the acetabulum to the fovea capitis, a small depression in the center of the head of the femur (Figure 9.13c).
- ***Acetabular labrum*** (as-ē-TAB-ū-lar LĀ-brum). Fibrocartilage rim attached to the margin of the acetabulum that increases the depth of the acetabulum. As a result, dislocation of the femur is rare (Figure 9.13c).
- ***Transverse ligament of the acetabulum***. Ligament that crosses over the inferior acetabulum to connect with the ligament of the head of the femur and the articular capsule (Figure 9.13c).

### Movements

The hip joint allows flexion, extension, abduction, adduction, circumduction, medial rotation, and lateral rotation of the thigh (see Figures 9.5–9.8). The extreme stability of the hip joint is related to the strong articular capsule and its accessory ligaments, the manner in which the femur fits into the acetabulum, and the muscles surrounding the joint. Although the shoulder and hip joints are both ball-and-socket joints, the hip joint does not have as wide a range of motion due to tension of the hip joint's strong ligaments.

## The Knee Joint

The **knee joint**, the largest and most complex joint of the body, is a modified hinge joint (Figure 9.14).

FIGURE 9.14 Right knee joint.

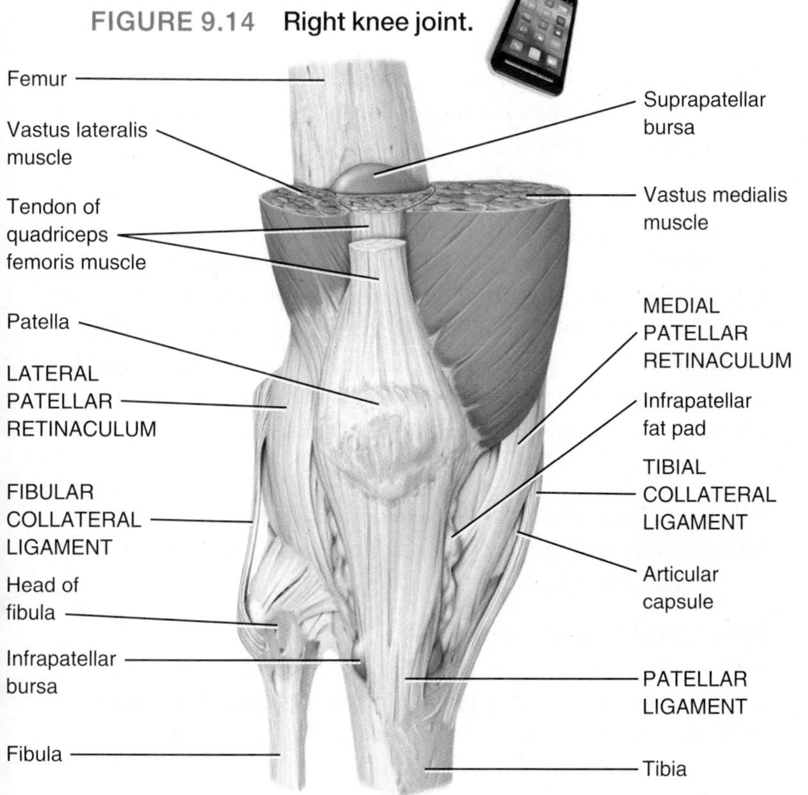


(a) Anterior superficial view

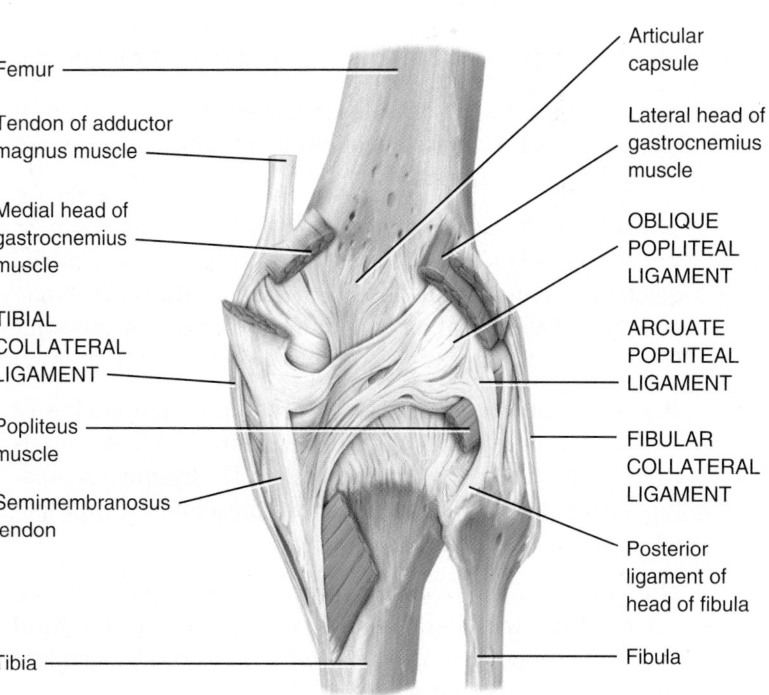


(b) Posterior deep view

(continues)

FIGURE 9.14 (continued)

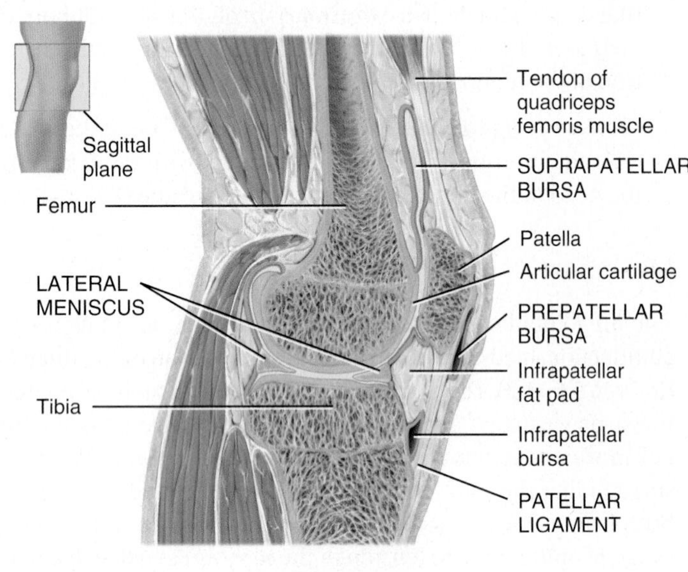


(c) Sagittal section

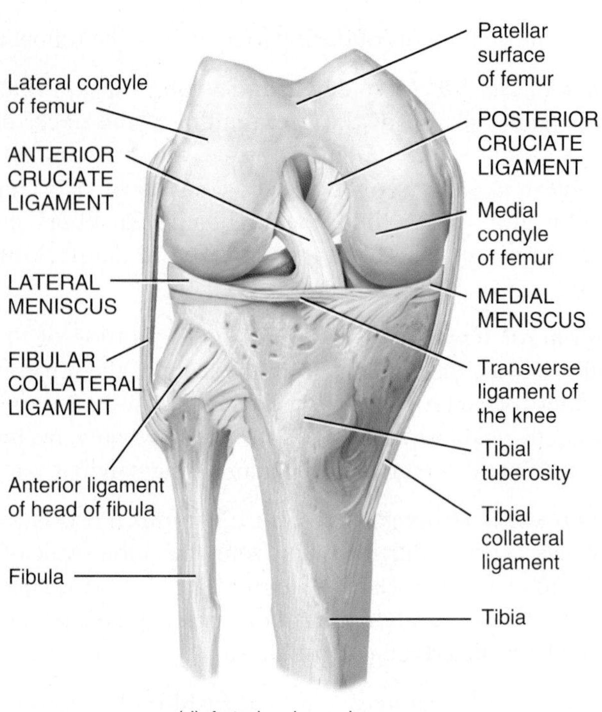


(d) Anterior deep view

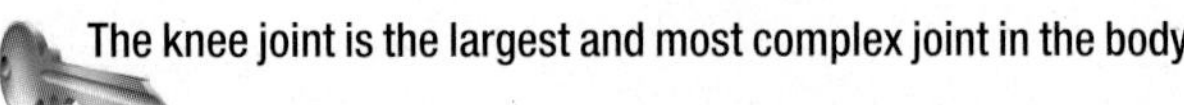

The knee joint is the largest and most complex joint in the body.


## Anatomical Components

Anatomical components of the knee joint include the following:

- *Articular capsule.* The capsule is weak and incomplete, but it is strengthened by muscle tendons and ligaments associated with the joint (Figure 9.14a, b).
- ***Medial and lateral patellar retinacula*** (ret′-i-NAK-ū-la; singular is *retinaculum*). Fused tendons of insertion of the quadriceps femoris muscle and the fascia lata (deep fascia of thigh) that strengthen the anterior surface of the joint (Figure 9.14a).
- ***Patellar ligament.*** Continuation of the insertion tendon of the quadriceps femoris (anterior thigh muscles) that extends from the patella to the tibial tuberosity. The ligament is separated from the synovial membrane of the joint by an infrapatellar fat pad (Figure 9.14a, c).
- ***Oblique popliteal ligament*** (pop-LIT-ē-al). Ligament that extends from the lateral condyle of the femur to the head and medial condyle of the tibia (Figure 9.14b). The ligament strengthens the posterior surface of the joint.
- ***Arcuate popliteal ligament.*** Extends from the lateral condyle of the femur to the head of the fibula. It strengthens the posterior surface of the joint (Figure 9.14b).
- ***Tibial collateral ligament.*** Ligament that extends from the medial condyle of the femur to the medial condyle of the tibia (Figure 9.14a, b, d). It strengthens the medial surface of the joint. The tibial collateral ligament is firmly attached to the medial meniscus. Because the tibial collateral ligament is firmly attached to the medial meniscus, tearing of the ligament frequently results in tearing of the meniscus and damage to the anterior cruciate ligament, described shortly.
- ***Fibular collateral ligament.*** Ligament that extends from the lateral condyle of the femur to the head of the fibula (Figure 9.14a, b, d). It strengthens the lateral aspect of the joint.
- ***Intracapsular ligaments*** (in′-tra-KAP-sū-lar). Ligaments within the articular capsule that connect the tibia and femur. The anterior and posterior **cruciate ligaments** (KROO-shē-āt = like a cross) are named based on their attachment sites on the tibia and because they cross one

## CLINICAL CONNECTION | *Knee Injuries*

The knee joint is the joint most vulnerable to damage because it is a mobile, weight-bearing joint and its stability depends almost entirely on its associated ligaments and muscles. Further, there is no complementary fit between the surfaces of the articulating bones. Following are several kinds of **knee injuries**. A **swollen knee** may occur immediately or hours after an injury. The initial swelling is due to escape of blood from damaged blood vessels adjacent to areas of injury, including rupture of the anterior cruciate ligament, damage to synovial membranes, torn menisci, fractures, or collateral ligament sprains. Delayed swelling is due to excessive production of synovial fluid, a condition commonly referred to as "water on the knee." The firm attachment of the tibial collateral ligament to the medial meniscus is clinically significant because tearing of the ligament typically also results in tearing of the meniscus. Such an injury may occur in sports such as football and rugby when the knee receives a blow from the lateral side while the foot is fixed on the ground. The force of the blow may also tear the anterior cruciate ligament, which is also connected to the anterior cruciate ligament. The term **"unhappy triad"** is applied to a knee injury that involves damage to the three components of the knee at the same time: the tibial collateral ligament, medial meniscus, and anterior cruciate ligament.

A **dislocated knee** refers to the displacement of the tibia relative to the femur. The most common type is dislocation anteriorly, resulting from hyperextension of the knee. A frequent consequence of a dislocated knee is damage to the popliteal artery.

If no surgery is required, treatment of knee injuries involves PRICE (protection, rest, ice, compression, and elevation) with some strengthening exercises and perhaps physical therapy.

another on their way to their destinations on the femur. The ***anterior cruciate ligament (ACL)*** extends posteriorly and laterally from a point *anterior* to the intercondylar area of the tibia to the medial surface of the lateral condyle of the femur (Figure 9.14d). The ACL limits hyperextension of the knee and prevents the anterior sliding of the tibia on the femur. This ligament is stretched or torn in about 70 percent of all serious knee injuries. The ***posterior cruciate ligament (PCL)*** extends anteriorly and medially from the posterior intercondylar area of the tibia to the lateral surface of the medial condyle of the femur (Figure 9.14d). The PCL prevents the posterior sliding of the tibia when the knee is flexed. This is important when walking down stairs or a steep incline.

- ***Articular discs (menisci)***. Two fibrocartilage discs between the tibial and femoral condyles help compensate for the irregular shapes of the bones and circulate synovial fluid. The ***medial meniscus*** is attached to the anterior intercondylar fossa of the tibia and to the posterior intercondylar fossa of the tibia between the attachments of the posterior cruciate ligament and lateral meniscus (Figure 9.14d). The ***lateral meniscus*** is attached anteriorly to the tibia and the anterior cruciate ligament, and posteriorly to the tibia and medial meniscus.
- **Bursae**. Several important bursae are associated with the knee to help relieve friction at the knee joint. For example, the **prepatellar bursa**, located between the patella and skin, is often damaged when you bump the anterior knee (Figure 9.14c).

### Movements

The knee joint allows flexion, extension, and slight medial rotation, and lateral rotation of the leg in the flexed position (see Figures 9.5f and 9.8c).

### ✓ CHECKPOINT

**18.** At the shoulder joint, which ligaments connect the scapula and humerus? Glenoid cavity and humerus?

**19.** Which structures provide the shoulder joint with stability?

**20.** At the elbow joint, which ligaments connect the humerus and ulna? Humerus and radius?

**21.** What movements can occur in the elbow joint?

**22.** At the hip joint, which ligaments connect the ilium and femur? Pubis and femur?

**23.** At the knee joint, which ligaments connect the anterior thigh muscle tendon and tibia? Lateral condyle of femur and fibula?

## Virtjean's Story

The sweet smell of baking cookies drifted out of the kitchen and down the hallways of Shady Acres. Two small children burst through the door of the kitchen, calling out, "Granny! Granny!"

Virtjean's granddaughter stood in the doorway. "Hi, Granny. We're finally here. You kids be careful. They were so excited to see you." She tipped her head back and inhaled deeply. "Mmm, you made cookies. Are they ready?"

"Yeah, are they ready, Granny? I love your cookies!" chimed in the children.

"Let's take a peek!" replied Virtjean, smiling as she turned toward the oven.

## EPILOGUE AND DISCUSSION

In this chapter we have seen that joints include any point of contact between two bones. These articulations are classified structurally and functionally, based on the type of movement they permit. The aging process, stress, and daily wear and tear can all have negative effects on joint function. Joint mobility is also reduced by diseases such as Virtjean's rheumatoid arthritis, which attacks the synovial membrane and articular cartilage, or osteoarthritis, which produces a progressive loss of articular cartilage and formation of new osseous tissue into obstructive bone spurs. Any number of conditions can alter joint flexibility and significantly impact our daily lives. An understanding of the structure and function of joints can assist in determining treatments and making adaptations to improve the quality of life. Perhaps in the future we will gain the knowledge to restore Virtjean's hands.

I. *How are synovial joints different from fibrous or cartilaginous joints? Which type of joint is most seriously affected by Virtjean's rheumatoid arthritis?*

J. *Virtjean has rheumatoid arthritis; the articular cartilage in her joints has been damaged and the synovial cavity is reduced. Explain how these conditions affect her joints.*

# Concept and Resource Summary

| Concept | Resources  |
|---|---|
| **Introduction**<br>1. A **joint**, or **articulation**, is the area where two or more bones meet, or where bone and cartilage, or bone and teeth, meet.<br>2. Most movements of the body occur at joints. | |
| **Concept 9.1** Joints are classified structurally and functionally.<br>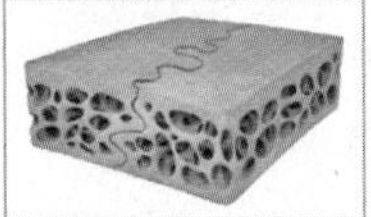<br>1. Structural classification of joints is based on (1) presence or absence of a synovial cavity and (2) the type of connective tissue binding the bones together.<br>2. There are three types of joints based on structure. **Fibrous joints** are held together by dense connective tissue and do not have a synovial cavity. **Cartilaginous joints** are held together by cartilage and do not have a synovial cavity. **Synovial joints** are held together by the dense connective tissue of an articular capsule and possess a synovial cavity.<br>3. Functional classification of joints is based on the degree of movement that they permit. There are three functional types of joints: immovable **synarthroses**; slightly movable **amphiarthroses**; and freely movable **diarthroses**. | Anatomy Overview—Joints |

Concept

Resources

## Concept 9.2 Fibrous joints lack a synovial cavity and are held together by dense connective tissue.

1. Fibrous joints include sutures, syndesmoses, and gomphoses.
2. A **suture** is composed of thin dense connective tissue. Sutures are synarthroses and are only found in the skull.
3. Some sutures, called **synostoses**, are present during childhood but eventually get replaced by osseous tissue in adulthood.
4. In a **syndesmosis**, the dense connective tissue of a ligament unites the bones and permits slight movement.
5. A **gomphosis** is a peg-in-a-socket joint composed of a bone held in its bony socket by a periodontal ligament. It is a synarthrosis; the only examples in the human body are the teeth.
6. **Interosseous membranes** are slightly movable joints found between the radius and ulna in the forearm and the tibia and fibula in the leg.

Anatomy Overview—Fibrous Joints
Concepts and Connections—Joint Classification

## Concept 9.3 Cartilaginous joints lack a synovial cavity and are held together by cartilage.

1. Cartilaginous joints include synchondroses and symphyses.
2. A **synchondrosis** is composed of hyaline cartilage and is a synarthrosis.
3. A **symphysis** has fibrocartilage connecting the bones at their articular cartilage surfaces; it is an amphiarthrosis.

Anatomy Overview—Cartilaginous Joints

## Concept 9.4 Articulating surfaces of bones at a synovial joint are covered with articular cartilage and enclosed within a synovial cavity.

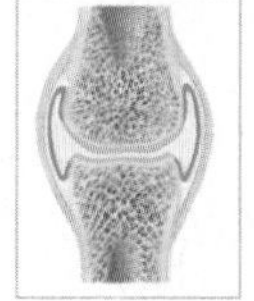

1. All synovial joints have a **synovial cavity** that allows the joint to function as a diarthrosis. The bones have a covering of **articular cartilage** on their articulating surfaces to reduce friction and absorb shock.
2. The synovial cavity is enclosed by a sleevelike **articular capsule** composed of an outer **fibrous membrane** that helps stabilize the joint and an inner **synovial membrane**.
3. The synovial membrane secretes a viscous, clear **synovial fluid** that forms a thin film over the articular capsule surfaces to reduce friction, absorb shock, supply oxygen and nutrients, and remove wastes.
4. **Accessory ligaments** located inside and outside the articular capsule are often found at synovial joints.
5. Certain synovial joints, like the knee, have pads of fibrocartilage, called **articular menisci**, lying between the articular surfaces of the bones and attached to the fibrous membrane. These discs stabilize the joint and direct the flow of synovial fluid to areas of greatest friction.
6. **Bursae** are sacs lined with a synovial membrane and filled with synovial fluid. They cushion adjacent body parts at certain joints and help alleviate friction between them. Tubelike bursae called **tendon sheaths** wrap around muscle tendons to prevent friction.

Anatomy Overview—Synovial Joints
Figure 9.3—Structure of a Typical Synovial Joint

Exercise—Judge the Joint
Concepts and Connections—Joint Classifications
Clinical Connection—Sprain and Strain
Clinical Connection—Rheumatism and Arthritis
Clinical Connection—Lyme Disease

## Concept 9.5 Synovial joint movement terminology indicates the direction of movement or the relationships of body parts during movement.

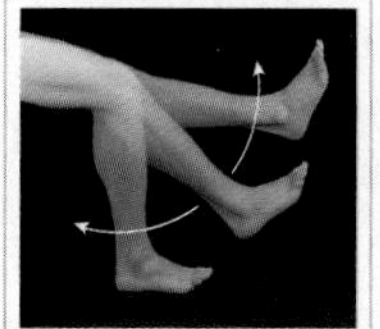

1. Movements at synovial joints are grouped into four categories: (1) gliding, (2) angular movements, (3) rotation, and (4) special movements.
2. **Gliding** occurs at plane joints where flat bone surfaces move back and forth and side to side in relation to one another without alteration of the angles between the bones.
3. **Angular movements** occur when the angle between articulating bones increases or decreases during movement. Angular movements include flexion, extension, lateral flexion, hyperextension, abduction, adduction, and circumduction.
4. **Flexion** is a decrease in the angle between articulating bones; **extension** is an increase in the angle between articulating bones. When flexion occurs laterally along the frontal plane, rather than along the sagittal plane, it is **lateral flexion**. When body parts at a joint undergo extension beyond their anatomical position, it is **hyperextension**.

Anatomy Overview—Movements Produced at Synovial Joints
Figure 9.5—Angular Movements at Synovial Joints—Flexion, Extension, Hyperextension, and Lateral Flexion

## Concept

5. **Abduction** is the movement of a bone away from the midline; **adduction** is the movement of a bone toward the midline.
6. **Circumduction** is movement of the distal end of a body part in a circle. It results from a sequence of movements: flexion, abduction, extension, and adduction.
7. **Rotation** involves a bone revolving around its own axis. In medial rotation a limb rotates toward the midline; in lateral rotation a limb rotates away from the midline.
8. **Special movements** occur only at certain joints. These include elevation, depression, protraction, retraction, inversion, eversion, dorsiflexion, plantar flexion, supination, pronation, and opposition.
9. Upward movement of a body part is **elevation**; downward movement of a body part is **depression**.
10. Movement of a body part anteriorly in the transverse plane is **protraction**; movement of a protracted body part back to the anatomical position is **retraction**.
11. Moving the soles of the feet medially is **inversion**; moving the soles laterally is **eversion**.
12. Bending of the foot at the ankle in the direction of its superior surface is termed **dorsiflexion**; bending of the foot at the ankle in the direction of its inferior surface is **plantar flexion**.
13. **Supination** is movement of the forearm resulting in the palm facing anteriorly; **pronation** is the movement of the forearm resulting in the palm facing posteriorly.
14. Movement of the thumb at the carpometacarpal joint across the palm to touch the fingertips is **opposition**.

### Concept 9.6 Synovial joints are described as plane, hinge, pivot, condyloid, saddle, or ball-and-socket.

1. Types of synovial joints are plane, hinge, pivot, condyloid, saddle, and ball-and-socket.
2. In a **plane joint** the articulating surfaces are flat, and the bones glide back and forth and side to side; they may also permit rotation.
3. In a **hinge joint**, the convex surface of one bone fits into the concave surface of another, and the motion is angular around one axis.
4. In a **pivot joint**, a round or pointed surface of one bone fits into a ring formed by another bone and a ligament, and movement is rotational.
5. In a **condyloid joint**, an oval projection of one bone fits into an oval cavity of another, and motion is angular around two axes; examples include the wrist joint and metacarpophalangeal joints of the second through fifth digits.
6. In a **saddle joint**, the articular surface of one bone is shaped like a saddle and the other bone fits into the saddle like a sitting rider.
7. In a **ball-and-socket joint**, the ball-shaped surface of one bone fits into the cuplike depression of another. Motion is around three axes.

### Concept 9.7 The shoulder, elbow, hip, and knee joints provide examples of synovial joint components, classifications, and movements.

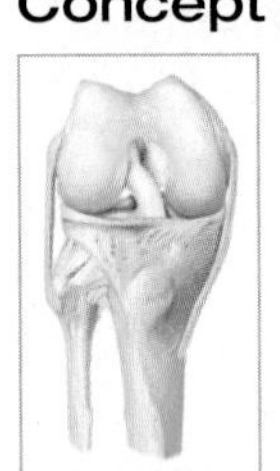

1. The **shoulder joint** is formed by the head of the humerus and the glenoid cavity of the scapula. Anterior and superior ligaments help reinforce the joint, and four bursae help reduce friction. Most joint stability comes from the rotator cuff muscles and their tendons that encircle the joint. This joint allows flexion, extension, abduction, adduction, medial rotation, lateral rotation, and circumduction.
2. The **elbow joint** is a hinge joint formed by the trochlea of the humerus, the trochlear notch of the ulna, and the head of the radius. It is stabilized by strong ligaments. This joint permits flexion and extension of the forearm.
3. The **hip joint** is a ball-and-socket joint formed by the head of the femur and the acetabulum of the hip bone. Its articular capsule is dense and strong, and reinforced by several strong ligaments. The acetabulum is a deep socket that secures articulation with the femur. Rotation and every angular movement are permitted; however, the hip joint does not have as wide a range of motion as the shoulder joint because of its strong ligaments.
4. The **knee joint** is the body's largest and most complex joint. The weak, incomplete articular capsule is strengthened by the muscle tendon of the anterior thigh muscles and other tendons and ligaments. The tibial and fibular collateral ligaments give the joint stability.

## Resources

Figure 9.6—Angular Movements at Synovial Joints—Abduction and Adduction

Figure 9.7—Angular Movements at Synovial Joints—Circumduction

Figure 9.8—Rotation at Synovial Joints

Exercise—Manage Movements

Anatomy Overview—Structural Subtypes of Synovial Joints

Anatomy Overview—Specific Synovial Joints

Anatomy Overview—Specific Synovial Diarthrotic Joints

Clinical Connection—Arthroplasty

Clinical Connection—Torn Cartilage and Arthroscopy

Temporomandibular Joint

Figure 9.14—Right Knee Joint

Clinical Connection—Dislocated Mandible

Clinical Connection—Autologous Chondrocyte Implantation

## Understanding the Concepts

1. What are the functional classifications of joints? Give an example of each type.
2. Functionally, why are sutures classified as synarthroses and syndesmoses as amphiarthroses?
3. Which types of cartilaginous joints are synarthroses? Which are amphiarthroses?
4. Maria is walking to class when she trips and falls, dislocating her shoulder. Which structures in the articular capsule have failed Maria?
5. Jamar likes to run but hates to stretch. Why will this eventually lead to an early retirement from regular running?
6. Jill likes to "crack" her knuckles, but her mother hates it and tells Jill that it can't be good for Jill's knuckles when she makes her bones snap into each other. What causes knuckles to crack? Is it damaging the articulating ends of Jill's bones?
7. After standing for several hours at a grocery store cash register, a checker's feet are sore and tired. While continuing to stand and check groceries, she attempts to relieve her foot discomfort. Which movements is the checker using when she raises her heels while rocking onto her toes? Raises her feet by rocking onto her heels?
8. Describe the movement that occurs at the knee joint when the anterior thigh muscles contract.
9. Which type of joint is found between vertebrae? How movable are these joints normally?
10. Why does the shoulder joint have more freedom of movement than any other joint of the body?
11. Why is the hip joint so stable? List all of the structures that contribute to its stability.
12. What movement occurs at the knee joint when the quadriceps femoris (anterior thigh muscles) contract?

CHAPTER

# 10

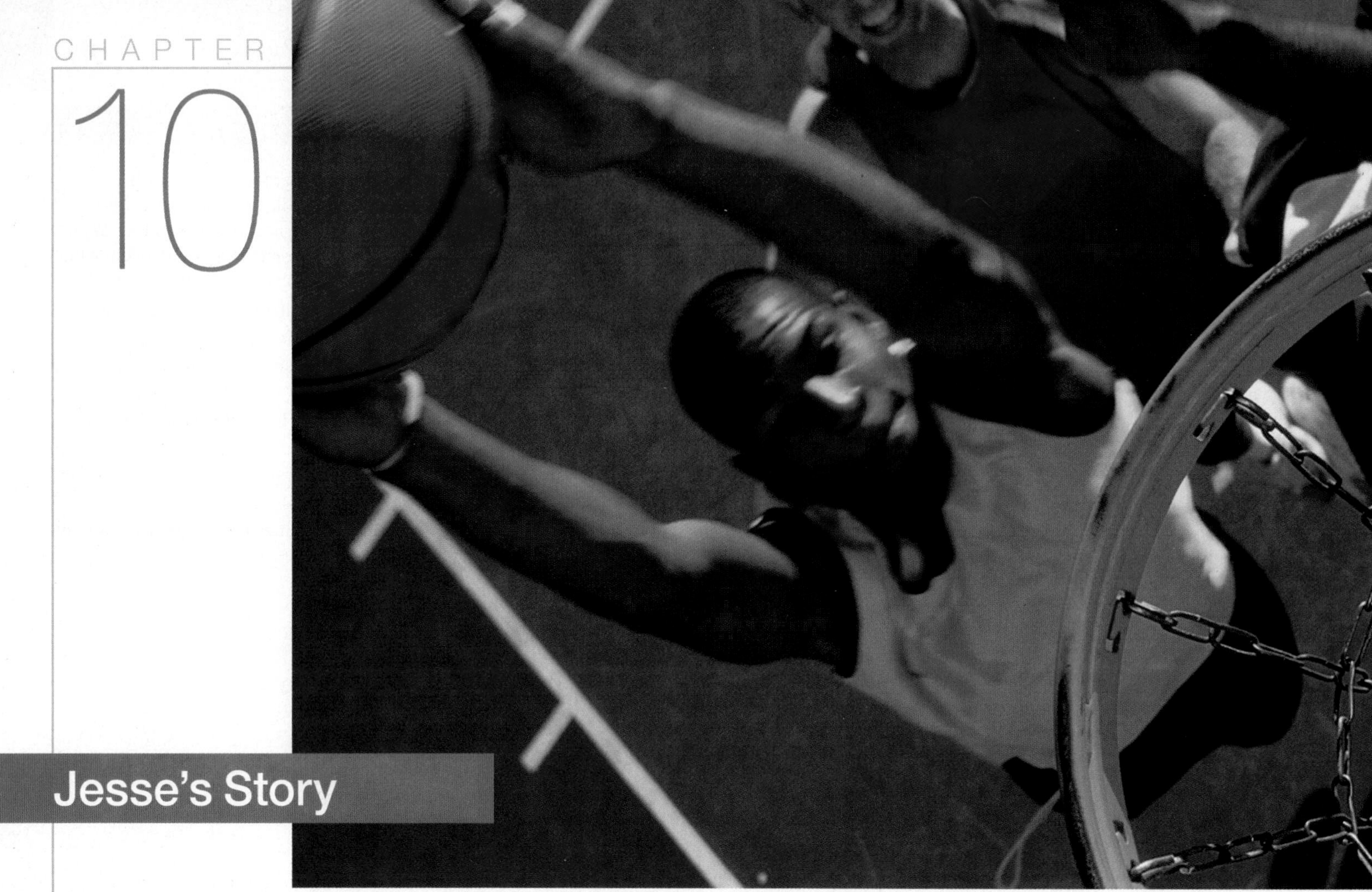

## Jesse's Story

Jesse moved down the basketball court with a determined look. His mind focused on his next move as he spun and drove forward to put the ball up. Jesse felt his feet leave the ground as he leaped upward. He grasped the basketball with two hands and brought them up, putting just the right spin on the ball with his fingers. But then a strange thing happened; he couldn't complete the move. His left arm would not obey his commands and the ball shot up past the backboard. "Oh man, what are you doing? What was that?" Jesse's friend Ronnie looked incredulously at Jesse. "What's the matter with you, man? If you can't make a shot like that, pass me the ball."

Jesse shot back an angry glance. "I was fouled. Look, if you're gonna get in my face, you can find someone else to play." Jesse turned and strode off the concrete playground court, not even slowing down as he grabbed his jacket and gym bag. He didn't stop jogging until he was several blocks away. He knew he wasn't really mad at Ronnie, he was mad at himself. He had always been good at sports; he played football, ran for the track team, and he loved basketball. But he had noticed lately that something wasn't right. He couldn't raise his left arm up over his head. It felt weaker the higher he tried to raise it. It had slowly been getting worse, and even increasing his weight lifting at the gym didn't seem to help. He had noticed, too, that his shoulder blade felt funny when he made certain movements, like it was coming loose. Even though he didn't want to admit it, he knew something was wrong. He wanted to find out what was happening; he knew that he couldn't make the varsity team in the fall if this kept getting worse. He would have to tell his mother.

When you think about muscles, what do you think of? Strength? Power? Professional athletes? We all have muscles that help us perform a variety of tasks, not just moving our bones to help sink a basket. In the next two chapters we will learn about what muscles are, what they do, and, in Jesse's case, the effects on daily life if they do not function properly.

# Muscle Tissue

## INTRODUCTION

**A**lthough bones provide leverage and form the framework of the body, they cannot move body parts by themselves. Motion results from the alternating contraction and relaxation of muscles, which make up 40–50 percent of total adult body weight. Your muscular strength reflects the primary function of muscle—the transformation of chemical energy into mechanical energy to produce body movements. In addition, muscle tissue stabilizes the body's position, regulates organ volume, generates heat to maintain normal body temperature, and propels fluids and food through various body systems.

## CONCEPTS

**10.1** Skeletal, cardiac, and smooth muscle tissues differ in location, structure, and function.

**10.2** Muscle tissue performs four functions and possesses four properties.

**10.3** Skeletal muscles are surrounded by connective tissues and are well supplied with nerves and blood vessels.

**10.4** Each skeletal muscle fiber is covered by a sarcolemma; each of its myofibrils is surrounded by sarcoplasmic reticulum and contains sarcomeres.

**10.5** The neuromuscular junction is where a muscle action potential is initiated.

**10.6** An action potential releases calcium ions that allow thick filaments to bind to and pull thin filaments toward the center of the sarcomere.

**10.7** Muscle tension is controlled by stimulation frequency and motor unit recruitment.

**10.8** Muscle fibers produce ATP from creatine phosphate, by anaerobic cellular respiration, and by aerobic cellular respiration.

**10.9** Skeletal muscle fibers are classified as slow oxidative fibers, fast oxidative–glycolytic fibers, or fast glycolytic fibers.

**10.10** Cardiac muscle tissue is found in the heart, and smooth muscle tissue is found in hollow internal structures.

# 10.1 Skeletal, cardiac, and smooth muscle tissues differ in location, structure, and function.

There are three types of **muscle tissue:** skeletal, cardiac, and smooth (see Table 4.8). Although the different types of muscle tissue share some properties, they differ from one another in their microscopic anatomy, location, and how they are controlled by the nervous and endocrine systems.

**Skeletal muscle tissue** is so named because the function of most skeletal muscles is to move bones of the skeleton. (A few skeletal muscles attach to and move the skin or other skeletal muscles.) Skeletal muscle tissue is referred to as *striated* because alternating light and dark protein bands (*striations*) are visible when the tissue is examined under a microscope (see Table 4.8). Skeletal muscle tissue works primarily in a *voluntary* manner; that is, its activity can be consciously (voluntarily) controlled by the somatic (voluntary) division of the nervous system. (Figure 12.2 depicts the divisions of the nervous system.) Most skeletal muscles also are controlled subconsciously to some extent. For example, your diaphragm continues to alternately contract and relax while you are asleep so that you don't stop breathing. Also, you do not need to consciously think about contracting the skeletal muscles that maintain your posture or stabilize body positions.

**Cardiac muscle tissue** is found only in the heart, where it forms most of the heart wall. Like skeletal muscle, cardiac muscle is *striated*, but its action is *involuntary*—its alternating contraction and relaxation cannot be consciously controlled (see Table 4.8). Rather, the heart beats because it has a pacemaker that initiates each contraction; this built-in (intrinsic) rhythm is called *autorhythmicity* (aw′-tō-rith-MISS-i-tē).

**Smooth muscle tissue** is located in the walls of hollow internal structures, such as blood vessels, airways, and most organs in the abdominopelvic cavity. It is also attached to hair follicles in the skin. Smooth muscle tissue gets its name because, under a microscope, this tissue lacks striations; hence, it appears *nonstriated* or *smooth* (see Table 4.8). The action of smooth muscle is usually *involuntary*, and, like cardiac muscle, some smooth muscle tissue, such as the muscles that propel food through your gastrointestinal tract, has autorhythmicity. Both cardiac muscle and smooth muscle are regulated by the autonomic (involuntary) division of the nervous system and by hormones released by endocrine glands.

✓ **CHECKPOINT**

1. Which types of muscle tissue can be consciously controlled? Which cannot be consciously controlled?
2. How does the appearance of smooth muscle differ from that of cardiac or skeletal muscle?

# 10.2 Muscle tissue performs four functions and possesses four properties.

## Functions of Muscle Tissue

Through sustained contraction or alternating contraction and relaxation, muscle tissue has four key functions:

- ***Produces body movements.*** Movements of the whole body, such as walking and running, and localized movements, such as grasping a pencil, keyboarding, or nodding the head, rely on the integrated functioning of skeletal muscles, bones, and joints.
- ***Stabilizes body positions.*** Skeletal muscle contractions stabilize joints and help maintain body positions, such as standing or sitting. Postural muscles contract continuously when you are awake; for example, sustained contractions of your neck muscles hold your head upright when you are listening intently to your anatomy and physiology lecture.
- ***Moves substances within the body.*** Sustained contractions of ringlike bands of smooth muscle called *sphincters* prevent outflow of the contents of a hollow organ. Temporary storage of food in the stomach or urine in the urinary bladder is possible because smooth muscle sphincters close off the outlets of these organs. Cardiac muscle contractions of the heart pump blood through blood vessels. Contraction and relaxation of smooth muscle in the walls of blood vessels help adjust their diameter and thus regulate the rate of blood flow. Smooth muscle contractions also move food and substances such as bile and enzymes through the gastrointestinal tract, push gametes (sperm and oocytes) through the passageways of the reproductive systems, and propel urine through the urinary system. Skeletal muscle contractions promote the flow of lymph and aid the return of blood to the heart.
- ***Generates heat.*** As muscle tissue contracts, it also produces heat. Much of the heat generated by muscle tissue is used to maintain normal body temperature. Involuntary contractions of skeletal muscles, known as *shivering*, can dramatically increase the rate of heat production.

## Properties of Muscle Tissue

Muscle tissue has four special properties that enable it to function and contribute to homeostasis:

- **Electrical excitability** (ek-sīt′-a-BIL-i-tē), a property of both muscle cells and neurons, is the ability to respond to certain stimuli by producing electrical signals called **action potentials** (**impulses**). The stimuli that trigger action potentials in muscle cells may be electrical signals arising in the muscle tissue itself, such as occurs in the heart's pacemaker, or chemical stimuli, such as neurotransmitters released by neurons, hormones distributed by the blood, or even local changes in pH.
- **Contractility** (kon′-trak-TIL-i-tē) is the ability of muscle tissue to shorten forcefully when stimulated by an action potential. When a skeletal muscle contracts, it generates *tension* (force of contraction) while pulling on its attachment points. If the tension generated is great enough to overcome the resistance of the object to be moved, the muscle shortens and movement occurs. An example is lifting a book off a table.
- **Extensibility** (ek-sten′-si-BIL-i-tē) is the ability of muscle tissue to stretch within limits, without being damaged. Normally, smooth muscle is subject to the greatest amount of stretching. For example, each time your stomach fills with food, the muscle in the wall is stretched.
- **Elasticity** (e-las-TIS-i-tē) is the ability of muscle tissue to return to its original length and shape after contraction or extension.

Skeletal muscle tissue is the focus of much of this chapter. Cardiac muscle tissue and smooth muscle tissue are described more briefly here. Cardiac muscle tissue is discussed in more detail in Chapter 19; smooth muscle tissue is included in Chapter 14, as well as in discussions of the various organs containing smooth muscle.

✓ **CHECKPOINT**

**3.** What are the general functions of muscle tissue?

**4.** What properties do all muscle tissues have in common?

**5.** Which property do muscle tissues and neurons share?

# 10.3 Skeletal muscles are surrounded by connective tissues and are well supplied with nerves and blood vessels.

Each of your skeletal muscles is a separate organ composed of hundreds to thousands of cells called **muscle fibers** because of their elongated shapes. Connective tissues surround skeletal muscle fibers and whole muscles, and blood vessels and nerves penetrate skeletal muscles (Figure 10.1). To understand how contraction of skeletal muscle can generate tension, you first need to understand its gross and microscopic anatomy.

## Connective Tissue Components

Connective tissue surrounds and protects muscle tissue. The **hypodermis** separates muscle from skin (see Figure 11.9c). It is composed of areolar connective tissue and adipose tissue, provides a pathway for nerves and blood and lymphatic vessels to enter and exit muscles, serves as an insulating layer that reduces heat loss, and protects muscles from physical trauma. **Fascia** (FASH-ē-a = bandage) is a sheet or broad band of dense connective tissue that supports and surrounds muscles and other organs of the body. Fascia holds together muscles with similar functions; allows free movement of muscles; carries nerves, blood vessels, and lymphatic vessels; and fills spaces between muscles.

Three layers of connective tissue extend from the fascia to further protect and strengthen skeletal muscle (Figure 10.1):

- **Epimysium** (ep-i-MĪZ-ē-um; *epi-* = upon), the outermost layer of dense connective tissue, encircles the entire muscle.
- **Perimysium** (per-i-MĪZ-ē-um; *peri-* = around) is also a layer of dense, irregular connective tissue, but it surrounds groups of 10 to 100 or more muscle fibers, separating them into bundles called **fascicles** (FAS-i-kuls = little bundles). Many fascicles are large enough to be seen with the naked eye. They give a cut of meat its characteristic "grain"; if you tear a piece of meat, it rips apart along the fascicles.
- **Endomysium** (en′-dō-MĪZ-ē-um; *endo-* = within) penetrates the interior of each fascicle and separates individual muscle fibers from one another. The endomysium is a thin sheath of areolar connective tissue.

The epimysium, perimysium, and endomysium may extend together beyond the muscle fibers to form a ropelike **tendon** that attaches a muscle to the periosteum of a bone. An example is the calcaneal (Achilles) tendon of the gastrocnemius (calf), which attaches the muscle to the calcaneus (see Figure 11.20d). When the connective tissue layers extend as a broad, flat sheet, it is called an **aponeurosis** (ap-ō-noo-RŌ-sis; *apo-* = from; *neur-* = a sinew). An example of an aponeurosis is the epicranial aponeurosis on top of the skull between the frontal and occipital bellies of the occipitofrontalis muscle (shown in Figure 11.3 a, c).

**FIGURE 10.1 Organization of skeletal muscle and its connective tissue coverings.**

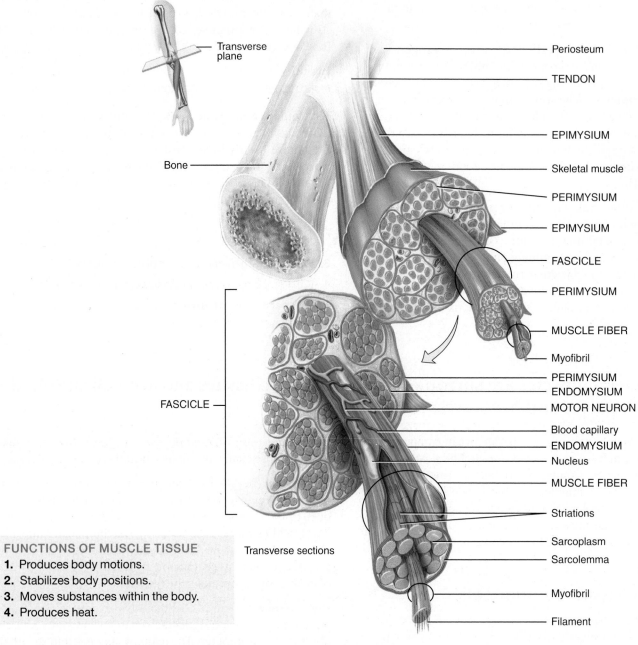

Transverse plane
Periosteum
TENDON
EPIMYSIUM
Bone
Skeletal muscle
PERIMYSIUM
EPIMYSIUM
FASCICLE
PERIMYSIUM
MUSCLE FIBER
Myofibril
PERIMYSIUM
ENDOMYSIUM
FASCICLE
MOTOR NEURON
Blood capillary
ENDOMYSIUM
Nucleus
MUSCLE FIBER
Striations
Sarcoplasm
Transverse sections
Sarcolemma
Myofibril
Filament

**FUNCTIONS OF MUSCLE TISSUE**

1. Produces body motions.
2. Stabilizes body positions.
3. Moves substances within the body.
4. Produces heat.

**A skeletal muscle consists of individual muscle fibers bundled into fascicles and surrounded by three connective tissue layers that are extensions of the fascia.**

## CLINICAL CONNECTION | *Fibromyalgia*

**Fibromyalgia** (fī-brō-mī-AL-jē-a; *algia* = painful condition) is a chronic, painful, nonarticular rheumatic disorder that affects the fibrous connective tissue components of muscles, tendons, and ligaments. A striking sign is pain that results from gentle pressure at specific "tender points." Even without pressure, there is pain, tenderness, and stiffness of muscles, tendons, and surrounding soft tissues. Besides muscle pain, those with fibromyalgia report severe fatigue, poor sleep, headaches, depression, irritable bowel syndrome, and inability to carry out their daily activities. There is no specific identifiable cause. Treatment consists of stress reduction, regular exercise, application of heat, gentle massage, physical therapy, medication for pain, and a low-dose antidepressant to help improve sleep.

## Nerve and Blood Supply

Skeletal muscles are well supplied with nerves and blood vessels. Generally, an artery and one or two veins accompany each nerve that penetrates a skeletal muscle. The neurons that stimulate muscle fibers to contract are called **motor neurons**. Each motor neuron has a threadlike process, called an **axon**, that extends from the brain or spinal cord to a group of skeletal muscle fibers (see Figure 10.6d). The axon of a motor neuron typically branches many times, each branch extending to a different skeletal muscle fiber.

Microscopic blood vessels called capillaries are plentiful in muscle tissue; each muscle fiber is in close contact with one or more capillaries (see Figure 10.6d). The blood capillaries bring in oxygen and nutrients and remove heat and the waste products of muscle metabolism. Especially during contraction, a muscle fiber synthesizes and uses considerable ATP (adenosine triphosphate). The generation of ATP requires oxygen, glucose, fatty acids, and other substances that are delivered to the muscle fiber in the blood.

✓ **CHECKPOINT**

6. What functions do fascia serve?
7. What is the structural difference between a tendon and an aponeurosis?

# 10.4 Each skeletal muscle fiber is covered by a sarcolemma; each of its myofibrils is surrounded by sarcoplasmic reticulum and contains sarcomeres.

The most important components of a skeletal muscle are the muscle fibers themselves. The diameter of a mature skeletal muscle fiber ranges from 10 to 100 $\mu$m.* The typical length of a mature skeletal muscle fiber is about 10 cm (4 in.), although some are as long as 30 cm (12 in.). During embryonic development, each skeletal muscle fiber arises from the fusion of a hundred or more small mesodermal cells called *myoblasts* (MĪ-ō-blasts) (Figure 10.2a). Hence, each mature skeletal muscle fiber has a hundred or more nuclei. Once fusion has occurred, the muscle fiber loses its ability to undergo cell division. Thus, the number of skeletal muscle fibers is set before you are born, and most of these cells last a lifetime.

The dramatic muscle growth that occurs after birth occurs by enlargement of existing muscle fibers, called **muscular hypertrophy** (hī-PER-trō-fē; *hyper-* = above or excessive; *-trophy* = nourishment), rather than by **muscular hyperplasia** (hi-per-PLĀ-zē-a; *-plasis* = molding), an increase in the number of fibers. Muscular hypertrophy is due to increased production of myofibrils (discussed shortly) and other organelles resulting from forceful, repetitive muscular activity, such as strength training. Because hypertrophied muscles contain more myofibrils, they are capable of more forceful contractions. During childhood, human growth hormone and other hormones stimulate an increase in the size of skeletal muscle fibers. The hormone testosterone (from the testes in males and in small amounts from other tissues, such as the ovaries in females) promotes further enlargement of muscle fibers.

**Muscular atrophy** (A-trō-fē; *a-* = without, *-trophy* = nourishment) is a decrease in the size and, therefore, strength of a muscle. Muscular atrophy is a result of progressive loss of myofibrils when muscles are not used or the nerve supply to a muscle is disrupted.

A few myoblasts do persist in mature skeletal muscle as *satellite cells* (Figure 10.2a). Satellite cells retain the capacity to fuse with one another or with damaged muscle fibers to regenerate functional muscle fibers. However, the number of new skeletal muscle fibers formed is not enough to compensate for significant skeletal muscle damage or degeneration. In such cases, skeletal muscle tissue undergoes **fibrosis**, the replacement of muscle fibers by fibrous scar tissue. For this reason, regeneration of skeletal muscle tissue is limited.

*One micrometer ($\mu$m) is $10^{-6}$ meter (1/25,000 in.).

## Sarcolemma, Transverse Tubules, and Sarcoplasm

The multiple nuclei of a skeletal muscle fiber are located just beneath the **sarcolemma** (sar′-kō-LEM-ma; *sarc-* = flesh; *-lemma* = sheath), the plasma membrane of a muscle fiber (Figure 10.2b, c). Thousands of tiny tunnel-like invaginations of the sarcolemma, called **transverse (T) tubules**, extend in toward the center of each muscle fiber. T tubules are open to the outside of the fiber and thus are filled with interstitial fluid. Muscle action potentials travel along the sarcolemma and through the T tubules, quickly spreading throughout the muscle fiber. This arrangement ensures that an action potential excites all parts of the muscle fiber at essentially the same instant.

The sarcolemma surrounds the **sarcoplasm** (SAR-kō-plazm), the cytoplasm of a muscle fiber. Sarcoplasm includes a substantial amount of glycogen, which is a large molecule composed of many glucose molecules (see Figure 2.15). Glycogen can be used for synthesis of ATP. In addition, the sarcoplasm contains a red-colored protein called **myoglobin** (mī-ō-GLŌB-in). This protein, found only in muscle, binds oxygen molecules that diffuse into muscle fibers from interstitial fluid. Myoglobin releases oxygen when mitochondria need it for ATP production. The mitochondria lie in rows throughout the muscle fiber, strategically close to the muscle proteins that use ATP during contraction so that ATP can be produced quickly as needed (Figure 10.2c).

## Myofibrils and Sarcoplasmic Reticulum

At high magnification the sarcoplasm appears stuffed with little parallel threads. These small structures are the **myofibrils** (mī-ō-FĪ-brils; *myo-* = muscle; *-fibrilla* = little fiber), the contractile

**FIGURE 10.2 Microscopic organization of skeletal muscle.** (a) During embryonic development, many myoblasts fuse to form one skeletal muscle fiber. (b-d) The sarcolemma of the fiber encloses sarcoplasm and myofibrils. Sarcoplasmic reticulum wraps around each myofibril. Thousands of transverse tubules invaginate from the sarcolemma toward the center of the muscle fiber.

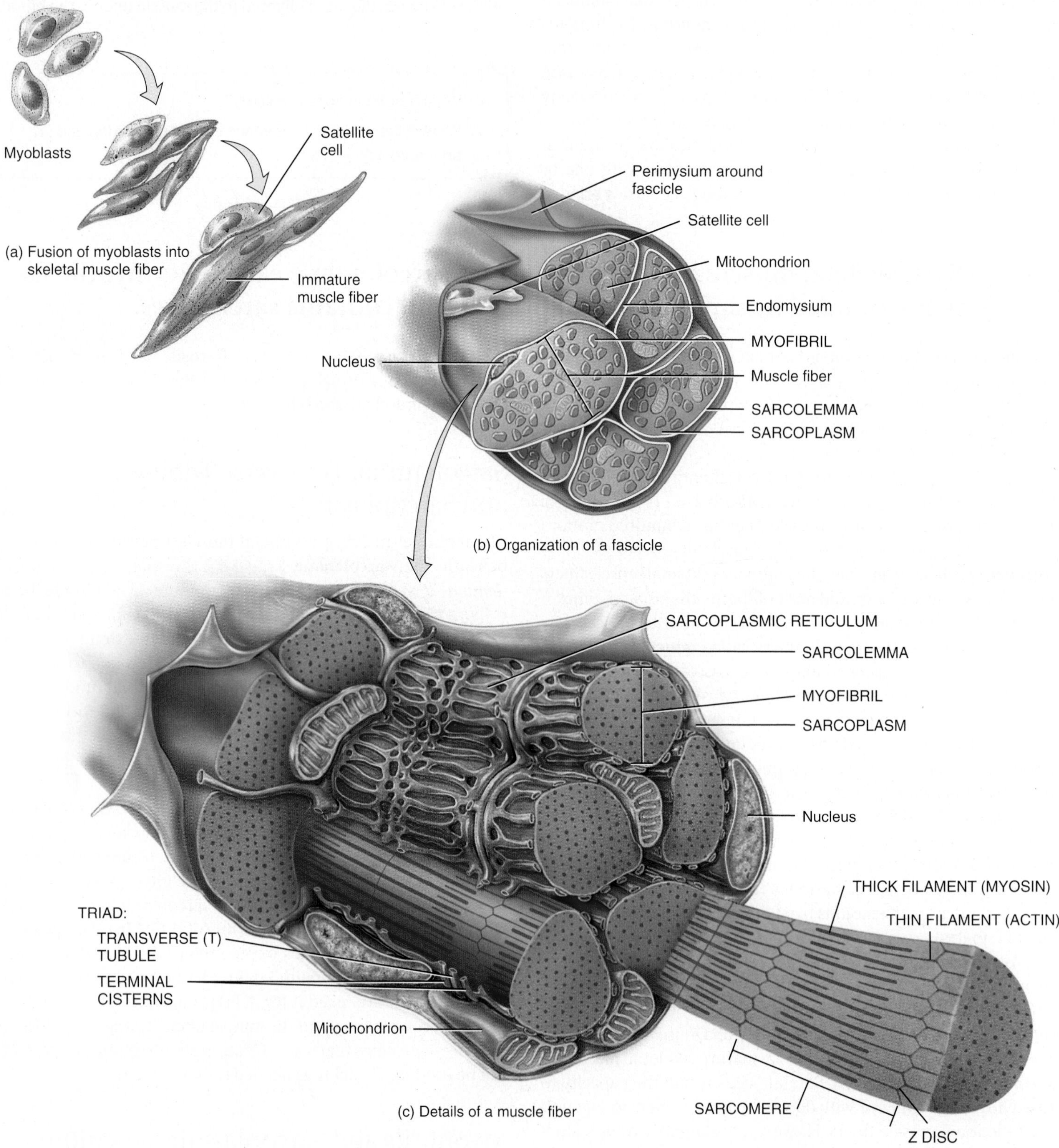

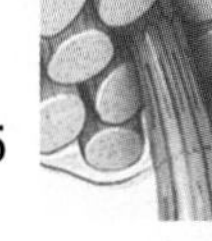

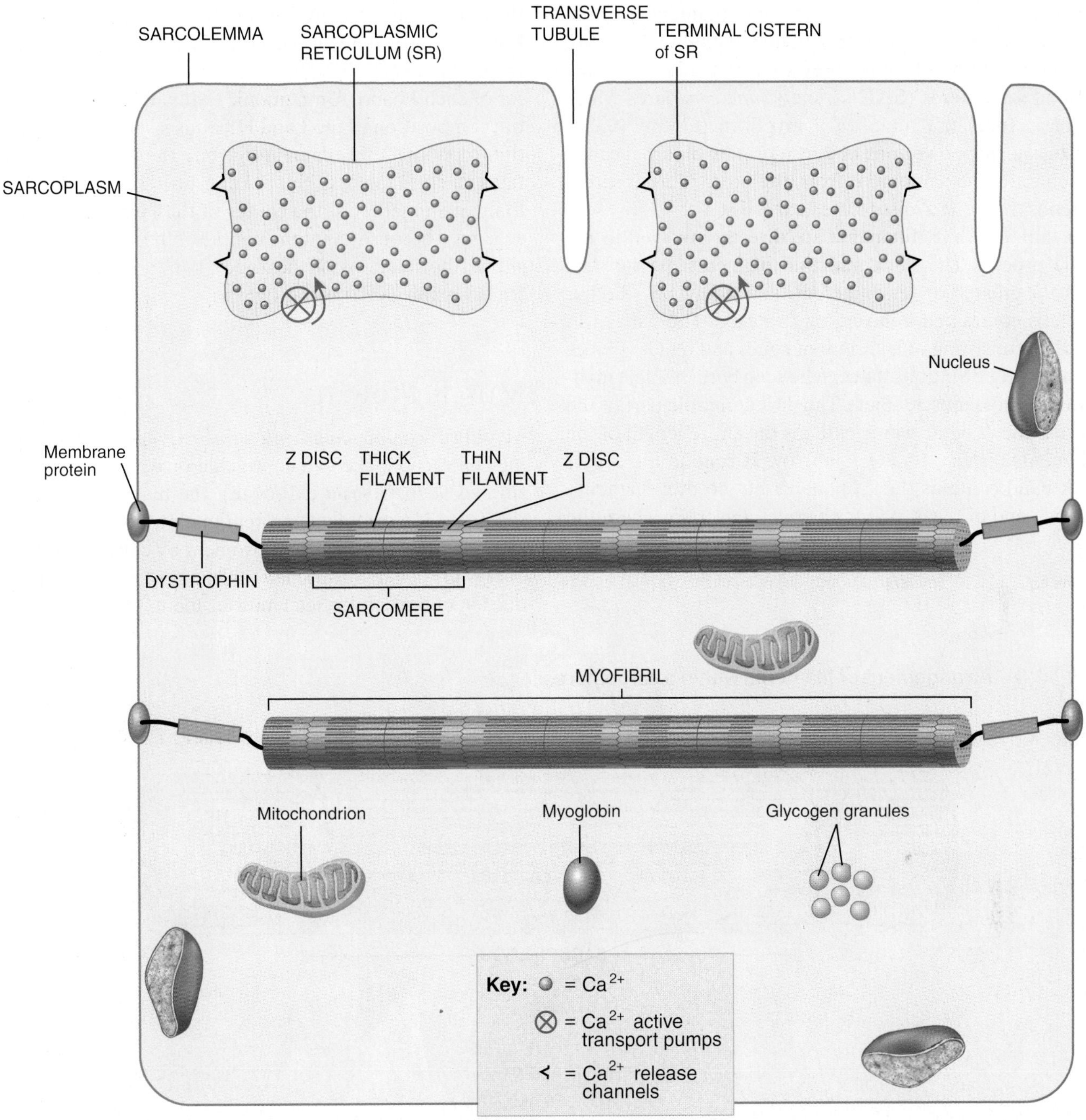

SARCOLEMMA
SARCOPLASMIC RETICULUM (SR)
TRANSVERSE TUBULE
TERMINAL CISTERN of SR
SARCOPLASM
Nucleus
Membrane protein
Z DISC
THICK FILAMENT
THIN FILAMENT
Z DISC
DYSTROPHIN
SARCOMERE
MYOFIBRIL
Mitochondrion
Myoglobin
Glycogen granules

**Key:** ● = $Ca^{2+}$
⊗ = $Ca^{2+}$ active transport pumps
< = $Ca^{2+}$ release channels

(d) Simplistic representation of a muscle fiber

**The contractile elements of muscle fibers, the myofibrils, contain overlapping thick and thin filaments.**

organelles of skeletal muscle (Figure 10.2c). Myofibrils are about 2 $\mu$m in diameter and extend the entire length of a muscle fiber. Their prominent striations make the entire muscle fiber appear striated.

A fluid-filled system of membranous sacs called the **sarcoplasmic reticulum (SR)** (sar′-kō-PLAZ-mik re-TIK-ū-lum) encircles each myofibril (Figure 10.2c). This elaborate system is similar to smooth endoplasmic reticulum in nonmuscular cells. Dilated end sacs of the sarcoplasmic reticulum called **terminal cisterns** (SIS-terns = reservoirs) butt against the T tubule from both sides. One T tubule and the two terminal cisterns on either side of it form a **triad** (*tri-* = three). In a relaxed muscle fiber, the sarcoplasmic reticulum stores calcium ions ($Ca^{2+}$). Release of $Ca^{2+}$ from the terminal cisterns of the sarcoplasmic reticulum triggers muscle contraction.

## Filaments and the Sarcomere

Within myofibrils are smaller protein structures, called *filaments* or *myofilaments* (Figure 10.2c). The diameter of the **thin**

**filaments** is about 8 nm*, and that of the **thick filaments** is about 16 nm. The filaments inside a myofibril do not extend the entire length of a muscle fiber. Instead, they are arranged in compartments called **sarcomeres** (SAR-kō-mers; *-mere* = part), which are the basic functional units of a myofibril (Figure 10.3a). Narrow, zigzag-shaped regions of dense protein material called **Z discs** separate one sarcomere from the next. Thus, a sarcomere extends from one Z disc to the next Z disc.

Both thin and thick filaments are directly involved in the contractile process. The thick and thin filaments overlap one another to a greater or lesser extent, depending on whether the muscle is contracted, relaxed, or stretched. The pattern of their overlap, consisting of a variety of zones and bands (Figure 10.3b), creates the striations that can be seen both in single myofibrils and in whole muscle fibers. The darker middle part of the sarcomere is the **A band**, which extends the entire length of the thick filaments (Figure 10.3b). A narrow **H zone** in the center of each A band contains thick filaments but no thin filaments. Toward each end of the A band is a *zone of overlap*, where the thick and thin filaments lie side by side. The **I band** is a lighter, less dense area that contains the rest of the thin filaments but no thick filaments (Figure 10.3b). A Z disc passes through the center of each I band. A mnemonic that will help you to remember the composition of the I and H bands is as follows: the letter I is thin (contains thin filaments), while the letter H is thick (contains thick filaments). Supporting proteins that hold the thick filaments together at the center of the H zone form the **M line**, so named because it is at the *middle* of the sarcomere. Figure 10.4 shows the relations of the zones, bands, and lines as seen in a transmission electron micrograph.

*One nanometer (nm) is $10^{-9}$ meter (0.001 $\mu$m); one micrometer ($\mu$m) = 1/25,000 of an inch.

## Muscle Proteins

Myofibrils contain *contractile proteins*, which generate force during contraction. The two contractile proteins in muscle are myosin and actin. **Myosin** (MĪ-ō-sin), the main component of thick filaments, is a contractile protein that pushes or pulls various cellular structures to achieve movement by converting the chemical energy in ATP to the mechanical energy of motion, that is, the production of force. In skeletal muscle, about 300 molecules of myosin

FIGURE 10.3 **Arrangement of filaments within a sarcomere.**

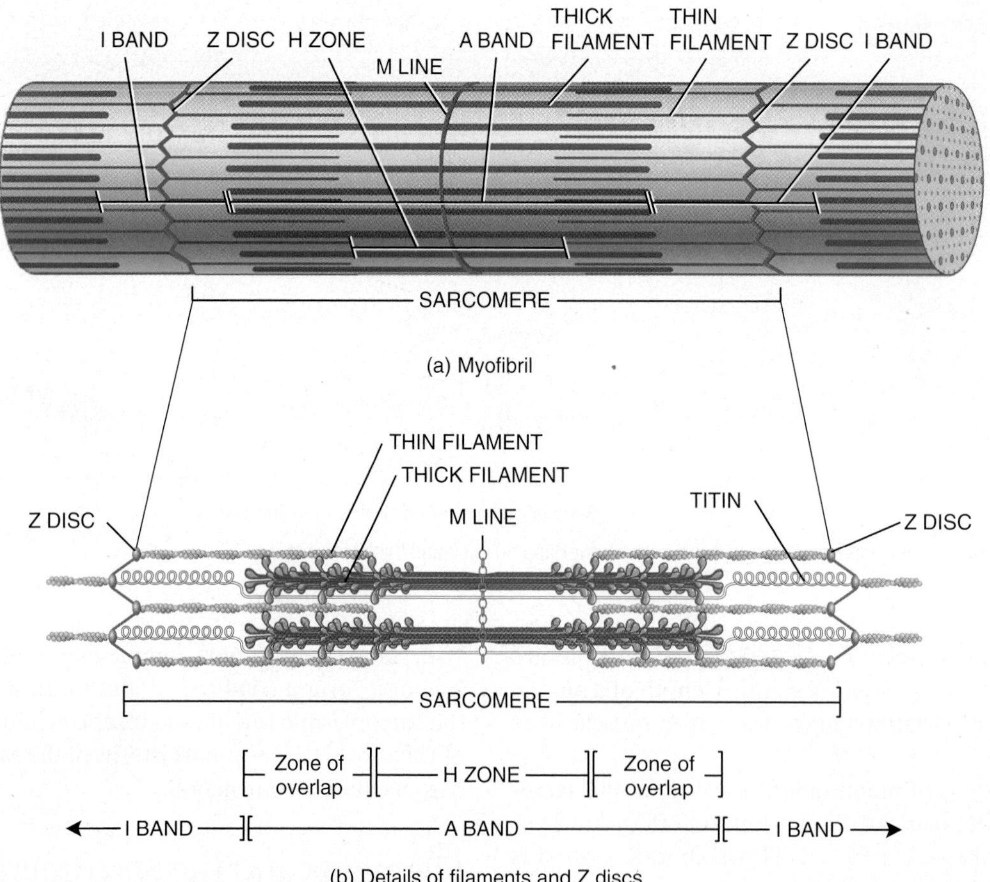


Myofibrils contain two types of contractile filaments: thick filaments and thin filaments.

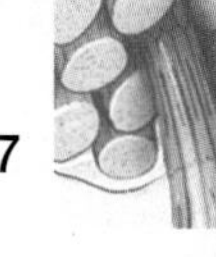

FIGURE 10.4 **Transmission electron micrograph showing the characteristic zones and bands of a sarcomere.**

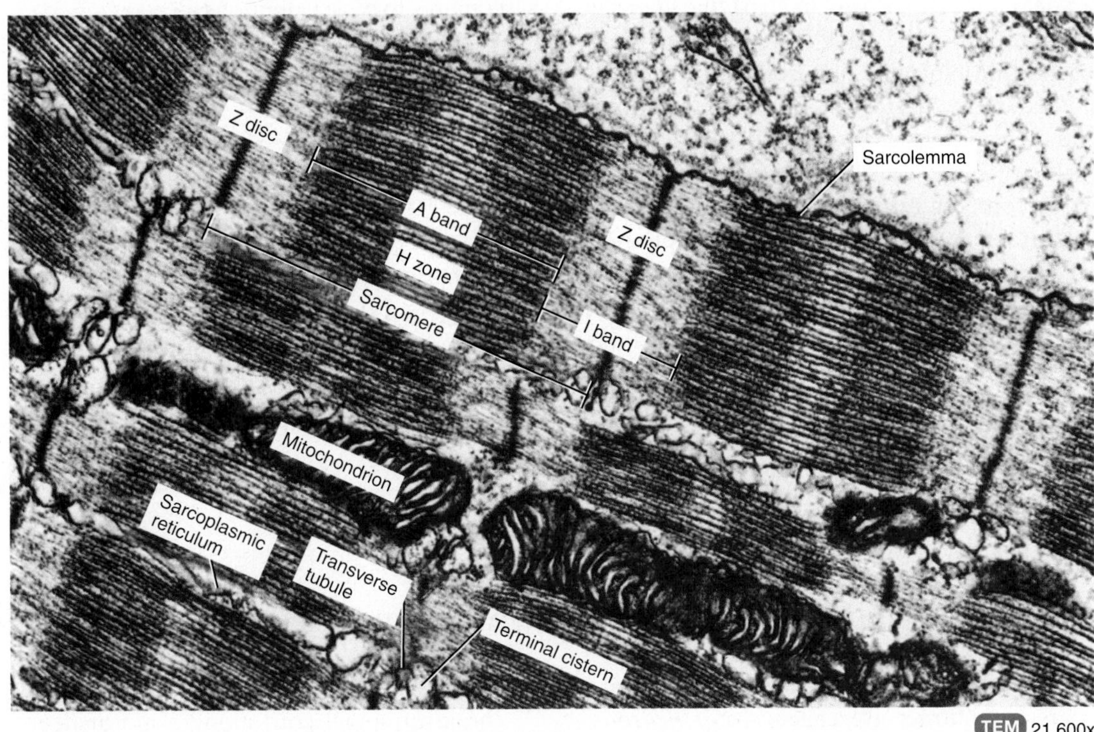


**The striations of skeletal muscle are alternating darker A bands and lighter I bands.**

form a single thick filament. Each myosin molecule is shaped like two golf clubs twisted together (Figure 10.5a). The *myosin tail* (twisted golf club handles) points toward the M line in the center of the sarcomere. Tails of neighboring myosin molecules lie parallel to one another, forming the shaft of the thick filament. The two projections of each myosin molecule (golf club heads) are called *myosin heads*. The heads project outward from the thick filament in a spiraling fashion, each head extending toward one of the six thin filaments that surround each thick filament.

Thin filaments are anchored to Z discs (see Figure 10.3b). Their main component is the contractile protein **actin** (AK-tin). Individual actin molecules join to form a thin filament that is twisted into a helix (Figure 10.5b). On each actin molecule is a *myosin-binding site*, where a myosin head can attach. Thin filaments also contain smaller amounts of two *regulatory proteins*, **tropomyosin** (trō-pō-MĪ-ō-sin) and **troponin** (TRŌ-pō-nin), which help switch the contraction process on and off. In relaxed muscle, myosin is blocked from binding to actin because strands of tropomyosin cover the myosin-binding sites on actin. The tropomyosin strand, in turn, is held in place by troponin.

Besides contractile and regulatory proteins, muscle contains about a dozen *structural proteins* that contribute to the alignment, stability, elasticity, and extensibility of myofibrils. One key structural protein is **titin** (*titan* = gigantic), the third

FIGURE 10.5 **Structure of thick and thin filaments.** (a) A thick filament contains myosin molecules (see enlargement). (b) Thin filaments contain actin, troponin, and tropomyosin.

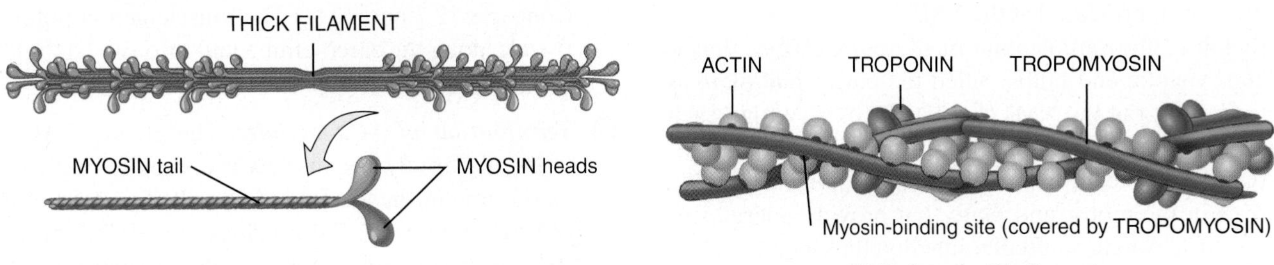


(a) One thick filament (above) and a myosin molecule (below)

(b) Portion of a thin filament

**Contractile proteins (myosin and actin) generate force during contraction; regulatory proteins (troponin and tropomyosin) help switch contraction on and off.**

most plentiful protein in skeletal muscle (after actin and myosin). Each titin filament connects a Z disc to an M line, thereby helping stabilize the position of the thick filament (see Figure 10.5b). The part of the titin molecule that extends from the Z disc is very elastic. Because it can stretch to at least four times its resting length and then spring back unharmed, titin accounts for much of the elasticity and extensibility of myofibrils. Titin probably helps the sarcomeres return to their resting length after a muscle has contracted or been stretched, may help prevent overextension of sarcomeres, and maintains the central location of the A bands.

**Dystrophin** (dis-TRŌ-fin) is a structural protein that links thin filaments of the sarcomere to integral membrane proteins of the sarcolemma, which are attached in turn to proteins in the connective tissue extracellular matrix that surrounds muscle fibers (see Figure 10.2d). Dystrophin is thought to help reinforce the sarcolemma and transmit the tension generated by the sarcomeres to muscle tendons.

✓ **CHECKPOINT**

**8.** What are the functions of T tubules, myoglobin, and sarcoplasmic reticulum, which are all unique to muscle fibers?

**9.** Arrange the following elements from smallest to largest: muscle fiber, thick filament, myofibril.

**10.** List the components of a myofibril found within the I band, the A band, and the H zone. Which components attach to the M line? To the Z disc?

**11.** What type of proteins compose thick filaments? Thin filaments?

## 10.5 The neuromuscular junction is the site where a muscle action potential is initiated.

As noted earlier in the chapter, the neurons that stimulate skeletal muscle fibers to contract are called *motor neurons*. Each motor neuron has a threadlike axon that extends from the brain or spinal cord to a group of skeletal muscle fibers. A muscle fiber contracts in response to muscle action potentials that arise at the **neuromuscular junction** (**NMJ**) (noo-rō-MUS-kū-lar), the synapse between a motor neuron and a skeletal muscle fiber (Figure 10.6a). A **synapse** is a region where communication occurs between a neuron and another cell—in this case, between a motor neuron and a skeletal muscle fiber. A small gap, called the *synaptic cleft*, separates the two cells. Because the cells do not physically touch, the action potential from the motor neuron cannot "jump the gap" to directly excite the muscle fiber. Instead, the motor neuron communicates with the muscle fiber indirectly, by releasing a chemical called a **neurotransmitter**.

At the NMJ, the end of the motor neuron, called the **axon terminal**, divides into a cluster of **synaptic end bulbs** (Figure 10.6a, b), the neural part of the NMJ. Suspended in the cytosol within each synaptic end bulb are membrane-enclosed sacs called **synaptic vesicles**. Inside each synaptic vesicle are thousands of molecules of **acetylcholine** (**ACh**) (as′-ē-til-KŌ-lēn), the neurotransmitter released at the NMJ.

The region of the sarcolemma of the muscle fiber that is opposite the synaptic end bulbs, called the **motor end plate**, is the muscle fiber part of the NMJ (Figure 10.6b, c). Within each motor end plate are **acetylcholine receptors** that bind specifically to ACh. These receptors are abundant in *junctional folds*, deep grooves in the motor end plate that provide a large surface area for ACh. A neuromuscular junction thus includes all of the synaptic end bulbs on one side of the synaptic cleft, plus the motor end plate of the muscle fiber on the other side.

An impulse (nerve action potential) excites a skeletal muscle fiber in the following way (Figure 10.6b, c):

1. ***Release of acetylcholine.*** An impulse travels from the brain or spinal cord along a motor neuron to the muscle fiber (Figure 10.6b). Arrival of the impulse at the synaptic end bulb stimulates voltage-gated channels to open. Because calcium ions ($Ca^{2+}$) are more concentrated in the extracellular fluid, $Ca^{2+}$ flows inward through the open channels. The entering $Ca^{2+}$ stimulates the synaptic vesicles to undergo exocytosis (Figure 10.6c). During exocytosis, the synaptic vesicles fuse with the motor neuron's plasma membrane, releasing ACh into the synaptic cleft. The ACh then diffuses across the synaptic cleft between the motor neuron and the motor end plate.

2. ***Activation of ACh receptors.*** Binding of two molecules of ACh to a receptor on the motor end plate opens an ion channel in the ACh receptor. Once the channel is open, small cations, most importantly $Na^+$, can flow across the membrane.

3. ***Generation of muscle action potential.*** The inflow of $Na^+$ (down its electrochemical gradient) makes the inside of the muscle fiber more positively charged. This change in the membrane potential triggers a muscle action potential. (The details of action potential generation are discussed in Concepts 12.7 and 12.8.) The muscle action potential then travels along the sarcolemma and into the T tubules, stimulating the contraction process.

4. ***Termination of ACh activity.*** The effect of ACh binding lasts only briefly because ACh is rapidly broken down by an enzyme in the synaptic cleft called **acetylcholinesterase** (**AChE**) (as′-ē-til-kō′-lin-ES-ter-ās). AChE breaks down ACh into acetyl and choline, products that cannot activate the ACh receptor.

Each impulse normally elicits one muscle action potential. If another impulse releases more acetylcholine, steps 2 and 3

FIGURE 10.6 **Structure of the neuromuscular junction, the synapse between a motor neuron and a skeletal muscle fiber.**

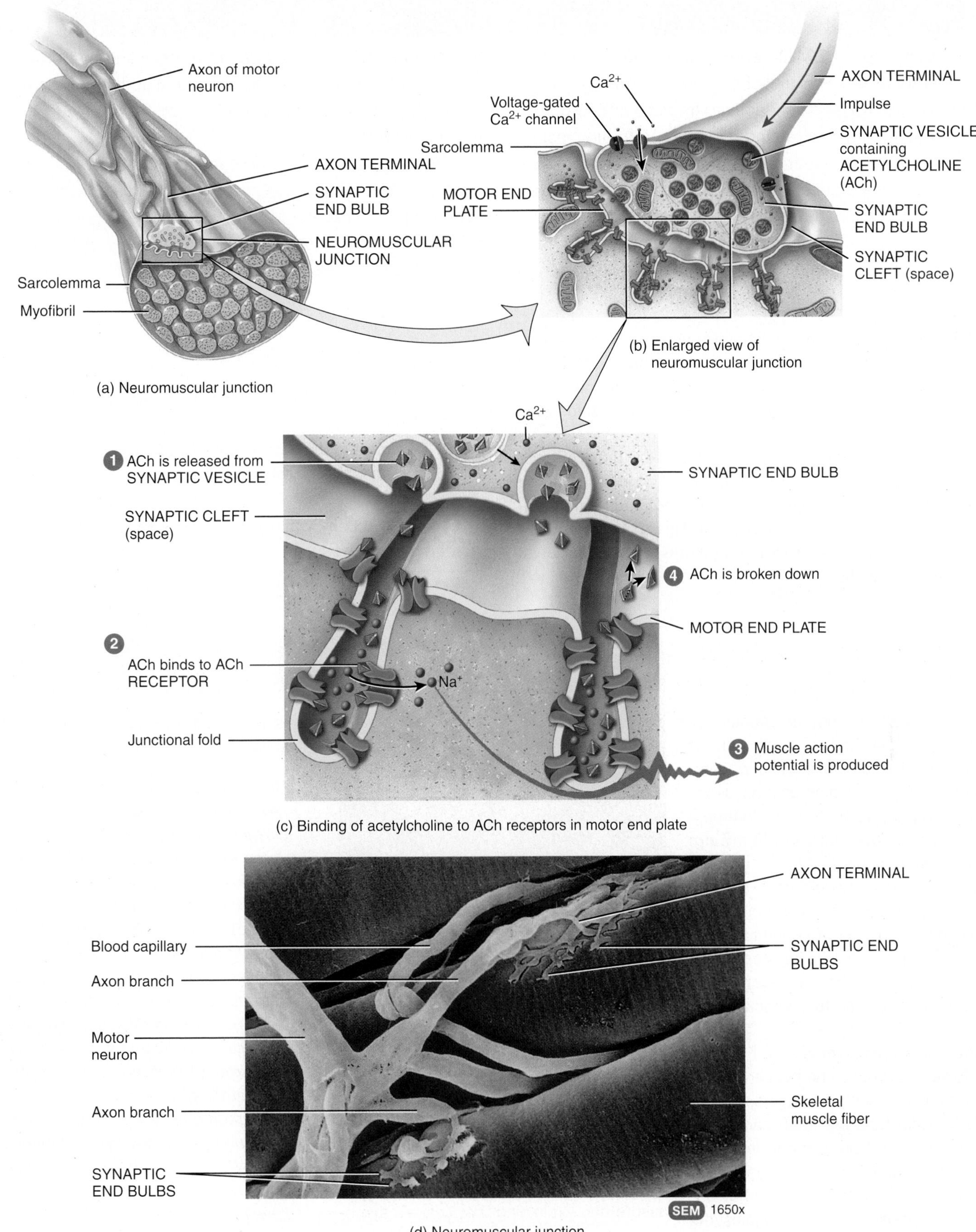


**Synaptic end bulbs at the tips of axon terminals contain synaptic vesicles filled with acetylcholine.**

## CLINICAL CONNECTION | *Electromyography*

**Electromyography (EMG)** (e-lek′-trō-mī-OG-ra-fē; *electro-* = electricity; *myo-* = muscle; *-graph* = to write) is a test that measures the electrical activity (muscle action potentials) in resting and contracting muscles. Normally, resting muscle produces no electrical activity; a slight contraction produces some electrical activity; and a more forceful contraction produces increased electrical activity. In the procedure, a ground electrode is placed over the muscle to be tested to eliminate background electrical activity. Then, a fine needle attached by wires to a recording instrument is inserted into the muscle. The electrical activity of the muscle is displayed as waves on an oscilloscope and heard through a loudspeaker.

EMG helps to determine if muscle weakness or paralysis is due to a malfunction of the muscle itself or the nerves supplying the muscle. EMG is also used to diagnose certain muscle disorders, such as muscular dystrophy, and to understand which muscles function during complex movements.

repeat. When impulses cease in the motor neuron, ACh is no longer released, and AChE rapidly breaks down the ACh already present in the synaptic cleft. The ion channels in the motor end plate close, and action potentials are no longer generated in the muscle fiber.

Because skeletal muscle fibers often are very long cells, the NMJ usually is located near the midpoint of a skeletal muscle fiber. Muscle action potentials arise at the NMJ and then propagate toward both ends of the muscle fiber. This arrangement permits nearly simultaneous activation (and thus contraction) of all parts of the muscle fiber.

### ✓ CHECKPOINT

**12.** Which portion of the sarcolemma contains acetylcholine receptors?

**13.** Describe the steps that occur for an action potential arriving at a synaptic end bulb to result in an action potential in a muscle fiber.

**14.** What happens to acetylcholine once it has bound to its receptor?

## RETURN TO **Jesse's Story**

As Jesse helped his mother unpack the groceries, he grabbed a large can of peaches. He tried to put it in the cupboard, but couldn't lift his arm high enough. Jesse set the heavy can back down. "Momma, I think I hurt my shoulder. I can't lift my arm up any higher than this."

As he waited in the doctor's examining room, Jesse thought about the night before, when he overheard his parents talking about his Uncle Walter, who had something called muscular dystrophy. "Come in," Jesse said in response to a knock on the door.

After introducing himself, Dr. Fitts had Jesse lower his arms as he pushed on the shoulder muscles with his fingers, checking the muscle tone. He then had Jesse raise his arms and bring them around in front. "Hmm, you have some winging of the scapula. Your shoulder is somewhat unstable. It's supposed to stay flat against the ribs, but sometimes the muscles weaken and it doesn't stay where it's supposed to." When Dr. Fitts asked Jesse to whistle, he could not pucker his lips. "I'm going to order some tests—a blood test, an electromyogram (EMG), a nerve conduction study, and a muscle biopsy."

As Jesse got dressed, Dr. Fitts spoke to his mother, indicating that Jesse's muscle weakness might be related to Uncle Walter's muscular dystrophy, because it is a genetic disorder. He did his best to reassure her, explaining that Jesse's symptoms indicated a less severe form of the disease.

A. *How could a reduced blood supply to the muscles of Jesse's shoulder explain his muscle weakness?*

B. *Jesse will have a biopsy (removal of a sample of living tissue for microscopic analysis) performed to assess the microscopic structure of his muscle; describe the cells and tissues that will be seen.*

C. *The pathology report on Jesse's muscle indicated the presence of "isolated small atrophic fibers intermixed with hypertrophic muscle fibers." What is muscle hypertrophy? What is muscle atrophy?*

D. *The pathology report also indicated that the sarcomeres "appear regular and show no signs of disarray or degeneration." What is a sarcomere, and what makes skeletal muscle tissue appear to have regular repeating striations?*

# 10.6 An action potential releases calcium ions that allow thick filaments to bind to and pull thin filaments toward the center of the sarcomere.

Once an action potential is propagated along the sarcolemma and into the T tubules, muscle contraction begins. The model describing the contraction of muscle is known as the **sliding filament mechanism**. Muscle contraction occurs because myosin heads attach to and "walk" along the thin filaments at both ends of a sarcomere, progressively pulling the thin filaments toward the M line (Figure 10.7). As a result, the thin filaments *slide* inward and meet at the center of a sarcomere. They may even move so far inward that their ends overlap (Figure 10.7c). As the thin filaments slide inward, the Z discs come closer together, and the sarcomere shortens. Note that the lengths of the individual thick and thin filaments do not change; it is the amount of overlap between them that changes. Shortening of the individual sarcomeres causes shortening of the whole muscle fiber, which in turn leads to shortening of the entire muscle.

## Excitation–Contraction Coupling

**Excitation–contraction coupling** refers to the steps that connect excitation (a muscle action potential propagating along the sarcolemma and into the T tubules) to contraction (sliding of the filaments). An increase in $Ca^{2+}$ concentration in the sarcoplasm starts muscle contraction; a decrease stops it. When a muscle fiber is relaxed, the concentration of $Ca^{2+}$ in its sarcoplasm is very low. However, a huge amount of $Ca^{2+}$ is stored inside the sarcoplasmic reticulum (Figure 10.8a). When a muscle action potential propagates along the sarcolemma and into the T tubules, it causes **$Ca^{2+}$ release channels** in the SR membrane to open (Figure 10.8b). As these channels open, $Ca^{2+}$ flows out of the SR into the sarcoplasm around the thick and thin filaments. The released $Ca^{2+}$ combines with troponin, moving the troponin–tropomyosin complex away from the myosin-binding sites on actin. Once these binding sites are exposed, myosin heads bind to them, and the contraction cycle begins.

## The Contraction Cycle

At the onset of contraction, the sarcoplasmic reticulum releases $Ca^{2+}$ into the sarcoplasm. The released $Ca^{2+}$ ions bind to troponin, which moves the troponin–tropomyosin complexes away from the myosin-binding sites on actin. Once the binding sites are uncovered, the contraction cycle begins. The **contraction**

FIGURE 10.7 **Sliding filament mechanism of muscle contraction, as it occurs in two adjacent sarcomeres.**

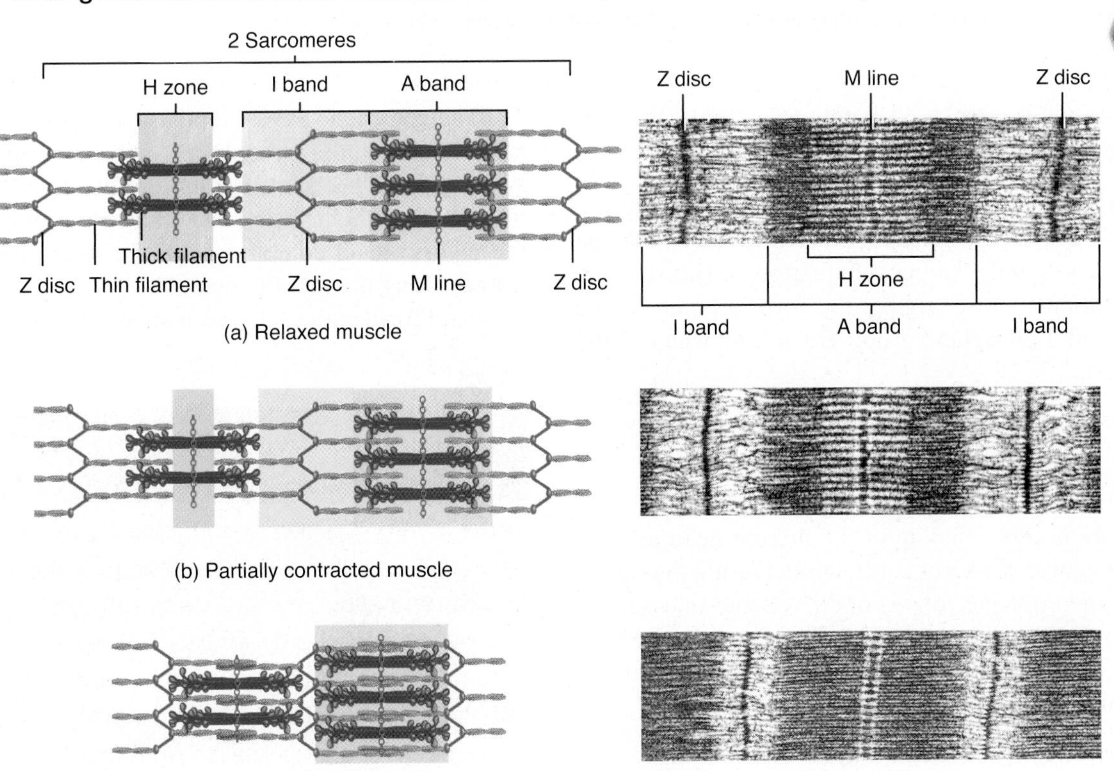


(a) Relaxed muscle

(b) Partially contracted muscle

(c) Maximally contracted muscle

**During muscle contractions, thin filaments move toward the M line of each sarcomere.**

**FIGURE 10.8 The role of $Ca^{2+}$ in the regulation of contraction by troponin and tropomyosin.** (a) During relaxation, the level of $Ca^{2+}$ in the sarcoplasm is low because active transport pumps force $Ca^{2+}$ into the sarcoplasmic reticulum (SR). (b) A muscle action potential propagating along a T tubule opens $Ca^{2+}$ release channels in the SR, calcium ions flow into the sarcoplasm, and contraction begins.

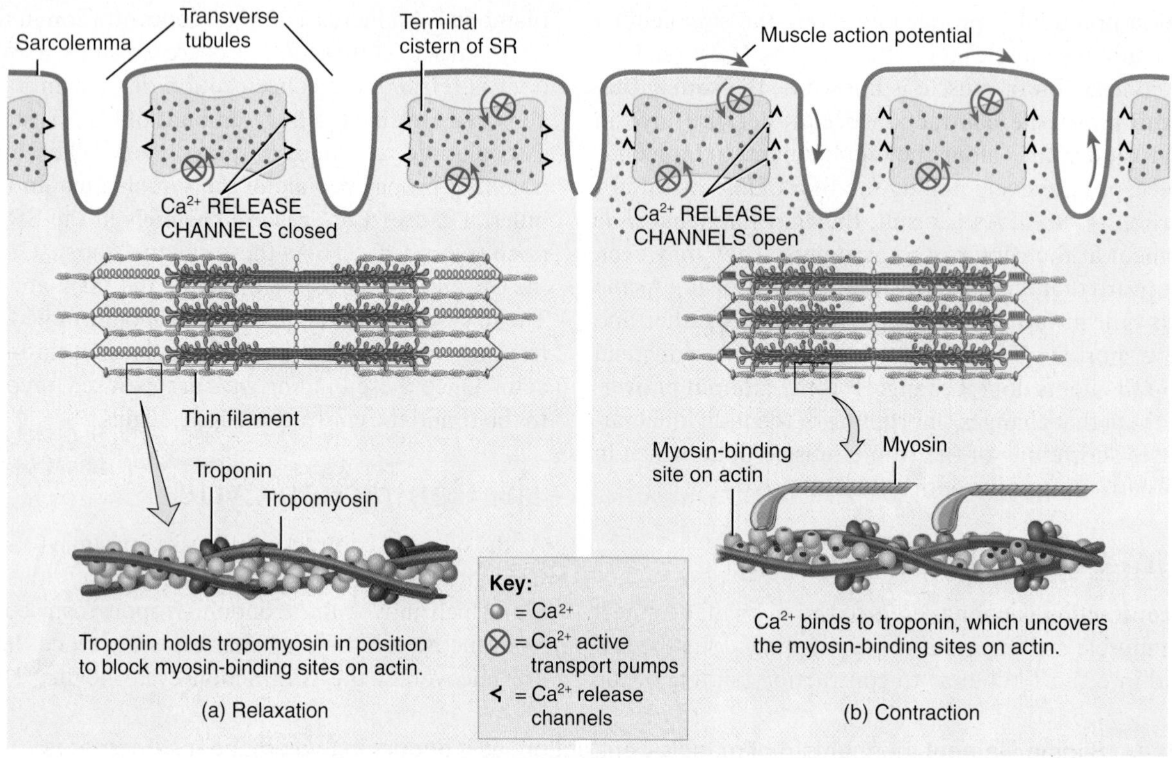


**An increase in the $Ca^{2+}$ level in the sarcoplasm starts the sliding of thin filaments; when the level of $Ca^{2+}$ in the sarcoplasm declines, sliding stops.**

**cycle**—the repeating sequence of events that causes the filaments to slide—consists of four steps (Figure 10.9):

1. ***ATP splits.*** The myosin head includes an ATP-binding site and ATPase, an enzyme that splits ATP into ADP (adenosine diphosphate) and P (a phosphate group). This splitting reaction reorients and energizes the myosin head. Notice that ADP and a phosphate group are still attached to the myosin head.
2. ***Myosin attaches to actin.*** The energized myosin head attaches to the myosin-binding site on actin and releases the phosphate group.
3. ***Power stroke occurs.*** Binding of the myosin head to actin triggers the power stroke of contraction. During the power stroke, the myosin head rotates or swivels and releases the ADP. The myosin head generates force as it rotates toward the center of the sarcomere, sliding the thin filament past the thick filament toward the M line.
4. ***Myosin detaches from actin.*** At the end of the power stroke, the myosin head remains firmly attached to actin until it binds another molecule of ATP. As ATP binds to the ATP-binding site on the myosin head, the myosin head detaches from actin.

The contraction cycle repeats for as long as ATP and $Ca^{2+}$ are available in the sarcoplasm. Splitting of ATP by the myosin ATPase again reorients the myosin head and transfers energy from ATP to the myosin head. Once reoriented and energized, the myosin head combines with the next myosin-binding site farther along the thin filament. The myosin heads keep rotating back and forth with each power stroke. Each power stroke pulls

## CLINICAL CONNECTION | *Rigor Mortis*

After death, cellular membranes become leaky. Calcium ions leak out of the sarcoplasmic reticulum into the sarcoplasm and allow myosin heads to bind to actin. ATP synthesis ceases shortly after breathing stops, however, so the myosin heads cannot detach from actin. The resulting condition, in which muscles are in a state of rigidity (cannot contract or stretch), is called **rigor mortis** (rigidity of death). Rigor mortis begins 3–4 hours after death and lasts about 24 hours; then it disappears as proteolytic enzymes from lysosomes digest the muscle proteins.

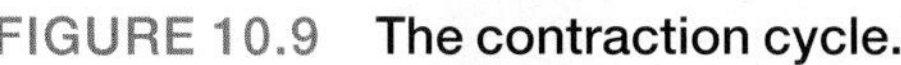

FIGURE 10.9 The contraction cycle.

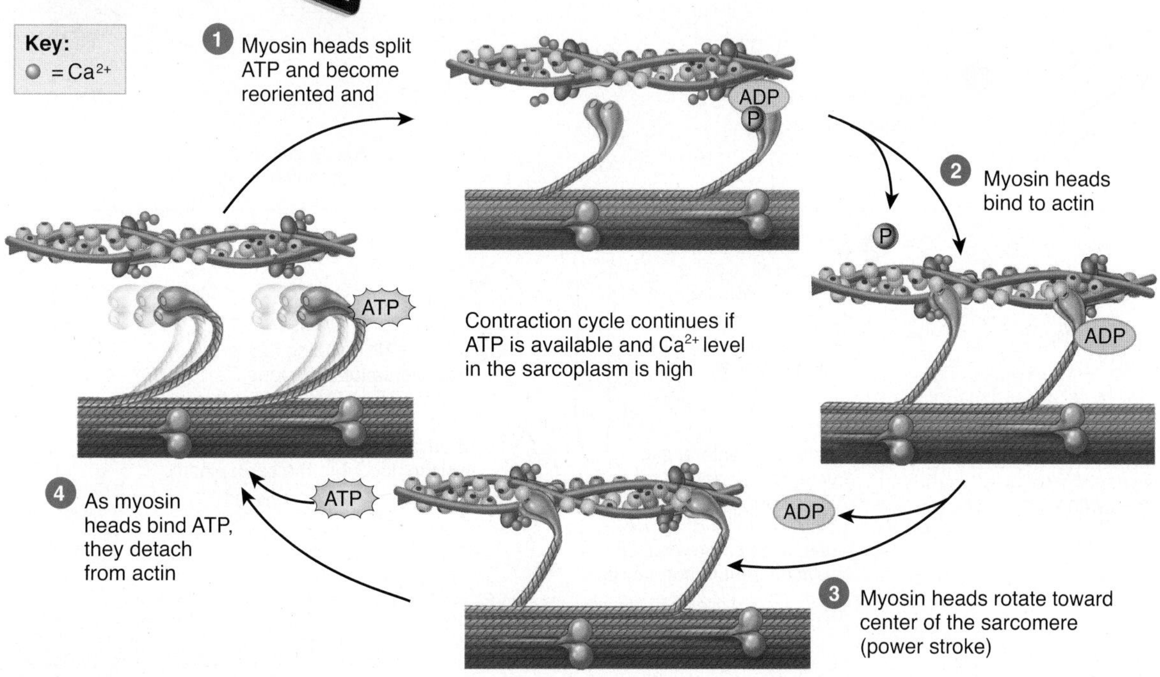
Key:
= $Ca^{2+}$
1 Myosin heads split ATP and become reoriented and
ADP
P
2 Myosin heads bind to actin
P
ATP
Contraction cycle continues if ATP is available and $Ca^{2+}$ level in the sarcoplasm is high
ADP
4 As myosin heads bind ATP, they detach from actin
ATP
ADP
3 Myosin heads rotate toward center of the sarcomere (power stroke)

**During the power stroke of contraction, myosin heads rotate and move the thin filaments past the thick filaments toward the center of the sarcomere.**

the thin filaments closer to the M line. Each of the 600 myosin heads in one thick filament repeatedly attach to actin, rotate, and detach about five times per second. At any one instant, some of the myosin heads are attached to actin and producing force, and others are detached and preparing to bind again.

The contraction cycle is analogous to walking on a foot-powered, nonmotorized treadmill. One foot (myosin head) strikes the treadmill belt (thin filament) and pushes it backward (toward the M line). Then the other foot comes down and imparts a second push. The belt (thin filament) moves smoothly while the walker (thick filament) remains stationary. Each myosin head progressively "walks" along a thin filament, coming closer to the Z disc with each "step," while the thin filament moves closer to the M line. Like the legs of a walker, the myosin head needs a constant supply of energy to keep going—one molecule of ATP for each contraction cycle.

As the contraction cycle continues, movement of myosin heads applies the force that draws the Z discs toward each other, and the sarcomere shortens. During a maximal muscle contraction, the sarcomere can shorten by as much as half the resting length. The Z discs, in turn, pull on neighboring sarcomeres, shortening the whole muscle fiber, which ultimately leads to shortening of the entire muscle. As the fibers of a skeletal muscle start to shorten, they first pull on their connective tissue coverings (endomysium, perimysium, and epimysium) and tendons. The coverings and tendons become taut, and the tension passed through the tendons pulls on the bones to which they are attached. The result is movement of a part of the body.

## Relaxation

Two changes permit a muscle fiber to relax after it has contracted. First, when impulses cease being transmitted by the motor neuron, ACh release stops and AChE rapidly breaks down the ACh already present in the synaptic cleft (see Figure 10.8). As ion channels close, muscle action potentials are no longer generated and propagated throughout the T tubules. The $Ca^{2+}$ release channels in the sarcoplasmic reticulum membrane close, stopping the release of $Ca^{2+}$ into the sarcoplasm.

Second, the sarcoplasmic reticulum membrane also contains **$Ca^{2+}$ active transport pumps** that use ATP to constantly move $Ca^{2+}$ from the sarcoplasm into the SR (see Figure 10.8a). With $Ca^{2+}$ release channels closed and the active transport pumps returning $Ca^{2+}$ back into the SR, the concentration of calcium ions in the sarcoplasm quickly decreases. Inside the SR, molecules of a calcium-binding protein, appropriately called **calsequestrin** (kal′-sē-KWES-trin), bind to the $Ca^{2+}$, enabling even more $Ca^{2+}$ to be stored (sequestered) within the SR. As the $Ca^{2+}$ level in the sarcoplasm drops, the troponin–tropomyosin complexes slide back over and cover the myosin-binding sites on actin. No longer attached to myosin heads, the thin filaments slip back to their relaxed positions, and the muscle fiber relaxes. As the fibers of a skeletal muscle lengthen, their connective tissue coverings and tendons become slack, and tendons decrease their pull on the bones to which they are attached.

Figure 10.10 summarizes the events that occur during contraction and relaxation in a skeletal muscle fiber.

**FIGURE 10.10 Summary of the events of contraction and relaxation in a skeletal muscle fiber.**

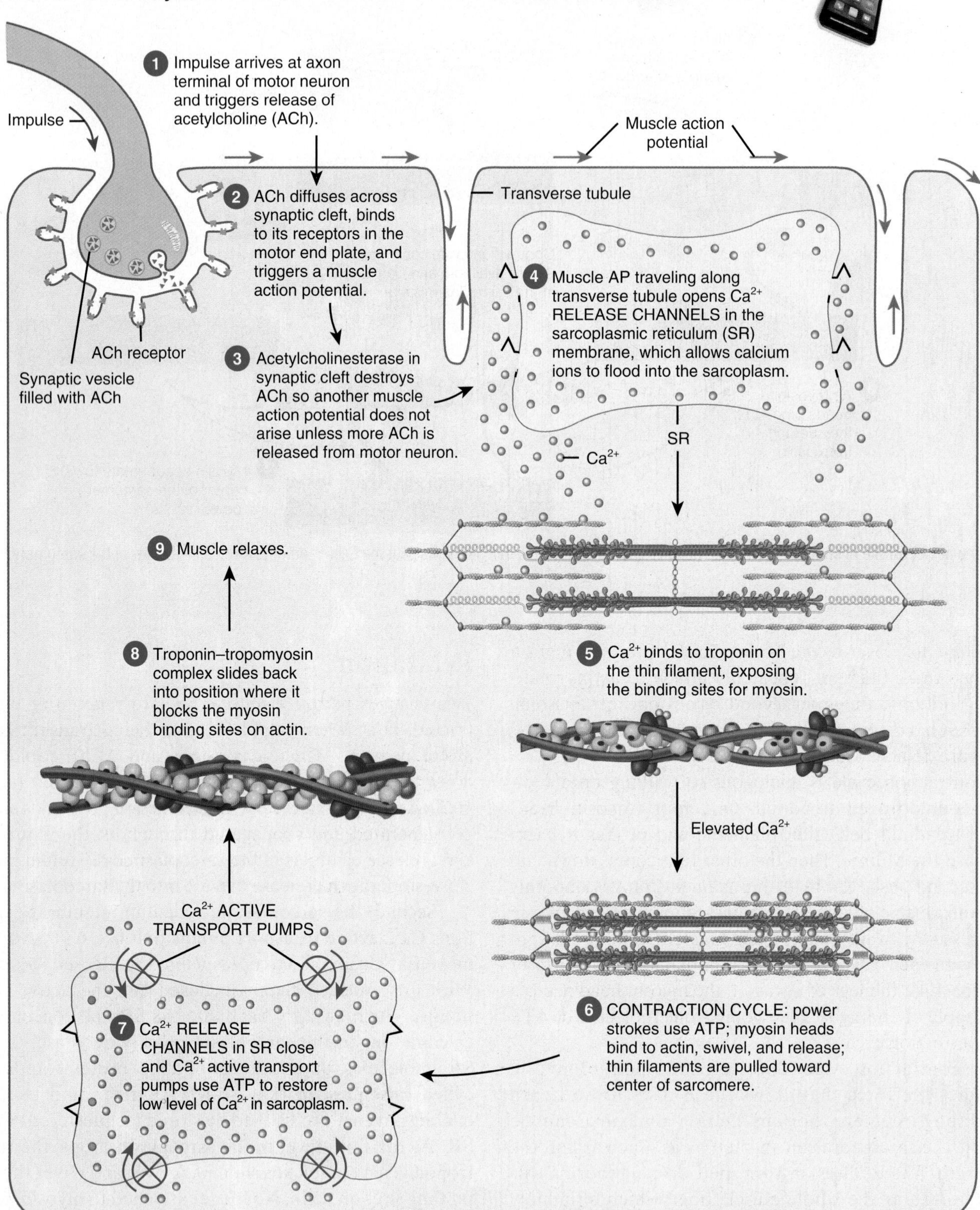

Impulse
1 Impulse arrives at axon terminal of motor neuron and triggers release of acetylcholine (ACh).
Muscle action potential
2 ACh diffuses across synaptic cleft, binds to its receptors in the motor end plate, and triggers a muscle action potential.
Transverse tubule
ACh receptor
Synaptic vesicle filled with ACh
3 Acetylcholinesterase in synaptic cleft destroys ACh so another muscle action potential does not arise unless more ACh is released from motor neuron.
4 Muscle AP traveling along transverse tubule opens Ca2+ RELEASE CHANNELS in the sarcoplasmic reticulum (SR) membrane, which allows calcium ions to flood into the sarcoplasm.
SR
Ca2+
9 Muscle relaxes.
8 Troponin–tropomyosin complex slides back into position where it blocks the myosin binding sites on actin.
5 Ca2+ binds to troponin on the thin filament, exposing the binding sites for myosin.
Elevated Ca2+
Ca2+ ACTIVE TRANSPORT PUMPS
7 Ca2+ RELEASE CHANNELS in SR close and Ca2+ active transport pumps use ATP to restore low level of Ca2+ in sarcoplasm.
6 CONTRACTION CYCLE: power strokes use ATP; myosin heads bind to actin, swivel, and release; thin filaments are pulled toward center of sarcomere.


**Acetylcholine released at the neuromuscular junction triggers a muscle action potential, which leads to muscle contraction.**

✓ CHECKPOINT

15. What happens to the distance between neighboring Z discs when thin filaments in a sarcomere slide toward the M line? When thin filaments slide away from the M line?
16. As each myosin head detaches from a thin filament, what stops the thin filament from sliding back to its original, relaxed position?
17. How does sarcomere shortening result in shortening of an entire muscle and movement of a bone?
18. How does a skeletal muscle fiber relax?
19. What are three functions of ATP in muscle contraction and relaxation?

# 10.7 Muscle tension is controlled by stimulation frequency and motor unit recruitment.

A single impulse in a motor neuron elicits a single muscle action potential in all muscle fibers with which it forms synapses. The contraction that results from a single muscle action potential has significantly smaller force than the maximum force or tension the fiber is capable of producing. The total tension that a single fiber can produce depends on the rate at which impulses arrive at the neuromuscular junction. The number of impulses per second is the *frequency of stimulation*. When considering the contraction of a whole muscle, the total tension it can produce depends on the number of muscle fibers that are contracting in unison.

## Motor Units

Even though each skeletal muscle fiber has only a single neuromuscular junction, the axon of a motor neuron branches out and forms neuromuscular junctions with many different muscle fibers. A **motor unit** consists of one motor neuron plus all the skeletal muscle fibers it stimulates (Figure 10.11). A single motor neuron makes contact with an average of 150 muscle fibers within a muscle, and all of the muscle fibers in one motor unit contract in unison. Typically, the muscle fibers of a motor unit are dispersed throughout a muscle rather than clustered together.

Muscles that control small, precise movements consist of many small motor units. For instance, muscles of the larynx (voice box) that control voice production have as few as two or three muscle fibers per motor unit, and muscles controlling eye movements may have 10 to 20 muscle fibers per motor unit. In contrast, skeletal muscles responsible for large, powerful movements, such as the biceps brachii muscle in the arm and the gastrocnemius muscle in the calf of the leg, have as many as 2000 to 3000 muscle fibers in some motor units. Stimulation of one motor neuron causes all of the muscle fibers in that motor unit to contract at the same time. Accordingly, the total strength of a muscle contraction depends, in part, on the size of the motor units and on the number of motor units that are activated at the same time.

FIGURE 10.11 **Motor units.** Shown are two motor neurons, each supplying the muscle fibers of its motor unit.

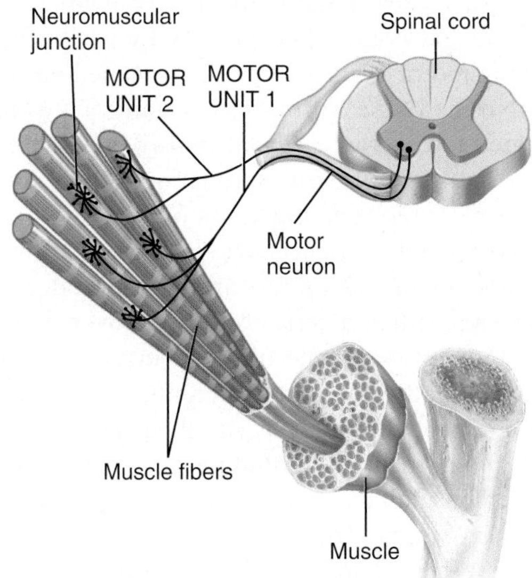


A motor unit consists of a motor neuron plus all the muscle fibers it stimulates.

## Twitch Contraction

A **twitch contraction** is the brief contraction of all the muscle fibers in a motor unit in response to a single impulse in its motor neuron. In the laboratory, a twitch can be produced by direct electrical stimulation of a motor neuron or its muscle fibers. The record of a muscle contraction, called a **myogram** (MĪ-ō-gram), is shown in Figure 10.12 (see also Figure 10.13a).

Note that a brief delay, called the **latent period**, occurs between application of the stimulus (time zero on the graph) and the

FIGURE 10.12 **Myogram of a twitch contraction.** The arrow indicates the time at which the stimulus occurred.

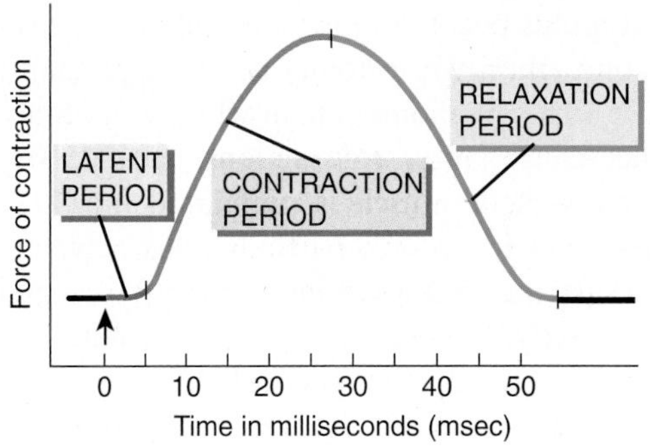


A myogram is a record of a muscle contraction.

**FIGURE 10.13 Myograms showing the effects of different frequencies of stimulation.** (a) Single twitch. (b) When a second stimulus occurs before the muscle has relaxed, wave summation occurs, and the second contraction is stronger than the first. (The solid black line indicates the force of contraction expected in a single twitch.) (c) In unfused tetanus, the curve looks jagged due to partial relaxation of the muscle between stimuli. (d) In fused tetanus, the contraction force is steady, sustained, and strong.

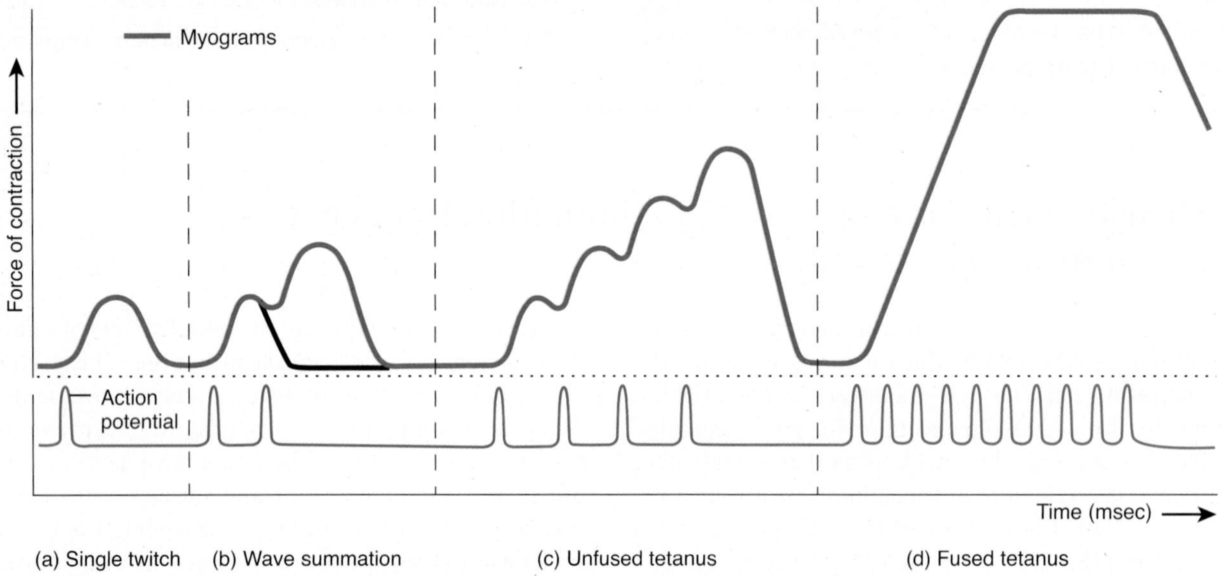


**Due to wave summation, the tension produced during a sustained contraction is greater than during a single twitch.**

beginning of contraction. During the latent period, the muscle action potential sweeps over the sarcolemma, and calcium ions are released from the SR. During the second phase, the **contraction period** (upward tracing), calcium ions bind to troponin, myosin-binding sites on actin are exposed, myosin heads attach to actin, and the resulting power strokes generate the force of contraction. During the third phase, the **relaxation period**, calcium ions are actively transported back into the SR; the level of $Ca^{2+}$ in the sarcoplasm decreases, myosin-binding sites are covered by tropomyosin, myosin heads detach from actin, power strokes cease, and the muscle fiber relaxes. The actual duration of these periods depends on the type of skeletal muscle fiber. Some fibers, such as the fast-twitch fibers that move the eyes (described shortly), have contraction periods as brief as 10 msec and equally brief relaxation periods. Others, such as the slow-twitch fibers that move the legs, have contraction and relaxation periods of about 100 msec each.

## Frequency of Stimulation

If a second stimulus occurs before a muscle fiber has relaxed, the second contraction will actually be stronger than the first (Figure 10.13b). This phenomenon, in which stimuli arriving in close sequence cause larger contractions, is called **wave summation**. When a skeletal muscle is stimulated at a rate of 20 to 30 times per second, it can only partially relax between stimuli. The result is a sustained but wavering contraction called **unfused tetanus** (*tetan-* = rigid, tense; Figure 10.13c). When a skeletal muscle fiber is stimulated at a higher rate of 80 to 100 times per second, it does not relax at all. The result is **fused tetanus**, a sustained contraction in which individual twitches cannot be detected (Figure 10.13d).

Wave summation and both kinds of tetanus result from the release of additional $Ca^{2+}$ from the sarcoplasmic reticulum by subsequent stimuli; the extra $Ca^{2+}$ adds to the $Ca^{2+}$ still in the sarcoplasm from the previous stimulus. Because of the $Ca^{2+}$ buildup in the sarcoplasm, contractions generated during fused tetanus are 5 to 10 times more forceful than those produced during a single twitch. Even so, smooth, sustained voluntary muscle contractions are achieved mainly by out-of-synchrony unfused tetanus in different motor units.

## Motor Unit Recruitment

The force of a muscle contraction becomes greater as more motor units are activated. The process by which the number of contracting motor units is increased is called **motor unit recruitment**. Normally, the various motor units of a whole muscle are stimulated to contract *asynchronously* (at different times): While some motor units are contracting, others are relaxed. Because alternately contracting motor units relieve one another, contraction of a whole muscle can be sustained for long periods.

Recruitment is one factor responsible for producing smooth movements rather than a series of jerky movements. As mentioned, the number of muscle fibers innervated by one motor neuron varies greatly. Precise movements, as occurs in muscles of the eye, are brought about by small changes in muscle contraction. Typically, the muscles that produce precise movements are composed of small motor units. In this way, when one motor unit is recruited or turned off, only slight changes occur in muscle tension. On the other hand, large motor units are active where large tension is needed and precision is less important, as occurs in muscles of the thigh and leg during locomotion.

## CLINICAL CONNECTION | *Aerobic Training Versus Strength Training*

Regular, repeated activities such as jogging or aerobic dancing increase the supply of oxygen-rich blood available to skeletal muscles for aerobic cellular respiration. By contrast, activities such as weight lifting rely more on anaerobic production of ATP through glycolysis. Such anaerobic activities stimulate synthesis of muscle proteins and result, over time, in increased muscle size (muscle hypertrophy). Athletes who engage in anaerobic training should have a diet that includes an adequate amount of protein. High protein intake will allow the body to synthesize muscle proteins and to increase muscle mass. As a result, aerobic training builds endurance for prolonged activities; in contrast, anaerobic training builds muscle strength for short-term feats. **Interval training** is a workout regimen that incorporates both types of training—for example, alternating sprints with jogging.

## Muscle Tone

Even at rest, a muscle exhibits **muscle tone** (*tonos* = tension), a small amount of tautness or tension in the muscle due to weak, involuntary contractions of its motor units. To sustain muscle tone, small groups of motor units are activated in a constantly shifting pattern, alternating from one group of motor units to the next. Muscle tone keeps skeletal muscles firm, but it does not result in a contraction strong enough to produce movement. For example, when you are awake, the tone of muscles in the back of the neck keep the head upright and prevent it from slumping forward on the chest, but they do not generate enough force to draw the head backward into hyperextension. Muscle tone is also important in smooth muscle tissue such as that found in the gastrointestinal tract, where the walls of the digestive organs maintain a steady pressure on their contents. The tone of smooth muscle tissue in the walls of blood vessels plays a crucial role in maintaining blood pressure.

Recall that a skeletal muscle contracts only after it is stimulated by motor neurons. Hence, muscle tone is established by the brain and spinal cord, which activate the muscle's motor neurons. When the motor neurons serving a skeletal muscle are damaged, the muscle becomes **flaccid** (FLAS-sid = flabby), a state of limpness in which muscle tone is lost.

## Isotonic and Isometric Contractions

Muscle contractions may be either isotonic or isometric. In an **isotonic contraction** (ī′-sō-TON-ik; *iso-* = equal; *-tonic* = tension), the tension (force of contraction) developed in the muscle remains constant while the muscle changes its length. Isotonic contractions are used for body movements and for moving objects. The two types of isotonic contractions are concentric and eccentric. In a **concentric isotonic contraction** (kon-SEN-trik), a muscle shortens and pulls on another structure, such as a tendon, to produce movement and to reduce the angle at a joint. Picking up a book off a table involves concentric isotonic contractions of the biceps brachii muscle in the arm (Figure 10.14a).

**FIGURE 10.14 Comparison between isotonic (concentric and eccentric) and isometric contractions.** Parts (a) and (b) show isotonic contraction of the biceps brachii muscle in the arm; part (c) shows isometric contraction of shoulder and arm muscles.

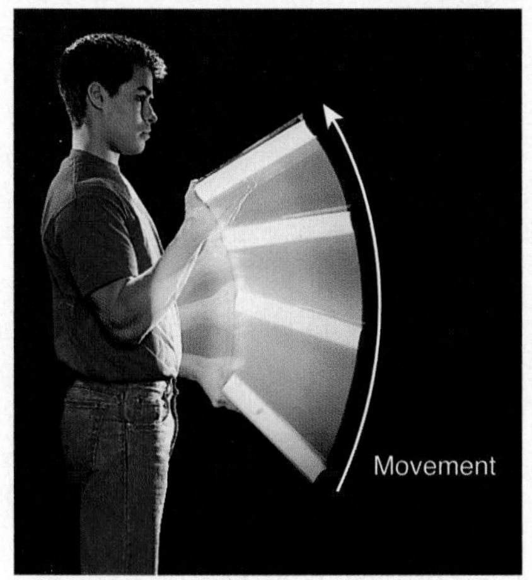


(a) Concentric contraction while picking up a book

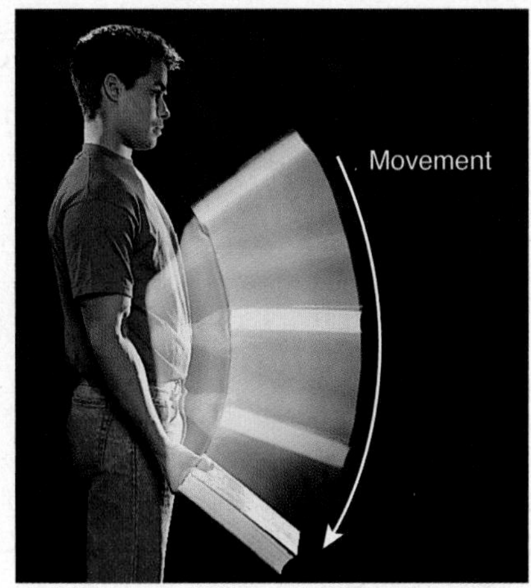


(b) Eccentric contraction while lowering a book

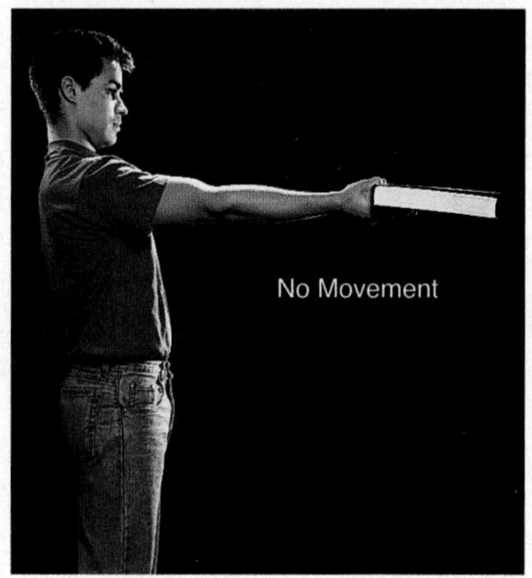


(c) Isometric contraction while holding a book steady

**In an isotonic contraction, tension remains constant as muscle length decreases or increases; in an isometric contraction, tension increases greatly without a change in muscle length.**

By contrast, as you lower the book to place it back on the table, the previously shortened biceps lengthens in a controlled manner while it continues to contract. When the length of a muscle increases during a contraction, the contraction is an **eccentric isotonic contraction** (ek-SEN-trik) (Figure 10.14b). During an eccentric isotonic contraction, the tension exerted by the myosin head resists the pull of gravity (the weight of the book, in this case) and slows the lengthening process.

In an **isometric contraction** (ī′-sō-MET-rik; *-metro* = measure), the tension generated is not enough to exceed the resistance of the object to be moved, and the muscle does not shorten. An example would be holding a book steady using an outstretched arm (Figure 10.14c). As the weight of the book pulls the arm downward, stretching the shoulder and arm muscles, isometric contraction of the shoulder and arm muscles counteracts the stretch. Isometric contractions are important for maintaining posture, for supporting objects in a fixed position, and for stabilizing some joints as other joints are moved. Most activities include both isotonic and isometric contractions.

### ✓ CHECKPOINT

**20.** What is a motor unit?

**21.** Which events occur during the latent period of a twitch contraction? During the relaxation period?

**22.** What is the difference between unfused tetanus and fused tetanus? Which type of stimulation can allow you to lift a heavier object?

**23.** How are calcium ions related to tetanus?

**24.** In what ways does motor unit recruitment assist muscle function?

## RETURN TO Jesse's Story

"Hi, Jesse, I'm Lisa. I'll be doing your nerve conduction and EMG studies today. We're going to measure two things—the electrical signals from your nerves that talk to your muscles and the electrical activity of the muscles themselves that make the muscles contract. I'm going to give you a small stimulus and record the response of your nerves. It's kind of like a static electricity shock."

"Okay, Jesse you did great. We're going to do the electromyogram next," said Lisa. "The EMG is for measuring the function of your muscles directly. The muscle receives an electrical signal from the nerve, then produces its own electrical signal which allows it to contract. These tests today help us figure out if it's your nerves or your muscles that aren't working properly. I need you to make the muscles contract. Raise your arm up. That's good. We have a pretty good signal."

Dr. Fitts read the results of Jesse's tests. His blood tests showed a normal serum creatine kinase but the left deltoid muscle biopsy revealed findings consistent with facioscapulohumeral muscular dystrophy (FSHD); isolated small atrophic fibers were present. His nerve conduction study and EMG reveal a myopathic process, most likely FSHD. Fitts sat back in his chair and picked up the phone.

"Hello, Mrs. Beatty?" he said.

"Yes, " she replied.

"This is Dr. Fitts. I have the results of Jesse's tests. The good news is that Jesse's condition is not life-threatening. He has what is called *facioscapulohumeral muscular dystrophy*. As I explained before, there are different types of muscular dystrophy. FSHD is generally less severe than the Duchenne muscular dystrophy that his uncle has. It may affect Jesse later in life, but he has a chance to be quite active. I think we can even help him with his shoulder weakness. We have a surgical procedure that will stabilize his scapula and allow a more normal range of motion."

Jesse didn't really understand all the talk about "autosomal dominant disorders" and "phenotypic expression." What he did understand was that his shoulder could be fixed with surgery, at least for now. They scheduled the surgery during the middle of his first semester.

E. *What is the name for the electrical signals that Lisa is measuring? What happens when the electrical signal from the nerve reaches the muscle fibers in Jesse's shoulder? How would the lack of proper electrical signals from the nervous system cause muscle weakness in Jesse's shoulder?*

F. *Lisa is measuring the electrical signals in both the nerve and the muscle; how are these two signals linked to initiate muscle contraction? As Jesse raises his arms, his muscles are actively contracting; which process couples muscle action potential to muscle contraction?*

G. *Based on the events of muscle stimulation and contraction that you have read about so far, with which events might problems occur that could lead to Jesse's muscle weakness? Could Jesse's muscles contract if the nerves that innervate those muscles were not functioning?*

H. *Jesse's nerve conduction test results reveal that impulses are traveling normally to the muscle. How does his loss of muscle fibers due to the muscular dystrophy affect the force generated by motor units in his atrophied muscles?*

# 10.8 Muscle fibers produce ATP from creatine phosphate, by anaerobic cellular respiration, and by aerobic cellular respiration.

## Production of ATP in Muscle Fibers

Unlike most cells of the body, skeletal muscle fibers often switch between virtual inactivity, when they are relaxed and using only a modest amount of ATP, and great activity, when they are contracting and using ATP at a rapid pace. A huge amount of ATP is needed to power the contraction cycle, to pump $Ca^{2+}$ back into the SR to achieve muscle relaxation, and for other metabolic reactions involved in muscle contraction. However, the ATP present inside muscle fibers is enough to power contraction for only a few seconds. If muscle contractions continue past that time, the muscle fibers must make more ATP. Fortunately, muscle fibers have three ways to produce ATP: (1) from creatine phosphate, (2) by anaerobic cellular respiration, and (3) by aerobic cellular respiration (Figure 10.15). All body cells make ATP through anaerobic and aerobic cellular respiration, but the use of creatine phosphate for ATP production is unique to muscle fibers. We consider the events of cellular respiration briefly here and then in further detail in Chapter 23.

### Creatine Phosphate

While muscle fibers are relaxed, they produce more ATP than they need for resting metabolism. The excess ATP is used to synthesize **creatine phosphate** (KRĒ-a-tēn), an energy-rich molecule that is found only in muscle fibers (Figure 10.15a). **Creatine** is

FIGURE 10.15 **Production of ATP for muscle contraction.** (a) Creatine phosphate is formed from ATP while the muscle is relaxed, then releases ATP as the muscle contracts. (b) Muscle glycogen is broken down into glucose, which glycolysis converts to pyruvic acid and ATP. Without sufficient oxygen, pyruvic acid is converted to lactic acid. (c) If adequate oxygen is available, mitochondria use pyruvic acid, fatty acids, and amino acids to produce additional ATP.

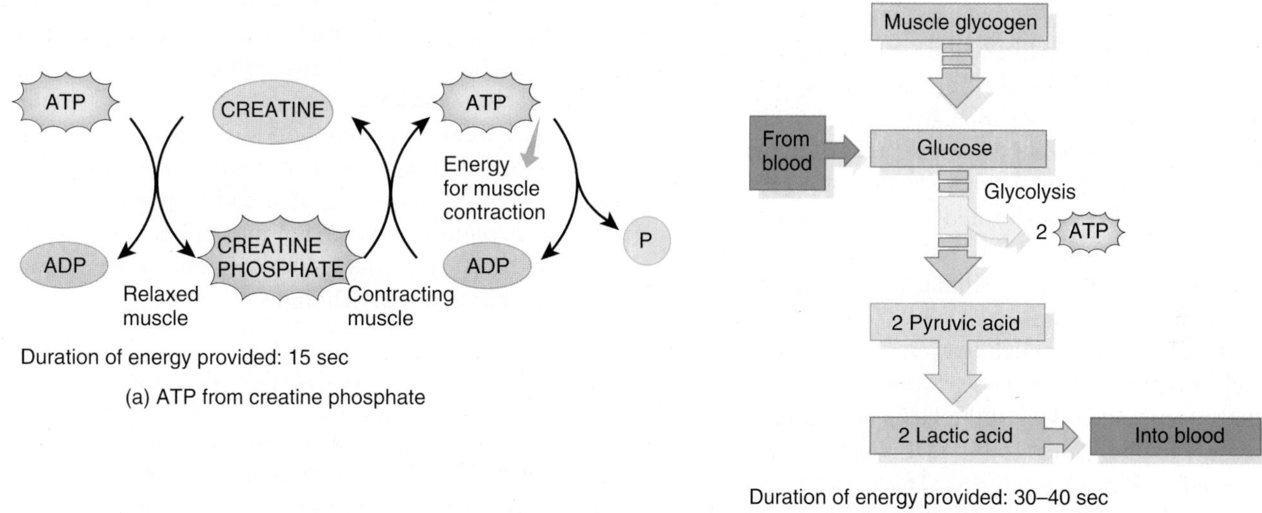


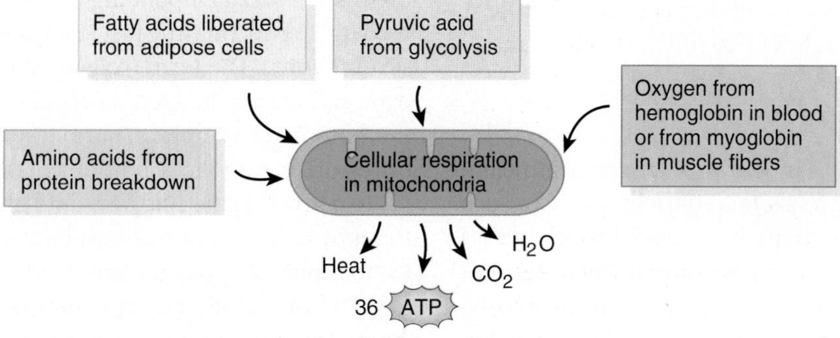


**During a long-term event such as a marathon race, most ATP is produced aerobically.**

synthesized in the liver, kidneys, and pancreas and then transported in the bloodstream to muscle fibers. The enzyme *creatine kinase* catalyzes the transfer of a high-energy phosphate group from ATP to creatine, forming creatine phosphate and ADP. Creatine phosphate is three to six times more plentiful than ATP in the sarcoplasm of a relaxed muscle fiber. When contraction begins and the ADP level starts to rise, creatine kinase catalyzes the transfer of a high-energy phosphate group from creatine phosphate back to ADP, quickly forming new ATP molecules. Creatine phosphate is the first source of energy when muscle contraction begins. Generating energy in a muscle fiber through anaerobic and aerobic cellular respiration takes more time than generating energy with creatine phosphate. Together, creatine phosphate and ATP provide enough energy for muscles to contract maximally for about 15 seconds. This amount of energy is sufficient for short bursts of activity, for example, to run a 100-meter dash.

### Anaerobic Cellular Respiration

**Anaerobic cellular respiration** (an′-a-RŌ-bik) is a series of ATP-producing reactions that do not require oxygen. When muscle activity continues past the 15-second mark and the supply of creatine phosphate within the muscle fibers is depleted, glucose is catabolized to generate ATP. Glucose passes easily from the blood into contracting muscle fibers and is also produced within muscle fibers by the breakdown of *glycogen* (Figure 10.15b). Then, a series of reactions known as glycolysis quickly breaks down each glucose molecule into two molecules of pyruvic acid. Glycolysis produces a net gain of two molecules of ATP.

Ordinarily, the pyruvic acid formed by glycolysis in the sarcoplasm enters the mitochondria. There, it enters a series of oxygen-requiring reactions called aerobic cellular respiration (described next) that produce a large amount of ATP. However, during periods of heavy exercise, not enough oxygen is available to muscle fibers. In such situations, anaerobic reactions convert most of the pyruvic acid to lactic acid. Most of the lactic acid diffuses out of the skeletal muscle fibers into the blood. Liver cells can convert some of the lactic acid back to glucose. This conversion has two benefits: providing new glucose molecules and reducing blood acidity. Glycolysis can provide enough energy for about 30 to 40 seconds of maximal muscle activity. Together, creatine phosphate and glycolysis can provide enough ATP to run a 400-meter race.

### Aerobic Cellular Respiration

During periods of rest or light to moderate exercise, a sufficient amount of oxygen is available to skeletal muscle fibers. In such cases, ATP used for muscular activity is produced from a series of oxygen-requiring reactions called **aerobic cellular respiration** (a-RŌ-bik). During this process, pyruvic acid from glycolysis enters the mitochondria, where it is completely oxidized in reactions that generate ATP, carbon dioxide, water, and heat (Figure 10.15c). Although aerobic cellular respiration is slower than glycolysis, one molecule of glucose yields about 36 molecules of ATP via aerobic cellular respiration. Compare this with the 2 molecules of ATP generated via anaerobic glycolysis.

Muscle tissue has two sources of oxygen: (1) oxygen that diffuses into muscle fibers from the blood and (2) oxygen released by myoglobin within muscle fibers. Both myoglobin (found only in muscle fibers) and hemoglobin (found only in red blood cells) bind oxygen when it is plentiful and release oxygen when it is scarce.

Aerobic cellular respiration provides enough ATP for prolonged activity as long as sufficient oxygen and nutrients are available. These nutrients include the pyruvic acid obtained from the glycolysis of glucose, fatty acids from the breakdown of triglycerides in adipose cells, and amino acids from the breakdown of proteins. In activities that last more than 10 minutes, aerobic cellular respiration provides most of the needed ATP. At the end of an endurance event such as a marathon race, nearly 100 percent of the ATP is being produced by aerobic cellular respiration.

## Muscle Fatigue

The inability of a muscle to contract forcefully after prolonged activity is called **muscle fatigue** (fa-TĒG). Even before actual muscle fatigue occurs, a person may experience tiredness and the desire to cease activity. This response, termed *central fatigue*, is caused by changes in the central nervous system (brain and spinal cord) and may be a protective mechanism to stop a person from exercising before muscle becomes damaged. Although the precise mechanisms that cause muscle fatigue are still unclear, several factors are thought to contribute. One is inadequate release of calcium ions from the SR, resulting in a decline of $Ca^{2+}$ level in the sarcoplasm. Other factors that contribute to muscle fatigue include depletion of creatine phosphate, insufficient oxygen, depletion of glycogen and other nutrients, buildup of lactic acid and ADP, and failure of impulses in the motor neuron to release enough acetylcholine.

## Oxygen Consumption after Exercise

During prolonged periods of muscle contraction, increases in breathing rate and blood flow enhance oxygen delivery to muscle tissue. After muscle contraction has stopped, heavy breathing continues for a while, and oxygen consumption remains above the resting level. Depending on the intensity of the exercise, the recovery period may be just a few minutes or it may last as long as several hours. The term **oxygen debt** refers to the added oxygen, over and above the oxygen consumed at rest, that is taken into the body after exercise. This extra oxygen is used to "pay back" or restore metabolic conditions to the resting level in three ways: (1) to convert lactic acid back into glycogen stores in the liver, (2) to resynthesize creatine phosphate and ATP in muscle fibers, and (3) to replace the oxygen removed from myoglobin.

The metabolic changes that occur *during exercise* can account for only some of the extra oxygen used *after exercise*. Only a small amount of glycogen is resynthesized from lactic acid. Instead, most glycogen stores are replenished much later from dietary carbohydrates. Much of the lactic acid that remains after exercise is converted back to pyruvic acid and used for ATP production via aerobic cellular respiration. Ongoing changes after exercise also boost oxygen use. First, the elevated body temperature after strenuous exercise increases the rate of chemical reactions throughout the body. Faster reactions use ATP more rapidly, and more oxygen is needed to produce ATP. Second, the heart and muscles used in breathing are still working harder than they were at rest, and thus they consume more ATP. Third, tissue repair processes are occurring at an increased pace. For these reasons, **recovery oxygen uptake** is a better term than oxygen debt for the elevated use of oxygen after exercise.

✓ CHECKPOINT

**25.** Which molecule is the only direct source of contraction energy for a muscle fiber?

**26.** Which ATP-producing reactions provide ATP during a 1000-meter run? Which occur within mitochondria?

**27.** What is muscle fatigue?

## 10.9 Skeletal muscle fibers are classified as slow oxidative fibers, fast oxidative–glycolytic fibers, or fast glycolytic fibers.

Skeletal muscle fibers vary in composition and function. For example, muscle fibers vary in their content of myoglobin, the red-colored protein that binds oxygen in muscle fibers. Those with a low myoglobin content appear pale and are called *white muscle fibers*. White muscle fibers are prevalent in the "white meat" in chicken breasts. Skeletal muscle fibers with a high myoglobin content have a darker appearance and are called *red muscle fibers*. Red muscle fibers are in the "dark meat" in chicken legs and thighs.

Skeletal muscle fibers contract and relax at different speeds; they also vary in which metabolic reactions are used to generate ATP and how quickly they fatigue. For example, a fiber is categorized as either slow or fast depending on how rapidly the ATPase in its myosin heads splits ATP. Based on these structural and functional characteristics, skeletal muscle fibers are classified into three main categories: (1) slow oxidative fibers, (2) fast oxidative–glycolytic fibers, and (3) fast glycolytic fibers.

### Slow Oxidative Fibers

**Slow oxidative (SO) fibers** are smallest in diameter and thus are the least powerful type of skeletal muscle fibers. They appear dark red because they contain large amounts of myoglobin and many blood capillaries. Because they have many mitochondria, SO fibers generate ATP mainly by aerobic cellular respiration, which is why they are called oxidative fibers. These fibers are said to be "slow" because the ATPase in the myosin heads splits ATP relatively slowly; thus, SO fibers have a slow speed of contraction. However, slow fibers are very resistant to fatigue and are capable of prolonged, sustained contractions for many hours. These slow-twitch, fatigue-resistant fibers are adapted for maintaining posture and for aerobic, endurance-type activities such as running a marathon.

### Fast Oxidative–Glycolytic Fibers

**Fast oxidative–glycolytic (FOG) fibers** are intermediate in diameter between the other two types of fibers. Like slow oxidative fibers, they contain large amounts of myoglobin and many blood capillaries. FOG fibers can generate considerable ATP by aerobic cellular respiration, which gives them a moderately high resistance to fatigue. Because their glycogen level is high, they also generate ATP by glycolysis. FOG fibers are "fast" because the ATPase in their myosin heads splits ATP faster than the myosin ATPase in SO fibers, which makes their speed of contraction faster. FOG fibers contribute to activities such as walking and sprinting.

### Fast Glycolytic Fibers

**Fast glycolytic (FG) fibers** are largest in diameter and contain the most myofibrils. Hence, they can generate the most powerful contractions. FG fibers are white fibers that have low myoglobin content and relatively few blood capillaries, but they contain large amounts of glycogen and generate ATP mainly by glycolysis. Due to their large size and their ability to use ATP at a fast rate, FG fibers contract strongly and quickly. These fast-twitch fibers are adapted for intense anaerobic movements of short duration, such as weight lifting or throwing a ball, but they fatigue quickly. Strength training programs that engage a person in activities requiring great strength for short periods increase the size, strength, and glycogen content of FG fibers. The FG fibers of a weight lifter may be 50 percent larger than those of a sedentary person or an endurance athlete. The increase in size is due to increased synthesis of muscle proteins. The result is muscle enlargement due to hypertrophy of the FG fibers.

### Distribution and Recruitment of Different Types of Fibers

Most skeletal muscles are a mixture of all three types of skeletal muscle fibers. The proportions vary somewhat, depending on the action of the muscle, the person's training regimen, and genetic factors. For example, the continually active postural muscles of the neck, back, and legs have a high proportion of SO fibers. Muscles of the shoulders and arms, in contrast, are not constantly active but are used briefly now and then to produce large amounts of tension, such as in lifting and throwing. These muscles have a high proportion of FG fibers. Leg muscles, which not only support the body but are also used for walking and running, have large numbers of both SO and FOG fibers.

Even though most skeletal muscles are a mixture of all three types of skeletal muscle fibers, the skeletal muscle fibers of any given motor unit are all of the same type. The different motor units in a muscle are recruited in a specific order, depending on need. For example, if weak contractions suffice to perform a task, only SO motor units are activated. If more force is needed, the motor units of FOG fibers are also recruited. Finally, if maximal force is required, motor units of FG fibers are also called into action. Activation of various motor units is controlled by the brain and spinal cord.

Table 10.1 summarizes the characteristics of the three types of skeletal muscle fibers.

✓ **CHECKPOINT**

**28.** Why are some skeletal muscle fibers classified as "fast" and others are said to be "slow"?

**TABLE 10.1**

Characteristics of the Three Types of Skeletal Muscle Fibers

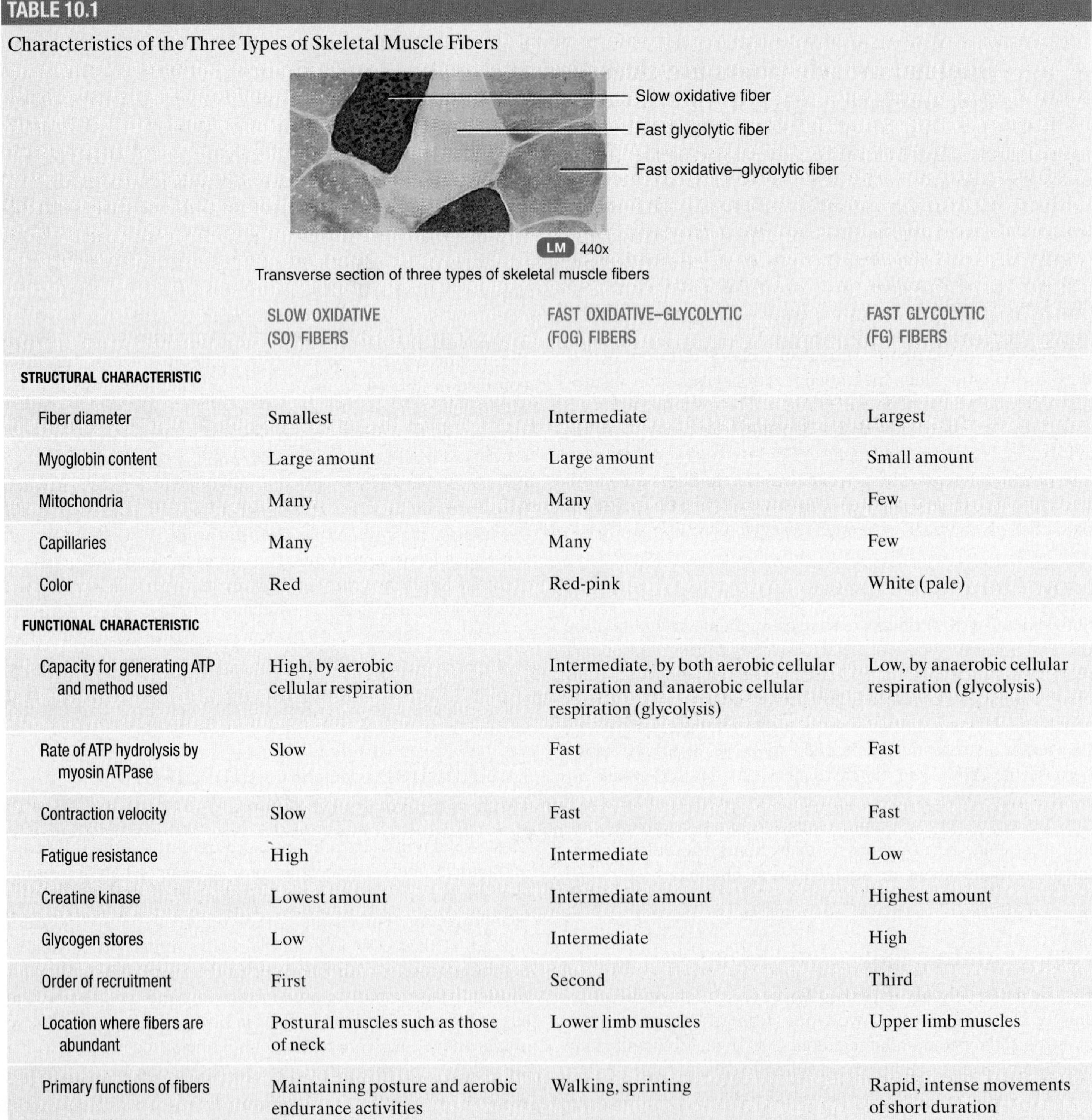


Transverse section of three types of skeletal muscle fibers

| | SLOW OXIDATIVE (SO) FIBERS | FAST OXIDATIVE–GLYCOLYTIC (FOG) FIBERS | FAST GLYCOLYTIC (FG) FIBERS |
|---|---|---|---|
| **STRUCTURAL CHARACTERISTIC** | | | |
| Fiber diameter | Smallest | Intermediate | Largest |
| Myoglobin content | Large amount | Large amount | Small amount |
| Mitochondria | Many | Many | Few |
| Capillaries | Many | Many | Few |
| Color | Red | Red-pink | White (pale) |
| **FUNCTIONAL CHARACTERISTIC** | | | |
| Capacity for generating ATP and method used | High, by aerobic cellular respiration | Intermediate, by both aerobic cellular respiration and anaerobic cellular respiration (glycolysis) | Low, by anaerobic cellular respiration (glycolysis) |
| Rate of ATP hydrolysis by myosin ATPase | Slow | Fast | Fast |
| Contraction velocity | Slow | Fast | Fast |
| Fatigue resistance | High | Intermediate | Low |
| Creatine kinase | Lowest amount | Intermediate amount | Highest amount |
| Glycogen stores | Low | Intermediate | High |
| Order of recruitment | First | Second | Third |
| Location where fibers are abundant | Postural muscles such as those of neck | Lower limb muscles | Upper limb muscles |
| Primary functions of fibers | Maintaining posture and aerobic endurance activities | Walking, sprinting | Rapid, intense movements of short duration |

# 10.10 Cardiac muscle tissue is found in the heart, and smooth muscle tissue is found in hollow internal structures.

## Cardiac Muscle Tissue

The principal tissue in the heart wall is **cardiac muscle tissue**. Compared with skeletal muscle fibers, cardiac muscle fibers are shorter in length and less circular in transverse section (Figure 10.16a). They also exhibit branching, which gives individual cardiac muscle fibers a "stair-step" appearance. Usually, one centrally located nucleus is present. Cardiac muscle fibers interconnect with one another by irregular transverse thickenings of

**FIGURE 10.16 Histology of cardiac muscle.** A photomicrograph of cardiac muscle tissue is shown in Table 4.8.

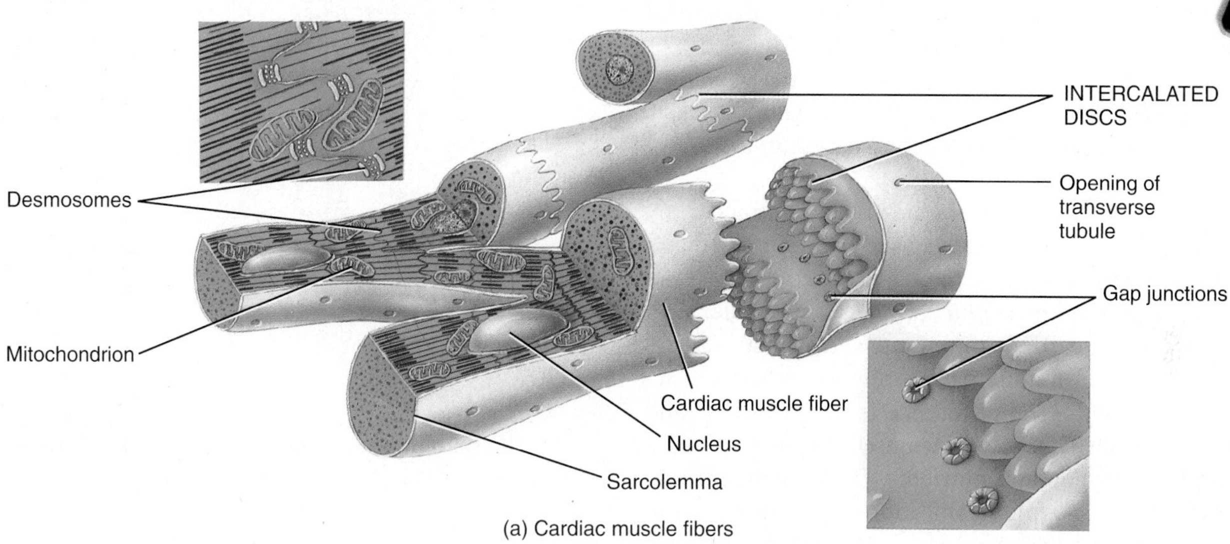


(a) Cardiac muscle fibers

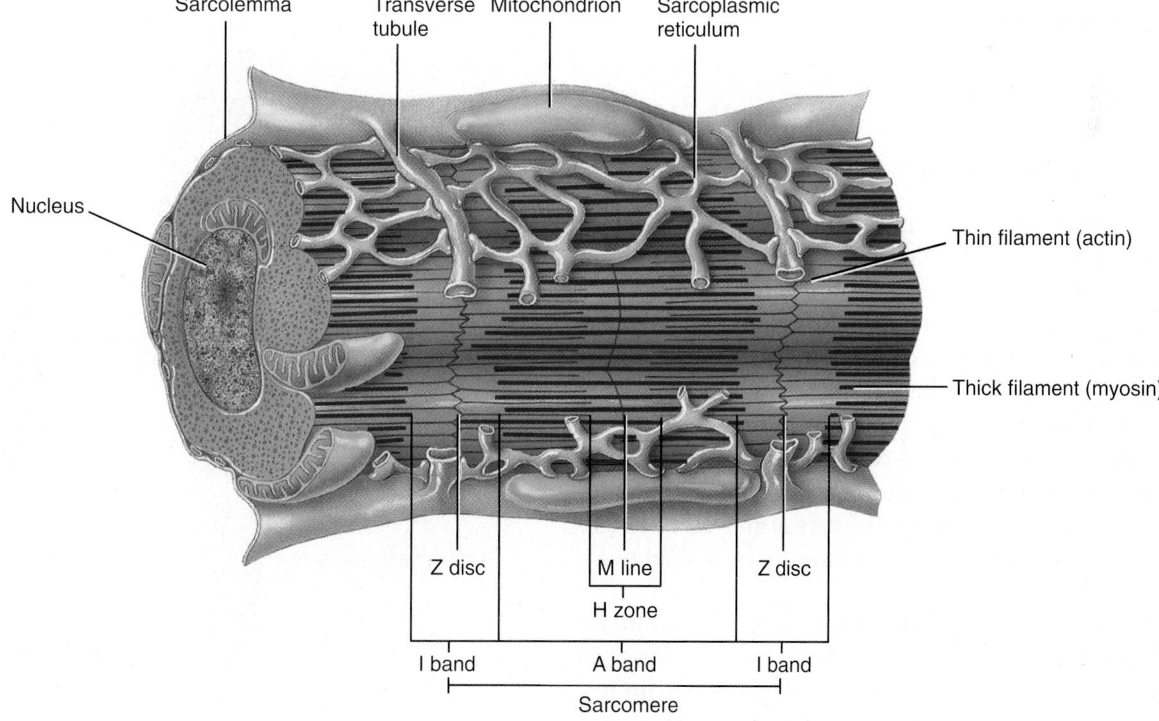


(b) Arrangement of components in a cardiac muscle fiber

Intercalated discs hold the fibers together and allow muscle action potentials to spread quickly from one cardiac muscle fiber to another.

the sarcolemma called **intercalated discs** (in-TER-ka-lāt-ed = inserted between). The discs contain *desmosomes,* which hold the fibers together, and *gap junctions*, which allow muscle action potentials to spread from one cardiac muscle fiber to another. Cardiac muscle fibers have the same arrangement of actin and myosin and the same bands, zones, and Z discs as skeletal muscle fibers (Figure 10.16b).

In response to a single action potential, cardiac muscle tissue remains contracted 10 to 15 times longer than skeletal muscle tissue. The long contraction is due to prolonged delivery of $Ca^{2+}$ into the sarcoplasm. In cardiac muscle fibers, $Ca^{2+}$ enters the sarcoplasm both from the sarcoplasmic reticulum (as in skeletal muscle fibers) and from the interstitial fluid that bathes the fibers. Because the channels in the sarcolemma that allow inflow of $Ca^{2+}$ from interstitial fluid stay open for a relatively long time, a cardiac muscle contraction lasts much longer than a skeletal muscle twitch.

We have seen that skeletal muscle tissue contracts only when stimulated by acetylcholine released by an impulse in a motor neuron. In contrast, cardiac muscle tissue contracts when stimulated by its own *autorhythmic muscle fibers*. Under normal resting conditions, cardiac muscle tissue contracts and relaxes about 75 times per minute. This continuous, rhythmic activity is a major physiological difference between cardiac and skeletal muscle tissue. Thus, cardiac muscle tissue requires a constant supply of oxygen and nutrients. The mitochondria in cardiac muscle fibers are larger and more numerous than in skeletal muscle fibers and produce most of the needed ATP via aerobic cellular respiration. In addition, cardiac muscle fibers can use lactic acid produced by skeletal muscle fibers to make ATP, a benefit during exercise.

## Smooth Muscle Tissue

Of the two types of smooth muscle tissue, the more common type is **visceral (single-unit) smooth muscle tissue** (Figure 10.17a). It is found in the skin and in part of the walls of small arteries and veins and of hollow organs such as the stomach, intestines, uterus, and urinary bladder. Like cardiac muscle tissue, visceral smooth muscle tissue is autorhythmic. The fibers connect to one another by gap junctions, forming a network through which muscle action potentials can spread. When a neurotransmitter, hormone, or autorhythmic signal stimulates one fiber, the muscle action potential is transmitted to neighboring fibers, which then contract in unison, as a single unit (hence the name).

The second type of smooth muscle tissue is **multiunit smooth muscle tissue** (Figure 10.17b). It consists of individual fibers, each with its own motor neuron terminals and with few gap junctions between neighboring fibers. Stimulation of one visceral muscle fiber causes contraction of many adjacent fibers, but stimulation of one multiunit fiber causes contraction

**FIGURE 10.17 Histology of smooth muscle tissue.** In (a), one autonomic motor neuron synapses with several visceral smooth muscle fibers. In (b), three autonomic motor neurons synapse with individual multiunit smooth muscle fibers. In (c), a smooth muscle fiber is shown in the relaxed state and the contracted state. A photomicrograph of smooth muscle tissue is shown in Table 4.8.

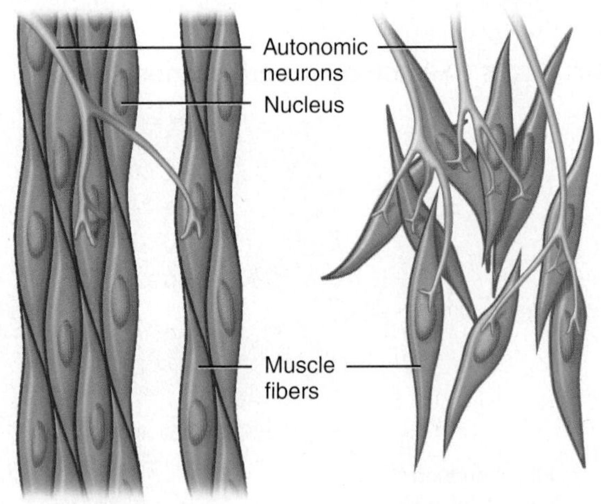


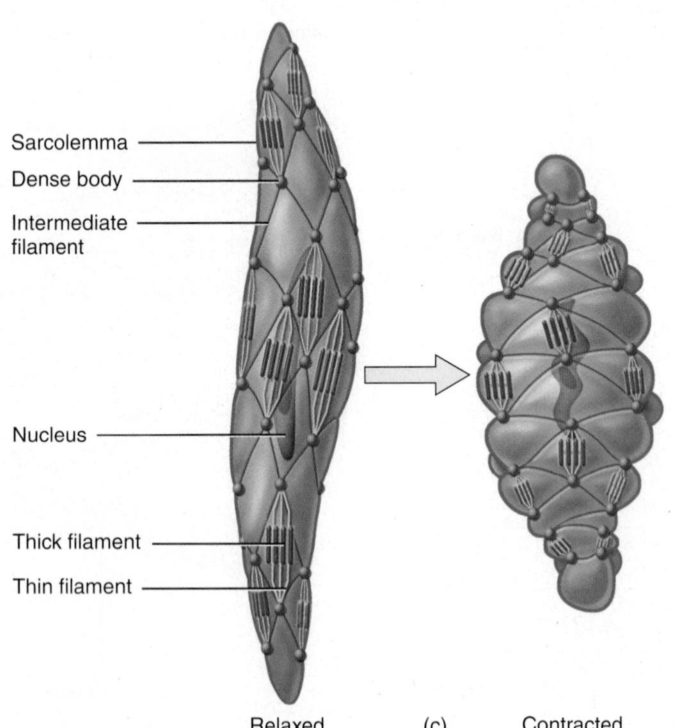


**Smooth muscle fibers have thick and thin filaments but no transverse tubules and little sarcoplasmic reticulum.**

of that fiber only. Multiunit smooth muscle tissue is found in the walls of large arteries, in airways to the lungs, in the arrector pili muscles that attach to hair follicles, and in the internal eye muscles.

Smooth muscle fibers are considerably smaller in length and diameter than skeletal muscle fibers and are tapered at both ends (Figure 10.17c). Within each fiber is a single, centrally located nucleus. The sarcoplasm of smooth muscle fibers contains both *thick filaments* and *thin filaments*, but they are not arranged in orderly sarcomeres as in striated muscle. Smooth muscle fibers also contain *intermediate filaments*, elements of the cytoskeleton. Because the various filaments have no regular pattern of overlap, smooth muscle lacks alternating dark and light bands and thus has a smooth, rather than striated, appearance. Bundles of intermediate filaments stretch between structures called *dense bodies*, which are functionally similar to the Z discs found in striated muscle (Figure 10.17c). Smooth muscle fibers also lack transverse tubules and have only scanty sarcoplasmic reticulum for storage of $Ca^{2+}$.

Although the principles of contraction are similar in all three types of muscle tissue, smooth muscle tissue exhibits some important physiological differences. During contraction, the sliding filament mechanism involving thick and thin filaments generates tension that is transmitted to intermediate filaments. These, in turn, pull on the dense bodies attached to the sarcolemma, causing a lengthwise shortening of the muscle fiber. As a smooth muscle fiber contracts, it rotates as a corkscrew turns. The fiber twists in a helix as it contracts, and rotates in the opposite direction as it relaxes (Figure 10.17c). Compared with contraction in a skeletal muscle fiber, contraction in a smooth muscle fiber starts more slowly and lasts much longer. In addition, smooth muscle can both shorten and stretch to a greater extent than other muscle types.

An increase in the concentration of $Ca^{2+}$ in the sarcoplasm of a smooth muscle initiates contraction, just as in striated muscle. However, the regulatory protein *calmodulin* (cal-MOD-ū-lin), rather than troponin, activates myosin head cycling in smooth muscle. Calcium ions flow into smooth muscle sarcoplasm from both the interstitial fluid and sarcoplasmic reticulum, but because there are no transverse tubules in smooth muscle fibers, it takes longer for $Ca^{2+}$ to reach the filaments in the center of the fiber and trigger the contractile process. This accounts, in part, for the slow onset of contraction of smooth muscle.

Calcium ions also move out of the muscle fiber slowly, which delays relaxation. The prolonged presence of $Ca^{2+}$ in the cytosol provides for *smooth muscle tone*, a state of continued partial contraction. This long-term muscle tone is important in the walls of blood vessels that maintain a steady pressure on blood, and in the walls of gastrointestinal tract organs that maintain a steady pressure on their contents.

Like cardiac muscle tissue, smooth muscle tissue is usually activated involuntarily. Most smooth muscle fibers contract or relax in response to action potentials from the autonomic nervous system. In addition, many smooth muscle fibers contract or relax in response to stretching, hormones, or local factors such as changes in pH, oxygen and carbon dioxide levels, temperature, and ion concentrations. For example, the hormone epinephrine causes relaxation of smooth muscle in the airways and in some blood vessel walls.

Table 10.2 summarizes the major characteristics of the three types of muscle tissue.

✓ **CHECKPOINT**

**29.** Of the three types of muscle tissue, which have contractile proteins organized into sarcomeres? Contain transverse tubules? Contain intermediate filaments? Contain dense bodies?

**30.** What are the functions of intercalated discs? In which muscle tissue type are they found?

**31.** What is the source of calcium ions for each of the three types of muscle tissue?

**32.** Which type of smooth muscle is found in the walls of hollow organs? Which is found in large blood vessels?

**33.** How does the speed of onset and duration of contraction in a smooth muscle fiber compare with that in a skeletal muscle fiber?

**TABLE 10.2**

Major Features of the Three Types of Muscle Tissue

| CHARACTERISTIC | SKELETAL MUSCLE | CARDIAC MUSCLE | SMOOTH MUSCLE |
|---|---|---|---|
| **Microscopic appearance and features** | Long cylindrical fiber with many peripherally located nuclei; striated  | Branched cylindrical fiber with one centrally located nucleus; intercalated discs join neighboring fibers; striated 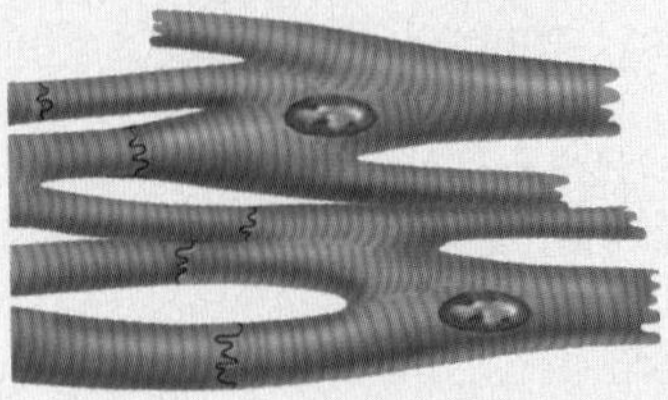 | Fiber is thickest in the middle, tapered at each end, and has one centrally positioned nucleus; not striated  |
| **Location** | Attached primarily to bones by tendons | Heart | Walls of hollow viscera, airways, blood vessels, iris and ciliary body of eye, arrector pili of hair follicles |
| **Fiber diameter** | Very large (10–100 μm) | Large (10–20 μm) | Small (5–10 μm) |
| **Connective tissue components** | Endomysium, perimysium, and epimysium | Endomysium | Endomysium |
| **Fiber length** | 100 μm–30 cm | 50–100 μm | 20–500 μm |
| **Contractile proteins organized into sarcomeres** | Yes | Yes | No |
| **Sarcoplasmic reticulum** | Abundant | Some | Scanty |
| **Transverse tubules present** | Yes, aligned with each A–I band junction | Yes, aligned with each Z disc | No |
| **Junctions between fibers** | None | Intercalated discs contain gap junctions and desmosomes | Gap junctions in visceral smooth muscle; none in multiunit smooth muscle |
| **Autorhythmicity** | No | Yes | Yes, in visceral smooth muscle |
| **Source of $Ca^{2+}$ for contraction** | Sarcoplasmic reticulum | Sarcoplasmic reticulum and interstitial fluid | Sarcoplasmic reticulum and interstitial fluid |
| **Speed of contraction** | Fast | Moderate | Slow |
| **Nervous control** | Mostly voluntary | Involuntary | Involuntary |
| **Contraction regulation by:** | Acetylcholine released by somatic motor neurons | Acetylcholine and norepinephrine released by autonomic motor neurons; several hormones | Acetylcholine and norepinephrine released by autonomic motor neurons; several hormones; local chemical changes; stretching |
| **Capacity for regeneration** | Limited, via satellite cells | Limited, under certain conditions | Considerable (compared with other muscle tissues, but limited compared with epithelium), via pericytes |

## Jesse's Story

## EPILOGUE AND DISCUSSION

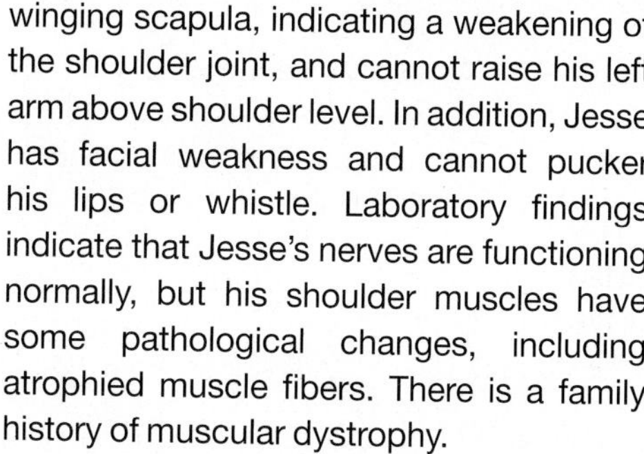

Jesse, a 16-year-old male athlete, presents to his sports medicine doctor complaining of arm weakness and shoulder discomfort. Upon examination, Jesse's doctor finds that Jesse has a winging scapula, indicating a weakening of the shoulder joint, and cannot raise his left arm above shoulder level. In addition, Jesse has facial weakness and cannot pucker his lips or whistle. Laboratory findings indicate that Jesse's nerves are functioning normally, but his shoulder muscles have some pathological changes, including atrophied muscle fibers. There is a family history of muscular dystrophy.

Based on the physical examination and laboratory results, the doctor diagnoses *facioscapulohumeral muscular dystrophy* (*FSHD*). This condition is characterized by muscle degeneration and weakness, like all muscular dystrophies. However, there are many types of muscular dystrophy. FSHD weakens muscles in the face, around the shoulder blades, and in the arm. The disease occurs with a frequency of 1 in 20,000, and symptoms usually appear during the first or second decade of life. The severity of the symptoms is worse the earlier the disease manifests itself. Unlike Duchenne muscular dystrophy, a more common and severe form of muscular dystrophy, FSHD does not affect life span. However, the severity of the disorder can vary from individual to individual. In Jesse's case, stabilizing the scapula by surgically fixing its medial borders to the ribs will probably result in improved abduction and flexion of the arms and improve range of motion in spite of the muscle weakness.

I. *What are the three types of muscle tissues found in the human body? Which of these was not functioning properly in Jesse's body? How were the functions of Jesse's muscle tissue impaired by his disorder?*

J. *Jesse had a nerve conduction study and an EMG. What was being tested?*

K. *Why was the doctor interested in checking nerve function if it was the muscles in Jesse's shoulder and arm that were weak?*

L. *Jesse plans on lifting weights and strength training as part of his recuperation; how will strengthening his shoulder help his quality of life?*

M. *Why is surgical fixation and stabilization of the scapula necessary in Jesse's case? How would Jesse's condition and its associated processes affect Jesse if they were left untreated?*

# Concept and Resource Summary

| Concept | Resources   |
|---|---|
| **Introduction**<br>1. The primary function of muscle is to change chemical energy into mechanical energy to produce body movement. | |
| 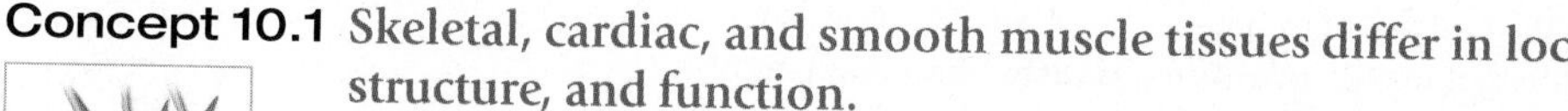 **Concept 10.1** Skeletal, cardiac, and smooth muscle tissues differ in location, structure, and function.<br><br>1. The three types of muscle tissue are skeletal, cardiac, and smooth. **Skeletal muscle tissue** is striated, under voluntary control, and functions to move bones of the skeleton. **Cardiac muscle tissue** is striated, under involuntary control, has autorhythmicity, and is only found in the heart. **Smooth muscle tissue** is nonstriated, under involuntary control, and is located in the walls of hollow internal structures. | Anatomy Overview—Muscle Tissue |

## Concept

## Resources

### Concept 10.2 Muscle tissue performs four functions and possesses four properties.

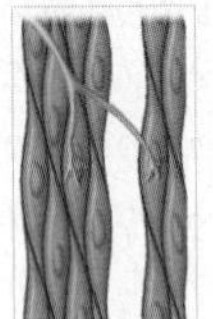

1. Muscle tissue has four important functions: producing body movements, stabilizing body positions, moving substances within the body, and producing heat.
2. Muscle tissue has four special properties: **electrical excitability**, **contractility**, **extensibility**, and **elasticity**.

### Concept 10.3 Skeletal muscles are surrounded by connective tissues and are well supplied with nerves and blood vessels.

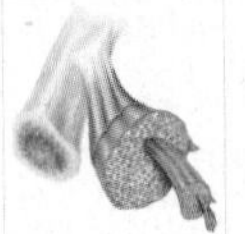

1. A skeletal muscle is an organ composed of elongated **muscle fibers** plus associated nerves, blood vessels, and connective tissues.
2. **Hypodermis** separates muscle from skin. **Fascia** unites muscles with similar functions, carries nerves and vessels, and fills spaces between muscles.
3. Three layers of connective tissue protect and strengthen skeletal muscle: the outermost layer encircles the muscle, and is the **epimysium**; **perimysium** bundles together groups of muscle fibers into **fascicles**; and the innermost **endomysium** surrounds each muscle fiber. These three layers may extend beyond the muscle to form a **tendon**, or a broad flat **aponeurosis**. Tendons and aponeuroses attach a muscle to bone.
4. One artery and one or two veins supply each skeletal muscle.
5. **Motor neurons** stimulate muscle fibers through processes called **axons**.

Resources:
- Anatomy Overview—Cross Section of Skeletal Muscle
- Animation—Contraction and Movement

### Concept 10.4 Each skeletal muscle fiber is covered by a sarcolemma; each of its myofibrils is surrounded by sarcoplasmic reticulum and contains sarcomeres.

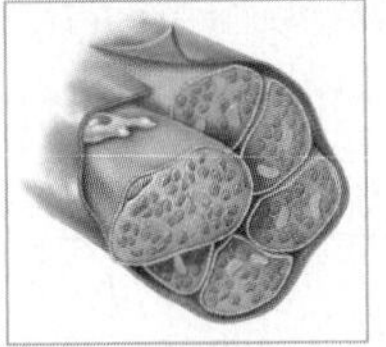

1. Embryonic development of skeletal muscle fibers arises from fusion of myoblasts into one elongated, multinucleate, amitotic muscle fiber. Some myoblasts persist as satellite cells in mature skeletal muscle that can fuse and regenerate muscle fibers in damaged tissue. Extensive damage involves **fibrosis**, the replacement of muscle fibers with scar tissue.
2. After birth, muscle growth occurs from enlargement of existing muscle fibers, termed **hypertrophy**, stimulated by human growth hormone, testosterone, and other hormones.
3. The **sarcolemma** has many invaginated, tunnel-like extensions called **transverse (T) tubules** that are also open to the cell's exterior.
4. The **sarcoplasm** contains glycogen, and a red protein, called **myoglobin**, that binds oxygen molecules for use in ATP synthesis.
5. The sarcoplasm is full of long contractile elements, called **myofibrils**, which give the fiber its striated appearance.
6. The **sarcoplasmic reticulum** (**SR**) wraps around each myofibril. The SR stores and releases $Ca^{2+}$. The dilated end sacs of the SR, called **terminal cisterns**, lie on each side of one T tubule, forming a **triad**.
7. Myofibrils contain **thin filaments** and **thick filaments** arranged in basic functional units called **sarcomeres**. Sarcomeres in a myofibril are separated from one another by **Z discs**.
8. A darker middle part of the sarcomere containing thick and thin overlapping filaments is the **A band**, at the center of which is a narrow **H zone** containing only thick myofilaments. **I bands** contain only thin filaments and are located near the Z discs at the ends of a sarcomere.
9. The contractile proteins, myosin and actin, are located inside myofibrils. **Myosin**, the main component of thick filaments, converts chemical energy in ATP to the mechanical energy of motion. **Actin** is the main component of thin filaments. Actin has a myosin-binding site where the myosin head can attach. Two regulatory proteins, **tropomyosin** and **troponin**, are also part of the thin filament, and help switch contraction on and off.
10. In addition to contractile and regulatory proteins, muscle contains structural proteins such as **titin** and **dystrophin**. Titin helps anchor a Z disc to the M line, and dystrophin links thin filaments of the sarcomere to the sarcolemma.

Resources:
- Anatomy Overview—Skeletal Muscle
- Animation—Muscle Cell Structures
- Clinical Connection—Exercise-induced Muscle Damage
- Clinical Connection—Muscular Dystrophy
- Clinical Connection—Creatine Supplementation
- Clinical Connection—Anabolic Steroids

**Concept** | **Resources**

## Concept 10.5 The neuromuscular junction is where a muscle action potential is initiated.

1. Muscle action potentials arise at the **neuromuscular junction**, the **synapse** between a motor neuron and a skeletal muscle fiber.
2. The synaptic end bulbs have **synaptic vesicles** containing the neurotransmitter **acetylcholine (ACh)**. The motor neuron releases ACh into the synaptic cleft; ACh binds to **acetylcholine receptors** on the **motor end plate** region of the sarcolemma of the muscle fiber.
3. Events in the excitation of a skeletal muscle fiber include the following: release of acetylcholine as impulse from the brain or spi nal cord reaches the synaptic end bulbs; activation of ACh receptors when ACh binds to a motor end plate ACh receptor, which opens ion channels; generation of a muscle action potential that stimulates a contraction; and breakdown of ACh by the enzyme **acetylcholinesterase** in the synaptic cleft.

Animation—Neuromuscular Junctions
Concepts and Connections—Events at the Neuromuscular Junction
Clinical Connection—Myasthenia Gravis

## Concept 10.6 An action potential releases calcium ions that allow thick filaments to bind to and pull thin filaments toward the center of the sarcomere.

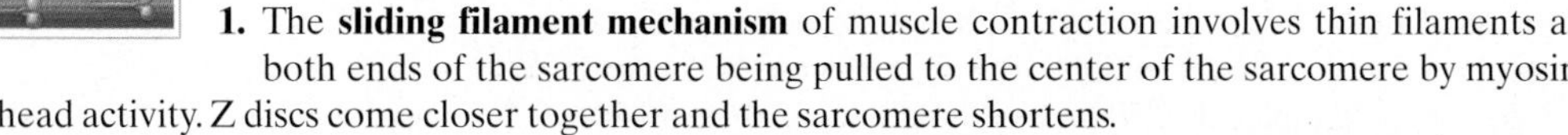

1. The **sliding filament mechanism** of muscle contraction involves thin filaments at both ends of the sarcomere being pulled to the center of the sarcomere by myosin head activity. Z discs come closer together and the sarcomere shortens.
2. When a muscle action potential propagates along the sarcolemma and T tubules, it opens **$Ca^{2+}$ release channels** in the SR membrane. $Ca^{2+}$ flows into the sarcoplasm and combines with troponin, moving the troponin–tropomyosin complex away from myosin-binding sites on actin. Myosin heads then bind to actin.
3. The **contraction cycle consists** of four steps: ATP splits in the myosin head to reorient and energize it; myosin attaches to actin when the energized myosin head attaches to the myosin-binding site on actin; the power stroke occurs when the energized myosin head releases the phosphate group, triggering ADP release and myosin rotation, which slides the thin filament toward the M line; and myosin detaches from actin as ATP binds to the myosin head.
4. ATPase splits ATP on the myosin head, and energy is transferred to the myosin head as it is energized and reoriented in position. The contraction cycle repeats as successive power strokes result in shortening of the sarcomeres. As Z discs pull on adjacent sarcomeres, the whole muscle fiber shortens, and, ultimately, the entire muscle shortens.
5. Cessation of impulses in the motor neuron stops ACh release. AChE breaks down any ACh in the synaptic cleft. Ion channels close, action potentials stop, and **$Ca^{2+}$ active transport pumps** use ATP to move $Ca^{2+}$ back into the SR. The troponin–tropomyosin complexes slide back to cover myosin-binding sites of actin, and thin filaments return to their relaxed positions.

Animation—Contraction of a Sarcomere
Figure 10.7—Sliding Filament Mechanism of Muscle Contraction

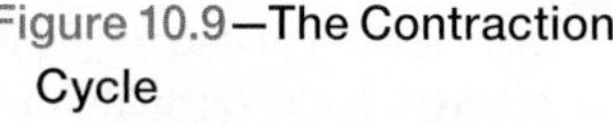

Figure 10.9—The Contraction Cycle
Figure 10.10—Summary of the Events of Contraction and Relaxation in a Skeletal Muscle Fiber
Exercise—Contraction Connections
Exercise—Muscle Car
Concepts and Connections—Excitation of Skeletal Muscle
Concepts and Connections—Skeletal Muscle Contraction Cycle

## Concept 10.7 Muscle tension is controlled by stimulation frequency and motor unit recruitment.

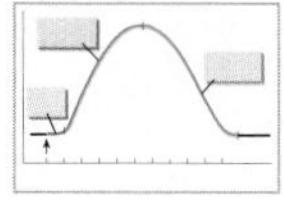

1. A single impulse in a motor neuron elicits a single muscle twitch contraction in all muscle fibers that it innervates. The frequency of stimulation governs the total tension that can be produced by a single muscle fiber. The total tension produced in a whole muscle depends on the number of fibers contracting in unison.
2. A **motor unit** is one motor neuron and all the skeletal muscle fibers the motor neuron stimulates.
3. Muscles controlling small, precise movements are innervated by many small motor units; muscles controlling large, powerful movements have fewer, large motor units. The size of a muscle's motor units and the number of motor units activated contribute to the contraction strength.
4. The response of a motor unit to a single impulse in its motor neuron is a **twitch contraction**. The three phases of a twitch are the **latent period** of cell events leading up to contraction; the **contraction period** of power strokes generating tension; and the **relaxation period**, during which the muscle is allowed to resume its original length.
5. Multiple stimuli that arrive before the muscle fiber has fully relaxed lead to **wave summation**. When the frequency of stimulation allows partial relaxation it is called **unfused tetanus**; rapid frequency of stimulation and sustained contraction is called **fused tetanus**.

Animation—Control of Muscle Tension
Exercise—Increase Muscle Tension
Clinical Connection—Abnormal Contraction of Skeletal Muscle
Clinical Connection—Hypotonia and Hypertonia

| Concept | Resources  |
|---|---|

6. **Motor unit recruitment** is the process of increasing the number of contracting motor units.
7. A muscle at rest exhibits **muscle tone**, a small amount of tension due to involuntary alternating contractions of a small number of its motor units that do not produce movement. Damaged motor neurons that cannot maintain muscle tone cause a muscle to become **flaccid**.
8. **Isotonic contractions** involve a change in muscle length without a change in its tension. There are two types: **concentric isotonic contractions** occur when the muscle shortens; **eccentric isotonic contractions** occur when the muscle lengthens.
9. An **isometric contraction** occurs when the load equals or exceeds the muscle tension, and the muscle does not lengthen or shorten.

## Concept 10.8 Muscle fibers produce ATP from creatine phosphate, by anaerobic cellular respiration, and by aerobic cellular respiration.

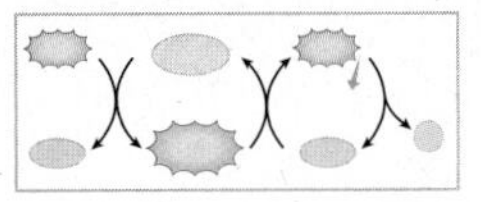

1. ATP is the only direct source of energy for muscle contraction. Muscle fibers have three ways to produce ATP: from creatine phosphate, by anaerobic cellular respiration, and by aerobic cellular respiration.
2. Muscle fibers break down excess ATP and transfer a phosphate group to **creatine**, forming **creatine phosphate** and ADP. During contraction, muscle fibers transfer the phosphate group from creatine phosphate to ADP, forming ATP.
3. A muscle at peak activity quickly depletes available ATP and creatine phosphate, and it will then catabolize glucose molecules from glycogen. Glycolysis is the initial pathway in glucose breakdown and yields two ATP molecules and two pyruvic acid molecules. When oxygen is unavailable, **anaerobic** reactions convert pyruvic acid to lactic acid. Blood removes lactic acid from skeletal muscle and carries much of it to the liver for reconversion to glucose.
4. When oxygen is available, the pyruvic acid molecules from glycolysis enter the mitochondria, where **aerobic cellular respiration** completely oxidizes each molecule of glucose to generate 36 ATP molecules.
5. **Muscle fatigue** is the inability of muscle to contract forcefully after prolonged activity.
6. Heavy breathing after prolonged muscle activity helps to repay the **oxygen debt**, more accurately referred to as **recovery oxygen uptake**. This increased oxygen intake helps to restore metabolic conditions to the resting level by conversion of lactic acid back to glycogen, resynthesis of creatine phosphate and ATP, and replacement of oxygen in muscle fibers.

Animation—Muscle Metabolism
Exercise—Fueling Contraction and Recovery
Concepts and Connections—Muscle Metabolism

## Concept 10.9 Skeletal muscle fibers are classified as slow oxidative fibers, fast oxidative–glycolytic fibers, or fast glycolytic fibers.

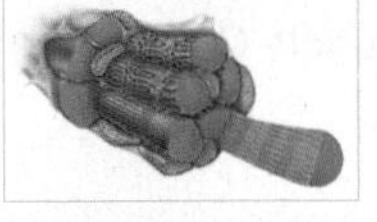

1. Skeletal muscle fibers with low myoglobin content appear pale and are called white muscle fibers; skeletal muscle fibers with high myoglobin content have a dark, reddish appearance and are called red muscle fibers.
2. Skeletal muscle fibers are classified as slow oxidative, fast oxidative–glycolytic, and fast glycolytic.
3. **Slow oxidative fibers** use aerobic cellular respiration, have a slow speed of contraction, and are fatigue-resistant. These fibers are beneficial for posture and endurance activities.
4. **Fast oxidative–glycolytic fibers** use aerobic cellular respiration and glycolysis, have fast speeds of contraction, and are moderately fatigue-resistant. These fibers are used for walking and sprinting.
5. **Fast glycolytic fibers** mainly use glycolysis, contract strongly and rapidly, and are adapted for intense bursts of anaerobic movements, but they fatigue rapidly.
6. Most skeletal muscles in the body are a mixture of all three types of muscle fibers. Training, genetics, and muscle action can slightly alter proportions of the fiber types.

Clinical Connection—Muscle Strain

## Concept 10.10 Cardiac muscle tissue is found in the heart, and smooth muscle tissue is found in internal hollow structures.

1. **Cardiac muscle tissue** is found in the wall of the heart. Cardiac muscle fibers have one nucleus and are branched and striated. Cardiac muscle fibers interconnect by **intercalated discs** containing desmosomes and gap junctions. Long, sustained contractions are supported by inflow of $Ca^{2+}$ into the sarcoplasm.

Anatomy Overview—Cardiac Muscle
Anatomy Overview—Smooth Muscle Tissue

| Concept | Resources |
|---|---|
| 2. Specialized cardiac muscle fibers have autorhythmicity. A constant supply of oxygen and nutrients is needed for the continuous contraction–relaxation cycle that occurs in cardiac muscle fibers.<br>3. There are two types of smooth muscle tissue: (1) **visceral (single-unit) smooth muscle tissue** is autorhythmic, and the fibers are connected by gap junctions allowing action potentials to spread throughout the network so that cells contract as a single unit; and (2) **multiunit smooth muscle tissue** acts independently, has few gap junctions, and lacks autorhythmicity.<br>4. **Smooth muscle fibers** have tapered ends, one central nucleus, are nonstriated, and lack sarcomeres. Intermediate filaments form bundles that stretch between dense bodies.<br>5. Thin and thick filaments of smooth muscle have a sliding mechanism that generates tension, resulting in lengthwise shortening of the fiber. Smooth muscle contraction starts slowly and lasts an extended time.<br>6. Smooth muscle is involuntary and responds to autonomic nervous system impulses, hormones, and local factors. | Figure 10.16—Histology of Cardiac Muscle<br>Figure 10.17—Histology of Smooth Muscle Tissue  |

## Understanding the Concepts

1. How is the muscle found in your heart different from the muscle that moves your arms?
2. What is the relationship between skeletal muscles and body temperature?
3. Which connective tissue coat surrounds groups of muscle fibers, separating them into bundles? What are these bundles called?
4. Describe the functions of titin and dystrophin.
5. What prevents an action potential arriving at the synaptic end bulbs of a motor neuron from being transferred directly to a muscle fiber?
6. Which of the numbered steps in Figure 10.10 are part of excitation–contraction coupling?
7. If the lengths of the thick and thin filaments do not change, how does a skeletal muscle fiber shorten during muscle contraction?
8. Would muscles of the fingers likely contain many small motor units or a few large ones? Is this also true of lower back muscles? Explain your answers.
9. Which type of contraction can allow you to make a sudden small jerk of your arm when startled? Maintain your neck's position while you are walking? Drag a chair to a table? Slowly sit down in the chair?
10. Why is the term recovery oxygen uptake more accurate than oxygen debt?
11. In what order are the various types of skeletal muscle fibers recruited when you sprint to make it to the bus stop?
12. What are the major functional differences between cardiac and skeletal muscle tissue? Between smooth and skeletal muscle tissue?

CHAPTER

# 11

## Stewart's Story

"Remember that muscles can only pull, they cannot push. For a joint to move properly, there needs to be muscles that pull it in different directions." Stewart follows along in the handout as Dr. Kendrick begins her lecture on muscle terminology. Stewart's class had been studying different body systems in anatomy and physiology for two months as part of the massage therapy program, but they were finally getting to the system that fascinated him most.

Stewart's interest in the muscular system had developed while playing high school sports. The coaching staff would refer to various muscle groups that the players were using, and Stewart could see how the different muscles worked together to perform some very intricate actions.

"We spent the last few days learning about muscle tissue and how sarcomere contraction occurs within the muscle fibers. Now we get to learn about the muscles at the organ level and how they relate to joints." As Dr. Kendrick continues the lecture on muscle attachments and muscle names, Stewart flips through the pages of the chapter. He works out at the gym regularly and is interested in learning more about the specific muscles that he exercises. He is amazed to realize the logic behind muscle names; he had always used the names but never put much thought into what they meant.

After the lecture, Stewart and Wally head to the student center for lunch. Stewart asks Wally what he thought about his revelation about muscle names.

"Yeah, it's interesting. We always talked about working out the *biceps*, but I never thought about it being a two-bellied muscle. It's also weird to realize that there's a *biceps* muscle on the back of the thigh."

"I know what you mean," Stewart replies, as they find an empty table at the food court. "I think it's wild that a muscle can have different functions depending on what action you're talking about. If you're talking about flexing the forearm, the *biceps brachii* is an agonist, but it's an antagonist to extension of the forearm."

"Well, it's certainly going to make studying the muscles interesting."

A solid understanding of the muscular system starts with a basic understanding of the organization of muscles at the organ level. As you study this chapter, pay attention to the action of each muscle and how it interacts with other muscles working on the same joint.

# The Muscular System

## INTRODUCTION

The voluntarily controlled skeletal muscle tissues of your body make up the **muscular system**. Almost 700 individual skeletal muscles, containing both muscle and connective tissues, are found in this important system. The primary function of most muscles is to produce movements of body parts. A few muscles function mainly to stabilize bones so that other skeletal muscles can execute a movement more effectively. This chapter presents many of the major skeletal muscles in the body, most of which are found on both the right and left sides. While learning the names of skeletal muscles, we will identify their attachment sites and the movements that they produce. Developing a working knowledge of these key aspects of skeletal muscle anatomy will enable you to understand how normal movements occur. This knowledge is especially crucial for professionals, such as those in the allied health and physical rehabilitation fields, who work with patients whose normal patterns of movement and physical mobility have been disrupted by physical trauma, surgery, or muscular paralysis.

## CONCEPTS

**11.1** Skeletal muscles produce movement when the insertion is pulled toward the origin.

**11.2** Skeletal muscles are named based on size, shape, action, location, or attachments.

**11.3** Muscles of the head produce facial expressions, eyeball movement, and assist in biting, chewing, swallowing, and speech.

**11.4** Muscles of the neck assist in swallowing and speech, and allow balance and movement of the head.

**11.5** Muscles of the abdomen protect the abdominal viscera, move the vertebral column, and assist breathing.

**11.6** Muscles of the pelvic floor and perineum support the pelvic viscera, function as sphincters, and assist in urination, erection, ejaculation, and defecation.

**11.7** Muscles inserting on the upper limb move and stabilize the pectoral girdle, and move the arm, forearm, and hand.

**11.8** Deep muscles of the back move the head and vertebral column.

**11.9** Muscles originating on the pelvic girdle or lower limb move the femur, leg, and foot.

# 11.1 Skeletal muscles produce movement when the insertion is pulled toward the origin.

## Muscle Attachment Sites: Origin and Insertion

As we have already mentioned, not all skeletal muscles produce movements. Those skeletal muscles that produce movements do so by exerting force on tendons, which in turn pull on bones or other structures (such as skin). Most muscles cross at least one joint and are usually attached to the articulating bones that form the joint (Figure 11.1a).

When a skeletal muscle contracts, it pulls one of the articulating bones toward the other. The two articulating bones usually do not move equally in response to contraction. One bone remains stationary or near its original position, either because other muscles stabilize that bone by contracting and pulling it in the opposite direction or because its structure makes it less movable. The attachment of a muscle's tendon to the stationary bone is called the **origin** (OR-i-jin). The attachment of the muscle's other tendon to the movable bone is called the **insertion** (in-SER-shun). A useful rule of thumb is that the origin is usually proximal and the insertion distal, especially in the limbs; the insertion is usually pulled toward the origin. The fleshy portion of the muscle between the tendons is called the **belly**. The **actions** of a muscle are the main movements that occur when the muscle contracts. A good analogy is a spring on a door. The part of the spring attached to the frame represents the origin, the part attached to the door is the insertion, the coils of the spring are the belly, and closing the door would be the action.

Muscles that move a body part often do not cover the moving part. Figure 11.1b shows that although one of the functions of the biceps brachii is to move the forearm, the belly of the muscle

**FIGURE 11.1 Relationship of skeletal muscles to bones.** (a) Muscles are attached to bones by tendons at their origin and insertion. (b) Skeletal muscles produce movements by pulling on bones. Bones serve as levers, and joints act as fulcrums for the levers. Here the lever–fulcrum principle is illustrated by the movement of the forearm. Note where the load (resistance) and effort are applied in this example.

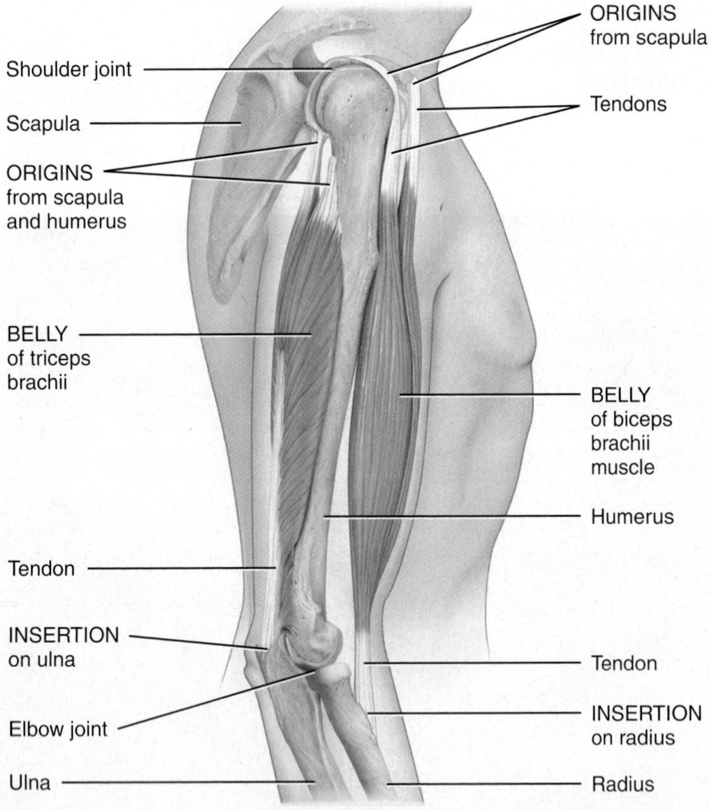

(a) Origin and insertion of a skeletal muscle

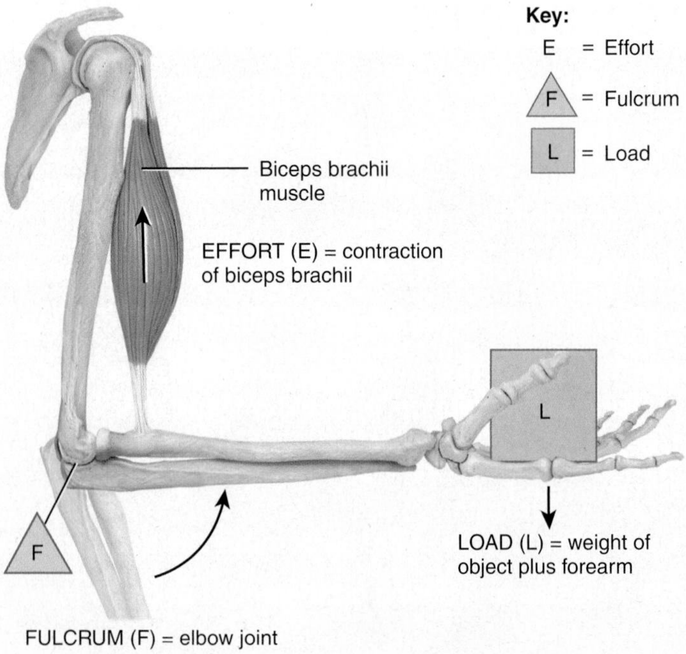

(b) Movement of forearm lifting a weight

In the limbs, the origin of a muscle is usually proximal and the insertion is usually distal.

lies over the humerus, not over the forearm. You will also see that muscles that cross two joints, such as the rectus femoris and sartorius of the thigh, have more complex actions than muscles that cross only one joint.

## Lever Systems and Leverage

A **lever** is a rigid structure that moves around a fixed point called the **fulcrum**. In producing movement, bones act as levers, and joints function as the fulcrums of these levers. A lever is acted on by two different forces: the **effort**, which causes movement, and the **load**, which resists movement. The effort is the force exerted by muscle contraction; the load is typically the weight of the body part that is moved plus any external object your body is moving. Motion occurs when the effort applied to the bone at the insertion exceeds the load. Consider the biceps brachii flexing the forearm at the elbow as an object is lifted (Figure 11.1b). When the forearm is raised, the elbow is the fulcrum. The weight of the forearm plus the weight of the object in the hand is the load. The force of contraction of the biceps brachii pulling the forearm up is the effort.

A lever operates at a *mechanical advantage* when the load is closer to the fulcrum and the effort is applied farther from the fulcrum; less effort is required to move the load. Conversely, a lever operates at a *mechanical disadvantage* when the load is farther from the fulcrum and the effort is closer to the fulcrum; more effort is required to move the load. For example, it is much easier to crush a piece of hard food (the load) with the teeth in the back of your mouth than with your front teeth because your back teeth are closer to the fulcrum (your jaw, or temporomandibular joint). Here is one more example you can try. Straighten out a paper clip. Now try to cut the paper clip with the tip of a pair of scissors (mechanical disadvantage) versus near the pivot of the scissors (mechanical advantage).

## Effects of Fascicle Arrangement

Recall from Chapter 10 that the skeletal muscle fibers (cells) within a muscle are arranged in bundles known as **fascicles** (FAS-i-kuls). Within a fascicle, all muscle fibers are parallel to one another. The fascicles, however, may form one of five patterns with respect to the tendons: parallel, fusiform (spindle-shaped, narrow toward the ends and wide in the middle), circular, triangular, or pennate (shaped like a feather) (Table 11.1).

Fascicular arrangement affects a muscle's power and range of motion. *Range of motion* refers to the range, measured in degrees of a circle, through which the bones of a joint can be moved. The longer and more parallel the fibers in a muscle, the greater is the range of motion it can produce. However, the contraction power of a muscle depends not on length but on its total cross-sectional area. Thicker muscles contain a greater number of fascicles, hence have more muscle fibers to produce a more forceful contraction. Fascicular arrangement often represents a compromise between power and range of motion. Pennate muscles, for instance, have a large number of short-fibered fascicles distributed over their tendons, giving them greater power but a smaller range of motion. In contrast, parallel muscles have comparatively fewer fascicles, but they have long fibers that extend the length of the muscle, so they have a greater range of motion but less power.

## Coordination among Muscles

Movements often are the result of several skeletal muscles acting as a group rather than individually. Most skeletal muscles are arranged in opposing pairs at joints—that is, flexors–extensors, abductors–adductors, and so on. Within opposing pairs, one muscle, called the **prime mover** or **agonist** (= leader), contracts to cause a desired action while the other muscle, the **antagonist** (*anti-* = against), stretches and yields to the effects of the prime mover. When you flex your elbow, the biceps brachii is the prime mover. While the biceps brachii is contracting, the triceps brachii (the antagonist), is relaxing (see Figure 11.1a). The antagonist and prime mover are usually located on opposite sides of the bone or joint, as is the case in this example.

With an opposing pair of muscles, the roles of the prime mover and antagonist can switch for different movements. For example, while extending the forearm at the elbow (i.e., lowering the load shown in Figure 11.1b), the triceps brachii becomes the prime mover and the biceps brachii is the antagonist. If a prime mover and its antagonist contract at the same time with equal force, there will be no movement.

Sometimes a prime mover crosses other joints before it reaches the joint at which its primary action occurs. The biceps brachii, for example, spans both the shoulder and elbow joints but has its primary action at the elbow joint. To prevent unwanted movements at intermediate joints or to otherwise aid the movement of the prime mover, muscles called **synergists** (SIN-er-gists; *syn-* = together; *-ergon* = work) contract and stabilize the intermediate joints. As an example, muscles that flex the fingers (prime movers) cross the carpal joints (intermediate joints). If movement at these intermediate joints was unrestrained, you would not be able to flex your fingers without flexing the wrist at the same time. Synergistic contraction of the wrist extensor muscles stabilizes the wrist joint and prevents unwanted movements, while the flexor muscles of the fingers contract to bring about the primary action, flexion of the fingers.

Some muscles in a group also act as **fixators**, stabilizing the origin of the prime mover so that the prime mover can act more efficiently. Fixators steady the proximal end of a limb while movements occur at the distal end. For example, the scapula in the pectoral (shoulder) girdle is a freely movable bone that serves as the origin for several muscles that move the arm. When

**TABLE 11.1**

Arrangement of Fascicles

**Parallel** Fascicles parallel to longitudinal axis of muscle; terminate at either end in flat tendons

*Example:* Stylohyoid muscle (see Figure 11.7)

**Fusiform** Fascicles nearly parallel to longitudinal axis of muscle; terminate in flat tendons; muscle tapers toward tendons, where diameter is less than at belly

*Example:* Digastric muscle (see Figure 11.7)

**Circular** Fascicles in concentric circular arrangements form sphincter muscles that enclose an orifice (opening)

*Example:* Orbicularis oculi muscle (see Figure 11.3)

**Triangular** Fascicles spread over broad area converge at thick central tendon; give muscle a triangular appearance

*Example:* Pectoralis major muscle (see Figure 11.2a)

**Pennate** Short fascicles in relation to total muscle length; tendon extends nearly entire length of muscle

**Unipennate** Fascicles are arranged on only one side of tendon

*Example:* Extensor digitorum longus muscle (see Figure 11.20a, b)

**Bipennate** Fascicles are arranged on both sides of centrally positioned tendons

*Example:* Rectus femoris muscle (see Figure 11.19a)

**Multipennate** Fascicles attach obliquely from many directions to several tendons

*Example:* Deltoid muscle (see Figure 11.9a)

the arm muscles contract, the scapula must be held steady. In abduction of the arm, the deltoid serves as the prime mover. As the deltoid pulls on the humerus to abduct the arm, fixators (pectoralis minor, trapezius, subclavius, serratus anterior, and others) hold the scapula firmly against the back of the chest (see Figure 11.14). Depending on the movement needed at the moment, muscles may act as prime movers, antagonists, synergists, or fixators.

**✓ CHECKPOINT**

1. How do skeletal muscles produce body movements by pulling on bones? (*Hint: Use the terms* origin, insertion, *and* belly.)
2. When you flex your knee, where are the fulcrum and load?
3. Would a narrow, long muscle likely have a large range of motion or a highly forceful contraction?

## 11.2 Skeletal muscles are named based on size, shape, action, location, or attachments.

The figures and tables that follow will assist you in learning the names of the principal skeletal muscles in various regions of the body. (Remember that there are nearly 700 individual muscles in the human body; we have not included them all here.) The selected skeletal muscles have been divided into groups according to the part of the body on which they act.

## CLINICAL CONNECTION | *Intramuscular Injections*

An **intramuscular (IM) injection** penetrates the skin and subcutaneous tissue to enter the muscle itself. Intramuscular injections are preferred when prompt absorption is desired, when larger doses than can be given subcutaneously are indicated, or when the drug is too irritating to give subcutaneously. The common sites for intramuscular injections include the gluteus medius muscle of the buttock (Figure 11.2b), lateral side of the thigh in the midportion of the vastus lateralis muscle (Figure 11.2a), and the deltoid muscle of the shoulder (Figure 11.2b). Muscles in these areas, especially the gluteal muscles in the buttock, are fairly thick, and absorption is promoted by their extensive blood supply. To avoid injury, intramuscular injections are given deep within the muscle, away from major nerves and blood vessels. Intramuscular injections have a faster speed of delivery than oral medications, but are slower than intravenous infusions.

In certain figures, superficial muscles cover deeper muscles. In other figures, superficial muscles have been removed or reflected (cut or pulled aside) to expose the deeper muscles. Within the figures, muscles that are named in all capital letters are specifically referred to in the corresponding tables. Additional related structures are also labeled in the figures for reference. Figure 11.2 provides an overview of principal superficial skeletal muscles. As you study groups of muscles, refer to Figure 11.2 to see how each group is related to the others.

The names of most of the skeletal muscles contain combinations of the word roots of their distinctive features. Learning the terms that refer to these features will help you remember the names of muscles. Such muscle features include the pattern of the muscle's fascicles; the size, shape, action, number of origins, and location of the muscle; and the sites of origin and insertion of the muscle. On the other hand, knowing the names of the muscles will give you clues to their features. Study Table 11.2 to become familiar with the terms used in muscle names.

To make it easier for you to learn and understand how skeletal muscles are named and to say their names, the tables include word roots and phonetic pronunciations. As you review each muscle in the table, notice the information provided next to its name. Once you have mastered the naming of the muscles, their actions will have more meaning and be easier to remember. The tables also provide the origin, insertion, actions, and innervation of each selected muscle. By learning the origin and insertion of a muscle, we can better understand its action. Employing common sense is often easier than memorization. As you think of a muscle's attachments, you can often reason through the movements, or the "actions," that muscle could produce when it contracts.

The innervation section of each table lists the nerve or nerves that cause contraction of each muscle. In general, cranial nerves, which arise from the lower parts of the brain, serve muscles in the head region. Spinal nerves, which arise from the spinal cord within the vertebral column, innervate muscles in the rest of the body. Cranial nerves are designated by both a name and a Roman numeral—for example, the facial (VII) nerve. Spinal nerves are numbered in groups according to the part of the spinal cord from which they arise: C = cervical (neck region), T = thoracic (chest region), L = lumbar (lower back region), and S = sacral (buttocks region). An example is T1, the first thoracic spinal nerve.

### ✓ CHECKPOINT

**4.** Using Figure 11.2, give an example of a muscle named for each of the following characteristics: direction of fibers, shape, action, size, origin and insertion, location, and number of tendons of origin.

**FIGURE 11.2 Principal superficial skeletal muscles.**

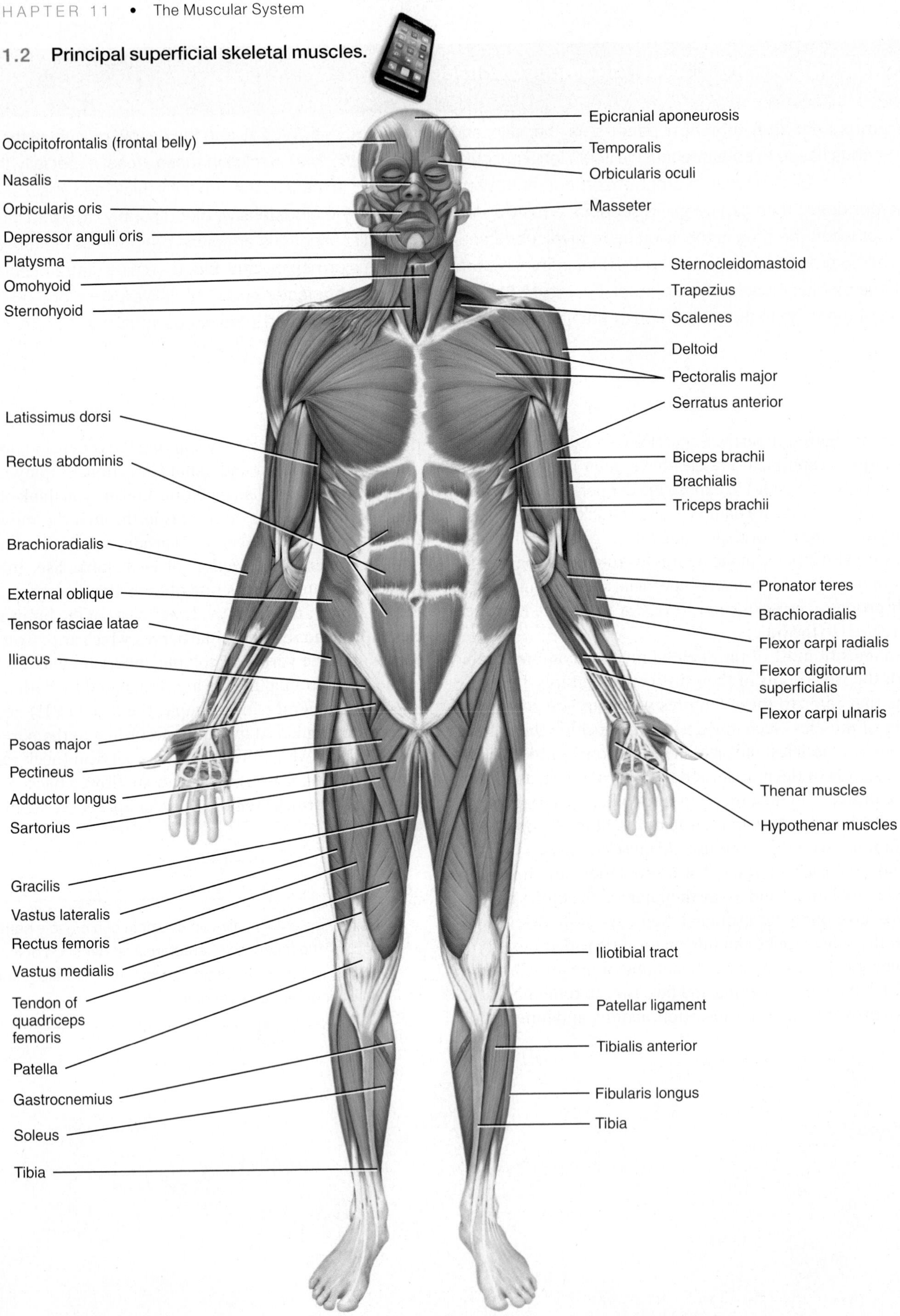

Epicranial aponeurosis
Occipitofrontalis (frontal belly)
Temporalis
Orbicularis oculi
Nasalis
Orbicularis oris
Masseter
Depressor anguli oris
Platysma
Sternocleidomastoid
Omohyoid
Trapezius
Sternohyoid
Scalenes
Deltoid
Pectoralis major
Serratus anterior
Latissimus dorsi
Rectus abdominis
Biceps brachii
Brachialis
Triceps brachii
Brachioradialis
External oblique
Pronator teres
Brachioradialis
Tensor fasciae latae
Flexor carpi radialis
Iliacus
Flexor digitorum superficialis
Flexor carpi ulnaris
Psoas major
Pectineus
Adductor longus
Thenar muscles
Sartorius
Hypothenar muscles
Gracilis
Vastus lateralis
Rectus femoris
Vastus medialis
Iliotibial tract
Tendon of quadriceps femoris
Patellar ligament
Patella
Tibialis anterior
Gastrocnemius
Fibularis longus
Soleus
Tibia
Tibia

(a) Anterior view

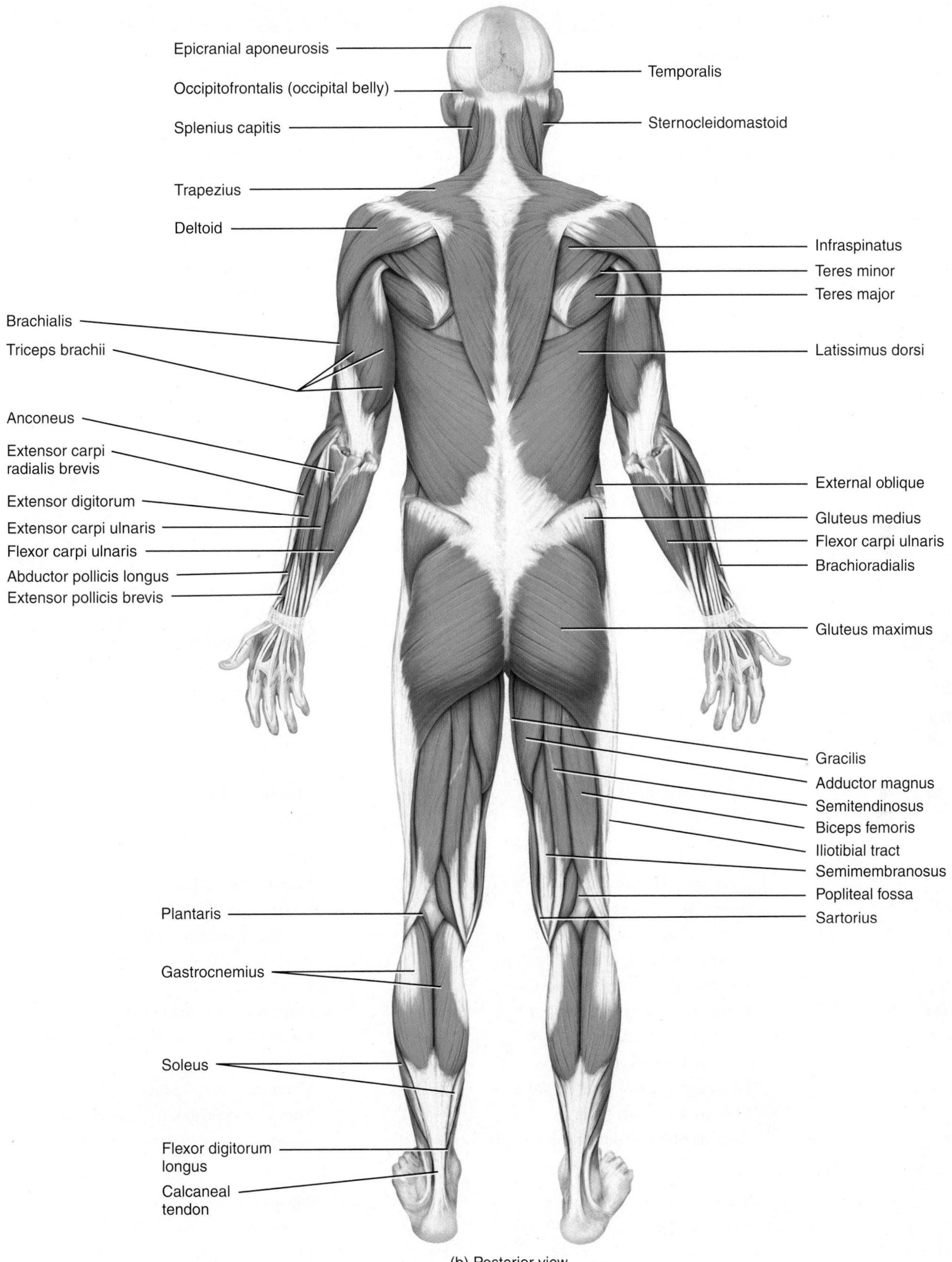


(b) Posterior view

**Most movements require several skeletal muscles acting in groups rather than individually.**

**TABLE 11.2**

Characteristics Used to Name Muscles

| NAME | MEANING | EXAMPLE | FIGURE |
|---|---|---|---|
| **Direction: Orientation of muscle fascicles relative to the body's midline** | | | |
| **Rectus** | Parallel to midline | Rectus abdominis | 11.9b |
| **Transverse** | Perpendicular to midline | Transversus abdominis | 11.9b |
| **Oblique** | Diagonal to midline | External oblique | 11.9c |
| **Size: Relative size of the muscle** | | | |
| **Maximus** | Largest | Gluteus maximus | 11.2b |
| **Medius** | Intermediate | Gluteus medius | 11.19c |
| **Minimus** | Smallest | Gluteus minimus | 11.19d |
| **Longus** | Long | Adductor longus | 11.19a |
| **Brevis** | Short | Adductor brevis | 11.19b |
| **Latissimus** | Widest | Latissimus dorsi | 11.14b |
| **Longissimus** | Longest | Longissimus capitis | 11.18a |
| **Magnus** | Large | Adductor magnus | 11.19a |
| **Major** | Larger | Pectoralis major | 11.9a |
| **Minor** | Smaller | Pectoralis minor | 11.13a |
| **Vastus** | Huge | Vastus lateralis | 11.19a |
| **Shape: Relative shape of the muscle** | | | |
| **Deltoid** | Triangular | Deltoid | 11.9a |
| **Trapezius** | Trapezoid | Trapezius | 11.2b |
| **Serratus** | Saw-toothed | Serratus anterior | 11.13b |
| **Rhomboid** | Diamond-shaped | Rhomboid major | 11.14c |
| **Orbicularis** | Circular | Orbicularis oculi | 11.3a |
| **Pectinate** | Comblike | Pectineus | 11.19a |
| **Piriformis** | Pear-shaped | Piriformis | 11.19d |
| **Platys** | Flat | Platysma | 11.3c |
| **Quadratus** | Square, four-sided | Quadratus femoris | 11.19d |
| **Gracilis** | Slender | Gracilis | 11.19a |
| **Action: Principal action of the muscle** | | | |
| **Flexor** | Decreases a joint angle | Flexor carpi radialis | 11.16a |
| **Extensor** | Increases a joint angle | Extensor carpi ulnaris | 11.16d |
| **Abductor** | Moves a bone away from the midline | Abductor pollicis longus | 11.16e |
| **Adductor** | Moves a bone closer to the midline | Adductor longus | 11.19a |
| **Levator** | Raises or elevates a body part | Levator scapulae | 11.13a |
| **Depressor** | Lowers or depresses a body part | Depressor labii inferioris | 11.3b |
| **Supinator** | Turns palm anteriorly | Supinator | 11.16b |
| **Pronator** | Turns palm posteriorly | Pronator teres | 11.16a |
| **Sphincter** | Decreases the size of an opening | External anal sphincter | 11.11 |
| **Tensor** | Makes a body part rigid | Tensor fasciae latae | 11.19a |
| **Rotator** | Rotates a bone around its longitudinal axis | Rotatore | 11.18a |
| **Number of Origins: Number of tendons of origin** | | | |
| **Biceps** | Two origins | Biceps brachii | 11.15a |
| **Triceps** | Three origins | Triceps brachii | 11.15b |
| **Quadriceps** | Four origins | Quadriceps femoris | 11.19a |
| **Location: Structure near which a muscle is found** | | | |
| *Example:* Temporalis, a muscle near the temporal bone | | | 11.3c |
| **Origin and Insertion: Sites where muscle originates and inserts** | | | |
| *Example:* Sternocleidomastoid, originating on the sternum and clavicle and inserting onto the mastoid process of the temporal bone | | | 11.2a |

# 11.3 Muscles of the head produce facial expressions, eyeball movement, and assist in biting, chewing, swallowing, and speech.

## Muscles of Facial Expression

The muscles of facial expression provide us with the ability to express a wide variety of emotions, including displeasure, surprise, fear, and happiness (Figure 11.3; Table 11.3). The muscles themselves lie within the hypodermis (connective tissue beneath the skin). They usually originate in the fascia or bones of the skull and insert into skin or other muscles. Because of their insertions, the muscles of facial expression move the skin rather than a joint when they contract.

FIGURE 11.3 Muscles of facial expression.

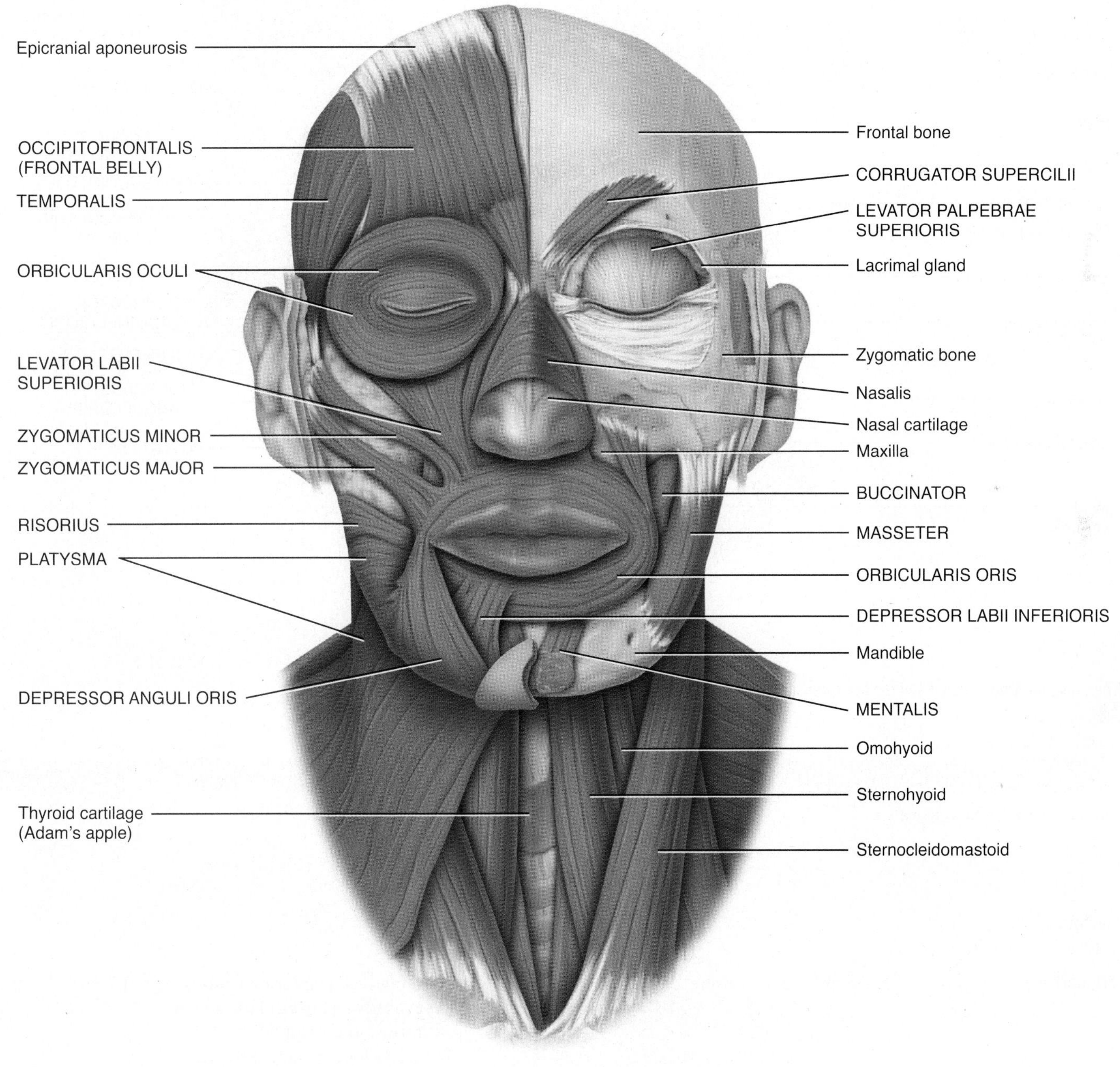


(a) Anterior superficial view (b) Anterior deep view

(continues)

FIGURE 11.3 (continued)

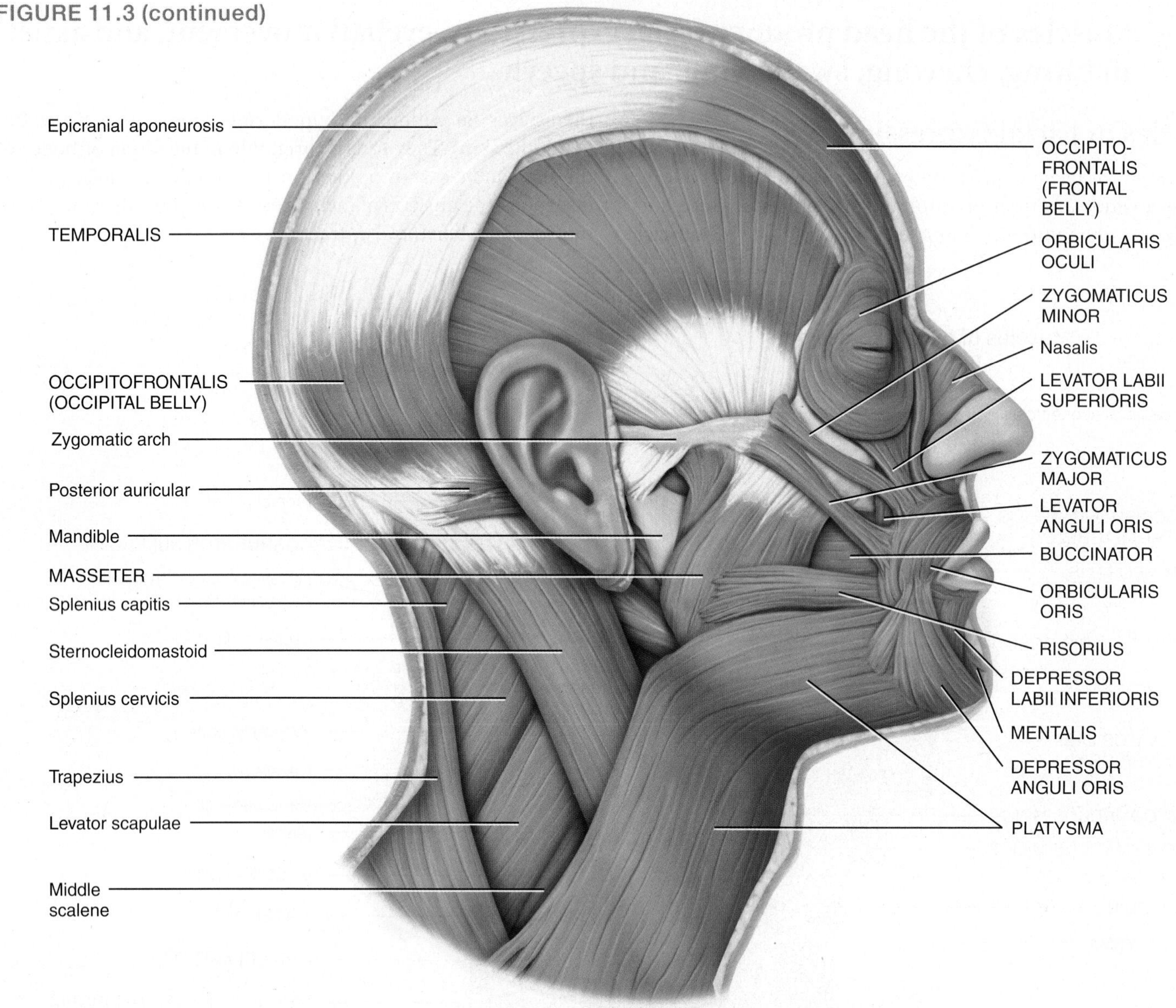


(c) Right lateral superficial view

**When they contract, muscles of facial expression move the skin rather than a joint.**

**TABLE 11.3**

Muscles of Facial Expression

| MUSCLE | ORIGIN | INSERTION | ACTION | INNERVATION |
|---|---|---|---|---|
| **Scalp Muscles** | | | | |
| **Occipitofrontalis** (ok-sip′-i-tō-frun-TĀ-lis) | | | | |
| **Frontal belly** | Epicranial aponeurosis | Skin superior to supraorbital margin | Draws scalp anteriorly, raises eyebrows, and wrinkles skin of forehead horizontally as in a look of surprise | Facial (VII) nerve |
| **Occipital belly** (*occipit-* = back of the head) | Occipital bone and mastoid process of temporal bone | Epicranial aponeurosis | Draws scalp posteriorly | Facial (VII) nerve |

| | | | | |
|---|---|---|---|---|
| **Mouth Muscles** | | | | |
| **Orbicularis oris** (or-bi′-kū-LAR-is OR-is; *orb-* = circular; *oris* = of the mouth) | Muscle fibers surrounding opening of mouth | Skin at corner of mouth | Closes and protrudes lips, as in kissing; compresses lips against teeth; and shapes lips during speech | Facial (VII) nerve |
| **Zygomaticus major** (zī-gō-MA-ti-kus; *zygomatic* = cheek bone; *major* = greater) | Zygomatic bone | Skin at angle of mouth and orbicularis oris | Draws angle of mouth superiorly and laterally, as in smiling | Facial (VII) nerve |
| **Zygomaticus minor** (*minor* = lesser) | Zygomatic bone | Upper lip | Raises (elevates) upper lip, exposing maxillary teeth | Facial (VII) nerve |
| **Levator labii superioris** (le-VĀ-tor LĀ-bē-ī soo-per′-ē-OR-is; *levator* = raises or elevates; *labii* = lip; *superioris* = upper) | Superior to infraorbital foramen of maxilla | Skin at angle of mouth and orbicularis oris | Raises upper lip | Facial (VII) nerve |
| **Depressor labii inferioris** (de-PRE-sor = depresses or lowers; *inferioris* = lower) | Mandible | Skin of lower lip | Depresses (lowers) lower lip | Facial (VII) nerve |
| **Depressor anguli oris** (*angul-* = angle or corner) | Mandible | Angle of mouth | Draws angle of mouth laterally and inferiorly, as in opening mouth | Facial (VII) nerve |
| **Levator anguli oris** | Inferior to infraorbital foramen | Skin of lower lip and orbicularis oris | Draws angle of mouth laterally and superiorly | Facial (VII) nerve |
| **Buccinator** (BUK-si-nā′-tor; *bucc-* = cheek) | Alveolar processes of maxilla and mandible and pterygomandibular raphe (fibrous band extending from the pterygoid process to the sphenoid bone to the mandible) | Orbicularis oris | Presses cheeks against teeth and lips, as in whistling, blowing, and sucking; draws corner of mouth laterally; and assists in mastication (chewing) by keeping food between the teeth (and not between teeth and cheeks) | Facial (VII) nerve |
| **Risorius** (ri-ZOR-ē-us; *risor-* = laughter) | Fascia over parotid (salivary) gland | Skin at angle of mouth | Draws angle of mouth laterally, as in grimacing | Facial (VII) nerve |
| **Mentalis** (men-TĀ-lis; *ment-* = the chin) | Mandible | Skin of chin | Elevates and protrudes lower lip and pulls skin of chin up, as in pouting | Facial (VII) nerve |
| **Neck Muscle** | | | | |
| **Platysma** (pla-TIZ-ma; *platys-* = flat, broad) | Fascia over deltoid and pectoralis major muscles | Mandible, muscle around angle of mouth, and skin of lower face | Draws outer part of lower lip inferiorly and posteriorly, as in pouting; depresses mandible | Facial (VII) nerve |
| **Orbit and Eyebrow Muscles** | | | | |
| **Orbicularis oculi** (OK-ū-lī = of the eye) | Medial wall of orbit | Circular path around orbit | Closes eye | Facial (VII) nerve |
| **Corrugator supercilii** (KOR-u-gā′-tor soo-per-SI-lē-ī; *corrugat-* = wrinkle; *supercilii* = of the eyebrow) | Medial end of superciliary arch of frontal bone | Skin of eyebrow | Draws eyebrow inferiorly and wrinkles skin of forehead vertically as in frowning | Facial (VII) nerve |
| **Levator palpebrae superioris** (PAL-pe-brē = eyelids) (see also Figure 11.4a) | Roof of orbit (lesser wing of sphenoid bone) | Skin of upper eyelid | Elevates upper eyelid (opens eye) | Oculomotor (III) nerve |

The **occipitofrontalis** is an unusual muscle in this group because it is made up of two parts: an anterior part called the **frontal belly**, which is superficial to the frontal bone, and a posterior part called the **occipital belly**, which is superficial to the occipital bone. The two muscular portions are held together by a strong *aponeurosis* (sheetlike tendon), the **epicranial aponeurosis** (ep-i-KRĀ-nē-al ap′-ō-noo-RŌ-sis), which covers the superior and lateral surfaces of the skull.

The **buccinator** forms the major muscular portion of the cheek. The buccinator is so named because it compresses the cheeks (*bucc-* = cheek) during blowing—for example, when a musician plays a wind instrument such as a trumpet. It functions in whistling, blowing, and sucking, and assists in chewing.

Among the noteworthy muscles in this group are those surrounding the orifices (openings) of the head, such as the eyes, nose, and mouth. These muscles function as **sphincters** (SFINGK-ters), which close the orifices, and *dilators* (DĪ-lā-tors), which open the orifices. For example, the **orbicularis oculi** closes the eye, and the **levator palpebrae superioris** opens it.

## Muscles That Move the Eyeballs

Muscles that move the eyeballs are called **extrinsic eye muscles** because they originate outside the eyeballs (in the orbit) and insert on its outer surface, the sclera ("white of the eye") (Figure 11.4; Table 11.4). The extrinsic eye muscles are some of the fastest contracting and most precisely controlled skeletal muscles in the body.

Three pairs of extrinsic eye muscles control movements of the eyeballs: (1) superior and inferior recti (singular is *rectus*), (2) lateral and medial recti, and (3) superior and inferior obliques. The four recti muscles (superior, inferior, lateral, and medial) arise from a common tendinous ring at the back of the orbit and insert into the sclera of the eyeball. The names of the four recti muscles clearly imply their actions. When the **superior rectus** pulls the eyeball, the eye looks up; as the **inferior rectus** contracts, the eye looks down. Contraction of the **lateral rectus** makes the eye rotate laterally, and when the **medial rectus** pulls, the eye moves medially.

The oblique muscles—superior and inferior—rotate the eyeball on its axis. The **superior oblique** originates posteriorly near the tendinous ring, passes anteriorly, and ends in a round tendon that extends through a pulleylike loop called the *trochlea* (= pulley), where it turns and inserts on the superior, lateral eyeball. When the superior oblique pulls the eyeball, the eye looks down and laterally. The **inferior oblique** originates on the maxilla at the anterior, medial floor of the orbit. It then passes posteriorly and laterally and inserts on the posterior, lateral eyeball. Because of this arrangement, contraction of the inferior oblique results in the eye looking up and laterally.

**FIGURE 11.4** Extrinsic muscles of the eyeball.

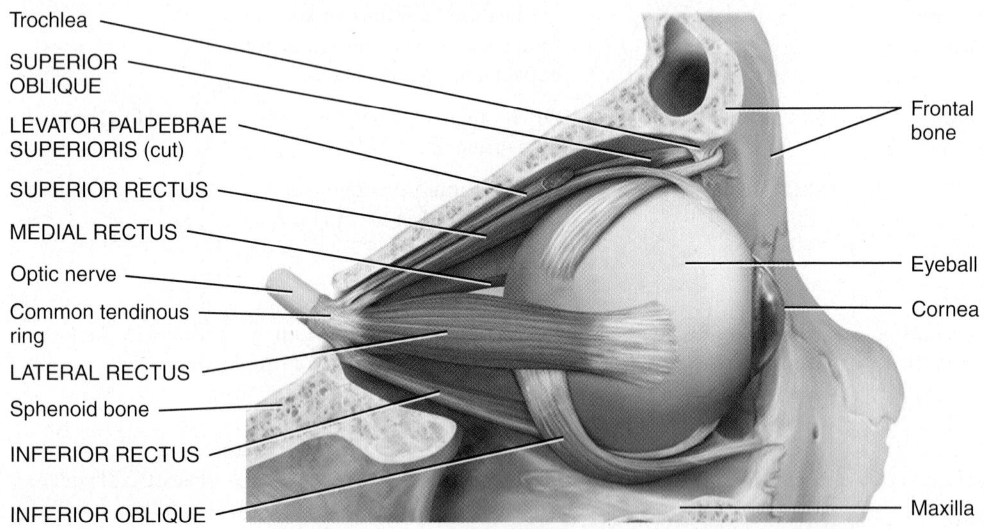


(a) Right lateral view of right eyeball

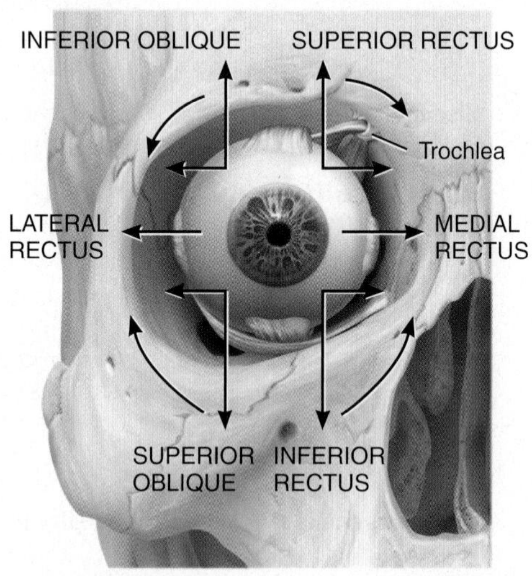


(b) Movements of right eyeball in response to contraction of extrinsic muscles

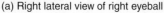

**The extrinsic muscles of the eyeball are among the fastest contracting and most precisely controlled skeletal muscles in the body.**

**TABLE 11.4**

Muscles That Move the Eyeballs—Extrinsic Eye Muscles

| MUSCLE | ORIGIN | INSERTION | ACTION | INNERVATION |
|---|---|---|---|---|
| **Superior rectus** (*rectus* = fascicles parallel to midline) | Common tendinous ring (attached to orbit around optic foramen) | Superior and central part of eyeball | Moves eyeball superiorly (elevation) and medially (adduction), and rotates it medially | Oculomotor (III) nerve |
| **Inferior rectus** | Same as above | Inferior and central part of eyeball | Moves eyeball inferiorly (depression) and medially (adduction), and rotates it medially | Oculomotor (III) nerve |
| **Lateral rectus** | Same as above | Lateral side of eyeball | Moves eyeball laterally (abduction) | Abducens (VI) nerve |
| **Medial rectus** | Same as above | Medial side of eyeball | Moves eyeball medially (adduction) | Oculomotor (III) nerve |
| **Superior oblique** (*oblique* = fascicles diagonal to midline) | Sphenoid bone, superior and medial to the tendinous ring in the orbit | Eyeball between superior and lateral recti; the muscle inserts into the superior and lateral surfaces of the eyeball via a tendon that passes through the trochlea | Moves eyeball inferiorly (depression) and laterally (abduction), and rotates it medially | Trochlear (IV) nerve |
| **Inferior oblique** | Maxilla in floor of orbit | Eyeball between inferior and lateral recti | Moves eyeball superiorly (elevation) and laterally (abduction), and rotates it laterally | Oculomotor (III) nerve |

## Muscles That Move the Mandible

The muscles that move the mandible (lower jawbone) at the temporomandibular joint (TMJ) are known as the *muscles of mastication* (chewing) (Figure 11.5; Table 11.5). These muscles also assist in speech. Of the four pairs of muscles involved in mastication, three are powerful closers of the jaw and account for the strength of the bite: **masseter**, **temporalis**, and **medial pterygoid**. Of these, the masseter is the strongest muscle of mastication. The **medial** and **lateral pterygoid** assist in mastication by moving the mandible from side to side to help grind food. Additionally, these muscles protract the mandible (thrust it forward).

## Muscles That Move the Tongue

The tongue is a highly mobile structure that is vital to digestive functions such as mastication, perception of taste, and deglutition (swallowing). It is also important in speech. The tongue's mobility is greatly aided by its suspension from the mandible, styloid process of the temporal bone, and hyoid bone.

**FIGURE 11.5 Muscles that move the mandible.**

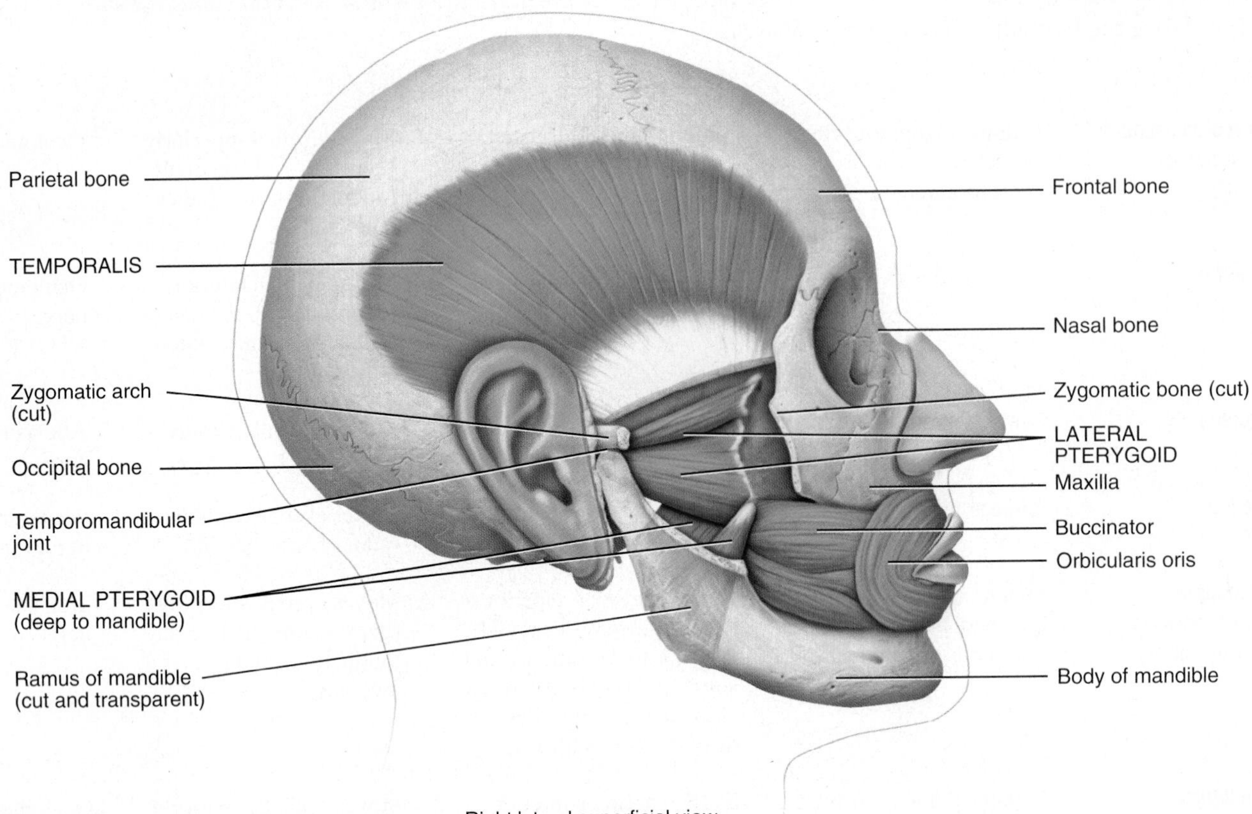


Right lateral superficial view

**The muscles that move the mandible are also known as muscles of mastication.**

**TABLE 11.5**

Muscles That Move the Mandible

| MUSCLE | ORIGIN | INSERTION | ACTION | INNERVATION |
|---|---|---|---|---|
| **Masseter** (MA-se-ter = a chewer) (see Figure 11.3c) | Maxilla and zygomatic arch | Angle and ramus of mandible | Elevates mandible, as in closing mouth | Mandibular division of trigeminal (V) nerve |
| **Temporalis** (tem′-pō-RĀ-lis; *tempor-* = time or temples) | Temporal bone | Coronoid process and ramus of mandible | Elevates and retracts mandible | Mandibular division of trigeminal (V) nerve |
| **Medial pterygoid** (TER-i-goyd; *medial* = closer to midline; *pterygoid* = like a wing) | Medial surface of lateral portion of pterygoid process of sphenoid bone; maxilla | Angle and ramus of mandible | Elevates and protracts (protrudes) mandible and moves mandible from side to side | Mandibular division of trigeminal (V) nerve |
| **Lateral pterygoid** *lateral* = farther from midline) | Greater wing and lateral surface of lateral portion of pterygoid process of sphenoid bone | Condyle of mandible; temporomandibular joint (TMJ) | Protracts mandible, depresses mandible as in opening mouth, and moves mandible from side to side | Mandibular division of trigeminal (V) nerve |

The tongue is divided into lateral halves by a median fibrous septum. The septum extends throughout the length of the tongue. Inferiorly, the septum attaches to the hyoid bone. Muscles of the tongue are of two principal types: extrinsic and intrinsic (Figure 11.6; Table 11.6). **Extrinsic tongue muscles** originate outside the tongue and insert into it. They move the entire tongue in various directions, such as anteriorly, posteriorly, and laterally. **Intrinsic tongue muscles** originate and insert within the tongue. These muscles alter the shape of the tongue rather than moving the entire tongue. The extrinsic and intrinsic muscles of the tongue insert into both lateral halves of the tongue. Only the extrinsic tongue muscles are considered in Table 11.6.

FIGURE 11.6 **Muscles that move the tongue.**

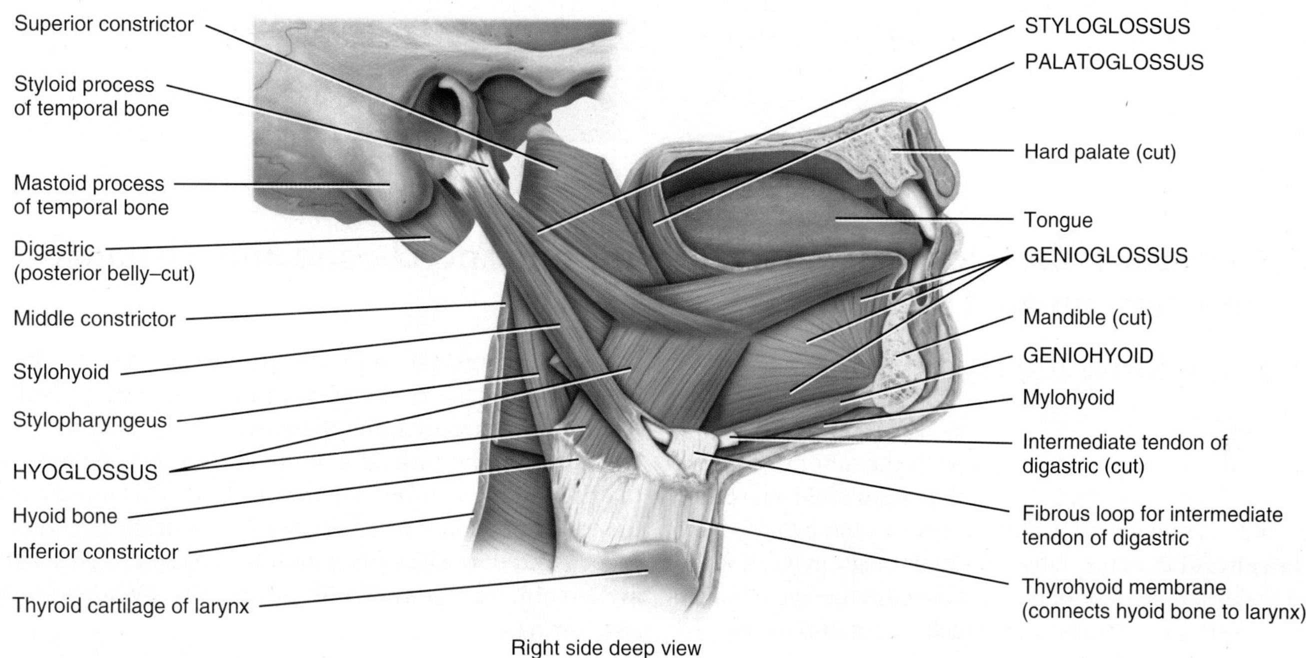


The extrinsic and intrinsic muscles of the tongue insert into both lateral halves of the tongue.

TABLE 11.6

Muscles That Move the Tongue—Extrinsic Tongue Muscles

| MUSCLE | ORIGIN | INSERTION | ACTION | INNERVATION |
|---|---|---|---|---|
| **Genioglossus** (jē′-nē-ō-GLOS-us; *genio-* = the chin; *-glossus* = tongue) | Mandible | Undersurface of tongue and hyoid bone | Depresses tongue and thrusts it anteriorly (protraction) | Hypoglossal (XII) nerve |
| **Styloglossus** (stī′-lō-GLOS-us; *stylō-* = stake or poles for styloid process of temporal bone) | Styloid process of temporal bone | Side and undersurface of tongue | Elevates tongue and draws it posteriorly (retraction) | Hypoglossal (XII) nerve |
| **Palatoglossus** (pal′-a-tō-GLOS-us; *palato-* = the roof of the mouth or palate) | Anterior surface of soft palate | Side of tongue | Elevates posterior portion of tongue and draws soft palate down on tongue | Pharyngeal plexus, which contains axons from both the vagus (X) and accessory (XI) nerves |
| **Hyoglossus** (hī′-ō-GLOS-us) | Greater horn and body of hyoid bone | Side of tongue | Depresses tongue and draws down its sides | Hypoglossal (XII) nerve |

When you study the extrinsic tongue muscles, you will notice that all of their names end in *-glossus*, meaning tongue. You will notice that the actions of the muscles are obvious, considering the positions of the mandible, styloid process, hyoid bone, and soft palate, which serve as origins for these muscles. For example, the **genioglossus** (origin: the mandible) pulls the tongue downward and forward, the **styloglossus** (origin: the styloid process) pulls the tongue upward and backward, the **hyoglossus** (origin: the hyoid bone) pulls the tongue downward and flattens it, and the **palatoglossus** (origin: the soft palate) raises the back portion of the tongue.

✓ **CHECKPOINT**

5. Why do the muscles of facial expression move the skin rather than a joint?
6. Which muscles of facial expression would you use to blow up a balloon? To squint? To pucker your lips? To pout?
7. Which extrinsic eye muscles contract as you gaze to your left without moving your head? Which ones relax?
8. Which extrinsic eye muscles cause elevation of the anterior eyeball?
9. What are the functions of the tongue?

## 11.4 Muscles of the neck assist in swallowing and speech, and allow balance and movement of the head.

### Muscles That Move the Hyoid Bone and Larynx

Two groups of muscles are associated with the anterior aspect of the neck (Figure 11.7; Table 11.7): the **suprahyoid muscles**, so named because they are located superior to the hyoid bone, and the **infrahyoid muscles**, named for their position inferior to the hyoid bone. Both groups of muscles stabilize the hyoid bone, allowing it to serve as a firm base on which the tongue can move.

As a group, the suprahyoid muscles elevate the hyoid bone, floor of the oral cavity, and tongue during swallowing. As its name suggests, the **digastric** (*di-* = two) has two bellies, anterior and posterior, united by an intermediate tendon that is held in position by a fibrous loop. This muscle elevates the hyoid bone and larynx (voice box) during swallowing and speech and depresses the mandible. Together, the **stylohyoid**, **mylohyoid**, and **geniohyoid** elevate the hyoid bone during swallowing.

FIGURE 11.7 Muscles of the anterior neck.

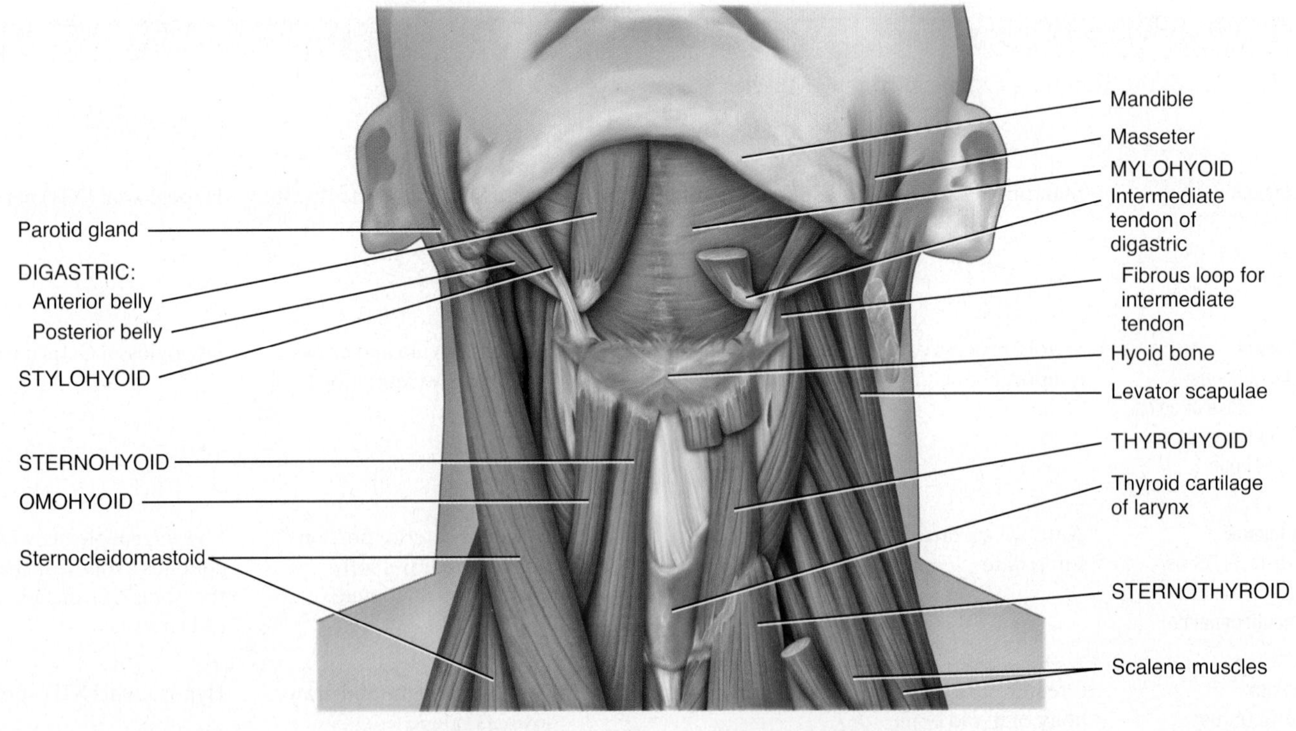

## TABLE 11.7

### Muscles That Move the Hyoid Bone and Larynx

| MUSCLE | ORIGIN | INSERTION | ACTION | INNERVATION |
|---|---|---|---|---|
| **Suprahyoid Muscles** | | | | |
| **Digastric** (dī′-GAS-trik; *di-* = two; *-gastr-* = belly) | Anterior belly from inner side of inferior border of mandible; posterior belly from temporal bone | Body of hyoid bone via an intermediate tendon | Elevates hyoid bone and depresses mandible, as in opening the mouth | Anterior belly: mandibular division of trigeminal (V) nerve<br>Posterior belly: facial (VII) nerve |
| **Stylohyoid** (stī′-lō-HĪ-oyd; *stylo-* = stake or pole, for styloid process of temporal bone; *-hyo-* = U-shaped, pertaining to hyoid bone) | Styloid process of temporal bone | Body of hyoid bone | Elevates hyoid bone and draws it posteriorly | Facial (VII) nerve |
| **Mylohyoid** (mī′-lō-HĪ-oyd) (*mylo-* = mill) | Inner surface of mandible | Body of hyoid bone | Elevates hyoid bone and floor of mouth and depresses mandible | Mandibular division of trigeminal (V) nerve |
| **Geniohyoid** (jē′-nē-ō-HĪ-oyd; *genio-* = chin) (see Figure 11.6) | Inner surface of mandible | Body of hyoid bone | Elevates hyoid bone, draws hyoid bone and tongue anteriorly, and depresses mandible | First cervical spinal nerve |
| **Infrahyoid Muscles** | | | | |
| **Omohyoid** (ō-mō-HĪ-oyd; *omo-* = relationship to the shoulder) | Superior border of scapula and superior transverse ligament | Body of hyoid bone | Depresses hyoid bone | Branches of spinal nerves C1–C3 |
| **Sternohyoid** (ster′-nō-HĪ-oyd; *sterno-* = sternum) | Medial end of clavicle and manubrium of sternum | Body of hyoid bone | Depresses hyoid bone | Branches of spinal nerves C1–C3 |
| **Sternothyroid** (ster′-nō-THĪ-royd; *-thyro-* = thyroid gland) | Manubrium of sternum | Thyroid cartilage of larynx | Depresses thyroid cartilage of larynx | Branches of spinal nerves C1–C3 |
| **Thyrohyoid** (thī′-rō-HĪ-oyd) | Thyroid cartilage of larynx | Greater horn of hyoid bone | Elevates thyroid cartilage and depresses hyoid bone | Branches of spinal nerves C1–C2 and descending hypoglossal (XII) nerve |

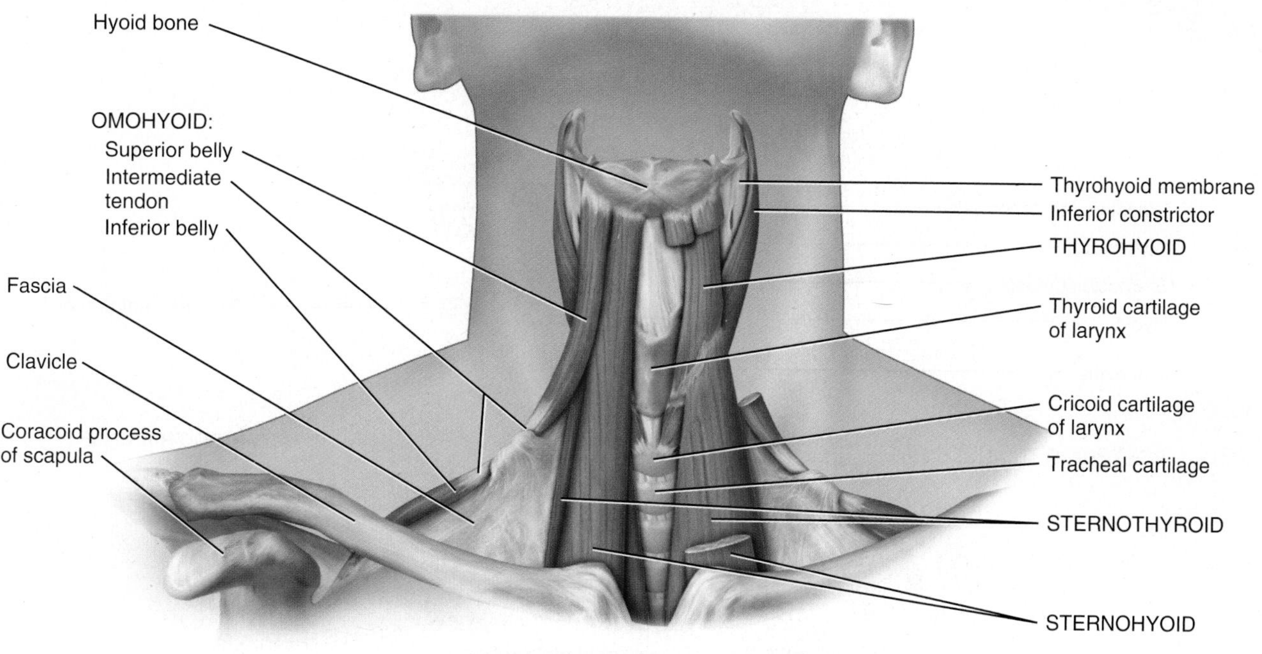


(c) Anterior superficial view (d) Anterior deep view

**The suprahyoid muscles elevate the hyoid bone, the floor of the oral cavity, and the tongue during swallowing.**

Most of the infrahyoid muscles depress the hyoid bone, and some move the thyroid cartilage (Adam's apple) of the larynx during swallowing and speech. The **omohyoid**, like the digastric, is composed of two bellies connected by an intermediate tendon. In this case, however, the two bellies are referred to as *superior* and *inferior,* rather than anterior and posterior. Together, the omohyoid, **sternohyoid**, and **thyrohyoid** depress the hyoid bone. In addition, the **sternothyroid** depresses the thyroid cartilage (Adam's apple) of the larynx to produce low sounds; the thyrohyoid elevates the thyroid cartilage to produce high sounds.

## Muscles That Move the Head

Balance and movement of the head on the vertebral column involve the action of several neck muscles (Figure 11.8; Table 11.8). Acting together (bilaterally), contraction of the **sternocleidomastoid** flexes

**FIGURE 11.8 Selected muscles of the lateral neck.**

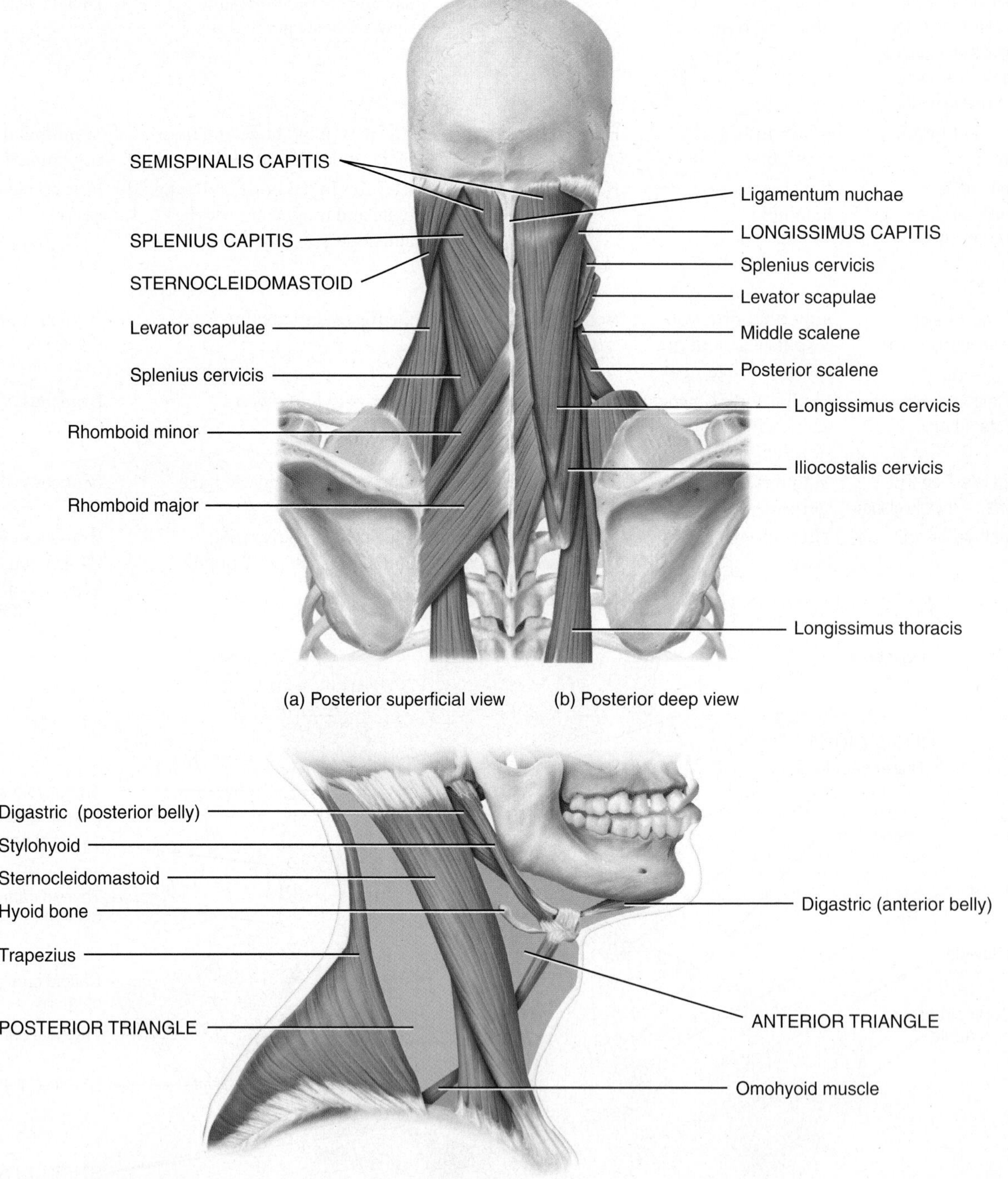


**Contraction of both sternocleidomastoids flexes the cervical curve of the vertebral column, which brings the anterior head toward the chest (see Figure 9.6a).**

the head at the neck. Bilateral contraction of each pair of **semispinalis capitis**, **splenius capitis**, and **longissimus capitis** muscles extends the head at the neck. However, when the above muscles contract unilaterally, their actions are quite different, involving primarily rotation of the head. For example, acting singly (unilaterally), each sternocleidomastoid laterally flexes the neck and rotates the head toward the opposite side.

The sternocleidomastoid is an important landmark that divides the neck into an **anterior triangle** and **posterior triangle** (Figure 11.8b). The triangles are important anatomically and surgically because of the lymph nodes, salivary glands, blood vessels, and nerves that lie within their boundaries.

**TABLE 11.8**

Muscles That Move the Head

| MUSCLE | ORIGIN | INSERTION | ACTION | INNERVATION |
|---|---|---|---|---|
| **Sternocleidomastoid** (ster′-nō-kli′-dō-MAS-toid; *sterno-* = breastbone; *cleido-* = clavicle; *mastoid* = mastoid process of temporal bone) | Sternum and clavicle | Mastoid process of temporal bone | Acting together (bilaterally), flex cervical portion of vertebral column, flex head, and elevate sternum during forced inhalation; acting singly (unilaterally), laterally extend and rotate head to side opposite contracting muscle | Accessory (XI) nerve |
| **Semispinalis capitis** (se′-mē-spi-NĀ-lis KAP-i-tis; *semi-* = half; *spine* = spinous process; *capit-* = head) | Transverse processes of C7–T7 and articular processes of C4–C6 | Occipital bone between superior and inferior nuchal lines | Acting together, extend head; acting singly, rotate head to side opposite contracting muscle | Cervical spinal nerves |
| **Splenius capitis** (SPLĒ-nē-us; *splenion-* = bandage) | Ligamentum nuchae and spinous processes of C7–T4 | Occipital bone and mastoid process of temporal bone | Acting together, extend head; acting singly, laterally flex and rotate head to same side as contracting muscle | Cervical spinal nerves |
| **Longissimus capitis** (lon-JIS-i-mus = longest) | Transverse processes of T1–T4 and articular processes of C4–C7 | Mastoid process of temporal bone | Acting together, extend head; acting singly, laterally flex and rotate head to same side as contracting muscle | Cervical spinal nerves |

✓ **CHECKPOINT**

**10.** What is the combined action of the suprahyoid and infrahyoid muscles?

**11.** What should happen to the hyoid bone when the suprahyoid muscles contract? When the infrahyoid muscles contract?

## RETURN TO Stewart's Story

Stewart exhales deeply as he finishes a repetition of bench presses at the gym. He stands and spots Wally as he begins his rep. Stewart catches his reflection in a wall mirror and thinks about that day's lecture on facial muscles. As he grins and bares his teeth he hears whistling and realizes that Wally is trying to get his attention.

"What on earth are you doing?" Wally asks. Stewart is embarrassed as he describes how he was reviewing some of the muscles from that day's lecture. "I just think it's amazing that there are minor/major pairs of some muscles, with the minor muscles always above the major muscle. I guess I was just practicing with the zygomaticus minor and zygomaticus major muscles." Wally shakes his head as he puts the weight back on the rack and stands up.

"All I know is that I was hungry after Dr. Kendrick talked about all of the muscles that move the mandible and tongue. I never thought about how busy that area of the body is when we talk and swallow." Wally palpates around his jaw and neck as he says, "You know, it was good to hear a lot of bony terms again. I knew it was important to learn all of those bony landmarks last month, but I'm glad that we'll be using the terms to talk about muscle attachments." Stewart nods as he looks at his neck in the mirror.

"I know what you mean. Some of these muscles will be challenging to learn, but it helps if you realize this big rope in my neck connects the sternum, clavicle, and mastoid process. That's going to make it a lot easier."

"Wally, when you injured your back in high school, did you ever learn what muscles were affected?" Wally shakes his head and instinctively rubs the middle of his back.

"No, I never thought to ask. I suppose we can dig out the book during study group tonight and figure it out. I definitely know the muscle actions that I couldn't do for a long time. That should give us an idea of which muscles were involved."

A. *In the opening story Stewart refers to the* biceps brachii *being both an agonist and an antagonist. What do the two terms mean? What muscle is the antagonist to the* biceps brachii?
B. *How does fascicle arrangement affect muscle function? Which fascicle arrangement would be best for the strong, short pulls required when Wally and Stewart do bench presses?*
C. *What does the name* zygomaticus minor *tell you about the muscle Stewart uses as he grimaces at his reflection?*
D. *What muscle is Stewart referring to as the "rope in the neck"? What actions does it perform?*

# 11.5 Muscles of the abdomen protect the abdominal viscera, move the vertebral column, and assist breathing.

## Muscles That Protect Abdominal Viscera and Move the Vertebral Column

The anterolateral abdominal wall includes four pairs of muscles: external oblique, internal oblique, transverse abdominis, and rectus abdominis (Figure 11.9; Table 11.9). From superficial to deep, the external oblique, internal oblique, and transverse abdominis form three layers of muscle around the abdomen. In each layer, the muscle fascicles extend in a different direction. This is a structural arrangement that affords considerable protection to the abdominal viscera, especially when the muscles have good tone. The **external oblique** is the superficial muscle. Its fascicles extend inferiorly and medially. The **internal oblique** is the intermediate muscle, with fascicles that extend at right angles to those of the external oblique. The **transverse abdominis** is the deep muscle, with most of its fascicles directed transversely around the abdominal wall.

The **rectus abdominis** is a long muscle that extends the entire length of the anterior abdominal wall, from the pubic crest and pubic symphysis to the cartilages of ribs 5–7 and the xiphoid process of the sternum. The anterior surface of the muscle is interrupted by three transverse fibrous bands of tissue called *tendinous intersections.*

Recall that an *aponeurosis* is a broad, flat tendon. The aponeuroses of the external oblique, internal oblique, and transverse abdominis form the *rectus sheaths*, which enclose the rectus abdominis. The sheaths meet at the midline to form the **linea**

FIGURE 11.9 **Muscles of the anterolateral abdominal wall.**

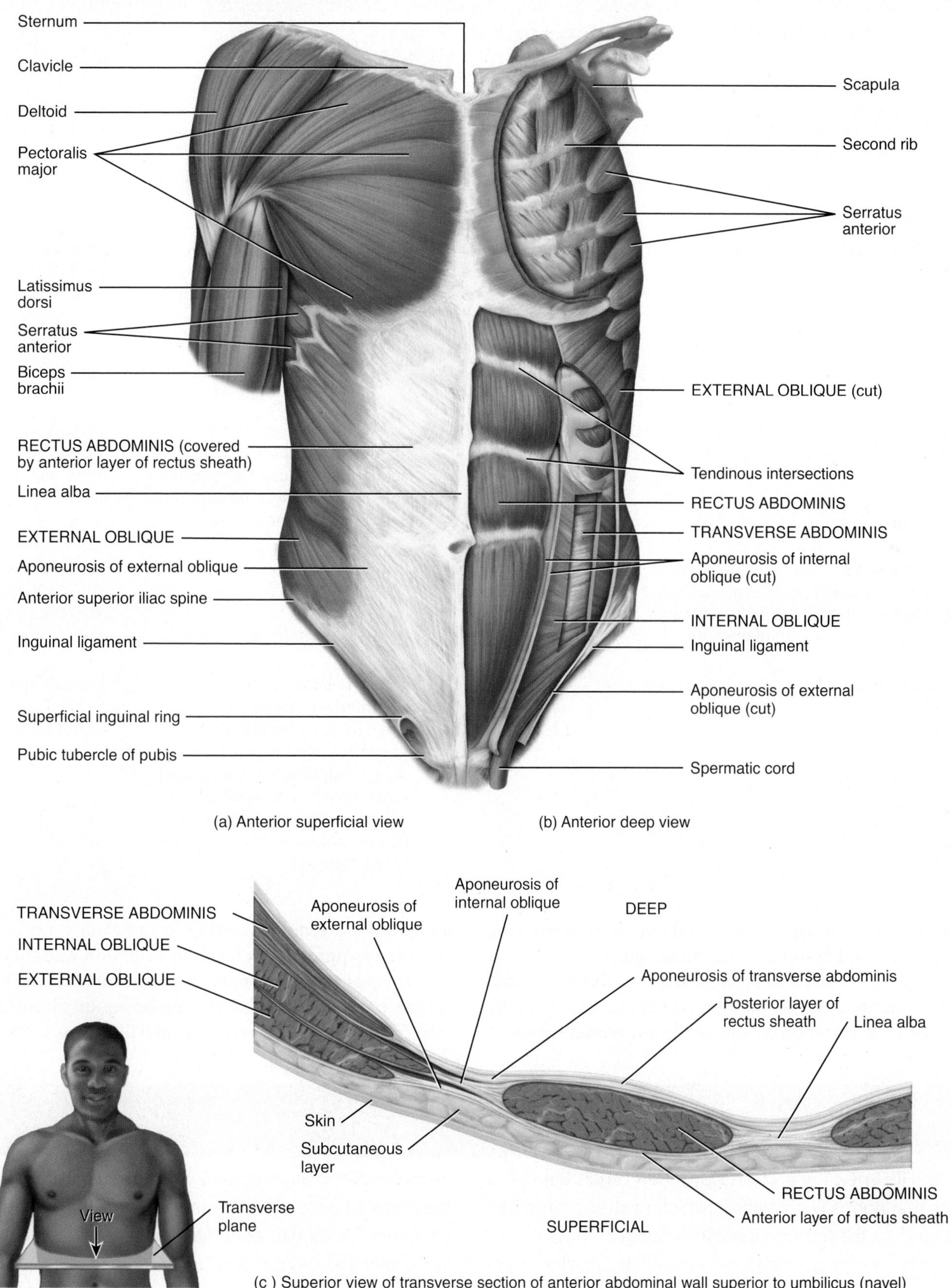


(c ) Superior view of transverse section of anterior abdominal wall superior to umbilicus (navel)

**The anterolateral abdominal muscles protect the abdominal viscera, move the vertebral column, and assist in forced expiration, defecation, urination, and childbirth.**

**TABLE 11.9**

Muscles That Protect Abdominal Viscera and Move the Vertebral Column

| MUSCLE | ORIGIN | INSERTION | ACTION | INNERVATION |
|---|---|---|---|---|
| **Rectus abdominis** (REK-tus ab-DOM-in-is; *rectus* = fascicles parallel to midline; *abdomin* = abdomen) | Pubic crest and pubic symphysis | Cartilage of fifth to seventh ribs and xiphoid process | Flexes vertebral column, especially lumbar portion, and compresses abdomen to aid in defecation, urination, forced exhalation, and childbirth | Thoracic spinal nerves T7–T12 |
| **External oblique** (ō-BLĒK; *external* = closer to surface; *oblique* = fascicles diagonal to midline) | Ribs 5–12 | Iliac crest and linea alba | Acting together (bilaterally), compress abdomen and flex vertebral column; acting singly (unilaterally), laterally flex vertebral column, especially lumbar portion, and rotate vertebral column | Thoracic spinal nerves T7–T12 and the iliohypogastric nerve |
| **Internal oblique** (*internal* = farther from surface) | Iliac crest, inguinal ligament, and thoracolumbar fascia | Cartilage of ribs 7–10 and linea alba | Acting together, compress abdomen and flex vertebral column; acting singly, laterally flex vertebral column, especially lumbar portion, and rotate vertebral column | Thoracic spinal nerves T8–T12, iliohypogastric nerve, and ilioinguinal nerve |
| **Transverse abdominis** (tranz-VERS = fascicles perpendicular to midline) | Iliac crest, inguinal ligament, lumbar fascia, and cartilages of ribs 5–10 | Xiphoid process, linea alba, and pubis | Compresses abdomen | Thoracic spinal nerves T8–T12, iliohypogastric nerve, and ilioinguinal nerve |
| **Quadratus lumborum** (kwod-RĀ-tus lum-BOR-um; *quad-* = four; *lumbo-* = lumbar region) (see Figure 11.10b) | Iliac crest and iliolumbar ligament | Inferior border of rib 12 and L1–L4 | Acting together, pull twelfth ribs inferiorly during forced exhalation, fix twelfth ribs to prevent their elevation during deep inhalation, and help extend lumbar portion of vertebral column; acting singly, laterally flex vertebral column, especially lumbar portion | Thoracic spinal nerve T12 and lumbar spinal nerves L1–L3 or L1–L4 |

**alba** (= white line), a tough, fibrous band that extends from the xiphoid process of the sternum to the pubic symphysis.

As a group, the muscles of the anterolateral abdominal wall help contain and protect the abdominal viscera; flex, laterally flex, and rotate the vertebral column at the intervertebral joints; compress the abdomen during forced exhalation; and produce the force required for defecation, urination, and childbirth.

The posterior abdominal wall is formed by the lumbar vertebrae, parts of the ilia of the hip bones, quadratus lumborum (see Figure 11.10b), psoas major, and iliacus (see Figure 11.19a).

## CLINICAL CONNECTION | *Inguinal Hernia*

A **hernia** (HER-nē-a) is a protrusion of an organ through a structure that normally contains it, which creates a lump that can be seen or felt through the skin's surface. The inguinal region is a weak area in the abdominal wall. It is often the site of an **inguinal hernia**, a rupture or separation of a portion of the inguinal area of the abdominal wall resulting in the protrusion of a part of the small intestine. Hernias are much more common in males than in females because the inguinal canals in males are larger to accommodate the spermatic cord and ilioinguinal nerve. Treatment of hernias most often involves surgery. The organ that protrudes is "tucked" back into the abdominal cavity and the defect in the abdominal muscles is repaired. In addition, a mesh is often applied to reinforce the area of weakness.

## Muscles Used in Breathing

The muscles described here alter the size of the thoracic cavity so that breathing can occur (Figure 11.10; Table 11.10). Inhalation (breathing in) occurs when the thoracic cavity increases in size, and exhalation (breathing out) occurs when the thoracic cavity decreases in size.

The dome-shaped **diaphragm** separates the thoracic and abdominal cavities. The diaphragm has a convex superior surface that forms the floor of the thoracic cavity (Figure 11.10c) and a concave, inferior surface that forms the roof of the abdominal cavity (Figure 11.10d). The muscular portion of the diaphragm is around the periphery of the muscle. The fibers of the muscular portion converge and insert into the *central tendon*, a strong aponeurosis located near the center of the muscle (Figure 11.10c, d). The central tendon fuses with the inferior surface of the pericardium (covering of the heart) and the pleurae (coverings of the lungs). The diaphragm has three major openings through which the aorta, esophagus, and inferior vena cava pass between the thorax and abdomen (Figure 11.10c, d).

The diaphragm is the most important muscle that powers breathing. During contraction, it depresses into a flatter shape, increasing thoracic cavity volume, which results in inhalation. As the diaphragm relaxes, it elevates back to the dome shape, decreasing thoracic cavity volume to produce exhalation. Movements of the diaphragm also help return blood from the abdomen to the heart. Together, the diaphragm and anterolateral abdominal muscles can be voluntarily contracted to help increase pressure in the abdomen to evacuate the pelvic

FIGURE 11.10 **Muscles used in breathing.**

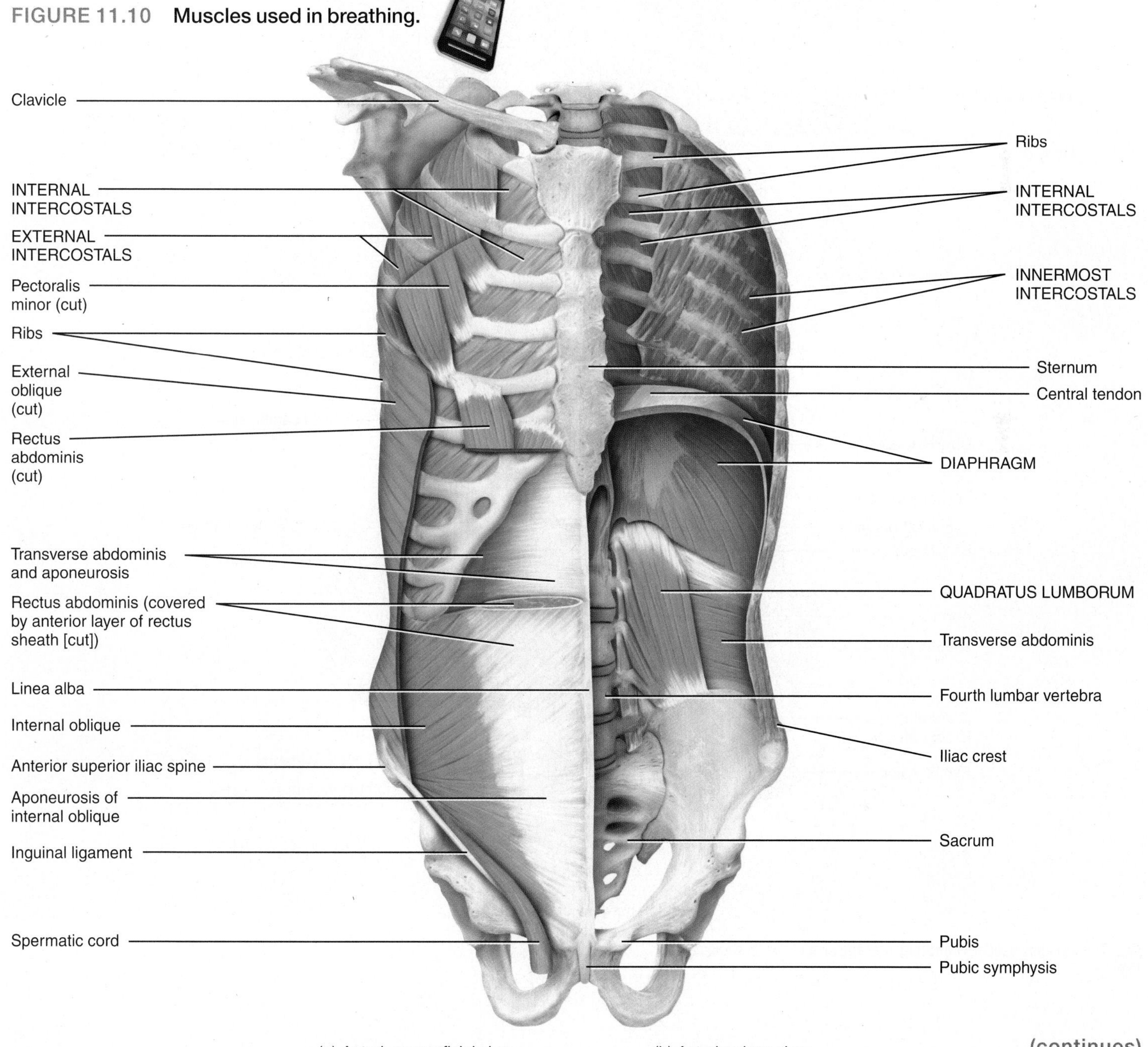


(a) Anterior superficial view (b) Anterior deep view

(continues)

**FIGURE 11.10 (continued)**

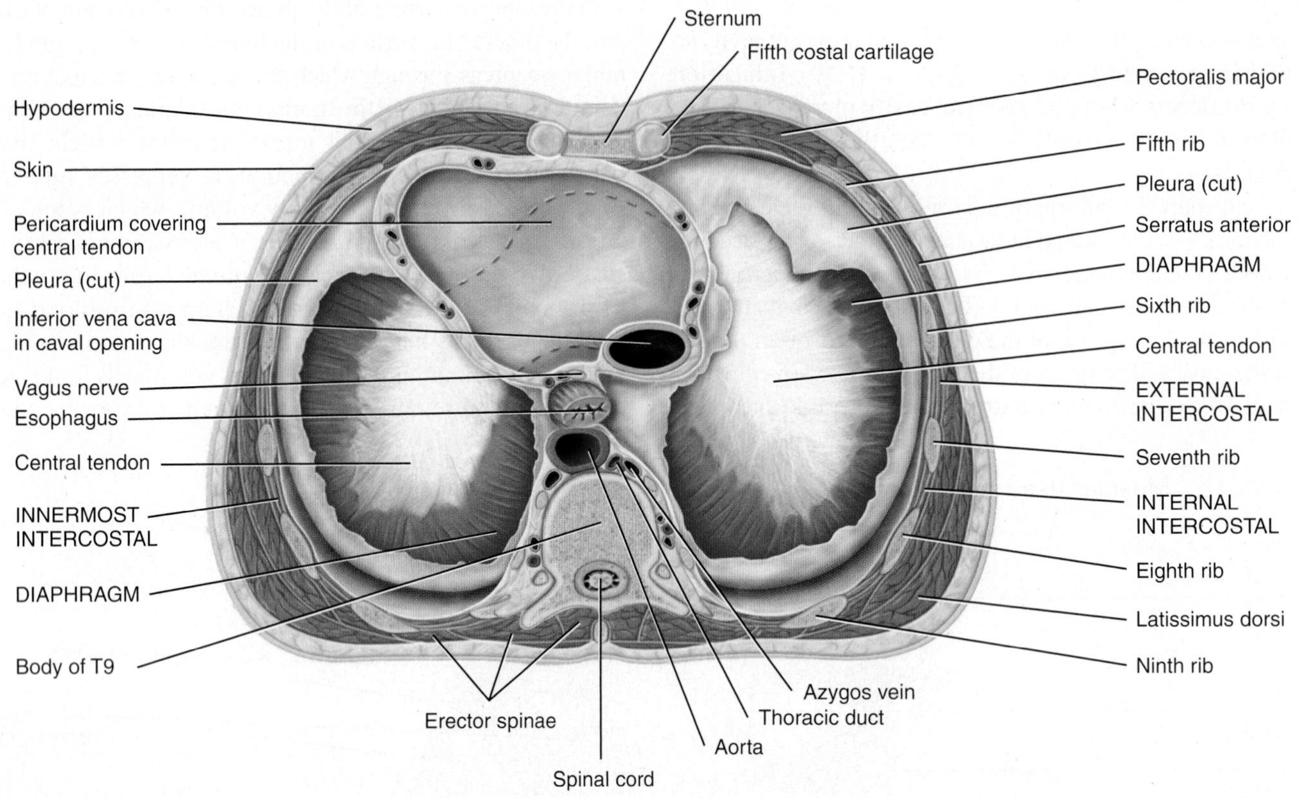


(c) Superior view of diaphragm

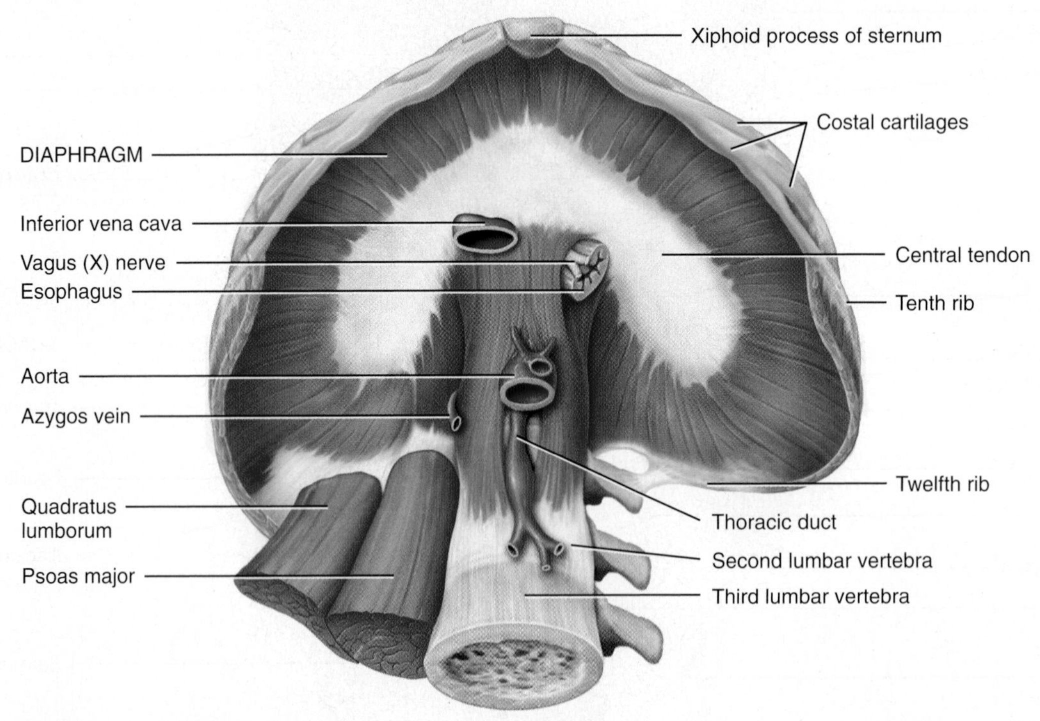


(d) Inferior view of diaphragm

**The muscles used in breathing alter the size of the thoracic cavity.**

**TABLE 11.10**

Muscles Used in Breathing

| MUSCLE | ORIGIN | INSERTION | ACTION | INNERVATION |
|---|---|---|---|---|
| **Diaphragm** (DĪ-a-fram; *dia-* = across; *-phragm* = wall) | Xiphoid process of the sternum, costal cartilages and adjacent portions of ribs 7–12, lumbar vertebrae and their intervertebral discs, and rib 12 | Central tendon | Contraction of the diaphragm causes it to flatten and increases the vertical dimension of the thoracic cavity, resulting in inhalation; relaxation of the diaphragm causes it to move superiorly and decreases the vertical dimension of the thoracic cavity, resulting in exhalation | Phrenic nerve, which contains axons from cervical spinal nerves (C3–C5) |
| **External intercostals** (in′-ter-KOS-tals; *external* = closer to surface; *inter-* = between; *-costa* = rib) | Inferior border of rib above | Superior border of rib below | Contraction elevates the ribs and increases the anteroposterior and lateral dimensions of the thoracic cavity, resulting in inhalation; relaxation depresses the ribs and decreases the anteroposterior and lateral dimensions of the thoracic cavity, resulting in exhalation | Thoracic spinal nerves T2–T12 |
| **Internal intercostals** (*internal* = further from surface) | Superior border of rib below | Inferior border of rib above | Contraction draws adjacent ribs together to further decrease the anteroposterior and lateral dimensions of the thoracic cavity during forced exhalation | Thoracic spinal nerves T2–T12 |

contents during defecation, urination, and childbirth. The increase in intra-abdominal pressure will also help support the vertebral column and prevent flexion during weight lifting. This greatly assists the back muscles in lifting a heavy weight.

Other muscles involved in breathing, called *intercostal muscles*, span the spaces between ribs. The **external intercostals** occupy the superficial layer, and their fibers run obliquely and anteriorly from the rib above to the rib below. They elevate the ribs during inhalation to help expand the thoracic cavity. The **internal intercostals** lie deep to the external intercostals. The fibers of these muscles are at right angles to fibers of the external intercostals and run obliquely and posteriorly from the rib below to the rib above. They draw adjacent ribs together during forced exhalation to help decrease the size of the thoracic cavity. The deepest muscle layer consists of the paired **innermost intercostal muscles**. These muscles extend in the same direction as the internal intercostals and may have the same role.

As you will see in Concept 22.3, the diaphragm and external intercostals are used during quiet inhalation and exhalation. However, during deep, forceful inhalation, as occurs during exercise or playing a wind instrument, the sternocleidomastoid, scalene, and pectoralis minor muscles are also used, and during deep, forceful exhalation, the external oblique, internal oblique, transverse abdominis, rectus abdominis, and internal intercostals are also used.

✓ **CHECKPOINT**

**12.** State the direction of the fascicles for the external oblique, internal oblique, transverse abdominis, and rectus abdominis.

**13.** Which abdominal muscle aids in urination?

**14.** Name the structures that pass through the diaphragm.

# 11.6 Muscles of the pelvic floor and perineum support the pelvic viscera, function as sphincters, and assist in urination, erection, ejaculation, and defecation.

## Muscles of the Pelvic Floor

The pelvic floor includes the coccygeus and the levator ani muscle group (Figure 11.11; Table 11.11). These muscles and their fascia are referred to as the **pelvic diaphragm**, which stretches from the pubis anteriorly, to the coccyx posteriorly, and from one lateral wall of the pelvis to the other. The anal canal and urethra pass through the pelvic diaphragm in both sexes, and the vagina goes through it in females.

**FIGURE 11.11 Muscles of the pelvic floor.**

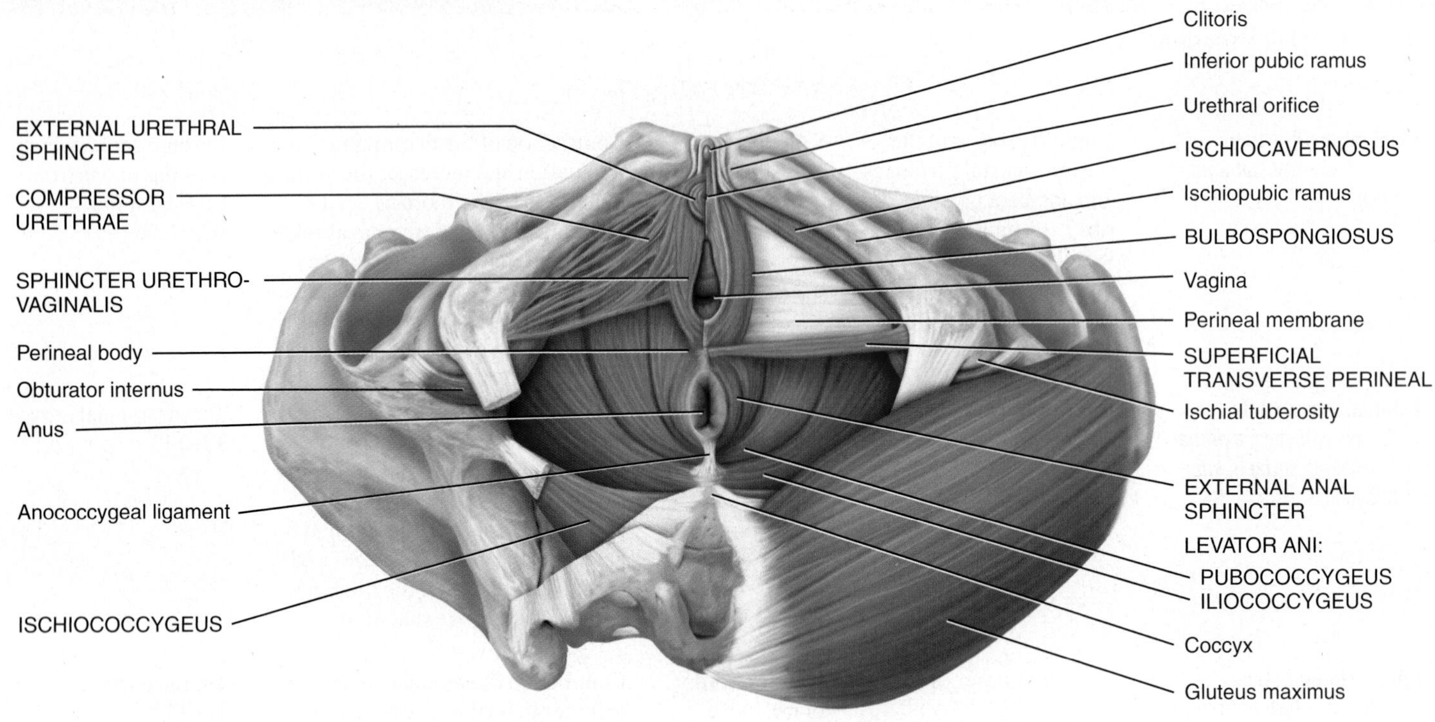


Inferior superficial view of a female perineum

**The pelvic diaphragm supports the pelvic viscera.**

**TABLE 11.11**

Muscles of the Pelvic Floor

| MUSCLE | ORIGIN | INSERTION | ACTION | INNERVATION |
|---|---|---|---|---|
| **Levator ani** (le-VĀ-tor Ā-nē; *levator* = raises; *ani* = anus) This muscle is divisible into two parts, the pubococcygeus muscle and the iliococcygeus muscle | | | | |
| **Pubococcygeus** (pū′-bō-kok-SIJ-ē-us; *pubo-* = pubis; *-coccygeus* = coccyx) | Pubis | Coccyx, urethra, anal canal, perineal body of the perineum (a wedge-shaped mass of fibrous tissue in the center of the perineum), and anococcygeal ligament (narrow fibrous band that extends from anus to coccyx) | Supports and maintains position of pelvic viscera; resists increase in intra-abdominal pressure during forced exhalation, coughing, vomiting, urination, and defecation; constricts anus, urethra, and vagina | Sacral spinal nerves S2–S4 |
| **Iliococcygeus** (il′-ē-ō-kok-SIJ-ē-us; *ilio-* = ilium) | Ischial spine | Coccyx | As above | Sacral spinal nerves S2–S4 |
| **Ischiococcygeus** (is′-kē-ō-kok-SIJ-ē-us) | Ischial spine | Lower sacrum and upper coccyx | Supports and maintains position of pelvic viscera; resists increase in intra-abdominal pressure during forced exhalation, coughing, vomiting, urination, and defecation; and pulls coccyx anteriorly following defecation or childbirth | Sacral spinal nerves S4–S5 |

The **levator ani** muscle group supports the pelvic viscera and resists the inferior thrust that accompanies increases in intra-abdominal pressure during functions such as forced exhalation, coughing, vomiting, urination, and defecation. The muscle also functions as a sphincter for the anal canal, urethra, and vagina. In addition to assisting the levator ani, the ischiococcygeus pulls the coccyx anteriorly after it has been pushed posteriorly during defecation or childbirth. Figure 11.11 shows these muscles in the female and Figure 11.12 illustrates them in the male.

## Muscles of the Perineum

The **perineum** is the region of the trunk inferior to the pelvic diaphragm. It is a diamond-shaped area that extends from the pubic symphysis anteriorly, to the coccyx posteriorly, and to the ischial tuberosities laterally. The female and the male perineums may be compared in Figures 11.11 and 11.12, respectively. A transverse line drawn between the ischial tuberosities divides the perineum into an anterior *urogenital triangle* that contains the external genitals and a posterior *anal triangle* that contains the anus (see Figure 25.21). Clinically, the perineum is very important to physicians who care for women during pregnancy and who treat disorders related to the female genital tract, urogenital organs, and the anorectal region.

The superficial muscles of the perineum help maintain erection of the penis in males and clitoris in females, and facilitate ejaculation in males (Table 11.12). The deep muscles of the perineum assist in urination and closing the vagina in females and urination and ejaculation in males. The **external anal sphincter** keeps the anal canal and anus closed except during defecation.

FIGURE 11.12 **Muscles of the perineum.**

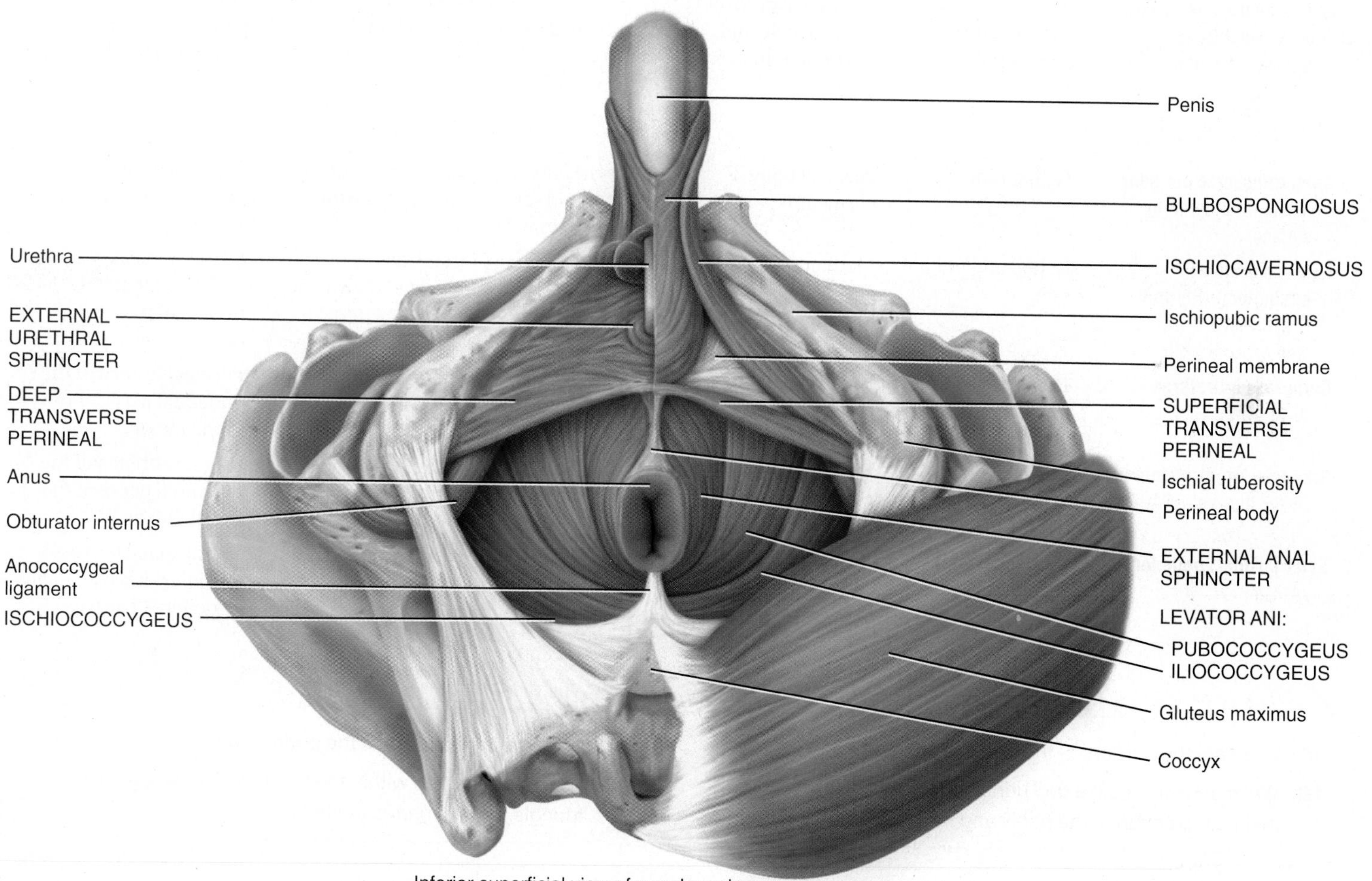


Inferior superficial view of a male perineum

**The muscles of the perineum assist in urination and help strengthen the pelvic floor in both females and males, and assist in ejaculation in males.**

**TABLE 11.12**

Muscles of the Perineum

| MUSCLE | ORIGIN | INSERTION | ACTION | INNERVATION |
|---|---|---|---|---|
| **Superficial Perineal Muscles** | | | | |
| **Superficial transverse perineal** (per-i-NĒ-al; *superficial* = closer to surface; *transverse* = across; *perineus* = perineum) | Ischial tuberosity | Perineal body of perineum | Stabilizes perineal body of perineum | Perineal branch of pudendal nerve of sacral plexus |
| **Bulbospongiosus** (bul′-bō-spon′-jē-Ō-sus; *bulb* = a bulb; *spongio-* = sponge) | Perineal body of perineum | Perineal membrane of deep muscles of perineum, corpus spongiosum of penis, and deep fascia on dorsum of penis in male; pubic arch and root and dorsum of clitoris in female | Helps expel urine during urination, helps propel semen along urethra, assists in erection of the penis in male; constricts vaginal orifice and assists in erection of clitoris in female | Perineal branch of pudendal nerve of sacral plexus |
| **Ischiocavernosus** (is′-kē-ō-ka′-ver-NŌ-sus; *ischio-* = the hip) | Ischial tuberosity and ischial and pubic rami | Corpus cavernosum of penis in male and clitoris in female | Maintains erection of penis in male and clitoris in female | Perineal branch of pudendal nerve of sacral plexus |
| **Deep Perineal Muscles** | | | | |
| **Deep transverse perineal** (*deep* = farther from surface) | Ischial rami | Perineal body of perineum | Helps expel last drops of urine and semen in male and urine in female | Perineal branch of pudendal nerve of sacral plexus |
| **External urethral sphincter** (ū-RĒ-thral SFINGK-ter) (see Figure 24.21) | Ischial and pubic rami | Median raphe in male and vaginal wall in female | Helps expel last drops of urine and semen in male and urine in female | Sacral spinal nerve S4 and inferior rectal branch of pudendal nerve |
| **Compressor urethrae** (ū-RĒ-thrē) | Ischiopubic ramus | Blends with partner on other side anterior to urethra | Serves as accessory sphincter of the urethra | Perineal branch of the pudendal nerve of the sacral plexus |
| **Sphincter urethrovaginalis** (ū-RĒ-thrō-vaj-i-NAL-is) | Perineal body | Blends with partner on other side anterior to urethra | Serves as a sphincter of urethra and facilitates closing of vagina | Perineal branch of the pudendal nerve of the sacral plexus |
| **External anal sphincter** (Ā-nal) | Anococcygeal ligament | Perineal body of perineum | Keeps anal canal and anus closed | Sacral spinal nerve S4 and inferior rectal branch of pudendal nerve |

✓ **CHECKPOINT**

**15.** What are the borders of the pelvic diaphragm?

**16.** Which muscles help expel urine and semen? Which maintain erection of the penis and clitoris?

**17.** What are the borders of the perineum?

**18.** What is contained within the borders of the urogenital triangle and the anal triangle?

# 11.7 Muscles inserting on the upper limb move and stabilize the pectoral girdle, and move the arm, forearm, and hand.

## Muscles That Move the Pectoral Girdle

The main action of the muscles that move the pectoral (shoulder) girdle (Figure 11.13; Table 11.13) is to stabilize the scapula so it can function as a steady origin for most of the muscles that move the humerus. Because scapular movements usually accompany humeral movements, the muscles also move the scapula to increase the range of motion of the humerus. For example, it would not be possible to raise your hand in class, which requires abducting your humerus past the horizontal, if the scapula did not follow the humerus by rotating upward.

Muscles that move the pectoral girdle can be classified into two groups based on their location in the thorax: anterior and posterior thoracic muscles. The **anterior thoracic muscles** are the subclavius, pectoralis minor, and serratus anterior. The **subclavius** extends from the clavicle to the first rib and steadies the clavicle during movements of the pectoral girdle. The **pectoralis minor** lies deep to the pectoralis major. In addition to its role in movements of the scapula, the pectoralis minor assists in forced inhalation. The **serratus anterior**, a large, fan-shaped muscle between the ribs and scapula, is named for the saw-toothed appearance of its origins on the ribs.

The **posterior thoracic muscles** are the trapezius, levator scapulae, rhomboid major, and rhomboid minor. The most superficial back muscle, the **trapezius**, extends from the skull and vertebral column medially to the pectoral girdle laterally. The two trapezius muscles form a trapezoid (diamond-shaped quadrangle)—hence its name. The **levator scapulae** lies in the posterior neck deep to the sternocleidomastoid and trapezoid. As its name suggests, one of its actions is to elevate the scapula. The **rhomboid major** and **rhomboid minor** lie deep to the trapezius and pass from the vertebrae to the scapula. They are named based on their shape—that is, a rhomboid (an oblique parallelogram). Both muscles are used when forcibly lowering the raised upper limbs, as in driving a stake with a sledgehammer.

FIGURE 11.13 **Muscles that move the pectoral girdle.**

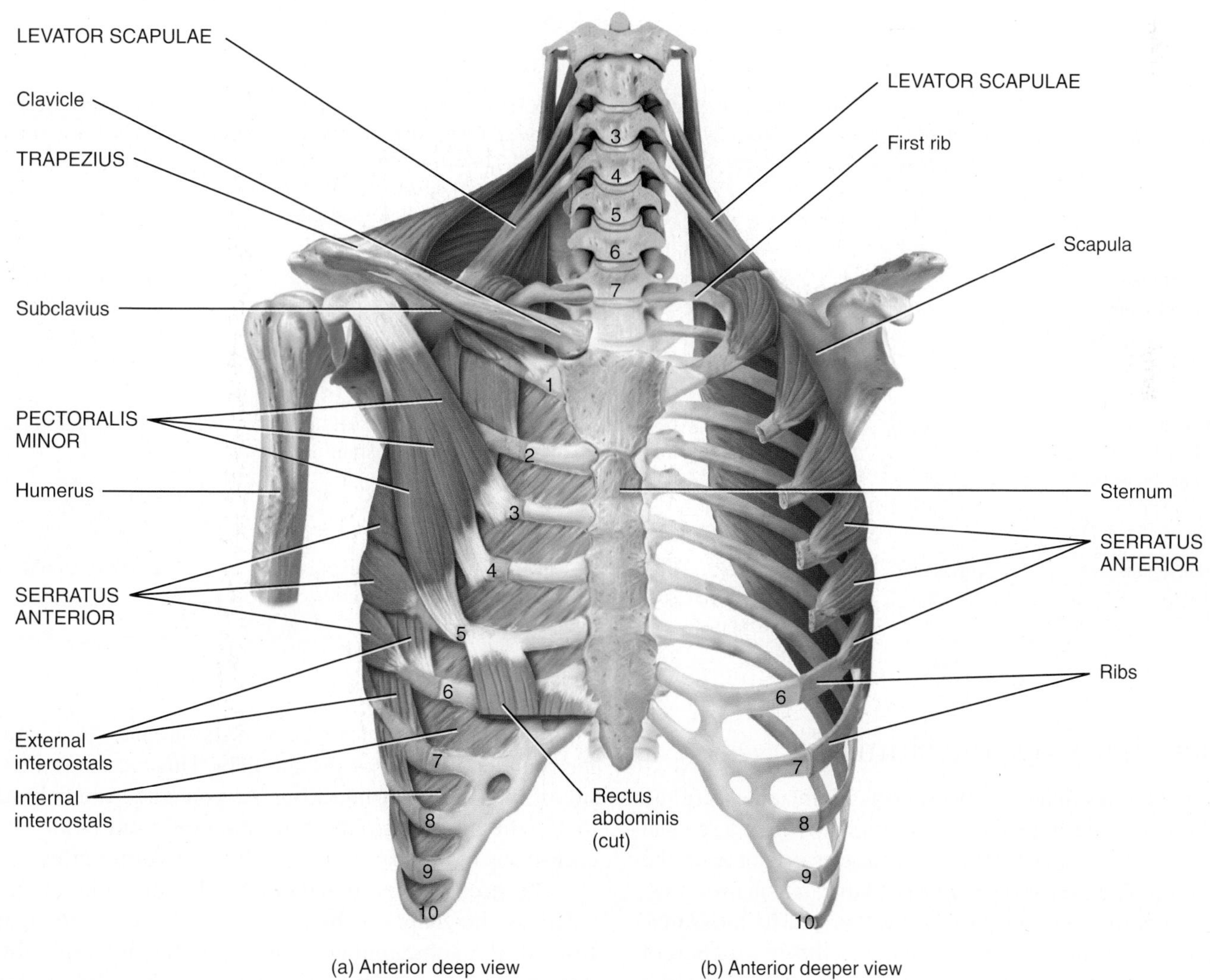


(a) Anterior deep view

(b) Anterior deeper view

**Muscles that move the pectoral girdle originate on the axial skeleton and insert on the clavicle or scapula.**

**TABLE 11.13**

Muscles That Move the Pectoral Girdle

| MUSCLE | ORIGIN | INSERTION | ACTION | INNERVATION |
|---|---|---|---|---|
| **Anterior Thoracic Muscles** | | | | |
| **Subclavius** (sub-KLĀ-vē-us; *sub-* = under; *-clavius* = clavicle | First rib | Clavicle | Depresses and moves clavicle anteriorly and helps stabilize pectoral girdle | Subclavian nerve |
| **Pectoralis minor** (pek′-tō-RĀ-lis; *pector-* = the breast, chest, thorax; *minor* = lesser) | Ribs 2–5, ribs 3–5, or ribs 2–4 | Coracoid process of scapula | Abducts scapula and rotates it downward; elevates third through fifth ribs during forced inhalation when scapula is fixed | Medial pectoral nerve |
| **Serratus anterior** (ser-Ā-tus = saw-toothed; *anterior* = front) | Ribs 1–8 or ribs 1–9 | Vertebral border and inferior angle of scapula | Abducts scapula and rotates it upward; elevates ribs when scapula is stabilized; known as "boxer's muscle" because it is important in horizontal arm movements such as punching and pushing | Long thoracic nerve |
| **Posterior Thoracic Muscles** | | | | |
| **Trapezius** (tra-PĒ-zē-us; = trapezoid-shaped) | Superior nuchal line of occipital bone, ligamentum nuchae, and spines of C7 and T1–T12 | Clavicle and acromion and spine of scapula | Superior fibers elevate scapula and can help extend head; middle fibers adduct scapula; inferior fibers depress scapula; superior and inferior fibers together rotate scapula upward; stabilizes scapula | Accessory (XI) nerve and cervical spinal nerves C3–C5 |
| **Levator scapulae** (le-VĀ-tor SKA-pū-lē; *levator* = raises; *scapulae* = of the scapula) | C1–C4 | Superior vertebral border of scapula | Elevates scapula and rotates it downward | Dorsal scapular nerve and cervical spinal nerves C3–C5 |
| **Rhomboid major** (rom-BOYD = rhomboid or diamond-shaped) (see Figure 11.14c) | Spines of T2–T5 | Vertebral border of scapula inferior to spine | Elevates and adducts scapula and rotates it downward; stabilizes scapula | Dorsal scapular nerve |
| **Rhomboid minor** (see Figure 11.14c) | Spines of C7 and T1 | Vertebral border of scapula superior to spine | Elevates and adducts scapula and rotates it downward; stabilizes scapula | Dorsal scapular nerve |

## Muscles That Move the Humerus

Of the nine muscles that cross the shoulder joint, all except the pectoralis major and latissimus dorsi originate on the scapula (Figure 11.14; Table 11.14). These two muscles originate on the axial skeleton. The **pectoralis major** is a large, fan-shaped muscle that covers the superior part of the thorax. The **latissimus dorsi** is a broad, triangular muscle located on the inferior part of the back. It is commonly called the "swimmer's muscle" because its many actions are used while swimming; consequently, many competitive swimmers have well-developed "lats." The **deltoid** is a thick, powerful muscle that covers the shoulder joint and forms the rounded contour of the shoulder. This muscle is a frequent site of intramuscular injections. As you study the deltoid, note that its fascicles originate from three different points and that each group of fascicles moves the humerus differently.

The strength and stability of the shoulder joint are not provided by the shape of the articulating bones or its ligaments. Instead, the **subscapularis**, **supraspinatus**, **infraspinatus**, and **teres minor** strengthen and stabilize the shoulder joint as they join the scapula to the humerus. Their tendons fuse together to form the *rotator cuff*, a nearly complete circle of tendons around

## CLINICAL CONNECTION | *Rotator Cuff Injury and Impingement Syndrome*

**Rotator cuff injury** is a strain or tear in the rotator cuff muscle and is common among baseball pitchers, volleyball players, racket sports players, and swimmers due to shoulder movements that involve vigorous circumduction. It also occurs as a result of wear and tear, aging, trauma, poor posture, improper lifting, and repetitive motions in certain jobs, such as placing items on a shelf above your head. Most often, there is tearing of the supraspinatus tendon or the rotator cuff. This tendon is especially predisposed to wear and tear because of its location between the head of the humerus and acromion of the scapula, which compresses the tendon during shoulder movements. Poor posture and poor body mechanics also increase compression of the supraspinatus muscle tendon.

One of the most common causes of shoulder pain and dysfunction in athletes is known as **impingement syndrome**, when the supraspinatus tendon is pinched between the head of the humerus and the acromion. The repetitive movement of the arm over the head that is common in baseball, overhead racquet sports, lifting weights over the head, spiking a volleyball, and swimming puts these athletes at risk. Impingement syndrome may also be caused by a direct blow or stretch injury. Continual pinching of the supraspinatus tendon as a result of overhead motions causes it to become inflamed and results in pain. If movement is continued despite the pain, the tendon may degenerate near the attachment to the humerus and ultimately may tear away from the bone (rotator cuff injury). Treatment consists of resting the injured tendons, strengthening the shoulder through exercise, massage therapy, and surgery if the injury is particularly severe. During surgery, an inflamed bursa may be removed, bone may be trimmed, and/or the coracoacromial ligament may be detached. Torn rotator cuff tendons may be trimmed and then reattached with sutures, anchors, or surgical tacks. These steps make more space, thus relieving pressure and allowing the arm to move freely.

FIGURE 11.14 **Muscles that move the humerus.**

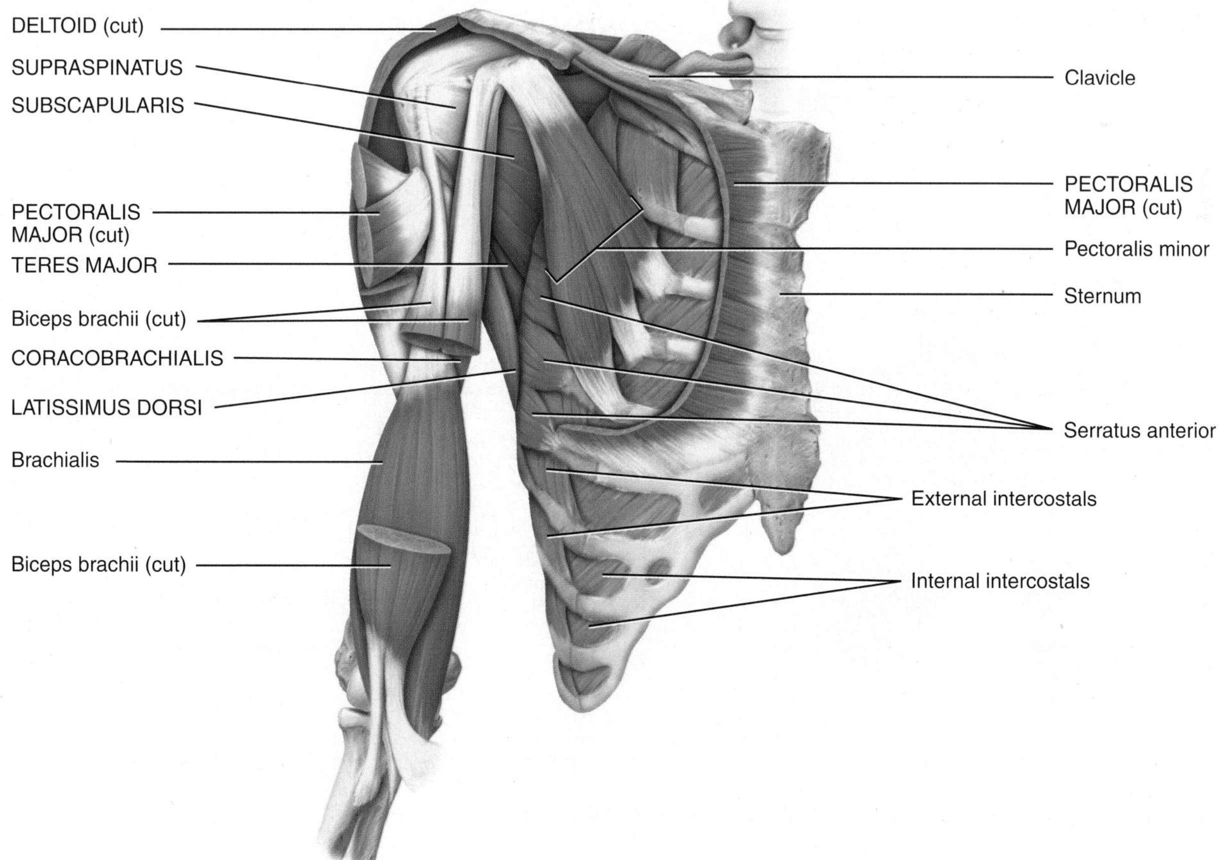


(a) Anterior deep view (the intact pectoralis major muscle is shown in Figure 11.9a)

**(continues)**

**FIGURE 11.14 (continued)**

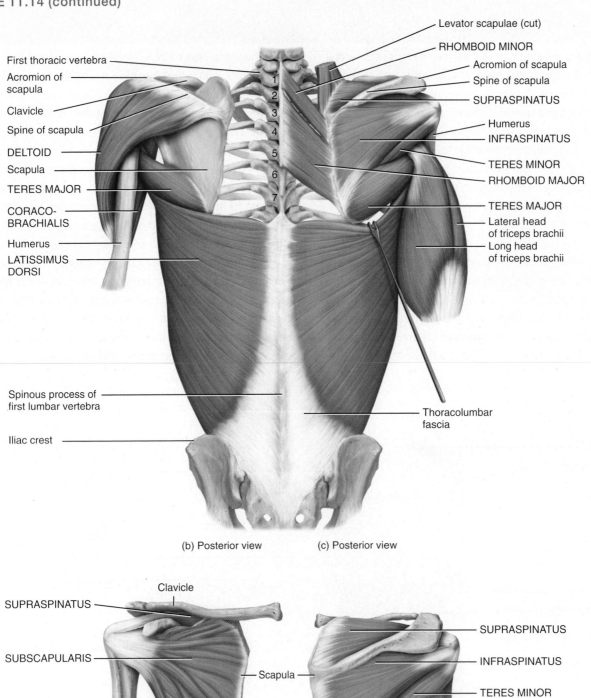

First thoracic vertebra
Acromion of scapula
Clavicle
Spine of scapula
DELTOID
Scapula
TERES MAJOR
CORACO-BRACHIALIS
Humerus
LATISSIMUS DORSI
Spinous process of first lumbar vertebra
Iliac crest
Levator scapulae (cut)
RHOMBOID MINOR
Acromion of scapula
Spine of scapula
SUPRASPINATUS
Humerus
INFRASPINATUS
TERES MINOR
RHOMBOID MAJOR
TERES MAJOR
Lateral head of triceps brachii
Long head of triceps brachii
Thoracolumbar fascia
1
2
3
4
5
6
7

(b) Posterior view (c) Posterior view

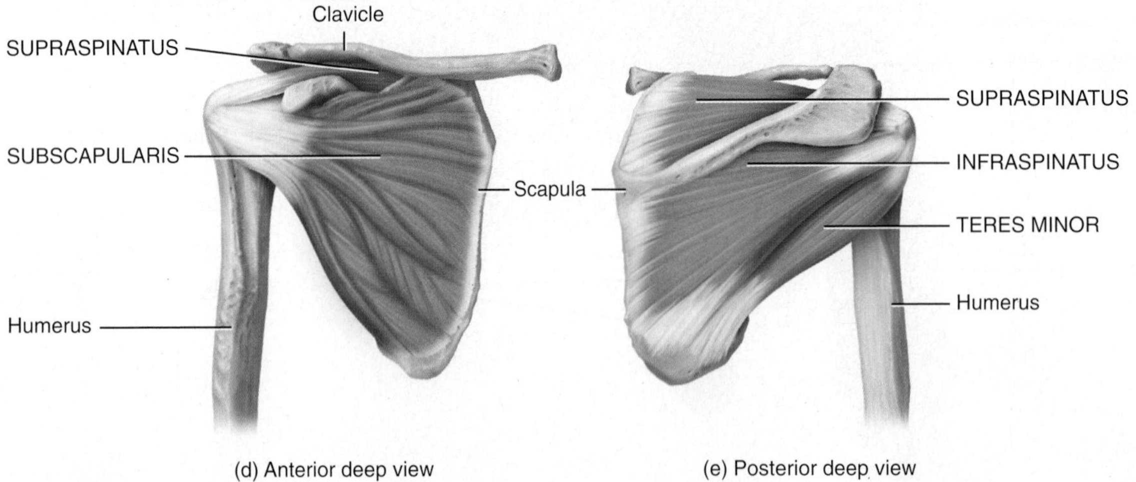


(d) Anterior deep view (e) Posterior deep view

**The strength and stability of the shoulder joint are provided by the tendons that form the rotator cuff.**

**TABLE 11.14**

Muscles That Move the Humerus

| MUSCLE | ORIGIN | INSERTION | ACTION | INNERVATION |
|---|---|---|---|---|
| **Axial Muscles That Move the Humerus** | | | | |
| **Pectoralis major** (pek′-tō-RĀ-lis; *pector-* = chest; *major* = larger) (see also Figure 11.9a) | Clavicle (clavicular head), sternum, and costal cartilages of ribs 2–6 and sometimes ribs 1–7 (sternocostal head) | Greater tubercle and lateral lip of the intertubercular sulcus of humerus | As a whole, adducts and medially rotates arm at shoulder joint; clavicular head flexes arm, and sternocostal head extends the flexed arm to side of trunk | Medical and lateral pectoral nerves |
| **Latissimus dorsi** (la-TIS-i-mus DOR-sī; *latissimus* = widest; *dorsi* = of the back) | Spines of T7–T12, lumbar vertebrae, crests of sacrum and ilium, ribs 9–12 | Intertubercular sulcus of humerus | Extends, adducts, and medially rotates arm at shoulder joint; draws arm inferiorly and posteriorly | Thoracodorsal nerve |
| **Scapular Muscles That Move the Humerus** | | | | |
| **Deltoid** (DEL-toyd = triangularly shaped) | Acromial extremity of clavicle (anterior fibers), acromion of scapula (lateral fibers), and spine of scapula (posterior fibers) | Deltoid tuberosity of humerus | Lateral fibers abduct arm at shoulder joint; anterior fibers flex and medially rotate arm at shoulder joint; posterior fibers extend and laterally rotate arm at shoulder joint | Axillary nerve |
| **Subscapularis** (sub-scap′-ū-LĀ-ris; *sub-* = below; *-scapularis* = scapula) | Subscapular fossa of scapula | Lesser tubercle of humerus | Medially rotates arm at shoulder joint | Upper and lower subscapular nerve |
| **Supraspinatus** (soo-pra-spī-NĀ-tus; *supra-* = above; *-spina-* = spine of the scapula) | Supraspinous fossa of scapula | Greater tubercle of humerus | Assists deltoid muscle in abducting arm at shoulder joint | Suprascapular nerve |
| **Infraspinatus** (in′-fra-spī-NĀ-tus; *infra-* = below) | Infraspinous fossa of scapula | Greater tubercle of humerus | Laterally rotates and adducts arm at shoulder joint | Suprascapular nerve |
| **Teres major** (TE-rēz; *teres* = long and round) | Inferior angle of scapula | Medial lip of intertubercular sulcus of humerus | Extends arm at shoulder joint and assists in adduction and medial rotation of arm at shoulder joint | Lower subscapular nerve |
| **Teres minor** | Inferior lateral border of scapula | Greater tubercle of humerus | Laterally rotates, extends, and adducts arm at shoulder joint | Axillary nerve |
| **Coracobrachialis** (kor′-a-kō-brā-kē-Ā-lis; *coraco-* = coracoid process of the scapula; *-brachi-* = arm) | Coracoid process of scapula | Middle of medial surface of shaft of humerus | Flexes and adducts arm at shoulder joint | Musculocutaneous nerve |

the shoulder joint, like the cuff on a shirtsleeve. The supraspinatus is especially subject to wear and tear because of its location between the head of the humerus and the acromion of the scapula, which compresses its tendon during shoulder movements, especially abduction of the arm.

## Muscles That Move the Radius and Ulna

Most of the muscles that move the radius and ulna (Figure 11.15; Table 11.15) are divided into **forearm flexors** and **forearm extensors**. Recall that the elbow joint is a hinge joint, capable only of flexion and extension. The biceps brachii, brachialis, and brachioradialis are flexors of the elbow joint; the triceps brachii and the anconeus are extensors.

The **biceps brachii** is the large muscle located on the anterior surface of the arm. As indicated by its name (*bi-* = two), it has two heads of origin (long and short), both from the scapula. Because the muscle spans both the shoulder and elbow joints, it flexes the arm at the shoulder joint, flexes the forearm at the elbow joint, and supinates the forearm at the radioulnar joints. The **brachialis** is deep to the biceps brachii. It is the most powerful flexor of the forearm at the elbow joint. For this reason, it is called the "workhorse" of the elbow flexors. The **brachioradialis** also flexes the forearm at the elbow joint, especially when a quick movement is required or when a weight is lifted slowly during flexion of the forearm.

The **triceps brachii** is the large muscle located on the posterior surface of the arm. It is the more powerful of the extensors of the forearm at the elbow joint. As its name implies (*tri-* = three), it has three heads of origin, one from the scapula (long head) and two from the humerus (lateral and medial heads). The **anconeus** is a small muscle located on the lateral part of the posterior elbow that assists the triceps brachii in extending the forearm at the elbow joint.

FIGURE 11.15 **Muscles that move the radius and ulna.**

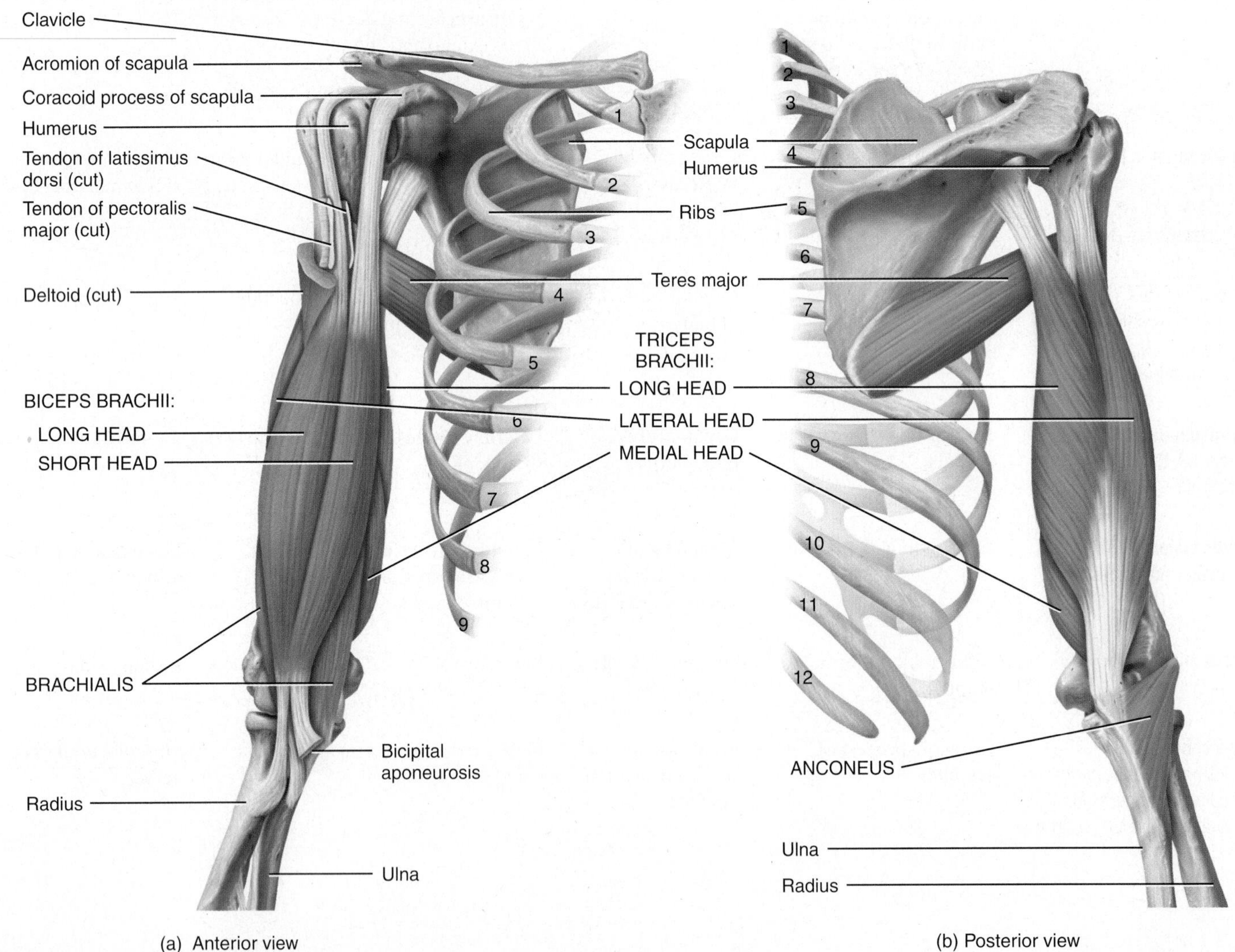


(a) Anterior view

(b) Posterior view

**The anterior arm muscles flex the forearm; the posterior arm muscles extend it.**

**TABLE 11.15**

Muscles That Move the Radius and Ulna

| MUSCLE | ORIGIN | INSERTION | ACTION | INNERVATION |
|---|---|---|---|---|
| **Forearm Flexors** | | | | |
| **Biceps brachii** (BĪ-ceps BRĀ-kē-ī; *biceps* = two heads of origin; *brachii* = arm) | *Long head:* tubercle above glenoid cavity of scapula (supraglenoid tubercle)<br>*Short head:* coracoid process of scapula | Radial tuberosity of radius and bicipital aponeurosis* | Flexes forearm at elbow joint, supinates forearm at radioulnar joints, and flexes arm at shoulder joint | Musculocutaneous nerve |
| **Brachialis** (brā-kē-Ā-lis) | Distal, anterior surface of humerus | Ulnar tuberosity and coronoid process of ulna | Flexes forearm at elbow joint | Musculocutaneous and radial nerves |
| **Brachioradialis** (brā′-kē-ō-rā-dē-Ā-lis; *-radi* = radius) (see Figure 11.16a) | Lateral border of distal end of humerus | Superior to styloid process of radius | Flexes forearm at elbow joint; supinates and pronates forearm at radioulnar joints to neutral position | Radial nerve |
| **Forearm Extensors** | | | | |
| **Triceps brachii** (TRĪ-ceps = three heads of origin) | *Long head:* infraglenoid tubercle, a projection inferior to glenoid cavity of scapula<br>*Lateral head:* lateral and posterior surface of humerus superior to radial groove<br>*Medial head:* entire posterior surface of humerus inferior to groove for radial nerve | Olecranon of ulna | Extends forearm at elbow joint and extends arm at shoulder joint | Radial nerve |
| **Anconeus** (an-KŌ-nē-us; *ancon-* = the elbow) (see also Figure 11.16d, e) | Lateral epicondyle of humerus | Olecranon and superior portion of shaft of ulna | Extends forearm at elbow joint | Radial nerve |
| **Forearm Pronators** (see Figure 11.16a) | | | | |
| **Pronator teres** (PRŌ-nā-tor TE-rēz; *pronator* = turns palm posteriorly) | Medial epicondyle of humerus and coronoid process of ulna | Midlateral surface of radius | Pronates forearm at radioulnar joints and weakly flexes forearm at elbow joint | Median nerve |
| **Pronator quadratus** (kwod-RĀ-tus = square, four-sided) | Distal portion of shaft of ulna | Distal portion of shaft of radius | Pronates forearm at radioulnar joints | Median nerve |
| **Forearm Supinator** (see Figure 11.16b) | | | | |
| **Supinator** (SOO-pi-nā-tor = turns palm anteriorly) | Lateral epicondyle of humerus and ridge near radial notch of ulna (supinator crest) | Lateral surface of proximal one-third of radius | Supinates forearm at radioulnar joints | Deep radial nerve |

*The **bicipital aponeurosis** is a broad aponeurosis from the tendon of insertion of the biceps brachii muscle that descends medially across the brachial artery and fuses with deep fascia over the forearm flexor muscles.

Some muscles that move the radius and ulna are concerned with supination and pronation of the forearm. The pronators, as suggested by their names, are the **pronator teres** and **pronator quadratus**. The supinator of the forearm is aptly named the **supinator**. You use the powerful action of the supinator when you twist a corkscrew or turn a screw with a screwdriver.

## Muscles That Move the Wrist, Hand, and Fingers

Muscles that move the wrist, hand, and fingers are located on the forearm and are many and varied (Figure 11.16; Table 11.16). Those muscles in this group that move the fingers in various ways are known as **extrinsic muscles of the hand** because they

**FIGURE 11.16 Muscles that move the wrist, hand, and fingers.**

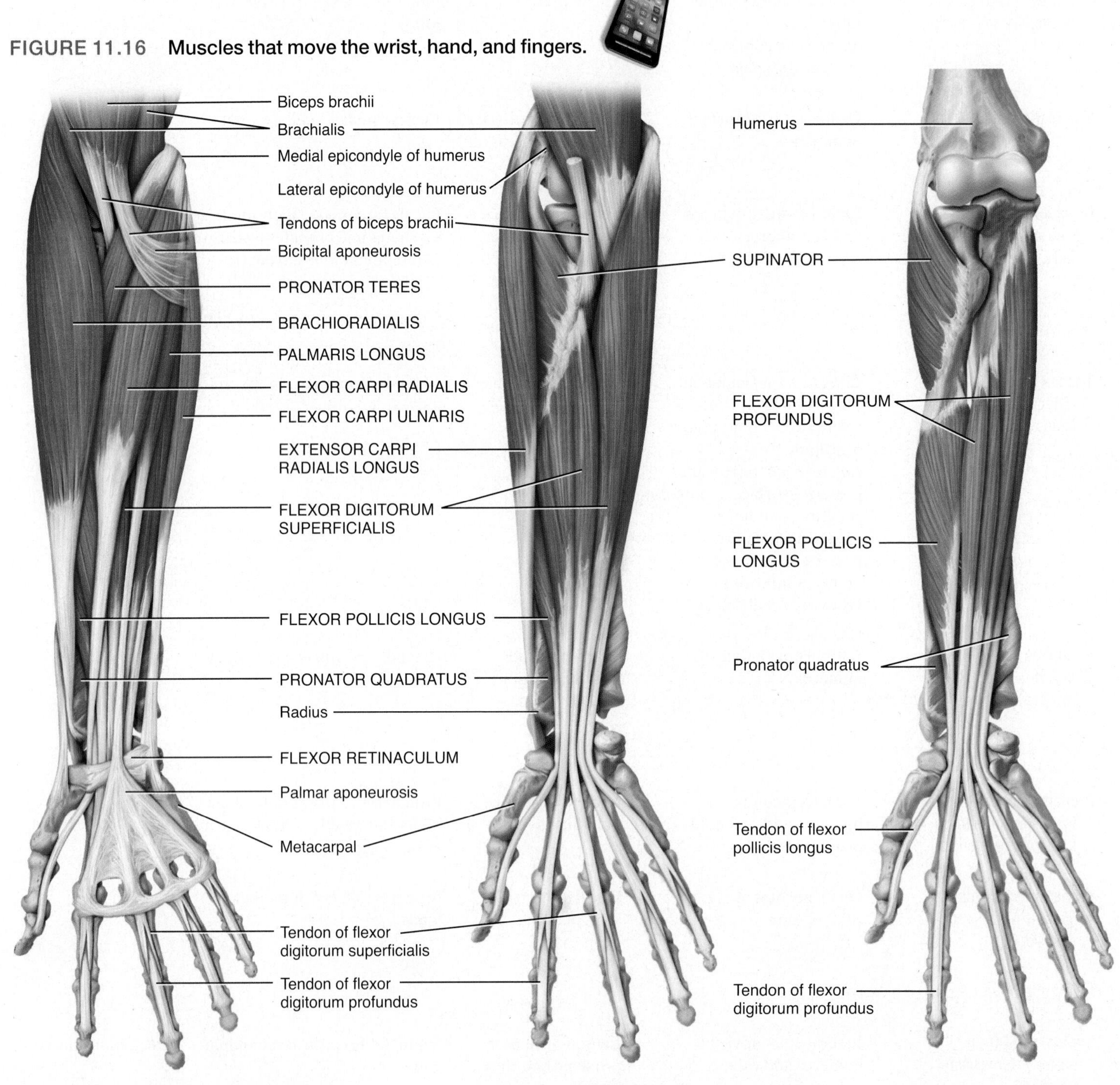


(a) Anterior superficial view

(b) Anterior intermediate view

(c) Anterior deep view

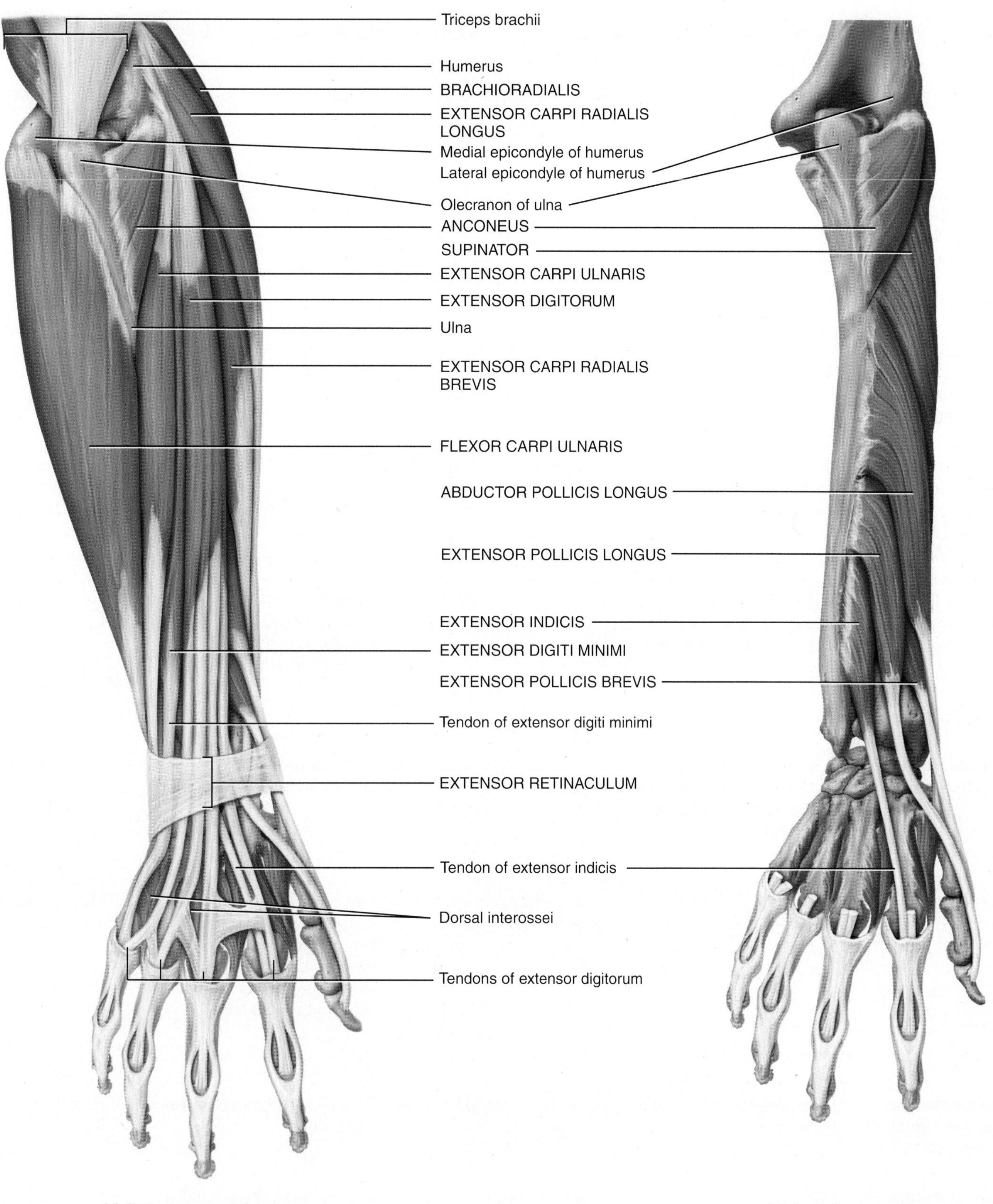


(d) Posterior superficial view

(e) Posterior deep view

The anterior forearm muscles function as flexors, and the posterior forearm muscles function as extensors.

**TABLE 11.16**

Muscles That Move the Wrist, Hand, and Fingers

| MUSCLE | ORIGIN | INSERTION | ACTION | INNERVATION |
|---|---|---|---|---|
| **Superficial Anterior (Flexor) Compartment of the Forearm** | | | | |
| **Flexor carpi radialis** (FLEK-sor KAR-pē rā′-dē-Ā-lis; *flexor* = decreases angle at joint; *carpi* = of the wrist; *radi-* = radius) | Medial epicondyle of humerus | Second and third metacarpals | Flexes and abducts hand (radial deviation) at wrist joint | Median nerve |
| **Palmaris longus** (pal-MA-ris LON-gus; *palma-* = palm; *longus* = long) | Medial epicondyle of humerus | Flexor retinaculum and palmar aponeurosis (fascia in center of palm) | Weakly flexes hand at wrist joint | Median nerve |
| **Flexor carpi ulnaris** (ul-NAR-is = ulna) | Medial epicondyle of humerus and superior posterior border of ulna | Pisiform, hamate, and base of fifth metacarpal | Flexes and adducts hand (ulnar deviation) at wrist joint | Ulnar nerve |
| **Flexor digitorum superficialis** (di-ji-TOR-um soo′-per-fish′-ē-Ā-lis; *digit* = finger or toe; *superficialis* = closer to surface) | Medial epicondyle of humerus, coronoid process of ulna, and a ridge along lateral margin of anterior surface (anterior oblique line) of radius | Middle phalanx of each finger* | Flexes middle phalanx of each finger at proximal interphalangeal joint, proximal phalanx of each finger at metacarpophalangeal joint, and hand at wrist joint | Median nerve |
| **Deep Anterior (Flexor) Compartment of the Forearm** | | | | |
| **Flexor pollicis longus** (POL-li-sis = thumb) | Anterior surface of radius and interosseous membrane (sheet of fibrous tissue that holds shafts of ulna and radius together) | Base of distal phalanx of thumb | Flexes distal phalanx of thumb at interphalangeal joint | Median nerve |
| **Flexor digitorum profundus** (prō-FUN-dus = deep) | Anterior medial surface of body of ulna | Base of distal phalanx of each finger | Flexes distal and middle phalanges of each finger at interphalangeal joints, proximal phalanx of each finger at metacarpophalangeal joint, and hand at wrist joint | Median and ulnar nerves |
| **Superficial Posterior (Extensor) Compartment of the Forearm** | | | | |
| **Extensor carpi radialis longus** (eks-TEN-sor = increases angle at joint) | Lateral supracondylar ridge of humerus | Second metacarpal | Extends and abducts hand at wrist joint | Radial nerve |
| **Extensor carpi radialis brevis** (BREV-is = short) | Lateral epicondyle of humerus | Third metacarpal | Extends and abducts hand at wrist joint | Radial nerve |

| | | | | |
|---|---|---|---|---|
| **Extensor digitorum** | Lateral epicondyle of humerus | Distal and middle phalanges of each finger | Extends distal and middle phalanges of each finger at interphalangeal joints, proximal phalanx of each finger at metacarpophalangeal joint, and hand at wrist joint | Radial nerve |
| **Extensor digiti minimi** (DIJ-i-tē Min-i-mē; *digit* = finger or toe; *minimi* = smallest) | Lateral epicondyle of humerus | Tendon of extensor digitorum on fifth phalanx | Extends proximal phalanx of little finger at metacarpophalangeal joint and hand at wrist joint | Deep radial nerve |
| **Extensor carpi ulnaris** | Lateral epicondyle of humerus and posterior border of ulna | Fifth metacarpal | Extends and adducts hand at wrist joint | Deep radial nerve |
| **Deep Posterior (Extensor) Compartment of the Forearm** | | | | |
| **Abductor pollicis longus** (ab-DUK-tor = moves part away from midline) | Posterior surface of middle of radius and ulna and interosseous membrane | First metacarpal | Abducts and extends thumb at carpometacarpal joint and abducts hand at wrist joint | Deep radial nerve |
| **Extensor pollicis brevis** | Posterior surface of middle of radius and interosseous membrane | Base of proximal phalanx of thumb | Extends proximal phalanx of thumb at metacarpophalangeal joint, first metacarpal of thumb at carpometacarpal joint, and hand at wrist joint | Deep radial nerve |
| **Extensor pollicis longus** | Posterior surface of middle of ulna and interosseous membrane | Base of distal phalanx of thumb | Extends distal phalanx of thumb at interphalangeal joint, first metacarpal of thumb at carpometacarpal joint, and abducts hand at wrist joint | Deep radial nerve |
| **Extensor indicis** (IN-di-kis; = index) | Posterior surface of ulna | Tendon of extensor digitorum of index finger | Extends distal and middle phalanges of index finger at interphalangeal joints, proximal phalanx of index finger at metacarpophalangeal joint, and hand at wrist joint | Deep radial nerve |

*Reminder: The thumb or pollex is the first finger and has two phalanges: proximal and distal. The remaining fingers are numbered II–V (2–5), and each has three phalanges: proximal, middle, and distal.

originate *outside* the hand and insert within it. As you will see, the names for the muscles give some indication of their origin, insertion, or action.

The anterior forearm muscles originate on the humerus, typically insert on the carpals, metacarpals, and phalanges, and function as flexors. The superficial anterior muscles in the forearm are arranged in the following order from lateral to medial: **flexor carpi radialis**, **palmaris longus**, and **flexor carpi ulnaris**. The **flexor digitorum superficialis** is deep to the other three muscles. The deep anterior muscles are arranged in the following order from lateral to medial: **flexor pollicis longus** (the only flexor of the distal phalanx of the thumb) and **flexor digitorum profundus** (ends in four tendons that insert into the distal phalanges of the fingers).

The posterior forearm muscles originate on the humerus, insert on the metacarpals and phalanges, and function as extensors. The superficial posterior muscles are arranged in the following order from lateral to medial: **extensor carpi radialis longus**, **extensor carpi radialis brevis**, **extensor digitorum** (divides into four tendons that insert into the middle and distal phalanges of the fingers), **extensor digiti minimi**, and the **extensor carpi**

**ulnaris.** The deep posterior muscles are arranged in the following order from lateral to medial: **abductor pollicis longus**, **extensor pollicis brevis**, **extensor pollicis longus**, and **extensor indicis**.

The tendons of the muscles of the forearm, which attach to the wrist or continue into the hand with blood vessels and nerves, are held close to bones by fascia. The tendons are also surrounded by tendon sheaths. At the wrist, the fascia is thickened into fibrous bands called *retinacula* (*retinacul-* = a holdfast). The *flexor retinaculum*, located over the palmar surface of the carpal bones, secures the tendons of the flexors of the fingers and wrist. The *extensor retinaculum*, located over the dorsal surface of the carpal bones, stabilizes tendons of the extensors of the wrist and fingers, which pass deep to it.

## Intrinsic Muscles of the Hand

The **intrinsic muscles of the hand** originate and insert *within* the hand (Figure 11.17; Table 11.17). In contrast to the extrinsic hand muscles, which create powerful but crude movements of the fingers, the intrinsic muscles produce the weak but intricate and precise movements of the fingers that characterize the human hand.

The intrinsic muscles of the hand are divided into three groups: **thenar**, **hypothenar**, and **intermediate**. The four thenar muscles act on the thumb and form the **thenar eminence**, the lateral rounded contour on the palm that is also called the ball of the thumb. The three hypothenar muscles act on the little finger and form the **hypothenar eminence**, the medial rounded contour on the palm that is also called the ball of the little finger. One of the hypothenar muscles, the **abductor digiti minimi**, is a powerful muscle that plays an important role in grasping an object with outspread fingers. The 11 intermediate muscles act on all of the fingers except the thumb. Two intermediate muscles, the **palmar interossei** and **dorsal interossei**, are located between the metacarpals and are important in abduction, adduction, flexion, and extension of the fingers, and in movements involved in skilled activities such as writing, typing, and playing a piano.

The functional importance of the hand is readily apparent when you consider that certain hand injuries can result in permanent disability. Most of the dexterity of the hand depends on movements of the thumb. The general activities of the hand are free motion, power grip (forcible movement of the fingers and thumb against the palm, as in squeezing), precision handling (a change in position of a handled object that requires exact control of finger and thumb positions, as in winding a watch or threading a needle), and pinch (compression between the thumb and index finger or between the thumb and first two fingers).

FIGURE 11.17 Intrinsic muscles of the hand.

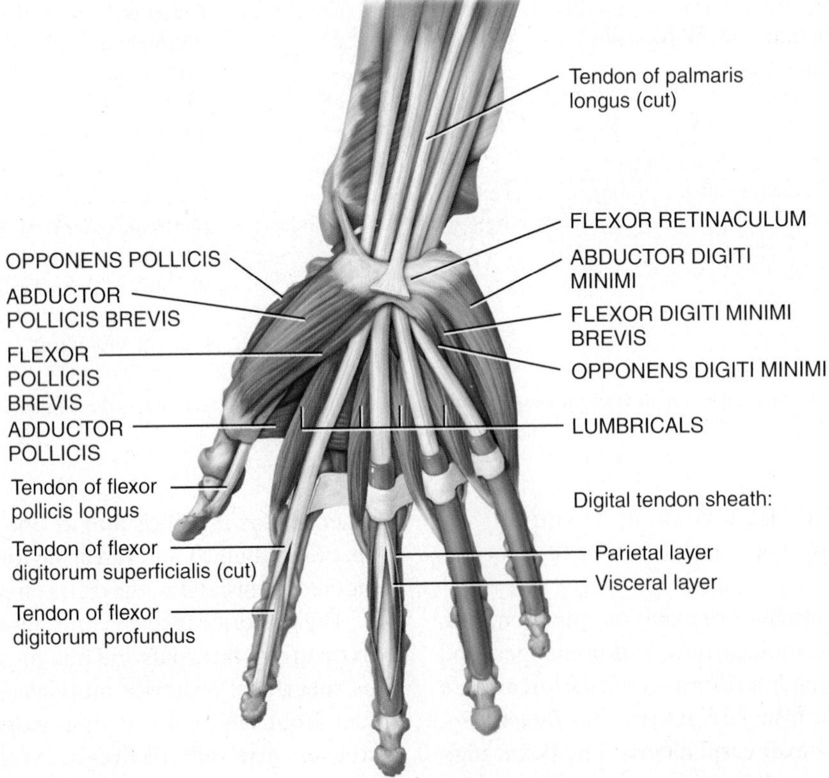


(a) Anterior superficial view

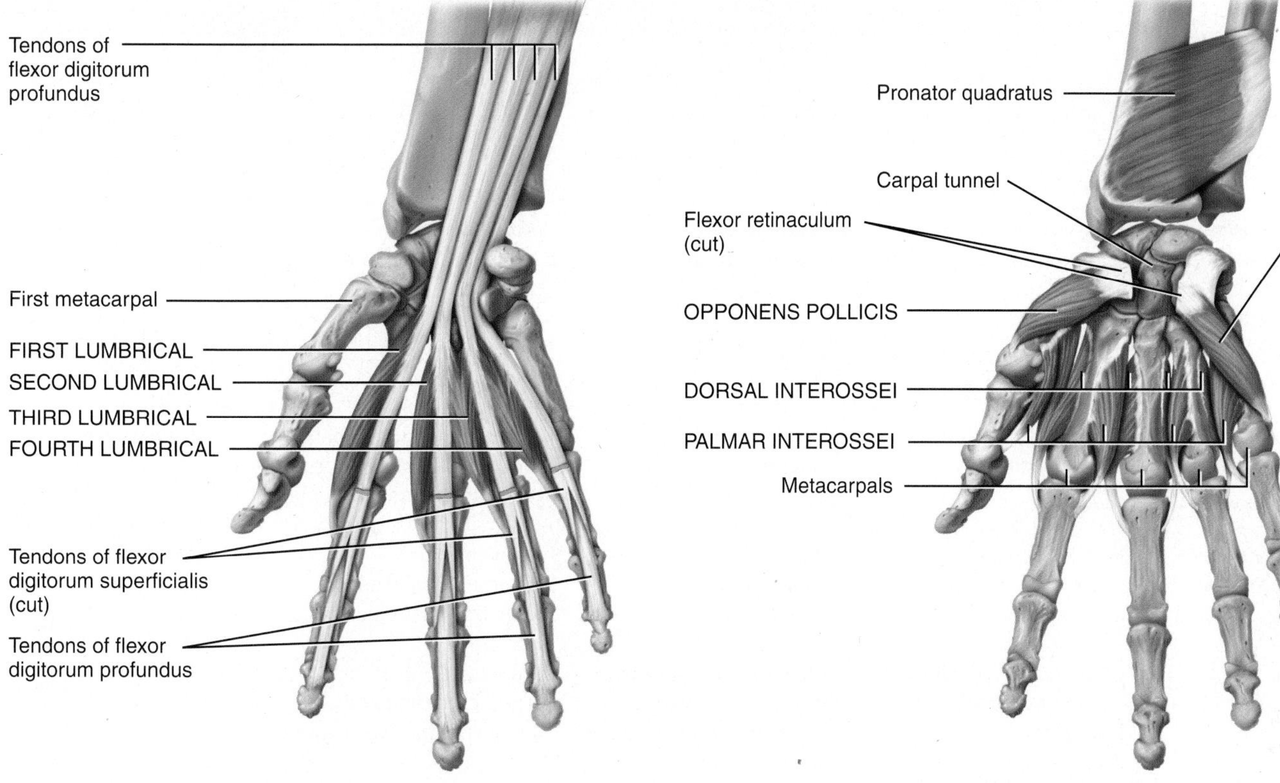


(b) Anterior intermediate view of lumbricals

(c) Anterior deep view

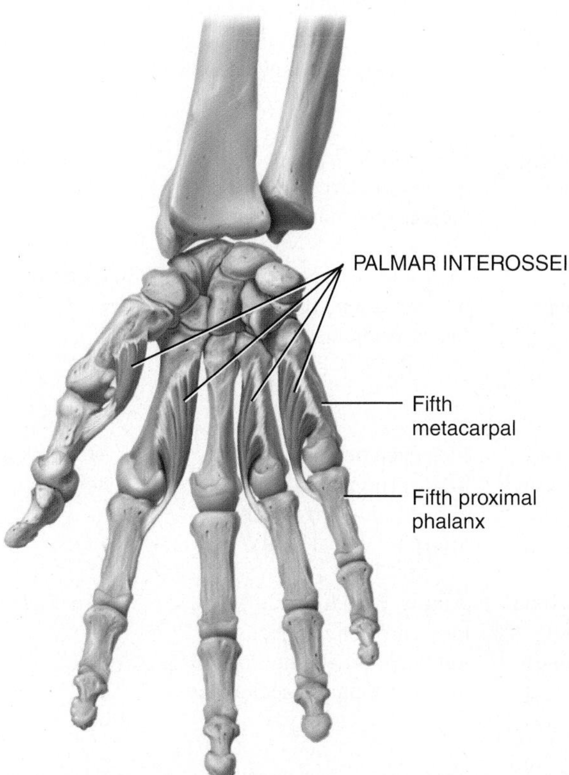


(d) Anterior deep view of palmar interossei

DORSAL INTEROSSEI

Fifth metacarpal

Fifth proximal phalanx

(e) Anterior deep view of dorsal interossei

**The intrinsic muscles of the hand produce the intricate and precise movements of the fingers that characterize the human hand.**

**TABLE 11.17**

Intrinsic Muscles of the Hand

| MUSCLE | ORIGIN | INSERTION | ACTION | INNERVATION |
|---|---|---|---|---|
| **Thenar (Lateral Aspect of Palm)** | | | | |
| **Abductor pollicis brevis** (ab-DUK-tor POL-li-sis BREV-is; *abductor* = moves part away from middle; *pollic-* = the thumb; *brevis* = short) | Flexor retinaculum, scaphoid, and trapezium | Lateral side of proximal phalanx of thumb | Abducts thumb at carpometacarpal joint | Median nerve |
| **Opponens pollicis** (op-PŌ-nenz = opposes) | Flexor retinaculum and trapezium | Lateral side of first metacarpal (thumb) | Moves thumb across palm to meet little finger (opposition) at the carpometacarpal joint | Median nerve |
| **Flexor pollicis brevis** (FLEK-sor = decreases angle at joint) | Flexor retinaculum, trapezium, capitate, and trapezoid | Lateral side of proximal phalanx of thumb | Flexes thumb at carpometacarpal and metacarpophalangeal joints | Median and ulnar nerves |
| **Adductor pollicis** (ad-DUK-tor = moves part toward midline) | *Oblique head:* capitate and metacarpals II and III *Transverse head*: metacarpal III | Medial side of proximal phalanx of thumb by a tendon containing a sesamoid bone | Adducts thumb at carpometacarpal and metacarpophalangeal joints | Ulnar nerve |
| **Hypothenar (Medial Aspect of Palm)** | | | | |
| **Abductor digiti minimi** (DIJ-i-tē MIN-i-mē; *digit-* = finger or toe; *minimi* = little) | Pisiform and tendon of flexor carpi ulnaris | Medial side of proximal phalanx of little finger | Abducts and flexes little finger at metacarpophalangeal joint | Ulnar nerve |
| **Flexor digiti minimi brevis** | Flexor retinaculum and hamate | Medial side of proximal phalanx of little finger | Flexes little finger at carpometacarpal and metacarpophalangeal joints | Ulnar nerve |
| **Opponens digiti minimi** | Flexor retinaculum and hamate | Medial side of fifth metacarpal (little finger) | Moves little finger across palm to meet thumb (opposition) at the carpometacarpal joint | Ulnar nerve |
| **Intermediate (Midpalmar)** | | | | |
| **Lumbricals** (LUM-bri-kals; *lumbric-* = earthworm) (four muscles) | Lateral sides of tendons and flexor digitorum profundus of each finger | Lateral sides of tendons of extensor digitorum on proximal phalanges of each finger | Flex each finger at metacarpophalangeal joints and extend each finger at interphalangeal joints | Median and ulnar nerves |
| **Palmar interossei** (PAL-mer in′-ter-OS-ē-ī *palmar* = palm; *inter-* = between; *-ossei* = bones) (three muscles) | Sides of shafts of metacarpals of all fingers (except the middle one) | Sides of bases of proximal phalanges of all fingers (except the middle one) | Adduct each finger at metacarpophalangeal joints; flex each finger at metacarpophalangeal joints | Ulnar nerve |
| **Dorsal interossei** (DOR-sal = back surface) (four muscles) | Adjacent sides of metacarpals | Proximal phalanx of each finger | Abduct fingers 2–4 at metacarpophalangeal joints; flex fingers 2–4 at metacarpophalangeal joints; and extend each finger at interphalangeal joints | Ulnar nerve |

## CLINICAL CONNECTION | *Carpal Tunnel Syndrome*

The **carpal tunnel** is a narrow passageway formed anteriorly by the flexor retinaculum and posteriorly by the carpal bones (Figure 11.17c). Through this tunnel pass the median nerve, the most superficial structure, and the long flexor tendons for the digits. Structures within the carpal tunnel, especially the median nerve, are vulnerable to compression, and the resulting condition is called **carpal tunnel syndrome**. Compression of the median nerve leads to sensory changes over the lateral side of the hand and muscle weakness in the thenar eminence. This results in pain, numbness, and tingling of the fingers. The condition may be caused by inflammation of the digital tendon sheaths, fluid retention, excessive exercise, infection, trauma, and/or repetitive activities that involve flexion of the wrist, such as keyboarding, cutting hair, or playing the piano. Treatment may involve the use of nonsteroidal anti-inflammatory drugs (such as ibuprofen or aspirin), wearing a wrist splint, corticosteroid injections, or surgery to cut the flexor retinaculum and release pressure on the median nerve.

### ✓ CHECKPOINT

**19.** Name the muscles that flex, extend, and adduct your arm at the shoulder joint.

**20.** The tendons of which muscles make up the rotator cuff?

**21.** Flex your arm at the elbow joint. Which muscles are contracting? Which muscles must relax so that you can flex your arm?

**22.** Name the muscles that supinate and pronate your forearm.

**23.** State the functions of the flexor retinaculum and extensor retinaculum.

**24.** Which extrinsic and intrinsic muscles of the hand are used to flex the thumb? To abduct the thumb?

**25.** Name the extrinsic and intrinsic muscles of the hand that flex the fingers and those that are used to adduct the fingers.

## 11.8 Deep muscles of the back move the head and vertebral column.

The muscles that move the vertebral column (spine) (Figure 11.18; Table 11.18) are quite complex because they have multiple origins and insertions and there is considerable overlap among them. The **splenius** muscles are attached to the sides and back of the neck. The two muscles in this group are named on the basis of their insertions: **splenius capitis** (head region) and **splenius cervicis** (cervical region). They extend, laterally flex, and rotate the head at the neck.

The **erector spinae** is the largest muscle mass of the back, forming a prominent bulge on either side of the vertebral column (Figure 11.18a). It is the chief extensor of the vertebral column. It is also important in controlling flexion, lateral flexion, and rotation of the vertebral column and in maintaining the lumbar curve, because the main mass of the muscle is in the lumbar region. The erector spinae consists of three groups of muscles: **iliocostalis group** (il′-ē-ō-kos-TĀ-lis), **longissimus group** (lon-JI-si-mus), and **spinalis group** (spi-NĀ-lis). The iliocostalis group is the most lateral, the longissimus group is intermediate, and the spinalis group is the most medial.

The **transversospinales**, so named because their fibers run from the transverse processes to the spinous processes of the vertebrae, extend, flex, and rotate the vertebral column, and rotate the head. The **segmental** muscles unite the spinous and transverse processes of consecutive vertebrae, and function primarily in stabilizing the vertebral column during its movements (Figure 11.18b). Because the **scalene** muscles assist in moving the vertebral column, they are also included in Table 11.18. The scalenes flex and rotate the head, and assist in deep inhalation (Figure 11.18c). Note in Table 11.9 that the rectus abdominis, external oblique, internal oblique, and quadratus lumborum also play roles in moving the vertebral column.

FIGURE 11.18 **Muscles that move the vertebral column.**

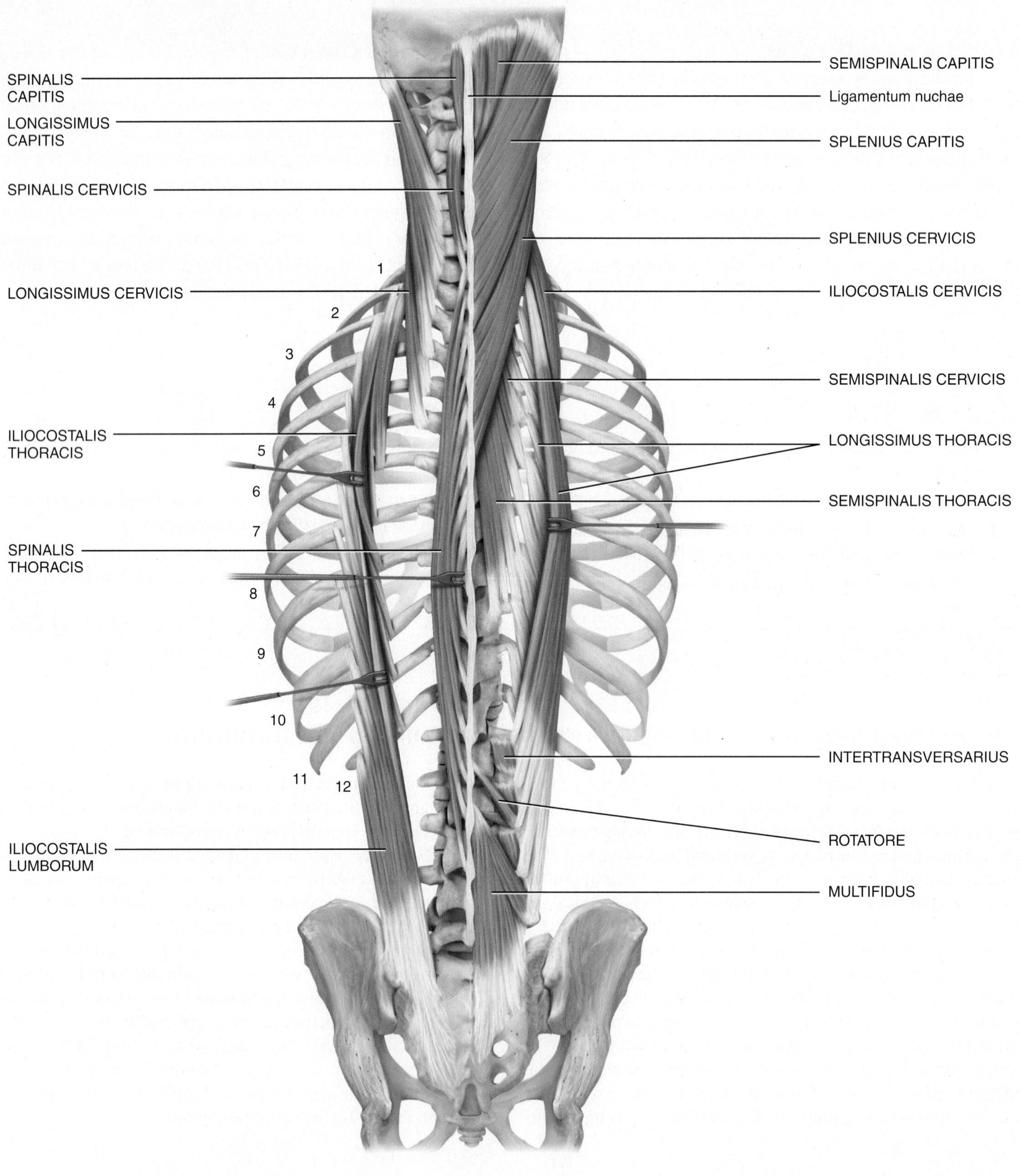


(a) Posterior view

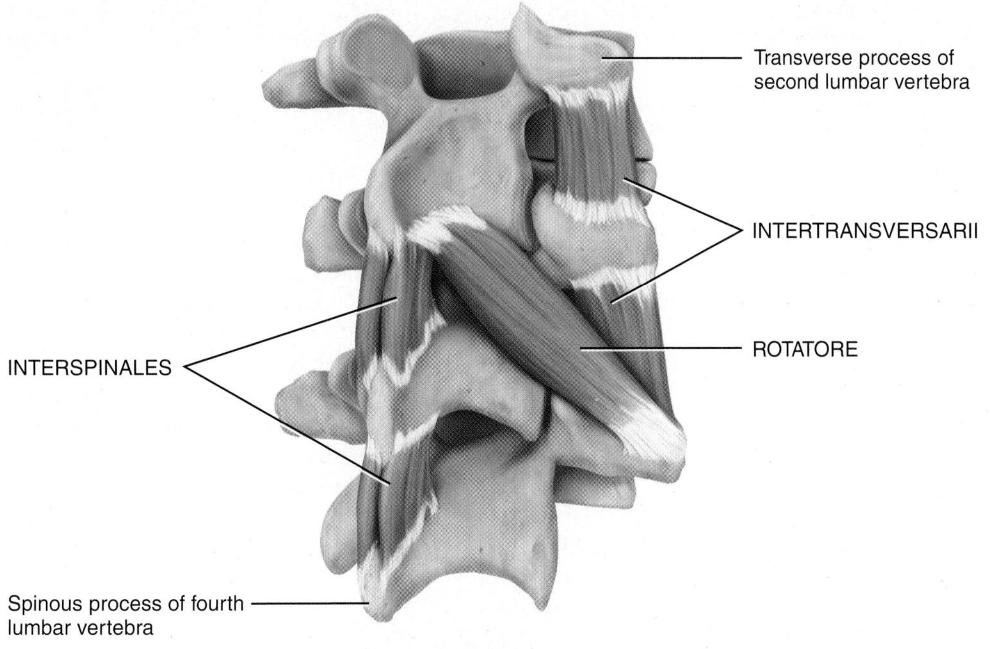


(b) Posterolateral view

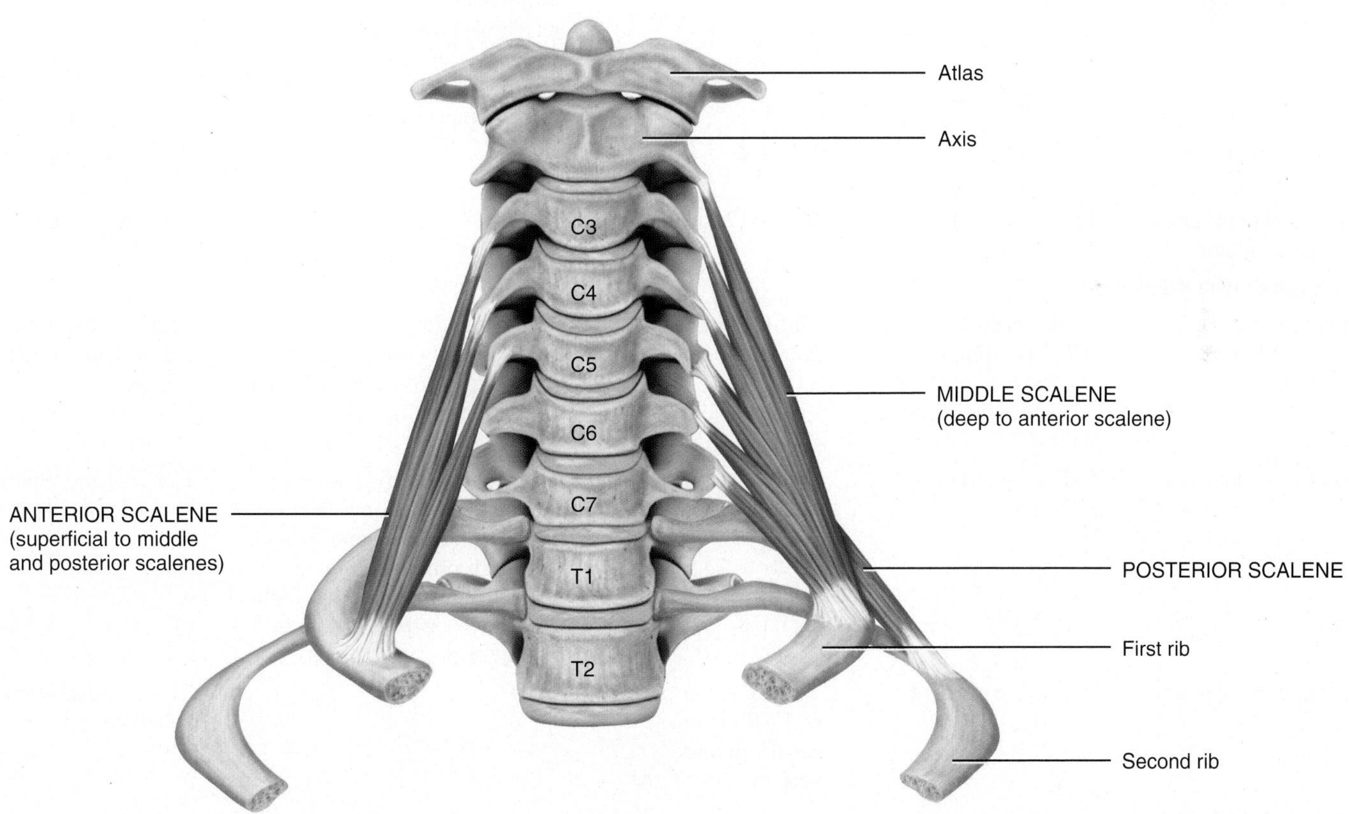


(c) Anterior view

**The erector spinae group (iliocostalis, longissimus, and spinalis muscles) is the chief extensor of the vertebral column.**

**TABLE 11.18**

Muscles That Move the Vertebral Column

| MUSCLE | ORIGIN | INSERTION | ACTION | INNERVATION |
|---|---|---|---|---|
| **Splenius (SPLĒ-nē-us)** | | | | |
| **Splenius capitis** (KAP-i-tis; *splenium* = bandage; *capit-* = head) | Ligamentum nuchae and spinous processes of C7–T4 | Occipital bone and mastoid process of temporal bone | Acting together (bilaterally), extend head; acting singly (unilaterally), laterally flex and rotate head to same side as contracting muscle | Middle cervical spinal nerves |
| **Splenius cervicis** (SER-vi-kis; *cervic-* = neck) | Spinous processes of T3–T6 | Transverse processes of C1–C2 or C1–C4 | Acting together, extend head; acting singly, laterally flex and rotate head to same side as contracting muscle | Inferior cervical spinal nerves |
| **Erector Spinae** (e-REK-tor SPI-nē) Consists of iliocostalis muscles (lateral), longissimus muscles (intermediate), and spinalis muscles (medial) | | | | |
| **Iliocostalis Group (Lateral)** | | | | |
| **Iliocostalis cervicis** (il′-ē-ō-kos-TĀL-is; *ilio-* = flank;-*costa-* = rib) | Ribs 1–6 | Transverse processes of C4–C6 | Acting together, muscles of each region (cervical, thoracic, and lumbar) extend and maintain erect posture of vertebral column of their respective regions; acting singly, laterally flex vertebral column of their respective regions | Cervical and thoracic spinal nerves |
| **Iliocostalis thoracis** (thō-RĀ-sis; = chest) | Ribs 7–12 | Ribs 1–6 | | Thoracic spinal nerves |
| **Iliocostalis lumborum** (lum-BOR-um) | Iliac crest | Ribs 7–12 | | Lumbar spinal nerves |
| **Longissimus Group (Intermediate)** | | | | |
| **Longissimus capitis** (lon-JIS-i-mus = longest) | Transverse processes of T1–T4 of superior four thoracic vertebrae and articular processes of C4–C7 | Mastoid process of temporal bone | Acting together, both longissimus capitis muscles extend head; acting singly, rotate head to same side as contracting muscle | Middle and inferior cervical and spinal nerves |
| **Longissimus cervicis** | Transverse processes of T4–T5 | Transverse processes of C2–C6 | Acting together, longissimus cervicis and both longissimus thoracis muscles extend vertebral column of their respective regions; acting singly, laterally flex vertebral column of their respective regions | Cervical and superior thoracic spinal nerves |
| **Longissimus thoracis** | Transverse processes of L1–L5 | Transverse processes of T1–T12 and superior lumbar vertebrae ribs 9–10 | | Thoracic and lumbar spinal nerves |
| **Spinalis Group (Medial)** | | | | |
| **Spinalis capitis** (spi-NĀ-lis = vertebral column) | Arises with semispinalis capitis | Occipital bone | Acting together, muscles of each region (cervical, thoracic, and lumbar) extend vertebral column of their respective regions | Cervical and superior thoracic spinal nerves |
| **Spinalis cervicis** | Ligamentum nuchae and spinous process of C7 | Spinous process of axis | | Inferior cervical and thoracic spinal nerves |
| **Spinalis thoracis** | Spinous processes of T10–12 | Spinous processes of T4–T8 (variable) | | Thoracic spinal nerves |

| Muscle | Origin | Insertion | Action | Innervation |
|---|---|---|---|---|
| **Transversospinales (trans-ver-sō-spi-NĀ-lēz)** | | | | |
| **Semispinalis capitis** (sem′-ē-spi-NĀ-lis; *semi-* = partially or one half) | Transverse processes of T1–6 or T1–7 and C7, and articular processes of C4–C6 | Occipital bone | Acting together, extend head; acting singly, rotate head to side opposite contracting muscle | Cervical and thoracic spinal nerves |
| **Semispinalis cervicis** | Transverse processes of T1–T5 or T1–T6 | Spinous processes of C1–C5 | Acting together, both semispinalis cervicis and both semispinalis thoracis muscles extend vertebral column of their respective regions; acting singly, rotate head to side opposite contracting muscle | Cervical and thoracic spinal nerves |
| **Semispinalis thoracis** | Transverse processes of T6–T10 | Spinous processes of T1–T4 and C6–C7 | | Thoracic spinal nerves |
| **Multifidus** (mul-TIF-i-dus; *multi-* = many; *-fid-* = segmented) | Sacrum, ilium, transverse processes of L1–L5, T1–T12, and C1–C4 | Spinous process of a more superior vertebra | Acting together, extend vertebral column; acting singly, laterally flex vertebral column and rotate head to side opposite contracting muscle | Cervical, thoracic, and lumbar spinal nerves |
| **Rotatores** (rō-ta-TŌ-rēz; singular is *rotatore* = to rotate) | Transverse processes of all vertebrae | Spinous process of vertebra superior to the one of origin | Acting together, extend vertebral column; acting singly, rotate vertebral column to side opposite contracting muscle | Cervical, thoracic, and lumbar spinal nerves |
| **Segmental (seg-MEN-tal)** | | | | |
| **Interspinales** (in-ter-spī-NĀ-lēz; *inter-* = between) | Superior surface of all spinous processes | Inferior surface of spinous process of vertebra superior to the one of origin | Acting together, extend vertebral column; acting singly, stabilize vertebral column during movement | Cervical, thoracic, and lumbar spinal nerves |
| **Intertransversarii** (in′-ter-trans-vers-AR-ē-ī; singular is *intertransversarius*) | Transverse processes of all vertebrae | Transverse process of vertebra superior to the one of origin | Acting together, extend vertebral column; acting singly, laterally flex vertebral column and stabilize it during movements | Cervical, thoracic, and lumbar spinal nerves |
| **Scalenes (SKĀ-lēnz)** | | | | |
| **Anterior scalene** (SKĀ-lēn; *anterior* = front; *scalene* = uneven) | Transverse processes of C3–C6 | Rib 1 | Acting together, right and left anterior scalene and middle scalene muscles flex head and elevate first ribs during deep inhalation; acting singly, laterally flex head and rotate head to side opposite contracting muscle | Cervical spinal nerves C5–C6 |
| **Middle scalene** | Transverse processes of C2–C7 | Rib 1 | | Cervical spinal nerves C3–C8 |
| **Posterior scalene** | Transverse processes of C4–C6 | Rib 2 | Acting together, flex head and elevate second ribs during deep inhalation; acting singly, laterally flex head and rotate head to side opposite contracting muscle | Cervical spinal nerves C6–C8 |

✓ **CHECKPOINT**

**26.** Which muscle groups constitute the erector spinae?

**27.** Name the muscles that laterally flex the head and those that laterally flex the vertebral column.

## RETURN TO Stewart's Story

"So, I think I must have done something to my *quadratus lumborum* on the left side. After I had my back injury, I couldn't laterally flex that side well and it hurt to breathe deeply." Wally puts down the atlas and stretches around his torso to demonstrate the motions described.

"Yeah, I think you're right." Stewart looks at a few more tables in the textbook and asks, "You mentioned how the doctors decided you hadn't damaged any of the intervertebral discs, but do you remember if you had any shoulder problems?"

Wally shakes his head. "No, the shoulders were fine. The problem seemed to be confined to the lumbar region. I suppose it could also be either of the obliques, but the pain seemed to be located more in the back than on the side."

"Well, if your shoulders weren't injured, you were lucky." They both look up as Jenny comes to the table, sits down to join them, and adds, "I damaged my rotator cuff playing baseball in high school and can still feel some of the aftereffects of the injury." Stewart asks her if she knows what part was injured. "Sure, it was the tendon of the subscapularis. That ruined my curve ball *and* my chances for a scholarship." Wally points at her palm. "Does that have anything to do with that scar in your hand?"

"No," Jenny replies, "this was from a wood-carving accident in junior high. I cut pretty deeply into the thenar eminence. I had to have a couple of surgeries to make sure each of the muscles healed properly, along with lots of physical therapy. Part of the physical therapy involved throwing a baseball, which got me more interested in the sport." Wally asked if she's ever had any wrist problems from the injury. "No, I've been lucky in that respect. I definitely use my hands and wrists a lot, but I've never had any carpal tunnel issues. What I usually have to watch are my lats or delts; they took a lot of abuse from my baseball days." That draws a strange look from Wally.

"Wait a minute, they work on the back of the shoulder. Why should they be a problem if you were throwing forward all of the time?"

"Think about it. I was propelling the ball forward, but I had to pull my shoulder back hard and fast to wind up those pitches. I was so focused on my limb muscles that I overused my shoulder muscles and now I'm paying for it." Wally nods slowly.

"Okay, I can see that now. They were counteracting the action of the pectoralis major and serratus anterior. Wow, it's amazing to think how much goes into a single motion."

E. *Wally mentions the obliques. Which oblique muscles would affect lumbar vertebral movement?*

F. *Where is the thenar eminence in Jenny's hand? What muscles comprise it?*

G. *What are the full names of the muscles Jenny refers to as "lats" and "delts"? What actions do they perform?*

# 11.9 Muscles originating on the pelvic girdle or lower limb move the femur, leg, and foot.

## Muscles That Move the Femur

While upper limb muscles are characterized by versatility of movement, lower limb muscles are larger and more powerful to provide stability, locomotion, and maintenance of posture. In addition, muscles of the lower limbs often cross two joints and act equally on both.

The majority of muscles that move the femur (thigh bone) (Figure 11.19; Table 11.19) originate on the pelvic girdle and insert on the femur. The **psoas major** and **iliacus** have a common insertion (lesser trochanter of the femur) and together are known as the **iliopsoas** (il′-ē-ō-SŌ-as). There are three gluteal muscles: gluteus maximus, gluteus medius, and gluteus minimus. The **gluteus maximus** is the largest and heaviest of the three muscles and is one of the largest muscles in the body. It is the chief extensor of the femur at the hip joint. The **gluteus medius** is mostly deep to the gluteus maximus and is a powerful abductor of the femur at the hip joint. It is a common site for an intramuscular injection. The **gluteus minimus** is the smallest of the gluteal muscles and lies deep to the gluteus medius. The **tensor fasciae latae** is located on the lateral surface of the thigh. Together, the tendons of the tensor fasciae latae and gluteus maximus form a structure called the *iliotibial tract* that inserts into the lateral condyle of the tibia. Three muscles on the medial thigh, the **adductor longus**, **adductor brevis**, and **adductor magnus**, adduct, flex, and medially rotate the femur at the hip joint.

**FIGURE 11.19 Muscles that move the femur.** 

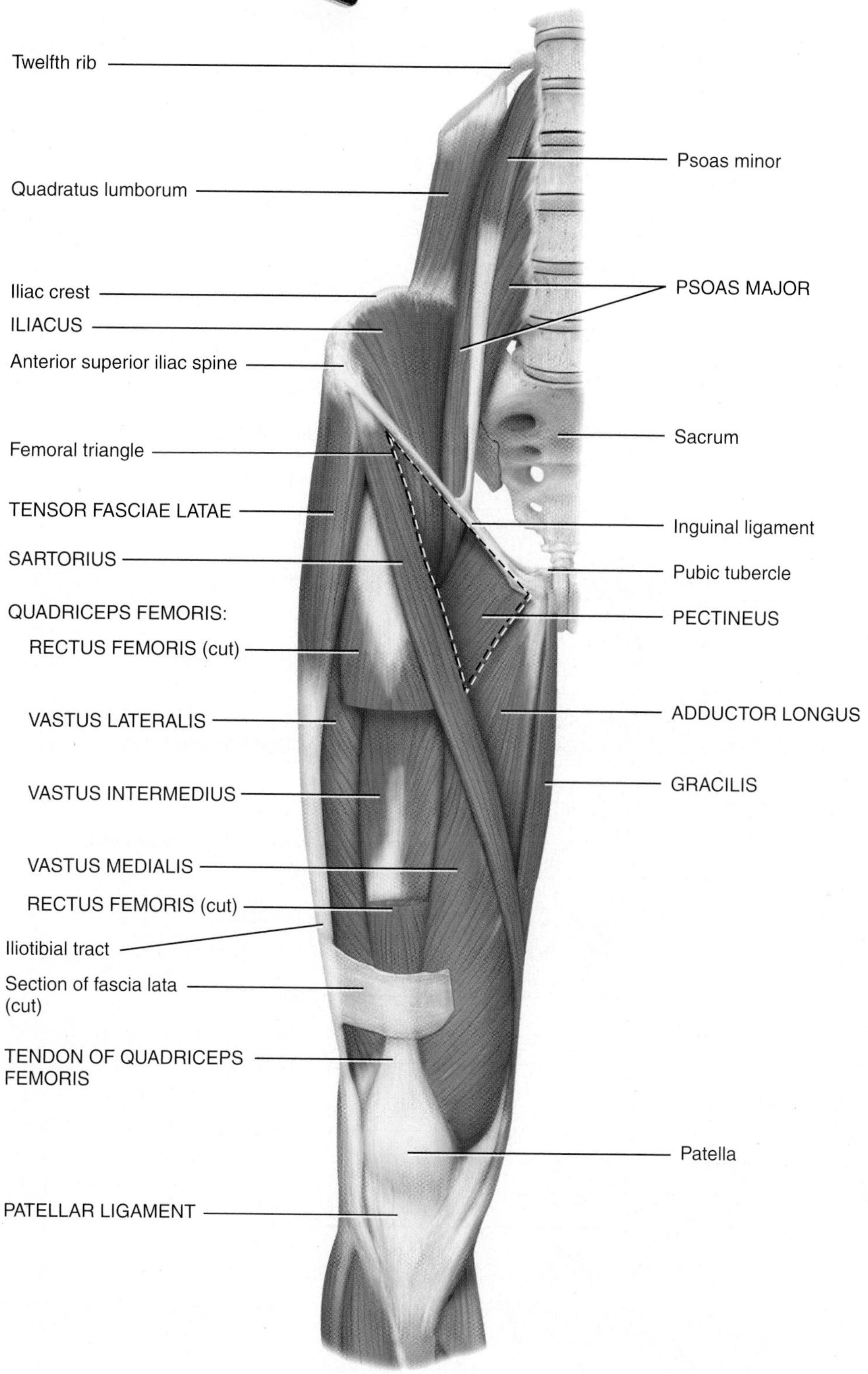

(a) Anterior superficial view (femoral triangle is indicated by a dashed line)

(continues)

FIGURE 11.19 (continued)

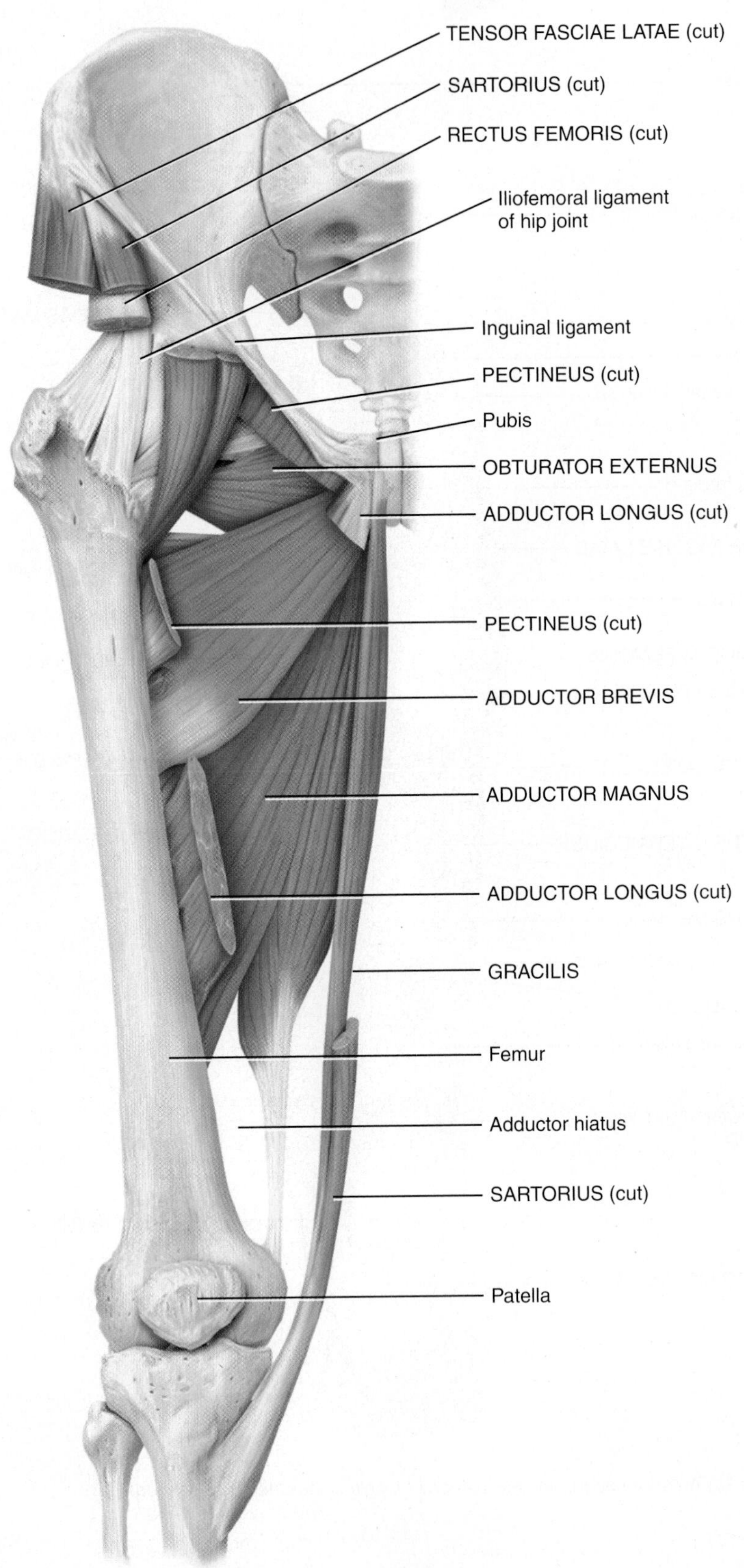


(b) Anterior deep view (femur rotated laterally)

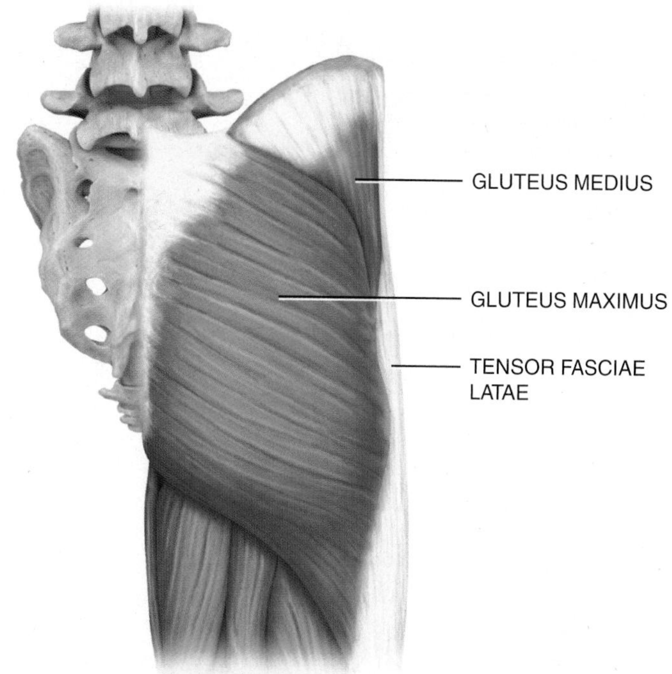


(c) Posterior superficial view

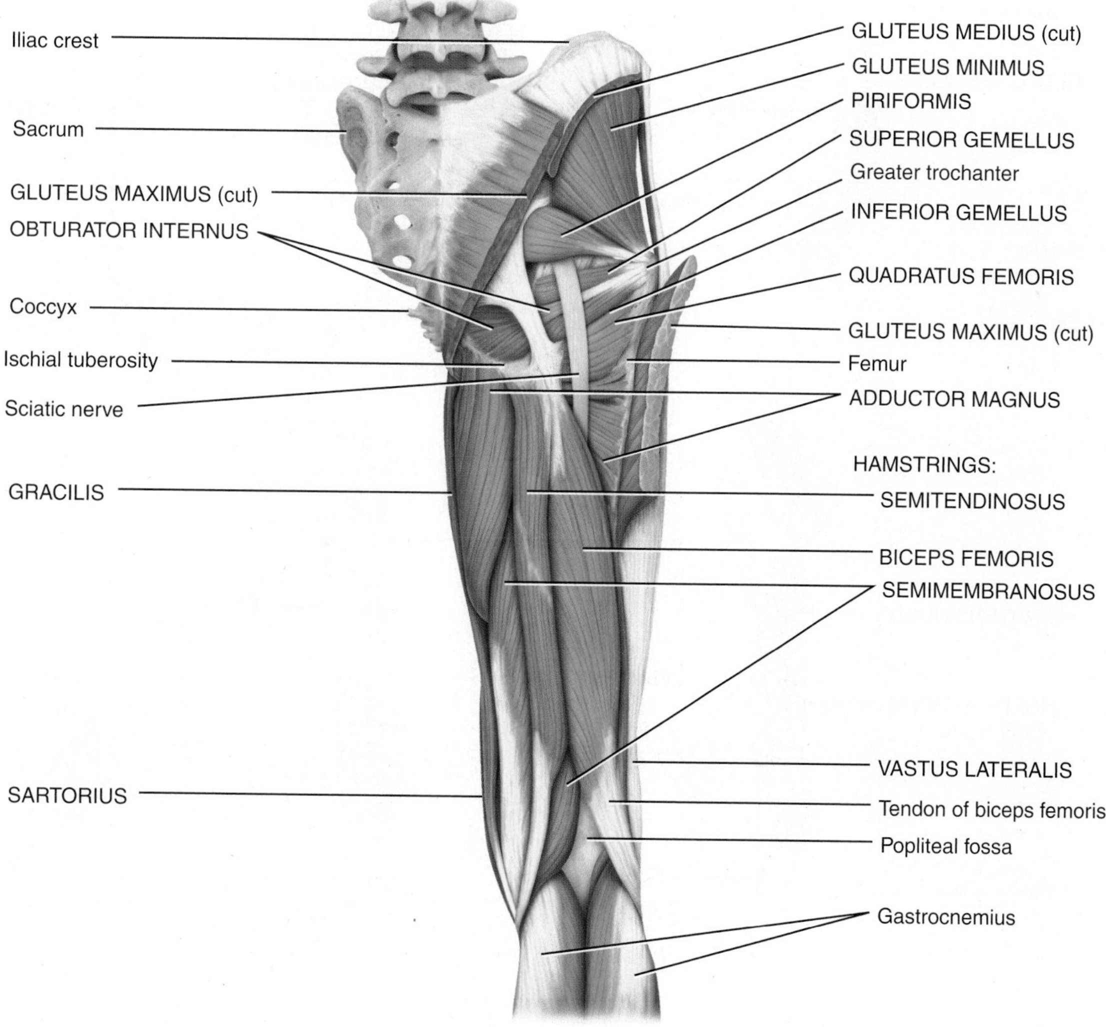


(d) Posterior superficial and deep view

(continues)

FIGURE 11.19 (continued)

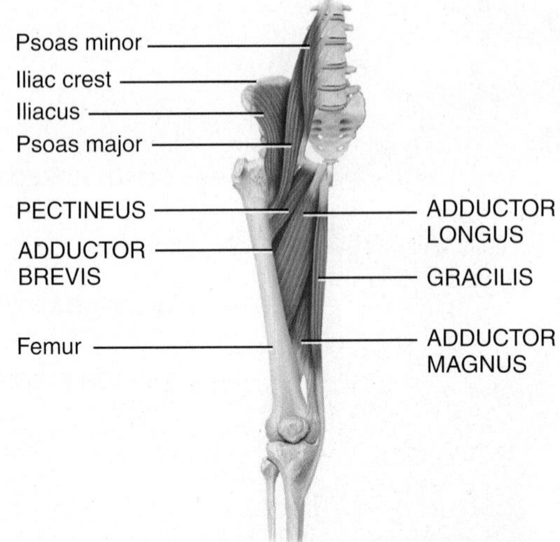


Anterior deep view

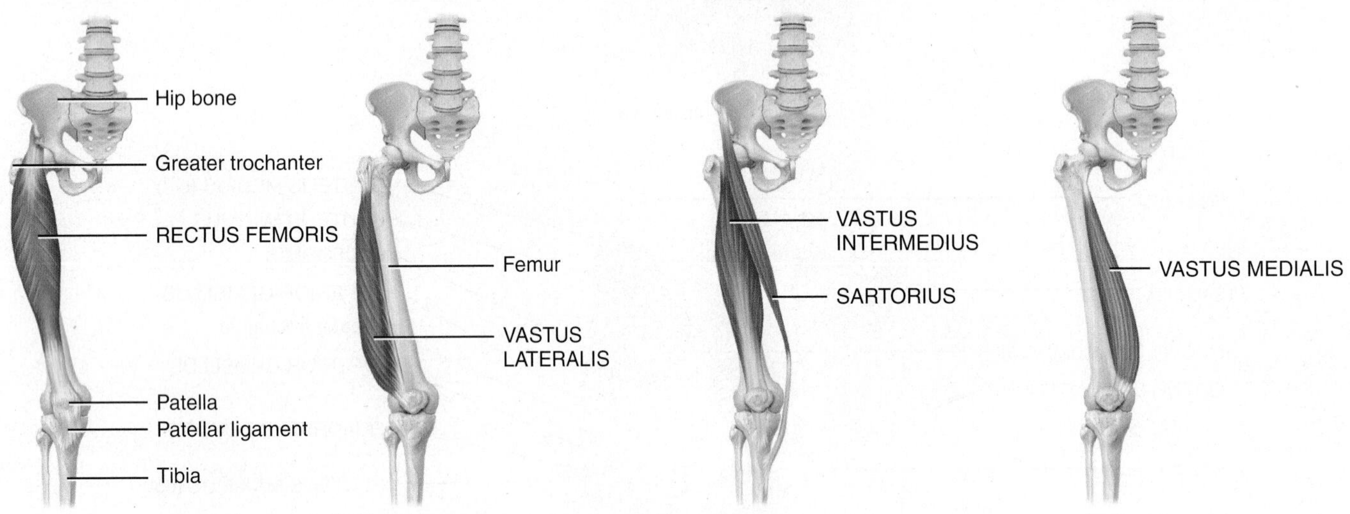


Anterior views

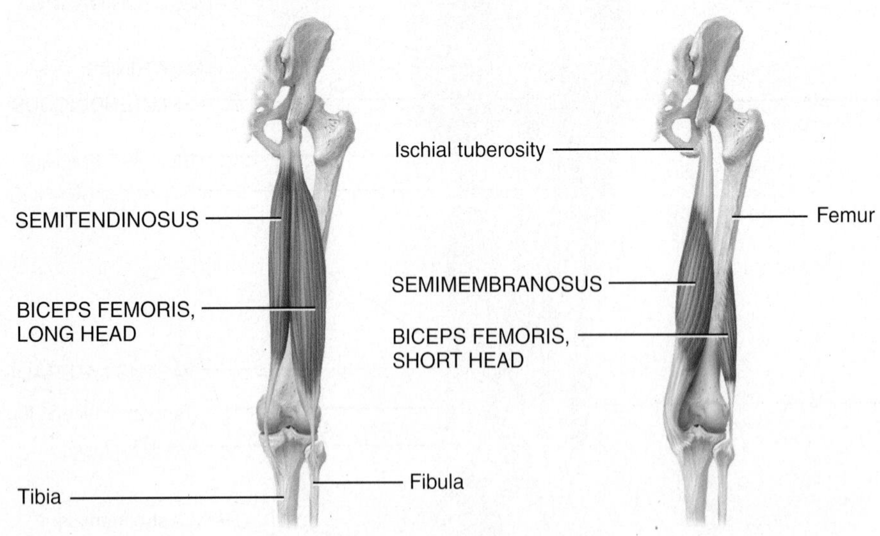


Posterior deep views

(e) Isolated muscles

**Most muscles that move the femur originate on the pelvic girdle and insert on the femur.**

**TABLE 11.19**

Muscles That Move the Femur

| MUSCLE | ORIGIN | INSERTION | ACTION | INNERVATION |
|---|---|---|---|---|
| **Iliopsoas** | | | | |
| **Psoas major** (SŌ-as; *psoa-* = a muscle of the loin) | Transverse processes and bodies of lumbar vertebrae | With iliacus into lesser trochanter of femur | Psoas major and iliacus muscles acting together flex thigh at hip joint, rotate thigh laterally, and flex trunk on hip as in sitting up from supine position | Lumbar spinal nerves L2–L3 |
| **Iliacus** (il′-ē-A-cus; *iliac-* = ilium) | Iliac fossa and sacrum | With psoas major into lesser trochanter of femur | | Femoral nerve |
| **Gluteus maximus** (GLOO-tē-us MAK-si-mus; *glute-* = rump or buttock; *maximus* = largest) | Iliac crest, sacrum, coccyx, and aponeurosis of sacrospinalis | Iliotibial tract of fascia lata and lateral part of linea aspera (gluteal tuberosity) under greater trochanter of femur | Extends thigh at hip joint and laterally rotates thigh | Inferior gluteal nerve |
| **Gluteus medius** (MĒ-de-us = middle) | Ilium | Greater trochanter of femur | Abducts thigh at hip joint and medially rotates thigh | Superior gluteal nerve |
| **Gluteus minimus** (MIN-i-mus = smallest) | Ilium | Greater trochanter of femur | Abducts thigh at hip joint and medially rotates thigh | Superior gluteal nerve |
| **Tensor fasciae latae** (TEN-sor FA-shē-ē LĀ-tē; *tensor* = makes tense; *fasciae* = of the band; *lat-* = wide) | Iliac crest | Tibia by way of the iliotibial tract | Flexes and abducts thigh at hip joint | Superior gluteal nerve |
| **Piriformis** (pir-i-FOR-mis; *piri-* = pear;-*form-* = shape) | Anterior sacrum | Superior border of greater trochanter of femur | Laterally rotates and abducts thigh at hip joint | Sacral spinal nerves S1 or S2, mainly S1 |
| **Obturator internus** (OB-too-rā′-tor in-TER-nus; *obturator* = obturator foramen; *intern-* = inside) | Inner surface of obturator foramen, pubis, and ischium | Medial surface of greater trochanter of femur | Laterally rotates and abducts thigh at hip joint | Nerve to obturator internus |
| **Obturator externus** (ex-TER-nus = outside) | Outer surface of obturator membrane | Deep depression inferior to greater trochanter (trochanteric fossa) of femur | Laterally rotates and abducts thigh at hip joint | Obturator nerve |
| **Superior gemellus** (jem-EL-lus; *superior* = above; *gemell-* = twins) | Ischial spine | Medial surface of greater trochanter of femur | Laterally rotates and abducts thigh at hip joint | Nerve to obturator internus |
| **Inferior gemellus** (*inferior* = below) | Ischial tuberosity | Medial surface of greater trochanter of femur | Laterally rotates and abducts thigh at hip joint | Nerve to quadratus femoris |

(continues)

**TABLE 11.19 CONTINUED**

Muscles That Move the Femur

| MUSCLE | ORIGIN | INSERTION | ACTION | INNERVATION |
|---|---|---|---|---|
| **Quadratus femoris** (kwod-RĀ-tus FEM-or-is; *quad-* = square, four-sided; *femoris* = femur) | Ischial tuberosity | Elevation superior to mid-portion of intertrochanteric crest (quadrate tubercle) on posterior femur | Laterally rotates and stabilizes hip joint | Nerve to quadratus femoris |
| **Adductor longus** (LONG-us; *adductor* = moves part closer to midline; *longus* = long) | Pubic crest and pubic symphysis | Linea aspera of femur | Adducts and flexes thigh at hip joint and laterally rotates thigh | Obturator nerve |
| **Adductor brevis** (BREV-is; = short) | Inferior ramus of pubis | Superior half of linea aspera of femur | Adducts and flexes thigh at hip joint and medially rotates thigh | Obturator nerve |
| **Adductor magnus** (MAG-nus = large) | Inferior ramus of pubis and ischium to ischial tuberosity | Linea aspera of femur | Adducts thigh at hip joint and laterally rotates thigh; anterior part flexes thigh at hip joint, and posterior part extends thigh at hip joint | Obturator and sciatic nerves |
| **Pectineus** (pek-TIN-ē-us; = a comb) | Superior ramus of pubis | Pectineal line of femur, between lesser trochanter and linea aspera | Flexes and adducts thigh at hip joint | Femoral nerve |

The *femoral triangle* is a space between inguinal ligament, adductor longus, and sartorius. Within the femoral triangle are the femoral nerve, femoral artery, femoral vein, and inguinal lymph nodes.

## Muscles That Move the Femur, Tibia, and Fibula

Medial muscles of the thigh adduct the femur at the hip joint (Figure 11.19; Table 11.20). The medial thigh muscles, the adductor magnus, adductor longus, adductor brevis, and pectineus, are included in Table 11.19 because they act on the femur. The **gracilis**, a long, straplike muscle on the medial aspect of the thigh and knee, not only adducts the thigh, but also flexes the leg at the knee joint. For this reason, it is listed in Table 11.20.

The muscles of the anterior aspect of the thigh extend the leg at the knee joint, and some also flex the thigh at the hip joint. The **quadriceps femoris** is the largest muscle in the body, covering most of the anterior surface and sides of the thigh. It has four distinct parts, usually described as four separate muscles: **rectus femoris** on the anterior aspect of the thigh, **vastus lateralis** on the lateral aspect of the thigh, **vastus medialis** on the medial aspect of the thigh, and **vastus intermedius** located deep to the rectus femoris between the vastus lateralis and vastus medialis. The common tendon for the four muscles is the *quadriceps tendon*, which inserts on the patella. The tendon continues below the patella as the *patellar ligament* and attaches to the tibial tuberosity. The **sartorius** is a long, narrow muscle that forms a band across the thigh from the ilium of the hip bone to the medial side of the tibia. The various movements it produces help effect the cross-legged sitting position in which the heel of one limb is placed on the knee of the opposite limb. It is known as the tailor's muscle because tailors often assume this cross-legged sitting position.

The muscles of the posterior thigh flex the leg and extend the thigh. The three posterior thigh muscles are collectively called the **hamstrings**: the **biceps femoris**, **semitendinosus**, and **semimembranosus**. The hamstrings are so named because butchers can hang hams for smoking by the tendons of these muscles, which are long and stringlike in the popliteal area. Because the hamstrings span two joints (hip and knee), they are able to both extend the thigh and flex the leg. The *popliteal fossa* is a diamond-shaped space on the posterior aspect of the knee bordered laterally by the tendons of the biceps femoris and medially by the tendons of the semitendinosus and semimembranosus.

## Muscles That Move the Foot and Toes

Muscles that move the foot and toes (Figure 11.20; Table 11.21) are located in the leg. Muscles of the anterior aspect of the leg dorsiflex the foot at the ankle joint. In addition to dorsiflexion, the **tibialis anterior**, easily palpated against the lateral surface of the tibia, inverts the foot, while the **fibularis tertius** everts the foot. The **extensor hallucis longus** and **extensor digitorum longus** dorsiflex the ankle and extend the toes. Analogous to the

## TABLE 11.20

Muscles That Move the Femur, Tibia, and Fibula

| MUSCLE | ORIGIN | INSERTION | ACTION | INNERVATION |
|---|---|---|---|---|
| **Medial (Adductor) Compartment of the Thigh** | | | | |
| **Adductor magnus** (MAG-nus)<br>**Adductor longus** (LONG-us)<br>**Adductor brevis** (BREV-is)<br>**Pectineus** (pek-TIN-ē-us) | See Table 11.19 | | | |
| **Gracilis** (gra-SIL-is = slender) | Body and inferior ramus of pubis | Medial surface of body of tibia | Adducts thigh at hip joint, medially rotates thigh, and flexes leg at knee joint | Obturator nerve |
| **Anterior (Extensor) Compartment of the Thigh** | | | | |
| **Quadriceps femoris** (KWOD-ri-ceps; *quadriceps* = four heads of origin; *femoris* = femur) | | | | |
| **Rectus femoris** (REK-tus = fascicles parallel to midline) | Anterior inferior iliac spine | Patella via quadriceps tendon and then tibial tuberosity via patellar ligament | All four heads extend leg at knee joint; rectus femoris muscle acting alone also flexes thigh at hip joint | Femoral nerve |
| **Vastus lateralis** (VAS-tus lat′-e-RĀ-lis; *vast* = huge; *lateralis* = lateral) | Greater trochanter and linea aspera of femur | | | |
| **Vastus medialis** (*medialis* = medial) | Linea aspera of femur | | | |
| **Vastus intermedius** (*intermedius* = middle) | Anterior and lateral surfaces of body of femur | | | |
| **Sartorius** (sar-TOR-ē-us; *sartor* = tailor; longest muscle in body) | Anterior superior iliac spine | Medial surface of body of tibia | Flexes leg at knee joint; flexes, abducts, and laterally rotates thigh at hip joint, thus crossing leg | Femoral nerve |
| **Posterior (Flexor) Compartment of the Thigh** | | | | |
| **Hamstrings** A collective designation for three separate muscles | | | | |
| **Biceps femoris** (BĪ-ceps = two heads of origin) | *Long head*: ischial tuberosity<br>*Short head*: linea aspera of femur | Head of fibula and lateral condyle of tibia | Flexes leg at knee joint and extends thigh at hip joint | Tibial and common peroneal nerves from sciatic nerve |
| **Semitendinosus** (sem′-ē-ten-di-NŌ-sus; *semi-* = half; *-tendo-* = tendon) | Ischial tuberosity | Proximal part of medial surface of shaft of tibia | Flexes leg at knee joint and extends thigh at hip joint | Tibial nerve from sciatic nerve |
| **Semimembranosus** (sem′-ē-mem-bra-NŌ-sus; *membran-* = membrane) | Ischial tuberosity | Medial condyle of tibia | Flexes leg at knee joint and extends thigh at hip joint | Tibial nerve from sciatic nerve |

wrist, the tendons of anterior leg muscles are held firmly to the ankle by thickenings of deep fascia called the *superior extensor retinaculum* and *inferior extensor retinaculum*.

The lateral leg contains two muscles that plantar flex and evert the foot at the ankle joint: the **fibularis longus** and **fibularis brevis**.

The superficial muscles of the posterior aspect of the leg are the gastrocnemius, soleus, and **plantaris**—the so-called calf muscles. The superficial muscles and most of the deep posterior leg muscles plantar flex the foot at the ankle joint. The large size of these muscles is directly related to the characteristic upright stance of humans. The **gastrocnemius** is the most superficial muscle and forms the prominence of the calf. The broad, flat **soleus**, which lies deep to the gastrocnemius, derives its name from its resemblance to a flat fish (sole). The superficial posterior leg muscles share a common tendon of insertion, the *calcaneal (Achilles) tendon*, which is the strongest tendon of the body. The calcaneal tendon inserts into the calcaneus, the heel bone of the ankle.

The deep muscles of the posterior leg are the popliteus, tibialis posterior, flexor digitorum longus, and flexor hallucis longus. The **popliteus** forms the floor of the popliteal fossa. The **tibialis**

FIGURE 11.20 **Muscles that move the foot and toes.**

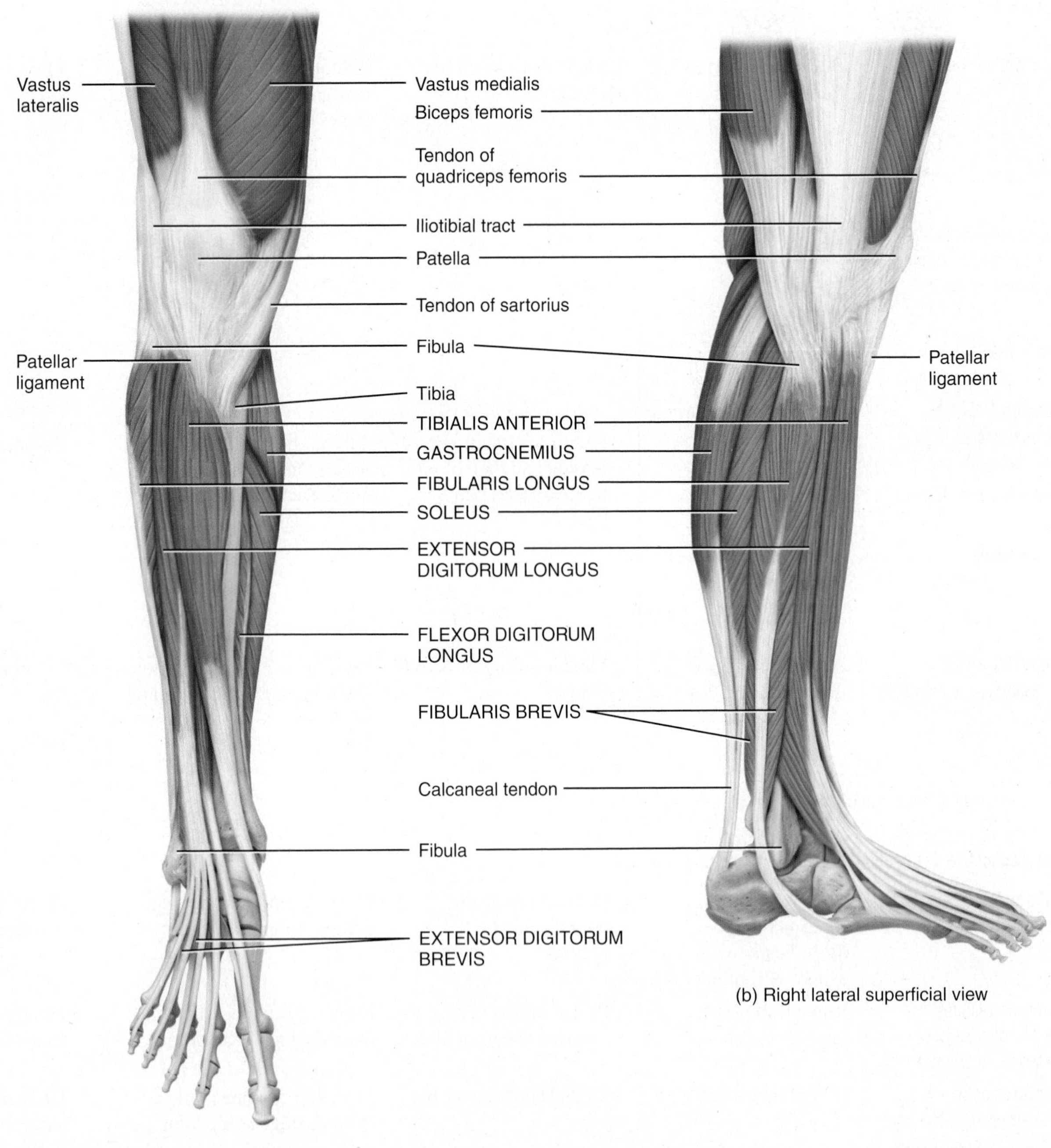


(a) Anterior superficial view

(b) Right lateral superficial view

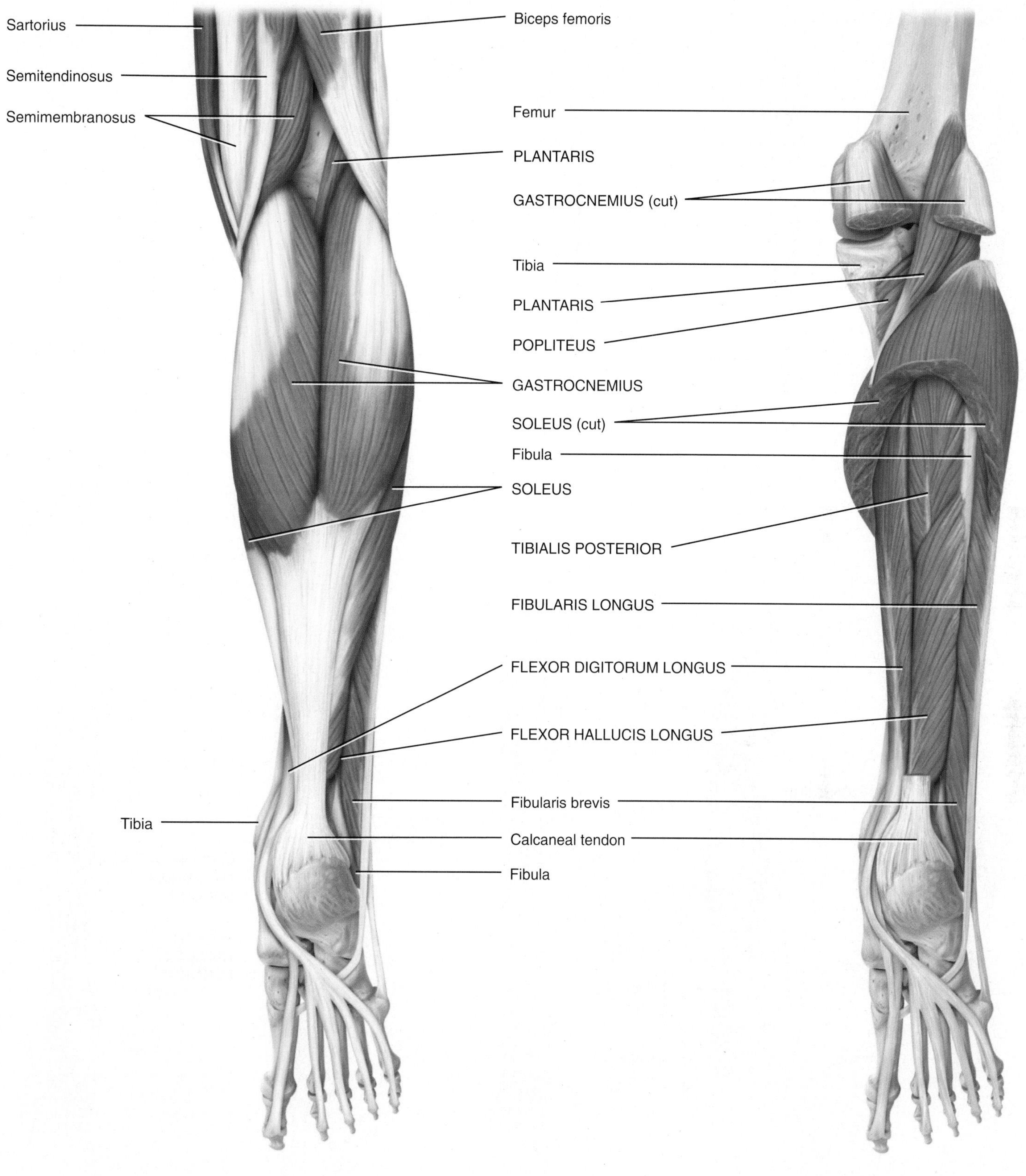


(c) Posterior superficial view

(d) Posterior deep view

(continues)

FIGURE 11.20 (continued)

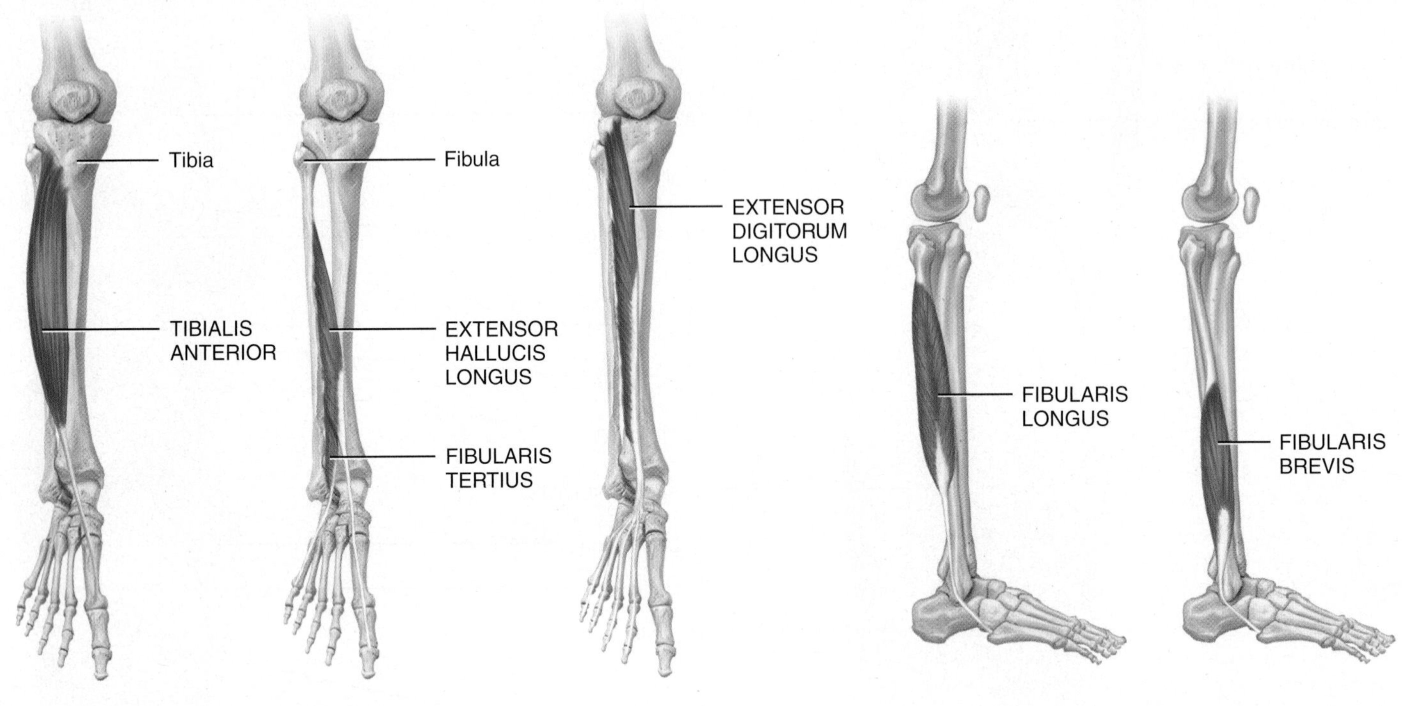


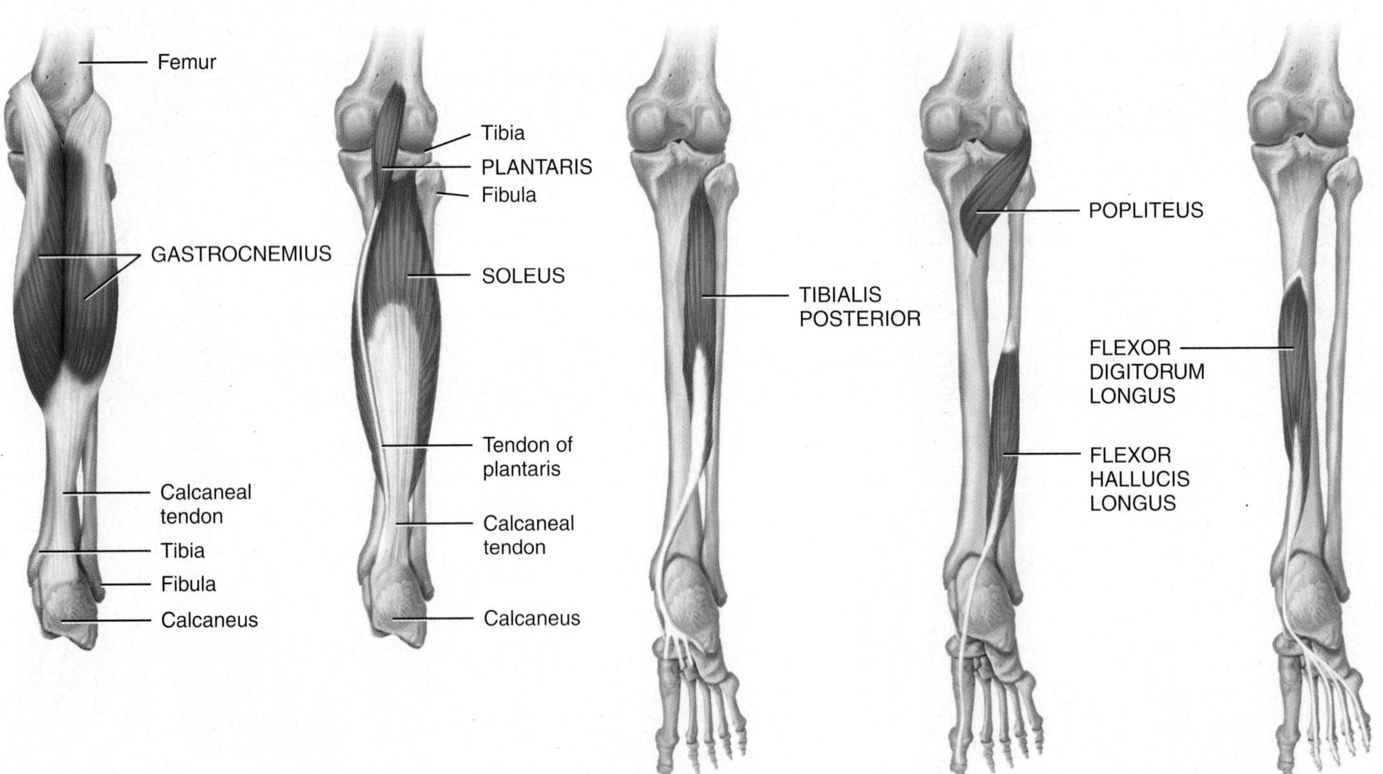


(e) Isolated muscles

**The superficial muscles of the posterior leg share a common tendon of insertion, the calcaneal (Achilles) tendon, which inserts into the calcaneus of the ankle.**

## TABLE 11.21

Muscles That Move the Foot and Toes

| MUSCLE | ORIGIN | INSERTION | ACTION | INNERVATION |
|---|---|---|---|---|
| **Anterior Compartment of the Leg** | | | | |
| **Tibialis anterior** (tib′-ē-Ā-lis = tibia; *anterior* = front) | Lateral condyle and body of tibia and interosseous membrane (sheet of fibrous tissue that holds shafts of tibia and fibula together) | Metatarsal I and first (medial) cuneiform | Dorsiflexes foot at ankle joint and inverts foot at intertarsal joints | Deep fibular (peroneal) nerve |
| **Extensor hallucis longus** (eks-TEN-sor HAL-ū-sis LON-gus; *extensor* = increases angle at joint; *halluc-* = hallux, or great toe; *longus* = long) | Anterior surface of fibula and interosseous membrane | Distal phalanx of great toe | Dorsiflexes foot at ankle joint and extends proximal phalanx of great toe at metatarsophalangeal joint | Deep fibular (peroneal) nerve |
| **Extensor digitorum longus** (di′-ji-TOR-um) | Lateral condyle of tibia, anterior surface of fibula, and interosseous membrane | Middle and distal phalanges of toes II–V* | Dorsiflexes foot at ankle joint and extends distal and middle phalanges of each toe at interphalangeal joints and proximal phalanx of each toe at metatarsophalangeal joint | Deep fibular (peroneal) nerve |
| **Fibularis (peroneus) tertius** (fib-ū-LĀ-ris TER-shus; *peron-* = fibula; *tertius* = third) | Distal third of fibula and interosseous membrane | Base of metatarsal V | Dorsiflexes foot at ankle joint and everts foot at intertarsal joints | Deep fibular (peroneal) nerve |
| **Lateral (Fibular) Compartment of the Leg** | | | | |
| **Fibularis (peroneus) longus** | Head and body of fibula and lateral condyle of tibia | Metatarsal I and first cuneiform | Plantar flexes foot at ankle joint and everts foot at intertarsal joints | Superficial fibular (peroneal) nerve |
| **Fibularis (peroneus) brevis** (BREV-is = short) | Body of fibula | Base of metatarsal V | Plantar flexes foot at ankle joint and everts foot at intertarsal joints | Superficial fibular (peroneal) nerve |
| **Superficial Posterior Compartment of the Leg** | | | | |
| **Gastrocnemius** (gas′-trok-NĒ-mē-us; *gastro-* = belly; *-cnem-* = leg) | Lateral and medial condyles of femur and capsule of knee | Calcaneus by way of calcaneal (Achilles) tendon | Plantar flexes foot at ankle joint and flexes leg at knee joint | Tibial nerve |
| **Soleus** (SŌ-lē-us; *sole* = a type of flat fish) | Head of fibula and medial border of tibia | Calcaneus by way of calcaneal (Achilles) tendon | Plantar flexes foot at ankle joint | Tibial nerve |
| **Plantaris** (plan-TĀR-is = the sole) | Femur superior to lateral condyle | Calcaneus by way of calcaneal (Achilles) tendon | Plantar flexes foot at ankle joint and flexes leg at knee joint | Tibial nerve |
| **Deep Posterior Compartment of the Leg** | | | | |
| **Popliteus** (pop-LIT-ē-us; *poplit-* = back of knee) | Lateral condyle of femur | Proximal tibia | Flexes leg at knee joint and medially rotates tibia to unlock the extended knee | Tibial nerve |
| **Tibialis posterior** (*posterior* = back) | Tibia, fibula, and interosseous membrane | Metatarsals II–IV; navicular; all three cuneiforms; and cuboid | Plantar flexes foot at ankle joint and inverts foot at intertarsal joints | Tibial nerve |
| **Flexor digitorum longus** (FLEK-sor = decreases angle at joint) | Posterior surface of tibia | Distal phalanges of toes II–V | Plantar flexes foot at ankle joint; flexes distal and middle phalanges of each toe at interphalangeal joints and proximal phalanx of each toe at metatarsophalangeal joint | Tibial nerve |
| **Flexor hallucis longus** | Inferior two-thirds of fibula | Distal phalanx of great toe | Plantar flexes foot at ankle joint; flexes distal phalanx of great toe at interphalangeal joint and proximal phalanx of great toe at metatarsophalangeal joint | Tibial nerve |

*Reminder: The great toe, or hallux, is the first toe and has two phalanges: proximal and distal. The remaining toes are numbered II–V (2–5), and each has three phalanges: proximal, middle, and distal.

**posterior** is the deepest muscle in the posterior compartment. It lies between the **flexor digitorum longus** and **flexor hallucis longus**, muscles that flex the toes.

## Intrinsic Muscles of the Foot

The muscles in Table 11.22 are termed **intrinsic foot muscles** because they originate and insert *within* the foot (Figure 11.21). The muscles of the hand are specialized for precise and intricate movements, but those of the foot are limited to support and locomotion. The intrinsic muscles of the foot primarily originate on the tarsal and metatarsal bones. They move the toes and contribute to the longitudinal arch of the foot. The fascia of the foot forms the *plantar aponeurosis*, which extends from the calcaneus to the phalanges of the toes. The plantar aponeurosis supports the longitudinal arch of the foot and encloses the flexor tendons of the foot.

FIGURE 11.21 **Intrinsic muscles of the foot.**

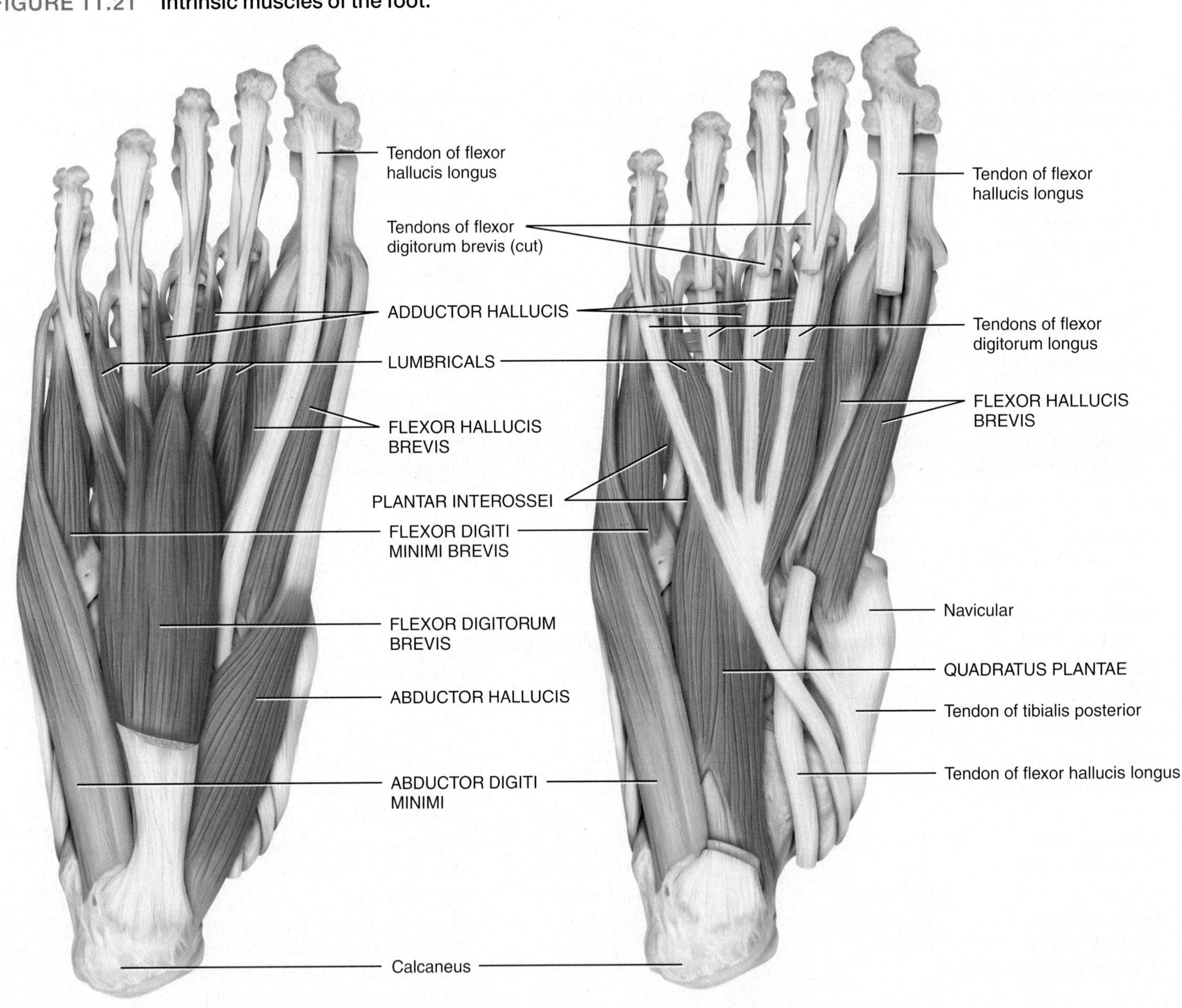


**The muscles of the hand are specialized for precise and intricate movements; those of the foot are limited to support and movement.**

**TABLE 11.22**

Intrinsic Muscles of the Foot

| MUSCLE | ORIGIN | INSERTION | ACTION | INNERVATION |
|---|---|---|---|---|
| **Dorsal** | | | | |
| **Extensor digitorum brevis** (eks-TEN-sor di-ji-TOR-um BREV-is; *extensor* = increases angle at joint; *digit* = finger or toe; *brevis* = short) (see Figure 11.20a) | Calcaneus and inferior extensor retinaculum | Tendons of extensor digitorum longus on toes II–IV and proximal phalanx of great toe* | Extends toes II–IV at interphalangeal joints | Deep fibular (peroneal) nerve |
| **Plantar** | | | | |
| **First Layer (most superficial)** | | | | |
| **Abductor hallucis** (ab-DUK-tor HAL-ū-sis; *abductor* = moves part away from midline; *hallucis* = hallux, or great toe) | Calcaneus, plantar aponeurosis, and flexor retinaculum | Medial side of proximal phalanx of great toe with the tendon of the flexor hallucis brevis | Abducts and flexes great toe at metatarsophalangeal joint | Medial plantar nerve |
| **Flexor digitorum brevis** (FLEK-sor = decreases angle at joint) | Calcaneus and plantar aponeurosis | Sides of middle phalanx of toes II–V | Flexes toes II–V at proximal interphalangeal and metatarsophalangeal joints | Medial plantar nerve |
| **Abductor digiti minimi** (DIJ-i-tē MIN-i-mē; *minimi* = little) | Calcaneus and plantar aponeurosis | Lateral side of proximal phalanx of little toe with the tendon of the flexor digiti minimi brevis | Abducts and flexes little toe at metatarsophalangeal joint | Lateral plantar nerve |
| **Second Layer** | | | | |
| **Quadratus plantae** (kwod-RĀ-tus plan-TĒ; *quad-* = square, four-sided; *planta* = the sole) | Calcaneus | Tendon of flexor digitorum longus | Assists flexor digitorum longus to flex toes II–V at interphalangeal and metatarsophalangeal joints | Lateral plantar nerve |
| **Lumbricals** (LUM-bri-kals; *lumbric-* = earthworm) | Tendons of flexor digitorum longus | Tendons of extensor digitorum longus on proximal phalanges of toes II–V | Extend toes II–V at interphalangeal joints and flex toes II–V at metatarsophalangeal joints | Medial and lateral plantar nerves |
| **Third Layer** | | | | |
| **Flexor hallucis brevis** | Cuboid and third (lateral) cuneiform | Medial and lateral sides of proximal phalanx of great toe via a tendon containing a sesamoid bone | Flexes great toe at metatarsophalangeal joint | Medial plantar nerve |
| **Adductor hallucis** (ad-DUK-tor = moves part toward midline) | Metatarsals 2–4, ligaments of 3–5 metatarsophalangeal joints, and tendon of peroneus longus | Lateral side of proximal phalanx of great toe | Adducts and flexes great toe at metatarsophalangeal joint | Lateral plantar nerve |
| **Flexor digiti minimi brevis** | Metatarsal 5 and tendon of peroneus longus | Lateral side of proximal phalanx of little toe | Flexes little toe at metatarsophalangeal joint | Lateral plantar nerve |
| **Fourth Layer (deepest)** | | | | |
| **Dorsal interossei** (DOR-sal in-ter-OS-ē-i; *dorsal* = back surface; *inter-* = between; *-ossei* = bones) (not illustrated) | Adjacent side of metatarsals | Proximal phalanges: both sides of toe II and lateral side of toes 3 and 4 | Abduct and flex toes II–IV at metatarsophalangeal joints and extend toes at interphalangeal joints | Lateral plantar nerve |
| **Plantar interossei** (PLAN-tar = relating to sole of foot) | Metatarsals 3–5 | Medial side of proximal phalanges of toes III–V | Adduct and flex proximal metatarsophalangeal joints and extend toes at interphalangeal joints | Lateral plantar nerve |

*The tendon that inserts into the proximal phalanx of the great toe, together with its belly, is often described as a separate muscle, the extensor hallucis brevis.

## ✓ CHECKPOINT

28. What are the principal differences between the muscles of the upper and lower limbs?
29. Which muscles extend the leg at the knee joint?
30. Name the muscles that are part of the iliopsoas, quadriceps femoris, and hamstrings.
31. Which muscle tendons form the medial and lateral borders of the popliteal fossa?
32. Name the muscles that dorsiflex the foot at the ankle joint and those that plantar flex it.
33. Which muscles flex the toes? Which extend the toes?
34. How do the intrinsic muscles of the hand and foot differ in function?

## Stewart's Story

## EPILOGUE AND DISCUSSION

Stewart and Wally are back in the gym, this time working out their lower limb muscles. Stewart is performing squats while Wally sits with a weight on his lap and is plantar flexing his feet. Stewart takes a break and gets a drink of water.

"It's funny how straightforward the lower limb lecture was, compared to the upper limb." Wally nods as he continues to lift the weight on his lap.

"Well, it makes sense when you realize how many more movements we perform with the upper limbs than with the lower limbs. Plus, we had already been through the upper limb muscles, so we had that information to apply to the lower limb muscles." Wally stands up to switch equipment with Stewart. As Stewart sits down with the weight on his lap, he jokes, "Well, at least you finally got to learn where that other *biceps* muscle is."

"Yeah, and I'm feeling it burn right now," Wally grunts as he lifts the weight on his shoulder through the squat. "It was neat that Dr. Kendrick answered all of our questions about weight-lifting muscles. She even gave us great advice for building the calf muscles." Stewart nods in agreement as he lifts the weight in his lap. In the seated position, he is working out the soleus muscle, found deep to the gastrocnemius muscle on the posterior calf.

As he feels the burn of the workout, Stewart is glad to have a better knowledge of the various muscles that move his body. Hearing Wally and Jenny's stories about injuries impresses upon him the idea of using his muscles intelligently so as to not damage them, so he knows how far he can push himself.

H. *Which lower limb biceps muscle are Stewart and Wally referring to? What other muscles are in the same group?*

I. *Why would Dr. Kendrick indicate that seated lifts build the calf muscles more than just squats?*

J. *What are the functions of the gluteus maximus and gluteus medius muscles? Which would be most sore from the squat exercises Stewart and Wally are doing?*

# Concept and Resource Summary

| Concept | Resources |
|---|---|

### Introduction

1. The **muscular system** is made up of the voluntarily controlled skeletal muscles of the body.
2. Most skeletal muscles produce movement of body parts; a few skeletal muscles stabilize bones during their movement.

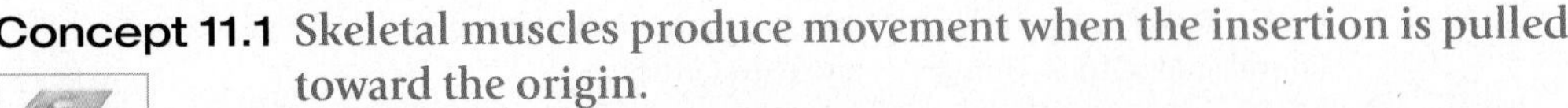

### Concept 11.1 Skeletal muscles produce movement when the insertion is pulled toward the origin.

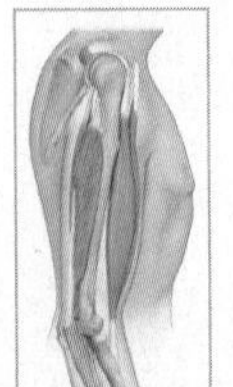

1. The contraction of skeletal muscles pulls on bones to cause movement. A muscle's **insertion** is its attachment on the movable bone that is typically moved toward its **origin**, which is its attachment on the immovable bone.
2. A **lever** is a rigid structure that moves around a fixed point called the **fulcrum**. **Effort** applied to the lever moves the **load**, or resistance. Bones act as levers, joints are the fulcrums, and skeletal muscles supply the effort.

**Resources:**

Anatomy Overview—Skeletal Muscle

Figure 11.1—Relationship of Skeletal Muscles to Bones

Concept | Resources

3. When a smaller effort can move a heavier load, the lever operates at a mechanical advantage. A lever operates at a mechanical disadvantage when a larger effort moves a lighter load.
4. The arrangements of **fascicles** include parallel, fusiform, circular, triangular, and pennate.
5. The muscle that accomplishes a desired action is termed the **prime mover (agonist)**, and the muscle that accomplishes the opposing action is the **antagonist**. **Synergists** assist the prime mover and stabilize intermediate joints to prevent unwanted movements. **Fixators** stabilize the origin of the prime mover to increase efficiency.

## Concept 11.2 Skeletal muscles are named based on size, shape, action, location, or attachments.

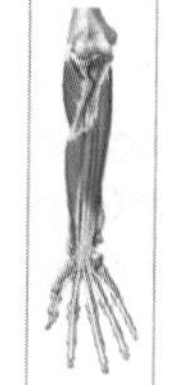

1. Skeletal muscle names are derived from specific features such as size, action, and attachments.

Figure 11.2—Principal Superficial Skeletal Muscles

## Concept 11.3 Muscles of the head produce facial expressions, eyeball movement, and assist in biting, chewing, swallowing, and speech.

1. Muscles of facial expression move the skin or muscles rather than bone when they contract. Around the eyes, nose, and mouth are sphincter muscles that encircle and close the orifices; dilator muscles open the orifices.
2. The **extrinsic eye muscles** are among the fastest contracting and most precisely controlled skeletal muscles in the body. They permit you to elevate, depress, abduct, adduct, and medially and laterally rotate the eyeballs.
3. Muscles for mastication (chewing) move the mandible. The **masseter**, **temporalis**, and **medial pterygoid** elevate the mandible for powerful closing of the jaw. The **medial** and **lateral pterygoids** move the mandible from side to side and protract it.
4. Tongue movements are important for mastication, swallowing, and speech. Extrinsic tongue muscles originate outside the tongue and insert into it to move it in various directions. Intrinsic tongue muscles originate and insert within the tongue to change its shape.

Anatomy Overview—Muscles of Facial Expression
Anatomy Overview—Muscles Moving Eyeballs
Anatomy Overview—Muscles for Speech, Swallowing, and Chewing

## Concept 11.4 Muscles of the neck assist in swallowing and speech, and allow balance and movement of the head.

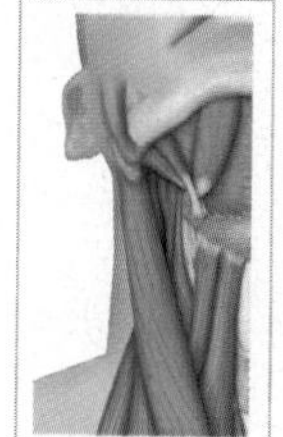

1. Two groups of muscles stabilize and move the hyoid bone, enabling it to serve as a firm base for attachment and action of the tongue. The **suprahyoid muscles** elevate the hyoid bone, oral cavity floor, and tongue during swallowing; the **infrahyoid muscles** depress the hyoid bone, and some move the larynx during swallowing and speech.
2. Several neck muscles balance and move the head on the vertebral column. Bilateral action of the sternocleidomastoids results in flexion of the head; unilateral action results in head rotation.

Anatomy Overview—Muscles That Move the Head
Anatomy Overview—Muscles for Speech, Swallowing, and Chewing

## Concept 11.5 Muscles of the abdomen protect the abdominal viscera, move the vertebral column, and assist breathing.

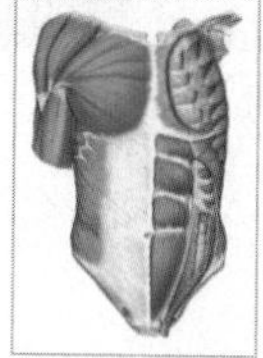

1. Muscles that act on the anterolateral abdominal wall help contain and protect the abdominal viscera, move the vertebral column, compress the abdomen, and produce the force required for defecation, urination, vomiting, and childbirth.
2. The **diaphragm**, a large, dome-shaped muscle that separates the thoracic and abdominal cavities, is the most important muscle for breathing. Contraction results in inhalation; relaxation results in exhalation.
3. The **external intercostals** elevate the ribs during inhalation. The **internal intercostals** help decrease the thoracic cavity volume during forced exhalation.

Anatomy Overview—Muscles of the Torso
Anatomy Overview—Muscles for Breathing
Figure 11.10—Muscles Used in Breathing

**Concept** | **Resources** 

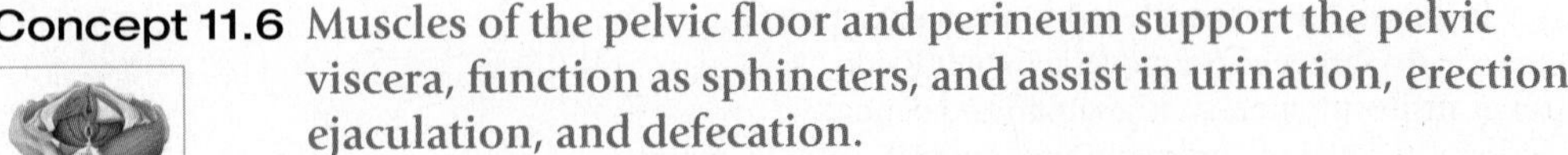

## Concept 11.6 Muscles of the pelvic floor and perineum support the pelvic viscera, function as sphincters, and assist in urination, erection, ejaculation, and defecation.

1. The **pelvic diaphragm** supports pelvic viscera, resists increases in intra-abdominal pressure, and acts as a sphincter for defecation, urination, and vaginal control.
2. The **perineum**, a diamond-shaped region inferior to the pelvic diaphragm, is important during childbirth. Muscles of the perineum also assist in erection of the penis and clitoris, ejaculation and urination, and defecation.

---

## Concept 11.7 Muscles inserting on the upper limb move and stabilize the pectoral girdle, and move the arm, forearm, and hand.

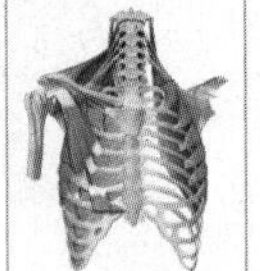

1. Muscles that move the pectoral girdle stabilize the scapula and facilitate its function as the origin for most of the muscles that move the humerus.
2. Seven of the nine muscles that cross the shoulder joint originate on the scapula; two originate on the axial skeleton. Tendons of several shoulder muscles form the rotator cuff, which encircles the shoulder joint to give it strength and stability.
3. The flexors of the elbow joint are the **biceps brachii**, **brachialis**, and **brachioradialis**. The extensors of the elbow joint are the **triceps brachii** and **anconeus**. The **pronator teres** and **pronator quadratus** pronate the forearm; the **supinator** allows supination of the forearm.
4. The numerous muscles that move the wrist, hand, and fingers are located on the forearm. The anterior forearm muscles act as flexors; the posterior forearm muscles act as extensors. The fascia at the wrist is thickened into fibrous bands called the flexor retinaculum and extensor retinaculum, which secure the tendons of certain forearm muscles.
5. **Intrinsic muscles of the hand** originate within the hand and provide us with the ability to grasp and manipulate objects precisely.

Resources:
- Anatomy Overview—Muscles That Move the Pectoral Girdle
- Anatomy Overview—Muscles That Move the Arm
- Anatomy Overview—Muscles That Move the Forearm
- Anatomy Overview—Muscles That Move the Hand
- Figure 11.16—Muscles That Move the Wrist, Hand, and Fingers
- Exercise—Shoulder Movement Target Practice

---

## Concept 11.8 Deep muscles of the back move the head and vertebral column.

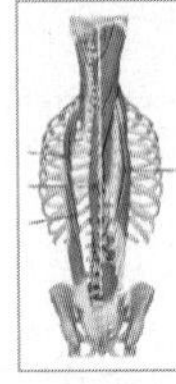

1. The **splenius** muscles attach to the sides and back of the neck for head extension, flexion, and rotation. The largest muscle mass of the back, the **erector spinae** group, consists of the **ilocostalis group**, **longissimus group**, and **spinalis group**. The erector spinae muscles are prime movers of the vertebral column, allowing back extension. The **transversospinales** and **segmental** muscles also function in vertebral column movements.

Resources:
- Anatomy Overview—Muscles of the Torso

---

## Concept 11.9 Muscles originating on the pelvic girdle or lower limb move the femur, leg, and foot.

1. Most muscles that move the thigh at the hip originate on the pelvic girdle and insert on the femur. The **iliopsoas** flexes the thigh. The **adductor longus**, **adductor brevis**, and **adductor magnus** adduct, medially rotate, and flex the thigh. The **gluteus medius** and **gluteus minimus** abduct the thigh at the hip. The **tensor fascia latae** flexes and abducts the thigh.
2. The **gracilis** is a long, straplike muscle that adducts the thigh and flexes the leg at the knee. The **quadriceps femoris** muscle group is a powerful extensor of the leg at the knee. The **hamstrings** muscle group on the posterior thigh flexes the leg at the knee and extends the thigh at the hip.
3. Muscles that move the foot at the ankle and toes are located in the leg. The **tibialis anterior**, **fibularis tertius**, **extensor hallucis longus**, and **extensor digitorum longus** dorsiflex the foot. The **tibialis posterior**, **fibularis longus**, and **fibularis brevis** plantar flex and evert the foot. The **gastrocnemius**, **soleus**, and **plantaris** also plantar flex the foot. The **flexor digitorum longus** and **flexor hallucis longus** flex the toes.
4. **Intrinsic foot muscles** originate and insert within the foot to move the toes and contribute to the longitudinal arch of the foot.

Resources:
- Anatomy Overview—Muscles That Move the Thigh
- Anatomy Overview—Muscles of the Foot
- Figure 11.19—Muscles That Move the Femur
- Exercise—Lower Limb Movement Target Practice
- Exercise—Manage Those Muscles

## Understanding the Concepts

1. A soccer player kicks a soccer ball down the field, flexing his thigh forward at the hip joint while keeping his knee joint locked in extension. What is the lever? The fulcrum? The effort? The load?
2. Two girls sit down on a playground seesaw, or teeter-totter. Both children weigh the same amount. Maria sits at the end of her side of the seesaw while Tonya sits halfway between the end and the fulcrum. Which girl will exert more force and be lowered to the ground?
3. What characteristics are used to name the quadriceps femoris? The adductor longus?
4. What would happen if an injury resulted in loss of nerve stimulation (innervation) to your masseter and temporalis?
5. When your physician says "Open your mouth, stick out your tongue, and say *ahh*" so she can examine the inside of your mouth for possible signs of infection, which muscles do you contract?
6. Which muscles do you contract to signify "yes"? Which do you contract to signify "no"?
7. Which muscles do you contract when you "suck in your tummy," thereby compressing the anterior abdominal wall? Which do you contract when you rotate the vertebral column?
8. Which muscles do you contract to increase the dimension of your thoracic cavity during quiet, normal inhalation? When you exhale?
9. Name the muscles you use to raise your shoulders and those you use to lower your shoulders.
10. Extend your hand at the wrist. Which muscles are contracting? Which muscles must relax so that you can extend your hand?
11. Which muscles do you exercise when performing sit-ups? Why might your ribs become sore from doing stomach exercises?
12. Which busy muscles flex, adduct, and medially rotate the femur at the hip joint as you rush to class? Which one flexes, abducts, and laterally rotates the thigh at the hip joint and flexes the leg at the knee joint?

CHAPTER

# 12

## Jennifer's Story

Jennifer looks at her watch and breathes a sigh of relief. The therapy department would close in fifteen minutes and she would be able to go home. As she watches her patient do his strengthening exercises Jennifer thinks about all of the things on her "to do" list. Jennifer says goodbye to Mr. Pacheco, sets up his next appointment, and reminds him to practice his exercises at home. Then she grabs her coat and rushes to the subway to take her train to Brookline.

Jennifer is fortunate enough to get a seat so she can relax on her 45-minute train ride. She feels more tired than usual and realizes that she has been feeling fatigued for weeks. She chalks it up to being a senior physical therapy student in her final semester of college with extensive clinical hours to complete.

After dinner, Jennifer opens her textbooks, anticipating several hours of study. She notices that her vision is a little blurry, and she is having trouble reading. She finally decides to go to bed and study in the morning when she is fresh.

Over the next few weeks, Jennifer experiences additional episodes of blurred vision, so she makes an appointment with her primary care provider to get a referral to an ophthalmologist (eye doctor), thinking she needs glasses. However, before her appointment she notices that she sometimes has trouble grasping her pen and even drops her favorite coffee cup one morning, smashing it to smithereens. She realizes that she does not have normal feeling in her right hand, and adds this new problem to her list to discuss with the doctor.

When Jennifer relates the details, Dr. Marino agrees to give her an ophthalmology referral but says he wants to examine her first. He tests her pupillary response and the movements of her eyeballs. He also checks the strength and reflexes in all of her limbs and even uses a cotton ball and a paperclip to test the level of sensation in her arms and legs. As he gives her the ophthalmology referral, he tells her that he also wants her to see a neurologist.

To understand Jennifer's case, you need to know how the nervous system receives information from the body, processes it, and formulates a response. Is Jennifer just tired? Does she need glasses? Why do you think she is having problems holding things in her hands? Read this chapter to learn the answers to these questions and discover the nature of Jennifer's problem.

# Introduction to the Nervous System

## INTRODUCTION

The nervous system and endocrine system share the responsibilities of maintaining the homeostasis of the human body. Both of these systems must sense changes in normal physiological set points (such as the maintenance of normal body temperature and blood pressure), integrate the information they are receiving, and respond by making changes that will keep controlled conditions within limits that maintain life. The nervous system regulates body activities by responding rapidly through impulses; the endocrine system responds more slowly, though no less effectively, by releasing hormones.

Very simply put, all you see, feel, think, and do is controlled by your nervous system. Because the nervous system is quite complex, we will consider different aspects of its structure and function in several related chapters. In this chapter we focus on the organization of the nervous system and the properties of the cells that make up nervous tissue—neurons (nerve cells) and neuroglia (cells that support the activities of neurons). In chapters that follow, we will examine the structure and functions of the brain and spinal cord (Chapter 13), and of the cranial nerves and spinal nerves (Chapter 14). Chapter 15 will discuss the somatic senses—touch, pressure, warmth, cold, pain, and others—and their sensory and motor pathways to explain how impulses pass into the spinal cord and brain or from the spinal cord and brain to muscles and glands. Our exploration of the nervous system concludes with an investigation of the special senses: smell, taste, vision, hearing, and equilibrium (Chapter 16).

## CONCEPTS

**12.1** The nervous system maintains homeostasis and integrates all body activities.

**12.2** The nervous system is organized into the central and peripheral nervous systems.

**12.3** Neurons are responsible for most of the unique functions of the nervous system.

**12.4** Neuroglia support, nourish, and protect neurons and maintain homeostasis.

**12.5** Neurons communicate with other cells.

**12.6** Graded potentials are the first response of a neuron to stimulation.

**12.7** The action potential is an all-or-none electrical signal.

**12.8** Action potentials propagate from the trigger zone to axon terminals.

**12.9** The synapse is a special junction between neurons.

**12.10** PNS neurons have a greater capacity for repair and regeneration than CNS neurons.

# 12.1 The nervous system maintains homeostasis and integrates all body activities.

With a mass of only 2 kg (4.5 lb), about 3 percent of the total body weight, the nervous system is one of the smallest and yet the most complex of the 11 body systems. The nervous system is responsible for all of your perceptions, behaviors, memories, and movements. To accomplish these functions, the nervous system carries out a complex array of tasks, such as sensing various smells, producing speech, remembering past events, providing signals that control body movements, and regulating the operation of internal organs. These diverse activities can be grouped into three basic functions:

- ***Sensory function.*** Sensory receptors *detect* internal stimuli, such as an increase in blood acidity, and external stimuli, such as a raindrop landing on your arm. This **sensory** information is then carried into the brain and spinal cord through cranial and spinal nerves.
- ***Integrative function.*** The nervous system *processes* sensory information by analyzing and storing some of it, and by making decisions for appropriate responses—an activity known as **integration**.
- ***Motor function.*** Once sensory information is integrated, the nervous system may *elicit an appropriate motor response* by activating **effectors** (muscles and glands) through cranial and spinal nerves. Stimulation of the effectors causes muscles to contract and glands to secrete.

The three basic functions of the nervous system occur, for example, when you answer your cell phone after hearing it ring. The sound of the ringing cell phone stimulates sensory receptors in your ears (sensory function). This auditory information is subsequently relayed into your brain where it is processed and the decision to answer the phone is made (integrative function). The brain then stimulates the contraction of specific muscles that will allow you to grab the phone and press the appropriate button to answer it (motor function).

✓ **CHECKPOINT**

1. How would each of the three functions of the nervous system be utilized when you realize it is time for lunch?

# 12.2 The nervous system is organized into the central and peripheral nervous systems.

The nervous system consists of an intricate network of billions of neurons and even more neuroglia and can be organized into two main subdivisions: the central nervous system and the peripheral nervous system (Figure 12.1).

## Central Nervous System

The **central nervous system (CNS)** consists of the brain and spinal cord (Figure 12.1). The **brain** is the part enclosed within the skull and contains about 100 billion neurons. The **spinal cord** is connected to the brain through the foramen magnum of the occipital bone and is encircled by the bones of the vertebral column. The spinal cord contains about 100 million neurons. The CNS processes many different kinds of incoming sensory information. It is also the source of thoughts, emotions, and memories. Most signals that stimulate muscles to contract and glands to secrete originate in the CNS.

## Peripheral Nervous System

The **peripheral nervous system (PNS)** (pe-RIF-e-ral) consists of all nervous tissue outside the CNS (Figure 12.1). Components of the PNS include nerves, ganglia, enteric plexuses, and sensory receptors. A **nerve** is a bundle of hundreds to thousands of axons plus associated connective tissue and blood vessels that lie outside the brain and spinal cord. **Cranial nerves** emerge from the brain and **spinal nerves** emerge from the spinal cord. Each nerve follows a defined path and serves a specific region of the body. **Ganglia** (GANG-lē-a = swelling or knot; singular is *ganglion*) are small clusters of nervous tissue, consisting primarily of neuron cell bodies, that are located outside of the brain and spinal cord. Ganglia are closely associated with cranial and spinal nerves. **Enteric plexuses** (PLEK-sus-ēz; singular is *plexus*) are extensive networks of neurons located in the walls of organs of the gastrointestinal tract. The neurons of these plexuses help regulate the digestive system (see Concept 23.1). **Sensory receptors** are nervous system structures that monitor changes in the external or internal environment. Examples of sensory receptors include touch receptors in the skin, photoreceptors in the eye, and olfactory receptors in the nose.

The PNS is divided into a **somatic nervous system (SNS)** (sō-MAT-ik; *somat-* = body), an **autonomic nervous system (ANS)** (aw'-tō-NOM-ik; *auto-* = self; *-nomic* = law), and an **enteric nervous system (ENS)** (en-TER-ik; *enteron-* = intestines) (Figure 12.2). The SNS consists of sensory neurons that convey information to the CNS from somatic receptors in the head, body wall, and limbs and from receptors for the special senses of vision, hearing, taste, and smell. The SNS also includes motor neurons that conduct impulses from the CNS to *skeletal muscles* only. Because these motor responses can be consciously controlled, the action of this part of the PNS is *voluntary*.

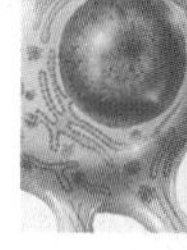

FIGURE 12.1 Major structures of the nervous system.

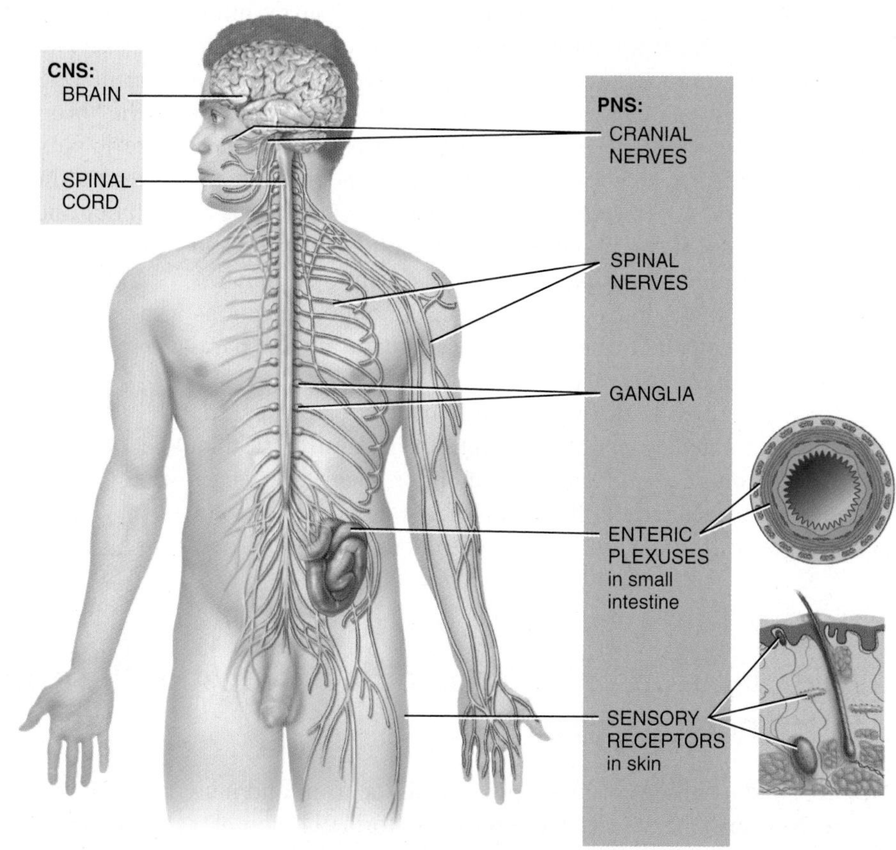


The central nervous system (CNS) consists of the brain and spinal cord; the peripheral nervous system (PNS) consists of all nervous tissue outside the CNS.

FIGURE 12.2 Organization of the nervous system. Blue boxes represent sensory components of the peripheral nervous system (PNS); red boxes represent motor components of the PNS; and green boxes represent effectors (muscles and glands).

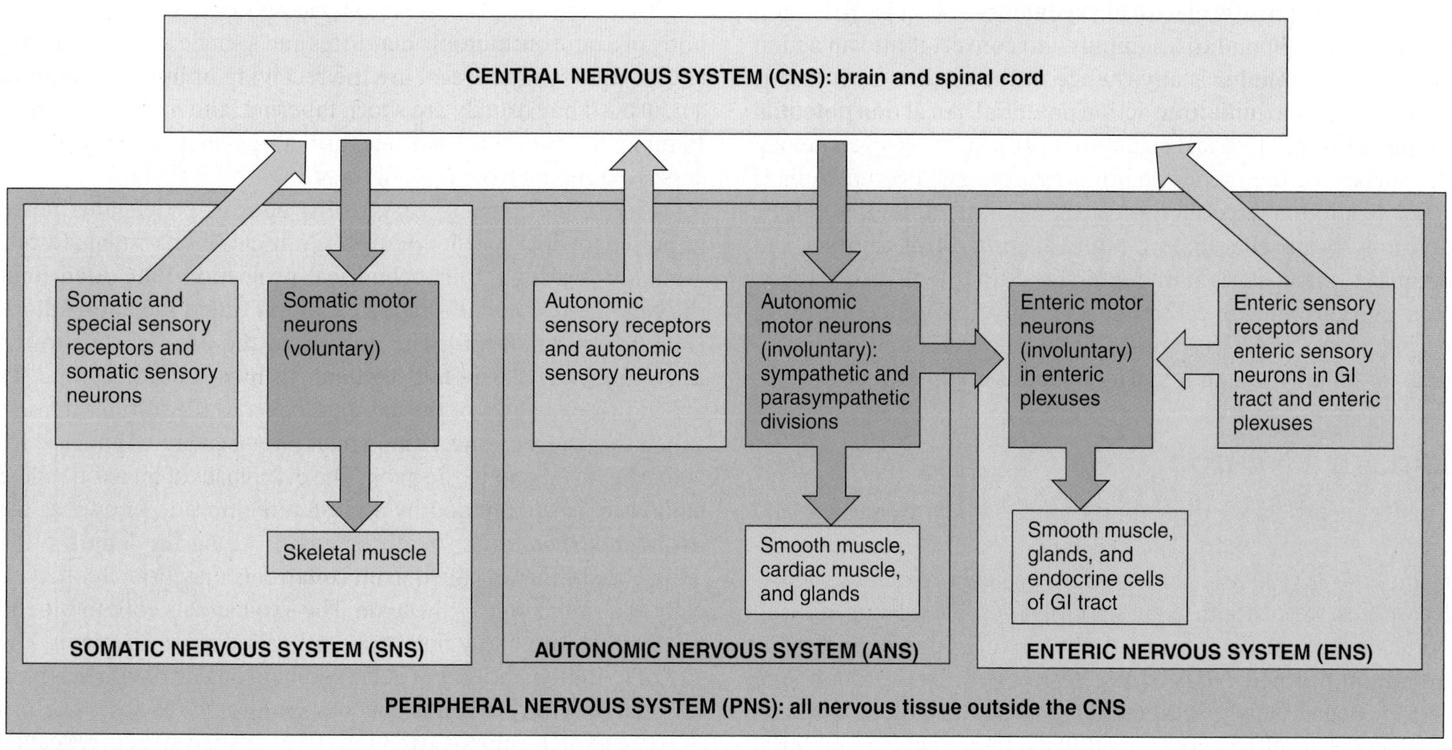


The SNS is under voluntary control; the ANS and ENS are under involuntary control.

The ANS consists of sensory neurons that convey information to the CNS from autonomic sensory receptors, located primarily in visceral organs such as the stomach and lungs. The ANS also consists of motor neurons that conduct nerve impulses from the CNS to *smooth muscle, cardiac muscle*, and *glands*. Because its motor responses are not normally under conscious control, the action of the ANS is *involuntary*. The motor part of the ANS consists of two branches, the **sympathetic division** and the **parasympathetic division**. With a few exceptions, effectors are stimulated by both divisions, usually with opposing actions. For example, sympathetic neurons increase heart rate, and parasympathetic neurons slow it down.

The ENS consists of over 100 million neurons in enteric plexuses that extend most of the length of the gastrointestinal (GI) tract. Enteric sensory neurons monitor chemical changes within the GI tract and the stretching of its walls. Enteric motor neurons govern contraction of GI tract smooth muscle to propel food through the GI tract, secretions of the GI tract organs such as acid secretion by the stomach, and activity of GI tract endocrine cells, which secrete hormones. The operation of the ENS, the "brain of the gut," is involuntary. Many neurons of the enteric plexuses function independently of the CNS to some extent, although they may communicate with the CNS via sympathetic and parasympathetic neurons.

✓ **CHECKPOINT**

2. Would a receptor more likely be located in your eyes, brain, or legs?
3. List the components of the CNS and the PNS.
4. What are the components and functions of the SNS, ANS, and ENS? Which subdivisions have voluntary actions? Involuntary actions?

## 12.3 Neurons are responsible for most of the unique functions of the nervous system.

Nervous tissue consists of two types of cells: neurons and neuroglia. **Neurons (nerve cells)** (NOO-rons) form the complex processing networks within the brain and spinal cord and also connect all regions of the body to the brain and spinal cord. Neurons carry out most of the unique functions of the nervous system, such as sensing, thinking, remembering, controlling muscle activity, and regulating glandular secretions. Like muscle cells, neurons possess **electrical excitability** (ek-sīt′-a-BIL-i-tē), the ability to respond to a stimulus and convert it into an action potential. A **stimulus** is any change in the environment that is strong enough to initiate an action potential. An **action potential (impulse)** is an electrical signal that propagates (travels) along the surface of the membrane of a neuron. As a result of their specialization, neurons have lost the ability to undergo mitotic divisions. **Neuroglia** support, nourish, and protect neurons, and maintain homeostasis in the interstitial fluid that bathes them. Neuroglia are smaller cells but they greatly outnumber neurons, perhaps by as much as 25 times. Unlike neurons, neuroglia continue to divide throughout an individual's lifetime.

### Parts of a Neuron

Most neurons have three parts: a cell body, dendrites, and an axon (Figure 12.3). The **cell body** contains a nucleus surrounded by cytoplasm that contains typical organelles such as lysosomes, mitochondria, and a Golgi complex. Neuronal cell bodies also contain prominent clusters of rough endoplasmic reticulum, termed **Nissl bodies** (NIS-el), where protein synthesis occurs. Newly synthesized proteins produced by Nissl bodies are used to replace cellular components, as material for growth of neurons, and to regenerate damaged axons in the PNS. The cytoskeleton includes both **neurofibrils** (noo-rō-FĪ-brils), composed of bundles of intermediate filaments that provide shape and support to the cell, and **microtubules** (mi′-krō-TOO-būls), which assist in moving materials between the cell body and axon. Most neurons cannot divide (are *amitotic*) because they lack the centrioles that are essential for mitosis to occur.

Two kinds of processes (extensions) emerge from the cell body of a neuron: multiple dendrites and a single axon. **Dendrites** (DEN-drīts = little trees) are the receiving or input portions of a neuron. They usually are short, tapering, and highly branched. In many neurons the dendrites form a tree-shaped array of processes extending from the cell body.

The single **axon** (= axis) of a neuron propagates nerve impulses toward another neuron, a muscle fiber, or a gland cell. An axon is a long, thin, cylindrical projection that often joins the cell body at a cone-shaped elevation called the **axon hillock** (HIL-lok = small hill). The portion of the axon closest to the axon hillock is the **initial segment**. In most neurons, impulses arise at the junction of the axon hillock and the initial segment, called the **trigger zone**, from which impulses are conducted toward the distal end of the axon. The cytoplasm of an axon, called **axoplasm**, is surrounded by a plasma membrane known as the **axolemma** (*lemma* = sheath or husk). Along the length of an axon, side branches called **axon collaterals** may branch off, typically at a right angle to the axon. The axon and its collaterals end by dividing into many fine processes called **axon terminals**. The site of communication between two neurons or between a neuron and an effector cell is called a **synapse** (SIN-aps). The tips of some axon terminals swell into bulb-shaped structures called **synaptic end bulbs**.

FIGURE 12.3 **Structure of a typical motor neuron and Schwann cell.** Arrows indicate the direction of information flow: dendrites → cell body → axon → axon terminals.

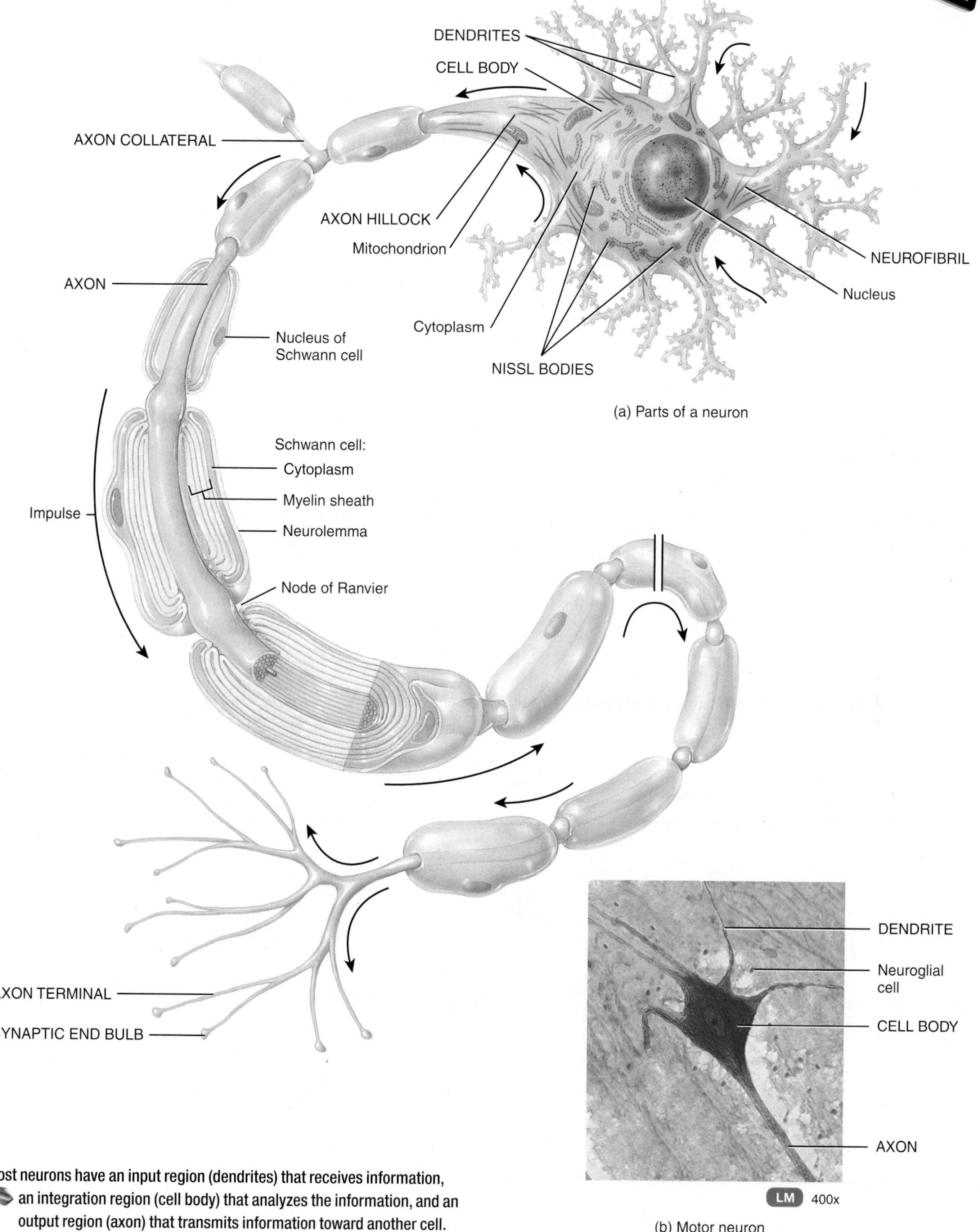


(a) Parts of a neuron

(b) Motor neuron

**Most neurons have an input region (dendrites) that receives information, an integration region (cell body) that analyzes the information, and an output region (axon) that transmits information toward another cell.**

The cell body is the portion of a neuron that synthesizes new cell products or recycles old ones. Newly synthesized substances are conveyed in the axoplasm from the cell body toward the axon terminals (**anterograde transport**; *antero-* = forward) for developing or regenerating axons and to replenish axoplasm in mature axons. Some substances are transported through axons by "motor" proteins that move substances along the surfaces of microtubules. Materials that form the membranes of the axolemma, synaptic end bulbs, and synaptic vesicles utilize microtubule transport. Some materials are transported from the axon terminal back to the cell body (**retrograde transport**; *retro-* = backward) to be degraded or recycled; others influence neuronal growth. Damage to an axon can impede the flow of axoplasm, leading to altered transmission of chemical and electrical signals.

## Structural Diversity and Classification of Neurons

Neurons display great diversity in size and shape. For example, neuron cell bodies range in diameter from 5 micrometers (μm) (slightly smaller than a red blood cell) up to 135 μm (barely large enough to see with the unaided eye). The pattern of dendritic branching is varied and distinctive for neurons in different parts of the nervous system. A few small neurons lack an axon, and many others have very short axons. The longest axons, however, are almost as long as a person is tall, extending from the toes to the lowest part of the brain.

Both structural and functional features are used to classify the various neurons in the body. Structurally, neurons are classified according to the number of processes extending from the cell body (Figure 12.4).

- **Multipolar neurons** usually have several dendrites and one axon (Figure 12.4a). Most neurons in the brain and spinal cord are of this type.
- **Bipolar neurons** have one main dendrite and one axon (Figure 12.4b). They are found in the retina of the eye, in the inner ear, and in the olfactory area (*olfact-* = to smell) of the brain.
- **Unipolar neurons** have dendrites and one axon that are fused together to form a continuous process that emerges from the cell body (Figure 12.4c). The dendrites of most unipolar neurons function as sensory receptors that detect a sensory stimulus such as touch, pressure, pain, or thermal stimuli (see Figure 12.9). The trigger zone for impulses in a unipolar neuron is at the junction of the dendrites and axon. The impulses then propagate toward the synaptic end bulbs. The cell bodies of most unipolar neurons are located in the ganglia of spinal and cranial nerves.

Functionally, neurons are classified according to the direction in which the impulse (action potential) is conveyed with respect to the CNS.

- **Sensory** or **afferent neurons** (AF-e-rent NOO-ronz; *af-* = toward; *-ferrent* = carried) either contain sensory receptors at

**FIGURE 12.4 Structural classification of neurons.** Breaks indicate that axons are longer than shown.

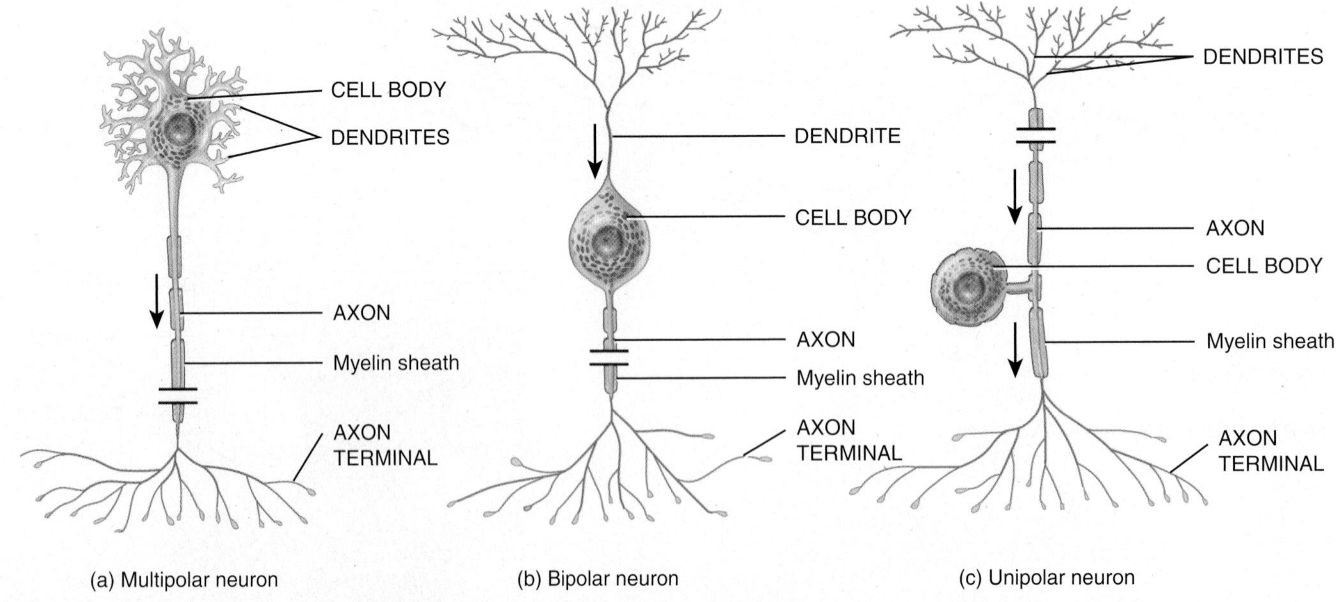


A multipolar neuron has many processes extending from the cell body, a bipolar neuron has two, and a unipolar neuron has one.

their distal ends (dendrites) (see blue neuron in Figure 12.9) or synapse with sensory receptors that are separate cells. Once an appropriate stimulus activates a sensory receptor, the sensory neuron forms an action potential in its axon and the action potential is conveyed *into* the CNS through cranial or spinal nerves. Most sensory neurons are unipolar in structure.

- **Motor** or **efferent neurons** (EF-er-ent; *ef-* = away from) convey action potentials away from the CNS to **effectors** (muscles and glands) in the periphery (PNS) through cranial or spinal nerves (see red neurons in Figure 12.9). Motor neurons are multipolar in structure.
- **Interneurons** are mainly located within the CNS between sensory and motor neurons (see purple neurons in Figure 12.9). Interneurons integrate (process) incoming sensory information from sensory neurons and then elicit a motor response by activating the appropriate motor neurons. Most interneurons are multipolar in structure.

✓ **CHECKPOINT**

5. Describe the parts of a neuron and the functions of each.
6. Name the location where a neuron and a muscle cell interact.
7. What roles do the dendrites, cell body, and axon play in communication of signals?
8. Describe three structural types of neurons.
9. What kind of neuron carries input toward your CNS? What kind of neuron carries input toward your PNS?

## 12.4 Neuroglia support, nourish, and protect neurons and maintain homeostasis.

**Neuroglia** (noo-RŌG-lē-a; *-glia* = glue), or simply **glia**, make up about half the volume of the CNS (Figure 12.5). Their name derives from the idea of early histologists that they were the "glue" that held nervous tissue together. We now know that neuroglia are not merely passive bystanders but rather active participants in the activities of nervous tissue. Generally, neuroglia are smaller than neurons, and they are 5 to 50 times more numerous (see Figure 12.3b). In contrast to neurons, glia do not generate or propagate action potentials, and they can readily multiply and divide. In cases of injury or disease, neuroglia multiply to fill in the spaces formerly occupied by neurons. Brain tumors derived from glia, called *gliomas* (glī-Ō-mas), tend to be highly malignant and to grow rapidly. Four of the six types of neuroglia—astrocytes, oligodendrocytes, microglia, and ependymal cells—are found only in the CNS. The remaining two types—Schwann cells and satellite cells—are present in the PNS.

### Neuroglia of the CNS

Neuroglia of the CNS include astrocytes, oligodendrocytes, microglia, and ependymal cells (Figure 12.5).

- **Astrocytes** (AS-trō-sīts; *astro-* = star; *-cyte* = cell), star-shaped cells with many processes, are the largest and most numerous of the neuroglia. The processes of astrocytes make contact with blood capillaries, neurons, and the *pia mater* (a thin membrane around the brain and spinal cord). Astrocytes cling to and support neurons. They help to maintain the appropriate chemical environment for the generation of impulses by providing nutrients to neurons, removing excess neurotransmitters (described shortly), and regulating the concentration of important ions. Astrocyte processes wrap around blood capillaries to inhibit movement of potentially harmful substances in blood, creating a *blood–brain barrier* that restricts the movement of substances between the blood and neurons of the CNS. Details of the blood–brain barrier are discussed in Concept 13.2.
- **Oligodendrocytes** (OL-i-gō-den′-drō-sīts; *oligo-* = few; *-dendro-* = tree) resemble astrocytes but are smaller and contain fewer processes. Oligodendrocyte processes are responsible for forming and maintaining the myelin sheath (described shortly) around CNS axons.
- **Microglial cells** or **microglia** (mī-KROG-lē-a; *micro-* = small) are small cells with slender processes that give off numerous spinelike projections. Microglia phagocytize microbes and damaged nervous tissue.
- **Ependymal cells** (ep-EN-de-mal; *epen-* = above; *-dym-* = garment) are cuboidal to columnar cells arranged in a single layer that have microvilli and cilia. These cells line the *ventricles* of the brain and *central canal* of the spinal cord (spaces filled with cerebrospinal fluid, which protects and nourishes the brain and spinal cord). Ependymal cells produce and assist in the circulation of cerebrospinal fluid. They also participate in the formation of the *blood–cerebrospinal fluid barrier* (see Concept 13.2).

**FIGURE 12.5 Neuroglia of the central nervous system.**

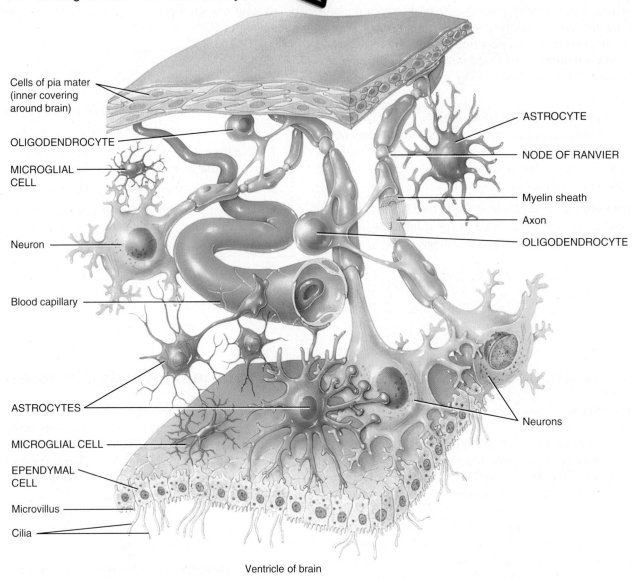



**Neuroglia of the CNS can be distinguished by size and the arrangement of their processes.**

## Neuroglia of the PNS

Neuroglia of the PNS completely surround axons and cell bodies. The two types of neuroglia in the PNS are Schwann cells and satellite cells (Figure 12.6).

- **Schwann cells** (SCHVON or SCHWON) form the myelin sheath around axons in the PNS. A Schwann cell can myelinate a single axon (Figure 12.6a) or enclose multiple unmyelinated axons (axons that lack a myelin sheath) (Figure 12.6b). Schwann cells participate in axon regeneration, which is more easily accomplished in the PNS than in the CNS.
- **Satellite cells** (SAT-i-līt) are flat cells surrounding the cell bodies of neurons of PNS ganglia (Figure 12.6c). (Recall that ganglia are collections of neuronal cell bodies outside the CNS.) In addition to providing structural support, satellite cells regulate the exchanges of materials between neuronal cell bodies and interstitial fluid.

## Myelination

The axons of most neurons are surrounded by a multilayered lipid and protein covering called the **myelin sheath** that electrically insulates them and increases the speed of impulse conduction. Axons with a myelin sheath are said to be **myelinated** (mī-e-li-NĀ-ted) (Figure 12.7a); axons without such a covering are said to be **unmyelinated** (Figure 12.7b). Schwann cells produce myelin sheaths around axons in the PNS; oligodendrocytes produce myelin sheaths around axons in the CNS.

FIGURE 12.6 **Neuroglia of the peripheral nervous system.**

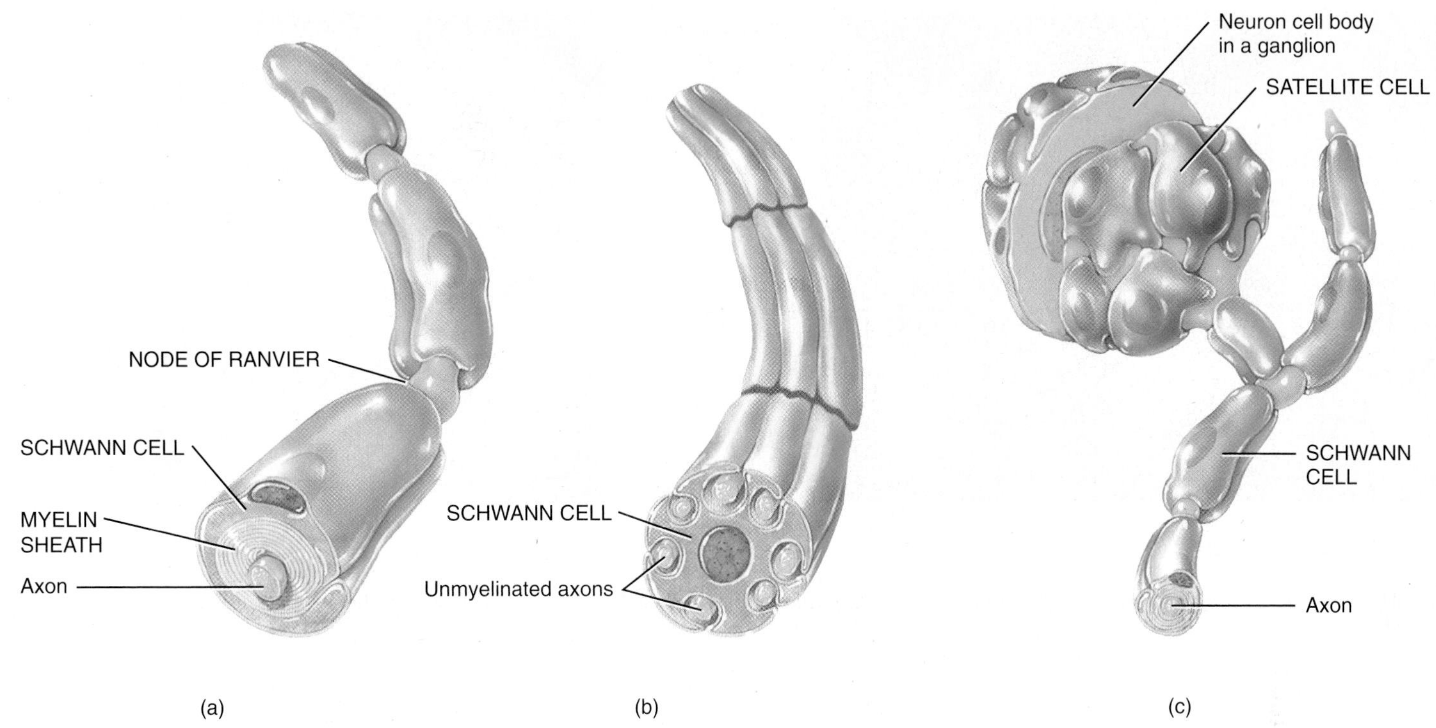


**Neuroglia of the PNS completely surround axons and cell bodies of neurons.**

FIGURE 12.7 **Myelinated and unmyelinated axons.**

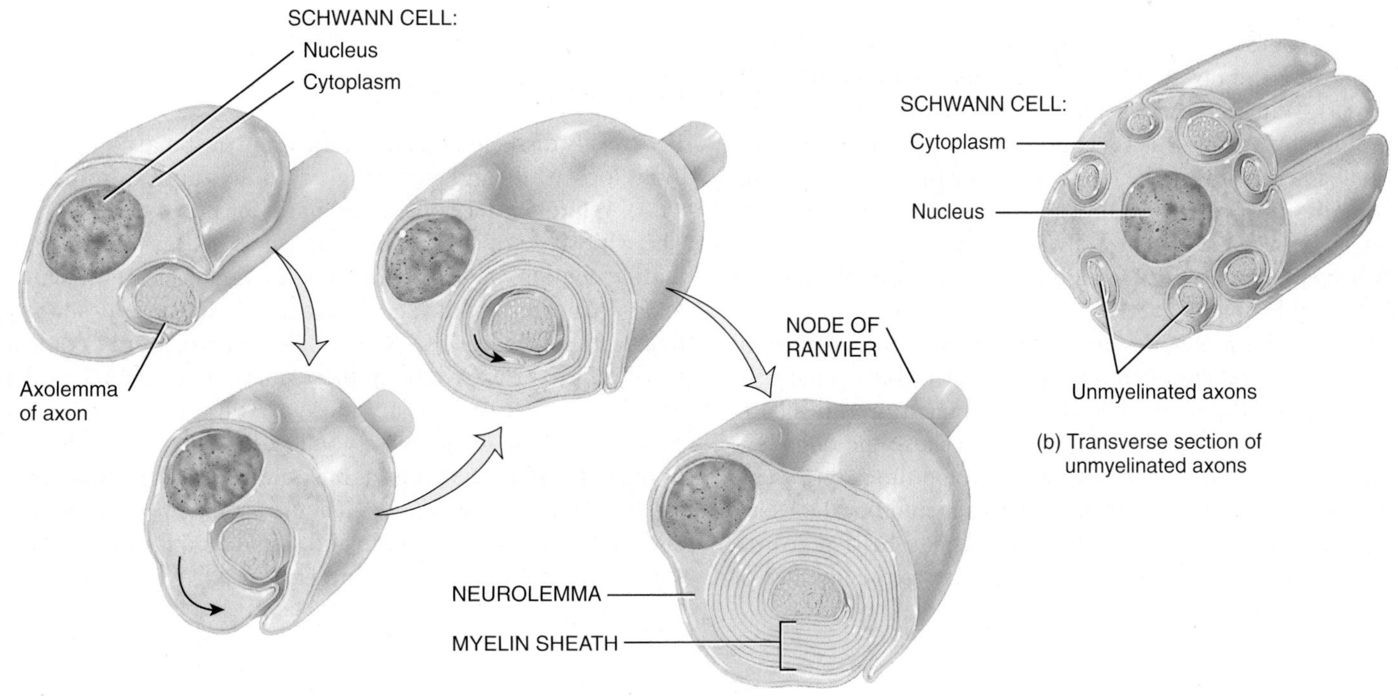


(a) Transverse sections of stages in the formation of a myelin sheath

(b) Transverse section of unmyelinated axons

(continues)

FIGURE 12.7 (continued)

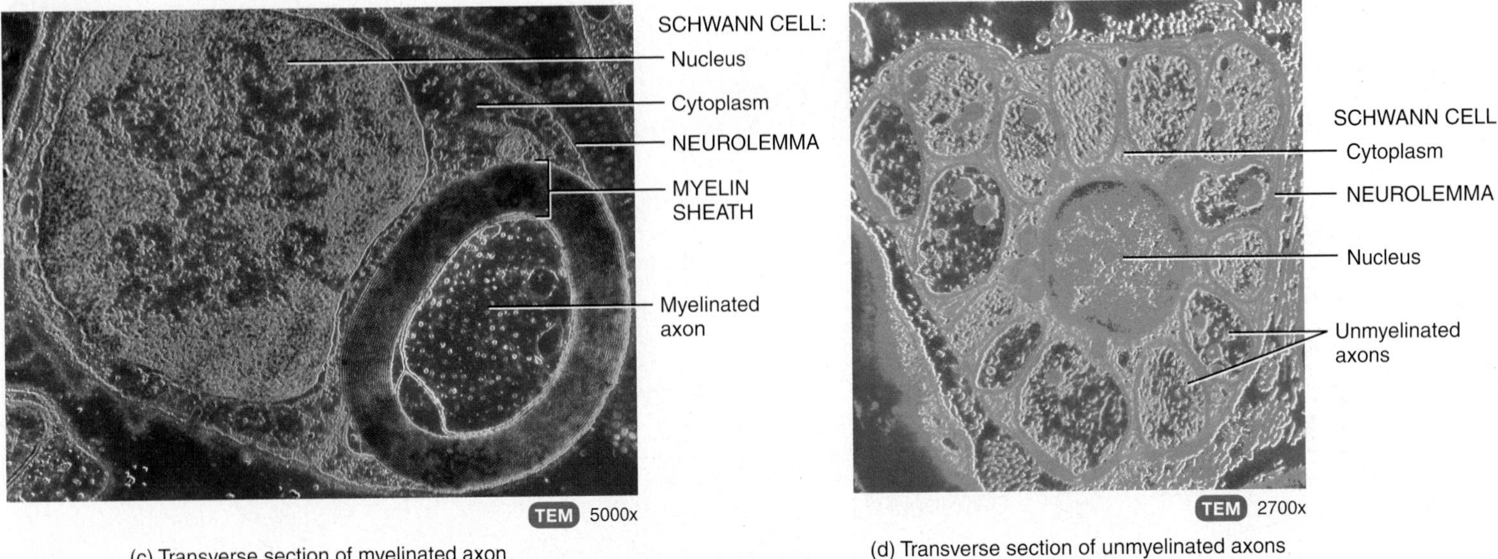


(c) Transverse section of myelinated axon

(d) Transverse section of unmyelinated axons

**Myelinated axons are surrounded by a myelin sheath produced by Schwann cells in the PNS or by oligodendrocytes in the CNS.**

Schwann cells begin to form myelin sheaths around axons during fetal development. Each Schwann cell wraps about 1 millimeter (0.04 in.) of a single axon's length by spiraling many times around the axon (Figure 12.7a). Eventually, multiple layers of Schwann cell plasma membrane surround the axon, with the Schwann cell's cytoplasm and nucleus forming the outermost layer. The inner portion, consisting of up to 100 layers of Schwann cell plasma membrane, is the myelin sheath. The outer, nucleated cytoplasmic layer of the Schwann cell enclosing the myelin sheath is the **neurolemma** (noo′-rō-LEM-ma). When an axon is injured, the neurolemma aids regeneration by forming a regeneration tube that guides and stimulates regrowth of the axon. Gaps in the myelin sheath, called **nodes of Ranvier** (RON-vē-ā), appear at intervals along the axon between adjacent Schwann cells (see Figure 12.3). Each Schwann cell wraps around one axon segment between two nodes of Ranvier.

In the CNS, an oligodendrocyte myelinates parts of several axons. Each oligodendrocyte extends about 15 broad, flat processes that spiral around CNS axons, forming a myelin sheath. A neurolemma is not present, however, because the oligodendrocyte cell body and nucleus do not envelop the axon. Nodes of Ranvier are present, but they are fewer in number than seen with PNS neurons. Axons in the CNS display little regrowth after injury. This is thought to be due, in part, to the absence of a neurolemma and in part to an inhibitory influence exerted by the oligodendrocytes on axon regrowth.

The amount of myelin increases from birth to maturity, and its presence greatly increases the speed of impulse conduction. An infant's responses to stimuli are neither as rapid nor as coordinated as those of an older child or an adult, in part because myelination is still in progress during infancy.

## Gray and White Matter

In a freshly dissected section of the brain or spinal cord, some regions look white and glistening, and others appear gray (Figure 12.8). **White matter** consists primarily of the myelinated axons. The whitish color of myelin gives white matter its name. **Gray matter** contains neuronal cell bodies, dendrites, unmyelinated axons, axon terminals, and neuroglia. It looks grayish, rather than white, because the Nissl bodies impart a gray color and there is little or no myelin in these areas. In the spinal cord, the white matter surrounds an inner core of gray matter that, depending on how imaginative you are, is shaped like a butterfly or the letter H. In the brain, a thin outer shell of gray matter covers the deeper white matter. The arrangement of gray matter and white matter in the brain and spinal cord is discussed more extensively in Concepts 13.3 and 13.9, respectively.

FIGURE 12.8 **Distribution of gray and white matter in the spinal cord and brain.**

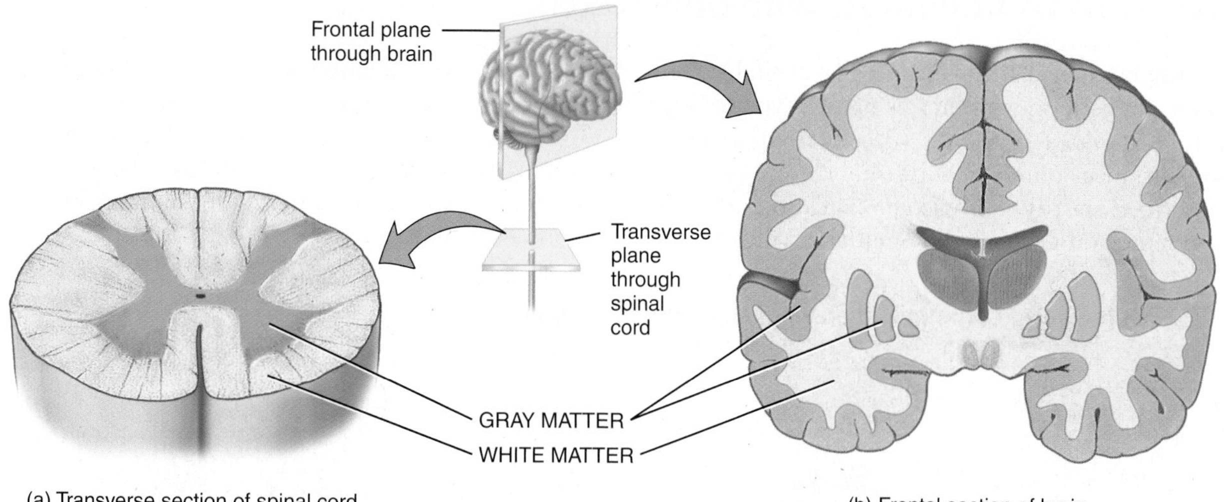


**White matter primarily consists of myelinated axons of many neurons. Gray matter consists of neuron cell bodies, dendrites, axon terminals, unmyelinated axons, and neuroglia.**

## ✓ CHECKPOINT

**10.** Which neuroglia could fight infection by phagocytizing bacterial cells and dead cells? Are these glial cells located within the CNS or PNS?

**11.** Which neuroglia help ensure a healthy environment for the nervous system? Are these glial cells located within the CNS or PNS?

**12.** Which neuroglia help move cerebrospinal fluid through the CNS?

**13.** What is the functional advantage of the myelin sheath?

**14.** Name two neuroglia that produce myelin sheaths. Are each of these glial cells located within the CNS or PNS?

**15.** Differentiate between the myelin sheath and neurolemma.

**16.** How do white matter and gray matter differ in structure?

## RETURN TO **Jennifer's Story**

Jennifer goes to see the ophthalmologist. Like Dr. Marino, Dr. Daniels tests Jennifer's pupillary response and movement of the eyeballs. Then she dilates the pupils of her eyes to examine the retina (the back of the eye). As she writes out a prescription for glasses, Dr. Daniels tells Jennifer that her vision is a little impaired and that the blurriness could be due to tired eyes from reading so much, but also suggests that Jennifer see a neurologist because she has some abnormal darting movements of her eyeballs called nystagmus.

At the neurologist's office, Jennifer tells her story and relays the information from Dr. Daniels about the nystagmus. Dr. Peters examines Jennifer's eyes and confirms that nystagmus is present. He starts his exam with the same tests that her primary care doctor had performed, adding an assessment of her cranial nerve function. Then he tests Jennifer's balance and finds that it is OK with her eyes open, but when she closes her eyes she starts to fall. When he recommends a magnetic resonance imaging (MRI) scan of her brain, Jennifer asks anxiously, " Do you think I have a brain tumor?" Dr. Peters reassures her, indicating that her symptoms could be due to a variety of conditions, and tells her that the MRI will help them find out what is causing them.

A. *Jennifer has blurred vision and abnormal feelings in her right hand. What kind of nerve fibers carry these messages to the brain?*

B. *What kind of nerve fibers control muscle function in the hand that would contribute to Jennifer's difficulty holding her pen?*

C. *Describe the components of the nervous system involved in letting Jennifer know that her vision is blurred, and that the sensation and weakness in her right hand and arm are abnormal.*

## 12.5 Neurons communicate with other cells.

Like muscle fibers, neurons are electrically excitable cells. They communicate with one another using two types of electrical signals: (1) *Graded potentials* (described shortly) are used for short-distance communication only. (2) *Action potentials* (also described shortly) allow communication over long distances within the body. An action potential in a muscle fiber is called a *muscle action potential*. When an action potential occurs in a neuron, it is called a *nerve action potential* (*impulse*). To understand the functions of graded potentials and action potentials, consider how the nervous system allows you to feel the smooth surface of a pen that you have picked up from a table (Figure 12.9):

FIGURE 12.9 **Overview of communication within the nervous system.**

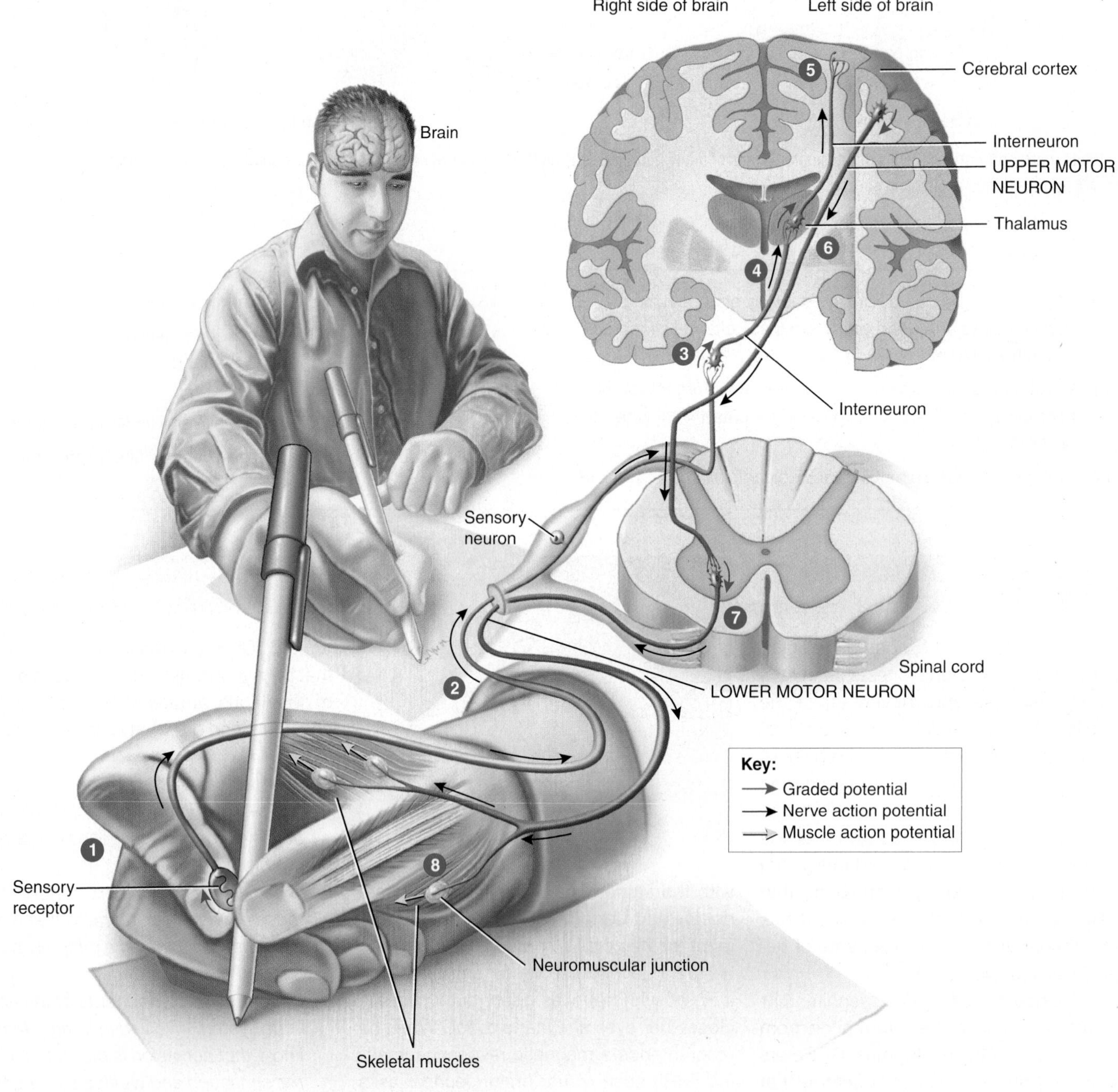


**Graded potentials and nerve and muscle action potentials are involved in the relay of sensory stimuli, integrative functions such as perception, and motor activities.**

1. As you touch the pen, a graded potential develops in a sensory receptor in the skin of the fingers.

2. The graded potential triggers the axon of the sensory neuron to form a nerve action potential, which travels along the axon into the CNS and ultimately causes the release of neurotransmitter at a synapse with an interneuron. A **neurotransmitter** (noo′-rō-trans′-MIT-ter) is a chemical released by a neuron that excites or inhibits other neurons or effector cells.

3. The neurotransmitter stimulates the interneuron to form a graded potential in its dendrites and cell body.

4. In response to the graded potential, the axon of the interneuron forms a nerve action potential. The nerve action potential travels along the axon, which results in neurotransmitter release at the next synapse with another interneuron.

5. This process of neurotransmitter release at a synapse followed by the formation of a graded potential, and then a nerve action potential, occurs over and over as interneurons in higher parts of the brain (such as the thalamus and cerebral cortex) are activated. Following activation of interneurons in the *cerebral cortex* (the outer part of the brain), you are now able to feel the smooth surface of the pen touch your fingers. Feeling the pen in your hand is an example of *perception*, the conscious awareness of a sensation, which is primarily a function of the cerebral cortex.

   Suppose that you want to use the pen to write a letter. The nervous system would respond in the following way (Figure 12.9):

6. A stimulus in the brain causes a graded potential to form in the dendrites and cell body of an **upper motor neuron**, a type of motor neuron that synapses with a lower motor neuron farther down in the CNS in order to contract a skeletal muscle. The graded potential subsequently causes a nerve action potential to occur in the axon of the upper motor neuron, leading to neurotransmitter release.

7. The neurotransmitter generates a graded potential in a **lower motor neuron**, a type of motor neuron that directly supplies skeletal muscle fibers. The graded potential triggers the formation of a nerve action potential and then release of the neurotransmitter at neuromuscular junctions formed with skeletal muscle fibers that control movements of the fingers.

8. The neurotransmitter stimulates the muscle fibers that control finger movements to form muscle action potentials. The muscle action potentials cause these muscle fibers to contract, allowing you to write with the pen.

The production of graded potentials and action potentials depends on two basic features of the plasma membrane of excitable cells: the existence of a resting membrane potential and the presence of specific types of ion channels. Like most other cells in the body, the plasma membrane of excitable cells exhibits a **membrane potential**, an electrical potential difference (voltage) across the plasma membrane. In excitable cells, this voltage is termed the **resting membrane potential**. The membrane potential is like voltage stored in a battery. If you connect the positive and negative terminals of a battery with a piece of wire, electrons will flow along the wire. This flow of charged particles is called **current**. In living cells, the flow of ions (rather than electrons) constitutes the electrical current.

## Ion Channels

Graded potentials and action potentials occur because the plasma membranes of neurons contain many different kinds of ion channels (Figure 12.10) that open or close in response to specific stimuli. Because the lipid bilayer of the plasma membrane is a good electrical insulator, the main paths for current to flow across the membrane are through the ion channels. When ion channels are open, they allow specific ions to move across the plasma membrane, down their **electrochemical gradients**—a concentration (chemical) difference plus an electrical difference. Recall that ions move from where they are more concentrated to where they are less concentrated (the chemical part of the gradient). Also, positively charged ions (cations) move toward a negatively charged area, and negatively charged ions (anions) move toward a positively charged area (the electrical part of the gradient). As ions move, they constitute a flow of electrical current that can change the membrane potential.

Ion channels open and close due to the presence of "gates." The gate is a part of the channel protein that can seal the channel pore shut or move aside to open the pore (see Figure 3.6). The electrical signals produced by neurons and muscle fibers rely on four types of ion channels:

- The gates of **leak channels** randomly alternate between open and closed positions (Figure 12.10a). Typically, plasma membranes have many more potassium ion ($K^+$) leak channels than sodium ion ($Na^+$) leak channels, and the $K^+$ leak channels are leakier than the $Na^+$ leak channels. Thus, the membrane's permeability to $K^+$ is much higher than its permeability to $Na^+$.
- A **ligand-gated channel** opens and closes in response to a specific **ligand**, a molecule that binds to a receptor (Figure 12.10b). A wide variety of ligands—including neurotransmitters, hormones, and ions—can open or close ligand-gated channels. The neurotransmitter acetylcholine, for example, opens cation channels that allow $Na^+$ and calcium ions ($Ca^{2+}$) to diffuse inward and $K^+$ to diffuse outward.
- A **mechanically gated channel** opens or closes in response to mechanical stimulation such as sound waves, touch, pressure, or tissue stretching (Figure 12.10c). The force distorts the channel from its resting position, opening the gate.
- A **voltage-gated channel** opens in response to a change in membrane potential (voltage) (Figure 12.10d). Voltage-gated channels participate in the generation and conduction of action potentials.

**FIGURE 12.10 Ion channels in the plasma membrane.** (a) Leak channels randomly open and close. (b) A chemical stimulus—here, the neurotransmitter acetylcholine—opens a ligand-gated channel. (c) A mechanical stimulus opens a mechanically gated channel. (d) A change in membrane potential opens voltage-gated $K^+$ channels during an action potential.

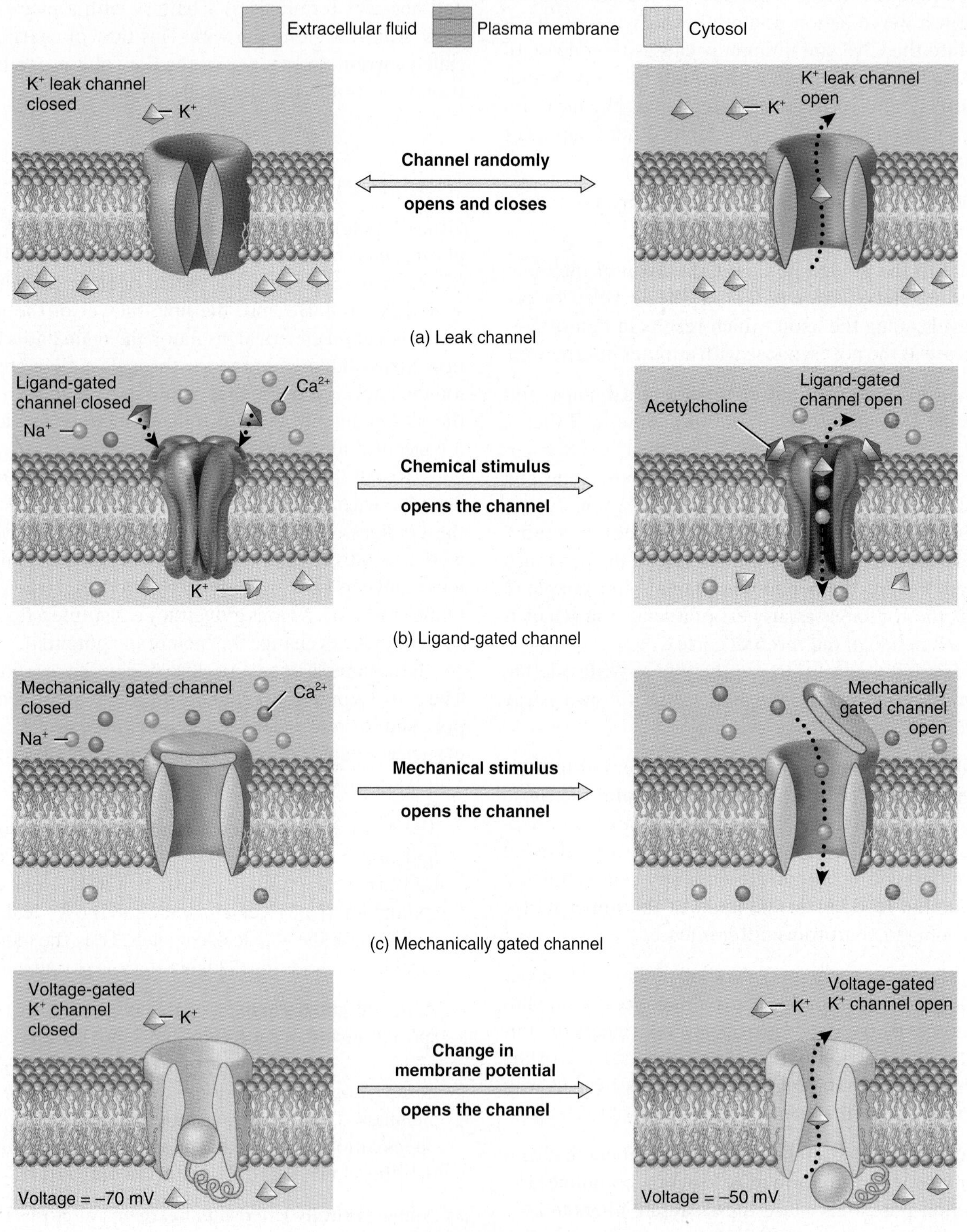


**When ion channels are open, specific ions can move across the plasma membrane down their electrochemical gradients.**

## Resting Membrane Potential

The resting membrane potential exists because of a small buildup of negative ions in the cytosol along the inside of the plasma membrane, and an equal buildup of positive ions in the extracellular fluid along the outside surface of the plasma membrane (Figure 12.11a). Such a separation of positive and negative electrical charges is a form of potential energy, which is measured in volts or millivolts (1 mV = 0.001 V). The greater the difference in charge across the plasma membrane, the larger is the membrane potential (voltage). Notice in Figure 12.11a that the buildup of charge occurs only very close to either side of the membrane. The cytosol or extracellular fluid elsewhere contains equal numbers of positive and negative charges and is electrically neutral.

**FIGURE 12.11 Resting membrane potential.** To measure resting membrane potential, the tip of the recording microelectrode is inserted inside the neuron, and the reference electrode is placed in the extracellular fluid. The electrodes are connected to a voltmeter that measures the difference in charge across the plasma membrane (in this case –70 mV inside relative to outside).

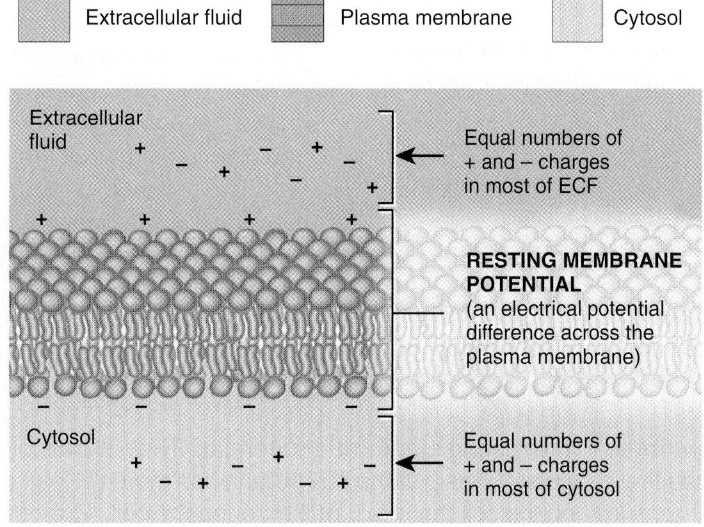


(a) Distribution of charges that produce the resting membrane potential of a neuron

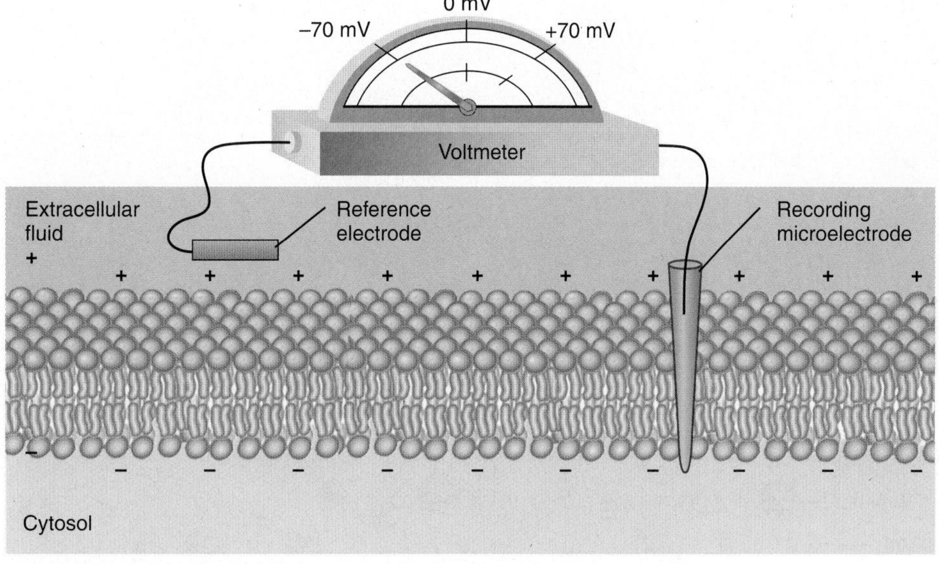


(b) Measurement of resting membrane potential of a neuron

**The resting membrane potential exists across the plasma membrane of an excitable cell under resting conditions.**

The resting membrane potential of a cell can be measured in the following way: The tip of a recording microelectrode is inserted inside the cell, and a reference electrode is placed outside the cell in the extracellular fluid. Electrodes are devices that conduct electrical charges. The recording microelectrode and the reference electrode are connected to an instrument known as a voltmeter, which detects the electrical difference (voltage) across the plasma membrane (Figure 12.11b). In neurons, the typical resting membrane potential is approximately –70 mV. The minus sign indicates that the cytosol side of the plasma membrane is negative relative to the outside. A cell that exhibits a membrane potential is said to be **polarized**.

The resting membrane potential arises from three major factors:

- ***Unequal distribution of ions across the plasma membrane.*** A major factor that contributes to the resting membrane potential is the unequal distributions of various ions in extracellular fluid and cytosol (Figure 12.12). Extracellular fluid is rich in $Na^+$ and chloride ions ($Cl^-$). In cytosol, however, there are large concentrations of $K^+$ and anions such as phosphates and amino acids. Because the plasma membrane typically has more $K^+$ leak channels than $Na^+$ leak channels, the number of potassium ions that diffuse down their concentration gradient out of the cell is greater than the number of sodium ions that diffuse down their concentration gradient into the cell. As more and more positive potassium ions exit, the cytosol side of the plasma membrane becomes increasingly negative and the extracellular fluid side of the plasma membrane becomes increasingly positive.
- ***Inability of most anions to leave the cell.*** Another factor contributes to the relative negativity of the cell interior: Most anions inside the cell are not free to leave (Figure 12.12). They cannot follow $K^+$ out of the cell because they are attached to nondiffusible molecules such as ATP and large proteins.
- ***Electrogenic nature of the sodium–potassium pump.*** Although there are only a few sodium leak channels, sodium ions do slowly diffuse inward, down their concentration gradient. Left unchecked, such inward leakage of $Na^+$ would eventually destroy the resting membrane potential. The small inward $Na^+$ leak and outward $K^+$ leak are offset by sodium–potassium pumps (Figure 12.12). These pumps help maintain the resting membrane potential by pumping out $Na^+$ as fast as it leaks in and, at the same time, returning $K^+$ to the cell interior. The sodium–potassium pumps expel three $Na^+$ for each two $K^+$ imported. Since these pumps remove more positive charges from the cell than they bring into the cell, they are *electrogenic*, which means they contribute to the negativity of the resting membrane potential.

FIGURE 12.12 **Factors that contribute to the resting membrane potential.** The cell interior is relatively negative and the extracellular fluid is relatively positive because (1) the plasma membrane has more $K^+$ leak channels (blue) than $Na^+$ leak channels (red), resulting in more $K^+$ ions leaving the cell than $Na^+$ ions entering the cell; (2) most anions are unable to leave the cell; (3) the electrogenic $Na^+$–$K^+$ pumps (purple) expel more positive ions (three $Na^+$) than they bring into the cell (two $K^+$).

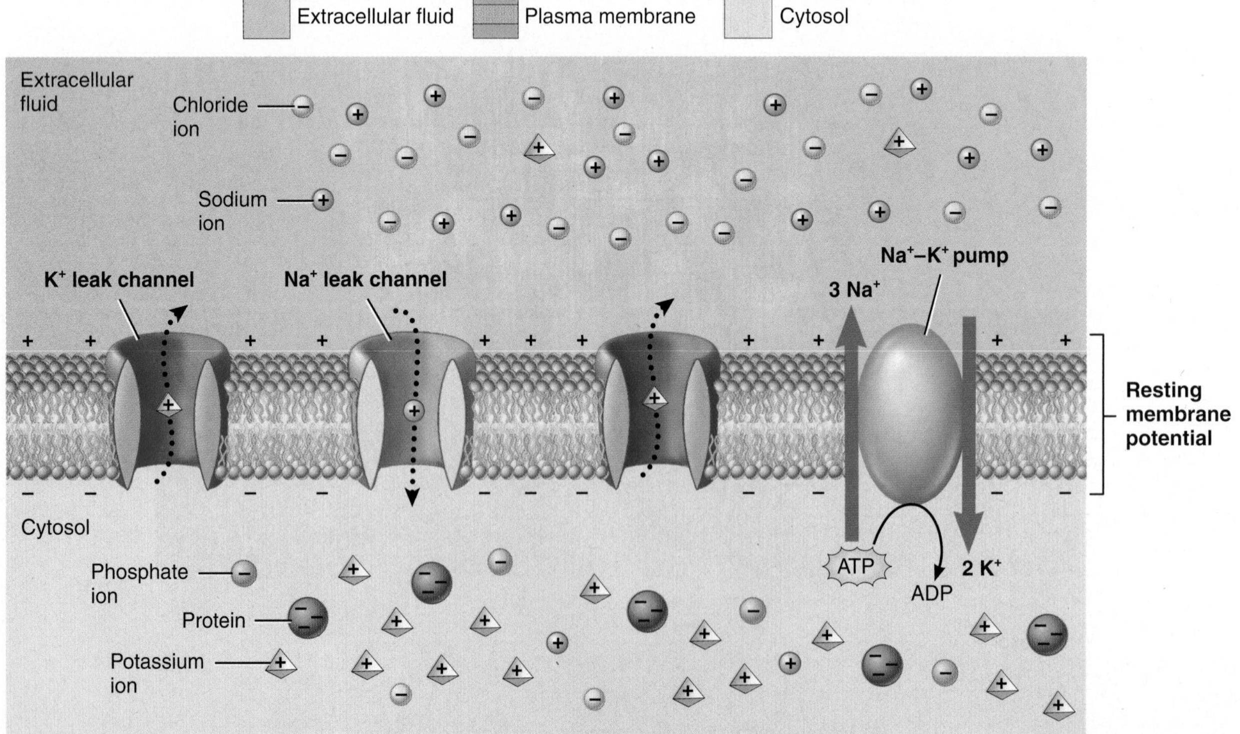


The resting membrane potential exists because the cytosol side of the plasma membrane has relatively more negative ions while the extracellular fluid has relatively more positive ions.

## ✓ CHECKPOINT

17. How does the cell membrane of a neuron allow the cell to store electrical charges?
18. What are the three stimuli that result in ion channels opening or closing? What happens when these channels open?
19. How are leak channels different from gated channels?
20. What two conditions allow maintenance of the resting membrane potential?

# 12.6 Graded potentials are the first response of a neuron to stimulation.

A **graded potential** (Figure 12.13a) is a small deviation from the membrane potential that makes the membrane either more polarized (inside more negative) or less polarized (inside less negative). When the response makes the membrane more polarized (inside more negative), it is termed a **hyperpolarizing graded potential** (hī-per-PŌ-lar-ī′-zing). When the response makes the membrane less polarized (inside less negative), it is termed a **depolarizing graded potential** (Figure 12.13b).

To say that these electrical signals are *graded* means that they vary in amplitude (size), depending on the strength of the stimulus (Figure 12.14). They are larger or smaller depending on how many ligand-gated or mechanically gated channels have opened (or closed) and how long each remains open. The opening or closing of these channels alters the flow of specific ions across the membrane, producing a flow of current that is localized, which means that it spreads along the plasma membrane for a short distance and then dies out. Because graded potentials die out within a few millimeters of their point of origin, they are useful for short-distance communication only. Although

FIGURE 12.13 **Graded potentials.** Most graded potentials occur in the dendrites and cell body (areas colored blue in the inset).

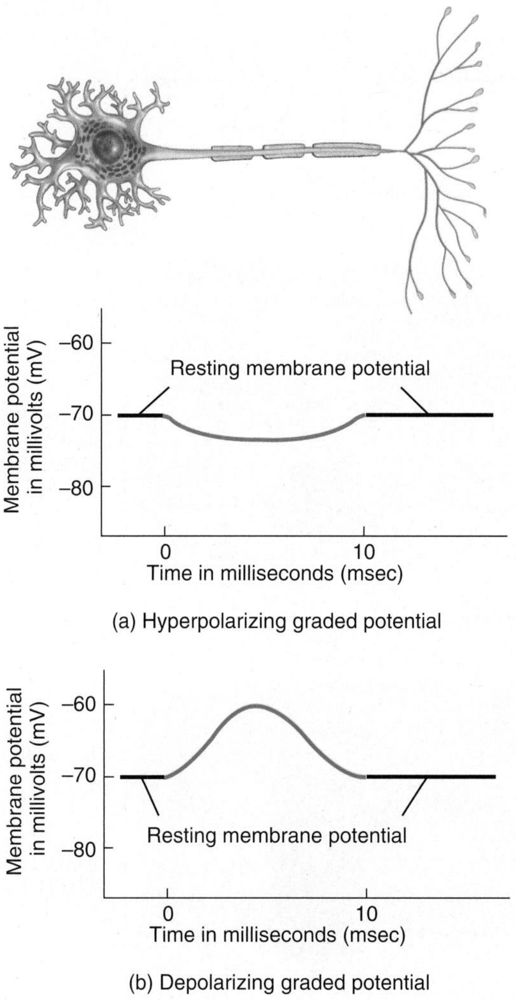

Membrane potential in millivolts (mV)
−60
−70
−80
Resting membrane potential
0
10
Time in milliseconds (msec)

(a) Hyperpolarizing graded potential

Membrane potential in millivolts (mV)
−60
−70
−80
Resting membrane potential
0
10
Time in milliseconds (msec)

(b) Depolarizing graded potential

**During a hyperpolarizing graded potential, the membrane polarization is more negative than the resting level; during a depolarizing graded potential, the membrane polarization is less negative than the resting level.**

FIGURE 12.14 **The graded nature of graded potentials.** As the stimulus strength increases, the amplitude (size) of each resulting depolarizing graded potential increases. Although not shown, a similar relationship exists between stimulus strength and the amplitude of a hyperpolarizing graded potential.

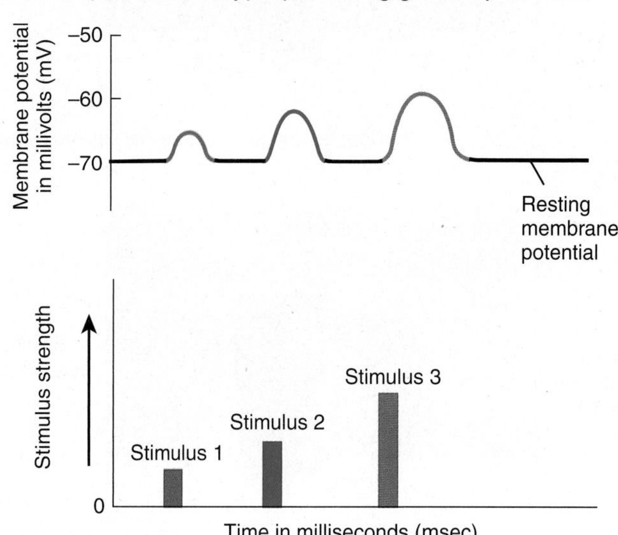


**The amplitude of a graded potential depends on the strength of the stimulus.**

graded potentials may only travel a short distance, they can generate another type of membrane potential, an action potential (described shortly) that travels the entire length of an axon.

A graded potential occurs when a stimulus causes ligand-gated or mechanically gated channels to open or close in an excitable cell's plasma membrane (Figure 12.15). Typically, ligand-gated and mechanically gated channels are present only in the dendrites and cell bodies of neurons. Hence, graded potentials occur mainly in the dendrites and cell body of a neuron.

Although an individual graded potential can die out as it spreads along the membrane, it can become stronger and last longer by summating with other graded potentials. **Summation** is the

FIGURE 12.15 **Generation of graded potentials.** (a) A mechanical stimulus opens a mechanically gated channel that allows cations to enter the cell; a depolarizing graded potential occurs as the cell interior becomes less negative than at rest. (b) A ligand stimulus (the neurotransmitter acetylcholine) opens a ligand-gated channel that allows the exchange of ions, and a depolarizing graded potential occurs as the cell interior becomes less negative than at rest. (c) A ligand stimulus (the neurotransmitter glycine) opens a ligand-gated channel that allows $Cl^-$ ions into the cell, and a hyperpolarizing graded potential occurs as the cell interior becomes more negative than at rest.

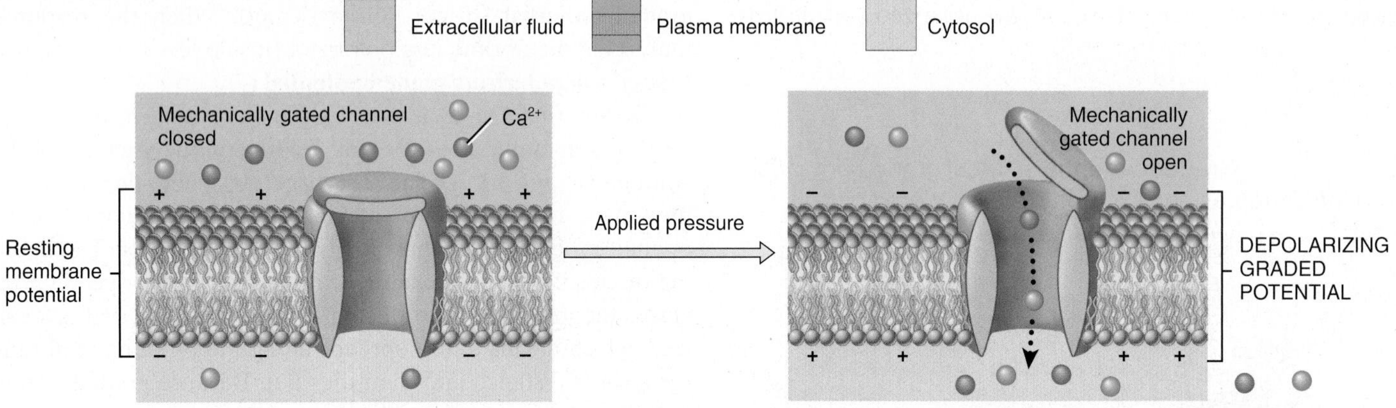


(a) Depolarizing graded potential caused by pressure, a mechanical stimulus

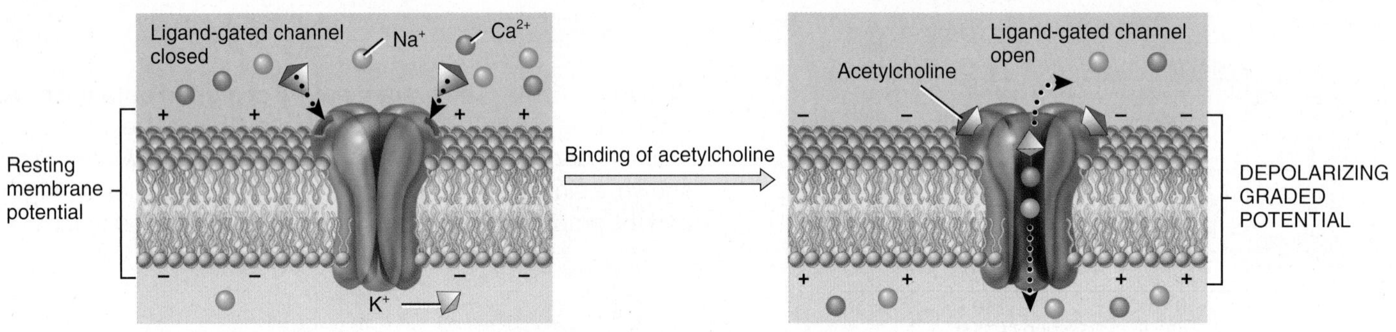


(b) Depolarizing graded potential caused by neurotransmitter acetylcholine, a ligand stimulus

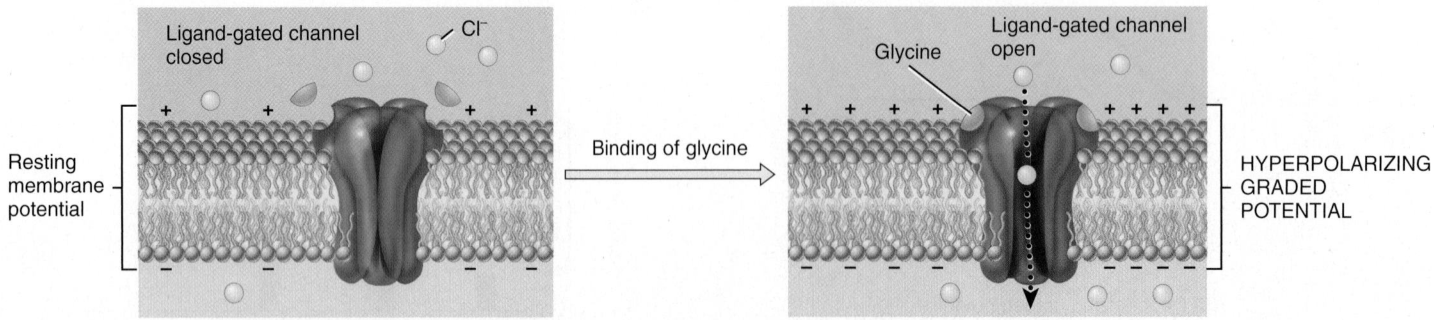


(c) Hyperpolarizing graded potential caused by neurotransmitter glycine, a ligand stimulus

**A graded potential forms in response to the opening of mechanically gated or ligand-gated channels.**

process by which graded potentials add together. If two depolarizing graded potentials summate, the net result is a larger depolarizing graded potential (Figure 12.16). If two hyperpolarizing graded potentials summate, the net result is a larger hyperpolarizing graded potential. If two equal but opposite graded potentials summate (one depolarizing and the other hyperpolarizing), then they cancel each other out and the overall graded potential disappears. You will learn more about the process of summation later in this chapter.

Graded potentials have different names depending on which type of stimulus causes them and where they occur. For example, when a graded potential occurs in a neuron in response to a neurotransmitter, it is called a *postsynaptic potential* (explained shortly). On the other hand, the graded potentials that occur in sensory receptors are termed *receptor potentials* (explained in Concept 15.2).

FIGURE 12.16 **Summation of graded potentials.** Summation of two depolarizing graded potentials happens when two stimuli occur very close together in time. The dotted lines represent the individual depolarizing graded potentials that would form if summation did not occur.

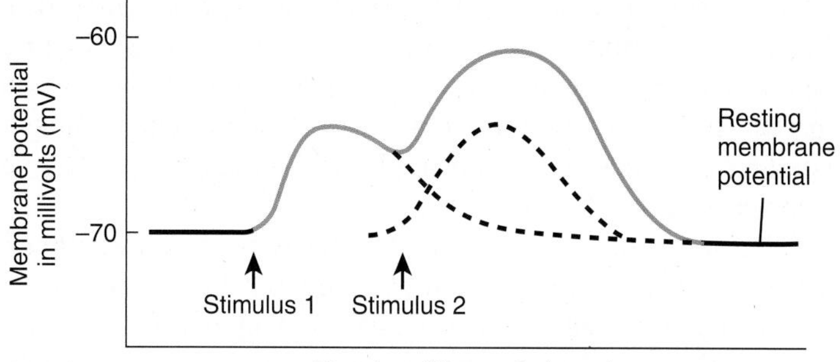


**Summation occurs when two or more graded potentials add together to become larger in amplitude.**

## ✓ CHECKPOINT

**21.** As more and more channels are opened, what happens to the magnitude of a graded potential?

**22.** What do we mean when we say graded potentials are "localized"?

**23.** In what neuronal structures do most graded potentials occur?

**24.** What types of ion channels produce graded potentials?

# 12.7 The action potential is an all-or-none electrical signal.

An **action potential**, or **impulse**, is a sequence of rapidly occurring events that briefly reverses the membrane potential and then eventually restores it to the resting state. An action potential has two main phases (Figure 12.17). During the **depolarizing**

FIGURE 12.17 **Action potential or impulse.** When a stimulus depolarizes the membrane to threshold (−55 mV), an action potential is generated. The action potential arises at the trigger zone of the neuron and then propagates along the axon plasma membrane to the axon terminals (region of neuron colored green).

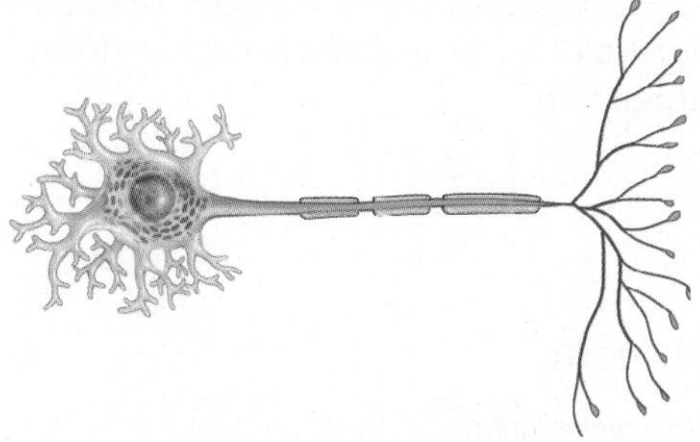

**An action potential consists of depolarizing and repolarizing phases.**

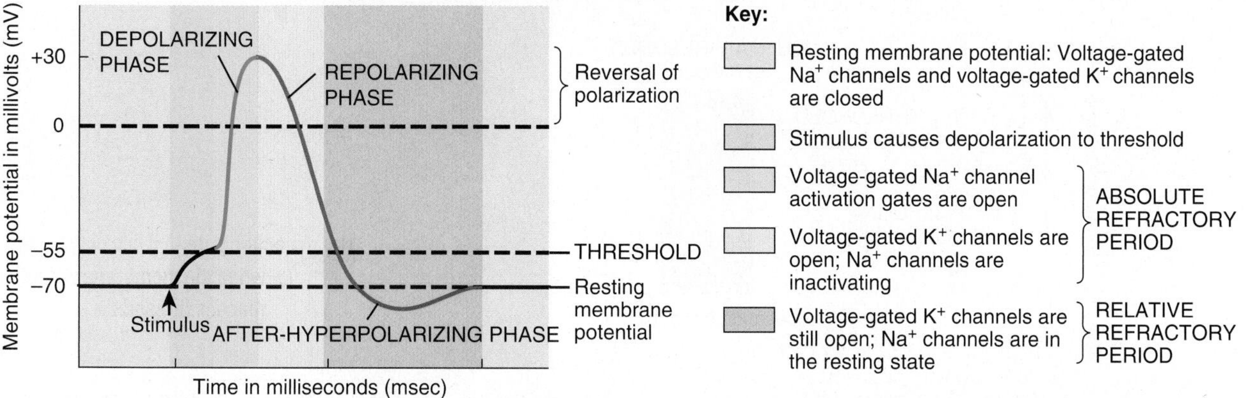

**phase**, the negative membrane potential becomes less negative, reaches zero, and then becomes positive. During the **repolarizing phase**, the membrane potential is restored to the resting state of −70 mV.

Two types of voltage-gated channels open and then close during an action potential. These channels are present mainly in the plasma membrane of the axon and axon terminals. The first channels to open, voltage-gated $Na^+$ channels, allow $Na^+$ to rush into the cell, resulting in the depolarizing phase. Then voltage-gated $K^+$ channels open, allowing $K^+$ to flow out, producing the repolarizing phase.

The generation of an action potential depends on whether a stimulus is able to bring the membrane potential to a certain level termed the **threshold** (about −55 mV in many neurons) (Figure 12.18). An action potential will not occur in response to a **subthreshold stimulus**, a weak depolarization that cannot bring the membrane potential to threshold. However, an action potential will occur in response to a **threshold stimulus**, a stimulus that is just strong enough to depolarize the membrane to threshold. Several action potentials will form in response to a **suprathreshold stimulus**, a stimulus that is strong enough to depolarize the membrane *above* threshold. Each of the action potentials caused by a suprathreshold stimulus has the same amplitude (size) as an action potential caused by a threshold stimulus. Therefore, once an action potential is generated, the amplitude of an action potential is always the same and does not depend on stimulus intensity. Instead, the greater the stimulus strength above threshold, the greater is the frequency of action potentials (the more often they occur).

As you have just learned, an action potential is generated in response to a threshold stimulus but does not form when there is a subthreshold stimulus. In other words, an action potential either occurs completely or it does not occur at all. This characteristic of an action potential is known as the **all-or-none principle**. The all-or-none principle of the action potential is similar to pushing the first domino in a long row of standing dominoes. When the push on the first domino is strong enough (when depolarization reaches threshold), that domino falls against the second domino, and the entire row topples (an action potential occurs). Stronger pushes on the first domino produce the identical effect—toppling of the entire row. Thus, pushing on the first domino produces an all-or-none event: The dominoes all fall or none fall.

## Depolarizing Phase

Depolarizing graded potentials originate in the dendrites or cell body of a neuron and then travel to the trigger zone. If the graded potential is able to depolarize the membrane to threshold, voltage-gated $Na^+$ channels open quickly. As voltage-gated $Na^+$ channels open, electrical and chemical gradients favor inward movement of $Na^+$. The resulting influx of $Na^+$ produces the depolarizing phase of the action potential as the membrane potential rises from the threshold level of −55 mV to +30 mV (see Figure 12.18). At the peak of the action potential, the cytosol side of the membrane is 30 mV more positive than the extracellular fluid side of the membrane.

Each voltage-gated $Na^+$ channel has an *activation gate* and an *inactivation gate*. In the *resting state* of a voltage-gated $Na^+$ channel, the inactivation gate is open, but the activation gate is closed (step **1** in Figure 12.19). As a result, $Na^+$ cannot move into the cell through these channels and the membrane potential is at −70 mV.

**FIGURE 12.18 Stimulus strength and action potential generation.** A subthreshold stimulus does not result in an action potential because it does not bring the membrane potential to threshold. An action potential does occur in response to a threshold stimulus, a stimulus just strong enough to depolarize the membrane to threshold. Several action potentials form in response to a suprathreshold stimulus, which depolarizes the membrane above threshold.

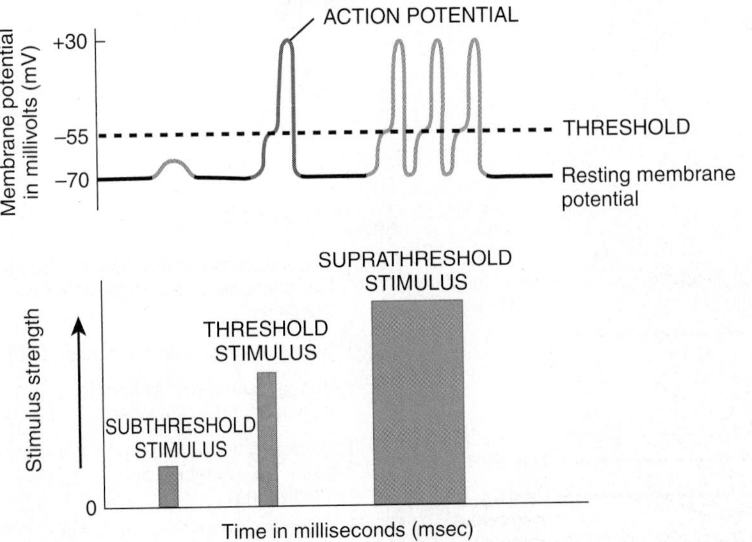


An action potential will only occur when the membrane potential reaches threshold.

**FIGURE 12.19 Changes in ion flow through voltage-gated channels during the depolarizing and repolarizing phases of an action potential.** Leak channels and sodium–potassium pumps are not shown.

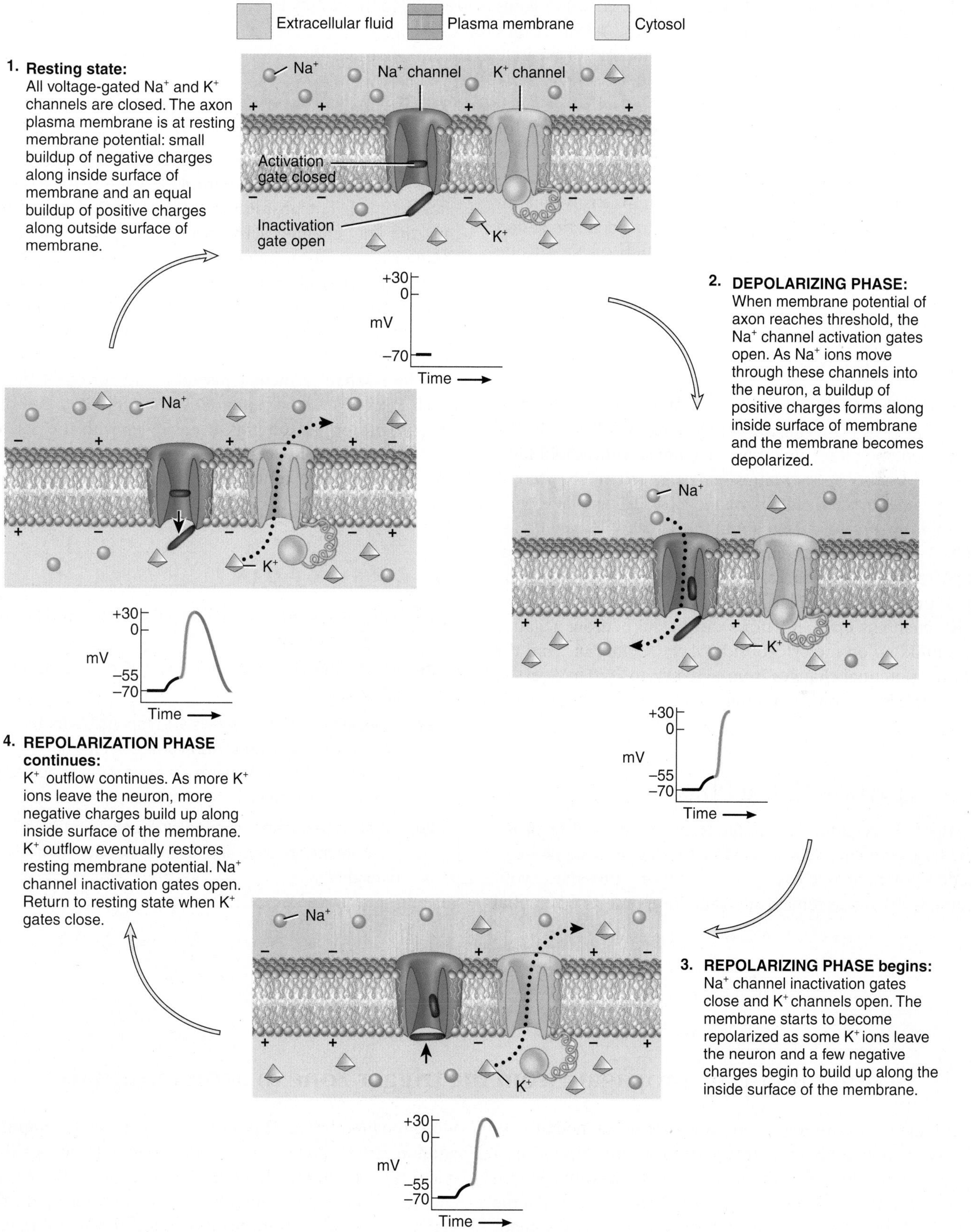

Extracellular fluid
Plasma membrane
Cytosol
1. **Resting state:**
All voltage-gated $Na^+$ and $K^+$ channels are closed. The axon plasma membrane is at resting membrane potential: small buildup of negative charges along inside surface of membrane and an equal buildup of positive charges along outside surface of membrane.
$Na^+$
$Na^+$ channel
$K^+$ channel
Activation gate closed
Inactivation gate open
$K^+$
+30
0
mV
–70
Time
2. **DEPOLARIZING PHASE:**
When membrane potential of axon reaches threshold, the $Na^+$ channel activation gates open. As $Na^+$ ions move through these channels into the neuron, a buildup of positive charges forms along inside surface of membrane and the membrane becomes depolarized.
$Na^+$
$K^+$
+30
0
mV
–55
–70
Time
3. **REPOLARIZING PHASE begins:**
$Na^+$ channel inactivation gates close and $K^+$ channels open. The membrane starts to become repolarized as some $K^+$ ions leave the neuron and a few negative charges begin to build up along the inside surface of the membrane.
$Na^+$
$K^+$
+30
0
mV
–55
–70
Time
4. **REPOLARIZATION PHASE continues:**
$K^+$ outflow continues. As more $K^+$ ions leave the neuron, more negative charges build up along inside surface of the membrane. $K^+$ outflow eventually restores resting membrane potential. $Na^+$ channel inactivation gates open. Return to resting state when $K^+$ gates close.
$Na^+$
$K^+$
+30
0
mV
–55
–70
Time

**Inflow of sodium ions ($Na^+$) causes the depolarizing phase and outflow of potassium ions ($K^+$) causes the repolarizing phase of an action potential.**

During the depolarizing phase, the resting membrane potential rises to threshold, when the activation gates in the $Na^+$ channel open. With both the activation and inactivation gates open, $Na^+$ influx begins (step **2** in Figure 12.19). As $Na^+$ rushes through the channels into the neuron, the buildup of positive charges on the cytosol side of the membrane increases the membrane potential to +30 mV.

Although the voltage-gated $Na^+$ channels are open for only a few ten-thousandths of a second, about 20,000 $Na^+$ ions flow across the membrane and change the membrane potential considerably. The sodium–potassium pumps easily bail out the 20,000 or so $Na^+$ that enter the cell during a single action potential and maintain the low concentration of $Na^+$ inside the cell.

## Repolarizing Phase

Shortly after the activation gates open, the inactivation gates close and $Na^+$ stops entering the neuron (step **3** in Figure 12.19). Besides opening voltage-gated $Na^+$ channels, a threshold-level depolarization also opens voltage-gated $K^+$ channels (steps **3** and **4** in Figure 12.19). Because voltage-gated $K^+$ channels open slowly, their opening occurs at about the same time the voltage-gated $Na^+$ channels are closing. The slower opening of voltage-gated $K^+$ channels and the closing of previously opened $Na^+$ channels produce the repolarizing phase of the action potential. With $K^+$ channels open, $K^+$ flows out of the neuron, negative charges build up on the cytosol side of the membrane, and the membrane potential changes from +30 mV to −70 mV, restoring the resting membrane potential.

## After-Hyperpolarizing Phase

While the voltage-gated $K^+$ channels are open, outflow of $K^+$ may be large enough to cause an **after-hyperpolarizing phase** of the action potential (see Figure 12.17). During this phase, voltage-gated $K^+$ channels remain open and the membrane potential becomes even more negative than the resting level of −70 mV. As the voltage-gated $K^+$ channels close, the membrane potential returns to the resting level.

## Refractory Period

The period of time after an action potential begins, during which an excitable cell cannot generate another action potential, is called the **refractory period** (rē-FRAK-tor-ē) (see key in Figure 12.17). The **absolute refractory period** occurs from the time the $Na^+$ channel activation gates open to when the $Na^+$ channel inactivation gates close (steps **2–3** in Figure 12.19). During this period, even a very strong stimulus cannot initiate a second action potential because inactivated $Na^+$ channels must return to the resting state before they can reopen (step **1** in Figure 12.19).

The **relative refractory period** is the period of time during which a second action potential can be initiated, but only by a larger-than-normal stimulus. It coincides with the period when the voltage-gated $K^+$ channels are still open after inactivated $Na^+$ channels have returned to their resting state (see key in Figure 12.17).

### ✓ CHECKPOINT

**25.** How are graded potentials and the threshold of an action potential related?

**26.** What is the all-or-none principle? Are graded potentials all-or-none?

**27.** What ions and ion channels are responsible for the depolarizing and repolarizing phases of the action potential? (Which channels are open during depolarization? During repolarization?)

**28.** What is the refractory period of an action potential? How are the absolute and relative refractory periods different?

# 12.8 Action potentials propagate from the trigger zone to axon terminals.

To communicate information from one part of the body to another, action potentials must travel from where they arise at a trigger zone to the axon terminals. In contrast to the graded potential, an action potential is not *decremental* (it does not die out). Instead, an action potential keeps its strength as it spreads along the membrane. This mode of travel is called **impulse propagation** (prop′-a-GĀ-shun), and it depends on positive feedback. As sodium ions flow into the neuron, voltage-gated $Na^+$ channels in adjacent segments of the membrane open. Thus, the action potential self-propagates along the membrane, rather like toppling that long row of dominoes by pushing over the first one in the line. In actuality, it is not the same action potential that propagates along the entire axon. Instead, the action potential regenerates over and over at adjacent regions of membrane from the trigger zone to the axon terminals. An action potential propagates in only one direction, toward the axon terminals (see leading edge of action potential in Figure 12.20), because any region of the axon that has just undergone an action potential must recover (experience the refractory period) before it is able to generate another action potential. Because action potentials can travel along a membrane without dying out, they function in communication over long distances.

## Continuous and Saltatory Conduction

The type of impulse propagation described so far is **continuous conduction**, which involves step-by-step depolarization and repolarization of each adjacent segment of the plasma membrane as ions flow through each voltage-gated channel along the membrane.

FIGURE 12.20 **Propagation of an action potential after it arises at the trigger zone.** Dotted lines indicate ionic current flow. Insert shows path of current flow. (a) In continuous conduction along an unmyelinated axon, ionic currents flow across each adjacent segment of the membrane. (b) In saltatory conduction along a myelinated axon, the action potential at the first node of Ranvier generates ionic currents in the cytosol and interstitial fluid that open voltage-gated $Na^+$ channels at the second node, and so on at each subsequent node.

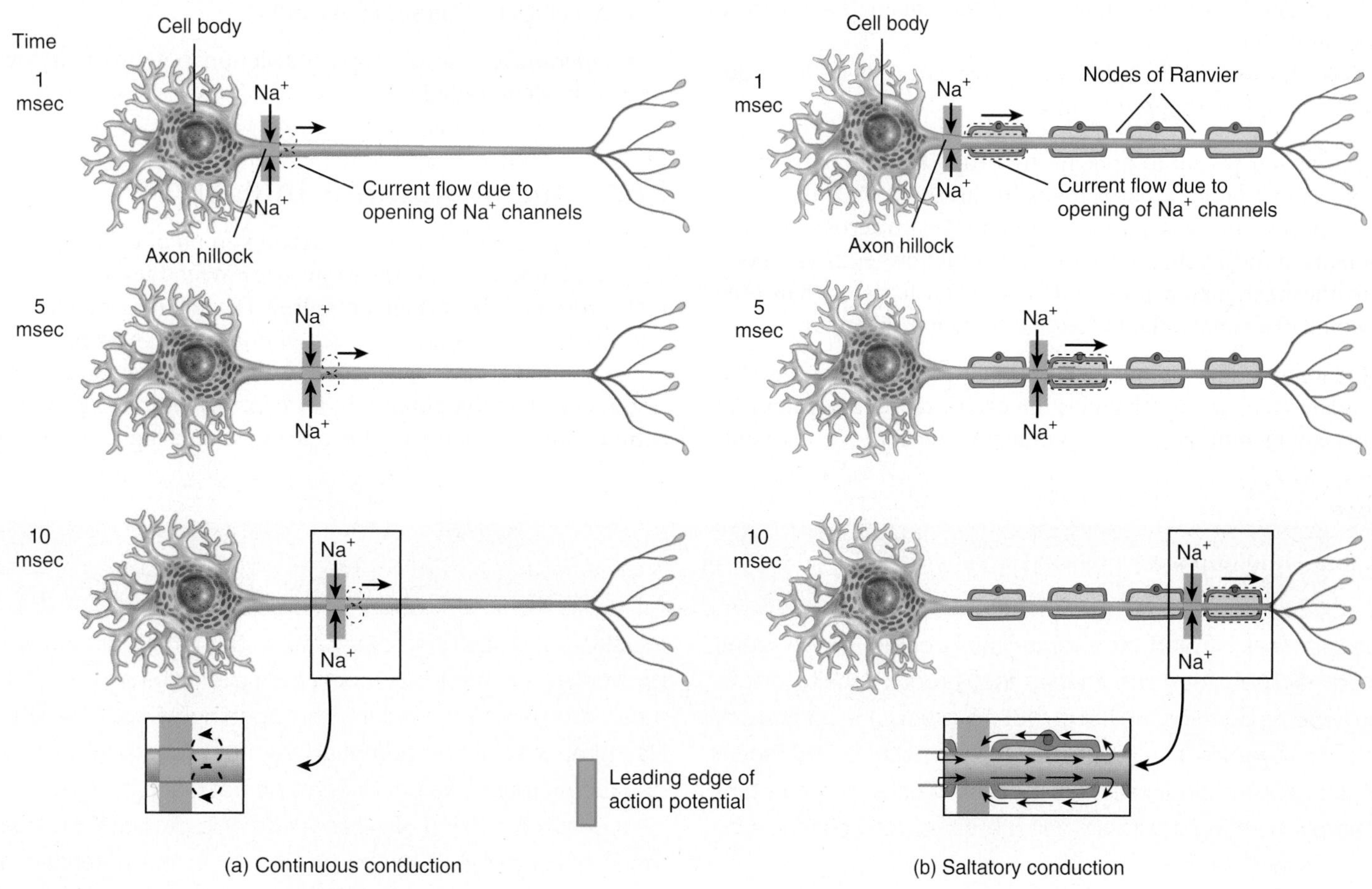


**Unmyelinated axons exhibit continuous conduction; myelinated axons exhibit saltatory conduction.**

Unmyelinated axons propagate action potentials by continuous conduction.

Action potentials propagate more rapidly along myelinated axons than along unmyelinated axons. If you compare parts a and b in Figure 12.20, you will see that the action potential propagates much farther along the myelinated axon in the same period of time. **Saltatory conduction** (SAL-ta-tō-rē; *saltat-* = leaping), the special mode of action potential propagation that occurs along myelinated axons, occurs because of the uneven distribution of voltage-gated channels. Voltage-gated channels are present primarily at the nodes of Ranvier (where there is no myelin sheath) rather than in regions where a myelin sheath covers the plasma membrane. Hence, current carried by $Na^+$ and $K^+$ flows across the membrane mainly at the nodes of Ranvier.

When an action potential propagates along a myelinated axon, an electric current flows from one node to the next through the extracellular fluid surrounding the myelin sheath and on through the cytosol. The action potential at the first node generates ionic currents that open voltage-gated $Na^+$ channels at the second node. The resulting ionic flow through the newly opened channels constitutes an action potential at the second node. Then the action potential at the second node generates an ionic current that opens voltage-gated $Na^+$ channels at the third node, and so on.

The flow of current across the membrane only at the nodes of Ranvier has two consequences:

- The action potential appears to "leap" from node to node as each nodal area depolarizes to threshold, thus the name "saltatory." Because an action potential leaps across long segments of the myelinated axon as current flows from one node to the next, action potentials travel much faster than they would in an unmyelinated axon of the same diameter.
- Opening a smaller number of channels only at the nodes, rather than many channels in each adjacent segment of plasma membrane, results in minimal inflow of $Na^+$ and outflow of $K^+$ each time an action potential occurs. Thus, less ATP is used by sodium–potassium pumps to maintain the low intracellular concentration of $Na^+$ and the low extracellular concentration of $K^+$.

## Factors That Affect the Speed of Propagation

The speed of propagation of an action potential is affected by three major factors: amount of myelination, axon diameter, and temperature.

- ***Amount of myelination.*** As you have just learned, action potentials propagate more rapidly along myelinated axons than along unmyelinated axons.
- ***Axon diameter.*** Larger-diameter axons propagate action potentials faster than smaller ones due to their larger surface areas. The largest-diameter axons (about 5–20 μm) propagate action potentials at speeds of 12 to 130 m/sec (27–280 mi/hr), medium-diameter axons (about 2–3 μm) propagate at speeds up to 15 m/sec (32 mi/hr), and action potentials transmitted in the smallest-diameter axons (0.5–1.5 μm) travel only at a speed of 0.5 to 22 m/sec (1–44 mi/hr).
- ***Temperature.*** Axons propagate action potentials at lower speeds when cooled.

## Encoding of Stimulus Intensity

How can your sensory systems detect stimuli of differing intensities if all impulses are the same size? Why does a light touch feel different from firmer pressure? The main answer to such questions is the *frequency of action potentials*—how often they are generated at the trigger zone. A light touch generates a low frequency of action potentials. By contrast, a firmer grip elicits action potentials that pass down the axon at a higher frequency.

### CLINICAL CONNECTION | *Neurotoxins and Local Anesthetics*

Certain shellfish and other organisms contain **neurotoxins** (noo′-rō-TOK-sins), substances that produce their poisonous effects by acting on the nervous system. One particularly lethal neurotoxin is tetrodotoxin (TTX), present in the viscera of Japanese pufferfish. TTX effectively blocks action potentials by inserting itself into voltage-gated $Na^+$ channels so they cannot open.

**Local anesthetics** are drugs that block pain and other somatic sensations. Examples include procaine (Novocaine®) and Lidocaine, which may be used to produce anesthesia in the skin during suturing of a gash, in the mouth during dental work, or in the lower body during childbirth. Like TTX, these drugs act by blocking the opening of voltage-gated $Na^+$ channels. Action potentials cannot propagate past the obstructed region, so pain signals do not reach the CNS.

Localized cooling of a nerve can also produce an anesthetic effect because axons propagate action potentials at lower speeds when cooled. The application of ice to injured tissue can reduce pain because propagation of the pain sensations along axons is partially blocked.

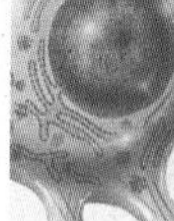

In addition to this "frequency code" for stimulus intensity, a second factor is the number of sensory neurons recruited (activated) by the stimulus. For example, a firm grip stimulates a larger number of pressure-sensitive neurons than does a light touch.

## Comparison of Electrical Signals Produced by Excitable Cells

You have seen that excitable cells—neurons and muscle fibers—produce two types of electrical signals: graded potentials and action potentials (impulses). One obvious difference between them is that graded potentials function only in short-distance communication, but the propagation of action potentials permits communication over long distances. Table 12.1 presents a summary of the differences between graded potentials and action potentials.

✓ **CHECKPOINT**

**29.** Why is saltatory conduction faster than continuous conduction of an action potential?

**30.** Why do large-diameter myelinated axons conduct action potentials more quickly than small-diameter unmyelinated axons?

**31.** How are stimuli of differing intensity detected as being different, such as warm soup versus hot soup?

**32.** Differentiate between graded potentials and action potentials.

**TABLE 12.1**

Comparison of Graded Potentials and Action Potentials in Neurons

| CHARACTERISTIC | GRADED POTENTIALS | ACTION POTENTIALS |
|---|---|---|
| **Origin** | Arise mainly in dendrites and cell body | Arise at trigger zones and propagate along the axon |
| **Types of channels** | Ligand-gated or mechanically gated ion channels | Voltage-gated channels for $Na^+$ and $K^+$ |
| **Conduction** | Not propagated; permit communication over short distances | Propagated; permit communication over longer distances |
| **Amplitude (size)** | Depending on strength of stimulus, varies from less than 1 mV to more than 50 mV | All-or-none (constant); typically about 100 mV |
| **Duration** | Typically longer, ranging from several milliseconds to several minutes | Shorter, ranging from 0.5 to 2 msec |
| **Polarity** | May be hyperpolarizing (inhibitory to generation of an action potential) or depolarizing (excitatory to generation of an action potential) | Always consists of depolarizing phase followed by repolarizing phase and return to resting membrane potential |
| **Refractory period** | Not present, thus summation can occur | Present, thus summation cannot occur |

## 12.9 The synapse is a special junction between neurons.

As you learned in Concept 12.3, a synapse is the site of communication between two neurons or between a neuron and an effector cell (muscle cell or glandular cell) through a series of events known as **synaptic transmission**.

In Concept 10.5 we examined the events occurring at the neuromuscular junction, the synapse between a somatic motor neuron and a skeletal muscle fiber (see Figure 10.6). Synapses between neurons operate in a similar fashion. Our focus in this chapter is on synaptic communication among the billions of neurons in the nervous system. Neurons filter, integrate, and process information using synaptic transmission. During learning, the structure and function of particular synapses change. For example, the changes in your synapses from studying will determine how well you do on your anatomy and physiology tests! Some diseases and neurological disorders result from disruptions of synaptic transmission. Synapses also are the sites of action for many therapeutic and addictive chemicals.

Most synapses are either **axodendritic** (ak′-sō-den-DRIT-ik = from axon to dendrite), **axosomatic** (ak′-sō-sō-MAT-ik = from axon to cell body), or **axoaxonic** (ak′-sō-ak-SON-ik = from axon to axon) (Figure 12.21). The neuron that carries an impulse toward a synapse is called the **presynaptic neuron** (*pre-* = before), and the neuron that carries an impulse away from a synapse is called the **postsynaptic neuron** (*post-* = after) (Figure 12.22). The two types of synapses, chemical and electrical, differ both structurally and functionally.

FIGURE 12.21 **Examples of synapses.** Arrows indicate the direction of information flow: presynaptic neuron → postsynaptic neuron. Presynaptic neurons usually synapse on the axon (axoaxonic; red), a dendrite (axodendritic; blue), or the cell body (axosomatic; green).

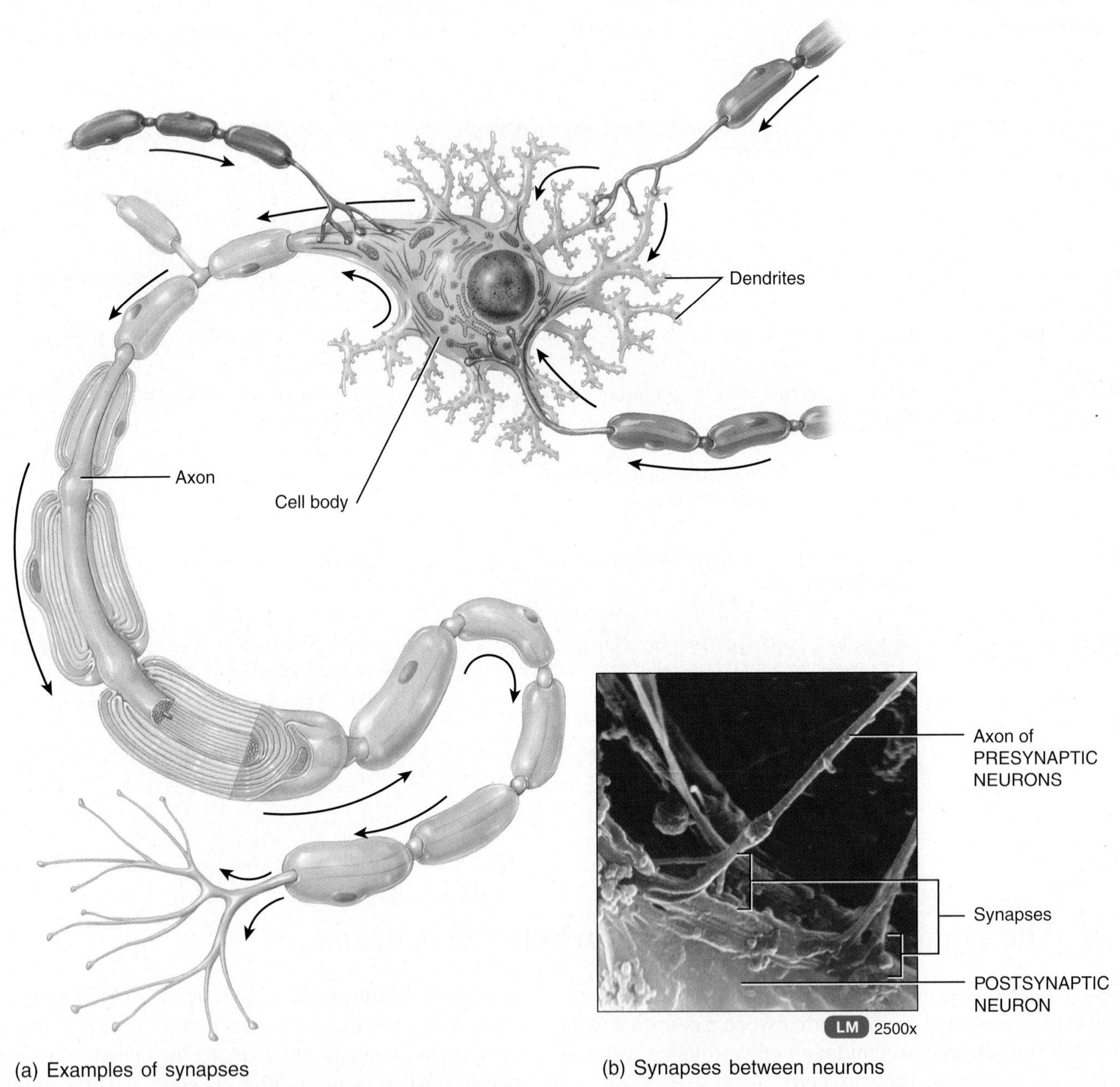


Neurons communicate with other neurons at synapses, which are junctions between one neuron and a second neuron or an effector cell.

## Chemical Synapses

Most synapses are **chemical synapses**. In a chemical synapse, an impulse in a presynaptic neuron causes the release if neurotransmitter molecules that produce an impulse in a postsynaptic neuron.

Although the plasma membranes of presynaptic and postsynaptic neurons are in close proximity at a chemical synapse, their plasma membranes do not touch. The neurons are separated by the **synaptic cleft**, a tiny space filled with interstitial fluid (Figure 12.22). Because impulses cannot conduct across the synaptic cleft, an alternative, indirect form of communication occurs between the presynaptic and postsynaptic neurons. In response to an impulse, the presynaptic neuron releases neurotransmitter molecules that diffuse through the fluid in the synaptic cleft and bind to receptors in the plasma membrane of the postsynaptic neuron. The postsynaptic neuron receives the chemical signal and, in turn, produces a **postsynaptic potential**, a type of graded potential. Thus, the presynaptic neuron converts an electrical signal (impulse) into a chemical signal (released neurotransmitter). The postsynaptic neuron receives the chemical signal and, in turn, generates an electrical signal (postsynaptic potential). The time required for these processes at a chemical synapse, a synaptic delay of about 0.5 msec, is the reason that chemical synapses relay signals more slowly than electrical synapses.

A typical chemical synapse transmits a signal as follows (Figure 12.22):

1. An impulse arrives at the synaptic end bulb of a presynaptic axon.
2. The depolarizing phase of the impulse opens **voltage-gated $Ca^{2+}$ channels** in the membrane of synaptic end bulbs. Because calcium ions are more concentrated in the extracellular fluid, $Ca^{2+}$ flows inward through the opened channels.
3. An increase in the concentration of $Ca^{2+}$ inside the presynaptic neuron triggers exocytosis of some of the synaptic vesicles.

FIGURE 12.22 **Signal transmission at a chemical synapse.** Through exocytosis of synaptic vesicles, a presynaptic neuron releases neurotransmitter molecules. After diffusing across the synaptic cleft, the neurotransmitter binds to receptors in the plasma membrane of the postsynaptic neuron and produces a postsynaptic potential.

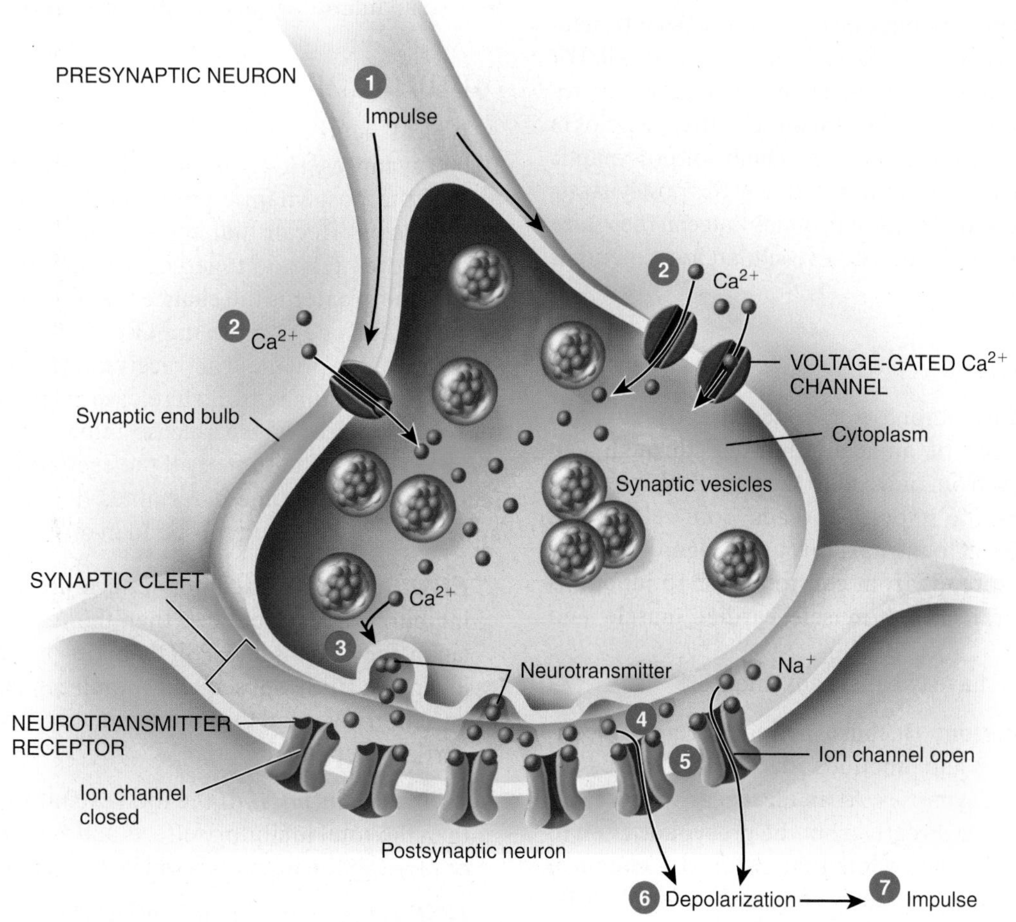


**At a chemical synapse, a presynaptic neuron converts an electrical signal (impulse) into a chemical signal (neurotransmitter release). The postsynaptic neuron then converts the chemical signal back into an electrical signal (postsynaptic potential).**

As vesicle membranes merge with the plasma membrane, neurotransmitter molecules within the synaptic vesicles are released into the synaptic cleft. Each synaptic vesicle contains several thousand molecules of neurotransmitter.

4. The neurotransmitter molecules diffuse across the synaptic cleft and bind to **neurotransmitter receptors** in the postsynaptic neuron's plasma membrane. The receptor shown in Figure 12.22 is part of a ligand-gated channel (see Figure 12.10b).
5. Binding of neurotransmitter molecules to their receptors on ligand-gated channels opens the channels and allows particular ions to flow across the membrane.
6. As ions flow through the opened channels, the voltage across the membrane changes. This change in membrane voltage is a postsynaptic potential. Depending on which ions the channels admit, the postsynaptic potential may be a depolarization (excitation) or hyperpolarization (inhibition). For example, opening of $Na^+$ channels allows inflow of $Na^+$, which causes depolarization. Opening of $K^+$ channels allows $K^+$ to move out, producing hyperpolarization.
7. When a depolarizing postsynaptic potential reaches threshold, it triggers an action potential in the axon of the postsynaptic neuron.

At most chemical synapses, only *one-way information transfer* can occur—from a presynaptic neuron to a postsynaptic neuron or to an effector, such as a muscle fiber or a gland cell. For example, synaptic transmission at a neuromuscular junction proceeds from a motor neuron to a skeletal muscle fiber (but not in the opposite direction). Only synaptic end bulbs of presynaptic neurons can release neurotransmitter, and only the postsynaptic neuron's membrane has the receptor proteins that can recognize and bind that neurotransmitter. As a result, action potentials move in one direction.

## Electrical Synapses

In an **electrical synapse**, impulses conduct directly between the plasma membranes of adjacent neurons through **gap junctions**. As ions flow from one cell to the next through these tunnel-like structures, the impulse spreads from cell to cell. As ions flow from one cell to the next through the gap junctions, the action potential spreads from cell to cell. Gap junctions are common in visceral smooth muscle, cardiac muscle, and the brain.

Electrical synapses have two main advantages:

- ***Faster communication.*** Because action potentials conduct directly through gap junctions, electrical synapses are faster than chemical synapses. At an electrical synapse, the action potential passes directly from the presynaptic cell to the postsynaptic cell. The events that occur at a chemical synapse take some time and delay communication slightly.
- ***Synchronization.*** Electrical synapses can synchronize (coordinate) the activity of a group of neurons or muscle fibers. Large numbers of neurons can produce action potentials in unison if they are connected by gap junctions. Synchronizing action potentials in the heart or in visceral smooth muscle coordinates contractions to produce a heartbeat or move food through the gastrointestinal tract.

## Excitatory and Inhibitory Postsynaptic Potentials

A neurotransmitter causes either an excitatory or an inhibitory graded potential. A neurotransmitter that depolarizes the postsynaptic membrane is excitatory because it brings the membrane closer to threshold (see Figure 12.13b). A depolarizing postsynaptic potential is called an **excitatory postsynaptic potential (EPSP)**. Although a single EPSP normally does not initiate an action potential, the postsynaptic neuron does become more excitable. Because it is partially depolarized, the postsynaptic neuron is more likely to reach threshold when the next EPSP occurs.

A neurotransmitter that causes *hyperpolarization* of the postsynaptic membrane (see Figure 12.13a) is inhibitory. During hyperpolarization, generation of an action potential is more difficult than usual because the membrane potential is more negative and thus even farther from threshold than in its resting state. A hyperpolarizing postsynaptic potential is termed an **inhibitory postsynaptic potential (IPSP)**.

## Summation of Postsynaptic Potentials

A typical neuron in the CNS receives input from 1000 to 10,000 synapses. Integration of these inputs involves **summation** of the postsynaptic potentials that form in the postsynaptic neuron. Recall that summation is the process by which graded potentials add together. The greater the summation of EPSPs, the greater is the chance that threshold will be reached. At threshold, one or more impulses (action potentials) arise. A single postsynaptic neuron receives input from many presynaptic neurons, some of which release excitatory neurotransmitters and some of which release inhibitory neurotransmitters (Figure 12.23). The sum of all the excitatory and inhibitory effects at any given time determines the effect on the postsynaptic neuron, which may respond in one of the following ways:

- ***EPSP.*** If the total excitatory effects are greater than the total inhibitory effects, but less than the threshold level of stimulation, the result is an EPSP that does not reach threshold. Following an EPSP, subsequent stimuli can more easily generate an action potential through summation because the neuron is partially depolarized.
- ***Action potential(s).*** If the total excitatory effects are greater than the total inhibitory effects and threshold is reached, one or more action potentials will be triggered.
- ***IPSP.*** If the total inhibitory effects are greater than the excitatory effects, the membrane hyperpolarizes (IPSP). The result is inhibition of the postsynaptic neuron and an inability to generate an action potential.

**FIGURE 12.23 Summation of postsynaptic potentials at the trigger zone of a postsynaptic neuron.** Presynaptic neurons 1, 3, and 5 release excitatory neurotransmitters (red dots) that generate excitatory postsynaptic potentials (EPSPs) (red arrows) in the membrane of a postsynaptic neuron. Presynaptic neurons 2 and 4 release inhibitory neurotransmitters (purple dots) that generate inhibitory postsynaptic potentials (IPSPs) (purple arrows) in the membrane of the postsynaptic neuron. The net summation of these EPSPs and IPSPs determines whether an action potential will be generated.



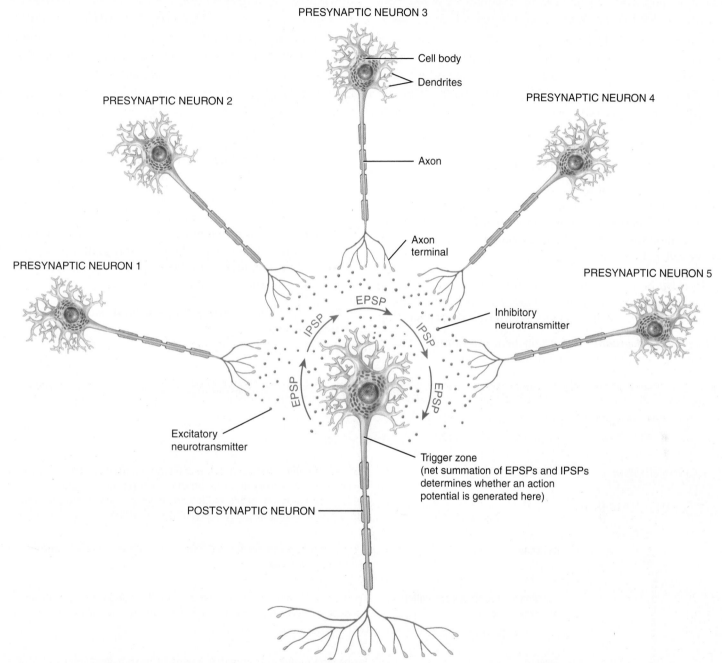

If the sum of all excitatory and inhibitory postsynaptic potentials is a depolarization that reaches threshold, then an action potential will occur at the trigger zone of a postsynaptic neuron.

## CLINICAL CONNECTION | *Strychnine Poisoning*

The importance of inhibitory neurons can be appreciated by observing what happens when their activity is blocked. Normally, inhibitory neurons in the spinal cord called *Renshaw cells* release the neurotransmitter glycine at inhibitory synapses with somatic motor neurons. This inhibitory input to their motor neurons prevents excessive contraction of skeletal muscles. **Strychnine** (STRIK-nīn) is a lethal poison that is mainly used as a pesticide to control rats, moles, gophers, and coyotes. When ingested, it binds to and blocks glycine receptors. The normal, delicate balance between excitation and inhibition in the CNS is disturbed, and motor neurons generate impulses without restraint. All skeletal muscles, including the diaphragm, contract fully and remain contracted. Because the diaphragm cannot relax, the victim cannot inhale, and suffocation results.

## Removal of Neurotransmitter

Removal of the neurotransmitter from the synaptic cleft is essential for normal synaptic function. If a neurotransmitter could linger in the synaptic cleft, it would influence the postsynaptic neuron, muscle fiber, or gland cell indefinitely. Neurotransmitter is removed in three ways:

- ***Diffusion.*** Some of the released neurotransmitter molecules diffuse away from the synaptic cleft. Once a neurotransmitter molecule is out of reach of its receptors, it can no longer exert an effect.
- ***Enzymatic degradation.*** Certain neurotransmitters are inactivated through enzymatic degradation. For example, the enzyme acetylcholinesterase breaks down acetylcholine in the synaptic cleft.
- ***Uptake by cells.*** Many neurotransmitters are actively transported back into the neuron that released them (reuptake). Others are transported into neighboring neuroglia (uptake). The neurons that release the neurotransmitter norepinephrine, for example, rapidly take up the norepinephrine and recycle it into new synaptic vesicles.

Table 12.2 summarizes the structural and functional elements of a neuron.

**TABLE 12.2**

Neuronal Structure and Function

| STRUCTURE | FUNCTIONS |
|---|---|
| **Dendrites** | Receive stimuli through activation of ligand-gated or mechanically gated ion channels; in sensory neurons, produce generator or receptor potentials; in motor neurons and interneurons, produce excitatory and inhibitory postsynaptic potentials (EPSPs and IPSPs) |
| **Cell body** | Receives stimuli and produces EPSPs and IPSPs through activation of ligand-gated ion channels |
| **Junction of axon hillock and initial segment of axon** | Trigger zone in many neurons; integrates EPSPs and IPSPs and, if sum is a depolarization that reaches threshold, initiates action potential (impulse) |
| **Axon** | Propagates impulses from initial segment (or from dendrites of sensory neurons) to axon terminals in a self-regenerating manner; impulse amplitude does not change as it propagates along the axon |
| **Axon terminals and synaptic end bulbs** | Inflow of $Ca^{2+}$ caused by depolarizing phase of impulse triggers exocytosis of neurotransmitter from synaptic vesicles |

Key:

- Plasma membrane includes chemically gated channels
- Plasma membrane includes voltage-gated $Na^+$ and $K^+$ channels
- Plasma membrane includes voltage-gated $Ca^{2+}$ channels

## ✓ CHECKPOINT

**33.** How do electrical and chemical synapses differ?

**34.** What sequence of events must occur for the release of neurotransmitter into the synaptic cleft?

**35.** Why is calcium required for the release of neurotransmitter at the synapse?

**36.** Explain the effects of excitatory and inhibitory postsynaptic potentials on the postsynaptic neuron.

## RETURN TO Jennifer's Story

Following her MRI, Jennifer goes back to the neurologist. She tells Dr. Peters that she has had some tremors in her left leg since she last saw him. He examines her again and finds the same abnormalities as before. He shows Jennifer the MRI and points out areas where the myelin sheaths of neurons in certain areas of her brain have deteriorated and formed scleroses (hardened scars or plaques). The destruction of the myelin sheaths short-circuits propagation of impulses, which is causing her symptoms. He gently explains that this pattern indicates that she has multiple sclerosis, a chronic neurological disorder of the central nervous system that she will have to cope with for the rest of her life.

Jennifer is devastated by the diagnosis, as she is about to graduate from Boston University and start her career in physical therapy. She asks Dr. Peters about her prognosis and available treatments. "Will I have any limitations?" The doctor tells her that the condition can either be progressive or relapsing/remitting, with alternating symptom-free periods and periods when the symptoms worsen. Dr. Peters gives Jennifer a prescription for an anti-inflammatory drug to try and slow the disease progression and relieve her current symptoms, and assures Jennifer that this is only one of a variety of medications available that they can try to treat her symptoms. But all Jennifer can think about is that there is no cure.

D. *Explain how the areas of demyelination cause Jennifer's symptoms.*

E. *Explain how a loss of myelin would influence $Na^+$–$K^+$ pump activity as related to the propagation of the action potential along the axon.*

F. *Would Jennifer's multiple sclerosis likely impact the transmission of action potentials from one neuron to another? Why or why not?*

# 12.10 PNS neurons have a greater capacity for repair and regeneration than CNS neurons.

Throughout your life, your nervous system exhibits **plasticity** (plas-TIS-i-tē), the capability to change based on experience. At the level of individual neurons, the changes that can occur include the sprouting of new dendrites, synthesis of new proteins, and changes in synaptic contacts with other neurons. Undoubtedly, both chemical and electrical signals drive the changes that occur. Despite plasticity, however, mammalian neurons have very limited powers of **regeneration**, the capability to replace or repair destroyed cells.

## Damage and Repair in the CNS

Little or no repair of damage to neurons occurs in the brain and spinal cord. Even when the cell body remains intact, a severed axon cannot be repaired or regrown. Although neurons are capable of arising in the hippocampus, an area of the brain that is crucial for learning, the nearly complete lack of regeneration in other regions of the brain and spinal cord seems to result from two factors. Axons in the CNS are myelinated by oligodendrocytes rather than Schwann cells, and this CNS myelin is one of the factors inhibiting regeneration of neurons. Also, after axonal damage, nearby astrocytes proliferate rapidly, forming scar tissue that is a physical barrier to regeneration. Thus, injury of the brain or spinal cord usually is permanent. Ongoing research seeks ways to improve the environment for existing spinal cord axons to bridge the injury gap. Scientists also are trying to find ways to stimulate dormant stem cells to replace neurons lost through damage or disease and to develop tissue-cultured neurons that can be used for transplantation purposes.

## Damage and Repair in the PNS

In the PNS, damaged dendrites and axons may be repaired if the cell body is intact, if the Schwann cells are functional, and if scar tissue formation does not occur too rapidly (Figure 12.24). Most nerves in the PNS consist of processes that are covered with a neurolemma. A person who injures axons of a nerve in an upper limb, for example, has a good chance of regaining nerve function.

When there is damage to an axon, changes usually occur both in the cell body of the affected neuron and in the portion of the axon distal to the site of injury. Following injury, the Nissl bodies break up into fine granular masses (Figure 12.24b). This alteration is called **chromatolysis** (krō-ma-TOL-i-sis; *chromato-* = color; *-lysis* = destruction). Within days the part of the axon distal to the damaged region breaks up into fragments and the myelin sheath deteriorates (Figure 12.24b). Even though the axon and myelin sheath degenerate, the neurolemma remains. Degeneration of the distal portion of the axon and myelin sheath is called **Wallerian degeneration** (waw-LĒR-ē-an).

Following chromatolysis, signs of recovery in the cell body become evident. Macrophages phagocytize the debris. Schwann cells on either side of the injured site multiply by mitosis, grow toward each other, and may form a **regeneration tube** across the injured area (Figure 12.24c). The tube guides growth of a new axon from the proximal area, across the injured area, into the distal area previously occupied by the original axon, and eventually toward receptors and effectors previously contacted by the neuron. Thus, some sensory and motor connections are reestablished and some functions restored. In time, the Schwann cells form a new myelin sheath.

FIGURE 12.24 **Damage and repair of a neuron in the PNS.**

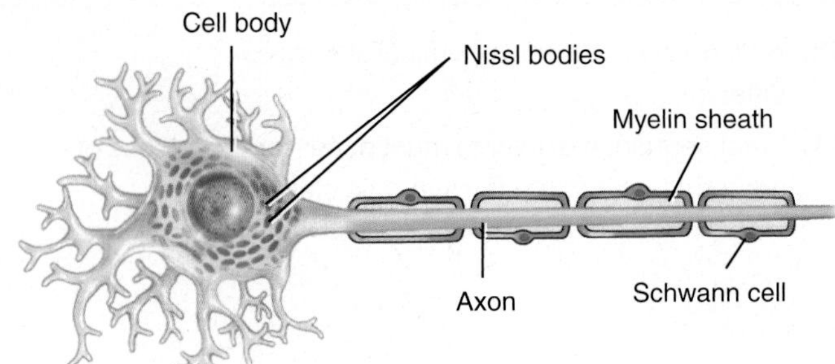


(a) Normal neuron

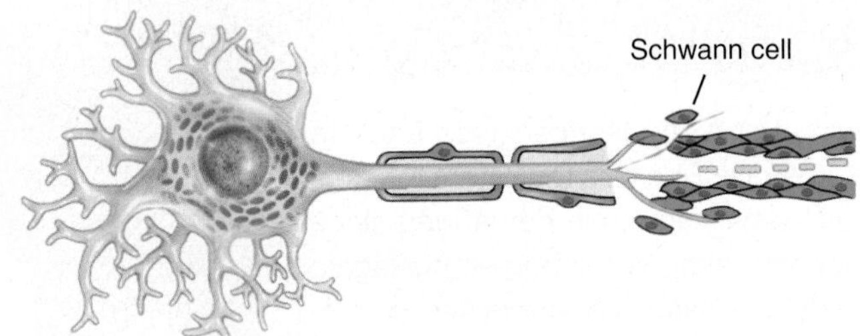


(b) Chromatolysis and Wallerian degeneration

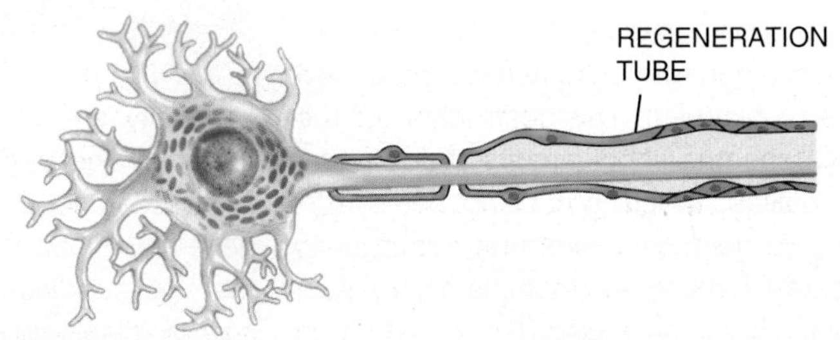


(c) Regeneration

**Myelinated axons in the peripheral nervous system may be repaired if the cell body remains intact and if Schwann cells remain active.**

### ✓ CHECKPOINT

**37.** Describe the repair process for a damaged PNS neuron.

**38.** How are Schwann cells involved in regeneration?

**39.** How is repair of nervous tissue in the PNS different from what occurs in the CNS?

## Jennifer's Story

### EPILOGUE AND DISCUSSION

Jennifer, a 22-year-old physical therapy student who has always been in good health, begins to have fatigue and blurred vision. This progresses to weakness in her hands and a tremor in her leg. She has no family history of any neurological conditions. She is given a prescription for glasses to wear when she reads her textbooks but this does not improve her visual problems. An MRI leads to a diagnosis of multiple sclerosis, a chronic, irreversible condition in which there is destruction of the myelin sheaths covering neurons throughout the central nervous system. In multiple sclerosis, myelin sheaths are destroyed by the patient's own immune system. Although the cause of this disease is unclear, it is understood that both genetic susceptibility and exposure to some environmental factor are thought to contribute.

In this chapter, you have learned how sensory neurons transmit information to the brain and spinal cord where the information is interpreted and a response generated that is transmitted down a motor neuron producing some action. You have also learned how and when the nervous system is able to repair itself. This chapter and those that follow (Chapters 13–16) are crucial to understanding the impact of a devastating diagnosis such as the one that Jennifer has just received.

G. *How would the outcome of Jennifer's disorder be different if multiple sclerosis was a disorder of the myelin sheath of the peripheral nervous system?*

# Concept and Resource Summary

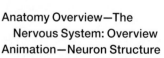

| Concept | Resources (WILEY PLUS) |
|---|---|
| **Introduction**<br>1. The nervous system, together with the endocrine system, governs homeostasis by sensing and responding to changes in normal physiological body set points.<br>2. The nervous system controls all body activities.<br>3. Neurons and neuroglia are the cells that form nervous tissue. | |
| 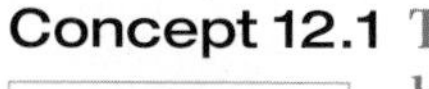 **Concept 12.1 The nervous system maintains homeostasis and integrates all body activities.**<br>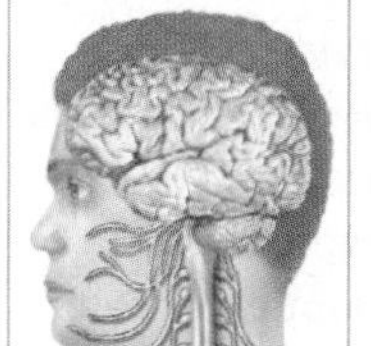<br>1. The nervous system integrates body activities through three basic functions.<br>2. The **sensory function** involves detecting internal and external stimuli.<br>3. The **integrative function** involves processing of sensory information.<br>4. The **motor function** occurs when the nervous system elicits a response by activating **effectors** such as muscles and glands. | Anatomy Overview—The Nervous System: Overview<br>Animation—Neuron Structure and Function |
| **Concept 12.2 The nervous system is organized into the central and peripheral nervous systems.**<br>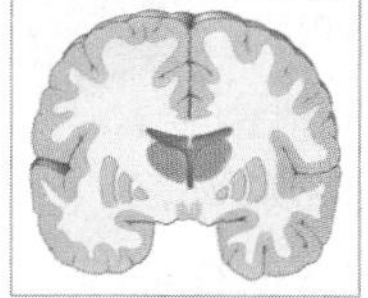<br>1. The **central nervous system (CNS)** consists of the **brain** and **spinal cord**. The CNS is responsible for integration of incoming sensory information, generation of motor commands, and thoughts and memories.<br>2. The **peripheral nervous system (PNS)** is composed of **cranial** and **spinal nerves**, **ganglia**, and **sensory receptors**. Subdivisions of the PNS include the **somatic nervous system (SNS)**, **autonomic nervous system (ANS)**, and **enteric nervous system (ENS)**.<br>3. The SNS consists of sensory neurons that conduct impulses from somatic and special sense receptors to the CNS. The SNS also has motor neurons that innervate skeletal muscles under voluntary control. | Animation—Structure and Function of the Nervous System: System Organization |

| Concept | Resources  |
|---|---|

4. The ANS consists of sensory neurons that conduct impulses from visceral organs to the CNS and motor neurons that convey impulses from the CNS to smooth muscle tissue, cardiac muscle tissue, and glands.
5. The ENS consists of neurons in enteric plexuses in the gastrointestinal (GI) tract that function somewhat independently of the ANS and CNS. The ENS monitors sensory changes in and controls operation of the GI tract.

## Concept 12.3 Neurons are responsible for most of the unique functions of the nervous system.

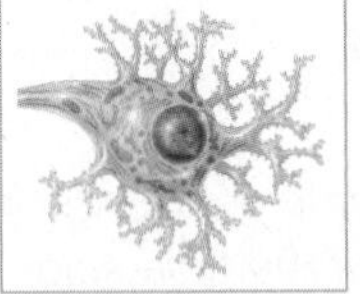

1. There are three parts to most **neurons**: cell body, dendrites, and axon.
2. The **cell body** houses the nucleus surrounded by cytoplasm containing organelles such as clusters of rough endoplasmic reticulum called **Nissl bodies**, lysosomes, mitochondria, Golgi complexes, and a cytoskeleton of neurofibrils and microtubules.
3. Multiple **dendrites** are processes that receive incoming impulses.
4. A single long, thin **axon** conducts impulses away from the cell body toward another neuron, muscle fiber, or gland cell.
5. The **axon hillock** is where the axon joins the cell body. **Axon collaterals** are side branches of an axon. The **axoplasm** is the axon's cytoplasm.
6. **Axon terminals** are numerous, fine processes at the ends of an axon and its collaterals.
7. Synaptic end bulbs are swollen tips of axon terminals where communication occurs between the neuron and other cells.
8. Newly synthesized cell products move by **anterograde transport** from the cell body toward the axon. Materials for recycling and degradation move from the axon to the cell body by **retrograde transport**.
9. Neurons can be classified structurally as **multipolar** (several dendrites and one axon), **bipolar** (one dendrite and one axon), or **unipolar** (one process arising from the cell body that branches into two axon-like processes).
10. Neurons are functionally classified as **sensory (afferent) neurons**, **motor (efferent) neurons**, and **interneurons**. Sensory neurons carry sensory information into the CNS. Motor neurons carry information out of the CNS to effectors (muscles and glands). Interneurons are located within the CNS and integrate (process) incoming sensory information from sensory neurons and then elicit a motor response by activating motor neurons.

**Resources:**

Anatomy Overview—Nerve

Anatomy Overview—Nervous Tissue: Neuron

Animation—Neuron Structure and Function

Figure 12.3—Structure of a Typical Motor Neuron and Schwann Cells

Exercise—Paint a Neuron

Clinical Connection—Tetanus

## Concept 12.4 Neuroglia support, nourish, and protect neurons and maintain homeostasis.

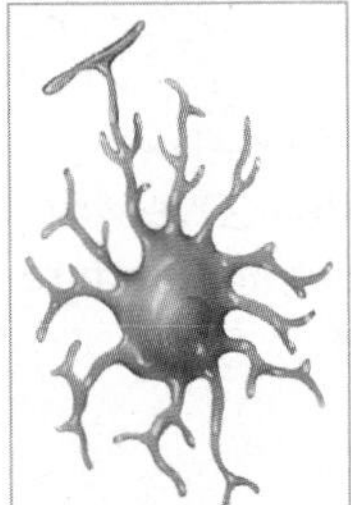

1. **Neuroglia** are smaller and more numerous than neurons. They can divide but cannot conduct impulses.
2. Neuroglia in the CNS include **astrocytes**, **oligodendrocytes**, **microglia**, and **ependymal cells**. Neuroglia in the PNS include Schwann cells and satellite cells.
3. The **myelin sheath** is a multilayered lipid and protein covering produced by layers of plasma membrane from neuroglia that wrap around **myelinated** axons.
4. In the PNS, Schwann cells form myelin sheaths; in the CNS, oligodendrocytes form myelin sheaths.
5. The unmyelinated gaps occurring at intervals along an axon are called the **nodes of Ranvier**.
6. The **neurolemma** is the outermost, nucleated layer of a Schwann cell that encloses the myelin sheath and functions in regeneration of damaged PNS axons.
7. The myelin sheath of CNS oligodendrocytes lacks a neurolemma. CNS axons show little regeneration.
8. Myelination increases from birth to maturity and functions to increase speed of impulse conduction along an axon.
9. Nervous system **white matter** appears white due to the presence of myelinated axons; the darker **gray matter** is composed of unmyelinated axons, cell bodies, and neuroglia.

**Resources:**

Anatomy Overview—Nerve

Anatomy Overview—Nervous Tissue: Neuroglia

Animation—Factors That Affect Conduction Rates: Myelination

Figure 12.5—Neuroglia of the Central Nervous System

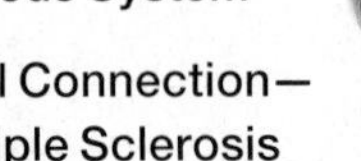

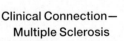

Clinical Connection—Multiple Sclerosis

Clinical Connection—Guillain-Barré Syndrome

Clinical Connection—Brain Tumors

| Concept | Resources |
|---|---|

## Concept 12.5 Neurons communicate with other cells.

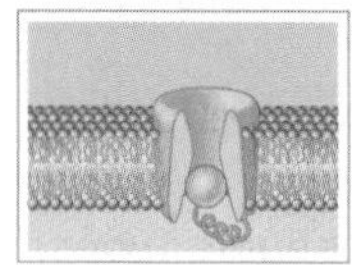

1. Neurons are electrically excitable cells that communicate by electrical signals called graded potentials over short distances or by action potentials over longer distances.
2. Nerve action potentials (impulses) occur in neurons; muscle action potentials occur in muscle fibers.
3. Graded potentials typically occur in sensory receptors, dendrites, and cell bodies, which allow axons to form action potentials resulting in release of **neurotransmitters**.
4. Excitable cell plasma membranes have a **resting membrane potential** due to an electrical voltage difference across the membrane established by specific ion channels.
5. Gated ion channels open or close in response to specific stimuli. When the channels open, the movement of cations and anions results in a flow of electrical current that can change the membrane potential.
6. There are four types of ion channels: **leak channels** have alternating open and closed gates; **ligand-gated channels** open and close due to a chemical stimulus in which **ligand** binds to a membrane receptor; **mechanically gated channels** open and close due to mechanical stimulation; and **voltage-gated channels** open when the membrane potential changes and participate in action potentials.
7. The resting membrane potential is more negative along the cytosol side of the plasma membrane and more positive along the extracellular fluid side of the membrane. The separation of these charges forms potential energy.
8. A cell with a membrane potential, such as a neuron, is **polarized**; neurons typically have a resting membrane potential of −70 mV.
9. Resting membrane potentials result from differences in distribution of charged ions across the plasma membrane, from a greater permeability to $K^+$ and $Cl^-$ compared to $Na^+$, and from action of the sodium–potassium pumps.

**Resources (WileyPLUS):**

Animation—Introduction to Membrane Potentials
Animation—Membrane Transport Proteins
Animation—Resting Membrane Potential
Figure 12.11—Resting Membrane Potential

Exercise—Keep the Resting Potential
Clinical Connection—Parkinson's Disease

## Concept 12.6 Graded potentials are the first response of a neuron to stimulation.

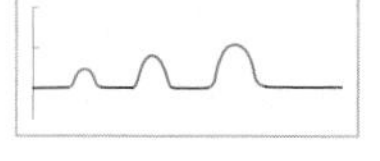

1. The opening or closing of ligand-gated and mechanically gated channels in response to a stimulus produces **graded potentials**.
2. **Depolarizing graded potentials** result in a less polarized membrane.
3. **Hyperpolarizing graded potentials** result in an even more polarized membrane.
4. The size of a graded potential varies with the strength of the stimulus.
5. Opening or closing of ion channels in graded potentials causes a localized flow of current along the membrane.

**Resources (WileyPLUS):**

Animation—Graded Potentials

## Concept 12.7 The action potential is an all-or-none electrical signal.

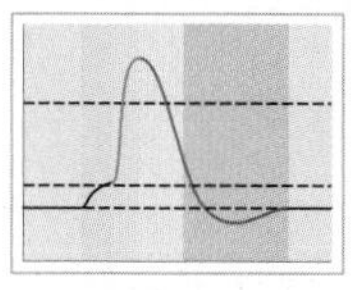

1. **Action potentials (impulses)** are rapid electrical events occurring in two phases: the depolarizing phase and the repolarizing phase.
2. Action potentials follow the **all-or-none principle**; once **threshold** depolarization occurs, voltage-gated channels open, and an action potential that is always the same size occurs.
3. During the **depolarizing phase**, voltage-gated $Na^+$ channels quickly open, $Na^+$ rushes into the cell, and the membrane potential becomes positive.
4. During the **repolarizing phase**, voltage-gated $K^+$ channels slowly open, $K^+$ flows out of the cell, and the membrane is repolarized.
5. Following an action potential there is a **refractory period** during which an excitable cell cannot generate another action potential.
6. The **absolute refractory period** is the time when even a very strong stimulus cannot generate a second action potential.
7. During a **relative refractory period**, a very strong stimulus can initiate a second action potential.

**Resources (WileyPLUS):**

Animation—Action Potentials
Figure 12.17—Action Potential

Exercise—Ruling the Gated Channels
Exercise—Define Stages of the Action Potential
Concepts and Connections—Membrane Potentials
Clinical Connection—Epilepsy

| Concept | Resources |
|---|---|

## Concept 12.8 Action potentials propagate from the trigger zone to axon terminals.

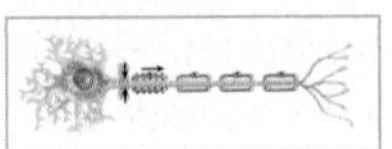

1. When $Na^+$ flows into open channels in one area of the membrane, voltage-gated $Na^+$ channels in adjacent segments of the membrane open, resulting in **impulse propagation** of the impulse toward the axon terminals.
2. **Continuous conduction** occurs in unmyelinated axons.
3. Myelinated axons undergo a more rapid **saltatory conduction** in which $Na^+$ and $K^+$ flow across the membrane only at the unmyelinated nodes of Ranvier as the impulse leaps from one node to another.
4. Impulses are conducted at a faster pace along larger-diameter axons.
5. Greater stimulus intensity, in addition to a greater number of recruited sensory neurons, leads to higher frequency of impulses and the ability to differentiate between stimuli.
6. Action potentials permit communication over long distances; graded potentials function only in short-distance communication. Their origins, amplitudes, duration, and types of channels also differ.

Animation—Propagation of Nerve Impulses
Exercise—Investigate Types of Impulse Conduction
Exercise—Speed Up Nerve Impulse Propagation

## Concept 12.9 The synapse is a special junction between neurons.

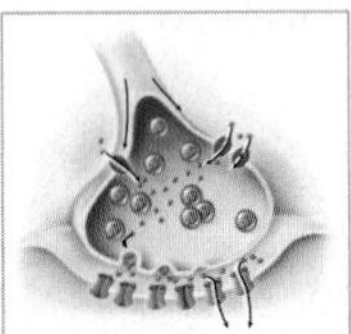

1. The synapse is the junction where communication occurs between two neurons or a neuron and an effector cell.
2. **Chemical synapses** conduct impulses from a neuron to another cell through a **synaptic cleft**. An impulse in a presynaptic axon opens **$Ca^{2+}$ channels** in synaptic end bulbs, leading to release of neurotransmitter that crosses the synaptic cleft and binds to postsynaptic neuron receptors, causing a **postsynaptic potential**.
3. **Electrical synapses** conduct impulses from a neuron to another cell through **gap junctions**. Electrical synapses allow faster communication and synchronization of activity of cell groups. They are common in visceral smooth muscle and cardiac muscle.
4. Some neurotransmitters will depolarize the postsynaptic neuron's membrane to produce an **excitatory postsynaptic potential (EPSP)**, bringing the membrane potential closer to threshold. Other neurotransmitters will hyperpolarize the postsynaptic neuron's membrane to produce an **inhibitory postsynaptic potential (IPSP)**, moving the membrane potential farther from threshold.
5. **Summation** of postsynaptic potentials determines whether the postsynaptic neuron will generate an action potential.
6. Neurotransmitter is removed from the synaptic cleft by diffusion, enzymatic degradation, or cellular uptake.

Animation—Events at the Synapse
Exercise—Are Your Synapses Working?
Exercise—Summation Smash
Concepts and Connections—Synaptic Summation
Additional Content—Neurotransmitters

## Concept 12.10 PNS neurons have a greater capacity for repair and regeneration than CNS neurons.

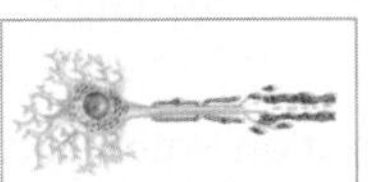

1. CNS neurons have little to no repair capacity.
2. Axons and dendrites that are associated with a neurolemma in the PNS may undergo repair if the cell body is intact, the Schwann cells are functional, and scar tissue formation does not occur too rapidly. Repair of PNS neurons begins with **Wallerian degeneration** of the distal region of the axon and myelin sheath.
3. A **regeneration tube** formed by Schwann cells directs passage of the regenerating axon to previously contacted receptors and effectors.

## Understanding the Concepts

1. Would you expect to find an effector in your ears, brain, or legs?
2. Which of the following are voluntary? Involuntary? Both? Explain each of your answers.
   - Moving your eyes across this page.
   - The tearing of your eyes when sand is blown into them.
3. What types of problems would result if your sensory neurons were damaged? If your interneurons were damaged? If your motor neurons were damaged?
4. What structural type of neuron would your eyes likely use to receive images of words on this page? What structural type of neuron would your brain likely use to determine what the words mean?
5. Your toothpaste tube is almost empty. Starting with the seam end, you tightly wrap the tube around a pencil until the last of the toothpaste is collected in the cap region of the tube. Assuming the toothpaste tube and pencil are analogous to components of nervous tissue, what structure would be represented by each of the following?
   - The cap region of the toothpaste tube containing the remaining toothpaste.
   - The tightly rolled component of the tube that is now empty of toothpaste.
   - The pencil.
6. What type of gated channel is activated by a touch on the arm?
7. What kind of graded potential describes a change in membrane potential from –70 to –60 mV? From –70 to –80 mV?
8. How would the arrival of hyperpolarizing graded potentials at the trigger zone of a neuron affect the generation of action potentials?
9. If the myelin sheath were destroyed or damaged, would you expect action potential propagation to speed up or slow down? Why?
10. Why can electrical synapses work in two directions, but chemical synapses are only able to transmit a signal in one direction?
11. What is the function of the regeneration tube in repair of neurons?

CHAPTER

# 13

## Annette's Story

Annette answers the phone late one evening to find her older son on the other end of the line.

"Hi, Mom," he says. "How are you doing? Would you like to have some company next week?"

"Why of course," she replies with a big smile. "I always look forward to seeing you. I hope you'll be bringing Jenny and the girls too."

Bill tells his mother that he'll be bringing his wife Jenny and daughters Samantha and Sophia. Annette and Bill discuss their plans and decide that Bill and his family will arrive on Monday and stay for four days.

After they hang up, Annette immediately starts making lists of things to do to prepare for the visit. She has to shop for all of their favorite foods, and plan some fun things to do with her granddaughters. With only three days to prepare, she needs to get moving. Maybe she should call Jeff to see if he wants to come visit with his brother, but decides against it, not sure she wants everyone at the house at the same time. After all, she has been told by Dr. Akers to take it easy and not to tire herself out.

Annette has been seeing Dr. Akers for three years because of tremors she was experiencing in her hands and forearms, difficulty starting and stopping movements when she walked, and muscular rigidity that was occurring when she tried to move her body in a certain way. At that time Dr. Akers diagnosed her with Parkinson's disease, a progressive movement disorder that primarily affects people over the age of 50. He prescribed levodopa/carbidopa, a medication that she must take every three hours during the day. Sometimes it partially relieves her symptoms, and sometimes it doesn't seem to do much at all. Recently, Dr. Akers placed her on a newer drug called ropinirole to make her brain respond as if it is receiving the neurotransmitter dopamine. Annette has seen no difference in her symptoms in spite of the dosage being increased on a regular basis.

Annette tries to stay upbeat about her condition, but it is hard sometimes. She considers talking to her family about it during their visit. But the thought of her children telling her what to do just doesn't sit well with her since she is only 70 years old. Until three years ago, she thought that she could expect to be independent for another 10 to 15 years or more.

What is happening to Annette? Is there anything else that can be done to help her?

# The Central Nervous System

## INTRODUCTION

Solving an equation, feeling hungry, laughing—the neural processes needed for each of these activities occurs in different regions of the brain, that portion of the central nervous system contained within the cranium. The brain is the control center for registering sensations, correlating them with one another and with stored information, making decisions, and taking actions. It is also the center for the intellect, emotions, behavior, and memory. But the brain encompasses yet a larger domain: with ideas that excite, artistry that dazzles, or rhetoric that mesmerizes, one person's thoughts and actions may influence and shape the lives of many others.

The spinal cord controls some of your most rapid reactions to environmental changes. If you pick up something hot, you may drop the object even before you are consciously aware of the extreme heat or pain. This is an example of a spinal cord reflex—a quick, automatic response to certain kinds of stimuli that involves neurons only in the spinal cord. The spinal cord also is the pathway for sensory impulses traveling to the brain and motor impulses traveling from the brain to skeletal muscles and other effectors.

This chapter explores how the brain and spinal cord are protected and nourished, functions occur in major regions of the central nervous system, and how together, the brain and spinal cord form the control center of the human body.

## CONCEPTS

**13.1** The CNS consists of the brain and spinal cord, and is protected by several structures.

**13.2** The CNS is nourished and protected by blood and cerebrospinal fluid.

**13.3** The cerebrum interprets sensory impulses, controls muscular movements, and functions in intellectual processes.

**13.4** The cerebral cortex can be divided functionally into sensory areas, motor areas, and association areas.

**13.5** The diencephalon includes the thalamus, hypothalamus, and pineal gland.

**13.6** The midbrain, pons, and medulla oblongata of the brain stem serve as a relay station and control center.

**13.7** The cerebellum coordinates movements and helps maintain normal muscle tone, posture, and balance.

**13.8** The limbic system controls emotions, behavior, and memory.

**13.9** The spinal cord receives sensory input and provides motor output through spinal nerves.

**13.10** The spinal cord conducts impulses between spinal nerves and the brain, and contains reflex pathways.

## 13.1 The CNS consists of the brain and spinal cord, and is protected by several structures.

The brain is continuous with the spinal cord and together they comprise the central nervous system (CNS). The brain consists of four major parts: cerebrum, diencephalon, brain stem, and cerebellum (Figure 13.1). The **cerebrum** (se-RĒ-brum = brain) is the largest part of the brain. Immediately below the cerebrum is the **diencephalon** (dī-en-SEF-a-lon; *di-* = through; *-encephalon* = brain), containing the thalamus and hypothalamus. The diencephalon provides a structural connection between the cerebrum and the brain stem. The **brain stem** consists of the midbrain, pons, and medulla oblongata (Figure 13.1b). Posterior to the brain stem is the **cerebellum** (ser′-e-BEL-um = little brain). The **spinal cord** extends inferiorly from the medulla oblongata to the level of the first or second lumbar vertebra (Figure 13.2).

FIGURE 13.1 **The brain.** The pituitary gland is discussed with the endocrine system in Concepts 17.3 and 17.4.

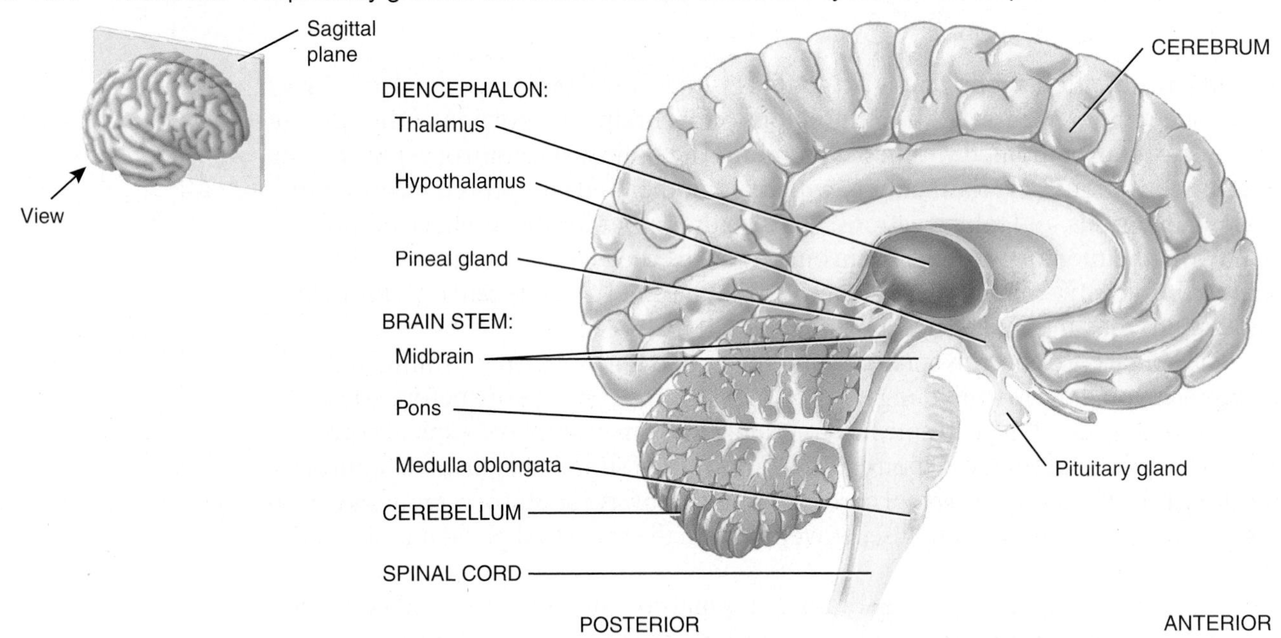


(a) Sagittal section, medial view



(b) Sagittal section, medial view

**The four principal parts of the brain are the cerebrum, diencephalon, brain stem, and cerebellum.**

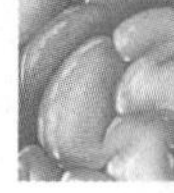

FIGURE 13.2 **External anatomy of the spinal cord and the spinal nerves.**

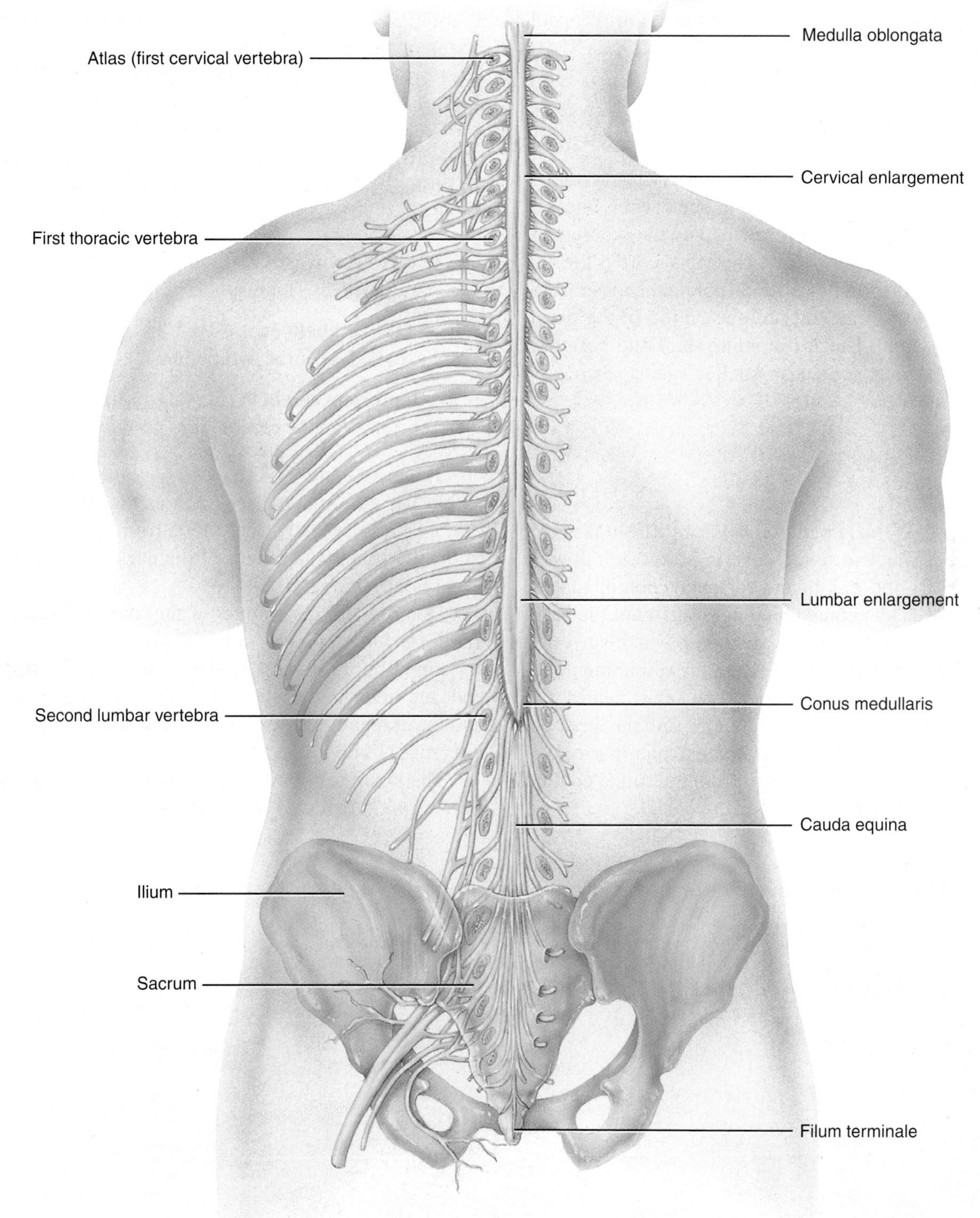

Medulla oblongata
Atlas (first cervical vertebra)
Cervical enlargement
First thoracic vertebra
Lumbar enlargement
Conus medullaris
Second lumbar vertebra
Cauda equina
Ilium
Sacrum
Filum terminale

Posterior view of entire spinal cord and portions of spinal nerves (unlabeled)

**The spinal cord extends from the medulla oblongata of the brain to the superior border of the second lumbar vertebra.**

Recall from Chapter 12 that the nervous tissue of the central nervous system is very delicate and does not respond well to injury or damage. Accordingly, this nervous tissue requires considerable protection.

## Skeletal Protection

The first layer of protection for the central nervous system is the hard bony skull and vertebral column. The brain is encased within the cranial cavity of the skull, formed by the securely interlocked cranial bones (Figure 13.3). The spinal cord is located within the vertebral canal of the vertebral column. The vertebral foramina of all the vertebrae, stacked one on top of the other, form the vertebral canal. The surrounding skull and vertebral column provide a strong, protective shelter against damaging blows or bumps. (see Figure 13.4c).

## Meninges

The **meninges** (me-NIN-jēz; singular is *meninx* [MEN-inks]) are three protective, connective tissue coverings that lie between the bony encasement of the skull and vertebral column and the brain and spinal cord. The **cranial meninges** surround the brain and are continuous with the **spinal meninges**, which surround the spinal cord. The cranial meninges and the spinal meninges have the same basic structure and bear the same names: the outer dura mater, the middle arachnoid mater, and the inner pia mater (Figures 13.3 and 13.4). All three spinal meninges cover the spinal nerves up to the point of exit from the spinal column through the intervertebral foramina.

The most superficial and strongest of the three meninges is the **dura mater** (DOO-ra MĀ-ter = tough mother), which is composed of dense, irregular connective tissue. The dura mater of the brain adheres directly to the periosteum of the interior surface of cranial bones. Within the vertebral canal, an **epidural space** (ep′-i-DOO-ral) exists between the dura mater and the wall of the vertebral canal. The spinal cord is protected not only by the strong dura mater, but also by a cushion of fat and connective tissue located within the epidural space. Three extensions of the dura mater separate portions of the brain:

- The **falx cerebri** (FALKS SER-e-brē; *falx* = sickle-shaped) separates the two hemispheres (sides) of the cerebrum.
- The **falx cerebelli** (cer-e-BEL-li) separates the two hemispheres (sides) of the cerebellum.
- The **tentorium cerebelli** (ten-TŌ-rē-um = tent) separates the cerebrum from the cerebellum.

The dura mater of the brain contains **dural sinuses**, spaces within the dura mater itself such as the **superior sagittal sinus** (Figure 13.3a), which drains blood from the brain and delivers it to the internal jugular veins of the neck. (A *sinus* is similar to a vein but has thinner walls.)

Deep to the dura mater is the **arachnoid mater** (a-RAK-noyd; *arachn-* = spider; *-oid* = similar to), an avascular covering so named because of its spider's web arrangement of delicate collagen fibers and some elastic fibers. Between the dura mater and the arachnoid mater is a thin **subdural space**, which contains interstitial fluid.

The innermost meningeal membrane is the **pia mater** (PĒ-a MĀ-ter; *pia* = delicate), a thin, transparent connective tissue layer

FIGURE 13.3 The protective coverings of the brain.

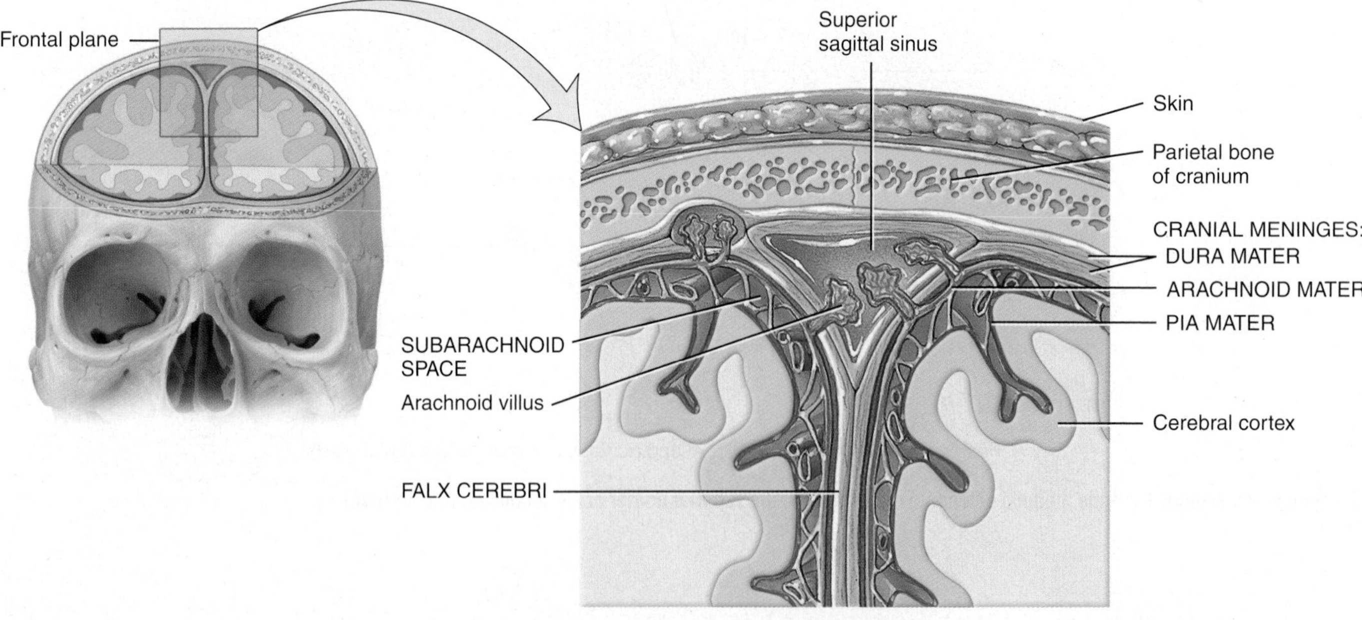


(a) Anterior view of frontal section through skull showing cranial meninges

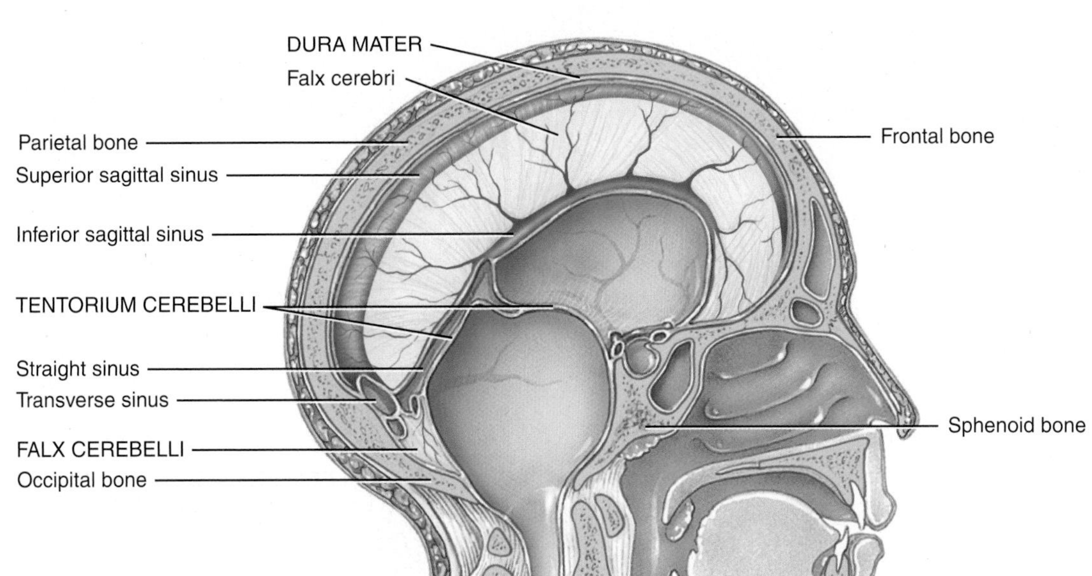


(b) Sagittal section of extensions of dura mater

**Cranial bones and the cranial meninges protect the brain.**

that adheres tightly to the surface of the brain and spinal cord. It contains interlacing bundles of collagen fibers and some fine elastic fibers. Within the pia mater are many blood vessels that supply oxygen and nutrients to the brain and spinal cord. Triangular-shaped membranous extensions of the pia mater called **denticulate ligaments** (den-TIK-ū-lāt = small tooth) project laterally from the spinal cord and fuse with the arachnoid mater and inner surface of the dura mater (see Figure 13.4a, b). Extending all along the entire length of the spinal cord, the denticulate ligaments protect the spinal cord against sudden displacement that could result in shock.

FIGURE 13.4 **Gross anatomy of the spinal cord and its protective coverings.**

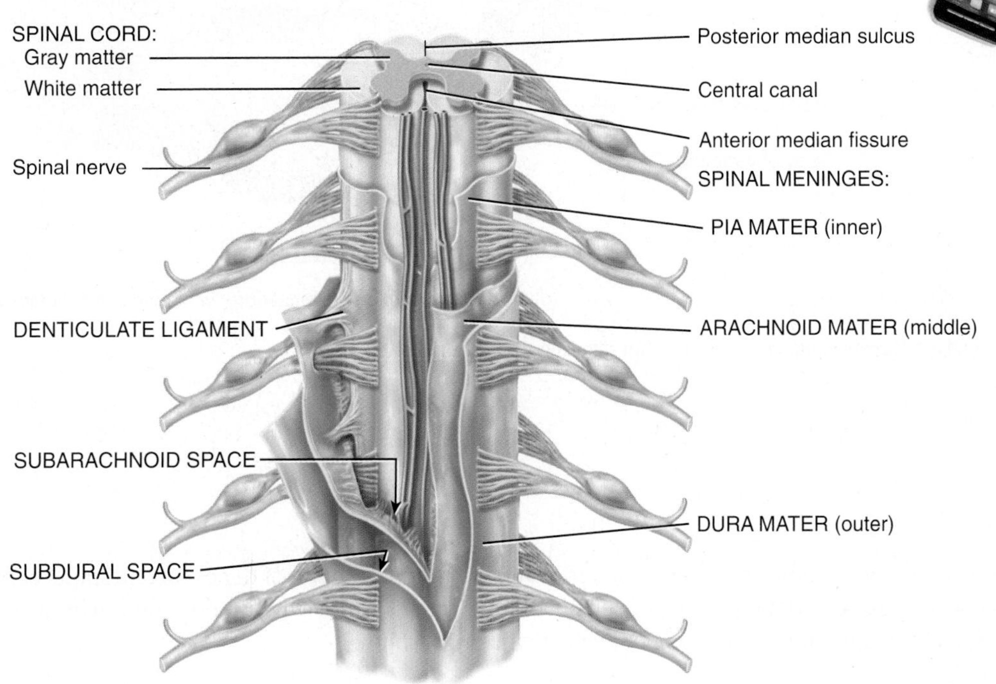


(a) Anterior view and transverse section through spinal cord

(continues)

FIGURE 13.4 (continued)

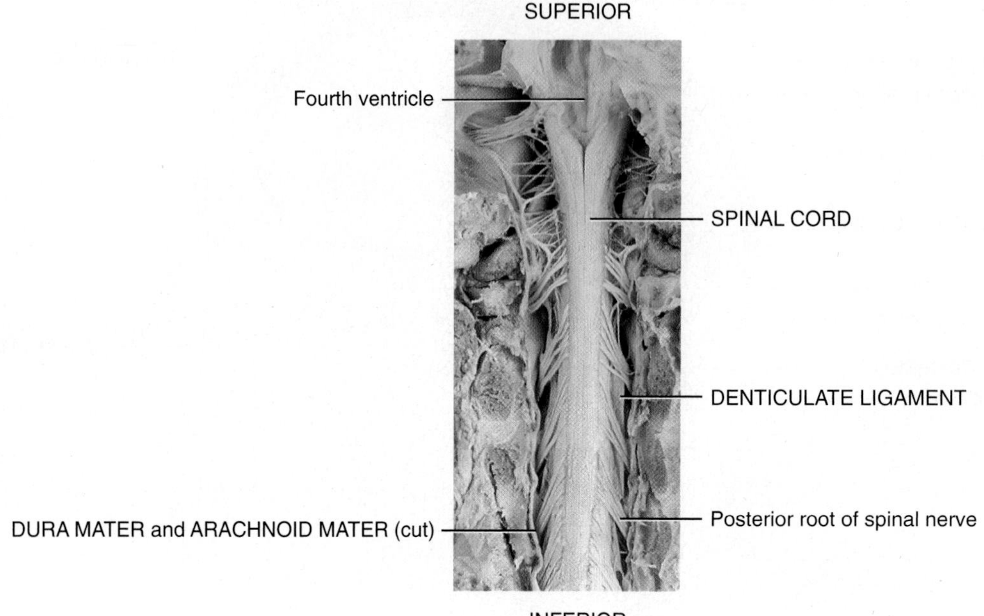


(b) Posterior view of cervical region of spinal cord

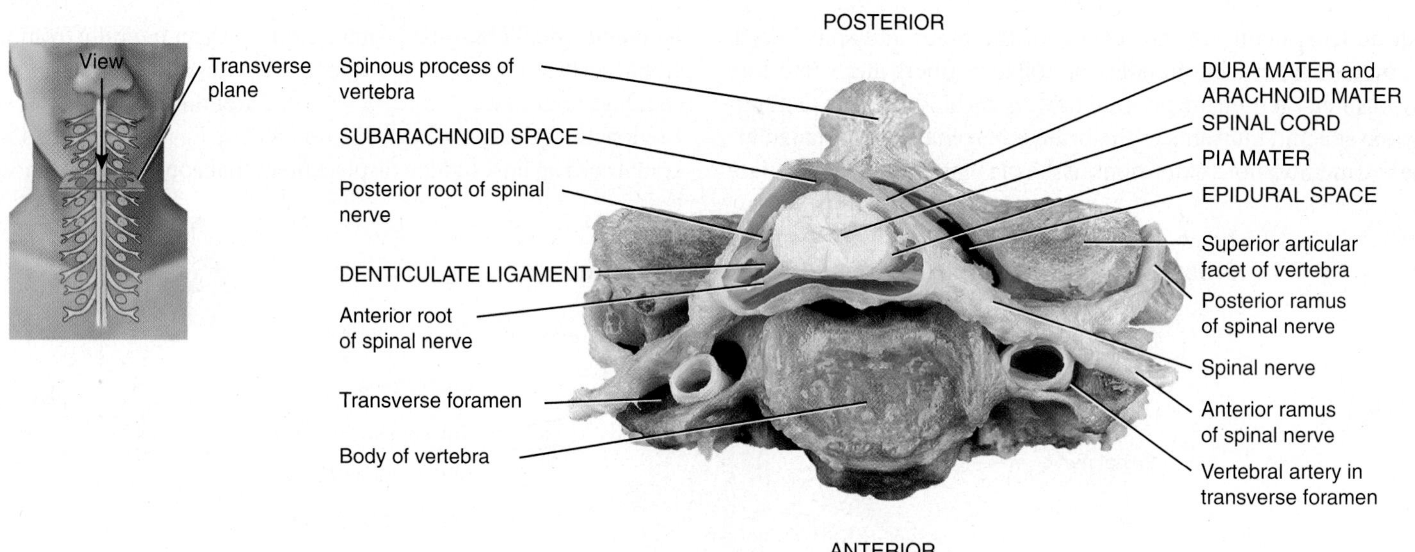


(c) Transverse section of spinal cord within a cervical vertebra

**Meninges are connective tissue coverings that surround the spinal cord and brain.**

## Cerebrospinal Fluid

Finally, a space between the arachnoid mater and the pia mater, called the **subarachnoid space**, contains cerebrospinal fluid. Cerebrospinal fluid is a buoyant liquid that suspends the delicate nervous tissue of the brain and spinal cord in a weightless environment while surrounding them with a shock-absorbing, hydraulic cushion.

✓ CHECKPOINT

1. Which of the four principal parts of the brain is the largest? Which part attaches to the spinal cord?
2. Where are the spinal meninges located? What are the three layers of the meninges, from superficial to deep?
3. Where are the epidural, subdural, and subarachnoid spaces located?
4. In which meningeal space does cerebrospinal fluid circulate?

# 13.2 The CNS is nourished and protected by blood and cerebrospinal fluid.

## Blood Flow to the Brain and Spinal Cord

Blood supplies oxygen and nutrients to the CNS. It flows to the brain mainly via the internal carotid and vertebral arteries (see Figure 20.18). Blood leaving the brain flows into the dural sinuses, which empty into the internal jugular veins (see Figure 20.23). Blood is supplied to the spinal cord by the posterior intercostal and lumbar arteries (see Figure 20.19). Blood leaves the spinal cord in the posterior intercostal and lumbar veins (see Figure 20.25).

In an adult, the brain represents only 2 percent of total body weight, but it consumes about 20 percent of the oxygen and glucose used by the body even when you are resting. Neurons synthesize ATP almost exclusively from glucose via reactions that use oxygen. When the activity of neurons and neuroglia increases in a particular region of the brain, blood flow to that area also increases. Even a brief slowing of brain blood flow may cause loss of consciousness, such as when you stand up too quickly after sitting for a long period of time. Typically, an interruption in blood flow for 1 or 2 minutes impairs neuronal function, and total deprivation of oxygen for about 4 minutes causes permanent injury. Because virtually no glucose is stored in the brain, the supply of glucose also must be continuous. If blood entering the brain has a low level of glucose, mental confusion, dizziness, convulsions, and loss of consciousness may occur. If the arterial supply to the brain is blocked, the resulting loss of blood flow can damage the brain, producing a cerebrovascular accident (CVA) or stroke. CVAs are the most common brain disorder. They affect 500,000 people each year in the United States and represent the third leading cause of death, behind heart attacks and cancer.

The **blood–brain barrier (BBB)** protects the CNS from harmful substances and pathogens by preventing passage of these substances from blood into the interstitial fluid of brain tissue. Once inside brain tissue, the cerebral arteries quickly divide into capillaries (microscopic blood vessels). Tight junctions seal together the endothelial cells of brain capillaries, which also are surrounded by a thick basement membrane. In addition, the processes of many astrocytes (one type of neuroglia) press up against the capillaries. The astrocytic processes secrete chemicals that maintain the permeability characteristics of the tight junctions, allowing selective passage of some substances from blood into brain tissue while inhibiting the passage of others. A few water-soluble substances, such as glucose, cross the BBB by active transport. Other substances, such as creatinine, urea, and most ions, cross the BBB very slowly. Still other substances—proteins and most antibiotic drugs—do not pass at all from the blood into brain tissue. However, lipid-soluble substances, such as oxygen, carbon dioxide, alcohol, and most anesthetic agents, are able to access brain tissue freely through the blood–brain barrier. Trauma, certain toxins, and inflammation can cause a breakdown of the blood–brain barrier.

## Cerebrospinal Fluid

**Cerebrospinal fluid (CSF)** is a clear, colorless liquid that protects the brain and spinal cord against chemical and physical injuries. It also carries oxygen, glucose, and other needed chemicals from the blood to neurons and neuroglia. CSF continuously circulates through cavities in the brain and spinal cord and around the brain and spinal cord in the subarachnoid space (between the arachnoid mater and pia mater). The total volume of CSF is 80 to 150 milliliters (3 to 5 oz) in an adult. CSF contains glucose, proteins, lactic acid, urea, cations ($Na^+$, $K^+$, $Ca^{2+}$, $Mg^{2+}$), and anions ($Cl^-$ and $HCO_3^-$); it also contains some white blood cells. The CSF has three basic functions:

- ***Mechanical protection.*** CSF serves as a shock-absorbing medium that protects the delicate tissues of the brain and spinal cord from jolts that would otherwise cause them to hit the bony walls of the cranial cavity and vertebral canal. The fluid also buoys the brain so that it "floats" in the cranial cavity.
- ***Chemical protection.*** CSF provides an optimal chemical environment for accurate neuronal signaling. Even slight changes in the ionic composition of CSF within the brain can seriously disrupt production of action potentials and postsynaptic potentials.
- ***Circulation.*** CSF is a medium for exchange of nutrients and waste products between the blood and adjacent nervous tissue.

## Formation of CSF in the Ventricles

**Figure 13.5** shows the four CSF-filled cavities within the brain, which are called **ventricles** (VEN-tri-kuls = little cavities). There is one **lateral ventricle** in each hemisphere of the cerebrum. (Think of them as ventricles 1 and 2.) Anteriorly, the lateral ventricles are separated by a thin membrane, the **septum pellucidum** (SEP-tum pe-LOO-si-dum; *pellucid* = transparent;

**FIGURE 13.5 Locations of ventricles within a "transparent" brain.** One interventricular foramen on each side connects a lateral ventricle to the third ventricle, and the cerebral aqueduct connects the third ventricle to the fourth ventricle.

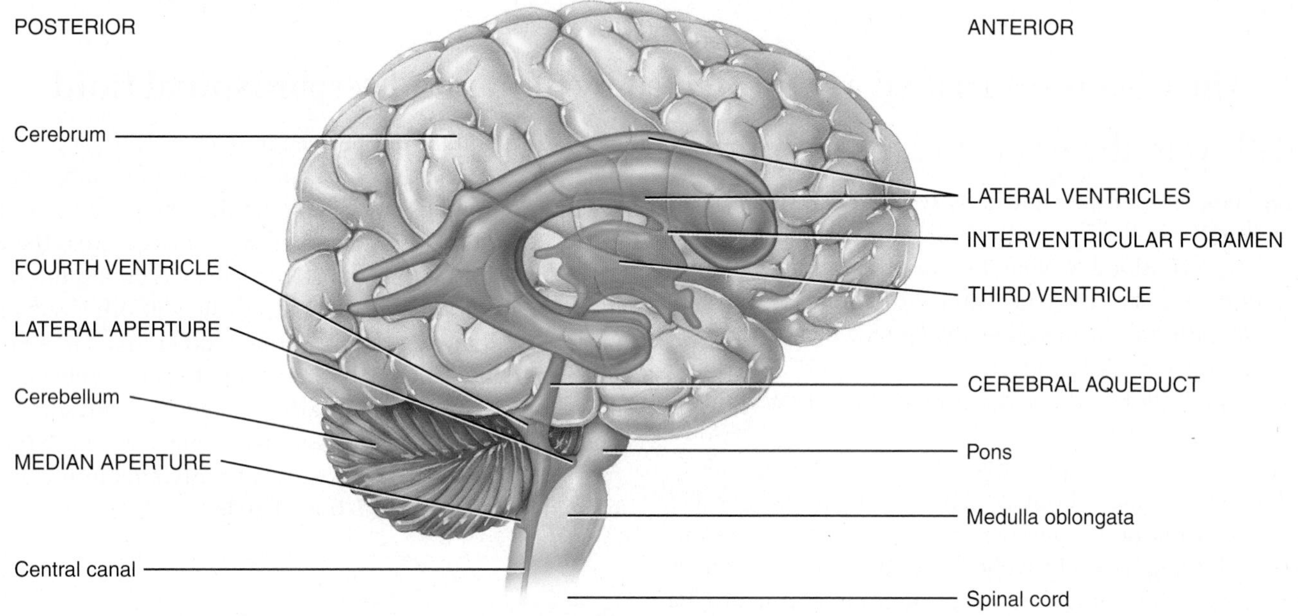


(a) Right lateral view of brain

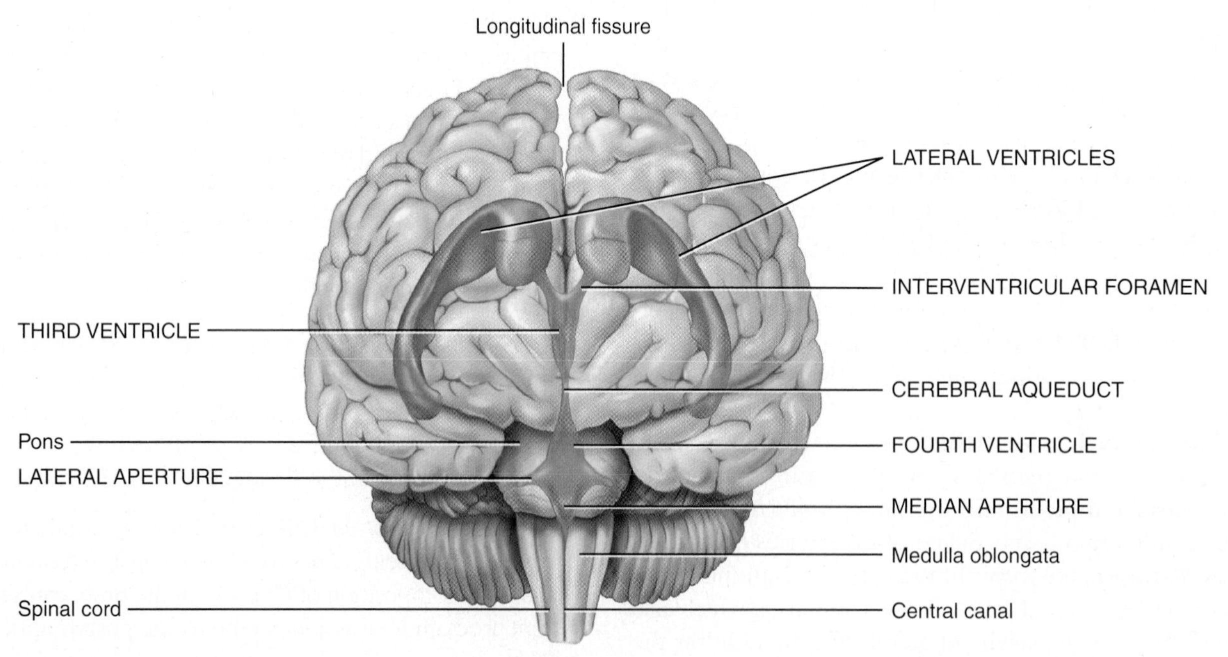


(b) Anterior view of brain

**Ventricles are cavities within the brain that are filled with cerebrospinal fluid.**

Figure 13.6a). The **third ventricle** is a narrow cavity along the midline superior to the hypothalamus and between the right and left halves of the thalamus. The **fourth ventricle** lies between the brain stem and the cerebellum.

The sites of CSF production are the **choroid plexuses** (KŌ-royd = membranelike), networks of capillaries in the walls of the ventricles. The capillaries are covered by ependymal cells (a type of neuroglia) joined by tight junctions. Selected substances (mostly water) filtered from the blood plasma are secreted by the ependymal cells to produce the cerebrospinal fluid. This secretory capacity is bidirectional and accounts for continuous production of CSF and transport of metabolites from the nervous tissue back to the blood. Because of the tight junctions between ependymal cells, materials entering CSF from choroid capillaries cannot leak between these cells; instead, they must pass through the ependymal cells. This **blood–cerebrospinal fluid barrier** permits certain substances to enter the CSF but excludes others, protecting the brain and spinal cord from potentially harmful blood-borne substances. A point of clarification: The blood–brain barrier controls passage of substances from *blood* into *interstitial fluid* of neural tissue, but the blood–cerebrospinal fluid barrier controls passage of substances from *blood* into *CSF*.

## Circulation of CSF

Ependymal cells line the ventricles of the brain and central canal of the spinal cord. Movement of their cilia assists in the flow of CSF. The CSF formed in the choroid plexuses of each lateral ventricle flows into the third ventricle through two narrow, oval openings, the **interventricular foramina** (in′-ter-ven-TRIK-ū-lar) (singular is *foramen*; Figure 13.6). Additional CSF is added by the choroid plexus in the roof of the third ventricle. The fluid then flows through the **cerebral aqueduct** (AK-we-dukt), which passes through the midbrain, into the fourth ventricle. The choroid plexus of the fourth ventricle contributes more fluid. CSF enters the subarachnoid space through three openings in the roof of the fourth ventricle: a single **median aperture** (AP-er-chur) and the paired **lateral apertures**, one on each side. CSF then circulates in the central canal of the spinal cord and in the subarachnoid space around the surface of the brain and spinal cord.

FIGURE 13.6 **Pathways of circulating cerebrospinal fluid.**

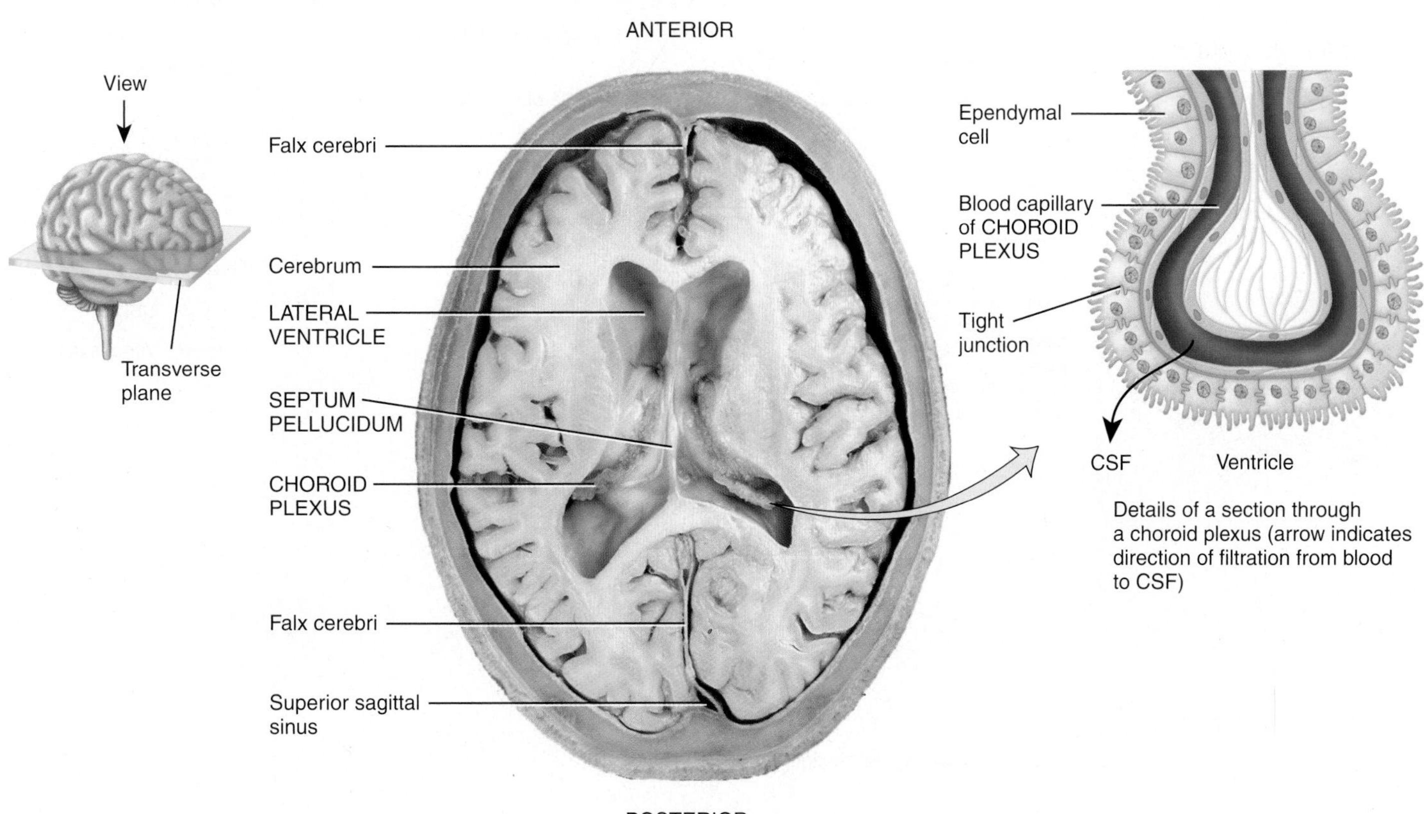


(a) Superior view of transverse section of brain showing choroid plexuses

(continues)

**FIGURE 13.6 (continued)**

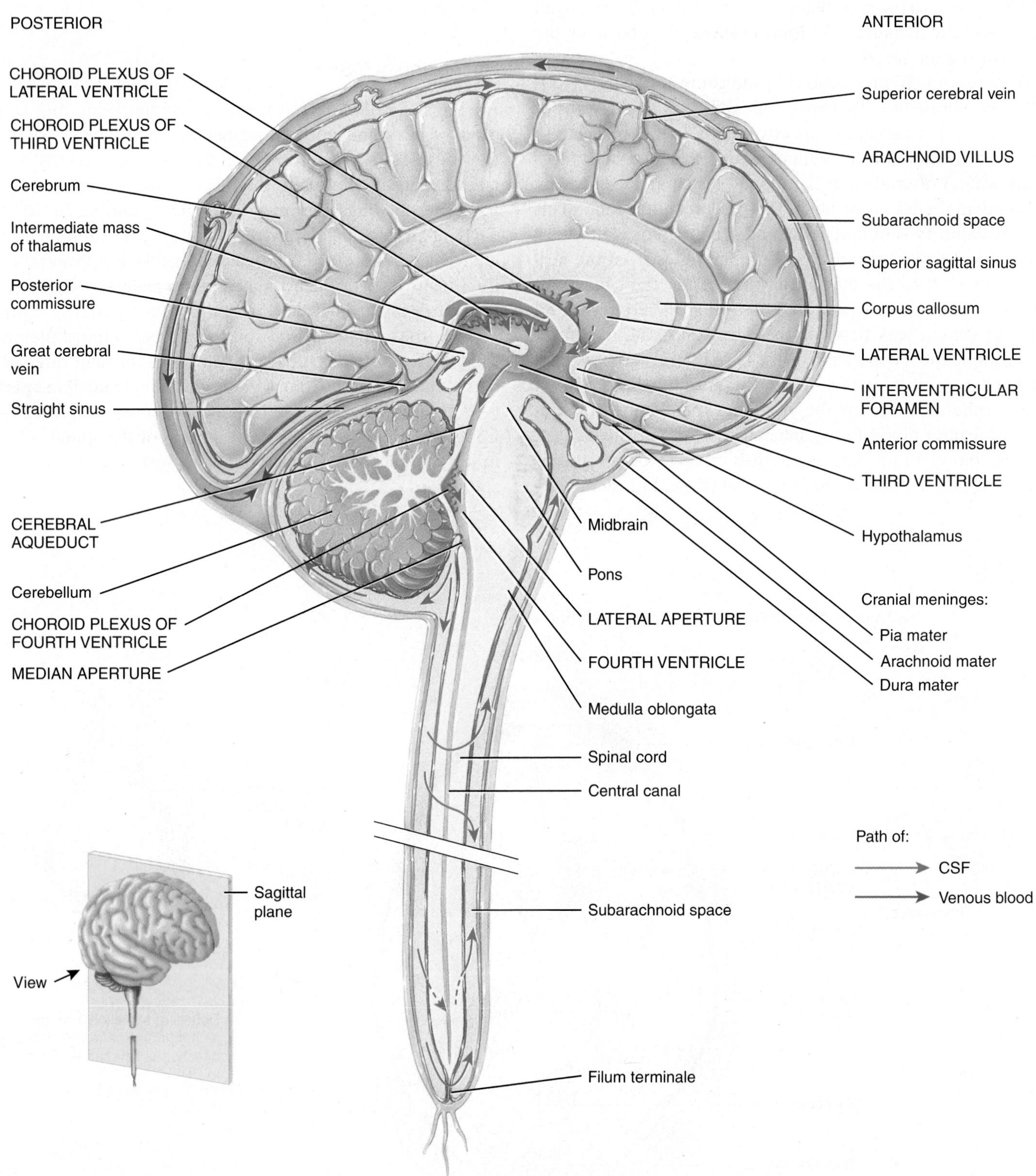
POSTERIOR
ANTERIOR
CHOROID PLEXUS OF LATERAL VENTRICLE
CHOROID PLEXUS OF THIRD VENTRICLE
Cerebrum
Intermediate mass of thalamus
Posterior commissure
Great cerebral vein
Straight sinus
CEREBRAL AQUEDUCT
Cerebellum
CHOROID PLEXUS OF FOURTH VENTRICLE
MEDIAN APERTURE
Superior cerebral vein
ARACHNOID VILLUS
Subarachnoid space
Superior sagittal sinus
Corpus callosum
LATERAL VENTRICLE
INTERVENTRICULAR FORAMEN
Anterior commissure
THIRD VENTRICLE
Hypothalamus
Midbrain
Pons
LATERAL APERTURE
FOURTH VENTRICLE
Medulla oblongata
Cranial meninges:
Pia mater
Arachnoid mater
Dura mater
Spinal cord
Central canal
Subarachnoid space
Filum terminale
Path of:
CSF
Venous blood
Sagittal plane
View

(b) Sagittal section of brain and spinal cord

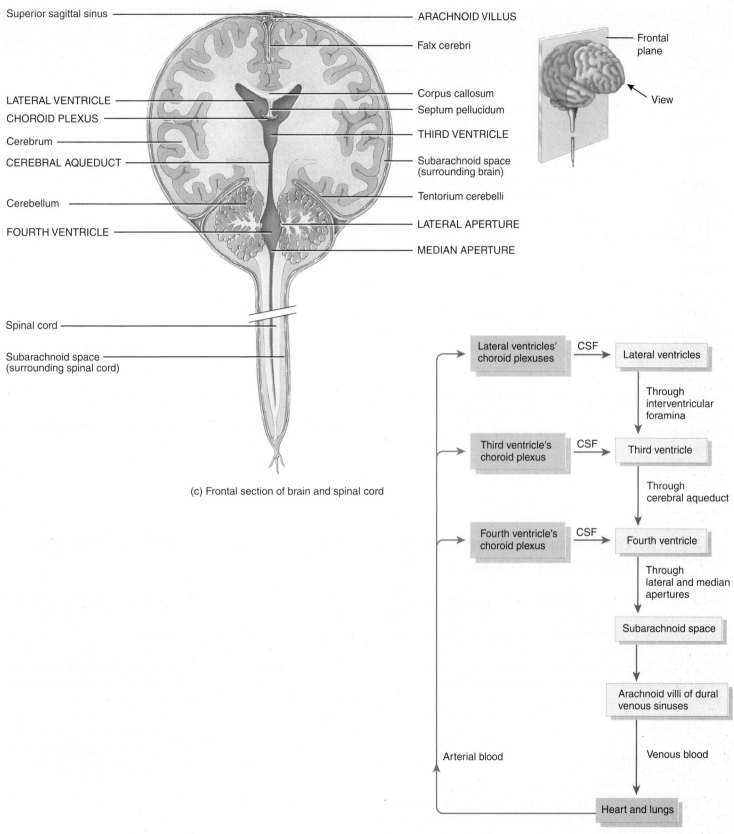

Superior sagittal sinus
ARACHNOID VILLUS
Falx cerebri
Frontal plane
View
LATERAL VENTRICLE
CHOROID PLEXUS
Cerebrum
CEREBRAL AQUEDUCT
Cerebellum
FOURTH VENTRICLE
Corpus callosum
Septum pellucidum
THIRD VENTRICLE
Subarachnoid space (surrounding brain)
Tentorium cerebelli
LATERAL APERTURE
MEDIAN APERTURE
Spinal cord
Subarachnoid space (surrounding spinal cord)

(c) Frontal section of brain and spinal cord

Lateral ventricles' choroid plexuses
CSF
Lateral ventricles
Through interventricular foramina
Third ventricle's choroid plexus
CSF
Third ventricle
Through cerebral aqueduct
Fourth ventricle's choroid plexus
CSF
Fourth ventricle
Through lateral and median apertures
Subarachnoid space
Arachnoid villi of dural venous sinuses
Venous blood
Heart and lungs
Arterial blood

(d) Summary of formation, circulation, and absorption of cerebrospinal fluid (CSF)

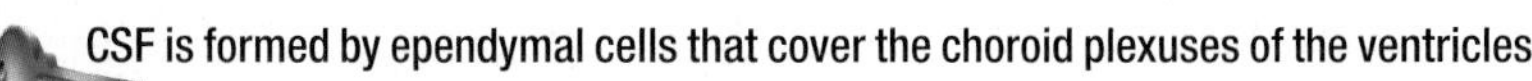

**CSF is formed by ependymal cells that cover the choroid plexuses of the ventricles.**

### CLINICAL CONNECTION | *Hydrocephalus*

Abnormalities in the brain—tumors, inflammation, or developmental malformations—can interfere with the circulation of CSF from the ventricles into the subarachnoid space. When excess CSF accumulates in the ventricles, the CSF pressure rises. Elevated CSF pressure causes a condition called **hydrocephalus** (hī′-drō-SEF-a-lus; *hydro-* = water; *cephal-* = head). The abnormal accumulation of CSF may be due to an obstruction to CSF flow or an abnormal rate of CSF production and/or reabsorption or an obstruction to its flow. In a baby whose fontanels have not yet closed, the head bulges due to the increased pressure. If the condition persists, the fluid buildup compresses and damages the delicate nervous tissue. Hydrocephalus is relieved by draining the excess CSF. In one procedure, called *endoscopic third ventriculostomy (ETV),* a neurosurgeon makes a hole in the floor of the third ventricle and the CSF drains directly into the subarachnoid space. In adults, hydrocephalus may occur after head injury, meningitis, or subarachnoid hemorrhage. Because the adult skull bones are fused together, this condition can quickly become life-threatening and requires immediate intervention.

CSF is gradually reabsorbed from the subarachnoid space into the blood through **arachnoid villi**, fingerlike extensions of the arachnoid mater that project into the dural sinuses, especially the superior sagittal sinus (see Figure 13.3a). Normally, CSF is reabsorbed as rapidly as it is formed by the choroid plexuses, at a rate of about 20 mL/hr (480 mL/day). Because the rates of formation and reabsorption are the same, the pressure of CSF normally is constant. For the same reason, the volume of CSF remains constant. Figure 13.6d summarizes the production and flow of CSF.

### ✓ CHECKPOINT

**5.** What vessels return blood from the head to the heart?

**6.** How does CSF protect the brain and spinal cord?

**7.** Which brain region is anterior to the fourth ventricle? Which is posterior to it?

**8.** What structures are the sites of CSF production, and where are they located?

**9.** Where is CSF reabsorbed into the blood?

## 13.3 The cerebrum interprets sensory impulses, controls muscular movements, and functions in intellectual processes.

The *cerebrum* is the "seat of intelligence." It provides us with the ability to read, write, and speak; to make calculations and compose music; and to remember the past, plan for the future, and imagine things that have never existed before. The cerebrum consists of an outer cerebral cortex, an internal region of cerebral white matter, and *nuclei*, functional clusters of neuronal cell bodies deep within the white matter.

### Cerebral Cortex

Gray matter, where neurons form synapses with one another, receives and integrates incoming and outgoing information. The **cerebral cortex** (*cortex* = rind or bark) is a region of gray matter that forms the outer rim of the cerebrum (Figure 13.7). Although only 2–4 mm (0.08–0.16 in.) thick, the cerebral cortex contains billions of neurons. During embryonic development, when brain size increases rapidly, the gray matter of the cortex enlarges much faster than the deeper white matter. As a result, the cortical region rolls and folds upon itself. The folds are called **gyri** (JĪ-rī = circles; singular is *gyrus*). The deepest grooves between folds are known as **fissures**; the shallower grooves between folds are termed **sulci** (SUL-sī = grooves; singular is *sulcus*). The most prominent fissure, the **longitudinal fissure**, separates the cerebrum into right and left halves called **cerebral hemispheres**. Within the longitudinal fissure between the cerebral hemispheres is the falx cerebri. The cerebral hemispheres are connected internally by the **corpus callosum** (kal-LŌ-sum; *corpus* = body; *callosum* = hard), a broad band of white matter containing axons that extend between the hemispheres (see Figure 13.6).

### Lobes of the Cerebrum

Each cerebral hemisphere can be further subdivided into several lobes. The lobes are named after the bones that cover them: frontal, parietal, temporal, and occipital lobes (Figure 13.7).

The **central sulcus** (SUL-kus) separates the **frontal lobe** from the **parietal lobe**. The **precentral gyrus** is located immediately anterior to the central sulcus, and the **postcentral gyrus** is located immediately posterior to the central sulcus. The **lateral cerebral sulcus** separates the **frontal lobe** from the **temporal lobe**. The **parieto-occipital sulcus** separates the **parietal lobe** from the **occipital lobe**. A fifth part of the cerebrum, the **insula**, cannot be seen at the surface of the brain because it lies within the lateral cerebral sulcus, deep to the parietal, frontal, and temporal lobes (Figure 13.7b, c).

FIGURE 13.7 **Cerebrum.** Because the insula cannot be seen externally, it has been projected to the surface in (b).

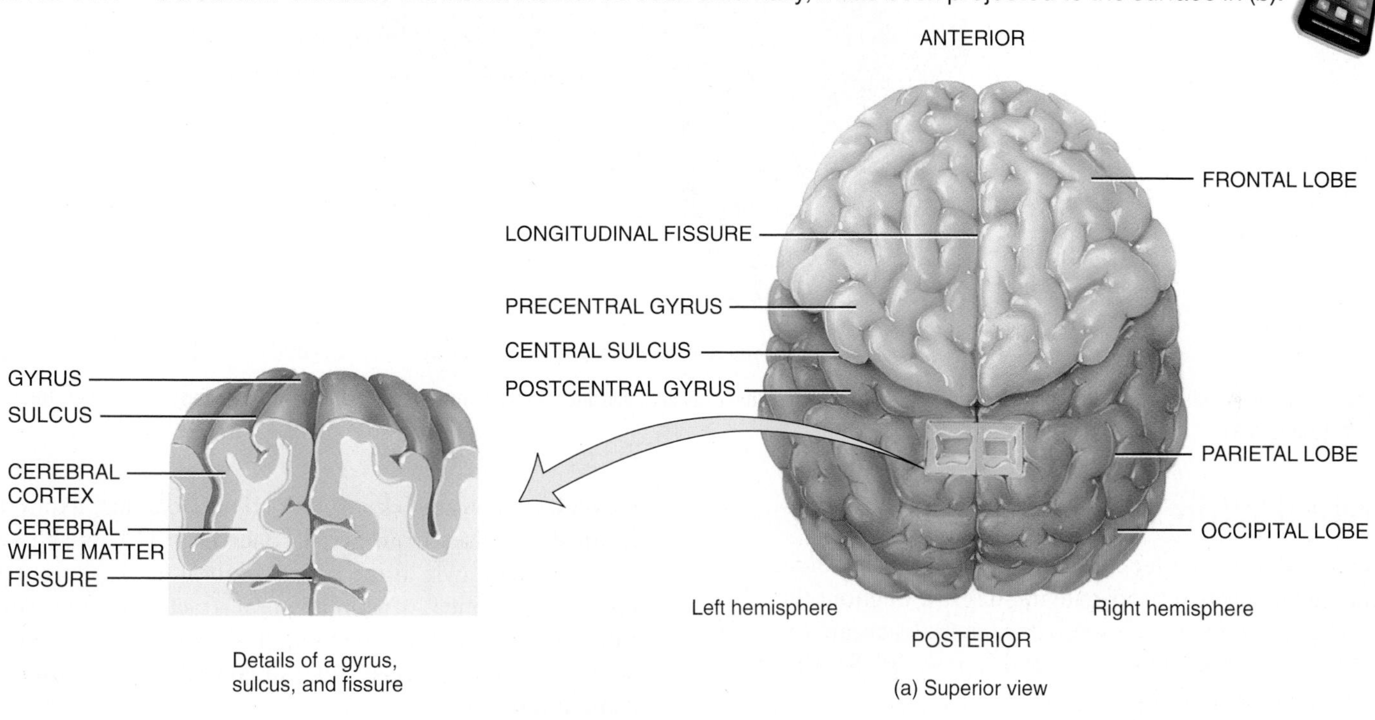


Details of a gyrus, sulcus, and fissure

(a) Superior view

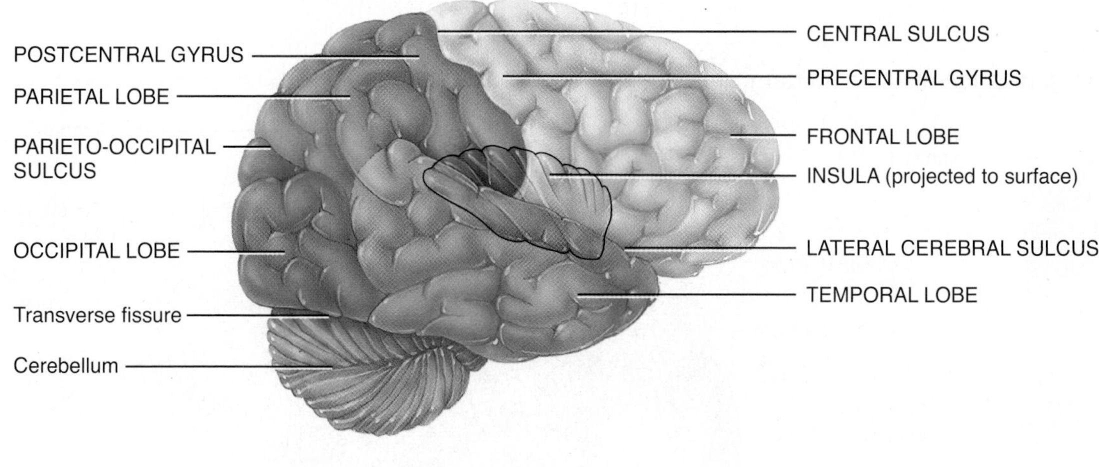


(b) Right lateral view

(continues)

FIGURE 13.7 (continued)

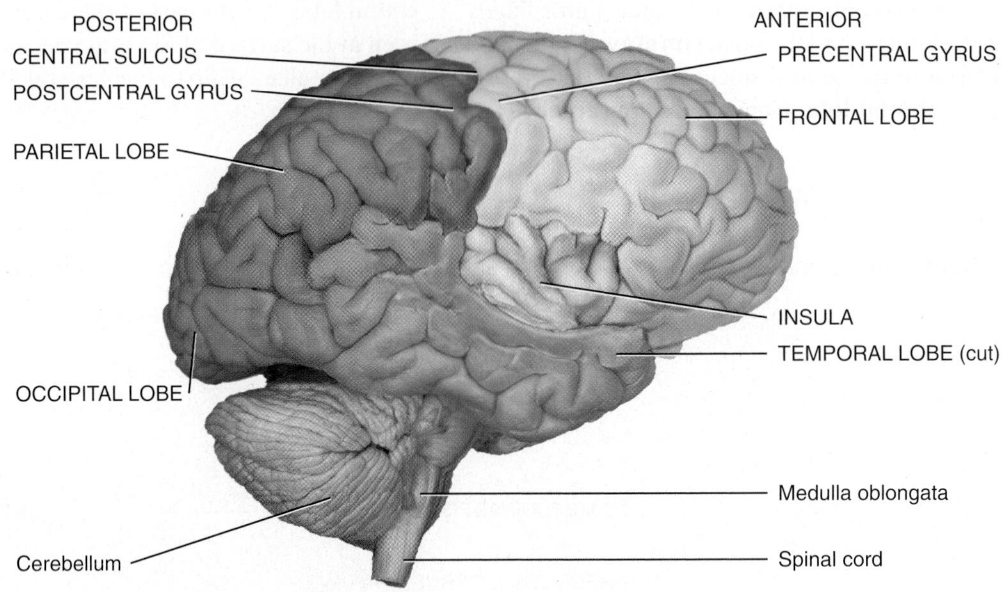


(c) Right lateral view with temporal lobe cut away

**The cerebrum provides us with the ability to read, write, and speak; to make calculations and compose music; to remember the past and plan for the future; and to create.**

## Cerebral White Matter

The **cerebral white matter** consists primarily of **tracts**, bundles of myelinated axons that propagate impulses throughout the CNS to allow communication between its regions. Tracts are the CNS equivalent of nerves (bundles of axons in the PNS). The three types of tracts are as follows (Figure 13.8):

- **Association tracts** contain axons that conduct impulses between gyri in the same hemisphere.
- **Commissural tracts** (kom′-i-SYŪR-al) contain axons that conduct impulses from gyri in one cerebral hemisphere to corresponding gyri in the other cerebral hemisphere. Three important groups of commissural tracts are the **corpus callosum** (the largest fiber bundle in the brain, containing about 300 million axons), **anterior commissure**, and **posterior commissure**.
- **Projection tracts** contain axons that conduct impulses from the cerebrum to lower parts of the CNS (thalamus, brain

FIGURE 13.8 Organization of white matter tracts of the left cerebral hemisphere.

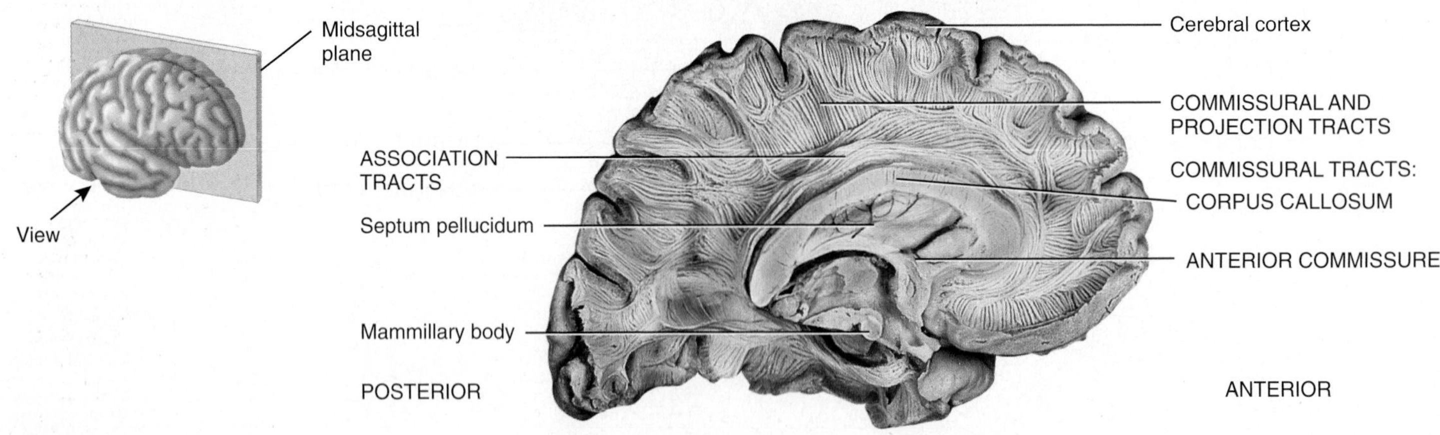


Medial view of tracts revealed by removing gray matter from a midsagittal section

**Association tracts, commissural tracts, and projection tracts form white matter areas in the cerebral hemispheres.**

stem, or spinal cord) or from lower parts of the CNS to the cerebrum. An example is the **internal capsule**, a thick band of white matter that contains both ascending and descending axons (see Figure 13.9b).

## Basal Nuclei

Deep within each cerebral hemisphere are three nuclei (masses of gray matter) that are collectively termed the **basal nuclei** (Figure 13.9). (Historically, these nuclei have been called the basal *ganglia.* However, this is a misnomer because a *ganglion* is an aggregate of neuronal cell bodies in the peripheral nervous system. While both terms still appear in the literature, we use *nuclei,* as this is the correct term as determined by the *Terminologia Anatomica,* the final say on correct anatomical terminology.)

Two of the basal nuclei are side by side, just lateral to the thalamus. They are the **globus pallidus** (GLŌ-bus PAL-i-dus; *globus* = ball; *pallidus* = pale), which is closer to the thalamus, and the **putamen** (pū-TĀ-men = shell), which is closer to the cerebral cortex. The third member of the basal nuclei is the **caudate**

FIGURE 13.9 **Basal nuclei.** The basal nuclei are shown in purple. In (a) the basal nuclei have been projected to the surface.

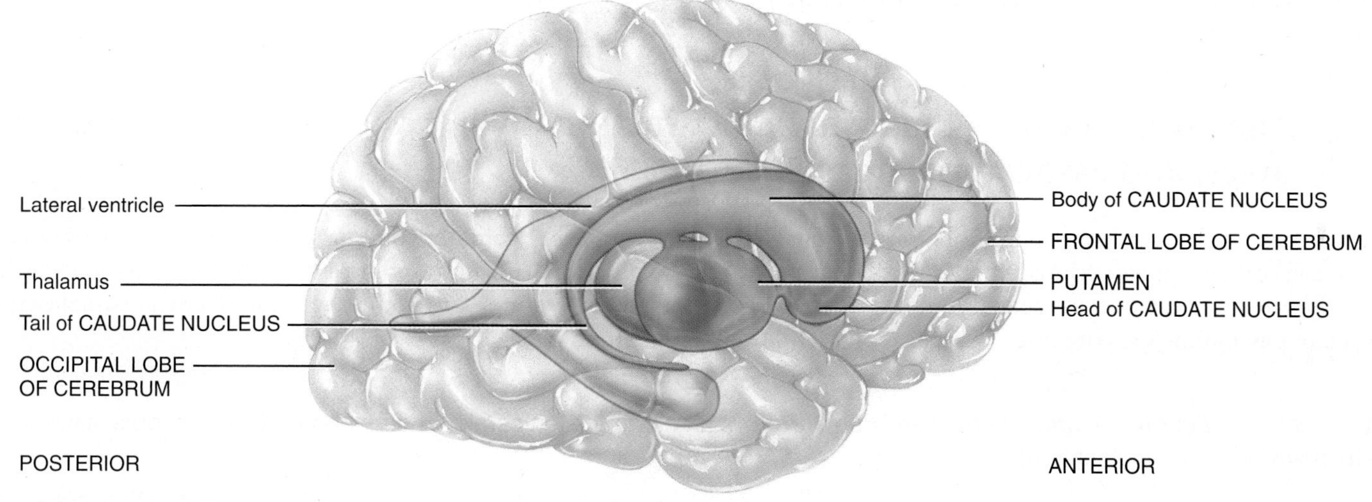


(a) Lateral view of right side of brain

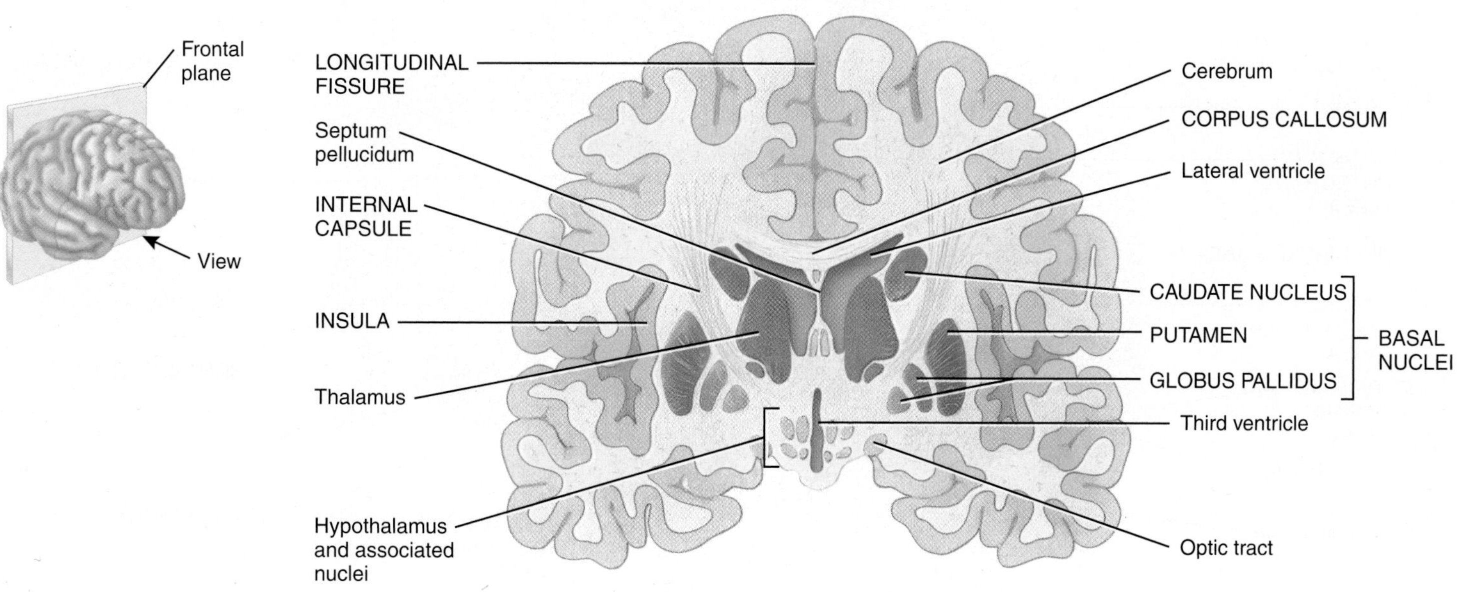


(b) Anterior view of frontal section

**The basal nuclei help initiate and terminate movements, suppress unwanted movements, and regulate muscle tone.**

**nucleus** (KAW-dāt; *caud-* = tail), which has a large "head" connected to a smaller "tail" by a long, comma-shaped "body."

The basal nuclei receive input from the cerebral cortex and provide output to motor portions of the cerebral cortex via nuclei of the thalamus. In addition, basal nuclei have extensive connections with one another. A major function of the basal nuclei is to help regulate initiation and termination of movements. The basal nuclei help regulate the muscle tone required for specific body movements. The basal nuclei also control subconscious contractions of skeletal muscles. Examples include automatic arm swings while walking and laughing in response to a funny joke. In addition, the basal nuclei help initiate and terminate some cognitive processes, such as attention, memory, and planning, Basal nuclei functions are described further in Concept 15.6.

See Table 13.2 for a summary of the functions of the cerebrum.

**✓ CHECKPOINT**

10. During development, does the gray matter or white matter enlarge more rapidly?
11. What are the brain folds, shallow grooves, and deep grooves called?
12. What deep groove separates the right and left cerebral hemispheres? What shallow grooves separate each lobe of the cerebrum from the next?
13. Which tracts carry impulses between gyri of the same hemisphere? Between gyri in opposite hemispheres? Between the cerebrum, thalamus, brain stem, and spinal cord?

## 13.4 The cerebral cortex can be divided functionally into sensory areas, motor areas, and association areas.

Specific types of sensory, motor, and integrative signals are processed in certain regions of the cerebral cortex (Figure 13.10). Generally, *sensory areas* receive sensory information and are involved in **perception**, the conscious awareness of a sensation; *motor areas* initiate voluntary movements, and *association areas* deal with more complex integrative functions such as memory, emotions, reasoning, will, judgment, personality traits, and intelligence.

FIGURE 13.10 **Functional areas of the cerebrum.** Broca's area is in the left cerebral hemisphere of most people; it is shown here to indicate its relative location.

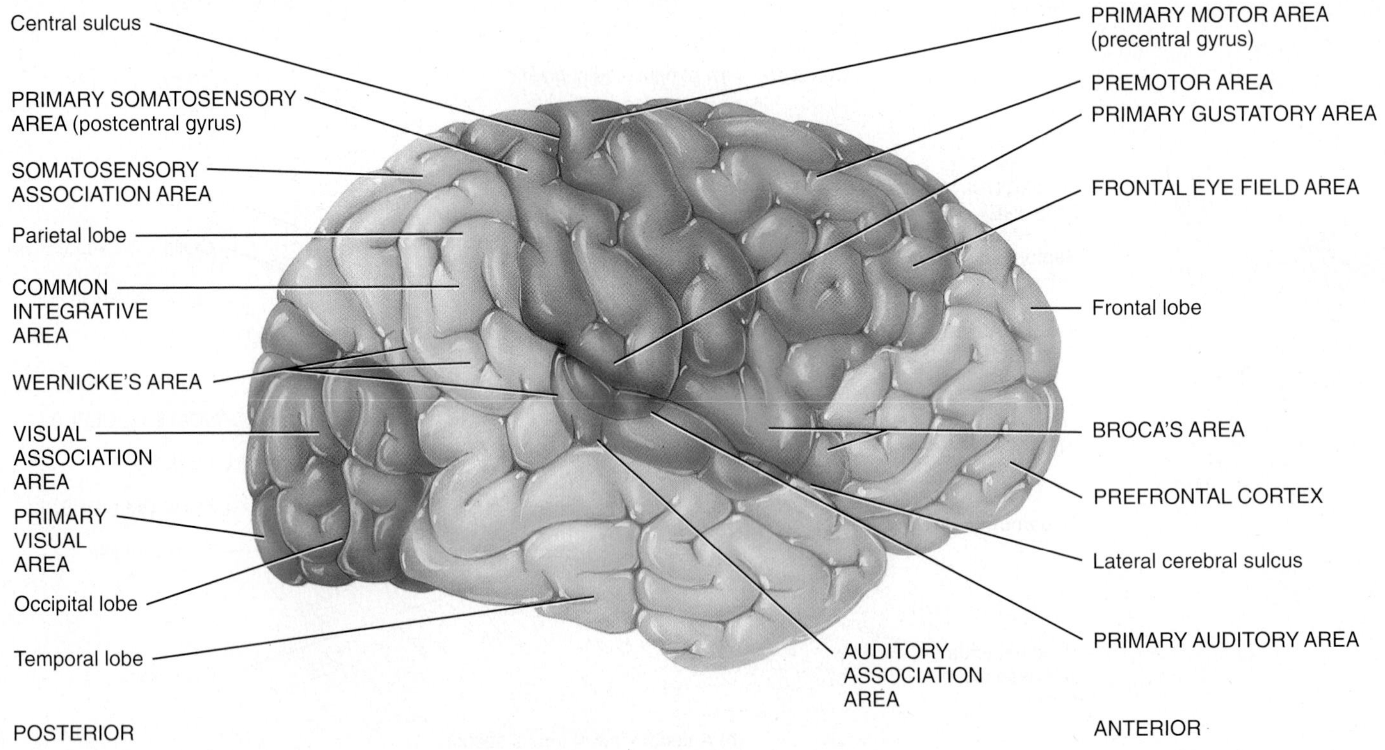


Lateral view of right cerebral hemisphere

**Particular areas of the cerebral cortex process sensory, motor, and integrative signals.**

## Sensory Areas

Sensory impulses arrive mainly in the posterior half of both cerebral hemispheres, in regions behind the central sulci. In the cerebral cortex, primary sensory areas receive sensory information from peripheral sensory receptors through lower regions of the brain. Sensory association areas often are adjacent to the primary areas. Sensory association areas integrate sensory experiences to generate meaningful patterns of recognition and awareness. For example, a person with damage in the *primary* visual area would be blind in at least part of his visual field, but a person with damage to a visual *association* area might see normally yet be unable to recognize ordinary objects such as a lamp or a toothbrush just by looking at them.

The following are some important sensory areas (Figure 13.10):

- The **primary somatosensory area** is located in each parietal lobe in the postcentral gyrus directly posterior to the central sulcus. The primary somatosensory area receives impulses for touch, pressure, vibration, itch, tickle, and temperature (coldness and warmth), pain, and proprioception (joint and muscle position). A "map" of the entire body is present in the primary somatosensory area: Each point within the area receives impulses from a specific part of the body (see Figure 15.6a). The size of the cortical area receiving impulses from a particular part of the body depends on the number of receptors present rather than on the size of the body part. For example, the region of the somatosensory area that receives impulses from the lips and fingertips is larger than the area that receives impulses from the thorax or hip. The primary somatosensory area allows you to localize the exact points of the body where somatic sensations originate, so that you know exactly where on your body to swat that mosquito.
- The **primary visual area** is located at the posterior tip of the occipital lobe, mainly on the medial surface (next to the longitudinal fissure). It receives visual information and is involved in visual perception.
- The **primary auditory area** is located in the superior part of the temporal lobe near the lateral cerebral sulcus. It receives information for sound and is involved in auditory perception.
- The **primary gustatory** area is located at the base of the postcentral gyrus superior to the lateral cerebral sulcus in the parietal cortex. It receives impulses for taste and is involved in taste perception and discrimination.
- The **primary olfactory area** is located in the temporal lobe on the medial aspect (and thus is not visible in Figure 13.10). It receives impulses for smell and is involved in the perception and discrimination of the various odors.

## Motor Areas

Motor output from the cerebral cortex flows mainly from the anterior part of each hemisphere. Among the most important motor areas are the following (Figure 13.10):

- The **primary motor area** is located in each frontal lobe in the precentral gyrus directly anterior to the central sulcus. As is true for the primary somatosensory area, a "map" of the entire body is present in the primary motor area: Each region within the area controls voluntary contractions of specific muscles or groups of muscles (see Figure 15.6b). Electrical stimulation of any point in the primary motor area causes contraction of specific skeletal muscle fibers on the opposite side of the body. Different muscles are represented unequally in the primary motor area. More cortical area is devoted to those muscles involved in skilled, complex, or delicate movement. For instance, the cortical region devoted to muscles that move the fingers is much larger than the region for muscles that move the toes, which is why most pianists play with their fingers.
- **Broca's area** (BRŌ-kaz) is located in the frontal lobe close to the lateral cerebral sulcus. Speaking and understanding language are complex activities that involve several sensory, association, and motor areas of the cortex. The planning and production of speech occurs in Broca's area, located in one frontal lobe—the *left* frontal lobe in 97 percent of the population. From Broca's area, impulses pass to the premotor area to control the muscles of the larynx, pharynx, and mouth for the specific, coordinated contractions that enable you to speak. Simultaneously, impulses propagate from Broca's area to the primary motor area to control the breathing muscles and to regulate the proper flow of air past the vocal cords. The coordinated contractions of your speech and breathing muscles enable you to speak your thoughts.

## Association Areas

The association areas of the cerebrum consist of large areas of all lobes of the cerebral cortex. Association areas often are adjacent to the primary sensory areas. They usually receive input both from the primary sensory areas and from other brain regions. Association areas integrate sensory experiences to generate meaningful patterns of recognition and awareness. Association areas are connected with one another by association tracts and include the following (Figure 13.10):

- The **somatosensory association area** is just posterior to and receives input from the primary somatosensory area, as well as from the thalamus and other parts of the brain. It integrates and interprets sensations. This area permits you to determine the exact shape and texture of an object by feeling it, to determine the orientation of one object with respect to another as

## CLINICAL CONNECTION | *Aphasia*

Much of what we know about language areas comes from studies of patients with language or speech disturbances that have resulted from brain damage. Broca's area, Wernicke's area, and other language areas are located in the left cerebral hemisphere of most people, regardless of whether they are left-handed or right-handed. Injury to language areas of the cerebral cortex results in **aphasia** (a-FĀ-zē-a; *a-* = without; *-phasia* = speech), an inability to use or comprehend words. Damage to Broca's speech area results in *nonfluent aphasia,* an inability to properly articulate or form words; people with nonfluent aphasia know what they wish to say but cannot speak. Damage to Wernicke's area, the common integrative area, or auditory association area results in *fluent aphasia,* characterized by faulty understanding of spoken or written words. A person experiencing this type of aphasia may fluently produce strings of words that have no meaning ("word salad"). For example, someone with fluent aphasia might say, "I rang car porch dinner light river pencil." The underlying deficit may be **word deafness** (an inability to understand spoken words), **word blindness** (an inability to understand written words), or both.

they are felt, and to sense the relationship of one body part to another. Another role of the somatosensory association area is the storage of memories of past somatic sensory experiences, enabling you to compare current sensations with previous experiences. For example, the somatosensory association area allows you to recognize objects such as a pencil and a paper clip simply by touching them.

- The **visual association area** is located in the occipital lobe anterior to the primary visual area. It receives sensory impulses from the primary visual area and the thalamus. It relates present and past visual experiences and is essential for recognizing and evaluating what is seen. For example, the visual association area allows you to recognize an object such as a spoon simply by looking at it.
- The **auditory association area** is located inferior and posterior to the primary auditory area in the temporal cortex. It allows you to recognize a particular sound as speech, music, or noise.
- **Wernicke's area** (VER-ni-kēz) is a broad region in the left temporal and parietal lobes. This area interprets the meaning of speech by recognizing spoken words. It is active as you translate words into thoughts. The regions in the *right* hemisphere that correspond to the locations of Broca's and Wernicke's areas in the left hemisphere also contribute to verbal communication by adding emotional content, such as anger or joy, to spoken words.
- The **common integrative area** is bordered by the somatosensory, visual, and auditory association areas. It integrates sensory interpretations from all sensory association areas, allowing the formation of thoughts based on a variety of sensory inputs. This area then transmits signals to other parts of the brain for the appropriate response to the sensory signals it has interpreted.
- The **prefrontal cortex** is an extensive area in the anterior portion of the frontal lobe that is well developed in primates, especially humans. This area has numerous connections with other areas of the cerebral cortex, thalamus, hypothalamus, limbic system, and cerebellum. The prefrontal cortex is concerned with the makeup of a person's personality, intellect, complex learning abilities, recall of information, initiative, judgment, foresight, reasoning, conscience, intuition, mood, planning for the future, and development of abstract ideas. A person with bilateral damage to the prefrontal cortices typically becomes rude, inconsiderate, incapable of accepting advice, moody, inattentive, less creative, unable to plan for the future, and incapable of anticipating the consequences of rash or reckless words or behavior.
- The **premotor area** is a motor association area that is immediately anterior to the primary motor area. The premotor area controls learned, skilled motor activities of a complex and sequential nature. It generates impulses that cause specific groups of muscles to contract in a specific sequence. Through repetition, the premotor area serves as a memory bank to store specific patterns of movements. For instance, while the primary motor area controls general movements of muscles of the hand, the premotor area directs muscles of the hand to tie a bow.
- The **frontal eye field area** in the frontal cortex controls voluntary scanning movements of the eyes—like those you just used in reading this sentence.

## Hemispheric Lateralization

Although the brain is almost symmetrical on its right and left sides, subtle anatomical differences between the two hemispheres exist. This asymmetry appears in the human fetus at about 30 weeks of gestation. Furthermore, although the

**TABLE 13.1**

Functional Differences between the Two Cerebral Hemispheres

| LEFT HEMISPHERE FUNCTIONS | RIGHT HEMISPHERE FUNCTIONS |
| --- | --- |
| Receives somatic sensory signals from and controls muscles on right side of body | Receives somatic sensory signals from and controls muscles on left side of body |
| Reasoning | Musical and artistic awareness |
| Numerical and scientific skills | Space and pattern perception |
| Ability to use and understand sign language | Recognition of faces and emotional content of facial expressions |
| Spoken and written language | Generating emotional content of language<br>Generating mental images to compare spatial relationships<br>Identifying and discriminating among odors |

two hemispheres share performance of many functions, each hemisphere also specializes in performing certain unique functions. This functional asymmetry is termed **hemispheric lateralization**.

In the most obvious example of hemispheric lateralization, the left hemisphere receives somatic sensory signals from and controls muscles on the right side of the body, and the right hemisphere receives sensory signals from and controls muscles on the left side of the body. In most people the left hemisphere is more important for reasoning, numerical and scientific skills, spoken and written language, and the ability to use and understand sign language. For example, patients with damage in the left hemisphere may be unable to properly articulate or form words. Conversely, the right hemisphere is more specialized for musical and artistic awareness; spatial and pattern perception; recognition of faces and emotional content of language; discrimination of different smells; and generating mental images of sight, sound, touch, taste, and smell to compare relationships among them. Patients with damage in the right hemisphere regions speak in a monotonous voice, having lost the ability to impart emotional inflection to what they say.

Despite some dramatic differences in functions of the two hemispheres, there is considerable variation from one person to another. Lateralization seems less pronounced in females than in males, both for language (left hemisphere) and for visual and spatial skills (right hemisphere). A possibly related observation is that females have, on average, a 12 percent larger anterior commissure and a broader posterior portion of the corpus callosum than males. Recall that both the anterior commissure and the corpus callosum are commissural tracts that provide communication between the two hemispheres.

Table 13.1 summarizes some of the functional differences between the two cerebral hemispheres.

**✓ CHECKPOINT**

**14.** What does the term "hemispheric lateralization" mean?

**15.** Which area of the cerebrum demonstrates hemispheric lateralization for the planning and production of speech?

## RETURN TO Annette's Story

"Ok, the guest rooms are prepared and dinner is planned. I am ready to sit down and relax," Annette says to herself. She plops down in the living room to wait for her guests to arrive and then realizes how much energy she has put into getting ready. She is tired but at the same time excited. She hasn't seen Bill in several months and it's been almost a year since she had seen Jenny and her granddaughters. "I bet the girls have really grown," she thinks. She stretches out on the sofa and closes her eyes.

"Wake up sleepyhead!" Bill says as he laughs at his mother's reaction. "Caught you goofing off again!" Annette struggles to sit up, her arms and legs being rather uncooperative at the moment. Bill is immediately concerned. "What's wrong, Mom? You usually just jump right up when I surprise you."

"Oh, I'm fine! It's just old age," she responds. "I just need to work out the kinks that set in during my nap." Annette hugs Bill, Jenny, and the girls, and starts right in with questions about school and extracurricular activities.

During dinner, Bill notices that Annette's hands tremble as she picks up her utensils, and that sometimes the food actually falls off the fork and she has to start over. Bill decides that Annette is not telling him everything, especially when he sees her leave food on her plate because it is too much trouble to keep trying to eat. He starts observing her very closely. After dinner, while helping Annette clean the kitchen, he asks about the medication bottle that he sees near the sink.

"Oh," replies Annette, "that's just something Dr. Akers gave me to help with my tremors. There's something I've been meaning to tell you. I have Parkinson's disease."

Bill fires off questions in rapid succession. "Why didn't you tell me? Are you good about taking your medicine? Have you told Jeff and Susan?"

"Oh, I didn't want to bother any of you. You all have your own lives, and it just . bothers me sometimes," says Annette. "I take my medicine regularly, but sometimes it works well and other times not so well," she admits.

He takes a deep breath and continues. "I know something about Parkinson's. I work with some scientists at the lab who are developing new protocols for Parkinson's patients. The disease affects people differently but it is progressive. I see that you have lost weight and don't seem to be responding consistently to your medication, so I'd like to go with you to see Dr. Akers while I am here. There are treatments other than medication that I'd like to discuss with him."

A. *Annette takes medication for her Parkinson's disease. What is a characteristic that the medication must have to cross the blood–brain barrier freely?*

B. *Which part of the brain is responsible for controlling the initiation and termination of movement?*

C. *What other symptom that Annette is having is also controlled by this region of the brain?*

D. *Annette has difficulty using her fork to move food from her plate to her mouth. Which area of her brain is not functioning properly to cause this symptom?*

# 13.5 The diencephalon includes the thalamus, hypothalamus, and pineal gland.

The *diencephalon* forms a central core of brain tissue that extends from the brain stem to the cerebrum and surrounds the third ventricle. It is involved in a wide variety of sensory and motor processing between higher and lower brain centers. Major regions of the diencephalon include the thalamus, hypothalamus, and pineal gland.

## Thalamus

The **thalamus** (THAL-a-mus = inner chamber) measures about 3 cm (1.2 in.) in length and makes up 80 percent of the diencephalon (Figure 13.11). It consists of paired oval masses of gray matter organized into nuclei, with interspersed tracts of white matter (Figures 13.1 and 13.9b). In most people, a bridge of gray matter called the **intermediate mass** crosses the third ventricle, joining the right and left halves of the thalamus (Figure 13.11). A vertical Y-shaped sheet of white matter called the **internal medullary lamina** divides the gray matter of the right and left sides of the thalamus. It consists of myelinated axons that enter and leave the various thalamic nuclei. Axons that connect the thalamus and cerebral cortex pass through the *internal capsule*, a thick band of white matter lateral to the thalamus (see Figure 13.9b).

The thalamus is the major relay station for most sensory impulses that reach the primary sensory areas of the cerebral

FIGURE 13.11 **Thalamus.** Note the position of the thalamus in (a) the lateral view, and in (b) the medial view. The various thalamic nuclei shown in (c) are correlated by color to the cortical regions in (a) and (b) to which they project.

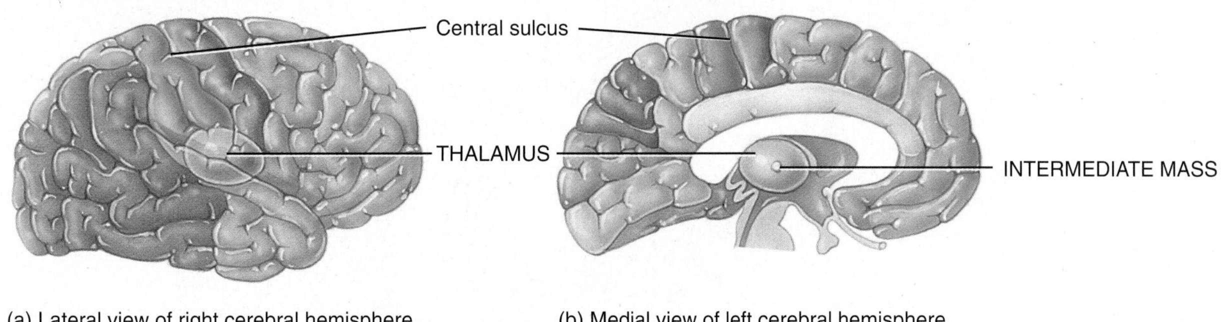


(a) Lateral view of right cerebral hemisphere

(b) Medial view of left cerebral hemisphere

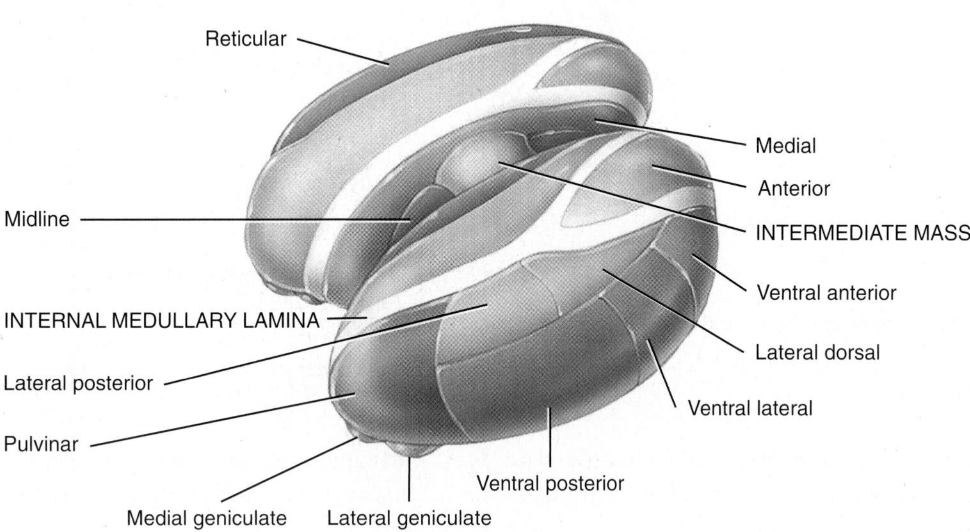


(c) Superolateral view of thalamus showing locations of thalamic nuclei (reticular nucleus is shown on left side only; all other nuclei are shown on right side)

**The thalamus is the main relay station for sensory impulses that reach the cerebral cortex from other parts of the brain and the spinal cord.**

cortex from the spinal cord and brain stem. In addition, the thalamus contributes to motor functions by transmitting information from the cerebellum and basal nuclei to the primary motor area of the cerebral cortex. It also relays impulses between different areas of the cerebrum. In addition to serving as a relay station, thalamic nuclei function along with other parts of the brain in movement control, emotions, maintenance of consciousness, pain perception, learning, memory, and cognition (thinking and knowing).

## Hypothalamus

The **hypothalamus** (hī-pō-THAL-a-mus; *hypo-* = under) is a small part of the diencephalon located inferior to the thalamus (see Figures 13.1 and 13.9b). It is composed of a dozen or so nuclei, as shown in Figure 13.12. Among these are the **mammillary bodies** (MAM-i-lar′-ē; *mammill-* = nipple-shaped), small, rounded nuclei projecting from the hypothalamus that serve as relay stations for reflexes related to the sense of smell. The stalklike **infundibulum** (in-fun-DIB-ū-lum = funnel) connects the pituitary gland to the hypothalamus.

The hypothalamus controls many body activities and is one of the major regulators of homeostasis. Sensory impulses related to both somatic and visceral senses arrive at the hypothalamus, as do impulses from receptors for vision, taste, and smell. Other receptors within the hypothalamus itself continually monitor conditions within the blood, including osmotic pressure, blood glucose level, certain hormone concentrations, and temperature. The hypothalamus has several very important connections with the pituitary gland and produces a variety of hormones, which are described in more detail in Concept 17.3. Some functions

**FIGURE 13.12 Hypothalamus.** Shown is a three-dimensional representation of hypothalamic nuclei located within the wall of the third ventricle.

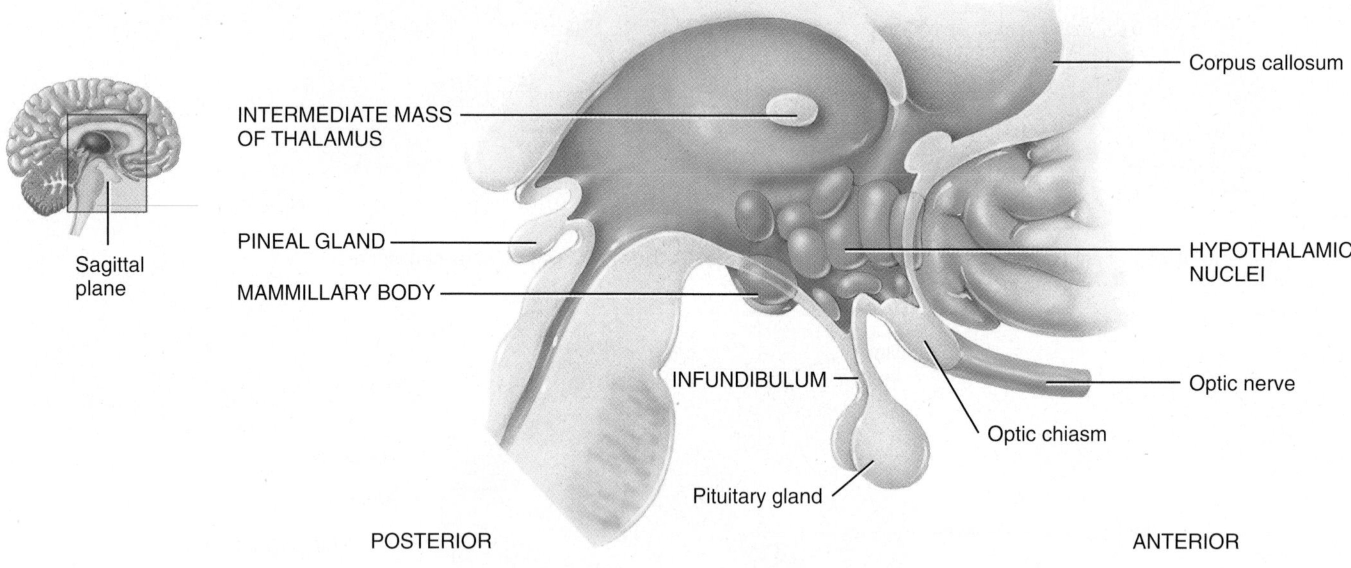


Sagittal section of brain showing hypothalamic nuclei

**The hypothalamus controls many body activities and is an important regulator of homeostasis.**

can be attributed to specific hypothalamic nuclei, but others are not so precisely localized. Important functions of the hypothalamus include the following:

- ***Control of the ANS***. The hypothalamus controls and integrates activities of the autonomic nervous system (ANS), which regulates contraction of smooth muscle and cardiac muscle and the secretions of many glands. Axons extend from the hypothalamus to sympathetic and parasympathetic nuclei in the brain stem and spinal cord. Through the ANS, the hypothalamus is a major regulator of visceral activities, including regulation of heart rate, movement of food through the gastrointestinal tract, and contraction of the urinary bladder.
- ***Production of hormones***. The hypothalamus produces several hormones and has two types of important connections with the pituitary gland, an endocrine gland located inferior to the hypothalamus (see Figures 13.1 and 13.12). First, hypothalamic hormones travel in the bloodstream through the infundibulum directly to the pituitary, where they stimulate or inhibit secretion of pituitary hormones. Second, axons extend from the hypothalamus through the infundibulum into the pituitary. The cell bodies of these neurons make hormones that are transported through their axons to the pituitary, where they are released.
- ***Regulation of emotional and behavioral patterns***. Together with the limbic system (described shortly), the hypothalamus participates in expressions of rage, aggression, pain, and pleasure, and the behavioral patterns related to sexual arousal.
- ***Regulation of eating and drinking***. The hypothalamus regulates food intake. It contains a **feeding center**, which promotes eating, and a **satiety center**, which causes a sensation of fullness and cessation of eating. The hypothalamus also contains a **thirst center**. When certain cells in the hypothalamus are stimulated by rising osmotic pressure of the extracellular fluid, they cause the sensation of thirst. The intake of water by drinking restores the osmotic pressure to normal, removing the stimulation and relieving the thirst.
- ***Control of body temperature***. The hypothalamus also functions as the body's *thermostat*. If the temperature of blood flowing through the hypothalamus is above normal, the hypothalamus directs the autonomic nervous system to stimulate activities that promote heat loss. When blood temperature is below normal, the hypothalamus generates impulses that promote heat production and retention.
- ***Regulation of circadian rhythms and states of consciousness***. The hypothalamus serves as the body's internal biological clock because it establishes **circadian rhythms** (ser-KĀ-dē-an), patterns of biological activity (such as the sleep–wake cycle) that occur on a daily schedule (cycle of about 24 hours). The hypothalamus receives input from the eyes (retina) and sends output to the reticular formation and the pineal gland.

## Pineal Gland

The **pineal gland** (PĪN-ē-al = pinecone-like) is about the size of a small pea and protrudes from the posterior midline of the third ventricle (Figure 13.12). The pineal gland secretes the hormone **melatonin**, which is thought to promote sleepiness as more melatonin is liberated during darkness than in light. When taken orally, melatonin also appears to contribute to the setting of the body's biological clock by inducing sleep and helping the body to adjust to jet lag.

The functions of the diencephalon are summarized in Table 13.2.

✓ **CHECKPOINT**

**16.** What structure usually connects the right and left halves of the thalamus?

**17.** Which part of the diencephalon contains the mammillary bodies? What is the function of the mammillary bodies?

**18.** What are the important functions of the hypothalamus?

**19.** What are the proposed functions of the pineal gland?

# 13.6 The midbrain, pons, and medulla oblongata of the brain stem serve as a relay station and control center.

The *brain stem* (Figure 13.13) is the part of the brain between the spinal cord and the diencephalon. It consists of three structurally and functionally connected regions: (1) midbrain, (2) pons, and (3) medulla oblongata. Extending through the brain stem is the reticular formation, a netlike region of interspersed gray and white matter. The midbrain, pons, and medulla contain both tracts and nuclei. Collectively, the structures of the brain stem act as relay centers for processing and controlling involuntary reflexes related to visual and auditory processing, eye movement, and regulation of autonomic functions including respiration, heart rate, blood pressure, and digestion.

## Midbrain

The **midbrain** extends from the pons to the diencephalon (see Figures 13.1, 13.6b, and 13.13) and is about 2.5 cm (1 in.) long. The cerebral aqueduct passes through the midbrain, connecting the third ventricle above with the fourth ventricle below.

FIGURE 13.13 **Inferior aspect of brain showing components of the brain stem.**

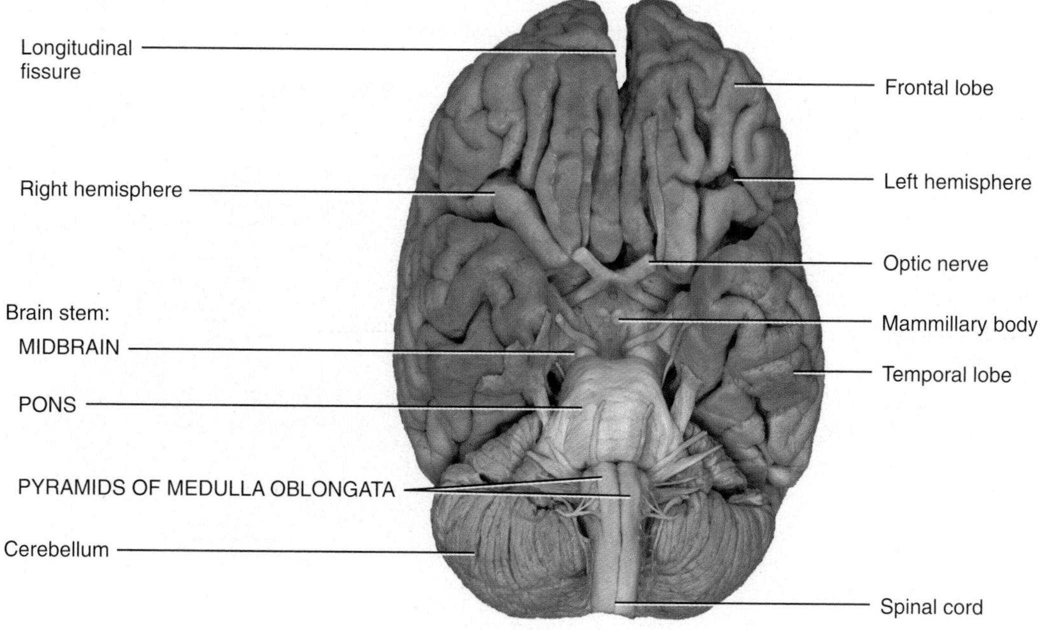


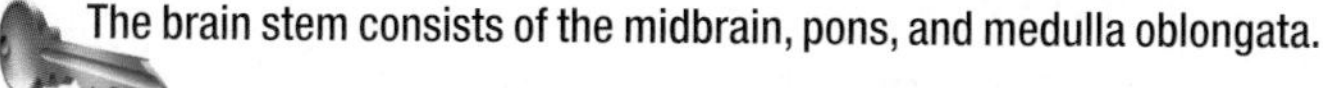
**The brain stem consists of the midbrain, pons, and medulla oblongata.**

The anterior part of the midbrain contains paired bundles of axons (tracts) known as the **cerebral peduncles** (pe-DUNG-kuls = little feet; see Figure 13.14). The cerebral peduncles consist of axons of motor (descending) tracts that conduct impulses from motor areas in the cerebral cortex to the medulla, pons, and spinal cord.

The posterior part of the midbrain contains four rounded elevations known as the **corpora quadrigemina** (kor-POR-a kwod-ri-JEM-i-nah). The two superior elevations are known as the **superior colliculi** (ko-LIK-ū-lī = little hills; singular is *colliculus;* Figure 13.14). They serve as reflex centers for certain visual activities. Through neural circuits from the retina

FIGURE 13.14 **Midbrain.** Cranial nerves, shown in yellow, will be discussed in Concept 14.2.

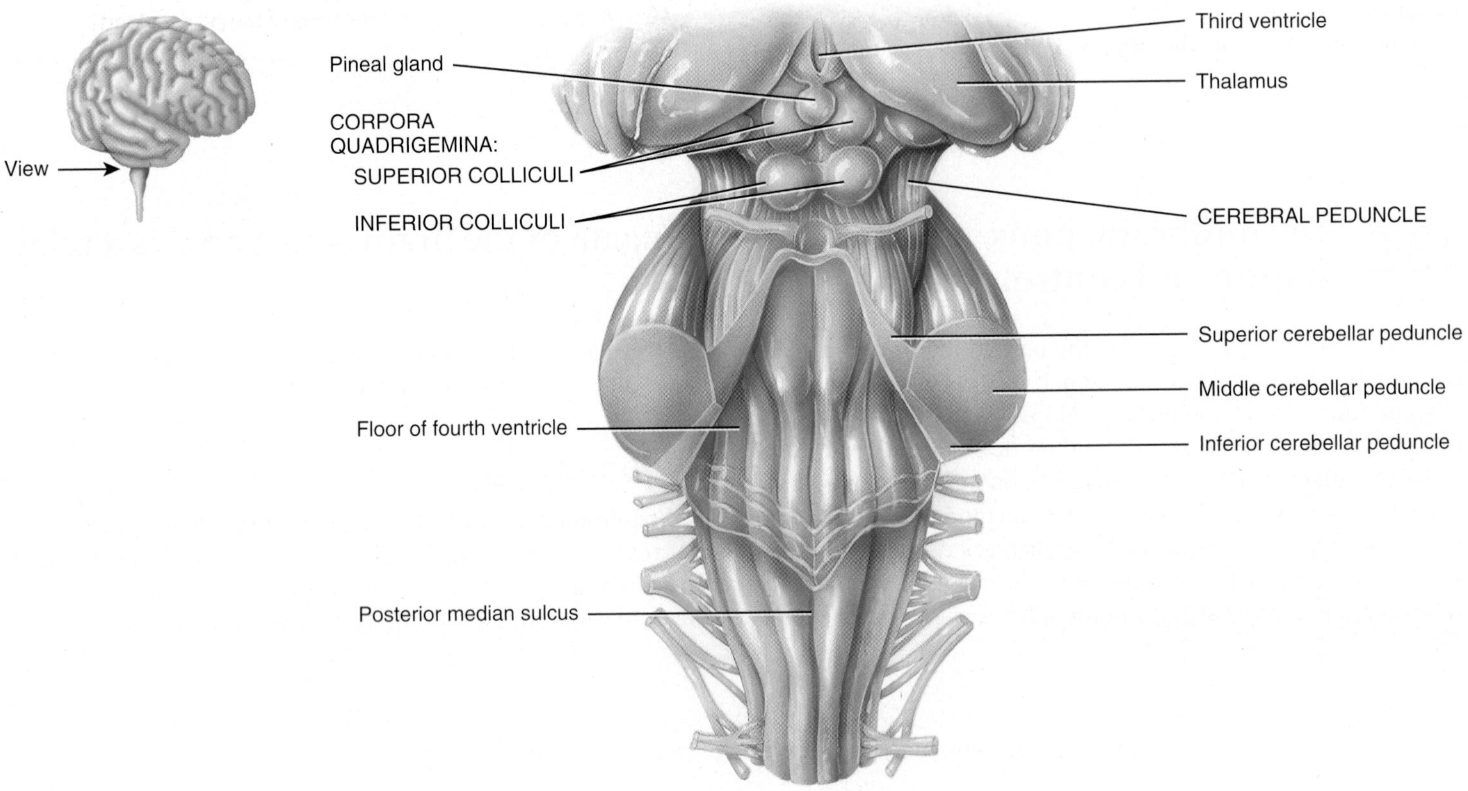

View
Pineal gland
CORPORA QUADRIGEMINA:
SUPERIOR COLLICULI
INFERIOR COLLICULI
Floor of fourth ventricle
Posterior median sulcus
Third ventricle
Thalamus
CEREBRAL PEDUNCLE
Superior cerebellar peduncle
Middle cerebellar peduncle
Inferior cerebellar peduncle

(a) Posterior view of midbrain in relation to brain stem

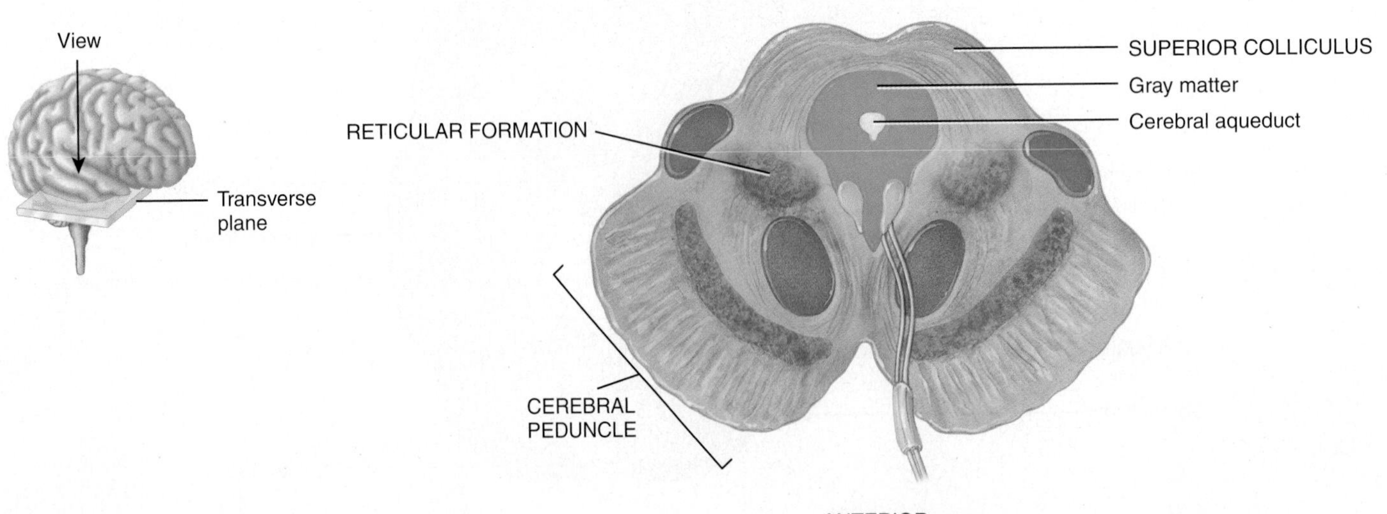


(b) Transverse section of midbrain

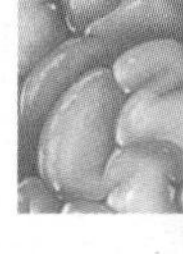



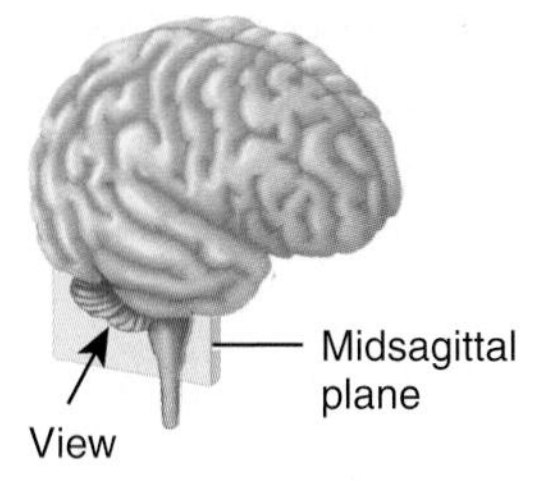


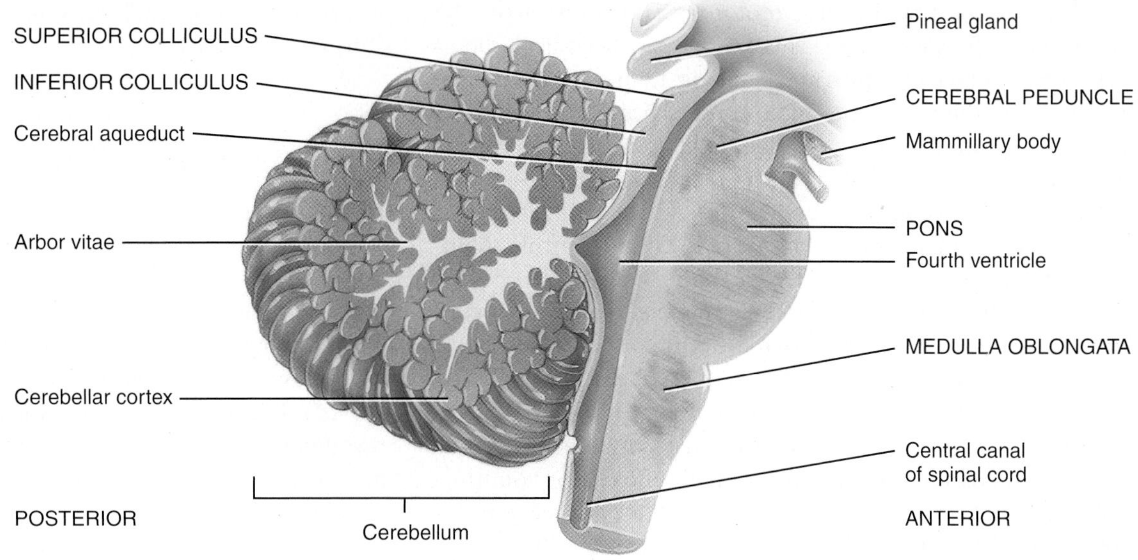


(c) Midsagittal section of cerebellum and brain stem

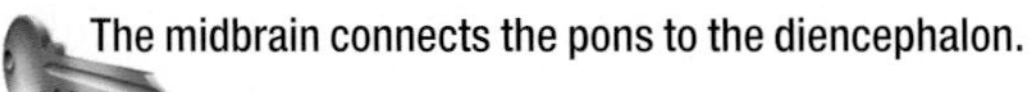


of the eye to the superior colliculi to the extrinsic eye muscles, visual stimuli elicit eye movements for tracking moving images (such as a moving car) and for scanning stationary images (as you are doing to read this sentence). The superior colliculi are also responsible for reflexes that govern movements of the head, eyes, and trunk in response to visual stimuli. The two inferior elevations, the **inferior colliculi**, are part of the auditory pathway, relaying impulses from the receptors for hearing in the ear to the brain. They also are reflex centers for the *startle reflex,* sudden movements of the head, eyes, and trunk that occur when you are surprised by a loud noise such as a gunshot.

The midbrain contains several nuclei. Neurons of nuclei that release dopamine extend to the basal nuclei to help control subconscious muscle activities. Loss of these neurons is associated with Parkinson's disease. Axons from the cerebellum and cerebral cortex form synapses in midbrain nuclei that help control muscular movements.

## Pons

The **pons** (= bridge) lies directly inferior to the midbrain and anterior to the cerebellum, and is about 2.5 cm (1 in.) long (see Figures 13.1, 13.13, and 13.14c). As its name implies, the pons is a bridge that connects parts of the brain with one another. These connections are provided by bundles of axons. Some axons of the pons connect the right and left sides of the cerebellum. Others are part of sensory (ascending) tracts and motor (descending) tracts. **Sensory (ascending) tracts** consist of axons that conduct sensory impulses toward higher brain areas. Tracts consisting of axons that carry motor impulses from higher brain areas are called **motor (descending) tracts**. Sensory and motor tracts of the brain are continuous with sensory and motor tracts in the spinal cord.

Signals for voluntary movements are relayed through the pons between the motor areas of the cortex (outer layer) of a cerebral hemisphere and the opposite hemisphere of the cerebellum. This complex circuitry plays an essential role in coordinating voluntary movements throughout the body. The **pneumotaxic area** (noo-mō-TAK-sik) and the **apneustic area** (ap-NOO-stik), shown in Figure 22.22, are nuclei in the pons that work with the medullary rhythmicity area of the medulla to help control breathing.

## Medulla Oblongata

The **medulla oblongata** (me-DOOL-la -ob′-long-GA-ta), or more simply the **medulla**, forms the inferior part of the brain stem and is a continuation of the spinal cord (see Figures 13.1, 13.13, and 13.14c). The medulla begins at the inferior border of the pons and extends to the foramen magnum, a distance of about 3 cm (1.2 in.).

Within the medulla's white matter are all sensory (ascending) tracts and motor (descending) tracts extending between the spinal cord and other parts of the brain. Some of the white matter forms bulges on the anterior aspect of the medulla.

These protrusions, called the **pyramids** (see Figures 13.13 and 13.15), are formed by the large motor tracts that pass from the cerebrum to the spinal cord. These motor tracts control voluntary movements of the limbs and trunk. Just superior to the junction of the medulla with the spinal cord, 90 percent of the axons in the left pyramid cross to the right side, and 90 percent of the axons in the right pyramid cross to the left side. This crossing is called the **decussation of pyramids** (deh-ku-SĀ-shun; *decuss-* = crossing) and explains why each side of the brain controls voluntary movements on the opposite side of the body.

The medulla also contains several nuclei that control vital body functions. The **cardiovascular center**, shown in Figure 20.11, regulates the rate and force of the heartbeat and the diameter of blood vessels. The **medullary rhythmicity area**, shown in Figure 22.22, adjusts the basic rhythm of breathing. Other nuclei in the medulla control reflexes for vomiting, coughing, sneezing, and hiccupping. Just lateral to each pyramid is an oval-shaped swelling called an **olive** (Figure 13.15). Neurons within the olives relay impulses to the cerebellum; these impulses provide instructions that the cerebellum uses to make adjustments to muscle activities as you learn new motor skills. Nuclei associated with sensations of touch, conscious proprioception, pressure, and vibration are also located in the medulla. Many ascending sensory axons form synapses in these nuclei, and postsynaptic neurons then relay the sensory information to the thalamus on the opposite side of the brain (see Figure 15.6a).

## Reticular Formation

In addition to the well-defined nuclei already described, much of the brain stem consists of small clusters of neuronal cell bodies (gray matter) interspersed among small bundles of myelinated axons (white matter). The broad region where white matter and gray matter exhibit a netlike arrangement is known as the **reticular formation** (re-TIK-ū-lar; *ret-* = net; Figure 13.14b). It extends from the inferior part of the diencephalon, throughout the brain stem, and into the superior part of the spinal cord. Neurons within the reticular formation have both ascending (sensory) and descending (motor) functions.

Part of the reticular formation, called the **reticular activating system (RAS)**, consists of sensory axons that project to the cerebral cortex (see Figure 15.9). The RAS helps maintain consciousness, a state of wakefulness in which an individual is fully alert, aware, and oriented. Visual and auditory stimuli and mental activities can stimulate the RAS to help maintain consciousness. The RAS is also active during **arousal** (awakening from

**FIGURE 13.15** **Internal anatomy of the medulla oblongata.**

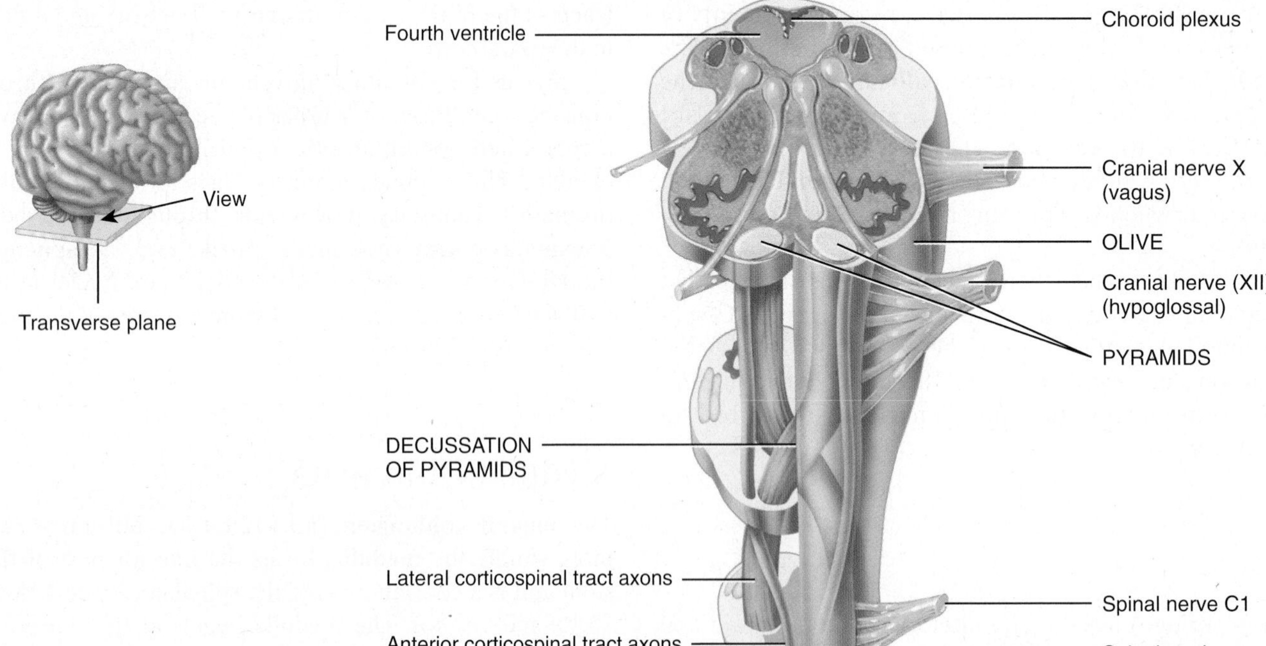


Transverse section and anterior surface of medulla oblongata

**The pyramids of the medulla contain large motor tracts that run from the cerebrum to the spinal cord.**

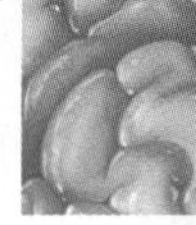

sleep) and helps maintain **attention** and *alertness.* The RAS also **prevents sensory overload** by filtering out insignificant sensory information so that it does not reach consciousness. Familiar or repeated impulses are ignored, while new or unusual impulses are passed on to the cerebral cortex. As you walk through a forest you may not notice the sound of a breeze stirring the overhead canopy of leaves, but you would be aware of a branch snapping behind you. Inactivation of the RAS produces **sleep**, a state of partial consciousness from which an individual can be aroused. Damage to the RAS, on the other hand, results in **coma**, a state of unconsciousness from which an individual cannot be aroused.

Even though the RAS receives input from the eyes, ears, and other sensory receptors, there is no input from receptors for the sense of smell; even strong odors may fail to cause arousal. People who die in house fires usually succumb to smoke inhalation without awakening. For this reason, all sleeping areas should have a nearby smoke detector that emits a loud alarm. A vibrating pillow or flashing light can serve the same purpose for those who are deaf or hard-of-hearing.

The reticular formation's main descending functions are to help regulate posture and *muscle tone*, the slight degree of involuntary contraction in normal resting muscles.

The functions of the brain stem are summarized in Table 13.2.

## ✓ CHECKPOINT

**20.** Where are the midbrain, pons, and medulla located relative to one another?

**21.** Which part of the brain stem contains the cerebral peduncles? What is their importance?

**22.** In which regions of the brain stem are the nuclei that help control breathing?

**23.** Define decussation of pyramids. What is the functional consequence of decussation of the pyramids?

**24.** What are important functions of the reticular formation?

## RETURN TO **Annette's Story**

Annette makes an appointment the next day, and Bill stays behind after his family returns home to go with her to see Dr. Akers. Bill gets right down to the purpose of the visit. "Dr. Akers, I have just learned that my mother has been taking levodopa for three years and ropinirole for a few months with only a partial response to her Parkinson's disease. She wants to live independently as long as possible and I don't think she can do this with medication alone. What do you know about deep brain stimulation (DBS)?"

Dr. Akers replied, "I've read about that but I have no experience with it. Annette will have to go to the university and see a neurosurgeon there. I'll give you a referral."

Bill and Annette next go to see Dr. Paynter at the university, who orders various diagnostic tests to determine whether Annette is a candidate for DBS. Dr. Paynter gives Annette the good news that she does qualify for DBS and explains that they can implant slender electrodes into the areas of Annette's brain that are causing her symptoms. The electrodes will be connected by thin wires to a battery-powered neurostimulator that will be implanted under her skin near her collarbone. The neurostimulator is similar to a heart pacemaker and will be programmed to send electrical impulses through the electrodes into her brain that block the abnormal electrical signals that are causing her symptoms. The doctor explains that the procedure is reversible if Annette feels that it has not helped her or a more effective procedure is developed in the future.

Annette quickly agrees to have the procedure done, thinking about all of the wonderful things she might be able to do with her grandchildren the next time they visit.

E. *What part of the brain transmits motor impulses from Annette's basal nuclei to her primary motor area?*

F. *Motor signals from Annette's primary motor area in her cerebral cortex must travel to the cerebellum so that she can pick up her fork. What part of the brain serves as a bridge from the cortex to the cerebellum?*

G. *Parkinson's disease is causing Annette's muscles to become increasingly rigid or hypertonic. Maintenance of normal muscle tone is controlled by which part of Annette's brain?*

## 13.7 The cerebellum coordinates movements and helps maintain normal muscle tone, posture, and balance.

The *cerebellum*, second only to the cerebrum in size, occupies the inferior and posterior aspects of the cranial cavity. Like the cerebrum, the cerebellum has a highly folded surface that greatly increases the surface area of its outer gray matter cortex, allowing for a greater number of neurons. The cerebellum accounts for about a tenth of the brain mass yet contains nearly half of the neurons in the brain. The cerebellum is posterior to the medulla and pons and inferior to the posterior portion of the cerebrum (see Figures 13.1 and 13.13). In superior or inferior views, the shape of the cerebellum resembles a butterfly. The central constricted area is the **vermis** (= worm), and the lateral "wings" or lobes are the **cerebellar hemispheres** (Figure 13.16). Each hemisphere consists of lobes separated by deep and distinct fissures. The **anterior lobe** and **posterior lobe** govern subconscious aspects of skeletal muscle movements. The **flocculonodular lobe** (flok-ū-lō-NOD-ū-lar; *flocculo-* = wool-like tuft) on the inferior surface contributes to equilibrium and balance.

The superficial layer of the cerebellum, called the **cerebellar cortex**, consists of gray matter in a series of slender, parallel folds. Deep to the gray matter are tracts of white matter called **arbor vitae** (AR-bor VĪ-tē = tree of life; see Figure 13.14c) that resemble branches of a tree. Even deeper, within the white matter, are the **cerebellar nuclei**, regions of gray matter that give rise to axons carrying impulses from the cerebellum to other brain centers and to the spinal cord.

Three paired **cerebellar peduncles** (pe-DUNG-kuls) attach the cerebellum to the brain stem (see Figures 13.14a and 13.16b). These bundles of white matter consist of axons that conduct impulses between the cerebellum and other parts of the brain. The **superior cerebellar peduncles** contain axons that extend from the cerebellum to the midbrain and thalamus; the **middle cerebellar peduncles** carry commands for voluntary movements from the pons (which receive input from motor areas of the cerebral cortex) into the cerebellum; and the **inferior cerebellar peduncles** carry sensory information into the cerebellum from the medulla, pons, and spinal cord.

The primary function of the cerebellum is to evaluate how well movements initiated by motor areas in the cerebrum are actually being carried out. When movements initiated by the

FIGURE 13.16 Cerebellum.

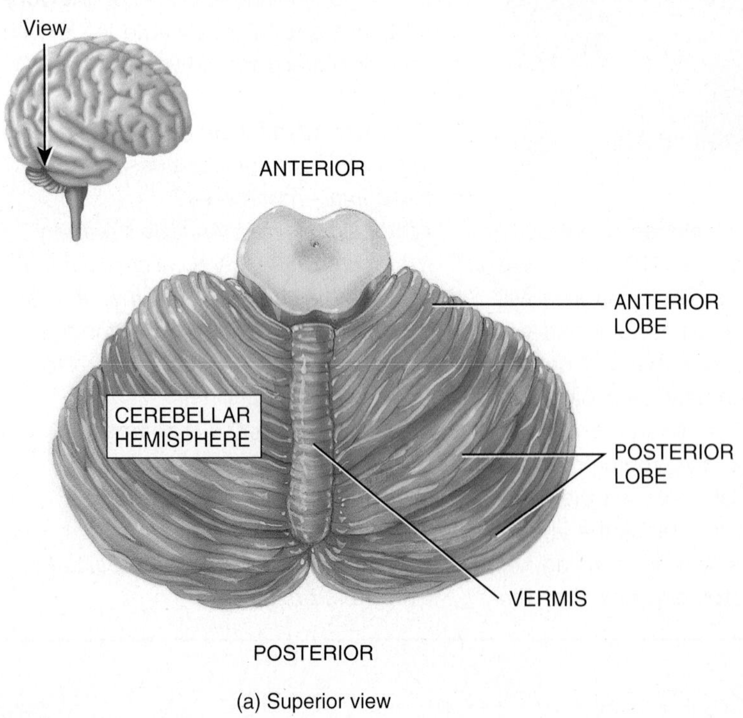


(a) Superior view

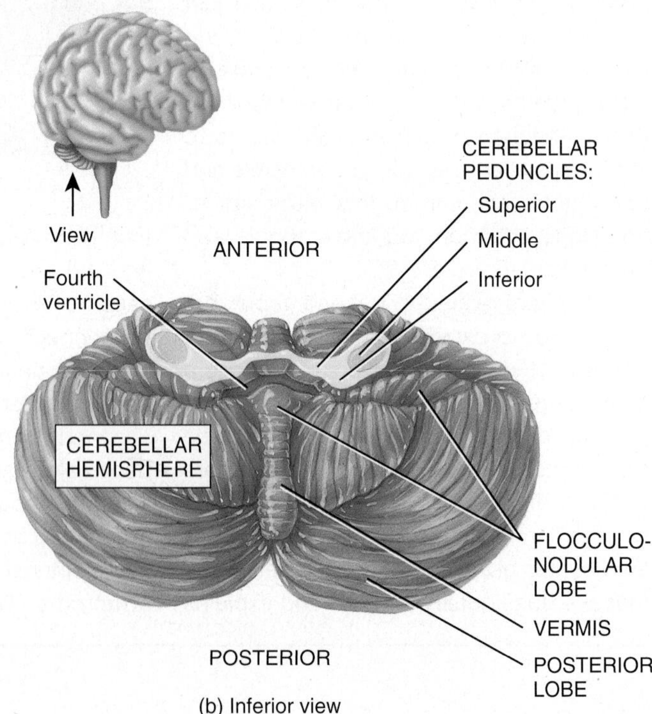


(b) Inferior view

The cerebellum coordinates skilled movements and regulates posture and balance.

cerebral motor areas are not being carried out correctly, the cerebellum detects the discrepancies. It then sends feedback signals to motor areas of the cerebral cortex via its connections to the thalamus. The feedback signals help correct the errors, smooth the movements, and coordinate complex sequences of skeletal muscle contractions. Besides coordinating skilled movements, the cerebellum is the main brain region that regulates posture and balance. These aspects of cerebellar function make possible all skilled muscular activities, from catching a baseball to dancing to speaking.

The functions of the cerebellum are summarized in Table 13.2.

✓ **CHECKPOINT**

**25.** How is the cerebellum externally similar to the cerebrum?

**26.** What are the functions of the three lobes of the cerebellum?

# 13.8 The limbic system controls emotions, behavior, and memory.

Encircling the upper part of the brain stem and the corpus callosum is a ring of structures on the inner border of the cerebrum and floor of the diencephalon that constitutes the **limbic system** (*limbic* = border; Figure 13.17). The limbic system is sometimes called the "emotional brain" because it plays a primary role in a range of emotions, including pain, pleasure, docility, affection, and anger. It also is involved in olfaction (smell) and memory.

FIGURE 13.17 **The limbic system.** Components of the limbic system are shown in green.

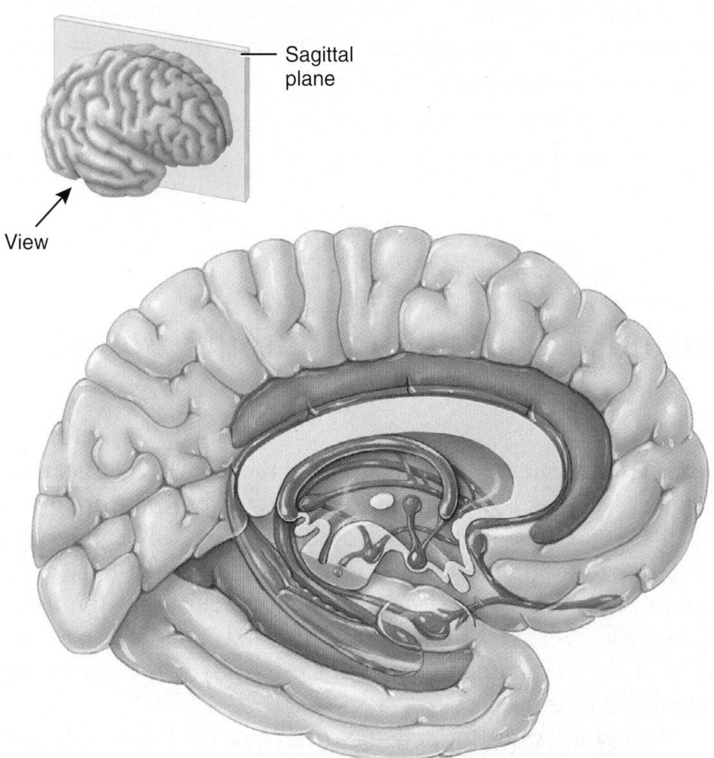


The limbic system governs emotional aspects of behavior.

Experiments have shown that when different areas of animals' limbic systems are stimulated, the animals' reactions indicate that they are experiencing intense pain or extreme pleasure. Stimulation of other limbic system areas in animals produces tameness and signs of affection. Stimulation of a cat's **amygdala** (a-MIG-da-la), one of the basal nuclei of the cerebrum, produces fear and a behavioral pattern called rage—the cat extends its claws, raises its tail, opens its eyes wide, hisses, and spits. By contrast, removal of the amygdala produces an animal that lacks fear and aggression. Likewise, a person whose amygdala is damaged fails to recognize fearful expressions in others or fails to express fear in situations where this emotion would normally be appropriate, for example, while being attacked by an animal.

Together with other parts of the cerebrum, the limbic system also functions in memory. People with damage to certain limbic system structures forget recent events and cannot commit anything to memory.

✓ **CHECKPOINT**

**27.** Where is the limbic system located?

## CLINICAL CONNECTION | *Brain Injuries*

**Brain injuries** are commonly associated with head trauma and result in part from displacement and distortion of neural tissue at the moment of impact. Additional tissue damage may occur when normal blood flow is restored after a period of ischemia (reduced blood flow). The sudden increase in oxygen level produces large numbers of *oxygen free radicals* (charged oxygen molecules with an unpaired electron). Brain cells recovering from the effects of a stroke or cardiac arrest also release free radicals. Free radicals cause damage by disrupting cellular DNA and enzymes and by altering plasma membrane permeability. Brain injuries can also result from *hypoxia* (cellular oxygen deficiency).

Various degrees of brain injury are described by specific terms. A **concussion** (kon-KU-shun) is an injury characterized by an abrupt, but temporary, loss of consciousness (from seconds to hours), disturbances of vision, and problems with equilibrium. It is caused by a blow to the head or the sudden stopping of a moving head (as in an automobile accident) and is the most common brain injury. A concussion produces no obvious bruising of the brain. Signs of a concussion are headache, drowsiness, nausea and/or vomiting, lack of concentration, confusion, or post-traumatic amnesia (memory loss).

A brain **contusion** (kon-TOO-zhun) is bruising due to trauma and includes the leakage of blood from microscopic vessels. It is usually associated with a concussion. In a contusion, the pia mater may be torn, allowing blood to enter the subarachnoid space. The area most commonly affected is the frontal lobe. A contusion usually results in an immediate loss of consciousness (generally lasting no longer than 5 minutes), loss of reflexes, transient cessation of respiration, and decreased blood pressure. Vital signs typically stabilize in a few seconds.

A **laceration** (las-er-Ā-shun) is a tear of the brain, usually from a skull fracture or a gunshot wound. A laceration results in rupture of large blood vessels, with bleeding into the brain and subarachnoid space. Consequences include *cerebral hematoma* (localized pool of blood, usually clotted, that swells against the brain tissue), edema, and increased intracranial pressure. If the blood clot is small enough, it may pose no major threat and may be absorbed. If the blood clot is large, it may require surgical removal. Swelling infringes on the limited space that the brain occupies in the cranial cavity. Swelling causes excruciating headaches. Brain tissue can also undergo *necrosis* (cellular death) due to the swelling; if the swelling is severe enough, the brain can herniate through the foramen magnum, resulting in death.

## 13.9 The spinal cord receives sensory input and provides motor output through spinal nerves.

### External Anatomy of the Spinal Cord

The *spinal cord* is roughly oval in shape, being flattened slightly anteriorly and posteriorly. In adults, the spinal cord extends from the medulla oblongata, the inferior part of the brain, to the level of the second lumbar vertebra (see Figure 13.2). During early childhood, both the spinal cord and the vertebral column grow longer as part of overall body growth. Elongation of the spinal cord stops around age 4 or 5, but growth of the vertebral column continues. Thus, the spinal cord does not extend the entire length of the adult vertebral column. The adult spinal cord is between 42 to 45 cm (16–18 in.) long and about 2 cm (0.75 in.) wide in the midthoracic region (see Figure 13.2).

When the spinal cord is viewed externally, two conspicuous enlargements can be seen. The superior enlargement, called the **cervical enlargement**, supplies nerves to and from the upper limbs. The inferior enlargement, called the **lumbar enlargement**, provides nerves to and from the lower limbs. Inferior to the lumbar enlargement, the spinal cord tapers to a conical portion referred to as the **conus medullaris** (KŌ-nus med-ū-LAR-is; *conus* = cone). Arising from the conus medullaris is the **filum terminale** (FĪ-lum ter-mi-NAL-ē = terminal filament), an extension of the pia mater that extends inferiorly and anchors the spinal cord to the coccyx.

**Spinal nerves** are the paths of communication between the spinal cord and specific regions of the body. The spinal cord appears to be segmented because the 31 pairs of spinal nerves emerge at regular intervals through intervertebral foramina of the vertebral column (see Figures 13.2 and 13.4c). For ease of discussion, each pair of spinal nerves is said to arise from a *spinal segment*. However, within the spinal cord there is no obvious segmentation.

Two bundles of axons, called **roots**, connect each spinal nerve to a segment of the cord (see Figures 13.4b, c and 13.18). The **posterior root** contains only sensory axons, which conduct impulses from sensory receptors in the skin, muscles, and internal organs into the central nervous system. Each posterior root also has a swelling, the **posterior root ganglion**, which contains the cell bodies of sensory neurons. The **anterior root** contains axons of motor neurons, which conduct impulses from the CNS to effectors (muscles and glands).

As spinal nerves branch from the spinal cord, they pass laterally to exit the spinal canal through the intervertebral foramina between adjacent vertebrae. However, because the spinal cord is shorter than the vertebral column, nerves that arise from the inferior part of the spinal cord do not leave the vertebral column at the same level as they exit from the cord. The roots of these lower spinal nerves angle inferiorly alongside the filum terminale in the vertebral canal like wisps of hair. Accordingly, the roots of these nerves are collectively named the **cauda equina** (KAW-da ē-KWĪ-na), meaning "horse's tail" (see Figure 13.2).

## Internal Anatomy of the Spinal Cord

The spinal cord white matter surrounds an inner core of gray matter (Figure 13.18). The white matter of the spinal cord consists primarily of bundles of myelinated axons of neurons. Two grooves penetrate the white matter of the spinal cord and divide it into right and left sides. The **anterior median fissure** is a wide groove on the anterior side. The **posterior median sulcus** is a narrow furrow on the posterior side. The gray matter of the spinal cord is shaped like the letter H or a butterfly; it consists of cell bodies of neurons, unmyelinated axons, dendrites, and neuroglia. The **gray commissure** (KOM-mi-shur) forms the crossbar of the H, connecting the gray matter on the right and left sides. In the center of the gray commissure is the **central canal**, a passageway filled with cerebrospinal fluid that extends the length of the spinal cord. At its superior end, the central canal is continuous with the fourth ventricle (a space that contains cerebrospinal fluid) in the medulla oblongata of the brain. Anterior to the gray commissure, the **anterior white commissure** connects the white matter of the right and left sides of the spinal cord.

The gray matter on each side of the spinal cord is subdivided into regions called **horns**. The **posterior gray horns** contain interneurons as well as axons of incoming sensory neurons. The **anterior gray horns** contain cell bodies of somatic motor neurons that provide impulses for skeletal muscle contractions. Between the posterior and anterior gray horns are the **lateral gray horns**. The lateral gray horns contain the cell bodies of autonomic motor neurons that regulate the activity of cardiac muscle, smooth muscle, and glands.

The white matter of the spinal cord, like the gray matter, is organized into regions. The anterior and posterior gray horns divide the white matter on each side into three broad areas called **columns:** (1) **anterior white columns**, (2) **posterior white columns**, and (3) **lateral white columns**. Within each column are sensory (ascending) and motor (descending) tracts that may extend long distances up or down the spinal cord.

FIGURE 13.18 **Internal anatomy of the spinal cord: the organization of gray matter and white matter.** For simplicity, dendrites are not shown. Blue, red, and green arrows in (a) indicate the direction of impulse propagation.

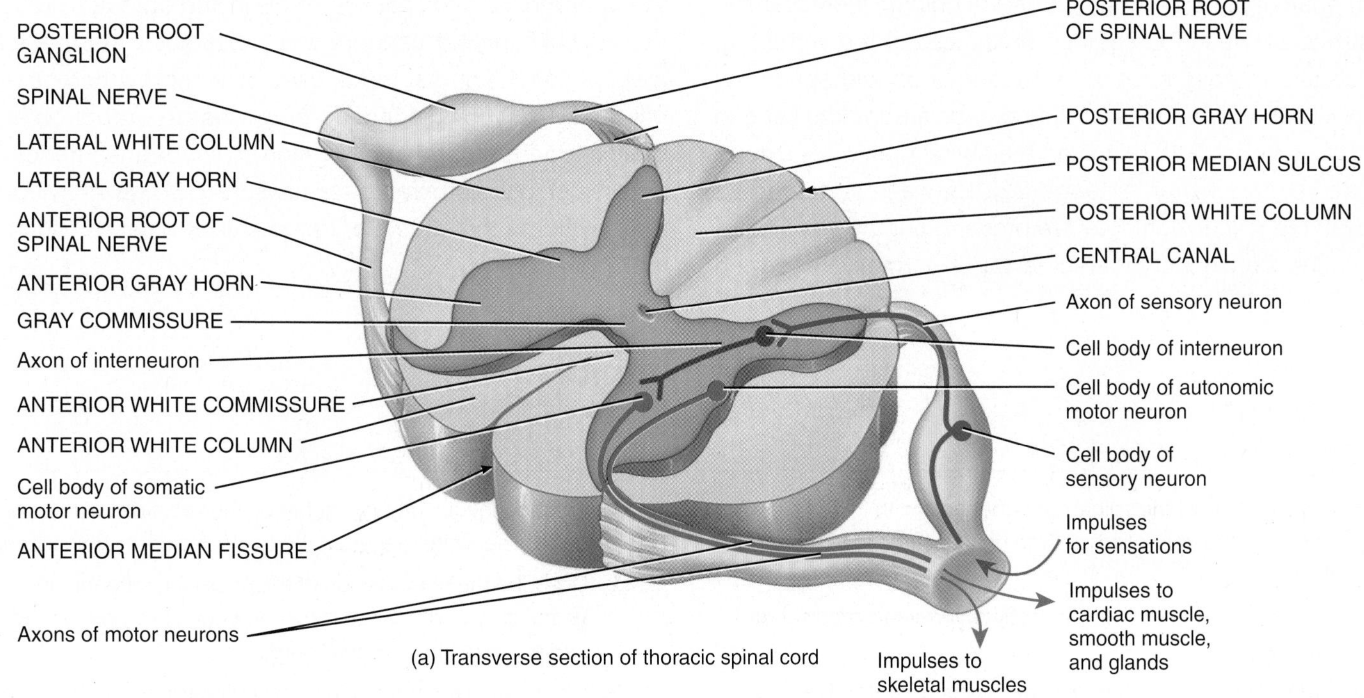

(a) Transverse section of thoracic spinal cord

(continues)

**FIGURE 13.18 (continued)**

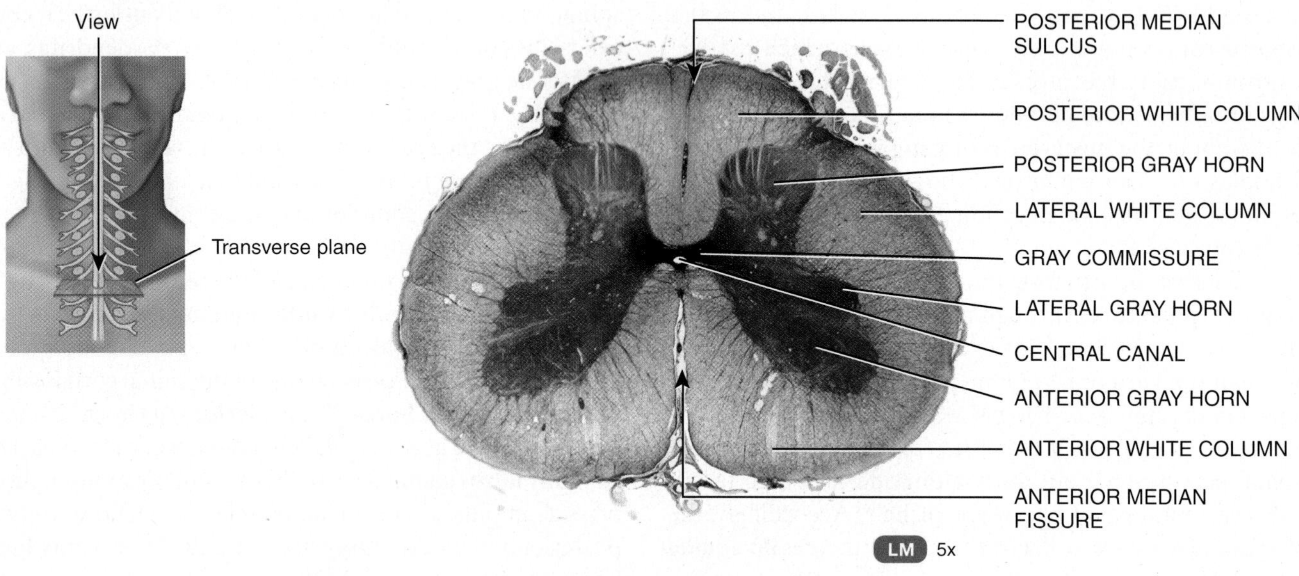


(b) Transverse section of thoracic spinal cord

**In the spinal cord, white matter surrounds the gray matter.**

## CLINICAL CONNECTION | *Spinal Tap*

In a **spinal tap (lumbar puncture),** a local anesthetic is given, and a long needle is inserted into the subarachnoid space. During this procedure, the patient lies on his or her side with the vertebral column flexed as in assuming the fetal position. Flexion of the vertebral column increases the distance between the spinous processes of the vertebrae, which allows easy access to the subarachnoid space. As you will soon learn, the spinal cord ends around the second lumbar vertebra (L2); however, the spinal meninges extend to the second sacral vertebra (S2). Between vertebrae L2 and S2 the spinal meninges are present, but the spinal cord is absent. Consequently, a spinal tap is normally performed in adults between vertebrae L3 and L4 or L4 and L5 because this region provides safe access to the subarachnoid space without the risk of damaging the spinal cord. (A line drawn across the highest points of the iliac crests, called the *supracristal line,* passes through the spinous process of the fourth lumbar vertebra and is used as a landmark for a spinal tap.) A spinal tap is used to withdraw cerebrospinal fluid (CSF) for diagnostic purposes; to introduce antibiotics, contrast media for myelography, or anesthetics; to administer chemotherapy; to measure CSF pressure; and/or to evaluate the effects of treatment for diseases such as meningitis.

## ✓ CHECKPOINT

**28.** What portion of the spinal cord connects with nerves of the upper limbs? With nerves of the lower limbs?

**29.** What are the conus medullaris, filum terminale, and cauda equina?

**30.** What structures are found in the posterior roots? The posterior root ganglia? The anterior roots?

**31.** What connects the gray matter on the two sides of the spinal cord? The white matter on the two sides of the spinal cord?

**32.** Describe the composition of the posterior, anterior, and lateral gray horns. What is the difference between a horn and a column in the spinal cord?

**33.** Where are ascending and descending tracts located in the spinal cord?

# 13.10 The spinal cord conducts impulses between spinal nerves and the brain, and contains reflex pathways.

The spinal cord has two principal functions in maintaining homeostasis: impulse propagation and information integration. Tracts within the *white matter* of the spinal cord serve as highways for impulse propagation. Sensory input travels along these tracts toward the brain, and motor output travels from the brain along these tracts toward skeletal muscles and other effector tissues. The *gray matter* of the spinal cord receives and integrates incoming and outgoing information.

## Sensory and Motor Tracts

As noted previously, the spinal cord conducts impulses along tracts. Often, the name of a tract indicates its position in the white matter and where it begins and ends. For example, the anterior corticospinal tract is located in the *anterior* white column; it begins in the *cerebral cortex* and ends in the *spinal cord*. Notice that the location of the axon terminals comes last in the name. This regularity in naming allows you to determine the direction of information flow along any tract named according to this convention. Thus, because the anterior corticospinal tract conveys impulses from the brain toward the spinal cord, it is a motor (descending) tract. Figure 13.19 highlights the major sensory and motor tracts in the spinal cord. Sensory and motor tracts are described in more detail in Concepts 15.4 and 15.6.

Impulses from sensory receptors propagate up the spinal cord to the brain along two main routes on each side: the spinothalamic tract and the posterior column. The **spinothalamic tract** (spī′-nō-tha-LAM-ik) conveys impulses for sensing pain, warmth, coolness, itching, tickling, deep pressure, and crude touch. The **posterior column** consists of tracts that convey impulses for discriminative touch, light pressure, vibration, and conscious proprioception (the awareness of the positions and movements of muscles, tendons, and joints).

Sensory input keeps the CNS informed of changes in the external and internal environments. The sensory information is integrated (processed) by interneurons in the spinal cord and brain. Responses to the integrative decisions are brought about by motor activities (muscular contractions and glandular secretions). The cerebral cortex plays a major role in controlling precise voluntary muscular movements. Other brain regions provide important integration for regulation of automatic movements. Motor output to skeletal muscles travels down the spinal cord in two types of descending pathways: direct and indirect. The **direct pathways** convey impulses that originate in the cerebral cortex and are destined to cause *voluntary* movements of skeletal muscles. **Indirect pathways** convey impulses from the brain stem to cause *automatic movements* that maintain skeletal muscle tone, sustain contraction of postural muscles, and play a major role in equilibrium by regulating muscle tone in response to movements of the head.

**FIGURE 13.19 Locations of major sensory and motor tracts, shown in a transverse section of the spinal cord.** Sensory tracts are indicated on one half and motor tracts on the other half of the cord, but actually all tracts are present on both sides.

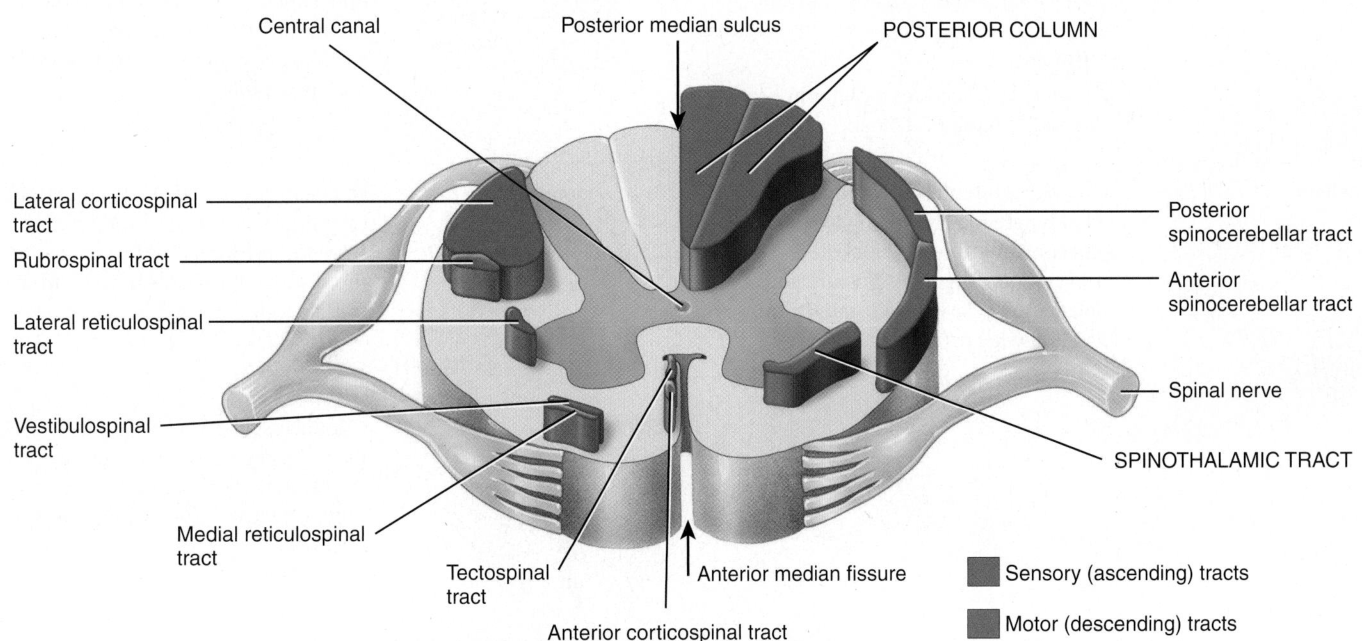


**The name of a tract often indicates its location in the white matter and where it begins and ends.**

## Spinal Reflexes

The spinal cord also serves as an integrating center for some reflexes. Reflexes are fast, involuntary responses to stimuli. When integration takes place in the spinal cord gray matter, the reflex is a **spinal reflex**. By contrast, if integration occurs in the brain rather than the spinal cord, the reflex is a **cranial reflex**. Reflexes will be described in more detail in Concept 14.4.

The functions of the spinal cord are summarized in Table 13.2.

**TABLE 13.2**

Functions of Principal Parts of the Central Nervous System

| PART | FUNCTIONS | PART | FUNCTIONS |
|---|---|---|---|
| **Cerebrum** 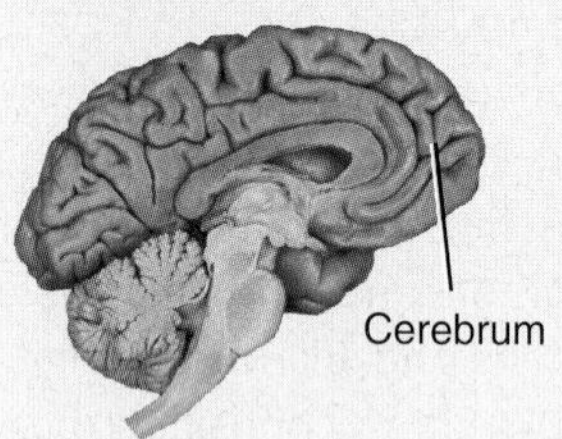  | Sensory areas interpret sensory impulses, motor areas control voluntary movement, and association areas function in emotional and intellectual processes. Basal nuclei coordinate automatic muscle movements and regulate muscle tone. Limbic system functions in emotional aspects of behavior related to survival. | **Brain stem** 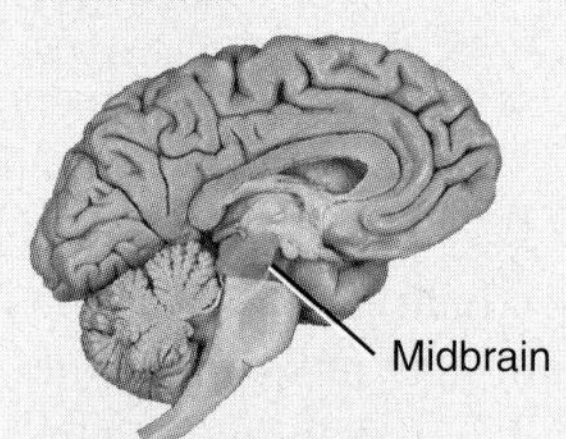  | *Midbrain:* Relays motor impulses from the cerebral cortex to the pons and sensory impulses from the spinal cord to the thalamus. Superior colliculi coordinate movements of the head, eyes, and trunk in response to visual stimuli, and the inferior colliculi coordinate movements of the head, eyes, and trunk in response to auditory stimuli. Contributes to control of movements. |
| **Diencephalon** 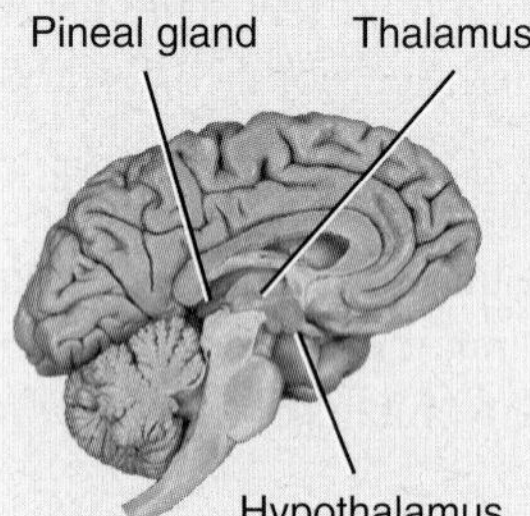  | *Thalamus:* Relays almost all sensory input to the cerebral cortex. Contributes to motor functions by transmitting information from the cerebellum and basal ganglia to the primary motor area of the cerebral cortex. Also plays a role in the maintenance of consciousness.<br>*Hypothalamus:* Controls and integrates activities of the autonomic nervous system and pituitary gland. Regulates emotional and behavioral patterns. Controls body temperature and regulates eating and drinking behavior. Helps maintain the waking state and establishes patterns of sleep. Produces hormones.<br>*Pineal gland:* Secretes melatonin, sets the body's biological clock. | 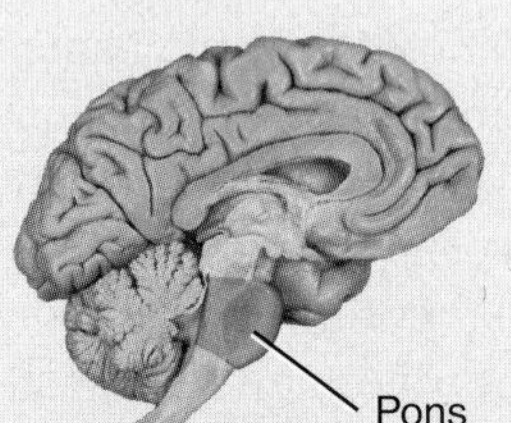  | *Pons:* Relays impulses between cerebral cortex and cerebellum and between the medulla and midbrain. Pneumotaxic and apneustic area, together with the medulla oblongata, help control breathing. |
| | | 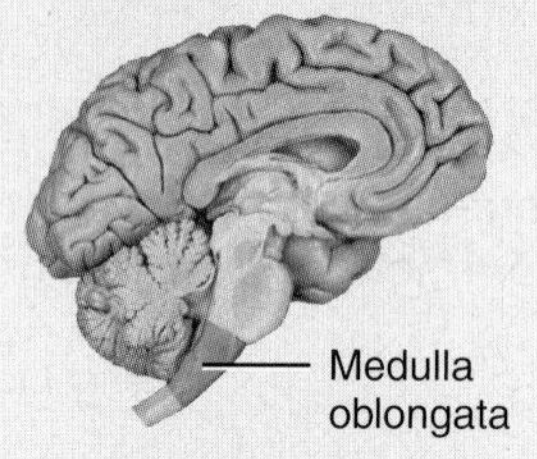  | *Medulla oblongata:* Relays motor and sensory impulses between other parts of the brain and the spinal cord. Vital centers regulate heartbeat, breathing (together with pneumotaxic and apneustic area of pons), and blood vessel diameter. Other centers coordinate swallowing, vomiting, coughing, sneezing, and hiccuping. |
| **Cerebellum** 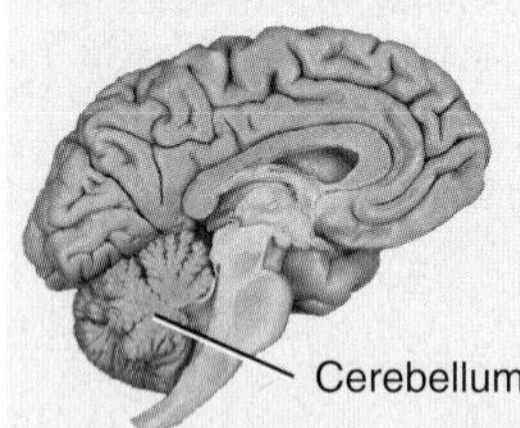  | Compares intended movements with what is actually happening to smooth and coordinate complex, skilled movements. Regulates posture and balance. May have a role in cognition and language processing. | 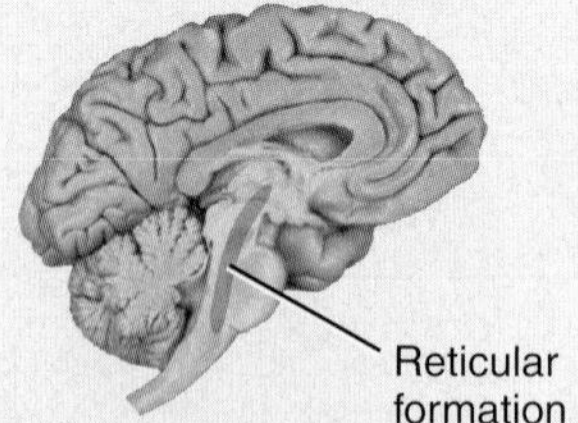  | *Reticular formation:* Helps maintain consciousness, causes awakening from sleep, filters repetitive sensory input, and contributes to regulation of muscle tone. |
| | | **Spinal cord** 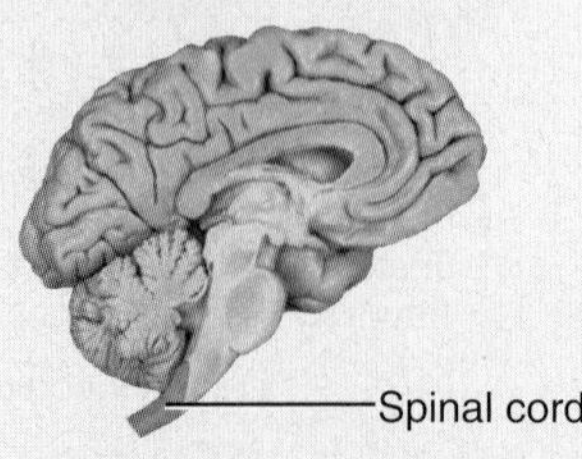  | Conducts sensory nerve impulses toward the brain and motor nerve impulses from the brain toward skeletal muscles and other effector tissues. Integrates spinal reflexes. |

## ✓ CHECKPOINT

**34.** Which component of the spinal cord white matter transmits information to your brain that your knee is flexed? Which components transmit information to your brain that your body feels cold?

**35.** Which kind of pathway would transmit motor instructions to consciously pick up a book? To maintain an upright posture while you sit in class?

## Annette's Story

## EPILOGUE AND DISCUSSION

Annette is very nervous as she is rolled into the operating room. There is plenty of scary equipment in the room and lots of people, but once she sees Dr. Paynter she feels reassured.

The nurse anesthetist introduces herself. "Hi, I'm Maggie. I'll be responsible for observing your vital signs throughout the procedure and will be available to answer any questions during the first part, when you'll be awake."

The operative team takes their time, double- and triple-checking to make sure they have the right patient and are all prepared for the correct procedure. Once it has been established that everything is in order, Dr. Paynter asks, "Annette, are you ready?" and she responds with a nod. Annette's head is placed in a stereotactic brace to keep it still during the procedure, then an MRI is performed to identify the area of the brain where Dr. Paynter needs to insert the electrodes. Dr. Paynter shaves a small area of Annette's scalp, injects it with local anesthetic, and drills a hole through the cranium and the attached outermost component of the meninges. She stimulates different parts of Annette's subthalamic area to find the location that initiates Annette's tremors to determine where the abnormal signals can be intercepted. Annette is awake during this procedure so her tremor activity can be observed. Once Dr. Paynter finds the target area, the electrode is secured in place and the head wound is closed.

About a week later Annette returns to the hospital for the second stage of the procedure. While Annette is under general anesthesia, Dr. Paynter implants the neurostimulator under the skin of her upper chest. The wires are passed beneath Annette's skin from the chest incision up through her neck, then scalp, and are finally attached to the electrodes coming out of her brain.

Following DBS surgery, Annette's neurostimulator is programmed to the level of stimulation that best negates her symptoms. Dr. Akers will monitor Annette's need for decreasing Parkinson's disease medication and Dr. Paynter will occasionally make slight adjustments to the stimulation parameters of her neurostimulator.

Annette's symptoms are much improved after undergoing the DBS procedure, and she is able to resume many of her previous activities. Annette notices that her hands do not shake a bit as she dials the phone to call Bill and thank him for the hundredth time for convincing her to seek further treatment and for recommending DBS.

H. *Which of the meninges did Dr. Paynter have to drill through during the procedure?*
I. *If Annette were to experience problems with balance, which area of the brain would be responsible?*
J. *How do nerve impulses for precise skeletal muscle movements travel from the brain to the spinal cord?*
K. *Which area of gray matter in Annette's spinal cord contains motor neurons responsible for skeletal muscle contractions?*

# Concept and Resource Summary

| Concept | Resources |
|---|---|
| **Introduction** | |
| 1. The central nervous system consists of the brain, the control center for thoughts, memories, and behaviors; and the spinal cord, the pathway for sensory impulses traveling to the brain and motor impulses traveling from the brain to skeletal muscles and other effectors. | |

## Concept 13.1 The CNS consists of the brain and spinal cord, and is protected by several structures.

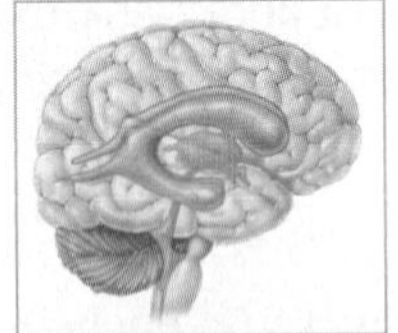

1. The four major parts of the brain are the **cerebrum**, the **diencephalon**, the **brain stem**, and the **cerebellum**.
2. The **spinal cord** extends from the medulla oblongata to the second lumbar vertebra.
3. The skull, vertebrae, meninges, and cerebrospinal fluid protect the CNS.
4. The **meninges** are three protective connective tissue coverings. The **dura mater**, the strongest meninx, adheres to the periosteum of the cranial bones, but it is separated from the vertebrae around the spinal cord by the **epidural space**. The middle **arachnoid mater** contains a web of delicate collagen and elastic fibers. The innermost meninx, the **pia mater**, is transparent, adheres tightly to the surface of the brain and spinal cord, and protects against shock and displacement of the spinal cord.
5. The **subarachnoid space** between the arachnoid mater and the pia mater contains cerebrospinal fluid.

**Resources:**
Anatomy Overview—The Brain
Anatomy Overview—The Spinal Cord
Figure 13.2—External Anatomy of the Spinal Cord and the Spinal Nerves
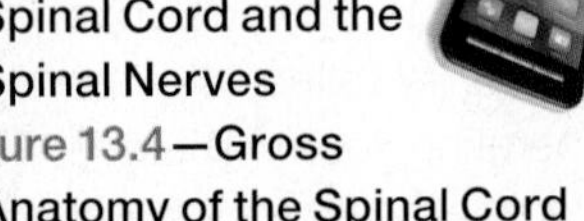
Figure 13.4—Gross Anatomy of the Spinal Cord and Its Protective Coverings
Clinical Connection–Meningitis

## Concept 13.2 The CNS is nourished and protected by blood and cerebrospinal fluid.

1. Blood from the internal carotid and vertebral arteries carries oxygen and nutrients to the brain; blood leaves through dural sinuses to the internal jugular veins.
2. Blood is carried to the spinal cord by the posterior intercostal and lumbar arteries; blood leaves by the posterior intercostal and lumbar veins.
3. The **blood–brain barrier** is created by tightly connected endothelial cells of CNS capillaries and astrocyte processes that press up against the capillaries. Glucose, oxygen, carbon dioxide, most anesthetics, and several other substances can cross the blood–brain barrier. Toxins, most antibiotics, and proteins cannot cross it.
4. **Cerebrospinal fluid** provides mechanical protection, chemical protection, and circulation of nutrients to the brain and spinal cord.
5. Cerebrospinal fluid is formed in the choroid plexuses and circulates through the **lateral ventricles**, **third ventricle**, **fourth ventricle**, **subarachnoid space**, and central canal. Ependymal cells use their cilia to maintain cerebrospinal fluid flow. Cerebrospinal fluid is returned to the blood across **arachnoid villi** that project into the dural sinuses.

**Resources:**
Anatomy Overview—The Nervous System: Overview
Clinical Connection—Breaching the Blood–Brain Barrier
Clinical Connection—Brain Tumors

## Concept 13.3 The cerebrum interprets sensory impulses, controls muscular movements, and functions in intellectual processes.

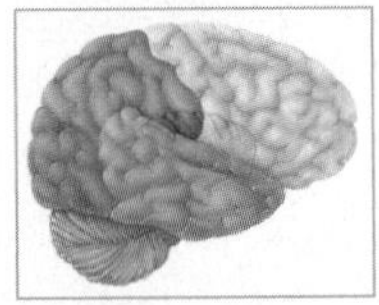

1. The cerebrum consists of an outer rim of gray matter called the **cerebral cortex**, deeper cerebral white matter, and gray matter nuclei. It has folds called **gyri**; grooves called **sulci**; and deep grooves called **fissures**. The **longitudinal fissure** separates the cerebrum into left and right **cerebral hemispheres** connected by the **corpus callosum**.
2. Each cerebral hemisphere is subdivided by sulci into four lobes: **frontal lobe**, **parietal lobe**, **temporal lobe**, and **occipital lobe**. A fifth deep region of the cerebrum, the **insula**, cannot be seen at the surface.
3. The **cerebral white matter** contains **tracts** of myelinated axons that propagate impulses for communication between CNS regions. There are three types of tracts: **association tracts** conduct impulses between gyri of the same hemisphere; **commisural tracts** conduct impulses between gyri of different hemispheres; and **projection tracts** conduct impulses between the cerebrum and lower CNS regions.
4. **Basal nuclei** located deep in each cerebral hemisphere include: the **globus pallidus**, the **putamen**, and the **caudate nucleus**. Basal nuclei help initiate and terminate movements, regulate muscle tone, control subconscious contractions of skeletal muscles, and influence some cognitive processes.

**Resources:**
Anatomy Overview—Cerebrum
Figure 13.7—Cerebrum

Exercise—Paint the Functional Areas of the Cerebral Cortex
Clinical Connection—Attention Deficit Disorder

| Concept | Resources |
|---|---|

## Concept 13.4 The cerebral cortex can be divided functionally into sensory areas, motor areas, and association areas.

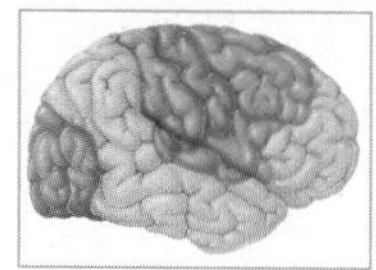

1. The sensory areas of the cerebral cortex allow perception of sensory information. The motor areas control the execution of voluntary movements. The association areas are concerned with more complex integrative functions such as memory, personality traits, and intelligence.
2. The **primary somatosensory area** receives impulses from somatic sensory receptors for touch, pressure, vibration, itch, tickle, temperature, pain, and proprioception and is involved in the perception of these sensations. The primary somatosensory area localizes the part of the body from which a sensation originates. The **primary visual area** receives visual information, the **primary auditory area** receives sound information, the **primary gustatory area** receives taste information, and the **primary olfactory area** receives smell information.
3. Motor areas include the **primary motor area**, which controls voluntary contractions of muscles, and **Broca's area**, which controls production of speech.
4. The **somatosensory association area** permits you to evaluate an object by touch and to sense the relationship of one body part to another. It also stores memories of past somatic sensory experiences.
5. The **visual association area** utilizes past visual experiences for recognizing and evaluating what is seen.
6. The **auditory association area** allows you to recognize a particular sound.
7. **Wernicke's area** interprets the meaning of speech by translating words into thoughts.
8. The **common integrative area** integrates sensory interpretations from the association areas, allowing thoughts based on sensory inputs.
9. The **prefrontal cortex** is concerned with personality, intellect, complex learning abilities, judgment, reasoning, conscience, intuition, and development of abstract ideas.
10. The **premotor area** controls sequential muscle contractions. It also serves as a memory bank for complex movements.
11. The **frontal eye field area** controls voluntary scanning movements of the eyes.
12. Each hemisphere has unique functions in a division of labor called **hemispheric lateralization**. Each hemisphere receives sensory signals from and controls movements of the opposite side of the body. The left hemisphere is more important for language, numerical and scientific skills, and reasoning. The right hemisphere is more important for musical and artistic awareness, spatial and pattern perception, recognition of faces, emotional content of language, identifying odors, and generating mental images of sight, sound, touch, taste and smell.

**Resources:**

Clinical Connection—Aphasia

Clinical Connection—Cerebral Palsy

## Concept 13.5 The diencephalon includes the thalamus, hypothalamus, and pineal gland.

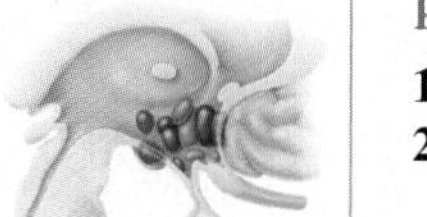

1. The thalamus, hypothalamus, and pineal gland are regions of the diencephalon.
2. The **thalamus** consists of paired oval masses of gray matter organized into nuclei that are connected by gray matter called the **intermediate mass**.
3. The thalamus relays and processes sensory impulses to the primary sensory areas of the cerebral cortex. It transmits input from the cerebellum to the primary motor area of the cerebral cortex. Together with other brain areas, it functions in movement control, emotions, consciousness, pain perception, learning, memory, and cognition.
4. The **hypothalamus** consists of many nuclei, including the **mammillary bodies**, which are relay stations for smell. The **infundibulum** connects the pituitary gland to the hypothalamus.
5. The hypothalamus is responsible for many body activities, including control of the autonomic nervous system, production of hormones, regulation of emotions and behavior, regulation of eating and drinking, control of body temperature, and regulation of circadian rhythms.
6. The **pineal gland** secretes the hormone **melatonin**, which is thought to promote sleep.

**Resources:**

Anatomy Overview—Diencephalon

Figure 13.12—Hypothalamus

Clinical Connection—Alzheimer's Disease

Clinical Connection—Transient Ischemic Attacks

Clinical Connection—Depression

| Concept | Resources  |
|---|---|

## Concept 13.6 The midbrain, pons, and medulla oblongata of the brain stem serve as a relay station and control center.

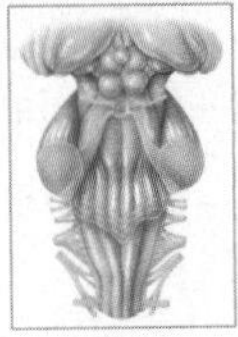

1. The three functional regions of the brain stem are the midbrain, the pons, and the medulla oblongata.
2. The tracts and nuclei of brain stem structures act as relay centers for processing and controlling involuntary reflexes for vision and hearing, and govern reflexes vital to life.
3. The **midbrain** contains anterior tracts called the **cerebral peduncles**; posteriorly, the **corpora quadrigemina** has **superior colliculi** for visual reflexes and **inferior colliculi** for auditory reflexes. Nuclei of the midbrain release dopamine for control of subconscious muscle activities.
4. The **pons** relays signals for voluntary movements from the cerebral cortex to the cerebellum. The **pneumotaxic area** and the **apneustic area** are pons nuclei that function with the medulla to control breathing. The pons contains **sensory (ascending) tracts** that conduct impulses to higher brain areas and **motor (descending) tracts** that conduct impulses from higher brain areas toward the spinal cord.
5. The **medulla oblongata** forms the inferior part of the brain stem and is a continuation of the spinal cord. Some of its white matter forms anterior bulges called the **pyramids** where there is a crossing of motor axons (**decussation of pyramids**) to the opposite side before continuing to the spinal cord. Nuclei of the medulla control vital body functions of heart rate, blood pressure, and the basic rhythm of breathing. **Olives**, lateral to each pyramid, relay motor impulses to the cerebellum while learning new motor skills.
6. The broad region of the brain stem where white matter and gray matter form a netlike arrangement is called the **reticular formation**. Part of the reticular formation, called the **reticular activating system**, helps arouse the body from sleep and maintain consciousness. Its motor function helps regulate posture and muscle tone.

Resources: Anatomy Overview—Brain Stem; Clinical Connection—Cerebrovascular Accident; Clinical Connection—Injury to the Medulla

## Concept 13.7 The cerebellum coordinates movements and helps maintain normal muscle tone, posture, and balance.

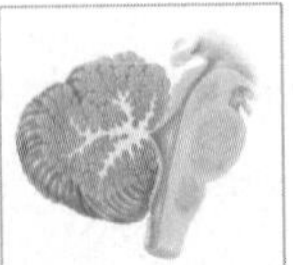

1. The cerebellum has a central area called the **vermis** and two lateral **cerebellar hemispheres**. The superficial **cerebellar cortex** consists of gray matter; deeper, branching white matter is the **arbor vitae**.
2. The cerebellum smoothes and coordinates the contractions of skeletal muscles. It also maintains posture and balance.

Resources: Anatomy Overview—Cerebellum; Clinical Connection—Ataxia

## Concept 13.8 The limbic system controls emotions, behavior, and memory.

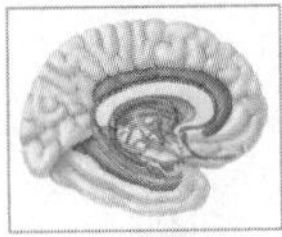

1. The **limbic system** includes components of the cerebrum and diencephalon. It encircles the upper brain stem and corpus callosum.
2. The limbic system plays a primary role in the emotional aspects of behavior and memory.

## Concept 13.9 The spinal cord receives sensory input and provides motor output through spinal nerves.

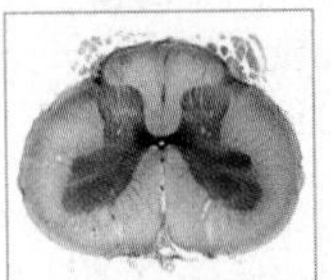

1. The spinal cord extends from the medulla oblongata to the second lumbar vertebra. It has **cervical** and **lumbar enlargements**, and an inferior **conus medullaris**. The **filum terminale** extends from the **conus medullaris** to anchor the spinal cord to the coccyx.
2. There are 31 pairs of **spinal nerves**, which function in communication between the spinal cord and peripheral nerves. Each spinal nerve connects to a segment of the spinal cord by two **roots**. The **posterior root** contains sensory axons and has a swelling, the **posterior root ganglion**, containing the cell bodies of sensory neurons. The **anterior root** contains axons of motor neurons.
3. The **cauda equina** is a collection of spinal nerve roots that hang inferiorly from the spinal cord in the vertebral canal.

Resources: Anatomy Overview—Spinal Nerves; Clinical Connection—Encephalitis; Clinical Connection—Spinal Tap

| Concept | Resources |
|---|---|

4. The H-shaped gray matter has a **central canal** that contains cerebrospinal fluid and is subdivided into **horns.** The **posterior gray horns** contain interneurons and axons of sensory neurons; the **anterior gray horns** contain cell bodies of somatic motor neurons; and the **lateral gray horns** contain cell bodies of autonomic motor neurons.
5. The white matter of the spinal cord is organized into **anterior white columns, posterior white columns,** and **lateral white columns** that contain **sensory (ascending) tracts** and **motor (descending) tracts.**

## Concept 13.10 The spinal cord conducts impulses between spinal nerves and the brain, and contains reflex pathways.

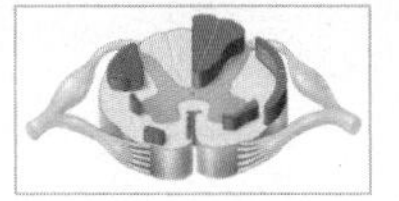

1. The spinal cord conducts impulses along sensory and motor tracts. Ascending impulses from sensory receptors travel along both sides of the spinal cord to the brain in the **spinothalamic tract** and the **posterior column.**
2. **Direct pathways** transmit impulses from the cerebral cortex to skeletal muscles for voluntary movements. **Indirect pathways** transmit impulses from the brain stem for automatic movements that regulate muscle tone, balance, and posture.
3. The spinal cord functions as an integration center for **spinal reflexes;** the brain integrates **cranial reflexes.**

Animation—Somatic Sensory and Motor Pathways
Exercise—Concentrate on Your Brain

Clinical Connection—Traumatic Spinal Cord Injuries

Clinical Connection—Poliomyelitis

## Understanding the Concepts

1. How does the skeleton protect the brain and spinal cord?
2. How does the blood–brain barrier protect the central nervous system?
3. What is the functional difference between the blood–brain barrier and the blood–cerebrospinal fluid barrier?
4. What structures must cerebrospinal fluid pass through to enter the dural sinuses? Are red blood cells and large proteins normally able to pass through these structures? Why or why not?
5. Describe the location, composition, and functions of the basal nuclei.
6. What area(s) of the cerebral cortex receives information about the colors in a flower? Controls flexion of your fingers? Instructs your fingers to play a familiar song on the piano? Distinguishes a baby's cry from a dog's bark? Interprets directions for installing software in a computer? Recognizes an object flying by as a basketball?
7. Why is the thalamus considered a "relay station" in the brain?
8. How does the hypothalamus influence the autonomic nervous system? Why is it considered part of both the nervous system and the endocrine system?
9. What functions are carried out by the superior colliculi? By the inferior colliculi?
10. Which structures contain the axons that carry information into and out of the cerebellum?
11. What are the functions of the limbic system?
12. How is the spinal cord partially divided into right and left sides?
13. Based on its name, what are the origin and destination of the spinothalamic tract? Is this a sensory tract or a motor tract?

CHAPTER

# 14

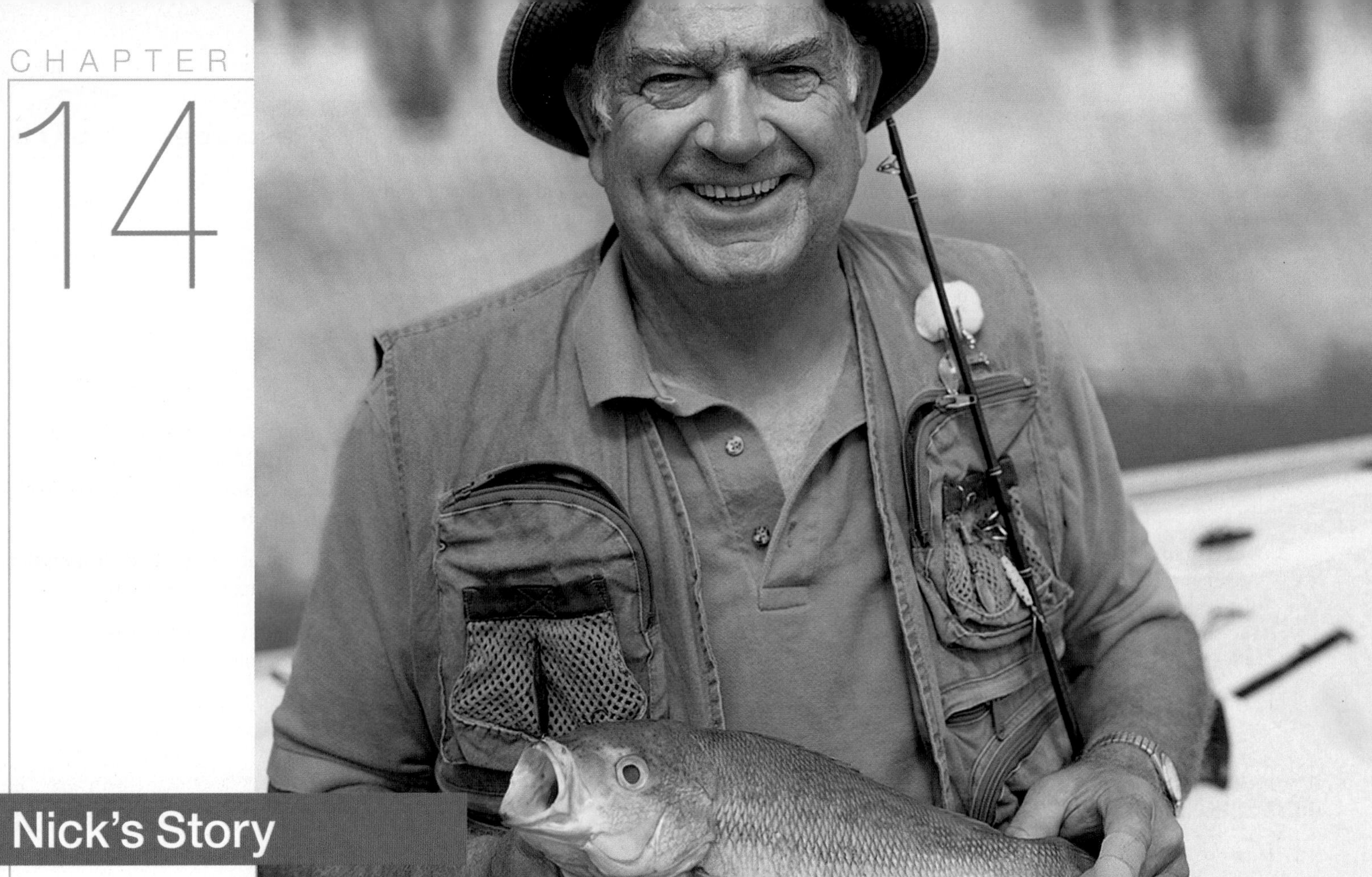

## Nick's Story

The door to the exam room opened slowly. "Well, hello, Nick, I haven't seen you in quite a while," said the doctor.

"Hi, Doc. Yeah, it's been a while; changed jobs again, and with my new insurance company's rules, well, I was afraid they wouldn't cover the cost of the visit." Nick shifted in his chair and began to stand up. He rose slowly, steadying himself on the arms of the chair.

"Steady there, Nick, let me help you over to the exam table."

"Sorry, Doc. I'm only fifty-six, but I feel like a hundred and fifty-six. I've been trying to keep the diabetes under control, but there's been a lot of stress in the last two years, what with my divorce and the company laying off its machinists. I've been feeling worse and worse. I got these weird pains in my feet, like burning and prickling, and I've been so clumsy lately. I get real dizzy when I sit down or stand up, which isn't too safe working in the shop, you know. I nearly broke my neck when I bent down to grab a tool the other day!"

The doctor quickly flipped through Nick's medical chart. "Hmm, it's been almost two years since you were here last, Nick. You know you really need to come in more often than that. Diabetes can lead to a lot of other health problems if you don't watch your blood sugars and get regular checkups."

"Oh, you docs are all the same, you just want our money," Nick said, winking at the doctor.

"Nick, you always get right to the point. I think you're just afraid of doctors!" They both laughed, but Dr. Jerouen was more concerned than he was ready to admit to Nick.

Nick is a Type 2 diabetic; his body is not able to regulate its blood sugar levels properly. This pathology causes very high blood sugar levels that can starve cells of the energy they need, damage the kidneys, and, in Nick's case, damage the nerves. As we follow Nick's case we will see that his neurological symptoms resulted from damage to his peripheral nerves. In order to understand his case, we need to know more about the peripheral nervous system. What are nerves? What information do they carry? How does damage to the peripheral nervous system affect the body?

# The Peripheral Nervous System

## INTRODUCTION

The **peripheral nervous system (PNS)** consists of nervous tissue found outside the brain and spinal cord, including nerves that exit the central nervous system and ganglia found throughout the body cavities. (**Ganglia** are collections of neuronal cell bodies outside the central nervous system.) The peripheral nervous system can be subdivided into three functional systems, the somatic nervous system, an autonomic nervous system, and an enteric nervous system.

The **somatic nervous system** consists of afferent nerves, sensory cranial and spinal nerves that connect somatic receptors to the central nervous system (CNS), and efferent nerves, motor cranial and spinal nerves that connect the CNS to skeletal muscles. Functionally, the somatic nervous system is associated with voluntary or conscious activities.

The **autonomic nervous system** includes autonomic sensory neurons, integrating centers within the CNS, and autonomic motor neurons. The autonomic nervous system usually operates without conscious control to regulate the activity of smooth muscle, cardiac muscle, and glands.

The **enteric nervous system** is a specialized network of nerves and ganglia within the wall of the gastrointestinal (GI) tract. The action of the enteric nervous system is involuntary to regulate GI tract activity. Although the enteric nervous system communicates with the CNS, it functions independently of both the autonomic and central nervous systems.

In this chapter, we explore the structure and function of the peripheral nervous system and its regulation by the CNS.

## CONCEPTS

**14.1** Nerves have three protective connective tissue coverings.

**14.2** Twelve pairs of cranial nerves distribute primarily to regions of the head and neck.

**14.3** Each spinal nerve branches into a posterior ramus, an anterior ramus, a meningeal branch, and rami communicantes.

**14.4** A reflex is produced by a reflex arc in response to a particular stimulus.

**14.5** The autonomic nervous system produces involuntary movements.

**14.6** The ANS includes preganglionic neurons, autonomic ganglia and plexuses, and postganglionic neurons.

**14.7** ANS neurons release acetylcholine or norepinephrine, resulting in excitation or inhibition.

**14.8** The sympathetic division supports vigorous physical activity; the parasympathetic division conserves body energy.

**14.9** Autonomic reflexes regulate controlled body conditions and are primarily integrated by the hypothalamus.

# 14.1 Nerves have three protective connective tissue coverings.

A **nerve** is a cordlike structure that consists of parallel bundles of axons and their associated neuroglial cells that lie outside the brain and spinal cord. Nerves emerging from the brain are called **cranial nerves**; those emerging from the spinal cord are called **spinal nerves**. Each cranial nerve and spinal nerve contains layers of protective connective tissue coverings (Figure 14.1). Individual axons, whether myelinated or unmyelinated, are wrapped in **endoneurium** (en′-dō-NOO-rē-um; *endo-* = within or inner; *-neurium* = nerve). Groups of axons with their endoneurium are held together in bundles called **fascicles**, each of which is wrapped in **perineurium** (per′-i-NOO-rē-um; *peri-* = around). The outermost covering over the entire nerve is the **epineurium** (ep′-i-NOO-rē-um; *epi-* = over). The dura mater of the CNS meninges fuses with the epineurium as the nerve passes through foramina of the skull or vertebral column. Note the presence of many blood vessels, which nourish the nerve (Figure 14.1b). You may recall from Chapter 10 that the connective tissue coverings of skeletal muscles—endomysium, perimysium, and epimysium—are similar in organization to those of nerves.

✓ **CHECKPOINT**

1. What is a nerve?
2. How are nerves protected?

FIGURE 14.1 **Organization and connective tissue coverings of a spinal nerve.**

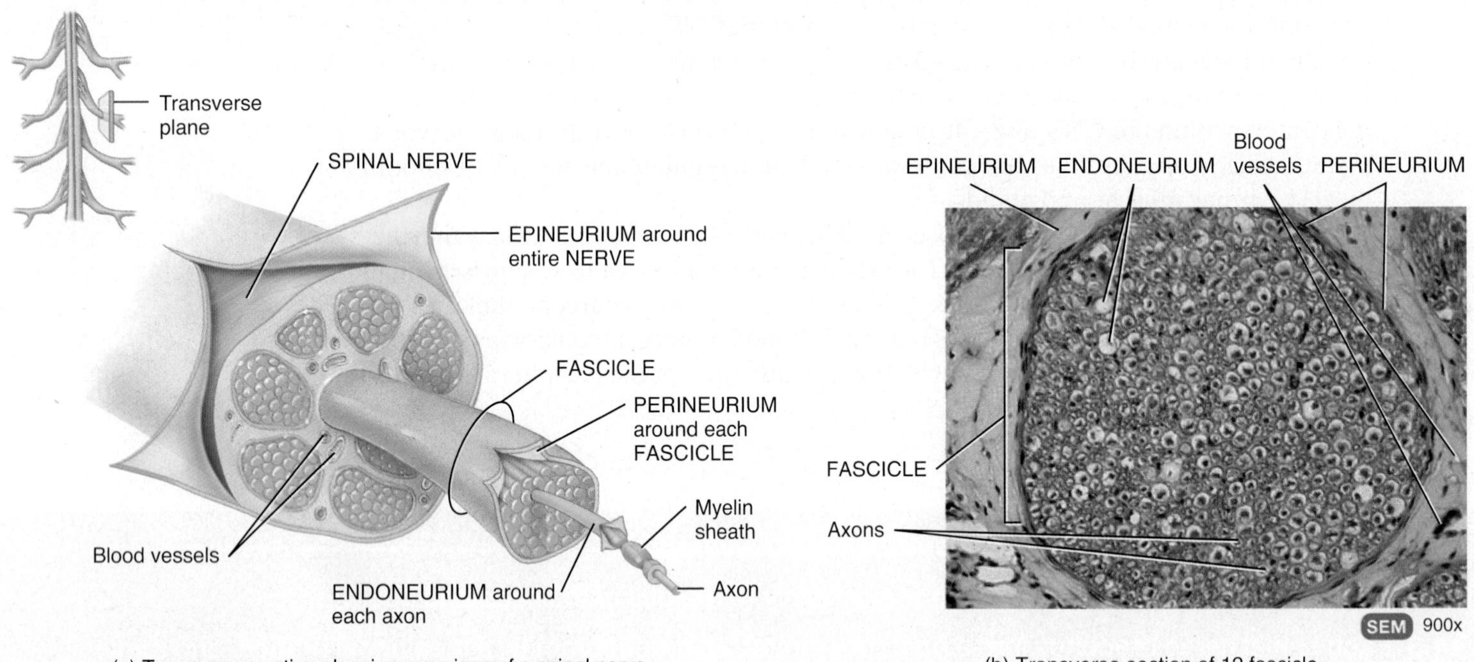


(a) Transverse section showing coverings of a spinal nerve

(b) Transverse section of 12 fascicle

**Three layers of connective tissue wrappings protect axons: endoneurium surrounds individual axons, perineurium surrounds bundles of axons (fascicles), and epineurium surrounds an entire nerve.**

## 14.2 Twelve pairs of cranial nerves distribute primarily to regions of the head and neck.

The 12 pairs of cranial nerves are so named because they arise from the brain inside the cranial cavity and pass through various foramina in the cranial bones. The cranial nerves and their branches are part of the peripheral nervous system. Each cranial nerve has both a number, designated by a Roman numeral, and a name (Figure 14.2). The numbers indicate the order, from anterior to posterior, in which the nerves arise from the brain. The names designate a nerve's distribution or function.

Three cranial nerves (I, II, and VIII) carry axons of sensory neurons and thus are called **sensory nerves**. Five cranial nerves (III, IV, VI, XI, and XII) are classified as **motor nerves** because they contain only axons of motor neurons. *Somatic motor*

FIGURE 14.2 Origins of cranial nerves.

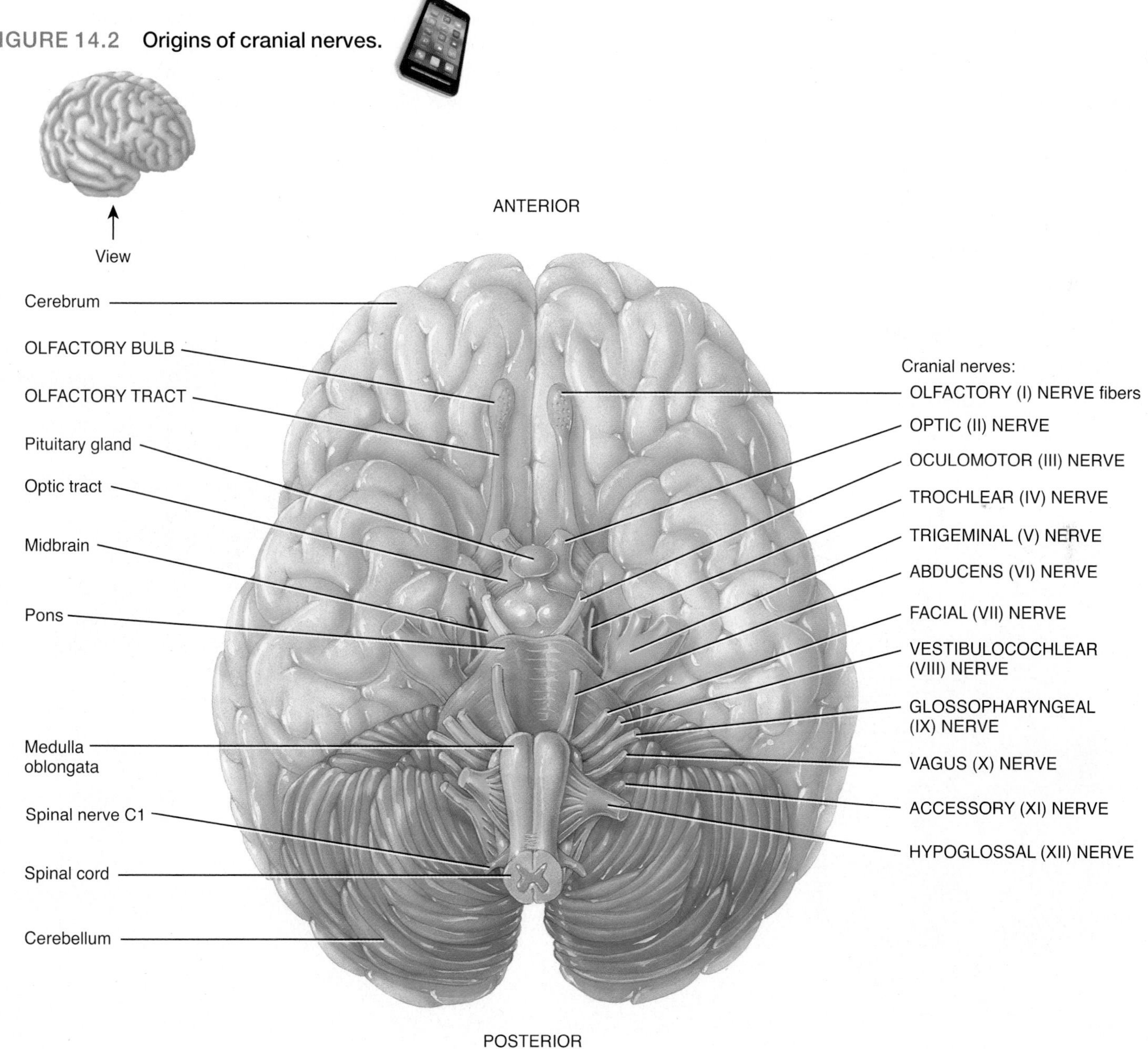


Inferior aspect of brain

There are 12 cranial nerves located on the inferior surface of the brain.

*axons* innervate skeletal muscles, and *autonomic motor axons*, part of the parasympathetic division of the autonomic nervous system, innervate glands, smooth muscle, and cardiac muscle. The cell bodies of sensory neurons are located in ganglia outside the brain; the cell bodies of motor neurons lie in nuclei within the brain. The remaining four cranial nerves (V, VII, IX, and X) are **mixed nerves**—they contain axons of both sensory neurons entering the brain and motor neurons leaving the brain.

Although the cranial nerves are mentioned singly in the following descriptions of their type, location, and function, remember that they are paired structures.

## Olfactory (I) Nerve

The **olfactory nerve** (ol-FAK-tō-rē; *olfact-* = to smell) is entirely sensory; it contains axons that conduct impulses for the sense of smell, or olfaction (Figure 14.3). The olfactory epithelium occupies the superior part of the nasal cavity, covering the inferior surface of the cribriform plate and extending down along the superior nasal concha. The olfactory receptors are located within the olfactory epithelium. Each has a single odor-sensitive knob-shaped dendrite projecting from one side of the cell body and an axon extending from the other side. Bundles of axons of olfactory receptors extend superiorly through the olfactory foramina in the cribriform plate of the ethmoid bone. On each side of the nose, these bundles of axons collectively form the right and left olfactory nerves.

Olfactory nerves end in the brain in paired masses of gray matter called the **olfactory bulbs**, two extensions of the brain that rest on the cribriform plate. Within the olfactory bulbs, the axon terminals of olfactory receptors form synapses with neurons in the olfactory pathway. The axons of these neurons make up the **olfactory tracts**, which extend posteriorly from the olfactory bulbs (see Figure 14.2). Axons in the olfactory tracts end in the primary olfactory area in the temporal lobe of the cerebral cortex.

**FIGURE 14.3 Olfactory (I) nerve.**

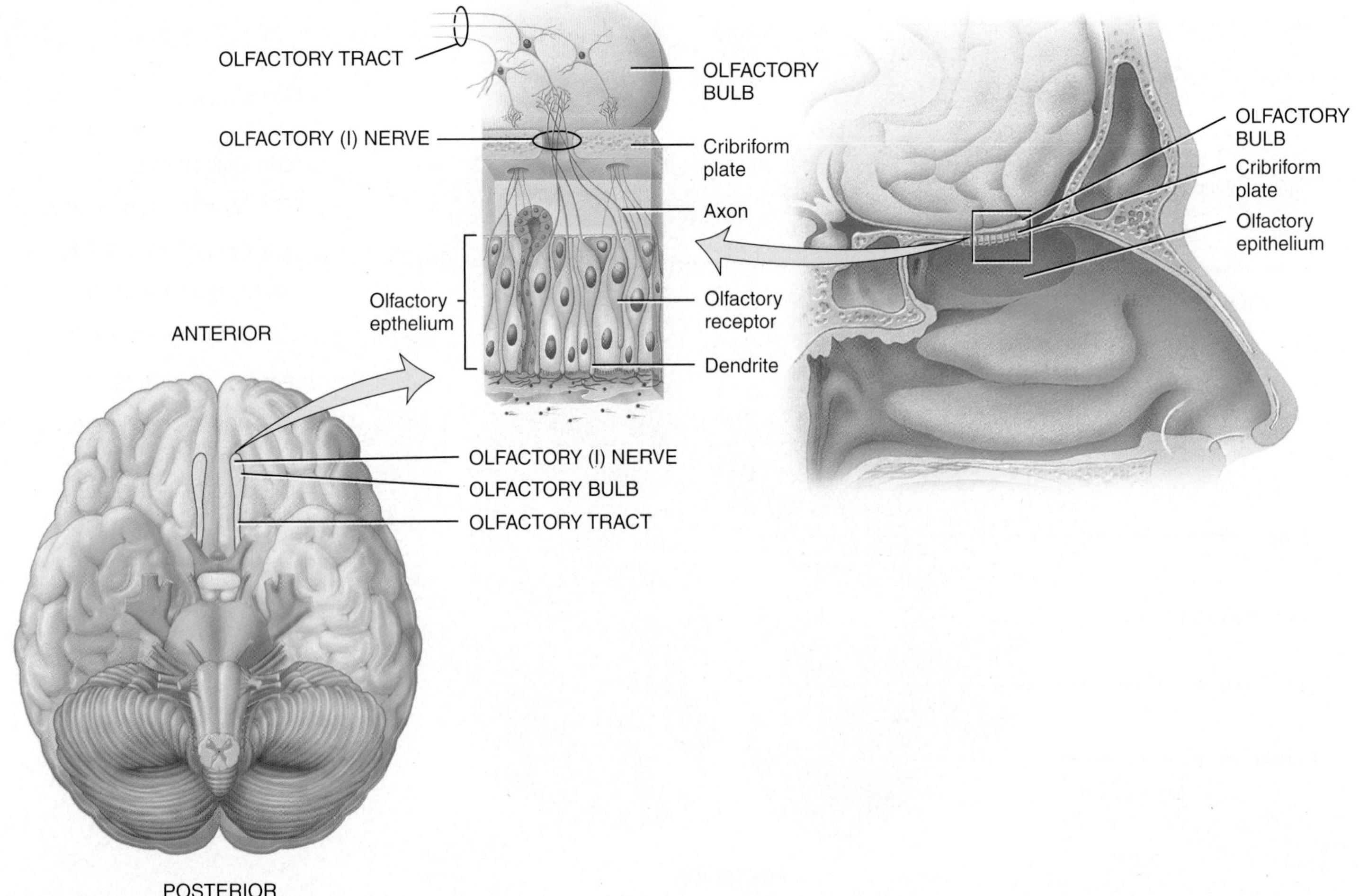


**The olfactory epithelium is located on the inferior surface of the cribriform plate and superior nasal conchae.**

## Optic (II) Nerve

The **optic nerve** (OP-tik; = eye, vision) is entirely sensory; it contains axons that conduct impulses for vision (Figure 14.4). In the retina of each eye, rods and cones initiate visual signals that are relayed via the optic nerves to the visual processing centers of the brain. The two optic nerves pass through the optic foramen of the orbit (see Figure 7.11), then merge to form the **optic chiasm** (KĪ-azm = cross, as in the letter X). Within the chiasm, axons from each eye may either continue on to the same side of the brain or cross to the opposite side for visual–spatial information processing. Posterior to the chiasm, the regrouped axons, some from each eye, form the **optic tracts**. Axons from the optic tracts project to the primary visual area of the occipital lobe in the cerebral cortex (see Figure 13.10).

**FIGURE 14.4** Optic (II) nerve.

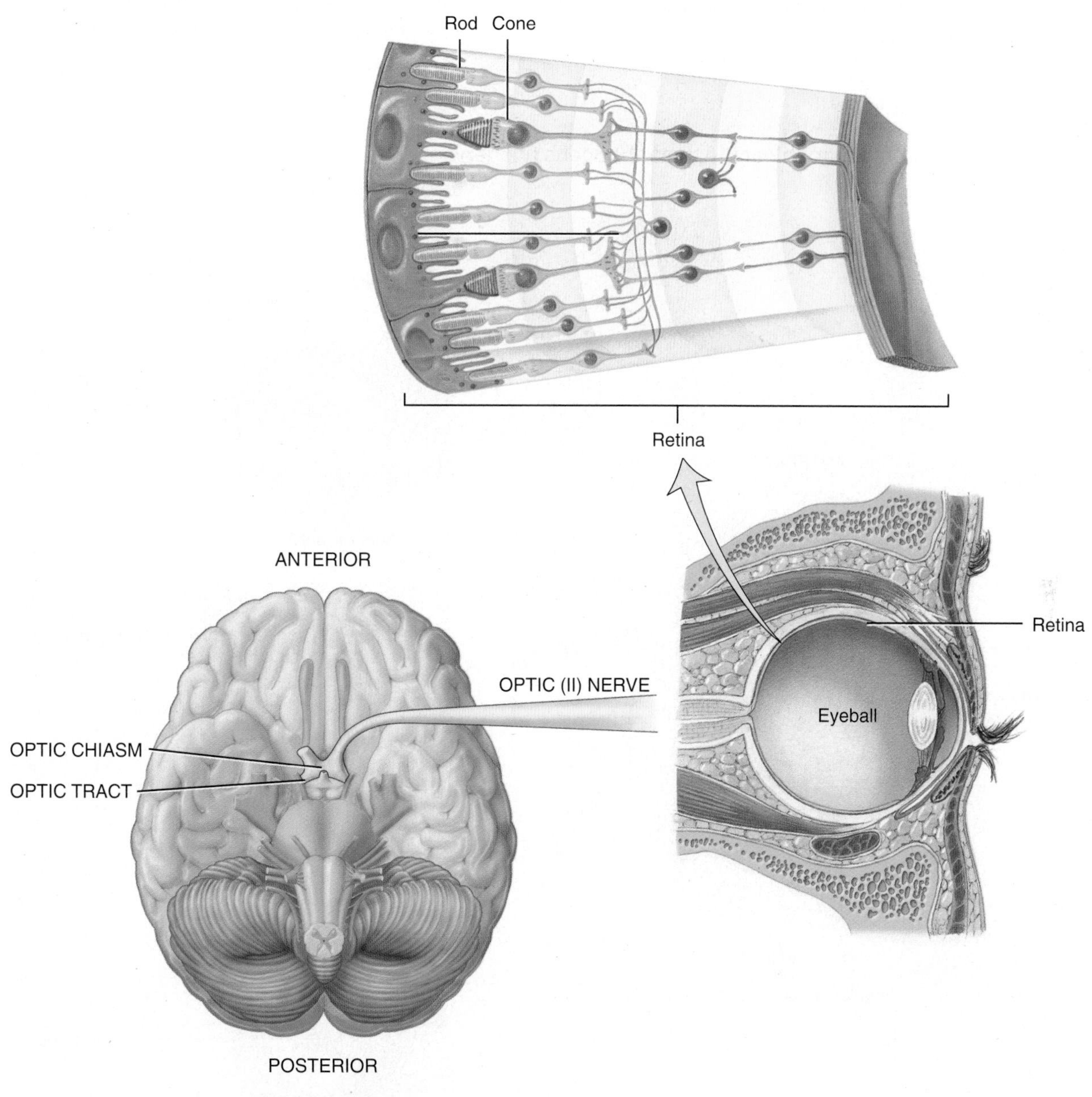


Visual signals are relayed from rods and cones in the retina to the primary visual area of the cerebral cortex.

## Oculomotor (III), Trochlear (IV), and Abducens (VI) Nerves

The oculomotor, trochlear, and abducens nerves are motor cranial nerves that pass through the superior orbital fissure into the orbit (see Figure 7.11) to control the muscles that move the eyeballs (Figure 14.5). They are all motor nerves that contain only motor axons as they exit the brain stem. Sensory axons from the extrinsic eyeball muscles initially travel through each of these nerves, but eventually enter the midbrain via the trigeminal nerve. These sensory axons convey impulses from the extrinsic eyeball muscles for **proprioception**, the nonvisual perception of the movements and position of the body.

The **oculomotor nerve** (ok′-ū-lō-MŌ-tor; *oculo-* = eye; *-motor* = a mover) extends from the midbrain near its junction with the pons (Figure 14.5a). The oculomotor nerve contains somatic motor axons that innervate extrinsic eyeball muscles (superior rectus, medial rectus, inferior rectus, and inferior oblique muscles) for movement of the eyeball, and the levator palpebrae superioris for raising the upper eyelid. Autonomic motor axons propagate to intrinsic eyeball muscles including the ciliary muscle to adjust the shape of the lens (see Figure 16.11), and the iris to adjust the size of the pupil (see Figure 16.9).

The **trochlear nerve** (TRŌK-lē-ar = a pulley) extends from the posterior side of the midbrain then wraps around the pons. These somatic motor axons innervate the superior oblique muscle,

FIGURE 14.5 **Oculomotor (III), trochlear (IV), and abducens (VI) nerves.**

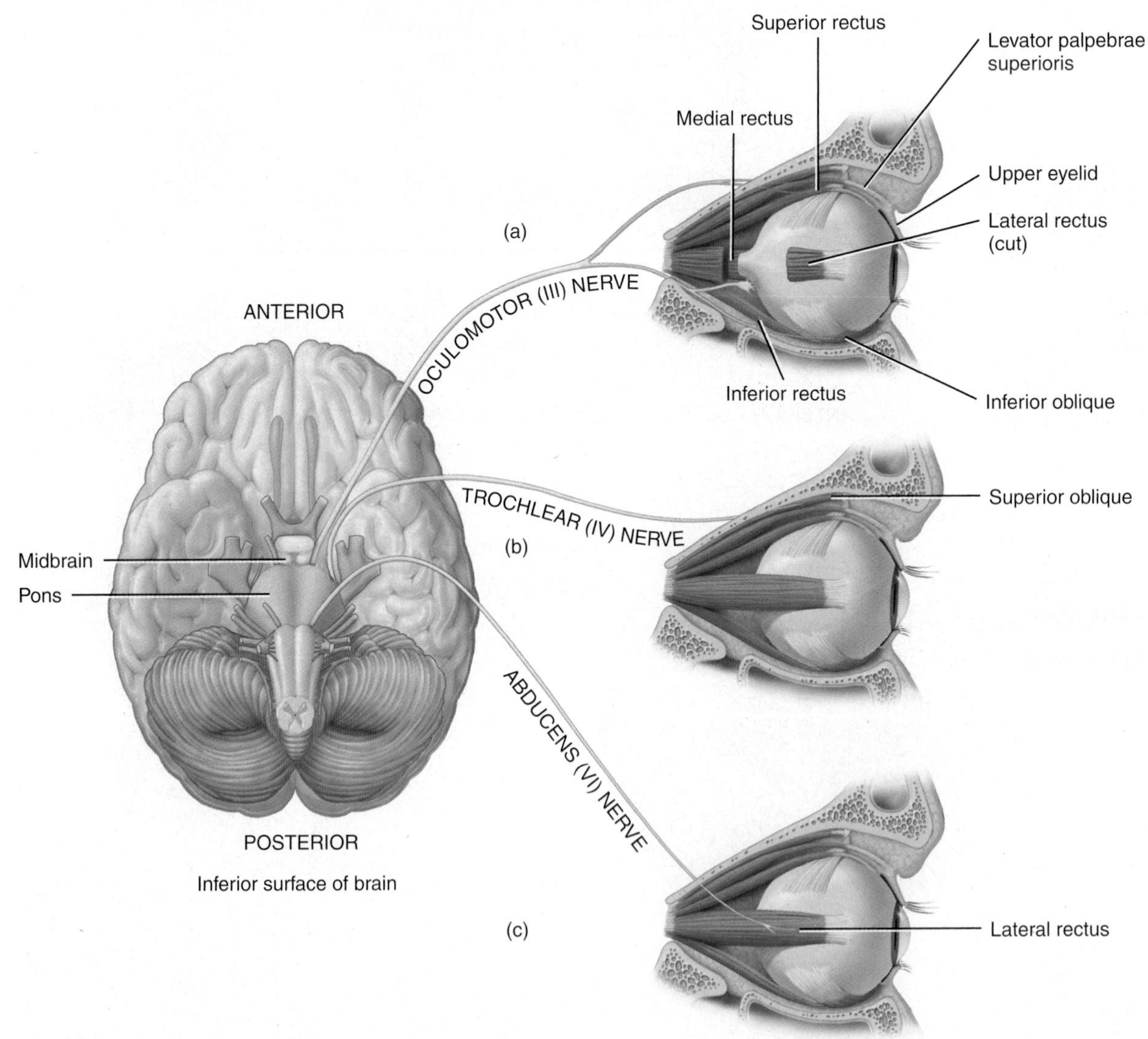


**The oculomotor nerve has the widest distribution to extrinsic eye muscles.**

another extrinsic eyeball muscle that controls movement of the eyeball (Figure 14.5b).

The **abducens nerve** (ab-DOO-senz; *ab-* = away; *-ducens* = to lead) originates in the pons (Figure 14.5c). Somatic motor axons extend from the pons to innervate the lateral rectus muscle of the eyeball, an extrinsic eyeball muscle controlling movement of the eyeball. The abducens nerve is so named because impulses cause abduction (lateral rotation) of the eyeball.

## Trigeminal (V) Nerve

The **trigeminal nerve** (trī-JEM-i-nal = triple, for its three branches), the largest of the cranial nerves, is a mixed cranial nerve that emerges on the lateral surface of the pons (Figure 14.6). As indicated by its name, the trigeminal nerve has three branches: ophthalmic, maxillary, and mandibular. The **ophthalmic nerve** (of-THAL- mik; *ophthalm-* = the eye) enters the orbit through the superior orbital fissure; the **maxillary nerve** (*maxilla* = upper jawbone) passes through the foramen rotundum; and the **mandibular nerve** (*mandibula* = lower jawbone) exits through the foramen ovale (see Figure 7.7a). Sensory axons in the trigeminal nerve carry impulses for touch, pain, and temperature sensations to the pons. The ophthalmic nerve contains sensory axons from the upper eyelid, eyeball, lacrimal glands (which secrete tears), nasal cavity, nose, forehead, and anterior scalp. The maxillary nerve includes sensory axons from the nose, palate, upper mouth, and lower eyelid. The mandibular nerve contains sensory axons from the anterior tongue (but not those for taste), cheek, skin over the jaw

FIGURE 14.6 Trigeminal (V) nerve.

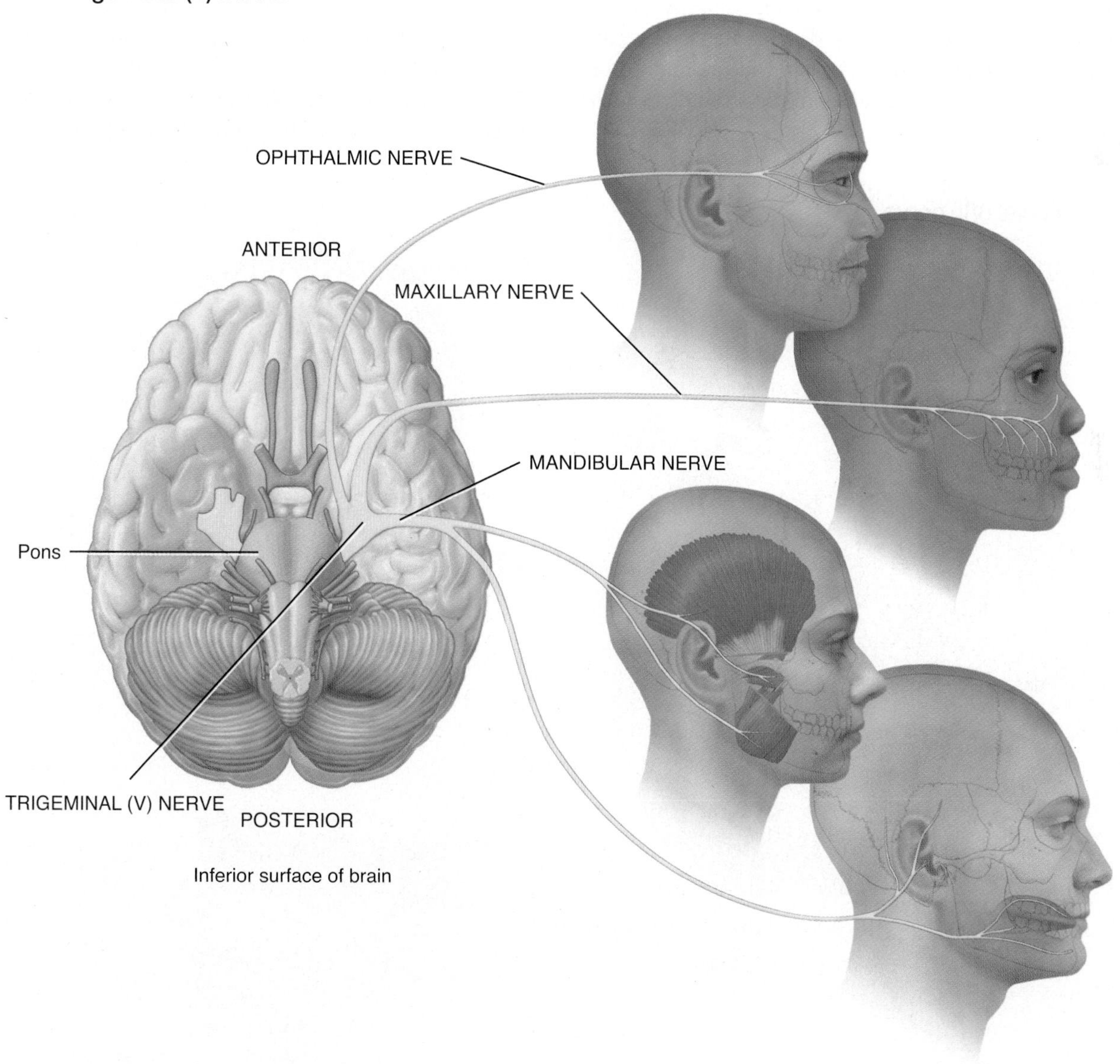


The three divisions of the trigeminal nerve leave the skull through the superior orbital fissure, foramen rotundum, and foramen ovale.

and side of the head, and lower mouth. The trigeminal nerve also contains sensory axons from proprioceptors (receptors that provide information regarding body position and movements) located in the muscles of mastication. Somatic motor axons of the trigeminal nerve supply muscles of mastication that control chewing movements.

## Facial (VII) Nerve

The **facial nerve** (FĀ-shal = face) is a mixed cranial nerve (Figure 14.7). Its sensory axons extend from the taste buds of the anterior tongue, pass through the stylomastoid foramen (see Figure 7.4) to the pons, and then continue on to the primary gustatory area in the parietal lobe of the cerebral cortex (see Figure 13.10). The sensory portion of the facial nerve also contains axons that relay touch, pain, and temperature sensations from the ear canal and proprioception from muscles of the face and scalp. Axons of somatic motor neurons arise in the pons and innervate facial, scalp, and neck muscles for facial expression. Axons of autonomic motor neurons innervate lacrimal glands and salivary glands (which secrete saliva).

## Vestibulocochlear (VIII) Nerve

The **vestibulocochlear nerve** (vest-tib-ū-lō-KOK-lē-ar; *vestibulo-* = small cavity; *-cochlear* = spiral, snail-like) is a sensory nerve and has two branches, the vestibular branch and the cochlear branch (Figure 14.8). The **vestibular branch** carries impulses for equilibrium from the semicircular canals, saccule, and utricle of the inner ear (see Figure 16.20) to the pons and cerebellum. The **cochlear branch** carries impulses for hearing from the spiral organ of the inner ear (see Figure 16.21b) to the primary auditory area in the temporal lobe of the cerebral cortex (see Figure 13.10).

## Glossopharyngeal (IX) Nerve

The **glossopharyngeal nerve** (glos′-Ō-fa-RIN-jē-al; *glosso-* = tongue; *-pharyngeal* = throat) is a mixed cranial nerve (Figure 14.9). Sensory axons arise from taste buds on the posterior tongue, from the external ear to convey touch, pain, and temperature sensations, and from proprioceptors in swallowing muscles. In the neck region, sensory axons carry impulses from baroreceptors in the carotid sinus that monitor blood pressure and chemoreceptors in the carotid bodies that monitor blood gas levels

**FIGURE 14.7** **Facial (VII) nerve.**

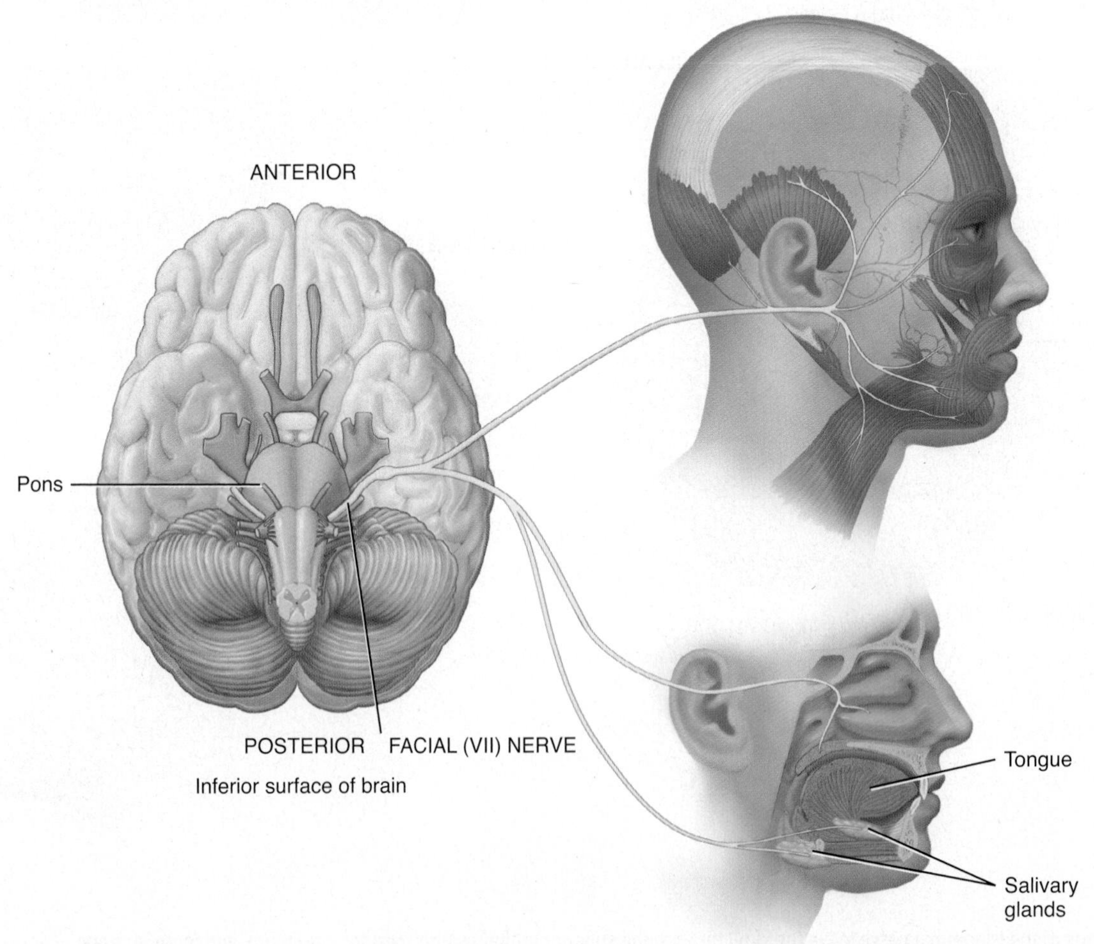


**The facial nerve causes contraction of the muscles of facial expression.**

FIGURE 14.8 Vestibulocochlear (VIII) nerve.

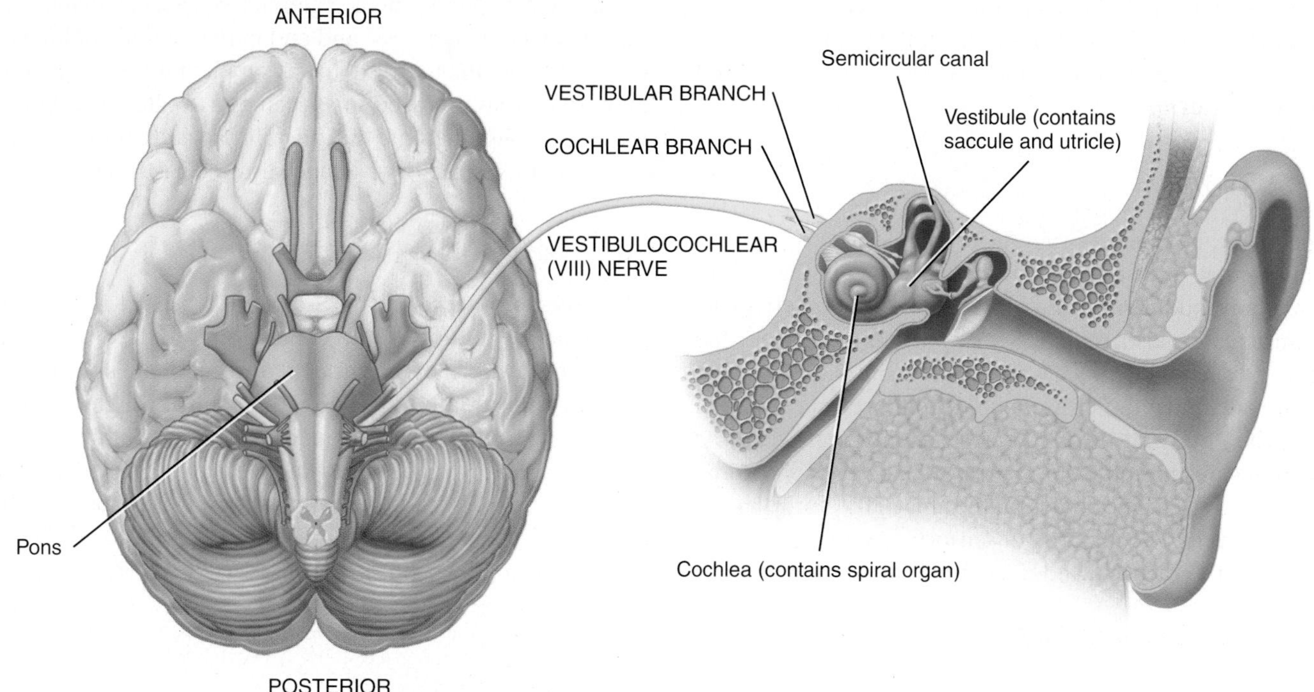


**The vestibular branch carries impulses for equilibrium, and the cochlear branch carries impulses for hearing.**

FIGURE 14.9 Glossopharyngeal (IX) nerve.

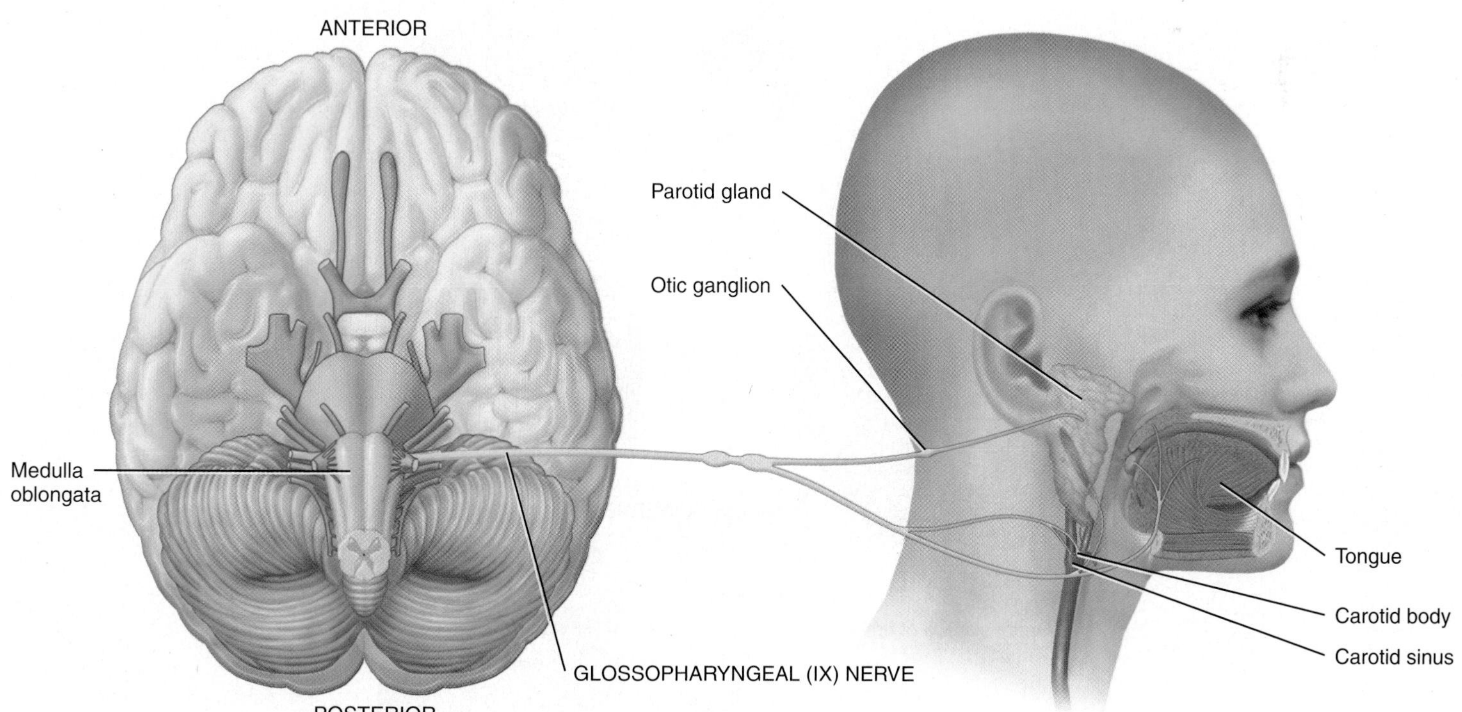


**Sensory axons in the glossopharyngeal nerve carry signals from the taste buds.**

(see Figure 14.10). Somatic motor axons arise in the medulla oblongata and pass through the jugular foramen (see Figure 7.7a). Somatic motor axons innervate a muscle of the pharynx used in swallowing. Autonomic motor neurons stimulate the parotid gland to secrete saliva.

## Vagus (X) Nerve

The **vagus nerve** (VĀ-gus = vagrant or wandering) is a mixed cranial nerve that is distributed from the head and neck into the thorax and abdomen (Figure 14.10). The nerve derives its name from its wide distribution. Sensory axons arise from the external ear to convey touch, pain, and temperature sensations, from taste buds in the throat, and from proprioceptors in muscles of the neck and throat. Sensory axons from receptors in the carotid sinus monitor blood pressure, and those from the carotid body and the aortic bodies (at the arch of the aorta) monitor blood gas levels. The majority of sensory neurons come from most organs of the thoracic and abdominal cavities that convey sensations (such as hunger, fullness, and discomfort) from these organs. Sensory axons in the vagus nerve pass through the jugular foramen (see Figure 7.7a) and end in the medulla oblongata and pons. The somatic motor neurons supply muscles of the pharynx and larynx that are used in speech and swallowing. Autonomic motor neurons innervate the lungs, heart, and smooth muscle and glands of the respiratory passageways and gastrointestinal tract.

## Accessory (XI) Nerve

The **accessory nerve** (ak-SES-ō-rē = assisting) is a motor cranial nerve (Figure 14.11). Its somatic motor axons arise in the anterior gray horn of the cervical portion of the spinal cord, exit the spinal cord laterally, enter the skull through the foramen magnum, and then exit through the jugular foramen (see Figure 7.7a) along with the vagus and glossopharyngeal nerves. The accessory nerve is traditionally considered a cranial nerve rather than spinal nerve because it passes through the skull. The accessory nerve conveys motor impulses to the sternocleidomastoid

**FIGURE 14.10 Vagus (X) nerve.**

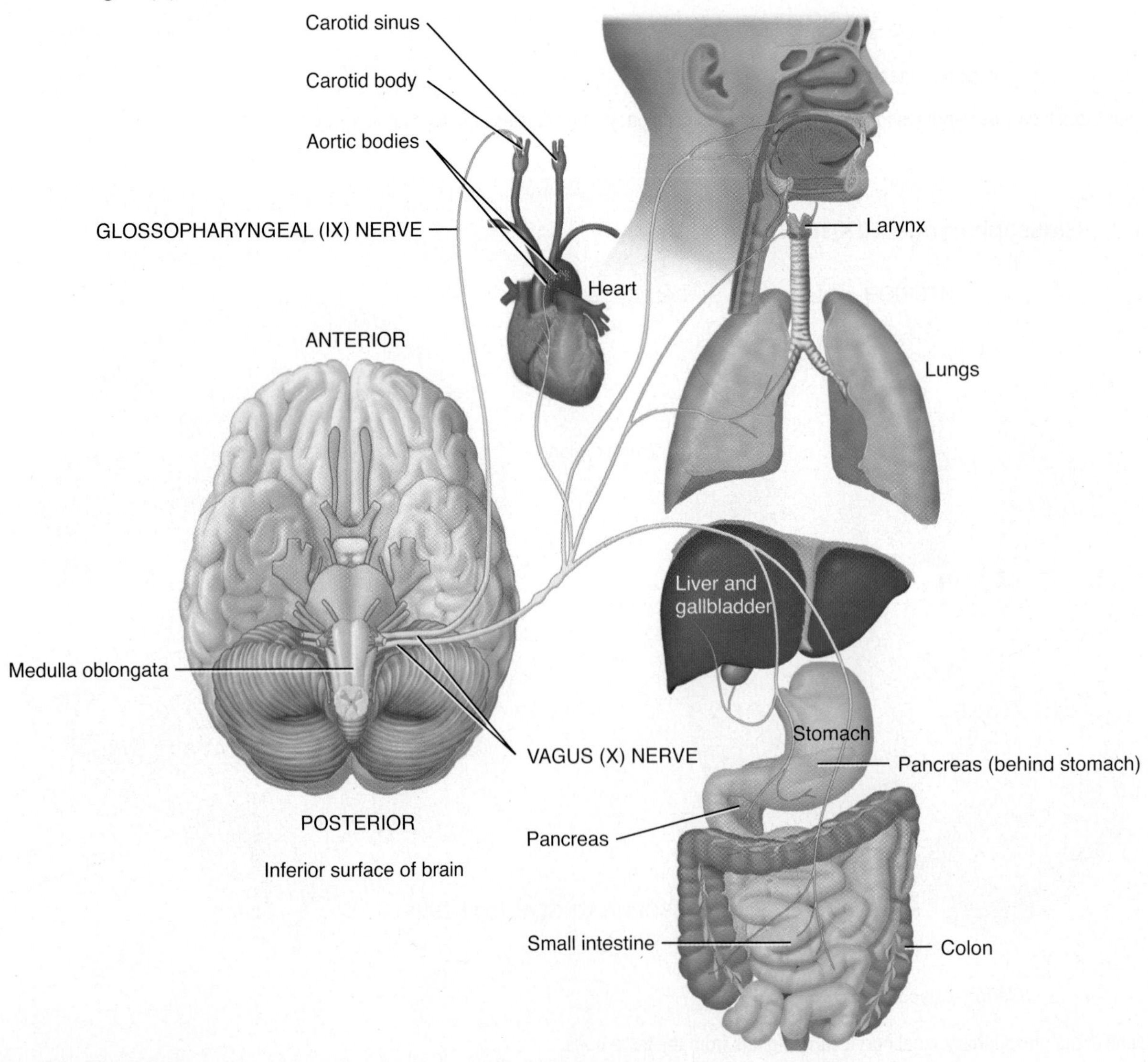


**The vagus nerve is widely distributed in the head, neck, thorax, and abdomen.**

FIGURE 14.11 Accessory (XI) nerve.

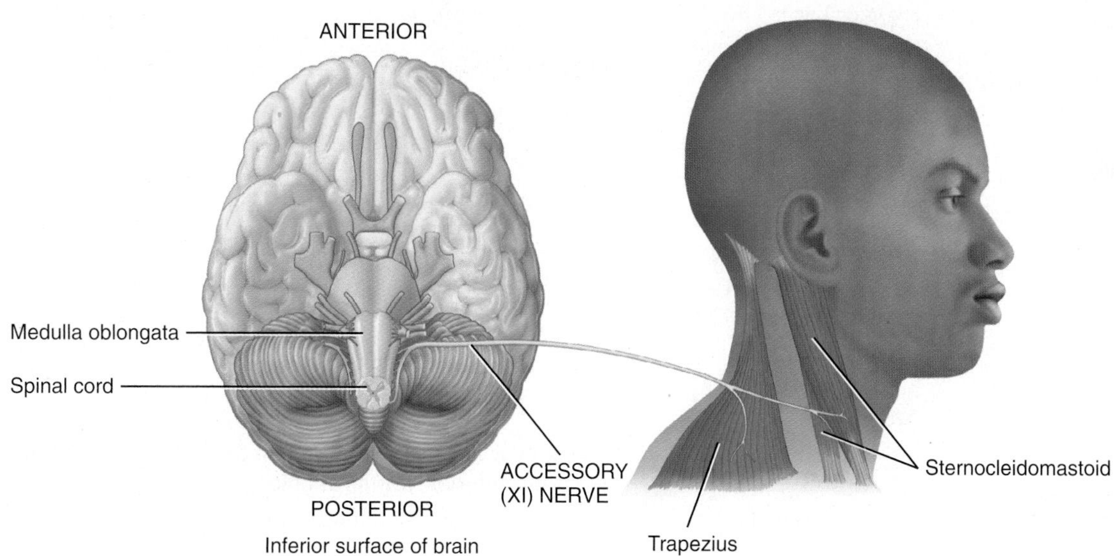


The accessory nerve exits the skull through the jugular foramen.

and trapezius muscles to coordinate head movements. Sensory axons from proprioceptors in the sternocleidomastoid and trapezius muscles begin their course toward the brain in the accessory nerve, but eventually leave the nerve to join nerves of the cervical plexus. From the cervical plexus the sensory axons enter the spinal cord, where they ascend to the medulla oblongata.

## Hypoglossal (XII) Nerve

The **hypoglossal nerve** (hī-pō-GLOS-al; *hypo-* = below; *-glossal* = tongue) is a motor cranial nerve (Figure 14.12). Somatic motor axons originate in the medulla oblongata and pass through the hypoglossal canal (see Figure 7.5) to supply tongue muscles with impulses for speech and swallowing. Sensory axons that originate from proprioceptors in the tongue muscles begin their course toward the brain in the hypoglossal nerve but leave the nerve to join cervical spinal nerves and end in the medulla oblongata.

Table 14.1 presents a summary of the components and principal functions of the cranial nerves, including a mnemonic to help you remember their names.

FIGURE 14.12 Hypoglossal (XII) nerve.

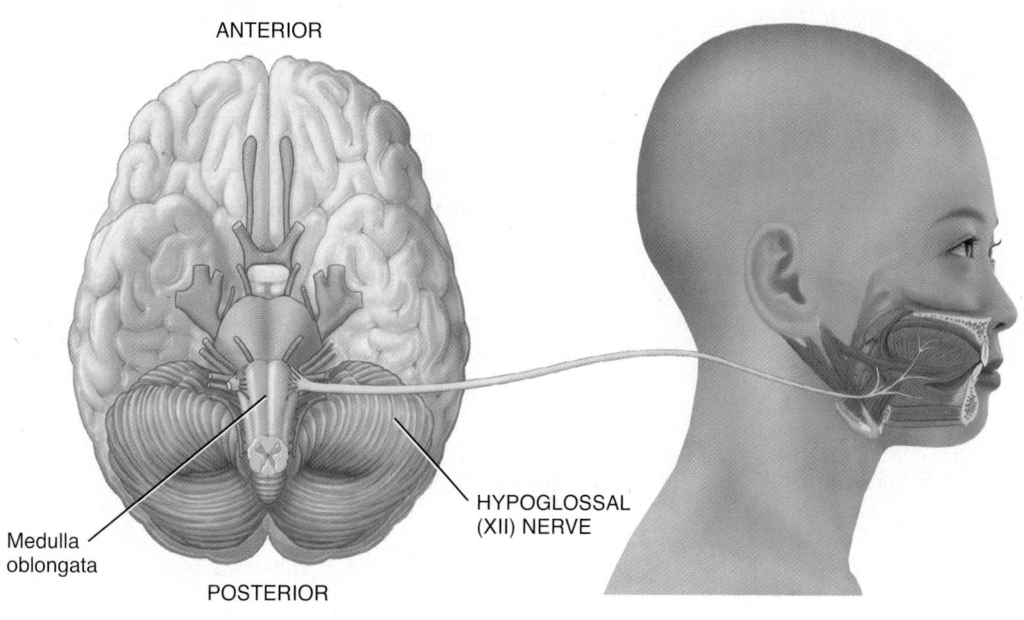


The hypoglossal nerve exits the skull through the hypoglossal canal.

**TABLE 14.1**

Summary of Cranial Nerves*

| CRANIAL NERVE | COMPONENTS | PRINCIPAL FUNCTIONS |
|---|---|---|
| **Olfactory (I)** | ***Special sensory*** | Olfaction (smell) |
| **Optic (II)** | ***Special sensory*** | Vision (sight) |
| **Oculomotor (III)** | ***Motor*** | |
| | Somatic | Movement of eyeballs and upper eyelid |
| | Autonomic | Adjusts lens for near vision (accommodation) |
| | | Constriction of pupil |
| **Trochlear (IV)** | ***Motor*** | |
| | Somatic | Movement of eyeballs |
| **Trigeminal (V)** | ***Mixed*** | |
| | Sensory | Touch, pain, and thermal sensations from scalp, face, and oral cavity (including teeth and anterior two-thirds of tongue) |
| | Motor (somatic) | Chewing and controls middle ear muscle |
| **Abducens (VI)** | ***Motor*** | |
| | Somatic | Movement of eyeballs |
| **Facial (VII)** | ***Mixed*** | |
| | Sensory | Taste from anterior two-thirds of tongue |
| | | Touch, pain, and thermal sensations from skin in external ear canal |
| | Motor (somatic) | Control of muscles of facial expression and middle ear muscle |
| | Motor (autonomic) | Secretion of tears and saliva |
| **Vestibulocochlear (VIII)** | ***Special sensory*** | Hearing and equilibrium |
| **Glossopharyngeal (IX)** | ***Mixed*** | |
| | Sensory | Taste from posterior one-third of tongue |
| | | Proprioception in some swallowing muscles |
| | | Monitors blood pressure and oxygen and carbon dioxide levels in blood |
| | | Touch, pain, and thermal sensations from skin of external ear and upper pharynx |
| | Motor (somatic) | Assists in swallowing |
| | Motor (autonomic) | Secretion of saliva |
| **Vagus (X)** | ***Mixed*** | |
| | Sensory | Taste from epiglottis |
| | | Proprioception from throat and voice box muscles |
| | | Monitors blood pressure and oxygen and carbon dioxide levels in blood |
| | | Touch, pain, and thermal sensations from skin of external ear |
| | | Sensations from thoracic and abdominal organs |
| | Motor (somatic) | Swallowing, vocalization, and coughing |
| | Motor (autonomic) | Motility and secretion of gastrointestinal organs |
| | | Constriction of respiratory passageways |
| | | Decreases heart rate |
| **Accessory (XI)** | ***Motor*** | |
| | Somatic | Movement of head and pectoral girdle |
| **Hypoglossal (XII)** | ***Motor*** | |
| | Somatic | Speech, manipulation of food, and swallowing |

*MNEMONIC FOR CRANIAL NERVES:

| **O**h | **O**h | **O**h | **T**o | **T**ouch | **A**nd | **F**eel | **V**ery | **G**reen | **V**egetables | **AH**! | |
|---|---|---|---|---|---|---|---|---|---|---|---|
| **O**lfactory | **O**ptic | **O**culomotor | **T**rochlear | **T**rigeminal | **A**bducens | **F**acial | **V**estibulocochlear | **G**lossopharyngeal | **V**agus | **A**ccessory | **H**ypoglossal |

## CLINICAL CONNECTION | *Dental Anesthesia*

The inferior alveolar nerve, a branch of the mandibular nerve, supplies all of the teeth in one-half of the mandible; it is often anesthetized in dental procedures. The same procedure will anesthetize the lower lip because the mental nerve is a branch of the inferior alveolar nerve. Because the lingual nerve runs very close to the inferior alveolar nerve near the mental foramen, it too is often anesthetized at the same time. For anesthesia to the upper teeth, the superior alveolar nerve endings, which are branches of the maxillary nerve, are blocked by inserting the needle beneath the mucous membrane. The anesthetic solution is then infiltrated slowly throughout the area of the roots of the teeth to be treated.

## ✓ CHECKPOINT

3. How are cranial nerves named and numbered? How many are there?
4. What is the difference between a mixed cranial nerve and a sensory cranial nerve?
5. How are somatic motor axons different from autonomic motor axons?
6. Where do the motor axons of the facial nerve originate?
7. Through which foramen does the glossopharyngeal nerve exit the skull?

## 14.3 Each spinal nerve branches into a posterior ramus, an anterior ramus, a meningeal branch, and rami communicantes.

Like the 12 pairs of cranial nerves, spinal nerves and the nerves that branch from them (Figure 14.13) are part of the peripheral nervous system. Spinal nerves connect the CNS to sensory receptors, muscles, and glands in all parts of the body. The 31 pairs of spinal nerves are named and numbered according to the region and level of the vertebral column from which they emerge (Figure 14.13). There are 8 pairs of *cervical nerves* (represented as C1–C8), 12 pairs of *thoracic nerves* (T1–T12), 5 pairs of *lumbar nerves* (L1–L5), 5 pairs of *sacral nerves* (S1–S5), and 1 pair of *coccygeal nerves* (Co1). The first cervical pair (C1) emerges between the occipital bone and the atlas (first cervical vertebra). All other spinal nerves emerge from the vertebral column through the intervertebral foramina between adjoining vertebrae.

Not all spinal cord segments are aligned with their corresponding vertebrae. Recall that the spinal cord ends near the level of the second lumbar vertebra, and that the roots of the lumbar, sacral, and coccygeal nerves descend at an angle to reach their respective foramina before emerging from the vertebral column. This arrangement constitutes the cauda equina.

As noted in Chapter 13, a typical spinal nerve has two connections to the cord: a **posterior root** and an **anterior root** (see Figure 13.18). The posterior and anterior roots unite to form a spinal nerve at the intervertebral foramen. Because the posterior root contains sensory axons and the anterior root contains motor axons, a spinal nerve is classified as a mixed nerve. The posterior root contains a **posterior root ganglion** in which cell bodies of sensory neurons are located.

**FIGURE 14.13** **External anatomy of the spinal cord and spinal nerves.**

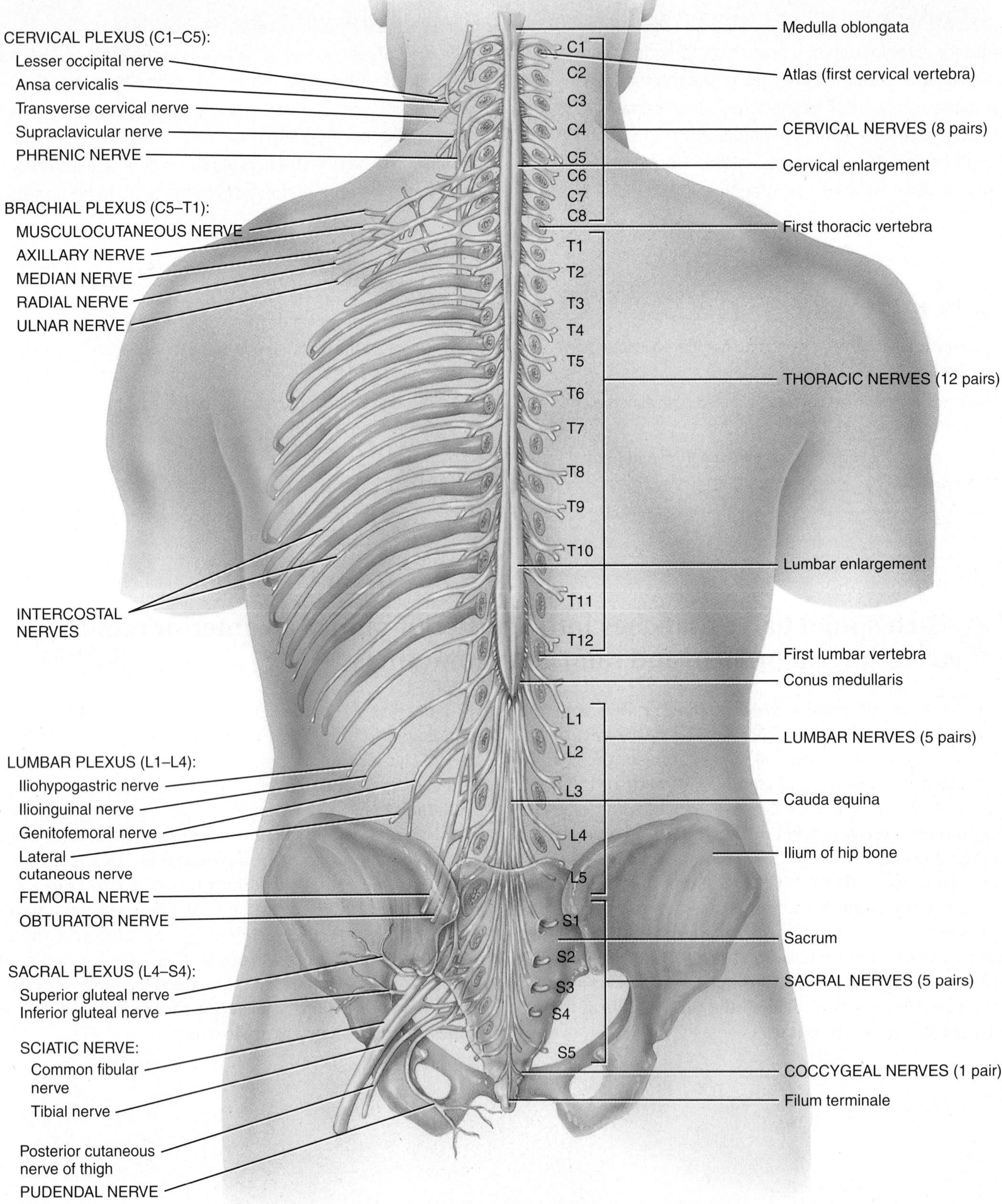


Posterior view of entire spinal cord and portions of spinal nerves

**The spinal nerves are named and numbered according to the region of the vertebral column from which they emerge.**

## Branches

A short distance after passing through its intervertebral foramen, a spinal nerve divides into several branches (Figure 14.14). These branches are known as **rami** (RĀ-mī = branches). The **posterior ramus** (RĀ-mus; singular form of *rami*) serves the muscles and skin of the posterior trunk of the body. The **anterior ramus** serves the muscles and structures of the upper and lower limbs and the skin of the lateral and anterior trunk. In addition to posterior and anterior rami, spinal nerves also give off a **meningeal branch** (me-NIN-jē′-al) that reenters the vertebral canal through the intervertebral foramen and supplies the vertebrae, spinal cord, and meninges. Other branches of a spinal nerve are the *rami communicantes* (kō-mū-ni-KAN-tēz), components of the autonomic nervous system, which will be discussed shortly.

## Plexuses

Axons from the anterior rami of spinal nerves, except for thoracic nerves T2–T12, do not go directly to the body structures they supply. Instead, they form networks on both the left and right sides of the body by joining with various numbers of axons from anterior rami of adjacent nerves. Such a network of axons is called a **plexus** (PLEK-sus = braid or network). Because axons of anterior rami recombine within each plexus, nerves emerging from a plexus contain axons of several spinal nerves. Since nerves to peripheral structures are supplied by several spinal nerves, total paralysis of a muscle is less likely when a single spinal cord segment is damaged. The principal plexuses are the cervical plexus, brachial plexus, lumbar plexus, and sacral plexus (see Figure 14.13). The nerves branching from the plexuses bear names that are often descriptive of the general regions they serve or the course they take. Each of the nerves, in turn, may have several branches named for the specific structures they innervate.

FIGURE 14.14 **Branches of a typical spinal nerve, shown in transverse section through the thoracic portion of the spinal cord.** See also Figure 13.4c.

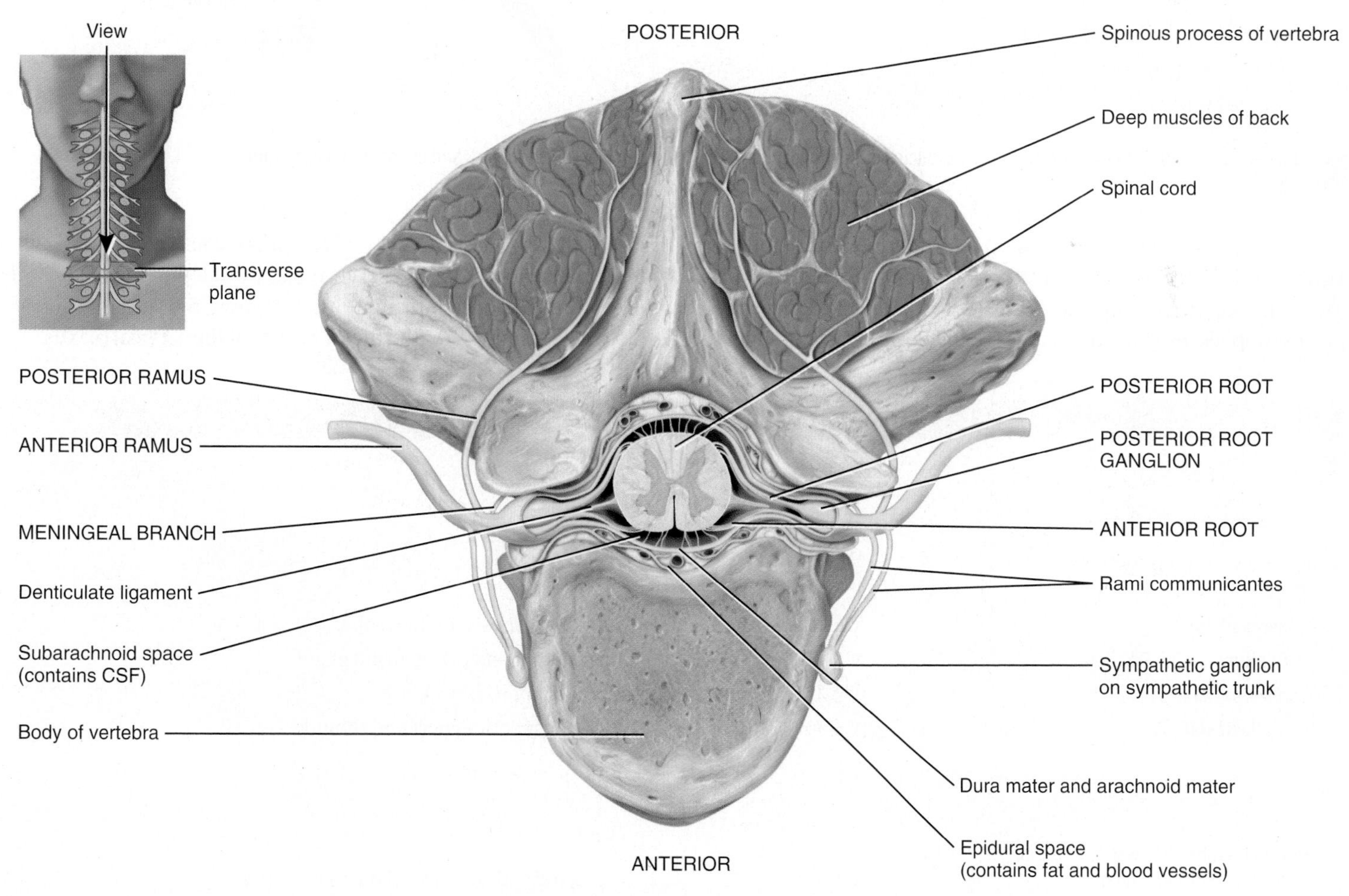


Transverse section of thoracic spinal cord

**The branches of a spinal nerve are the posterior ramus, anterior ramus, meningeal branch, and rami communicantes.**

FIGURE 14.15 Cervical plexus in anterior view.

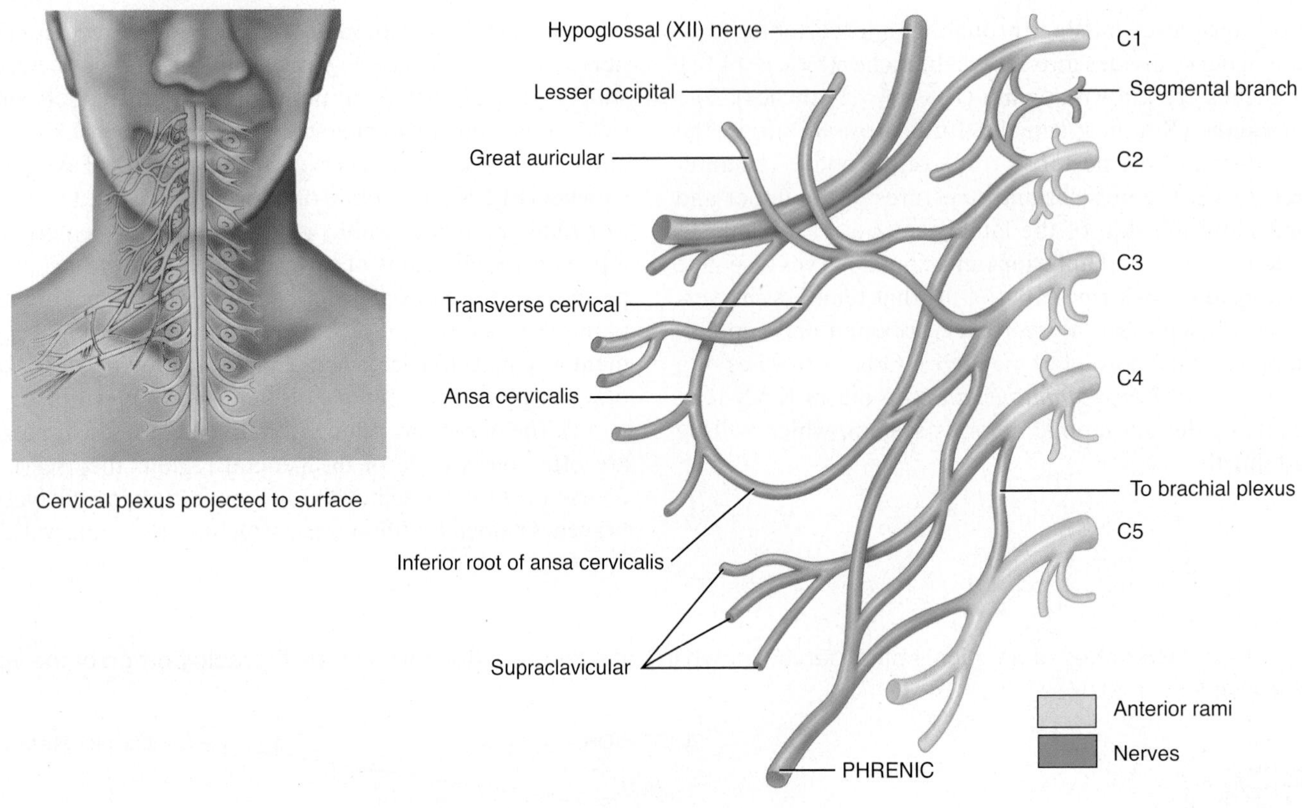


Cervical plexus projected to surface

Origin of cervical plexus

**The cervical plexus innervates the skin and muscles of the head, neck, superior portion of the shoulders and chest, and diaphragm.**

The **cervical plexus** (SER-vi-kul) is formed by the first four cervical nerves (C1–C4), with contributions from C5 (Figure 14.15). There is a cervical plexus on each side of the neck. The cervical plexus supplies the skin and muscles of the head, neck, and superior part of the shoulders and chest. The **phrenic nerve** is an important nerve that arises from the cervical plexus. The phrenic nerve stimulates the diaphragm, a major respiratory muscle, to contract. Table 14.2 summarizes the major nerves of the cervical plexus.

**TABLE 14.2**

Cervical Plexus

| NERVE | ORIGIN | DISTRIBUTION |
|---|---|---|
| **Superficial (Sensory) Branches** | | |
| **Lesser occipital** | C2 | Skin of scalp posterior and superior to ear |
| **Great auricular** (aw-RIK-ū-lar) | C2–C3 | Skin over ear and over parotid glands |
| **Transverse cervical** | C2–C3 | Skin over anterior neck |
| **Supraclavicular** | C3–C4 | Skin over superior chest and shoulder |
| **Deep (Largely Motor) Branches** | | |
| **Ansa cervicalis** (AN-sa ser-vi-KAL-is) (superior and inferior roots) | C1–C3 | Infrahyoid and geniohyoid muscles of neck |
| **Phrenic** (FREN-ik) | C3–C5 | Diaphragm |
| **Segmental branches** | C1–C5 | Deep muscles of neck, levator scapulae, and middle scalene muscles |

Spinal nerves C5–C8 and T1 form the **brachial plexus** (BRĀ-kē-al; Figure 14.16a). It passes above the first rib posterior to the clavicle and then enters the axilla (Figure 14.16b). The brachial plexus provides almost the entire nerve supply of the shoulders and upper limbs. Since the brachial plexus is so complex, an explanation of its various parts is helpful. The anterior rami of several spinal nerves in the plexus unite to form *trunks*; the trunks divide into *divisions*, which unite to form *cords*. Five important nerves branch from the cords of the brachial plexus:

1. The **axillary nerve** innervates the shoulder.
2. The **musculocutaneous nerve** innervates the anterior arm.
3. The **radial nerve** innervates the posterior arm and forearm.
4. The **median nerve** supplies the anterior forearm and some of the hand.
5. The **ulnar nerve** innervates the medial forearm and most of the hand. Table 14.3 summarizes the major nerves of the brachial plexus.

FIGURE 14.16 **Brachial plexus in anterior view.**

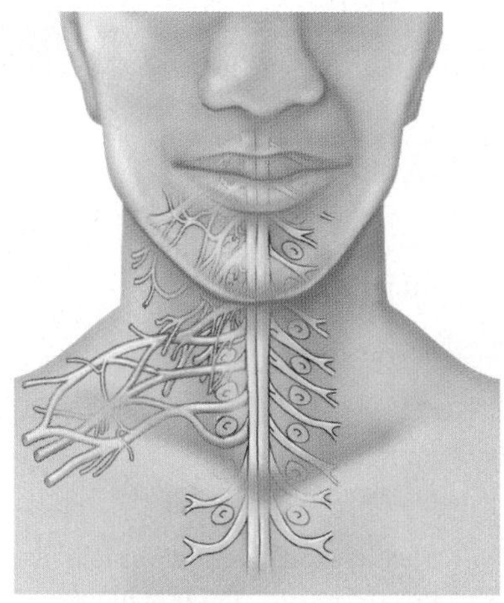

Brachial plexus projected to surface

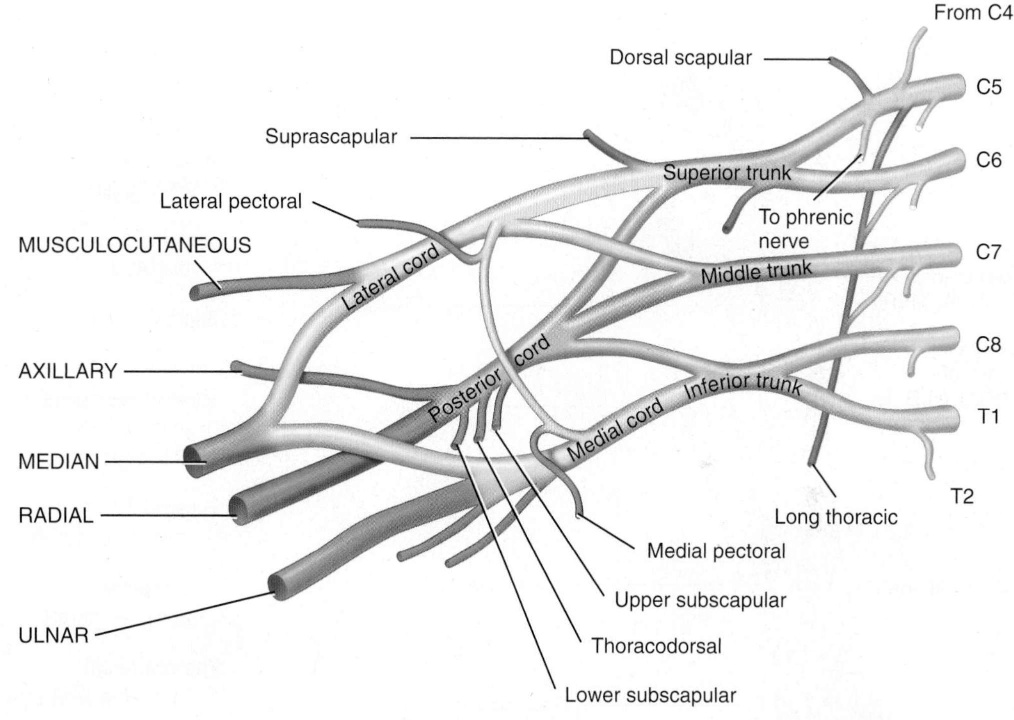


(a) Origin of brachial plexus

MNEMONIC for subunits of brachial plexus:
**R**isk **T**akers **D**on't **C**autiously **B**ehave.
**R**oots, **T**runks, **D**ivisions, **C**ords, **B**ranches

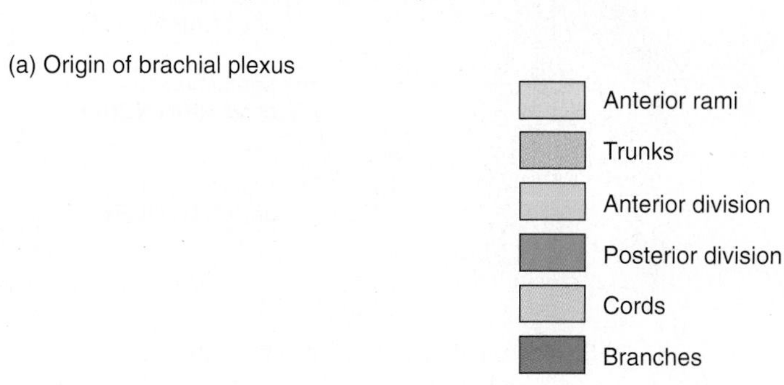


(continues)

FIGURE 14.16 (continued)

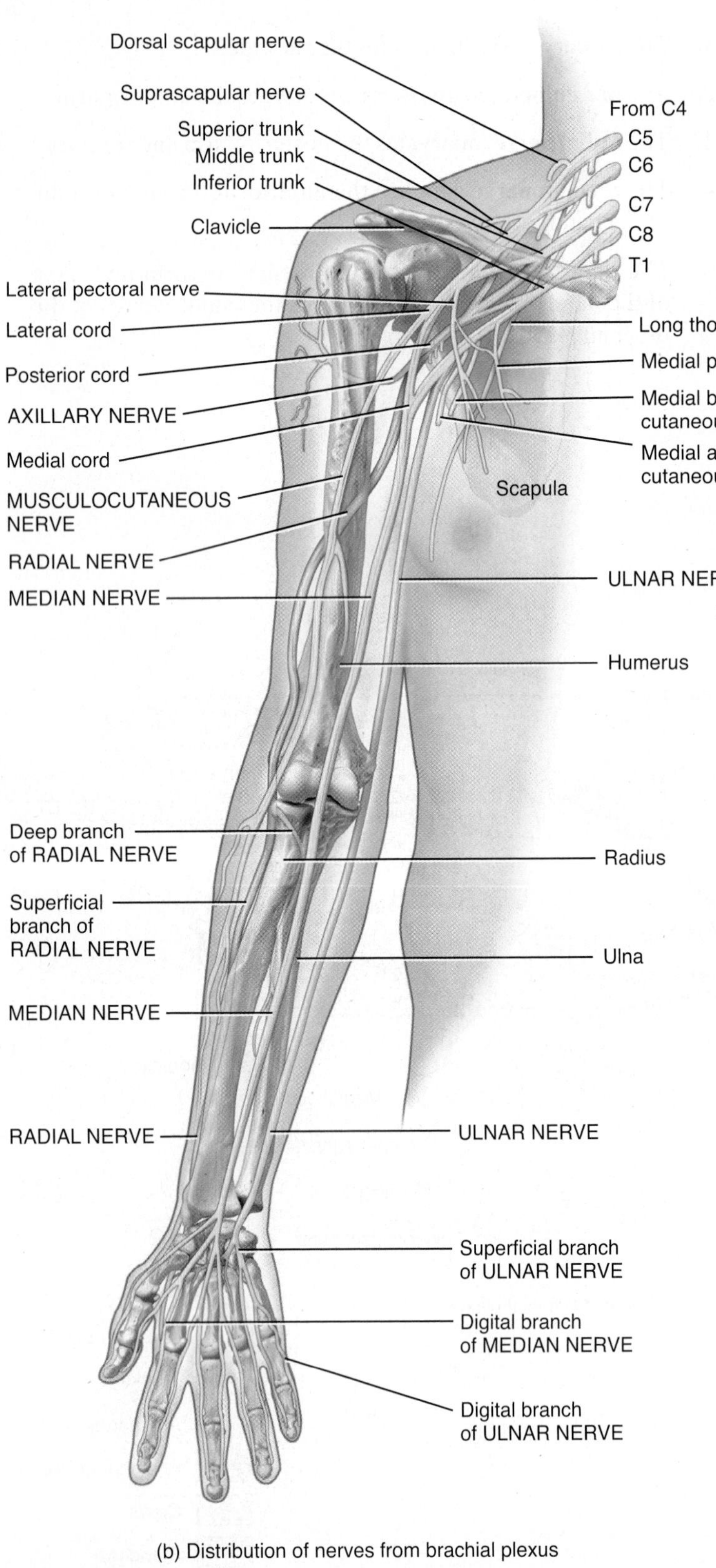


(b) Distribution of nerves from brachial plexus

**The brachial plexus innervates the shoulders and upper limbs.**

**TABLE 14.3**

Brachial Plexus

| NERVE | ORIGIN | DISTRIBUTION |
|---|---|---|
| **Dorsal scapular** (SKAP-ū-lar) | C5 | Levator scapulae and rhomboid muscles |
| **Long thoracic** (thor-AS-ik) | C5–C7 | Serratus anterior muscle |
| **Suprascapular** | C5–C6 | Supraspinatus and infraspinatus muscles |
| **Musculocutaneous** (mus′-kū-lo-kū-TĀN-ē-us) | C5–C7 | Flexors of arm (coracobrachialis, biceps brachii, and brachialis muscles); skin over lateral forearm |
| **Pectoral** (PEK-to-ral) (medial and lateral) | C5–T1 | Pectoralis major and pectoralis minor muscles |
| **Subscapular** (upper and lower) | C5–C6 | Subscapularis and teres major muscles |
| **Thoracodorsal** (thor′-ak-ō-DŌR-sal) | C6–C8 | Latissimus dorsi muscle |
| **Axillary** (AK-si-lar-ē) | C5–C6 | Deltoid and teres minor muscles; skin over deltoid and superior posterior arm |
| **Median** | C5–T1 | Flexors of forearm, except flexor carpi ulnaris and some muscles of the hand; skin of lateral two-thirds of palm of hand and fingers |
| **Radial** | C5–T1 | Triceps brachii and other extensor muscles of arm and extensor muscles of forearm; skin of posterior arm and forearm, lateral two-thirds of dorsum of hand, and fingers over proximal and middle phalanges |
| **Ulnar** | C8–T1 | Flexor carpi ulnaris, flexor digitorum profundus, and most muscles of the hand; skin over medial hand |

Spinal nerves L1–L4 form the **lumbar plexus** (LUM-bar) (Figure 14.17a). Unlike the brachial plexus, there is minimal intermingling of fibers in the lumbar plexus. The lumbar plexus supplies the anterior and lateral abdominal wall, external genitals, and part of the lower limbs (Figure 14.17b). Some important nerves arising from this plexus are the **femoral nerve** innervating the anterior and medial thigh and medial leg, and the **obturator nerve** innervating the medial thigh. Table 14.4 summarizes the major nerves of the lumbar plexus.

FIGURE 14.17 **Lumbar plexus in anterior view.**

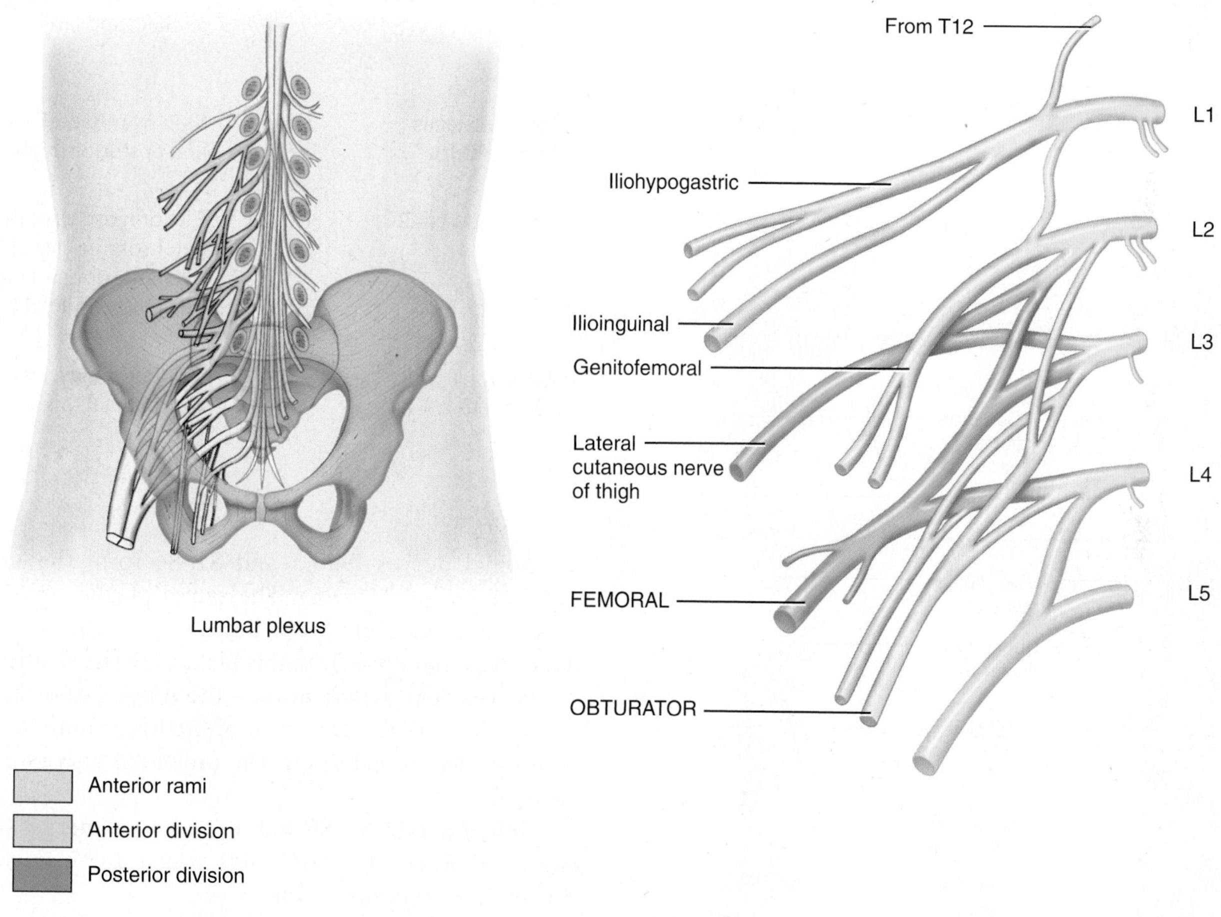


(a) Origin of lumbar plexus

(continues)

FIGURE 14.17 (continued)

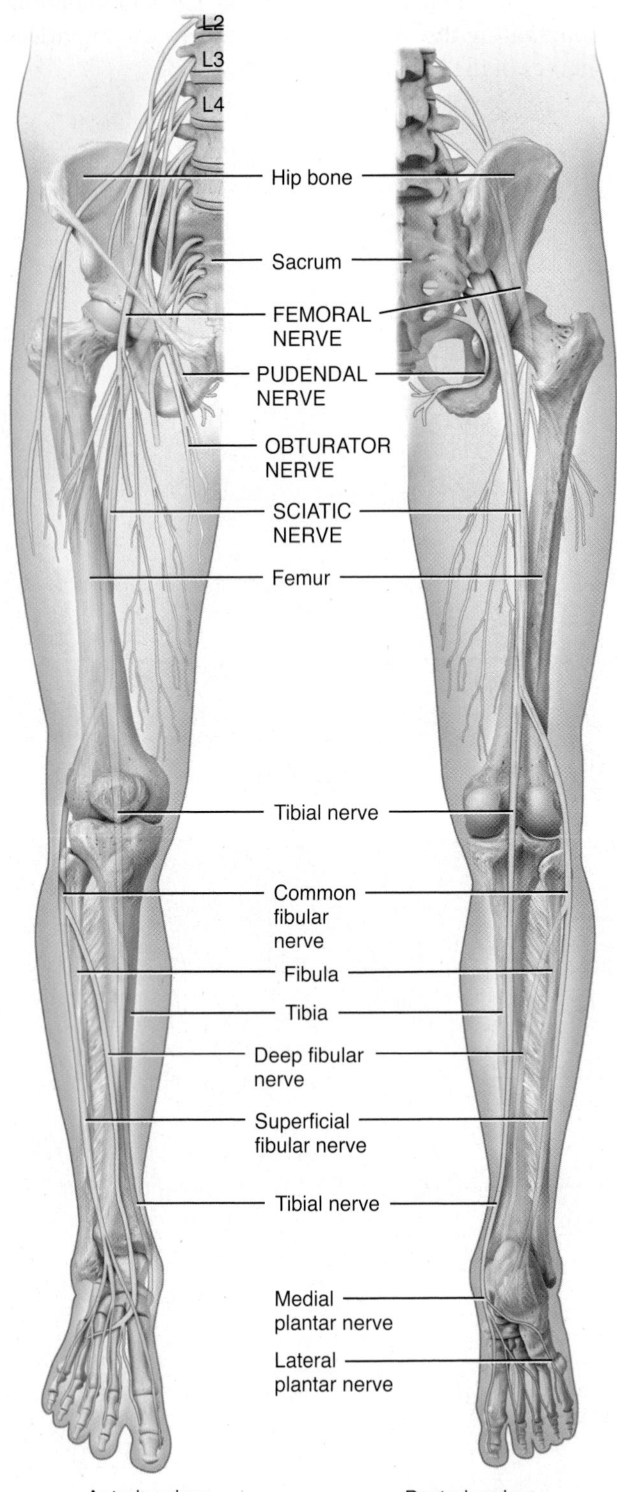


(b) Distribution of nerves from lumbar and sacral plexuses

**The lumbar plexus innervates the anterolateral abdominal wall, external genitals, and part of the lower limbs.**

**TABLE 14.4**

Lumbar Plexus

| NERVE | ORIGIN | DISTRIBUTION |
|---|---|---|
| **Iliohypogastric** (il′-ē-ō-hī-pō-GAS-trik) | L1 | Abdominal muscles; skin over inferior abdomen and buttocks |
| **Ilioinguinal** (il′-ē-ō-IN-gwi-nal) | L1 | Abdominal muscles (with iliohypogastric); skin of superior medial thigh, penis and scrotum in male, and labia majora and mons pubis in female |
| **Genitofemoral** (jen′-i-tō-FEM-or-al) | L1–L2 | Cremaster muscle; skin over middle anterior thigh, scrotum in male, and labia majora in female |
| **Lateral cutaneous nerve of thigh** | L2–L3 | Skin over lateral, anterior, and posterior thigh |
| **Femoral** (FEM-or-al) | L2–L4 | Flexor muscles of thigh and extensor muscles of leg; skin over anterior and medial thigh and medial side of leg and foot |
| **Obturator** (OB-too-rā-tor) | L2–L4 | Adductor muscles of thigh; skin over medial thigh |

Spinal nerves L4–L5 and S1–S4 form the **sacral plexus** (SĀ-kral) (Figure 14.18). The sacral plexus supplies the buttocks, perineum, and lower limbs (see Figure 14.17b). Among the nerves that arise from this plexus are the sciatic and pudendal nerves. The **sciatic nerve**—the longest and thickest nerve in the body—innervates most of the lower limb, except for the anterior and medial thigh. The **pudendal nerve** innervates the perineum.

Spinal nerves S4–S5 and the coccygeal nerves form a small **coccygeal plexus** (kok-SIG-ē-al), which supplies a small area of skin in the coccygeal region.

Table 14.5 summarizes the major nerves of the sacral plexus.

## Intercostal Nerves

The anterior rami of spinal nerves T2–T12 do not enter into the formation of plexuses and are known as **intercostal nerves** (see Figure 14.13). These nerves and their branches directly innervate the structures they supply: the intercostal muscles between ribs, and the muscles and skin of the thorax and abdominal wall.

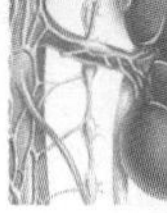

FIGURE 14.18 **Sacral and coccygeal plexuses in anterior view.** The distribution of the nerves of the sacral plexus is shown in Figure 14.17b.

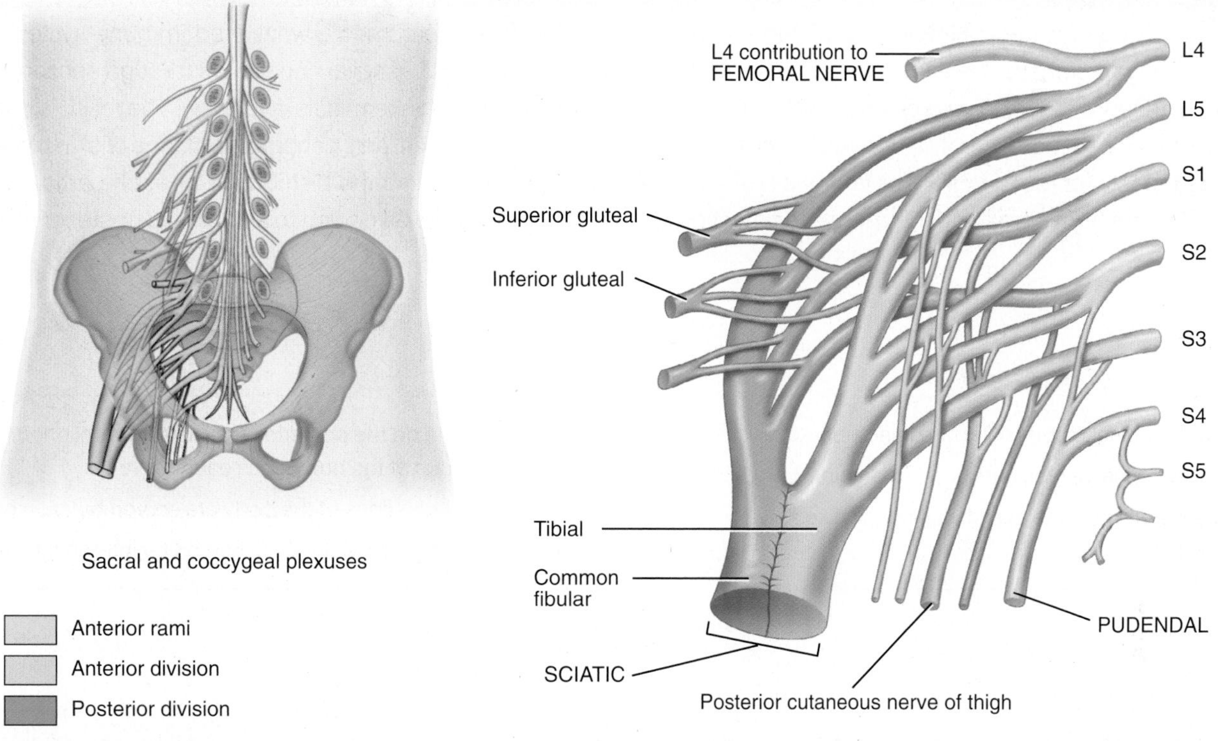


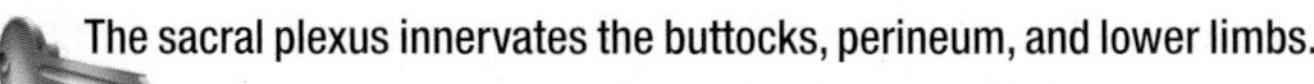

**The sacral plexus innervates the buttocks, perineum, and lower limbs.**

**TABLE 14.5**

Sacral Plexus

| NERVE | ORIGIN | DISTRIBUTION |
|---|---|---|
| **Superior gluteal** (GLOO-tē-al) | L4–L5 and S1 | Gluteus minimus and gluteus medius muscles, and tensor fasciae latae |
| **Inferior gluteal** | L5–S2 | Gluteus maximus muscle |
| **Posterior cutaneous nerve of thigh** | S1–S3 | Skin over anal region, inferior lateral aspect of buttocks, superior posterior aspect of thigh, superior part of calf, scrotum in male, and labia majora in female |
| **Sciatic** (sī-AT-ik) | L4–S3 | Hamstring and adductor magnus muscles; composed of two nerves bound together by common sheath that splits, usually at the knee |
| **Tibial** (TIB-ē-al) (including medial plantar branches) | L4–S3 | Gastrocnemius, plantaris, soleus, popliteus, tibialis posterior muscles, flexors of toes, intrinsic muscles of the foot; skin over posterior leg and plantar surface of foot |
| **Common fibular** (FIB-ū-lar) (including superficial and deep fibular branches) | L4–S2 | Fibularis muscles (brevis, longus, tertius), tibialis anterior muscles, and extensor muscles of toes; skin over anterior leg and dorsum of foot |
| **Pudendal** (pū-DEN-dal) | S2–S4 | Muscles of perineum; skin of penis and scrotum in male and clitoris, labia majora, labia minora, and vagina in female |

## CLINICAL CONNECTION | *Shingles*

**Shingles**, or **herpes zoster** (HER-pēz ZOS-ter), is an acute infection of the peripheral nervous system caused by the varicella-zoster virus that also causes chickenpox. As a person recovers from chickenpox, the virus retreats to sensory neuron cell bodies in posterior root ganglia. The immune system usually prevents the virus from spreading. From time to time, however, the virus overcomes a weakened immune system, leaves the posterior root ganglia, and travels through sensory neurons toward their axon terminals in the skin. The result is pain, discoloration of the skin, and a characteristic line of skin blisters. The line of blisters marks the distribution of each particular cutaneous sensory nerve belonging to an infected posterior root ganglion.

## ✓ CHECKPOINT

8. How are spinal nerves named and numbered?
9. How do spinal nerves connect to the spinal cord?
10. Why are all spinal nerves classified as mixed nerves?
11. Where do the posterior and anterior rami originate and what regions of the body does each serve?
12. Which regions of the body are served by each plexus?
13. Which body regions are served by intercostal nerves?

# 14.4 A reflex is produced by a reflex arc in response to a particular stimulus.

## Reflexes and Reflex Arcs

A **reflex** is a fast, involuntary, unplanned sequence of actions that occurs in response to a particular stimulus. Some reflexes are present from birth; for example, the reflex involved in pulling your hand away from a hot surface is intrinsic to your nervous system. Other reflexes are learned or acquired. For instance, you learn many reflexes while acquiring driving expertise. Slamming on the brakes in an emergency is one example.

The brain stem and spinal cord serve as integrating centers for reflexive information. When integration occurs in the gray matter of the brain stem, the reflex is a **cranial reflex**. An example of a cranial reflex is the tracking movements of your eyes as you read this sentence. If the spinal cord gray matter is the site of integration, the reflex is a **spinal reflex**. For example, jerking your bare foot up from hot pavement is a spinal reflex.

You are probably most aware of **somatic reflexes**, which involve contraction of skeletal muscles. Equally important, however, are the **autonomic reflexes**, which generally are not consciously perceived. They involve responses of smooth muscle, cardiac muscle, and glands. Body functions such as heart rate, digestion, urination, and defecation are controlled by the autonomic nervous system through autonomic reflexes.

Impulses propagating into, through, and out of the CNS follow specific pathways, depending on the kind of information, its origin, and its destination. The pathway followed by impulses that produce a reflex is called a **reflex arc**. A reflex arc includes the following five functional components (Figure 14.19):

1. **Sensory receptor**. The distal end (dendrite) of a sensory neuron or an associated sensory structure serves as a sensory receptor. It responds to a specific **stimulus**—a change in the internal or external environment—by triggering one or more impulses in the sensory neuron.
2. **Sensory neuron**. The impulses propagate from the sensory receptor along the sensory neuron to the axon terminals in the gray matter of the spinal cord or brain stem.
3. **Integrating center**. One or more regions of gray matter within the CNS act as an integrating center. In the simplest type of reflex, the integrating center is a single synapse between a sensory neuron and a motor neuron. A reflex pathway having only one synapse in the integrating center is termed a **monosynaptic reflex arc** (mon′-ō-si-NAP-tik; *mono-* = one). More often, the integrating center includes one or more interneurons, which may relay impulses to other interneurons as well as to a motor neuron. A **polysynaptic reflex arc** (*poly-* = many) involves more than two types of neurons and more than one CNS synapse.

FIGURE 14.19 **General components of a reflex arc.** The arrows show the direction of impulse propagation.

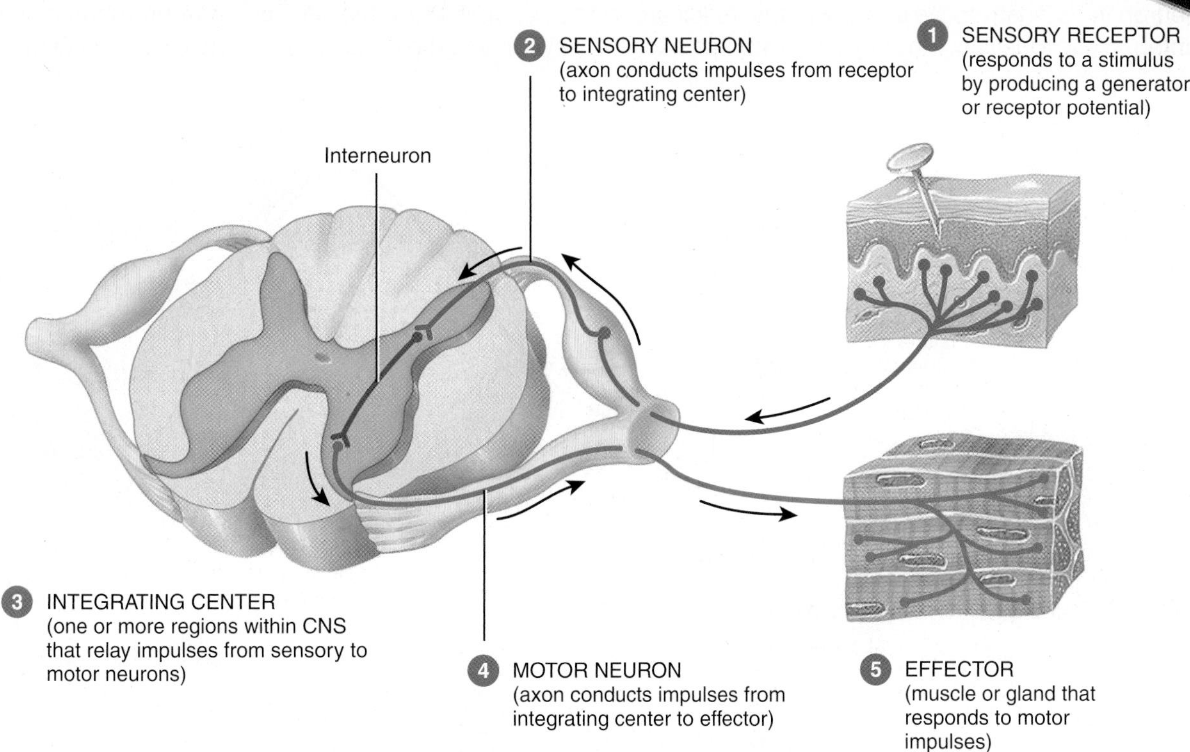
2 SENSORY NEURON
(axon conducts impulses from receptor to integrating center)
1 SENSORY RECEPTOR
(responds to a stimulus by producing a generator or receptor potential)
Interneuron
3 INTEGRATING CENTER
(one or more regions within CNS that relay impulses from sensory to motor neurons)
4 MOTOR NEURON
(axon conducts impulses from integrating center to effector)
5 EFFECTOR
(muscle or gland that responds to motor impulses)

**A reflex is a fast, predictable sequence of involuntary actions that occur in response to certain changes in the environment.**

4. **Motor neuron**. Impulses triggered by the integrating center propagate out of the CNS along a motor neuron to the part of the body that will respond.
5. **Effector**. The part of the body that responds to the motor impulse, such as a muscle or gland, is the effector. Its action is called a reflex. If the effector is skeletal muscle, the reflex is a somatic reflex. If the effector is smooth muscle, cardiac muscle, or a gland, the reflex is an autonomic reflex.

Because reflexes are normally so predictable, they provide useful information about the health of the nervous system and can greatly aid diagnosis of disease. Damage or disease anywhere along its reflex arc can cause a reflex to be absent or abnormal. For example, tapping the patellar ligament normally causes reflex extension of the knee joint. Absence of the patellar reflex could indicate damage to the sensory or motor neurons, or a spinal cord injury in the lumbar region. Somatic reflexes generally can be tested simply by tapping or stroking the body surface.

## The Stretch Reflex

A **stretch reflex** causes contraction of a skeletal muscle (the effector) in response to stretching of the muscle. This type of reflex is a *monosynaptic reflex arc* with activation of a single sensory neuron that synapses in the CNS with a single motor neuron. Stretch reflexes can be elicited by tapping on tendons attached to muscles at the elbow, wrist, knee, and ankle joints. An example of a stretch reflex is the patellar reflex (knee jerk), which is described in the Clinical Connection entitled Reflexes and Diagnosis later in the chapter.

A stretch reflex operates as follows (Figure 14.20):

1. Slight stretching of a muscle stimulates sensory receptors in the muscle called **muscle spindles** (shown in more detail in Figure 15.4). Muscle spindles monitor changes in the length of the muscle.
2. In response to being stretched, a muscle spindle generates one or more impulses that propagate along a somatic sensory neuron that extends through the posterior root of the spinal nerve and into the spinal cord.
3. In the spinal cord gray matter (integrating center), the sensory neuron synapses with and activates a motor neuron.
4. Impulses arise in the motor neuron and propagate along its axon, which extends from the spinal cord into the anterior root and through peripheral nerves to the stimulated muscle.

**FIGURE 14.20 Stretch reflex.** This monosynaptic reflex arc has only one synapse in the CNS—between the sensory neuron and motor neuron. Also illustrated is a polysynaptic reflex arc, with two synapses in the CNS and one interneuron, which innervates antagonistic muscles. Plus signs (+) indicate excitatory synapses; the minus sign (−) indicates an inhibitory synapse.

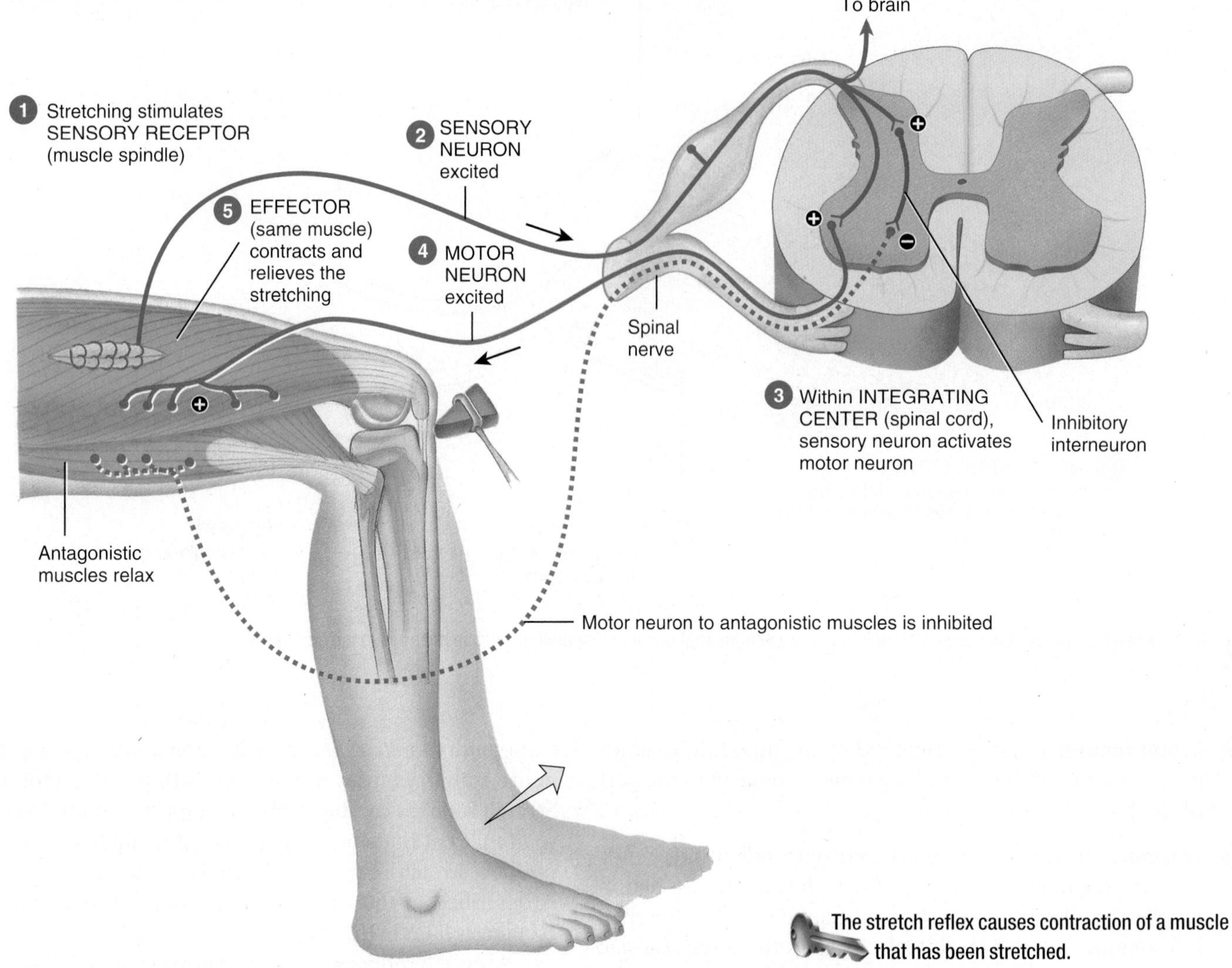
To brain
1 Stretching stimulates SENSORY RECEPTOR (muscle spindle)
2 SENSORY NEURON excited
5 EFFECTOR (same muscle) contracts and relieves the stretching
4 MOTOR NEURON excited
Spinal nerve
3 Within INTEGRATING CENTER (spinal cord), sensory neuron activates motor neuron
Inhibitory interneuron
Antagonistic muscles relax
Motor neuron to antagonistic muscles is inhibited

The stretch reflex causes contraction of a muscle that has been stretched.

5 Acetylcholine is released at the neuromuscular junctions between the axon terminals of the motor neuron and skeletal muscle fibers of the stretched muscle. Acetylcholine triggers action potentials in the stretched muscle (effector), and the muscle contracts. Thus, muscle stretch is followed by muscle contraction, which relieves the stretching.

In the reflex arc just described, sensory impulses enter the spinal cord on the same side from which motor impulses leave it. This arrangement is called an **ipsilateral reflex** (ip′-si-LAT-er-al = same side). All monosynaptic reflexes are ipsilateral.

By adjusting how vigorously a muscle spindle responds to stretching, the brain sets an overall level of **muscle tone**, which is the small degree of contraction present while the muscle is at rest. Because the stimulus for the stretch reflex is stretching of muscle, this reflex helps avert injury by preventing overstretching of muscles.

Although the stretch reflex pathway itself is monosynaptic (just two neurons and one synapse), a *polysynaptic reflex arc* to the antagonistic muscles operates at the same time. This arc involves three neurons and two synapses. An axon collateral (branch) from the sensory neuron also synapses with an inhibitory interneuron in the integrating center. In turn, the interneuron synapses with and inhibits a motor neuron that normally excites the antagonistic muscles (dotted line in Figure 14.20). Thus, when the stretched muscle contracts during a stretch reflex, antagonistic muscles that oppose the contraction simultaneously relax. This type of innervation that simultaneously results in contraction of one muscle and relaxation of its antagonistic muscles, is termed **reciprocal innervation** (rē-SIP-rō′-kal in′-er-VĀ-shun). Reciprocal innervation prevents conflict between opposing muscles and is vital in coordinating body movements.

Axon collaterals of the sensory neuron also relay impulses to the brain about the state of stretch or contraction of skeletal muscles, enabling the brain to coordinate muscular movements. The impulses that pass to the brain also allow conscious awareness that the reflex has occurred.

The stretch reflex can also help maintain posture. For example, if a standing person begins to lean forward, the gastrocnemius and other calf muscles are stretched. Consequently, stretch reflexes are initiated in these muscles, which cause them to contract and reestablish the body's upright posture.

## The Flexor Reflex

A polysynaptic reflex arc results when, for instance, you step on a tack. In response to such a painful stimulus, you immediately withdraw your leg. This reflex, called the **flexor** or **withdrawal reflex**, operates as follows (Figure 14.21):

1. Stepping on a tack stimulates a pain-sensitive sensory neuron (pain receptor).
2. This stimulated sensory neuron generates impulses, which propagate into the spinal cord.
3. Within the spinal cord (integrating center), the sensory neuron activates interneurons that extend to several spinal cord segments.

FIGURE 14.21 **Flexor (withdrawal) reflex.** This reflex arc is polysynaptic and ipsilateral. Plus signs (+) indicate excitatory synapses.

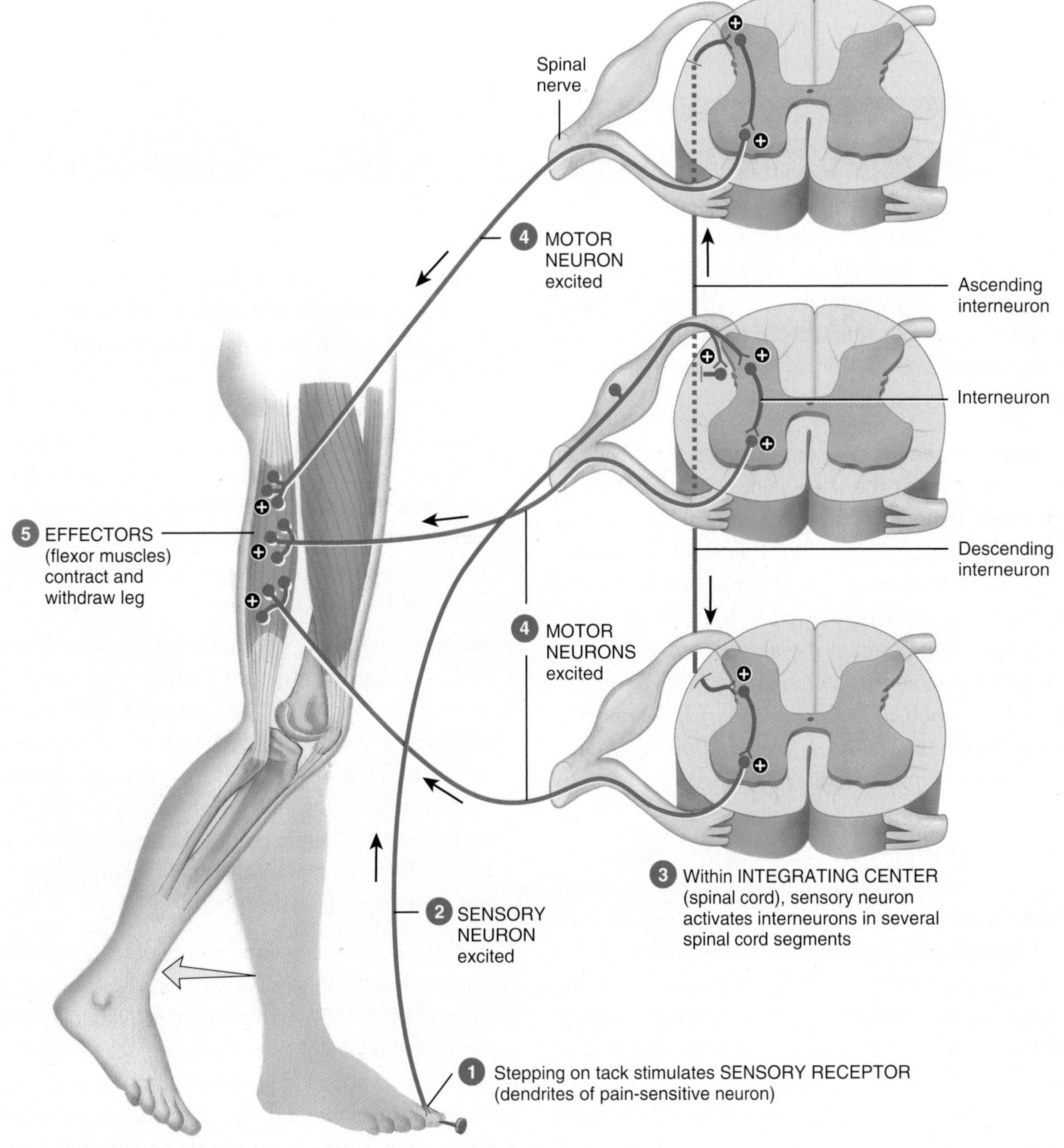
Spinal nerve
4 MOTOR NEURON excited
Ascending interneuron
Interneuron
Descending interneuron
5 EFFECTORS (flexor muscles) contract and withdraw leg
4 MOTOR NEURONS excited
3 Within INTEGRATING CENTER (spinal cord), sensory neuron activates interneurons in several spinal cord segments
2 SENSORY NEURON excited
1 Stepping on tack stimulates SENSORY RECEPTOR (dendrites of pain-sensitive neuron)

The flexor reflex causes withdrawal of a part of the body in response to a painful stimulus.

4. The interneurons activate several motor neurons. The stimulated motor neurons generate impulses, which propagate toward the axon terminals.

5. Acetylcholine released by the motor neurons causes the flexor muscles in the thigh (effectors) to contract, producing withdrawal of the leg. This reflex is protective because contraction of flexor muscles moves a limb away from the source of a possibly damaging stimulus.

The flexor reflex, like the stretch reflex, is ipsilateral—the incoming and outgoing impulses propagate into and out of the same side of the spinal cord. The flexor reflex also illustrates another feature of polysynaptic reflex arcs. Moving your entire lower or upper limb away from a painful stimulus involves contraction of more than one muscle group. Hence, several motor neurons must simultaneously convey impulses to several limb muscles. Because impulses from one sensory neuron ascend and descend in the spinal cord and activate interneurons in several segments of the spinal cord, this type of reflex is called an **intersegmental reflex arc** (in′-ter-seg-MEN-tal; *inter-* = between). Through intersegmental reflex arcs, a single sensory neuron can activate several motor neurons, thereby stimulating more than one effector. The monosynaptic stretch reflex, in contrast, involves muscles receiving impulses from one spinal cord segment only.

As in the stretch reflex, reciprocal innervation occurs in the flexor reflex. When the flexor muscles of a painfully stimulated

## CLINICAL CONNECTION | *Reflexes and Diagnosis*

Reflexes are often used for diagnosing disorders of the nervous system and locating injured tissue. If a reflex ceases to function or functions abnormally, the physician may suspect that the damage lies somewhere along a particular conduction pathway. Many somatic reflexes can be tested simply by tapping or stroking the body. Among the somatic reflexes of clinical significance are the following:

- **Patellar reflex (knee jerk).** This stretch reflex involves extension of the leg at the knee joint by contraction of the quadriceps femoris muscle in response to tapping the patellar ligament (Figure 14.20). This reflex is blocked by damage to the sensory or motor nerves supplying the muscle or to the integrating centers in the second, third, or fourth lumbar segments of the spinal cord. It is often absent in people with chronic diabetes mellitus or neurosyphilis, both of which cause degeneration of nerves. It is exaggerated in disease or injury involving certain motor tracts descending from the higher centers of the brain to the spinal cord.
- **Achilles reflex (a-KIL-ēz) (ankle jerk).** This stretch reflex involves plantar flexion of the foot by contraction of the gastrocnemius and soleus muscles in response to tapping the calcaneal (Achilles) tendon. Absence of the Achilles reflex indicates damage to the nerves supplying the posterior leg muscles or to neurons in the lumbosacral region of the spinal cord. This reflex may also disappear in people with chronic diabetes, neurosyphilis, alcoholism, and subarachnoid hemorrhages. An exaggerated Achilles reflex indicates cervical cord compression or a lesion of the motor tracts of the first or second sacral segments of the cord.
- **Babinski sign (ba-BIN-skē).** This reflex results from gentle stroking of the lateral outer margin of the sole. The great toe extends, with or without a lateral fanning of the other toes. This phenomenon normally occurs in children under 1½ years of age and is due to incomplete myelination of fibers in the corticospinal tract. A positive Babinski sign after age 1½ is abnormal and indicates an interruption of the corticospinal tract as the result of a lesion of the tract, usually in the upper portion. The normal response after age 1½ is the **plantar flexion reflex,** or **negative Babinski**—a curling under of all the toes.
- **Abdominal reflex.** This reflex involves contraction of the muscles that compress the abdominal wall in response to stroking the side of the abdomen. The response is an abdominal muscle contraction that causes the umbilicus to move in the direction of the stimulus. Absence of this reflex is associated with lesions of the corticospinal tracts. It may also be absent because of lesions of the peripheral nerves, lesions of integrating centers in the thoracic part of the cord, or multiple sclerosis.

Most autonomic reflexes are not practical diagnostic tools because it is difficult to stimulate visceral effectors, which are deep inside the body. An exception is the pupillary light reflex, in which the pupils of both eyes decrease in diameter when either eye is exposed to light. Because the reflex arc includes synapses in lower parts of the brain, the **absence of a normal pupillary light reflex** may indicate brain damage or injury.

lower limb are contracting, the extensor muscles of the same limb are relaxing to some degree. If both sets of muscles contracted at the same time, the two sets of muscles would pull on the bones in opposite directions, which might immobilize the limb. Because of reciprocal innervation, however, one set of muscles contracts while the other relaxes. Reflexes involving intersegmental reflex arcs along with reciprocal innervation of antagonistic muscles result in coordinated movements of the body away from dangerous situations.

✓ **CHECKPOINT**

**14.** How are cranial and spinal reflexes different?

**15.** How are somatic and autonomic reflexes different?

**16.** List the components of a reflex arc in order.

**17.** Define each of the following terms in relation to reflex arcs: monosynaptic, polysynaptic, reciprocal innervation, and intersegmental.

## RETURN TO **Nick's Story**

"Well, Nick," said Dr. Jerouen. "You were diagnosed with Type 2 diabetes mellitus eight years ago. You've made pretty regular visits up until now, with no complaints. When did these problems start?"

"I've been having these prickling sensations, in both my feet, since March or April. I lose my balance sometimes when I'm getting in and out of my boat. And my eyes get funny once in a while."

Dr. Jerouen interrupted. "So this has been progressing for about a year. Have you been monitoring your blood glucose levels regularly?"

"I haven't been real good about it," admitted Nick.

"You may have developed diabetic neuropathy, nerve damage from not controlling your blood sugar. I'm going to run some blood tests and refer you to a neurologist to check your nerve function."

About a week later, Nick went to the neurologist.

"Hi, Nick, I'm Dr. Meyerson. Tell me a little bit about what's been happening." As Nick began recounting his symptoms, Dr. Meyerson worked quickly and efficiently. She closely observed Nick's face, looking for symmetry, noting his facial muscle tone, and listening to the sound of his voice and the articulation of his words. "Smile. Now stick out your tongue. Wrinkle your forehead. Good. I'm going to shine a light into your eyes. Hmmm. Follow the end of this pen with your eyes as I move it around in front of you. Good. Have you had any loss of hearing or problems with your vision?" she asked.

"I have trouble focusing sometimes, and my hearing is always bad," said Nick in a slightly hoarse voice.

"Nick, I'm going to tap you with this hammer to check your reflexes," said Dr. Meyerson. "So, Nick? Do you have any hobbies?" Dr. Meyerson was only half listening to him talk as she briskly tapped Nick's patellar tendon several times and observed the response. She noticed that Nick's knee reflex was somewhat weak. Moving up to Nick's arms, the doctor tapped the back of his arm, then the inside of the elbow.

Using the blunt end of the reflex hammer, Dr. Meyerson ran the tool up the bottom of Nick's foot from the outside of the heel to the ball of his foot. She observed his toes. "Did you feel that?" she asked Nick.

"Not really. Can't feel much in the feet these days," admitted Nick.

"Nick, I'm going to put pressure on your foot. You resist me as hard as you can." Placing her hand on top of Nick's foot, Dr. Meyerson pushed down. "Nick, hold your foot up as I try to push it down. Hmm, you've lost quite a bit of muscle strength and sensation, and your reflexes are somewhat weak. Your tests suggest some peripheral nerve damage. There's not much we can do for the neuropathy except try to get your diabetes under control, to prevent further nerve damage. If your pain increases, there are a few drugs we can try. Come back and see me in six months. But if your symptoms get worse or you begin having trouble with indigestion, vomiting, or constipation, you should come in sooner. It could be a sign of autonomic involvement."

"Auto *what?*" asked Nick.

"Your autonomic nervous system automatically controls things like digestion and blood pressure. But the nerves that carry those signals can be damaged just like your somatic nervous system, the part of your nervous system that voluntarily directs your muscles to move your arms and legs."

A. *Which symptoms that Nick has described so far are relevant to the nervous system? Are his symptoms sensory, motor, or both?*

B. *Do you think the symptoms Nick describes are likely caused by peripheral nerve damage? Could they be caused by damage to the central nervous system?*

C. *Diabetic neuropathies damage peripheral nerves. Which component of the reflex arc is most likely to be damaged in Nick's situation?*

## 14.5 The autonomic nervous system produces involuntary movements.

Like the somatic nervous system, the **autonomic nervous system (ANS)** operates via reflex arcs. A continual flow of impulses propagates from autonomic sensory neurons in visceral organs and blood vessels into integrating centers in the CNS. Then, impulses in autonomic motor neurons propagate to various effector tissues, thereby regulating the activity of smooth muscle, cardiac muscle, and glands.

The somatic nervous system includes both sensory and motor neurons. Sensory neurons convey input from receptors for the **special senses** (vision, hearing, taste, smell, and equilibrium) and from receptors for **somatic senses** (pain, temperature, tactile sensations, and proprioceptive sensations). All of these sensations normally are consciously perceived. In turn, somatic motor neurons innervate skeletal muscles—the effectors of the somatic nervous system—and produce voluntary movements. When a somatic motor neuron stimulates a skeletal muscle, the muscle contracts; the effect is always excitation. If somatic motor neurons cease to stimulate a muscle, the result is a paralyzed, limp muscle that has no muscle tone. Although we are generally not conscious of breathing, the muscles that generate respiratory movements are skeletal muscles controlled by somatic motor neurons. If the respiratory motor neurons become inactive, breathing stops.

The main input to the ANS comes from **autonomic sensory neurons**. Mostly, these neurons are associated with **interoceptors** (IN-ter-ō-sep′-tors), sensory receptors located in blood vessels, visceral organs, muscles, and the nervous system that monitor the internal environment of the body. Examples of interoceptors are chemoreceptors that monitor blood $CO_2$ level and mechanoreceptors that detect the degree of stretch in the walls of organs or blood vessels. Unlike sensory signals triggered by a flower's perfume, a beautiful painting, or a delicious meal, sensory signals from interoceptors are not consciously perceived most of the time. On the other hand, intense activation of interoceptors may produce conscious sensations. Two examples of perceived visceral sensations are pain sensations from damaged viscera and angina pectoris (chest pain) from inadequate blood flow to the heart. The ANS can also be influenced by sensory input from somatic sensory and special sensory neurons. For example, severe pain from an injured foot can produce dramatic changes in the autonomic regulation of heart rate, breathing rate, and gastrointestinal tract activity.

**Autonomic motor neurons** regulate visceral activities by either increasing (exciting) or decreasing (inhibiting) ongoing activities in their effector tissues (cardiac muscle, smooth muscle, and glands). Changes in the diameter of the pupils, dilation and constriction of blood vessels, and adjustment of the rate and force of the heartbeat are examples of autonomic motor responses. Unlike skeletal muscle, tissues innervated by the ANS often function to some extent even if their nerve supply is damaged. For example, the heart continues to beat when it is removed for transplantation into another person, smooth muscle in the lining of the gastrointestinal tract contracts rhythmically on its own, and glands produce some secretions in the absence of ANS control.

The ANS usually operates without conscious control and was originally named *autonomic* because it was thought to function autonomously, or in a self-governing manner, without control by the CNS. However, we now know that the hypothalamus and brain stem do regulate ANS activity. Most autonomic responses cannot be consciously altered to any great degree. You probably cannot voluntarily slow your heartbeat to half its normal rate. For this reason, some autonomic responses are the basis for polygraph ("lie detector") tests. However, practitioners of yoga or other meditation techniques may learn how to regulate at least some of their autonomic activities through long practice. **Biofeedback**, in which monitoring devices display information about a body function such as heart rate or blood pressure, enhances the ability to learn such conscious control. Signals from the general somatic and special senses, acting via the limbic system, also influence responses of autonomic motor neurons. Seeing a bike about to hit you, hearing the squealing brakes of a nearby car, or being grabbed by an attacker would all increase the rate and force of your heartbeat.

The axon of a single, myelinated somatic motor neuron extends from the CNS all the way to the skeletal muscle fibers in its motor unit (Figure 14.22a). By contrast, most autonomic motor pathways consist of two motor neurons in series, that is, one following the other (Figure 14.22b). The first autonomic motor neuron (**preganglionic neuron**) has its cell body in the CNS; its myelinated axon extends from the CNS to an **autonomic ganglion**. (Recall that a *ganglion* is a collection of neuronal cell bodies in the PNS.) The cell body of the second neuron (**postganglionic neuron**) is also in that same autonomic ganglion; its unmyelinated axon extends directly from the ganglion to the effector (smooth muscle, cardiac muscle, or a gland). Alternatively, in some autonomic pathways, the first motor neuron extends to specialized cells called *chromaffin cells* in the adrenal medullae (inner portion of the adrenal glands) rather than an autonomic ganglion. All somatic motor neurons release only acetylcholine as their neurotransmitter, but autonomic motor neurons may release acetylcholine or norepinephrine.

The motor portion of the ANS has two principal branches: the **sympathetic division** and the **parasympathetic division**. Most

**FIGURE 14.22 Motor neuron pathways in the (a) somatic nervous system and (b) autonomic nervous system.** Note that somatic motor neurons release acetylcholine (ACh); autonomic motor neurons release either ACh or norepinephrine (NE).

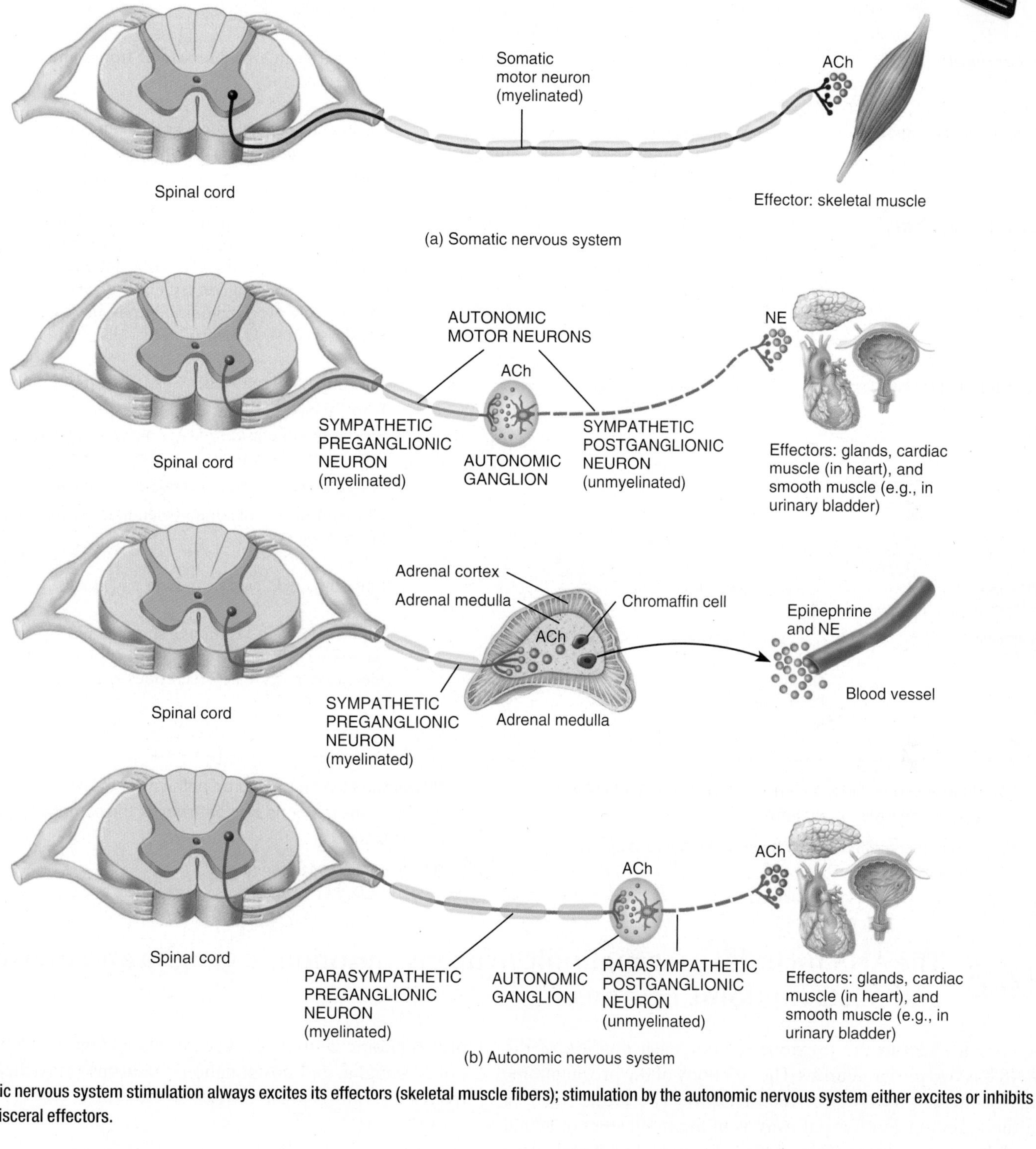

Somatic motor neuron (myelinated)
ACh
Spinal cord
Effector: skeletal muscle

(a) Somatic nervous system

AUTONOMIC MOTOR NEURONS
ACh
NE
Spinal cord
SYMPATHETIC PREGANGLIONIC NEURON (myelinated)
AUTONOMIC GANGLION
SYMPATHETIC POSTGANGLIONIC NEURON (unmyelinated)
Effectors: glands, cardiac muscle (in heart), and smooth muscle (e.g., in urinary bladder)

Adrenal cortex
Adrenal medulla
Chromaffin cell
ACh
Epinephrine and NE
Blood vessel
Spinal cord
SYMPATHETIC PREGANGLIONIC NEURON (myelinated)
Adrenal medulla

ACh
ACh
Spinal cord
PARASYMPATHETIC PREGANGLIONIC NEURON (myelinated)
AUTONOMIC GANGLION
PARASYMPATHETIC POSTGANGLIONIC NEURON (unmyelinated)
Effectors: glands, cardiac muscle (in heart), and smooth muscle (e.g., in urinary bladder)

(b) Autonomic nervous system

**Somatic nervous system stimulation always excites its effectors (skeletal muscle fibers); stimulation by the autonomic nervous system either excites or inhibits visceral effectors.**

organs receive **dual innervation**; that is, they receive impulses from both sympathetic and parasympathetic motor neurons. In some organs, impulses from one division of the ANS stimulate the organ to increase its activity (excitation), and impulses from the other division decrease the organ's activity (inhibition). For example, an increased rate of impulses from the *sympathetic* division *increases* heart rate, and an increased rate of impulses from the *parasympathetic* division *decreases* heart rate.

Table 14.6 compares the somatic and autonomic nervous systems.

**TABLE 14.6**

Comparison of the Somatic and Autonomic Nervous Systems

| | SOMATIC NERVOUS SYSTEM | AUTONOMIC NERVOUS SYSTEM |
|---|---|---|
| **Sensory input** | From somatic senses and special senses | Mainly from interoceptors; some from somatic senses and special senses |
| **Control of motor output** | Voluntary control from cerebral cortex, with contributions from basal ganglia, cerebellum, brain stem, and spinal cord | Involuntary control from hypothalamus, limbic system, brain stem, and spinal cord; limited control from cerebral cortex |
| **Motor neuron pathway** | One-neuron pathway: Somatic motor neurons extending from CNS synapse directly with effector | Usually two-neuron pathway: Preganglionic neurons extending from CNS synapse with postganglionic neurons in autonomic ganglion, and postganglionic neurons extending from ganglion synapse with visceral effector; alternatively, preganglionic neurons may extend from CNS to synapse with chromaffin cells of adrenal medullae |
| **Neurotransmitters and hormones** | All somatic motor neurons release ACh | All sympathetic and parasympathetic preganglionic neurons release ACh<br>Most sympathetic postganglionic neurons release NE; those to most sweat glands release ACh<br>All parasympathetic postganglionic neurons release ACh<br>Chromaffin cells of adrenal medullae release epinephrine and norepinephrine. |
| **Effectors** | Skeletal muscle | Smooth muscle, cardiac muscle, and glands |
| **Responses** | Contraction of skeletal muscle | Contraction or relaxation of smooth muscle; increased or decreased rate and force of contraction of cardiac muscle; increased or decreased secretions of glands |

✓ **CHECKPOINT**

**18.** What are the main input and output components of the autonomic nervous system?

**19.** What are the effectors of the somatic nervous system? The ANS?

**20.** How is the motor neuron pathway of the ANS different from the motor neuron pathway of the somatic nervous system?

**21.** What is dual innervation?

## 14.6 The ANS includes preganglionic neurons, autonomic ganglia and plexuses, and postganglionic neurons.

As you learned in the previous section, each division of the ANS has two motor neurons. The cell body of the preganglionic neuron (see Figure 14.22b), the first of the two motor neurons in the autonomic motor pathway, is in the brain stem or spinal cord. Its axon exits the CNS as part of a cranial or spinal nerve and extends to an autonomic ganglion. There it synapses with a postganglionic neuron, the second neuron in the autonomic motor pathway (see Figure 14.22b). Notice that the postganglionic neuron lies entirely outside the CNS. Its cell body and dendrites are located in an autonomic ganglion, where it forms synapses with one or more preganglionic axons. The axon of the postganglionic neuron terminates in a visceral effector. Thus, preganglionic neurons convey impulses from the CNS to autonomic ganglia, and postganglionic neurons relay the impulses from autonomic ganglia to visceral effectors.

### Preganglionic Neurons

In the sympathetic division, cell bodies of preganglionic neurons are located in the gray matter in the thoracic and first two or three lumbar segments of the spinal cord (Figure 14.23). Cell bodies of preganglionic neurons of the parasympathetic division are located in the brain stem and sacral segments of the spinal cord (Figure 14.24).

**FIGURE 14.23 Structure of the sympathetic division of the autonomic nervous system.** Solid lines represent preganglionic axons; dashed lines represent postganglionic axons. Although the innervated structures are shown only for one side of the body for diagrammatic purposes, the sympathetic division actually innervates tissues and organs on both sides.

**SYMPATHETIC DIVISION (thoracolumbar)**

**Key:**
Preganglionic neurons
Postganglionic neurons

Distributed primarily to smooth muscle of blood vessels of these organs:

Brain
Spinal cord
C1
C2
C3
C4
C5
C6
C7
C8
T1
T2
T3
T4
T5
T6
T7
T8
T9
T10
T11
T12
L1
L2
L3
L4
L5
S1
S2
S3
S4
S5
Coccygeal (fused together)

**Superior cervical ganglion**
**Middle cervical ganglion**
**Inferior cervical ganglion**
**CARDIAC PLEXUS**
**PULMONARY PLEXUS**
**Greater splanchnic nerve**
**Celiac ganglion**
**Aorticorenal ganglion**
**Lesser splanchnic nerve**
**Least splanchnic nerve**
**Superior mesenteric ganglion**
**Renal ganglion**
**Lumbar splanchnic nerve**
**Inferior mesenteric ganglion**
**PREVERTEBRAL GANGLIA**

Skin
Sweat gland
Hair follicle smooth muscle
Blood vessels (each sympathetic trunk innervates skin and viscera)

**SYMPATHETIC TRUNK GANGLIA** (on both sides)

Eye
Pineal gland
Lacrimal gland
Mucous membrane of nose and palate
Parotid gland
Sublingual and submandibular glands
Heart
Atrial muscle fibers
SA/AV nodes
Ventricular muscle fibers
Trachea
Bronchi
Lungs
Liver, gallbladder, and bile ducts
Stomach
Spleen
Pancreas
Transverse colon
Small intestine
Descending colon
Ascending colon
Sigmoid colon
Rectum
Adrenal gland
Kidney
Ureter
Urinary bladder
External genitals
Uterus

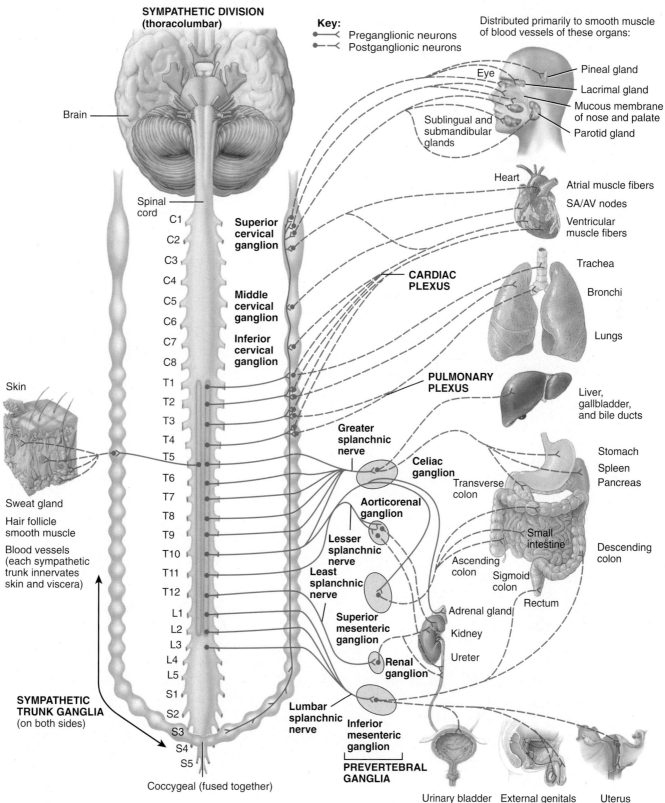

Cell bodies of sympathetic preganglionic neurons are located in the gray matter in the thoracic and lumbar segments of the spinal cord.

**FIGURE 14.24 Structure of the parasympathetic division of the autonomic nervous system.** Solid lines represent preganglionic axons; dashed lines represent postganglionic axons. Although the innervated structures are shown only for one side of the body for diagrammatic purposes, the parasympathetic division actually innervates tissues and organs on both sides.

**PARASYMPATHETIC DIVISION (craniosacral)**

**Key:**
Preganglionic neurons
Postganglionic neurons

**TERMINAL GANGLIA**

Brain
CN III
CN VII
CN IX
CN X
**Ciliary ganglion**
**Pterygopalatine ganglion**
**Submandibular ganglion**
**Otic ganglion**
Spinal cord
C1
C2
C3
C4
C5
C6
C7
C8
T1
T2
T3
T4
T5
T6
T7
T8
T9
T10
T11
T12
L1
L2
L3
L4
L5
S1
S2
S3
S4
S5
Coccygeal
**PELVIC SPLANCHNIC NERVES**
Eye
Lacrimal gland
Mucous membrane of nose and palate
Parotid gland
Sublingual and submandibular glands
Heart
Atrial muscle fibers
SA/AV nodes
Larynx
Trachea
Bronchi
Lungs
Liver, gallbladder, and bile ducts
Stomach
Pancreas
Transverse colon
Small intestine
Descending colon
Ascending colon
Sigmoid colon
Rectum
Ureter
Urinary bladder
External genitals
Uterus

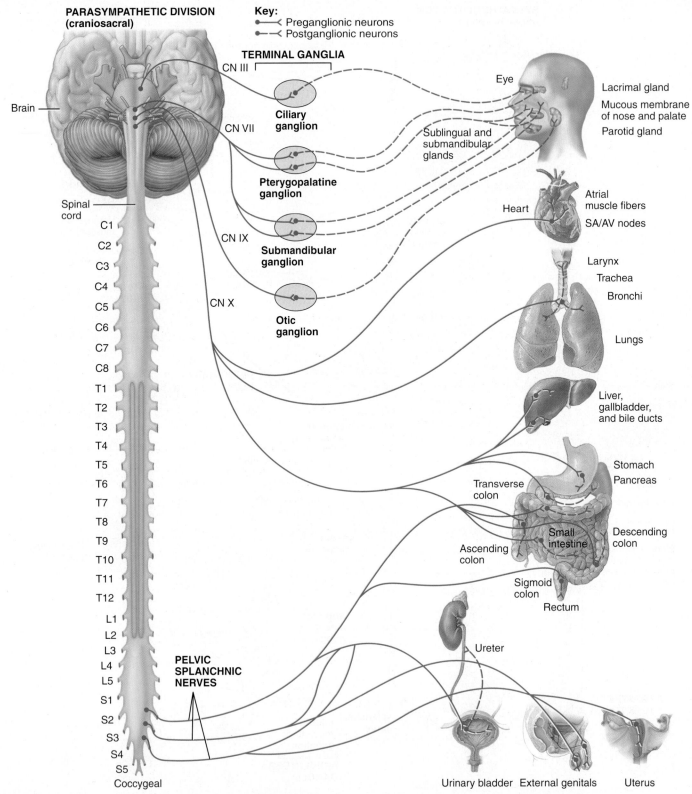

Cell bodies of parasympathetic preganglionic neurons are located in the brain stem and in the gray matter in the sacral segments of the spinal cord.

## Autonomic Ganglia

There are two major groups of autonomic ganglia: (1) **sympathetic ganglia**, which are components of the sympathetic division of the ANS, and (2) **parasympathetic ganglia**, which are components of the parasympathetic division of the ANS.

Sympathetic ganglia are the sites of synapses between sympathetic preganglionic and postganglionic neurons. There are two major types of sympathetic ganglia: sympathetic trunk ganglia and prevertebral ganglia. **Sympathetic trunk ganglia** lie in a vertical row on either side of the vertebral column (Figure 14.23). These ganglia extend from the base of the skull to the coccyx. Postganglionic axons from sympathetic trunk ganglia primarily innervate organs above the diaphragm such as the head, neck, shoulders, and heart. Because the sympathetic trunk ganglia are near the spinal cord, most sympathetic preganglionic axons are short and most sympathetic postganglionic axons are long. The second group of sympathetic ganglia, the **prevertebral ganglia**, lies in the abdomen and anterior to the vertebral column. In general, postganglionic axons from prevertebral ganglia innervate organs below the diaphragm.

Preganglionic neurons of the parasympathetic division synapse with postganglionic neurons in **terminal ganglia**. Most of these ganglia are located close to or actually within the wall of their target organ (Figure 14.24). Because terminal ganglia are located either close to or in the wall of the visceral organ, parasympathetic preganglionic axons are long, in contrast to parasympathetic postganglionic axons, which are short.

## Postganglionic Neurons

Once axons of sympathetic preganglionic neurons pass to sympathetic trunk ganglia, they may connect with postganglionic neurons in one of the following ways (Figure 14.25):

1. An axon may synapse with postganglionic neurons in the ganglion it first reaches.

FIGURE 14.25 **Types of connections between ganglia and postganglionic neurons in the sympathetic division of the ANS.** Also illustrated are the rami communicantes.

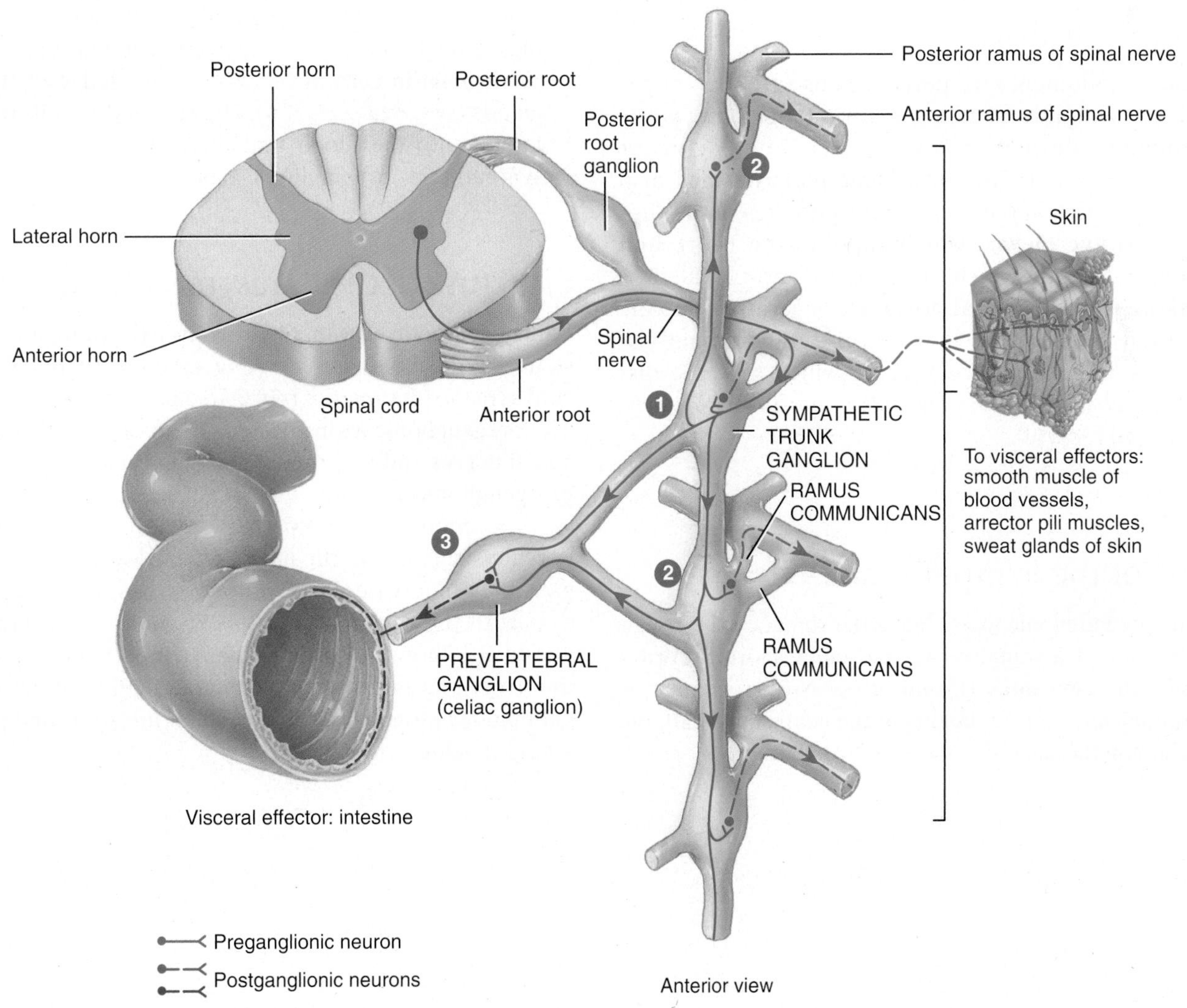


Sympathetic ganglia lie in two chains on either side of the vertebral column (sympathetic trunk ganglia) and anterior to the vertebral column (prevertebral ganglia).

2. An axon may ascend or descend to a higher or lower ganglion before synapsing with postganglionic neurons.

3. An axon may continue, without synapsing, through the sympathetic trunk ganglion to end at a prevertebral ganglion where it synapses with postganglionic neurons.

A single sympathetic preganglionic fiber has many axon collaterals (branches) and may synapse with 20 or more postganglionic neurons. This pattern of projection is an example of divergence and helps explain why many sympathetic responses affect almost the entire body simultaneously. After exiting their ganglia, the postganglionic axons typically terminate in several visceral effectors (see Figure 14.23).

Parasympathetic preganglionic axons extend to terminal ganglia near or within a visceral effector (see Figure 14.24). In the terminal ganglion, the presynaptic neuron usually synapses with only four or five postsynaptic neurons, all of which supply a single visceral effector, allowing parasympathetic responses to be localized to a single effector.

## Autonomic Plexuses

In the thorax, abdomen, and pelvis, axons of both sympathetic and parasympathetic neurons form tangled networks called **autonomic plexuses**, many of which lie along major arteries (Figure 14.26). The autonomic plexuses may also contain sympathetic ganglia. The major plexuses in the thorax are the **cardiac plexus**, which supplies the heart, and the **pulmonary plexus**, which supplies the bronchial tree. Located along the abdominal aorta are several autonomic plexuses; often the plexuses that innervate abdominal viscera are named after the artery along which they are distributed (e.g., *celiac, superior mesenteric, inferior mesenteric, hypogastric,* and *renal*).

## Structure of the Sympathetic Division

Sympathetic preganglionic axons leave the spinal cord through the anterior root of a spinal nerve and enter short pathways called **rami communicantes** (kō-mū-ni-KAN-tē-z; singular is **ramus communicans**) before passing to the nearest sympathetic trunk ganglion on the same side (see Figure 14.25).

Axons leave the sympathetic trunk in three possible ways: (1) They can enter spinal nerves; (2) they can form sympathetic nerves; and (3) they can form splanchnic nerves.

- Some sympathetic postganglionic axons leave the sympathetic trunk ganglia by passing through short rami communicantes to spinal nerves, where they provide sympathetic innervation to visceral effectors (see Figure 14.25).
- Sympathetic postganglionic axons may leave the sympathetic trunk ganglia by forming **sympathetic nerves** that extend to visceral effectors in the thoracic cavity. The postganglionic axons may enter the cardiac plexus to provide sympathetic innervation to the heart or enter the pulmonary plexus to supply the smooth muscle of the bronchi and bronchioles of the lungs (see Figure 14.23).
- Recall that some sympathetic preganglionic axons pass through a sympathetic trunk ganglion without terminating in it. Beyond the sympathetic trunk ganglia, they form nerves known as **splanchnic nerves** (SPLANK-nik; see Figure 14.23), most of which extend to outlying prevertebral ganglia that supply the organs of the abdominopelvic cavity. Some sympathetic preganglionic axons pass, without synapsing, through the sympathetic trunk ganglia, and then extend to **chromaffin cells** in the adrenal medullae of the adrenal glands (see Figure 14.22b). The chromaffin cells release hormones into the blood that intensify responses elicited by sympathetic postganglionic axons.

## Structure of the Parasympathetic Division

Parasympathetic preganglionic axons exit from either the brain stem as part of a cranial nerve or sacral segments of the spinal cord as part of the anterior root of a spinal nerve (see Figure 14.24). The preganglionic axons of both the cranial nerves and sacral spinal nerves end in terminal ganglia, where they synapse with postganglionic neurons.

The cranial parasympathetic axons have ganglia that innervate structures in the head and are located close to the organs they innervate. Preganglionic axons that leave the brain as part of the vagus (X) nerves extend to many terminal ganglia in the thorax to provide parasympathetic innervation to the heart and lungs, and to the abdomen to supply the liver, gallbladder, stomach, pancreas, small intestine, and part of the large intestine.

FIGURE 14.26 **Autonomic plexuses in the thorax, abdomen, and pelvis.**

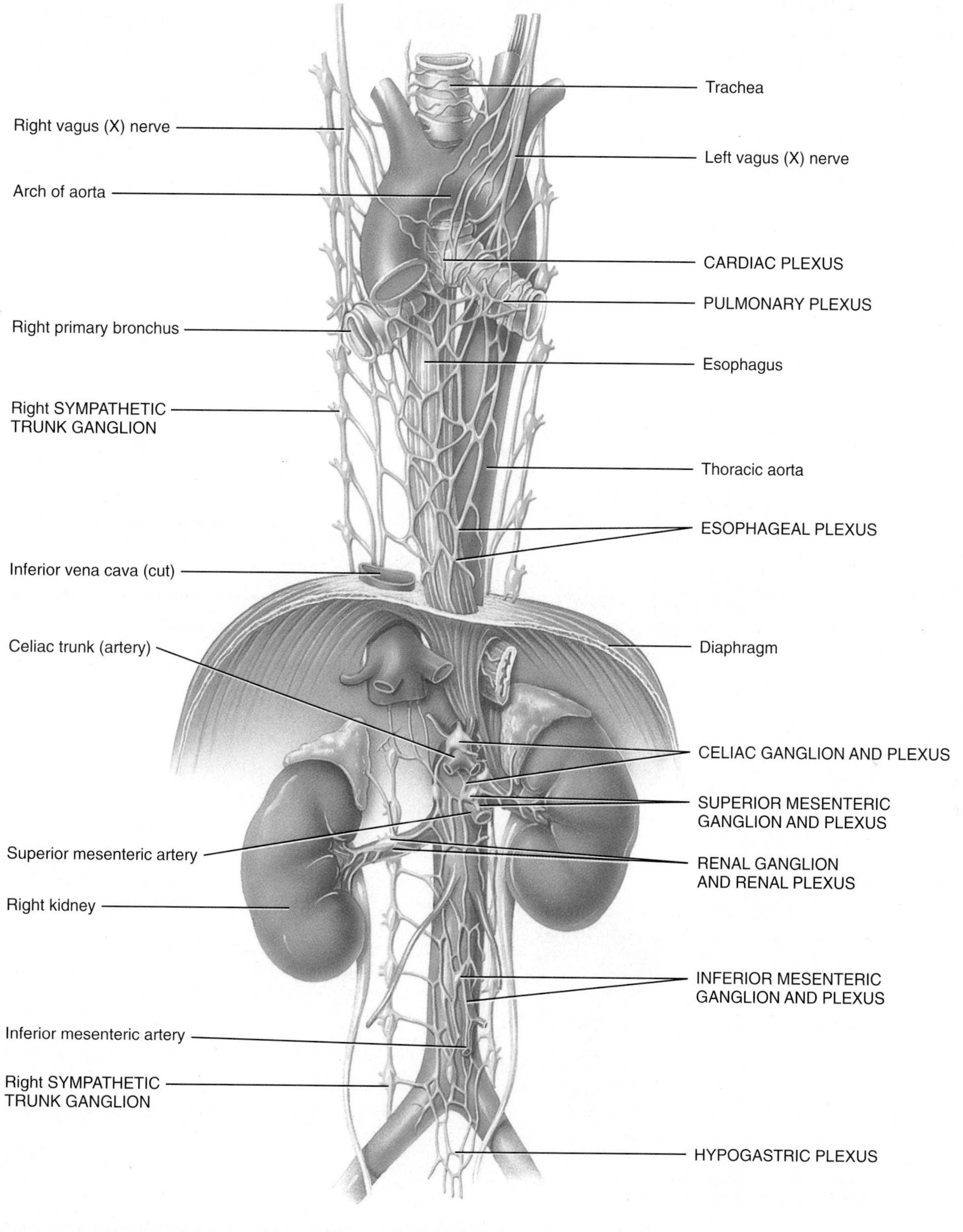


Anterior view

**An autonomic plexus is a network of sympathetic and parasympathetic axons and may also include sympathetic ganglia.**

Parasympathetic preganglionic axons course through the sacral spinal nerves, then branch off these nerves to form **pelvic splanchnic nerves** (Figure 14.27). Pelvic splanchnic nerves synapse with parasympathetic postganglionic neurons located in terminal ganglia in the walls of the innervated viscera. From the terminal ganglia, parasympathetic postganglionic axons innervate smooth muscle and glands in the walls of the colon, ureters, urinary bladder, and reproductive organs.

## Structure of the Enteric Nervous System

Although the enteric nervous system (en-TER-ik) and autonomic nervous system are considered separate, independent functional divisions of the peripheral nervous system, enteric components are associated with ANS structures.

To understand and appreciate the enteric nervous system, it is important to realize that the gastrointestinal tract, like the surface of the body, forms an extensive area of contact with

FIGURE 14.27 **Pelvic splanchnic nerves.**

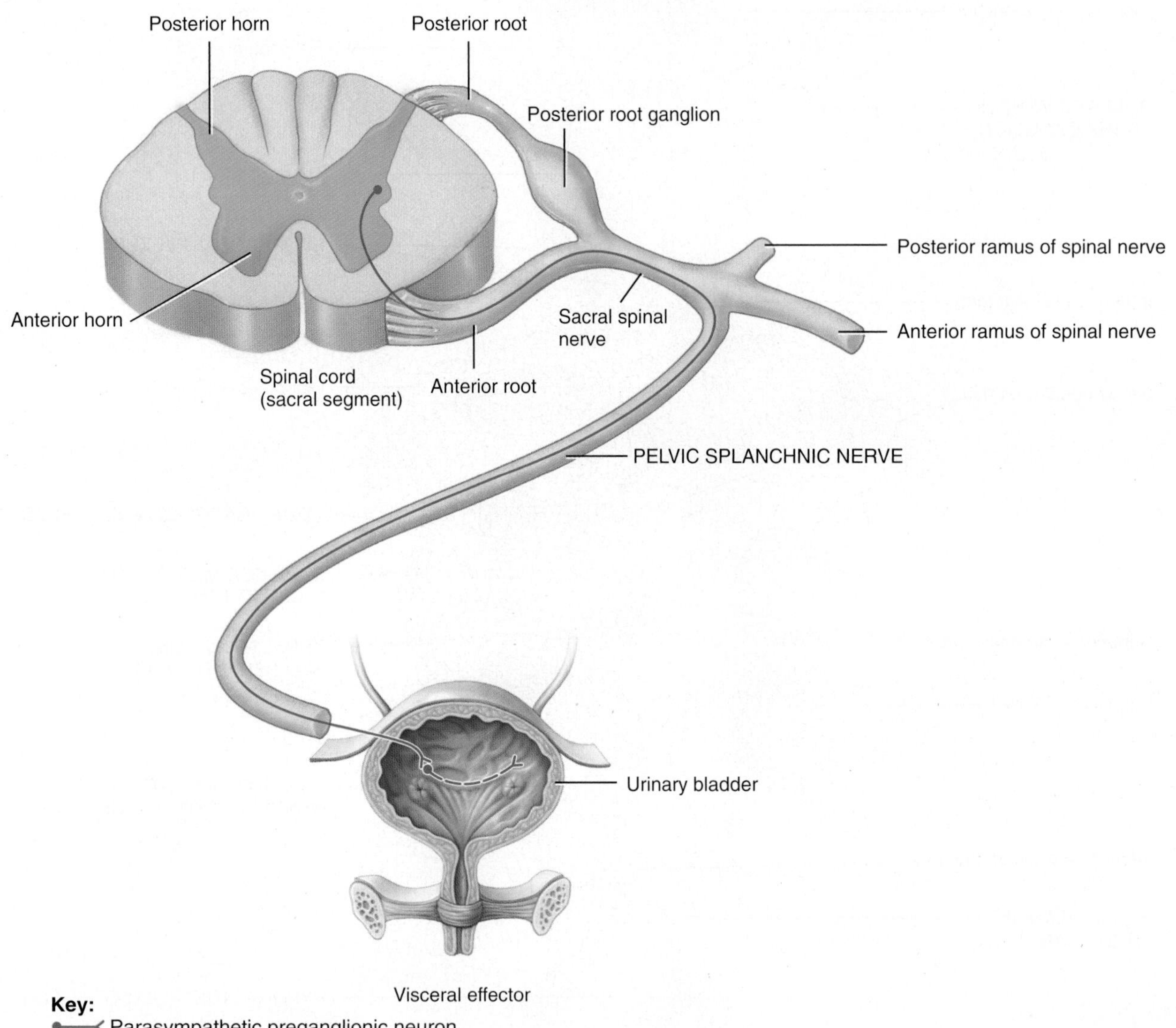


**Parasympathetic preganglionic axons extend through pelvic splanchnic nerves to parasympathetic postganglionic neurons in terminal ganglia within the walls of pelvic viscera.**

the environment. Although this gastrointestinal environment is inside the body, it is still considered part of the external environment. Just as the surface of the body must respond to important environmental stimuli in order to function properly, the surface of the gastrointestinal tract must respond to surrounding stimuli to generate proper homeostatic controls. In fact, these responses and controls are so important that the gastrointestinal tract has its own nervous system with intrinsic input, processing, and output. The enteric nervous system can and does function independent of central nervous system activity, but can also receive controlling input from the central nervous system.

The enteric nervous system is the specialized network of nerves and ganglia forming a complex, integrated neuronal network within the wall of the gastrointestinal tract, pancreas, and gallbladder. This incredible nerve network contains in the neighborhood of 100 million neurons, approximately the same number as the spinal cord, and is capable of continued function without input from the central nervous system. The enteric network of nerves and ganglia contains sensory neurons capable of monitoring tension in the intestinal wall and accessing the composition of the intestinal contents. These sensory neurons relay their input signals to interneurons within the enteric ganglia. The interneurons establish an integrative network that processes the incoming signals and generates regulatory output signals to motor neurons throughout plexuses within the wall of the digestive organs. The motor neurons carry the output signals to the smooth muscle and glands of the gastrointestinal tract to exert control over its motility (movement) and secretory activities.

Most of the nerve fibers that innervate the digestive organs arise from two plexuses within the enteric nervous system. The largest, the **myenteric plexus** (mī-en-TER-ik), is positioned between the outer longitudinal and circular muscle layers from the upper esophagus to the anus. The myenteric plexus communicates extensively with a somewhat smaller plexus, the **submucous plexus**, which occupies the gut wall between the circular muscle layer and the muscularis mucosae (see Concept 23.1) and runs from the stomach to the anus. Neurons emerge from the ganglia of these two plexuses to form smaller plexuses around blood vessels and within the muscle layers and mucosa of the gut wall. It is this system of nerves that makes possible the normal motility and secretory functions of the gastrointestinal tract.

### ✓ CHECKPOINT

**22.** What is the functional difference between preganglionic and postganglionic neurons?

**23.** Contrast the locations of sympathetic trunk ganglia, prevertebral ganglia, and terminal ganglia.

**24.** Which ganglia are associated with the sympathetic division? Parasympathetic division?

**25.** Which division, sympathetic or parasympathetic, has longer preganglionic axons? Why?

**26.** What are the functions of the enteric nervous system?

## 14.7 ANS neurons release acetylcholine or norepinephrine, resulting in excitation or inhibition.

Autonomic neurons release neurotransmitters at synapses between neurons (preganglionic to postganglionic) and at synapses between neurons and autonomic effectors. Based on the neurotransmitter they produce and release, autonomic neurons are classified as either cholinergic or adrenergic. Autonomic neurotransmitters exert their effects by binding to specific receptors located in the plasma membrane of a postsynaptic neuron or effector cell.

### Cholinergic Neurons and Receptors

**Cholinergic neurons** (kō-lin-ER-jik) release the neurotransmitter **acetylcholine (ACh)**. Cholinergic neurons include (1) all sympathetic and parasympathetic preganglionic neurons, (2) sympathetic postganglionic neurons that innervate most sweat glands, and (3) all parasympathetic postganglionic neurons (Figure 14.28).

ACh is stored in synaptic vesicles and released by exocytosis. It then diffuses across the synaptic cleft and binds with specific **cholinergic receptors** on the *postsynaptic* plasma membrane. The two types of cholinergic receptors are nicotinic receptors and muscarinic receptors. **Nicotinic receptors** (nik′-ō-TIN-ik) are present in the plasma membrane of dendrites and cell bodies of both sympathetic and parasympathetic postganglionic neurons (Figure 14.28a, b), the plasma membranes of chromaffin cells of the adrenal medullae, and in the motor end plate at the neuromuscular junction. They are so named because nicotine mimics the action of ACh by binding to these receptors. (Nicotine, a natural substance in tobacco leaves, is not a naturally occurring substance in humans and is not normally present in nonsmokers.) **Muscarinic receptors** (mus′-ka-RIN-ik) are present on all effectors (smooth muscle, cardiac muscle, and glands) innervated by parasympathetic postganglionic axons and on most sweat glands (Figure 14.28b, c). These receptors are so named because a mushroom poison called muscarine mimics the actions of ACh by binding to them. Nicotine does not activate muscarinic receptors, and muscarine does not activate nicotinic receptors, but ACh does activate both of these types of cholinergic receptors.

Activation of nicotinic receptors by ACh causes depolarization and thus excitation of the postsynaptic cell, which can be a postganglionic neuron, an autonomic effector, or a skeletal muscle fiber. Activation of muscarinic receptors by ACh sometimes causes depolarization (excitation) and sometimes causes hyperpolarization (inhibition), depending on the target cell. For example, binding of ACh to muscarinic receptors inhibits (relaxes) smooth muscle sphincters in the gastrointestinal tract but excites (contracts) smooth muscle fibers in the the iris of the eye. Because acetylcholine is quickly inactivated by the enzyme **acetylcholinesterase**, effects triggered by cholinergic neurons are brief.

**FIGURE 14.28 Cholinergic neurons and adrenergic neurons in the sympathetic and parasympathetic divisions.**

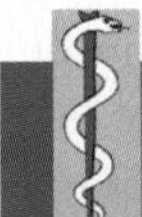

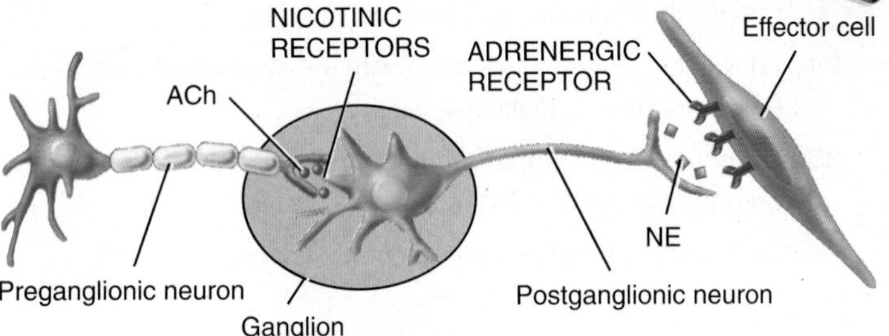


(a) Sympathetic division–innervation to most effector tissues

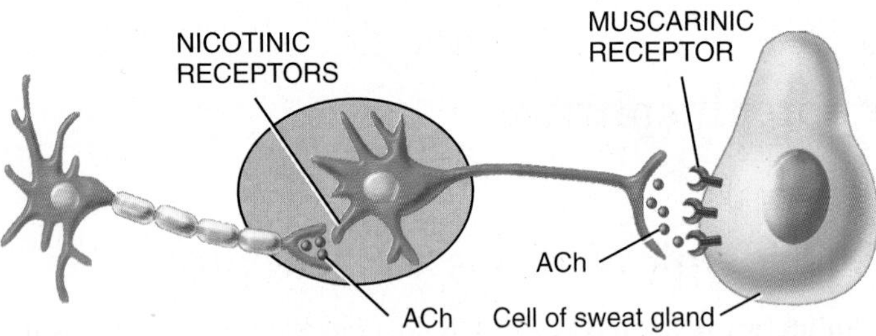


(b) Sympathetic division–innervation to most sweat glands

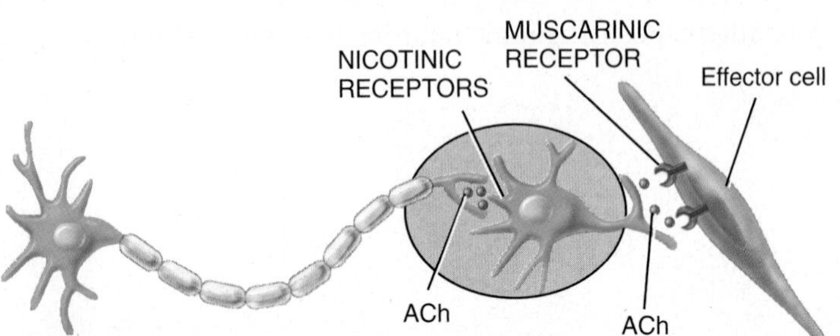


(c) Parasympathetic division

**Cholinergic neurons release acetylcholine; adrenergic neurons release norepinephrine.**

## Adrenergic Neurons and Receptors

**Adrenergic neurons** (ad′-ren-ER-jik) release **norepinephrine (NE)**, also known as noradrenaline (Figure 14.28a). Most sympathetic postganglionic neurons are adrenergic. Like ACh,

### CLINICAL CONNECTION | *Raynaud Phenomenon*

In **Raynaud phenomenon** (rā-NŌ) the digits (fingers and toes) become ischemic (lack blood) after exposure to cold or with emotional stress. The condition is due to excessive sympathetic stimulation of smooth muscle in the arterioles of the digits and a heightened response to stimuli that cause vasoconstriction. When arterioles in the digits vasoconstrict in response to sympathetic stimulation, blood flow is greatly diminished. As a result, the digits may blanch (look white due to blockage of blood flow) or become cyanotic (look blue due to deoxygenated blood in capillaries). In extreme cases, the digits may become necrotic from lack of oxygen and nutrients. With rewarming after cold exposure, the arterioles may dilate, causing the fingers and toes to look red. Many patients with Raynaud phenomenon have low blood pressure. Some have increased numbers of α-adrenergic receptors. Raynaud is most common in young women and occurs more often in cold climates. Patients with Raynaud phenomenon should avoid exposure to cold, wear warm clothing, and keep the hands and feet warm. Drugs used to treat Raynaud include nifedipine, a calcium channel blocker that relaxes vascular smooth muscle, and prazosin, which relaxes smooth muscle by blocking alpha receptors. Smoking and the use of alcohol or illicit drugs can exacerbate the symptoms of this condition.

NE is synthesized and stored in synaptic vesicles and released by exocytosis. Molecules of NE diffuse across the synaptic cleft and bind to specific adrenergic receptors on the postsynaptic membrane, causing either excitation or inhibition of the effector cell.

**Adrenergic receptors** bind both norepinephrine and epinephrine. Norepinephrine can be released either as a neurotransmitter by sympathetic postganglionic neurons, or as a hormone into the blood by chromaffin cells of the adrenal medullae; epinephrine is released as a hormone. The two main types of adrenergic receptors are **alpha ($\alpha$) receptors** and **beta ($\beta$) receptors**, which are found on visceral effectors innervated by most sympathetic postganglionic axons. These receptors are further classified into subtypes— $\alpha_1$, $\alpha_2$, $\beta_1$, $\beta_2$, and $\beta_3$—based on the specific responses they elicit and by their selective binding of drugs that activate or block them.

The activity of norepinephrine at a synapse is terminated either when NE is taken up by the axon that released it or when NE is inactivated by enzymes. Compared to ACh, norepinephrine lingers in the synaptic cleft for a longer time. Thus, effects triggered by adrenergic neurons typically are longer lasting than those triggered by cholinergic neurons.

Table 14.7 describes the locations of cholinergic and adrenergic receptors and summarizes the responses that occur when each type of receptor is activated.

**TABLE 14.7**

Locations and Responses of Adrenergic and Cholinergic Receptors

| TYPE OF RECEPTOR | MAJOR LOCATIONS | ACTIVATING NEUROTRANSMITTER | EFFECTS OF RECEPTOR ACTIVATION |
|---|---|---|---|
| **Cholinergic** | | | |
| **Nicotinic** | Sympathetic and parasympathetic postganglionic neurons | Acetylcholine | Impulses in postsynaptic neurons |
| | Skeletal muscles | | Contraction |
| **Muscarinic** | Effectors innervated by parasympathetic postganglionic neurons | Acetylcholine | Excitation or inhibition of postsynaptic cell |
| | Sweat glands | | Increased sweating |
| **Adrenergic** | | | |
| $\alpha_1$ | Blood vessels serving salivary glands, skin, mucosa, kidneys, and abdominal viscera | Norepinephrine and epinephrine | Vasoconstriction |
| | Sphincter muscles of stomach and urinary bladder | | Closing of sphincters |
| | Salivary glands | | Increased secretion |
| $\alpha_2$ | Blood vessels | Norepinephrine and epinephrine | Vasodilation |
| | Pancreas | | Decreased secretion of insulin and digestive enzymes |
| | Platelets in blood | | Aggregation to promote blood clotting |
| $\beta_1$ | Heart | Norepinephrine and epinephrine | Increased force and rate of contraction |
| | Kidneys | | Secretion of renin |
| | Adipose cells | | Breakdown of triglycerides to fatty acids |
| $\beta_2$ | Lungs | Norepinephrine and epinephrine | Dilation of airways |
| | Blood vessels serving heart, skeletal muscle, adipose tissue, and liver | | Vasodilation |
| | Digestive and urinary organs | | Relaxation of organ walls |
| | Liver | | Breakdown of glycogen to glucose |
| $\beta_3$ | Adipose tissue | Norepinephrine and epinephrine | Heat production |

## ✓ CHECKPOINT

**27.** Define cholinergic and adrenergic neurons.

**28.** Which neurotransmitters bind to cholinergic receptors and which bind to adrenergic receptors?

**29.** Which autonomic neurons produce a shorter effect, cholinergic or adrenergic? Why?

## RETURN TO **Nick's Story**

As Nick stood up he swayed a bit and placed one hand on the table and the other on his head. "Hey, Nick, are you feeling alright?" his friend Tony asked.

"I get a little dizzy sometimes when I stand up. It's nothing. Let's just get to the lake. I want to catch some fish," said Nick.

"Alright," agreed Tony. "Just don't pass out on me. You're too big to carry back to town on my shoulders."

As they drove down the narrow dirt road, Tony asked, "So, Nick, you been feeling alright?"

Nick shook his head. "My stomach's been giving me an awful time. I've been throwing up, can't keep anything down. What I do keep down, well, it doesn't come out. I've been stopped up for weeks. I should probably go see the doctor. I was supposed to go about 3 months ago, but I just didn't have time. Here we are. Let's get the boat in the water."

As Nick got out of the truck his head felt light, and it felt like someone was grabbing his scalp. "Tony, something bad's going to happen." As Tony turned, he saw Nick fall to the ground.

Nick's eyes opened slowly. "Nick, wake up. Can you hear me?" asked Dr. Meyerson.

"Where am I?" Nick said hoarsely.

"You're in the intensive care unit. You've been unconscious for several hours." The doctor shined a light in Nick's eyes. His pupils constricted slowly. "Hmm," she said. She turned his head quickly to the right and then to the left, noting the movement of Nick's eyes. Dr. Meyerson pinched Nick's index finger hard. "Can you feel that?" she asked Nick.

"I can feel you squeezing my finger," he said.

Pulling out her reflex hammer, the doctor tapped Nick's biceps tendon, then his patellar tendon.

"What happened to me, anyway?" asked Nick.

"You passed out. Your blood sugars were sky high. You're lucky your friend got you here fast. We can talk more tomorrow, Nick. Just rest for now," advised Dr. Meyerson.

The doctor stopped by the nurse's station. "How's he doing?" asked the nurse.

"He's got some substantial neuropathy; pupillary reflex is slow, and pain sensation is pretty limited. Were you here when he came in yesterday?"

Yolanda sighed. "Yes, and it wasn't pretty. He was sweating buckets—rapid heart rate, elevated blood pressure, rapid respiration. He responded quickly to treatment. I guess he hasn't been taking care of his diabetes. Blood sugars were really high, and he's been having a lot of gastrointestinal problems—vomiting, constipation, things like that."

D. *Which division of the autonomic nervous system would be affected and would be causing Nick's GI tract symptoms?*

E. *Nick's light-headedness is caused by a condition known as orthostatic hypotension, a rapid drop in blood pressure upon standing up. Based on what you have learned so far, how does the autonomic nervous system control blood pressure?*

F. *After becoming comatose, Nick was sweating profusely, and had rapid heart and respiratory rates and elevated blood pressure. Which area of the brain interacts with the autonomic nervous system during physical stress to initiate these responses?*

G. *Nick has digestive symptoms indicating reduced gastrointestinal mobility. What autonomic receptors regulate closing of sphincters and relaxation of organ walls?*

# 14.8 The sympathetic division supports vigorous physical activity; the parasympathetic division conserves body energy.

As noted earlier, most body organs receive innervation from both divisions of the ANS, which typically work in opposition to one another. The balance between sympathetic and parasympathetic activity, called **autonomic tone**, is regulated by the hypothalamus. Typically, the hypothalamus turns up sympathetic tone at the same time it turns down parasympathetic tone, and vice versa. The two divisions can affect body organs differently because their postganglionic neurons release different neurotransmitters and because the effector organs possess different adrenergic and cholinergic receptors. A few structures receive only sympathetic innervation—sweat glands, arrector pili muscles attached to hair follicles in the skin, the kidneys, the spleen, most blood vessels, and the adrenal medullae (see Figure 14.23). In these structures there is no opposition from the parasympathetic division. Still, an increase in sympathetic tone has one effect, and a decrease in sympathetic tone produces the opposite effect. Although both the sympathetic and parasympathetic divisions are concerned with maintaining health, they do so in dramatically different ways.

## Sympathetic Responses

During physical activity or emotional stress, the sympathetic division dominates the parasympathetic division. High sympathetic tone favors body functions that can support vigorous physical activity and rapid production of ATP. At the same time, the sympathetic division reduces body functions that favor the storage of energy. Besides physical exertion, various emotions—such as fear, embarrassment, or rage—stimulate the sympathetic division. Visualizing body changes that occur during "E situations" such as exercise, emergency, excitement, and embarrassment will help you remember most of the sympathetic responses. Sympathetic activities result in increased alertness and metabolic activities in order to prepare the body for an emergency situation. Activation of the sympathetic division and release of hormones by the adrenal medullae set in motion a series of physiological responses collectively called the **fight-or-flight response**, which includes the following effects:

- Pupils of the eyes dilate.
- Heart rate, force of heart contraction, and blood pressure increase.
- Airways dilate, allowing faster movement of air into and out of the lungs.
- Blood vessels supplying the kidneys and gastrointestinal tract constrict, which decreases blood flow through these tissues. The result is a slowing of urine formation and digestive activities, which are not essential during exercise.
- Blood vessels supplying organs involved in exercise or fighting off danger—skeletal muscles, cardiac muscle, liver, and adipose tissue—dilate, allowing greater blood flow through these tissues.
- The liver increases breakdown of glycogen to glucose, and adipose tissue increases breakdown of triglycerides to fatty acids and glycerol, raising blood levels of these molecules for greater ATP production.
- Processes that are not essential for meeting the stressful situation are inhibited. For example, muscular movements of the gastrointestinal tract and digestive secretions slow down or even stop.

The effects of sympathetic stimulation are longer lasting and more widespread than the effects of parasympathetic stimulation for three reasons:

- Sympathetic postganglionic axons diverge more extensively; many tissues are activated simultaneously.
- Acetylcholinesterase quickly inactivates acetylcholine, but norepinephrine lingers in the synaptic cleft.
- Epinephrine and norepinephrine secreted from the adrenal medulla intensify and prolong the responses caused by NE liberated from sympathetic postganglionic axons.

## Parasympathetic Responses

In contrast to the *fight-or-flight* activities of the sympathetic division, the parasympathetic division enhances *rest-and-digest* activities. Parasympathetic responses support body functions that conserve and restore body energy during times of rest and recovery. In the quiet intervals between periods of exercise, parasympathetic impulses to the digestive glands and the smooth muscle of the gastrointestinal tract predominate over sympathetic impulses. This allows energy-supplying food to be digested and absorbed. At the same time, parasympathetic responses reduce body functions that support physical activity.

The acronym *SLUDD* can be helpful in remembering five parasympathetic responses. It stands for salivation (S), lacrimation (L), urination (U), digestion (D), and defecation (D). All of these activities are stimulated mainly by the parasympathetic division. Besides the increasing SLUDD responses, other important parasympathetic responses are "three decreases": decreased heart rate, decreased diameter of airways, and decreased diameter (constriction) of the pupils.

Table 14.8 compares the structural and functional features of the sympathetic and parasympathetic divisions of the ANS. Table 14.9 lists the responses of glands, cardiac muscle, and smooth muscle to stimulation by the sympathetic and parasympathetic divisions of the ANS.

**TABLE 14.8**

Comparison of Sympathetic and Parasympathetic Divisions of the ANS

| | SYMPATHETIC | PARASYMPATHETIC |
|---|---|---|
| **Origin of preganglionic neuron** | Spinal cord segments T1–L2 | Cranial nerves exiting brain stem (III, VII, IX, and X) and spinal cord segments S2–S4 |
| **Associated ganglia** | Sympathetic trunk ganglia and prevertebral ganglia | Terminal ganglia |
| **Ganglia locations** | Close to CNS and distant from visceral effectors | Typically near or within wall of visceral effectors |
| **Axon length and divergence** | Short preganglionic neuron axons; long postganglionic neuron axons that pass to many visceral effectors | Long preganglionic neuron axons; short postganglionic neuron axons that pass to a single visceral effector |
| **Rami communicantes** | Present | Absent |
| **Neurotransmitters** | Preganglionic neurons release acetylcholine (ACh); most postganglionic neurons release norepinephrine (NE); postganglionic neurons that innervate most sweat glands and some blood vessels in skeletal muscle release ACh | Preganglionic neurons and postganglionic neurons release ACh |
| **Physiological effects** | Fight-or-flight responses to prepare body for intense physical activity | Rest-and-digest responses to conserve and restore energy |

**TABLE 14.9**

Effects of Sympathetic and Parasympathetic Divisions of the ANS

| VISCERAL EFFECTOR | EFFECT OF SYMPATHETIC STIMULATION | EFFECT OF PARASYMPATHETIC STIMULATION |
|---|---|---|
| **Glands** | | |
| **Adipose tissue*** | Release of fatty acids into blood | No known effect |
| **Adrenal medullae** | Secretion of epinephrine and norepinephrine | No known effect |
| **Gastric glands** | Inhibits secretion of gastric juice | Stimulates secretion of gastric juice |
| **Intestinal glands** | Inhibits secretion of intestinal juice | Stimulates secretion of intestinal juice |
| **Kidneys*** | Secretion of renin | No known effect |
| **Lacrimal glands** | No known effect | Stimulates secretion of tears |
| **Liver*** | Glucose secretion; decreased bile secretion | Glycogen synthesis: increased bile secretion |
| **Pancreas** | Inhibits secretion of digestive enzymes and insulin; stimulates secretion of glucagon | Stimulates secretion of digestive enzymes and insulin |
| **Pineal gland** | Increases secretion of melatonin | No known effect |
| **Posterior pituitary** | Secretion of antidiuretic hormone (ADH) | No known effect |
| **Salivary glands** | Inhibits secretion of saliva | Stimulates secretion of saliva |
| **Sweat glands** | Increases sweating | No known effect |
| **Cardiac (heart) muscle** | Increased heart rate and force of contractions | Decreased heart rate and force of contractions |

| | | |
|---|---|---|
| **Smooth Muscle** | | |
| **Ciliary muscle of eye** | Relaxation for distant vision | Contraction for close vision |
| **Gallbladder and ducts** | Storage of bile | Release of bile into small intestine |
| **Hair follicles, arrector pili muscle** | Erection of hairs | No known effect |
| **Iris of eye** | Dilation of pupil | Constriction of pupil |
| **Lungs, bronchial muscle** | Airway dilation | Airway constriction |
| **Sex organs** | Ejaculation of semen (males) | Erection of clitoris (females) and penis (males) |
| **Spleen** | Discharge of stored blood into general circulation | No known effect |
| **Stomach and intestines** | Decreased motility: contraction of sphincters | Increased motility: relaxation of sphincters |
| **Urinary bladder** | Relaxation of muscular wall; contraction of sphincter | Contraction of muscular wall: relaxation of sphincter |
| **Uterus** | Inhibits contraction in nonpregnant woman; promotes contraction in pregnant woman | Minimal effect |
| **Vascular Smooth Muscle** | | |
| **Abdominal viscera arterioles** | Vasoconstriction | No known effect |
| **Coronary (heart) arterioles** | Vasodilation | Minimal effect |
| **Kidney arterioles** | Decreased production of urine | No known effect |
| **Skeletal muscle arterioles** | Vasodilation | No known effect |

*Grouped with glands because they release substances into the blood.

✓ **CHECKPOINT**

**30.** Which part of the brain controls the balance between sympathetic and parasympathetic tone?

**31.** What happens during the fight-or-flight response? Which division of the ANS is responsible?

**32.** What happens during the rest-and-digest response of the parasympathetic division of the ANS?

**33.** Describe the sympathetic response in a frightening situation for each of the following body parts: iris of eye, lungs, urinary bladder, stomach, intestines, heart, arterioles of the abdominal viscera, and arterioles of skeletal muscles.

**34.** Give three examples of the opposite effects of the sympathetic and parasympathetic divisions of the autonomic nervous system.

# 14.9 Autonomic reflexes regulate controlled body conditions and are primarily integrated by the hypothalamus.

## Autonomic Reflexes

**Autonomic reflexes** are responses that occur when impulses pass through an autonomic reflex arc. These reflexes play a key role in regulating controlled conditions in the body such as *blood pressure*, by adjusting heart rate, force of ventricular contraction, and blood vessel diameter; *digestion*, by adjusting the motility and muscle tone of the gastrointestinal tract; and *defecation and urination*, by regulating the opening and closing of sphincters.

The components of an autonomic reflex arc are as follows:

- ***Receptor***. Like the receptor in a somatic reflex arc (see Figure 14.19), the receptor in an autonomic reflex arc is the distal end of a sensory neuron, which responds to a stimulus and produces a change that will ultimately trigger impulses.
- ***Sensory neuron***. Conducts impulses from receptors to the CNS.
- ***Integrating center***. Interneurons within the CNS relay signals from sensory neurons to motor neurons. Integrating centers for most autonomic reflexes are located in the hypothalamus and brain stem. Some autonomic reflexes, such as those for urination and defecation, have integrating centers in the spinal cord.
- ***Motor neurons***. Impulses triggered by the integrating center propagate out of the CNS along motor neurons to an effector. In an autonomic reflex arc, two motor neurons connect

the CNS to an effector: The preganglionic neuron conducts motor impulses from the CNS to an autonomic ganglion, and the postganglionic neuron conducts motor impulses from an autonomic ganglion to an effector (see Figure 14.22b).

- ***Effector***. In an autonomic reflex arc, the effectors are smooth muscle, cardiac muscle, and glands, and the reflex is called an autonomic reflex.

## Autonomic Control by Higher Centers

Normally, we are not aware of muscular contractions of our digestive organs, heartbeat, changes in the diameter of our blood vessels, and pupil dilation and constriction because the integrating centers for these autonomic responses are in the spinal cord or brain stem. Somatic or autonomic sensory neurons deliver input to these centers, and autonomic motor neurons provide output that adjusts activity in the visceral effector, usually without our conscious perception.

The hypothalamus is the major control and integration center of the ANS. The hypothalamus receives sensory input related to visceral functions, smell, and taste, as well as input related to changes in temperature, osmotic pressure, and levels of various substances in blood. It also receives input from the limbic system relating to emotions. Output from the hypothalamus influences autonomic centers in both the brain stem (such as the cardiovascular, salivation, swallowing, and vomiting centers) and in the spinal cord (such as the defecation and urination reflex centers).

Anatomically, the hypothalamus is connected to both the sympathetic and parasympathetic divisions of the ANS by axons that form tracts from the hypothalamus to sympathetic and parasympathetic neurons in the brain stem and spinal cord through relays in the reticular formation. The posterior and lateral parts of the hypothalamus control the sympathetic division. Stimulation of these areas produces an increase in heart rate and force of contraction, a rise in blood pressure due to constriction of blood vessels, an increase in body temperature, dilation of the pupils, and inhibition of the gastrointestinal tract. In contrast, the anterior and medial parts of the hypothalamus control the parasympathetic division. Stimulation of these areas results in a decrease in heart rate, lowering of blood pressure, constriction of the pupils, and increased secretion and motility of the gastrointestinal tract.

### ✓ CHECKPOINT

**35.** Give three examples of controlled conditions in the body that are kept in homeostatic balance by autonomic reflexes.

**36.** How is the hypothalamus anatomically connected to the autonomic nervous system?

## Nick's Story

## EPILOGUE AND DISCUSSION

From Nick's story and the information in this chapter, you have learned that the peripheral nervous system consists of somatic and autonomic divisions, both of which can be affected by disturbances in body homeostasis. In this case, Nick's diabetes mellitus led to a chronic condition of elevated blood glucose (regulation of blood glucose by the endocrine system will be discussed further in Chapter 17). Chronically elevated glucose levels damage peripheral nerves. Small unmyelinated fibers in the distal lower extremities are usually affected first, with a resulting loss of pain and temperature sensation. Large myelinated fibers can also be affected, causing further sensory and proprioceptive sensation loss.

Typically, symptoms like Nick's do not resolve themselves but can be controlled as the blood sugar levels are normalized. In extreme cases of uncontrolled diabetes mellitus, significant somatic and autonomic involvement and even coma can occur. There is no cure for diabetic polyneuropathy, but it is treatable by controlling blood glucose levels and by continued monitoring of the diabetic condition. It is important to rule out other possible causes of peripheral neuropathy before diagnosing diabetic polyneuropathy, since treatment methods may vary.

H. *Why would the term* polyneuropathy *be appropriate for the symptoms Nick was experiencing? (Hint:* Poly *means "many.")*

I. *What symptoms noted by Nick's primary care physician indicated a polyneuropathy?*

J. *Why are Nick's generalized symptoms more indicative of a peripheral polyneuropathy than a central nervous system lesion to the brain or spinal cord?*

K. *Which of Nick's symptoms were related to somatic reflexes? Which were related to autonomic reflexes?*

# Concept and Resource Summary

**Concept** | **Resources**

## Introduction

1. The **peripheral nervous system** consists of nervous tissue outside the brain and spinal cord, and is subdivided into the somatic, autonomic, and enteric nervous systems.
2. The **somatic nervous system** is involved in voluntary skeletal muscle activities.
3. The **autonomic nervous system** regulates the activities of smooth muscle, cardiac muscle, and glands.
4. The **enteric nervous system** regulates GI tract activity.

## Concept 14.1 Nerves have three protective connective tissue coverings.

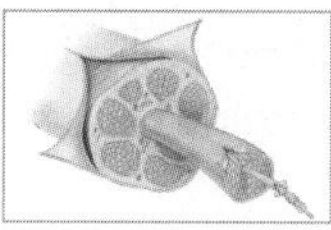

1. A **nerve** is a bundle of axons outside the CNS, along with its associated connective tissues and blood vessels. **Cranial nerves** arise from the brain and **spinal nerves** arise from the spinal cord. Each axon in a nerve is wrapped in **endoneurium**. Groups of axons are then wrapped by **perineurium** into bundles called **fascicles**. The entire nerve is covered with **epineurium**.

Resources: Anatomy Overview—Nerves

## Concept 14.2 Twelve pairs of cranial nerves distribute primarily to regions of the head and neck.

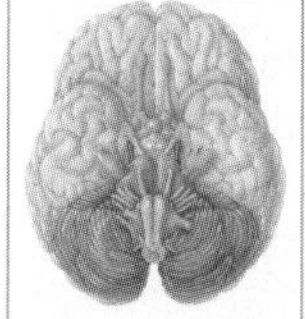

1. The 12 pairs of cranial nerves are numbered from anterior to posterior in order of their emergence from the brain and spinal cord.
2. Cranial nerves are considered **sensory nerves**, **motor nerves**, or **mixed nerves** containing axons of both sensory and motor neurons.
3. The **olfactory (I) nerve** is sensory and functions in the sense of smell.
4. The **optic (II) nerve** is sensory and conducts visual impulses. The paired nerves form the **optic chiasm**, in which visual information may continue to the same side of the brain or cross to the opposite side.
5. The **oculomotor (III) nerve** is a motor nerve stimulating eyeball and eyelid movement, and change in lens shape and pupil size.
6. The **trochlear (IV) nerve** is a motor nerve that stimulates eyeball movement.
7. The **trigeminal (V) nerve** is a mixed nerve that carries sensory impulses from the scalp, face, and mouth for touch, pain, temperature sensations, and proprioception. Motor impulses stimulate chewing movements.
8. The **abducens (VI) nerve** is a motor nerve that stimulates eyeball movement.
9. The **facial (VII) nerve** is a mixed nerve that transmits sensory impulses for taste from taste buds of the tongue and for proprioception from the face and scalp. Motor impulses stimulate facial expression and secretion by lacrimal glands and salivary glands.
10. The **vestibulocochlear (VIII) nerve** is a sensory nerve with two branches serving the inner ear. The **vestibular nerve** conveys impulses for equilibrium; the **cochlear nerve** conveys impulses for hearing.
11. The **glossopharyngeal (IX) nerve** is a mixed nerve that transmits sensory impulses from the tongue for taste and somatic sensations, blood pressure and blood gas levels from sensory receptors, and proprioception from swallowing muscles. Motor impulses stimulate the pharynx for swallowing and speech.
12. The **vagus (X) nerve** is a mixed nerve that transmits sensory impulses for taste from taste buds of the epiglottis and pharynx, proprioception from muscles of the neck and throat, blood pressure and respiratory functions from sensory receptors, and sensations from most thoracic and abdominal organs. Motor impulses stimulate speech, swallowing, smooth muscle contraction of respiratory and abdominal organs, and secretion of digestive fluids.
13. The **accessory (XI) nerve** is a motor nerve that transmits impulses for head movement.
14. The **hypoglossal (XII) nerve** is a motor nerve that transmits impulses for speech and swallowing.

Resources:

Anatomy Overview—Cranial Nerves

Figure 14.2—Origins of Cranial Nerves

Exercise—Cranial Nerve Target Practice—Sensory Functions

Exercise—Cranial Nerve Target Practice—Motor Functions

Clinical Connection—Anosmia

Clinical Connection—Anopia

Clinical Connection—Strabismus, Ptosis, and Diplopia

Clinical Connection—Trigeminal Neuralgia

Clinical Connection—Bell's Palsy

Clinical Connection—Vertigo, Ataxia, and Nystagmus

Clinical Connection—Dysphagia, Aptyalia, and Ageusia

Clinical Connection—Vagal Paralysis, Dysphagia, and Tachycardia

Clinical Connection—Paralysis of the Sternocleidomastoid and Trapezius Muscles

Clinical Connection—Dysarthria and Dysphagia

| Concept | Resources |
|---|---|

## Concept 14.3 Each spinal nerve branches into a posterior ramus, an anterior ramus, a meningeal branch, and rami communicantes.

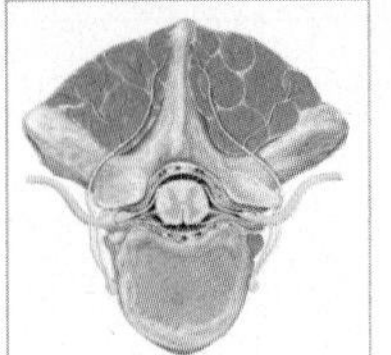

1. The 31 pairs of spinal nerves are numbered according to where they emerge from the vertebral column, beginning with the first cervical pair that emerges between the atlas and the occipital bone. There are eight pairs of cervical nerves, twelve pairs of thoracic nerves, five pairs of lumbar nerves, five pairs of sacral nerves, and one pair of coccygeal nerves.
2. Spinal nerves connect the CNS to sensory receptors, muscles, and glands. Because spinal nerves arise from the union of posterior (sensory) and anterior (motor) roots of the spinal cord, all spinal nerves are mixed nerves.
3. After a spinal nerve passes through its intervertebral foramen, the nerve divides into branches. Branches include the **posterior rami**, serving muscles and skin of the posterior trunk; the **anterior rami**, serving the muscles and structures of limbs and the skin of the lateral and anterior trunk; **meningeal branches**, which supply the vertebrae, spinal cord, and meninges; and the rami communicantes of the autonomic nervous system.
4. Except for T2–T12, anterior rami do not go directly to body structures, but instead form **plexuses** with axons of adjacent ventral rami. Nerves emerge from each plexus.
5. The **cervical plexus** serves the skin and muscles of the head, neck, and superior shoulders and chest. The **brachial plexus** innervates the shoulders and upper limbs. The **lumbar plexus** innervates the anterior and lateral abdominal wall, external genitals, and part of the lower limbs. The **sacral plexus** supplies the buttocks, perineum, and lower limbs. The coccygeal plexus supplies the coccygeal region.
6. Anterior rami of T2–T12 do not form plexuses and are called the **intercostal nerves**. Intercostal nerves innervate intercostal muscles and muscles and skin of the thorax and abdominal wall.

**Resources:**

Anatomy Overview—Spinal Nerves

Figure 14.13—External Anatomy of the Spinal Cord and Spinal Nerves

Clinical Connection—Injuries to the Phrenic Nerves

Clinical Connection—Injuries to Nerves Emerging from the Brachial Plexus

Clinical Connection—Injuries to the Lumbar Plexus

Clinical Connection—Injury to the Sciatic Nerve

Clinical Connection—Spinal Cord Compression

## Concept 14.4 A reflex is produced by a reflex arc in response to a particular stimulus.

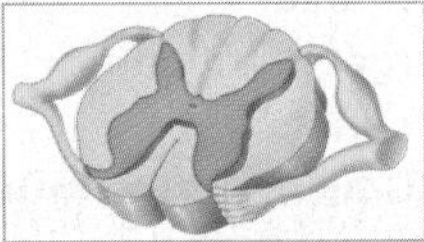

1. A **reflex** is a fast, involuntary response to a stimulus.
2. Integrating centers for reflexive information are the gray matter of the brain stem in a **cranial reflex** and spinal cord gray matter in a **spinal reflex**.
3. **Somatic reflexes** involve motor responses by skeletal muscles; **autonomic reflexes** involve smooth muscle, cardiac muscle, and glands as the effectors.
4. A **reflex arc** has five functional components in the sequential pathway of impulse propagation in response to a **stimulus**: (1) **sensory receptor**, (2) **sensory neuron**, (3) **integrating center**, (4) **motor neuron**, and (5) **effector**.
5. A **monosynaptic reflex arc** has one synapse in the CNS, and a **polysynaptic reflex arc** has more than one CNS synapse.
6. A **stretch reflex** begins when stretching of a muscle stimulates **muscle spindles**, and the monosynaptic reflex response is contraction of the stretched muscle. There is simultaneous relaxation of antagonist muscles by a polysynaptic reflex arc governed by **reciprocal innervation**. Stretch reflexes function in maintaining muscle tone and posture.
7. The stretch reflex is an example of an **ipsilateral reflex** because sensory impulses enter the spinal cord on the same side as motor impulses leave it.
8. The **flexor** or **withdrawal reflex** is a polysynaptic reflex arc initiated by a painful stimulus, in which the response of flexor leg muscle contraction causes withdrawal of the leg. The response of several muscle groups through activation of interneurons in several segments of the spinal cord is called an **intersegmental reflex arc**.

**Resources:**

Animation—Reflex Arcs

Animation—Reflexes

Figure 14.19—General Components of a Reflex Arc

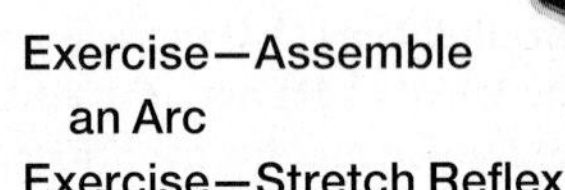

Exercise—Assemble an Arc

Exercise—Stretch Reflex

| Concept | Resources |
|---|---|

## Concept 14.5 The autonomic nervous system produces involuntary movements.

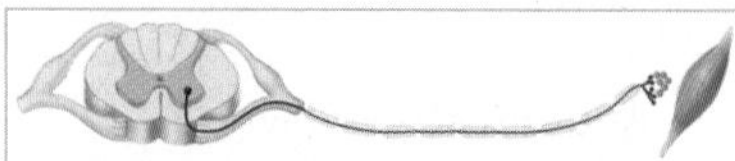

1. The somatic nervous system is under conscious control. It has sensory neurons conveying input from receptors for **special senses** and **somatic senses**. Somatic motor neurons innervate skeletal muscle effectors, which always results in excitation for muscle contraction.
2. The **autonomic nervous system (ANS)** involves **autonomic sensory neurons** conveying input from **interoceptors** in viscera and blood vessels. **Autonomic motor neurons** carry motor impulses to cardiac muscle, smooth muscle, and glands.
3. The ANS usually operates without conscious control. The hypothalamus and brain stem regulate ANS ac tivity.
4. The ANS has two principal branches: the **sympathetic division** and the **parasympathetic division**.
5. The somatic nervous system has one motor neuron from the CNS to the skeletal muscle. The sympathetic and parasympathetic ANS pathways have two motor neurons in a chain. The first motor neuron has its cell body in the CNS, and its myelinated axon extends from the CNS to an autonomic ganglion that contains the cell body of the second motor neuron. The unmyelinated axon of the second motor neuron extends to an effector.
6. All somatic motor neurons release acetylcholine; autonomic motor neurons release acetylcholine or norepinephrine.
7. Most organs receive **dual innervation** from the sympathetic and parasympathetic divisions of the ANS. Generally, these two divisions transmit opposing signals to the organ that they innervate.

**Resources:**

Anatomy Overview—Visceral Receptors

Anatomy Overview—Visceral Effectors

Figure 14.22—Motor Neuron Pathways in the Somatic Nervous System and Autonomic Nervous System

## Concept 14.6 The ANS includes preganglionic neurons, autonomic ganglia and plexuses, and postganglionic neurons.

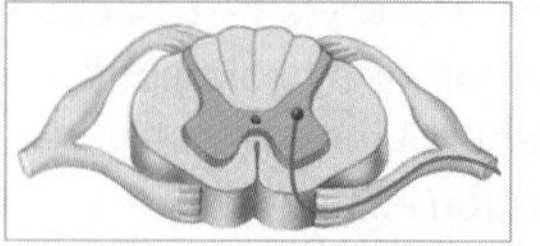

1. The ANS pathway involves a preganglionic neuron with its cell body in the CNS and a myelinated axon extending to the autonomic ganglion, where it synapses with the postganglionic neuron. The cell body and dendrites of this second motor neuron are in the autonomic ganglion, and its unmyelinated axon extends to a visceral effector.
2. In the sympathetic division, cell bodies of preganglionic neurons are in thoracic and lumbar spinal cord segments; in the parasympathetic division, preganglionic neuron cell bodies are in the brain stem and sacral spinal cord segments.
3. **Sympathetic trunk ganglia** extend on each side of the vertebral column, and their postganglionic axons typically innervate organs above the diaphragm. **Prevertebral ganglia** lie anterior to the vertebral column, and their postganglionic axons typically innervate organs below the diaphragm.
4. Preganglionic axons of the parasympathetic division extend from the CNS to synapse with postganglionic neurons in **terminal ganglia** located close to visceral organs.
5. In the thorax, abdomen, and pelvis, sympathetic and parasympathetic neurons form **autonomic plexuses** adjacent to major arteries. These plexuses may also contain sympathetic ganglia.
6. Sympathetic preganglionic axons leave the spinal cord, travel through **rami communicantes**, enter sympathetic trunk ganglia, and connect to postganglionic neurons; the postganglionic neurons then continue in spinal nerves to effectors.
7. Parasympathetic preganglionic axons leave the brain stem and spinal cord, enter terminal ganglia located near or in visceral organs, and connect to postganglionic neurons supplying local effectors. Sacral parasympathetic preganglionic neurons travel through **pelvic splanchnic nerves** to reach terminal ganglia.
8. Enteric nerves and ganglia control gastrointestinal tract motility and secretory activities. The **myenteric** and **submucous plexuses** innervate most of the digestive organs.

**Resources:**

Anatomy Overview—Organization of the ANS

Animation—ANS: Motor Pathways

Figure 14.23—Structure of the Sympathetic Division of the Autonomic Nervous System

Figure 14.24—Structure of the Parasympathetic Division of the Autonomic Nervous System

Exercise—Assemble the Structures of the ANS

Clinical Connection–Horner's Syndrome

| Concept | Resources |
|---|---|

## Concept 14.7 ANS neurons release acetylcholine or norepinephrine, resulting in excitation or inhibition.

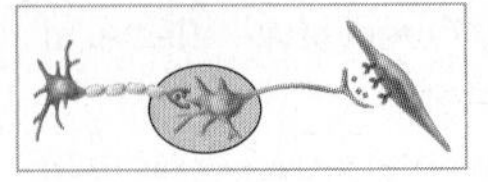

1. In an autonomic pathway, neurotransmitters are released at synapses between preganglionic and postganglionic neurons, and between postganglionic neurons and effectors.
2. **Cholinergic neurons** release **acetylcholine (ACh)**. Cholinergic neurons include all sympathetic and parasympathetic preganglionic neurons, sympathetic postganglionic neurons that innervate sweat glands, and all parasympathetic postganglionic neurons.
3. ACh binds to **cholinergic receptors** on postsynaptic plasma membranes. There are two types of cholinergic receptors: **nicotinic receptors** on dendrites and cell bodies of sympathetic and parasympathetic postganglionic neurons, and in the motor end plate at the neuromuscular junction; and **muscarinic receptors** on all effectors innervated by parasympathetic postganglionic axons, and most sweat glands.
4. ANS **adrenergic neurons** release **norepinephrine (NE)**. Most sympathetic postganglionic neurons are adrenergic.
5. Adrenergic receptors bind both norepinephrine, released by sympathetic postganglionic neurons, and epinephrine, an adrenal hormone. The two adrenergic receptors are **alpha (α) receptors** and **beta (β) receptors** found on visceral effectors.

**Resources:**
- Anatomy Overview—Neurotransmitters
- Animation—The ANS: Types of Neurotransmitters and Neurons
- Figure 14.28—Cholinergic Neurons and Adrenergic Neurons in the Sympathetic and Parasympathetic Divisions

## Concept 14.8 The sympathetic division supports vigorous physical activity; the parasympathetic division conserves body energy.

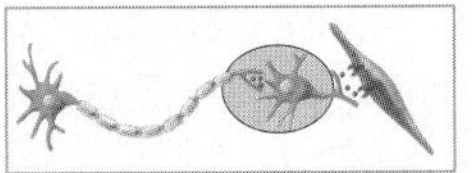

1. The balance between sympathetic and parasympathetic activity is regulated by the hypothalamus.
2. Although most body structures have dual innervation by the ANS, some structures receive only sympathetic innervation. A decrease in sympathetic tone produces an opposing effect on these structures.
3. High sympathetic tone supports vigorous physical activity, rapid ATP production, and decreased energy storage. Activation of the sympathetic division gives rise to a series of physiological responses called the **fight-or-flight response**.
4. Parasympathetic responses support rest-and-digest activities such as conservation and restoration of body energy during rest and recovery periods, and increased gastrointestinal tract activity. The acronym SLUDD represents parasympathetic responses: salivation, lacrimation, urination, digestion, and defecation. Parasympathetic responses also include decreased heart rate, decreased diameter of airways, and decreased diameter of pupils.

**Resources:**
- Anatomy Overview—Effectors
- Animation—Physiological Effects of the ANS
- Animation—The Alarm Reaction
- Exercise—Sort ANS Functions
- Exercise—What Is Your ANS Status?
- Clinical Connection—Reflex Sympathetic Dysreflexia

## Concept 14.9 Autonomic reflexes regulate controlled body conditions and are primarily integrated by the hypothalamus.

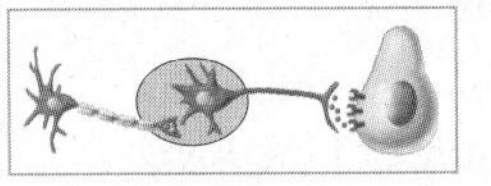

1. **Autonomic reflexes** regulate controlled body conditions including blood pressure, digestion, defecation, and urination.
2. The components of a typical autonomic reflex arc are as follows: **receptor**, **sensory neuron**, **integrating center** in the hypothalamus and brain stem, **motor neurons** arranged in a two-motor-neuron chain of the ANS, and a visceral **effector** such as smooth muscle, cardiac muscle, and glands.
3. The hypothalamus is the major control and integration center of the ANS. It is connected to both the sympathetic and the parasympathetic divisions by tracts that relay through the reticular formation of the brain stem.

**Resources:**
- Anatomy Overview—The ANS Control Centers
- Concepts and Connections—Nervous System Negative Feedback Loop
- Clinical Connection—Autonomic Dysreflexia

## Understanding the Concepts

1. Which part of a nerve does each connective tissue covering surround?
2. How does the accessory nerve differ from the other cranial nerves?
3. What important loss of function could reveal damage to each cranial nerve?
4. Why could severing of the spinal cord in the cervical region cause respiratory arrest?
5. Injury of which nerve could cause paralysis of the biceps brachii muscle? The triceps brachii muscle? From which plexus do they branch?
6. Injury to which nerve could cause pain that extends down the anterior lower limb? Down the posterior lower limb?
7. How are the intercostal nerves different from other spinal nerves?
8. Which branch of the nervous system includes all integrating centers for reflexes?
9. How is the response of ANS effectors different from the response of somatic nervous system effectors?
10. Why can the sympathetic division produce simultaneous effects throughout the body, but parasympathetic effects typically are localized to specific organs?
11. What types of effector tissues contain muscarinic receptors?
12. Why is the parasympathetic division of the ANS sometimes referred to as an energy conservation or restoration system?
13. How does an autonomic reflex arc differ from a somatic reflex arc?

CHAPTER

# 15

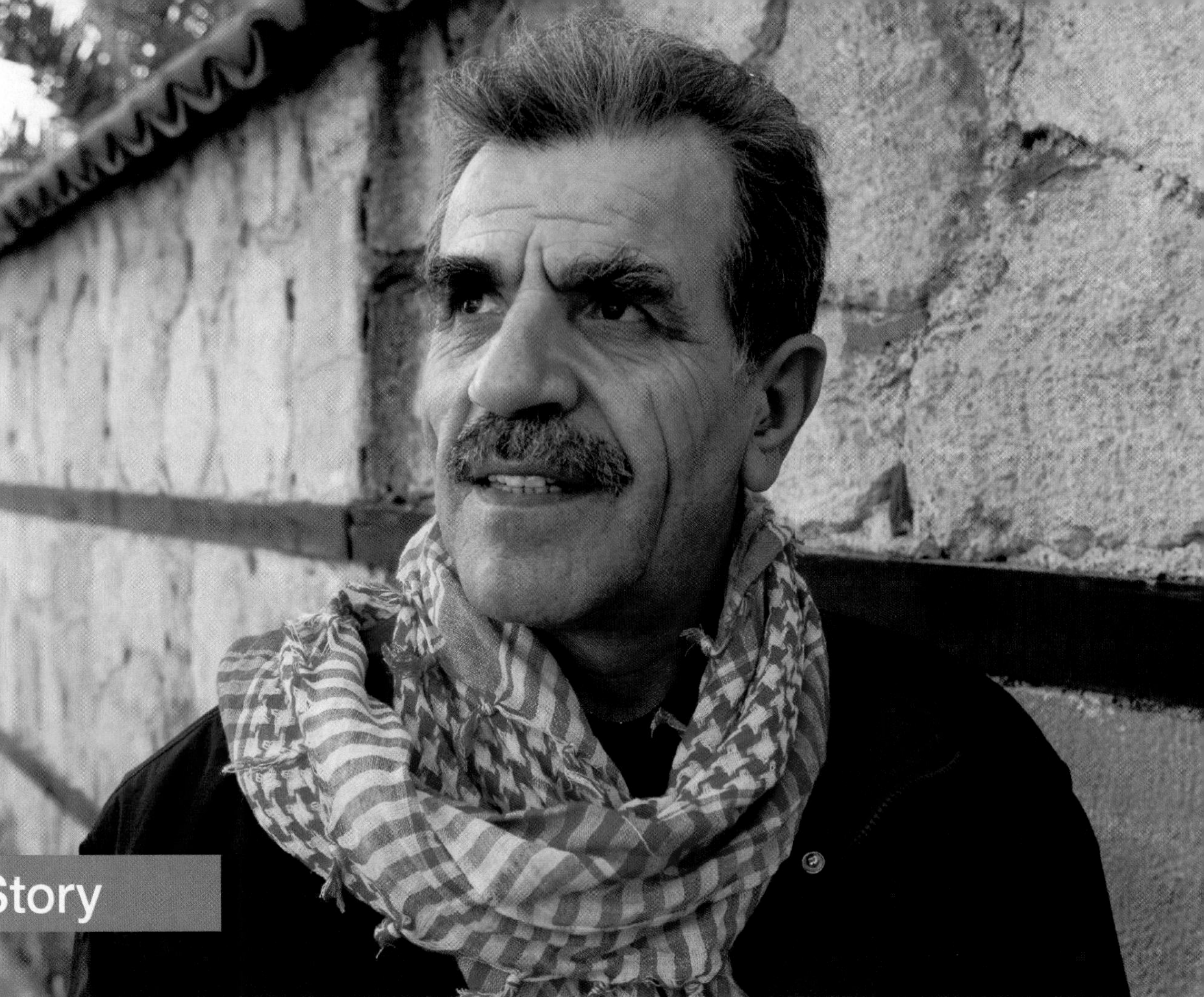

## Mustafa's Story

Mustafa strolls around his village, greeting his neighbors and the shopkeepers as they reposition their wares to attract customers.

"I have a new grandson," announces Ahmed.

"Congratulations! May he have a long life!" respond all those within hearing distance.

A group of merchants is discussing the latest political news of the day and wondering about the effects on their businesses. This is Mustafa's favorite time of day—a time to sustain friendships and learn the village news to discuss with his family during dinner.

Mustafa returns home to find that dinner has been prepared and is being brought to the patio. The pleasant smell of jasmine wafts over the wall from the public garden next door. Mustafa's wife, Khalia, places the platter of roast lamb on the table while his daughters and daughters-in-law follow with the rice, vegetables, sauces, and bread. A large pitcher of lemonade is carried out and glasses are filled. Everything looks and smells delicious. Soon the entire family is on the patio enjoying the food, one another's company, and the bits of news that Mustafa has brought home.

The evening meal in Mustafa's home is a time to relax after a busy day. When the meal is finished, the children run and play in the yard, hoping their mothers will linger at the table to talk, delaying their bedtime.

Mustafa starts to tell his family what he heard at the village. "Today, I ran into Ahmed and he told me that he has a. . . ." Suddenly Mustafa's words become garbled and his eyes widen in fright as he slumps to one side in his chair. Khalia rushes to him and says, "Mustafa, what is wrong? How can I help you?"

Two of his sons try to help him stand up, but Mustafa cannot walk. His right side will not support his weight and is flaccid. He becomes combative, trying to fight off his sons, but he can only use his left arm and leg. The entire right side of his body drags. His sons pick him up, carry him into his bedroom, and place him gently on the bed. One son loosens Mustafa's clothing, while the other calls the village ambulance to come and take him to the hospital.

What is happening to Mustafa? Will he get better? Read this chapter to see how disruptions in the central nervous system such as the one Mustafa has just experienced can affect the entire body.

# Sensory, Motor, and Integrative Systems

## INTRODUCTION

In the three previous chapters we have explored how the nervous system is organized and how it processes information and responds to the environment. In order for such responses to occur, the body must be able to sense the internal and external environment and convey these sensations to the appropriate processing areas of the central nervous system (CNS). As sensory impulses reach the CNS, they become part of a large pool of sensory input. However, not every bit of input to the CNS elicits a response. Rather, each piece of incoming information is combined with other arriving and previously stored information in a process called *integration*. Integration occurs at many sites along pathways in the CNS, such as the spinal cord, brain stem, cerebellum, basal nuclei, and cerebral cortex. The motor responses that govern muscle contraction are modified at several of these CNS levels to produce appropriate motor responses to sensory input.

Disruption of any of the sensory, motor, or integrative structures or pathways can cause significant disturbances in homeostasis. By learning about the functional role of each of these components, it will be easier to understand the disease processes associated with them.

In this chapter we learn more about how information is sensed, the pathways along which information travels, and how sensory information is perceived, modified, and integrated to produce appropriate motor responses. We also introduce two complex integrative functions of the brain: (1) wakefulness and sleep and (2) learning and memory.

## CONCEPTS

**15.1** Sensations arise as a result of stimulation, transduction, generation, and integration.

**15.2** Sensory receptors can be classified structurally, functionally, or by the type of stimulus detected.

**15.3** Somatic sensations include tactile sensations, thermal sensations, pain, and proprioception.

**15.4** Somatic sensory pathways relay information from sensory receptors to the cerebral cortex and cerebellum.

**15.5** The somatosensory and primary motor areas of the cerebral cortex unequally serve different body regions.

**15.6** Somatic motor pathways carry impulses from the brain to effectors.

**15.7** Wakefulness and memory are integrative functions of the brain.

# 15.1 Sensations arise as a result of stimulation, transduction, generation, and integration.

In its broadest definition, **sensation** is the conscious or subconscious awareness of changes in the external or internal environment. The nature of the sensation and the type of reaction generated vary according to the ultimate destination of impulses that convey sensory information to the CNS. Sensory impulses relayed to the spinal cord may serve as input for spinal reflexes, such as the stretch reflex you learned about in Chapter 14. Sensory impulses that reach the lower brain stem elicit more complex reflexes, such as changes in heart rate or breathing rate. When sensory impulses reach the cerebral cortex, we become consciously aware of the sensory stimuli and can precisely locate and identify specific sensations such as touch, pain, hearing, or taste.

**Perception**, the conscious awareness and interpretation of sensations, is primarily a function of the cerebral cortex. We have no perception of some sensory information because it never reaches the cerebral cortex. For example, certain sensory receptors constantly monitor the pressure of blood in blood vessels. Because the impulses conveying blood pressure information propagate to the cardiovascular center in the medulla oblongata rather than to the cerebral cortex, blood pressure is not consciously perceived.

## Sensory Modalities

Each unique type of sensation—such as touch, pain, vision, or hearing—is called a **sensory modality** (mō-DAL-i-tē). A given sensory neuron carries information for only one sensory modality. Neurons relaying impulses for touch to the somatosensory area of the cerebral cortex do not transmit impulses for pain. Likewise, impulses from the eyes are perceived as sight, and those from the ears are perceived as sounds.

The different sensory modalities can be grouped into two classes: general senses and special senses.

- The **general senses** refer to both somatic senses and visceral senses. **Somatic senses** (*somat-* = of the body) include tactile sensations (touch, pressure, vibration, itch, and tickle), thermal sensations (warm and cold), pain sensations, and proprioceptive sensations. Proprioceptive sensations allow perception of both the static (nonmoving) positions and movements of our head and limbs. **Visceral senses** provide information about conditions within internal organs.
- The **special senses** include the sensory modalities of smell, taste, vision, hearing, and equilibrium or balance.

In this chapter we discuss the somatic senses and visceral pain. The special senses will be the focus of Chapter 16. Visceral senses were introduced in Chapter 14 and will be discussed in association with individual organs in later chapters.

## The Process of Sensation

The process of sensation begins in a **sensory receptor**, which can be either a specialized cell or the dendrites of a sensory neuron. A given sensory receptor responds to only one particular kind of **stimulus**, a change in the environment that can activate certain sensory receptors. That sensory receptor responds only weakly or not at all to other stimuli. This characteristic of sensory receptors is known as **selectivity**.

For a sensation to arise, the following four events typically occur:

1. ***Stimulation of the sensory receptor.*** An appropriate stimulus must occur within the sensory receptor's *receptive field*, that is, the body region where stimulation activates the receptor and produces a response.
2. ***Transduction of the stimulus.*** A sensory receptor *transduces* (converts) energy from a stimulus into a graded potential. For example, air molecules that we can smell contain chemical energy that is transduced by olfactory (smell) receptors in our nose into electrical energy in the form of a graded potential. Recall that graded potentials vary in amplitude (size), depending on the strength of the stimulus that causes them, and are not propagated. (See Concept 12.8 to review the differences between graded potentials and action potentials.) Each type of sensory receptor exhibits selectivity; it can transduce only one kind of stimulus.
3. ***Generation of impulses.*** When a graded potential in a sensory neuron reaches threshold, it triggers one or more impulses, which then propagate toward the CNS. Sensory neurons that conduct impulses from the peripheral nervous system (PNS) into the CNS are called **first-order neurons**.
4. ***Integration of sensory input.*** A particular region of the CNS receives and integrates the sensory impulses. Conscious sensations or perceptions are integrated in the cerebral cortex. You seem to see with your eyes, hear with your ears, and feel pain in an injured part of your body because sensory impulses from each part of the body arrive in a specific region of the cerebral cortex, which interprets the sensation as coming from the stimulated sensory receptors.

### ✓ CHECKPOINT

1. What is a sensory modality?
2. Distinguish between general senses and special senses.
3. What is receptor selectivity?
4. What four events must happen for a sensation to occur?
5. What is transduction?

# 15.2 Sensory receptors can be classified structurally, functionally, or by the type of stimulus detected.

## Types of Sensory Receptors

Sensory receptors can be grouped into different classes by structure, location of the receptor or origin of the stimulus, or by the type of stimulus detected.

### Microscopic Structure

Structurally, sensory receptors may be free nerve endings or encapsulated nerve endings of first-order sensory neurons, or separate cells that synapse with first-order sensory neurons.

- **Free nerve endings** are bare dendrites; they lack any structural specializations that can be seen under a light microscope (Figure 15.1a). Receptors for pain, temperature, tickle, itch, and some touch sensations are free nerve endings.
- **Encapsulated nerve endings** are dendrites enclosed in a connective tissue capsule (Figure 15.1b). Receptors for pressure, vibration, and some touch sensations are encapsulated nerve endings.
- **Separate cells** may synapse with sensory neurons (Figure 15.1c). Sensory receptors for some special senses are specialized, separate cells. These include *hair cells* for hearing and equilibrium

FIGURE 15.1 **Types of sensory receptors and their relationship to first-order sensory neurons.** (a) Free nerve endings, in this case a cold-sensitive receptor, are bare dendrites of first-order neurons. (b) An encapsulated nerve ending, in this case a pressure-sensitive receptor, are enclosed dendrites of first-order neurons. (c) A separate receptor cell—here, a gustatory (taste) receptor—synapses with a first-order neuron.

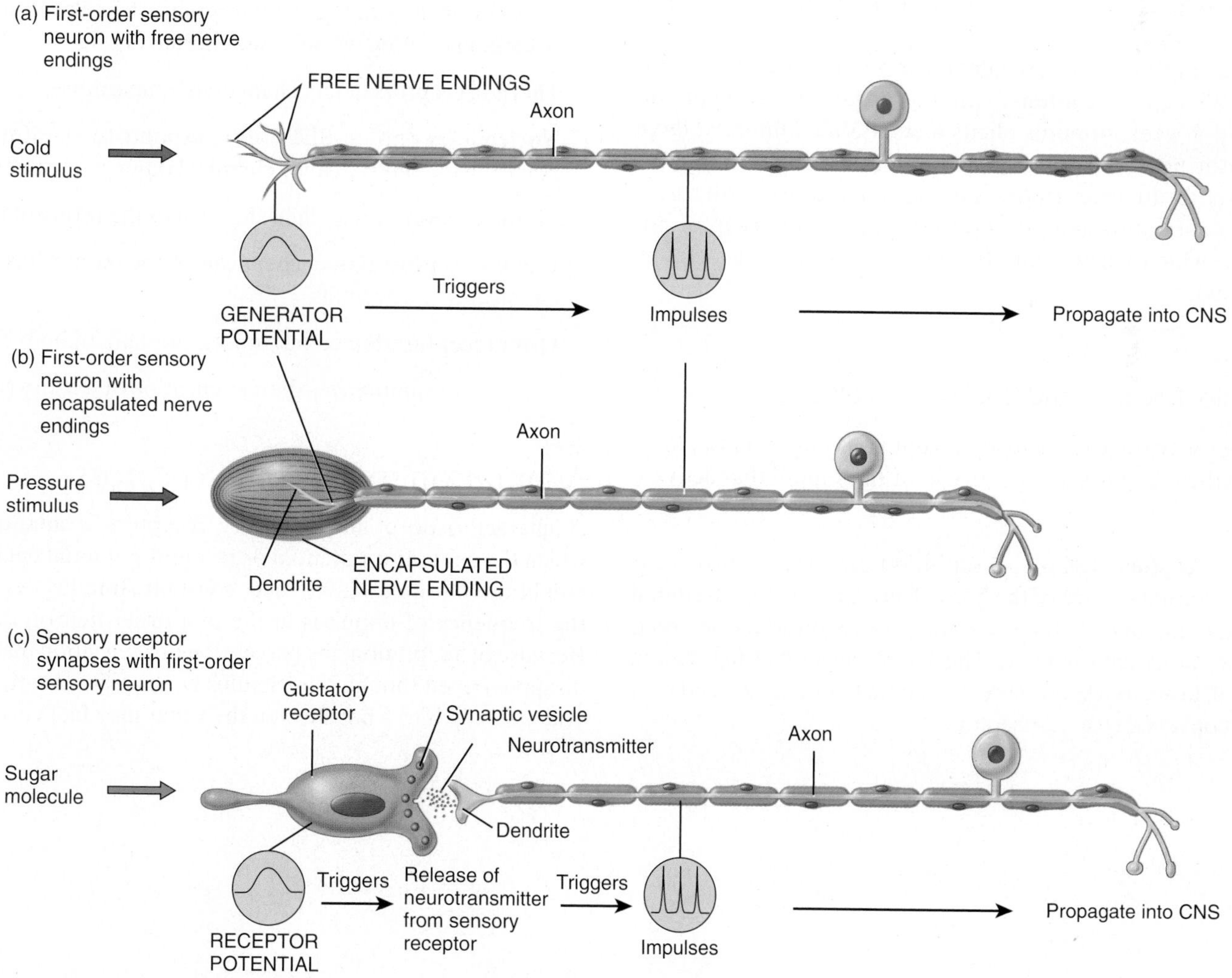

**Free and encapsulated nerve endings trigger impulses in the same first-order neurons. Separate sensory receptors release a neurotransmitter that triggers impulses in a first-order neuron.**

in the inner ear, *gustatory receptor cells* in taste buds, and *photoreceptors* in the retina of the eye for vision. You will learn more about separate receptor cells for the special senses in Chapter 16.

Sensory receptors produce two kinds of graded potentials—generator potentials and receptor potentials—in response to a stimulus. When stimulated, the dendrites of free nerve endings and encapsulated nerve endings produce **generator potentials** (Figure 15.1a, b). When a generator potential is large enough to reach threshold, it triggers one or more action potentials in the axon of a first-order sensory neuron. The resulting impulse propagates along the axon into the CNS. Thus, generator potentials generate action potentials.

By contrast, sensory receptors that are separate cells produce graded potentials termed **receptor potentials**. Receptor potentials trigger release of neurotransmitter from the sensory receptor (Figure 15.1c). The neurotransmitter molecules diffuse across the synaptic cleft and produce a postsynaptic potential in the first-order neuron. The postsynaptic potential may trigger one or more action potentials, which propagate as impulses along the axon of the first-order neuron into the CNS.

The amplitude of both generator potentials and receptor potentials varies; an intense stimulus produces a large potential, and a weak stimulus elicits a small one. Similarly, large generator potentials or receptor potentials trigger impulses at higher frequencies (more impulses) in the first-order neuron, in contrast to small generator potentials or receptor potentials, which trigger impulses at lower frequencies (fewer impulses).

### Receptor Location and Origin of Stimulus

Another way to group sensory receptors is based on the location of the receptors and the origin of the stimuli that activate them.

- **Exteroceptors** (EKS-ter-ō-sep′-tors) are located at or near the external surface of the body. They are sensitive to stimuli originating outside the body and provide information about the *external* environment. The sensations of hearing, vision, smell, taste, touch, pressure, vibration, temperature, and pain are conveyed by exteroceptors.
- **Interoceptors** (IN-ter-ō-sep′-tors) are located in blood vessels, visceral organs, muscles, and the nervous system. Interoceptors monitor conditions in the *internal* environment. The impulses produced by interoceptors usually are not consciously perceived.
- **Proprioceptors** (PRŌ-prē-ō-sep′-tors; *proprio-* = one's own) are located in muscles, tendons, joints, and the inner ear. They provide information about body position, muscle length and tension, and the position and movement of your joints.

### Type of Stimulus Detected

A third way to group sensory receptors is according to the type of stimulus they detect. Most stimuli are in the form of mechanical energy, such as sound waves or pressure changes; electromagnetic energy, such as light or heat; or chemical energy, such as in a molecule of glucose.

- **Mechanoreceptors** are sensitive to mechanical stimuli such as the deformation, stretching, or bending of cells. Mechanoreceptors provide sensations of touch, pressure, vibration, proprioception, hearing, and equilibrium. They also monitor the stretching of blood vessels and internal organs.
- **Thermoreceptors** detect changes in temperature.
- **Nociceptors** (nō′-si-SEP-tors) respond to painful stimuli resulting from physical or chemical damage to tissue.
- **Photoreceptors** detect light that strikes the retina of the eye.
- **Chemoreceptors** detect chemicals in the mouth (taste), nose (smell), and body fluids.
- **Osmoreceptors** detect the osmotic pressure of body fluids.

Table 15.1 summarizes the classification of sensory receptors.

## Adaptation in Sensory Receptors

A characteristic of most sensory receptors is **adaptation**, in which the generator potential or receptor potential decreases in amplitude during a maintained, constant stimulus. As a result, the frequency of impulses in the first-order neuron decreases. Because of adaptation, the perception of a sensation may fade or disappear even though the stimulus persists. For example, when you first step into a hot shower, the water may feel very hot, but

## TABLE 15.1

Classification of Sensory Receptors

| BASIS OF CLASSIFICATION | DESCRIPTION |
|---|---|
| **MICROSCOPIC STRUCTURE** | |
| Free nerve endings | Bare dendrites associated with pain, thermal, tickle, itch, and some touch sensations |
| Encapsulated nerve endings | Dendrites enclosed in connective tissue capsule for pressure, vibration, and some touch sensations |
| Separate cells | Receptor cells synapse with first-order sensory neurons; located in retina of eye (photoreceptors), inner ear (hair cells), and taste buds of tongue (gustatory receptor cells) |
| **RECEPTOR LOCATION AND ACTIVATING STIMULI** | |
| Exteroceptors | Located at or near body surface; sensitive to stimuli originating outside body; provide information about external environment; convey visual, smell, taste, touch, pressure, vibration, thermal, and pain sensations |
| Interoceptors | Located in blood vessels, visceral organs, and nervous system; provide information about internal environment; impulses usually are not consciously perceived but occasionally may be felt as pain or pressure |
| Proprioceptors | Located in muscles, tendons, joints, and inner ear; provide information about body position, muscle length and tension, position and motion of joints, and equilibrium (balance) |
| **TYPE OF STIMULUS DETECTED** | |
| Mechanoreceptors | Detect mechanical stimuli; provide sensations of touch, pressure, vibration, proprioception, and hearing and equilibrium; also monitor stretching of blood vessels and internal organs |
| Thermoreceptors | Detect changes in temperature |
| Nociceptors | Respond to painful stimuli resulting from physical or chemical damage to tissue |
| Photoreceptors | Detect light that strikes the retina of the eye |
| Chemoreceptors | Detect chemicals in mouth (taste), nose (smell), and body fluids |
| Osmoreceptors | Sense osmotic pressure of body fluids |

soon the sensation decreases to one of comfortable warmth even though the stimulus (the high temperature of the water) does not change.

Receptors vary in how quickly they adapt. **Rapidly adapting receptors** adapt very quickly. They are specialized for signaling *changes* in a stimulus. Receptors associated with pressure, touch, and smell are rapidly adapting. **Slowly adapting receptors**, by contrast, adapt slowly and continue to trigger impulses as long as the stimulus persists. Slowly adapting receptors monitor stimuli associated with pain, body position, and chemical composition of the blood.

### ✓ CHECKPOINT

6. How do sensory receptors differ structurally?
7. What is adaptation? How does it occur?
8. What is the difference between rapidly adapting and slowly adapting receptors?

## RETURN TO Mustafa's Story

The ambulance transports Mustafa to the local hospital. His sons and Khalia follow in their car. At the hospital Mustafa is taken immediately to an examining room where the nurse starts an intravenous line and inserts a catheter into his bladder. Dr. Nasri, noting the flaccidity of Mustafa's right side, recognizes that Mustafa has experienced a cerebrovascular accident, more commonly known as a stroke or "brain attack." Dr. Nasri uses his ophthalmoscope to look into Mustafa's eyes. He says to Mustafa, "Open your mouth and stick out your tongue." Mustafa does not follow this command, so Dr. Nasri starts to check the reflexes in his upper and lower limbs. There is no response when Dr. Nasri taps the right upper and lower limbs with his reflex hammer, but the left upper and lower limbs both jerk when the hammer is applied. Dr. Nasri orders a computerized tomography (CT) scan of Mustafa's head to find out where the brain attack has occurred and if it is from a blood clot in a vessel traveling to his head or from a ruptured artery in his head. While waiting for the CT, Dr. Nasri takes a safety pin from his pocket and gently touches Mustafa's upper and lower limbs with it. Mustafa has no response when his right limbs are touched, but he becomes combative again and vocalizes sounds that cannot be understood when his left limbs are touched. Dr. Nasri takes an ice cube and holds it to Mustafa's right side and left side. Again, on the right there is no response, but Mustafa withdraws his left upper limb quickly from the cold, wet ice cube.

Dr. Nasri speaks to Mustafa's family. "Mustafa has had a stroke. Since you were able to get him here so quickly, I have ordered a CT scan to see if he is a candidate to receive thrombolytic therapy. If the cause of his stroke is a blood clot, I will administer a drug into Mustafa's vein. It will travel to the brain and partially dissolve the clot. This therapy will improve his chances for a full recovery."

A. *Mustafa has no sensation on the right side of his body. Based on microscopic structure, which type of sensory receptor is responsible for transmitting the sensation of cold from the ice cube?*

B. *Mustafa is unable to balance himself. Based on receptor location, which type of sensory receptor is responsible for providing information about his body position?*

C. *Mustafa's right side lacks sensations when Dr. Nasri pricked him with the pin or touched him with ice. Based on receptor location, which type of sensory receptor is not functioning?*

D. *Mustafa's symptoms can also be categorized according to the type of stimuli that the receptors detect. Which of the following receptor types can be correlated with the symptoms that Mustafa is experiencing? Be specific in your answers.*

*Chemoreceptors* *Mechanoreceptors*
*Nociceptors* *Osmoreceptors*
*Photoreceptors* *Thermoreceptors*

# 15.3 Somatic sensations include tactile sensations, thermal sensations, pain, and proprioception.

Somatic sensations arise from stimulation of sensory receptors embedded in the skin or hypodermis; in mucous membranes of the mouth, vagina, and anus; in muscles, tendons, and joints; and in the inner ear. The sensory receptors for somatic sensations are distributed unevenly—some parts of the body surface are densely populated with receptors, and others contain only a few. The areas with the highest density of somatic sensory receptors are the tip of the tongue, the lips, and the fingertips. Somatic sensations that arise from stimulating the skin surface are **cutaneous sensations** (kū-TĀ-nē-us; *cutane-* = skin). There are four modalities of somatic sensations: tactile, thermal, pain, and proprioceptive.

## Tactile Sensations

The **tactile sensations** (TAK-tīl; *tact-* = touch) include touch, pressure, vibration, itch, and tickle. Encapsulated mechanoreceptors mediate sensations of touch, pressure, and vibration. Other tactile sensations, such as itch and tickle sensations, are detected by free nerve endings. Tactile receptors in the skin or hypodermis include corpuscles of touch, hair root plexuses, tactile discs, Ruffini corpuscles, lamellated corpuscles, and free nerve endings (Figure 15.2).

### Touch

Sensations of **touch** generally result from stimulation of tactile receptors in the skin or hypodermis. There are two types of rapidly adapting touch receptors. **Corpuscles of touch** or *Meissner corpuscles* (MĪS-ner) are touch receptors located in the dermal papillae of hairless skin. Each corpuscle is an egg-shaped mass of dendrites enclosed by a capsule of connective tissue. Because corpuscles of touch are rapidly adapting receptors, they generate impulses mainly at the onset of a touch. They are abundant in the fingertips, hands, eyelids, tip of the tongue, lips, nipples,

FIGURE 15.2 **Structure and location of sensory receptors in the skin and hypodermis.**

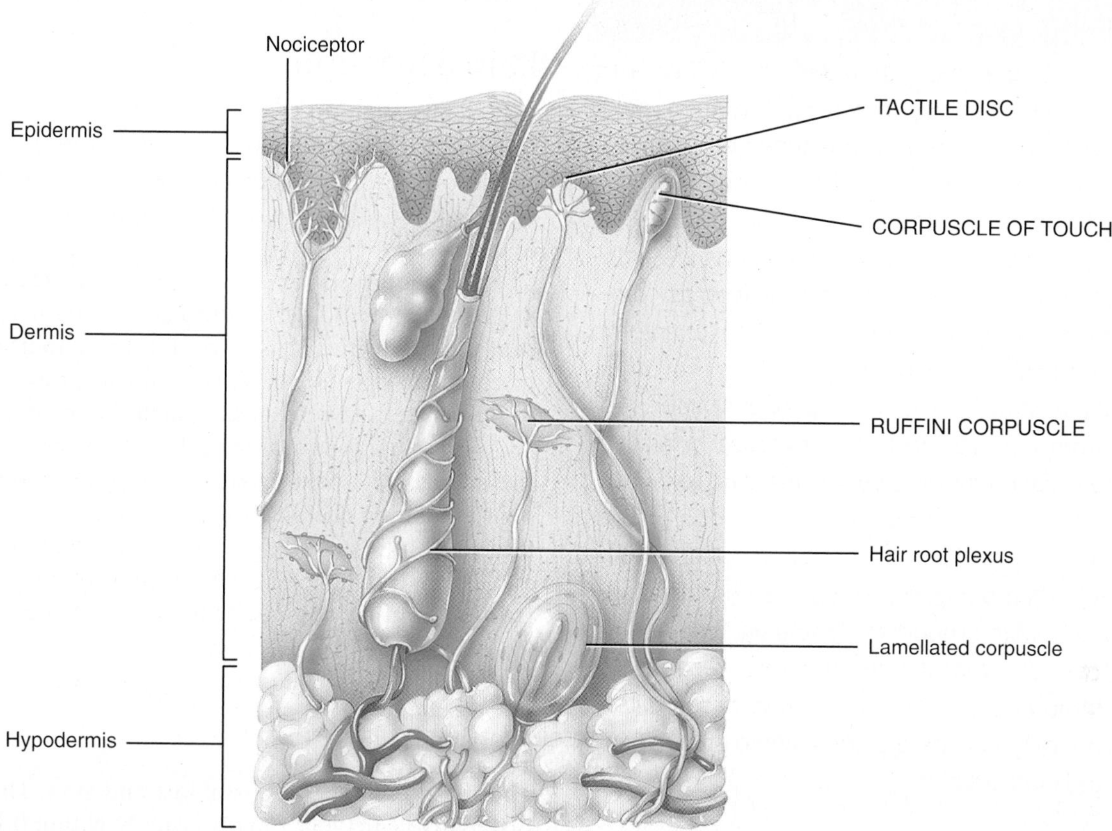


**The skin contains sensory receptors for somatic sensations of touch, pressure, vibration, warmth, cold, and pain.**

soles, clitoris, and tip of the penis. **Hair root plexuses** are rapidly adapting touch receptors found in hairy skin; they consist of free nerve endings wrapped around hair follicles. Hair root plexuses detect movements on the skin surface that disturb hairs. For example, an insect landing on a hair causes movement of the hair shaft, which in turn stimulates the free nerve endings.

There are also two types of slowly adapting touch receptors. **Tactile discs**, also known as *Merkel discs*, are saucer-shaped, flattened free nerve endings that make contact with tactile (Merkel) cells of the stratum basale (see Figure 5.2a). These touch receptors are plentiful in the fingertips, hands, lips, and external genitalia. **Ruffini corpuscles** are elongated, encapsulated receptors located deep in the dermis, and in ligaments and tendons. Present in the hands and abundant on the soles, they are most sensitive to stretching that occurs as digits or limbs are moved.

## Pressure

**Pressure** is a sustained sensation that is felt over a larger area than touch. It occurs when deeper tissues are compressed. Receptors that contribute to sensations of pressure include corpuscles of touch, tactile discs, and lamellated corpuscles. A **lamellated corpuscle**, or *pacinian corpuscle* (pa-SIN-ē-an), is a large oval structure composed of a multilayered connective tissue capsule that encloses a dendrite. Like corpuscles of touch, lamellated corpuscles adapt rapidly. They are widely distributed in the body: in the dermis and hypodermis; in tissues that underlie mucous and serous membranes; around joints, tendons, and muscles; in the periosteum; and in the mammary glands, external genitalia, and some viscera.

## Vibration

Sensations of **vibration** result from rapidly repetitive sensory signals from tactile receptors. The receptors for vibration sensations are corpuscles of touch and lamellated corpuscles. Corpuscles of touch can detect lower frequency vibrations; lamellated corpuscles detect higher frequency vibrations.

## Itch

The **itch** sensation results from stimulation of free nerve endings by certain chemicals, such as bradykinin, often because of a local inflammatory response.

### CLINICAL CONNECTION | *Phantom Limb Sensation*

Patients who have had a limb amputated may still experience sensations such as itching, pressure, tingling, or pain as if the limb were still there. This phenomenon is called **phantom limb sensation**. Although the limb has been removed, severed endings of sensory axons are still present in the remaining stump. If these severed endings are activated, the cerebral cortex interprets the sensation as coming form the sensory receptors in the nonexisting (phantom) limb. Another explanation for phantom limb sensation is that the area of the cerebral cortex that previously received sensory input from the missing limb undergoes extensive functional reorganization that allows it to respond to stimuli from another body part. The remodeling of this cortical area is thought to give rise to false sensory perceptions from the missing limb. Phantom limb pain can be very distressing to an amputee. Many report that the pain is severe or extremely intense, and that it often does not respond to traditional pain medication therapy. In such cases, alternative treatments may include electrical nerve stimulation, acupuncture, and biofeedback.

### Tickle

Free nerve endings are thought to mediate the **tickle** sensation. This intriguing sensation typically arises only when someone else touches you, not when you touch yourself. The explanation of this puzzle seems to lie in the impulses that conduct to and from the cerebellum when you are moving your fingers and touching yourself, which do not occur when someone else is tickling you.

## Thermal Sensations

**Thermoreceptors** are free nerve endings that have receptive fields on the skin surface. Two distinct **thermal sensations**—coldness and warmth—are detected by different thermoreceptors. Temperatures between 10°C and 40°C (50–105°F) activate **cold receptors** located in the epidermis. Temperatures between 32°C and 48°C (90–118°F) are stimulated by **warm receptors** located in the dermis. Cold and warm receptors both adapt rapidly at the onset of a stimulus but continue to generate impulses at a lower frequency throughout a prolonged stimulus. Temperatures below 10°C and above 48°C primarily stimulate nociceptors (pain receptors), rather than thermoreceptors, producing painful sensations.

## Pain Sensations

Pain is indispensable for survival. It serves a protective function by signaling the presence of harmful, tissue-damaging conditions. From a medical standpoint, describing the type and location of pain may help pinpoint the underlying cause of disease.

**Nociceptors** (*noci-* = harmful), the receptors for pain, are free nerve endings found in every tissue of the body except the brain (see Figure 15.2). Intense thermal, mechanical, or chemical stimuli can activate nociceptors. Tissue irritation or injury releases chemicals such as prostaglandins, kinins, and potassium ions that stimulate nociceptors. Pain may persist even after a pain-producing stimulus is removed because pain-mediating chemicals linger, and because nociceptors exhibit very little adaptation. Conditions that elicit pain include excessive stretching of a structure, prolonged muscular contractions, muscle spasms, or ischemia (inadequate blood flow to an organ).

### Types of Pain

There are two types of pain: fast and slow. The perception of **fast pain** occurs very rapidly, usually within 0.1 second after a stimulus is applied. This type of pain is also known as acute, sharp, or pricking pain. The pain felt from a needle puncture or knife cut to the skin is fast pain. The perception of **slow pain**, by contrast, begins a second or more after a stimulus is applied. It then gradually increases in intensity over a period of several seconds or minutes. This type of pain is also referred to as chronic, burning, aching, or throbbing pain, like that associated with a toothache. You can perceive the difference in these two types of pain when you stub your toe; first you feel the sharp sensation of fast pain, followed by the slower, aching sensation of slow pain.

Pain that arises from stimulation of receptors in the skin is called **superficial somatic pain**. Stimulation of receptors in skeletal muscles, joints, tendons, and fascia causes **deep somatic pain**. **Visceral pain** results from stimulation of nociceptors in visceral organs. Diffuse stimulation of visceral nociceptors might result from distension or ischemia of an internal organ. For example, a kidney stone might cause severe pain by obstructing and distending a ureter.

### Localization of Pain

Fast pain is very precisely localized to the stimulated area. For example, if someone pricks you with a pin, you know exactly which part of your body was stimulated. Somatic slow pain has more diffuse localization (involves large areas); it usually appears to come from a larger area of the skin. In some instances of visceral slow pain, the affected area is where the pain is felt. However, in many instances of visceral pain, the pain is felt in or just deep to the skin that overlies the stimulated organ or in a surface area far from the stimulated organ. This phenomenon is called **referred pain**. Figure 15.3 shows skin regions to which visceral pain may be referred. In general, the visceral organ involved and the area to which the pain is referred are served by the same segment of the spinal cord. For example, sensory fibers from the heart, the skin over the heart, and the skin along the medial aspect of the left arm enter the same spinal cord segments, segments T1 to T5. Thus, the pain of a heart attack typically is felt in the skin over the heart and along the left arm.

## Proprioceptive Sensations

**Proprioceptive sensations** allow us to know where our head and limbs are located and how they are moving even if we are not looking at them, so that we can walk, type, or dress without using our eyes. Proprioceptive sensations arise in receptors called **proprioceptors**. Proprioceptors embedded in muscles and tendons inform us of the degree to which muscles are contracted, the amount of tension on tendons, and the positions of joints. Hair cells of the inner ear are proprioceptors for balance and equilibrium that monitor the orientation of the head relative to gravity and head position during movements. Because proprioceptors adapt only slightly, the brain continually receives impulses related to the position of different body parts and makes adjustments to ensure coordination.

Proprioceptors also allow us to estimate the weight of objects and determine the muscular effort needed to perform a task. For example, as you pick up an object you quickly realize how heavy it is, and you then exert the correct amount of effort needed to lift it. We shall now discuss three types of proprioceptors: muscle spindles within skeletal muscles, tendon organs within tendons, and joint kinesthetic receptors within synovial joint capsules.

### Muscle Spindles

Muscle spindles are the proprioceptors in skeletal muscles that monitor changes in the length of skeletal muscles and participate in stretch reflexes (shown in Figure 14.20). By adjusting how vigorously a muscle spindle responds to stretching of a skeletal muscle, the brain sets an overall level of **muscle tone**, the small degree of contraction that is present while the muscle is at rest.

Each **muscle spindle** consists of several slowly adapting sensory nerve endings that wrap around specialized muscle fibers called **intrafusal muscle fibers** (in′-tra-FŪ-sal = within a spindle;

FIGURE 15.3 **Distribution of referred pain.** The colored parts of the diagrams indicate skin areas to which visceral pain is referred.

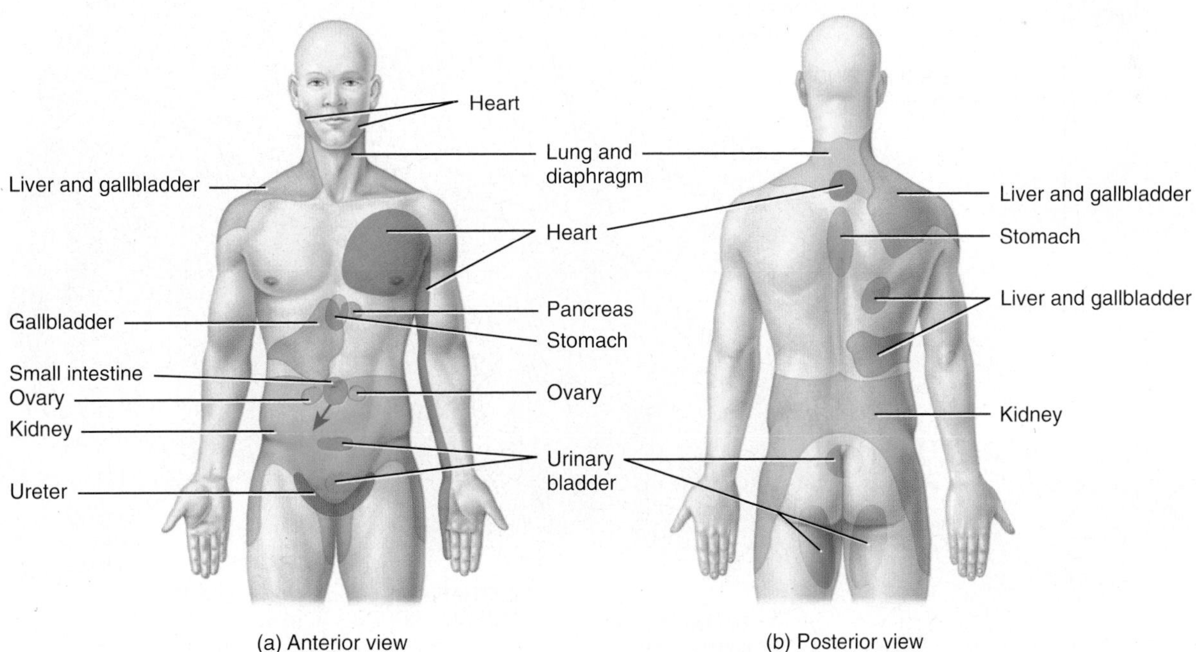


**Nociceptors are present in almost every tissue of the body.**

Figure 15.4). A connective tissue capsule encloses the sensory nerve endings and intrafusal fibers and anchors the spindle to the endomysium and perimysium. Muscle spindles are interspersed among most skeletal muscle fibers and aligned parallel to them. In muscles that produce finely controlled movements, such as those of the fingers or eyes, muscle spindles are plentiful. Muscles involved in coarser but more forceful movements, like the quadriceps femoris and hamstring muscles of the thigh, have fewer muscle spindles.

The main function of muscle spindles is to measure *muscle length*—how much a muscle is being stretched. Stretching of the intrafusal muscle fibers stimulates the sensory nerve endings. The resulting impulses propagate into the CNS. Information from muscle spindles arrives at the somatic sensory areas of the cerebral cortex, allowing conscious perception of limb positions and movements. At the same time, impulses from muscle spindles pass to the cerebellum, where the input is used to coordinate muscle contractions.

In addition to their sensory nerve endings that wrap around the central portion of intrafusal fibers, muscle spindles contain motor neurons, called **gamma motor neurons**, that terminate near both ends of the intrafusal fibers. Gamma motor neurons adjust the tension in a muscle spindle to variations in the length of the muscle. For example, when a muscle shortens, gamma motor neurons stimulate the intrafusal fibers to contract slightly. This keeps the intrafusal fibers taut and maintains the sensitivity of the muscle spindle to stretching of the muscle. As the

**FIGURE 15.4 Two types of proprioceptors: a muscle spindle and a tendon organ.** In muscle spindles, which monitor changes in skeletal muscle length, sensory nerve endings wrap around intrafusal muscle fibers. In tendon organs, which monitor the force of muscle contraction, sensory nerve endings are activated by increasing tension on a tendon.

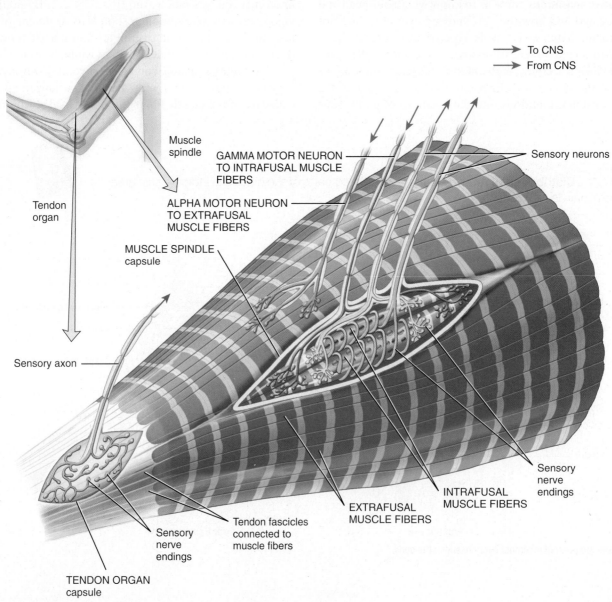

Proprioceptors provide information about body position and movement.

frequency of impulses in its gamma motor neuron increases, a muscle spindle becomes more sensitive to stretching.

Surrounding muscle spindles are ordinary skeletal muscle fibers, called **extrafusal muscle fibers** (*extrafusal* = outside a spindle), which are innervated by **alpha motor neurons**. During the stretch reflex, impulses in muscle spindle sensory axons propagate into the spinal cord and brain stem, where they activate alpha motor neurons that are connected to extrafusal muscle fibers in the same muscle. In this way, activation of its muscle spindles causes contraction of a skeletal muscle, which relieves the stretching.

## Tendon Organs

Tendon organs are located at the junction of a tendon and a muscle. By initiating reflexes, tendon organs protect tendons and their associated muscles from damage due to excessive tension. When a muscle contracts, it exerts a force (tension) that pulls the points of attachment located at either end of the muscle toward each other. Each **tendon organ** consists of a thin capsule of connective tissue that encloses a few tendon fascicles (bundles of collagen fibers) (see Figure 15.4). Sensory nerve endings entwine among and around the collagen fibers of the tendon organ. As a muscle contracts, the tendon organs generate nerve impulses that propagate into the CNS, providing information about changes in muscle tension. The resulting reflex activity decreases muscle tension by causing muscle relaxation.

## Joint Kinesthetic Receptors

Several types of **joint kinesthetic receptors** (kin′-es-THET-ik; *kin-* = motion), associated with the articular (joint) capsules of synovial joints, sense joint position and movement. Free nerve endings and Ruffini corpuscles within the articular capsules respond to pressure. Lamellated corpuscles in the connective tissue outside articular capsules respond to changes in joint movement. Ligaments of joints contain receptors similar to tendon organs that adjust tension of adjacent muscles when excessive strain is placed on the joint.

Table 15.2 summarizes the types of somatic sensory receptors and the sensations they convey.

**TABLE 15.2**

Summary of Receptors for Somatic Sensations

| RECEPTOR TYPE | RECEPTOR STRUCTURE AND LOCATION | SENSATIONS | ADAPTATION RATE |
|---|---|---|---|
| **TACTILE RECEPTORS** | | | |
| Corpuscles of touch (Meissner corpuscles) | Capsule surrounds mass of dendrites in dermal papillae of hairless skin | Touch, pressure, and slow vibrations | Rapid |
| Hair root plexuses | Free nerve endings wrapped around hair follicles in skin | Touch | Rapid |
| Tactile (Merkel) discs | Saucer-shaped free nerve endings make contact with tactile cells in epidermis | Touch and pressure | Slow |
| Ruffini corpuscles | Elongated capsule surrounds dendrites deep in dermis and in ligaments and tendons | Stretching of skin | Slow |
| Lamellated corpuscles | Oval, layered capsule surrounds dendrites; present in dermis and subcutaneous layer, submucosal tissues, joints, periosteum, and some viscera | Pressure and fast vibrations | Rapid |
| Itch and tickle receptors | Free nerve endings in skin and mucous membranes | Itching and tickling | Both slow and rapid |
| **THERMORECEPTORS** | | | |
| Warm receptors and cold receptors | Free nerve endings in skin and mucous membranes of mouth, vagina, and anus | Warmth or cold | Initially rapid, then slow |
| **PAIN RECEPTORS** | | | |
| Nociceptors | Free nerve endings in every body tissue except brain | Pain | Slow |
| **PROPRIOCEPTORS** | | | |
| Muscle spindles | Sensory nerve endings wrap around central area of encapsulated intrafusal muscle fibers within most skeletal muscles | Muscle length | Slow |
| Tendon organs | Capsule encloses collagen fibers and sensory nerve endings at junction of tendon and muscle | Muscle tension | Slow |
| Joint kinesthetic receptors | Corpuscles of touch, Ruffini corpuscles, tendon organs, and free nerve endings | Joint position and movement | Rapid |

✓ **CHECKPOINT**

9. Which somatic sensory receptors adapt slowly, and which adapt rapidly?
10. Why do we need pain receptors? How are they activated?
11. Distinguish among superficial somatic pain, deep somatic pain, and visceral pain.
12. What aspects of muscle function are monitored by muscle spindles and tendon organs?
13. What is the function of joint kinesthetic receptors?

## 15.4 Somatic sensory pathways relay information from sensory receptors to the cerebral cortex and cerebellum.

The sensory and motor pathways of the body provide routes for input into the brain and spinal cord and output to targeted organs for responses, such as muscle contraction. **Somatic sensory pathways** relay information from somatic sensory receptors to the primary somatosensory area in the cerebral cortex and to the cerebellum. These ascending pathways consist of thousands of sets of three neurons: a first-order neuron, a second-order neuron, and a third-order neuron.

1. **First-order neurons** conduct impulses from somatic receptors into the brain stem or spinal cord. From the face, mouth, teeth, and eyes, somatic sensory impulses propagate along *cranial nerves* into the brain stem. From the neck, trunk, limbs, and posterior head, somatic sensory impulses propagate along *spinal nerves* into the spinal cord.
2. **Second-order neurons** conduct impulses from the brain stem and spinal cord to the thalamus. Axons of second-order neurons *decussate* (cross over to the opposite side) in the brain stem or spinal cord before ascending to the thalamus. Thus, all somatic sensory information from one side of the body reaches the thalamus on the opposite side.
3. **Third-order neurons** conduct impulses from the thalamus to the primary somatosensory area of the cerebral cortex on the same side.

Regions within the CNS where neurons synapse with other neurons that are a part of a particular pathway are known as **relay stations** because neural signals are being relayed from one region of the CNS to another. For example, the neurons of many sensory pathways synapse with neurons in the thalamus; therefore, the thalamus functions as a major relay station. In addition to the thalamus, many other regions of the CNS, including the spinal cord and brain stem, can function as relay stations.

Somatic sensory impulses may ascend to the cerebral cortex through the posterior column–medial lemniscus pathway or the spinothalamic pathways. Somatic sensory impulses reach the cerebellum via the spinocerebellar tracts.

### Posterior Column–Medial Lemniscus Pathway to the Cerebral Cortex

Impulses for most touch, pressure, vibration, and proprioception ascend to the cerebral cortex along the **posterior column–medial lemniscus pathway** (lem-NIS-kus = ribbon) (Figure 15.5a). The name of the pathway comes from the names of two white matter tracts that convey the impulses: the posterior column of the spinal cord and the medial lemniscus of the brain stem.

First-order neurons extend from sensory receptors in the limbs, trunk, neck, and posterior head into the spinal cord and ascend to the medulla oblongata on the same side of the body. The cell bodies of these first-order neurons are in the posterior root ganglia of spinal nerves. In the spinal cord, their axons form the **posterior columns** (white matter located on the posterior side of the spinal cord) (see Figure 13.18) and travel superiorly to synapse with second-order neurons in nuclei of the medulla oblongata. The axons of second-order neurons cross to the opposite side of the medulla oblongata and enter the **medial lemniscus**, a thin, ribbonlike projection tract that extends from the medulla oblongata and travels superiorly to the thalamus. In the thalamus, the axons of second-order neurons synapse with third-order neurons, which project their axons to the primary somatosensory area of the cerebral cortex.

### Spinothalamic Pathway to the Cerebral Cortex

Impulses for pain, temperature, itch, and tickle from the limbs, trunk, neck, and posterior head ascend to the cerebral cortex along the **spinothalamic pathway** (spī-nō-tha-LAM-ik). The name of this pathway comes from the anterior and lateral white tracts. Like the posterior column–medial lemniscus pathway, the spinothalamic pathway is composed of three-neuron sets (Figure 15.5b). The first-order neurons connect receptors in the limbs, trunk, neck, or posterior head with the spinal cord. The

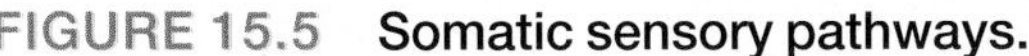

FIGURE 15.5 Somatic sensory pathways.

(a) Posterior column–medial lemniscus pathway

(b) Spinothalamic pathways

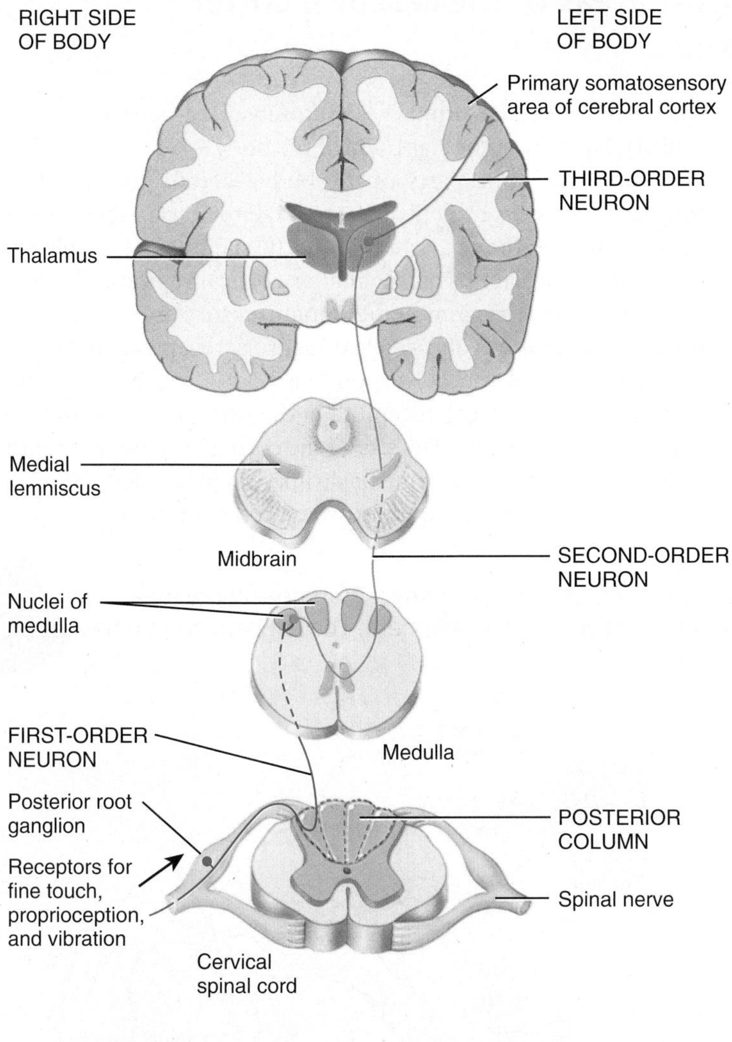

**Impulses propagate along sets of first-order, second-order, and third-order neurons to the primary somatosensory area of the cerebral cortex.**

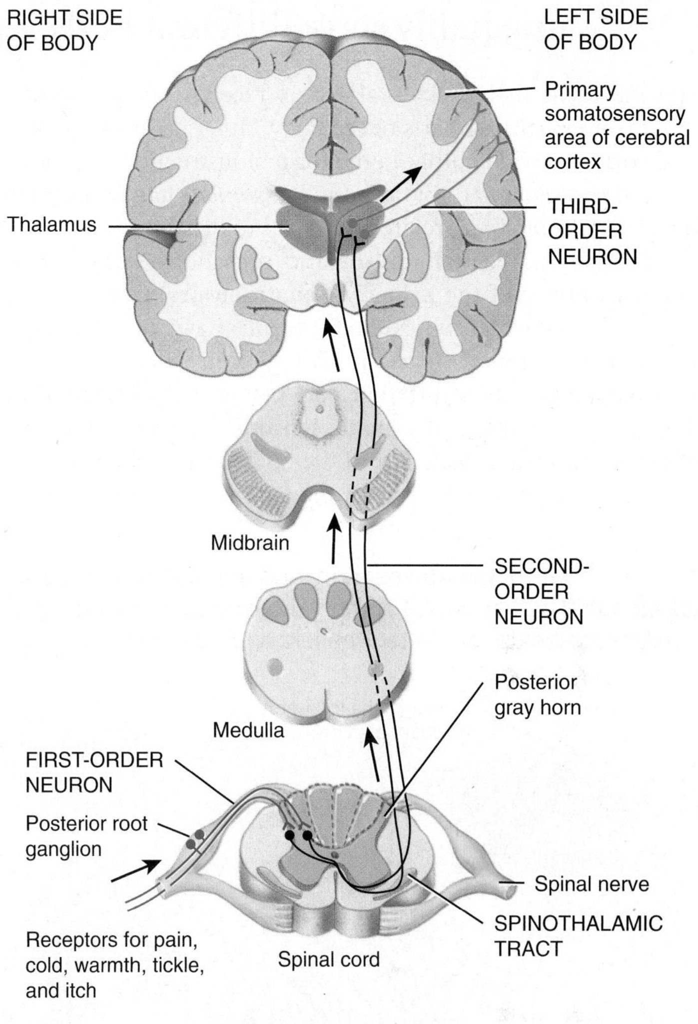

cell bodies of the first-order neurons are in the posterior root ganglia. The axons of the first-order neurons synapse with second-order neurons in the posterior gray horn of the spinal cord. The axons of second-order neurons cross to the opposite side of the spinal cord, then travel superiorly to the brain stem as the **spinothalamic tract** and end in the thalamus, where they synapse with third-order neurons. The axons of the third-order neurons project to the primary somatosensory area of the cerebral cortex on the same side as the thalamus.

## Somatic Sensory Pathways to the Cerebellum

Two tracts in the spinal cord—the **posterior spinocerebellar tract** (spī-nō-ser-e-BEL-ar) and the **anterior spinocerebellar tract**—are the major routes proprioceptive impulses take to reach the cerebellum (see Figure 13.19). Although they are not consciously perceived, sensory impulses conveyed to the cerebellum along these two pathways are critical for posture, balance, and coordination of skilled movements.

### ✓ CHECKPOINT

14. Which sensations could be lost if the posterior column–medial lemniscus pathway is damaged?
15. State three differences between the posterior column–medial lemniscus pathway and the spinothalamic pathway.
16. What type of sensory information is carried in the spinocerebellar pathways, and what is its usefulness?

## 15.5 The somatosensory and primary motor areas of the cerebral cortex unequally serve different body regions.

Specific areas of the cerebral cortex receive somatic sensory input from particular parts of the body. Other areas of the cerebral cortex provide output in the form of instructions for movement of particular parts of the body. The *somatic sensory map* and the *somatic motor map* relate body parts to these cortical areas.

Precise localization of somatic sensations occurs when impulses arrive at the **primary somatosensory area** (see Figure 13.10), which occupies the postcentral gyri of the parietal lobes of the cerebral cortex. Each region in this area receives sensory input from a different part of the body. Figure 15.6a maps the destination of somatic sensory signals coming from different parts of the left side of the body to the somatosensory area of the right cerebral hemisphere. The left cerebral hemisphere has a similar primary somatosensory area that receives sensory input from the right side of the body.

Note that some parts of the body—chiefly the lips, face, tongue, and hand—provide input to large regions in the somatosensory area. Other parts of the body, such as the trunk and lower limbs, project to much smaller cortical regions. The relative sizes of these regions in the somatosensory area are proportional to the number of specialized sensory receptors, and hence the sensitivity, of the corresponding body part. For example, there are many sensory receptors in the skin of the lips but few in the skin of the trunk. The size of the cortical region that represents a body part may expand or shrink somewhat, depending on the quantity of sensory impulses received from that body part.

**FIGURE 15.6 Somatic sensory and somatic motor maps in the cerebral cortex.** (a) Primary somatosensory area (postcentral gyrus) and (b) primary motor area (precentral gyrus) of the right cerebral hemisphere. The left hemisphere has similar representation. (After Penfield and Rasmussen.)

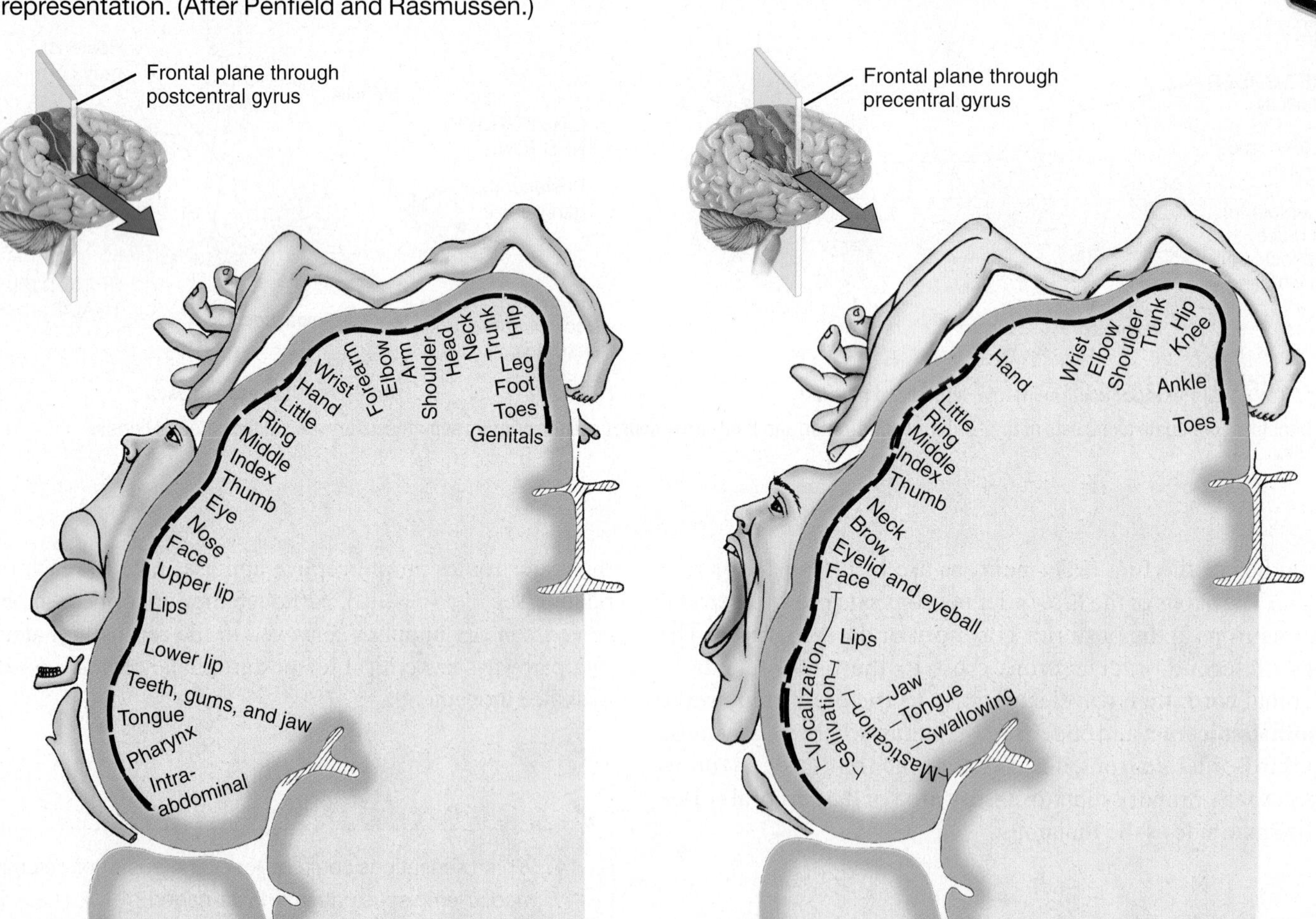


(a) Frontal section of primary somatosensory area in right cerebral hemisphere

(b) Frontal section of primary motor area in right cerebral hemisphere

**Each point on the body surface maps to a specific region in both the primary somatosensory area and the primary motor area.**

For example, people who learn to read Braille eventually have a larger cortical region in the somatosensory area to represent the fingertips.

Control of body movements occurs in several regions of the brain. The **primary motor area** (see Figure 13.10), located in the precentral gyrus of the frontal lobe (Figure 15.6b) of the cerebral cortex, is a major control region for the execution of voluntary movements. The adjacent **premotor area** also contributes axons to the descending motor pathways. As is true for somatic sensory representation in the somatosensory area, different muscles are represented unequally in the primary motor area. More cortical area is devoted to those muscles involved in skilled, complex, or delicate movements. Muscles in the hand, lips, tongue, and vocal cords have large representations; the trunk has a much smaller representation.

### ✓ CHECKPOINT

**17.** Which body parts have the largest representation in the primary somatosensory areas?

**18.** Which body parts have the largest representation in the primary motor areas?

**19.** Why are certain body regions represented by larger portions of the primary motor area than others?

## RETURN TO **Mustafa's Story**

Mustafa returns from the CT scan. His family is anxious to see him but feel distraught when Mustafa still does not seem to recognize them. The nurse reassures them that this is normal after a stroke, because of swelling of the brain. Dr. Nasri enters the room and smiles. "Good news. Mustafa has a blood clot in an artery in his head affecting the blood supply to the left side of the brain. Since it has been less than three hours since he first experienced symptoms, we can give him a drug to try and dissolve the clot. The drug is called tissue plasminogen activator, tPA for short. I will need Khalia to sign a consent form for the treatment since Mustafa is confused and I am not sure how much he understands what is happening."

Khalia discusses the procedure with her children, and they all agree that they want Mustafa to have the treatment. Khalia signs the consent form.

Dr. Nasri administers the tPA slowly and carefully monitors Mustafa's vital signs and responses. After administration of the tPA is complete and Mustafa is stable, he is moved to a room in the stroke unit for further observation. Khalia is allowed to accompany him. The nurses tell Mustafa and Khalia that they will be monitoring his neurological status every hour to document any change.

"How do you check his neurological status?" Khalia asks.

"I test his orientation by seeing if he can tell me who he is, where he is, and what time of day it is. Then I check the pupils of his eyes to see if they react to light in the same way. I also ask him to squeeze my hands with his to see if the strength in both arms is the same. If I find changes, there are other tests that I can do to narrow down his neurological deficits. Hopefully, with the tPA I will see improvement each time I assess him."

E. *Mustafa has a clot in an artery on the left side of his brain. Why has he lost function on the right side of his body?*

F. *The nurse uses a serrated wheel to test for the presence of sensation on the surface of Mustafa's body. Which part of his body has the most sensory nerve endings that would respond to the pressure of the wheel?*

G. *The nurse slides the handle of a reflex hammer along the palms of Mustafa's hands. Which type of sensory receptor is being stimulated by the touch of the handle?*

H. *Dr. Nasri uses a tuning fork to test for vibratory sensation, a more subtle sensation than touch. Which types of sensory receptors are stimulated by vibrations?*

## 15.6 Somatic motor pathways carry impulses from the brain to effectors.

Neurons in the brain and spinal cord orchestrate all voluntary and involuntary movements. Ultimately, all excitatory and inhibitory signals that control movement converge on motor neurons that extend out of the brain stem and spinal cord to innervate skeletal muscles in the body. These neurons, also known as **lower motor neurons**, have their cell bodies in the brain stem and spinal cord. From the brain stem, axons of lower motor neurons extend though *cranial nerves* to innervate skeletal muscles of the head. From the spinal cord, axons extend through *spinal nerves* to innervate skeletal muscles of the limbs and trunk.

Other neurons in the brain and spinal cord participate in control of movement by providing input to lower motor neurons:

- ***Local circuit neurons.*** Input arrives at lower motor neurons from nearby interneurons called **local circuit neurons**. Local circuit neurons receive input from somatic sensory receptors, such as muscle spindles, as well as from higher centers in the brain. They help coordinate rhythmic activity in specific muscle groups, such as alternating flexion and extension of the lower limbs during walking.
- ***Upper motor neurons.*** Upper motor neurons originating in the brain stem help maintain muscle tone, posture, and balance. Both local circuit neurons and lower motor neurons receive input from **upper motor neurons**. Most upper motor neurons synapse with local circuit neurons, which in turn synapse with lower motor neurons. Upper motor neurons from the cerebral cortex are essential for executing voluntary movements. Upper motor neurons from the brain stem help maintain muscle tone, posture, and balance.
- ***Basal nuclei neurons.*** Neurons of the basal nuclei assist movement by providing input to upper motor neurons. Basal nuclei neurons help initiate and terminate movements, suppress unwanted movements, and establish a normal level of muscle tone.
- ***Cerebellar neurons.*** Neurons of the cerebellum also aid movement by controlling the activity of upper motor neurons. A prime function of the cerebellum is to monitor differences between intended movements and movements actually performed. Then it issues commands to upper motor neurons to reduce errors in movement. The cerebellum thus coordinates body movements and helps maintain normal posture and balance.

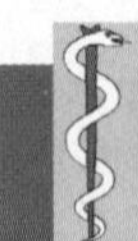

## CLINICAL CONNECTION | *Paralysis*

Damage or disease of *lower* motor neurons produces **flaccid paralysis** (FLAK-sid) of muscles on the same side of the body. There is neither voluntary nor reflex action of the innervated muscle fibers, muscle tone is decreased or lost, and the muscle remains limp or flaccid. Injury or disease of *upper* motor neurons in the cerebral cortex removes inhibitory influences that some of these neurons have on lower motor neurons, which causes **spastic paralysis** of muscles on the opposite side of the body. In this condition muscle tone is increased, reflexes are exaggerated, and pathological reflexes such as the Babinski sign appear (described in the Chapter 14 Clinical Connection entitled Reflexes and Diagnosis; see Concept 14.4).

## Organization of Upper Motor Neuron Pathways

The axons of upper motor neurons extend from the brain to lower motor neurons via two types of descending somatic motor pathways—direct and indirect. Direct motor pathways provide input to lower motor neurons via axons that extend directly from the cerebral cortex. Indirect motor pathways provide input to lower motor neurons from motor centers in the basal nuclei, cerebellum, and cerebral cortex. Direct and indirect pathways both govern generation of impulses in the lower motor neurons, the neurons that stimulate contraction of skeletal muscles.

### Direct Motor Pathways

Impulses for voluntary movements propagate from the cerebral cortex to lower motor neurons via the direct motor pathways (Figure 15.7). The **direct motor pathways** consist of axons that descend from upper motor neurons located in the primary motor area of the cerebral cortex.

The **corticospinal pathways** (kor′-ti-kō-SPĪ-nal) are major tracts that conduct impulses controlling muscles of the limbs and trunk. While descending from the primary motor area, most corticospinal axons *decussate* (cross over) to the opposite side of the medulla oblongata and then descend into the spinal cord where they synapse with a lower motor neuron (Figure 15.7). Thus, the right cerebral cortex controls muscles on the left side of the body, and the left cerebral cortex controls muscles on the right side of the body. There are two types of corticospinal tracts: the lateral corticospinal tract and the anterior corticospinal tract.

Corticospinal axons that decussate in the medulla oblongata form the **lateral corticospinal tract** in the lateral white columns of the spinal cord (see Figure 13.19). These axons synapse with lower motor neurons in the spinal cord (Figure 15.7a). Axons of these lower motor neurons exit the cord in the anterior roots of spinal nerves and terminate in skeletal muscles that control precise, agile, and highly skilled movements of the hands and feet. Examples include the movements needed to button a shirt or play the piano.

Corticospinal axons that do not decussate in the medulla oblongata form the **anterior corticospinal tract** in the anterior white columns of the spinal cord (see Figure 13.19 and Figure 15.7b). At each spinal cord level, some of these axons cross via the anterior white commissure (see Figure 13.18). Then they synapse with lower motor neurons in the anterior gray horn. Axons of these lower motor neurons exit the spinal cord in the anterior roots of spinal nerves and terminate in skeletal muscles that control movements of the trunk and proximal parts of the limbs.

In addition to the corticospinal tracts, another direct motor pathway is the **corticobulbar tract** (kor′-ti-kō-BUL-bar) which conducts impulses for the control of skeletal muscles in the head. Some of the axons of the corticobulbar tract decussate,

FIGURE 15.7 **Somatic motor pathways.** Shown here are the two most direct pathways whereby signals initiated by the primary motor area in one hemisphere control skeletal muscles on the opposite side of the body.

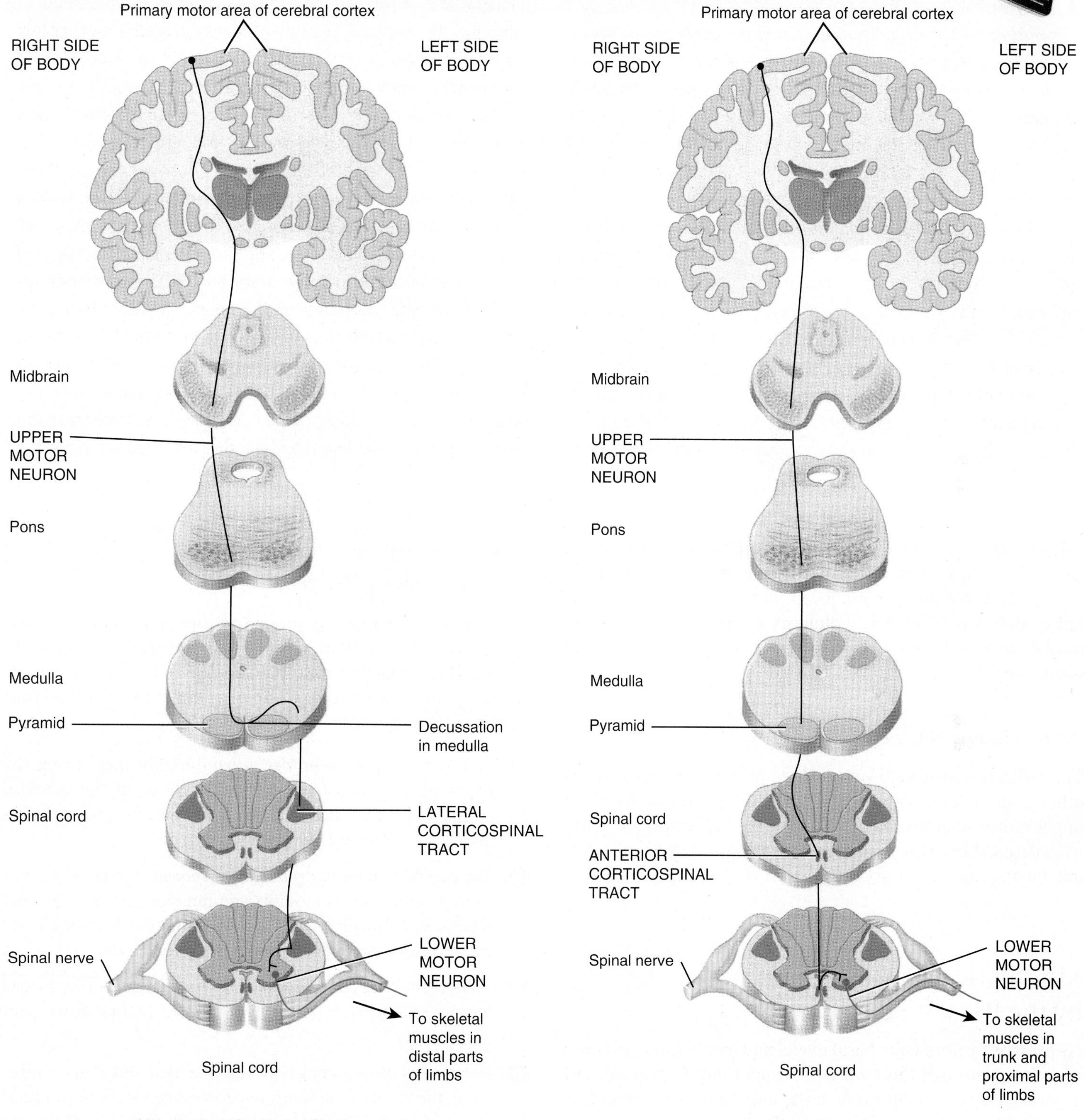


(a) Lateral corticospinal pathway

(b) Anterior corticospinal pathway

**Direct motor pathways convey impulses that result in precise, voluntary movements.**

## CLINICAL CONNECTION | *Disorders of the Basal Nuclei*

**Disorders of the basal nuclei** can affect body movements, cognition, and behavior. Uncontrollable shaking (tremor) and muscle rigidity (stiffness) are hallmark signs of **Parkinson's disease** (**PD**). (See the Chapter 15 WileyPLUS Clinical Connection entitled Parkinson's Disease and the Chapter 13 case story.) In this disorder, dopamine-releasing neurons in the midbrain and basal nuclei degenerate.

**Huntington disease** (**HD**) is an inherited disorder in which the basal nuclei degenerate. A key sign of HD is **chorea** (KŌ-rē-a = a dance), in which rapid, jerky movements occur involuntarily and without purpose. Progressive mental deterioration also occurs. Symptoms of HD often do not appear until age 30 or 40. Death occurs 10 to 20 years after symptoms first appear.

**Tourette syndrome** is a disorder that is characterized by involuntary body movements (motor tics) and the use of inappropriate or unnecessary sounds or words (vocal tics). Although the cause is unknown, research suggests that this disorder involves a dysfunction of the cognitive neural circuits between the basal nuclei and the prefrontal cortex.

Some psychiatric disorders, such as schizophrenia and obsessive-compulsive disorder, are thought to involve dysfunction of the behavioral neural circuits between the basal nuclei and the limbic system. In **schizophrenia**, excess dopamine activity in the brain causes a person to experience delusions, distortions of reality, paranoia, and hallucinations. People who have **obsessive-compulsive disorder** (**OCD**) experience repetitive thoughts (obsessions) that cause repetitive behaviors (compulsions) that they feel obligated to perform. For example, a person with OCD might have repetitive thoughts about someone breaking into the house; these thoughts might drive that person to check the doors of the house over and over again (for minutes or hours at a time) to make sure that they are locked.

and others do not, as they descend from the cerebral cortex and then synapse with lower motor neurons of cranial nerves exiting the brain stem. These lower motor neurons convey impulses that control precise, voluntary movements of the eyes, tongue, and neck, plus chewing, facial expression, speech, and swallowing.

### Indirect Motor Pathways

The **indirect motor pathways** include all somatic motor tracts other than the corticospinal and corticobulbar tracts. Axons of upper motor neurons that give rise to the indirect motor pathways descend from the brain stem into spinal cord tracts that terminate on lower motor neurons.

## Modulation of Movement by the Basal Nuclei

As previously noted, the basal nuclei and cerebellum influence movement through their effects on upper motor neurons. The basal nuclei play a major role in the initiation and termination of movements. They receive input from sensory, association, and motor areas of the cerebral cortex, then send feedback signals through the thalamus to the motor areas of the cerebral cortex that appear to initiate and terminate movements. The basal nuclei send impulses into the thalamus and midbrain to suppress unwanted movements, and into the brain stem to reduce muscle tone.

## Modulation of Movement by the Cerebellum

In addition to maintaining proper posture and balance, the cerebellum is active in both learning and performing rapid, coordinated, highly skilled movements such as hitting a golf ball, speaking, and swimming. Cerebellar function involves four activities (Figure 15.8):

1. The cerebellum *monitors intentions for movement* by receiving impulses from the motor areas of the cerebral cortex and basal nuclei via the pons regarding what movements are planned (red arrows).
2. The cerebellum *monitors actual movement* by receiving input from proprioceptors in joints and muscles that reveals what is actually happening (blue arrows). Impulses from the inner ear and from the eyes also enter the cerebellum.
3. The cerebellum *compares the command signals* (intentions for movement) *with sensory information* (actual movement performed).
4. If there is a discrepancy between intended and actual movement, the cerebellum *sends out corrective feedback* to upper motor neurons. This information travels via the thalamus to the cerebral cortex and goes directly to brain stem motor centers (green arrows). As movements occur, the cerebellum continuously provides error corrections to upper motor neurons, which decreases errors and smoothes motions. The cerebellum also contributes over longer periods to the learning of new motor skills.

FIGURE 15.8 **Input to and output from the cerebellum.**

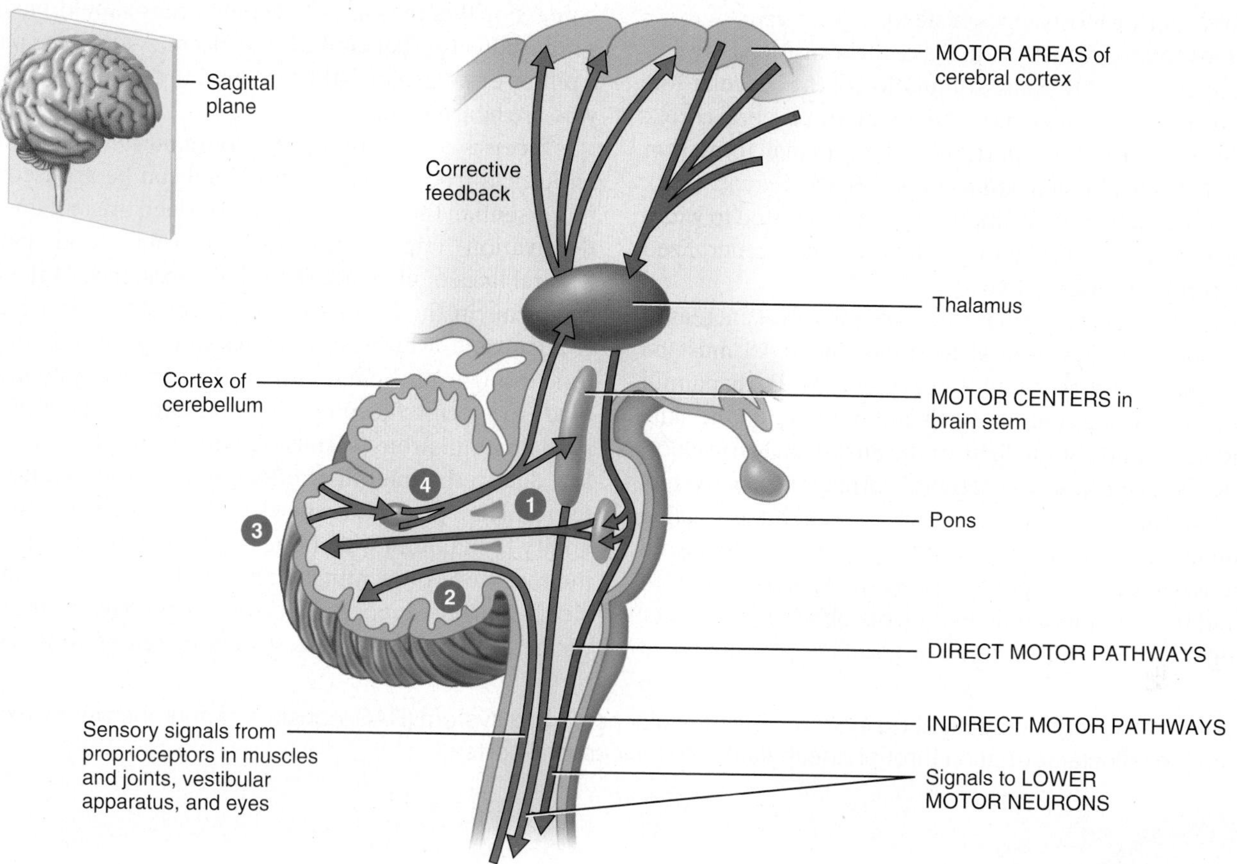


**The cerebellum coordinates and smoothes contractions of skeletal muscles during skilled movements.**

Skilled activities such as volleyball provide good examples of the contribution of the cerebellum to movement. To block a spike, you must bring your arms forward just far enough to make solid contact. How do you stop at exactly the right point? Before you even hit the ball, the cerebellum has sent impulses to the cerebral cortex and basal nuclei informing them where your swing must stop. In response to impulses from the cerebellum, the cortex and basal nuclei transmit motor impulses to opposing body muscles to stop the swing.

✓ **CHECKPOINT**

**20.** What is the function of lower motor neurons?

**21.** Summarize the functions of neurons that provide input to lower motor neurons.

**22.** Why are the two main somatic motor pathways named "direct" and "indirect"?

# 15.7 Wakefulness and memory are integrative functions of the brain.

We turn now to a fascinating, though incompletely understood, function of the cerebrum: integration, the processing of sensory information by analyzing and storing it and making decisions for various responses. The **integrative functions** include cerebral activities such as sleep and wakefulness, learning and memory, and emotional responses. (The role of the limbic system in emotional behavior was discussed in Concept 13.8.)

## Wakefulness and Sleep

Humans sleep and awaken in a 24-hour cycle called a **circadian rhythm** (ser-KĀ-dē-an; *circa-* = about; *-dia-* = a day) that is established by the hypothalamus. A person who is awake is in a state of readiness and is able to react consciously to various stimuli. Electroencephalogram (EEG) recordings show that the cerebral cortex is very active during wakefulness; fewer impulses arise during most stages of sleep.

How does your nervous system make the transition between wakefulness and sleep? Because stimulation of the reticular formation in the brain stem increases activity of the cerebral cortex to maintain consciousness, a portion of the reticular formation is known as the **reticular activating system (RAS)** (Figure 15.9). When this area is active, many impulses are transmitted to widespread areas of the cerebral cortex. The effect is a generalized increase in cerebral cortex activity.

**Arousal**, or awakening from sleep, also involves increased activity in the RAS. For arousal to occur, the RAS must be stimulated. Many sensory stimuli can activate the RAS: painful stimuli detected by nociceptors, touch and pressure on the skin, movement of the limbs, bright light, or the buzz of an alarm clock. Once the RAS is activated, the cerebral cortex is also activated, and arousal occurs. The result is a state of wakefulness called **consciousness** (KON-shus-nes). Notice in Figure 15.9 that even though the RAS receives input from somatic sensory receptors, the eyes, and the ears, there is no input from olfactory receptors; even strong odors may fail to cause arousal. People who die in house fires usually succumb to smoke inhalation without awakening. For this reason, all sleeping areas should have a nearby smoke detector that emits a loud alarm. A vibrating pad beneath a pillow or a flashing light can serve the same purpose for those who are hearing impaired.

**Sleep** is a state of altered consciousness or partial unconsciousness from which an individual can be aroused. Although it is essential, the exact functions of sleep are still unclear. Sleep deprivation impairs attention, learning, and performance. Several lines of evidence suggest the existence of sleep-inducing chemicals in the brain. One apparent sleep-inducing chemical is adenosine, which accumulates during periods of high ATP (adenosine triphosphate) use by the nervous system. Adenosine binds to specific receptors that inhibit neurons of the RAS that participate in arousal. Thus, activity in the RAS during sleep is low due to the inhibitory effect of adenosine. Caffeine (in coffee) and theophylline (in black tea)—substances known for their ability to maintain wakefulness—bind to and block the adenosine receptors, preventing adenosine from binding and inducing sleep. During sleep, activity in the parasympathetic division of the autonomic nervous system increases while sympathetic

**FIGURE 15.9 The reticular activating system.** The reticular activating system (RAS) consists of neurons whose axons project from the reticular formation through the thalamus to the cerebral cortex.

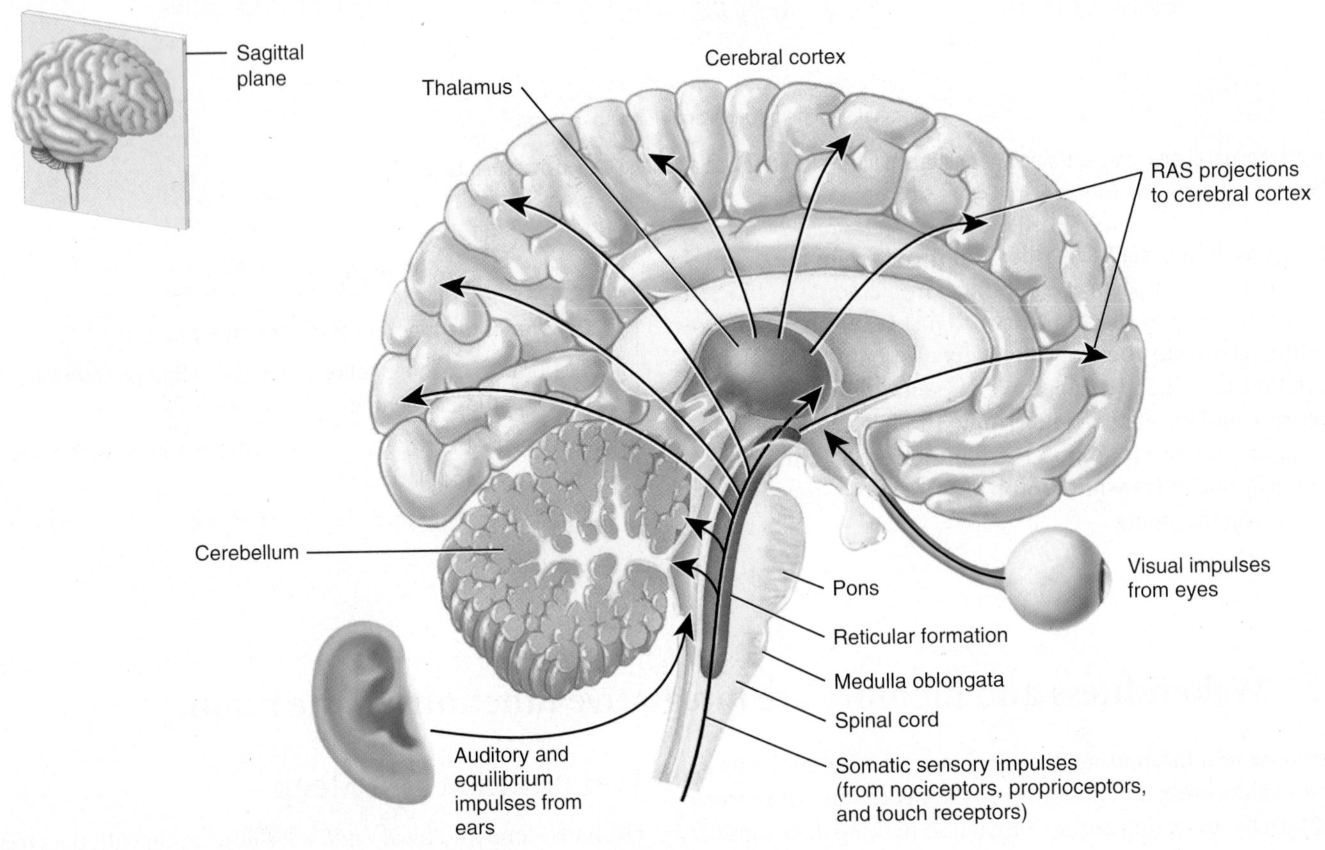


Increased activity of the RAS causes awakening from sleep (arousal).

## CLINICAL CONNECTION | *Sleep Disorders*

Sleep disorders affect over 70 million Americans each year. Common sleep disorders include insomnia, sleep apnea, and narcolepsy. A person with **insomnia** (in-SOM-nē-a) has difficulty in falling asleep or staying asleep. Possible causes of insomnia include stress, excessive caffeine intake, disruption of circadian rhythms (for example, working the night shift instead of the day shift at your job), and depression. **Sleep apnea** (AP-nē-a) is a disorder in which a person repeatedly stops breathing for 10 or more seconds while sleeping. Most often, it occurs because a loss of muscle tone in pharyngeal muscles allows the airway to collapse. **Narcolepsy** (NAR-kō-lep-sē) is a condition in which sleep cannot be inhibited during waking periods. As a result, involuntary periods of sleep that last about 15 minutes occur throughout the day. Recent studies have revealed that people with narcolepsy have a deficiency of the neuropeptide *orexin*, which is also known as **hypocretin**. Orexin is released from certain neurons of the hypothalamus and has a role in promoting wakefulness.

activity decreases. For example, heart rate and blood pressure decrease during sleep.

## Learning and Memory

Without memory, we would repeat mistakes and be unable to learn. Similarly, we would not be able to repeat our successes or accomplishments, except by chance. Although both learning and memory have been studied extensively, we still have no completely satisfactory explanation for how we remember information or events. However, we do have some understanding about how information is stored, and it is clear that there are different categories of memory.

**Learning** is the ability to acquire new information or skills through instruction or experience. **Memory** is the process by which information acquired through learning is stored and retrieved. For an experience to become part of memory, it must produce structural and functional changes in the brain. This capability for change involves changes in individual neurons—for example, synthesis of different proteins or sprouting of new dendrites—as well as changes in the strengths of synaptic connections among neurons. The parts of the brain known to be involved with memory include the association areas of the cerebral cortex, thalamus and hypothalamus, and parts of the limbic system. If a particular body part is used more intensively or in a newly learned activity, such as reading Braille, the cortical areas devoted to that body part gradually expand.

Memory occurs in stages over a period of time. **Immediate memory** is the ability to recall ongoing experiences for a few seconds. It provides a perspective to the present time that allows us to know where we are and what we are doing. **Short-term memory** is the temporary ability to recall a few pieces of information for seconds to minutes. One example is when you look up an unfamiliar telephone number, cross the room to the phone, and then dial the new number. If the number has no special significance, it is usually forgotten within a few seconds. Some evidence supports the notion that short-term memory depends more on electrical and chemical events in the brain than on structural changes, such as the formation of new synapses.

Information in short-term memory may later be transformed into a more permanent type of memory, called **long-term memory**, which lasts from days to years. If you use that new telephone number often enough, it becomes part of long-term memory. Information in long-term memory usually can be retrieved for use whenever needed. Long-term memories for information that can be expressed by language, such as a telephone number, apparently are stored in wide regions of the cerebral cortex. Memories for motor skills, such as how to serve a tennis ball, are stored in the basal nuclei and cerebellum as well as in the cerebral cortex.

Although the brain receives many stimuli, we pay attention to only a few of them at a time. It has been estimated that only 1 percent of all information that comes to our consciousness is stored as long-term memory. Moreover, much of what goes into long-term memory is eventually forgotten. Memory does not record every detail as if it were magnetic tape. Even when details are lost, we can often explain the idea or concept using our own words and ways of viewing things.

Several conditions that inhibit the electrical activity of the brain, such as anesthesia, coma, electroconvulsive therapy, and ischemia of the brain, disrupt retention of recently acquired information without altering previously established long-term memories. People who suffer retrograde amnesia cannot remember anything that occurred during the 30 minutes or so before the amnesia developed. As a person recovers from amnesia, the most recent memories return last.

Anatomical changes occur in neurons when they are stimulated. For example, electron micrographs of neurons subjected to prolonged, intense activity reveal an increase in the number of axon terminals and enlargement of synaptic end bulbs in presynaptic neurons, as well as an increase in the number of dendritic branches in postsynaptic neurons. Opposite changes occur when neurons are inactive. For example, the cerebral cortex in the visual area of animals that have lost their eyesight becomes thinner.

### ✓ CHECKPOINT

**23.** How is wakefulness related to the reticular activating system?

**24.** Define memory. What are the three kinds of memory?

## Mustafa's Story — EPILOGUE AND DISCUSSION

Mustafa is a 65-year-old man who has experienced a cerebrovascular accident, commonly referred to as a brain attack or stroke, from a blood clot lodging in an artery in the left side of his brain. This has caused him to be aphasic (unable to articulate words and thoughts) and to have paralysis on the right side of his body. Fortunately, his family is able to get him to the hospital within the three-hour window for receiving tPA, a potent clot-busting drug. Following the administration of tPA, Mustafa is hospitalized for a few days, and some of his motor function and speech capabilities return. However, Mustafa continues to have difficulty with his balance. The right side of his body is weak and he has difficulty expressing himself, although he has no trouble understanding what is said to him.

Mustafa is transferred to a rehabilitation facility where he will receive intensive physical and speech therapy for the next three to four weeks. The therapists explain to Khalia that Mustafa's brain has forgotten how to put words together and how to coordinate his balance. He is going to have to learn to talk and walk again. They encourage Khalia to sit in on the training sessions so she can reinforce the exercises when Mustafa returns home.

I. *Mustafa increasingly feels sensory stimuli to his right side. Trace the path of a touch stimulus from his right upper limb to the somatosensory area of the cerebral cortex.*

J. *If Mustafa experiences pain after his physical therapy sessions, how will that sensation reach his somatosensory area?*

K. *Mustafa is receiving physical therapy to help him regain his balance and his proprioceptive capability. Which part of the brain is active in monitoring and correcting balance?*

L. *Mustafa is receiving speech therapy to help him learn to speak again. Has he forgotten every word he ever knew? Explain your answer.*

# Concept and Resource Summary

| Concept | Resources |
|---|---|

### Introduction

1. The nervous system must sense the internal and external environments, and convey the sensations to the CNS for modification and integration with other incoming or stored information in order to determine appropriate motor responses.

### Concept 15.1 Sensations arise as a result of stimulation, transduction, generation, and integration.

1. **Sensation** is the conscious or subconscious detection of changes in the external or internal environment.
2. **Perception** is the conscious awareness and interpretation of sensations.
3. A **sensory modality** is one unique type of sensation. A sensory receptor has **selectivity**, in that it responds vigorously to only one kind of **stimulus**.
4. Sensory modalities are grouped into two classes: general senses and special senses. **General senses** include **somatic senses** and **visceral senses**. Somatic sensory modalities include tactile, thermal, pain, and proprioceptive sensations. Visceral sensations mediated by the autonomic nervous system provide information about conditions within internal organs. **Special senses** include the sensory modalities of smell, taste, vision, hearing, and equilibrium.
5. The process of sensation involves a sequence of four events: (1) stimulation of a sensory receptor, (2) transduction of the stimulus, (3) generation of impulses, and (4) integration of sensory input.

## Concept

## Resources

### Concept 15.2 Sensory receptors can be classified structurally, functionally, or by the type of stimulus detected.

1. There are three types of sensory receptors based on structure: **free nerve endings** are bare dendrites; **encapsulated nerve endings** are dendrites enclosed in connective tissue; and **separate cells**.
2. Graded potentials have amplitudes that vary with the intensity of the stimulus. Two types are generator potentials and receptor potentials. Free nerve endings and encapsulated nerve endings produce **generator potentials**. Separate cells produce **receptor potentials**.
3. Sensory receptors can be grouped based on location of receptors and origin of the stimuli. **Exteroceptors** monitor the external environment. **Interoceptors** monitor the internal environment. **Proprioceptors** provide information about body position and movement.
4. Sensory receptors can also be distinguished by the type of stimulus they detect, including electromagnetic energy, mechanical energy, and chemical energy.
5. Adaptation is a decrease in sensitivity during a constant stimulus. There are **rapidly adapting receptors** and **slowly adapting receptors**.

Anatomy Overview—Chemoreceptors
Anatomy Overview—Baroreceptors
Anatomy Overview—Proprioceptors

### Concept 15.3 Somatic sensations include tactile sensations, thermal sensations, pain, and proprioception.

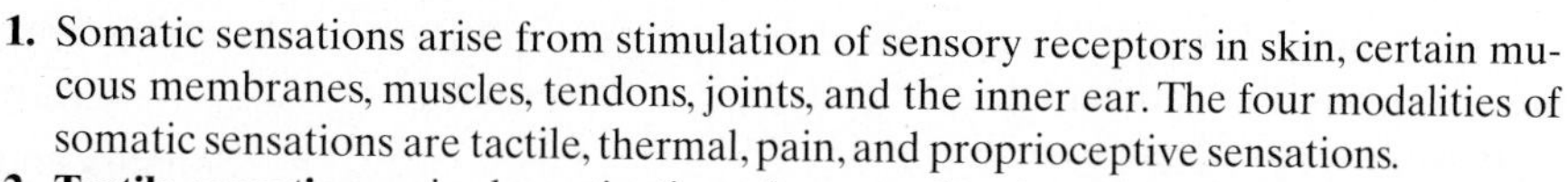

1. Somatic sensations arise from stimulation of sensory receptors in skin, certain mucous membranes, muscles, tendons, joints, and the inner ear. The four modalities of somatic sensations are tactile, thermal, pain, and proprioceptive sensations.
2. **Tactile sensations** arise by activation of encapsulated mechanoreceptors for touch, pressure, and vibration; and by activation of free nerve endings for itch and tickle sensations.
3. Rapidly adapting **touch** receptors include **corpuscles of touch** in hairless skin and **hair root plexuses** in hairy skin. Slowly adapting touch receptors include **tactile discs** in the epidermis of the skin and **Ruffini corpuscles** in the dermis of the skin and in ligaments and tendons.
4. **Pressure receptors** are widely distributed in the body, and include corpuscles of touch, tactile discs, and **lamellated corpuscles**.
5. **Vibration** receptors are corpuscles of touch and lamellated corpuscles.
6. **Itch** and **tickle** sensations result from stimulation of free nerve endings.
7. **Thermoreceptors** are free nerve endings on the skin that detect coldness and warmth. **Cold receptors** are in the epidermis, and **warm receptors** are in the dermis.
8. Receptors for pain are free nerve endings, called **nociceptors**, and are found in every body tissue except the brain. **Fast pain** is known as acute, sharp, or pricking pain. **Slow pain** is known as chronic, burning, aching, or throbbing pain. Depending on the location of the receptors, pain can be **superficial somatic pain**, **deep somatic pain**, or **visceral pain**. **Referred pain**, in a surface away from the affected organ, can also be experienced with visceral pain.
9. **Proprioceptive sensations** allow us to be aware of the movements and position of the head and limbs. **Proprioceptors** embedded in muscles and tendons convey to the brain the amount of tension and stretch in muscles and joints. **Muscle spindles**, composed of specialized **intrafusal muscles fibers** and nerve endings, are proprioceptors that facilitate the setting of overall **muscle tone** by the brain. The function of muscle spindles is to measure muscle length.
10. **Tendon organs** give information about changes in muscle tension.
11. **Joint kinesthetic receptors** sense joint position and movement.

Anatomy Overview—Sensory Receptors of the Skin

Figure 15.3—Distribution of Referred Pain
Figure 15.4—Two Types of Proprioceptors: A Muscle Spindle and a Tendon Organ

Clinical Connection—Analgesia: Relief from Pain

## Concept

## Resources

### Concept 15.4 Somatic sensory pathways relay information from sensory receptors to the cerebral cortex and cerebellum.

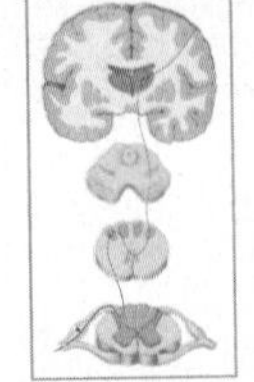

1. **Somatic sensory pathways** relay information from somatic sensory receptors to the primary somatosensory area in the cerebral cortex and to the cerebellum. **First-order neurons** conduct impulses from somatic receptors via nerves into the brain stem or spinal cord. **Second-order neurons** conduct impulses from where they decussate in the brain stem or spinal cord to the thalamus. **Third-order neurons** conduct impulses from the thalamus to the primary somatosensory area of the cortex on the same side.
2. The **posterior column–medial lemniscus pathway** conducts impulses for conscious proprioception, touch, and vibratory sensations to the primary somatosensory area of the cerebral cortex.
3. The **spinothalamic pathway** relays impulses for pain, temperature, tickle, and itch sensations to the primary somatosensory area of the cerebral cortex.
4. The **posterior** and **anterior spinocerebellar tracts** of the spinal cord are the major routes for proprioceptive impulses to travel to the cerebellum. These are critical for posture, balance, and coordination of skilled movements.

Animation—Somatic Sensory Pathways

Figure 15.5—Somatic Sensory Pathways

### Concept 15.5 The somatosensory and primary motor areas of the cerebral cortex unequally serve different body regions.

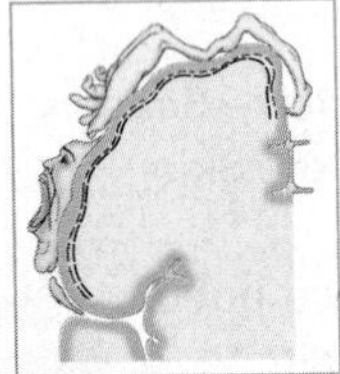

1. Precise localization of somatic sensations occurs when impulses arrive at the **primary somatosensory area** in the parietal lobes. Each region of the **primary somatosensory** area receives sensory input from different body areas.
2. Voluntary body movements are controlled principally in the **primary motor area** in the frontal lobe. The size of the cortical area governing a muscle is proportional to the number of motor units in the muscle.

Anatomy Overview—Cerebrum

Figure 15.6—Somatic Sensory and Somatic Motor Maps in the Cerebral Cortex

### Concept 15.6 Somatic motor pathways carry impulses from the brain to effectors.

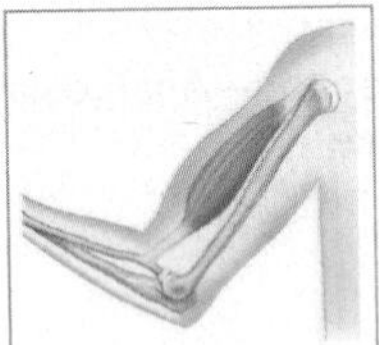

1. Axons of **lower motor neurons (LMNs)** extend from cranial and spinal nerves to skeletal muscles.
2. Neurons in the brain and spinal cord give input to lower motor neurons. **Local circuit neurons** help coordinate rhythmic activity in specific muscle groups. **Upper motor neurons** from the cerebral cortex give input for voluntary movements. **Basal nuclei neurons** help initiate and terminate movements, suppress unwanted movements, and establish normal muscle tone. **Cerebellar neurons** control activity of **upper motor neurons** to reduce movement errors.
3. **Direct motor pathways** have axons that extend from the cerebral cortex to **lower motor neurons**. **Indirect motor pathways** provide input to **lower motor neurons** from motor centers in the brain stem.
4. Three tracts are direct motor pathways: **lateral corticospinal tracts** control precise, skilled movements of the hands and feet; **anterior corticospinal tracts** control movements of the trunk and proximal limbs; and **corticobulbar tracts** control skeletal muscles of the head.
5. **Indirect motor pathways** include all somatic motor pathways other than the corticospinal and corticobulbar tracts.
6. Basal nuclei receive input from sensory, association, and motor areas of the cerebral cortex, then respond with feedback signals to the motor cortex. Basal nuclei suppress unwanted movements and influence muscle tone.
7. The cerebellum monitors intentions for movements, monitors actual movements, compares them, and sends out corrective signals to **upper motor neurons**.

Animation—Somatic Motor Pathways

Figure 15.7—Somatic Motor Pathways

Clinical Connection—Parkinson's Disease

Clinical Connection—Cerebral Palsy

| Concept | Resources |
|---|---|
| **Concept 15.7** Wakefulness and memory are integrative functions of the brain. | |
| 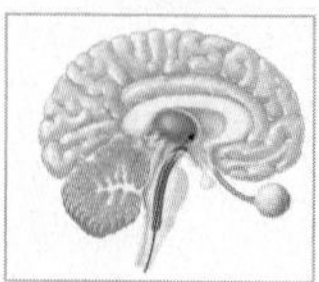 1. **Wakefulness** and **sleep** are integrative functions that are controlled by the **reticular activating system** (**RAS**) in the brain stem. The reticular activating system sends impulses to the cerebral cortex that increase its activity. **Arousal** from sleep involves increased RAS activity. The RAS can be activated by many different sensory stimuli, which then activate the cerebral cortex, resulting in **consciousness**. The RAS is inhibited during sleep. | Clinical Connection—Coma |
| 2. **Learning** is the ability to acquire new information or skills. | |
| 3. **Memory**, the ability to store and recall information or skills, involves structural changes in neurons and synapses within the brain. The three types are immediate, short-term, and long-term memory. | |

## Understanding the Concepts

1. How is a sensation different from a perception?
2. How are generator potentials and receptor potentials similar? How do they differ?
3. Which body parts would be more sensitive, those with a higher density of somatic sensory receptors or those with a lower density? Explain your answer.
4. What is referred pain, and how is it useful in diagnosing internal disorders?
5. What types of sensory deficits could be produced by damage to the right lateral spinothalamic tract?
6. Why would you expect the face to be represented by a larger portion of the somatosensory area of the cerebral cortex than a physically larger area such as the trunk?
7. What is the significance of the genitals being represented in the primary somatosensory area but not in the primary motor area of the cerebral cortex? (*Hint:* Genital tissues contain smooth muscle but no skeletal muscle tissue.)
8. What movement-related differences would you expect to see with damage to the basal nuclei versus damage to the cerebellum?
9. Why should every sleeping room have a smoke detector?

CHAPTER

16

# Dan's Story

Dan rolled over in bed. The muscles in his back ached and he could feel the pain in his knees again, but it had been worth it. He played over the events of the last run in his head—the old wooden building, the steep second-story stairs, the screams of the terrified elderly woman. He didn't think about the danger when it was happening, he just went in. He was proud of being a volunteer firefighter.

He heard a soft giggle and felt tickling on his bare back. He knew it was little Danny tickling him, so he rolled over and grabbed the toddler, pulling him into bed. Dan could smell the sweet odor of dried milk and sugar-coated cereal on Danny's pajamas and fresh coffee brewing in the kitchen. He thought, "I'm lucky to smell anything at all after breathing in all that smoke last night."

"Dan! We have to get going to the beach before it gets too crowded," Dan's wife Pam called from the kitchen.

"Go tell Mommy I'll be there in a minute," said Dan.

Dan swung his feet over the edge of the bed. He reached for the window shade, but his hand slipped. The spring-loaded roll shade flung open with a snap, flooding the room with the bright morning sun and causing Dan to squint. "No clouds today, I guess," he thought to himself. He reached for a pair of sunglasses on the nightstand. Dan walked into the kitchen still wearing the sunglasses.

The phone rang suddenly. Dan picked it up. "Three alarms, huh? Alright. Bye." Dan turned to his wife and said, "There's been a fire at the chemical plant down by the county line; they need everyone they can get. If that place explodes, it will take out half the town. I'm sorry. We'll get to the beach another day, but I have to go."

Consider the sensations experienced by Dan. Like many of us, he takes his special senses for granted. As we will learn, the special senses are not only important for sensing our environment, but also for enjoying the richness of life.

# The Special Senses

## INTRODUCTION

Recall from Chapter 15 that the general senses include somatic sensations (tactile, thermal, pain, and proprioceptive) and visceral sensations. As you learned in that chapter, receptors for the general senses are scattered throughout the body and are relatively simple in structure. Receptors for the special senses—smell, taste, vision, hearing, and equilibrium—are anatomically distinct from one another and are concentrated in complex sensory organs of the head, such as the eyes and ears. Neural pathways for the special senses are also more complex than those of the general senses.

Familiar sensations that you take for granted, such as pleasant smells, the sound of laughter, the taste of your favorite food, and the ability to see the face of someone you love or the beauty of a sunrise, are mediated by the special senses. In this chapter we examine the structure of the special sense organs, how they receive and respond to stimuli, and the pathways involved in conveying their information to the central nervous system.

## CONCEPTS

**16.1** Impulses for smell propagate along the olfactory nerve to the brain.

**16.2** Impulses for taste propagate along the facial, glossopharyngeal, and vagus nerves to the brain.

**16.3** The eye is protected by eyelids, eyelashes, eyebrows, and a lacrimal apparatus.

**16.4** The eye is constructed of three layers.

**16.5** Image formation involves refraction of light rays, accommodation, pupil constriction, and convergence.

**16.6** The neural pathway of light is photoreceptors → bipolar cells → ganglion cells → optic nerve → primary visual cortex.

**16.7** The ear is divided into external, middle, and internal regions.

**16.8** The pathway of sound is tympanic membrane → ossicles → oval window → cochlea → vestibulocochlear nerve → primary auditory cortex.

**16.9** Impulses for equilibrium propagate along the vestibulocochlear nerve to the brain.

# 16.1 Impulses for smell propagate along the olfactory nerve to the brain.

Both smell and taste are chemical senses; the sensations arise from the interaction of molecules with smell or taste receptors. Because impulses for smell and taste propagate to the limbic system (and to higher cortical areas as well), certain odors and tastes can evoke strong emotional responses or a flood of memories.

## Anatomy of the Olfactory Epithelium

The nose contains between 10 and 100 million receptors for the sense of smell, or **olfaction** (ol-FAK-shun; *olfact-* = smell), contained within an area called the olfactory epithelium. The total area of the **olfactory epithelium** is 5 cm$^2$ (a little less than 1 in.$^2$). The olfactory epithelium occupies the superior part of the nasal cavity, covering the inferior surface of the cribriform plate of the ethmoid bone and extending along the superior nasal concha (Figure 16.1a). It consists of three types of cells: olfactory receptor cells, supporting cells, and basal cells (Figure 16.1b, c).

**Olfactory receptor cells** are the first-order neurons of the olfactory pathway. They are bipolar neurons with an exposed dendrite and an axon projecting through a foramen of the cribriform plate of the ethmoid bone and ending in the olfactory bulb. The parts of the olfactory receptor cell that respond to inhaled chemicals are the **olfactory hairs**, nonmotile cilia that project from the dendrite. **Odorants** are chemicals that can stimulate the olfactory hairs and, therefore, be detected as odors. Olfactory receptors respond to the chemical stimulation of an odorant molecule by producing a generator potential, thus initiating the olfactory response.

**Supporting cells** are columnar epithelial cells of the mucous membrane lining the nose. They provide physical support, nourishment, and electrical insulation for the olfactory receptor cells, and they help detoxify chemicals that come in contact with the

FIGURE 16.1 **Olfactory epithelium and olfactory receptor cells.** (a) Location of olfactory epithelium in the nasal cavity. (b) Axons of olfactory receptor cells extend through the cribriform plate and terminate in the olfactory bulb. (c) Histology of the olfactory epithelium.

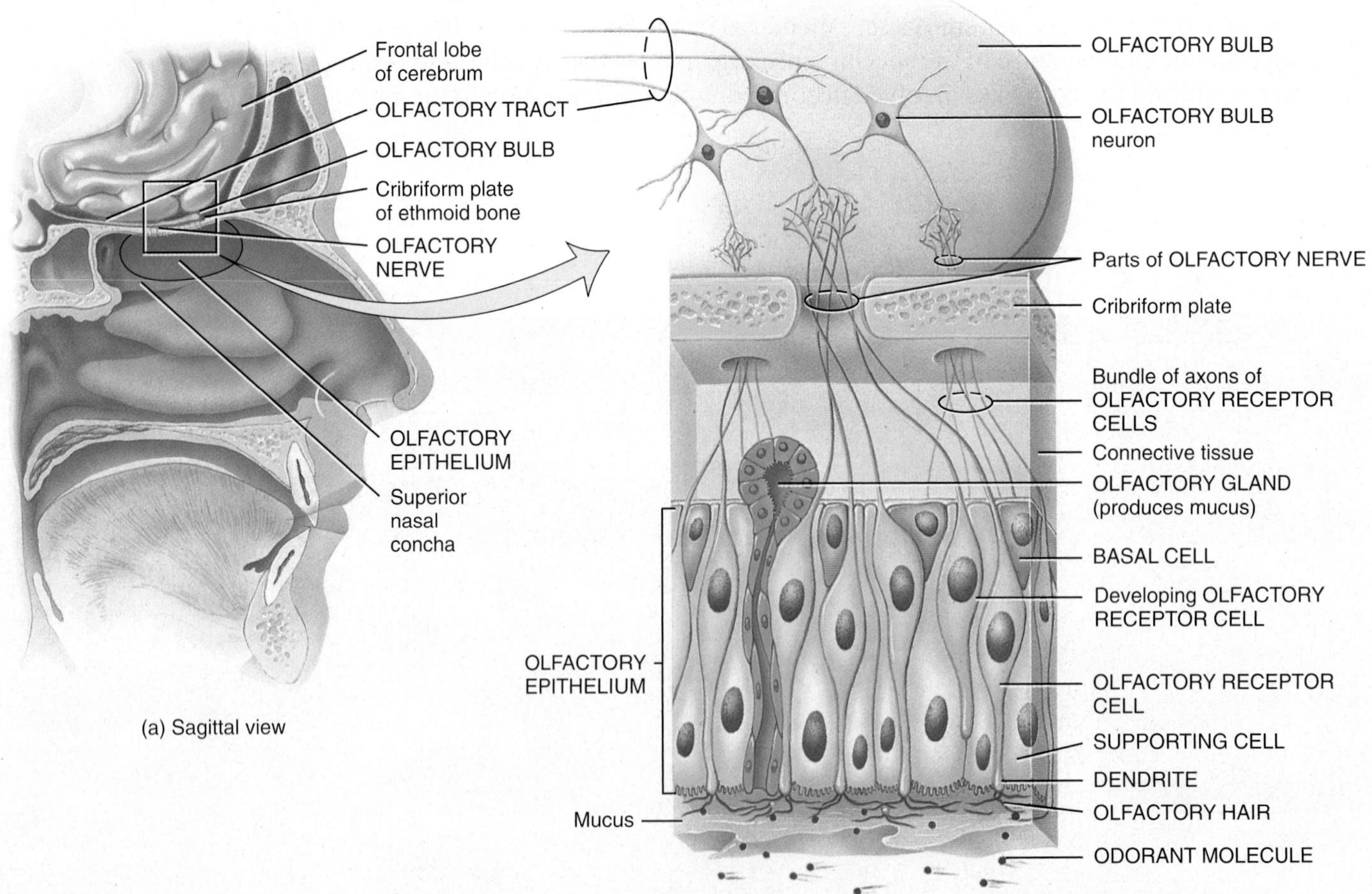

olfactory epithelium. **Basal cells** are stem cells that lie between the bases of the supporting cells and continually undergo cell division to produce new olfactory receptor cells, which live for only a month or so before being replaced. This process is remarkable considering that olfactory receptors are neurons, and, as you have already learned, mature neurons are generally not replaced.

Within the connective tissue that supports the olfactory epithelium are **olfactory glands**, which produce mucus that is carried to the surface of the epithelium by ducts. The secretion moistens the surface of the olfactory epithelium and dissolves odorants, allowing them to interact with olfactory hairs. Both supporting cells of the nasal epithelium and olfactory glands are innervated by branches of the facial nerve, which can be stimulated by certain chemicals. Impulses in the facial nerve in turn stimulate the lacrimal glands in the eyes and nasal mucous glands. The result is tears and a runny nose after inhaling substances such as pepper, onion, or the vapors of household ammonia.

## Physiology of Olfaction

Many attempts have been made to distinguish among and classify "primary" sensations of smell. Genetic evidence now suggests the existence of hundreds of primary odors. Our ability to recognize about 10,000 different odors probably depends on patterns of activity in the brain that arise from activation of many different combinations of olfactory receptor cells.

Olfactory receptor cells react to odorant molecules in the same way that most sensory receptors react to their specific stimuli: A depolarizing generator potential develops and triggers one or more action potentials. How the generator potential arises in olfactory receptors is known in some cases. Many odorants bind to receptor proteins, called *odorant binding proteins*, on the plasma membrane and activate the enzyme adenylate cyclase, which synthesizes cyclic adenosine monophosphate (cAMP) (Figure 16.2). The result is the following chain of events: cAMP opens sodium ion ($Na^+$) channels → $Na^+$ flows into olfactory receptor cell → depolarizing generator potential is generated → action potential is triggered → impulses are propagated along axon of olfactory receptor cell.

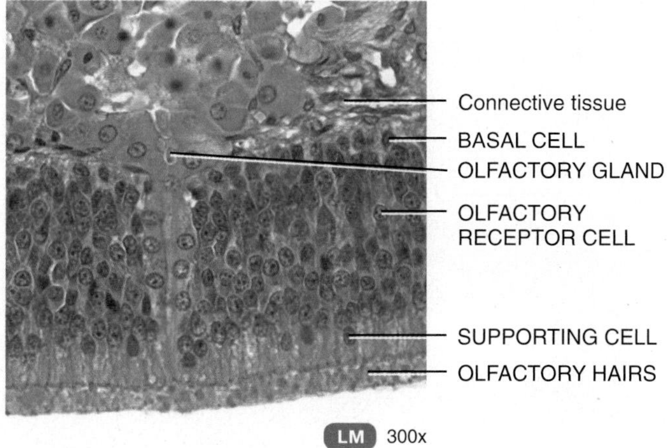


(c) Histology of olfactory epithelium

**The olfactory epithelium consists of olfactory receptor cells, supporting cells, and basal cells.**

FIGURE 16.2 **Olfactory transduction.** Binding of the odorant molecule to the plasma protein receptor activates adenylate cyclase, resulting in the production of cyclic AMP. Cyclic AMP (cAMP) opens sodium channels and $Na^+$ enters the olfactory receptor cell. The resulting depolarization may generate an action potential, which is transmitted to the axon terminals.

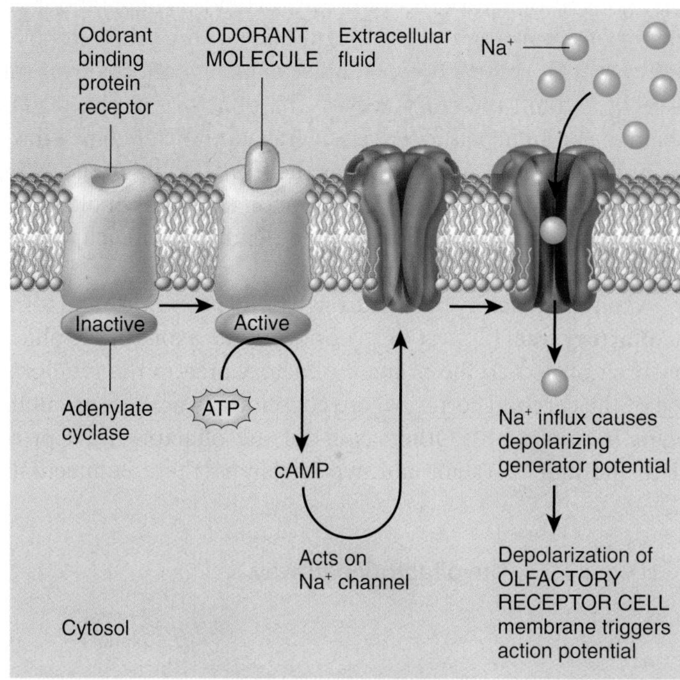


**Odorants can produce depolarizing generator potentials, which can lead to action potentials.**

## Odor Thresholds and Adaptation

Olfaction, like all of the special senses, has a low threshold. Only a few molecules of certain substances need be present in air to be perceived as an odor. A good example is the chemical methyl mercaptan, which smells like rotten cabbage and can be detected in concentrations as low as 1/25 billionth of a milligram per milliliter of air. Because the natural gas used for cooking and heating is odorless but lethal and potentially explosive if it accumulates, a small amount of methyl mercaptan is added to natural gas to provide olfactory warning of gas leaks.

*Adaptation* (decreasing sensitivity) to odors occurs rapidly. Olfactory receptors adapt by about 50 percent in the first second or so after stimulation but adapt very slowly thereafter. Still, complete insensitivity to certain strong odors occurs in about a minute after exposure. Apparently, reduced sensitivity involves an adaptation process in the central nervous system as well.

## The Olfactory Pathway

On each side of the nose, bundles of the slender, unmyelinated axons of olfactory receptor cells extend through about 20 olfactory foramina in the cribriform plate of the ethmoid bone (Figure 16.1b). These 40 or so bundles of axons collectively form the right and left **olfactory nerves**. The olfactory nerves terminate in the brain in paired masses of gray matter called the **olfactory bulbs**, which are located below the frontal lobes of the cerebrum. Within the olfactory bulbs, the axon terminals of olfactory receptor cells form synapses with the dendrites and cell bodies of olfactory bulb neurons.

Axons of olfactory bulb neurons extend posteriorly and form the **olfactory tract** (Figure 16.1b). Some of the axons of the olfactory tract project to the primary olfactory area in the temporal lobe of the cerebral cortex, where conscious awareness of smells begins (Figure 16.3). Other axons of the olfactory tract project to the limbic system and hypothalamus; these connections account for our emotional and memory-evoked responses to odors. Examples include sexual excitement upon smelling a certain perfume, nausea upon smelling a food that once made you violently ill, or an odor-evoked memory of a childhood experience.

From the primary olfactory area, pathways also project to the frontal lobe, an important region for odor identification and discrimination. People who suffer damage to this area have been reported to have difficulty identifying different odors.

### ✓ CHECKPOINT

1. What is the function of the supporting cells of the nasal epithelium?
2. How do basal cells contribute to olfaction?
3. Olfactory receptor cells are unique among neurons because they can undergo what process?
4. Why must the olfactory epithelium have a coating of mucus?
5. How long would it take for your olfactory receptor cells to adapt to the smell of something very rotten?

FIGURE 16.3 **The olfactory pathway.**

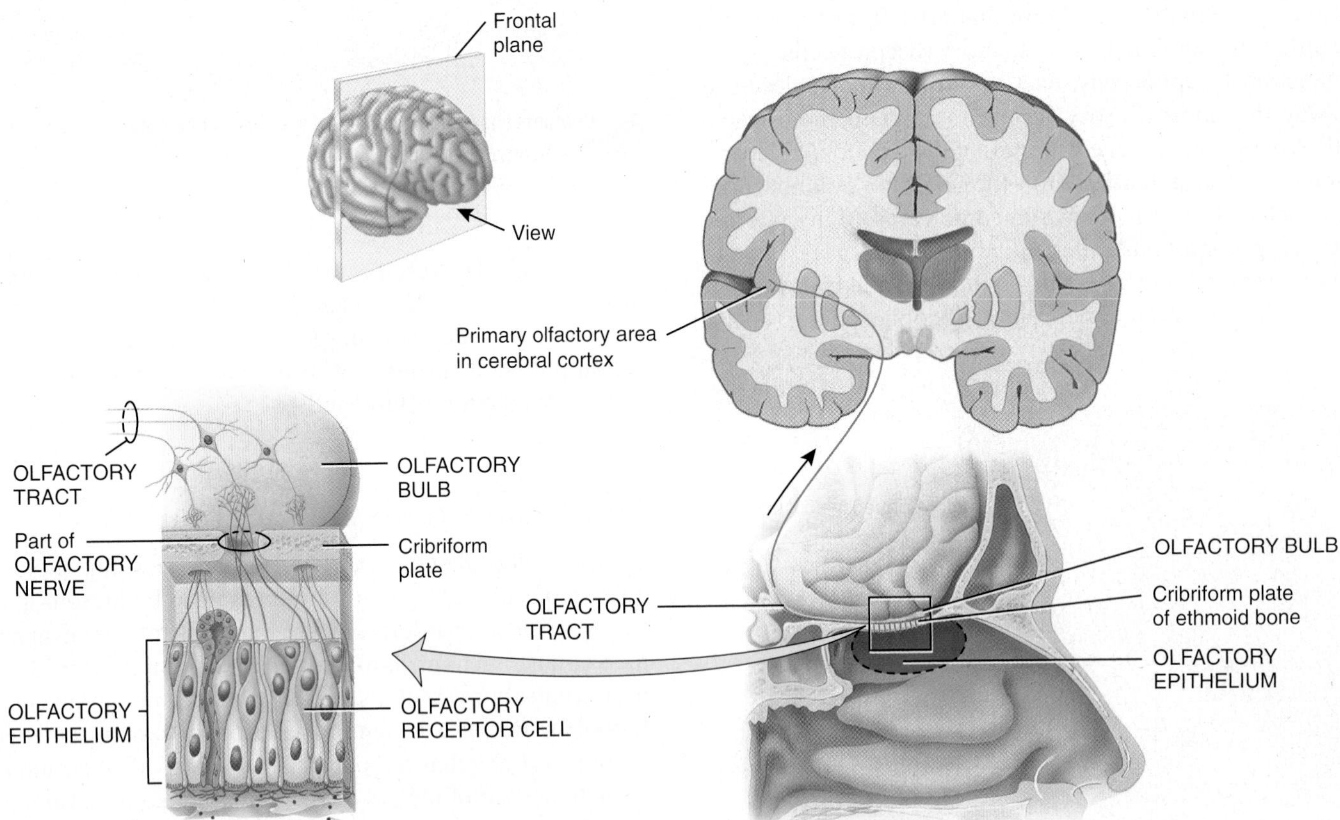


**From the olfactory epithelium, olfactory impulses are propagated along the olfactory bulb and tract to the primary olfactory area of the cerebral cortex.**

# 16.2 Impulses for taste propagate along the facial, glossopharyngeal, and vagus nerves to the brain.

Taste or **gustation** (gus-TĀ-shun; *gust-* = taste), like olfaction, is a chemical sense. However, taste is much simpler than olfaction in that only five primary tastes are considered distinguishable: *sour, sweet, bitter, salty,* and *umami* (ū-MAM-ē). The umami taste, more recently discovered than the other tastes, was first reported by Japanese scientists and is described as "meaty" or "savory." Umami is believed to arise from taste receptors that are stimulated by monosodium glutamate (MSG), a substance naturally present in many foods and added to others as a flavor enhancer. All other flavors, such as chocolate, pepper, and coffee, are combinations of the five primary tastes, plus accompanying olfactory and tactile (touch) sensations. Odors from food can pass upward from the mouth into the nasal cavity, where they stimulate olfactory receptor cells. Because olfaction is much more sensitive than taste, a given concentration of a food substance may stimulate the olfactory system thousands of times more strongly than it stimulates the gustatory system. When you have a cold or are suffering from allergies and cannot taste your food, it is actually olfaction that is blocked, not taste.

## Anatomy of Taste Buds and Papillae

The receptors for sensations of taste are located in the taste buds (Figure 16.4). Most of the nearly 10,000 taste buds of a young adult are on the tongue, but some are found on the soft palate (posterior portion of the roof of the mouth), pharynx (throat), and epiglottis (cartilage lid over voice box). The number of taste buds declines with age.

FIGURE 16.4 **The relationship of gustatory receptor cells in taste buds to tongue papillae.**

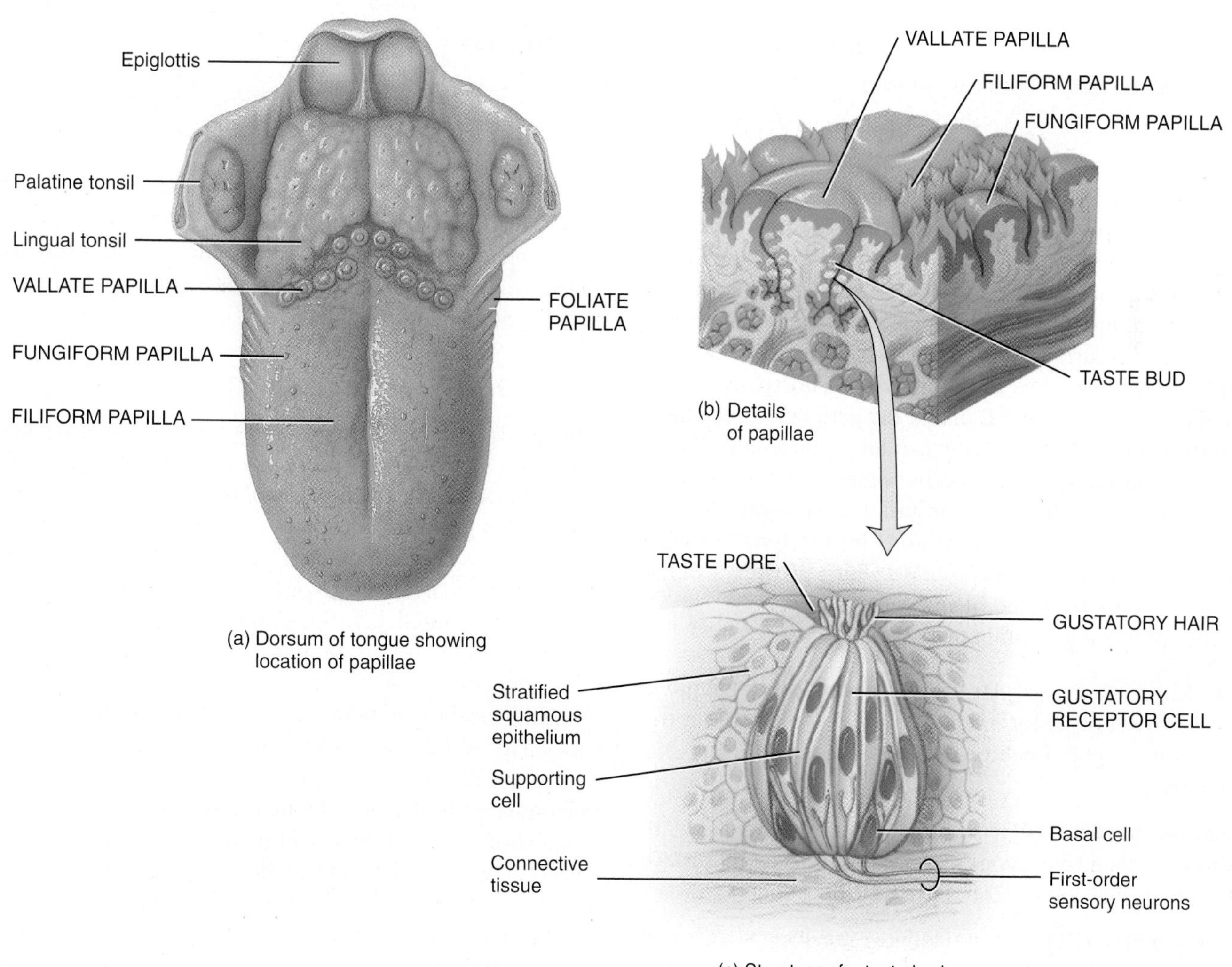


(continues)

FIGURE 16.4 (continued)

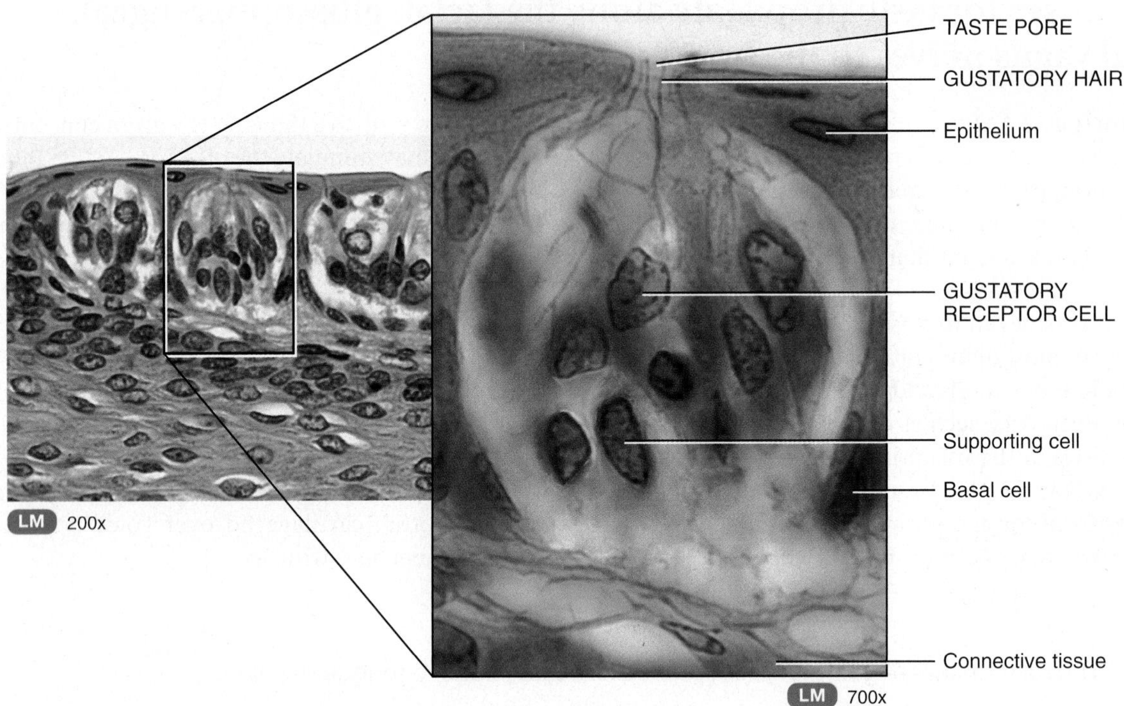


(d) Histology of a taste bud from a vallate papilla

Gustatory receptor cells are located in taste buds.

Each **taste bud** is an oval body consisting of three kinds of epithelial cells: supporting cells, gustatory receptor cells, and basal cells (Figure 16.4c, d). The *supporting cells* surround the **gustatory receptor cells** (GUS-ta-tō-rē). A single, long microvillus, called a **gustatory hair**, projects from each gustatory receptor cell to the external surface through the **taste pore**, an opening in the taste bud. Each gustatory receptor cell has a life span of about 10 days. *Basal cells*, stem cells found at the periphery of the taste bud near the connective tissue layer, produce supporting cells. At their base, the gustatory receptor cells synapse with dendrites of the first-order neurons that begin the gustatory pathway.

Taste buds are found in elevations on the tongue called **papillae** (pa-PIL-ē; singular is *papilla*), which provide a rough texture to the upper surface of the tongue (Figure 16.4a, b). Three types of papillae contain taste buds:

- About 12 very large, circular **vallate (circumvallate) papillae** (VAL-āt = wall-like) form an inverted V-shaped row at the back of the tongue. Each of these papillae houses 100–300 taste buds.
- **Fungiform papillae** (FUN-ji-form = mushroomlike) are mushroom-shaped elevations scattered over the entire surface of the tongue that contain about five taste buds each.
- **Foliate papillae** (FŌ-lē-āt = leaflike) are located in small trenches on the lateral margins of the tongue, but most of their taste buds degenerate in early childhood.

In addition, the entire surface of the tongue has **filiform papillae** (FIL-i-form = threadlike). These pointed, threadlike structures contain tactile receptors but no taste buds. They increase friction between the tongue and food, making it easier for the tongue to move food in the oral cavity.

## Physiology of Gustation

Chemicals that stimulate gustatory receptor cells are known as **tastants**. Once a tastant is dissolved in saliva, it can make contact with the plasma membrane of the gustatory hairs, which are the sites on gustatory receptor cells that respond to tastants. The result is a receptor potential that stimulates release of neurotransmitter molecules from the gustatory receptor cell. In turn, the liberated neurotransmitter molecules trigger action potentials in the first-order neurons that synapse with gustatory receptor cells.

The receptor potential arises differently for different tastants. The sodium ions ($Na^+$) in a salty food enter gustatory receptor cells via $Na^+$ channels in the plasma membrane. The accumulation of $Na^+$ inside the receptor cell causes depolarization, which leads to release of neurotransmitter. The hydrogen ions ($H^+$) in sour tastants may flow into gustatory receptor cells via $H^+$ channels, and produce a depolarization that leads to release of neurotransmitter. Other tastants, responsible for stimulating sweet, bitter, and umami tastes, do not themselves enter gustatory receptor cells. Rather, they bind to receptor proteins

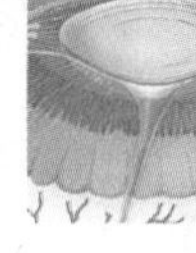

on the plasma membrane that activate several different chemicals inside the gustatory receptor cell. Again, the result is the same—release of neurotransmitter.

If all tastants cause release of neurotransmitter from many gustatory receptor cells, why do foods taste different? The answer to this question is thought to lie in the patterns of impulses in groups of first-order gustatory neurons that synapse with the gustatory receptor cells. Different tastes elicit activation of different groups of gustatory neurons. In addition, although each individual gustatory receptor cell responds to more than one of the five primary tastes, it may respond more strongly to some tastants than to others.

## Taste Thresholds and Adaptation

The threshold for taste varies for each of the primary tastes. The threshold for bitter substances, such as quinine, is lowest. Because poisonous substances often are bitter, the low threshold (or high sensitivity) may have a protective function. The threshold for sour substances, such as lemon, as measured by using hydrochloric acid, is somewhat higher. The thresholds for salty substances, represented by sodium chloride, and for sweet substances, as measured by using sucrose, are higher than those for bitter or sour substances.

Complete adaptation to a specific taste can occur in 1–5 minutes of continuous stimulation. Taste adaptation is due to changes that occur in the gustatory receptor cells, in olfactory receptor cells, and in neurons of the gustatory pathway in the CNS.

## The Gustatory Pathway

Three cranial nerves contain axons of gustatory neurons that innervate the taste buds. The facial nerve serves taste buds in the anterior two-thirds of the tongue, the glossopharyngeal nerve serves taste buds in the posterior one-third of the tongue, and the vagus nerve serves taste buds in the throat and epiglottis (Figure 16.5). From taste buds, impulses propagate along these cranial nerves to the medulla oblongata. From the medulla oblongata, some axons carrying taste signals project to the limbic system and the hypothalamus; others project to the thalamus. Taste signals that project from the thalamus to the primary gustatory area in the parietal lobe of the cerebral cortex (see Figure 13.10) give rise to the conscious perception of taste.

**FIGURE 16.5 The gustatory pathway.**

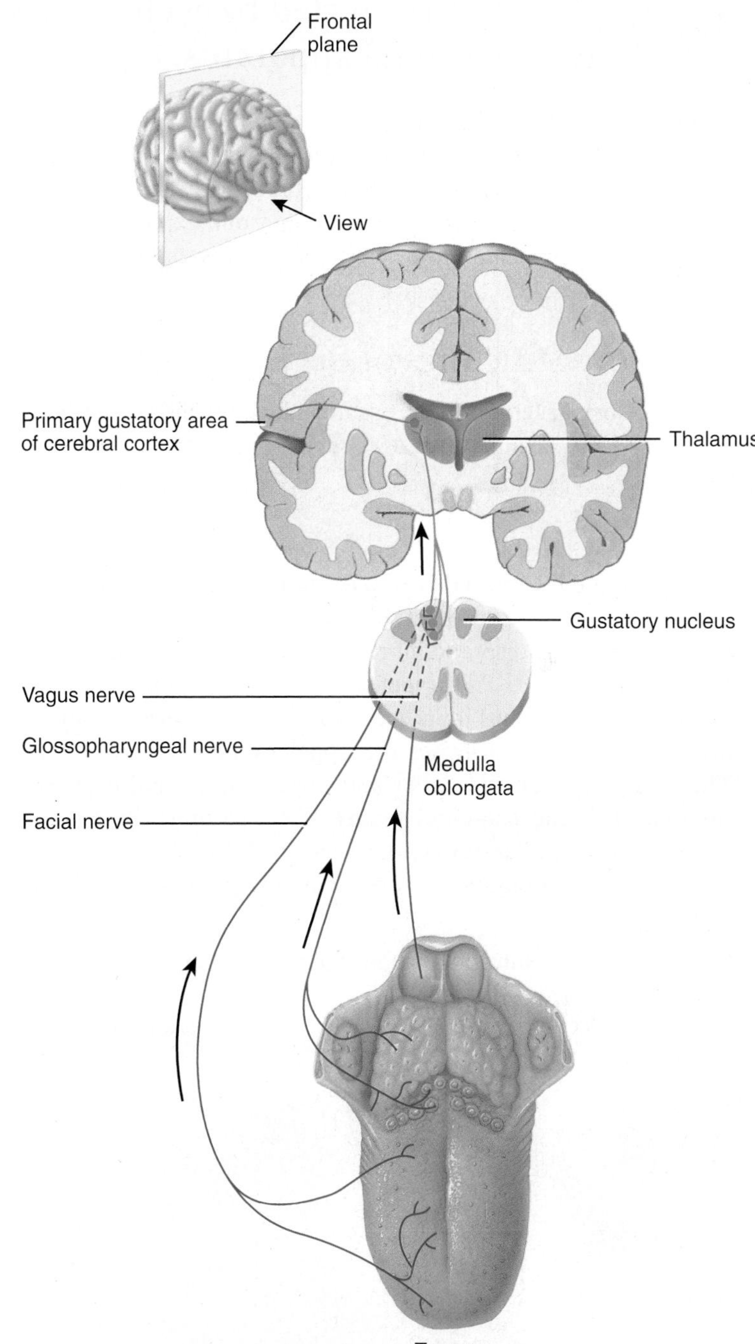


**When gustatory impulses arrive at the primary gustatory area of the cerebral cortex, we consciously perceive taste.**

### ✓ CHECKPOINT

6. For each of the primary tastes, give an example of a food that strongly represents that taste.
7. Where on the tongue is each of the four types of papillae located?
8. What is the function of supporting cells in taste buds?
9. Sensory receptors can be classified according to the type of stimulus they detect. Which type of receptors are olfactory and gustatory receptor cells (see Concept 15.2)?
10. What is the sequence of events from the binding of a tastant molecule to a gustatory hair to the generation of an action potential in a first-order gustatory neuron?

# 16.3 The eye is protected by eyelids, eyelashes, eyebrows, and a lacrimal apparatus.

**Vision** or sight is extremely important to human survival. More than half of the sensory receptors in the human body are located in the eyes, and a large part of the cerebral cortex is devoted to processing visual information. We begin our exploration of the eyes by introducing the accessory structures that protect or assist their function.

## Accessory Structures of the Eye

The **accessory structures** of the eye include the eyelids, eyelashes, eyebrows, lacrimal apparatus, and extrinsic eye muscles.

### Eyelids

The upper and lower **eyelids**, or **palpebrae** (PAL-pe-brē), shade the eyes during sleep, protect the eyes from excessive light and foreign objects, and spread lubricating secretions over the eyeballs (Figure 16.6). The upper eyelid is more movable than the lower and contains in its superior region the **levator palpebrae superioris**, which raises the upper eyelid (Figure 16.7a). The space between the upper and lower eyelids that exposes the eyeball is the **palpebral fissure** (PAL-pe-bral). Its angles are known as the **lateral commissure** (KOM-i-shur), which is closer to the temporal bone, and the **medial commissure**, which is more proximal to the nasal bone. In the medial commissure is a small, reddish elevation, the **lacrimal caruncle** (KAR-ung-kul), which contains sebaceous (oil) glands and sudoriferous (sweat) glands. The whitish material that sometimes collects in the medial commissure comes from these glands.

From superficial to deep, each eyelid consists of skin overlying the orbicularis oculi, meibomian glands, and conjunctiva (Figure 16.7a). The **orbicularis oculi** muscle closes the eyelid. Embedded in each eyelid is a row of modified sebaceous glands, known as **meibomian glands** (mī-BŌ-mē-an), which secrete a fluid that helps keep the eyelids from adhering to each other. The **conjunctiva** (kon′-junk-TĪ-va) is a thin, protective mucous membrane. The **palpebral conjunctiva** lines the inner aspect of each eyelid. The **bulbar conjunctiva** passes from the eyelids onto the anterior surface of the eyeball, where it covers the sclera (the "white" of the eye) but not the cornea, which is a transparent region that forms the outer anterior surface of the eyeball. Both the sclera and the cornea are discussed in more detail shortly. Dilation and congestion of the blood vessels of the bulbar conjunctiva due to local irritation or infection are the cause of *bloodshot eyes*.

**FIGURE 16.6** Surface anatomy of the right eye.

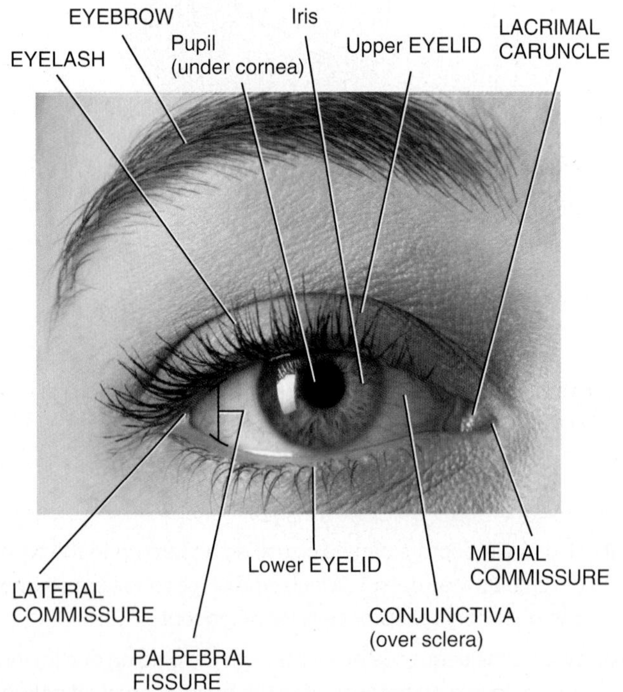


The eye is protected by the eyebrow, eyelashes, and eyelids.

### Eyelashes and Eyebrows

The **eyelashes** project from the border of each eyelid, and the **eyebrows** arch transversely above the upper eyelids. Eyelashes and eyebrows help protect the eyeballs from foreign objects, perspiration, and the direct rays of the sun.

### The Lacrimal Apparatus

The **lacrimal apparatus** (LAK-ri-mal; *lacrim-* = tears) is a group of structures that produces and drains **lacrimal fluid**, or **tears**. The **lacrimal glands**, each about the size and shape of an almond, secrete lacrimal fluid, which drains into **lacrimal ducts** that empty tears onto the surface of the conjunctiva of the upper lid (Figure 16.7b). From there, the tears pass medially over the anterior surface of the eye to enter two small openings called **lacrimal puncta**. Tears then pass into two ducts, the

**CLINICAL CONNECTION | *Conjunctivitis (pinkeye)***

**Conjunctivitis**, inflammation of the conjunctiva, is caused by bacteria such as pneumococci, staphylococci, or *Hemophilus influenzae.* It is very contagious and more common in children. Conjunctivitis may also be caused by irritants such as dust, smoke, or pollutants in the air, in which case it is not contagious.

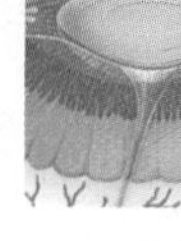

FIGURE 16.7 Accessory structures of the eye.

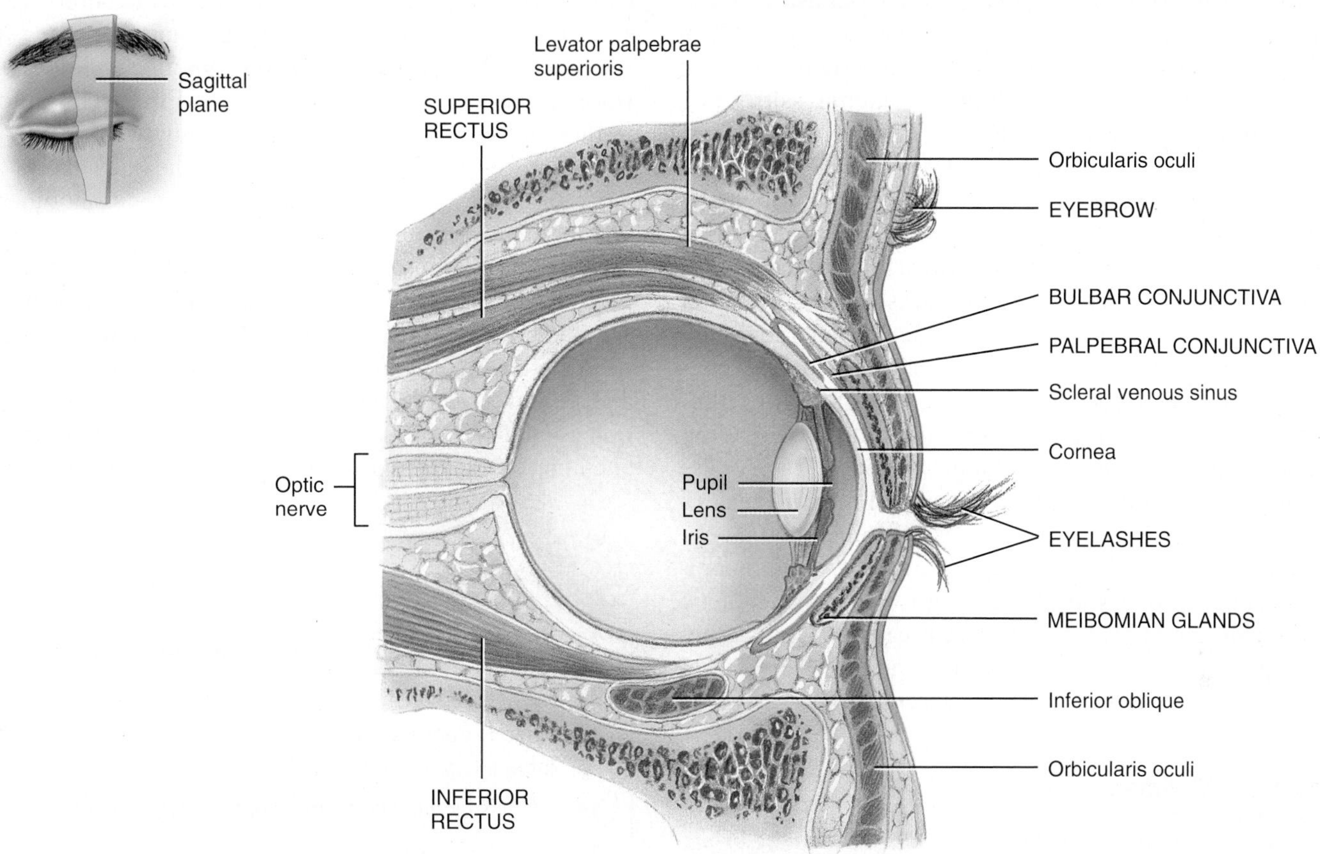


(a) Sagittal section of eye and its accessory structures

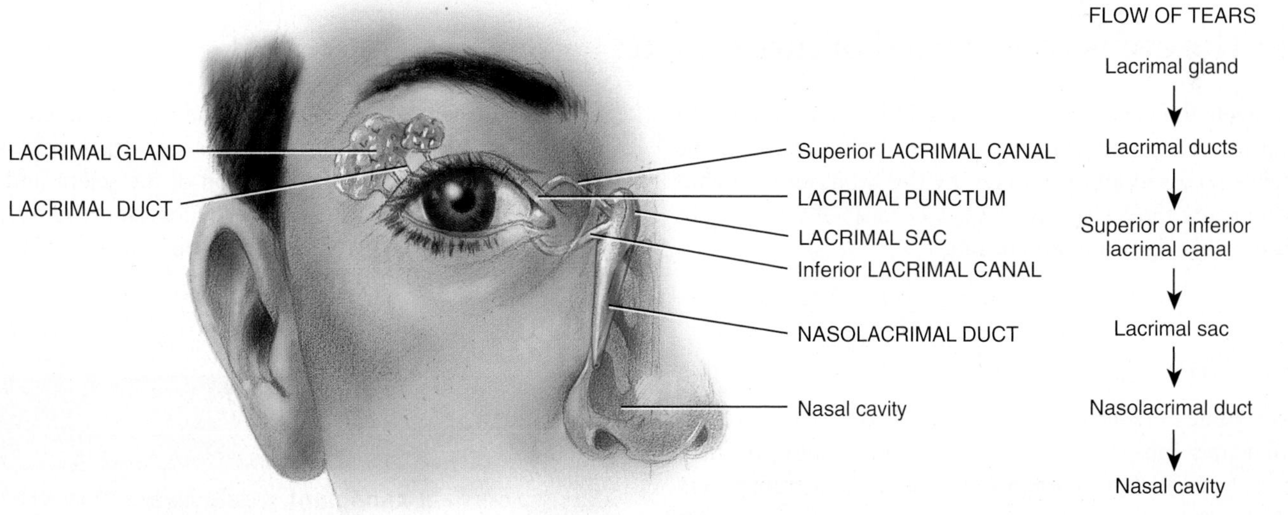


(b) Anterior view of lacrimal apparatus

**Accessory structures of the eye include the eyelids, eyelashes, eyebrows, the lacrimal apparatus, and extrinsic eye muscles.**

**lacrimal canals**, which lead into the **lacrimal sac**. As tears fill the lacrimal sac, they are pumped into the **nasolacrimal duct** by the blinking muscular action of the orbicularis oculi (see Figure 11.3). The nasolacrimal duct carries tears into the nasal cavity.

Lacrimal fluid is a watery solution containing salts, some mucus, and **lysozyme**, a protective bactericidal enzyme. The fluid protects, cleans, lubricates, and moistens the eyeball. After being secreted by the lacrimal gland, lacrimal fluid is spread medially over the surface of the eyeball by the blinking of the eyelids. Each gland produces about 1 mL of lacrimal fluid per day.

Normally, tears are cleared away as fast as they are produced, either by evaporation or by passing into the lacrimal canals and then into the nasal cavity. If an irritating substance makes contact with the conjunctiva, however, the lacrimal glands are stimulated to oversecrete, and tears accumulate (watery eyes) as the tears dilute and wash away the irritating substance. Watery eyes also occur when an inflammation of the nasal mucosa, such as occurs with a cold, obstructs the nasolacrimal ducts and blocks drainage of tears. Only humans express emotions, both happiness and sadness, by **crying**. In response to parasympathetic stimulation, the lacrimal glands produce excessive lacrimal fluid that may spill over the edges of the eyelids and even fill the nasal cavity with fluid. This is how crying produces a runny nose.

### Extrinsic Eye Muscles

The eyes sit in the bony depressions of the skull called the *orbits*. The extrinsic eye muscles extend from the walls of the orbit of the skull to the sclera (white) of the eyeball. These muscles are capable of moving the eye in almost any direction. Six extrinsic eye muscles move each eyeball: the **superior rectus**, **inferior rectus**, **lateral rectus**, **medial rectus**, **superior oblique**, and **inferior oblique** (see Figures 11.4, 16.7a, and 16.8). The motor units in these muscles are small and serve only two or three muscle fibers—fewer than in any other part of the body except the larynx (voice box). Such small motor units permit smooth, precise, and rapid movements of the eyes. As indicated in Figure 11.4b, the extrinsic eye muscles move the eyeball laterally, medially, superiorly, and inferiorly. For example, looking to the right requires simultaneous contraction of the right eye's lateral rectus and left eye's medial rectus and relaxation of the left eye's lateral rectus and right eye's medial rectus. The oblique muscles preserve rotational stability of the eyeball. Neurons in the brain stem and cerebellum coordinate and synchronize the movements of the eyes.

**✓ CHECKPOINT**

**11.** Give the name and function of each muscle of the eyelids.

**12.** Why do eyelids of healthy eyes not stick together?

**13.** Which structure shown in Figure 16.7 is continuous with the inner lining of the eyelids?

**14.** What is lacrimal fluid, and what are its functions?

**15.** Why does your nose run when you cry?

## 16.4 The eye is constructed of three layers.

The adult **eyeball** measures about 2.5 cm (1 in.) in diameter. Of its total surface area, only the anterior one-sixth is exposed; the remainder is recessed and protected by the orbit, into which it fits. Anatomically, the wall of the eyeball consists of three layers: fibrous tunic, vascular tunic, and retina.

### Fibrous Tunic

The **fibrous tunic** (TOO-nik), the superficial layer of the eyeball, is avascular and consists of the anterior cornea and posterior sclera (Figure 16.8). The **cornea** (KOR-nē-a) is a transparent coat that covers the colored iris. Because it is curved, the cornea helps focus light onto the retina. The cornea is composed primarily of dense connective tissue, but the many layers of collagen fibers are arranged to allow transmission of light. Since the central part of the cornea receives oxygen from the outside air, contact lenses that are worn for long periods of time must be permeable to permit oxygen to pass through them. The **sclera** (SKLE-ra; *scler-* = hard), the "white" of the eye, is a layer of dense connective tissue. The sclera covers the entire eyeball except the cornea; it gives shape to the eyeball, makes it more rigid, protects its inner parts, and is an attachment site for extrinsic eye muscles. At the junction of the sclera and cornea is an opening known as the **scleral venous sinus**. A fluid called *aqueous humor* drains into this sinus (see Figure 16.11).

**CLINICAL CONNECTION | *Corneal Transplant***

During a **corneal transplant**, a defective cornea is removed and a donor cornea of similar diameter is sewn in. It is the most common and most successful transplant operation. Since the cornea is avascular, antibodies in the blood that might cause rejection do not enter the transplanted tissue, and rejection rarely occurs. The shortage of donor corneas has been partially overcome by the development of artificial corneas made of plastic.

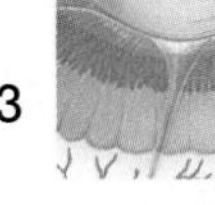



FIGURE 16.8 **Anatomy of the eyeball.**

Transverse plane

ANTERIOR CAVITY (contains AQUEOUS HUMOR):
ANTERIOR CHAMBER
POSTERIOR CHAMBER
SCLERAL VENOUS SINUS
Light
Visual axis
CORNEA
PUPIL
IRIS
LENS
ZONULAR FIBERS
Bulbar conjunctiva
Lacrimal sac
CILIARY BODY:
CILIARY MUSCLE
CILIARY PROCESS
ORA SERRATA
RETINA
CHOROID
SCLERA
Medial rectus
Lateral rectus
VITREOUS CHAMBER (contains VITREOUS BODY)
MEDIAL
LATERAL
CENTRAL RETINAL ARTERY AND VEIN
OPTIC NERVE
OPTIC DISC
FOVEA CENTRALIS

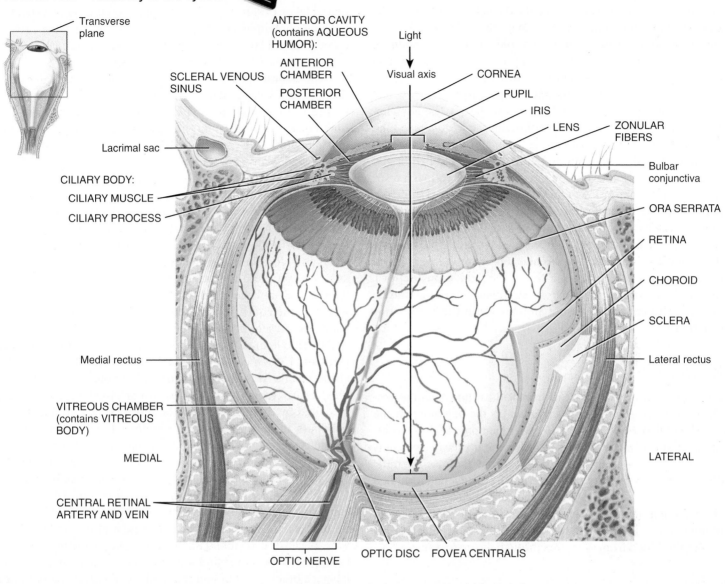

Superior view of transverse section of right eyeball

**The wall of the eyeball consists of three layers: the fibrous tunic, the vascular tunic, and the retina.**

## Vascular Tunic

The **vascular tunic** is the middle layer of the eyeball and has three parts: choroid, ciliary body, and iris (Figure 16.8). The highly vascularized **choroid** (KŌ-royd), which is the posterior portion of the vascular tunic, lines most of the internal surface of the sclera. Its numerous blood vessels provide nutrients to the retina and, because of its dark pigmentation, absorbs stray light rays, which prevents reflection and scattering of light within the eyeball. As a result, images cast on the retina remain sharp and clear.

In the anterior portion of the vascular tunic, the choroid becomes the **ciliary body** (SIL-ē-ar′-ē). It extends from the **ora serrata** (Ō-ra ser-RĀ-ta), the jagged anterior margin of the retina, to a point just posterior to the junction of the sclera and cornea. The ciliary body consists of the ciliary processes and the ciliary muscle. The **ciliary processes** are protrusions or folds on the internal surface of the ciliary body. They contain capillaries that secrete aqueous humor. Extending from the ciliary process are **zonular fibers (suspensory ligaments)** that attach to the lens. The **ciliary muscle** is a circular band of smooth muscle. Contraction or relaxation of the ciliary muscle changes the tightness of the zonular fibers, which alters the shape of the lens, adapting it for near or far vision.

The **iris**, the colored portion of the eyeball, is shaped like a flattened donut. It is suspended between the cornea and the lens and is attached at its outer margin to the ciliary processes. It consists of melanin-producing melanocytes and circular and radial smooth muscle fibers that regulate the diameter of the pupil. The amount of melanin in the iris determines the eye color. The eyes appear brown to black when the iris contains a large amount of melanin, blue when its melanin concentration is very low, and green when its melanin concentration is moderate.

The iris regulates the amount of light entering the eyeball through the **pupil**, the hole in the center of the iris. The pupil appears black because, as you look through the lens, you see the heavily pigmented back of the eye (the choroid). However, if bright light is directed into the pupil, the reflected light is red because of the blood vessels on the surface of the retina. It is for this reason that a person's eyes sometimes appear red in a photograph when the flash is directed into the pupil. Autonomic reflexes regulate pupil diameter in response to light levels (Figure 16.9). When bright light stimulates the eye, parasympathetic neurons stimulate the **circular muscle** of the iris to contract, causing a decrease in the size of the pupil (constriction). In dim light, sympathetic neurons stimulate the **radial muscle** of the iris to contract, causing an increase in the pupil's size (dilation).

## Retina

The third and inner layer of the eyeball, the **retina**, lines the posterior three-quarters of the eyeball and is the beginning of the visual pathway (see Figure 16.8). (A mnemonic that might assist your study of the layers present in the posterior eyeball is "Scary Cadavers—Run!" for *S*clera, *C*horoid, *R*etina.) The **optic disc** is the site where the **optic nerve** exits the eyeball. Bundled together with the optic nerve are the **central retinal artery** and **central retinal vein**. Branches of the central retinal artery fan out to nourish the anterior surface of the retina; the **central retinal vein** drains blood from the retina through the optic disc.

**FIGURE 16.9 Responses of the pupil to light of varying brightness.**

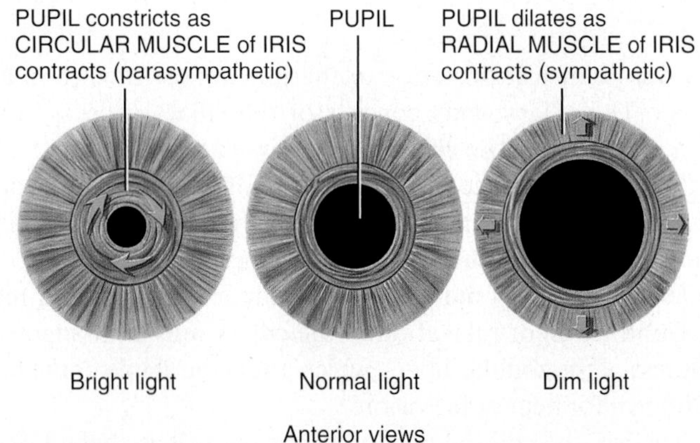
PUPIL constricts as CIRCULAR MUSCLE of IRIS contracts (parasympathetic)
PUPIL
PUPIL dilates as RADIAL MUSCLE of IRIS contracts (sympathetic)
Bright light
Normal light
Dim light
Anterior views

**Contraction of the circular muscle constricts the pupil; contraction of the radial muscle dilates the pupil.**

The retina consists of an outer pigmented layer and an inner neural layer (Figure 16.10b). The **pigmented layer** is a sheet of melanin-containing epithelial cells located between the choroid and the neural layer. The melanin in the pigmented layer of the retina, as in the choroid, helps to absorb stray light rays.

The **neural layer** is a multilayered outgrowth of the brain that processes visual data before sending impulses into axons that form the optic nerve. The neural layer contains three distinct layers of retinal neurons—**photoreceptors**, **bipolar cells**, and **ganglion cells**—separated by two zones, the *outer* and *inner synaptic layers*, where synaptic contacts are made (Figure 16.10a, c). Note that light passes through the ganglion and bipolar cell layers before it reaches the photoreceptors.

Photoreceptors are specialized cells that begin the process of converting light rays to impulses. There are two types of photoreceptors: rods and cones. Each retina has about 6 million cones and 120 million rods. **Rods** allow us to see in dim light, such as moonlight. Because rods do not provide color vision, in dim light we see only black, white, and shades of gray. Brighter lights stimulate the **cones**, which produce color vision. Three types of cones are present in the retina: (1) *blue cones*, which are sensitive to blue light, (2) *green cones*, which are sensitive to green light, and (3) *red cones*, which are sensitive to red light. Color vision results from the stimulation of various combinations of these three types of cones. Most of our visual experiences are mediated by the cone system, the loss of which produces legal blindness. In contrast, a person who loses rod vision mainly has difficulty seeing in dim light and thus should not drive at night.

From photoreceptors, visual information flows to bipolar cells through the outer synaptic layer, to bipolar cells, and then through the inner synaptic layer to ganglion cells. The axons of ganglion cells extend posteriorly to the optic disc and exit the eyeball as the optic nerve. The optic disc is also called the *blind spot*. Because it contains no rods or cones, we cannot see an image that strikes the blind spot. Normally, you are not aware of having a blind spot, but you can easily demonstrate its presence. Hold this page about 20 in. from your face with the cross shown below directly in front of your right eye. You should be able to see the cross and the square when you close your left eye. Now, keeping the left eye closed, slowly bring the page closer to your face while keeping the right eye on the cross. At a certain distance the square will disappear because its image falls on the blind spot.

+ ■

The **fovea centralis** (FŌ-vē-a) (see Figure 16.8) is a small depression at the center of the posterior portion of the retina, at the visual axis of the eye. The fovea centralis contains only cones. In addition, the layers of bipolar and ganglion cells, which scatter light to some extent, do not cover the cones here; these layers are displaced to the periphery of the fovea centralis. As a result, the fovea centralis is the area of highest **visual acuity** (a-KŪ-i-tē) (sharpness of vision). A main reason that you move your head and eyes while looking at something is to place images of interest on your fovea centralis—as you do to read the words

FIGURE 16.10 **Microscopic structure of the retina.** The downward blue arrow to the right in part (a) indicates the direction of the visual signals passing through the neural layer of the retina, and the upward yellow arrow shows the path of light. Eventually, impulses arise in ganglion cells and propagate along their axons, which make up the optic nerve.

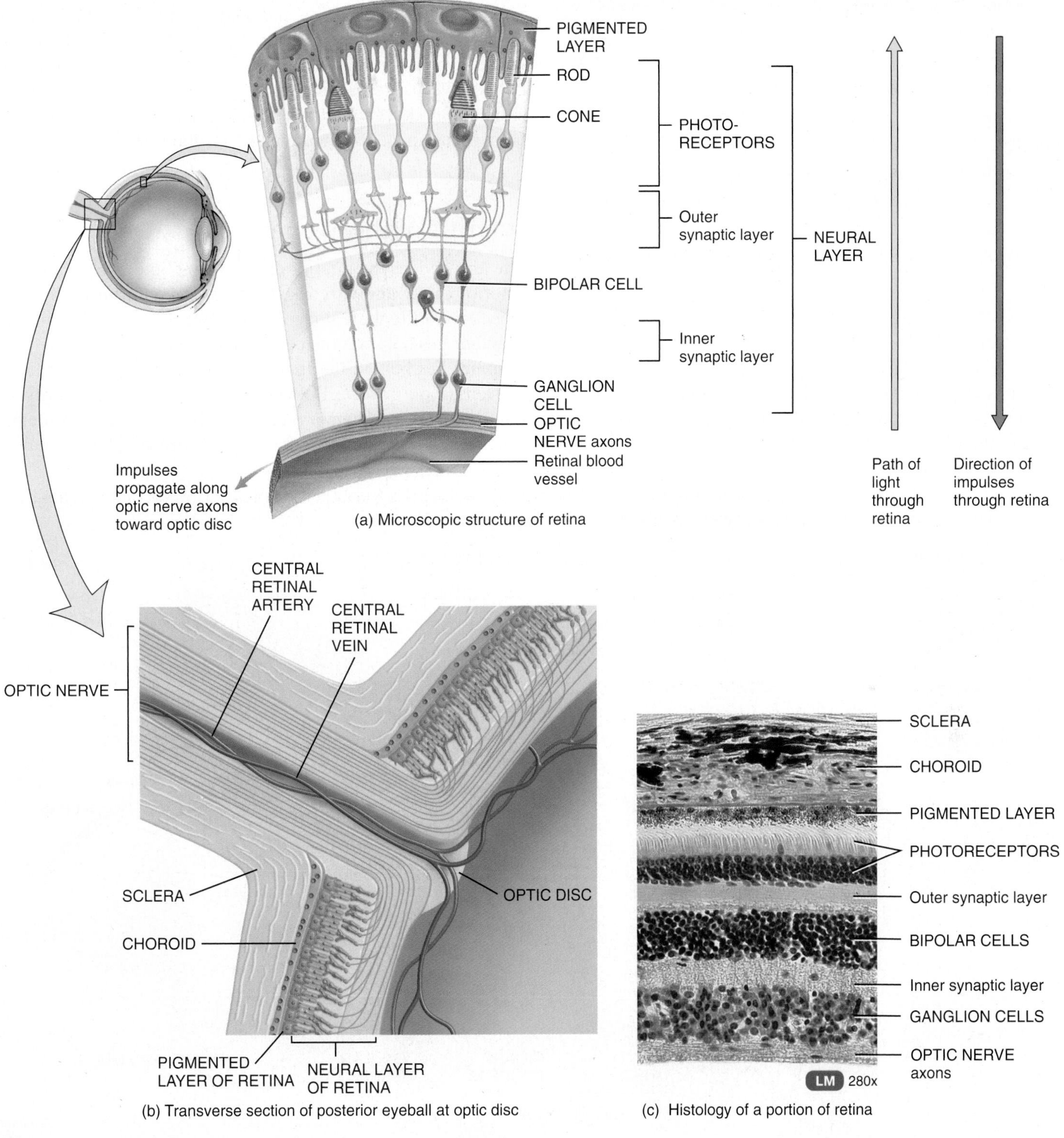


(a) Microscopic structure of retina

(b) Transverse section of posterior eyeball at optic disc

(c) Histology of a portion of retina

**In the retina, visual signals pass from photoreceptors to bipolar cells to ganglion cells.**

## CLINICAL CONNECTION | *Detached Retina*

A **detached retina** may occur due to trauma, such as a blow to the head, in various eye disorders, or as a result of age-related degeneration. The detachment occurs between the neural portion of the retina and the pigment epithelium. Fluid accumulates between these layers, forcing the thin, pliable retina to billow outward. The result is distorted vision and blindness in the corresponding field of vision. The retina may be reattached by laser surgery or cryosurgery (localized application of extreme cold). Reattachment must be accomplished quickly to avoid permanent damage to the retina.

in this sentence! Rods are absent from the fovea centralis and are more plentiful toward the periphery of the retina. Because rod vision is more light-sensitive than cone vision, you can see a faint object (such as a dim star) better if you gaze slightly to one side rather than looking directly at it.

## Lens

Posterior to the pupil and iris, within the cavity of the eyeball, is the **lens** (see Figure 16.8). The cells of the lens are arranged like the layers of an onion and contain proteins called **crystallins** (KRIS-ta-lins) that make up the refractive media of the lens, which normally is perfectly transparent. It is enclosed by a clear connective tissue capsule and held in position by encircling zonular fibers, which attach to the ciliary processes. The lens helps focus images on the retina to facilitate clear vision.

## Interior of the Eyeball

The lens divides the interior of the eyeball into two cavities: the anterior cavity and vitreous chamber. The **anterior cavity**—the space anterior to the lens—consists of two chambers. The **anterior chamber** lies between the cornea and the iris, and the **posterior chamber** lies behind the iris and in front of the zonular fibers and lens (Figure 16.11). Both chambers of the anterior cavity are filled with **aqueous humor** (Ā-kwē-us HŪ-mor; *aqua* = water), a watery fluid that nourishes the lens and cornea. Aqueous humor

**FIGURE 16.11 The anterior and posterior chambers of the eye.** The section is through the anterior portion of the eyeball at the junction of the cornea and sclera. Arrows indicate the flow of aqueous humor.

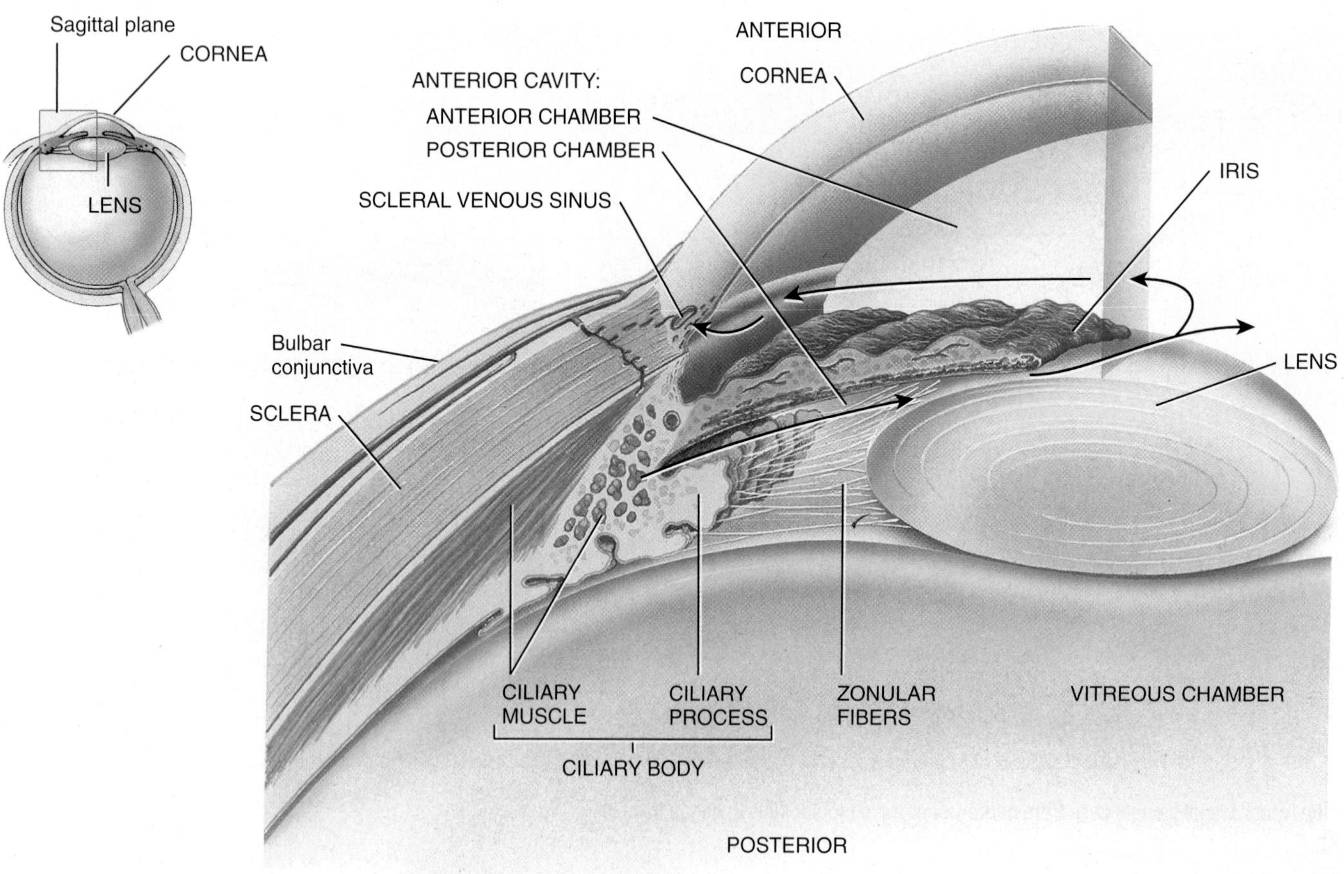


The iris separates the anterior and posterior chambers of the eye.

continually filters out of blood capillaries in the ciliary processes of the ciliary body and enters the posterior chamber. It then flows forward between the iris and the lens, through the pupil, and into the anterior chamber. From the anterior chamber, aqueous humor drains into the scleral venous sinus and then into the blood. Normally, aqueous humor is completely replaced about every 90 minutes.

The larger, posterior cavity of the eyeball is the **vitreous chamber** (VIT-rē-us), which lies between the lens and the retina (Figure 16.11). Within the vitreous chamber is the **vitreous body**, a jellylike substance that holds the retina flush against the choroid, giving the retina an even surface for the reception of clear images. Unlike the aqueous humor, the vitreous body does not undergo constant replacement. It is formed during embryonic life and is not replaced thereafter. The vitreous body contains phagocytic cells that remove debris, keeping this part of the eye clear for unobstructed vision.

The pressure in the eye, called **intraocular pressure**, is produced mainly by the aqueous humor and partly by the vitreous body; normally it is about 16 mm Hg (millimeters of mercury). The intraocular pressure maintains the shape of the eyeball and prevents it from collapsing.

Table 16.1 summarizes the structures associated with the eyeball.

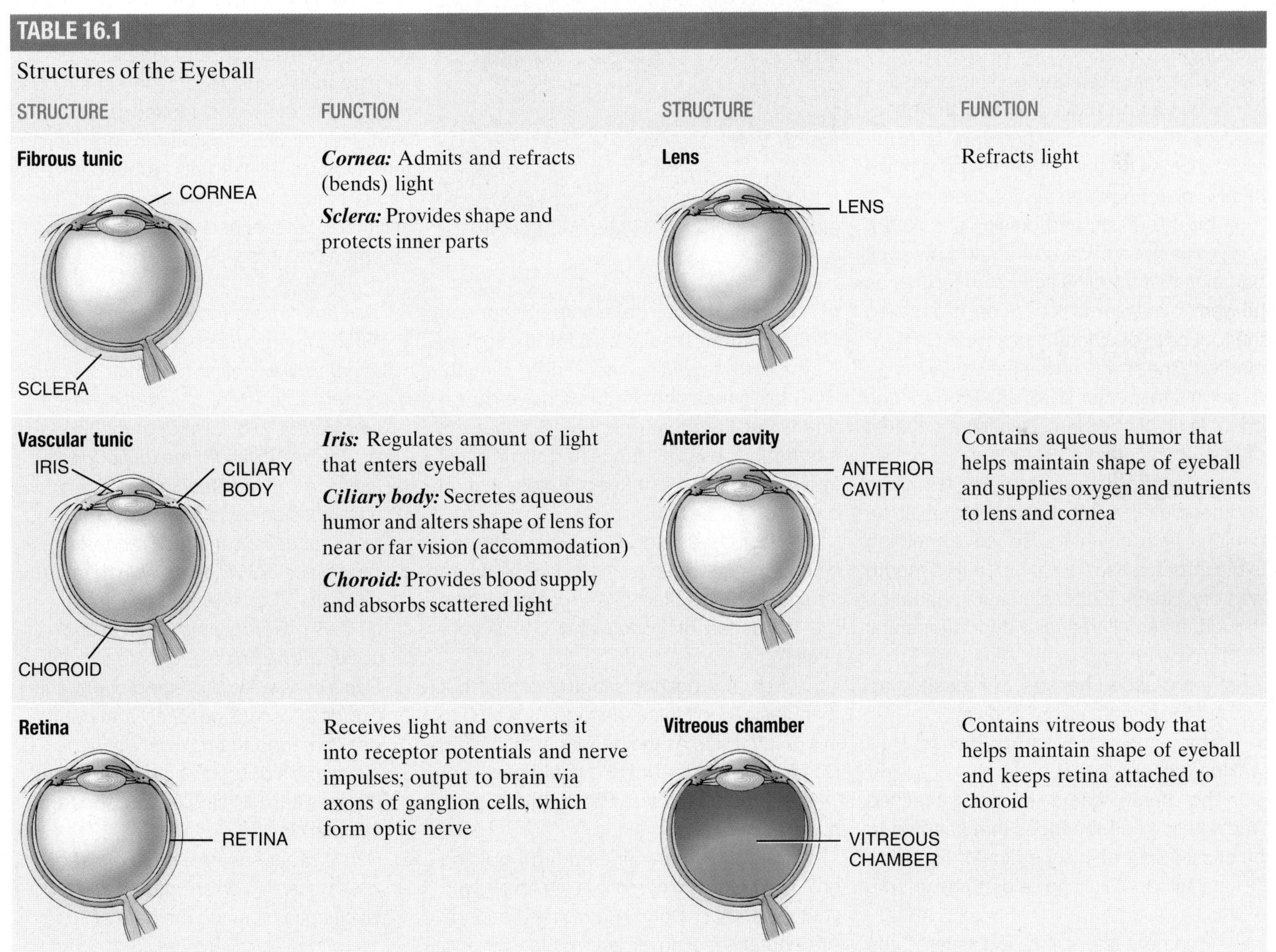


**TABLE 16.1**

Structures of the Eyeball

| STRUCTURE | FUNCTION | STRUCTURE | FUNCTION |
|---|---|---|---|
| **Fibrous tunic** | ***Cornea:*** Admits and refracts (bends) light<br>***Sclera:*** Provides shape and protects inner parts | **Lens** | Refracts light |
| **Vascular tunic** | ***Iris:*** Regulates amount of light that enters eyeball<br>***Ciliary body:*** Secretes aqueous humor and alters shape of lens for near or far vision (accommodation)<br>***Choroid:*** Provides blood supply and absorbs scattered light | **Anterior cavity** | Contains aqueous humor that helps maintain shape of eyeball and supplies oxygen and nutrients to lens and cornea |
| **Retina** | Receives light and converts it into receptor potentials and nerve impulses; output to brain via axons of ganglion cells, which form optic nerve | **Vitreous chamber** | Contains vitreous body that helps maintain shape of eyeball and keeps retina attached to choroid |

## ✓ CHECKPOINT

16. What are the components of the fibrous tunic and vascular tunic?
17. How does melanin assist vision?
18. Which part of the retina produces the sharpest vision when light falls on it?
19. What is the function of the aqueous humor? The vitreous body?
20. What separates the anterior and posterior chambers of the eyeball? The anterior cavity from the vitreous chamber?
21. What is the circulation path of aqueous humor?

## RETURN TO **Dan's Story**

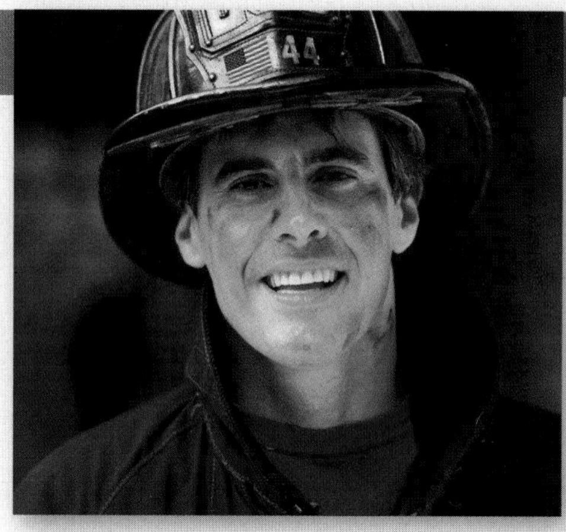

As Dan swung in behind the parked fire engine, he noticed that the putrid metallic chemical odor seemed less intense, even though he was closer to the fire. Dan quickly put on his gear, then grabbed his breathing apparatus as he moved toward the building. As he entered the building, Dan felt resistance as he tried to inhale; probably a sticky valve on his respirator. He stopped, took off the mask, and quickly checked it again. His nose began to run and his eyes began to tear from the acrid smoke. Just as he went to put his mask back on, he smelled the foul odor of a natural gas leak. He fumbled with the straps on the head gear, trying to get his respirator back on. Then a large bright flash blinded him. Dan felt a moment of pain, then nothing.

Dan woke with a start. He felt dizzy and couldn't see. Most disturbing of all, he couldn't hear anything. He felt something in his throat, a hard plastic tube. His mouth was dry. He couldn't taste anything, but he thought he smelled a faint odor of hospital antiseptic cleaner.

Pam grasped his arm. He reached up and felt the bandages on his eyes.

"Someone come quickly, please! He's waking up!" shouted Pam.

The nurse came in and checked Dan's IV line and the tubes that seemed to be sticking out of every orifice of his body. "I'll call the doctor," she said as she left the room.

A moment later Dr. Burke entered the room.

"Dr. Burke, can he hear us?" asked Pam.

"Well, Dan was lucky. His equipment spared his face and body from the shrapnel of the blast. He may have some temporary blindness from the flash of the explosion. It's hard to say with his hearing, though. Both his tympanic membranes were ruptured. But I'm most concerned about the internal injuries to his lungs and some of his internal organs. We'll keep a close watch on him."

A few days later, Dr. Burke returned with some good news. "Hi, Dan. How are you doing?" he asked. "We're going to take the bandages off today and check your eyes."

"Well, my hearing's better today, but thankfully, I can't taste the food," joked Dan in a raspy voice. As the bandages came off, Dan felt a rush of relief. He could see—not well, but he could see! His eyes felt sore, puffy, and watery.

"Dan, how many fingers am I holding up? Two? Good. Now follow my finger with your eyes, that's good. How's the hearing? Are you still feeling dizzy?" Dan was only half listening to the doctor's questions as his eyes scanned the room.

Later, Dan sat in his hospital room trying to watch the local news. He had been in the hospital for over a week now and was feeling better in spite of the constant pain. His internal injuries hadn't been serious. The television appeared blurry, and he had trouble focusing on the screen. Dan rubbed his eyes and looked up. He saw what looked like a curtain falling over the television. He blinked and rubbed his eyes again, trying to clear his vision, but it was still there. He reached for the call button to summon the nurse.

A. *Dan has suffered a blast injury. Although rare, these are devastating events involving injury to multiple organ systems. The odor of which chemical was detected by Dan prior to the explosion? Why did the foul odor cause a strong sense of fear in Dan?*

B. *If Dan has damage to his gustatory receptor cells, can these receptor cells be replaced? Will he be able to taste normally? Explain your answers.*

C. *If Dan's olfactory epithelium is damaged, will he be able to taste normally?*

D. *Dan's eyes were protected during the explosion, so it's not likely that debris has damaged them. What might be causing his blurred vision? Why might his eyes hurt? Which membranes are inflamed?*

E. *Dan is experiencing a rapid onset of visual loss. Which structures could be involved in blocking the path of light to his retina? Is it likely that such a rapid loss of vision is due to blockage of light entering the eye, or is something wrong with the retina?*

# 16.5 Image formation involves refraction of light rays, accommodation, pupil constriction, and convergence.

In some ways the eye is like a camera: Its optical elements focus an image of some object on a light-sensitive "film"—the retina—while ensuring the correct amount of light to make the proper "exposure." To understand how the eye forms clear images of objects on the retina, we must examine three processes: (1) the refraction or bending of light by the lens and cornea; (2) accommodation, the change in shape of the lens; and (3) constriction or narrowing of the pupil.

## Refraction of Light Rays

When light rays traveling through a transparent substance (such as air) pass into a second transparent substance with a different density (such as water), they bend at the junction between the two substances. This bending is called **refraction** (rē-FRAK-shun) (Figure 16.12a). As light rays enter the eye, they are refracted at the anterior and posterior surfaces of the cornea. Both surfaces of the lens of the eye further refract the light rays so they come into exact focus on the retina.

Images focused on the retina are inverted, or upside-down (Figure 16.12b, c). They also undergo right-to-left reversal; that is, light from the right side of an object strikes the left side of the retina, and vice versa. The reason the world does not look inverted and reversed is that the brain "learns" early in life to coordinate visual images with the orientations of objects. The brain stores the inverted and reversed images we acquired when we first reached for and touched objects and interprets those visual images as being correctly oriented in space.

About 75 percent of the total refraction of light occurs at the cornea. The lens provides the remaining 25 percent of focusing power and also changes the focus to view near or distant objects. When an object is 6 m (20 ft) or more away from the viewer, the light rays reflected from the object are nearly parallel to one another (Figure 16.12b). The lens must bend these parallel rays just enough so that they fall exactly focused on the fovea centralis, where vision is sharpest. Because light rays that are reflected from objects closer than 6 m (20 ft) are divergent rather than parallel (Figure 16.12c), the rays must be refracted more if they are to be focused on the retina. This additional refraction is accomplished through the process called accommodation.

FIGURE 16.12 **Refraction of light rays.** (a) Refraction is the bending of light rays at the junction of two transparent substances with different densities. (b) The cornea and lens refract light rays from distant objects so the image is focused on the retina. (c) In accommodation, the lens becomes more spherical, which increases the refraction of light.

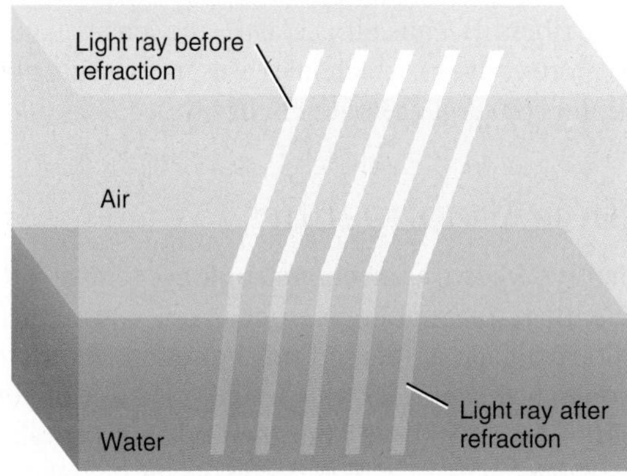


(a) Refraction of light rays

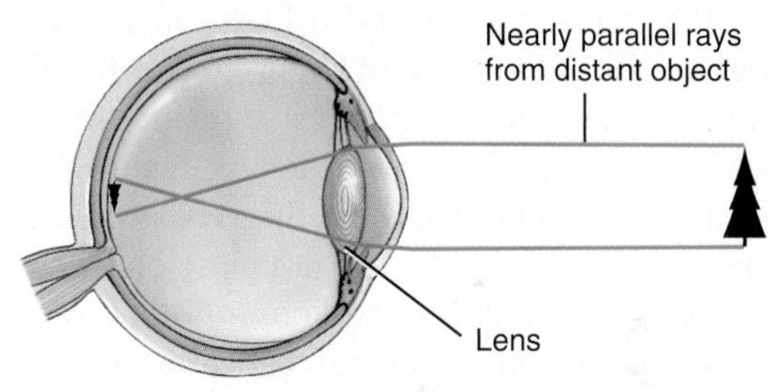


(b) Viewing distant object

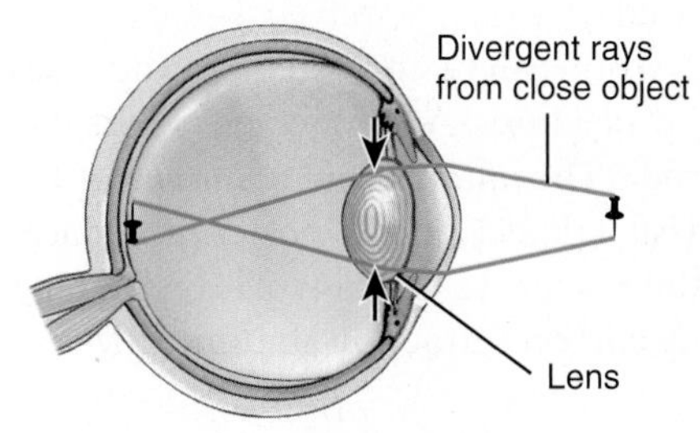


(c) Accommodation

**Images focused on the retina are inverted and left-to-right reversed.**

## Accommodation and the Near Point of Vision

A surface that curves outward, like the surface of a ball, is said to be *convex*. When the surface of a lens is convex, that lens will refract incoming light rays toward each other, so that they eventually intersect. If the surface of a lens curves inward, like the inside of a hollow ball, the lens is said to be *concave* and causes light rays to refract away from one another. The lens of the eye is convex on both its anterior and posterior surfaces, and its focusing power increases as its curvature becomes greater. When the eye is focusing on a close object, the lens becomes more curved, causing greater refraction of the light rays. This increase in the curvature of the lens for near vision is called **accommodation** (a-kom-a-DĀ-shun) (Figure 16.12c). The **near point of vision** is the minimum distance from the eye that an object can be clearly focused with maximum accommodation. This distance is about 10 cm (4 in.) in a young adult.

How does accommodation occur? When you are viewing distant objects, the ciliary muscle is relaxed and the lens is flatter because it is stretched in all directions by taut zonular fibers (Figure 16.12b). When you view a close object, the ciliary muscle contracts. As the muscle contracts, the ciliary muscle is pulled toward the lens, which pulls the ciliary process forward toward the lens (Figure 16.12c). This action releases tension on the lens and zonular fibers. Because it is elastic, the lens becomes more spherical (more convex), which increases its focusing power and causes greater convergence of the light rays.

## Refraction Abnormalities

The normal eye, known as an **emmetropic eye** (em′-e-TROP-ik), can sufficiently refract light rays from an object 6 m (20 ft) away so that a clear image is focused on the retina (Figure 16.13a). Many people, however, lack this ability because of refraction abnormalities. Among these abnormalities are **myopia** (mī-Ō-pē-a), or *nearsightedness*, which occurs when the eyeball is too long relative to the focusing power of the cornea and lens, or when the lens is thicker than normal, so an image converges in front of the retina. Myopic individuals can see close objects clearly, but not distant objects. In **hyperopia** (hī-per-Ō-pē-a), or *farsightedness*, the eyeball length is short relative to the focusing power of the cornea and lens, or the lens is thinner than normal, so an image converges behind the retina. Hyperopic individuals can see distant objects clearly, but not close ones. Figure 16.13 illustrates these conditions and explains how they are corrected. Another refraction abnormality is **astigmatism** (a-STIG-ma-tizm), in which either the cornea or the lens has an irregular curvature. As a result, parts of the image are out of focus and vision is blurred or distorted.

Most errors of vision can be corrected by eyeglasses, contact lenses, or surgical procedures. A contact lens floats on a film of tears over the cornea. The anterior outer surface of the contact lens corrects the visual defect, and its posterior surface matches the curvature of the cornea. LASIK involves reshaping the cornea to permanently correct refraction abnormalities.

**FIGURE 16.13 Refraction abnormalities in the eyeball and their correction.** (a) Normal (emmetropic) eye. (b) In the nearsighted or myopic eye, the image is focused in front of the retina. The condition may result from an elongated eyeball or thickened lens. (c) Myopia can be corrected by use of a concave lens that diverges entering light rays so that they come into focus directly on the retina. (d) In the farsighted or hyperopic eye, the image is focused behind the retina. The condition results from a shortened eyeball or a thin lens. (e) Hyperopia can be corrected by a convex lens that converges entering light rays so that they focus directly on the retina.

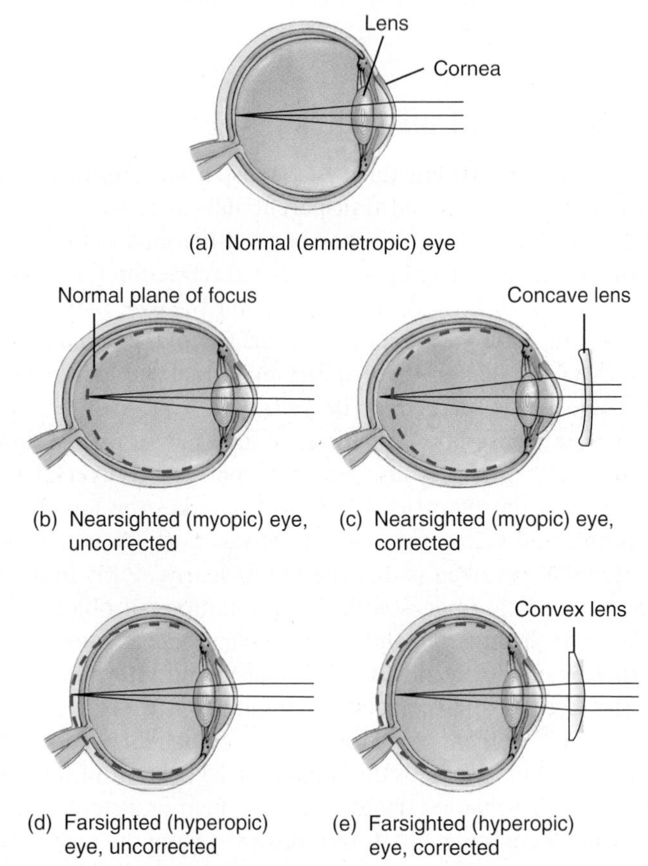


**In myopia (nearsightedness), distant objects can't be seen clearly; in hyperopia (farsightedness), nearby objects can't be seen clearly.**

## Constriction of the Pupil

The circular muscle fibers of the iris also have a role in the formation of clear retinal images. Part of the accommodation mechanism consists of the contraction of the circular muscle of the iris to constrict the pupil (see Figure 16.9). **Constriction of the pupil** is a narrowing of the diameter of the hole through which light enters the eye due to the contraction of the circular muscles of the iris. This autonomic reflex occurs simultaneously with accommodation for near objects and prevents light rays from entering the eye through the periphery of the lens. Light rays entering at the periphery would not be brought to focus on the retina and would result in blurred vision. The pupil, as noted earlier, also constricts in bright light.

## Convergence

Because of the position of their eyes in their heads, many animals, such as horses and goats, see one set of objects off to the left through one eye and an entirely different set of objects off to the right through the other. In humans, both eyes focus on only one set of objects—a characteristic called **binocular vision**. This feature of our visual system allows the perception of depth and an appreciation of the three-dimensional nature of objects. Binocular vision occurs when light rays from an object strike corresponding points on the two retinas. When we stare straight ahead at a distant object, the incoming light rays are aimed directly at both pupils and are refracted to comparable spots on the retinas of both eyes. As we move closer to an object, however, the eyes must rotate medially if the light rays from the object are to strike the same points on both retinas. The term **convergence** refers to this medial movement of the two eyeballs so that both are directed toward the object being viewed, for example, tracking a pencil moving toward your eyes. The nearer the object, the greater the degree of convergence needed to maintain binocular vision. The coordinated action of the extrinsic eye muscles brings about convergence.

✓ **CHECKPOINT**

**22.** What is refraction? Which components of the eye are primarily responsible for refracting light?

**23.** If you look at the horizon to determine where you are, then look down to read a map, what process must your eyes accomplish to keep your vision focused?

**24.** Which sequence of events occurs when you look at a distant object? A close object?

**25.** What is convergence? Why is it important for human vision?

# 16.6 The neural pathway of light is photoreceptors → bipolar cells → ganglion cells → optic nerve → primary visual cortex.

## Photoreceptors and Photopigments

Rods and cones were named for differences in the appearance of the *outer segment*, the distal end of the photoreceptor next to the pigmented layer. The outer segments of rods are cylindrical or rod-shaped; those of cones are tapered or cone-shaped (Figure 16.14). Transduction of light energy into a receptor potential occurs in the outer segment of the photoreceptor. The photopigments are proteins in the plasma membrane of the outer segment. In cones the plasma membrane is folded back and forth in a pleated fashion, but in rods the pleats pinch off from the plasma membrane to form discs. The outer segment of each rod contains a stack of about 1000 discs, piled up like coins inside a wrapper. Photoreceptor outer segments are renewed at an astonishingly rapid pace. In rods, one to three new discs are added to the base of the outer segment every hour while old discs slough off at the tip and are phagocytized by epithelial cells of the pigmented layer of the retina.

The *inner segment* of a rod or cone contains the cell nucleus, Golgi complex, and mitochondria. At its proximal end, the photoreceptor expands into bulblike synaptic terminals filled with synaptic vesicles.

The first step in visual transduction is absorption of light by a **photopigment**, a colored protein that undergoes structural changes when it absorbs light, in the outer segment of a photoreceptor. Light absorption initiates the events that lead to the production of a receptor potential. The photopigment in rods is **rhodopsin** (rō-DOP-sin; *rhod-* = rose; *-opsin* = related to vision). Three different **cone photopigments** are present in the retina, one in each of the three types of cones. Color vision results from different colors of light selectively activating the different cone photopigments.

Photopigments contain two parts: a glycoprotein known as **opsin** and a derivative of vitamin A called **retinal**. Vitamin A derivatives are formed from carotene, the plant pigment that gives carrots their orange color. Good vision depends on adequate dietary intake of carotene-rich vegetables such as carrots, spinach,

FIGURE 16.14 **Structure of rod and cone photoreceptors.**

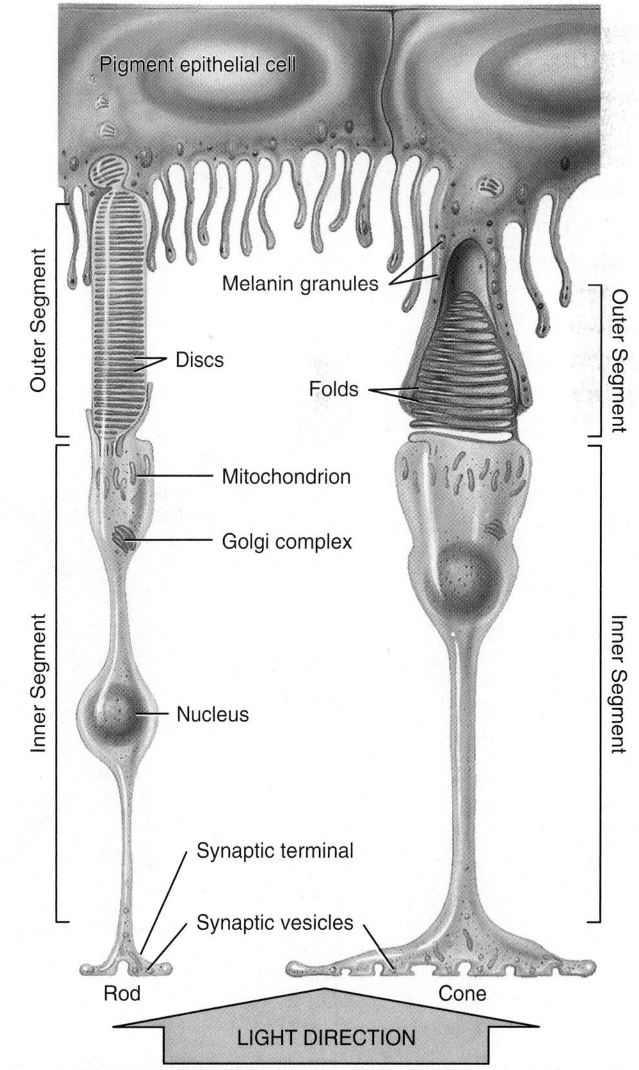


**Transduction of light energy into receptor potentials occurs in the outer segments of rods and cones.**

broccoli, and yellow squash, or other foods that contain vitamin A, such as liver. Retinal is the light-absorbing part of photopigments. In the human retina, there are four different opsins, one in each type of cone and one in the rods (rhodopsin). Small molecular variations of the different opsins permit the rods and cones to absorb different colors (wavelengths) of incoming light.

Photopigments respond to light in the following cyclical process (Figure 16.15):

1. In darkness, retinal has a bent shape, called *cis*-retinal, which fits snugly into the opsin portion of the photopigment. When *cis*-retinal absorbs a photon of light, it straightens out to a shape called *trans*-retinal. This *cis*-to-*trans* conversion is called **isomerization** (ī-som′-er-i-ZĀ-shun) and is the first step in visual transduction. After retinal isomerizes, several chemical changes occur that generate a receptor potential.
2. In about a minute, *trans*-retinal completely separates from opsin. The final products look colorless, so this part of the cycle is termed **bleaching** of photopigment.
3. An enzyme called **retinal isomerase** converts *trans*-retinal back to *cis*-retinal.
4. The *cis*-retinal then can bind to opsin, reforming a functional photopigment. This part of the cycle—resynthesis of a photopigment—is called **regeneration**.

FIGURE 16.15 **The cyclical bleaching and regeneration of photopigment.** Yellow section of cycle indicates light exposure. Blue arrows indicate bleaching steps; black arrows indicate regeneration steps.

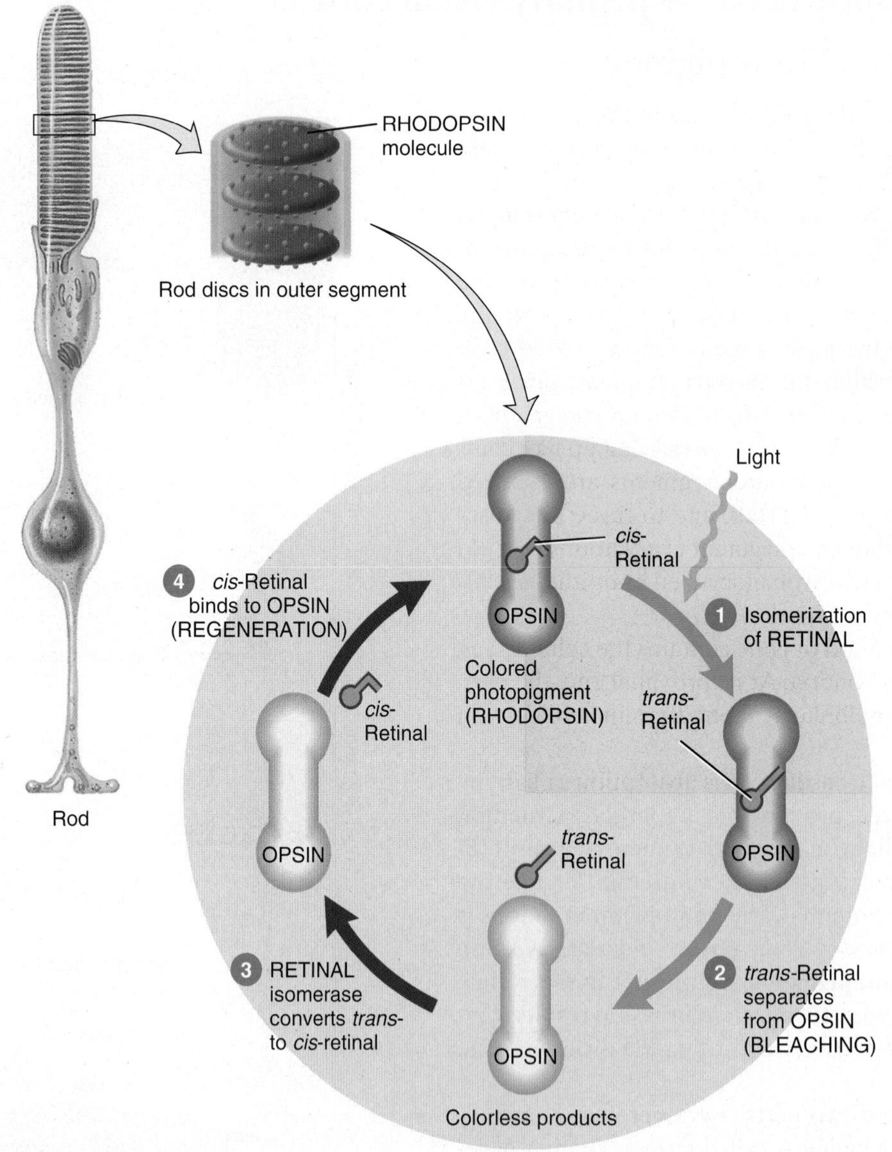


Retinal, a derivative of vitamin A, is the light-absorbing component of photopigments.

Bleaching of photopigment occurs when you continue to see an image after a camera flashbulb has flashed at you. When your retinas are exposed to a bright light, the photopigments can't respond to further light stimulation until regeneration has occurred. Until the photoreceptors recover, your retinas respond to an afterimage, a "ghost" image of the bright light.

## Light and Dark Adaptation

When you emerge from dark surroundings (say, a tunnel) into the sunshine, **light adaptation** occurs—your visual system adjusts in seconds to the brighter environment by decreasing its sensitivity. On the other hand, when you enter a darkened room such as a movie theater, your visual system undergoes **dark adaptation**—its sensitivity increases slowly over many minutes.

As the light level increases, more and more photopigment is bleached, assisting light adaptation. While light is bleaching some photopigment molecules, however, others are being regenerated. In daylight, regeneration of rhodopsin cannot keep up with the bleaching process, so rods contribute little to daylight vision. In contrast, cone photopigments regenerate rapidly enough that some of the *cis* form is always present, even in very bright light.

If the light level decreases abruptly, sensitivity increases. In complete darkness, full regeneration of cone photopigments occurs during the first 8 minutes of dark adaptation. Rhodopsin regenerates more slowly, and our visual sensitivity increases until even a single photon (the smallest unit of light) can be detected. In that situation, barely perceptible light appears gray-white, regardless of its color. At very low light levels, such as starlight, objects appear as shades of gray because only the rods are functioning.

## Release of Neurotransmitter by Photoreceptors

The absorption of light and isomerization of retinal initiates chemical changes in the photoreceptor that lead to production of a receptor potential. To understand how the receptor potential arises, however, we first need to examine the operation of photoreceptors in the *absence* of light. In darkness, sodium ions ($Na^+$) flow into photoreceptor outer segments through $Na^+$ channels that are held open by **cyclic GMP (guanosine monophosphate)** or **cGMP** (Figure 16.16a). The inflow of $Na^+$ partially depolarizes the photoreceptor. As a result, in darkness the membrane potential of a photoreceptor is about −30 mV. This is much closer to zero than a typical neuron's resting membrane potential of −70 mV. The depolarization during darkness triggers continual release of the neurotransmitter **glutamate** at the synaptic terminals. At synapses between photoreceptors and bipolar cells, glutamate is an inhibitory neurotransmitter. Glutamate triggers inhibitory postsynaptic potentials (IPSPs) that hyperpolarize the bipolar cells and prevent them from sending signals to the ganglion cells.

FIGURE 16.16 **Operation of rod photoreceptors.**

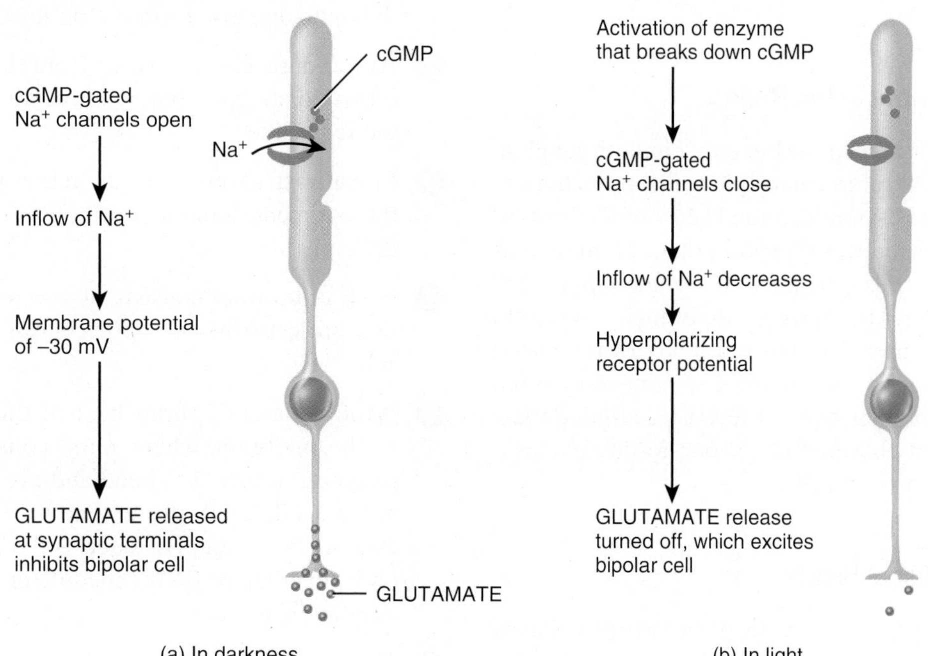


**Light causes a hyperpolarizing receptor potential in photoreceptors, which decreases release of an inhibitory neurotransmitter (glutamate).**

## CLINICAL CONNECTION | *Color Blindness and Night Blindness*

Most forms of **color blindness**, an inherited inability to distinguish between certain colors, result from the absence or deficiency of one of the three types of cones. The most common type is *red-green color blindness*, in which red cones or green cones are missing. As a result, the person cannot distinguish between the colors red and green. Prolonged vitamin A deficiency and the resulting below-normal amount of rhodopsin may cause **night blindness** or **nyctalopia** (nik′-ta-LŌ-pē-a), an inability to see well at low light levels.

When light strikes the retina, enzymes are activated that break down cGMP. As a result, some cGMP-gated $Na^+$ channels close, $Na^+$ inflow decreases, and the membrane potential becomes more negative, approaching −70 mV (Figure 16.16b). This sequence of events produces a hyperpolarizing receptor potential in the photoreceptor that decreases the release of glutamate. As less inhibitory neurotransmitter is released by the photoreceptors, the bipolar cells generate more receptor potentials. Thus, light *excites* the bipolar cells that synapse with rods by turning *off* the release of an inhibitory neurotransmitter. The excited bipolar cells subsequently stimulate the ganglion cells to form action potentials.

# The Visual Pathway

Visual signals in the retina undergo considerable processing at synapses among the various types of retinal neurons (see Figure 16.10 and related discussion). Then, axons of the ganglion cells provide output from the retina to the brain, exiting the eyeball as the optic nerve.

## Processing of Visual Input in the Retina

There are 126 million photoreceptors but only 1 million ganglion cells in the human eye. Therefore, many photoreceptors tend to converge on individual postsynaptic retinal neurons. Light stimulation of rods and cones induces receptor potentials in bipolar cells. Between 6 and 600 rods synapse with a single bipolar cell in the outer synaptic layer. The convergence of many rods onto a single bipolar cell increases the light sensitivity of rod vision but slightly blurs the image that is perceived. A cone more often synapses with just one bipolar cell. Cone vision, although less sensitive, has higher acuity because of the one-to-one synapses between cones and their bipolar cells.

## Visual Pathway and Visual Fields

The axons within the optic nerve pass through the **optic chiasm** (kī-AZ-m = a crossover, as in the letter X), a crossing point of the optic nerves (Figure 16.17a, b). Some axons cross to the opposite side, but others remain uncrossed. After passing through the optic chiasm, the axons, now part of the **optic tract**, enter the brain and terminate in the thalamus. Here they synapse with neurons whose axons form the **optic radiations**, which project to the primary visual areas in the occipital lobes of the cerebral cortex (see Figure 13.10), and visual perception begins.

Everything that can be seen by one eye is that eye's **visual field**. Because our two eyes are located anteriorly in our head, their visual fields overlap considerably (Figure 16.17b). We have binocular vision due to the large region where the visual fields of the two eyes overlap. The visual field of each eye is divided into two regions: the nasal or central half and the temporal or peripheral half (Figure 16.17c, d). For each eye, light rays from an object in the nasal half of the visual field fall on the temporal half of the retina, and light rays from an object in the temporal half of the visual field fall on the nasal half of the retina. Visual information from the *right* half of each visual field is conveyed to the *left* side of the brain, and visual information from the *left* half of each visual field is conveyed to the *right* side of the brain, as follows (Figure 16.17c, d):

1. The axons of ganglion cells in one eye exit the eyeball at the optic disc and form the optic nerve on that side.
2. At the optic chiasm, axons from the temporal half of each retina do not cross but continue directly to the thalamus on the same side.
3. In contrast, axons from the nasal half of each retina cross the optic chiasm and continue to the opposite side of the thalamus.
4. Each optic tract consists of crossed and uncrossed axons that project from the optic chiasm to the thalamus on one side.
5. Axon collaterals (branches) of the ganglion cells project to the midbrain, where pupil constriction is regulated in response to light and head and eye movements are coordinated. Collaterals also extend to the hypothalamus, which establishes patterns of sleep and other activities that occur on a circadian or daily schedule in response to intervals of light and darkness.
6. The axons of thalamic neurons form the optic radiations as they project from the thalamus to the primary visual area of the cerebral cortex on the same side.

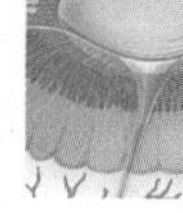

FIGURE 16.17 **The visual pathway.** (a) Partial dissection of the brain reveals the optic radiations (axons extending from the thalamus to the occipital lobe). (b) An object in the binocular visual field can be seen with both eyes. In (c) and (d), note that information from the right side of the visual field of each eye projects to the left side of the brain, and information from the left side of the visual field of each eye projects to the right side of the brain.

View

ANTERIOR

Visual pathway:

Optic nerve
↓
OPTIC CHIASM
↓
OPTIC TRACT
↓
Thalamus
↓
OPTIC RADIATIONS
↓
Primary visual area of cerebral cortex

POSTERIOR

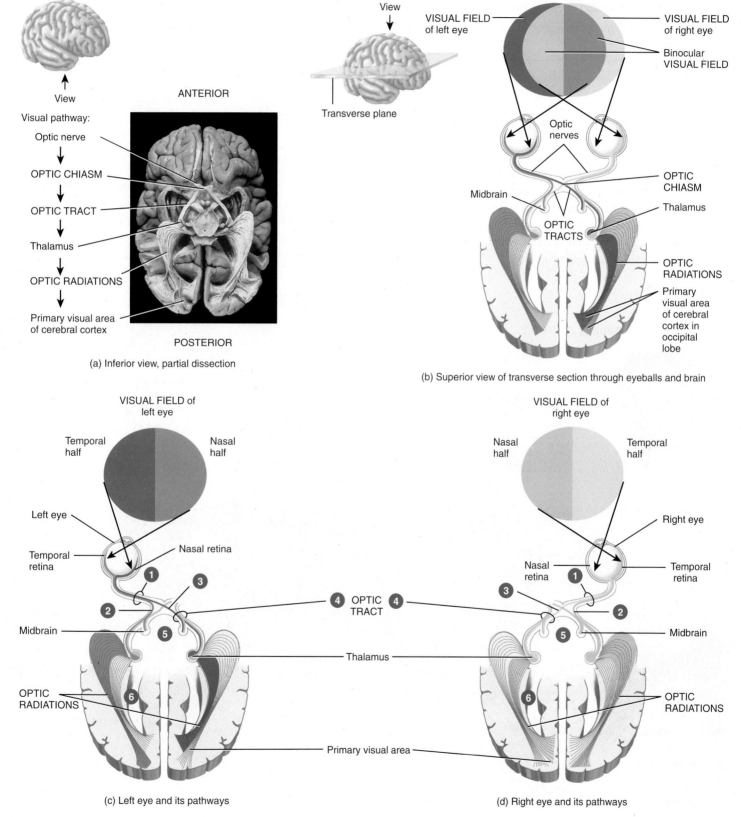

(a) Inferior view, partial dissection

View

Transverse plane

VISUAL FIELD of left eye

VISUAL FIELD of right eye

Binocular VISUAL FIELD

Optic nerves

OPTIC CHIASM

Midbrain

Thalamus

OPTIC TRACTS

OPTIC RADIATIONS

Primary visual area of cerebral cortex in occipital lobe

(b) Superior view of transverse section through eyeballs and brain

VISUAL FIELD of left eye

Temporal half

Nasal half

Left eye

Temporal retina

Nasal retina

1

2

3

4 OPTIC TRACT 4

Midbrain

5

Thalamus

OPTIC RADIATIONS

6

Primary visual area

(c) Left eye and its pathways

VISUAL FIELD of right eye

Nasal half

Temporal half

Right eye

Nasal retina

Temporal retina

1

2

3

Midbrain

5

OPTIC RADIATIONS

6

(d) Right eye and its pathways

**The axons of ganglion cells in the temporal half of each retina extend to the thalamus on the same side; the axons of ganglion cells in the nasal half of each retina extend to the thalamus on the opposite side.**

**✓ CHECKPOINT**

**26.** What are the functional similarities between rods and cones?

**27.** What is the conversion of *cis*-retinal to *trans*-retinal called? What is the conversion of *trans*-retinal to *cis*-retinal called?

**28.** How do photopigments respond to light and recover in darkness?

**29.** Why does a decreased release of glutamate by photoreceptors generate a receptor potential in bipolar cells?

**30.** What is the composition of the optic nerve?

**31.** Light rays from an object in the temporal half of the visual field strike which half of the retina?

# 16.7 The ear is divided into external, middle, and internal regions.

Hearing is the ability to perceive sounds. The ear is an engineering marvel because its sensory receptors can transduce sound vibrations with amplitudes as small as the diameter of an atom of gold (0.3 nm) into electrical signals 1000 times faster than photoreceptors can respond to light. The ear also contains receptors for equilibrium, the sense that helps you maintain your balance and be aware of your orientation in space.

## Anatomy of the Ear

The ear is divided into three main regions: the external ear, which collects sound waves and channels them inward; the middle ear, which conveys sound vibrations to the oval window; and the internal ear, which houses the receptors for hearing and equilibrium (Figure 16.18).

FIGURE 16.18 Anatomy of the ear.

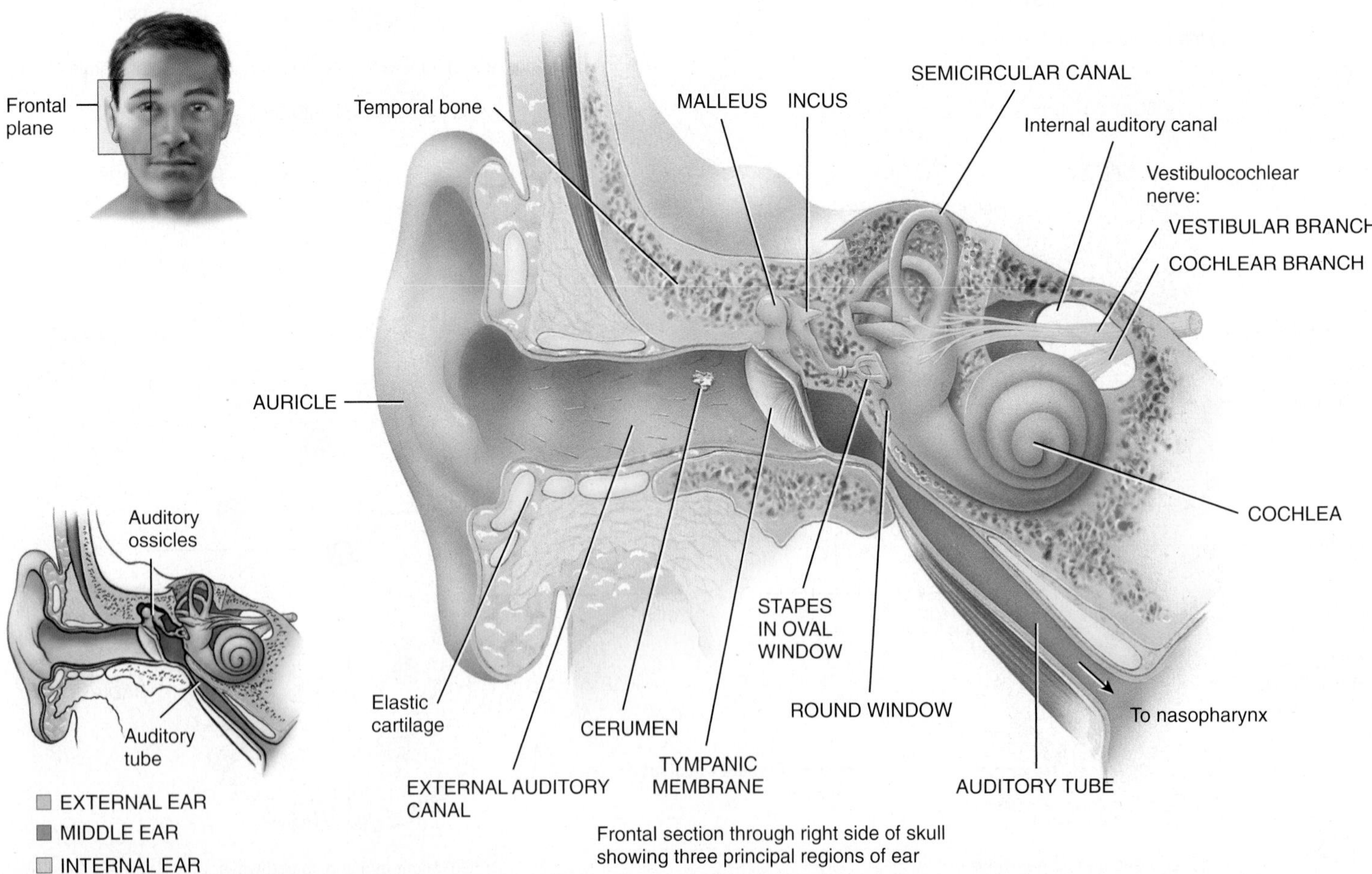


Frontal section through right side of skull showing three principal regions of ear

**The external ear collects sound waves, which are transmitted through the middle ear to the internal ear, where the receptors for hearing and equilibrium are located.**

## External (Outer) Ear

The **external (outer) ear** consists of the auricle, external auditory canal, and tympanic membrane. The **auricle** (AW-ri-kul) is a flap of elastic cartilage shaped like the flared end of a trumpet and covered by skin. Ligaments and muscles attach the auricle to the head. The **external auditory canal** (*audit-* = hearing) is a curved tube about 2.5 cm (1 in.) long that lies in the temporal bone and leads from the auricle to the tympanic membrane. The **tympanic membrane** (tim-PAN-ik; *tympan-* = a drum), or **eardrum**, is a thin, semitransparent partition between the external auditory canal and middle ear. The tympanic membrane is covered by skin on the side facing the external auditory canal and mucus membrane on the internal surface. Tearing of the tympanic membrane is called a *perforated eardrum*. It may be due to pressure from a cotton swab, trauma, or a middle ear infection. A perforated eardrum usually heals within a month.

Near the exterior opening, the external auditory canal contains a few hairs and specialized sweat glands called **ceruminous glands** (se-ROO-mi-nus) that secrete earwax or **cerumen** (se-ROO-men). The combination of hairs and cerumen helps prevent dust and foreign objects from entering the ear. Cerumen also prevents damage to the delicate skin of the external ear canal by water and insects. Cerumen usually dries up and falls out of the ear canal. However, some people produce a large amount of cerumen, which can become impacted and can muffle incoming sounds. The treatment for *impacted cerumen* is usually periodic ear irrigation or removal of earwax by trained medical personnel.

## Middle Ear

The **middle ear** is a small, air-filled cavity in the petrous portion of the temporal bone (Figure 16.19). It is separated from the external ear by the tympanic membrane and from the internal ear by a thin bony partition that contains two small membrane-covered openings: the oval window and the round window. Extending across the middle ear and attached to it by ligaments are the three smallest bones in the body, the **auditory ossicles** (OS-si-kuls) which are connected by synovial joints. The bones, named for their shapes, are the malleus, incus, and stapes—commonly called the hammer, anvil, and stirrup, respectively. The auditory ossicles are connected to one another by synovial joints. The "handle" of the **malleus** (MAL-ē-us) attaches to the internal surface of the tympanic membrane. The head of the

**FIGURE 16.19 The right middle ear containing the auditory ossicles.**

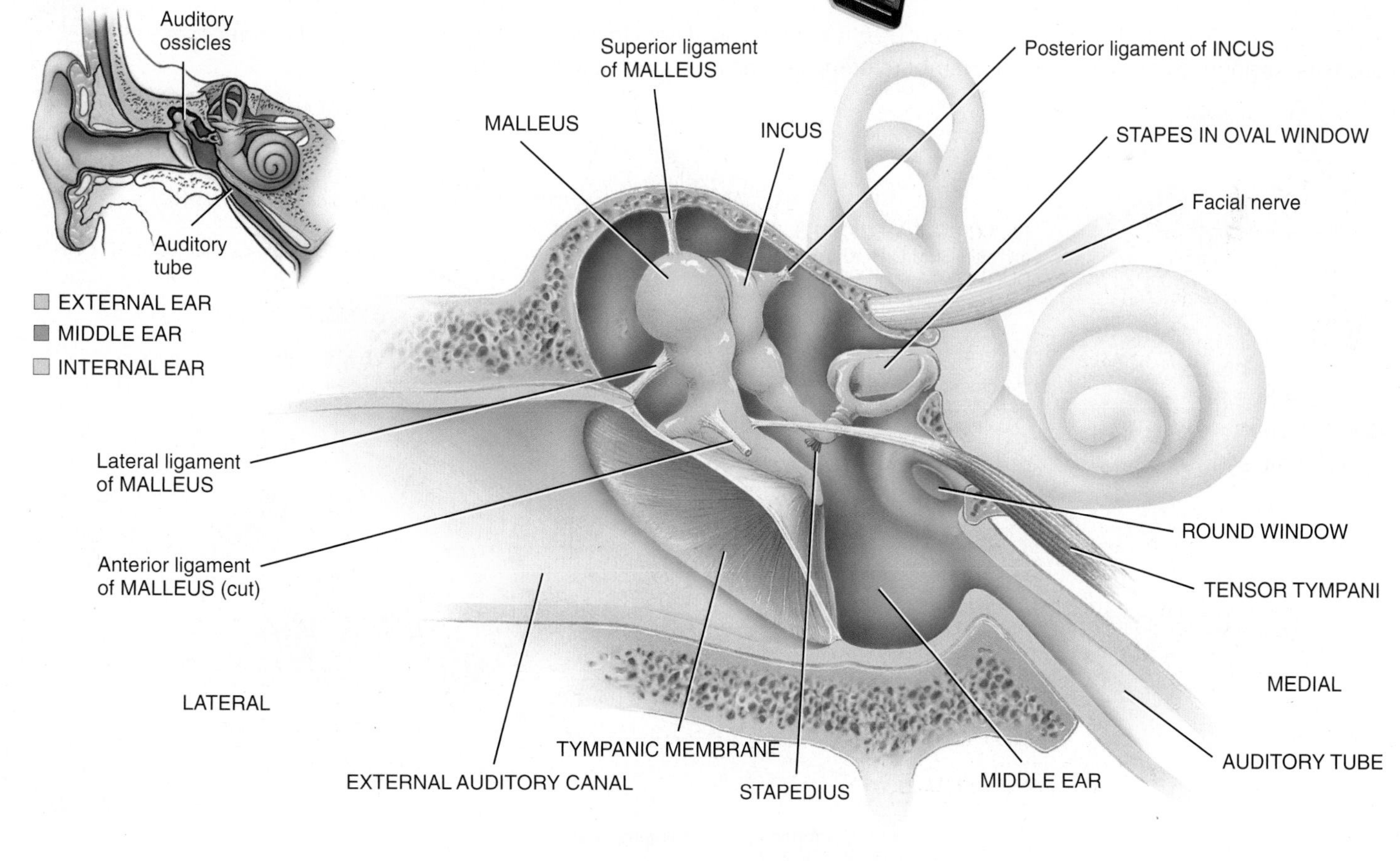

Auditory ossicles
Auditory tube
EXTERNAL EAR
MIDDLE EAR
INTERNAL EAR
Superior ligament of MALLEUS
MALLEUS
INCUS
Posterior ligament of INCUS
STAPES IN OVAL WINDOW
Facial nerve
Lateral ligament of MALLEUS
Anterior ligament of MALLEUS (cut)
ROUND WINDOW
TENSOR TYMPANI
MEDIAL
LATERAL
TYMPANIC MEMBRANE
EXTERNAL AUDITORY CANAL
STAPEDIUS
MIDDLE EAR
AUDITORY TUBE

Frontal section showing location of auditory ossicles

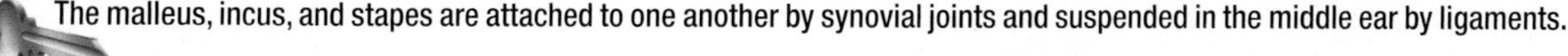

**The malleus, incus, and stapes are attached to one another by synovial joints and suspended in the middle ear by ligaments.**

malleus articulates with the body of the incus. The **incus** (ING-kus), the middle bone in the series, articulates with the head of the stapes. The base or footplate of the **stapes** (STĀ-pēz) fits into the **oval window**. Directly below the oval window is another membrane-covered opening, the **round window**.

Besides the ligaments, two tiny skeletal muscles also attach to the auditory ossicles. The **tensor tympani** (TIM-pan-ē) limits movement and increases tension on the tympanic membrane to prevent damage to the internal ear from loud noises. The **stapedius** (sta-PĒ-de-us) is the smallest skeletal muscle in the human body. By dampening large vibrations of the stapes due to loud noises, it protects the oval window. Because it takes a fraction of a second for the tensor tympani and stapedius muscles to contract, they help protect the internal ear from prolonged loud noises, but not from sudden, brief ones such as a gunshot.

The anterior wall of the middle ear contains the opening of the **auditory tube**, commonly known as the **eustachian tube** (ū′-STĀ-kē-an). The auditory tube connects the middle ear with the nasopharynx (superior portion of the throat). It is normally closed at its medial (pharyngeal) end. During swallowing and yawning, it opens, allowing air to enter or leave the middle ear until the pressure in the middle ear equals the atmospheric pressure. Most of us have experienced our ears popping as the pressures equalize. When the pressures are balanced, the tympanic membrane vibrates freely as sound waves strike it. If the pressure is not equalized, intense pain, hearing impairment, ringing in the ears, and vertigo could develop. The auditory tube also is a route for pathogens to travel from the nose and throat to the middle ear, causing the most common type of ear infection (see the Clinical Connection entitled Otitis Media in your WileyPLUS resources).

### Internal (Inner) Ear

The **internal (inner) ear** is a complicated series of canals (Figure 16.20). Structurally, it consists of two main divisions: an outer bony labyrinth that encloses an inner membranous labyrinth. The **bony labyrinth** is a series of cavities in the petrous portion of the temporal bone divided into three areas: (1) the semicircular canals and (2) the vestibule, both of which contain receptors for equilibrium, and (3) the cochlea, which contains receptors for hearing. The bony labyrinth contains **perilymph**. This fluid, which is chemically similar to cerebrospinal fluid, surrounds the **membranous labyrinth**, a series of sacs and tubes inside the bony labyrinth and having the same general form as the bony labyrinth. The membranous labyrinth contains **endolymph**.

The **vestibule** (VES-ti-būl) is the oval central portion of the bony labyrinth. The membranous labyrinth in the vestibule

**FIGURE 16.20 The right internal ear.** The outer, cream-colored area is part of the bony labyrinth; the inner, pink-colored and orange-colored fluid-filled chambers compose the membranous labyrinth.

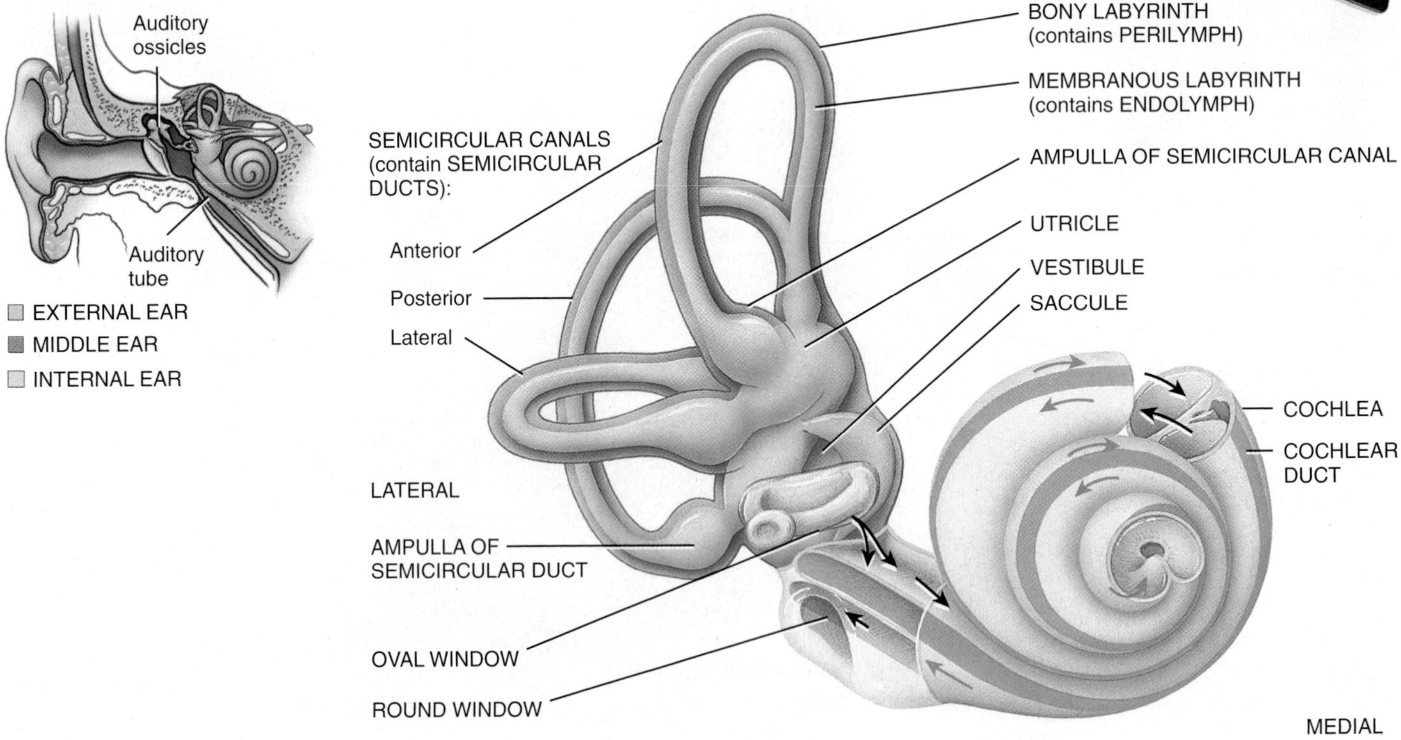

Components of right internal ear

**The bony labyrinth contains perilymph, and the membranous labyrinth contains endolymph.**

consists of two sacs called the **utricle** (Ū-tri-kul = little bag) and the **saccule** (SAK-ūl = little sac), which are connected by a small duct. Projecting superiorly and posteriorly from the vestibule are the three bony **semicircular canals**, each of which lies at approximately right angles to the other two. Based on their positions, they are named the anterior, posterior, and lateral semicircular canals. The anterior and posterior semicircular canals are vertically oriented; the lateral canal is horizontally oriented. At one end of each canal is a swollen enlargement called the **ampulla** (am-PUL-la = saclike duct). The portions of the membranous labyrinth that lie inside the bony semicircular canals are called the **semicircular ducts**. These structures connect with the utricle of the vestibule. The **vestibular branch** (ves-TIB-ū-lar) of the vestibulocochlear nerve transmits impulses for equilibrium from the vestibule and semicircular canals. Cell bodies of the sensory neurons are located in the **vestibular ganglia** (see Figure 16.21b).

Anterior to the vestibule is the **cochlea** (KOK-lē-a = snail-shaped), a bony spiral canal (Figure 16.21a) that resembles a snail's shell and makes almost three turns around a central bony hub. Sections through the cochlea reveal that it is divided into three channels: cochlear duct, scala vestibuli, and scala tympani (Figure 16.21a–c). The **cochlear duct** (KOK-lē-ar) is a continuation of the membranous labyrinth into the cochlea; it is filled with endolymph. The channel above the cochlear duct is the **scala vestibuli**, which ends at the oval window. The channel below is the **scala tympani**, which ends at the round window. Both the scala vestibuli and scala tympani are part of the bony labyrinth of the cochlea; therefore, these chambers are filled with perilymph. The scala vestibuli and scala tympani are completely separated by the cochlear duct, except for an opening at the apex of the cochlea, the **helicotrema** (hel-i-kō-TRĒ-ma; see Figure 16.22). The cochlea adjoins the wall of the vestibule, into which the scala vestibuli opens. The perilymph in the vestibule is continuous with that of the scala vestibuli.

The **vestibular membrane** separates the cochlear duct from the scala vestibuli, and the **basilar membrane** (BĀS-i-lar) separates the cochlear duct from the scala tympani. Resting on the basilar membrane is the **spiral organ** (Figure 16.21c, d). The spiral organ is a coiled sheet of epithelial cells, including supporting cells and about 16,000 **hair cells**, which are the receptors for hearing. At the apical tip of each hair cell is a **hair bundle** consisting of 40–80 long, hairlike microvilli that extend into the endolymph of the cochlear duct and are arranged in several rows of graded height. At their basal ends, hair cells synapse with sensory neurons from the **cochlear branch** of the vestibulocochlear nerve. Cell bodies of the sensory neurons are located in the **spiral ganglion** (Figure 16.21b, c). The **tectorial membrane** (tek-TŌ-rē-al; *tector-* = covering), a flexible gelatinous membrane, projects over and comes in contact with the hair cells of the spiral organ (Figure 16.21d).

**FIGURE 16.21 Semicircular canals, vestibule, and cochlea of the right ear.** Note that the cochlea makes nearly three complete turns.

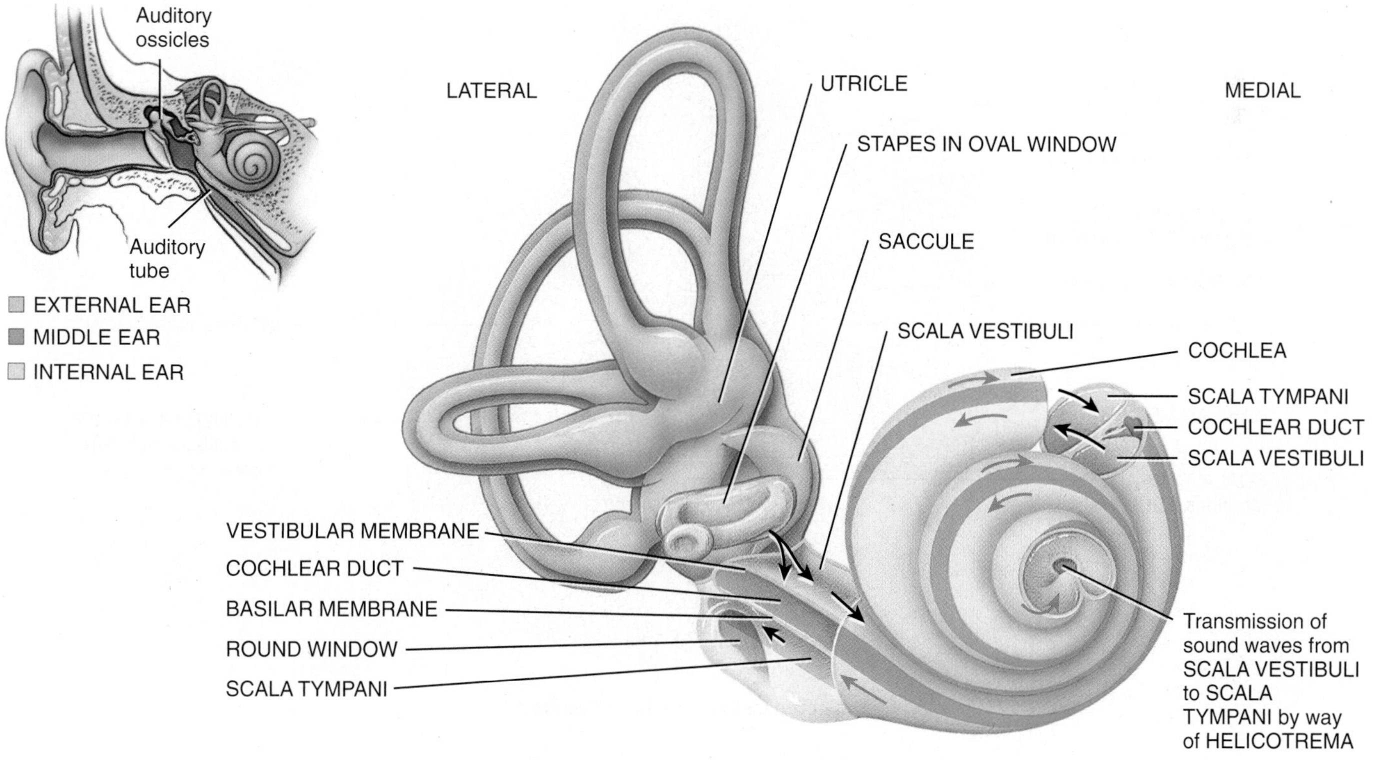


(a) Sections through cochlea

(continues)

FIGURE 16.21 (continued)

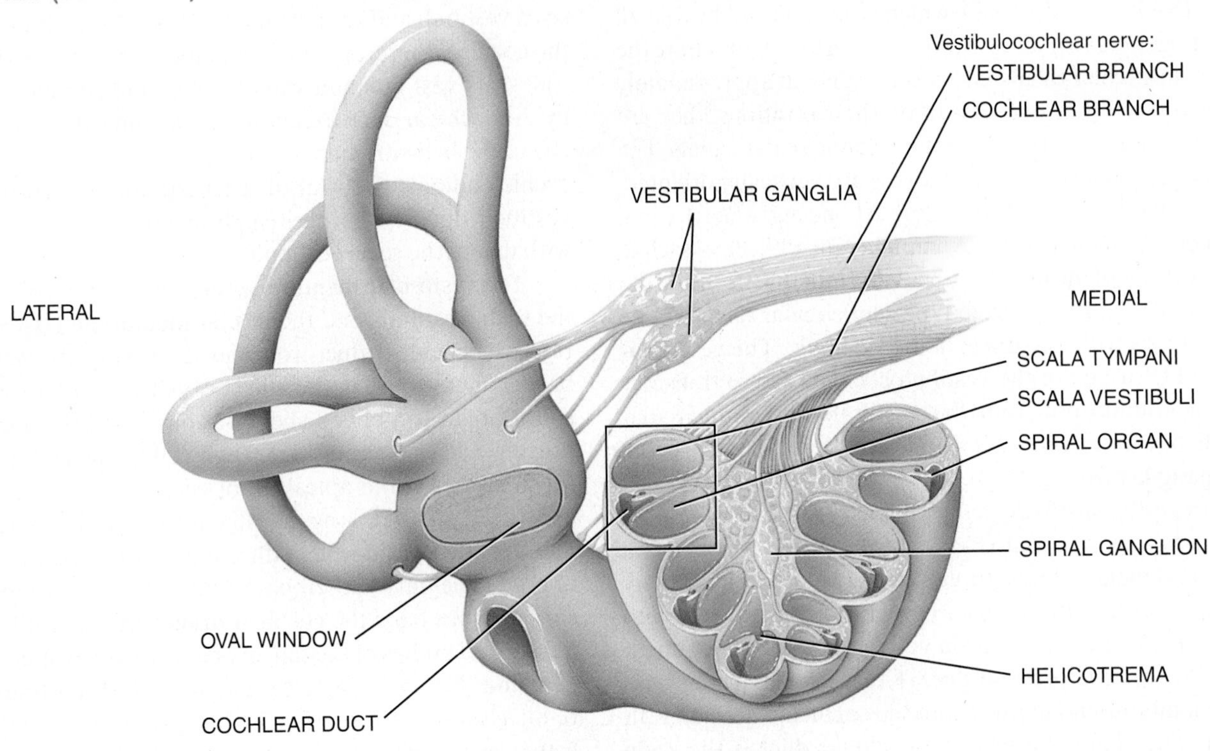


(b) Components of vestibulocochlear nerve

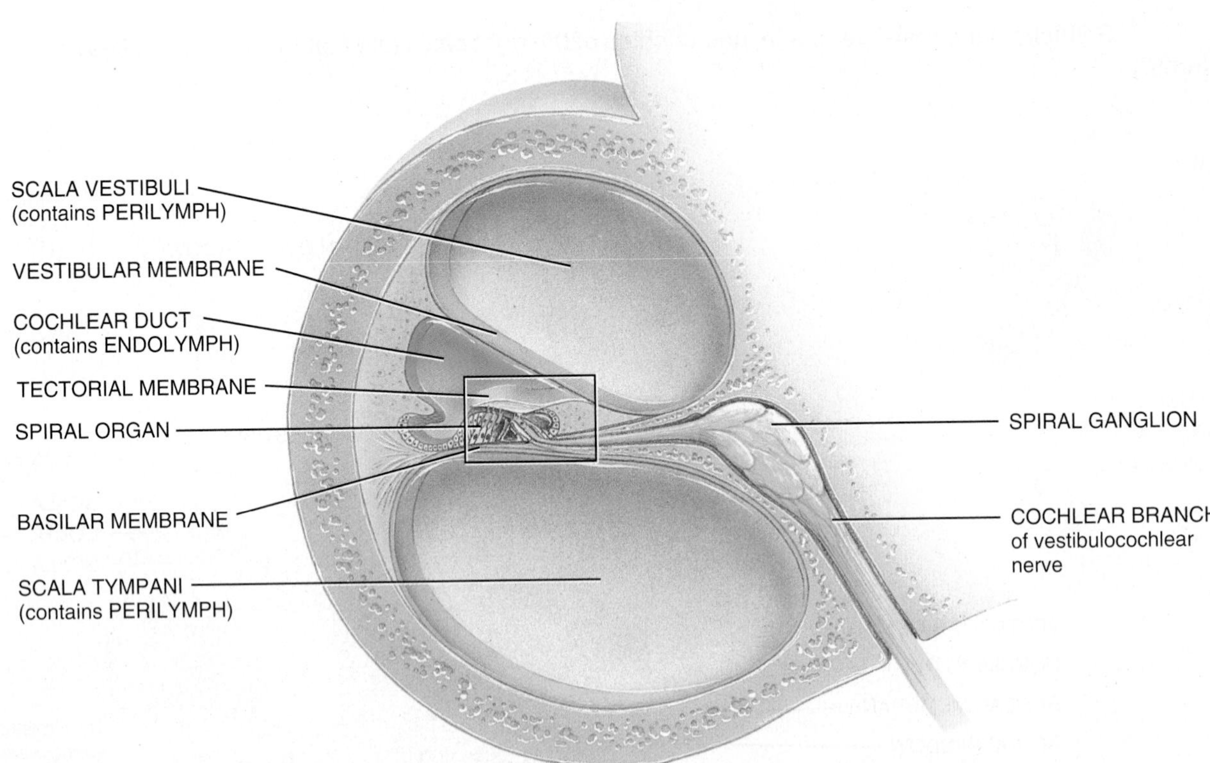


(c) Section through one turn of cochlea

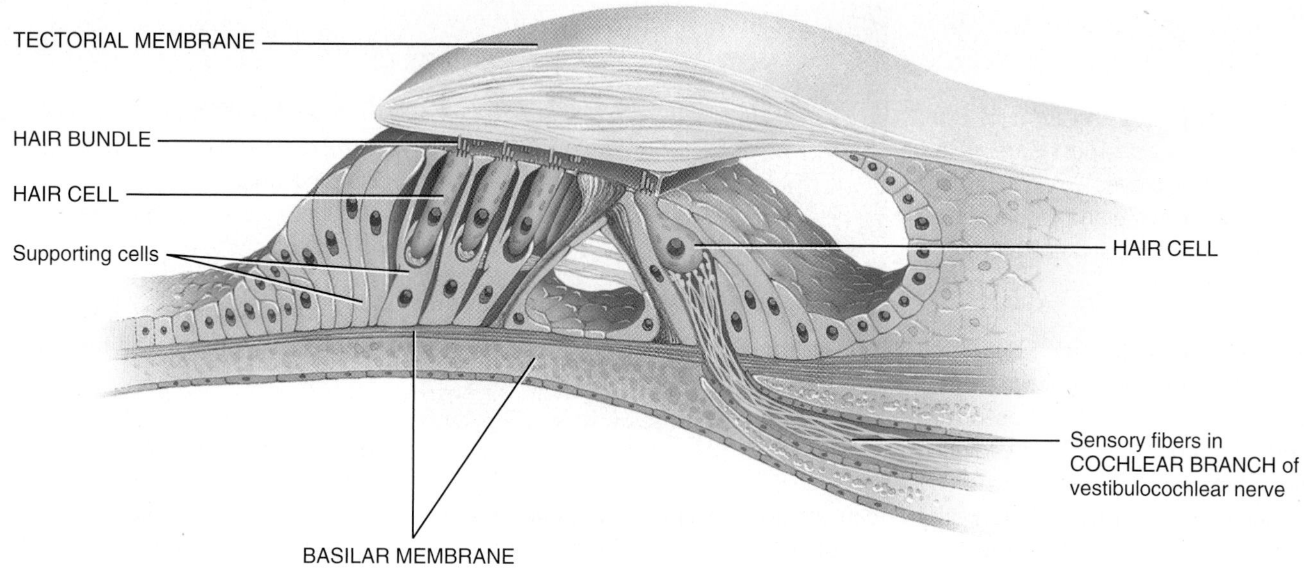


(d) Enlargement of spiral organ

**The three channels in the cochlea are the scala vestibuli, the scala tympani, and the cochlear duct.**

## ✓ CHECKPOINT

**32.** If you lost your auricles (common in severe burns of the head), how would your hearing be affected?

**33.** Which structures separate the middle ear from the external ear? From the internal ear?

**34.** List the auditory ossicles in order of sound wave transmission through the middle ear. For each ossicle, name two structures to which it is attached.

**35.** What are the three subdivisions of the bony labyrinth?

**36.** Which component of the cochlea contains the receptors for hearing?

**37.** Which nerve innervates the internal ear? Which branch of that nerve transmits equilibrium signals? Which branch transmits auditory signals?

## RETURN TO Dan's Story

Dr. Burke held a penlight in front of Dan, watching the pupils constrict as he swept the light toward Dan's eyes.

"Stay focused on the end of this pen as I move it," requested Dr. Burke. He brought the pen in toward Dan's face, then pulled away, then back in.

"It's most likely you have a detached retina, probably from the impact of the explosion. A layer of your retina has pulled away from the inner lining of the eye. It might have happened anyway, but these things usually don't occur in healthy 30-year-old men. The ophthalmologist said you had some abnormalities with your eyes: high intraocular pressure, some small hemorrhages in the vasculature of your retinas, and poor accommodation. But this problem won't just heal by itself," said Dr. Burke.

Dr. Burke consulted with an eye surgeon, who visited Dan shortly thereafter. Reattaching the retina required delicate laser surgery. The detachment would disrupt the supply of blood to the retina, which could lead to permanent blindness if the retina wasn't reattached. During the laser procedure, his eye was numbed, then a gas bubble was injected into the eye. After the procedure Dan could see, but his vision was blurry in the eye that had been operated on. It was explained to Dan that the retina had only detached at the perimeter. It then slid down over his fovea centralis, the part of his retina needed for good vision.

Pam sat next to Dan's hospital bed. "Okay, now that they repaired your eye, what did they say about the hearing loss?" she asked.

"Well, I can hear better out of the right ear, but I am pretty much deaf in the left. The audiologist said my eardrum wasn't quite as damaged on the right side. They said my eardrum might grow back, but I might need surgery on it, too."

Pam searched Dan's face. "But will you be able to hear like before? Is there any long-term damage?" she asked.

"The internal ear might be damaged. The doc said that was more serious, less likely to heal," explained Dan.

"Hi, Pam. Hi, Dan, how're you doing today? I hear they're letting you go soon," Captain Donald called from the doorway.

"Hey, Cap! Come on in!" Dan replied.

"Are you still dizzy all the time?" asked Captain Donald.

"It comes and goes," responded Dan.

"How are the ears?" asked Donald.

"What, what?" Dan joked, laughing. "Well, 50 percent loss in the right, 75 percent in the left. But the eardrums are healing."

"Dan, you know I can't let you back on the fire department if you're not at 100 percent. Someone could get hurt."

Dan looked at Captain Donald. He knew he was right, but it was hard news to hear.

F. *A "detached retina" occurs when the neural layer of the retina separates from the pigmented layer. The usual cause is a sudden impact to the eye. If this occurred at the fovea centralis, would Dan lose all of his vision from that eye? Why or why not?*

G. *Which of Dan's middle ear structures would have served a protective function during the explosion?*

H. *Why would damage to Dan's internal ear be less likely to heal than damage to his tympanic membrane? Which components of his internal ear may be damaged?*

I. *Why would a firefighter need good vision, sight, and hearing in addition to a high level of fitness?*

## 16.8 The pathway of sound is tympanic membrane → ossicles → oval window → cochlea → vestibulocochlear nerve → primary auditory cortex.

### The Nature of Sound Waves

In order to understand the physiology of hearing, it is necessary to learn something about its input, which occurs in the form of sound waves. **Sound waves** are alternating high- and low-pressure regions traveling in the same direction through some medium (such as air). They originate from a vibrating object in much the same way that ripples arise and travel over the surface of a pond when you toss a stone into it. The number of sound waves that arrive in a given time is defined as *frequency* of a sound and is measured in *hertz* (Hz; 1 Hz = 1 cycle per second). We interpret different sound frequencies as differences in pitch. The higher the frequency of vibration, the higher is the pitch. The entire audible range of the human ear extends from 20 to 20,000 Hz. Sounds of speech primarily contain frequencies between 100 and 3000 Hz, and the "high C" sung by a soprano has a dominant frequency at 1048 Hz. The sounds from a jet plane several miles away range from 20 to 100 Hz.

The larger the *intensity* (size or amplitude) of the vibration, the *louder* is the sound. Sound intensity is measured in units called decibels (dB). The hearing threshold—the point at which an average young adult can just distinguish sound from silence—is defined as 0 dB. Rustling leaves have a decibel level of 15; whispered speech, 30; normal conversation, 60; a vacuum cleaner, 75; shouting, 80; and a nearby motorcycle or jackhammer, 90. Sound becomes uncomfortable to a normal ear at about 120 dB, and painful above 140 dB.

## CLINICAL CONNECTION | *Loud Sounds and Hair Cell Damage*

Exposure to loud music and the engine roar of jet planes, revved-up motorcycles, lawn mowers, and vacuum cleaners damages hair cells of the cochlea. Because prolonged noise exposure causes hearing loss, employers in the United States must require workers to use hearing protectors when occupational noise levels exceed 90 dB. Rock concerts and even inexpensive headphones can easily produce sounds over 110 dB. Continued exposure to high-intensity sounds is one cause of **deafness**, a significant or total hearing loss. The louder the sounds, the more rapid is the hearing loss. Deafness usually begins with loss of sensitivity for high-pitched sounds. If you are listening to music through headphones and bystanders can hear it, the dB level is in the damaging range. Most people fail to notice their progressive hearing loss until destruction is extensive and they begin having difficulty understanding speech. Wearing earplugs with a noise-reduction rating of 30 dB while engaging in noisy activities can protect the sensitivity of your ears.

## Physiology of Hearing

The following events are involved in hearing (Figure 16.22):

1. The auricle directs sound waves into the external auditory canal.
2. When sound waves strike the tympanic membrane, the alternating high and low pressure of the air causes the tympanic membrane to vibrate back and forth. The tympanic membrane vibrates slowly in response to low-frequency (low-pitched) sounds and rapidly in response to high-frequency (high-pitched) sounds. It vibrates more forcefully in response to higher intensity (louder) sounds, more gently in response to lower intensity (quieter) sounds.
3. The central area of the tympanic membrane connects to the malleus, which also starts to vibrate. The vibration is transmitted from the malleus to the incus, and then to the stapes.
4. As the stapes moves back and forth, its oval-shaped footplate vibrates in the oval window. The vibrations at the oval window are about 20 times more vigorous than the

FIGURE 16.22 **Events in the stimulation of auditory receptors in the right ear.** The cochlea has been uncoiled to more easily visualize the transmission of sound waves and their distortion of the vestibular and basilar membranes of the cochlear duct.

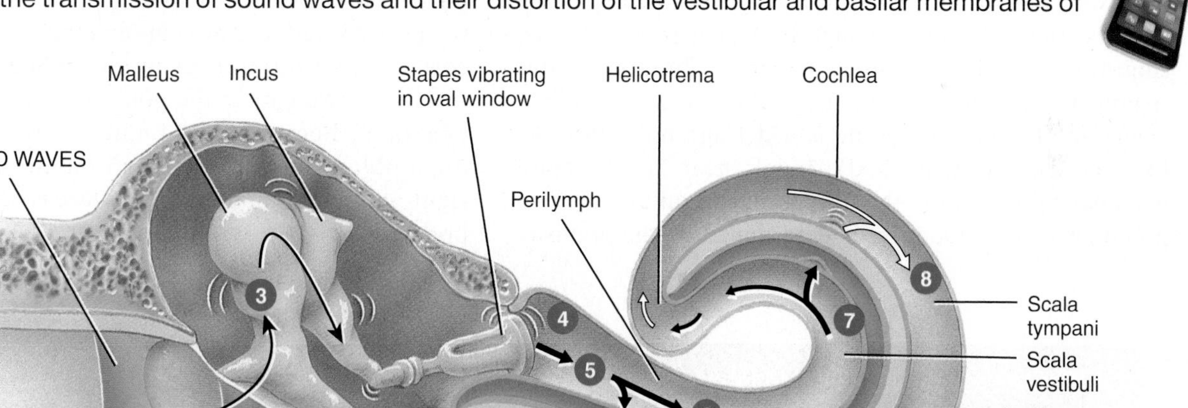


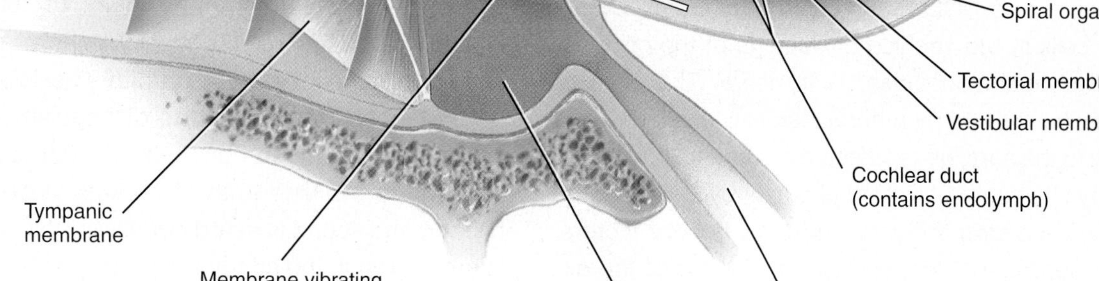


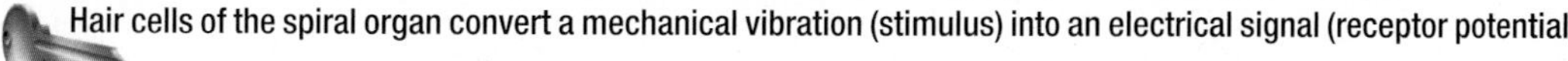

**Hair cells of the spiral organ convert a mechanical vibration (stimulus) into an electrical signal (receptor potential).**

vibrations at the tympanic membrane because the auditory ossicles efficiently transmit small vibrations, spread over a large surface area (tympanic membrane), into larger vibrations of a smaller surface (oval window).

5. The vibrations of the stapes at the oval window sets up fluid pressure waves in the cochlea by pushing on the perilymph of the scala vestibuli.

6. Pressure waves are transmitted from the scala vestibuli to the scala tympani and eventually to the round window, where the pressure waves are finally absorbed as the flexible membrane of the round window bulges outward into the middle ear. (See 9 in the figure.)

7. The pressure waves travel through the perilymph of the scala vestibuli, then the vestibular membrane, and then move into the endolymph inside the cochlear duct.

8. The pressure waves in the endolymph cause the basilar membrane to vibrate, which moves the hair cells of the spiral organ against the tectorial membrane. This leads to bending of the hair cell microvilli, which produces receptor potentials that ultimately lead to the generation of impulses.

9. Sound waves of various frequencies cause certain regions of the basilar membrane to vibrate more intensely than other regions. Each segment of the basilar membrane is "tuned" for a particular pitch. Because the membrane is narrower and stiffer at the base of the cochlea (portion closer to the oval window), high-frequency (high-pitched) sounds near 20,000 Hz induce maximal vibrations in this region. Toward the apex of the cochlea near the helicotrema, the basilar membrane is wider and more flexible; low-frequency (low-pitched) sounds near 20 Hz cause maximal vibration of the basilar membrane there. As noted previously, loudness is determined by the intensity of sound waves. High-intensity sound waves cause larger vibrations of the basilar membrane, which leads to a higher frequency of impulses reaching the brain. Louder sounds also may stimulate a larger number of hair cells.

The hair cells convert mechanical vibrations into electrical signals. As the basilar membrane vibrates, the hair bundles at the apex of the hair cell bend back and forth. Bending of the hair bundles stimulates the hair cells, producing a depolarizing receptor potential that triggers the release of a neurotransmitter, which is probably glutamate. The released neurotransmitter generates impulses in the sensory neurons that synapse with the base of the hair cells. As more neurotransmitter is released, the frequency of impulses in the sensory neurons increases.

## The Auditory Pathway

Bending of the microvilli of the hair cells in the spiral organ causes the release of a neurotransmitter, which generates impulses in the sensory neurons that innervate the hair cells. These sensory neurons form the cochlear branch of the vestibulocochlear nerve (Figure 16.23). Their axons synapse in nuclei in the medulla oblongata on the same side. From there, axons ascend directly to the midbrain or cross to the opposite side and terminate either in nuclei of the pons or in the midbrain. From the pons, axons ascend to the midbrain in a tract called the **lateral meniscus**. From the midbrain, impulses are conveyed to nuclei in the thalamus, and finally to the primary auditory area in the temporal lobe of the cerebral cortex. Slight differences in the timing of impulses arriving from your two ears at the pons allow you to locate the source of a sound. Because many auditory axons cross over in the medulla oblongata while others remain on the same side, the right and left primary auditory areas receive impulses from both ears.

### CLINICAL CONNECTION | *Deafness*

**Deafness** is significant or total hearing loss. **Sensorineural deafness** (sen′-sō-rē-NOO-ral) is caused by either impairment of hair cells in the cochlea or damage of the cochlear branch of the vestibulocochlear (VIII) nerve. This type of deafness may be caused by atherosclerosis, which reduces blood supply to the ears; by repeated exposure to loud noise, which destroys hair cells of the spiral organ; by certain drugs such as aspirin and streptomycin; and/or by genetic factors. **Conduction deafness** is caused by impairment of the external and middle ear mechanisms for transmitting sounds to the cochlea. Causes of conduction deafness include otosclerosis, the deposition of new bone around the oval window; impacted cerumen; injury to the eardrum; and aging, which often results in thickening of the eardrum and stiffening of the joints of the auditory ossicles. A hearing test called *Weber's test* is used to distinguish between sensorineural and conduction deafness. In the test, the stem of a vibrating fork is held to the forehead. In people with normal hearing, the sound is heard equally in both ears. If the sound is heard best in the affected ear, the deafness is probably of the conduction type; if the sound is heard best in the normal ear, it is probably of the sensorineural type.

FIGURE 16.23 The auditory pathway.

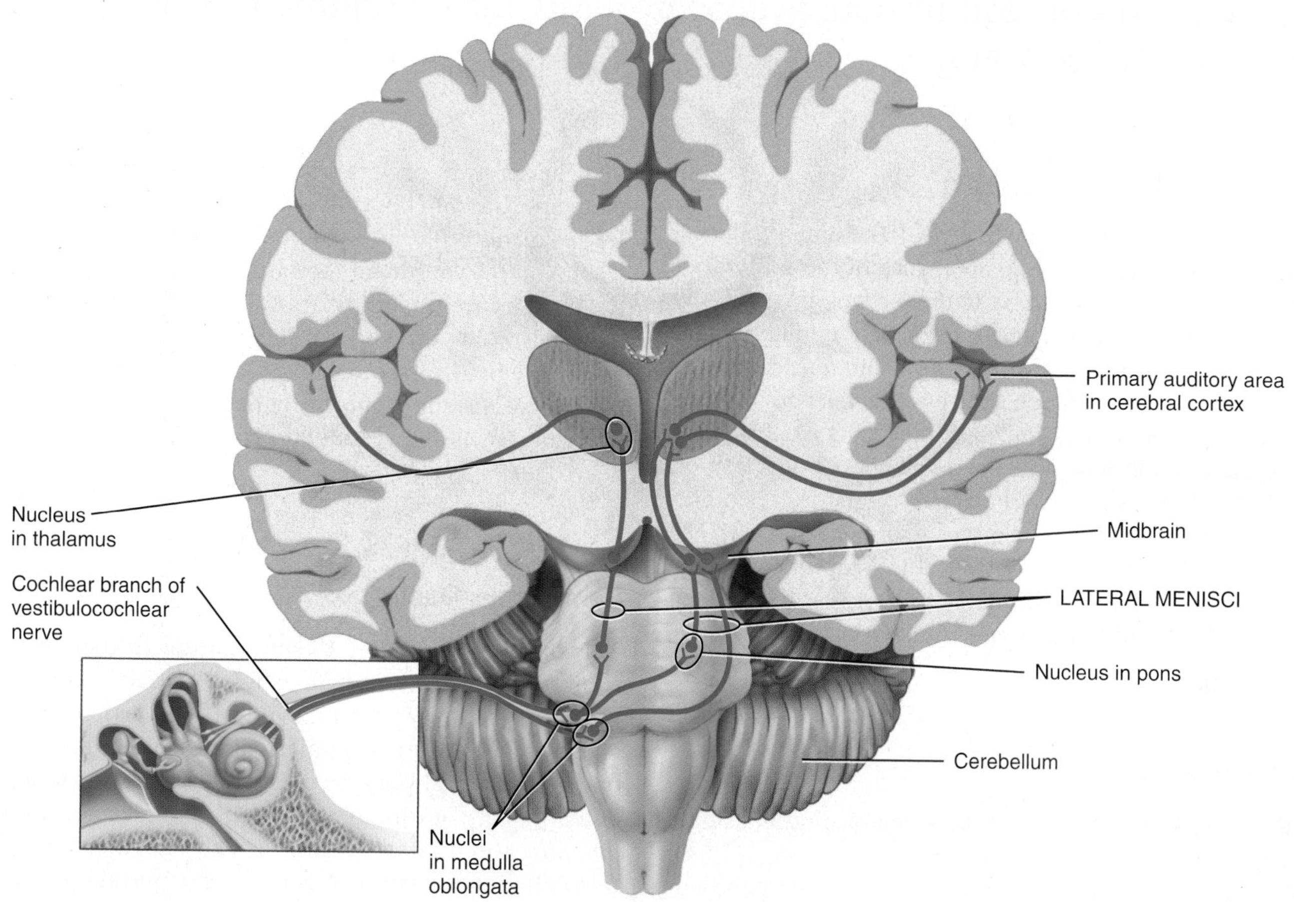


**From the cochlea, auditory impulses are propagated along the cochlear branch of the vestibulocochlear nerve and then to the brain stem, thalamus, and cerebral cortex.**

## ✓ CHECKPOINT

**38.** Would strongly blowing through a whistle produce sound waves of high or low frequency? High or low intensity?

**39.** How are sound waves transmitted from the auricles to the spiral organ?

**40.** How does the tympanic membrane respond to lower intensity sounds? To low-frequency sounds?

**41.** If fluid waves in the cochlea bounced back and forth, sounds would echo inside your cochleas. What stops fluid waves from traveling more than once through the cochlea?

**42.** What is the pathway for auditory impulses from the cochlea to the cerebral cortex?

# 16.9 Impulses for equilibrium propagate along the vestibulocochlear nerve to the brain.

## Physiology of Equilibrium

There are two types of **equilibrium** (ē-kwi-LIB-rē-um) or balance. **Static equilibrium** refers to the maintenance of the position of the body (mainly the head) relative to the force of gravity. Body movements that stimulate the receptors for static equilibrium include tilting the head forward or backward and *linear* (straight line) acceleration or deceleration, such as when the body is being moved in an elevator or in a plane that is speeding up or slowing down. **Dynamic equilibrium** is the maintenance of body position (mainly the head) in response to *rotational* (turning) acceleration or deceleration, such as when you turn your head or spin your body around while dancing. Collectively, the receptor organs for equilibrium are called the **vestibular apparatus** (ves-TIB-ū-lar); these include the saccule, utricle, and semicircular ducts.

### Saccule and Utricle

The walls of both the utricle and the saccule contain a small, thickened region called a **macula** (MAK-ū-la; Figure 16.24). The two maculae (plural; MAK-ū-lē), which are perpendicular to one another, are the receptors for static equilibrium. They provide sensory information on the position of the head in space, essential to maintaining appropriate posture and balance. The maculae also detect linear acceleration and deceleration—for example, the sensations you feel while in a car that is speeding up or slowing down.

The maculae contain hair cells, sensory receptors that are surrounded by columnar supporting cells. Hair cells have on their surface hair bundles that consist of 40–80 microvilli. The hair bundles of the hair cells project into a thick, gelatinous layer called the **otolithic membrane** (ō-tō-LITH-ik). Calcium carbonate crystals, called **otoliths** (Ō-tō-liths; *oto-* = ear; *-liths* = stones), extend over the surface of the otolithic membrane.

The densely packed otoliths add weight to the otolithic membrane, amplifying the pull of gravity during movements. Because the otolithic membrane sits on top of the macula, when you tilt your head forward, the otolithic membrane and the otoliths are pulled by gravity and slide downhill over the hair cells in the direction of the tilt, bending the hair bundles. The movement of the hair bundles produces receptor potentials, releasing neurotransmitter that stimulates impulses in sensory neurons in the vestibular branch of the vestibulocochlear nerve (see Figure 16.21b).

### Semicircular Ducts

The three semicircular ducts function in dynamic equilibrium. The ducts lie at right angles to one another in three planes of space (Figure 16.25): The two vertical ducts are the anterior and posterior semicircular ducts, and the horizontal one is the lateral semicircular duct. This positioning permits detection of rotational acceleration or deceleration. In the *ampulla*, the dilated portion of each duct, is a small elevation called the **crista** (KRIS-ta = *crest*; plural is *cristae*). Each crista contains a group of hair cells and supporting cells. Covering the crista is a mass of gelatinous material called the **cupula** (KŪ-pū-la). Hair bundles project from the hair cells into the cupula.

When your head moves along the plane of one of the semicircular ducts, the enclosed endolymph moves through the semicircular duct, pushing on the cupula and bending the hair bundles of the hair cells. The three different planes of the three semicircular ducts in each ear are able to respond to virtually any rotational movement of the head. Bending of the hair bundles produces receptor potentials. In turn, the receptor potentials lead to impulses that pass along the vestibular branch of the vestibulocochlear nerve (see Figure 16.21b).

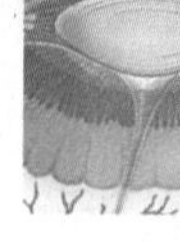



FIGURE 16.24 **Location and structure of the maculae of the right ear.**

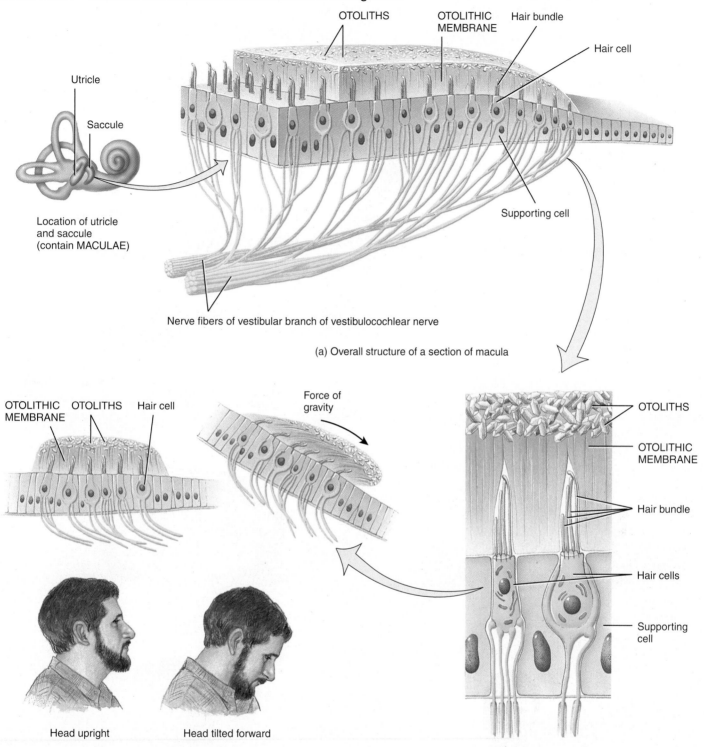

Utricle

Saccule

Location of utricle
and saccule
(contain MACULAE)

OTOLITHS

OTOLITHIC
MEMBRANE

Hair bundle

Hair cell

Supporting cell

Nerve fibers of vestibular branch of vestibulocochlear nerve

(a) Overall structure of a section of macula

OTOLITHIC
MEMBRANE

OTOLITHS

Hair cell

Force of
gravity

OTOLITHS

OTOLITHIC
MEMBRANE

Hair bundle

Hair cells

Supporting
cell

Head upright

Head tilted forward

(c) Position of macula with head upright (left) and tilted forward (right)

(b) Details of two hair cells

**The maculae of the utricle and saccule are the sensory receptors for static equilibrium.**

FIGURE 16.25 **Location and structure of the semicircular ducts of the right ear.**

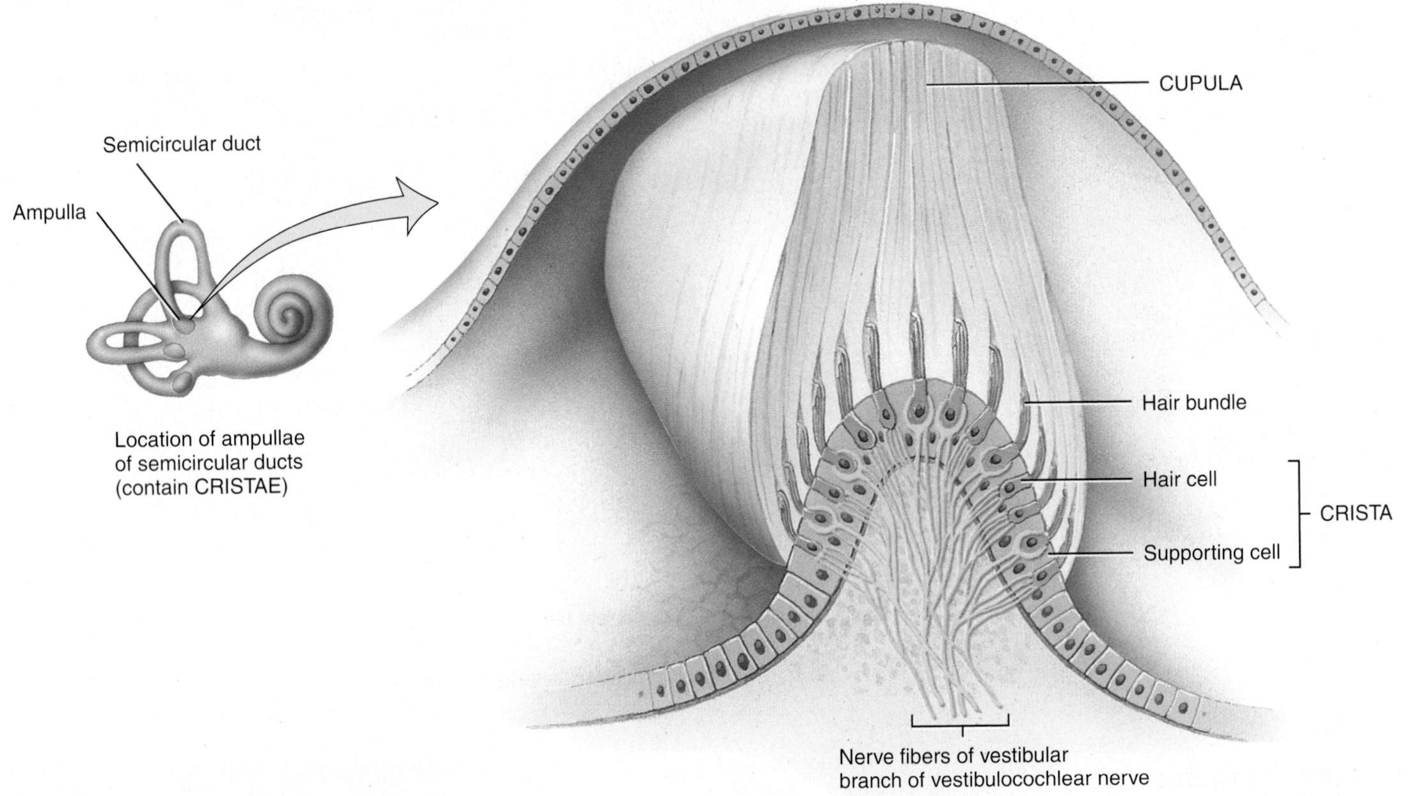


(a) Details of a crista

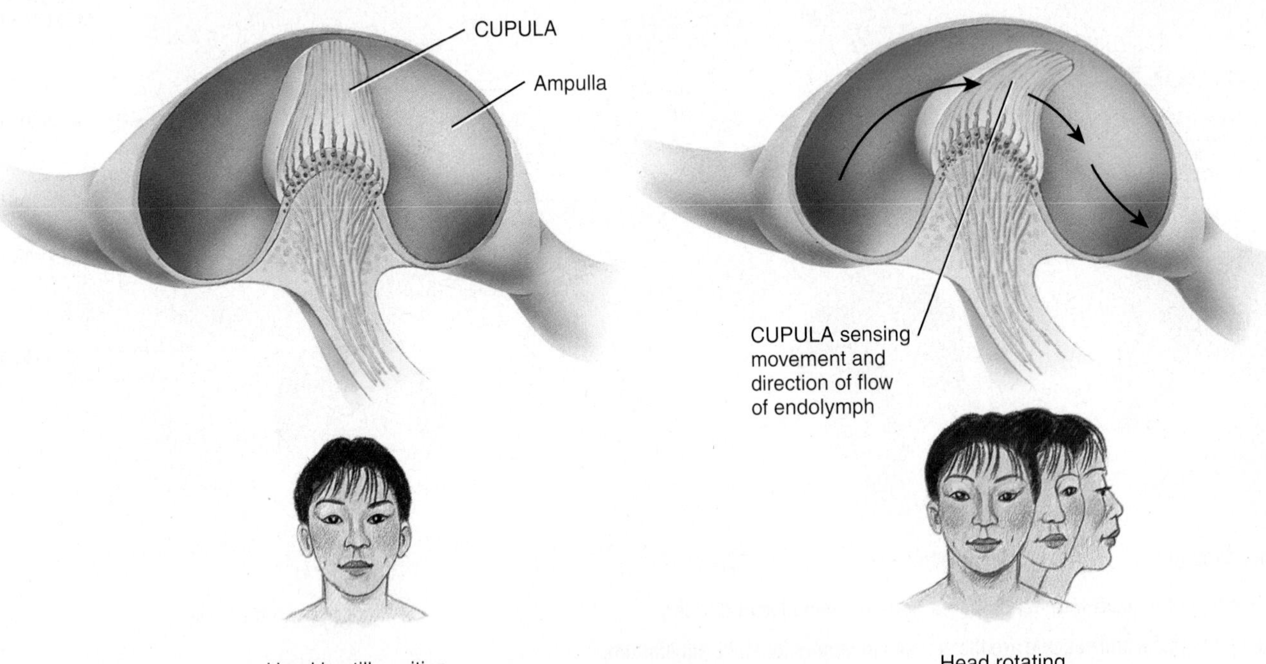


(b) Position of a cupula with head in still position (left) and when head rotates (right)

**The positions of the semicircular ducts permit detection of the rotational movements of dynamic equilibrium.**

## Equilibrium Pathways

Bending of hair bundles of the hair cells in the utricle, saccule, or semicircular ducts causes the release of a neurotransmitter (probably glutamate), which generates impulses in the sensory neurons that innervate the hair cells. Impulses pass along the axons of these neurons, which form the vestibular branch of the vestibulocochlear nerve (Figure 16.26). Most of these axons synapse with sensory neurons in nuclei in the medulla oblongata and pons. The remaining axons terminate in the cerebellum.

The medulla oblongata and pons integrate information arriving from the utricle, saccule, semicircular ducts, eyes, and somatic receptors, especially proprioceptors in the neck muscles that indicate the position of the head, and then send commands to the following:

- **The oculomotor, trochlear, and abducens nerves**. Control movements of the eyes coupled with those of the head to help maintain focus on the visual field.
- **The accessory nerves**. Help control head and neck movements to assist in maintaining equilibrium.
- **The vestibulospinal tract**. Conveys impulses down the spinal cord to maintain muscle tone in skeletal muscles to help maintain equilibrium.
- **The thalamus and the vestibular area**. The thalamus conveys impulses to the vestibular area, part of the primary somatosensory area in the parietal lobe of the cerebral cortex (see Figure 13.10), which provides us with the conscious awareness of the position and movements of the head.

The cerebellum also plays a key role in maintaining equilibrium. In response to input from the utricle, saccule, and semicircular ducts, the cerebellum continuously sends impulses to the motor areas of the cerebrum. This feedback allows correction of signals from the motor cortex to specific skeletal muscles to smooth movements and coordinate complex sequences of muscle contractions to help maintain equilibrium.

Table 16.2 summarizes the structures of the ear related to hearing and equilibrium.

FIGURE 16.26 **The equilibrium pathway.**

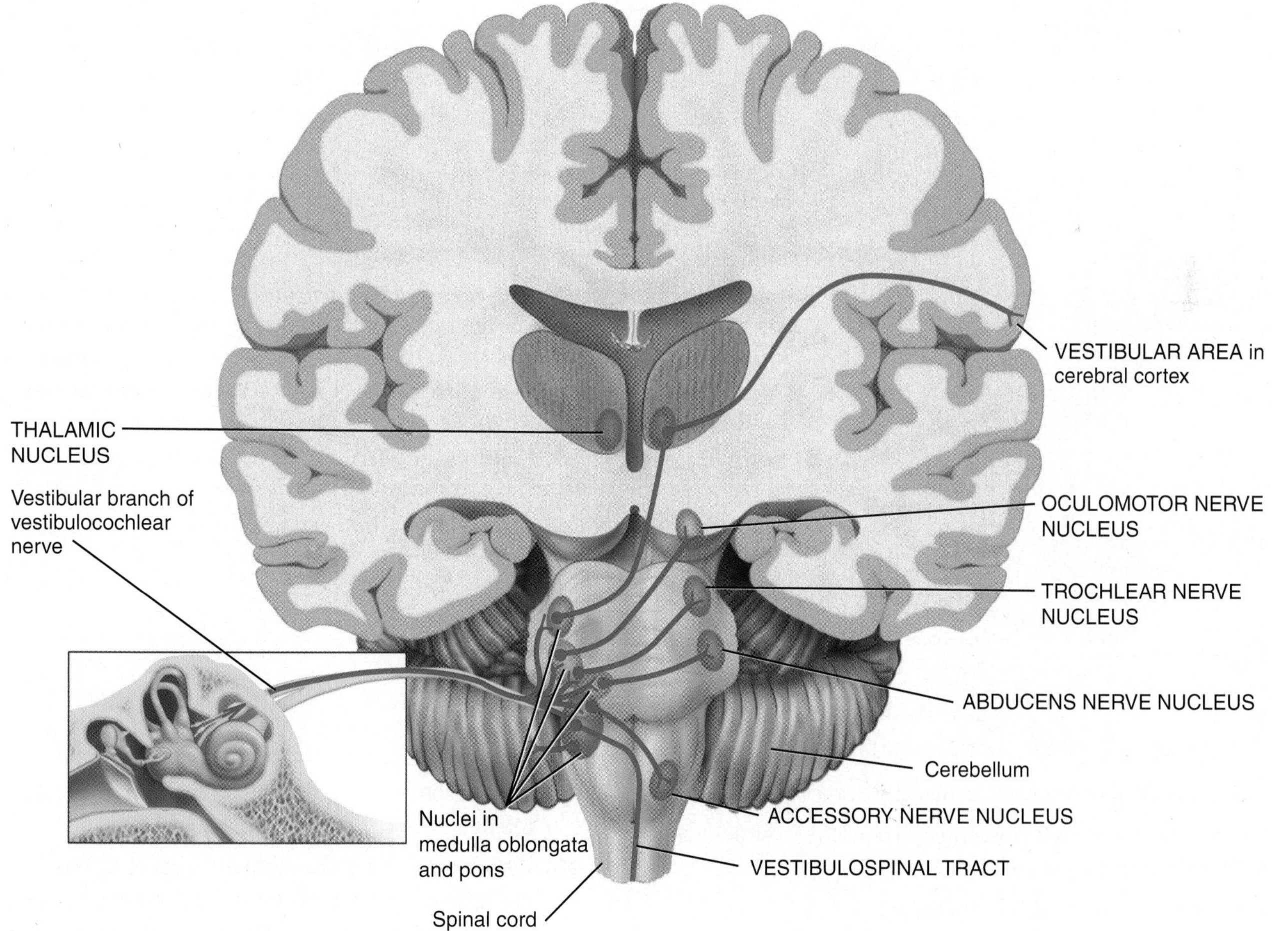


**From the semicircular ducts, utricle, and saccule, equilibrium impulses are propagated along the vestibular branch of the vestibulocochlear nerve to the brain stem, cerebellum, thalamus, and cerebral cortex.**

**TABLE 16.2**

Structures of the Ear

| REGIONS OF THE EAR AND KEY STRUCTURES | FUNCTION |
|---|---|
| **External (outer) ear** 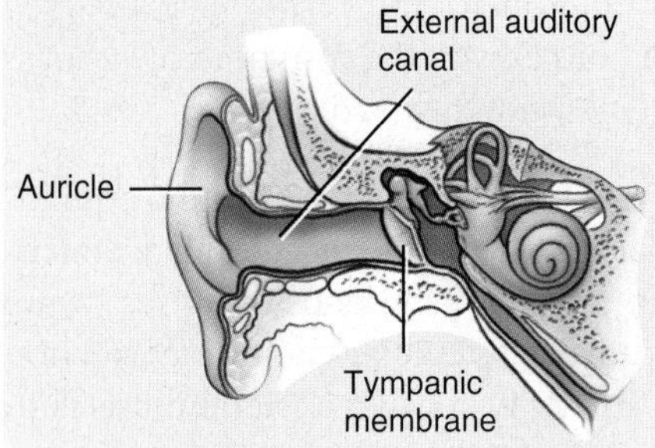  | ***Auricle:*** Collects sound waves<br>***External auditory canal:*** Directs sound waves to eardrum<br>***Tympanic membrane:*** Sound waves cause it to vibrate, which in turn causes malleus to vibrate |
| **Middle ear** 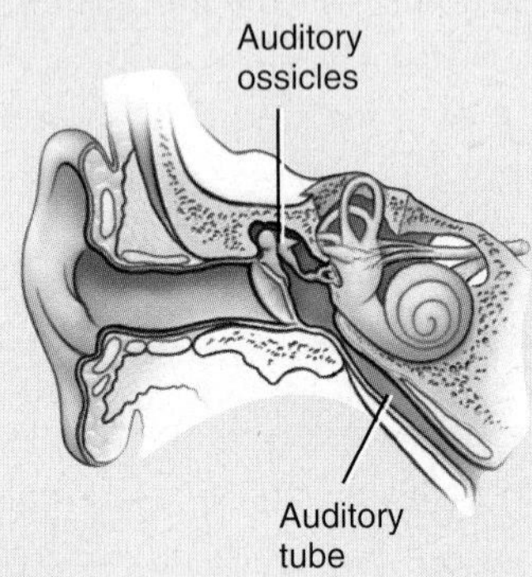  | ***Auditory ossicles:*** Transmit and amplify vibrations from tympanic membrane to oval window<br>***Auditory tube:*** Equalizes air pressure on both sides of tympanic membrane |
| **Internal (inner) ear** 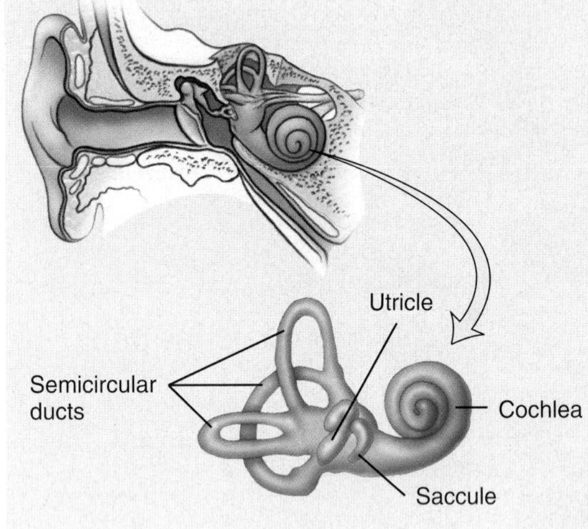  | ***Cochlea:*** Contains a series of fluids, channels, and membranes that transmit vibrations to spiral organ; the organ of hearing; hair cells in spiral organ produce receptor potentials, which elicit nerve impulses in cochlear branch of vestibulocochlear nerve<br>***Vestibular apparatus:*** Includes semicircular ducts, utricle, and saccule, which generate nerve impulses that propagate along vestibular branch of vestibulocochlear nerve<br>***Semicircular ducts:*** Contain cristae, site of hair cells for dynamic equilibrium (maintenance of body position, mainly the head, in response to rotational acceleration and deceleration movements)<br>***Utricle:*** Contains macula, site of hair cells for static equilibrium (maintenance of body position, mainly the head, relative to force of gravity)<br>***Saccule:*** Contains macula, site of hair cells for static equilibrium (maintenance of body position, mainly the head, relative to force of gravity) |

## ✓ CHECKPOINT

**43.** What is the difference between static and dynamic equilibrium?

**44.** How are the receptors for hearing and equilibrium structurally similar? How do they differ?

**45.** What function is served by the otoliths?

**46.** Why do we need three, rather than one or two, semicircular canals?

**47.** With which type of equilibrium are the semicircular ducts, the utricle, and the saccule associated?

**48.** Where do axons in the vestibular branch of the vestibulocochlear nerve terminate? What purposes are served by the transmission of equilibrium input to these locations?

## Dan's Story — EPILOGUE AND DISCUSSION

Dan is a healthy 30-year-old male who has suffered a blast injury from a natural gas explosion. The explosion has caused significant damage to his special senses. Dan initially had swelling and bruising, and possibly a concussion from the blast. His vision has been disrupted, and he suffered a detached retina, which was surgically repaired. Immediately after his accident, his sense of smell was abnormal, and he had a reduced ability to taste. Hearing loss is significant due to damage to middle ear components that conduct sound to the internal ear, and possibly due to damage to the hair cells within the cochlea. Dan is also experiencing chronic vertigo, a feeling of dizziness due to blast shock damage to the vestibular apparatus.

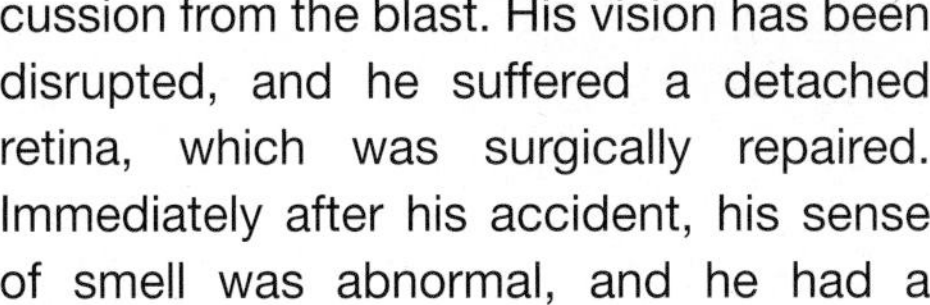

Dan was fortunate that he was wearing protective equipment when he was exposed to the explosion. Typically, the predominant post-explosion injuries suffered in such a blast involve penetrating injuries and blunt trauma.

J. *Which structures of Dan's internal ear could have been damaged to produce his continued sensations of dizziness?*

K. *How could the explosion have damaged hair cells in Dan's vestibular apparatus?*

L. *Is it likely that Dan's senses of taste and smell will return? Why or why not?*

M. *During Dan's entry into the burning building, how did his special senses warn him of impending danger?*

# Concept and Resource Summary

| Concept | Resources |
|---|---|

## Introduction

1. The **special senses** include smell, taste, vision, hearing, and equilibrium. Receptors for these senses are located in complex sensory organs of the head.

## Concept 16.1 Impulses for smell propagate along the olfactory nerve to the brain.

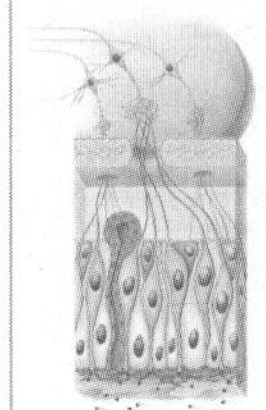

1. The receptors for **olfaction**, **olfactory receptor cells**, are in the **olfactory epithelium** in the superior area of the nasal cavity.
2. **Olfactory hairs** respond to inhaled chemicals, or **odorants**. **Supporting cells** detoxify chemicals, electrically insulate receptors, and provide support and nourishment. **Basal cells** continually divide to produce new olfactory receptor cells. **Olfactory glands** produce mucus to dissolve odorants.
3. Olfactory receptor cells are excited by odorant molecules binding to odorant binding proteins. This results in production of a generator potential, which may lead to an action potential.
4. Olfaction has a low threshold; only a few molecules are needed to be perceived as an odor. Adaptation to odors occurs rapidly.
5. Bundles of axons of olfactory receptor cells pass through foramina in the cribriform plate of the ethmoid bone and form the left and right **olfactory nerves**, which convey impulses of smells to the **olfactory bulbs**, **olfactory tracts**, limbic system, and cerebral cortex (temporal and frontal lobes).

**Resources:**

Anatomy Overview—Chemoreceptors
Anatomy Overview—Olfactory Nerve
Anatomy Overview—The Special Senses

Clinical Connection—Hyposmia

## Concept

## Resources 

### Concept 16.2 Impulses for taste propagate along the facial, glossopharyngeal, and vagus nerves to the brain.

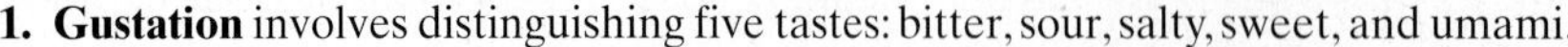

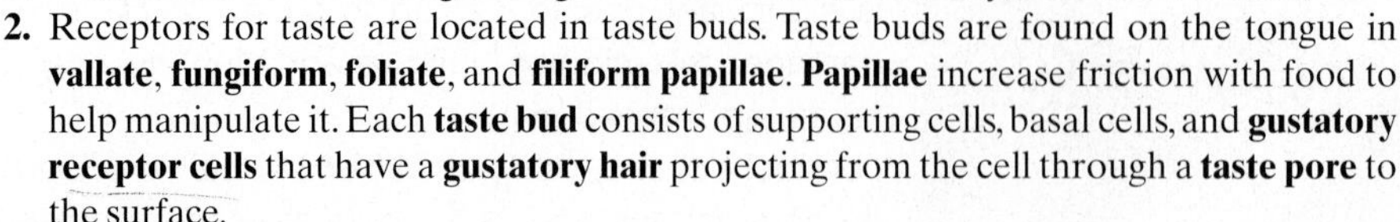

1. **Gustation** involves distinguishing five tastes: bitter, sour, salty, sweet, and umami.
2. Receptors for taste are located in taste buds. Taste buds are found on the tongue in **vallate, fungiform, foliate**, and **filiform papillae. Papillae** increase friction with food to help manipulate it. Each **taste bud** consists of supporting cells, basal cells, and **gustatory receptor cells** that have a **gustatory hair** projecting from the cell through a **taste pore** to the surface.
3. **Tastants** dissolve in saliva and bind to gustatory hairs, stimulating the receptor cells to have a receptor potential that stimulates the release of neurotransmitter, which can generate action potentials in first-order neurons.
4. The threshold for each of the primary tastes varies, and adaptation to taste occurs quickly.
5. Three cranial nerves propagate gustation impulses to the medulla oblongata; some then project to the limbic system, hypothalamus, thalamus, and cerebral cortex (parietal lobe).

Anatomy Overview—Chemoreceptors
Anatomy Overview—Facial Nerve
Anatomy Overview—Glossopharyngeal Nerve
Anatomy Overview—Vagus Nerve

Clinical Connection—Taste Aversion

### Concept 16.3 The eye is protected by eyelids, eyelashes, eyebrows, and a lacrimal apparatus.

1. The **accessory structures** of the eye include eyelids, eyelashes, eyebrows, lacrimal apparatus, and extrinsic eye muscles.
2. The **eyelids** shade and protect the eyes, and spread lubricants over the eyeballs. Each eyelid contains the **orbicularis oculi, meibomian glands**, and **conjunctiva**.
3. The **eyelashes** and **eyebrows** help protect the eyeballs.
4. The **lacrimal apparatus** includes the **lacrimal glands, lacrimal ducts, lacrimal puncta, lacrimal canals, lacrimal sac**, and **nasolacrimal duct**. This group of structures produce and drain **lacrimal fluid**. Tears protect, clean, and lubricate the eyeball.
5. Six extrinsic muscles move each eyeball: the **superior rectus, inferior rectus, lateral rectus, medial rectus, superior oblique**, and **inferior oblique**.

Anatomy Overview—Oculomotor Nerve
Anatomy Overview—Trochlear Nerve
Anatomy Overview—Abducens Nerve

### Concept 16.4 The eye is constructed of three layers.

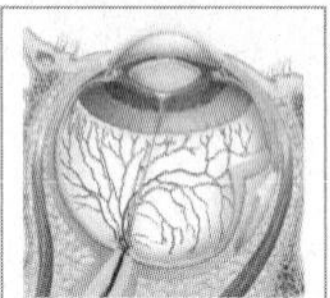

1. The wall of the eyeball has three layers: the fibrous tunic, vascular tunic, and retina.
2. The superficial **fibrous tunic** consists of the posterior, tough, protective, white **sclera**; and the anterior, curved, transparent **cornea**. The **scleral venous sinus** is at the junction of the sclera and cornea, and functions to drain aqueous humor.
3. The **vascular tunic** is between the fibrous tunic and retina. The vascular **choroid** supplies nutrients to the retina and absorbs scattered light rays. The **ciliary body** consisting of the **ciliary processes** and **ciliary muscle** that secretes aqueous humor and controls the shape of the lens, and the **iris**, which regulates the diameter of its central opening, the **pupil**.
4. The **retina** is the innermost tunic. The outer **pigmented layer** prevents scattering of light rays. The **neural layer** contains **photoreceptors, bipolar cells**, and **ganglion cells**. The **rods** are photoreceptors that are stimulated by even low light and allow night or dim vision. **Cones** are photoreceptors that are stimulated by bright light and produce color vision.
5. Visual information passes from photoreceptors to bipolar cells to ganglion cells. Axons of ganglion cells extend to the **optic disc** and exit the eyeball as the **optic nerve**. The optic disc is called the blind spot because it contains no rods or cones.
6. The central, posterior portion of the retina contains only cones. This area, the **fovea centralis**, is the area of highest **visual acuity**.
7. The **lens** is posterior to the pupil and iris, and suspended in the eyeball cavity by the zonular fibers attached to the ciliary processes. It is transparent and contains proteins called **crystallins**. The lens functions to focus images on the retina.
8. The lens divides the interior of the eyeball into the anterior cavity and the vitreous chamber. The **anterior cavity**, the space anterior to the lens, contains **anterior** and **posterior chambers** filled with **aqueous humor**. The larger **vitreous chamber**, between the lens and the retina, contains the **vitreous body**.

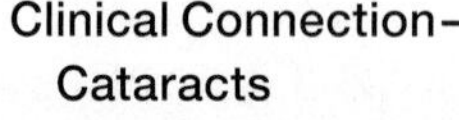

Figure 16.8—Anatomy of the Eyeball

Clinical Connection—Cataracts

Clinical Connection—Age-related Macular Disease

Clinical Connection—Glaucoma

| Concept | Resources |
|---|---|

## Concept 16.5 Image formation involves refraction of light rays, accommodation, pupil constriction, and convergence.

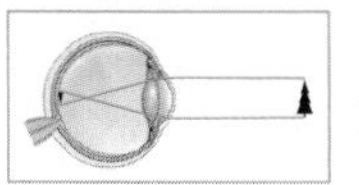

1. Light rays entering the eye undergo **refraction**, the bending of light rays at each surface of the cornea and the lens, which focuses an inverted image on the fovea centralis of the retina.
2. The anterior and posterior lens surfaces are convex, and increasing the lens curvature increases its focusing power. **Accommodation** is an increase in the curvature of the lens for near vision.
3. An **emmetropic eye** is a normal eye that can refract light rays so that a clear image is focused on the retina. **Myopia** (nearsightedness) is the inability to focus properly on distant objects. **Hyperopia** (farsightedness) results in the inability to focus on nearby objects. **Astigmatism** occurs when there is an irregular curvature to the cornea or lens that causes parts of an image to be out of focus or distorted.
4. **Constriction of the pupil** is an autonomic reflex that prevents light rays from entering at the periphery of the lens, which causes blurred vision.
5. Human eyes have **binocular vision** to allow depth perception. When objects move closer, the eyes must move medially and undergo **convergence**, so that light rays from the object can strike the same points on both eyes.

**Resources:**
Clinical Connection—Presbyopia
Clinical Connection—LASIK

## Concept 16.6 The neural pathway of light is photoreceptors → bipolar cells → ganglion cells → optic nerve → primary visual cortex.

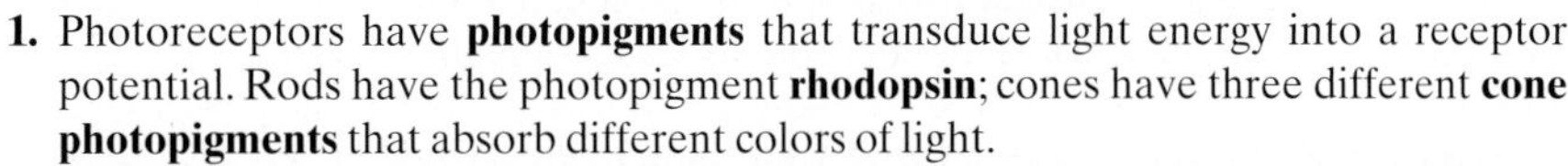

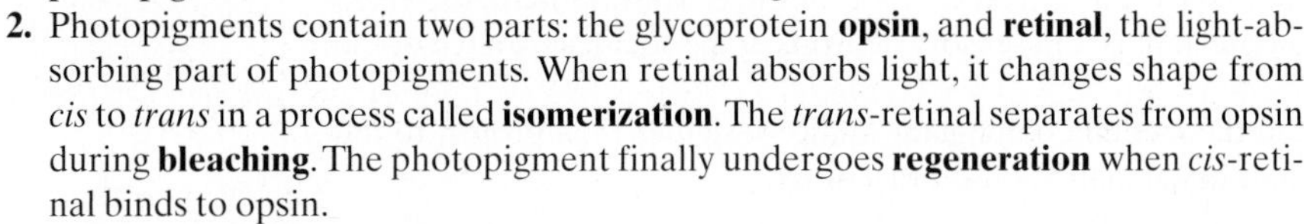

1. Photoreceptors have **photopigments** that transduce light energy into a receptor potential. Rods have the photopigment **rhodopsin**; cones have three different **cone photopigments** that absorb different colors of light.
2. Photopigments contain two parts: the glycoprotein **opsin**, and **retinal**, the light-absorbing part of photopigments. When retinal absorbs light, it changes shape from *cis* to *trans* in a process called **isomerization**. The *trans*-retinal separates from opsin during **bleaching**. The photopigment finally undergoes **regeneration** when *cis*-retinal binds to opsin.
3. **Light adaptation** occurs when photoreceptors adjust to a brighter environment by decreasing their sensitivity as increasing amounts of photopigment are bleached.
4. **Dark adaptation** occurs when photoreceptors increase sensitivity by increasing photopigment regeneration. Only rods function at low light intensity.
5. Light striking the retina activates enzymes that lead to a receptor potential in the photoreceptors, then receptor potentials in bipolar cells that synapse on photoreceptors.
6. Bipolar cells transmit receptor potentials to ganglion cells, which generate action potentials. Ganglion cell axons exit the eyeball as the optic nerve. When optic nerve axons pass through the **optic chiasm**, they either cross to the opposite side or continue straight ahead, forming the **optic tract** that enters the thalamus. **Optic radiations** allow for projection to the primary visual areas of the cerebral cortex.
7. The eye's **visual field** is everything that can be seen by one eye.

**Resources:**
Anatomy Overview—Optic Nerve
Figure 16.17—The Visual Pathway

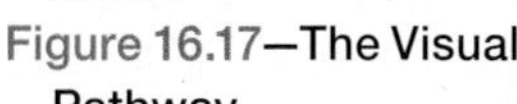

## Concept 16.7 The ear is divided into external, middle, and internal regions.

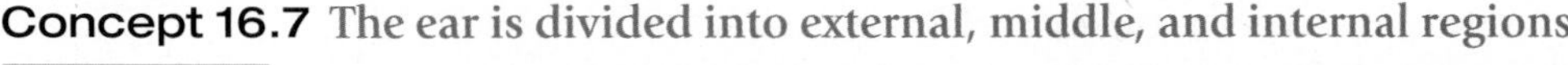

1. The external ear collects and channels sound waves inward; the middle ear conveys sound vibrations to the oval window; and the internal ear has the receptors for hearing and equilibrium.
2. The **external ear** consists of three structures: the **auricle** collects sound waves; the **external auditory canal** carries sound waves from the auricle to the tympanic membrane; and the **tympanic membrane** separates the external auditory canal from the middle ear.
3. The **middle ear** consists of the **auditory ossicles**, **oval window**, **round window**, and **auditory tube**, which equalizes atmospheric pressure.
4. The **internal ear** consists of the **bony labyrinth** and the **membranous labyrinth**. The bony labyrinth is divided into the semicircular canals and the vestibule, both of which contain receptors for equilibrium; and the cochlea, which contains receptors for hearing.

**Resources:**
Figure 16.19—The Right Middle Ear Containing the Auditory Ossicles
Figure 16.20—The Right Internal Ear
Clinical Connection—Otitis Media

| Concept | Resources |
|---|---|

5. The **vestibule** contains two sacs, the **utricle** and **saccule**. Three bony **semicircular canals** project from the vestibule and contain the **semicircular ducts** that communicate with the utricle. The vestibule and semicircular canals transmit impulses to the **vestibular branch** of the vestibulocochlear nerve.
6. The spiral-shaped **cochlea** has three interior channels: the **scala vestibuli**, **scala tympani**, and **cochlear duct**. The **spiral organ** contains **hair cells**, the receptors for hearing. The basal ends of hair cells synapse with sensory neurons from the **cochlear branch** of the vestibulocochlear nerve.

## Concept 16.8 The pathway of sound is tympanic membrane → ossicles → oval window → cochlea → vestibulocochlear nerve → primary auditory cortex.

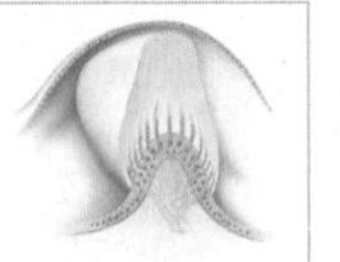

1. **Sound waves** are alternating high- and low-pressure regions traveling in a medium.
2. The number of pressure waves that arrive in a given time is **frequency** of a sound; different sound frequencies are interpreted as differences in pitch. Sound volume is its **intensity** (size or amplitude).
3. Sound waves entering the external auditory canal cause vibrations of the tympanic membrane. The ossicles vibrate in turn. Sound waves enter the external auditory canal, strike the tympanic membrane, pass through the ossicles, strike the oval window, set up waves in the perilymph, strike the vestibular membrane and scala tympani, set up waves in the endolymph, vibrate the basilar membrane, and then stimulate hair cells on the spiral organ.
4. Low-frequency sounds cause vibrations in the basilar membrane toward the apex of the cochlea, and high-frequency sounds induce vibrations of the base of the cochlea. Louder sounds lead to a higher frequency of impulses leaving the spiral organ.
5. As the microvilli of hair cells bend, they convert mechanical vibrations to receptor potentials, which releases neurotransmitter that can initiate impulses in sensory neurons in the cochlear branch of the vestibulocochlear nerve.
6. Cochlear branch neurons terminate in the medulla oblongata. Auditory signals then project to the thalamus and finally to the primary auditory area in the cerebral cortex.

Resources:

Anatomy Overview—Vestibulocochlear Nerve

Figure 16.22—Events in the Stimulation of Auditory Receptors in the Right Ear

Clinical Connection—Cochlear Implants

## Concept 16.9 Impulses for equilibrium propagate along the vestibulocochlear nerve to the brain.

1. **Static equilibrium** is the orientation of the body relative to the pull of gravity. The **maculae** of the utricle and saccule are the sense organs for static equilibrium. Body movements that stimulate the receptors for static equilibrium include tilting the head forward or backward and linear acceleration or deceleration.
2. The receptor organs for equilibrium, the **vestibular apparatus**, consist of the saccule, utricle, and semicircular ducts.
3. The maculae contain hair cells, which act as the sensory receptors. Hair cells have hair bundles consisting of microvilli. Resting on the hair cells is the **otolithic membrane** covered by a layer of **otoliths**, which amplify the pull of gravity during movements.
4. **Dynamic equilibrium** is the maintenance of body position in response to rotational movements. The **cristae** in the three semicircular ducts are the main sense organs of dynamic equilibrium.
5. Each crista contains hair cells with hair bundles that project into the **cupula**, a gelatinous mass. When the head moves, endolymph moves through the ducts, bending the hair bundles and producing receptor potentials in the hair cells. The receptor potentials result in impulses in the vestibular branch of the vestibulocochlear nerve.
6. Most vestibular branch axons terminate in the medulla oblongata and pons; remaining axons terminate in the cerebellum.
7. The cerebellum receives sensory information from all structures of the vestibular apparatus and plays a key role in maintaining static and dynamic equilibrium by making corrective adjustments to skeletal muscle movements.

Resources:

Anatomy Overview—Vestibulocochlear Nerve

Clinical Connection—Motion Sickness

Clinical Connection—Ménière's Disease

## Understanding the Concepts

1. Describe the sequence of events from the binding of an odorant molecule to an olfactory hair to the arrival of an impulse in an olfactory bulb.
2. How do the receptor cells for olfaction and gustation differ in structure and function? How do the olfactory and gustatory pathways differ?
3. Why is the low threshold of bitter substances a survival advantage?
4. What protective functions would be compromised if a person lost his or her eyelashes and eyebrows?
5. Which division of the autonomic nervous system causes constriction of pupils? Which causes dilation of pupils?
6. Through which structures does light pass as it travels through the neural layer of the retina? How does visual information pass from photoreceptors to the optic nerve?
7. Where is aqueous humor produced, what is its circulation path, and where does it drain from the eyeball?
8. Which characteristics allow the lens to transmit light? Why are we unable to see an image that strikes the blind spot?
9. Which refraction abnormality do you likely have if both near and far objects are out of focus?
10. Describe the pathway of impulses triggered by an object in the nasal half of the visual field of the left eye to the primary visual area of the cortex.
11. Why do small particles, such as dust from a dirt trail, normally not travel to the interior of your external auditory canal?
12. An infection of the throat can lead to a secondary infection of the middle ear (otitis media). How are these structures connected?
13. How is the human ear able to identify the origin of a sound?
14. Why might loud noises cause dizziness?
15. Which part of the basilar membrane vibrates most vigorously in response to banging on a bass drum (low-frequency sounds)? How would the basilar membrane respond to someone quietly whispering (low-intensity sounds)?
16. If you were blindfolded, would maculae or cristae allow you to sense that you were hanging upside down? Where are maculae and cristae found?

CHAPTER

# 17

## Lisa's Story

Mrs. Vasquez has noticed that her daughter Lisa has been tired and nauseous recently. When she picks Lisa up from basketball practice at the high school, she asks her how she is feeling.

"I don't understand it, Mom. My belly hurts all the time. No matter what I eat, it doesn't get any better. The coach gave me a power drink during practice, which helped, but I still feel sick after such a hard workout." After Lisa reminds her mom that she'd even vomited a few times, they agree to see their family physician the next day.

The following day, Dr. Carson looks over Lisa's file as she and her mother describe her symptoms. "Lisa, we gave you a complete physical when your family moved here from Panama three years ago and there wasn't anything wrong at that time. Has anything changed since then?" Lisa mentions that she has finally started having menstrual periods, a few years after most of her friends. She tells Dr. Carson that she's always thirsty and wonders if there might be a connection.

"I'm not sure. You've told me how you've had your tap water tested and found nothing, and we can't find any evidence of an infection. Let me look over your lab results, do some research, and we'll meet again to figure this out."

On their next visit, Dr. Carson suggests that Lisa might have a hormone imbalance. "The glands in your body produce hormones for a variety of functions. If one of the glands isn't making the correct amount of hormone, that could result in some of the symptoms you're experiencing."

Dr. Carson refers them to Dr. Badeer, an endocrinologist, to identify any hormonal imbalances Lisa might have and determine a treatment plan. In the meantime, he suggests that they keep a record of any additional symptoms and pass that information along to Dr. Badeer.

The endocrine glands can affect target tissues throughout the body. As we shall see in Lisa's case, hormone imbalances can have subtle yet significant effects on various systems in the body.

# The Endocrine System

## INTRODUCTION

As girls and boys enter puberty, they start to develop striking differences in physical appearance and behavior. Perhaps no other period in life so dramatically demonstrates the impact of the endocrine system in directing development and regulating body functions. The endocrine system controls body activities by releasing hormones. In girls, estrogens promote accumulation of adipose tissue in the breasts and hips, sculpting a feminine shape. At the same time, or a little later, an increasing level of testosterone in boys begins to help build muscle mass and enlarge the vocal cords, producing a lower-pitched voice. These changes are just a few examples of the powerful influence of hormones. Less dramatically, perhaps, multitudes of hormones help maintain homeostasis on a daily basis. They regulate the activity of smooth muscle, cardiac muscle, and some glands; alter metabolism; spur growth and development; influence reproductive processes; and participate in the establishment of circadian (daily) rhythms established by the hypothalamus.

## CONCEPTS

**17.1** The nervous and endocrine systems function together to regulate body activities.

**17.2** The secretion of hormones is regulated by the nervous system, chemical changes in the blood, and other hormones.

**17.3** The hypothalamus regulates anterior pituitary hormone secretion of seven important hormones.

**17.4** Oxytocin and antidiuretic hormone originate in the hypothalamus and are stored in the posterior pituitary.

**17.5** The thyroid gland secretes thyroxine, triiodothyronine, and calcitonin.

**17.6** The parathyroid glands secrete parathyroid hormone, which regulates calcium, magnesium, and phosphate ion levels.

**17.7** The adrenal glands are structurally and functionally two independent endocrine glands.

**17.8** The pancreatic islets regulate blood glucose level by secreting glucagon and insulin.

**17.9** The ovaries produce estrogens, progesterone, and inhibin; the testes produce testosterone and inhibin.

**17.10** The pineal gland, thymus, and other organs also secrete hormones.

# 17.1 The nervous and endocrine systems function together to regulate body activities.

The nervous and endocrine systems act together to coordinate functions of all body systems. Recall that the nervous system acts through nerve impulses (action potentials) conducted along axons of neurons. At synapses, impulses trigger the release of mediator molecules called *neurotransmitters* (shown in Figure 12.22). The endocrine system also controls body activities by releasing mediator molecules called *hormones*, but the means of control of the two systems are very different.

A **hormone** (*hormon* = to excite or get moving) is a molecule that is released in one part of the body but regulates the activity of cells in other parts of the body. Most hormones enter interstitial fluid and then the bloodstream. The circulating blood delivers hormones to cells throughout the body. Both neurotransmitters and hormones exert their effects by binding to receptors on or in their "target" cells. Several mediator molecules act as both neurotransmitters and hormones. One example is norepinephrine, which is released as a neurotransmitter by sympathetic postganglionic neurons and as a hormone by cells of the adrenal medullae.

Responses produced by the endocrine system often are slower than responses initiated by the nervous system. Although some hormones act within seconds, most take several minutes or more to cause a response. When the glucose level in your blood rises after ingesting a sugary treat, insulin is secreted from the pancreas, travels through the bloodstream, and stimulates uptake of glucose by body cells; after several minutes, the blood glucose level returns to normal. In contrast, reflex activity of your nervous system moves your hand away from a hot pan within milliseconds. The effects of nervous system activation are generally briefer than those of endocrine system activation. The nervous system acts on specific muscles to contract or relax and on glands to secrete more or less of their product. The influence of the endocrine system is much broader; it helps regulate virtually all types of body cells.

You will have numerous opportunities to see how the nervous and endocrine systems function together as an interlocking "supersystem." For example, when you are in a life-threatening situation, the autonomic nervous system transmits sympathetic impulses to the adrenal glands, stimulating them to release hormones that will increase the level of glucose in the blood. This ensures that the brain, skeletal muscles, and heart have sufficient fuel to assist in fighting off an attacker or fleeing from danger.

Table 17.1 compares the characteristics of the nervous and endocrine systems. In the nervous system chapters you learned that the nervous system serves as a control and integration center for your organ systems. In this chapter, we focus on the major endocrine glands and explore how hormones also govern body activities.

## Endocrine Glands

Recall from Chapter 4 that the body contains two kinds of glands: exocrine glands and endocrine glands. **Exocrine glands** (*exo-* = outside) secrete their products into ducts that carry the secretions into body cavities, into the lumen of an organ, or to the outer surface of the body. Exocrine glands include sudoriferous (sweat), sebaceous (oil), mucous, and digestive glands.

**TABLE 17.1**

Comparison of Control by the Nervous and Endocrine Systems

| CHARACTERISTIC | NERVOUS SYSTEM | ENDOCRINE SYSTEM |
|---|---|---|
| **Mediator molecules** | Neurotransmitters released locally in response to impulses | Hormones delivered to tissues throughout body by blood |
| **Site of mediator action** | Close to site of release, at synapse; binds to receptors in postsynaptic membrane | Far from site of release (usually); binds to receptors on or in target cells |
| **Types of target cells** | Muscle (smooth, cardiac, and skeletal) cells, gland cells, other neurons | Cells throughout body |
| **Time to onset of action** | Typically within milliseconds (thousandths of a second) | Seconds to hours or days |
| **Duration of action** | Generally briefer (milliseconds) | Generally longer (seconds to days) |

**Endocrine glands** (*endo-* = within) secrete their products (hormones) into the interstitial fluid surrounding the secretory cells rather than into ducts. From the interstitial fluid, hormones diffuse into capillaries and blood carries them to target cells throughout the body. Most hormones are required in very small amounts, so circulating levels typically are low. As they approach the target cells, hormones diffuse from blood into the interstitial fluid, where they interact with their target cells.

The endocrine glands include the pituitary, thyroid, parathyroid, adrenal, and pineal glands (Figure 17.1). In addition, several organs and tissues do not function exclusively as endocrine glands but contain cells that secrete hormones. These include the hypothalamus, thymus, pancreas, ovaries, testes, kidneys, stomach, liver, small intestine, skin, heart, adipose tissue, and placenta. Taken together, all endocrine glands and hormone-secreting cells constitute the **endocrine system**.

FIGURE 17.1 **Location of many endocrine glands.** Also shown are other organs that contain endocrine cells and associated structures.

Thyroid gland
Trachea
Parathyroid glands
**Parathyroid glands** (behind thyroid glands)
Pineal gland
Hypothalamus
Pituitary gland
Thyroid gland
Trachea
Skin
Thymus
Lung
Heart
Liver
Stomach
Kidney
Adrenal glands
Pancreas
Small intestine
Uterus
Ovary
Female
Scrotum
Testes
Male

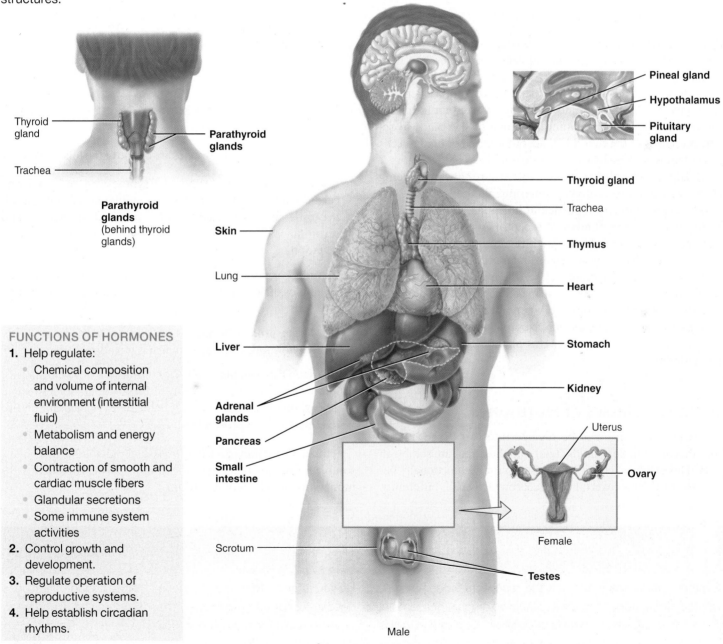

**FUNCTIONS OF HORMONES**

1. Help regulate:
   - Chemical composition and volume of internal environment (interstitial fluid)
   - Metabolism and energy balance
   - Contraction of smooth and cardiac muscle fibers
   - Glandular secretions
   - Some immune system activities
2. Control growth and development.
3. Regulate operation of reproductive systems.
4. Help establish circadian rhythms.

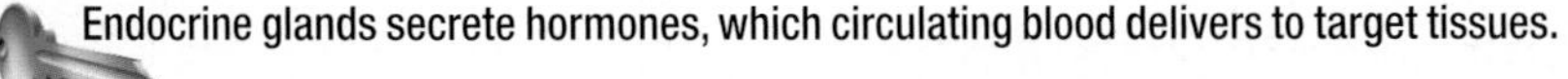

Endocrine glands secrete hormones, which circulating blood delivers to target tissues.

✓ CHECKPOINT

1. Why is the release of acetylcholine at the neuromuscular junction not considered a "hormonal event"?
2. Which system produces a quicker response, the nervous system or the endocrine system? Which system produces a longer-acting response?
3. How does a sudoriferous (sweat) gland differ from an adrenal gland?

# 17.2 The secretion of hormones is regulated by the nervous system, chemical changes in the blood, and other hormones.

## The Role of Hormone Receptors

Although a given hormone travels throughout the body in the blood, it affects only specific **target cells**. Hormones, like neurotransmitters, influence their target cells by chemically binding to specific protein **receptors**. Only the target cells for a given hormone have receptors that bind to and recognize that hormone. For example, thyroid-stimulating hormone (TSH) binds to receptors on cells of the thyroid gland, but it does not bind to cells of the ovaries because ovarian cells do not have TSH receptors.

Receptors, like other cellular proteins, are constantly being synthesized and broken down. Generally, a target cell has 2000 to 100,000 receptors for a particular hormone. If a hormone is present in excess, the number of receptors on target cells may decrease—an effect called **down-regulation**. For example, when certain cells of the testes are exposed to a high concentration of luteinizing hormone (LH), the number of LH receptors decreases. Down-regulation makes a target cell *less sensitive* to a hormone. In contrast, when a hormone is deficient, the number of LH receptors increases to make the target tissue *more sensitive* to the hormone. This phenomenon is known as **up-regulation.**

## Chemical Classes of Hormones

Chemically, hormones can be divided into two broad classes: those that are soluble in lipids (fats), and those that are soluble in water. This chemical classification is also useful functionally because the two classes exert their effects on target cells differently.

The lipid-soluble hormones include the following:

- **Steroid hormones** are derived from cholesterol.
- **Thyroid hormones** are synthesized by attaching iodine to the amino acid tyrosine.
- **Nitric oxide** is a gas that serves as both a hormone and a neurotransmitter.

Most water-soluble hormones are of one of the following types:

- **Amine hormones** (a-MĒN) are synthesized by modifying certain amino acids. For instance, epinephrine and norepinephrine are synthesized from the amino acid tyrosine, and serotonin and melatonin are derived from the amino acid tryptophan.
- **Peptide hormones** consist of chains of 3 to 49 amino acids. Examples of peptide hormones are antidiuretic hormone and oxytocin.
- The larger **protein hormones** include 50 to 200 amino acids. Protein hormones include human growth hormone and insulin.
- **Eicosanoid hormones** (ī-KŌ-sa-noid; *eicos-* = twenty forms; *-oid* = resembling) are derived from arachidonic acid, a 20-carbon fatty acid. The two major types of eicosanoids are leukotrienes and prostaglandins.

Table 17.2 summarizes the classes of lipid-soluble and water-soluble hormones. It also provides an overview of the major hormones and their sites of secretion.

### CLINICAL CONNECTION | *Blocking Hormone Receptors*

Synthetic hormones that **block the receptors** for some naturally occurring hormones are available as drugs. For example, RU486 (mifepristone), which is used to induce abortion, binds to the receptors for progesterone (a female sex hormone) and prevents progesterone from exerting its normal effect, in this case preparing the lining of the uterus for implantation. When RU486 is given to a pregnant woman, the uterine conditions needed for nurturing an embryo are not maintained, embryonic development stops, and the embryo is sloughed off along with the uterine lining. This example illustrates an important endocrine principle: If a hormone is prevented from interacting with its receptors, the hormone cannot perform its normal functions.

**TABLE 17.2**

Hormones by Chemical Class

| CHEMICAL CLASS | HORMONES | SITE OF SECRETION |
|---|---|---|
| **LIPID-SOLUBLE** | | |
| **Steroid hormones** | Aldosterone, cortisol, androgens | Adrenal cortex |
| 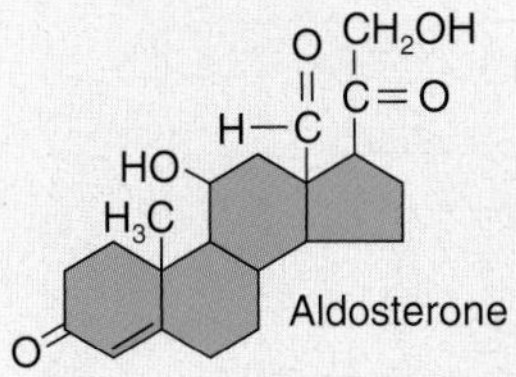 | Calcitriol | Kidneys |
| | Testosterone | Testes |
| (Aldosterone) | Estrogens, progesterone | Ovaries |
| **Thyroid hormones** 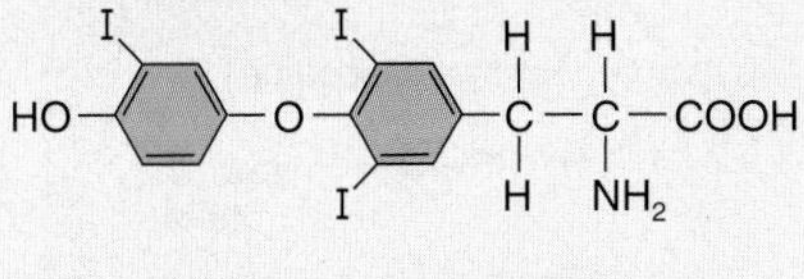 (Triiodothyronine ($T_3$)) | $T_3$ (triiodothyronine), $T_4$ (thyroxine) | Thyroid gland (follicular cells) |
| **Gas** | Nitric oxide (NO) | Endothelial cells lining blood vessels |
| **WATER-SOLUBLE** | | |
| **Amines** | Epinephrine, norepinephrine (catecholamines) | Adrenal medulla |
| 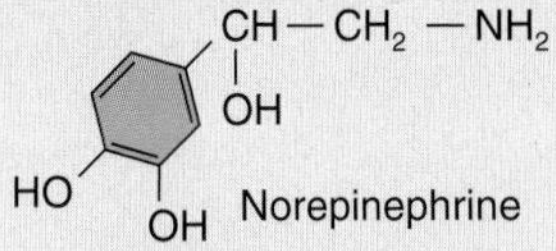 | Melatonin | Pineal gland |
| | Histamine | Mast cells in connective tissues |
| (Norepinephrine) | Serotonin | Platelets in blood |
| **Peptides and proteins** | All hypothalamic releasing and inhibiting hormones | Hypothalamus |
| 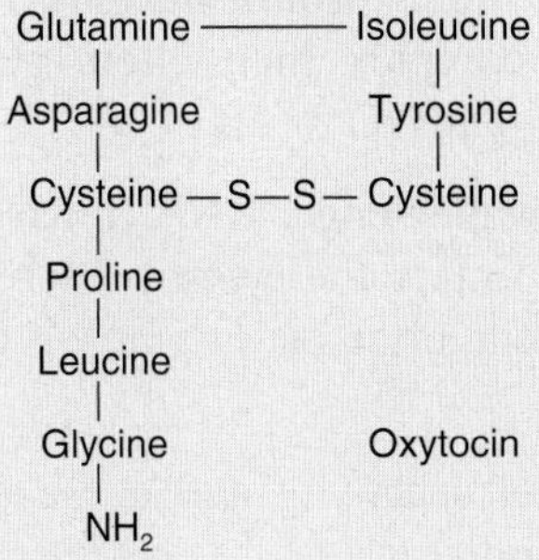 | Oxytocin, antidiuretic hormone | Posterior pituitary |
| | Human growth hormone, thyroid-stimulating hormone, adrenocorticotropic hormone, follicle-stimulating hormone, luteinizing hormone, prolactin, melanocyte-stimulating hormone | Anterior pituitary |
| | Insulin, glucagon, somatostatin, pancreatic polypeptide | Pancreas |
| | Parathyroid hormone | Parathyroid glands |
| (Oxytocin: Glutamine–Isoleucine–Tyrosine–Cysteine–S–S–Cysteine–Asparagine–Glutamine; Cysteine–Proline–Leucine–Glycine–$NH_2$) | Calcitonin | Thyroid gland (parafollicular cells) |
| | Gastrin, secretin, cholecystokinin, GIP (glucose-dependent insulinotropic peptide) | Stomach and small intestine (enteroendocrine cells) |
| | Erythropoietin | Kidneys |
| | Leptin | Adipose tissue |
| **Eicosanoids** 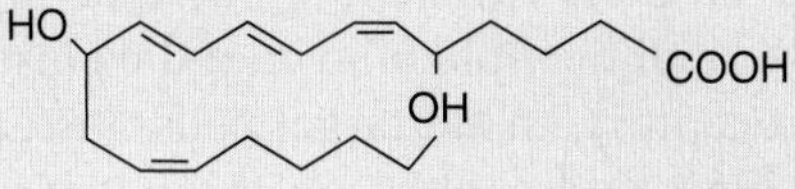 (A leukotriene ($LTB_4$)) | Prostaglandins, leukotrienes | All cells except red blood cells |

## Hormone Transport in the Blood

Most water-soluble hormone molecules circulate in the watery blood plasma in a "free" form (not attached to other molecules), but most lipid-soluble hormone molecules are bound to **transport proteins**. The transport proteins, which are synthesized by cells in the liver, have three functions:

- They make lipid-soluble hormones temporarily water-soluble, thus increasing their solubility in blood.
- They retard passage of small hormone molecules through the filtering mechanism in the kidneys, thus slowing the rate of hormone loss in the urine.
- They provide a ready reserve of hormone, already present in the bloodstream.

In general, less than 10 percent of lipid-soluble hormone molecules are not bound to a transport protein. This **free fraction** diffuses out of capillaries, binds to receptors, and triggers hormonal responses. As free hormone molecules leave the blood and bind to their receptors, transport proteins release new ones to replenish the free fraction.

FIGURE 17.2 **Mechanism of action of lipid-soluble hormones.**

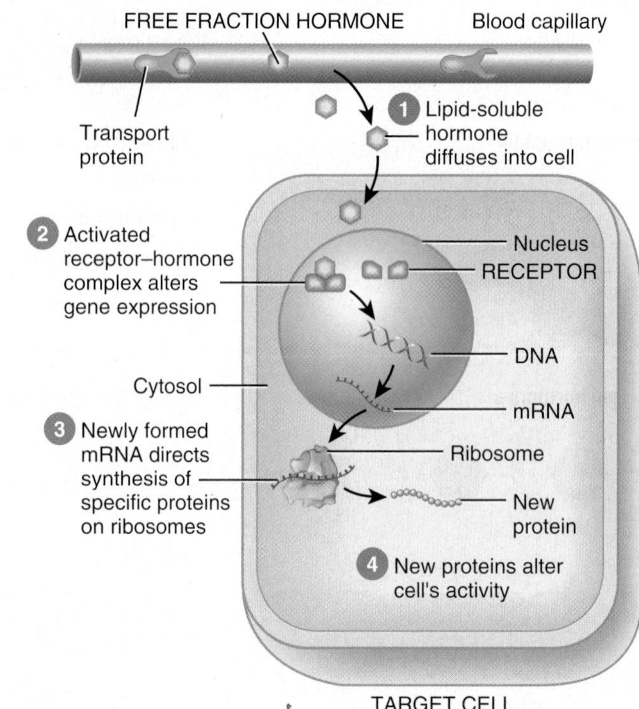


Lipid-soluble hormones bind to receptors inside target cells.

## Mechanism of Hormone Action

The response to a hormone depends on both the hormone and the target cell. Various target cells respond differently to the same hormone. Insulin, for example, stimulates synthesis of glycogen in liver cells but synthesis of triglycerides in adipose cells.

The response to a hormone is not always the synthesis of new molecules, as is the case for insulin. Hormones may also change the permeability of the plasma membrane, stimulate transport of a substance into or out of the target cells, alter the rate of a metabolic reaction, or cause contraction of smooth muscle or cardiac muscle. Before a hormone can set in motion a cellular response, the hormone must first "announce its arrival" to a target cell by binding to its receptors. The receptors for lipid-soluble hormones are located inside target cells; the receptors for water-soluble hormones are part of the plasma membrane of target cells.

### Action of Lipid-soluble Hormones

Lipid-soluble hormones bind to their receptors *within* target cells. Their mechanism of action is as follows (Figure 17.2):

1. A lipid-soluble hormone molecule detaches from its transport protein in the bloodstream, then diffuses from the blood through interstitial fluid, and through the lipid bilayer of the plasma membrane into a cell.
2. If the cell is a target cell, the hormone binds to and activates receptors located within the cytosol or nucleus. The activated receptor–hormone complex then alters gene expression by turning specific genes on or off.
3. As the DNA is transcribed, new messenger RNA (mRNA) forms, leaves the nucleus, and enters the cytosol. There it directs synthesis of a new protein, often an enzyme, on the ribosomes.
4. The new proteins alter the cell's activity and cause the responses typical of that hormone.

### Action of Water-soluble Hormones

Because most amino acid–based hormones are not lipid-soluble, they cannot diffuse through the lipid bilayer of the plasma membrane and bind to receptors inside target cells. Instead, water-soluble hormones bind to receptors on the *outside* of the target cell. When a water-soluble hormone binds to its receptor at the surface of the plasma membrane, it acts as the **first messenger**. The hormone then stimulates production of a **second messenger** inside the cell, where specific hormone-stimulated responses take place. One common second messenger is **cyclic AMP (cAMP)**.

The action of a typical water-soluble hormone occurs as follows (Figure 17.3):

1. A water-soluble hormone (the first messenger) diffuses from the blood through interstitial fluid and then binds to its receptor on the exterior surface of a target cell's plasma membrane.
2. As a result of the binding, the enzyme **adenylate cyclase** (a-DEN-i-lāt SĪ-klās) converts ATP into cAMP. Because the enzyme's active site is on the inner surface of the plasma membrane, this reaction occurs in the cytosol of the cell.
3. Cyclic AMP (the second messenger) activates several enzymes.
4. Activated enzymes catalyze reactions that produce physiological responses.
5. After a brief period, cAMP is inactivated by an enzyme called **phosphodiesterase** (fos′-fō-dī-ES-ter′-ās). Thus, the cell's response is turned off unless new hormone molecules continue to bind to their receptors on the plasma membrane.

Hormones that bind to plasma membrane receptors can induce their effects at very low concentrations because they initiate a chain reaction, each step of which amplifies the initial effect. For example, the binding of a single molecule of epinephrine (first messenger) to its receptor on a liver cell may activate a hundred or so adenylate cyclase molecules. If each molecule of adenylate cyclase produces as few as 1000 cyclic AMP, then 100,000 of these second messengers will be liberated inside the cell. Each molecule of cyclic AMP may activate an enzyme molecule, which in turn can act on hundreds or thousands of substrate molecules. In the breakdown of glycogen, binding of a single molecule of epinephrine to its receptor results in the breakdown of millions of glycogen molecules into glucose monomers.

FIGURE 17.3 **Mechanism of action of water-soluble hormones.**

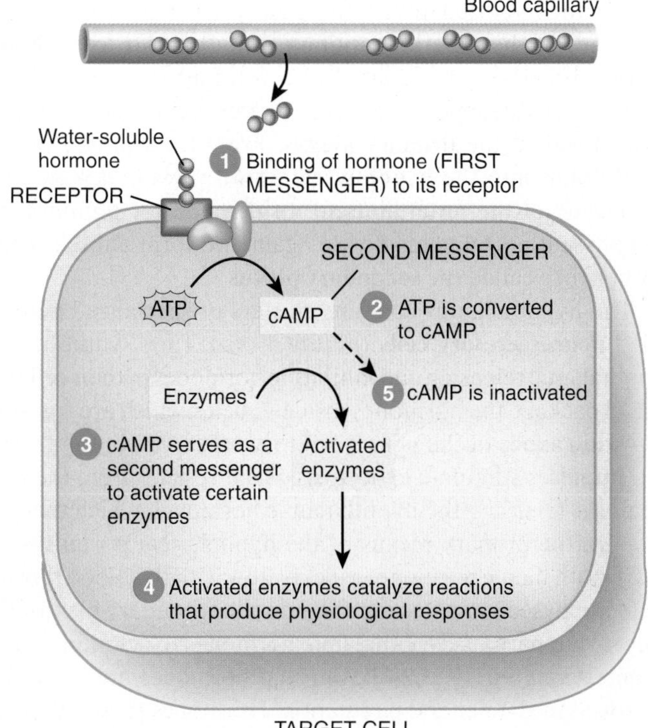


**Water-soluble hormones bind to receptors embedded in the plasma membrane of target cells.**

### Hormone Interactions

The responsiveness of a target cell to a hormone depends on (1) the hormone's concentration in the blood, (2) the abundance of the target cell's hormone receptors, and (3) influences exerted by other hormones. A target cell responds more vigorously when the level of a hormone rises or when the target cell has more receptors (up-regulation). In addition, the actions of some hormones on target cells require a simultaneous or recent exposure to a second hormone to exert their full effects. In such cases, the second hormone is said to have a **permissive effect**. For example, epinephrine alone only weakly stimulates *lipolysis* (the breakdown of triglycerides), but when small amounts of thyroid hormones are present, the same amount of epinephrine stimulates lipolysis more powerfully. In some cases the permissive hormone increases the number of receptors for the other hormone, and in others it promotes the synthesis of an enzyme required for the expression of the other hormone's effects.

When the effect of two hormones acting together is greater than the effect of each hormone acting alone, the two hormones are said to have a **synergistic effect**. For example, normal development of oocytes in the ovaries requires both follicle-stimulating hormone from the anterior pituitary and estrogens from the ovaries. Neither hormone alone is sufficient for normal oocyte development.

When one hormone opposes the actions of another hormone, the two hormones are said to have **antagonistic effects**. An example of an antagonistic pair of hormones is insulin (which stimulates the liver to synthesize glycogen) and glucagon (which stimulates the liver to break down glycogen).

## Control of Hormone Secretion

The release of most hormones occurs in short bursts, with little or no secretion between bursts. When stimulated, an endocrine gland will release its hormone in more frequent bursts, thus increasing the concentration of the hormone in the blood. In the absence of stimulation, hormone release is inhibited, and the blood level of the hormone decreases. Regulation of hormone secretion normally prevents overproduction or underproduction of any given hormone.

Hormone secretion is regulated in three ways:

- ***Signals from the nervous system.*** Impulses to the adrenal medullae alter the release of epinephrine.

- ***Chemical changes in the blood.*** Blood $Ca^{2+}$ level regulates the secretion of parathyroid hormone.
- ***Other hormones.*** Adrenocorticotropic hormone stimulates the release of cortisol by the adrenal cortex.

Most systems that regulate secretion of hormones work by negative feedback (see Figure 1.3), but a few operate by positive feedback (see Figure 1.4). For example, during childbirth, the hormone oxytocin stimulates contractions of the uterus, and uterine contractions, in turn, stimulate more oxytocin release, a positive feedback loop that is only broken by the birth of the baby.

Now that you have a general understanding of the roles of hormones in the endocrine system, we next discuss the various endocrine glands and the hormones they secrete.

✓ **CHECKPOINT**

4. Why would follicle-stimulating hormone act on the ovaries but not on the thyroid?
5. Identify the chemical classes of hormones.
6. Where do lipid-soluble hormones bind to their target cells? Where do water-soluble hormones bind to their target cells?
7. Why is cyclic AMP considered a "second messenger"?
8. What factors determine the responsiveness of a target cell to a hormone?
9. What are the three influences on hormone secretion?

# 17.3 The hypothalamus regulates anterior pituitary hormone secretion of seven important hormones.

## The Hypothalamus

For many years, the **pituitary gland** (pi-TOO-i-tār-ē), or **hypophysis** (hī-POF-i-sis), was called the "master" endocrine gland because it secretes several hormones that control other endocrine glands. We now know that the pituitary gland itself has a master—the **hypothalamus** (see Figure 17.1). This small region of the brain below the thalamus is the major integrating link between the nervous and endocrine systems. Cells in the hypothalamus synthesize at least nine different hormones, and the pituitary gland secretes seven. Together, the hypothalamic and pituitary hormones play important roles in the regulation of virtually all aspects of growth, development, metabolism, and homeostasis.

## The Pituitary Gland

The pituitary gland is a pea-shaped structure measuring 1–1.5 cm (0.5 in.) in diameter that lies in the sella turcica of the sphenoid bone. It attaches to the hypothalamus by a stalk, the **infundibulum** (in′-fun-DIB-ū-lum; = = a funnel; Figure 17.4). The pituitary gland has two anatomically and functionally separate portions. The glandular anterior lobe, called the anterior pituitary, accounts for about 75 percent of the total weight of the pituitary gland and produces hormones that regulate a wide range of bodily activities, from growth to reproduction. The posterior pituitary is composed of nervous tissue that extends down from the hypothalamus as a posterior lobe. As you will see shortly, the posterior pituitary does not produce any hormones, but it does release hormones that are synthesized by the hypothalamus.

### Anterior Pituitary

The release of hormones from the **anterior pituitary**, or **adenohypophysis** (ad′-e-nō-hī-POF-i-sis; *adeno-* = gland; *hypophysis* = undergrowth) is influenced by hormones from the hypothalamus. These hypothalamic hormones are an important link between the nervous and endocrine systems.

Hypothalamic hormones that release or inhibit anterior pituitary hormones reach the anterior pituitary through a portal system. Usually, blood passes from a capillary to a vein, and then back to the heart. In a *portal system*, blood flows from one capillary network into a portal vein, and then into a second capillary network before returning to the heart. In the **hypophyseal portal system** (hī′-pō-FIZ-ē-al; Figure 17.4a), blood flows from capillaries in the hypothalamus into portal veins that carry blood to capillaries of the anterior pituitary. The **superior hypophyseal arteries**, branches of the internal carotid arteries, bring blood into the hypothalamus, where the arteries divide into a capillary network called the **primary plexus**. From the primary plexus, blood drains into the **hypophyseal portal veins** that pass down the outside of the infundibulum. In the anterior pituitary, the hypophyseal portal veins divide again and form another capillary network called the **secondary plexus**.

The hypothalamus contains clusters of specialized neurons called **neurosecretory cells** (Figure 17.4b). They synthesize the hypothalamic releasing and inhibiting hormones in their cell bodies and package the hormones inside vesicles, which are transmitted within axons of the neurosecretory cells to their axon terminals. Impulses stimulate exocytosis of the vesicles from the axon terminals, releasing the hypothalamic hormones, which then diffuse into the primary plexus of the hypophyseal portal system. The hypothalamic hormones quickly flow with the blood through the hypophyseal portal veins and into the secondary plexus. This direct route permits hypothalamic hormones to act immediately on anterior pituitary cells, before the hormones are diluted or destroyed in the general circulation. Hormones secreted by anterior pituitary cells pass into the secondary plexus capillaries, which drain into the **anterior hypophyseal veins** and out into the general circulation. Anterior pituitary hormones then travel to target tissues throughout the body.

FIGURE 17.4 **Hypothalamus and pituitary gland, and their blood supply.** Releasing and inhibiting hormones synthesized by hypothalamic neurons diffuse from the axon terminals into capillaries of the primary plexus of the hypophyseal portal system, where they are carried by the hypophyseal portal veins to the secondary plexus of the hypophyseal portal system for distribution to target cells in the anterior pituitary.

INFUNDIBULUM
Posterior pituitary
ANTERIOR PITUITARY

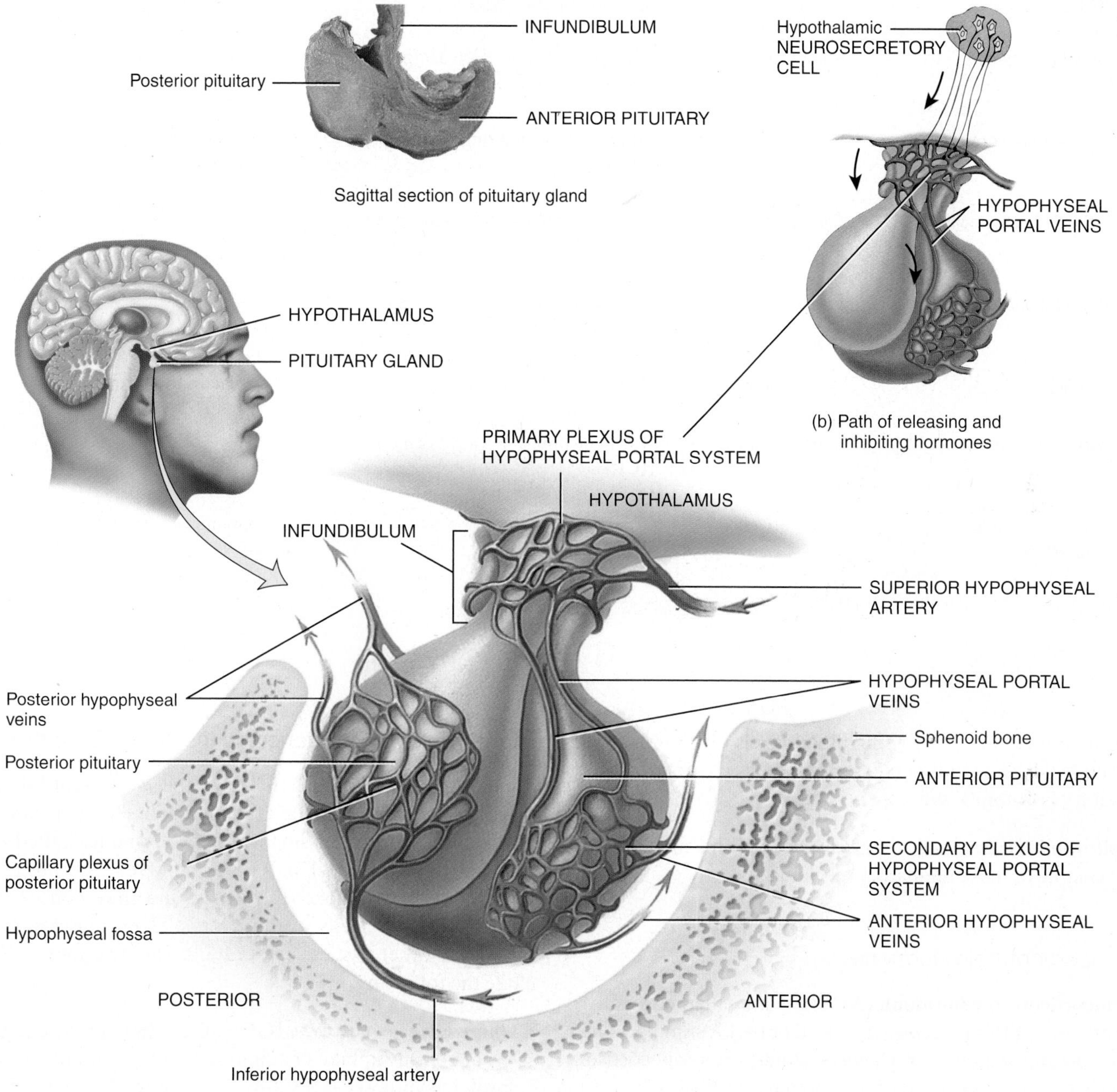

Sagittal section of pituitary gland

Hypothalamic
NEUROSECRETORY
CELL
HYPOPHYSEAL
PORTAL VEINS

(b) Path of releasing and inhibiting hormones

HYPOTHALAMUS
PITUITARY GLAND
PRIMARY PLEXUS OF
HYPOPHYSEAL PORTAL SYSTEM
HYPOTHALAMUS
INFUNDIBULUM
SUPERIOR HYPOPHYSEAL
ARTERY
Posterior hypophyseal
veins
HYPOPHYSEAL PORTAL
VEINS
Sphenoid bone
Posterior pituitary
ANTERIOR PITUITARY
Capillary plexus of
posterior pituitary
SECONDARY PLEXUS OF
HYPOPHYSEAL PORTAL
SYSTEM
ANTERIOR HYPOPHYSEAL
VEINS
Hypophyseal fossa
POSTERIOR
ANTERIOR
Inferior hypophyseal artery

(a) Relationship of hypothalamus to pituitary gland

**Hypothalamic hormones are an important link between the nervous and endocrine systems.**

Those anterior pituitary hormones that influence another endocrine gland are called **tropic hormones** (TRŌ-pik) or **tropins**. Several of the anterior pituitary hormones are tropins. The two gonadotropins, follicle-stimulating hormone and luteinizing hormone, regulate the functions of the gonads (ovaries and testes). Thyrotropin stimulates the thyroid gland, and corticotropin acts on the cortex of the adrenal gland.

## Control of Secretion by the Anterior Pituitary

Secretion of anterior pituitary hormones is regulated in two ways:

- Neurosecretory cells in the hypothalamus secrete **releasing hormones**, which stimulate secretion of anterior pituitary hormones, and **inhibiting hormones**, which suppress secretion of anterior pituitary hormones (Table 17.3).

**TABLE 17.3**

Hormones of the Anterior Pituitary

| HORMONE | SECRETED BY | HYPOTHALAMIC-RELEASING HORMONE (STIMULATES SECRETION) | HYPOTHALAMIC-INHIBITING HORMONE (SUPPRESSES SECRETION) |
|---|---|---|---|
| **Human growth hormone (hGH)** | Somatotrophs | Growth hormone–releasing hormone (GHRH) | Growth hormone–inhibiting hormone (GHIH) |
| **Thyroid-stimulating hormone (TSH)** | Thyrotrophs | Thyrotropin-releasing hormone (TRH) | Growth hormone–inhibiting hormone (GHIH) |
| **Follicle-stimulating hormone (FSH)** | Gonadotrophs | Gonadotropin-releasing hormone (GnRH) | — |
| **Luteinizing hormone (LH)** | Gonadotrophs | Gonadotropin-releasing hormone (GnRH) | — |
| **Prolactin (PRL)** | Lactotrophs | Prolactin-releasing hormone (PRH)* | Prolactin-inhibiting hormone (PIH) |
| **Adrenocorticotropic hormone (ACTH)** | Corticotrophs | Corticotropin-releasing hormone (CRH) | — |
| **Melanocyte-stimulating hormone (MSH)** | Corticotrophs | Corticotropin-releasing hormone (CRH) | Dopamine |

*Thought to exist, but exact nature is uncertain.

- Negative feedback from rising blood levels of hormones released by target glands decreases secretions of some anterior pituitary hormones (see Figure 17.5).

The following discussion of adrenocorticotropic hormone serves as an example of these regulatory mechanisms.

### Adrenocorticotropic Hormone

**Adrenocorticotropic hormone (ACTH)**, also called corticotropin (kor′-ti-kō-TRŌ-pin; *cortico-* = rind or bark), stimulates the production and secretion of glucocorticoids, primarily cortisol, by the cortex (outer portion) of the adrenal glands (Figure 17.5). **Corticotropin-releasing hormone (CRH)** released from the hypothalamus stimulates the anterior pituitary to secrete ACTH. As the blood level of cortisol rises, secretion of both ACTH and corticotropin-releasing hormone drops due to negative feedback suppression of the anterior pituitary and hypothalamus. Stress-related stimuli, such as low blood glucose or physical trauma, and interleukin-1, a substance produced by macrophages, also stimulate release of ACTH. Glucocorticoids inhibit both corticotropin-releasing hormone and ACTH release via negative feedback (see Figure 17.16). There is no corticotropin-inhibiting hormone.

### Human Growth Hormone and Insulinlike Growth Factors

**Human growth hormone (hGH)** is the most plentiful anterior pituitary hormone. The main function of hGH is to promote synthesis and secretion of small protein hormones called **insulin-like growth factors (IGFs)**. IGFs are so named because some of their actions are similar to those of insulin. In response to human growth hormone, cells in the liver, skeletal muscles, cartilage, bones, and other tissues secrete IGFs. The functions of IGFs include the following:

- ***Increasing growth rate.*** IGFs cause cells to grow and multiply by increasing uptake of amino acids into cells and accelerating protein synthesis. IGFs also decrease the breakdown of proteins and the use of amino acids for ATP production. Due to these effects of the IGFs, hGH increases the growth rate of the skeleton and skeletal muscles during childhood and the teenage years. In adults, hGH and IGFs help maintain muscle and bone mass and promote healing of injuries and tissue repair.
- ***Increasing lipolysis.*** IGFs enhance lipolysis in adipose tissue, which results in increased use of the released fatty acids for ATP production by body cells.
- ***Elevating blood glucose.*** IGFs and hGH influence carbohydrate metabolism by decreasing glucose uptake by most body

FIGURE 17.5 **Negative feedback regulation of hypothalamic and anterior pituitary secretion.** Solid green arrows show stimulation of secretions; dashed red lines show inhibition of secretion by negative feedback.

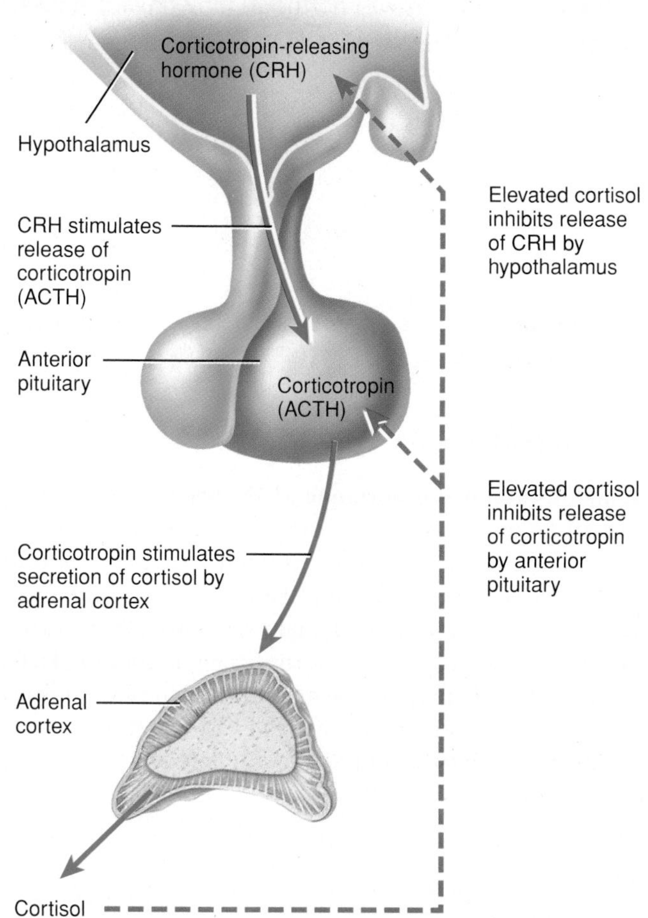


**Cortisol secreted by the adrenal cortex suppresses secretion of CRH and ACTH.**

FIGURE 17.6 **Effects of human growth hormone (hGH) and insulinlike growth factors (IGFs).** Dashed lines indicate inhibition.

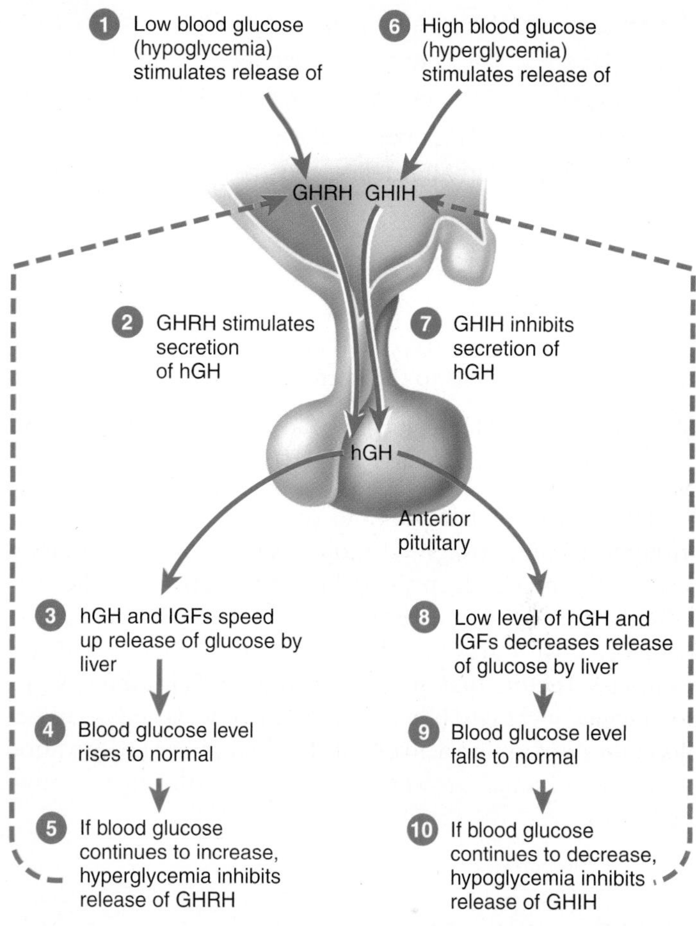


**Secretion of hGH is stimulated by growth hormone–releasing hormone (GHRH) and inhibited by growth hormone–inhibiting hormone (GHIH).**

cells, which decreases their use of glucose for ATP production. This action spares glucose so that it is available to neurons for ATP production in times of glucose scarcity. IGFs and hGH may also stimulate the liver to release glucose into the blood, thereby raising blood glucose levels.

The anterior pituitary releases hGH in bursts that occur every few hours, especially during sleep. Two hypothalamic hormones control secretion of hGH: **growth hormone–releasing hormone (GHRH)** promotes secretion of human growth hormone, and **growth hormone–inhibiting hormone (GHIH)** suppresses it. Blood glucose level is a major regulator of GHRH and GHIH secretion (Figure 17.6):

1. **Hypoglycemia** (hī′-po-glī-SĒ-mē-a), an abnormally low blood glucose level, stimulates the hypothalamus to secrete GHRH, which flows through the hypophyseal portal veins to the anterior pituitary.
2. GHRH stimulates the anterior pituitary to release human growth hormone.
3. Human growth hormone stimulates secretion of IGFs, which speed up breakdown of liver glycogen into glucose, causing glucose to enter the blood more rapidly.
4. As a result, blood glucose rises to the normal level.
5. An increase in blood glucose above the normal level inhibits release of GHRH.
6. **Hyperglycemia** (hī′-per-glī-SĒ-mē-a), an abnormally high blood glucose concentration, stimulates the hypothalamus to secrete GHIH.
7. On reaching the anterior pituitary in portal blood, GHIH inhibits secretion of human growth hormone.
8. A low level of human growth hormone and IGFs slows breakdown of glycogen in the liver, and glucose is released into the blood more slowly.

## CLINICAL CONNECTION | *Diabetogenic Effect of hGH*

One symptom of excess human growth hormone (hGH) is hyperglycemia. Persistent hyperglycemia in turn stimulates the pancreas to secrete insulin continually. Such excessive stimulation, if it lasts for weeks or months, may cause "beta-cell burnout," a greatly decreased capacity of pancreatic beta cells to synthesize and secrete insulin. Thus, excess secretion of human growth hormone may have a **diabetogenic effect** (dī′-a-bet′-ō-JEN-ik); that is, it causes diabetes mellitus (lack of insulin activity).

9 Blood glucose falls to the normal level.

10 A decrease in blood glucose below the normal level (hypoglycemia) inhibits release of GHIH.

Other stimuli that promote secretion of human growth hormone include decreased fatty acids and increased amino acids in the blood; deep sleep; increased activity of the sympathetic division of the autonomic nervous system, such as might occur with stress or vigorous physical exercise; and other hormones. Factors that inhibit human growth hormone secretion include increased blood levels of fatty acids and decreased blood levels of amino acids, light sleep, emotional deprivation, obesity, and human growth hormone itself (through negative feedback).

### Thyroid-stimulating Hormone

**Thyroid-stimulating hormone (TSH)**, or **thyrotropin**, stimulates the thyroid gland to synthesize and secrete the two thyroid hormones, triiodothyronine ($T_3$) and thyroxine ($T_4$) (see Concept 17.5). **Thyrotropin-releasing hormone (TRH)** from the hypothalamus controls TSH secretion. Release of TRH, in turn, depends on blood levels of $T_3$ and $T_4$: high levels of $T_3$ and $T_4$ inhibit secretion of TRH via negative feedback. There is no thyrotropin-inhibiting hormone. The release of TRH is explained later in this chapter (see Figure 17.11).

### Follicle-stimulating Hormone

In females, the ovaries are the targets for **follicle-stimulating hormone (FSH)**. Each month FSH initiates the development of several ovarian follicles, saclike arrangements of secretory cells that surround a developing oocyte (future ovum). FSH also stimulates the ovarian follicles to secrete estrogens (female sex hormones). In males, FSH stimulates sperm production in the testes. **Gonadotropin-releasing hormone (GnRH)** from the hypothalamus stimulates FSH release. Release of GnRH and FSH is suppressed by estrogens (female sex hormones) in females and by testosterone (the principal male sex hormone) in males through negative feedback systems. There is no gonadotropin-inhibiting hormone.

### Luteinizing Hormone

In females, **luteinizing hormone (LH)** triggers **ovulation**, the release of an oocyte from an ovary. After ovulation, LH stimulates secretion of progesterone (another female sex hormone) by the ovary. Together, FSH and LH also stimulate the ovary to secrete estrogens (additional female sex hormones). Estrogens and progesterone prepare the uterus for implantation of a fertilized ovum and help prepare the mammary glands for milk secretion. In males, LH stimulates the testes to secrete testosterone. Secretion of LH, like that of FSH, is stimulated by gonadotropin-releasing hormone (GnRH).

### Prolactin

**Prolactin** helps initiate and maintain milk production by the mammary glands. Once the mammary glands have been primed by other hormones (estrogens, progesterone, glucocorticoids, human growth hormone, thyroxine, and insulin) exerting permissive effects, prolactin brings about milk production. Ejection of milk from the mammary glands depends on the hormone oxytocin, which is released from the posterior pituitary. These two hormones act synergistically. Together, milk secretion and ejection constitute *lactation*.

The hypothalamus secretes both inhibitory and excitatory hormones that regulate prolactin secretion. In females, **prolactin-inhibiting hormone (PIH)** inhibits the release of prolactin from the anterior pituitary most of the time. Each month, just before menstruation begins, the secretion of PIH diminishes and the blood level of prolactin rises, but not enough to stimulate milk production. Breast tenderness just before menstruation may be caused by elevated prolactin. As the menstrual cycle begins anew, PIH is again secreted and the prolactin level drops. During pregnancy, the prolactin level rises, stimulated by **prolactin-releasing hormone** from the hypothalamus. The sucking action of a nursing infant causes a reduction in hypothalamic secretion of PIH, and therefore an increase in prolactin secretion.

The function of prolactin in males is not known, but its hypersecretion causes erectile dysfunction or impotence (the inability to have an erection of the penis).

## Melanocyte-stimulating Hormone

**Melanocyte-stimulating hormone (MSH)** increases skin pigmentation in amphibians by stimulating the dispersion of melanin granules in melanocytes. Its exact role in humans is unknown, but the presence of MSH receptors in the brain suggests it may influence brain activity. Excessive levels of CRH can stimulate MSH release, and prolactin-inhibiting hormone (PIH) inhibits MSH release.

Table 17.4 summarizes the principal actions of the anterior pituitary hormones.

### ✓ CHECKPOINT

10. What is the functional importance of the hypophyseal portal veins?
11. Which anterior pituitary hormones stimulate a second endocrine gland to release a hormone?
12. You have noticed your neighbor's 8-year-old child is the same height as her 5-year-old cousin. This child could be deficient in which hormone?
13. If a person has a pituitary tumor that secretes a large amount of hGH and the tumor cells are not responsive to regulation by GHRH and GHIH, will hyperglycemia or hypoglycemia be more likely? Explain your answer.

**TABLE 17.4**

Principal Actions of Anterior Pituitary Hormones

| HORMONE | TARGET TISSUES | PRINCIPAL ACTIONS |
|---|---|---|
| **Human growth hormone (hGH)** | Liver (and other tissues) | Stimulates liver, muscle, cartilage, bone, and other tissues to synthesize and secrete insulinlike growth factors (IGFs); IGFs promote growth of body cells, protein synthesis, tissue repair, lipolysis, and elevation of blood glucose concentration |
| **Thyroid-stimulating hormone (TSH)** | Thyroid gland | Stimulates synthesis and secretion of thyroid hormones by thyroid gland |
| **Follicle-stimulating hormone (FSH)** | Ovaries; Testes | In females, initiates development of oocytes and induces ovarian secretion of estrogens; in males, stimulates testes to produce sperm |
| **Luteinizing hormone (LH)** | Ovaries; Testes | In females, stimulates secretion of estrogens and progesterone, ovulation, and formation of corpus luteum; in males, stimulates testes to produce testosterone |
| **Prolactin (PRL)** | Mammary glands | Together with other hormones, promotes milk production by mammary glands |
| **Adrenocorticotropic hormone (ACTH)** | Adrenal cortex | Stimulates secretion of glucocorticoids (mainly cortisol) by adrenal cortex |
| **Melanocyte-stimulating hormone (MSH)** | Brain | Exact role in humans is unknown but may influence brain activity; when present in excess, can cause darkening of skin |

## 17.4 Oxytocin and antidiuretic hormone originate in the hypothalamus and are stored in the posterior pituitary.

Although the **posterior pituitary**, or **neurohypophysis** (noo′-rō-hī-POF-i-sis; *neuro-* = nerve; *hypophysis* = undergrowth), does not *synthesize* hormones, it does *store* and *release* two hormones. The posterior pituitary contains axon terminals of more than 10,000 hypothalamic neurosecretory cells. The cell bodies of the neurosecretory cells are in the hypothalamus; their axons begin in the hypothalamus and end near capillaries in the posterior pituitary (Figure 17.7). Different neurosecretory cells produce two hormones: oxytocin and antidiuretic hormone.

After their production in the cell bodies of neurosecretory cells, oxytocin and antidiuretic hormone are packed into vesicles and transported to the axon terminals in the posterior pituitary, where they are stored until impulses trigger exocytosis and release of the hormone. As the vesicles undergo exocytosis, the hormone within them is released and diffuses into the capillaries of the posterior pituitary.

Blood is supplied to the posterior pituitary by the **inferior hypophyseal arteries** (see Figure 17.4), which branch from the internal carotid arteries. In the posterior pituitary, the inferior hypophyseal arteries drain into the **capillary plexus** of the posterior pituitary, a capillary network that receives oxytocin and antidiuretic hormone secreted from the axon terminals of the neurosecretory cells. From this plexus, hormones pass into the **posterior hypophyseal veins** for distribution to target cells.

### Oxytocin

During and after delivery of a baby, **oxytocin (OT)** (ok′-sē-TŌ-sin; *oxytoc-* = quick birth) affects two target tissues: the mother's uterus and breasts. During delivery, stretching of the cervix of the uterus stimulates the release of oxytocin which, in turn, enhances contraction of smooth muscle cells in the wall of the uterus (see Figure 1.4). Initial uterine contractions push the baby's head into the cervix, the lowest part of the uterus. After delivery, OT stimulates milk ejection ("letdown") from the mammary glands in response to the mechanical stimulus provided by a suckling infant during nursing.

The function of oxytocin in males and in nonpregnant females is not clear. Experiments with animals have suggested that its actions within the brain foster parental caretaking behavior toward young offspring. OT may also be responsible, in part, for the feelings of sexual pleasure during and after intercourse.

FIGURE 17.7 **Relationship between hypothalamus and posterior pituitary.** Hormone molecules synthesized in the cell bodies of hypothalamic neurosecretory cells move down to their axon terminals in the posterior pituitary. Impulses trigger exocytosis of the vesicles, thereby releasing the hormone molecules.

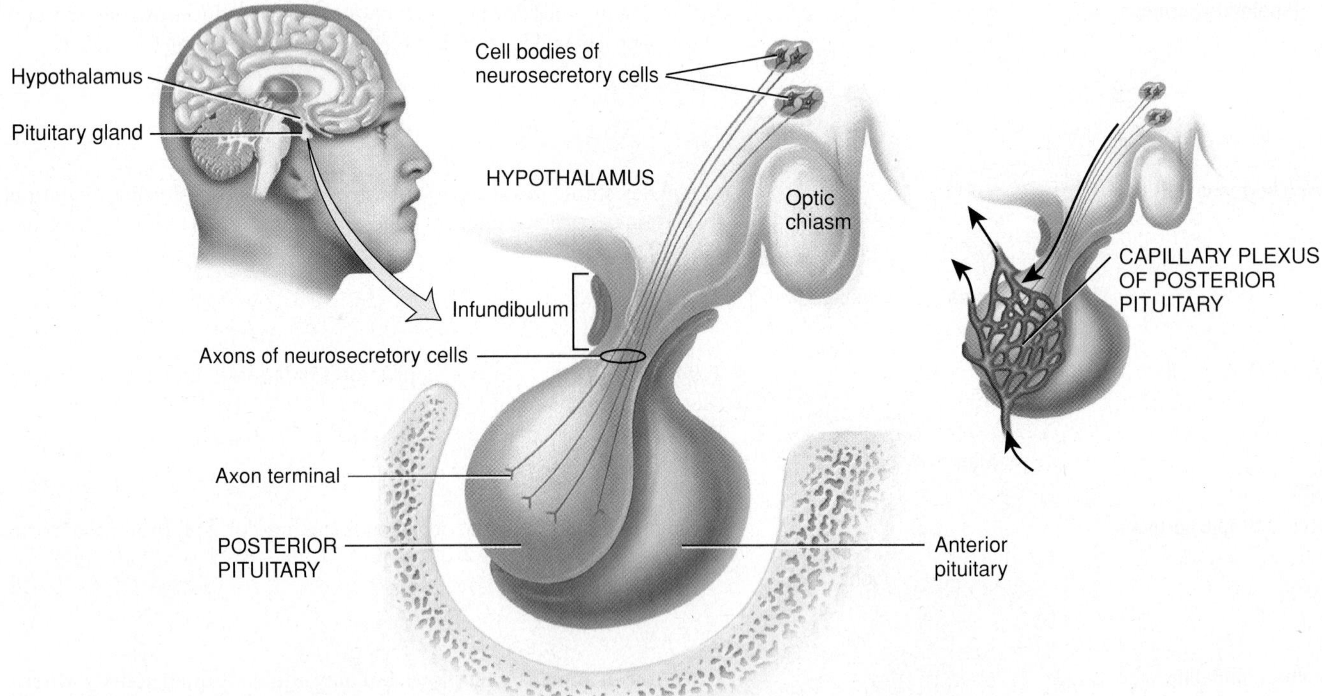


**Oxytocin and antidiuretic hormone are synthesized in the hypothalamus and released into capillaries in the posterior pituitary.**

## CLINICAL CONNECTION | *Oxytocin and Childbirth*

Years before oxytocin was discovered, it was common practice in midwifery to let a first-born twin nurse at the mother's breast to speed the birth of the second child. Now we know why this practice is helpful—it stimulates the release of oxytocin. Even after a single birth, nursing promotes expulsion of the placenta (afterbirth) and helps the uterus regain its smaller size. Synthetic OT (Pitocin) often is given to induce labor or to increase uterine tone and control hemorrhage just after giving birth.

## Antidiuretic Hormone

As its name implies, **antidiuretic hormone (ADH)** (an-tī-dī-ū-RET-ik; *anti-* = against; *-diuretic* = increases urine production) decreases urine production. ADH causes the kidneys to return more water to the blood, thus decreasing urine volume. In the absence of ADH, urine output increases more than tenfold, from the normal 1 to 2 liters to about 20 liters a day. ADH also decreases the water lost through sweating and causes constriction of arterioles, which increases blood pressure. This hormone's other name, **vasopressin** (vā-sō-PRES-in; *vaso-* = blood; *-pressus* = to press), reflects this effect on blood pressure.

The amount of ADH secreted varies with blood osmotic pressure. Blood osmotic pressure is proportional to the concentration of solutes in the blood plasma. In other words, concentrated blood has a high osmotic pressure, while dilute blood plasma has a high osmotic pressure. Figure 17.8 shows regulation of ADH secretion and the actions of ADH:

1. High blood osmotic pressure—due to dehydration or a decline in blood volume because of hemorrhage, diarrhea, or excessive sweating—stimulates **osmoreceptors**, neurons in the hypothalamus that monitor blood osmotic pressure.
2. Osmoreceptors activate the hypothalamic neurosecretory cells that synthesize and release ADH.
3. When neurosecretory cells receive excitatory input from the osmoreceptors, they generate impulses that travel through the axons of neurosecretory cells to the axon terminals in the posterior pituitary, where they cause exocytosis of ADH-containing vesicles. The liberated ADH diffuses into capillaries of the posterior pituitary where it is carried through the bloodstream to target tissues.
4. ADH stimulates the kidneys to retain more water, which decreases urine output. Sudoriferous (sweat) glands respond to ADH by decreasing their secretory activity, which lowers the rate of water loss by perspiration from the skin. Smooth muscle in the walls of arterioles (small arteries) contracts in response to the high level of ADH, which constricts (narrows) the lumen of these blood vessels and increases blood pressure.
5. Low blood osmotic pressure inhibits the osmoreceptors.
6. Inhibition of osmoreceptors reduces or stops ADH secretion. The kidneys then retain less water by forming a larger volume of urine, secretory activity of sweat glands increases, and arterioles dilate. The blood volume and osmotic pressure of body fluids return to normal.

**FIGURE 17.8 Regulation of secretion and actions of antidiuretic hormone (ADH).**

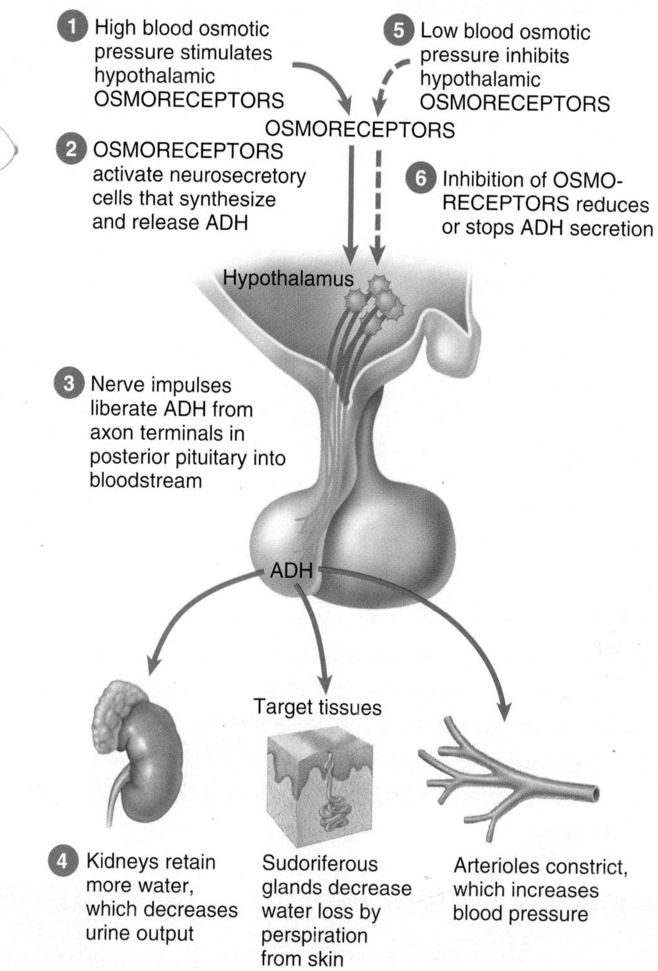
1 High blood osmotic pressure stimulates hypothalamic OSMORECEPTORS
5 Low blood osmotic pressure inhibits hypothalamic OSMORECEPTORS
OSMORECEPTORS
2 OSMORECEPTORS activate neurosecretory cells that synthesize and release ADH
6 Inhibition of OSMORECEPTORS reduces or stops ADH secretion
Hypothalamus
3 Nerve impulses liberate ADH from axon terminals in posterior pituitary into bloodstream
ADH
Target tissues
4 Kidneys retain more water, which decreases urine output
Sudoriferous glands decrease water loss by perspiration from skin
Arterioles constrict, which increases blood pressure

ADH acts to retain body water and increase blood pressure.

Secretion of ADH can also be altered in other ways. Pain, stress, trauma, anxiety, acetylcholine, nicotine, and drugs such as morphine, tranquilizers, and some anesthetics stimulate ADH secretion, leading to decreased urination. Drinking alcohol often causes frequent and copious urination because the alcohol inhibits secretion of ADH. The resulting dehydration may cause both the thirst and the headache typical of a hangover.

Table 17.5 lists the posterior pituitary hormones, control of their secretion, and their principal actions.

**TABLE 17.5**

Posterior Pituitary Hormones

| HORMONE AND TARGET TISSUES | CONTROL OF SECRETION | PRINCIPAL ACTIONS |
|---|---|---|
| **Oxytocin (OT)** 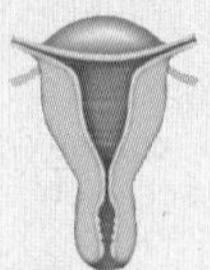 Uterus 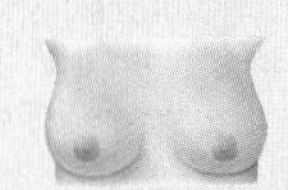 Mammary glands | Neurosecretory cells of hypothalamus secrete OT in response to uterine distension and stimulation of nipples | Stimulates contraction of smooth muscle cells of uterus during childbirth; stimulates contraction of myoepithelial cells in mammary glands to cause milk ejection |
| **Antidiuretic hormone (ADH) or vasopressin** 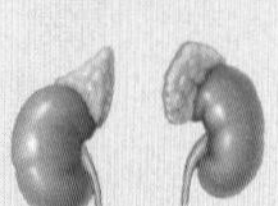 Kidneys 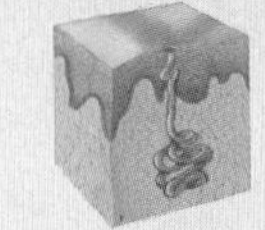 Sudoriferous glands 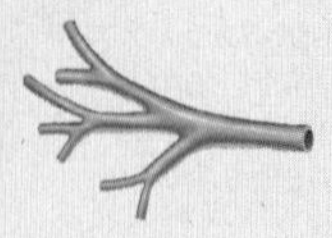 Arterioles | Neurosecretory cells of hypothalamus secrete ADH in response to elevated blood osmotic pressure, dehydration, loss of blood volume, pain, or stress; inhibitors of ADH secretion include low blood osmotic pressure, high blood volume, and alcohol | Conserves body water by decreasing urine volume; decreases water loss through perspiration; raises blood pressure by constricting arterioles |

✓ **CHECKPOINT**

**14.** Where are the hormones secreted by the posterior pituitary produced? How do OT and ADH get from their production sites to the capillaries of the posterior pituitary?

**15.** Functionally, how are the axons of hypothalamic neurosecretory cells and the hypophyseal portal veins similar?

**16.** What feedback stimuli influence the release of OT?

**17.** What effect would drinking a large volume of water have on the osmotic pressure of your blood? The level of ADH?

## RETURN TO Lisa's Story

Dr. Badeer shows Lisa and her mother an illustration of the endocrine glands of the body and explains the hormones made by each.

"Your anterior pituitary produces a number of hormones that regulate other glands in the body, and we have tests that can detect imbalances of those hormones in your bloodstream. Your body temperature and heart rate are normal, which suggests that we are not seeing a problem with your thyroid gland. Blood calcium levels are normal, which rules out a problem with either the thyroid or the parathyroid gland. Blood glucose levels are normal, which tends to rule out the pancreas."

Mrs. Vasquez asks Dr. Badeer about her daughter's constant thirst. "That is called polydipsia. It could indicate a problem with the adrenal gland or hypothalamus. Lisa, have you ever experienced lightheadedness or muscle cramps?"

Lisa confirms that she often feels dizzy when she stands up too quickly, but lately she's felt tired more often. "I just assumed that it was from playing on the basketball team."

Mrs. Vasquez mentions how they had been reading up on endocrine disorders following their visit with Dr. Carson. "We read something about how the adrenal gland makes hormones that affect blood volume and metabolism. We found a disorder called Addison's disease. Could that be causing Lisa's problems?" Dr. Badeer explains how Addison's disease presents with decreased production of glucocorticoids and mineralocorticoids, which could cause several of Lisa's symptoms.

"However, I do not think it would explain the late start of her menstrual periods. Lisa, can you tell me how often you menstruate and how much is released?" From Lisa's answers, Dr. Badeer surmises that she is demonstrating oligomenorrhea, a reduced amount of menstrual flow.

Dr. Badeer notes that Lisa has a fair amount of body hair and asks whether that has changed since she started menstruating. "No, I've always had this much body hair. Is that important?" Dr. Badeer explains how it could indicate an increased production of androgens from the adrenal gland.

"I am going to schedule you for those blood tests so that we can examine your hormone levels. That should help us determine exactly what is imbalanced and how to correct it."

A. *If Lisa's anterior pituitary gland isn't producing enough hormones, how would that affect the target glands?*

B. *If Addison's disease is a problem with the adrenal cortex, how could that affect the hypothalamus and the anterior pituitary?*

## 17.5 The thyroid gland secretes thyroxine, triiodothyronine, and calcitonin.

The butterfly-shaped **thyroid gland** is located just inferior to the larynx (voice box). It is composed of right and left **lobes**, one on either side of the trachea, that are connected by an **isthmus** (IS-mus = a narrow passage) that lies anterior to the trachea (Figure 17.9a, d). About 30 percent of thyroid glands have a small third lobe, called the *pyramidal lobe*, that extends superiorly from the isthmus.

FIGURE 17.9 The thyroid gland.

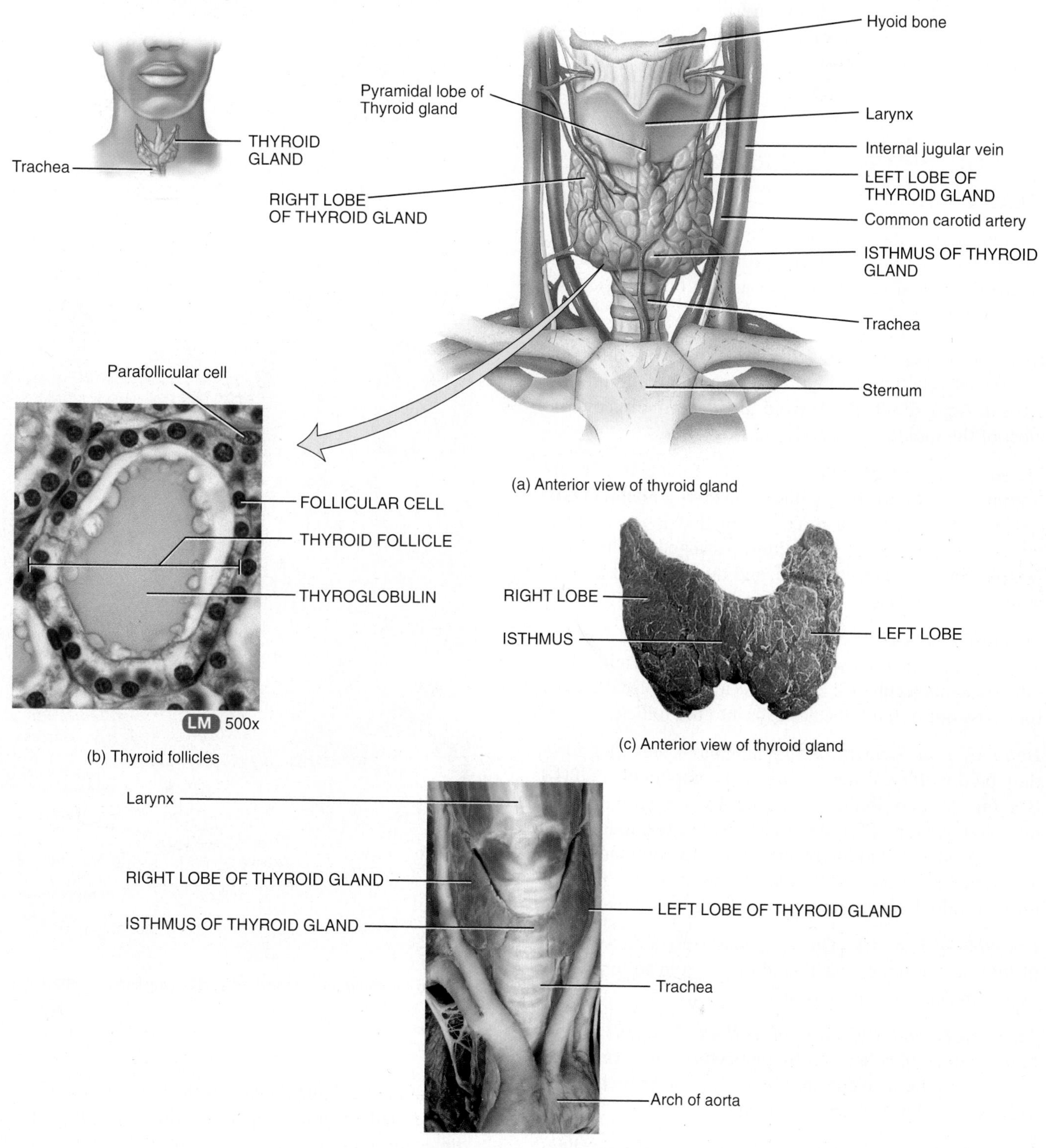


(a) Anterior view of thyroid gland

(b) Thyroid follicles

(c) Anterior view of thyroid gland

(d) Anterior view

Thyroid hormones regulate oxygen use and basal metabolic rate, cellular metabolism, and growth and development.

Microscopic spherical sacs called **thyroid follicles** (Figure 17.9b) make up most of the thyroid gland. The wall of each follicle consists primarily of glandular cells called **follicular cells** (fō-LIK-ū-lar). The follicular cells produce two hormones: **triiodothyronine** (trī′-ī-ō-dō-THĪ-rō-nēn), or $\mathbf{T_3}$, because it contains three atoms of iodine, and **thyroxine** (thī-ROK-sēn), or $\mathbf{T_4}$, which contains four atoms of iodine. $T_3$ and $T_4$ together are also known as **thyroid hormones**. A few cells called **parafollicular cells** (par′-a-fō-LIK-ū-lar) lie between follicles (Figure 17.9b). They produce the hormone **calcitonin** (kal-si-TŌ-nin), which helps regulate calcium homeostasis.

## Formation, Storage, and Release of Thyroid Hormones

The thyroid gland is the only endocrine gland that stores its secretory product in large quantity—normally about a 100-day supply. Synthesis and secretion of $T_3$ and $T_4$ occur as follows (Figure 17.10):

1. ***Iodide trapping.*** Thyroid follicular cells trap iodide ions ($I^-$) by actively transporting them from the blood into the cytosol. As a result, the thyroid gland normally contains most of the iodide in the body.
2. ***Synthesis of thyroglobulin.*** While the follicular cells are trapping $I^-$, they are also synthesizing **thyroglobulin (TGB)** (thī′-rō-GLOB-ū-lin), a glycoprotein that is produced in the rough endoplasmic reticulum, packaged in the Golgi complex into secretory vesicles, and then secreted into the lumen of the follicle.
3. ***Oxidation of iodide.*** Negatively charged **iodide ions** undergo oxidation (removal of electrons), transforming them into **iodine molecules** ($2I^- \rightarrow I_2$). As iodine molecules form, they pass out of the follicular cells into the follicle.
4. ***Iodination of tyrosine.*** As iodine molecules ($I_2$) form, they bind with **tyrosines**, amino acid components of TGB. Binding of one iodine molecule to a tyrosine yields monoiodotyrosine ($T_1$), and attachment of two iodine molecules produces diiodotyrosine ($T_2$). The combined TGB and attached iodine molecules forms a sticky material termed **colloid** that accumulates in the lumen of the follicle.
5. ***Coupling of $T_1$ and $T_2$.*** During the last step in the synthesis of thyroid hormone, one $T_1$ and one $T_2$ join to form $T_3$ or two $T_2$ molecules join to form $T_4$.
6. ***Pinocytosis and digestion of colloid.*** Droplets of colloid reenter follicular cells by pinocytosis and merge with lysosomes. Digestive enzymes in the lysosomes break down TGB, separating molecules of $T_3$ and $T_4$.
7. ***Secretion of thyroid hormones.*** Because $T_3$ and $T_4$ are lipid-soluble, they diffuse through the plasma membrane into interstitial fluid and then enter the bloodstream. $T_4$ normally is secreted in greater quantity than $T_3$, but $T_3$ is several times more potent. In addition, after $T_4$ enters a body cell, most of it is converted to $T_3$ by removal of one iodine.

**FIGURE 17.10 Steps in the synthesis and secretion of thyroid hormones.** $I^-$ = iodide, $I_2$ = iodine, TGB = thyroglobulin, TBG = thyroxine-binding globulin.

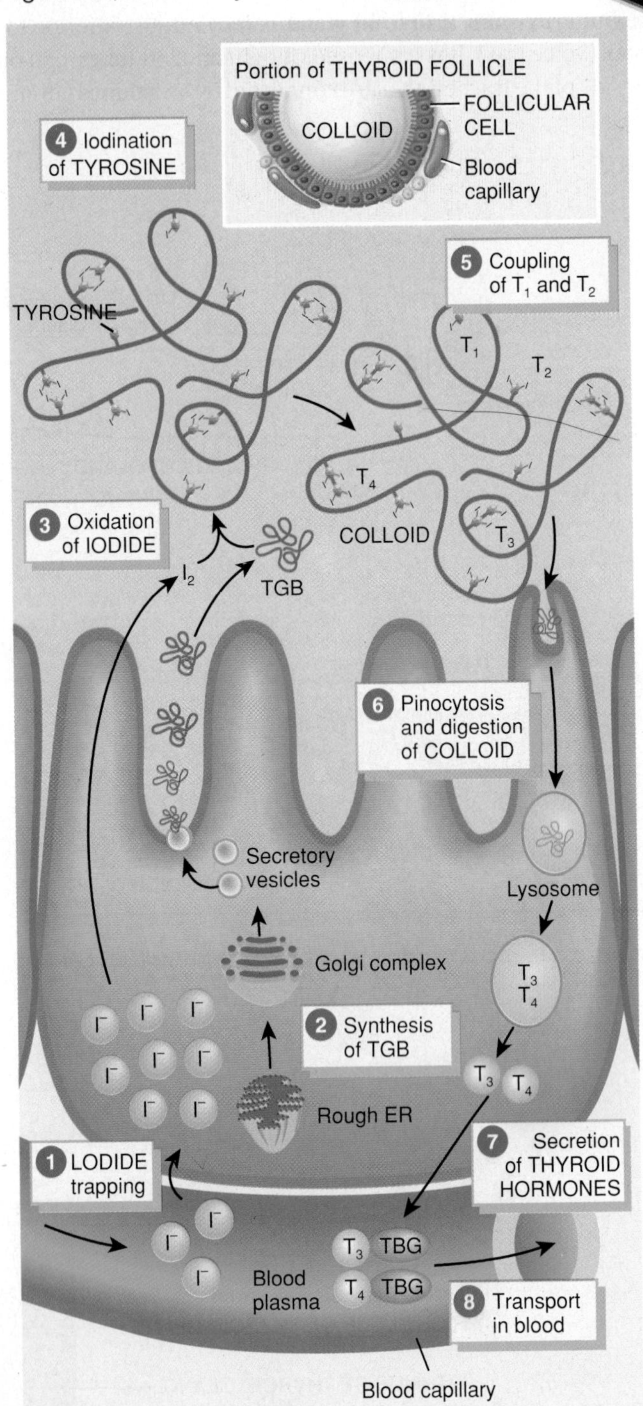


Thyroid hormones are synthesized by attaching iodine atoms to tyrosine.

8 ***Transport in the blood.*** Most $T_3$ and $T_4$ travel through the bloodstream bound to the transport protein **thyroxine-binding globulin (TBG)**.

## Actions of Thyroid Hormones

Because most body cells have receptors for thyroid hormones, $T_3$ and $T_4$ exert their effects throughout the body. Thyroid hormones:

- Increase **basal metabolic rate (BMR)**, the rate of oxygen consumption under standard or basal conditions (awake, at rest, and fasting), by stimulating the use of cellular oxygen to produce ATP. When BMR increases, cellular metabolism of carbohydrates, lipids, and proteins also increases.
- Stimulate cells to use large amounts of ATP. As cells produce and use more ATP, more heat is given off, and body temperature rises. This phenomenon is called the **calorigenic effect** (ka-lor′-i-JEN-ik). In this way, thyroid hormones play an important role in the maintenance of normal body temperature. Normal mammals can survive in freezing temperatures, but those whose thyroid glands have been removed cannot.
- Stimulate protein synthesis and increase the use of glucose and fatty acids for ATP production.
- Increase the breakdown of triglycerides and enhance cholesterol excretion, thus reducing blood cholesterol level.
- Enhance some actions of norepinephrine and epinephrine. For this reason, symptoms of hyperthyroidism (above normal secretion of thyroid hormones) include increased heart rate, more forceful heartbeats, and increased blood pressure.
- Stimulate body growth, particularly the growth of the nervous and skeletal systems. Deficiency of thyroid hormones during fetal development, infancy, or childhood causes severe mental retardation and stunted bone growth.

## Control of Thyroid Hormone Secretion

Thyrotropin-releasing hormone (TRH) from the hypothalamus and thyroid-stimulating hormone (TSH) from the anterior pituitary stimulate synthesis and release of thyroid hormones, as shown in Figure 17.11:

1 Low blood levels of $T_3$ and $T_4$ or low metabolic rate stimulate the hypothalamus to secrete TRH.

2 TRH flows through the hypophyseal portal veins to the anterior pituitary, where it stimulates secretion of TSH.

3 TSH stimulates $T_3$ and $T_4$ synthesis and secretion by the thyroid follicles.

4 The thyroid follicles release $T_3$ and $T_4$ into the blood until the metabolic rate returns to normal.

**FIGURE 17.11 Regulation of secretion of thyroid hormones.** TRH = thyrotropin-releasing hormone, TSH = thyroid-stimulating hormone, $T_3$ = triiodothyronine, and $T_4$ = thyroxine.

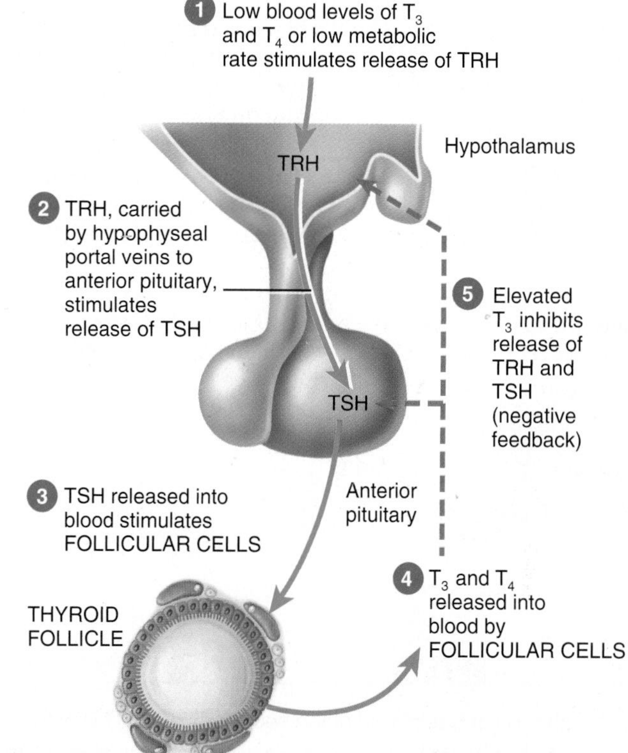


TSH promotes release of thyroid hormones ($T_3$ and $T_4$) by the thyroid gland.

5 An elevated level of $T_3$ inhibits release of TRH and TSH (negative feedback inhibition).

Conditions that increase ATP demand—a cold environment, hypoglycemia, high altitude, and pregnancy—also increase the secretion of the thyroid hormones.

## Calcitonin

The hormone produced by the **parafollicular cells** of the thyroid gland (see Figure 17.9b) is **calcitonin** (kal-si-TŌ-nin). Calcitonin decreases the level of calcium in the blood by inhibiting the action of osteoclasts, the cells that break down bone extracellular matrix. The secretion of calcitonin is controlled by a negative feedback system (see Figure 17.13). When its blood level is high, calcitonin lowers the amount of blood calcium and phosphates by inhibiting the breakdown of bone extracellular matrix by osteoclasts and by accelerating uptake of calcium and phosphates into bone extracellular matrix.

Table 17.6 summarizes the hormones produced by the thyroid gland, control of their secretion, and their principal actions.

**TABLE 17.6**

Thyroid Gland Hormones

| HORMONE AND SOURCE | CONTROL OF SECRETION | PRINCIPAL ACTIONS |
|---|---|---|
| **$T_3$ (triiodothyronine)** and **$T_4$ (thyroxine)** or thyroid hormones from follicular cells 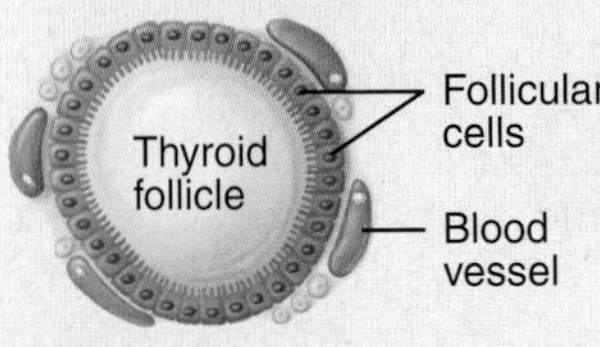  | Secretion is increased by thyrotropin-releasing hormone (TRH), which stimulates release of thyroid-stimulating hormone (TSH) in response to low thyroid hormone levels, low metabolic rate, cold, pregnancy, and high altitudes; TRH and TSH secretions are inhibited in response to high thyroid hormone levels; high iodine level suppresses $T_3/T_4$ secretion | Increase basal metabolic rate; stimulate synthesis of proteins; increase use of glucose and fatty acids for ATP production; increase lipolysis; enhance cholesterol excretion; accelerate body growth; contribute to development of nervous system |
| **Calcitonin (CT)** from parafollicular cells 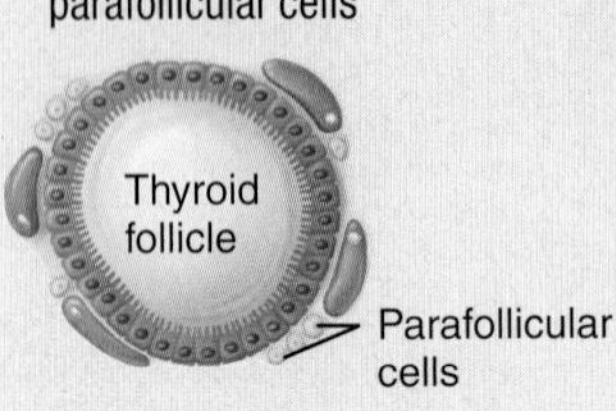  | High blood calcium levels stimulate secretion; low blood calcium levels inhibit secretion | Lowers blood levels of calcium and phosphates by inhibiting bone resorption by osteoclasts and by accelerating uptake of calcium and phosphates into bone extracellular matrix |

✓ **CHECKPOINT**

**18.** Which cells of the thyroid gland secrete $T_3$ and $T_4$? Which secrete calcitonin? Which of these hormones are also called thyroid hormones?

**19.** How are the thyroid hormones synthesized, stored, and transported in the bloodstream?

**20.** How is the secretion of $T_3$ and $T_4$ regulated?

**21.** Explain how blood levels of $T_3/T_4$, TSH, and TRH would change in a laboratory animal that has undergone a thyroidectomy (complete removal of its thyroid gland).

**22.** Why might hypothyroidism (low thyroid hormone level) in an infant result in mental retardation?

**23.** What is the primary target for calcitonin? What is the action of calcitonin?

# 17.6 The parathyroid glands secrete parathyroid hormone, which regulates calcium, magnesium, and phosphate ion levels.

Partially embedded in the posterior surface of the lateral lobes of the thyroid gland are several small, round masses of tissue called the **parathyroid glands** (*para-* = beside). There usually are four parathyroid glands, one superior and one inferior gland attached to each lateral thyroid lobe (Figure 17.12a, d).

Microscopically, the parathyroid glands contain two kinds of epithelial cells (Figure 17.12b, c). The more numerous **chief cells** produce parathyroid hormone. The function of the other kind of cell, called an *oxyphil cell*, is not known.

## Parathyroid Hormone

**Parathyroid hormone (PTH)** is the major regulator of the levels of calcium ($Ca^{2+}$), magnesium ($Mg^{2+}$), and phosphate ($HPO_4^{2-}$) ions in the blood. The specific action of PTH is to stimulate osteoclast activity. As osteoclasts increase their digestion of the bone extracellular matrix, $Ca^{2+}$ and $HPO_4^{2-}$ are released into the blood. PTH also acts on the kidneys during urine formation from blood by decreasing $Ca^{2+}$ and $Mg^{2+}$ urinary loss while increasing $HPO_4^{2-}$ loss into the urine. Because more $HPO_4^{2-}$ is lost in the urine than is gained from the bones, PTH decreases blood $HPO_4^{2-}$ level and increases blood $Ca^{2+}$ and $Mg^{2+}$ levels. A third effect of PTH on the kidneys is to promote formation of the hormone **calcitriol** (kal′-si-TRĪ-ol), the active form of vitamin D. Calcitriol increases the rate of $Ca^{2+}$, $HPO_4^{2-}$, and $Mg^{2+}$ absorption from food within the gastrointestinal tract into the blood.

The blood calcium level directly controls the secretion of both *calcitonin* and parathyroid hormone through negative

FIGURE 17.12 **The parathyroid glands.**



(a) Posterior view



(b) Parathyroid gland



(c) Portion of thyroid gland (left) and parathyroid gland (right)



(d) Posterior view of parathyroid glands

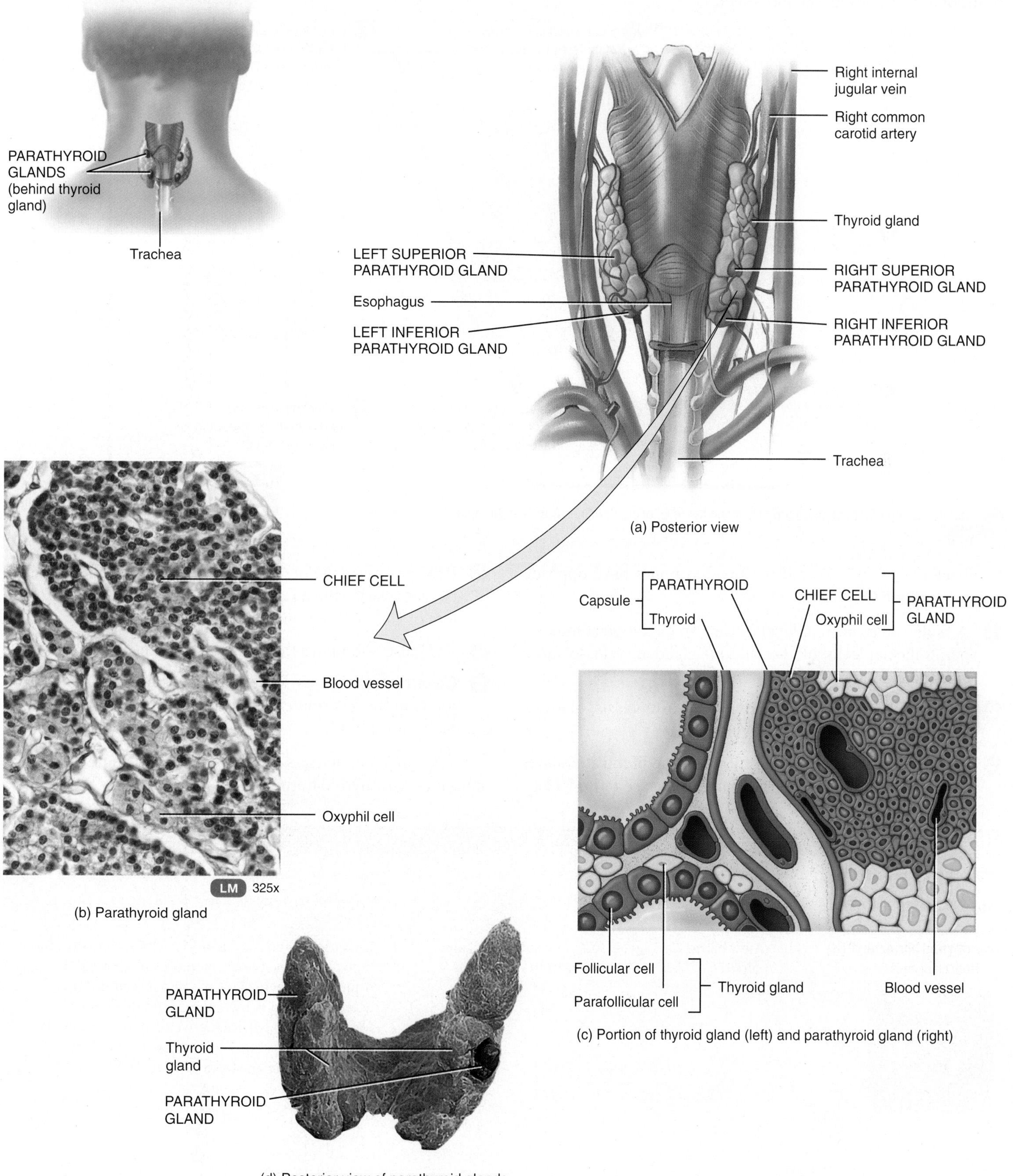

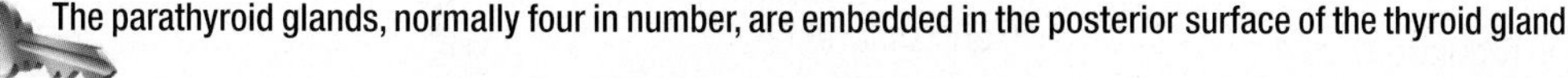

**The parathyroid glands, normally four in number, are embedded in the posterior surface of the thyroid gland.**

FIGURE 17.13 **The roles of calcitonin (green arrows), parathyroid hormone (blue arrows), and calcitriol (orange arrows) in homeostasis of blood calcium level.**

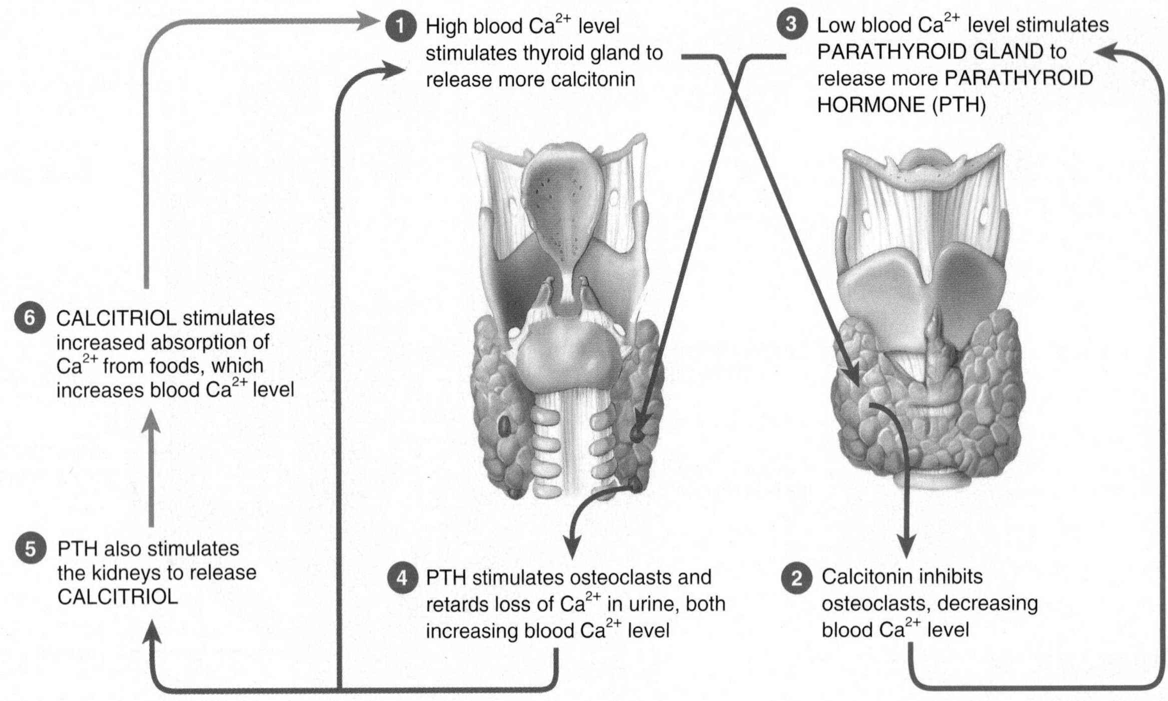

1 High blood $Ca^{2+}$ level stimulates thyroid gland to release more calcitonin
3 Low blood $Ca^{2+}$ level stimulates PARATHYROID GLAND to release more PARATHYROID HORMONE (PTH)
6 CALCITRIOL stimulates increased absorption of $Ca^{2+}$ from foods, which increases blood $Ca^{2+}$ level
5 PTH also stimulates the kidneys to release CALCITRIOL
4 PTH stimulates osteoclasts and retards loss of $Ca^{2+}$ in urine, both increasing blood $Ca^{2+}$ level
2 Calcitonin inhibits osteoclasts, decreasing blood $Ca^{2+}$ level

**Calcitonin and PTH have opposite effects on the level of calcium ions ($Ca^{2+}$) in the blood.**

feedback (Figure 17.13), and the two hormones have opposite effects on blood $Ca^{2+}$ level:

1. A higher-than-normal level of $Ca^{2+}$ in the blood stimulates parafollicular cells of the thyroid gland to release more calcitonin.
2. Calcitonin inhibits the activity of osteoclasts, thereby decreasing the blood $Ca^{2+}$ level.
3. A lower-than-normal level of $Ca^{2+}$ in the blood stimulates chief cells of the parathyroid gland to release more PTH.
4. PTH promotes the breakdown of bone extracellular matrix by osteoclasts, which releases $Ca^{2+}$ into the blood, and slows loss of $Ca^{2+}$ in the urine, raising the blood level of $Ca^{2+}$.
5. PTH also stimulates the kidneys to synthesize calcitriol.
6. Calcitriol stimulates increased absorption of $Ca^{2+}$ from foods in the gastrointestinal tract, which helps increase the blood level of $Ca^{2+}$.

Table 17.7 summarizes control of secretion and the principal actions of parathyroid hormone.

**TABLE 17.7**

Parathyroid Gland Hormone

| HORMONE AND SOURCE | CONTROL OF SECRETION | PRINCIPAL ACTIONS |
|---|---|---|
| **Parathyroid hormone (PTH)** from chief cells<br>Chief cell | Low blood $Ca^{2+}$ levels stimulate secretion; high blood $Ca^{2+}$ levels inhibit secretion | Increases blood $Ca^{2+}$ and $Mg^{2+}$ levels and decreases blood phosphate level; increases bone breakdown by osteoclasts; decreases $Ca^{2+}$ and increases $HPO_4^{2-}$ excretion in urine by kidneys; and promotes formation of calcitriol (active form of vitamin D), which increases rate of dietary $Ca^{2+}$ and $Mg^{2+}$ absorption from gastrointestinal tract |

✓ **CHECKPOINT**

**24.** What are the primary target tissues for PTH and calcitriol?

**25.** How is secretion of parathyroid hormone regulated?

**26.** In which ways are the actions of PTH and calcitriol similar and different?

# 17.7 The adrenal glands are structurally and functionally two independent endocrine glands.

The paired **adrenal glands**, or **suprarenal glands** (*supra-* = above; *-renal* = kidney), one of which lies superior to each kidney, have a flattened pyramidal shape (Figure 17.14a, c). Each adrenal gland has two structurally and functionally distinct regions: a large, peripherally located **adrenal cortex**, comprising 80–90 percent of the gland, and a small, centrally located **adrenal medulla** (Figure 17.14b). A

FIGURE 17.14 The adrenal (suprarenal) glands.

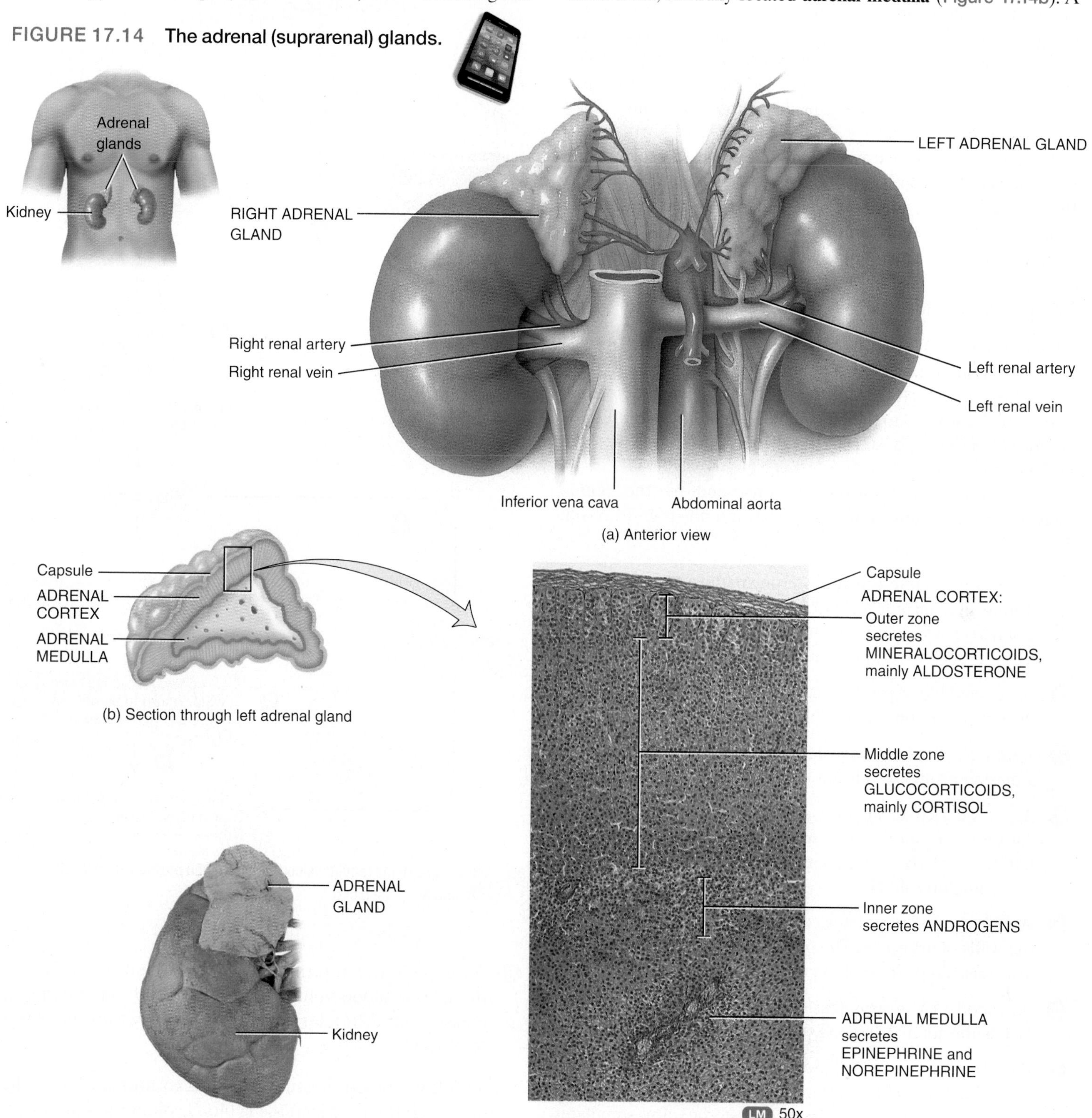


(a) Anterior view

(b) Section through left adrenal gland

(c) Anterior view of adrenal gland and kidney

(d) Subdivisions of adrenal gland

The adrenal cortex secretes steroid hormones that are essential for life; the adrenal medulla secretes norepinephrine and epinephrine.

connective tissue capsule covers the gland. The adrenal glands, like the thyroid gland, are highly vascularized.

## Adrenal Cortex

The adrenal cortex produces steroid hormones that are essential for life. Complete loss of adrenocortical hormones leads to death due to dehydration and electrolyte imbalances in a few days to a week, unless hormone replacement therapy begins promptly. The adrenal cortex is subdivided into three zones, each of which secretes different hormones (Figure 17.14d). The outer zone, just deep to the connective tissue capsule, secretes hormones called **mineralocorticoids** (min′-er-al-ō-KOR-ti-koyds) because they affect mineral homeostasis. The middle zone secretes mainly **glucocorticoids** (gloo′-kō-KOR-ti-koyds), so named because they affect glucose homeostasis. The inner zone synthesizes small amounts of weak **androgens** (*andro-* = a man), steroid hormones that have masculinizing effects.

### Mineralocorticoids

**Aldosterone** (al-DOS-ter-ōn) is the major mineralocorticoid. It regulates homeostasis of two mineral ions, sodium ions ($Na^+$) and potassium ions ($K^+$). Aldosterone increases reabsorption of $Na^+$ from the urine into the blood, and it stimulates excretion of $K^+$ into the urine. Aldosterone also helps adjust blood pressure and blood volume.

Secretion of aldosterone is controlled by the **renin–angiotensin–aldosterone pathway** (RĒ-nin an′-jē-ō-TEN-sin; Figure 17.15):

1. The renin–angiotensin–aldosterone pathway is activated by stimuli that include dehydration, $Na^+$ deficiency, or hemorrhage. These conditions decrease blood volume, which leads to decreased blood pressure.
2. Decreased blood pressure stimulates the kidneys to secrete the enzyme **renin** into the blood.
3. Renin converts **angiotensinogen** (an′-jē-ō-ten-SIN-ō-jen), a plasma protein produced by the liver, into **angiotensin I.**
4. As blood rich in angiotensin I flows through the lungs, another enzyme called **angiotensin-converting enzyme (ACE)** converts inactive angiotensin I into the active hormone **angiotensin II.**
5. Angiotensin II stimulates contraction of smooth muscle in the walls of arterioles. The resulting vasoconstriction of the arterioles helps raise blood pressure to normal.
6. As blood rich in angiotensin II flows through the adrenal glands, the adrenal cortex is stimulated to secrete aldosterone.
7. Blood containing an increased level of aldosterone circulates to the kidneys where it promotes the movement of $Na^+$ and water from urine to the blood. Aldosterone also stimulates the kidneys to increase excretion of $K^+$ into the urine.
8. As more water returns to the blood (and less is lost in the urine), blood volume increases. As blood volume increases and arterioles constrict, blood pressure increases to normal.

Therefore, blood pressure increases to normal due to the aldosterone-influenced increase in blood volume and the angiotensin II-influenced vasoconstriction of arterioles.

**FIGURE 17.15 The renin–angiotensin–aldosterone pathway.**

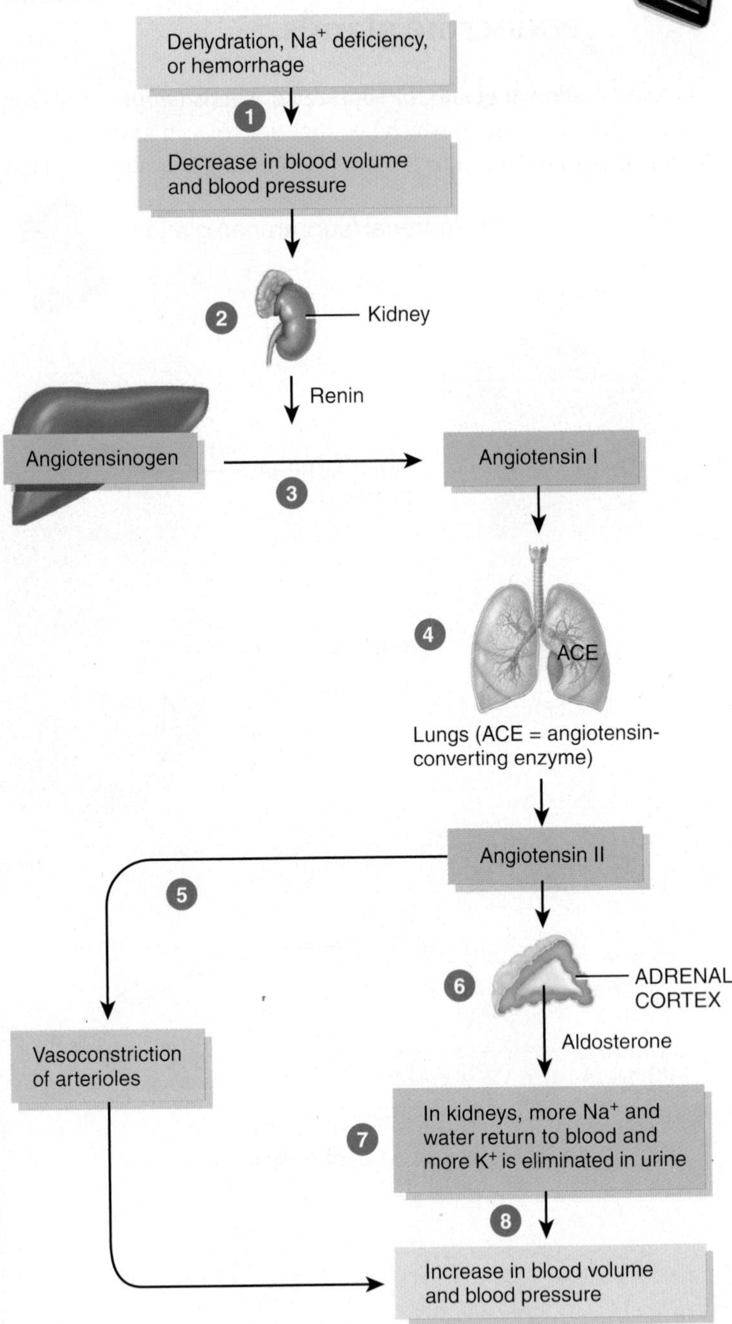


Aldosterone helps regulate blood volume, blood pressure, and levels of $Na^+$ and $K^+$ in the blood.

## Glucocorticoids

The glucocorticoids, which regulate metabolism and resistance to stress, include **cortisol** (KOR-ti-sol), **hydrocortisone**, **corticosterone** (kor′-ti-KOS-ter-on), and **cortisone** (KOR-ti-sōn). Of the three, cortisol is the most abundant, accounting for about 95 percent of glucocorticoid activity.

Control of glucocorticoid secretion is by a typical negative feedback system (Figure 17.16; see also Figure 17.5):

1. Low blood levels of glucocorticoids, mainly cortisol, stimulate neurosecretory cells in the hypothalamus to secrete corticotropin-releasing hormone (CRH).
2. CRH travels through the infundibulum to the anterior pituitary, where it promotes the release of corticotropin (ACTH) from the anterior pituitary.

FIGURE 17.16 **Negative feedback regulation of glucocorticoid secretion.**

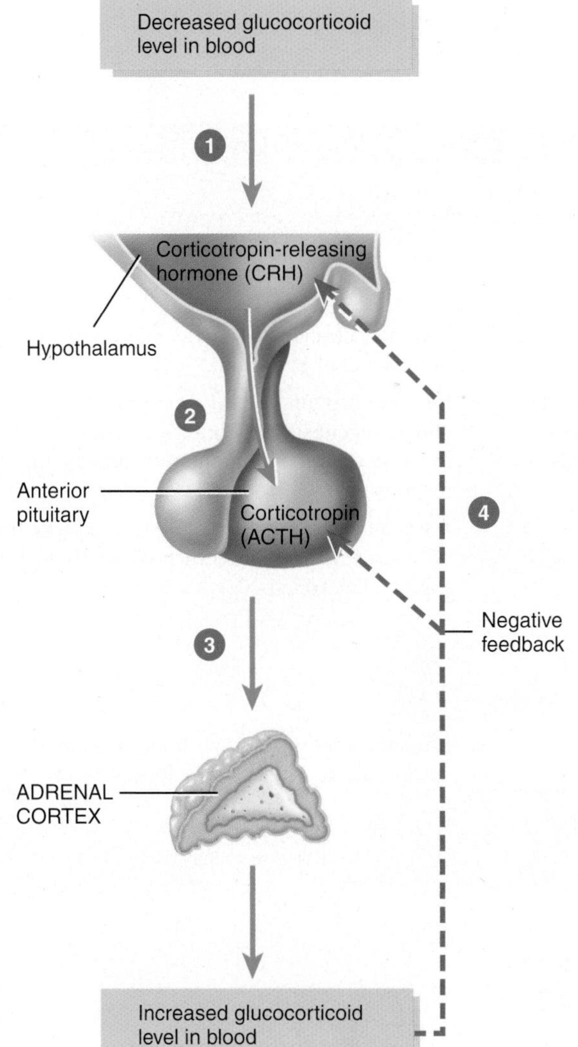


**Low levels of glucocorticoids stimulate the release of ACTH; ACTH in turn stimulates glucocorticoid secretion by the adrenal cortex.**

3. ACTH flows in the blood to the adrenal cortex, where it stimulates glucocorticoid secretion.
4. As the level of glucocorticoid rises, it exerts negative feedback inhibition both on the anterior pituitary to reduce release of ACTH and on the hypothalamus to reduce release of CRH.

Glucocorticoids have the following effects:

- ***Protein breakdown.*** Glucocorticoids increase the rate of protein breakdown, mainly in muscle fibers, and thus increase amino acid concentrations in the blood. These amino acids can be used by body cells for synthesis of new proteins or for ATP production.
- ***Glucose formation.*** Upon stimulation by glucocorticoids, liver cells may convert certain amino acids or lactic acid to glucose, which cells can use for ATP production. Such conversion of a substance other than glycogen or another monosaccharide into glucose is called **gluconeogenesis** (gloo′-ko-nē-ō-JEN-e-sis).
- ***Breakdown of triglycerides.*** Glucocorticoids stimulate the breakdown of triglycerides in adipose tissue. The fatty acids released into the blood can be used for ATP production by many body cells.
- ***Resistance to stress.*** Glucocorticoids work in many ways to provide resistance to stress. The additional amino acids, glucose, and fatty acids provide tissues with an increased source of ATP to combat a range of stresses, including exercise, fasting, fright, temperature extremes, high altitude, bleeding, infection, surgery, trauma, and disease.
- ***Anti-inflammatory effects.*** Glucocorticoids inhibit white blood cells that participate in inflammatory responses. Unfortunately, glucocorticoids also retard tissue repair, which slows wound healing. Although high doses can cause severe mental disturbances, synthetic glucocorticoids are very useful in the treatment of chronic inflammatory disorders such as rheumatoid arthritis.
- ***Depression of immune responses.*** High doses of glucocorticoids depress immune responses. For this reason, glucocorticoids are prescribed for organ transplant recipients to retard tissue rejection by the immune system.

## Androgens

In both males and females, the adrenal cortex secretes small amounts of weak androgens. After puberty in males, the androgen testosterone is also released in much greater quantity by the testes. Thus, the relative amount of androgens secreted by the adrenal gland in males is usually so low that their effects are insignificant. In females, however, adrenal androgens play important roles. They contribute to libido (sex drive) and are converted into estrogens (feminizing sex steroids) by other body tissues. After menopause, when ovarian secretion of estrogens ceases, all female estrogens come from conversion of adrenal androgens. Adrenal androgens also stimulate growth of axillary and pubic hair in boys and girls and contribute to growth spurts before puberty. Although control of adrenal androgen secretion is not fully understood, the main hormone that stimulates its secretion is ACTH.

## Adrenal Medulla

The inner region of the adrenal gland, the adrenal medulla, is a modified sympathetic ganglion of the autonomic nervous system. It develops from the same embryonic tissue as all other sympathetic ganglia, but its neurons, which lack axons, form clusters around large blood vessels. Rather than releasing a neurotransmitter, the neurons of the adrenal medulla secrete hormones. Because the autonomic nervous system exerts direct control over the adrenal medulla through sympathetic innervation, hormone release can occur very quickly.

The two major hormones synthesized by the adrenal medulla are **epinephrine** (ep′-i-NEF-rin) and **norepinephrine (NE)**, also called **adrenaline** and **noradrenaline**. The hormones of the adrenal medulla intensify sympathetic responses that occur in other parts of the body. In stressful situations and during exercise, sympathetic impulses initiated by the hypothalamus stimulate the adrenal medulla to secrete epinephrine and norepinephrine. These two hormones greatly augment the fight-or-flight response that was described in Concept 14.8. By increasing heart rate and force of contraction, epinephrine and norepinephrine increase the pumping output of the heart, which increases blood pressure. They also increase blood flow to the heart, liver, skeletal muscles, and adipose tissue; dilate airways to the lungs; and increase blood levels of glucose and fatty acids.

Table 17.8 summarizes the hormones produced by the adrenal glands, control of their secretion, and their principal actions.

**TABLE 17.8**

Adrenal Gland Hormones

| HORMONES AND SOURCE | CONTROL OF SECRETION | PRINCIPAL ACTIONS |
| --- | --- | --- |
| **ADRENAL CORTEX HORMONES** | | |
| **Mineralocorticoids (mainly aldosterone)** | Increased blood $K^+$ level and angiotensin II stimulate secretion | Increase blood levels of $Na^+$ and water; decrease blood level of $K^+$ |
| **Glucocorticoids (mainly cortisol)** | ACTH stimulates release; corticotropin-releasing hormone promotes ACTH secretion in response to stress and low blood levels of glucocorticoids | Increase protein breakdown, stimulate gluconeogenesis and lipolysis, provide resistance to stress, dampen inflammation, depress immune responses |
| **Androgens**<br>Adrenal cortex | ACTH stimulates secretion | Assist in early growth of axillary and pubic hair in both sexes; in females, contribute to libido and are source of estrogens after menopause |
| **ADRENAL MEDULLA HORMONES** | | |
| **Epinephrine** and **norepinephrine**<br>Adrenal medulla | Sympathetic nervous system stimulation in response to stress | Enhance effects of sympathetic division of autonomic nervous system during stress |

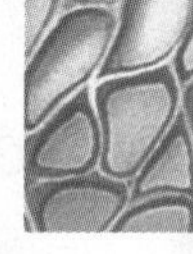

## CLINICAL CONNECTION | *Congenital Adrenal Hyperplasia*

**Congenital adrenal hyperplasia (CAH)** (hī-per-PLĀ-zhē-a) is a genetic disorder in which one or more enzymes needed for synthesis of cortisol are absent. Because the cortisol level is low, secretion of ACTH by the anterior pituitary is high due to lack of negative feedback inhibition. ACTH in turn stimulates growth and secretory activity of the adrenal cortex. As a result, both adrenal glands are enlarged. However, certain steps leading to synthesis of cortisol are blocked. Thus, precursor molecules accumulate, and some of these are weak androgens that can undergo conversion to testosterone. The result is **virilism** (VIR-i-lizm), or masculinization. In a female, virile characteristics include growth of a beard, development of a much deeper voice and a masculine distribution of body hair, growth of the clitoris so it may resemble a penis, atrophy of the breasts, and increased muscularity that produces a masculine physique. In prepubertal males, the syndrome causes the same characteristics as in females, plus rapid development of the male sexual organs and emergence of male sexual desires. In adult males, the virilizing effects of CAH are usually completely obscured by the normal virilizing effects of the testosterone secreted by the testes. As a result, CAH is often difficult to diagnose in adult males. Treatment involves cortisol therapy, which inhibits ACTH secretion and thus reduces production of adrenal androgens.

### ✓ CHECKPOINT

**27.** In which two ways can angiotensin II increase blood pressure? What are its target tissues for each mechanism?

**28.** Which stimulus initiates the renin–angiotensin–aldosterone pathway? Which stimulus turns the pathway off? Is this a negative or positive feedback loop?

**29.** How do the hypothalamus and anterior pituitary influence glucocorticoid secretion?

**30.** How do glucocorticoids influence ATP production? How does this influence your resistance to stress?

**31.** Which chemical does the adrenal medulla and the autonomic nervous system both produce?

**32.** Complete removal of the adrenal glands would eliminate the release of which hormones?

## 17.8 The pancreatic islets regulate blood glucose level by secreting glucagon and insulin.

The **pancreas** (*pan-* = all; *-creas* = flesh) is both an endocrine gland and an exocrine gland (Figure 17.17). We discuss its endocrine functions in this chapter and its exocrine functions in the digestive system in Concept 23.5. The pancreas is a flattened organ that is located in the curve of the duodenum, the first part of the small intestine (Figure 17.17a). Roughly 99 percent of the pancreas consists of clusters of exocrine cells called **acini** (AS-i-nī). The acini produce digestive enzymes, which flow into the gastrointestinal tract through a network of ducts. Scattered among the acini are 1–2 million tiny clusters of endocrine tissue called **pancreatic islets** (Ī-lets) or **islets of Langerhans** (LAHNG-er-hanz; Figure 17.17b, c). Abundant capillaries serve both the exocrine and endocrine portions of the pancreas.

**FIGURE 17.17 The pancreas.**

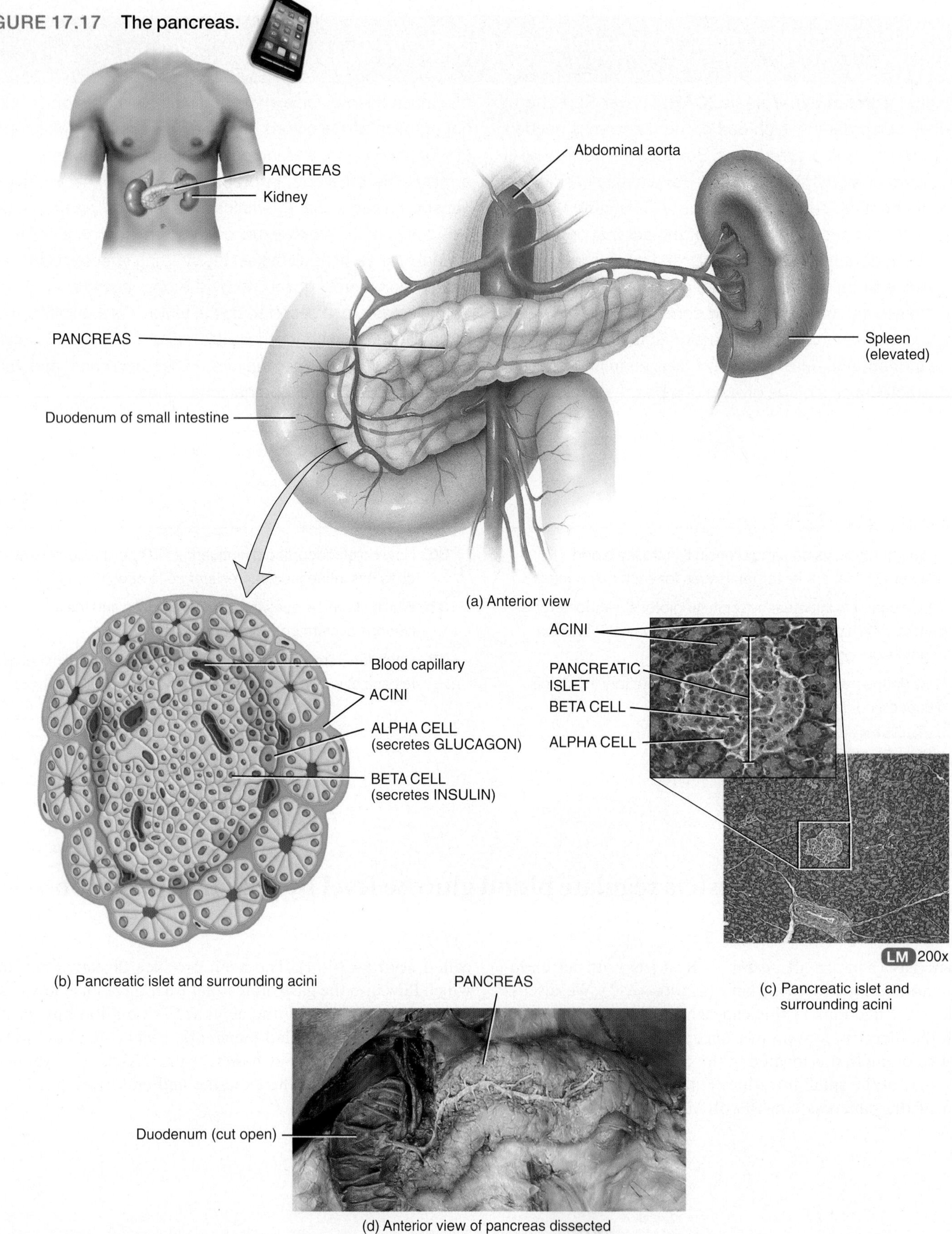


(a) Anterior view

(b) Pancreatic islet and surrounding acini

(c) Pancreatic islet and surrounding acini

(d) Anterior view of pancreas dissected to reveal pancreatic duct

**Pancreatic hormones regulate blood glucose level.**

## Regulation of Glucagon and Insulin Secretion

Some of the pancreatic islet cells, the **alpha cells**, secrete the hormone **glucagon** (GLOO-ka-gon). Other neighboring islet cells, the **beta cells**, secrete **insulin** (IN-soo-lin). The principal action of glucagon is to increase blood glucose level when it falls below normal. Insulin, on the other hand, helps lower blood glucose level when it is too high. The level of blood glucose controls secretion of glucagon and insulin via negative feedback. Figure 17.18 shows the conditions that stimulate the pancreatic islets to secrete their hormones, the ways in which glucagon and insulin produce their effects on blood glucose level, and negative feedback control of hormone secretion.

1. Low blood glucose level (hypoglycemia) stimulates secretion of glucagon from alpha cells of the pancreatic islets.
2. Glucagon acts on liver cells to accelerate the breakdown of glycogen into glucose and formation of glucose from lactic acid and certain amino acids.
3. As a result, the liver releases glucose into the blood more rapidly, and blood glucose level rises.
4. If blood glucose continues to rise, high blood glucose level (hyperglycemia) inhibits release of glucagon by alpha cells (negative feedback).
5. At the same time, high blood glucose (hyperglycemia) stimulates secretion of insulin by beta cells of the pancreatic islets.
6. Insulin acts on body cells to increase uptake of glucose, accelerate synthesis of glycogen from glucose, increase uptake of amino acids by cells and increase protein synthesis, and increase fatty acid synthesis.
7. As glucose enters and is utilized by cells, blood glucose level falls.
8. If blood glucose level drops below normal, low blood glucose inhibits release of insulin by beta cells (negative feedback) and stimulates release of glucagon.

Although blood glucose level is the most important regulator of glucagon and insulin, these hormones are also regulated by other hormones, the autonomic nervous system, and your diet. Human growth hormone and adrenocorticotropic hormone indirectly stimulate insulin secretion because they elevate blood glucose level. Digestion and absorption of food containing *both* carbohydrates and proteins stimulates insulin release. In contrast, the high blood levels of amino acids following meals containing mainly proteins stimulates glucagon secretion. The sympathetic division of the ANS stimulates secretion of glucagon during exercise.

Table 17.9 summarizes the hormones produced by the pancreas, control of their secretion, and their principal actions.

**FIGURE 17.18 Regulation of blood glucose level by negative feedback systems involving glucagon (blue arrows) and insulin (orange arrows).**

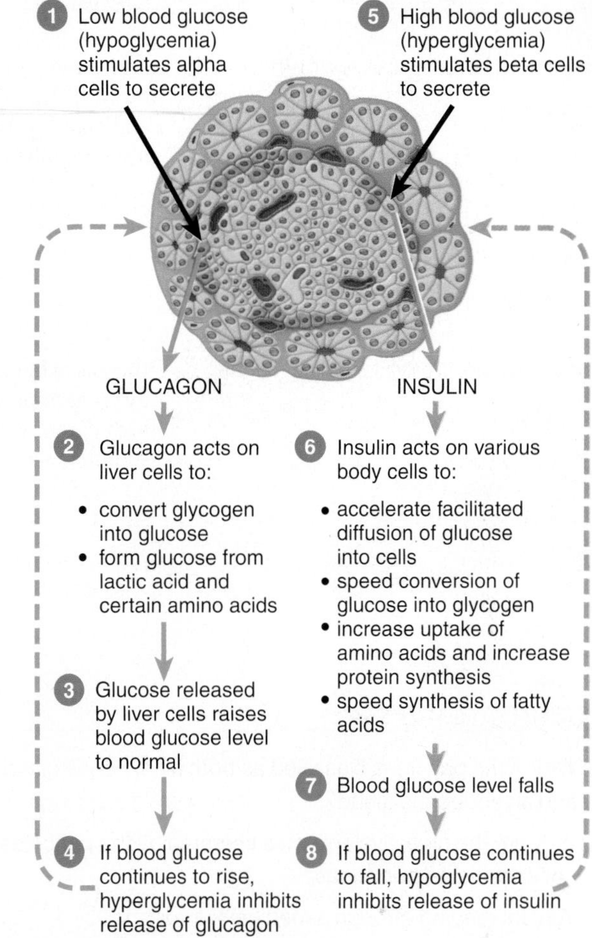
1 Low blood glucose (hypoglycemia) stimulates alpha cells to secrete
5 High blood glucose (hyperglycemia) stimulates beta cells to secrete
GLUCAGON
INSULIN
2 Glucagon acts on liver cells to:
• convert glycogen into glucose
• form glucose from lactic acid and certain amino acids
3 Glucose released by liver cells raises blood glucose level to normal
4 If blood glucose continues to rise, hyperglycemia inhibits release of glucagon
6 Insulin acts on various body cells to:
• accelerate facilitated diffusion of glucose into cells
• speed conversion of glucose into glycogen
• increase uptake of amino acids and increase protein synthesis
• speed synthesis of fatty acids
7 Blood glucose level falls
8 If blood glucose continues to fall, hypoglycemia inhibits release of insulin

**Low blood glucose stimulates secretion of glucagon; high blood glucose stimulates secretion of insulin.**

**TABLE 17.9**

Selected Pancreatic Islet Hormones

| HORMONE AND SOURCE | CONTROL OF SECRETION | PRINCIPAL ACTIONS |
|---|---|---|
| **Glucagon** from alpha cells of pancreatic islets 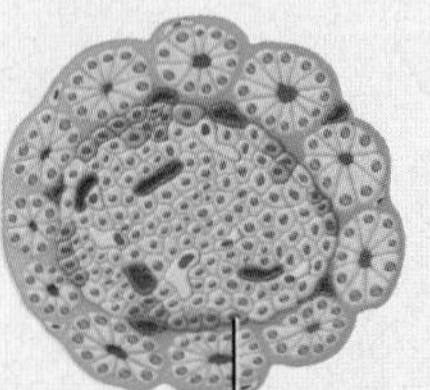  | Decreased blood levels of glucose, stimulation by sympathetic division of ANS during exercise, and mainly protein meals stimulate secretion; insulin inhibits secretion | Raises blood glucose levels by accelerating breakdown of glycogen into glucose in liver, converting other nutrients into glucose in liver, and releasing glucose into the blood |
| **Insulin** from beta cells of pancreatic islets 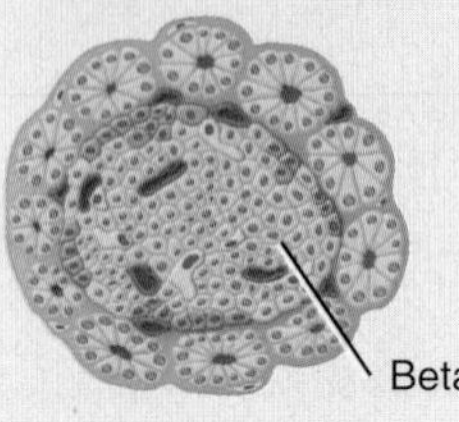  | Increased blood levels of glucose, stimulation by parasympathetic division of ANS following high-carbohydrate meals, hGH, and ACTH stimulate secretion | Lowers blood glucose levels by accelerating transport of glucose into cells, converting glucose into glycogen, and stimulating protein and fatty acid synthesis |

## ✓ CHECKPOINT

**33.** Why is the pancreas classified as both an exocrine gland and an endocrine gland?

**34.** Why are the digestive enzymes secreted by the pancreas not considered hormones?

**35.** You are reading through a medical journal and encounter an article entitled "Clinical Implications of Beta-Cell Tumor Hypersecretion." What is the article about?

**36.** You've just eaten a big meal and are feeling very relaxed. Which division of the autonomic nervous system interacts with the pancreas to regulate glucose level after the meal? The secretion of which pancreatic islet hormone is stimulated?

**37.** You're swimming laps in the school pool. Which division of the autonomic nervous system interacts with the pancreas to regulate glucose level while you swim? The secretion of which pancreatic islet hormone is stimulated?

## RETURN TO **Lisa's Story**

Lisa and her mother return to Dr. Badeer's office when her blood work results are ready.

"Lisa, your test results are consistent with congenital adrenal hyperplasia. Your adrenal glands are not making enough glucocorticoids or mineralocorticoids, which could explain the fatigue and dizziness. Without enough aldosterone, one of the mineralocorticoids, you can easily become dehydrated and that could explain your constant thirst."

Mrs. Vasquez asks if that explains her daughter's excess body hair. "Yes, I believe so. One of the tests shows that Lisa is missing an enzyme, 21-hydroxylase. This enzyme is necessary to create cortisol and aldosterone. Without these hormones, her hypothalamus and anterior pituitary gland—which normally regulate the blood levels of these hormones—would send signals to stimulate growth within the adrenal gland in order to make more of the necessary hormone. However, it would also stimulate an increase in the production of a group of hormones called androgens. These would produce a condition called hirsutism, which can present as excess body hair. The hormone imbalance could also interfere with the onset of her menstrual periods."

Lisa asks what can be done to fix her problem. "I want to start you on cortisol

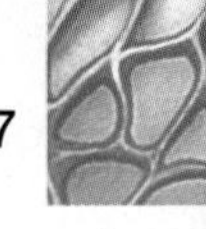



and aldosterone treatments. That should help with the fatigue and water issues. As these hormones begin to send correct messages to the hypothalamus and anterior pituitary, that should help balance your sex hormone levels and stabilize the menstrual irregularities."

Lisa begins hormone supplemental treatments and shows a steady reduction in her thirst and fatigue symptoms over the next few months. Within a year, her menstrual periods develop a more regular pattern.

C. *How would cortisol treatments reduce the production of androgens?*
D. *If Dr. Badeer wanted to perform an ultrasound to determine the relative size of the adrenal glands, where in Lisa's body would he position the instrument?*
E. *If, instead of reduced adrenal hormones, Lisa had a tumor that produced too much of the adrenal hormones, what would have been the signs and symptoms?*
F. *What other endocrine glands have hormones that could cause Lisa's fatigue, excessive thirst, stomach pain, and hirsutism?*

## 17.9 The ovaries produce estrogens, progesterone, and inhibin; the testes produce testosterone and inhibin.

**Gonads** are the organs that produce gametes—oocytes in females and sperm in males. In addition to their reproductive function, the gonads secrete hormones.

The female gonads, the **ovaries**, paired organs located in the female pelvic cavity, produce the female sex hormones **estrogens** (there are at least six known estrogens) and **progesterone**. These female sex hormones, along with follicle-stimulating hormone (FSH) and luteinizing hormone from the anterior pituitary, regulate the female reproductive cycle, maintain pregnancy, and prepare the mammary glands for lactation. They also promote enlargement of the breasts and widening of the hips at puberty, and help maintain these female secondary sex characteristics. The ovaries also produce **inhibin**, a hormone that inhibits secretion of FSH. During pregnancy, the ovaries and placenta produce a hormone called **relaxin**, which increases the flexibility of the pubic symphysis during pregnancy and helps dilate the uterine cervix during labor and delivery. These actions help ease the baby's passage by enlarging the birth canal.

The male gonads, the **testes**, are oval glands that lie in the scrotum. The testes produce and secrete **testosterone**, the primary **androgen** or male sex hormone. Testosterone stimulates descent of the testes before birth, regulates production of sperm, and stimulates development and maintenance of masculine secondary sex characteristics such as beard growth and deepening of the voice. The testes also produce inhibin, which inhibits FSH secretion.

A detailed discussion of the ovaries, testes, and the hormones they produce will be presented in Chapter 25.

Table 17.10 summarizes the hormones produced by the ovaries and testes and their principal actions.

**TABLE 17.10**

Hormones of the Ovaries and Testes

| HORMONE | PRINCIPAL ACTIONS |
|---|---|
| **OVARIAN HORMONES** | |
| **Estrogens and progesterone** (Ovaries) | Together with gonadotropic hormones of anterior pituitary, regulate female reproductive cycle, maintain pregnancy, prepare mammary glands for lactation, and promote development and maintenance of female secondary sex characteristics |
| **Relaxin** | Increases flexibility of pubic symphysis during pregnancy; helps dilate uterine cervix during labor and delivery |
| **Inhibin** | Inhibits secretion of FSH from anterior pituitary |
| **TESTICULAR HORMONES** | |
| **Testosterone** (Testes) | Stimulates descent of testes before birth; regulates sperm production; promotes development and maintenance of male secondary sex characteristics |
| **Inhibin** | Inhibits secretion of FSH from anterior pituitary |

✓ **CHECKPOINT**

**38.** What two endocrine glands produce androgens?

**39.** Where do androgens in females come from?

**40.** What is the target tissue of inhibin?

# 17.10 The pineal gland, thymus, and other organs also secrete hormones.

## The Pineal Gland

The **pineal gland** (PĪ N-ē-al = pinecone shape) is a small endocrine gland attached to the roof of the third ventricle of the brain at the midline (see Figure 17.1). The gland consists of masses of secretory cells and supportive neuroglia.

The pineal gland secretes **melatonin**; this hormone appears to contribute to the setting of the body's biological clock, which is controlled by the hypothalamus. As more melatonin is liberated during darkness than in light, this hormone is thought to promote sleepiness. In response to visual input from the eyes (retina), the hypothalamus stimulates sympathetic neurons, which in turn stimulate the pineal gland to secrete melatonin in a rhythmic pattern, with low levels of melatonin secreted during the day and significantly higher levels secreted at night. During sleep, the level of melatonin in the bloodstream increases tenfold and then declines to a low level again before awakening. Small doses of melatonin given orally can induce sleep and reset daily rhythms, which might benefit workers whose shifts alternate between daylight and nighttime hours.

In animals that breed during specific seasons, melatonin inhibits reproductive functions outside the breeding season, but it is unclear whether melatonin influences human reproductive function. The level of melatonin is higher in children and declines with age into adulthood, but there is no evidence that changes in melatonin secretion correlate with the onset of puberty and sexual maturation. Nevertheless, because melatonin causes atrophy of the gonads in several animal species, the possibility of adverse effects on human reproduction must be studied before its use to reset daily rhythms can be recommended.

### CLINICAL CONNECTION | *Seasonal Affective Disorder and Jet Lag*

**Seasonal affective disorder (SAD)** is a type of depression that afflicts some people during the winter months, when day length is short. It is thought to be due, in part, to overproduction of melatonin. Full-spectrum bright-light therapy—repeated doses of several hours of exposure to artificial light as bright as sunlight—provides relief for some people. Three to six hours of exposure to bright light also appears to speed recovery from jet lag, the fatigue suffered by travelers who quickly cross several time zones.

## The Thymus

The **thymus** is located behind the sternum between the lungs. Because of its role in immunity, the details of the structure and functions of the thymus are discussed in Chapter 21. The hormones produced by the thymus—**thymosin**, **thymic humoral factor**, **thymic factor**, and **thymopoietin** (thī-mō-poy-Ē-tin)—promote the maturation of T cells (a type of white blood cell that destroys microbes and foreign substances) and may retard the aging process.

## Hormones from Other Endocrine Tissues and Organs

As you learned at the beginning of this chapter, cells in organs other than those usually classified as endocrine glands have an endocrine function and secrete hormones. You learned about several of these in this chapter: the hypothalamus, pancreas, ovaries, testes, and thymus. Table 17.11 provides an overview of these organs and tissues and their hormones and actions.

## Eicosanoids

Two families of eicosanoid molecules—the **prostaglandins** (pros′-ta-GLAN-dins) and the **leukotrienes** (loo-kō-TRĪ-ēns)—are found in virtually all body cells except red blood cells, where they respond to chemical or mechanical stimuli. They are synthesized by clipping a 20-carbon fatty acid, called **arachidonic acid** (a-rak-i-DON-ik), from membrane phospholipid molecules. From arachidonic acid, different enzymatic reactions produce the various prostaglandins or leukotrienes.

To exert their effects, eicosanoids bind to receptors on target-cell plasma membranes and stimulate or inhibit the synthesis of second messengers such as cyclic AMP. The prostaglandins alter smooth muscle contraction, glandular secretions, blood flow, reproductive processes, platelet function, respiration, impulse transmission, lipid metabolism, and immune responses. They also have roles in promoting inflammation and fever, and in intensifying pain. Leukotrienes stimulate chemotaxis (attraction to a chemical stimulus) of white blood cells and mediate inflammation.

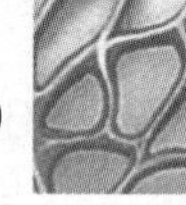



**TABLE 17.11**

Hormones Produced by Other Organs and Tissues That Contain Endocrine Cells

| HORMONE | PRINCIPAL ACTIONS |
|---|---|
| **GASTROINTESTINAL TRACT** | |
| **Gastrin** | Promotes secretion of gastric juice; increases movements of the stomach |
| **Glucose-dependent insulinotropic peptide (GIP)** | Stimulates release of insulin by pancreatic beta cells |
| **Secretin** | Stimulates secretion of pancreatic juice and bile |
| **Cholecystokinin (CCK)** | Stimulates secretion of pancreatic juice; regulates release of bile from gallbladder; causes feeling of fullness after eating |
| **PLACENTA** | |
| **Human chorionic gonadotropin (hCG)** | Stimulates corpus luteum in ovary to continue production of estrogens and progesterone to maintain pregnancy |
| **Estrogens and progesterone** | Maintain pregnancy; help prepare mammary glands to secrete milk |
| **Human chorionic somatomammotropin (hCS)** | Stimulates development of mammary glands for lactation |
| **KIDNEYS** | |
| **Renin** | Part of reaction sequence that raises blood pressure by bringing about vasoconstriction and secretion of aldosterone |
| **Erythropoietin (EPO)** | Increases rate of red blood cell formation |
| **Calcitriol* (active form of vitamin D)** | Aids in absorption of dietary calcium and phosphorus |
| **HEART** | |
| **Atrial natriuretic peptide (ANP)** | Decreases blood pressure |
| **ADIPOSE TISSUE** | |
| **Leptin** | Suppresses appetite; may increase FSH and LH activity |

*Synthesis begins in the skin, continues in the liver, and ends in the kidneys.

### ✓ CHECKPOINT

**41.** Why might world travelers flying across several time zones be tempted to try small doses of melatonin supplements?

**42.** How is the autonomic nervous system related to the pineal gland?

**43.** How do thymic hormones play a role in immunity?

## Lisa's Story — EPILOGUE AND DISCUSSION

Lisa is a 15-year-old girl who presents to her family's physician with abdominal discomfort, polydipsia (frequent thirst), and oligomenorrhea (reduced menstruation). A referral to an endocrinologist uncovers the presence of hirsutism (excess body hair) and blood tests discover a lack of the enzyme 21-hydroxylase. These facts suggest that Lisa suffers from a form of congenital adrenal hyperplasia, which shows decreased production of glucocorticoids and mineralocorticoids but an increase in androgens. Hormone treatments of the missing hormones eventually relieve her symptoms.

G. *Lack of cortisol stimulates the hypothalamus to increase its release of CRH. What effects would result from this CRH increase?*

H. *If Lisa had not received hormone treatments before the end of puberty, what could have been the long-term effects of her condition?*

I. *Why did Dr. Badeer not suspect the problem to be associated with the ovaries and sex hormones?*

# Concept and Resource Summary

| Concept | Resources |
|---|---|
| **Introduction**<br>1. The endocrine system releases hormones to control body activities and help maintain homeostasis. | |
| **Concept 17.1 The nervous and endocrine systems function together to regulate body activities.**<br>1. A **hormone** is a molecule that is released in one part of the body but regulates the activity of cells in other parts of the body.<br>2. In contrast to the rapid communication and control associated with the nervous system, the endocrine system responses are slower and more sustained.<br>3. **Exocrine gland** secretions enter ducts that carry the secretions to body surfaces or into cavities. **Endocrine glands** secrete hormones into interstitial fluid to diffuse into the blood where they circulate to target tissues.<br>4. The **endocrine system** comprises all of the endocrine glands and hormone-secreting cells. | Anatomy Overview—The Endocrine System |
| **Concept 17.2 The secretion of hormones is regulated by the nervous system, chemical changes in the blood, and other hormones.**<br>1. Hormones affect only **target cells** that have specific **receptors** to bind to a given hormone. The number of hormone receptors may decrease (**down-regulation**) or increase (**up-regulation**).<br>2. There are two classes of hormones: lipid-soluble and water-soluble.<br>3. Water-soluble hormones circulate in blood plasma unattached to plasma proteins; most lipid-soluble hormones circulate attached to transport proteins. **Transport proteins** increase blood solubility of lipid-soluble hormones, prevent loss of small hormone molecules to urine, and provide a ready reserve of hormones in the blood.<br>4. Hormones can stimulate synthesis of molecules, alterations in membrane permeability, stimulation of membrane transport, alterations in metabolic rate, and contraction of smooth and cardiac muscle.<br>5. Three factors influence the responsiveness of a target cell to a hormone: the hormone's concentration, abundance of target cell receptors, and influences exerted by other hormones.<br>6. Hormone secretion is regulated by nervous system signals, chemical changes in the blood, and other hormones. | Anatomy Overview—Lipid-soluble Hormones<br>Anatomy Overview—Water-soluble Hormones<br>Anatomy Overview—Local Hormones<br>Animation—Introduction to Hormonal Regulation, Secretion, and Concentration<br>Animation—Mechanisms of Hormone Action<br>Animation—Introduction to Hormone Feedback Loops<br>Clinical Connection—Administering Hormones |
| **Concept 17.3 The hypothalamus regulates anterior pituitary secretion of seven important hormones.**<br>1. The **hypothalamus** is the major integrating link between the nervous and endocrine systems. Although the **pituitary gland** secretes several hormones that control other endocrine glands, it is also controlled by the hypothalamus.<br>2. The pituitary gland sits in the sella turcica of the sphenoid bone and is attached by a stalk, the **infundibulum**, to the hypothalamus. The pituitary gland is divided into the anterior pituitary and the posterior pituitary.<br>3. Secretion of **anterior pituitary** hormones is stimulated by **releasing hormones** and suppressed by **inhibiting hormones** from the hypothalamus. These hypothalamic hormones reach the anterior pituitary through the **hypophyseal portal system** that links the hypothalamus and the anterior pituitary.<br>4. Several anterior pituitary hormones are **tropic hormones**, hormones that influence the secretion of other endocrine glands.<br>5. Anterior pituitary hormone secretion is regulated by two factors: releasing and inhibiting hormones of the hypothalamus, and negative feedback from rising blood levels of target gland hormones.<br>6. **Adrenocorticotropic hormone (ACTH)** controls the secretion of glucocorticoids such as cortisol from the adrenal cortex. **Corticotropin-releasing hormone (CRH)** stimulates secretion of ACTH. Glucocorticoids cause inhibition of CRH and ACTH release. | Anatomy Overview—The Hypothalamus and Pituitary Gland<br>Anatomy Overview—Hormones of the Anterior Pituitary Gland<br>Anatomy Overview—Anterior Pituitary Reproductive Hormones<br>Animation—GHRH/hGH<br>Animation—ACTH/Cortisol: Glycogenolysis<br>Animation—TRH/TSH: Production<br>Animation—hGH: Growth and Development |

| Concept | Resources |
|---|---|

7. **Human growth hormone (hGH)** promotes synthesis and secretion of **insulinlike growth factors**. IGFs stimulate body growth and repair. **Growth hormone–releasing hormone (GHRH)** and **growth hormone–inhibiting hormone (GHIH)** control hGH secretion. Blood glucose level is a major regulator of GHRH and GHIH secretion.
8. **Thyroid-stimulating hormone** stimulates the thyroid gland to secrete triiodothyronine and thyroxine. Its secretion is stimulated by **thyrotropin-releasing hormone**.
9. **Follicle-stimulating hormone (FSH)** targets the ovaries for monthly development of several ovarian follicles that surround a developing oocyte. FSH also stimulates the testes to produce sperm. Its secretion is stimulated by gonadotropin-releasing hormone.
10. **Luteinizing hormone (LH)** triggers **ovulation** and progesterone secretion by the ovary. With FSH, LH stimulates secretion of estrogens by the ovaries. In males, LH stimulates secretion of testosterone by the testes. Its secretion is stimulated by gonadotropin-releasing hormone.
11. **Prolactin** initiates milk production by mammary glands. Secretion of PRL is stimulated by **prolactin-releasing hormone** and inhibited by **prolactin-inhibiting hormone**.
12. Although there are **melanocyte-stimulating hormone** receptors in the brain, its function in humans is unknown. CRH stimulates its release and PIH inhibits its secretion.

Animation—hGH: Glycogenolysis and Lipolysis

Figure 17.4—Hypothalamus and Pituitary Gland, and their Blood Supply

Clinical Connection—Pituitary Dwarfism, Giantism, and Acromegaly

Exercise—Hormone Actions

## Concept 17.4 Oxytocin and antidiuretic hormone originate in the hypothalalmus and are stored in the posterior pituitary.

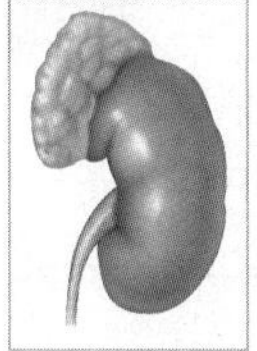

1. The **posterior pituitary** stores and releases two hypothalamic hormones, oxytocin and antidiuretic hormone, which are produced by hypothalamic neurosecretory cells. Impulses trigger their release.
2. **Oxytocin (OT)** enhances smooth muscle contractions in the uterine wall to facilitate labor and delivery, and stimulates milk ejection from mammary glands after delivery. Release of OT is stimulated by uterine stretching and suckling during nursing.
3. **Antidiuretic hormone (ADH)** causes the kidneys to return more water to the blood while decreasing urine volume, decreases water loss through sweating, and causes constriction of arterioles to increase blood pressure. High blood osmotic pressure stimulates secretion of ADH; low blood osmotic pressure inhibits ADH secretion.

Anatomy Overview—The Hypothalamus
Anatomy Overview—Hormones of the Hypothalamus
Anatomy Overview—Hypothalamic Reproductive Hormones
Animation—Antidiuretic Hormone

Clinical Connection—Diabetes Insipidus

## Concept 17.5 The thyroid gland secretes thyroxine, triiodothyronine, and calcitonin.

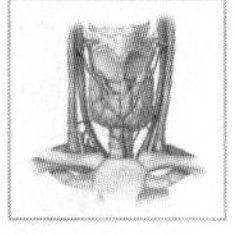

1. The **thyroid gland** is located inferior to the larynx and anterior to the trachea. Internally, the gland consists of **thyroid follicles** with **follicular cells** that produce two **thyroid hormones**: **triiodothyronine ($T_3$)** and **thyroxine ($T_4$)**, and **parafollicular cells** that produce **calcitonin**.
2. The thyroid gland is the only endocrine gland that stores its secretory products in large supply.
3. Thyroid hormones are synthesized from **iodine** and **tyrosine** within **thyroglobulin**. They are transported in the blood bound mostly to **thyroxine-binding globulin**.
4. Thyroid hormones regulate oxygen use and metabolic rate, cellular metabolism, and growth and development.
5. Secretion is controlled by thyrotropin-releasing hormone from the hypothalamus and thyroid-stimulating hormone from the anterior pituitary.
6. **Calcitonin (CT)** lowers the blood levels of calcium and phosphates, and promotes their uptake into bone extracellular matrix. Secretion of CT is controlled by the calcium level in the blood.

Anatomy Overview—The Thyroid
Anatomy Overview—Thyroid Hormones
Animation—TRH/TSH
Animation—Calcitonin
Animation—Thyroid Hormones and Glucose and Lipid Catabolism

Figure 17.10—Steps in the Synthesis and Secretion of Thyroid Hormones

Exercise—Calcium Homeostasis

Clinical Connection—Thyroid Gland Disorders

## Concept 17.6 The parathyroid glands secrete parathyroid hormone, which regulates calcium, magnesium, and phosphate ion levels.

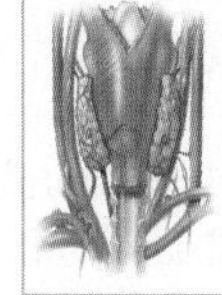

1. The **parathyroid glands** are embedded in the posterior surface of the thyroid gland. The **chief cells** produce **parathyroid hormone (PTH)**.
2. PTH regulates blood levels of calcium ($Ca^{2+}$), magnesium ($Mg^{2+}$), and phosphate ($HPO_4^{2-}$) ions.
3. When blood calcium level is low, PTH secretion is stimulated, causing increased $Ca^{2+}$ and $HPO_4^{2-}$ release into the blood from bones.

Anatomy Overview—Parathyroid Glands
Anatomy Overview—Hormones of the Parathyroid
Animation—Parathyroid Hormone

| Concept | Resources |
|---|---|
| 4. PTH also decreases the loss of $Ca^{2+}$ and $Mg^{2+}$ from the blood to urine and increases the loss of $HPO_4^{2-}$ from blood to urine.<br>5. PTH stimulates synthesis of **calcitriol** by the kidneys. Calcitriol increases absorption of $Ca^{2+}$, $HPO_4^{2-}$, and $Mg^{2+}$ from ingested foods into the blood. | Exercise—Calcium Homeostasis<br>Clinical Connection—Parathyroid Gland Disorders |

## Concept 17.7 The adrenal glands are structurally and functionally two independent endocrine glands.

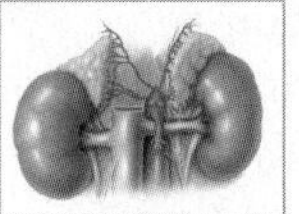

1. The paired **adrenal glands** lie superior to each kidney. Each adrenal gland has an outer **adrenal cortex** and inner **adrenal medulla**.
2. The adrenal cortex secretes **mineralocorticoids**, **glucocorticoids**, and weak **androgens**.
3. The major mineralocorticoid, **aldosterone**, increases kidney reabsorption of $Na^+$ and water from urine to the blood, increases excretion of $K^+$ into the urine, and helps adjust blood pressure and blood volume. Secretion of aldosterone is controlled by the **renin–angiotensin–aldosterone pathway**.
4. The glucocorticoids (mainly **cortisol**) regulate metabolism and resistance to stress. Secretion of glucocorticoids is initiated by **corticotropin-releasing hormone** from the hypothalamus, which stimulates release of ACTH from the anterior pituitary. Glucocorticoids increase protein breakdown, enhance glucose formation, break down triglycerides, inhibit inflammation, and depress immune system responses.
5. Androgens stimulate growth of axillary and pubic hair, aid prepubertal growth spurts, and contribute to libido in females.
6. The adrenal medulla secretes **epinephrine** and **norepinephrine**. These hormones augment the sympathetic nervous system fight-or-flight response.

**Resources:**
Anatomy Overview—The Adrenal Glands
Anatomy Overview—Hormones of the Adrenal Glands
Animation—ACTH/Cortisol
Animation—Epinephrine/NE
Animation—Cortisol 
Figure 17.14—The Adrenal (Suprarenal) Glands
Figure 17.15—The Renin–Angiotensin–Aldosterone Pathway
Clinical Connection—Adrenal Gland Disorders

## Concept 17.8 The pancreatic islets regulate blood glucose levels by secreting glucagon and insulin.

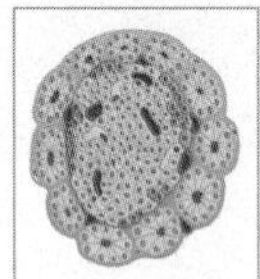

1. The **pancreas**, which is located in the abdomen near the duodenum of the small intestine, is both an endocrine and an exocrine gland. Exocrine cells produce digestive enzymes. The pancreatic endocrine tissue is referred to as **pancreatic islets** and is composed of **alpha cells** that secrete **glucagon** and **beta cells** that secrete **insulin**.
2. Low blood glucose level causes secretion of glucagon, which acts on liver cells to release glucose into the blood. High blood glucose level causes secretion of insulin, which acts on cells to increase diffusion of glucose into cells. Insulin also increases protein and fatty acid synthesis.
3. In addition to blood glucose level, the autonomic nervous system, diet, and other hormones can also stimulate glucagon and insulin release.

**Resources:**
Anatomy Overview—The Pancreas
Anatomy Overview—Hormones of the Pancreas
Animation—Glucagon
Animation—Insulin 
Figure 17.17—The Pancreas
Figure 17.18—Regulation of Blood Glucose Level by Negative Feedback Systems Involving Glucagon and Insulin
Exercise—Glucose Regulation Feedback Loop
Concepts and Connections—Blood Glucose Regulation
Clinical Connection—Pancreatic Islet Disorders

## Concept 17.9 The ovaries produce estrogens, progesterone, and inhibin; the testes produce testosterone and inhibin.

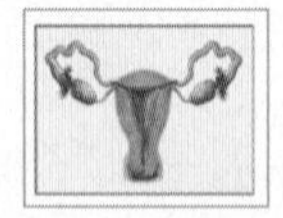

1. The paired **ovaries** are located in the pelvic cavity and produce **estrogens** and **progesterone**. Along with FSH and LH, these hormones regulate the female reproductive cycle, maintain pregnancy, prepare the mammary glands for lactation, and promote development and maintenance of female secondary sex characteristics.
2. The ovaries also produce **inhibin**, a hormone that inhibits FSH secretion.

**Resources:**
Anatomy Overview—The Ovaries
Anatomy Overview—Ovarian Hormones
Anatomy Overview—The Testes
Anatomy Overview—Testicular Hormones

## Concept

3. During pregnancy, the ovaries and placenta produce the hormone **relaxin**, which increases the flexibility of the pubic symphysis and helps dilate the cervix.
4. The paired **testes** are located in the scrotum and produce **testosterone**, the primary male sex hormone or **androgen**, which regulates sperm production and stimulates development and maintenance of masculine secondary sex characteristics.
5. The testes also produce **inhibin**, which inhibits FSH secretion.

## Resources 

Animation—Hormonal Control of Male Reproductive Function
Animation—Hormonal Regulation of Female Reproductive System
Exercise—Match the Male Hormones
Exercise—Match the Female Hormones

### Concept 17.10 The pineal gland, thymus, and other organs also secrete hormones.

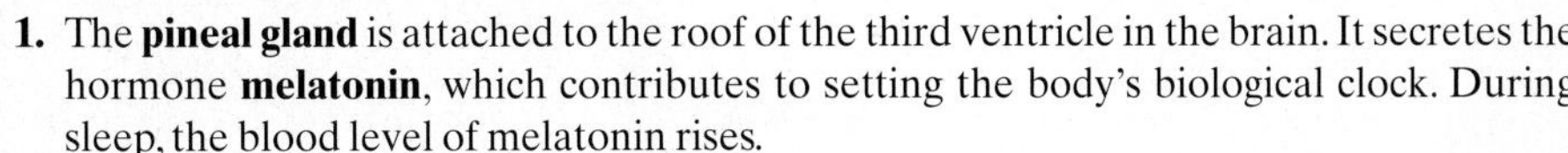

1. The **pineal gland** is attached to the roof of the third ventricle in the brain. It secretes the hormone **melatonin**, which contributes to setting the body's biological clock. During sleep, the blood level of melatonin rises.
2. The **thymus** secretes several hormones related to immunity by promoting the maturation of T cells.
3. Body tissues other than those normally classified as endocrine glands contain endocrine tissue and secrete hormones, including the gastrointestinal tract, placenta, kidneys, skin, and heart.
4. **Prostaglandins** and **leukotrienes** are eicosanoids that act in most body tissues to stimulate or inhibit the synthesis of second messengers such as cyclic AMP.

Anatomy Overview—The Endocrine System
Anatomy Overview—The Pineal Gland
Animation—The General Adaptation Syndrome
Exercise—Concentrate on the Endocrine System
Exercise—Produce That Hormone
Exercise—Hormone Actions
Concepts and Connections—Endocrine System Negative Feedback Loop

Clinical Connection—Nonsteroidal Anti-inflammatory Drugs

## Understanding the Concepts

1. Why are organs such as the kidneys, stomach, heart, and skin considered to be part of the endocrine system?
2. A report on a blood hormone concentration states that the "free fraction" levels were 0.6 micrograms per deciliter of blood. Is this level an accurate reflection of the actual total concentration of the hormone in the blood? Why or why not?
3. How do lipid-soluble hormones influence target cells to produce new molecules? Which type of molecule is produced?
4. What are the differences among permissive effects, synergistic effects, and antagonistic effects of hormones?
5. How do hypothalamic releasing and inhibiting hormones influence secretions of the anterior pituitary?
6. Sometimes during difficult deliveries or following long labor a synthetic form of oxytocin is given to the mother. Why?
7. Iodized salt is common throughout the United States. Why might adding iodine to salt be a good idea?
8. A fellow student with hyperthyroidism complains of feeling hot all the time; explain why.
9. If you received a thyroidectomy, should you be concerned about your skeletal system? Why or why not?
10. What is the benefit of adding vitamin D to various food products such as milk?
11. Would a drug that blocks the action of angiotensin-converting enzyme be used to raise or lower blood pressure? Where does this enzyme function?
12. Why are high-dose glucocorticoids sometimes used following tissue transplant surgery?
13. If a female has an adrenal tumor that causes masculinization, is the tumor located in the adrenal cortex or adrenal medulla? Which hormone class is being produced in excess?
14. You have fallen asleep in class and wake up in time to hear the instructor say that the adrenal gland is composed of nervous tissue. You initially are confused because you thought the lecture was about the endocrine system, but then you realize that the instructor is partly right. Why?
15. How is the release of glucagon and insulin be regulated if there were no hypothalamic releasing or inhibiting factors present in the body?
16. Which two hormones are involved in labor and delivery? (*Hint:* One is a posterior pituitary hormone, the other is an ovarian hormone.)

CHAPTER

# 18

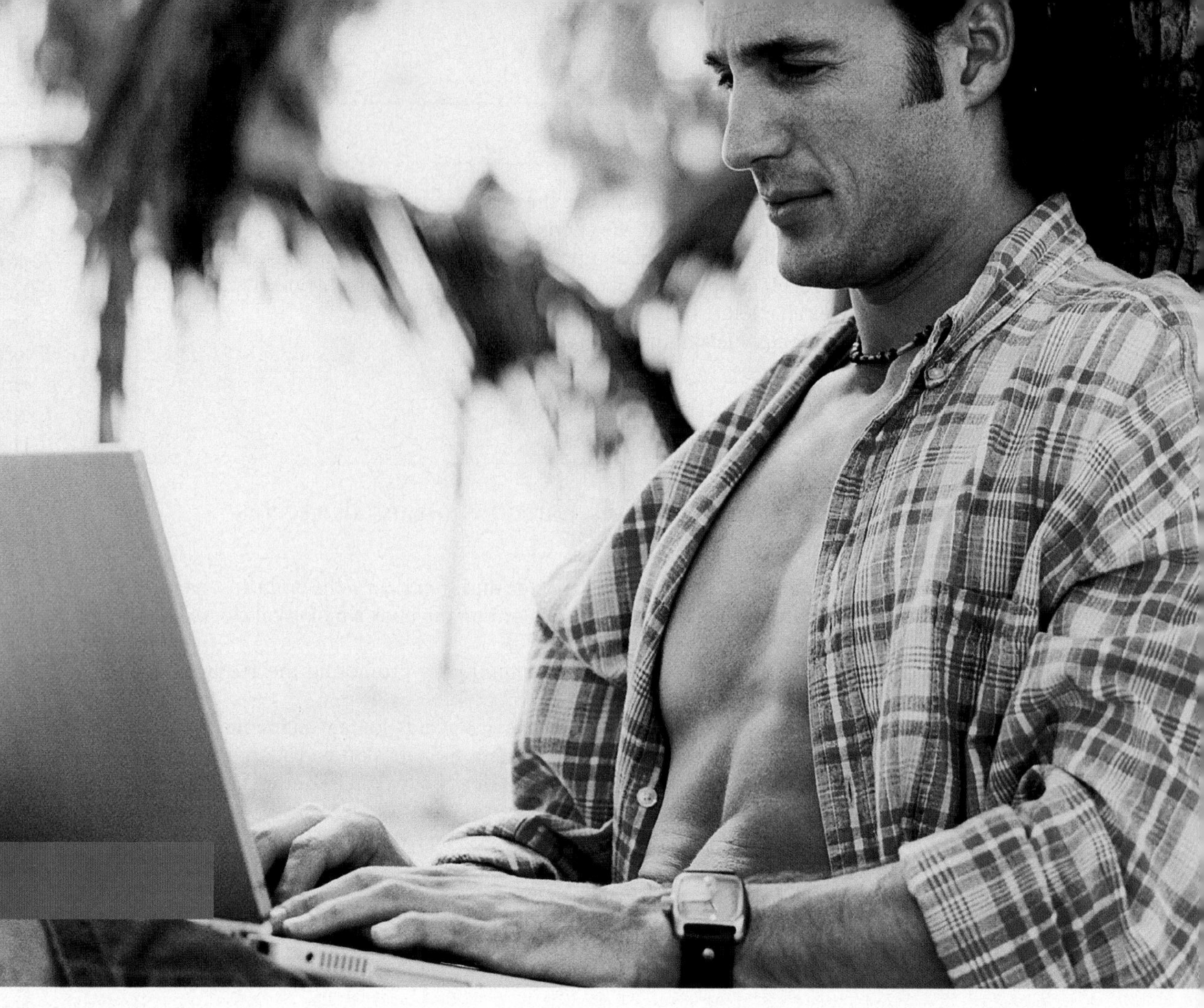

## Ed's Story

Ed leaned back in his chair, reaching for his margarita. A warm Gulf breeze, the first of the early evening, moved through his hair. He had been coming to Belize for over 10 years now and his work with the government park service had been very successful. Documenting the local flora and fauna had been a daunting but satisfying task for the ecologist, who at 46 had finally settled into his mid-career stride. As he sat enjoying the sunset, he swatted at the occasional insect. Tomorrow he and his team of students would be heading back to the States. It had been a particularly hot, wet season, and the mosquitoes had been intense in the national park where they had been working. He felt wiped out, which was not unusual for having finished a month of intensive fieldwork with students. As he sat, his eyelids felt heavy and his eyes began to close. The high-pitched whine of the mosquito's wings didn't disturb Ed as he dozed. The *Anopheles* mosquito lit on his arm and began its work. It probed Ed's skin, finding just the right spot with its sharp-ended proboscis, then pierced through the epithelial barrier. The mosquito began pumping salivary enzymes into the dermis to prevent the blood from coagulating. Her meal of blood would require only seconds to complete, yet it was still enough time for the protozoan hitchhikers living in her salivary glands to be injected into Ed's arm. The malarial sporozoites of *Plasmodium falciparum* squirmed inside the dermis, penetrating the superficial capillaries of Ed's skin to begin their journey to his liver.

Malaria is a very serious and sometimes fatal parasitic disease affecting the blood. Although it is rare in the United States and Europe, malaria is a leading cause of death worldwide and is endemic to (constantly present in) certain regions of the world. The vast majority of malaria cases in the United States (about 1200 each year) occur in individuals who have traveled to high-risk areas and in immigrants from such regions. Ed's experience will illustrate several concepts related to the function of blood and demonstrate that blood is a complex tissue that is intimately involved in maintaining the body's homeostasis.

# The Cardiovascular System: The Blood

## INTRODUCTION

The **cardiovascular system** (*cardio-* = heart; *-vascular* = blood or blood vessels) consists of three interrelated components: blood, the heart, and blood vessels. The focus of this chapter is blood; the next two chapters will explore the heart and blood vessels, respectively. Blood transports various substances, helps regulate several life processes, and affords protection against disease. For all of its similarities in origin, composition, and functions, blood is as unique from one person to another as are skin, bone, and hair. Despite these differences, blood is the most easily and widely shared of human tissues, saving many thousands of lives every year through blood transfusions. Health-care professionals routinely examine and analyze the blood through various tests when trying to determine the cause of different diseases.

Most cells of the body cannot move around to obtain oxygen and nutrients and get rid of carbon dioxide and other wastes. Instead, these needs are met by two fluids: blood and interstitial fluid. **Blood** is a connective tissue consisting of plasma in which various cells and cell fragments are suspended. While plasma is the extracellular fluid within blood, **interstitial fluid** is the extracellular fluid that directly bathes body cells and is constantly renewed by the blood. Blood transports oxygen from the lungs and nutrients from the gastrointestinal tract to cells throughout the body. The oxygen and nutrients diffuse from the blood into the interstitial fluid and then into body cells. Carbon dioxide and other wastes move in the reverse direction, from the body cells into the interstitial fluid, and then into the blood. Blood then transports the wastes to various organs—the lungs, kidneys, skin, and digestive system—for elimination from the body.

In order for blood to reach all cells, it must be moved throughout the body. The *heart* is the pump that circulates the blood (see Chapter 19). *Blood vessels* convey blood from the heart to body cells (via arteries) and from body cells back to the heart (via veins) (see Chapter 20).

## CONCEPTS

**18.1** Blood contains plasma and formed elements and transports essential substances through the body.
**18.2** Hemopoiesis is the production of formed elements.
**18.3** Mature red blood cells are biconcave cells containing hemoglobin.
**18.4** Red blood cells have a life cycle of 120 days.
**18.5** Erythropoiesis is the process of red blood cell formation.
**18.6** Blood is categorized into groups based on surface antigens.
**18.7** White blood cells combat inflammation and infection.
**18.8** Platelets reduce blood loss from damaged vessels.
**18.9** Hemostasis is the sequence of events that stops bleeding from a damaged blood vessel.

# 18.1 Blood contains plasma and formed elements and transports essential substances through the body.

## Functions of Blood

**Blood**, a liquid connective tissue, has three general functions:

- ***Transportation***. Blood transports oxygen from the lungs to cells throughout the body and carbon dioxide from body cells to the lungs for exhalation. It also carries nutrients from the gastrointestinal tract to body cells, hormones from endocrine glands to other body cells, and heat and waste products away from cells to various organs for elimination from the body.
- ***Regulation***. Circulating blood helps maintain homeostasis of all body fluids. Blood helps regulate pH through the use of buffers (chemicals that convert strong acids or bases into weak ones). It also helps adjust body temperature through the heat-absorbing and coolant properties of the water in blood and blood's variable rate of flow through the skin, where excess heat can be lost from the blood to the environment. Also, blood osmotic pressure influences the water content of cells, mainly through interactions of dissolved ions and proteins.
- ***Protection***. Blood can clot (become gel-like) in response to an injury, which protects against its excessive loss from the cardiovascular system. Blood's white blood cells protect against disease by carrying on phagocytosis. Several types of blood proteins, including antibodies, interferons, and complement, also help protect against disease.

## Physical Characteristics of Blood

Blood is denser and more viscous (thicker) than water, which is part of the reason it flows more slowly than water. The temperature of blood is 38°C (100.4°F), about 1°C higher than oral or rectal body temperature, and it has a slightly alkaline pH ranging from 7.35 to 7.45. The color of blood varies with its oxygen content. When saturated with oxygen, it is bright red. When blood has a low oxygen content, it is dark red. Blood constitutes about 8 percent of the total body weight. The blood volume is 5 to 6 liters (1.5 gal) in an average-sized adult male and 4 to 5 liters (1.2 gal) in an average-sized adult female. The difference in volume is due to differences in body size. Several hormones, regulated by negative feedback, ensure that blood volume and osmotic pressure remain relatively constant.

## Components of Blood

Whole blood is composed of **plasma**, a liquid extracellular fluid that contains dissolved substances, and formed elements, which include cells and cell fragments. If a sample of blood is centrifuged (spun at high speed) in a small glass tube, the formed elements (which are more dense) sink to the bottom of the tube, and the plasma (which is less dense) forms a layer on top (Figure 18.1a). Blood is about 45 percent formed elements and about 55 percent plasma. Normally, more than 99 percent of the formed elements are cells named for their red color—red blood cells (RBCs). Pale, colorless white blood cells (WBCs) and platelets comprise less than 1 percent of the formed elements. Because they are less dense than red blood cells but denser than plasma, WBCs and platelets form a very thin layer, called the *buffy coat*, between the packed RBCs and plasma in centrifuged blood. Figure 18.1b shows the composition of plasma and the numbers of the various types of formed elements in blood.

### Plasma

When the formed elements are removed from blood, the straw-colored plasma remains. Plasma is about 91.5 percent water and 8.5 percent solutes, most of which are proteins. Some of the proteins in plasma are also found elsewhere in the body, but those confined to blood, called **plasma proteins**, are synthesized

### CLINICAL CONNECTION | *Withdrawing Blood*

**Blood samples** for laboratory testing may be obtained in several ways. The most common procedure is **venipuncture** (ven′-i-PUNK-chur), withdrawal of blood from a vein using a needle and collecting tube, which contains various additives. A tourniquet is wrapped around the arm above the venipuncture site, which causes blood to accumulate in the vein. This increased blood volume makes the vein stand out. Opening and closing the fist further causes it to stand out, making the venipuncture more successful. A common site for venipuncture is the median cubital vein anterior to the elbow (see Figure 20.24c). Another method of withdrawing blood is through a **finger** or **heel stick**. Diabetic patients who monitor their daily blood sugar typically perform a finger stick, and it is often used for drawing blood from infants and children. In an **arterial stick**, blood is withdrawn from an artery; this test is used to determine the level of oxygen in oxygenated blood.

FIGURE 18.1 **Components of blood in a normal adult.**

PLASMA (55%)
Buffy coat, composed of WHITE BLOOD CELLS and PLATELETS
RED BLOOD CELLS (45%)

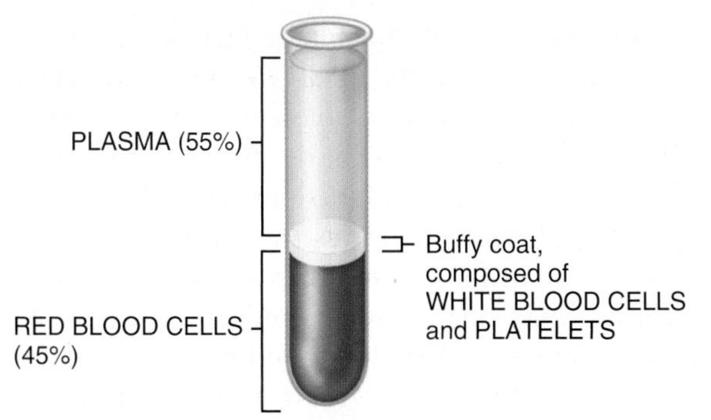

(a) Appearance of centrifuged blood

**FUNCTIONS OF BLOOD**

1. Transports oxygen, carbon dioxide, nutrients, hormones, heat, and wastes.
2. Regulates pH, body temperature, and water content of cells.
3. Protects against blood loss through clotting and against disease through phagocytic white blood cells and antibodies.

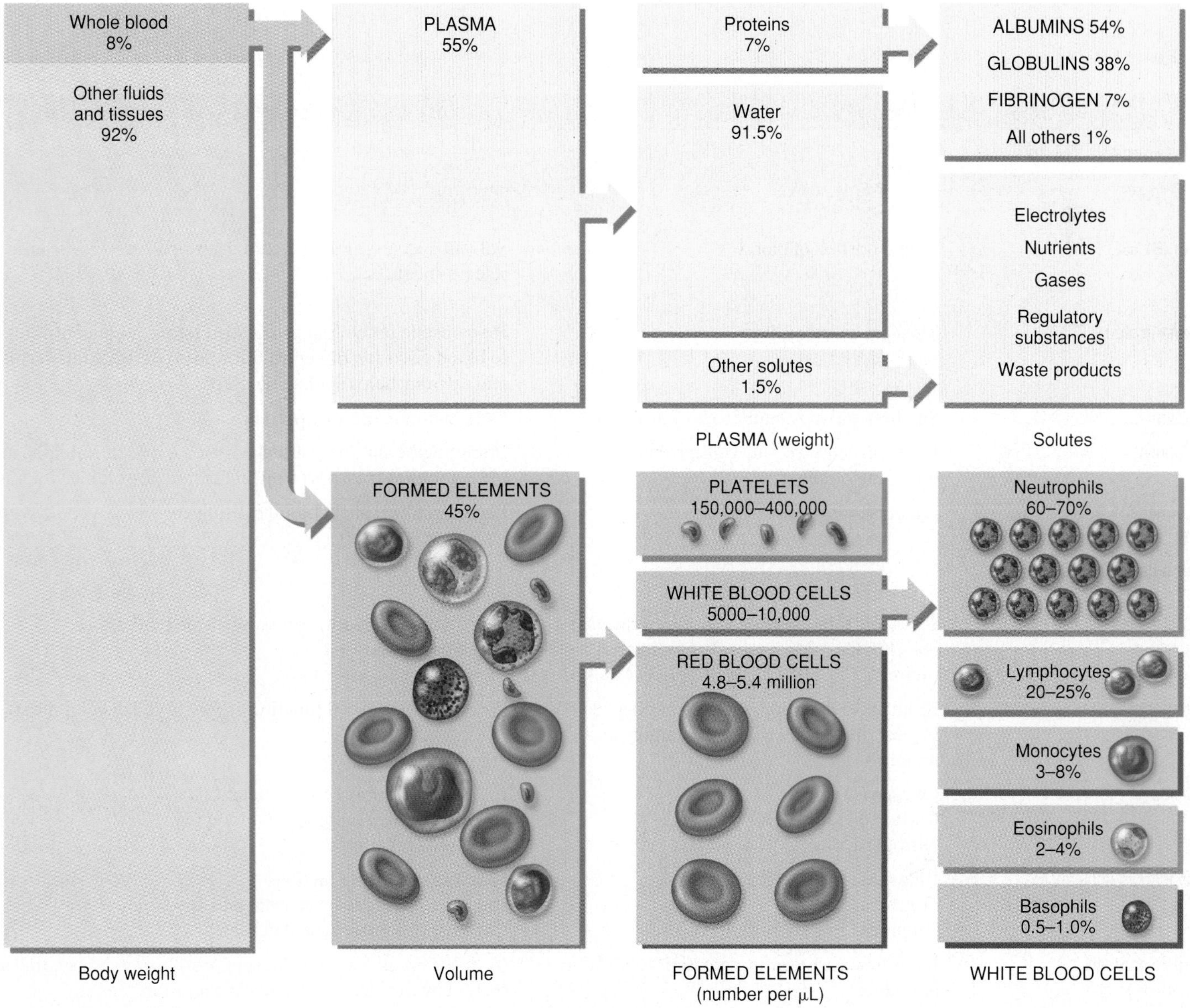

Whole blood 8%
Other fluids and tissues 92%
PLASMA 55%
Proteins 7%
Water 91.5%
Other solutes 1.5%
ALBUMINS 54%
GLOBULINS 38%
FIBRINOGEN 7%
All others 1%
Electrolytes
Nutrients
Gases
Regulatory substances
Waste products
PLASMA (weight)
Solutes
FORMED ELEMENTS 45%
PLATELETS 150,000–400,000
WHITE BLOOD CELLS 5000–10,000
RED BLOOD CELLS 4.8–5.4 million
Neutrophils 60–70%
Lymphocytes 20–25%
Monocytes 3–8%
Eosinophils 2–4%
Basophils 0.5–1.0%
Body weight
Volume
FORMED ELEMENTS (number per μL)
WHITE BLOOD CELLS

(b) Components of blood

**Blood is a connective tissue that consists of plasma (liquid) plus formed elements (red blood cells, white blood cells, and platelets).**

mainly by liver cells. The most plentiful plasma proteins are the **albumins** (al′-BŪ-mins), which account for about 54 percent of all plasma proteins. Among other functions, albumins help maintain proper blood osmotic pressure, which is an important factor in the exchange of fluids across capillary walls. **Globulins** (GLOB-ū-lins), which compose 38 percent of plasma proteins, include **antibodies**, or **immunoglobulins** (im′-ū-nō-GLOB-ū-lins), defensive proteins produced during certain immune responses. **Fibrinogen** (fī-BRIN-ō-jen) makes up about 7 percent of plasma proteins and is a key protein in the formation of blood clots. Other solutes in plasma include electrolytes, nutrients, gases, regulatory substances such as enzymes and hormones, and waste products.

Table 18.1 describes the chemical composition of plasma.

## Formed Elements

The **formed elements** of the blood include **red blood cells (RBCs)**, **white blood cells (WBCs)**, and **platelets** (Figure 18.2). RBCs and WBCs are whole cells, but platelets are cell fragments. RBCs and platelets have just a few roles, but WBCs have a number of specialized functions. Several distinct types of WBCs—neutrophils, lymphocytes, monocytes, eosinophils, and basophils—each with a unique microscopic appearance, carry out these functions, which are discussed later in Concept 18.7.

The percentage of total blood volume occupied by RBCs is called the **hematocrit** (hē-MAT-ō-krit). For example, a hematocrit of 40 percent indicates that 40 percent of the volume

**TABLE 18.1**

Substances in Plasma

| CONSTITUENT | DESCRIPTION | FUNCTION |
|---|---|---|
| **Water (91.5%)** | Liquid portion of blood | Solvent and suspending medium; absorbs, transports, and releases heat |
| **Plasma proteins (7%)** | Most produced by liver | Responsible for colloid osmotic pressure; major contributors to blood viscosity; transport hormones (steroid), fatty acids, and calcium; help regulate blood pH |
| Albumins | Smallest and most numerous of proteins | Help maintain osmotic pressure |
| Globulins | Large proteins (plasma cells produce immunoglobulins) | Immunoglobulins help attack viruses and bacteria; alpha and beta globulins transport iron, lipids, and fat-soluble vitamins |
| Fibrinogen | Large protein | Plays essential role in blood clotting |
| **Other solutes (1.5%)** | | |
| Electrolytes | Inorganic salts; positively charged (cations) $Na^+$, $K^+$, $Ca^{2+}$, $Mg^{2+}$; negatively charged (anions) $Cl^-$, $HPO_4^{2-}$, $SO_4^{2-}$, $HCO_3^-$ | Help maintain osmotic pressure and essential roles in cell functions |
| Nutrients | Products of digestion, such as amino acids, glucose, fatty acids, glycerol, vitamins, and minerals | Essential roles in cell functions, growth, and development |
| Gases | Oxygen ($O_2$)<br>Carbon dioxide ($CO_2$)<br>Nitrogen ($N_2$) | Important in many cellular functions<br>Involved in the regulation of blood pH<br>No known function |
| Regulatory substances | Enzymes<br>Hormones<br>Vitamins | Catalyze chemical reactions<br>Regulate metabolism, growth, and development<br>Cofactors for enzymatic reactions |
| Waste products | Urea, uric acid, creatine, creatinine, bilirubin, ammonia | Most are breakdown products of protein metabolism that are carried by the blood to organs of excretion |

FIGURE 18.2 **Formed elements of blood.**

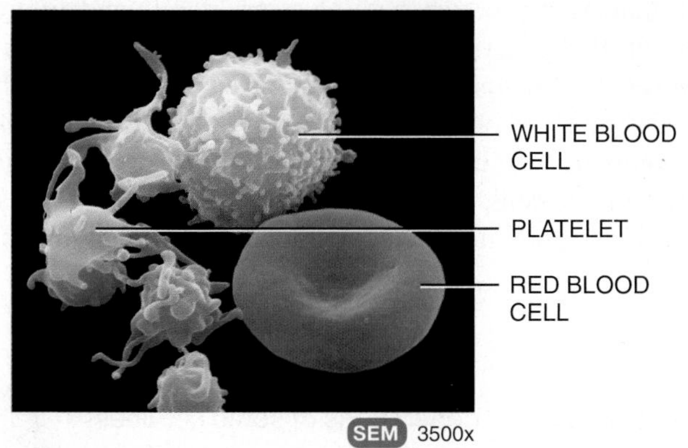


(a) Scanning electron micrograph

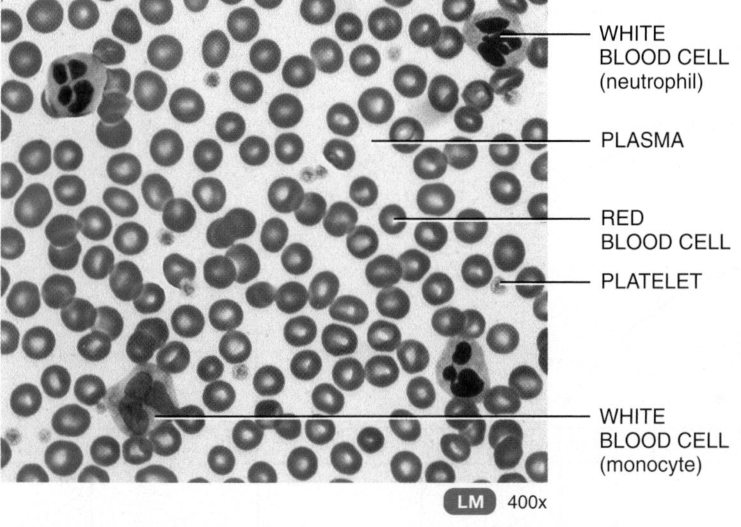


(b) Blood smear (thin film of blood spread on glass slide)

**The formed elements of blood are red blood cells, white blood cells, and platelets.**

of blood is composed of RBCs. The normal range of hematocrit for adult females is about 38–46 percent; for adult males it is about 40–54 percent. The hormone testosterone, present in much higher concentration in males than in females, stimulates synthesis of **erythropoietin** (EPO) (e-rith′-rō-POY-ē-tin), the hormone that stimulates production of RBCs. Thus, testosterone contributes to higher hematocrits in males. Lower values in women during their reproductive years also may be due to excessive loss of blood during menstruation. A significant drop in the hematocrit indicates *anemia* (a-NĒ-mē-a), a lower-than-normal number of RBCs. In *polycythemia* (pol′-ē-si-THĒ-mē-a), the percentage of RBCs is abnormally high, and the hematocrit may be 65 percent or higher. This raises the viscosity of blood, making the blood more difficult for the heart to pump. Increased viscosity also contributes to high blood pressure and increased risk of stroke. Polycythemia may be caused by conditions such as an abnormal increase in RBC production, tissue hypoxia (low oxygen levels), dehydration, and blood doping or the use of EPO by athletes.

### ✓ CHECKPOINT

1. List the substances that blood transports.
2. Which processes does blood help regulate?
3. Which formed elements of blood are most numerous?
4. What is the buffy coat?
5. List the functions of plasma proteins.
6. Which formed elements of the blood are cell fragments?

## 18.2 Hemopoiesis is the production of formed elements.

Although some lymphocytes have a lifetime measured in years, most formed elements of the blood last only hours, days, or weeks, and must be replaced continually. Negative feedback systems regulate the total number of RBCs and platelets in circulation, and their numbers normally remain steady. However, the abundance of the different types of WBCs varies in response to challenges by invading pathogens and other foreign antigens.

The process by which the formed elements of blood develop is called **hemopoiesis** (hĒ-mō-poy-Ē-sis; *-poiesis* = making) or **hematopoiesis** (hem′-a-tō-poy-Ē-sis). Before birth, hemopoiesis first occurs in the yolk sac of an embryo and later in the liver, spleen, thymus, and lymph nodes of a fetus. Red bone marrow becomes the primary site of hemopoiesis in the last three months before birth, and continues as the source of formed elements after birth and throughout life.

**Red bone marrow** is a highly vascularized connective tissue located in the microscopic spaces between trabeculae of spongy bone tissue. It is present chiefly in bones of the axial skeleton, pectoral and pelvic girdles, and the proximal epiphyses of the humerus and femur. Red bone marrow cells contains **pluripotent stem cells** (ploo-RIP-ō-tent; *pluri-* = several), cells that have the ability to develop into many different types of cells (Figure 18.3). In newborns, all bone marrow is red and thus active in blood cell production. As an individual ages, the rate of blood cell formation decreases; the red bone marrow in the medullary (marrow) cavity of long bones becomes inactive and is replaced by yellow bone marrow, which consists largely of fat cells. Under certain conditions, such as severe bleeding, yellow bone marrow can revert to red bone marrow to assist in blood cell production.

In order to form blood cells, pluripotent stem cells in red bone marrow reproduce themselves, proliferate, and differentiate into cells that give rise to two further types of stem cells; *myeloid stem cells* and *lymphoid stem cells* have the capacity to develop into several types of cells (Figure 18.3). Myeloid stem cells begin and complete their development in red bone marrow and give rise to red blood cells, platelets, mast cells, eosinophils, basophils, neutrophils, and monocytes. Lymphoid stem cells begin their

FIGURE 18.3 **Origin, development, and structure of formed elements.** Some of the generations of some cell lines have been omitted.

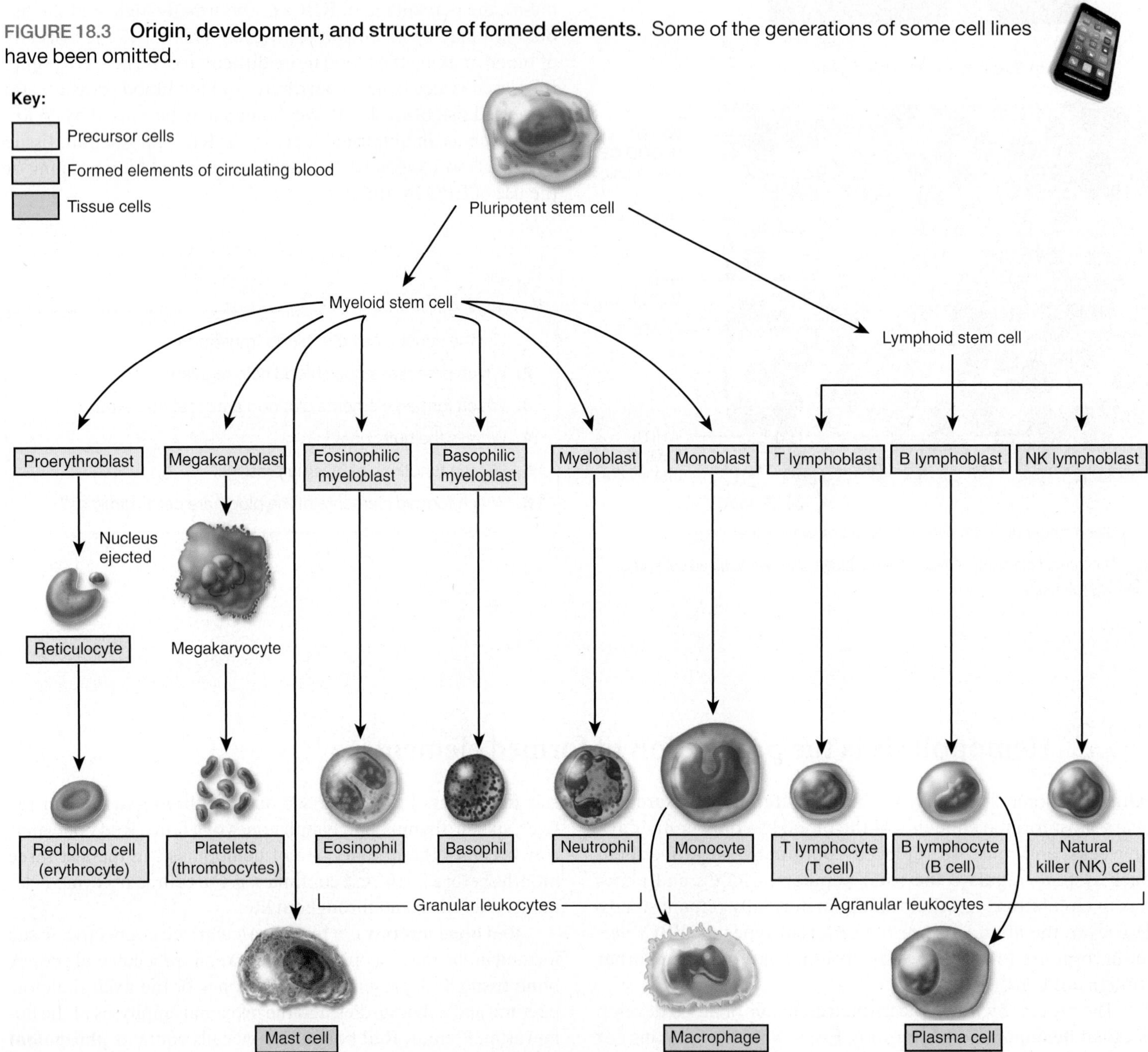


**Formed element production is called hemopoiesis and occurs mainly in red bone marrow after birth.**

development in red bone marrow but complete it in lymphatic tissues; they give rise to lymphocytes and natural killer (NK) cells. After blood cells form, they enter the bloodstream in blood vessels leaving the bone. With the exception of lymphocytes, formed elements do not divide once they leave red bone marrow.

Myeloid and lymphoid stem cells give rise to **precursor cells**, also known as **blasts**. Precursor cells are committed to giving rise to more specific elements of blood. Over several cell divisions they develop into the actual formed elements of blood. For example, monoblasts develop into monocytes, eosinophilic myeloblasts develop into eosinophils, and so on.

Several hormones called **hemopoietic growth factors** (hē-mō-poy-ET-ik) regulate the differentiation and proliferation of particular formed elements. **Erythropoietin (EPO)** (e-rith′-rō-POY-e-tin) produced by the kidneys increases the number of red blood cell precursors. **Thrombopoietin** (throm′-bō-POY-ē-tin) is a hormone produced by the liver that stimulates the formation of platelets from megakaryocytes. Some **cytokines** (SĪ-tō-kīns), small glycoprotein hormones typically produced by cells such as red bone marrow cells and macrophages, stimulate white blood cell formation.

✓ **CHECKPOINT**

**7.** What is the primary site for the development of formed elements after birth?

**8.** What is the unique ability of pluripotent stem cells?

## 18.3 Mature red blood cells are biconcave cells containing hemoglobin.

Red blood cells (RBCs) contain the oxygen-carrying protein **hemoglobin**, which is a pigment that gives whole blood its red color. A healthy adult male has about 5.4 million red blood cells per microliter (μL) of blood, and a healthy adult female has about 4.8 million. (One drop of blood is about 50 μL.) To maintain normal numbers of RBCs, new cells must enter the circulation at the astonishing rate of at least 2 million per second, a pace that balances the equally high rate of RBC destruction.

### RBC Anatomy

**Erythrocytes** (e-RITH-rō-sīts; *erythro-* = red; *-cyte* = cell), or *mature* red blood cells (RBCs), are biconcave (concave on both sides) discs with a simple structure; they lack a nucleus and other organelles and can neither reproduce nor carry on extensive metabolic activities (Figure 18.4a). The cytosol of erythrocytes contains hemoglobin molecules, which were synthesized before loss of the nucleus during RBC production. Essentially, erythrocytes consist of a plasma membrane enclosing a cytosol rich with hemoglobin. The RBC plasma membrane is both strong and flexible, which allows the cells to deform without rupturing as they squeeze through narrow capillaries (see Figure 20.1e). As you will see later, certain glycolipids in the plasma membrane of RBCs are antigens that account for the various blood groups such as the ABO and Rh groups.

### RBC Physiology

Red blood cells are highly specialized for oxygen transport. Because erythrocytes have no nucleus, all of their internal space is available to transport oxygen molecules. Since erythrocytes lack mitochondria and generate ATP anaerobically (without oxygen), they do not use up any of the oxygen they transport. Even the shape of an erythrocyte facilitates its function. A biconcave disc has a much greater surface area for the diffusion of gas molecules into and out of the erythrocyte than, say, a sphere or a cube.

Each erythrocyte contains about 280 million hemoglobin molecules (Figure 18.4b). A hemoglobin molecule consists of a protein called **globin**, composed of four polypeptide chains; a ringlike nonprotein pigment called a **heme** is bound to each of the four chains. At the center of the heme ring is an iron ion ($Fe^{2+}$) (Figure 18.4c) that can combine reversibly with one oxygen molecule, allowing each hemoglobin molecule to bind four oxygen molecules. Each oxygen molecule picked up from the lungs is transported bound to an iron ion within the heme ring. As blood flows through tissue capillaries, hemoglobin releases

**FIGURE 18.4 Shapes of an erythrocyte and a hemoglobin molecule, and structure of a heme group.** In (b), note that each of the four polypetide chains (blue) of a hemoglobin molecule has one heme group (gold), which contains an iron ion, $Fe^{2+}$ (shown in red).

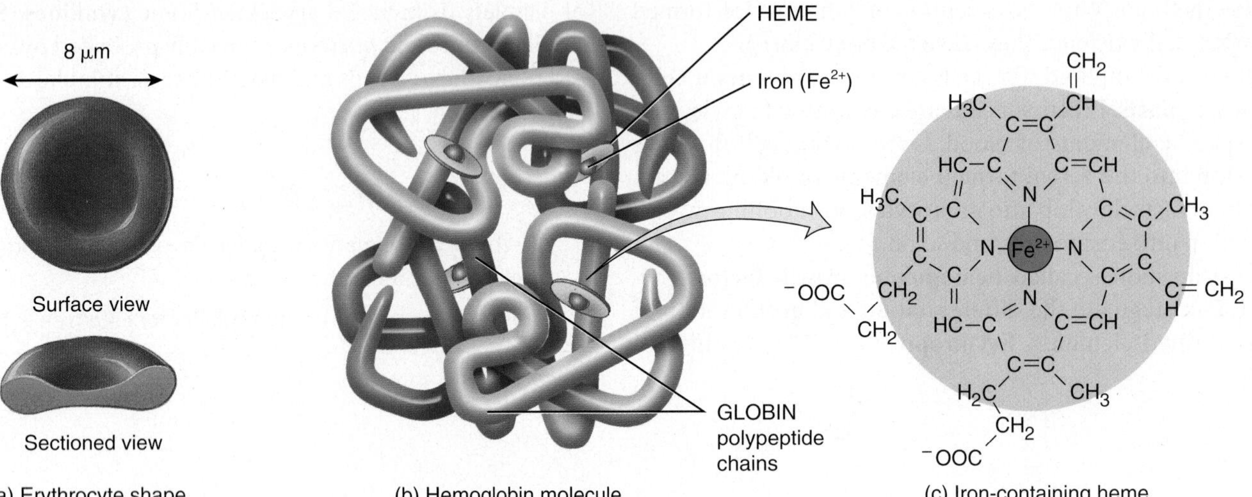


**The iron portion of a heme group binds oxygen for transport by hemoglobin.**

oxygen, which diffuses first into the interstitial fluid and then into cells.

Hemoglobin also transports about 23 percent of the total carbon dioxide, a waste product of metabolism. (The remaining carbon dioxide is dissolved in plasma or carried as bicarbonate ions.) Blood flowing through tissue capillaries picks up carbon dioxide, some of which combines with amino acids in the globin part of hemoglobin. As blood flows through the lungs, the carbon dioxide is released from hemoglobin and then exhaled (see Chapter 22).

See **Table 18.5** for a summary of the number, characteristics, and functions of RBCs.

**✓ CHECKPOINT**

**9.** Which characteristics of red blood cells facilitate oxygen transport?

**10.** How many molecules of $O_2$ can one hemoglobin molecule transport?

**11.** Which part of the hemoglobin molecule carries oxygen? Which part carries carbon dioxide?

## 18.4 Red blood cells have a life cycle of 120 days.

Red blood cells live only about 120 days because of the wear and tear their plasma membranes undergo as they squeeze through blood capillaries. Without a nucleus and other organelles, RBCs cannot synthesize new components to replace damaged ones. The plasma membrane becomes more fragile with age, and the cells are more likely to burst, especially as they squeeze through narrow channels in the spleen. Ruptured red blood cells are removed from circulation and destroyed by phagocytic macrophages in the spleen and liver, and the breakdown products are recycled, as follows (**Figure 18.5**):

1. Macrophages in the spleen, liver, and red bone marrow phagocytize ruptured and worn-out red blood cells, splitting apart the globin and heme portions of hemoglobin.
2. Globin is broken down into amino acids, which can be reused by body cells to synthesize other proteins.

FIGURE 18.5 Formation and destruction of red blood cells and recycling of hemoglobin components.

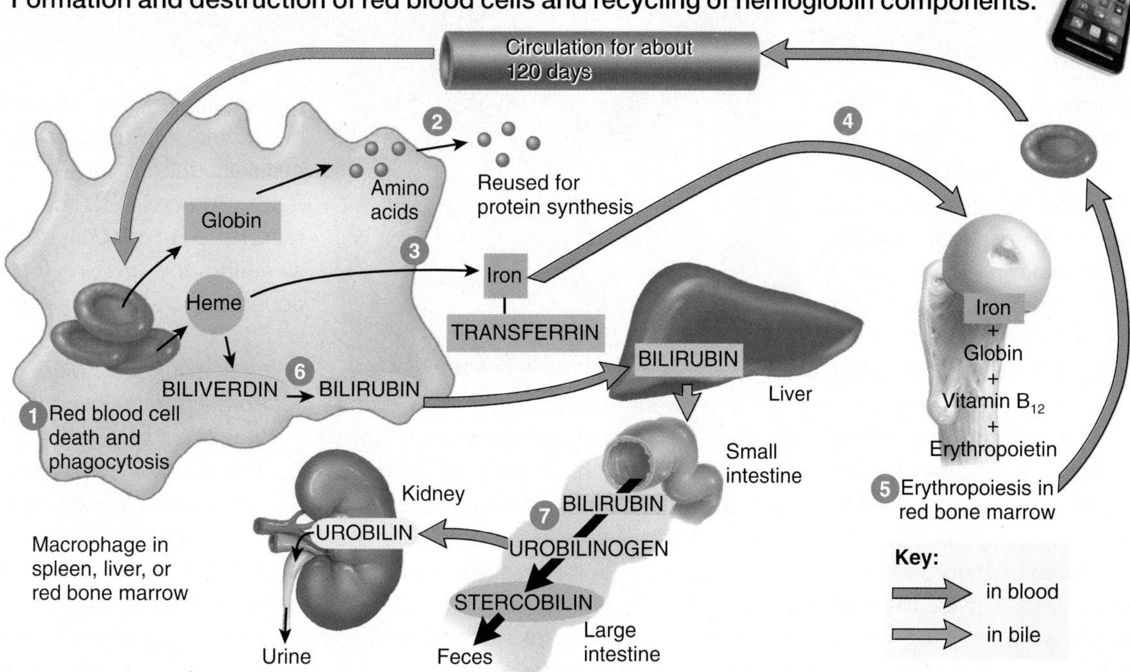


The rate of RBC formation by red bone marrow normally equals the rate of RBC destruction by macrophages.

3. Iron removed from the heme portion associates with the plasma protein **transferrin** (trans-FER-in; *trans-* = across; *-ferr-* = iron), a transporter for iron in the bloodstream.

4. The iron–transferrin complex is then carried to red bone marrow, where RBC precursor cells use it in hemoglobin synthesis. Iron is needed for the heme portion of the hemoglobin molecule, and amino acids are needed for the globin portion. Vitamin $B_{12}$ is also needed for synthesis of hemoglobin.

5. Within red bone marrow, erythropoiesis (red blood cell production) releases red blood cells into the circulation.

6. When iron is removed from heme, the non-iron portion of heme is converted to **biliverdin** (bil′-i-VER-din), a green pigment, and then into **bilirubin** (bil′-i-ROO-bin), a yellow-orange pigment. Bilirubin enters the blood and is transported to the liver. Within the liver, bilirubin is secreted by liver cells into bile, which passes into the small intestine and then into the large intestine.

7. In the large intestine, bacteria convert bilirubin into **urobilinogen** (ūr-ō-bī-LIN-ō-jen). Some urobilinogen is absorbed back into the blood, is converted to a yellow pigment called **urobilin** (ūr-ō-BĪ-lin), and excreted in urine. Most urobilinogen is eliminated in feces in the form of a brown pigment called **stercobilin** (ster′-kō-BĪ-lin), which gives feces its characteristic color.

✓ CHECKPOINT

**12.** What is the function of transferrin?

**13.** As red blood cells die, how are the components of hemoglobin recycled or excreted?

## RETURN TO Ed's Story

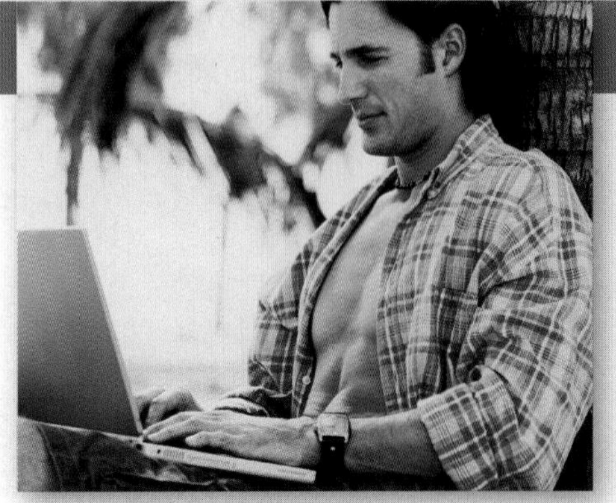

As Ed sat on the plane, he had no idea that his immune system was on alert. It had been 24 hours since the intruder had entered his bloodstream. The microscopic sporozoites had hitched a ride on the stream of Ed's life-giving blood. As they traveled, they collided with red blood cells, white blood cells, and proteins. The sporozoite surfaces were covered with alien proteins that Ed's body didn't recognize. A few were destroyed by white blood cells, but most found their way to the liver. Proteolytic enzymes released by the sporozoites started eating away at the cell surface while microscopic molecular motors inside the parasites pushed them into Ed's liver cells. Once hidden inside the cells of Ed's own body, the sporozoites would transform and multiply in relative safety.

Ed has been home now for four weeks. He went to bed early one night, feeling a bit nauseous and achy. By 3 o'clock the next morning, he was tossing and turning in his bed, and sweat soaked the sheets. Inside his body the next stage in the protozoa's life cycle had begun. Thousands of transformed *Plasmodium* merozoites now burst forth from the liver and traveled into the bloodstream. The merozoites bound to the surface of red blood cells and invaded the interior. Once inside, the merozoites hijacked the RBCs' protein-manufacturing centers, robbing them of vital nutrients and masking the surface of the infected cells with rogue proteins to prevent detection by Ed's immune system. As the parasites grew and multiplied, the infected RBCs swelled and finally ruptured, releasing toxins and cellular debris into Ed's bloodstream. Each ruptured RBC meant more potential merozoites to infect more RBCs. Ed awoke sweating and gasping for air. His muscles ached as if he had run a race.

A. *How would Ed's blood help protect him from a foreign invader such as the one now in his system?*
B. *The sporozoites have traveled to Ed's liver. How is the function of hepatocytes related to blood?*
C. *Ed's red blood cells are being destroyed by the parasite that infects him. What is the process of blood cell formation that replenishes lost blood cells? Where in his body would new red blood cells be produced?*
D. *Why would Ed's RBCs be a good host for a parasitic protozoan?*

## 18.5 Erythropoiesis is the process of red blood cell formation.

As you have already learned, the production of the formed elements of blood is called hemopoiesis. The specific production of RBCs is termed **erythropoiesis** (e-rith′-rō-poy-Ē-sis). Erythropoiesis starts in the red bone marrow with a precursor cell called a proerythroblast (see Figure 18.3). The proerythroblast divides several times, producing cells that begin to synthesize hemoglobin. Ultimately, a cell near the end of the development sequence ejects its nucleus and becomes a **reticulocyte** (re-TIK-ū-lō-sit). Loss of the nucleus causes the center of the cell to indent, producing the RBC's distinctive biconcave shape. Reticulocytes, which are about 34 percent hemoglobin and retain some mitochondria, ribosomes, and endoplasmic reticulum, pass from red bone marrow into the bloodstream. Reticulocytes develop into erythrocytes (mature RBCs) within 1 or 2 days after their release from red bone marrow. Determining the percentage of reticulocytes compared to all circulating RBCs indicates the rate of erythropoiesis. (See the Clinical Connection entitled Reticulocyte Count later in this section.)

Normally, erythropoiesis and red blood cell destruction proceed at roughly the same pace. If the oxygen-carrying capacity of the blood falls because erythropoiesis is not keeping up with RBC destruction, a negative feedback system increases RBC production (Figure 18.6). The controlled condition in this particular feedback loop is the amount of oxygen delivered to the kidneys (and thus to body tissues in general). Cellular oxygen deficiency, called **hypoxia** (hī-POKS-ē-a), may occur if too little oxygen enters the blood. For example, the lower oxygen content of air at high altitudes reduces the level of oxygen in the blood. Oxygen delivery may also decrease due to anemia, which can result from many causes; lack of iron, certain amino acids, and lack of vitamin $B_{12}$ are but a few. (See the Clinical Connection entitled Anemia in your WileyPLUS resources, which are listed in the summary at the end of this chapter.) Circulatory problems that reduce blood flow to tissues may also reduce oxygen delivery. Whatever the cause, hypoxia stimulates the kidneys to increase the release of the hormone erythropoietin.

FIGURE 18.6 **Negative feedback regulation of erythropoiesis (red blood cell formation).**

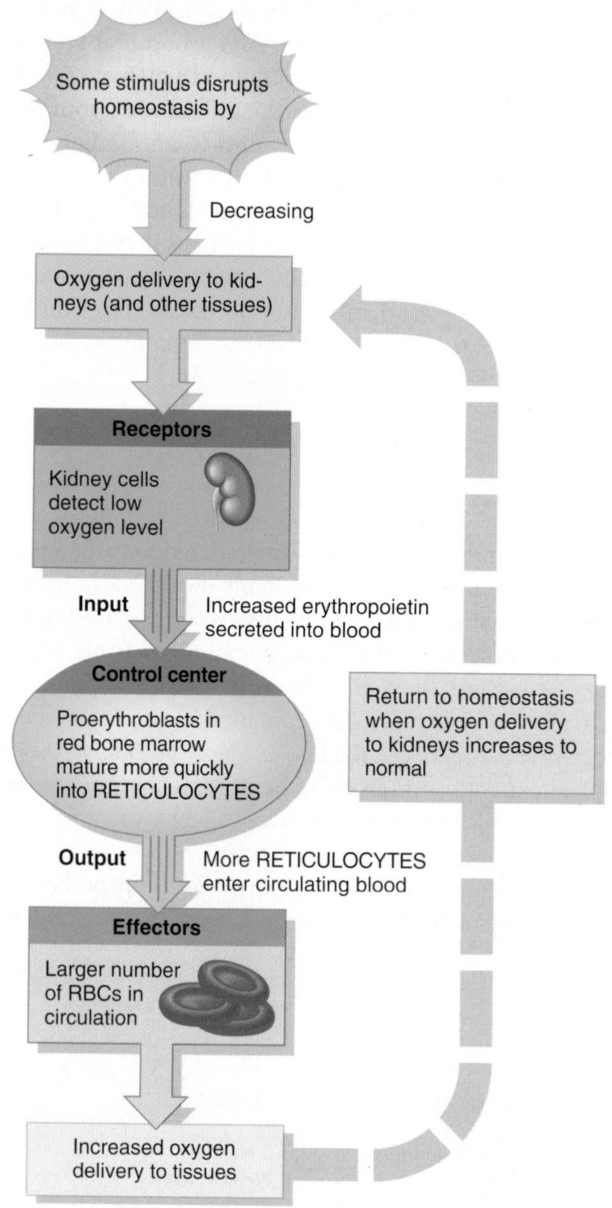


**The main stimulus for erythropoiesis is when kidney cells detect low oxygen levels in the blood.**

Erythropoietin circulates through the blood to the red bone marrow, where it speeds the development of proerythroblasts into reticulocytes. As the number of circulating RBCs increases, more oxygen can be delivered to body tissues.

**CLINICAL CONNECTION | *Reticulocyte Count***

The rate of erythropoiesis is measured by a **reticulocyte count.** Normally, a little less than 1 percent of the oldest RBCs are replaced by newcomer reticulocytes on any given day. It then takes one to two days for the reticulocytes to lose the last vestiges of endoplasmic reticulum and become erythrocytes. Thus, reticulocytes account for about 0.5–1.5 percent of all RBCs in a normal blood sample. A low "retic" count in a person who is anemic might indicate a shortage of erythropoietin (EPO) or an inability of the red bone marrow to respond to EPO, perhaps because of a nutritional deficiency or leukemia. A high "retic" count might indicate a good red bone marrow response to previous blood loss or to iron therapy in someone who had been iron-deficient. It could also point to illegal use of Epoetin alfa by an athlete. (See Clinical Connection on Blood Doping in your WileyPLUS Resources, which are listed in the summary at the end of this chapter.)

**✓ CHECKPOINT**

**14.** Which forms of red blood cells are found in circulating blood?

**15.** A hematocrit indicates the percentage of the blood volume occupied by red blood cells. How might your hematocrit change if you moved from a town at sea level to a high mountain village where the air contains less oxygen?

## 18.6 Blood is categorized into groups based on surface antigens.

The surfaces of erythrocytes contain a genetically determined assortment of **antigens** composed of glycoproteins and glycolipids. These antigens, called **agglutinogens** (ag′-loo-TIN-ō-jens), occur in characteristic combinations. Based on the presence or absence of various antigens, blood is categorized into different **blood groups**. Within a given blood group, there may be two or more different **blood types**. There are at least 24 blood groups, and more than 100 antigens that can be detected on the surface of red blood cells. Here we discuss two major blood groups: ABO and Rh. The incidence of ABO and Rh blood types varies among different population groups, as indicated in Table 18.2.

**TABLE 18.2**

Blood Types in the United States

| POPULATION GROUP | BLOOD TYPE (PERCENTAGE) | | | | |
|---|---|---|---|---|---|
| | O | A | B | AB | $RH^+$ |
| **European-American** | 45 | 40 | 11 | 4 | 85 |
| **African-American** | 49 | 27 | 20 | 4 | 95 |
| **Korean-American** | 32 | 28 | 30 | 10 | 100 |
| **Japanese-American** | 31 | 38 | 21 | 10 | 100 |
| **Chinese-American** | 42 | 27 | 25 | 6 | 100 |
| **Native American** | 79 | 16 | 4 | 1 | 100 |

## ABO Blood Group

The **ABO blood group** is based on two antigens called A and B (Figure 18.7). People whose RBCs display only antigen A have **type A** blood. Those who have only antigen B are **type B**. Individuals who have both A and B antigens are **type AB**, and those who have neither antigen A nor antigen B are **type O**.

In addition to antigens on the surfaces of RBCs, plasma usually contains antibodies called **agglutinins** (a-GLOO-ti-nins) that react with the A or B antigens if the two are mixed. These are the **anti-A antibody**, which reacts with antigen A, and the **anti-B antibody**, which reacts with antigen B. The antibodies present in each of the four ABO blood types are shown in Figure 18.7. You do not have antibodies that react with the antigens of your own RBCs, but most likely you do have antibodies for any antigens of the ABO group that your RBCs lack. For example, if you have type B blood, you have B antigens on your RBCs, and you have anti-A antibodies in your plasma. Fortunately you would not have anti-B antibodies in your plasma, or they would attack your RBCs. Although agglutinins start to appear in the blood within a few months after birth, the reason for their presence is not clear. Perhaps they are formed in response to bacteria that normally inhabit the gastrointestinal tract. Because anti-A and anti-B antibodies are large molecules that do not cross the placenta, ABO incompatibility between a mother and her fetus rarely causes problems.

## Transfusions

Despite the differences in RBC antigens, blood is the most easily shared of human tissues, saving many thousands of lives every year through transfusions. A **transfusion** (trans-FŪ-shun) is the transfer of whole blood or blood components (only red blood cells or plasma) into the bloodstream. A transfusion is most often given to alleviate anemia, to increase blood volume (for example, after a severe *hemorrhage*, the loss of a large amount of blood), or to improve immunity. However, the normal components of one person's RBC plasma membrane can trigger damaging antigen–antibody responses in a transfusion recipient.

In an incompatible blood transfusion, antibodies in the *recipient's* plasma bind to the antigens on the *donated* RBCs,

**FIGURE 18.7** **Antigens and antibodies involved in the ABO blood groups.**

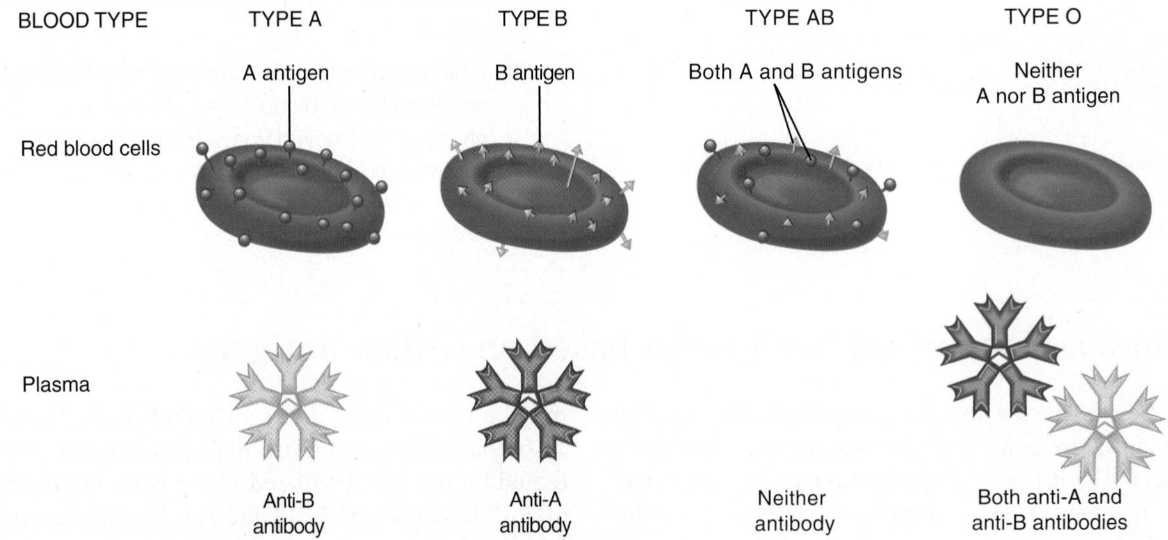


**The antibodies in your plasma do not react with the antigens on your red blood cells.**

which causes **agglutination** (a-gloo-ti-NĀ-shun), or clumping, of the donated RBCs. Note that the *recipient's* RBCs do not usually clump following blood transfusion because antibodies in the *donor's* plasma are so diluted in the recipient's bloodstream. Clumping of RBCs can block small blood vessels, preventing blood flow to essential tissues. Agglutination is an antigen–antibody response in which RBCs become cross-linked to one another. (Note that agglutination is not the same as blood clotting, a process that is described later in Concept 18.9.) Formation of these antigen–antibody complexes causes **hemolysis** (hē-MOL-i-sis) or rupture of the RBCs and the release of hemoglobin into the plasma. The liberated hemoglobin may cause kidney damage by clogging the filtration membranes. If the kidneys stop working entirely (acute renal failure), the individual may die.

As an example of an incompatible blood transfusion, consider what happens if a person with type A blood receives a transfusion of type B blood. The recipient's blood (type A) contains A antigens on the red blood cells and anti-B antibodies in the plasma. The donor's blood (type B) contains B antigens and anti-A antibodies. In this situation, two things can happen. First, the anti-B antibodies in the recipient's plasma can bind to the B antigens on the donor's RBCs, causing agglutination and hemolysis of the donated RBCs. Second, the anti-A antibodies in the donor's plasma can bind to the A antigens on the recipient's RBCs. As noted above, the second reaction is usually not serious because the donor's anti-A antibodies become so diluted in the recipient's plasma that they do not cause any significant agglutination and hemolysis of the recipient's RBCs. Table 18.3 summarizes the interactions of the four blood types of the ABO system.

**TABLE 18.3**

ABO Blood Group Interactions

| | BLOOD TYPE | | | |
|---|---|---|---|---|
| **CHARACTERISTIC** | **A** | **B** | **AB** | **O** |
| **Antigen (agglutinogen) on RBC** | A | B | Both A and B | Neither A nor B |
| **Antibody (agglutinin) in plasma** | Anti-B | Anti-A | Neither anti-A nor anti-B | Both anti-A and anti-B |
| **Compatible donor blood types (no hemolysis)** | A, O | B, O | A, B, AB, O | O |
| **Incompatible donor blood types (hemolysis)** | B, AB | A, AB | — | A, B, AB |

People with type AB blood do not have anti-A or anti-B antibodies in their plasma. They are sometimes called *universal recipients* because theoretically they can receive blood from donors of all four ABO blood types. They have no antibodies to attack antigens on donated RBCs (Table 18.3). People with type O blood have neither A nor B antigens on their RBCs and are sometimes called *universal donors*. Because there are no antigens on their RBCs for antibodies to attack, people with type O blood can theoretically donate blood to all four ABO blood types. Type O persons requiring blood may receive only type O blood, as they have antibodies to both A and B antigens in their plasma (Table 18.3). In practice, use of the terms *universal recipient* and *universal donor* is misleading and dangerous. Blood contains antigens and antibodies other than those associated with the ABO system that can also cause transfusion problems. Thus, blood should always be carefully cross-matched or screened before transfusion.

In about 80 percent of the population, soluble antigens of the ABO type appear in saliva and other body fluids, in which case blood type can be identified from a sample of saliva.

## Rh Blood Group

The **Rh blood group** is so named because the Rh antigen was discovered in the blood of the *Rhesus* monkey. People whose RBCs have Rh antigens are designated $Rh^+$ (Rh positive); those who lack Rh antigens are designated $Rh^-$ (Rh negative). Table 18.2 shows the incidence of $Rh^+$ and $Rh^-$ individuals in various populations. Normally, plasma does not contain anti-Rh antibodies. If an $Rh^-$ person receives an $Rh^+$ blood transfusion, however, the immune system starts to make anti-Rh antibodies that will remain in the blood. If a second transfusion of $Rh^+$ blood is given later, the previously formed anti-Rh antibodies will cause agglutination and hemolysis of the RBCs in the donated blood, and a severe reaction may occur. This can happen through an incompatible blood transfusion, by sharing hypodermic needles, or when a pregnant $Rh^-$ woman is carrying an $Rh^+$ fetus; the latter results in a condition called *hemolytic disease of the newborn* and is described in a Clinical Connection of that name in your WileyPLUS resources listed at the end of this chapter.

## Typing and Cross-matching Blood for Transfusion

During the ABO blood typing procedure, single drops of blood are mixed with different *antisera*, solutions that contain antibodies (Figure 18.8). One drop of blood is mixed with anti-A serum, which contains anti-A antibodies that will agglutinate red blood cells that possess A antigens. Another drop is mixed with anti-B serum, which contains anti-B antibodies that will agglutinate red blood cells that possess B antigens. If the red blood cells agglutinate only when mixed with anti-A serum, the blood is type A. If the red blood cells agglutinate only when mixed with anti-B serum, the blood is type B. The blood is type AB if both drops agglutinate; if neither drop agglutinates, the blood is type O.

In the procedure for determining Rh factor, a drop of blood is mixed with antiserum containing antibodies that will agglutinate RBCs displaying Rh antigens. If the blood agglutinates, it is $Rh^+$; no agglutination indicates $Rh^-$.

To avoid blood-type mismatches, laboratory technicians first type the patient's blood and then either cross-match it to potential donor blood or screen it for the presence of antibodies. Once the patient's blood type is known, donor blood of the same ABO and Rh type is selected. In a **cross-match**, the possible donor RBCs are mixed with the recipient's serum. If agglutination does not occur, the recipient does not have antibodies that will attack the donor RBCs. Alternatively, the recipient's serum can be **screened** against a test panel of RBCs having antigens known to cause blood transfusion reactions to detect any antibodies that may be present.

FIGURE 18.8 **ABO blood typing.**

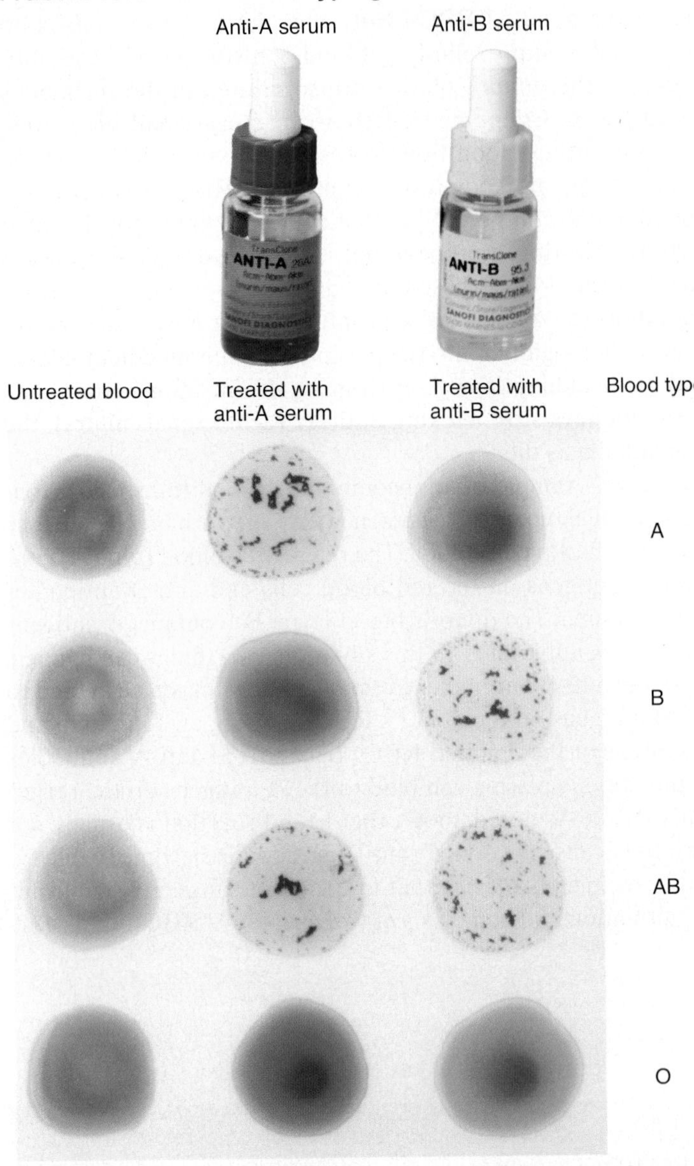


**In the procedure for ABO blood typing, blood is mixed with anti-A serum and anti-B serum.**

### ✓ CHECKPOINT

**16.** Which ABO antigens are found in type AB blood? In type O blood?

**17.** Where in blood are antibodies to the A or B antigens found?

**18.** Which antibodies are found in type A blood? In type AB blood? In type O blood?

**19.** Which blood type can be given to people with type AB blood?

**20.** What is hemolysis, and how can it occur after a mismatched Rh blood transfusion?

## 18.7 White blood cells combat inflammation and infection.

Unlike red blood cells, white blood cells (WBCs), or **leukocytes** (LOO-kō-sīts; *leuko-* = white), have nuclei and a full complement of other organelles but do not contain hemoglobin (Figure 18.9). WBCs are classified as granular or agranular, depending on whether they contain cytoplasmic granules (vesicles) made visible by staining when viewed through a light microscope. *Granular leukocytes* include neutrophils, eosinophils, and basophils; *agranular leukocytes* include lymphocytes and monocytes. As shown in Figure 18.3, granular leukocytes and monocytes develop from myeloid stem cells. In contrast, lymphocytes develop from lymphoid stem cells.

### WBC Types

#### Granular Leukocytes

After staining, each of the three types of granular leukocytes displays conspicuous granules with distinctive coloration that can be recognized under a light microscope. Granular leukocytes can be distinguished as follows:

- **Neutrophil** (NOO-trō-fil). The granules of a neutrophil are smaller, evenly distributed, and pale lilac in color (Figure 18.9a). Because the granules do not strongly attract either the acidic

FIGURE 18.9 Types of white blood cells.

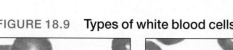

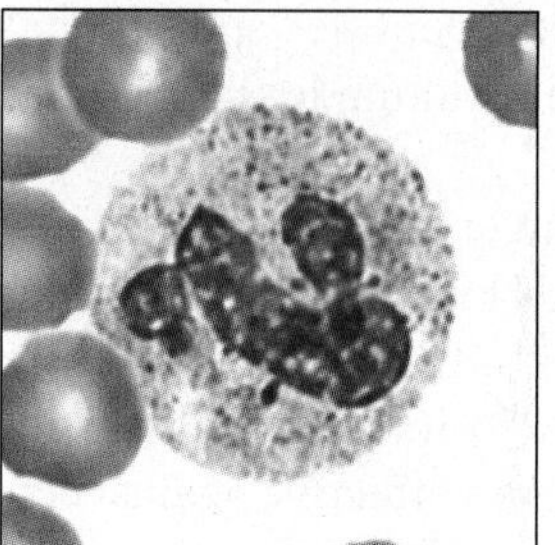

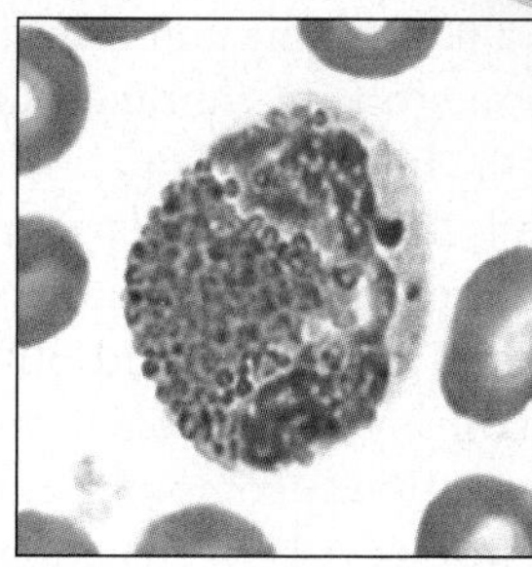

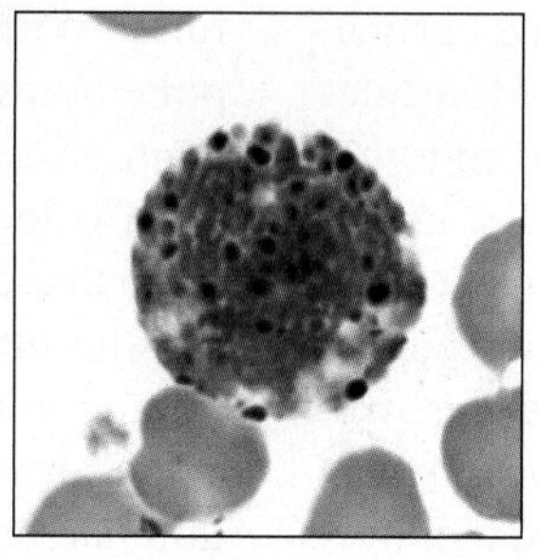

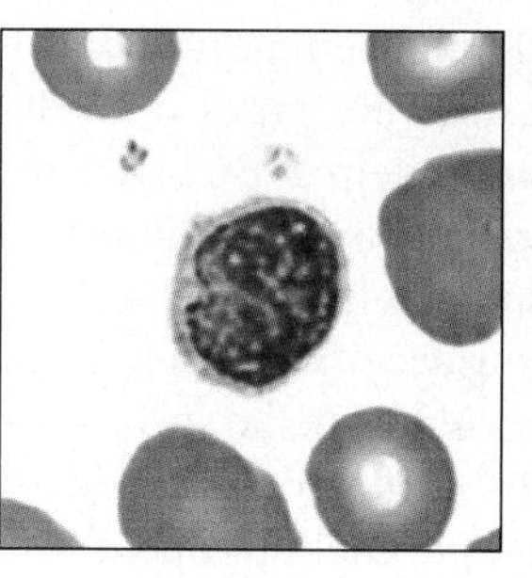

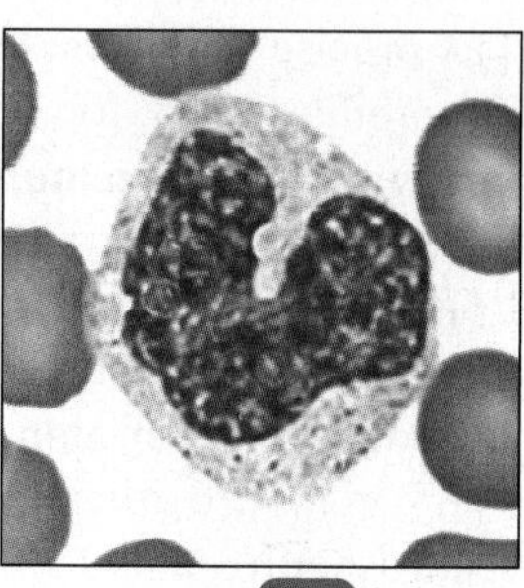


(a) Neutrophil (b) Eosinophil (c) Basophil (d) Lymphocyte (e) Monocyte

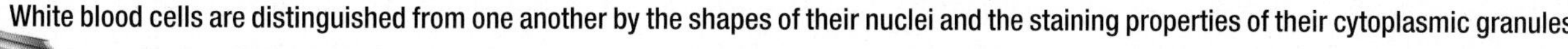

White blood cells are distinguished from one another by the shapes of their nuclei and the staining properties of their cytoplasmic granules.

(red) or basic (blue) stain, these WBCs are *neutrophilic* (= neutral-loving). The nucleus has two to five lobes, connected by very thin strands of nuclear material. As the cells age, the number of nuclear lobes increases. Because older neutrophils thus have several differently shaped nuclear lobes, they are often called *polymorphonuclear leukocytes (PMNs;* poly = many).

- **Eosinophil** (ē-ō-SIN-ō-fil). The large, uniform-sized granules within an eosinophil are *eosinophilic* (= eosin-loving)—they stain red-orange with acidic dyes (Figure 18.9b). The granules usually do not cover or obscure the nucleus, which most often has two or three lobes connected by a thin or thick strand of nuclear material.

- **Basophil** (BĀ-sō-fil). The round, variable-sized granules of a basophil are *basophilic* (= basic-loving)—they stain blue-purple with basic dyes (Figure 18.9c). The granules commonly obscure the nucleus, which has two lobes.

### Agranular Leukocytes

Even though so-called agranular leukocytes possess cytoplasmic granules, the granules are not visible under a light microscope because of their small size and poor staining qualities. Characteristics of agranular leukocytes are as follows:

- **Lymphocyte** (LIM-fō-sīt). The nucleus of a lymphocyte stains dark and is round or slightly indented (Figure 18.9d). The cytoplasm stains sky blue and forms a rim around the nucleus. The larger the cell, the more cytoplasm is visible. Lymphocytes may be as small as 6–9 μm in diameter or as large as 10–14 μm in diameter. Although the functional significance of the size difference between small and large lymphocytes is unclear, the distinction is still clinically useful because an increase in the number of large lymphocytes has diagnostic significance in acute viral infections and in some immunodeficiency diseases.

- **Monocyte** (MON-ō-sīt′). The nucleus of a monocyte is usually kidney-shaped or horseshoe-shaped, and the cytoplasm is blue-gray and has a foamy appearance (Figure 18.9e). The cytoplasm's color and appearance are due to very fine *azurophilic granules* (az′-ū-rō-FIL-ik; *azur* = blue; *philic* = loving), which are lysosomes. Blood is merely a conduit for monocytes, which migrate from the bloodstream into the tissues, where they enlarge and differentiate into **macrophages** (MAK-rō-fā-jez = large eaters). Some become **fixed macrophages**, which means they reside in a particular tissue; examples are alveolar macrophages in the lungs or macrophages in the spleen. Others become **wandering macrophages**, which roam the tissues and gather at sites of infection or inflammation.

## WBC Functions

The skin and mucous membranes of the body are continuously exposed to microscopic organisms, such as bacteria, some of which are capable of invading deeper tissues and causing disease. Once pathogens enter the body, the general function of white blood cells is to combat them by phagocytosis or immune responses. To accomplish these tasks, many WBCs leave the bloodstream and collect at sites of pathogen invasion or inflammation. Once granular leukocytes and monocytes leave the bloodstream to fight injury or infection, they never return to it. Lymphocytes, on the other hand, continually recirculate—from blood to interstitial spaces of tissues to lymphatic fluid and back to blood. Only 2 percent of the total lymphocyte population is circulating in the blood at any given time; the rest are in lymphatic fluid and organs such as the skin, lungs, lymph nodes, and spleen.

RBCs are contained within the bloodstream, but WBCs are able to leave the bloodstream by a process termed **emigration** (em′-i-GRĀ-shun; *e-* = out; *-migra-* = wander), formerly called *diapedesis* (dī-a-pe-DĒ-sis). During emigration, WBCs roll along the endothelium that forms blood capillary walls, stick to it, and then squeeze between the endothelial cells (Figure 18.10). The precise signals that stimulate emigration through a particular capillary vary for the different types of WBCs. Molecules known as **adhesion molecules** help WBCs stick to the endothelium. For example, in response to nearby injury and inflammation, endothelial cells display adhesion molecules called *selectins*. Selectins stick to the surface of neutrophils, causing them to slow down and roll along the endothelial surface. On the neutrophil surface are other adhesion molecules called *integrins*, which secure neutrophils to the endothelium and assist their movement through the capillary wall and into the interstitial fluid of the injured tissue.

Neutrophils and macrophages are active in **phagocytosis** (fag′-ō-sī-TŌ-sis); they can ingest bacteria and dispose of dead matter (see Figure 3.12). Several different chemicals released by microbes and inflamed tissues attract phagocytes, a phenomenon called **chemotaxis** (kē-mō-TAK-sis).

Among WBCs, neutrophils respond most quickly to tissue destruction by bacteria. After engulfing a pathogen during phagocytosis, a neutrophil unleashes several destructive chemicals to destroy the ingested pathogen. These chemicals include the enzyme **lysozyme** (LĪ-sō-zīm′), which destroys certain bacteria, and strong **oxidants**, such as hydrogen peroxide ($H_2O_2$) and the hypochlorite anion ($OCl^-$), which is similar to household bleach. Neutrophils also contain **defensins**, proteins that exhibit a broad range of antibiotic activity against bacteria and fungi. Within a neutrophil, defensins form peptide "spears" that poke holes in microbe membranes; the resulting loss of cellular contents kills the invader.

FIGURE 18.10 **Emigration of white blood cells.**

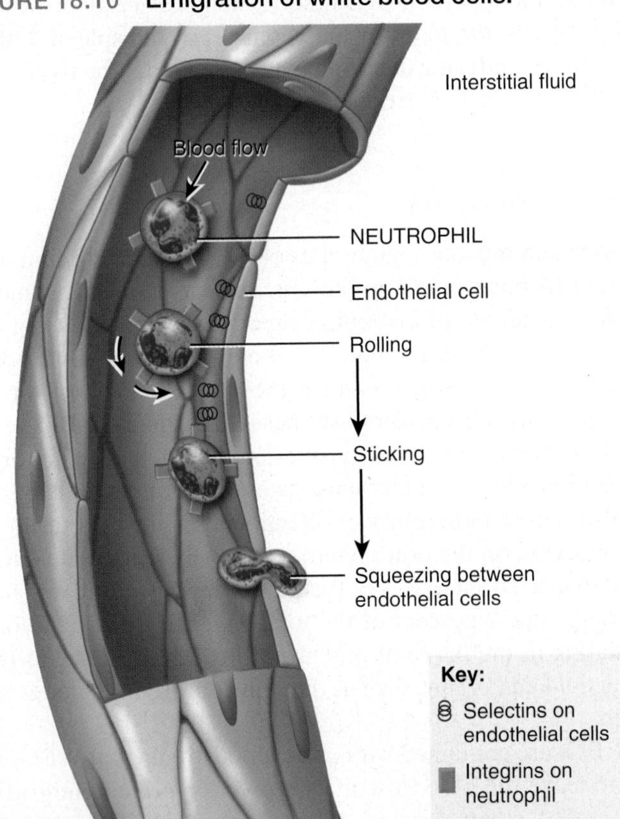


**Adhesion molecules (selectins and integrins) assist the emigration of WBCs from the bloodstream into interstitial fluid.**

Eosinophils leave the capillaries and enter tissue fluid. They are believed to release enzymes, such as histaminase, that combat the effects of histamine and other substances involved in inflammation during allergic reactions. Eosinophils also phagocytize antigen–antibody complexes and are effective against certain parasitic worms. A high eosinophil count often indicates an allergic condition or a parasitic infection.

At sites of inflammation, basophils leave capillaries; enter tissues; and then release heparin, histamine, and serotonin. These substances intensify the inflammatory reaction and are involved in hypersensitivity (allergic) reactions. Like basophils, *mast cells* release substances involved in inflammation, including heparin and histamine. Mast cells are widely dispersed in the body, particularly in connective tissues of the skin and mucous membranes of the respiratory and gastrointestinal tracts.

Lymphocytes are soldiers on the front lines of immune system battles (described in detail in Concept 21.6). Most lymphocytes continually move among lymphoid tissues, lymph, and blood, spending only a few hours at a time in blood. Thus, only a small proportion of the total lymphocytes are present in the blood at any given time. Three main types of lymphocytes are B cells, T cells, and natural killer cells. B cells are particularly effective in destroying bacteria and inactivating their toxins. T cells attack viruses, fungi, transplanted cells, cancer cells, and some bacteria, and are responsible for transfusion reactions, allergies, and the rejection of transplanted organs. Immune responses carried out by both B cells and T cells help combat infection and provide protection against some diseases. Natural killer cells attack a wide variety of infectious microbes and certain tumor cells.

Monocytes take longer to reach a site of infection than neutrophils, but they arrive in larger numbers and destroy more microbes. Upon their arrival, monocytes enlarge and differentiate into wandering macrophages, which clean up cellular debris and microbes by phagocytosis after an infection.

## WBC Life Span

Microbes have continuous access to the body through the mouth, nose, and pores of the skin. Furthermore, many cells age and die daily, and their remains must be removed. However, a WBC can phagocytize only a certain amount of material before the ingested material interferes with the WBC's own metabolic activities. In a healthy body, some WBCs, especially lymphocytes, can live for several months or years, but most live only a few days. During a period of infection, phagocytic WBCs may live only a few hours.

RBCs are far more numerous than WBCs and outnumber WBCs by about 700:1. There are normally about 5000–10,000

WBCs per μL of blood. An abnormally low level of white blood cells (below 5000/mL) is termed **leukopenia** (loo-kō-PĒ-nē-a). It is never beneficial and may be caused by radiation, shock, and certain chemotherapeutic agents. **Leukocytosis** (loo′-kō-sī-TŌ-sis), an increase in the number of WBCs above 10,000/μL, is a normal, protective response to stresses such as invading microbes, strenuous exercise, anesthesia, and surgery. Leukocytosis usually indicates an inflammation or infection. A physician may order a **differential white blood cell count**, a count of each of the five types of WBCs, to detect infection or inflammation, determine the effects of possible poisoning by chemicals or drugs, monitor blood disorders (such as leukemia) and the effects of chemotherapy, or detect allergic reactions and parasitic infections. Differential white blood cell counts measure the number of each kind of WBC in a sample of 100 WBCs. Because each type of white blood cell plays a different role, determining the *percentage* of each type in the blood assists in diagnosing the condition.

Table 18.4 lists the significance of both high and low WBC counts. See Table 18.5 for a summary of the numbers, characteristics, and functions of the various WBCs.

**TABLE 18.4**

Significance of High and Low White Blood Cell Counts

| WBC TYPE | HIGH COUNT MAY INDICATE | LOW COUNT MAY INDICATE |
|---|---|---|
| **Neutrophils** | Bacterial infection, burns, stress, inflammation | Radiation exposure, drug toxicity, vitamin $B_{12}$ deficiency, systemic lupus erythematosus (SLE) |
| **Lymphocytes** | Viral infections, some leukemias | Prolonged illness, immunosuppression, treatment with cortisol |
| **Monocytes** | Viral or fungal infections, tuberculosis, some leukemias, other chronic diseases | Bone marrow suppression, treatment with cortisol |
| **Eosinophils** | Allergic reactions, parasitic infections, autoimmune diseases | Drug toxicity, stress |
| **Basophils** | Allergic reactions, leukemias, cancers, hypothyroidism | Pregnancy, ovulation, stress, hyperthyroidism |

✓ **CHECKPOINT**

**21.** Which WBCs are called agranular leukocytes? Why?

**22.** How is the circulation of lymphocytes within the body different from other WBC's?

**23.** What is the importance of emigration, chemotaxis, and phagocytosis in fighting bacterial invaders?

**24.** How do neutrophils utilize chemicals to kill ingested pathogens?

**25.** What is a differential white blood cell count?

## 18.8 Platelets reduce blood loss from damaged vessels.

In addition to developing into RBCs and WBCs, hemopoietic stem cells also differentiate into cells that produce platelets. Under the influence of the hormone **thrombopoietin** (throm′-bō-POY-e-tin), myeloid stem cells develop into precursor cells called *megakaryoblasts* (see Figure 18.3). Megakaryoblasts transform into megakaryocytes, huge cells that splinter into 2000–3000 fragments in the red bone marrow and then enter the bloodstream. Each fragment, enclosed by a piece of the plasma membrane, is a platelet, or **thrombocyte**. Platelets are disc-shaped and exhibit many vesicles but no nucleus. When blood vessels are damaged, platelets help stop blood loss by forming a platelet plug that fills the gap in the blood vessel wall. Their vesicles also contain chemicals that, once released, promote blood clotting. Platelets have a short life span, normally just 5 to 9 days. Aged and dead platelets are removed by fixed macrophages in the spleen and liver.

### CLINICAL CONNECTION | *Complete Blood Count*

**A complete blood count (CBC)** is a very valuable test that screens for anemia and various infections. Usually included are counts of RBCs, WBCs, and platelets per μL of whole blood; hematocrit; and differential white blood cell count. The amount of hemoglobin in grams per milliliter of blood also is determined. Normal hemoglobin ranges are as follows: infants, 14–20 g/100 mL of blood; adult females, 12–16 g/100 mL of blood; and adult males, 13.5–18 g/100 mL of blood.

The numbers, characteristics, and functions of platelets, along with those of RBCs and WBCs, are summarized in Table 18.5.

✓ **CHECKPOINT**

**26.** What functions are served by platelets?

**TABLE 18.5**

Formed Elements in Blood

| NAME AND APPEARANCE | NUMBER | CHARACTERISTICS* | FUNCTIONS |
|---|---|---|---|
| **Red Blood Cells (RBCs) or Erythrocytes** | 4.8 million/μL in females; 5.4 million/μL in males | 7–8 μm diameter, biconcave discs, without nuclei; live for about 120 days | Hemoglobin within RBCs transports most oxygen and part of carbon dioxide in blood |
| **White Blood Cells (WBCs) or Leukocytes** | 5000–10,000/μL | Most live for a few hours to a few days† | Combat pathogens and other foreign substances that enter body |
| **Granular leukocytes** | | | |
| Neutrophils | 60–70% of all WBCs | 10–12 μm diameter; nucleus has 2–5 lobes connected by thin strands of chromatin; cytoplasm has very fine, pale lilac granules | Phagocytosis.; destruction of bacteria with lysozyme, defensins, and strong oxidants, such as superoxide anion, hydrogen peroxide, and hypochlorite anion |
| Eosinophils | 2–4% of all WBCs | 10–12 μm diameter; nucleus usually has 2 lobes connected by thick strand of chromatin; large, red-orange granules fill cytoplasm | Combat effects of histamine in allergic reactions, phagocytize antigen–antibody complexes, and destroy certain parasitic worms |
| Basophils | 0.5–1% of all WBCs | 8–10 μm diameter; nucleus has 2 lobes; large cytoplasmic granules appear deep blue-purple | Liberate heparin, histamine, and serotonin in allergic reactions that intensify overall inflammatory response |
| **Agranular leukocytes** | | | |
| Lymphocytes (T cells, B cells, and natural killer cells) | 20–25% of all WBCs | Small lymphocytes are 6–9 μm in diameter; large lymphocytes are 10–14 μm in diameter; nucleus is round or slightly indented; cytoplasm forms rim around nucleus that looks sky blue; the larger the cell, the more cytoplasm is visible | Mediate immune responses, including antigen–antibody reactions; B cells develop into plasma cells, which secrete antibodies; T cells attack invading viruses, cancer cells, and transplanted tissue cells; natural killer cells attack wide variety of infectious microbes and certain spontaneously arising tumor cells |
| Monocytes | 3–8% of all WBCs | 12–20 μm diameter; nucleus is kidney- or horseshoe-shaped; cytoplasm is blue-gray and appears foamy | Phagocytosis (after transforming into fixed or wandering macrophages) |
| **Platelets (thrombocytes)** | 150,000–400,000/μL | 2–4 μm diameter cell fragments that live for 5–9 days; contain many vesicles but no nucleus | Form platelet plug in hemostasis; release chemicals that promote vascular spasm and blood clotting |

*Colors are those seen when using Wright's stain.

†Some lymphocytes, called T and B memory cells, can live for many years once they are established.

## RETURN TO Ed's Story

"So, what seems to be troubling you, Ed?" asked the doctor as he reviewed Ed's chart.

"Jerry, I'm sore and achy all over, I've been having headaches, hot and cold spells, night sweats; I wake up soaking wet! I just got back from Central America about four weeks ago. I usually get a cold or mild flu when I travel, but this is the worst I've had. On top of everything else, I've been having diarrhea, too."

Jerry nodded and asked, "Were you up-to-date on your shots before you left?"

"I've been going to Belize with my students for 10 years, so I keep up-to-date on the shots and I got chloroquine for malaria prophylaxis. I was supposed to take 300 milligrams weekly, but I got busy. I might have missed a few doses."

"Well, you are definitely a candidate for a malarial infection. We'd better check to be sure. Malaria is nothing to take lightly; it can cause severe anemia, kidney and liver damage, and seizures. We'll do a blood smear and I'll also draw some more blood for lab analysis."

Jerry returned once the lab work was back. "Ed, your blood smear was positive for malarial parisitemia. The blood work suggests your kidneys have sustained some damage; hematocrit is low, bilirubin levels are elevated, blood glucose is low. I'd like to admit you to the hospital right away. We'll need to start treatment immediately."

"I thought it was just the flu," Ed said weakly, then reached for his cell phone to call his wife.

Jerry left the room and walked to the front desk. "Linda, we need to get this guy admitted right away. He's got malaria. His kidneys are shutting down and he may need dialysis. We'll need to get him started right away on IV quinidine, for the malaria. And try to get hold of nephrology. We're going to need a consult on this one."

As he lay in the hospital bed, Ed heard two voices in a hushed but serious conversation just outside his door. "I don't think quinidine is a good choice. His platelet count is already low and quinidine will cause further thrombocytopenia. We could try quinine instead to address the malaria, without increasing his risk of internal bleeding. At any rate, we'll have to get him on dialysis right away. His sodium levels are way down and his potassium and bilirubin levels are sky high."

Jerry replied, "I'll start him on 10 mg/kg quinine dihydrochloride in 5% dextrose. I don't think he could handle oral medication right now. He's been vomiting quite a bit and says he can't keep food or drink down. The dextrose should also help with his hypoglycemia."

Jerry entered Ed's room. "Hi, Ed. How are you feeling?" he asked.

"I'm feeling okay, but I'm nauseous and stiff, and it feels warm in here. Is everything okay? I heard you talking in the hallway."

"The nephrologist and I were just consulting on your case. Your platelet count is quite low and your kidneys have temporarily shut down. We'll get you on medication for the malaria and we'll have to put you on dialysis for a while until your kidneys begin to work on their own again."

Ed pushed himself up in the bed and asked, "Shouldn't the chloroquine have prevented this, Doc? I only missed a few doses."

"Well, *Plasmodium falciparum* is chloroquine-resistant, Ed. Just like the excessive use of antibiotics has led to antibiotic-resistant microbes, certain species of *Plasmodium* have developed resistance. There's still not a lot known about how *Plasmodium* is so successful in fighting off the human immune system, but it does. Your white blood cell count actually was high, but this bug has a knack for hiding in cells and altering the recognition of infected red cells by the immune system."

E. *If erythropoiesis cannot keep up with the pace of destruction of RBCs, what will happen to the oxygen-carrying capacity of Ed's blood?*

F. *Jaundice is a condition characterized by a yellowish color to the skin. Which pigments are produced from the breakdown of blood that might cause Ed to appear jaundiced?*

G. *Why would Ed potentially need a transfusion of blood?*

H. *Ed's kidneys have been damaged as they receive the released hemoglobin and RBC fragments. How is the kidney involved in blood cell production?*

I. *Which type of WBC phagocytizes damaged tissues and pathogens and may have been active early in Ed's infection (at the site of the bite where the inflammation was occurring)?*

# 18.9 Hemostasis is the sequence of events that stops bleeding from a damaged blood vessel.

**Hemostasis** (hē-mō-STĀ-sis; *-stasis* = standing still), not to be confused with the very similar term *homeostasis*, is a sequence of responses that stops bleeding when blood vessels are injured. When blood vessels are damaged or ruptured, the hemostatic response must be quick, localized to the region of damage, and carefully controlled in order to be effective. Three mechanisms reduce loss of blood from blood vessels: (1) vascular spasm, (2) platelet plug formation, and (3) blood clotting (coagulation). When successful, hemostasis prevents **hemorrhage** (HEM-or-ij; *-rhage* = burst forth), the loss of a large amount of blood from the vessels. Hemostasis can prevent hemorrhage from smaller blood vessels, but extensive hemorrhage from larger vessels usually requires medical intervention.

## Vascular Spasm

When arteries or arterioles are damaged, the smooth muscle in their walls contracts immediately, a response called **vascular spasm**. Vascular spasm reduces blood loss for several minutes to several hours, during which time the other hemostatic mechanisms begin to operate. The spasm is probably caused by damage to the smooth muscle, by substances released from activated platelets, and by reflexes initiated by pain receptors.

## Platelet Plug Formation

When platelets come into contact with parts of a damaged blood vessel, their characteristics change drastically and they quickly come together to form a platelet plug that helps fill the gap in the injured blood vessel wall. Platelet plug formation occurs as follows (Figure 18.11):

1. Initially, platelets contact and stick to parts of a damaged blood vessel, such as collagen fibers of the connective tissue underlying the damaged endothelial cells. This process is called **platelet adhesion**.
2. Due to adhesion, the platelets become activated, and their characteristics change dramatically. They extend many projections that enable them to contact and interact with one another, and they begin to liberate the contents of their vesicles. This phase is called the **platelet release reaction**. Liberated chemicals activate nearby platelets and sustain the vascular spasm, which decreases blood flow through the injured vessel.
3. The release of platelet chemicals makes other platelets in the area sticky, and the stickiness of the newly recruited and activated platelets causes them to stick to the originally activated platelets. This gathering of platelets is called **platelet aggregation**. Eventually, the accumulation and attachment of large numbers of platelets form a mass called a **platelet plug**.

A platelet plug is very effective in preventing blood loss in a small vessel. Although initially the platelet plug is loose, it becomes quite tight when reinforced by fibrin threads formed during blood clotting (see Figure 18.12). A platelet plug can stop blood loss completely if the hole in a blood vessel is not too large.

**FIGURE 18.11 Platelet plug formation.**

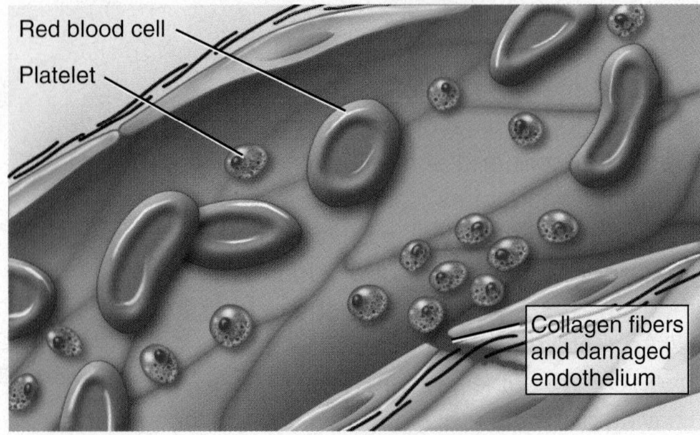


1 Platelet adhesion

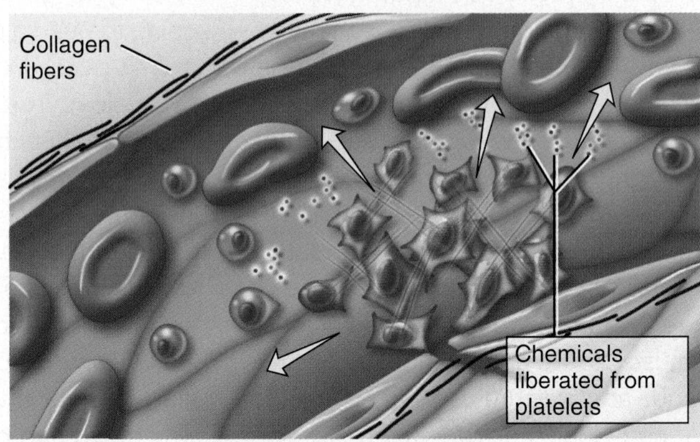


2 Platelet release reaction

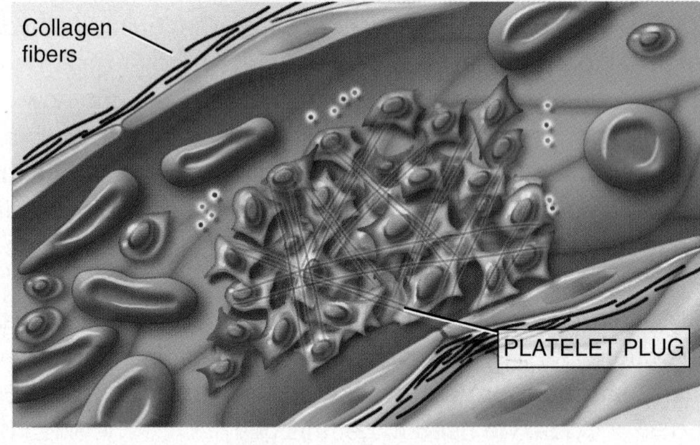


3 Platelet aggregation

**A platelet plug can stop blood loss completely if the hole in a blood vessel is small.**

## Blood Clotting

Normally, blood remains in its liquid form as long as it stays within its vessels. If it is withdrawn from the body, however, it thickens and forms a gel. Eventually, the gel separates from the liquid. The straw-colored liquid, called **serum**, is simply plasma minus the clotting proteins. The gel is called a **clot** and consists of a network of insoluble protein fibers called **fibrin** in which the formed elements of blood are trapped (Figure 18.12).

The process of gel formation, called **clotting** or **coagulation** (kō-ag-ū-LĀ-shun), is a series of chemical reactions that culminates in the formation of fibrin threads. If blood clots too easily, the result can be **thrombosis**—clotting in an undamaged blood vessel. If the blood takes too long to clot, hemorrhage can result.

Clotting is a complex cascade of enzymatic reactions in which various chemicals known as **clotting factors** (calcium ions, enzymes, and molecules associated with platelets or released from damaged tissues) activate one another in a fixed sequence. Finally, the series of chemical reactions forms a network of insoluble protein fibers. Clotting can be divided into three stages (Figure 18.13):

1. Two pathways, called the extrinsic pathway (Figure 18.13a) and the intrinsic pathway (Figure 18.13b), which will be

FIGURE 18.12 **Blood clot formation.** Notice the platelet and red blood cells entrapped in fibrin threads.

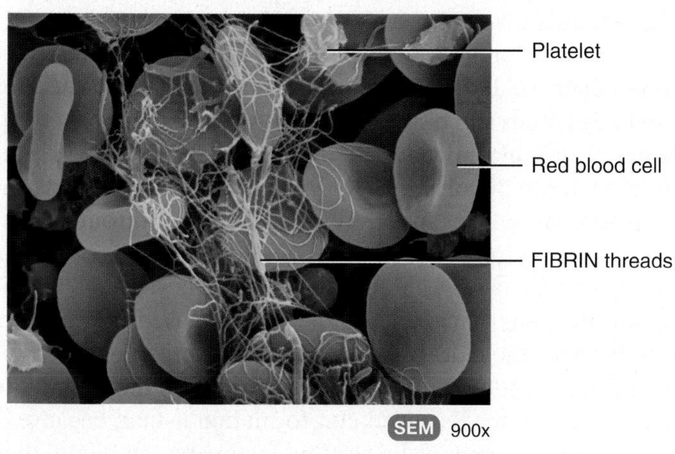


(a) Early stage

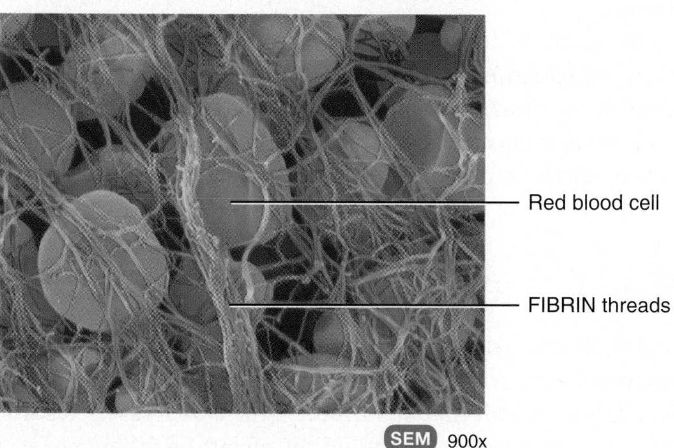


(b) Last stage showing red blood cells trapped in fibrin threads

A blood clot is a gel that contains formed elements of the blood entangled in fibrin threads.

FIGURE 18.13 **Blood clotting.** Green arrows represent positive feedback.

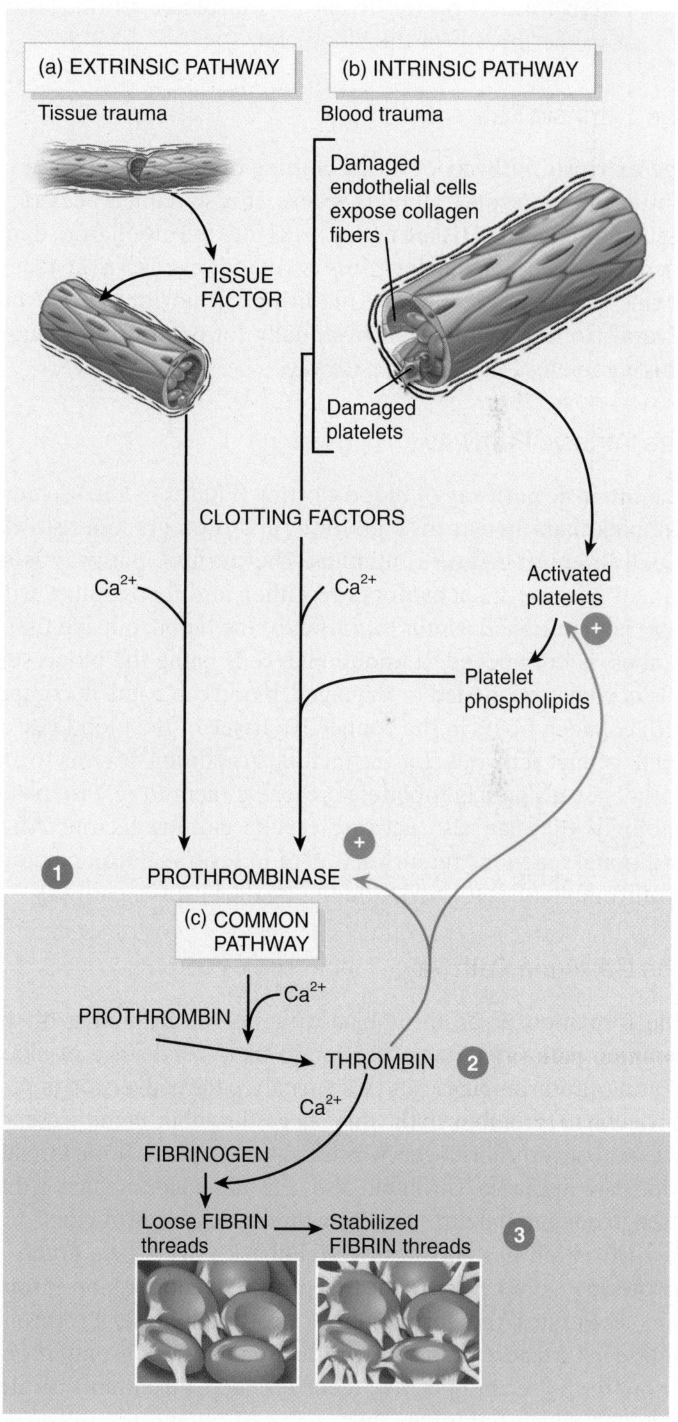


During blood clotting, the clotting factors activate one another, resulting in a cascade of reactions that includes positive feedback cycles.

described shortly, lead to the formation of the active enzyme **prothrombinase**. Once prothrombinase is formed, the steps involved in the next two stages of clotting are the same for both the extrinsic and intrinsic pathways, and together these two stages are referred to as the common pathway.

2 Prothrombinase converts **prothrombin** (a plasma protein formed by the liver) into the enzyme **thrombin**.

3 Thrombin converts soluble **fibrinogen** (another plasma protein formed by the liver) into insoluble fibrin. Fibrin forms the threads of the blood clot.

### The Extrinsic Pathway

The **extrinsic pathway** of blood clotting occurs rapidly, within a matter of seconds if trauma is severe. It is so named because a tissue protein called **tissue factor** leaks into the blood from damaged tissue cells *outside (extrinsic to)* blood vessels. Tissue factor begins a sequence of reactions requiring calcium ions ($Ca^{2+}$) and several clotting factors that eventually forms prothrombinase. This completes the extrinsic pathway.

### The Intrinsic Pathway

The **intrinsic pathway** of blood clotting (Figure 18.13b) is more complex than the extrinsic pathway, and it occurs more slowly, usually requiring several minutes. The intrinsic pathway is so named because its activators are either in direct contact with blood or contained *within (intrinsic to)* the blood; outside tissue damage is not needed. If endothelial cells lining the blood vessels become roughened or damaged, blood can come in contact with collagen fibers in the connective tissue of the blood vessel. Such contact activates clotting factors. In addition, trauma to endothelial cells damages platelets, causing them to release phospholipids that can also activate certain clotting factors. After additional reactions requiring $Ca^{2+}$ and several clotting factors, prothrombinase is formed, completing the intrinsic pathway.

### The Common Pathway

The formation of prothrombinase marks the beginning of the **common pathway** (Figure 18.13c). In the second stage of blood clotting, prothrombinase and $Ca^{2+}$ catalyze the conversion of prothrombin to thrombin. In the third stage, thrombin, in the presence of $Ca^{2+}$, converts fibrinogen, which is soluble, to loose fibrin threads, which are insoluble. Thrombin also activates a clotting factor that strengthens and stabilizes the fibrin threads into a sturdy clot.

Thrombin has two positive feedback effects (see green + arrows in Figure 18.13). In the first positive feedback loop, thrombin accelerates the formation of prothrombinase. Prothrombinase in turn accelerates the production of more thrombin, and so on. In the second positive feedback loop, thrombin activates platelets, which reinforces their aggregation and the release of platelet phospholipids. As the positive feedback loop continues, the fibrin clot grows.

### Clot Retraction and Blood Vessel Repair

Once a clot is formed, it plugs the ruptured area of the blood vessel and thus stops blood loss. **Clot retraction** is the consolidation or tightening of the fibrin clot. Platelets trapped around the fibrin threads contract, pulling on the fibrin threads and tightening the clot. Because the fibrin threads are attached to the damaged surfaces of the blood vessel, as the clot retracts, it pulls the edges of the damaged blood vessel closer together, decreasing the risk of further damage. Permanent repair of the blood vessel can then take place. In time, fibroblasts form connective tissue in the ruptured area, and new endothelial cells repair the blood vessel lining.

## Hemostatic Control Mechanisms

Even when blood vessels are not damaged, small clots begin to form many times a day, often at a site of minor roughness inside a blood vessel. Because blood clotting involves positive feedback cycles, a clot has a tendency to enlarge and potentially block blood flow through undamaged vessels. Usually, small, inappropriate clots dissolve in a process called **fibrinolysis** (fī′-bri-NOL-i-sis). When a clot is formed, an inactive plasma enzyme called **plasminogen** (plaz-MIN-ō-jen) is incorporated into the clot. Both body tissues and blood contain substances that can activate plasminogen to **plasmin**, an active plasma enzyme. Once plasmin is formed, it can dissolve the clot by digesting fibrin threads and inactivating substances such as fibrinogen and prothrombin.

Clot formation normally remains localized at the site of blood vessel damage. A clot does not extend beyond the wound site into the general circulation, in part because fibrin absorbs thrombin into the clot. This helps stop the spread of thrombin into the blood and thus inhibits clotting except at the wound site. Another reason for localized clot formation is that, because of the dispersal of some of the clotting factors by the blood, their concentrations are not high enough to bring about widespread clotting.

In addition, substances that suppress or prevent clotting, called **anticoagulants** (an′-tī-kō-AG-ū-lants), are present in blood. Most block the action of clotting factors. For example, basophils and mast cells produce *heparin*, an anticoagulant that helps block the action of thrombin.

## Clotting in Blood Vessels

Despite fibrinolysis and the action of anticoagulants, blood clots sometimes form within blood vessels. Such clots may be initiated by roughened endothelial surfaces of a blood vessel resulting from atherosclerosis (accumulation of fatty substances on arterial walls), trauma, or infection. These conditions induce adhesion of platelets. Clots may also form in blood vessels when blood flows too slowly, allowing clotting factors to accumulate locally in high enough concentrations to form a clot.

Clotting in an unbroken blood vessel is called **thrombosis** (throm-BŌ-sis; *thromb-* = clot; *-osis* = a condition of). The clot itself, called a **thrombus** (THROM-bus), may dissolve spontaneously. If it remains intact, however, the thrombus may become dislodged and be swept away in the blood. A blood clot, bubble of air, fat from broken bones, or a piece of debris transported by the bloodstream is called an **embolus** (*em-* = in; *-bolus* = a mass; plural is *emboli*). An embolus that breaks away from an arterial wall may lodge in a smaller-diameter artery downstream and block blood flow to a vital organ. If it blocks blood flow to the brain, kidney, or heart, the embolus can cause a stroke, kidney failure, or heart attack, respectively. When an embolus lodges in the lungs, the condition is called a **pulmonary embolism**. Massive emboli in the lungs may result in heart failure and death in a few minutes or hours.

### CLINICAL CONNECTION | *Anticoagulants*

Patients who are at increased risk of forming blood clots may receive anticoagulants. Examples are heparin or warfarin. Heparin is often administered during hemodialysis and open-heart surgery. **Warfarin (Coumadin®)** acts as an antagonist to vitamin K and thus blocks synthesis of four clotting factors. Warfarin is slower acting than heparin. To prevent clotting in donated blood, blood banks and laboratories often add substances that remove $Ca^{2+}$; examples are EDTA (ethylene diamine tetraacetic acid) and CPD (citrate phosphate dextrose).

### ✓ CHECKPOINT

**27.** Describe how vascular spasm and platelet plug formation occur.

**28.** What is the outcome of the first stage of clotting?

**29.** Why is the clotting mechanism described as a positive feedback cycle?

**30.** Why is adequate dietary calcium important to hemostasis?

**31.** How does clot retraction assist in blood vessel repair?

## Ed's Story — EPILOGUE AND DISCUSSION

Ed, a 46-year-old ecologist, returns from a field research excursion to Belize. Approximately four weeks after his return, he seeks medical attention for flu-like symptoms: fever, chills, sweating, nausea and vomiting, and abdominal pain. He admits to poor compliance with his malaria chemoprophylaxis (a chemical that prevents disease). A blood smear at the time of his exam indicates the presence of *Plasmodium falciparum*. His laboratory blood tests reveal he has thrombocytopenia (low platelet count), leukocytosis (high WBC count), hypoglycemia (low blood glucose level), hyponatremia (low blood sodium level), hyperkalemia (elevated potassium level), and elevated creatinine (waste product of muscle metabolism). A diagnosis of acute malaria with hemolytic anemia and acute renal failure requiring dialysis is made.

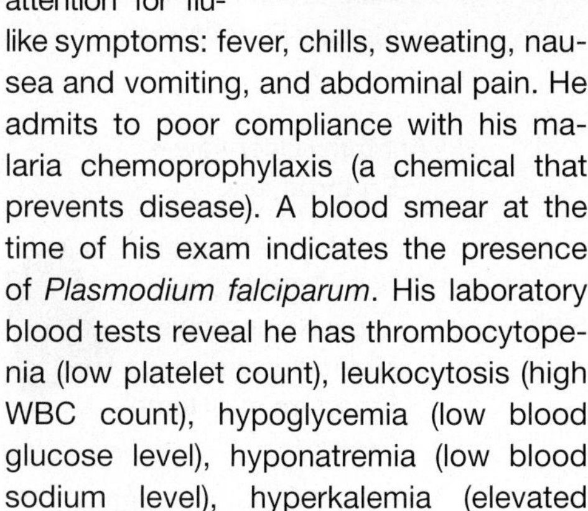

Ed is fortunate that his physician did not misdiagnose this illness as a common flu. The most revealing clue was that Ed had traveled to a region of the world where malaria is endemic. The malarial parasite is one of four species of *Plasmodium,* a protozoan with a multistage life cycle. The majority of the pest's time in humans is spent multiplying inside RBCs; when the number of creatures is great enough, the RBCs lyse, discharging organisms that infect more RBCs. Since the organisms live inside RBCs and multiply rapidly, WBCs have difficulty mounting a significant immune response. A regular pattern of fevers and chills following a 48-hour cycle is observed in most patients with this condition. The fever spikes correlate with the release of toxins into the bloodstream from massive RBC hemolysis.

J. *Which organ systems were involved in Ed's case?*

K. *How would the hemolysis of Ed's RBCs affect the function of his blood?*

L. *Why would you expect Ed to have a high eosinophil count?*

M. *Why would Ed's low platelet count concern his physicians? What are the three hemostatic mechanisms that normally occur in a healthy person?*

N. *If Ed's platelet count is low, which hormone could be administered to stimulate platelet production?*

# Concept and Resource Summary

**Concept** | **Resources**

## Introduction

1. The **cardiovascular system** consists of the blood, the heart, and blood vessels.
2. **Blood** is a connective tissue composed of plasma (liquid portion) and formed elements (cells and cell fragments).

## Concept 18.1 Blood contains plasma and formed elements and transports essential substances through the body.

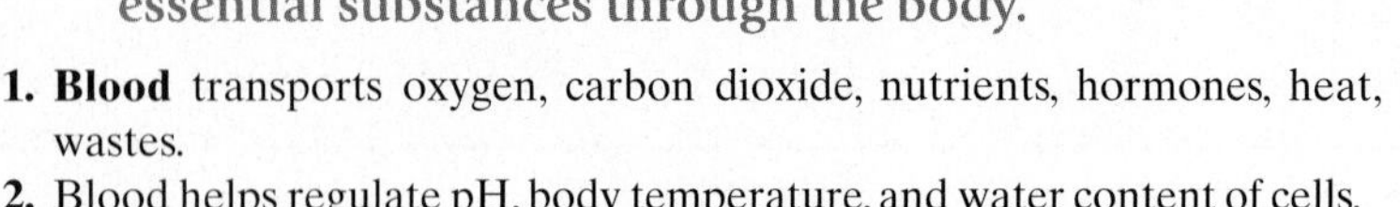

1. **Blood** transports oxygen, carbon dioxide, nutrients, hormones, heat, and metabolic wastes.
2. Blood helps regulate pH, body temperature, and water content of cells.
3. It provides protection through clotting and by combating toxins and microbes through certain phagocytic white blood cells or specialized blood plasma proteins.
4. Blood is viscous and slightly alkaline, ranging from pH 7.35 to 7.45 with a temperature of 38°C (100.4°F). Its volume is 4–6 liters.
5. Whole blood is composed of plasma, a watery liquid containing dissolved substances, and formed elements, which are cells and cell fragments.
6. **Plasma** consists of 91.5 percent water and 8.5 percent solutes. Principal solutes include proteins (albumins, globulins, fibrinogen), nutrients, vitamins, hormones, respiratory gases, electrolytes, and waste products.
7. The **formed elements** include red blood cells (RBCs), white blood cells (WBCs), and platelets.
8. A **hematocrit** is the percentage of total blood volume occupied by RBCs.

## Concept 18.2 Hemopoiesis is the production of formed elements.

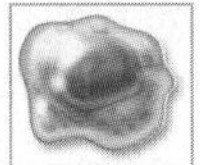

1. Hemopoiesis is the formation of blood cells from hemopoietic stem cells in red bone marrow.
2. The **pluripotent stem cells** of the red bone marrow give rise to lymphoid stem cells and myeloid stem cells.
3. Myeloid stem cells differentiate into RBCs, platelets, mast cells, granulocytes, and monocytes. Lymphoid stem cells start differentiating into lymphocytes in red bone marrow, and then complete development in lymphatic tissues.

**Resources:**

Anatomy Overview—Erythrocytes

Figure 18.3—Origin, Development, and Structure of Formed Elements

Clinical Connection—Bone Marrow Examination

## Concept 18.3 Mature red blood cells are biconcave cells containing hemoglobin.

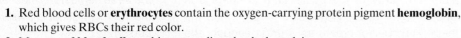

1. Red blood cells or **erythrocytes** contain the oxygen-carrying protein pigment **hemoglobin**, which gives RBCs their red color.
2. Mature **red blood cells** are biconcave discs that lack nuclei.
3. A hemoglobin molecule consists of the protein **globin** composed of four polypeptide chains; bound to each chain is a ringlike, nonprotein pigment called **heme**. Heme contains iron that binds to oxygen. About 23 percent of the carbon dioxide transported by the blood combines with globin.

**Resources:**

Anatomy Overview—Erythrocytes

Figure 18.4—Shapes of an Erythrocyte and a Hemoglobin Molecule, and Structure of a Heme Group

Clinical Connection—Sickle Cell Disease

| Concept | Resources |
|---|---|

## Concept 18.4 Red blood cells have a life cycle of 120 days.

1. The life span of a red blood cell is about 120 days. Fragile, old, or damaged RBCs are destroyed by phagocytic macrophages.
2. As macrophages phagocytize RBCs, their hemoglobin is broken down and recycled. Amino acids from globin are used to make proteins. Iron from heme is used to synthesize hemoglobin for new RBCs. The non-iron portion of heme is eventually converted to **bilirubin** and secreted into bile that passes into the intestines.

Figure 18.5—Formation and Destruction of Red Blood Cells and Recycling of Hemoglobin Components

Clinical Connection—Medical Uses of Hemopoietic Growth Factors

Clinical Connection—Iron Overload and Tissue Damage

## Concept 18.5 Erythropoiesis is the process of red blood cell formation.

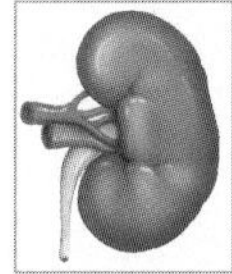

1. The production of red blood cells, termed **erythropoiesis**, begins in the red bone marrow. **Reticulocytes** enter the bloodstream where, within 1 to 2 days, they become mature erythrocytes.
2. Normally, erythropoiesis occurs at the same pace as red blood cell destruction.
3. Erythropoiesis is stimulated by **hypoxia**, which results in the release of erythropoietin by the kidneys.

Animation—Erythropoietin

Clinical Connection—Anemia

Clinical Connection—Blood Doping

## Concept 18.6 Blood is categorized into groups based on surface antigens.

1. Genetically determined **agglutinogens** are found on the surfaces of RBCs. These surface antigens are the basis for different **blood groups** and **blood types**.
2. The **ABO blood group** is based on the presence or absence of antigen A and antigen B on the RBC surface. Plasma typically contains **anti-A antibodies** or **anti-B antibodies**, called **agglutinins**, directed against A or B antigens, respectively.
3. The transfer of whole blood or blood components into the bloodstream is called a **transfusion**. In an incompatible transfusion, the recipient's antibodies bind to antigens on the donated RBCs, which cause RBC clumping, or **agglutination**.
4. The **Rh blood group** is based on the Rh antigen. RBCs with Rh antigen are classified as $Rh^+$, and those without the Rh antigen are $Rh^-$.
5. Before blood is transfused, a recipient's blood is typed and then either **cross-matched** to potential donor blood or **screened** for the presence of antibodies.

Clinical Connection—Hemolytic Disease of the Newborn

## Concept 18.7 White blood cells combat inflammation and infection.

1. White blood cells or **leukocytes** have a nucleus, and do not have hemoglobin.
2. The granular leukocytes have visual differences under a light microscope. **Eosinophils** have large, uniform granules stained red-orange and a nucleus with two or three lobes. **Basophils** have variable-sized granules stained blue-purple that obscure the two-lobed nucleus. **Neutrophils** have small, pale lilac granules. The nucleus has two to five lobes.
3. The agranular leukocytes can be distinguished. **Lymphocytes** have a sky blue cytoplasm with a dark, round, slightly indented nucleus. **Monocytes** have a blue-gray cytoplasm with a foamy appearance and a kidney-shaped nucleus. Monocytes that migrate into tissues enlarge into **macrophages**.
4. WBCs can leave the bloodstream by **emigration** to move to tissue sites of pathogen invasion or inflammation.

Anatomy Overview—White Blood Cells

Figure 18.9—Types of White Blood Cells

| Concept | Resources  |
|---|---|

5. **Chemotaxis** occurs when pathogens and inflamed tissues release chemicals that attract phagocytic neutrophils and macrophages. These cells ingest and dispose of pathogens and dead matter during **phagocytosis**. Neutrophils respond most rapidly to an infection site. Monocytes arrive later at an infection site but arrive in large numbers.
6. Eosinophils stop the effects of histamine and other mediators of inflammation in allergic reactions. They also attack parasitic worms.
7. Basophils release heparin, histamine, and serotonin, which intensify the inflammatory reaction.
8. Lymphocytes function in immune responses. B cells destroy bacteria. T cells combat viruses, fungi, transplanted cells, cancer cells, and some bacteria. Natural killer cells attack various microbes and tumor cells.
9. Most WBCs live only a few hours to a few days. A **differential white blood cell count** measures the number of each type of WBC in a sample of 100 WBCs.

Concepts and Connections—Blood
Clinical Connection—Leukemia

## Concept 18.8 Platelets reduce blood loss from damaged vessels.

1. The hormone **thrombopoietin** stimulates myeloid stem cells to develop into megakaryocytes. These huge cells splinter into 2000–3000 fragments and enter the bloodstream. Each fragment is a platelet or **thrombocyte**.
2. Platelets lack a nucleus and have chemicals that promote blood clotting. Platelets stop blood loss in damaged vessels by forming a platelet plug in the vessel wall.

Anatomy Overview—Platelets
Concepts and Connections—Blood
Clinical Connection—Stem Cell Transplants from Bone Marrow and Cord Blood

## Concept 18.9 Hemostasis is the sequence of events that stops bleeding from a damaged blood vessel.

1. **Hemostasis** is a sequence of responses to stop blood loss from a damaged blood vessel. Three mechanisms are involved: vascular spasm, platelet plug formation, and blood clotting.
2. **Vascular spasms** of the smooth muscle in the wall of a damaged vessel help reduce blood loss.
3. Platelets that come into contact with damaged blood vessels aggregate into a **platelet plug** to stop bleeding.
4. A **clot** is a network of insoluble protein fibers (**fibrin**) in which formed elements of blood are trapped. **Clotting** is a cascade of reactions involving **clotting factors** that activate one another. Stages of clotting include the following: formation of **prothrombinase**, conversion of **prothrombin** into **thrombin**, and conversion by thrombin of soluble **fibrinogen** into insoluble **fibrin**.
5. Clotting is initiated by the interplay of the **extrinsic** and **intrinsic pathways** of blood clotting. Because the next two stages of clotting are the same for both pathways, they are collectively called the **common pathway**.
6. The clot plugs the damaged area of the blood vessel and undergoes **clot retraction**, which pulls the blood vessel edges closer together.
7. Clotting in an unbroken blood vessel is called **thrombosis**; the clot is called a **thrombus**. A dislodged thrombus can be swept away in the blood and is then called an **embolus**.

Figure 18.13—Blood Clotting

Concepts and Connections—Blood
Clinical Connection—Hemophilia
Clinical Connection—Aspirin and Thrombolytic Agents

## Understanding the Concepts

1. How does the volume of blood in your body compare to the volume of fluid in a two-liter bottle of soda?
2. What is the significance of a lower-than-normal hematocrit? A higher-than-normal hematocrit? What might be the effect of a bacterial infection on the hematocrit?
3. How is the development of lymphocytes unique when compared with the development of the other formed elements?
4. How are red blood cells able to squeeze through capillaries that are smaller in diameter than they are?
5. Why do red blood cells live for only about 120 days?
6. What is erythropoiesis? Which factors speed up and slow down erythropoiesis?
7. What would happen if a person with type B blood were given a transfusion of type O blood?
8. During an anatomy and physiology exam you are asked to view white blood cells in prepared slides of standard human blood smears. Based on the observations below, what is the name and function of each WBC?
   - WBC has a round nucleus surrounded by a blue halo of cytoplasm with no visible granules.
   - WBC contains dense blue-purple granules that hide the nucleus.
   - WBC has a U-shaped nucleus and a bluish, foamy cytoplasm with no visible granules.
   - WBC contains small, pale lilac granules and a four-lobed nucleus.
   - WBC contains red-orange granules and a two-lobed nucleus.
9. Why would the level of leukocytes be higher in an individual who has been infected with a parasitic disease?
10. In regions where malaria is endemic, some people build up immune resistance to the malaria pathogen. Which WBCs are responsible for the immune response against pathogens? How do they function?
11. Why are platelets not called blood cells?
12. What is the function of prothrombinase and thrombin in clotting? Explain how the extrinsic and intrinsic pathways of blood clotting differ.

CHAPTER

19

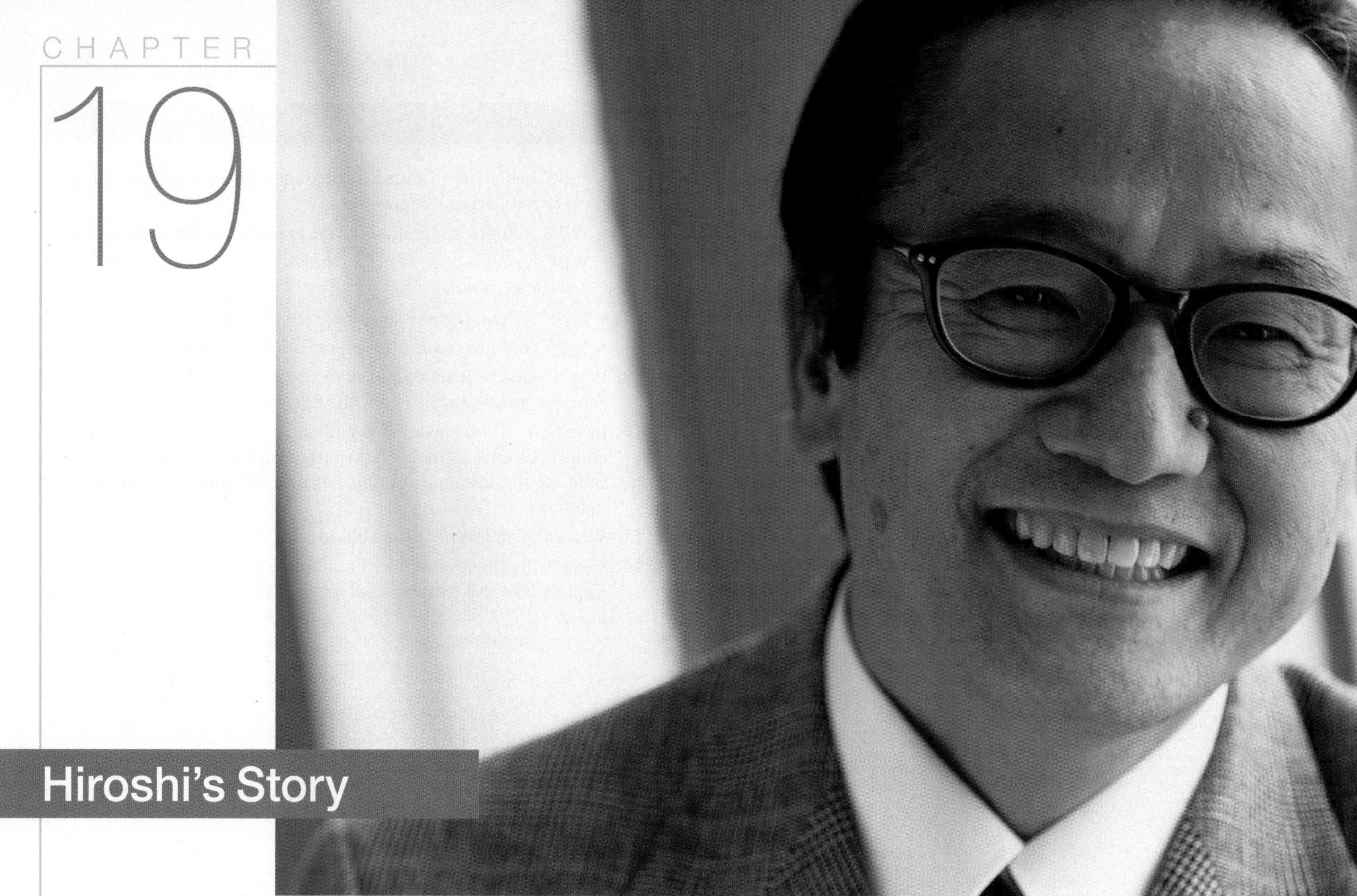

# Hiroshi's Story

Hiroshi's business meeting ran late, and he is rushing through Heathrow Airport to make sure he's in time to check his luggage for his flight home to Chicago. He arrives at the ticket counter to face a long, slow-moving line, but he finally makes it to the front and gets his boarding pass. The clerk tells him to hurry: he only has thirty minutes to make it through security and to his gate for boarding.

Without his luggage Hiroshi is able to move more quickly, but the security line, as usual, is long. He places his laptop in a bin and pushes it toward the screener. He walks through the metal detector, collects his computer, and runs down the concourse to make his flight. By the time he gets in the boarding line, he is gasping for air, and thinks to himself, "I am really out of shape. I should have been able to run from security to the gate without getting this short of breath."

Hiroshi boards the plane, finds his window seat, and collapses into it, wishing he had a bottle of water to drink. He leans his head against the window, watching the last of the luggage being loaded into the belly of the plane. His breathing is a little easier, but his chest feels tight and he is experiencing a burning sensation. He decides to take one of the pills that his doctor prescribed to help him relax while flying, but it will have to wait until the flight attendant offers him something to drink. In the meantime, he closes his eyes and tries to focus on the meditation techniques he has learned to ease himself into a more relaxed state.

Hiroshi is a 49-year-old man with a high-stress engineering job. He has no personal history of health problems, but his grandfather died of a sudden heart attack at the age of 70 and his father died of the same condition at the age of 60. Hiroshi is having a heart attack. What causes a heart attack, and what are its complications? You will learn the answers to these questions and others in this chapter on the circulatory system, which emphasizes the crucial role of the heart in maintaining the supply of oxygen to the body.

# The Cardiovascular System: The Heart

## INTRODUCTION

In the last chapter we learned that the **cardiovascular system** consists of the blood, the heart, and blood vessels. We also examined the composition and functions of blood. For blood to reach body cells and exchange materials with them, it must constantly be pumped by the heart through the body's blood vessels. The heart beats about 100,000 times every day, which adds up to about 35 million beats in a year and approximately 2.5 billion times in an average lifetime. The left side of the heart pumps blood through an estimated 100,000 km (60,000 mi) of blood vessels, which is equivalent to traveling around the earth's equator about 3 times. The right side of the heart pumps blood through the lungs, so that it can pick up oxygen and unload carbon dioxide. Even while you are sleeping, your heart pumps 30 times its own weight each minute, which amounts to about 5 liters (5.3 qt) to the lungs and the same volume to the rest of the body. At this rate, the heart pumps more than 14,000 liters (3600 gal) of blood in a day or 10 million liters (2.6 million gal) in a year. However, you don't spend all of your time sleeping, and your heart pumps more vigorously when you are active; the actual blood volume the heart pumps in a single day is even larger. This chapter explores the structure of the heart and the unique properties that permit it to pump for a lifetime without a moment of rest.

## CONCEPTS

**19.1** The heart is located in the mediastinum and has a muscular wall covered by pericardium.

**19.2** The heart has four chambers, two upper atria and two lower ventricles.

**19.3** Heart valves ensure one-way flow of blood.

**19.4** The heart pumps blood to the lungs for oxygenation, then pumps oxygen-rich blood throughout the body.

**19.5** The cardiac conduction system coordinates heart contractions for effective pumping.

**19.6** The electrocardiogram is a record of electrical activity associated with each heartbeat.

**19.7** The cardiac cycle represents all of the events associated with one heartbeat.

**19.8** Cardiac output is the blood volume ejected by a ventricle each minute.

# 19.1 The heart is located in the mediastinum and has a muscular wall covered by pericardium.

## Location of the Heart

For all its might, the **heart** is relatively small, roughly the same size as your closed fist—about 12 cm (5 in.) long, 9 cm (3.5 in.) wide at its broadest point, and 6 cm (2.5 in.) thick. The heart is enclosed in the **mediastinum** (mē′-dē-a-STĪ-num), an anatomical region between the lungs that extends from the sternum to the vertebral column, and between the lungs (Figure 19.1a, b). The heart is located near the midline of the thoracic cavity. Recall that the midline is an imaginary vertical line that divides the body into unequal left and right sides. About two-thirds of the mass of the heart lies to the left of the body's midline. You can visualize the heart as a cone lying on its side. The pointed **apex** is formed by the tip of the left ventricle (a lower chamber of the heart) and rests on the diaphragm. The apex is directed anteriorly, inferiorly, and to the left. The **base** of the heart, the broad superior portion opposite the apex, is formed by the atria (upper chambers of the heart). The major blood vessels of the heart enter and exit at the base.

## Pericardium

The membrane that surrounds and protects the heart is the **pericardium** (per′-i-KAR-dē-um; *peri-* = around). It confines the heart to its position in the mediastinum, while allowing sufficient freedom of movement for vigorous and rapid contraction. The pericardium consists of two principal parts: the fibrous pericardium and the

**FIGURE 19.1 Position of the heart and associated structures in the mediastinum (dashed outline).** In this and subsequent illustrations, blood vessels that carry oxygenated blood (which looks bright red) are colored red, and those that carry deoxygenated blood (which looks dark red) are colored blue.

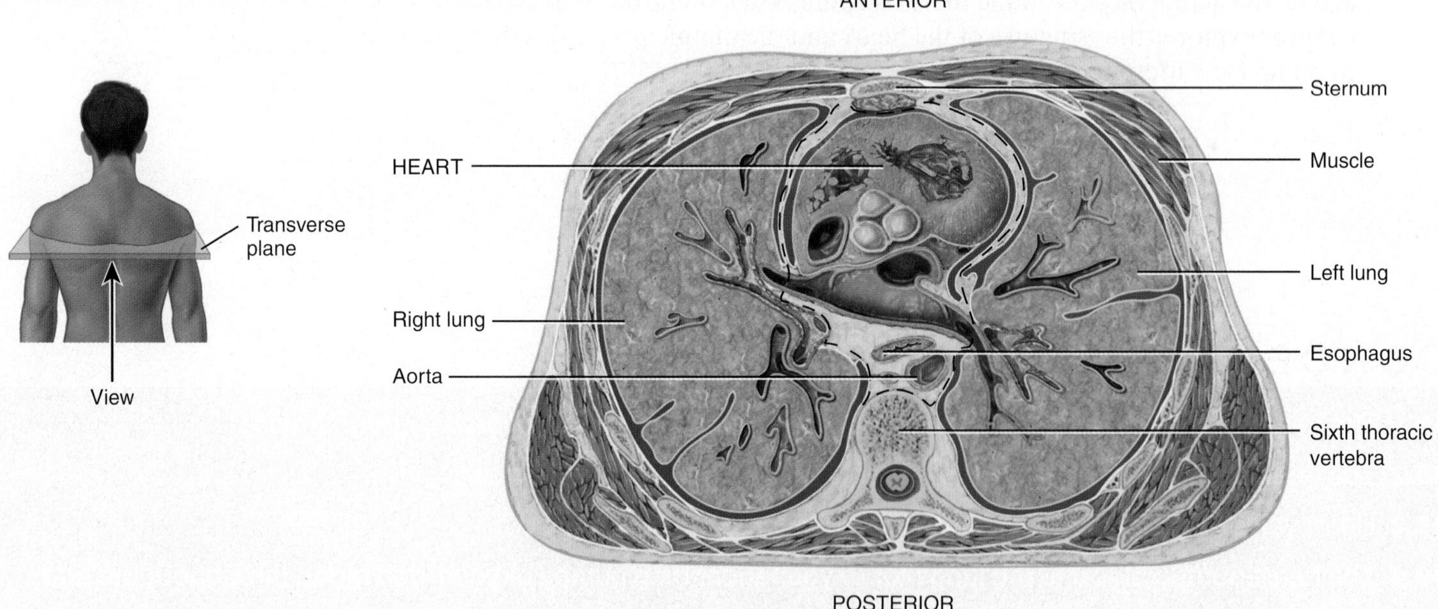


(a) Inferior view of transverse section of thoracic cavity showing heart in mediastinum

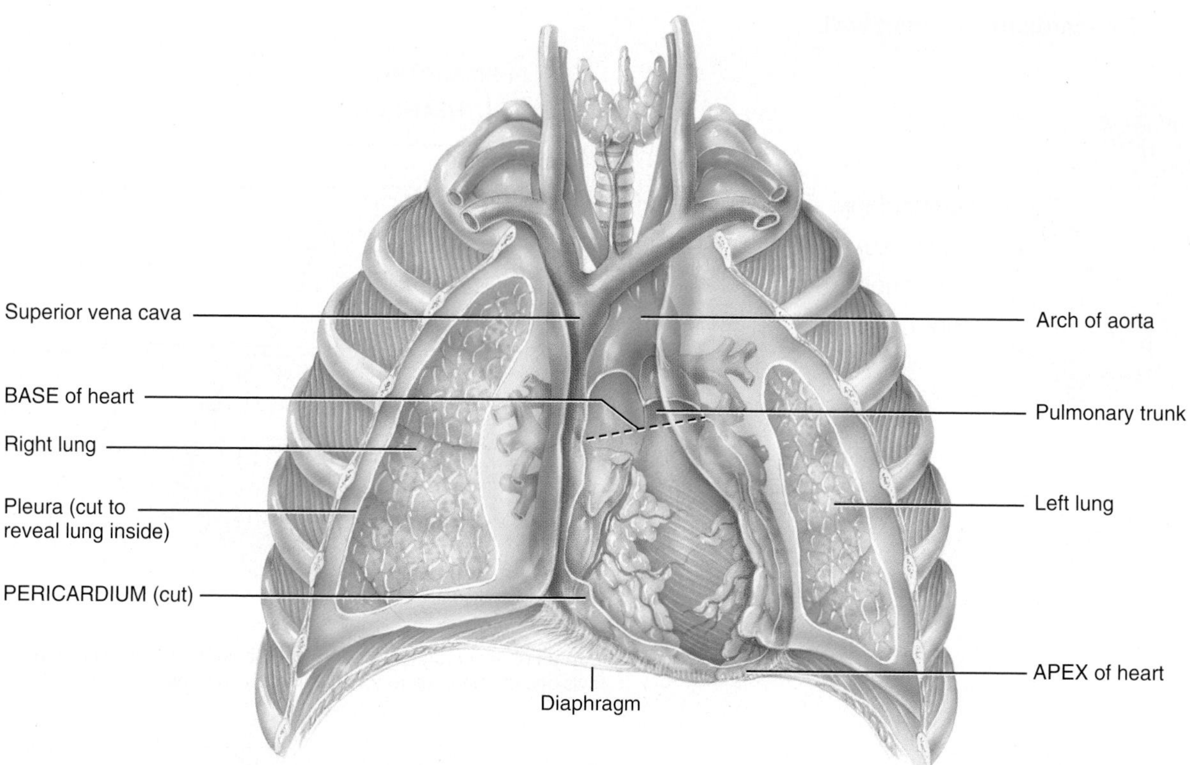


(b) Anterior view of heart in thoracic cavity

**The heart is located between the lungs, with about two-thirds of its mass to the left of the midline.**

serous pericardium (Figure 19.2a). The superficial **fibrous pericardium** is composed of tough, dense irregular connective tissue. It prevents overstretching of the heart, provides protection, and anchors the heart in the mediastinum.

The deeper **serous pericardium** is a thinner, more delicate membrane that forms a double layer around the heart (Figure 19.2a, b). The outer **parietal layer** of the serous pericardium is fused to the fibrous pericardium, and the inner **visceral layer**, also called the **epicardium** (ep′-i-KAR-dē-um; *epi-* = on top of), is one of the layers of the heart wall and adheres tightly to the surface of the heart. Between the parietal and visceral layers of the serous pericardium is a **pericardial cavity** containing **pericardial fluid**, a slippery, lubricating secretion of the pericardial membranes that reduces friction between the serous pericardial membranes as the heart moves.

## Layers of the Heart Wall

The wall of the heart (Figure 19.2a) is composed of three layers: epicardium (external layer), myocardium (middle layer), and endocardium (inner layer). The epicardium is composed of two tissue layers. As noted earlier, the **epicardium** is also called the visceral layer of serous pericardium. This thin, transparent outer layer of the heart wall is composed of mesothelium and connective tissue that imparts a smooth, slippery texture to the outermost surface of the heart. The epicardium contains blood vessels and lymphatic vessels that supply the myocardium.

The middle **myocardium** (mī′-ō-KAR-dē-um; *myo-* = muscle) makes up approximately 95 percent of the heart wall. It is composed of cardiac muscle tissue and is responsible for its pumping action. The cardiac muscle fibers are organized in bundles that swirl diagonally around the heart and generate the strong pumping actions of the heart (Figure 19.2c). Although cardiac muscle fibers are striated like skeletal muscle fibers, they are involuntary like smooth muscle fibers (see Concept 4.7 for a discussion of muscle histology).

The innermost **endocardium** (en′-dō-KAR-dē-um; *endo-* = within) is a thin layer of endothelium and connective tissue that provides a smooth lining for the heart chambers and covers the heart valves (Figure 19.2a). The smooth endothelial lining minimizes the surface friction as blood passes through the heart. The endocardium is continuous with the endothelial lining of the large blood vessels attached to the heart.

FIGURE 19.2 **Pericardium and heart wall.**

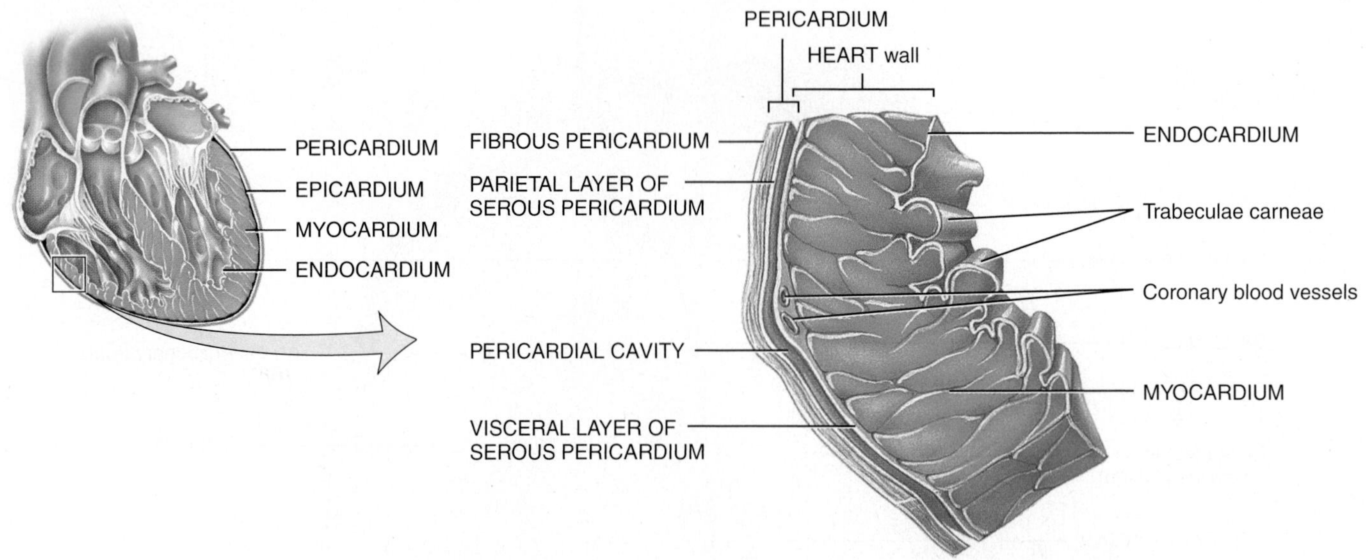


(a) Portion of pericardium and right ventricular heart wall showing divisions of pericardium and layers of heart wall

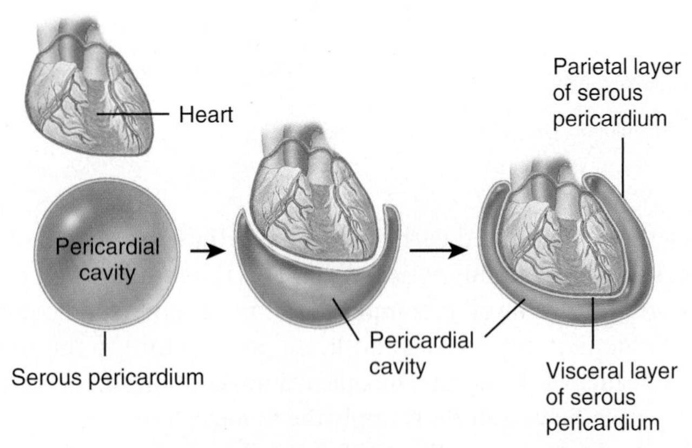


(b) Simplified relationship of serous pericardium to heart

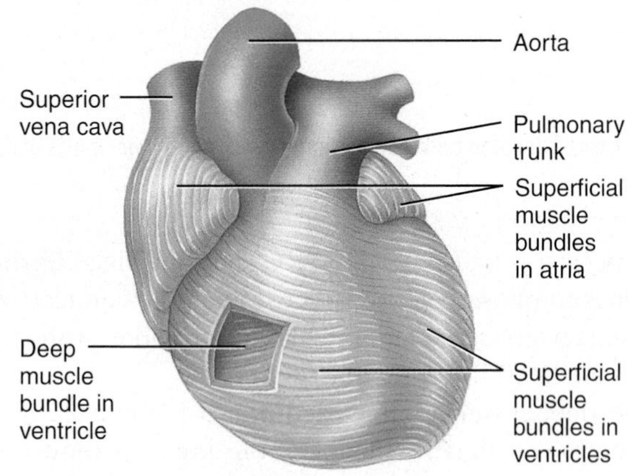


(c) Cardiac muscle bundles of myocardium

**The pericardium is a triple-layered sac that surrounds and protects the heart.**

## CLINICAL CONNECTION | *Cardiopulmonary Resuscitation*

Because the heart lies between two rigid structures—the vertebral column and the sternum (Figure 19.1a)—external pressure on the chest (compression) can be used to force blood out of the heart and into the circulation. In cases in which the heart suddenly stops beating, **cardiopulmonary resuscitation** (**CPR**) (kar-dē-ō-PUL-mo-nar′-ē rē-sus-i-TĀ-shun)—properly applied cardiac compressions, performed with artificial ventilation of the lungs via mouth-to-mouth respiration—saves lives. CPR keeps oxygenated blood circulating until the heart can be restarted.

Researchers have found that chest compressions alone are equally as effective as, if not better than, traditional CPR with lung ventilation. This is good news because it is easier for an emergency dispatcher to give instructions limited to chest compressions to frightened, nonmedical bystanders. As public fear of contracting contagious diseases such as hepatitis, HIV, and tuberculosis continues to rise, bystanders are much more likely to perform chest compressions alone than treatment involving mouth-to-mouth rescue breathing.

✓ CHECKPOINT

1. How is the heart positioned in the mediastinum?
2. What is the function of pericardial fluid?
3. From most superficial to deepest, what are the layers of the pericardium and heart wall?

## 19.2 The heart has four chambers, two upper atria and two lower ventricles.

The heart contains four chambers. The two upper receiving chambers are the **atria** (= entry halls or chambers), and the two lower pumping chambers are the **ventricles** (= little bellies). The paired atria receive blood from *veins*, blood vessels returning blood to the heart, while the ventricles eject the blood from the heart into blood vessels called *arteries*. On the anterior surface of each atrium is a wrinkled pouchlike structure called an **auricle** (OR-i-kul; *auri-* = ear), so named because of its passing resemblance to a dog's ear (Figure 19.3). Each auricle slightly increases the capacity of an atrium so that it can hold a greater volume of blood. Also on the surface of the heart are a series of grooves, called *sulci* (SUL-sī), that contain coronary blood vessels and a variable amount of fat. Each sulcus (SUL-kus; singular) marks the external boundary between two chambers of the heart. The **coronary sulcus** (*coron-* = resembling a crown) encircles most of the heart and marks the external boundary between the superior atria and inferior ventricles. The **anterior interventricular sulcus** (in′-ter-ven-TRIK-ū-lar) is a shallow groove on the anterior surface of the heart that marks the external boundary between the right and left ventricles. This sulcus continues around to the posterior surface of the heart as the **posterior interventricular sulcus**, which marks the external boundary between the ventricles on the posterior aspect of the heart.

FIGURE 19.3 Structure of the heart: surface anatomy.

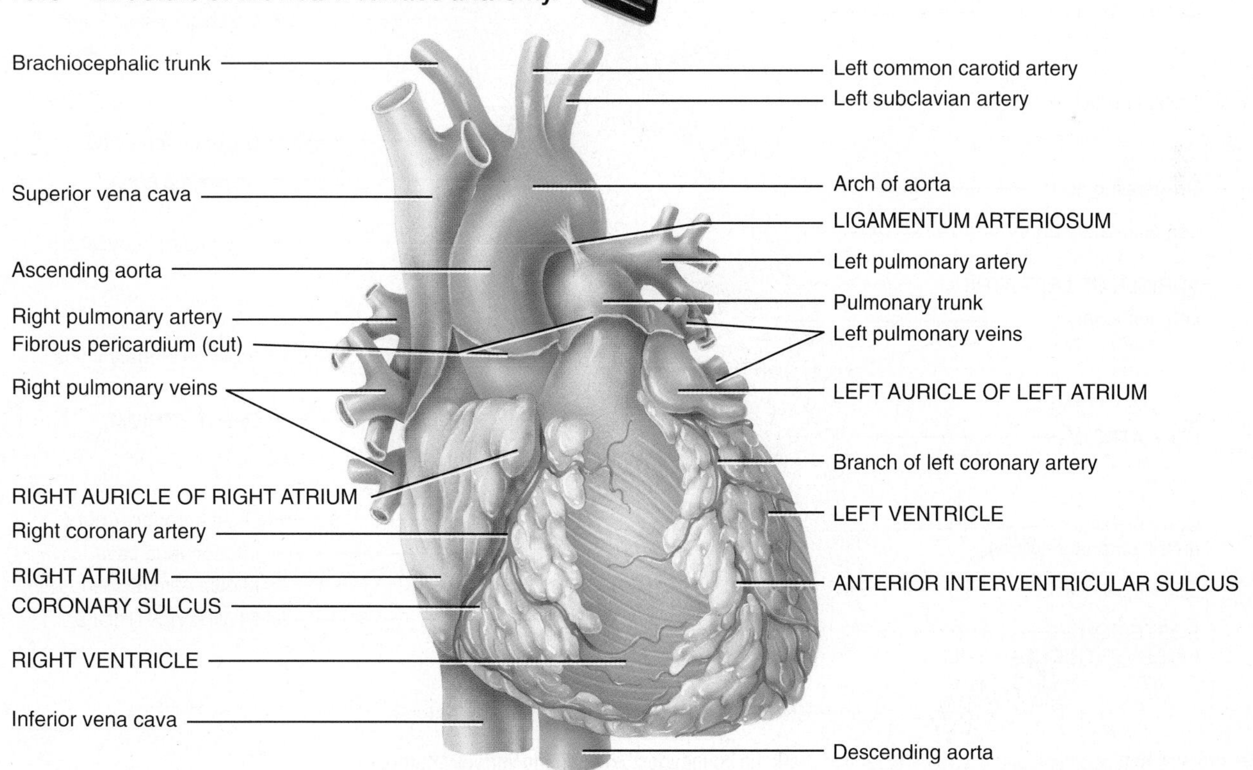


(a) Anterior external view showing surface features

(continues)

**FIGURE 19.3 (continued)**

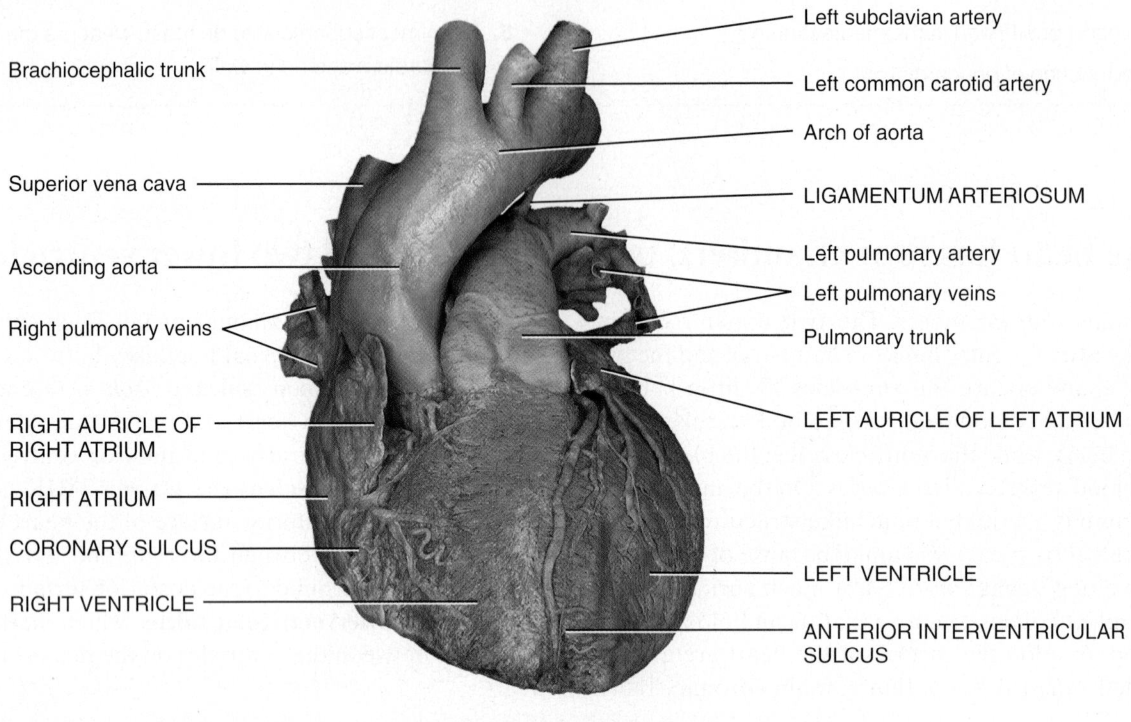


(b) Anterior external view

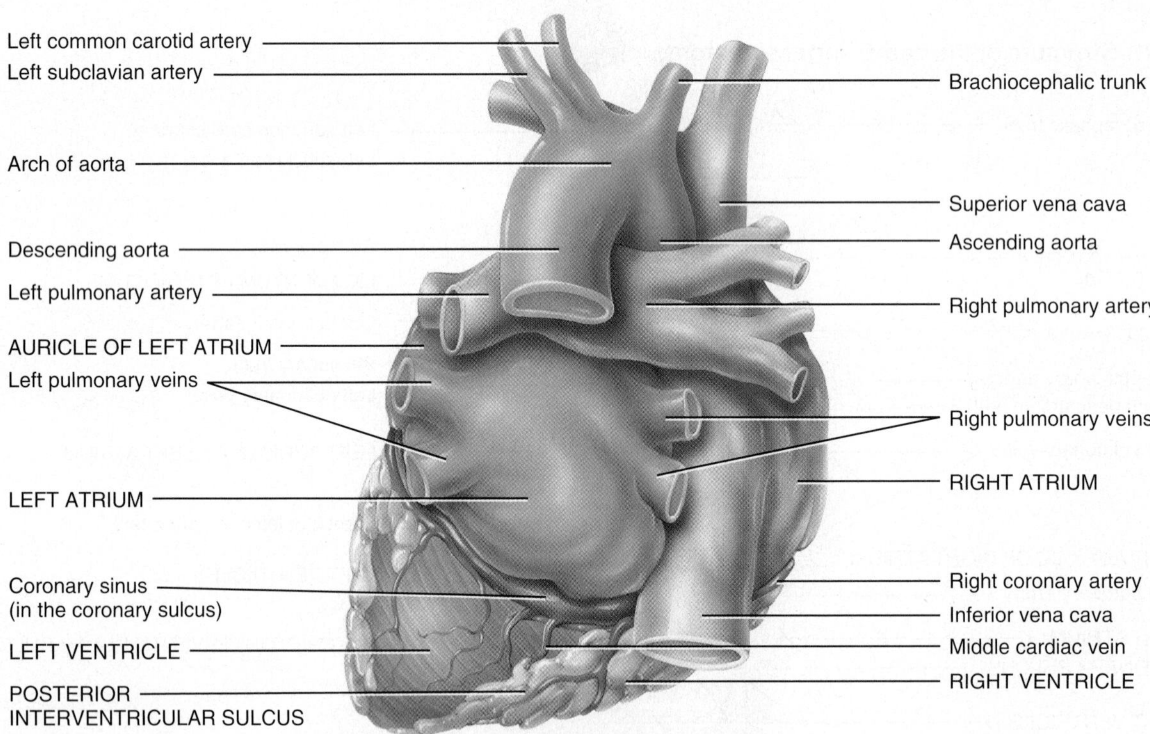


(c) Posterior external view showing surface features

**Sulci are grooves that contain blood vessels and fat and mark the boundaries between the various chambers.**

## Right Atrium

The **right atrium** receives blood from three veins: *superior vena cava, inferior vena cava*, and *coronary sinus* (Figure 19.4). The posterior wall of the right atrium is smooth, but the anterior wall is rough due to muscular ridges called **pectinate muscles** (PEK-tin-āt; *pectin* = comb), which also extend into the auricle (Figure 19.4b). Between the right atrium and left atrium is a thin partition called the **interatrial septum** (*inter-* = between; *-septum* = a dividing wall or partition). A prominent feature of this septum is an oval depression called the **fossa ovalis**, the remnant of the *foramen ovale*, an opening in the interatrial septum of the fetal heart that normally closes soon after birth. Blood passes from the right atrium into the right ventricle through the **tricuspid valve** (trī-KUS-pid; *tri-* = three; *-cuspid* = point), so named because it consists of three flaps or **cusps** (Figure 19.4a).

## Right Ventricle

The **right ventricle** forms most of the anterior surface of the heart (Figure 19.3a). The interior of the right ventricle contains a series of ridges formed by raised bundles of cardiac muscle fibers called **trabeculae carneae** (tra-BEK-ū-lē KAR-nē-ē; *trabeculae* = little beams; *carneae* = fleshy; Figure 19.4). The cusps of the tricuspid valve are connected to tendonlike cords, the **chordae tendineae** (KOR-dē ten-DI-nē-ē; *chord-* = cord; *tend-* = tendon), which in turn are connected to cone-shaped trabeculae carneae called **papillary muscles** (*papill-* = nipple). The right ventricle is separated from the left ventricle by a partition called the **interventricular septum**. Blood passes from the right ventricle through the **pulmonary valve** into a large artery, the *pulmonary trunk*, which carries blood to the lungs for oxygenation. Arteries always take blood away from the heart.

## Left Atrium

The **left atrium** receives oxygenated blood from the lungs through four *pulmonary veins* (Figure 19.4). Like the right atrium, the inside of the left atrium has a smooth posterior wall. Because the ridged pectinate muscles are confined to the auricle of the left atrium, the anterior wall of the left atrium is smooth. Blood passes from the left atrium into the left ventricle through

FIGURE 19.4 **Structure of the heart: internal anatomy.**

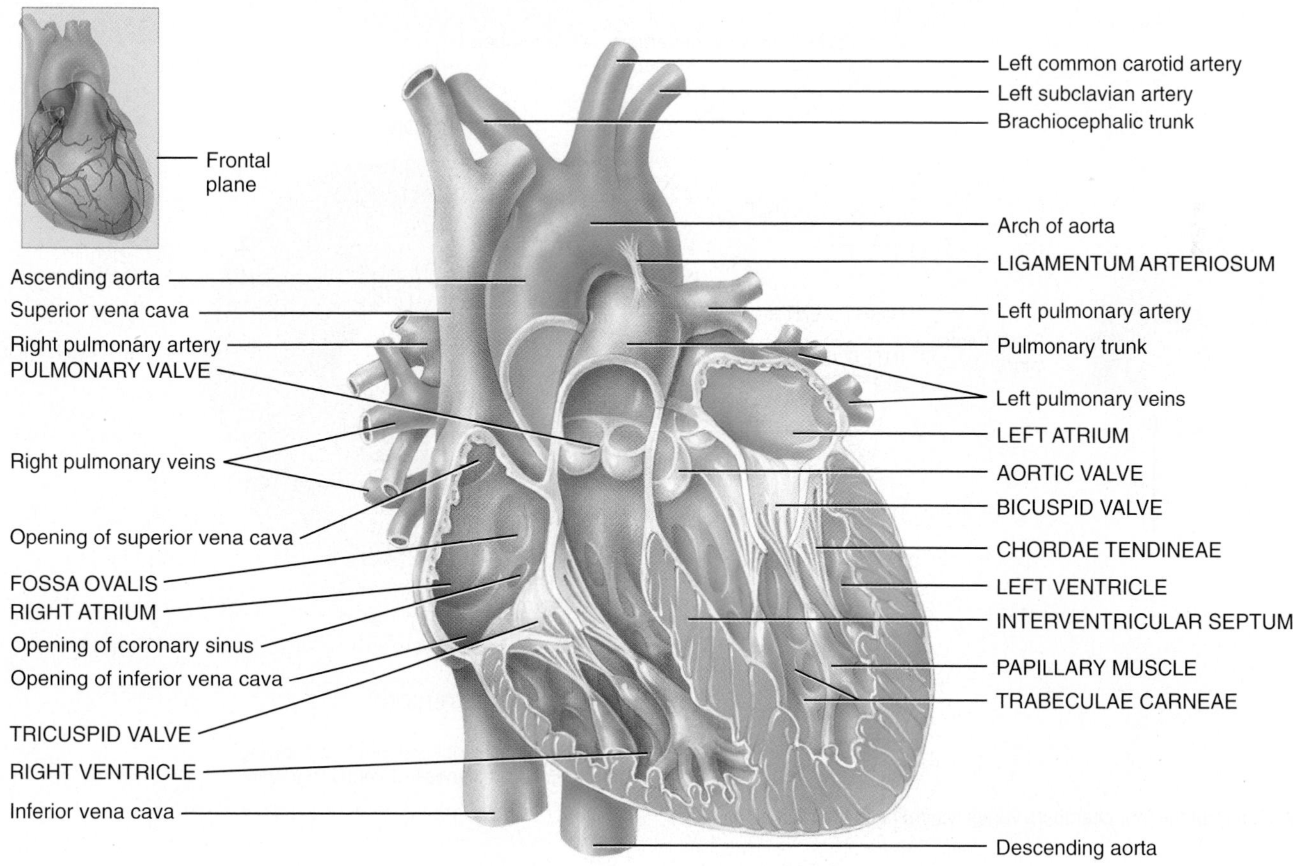


(a) Anterior view of frontal section showing internal anatomy

(continues)

FIGURE 19.4 (continued)

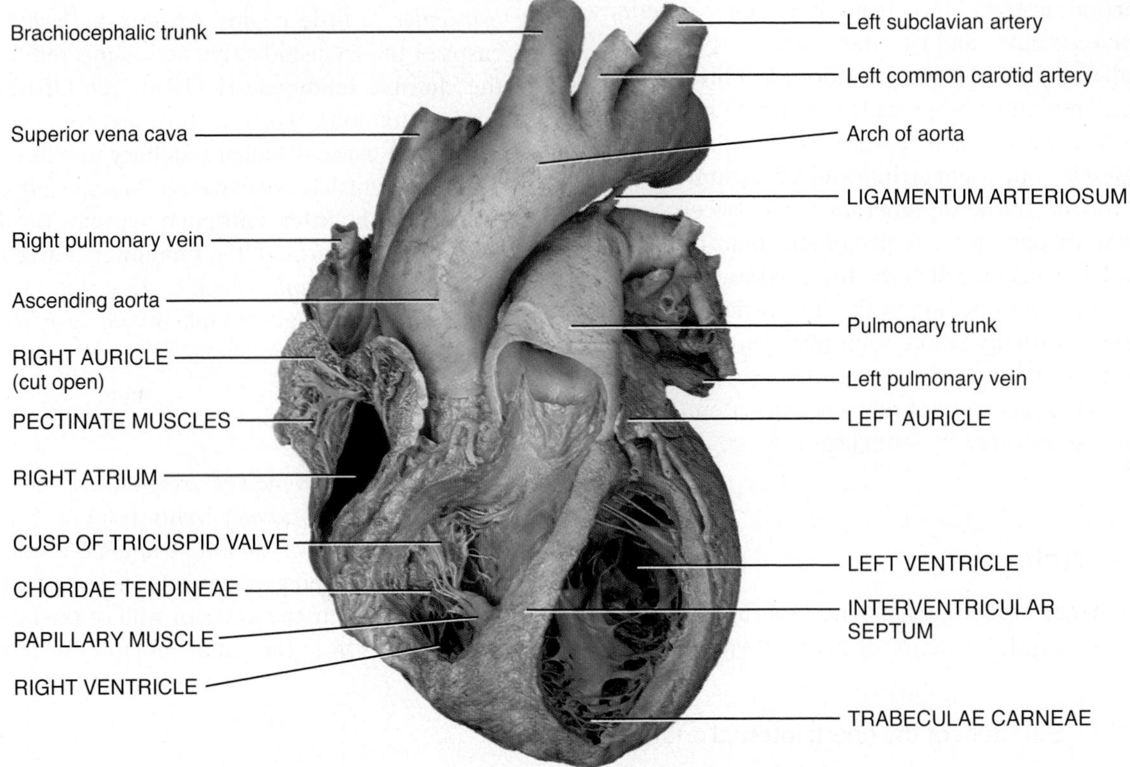


(b) Anterior view of partially sectioned heart

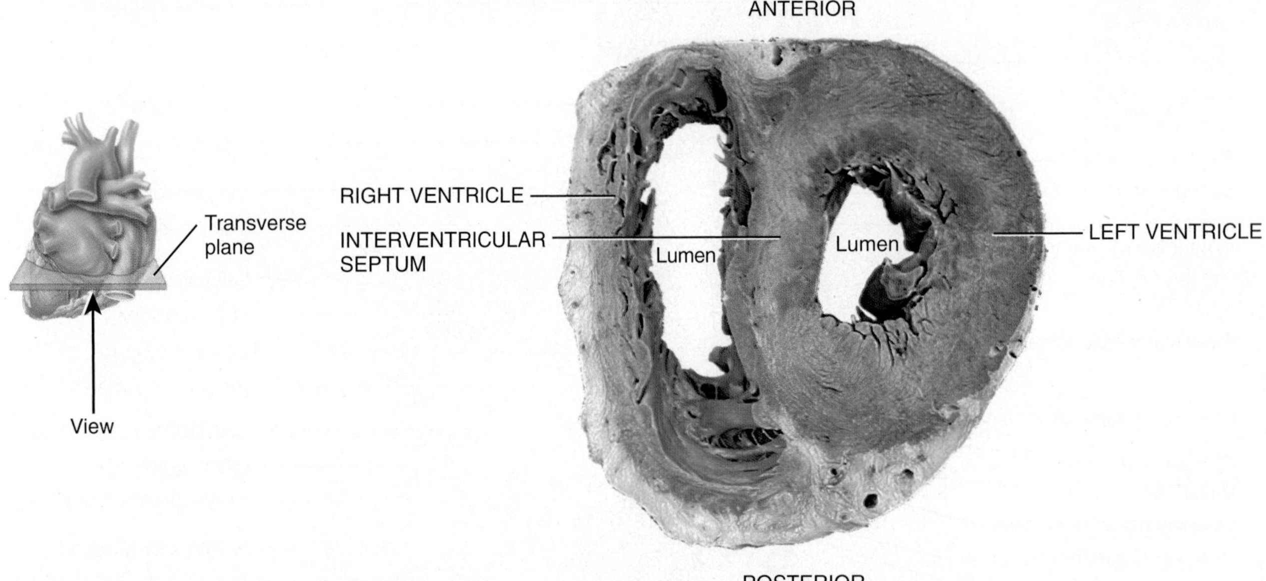


(c) Inferior view of transverse section showing differences in thickness of ventricular walls

The thickness of the four chambers varies according to their functions.

the **bicuspid (mitral) valve** (*bi-* = two), which, as its name implies, has two cusps. The term *mitral* refers to its resemblance to a bishop's miter (hat), which is two-sided.

## Left Ventricle

The **left ventricle** is the thickest chamber of the heart, averaging 10–15 mm (0.4–0.6 in.), and forms the apex of the heart (see Figure 19.1b). Like the right ventricle, the left ventricle contains trabeculae carneae and has chordae tendineae that anchor the cusps of the bicuspid valve to papillary muscles (Figure 19.4). Blood passes from the left ventricle through the **aortic valve** into the largest artery of the body, the *ascending aorta* (*aorte* = to suspend, because the aorta once was believed to lift up the heart). Some of the blood in the aorta flows into the *coronary arteries*, which branch from the ascending aorta and carry blood to the heart wall; the remainder of the blood passes into the *arch of the aorta* and *descending aorta*. Branches of the arch of the aorta and descending aorta carry blood throughout the body.

During fetal life, a temporary blood vessel, called the *ductus arteriosus*, shunts blood from the pulmonary trunk into the aorta to bypass the nonfunctioning fetal lungs. The ductus arteriosus normally closes shortly after birth, leaving a remnant known as the **ligamentum arteriosum** (lig′-a-MEN-tum ar-ter-ē-Ō-sum), which connects the arch of the aorta and pulmonary trunk (Figure 19.4a).

## Myocardial Thickness and Function

The thickness of the myocardium of the four chambers varies according to the amount of work each chamber has to perform. The walls of the atria are thin compared to those of the ventricles because the atria need only enough cardiac muscle tissue to deliver blood into the ventricles (Figure 19.4a). Since the ventricles pump blood greater distances, their walls are thicker. Although the right and left ventricles act as two separate pumps that simultaneously eject equal volumes of blood, the right ventricle has a much smaller workload. It pumps blood only a short distance to the lungs (pulmonary circulation). The left ventricle pumps blood great distances to all other parts of the body (systemic circulation). Therefore, the left ventricle works much harder than the right ventricle to maintain the same rate of blood flow; therefore, the muscular wall of the left ventricle is considerably thicker than the wall of the right ventricle (Figure 19.4c).

✓ **CHECKPOINT**

4. What function do the auricles serve?
5. The coronary sulcus forms a boundary between which chambers of the heart?
6. The anterior interventricular sulcus forms a boundary between which chambers of the heart?
7. What structure separates the atria? The ventricles?

# 19.3 Heart valves ensure one-way flow of blood.

As each chamber of the heart contracts, it pushes a volume of blood into a ventricle or out of the heart into an artery. Blood flows through the heart from areas of higher blood pressure to areas of lower blood pressure. As the walls of each chamber contract and relax, resulting pressure differences across the heart valves force valves to open and close. Each of the four valves helps to ensure the one-way flow of blood by opening to let blood through and then closing to prevent its backflow.

## Operation of the Atrioventricular Valves

Because they are located between an atrium and a ventricle, the tricuspid and bicuspid valves are also called the right and left **atrioventricular (AV) valves** (ā′-trē-ō-ven-TRIK-ū-lar). When an AV valve is open, the rounded ends of the cusps project into the ventricle (Figure 19.5a, d). When the ventricles are relaxed, the papillary muscles are relaxed, the chordae tendineae are slack, and blood moves from a higher pressure in the atria to a lower pressure in the ventricles through open AV valves. When the ventricles contract, the pressure of the ventricular blood drives the cusps upward until their edges meet and close the valve opening (Figure 19.5b, e). At the same time, the papillary muscles contract, which pulls on and tightens the chordae tendineae. This prevents the valve cusps from pushing up into the atria in response to the high ventricular pressure. If the AV valves or chordae tendineae were to become damaged, blood might flow back into the atria when the ventricles contract.

## Operation of the Semilunar Valves

The aortic and pulmonary valves are known as the left and right **semilunar (SL) valves** (sem′-ī-LOO-nar; *semi-* = half; *lunar* = moon-shaped) because they are made up of three crescent moon-shaped cusps (Figure 19.5d). The semilunar valves allow ejection of blood from the heart into arteries but prevent arterial blood from flowing back into the ventricles. When the walls of the ventricles contract, the blood pressure in the ventricles exceeds the pressure in the arteries and forces the SL valves to

**FIGURE 19.5 Responses of the valves to the pumping of the heart.**

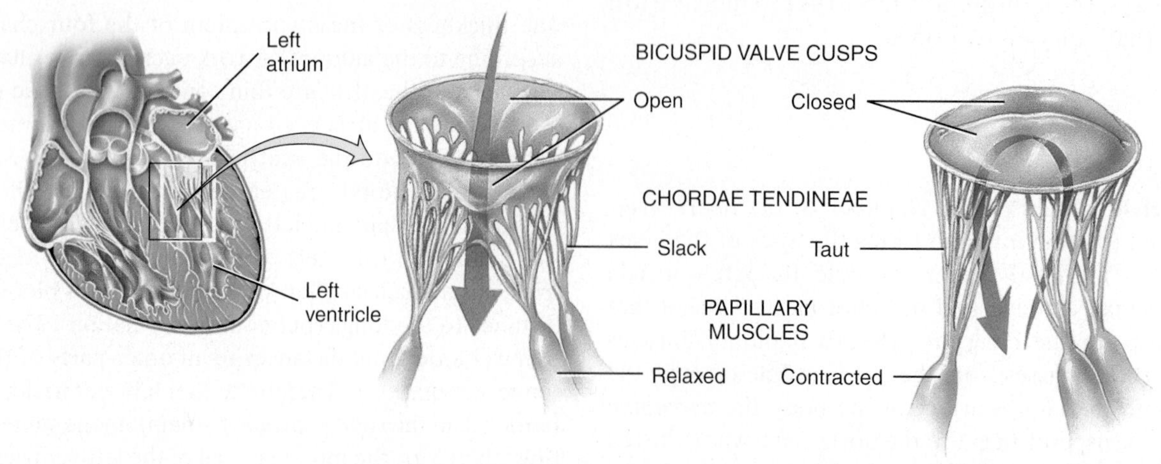


(a) Bicuspid valve open

(b) Bicuspid valve closed

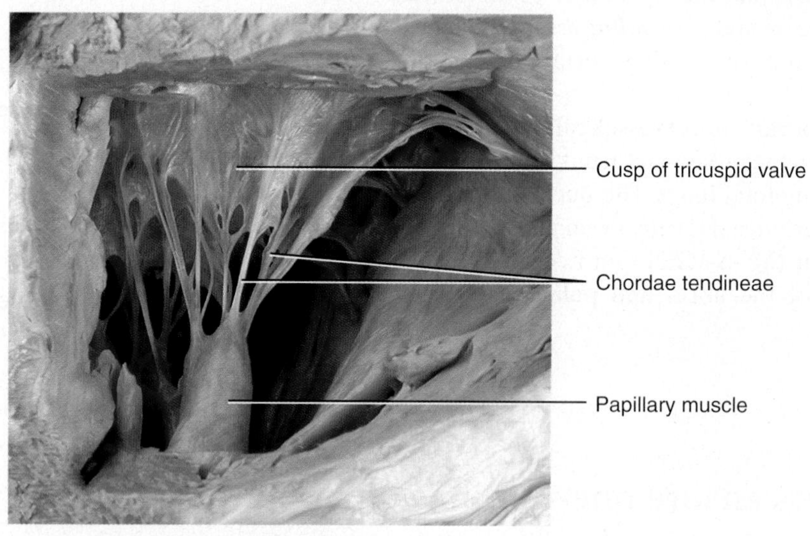


(c) Tricuspid valve open

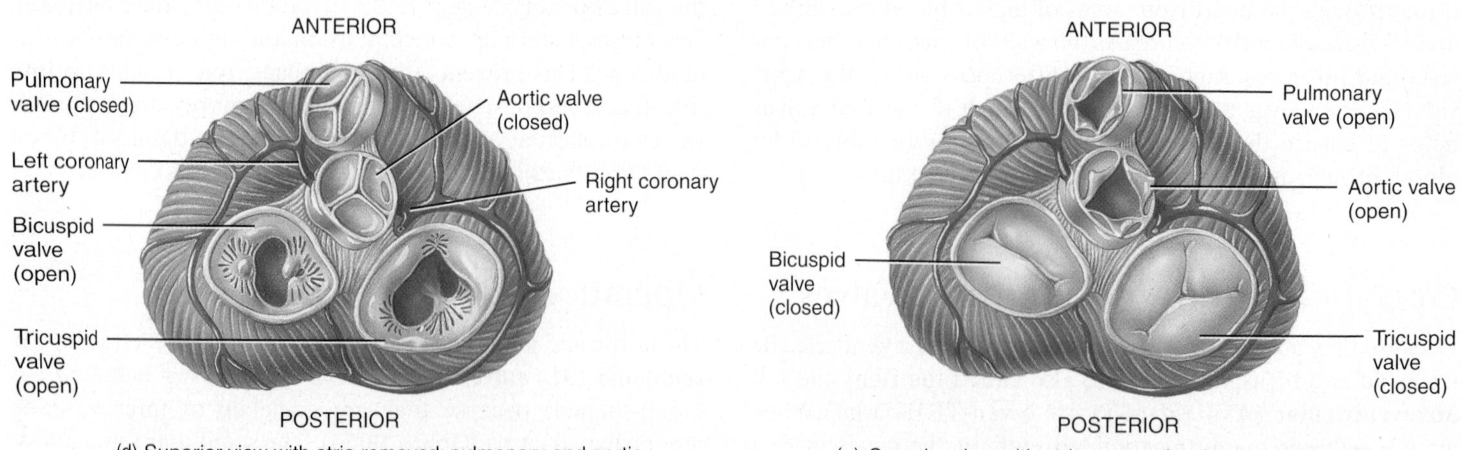


(d) Superior view with atria removed: pulmonary and aortic valves closed, bicuspid and tricuspid valves open

(e) Superior view with atria removed: pulmonary and aortic valves open, bicuspid and tricuspid valves closed

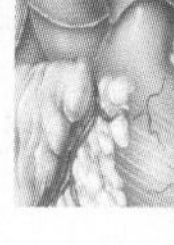

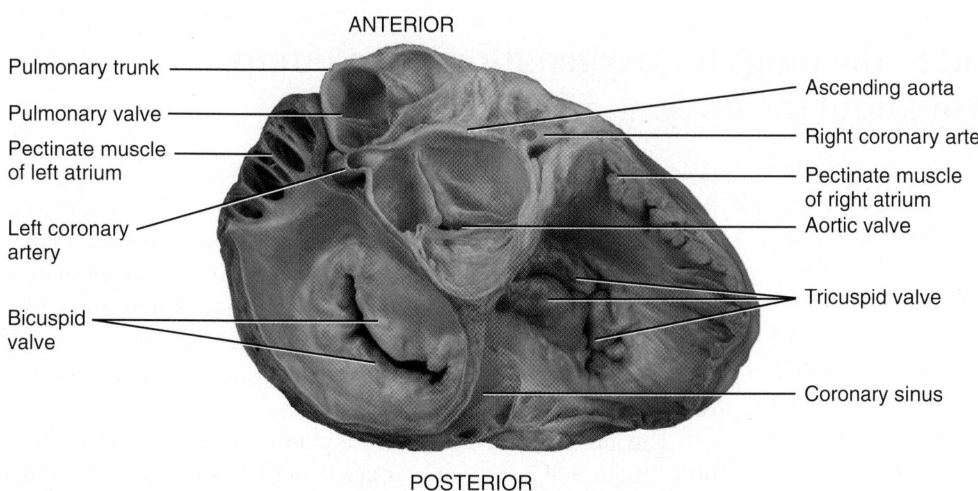


(f) Superior view of atrioventricular and semilunar valves

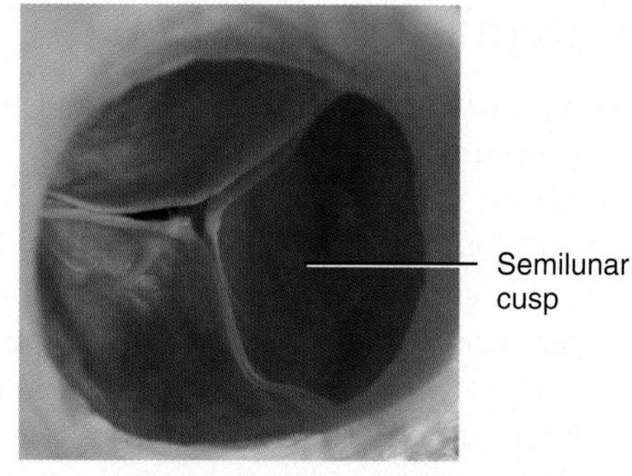


(g) Superior view of aortic valve

**Heart valves open and close in response to pressure changes as the heart contracts and relaxes.**

open, permitting ejection of blood through the SL valves into the pulmonary trunk and aorta (Figure 19.5e). As the ventricles relax, blood starts to flow back toward the heart. This backflowing blood fills the valve cusps, which causes the semilunar valves to contact each other tightly and close the opening between the ventricle and artery (Figure 19.5d).

Surprisingly perhaps, there are no valves between the venae cavae and the right atrium or between the pulmonary veins and the left atrium. As the atria contract, a small amount of blood does flow backward into these vessels. However, backflow is minimized because contracting atria compress and nearly collapse the openings from these veins.

### ✓ CHECKPOINT

**8.** What is the main force that causes blood to flow through the heart and the heart valves to open and close?

**9.** Why does blood ejected into the pulmonary trunk or aorta normally not flow back through the semilunar valve?

## CLINICAL CONNECTION | *Heart Valve Disorders*

When heart valves operate normally, they open fully and close completely at the proper times. A narrowing of a heart valve opening that restricts blood flow is known as **stenosis** (ste-NŌ-sis = a narrowing of an opening); failure of a valve to close completely is termed **insufficiency** (in′-su-FISH-en-sē) or **incompetence**. In **mitral stenosis**, scar formation or a congenital defect causes narrowing of the mitral valve. One cause of **mitral insufficiency**, in which there is backflow of blood from the left ventricle into the left atrium, is **mitral valve prolapse (MVP)**. In MVP one or both cusps of the mitral valve protrude into the left atrium during ventricular contraction. Mitral valve prolapse is one of the most common valvular disorders, affecting as much as 30 percent of the population. It is more prevalent in women than in men, and does not always pose a serious threat. In **aortic stenosis** the aortic valve is narrowed, and in **aortic insufficiency** there is backflow of blood from the aorta into the left ventricle.

Certain infectious diseases can damage or destroy the heart valves. One example is **rheumatic fever**, an acute systemic inflammatory disease that usually occurs after a streptococcal infection of the throat. The bacteria trigger an immune response in which antibodies produced to destroy the bacteria instead attack and inflame the connective tissues in joints, heart valves, and other organs. Even though rheumatic fever may weaken the entire heart wall, most often it damages the mitral and aortic valves.

If daily activities are affected by symptoms and if a heart valve cannot be repaired surgically, then the valve must be replaced. Tissue valves may be provided by human donors or pigs; sometimes, mechanical replacements are used. In any case, valve replacement involves open heart surgery. The aortic valve is the most commonly replaced heart valve.

# 19.4 The heart pumps blood to the lungs for oxygenation, then pumps oxygen-rich blood throughout the body.

## Systemic and Pulmonary Circulations

With each beat, the heart pumps blood into two closed circuits—the **pulmonary circulation** (*pulmon-* = lung) carrying blood to the air sacs (alveoli) of the lungs and the **systemic circulation** carrying blood to the rest of the body. The two circuits are arranged in series so that the output of one becomes the input of the other, as would happen if you attached two garden hoses (see Figure 20.14). Figure 19.6 shows the route of blood flow through the heart and the pulmonary and systemic circulations.

The right side of the heart is the pump for pulmonary circulation. The right atrium receives all of the dark-red *deoxygenated* (oxygen-poor) blood returning from the systemic circulation and then delivers it into the right ventricle, which pumps the deoxygenated blood into the *pulmonary trunk*. The pulmonary trunk divides into *pulmonary arteries* that carry blood to the lungs. (Arteries carry blood away from the heart—*a*rtery = *a*way.) In the lungs, blood flows through progressively smaller pulmonary arteries, and finally flows into extensive beds of *pulmonary capillaries*. In the thin-walled pulmonary capillaries, deoxygenated blood unloads $CO_2$ (carbon dioxide), which is exhaled, and picks up $O_2$ (oxygen) from inhaled air. The freshly oxygenated blood then flows through *pulmonary veins* to the left atrium. (Veins return blood to the heart.)

The left side of the heart is the pump for the systemic circulation. Bright red *oxygenated* (oxygen-rich) blood from the lungs entering the left atrium passes into the left ventricle. The left ventricle

**FIGURE 19.6 Systemic and pulmonary circulations.**

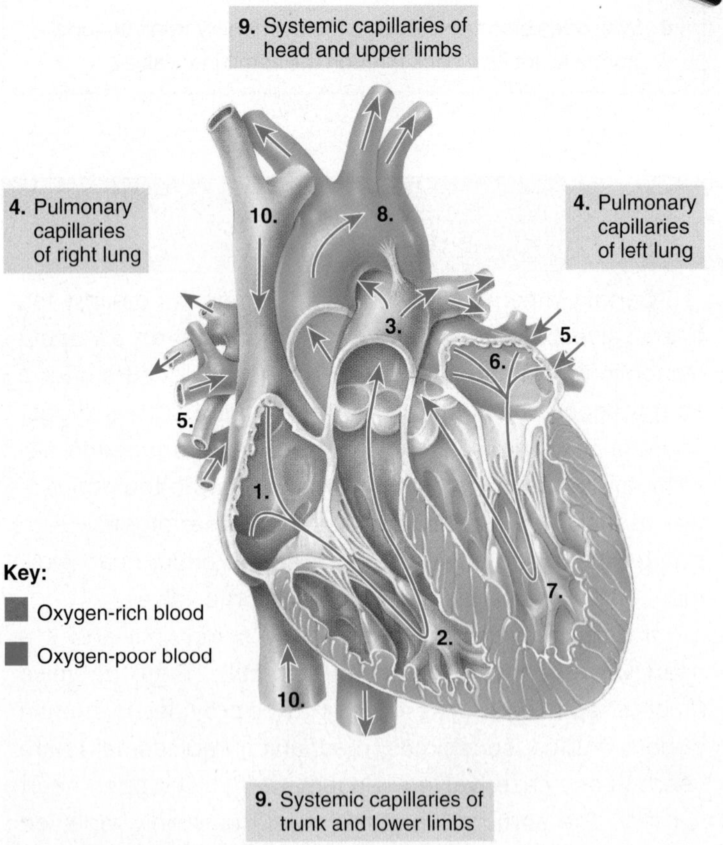

(a) Blood flow through heart

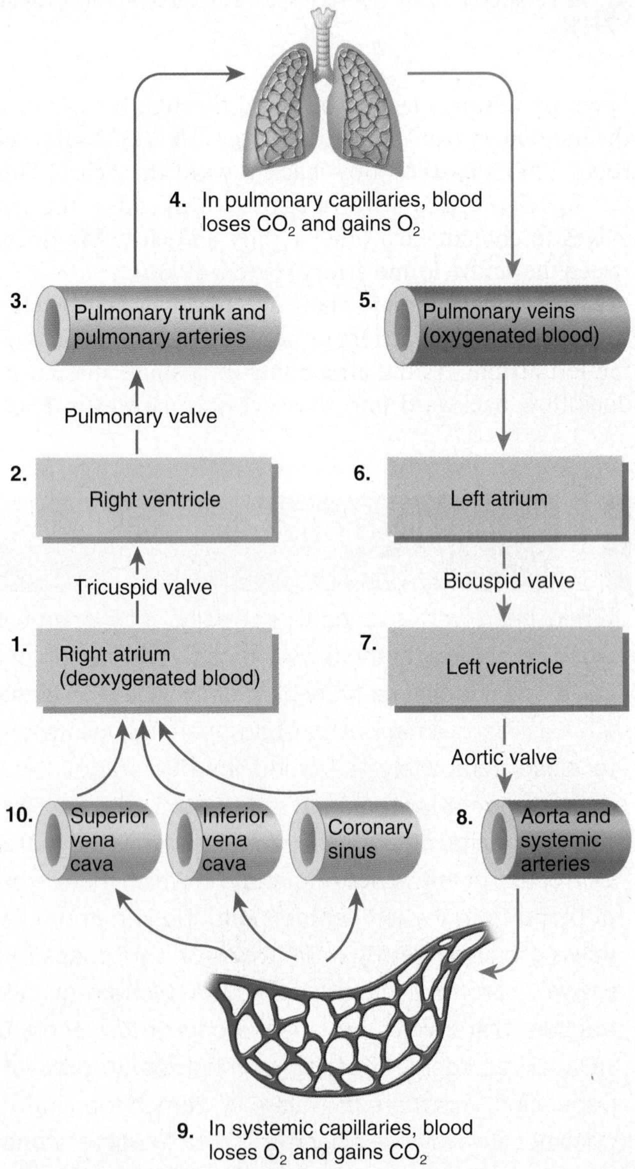

(b) Blood flow through pulmonary and systemic circulations

**The right side of the heart pumps deoxygenated blood into the pulmonary circulation to the lungs; the left side of the heart pumps freshly oxygenated blood into the systemic circulation to all body tissues except the air sacs of the lungs.**

ejects blood into the *aorta* (see Figure 19.4a), From the aorta, blood enters progressively smaller *systemic arteries* that carry it to all organs of the body—except for the air sacs (alveoli) of the lungs, which are supplied by the pulmonary circulation. In systemic tissues, progressively smaller arteries finally lead into *systemic capillaries*. Exchange of nutrients and gases occurs across the thin systemic capillary walls. Blood unloads $O_2$ and picks up $CO_2$ from body cells. The deoxygenated blood flows from systemic tissues to the right atrium through three veins, the superior vena cava, inferior vena cava, and coronary sinus (see Figures 19.3c and 19.4a).

## Coronary Circulation

Nutrients are not able to diffuse quickly enough from blood within the chambers of the heart to supply all of the layers of cells that make up the heart wall. For this reason, the wall of the myocardium has its own network of blood vessels. The flow of blood through the many vessels that pierce the myocardium is called the **coronary (cardiac) circulation** (*coron-* = crown) because these arteries encircle the heart like a crown encircles the head (see Figure 19.7a).

### Coronary Arteries

The principal coronary vessels, the **left** and **right coronary arteries**, branch from the ascending aorta and supply oxygenated blood to the myocardium (Figure 19.7a). The **left coronary artery** passes inferior to the left auricle and divides into the anterior interventricular and circumflex branches. The **anterior interventricular branch**, or *left anterior descending (LAD) artery*, is in the anterior interventricular sulcus and supplies oxygenated blood to the walls of both ventricles. The **circumflex branch** (SER-kum-fleks) lies in the coronary sulcus and distributes oxygenated blood to the walls of the left atrium and left ventricle.

The **right coronary artery** supplies small branches to the right atrium. It continues inferior to the right auricle and ultimately divides into the posterior interventricular and marginal branches (Figure 19.7a). The **posterior interventricular branch** follows the posterior interventricular sulcus and supplies the walls of the two ventricles with oxygenated blood. The **marginal branch** extends beyond the coronary sulcus to run along the right margin of the heart and transports oxygenated blood to the myocardium of the right ventricle.

FIGURE 19.7 **Coronary circulation.** The views of the heart from the anterior aspect in (a) and (b) are drawn as if the heart were transparent to reveal blood vessels on the posterior aspect.

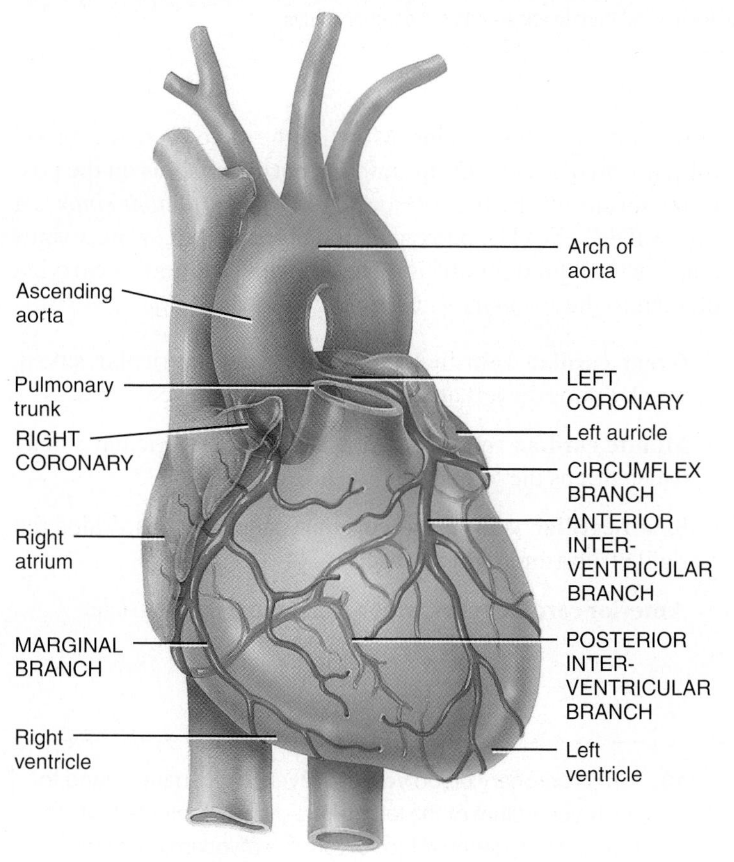


(a) Anterior view of coronary arteries

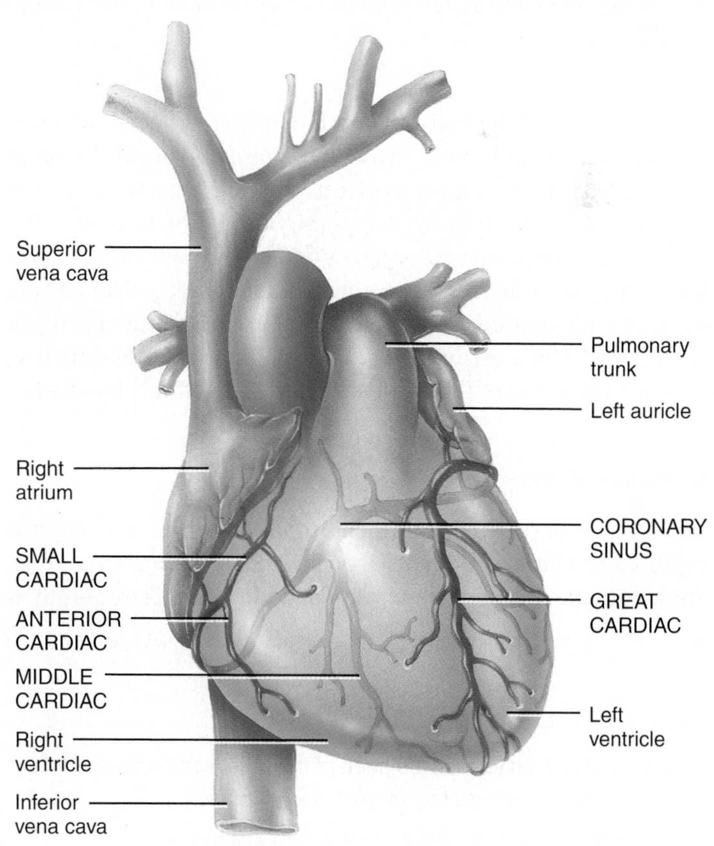


(b) Anterior view of coronary veins

(continues)

FIGURE 19.7 (continued)

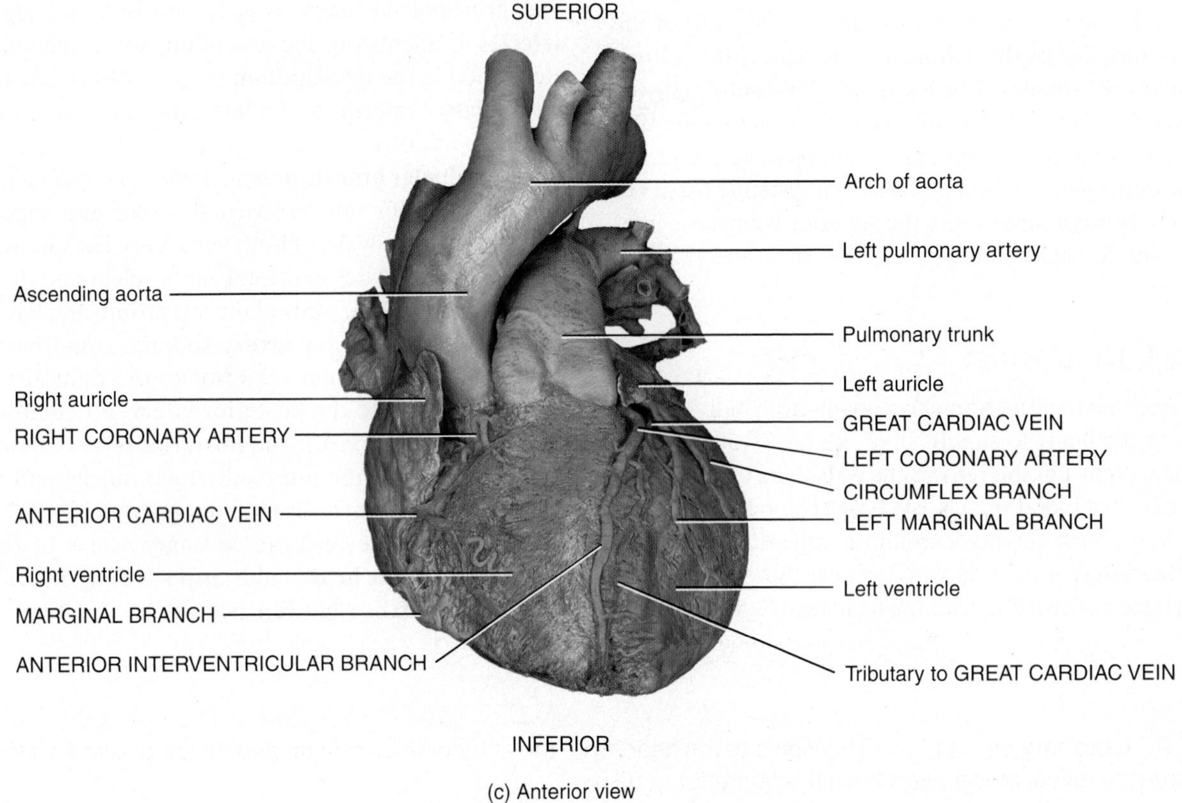


(c) Anterior view

**The right and left coronary arteries deliver blood to the heart; the coronary veins drain blood from the heart into the coronary sinus.**

Most parts of the body receive blood from branches of more than one artery, and where two or more arteries supply the same body region, they usually connect. These connections, called **anastomoses** (a-nas′-tō-MŌ-sēs), provide alternate routes for blood to reach a particular organ or tissue. The myocardium contains many anastomoses that connect branches of coronary arteries to provide detours for arterial blood if a main route becomes obstructed. Thus, cardiac muscle tissue may receive sufficient oxygen even if one of the coronary arteries is partially blocked.

## Coronary Veins

After blood passes through the coronary arteries, it flows into capillaries, where it delivers oxygen and nutrients to cardiac muscle fibers, collects their carbon dioxide and wastes, and then moves into coronary veins. Most of the deoxygenated blood from the myocardium drains into the **coronary sinus** on the posterior surface of the heart (Figure 19.7b). (A *vascular sinus* is a thin-walled vein.) The deoxygenated blood in the coronary sinus empties into the right atrium. The principal tributaries carrying blood into the coronary sinus are the following (Figure 19.7b):

- **Great cardiac vein** in the anterior interventricular sulcus, which drains the left atrium and left and right ventricles.
- **Middle cardiac vein** in the posterior interventricular sulcus, which drains the left and right ventricles.
- **Small cardiac vein** in the coronary sulcus, which drains the right atrium and right ventricle.
- **Anterior cardiac veins**, which drain the right ventricle.

### ✓ CHECKPOINT

10. From which vessels do each of the atria receive blood? Into which vessels do the ventricles eject blood?
11. Which blood vessels that enter and exit the heart carry oxygenated blood? Which carry deoxygenated blood?
12. Which blood vessel supplies blood to the coronary arteries?
13. Which coronary blood vessel delivers oxygenated blood to the myocardium of the left atrium and left ventricle? Which drains deoxygenated blood from the myocardium of the left atrium and left ventricle? Which drains deoxygenated blood from the myocardium of the right atrium and right ventricle?

## CLINICAL CONNECTION | *Myocardial Ischemia and Infarction*

Partial obstruction of blood flow in the coronary arteries may cause **myocardial ischemia** (is-KĒ-mē-a; *ische-* = to obstruct; *-emia* = in the blood), a condition of reduced blood flow to the myocardium. Usually, ischemia causes **hypoxia** (hī-POKS-ē-a = reduced oxygen supply), which may weaken cells without killing them. **Angina pectoris** (an-JĪ-na, or AN-ji-na, PEK-to-ris), which literally means "strangled chest," is a severe pain that usually accompanies myocardial ischemia. Typically, sufferers describe it as tightness or a squeezing sensation, as though the chest were in a vise. The pain associated with angina pectoris is often referred to the neck, chin, or down the left arm to the elbow. **Silent myocardial ischemia**, ischemic episodes without pain, is particularly dangerous because the person has no forewarning of an impending heart attack.

A complete obstruction to blood flow in a coronary artery may result in a **myocardial infarction (MI)** (in-FARK-shun), commonly called a *heart attack. Infarction* means the death of an area of tissue because of interrupted blood supply. Because the heart tissue distal to the obstruction dies and is replaced by noncontractile scar tissue, the heart muscle loses some of its strength. Depending on the size and location of the infarcted (dead) area, an infarction may disrupt the cardiac conduction system of the heart and cause sudden death by triggering ventricular fibrillation. Fibrillation (fi-bri-LĀ-shun) refers to rapid, uncoordinated heartbeats. Treatment for a myocardial infarction may involve injection of a thrombolytic (clot-dissolving) agent such as streptokinase or t-PA, plus heparin (an anticoagulant), or performing coronary angioplasty or coronary artery bypass grafting. Fortunately, heart muscle can remain alive in a resting person if it receives as little as 10–15 percent of its normal blood supply.

## RETURN TO **Hiroshi's Story**

Once the plane is airborne, the flight attendant offers drinks. Hiroshi asks for a cup of hot tea, and takes a Xanax tablet to relax, as the meditation techniques have not worked so far. He feels nauseated and doesn't understand why he is sweating so much. He declines the meal, turns up the air at his seat, and sleeps fitfully until the plane lands in Chicago. He slowly makes his way through O'Hare Airport to the baggage claim area, feeling even more short of breath than he did in London. He gets his luggage and sits down to rest before going outside to be picked up by his wife, Takara.

The next thing Hiroshi knows, Takara is standing in front of him saying, "Hiroshi, I was waiting for you outside. When you did not come out, I parked in the garage and came in. Why are you just sitting here? You look awful. You are gray and sweating. What's wrong?"

"I don't know," replies Hiroshi. "I haven't felt good since I arrived at the airport in London. Maybe I'm coming down with the flu. I am really short of breath," he says as he begins to cough.

Takara says, "I don't like the way you look. Come, get into the car. I am going to take you to the emergency room."

"No, just take me home so I can go to bed. I am really tired," argues Hiroshi.

"Absolutely not! We are going to the hospital," Takara responds firmly.

As Takara provides insurance information to the admitting clerk at the hospital, the nurse measures Hiroshi's vital signs, listens to his heart and lungs, hooks him up to a cardiac monitor and nasal oxygen, and starts an intravenous line. The nurse calls Dr. Morris, reporting that Hiroshi has an irregular heart rate, an extra heart sound known as a gallop rhythm, and fluid in his lungs. While waiting for Dr. Morris to arrive, the nurse draws some blood and sends it to the lab.

Dr. Morris arrives, and after greeting Hiroshi and Takara he looks at the readings on the monitor and listens to Hiroshi's heart and lungs to confirm the nurse's findings. As he places his stethoscope around his neck, he tells the couple that Hiroshi has experienced a myocardial infarction, or heart attack, while on the plane and that Hiroshi's left coronary artery is blocked. "I need to perform a cardiac catheterization to place a stent around the blockage, but because the injury occurred several hours ago, some of Hiroshi's myocardial tissue may already have died. The cardiac catheterization will help us determine the status of his cardiac function. In the meantime, the nurse will give Hiroshi some intravenous medications, morphine to ease his breathing and help him relax, and Lasix, a diuretic to remove the excess fluid from his body."

A. *Hiroshi's left coronary artery is blocked. What effect will this have on the myocardium of his heart?*

B. *Why would the inability of Hiroshi's left ventricle to effectively pump blood cause him to experience shortness of breath and peripheral swelling from excess fluid (edema)?*

# 19.5 The cardiac conduction system coordinates heart contractions for effective pumping.

## Cardiac Muscle Tissue

Compared with skeletal muscle fibers, cardiac muscle fibers are shorter in length (see Figure 10.16). They also exhibit branching, which gives individual cardiac muscle fibers a "stair-step" appearance (see Table 4.8). The ends of cardiac muscle fibers connect to neighboring fibers by irregular transverse thickenings of the sarcolemma called **intercalated discs** (in-TER-kā-lāt-ed; *intercalat-* = to insert between). The discs contain **desmosomes**, which hold the fibers together, and **gap junctions**, which allow muscle action potentials to conduct from one cardiac muscle fiber to another. Gap junctions allow the entire myocardium of the atria or the ventricles to contract as a single, coordinated unit.

Cardiac muscle fibers have the same arrangement of thin and thick filaments, and the same bands, zones, and Z discs, as skeletal muscle fibers. Mitochondria are larger and more numerous in cardiac muscle fibers than in skeletal muscle fibers. The sarcoplasmic reticulum of cardiac muscle fibers is somewhat smaller than the SR of skeletal muscle fibers. As a result, cardiac muscle has a smaller intracellular reserve of $Ca^{2+}$.

## Autorhythmic Fibers: The Cardiac Conduction System

An inherent and rhythmical electrical activity is the reason for the heart's lifelong beat. The source of this electrical activity is a network of specialized cardiac muscle fibers called **autorhythmic fibers** (aw′-tō-RITH-mik; *auto-* = self) because they are self-excitable; they repeatedly and rhythmically generate action potentials that trigger heart contractions. Autorhythmic fibers continue to stimulate a heart to beat even after it is removed from the body—for example, for transplant into another person—and all of its nerves have been cut. The remaining "working" myocardial fibers, **contractile fibers**, provide the powerful contractions that propel blood.

Autorhythmic fibers have two important functions:

- They act as a **pacemaker**, setting the rhythm of electrical excitation that causes contraction of the heart.
- They form the **cardiac conduction system**, which provides a path for each cycle of cardiac excitation to progress through the heart. The cardiac conduction system ensures that cardiac chambers are stimulated to contract in a coordinated manner, which makes the heart an effective pump.

Cardiac action potentials propagate through the cardiac conduction system in the following sequence (Figure 19.8a):

1. Cardiac excitation normally begins in the **sinoatrial (SA) node**, located in the right atrial wall just inferior to the opening of the superior vena cava. SA node cells do not have a stable resting potential. Rather, they repeatedly depolarize to threshold spontaneously. The spontaneous depolarization is a **pacemaker potential**. When the pacemaker potential reaches threshold, it triggers an action potential (Figure 19.8b). Each action potential from the SA node propagates throughout both atria via gap junctions in the intercalated discs of the atrial muscle fibers. Following the action potential, the two atria contract at the same time.
2. By propagating along atrial muscle fibers, the action potential reaches the **atrioventricular (AV) node**, located in the interatrial septum, just anterior to the opening of the coronary sinus (Figure 19.8a).
3. From the AV node, the action potential enters the **atrioventricular (AV) bundle** (also known as the **bundle of His**, pronounced HIZ). This bundle is the only site where action potentials can conduct from the atria to the ventricles. (Elsewhere, dense connective tissue electrically insulates the atria from the ventricles.)
4. After propagating along the AV bundle, the action potential enters both the **right** and **left bundle branches** that course through the interventricular septum toward the apex of the heart.
5. Finally, **Purkinje fibers** (pur′-KIN-jē) conduct the action potential from the apex of the heart upward to the remainder of the ventricular myocardium. Then the ventricles contract superiorly from the apex, pushing blood upward toward the semilunar valves.

The SA node initiates action potentials about 100 times per minute, faster than those of any other autorhythmic fibers. Action potentials from the SA node spread through the cardiac conduction system and stimulate other regions before those regions are able to generate their own action potentials. Thus, the SA node sets the rhythm for contraction of the heart—it acts as the *natural pacemaker* of the heart. Impulses from the autonomic nervous system (ANS) and blood-borne hormones (such as epinephrine) *modify the timing and strength* of each heartbeat, but they *do not establish the fundamental rhythm*. For example, in a person at rest, acetylcholine released by the parasympathetic division of the ANS typically slows SA node pacing to about 75 action potentials per minute, causing 75 heartbeats per minute.

## Contraction of Contractile Fibers

The action potential initiated by the SA node travels along the cardiac conduction system and spreads out to excite the

FIGURE 19.8 **The cardiac conduction system.** Autorhythmic fibers in the SA node, located in the right atrial wall (a), act as the heart's pacemaker, initiating cardiac action potentials (b) that cause contraction of the heart's chambers.

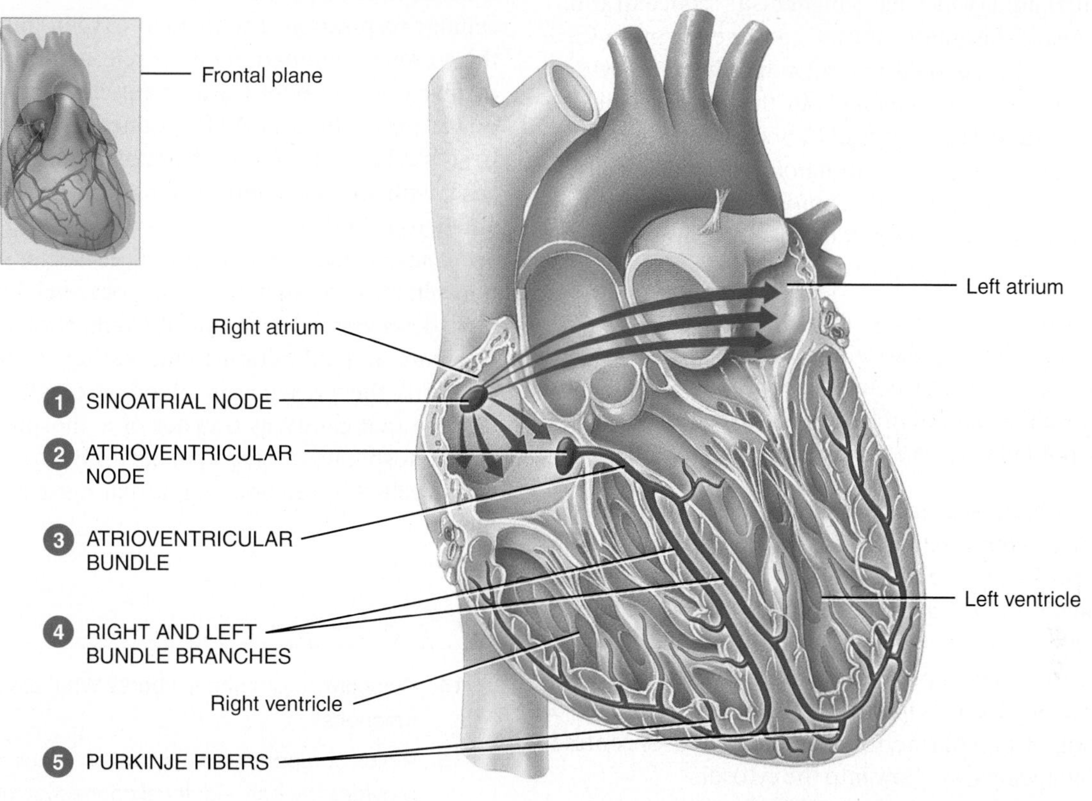


(a) Anterior view of frontal section

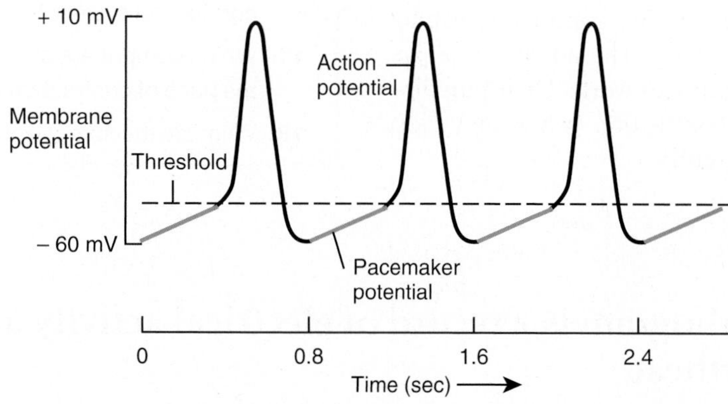


(b) Pacemaker potentials (green) and action potentials (black) in autorhythmic fibers of SA node

**The cardiac conduction system ensures that cardiac chambers contract in a coordinated manner.**

**contractile fibers** of the atria and ventricles. An action potential occurs in a contractile fiber as follows:

1. ***Depolarization.*** When a contractile fiber is brought to threshold by an action potential from neighboring fibers, its $Na^+$ channels open. $Na^+$ flows from the interstitial fluid into the contractile fiber because the cytosol of contractile fibers is electrically more negative than interstitial fluid, and because the $Na^+$ concentration is higher in the interstitial fluid. This inflow of $Na^+$ produces a **depolarization** (dē′-pō-lar-i-ZĀ-shun) that raises the contractile fiber's membrane potential (the cytosol side of the membrane becomes less negative than the extracellular fluid side).
2. ***Plateau.*** The next phase of an action potential in a contractile fiber is the **plateau**, a period of maintained depolarization.

It is due in part to opening of $Ca^{2+}$ channels in the sarcolemma. When these channels open, calcium ions move from the interstitial fluid (which has a higher $Ca^{2+}$ concentration) into the cytosol. This inflow of $Ca^{2+}$ causes even more $Ca^{2+}$ to pour out of the sarcoplasmic reticulum into the cytosol through additional $Ca^{2+}$ channels in the sarcoplasmic reticulum membrane. The increased $Ca^{2+}$ concentration in the cytosol ultimately triggers contraction. The plateau phase lasts for about 0.25 seconds. By comparison, depolarization in a neuron or skeletal muscle fiber is much briefer, about 0.001 seconds, because it lacks a plateau phase.

3. ***Repolarization.*** The recovery of the resting membrane potential during the **repolarization** (rē′-pō-lar-i-ZĀ-shun) phase of a cardiac action potential occurs as $K^+$ channels open. The resulting outflow of $K^+$ lowers the contractile fiber's membrane potential to the resting membrane potential.

The mechanism of contraction is similar in cardiac and skeletal muscle: The electrical activity (action potential) leads to the mechanical response (contraction). As $Ca^{2+}$ concentration rises inside a contractile fiber, $Ca^{2+}$ binds to troponin, which allows the thick and thin filaments to slide past one another, and contractile fiber tension develops. Substances that alter the movement of $Ca^{2+}$ through $Ca^{2+}$ channels influence the strength of heart contractions. Epinephrine, for example, increases contraction force by enhancing $Ca^{2+}$ flow into the cytosol.

In muscle, the **refractory period** (rē-FRAK-tō-rē) is the time interval during which a second contraction cannot be triggered. The refractory period of a cardiac muscle fiber lasts longer than the contraction itself. As a result, another contraction cannot begin until relaxation is well underway. The advantage is apparent if you consider how the ventricles work. Their pumping function depends on alternating contraction (when they eject blood) and relaxation (when they refill).

## ATP Production in Cardiac Muscle

Cardiac muscle produces most of the ATP it needs by aerobic cellular respiration. The needed oxygen diffuses from blood in the coronary circulation and is released from myoglobin inside cardiac muscle fibers. Cardiac muscle fibers use several fuels to power mitochondrial ATP production. In a person at rest, the heart's ATP comes mainly from oxidation of fatty acids and glucose, with smaller contributions from lactic acid, amino acids, and ketone bodies.

Cardiac muscle also produces some ATP from creatine phosphate. One sign that a myocardial infarction (heart attack, described in the Clinical Connection entitled Myocardial Ischemia and Infarction found earlier in the chapter) has occurred is the presence in blood of creatine kinase (CK), the enzyme that catalyzes transfer of a phosphate group from creatine phosphate to ADP to make ATP. Normally, CK is confined within cells. Injured or dying cardiac muscle releases CK into the blood.

✓ **CHECKPOINT**

14. What are autorhythmic fibers? What are their two functions?
15. Which component of the cardiac conduction system provides the only electrical connection between the atria and the ventricles?
16. What is the path of an action potential through the cardiac conduction system?
17. Which component of the cardiac conduction system sets the pace of contraction in a normal heart?
18. What are the sources of energy for cardiac muscle fibers?

# 19.6 The electrocardiogram is a record of electrical activity associated with each heartbeat.

## Electrocardiogram

As action potentials propagate through the heart, they generate electrical currents that can be detected by electrodes placed on the body's surface. A recording of the electrical changes that accompany the heartbeat is called an **electrocardiogram** (e-lek′-trō-KAR-dē-ō-gram), which is abbreviated as either **ECG** or **EKG** (from the German word *Electrokardiogram*). The ECG is a composite record of action potentials produced by cardiac muscle fibers during each heartbeat. The instrument used to record the changes is an **electrocardiograph**.

In clinical practice, electrodes are positioned on the arms and legs (limb leads) and at six positions on the chest (chest leads) to record the ECG. The electrocardiograph amplifies the heart's electrical signals and produces 12 different tracings from different combinations of limb and chest leads. Each limb and chest electrode records slightly different electrical activity because of the difference in its position relative to the heart.

In a typical record, three clearly recognizable waves appear with each heartbeat (Figure 19.9). The first, called the **P wave**, is a small upward deflection on the ECG; it represents **atrial depolarization**, the depolarizing (electrical excitation) phase of the cardiac action potential as it spreads from the SA node through contractile fibers in both atria leading to their contraction. The second wave, called the **QRS complex**, begins as a downward deflection (Q); continues as a large, upright, triangular wave (R);

FIGURE 19.9 **Normal electrocardiogram (ECG) of a single heartbeat.** P wave = atrial depolarization; QRS complex = onset of ventricular depolarization; T wave = ventricular repolarization.

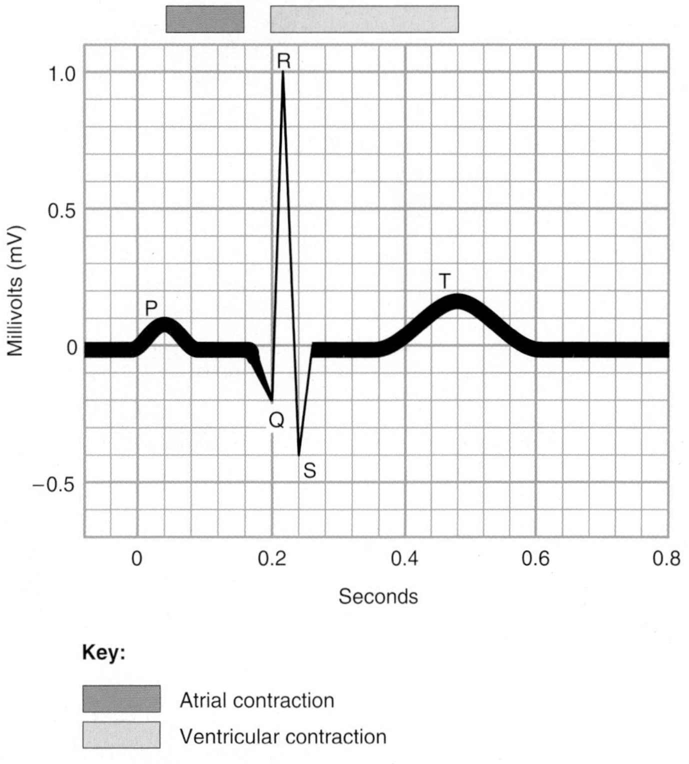


**An electrocardiogram is a recording of the electrical activity that initiates each heartbeat.**

and ends as a downward wave (S). The QRS complex represents the onset of **ventricular depolarization**, as the cardiac action potential spreads through ventricular contractile fibers. Shortly after the QRS complex begins, the ventricles start to contract. The third wave is an upward deflection called the **T wave**; it indicates **ventricular repolarization** and occurs just as the ventricles are starting to relax.

Comparing the size and duration of waves of an ECG with one another and with normal ECG records can be useful in diagnosing abnormal cardiac rhythms and conduction patterns, determining whether the heart is damaged or enlarged, determining the cause of chest pain, and in following the course of recovery from a heart attack. For example, larger P waves indicate enlargement of an atrium; an enlarged Q wave may indicate a myocardial infarction; and an enlarged R wave generally indicates enlarged ventricles. The T wave is flatter than normal when the heart muscle is receiving insufficient oxygen—as, for example, in coronary artery disease. The T wave may be elevated when blood $K^+$ levels are high. The time span between the P wave and the beginning of the QRS complex lengthens when the action potential is forced to detour around scar tissue caused by disorders such as coronary artery disease or rheumatic fever.

## Correlation of ECG Waves with Heart Activity

As we have seen, the atria and ventricles depolarize and then contract at different times because the cardiac conduction system routes cardiac action potentials along a specific pathway. The term **systole** (SIS-tō-lē = contraction) refers to the phase of contraction; the phase of relaxation is **diastole** (dī-AS-tō-lē = dilation or expansion). The ECG waves predict the timing of atrial and ventricular systole and diastole. The sequence of systole and diastole, as they relate to the waves of an ECG, is as follows (Figure 19.10):

1. A cardiac action potential arises in the SA node. It propagates throughout the atrial muscle and down to the AV node. As the atrial contractile fibers depolarize, the P wave appears in the ECG.
2. Contraction of atrial contractile fibers (*atrial systole*) begins after the P wave appears.
3. The action potential enters the AV bundle, where it is propagated through the bundle branches, Purkinje fibers,

**FIGURE 19.10 Timing and route of action potential depolarization through the heart.** Green indicates depolarization and red indicates repolarization.

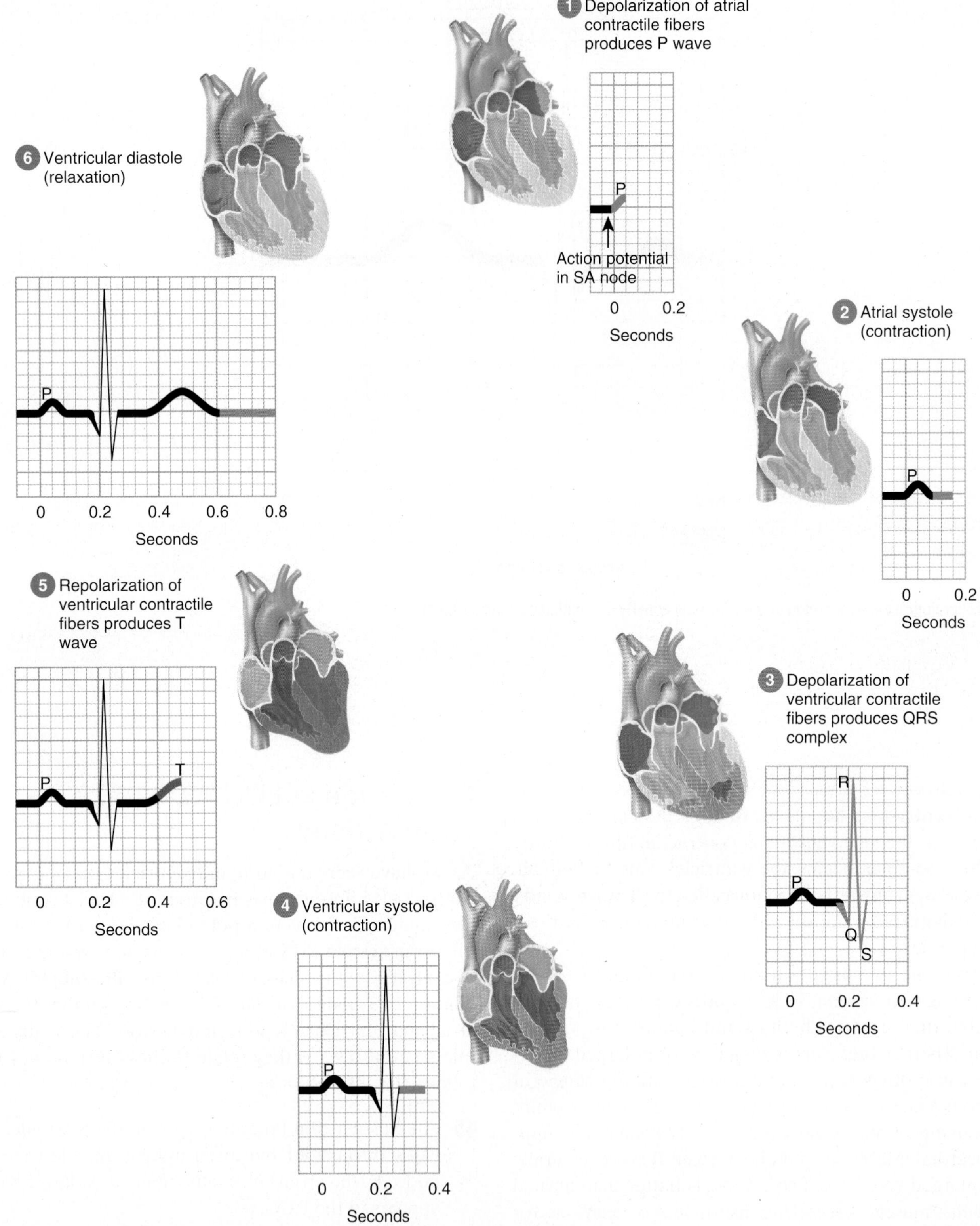


**Depolarization causes contraction and repolarization causes relaxation of cardiac muscle fibers.**

and the entire ventricular myocardium. Depolarization progresses down the interventricular septum, then upward and outward from the apex, producing the QRS complex. At the same time, atrial repolarization is occurring, but it is not usually evident in an ECG because the larger QRS complex masks it. As the atria repolarize, they relax (*atrial diastole*).

4 Contraction of ventricular contractile fibers (*ventricular systole*) begins after the QRS complex appears. As contraction proceeds from the apex toward the base of the heart, blood is squeezed upward toward the semilunar valves.

5 Repolarization of ventricular contractile fibers begins at the apex and spreads throughout the ventricular myocardium. This produces the T wave in the ECG.

6 After the T wave begins, the ventricles relax (*ventricular diastole*).

Briefly, contractile fibers in both the atria and ventricles are relaxed. Then the P wave appears again in the ECG, the atria begin to contract, and the cycle repeats. Events in the heart occur in cycles that repeat for as long as you live. Next, we will see how the pressure changes associated with relaxation and contraction of the heart chambers allow the heart to alternately fill with blood and then eject blood into the aorta and pulmonary trunk.

✓ **CHECKPOINT**

**19.** What is an electrocardiogram?

**20.** How does each ECG wave relate to the phases of an action potential?

**21.** What is the diagnostic significance of the ECG?

## 19.7 The cardiac cycle represents all of the events associated with one heartbeat.

The **cardiac cycle** comprises all of the events associated with a single heartbeat. During the cardiac cycle, the two atria contract (atrial systole) while the two ventricles relax (ventricular diastole). Then while the two ventricles contract (ventricular systole), the two atria relax (atrial diastole) (see Figure 19.10).

### Heart Sounds during the Cardiac Cycle

**Auscultation** (aws-kul-TĀ-shun; *ausculta-* = listening) is the act of listening to sounds within the body, and it is usually done with a stethoscope. The sound of the heartbeat comes primarily from turbulence in blood flow created by the closure of the heart valves. Smoothly flowing blood is silent. Recall the sounds made by whitewater rapids or a waterfall as compared with the silence of a smoothly flowing river. During each cardiac cycle, there are four **heart sounds**, but in a normal heart only the first and second heart sounds are loud enough to be heard by listening through a stethoscope. See Figure 19.12c for the timing of heart sounds relative to other events in the cardiac cycle.

The first sound (S1), which can be described as a **lubb** sound, is louder and a bit longer than the second sound; it is caused by blood turbulence due to closure of the atrioventricular (AV) valves as the ventricles contract (ventricular systole), squeezing blood back against the AV valve cusps. The second sound (S2), which is shorter and not as loud as the first, can be described as a **dupp** sound; it is caused by blood turbulence due to closure of the semilunar (SL) valves as the ventricles relax (ventricular diastole) and arterial blood pushes back against the SL valve cusps.

Although S1 and S2 are associated with the closure of valves, they are best heard at the surface of the chest in locations that are slightly different from the locations of the valves (Figure 19.11). This is because the sound is carried away from the valves by blood flow. Normally not loud enough to be heard, S3 is due to blood turbulence as the ventricles fill with blood, and S4 is due to blood turbulence during atrial systole.

FIGURE 19.11 **Heart sounds.** Location of valves (purple) and auscultation sites (red) for heart sounds.

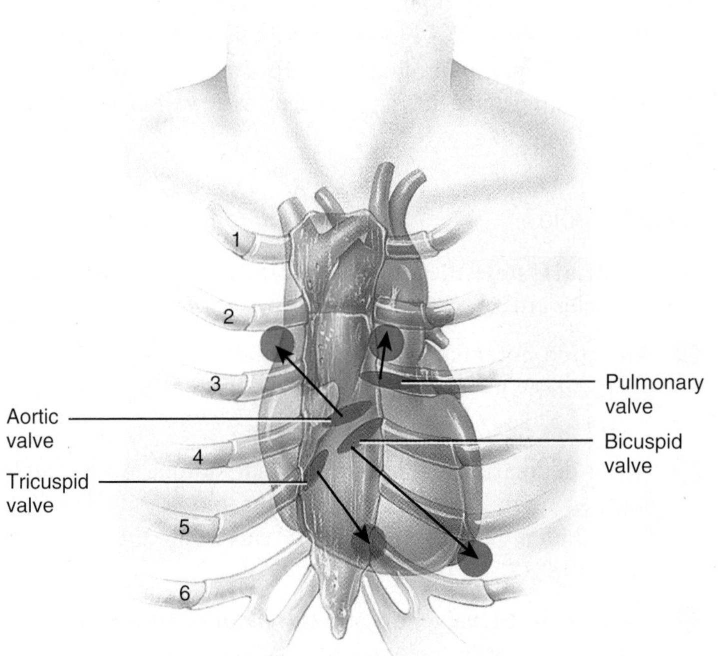


Anterior view of heart valve locations and auscultation sites

**Listening to sounds within the body is called auscultation; it is usually done with a stethoscope.**

## CLINICAL CONNECTION | *Heart Murmurs*

Heart sounds provide valuable information about the mechanical operation of the heart. A **heart murmur** is an abnormal sound consisting of a clicking, rushing, or gurgling noise that (1) is heard before, between, or after the normal heart sounds, or (2) may mask the normal heart sounds. Heart murmurs in children are extremely common and usually do not represent a health condition. Murmurs are most frequently discovered in children between the ages of two and four. These types of heart murmurs are referred to as *innocent* or *functional heart murmurs;* they often subside or disappear with growth. Although some heart murmurs in adults are innocent, most often an adult murmur indicates a valve disorder. When a heart valve exhibits *stenosis* (narrowing of an opening), the heart murmur is heard while the valve should be fully open but is not. For example, mitral stenosis produces a murmur as the ventricles relax, between S2 and the next S1 (see Clinical Connection entitled Heart Valve Disorders earlier in this chapter). In contrast, an incompetent heart valve causes a murmur to appear when the valve should be fully closed but is not. So, a murmur due to *mitral insufficiency* (also included in the Clinical Connection on Heart Valve Disorders) occurs during ventricular systole, between S1 and S2.

# Pressure and Volume Changes during the Cardiac Cycle

In each cardiac cycle, alternating contraction of the atria and ventricles forces blood from areas of higher pressure to areas of lower pressure. As each chamber of the heart contracts, blood pressure within it increases, pushing blood into a ventricle or out of the heart into an artery. Figure 19.12 shows the relationship between the heart's electrical signals (ECG) and changes in atrial pressure, ventricular pressure, aortic pressure, and ventricular volume during the cardiac cycle. The pressures given in Figure 19.12 apply to the left side of the heart; pressures on the right side are considerably lower. However, each ventricle expels the same volume of blood per beat, and the same pattern exists for both pumping chambers. To examine and correlate the events taking place during a cardiac cycle, we will begin with atrial systole.

## Atrial Systole

During **atrial systole**, the atria are contracting. At the same time, the ventricles are relaxed.

1. An action potential from the SA node results in atrial depolarization, marked by the P wave in the ECG shown in Figure 19.12a.

2. Atrial depolarization produces atrial systole. As the atria contract, they exert pressure on the blood within, which forces blood through the open AV valves into the ventricles (Figure 19.12b).

3. Because the SL valves are still closed, the ventricles fill with blood. Atrial systole contributes a final 25 mL of blood to the volume already in each ventricle (about 105 mL). The end of atrial systole is also the end of ventricular diastole (relaxation). Thus, each ventricle contains about 130 mL at the end of ventricular diastole. This blood volume is called the **end-diastolic volume (EDV)** (Figure 19.12d). At the end of ventricular diastole, the AV valves are still open and the SL valves are still closed.

## Ventricular Systole

During **ventricular systole**, the ventricles are contracting. At the same time, the atria are relaxed in **atrial diastole**.

4. The QRS complex in the ECG marks the onset of ventricular depolarization (Figure 19.12a).

5. Ventricular depolarization leads to **ventricular systole**. As ventricular systole begins, pressure rises inside the ventricles and pushes blood up against the AV valves, forcing them shut (Figure 19.12b). As the AV valves slam shut, resulting blood turbulence produces the first audible heart sound (S1). For a split second, both the SL and AV valves are closed. This is the period of **isovolumetric contraction** (ī-sō-VOL-ū-met′-rik; *iso-* = same). During this interval, the ventricles contract, yet because all four valves are closed, blood volume in the ventricles remains the same (*isovolumic*).

6. As the ventricles continue to contract, pressure inside the closed chambers rises sharply. When ventricular pressure rises above the pressure in the aorta and pulmonary trunk, both SL valves are forced open, and blood is ejected from the heart (Figure 19.12b). The period when the SL valves are open is termed **ventricular ejection**. During this period, the ventricles are contracting so strongly that pressure within the chambers continues to rise even as blood is ejected through the SL valves.

7. In the resting body, the volume of blood ejected from each ventricle during ventricular systole is about 70 mL (a little more than 2 oz). The volume of blood remaining in each ventricle at the end of systole, about 60 mL, is the **end-systolic volume (ESV)** (Figure 19.12d).

FIGURE 19.12 **Cardiac cycle.** (a) ECG. (b) Changes in left atrial pressure (green line), left ventricular pressure (blue line), and aortic pressure (red line) as they relate to the opening and closing of heart valves. (c) Heart sounds. (d) Changes in left ventricular volume. (e) Phases of the cardiac cycle.

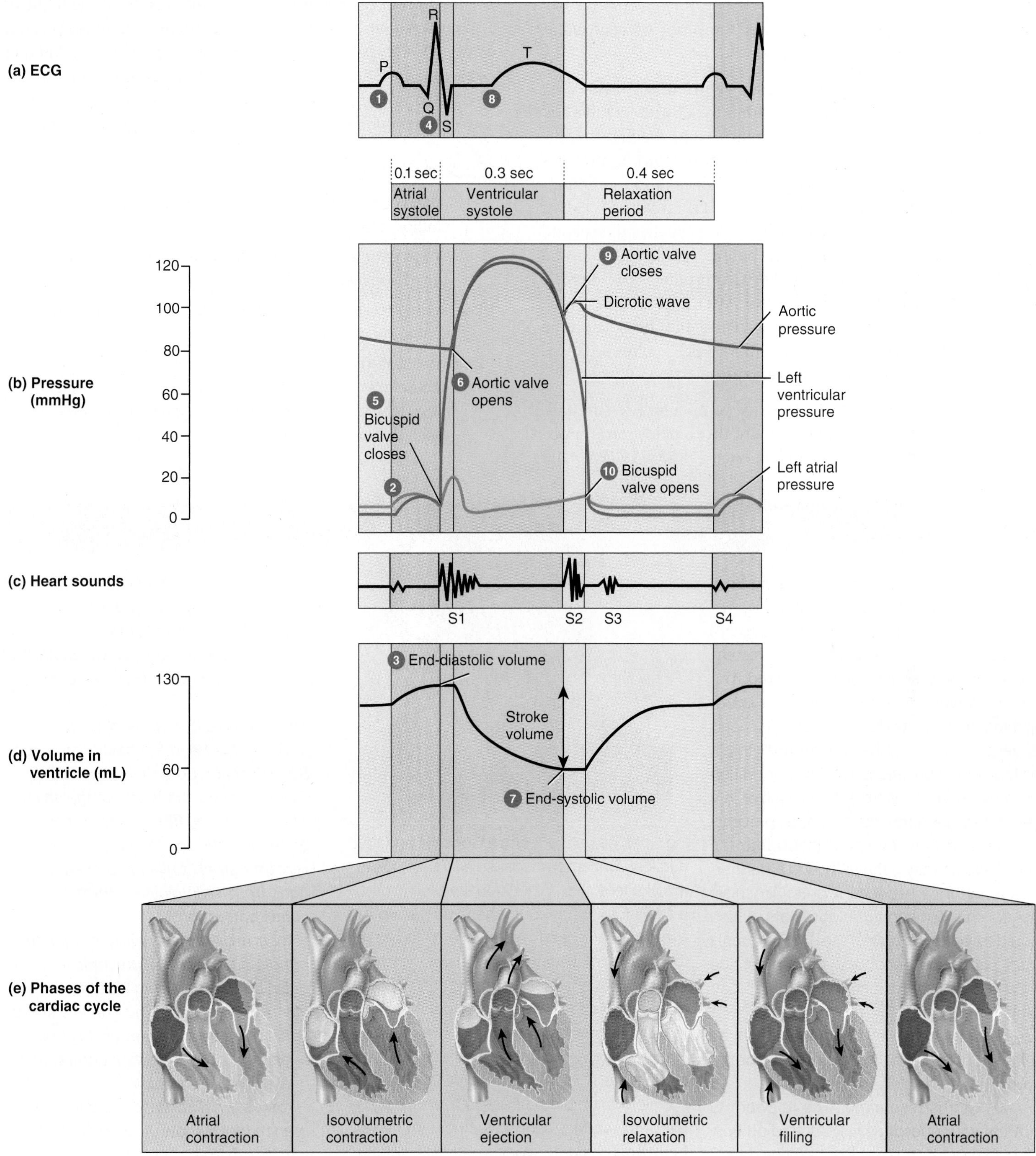


In each cardiac cycle, blood flows from areas of higher pressure to areas of lower pressure.

### Atrial and Ventricular Diastole

Immediately after ventricular systole, all four chambers are relaxed.

8 The T wave in the ECG marks the onset of ventricular repolarization (Figure 19.12a).

9 Ventricular repolarization causes **ventricular diastole**. As the ventricles relax, pressure within the chambers falls, and blood in the aorta and pulmonary trunk begins to flow backward toward the regions of lower pressure in the ventricles. As this back-flowing blood catches in the semilunar cusps, the SL valves close (Figure 19.12b). As the SL valves slam shut, resulting blood turbulence produces the second audible heart sound (S2). Rebound of blood from the closed cusps of the aortic valve also produces the **dicrotic wave** on the aortic pressure curve. After the SL valves close, there is a brief interval when ventricular blood volume does not change because all four valves are closed. This is the period of **isovolumetric relaxation**.

10 As the ventricles continue to relax, ventricular pressure falls quickly. When ventricular pressure drops below atrial pressure, the AV valves are forced open (Figure 19.12b), and **ventricular filling** begins. Blood that has been flowing into and building up in the atria during ventricular systole then rushes rapidly into the ventricles. Filling of the ventricles with blood produces blood turbulence that results in the S3 heart sound. About 75 percent of the ventricular filling occurs after the AV valves open and before the atria contract. Another cardiac cycle begins as atrial depolarization, signaled by the P wave in the ECG, causes atrial systole. As atria contract, resulting blood turbulence produces the S4 heart sound.

### ✓ CHECKPOINT

**22.** Which heart events can usually be heard through a stethoscope?

**23.** Which term is used for the contraction phase of the cardiac cycle? The relaxation phase?

**24.** In which phase of the cardiac cycle are the atria during ventricular systole?

**25.** What is the name for the volume of blood that remains in each ventricle at the end of ventricular diastole? What is the name for the volume of blood that remains in each ventricle at the end of ventricular systole?

## RETURN TO Hiroshi's Story

Following the cardiac catheterization, Dr. Morris returns to talk with Takara.

"I was able to place the stent to restore circulation to Hiroshi's heart. Unfortunately, some damage could not be repaired, but his blood work shows that his heart is beginning to heal itself. However, his heart is very irritated and he has an arrhythmia called atrial fibrillation. This is a condition in which atrial myocardial fibers quiver in a very rapid, uncoordinated manner, preventing his atria from pumping blood into the ventricles as they should. This is a dangerous arrhythmia because it puts Hiroshi at risk for the formation of blood clots, which could travel to his brain if the atria suddenly start contracting normally. I will prescribe an anticoagulant medication to keep blood clots from forming."

Takara asks, "How long will Hiroshi be in the hospital? Will he be able to return to work?"

Dr. Morris responds, "The answers to both of your questions will depend on how Hiroshi progresses over the next few days."

Hiroshi begins to feel better and wants to go home, but he still has trouble breathing when he gets out of bed to walk, and the nurse is concerned because his pulse rate is 45. She tells him, "When you are lying in bed, a rate of 45 is fine, but when you get up to exercise, your heart rate needs to increase to provide you with the oxygen you need for exertion. Your heart is not doing this so I am going to have to let Dr. Morris know."

Dr. Morris returns to see Hiroshi. After looking at the most recent electrocardiogram, he tells Hiroshi, "You have a complete bundle branch block; there is a blockage in the pathway that sends electrical impulses from your atria to your ventricles, which slows the pumping action of your ventricles. I will put you on this afternoon's schedule for the insertion of a pacemaker into your chest wall. The pacemaker will adjust your heart rate to meet the demands of your body. You will experience less shortness of breath, and you'll feel better when you get out of bed."

C. *The sinoatrial (SA) node, the pacemaker of the heart, normally generates 80-100 action potentials per minute. Given what you know about the ability of autorhythmic fibers to spontaneously generate action potentials, how might Hiroshi's atrial fibrillation prevent the SA node from maintaining a normal rate of atrial contraction?*

D. *Hiroshi is diagnosed with a complete bundle branch block, preventing impulse conduction from his atrioventricular (AV) bundle and the Purkinje fibers of his ventricles. How does this blockage affect coordination of contractions of the heart chambers?*

E. *Hiroshi's atrial fibrillation and bundle branch block show up on his ECG tracing. Which components of his ECG tracing might be different than those from a normal ECG?*

# 19.8 Cardiac output is the blood volume ejected by a ventricle each minute.

Although the heart's autorhythmic fibers enable it to beat independently, its operation is governed by events occurring throughout the body. All body cells must receive a certain amount of oxygenated blood each minute to maintain health and life. When cells are metabolically active, as during exercise, they take up even more oxygen from the blood, and the heart is stimulated to work harder. During periods of rest, cellular metabolic need is reduced, and the workload of the heart decreases.

**Cardiac output (CO)** is the volume of blood ejected by a ventricle each minute into the aorta or pulmonary trunk. Cardiac output equals the **stroke volume (SV)**, the volume of blood ejected by the ventricle during each contraction, multiplied by the **heart rate (HR)**, the number of heartbeats per minute:

$$\underset{(\text{mL/min})}{\text{CO}} = \underset{(\text{mL/beat})}{\text{SV}} \times \underset{(\text{beats/min})}{\text{HR}}$$

In a typical resting adult male, stroke volume averages 70 mL/beat, and heart rate is about 75 beats/min. Thus the average cardiac output in a resting adult is

$$\text{CO} = 70 \text{ mL/beat} \times 75 \text{ beats/min} = 5250 \text{ mL/min } (5.25 \text{ L/min})$$

This volume is close to the total blood volume, which is about 5 liters in a typical adult. Thus, your entire blood volume flows through your pulmonary and systemic circulations each minute. When body tissues use more or less oxygen, cardiac output changes to meet the need. Factors that increase stroke volume or make the heart beat faster normally increase cardiac output. During intense exercise, for example, stroke volume may increase to 130 mL/beat and heart rate may accelerate to 150 beats/min, resulting in a cardiac output of 19.5 L/min.

## Regulation of Stroke Volume

Although some blood is always left in the ventricles at the end of their contraction, a healthy heart pumps out the entire volume of blood that has entered its chambers during the previous diastole. In other words, the more blood that returns to the heart during diastole, the more blood that is ejected during the next systole. Three factors regulate stroke volume and ensure that the left and right ventricles pump equal volumes of blood:

- ***Preload.*** The **preload** is the degree to which the heart is stretched before it contracts. Preload can be compared to the stretching of a rubber band. The more the rubber band is stretched, the more forcefully it will snap back. Within limits, the more the heart is stretched as it fills during diastole, the greater is the force of contraction during systole. This relationship is known as the **Frank–Starling law of the heart**. The Frank–Starling law of the heart equalizes the output of the right and left ventricles and keeps the same volume of blood flowing to both the systemic and pulmonary circulations. For example, if the left side of the heart pumps a little more blood than the right side, a larger volume of blood returns to the right ventricle. The right ventricle contracts more forcefully on the next beat, bringing the two sides of the heart back into balance.
- ***Contractility.*** The second factor that influences stroke volume is myocardial **contractility**, the strength of contraction of individual ventricular muscle fibers. Even at a constant degree of stretch, the heart can contract more or less forcefully when certain substances are present. Stimulation by the sympathetic division of the autonomic nervous system (ANS), hormones such as epinephrine and norepinephrine,

### CLINICAL CONNECTION | *Congestive Heart Failure*

In **congestive heart failure (CHF),** there is a loss of pumping efficiency by the heart. Causes of CHF include coronary artery disease (see Chapter 19 WileyPLUS Clinical Connection entitled Coronary Artery Disease), congenital defects, long-term high blood pressure (which increases the afterload), myocardial infarctions (regions of dead heart tissue due to a previous heart attack, which decrease pumping efficiency), and valve disorders. As the pump becomes less effective, more blood remains in the ventricles at the end of each cycle, and gradually the end-diastolic volume (preload) increases. Initially, increased preload may promote increased force of contraction (the Frank–Starling law of the heart), but as the preload increases further, the heart is overstretched and contracts less forcefully. The result is a potentially lethal positive feedback loop: Less-effective pumping leads to even lower pumping capability.

Often, one side of the heart starts to fail before the other. If the left ventricle fails first, it can't pump out all of the blood it receives. As a result, blood backs up in the lungs and causes *pulmonary edema,* fluid accumulation in the lungs that can cause suffocation if left untreated. If the right ventricle fails first, blood backs up in the systemic veins and, over time, the kidneys cause an increase in blood volume. In this case, the resulting *peripheral edema* usually is most noticeable in the feet and ankles.

and increased $Ca^{2+}$ level in the interstitial fluid increase the force of contraction of cardiac muscle fibers. In contrast, inhibition of the sympathetic division of the ANS, some anesthetics, and increased $K^{+}$ levels in the extracellular fluid decrease contractility. *Calcium channel blockers* are drugs that reduce $Ca^{2+}$ inflow, thereby decreasing the strength of the heartbeat.

- ***Afterload.*** Ejection of blood from the heart begins when pressure in the right ventricle exceeds the pressure in the pulmonary trunk, and when the pressure in the left ventricle exceeds the pressure in the aorta. At that point, the higher pressure in the ventricles causes blood to push the semilunar valves open. The amount of pressure the contracting ventricles must produce to force open the SL valves is termed the **afterload**. When the required pressure is higher than normal, the valves open later than normal, stroke volume decreases, and more blood remains in the ventricles at the end of systole. Conditions that can increase afterload include hypertension (elevated blood pressure) and narrowing of arteries by atherosclerosis (see the Chapter 19 WileyPLUS Clinical Connection entitled Coronary Artery Disease).

## Regulation of Heart Rate

As you have just learned, cardiac output depends on both heart rate and stroke volume. Adjustments in heart rate are important in the short-term control of cardiac output and blood pressure. Without external input, the sinoatrial node would set a constant heart rate of about 100 beats/min. However, tissues require different blood flow volumes under different conditions. During exercise, for example, cardiac output rises to supply working tissues with increased amounts of oxygen and nutrients. Stroke volume may fall if the ventricular myocardium is damaged or if bleeding reduces blood volume. In these cases, adequate cardiac output is maintained by increasing the heart rate. Among the several factors that contribute to regulation of heart rate, the most important are the autonomic nervous system and the hormones epinephrine and norepinephrine.

### Autonomic Regulation of Heart Rate

Nervous system regulation of the heart originates in the **cardiovascular center** in the medulla oblongata. This region of the brain stem receives input from a variety of sensory receptors and from higher brain centers, such as the limbic system and cerebral cortex. In response, the cardiovascular center increases or decreases the frequency of impulses sent to the heart through both the sympathetic and parasympathetic branches of the ANS (Figure 19.13).

Even before physical activity begins, especially in competitive situations, heart rate may climb. This anticipatory increase occurs because the limbic system sends impulses to the cardiovascular center in the medulla oblongata. As physical activity begins, **proprioceptors** that are monitoring the position of limbs and muscles send an increased frequency of impulses to the cardiovascular center. Proprioceptor input is a major stimulus for

FIGURE 19.13 **Nervous system control of the heart.**

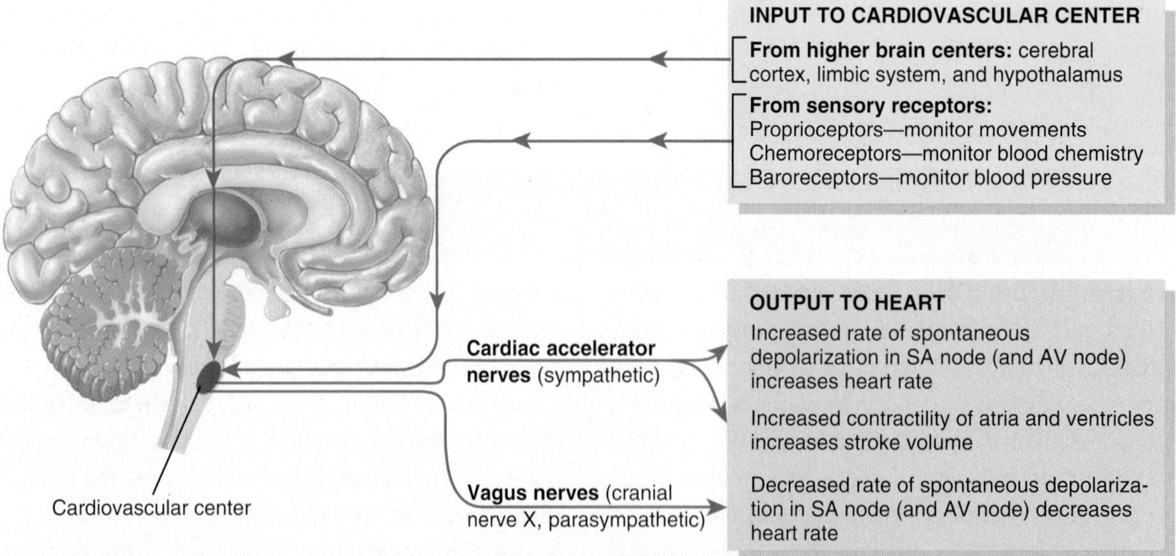


**The cardiovascular center in the medulla oblongata controls both sympathetic and parasympathetic nerves that innervate the heart.**

the quick rise in heart rate that occurs at the onset of physical activity. Other sensory receptors that provide input to the cardiovascular center include **chemoreceptors**, which monitor chemical changes in the blood, and **baroreceptors**, which detect changes in blood pressure by monitoring the stretching of major arteries and veins caused by the pressure of the blood flowing through them. The role of baroreceptors in the regulation of blood pressure is discussed in detail in Concept 20.6.

Sympathetic neurons extend from the cardiovascular center to the heart in **cardiac accelerator nerves**. They innervate the SA node, AV node, and most portions of the myocardium. Impulses in the cardiac accelerator nerves trigger the release of norepinephrine. Norepinephrine speeds up spontaneous depolarization at SA and AV nodes, so that these pacemakers fire impulses more rapidly and heart rate increases. Norepinephrine also enhances $Ca^{2+}$ entry through $Ca^{2+}$ channels, thereby increasing contractility. As a result, a greater volume of blood is ejected during systole.

Also arising from the cardiovascular center are parasympathetic neurons that reach the heart via the right and left **vagus nerves**. These parasympathetic neurons terminate in the SA node, AV node, and atrial myocardium. The neurotransmitter they release—acetylcholine—decreases heart rate by slowing the rate of spontaneous depolarization in autorhythmic fibers.

A continually shifting balance exists between sympathetic and parasympathetic stimulation of the heart. At rest, parasympathetic stimulation predominates. The resting heart rate—about 75 beats/min—is usually lower than the autorhythmic rate of the SA node (about 100 beats/min).

## Chemical Regulation of Heart Rate

Certain chemicals influence both the basic physiology of cardiac muscle and the heart rate. A number of hormones and cations have major effects on the heart:

- ***Hormones.*** Epinephrine and norepinephrine (from the adrenal medullae) increase both heart rate and contractility. Exercise, stress, and excitement stimulate the adrenal medullae to release more epinephrine and norepinephrine. Thyroid hormones also enhance cardiac contractility and increase heart rate. One sign of hyperthyroidism (excessive thyroid hormone) is **tachycardia** (tak′-i-KAR-dē-a), an elevated resting heart rate.
- ***Cations.*** Differences between intracellular and extracellular concentrations of several cations (for example, $Na^+$ and $K^+$) are crucial for the production of action potentials in all nerve and muscle fibers, so it is not surprising that ionic imbalances can quickly compromise the pumping effectiveness of the heart. Elevated blood levels of $Na^+$ blocks $Ca^{2+}$ inflow into cardiac muscle fibers, decreasing the force of contraction. Excess $K^+$ in the blood decreases heart rate by blocking generation of action potentials. A moderate increase in interstitial $Ca^{2+}$ level increases heart rate and contraction force.

## Other Factors in Heart Rate Regulation

Age, gender, physical fitness, and body temperature also influence resting heart rate. A newborn baby is likely to have a resting heart rate of over 120 beats/min; the rate then gradually declines throughout life. Adult females often have slightly higher resting heart rates than adult males, although regular exercise tends to bring resting heart rate down in both sexes. A physically fit person may even exhibit **bradycardia** (brad-ē-KAR-dē-a; *bradys-* = slow), a resting heart rate under 50 beats/min. This is a beneficial effect of endurance-type training because a slowly beating heart is more energy efficient than one that beats more rapidly.

Increased body temperature, such as occurs during fever or strenuous exercise, causes the SA node to discharge impulses more quickly, thereby increasing heart rate. Decreased body temperature decreases heart rate and force of contraction. During surgical repair of certain heart abnormalities, it is helpful to slow a patient's heart rate by hypothermia (hī′-pō-THER-mē-a), deliberate cooling of the body to a low core temperature. Hypothermia also slows metabolism, which reduces the oxygen needs of the tissues, allowing the heart and brain to withstand short periods of interrupted or reduced blood flow during the procedure.

### ✓ CHECKPOINT

26. What is cardiac output, and how is it calculated?
27. What is stroke volume? Which factors regulate stroke volume?
28. Where in the central nervous system is heart rate adjusted?
29. Which sensory receptors provide feedback to the cardiovascular center? Which specific input does each receptor provide?
30. How do hormones alter heart activity?

## Hiroshi's Story

## EPILOGUE AND DISCUSSION

Hiroshi experienced a heart attack while on a plane from London to Chicago. His symptoms were atypical, with none of the crushing chest pain that most men experience during a myocardial infarction. The length of time that passed between the heart attack and treatment caused permanent damage to Hiroshi's heart. A stent was inserted, which diminished further damage. Hiroshi experienced atrial fibrillation, an arrhythmia of the atria, and a complete bundle branch block so that no atrial impulses could get to the ventricles. Because Hiroshi's ventricles contracted at such a slow rate that he could not resume his normal activities, he had a pacemaker inserted into his chest wall and connected to his ventricles. The pacemaker will maintain a steady heart rate and will increase his heart rate when Hiroshi needs more oxygen sent to the organs of his body. Although the pacemaker will alleviate fatigue, the ventricles of Hiroshi's heart have been damaged to the extent that they are no longer able to meet the cardiac output demands of his heart, even with medications to stimulate contraction. Fortunately, Hiroshi's vital signs have remained within normal limits, so he does not have to worry about hypertension at this time. He will be given medication to control his atrial fibrillation or he may undergo atrial fibrillation ablation, which uses a laser to create a small scar in the atrial conduction pathway that prevents the spread of abnormal electrical impulses through the atria. Hiroshi will need the pacemaker for the rest of his life and, depending on the extent of the damage to his heart, Hiroshi may eventually need a heart transplant.

F. *Hiroshi's left ventricle does not empty with each contraction, causing blood to back up through his circulatory system. How would this affect his preload and afterload?*

G. *Hiroshi experienced shortness of breath prior to his hospitalization. He was given a diuretic that promotes $Na^+$ excretion in urine. How could a decrease in the circulating level of $Na^+$ affect contractility of the heart?*

# Concept and Resource Summary

| Concept | Resources |
|---|---|
| **Introduction**<br>1. The heart pumps blood to all body cells. | |
| **Concept 19.1 The heart is located in the mediastinum and has a muscular wall covered by pericardium.**<br>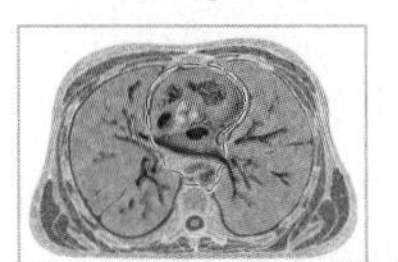<br>1. The **heart** is located in the **mediastinum** of the thoracic cavity, where it rests on the diaphragm. It is situated obliquely, with the **apex** (pointed end) directed inferiorly to the left and the **base** positioned superiorly to the right.<br>2. The heart is surrounded by a protective membrane, the **pericardium**, consisting of a tough superficial **fibrous pericardium** and a deeper more delicate **serous pericardium** that forms a double layer around the heart, an outer **parietal layer** and an inner **visceral layer**. A thin, slippery, lubricating **pericardial fluid** is found in the **pericardial cavity** between the parietal and visceral layers of the serous pericardium.<br>3. The three layers of the heart wall from superficial to deep are the **epicardium** (visceral layer of serous pericardium), the **myocardium** (cardiac muscle tissue), and the **endocardium** (endothelium overlying connective tissue). The endocardium also covers the valves of the heart. | Anatomy Overview—Cardiac Muscle<br>Clinical Connection—Pericarditis<br>Clinical Connection—Myocarditis and Endocarditis |

| Concept | Resources |
|---|---|

## Concept 19.2 The heart has four chambers, two upper atria and two lower ventricles.

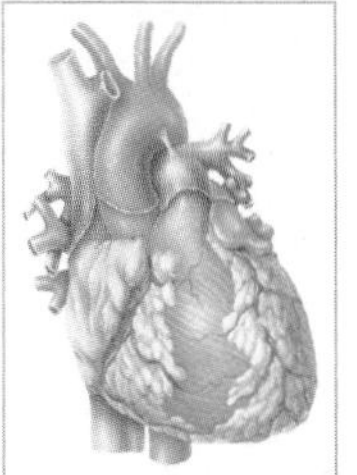

1. There is a wrinkled, pouchlike **auricle** on the anterior surface of each atrium that increases blood volume capacity.
2. The **coronary sulcus** encircles the heart between the superior atria and inferior ventricles. The **anterior interventricular sulcus** marks the anterior boundary between the ventricles, while the **posterior interventricular sulcus** is the posterior boundary between the ventricles.
3. The **right atrium** receives blood from the superior vena cava, inferior vena cava, and coronary sinus. It is separated internally from the left atrium by the **interatrial septum**, which contains the **fossa ovalis**. Blood exits the right atrium through the **tricuspid valve**.
4. The **right ventricle** receives blood from the right atrium. It is separated internally from the left ventricle by the **interventricular septum** and pumps blood through the **pulmonary valve** into the pulmonary trunk that carries blood to the lungs. The cusps of the tricuspid valve connect to **chordae tendineae** that in turn are connected to cone-shaped **papillary muscles** of the ventricular wall.
5. The **left atrium** receives blood from the lungs through pulmonary veins. Blood from the left atrium passes through the **bicuspid (mitral) valve** as it flows into the left ventricle.
6. The **left ventricle** contains chordae tendineae that anchor the **cusps** of the bicuspid valve to papillary muscles. Blood exiting the left ventricle passes through the **aortic valve** and into the ascending aorta, which carries blood to the heart wall and to the rest of the body.
7. The thickness of the myocardium of the four heart chambers varies according to each chamber's function. The left ventricle, with the highest workload, has the thickest wall because it must generate contractions that pump blood to all parts of the body.

Anatomy Overview—Heart Structures
Figure 19.3—Structure of the Heart: Surface Anatomy
Exercise—Paint the Heart
Clinical Connection—Congenital Heart Defects

## Concept 19.3 Heart valves ensure one-way flow of blood.

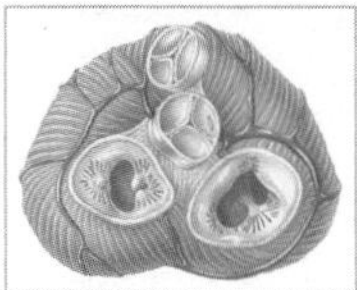

1. Blood flows through the heart from areas of higher pressure to areas of lower pressure. Contraction of the walls of a chamber increases pressure of the blood within the chambers.
2. Heart valves open and close in response to pressure differences across the valves.
3. The location of the bicuspid and tricuspid valves between the atria and ventricles has given them the name **atrioventricular (AV) valves**. Contraction of an atrium increases atrial blood pressure above ventricular pressure, causing blood to push open the AV valve and flow into the ventricle.
4. The **semilunar (SL) valves** are the aortic valve, at the entrance to the aorta, and the pulmonary valve, at the entrance to the pulmonary trunk. Contraction of a ventricle increases ventricular blood pressure above arterial blood pressure, which opens the SL valves and ejects blood into the pulmonary trunk and aorta.

Anatomy Overview—Heart Structures
Exercise—Paint the Heart

## Concept 19.4 The heart pumps blood to the lungs for oxygenation, then pumps oxygen-rich blood throughout the body.

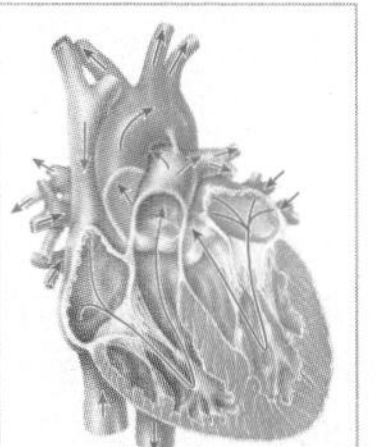

1. The right side of the heart is the pump for **pulmonary circulation**, the circulation of deoxygenated blood through the lungs where blood unloads $CO_2$ and picks up $O_2$. The right ventricle ejects blood into the pulmonary trunk, and blood then flows into pulmonary arteries, pulmonary capillaries, and pulmonary veins, which carry it back to the left atrium.
2. The left side of the heart is the pump for **systemic circulation**, the circulation of oxygenated blood throughout the body except for the air sacs of the lungs. The left ventricle ejects blood into the aorta, and blood then flows into systemic arteries, systemic capillaries, and finally enters the right atrium through three veins, the superior vena cava, inferior vena cava, and coronary sinus.
3. The wall of the heart has its own blood vessels called the **coronary (cardiac) circulation** that delivers blood to and from the myocardium. The main arteries of the coronary circulation are the **left** and **right coronary arteries**; the main veins are the **cardiac veins** and the **coronary sinus**.

Anatomy Overview—Heart Structures
Anatomy Overview—Pulmonary Circulation
Figure 19.6—Systemic and Pulmonary Circulations
Exercise–Drag and Drop Blood Flow
Clinical Connection—Coronary Artery Disease

**Concept** | **Resources** 

4. The principal coronary vessels are the **left** and **right coronary arteries**, which arise from the ascending aorta. The **left coronary artery** divides into the **anterior interventricular branch** serving both ventricles, and the **circumflex branch** serving the left atrium and ventricle. The **right coronary artery** branches to supply the right atrium. Branches of the right coronary artery form the **posterior interventricular branch** serving walls of both ventricles, and the **marginal branch** supplying the right ventricle.
5. Arteries of the coronary circulation have many **anastomoses** (connections) that allow detours for arterial blood if a main route becomes obstructed.

---

## Concept 19.5 The cardiac conduction system coordinates heart contractions for effective pumping.

1. Cardiac muscle fibers are connected end-to-end via **intercalated discs. Desmosomes** in the discs provide strength, and **gap junctions** allow muscle action potentials to conduct from one muscle fiber to its neighbors.
2. **Autorhythmic fibers** form the cardiac conduction system, cardiac muscle fibers that spontaneously depolarize and generate action potentials. The powerful contractions of **contractile fibers**, which comprise the majority of myocardial fibers, propel blood throughout the body.
3. Cardiac conduction system action potentials occur as follows: Excitation is initiated at the heart's **pacemaker**, the **sinoatrial (SA) node**, and an action potential propagates throughout both atria, the **atrioventricular (AV) node**, the **atrioventricular (AV) bundle**, the **right** and **left bundle branches**, and finally, the **Purkinje fibers**. The Purkinje fibers conduct the action potential from the apex upward to the remainder of the ventricular myocardium.
4. Phases of an action potential in a ventricular contractile fiber include **depolarization**, **plateau**, and **repolarization**. Cardiac muscle tissue has a long **refractory period**, which allows chambers to refill with blood.
5. Cardiac muscle tissue produces most ATP from aerobic cellular respiration.

Resources:
Anatomy Overview—Heart Structures
Animation—Cardiac Conduction
Figure 19.8—The Conduction System of the Heart
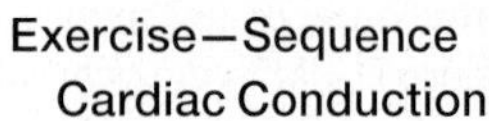
Exercise—Sequence Cardiac Conduction
Clinical Connection—Regeneration of Heart Cells
Clinical Connection—Artificial Pacemakers

---

## Concept 19.6 The electrocardiogram is a record of electrical activity associated with each heartbeat.

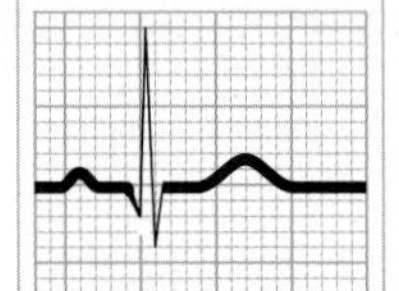

1. An **electrocardiogram** (**ECG** or **EKG**) is a recording of the electrical changes that accompany each heartbeat. The ECG is a composite of all the action potentials produced by cardiac muscle fibers during each heartbeat.
2. A normal ECG consists of a **P wave** (**atrial depolarization**), a **QRS complex** (onset of **ventricular depolarization**), and a **T wave** (**ventricular repolarization**).
3. ECG waves correlate to atrial and ventricular relaxation, or **diastole**, and contraction, or **systole**. The sequence of systole and diastole related to the ECG is as follows: (1) depolarization of the atria is represented by the P wave; (2) atrial systole occurs; (3) ventricular depolarization is seen as the QRS complex; (4) ventricular systole begins; (5) repolarization of the ventricles produces the T wave; and (6) ventricular diastole occurs.

Resources:
Animation—Cardiac Cycle and ECG
Exercise—ECG Jigsaw Puzzle
Clinical Connection—Arrhythmias

---

## Concept 19.7 The cardiac cycle represents all of the events associated with one heartbeat.

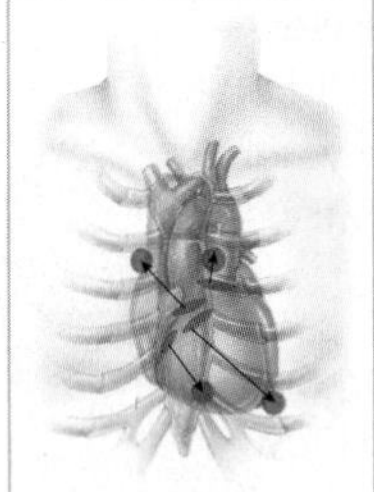

1. A **cardiac cycle** consists of the systole (contraction) and diastole (relaxation) of both atria, plus the systole and diastole of both ventricles.
2. Using a stethoscope, the first two of the four **heart sounds** of a cardiac cycle are loud enough to be heard. S1, the first heart sound (**lubb**), is caused by blood turbulence associated with the closing of the atrioventricular valves. S2, the second sound (**dupp**), is caused by blood turbulence associated with the closing of semilunar valves.
3. The heart's electrical signals cause alternating contraction and relaxation of the atria and ventricles. Contraction of a heart chamber causes an increase in the blood pressure within the chamber, which pushes blood to areas of lower pressure.

Resources:
Animation—Cardiac Cycle
Concepts and Connections—Cardiac Cycle
Exercise—Cardiac Cycle

4. The events during a cardiac cycle include the following: (1) atrial depolarization and P wave occur; (2) **atrial systole** forces blood across open AV valves into relaxed ventricles; (3) ventricular filling occurs; (4) QRS complex signals the onset of ventricular depolarization; (5) **atrial diastole** and **ventricular systole** close the AV valves; (6) increasing pressure inside the ventricles opens the SL valves, resulting in **ventricular ejection**; (7) at rest, 70 mL of blood is ejected from each ventricle; (8) T wave indicates ventricular repolarization; (9) **ventricular diastole** decreases ventricular pressure, causing SL valves to close; and (10) as ventricular pressure continues to drop, AV valves open, and **ventricular filling** begins.

## Concept 19.8 Cardiac output is the blood volume ejected by a ventricle each minute.

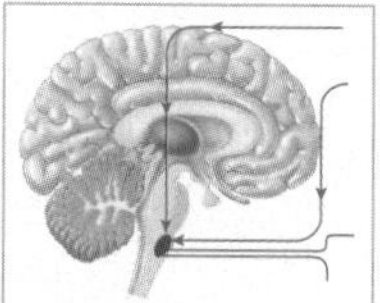

1. **Cardiac output** is the volume of blood ejected each minute from the left ventricle into the aorta or by the right ventricle into the pulmonary trunk.
2. **Stroke volume** is the volume of blood ejected by the ventricle with each contraction (beat). **Heart rate** is the number of heartbeats per minute.
3. Stroke volume is related to **preload** (stretch on the heart before it contracts), **contractility** (forcefulness of contraction), and **afterload** (pressure that must be exceeded before the SL valves open and ventricular ejection begins). According to the **Frank-Starling law of the heart**, greater stretch of ventricular muscle fibers during diastole leads to greater force of contraction during systole.
4. The **cardiovascular center** in the medulla oblongata is the origin of nervous system regulation of the heart. This region receives input from **proprioceptors**, **chemoreceptors**, and **baroreceptors**, and from the limbic system and cerebral cortex.
5. Sympathetic neurons arising from the cardiovascular center release norepinephrine to increase heart rate and force of contraction.
6. Parasympathetic neurons from the cardiovascular center release acetylcholine to decrease heart rate.
7. Epinephrine, norepinephrine, and thyroid hormones increase heart rate and contractility. Elevated $Na^+$ or $K^+$ decreases heart rate and contractility, while elevated $Ca^{2+}$ increases heart rate and contractility.
8. Other factors that influence heart rate include age, gender, physical fitness, and body temperature.

Animation—Cardiac Output Factors
Exercise—Cardiac Ouput
Concepts and Connections—Cardiac Output

Clinical Connection—Help for Failing Hearts

## Understanding the Concepts

1. Which layer of the pericardium is both a part of the pericardium and a part of the heart wall?
2. Which of the four heart chambers has the thickest wall? Why?
3. As the ventricles contract, what prevents the atrioventricular valves from swinging upward into the atria?
4. Describe the sequence of heart chambers, heart valves, and blood vessels a drop of blood encounters from the time it flows into the right atrium until it reaches the aorta. When during its journey does the drop of blood contain the most oxygen? The least oxygen?
5. How does blood flowing through the heart chambers eventually reach the myocardium?
6. Why do cardiac muscle fibers need a long refractory period?
7. How does each ECG wave relate to contraction and relaxation of the atria and ventricles?
8. Describe the status of each of the four heart valves during ventricular filling. How would a defect in the atrioventricular valve affect ventricular filling?
9. Why must ventricular pressure be greater than arterial pressure during ventricular ejection?
10. Which events cause the AV valves to open and shut? Which events make the SL valves open and shut?
11. Will stroke volume increase or decrease in each of the following situations? Explain each of your answers.
    - Your blood pressure rises when you're angry.
    - An anesthetic decreases the strength of contraction of your ventricles.
    - When you are exercising, contraction of skeletal muscles returns more blood to the heart.
12. What mechanism does the cardiovascular center use to speed up or slow down heart rate?

CHAPTER

# 20

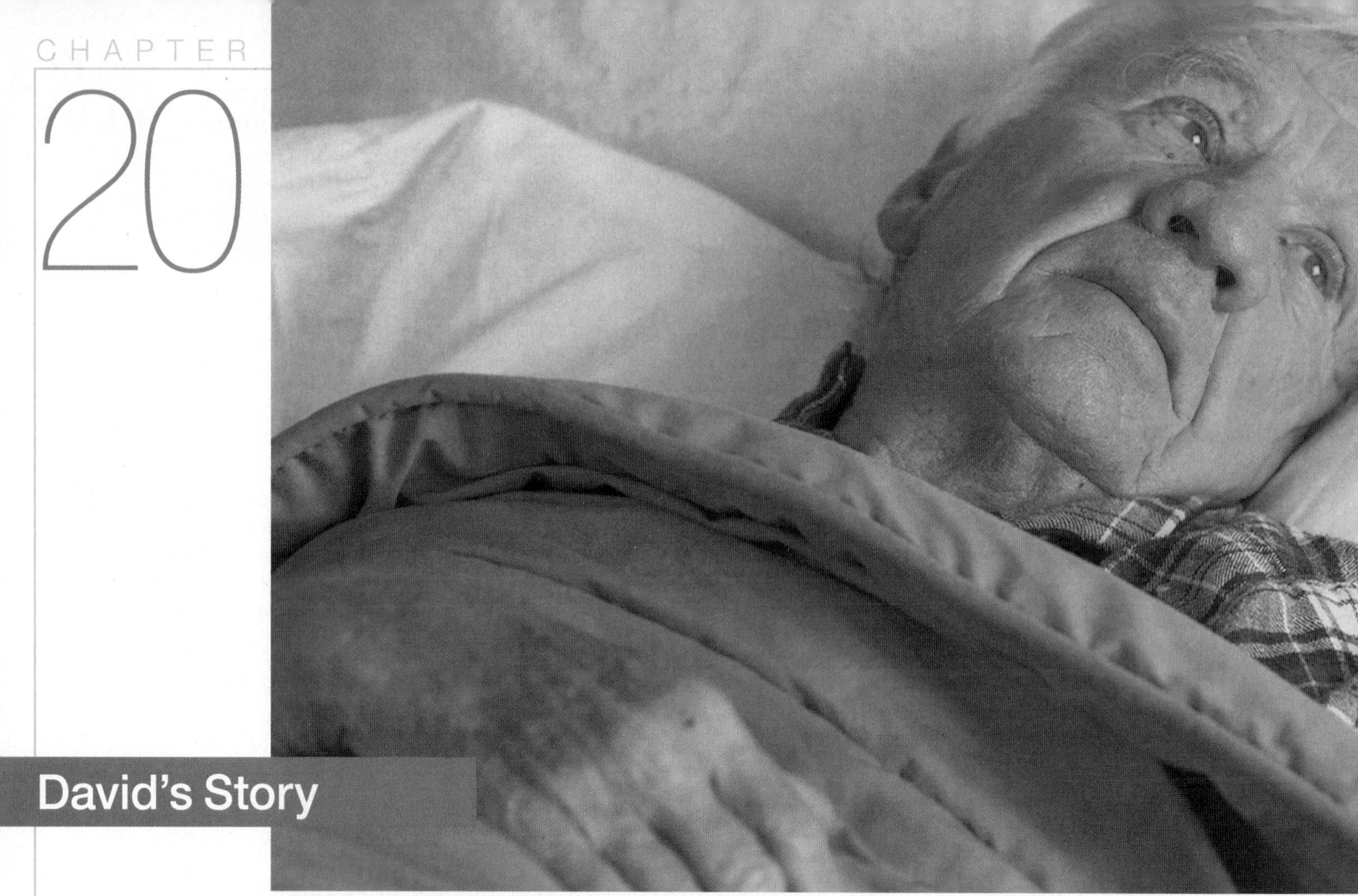

## David's Story

Sandy got up early to get ready for school. As she made coffee in the small kitchenette of the apartment where she and her grandfather lived, she heard him coughing. "Grandpa David, it's time to get up," she called as she finished putting water in the coffeepot.

"I'm just going to stay in here awhile. I've got a headache," replied David.

Sandy thought it was odd. Her grandfather usually got up early, much earlier than she did. She walked down the hallway and knocked on his door. "Grandpa, are you all right?" she asked as she entered his room.

"Just leave me alone. I'm fine, just a little under the weather today," he answered.

She walked over to him and put her hand on his head. His pale skin was cool and clammy, and a strange mottled rash covered his face and neck. She noticed that he was breathing rapidly, and felt his pulse racing in the artery that ran along his temple. "Grandpa David, I think you're really sick. Maybe you should go to the doctor."

David grabbed his blanket and turned away from her. "It's just the flu. I threw up last night. I'll be fine. I just need some rest. Now go to school. Tell your Grandma to come in here. I want to talk to her."

Sandy stared at him and said, "Grandpa, she's not here." Sandy wasn't sure what to do. Her grandmother had been dead for 10 years. She felt scared. Something was not right—Grandpa David was sick, and confused. "Grandpa David, I'm going to see if Mrs. Mahoney can drive us to the clinic."

The cardiovascular system contributes to homeostasis of other body systems by transporting and distributing blood throughout the body, delivering oxygen and nutrients, and carrying away wastes. Defects in the functioning of the cardiovascular system can have profound effects on the overall health of the body. Such defects can take many forms, and, as we will see in David's case, failure of the cardiovascular system to deliver enough oxygen and nutrients to the body can have significant consequences.

# The Cardiovascular System: Blood Vessels

## INTRODUCTION

Blood vessels transport and distribute blood throughout the body to deliver materials (such as oxygen, nutrients, and hormones) and carry away wastes. Blood vessels form a closed system of tubes that carries blood away from the heart, transports it to the tissues of the body where nutrients and wastes are exchanged, and then returns it to the heart. Blood is pumped through an estimated 100,000 km (60,000 mi) of blood vessels by the left side of the heart, while the right side of the heart pumps blood through pulmonary blood vessels that allow blood to pick up oxygen and unload carbon dioxide from the lungs. In Chapters 18 and 19 we described the composition and functions of blood and the structure and function of the heart. In this chapter we examine the structure and functions of the various types of blood vessels (arteries, arterioles, capillaries, venules, and veins); the forces involved in circulating blood throughout the [illegible] constitute the major circulatory routes.

## CONCEPTS

**20.1** Most blood vessel walls have three distinct tissue la

**20.2** Blood ejected from the heart flows through elastic a

**20.3** Capillaries are microscopic blood vessels that funct

**20.4** Venules and veins return blood to the heart.

**20.5** Blood flows from regions of higher pressure to thos

**20.6** Blood pressure is regulated by neural and hormonal

**20.7** Measurement of the pulse and blood pressure helps

**20.8** The two main circulatory routes are the pulmonary c

**20.9** Systemic arteries carry blood from the heart to all b

**20.10** Systemic veins return blood to the heart from all bo

Endothelium

## 20.1 Most blood vessel walls have three distinct tissue layers.

The walls of blood vessels, except the smallest vessels, consist of three layers, or tunics, of different tissues. The three structural layers from innermost to outermost are the *tunica interna*, *tunica media*, and *tunica externa* (Figure 20.1). Modifications of this basic design account for the structural and functional differences among the various blood vessel types. These structural variations correlate to the differences in function that occur throughout the cardiovascular system.

### Tunica Interna

The **tunica interna** (TOO-ni-ka; *tunic* = garment or *coat; interna* = innermost) forms the thin inner lining of a blood vessel and is in direct contact with blood as it flows through the **lumen** (LOO-men), or interior opening, of the vessel (Figure 20.1a, b). Its innermost layer, the *endothelium*, is a single layer of flattened cells that lines the inner surface of the entire cardiovascular system (heart and blood vessels). Endothelial cells have a smooth luminal surface that facilitates efficient blood flow by reducing surface friction, secrete chemical mediators that influence vessel contraction, and assist with capillary permeability. Deep to the endothelium is a *basement membrane*. It anchors the endothelium to the underlying connective tissue and provides a physical support base that imparts significant tensile strength and resilience for stretching and recoil. The outermost part of the tunica interna is the *internal elastic lamina* (*lamina* = thin plate; plural is *laminae*), a thin sheet of elastic fibers with windowlike openings that give it the look of Swiss cheese. These openings facilitate diffusion of materials through the tunica interna to the thicker tunica media.

**FIGURE 20.1 Comparative structure of blood vessels.** The capillary in (c) is enlarged relative to the structures shown in parts (a) and (b).

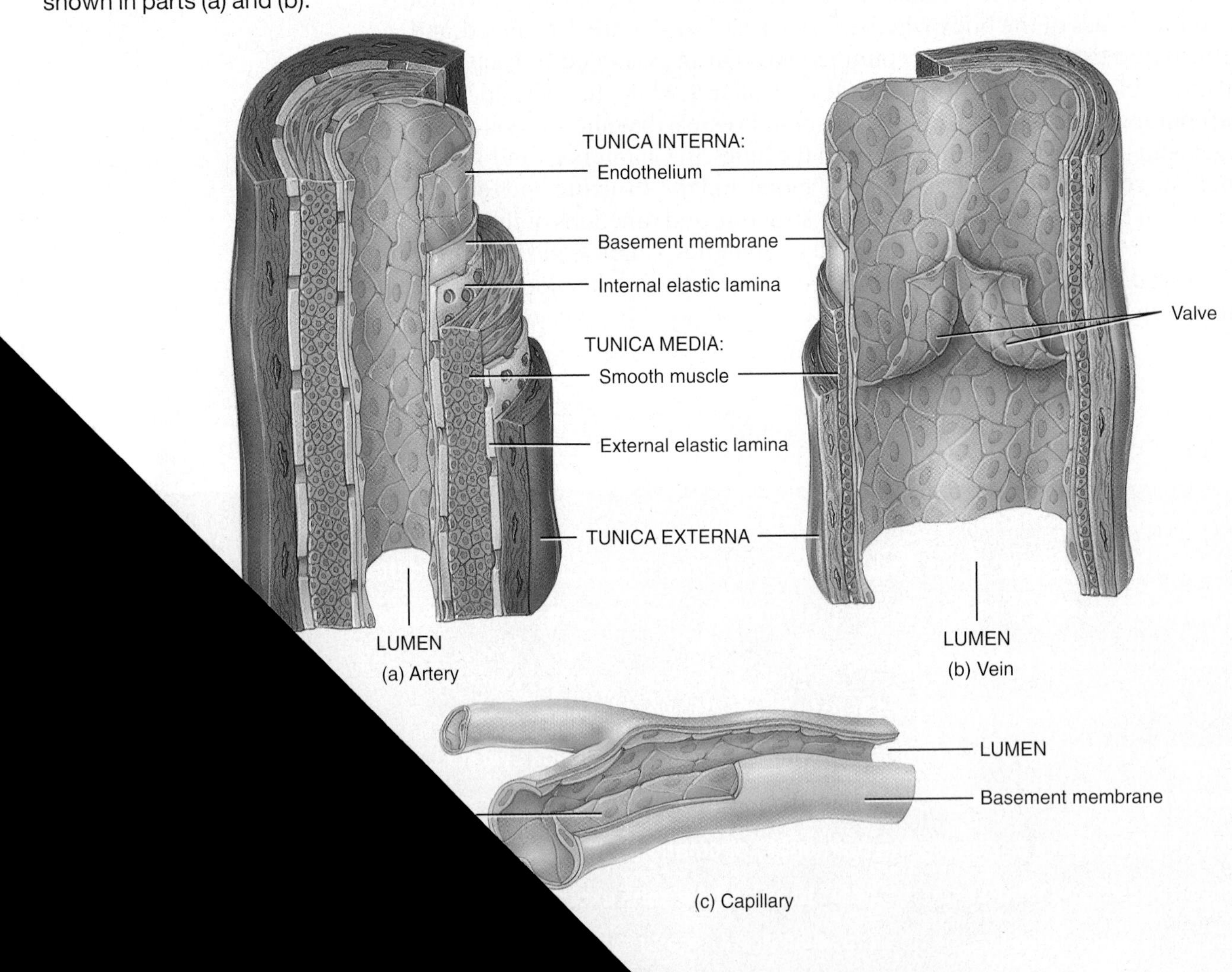

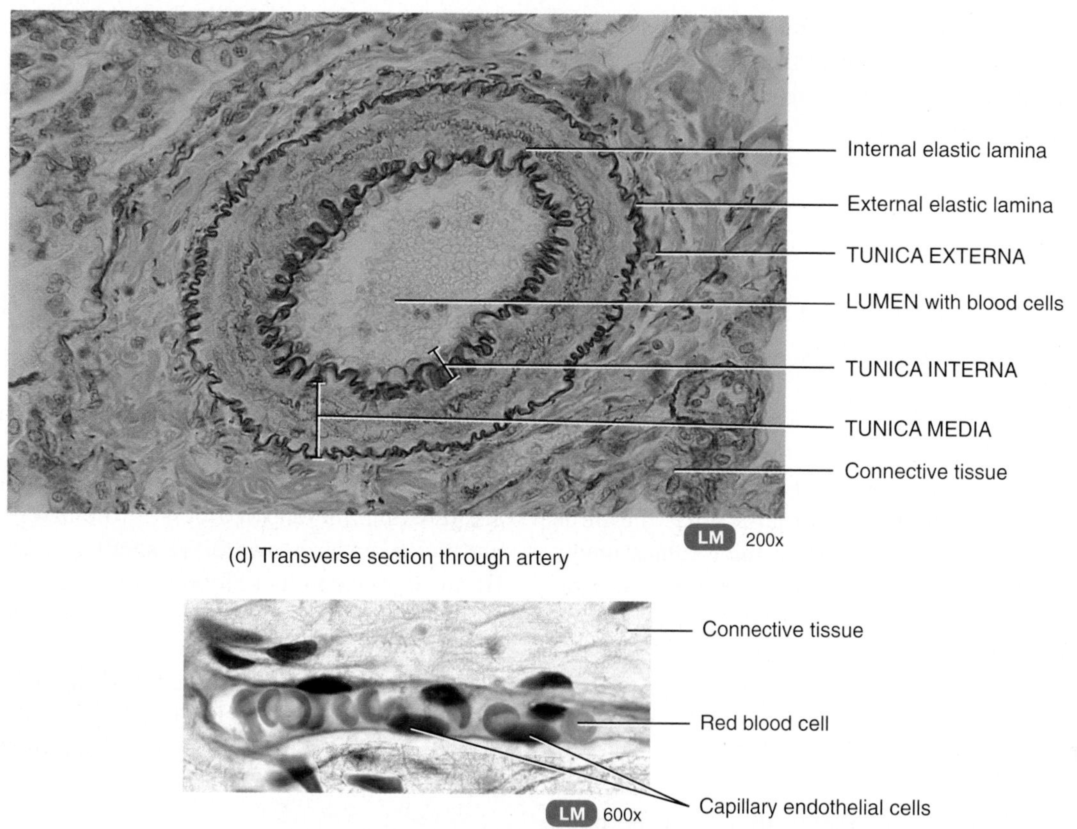


(d) Transverse section through artery

(e) Red blood cells passing through capillary

**Arteries carry blood from the heart to tissues; veins carry blood from tissues back to the heart.**

## Tunica Media

The **tunica media** (*media* = middle) is a relatively thick layer comprised mainly of substantial amounts of elastic fibers and smooth muscle cells (Figure 20.1a, b). The elastic fibers allow vessels to stretch and recoil under the applied pressure of blood. The primary role of the smooth muscle cells, which extend circularly around the lumen like a ring encircles your finger, is to regulate the diameter of the lumen. Sympathetic axons of the autonomic nervous system innervate the smooth muscle of the tunica media. An increase in sympathetic stimulation typically stimulates the smooth muscle to contract, squeezing the vessel wall and narrowing the lumen. Such a decrease in the diameter of the lumen of a blood vessel is called **vasoconstriction** (vā-sō-kon-STRIK-shun). In contrast, when sympathetic stimulation decreases, smooth muscle fibers relax. The resulting increase in lumen diameter is called **vasodilation** (vā-sō-dī-LĀ-shun). As you will learn in more detail shortly, the rate of blood flow through different parts of the body is regulated by the extent of smooth muscle contraction in the walls of particular vessels. Furthermore, the extent of smooth muscle contraction in particular vessel types is crucial in the regulation of blood pressure. In addition to regulating blood flow and blood pressure, smooth muscle of the tunica media contracts when an artery or arteriole is damaged (in a process called *vascular spasm*) to help limit loss of blood through the injured vessel if the vessel is small.

The tunica media is the most variable of the tunics. As you study the different types of blood vessels in the remainder of this chapter, you will see that the structural differences in this layer account for the many variations in function among the different vessel types. Separating the tunica media from the tunica externa is a network of elastic fibers, the *external elastic lamina*, which is part of the tunica media.

## Tunica Externa

The outer covering of a blood vessel, the **tunica externa** (*externa* = outermost), consists of elastic and collagen fibers (Figure 20.1a, b). The tunica externa contains numerous nerves and, especially in larger vessels, tiny blood vessels that supply blood to the tissue of the vessel wall. The tunica externa also helps anchor the vessel to surrounding tissues.

Next we investigate the types of blood vessels: arteries, arterioles, capillaries, venules, and veins.

✓ CHECKPOINT

1. What is the function of elastic fibers and smooth muscle in the tunica media of arteries?

## 20.2 Blood ejected from the heart flows through elastic arteries, muscular arteries, and then arterioles.

Because **arteries** (AR-ter-ēz; *ar-* = air; *-ter-* = to carry) were found empty at death, in ancient times they were thought to contain only air. Arteries carry blood *away from the heart* (mnemonic: ***a**rteries* = ***a**way*). The wall of an artery has the three layers of a typical blood vessel, but has a thick tunica media (see Figure 20.1a). Due to their plentiful elastic fibers, arteries normally have walls that stretch easily or expand without tearing in response to a small increase in blood pressure.

### Elastic Arteries

**Elastic arteries** are the largest diameter arteries in the body, ranging from the garden hose–sized aorta and pulmonary trunk to the finger-sized branches of the aorta. Measuring approximately one-tenth of the vessel's total diameter, their walls are relatively thin compared to the overall size of the vessel. These vessels are characterized by a thick tunica media that is dominated by elastic fibers. Elastic arteries perform an important function: They help propel blood onward while the ventricles are relaxing. As blood is ejected from the heart into elastic arteries, their highly elastic walls stretch, easily accommodating the surge of blood. As they stretch, the elastic fibers momentarily store mechanical energy, functioning as a **pressure reservoir** (REZ-er-vwa) (Figure 20.2a). Then, while the ventricles are relaxing, the elastic fibers in the artery walls recoil, forcing blood onward toward the smaller arteries (Figure 20.2b). Because they conduct blood from the heart to muscular arteries, elastic arteries are also called *conducting arteries*. Elastic arteries include the two major trunks that exit the heart (the aorta and the pulmonary trunk), along with the aorta's major initial branches, such as the brachiocephalic trunk, subclavian artery, common carotid artery, and common iliac arteries (see Figure 20.16b).

### Muscular Arteries

Large elastic arteries divide into medium-sized **muscular arteries** that branch out into the various regions of the body. The tunica media of muscular arteries contains more smooth muscle and fewer elastic fibers than elastic arteries. The large amount of smooth muscle, approximately three-quarters of the total mass, makes the walls of muscular arteries relatively thick. Because

FIGURE 20.2 **Pressure reservoir function of elastic arteries.**

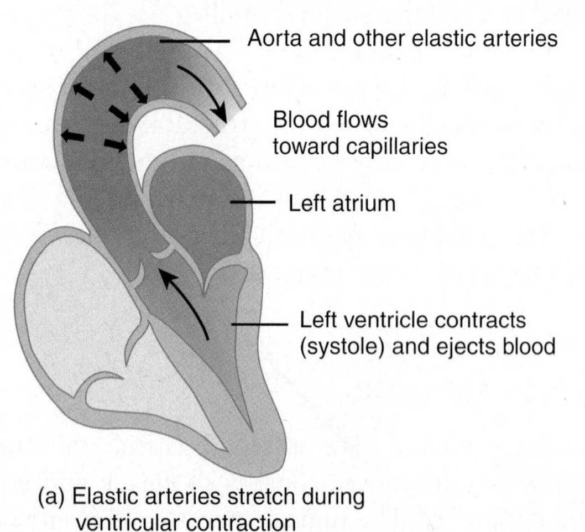


(a) Elastic arteries stretch during ventricular contraction

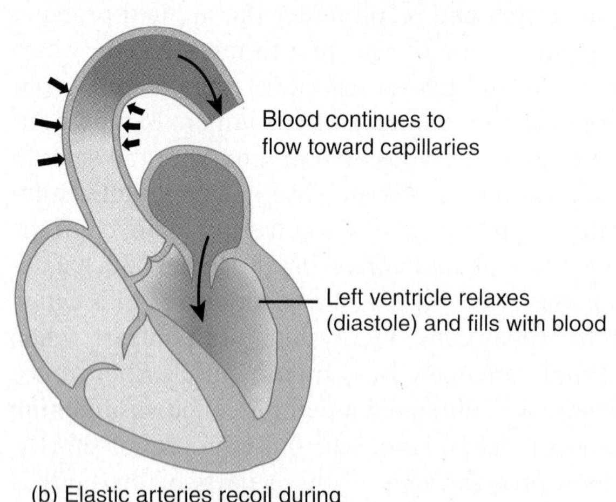


(b) Elastic arteries recoil during ventricular relaxation

**Recoil of elastic arteries keeps blood flowing during ventricular relaxation (diastole).**

of the reduced amount of elastic tissue in the walls of muscular arteries, these vessels do not have the ability to recoil and help propel the blood like the elastic arteries. Instead, muscular arteries are capable of greater vasoconstriction and vasodilation to adjust the rate of blood flow. Muscular arteries span a range of sizes from the pencil-sized femoral and axillary arteries to string-sized arteries that enter organs. Compared to elastic arteries, the walls of muscular arteries comprise a larger percentage (approximately 25 percent) of the total vessel diameter. Because the muscular arteries continue to branch and ultimately distribute blood to each of the various regions of the body, they are called *distributing arteries*. Examples include the brachial artery in the arm and radial artery in the forearm (see Figure 20.18b).

## Anastomoses

Most tissues of the body receive blood from more than one artery. The union of the branches of two or more arteries supplying the same body region is called an **anastomosis** (a-nas-tō-MŌ-sis = connecting; plural is *anastomoses*). Anastomoses between arteries provide alternative routes for blood to reach a tissue or organ. If blood flow stops for a short time when normal movements compress a vessel, or if a vessel is blocked by disease, injury, or surgery, then circulation to a part of the body is not necessarily stopped. The alternative route of blood flow to a body part through an anastomosis is known as **collateral circulation**. Anastomoses may also occur between veins and between arterioles and venules. Arteries that do not anastomose are known as **end arteries**. Obstruction of an end artery interrupts the blood supply to a whole segment of an organ, producing *necrosis* (death) of that segment. Alternative blood routes may also be provided by vessels that do not anastomose yet supply the same region of the body.

## Arterioles

Muscular arteries branch into small arteries, which in turn divide into still smaller arteries called **arterioles** (ar-TĒR-ē-ŌIs). Literally meaning small arteries, arterioles are abundant microscopic vessels that regulate the flow of blood into the capillary networks of the body's tissues (Figure 20.3). The wall thickness of arterioles is about one-half of the total vessel diameter. Arterioles have a thin tunica interna with a thin, *fenestrated*

FIGURE 20.3 **Arterioles, capillaries, and venules.** Precapillary sphincters regulate the flow of blood through capillary beds.

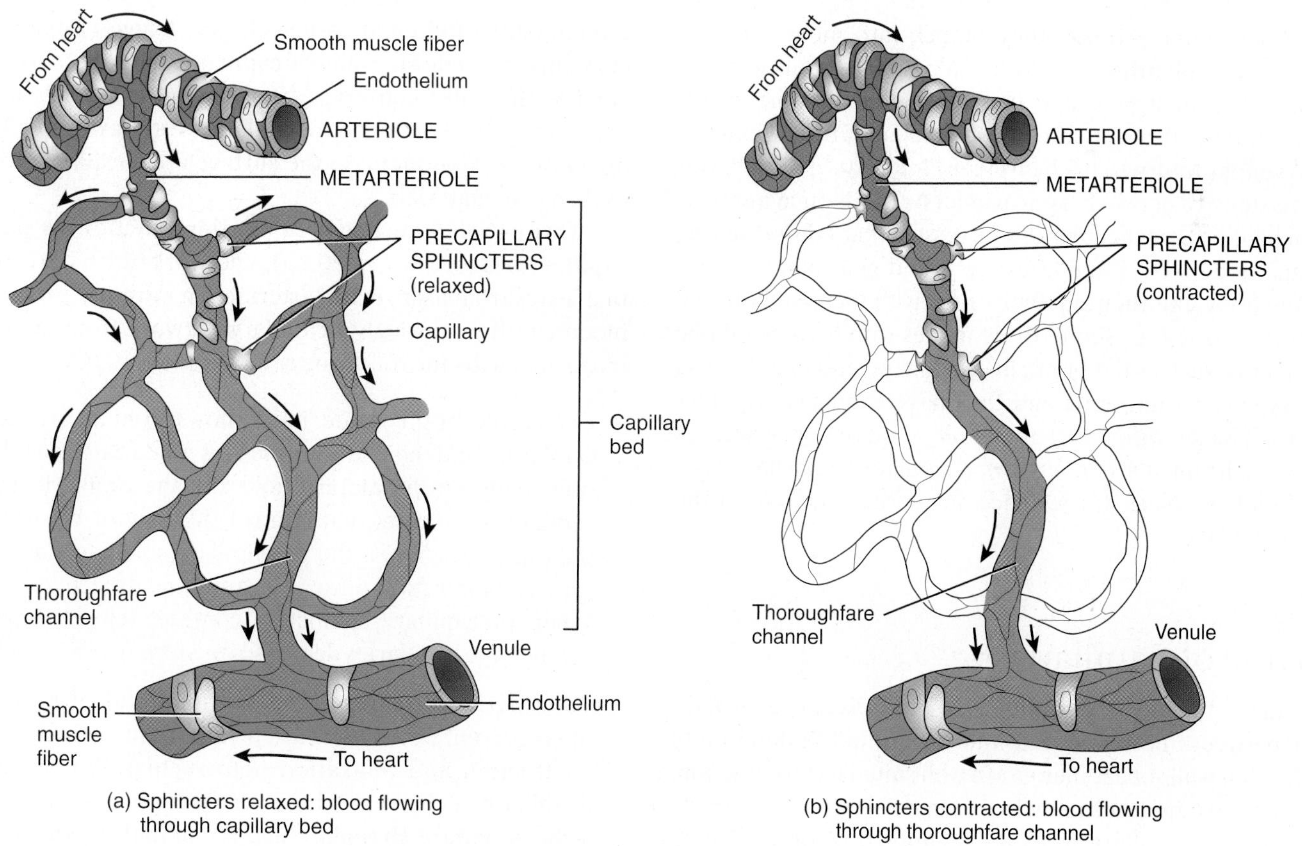


(a) Sphincters relaxed: blood flowing through capillary bed

(b) Sphincters contracted: blood flowing through thoroughfare channel

**In capillaries, nutrients, gases, and wastes are exchanged between the blood and interstitial fluid.**

(with small pores) internal elastic lamina. Arterioles have a tunica media that consists of one to two layers of smooth muscle cells having a circular orientation in the vessel wall. The terminal end of the arteriole, the region called the **metarteriole** (met′-ar-TĒR-ē-ōl; *meta-* = after), tapers toward the capillary junction. At the metarteriole–capillary junction, the distal-most muscle cell forms the **precapillary sphincter** (SFINGK-ter = to bind tight), which monitors the blood flow into the capillary. The tunica externa of the arteriole contains abundant sympathetic nerves. This sympathetic nerve supply, along with the actions of chemical mediators, can alter the diameter of arterioles to vary the flow of blood through these vessels.

Arterioles play a key role in regulating blood flow from arteries into capillaries by regulating *resistance*, the opposition to blood flow. In a blood vessel, resistance is due mainly to friction between blood and the inner walls of blood vessels. When the blood vessel diameter is smaller, the friction is greater, so there is more resistance to blood flow. Contraction of the smooth muscle of an arteriole causes vasoconstriction, which increases resistance even more and decreases blood flow into capillaries supplied by that arteriole. By contrast, relaxation of the smooth muscle of an arteriole causes vasodilation, which decreases resistance and increases blood flow into capillaries. Changes in the diameter of arterioles can also affect blood pressure: vasoconstriction of arterioles increases blood pressure, and vasodilation of arterioles decreases blood pressure.

✓ **CHECKPOINT**

2. How do elastic arteries and muscular arteries differ in structure?
3. Contrast the functions of elastic arteries, muscular arteries, and arterioles.
4. What is the relationship between anastomoses and collateral circulation?

# 20.3 Capillaries are microscopic blood vessels that function in exchange between blood and interstitial fluid.

As arterioles enter a tissue, they branch into numerous tiny vessels called **capillaries** (KAP-i-lar′-ēz; *capillus* = little hair). Capillaries, the smallest of blood vessels, are microscopic vessels that form the "U-turns" that connect arterial outflow from the heart to venous return to the heart (see Figure 20.3). Capillaries form an extensive network, approximately 20 billion in number, of branched, interconnecting blood vessels that course among individual body cells. Capillaries are found near almost every cell in the body, but their number varies with the metabolic activity of the tissue they serve. Body tissues with high metabolic requirements, such as the brain, liver, kidneys, and muscles, use more oxygen and nutrients and thus have extensive capillary networks. Tissues with lower metabolic requirements, such as tendons and ligaments, contain fewer capillaries. Capillaries are absent in a few tissues, for instance, the cornea and lens of the eye, and cartilage.

## Structure of Capillaries

The primary function of capillaries is the exchange of substances between the blood and interstitial fluid bathing body cells. The thin walls of capillaries are well suited to this function. Capillaries have no tunica media or tunica externa (see Figure 20.1c). Because capillary walls are composed of only a single layer of endothelial cells and a basement membrane, a substance in the blood only needs to pass through one cell layer to reach the interstitial fluid and tissue cells. Exchange of materials occurs only through the walls of capillaries and the beginning of venules; the walls of arteries, arterioles, most venules, and veins present too thick a barrier. Capillaries form extensive branching networks that increase the surface area available for rapid exchange of materials.

Throughout the body, capillaries function as part of a **capillary bed** (see Figure 20.3a), a network of 10–100 capillaries that arises from a single metarteriole. In most parts of the body, blood can flow through a capillary network from an arteriole into a venule by the following routes:

- *Capillaries.* In this route, blood flows from an arteriole into capillaries and then into venules. As noted earlier, at the junctions between the metarteriole and the capillaries are precapillary sphincters that control the flow of blood through the capillaries. When the precapillary sphincters are relaxed (open), blood flows into the capillaries (see Figure 20.3a). When precapillary sphincters contract (close), blood flow through the capillaries decreases or stops (see Figure 20.3b).
- *Thoroughfare channel.* The proximal end of a metarteriole is surrounded by scattered smooth muscle fibers whose contraction and relaxation help regulate blood flow. The distal end of the vessel has no smooth muscle and is called a **thoroughfare channel**. Such a channel provides a direct route for blood from an arteriole to a venule, thus bypassing capillaries.

## Types of Capillaries

There are three different types of capillaries: continuous capillaries, fenestrated capillaries, and sinusoids (Figure 20.4). Most capillaries are **continuous capillaries**, in which the plasma membranes of endothelial cells form a continuous tube that is interrupted only by *intercellular clefts*, gaps between neighboring endothelial cells (Figure 20.4a). Most areas of the brain contain continuous capillaries, which allow only a few substances to move across their walls. Because of the "tight" nature of these capillary walls, the resulting blockade to movement of materials into and out of brain capillaries is known as the *blood–brain barrier* (see Concept 13.2).

Some capillaries are **fenestrated capillaries** (fen′-es-TRĀ-ted; *fenestr-* = window). The plasma membranes of the endothelial cells in these capillaries contain many *fenestrations*, small pores or holes (Figure 20.4b). The many fenestrations allow these capillaries greater exchange of materials. Fenestrated capillaries are found in the kidneys, where filtration of blood actively occurs.

**Sinusoids** (SĪ-nū-soyds; *sinus* = curve) have endothelial cells with large fenestrations, an incomplete or absent basement membrane, and very large intercellular clefts (Figure 20.4c). Sinusoid walls are so leaky that large protein molecules and even blood cells can pass from a tissue into the bloodstream. For example, newly formed blood cells enter the bloodstream through the sinusoids of red bone marrow.

Usually blood passes from the heart and then (in sequence) through arteries, arterioles, capillaries, venules, and veins and then back to the heart. In some parts of the body, however, blood passes from one capillary network into another through a vein called a portal vein. Such a circulation of blood is called a **portal system**.

FIGURE 20.4 **Types of capillaries.**

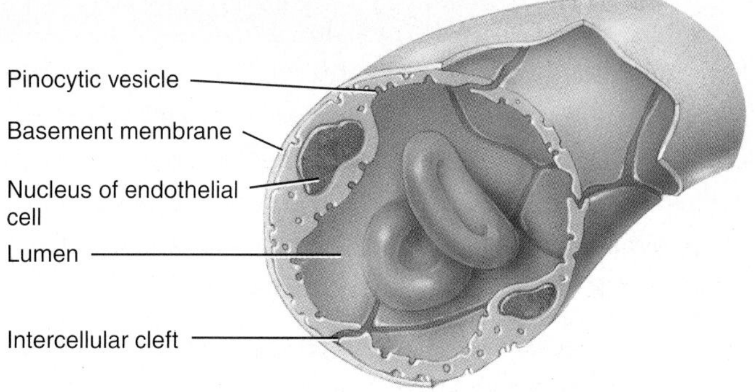


(a) Continuous capillary formed by endothelial cells

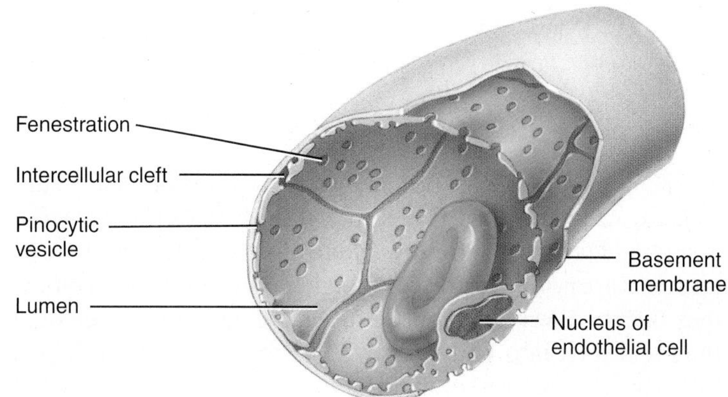


(b) Fenestrated capillary

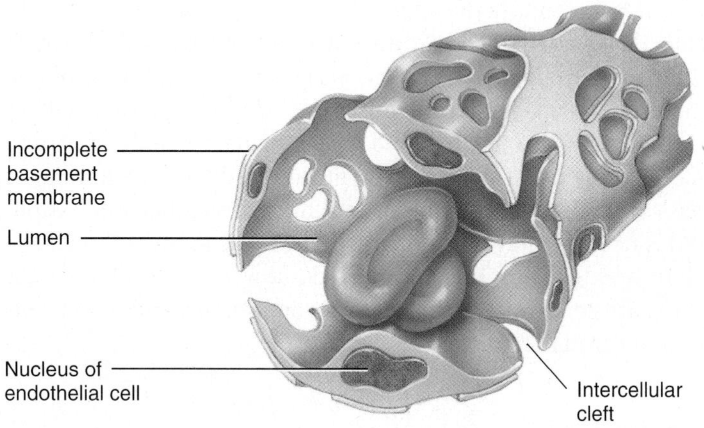


(c) Sinusoid

**Capillaries are microscopic blood vessels that connect arterioles and venules.**

## Autoregulation of Capillary Blood Flow

In each capillary bed, local changes can regulate vasodilation and vasoconstriction. When tissue cells release vasodilators, nearby arterioles dilate and precapillary sphincters relax. As a result, blood flow into the capillary networks increases, and oxygen delivery to the tissue rises. Vasoconstrictors have the opposite effect. The ability of a tissue to automatically adjust its blood flow to match its metabolic demands is called **autoregulation** (aw′-tō-reg′-ū-LĀ-shun). In tissues such as the heart and skeletal muscle, where the demand for oxygen and nutrients and for the removal of wastes can increase as much as tenfold during physical activity, autoregulation is an important contributor to increased blood flow through the tissue. Autoregulation also controls regional blood flow in the brain; blood distribution to various parts of the brain changes dramatically for different mental and physical activities. During a conversation, for example, blood flow increases to your motor speech areas when you are talking and increases to the auditory areas when you are listening.

## Capillary Exchange

The mission of the entire cardiovascular system is to keep blood flowing through capillaries to allow **capillary exchange**, the movement of substances between blood and interstitial fluid. The 7 percent of the blood in systemic capillaries at any given time is continually exchanging materials with interstitial fluid.

## CLINICAL CONNECTION | *Edema*

If filtration greatly exceeds reabsorption, the result is **edema** (e-DĒ-ma = swelling), an abnormal increase in interstitial fluid volume. Edema is not usually detectable in tissues until interstitial fluid volume has risen to 30 percent above normal. Edema can result from either excess filtration or inadequate reabsorption.

Two situations may cause excess filtration:

- *Increased capillary blood pressure* causes more fluid to be filtered from capillaries.
- *Increased permeability of capillaries* raises interstitial fluid osmotic pressure by allowing some plasma proteins to escape. Such leakiness may be caused by the destructive effects of chemical, bacterial, thermal, or mechanical agents on capillary walls.

One situation commonly causes inadequate reabsorption:

- *Decreased concentration of plasma proteins* lowers the blood colloid osmotic pressure. Inadequate synthesis or dietary intake or loss of plasma proteins is associated with liver disease, burns, malnutrition, and kidney disease.

### Diffusion

Most capillary exchange occurs by simple diffusion. Many substances, such as oxygen ($O_2$), carbon dioxide ($CO_2$), and glucose enter and leave capillaries by simple diffusion. Because $O_2$ and nutrients are normally present in higher concentrations in blood, they diffuse down their concentration gradients into interstitial fluid and then into body cells. $CO_2$ and other wastes released by body cells are present in higher concentrations in interstitial fluid, so they diffuse into blood.

Substances can cross the walls of a capillary by diffusing through the intercellular clefts or fenestrations or by diffusing through the endothelial cells (Figure 20.4). Water-soluble substances, such as glucose, diffuse across capillary walls through intercellular clefts or fenestrations (Figure 20.4). Lipid-soluble materials, such as steroid hormones, diffuse across capillary walls directly through the lipid bilayer of endothelial cell plasma membranes. Most plasma proteins and red blood cells cannot pass through capillary walls of continuous and fenestrated capillaries because they are too large to fit through the intercellular clefts and fenestrations. In sinusoids, however, the intercellular clefts are so large that they allow even proteins and blood cells to pass through their walls.

### Transcytosis

A small quantity of material crosses capillary walls by **transcytosis** (tranz′-sī-TŌ-sis; *trans-* = across; *-cyt-* = cell; *-osis* = process). In this process, substances in blood plasma become enclosed within tiny pinocytic vesicles that first enter endothelial cells by endocytosis, then move across the cell and exit on the other side by exocytosis. This method of transport is important mainly for large, lipid-insoluble molecules that cannot cross capillary walls in any other way. For example, certain antibodies (protein molecules) pass from the maternal circulation into the fetal circulation by transcytosis.

### Influences on Capillary Exchange

Because of the small diameter of capillaries, blood flows more slowly through them than through larger blood vessels. The slow flow aids the exchange of substances into and out of capillaries. The exchange of gases, nutrients, hormones, and metabolic waste products between capillary blood and interstitial fluid occurs due to pressure differences. **Capillary blood pressure**, the pressure of blood against the walls of capillaries, "pushes" fluid out of capillary blood into interstitial fluid. An opposing pressure, termed **blood colloid osmotic pressure**, "pulls" fluid into capillary blood. (Recall that an osmotic pressure is the pressure of a fluid due to its solute concentration. The higher the solute concentration, the greater is the osmotic pressure.) Most solutes exchange freely and so are present in nearly equal concentrations in capillary blood and interstitial fluid. Plasma proteins, however, are too large to pass through either fenestrations or gaps between endothelial cells. The presence of proteins in plasma and their virtual absence in interstitial fluid gives blood the higher osmotic pressure. Blood colloid osmotic pressure is osmotic pressure due mainly to plasma proteins.

If the pressures that push fluid out of capillaries exceed the pressures that pull fluid into capillaries, fluid will move *out of* capillary blood *into* the surrounding interstitial fluid, a movement called **filtration** (Figure 20.5). If, however, the pressures that push fluid out of interstitial spaces into capillaries exceed the pressures that pull fluid out of capillaries, then fluid will move *out of* interstitial fluid *into* capillary blood, a process termed **reabsorption**.

At the arterial end of a capillary, capillary blood pressure is higher than blood colloid osmotic pressure. Thus, water and solutes are filtered out of the capillary blood into the surrounding interstitial fluid. Capillary blood pressure decreases progressively as blood flows along a capillary. At the venous end of a capillary, blood pressure drops below blood colloid osmotic pressure. Then, water and solutes are reabsorbed from interstitial fluid

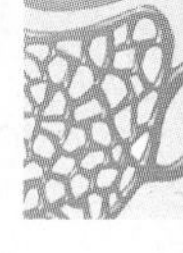

FIGURE 20.5 **Capillary exchange.**

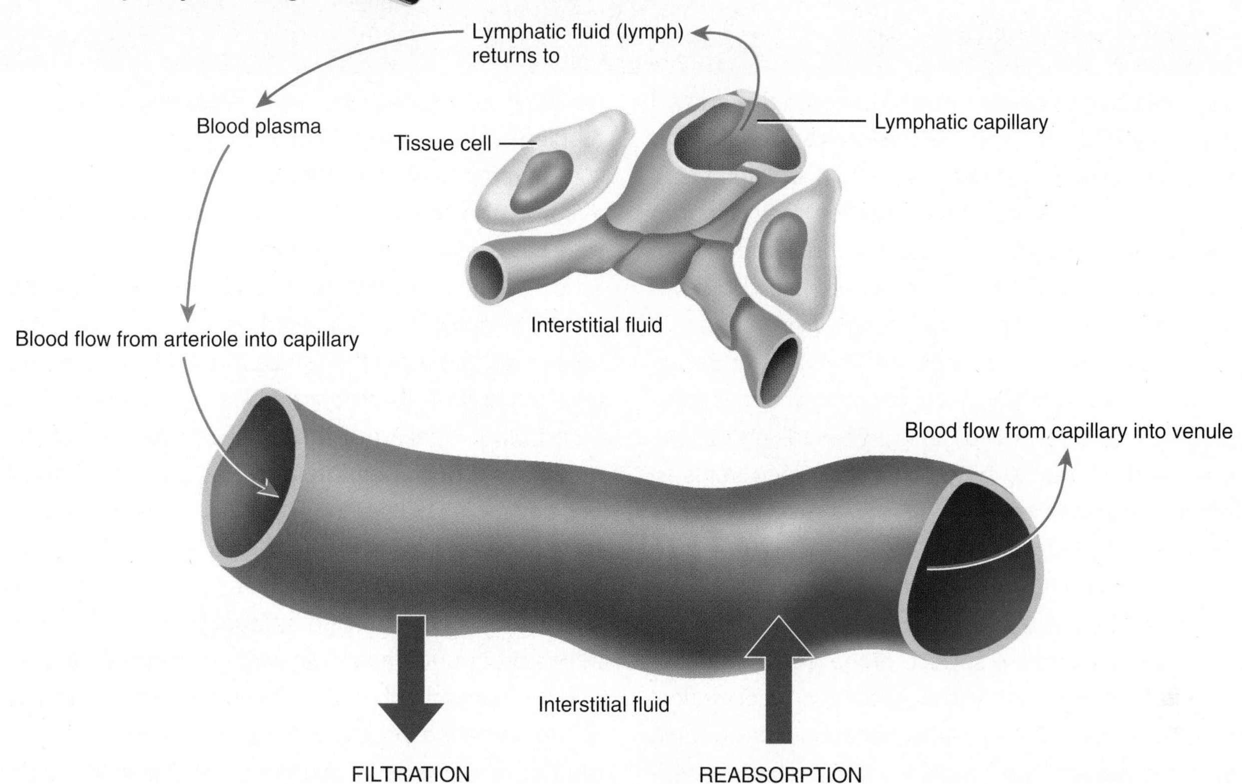


**Capillary blood pressure pushes fluid out of capillaries (filtration); blood colloid osmotic pressure pulls fluid into capillaries (reabsorption).**

into the capillary blood. On average, about 85 percent of the fluid filtered out of capillaries is reabsorbed back into capillaries. The excess filtered fluid and the few plasma proteins that do escape from blood into interstitial fluid enter lymphatic capillaries (Figure 20.5) and eventually are returned by the lymphatic system to the cardiovascular system.

✓ **CHECKPOINT**

5. How do materials move through capillary walls?
6. How is blood flow directed through a capillary bed?
7. Contrast the structure and function of the three types of capillaries.
8. What is autoregulation?
9. State the ways substances enter and leave blood plasma.
10. What happens to excess filtered fluid and proteins that are not reabsorbed back into capillaries?
11. How do capillary blood pressure and blood colloid osmotic pressure determine fluid movement across the walls of capillaries?

## 20.4 Venules and veins return blood to the heart.

### Venules

Groups of capillaries within a tissue reunite to form small vessels called **venules** (VEN-ūls = little veins). Unlike their thick-walled arterial counterparts, venules and veins have thin walls that do not readily maintain their shape. Venules drain capillary blood and begin the return flow of blood back toward the heart (see Figure 20.3). The smallest venules, those closest to the capillaries, have loosely organized intercellular junctions and thus are very porous. Venules function as sites of exchange of nutrients and wastes and movement of white blood cells from the bloodstream into an inflamed or infected tissue. As venules continue to enlarge, they acquire thicker walls across which exchanges with the interstitial fluid can no longer occur.

## CLINICAL CONNECTION | *Varicose Veins*

Leaky venous valves can cause veins to become dilated and twisted in appearance, a condition called **varicose veins** (VAR-i-kōs) or **varices** (VAR-i-sēz; *varic-* = a swollen vein). The singular is *varix* (VAR-iks). The condition may occur in the veins of almost any body part, but it is most common in the esophagus, anal canal, and superficial veins of the lower limbs. Those in the lower limbs can range from cosmetic problems to serious medical conditions. The valvular defect may be congenital or may result from mechanical stress (prolonged standing or pregnancy) or aging. The leaking venous valves allow the backflow of blood from the deep veins to the less efficient superficial veins, where the blood pools. This creates pressure that distends the vein and allows fluid to leak into surrounding tissue. As a result, the affected vein and the tissue around it may become inflamed and painfully tender. Veins close to the surface of the legs, especially the saphenous vein, are highly susceptible to varicosities; deeper veins are not as vulnerable because surrounding skeletal muscles prevent their walls from stretching excessively. Varicose veins in the anal canal are referred to as *hemorrhoids* (HEM-ō-royds). Esophageal varices result from dilated veins in the walls of the lower part of the esophagus and sometimes the upper part of the stomach. Bleeding esophageal varices are life-threatening and are usually a result of chronic liver disease.

Several treatment options are available for varicose veins in the lower limbs. *Elastic stockings* (support hose) may be used for individuals with mild symptoms or for whom other options are not recommended. *Sclerotherapy* (sklerō-THER-a-pē) involves injection of a solution into varicose veins that damages the tunica interna by producing a harmless superficial *thrombophlebitis* (inflammation involving a blood clot). Healing of the damaged part leads to scar formation that occludes the vein. *Radiofrequency endovenous occlusion* (ō-KLOO-zhun) involves the application of radiofrequency energy to heat up and close off varicose veins. *Laser occlusion* uses laser therapy to shut down veins. In a surgical procedure called *stripping*, veins may be removed. In this procedure, a flexible wire is threaded through the vein and then pulled out to strip (remove) it from the body.

In some individuals the superficial veins can be seen as blue-colored tubes passing under the skin. While the venous blood is a deep dark red, the veins appear blue because their thin walls and the tissue of the skin absorb the red-light wavelengths, allowing the blue light to pass through the surface to our eyes where we see them as blue.

## Veins

Venules merge to form progressively larger blood vessels called veins. **Veins** (VĀNZ) are the blood vessels that convey blood from the tissues *back to the heart.* Veins have very thin walls relative to their total diameter (average thickness is less than one-tenth of the vessel diameter). Although veins are composed of essentially the same three layers as arteries, the relative thicknesses of the layers are different (see Figure 20.1b). The tunica interna and tunica media of veins is thinner than that of arteries. The tunica externa of veins is the thickest layer and consists of collagen and elastic fibers. Because they lack the external or internal elastic laminae found in arteries, veins are distensible enough to adapt to variations in the volume and pressure of blood passing through them, but are not structured to withstand high pressure. The lumen of a vein is larger than that of a comparable artery, and their thinner walls often result in veins appearing collapsed (flattened) when sectioned.

By the time blood leaves capillary beds and moves into veins, it has lost a great deal of pressure. Therefore, the average blood pressure in veins is considerably lower than in arteries. The difference in pressure can be noticed when blood flows from a cut vessel. Blood leaves a cut vein in an even, slow flow but spurts rapidly from a cut artery. Most of the structural differences between arteries and veins reflect this pressure difference. For example, the walls of veins are not as strong as those of arteries.

Many veins, especially those in the limbs, contain **valves**, thin folds of tunica interna that form flaplike cusps (see Figure 20.1b). The valve cusps project into the lumen, pointing toward the heart (Figure 20.6). The low blood pressure in veins allows blood returning to the heart to slow and even back up; the valves aid in venous return by preventing the backflow of blood.

A **vascular sinus** is a vein with a thin endothelial wall that has no smooth muscle to alter its diameter. In a vascular sinus, the surrounding dense connective tissue replaces the tunica media and tunica externa in providing support. For example,

FIGURE 20.6 **Venous valves.**

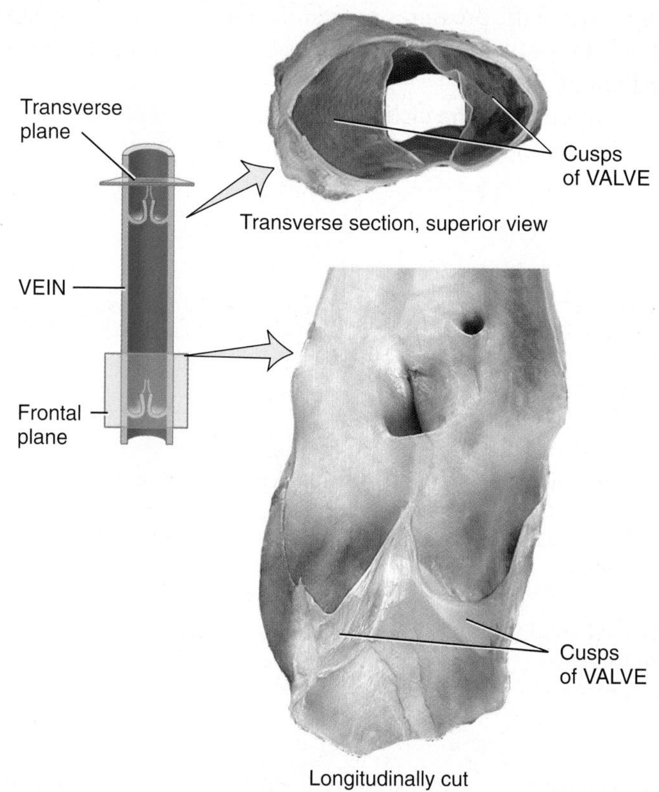


Valves in veins allow blood to flow in one direction only—toward the heart.

the coronary sinus of the heart is a vascular sinus that conveys deoxygenated blood from the heart wall to the right atrium (see Figure 19.3c).

## Venous Return

**Venous return**, the volume of blood flowing back to the heart through the systemic veins, is primarily due to the blood pressure generated by contractions of the heart's left ventricle. Blood pressure is measured in millimeters of mercury and abbreviated mm Hg. Although small, the blood pressure difference from venules (averaging about 16 mm Hg) to the right atrium (near 0 mm Hg) normally is sufficient to cause venous return to the heart. For example, when you stand up at the end of an anatomy and physiology lecture, the pressure pushing blood up the veins in your lower limbs is barely enough to overcome the force of gravity pushing it back down. Besides blood pressure generated by the heart, two other mechanisms help boost venous return by "pumping" blood from the lower body back to the heart: (1) the skeletal muscle pump and (2) the respiratory pump. Both pumps depend on the presence of valves in veins.

The **skeletal muscle pump** operates as follows (Figure 20.7):

1. While you are standing at rest, both the venous valve closer to the heart and the one farther from the heart in this part of the leg are open, and blood flows upward toward the heart.
2. Contraction of leg muscles, such as when you stand on tiptoes or take a step, compresses the vein. The compression pushes blood through the valve closer to the heart, an action called *milking*. At the same time, the valve farther from the heart in the uncompressed segment of the vein closes as some blood is pushed back against it. People who are immobilized through injury or disease lack these contractions of leg muscles. As a result, their venous return is slower and they may develop circulation problems.
3. Just after muscle relaxation, pressure falls in the previously compressed section of vein, which causes the valve closer to the heart to close. The valve farther from the heart now opens because blood pressure in the foot is higher than in the leg, and the vein fills with blood from the foot.

FIGURE 20.7 **Action of the skeletal muscle pump in returning blood to the heart.**

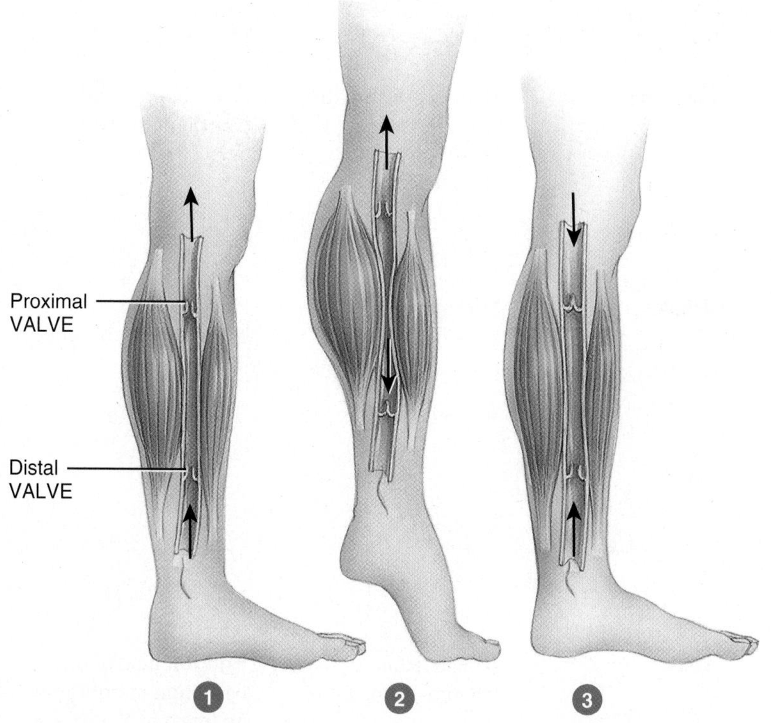


Milking refers to skeletal muscle contractions that drive venous blood toward the heart.

The **respiratory pump** is also based on alternating compression and decompression of veins. During inhalation the diaphragm moves downward, which causes a decrease in pressure in the thoracic cavity and an increase in pressure in the abdominal cavity. As a result, abdominal veins are compressed, and a greater volume of blood moves from the compressed abdominal veins into the decompressed thoracic veins and then into the right atrium. When the pressures reverse during exhalation, the valves in the veins prevent backflow of blood from the thoracic veins to the abdominal veins.

A summary of the distinguishing features of blood vessels is presented in Table 20.1.

**TABLE 20.1**

Distinguishing Features of Blood Vessels

| BLOOD VESSEL | SIZE | TUNICA INTERNA | TUNICA MEDIA | TUNICA EXTERNA | FUNCTION |
|---|---|---|---|---|---|
| **Elastic arteries** | Largest arteries in the body | Well-defined internal elastic lamina | Thick and dominated by elastic fibers; well-defined external elastic lamina | Thinner than tunica media | Conduct blood from heart to muscular arteries |
| **Muscular arteries** | Medium-sized arteries | Well-defined internal elastic lamina | Thick and dominated by smooth muscle; thin external elastic lamina | Thicker than tunica media | Distribute blood to arterioles |
| **Arterioles** | Microscopic (15–300 μm in diameter) | Thin with a fenestrated internal elastic lamina that disappears distally | One or two layers of circularly oriented smooth muscle; distal-most smooth muscle cell forms a precapillary sphincter | Loose collagenous connective tissue and sympathetic nerves | Deliver blood to capillaries and help regulate blood flow from arteries to capillaries |
| **Capillaries** | Microscopic; smallest blood vessels (5–10 μm in diameter) | Endothelium and basement membrane | None | None | Permit exchange of nutrients and wastes between blood and interstitial fluid; distribute blood to venules |
| **Venules** | Microscopic (10–200 μm in diameter) | Endothelium and basement membrane | Proximally none; distally one or two layers of circularly oriented smooth muscle | Sparse | Permit exchange of nutrients and wastes between blood and interstitial fluid and function in white blood cell emigration; reservoirs for accumulating large volumes of blood; pass blood into vein |
| **Veins** | Range from 0.5 mm–3 cm in diameter | Endothelium and basement membrane; no internal elastic lamina; contain valves; lumen much larger than in accompanying artery | Much thinner than in arteries; no external elastic lamina | Thickest of the three layers | Return blood to heart, facilitated by valves in veins in limbs |

## CLINICAL CONNECTION | *Syncope*

**Syncope** (SIN-kō-pē), or fainting, is a sudden, temporary loss of consciousness that is not due to head trauma, followed by spontaneous recovery. It is most commonly due to cerebral ischemia, lack of sufficient blood flow to the brain. Syncope may occur for several reasons:

- *Vasodepressor syncope* is due to sudden emotional stress or real, threatened, or fantasized injury.
- *Situational syncope* is caused by pressure stress associated with urination, defecation, or severe coughing.
- *Drug-induced syncope* may be caused by drugs such as antihypertensives, diuretics, vasodilators, and tranquilizers.
- *Orthostatic hypotension*, an excessive decrease in blood pressure that occurs upon standing up, may cause fainting.

## Blood Distribution

The largest portion of your blood volume at rest—about 64 percent—is in systemic veins and venules (Figure 20.8). Systemic arteries and arterioles hold about 13 percent of the blood volume, pulmonary blood vessels hold about 9 percent, systemic capillaries hold about 7 percent, and the heart chambers hold about 7 percent. Because systemic veins and venules contain a large percentage of the blood volume, they function as **blood reservoirs** from which blood can be diverted quickly if the need arises. For example, during increased muscular activity, the cardiovascular center in the brain stem sends more sympathetic impulses to veins. As a result, veins constrict, reducing the volume of blood in reservoirs and allowing a greater blood volume to flow to skeletal muscles, where it is needed most. A similar mechanism operates in cases of hemorrhage; when blood volume and pressure decrease, veins constrict to help counteract the drop in blood pressure. Among the principal blood reservoirs are the veins of the abdominal organs (especially the liver and spleen) and the veins of the skin.

FIGURE 20.8 **Blood distribution in the cardiovascular system at rest.**

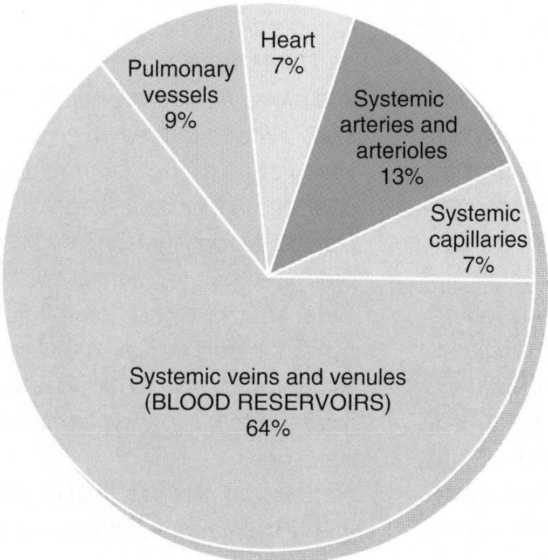


**Because systemic veins and venules contain more than half the total blood volume, they are called blood reservoirs.**

### ✓ CHECKPOINT

12. Which vessel—the femoral artery or the femoral vein—has a thicker wall? Which has a wider lumen?
13. What are the primary structural and functional differences between arteries, capillaries, and veins?
14. Explain how venous blood is returned to the heart.
15. Referencing Figure 20.8, if your total blood volume is 5 liters, what volume is in your venules and veins right now? In your capillaries?

## RETURN TO David's Story

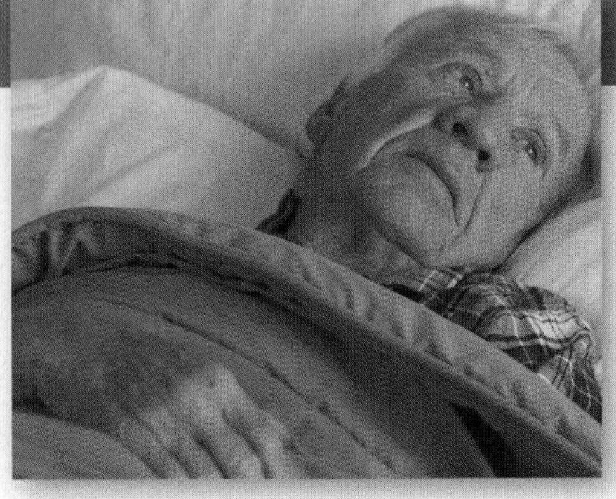

Sandy left her grandfather to get her neighbor, Mrs. Mahoney. Mrs. Mahoney knocked on the door. "David, you don't look so hot. How long has this been going on?" She felt his forehead. He felt cool, he looked pale, and his hands felt cold. "I think Sandy is right. You need to get to the clinic. Better to be safe than sorry."

Reluctantly, David went to the clinic with Sandy and Mrs. Mahoney. The nurse found David's blood pressure to be quite low, only 90/60. His pulse was a rapid 130, and his respirations were rapid. "David, how long have you been feeling poorly?" she asked.

"About the last week or so I've been really tired. I've been helping a friend tear down his garage."

"Is that how you got those scrapes on your arms?" the nurse asked. She looked at David's arm; one of the wounds was red and puffy, possibly infected. "How long ago did that happen, David?"

"About two weeks ago," he replied.

She made a note of the accident on his chart. "David, have you been having any trouble urinating in the last few weeks?"

"Yeah," he said, looking down at his hands. "But at my age, that's nothing new. And I did have a fever about a week ago, but it went away."

Dr. Rolandi saw the nurse exit the exam room. "So who's in room three, Gail?"

"David, possible septic shock; he's got a nasty wound on his arm he didn't take care of. His blood pressure is quite low, respirations are rapid, temperature is below normal, hasn't been urinating. He's complaining of nausea, vomiting, and had a fever last week."

"Thanks," said Dr. Rolandi. The doctor asked David a few more questions, palpated his abdomen, and listened to his heart. "David, I'd like to admit you to the hospital for treatment. All of your symptoms point to septicemia and impending septic shock, probably from that injury on your arm. Sometimes, if the conditions are right, a bacterial infection can overwhelm the body's immune system. The infection itself doesn't even have to be that bad. It's the toxins, or poisons, the bacteria produce that can cause the body to react by going into septic shock. Your capillaries get leaky, you begin to lose fluid into your tissues, and your blood volume drops, which makes your blood pressure drop. When your blood pressure drops, your heart has to work harder to circulate blood throughout your body. If the condition is left untreated, your lungs and kidneys can fail."

The doctor left the exam room to find the nurse. "Gail, we need to admit David, in exam room three. Let's get him started on IV saline and oxygen right away, and I'll be ordering an antibiotic. Put a rush on the labs. We'll need arterial blood gases, complete blood cell count, and a blood culture. Get a culture of that arm wound, too."

Gail nodded. "Septicemia?" she asked.

"Most likely. We want to avoid septic shock," replied Dr. Rolandi. "His capillary refill times were delayed, and he has some of the symptoms. He's lucky his granddaughter brought him in."

Sandy and Mrs. Mahoney approached. "Doctor, how's Grandpa?" Sandy asked.

"Not well, I'm sorry to say. He must be admitted to the hospital. His infected wound may have caused septicemia. Toxins from the bacteria have produced an inflammatory response throughout his body. We need to get his blood volume up, which will increase his blood pressure. Right now, his blood is pooling in his capillary beds and he's losing plasma proteins as the capillaries become more leaky. This has led to what we call circulatory insufficiency; there's not enough blood pressure to keep his blood moving throughout his body as it should."

A. *David's skin is cool to the touch, pale, and blotchy. Which blood vessels regulate blood flow to the capillaries of the skin? What changes in the peripheral vasculature of the skin might cause the changes in its appearance?*
B. *David's blood pressure is low. Which blood vessels regulate blood pressure?*
C. *What structures control the flow of David's blood into his capillary beds?*
D. *How would David's blood reservoirs respond to his low systemic blood pressure?*
E. *How would the loss of plasma proteins affect capillary exchange in David's tissues?*

## 20.5 Blood flows from regions of higher pressure to those of lower pressure.

**Blood flow** is the volume of blood that flows through any tissue in a given time period (in mL/min). *Total blood flow*, the volume of blood that circulates through systemic (or pulmonary) blood vessels each minute, is equal to the cardiac output (see Concept 19.8). How the blood exiting the heart becomes distributed to various body tissues depends on the *blood pressure difference* that drives blood flow through a tissue, and the *resistance* to blood flow by specific blood vessels. Blood flows from regions of higher pressure to regions of lower pressure; the greater the pressure difference, the greater the blood flow. Conversely, the higher the resistance, the smaller the blood flow.

## Blood Pressure

Contraction of the ventricles generates **blood pressure (BP)**, the pressure exerted by blood on the walls of a blood vessel. BP is highest in the aorta and large systemic arteries, where in a resting young adult it rises to about 110 mm Hg during systole (ventricular contraction) and drops to about 70 mm Hg during diastole (ventricular relaxation). **Systolic blood pressure** (sis-TOL-ik) is the highest pressure attained in arteries during systole, and **diastolic blood pressure** (dī-a-STOL-ik) is the lowest arterial pressure during diastole (Figure 20.9). As blood leaves the aorta and flows through the systemic circulation, its pressure falls progressively as the distance from the left ventricle increases. Blood pressure decreases to about 35 mm Hg as blood passes from systemic arteries through systemic arterioles and into capillaries, where the pressure fluctuations disappear. At the venous end of capillaries, blood pressure has dropped to about 16 mm Hg. Blood pressure continues to drop as blood enters systemic venules and then veins, because these vessels are farthest from the left ventricle. Finally, blood pressure reaches 0 mm Hg as blood enters the right atrium.

Blood pressure is determined by cardiac output (discussed in Concept 19.8), blood volume, and vascular resistance (described shortly). Blood pressure depends in part on the total volume of blood in the cardiovascular system. The normal volume of blood in an adult is about 5 liters (5.3 qt). Any decrease in this volume, as from hemorrhage, decreases the amount of blood that is circulated through the arteries. A modest decrease can be compensated for by homeostatic mechanisms that help maintain blood pressure, but if the decrease in blood volume is greater than 10 percent of total blood volume, blood pressure drops. Conversely, anything that increases blood volume, such as water retention in the body, tends to increase blood pressure.

**FIGURE 20.9 Blood pressure in various parts of the cardiovascular system.** The dashed line is the mean (average) blood pressure in the aorta, arteries, and arterioles.

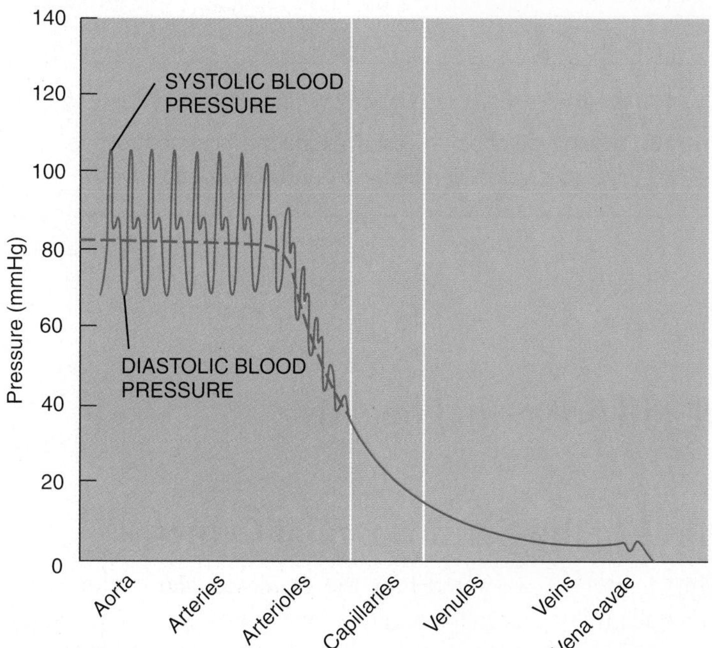


**Blood pressure falls progressively as blood flows from systemic arteries through capillaries and back to the right atrium.**

## Vascular Resistance

**Vascular resistance** is opposition to blood flow due to friction between blood and the walls of blood vessels. As vascular resistance increases, blood flow decreases. An increase in vascular resistance increases blood pressure; a decrease in vascular resistance has the opposite effect. Vascular resistance depends on the size of the blood vessel lumen, blood viscosity, and the total blood vessel length.

- ***Size of the lumen.*** The smaller the lumen of a blood vessel, the greater is its resistance to blood flow. Vasoconstriction narrows the lumen, and vasodilation widens it. Normally, moment-to-moment fluctuations in blood flow through a given tissue are due to vasoconstriction and vasodilation of the tissue's arterioles. As arterioles dilate, resistance decreases, and blood pressure falls. As arterioles constrict, resistance increases, and blood pressure rises.
- ***Blood viscosity.*** The **viscosity** (vis-KOS-i-tē = thickness) of blood depends mostly on the ratio of red blood cells to plasma (fluid) volume, and to a smaller extent on the concentration of proteins in plasma. The higher the blood's viscosity, the higher the resistance. Thus, any condition that increases the viscosity of blood, such as dehydration or *polycythemia* (an unusually high number of red blood cells), increases blood pressure. A depletion of red blood cells or plasma proteins, due to anemia or hemorrhage, decreases viscosity and thus decreases blood pressure.
- ***Total blood vessel length.*** Resistance to blood flow through a vessel is directly proportional to the length of the blood vessel. The longer a blood vessel, the greater is the contact between the vessel wall and the blood, which increases resistance. Obese people often have hypertension (elevated blood pressure) because the additional blood vessels in their adipose tissue increase their total blood vessel length. An estimated 650 km (about 400 miles) of additional blood vessels develop for each extra kilogram (2.2 lb) of fat, one reason why overweight individuals may have hypertension (elevated blood pressure).

**Systemic vascular resistance (SVR)**, also known as *total peripheral resistance*, refers to all of the vascular resistances offered by systemic blood vessels. The diameters of arteries and veins are large, so their resistance is small because most of the blood does not come into physical contact with the walls of the blood vessel. The smallest vessels—arterioles, capillaries, and venules—contribute the most resistance. A major function of arterioles is to control SVR—and therefore blood pressure and blood flow to particular tissues—by changing their diameters. Arterioles need to vasodilate or vasoconstrict only slightly to have a large effect on SVR.

**FIGURE 20.10 Summary of factors that increase blood pressure.** Changes noted within green boxes increase cardiac output; changes noted within blue boxes increase systemic vascular resistance.

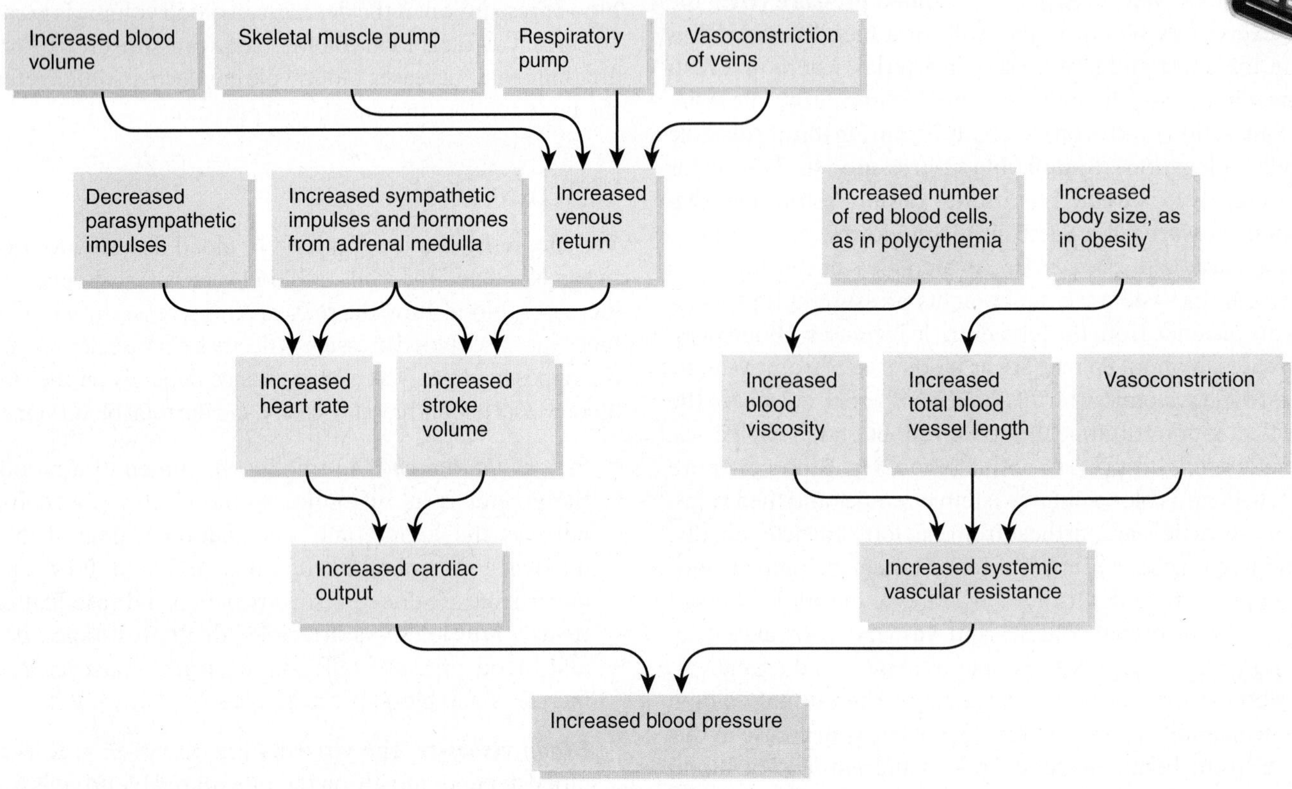


**Increases in cardiac output and increases in systemic vascular resistance will increase blood pressure.**

Figure 20.10 summarizes several factors that increase blood pressure.

✓ **CHECKPOINT**

16. Which activity is occurring in the heart to produce systolic blood pressure? Which heart activity is occurring during diastolic blood pressure?
17. What is vascular resistance? What factors contribute to it?
18. Which type of blood vessel exerts the most control over systemic vascular resistance, and how is it achieved?

# 20.6 Blood pressure is regulated by neural and hormonal negative feedback systems.

Several interconnected negative feedback systems control blood pressure by adjusting cardiac output, blood volume, and systemic vascular resistance. Some systems allow rapid adjustments to cope with sudden changes, such as the drop in blood pressure in the brain that occurs when you get out of bed; others act more slowly to provide long-term regulation of blood pressure.

## Role of the Cardiovascular Center

In Chapter 19, we noted how the **cardiovascular center** in the medulla oblongata helps regulate heart rate and stroke volume. The cardiovascular center also controls the neural and hormonal negative feedback systems that regulate blood pressure and blood flow to specific tissues.

FIGURE 20.11 **The cardiovascular center.** The cardiovascular center receives input from higher brain centers, proprioceptors, baroreceptors, and chemoreceptors. It then provides output to the sympathetic and parasympathetic divisions of the autonomic nervous system.

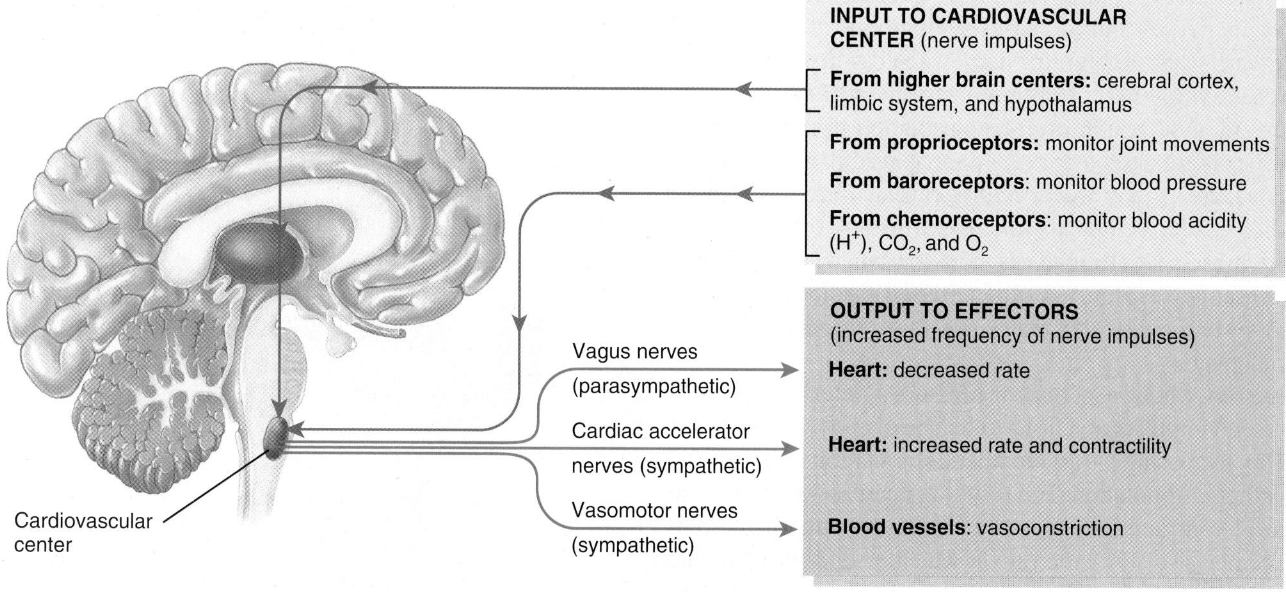


**The cardiovascular center is the main region for nervous system regulation of the heart and blood vessels.**

The cardiovascular center receives input both from higher brain regions and from sensory receptors (Figure 20.11). Nerve impulses descend from the cerebral cortex, limbic system, and hypothalamus to affect the cardiovascular center. For example, even before you start to run a race, your heart rate may increase due to nerve impulses conveyed from the limbic system to the cardiovascular center. If your body temperature rises during a race, the hypothalamus sends nerve impulses to the cardiovascular center. The resulting vasodilation of skin blood vessels allows heat to dissipate more rapidly from the surface of the skin. The three main types of sensory receptors that provide input to the cardiovascular center are proprioceptors, baroreceptors, and chemoreceptors. *Proprioceptors* (prō′-prē-ō-SEP-tors) monitor movements of joints and muscles, and provide input to the cardiovascular center during physical activity. Their activity accounts for the rapid increase in heart rate at the beginning of exercise. *Baroreceptors* (bar′-ō-rē-SEP-tors) monitor changes in pressure and stretch in the walls of blood vessels, and *chemoreceptors* (kē′-mō-rē-SEP-tors) monitor the concentration of various chemicals in the blood.

Output from the cardiovascular center flows along sympathetic and parasympathetic neurons of the autonomic nervous system (Figure 20.11). Opposing sympathetic (stimulatory) and parasympathetic (inhibitory) innervation control the heart. Parasympathetic stimulation, conveyed along the **vagus nerves**, decreases heart rate, while sympathetic impulses in the **cardiac accelerator nerves** increase both the rate and force of contraction of the heart.

The cardiovascular center also sends impulses continually to smooth muscle in blood vessel walls via **vasomotor nerves** (vā-sō-MŌ-tor). The result is a moderate state of vasoconstriction, called **vasomotor tone**, that sets the resting level of systemic vascular resistance. Sympathetic stimulation of most veins results in their constriction and movement of blood out of venous blood reservoirs, which increases blood pressure.

## Neural Regulation of Blood Pressure

The nervous system regulates blood pressure via negative feedback loops that occur as two types of reflexes: baroreceptor reflexes and chemoreceptor reflexes.

### Baroreceptor Reflexes

**Baroreceptors** (pressure receptors) are located in the aorta, internal carotid arteries (arteries in the neck that supply blood to the brain), and other large arteries in the neck and chest. They send impulses to the cardiovascular center to help regulate blood pressure.

The right and left internal carotid arteries supply blood to the brain (see Figure 20.16). The *carotid sinuses* are small widenings of the internal carotid arteries just above the point where they branch from the common carotid arteries. Baroreceptors are located in the wall of the carotid sinuses. When blood pressure rises, it stretches the wall of the carotid sinus, stimulating the baroreceptors. The stretched baroreceptors increase their transmission of nerve impulses to the cardiovascular center, which initiates reflex activity to lower blood pressure to the brain. In a similar fashion, baroreceptors in the wall of the arch of the aorta (see Figure 20.16) initiate reflex activity via the cardiovascular center that regulates systemic blood pressure.

When blood pressure falls, the baroreceptors are stretched less, and they send nerve impulses at a slower rate to the cardiovascular center (Figure 20.12). In response, the cardiovascular center decreases parasympathetic stimulation of the heart via the vagus nerves and increases sympathetic stimulation of the heart via cardiac accelerator nerves. The cardiovascular center increases sympathetic stimulation to the adrenal medullae to increase their secretion of epinephrine and norepinephrine, both of which increase heart rate and force of contraction. As the heart beats faster and more forcefully, cardiac output increases. The cardiovascular center also increases stimulation of vascular smooth muscle resulting in vasoconstriction, which increases systemic vascular resistance. Both increased cardiac output and increased systemic vascular resistance raise blood pressure to the normal level.

Conversely, when an increase in pressure is detected, the baroreceptors send impulses at a faster rate. The cardiovascular center responds by increasing parasympathetic stimulation and decreasing sympathetic stimulation. The resulting decreases in heart rate and force of contraction reduce the cardiac output. The cardiovascular center also slows the rate at which it sends sympathetic impulses along vasomotor neurons that normally cause vasoconstriction. The resulting vasodilation lowers systemic vascular resistance. Decreased cardiac output and decreased systemic vascular resistance both lower systemic blood pressure to the normal level.

Moving from a prone (lying down) to an erect position decreases blood pressure and blood flow in the head and upper part of the body. The baroreceptor reflexes, however, quickly counteract the drop in pressure. Sometimes these reflexes operate more slowly than normal, especially in the elderly, in which case a person can faint due to reduced brain blood flow upon standing up too quickly.

### Chemoreceptor Reflexes

**Chemoreceptors**, sensory receptors that monitor the chemical composition of blood, are located close to the baroreceptors of the carotid sinus and arch of the aorta in small structures called **carotid bodies** and **aortic bodies**, respectively. These chemoreceptors detect changes in blood levels of $O_2$, $CO_2$, and $H^+$. When these chemoreceptors are stimulated by a drop in blood $O_2$ level (*hypoxia*), an increase in blood $CO_2$ level (*hypercapnia*), or acidic blood (an increase in $H^+$ concentration, or *acidosis*), they send impulses to the cardiovascular center. In response, the cardiovascular center increases sympathetic stimulation to arterioles and veins, producing vasoconstriction and an increase in blood pressure to drive gas exchange through the lungs.

FIGURE 20.12 **Negative feedback regulation of blood pressure via baroreceptor reflexes.**

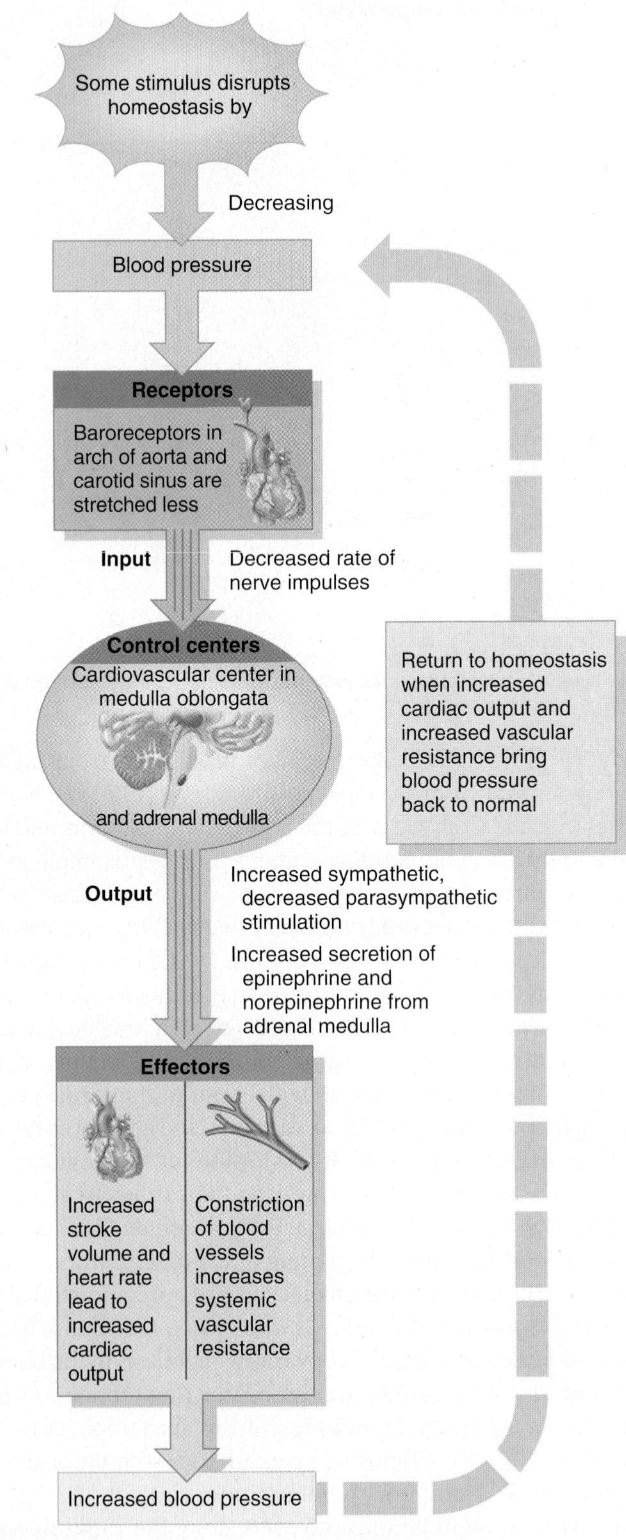
Some stimulus disrupts homeostasis by
Decreasing
Blood pressure
Receptors
Baroreceptors in arch of aorta and carotid sinus are stretched less
Input
Decreased rate of nerve impulses
Control centers
Cardiovascular center in medulla oblongata
and adrenal medulla
Output
Increased sympathetic, decreased parasympathetic stimulation
Increased secretion of epinephrine and norepinephrine from adrenal medulla
Effectors
Increased stroke volume and heart rate lead to increased cardiac output
Constriction of blood vessels increases systemic vascular resistance
Increased blood pressure
Return to homeostasis when increased cardiac output and increased vascular resistance bring blood pressure back to normal

**When blood pressure decreases, heart rate increases.**

## Hormonal Regulation of Blood Pressure

Several hormones help regulate blood pressure and blood flow by altering cardiac output, changing systemic vascular resistance, or adjusting the total blood volume:

- ***Renin–angiotensin–aldosterone system*** (an′-jē-ō-TEN-sin). When blood volume falls or blood flow to the kidneys

## CLINICAL CONNECTION | *Carotid Sinus Massage and Carotid Sinus Syncope*

Because the carotid sinus is close to the anterior surface of the neck, it is possible to stimulate the baroreceptors there by putting pressure on the neck. Physicians sometimes use **carotid sinus massage**, which involves carefully massaging the neck over the carotid sinus, to slow heart rate in a person who has paroxysmal superventricular tachycardia, a type of tachycardia that originates in the atria. Anything that stretches or puts pressure on the carotid sinus, such as hyperextension of the head, tight collars, or carrying heavy shoulder loads, may also slow heart rate and can cause **carotid sinus syncope** (SIN-kō-pē), fainting due to inappropriate stimulation of the carotid sinus baroreceptors.

decreases, the kidneys secrete **renin** into the bloodstream. Renin helps activate the hormone **angiotensin II**, which raises blood pressure in two ways. Angiotensin II causes vasoconstriction; it raises blood pressure by increasing systemic vascular resistance. Second, it stimulates secretion of **aldosterone**, which increases reabsorption of sodium ions and water by the kidneys. As water is reabsorbed into the blood, total blood volume increases, which increases blood pressure.

- ***Epinephrine and norepinephrine***. In response to sympathetic stimulation, the adrenal medulla releases epinephrine and norepinephrine. These hormones increase cardiac output by increasing the rate and force of heart contractions, and cause vasoconstriction of some arterioles and veins. The resulting increase in cardiac output and vasoconstriction raises blood pressure. Epinephrine also produces vasodilation of arterioles in skeletal muscles, which helps increase blood flow to muscles during exercise.
- ***Antidiuretic hormone***. Antidiuretic hormone is produced by the hypothalamus and released from the posterior pituitary in response to dehydration or decreased blood volume. Among other actions, antidiuretic hormone causes vasoconstriction, which increases blood pressure.
- ***Atrial natriuretic peptide***. Released by cells in the atria of the heart, atrial natriuretic peptide lowers blood pressure by causing vasodilation and by promoting the loss of salt and water in the urine, which reduces blood volume.

Table 20.2 summarizes the regulation of blood pressure by hormones.

**TABLE 20.2**

Blood Pressure Regulation by Hormones

| HORMONE | PRINCIPAL ACTION INFLUENCING BLOOD PRESSURE | EFFECT ON BLOOD PRESSURE |
|---|---|---|
| **CARDIAC OUTPUT** | | |
| **Norepinephrine, epinephrine** | Increased heart rate and contractility | Increase |
| **SYSTEMIC VASCULAR RESISTANCE** | | |
| **Angiotensin I, antidiuretic hormone, norepinephrine, epinephrine** | Vasoconstriction | Increase |
| **Atrial natriuretic peptide, epinephrine** | Vasodilation | Decrease |
| **BLOOD VOLUME** | | |
| **Aldosterone, antidiuretic hormone** | Increased blood volume | Increase |
| **Atrial natriuretic peptide** | Decreased blood volume | Decrease |

## ✓ CHECKPOINT

**19.** What are the principal inputs to and outputs from the cardiovascular center?

**20.** What types of effector tissues are regulated by the cardiovascular center?

**21.** Following a drop in blood pressure, how does the baroreceptor reflex return blood pressure to a normal level?

**22.** What is the role of chemoreceptors in the regulation of blood pressure?

**23.** How do hormones regulate blood pressure?

# 20.7 Measurement of the pulse and blood pressure helps assess cardiovascular system function.

## Pulse

The alternate expansion and recoil of elastic arteries after each contraction of the left ventricle creates a traveling pressure wave that is called the **pulse**. The pulse is strongest in the arteries closest to the heart, becomes weaker in arterioles, and disappears altogether in capillaries. The pulse may be felt in any artery that lies near the surface of the body that can be compressed against a bone or other firm structure. Table 20.3 depicts some common pulse points.

The pulse rate normally is the same as the heart rate, about 70 to 80 beats per minute at rest. **Tachycardia** (tak′-i-KAR-dē-a; *tachy-* = fast) is a rapid resting heart or pulse rate over 100 beats/min. **Bradycardia** (brād′-i-KAR-dē-a; *brady-* = slow) is a slow resting heart or pulse rate under 50 beats/min. Endurance-trained athletes normally exhibit bradycardia.

## Measuring Blood Pressure

In clinical use, the term *blood pressure* usually refers to the pressure in arteries generated by the left ventricle during systole and the pressure remaining in the arteries when the ventricle is in diastole. Blood pressure is usually measured in the brachial artery in the left arm (Table 20.3). The device used to measure blood pressure is a **sphygmomanometer** (sfig′-mō-ma-NOM-e-ter; *sphygmo-* = pulse; *-manometer* = instrument used to measure

**TABLE 20.3**

Pulse Points

| STRUCTURE | LOCATION | STRUCTURE | LOCATION |
|---|---|---|---|
| **Superficial temporal artery** | Medial to ear | **Femoral artery** | Inferior to inguinal ligament |
| **Facial artery** | Mandible (lower jawbone) on line with corners of mouth | **Popliteal artery** | Posterior to knee |
| **Common carotid artery** | Lateral to larynx (voice box) | **Radial artery** | Distal aspect of wrist |
| **Brachial artery** | Medial side of biceps brachii muscle | **Dorsal artery of foot** | Superior to instep of foot |

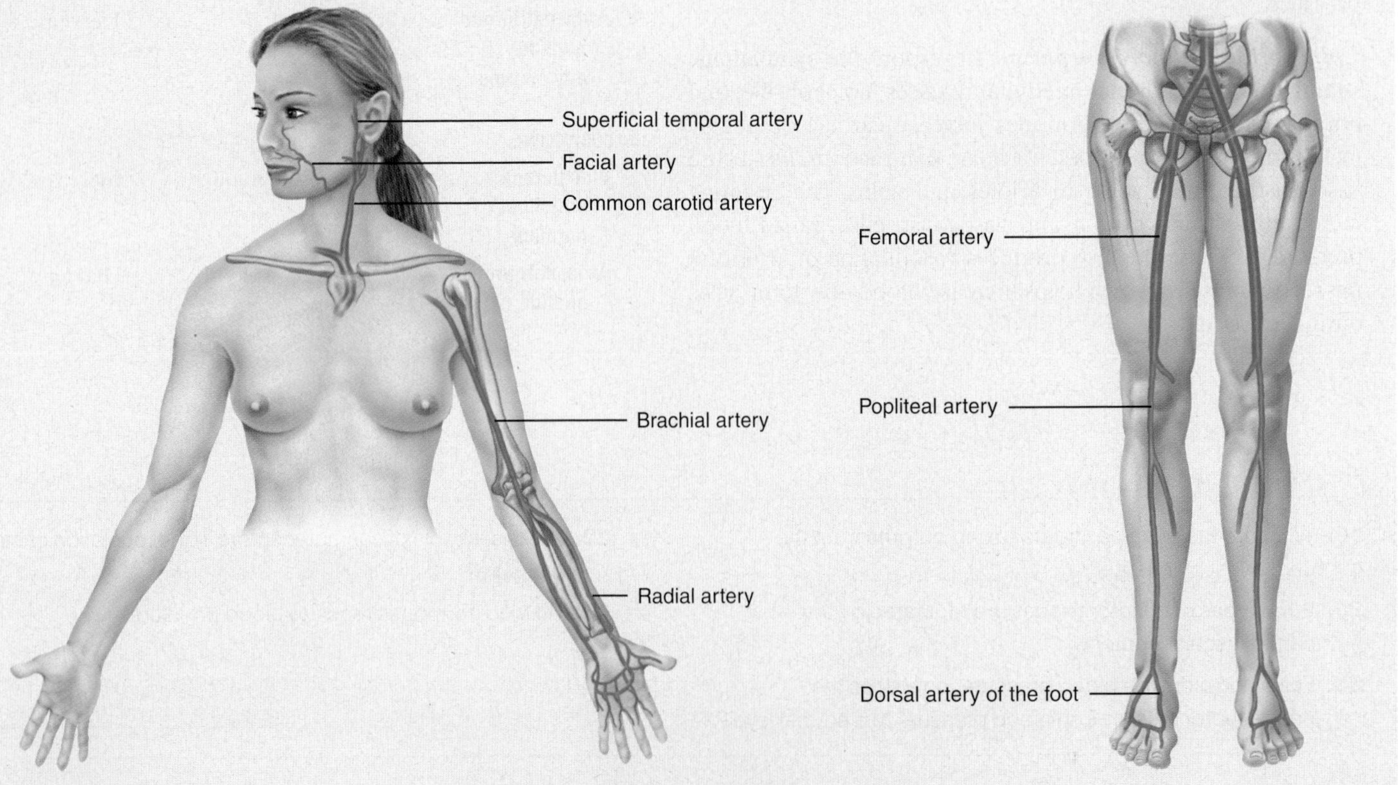

pressure). It consists of a rubber cuff connected to a rubber bulb that is used to inflate the cuff, and a meter that registers the pressure in the cuff. With the arm resting on a table so that it is about the same level as the heart, the cuff of the sphygmomanometer is wrapped around a bared arm. The cuff is inflated by squeezing the bulb until the brachial artery is compressed and blood flow stops, about 30 mm Hg higher than the person's usual systolic pressure. The technician places a stethoscope below the cuff on the brachial artery and slowly deflates the cuff. When the cuff is deflated enough to allow the artery to open, a spurt of blood passes through, resulting in the first sound heard through the stethoscope. This sound corresponds to *systolic blood pressure*—the force with which blood is pushing on arterial walls just after ventricular contraction (Figure 20.13). As the cuff is deflated further, the sounds suddenly become too faint to be heard through the stethoscope. This level, called the *diastolic blood pressure*, represents the force exerted by the blood remaining in arteries during ventricular relaxation. At pressures below diastolic blood pressure, sounds disappear altogether. The various sounds that are heard while taking blood pressure are called **Korotkoff sounds** (kō-ROT-kof).

The normal blood pressure of an adult male is less than 120 mm Hg systolic and less than 80 mm Hg diastolic. For example, 110 systolic and 70 diastolic (written as 110/70 and spoken aloud as "110 over 70") is *less* than 120/80, so would be considered a normal blood pressure. In young adult females, the pressures are 8 to 10 mm Hg less. People who exercise regularly and are in good physical condition may have even lower blood pressures. Thus, blood pressure slightly lower than 120/80 may be a sign of good health and fitness.

FIGURE 20.13 **Relationship of blood pressure changes to cuff pressure.**

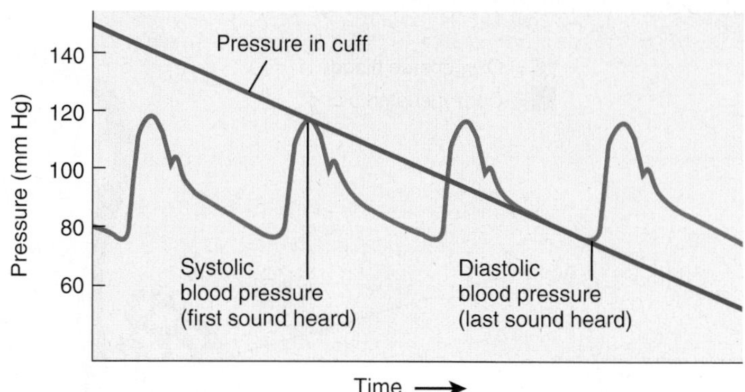


As the cuff is deflated, sounds first occur at a blood pressure level referred to as the systolic blood pressure; the sounds suddenly become faint at a blood pressure level referred to as the diastolic blood pressure.

The difference between systolic and diastolic pressure is called **pulse pressure**. This pressure, normally about 40 mm Hg, provides information about the condition of the cardiovascular system. For example, conditions such as atherosclerosis greatly increase pulse pressure.

**✓ CHECKPOINT**

**24.** Explain the terms tachycardia and bradycardia.

**25.** How are systolic and diastolic blood pressures measured with a sphygmomanometer?

**26.** If a blood pressure is reported as "142 over 95," what are the diastolic and systolic pressures?

## 20.8 The two main circulatory routes are the pulmonary circulation and the systemic circulation.

Blood vessels are organized into **circulatory routes** that carry blood throughout the body. Figure 20.14 shows the circulatory routes for blood flow. The routes are parallel—in most cases a portion of the cardiac output flows separately to each tissue of the body. Thus, each organ receives its own supply of freshly oxygenated blood. The two main circulatory routes are the pulmonary circulation and the systemic circulation.

### Pulmonary Circulation

The **pulmonary circulation** (pul′-mo-NAR-ē; *pulmo-* = lung) carries dark red deoxygenated blood from the right ventricle to the air sacs (alveoli) within the lungs and returns bright red oxygenated blood from the air sacs to the left atrium

**FIGURE 20.14 Circulatory routes.** Short black arrows indicate the pulmonary circulation (detailed in Figure 20.15), long black arrows show the systemic circulation (detailed in Figures 20.16–20.27), and red arrows outline the hepatic portal circulation (detailed in Figure 20.28).

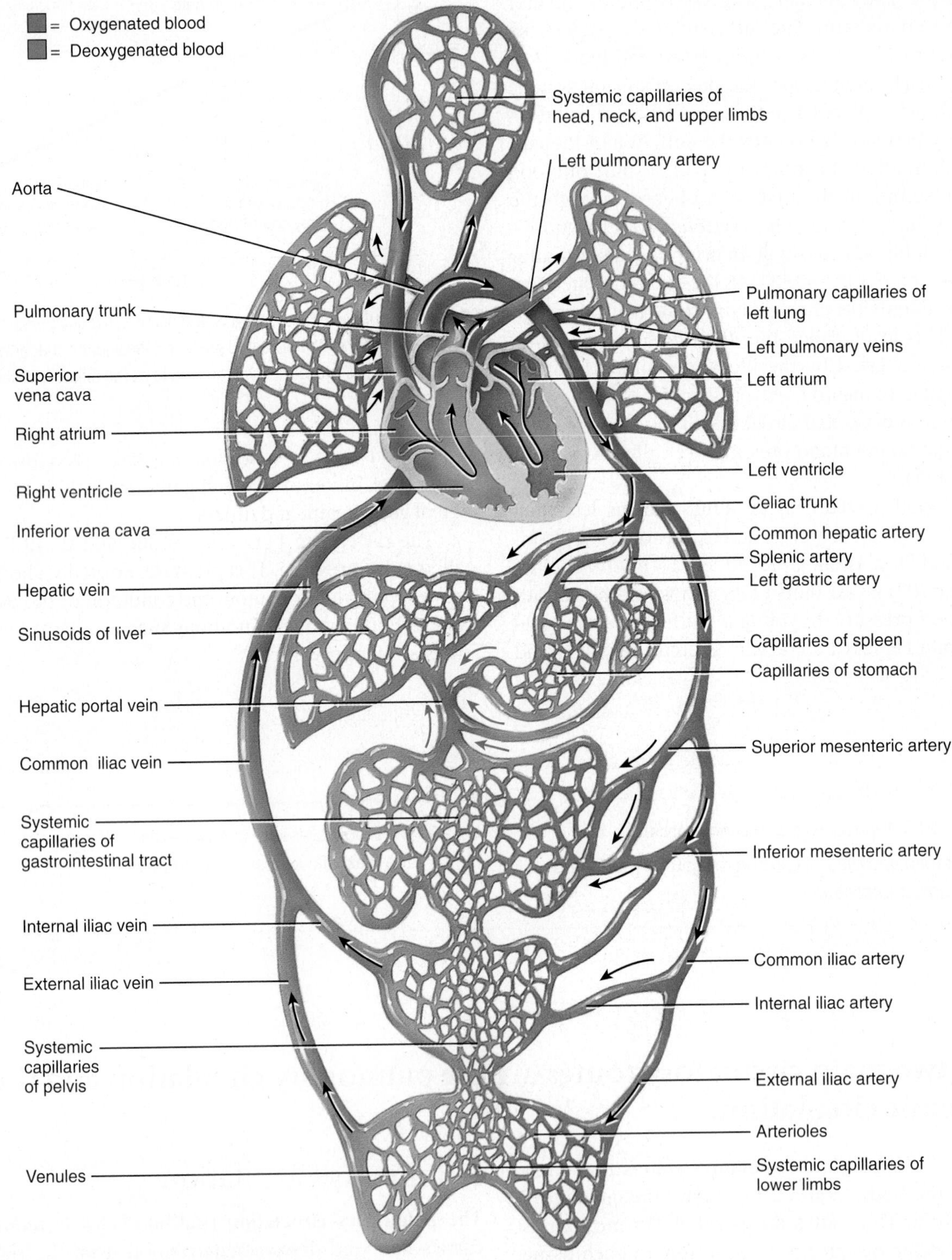


**Blood vessels are organized into circulatory routes that deliver blood to tissues of the body.**

(Figure 20.15). The **pulmonary trunk** emerges from the right ventricle and passes superiorly, posteriorly, and to the left. It then divides into two branches: the **right pulmonary artery** to the right lung and the **left pulmonary artery** to the left lung. After birth, the pulmonary arteries are the only arteries that carry deoxygenated blood. On entering the lungs, the branches divide and subdivide until finally they form capillaries around the air sacs (alveoli) within the lungs. $CO_2$ passes from the capillary blood into the air sacs and is exhaled. Inhaled $O_2$ passes from the air within the lungs into the capillary blood. The pulmonary capillaries unite to form

FIGURE 20.15 Pulmonary circulation.

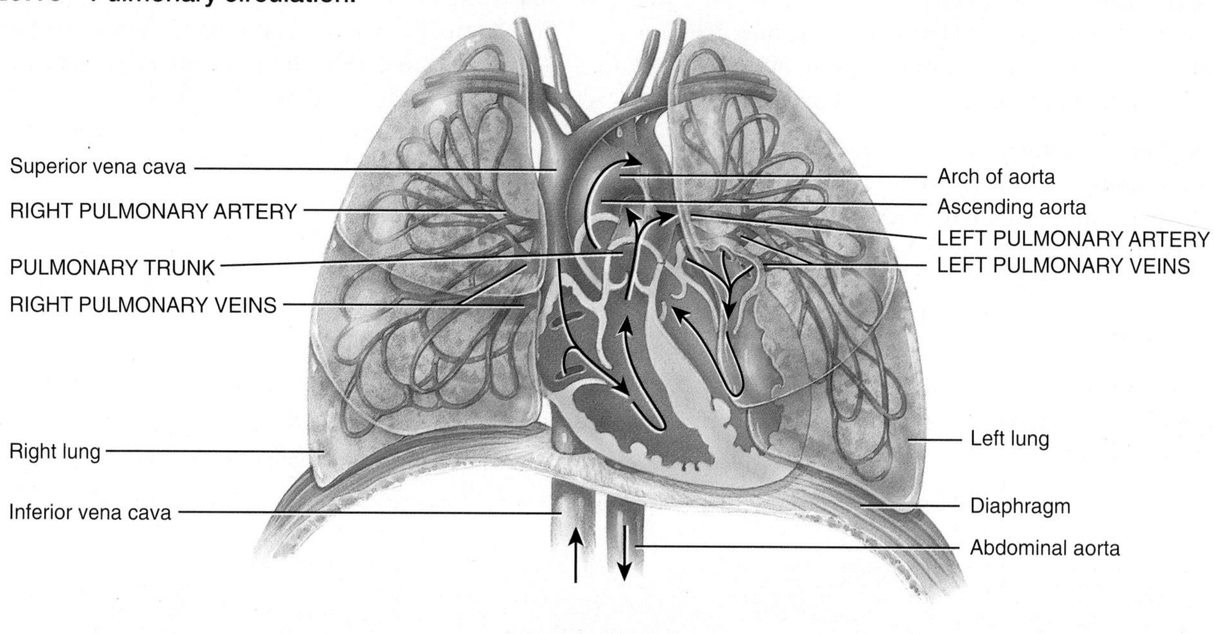


(a) Anterior view

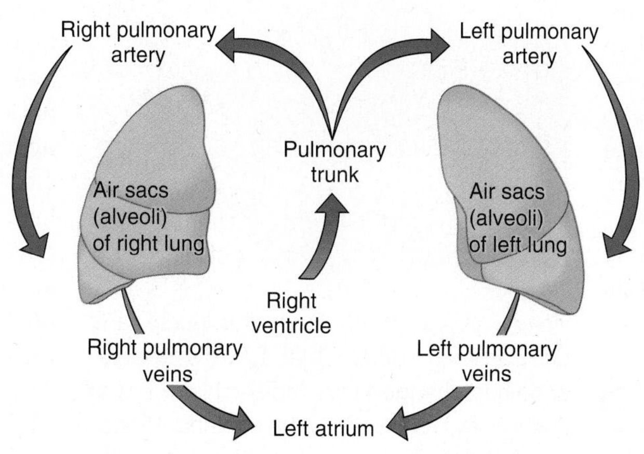


(b) Scheme of pulmonary circulation

**The pulmonary circulation brings deoxygenated blood from the right ventricle to the lungs and returns oxygenated blood from the lungs to the left atrium.**

venules and eventually **pulmonary veins**, which exit the lungs and carry the oxygenated blood to the left atrium. Two left and two right pulmonary veins enter the left atrium. After birth, the pulmonary veins are the only veins that carry oxygenated blood.

## Systemic Circulation

Blood leaving the aorta and flowing through the systemic arteries is oxygenated and a bright red color (see Figure 20.14). As blood flows through capillaries, it loses some of its oxygen and picks up carbon dioxide, becoming a dark red color. The **systemic circulation** includes all of the arteries and arterioles that carry bright red oxygenated blood from the left ventricle to systemic capillaries, plus the veins and venules that return dark red deoxygenated blood to the right atrium after flowing through body organs.

Subdivisions of the systemic circulation include the **coronary circulation** (see Figure 19.7), which supplies the myocardium of the heart; **cerebral circulation**, which supplies the brain (see Figure 20.18c); and the hepatic portal circulation, which extends from the gastrointestinal tract to the liver and is described later in the chapter (see Figure 20.28).

In the figures and tables that follow, the principal arteries and veins of the systemic circulation are described and illustrated to assist you in learning their names and locations. Figure 20.16a shows an overview of the major arteries, and Figure 20.22 shows an overview of the major veins. As you study the various blood vessels, refer to these two figures to see the relationships of the blood vessels under consideration to other regions of the body.

The selected blood vessels are organized according to regions of the body. Each presentation contains the following information:

- **An overview**. This information provides a general orientation to the blood vessels under consideration, with emphasis on how the blood vessels are organized into various regions as well as distinguishing and/or interesting features of the blood vessels.

- **Blood vessel names**. Students often have difficulty with the pronunciations and meanings of blood vessel names. To learn them more easily, study the phonetic pronunciations and word derivations that indicate how they get their names.
- **Region supplied or drained**. For each artery listed, there is a description of parts of the body that receive blood from the vessel. For each vein listed, there is a description of parts of the body that are drained by the vessel.
- **Illustrations**. The drawings and photographs include illustrations of blood vessels under consideration and flow diagrams to indicate the patterns of blood distribution or drainage. Oxygenated blood is shown in red and deoxygenated blood is shown in blue. Cadaver photographs are also included to provide more realistic views of selected blood vessels.

## ✓ CHECKPOINT

**27.** In an adult, which are the only arteries that carry deoxygenated blood? Which are the only veins that carry oxygenated blood?

**28.** Why is less blood pressure needed in pulmonary arteries than systemic arteries?

**29.** What is the main function of the systemic circulation?

## RETURN TO David's Story

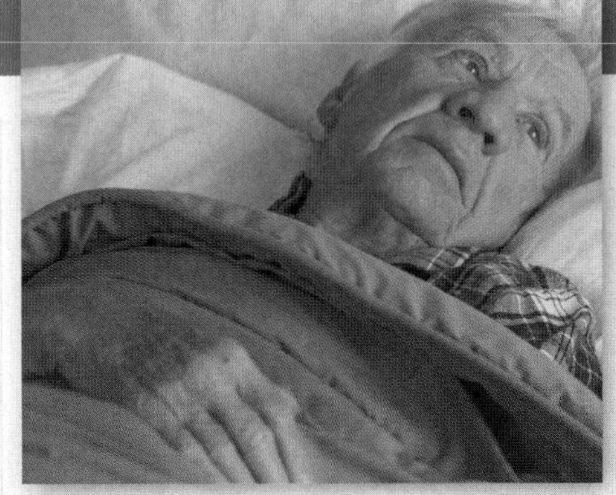

David finally consented to being admitted to the hospital. The staff started him on intravenous saline and an antibiotic, had given him oxygen, and were watching him closely—taking his blood pressure often, listening to his lungs and heart, and monitoring his urine output. "Why do you keep coming in here and bothering me?" David grumbled.

The nurse sighed. "David, we have to keep a close eye on you. We want to be sure the fluid in your tissues is returning to your bloodstream. If your lungs fill up with fluid, you won't be able to breathe. Your blood pressure is better now; it's up to one hundred thirty-five over eighty. And you're producing urine again, which means your kidneys are working. You're getting rid of the excess fluids."

"Hello, David," said Dr. Barbano, the physician who had taken over David's case when he was admitted. "How are you feeling?"

"Okay, better than before. I had felt so out of breath and weak," answered David.

"Well, your blood oxygen level was down and the pH of your blood was abnormal, too. Your body responds to changes in blood flow, pressure, pH, oxygen level, and carbon dioxide level, and tries to keep the body in balance. That's what we're helping your body do now, and we're also trying to get rid of that excess fluid to prevent any damage to your organs." Another nurse came in to check his blood pressure while Dr. Barbano was still speaking to David.

"Here we go again," said David. He laughed, but then the laugh turned into a violent coughing fit. Both the nurse and Dr. Barbano noticed that David's appearance had changed very rapidly; his skin was flushed and reddened as he strained. Blood-red frothy sputum came from David's mouth.

"Doctor, his oxygen saturations are dipping into the eighties," said the nurse. The doctor took out his stethoscope and listened to David's lungs and heart.

"Get him back on oxygen. He may have a pulmonary embolism. Did you notice any swelling in the extremities?" The nurse pulled back the sheet that covered David's legs and they saw that his right calf was swollen.

So much was going on that Sandy didn't understand. She sat in a waiting room, twisting a tissue around her fingers and hoping to hear something about this latest crisis. The nurse finally came in to talk to her. "Sandy, your grandfather is doing okay. He can't see you right now, though. The doctor had to order some tests and had us put in an airway to help him breathe."

"He's going to be okay, though, isn't he?" asked Sandy through her tears.

The nurse reached out and took Sandy's hand. "He's been through a rough time, Sandy. You have to hope for the best. A clot formed in one of the veins in his leg and part of it broke off and traveled up to his heart and then into his lungs. The clot is blocking blood from getting to the lung, so the blood isn't carrying enough oxygen to your grandfather's body. The airway is allowing us to give him oxygen until Dr. Barbano decides how to treat the clot."

F. *David has lost blood volume, resulting in a drop in blood pressure. His autonomic nervous system countered the dropping blood pressure by stimulated vasoconstriction. How does vasoconstriction increase blood pressure?*

G. *If Sandy had not taken Grandpa David to the hospital, how might his inactivity have caused his condition to become worse?*

H. *Which structures in David's cardiovascular system would respond to low blood oxygen and decreased blood pH? What would the response be?*

I. *Because David's blood volume had dropped, blood flow to his kidneys had been reduced. What would be the hormonal response of the kidneys to these two changes?*

J. *Why would a blood clot in David's leg travel to his lungs?*

# 20.9 Systemic arteries carry blood from the heart to all body organs except the lungs.

Freshly oxygenated blood flows into the left atrium from the pulmonary circulation. Contractions of the left ventricle then eject the oxygenated blood into the aorta. Branching from the aorta are all systemic arteries, which carry oxygen, nutrients, and hormones to body tissues.

## The Aorta and Its Branches

The **aorta** (ā-OR-ta = to lift up; Figure 20.16; Table 20.4) is the largest artery of the body, with a diameter of 2–3 cm (about 1 in.). Its four principal divisions are the ascending aorta, arch

FIGURE 20.16 **Aorta and its principal branches.**



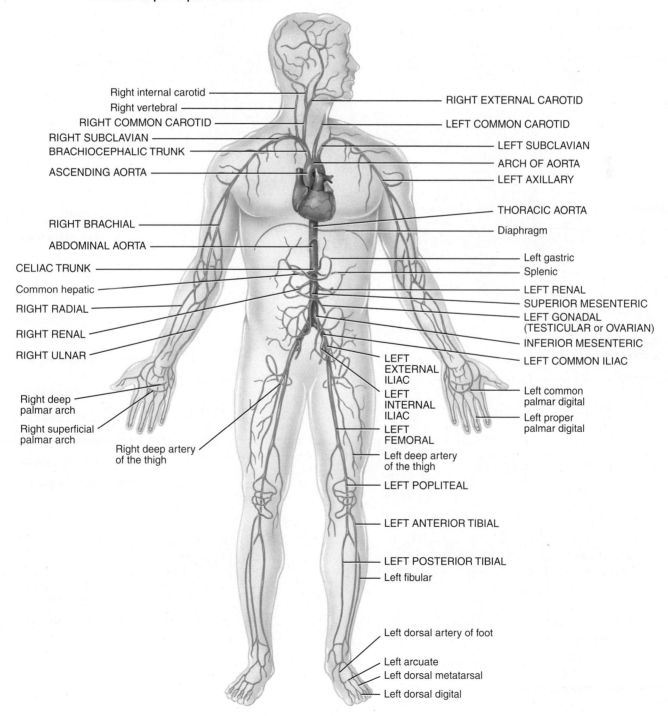

(a) Overall anterior view of principal branches of aorta

(continues)

FIGURE 20.16 (continued)

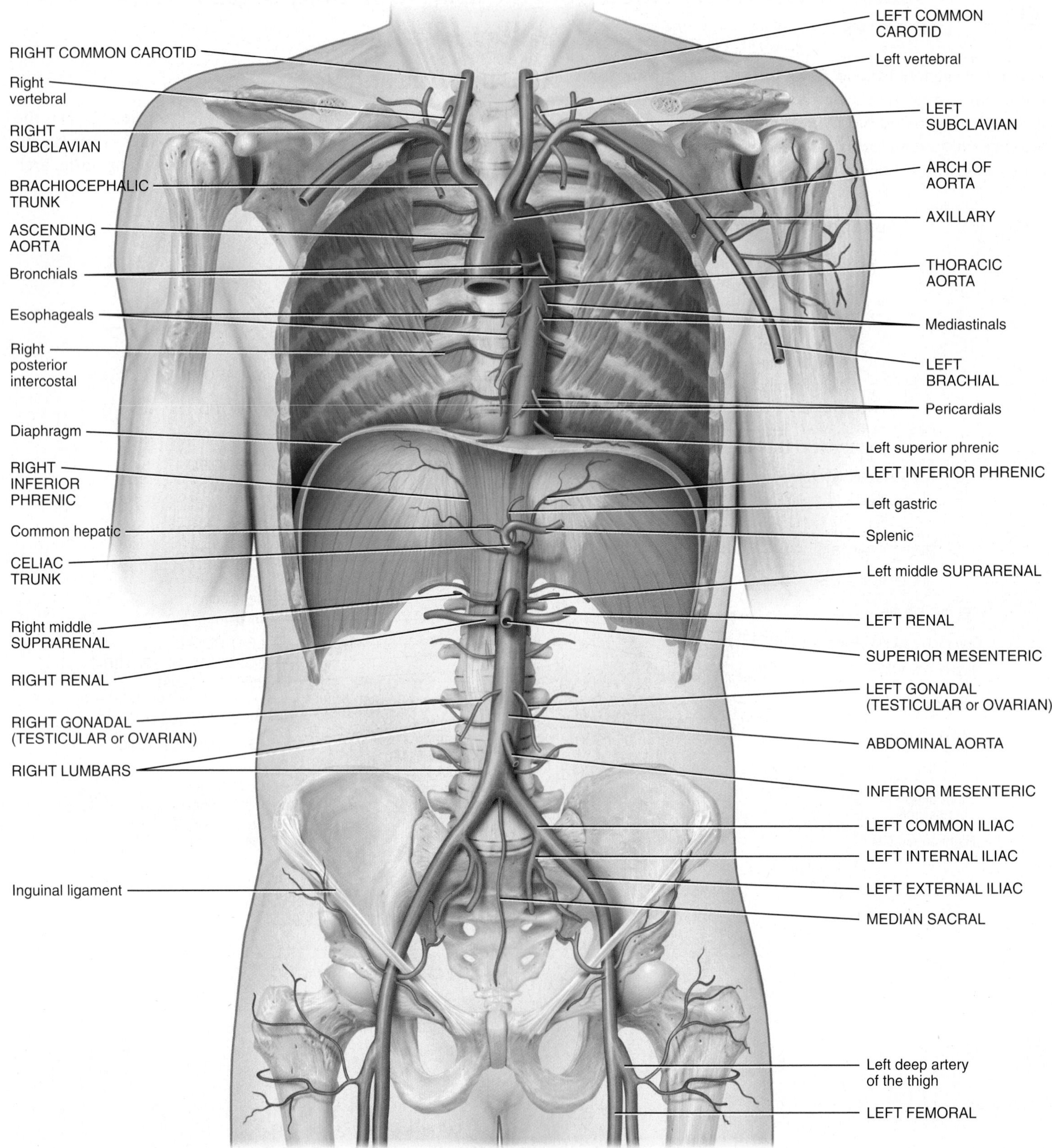


(b) Detailed anterior view of principal branches of aorta

**All systemic arteries branch from the aorta.**

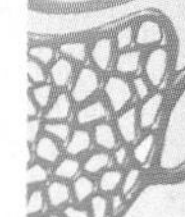

**TABLE 20.4**

The Aorta and Its Branches

| DIVISION AND BRANCHES | REGION SUPPLIED |
|---|---|
| **ASCENDING AORTA** | |
| **Right and left coronary arteries** | Heart |
| **ARCH OF THE AORTA** | |
| **Brachiocephalic trunk** (brā′-kē-ō-se-FAL-ik) | |
| **Right common carotid artery** (ka-ROT-id) | Right side of head and neck |
| **Right subclavian artery** (sub-KLĀ-vē-an) | Right upper limb |
| **Left common carotid artery** | Left side of head and neck |
| **Left subclavian artery** | Left upper limb |
| **THORACIC AORTA (*THORAC-* = CHEST)** | |
| **Pericardial arteries** (per-Ī-KAR-dē-al) | Pericardium |
| **Bronchial arteries** (BRONG-kē-al) | Bronchi of lungs |
| **Esophageal arteries** (e-sof′-a-JĒ-al) | Esophagus |
| **Mediastinal arteries** (mē′-dē-as-TĪ-nal) | Structures in mediastinum |
| **Posterior intercostal arteries** (in′-ter-KOS-tal) | Intercostal and chest muscles |
| **Subcostal arteries** (sub-KOS-tal) | Same as posterior intercostals |
| **Superior phrenic arteries** (FREN-ik) | Superior and posterior surfaces of diaphragm |
| **ABDOMINAL AORTA** | |
| **Inferior phrenic arteries** (FREN-ik) | Inferior surface of diaphragm |
| **Celiac trunk** (SĒ-lē-ak) | |
| **Common hepatic artery** (he-PAT-ik) | Liver |
| **Left gastric artery** (GAS-trik) | Stomach and esophagus |
| **Splenic artery** (SPLĒN-ik) | Spleen, pancreas, and stomach |
| **Superior mesenteric artery** (MES-en-ter′-ik) | Small intestine, cecum, ascending and transverse colons, and pancreas |
| **Suprarenal arteries** (soo-pre-RĒ-nal) | Adrenal (suprarenal) glands |
| **Renal arteries** (RĒ-nal) | Kidneys |
| **Gonadal arteries** (gō-NAD-al) | |
| **Testicular arteries** (tes-TIK-ū-lar) | Testes (male) |
| **Ovarian arteries** (ō-VAR-ē-an) | Ovaries (female) |
| **Inferior mesenteric artery** | Transverse, descending, and sigmoid colons, rectum |
| **Common iliac arteries** (IL-ē-ak) | |
| **External iliac arteries** | Lower limbs |
| **Internal iliac arteries** | Uterus (female), prostate (male), muscles of buttocks, and urinary bladder |

of the aorta, thoracic aorta, and abdominal aorta. The portion of the aorta that emerges from the left ventricle posterior to the pulmonary trunk is the **ascending aorta**. The ascending aorta gives off two coronary artery branches that supply the myocardium of the heart. The ascending aorta turns to the left, forming the **arch of the aorta**, then descends, following closely to the vertebral bodies, and passes through the diaphragm. The section of the aorta between the arch of the aorta and the diaphragm is called the **thoracic aorta**. As the aorta continues to descend below the diaphragm, it follows closely the vertebral column and divides into the two common iliac arteries, which carry blood to the lower limbs. The section between the diaphragm and the common iliac arteries is the **abdominal aorta**. Each division of the aorta gives off arteries that branch into distributing arteries that lead to various organs. Within the organs, the arteries divide into arterioles and then into capillaries that service the systemic tissues (all tissues except the alveoli of the lungs).

✓ **CHECKPOINT**

30. What are the names of the four divisions of the aorta, in order of blood flow?
31. Which general regions do each of the four principal divisions of the aorta supply?

## Ascending Aorta

The ascending aorta (see Figure 19.4) begins at the aortic valve posterior to the pulmonary trunk and right auricle. It is directed superiorly, slightly anteriorly, and to the right, where it becomes the arch of the aorta (Figure 20.16b). The right and left **coronary arteries** (*coron-* = crown) arise from the ascending aorta just superior to the aortic semilunar valve (see Figures 19.7 and 20.17). They form a crownlike ring around the heart, giving off branches to the myocardium. The **posterior interventricular branch** (in-ter-ven-TRIK-ū-lar; *inter-* = between) of the right coronary artery supplies both ventricles, and the **marginal branch** supplies the right ventricle. The **anterior interventricular branch**, also known as the **left anterior descending branch**, of the left coronary artery supplies both ventricles, and the **circumflex branch** (SER-kum-flex; *circum-* = around; *-flex* = to bend) supplies the left atrium and left ventricle.

## The Arch of the Aorta

The arch of the aorta (Figure 20.18; Table 20.5) is the continuation of the ascending aorta. It emerges from the pericardium posterior to the sternum. The arch is directed superiorly and posteriorly to the left and then inferiorly, where it becomes the thoracic aorta. Three major arteries branch from the superior aspect of the arch of the aorta: the brachiocephalic trunk, the left common carotid, and the left subclavian. The first and largest branch from the arch of the aorta is the **brachiocephalic trunk** (brā′-kē-Ō-se-FAL-ik; *brachio-* = arm; *-cephalic* = head). It extends superiorly, bending slightly to the right, and divides to form the right subclavian artery and right common carotid artery.

FIGURE 20.17 **Distribution of ascending aorta.**

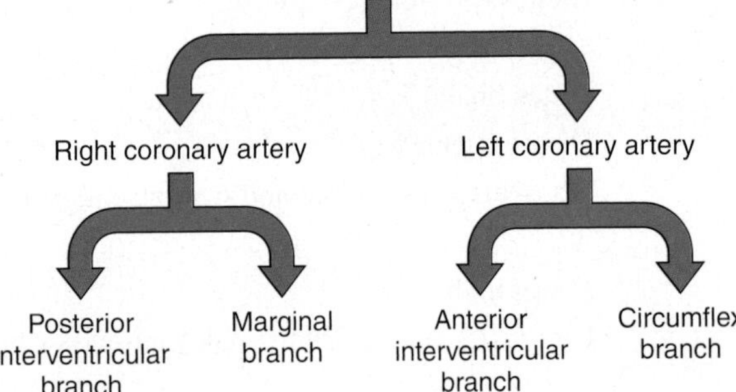


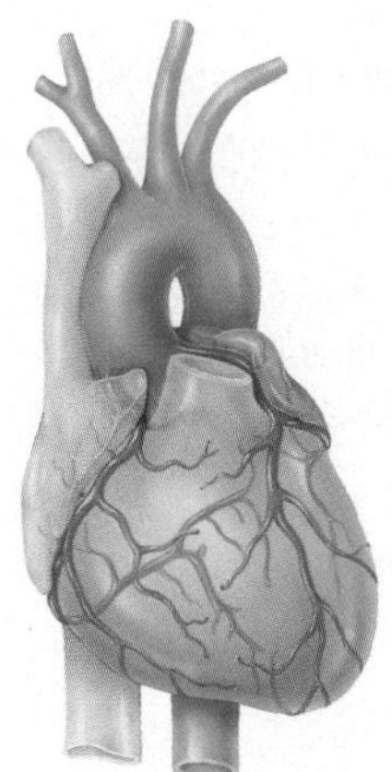

**The coronary arteries deliver blood to the myocardium of the heart.**

FIGURE 20.18 Arch of the aorta and its branches.

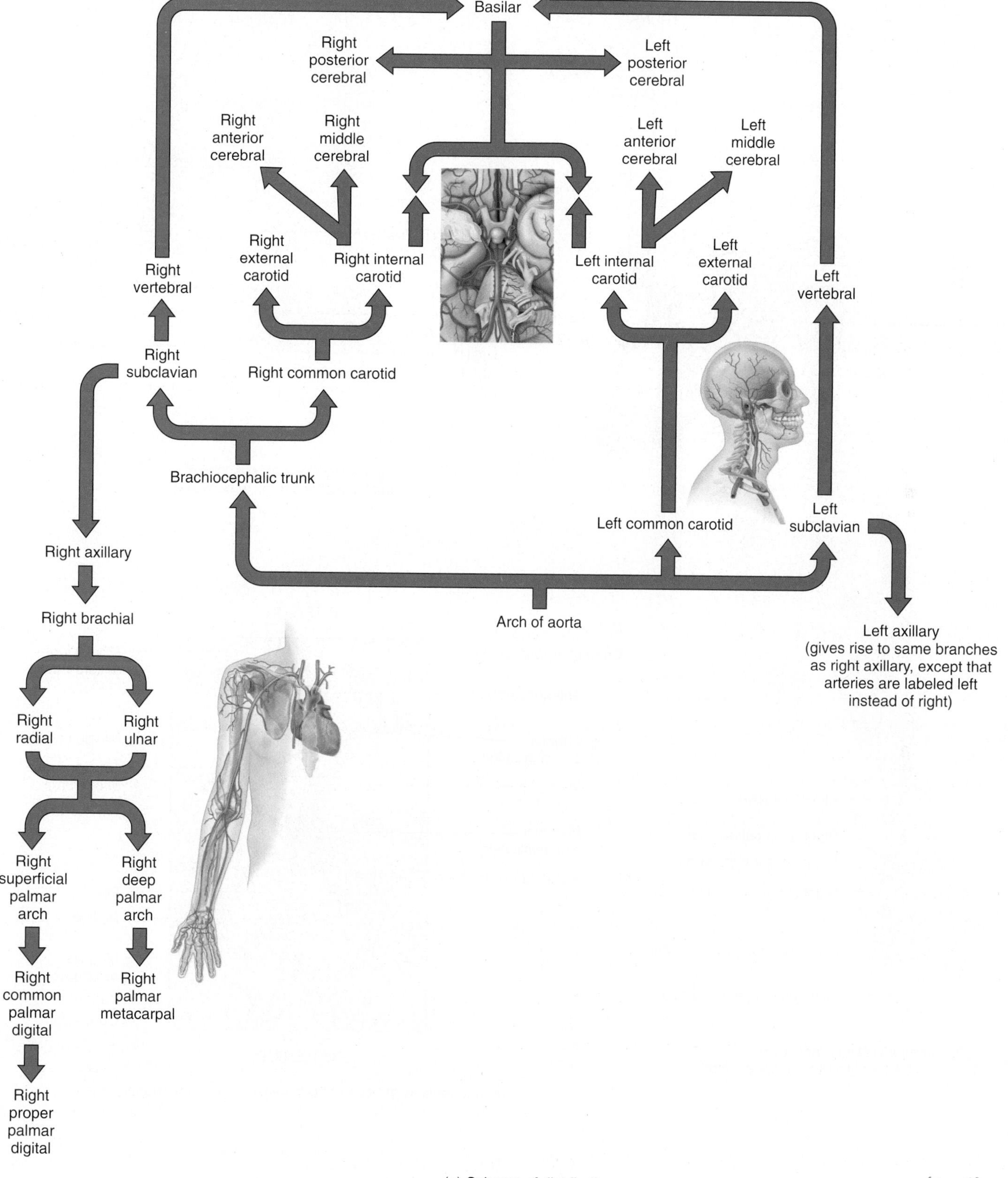


(a) Scheme of distribution

(continues)

FIGURE 20.18 (continued)

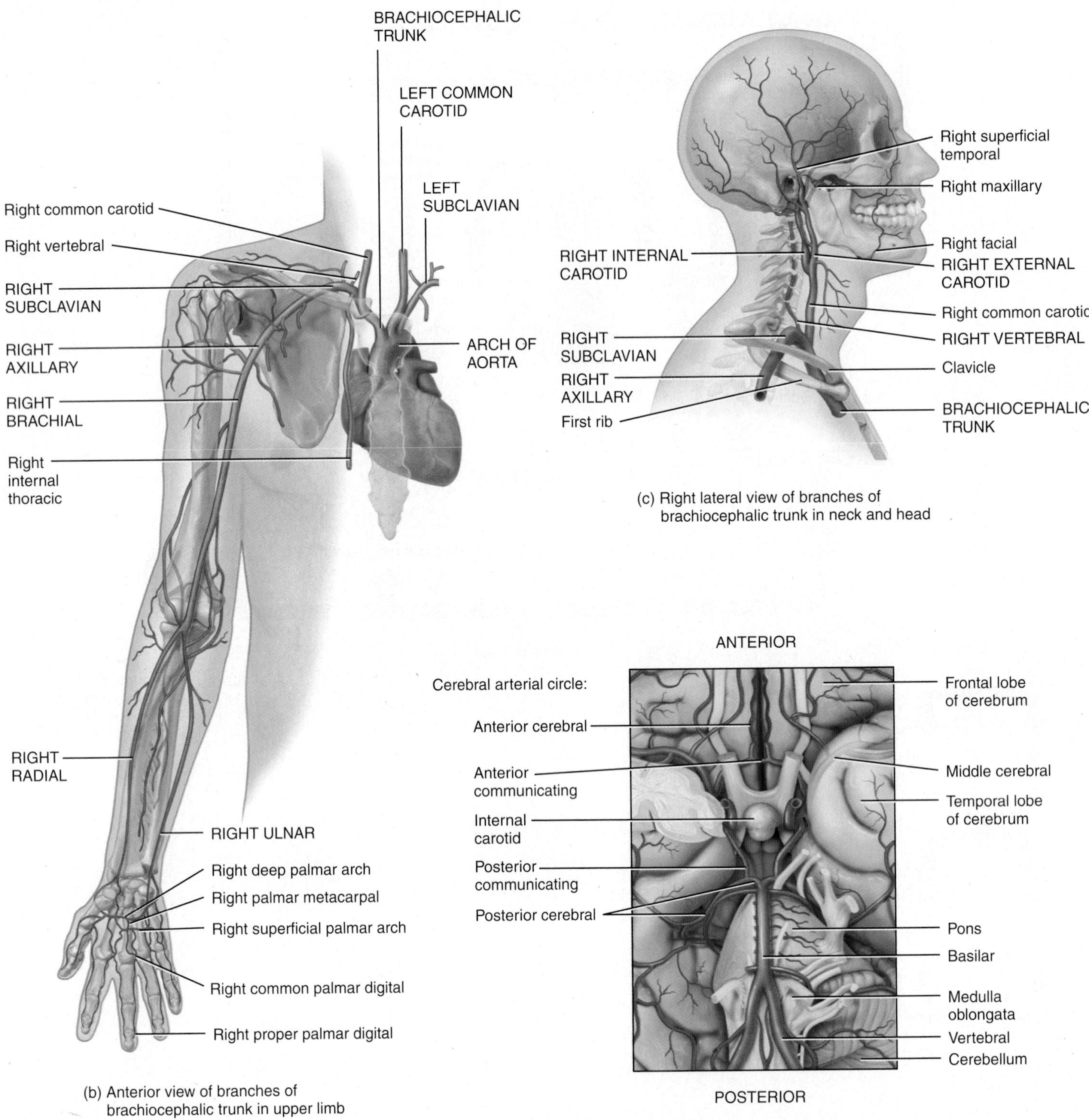


(b) Anterior view of branches of brachiocephalic trunk in upper limb

(c) Right lateral view of branches of brachiocephalic trunk in neck and head

(d) Inferior view of base of brain showing cerebral arterial circle

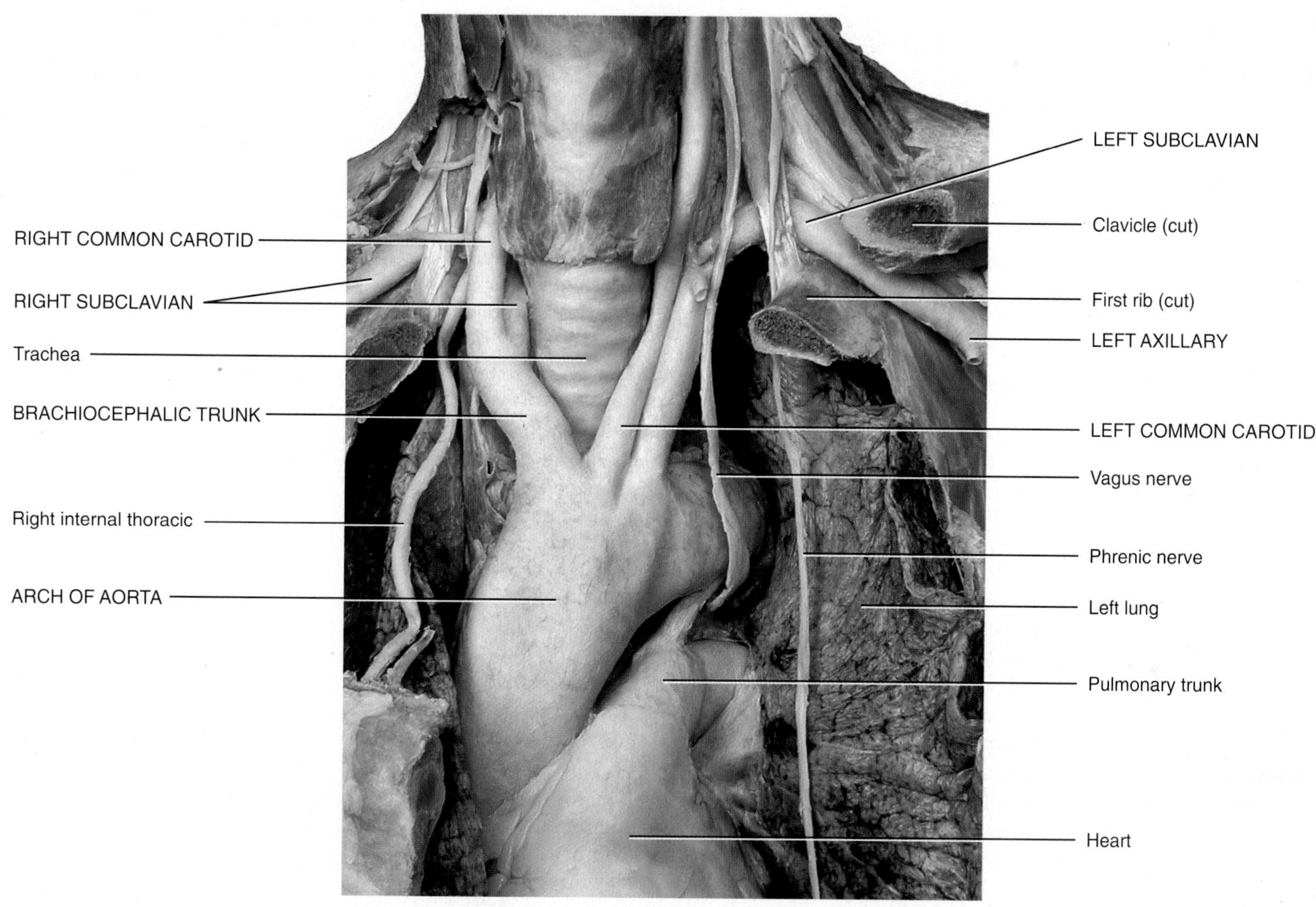


(e) Anterior view of branches of arch of aorta

**The arch of the aorta is the continuation of the ascending aorta.**

**TABLE 20.5**

The Arch of the Aorta

| BRANCH | DESCRIPTION AND BRANCHES | REGIONS SUPPLIED |
|---|---|---|
| **Brachiocephalic trunk** | First branch of arch of the aorta; divides to form right subclavian artery and right common carotid artery (Figure 20.18b) | Head, neck, upper limb, and thoracic wall |
| **Right subclavian artery*** (sub-KLĀ-ve-an) | Extends from brachiocephalic artery to inferior border of first rib; gives rise to a number of branches at base of neck | Brain, spinal cord, neck, shoulder, thoracic muscle wall, and scapular muscles |
| **Internal thoracic (mammary) artery** (thor-AS-ik; *thorac-* = chest) | Arises from first part of subclavian artery and descends posterior to costal cartilages of superior six rib just lateral to sternum; terminates at sixth intercostal space by bifurcating (branching into two arteries) and sends branches into intercostal spaces<br>**Clinical note:** In **coronary artery bypass grafting**, if only a single vessel is obstructed, the internal thoracic (usually the left) is used to create the bypass. The upper end of the artery is left attached to the subclavian artery and the cut end is connected to the coronary artery at a point distal to the blockage. The lower end of the internal thoracic is tied off. Artery grafts are preferred over vein grafts because arteries can withstand the greater pressure of blood flowing through coronary arteries and are less likely to become obstructed over time. | Anterior thoracic wall |

(continues)

**TABLE 20.5 CONTINUED**

The Arch of the Aorta

| BRANCH | DESCRIPTION AND BRANCHES | REGIONS SUPPLIED |
|---|---|---|
| **Vertebral artery** (VER-te-bral) | Major branch to brain of right subclavian artery before it passes into axilla (Figure 20.18c); ascends through neck, passes through transverse foramina of cervical vertebrae, and enters skull via foramen magnum to reach inferior surface of brain; unites with left vertebral artery to form **basilar artery** (BAS-i-lar); basilar artery passes along midline of anterior aspect of brain stem and gives off several branches (**posterior cerebral** and **cerebellar arteries**) | Posterior portion of cerebrum, cerebellum, pons, and inner ear |
| **Axillary artery*** (AK-sil-ar-ē = armpit) | Continuation of right subclavian artery into axilla; begins where subclavian artery passes inferior border of first rib and ends as it crosses distal margin of teres major muscle; gives rise to numerous branches in axilla | Thoracic, shoulder, and scapular muscles and humerus |
| **Brachial artery*** (BRĀ-kē-al = arm) | Continuation of axillary artery into arm; begins at distal border of teres major muscle and terminates by bifurcating into radial and ulnar arteries just distal to bend of elbow; superficial and palpable along medial side of arm; as it descends toward elbow it curves laterally and passes through cubital fossa, a triangular depression anterior to elbow, where you can easily detect pulse of brachial artery and listen to various sounds when taking a person's blood pressure<br><br>**Clinical note: Blood pressure** is usually measured in the brachial artery. In order to control hemorrhage, the best place to compress the brachial artery is near the middle of the arm where it is superficial and easily pressed against the humerus. | Muscles of arm, humerus, elbow joint |
| **Radial artery** (RĀ-dē-al = radius) | Smaller branch of brachial bifurcation; a direct continuation of brachial artery; passes along lateral (radial) aspect of forearm and enters wrist where it bifurcates into superficial and deep branches that anastomose with corresponding branches of ulnar artery to form palmar arches of hand; makes contact with distal end of radius at wrist, where it is covered only by fascia and skin<br><br>**Clinical note:** Because of its superficial location at this point, it is a common site for measuring **radial pulse**. | Major blood source to muscles of posterior compartment of forearm |
| **Ulnar artery** (UL-nar = ulna) | Larger branch of brachial artery passes along medial (ulnar) aspect of forearm and then into wrist, where it branches into superficial and deep branches that enter hand; these branches anastomose with corresponding branches of radial artery to form palmar arches of hand | Major blood source to muscles of anterior compartment of forearm |
| **Superficial palmar arch** (*palma* = palm) | Formed mainly by superficial branch of ulnar artery, with contribution from superficial branch of radial artery; superficial to long flexor tendons of fingers and extends across palm at bases of metacarpals; gives rise to **common palmar digital arteries**, each of which divides into **proper palmar digital arteries** | Muscles, bones, joints, and skin of palm and fingers |
| **Deep palmar arch** | Arises mainly from deep branch of radial artery, but receives contribution from deep branch of ulnar artery; deep to long flexor tendons of fingers and extends across palm just distal to bases of metacarpals; gives rise to **palmar metacarpal arteries**, which anastomose with common palmar digital arteries from superficial arch | Muscles, bones, and joints of palm and fingers |
| **Right common carotid** | Begins at bifurcation of brachiocephalic trunk, posterior to right sternoclavicular joint; passes superiorly into neck to supply structures in head (Figure 20.18c); divides into right external and right internal carotid arteries at superior border of larynx (voice box)<br><br>**Clinical note: Pulse** may be detected in the common carotid artery, just lateral to the larynx. It is convenient to detect a carotid pulse when exercising or when administering cardiopulmonary resuscitation. | Head and neck |

| BRANCH | DESCRIPTION AND BRANCHES | REGIONS SUPPLIED |
|---|---|---|
| **External carotid artery** | Begins at superior border of larynx and terminates near temporomandibular joint of parotid gland, where it divides into two branches: superficial temporal and maxillary arteries<br>**Clinical note:** The **carotid pulse** can be detected in the external carotid artery just anterior to the sternocleidomastoid muscle at the superior border of the larynx. | Major blood source to all structures of head except brain<br>Supplies skin, connective tissues, muscles, bones, joints, dura and arachnoid mater in head and supplies much of neck anatomy |
| **Internal carotid artery** | Arises from common carotid artery; enters cranial cavity through carotid foramen in temporal bone and emerges in cranial cavity near base of sella turcica of sphenoid bone; gives rise to numerous branches inside cranial cavity and terminates as anterior cerebral arteries; **anterior cerebral artery** passes forward toward frontal lobe of cerebrum and **middle cerebral artery** passes laterally between temporal and parietal lobes of cerebrum; inside cranium (Figure 20.18d), anastomoses of left and right internal carotid arteries via anterior communicating artery between two anterior cerebral arteries, along with internal carotid–basilar artery anastomoses, form an arrangement of blood vessels at base of brain called **cerebral arterial circle (circle of Willis)** (Figure 20.18d); internal carotid-basilar anastomosis occurs where **posterior communicating arteries** arising from internal carotid artery anastomose with posterior cerebral arteries from basilar artery to link internal carotid blood supply with vertebral blood supply; cerebral arterial circle equalizes blood pressure to brain and provides alternate routes for blood flow to brain, should arteries become damaged | Eyeball and other orbital structures, ear, and parts of nose and nasal cavity<br>Frontal, temporal, parietal lobes of the cerebrum of brain, pituitary gland, and pia mater |
| **Left common carotid artery** | Arises as second branch of arch of the aorta and ascends through mediastinum to enter neck deep to clavicle, then follows similar path to right common carotid artery | Distribution similar to right common carotid artery |
| **Left subclavian artery** | Arises as third and final branch of arch of the aorta; passes superior and lateral through mediastinum and deep to clavicle at base of neck as it courses toward upper limb; has similar course to right subclavian artery after leaving mediastinum | Distribution similar to right subclavian artery |

*This is an example of the practice of giving the same vessel different names as it passes through different regions. See the axillary and brachial arteries.

The second branch from the arch of the aorta is the **left common carotid artery** (ka-ROT-id), which divides into the same branches with the same names as the right common carotid artery. The third branch from the arch of the aorta is the **left subclavian artery** (sub-KLĀ-vē-an), which distributes blood to the left vertebral artery and vessels of the left upper limb. Arteries branching from the left subclavian artery are similar in distribution and name to those branching from the right subclavian artery.

## Thoracic Aorta

The thoracic aorta (Table 20.6; Figure 20.19) is a continuation of the arch of the aorta. It lies to the left of the vertebral column. As it descends, it passes through the diaphragm and becomes the abdominal aorta as it enters the abdominal cavity. Along its course, the thoracic aorta sends off numerous small arteries, **visceral branches** (VIS-er-al) to viscera and **parietal branches** (pa-RĪ-e-tal) to body wall structures.

**TABLE 20.6**

Thoracic Aorta

| BRANCH | DESCRIPTION AND BRANCHES | REGIONS SUPPLIED |
|---|---|---|
| **VISCERAL BRANCHES** | | |
| **Pericardial arteries** (per′-i-KAR-dē-al; *peri-* = around; *-cardia* = heart) | Two to three small arteries that arise from variable levels of thoracic aorta and pass forward to pericardial sac surrounding heart | Tissues of pericardial sac |
| **Bronchial arteries** (BRONG-kē-al = windpipe) | Arise from thoracic aorta or one of its branches; right bronchial artery typically arises from third posterior intercostal artery; two left bronchial arteries arise from upper end of thoracic aorta; all follow bronchial tree into lungs | Supply tissues of bronchial tree and surrounding lung tissue down to level of alveolar ducts |
| **Esophageal arteries** (e-sof′-a-JĒ-al; *eso-* = to carry; *phage* = food) | Four to five arteries that arise from anterior surface of thoracic aorta and pass forward to branch onto esophagus | All tissues of esophagus |
| **Mediastinal arteries** (mē-dē-as-TĪ-nal) | Arise from various points on thoracic aorta | Assorted tissues within mediastinum, primarily connective tissue and lymph nodes |
| **PARIETAL BRANCHES** | | |
| **Posterior intercostal arteries** (in′ter-KOS-tal; *inter-* = between; *-costa* = rib) | Typically, nine pairs of arteries that arise from posterolateral aspect on each side of thoracic aorta; each passes laterally and then anteriorly through intercostal space, where they will eventually anastomose with anterior branches from internal thoracic arteries | Skin, muscles, and ribs of thoracic wall; thoracic vertebrae, meninges, and spinal cord; mammary glands |
| **Subcostal arteries** (sub-KOS-tal; *sub-* = under) | The lowest segmental branches of thoracic aorta; one on each side passes into thoracic body wall inferior to twelfth rib and courses forward into upper abdominal region of body wall | Skin, muscles, and ribs; twelfth thoracic vertebra, meninges, and spinal cord |
| **Superior phrenic arteries** (FREN-ik = pertaining to diaphragm) | Arise from lower end of thoracic aorta and pass onto superior surface of diaphragm | Diaphragm muscle and pleura covering diaphragm |

**FIGURE 20.19** **Thoracic aorta and abdominal aorta and their principal branches.**

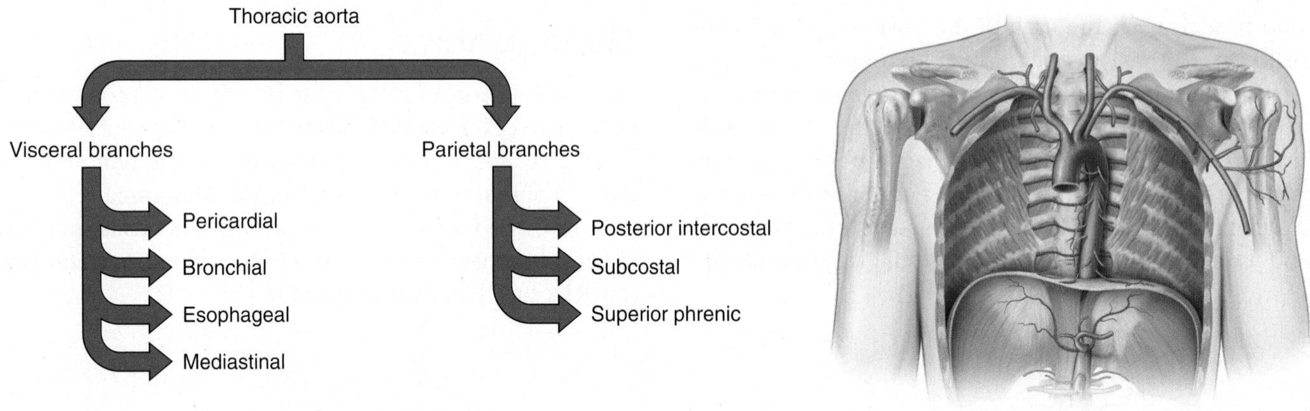


(a) Scheme of distribution

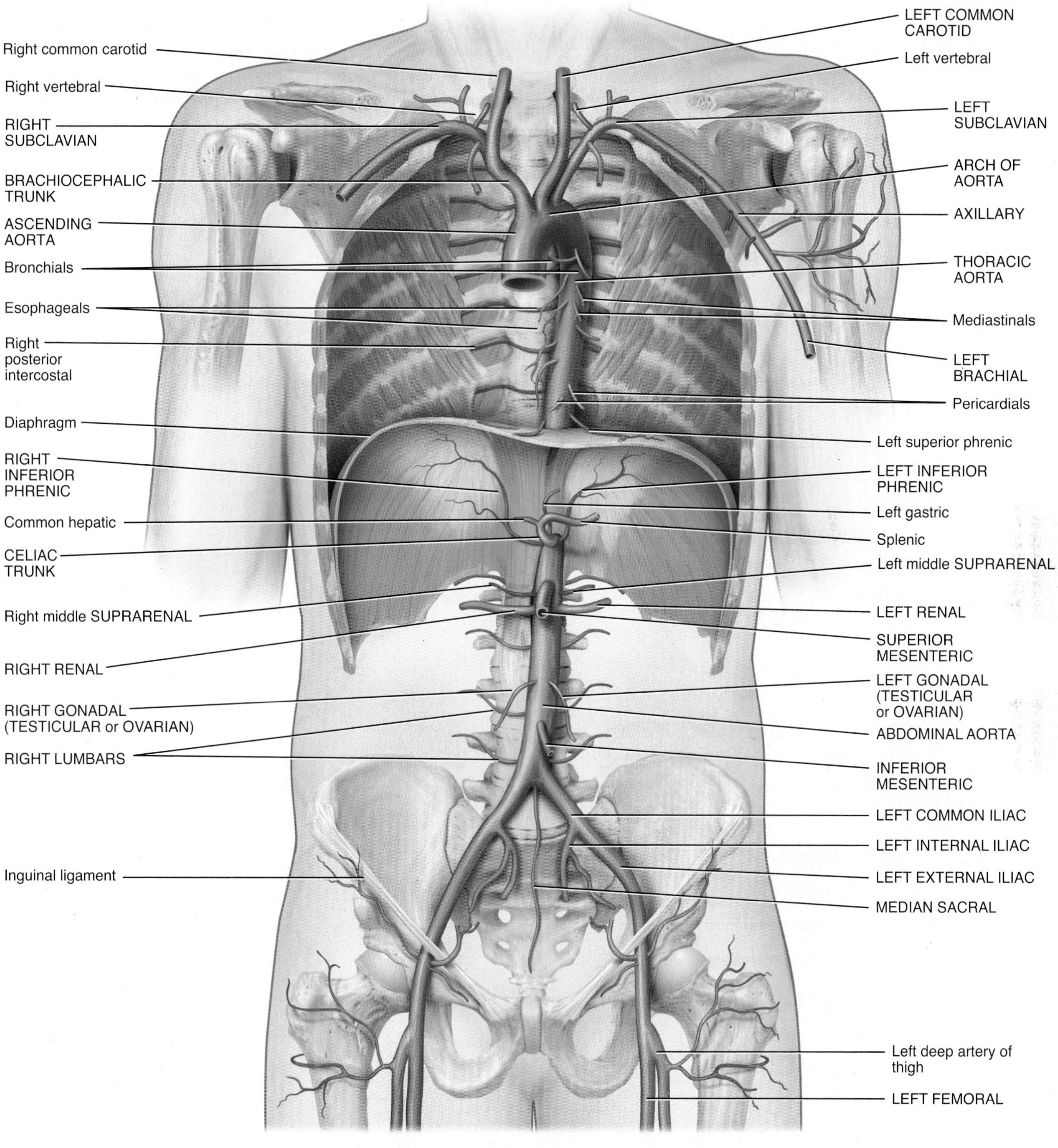

Right common carotid
Right vertebral
RIGHT SUBCLAVIAN
BRACHIOCEPHALIC TRUNK
ASCENDING AORTA
Bronchials
Esophageals
Right posterior intercostal
Diaphragm
RIGHT INFERIOR PHRENIC
Common hepatic
CELIAC TRUNK
Right middle SUPRARENAL
RIGHT RENAL
RIGHT GONADAL (TESTICULAR or OVARIAN)
RIGHT LUMBARS
Inguinal ligament
LEFT COMMON CAROTID
Left vertebral
LEFT SUBCLAVIAN
ARCH OF AORTA
AXILLARY
THORACIC AORTA
Mediastinals
LEFT BRACHIAL
Pericardials
Left superior phrenic
LEFT INFERIOR PHRENIC
Left gastric
Splenic
Left middle SUPRARENAL
LEFT RENAL
SUPERIOR MESENTERIC
LEFT GONADAL (TESTICULAR or OVARIAN)
ABDOMINAL AORTA
INFERIOR MESENTERIC
LEFT COMMON ILIAC
LEFT INTERNAL ILIAC
LEFT EXTERNAL ILIAC
MEDIAN SACRAL
Left deep artery of thigh
LEFT FEMORAL

(b) Detailed anterior view of principal branches of aorta

**The thoracic aorta is the continuation of the ascending aorta.**

## ✓ CHECKPOINT

**32.** What are the three major branches of the arch of the aorta in order of origination? Which regions do they supply?

**33.** Which structure is the anatomical border between the thoracic and abdominal aorta?

## Abdominal Aorta

The abdominal aorta (Figure 20.20; Table 20.7) is the continuation of the thoracic aorta after it passes through the diaphragm. It begins as the aorta penetrates the diaphragm and ends where it divides into the right and left common iliac arteries. The abdominal aorta lies anterior to the vertebral column. As with the thoracic aorta, the abdominal aorta gives off visceral and parietal branches. The paired visceral branches arise from the lateral surfaces of the aorta and include the **suprarenal, renal**, and **gonadal arteries**. The unpaired visceral branches arise from the anterior surface of the aorta and include the **celiac trunk**, the **superior mesenteric artery**, and the **inferior mesenteric artery** (see Figure 20.19b). The parietal branches arise from the posterior surfaces of the aorta: the paired **inferior phrenic** and **lumbar arteries**, and the unpaired **median sacral artery**.

**FIGURE 20.20** **Abdominal aorta and principal branches.**

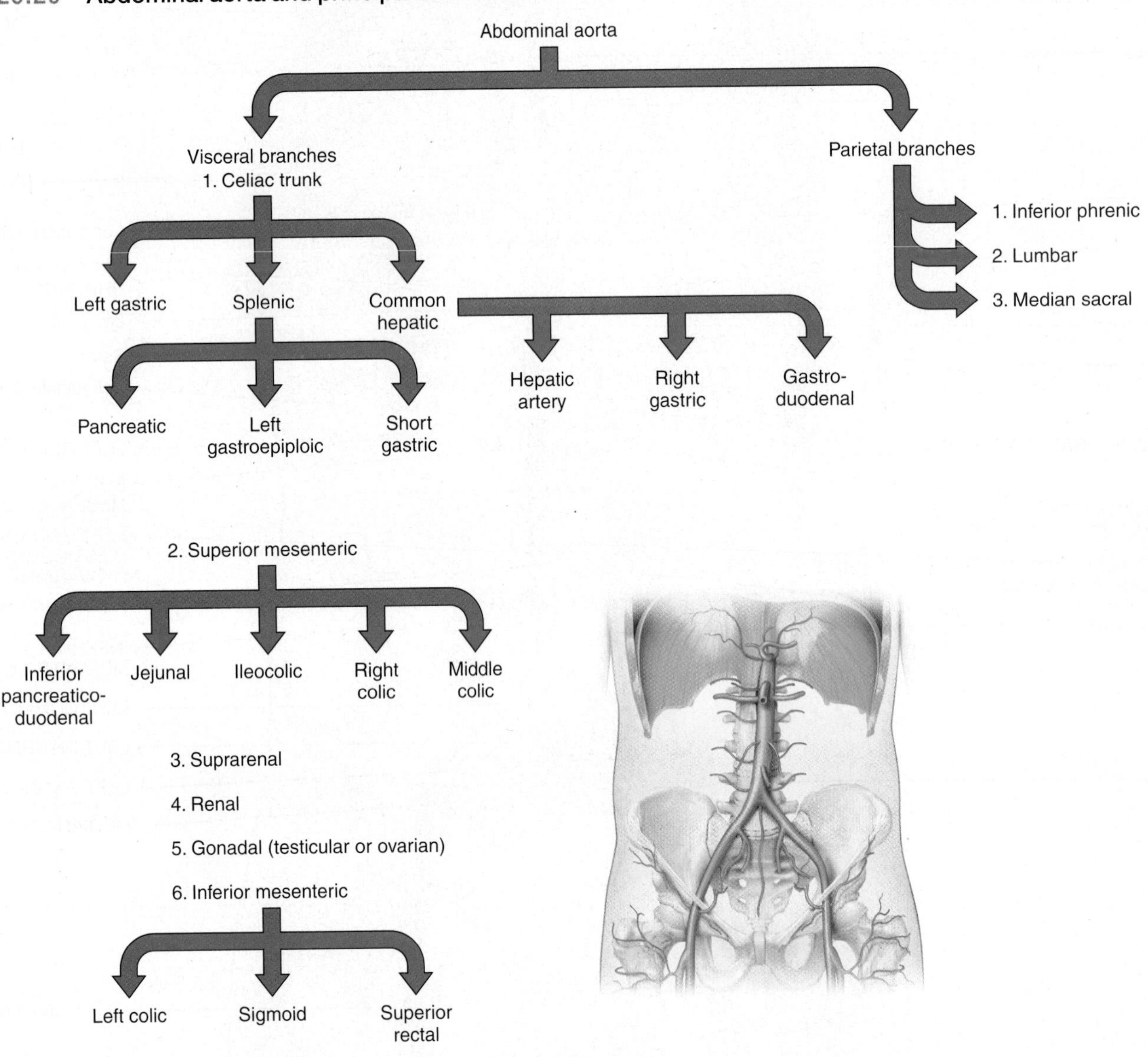


(a) Scheme of distribution

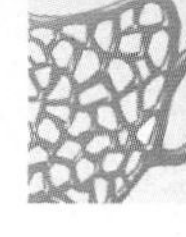

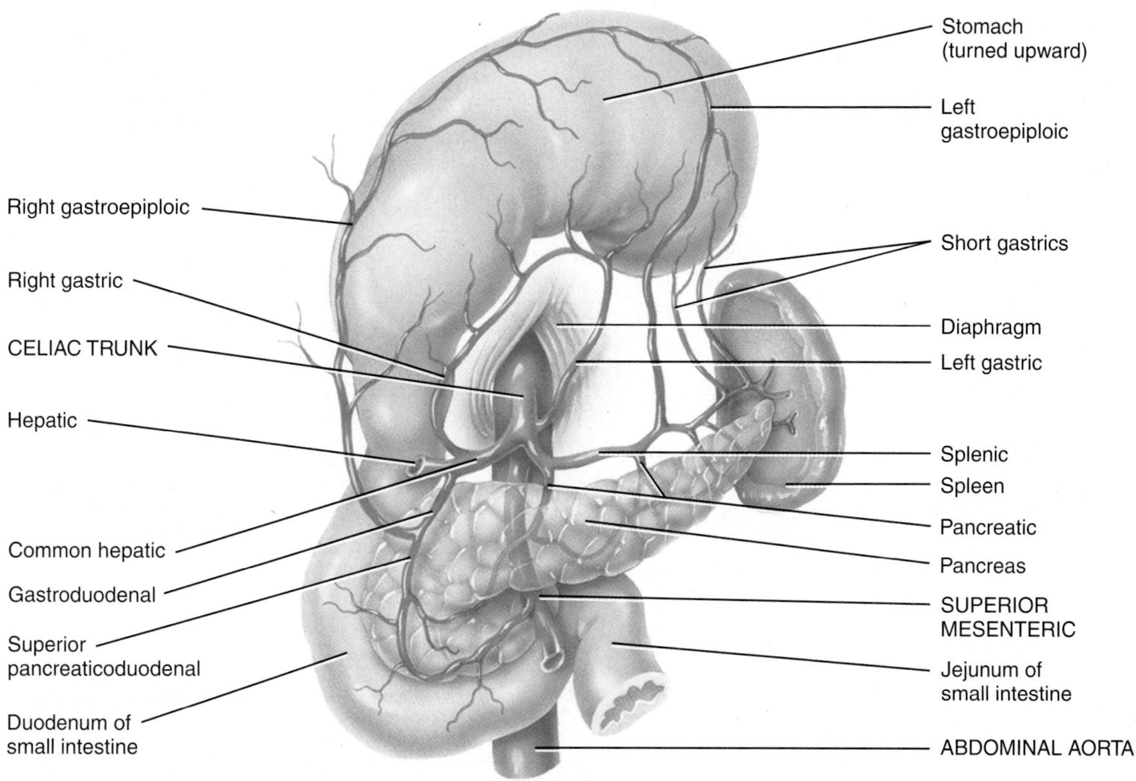


(b) Anterior view of celiac trunk and its branches

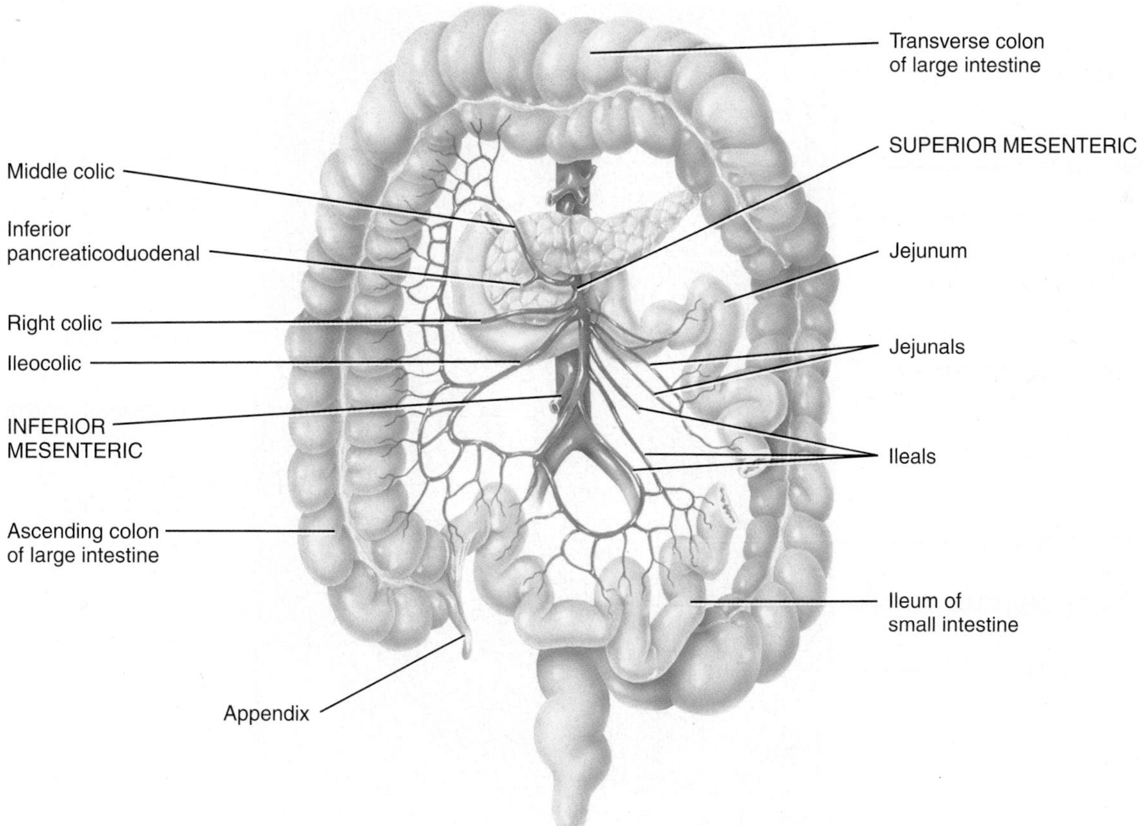


(c) Anterior view of superior mesenteric artery and its branches

(continues)

**FIGURE 20.20 (continued)**

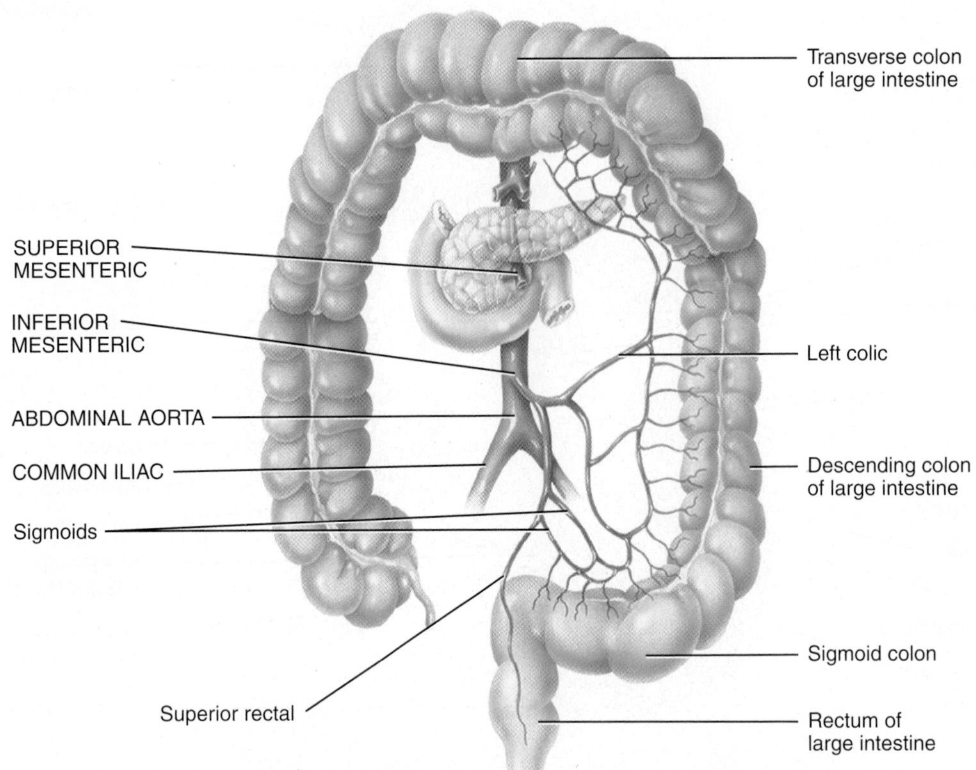


(d) Anterior view of inferior mesenteric artery and its branches

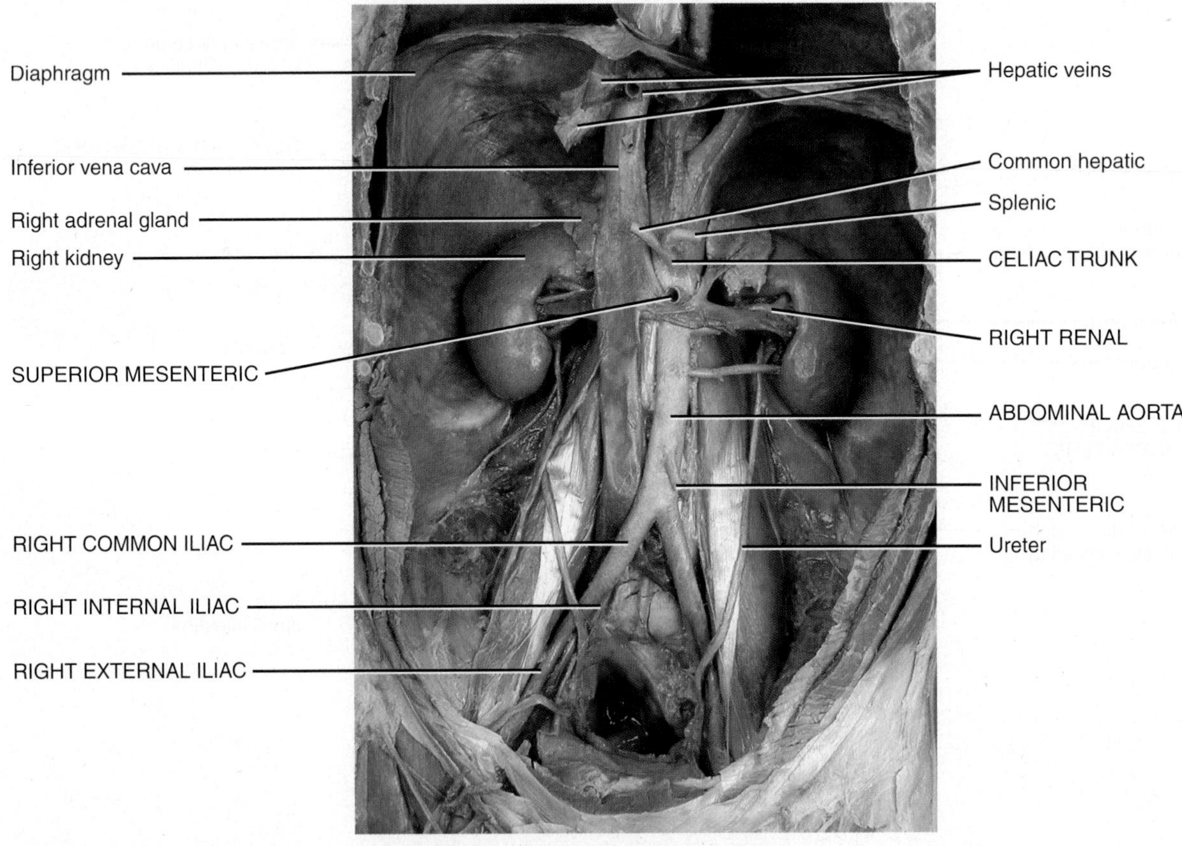


(e) Anterior view of arteries of abdomen and pelvis

**The abdominal aorta is the continuation of the thoracic aorta.**

**TABLE 20.7**

Abdominal Aorta

| BRANCH | DESCRIPTION AND BRANCHES | REGIONS SUPPLIED |
|---|---|---|
| **UNPAIRED VISCERAL BRANCHES** | | |
| **Celiac trunk (artery)** (SĒ-lē-ak) | First visceral branch of aorta inferior to diaphragm; arises from abdominal aorta at level of twelfth thoracic vertebra as aorta passes through hiatus in diaphragm; divides into three branches: left gastric, splenic, and common hepatic arteries (Figure 20.20b) | Supplies all organs of gastrointestinal tract that arise from embryonic foregut, that is, from abdominal part of esophagus to duodenum, and also spleen |
| | • **Left gastric artery** (GAS-trik = stomach). Smallest of three celiac branches arises superiorly to left toward esophagus and then turns to follow lesser curvature of stomach. On lesser curvature of stomach it anastomoses with right gastric artery. | Abdominal part of esophagus and lesser curvature of stomach |
| | • **Splenic artery** (SPLEN-ik = spleen). Largest branch of celiac trunk arises from left side of celiac trunk distal to left gastric artery, and passes horizontally to left along pancreas. Before reaching spleen, it gives rise to three named arteries: | Spleen, pancreas, fundus and greater curvature of stomach, and greater omentum |
| | ▪ **Pancreatic arteries** (pan-krē-AT-ik), a series of small arteries that arise from splenic and descend into tissue of pancreas | Pancreas |
| | ▪ **Left gastroepiploic (gastro-omental) artery** (gas′-trō-ep′-i-PLŌ-ik; *epiplo-* = omentum) arises from terminal end of splenic artery and passes from left to right along greater curvature of stomach | Greater curvature of stomach and greater omentum |
| | ▪ **Short gastric arteries** arise from terminal end of splenic artery and pass onto fundus of stomach | Fundus of stomach |
| | • **Common hepatic artery** (he-PAT-ik = liver). Intermediate in size between left gastric and splenic arteries; arises from right side of celiac trunk and gives rise to three arteries: | Liver, gallbladder, lesser omentum, stomach, pancreas, and duodenum |
| | ▪ **Hepatic artery** branches from common hepatic artery and ascends along bile ducts into liver and gallbladder | Liver, gallbladder, and lesser omentum |
| | ▪ **Right gastric artery** arises from common hepatic artery and curves back to left along lesser curvature of stomach, where it anastomoses with left gastric artery | Stomach and lesser omentum |
| | ▪ **Gastroduodenal artery** (gas′-trō-doo′-ō-DĒ-nal) passes inferiorly toward stomach and duodenum and sends branches along greater curvature of stomach | Stomach, duodenum, and pancreas |
| **Superior mesenteric artery** (MES-en-ter′-ik; *meso-* = middle; *-enteric* = pertaining to intestines) | Arises from anterior surface of abdominal aorta about 1 cm inferior to celiac trunk at level of first lumbar vertebra (Figure 20.20c); extends inferiorly and anteriorly between layers of mesentery (portion of peritoneum that attaches small intestine to posterior abdominal wall); it anastomoses extensively and has five branches: | Supplies all organs of gastrointestinal tract that arise from embryonic midgut, that is, from duodenum to transverse colon |
| | • **Inferior pancreaticoduodenal artery** (pan′-krē-at′-i-kō-doo′-ō-DĒ-nal). Passes superiorly and to right toward head of pancreas and duodenum. | Pancreas and duodenum |
| | • **Jejunal** (je-JOO-nal) and **ileal arteries** (IL-ē-al). Spread through mesentery to pass to loops of jejunum and ileum (small intestine). | Jejunum and ileum, which is majority of small intestine |
| | • **Ileocolic artery** (il′-ē-ō-KOL-ik). Passes inferiorly and laterally toward right side toward terminal part of ileum, cecum, appendix, and first part of ascending colon. | Terminal part of ileum, cecum, appendix, and first part of ascending colon |

(continues)

## TABLE 20.7 CONTINUED

### Abdominal Aorta

| BRANCH | DESCRIPTION AND BRANCHES | REGIONS SUPPLIED |
|---|---|---|
| | • **Right colic artery** (KOL-ik). Passes laterally to right toward ascending colon. | Ascending colon and first part of transverse colon |
| | • **Middle colic artery**. Ascends slightly to right toward transverse colon. | Most of transverse colon |
| **Inferior mesenteric artery** | Arises from anterior aspect of abdominal aorta at level of third lumbar vertebra and then passes inferiorly to left of aorta (Figure 20.20d). It anastomoses extensively and has three branches: | Supplies all organs of gastrointestinal tract that arise from embryonic hindgut from transverse colon to rectum |
| | • **Left colic artery.** Ascends laterally to left toward distal end of transverse colon and descending colon. | End of transverse colon and descending colon |
| | • **Sigmoid arteries** (SIG-moyd). Descend laterally to left toward sigmoid colon. | Sigmoid colon |
| | • **Superior rectal artery** (REK-tal). Passes inferiorly to superior part of rectum. | Upper part of rectum |
| **Suprarenal arteries** (soo′-pra-RĒ-nal; *supra-* = above; *-ren-* = kidney) | There are typically three pairs (superior, middle, and inferior), but only middle pair originates directly from abdominal aorta (see Figure 20.19b); middle suprarenal arteries arise from abdominal aorta at level of first lumbar vertebra at or superior to renal arteries; superior suprarenal arteries arise from inferior phrenic arteries; inferior suprarenal arteries originate from renal arteries | Suprarenal (adrenal) glands |
| **Renal arteries** (RĒ-nal; *ren* = kidney) | Right and left **renal arteries** usually arise from lateral aspects of abdominal aorta at superior border of second lumbar vertebra, about 1 cm inferior to superior mesenteric artery (see Figure 20.19b); right renal artery, which is longer than left, arises slightly lower than left and passes posterior to right renal vein and inferior vena cava; left renal artery is posterior to left renal vein and is crossed by inferior mesenteric vein | All tissues of kidneys |
| **Gonadal arteries** (gō-NAD-al; *gon-* = seed) [**testicular** (tes-TIK-ū-lar) or **ovarian** (ō-VAR-ē-an)] | Arise from anterior aspect of abdominal aorta at level of second lumbar vertebra just inferior to renal arteries (see Figure 21.20); in males, specifically referred to as **testicular arteries** and descend along posterior abdominal wall to pass through inguinal canal and descend into scrotum; in females, called **ovarian arteries** and are much shorter than testicular arteries and remain within abdominal cavity | Males: testis, epididymis, ductus deferens, and ureters<br>Females: ovaries, uterine (fallopian) tubes, and ureters |
| **UNPAIRED PARIETAL BRANCH** | | |
| **Median sacral artery** (SĀ-kral = pertaining to sacrum) | Arises from posterior surface of abdominal aorta about 1 cm superior to *bifurcation* (division into two branches) of aorta into right and left common iliac arteries (see Figure 20.19b) | Sacrum, coccyx, sacral spinal nerves, and piriformis muscle |
| **PAIRED PARIETAL BRANCH** | | |
| **Inferior phrenic arteries** (FREN-ik = pertaining to diaphragm) | First paired branches of abdominal aorta; arise immediately superior to origin of celiac trunk (see Figure 20.19b; may also arise from renal arteries) | Diaphragm and suprarenal (adrenal) glands |
| **Lumbar arteries** (LUM-bar = pertaining to loin) | Four pairs arise from posterolateral surface of abdominal aorta in a way similar to posterior intercostal arteries of thorax (see Figure 20.19b); pass laterally into abdominal muscle wall and curve toward anterior aspect of wall | Lumbar vertebrae, spinal cord and meninges, skin and muscles of posterior and lateral part of abdominal wall |

## ✓ CHECKPOINT

34. What are the paired and unpaired visceral branches of the abdominal aorta? What is the general region that each branch supplies?
35. What are the paired and unpaired parietal branches of the abdominal aorta? What is the general region that each branch supplies?

## Arteries of the Pelvis and Lower Limbs

The abdominal aorta ends by dividing into the right and left **common iliac arteries** to supply the pelvis and lower limbs (Figure 20.21; Table 20.8). The common iliac arteries, in turn, divide into the **internal** and **external iliac arteries**. In sequence, the external iliacs become the **femoral arteries** in the thighs, the **popliteal arteries** posterior to the knee, and the **anterior** and **posterior tibial arteries** in the legs.

FIGURE 20.21 **Arteries of the pelvis and right lower limb.**

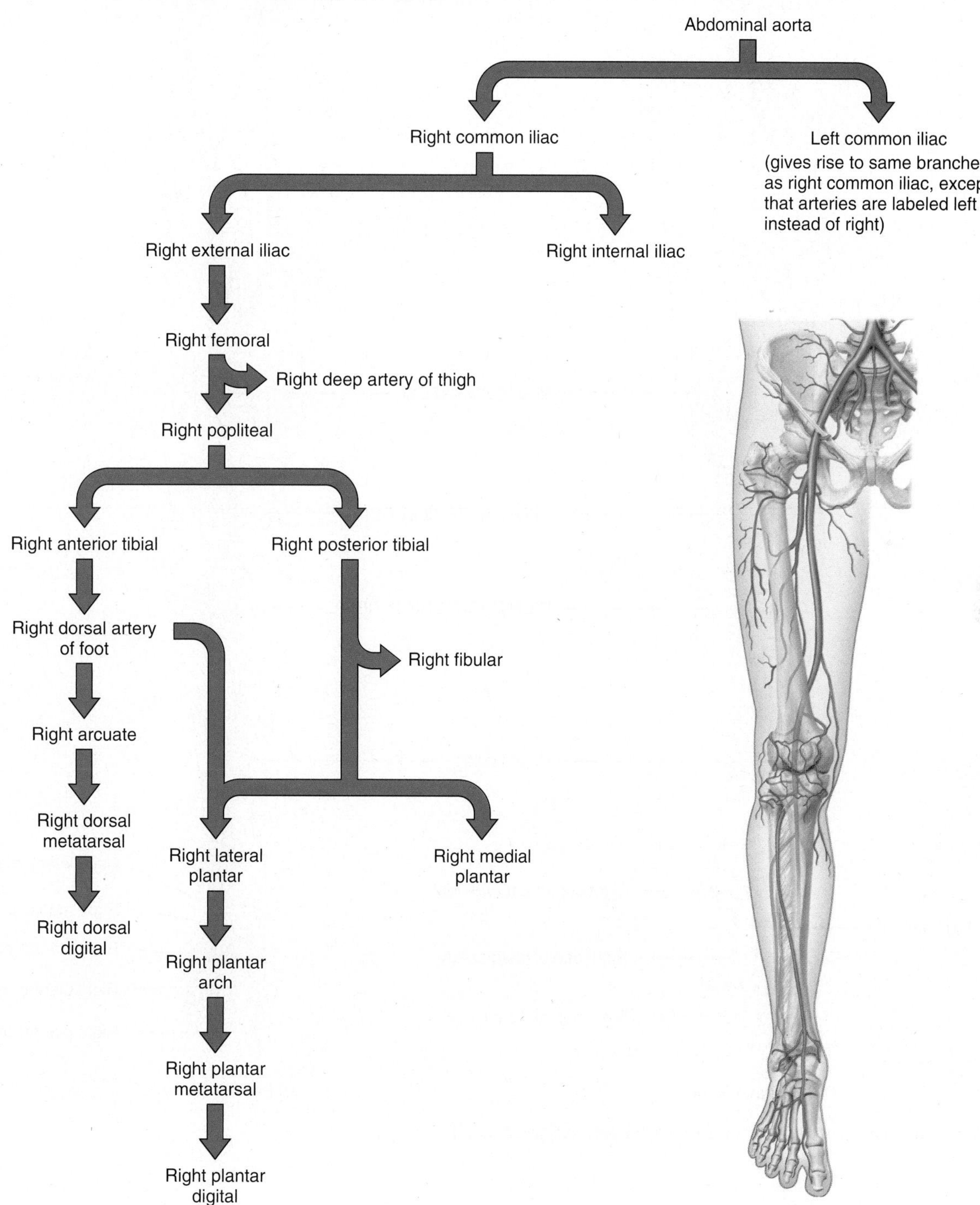


(a) Scheme of distribution

(continues)

**FIGURE 20.21 (continued)**

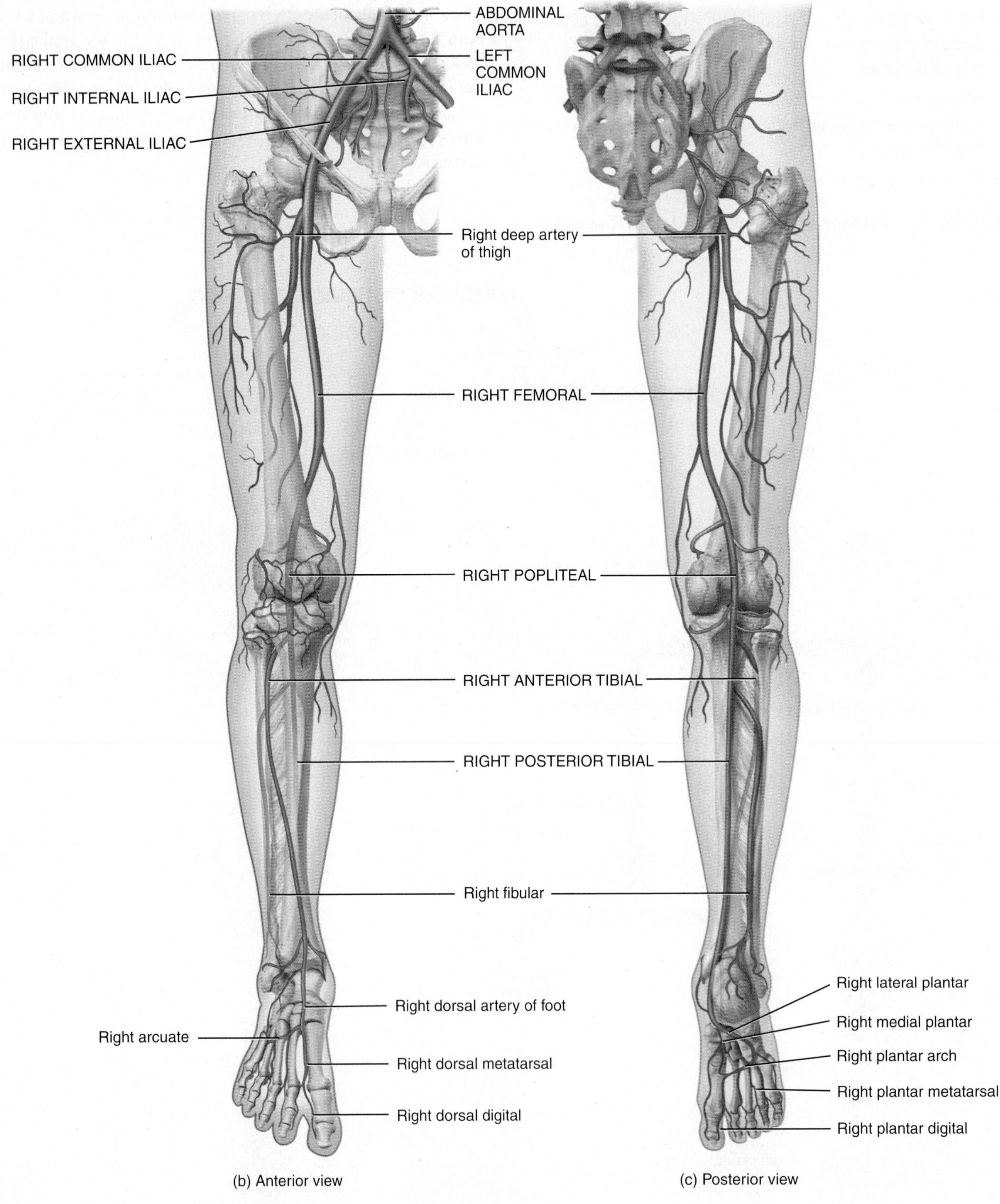


(b) Anterior view

(c) Posterior view

**The internal iliac arteries carry most of the blood supply to the pelvic viscera and wall.**

**TABLE 20.8**

Arteries of the Pelvis and Lower Limbs

| BRANCH | DESCRIPTION AND BRANCHES | REGIONS SUPPLIED |
|---|---|---|
| **Common iliac arteries** (IL-ē-ak = pertaining to ilium) | Arise from abdominal aorta at about level of fourth lumbar vertebra; each common iliac artery passes inferiorly and slightly laterally for about 5 cm (2 in.) and gives rise to two branches: internal and external iliac arteries | Pelvic muscle wall, pelvic organs, external genitals, and lower limbs |
| **Internal iliac arteries** | Primary arteries of pelvis; begin at bifurcation (division into two branches) of common iliac arteries anterior to sacroiliac joint at level of lumbosacral intervertebral disc; pass posteriorly as they descend into pelvis and divide into anterior and posterior divisions | Pelvic muscle wall, pelvic organs, buttocks, external genitals, and medial muscles of thigh |
| **External iliac arteries** | Larger than internal iliac arteries and begin at bifurcation of common iliac arteries; descend along medial border of psoas major muscles following pelvic brim, pass posterior to midportion of inguinal ligaments, and become femoral arteries as they pass beneath inguinal ligament and enter thigh | Lower abdominal wall, cremaster muscle in males and round ligament of uterus in females, and lower limb |
| **Femoral arteries** (FEM-o-ral = pertaining to thigh) | Continuations of external iliac arteries as they enter thigh; in *femoral triangle* of upper thighs they are superficial along with femoral vein and nerve and deep inguinal lymph nodes (see Figure 11.19a); pass beneath sartorius muscle as they descend along anteromedial aspects of thighs and follow its course to distal end of thigh where they pass through opening in tendon of adductor magnus muscle to end at posterior aspect of knee, where they become popliteal arteries<br><br>**Clinical note**: In **cardiac catheterization**, a catheter is inserted through a blood vessel and advanced into the major vessels to access a heart chamber. A catheter often contains a measuring instrument or other device at its tip. To reach the left side of the heart, the catheter is inserted into the femoral artery and passed into the aorta to the coronary arteries or heart chamber. | Muscles of thigh (quadriceps, adductors, and hamstrings), femur, and ligaments and tendons around knee joint |
| **Popliteal arteries** (pop′-li-TĒ-al = posterior surface of knee) | Continuation of femoral arteries through popliteal fossa (space behind knee); descend to inferior border of popliteus muscles, where they divide into anterior and posterior tibial arteries | Muscles of distal thigh, skin of knee region, muscles of proximal leg, knee joint, femur, patella, tibia, and fibula |
| **Anterior tibial arteries** (TIB-ē-al = pertaining to shin) | Descend from bifurcation of popliteal arteries at distal border of popliteus muscles; smaller than posterior tibial arteries; pass over interosseous membrane of tibia and fibula to descend through anterior muscle compartment of leg; become **dorsal arteries of foot (dorsalis pedis arteries)** at ankles; on dorsum of feet, dorsal arteries of foot give off transverse branch at first medial cuneiform bone called **arcuate arteries** (*arcuat-* = bowed) that run laterally over bases of metatarsals; from arcuate arteries branch **dorsal metatarsal arteries**, which course along metatarsal bones; dorsal metatarsal arteries terminate by dividing into **dorsal digital arteries**, which pass into toes | Tibia, fibula, anterior muscles of leg, dorsal muscles of foot, tarsal bones, metatarsal bones, and phalanges |

(continues)

**TABLE 20.8 CONTINUED**

Arteries of the Pelvis and Lower Limbs

| BRANCH | DESCRIPTION AND BRANCHES | REGIONS SUPPLIED |
|---|---|---|
| **Posterior tibial arteries** | Direct continuations of popliteal arteries, descend from bifurcation of popliteal arteries; pass down posterior muscular compartment of legs deep to soleus muscles; pass posterior to medial malleolus at distal end of leg and curve forward toward plantar surface of feet; pass deep to flexor retinaculum on medial side of feet and terminate by branching into medial and lateral plantar arteries; give rise to **fibular (peroneal) arteries** in middle of leg, which course laterally as they descend into lateral compartment of leg; **medial plantar arteries** (PLAN-tar = sole) pass along medial side of sole and **lateral plantar arteries** angle toward lateral side of sole and unite with branch of dorsal arteries of foot to form **plantar arch**; arch begins at base of fifth metatarsal and extends medially across metatarsals; as arch crosses foot, it gives off **plantar metatarsal arteries**, which course along plantar surface of metatarsal bones; these arteries terminate by dividing into **plantar digital arteries** that pass into toes | Posterior and lateral muscle compartments of leg, plantar muscles of foot, tibia, fibula, tarsal, metatarsal, and phalangeal bones |

✓ **CHECKPOINT**

**36.** What are the two major branches of the common iliac arteries? Which general regions do those arteries supply?

## 20.10 Systemic veins return blood to the heart from all body organs except the lungs.

As veins drain blood away from various parts of the body, they are removing carbon dioxide and other wastes and heat from the tissues.

For the most part, arteries are deep; veins (Figure 20.22; Table 20.9) may be superficial or deep. Superficial veins are located just beneath the skin and can be seen easily. Superficial veins are clinically important as sites for withdrawing blood or giving injections. Deep veins generally travel alongside arteries and usually bear the same name. Because there are no large superficial arteries, the names of superficial veins do not correspond to those of arteries. Arteries usually follow definite pathways; veins are more difficult to follow because they connect in irregular networks in which many tributaries merge to form a large vein.

Although only one systemic artery, the aorta, takes oxygenated blood away from the heart (left ventricle), three systemic veins [the **coronary sinus**, **superior vena cava** (KĀ-va), and **inferior vena cava**] return deoxygenated blood to the right atrium of the heart (Figure 20.22; Table 20.9). The coronary sinus receives blood from the cardiac veins draining the heart wall. The superior vena cava receives blood from other veins superior to the diaphragm, except the air sacs (alveoli) of the lungs. The inferior vena cava receives blood from veins inferior to the diaphragm.

FIGURE 20.22 **Principal veins.**

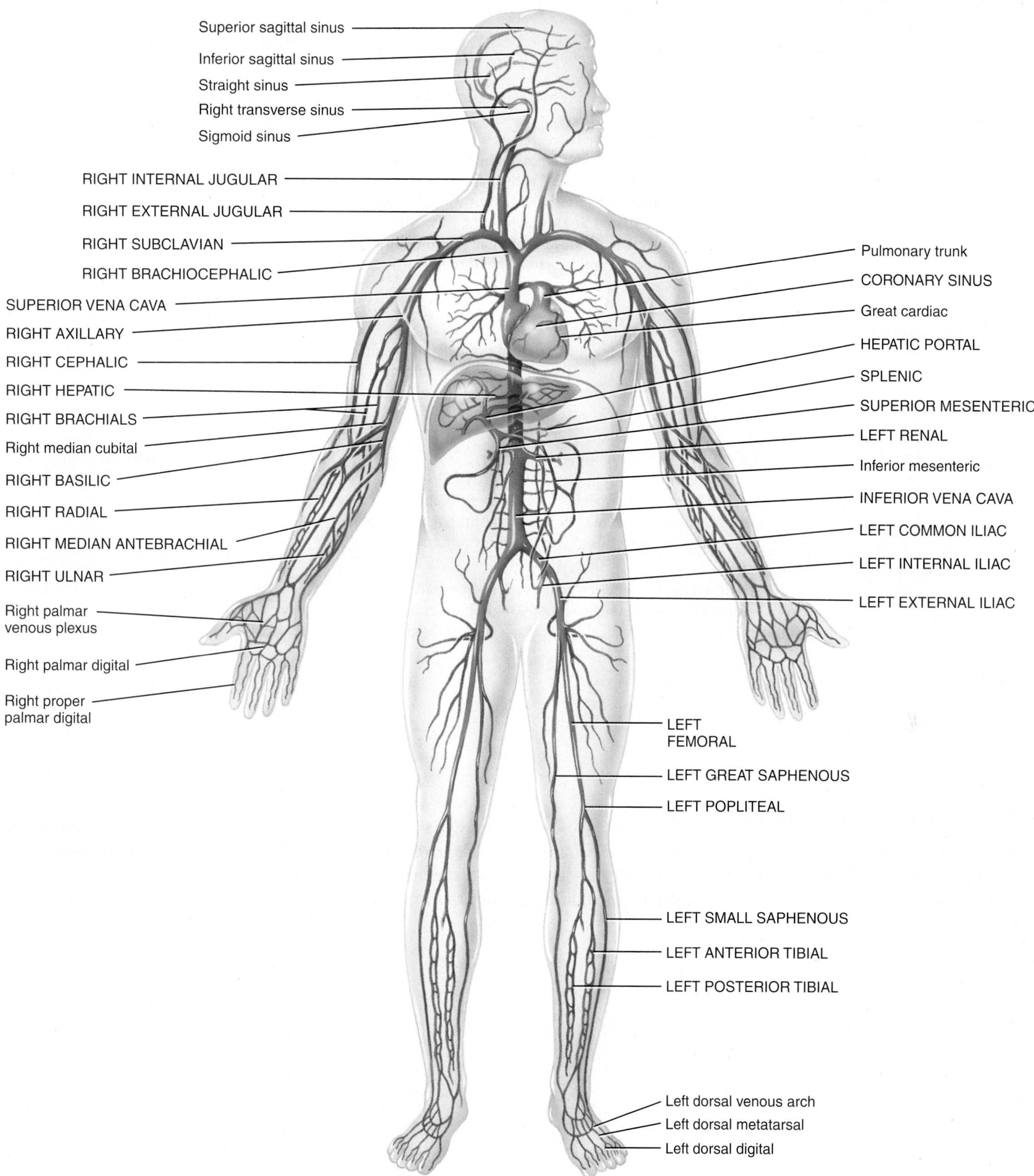


Overall anterior view of principal veins

**Deoxygenated blood returns to the heart via the superior vena cava, inferior vena cava, and coronary sinus.**

**TABLE 20.9**

Veins of the Systemic Circulation

| VEINS | DESCRIPTION AND TRIBUTARIES | REGIONS DRAINED |
|---|---|---|
| **Coronary sinus** (KOR-ō-nar-ē; *corona* = crown) | Main vein of heart; receives almost all venous blood from myocardium; located in coronary sulcus (see Figure 19.3c) on posterior aspect of heart and opens into right atrium between orifice of inferior vena cava and tricuspid valve; wide venous channel into which three veins drain; receives **great cardiac vein** (from anterior interventricular sulcus) into its left end, and **middle cardiac vein** (from posterior interventricular sulcus) and **small cardiac vein** into its right end; several **anterior cardiac veins** drain directly into right atrium | All tissues of heart |
| **Superior vena cava (SVC)** (VĒ-na KĀ-va; *vena* = vein; *cava* = cavelike) | About 7.5 cm (3 in.) long and 2 cm (1 in.) in diameter; empties its blood into superior part of right atrium; begins posterior to right first costal cartilage by union of right and left brachiocephalic veins and ends at level of right third costal cartilage, where it enters right atrium | Head, neck, upper limbs, and thorax |
| **Inferior vena cava (IVC)** | Largest vein in body, about 3.5 cm (1.4 in.) in diameter; begins anterior to fifth lumbar vertebra by union of common iliac veins, ascends behind peritoneum to right of midline, pierces caval opening of diaphragm at level of eighth thoracic vertebra, and enters inferior part of right atrium<br>**Clinical note:** The inferior vena cava is commonly **compressed during the later stages of pregnancy** by the enlarging uterus, producing edema of the ankles and feet and temporary varicose veins. | Abdomen, pelvis, and lower limbs |

## Veins of the Head and Neck

Most blood draining from the head passes into three pairs of veins: the **internal jugular** (JUG-ū-lar), **external jugular**, and **vertebral veins** (Figure 20.23; Table 20.10). Within the cranial cavity of the skull, all veins drain into dural venous sinuses and then into the internal jugular veins. **Dural venous sinuses** are venous channels between layers of the dura mater that surrounds the brain.

FIGURE 20.23 **Principal veins of the head and neck.**

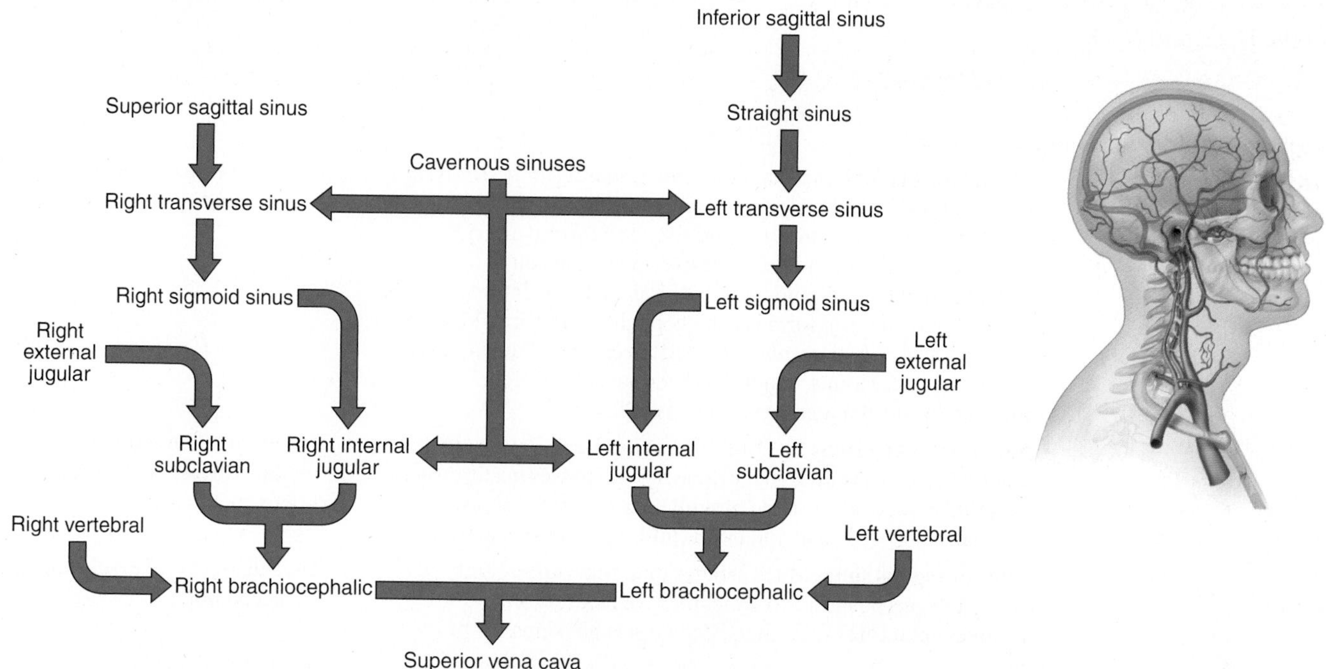


(a) Scheme of drainage

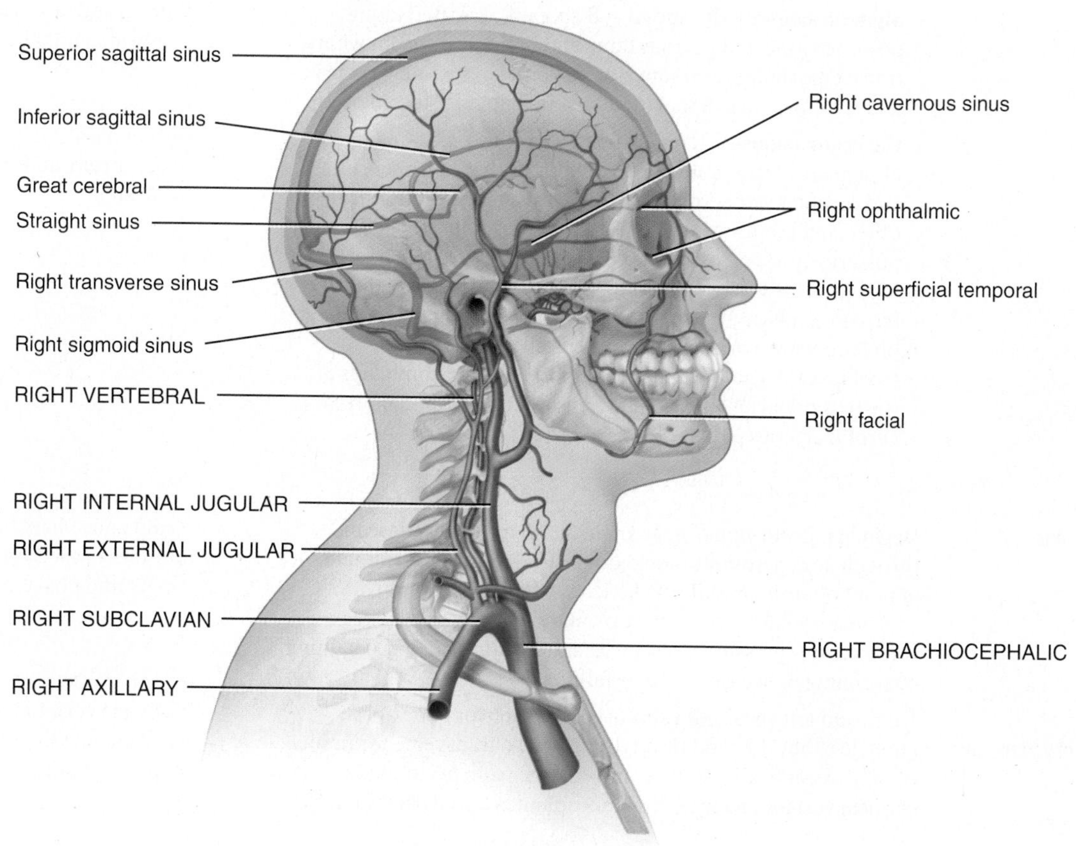


(b) Right lateral view

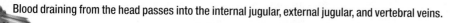

**Blood draining from the head passes into the internal jugular, external jugular, and vertebral veins.**

**TABLE 20.10**

Veins of the Head and Neck

| VEINS | DESCRIPTION AND TRIBUTARIES | REGIONS DRAINED |
|---|---|---|
| **Brachiocephalic veins** | (See Table 20.12) | |
| **Internal jugular veins** (JUG-ū-lar = throat) | Begin at base of cranium as sigmoid sinus and inferior petrosal sinus converge at opening of the jugular foramen; descend within carotid sheath lateral to internal and common carotid arteries, deep to sternocleidomastoid muscles; receive numerous tributaries from the face and neck; internal jugular veins anastomose with subclavian veins to form brachiocephalic veins (brā′ -kē-ō-se-FAL-ik; *brachi-* = arm; *-cephal-* = head) deep and slightly lateral to sternoclavicular joints; major dural venous sinuses that contribute to internal jugular vein are as follows: | Brain, meninges, bones of cranium, muscles and tissues of face and neck |
| | • **Superior sagittal sinus** (SAJ-i-tal = arrow). Begins at frontal bone, where it receives vein from nasal cavity, and passes posteriorly to occipital bone along midline of skull deep to sagittal sinus. It usually angles to right and drains into right transverse sinus. | Nasal cavity; superior, lateral, and medial aspects of cerebrum; skull bones; meninges |
| | • **Inferior sagittal sinus**. Much smaller than superior sagittal sinus, it begins posterior to attachment of falx cerebri and receives great cerebral vein to become straight sinus. | Medial aspects of cerebrum and diencephalon |
| | • **Straight sinus**. Runs in tentorium cerebelli and is formed by union of inferior sagittal sinus and great cerebral vein. It typically drains into left transverse sinus. | Medial and inferior aspects of cerebrum and the cerebellum |
| | • **Sigmoid sinuses** (SIG-moyd = S-shaped). Located along posterior aspect of petrous temporal bone. They begin where transverse sinuses and superior petrosal sinuses anastomose and terminate in internal jugular vein at jugular foramen. | Lateral and posterior aspect of cerebrum and the cerebellum |
| | • **Cavernous sinuses** (KAV-er-nus = cavelike). Located on either side of sphenoid bone. Ophthalmic veins from orbits and cerebral veins from cerebral hemispheres, along with other small sinuses, empty into cavernous sinuses. They drain posteriorly to petrosal sinuses to eventually return to internal jugular veins. Cavernous sinuses are unique because they have major blood vessels and nerves passing through them on their way to orbit and face. Oculomotor (III) nerve, trochlear (IV) nerve, ophthalmic and maxillary branches of the trigeminal (V) nerve, abducens (VI) nerve, and internal carotid arteries pass through cavernous sinuses. | Orbits, nasal cavity, frontal regions of cerebrum, and superior aspect of brain stem |
| **Subclavian veins** | (See Table 20.11) | |
| **External jugular veins** | Begin in parotid glands near angle of the mandible; descend through neck across sternocleidomastoid muscles; terminate at point opposite middle of clavicles, where they empty into subclavian veins; become very prominent along side of neck when venous pressure rises, for example, during heavy coughing or straining or in cases of heart failure | Scalp and skin of head and neck, muscles of face and neck, and oral cavity and pharynx |
| **Vertebral veins** (VER-te-bral = of vertebrae) | Right and left **vertebral veins** originate inferior to occipital condyles; they descend through successive transverse foramina of first six cervical vertebrae and emerge from foramina of sixth cervical vertebra to enter brachiocephalic veins in root of neck | Cervical vertebrae, cervical spinal cord and meninges, and some deep muscles in neck |

**✓ CHECKPOINT**

**37.** Which general regions of the body are drained by the superior vena cava and the inferior vena cava?

**38.** Which vessels drain into the coronary sinus?

**39.** Into which veins in the neck does all venous blood from the head and neck drain?

**40.** Which general areas are drained by the internal jugular, external jugular, and vertebral veins?

## Veins of the Upper Limbs

Both superficial and deep veins return blood from the upper limbs to the heart (Figure 20.24; Table 20.11). **Superficial veins** are located just deep to the skin and are often visible. They anastomose extensively with one another and with deep veins, and they do not accompany arteries. Superficial veins are larger than deep veins and return most of the blood from the upper limbs. **Deep veins** are located deep in the body. They usually accompany arteries and have the same names as the corresponding arteries. Both superficial and deep veins have valves, but valves are more numerous in the deep veins.

FIGURE 20.24 **Principal veins of the right upper limb.**

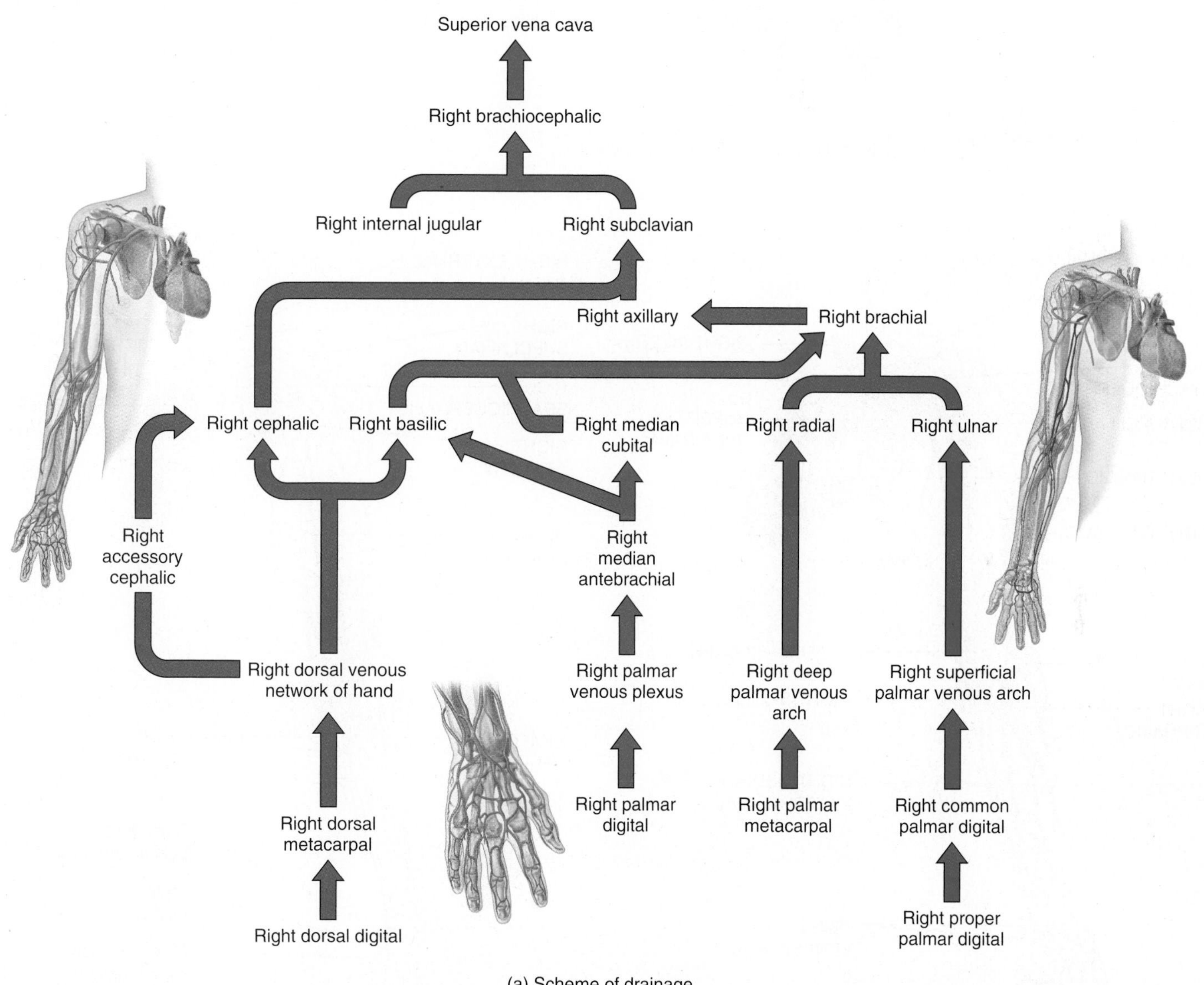


(a) Scheme of drainage

(continues)

**FIGURE 20.24 (continued)**

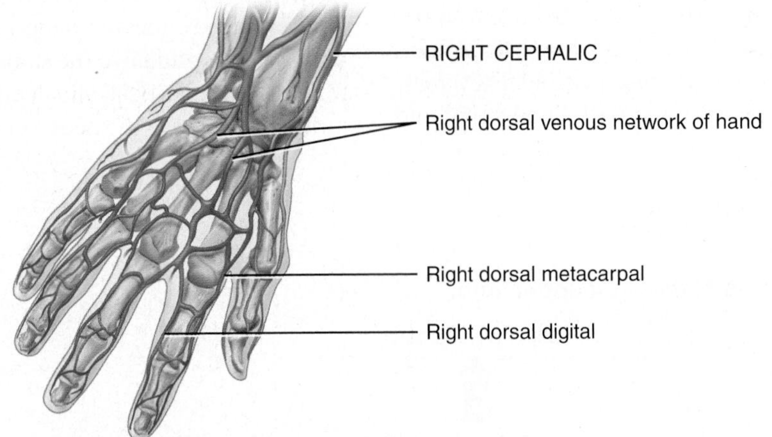


(b) Posterior view of superficial veins of hand



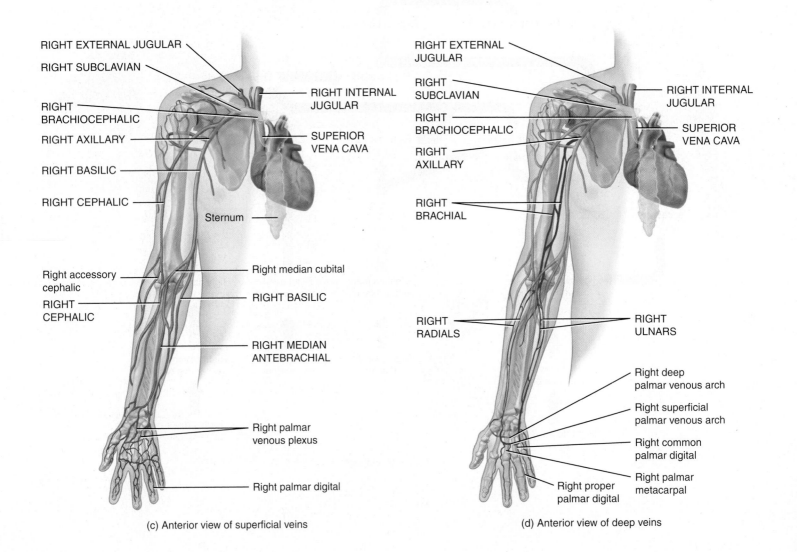

(c) Anterior view of superficial veins



(d) Anterior view of deep veins

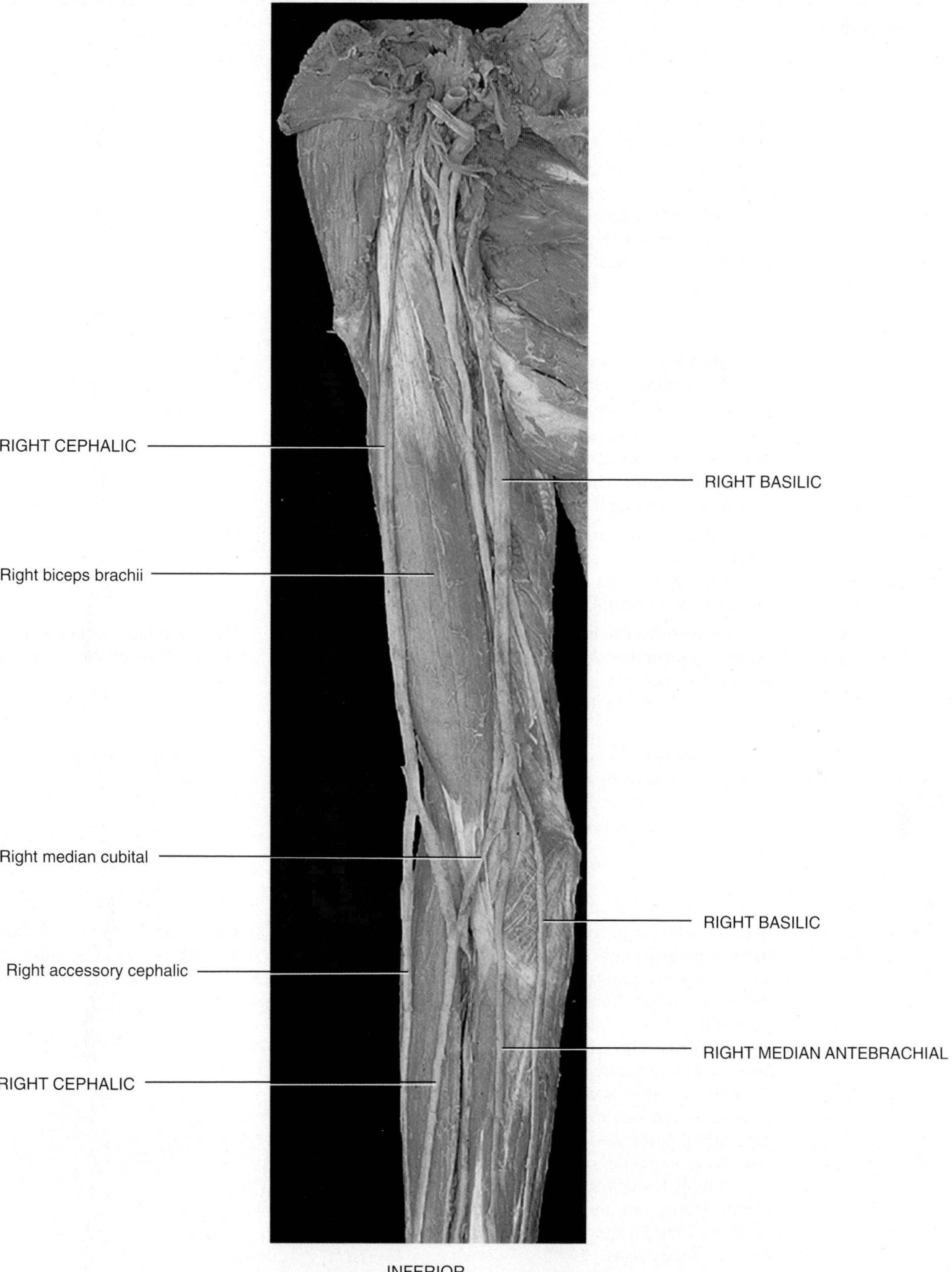


(e) Anterior view of superficial veins of arm and forearm

**Superficial veins of the upper limb anastomose extensively.**

## TABLE 20.11

### Veins of the Upper Limb

| VEINS | DESCRIPTION AND TRIBUTARIES | REGIONS DRAINED |
|---|---|---|
| **DEEP VEINS** | | |
| **Brachiocephalic veins** | (See Table 20.12) | |
| **Subclavian veins** (sub-KLĀ-vē-an; *sub-* = under; *-clavian* = pertaining to clavicle) | Continuations of axillary veins; pass over first rib deep to clavicle to terminate at sternal end of clavicle, where they unite with internal jugular veins to form brachiocephalic veins; thoracic duct of lymphatic system delivers lymph into junction between left subclavian and left internal jugular veins; right lymphatic duct delivers lymph into junction between right subclavian and right internal jugular veins (see Figure 21.3)<br>**Clinical note**: In a procedure called **central line placement**, the right subclavian vein is frequently used to administer nutrients and medication and measure venous pressure. | Skin, muscles, bones of arms, shoulders, neck, and superior thoracic wall |
| **Axillary veins** (AK-sil-ār-ē; *axilla* = armpit) | Arise as brachial veins and basilic veins unite near base of axilla (armpit); ascend to outer borders of first ribs, where they become subclavian veins; receive numerous tributaries in axilla that correspond to branches of axillary arteries | Skin, muscles, bones of arm, axilla, shoulder, and superolateral chest wall |
| **Brachial veins** (BRĀ-kē-al; *brachi-* = arm) | Accompany brachial arteries; begin in anterior aspect of elbow region where radial and ulnar veins join one another; as they ascend through arm, basilic veins join them to form axillary vein near distal border of teres major muscle | Muscles and bones of elbow and brachial regions |
| **Ulnar veins** (UL-nar = pertaining to ulna) | Begin at **superficial palmar venous arches**, which drain **common palmar digital veins** and **proper palmar digital veins** in fingers; course along medial aspect of forearms, pass alongside ulnar arteries, and join with radial veins to form brachial veins | Muscles, bones, and skin of hand, and muscles of medial aspect of forearm |
| **Radial veins** (RĀ-dē-al = pertaining to radius) | Begin at **deep palmar venous arches** (Figure 20.24d), which drain **palmar metacarpal veins** in palms; drain lateral aspects of forearms and pass alongside radial arteries; unite with ulnar veins to form brachial veins just inferior to elbow joint | Muscles and bones of lateral hand and forearm |
| **SUPERFICIAL VEINS** | | |
| **Cephalic veins** (se-FAL-ik = pertaining to head) | Begin on lateral aspect of **dorsal venous networks of hands (dorsal venous arches)**; networks of veins on dorsum of hands formed by dorsal metacarpal veins (Figure 20.24b); these veins in turn drain **dorsal digital veins**, which pass along sides of fingers; arch around radial side of forearms to anterior surface and ascend through entire limbs along anterolateral surface; end where they join axillary veins, just inferior to clavicles; **accessory cephalic veins** originate either from venous plexus on dorsum of forearms or from medial aspects of dorsal venous networks of hands, and unite with cephalic veins just inferior to elbow; after receiving median cubital veins, basilic veins continue ascending until they reach middle of arm; there they penetrate tissues deeply and run alongside brachial arteries until they join with brachial veins to form axillary veins | Integument and superficial muscles of lateral aspect of upper limb |

| | | |
|---|---|---|
| **Basilic veins** (ba-SIL-ik = royal, of prime importance) | Begin on medial aspects of dorsal venous networks of hands and ascend along posteromedial surface of forearm and anteromedial surface of arm (Figure 20.24c); connected to cephalic veins anterior to elbow by **median cubital veins** (*cubital* = pertaining to elbow)<br>**Clinical note**: If veins must be **punctured** for an injection, transfusion, or removal of a blood sample, the median cubital veins are preferred | Integument and superficial muscles of medial aspect of upper limb |
| **Median antebrachial veins (median veins of forearm)** (an′-tē-BRĀ-kē-al; *ante-* = before, in front of; *brachi-* = arm) | Begin in **palmar venous plexuses**, networks of veins on palms; drain **palmar digital veins** in fingers; ascend anteriorly in forearms to join basilic or median cubital veins, sometimes both | Integument and superficial muscles of palm and anterior aspect of upper limb |

✓ **CHECKPOINT**

**41.** Which regions of the upper limb are drained by the cephalic, basilic, median antebrachial, radial, and ulnar veins?

**42.** Which vein provides collateral circulation between the cephalic and basilic veins?

**43.** From which vein in the upper limb is a blood sample often taken?

**44.** Into which veins does lymph drain from the lymphatic system into the circulatory system?

## Veins of the Thorax

Although the brachiocephalic veins drain some portions of the thorax, most thoracic structures are drained by a network of veins, called the **azygos system** (AZ-i-gus or a-ZĪ-gus), that runs on either side of the vertebral column (Figure 20.25; Table 20.12). The system consists of three veins—the **azygos**, **hemiazygos**, and **accessory hemiazygos veins**—that show considerable variation in origin, course, tributaries, anastomoses, and termination. Ultimately they empty into the superior vena cava.

In addition to collecting blood from the thorax and abdominal wall, the azygos system may serve as a bypass for the inferior vena cava, which drains blood from the lower body. Several small veins directly link the azygos system with the inferior vena cava. Larger veins that drain the lower limbs and abdomen also connect into the azygos system. If the inferior vena cava or hepatic portal vein becomes obstructed, blood that typically passes through the inferior vena cava can detour into the azygos system to return blood from the lower body to the superior vena cava.

**FIGURE 20.25** **Principal veins of the thorax, abdomen, and pelvis.**

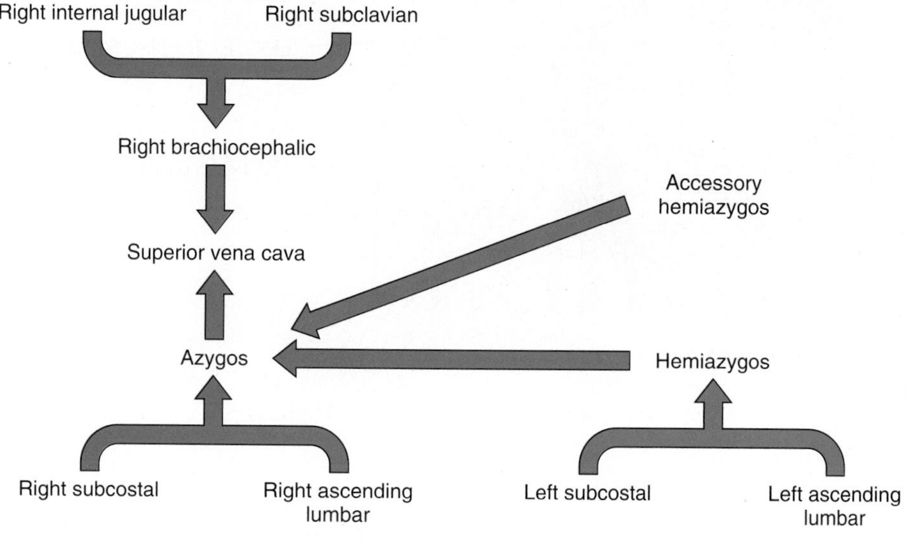


(a) Scheme of drainage

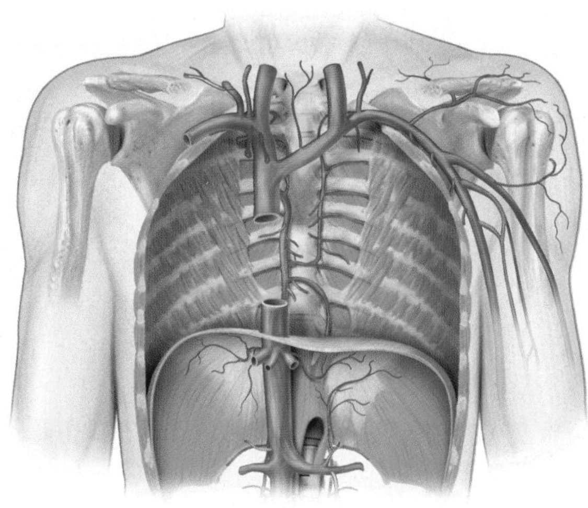

(continues)

FIGURE 20.25 (continued)

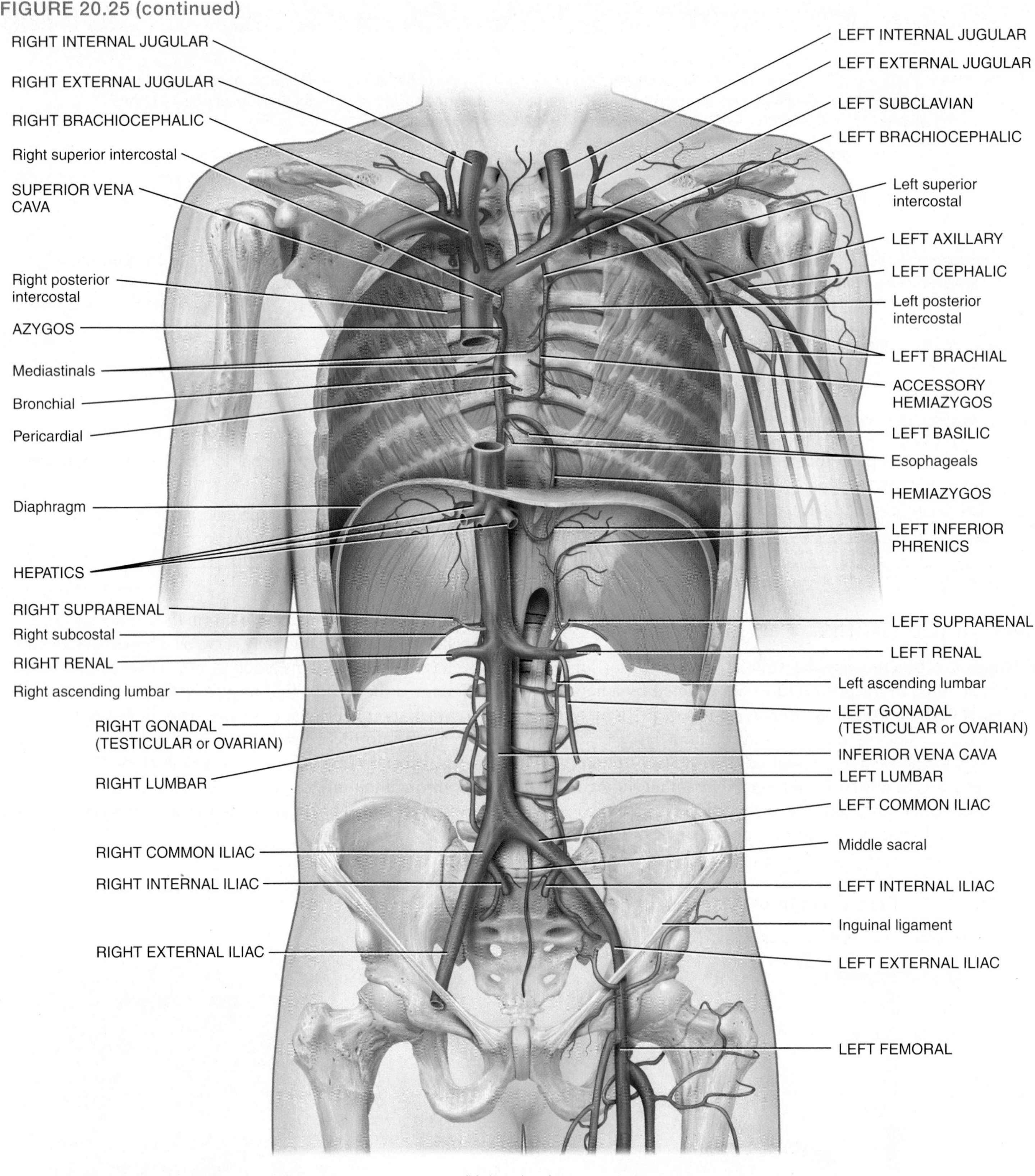


(b) Anterior view

**Most thoracic structures are drained by the azygos system of veins.**

**TABLE 20.12**

Veins of the Thorax

| VEINS | DESCRIPTION AND TRIBUTARIES | REGIONS DRAINED |
|---|---|---|
| **Brachiocephalic veins** (brā′-kē-ō-se-FAL-ik; *brachio-* = arm; *-cephalic* = pertaining to head) | Form by union of subclavian and internal jugular veins; two brachiocephalic veins unite to form superior vena cava; because superior vena cava is to right of body's midline, left brachiocephalic vein is longer than right; right brachiocephalic vein is anterior and to right of brachiocephalic trunk and follows more vertical course; left brachiocephalic vein is anterior to brachiocephalic trunk, left common carotid and left subclavian arteries, trachea, left vagus (X) nerve, and phrenic nerve; it approaches a more horizontal position as it passes from left to right | Head, neck, upper limbs, mammary glands, and superior thorax |
| **Azygos vein** (az-Ī-gus = unpaired) | An unpaired vein that is anterior to vertebral column, slightly to right of midline; usually begins at junction of right ascending lumbar and right subcostal veins near diaphragm; arches over root of right lung at level of fourth thoracic vertebra to end in superior vena cava; receives the following tributaries: **right posterior intercostal**, **hemiazygos**, **accessory hemiazygos**, **esophageal**, **mediastinal**, **pericardial**, and **bronchial veins** | Right side of thoracic wall, thoracic viscera, and posterior abdominal wall |
| **Hemiazygos vein** (HEM-ē-ā-zī-gus; *hemi-* = half) | Anterior to vertebral column and slightly to left of midline; often begins at junction of left ascending lumbar and left subcostal veins; terminates by joining azygos vein at about level of ninth thoracic vertebra; receives following tributaries: ninth through eleventh **left posterior intercostal**, **esophageal**, **mediastinal**, and sometimes **accessory hemiazygos veins** | Left side of lower thoracic wall, thoracic viscera, and left posterior abdominal wall |
| **Accessory hemiazygos vein** | Anterior to vertebral column and to left of midline; begins at fourth or fifth intercostal space and descends from fifth to eighth thoracic vertebra or ends in hemiazygos vein; terminates by joining azygos vein at about level of eighth thoracic vertebra; receives the following tributaries: fourth through eighth **left posterior intercostal veins** (first through third posterior intercostal veins drain into left brachiocephalic vein), **left bronchial**, and **mediastinal veins** | Left side of upper thoracic wall and thoracic viscera |

✓ **CHECKPOINT**

**45.** What is the importance of the azygos system relative to the inferior vena cava?

**46.** Which components of the azygos system drain the left side of the thorax? Into which vein do they drain?

## Veins of the Abdomen and Pelvis

Blood from the abdominal and pelvic viscera and abdominal wall returns to the heart through the inferior vena cava (Figure 20.26; Table 20.13). Many small veins enter the inferior vena cava. Most carry return flow from parietal branches of the abdominal aorta, and their names correspond to the names of the arteries (see also Figure 20.25b).

The inferior vena cava does not receive veins directly from the gastrointestinal tract, spleen, pancreas, and gallbladder. These organs pass their blood into a common vein, the hepatic portal vein, which delivers the blood to the liver. The superior mesenteric and splenic veins unite to form the hepatic portal vein (see Figure 20.28b). This special flow of venous blood, called the hepatic portal circulation, is described shortly. After passing through the liver for processing, blood drains into the hepatic veins, which empty into the inferior vena cava.

FIGURE 20.26 **Drainage into the inferior vena cava.**

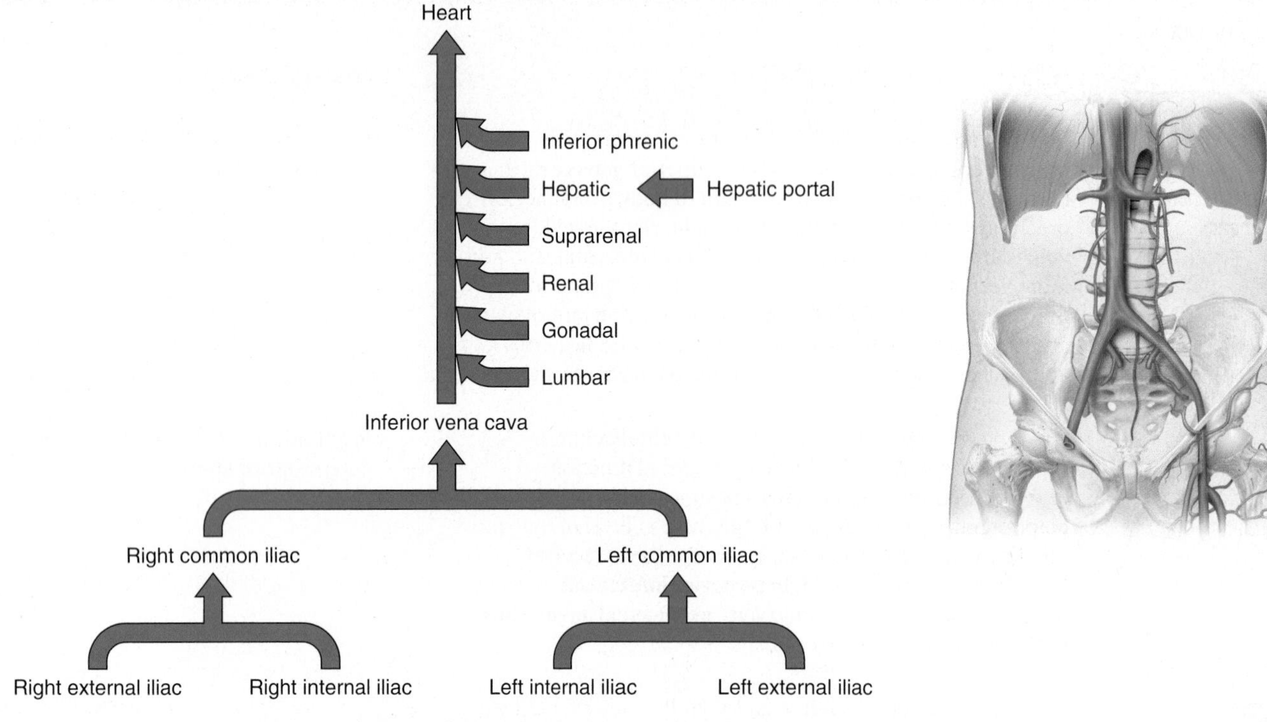


Scheme of drainage

**The hepatic portal vein delivers blood to the liver from the gastrointestinal tract, spleen, pancreas, and gallbladder.**

**TABLE 20.13**

Veins of the Abdomen and Pelvis

| VEINS | DESCRIPTION AND TRIBUTARIES | REGIONS DRAINED |
|---|---|---|
| **Inferior vena cava** | (See Table 20.9) | |
| **Inferior phrenic veins** (FREN-ik = pertaining to diaphragm) | Arise on inferior surface of diaphragm; left inferior phrenic vein usually sends one tributary to left suprarenal vein, which empties into left renal vein, and another tributary into inferior vena cava; right inferior phrenic vein empties into inferior vena cava | Inferior surface of diaphragm and adjoining peritoneal tissues |
| **Hepatic veins** (he-PAT-ik = pertaining to liver) | Typically two or three in number; drain sinusoidal capillaries of liver; capillaries of liver receive venous blood from capillaries of gastrointestinal organs via hepatic portal vein; **hepatic portal vein** receives the following tributaries from gastrointestinal organs: | |
| | • **Left gastric vein**. Arises from left side of lesser curvature of stomach and joins left side of hepatic portal vein in lesser omentum. | Terminal esophagus, stomach, liver, gallbladder, spleen, pancreas, small intestine, and large intestine |
| | • **Right gastric vein**. Arises from right aspect of lesser curvature of stomach and joins hepatic portal vein on its anterior surface within lesser omentum. | Lesser curvature of stomach, abdominal portion of esophagus, stomach, and duodenum |
| | • **Splenic vein**. Arises in spleen and crosses abdomen transversely posterior to stomach to anastomose with superior mesenteric vein to form hepatic portal vein. It receives **inferior mesenteric vein** near its junction with hepatic portal vein. | Spleen, fundus and greater curvature of stomach, pancreas, greater omentum, descending colon, sigmoid colon, and rectum |
| | • **Superior mesenteric vein**. Arises from numerous tributaries from most of small intestine and first half of large intestine and ascends to join splenic vein to form hepatic portal vein. | Duodenum, jejunum, ileum, cecum, appendix, ascending colon, and transverse colon |

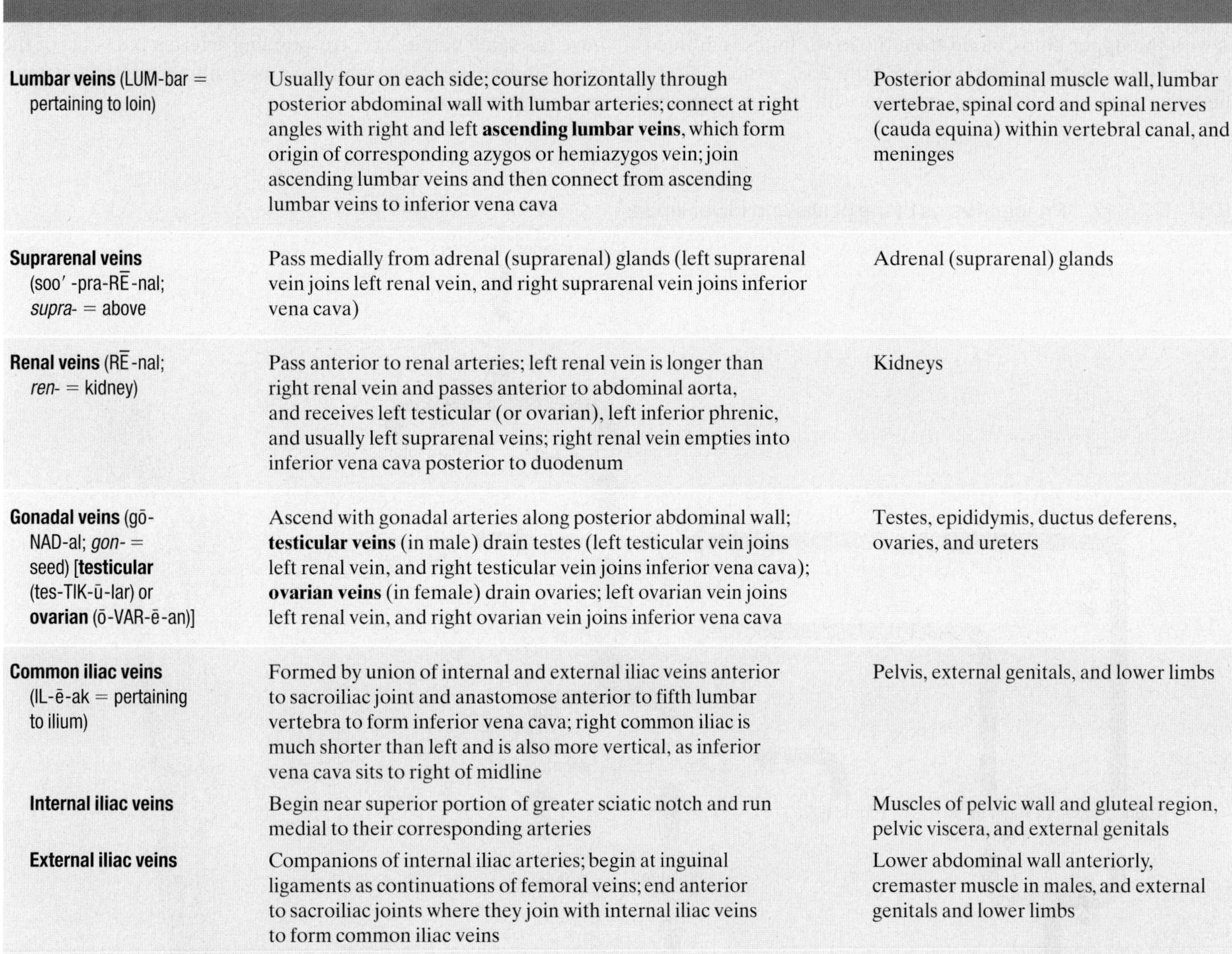

| | | |
|---|---|---|
| **Lumbar veins** (LUM-bar = pertaining to loin) | Usually four on each side; course horizontally through posterior abdominal wall with lumbar arteries; connect at right angles with right and left **ascending lumbar veins**, which form origin of corresponding azygos or hemiazygos vein; join ascending lumbar veins and then connect from ascending lumbar veins to inferior vena cava | Posterior abdominal muscle wall, lumbar vertebrae, spinal cord and spinal nerves (cauda equina) within vertebral canal, and meninges |
| **Suprarenal veins** (soo′ -pra-RĒ-nal; *supra-* = above | Pass medially from adrenal (suprarenal) glands (left suprarenal vein joins left renal vein, and right suprarenal vein joins inferior vena cava) | Adrenal (suprarenal) glands |
| **Renal veins** (RĒ-nal; *ren-* = kidney) | Pass anterior to renal arteries; left renal vein is longer than right renal vein and passes anterior to abdominal aorta, and receives left testicular (or ovarian), left inferior phrenic, and usually left suprarenal veins; right renal vein empties into inferior vena cava posterior to duodenum | Kidneys |
| **Gonadal veins** (gō-NAD-al; *gon-* = seed) [**testicular** (tes-TIK-ū-lar) or **ovarian** (ō-VAR-ē-an)] | Ascend with gonadal arteries along posterior abdominal wall; **testicular veins** (in male) drain testes (left testicular vein joins left renal vein, and right testicular vein joins inferior vena cava); **ovarian veins** (in female) drain ovaries; left ovarian vein joins left renal vein, and right ovarian vein joins inferior vena cava | Testes, epididymis, ductus deferens, ovaries, and ureters |
| **Common iliac veins** (IL-ē-ak = pertaining to ilium) | Formed by union of internal and external iliac veins anterior to sacroiliac joint and anastomose anterior to fifth lumbar vertebra to form inferior vena cava; right common iliac is much shorter than left and is also more vertical, as inferior vena cava sits to right of midline | Pelvis, external genitals, and lower limbs |
| **Internal iliac veins** | Begin near superior portion of greater sciatic notch and run medial to their corresponding arteries | Muscles of pelvic wall and gluteal region, pelvic viscera, and external genitals |
| **External iliac veins** | Companions of internal iliac arteries; begin at inguinal ligaments as continuations of femoral veins; end anterior to sacroiliac joints where they join with internal iliac veins to form common iliac veins | Lower abdominal wall anteriorly, cremaster muscle in males, and external genitals and lower limbs |

## ✓ CHECKPOINT

**47.** Which vein returns blood from the abdominopelvic viscera to the heart?

**48.** Which structures are drained by the lumbar, gonadal, renal, suprarenal, inferior phrenic, and hepatic veins?

## Veins of the Lower Limbs

As with the upper limbs, blood from the lower limbs is drained by both superficial and deep veins (Figure 20.27; Table 20.14). The superficial veins often anastomose with one another and with deep veins along their length. Deep veins, for the most part, have the same names as corresponding arteries. All veins of the lower limbs have valves, which are more numerous than in veins of the upper limbs.

FIGURE 20.27 **Principal veins of the pelvis and lower limbs.**

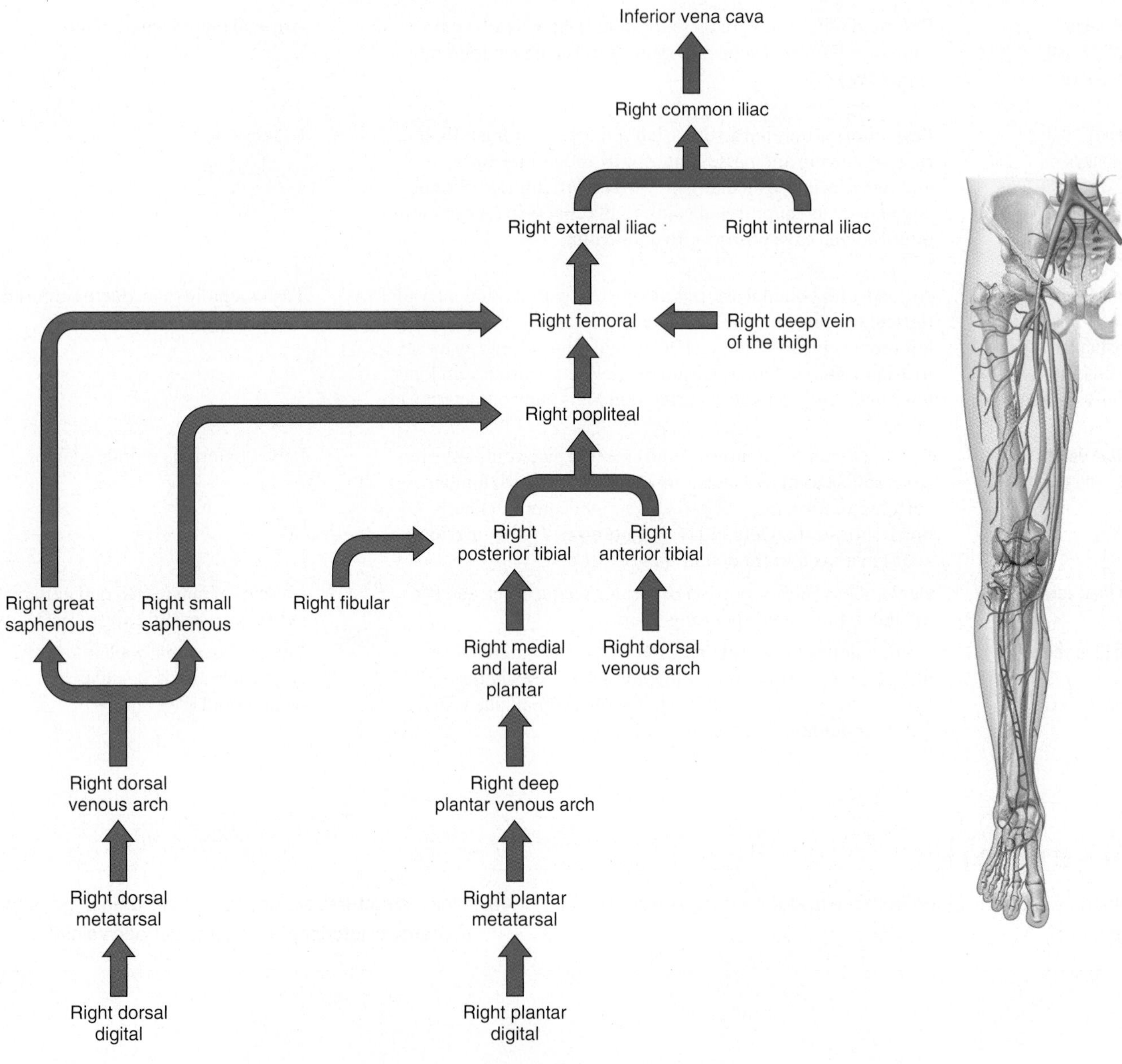


(a) Scheme of drainage

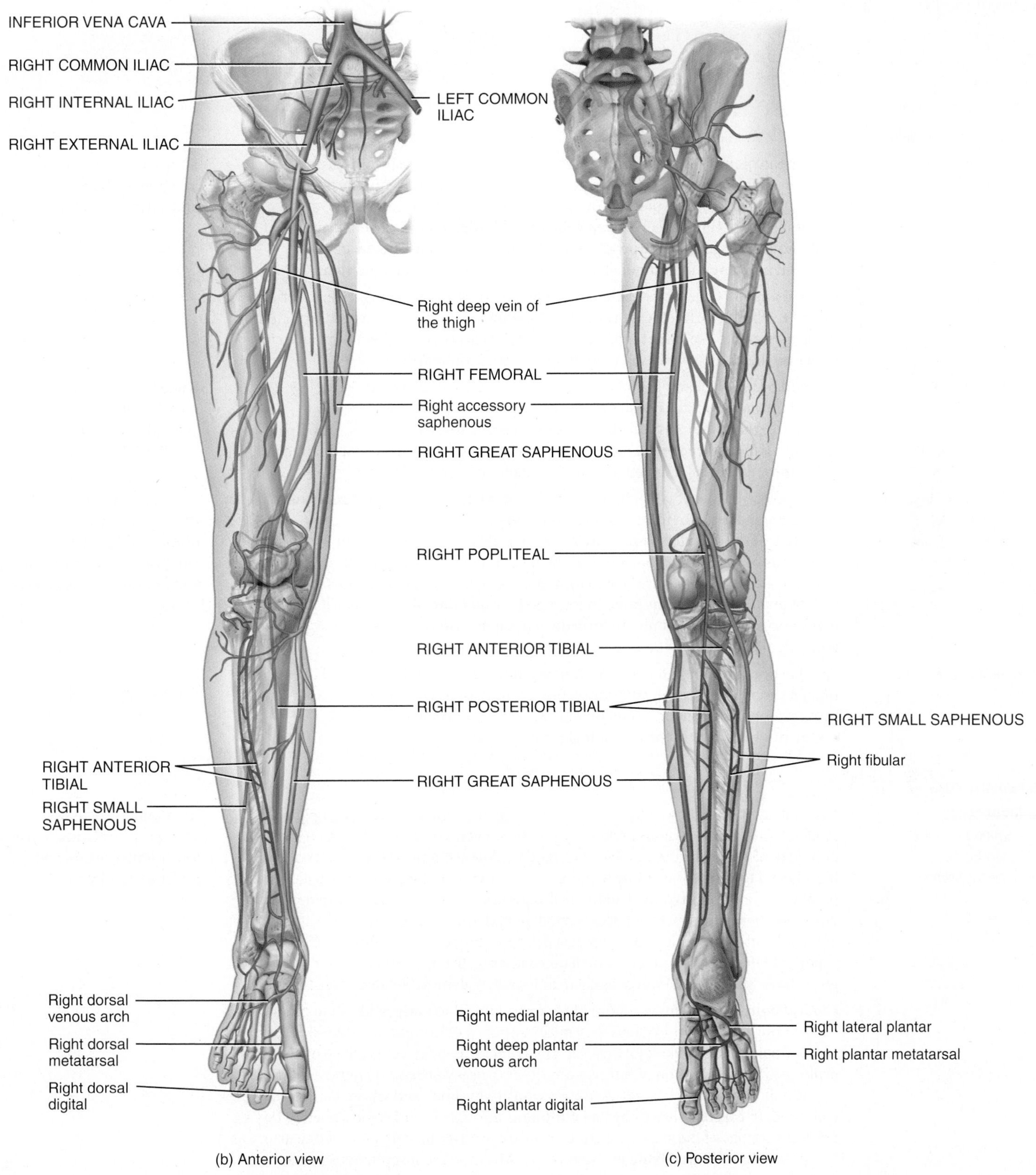


(b) Anterior view

(c) Posterior view

**Deep veins usually bear the names of their companion arteries.**

**TABLE 20.14**

Veins of the Lower Limbs

| VEINS | DESCRIPTION AND TRIBUTARIES | REGIONS DRAINED |
|---|---|---|
| **DEEP VEINS** | | |
| **Common iliac veins** | (See Table 20.13) | |
| **External iliac veins** | (See Table 20.13) | |
| **Femoral veins** (FEM-o-ral) | Accompany femoral arteries and are continuations of popliteal veins just superior to knee where veins pass through opening in adductor magnus muscle; ascend deep to sartorius muscle and emerge from beneath muscle in femoral triangle at proximal end of thigh; receive **deep veins of thigh (deep femoral veins)** and great saphenous veins just before penetrating abdominal wall; pass below inguinal ligament and enter abdominopelvic region to become external iliac veins | Skin, lymph nodes, muscles, and bones of thigh, and external genitals |
| | **Clinical note**: In order to take **blood samples** or **pressure recordings** from the right side of the heart, a catheter is inserted into the femoral vein as it passes through the femoral triangle. The catheter passes through the external and common iliac veins, then into the inferior vena cava, and finally into the right atrium. | |
| **Popliteal veins** (pop′-li-TĒ-al = pertaining to hollow behind knee) | Formed by union of anterior and posterior tibial veins at proximal end of leg; ascend through popliteal fossa with popliteal arteries and tibial nerve; terminate where they pass through window in adductor magnus muscle and pass to front of knee to become femoral veins; also receive blood from small saphenous veins and tributaries that correspond to branches of popliteal artery | Knee joint and skin, muscles, and bones around knee joint |
| **Posterior tibial veins** (TIB-ē-al) | Begin posterior to medial malleolus at union of **medial** and **lateral plantar veins** from plantar surface of foot; ascend through leg with posterior tibial artery and tibial nerve deep to soleus muscle; join posterior tibial veins about two-thirds of way up leg; join anterior tibial veins near top and lateral aspects of interosseous membrane of leg to form popliteal veins; on plantar surface of foot **plantar digital veins** unite to form **plantar metatarsal veins**, which parallel metatarsals, and in turn unite to form **deep plantar venous arches**; medial and lateral plantar veins emerge from deep plantar venous arches | Skin, muscles, and bones on plantar surface of foot, and skin, muscles, and bones from posterior and lateral aspects of leg |
| **Anterior tibial veins** | Arise in dorsal venous arch and accompany anterior tibial artery; ascend deep to tibialis anterior muscle on anterior surface of interosseous membrane; pass through opening at superior end of interosseous membrane to join posterior tibial veins to form popliteal veins | Dorsal surface of foot, ankle joint, anterior aspect of leg, knee joint, and tibiofibular joint |
| **SUPERFICIAL VEINS** | | |
| **Great (long) saphenous veins** (sa-FĒ-nus = clearly visible) | Longest veins in body; ascend from foot to groin in subcutaneous layer; begin at medial end of dorsal venous arches of foot; **dorsal venous arches** (VĒ-nus) are networks of veins on dorsum of foot formed by **dorsal digital veins**, which collect blood from toes, and then unite in pairs to form **dorsal metatarsal veins**, which parallel metatarsals; as dorsal metatarsal veins approach foot, they combine to form dorsal venous arches; pass anterior to medial malleolus of tibia and then superiorly along medial aspect of leg and thigh just deep to skin; receive tributaries from superficial tissues and connect with deep veins as well; empty into femoral veins at groin; have from 10 to 20 valves along their length, with more located in leg than thigh | Integumentary tissues and superficial muscles of lower limbs, groin, and lower abdominal wall |
|  | **Clinical note**: These veins are more likely to be subject to **varicosities** than other veins in the lower limbs because they must support a long column of blood and are not well supported by skeletal muscles. The great saphenous veins are often used for prolonged administration of intravenous fluids. This is particularly important in very young children and in patients of any age who are in shock and whose veins are collapsed. In **coronary artery bypass grafting**, if multiple blood vessels need to be grafted, sections of the great saphenous vein are used along with at least one artery as a graft (see first **Clinical Note** in Table 20.5). After the great saphenous vein is removed and divided into sections, the sections are used to bypass the blockages. The vein grafts are reversed so that the valves do not obstruct the flow of blood. | |
| **Small saphenous veins** | Begin at lateral aspect of dorsal venous arches of foot; pass posterior to lateral malleolus of fibula and ascend deep to skin along posterior aspect of leg; empty into popliteal veins in popliteal fossa, posterior to knee; have from 9 to 12 valves; may communicate with great saphenous veins in proximal thigh | Integumentary tissues and superficial muscles of foot and posterior aspect of leg |

✓ CHECKPOINT

**49.** Which vein drains the medial side of the entire lower limb? The anterior leg? The knee joint?

**50.** What is the clinical importance of the great saphenous veins?

## Hepatic Portal Circulation

The **hepatic portal circulation** (hep′-A-tik) carries venous blood from the gastrointestinal organs and spleen to the liver (Figure 20.28). A vein that carries blood from one capillary network to another is called a *portal vein*. **The hepatic portal vein** (*hepat-* = liver) receives blood from capillaries of gastrointestinal organs and the spleen and delivers it to the sinusoids of the liver. After a meal, hepatic portal blood is rich in nutrients absorbed from the gastrointestinal tract. The liver stores some of them and modifies others before they pass into the general circulation. For example, the liver converts glucose into glycogen for storage, reducing blood glucose level shortly after a meal. The liver also detoxifies harmful substances, such as alcohol, that have been absorbed from the gastrointestinal tract and destroys bacteria by phagocytosis.

The splenic and superior mesenteric veins unite to form the hepatic portal vein. The **splenic vein** drains blood from the spleen, stomach, pancreas, and portions of the large intestine. The **superior mesenteric vein** (mes′-en-TER-ik) drains blood from the small intestine and portions of the large intestine, stomach, and pancreas. The *inferior mesenteric vein*, which

FIGURE 20.28 **Hepatic portal circulation.** A schematic diagram of blood flow through the liver, including arterial circulation, is shown in (a); deoxygenated blood is indicated in blue, oxygenated blood in red.

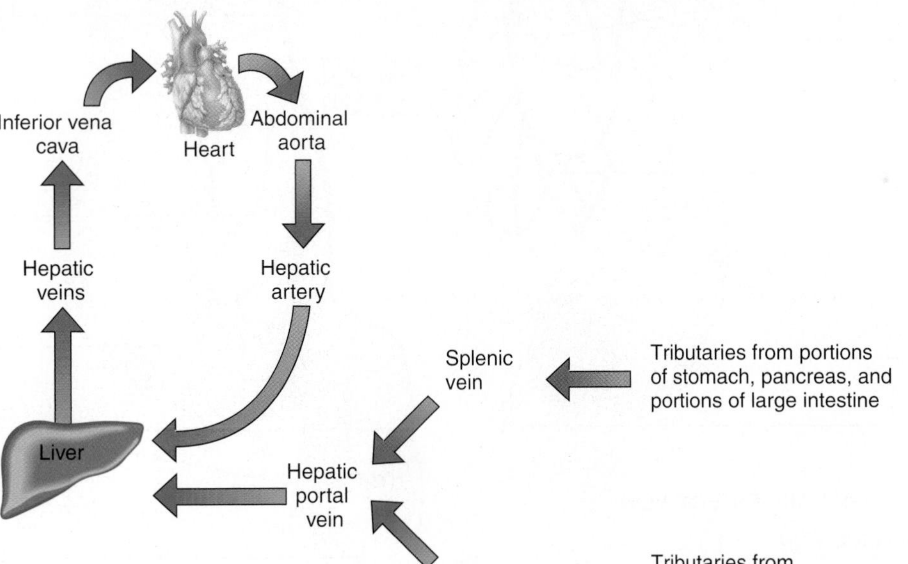


(a) Scheme of principal blood vessels of hepatic portal circulation and arterial supply and venous drainage of liver

(continues)

**FIGURE 20.28 (continued)**

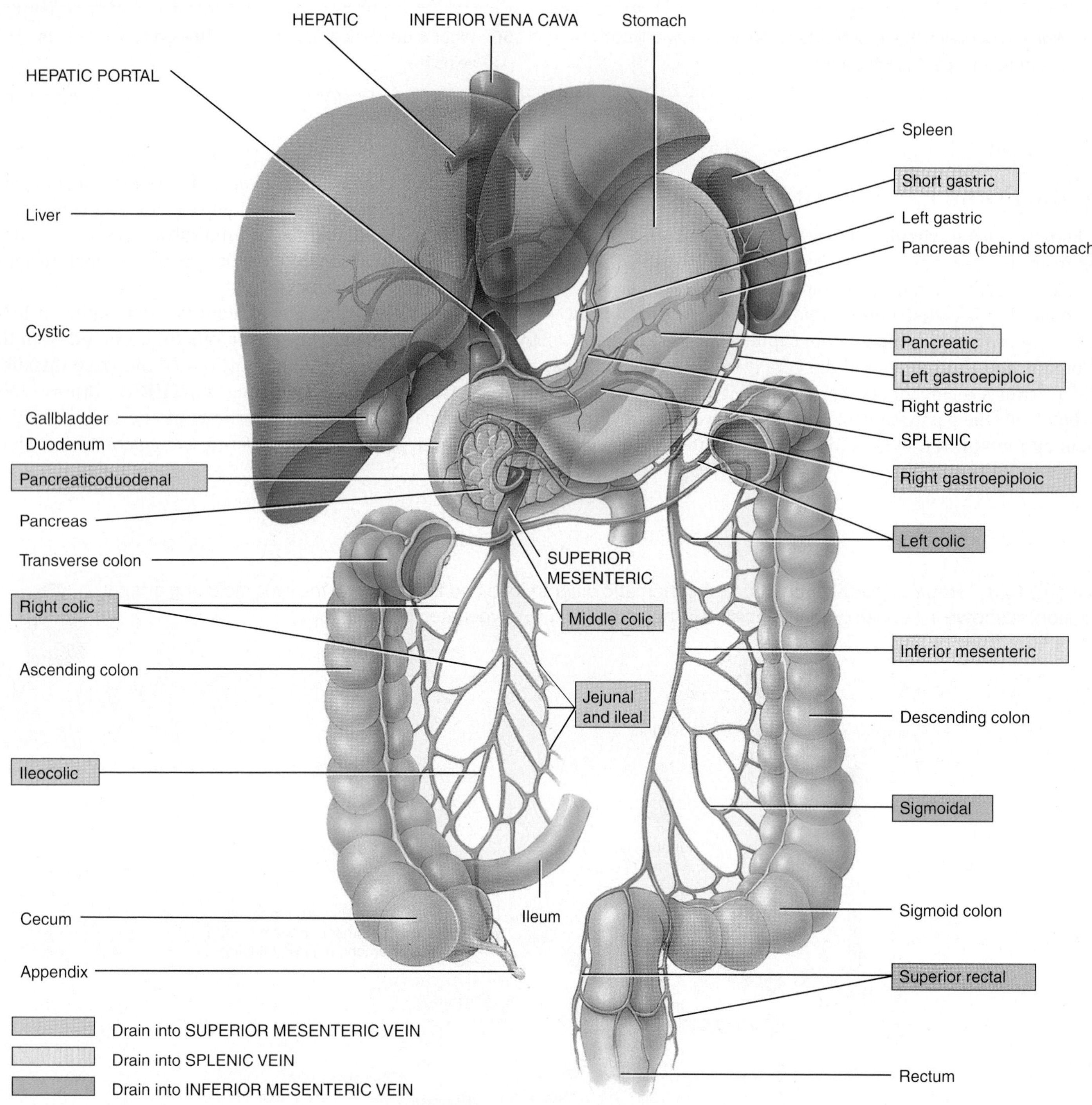


(b) Anterior view of veins draining into hepatic portal vein

**The hepatic portal circulation delivers venous blood from the organs of the gastrointestinal tract and spleen to the liver.**

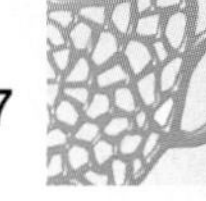

passes into the splenic vein, drains portions of the large intestine. The *right* and *left gastric veins*, which open directly into the hepatic portal vein, drain the stomach. The *cystic vein*, which also opens into the hepatic portal vein, drains the gallbladder.

At the same time the liver is receiving nutrient-rich but deoxygenated blood via the hepatic portal vein, it is also receiving oxygenated blood from the hepatic artery, a branch of the celiac trunk. The oxygenated blood mixes with the deoxygenated blood in sinusoids. Eventually, blood leaves the sinusoids of the liver through the **hepatic veins**, which drain into the inferior vena cava.

## ✓ CHECKPOINT

**51.** Which vein carries nutrient-rich blood to the liver? Which artery carries oxygen-rich blood to the liver?

**52.** Which veins carry blood away from the liver?

## David's Story

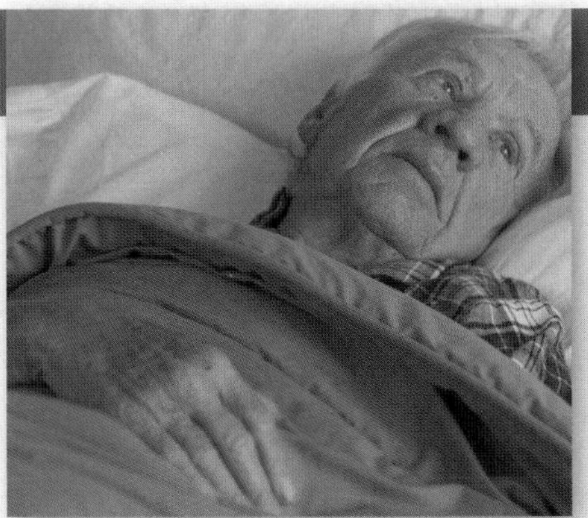

## EPILOGUE AND DISCUSSION

David, a retired man formerly in good health, is taken to the clinic complaining of fatigue, nausea, and vomiting. He denies that anything is wrong, but upon examination it is found that he is confused, has cool, clammy skin, rapid pulse and respirations, and is hypotensive (blood pressure = 90/60). He has not been producing a normal volume of urine, and indicates that he sustained an injury to his arm two weeks prior and had a fever one week prior to his exam. The wound appears infected. Palpation of the abdomen reveals enlargement of the liver, and capillary refill times are delayed. A diagnosis of septicemia with impending septic shock is made, and David is admitted to the hospital immediately. David's low blood pressure is due to reduced blood volume caused by damage to the capillaries initiated by bacterial toxins. The capillaries have become leaky, and plasma proteins and fluid have left the circulatory system and are infiltrating his tissues. David is placed on fluid replacement therapy to increase his blood volume, an antibiotic to address the potential infection, and oxygen to increase his blood oxygen level.

Following admission, David experiences a deep vein thrombosis in which fragments form an embolus that travels the venous circulatory system to David's lungs. The pulmonary embolism induces an acute respiratory crisis; David's blood oxygen level drops rapidly, and respiratory distress occurs. The treatment options available to David's doctor depend on the severity of the pulmonary artery obstruction. Treatment goals typically focus on dealing with acute symptoms and preventing further pulmonary emboli. In severe cases, surgery or treatment with clot-busting drugs may be tried. The overall mortality rate for pulmonary embolism is around 15 percent. In milder cases, such as David's, individuals can survive with pulmonary emboli, although with reduced oxygen exchange at the lung.

K. *Why would the loss of plasma proteins cause fluid to leave the capillaries and enter David's tissues?*

L. *David's low blood pressure has resulted from loss of blood volume. How did his body respond to the blood volume decrease?*

M. *How did the cardiovascular center of David's brain respond to his condition?*

N. *David's condition led to the formation of a pulmonary embolism, a blood clot that is blocking blood flow to his lungs. Why would this reduce his blood oxygen? Which pathway does blood travel to get to and from the lungs? How might this affect David's cardiac output?*

# Concept and Resource Summary

| Concept | Resources |
|---|---|
| **Introduction** | |
| 1. Blood vessels carry blood to all body tissues and then return it to the heart. | |

## Concept 20.1 Most blood vessel walls have three distinct tissue layers.

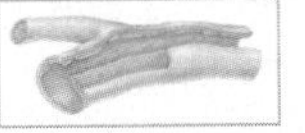

1. Blood vessel walls consist of an innermost **tunica interna**, which surrounds the **lumen** where blood flows, the middle **tunica media**, and the outermost **tunica externa**.
2. Sympathetic fibers of the autonomic nervous system regulate contraction of smooth muscle fibers of the tunica media. An increase in sympathetic stimulation produces **vasoconstriction** while a decrease causes **vasodilation**.
3. The extent of smooth muscle contraction helps regulate blood flow and blood pressure.
4. Contraction of the tunica media can decrease blood loss through a damaged artery or arteriole.

**Resources:**
Figure 20.1—Comparative Structure of Blood Vessels

## Concept 20.2 Blood ejected from the heart flows through elastic arteries, muscular arteries, and then arterioles.

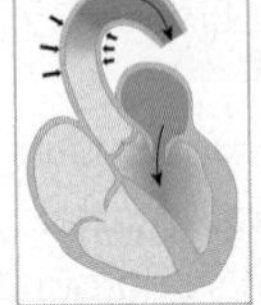

1. **Arteries** carry blood away from the heart.
2. **Elastic arteries** are those close to the heart and have highly elastic walls to accommodate the surge of blood and propel blood onward when the ventricles are relaxing.
3. **Muscular arteries** have lots of smooth muscle in the tunica media for greater vasoconstriction and vasodilation to adjust blood flow to various regions of the body.
4. Many blood vessels **anastomose**, forming **collateral circulation** routes.
5. **Arterioles** deliver blood to capillaries. Through constriction and dilation, arterioles regulate blood flow from arteries into capillaries and help alter blood pressure.

**Resources:**
Anatomy Overview—Arteries and Arterioles

## Concept 20.3 Capillaries are microscopic blood vessels that function in exchange between blood and interstitial fluid.

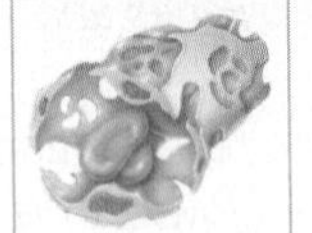

1. **Capillaries** are microscopic blood vessels through which materials are exchanged between blood and the interstitial fluid bathing tissue cells.
2. Precapillary sphincters regulate blood flow through a **capillary bed**.
3. **Continuous capillaries** lack fenestrations and allow only limited diffusion. **Fenestrated capillaries** have many fenestrations that allow quick fluid and larger exchange. **Sinusoids** have large fenestrations, an incomplete basement membrane, and are quite leaky.
4. **Autoregulation** is local, automatic adjustments of blood flow through a capillary bed in response to metabolic needs of a tissue.
5. During **capillary exchange**, substances enter and leave capillaries by simple diffusion, or **transcytosis**, within vesicles. **Capillary blood pressure** "pushes" fluids out of capillaries into interstitial fluid (**filtration**). **Blood colloid osmotic pressure** "pulls" fluid into capillaries from interstitial fluid (**reabsorption**).

**Resources:**
Anatomy Overview—Capillaries
Animation—Capillary Exchange
Figure 20.5—Capillary Exchange

Exercise—Capillary Exchange Pick 'em
Concepts and Connections—Capillary Exchange

## Concept 20.4 Venules and veins return blood to the heart.

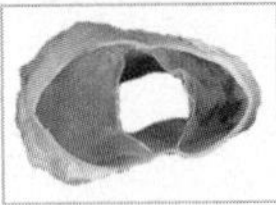

1. **Venules** are small vessels that collect blood from capillaries and merge to form veins.
2. **Veins** have thinner tunica interna and tunica media than arteries. Veins have **valves** to prevent backflow of blood. A **vascular sinus** is a vein with very thin walls.
3. **Venous return**, the volume of blood flowing back to the heart through systemic veins, occurs due to the pumping action of the heart, aided by skeletal muscle contractions (the **skeletal muscle pump**), and pressure changes associated with breathing (the **respiratory pump**).
4. Systemic veins are collectively called **blood reservoirs** because they hold a large volume of blood. If the need arises, this blood can be shifted into other blood vessels through vasoconstriction of veins.

**Resources:**
Anatomy Overview—Veins and Venules
Exercise—Vein Archery

**Concept** | **Resources** WILEY PLUS

## Concept 20.5 Blood flows from regions of higher pressure to those of lower pressure.

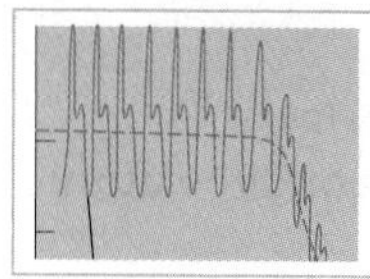

1. **Blood flow** is the volume of blood flowing through a tissue in a given period. Total blood flow, the volume of blood flowing through all systemic blood vessels each minute, is equal to the cardiac output. Blood flow is greater where there is a greater blood pressure gradient and less where there is higher resistance.
2. Ventricular contraction generates **blood pressure**, which is the pressure exerted by blood on blood vessel walls. **Systolic blood pressure** is the highest pressure during systole, and **diastolic blood pressure** is the lowest pressure during diastole. Systemic blood pressure is highest in the aorta and lowest in the vena cava.
3. Blood pressure is influenced by cardiac output, blood volume, and vascular resistance.
4. **Vascular resistance** depends on the size of the blood vessel lumen, blood **viscosity**, and total blood vessel length.
5. **Systemic vascular resistance** refers to all of the vascular resistances from all systemic blood vessels.

Animation—MABP
Animation—Vascular Regulation
Animation—Lymph Flow
Figure 20.10—Summary of Factors That Increase Blood Pressure
Concepts and Connections—Blood Flow
Clinical Connection—Hypertension

## Concept 20.6 Blood pressure is regulated by neural and hormonal negative feedback systems.

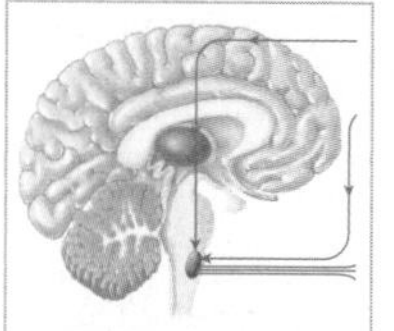

1. The **cardiovascular center** in the medulla oblongata regulates heart rate, heart contractility, and blood vessel diameter.
2. The cardiovascular center receives input from higher regions, proprioceptors, baroreceptors, and chemoreceptors.
3. Parasympathetic impulses from the cardiovascular center decrease heart rate; sympathetic impulses increase heart rate and contractility.
4. **Baroreceptors** monitor blood pressure and initiate reflexes that help regulate blood pressure in the brain.
5. **Chemoreceptors** monitor blood levels of oxygen, carbon dioxide, and hydrogen ions and initiate reflexes that help regulate general systemic blood pressure.
6. Hormones help regulate blood pressure. **Angiotensin II** and **antidiuretic hormone** cause vasoconstriction. **Aldosterone** increases blood volume. **Epinephrine** and **norepinephrine** increase cardiac output and cause vasoconstriction. **Atrial natriuretic peptide** causes vasodilation.

Animation—Regulating Blood Pressure
Exercise—Regulate BP with Hormones
Exercise—Regulate BP with Nerve Impulses
Concepts and Connections—Blood Pressure Regulation

## Concept 20.7 Measurement of the pulse and blood pressure helps assess cardiovascular system function.

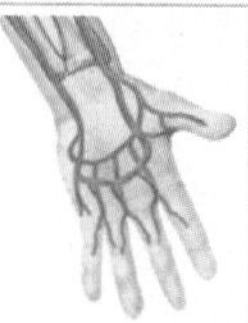

1. **Pulse** is the alternate expansion and recoil of an artery wall after each contraction of the left ventricle. Pulse can be felt in any artery that lies near the body surface or over a hard tissue.
2. A **sphygmomanometer** measures blood pressure. The first **Korotkoff sound** that is heard through the stethoscope corresponds to systolic blood pressure, the arteriole blood pressure just after ventricular contraction. The cuff pressure when the Korotkoff sounds are no longer heard is the diastolic blood pressure, the arteriole pressure during ventricular relaxation.
3. **Pulse pressure** is the difference between systolic and diastolic blood pressure.

Animation—MABP
Clinical Connection—Shock and Homeostasis

| Concept | Resources |
|---|---|

## Concept 20.8 The two main circulatory routes are the pulmonary circulation and the systemic circulation.

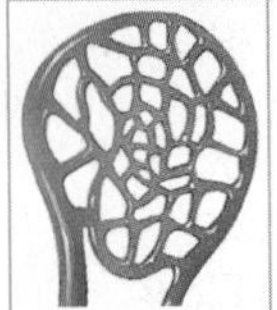

1. Blood vessels are organized into two main **circulatory routes**: the pulmonary circulation and the systemic circulation.
2. The **pulmonary circulation** carries deoxygenated blood from the right ventricle to the air sacs of the lungs and returns oxygenated blood from the air sacs to the left atrium.
3. The **systemic circulation** carries oxygenated blood from the left ventricle to systemic capillaries throughout the body, and returns the deoxygenated blood to the right atrium.

Anatomy Overview—Comparison of Circulatory Routes
Anatomy Overview—Pulmonary Circulation

## Concept 20.9 Systemic arteries carry blood from the heart to all body organs except the lungs.

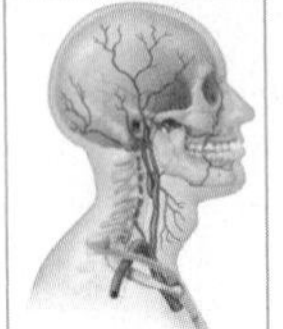

1. The **aorta** is divided into the **ascending aorta**, **arch of the aorta**, **thoracic aorta**, and **abdominal aorta**.
2. The ascending aorta begins at the heart and gives rise to the right and left **coronary arteries** serving the myocardium of the heart.
3. The ascending aorta continues as the arch of the aorta, which has three major branches: the **brachiocephalic trunk**, the **left common carotid artery**, and the **left subclavian artery**.
4. The thoracic aorta is a continuation of the arch of the aorta and gives off branches that supply thoracic organs. As it passes through the diaphragm to enter the abdominal cavity, it becomes the abdominal aorta.
5. The abdominal aorta gives off branches that supply the abdominal organs.
6. The **common iliac arteries** branch from the abdominal aorta to divide into several arteries that supply the pelvis and lower limbs.

Anatomy Overview—Comparison of Circulatory Routes
Exercise—Artery Archery

## Concept 20.10 Systemic veins return blood to the heart from all body organs except the lungs.

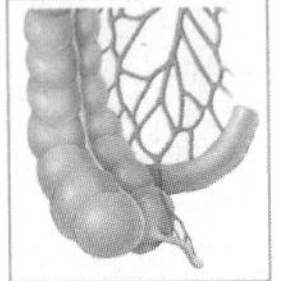

1. All veins of the systemic circulation drain into the **superior vena cava**, **inferior vena cava**, or the **coronary sinus**, which in turn empty into the right atrium. The superior vena cava receives blood from veins above the diaphragm, the inferior vena cava receives blood from veins below the diaphragm, and the coronary sinus drains the heart wall.
2. The **internal jugular**, **external jugular**, and **vertebral veins** drain the head. The **dural venous sinuses** receive blood from all veins within the brain and drain it into the internal jugular veins.
3. The upper limbs have large, **superficial veins** that anastomose with one another and with **deep veins**. Deep veins usually accompany arteries.
4. The **azygos system**, composed of the **azygos**, **hemiazygos**, and **accessory hemiazygos veins**, drains most of the thorax. They ultimately drain into the superior vena cava.
5. Blood from the gastrointestinal tract, spleen, pancreas, and gallbladder passes into the hepatic portal vein for transport to the liver for processing before emptying into the inferior vena cava.
6. The lower limb has both superficial and deep veins.
7. The **hepatic portal circulation** directs venous blood from the gastrointestinal organs and spleen into the **hepatic portal vein** of the liver before it returns to the heart. This portal system enables the liver to utilize nutrients and detoxify harmful substances in the blood. The **splenic vein** drains the spleen, stomach, and large intestine. The **superior mesenteric vein** drains the small intestine, large intestine, stomach, and pancreas. The **hepatic veins** drain the liver.

Figure 20.28—Hepatic Portal Circulation

Exercise—Vein Archery

## Understanding the Concepts

1. What happens to the diameter of the lumen when a blood vessel vasoconstricts? What happens when it vasodilates?
2. In atherosclerosis, the walls of elastic arteries become less compliant (stiffer and, therefore, less able to stretch). What effect does reduced compliance have on the pressure reservoir function of arteries?
3. Why do metabolically active tissues have extensive capillary networks?
4. A person who has liver failure cannot synthesize the normal amount of plasma proteins. How does a deficit of plasma proteins affect blood colloid osmotic pressure? What is the effect on capillary filtration and reabsorption?
5. Why are valves more important in arm veins and leg veins than in neck veins?
6. While exercising, your cardiac output increases. Does your total blood flow increase or decrease? Explain your answer.
7. What happens to blood pressure when a blood vessel's distance from the left ventricle increases? When you're dehydrated? When dieting decreases adipose tissue volume?
8. Would the negative feedback cycle illustrated in Figure 20.12 happen when you lie down or when you stand up? Explain your answer.
9. What causes the pulse?
10. Why does blood change from a bright red color to a dark red color as it travels through the systemic circulation?
11. Which branches of the coronary arteries supply the left ventricle? Why does the left ventricle have such an extensive arterial blood supply?
12. What is the main function of the hepatic portal circulation?

CHAPTER

# 21

## Marlene's Story

It had been seven weeks since her surgery and Marlene was still feeling the effects of the lumpectomy, axillary dissection (removal of armpit lymph nodes), and radiation therapy. When the doctor had told her she had breast cancer, Marlene's world was turned upside down; she was only 57 years old, healthy and active. Her initial shock and numbness had given way to a determination to "beat this thing." The doctor had said the prognosis was encouraging. Several lymph nodes were removed from Marlene's armpit at the time of the lumpectomy to determine if the cancer present in the bean-sized lump had spread to the lymphatic system and potentially to the rest of her body. Thankfully it had not. She was tired and disoriented after the radiation treatments; even now she was only beginning to recover. She felt good enough this evening to wash the dishes herself, rather than having her husband do them. The surgery that removed her axillary nodes had left Marlene with a nagging pinching sensation in her armpit and numb areas of skin near the scar, but overall she felt satisfied with the result. The swelling in her arm and hand did bother her, however. She didn't think much about the swelling at first, as the surgeon indicated it might occur if lymphatic vessels are injured during surgery. It had started several days ago, but over the last few days her swollen arm and hand had really become bothersome.

One function of the lymphatic system is to help return interstitial fluid to the venous circulation. Removal of lymphatic tissues, such as the lymph nodes that were excised during Marlene's lumpectomy, can compromise fluid drainage. In Marlene's case, inadequate drainage of her upper limb is producing tissue swelling, also referred to as *edema*.

# The Lymphatic System and Immunity

## INTRODUCTION

Despite constant exposure to a variety of **pathogens** (PATH-ō-jens), disease-producing microbes such as bacteria and viruses, cuts and bumps, exposure to ultraviolet rays in sunlight, chemical toxins, and minor burns, most people remain healthy. **Immunity** (i-MŪ-ni-tē) or **resistance** is the ability to ward off damage or disease through our defenses. Vulnerability, or lack of resistance, is termed **susceptibility**.

**Innate (nonspecific) immunity** refers to defenses that are always present and available to provide rapid responses to protect us against all types of disease. Among the components of innate immunity are the first line of defense (physical and chemical barriers of the skin and mucous membranes) and the second line of defense (antimicrobial proteins, natural killer cells, phagocytes, inflammation, and fever).

**Adaptive (specific) immunity** involves specific recognition of a microbe once it has breached the innate immunity defenses. Adaptive immunity develops a *specific* response to a *specific* microbe.

The body system responsible for adaptive immunity (and some aspects of innate immunity) is the lymphatic system. Physiologically, the cells and tissues that carry out immune responses are sometimes referred to as the **immune system**. This chapter will explore the mechanisms that provide defenses against intruders and promote repair of damaged body tissues.

## CONCEPTS

**21.1** The lymphatic system drains interstitial fluid, transports dietary lipids, and protects against invasion.

**21.2** Lymph flows through lymphatic capillaries, lymphatic vessels, and lymph nodes.

**21.3** The lymphatic organs and tissues include the thymus, lymph nodes, spleen, and lymphatic follicles.

**21.4** Innate immunity includes external physical and chemical barriers and various internal defenses.

**21.5** The complement system destroys microbes through phagocytosis, cytolysis, and inflammation.

**21.6** Adaptive immunity involves the production of a specific lymphocyte or antibody against a specific antigen.

**21.7** In cell-mediated immunity, cytotoxic T cells directly attack target cells.

**21.8** In antibody-mediated immunity, antibodies specifically target a particular antigen.

**21.9** Immunological memory results in a more intense secondary response to an antigen.

## 21.1 The lymphatic system drains interstitial fluid, transports dietary lipids, and protects against invasion.

The **lymphatic system** (lim-FAT-ik) consists of a fluid called lymph, vessels called lymphatic vessels that transport the lymph, a number of structures and organs containing lymphatic tissue, and red bone marrow (Figure 21.1). The lymphatic system assists in circulating body fluids and helps defend the body against disease-causing agents. Most components of blood plasma filter

FIGURE 21.1 **Components of the lymphatic system.**

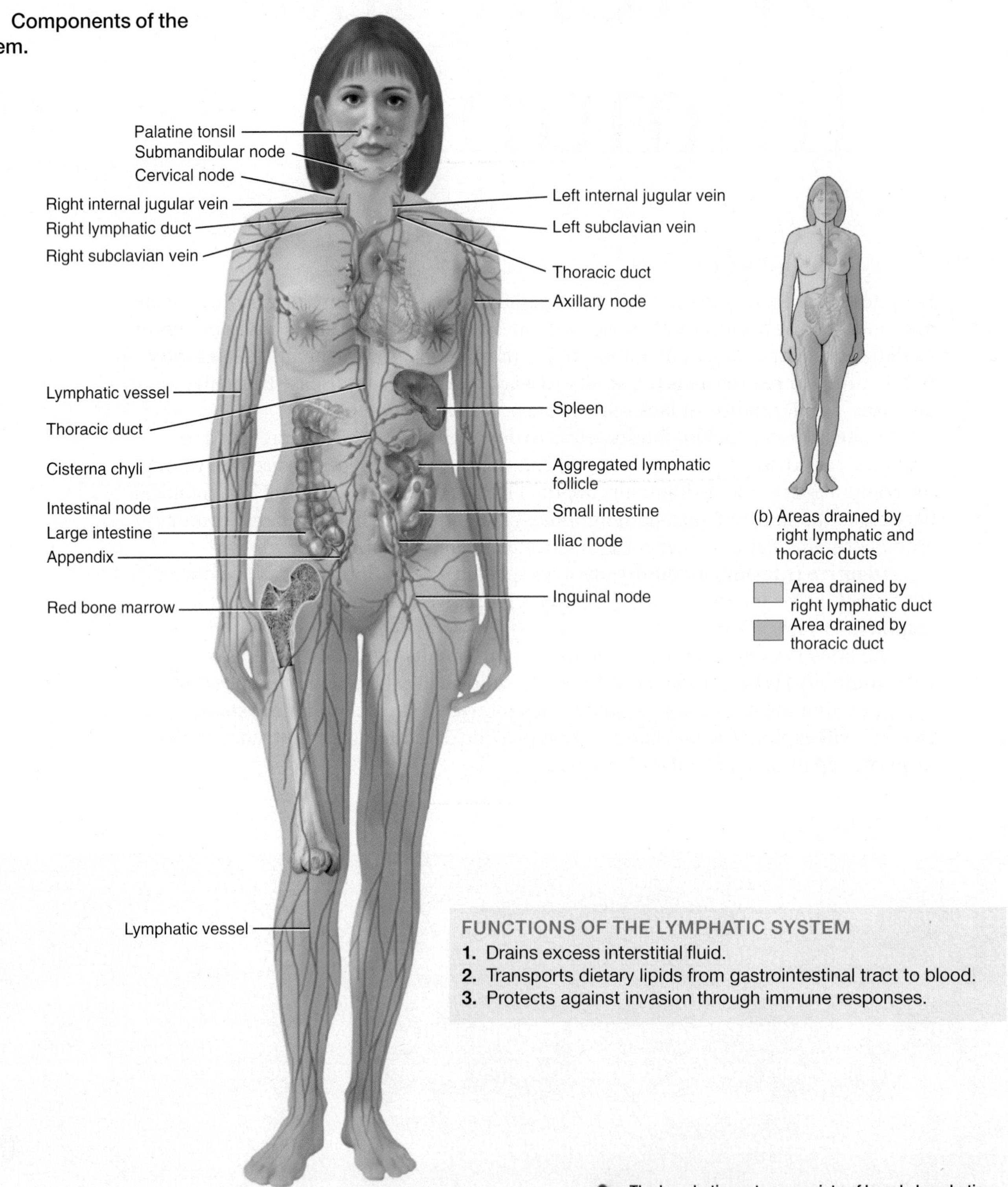


(a) Anterior view of principal components of lymphatic system

The lymphatic system consists of lymph, lymphatic vessels, lymphatic tissues, and red bone marrow.

across blood capillary walls to form interstitial fluid, the fluid that surrounds the cells of body tissues. After interstitial fluid passes into lymphatic vessels, it is called **lymph** (LIMF = clear fluid). The major difference between interstitial fluid and lymph is location: Interstitial fluid is found between cells, and lymph is located within lymphatic vessels and lymphatic tissue.

**Lymphatic tissue** is a specialized form of reticular connective tissue (see Table 4.2) that contains large numbers of lymphocytes. Recall from Chapter 18 that lymphocytes are white blood cells. Two types of lymphocytes participate in adaptive immune responses: B lymphocytes (B cells) and T lymphocytes (T cells).

The lymphatic system has three primary functions:

- ***Drains excess interstitial fluid.*** Lymphatic vessels drain excess interstitial fluid from tissue spaces and return it to the blood.
- ***Transports dietary lipids.*** Lymphatic vessels transport lipids and lipid-soluble vitamins (A, D, E, and K) absorbed by the gastrointestinal tract.
- ***Carries out immune responses.*** Lymphatic tissue initiates highly specific responses directed against particular microbes or abnormal cells.

✓ **CHECKPOINT**

1. What is lymphatic tissue?
2. What is lymph?

## 21.2 Lymph flows through lymphatic capillaries, lymphatic vessels, and lymph nodes.

Lymphatic drainage begins at lymphatic capillaries. These tiny vessels, which are located in the spaces between cells, are closed at one end (Figure 21.2). Just as blood capillaries converge to form venules and then veins, lymphatic capillaries unite to form larger **lymphatic vessels** (Figure 21.1), which resemble veins in structure but have thinner walls and more valves. At intervals along the lymphatic vessels, lymph flows through **lymph nodes**, encapsulated bean-shaped organs consisting of masses of B cells and T cells (which you will learn more about in Concept 21.6). In the skin, lymphatic vessels lie in the subcutaneous tissue and generally follow the same route as veins; lymphatic vessels of the viscera generally follow arteries, forming plexuses (networks) around them. Tissues that lack lymphatic capillaries include avascular tissues (such as cartilage, the epidermis, and the cornea of the eye), the central nervous system, portions of the spleen, and red bone marrow.

### Lymphatic Capillaries

Lymphatic capillaries have greater permeability than blood capillaries and thus can absorb large molecules such as proteins and lipids. **Lymphatic capillaries** also have a unique one-way structure that permits interstitial fluid to flow into them but not

FIGURE 21.2 Lymphatic capillaries.

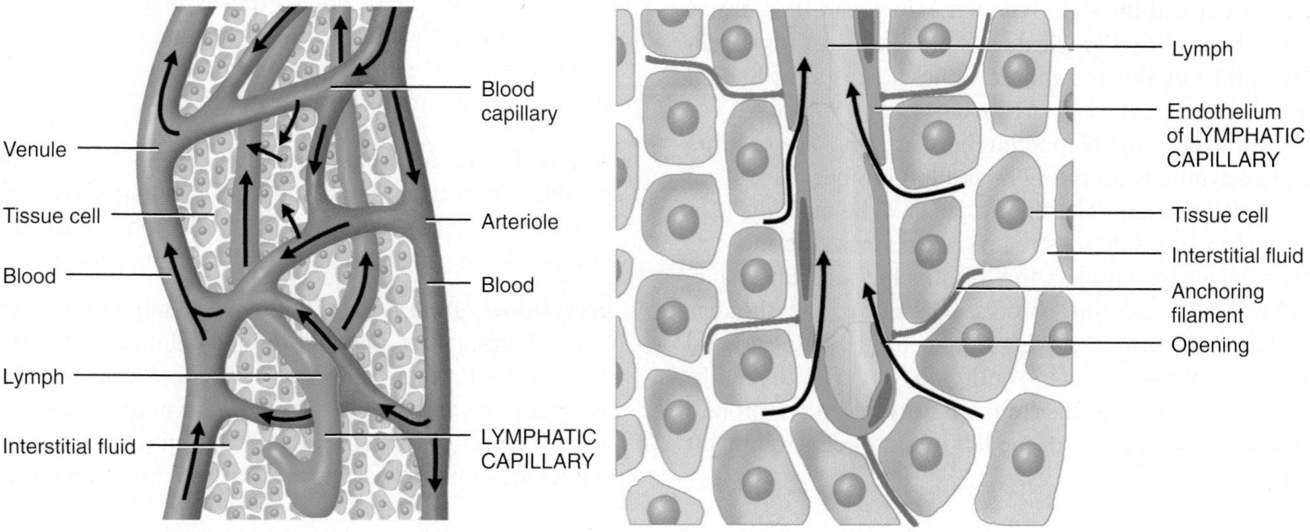


(a) Relationship of lymphatic capillaries to tissue cells and blood capillaries

(b) Details of lymphatic capillary

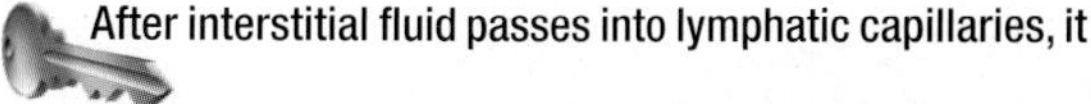

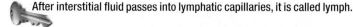

After interstitial fluid passes into lymphatic capillaries, it is called lymph.

out. The ends of endothelial cells that make up the wall of a lymphatic capillary overlap (Figure 21.2b). When pressure is greater in the interstitial fluid than in lymph, the cells separate slightly, like the opening of a one-way swinging door, and interstitial fluid enters the lymphatic capillary. When pressure is greater inside the lymphatic capillary, the cells adhere more closely, and lymph cannot escape back into interstitial fluid. Attached to the lymphatic capillaries are *anchoring filaments,* which contain elastic fibers. They extend out from the lymphatic capillary, attaching lymphatic endothelial cells to surrounding tissues. When excess interstitial fluid accumulates and causes tissue swelling (edema), the anchoring filaments are pulled, making the openings between cells even larger so that more fluid can flow into the lymphatic capillary.

In the small intestine, specialized lymphatic capillaries called **lacteals** (LAK-tē-als; *lact-* = milky) absorb dietary lipids, transfer them to lymphatic vessels, and ultimately into the blood (see Figure 23.21).

## Lymph Trunks and Ducts

As you have already learned, lymph passes from lymphatic capillaries into lymphatic vessels and then through lymph nodes. Lymphatic vessels pass lymph, node to node, through sequential chains of lymph nodes. Upon exiting a chain of nodes, lymphatic vessels in a particular region of the body unite to form **lymph trunks** (Figure 21.3). The **lumbar trunks** drain lymph from the lower limbs, pelvis, kidneys, adrenal glands, and abdominal wall. The **intestinal trunk** drains lymph from the stomach, intestines, pancreas, spleen, and liver. The **bronchomediastinal trunks** (brong-kō-mē′-dē-as-TĪ-nal) drain lymph from the thoracic wall, lung, and heart. The **subclavian trunks** drain the upper limbs. The **jugular trunks** drain the head and neck.

Lymph passes from lymph trunks into two main channels, the thoracic duct and the right lymphatic duct, and then drains into venous blood. The **thoracic duct** is the main duct for the return of lymph to blood. The thoracic duct begins as a dilation called the **cisterna chyli** (sis-TER-na KĪ-lē; *cisterna* = cavity or reservoir) anterior to the second lumbar vertebra. The cisterna chyli receives lymph from the right and left lumbar trunks and from the intestinal trunk. The thoracic duct is the main duct for return of lymph to the bloodstream. In the neck, the thoracic duct also receives lymph from the left jugular, left subclavian, and left bronchomediastinal trunks. Therefore, the thoracic duct receives lymph from the left side of the head, neck, and chest; the left upper limb; and the entire body inferior to the ribs (see Figure 21.1b). The thoracic duct in turn drains lymph into venous blood at the junction of the left internal jugular and left subclavian veins.

The **right lymphatic duct** (Figure 21.3) is only about 1.2 cm (0.5 in.) long and receives lymph from the right jugular, right subclavian, and right bronchomediastinal trunks. Thus, the right lymphatic duct receives lymph from the upper right side of the body (see Figure 21.1b). From the right lymphatic duct, lymph drains into venous blood at the junction of the right internal jugular and right subclavian veins.

## Formation and Flow of Lymph

Most components of blood plasma, such as nutrients and hormones, filter freely through the capillary walls to form interstitial fluid, but more fluid filters out of blood capillaries than returns to them by reabsorption (see Figure 20.5). The excess filtered fluid—about 3 liters per day—drains into lymphatic vessels and becomes lymph. Because most plasma proteins are too large to leave blood vessels, interstitial fluid contains only a small amount of protein. Proteins that do leave blood plasma cannot return to the blood by diffusion because the concentration gradient (high level of proteins inside blood capillaries, low level outside) opposes such movement. The proteins can, however, move readily through the more permeable lymphatic capillaries into lymph. Thus, an important function of lymphatic vessels is to return lost plasma proteins to the bloodstream.

Like veins, lymphatic vessels contain valves, which ensure the one-way movement of lymph. As noted previously, lymph drains into venous blood through the right lymphatic duct and the thoracic duct at the junction of the internal jugular and subclavian veins (Figure 21.3). Thus, the sequence of fluid flow is blood capillaries (blood) → interstitial spaces (interstitial fluid) → lymphatic capillaries (lymph) → lymphatic vessels (lymph) → lymph nodes (lymph) → lymphatic vessels (lymph) → lymphatic ducts (lymph) → junction of the internal jugular and subclavian veins (blood). Figure 21.4 illustrates this sequence, along with the relationship of the lymphatic and cardiovascular systems. Both systems form a very efficient circulatory system.

The same two "pumps" that aid the return of venous blood to the heart maintain the flow of lymph:

- ***Skeletal muscle pump.*** The "milking" action of skeletal muscle contractions (see Figure 20.7) compresses lymphatic vessels (as well as veins) and forces lymph toward the junction of the internal jugular and subclavian veins.
- ***Respiratory pump.*** Lymph flow is also maintained by pressure changes that occur during inhalation (breathing in). Lymph flows from the abdominal region, where the pressure is higher, toward the thoracic region, where it is lower. When the pressures reverse during exhalation (breathing out), the valves in lymphatic vessels prevent backflow of lymph.

**FIGURE 21.3** Routes for drainage of lymph from lymph trunks into the thoracic and right lymphatic ducts.

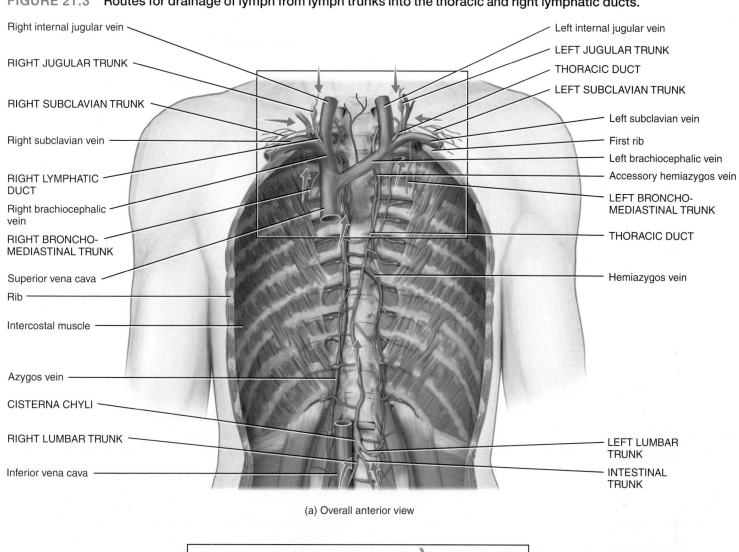

Right internal jugular vein
RIGHT JUGULAR TRUNK
RIGHT SUBCLAVIAN TRUNK
Right subclavian vein
RIGHT LYMPHATIC DUCT
Right brachiocephalic vein
RIGHT BRONCHO-MEDIASTINAL TRUNK
Superior vena cava
Rib
Intercostal muscle
Azygos vein
CISTERNA CHYLI
RIGHT LUMBAR TRUNK
Inferior vena cava
Left internal jugular vein
LEFT JUGULAR TRUNK
THORACIC DUCT
LEFT SUBCLAVIAN TRUNK
Left subclavian vein
First rib
Left brachiocephalic vein
Accessory hemiazygos vein
LEFT BRONCHO-MEDIASTINAL TRUNK
THORACIC DUCT
Hemiazygos vein
LEFT LUMBAR TRUNK
INTESTINAL TRUNK

(a) Overall anterior view

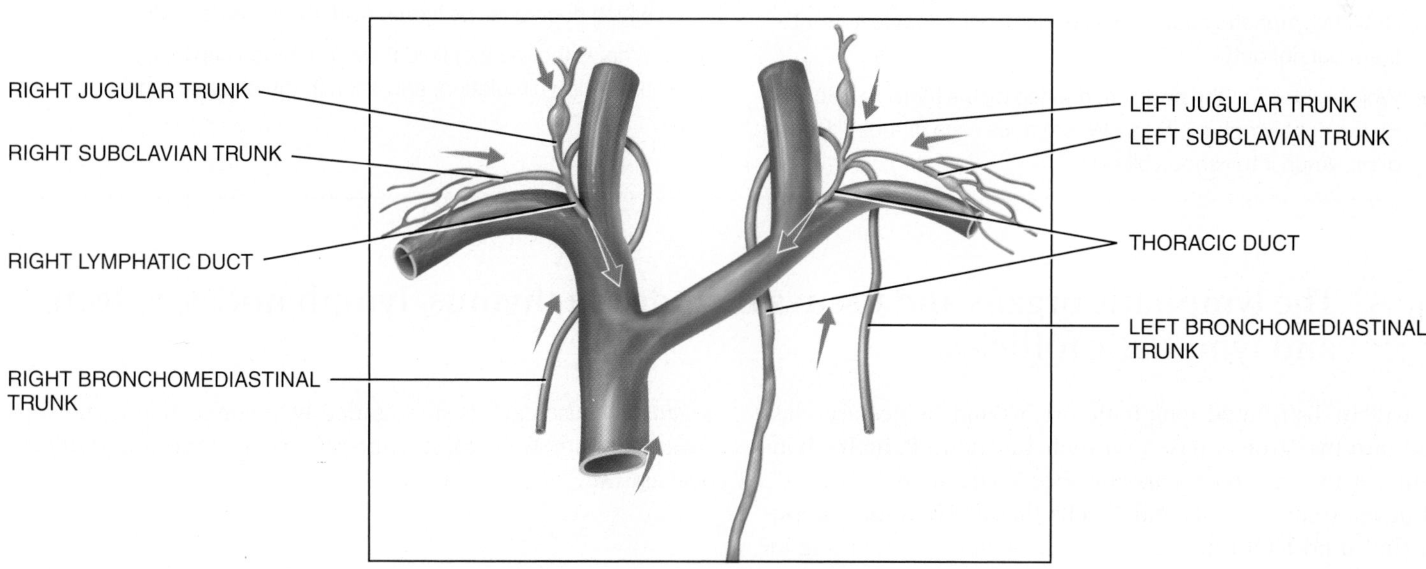


(b) Detailed anterior view of thoracic and right lymphatic ducts

**Lymph returns to the bloodstream through the thoracic duct and right lymphatic duct.**

FIGURE 21.4 **Relationship of the lymphatic system to the cardiovascular system.** The arrows indicate the direction of flow of lymph and blood.

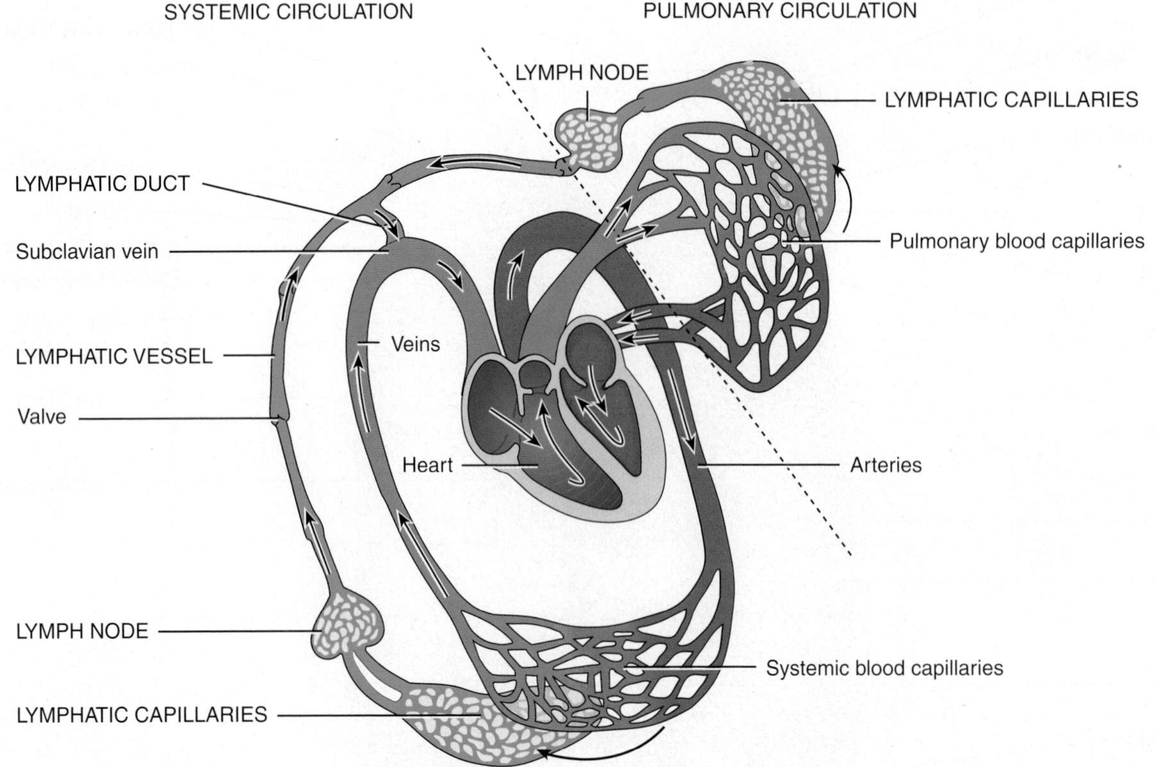


**The sequence of fluid flow is blood capillaries (blood) → interstitial spaces (interstitial fluid) → lymphatic capillaries (lymph) → lymphatic vessels (lymph) → lymph nodes (lymph) → lymphatic vessels (lymph) → lymphatic ducts (lymph) → junction of the internal jugular and subclavian veins (blood).**

✓ **CHECKPOINT**

3. How do lymphatic vessels differ in structure from veins?
4. How do lymphatic capillaries allow interstitial fluid to flow into them but not out?
5. Which regions of the body are drained by the thoracic duct? By the right lymphatic duct? Where does each of these ducts drain lymph into venous blood?
6. Which lymphatic vessels empty into the cisterna chyli, and which duct receives lymph from the cisterna chyli?
7. What is the route of fluid flow from blood capillaries, through the lymph circulation, and back to the bloodstream?

## 21.3 The lymphatic organs and tissues include the thymus, lymph nodes, spleen, and lymphatic follicles.

The widely distributed lymphatic organs and tissues are classified into two groups based on their functions. **Primary lymphatic organs** and **tissues** are the sites where stem cells divide and develop into B cells and T cells that are **immunocompetent** (im′-ū-nō-KOM-pe-tent), that is, capable of mounting an immune response. The primary lymphatic organs and tissues are the thymus and the red bone marrow (in flat bones and the epiphyses of long bones of adults). The **secondary lymphatic organs** and **tissues** are the sites where most immune responses occur. They include lymph nodes, the spleen, and lymphatic follicles. The thymus, lymph nodes, and spleen are considered organs because each is surrounded by a connective tissue capsule; lymphatic follicles, in contrast, are not considered organs because they lack a capsule.

### Thymus

The **thymus** is a bilobed organ covered by a connective tissue capsule. It is located in the mediastinum between the sternum and the aorta, and superior to the heart (Figure 21.5). The thymus contains large numbers of T cells and scattered dendritic

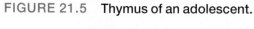

FIGURE 21.5 Thymus of an adolescent.

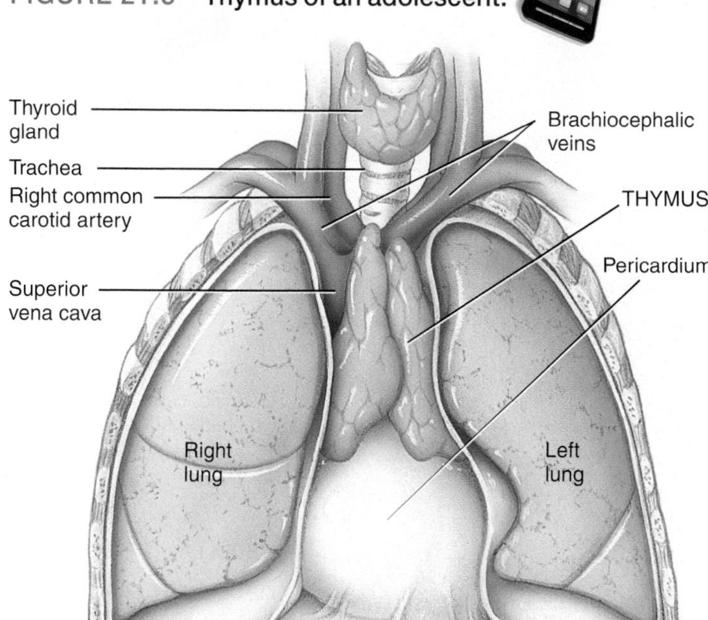


The bilobed thymus is largest at puberty and then atrophies with age.

cells, epithelial cells, and macrophages. Immature T cells migrate from red bone marrow to the thymus, where they multiply and begin to mature. **Dendritic cells** (den-DRIT-ik; *dendr-* = a tree), which are derived from monocytes and so named because they have long, branched projections that resemble the dendrites of a neuron, assist in the maturation process. As discussed shortly, dendritic cells in other parts of the body, such as lymph nodes, play another key role in immune responses. Each of the specialized **epithelial cells** has several long processes that surround and serve as a framework for as many as 50 T cells. These epithelial cells help "educate" the immature T cells to distinguish *self* (your own) molecules from *foreign* (from outside your body) molecules. Additionally, they produce hormones that are thought to aid in the maturation of T cells. Only about 2 percent of T cells developing in the thymus "graduate" into mature T cells. The remaining cells die through *apoptosis* (programmed cell death). Thymic **macrophages** help clear out the debris of dead and dying cells. Mature T cells leave the thymus via the blood and are carried to lymph nodes, the spleen, and other lymphatic tissues, where they populate parts of these organs and tissues.

In infants, the thymus is large, with a mass of about 70 g (2.3 oz). After puberty, adipose and areolar connective tissues begin to replace the thymic tissue. By the time a person reaches maturity, the gland has atrophied considerably, and in old age it may weigh only 3 g (0.1 oz). Before the thymus atrophies, it populates the secondary lymphatic organs and tissues with T cells. However, some T cells continue to proliferate in the thymus throughout an individual's lifetime.

## Lymph Nodes

Located along lymphatic vessels are about 600 bean-shaped lymph nodes. They are scattered throughout the body, both superficially and deep, and usually occur in groups (see Figure 21.1a). Large groups of lymph nodes are present near the mammary glands and in the axillae and groin.

A capsule of dense connective tissues covers and extends into the lymph node (Figure 21.6). The capsular extensions, called **trabeculae** (trā-BEK-ū-lē), divide the node into compartments. Internal to the capsule is a supporting network of reticular fibers. A lymph node is divided into a superficial **cortex** and a deep **medulla**. Within the cortex are aggregates of dividing B cells called **lymphatic nodules**. The centers of many lymphatic nodules contain a region of light-staining cells called a *germinal center*. In the germinal center B cells proliferate and develop into antibody-producing plasma cells or develop into memory B cells (described later in Concept 21.8).

Lymph flows through a node in one direction only (Figure 21.6a). It enters through **afferent lymphatic vessels** (AF-er-ent; *afferent* = to carry toward), which penetrate the surface of the node at several points. The afferent lymphatic vessels contain valves that open toward the center of the node, directing the lymph *inward*. Within the node, lymph enters **sinuses**, a series of irregular channels that contain branching reticular fibers. Embedded in this network of reticular fibers are lymphocytes and macrophages. From the afferent lymphatic vessels, lymph flows into the **subcapsular sinus** (sub-KAP-soo-lar) immediately beneath the capsule. From the subscapular sinus the lymph flows through **trabecular sinuses** (tra-BEK-ū-lar), which extend through the cortex parallel to the trabeculae, and into **medullary sinuses**, which extend through the medulla. The medullary sinuses drain into one or two **efferent lymphatic vessels** (EF-er-ent; *efferent* = to carry away). They contain valves that open away from the center of the lymph node to convey lymph *out* of the node. Efferent lymphatic vessels emerge from one side of the lymph node at a depression called a **hilum** (HĪ-lum).

Lymph nodes function as a filter. As lymph enters one end of a lymph node, foreign substances are trapped by the reticular fibers within the sinuses of the lymph node. Macrophages destroy some foreign substances by phagocytosis, while lymphocytes destroy others by immune responses. The filtered lymph then leaves the other end of the lymph node. Since there are many afferent lymphatic vessels that bring lymph into a lymph node, and only one or two efferent lymphatic vessels that transport lymph out of a lymph node, the slow flow of lymph within the lymph nodes allows sufficient time for lymph to be filtered. Additionally, all lymph flows through multiple lymph nodes on its path through the lymph vessels. This exposes the lymph to multiple filtering events before returning to the blood.

**FIGURE 21.6 Structure of a lymph node.** Arrows indicate direction of lymph flow through a lymph node.

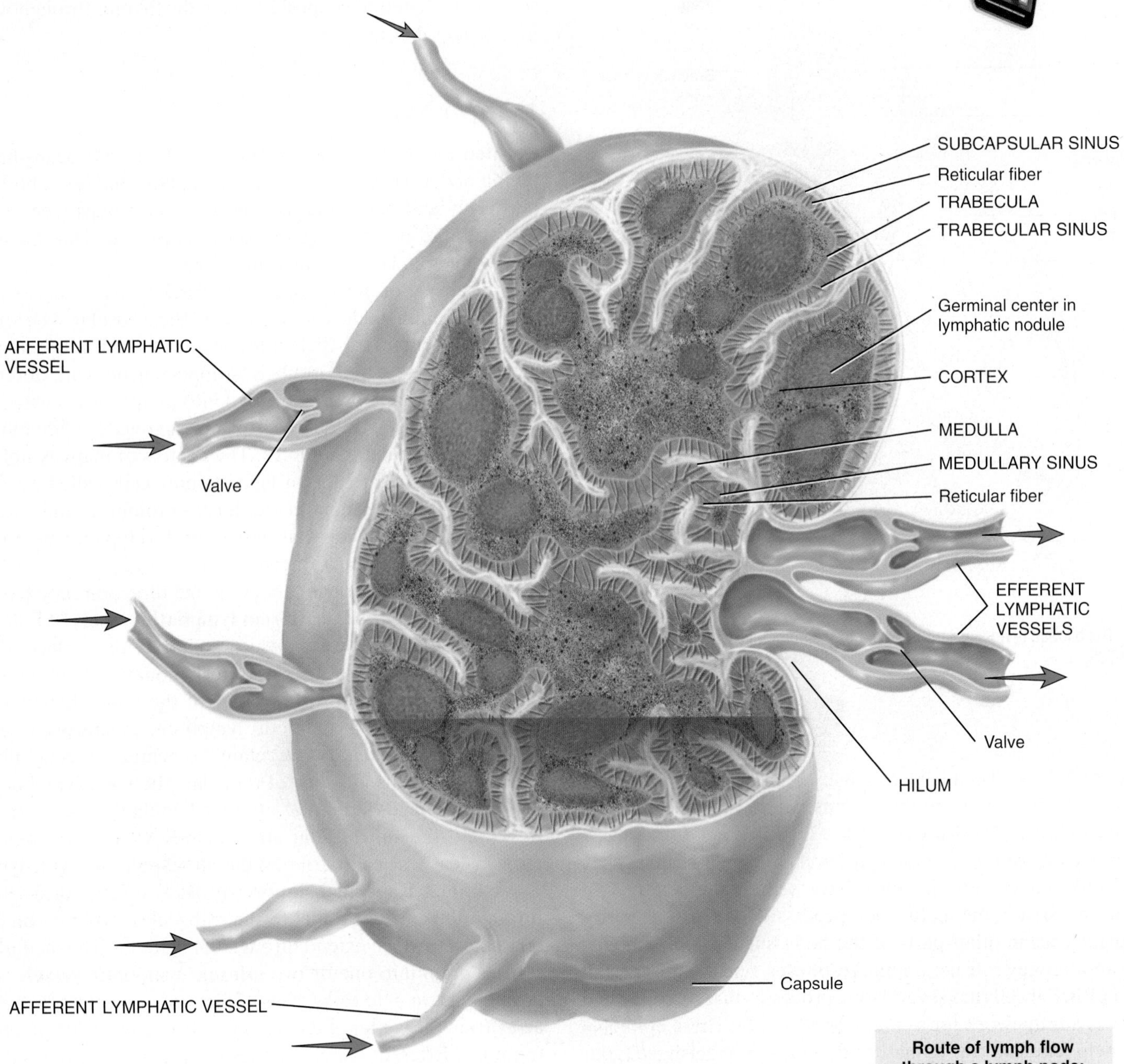

(a) Partially sectioned lymph node

**Route of lymph flow through a lymph node:**
Afferent lymphatic vessel
↓
Subcapsular sinus
↓
Trabecular sinus
↓
Medullary sinus
↓
Efferent lymphatic vessel

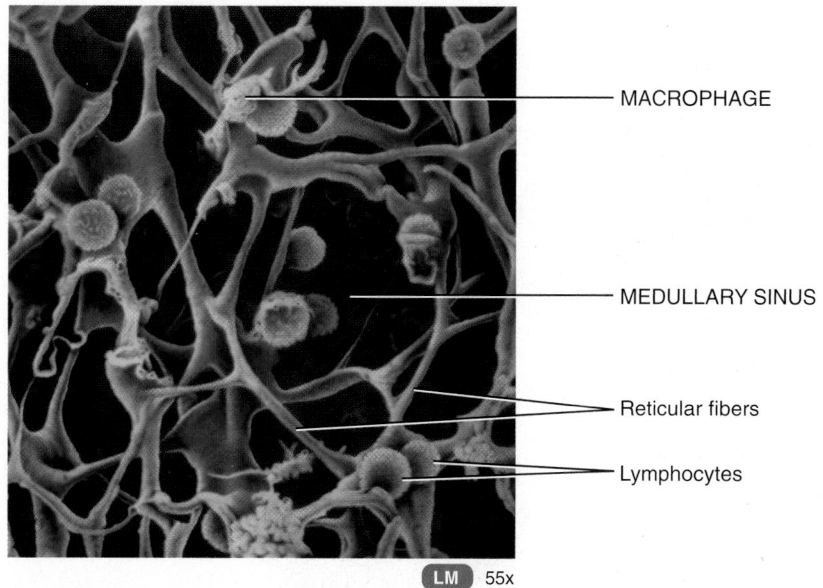


(b) Reticular connective tissue in lymph node

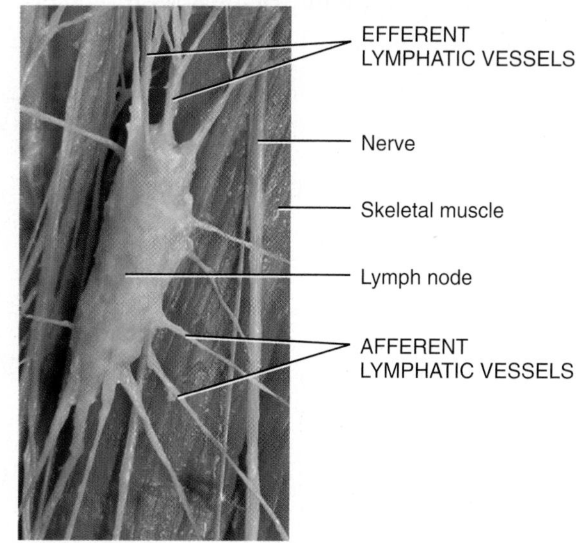


(c) Anterior view of inguinal lymph node

**Lymph nodes are present throughout the body, usually clustered in groups.**

## Spleen

The **spleen** is the largest single mass of lymphatic tissue in the body, measuring about 12 cm (5 in.) in length (see Figure 21.1a). It is located on the left side of the abdominal cavity between the stomach and diaphragm. Neighboring organs, the stomach, kidney, and colon, make indentations, *impressions*, in the visceral surface of the spleen (Figure 21.7a). Like lymph nodes, the spleen has a hilum. The *splenic artery* and *splenic vein* pass through the hilum.

A capsule of dense connective tissue surrounds the spleen. Supportive trabeculae extend inward from the capsule. The spleen contains two types of tissue called white pulp and red pulp (Figure 21.7b, c). **White pulp** is lymphatic tissue, consisting mostly of lymphocytes and macrophages arranged around branches of the splenic artery called **central arteries**. The **red pulp** consists of blood-filled **venous sinuses** and strands of splenic tissue called **splenic cords**. Splenic cords consist of red blood cells, macrophages, lymphocytes, plasma cells, and granulocytes. Veins are closely associated with the red pulp.

Blood flowing into the spleen through the splenic artery enters the central arteries of the white pulp. Within the white pulp, B cells and T cells carry out immune responses while spleen macrophages destroy blood-borne pathogens by phagocytosis. Within the red pulp, the spleen performs three functions related to blood cells:

- ***Removal of aged and defective formed elements.*** Macrophages in the spleen remove worn-out or defective blood cells and platelets.
- ***Platelet storage.*** Up to one-third of the body's supply of platelets are stored in the spleen.
- ***Blood cell production.*** The spleen produces blood cells (hemopoiesis) during fetal life.

### CLINICAL CONNECTION | *Ruptured Spleen*

The spleen is the organ most often damaged in cases of abdominal trauma. Severe blows over the inferior left chest or superior abdomen can fracture the protecting ribs. Such crushing injury may result in a **ruptured spleen**, which causes significant hemorrhage and shock. Prompt removal of the spleen, called a **splenectomy** (splē-NEK-tō-mē), is needed to prevent death due to bleeding. Other structures, particularly red bone marrow and the liver, can take over some functions normally carried out by the spleen. Immune functions, however, decrease in the absence of a spleen. The spleen's absence also places the patient at higher risk for **sepsis** (a blood infection) due to loss of the filtering and phagocytic functions of the spleen. To reduce the risk of sepsis, patients who have undergone a splenectomy take prophylactic (preventive) antibiotics before any invasive procedures.

**FIGURE 21.7 Structure of the spleen.**

SUPERIOR
POSTERIOR
ANTERIOR
Splenic artery
Gastric impression
Splenic vein
Colic impression
Hilum
Renal impression

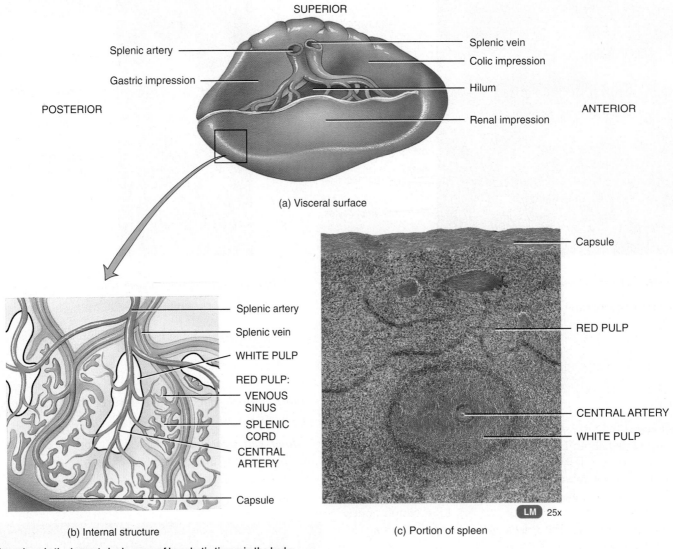

(a) Visceral surface

Splenic artery
Splenic vein
WHITE PULP
RED PULP:
VENOUS SINUS
SPLENIC CORD
CENTRAL ARTERY
Capsule

(b) Internal structure

Capsule
RED PULP
CENTRAL ARTERY
WHITE PULP
LM 25x

(c) Portion of spleen

**The spleen is the largest single mass of lymphatic tissue in the body.**

## Lymphatic Follicles

**Lymphatic follicles** are masses of lymphatic tissue that are not surrounded by a capsule. Because they are scattered throughout the mucous membranes (mucosa) lining the respiratory airways and the gastrointestinal, urinary, and reproductive tracts, lymphatic nodules in these areas are also referred to as **mucosa-associated lymphatic tissue (MALT)**. MALT carries out immune responses by intercepting pathogens and potentially harmful substances that are constantly entering the body through passageways that open to the exterior.

### CLINICAL CONNECTION | *Tonsillitis*

**Tonsillitis** is an infection or inflammation of the tonsils. Most often it is caused by a virus, but it may also be caused by the same bacteria that cause strep throat. The principal symptom of tonsillitis is a sore throat. Fever, swollen lymph nodes, nasal congestion, difficulty in swallowing, and headache may also occur. Tonsillitis of viral origin usually resolves on its own. Bacterial tonsillitis is typically treated with antibiotics. **Tonsillectomy** (ton-si-LEK-tō-mē; *ectomy* = incision), the removal of a tonsil, may be indicated for individuals who do not respond to other treatments. Such individuals usually have tonsillitis lasting for more than three months (despite medication), obstructed air pathways, and difficulty in swallowing and talking. It appears that tonsillectomy does not interfere with a person's response to subsequent infections.

Although many lymphatic follicles are small and solitary, some occur as large aggregations in specific parts of the body. Among these are the tonsils in the pharyngeal region and the aggregated lymphatic follicles (Peyer's patches) in the ileum of the small intestine (see Figure 21.1a). Aggregations of lymphatic follicles also occur in the appendix. Usually there are five **tonsils**, which form a ring at the junction of the oral cavity, nasal cavity, and pharynx (see Figure 22.2b). The tonsils are strategically positioned to participate in immune responses against inhaled or ingested foreign substances. The single **pharyngeal tonsil** (fa-RIN-jē-al), or **adenoid**, is embedded in the posterior wall of the nasopharynx. The two **palatine tonsils** (PAL-a-tīn) lie at the posterior region of the oral cavity, one on either side; these are the tonsils commonly removed in a tonsillectomy. The paired **lingual tonsils** (LIN-gwal), located at the base of the tongue, may also require removal during a tonsillectomy.

✓ CHECKPOINT

8. What is the role of the thymus in immunity?
9. What happens to the size of the thymus as you age?
10. Through what structures does lymph flow as it travels through a lymph node?
11. What happens to foreign substances that enter a lymph node in lymph?
12. What are the functions of the spleen and tonsils?

## 21.4 Innate immunity includes external physical and chemical barriers and various internal defenses.

As noted in the introduction to this chapter, innate (nonspecific) immunity does not involve specific recognition of a microbe and acts against all microbes in the same way. Innate immune responses represent immunity's early warning system and are designed to prevent microbes from gaining access to body tissues and to help eliminate those that do gain access. Innate immunity includes the external physical and chemical barriers provided by the skin and mucous membranes. It also includes various internal defenses, such as antimicrobial proteins, natural killer cells, phagocytes, inflammation, and fever.

### First Line of Defense: Skin and Mucous Membranes

The skin and mucous membranes of the body are the first line of defense against pathogens. These structures provide physical and chemical barriers that discourage pathogens and foreign substances from penetrating the body and causing disease.

With its many layers of closely packed, keratinized cells, the outer epithelial layer of the skin—the *epidermis*—provides a formidable physical barrier to the entrance of microbes (see Figure 5.1). In addition, continual shedding of epidermal cells helps remove microbes at the skin surface. Bacteria rarely penetrate the intact surface of healthy epidermis. However, if cuts, burns, or punctures break the surface, pathogens can penetrate the epidermis and invade adjacent tissues or circulate in the blood to other parts of the body.

The epithelial layer of *mucous membranes*, which line body cavities, secretes a fluid called *mucus* that lubricates and moistens the cavity surface. Because mucus is sticky, it traps many microbes and foreign substances. The mucous membrane of the nose has mucus-coated *hairs* that trap and filter microbes, dust, and pollutants from inhaled air. The mucous membrane of the upper respiratory tract contains *cilia*, microscopic hairlike projections on the surface of the epithelial cells. The waving action of cilia propels inhaled dust and microbes that have become trapped in mucus toward the throat. *Coughing* and *sneezing* accelerate movement of mucus and its entrapped pathogens out of the body. *Swallowing* mucus sends pathogens to the stomach, where gastric juice destroys them.

The cleansing of the urethra by the *flow of urine* retards microbial colonization of the urinary system. *Vaginal secretions* likewise move microbes out of the body in females. *Defecation* and *vomiting* also expel microbes. For example, in response to some microbial toxins, the smooth muscle of the lower gastrointestinal tract contracts vigorously; the resulting diarrhea rapidly expels many of the microbes.

Certain chemicals also contribute to the resistance of the skin and mucous membranes to microbial invasion. Sebaceous (oil) glands of the skin secrete an oily substance called *sebum*. Sebum forms a protective film over the surface of the skin that inhibits the growth of certain pathogenic bacteria and fungi. Bacterial growth is inhibited by the acidity of the skin, caused in part by the secretion of fatty acids and lactic acid. *Perspiration* helps flush microbes from the surface of the skin. *Gastric juice*, produced by the glands of the stomach, is a mixture of hydrochloric acid, enzymes, and mucus. The strong acidity of gastric juice destroys many bacteria and most bacterial toxins. *Vaginal secretions* also are slightly acidic, which discourages bacterial growth.

Other fluids produced by various organs also help protect epithelial surfaces of the skin and mucous membranes. The *lacrimal apparatus* (LAK-ri-mal) of the eyes (see Figure 16.7b) produces and drains away *tears* in response to irritants. Blinking spreads tears over the surface of the eyeball, and the continual washing action of tears helps to dilute microbes and keep them from settling on the surface of the eye. Tears also contain lysozyme (LĪ-sō-zīm), an enzyme capable of breaking down the cell walls of certain bacteria. Lysozyme is also present in saliva, perspiration, nasal secretions, and tissue fluids. *Saliva*, produced by the salivary glands, washes microbes from the surfaces of the teeth and from the mucous membrane of the mouth, much as tears wash the eyes. The flow of saliva reduces colonization of the mouth by microbes.

## Second Line of Defense: Internal Defenses

Although the physical and chemical barriers of the skin and mucous membranes are very effective in preventing invasion by pathogens, they may be broken by injuries or everyday activities such as brushing the teeth or shaving. Any pathogens that get past the surface barriers encounter a second line of defense, which includes the following: natural killer cells, phagocytes, inflammation, fever, and antimicrobial proteins.

### Natural Killer Cells and Phagocytes

About 5–10 percent of lymphocytes in the blood are **natural killer (NK) cells**. They are also present in the spleen, lymph nodes, and red bone marrow. NK cells have the ability to kill a wide variety of infected body cells and certain tumor cells.

NK cells attack any body cells that display abnormal or unusual plasma membrane proteins. The binding of NK cells to a target cell, such as an infected human cell, causes the NK cells to release toxic granules. Some granules contain a protein called **perforin** (PER-for-in) that inserts into the plasma membrane of the target cell and creates channels (perforations) in the membrane. As a result, extracellular fluid flows into the target cell and the cell bursts, a process called **cytolysis** (sī-TOL-i-sis; *cyto-* = cell; *-lysis* = loosening). Other granules of NK cells release **granzymes** (GRAN-zīms), protein-digesting enzymes that induce the target cell to undergo **apoptosis**, or self-destruction. This type of attack kills the infected target cell, but not the microbes inside the cell. The released microbes can be destroyed by phagocytes.

**Phagocytes** (FAG-ō-sīts; *phago-* = eat; *-cytes* = cells) are specialized cells that perform **phagocytosis** (fag-ō-sī-TŌ-sis; *-osis* = process), the ingestion of microbes or other particles such as cellular debris. The two major types of phagocytes are **neutrophils** and **macrophages** (MAK-rō-fā-jez). When an infection occurs, neutrophils and monocytes migrate to the infected area. During this migration, the monocytes enlarge and develop into actively phagocytic cells called **wandering macrophages**. Other macrophages, called **fixed macrophages**, stand guard in specific tissues, including the skin, liver, lungs, brain, spleen, lymph nodes, and red bone marrow.

Phagocytosis occurs in five phases: chemotaxis, adherence, ingestion, digestion, and killing (Figure 21.8):

1. ***Chemotaxis.*** Phagocytosis begins with **chemotaxis** (kē′-mō-TAK-sis), chemical attraction of phagocytes to a site of damage. For example, an invading microbe or damaged tissue cell might release chemicals that attract phagocytes.
2. ***Adherence*** (ad-HĒR-ens). Attachment of the phagocyte to the microbe or other foreign material is termed **adherence**.

FIGURE 21.8 Phagocytosis of a microbe.

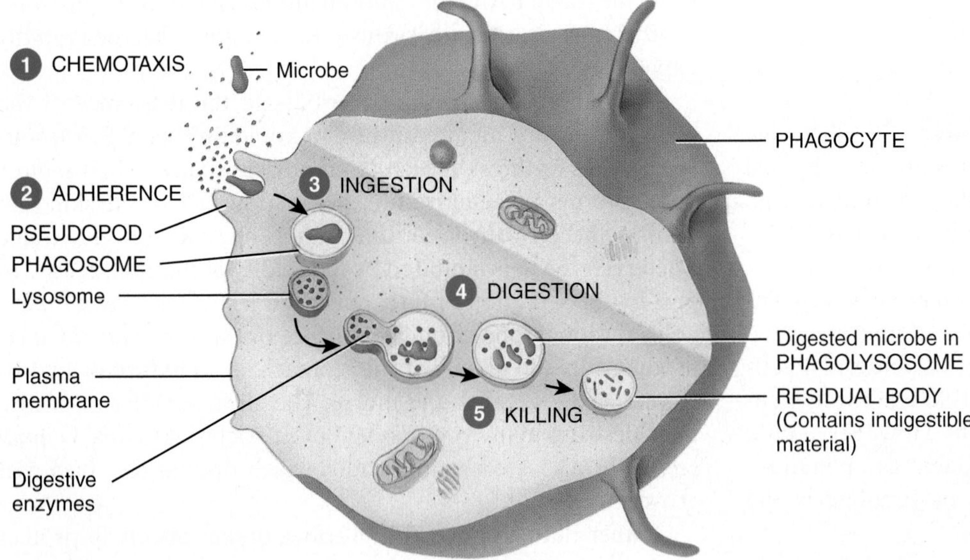


(a) Phases of phagocytosis

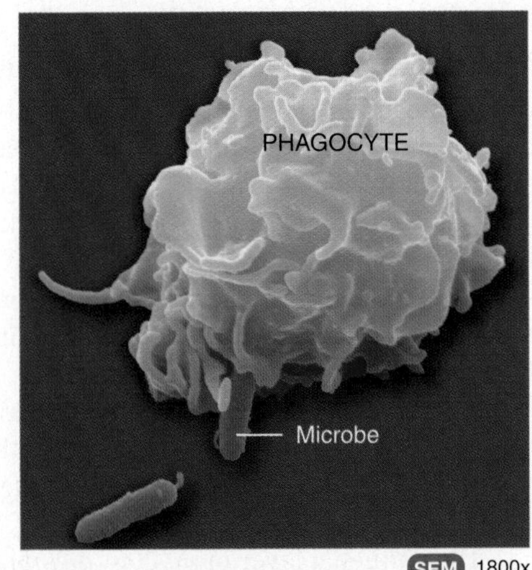


(b) Phagocyte (white blood cell) engulfing microbe.

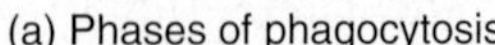

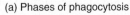

The major types of phagocytes are neutrophils and macrophages.

❸ ***Ingestion.*** The plasma membrane of the phagocyte extends projections, called **pseudopods** (SOO-dō-pods), which engulf the microbe in a process called **ingestion**. As the pseudopods meet, they fuse, surrounding the microorganism within a sac called a **phagosome** (FAG-ō-sōm).

❹ ***Digestion.*** The phagosome enters the cytoplasm and fuses with lysosomes to form a **phagolysosome** (fag-ō-LĪ-sō-sōm). Within the phagolysosome, the lysosome contributes lysozyme and other digestive enzymes that result in **digestion** of the microbe.

❺ ***Killing.*** The chemical onslaught by lysosomal enzymes quickly digests many types of microbes, **killing** them. Any materials that cannot be digested further remain in structures called **residual bodies**.

FIGURE 21.9 **Inflammation.**

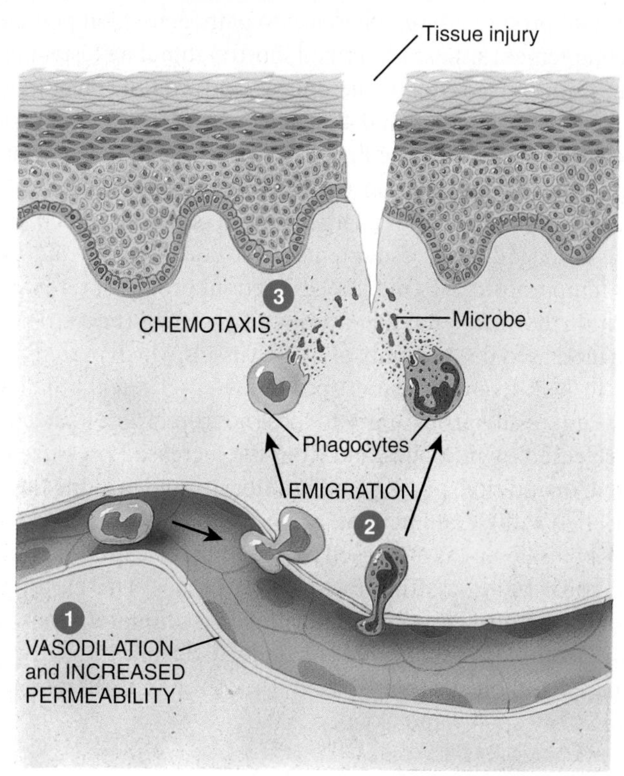


Phagocytes migrate from blood to site of tissue injury

**The three stages of inflammation are vasodilation and increased permeability of blood vessels, phagocyte emigration, and chemotaxis and microbial attack.**

## Inflammation

**Inflammation** is an innate immune response of the body to tissue damage. The conditions that may produce inflammation include pathogens, abrasions, chemical irritations, distortion or disturbances of cells, and extreme temperatures. The four characteristic signs and symptoms of inflammation are *redness, heat, swelling,* and *pain*. Inflammation can also cause a *loss of function* in the injured area (for example, the inability to detect sensations), depending on the site and extent of the injury.

Inflammation is an attempt to dispose of microbes, toxins, or foreign material at the site of injury, to prevent their spread to other tissues, and to prepare the site for tissue repair in an attempt to restore tissue homeostasis. The inflammatory response involves three basic activities (Figure 21.9):

❶ ***Vasodilation and increased permeability of local blood vessels.*** In a region of tissue injury, injured tissue cells, white blood cells, and platelets release **histamine** and other substances that produce two immediate changes in blood vessels: **vasodilation** (increase in diameter) of local arterioles and increased permeability of local capillaries. Vasodilation allows more blood to flow through the damaged area and helps remove microbial toxins and dead cells. **Increased permeability** means that substances normally retained in blood are permitted to pass out of the blood vessels, allowing defensive substances such as antibodies and clotting factors to enter the injured area from the blood.

❷ ***Emigration of phagocytes.*** Within an hour after the inflammatory process starts, phagocytes appear at the site of injury. As large amounts of blood accumulate, neutrophils begin to stick to the inner surfaces of blood vessels. The neutrophils then begin to squeeze through the wall of the blood vessel to reach the damaged area, a process called **emigration** (em′-i-GRĀ-shun). Drawn by chemotaxis to the damaged area, neutrophils attempt to destroy the invading microbes by phagocytosis. A steady stream of neutrophils is ensured by the production and release of additional cells from red bone marrow.

❸ ***Chemotaxis and microbial attack.*** Although neutrophils predominate in the early stages of infection, they die off rapidly. As the inflammatory response continues, monocytes follow the neutrophils by chemotaxis into the infected area. Once in the tissue, monocytes transform into wandering macrophages. True to their name, macrophages are much more potent phagocytes than neutrophils. They are large enough to engulf damaged tissue, worn-out neutrophils, and invading microbes. Eventually, macrophages also die. Within a few days, a pocket of dead phagocytes and damaged tissue forms; this collection of dead cells and fluid is called **pus**. Pus formation occurs in most inflammatory responses and usually continues until the infection subsides. Pus may eventually reach the surface of the body, drain into an internal cavity and be dispersed, or accumulate at the infection site and gradually be absorbed.

Besides histamine, other substances contribute to the inflammatory response. **Prostaglandins** (pros′-ta-GLAN-dins) released by damaged cells, intensify the effects of histamine and may stimulate the emigration of phagocytes through capillary walls. **Leukotrienes** (loo′-kō-TRĪ-ēns) produced by white blood cells increase capillary permeability, chemotactically attract phagocytes, and promote their adherence to pathogens. Components of the complement system (described shortly) stimulate histamine release, attract phagocytes by chemotaxis, and promote phagocytosis.

It's easy to understand the signs and symptoms of inflammation (redness, heat, swelling, and pain) when you consider the events that occur. Dilation of arterioles and increased permeability of capillaries allow a large amount of blood to accumulate in the damaged area, resulting in local redness and heat. As the local temperature rises, metabolic reactions proceed more rapidly, which further heats the damaged area. Swelling (edema) results from increased permeability of blood vessels, which permits more fluid to leak from capillaries into tissue spaces of the damaged area. Pain results from injury to local neurons, from toxic chemicals released by microbes, and from the increased pressure from edema. In addition, prostaglandins intensify and prolong the pain associated with inflammation.

Repair begins as increased permeability of capillaries allows leakage of blood-clotting factors into tissues. The clotting sequence is set into motion, and fibrinogen is ultimately converted to an insoluble, thick mesh of fibrin threads that traps invading microbes and blocks their spread. Tissue repair is discussed further in Concept 4.9 and Concept 5.8.

### Fever

While inflammation elevates temperature *locally* at the site of damage, **fever** is an abnormally high *systemic* (whole body) temperature that occurs when the hypothalamic thermostat is reset. It commonly occurs during infection and can accompany inflammation. Many bacterial toxins elevate body temperature, sometimes by triggering release of fever-causing cytokines (described in Concept 21.6) from macrophages. Elevated body temperature intensifies the effects of interferons, inhibits the growth of some microbes, and speeds up body reactions that aid repair.

### Antimicrobial Proteins

Various body fluids contain **antimicrobial proteins** that discourage microbial growth; the most important of these are interferons and proteins that make up the complement system.

**Interferons** (in′-ter-FĒR-ons) are proteins produced by cells infected with viruses. Once released by virus-infected cells, interferons diffuse to uninfected neighboring cells, where they stimulate synthesis of antiviral proteins that block viral replication. Although interferons do not prevent viruses from penetrating host cells, they do stop replication. Viruses can cause disease only if they can replicate within body cells.

The **complement system** is a group of normally inactive proteins in blood plasma and on plasma membranes. When activated, these proteins "complement" or enhance certain immune reactions. The complement system causes cytolysis (bursting) of microbes, promotes phagocytosis, and contributes to inflammation.

Table 21.1 summarizes the components of innate immunity.

**TABLE 21.1**

Innate Defenses

| COMPONENT | FUNCTIONS |
|---|---|
| **FIRST LINE OF DEFENSE: SKIN AND MUCOUS MEMBRANES** | |
| **Physical Factors** | |
| **Epidermis of skin** | Forms physical barrier to entrance of microbes |
| **Mucous membranes** | Inhibit entrance of many microbes, but not as effective as intact skin |
| **Mucus** | Traps microbes in respiratory and gastrointestinal tracts |
| **Hairs** | Filter out microbes and dust in nose |
| **Cilia** | Together with mucus, trap and remove microbes and dust from upper respiratory tract |
| **Lacrimal apparatus** | Tears dilute and wash away irritating substances and microbes |
| **Saliva** | Washes microbes from surfaces of teeth and mucous membranes of mouth |
| **Urine** | Washes microbes from urethra |
| **Defecation and vomiting** | Expel microbes from body |
| **Chemical Factors** | |
| **Sebum** | Forms protective acidic film over skin surface that inhibits growth of many microbes |
| **Lysozyme** | Antimicrobial substance in perspiration, tears, saliva, nasal secretions, and tissue fluids |
| **Gastric juice** | Destroys bacteria and most toxins in stomach |
| **Vaginal secretions** | Slight acidity discourages bacterial growth; flush microbes out of vagina |
| **SECOND LINE OF DEFENSE: INTERNAL DEFENSES** | |
| **Antimicrobial Proteins** | |
| **Interferons** | Protect uninfected host cells from viral infection |
| **Complement system** | Causes cytolysis of microbes; promotes phagocytosis; contributes to inflammation |
| **Natural killer (NK) cells** | Kill infected target cells by releasing granules that contain perforin and granzymes; phagocytes then kill released microbes |
| **Phagocytes** | Ingest foreign particulate matter |
| **Inflammation** | Confines and destroys microbes; initiates tissue repair |
| **Fever** | Intensifies effects of interferons; inhibits growth of some microbes; speeds up body reactions that aid repair |

## CLINICAL CONNECTION | *Abscesses and Ulcers*

If pus cannot drain out of an inflamed region, the result is an **abscess**—an excessive accumulation of pus in a confined space. Common examples are pimples and boils. When superficial inflamed tissue sloughs off the surface of an organ or tissue, the resulting open sore is called an **ulcer**. People with poor circulation—for instance, diabetics with advanced atherosclerosis—are susceptible to ulcers in the tissues of their legs. These ulcers, which are called stasis ulcers, develop because of poor oxygen and nutrient supply to tissues that then become very susceptible to even a very mild injury or an infection.

## ✓ CHECKPOINT

**13.** What is innate immunity?

**14.** Name the physical and chemical factors that provide protection from disease in the skin and mucous membranes.

**15.** Which internal defenses provide protection against microbes that penetrate the skin and mucous membranes?

**16.** What is the name for the antimicrobial proteins that prevent viral replication?

**17.** Which type of white blood cells are natural killer cells? Which type are phagocytes?

**18.** Describe the phases of phagocytosis.

**19.** How are microbes killed following their ingestion by a phagocyte?

## RETURN TO Marlene's Story

As she reached into the tepid dishwater, Marlene felt a dull burning pain. A sharp knife at the bottom of the dishpan had opened a gash between her thumb and forefinger. "Francis!" she called out to her husband. Francis came rushing in.

"Marlene, what's wrong?" he asked.

"I cut myself. Is it bad?"

Francis took her hand. "It's not too deep. May need a few stitches, though. Let's clean it up and get something on it, then we'll go to the urgent care center."

"If it's not that bad, let's just put a bandage on it. I am so tired of doctors and nurses and hospitals," replied Marlene.

Francis carefully washed the wound and placed a clean cotton pad over it. "Your hand seems so swollen, Marlene," he remarked.

Three days later, Marlene noticed that the swelling was worse, and red streaks extended from her hand up her forearm. She gently lifted the bandage. Pus oozed from the wound, and its margins (edges) were red and inflamed. She reached into her medicine cabinet for a bottle of hydrogen peroxide but couldn't open it. "Francis? Could you open this for me?" He looked at her hand.

"Marlene, your hand looks worse. Let me take you in to have it looked at."

"Okay," she sighed.

It didn't take the doctor at the urgent care center long to diagnose the problem. "You'll need to go on antibiotics right away. Your cut is infected and the red streaks on your arm indicate *lymphangitis*, inflammation of the lymphatic vessels. The swelling you experienced before you cut your hand was probably caused by a blockage in the lymphatic system, secondary to your surgery. Radiation therapy can cause localized inflammation. You should follow up with your oncologist about that. Right now we need to focus on treating the infection. Strep bacteria are very common on the skin and in the environment. When they get into cuts and scrapes, they can multiply rapidly and overwhelm the immune system. I am most concerned with treating the infection so we can prevent it from spreading."

A. *Marlene has had lymph nodes removed from her armpit. Why would removal of lymph nodes cause swelling in her upper limb?*

B. *Because Marlene was tired following the radiation therapy, she was not moving around a lot. How might this be related to the lymph fluid buildup in her arm?*

C. *Which elements of innate immunity were at work in Marlene's body when she cut her hand?*

## 21.5 The complement system destroys microbes through phagocytosis, cytolysis, and inflammation.

The **complement system** (KOM-ple-ment) is a group of over 30 proteins produced by the liver and found circulating in blood plasma and within tissues throughout the body. Most complement proteins are designated by an uppercase letter *C*, numbered C1 through C9, named for the order in which they were discovered. The C1–C9 complement proteins are normally inactive; they become activated only when split by enzymes into active fragments, which are indicated by lowercase letters *a* and *b*. For example, inactive complement protein C3 is split into the activated fragments C3a and C3b. The active fragments carry out the destructive actions of the C1–C9 complement proteins.

Complement proteins act in a *cascade*—one reaction triggers another reaction, which in turn triggers another reaction, and so on. With each succeeding reaction, more and more product is formed so that the net effect of complement is amplified many times. The complement cascade of reactions begins with activation of C3. Once activated, C3 starts a sequence of reactions that destroys microbes by causing phagocytosis, cytolysis, and inflammation (Figure 21.10):

1. ***Activation.*** Inactivated C3 splits into activated C3a and C3b.
2. ***Opsonization.*** C3b binds to the surface of a microbe serving as an attachment site for phagocytes. Thus C3b enhances phagocytosis by coating a microbe, a process called **opsonization** (op-sō-ni-ZĀ-shun). Opsonization promotes attachment of a phagocyte to a microbe.

FIGURE 21.10 **Complement activation.** (Adapted from Tortora, Funke, and Case, *Microbiology: An Introduction, Eighth Edition,* Figure 16.10, Pearson Benjamin-Cummings, 2004.)

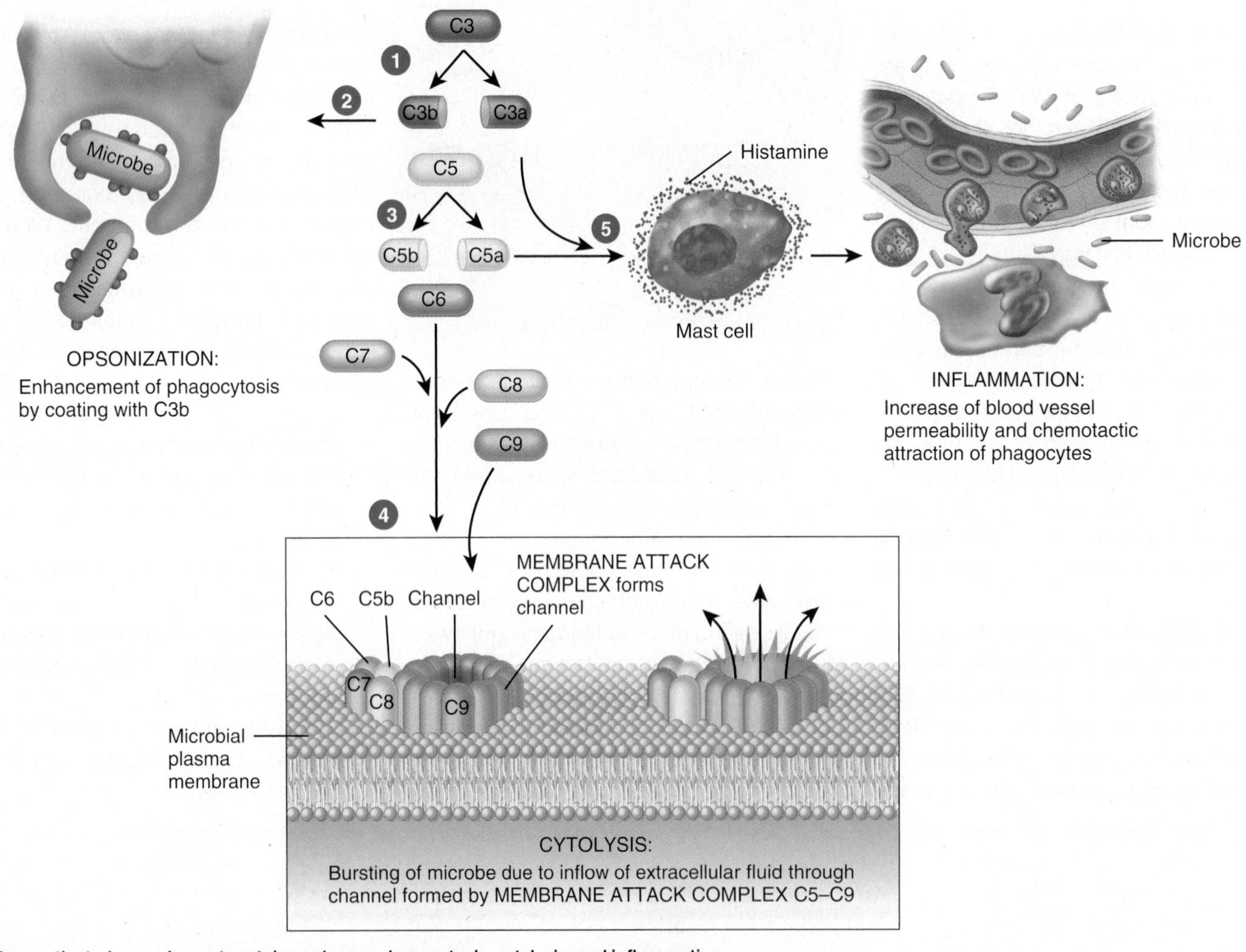


When activated, complement proteins enhance phagocytosis, cytolysis, and inflammation.

3 ***Cytolysis.*** C3b also initiates a series of reactions that bring about cytolysis. C3b splits C5 into activated C5a and C5b. The C5b fragment then binds to C6 and C7, which attach to the plasma membrane of an invading microbe.

4 ***Formation of membrane attack complex.*** C8 and C9 molecules join C5b, C6, and C7 to form a cylinder-shaped **membrane attack complex**, which inserts into the microbial plasma membrane. The membrane attack complex creates a channel through the plasma membrane that results in cytolysis, the bursting of the microbial cells due to the inflow of extracellular fluid through the channels.

5 ***Inflammation and chemotaxis.*** C3a and C5a bind to mast cells, connective tissue cells that produce histamine, and cause them to release it, which increases blood vessel permeability during inflammation. C5a also attracts phagocytes to the site of inflammation (chemotaxis, the first step in phagocytosis; see Figure 21.8).

Once complement is activated, proteins in blood and on body cells break down activated C3. In this way, its destructive capabilities cease very quickly so that damage to body cells is minimized.

✓ **CHECKPOINT**

**20.** Where are complement proteins produced?

**21.** How does the complement system destroy microbes?

**22.** How is the complement system activated?

# 21.6 Adaptive immunity involves the production of a specific lymphocyte or antibody against a specific antigen.

The various aspects of innate immunity have one thing in common: They are not specifically directed against a particular type of invader. The ability of the body to defend itself against *specific* invading pathogens and foreign tissues that have breached the innate immunity defenses is called adaptive (specific) immunity. Substances that are recognized as foreign (nonself) and provoke immune responses are called **antigens** (AN-ti-jens). Microbes and pollen are examples of antigens. Two properties distinguish adaptive immunity from innate immunity: (1) *specificity* for particular foreign molecules (antigens), which also involves distinguishing self from nonself molecules, and (2) *memory* for most previously encountered antigens so that a second encounter prompts an even more rapid and vigorous response.

## Maturation of T Cells and B Cells

Adaptive immunity involves two types of lymphocytes: **B cells** and **T cells**. Both develop from stem cells that originate in red bone marrow (see Figure 18.3). B cells complete their development in red bone marrow, a process that continues throughout life. Immature T cells migrate from red bone marrow into the thymus, where they mature (Figure 21.11). Most T cells arise before puberty, but they continue to mature and leave the thymus throughout life. B cells and T cells are named based on where they mature. Although in humans B cells mature in red bone marrow, in birds B cells mature in an organ called the *bursa of Fabricius.* This organ is not present in humans, but the term *B cell* is still used with the letter *B* standing for *bursa equivalent,* which is red bone marrow in humans. T cells are so named because they mature in the *thymus* gland.

Before T cells leave the thymus or B cells leave red bone marrow, they become immunocompetent, able to carry out adaptive immune responses. Immunocompetence (im′-ū-nō-KOM-pe-tens) occurs as B cells and T cells make several distinctive proteins that they insert into their plasma membranes. Some of these proteins function as **antigen receptors**—molecules capable of recognizing specific antigens (Figure 21.11).

There are two major types of mature T cells that exit the thymus: helper T cells and cytotoxic T cells (Figure 21.11). In addition to containing antigen receptors, the plasma membranes of helper T cells include a protein called CD4, and the plasma membranes of cytotoxic T cells include a protein called CD8. As you will see in Concept 21.7, these two types of T cells have very different functions.

## Types of Adaptive Immunity

There are two types of adaptive immunity: cell-mediated immunity and antibody-mediated immunity. Antigens trigger both types of adaptive immunity.

In **cell-mediated immunity**, cytotoxic T cells directly attack invading antigens. Cell-mediated immunity is particularly effective against intracellular pathogens, which include any viruses, bacteria, or fungi that are inside cells; some cancer cells; and foreign tissue transplants. Thus, cell-mediated immunity always involves cells attacking cells. In **antibody-mediated immunity**, B cells transform into plasma cells, which synthesize and secrete specific **antibodies**. A given antibody can bind to and inactivate a specific antigen. Antibody-mediated immunity works mainly against extracellular pathogens, which include viruses, bacteria,

FIGURE 21.11 **B cells and immature T cells arise from stem cells in red bone marrow.**

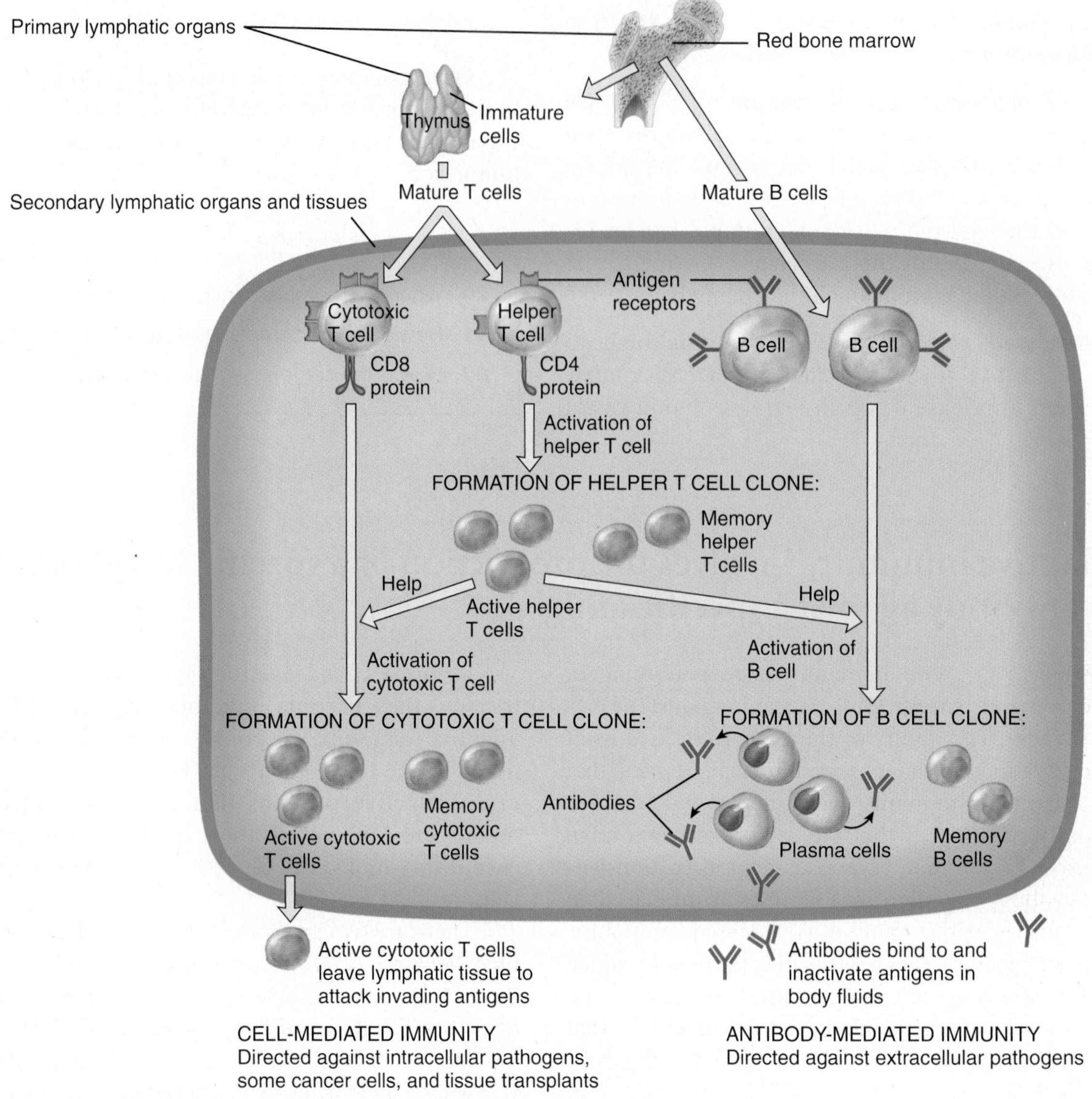


**The two types of adaptive immunity are cell-mediated immunity and antibody-mediated immunity.**

or fungi that are in body fluids outside cells. Helper T cells aid the immune responses of both cell-mediated immunity and antibody-mediated immunity.

When a specific antigen invades the body, there are usually many copies of that antigen spread throughout the body's tissues and fluids. Copies of the antigen that enter body cells provoke a cell-mediated immune response by cytotoxic T cells, while copies of the antigen in extracellular fluid produce an antibody-mediated immune response by B cells. Cell-mediated immune responses and antibody-mediated immune responses often work together to rid the body of numerous copies of an antigen.

## Clonal Selection

As you just learned, when a specific antigen invades the body, there are usually many copies of that antigen located throughout the body's tissues and fluids. The numerous copies of the antigen initially outnumber the small group of helper T cells, cytotoxic T cells, and B cells with the correct antigen receptors to respond to that antigen. Therefore, once each of these lymphocytes encounters a copy of the antigen, it undergoes clonal selection. **Clonal selection** is the process by which a lymphocyte divides and *differentiates* (forms more highly specialized cells) in response to contact with a specific antigen. The result of clonal selection is the formation of a population of identical cells, called a **clone**, which can recognize the same specific antigen as the original lymphocyte (Figure 21.11). Before the first exposure to a given antigen, only a few lymphocytes are able to recognize it, but once clonal selection occurs, there are thousands of lymphocytes that can respond to that antigen. The swollen tonsils or lymph nodes in your neck you experienced the last time you were sick were probably caused by clonal selection of lymphocytes participating in an immune response.

When a lymphocyte undergoes clonal selection, the resulting clone includes two major types of cells: effector cells and memory cells. **Effector cells** carry out immune responses that ultimately destroy or inactivate the antigen. Most effector cells eventually die after the immune response has been completed. **Memory cells** do not actively participate in the initial immune response to the antigen. However, if the same antigen enters the body again in the future, memory cells are available to initiate a far swifter reaction than occurred during the first invasion. Memory cells respond to the antigen by proliferating and differentiating into more effector cells and more memory cells. Consequently, the second response to the antigen is usually so fast and so vigorous that the antigen is destroyed before any signs or symptoms of disease can occur. Most memory cells do not die at the end of an immune response. Instead, they have long life spans (often lasting for decades). The functions of effector cells and memory cells are described in more detail in Concepts 21.7 and 21.8.

## Antigens and Antigen Receptors

Antigens are able to provoke an immune response by stimulating the production of specific antibodies and/or the proliferation of specific T cells. The term *antigen* derives from its function as an *anti*body *gen*erator.

Entire microbes or parts of microbes may act as antigens. Components of bacterial structures such as flagella and cell walls are antigenic, as are bacterial toxins. Nonmicrobial examples of antigens include components of pollen, egg white, incompatible blood cells, and transplanted tissues and organs. The huge variety of antigens in the environment provides myriad opportunities for provoking immune responses. Typically, large antigen molecules have small parts called **epitopes** (EP-i-tōps) that trigger immune responses (Figure 21.12). Most antigens have many epitopes, each of which induces production of a specific antibody or activates a specific T cell.

Antigens are large, complex molecules. Most often, they are proteins. Large molecules that have simple, repeating subunits—for example, cellulose and most plastics—are not usually antigenic. This is why plastic materials can be used in artificial heart valves or joints. A small molecule can stimulate an immune response if it is attached to a larger carrier molecule. An example is the small lipid toxin in poison ivy, which triggers an immune response after combining with a body protein.

FIGURE 21.12 **Epitopes.**

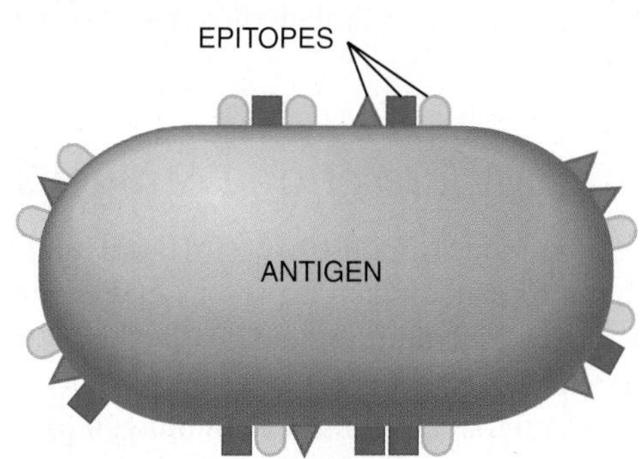


**Most antigens have several epitopes that induce the production of different antibodies or activate different T cells.**

Likewise, penicillin may stimulate an immune response when it combines with body proteins.

An amazing feature of the human immune system is its ability to recognize and bind to at least a billion ($10^9$) different epitopes. Before a particular antigen ever enters the body, T cells and B cells that can recognize and respond to that intruder are ready and waiting. Cells of the immune system can even recognize artificially made molecules that do not exist in nature. B cells and T cells can recognize so many epitopes because they have an equally large diversity of antigen receptors.

Antigens that get past the innate defenses generally meet their demise by entering lymphatic tissue. Most antigens that enter the bloodstream (for example, through an injured blood vessel) are trapped as they flow through the spleen. Antigens that penetrate the skin enter lymphatic vessels and lodge in lymph nodes. Antigens that penetrate mucous membranes (for example, through the nose) are entrapped by mucosa-associated lymphatic tissue (MALT).

## Major Histocompatibility Complex Molecules

Located in the plasma membrane of body cells are "self-antigens" known as **major histocompatibility complex (MHC) molecules** (his′-tō-kom-pat′-i-BIL-i-tē). Unless you have an identical twin, your MHC molecules are unique. Thousands to several hundred thousand MHC molecules mark the surface of each of your body cells except red blood cells. Although MHC molecules are the reason that tissues may be rejected when they are transplanted from one person to another, their normal function is to help T cells distinguish your own cells from foreign cells. Such recognition is an important first step in any adaptive immune response.

As a rule, antigens are foreign substances; they are not usually part of our body tissues. However, sometimes the immune system fails to distinguish "friend" (self) from "foe" (nonself).

The result is an autoimmune disorder in which our own molecules or cells are attacked as though they were foreign.

## Processing and Presenting Antigens

For an adaptive immune response to occur, B cells and T cells must recognize that a foreign antigen is present. B cells can recognize and bind to antigens in lymph, interstitial fluid, or blood plasma. T cells only recognize fragments of antigenic proteins that are processed and presented in a certain way. In **antigen processing**, antigenic proteins are broken down into peptide fragments that then associate with MHC molecules. Next, the antigen–MHC complex is inserted into the plasma membrane of a body cell. The insertion of the complex into the plasma membrane is called **antigen presentation**. When a fragment comes from a *self-protein*, T cells ignore the antigen–MHC complex. However, if the peptide fragment comes from a *foreign protein,* T cells recognize the antigen–MHC complex as an intruder, and an immune response takes place.

### Processing of Exogenous Antigens

Foreign antigens that are present in fluids *outside* body cells are termed **exogenous antigens** (eks-OJ-e-nus). They include intruders such as bacteria, parasitic worms, inhaled pollen, and viruses that have not yet infected a body cell. A special class of cells called **antigen-presenting cells (APCs)** process and present exogenous antigens. APCs include dendritic cells, macrophages, and B cells. They are strategically located in places where antigens are likely to penetrate our innate defenses and enter the body, such as the epidermis of the skin; mucous membranes that line the respiratory, gastrointestinal, urinary, and reproductive tracts; and lymph nodes.

The steps in the processing and presenting of an exogenous antigen by an APC occur as follows (Figure 21.13):

1. ***Phagocytosis of antigen.*** Antigen-presenting cells ingest exogenous antigens via phagocytosis. Ingestion could occur almost anywhere in the body that invaders, such as microbes, have penetrated the innate defenses.

FIGURE 21.13 **Processing and presenting of exogenous antigen by an antigen-presenting cell.**

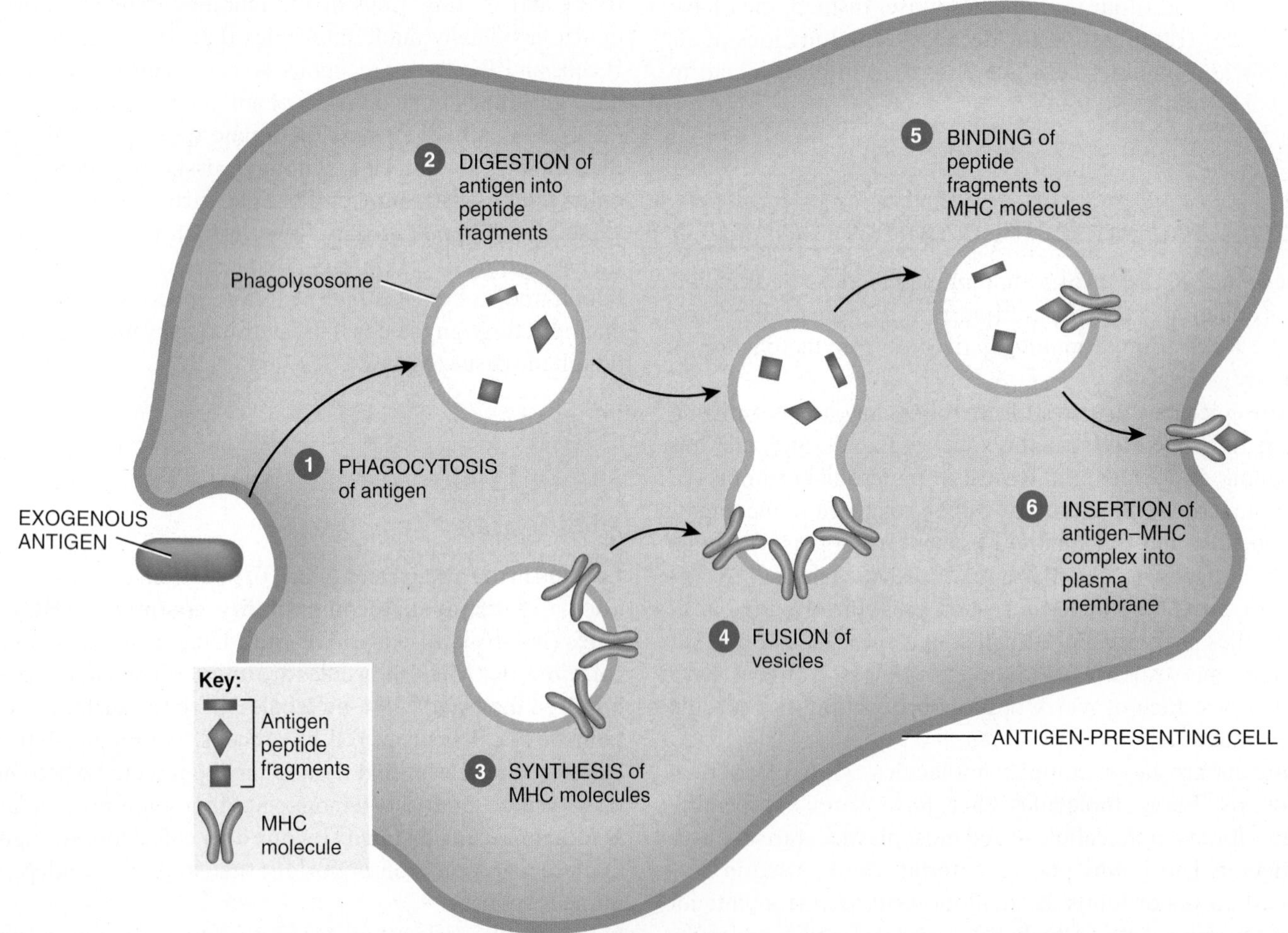


**Fragments of exogenous antigens are processed and then presented with MHC molecules on the surface of an antigen-presenting cell.**

2. ***Digestion of antigen into peptide fragments.*** Within the phagolysosome, protein-digesting enzymes split large antigens into short peptide fragments.

3. ***Synthesis of MHC molecules.*** At the same time the antigen is being digested, the APC synthesizes MHC molecules and packages them into vesicles.

4. ***Fusion of vesicles.*** The vesicles containing antigen peptide fragments and MHC molecules merge and fuse.

5. ***Binding of peptide fragments to MHC molecules.*** After fusion of the two vesicles, antigen peptide fragments bind to MHC molecules, forming **antigen–MHC complexes**.

6. ***Insertion of antigen–MHC complexes into plasma membrane.*** The combined vesicle that contains antigen–MHC complexes undergoes exocytosis. As a result, antigen–MHC complexes are inserted into the plasma membrane.

After processing an exogenous antigen, the APC migrates to lymphatic tissue to present the antigen to T cells. Within lymphatic tissue, a small number of T cells that have compatibly shaped receptors recognize and bind to the antigen–MHC complex, triggering an adaptive immune response. The presentation of exogenous antigen together with MHC molecules by APCs informs T cells that intruders are present in the body and that combative action should begin.

### Processing of Endogenous Antigens

Foreign antigens that are present inside body cells are termed **endogenous antigens** (en-DOJ-e-nus). Such antigens may be viral proteins produced after a virus infects the cell and takes over the cell's metabolic machinery, toxins produced from intracellular bacteria, or abnormal proteins synthesized by a cancerous cell. Within infected cells, fragments of endogenous antigens bind with MHC molecules. The resulting endogenous antigen–MHC complex then moves to the surface of the cell, where it is inserted into the plasma membrane. The display of an endogenous antigen bound to an MHC molecule signals that a cell has been infected and needs help.

## Cytokines

**Cytokines** (SĪ-tō-kīns) are small protein hormones that stimulate or inhibit many normal cell functions in the immune system, such as cell growth and differentiation. Lymphocytes and antigen-presenting cells secrete cytokines, as do a variety of other cells. Cytokines include interferons and interleukins. Some cytokines stimulate proliferation of blood cells in red bone marrow. Others regulate activities of cells involved in innate defenses or adaptive immune responses.

✓ **CHECKPOINT**

**23.** What is meant when we say a cell is immunocompetent, and which body cells can become immunocompetent?

**24.** Where do T cells and B cells develop?

**25.** What are the functions of cell-mediated immunity and antibody-mediated immunity?

**26.** How do antigens induce an immune response? Use the terms *antigen processing* and *antigen presentation* in your answer.

**27.** What is the normal function of major histocompatibility complex molecules?

**28.** Give three examples of endogenous antigens.

# 21.7 In cell-mediated immunity, cytotoxic T cells directly attack target cells.

The presentation of an antigen–MHC complex by APCs informs T cells that intruders are present in the body and that combative action should begin. A cell-mediated immune response begins with *activation* of a small number of T cells by a specific antigen. Once a T cell has been activated, it undergoes clonal selection. Recall that clonal selection is the process by which a lymphocyte divides and differentiates in response to a specific antigen. The resulting clone of cells can recognize the same antigen as the original lymphocyte (see Figure 21.11). Some of the cells of a T cell clone become effector cells, while other cells of the clone become memory cells. The effector cells of a T cell clone carry out immune responses that ultimately result in *elimination* of the intruder.

## Activation of T Cells

At any given time, most T cells are inactive. Antigen receptors on the surface of T cells, called **T cell receptors (TCRs)**, recognize and bind to specific foreign antigen fragments that are presented in antigen–MHC complexes. There are millions of different T cells; each has its own unique receptors that can recognize a specific antigen–MHC complex. When an antigen enters the body, only a few T cells have receptors that can recognize and bind to that specific antigen–MHC complex (antigen recognition). Antigen recognition also involves other surface proteins on T cells, the CD4 or CD8 proteins. These proteins help maintain coupling of the antigen–MHC complexes with the T cell receptor.

Binding of a T cell to an antigen–MHC complex is the *first signal* in activation of a T cell. A T cell becomes activated only if it binds to the foreign antigen and at the same time receives a *second signal*, a process known as **costimulation**. Most costimulators are cytokines, such as **interleukin-2 (IL-2)**.

The need for two signals to activate a T cell is a little like starting and driving a car: When you insert the correct key (antigen) in the ignition (T cell receptor) and turn it, the car starts (recognition of specific antigen), but it cannot move forward until you move the gear shift into drive (costimulation). The need for costimulation may prevent immune responses from occurring accidentally. Different costimulators affect the activated T cell in different ways, just as shifting a car into reverse has a different effect than shifting it into drive. Once a T cell has received these two signals (antigen recognition and costimulation), it is *activated*. An activated T cell subsequently undergoes clonal selection.

### Activation and Clonal Selection of Helper T Cells

During T cell development, two major types of mature T cells exit the thymus: helper T cells and cytotoxic T cells (see Figure 21.11). Inactive **helper T cells** recognize MHC molecules at the surface of an APC (Figure 21.14). With the aid of the CD4 protein, the helper T cell and APC interact with each other, costimulation occurs, and the helper T cell becomes activated.

Once activated, the helper T cell undergoes clonal selection (Figure 21.14). The result is the formation of a clone of helper T cells that consists of active helper T cells and memory helper T cells. Within hours after costimulation, **active helper T cells** start secreting a variety of cytokines. One important cytokine produced by helper T cells is interleukin-2 (IL-2), which is needed for virtually all immune responses and is the prime trigger of T cell proliferation. IL-2 can act as a costimulator for resting T cells, and it enhances activation and proliferation of T cells, B cells, and natural killer cells.

The **memory helper T cells** of a helper T cell clone are not active cells. However, if the same antigen enters the body again in the future, memory helper T cells can quickly proliferate and differentiate into more active helper T cells and more memory helper T cells.

**FIGURE 21.14 Activation and clonal selection of a helper T cell.**

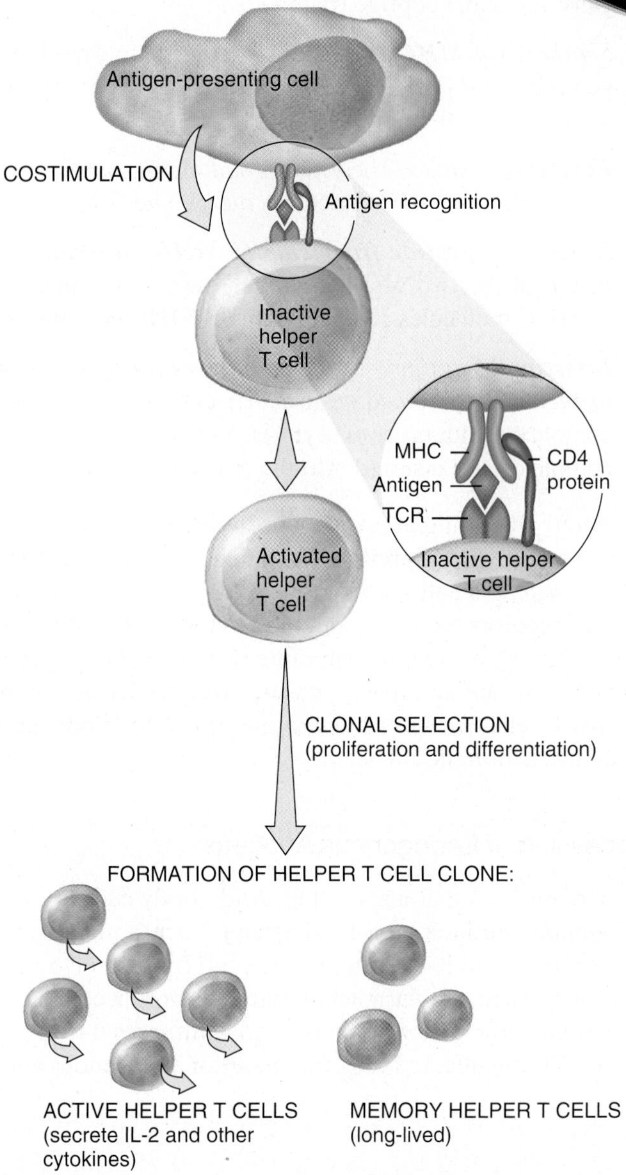

Antigen-presenting cell
COSTIMULATION
Antigen recognition
Inactive helper T cell
MHC
Antigen
TCR
CD4 protein
Inactive helper T cell
Activated helper T cell
CLONAL SELECTION (proliferation and differentiation)
FORMATION OF HELPER T CELL CLONE:
ACTIVE HELPER T CELLS (secrete IL-2 and other cytokines)
MEMORY HELPER T CELLS (long-lived)

**Once a helper T cell is activated, it forms a clone of active and memory helper T cells.**

### Activation and Clonal Selection of Cytotoxic T Cells

Most T cells that display CD8 develop into **cytotoxic T cells** (sī′-tō-TOK-sik). Cytotoxic T cells recognize foreign antigens combined with MHC molecules on the surfaces of body cells infected by microbes, some tumor cells, and cells of a tissue transplant (Figure 21.15). The plasma membranes of cytotoxic T cells have the CD8 protein. Antigen recognition occurs as the T cell receptor and CD8 protein interact with the MHC molecule. Following antigenic recognition, costimulation occurs with interleukin-2 or other cytokines produced by helper T cells.

Once activated, the cytotoxic T cell undergoes clonal selection. The result is the formation of a clone of cytotoxic T cells that consists of active cytotoxic T cells and memory cytotoxic T cells. **Active cytotoxic T cells** are the effector cells of a T cell clone; they attack other body cells that have been infected with the antigen. **Memory cytotoxic T cells** do not attack infected body cells. Instead, they can quickly proliferate and differentiate into more active cytotoxic T cells and more memory cytotoxic T cells if the same antigen enters the body at a future time.

FIGURE 21.15 **Activation and clonal selection of a cytotoxic T cell.**

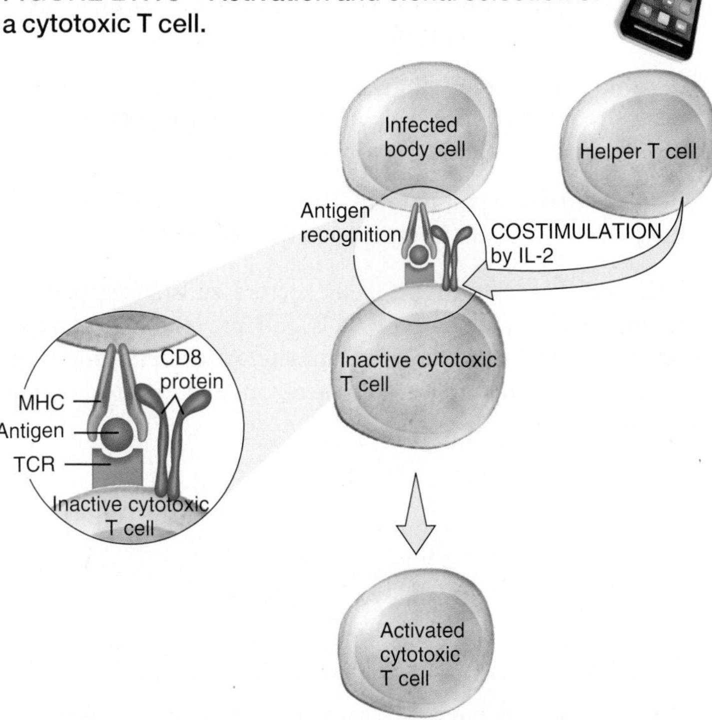


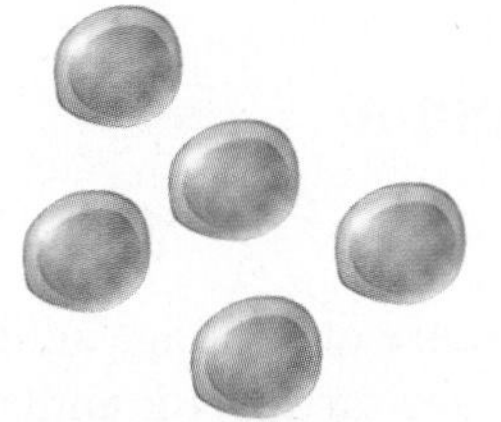


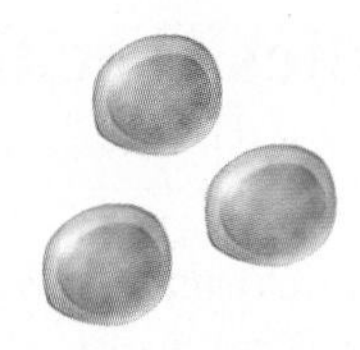


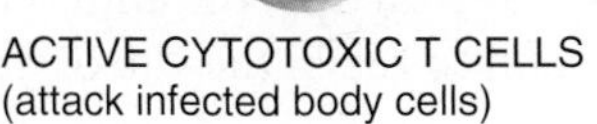

**Once a cytotoxic T cell is activated, it forms a clone of active and memory cytotoxic T cells.**

## Elimination of Invaders

Cytotoxic T cells are the soldiers that march forth to do battle with foreign invaders in cell-mediated immune responses. The name "cytotoxic" reflects their function—killing cells. They leave lymphatic organs and tissues and migrate to seek out and destroy infected target cells, cancer cells, and transplanted cells (Figure 21.16). Cytotoxic T cells recognize and attach to target cells. Then, the cytotoxic T cells deliver a "lethal hit" that kills the target cells.

FIGURE 21.16 **Activity of cytotoxic T cells.** After delivering a "lethal hit," a cytotoxic T cell can detach and attack another infected target cell displaying the same antigen.

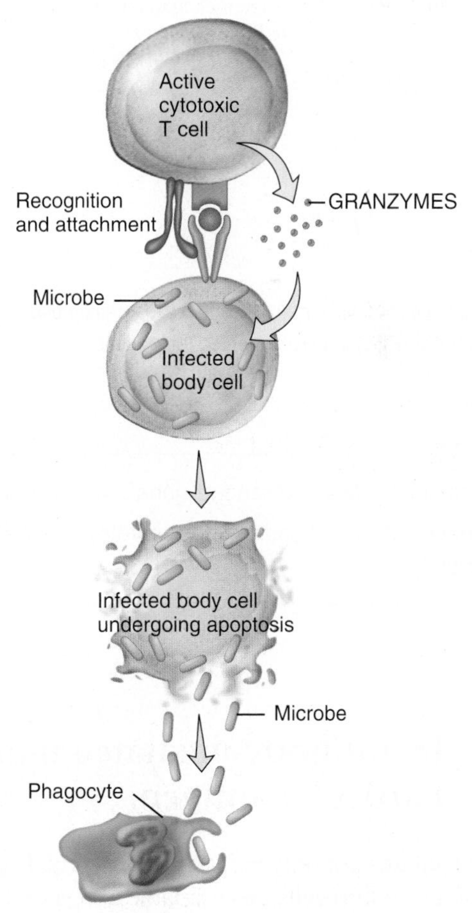


(a) Cytotoxic T cell destruction of infected cell by release of granzymes that cause apoptosis; released microbes are destroyed by phagocyte

(continues)

FIGURE 21.16 (continued)

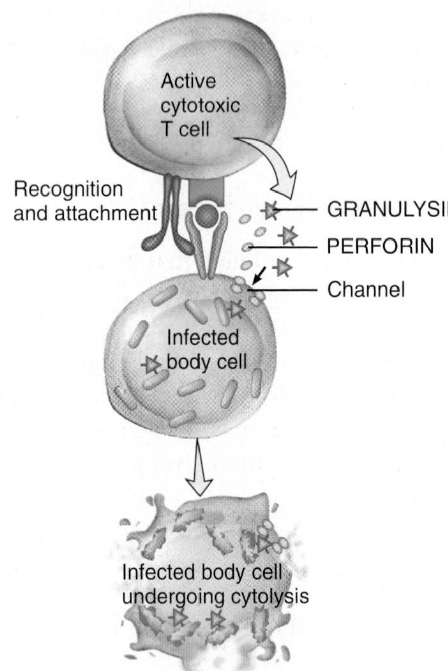


(b) Cytotoxic T cell destruction of infected cell by release of perforins that cause cytolysis; microbes are destroyed by granulysin

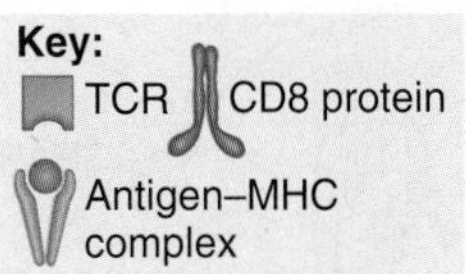


**Cytotoxic T cells release granzymes that trigger apoptosis and perforin that triggers cytolysis of infected target cells.**

Cytotoxic T cells kill infected target body cells much like natural killer cells do. The major difference is that cytotoxic T cells have receptors specific for a particular microbe and thus kill only target body cells infected with *one* particular type of microbe. In contrast, natural killer cells can destroy a wide variety of microbe-infected body cells.

After binding to infected target cells that have microbial antigens displayed on their surface, cytotoxic T cells kill the infected cell with various lethal substances. Cytotoxic T cells release the following substances:

- ***Granzymes.*** Granzymes trigger *apoptosis* in the infected cell (Figure 21.16a). Once the infected cell is destroyed, phagocytes kill the released microbes.
- ***Perforin.*** Perforin creates channels in the plasma membrane of the infected cell (Figure 21.16b). As extracellular fluid flows into the target cell, the infected cell bursts.
- ***Granulysin.*** **Granulysin** (gran′-ū-LĪ-sin) passes into the target cell through perforin channels then destroys the microbes by creating holes in their plasma membranes.
- ***Lymphotoxin.*** **Lymphotoxin** (lim′-fo-TOK-sin) causes the target cell's DNA to fragment and the cell to die.

In addition, cytotoxic T cells secrete substances that attract and activate phagocytic cells and prevent migration of phagocytes away from the infection site. After detaching from a target cell, a cytotoxic T cell can seek out and destroy another target cell.

### ✓ CHECKPOINT

**29.** What are the first and second signals in activation of a T cell?

**30.** What are the functions of helper, cytotoxic, and memory T cells?

**31.** Besides cells infected by microbes, which other types of cells do cytotoxic T cells attack?

**32.** How do cytotoxic T cells kill infected target cells?

## 21.8 In antibody-mediated immunity, antibodies specifically target a particular antigen.

The body contains not only millions of different T cells but also millions of different B cells, each capable of responding to a specific antigen. Cytotoxic T cells leave lymphatic tissues to seek out and destroy a foreign antigen, but B cells stay put within a lymphatic organ or tissue. In the presence of a foreign antigen, a specific B cell forms plasma cells that secrete specific antibodies, which circulate in the lymph and blood to reach the site of invasion and disable the antigen.

## Activation and Clonal Selection of B Cells

During activation of a B cell, an antigen binds to **B cell receptors** (Figure 21.17). B cell antigen receptors are chemically similar to the antibodies that eventually are secreted by the plasma cells. B cells can respond to an unprocessed antigen, but their response is much more intense when they process the antigen first. Antigen processing begins as the B cell takes in the antigen, breaks it down into peptide fragments, combines it with MHC molecules, and moves the antigen–MHC complex to the B cell plasma membrane. Helper T cells then recognize the antigen–MHC complex and deliver the costimulation needed for B cell division and differentiation. The helper T cell produces interleukin-2 and other cytokines, which function as costimulators to activate B cells.

Once activated, a B cell undergoes clonal selection, forming a clone of plasma cells and memory B cells (Figure 21.17). **Plasma cells** are the effector cells of a B cell clone; they secrete specific antibodies. A few days after exposure to an antigen, a plasma cell secretes hundreds of millions of antibodies each day for about 4 or 5 days, until the plasma cell dies. Most antibodies travel in lymph and blood to the invasion site. Interleukins produced by helper T cells enhance B cell division into plasma cells and secretion of antibodies by plasma cells. **Memory B cells** do not secrete antibodies. Instead, they can quickly divide and differentiate into more plasma cells and more memory B cells should the same antigen reappear at a future time.

Different antigens stimulate different B cells to develop into plasma cells and their accompanying memory B cells. All of the B cells of a particular clone are capable of secreting only one type of antibody, which is identical to the antigen receptor displayed by the B cell that first responded.

FIGURE 21.17 **Activation and clonal selection of B cells.** Plasma cells are actually much larger than B cells.

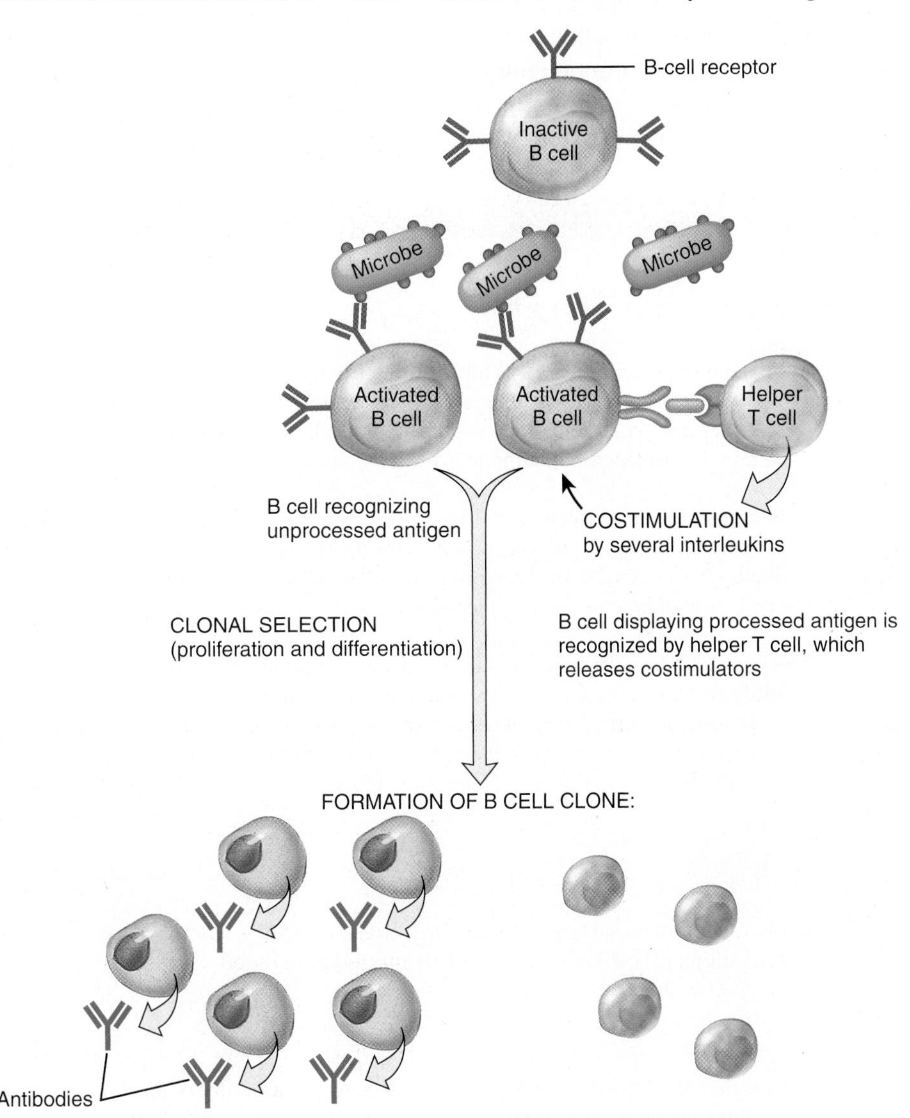


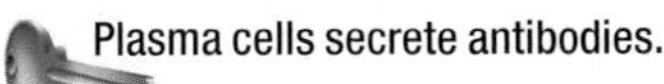
Plasma cells secrete antibodies.

## CLINICAL CONNECTION | *Infectious Mononucleosis*

**Infectious mononucleosis** (mon′-ō-noo-klē-Ō-sis) or "mono" is a contagious disease caused by the *Epstein–Barr virus* (*EBV*). It occurs mainly in children and young adults, and more often in females than in males. The virus most commonly enters the body through intimate oral contact such as kissing, which accounts for its common name, the "kissing disease." EBV then multiplies in lymphatic tissues and filters into the blood, where it infects and multiplies in B cells, the primary host cells. Because of this infection, the B cells become so enlarged and abnormal in appearance that they resemble monocytes, the primary reason for the term *mononucleosis*. In addition to an elevated white blood cell count with an abnormally high percentage of lymphocytes, signs and symptoms include fatigue, headache, dizziness, sore throat, enlarged and tender lymph nodes, and fever. There is no cure for infectious mononucleosis, but the disease usually runs its course in a few weeks.

## Antibodies

An antibody can combine specifically with the epitope on the antigen that triggered the antibody's production. The antibody's structure matches its antigen much as a lock accepts a specific key. Antibodies belong to a group of glycoproteins called globulins, and for this reason they are also known as **immunoglobulins (Igs)** (im′-ū-nō-GLOB-ū-lins). Antibodies assume either a T shape or a Y shape, with each of the two "arms" serving as an **antigen-binding site**, the part of the antibody that recognizes and attaches specifically to a particular antigen (Table 21.2). Because antibodies have two antigen-binding sites, they are able to simultaneously bind to epitopes that are some distance apart—for example, on the surfaces of two different microbes.

The five classes of antibodies are designated IgG, IgA, IgM, IgD, and IgE. Each class has a distinct chemical structure and a specific biological role. Because they appear first and are relatively short-lived, IgM antibodies indicate a recent invasion. In a sick

**TABLE 21.2**

Classes of Immunoglobulins

| NAME AND STRUCTURE | CHARACTERISTICS AND FUNCTIONS |
|---|---|
| **IgG** | Most abundant, about 80% of all antibodies in blood; found in blood, lymph, and intestines; monomer (one-unit) structure; protects against bacteria and viruses by enhancing phagocytosis, neutralizing toxins, and triggering complement system; is the only class of antibody to cross placenta from mother to fetus, conferring considerable immune protection in newborns |
| **IgA** | Found mainly in sweat, tears, saliva, mucus, breast milk, and gastrointestinal secretions; smaller quantities are present in blood and lymph; makes up 10–15% of all antibodies in blood; occurs as monomers and dimers (two units); levels decrease during stress, lowering resistance to infection; provides localized protection of mucous membranes against bacteria and viruses |
| **IgM** | About 5–10% of all antibodies in blood; also found in lymph; occurs as pentamers (five units); first antibody class to be secreted by plasma cells after initial exposure to any antigen; activates complement and causes agglutination and lysis of microbes; also present as monomers on surfaces of B cells, where they serve as antigen receptors; in blood plasma, anti-A and anti-B antibodies of ABO blood group, which bind to A and B antigens during incompatible blood transfusions, are also IgM antibodies (see Figure 18.7) |
| **IgD** | Mainly found on surfaces of B cells as antigen receptors, where it occurs as monomers; involved in activation of B cells; about 0.2% of all antibodies in blood |
| **IgE** | Less than 0.1% of all antibodies in blood; occurs as monomers; located on mast cells and basophils; involved in allergic and hypersensitivity reactions; provides protection against parasitic worms |

patient, the responsible pathogen may be suggested by the presence of high levels of IgM specific to a particular organism. Resistance of the fetus and newborn baby to infection stems mainly from maternal IgG antibodies that cross the placenta before birth and IgA antibodies in breast milk after birth. Table 21.2 summarizes the structures and functions of the five classes of antibodies.

## Antibody Actions

Antibodies form antigen–antibody complexes with the antigen that initiated the antibody production. Antigen–antibody binding disables antigens in various ways. Actions of the five classes of antibodies differ somewhat but include the following:

- ***Neutralize antigen.*** The binding of an antibody with its antigen neutralizes, or blocks, some bacterial toxins and prevents attachment of some viruses to body cells.
- ***Immobilize bacteria.*** If antibodies form against antigens on the cilia or flagella of motile bacteria, the antigen–antibody complex may cause some bacteria to lose their motility, which limits their spread into nearby tissues.
- ***Agglutinate and precipitate antigen.*** Because antibodies have two antigen binding sites, antigen–antibody binding may connect pathogens to one another, causing agglutination (clumping together). Phagocytic cells ingest agglutinated microbes more readily.
- ***Activate complement.*** Antigen–antibody complexes activate complement proteins (see Concept 21.5).
- ***Enhance phagocytosis.*** Once antigens have bound to an antibody, the antibody acts as a "flag" that attracts phagocytes. Antibodies enhance the activity of phagocytes by causing agglutination and precipitation, by activating complement, and by coating microbes so that they are more susceptible to phagocytosis.

Table 21.3 summarizes the activities of cells involved in adaptive immune responses.

**TABLE 21.3**

Functions of Cells Participating in Adaptive Immune Responses

| CELL | FUNCTIONS |
|---|---|
| **ANTIGEN-PRESENTING CELLS (APCs)** | |
| **Macrophage** | Processing and presentation of foreign antigens to T cells; secretion of interleukin-1, which stimulates secretion of interleukin-2 by helper T cells and induces proliferation of B cells; secretion of interferons that stimulate T cell growth |
| **Dendritic cell** | Processes and presents antigen to T cells and B cells; found in mucous membranes, skin, and lymph nodes |
| **B cell** | Processes and presents antigen to helper T cells |
| **LYMPHOCYTES** | |
| **Cytotoxic T cell** | Kills host target cells by releasing granzymes that induce apoptosis, perforin that forms channels to cause cytolysis, granulysin that destroys microbes, lymphotoxin that destroys target cell DNA, gamma-interferon that attracts macrophages and increases their phagocytic activity, and macrophage migration inhibition factor that prevents macrophage migration from site of infection |
| **Helper T cell** | Cooperates with B cells to amplify antibody production by plasma cells and secretes interleukin-2, which stimulates proliferation of T cells and B cells |
| **Memory T cell** | Remains in lymphatic tissue and recognizes original invading antigens, even years after first encounter |
| **B cell** | Differentiates into antibody-producing plasma cell |
| **Plasma cell** | Descendant of B cell that produces and secretes antibodies |
| **Memory B cell** | Descendant of B cell that remains after immune response and is ready to respond rapidly and forcefully should the same antigen enter body in future |

✓ **CHECKPOINT**

**33.** Which type of T cell participates in both cell-mediated and antibody-mediated immune responses?

**34.** Outline the steps in antibody production.

**35.** How many antigens can bind to one antibody molecule?

**36.** How do the five classes of antibodies differ in function?

**37.** How do antibodies disable antigens?

## CLINICAL CONNECTION | *Monoclonal Antibodies*

The antibodies produced against a given antigen by plasma cells can be harvested from an individual's blood. However, because an antigen typically has many epitopes, several different clones of plasma cells produce different antibodies against the antigen. If a single plasma cell could be isolated and induced to proliferate into a clone of identical plasma cells, then a large quantity of identical antibodies could be produced. Unfortunately, lymphocytes and plasma cells are difficult to grow in culture, so scientists sidestepped this difficulty by fusing B cells with tumor cells that grow easily and proliferate endlessly. The resulting hybrid cell is called a **hybridoma** (hī-bri-DŌ-ma). Hybridomas are long-term sources of large quantities of pure, identical antibodies, called **monoclonal antibodies (MAbs)** (mon′-ō-KLŌ-nal) because they come from a single clone of identical cells. One clinical use of monoclonal antibodies is for measuring levels of a drug in a patient's blood. Other uses include the diagnosis of strep throat, pregnancy, allergies, and diseases such as hepatitis, rabies, and some sexually transmitted diseases. MAbs have also been used to detect cancer at an early stage and to ascertain the extent of metastasis. They may also be useful in preparing vaccines to counteract the rejection associated with transplants, to treat autoimmune diseases, and perhaps to treat AIDS.

## RETURN TO **Marlene's Story**

"It was just a small cut. How could it have gotten so infected?" Francis asked the doctor.

The doctor responded, "The immune system relies on two basic types of defenses, innate and adaptive. The body's innate or nonspecific response is to send white blood cells to the site of an injury. Certain white blood cells will recognize anything that's foreign to the body and begin the attack. Lymphatic vessels help carry foreign substances to the lymphatic organs, where they are presented to other white blood cells called T cells and B cells that then mount a second, more specific attack. Marlene, in your case, the swelling of your arm and preexisting inflammation from the surgery put you at a higher risk of infection since some of your lymph nodes were missing and lymphatic fluid couldn't return effectively to your lymphatic organs. I'd like to start you on high-dose penicillin; that should help reduce the infection and give your immune system a chance to fight off this bug more efficiently."

After consulting with Marlene's oncologist, the internist from the urgent care center admitted her to the hospital for several days, to make sure the infection could be monitored, and ordered an intravenous antibiotic. After a few days, the red streaks on her arm faded, and her temperature, which had spiked at 102°F, had fallen back to near normal. Finally, they let her return home.

"How are you feeling today?" Francis asked.

"My arm is still swollen. It hurts, and my throat feels swollen, like I'm getting a cold. After all this, I don't know how anything could survive in my body."

Francis smiled at her. "The therapist told me that once your infection was under control, they could help you with the swelling in your arm. Did you make an appointment?"

"Yes, I see her in 2 weeks. She said that massage would help reduce the swelling and mentioned something about a special compression garment."

Marlene made the trip to the physical therapist's office as scheduled. "I suggest a course of self-massage. I'd also like to fit you for a compression garment, which should help reduce the swelling if you're going to be inactive for any length of time. Keeping the swelling under control reduces the risk of infection and will improve the function of your arm and hand. Any questions?" the therapist asked.

"What caused the swelling in the first place?" asked Marlene.

"With the type of surgery you had, you'll always be at risk for lymphedema, the accumulation of fluid in lymphatic vessels. Accumulated fluids are normally returned to the bloodstream by lymphatic vessels; but in your case, swelling has occurred because those vessels are blocked."

D. *Which lymphatic organs in Marlene's body would normally trap the antigens and allow an adaptive immune response to be mounted?*

E. *Why might the lymph nodes in Marlene's throat be enlarged?*

F. *How would massage help reduce the fluid buildup in Marlene's arm? How would improving the flow of lymph reduce Marlene's future chances of getting a similar infection and lymphangitis?*

# 21.9 Immunological memory results in a more intense secondary response to an antigen.

A hallmark of immune responses is memory for specific antigens that have triggered immune responses in the past. **Immunological memory** is due to the presence of long-lasting antibodies and very long-lived memory cells that arise during clonal selection of antigen-stimulated B cells and T cells.

Immune responses, whether cell-mediated or antibody-mediated, are much quicker and more intense after a second or subsequent exposure to an antigen than after the first exposure. Initially, only a few cells have the correct antigen receptors to respond, and the immune response may take several days to build to maximum intensity. Because thousands of memory cells exist after an initial encounter with an antigen, within hours of the next appearance of the antigen memory cells can divide and differentiate into helper T cells, cytotoxic T cells, or plasma cells.

One measure of immunological memory is *antibody titer* (TĪ-ter), the amount of antibody in blood serum. After an initial contact with an antigen, no antibodies are present for several days (Figure 21.18). Then, the levels of antibodies slowly increase, first IgM and then IgG, followed by a gradual decline in antibody titer. This is the **primary response**.

Memory cells may remain for decades. Every new encounter with the same antigen results in a rapid division of memory cells. After subsequent encounters, the antibody titer is far greater than during a primary response and consists mainly of IgG antibodies. This accelerated, more intense response is called the **secondary response**. Antibodies produced during a secondary response have an even higher affinity for the antigen than those produced during a primary response, and thus they are more successful in disposing of it.

Primary and secondary responses occur during microbial infection. When you recover from an infection without taking antimicrobial drugs, it is usually because of the primary response. If the same microbe infects you later, the secondary response could be so swift that the microbes are destroyed before you exhibit any signs or symptoms of infection.

Immunological memory provides the basis for immunization by vaccination against certain diseases, such as polio. When you receive the vaccine, which may contain weakened or killed whole microbes or portions of microbes, your B cells and T cells are activated to produce a primary response. Should you subsequently encounter the living pathogen as an infecting microbe, waiting memory cells initiate a secondary response.

Table 21.4 summarizes the various ways to acquire adaptive immunity.

**FIGURE 21.18 Secretion of antibodies.** The primary response (after first exposure) is milder than the secondary response (after second exposure) to a given antigen.

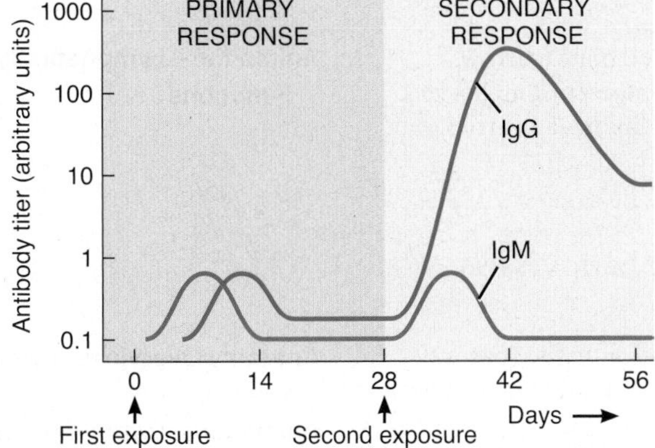


Immunological memory is the basis for successful immunization by vaccination.

**TABLE 21.4**

Ways to Acquire Adaptive Immunity

| METHOD | DESCRIPTION |
|---|---|
| **Naturally acquired active immunity** | Following exposure to a microbe, antigen recognition by B cells and T cells and costimulation lead to formation of antibody-secreting plasma cells, cytotoxic T cells, and B and T memory cells |
| **Naturally acquired passive immunity** | IgG antibodies are transferred from mother to fetus across placenta, or IgA antibodies are transferred from mother to baby in milk during breast-feeding |
| **Artificially acquired active immunity** | Antigens introduced during vaccination stimulate cell-mediated and antibody-mediated immune responses, leading to production of memory cells; antigens are pretreated to be immunogenic but not pathogenic (they will trigger an immune response but not cause significant illness) |
| **Artificially acquired passive immunity** | Intravenous injection of immunoglobulins (antibodies) |

## ✓ CHECKPOINT

**38.** What role do memory cells play in immunological memory?

**39.** How is the secondary response to an antigen different from the primary response?

**40.** Which type of antibody responds most strongly during the secondary response?

## Marlene's Story — EPILOGUE AND DISCUSSION

Marlene is a 57-year-old woman recovering from a breast lumpectomy and lymph node removal due to breast cancer treatment, and subsequent lymphangitis from a cut to the hand that becomes infected. At an urgent care center she presents with a fever and infection. She is admitted to the hospital, where she is placed on a course of intravenous antibiotics. Following resolution of the infection, she consults with a physical therapist, who suggests treatment to reduce swelling and discomfort due to lymph node removal.

G. *Why would reduced lymph return have increased Marlene's risk of localized infection?*

H. *How would removal of her lymph nodes have reduced Marlene's ability to fight infection?*

I. *If Marlene suffered a cut on the side of her body opposite the surgery, would her body's response have been different? Why or why not?*

# Concept and Resource Summary

| Concept | Resources (WILEY PLUS) |
|---|---|
| **Introduction**<br>**1.** The ability to ward off disease is called **immunity** (**resistance**). Lack of resistance is called **susceptibility**.<br>**2.** **Innate (nonspecific) immunity** is always present to provide rapid responses against disease.<br>**3.** **Adaptive (specific) immunity** develops in response to specific **pathogens** when they have surpassed innate immunity defenses. | |
| **Concept 21.1 The lymphatic system drains interstitial fluid, transports dietary lipids, and protects against invasion.**<br>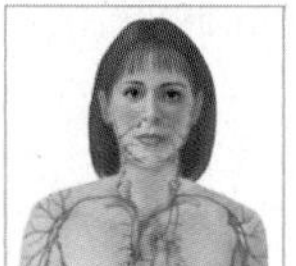<br>**1.** The **lymphatic system** consists of **lymph**, lymphatic tissues, and red bone marrow.<br>**2.** **Lymphatic tissue** is specialized reticular tissue containing large numbers of lymphocytes.<br>**3.** The lymphatic system drains excess interstitial fluid, transports lipids, and carries out immune responses against invasion. | Animation—Lymphatic System Functions |
| **Concept 21.2 Lymph flows through lymphatic capillaries, lymphatic vessels, and lymph nodes.**<br>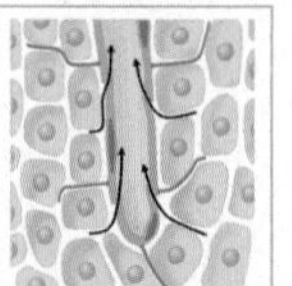<br>**1.** Lymphatic drainage begins at closed-ended **lymphatic capillaries** in tissue spaces between cells. The structure of lymphatic capillaries allows interstitial fluid to flow in but not out.<br>**2.** Lymph capillaries merge to form larger vessels, called **lymphatic vessels**, which convey lymph into and out of **lymph nodes**.<br>**3.** The route of lymph flow is from lymph capillaries to lymphatic vessels, to **lymph trunks**, then to the **thoracic duct** and **right lymphatic duct**.<br>**4.** Lymph flows due to the "milking action" of skeletal muscle contractions and pressure changes that occur during respiratory movements. Valves in lymphatic vessels aid lymph flow by preventing backflow of lymph. | Anatomy Overview—Lymphatic Vessels<br>Animation—Lymph Formation and Flow<br>Exercise—Lymphatic Highway<br>Clinical Connection—Metastasis Through Lymphatic Vessels |

| Concept | Resources |
|---|---|

## Concept 21.3 The lymphatic organs and tissues include the thymus, lymph nodes, spleen, and lymphatic follicles.

1. **Primary lymphatic organs** and **tissues** are the sites where stem cells divide and develop into **immunocompetent** B cells and T cells. They are the red bone marrow and the thymus.
2. **Secondary lymphatic organs** and **tissues** are the sites where most immune responses occur. They include lymph nodes, the spleen, and lymphatic follicles.
3. The **thymus** lies posterior to the sternum and superior to the heart. It is the site of T cell maturation.
4. **Lymph nodes** are located along lymphatic vessels. Lymph enters lymph nodes through **afferent lymphatic vessels**, is filtered, and exits through **efferent lymphatic vessels**.
5. The **spleen** is the largest mass of lymphatic tissue in the body. It is a site where B cells divide into plasma cells and macrophages phagocytize worn-out red blood cells and platelets. Within the spleen, B cells and T cells carry out immune functions and macrophages phagocytize blood-borne pathogens and worn-out blood cells.
6. **Lymphatic follicles** are concentrations of lymphatic tissue that are not surrounded by a capsule. They are scattered throughout the mucosa of the gastrointestinal, respiratory, urinary, and reproductive tracts and are termed **mucosa-associated lymphatic tissue** (MALT). **Tonsils** are examples of lymphatic follicles.

Anatomy Overview—The Spleen and Lymph Nodes
Anatomy Overview—The Thymus

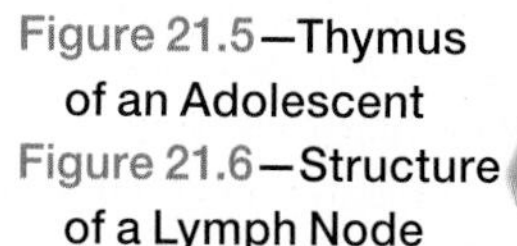

Figure 21.5—Thymus of an Adolescent
Figure 21.6—Structure of a Lymph Node

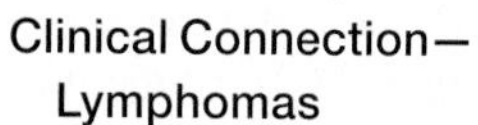

Clinical Connection—Lymphomas

## Concept 21.4 Innate immunity includes external physical and chemical barriers and various internal defenses.

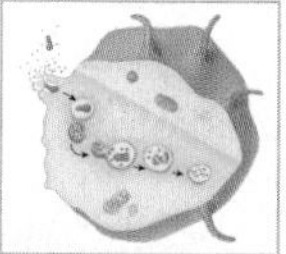

1. Innate (nonspecific) immunity includes physical factors, chemical factors, natural killer cells, phagocytes, inflammation, fever, and antimicrobial proteins.
2. The skin and mucous membranes are the first line of defense against entry of pathogens. The closely packed keratinized cells of the epidermis are a physical barrier to invasion. Mucous membranes have mucus that traps microbes and cilia that sweep microbes away from the lungs. Tears, saliva, urine, and vaginal secretions flush microbes from the body. Defecation and vomiting expel microbes. The acidity of gastric juice and vaginal secretions discourages bacterial growth.
3. Pathogens that get past the first line of defense are attacked by internal defenses. **Natural killer cells** and **phagocytes** attack and kill pathogens and defective cells in the body.
4. **Inflammation** aids disposal of microbes, toxins, or foreign material at the site of an injury and prepares the site for tissue repair.
5. **Fever** intensifies the antiviral effects of interferons, inhibits growth of some microbes, and speeds up body reactions that aid repair.
6. **Antimicrobial proteins** (**interferons** and **complement**) inhibit microbial growth.

Anatomy Overview—The Integument and Disease Resistance
Animation—Introduction to Disease Resistance
Animation—Nonspecific Disease Resistance
Exercise—Integument vs. Disease
Concepts and Connections—Nonspecific Disease Resistance
Concepts and Connections—Role of Integument in Disease Resistance

Clinical Connection—Microbial Evasion of Phagocytosis

## Concept 21.5 The complement system destroys microbes through phagocytosis, cytolysis, and inflammation.

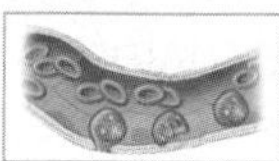

1. The **complement system** is a group of proteins that "complement" immune responses and help clear antigens from the body. Once activated, complement proteins bring about phagocytosis, cytolysis, and inflammation.

Animation—Complement Proteins
Concepts and Connections—Complement Proteins

## Concept 21.6 Adaptive immunity involves the production of a specific lymphocyte or antibody against a specific antigen.

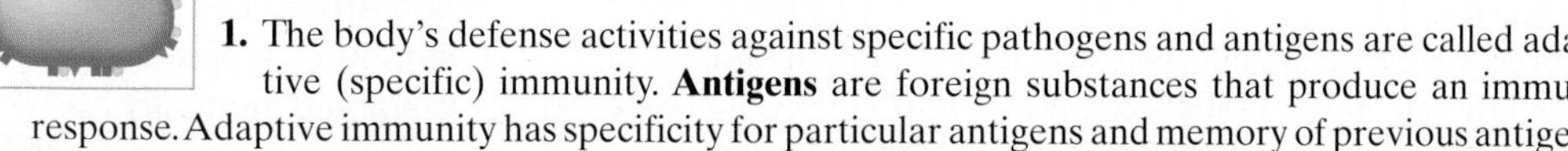

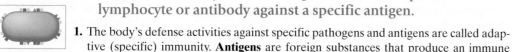

1. The body's defense activities against specific pathogens and antigens are called adaptive (specific) immunity. **Antigens** are foreign substances that produce an immune response. Adaptive immunity has specificity for particular antigens and memory of previous antigens.
2. **B cells** and **T cells** arise from stem cells in red bone marrow. B cells mature in red bone marrow; T cells mature in the thymus.
3. B cells and T cells become immunocompetent, capable of carrying out adaptive immune responses by inserting antigen receptors into their plasma membranes. **Antigen receptors** are molecules that are capable of recognizing specific antigens.

Anatomy Overview—B Lymphocytes
Anatomy Overview—T Lymphocytes
Anatomy Overview—Antigens and Antibodies

## Concept

4. In **cell-mediated immunity**, cytotoxic T cells directly attack invading antigens and are effective against intracellular pathogens, cancer cells, and transplanted tissues. In **antibody-mediated immunity**, B cells become plasma cells that secrete **antibodies** used against extracellular pathogens.
5. **Clonal selection** is the process by which a lymphocyte divides and differentiates into a **clone** of cells that can recognize the same specific antigen as the original lymphocyte. The clone includes effector cells and memory cells. **Effector cells** destroy or inactivate the antigen. **Memory cells** do not actively participate in the initial immune response, but if the antigen reappears in the body in the future, memory cells quickly divide and differentiate into more effector and memory cells.
6. Antigens stimulate the production of specific antibodies. Immune responses are triggered by **epitopes**, particles on the surface of antigens.
7. "Self-antigens" called **major histocompatibility complex (MHC) molecules** are unique to each person's body cells. All cells except red blood cells display MHC molecules.
8. **Antigen-presenting cells** process **exogenous antigens** (present outside body cells), which results in an **antigen–MHC complex** being displayed on the cell's plasma membrane. The exogenous antigen–MHC complexes trigger adaptive immune responses in T cells.
9. **Endogenous antigens** (present inside body cells) bind with MHC molecules. The endogenous antigen–MHC complexes insert in the plasma membrane, which signals that the cell is infected.
10. **Cytokines** are small hormones that stimulate or inhibit many normal cell functions such as growth and differentiation. Other cytokines regulate immune responses.

## Resources

Animation—Introduction to Disease Resistance
Animation—Introduction to Antibody-mediated Immunity
Animation—Introduction to Cell-mediated Immunity
Concepts and Connections—Specific Disease Resistance

Clinical Connection—Cytokine Therapy

---

### Concept 21.7 In cell-mediated immunity, cytotoxic T cells directly attack target cells.

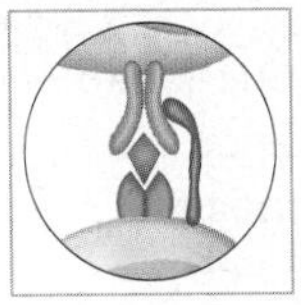

1. A cell-mediated immune response begins with activation of a small number of T cells by a specific antigen and ultimately results in elimination of the intruder.
2. During the activation process, **T cell receptors** recognize and bind to antigen fragments associated with MHC molecules on the surface of a body cell. Activation of T cells also requires **costimulation**, usually by **interleukin-2**. Once a T cell has been activated, it undergoes clonal selection.
3. **Helper T cells** display the CD4 protein and are activated by MHC molecules presented by an APC. Once activated, helper T cells undergo clonal selection, then secrete several cytokines, most importantly interleukin-2, which acts as a costimulator for other T cells, B cells, and natural killer cells.
4. **Cytotoxic T cells** display the CD8 protein and are activated by MHC molecules on the surfaces of body cells infected by microbes, tumor cells, and transplanted tissue cells. Once activated, cytotoxic T cells undergo clonal selection.
5. **Active cytotoxic T cells** kill infected cells by releasing granzymes, which cause target cell apoptosis (phagocytes then kill the released microbes); perforin, which causes target cell cytolysis; and **granulysin**, which enters the target cell to destroy microbes.

Anatomy Overview—T Lymphocytes
Animation—Cell-mediated Immunity
Figure 21.14—Activation and Clonal Selection of a Helper T Cell 
Figure 21.15—Activation and Clonal Selection of a Cytotoxic T Cell
Exercise—Activate and Attack
Concepts and Connections—Specific Disease Resistance

Clinical Connection—Immunological Surveillance

Clinical Connection—Graft Rejection and Tissue Typing

Clinical Connection—AIDS: Acquired Immunodeficiency Syndrome

---

### Concept 21.8 In antibody-mediated immunity, antibodies specifically target a particular antigen.

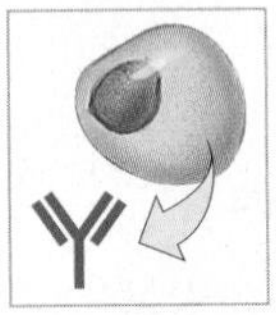

1. An antibody-mediated immune response begins when a specific antigen binds to **B cell receptors**, activating the B cell.
2. B cells can respond to unprocessed antigens, but their response is more intense when they process the antigen by inserting antigen–MHC complexes into the B cell plasma membrane. Helper T cells secrete cytokines that provide costimulation for activation of B cells.
3. Once activated, a B cell undergoes clonal selection, forming a clone of **plasma cells** that secrete antibodies and **memory B cells** to quickly produce more plasma and memory B cells should the same antigen ever reappear.

Anatomy Overview—B Lymphocytes
Anatomy Overview—Antigens and Antibodies
Animation—Antibody-mediated Immunity
Figure 21.17—Activation and Clonal Selection of B Cells 
Exercise—Activate and Attack

**Concept**

4. An antibody is a protein that combines specifically with the antigen that triggered its production. Based on chemistry and structure, antibodies are grouped into five principal classes, each with specific biological roles.
5. Antibodies form antigen–antibody complexes with their complementary antigens, which disables antigens by blocking attachment of the antigen to body cells, immobilizing bacteria, agglutinating and precipitating antigen, activating complement, and enhancing phagocytosis.

**Resources**

Exercise—Antibody Ambush
Concepts and Connections—Specific Disease Resistance
Clinical Connection—Autoimmune Diseases
Clinical Connection—Severe Combined Immunodeficiency Disease
Clinical Connection—Cancer Immunology
Clinical Connection—Systemic Lupus Erythematosus

### Concept 21.9 Immunological memory results in a more intense secondary response to an antigen.

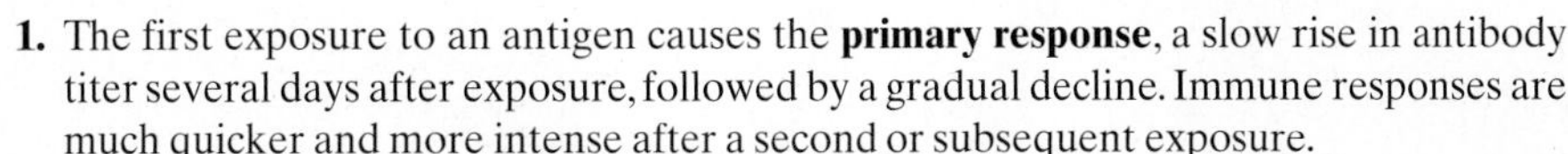

1. The first exposure to an antigen causes the **primary response**, a slow rise in antibody titer several days after exposure, followed by a gradual decline. Immune responses are much quicker and more intense after a second or subsequent exposure.
2. The **secondary response** is an accelerated response to an antigen after a subsequent exposure. Memory B cells and memory T cells that remain in the body produce the secondary response.

Animation—Primary and Secondary Infections
Clinical Connection—Allergic Reactions

## Understanding the Concepts

1. What functions are served by the lymphatic system?
2. Describe how running, inhalation, and exhalation each influence the flow of lymph.
3. Why would the thymus be considered a primary lymphatic organ?
4. Contrast the functions of natural killer cells and phagocytes.
5. What are the functions of inflammation?
6. What causes each of the following signs and symptoms of inflammation: *redness*, *heat*, *swelling*, and *pain*?
7. What purpose is served by fever?
8. What is adaptive immunity? How does it compare with innate immunity?
9. What is the relationship between an antigen and an antibody?
10. Why is it a survival advantage to have APCs in the respiratory tract?
11. How do the CD4 and CD8 proteins assist T cell activation?
12. How many different types of antibodies will be secreted by plasma cells derived from the same clone? Explain your answer.
13. How are cell-mediated and antibody-mediated immune responses similar? How do they differ?
14. According to the graph in Figure 21.18, how much more IgG is circulating in the blood in the secondary response than in the primary response? (*Hint*: Notice that each mark on the antibody titer axis represents a 10-fold increase.)

CHAPTER

# 22

## Cari's Story

Cari had gone to bed early. Her throat had started to feel sore, the same feeling she always got when she was getting a cold. The tickle at the back of her throat wasn't that unusual; she coughed a lot, having been a pack-a-day smoker for 20 years. But this was different from her smoker's cough; this was turning into another cold. She tended to have a lot of them. She woke up after a fitful night of sleep with a pounding headache and perspiration-soaked bed sheets. The back of her throat was dry, and her shoulders ached. She reached for a tissue and blew her nose.

"What's the matter, did you catch another cold?" asked her husband Ken.

"It's just a summer cold, maybe allergies. I'll take some antihistamines and vitamin C," she replied.

"You look pretty bad. Maybe you should call in sick today and see the doctor," advised Ken.

"I just don't have time, Ken. I've got to get the Jenson house finished by September and there have already been delays with the subcontractors." Cari got out of bed, retrieving her cigarettes and lighter from the nightstand. She slid open the patio door, walked out onto the deck, and lit up a cigarette. Ken followed her.

"Cari, maybe this is a good time to quit; you know it's been a year for me now."

Respiratory tract infections can involve the upper or lower respiratory tracts. The respiratory system is susceptible to infectious microorganisms of several types. An individual's response to a respiratory tract infection depends on the type of infectious organism, and the person's age and overall health status. In Cari's case, a long history of smoking has led to chronic respiratory tract infections. As we will see in this chapter, frequent colds are not the only problem caused by her smoking habit.

# The Respiratory System

## INTRODUCTION

Your body's cells continually use oxygen ($O_2$) for the metabolic reactions that release energy from nutrient molecules and produce ATP. At the same time, these reactions release carbon dioxide ($CO_2$). Because an excessive amount of $CO_2$ produces acidity that can be toxic to cells, excess $CO_2$ must be eliminated quickly and efficiently. The cardiovascular and respiratory systems function together to supply $O_2$ to cells and eliminate $CO_2$. The cardiovascular system transports blood containing those gases between the lungs and body cells. The **respiratory system** (RES-pi-ra-tōr-ē) provides for gas exchange—intake of $O_2$ and elimination of $CO_2$. Failure of the exchange of either gas disrupts homeostasis by causing rapid death of cells from oxygen starvation and buildup of waste products. The extensive area of contact between capillary blood vessels of the respiratory system and the external environment allows the body to constantly replenish gases in the internal fluid environment that surrounds every body cell. In addition to functioning in gas exchange, the respiratory system participates in regulating blood pH, contains receptors for the sense of smell, filters inspired air, produces sounds, and rids the body of some water and heat in exhaled air.

## CONCEPTS

**22.1** Inhaled air travels in the upper respiratory system through the nasal cavities and then through the pharynx.

**22.2** Inhaled air travels in the lower respiratory system from the larynx to alveoli.

**22.3** Inhalation and exhalation result from pressure changes caused by muscle contraction and relaxation.

**22.4** Lung volumes and capacities are measured to determine the respiratory status of an individual.

**22.5** Oxygen and carbon dioxide diffusion is based on partial pressure gradients and solubility.

**22.6** Respiration occurs between alveoli and pulmonary capillaries and between systemic capillaries and tissue cells.

**22.7** Oxygen is primarily transported attached to hemoglobin, while carbon dioxide is transported in three different ways.

**22.8** The basic rhythm of respiration is controlled by the respiratory center in the brain stem.

**22.9** Respiration may be modified by cortical influences, chemical stimuli, proprioceptor input, and the inflation reflex.

**22.10** Acid–base balance is maintained by controlling the $H^+$ concentration of body fluids.

# 22.1 Inhaled air travels in the upper respiratory system through the nasal cavities and then through the pharynx.

The **respiratory system** consists of the nose, pharynx (throat), larynx (voice box), trachea (windpipe), bronchi, and lungs (Figure 22.1). Its parts can be classified according to either structure or function. *Structurally,* the respiratory system consists of two parts:

- The **upper respiratory system** includes the nose, pharynx, and associated structures.
- The **lower respiratory system** includes the larynx, trachea, bronchi, and lungs.

*Functionally*, the respiratory system also consists of two parts:

- The **conducting zone** consists of the nose, nasal cavity, pharynx, larynx, trachea, bronchi, bronchioles, and terminal bronchioles—cavities and tubes that filter, warm, and moisten air and conduct it into the lungs.
- The **respiratory zone** consists of the respiratory bronchioles, alveolar ducts, alveolar sacs, and alveoli—structures that are the main sites of gas exchange between air and blood.

FIGURE 22.1 **Structures of the respiratory system.**

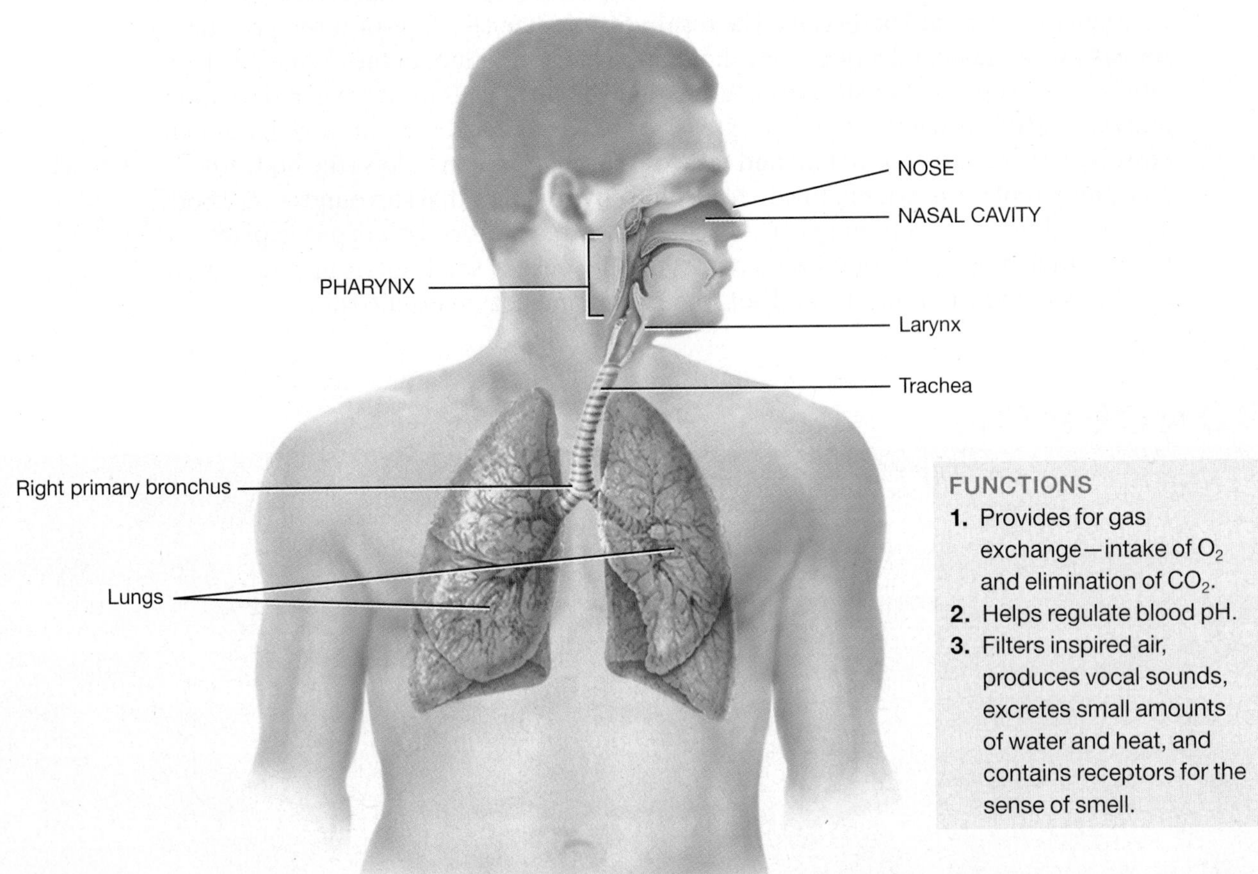
NOSE
NASAL CAVITY
PHARYNX
Larynx
Trachea
Right primary bronchus
Lungs
FUNCTIONS
1. Provides for gas exchange—intake of $O_2$ and elimination of $CO_2$.
2. Helps regulate blood pH.
3. Filters inspired air, produces vocal sounds, excretes small amounts of water and heat, and contains receptors for the sense of smell.

(a) Anterior view showing organs of respiration

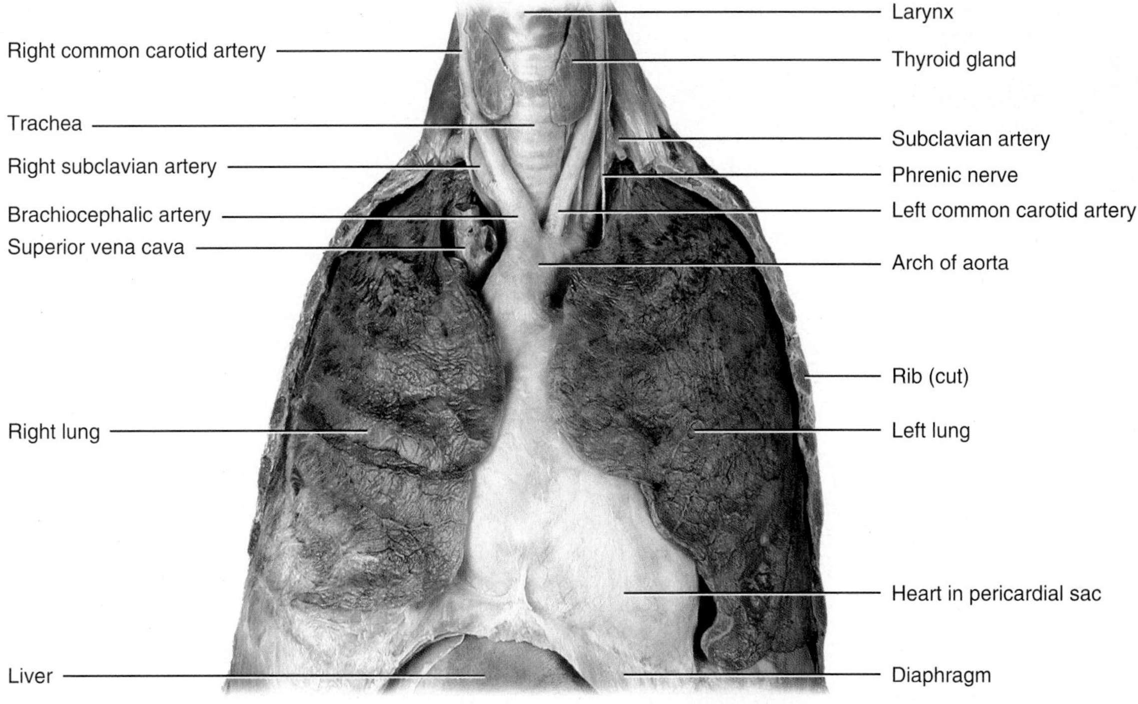


(b) Anterior view of lungs and heart after removal of anterolateral thoracic wall and pleura

**The upper respiratory system includes the nose and pharynx; the lower respiratory system includes the larynx, trachea, bronchi, and lungs.**

## Nose

The **nose** is a specialized organ at the entrance of the respiratory system that is divided into an external portion and an internal nasal cavity. The **external nose** is the portion visible on the face and consists of a supporting framework of bone and cartilage covered with skin and lined by mucous membrane (Figure 22.2a). The frontal bone, nasal bones, and maxillae form the *bony framework* of the external nose. The *cartilaginous framework* of the external nose consists of the **septal cartilage**, which forms the anterior portion of the nasal septum; the **lateral nasal cartilages** inferior to the nasal bones; and the **alar cartilages** (Ā-lar), which form a portion of the walls of the nostrils. Because it consists of pliable hyaline cartilage, the cartilaginous framework is somewhat flexible. The external nose has two openings called the **external nares** (NĀ-rez; singular is *naris*), or **nostrils**.

FIGURE 22.2 **Respiratory structures in the head and neck.**

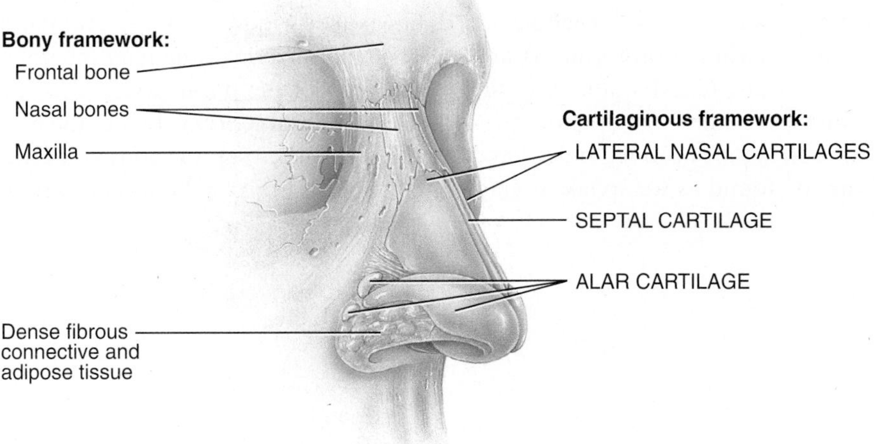


(a) Anterolateral view of external nose showing bony and cartilaginous framework

(continues)

FIGURE 22.2 (continued)

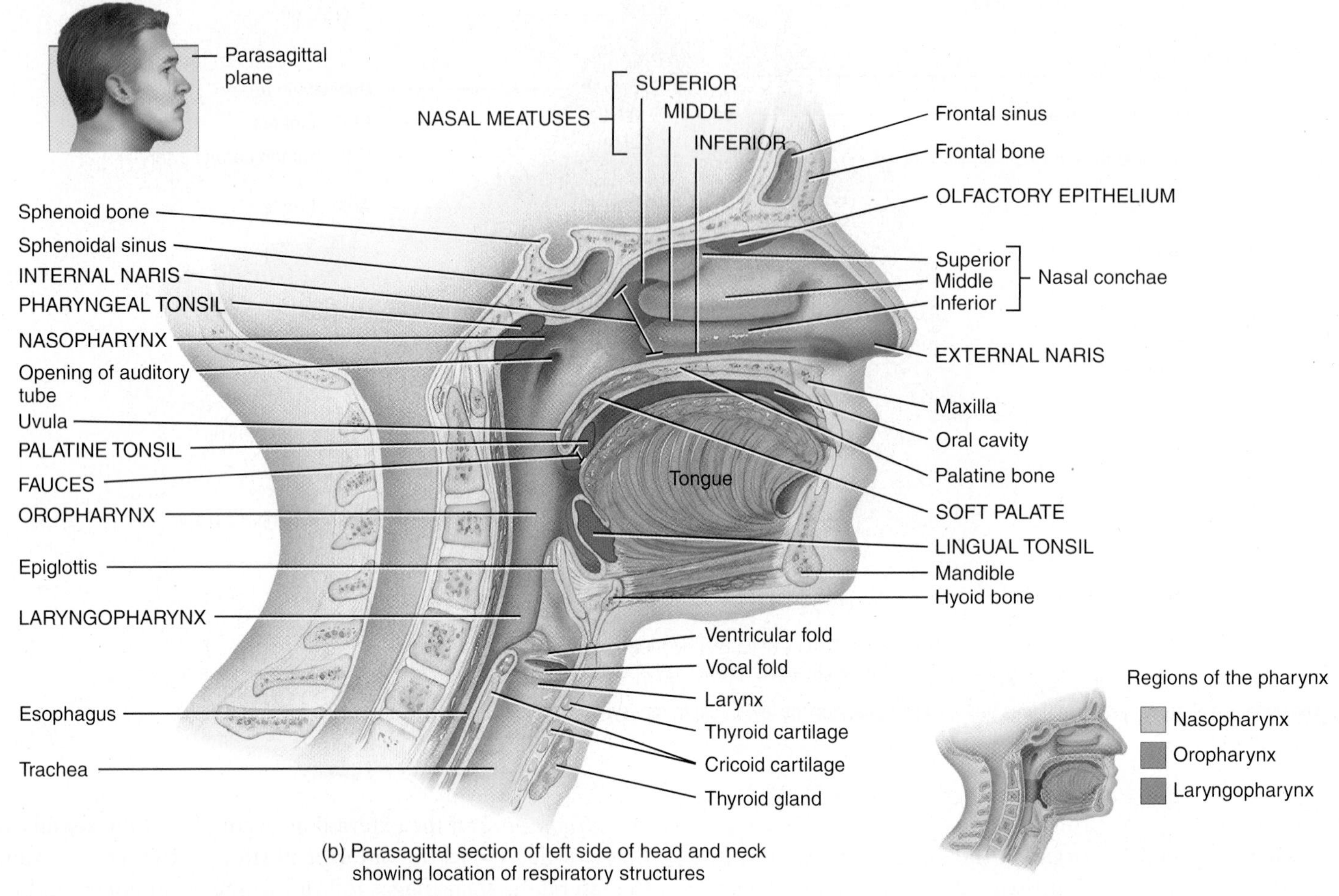


(b) Parasagittal section of left side of head and neck showing location of respiratory structures

**As air passes through the nose, it is warmed, filtered, and moistened, and olfaction occurs.**

The **nasal cavity** is a large space inside the skull that lies inferior to the nasal bone and superior to the oral cavity (Figure 22.2b). Anteriorly, the nasal cavity merges with the external nose, and posteriorly it communicates with the pharynx through two openings called the **internal nares**. Ducts from the *paranasal sinuses* (which drain mucus) and the *nasolacrimal ducts* (which drain tears) open into the nasal cavity. The **paranasal sinuses** are mucous membrane–lined cavities in certain skull bones that produce mucus and serve as resonating chambers for sound as we speak or sing. *Resonance* refers to prolonging, amplifying, or modifying a sound by vibration.

The lateral walls of the nasal cavity are formed by the ethmoid, maxillae, lacrimal, palatine, and inferior nasal conchae bones (see Figure 7.8); the ethmoid bone also forms the roof. The palatine bones and palatine processes of the maxillae form the floor of the nasal cavity (see Figure 7.9). A vertical partition, the **nasal septum**, divides the nasal cavity into right and left sides. The nasal septum consists of the vomer, the perpendicular plate of the ethmoid bone, and cartilage (see Figure 7.9).

Three shelves formed by projections of the superior, middle, and inferior nasal conchae extend out of each lateral wall of the nasal cavity. The conchae subdivide each side of the nasal cavity into a series of groovelike passageways—the **superior**, **middle**, and **inferior nasal meatuses** (mē-Ā-tus-ez = openings or passages; singular is *meatus*). The nasal cavity and conchae are lined with a mucous membrane pseudostratified ciliated columnar epithelium with numerous goblet cells (see Table 4.1). The arrangement of conchae and meatuses increases surface area in the internal nose and prevents dehydration by trapping water droplets during exhalation. Olfactory receptors lie in a region of the membrane lining the superior nasal conchae and adjacent nasal septum called the **olfactory epithelium**.

The interior structures of the nose are specialized for three basic functions: (1) filtering, warming, and moistening incoming air; (2) detecting olfactory (smell) stimuli; and (3) modifying the vibrations of speech sounds. When air enters the nostrils, it passes coarse hairs that trap large dust particles. As inhaled air whirls around the conchae and meatuses, it is warmed by blood circulating in abundant capillaries. Mucus secreted by the goblet cells moistens the air and traps dust particles. Drainage from the nasolacrimal ducts and paranasal sinuses also helps moisten the air. Cilia move the dust-laden mucus toward the pharynx, at which point it can be swallowed or spit out, thus removing particles from the respiratory tract.

## Pharynx

The **pharynx** (FAIR-inks), or throat, is a funnel-shaped tube about 13 cm (5 in.) long that starts at the internal nares and extends to the larynx (voice box) (Figure 22.2). The pharynx lies posterior to the nasal and oral cavities and anterior to the cervical (neck) vertebrae. Its wall is composed of skeletal muscles and is lined with a mucous membrane. Contraction of the skeletal muscles assists swallowing. The pharynx can be divided into three regions—the nasopharynx, oropharynx, and laryngopharynx (see the lower orientation diagram in Figure 22.2).

The superior portion of the pharynx, called the **nasopharynx**, lies posterior to the nasal cavity and extends to the soft palate. The **soft palate**, the posterior roof of the mouth between the nasopharynx and oropharynx, has five openings: two internal nares, two openings that lead into the *auditory (eustachian) tubes*, and the opening into the oropharynx. Air and dust-laden mucus from the nasal cavity enters the nasopharynx through the internal nares. Cilia, of the pseudostratified columnar epithelium lining the nasopharynx, move the mucus down toward the most inferior part of the pharynx. The nasopharynx also exchanges small amounts of air with the auditory tubes to equalize air pressure between the pharynx and the middle ear.

The intermediate portion of the pharynx, the **oropharynx**, lies posterior to the oral cavity and extends from the soft palate inferiorly to the level of the hyoid bone. It has only one opening, the **fauces** (FAW-sēz = throat), the opening from the mouth. The oropharynx has both respiratory and digestive functions, serving as a common passageway for air, food, and drink. Because the oropharynx is subject to abrasion by food particles, it is lined with stratified squamous epithelium.

The inferior portion of the pharynx, the **laryngopharynx** (la-rin′-gō-FAIR-inks), begins at the level of the hyoid bone. At its inferior end it opens into the esophagus (food tube) posteriorly and the larynx anteriorly. Like the oropharynx, the laryngopharynx is both a respiratory and a digestive pathway and is lined with stratified squamous epithelium.

The pharynx functions as a passageway for air and food, provides a resonating chamber for speech sounds, and houses the tonsils, which participate in immune responses against foreign invaders. The five tonsils are strategically positioned in the pharynx to trap and destroy inhaled or ingested pathogens. The posterior wall of the nasopharynx contains the single **pharyngeal tonsil** (fa-RIN-jē-al), or **adenoid**. Two pairs of tonsils, the **palatine** and **lingual tonsils**, are found in the oropharynx.

### ✓ CHECKPOINT

1. What is the path taken by air molecules into and through the nose?
2. List the functions of the nose.
3. Name the openings into and out of the nose.
4. What are the functions of the paranasal sinuses?
5. How does the nose moisten and clean inhaled air before it enters the rest of the respiratory tract?
6. What are the superior and inferior borders of the pharynx?

# 22.2 Inhaled air travels in the lower respiratory system from the larynx to alveoli.

The larynx, trachea, bronchi, and lungs are structurally considered the **lower respiratory system**.

## Larynx

The **larynx** (LAIR-inks), or voice box, is a short passageway that connects the laryngopharynx with the trachea (see Figure 22.2b). It lies in the midline of the neck anterior to the esophagus and the fourth through sixth cervical vertebrae. The lining of the larynx superior to the vocal folds is stratified squamous epithelium. The lining of the larynx inferior to the vocal folds is pseudostratified columnar epithelium with goblet cells. The mucus produced by the goblet cells helps trap dust not removed in the upper passages. The cilia in the upper respiratory tract move mucus and trapped particles *down* toward the pharynx; the cilia in the lower respiratory tract move them *up* toward the pharynx. Once in the pharynx, the mucus-trapped dust particles can be swallowed or spit out. The larynx is composed of cartilaginous walls that are held in position by ligaments and skeletal muscles.

The **epiglottis** (*epi-* = over; -glottis = tongue) is a large, leaf-shaped piece of elastic cartilage that is covered with epithelium (see Figure 22.2b). The "stem" of the epiglottis is the tapered inferior portion that is attached to the anterior rim of the thyroid cartilage. The broad superior "leaf" portion of the epiglottis is unattached and is free to move up and down like a trap door. The **glottis** consists of the vocal folds (a pair of folds of mucous membrane in the larynx) and the opening between them (see Figure 22.4a). During swallowing, the pharynx and larynx rise. Elevation of the pharynx widens it to receive food or drink; elevation of the larynx causes the epiglottis to move down and form a lid over the glottis, closing it off. Closing of the larynx by the epiglottis during swallowing routes liquids and foods into the esophagus while keeping them out of the larynx and airways. When small particles of dust, smoke, food, or liquids pass into the larynx, a cough reflex occurs, usually expelling the material.

The **thyroid cartilage** (Adam's apple) consists of hyaline cartilage that forms the anterior wall of the larynx and gives it a triangular shape (Figure 22.3). It is usually larger in males than in females due to the influence of male sex hormones on its growth during puberty.

The **cricoid cartilage** (KRĪ-koyd = ringlike) is a ring of hyaline cartilage that forms the inferior wall of the larynx. The

FIGURE 22.3 Larynx.

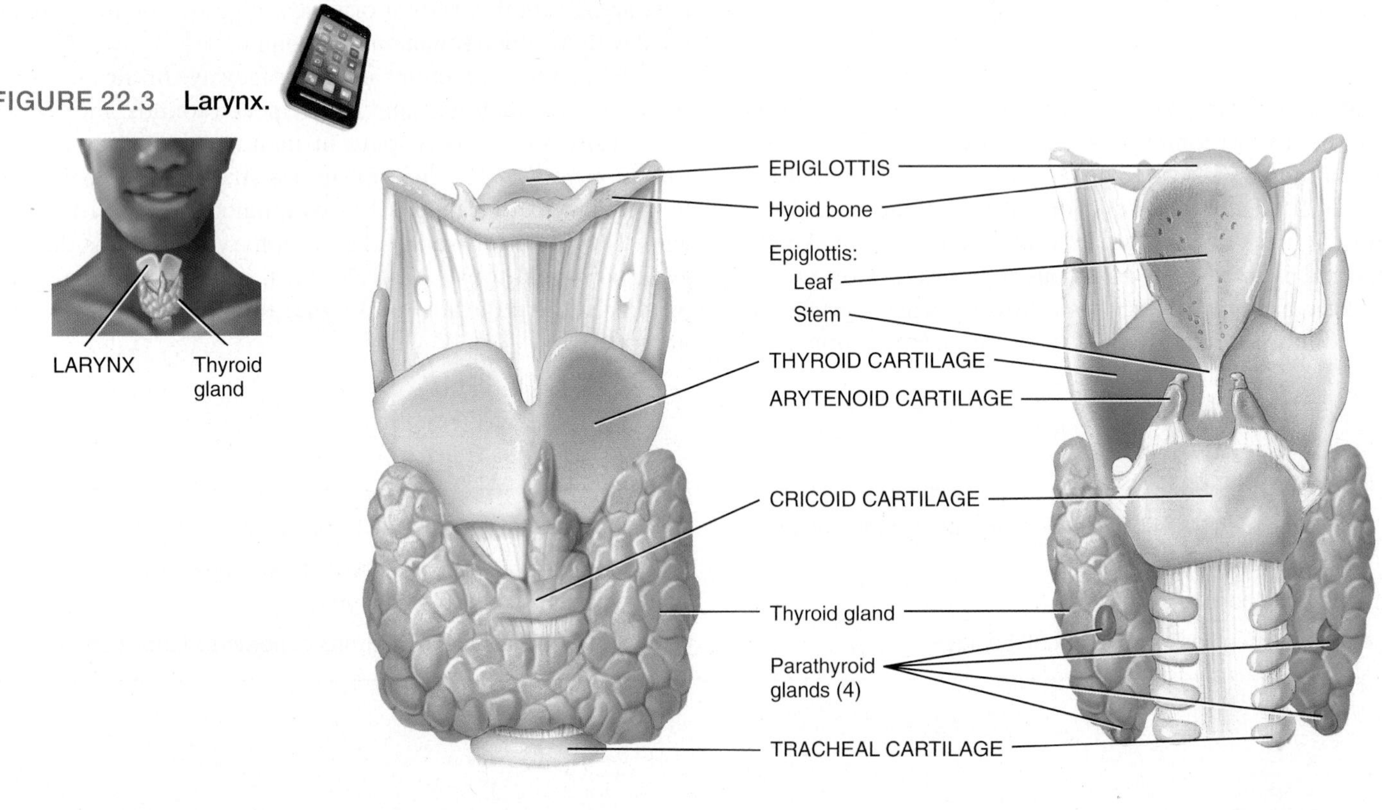

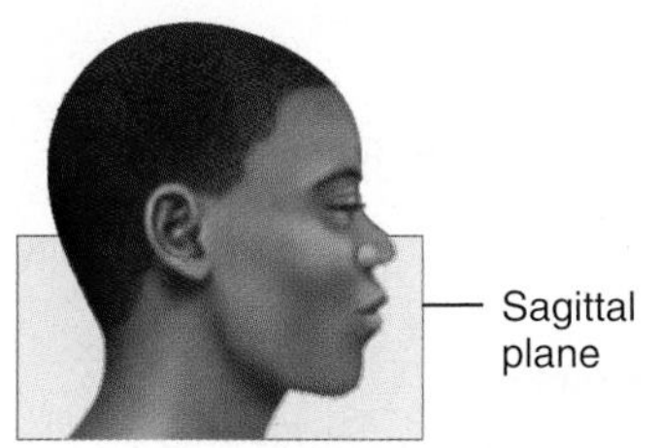


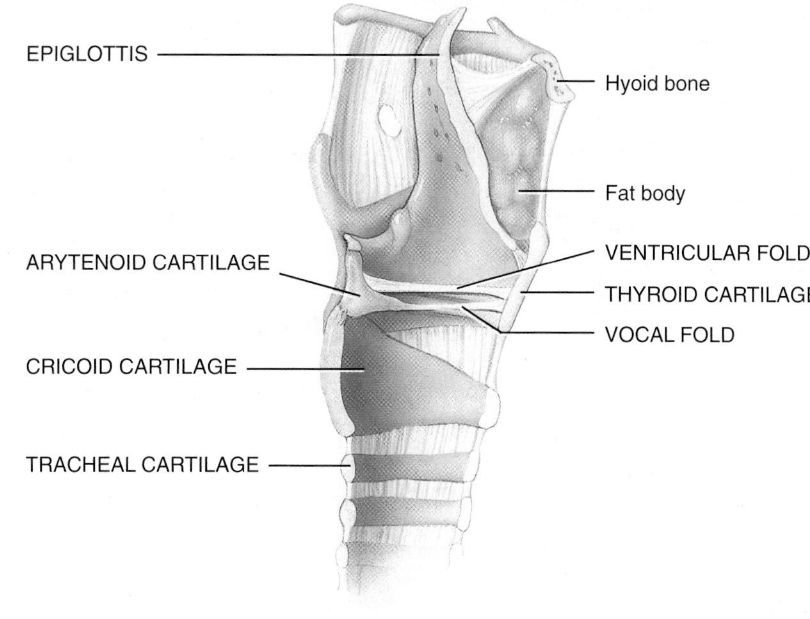


(c) Sagittal section

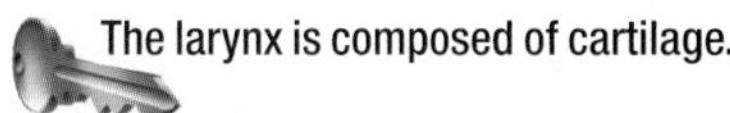


cricoid cartilage is connected superiorly to the thyroid cartilage and inferiorly to the first ring of cartilage of the trachea. The cricoid cartilage is the landmark for making an emergency airway, called a tracheotomy (see Clinical Connection entitled Tracheotomy and Intubation later in the chapter).

The paired **arytenoid cartilages** (ar′-i-TĒ-noyd = ladlelike) are triangular pieces of mostly hyaline cartilage located at the posterior, superior border of the cricoid cartilage. They form synovial joints with the cricoid cartilage and have a wide range of mobility important for voice production.

## The Structures of Voice Production

The mucous membrane of the larynx forms two pairs of folds (Figure 22.3c): a superior pair called the **ventricular folds (false vocal cords)** and an inferior pair called the **vocal folds (true vocal cords)**. While the ventricular folds do not function in voice production, they do have other important functional roles. When the ventricular folds are brought together, they hold the breath against pressure in the thoracic cavity, such as might occur when you strain to lift a heavy object.

The vocal folds produce sounds during speaking and singing. Deep to the mucous membrane of the vocal folds, bands of elastic ligaments are stretched between pieces of the rigid cartilages of the larynx like the strings on a guitar. Laryngeal muscles attach to both the rigid cartilage and the vocal folds. When the muscles contract they move the cartilages, which pulls the elastic ligaments tight, and this stretches the vocal folds out into the air passageway, narrowing the glottis. Contracting and relaxing the muscles varies the tension in the vocal folds, much like loosening or tightening a guitar string. Air passing through the larynx vibrates the folds and produces sound (phonation) by setting up sound waves in the column of air in the pharynx, nose, and mouth. The variation in the pitch of the sound is related to the tension in the vocal folds. The greater the air pressure, the louder the sound produced by the vibrating vocal folds.

When the muscles of the larynx contract, they pull on the arytenoid cartilages, which causes the cartilages to pivot and slide. Contraction of various laryngeal muscles could, for example, move the vocal folds apart, thereby opening the glottis (Figure 22.4a), bring the vocal folds together to close the glottis (Figure 22.4b), or elongate (and place tension on) the vocal folds. Pitch is controlled by the tension on the vocal folds. If they are pulled taut by the laryngeal muscles, they vibrate more rapidly and a higher pitch results. Decreasing the muscular tension on the vocal folds causes them to vibrate more slowly and produce lower-pitch sounds. Due to the influence of male sex hormones, vocal folds are usually thicker and longer in males than in females, and therefore they vibrate more slowly. This is why a man's voice generally has a lower range of pitch than a woman's voice.

**FIGURE 22.4 Movement of the vocal folds.**

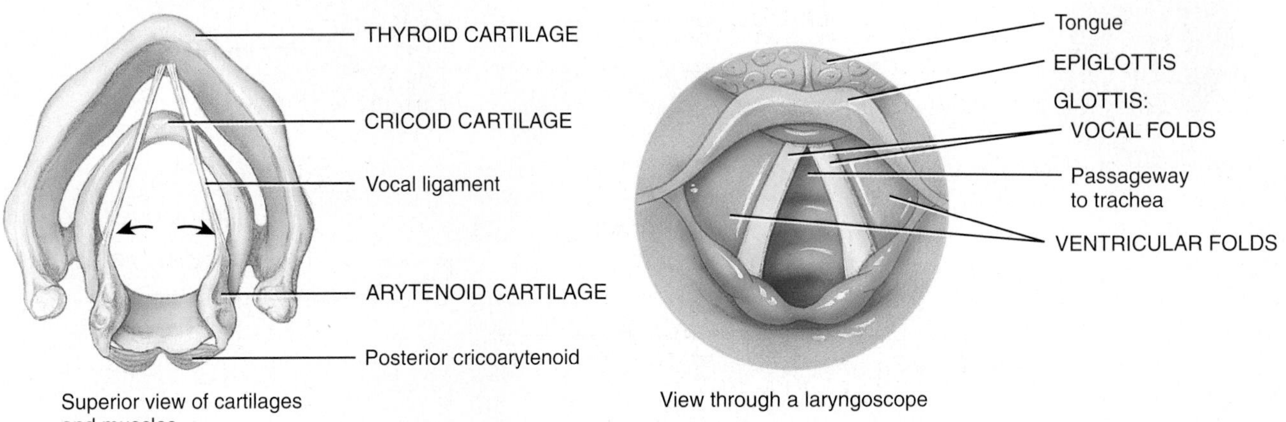


(a) Movement of vocal folds apart (abduction)

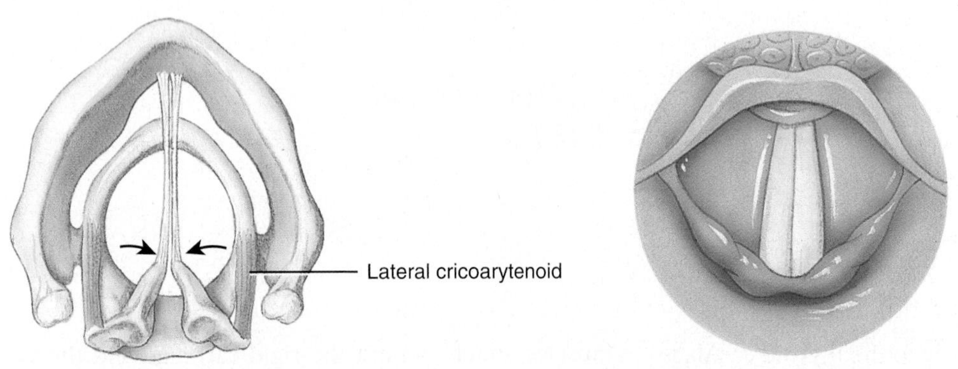


(b) Movement of vocal folds together (adduction)

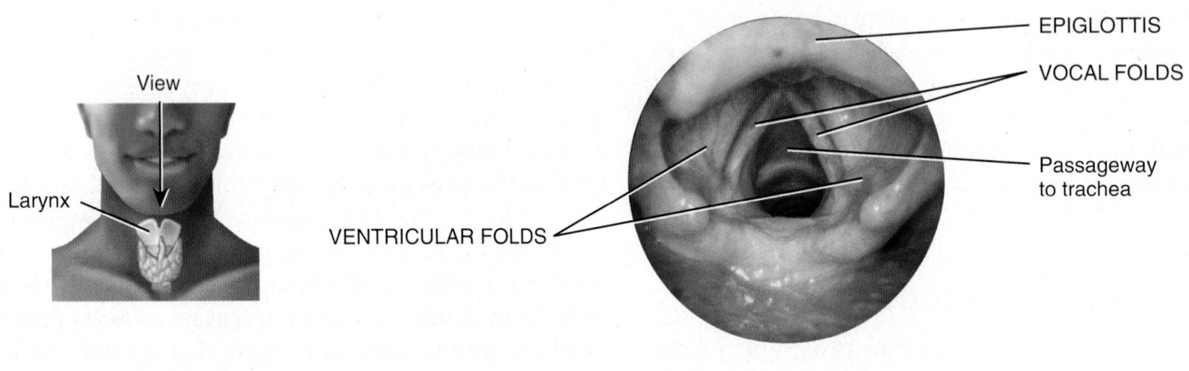


(c) Superior view

**The glottis consists of the vocal folds and the space between them.**

Sound originates from the vibration of the vocal folds, but other structures are necessary for converting the sound into recognizable speech. The pharynx, mouth, nasal cavity, and paranasal sinuses all act as resonating chambers that give the voice its human and individual quality. We produce the vowel sounds by constricting and relaxing the muscles in the wall of the pharynx. Muscles of the face, tongue, and lips help us enunciate words.

Whispering is accomplished by closing all but the posterior portion of the glottis. Because the vocal folds do not vibrate during whispering, there is no pitch to this form of speech. However, we can still produce intelligible speech while whispering by changing the shape of the oral cavity as we enunciate. As the size of the oral cavity changes, its resonance qualities change, which imparts a vowel-like pitch to the air as it rushes toward the lips.

## CLINICAL CONNECTION | *Tracheotomy and Intubation*

Several conditions may block airflow by obstructing the trachea. The rings of cartilage that support the trachea may be accidentally crushed, the mucous membrane may become inflamed and swell so much that it closes off the passageway, excess mucus secreted by inflamed membranes may clog the lower respiratory passages, a large object may be aspirated (breathed in), or a cancerous tumor may protrude into the airway. Two methods are used to reestablish airflow past a tracheal obstruction. If the obstruction is above the level of the larynx, a **tracheotomy** (trā-kē-O-tō-mē) may be performed. In this procedure, also called a *tracheostomy*, a skin incision is followed by a short longitudinal incision into the trachea below the cricoid cartilage. A tracheal tube is then inserted to create an emergency air passageway. The second method is **intubation** (in′-too-BĀ-shun), in which a tube is inserted into the mouth or nose and passed inferiorly through the larynx and trachea. The firm wall of the tube pushes aside any flexible obstruction, and the lumen of the tube provides a passageway for air; any mucus clogging the trachea can be suctioned out through the tube.

## Trachea

The **trachea** (TRĀ-kē-a = sturdy), or windpipe, is a tubular passageway for air that is about 12 cm (5 in.) long. It is located anterior to the esophagus (Figure 22.5) and extends from the larynx to the superior border of the fifth thoracic vertebra, where it divides into right and left primary bronchi (see Figure 22.6).

The wall of the trachea is lined with mucous membrane composed of pseudostratified columnar epithelium; its cilia and mucus-secreting goblet cells provide the same protection against dust as the membranes lining the nasal cavity and larynx. The 16–20 incomplete, horizontal rings of hyaline cartilage resemble the letter C and are stacked one on top of another. They may be felt through the skin inferior to the larynx. The open part of each C-shaped cartilage ring faces posteriorly toward the esophagus (Figure 22.5) and is spanned by a *fibromuscular membrane.* Within this membrane are transverse smooth muscle fibers, a muscle called the **trachealis** (trā-kē-A-lis), and elastic connective tissue that allows the diameter of the trachea to change subtly during inhalation and exhalation, which is important in maintaining

**FIGURE 22.5 Location of the trachea in relation to the esophagus.**

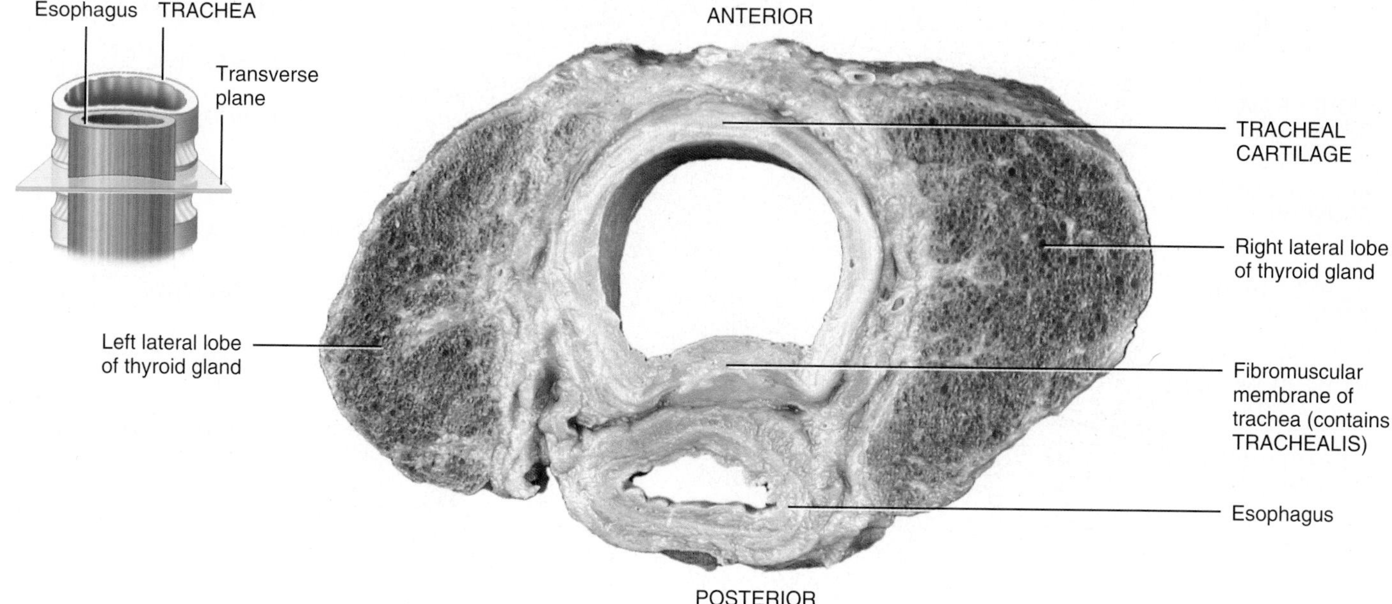


Superior view of transverse section of thyroid gland, trachea, and esophagus

**The trachea is anterior to the esophagus and extends from the larynx to the primary bronchi.**

efficient airflow. The C-shaped cartilage rings provide a semirigid support so that the tracheal wall does not collapse inward (especially during inhalation) and obstruct the air passageway.

## Bronchi

The trachea divides into a **right primary bronchus** (BRON-kus = windpipe), which goes into the right lung, and a **left primary bronchus**, which goes into the left lung (Figure 22.6). The right primary bronchus is more vertical and shorter than the left. As a result, an aspirated object is more likely to enter and lodge in the right primary bronchus than the left. Like the trachea, the primary **bronchi** (BRON-kē) contain incomplete rings of cartilage and are lined by pseudostratified columnar epithelium.

At the point where the trachea divides into right and left primary bronchi, an internal ridge called the **carina** (ka-RĪ-na = keel of a boat) is formed by a projection of the last tracheal cartilage. The mucous membrane of the carina is one of the most sensitive areas of the entire larynx and trachea for triggering a cough reflex.

On entering the lungs, the primary bronchi divide to form smaller bronchi—the **secondary (lobar) bronchi**, one for each lobe of the lung. (The right lung has three lobes; the left lung has two.) The secondary bronchi continue to branch, forming still smaller bronchi, called **tertiary (segmental) bronchi** (TER-shē-e-rē), that divide into **bronchioles**. Bronchioles in turn branch repeatedly, and the smallest passageways branch into even smaller tubes called **terminal bronchioles**. The terminal bronchioles represent the end of the conducting zone of the

FIGURE 22.6 **Branching of airways from the trachea: the bronchial tree.**

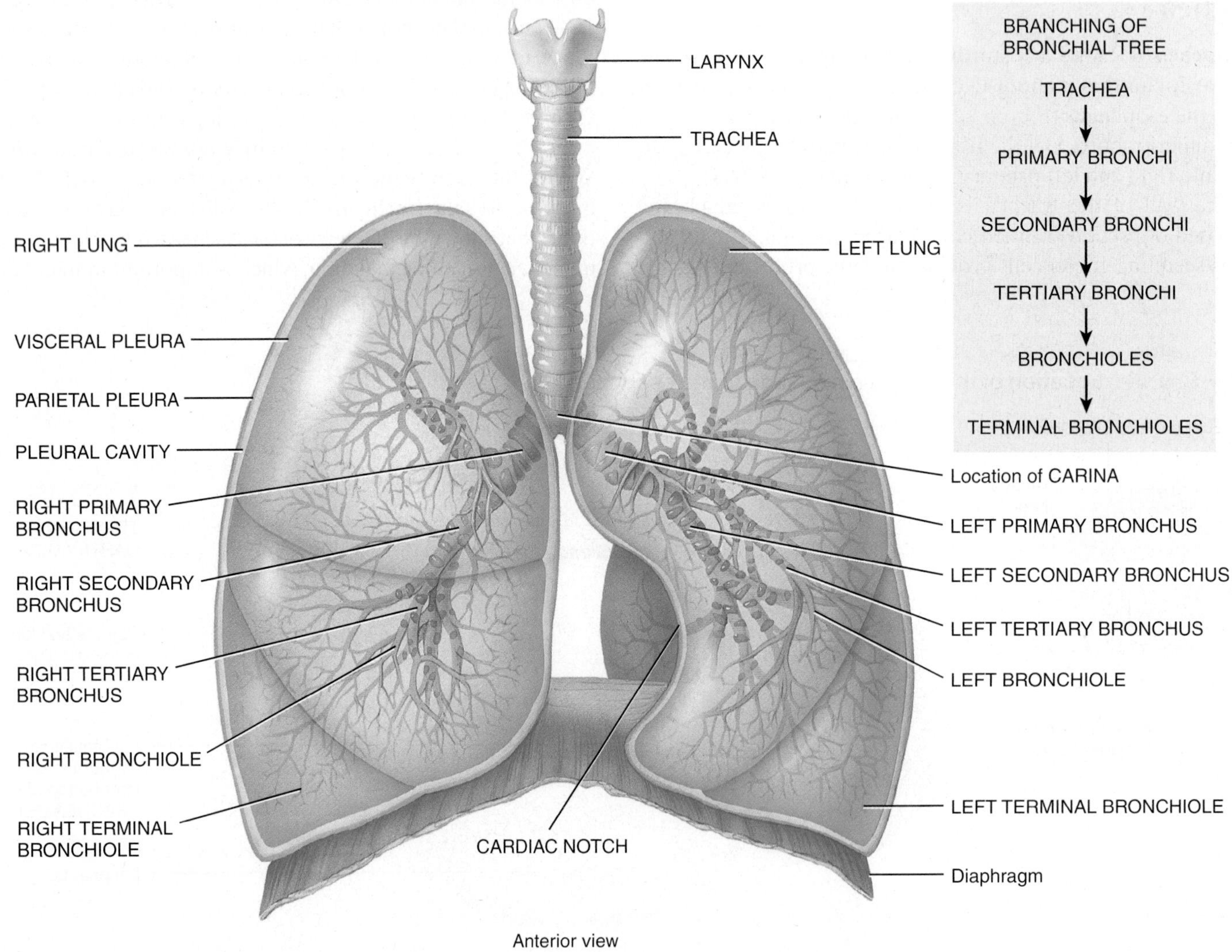


**The bronchial tree consists of airways that begin at the trachea and end at the terminal bronchioles.**

respiratory system. Because this extensive branching from the trachea through the terminal bronchioles resembles an upside-down tree, it is commonly referred to as the **bronchial tree**.

As the branching becomes more extensive in the bronchial tree, structural changes occur:

- The mucous membrane in the bronchial tree changes from pseudostratified columnar epithelium in the bronchi to ciliated simple columnar epithelium with some goblet cells (see Table 4.1) in larger bronchioles, to simple cuboidal epithelium (see Table 4.1) in smaller and terminal bronchioles. Recall that mucus produced by goblet cells traps inhaled particles, and cilia move the mucus and trapped particles toward the pharynx for removal. In regions where nonciliated simple cuboidal epithelium is present, inhaled particles are removed by macrophages.
- Plates of cartilage gradually replace the incomplete rings of cartilage in primary bronchi and finally disappear in the distal bronchioles.
- As the amount of cartilage decreases, the amount of smooth muscle increases. Smooth muscle encircles the lumen in spiral bands. However, because there is no supporting cartilage, muscle spasms can close off the airways. This is what happens during an asthma attack, which can be a life-threatening situation.

## Lungs

The **lungs** (= lightweights, because they float) are paired spongy, cone-shaped organs in the thoracic cavity (see Figure 22.1b). They are separated from each other by the heart and other structures in the *mediastinum,* which divides the thoracic cavity into two anatomically distinct chambers. As a result, if trauma causes one lung to collapse, the other may remain expanded. Two layers of serous membrane enclose and protect each lung. The superficial layer, called the **parietal pleura** (PLOOR-a; *pleur-* = side), lines the wall of the thoracic cavity; the deep layer, the **visceral pleura**, covers the lungs themselves (Figure 22.7). Between the visceral and parietal pleurae is a narrow space, the **pleural cavity**, which contains a lubricating fluid secreted by the membranes. This *pleural fluid* reduces friction between the membranes, allowing them to slide easily over one another during breathing. Pleural fluid also causes the two membranes to adhere to one another just as a film of water causes two glass microscope slides to stick together, a phenomenon called surface tension. Separate pleural cavities surround the left and right lungs.

FIGURE 22.7 Relationship of the pleural membranes to the lungs.

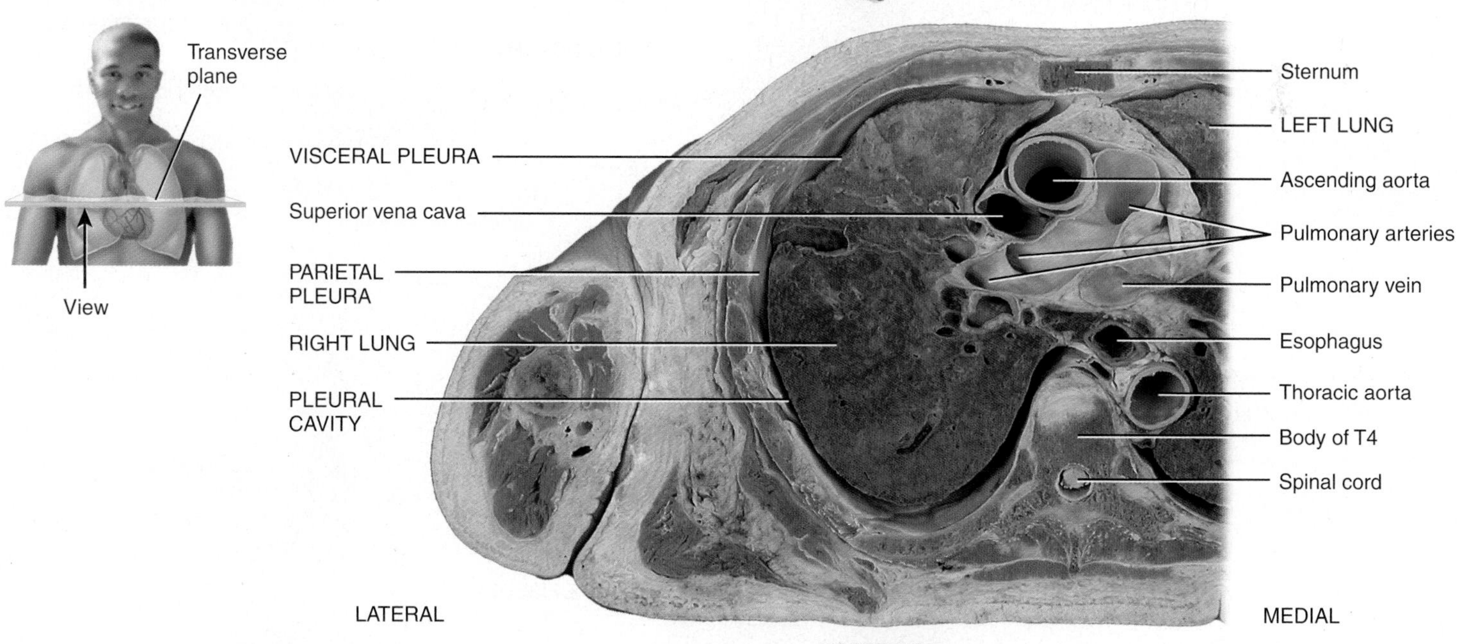


Inferior view of transverse section through thoracic cavity showing pleural cavity and pleural membranes

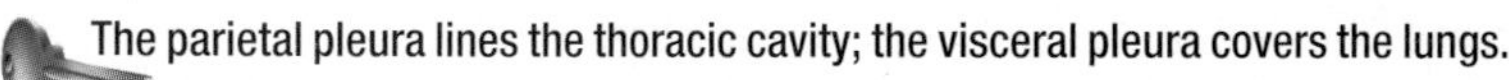

The parietal pleura lines the thoracic cavity; the visceral pleura covers the lungs.

The lungs lie against the ribs and extend from just slightly superior to the clavicles to the diaphragm (Figure 22.8a). The broad inferior portion of the lung, the **base**, is concave and fits over the convex area of the diaphragm. The narrow superior portion of the lung is the **apex**. The medial surface of each lung contains a region, the **hilum**, through which bronchi, pulmonary blood vessels, lymphatic vessels, and nerves enter and exit (Figure 22.8e). The left lung also contains a medial indentation, the **cardiac notch**, in which the heart lies. Due to the space occupied by the heart, the left lung is about 10 percent smaller than the right lung.

FIGURE 22.8 **Surface anatomy of the lungs.**

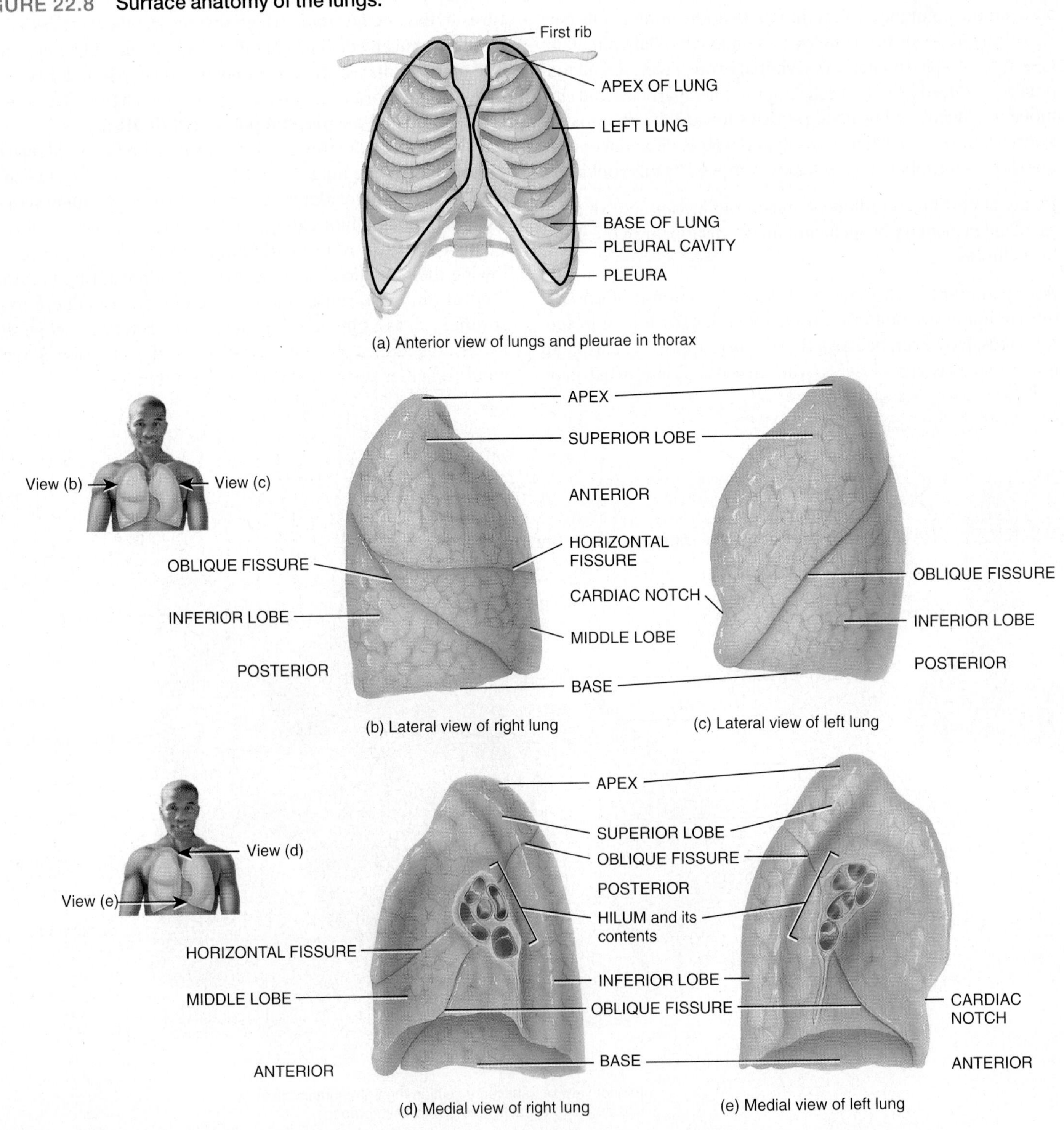


The oblique fissure divides the left lung into two lobes. The oblique and horizontal fissures divide the right lung into three lobes.

**CLINICAL CONNECTION | *Abdominal Thrust (Heimlich) Maneuver***

The abdominal thrust maneuver is a first-aid procedure designed to clear the airways of obstructing objects. It is performed by applying a quick upward thrust between the navel and costal margin that causes sudden elevation of the diaphragm and forceful, rapid expulsion of air in the lungs; this action forces air out the trachea to eject the obstructing object. The abdominal thrust maneuver is also used to expel water from the lungs of near-drowning victims before resuscitation is begun.

Although the right lung is thicker and broader, it is also somewhat shorter than the left lung because the diaphragm is higher on the right side, accommodating the liver that lies inferior to it.

### Lobes, Fissures, and Lobules

Deep grooves called fissures divide each lung into lobes (Figure 22.8b–e). Both lungs have an **oblique fissure**, which extends inferiorly and anteriorly; the right lung also has a **horizontal fissure**. The oblique fissure divides the left lung into **superior** and **inferior lobes**. The oblique and horizontal fissures divide the right lung into **superior, middle**, and **inferior lobes**.

Each lobe receives its own secondary (lobar) bronchus. Thus, the right primary bronchus gives rise to three secondary (lobar) bronchi called the **superior, middle**, and **inferior secondary (lobar) bronchi**, and the left primary bronchus gives rise to **superior** and **inferior secondary (lobar) bronchi**. As noted above, within the lung, the secondary bronchi give rise to the tertiary (segmental) bronchi. The segment of lung tissue that each tertiary bronchus supplies is called a **bronchopulmonary segment** (brong′-kō-PUL-mō-nar′-ē). Bronchial and pulmonary disorders (such as tumors or abscesses) that are localized in a bronchopulmonary segment may be surgically removed without seriously disrupting the surrounding lung tissue.

Each bronchopulmonary segment of the lungs has many small compartments called **lobules**; each lobule is wrapped in elastic connective tissue and contains a lymphatic vessel, an arteriole, a venule, and a branch from a terminal bronchiole (Figure 22.9). Terminal bronchioles subdivide into microscopic branches called **respiratory bronchioles** that have alveoli

**FIGURE 22.9** Microscopic anatomy of a lobule of the lungs.

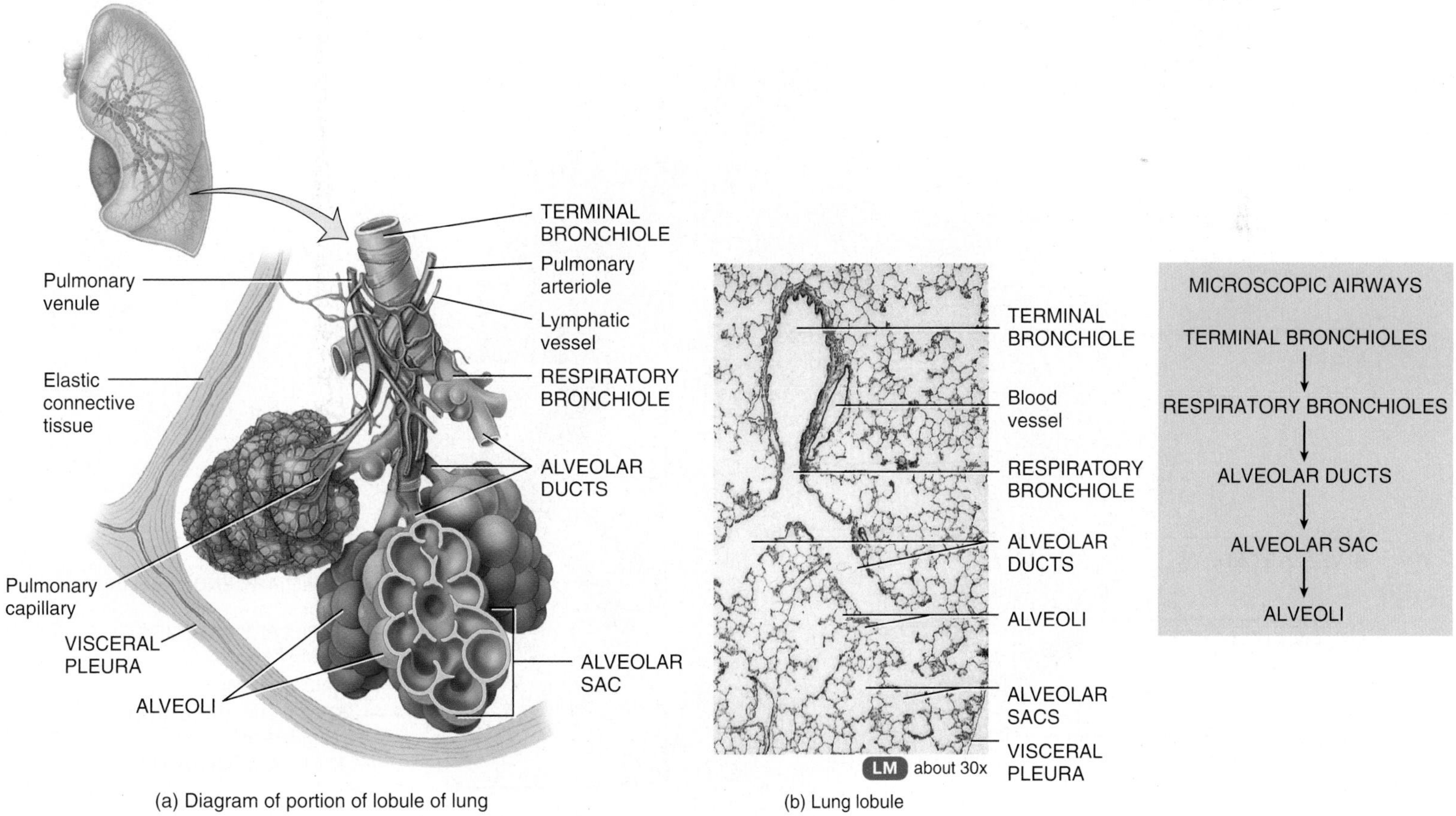

TERMINAL BRONCHIOLE
Pulmonary venule
Pulmonary arteriole
Lymphatic vessel
Elastic connective tissue
RESPIRATORY BRONCHIOLE
ALVEOLAR DUCTS
Pulmonary capillary
VISCERAL PLEURA
ALVEOLI
ALVEOLAR SAC

(a) Diagram of portion of lobule of lung

TERMINAL BRONCHIOLE
Blood vessel
RESPIRATORY BRONCHIOLE
ALVEOLAR DUCTS
ALVEOLI
ALVEOLAR SACS
VISCERAL PLEURA
LM about 30x

(b) Lung lobule

MICROSCOPIC AIRWAYS
TERMINAL BRONCHIOLES
↓
RESPIRATORY BRONCHIOLES
↓
ALVEOLAR DUCTS
↓
ALVEOLAR SAC
↓
ALVEOLI

Alveolar sacs consist of two or more alveoli that share a common opening into an alveolar duct.

(described shortly) budding from their walls. Because alveoli participate in gas exchange, respiratory bronchioles are considered the first part of the respiratory zone of the respiratory system. Respiratory bronchioles in turn subdivide into several **alveolar ducts** (al-VĒ-ō-lar).

## Alveoli

Around the circumference of the alveolar ducts are numerous alveoli and alveolar sacs. An **alveolus** (al-VĒ-ō-lus) is a cup-shaped bulge lined by simple squamous epithelium and supported by a thin elastic basement membrane; an **alveolar sac** consists of two or more alveoli that share a common opening to an alveolar duct (Figure 22.9). The walls of alveoli consist of two types of alveolar epithelial cells (Figure 22.10). The more numerous **type I alveolar cells** are simple squamous epithelial cells that form a nearly continuous lining of the alveolar wall. The thin, type I alveolar cells are the main sites of gas exchange. **Type II alveolar cells**, also called **septal cells**, are fewer in number and are found between type I alveolar cells. Type II alveolar cells are rounded epithelial cells that secrete alveolar fluid, which keeps the surface between the cells and the air moist. Included in the alveolar fluid is **surfactant** (sur-FAK-tant), a complex mixture of phospholipids and lipoproteins. Surfactant lowers the surface tension of alveolar fluid, which reduces the tendency of alveoli to collapse (described later). Also present are **alveolar macrophages** (*dust cells*), wandering phagocytes that remove fine dust particles and other debris from the alveolar spaces. On the outer surface of each alveolar sac an arteriole and venule disperse into a lush network of blood capillaries (Figure 22.9a).

The exchange of $O_2$ and $CO_2$ between the air spaces in the lungs and the blood takes place by diffusion across the alveolar and capillary walls, which together form the

**FIGURE 22.10 Structural components of an alveolus.** The respiratory membrane consists of a layer of type I and type II alveolar cells, an epithelial basement membrane, a capillary basement membrane, and the capillary endothelium.

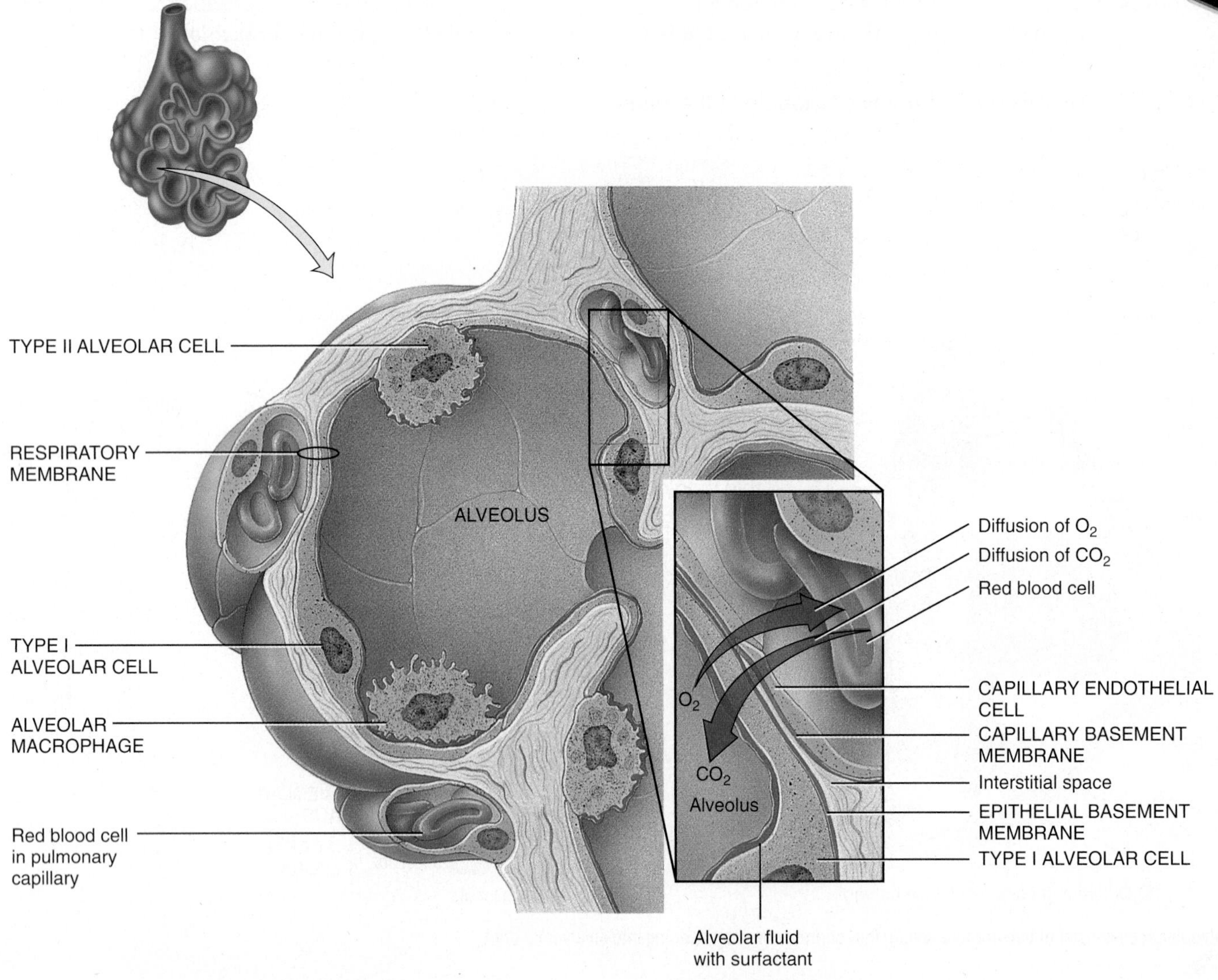


(a) Section through an alveolus showing cellular components

(b) Details of respiratory membrane

**respiratory membrane**. Extending from the alveolar air space to blood plasma, the respiratory membrane consists of four layers (Figure 22.10b):

- A layer of type I and type II alveolar cells that constitute the alveolar wall.
- An **epithelial basement membrane** underlying the alveolar wall.
- A **capillary basement membrane**.
- The **capillary endothelium**.

Despite having several layers, the respiratory membrane is very thin—only 0.5 μm thick—to allow rapid diffusion of gases. It has been estimated that the lungs contain 300 million alveoli, providing an immense surface area of 70 $m^2$ (750 $ft^2$)—about the size of a racquetball court—for gas exchange.

## ✓ CHECKPOINT

7. How does the epiglottis prevent aspiration of foods and liquids?
8. What is the main function of the vocal folds?
9. What is the benefit of not having cartilage between the trachea and the esophagus? What is the purpose of the solid C-shaped cartilage rings?
10. What is the path taken by air molecules into and through the bronchial tree?
11. Name the locations of the parietal pleura, visceral pleura, and pleural cavity.
12. What is the path taken by air molecules as they travel from terminal bronchioles to alveoli?

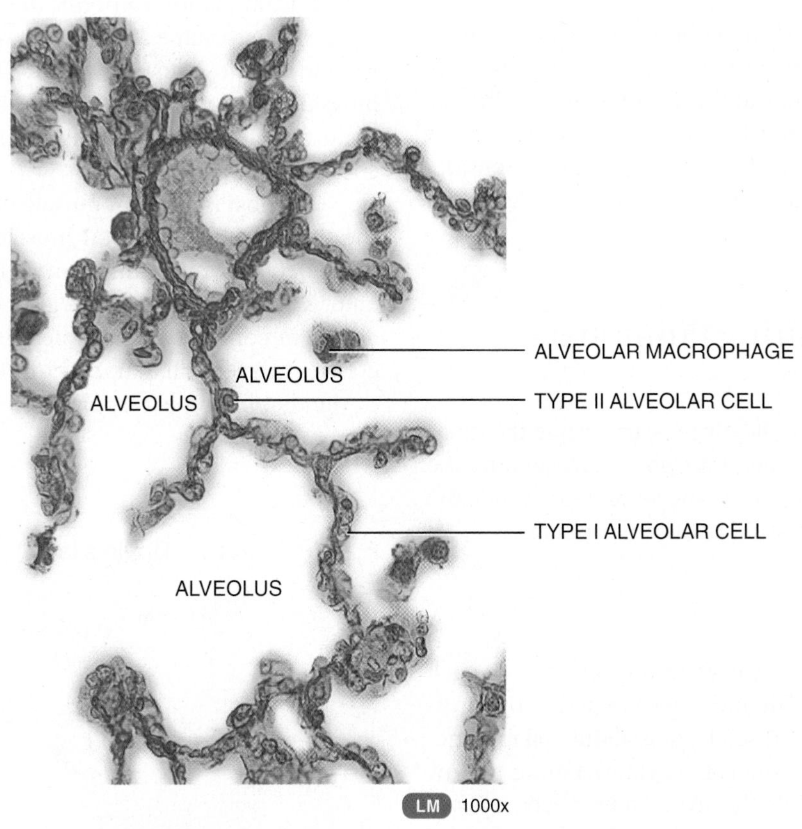


(c) Details of several alveoli

**The exchange of respiratory gases occurs by diffusion across the respiratory membrane.**

# 22.3 Inhalation and exhalation result from pressure changes caused by muscle contraction and relaxation.

The process of gas exchange in the body, called **respiration**, has three basic steps:

1. **Pulmonary ventilation** (*pulmon-* = lung), or **breathing**, is the inhalation (inflow) and exhalation (outflow) of air and involves the exchange of air between the atmosphere and the alveoli of the lungs.
2. **External respiration** is the *pulmonary* exchange of gases across the respiratory membrane between the alveoli of the lungs and the blood in pulmonary capillaries. In this process, pulmonary capillary blood gains $O_2$ and loses $CO_2$.
3. **Internal respiration** is the *tissue* exchange of gases between blood in systemic capillaries and tissue cells. During internal respiration, the blood loses $O_2$ and gains $CO_2$. Within cells, the metabolic reactions that consume $O_2$ and give off $CO_2$ during the production of ATP are termed *cellular respiration* (discussed in Concepts 10.8 and 23.9).

In pulmonary ventilation, air flows between the atmosphere and the alveoli of the lungs because of alternating pressure differences created by contraction and relaxation of respiratory muscles. The rate of airflow and the amount of effort needed for breathing are also influenced by alveolar surface tension, compliance of the lungs, and airway resistance (each described shortly).

## Pressure Changes during Pulmonary Ventilation

Air moves into the lungs when the air pressure inside the lungs is less than the air pressure in the atmosphere surrounding the body. Air moves out of the lungs when the air pressure inside the lungs is greater than the air pressure in the atmosphere.

### Inhalation

Breathing in is called **inhalation (inspiration)**. Just before each inhalation, the air pressure inside the lungs is equal to the pressure of the atmosphere, which at sea level is about 760 millimeters of mercury (mmHg), or 1 atmosphere (atm). For air to flow into the lungs, the pressure inside the alveoli must become lower than the atmospheric pressure. This condition is achieved by increasing the size of the lungs.

The pressure of a gas in a closed container is inversely proportional to the volume of the container. This means that if the size of a closed container is increased, the pressure the gas exerts on the walls of the container decreases, and that if the size of the container is decreased, then the pressure exerted on the walls increases. This inverse relationship between volume and pressure, called **Boyle's law**, may be demonstrated as follows (Figure 22.11): Suppose we place a gas in a cylinder that has a movable piston and a pressure gauge, and suppose that the initial pressure created by the gas molecules striking the wall of the container is 1 atm. If the piston is pushed down, the gas molecules are squeezed into a smaller volume; now with less wall area, each segment of the exposed wall is struck more often by the gas molecules, creating greater gas pressure. The gauge shows that the pressure doubles as the gas is compressed to half its original volume. In other words, the same number of molecules in half the volume produces twice the pressure. Conversely, if the piston is raised to increase the volume, the gas molecules spread farther apart and the pressure decreases. Thus, the pressure of a gas varies inversely with volume.

Differences in pressure caused by changes in lung volume force air into our lungs when we inhale and out when we exhale. For inhalation to occur, the lungs must expand, which increases lung volume and thus decreases the pressure in the lungs to below atmospheric pressure. The first step in expanding the lungs during normal quiet inhalation involves contraction of the main muscles of inhalation, the diaphragm and external intercostals (Figure 22.12).

The most important muscle of inhalation is the diaphragm, the dome-shaped skeletal muscle that forms the floor of the thoracic cavity. It is innervated by the phrenic nerves, which emerge from the cervical region of the spinal cord. As the diaphragm contracts, it becomes flatter and descends, which increases the vertical diameter of the thoracic cavity. Contraction of the diaphragm is responsible for about 75 percent of the air

FIGURE 22.11 **Boyle's law.**

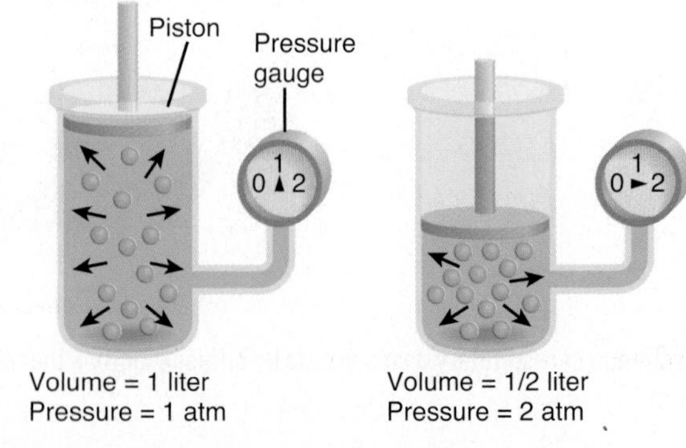


**The volume of a gas varies inversely with its pressure.**

FIGURE 22.12 **Muscles of inhalation and exhalation and their actions.** The pectoralis minor (not shown here) is illustrated in Figure 11.13a.

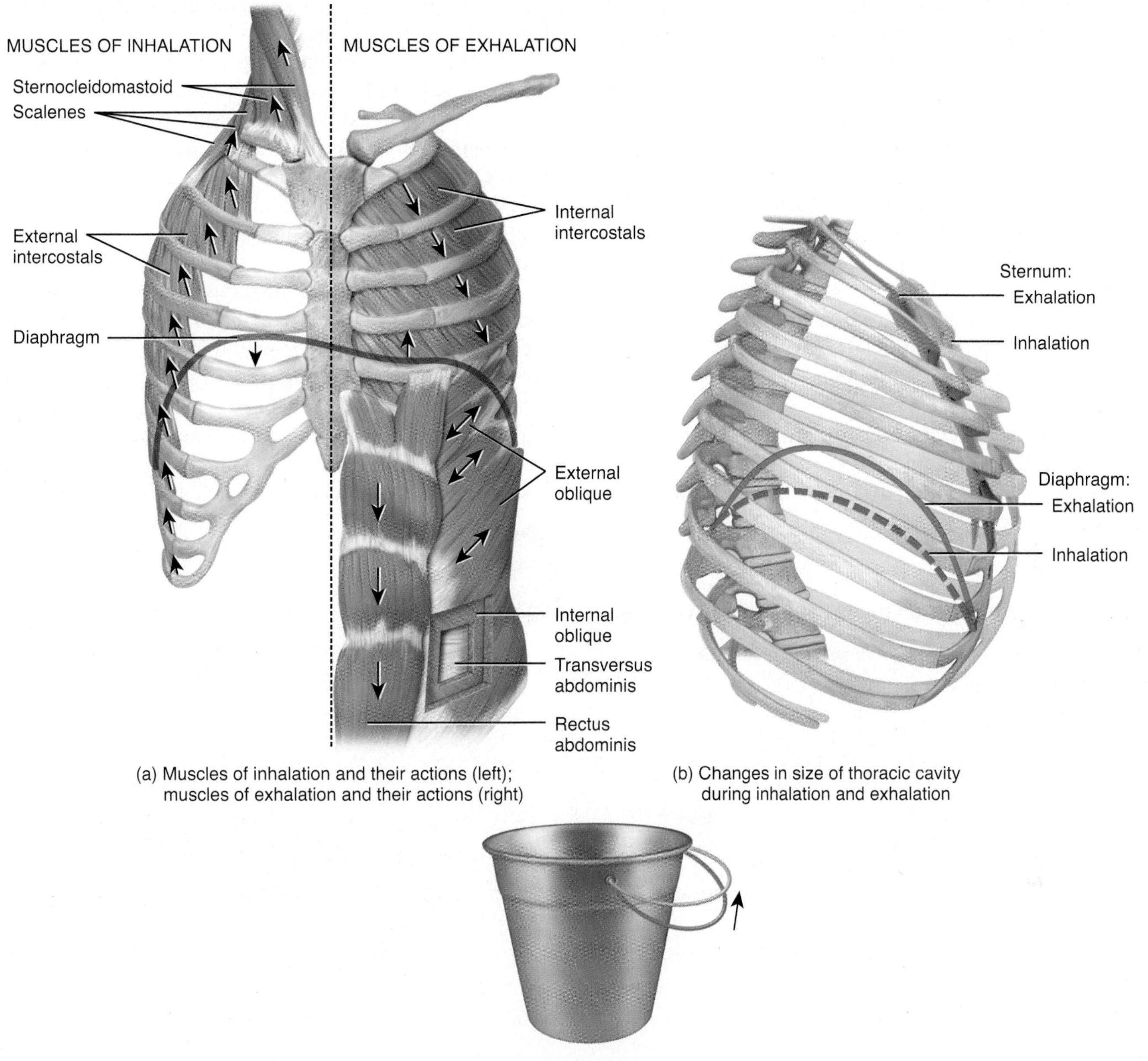


(a) Muscles of inhalation and their actions (left); muscles of exhalation and their actions (right)

(b) Changes in size of thoracic cavity during inhalation and exhalation

(c) During inhalation, the ribs move upward and outward like the handle on a bucket

**During quiet inhalation, the diaphragm and external intercostals contract, the lungs expand, and air moves into the lungs. During quiet exhalation, the diaphragm relaxes and the lungs recoil inward, forcing air out of the lungs.**

that enters the lungs during quiet breathing. Advanced pregnancy, excessive obesity, or confining abdominal clothing can prevent complete descent of the diaphragm and may cause shortness of breath.

The next most important muscles of inhalation are the external intercostals. When these muscles contract, they pull the ribs upward and outward. As a result, there is an increase in the anteroposterior and lateral diameters of the thoracic cavity. Contraction of the external intercostals is responsible for about 25 percent of the air that enters the lungs during normal quiet breathing.

At rest just before an inhalation, the air pressure inside

the lungs is the same as the pressure of the atmosphere, 760 mmHg (Figure 22.13). During quiet inhalations, the pressure in the pleural cavity between the visceral and parietal pleurae, called **intrapleural pressure**, is lower than the atmospheric pressure. Just before inhalation, it is about 4 mmHg less than the atmospheric pressure, or about 756 mmHg at an atmospheric pressure of 760 mmHg. As the diaphragm and external intercostals contract and the overall size of the thoracic cavity increases, the volume of the pleural cavity also increases, which causes intrapleural pressure to decrease from 756 to about 754 mmHg. During expansion of the thorax, the parietal and visceral pleurae normally adhere tightly because of the subatmospheric pressure between them and because of the surface tension created by their moist adjoining surfaces. As the thoracic cavity expands, the parietal pleura lining the cavity is pulled outward in all directions, and the visceral pleura and lungs are pulled along with it.

Maintenance of a subatmospheric pressure in the pleural cavity is vital to the functioning of the lungs because it helps keep the alveoli slightly inflated. Alveoli are so elastic that at the end of an exhalation they recoil inward and tend to collapse on themselves like the walls of a deflated balloon. The "suction" created by the slightly lower pressure in the pleural cavities helps prevent lung collapse.

As the volume of the lungs increases, the pressure inside the lungs, called the **alveolar pressure**, drops from 760 to 758 mmHg. A pressure difference is thus established between the atmosphere and the alveoli. Because air always flows to a region of lower pressure, inhalation takes place as air flows from the atmosphere (higher pressure) into the lungs (lower pressure). Air continues to flow into the lungs as long as a pressure difference exists.

During deep, forceful inhalations, accessory muscles of inspiration also participate to further increase the size of the thoracic cavity, allowing greater airflow into the lungs than can occur during quiet inhalation (Figure 22.12a). The accessory muscles are so named because they make little, if any, contribution during normal quiet inhalation, but during forced ventilation, such as occurs during exercise, they may contract vigorously. The accessory muscles of inhalation include the sternocleidomastoids,

FIGURE 22.13 **Pressure changes in pulmonary ventilation.**

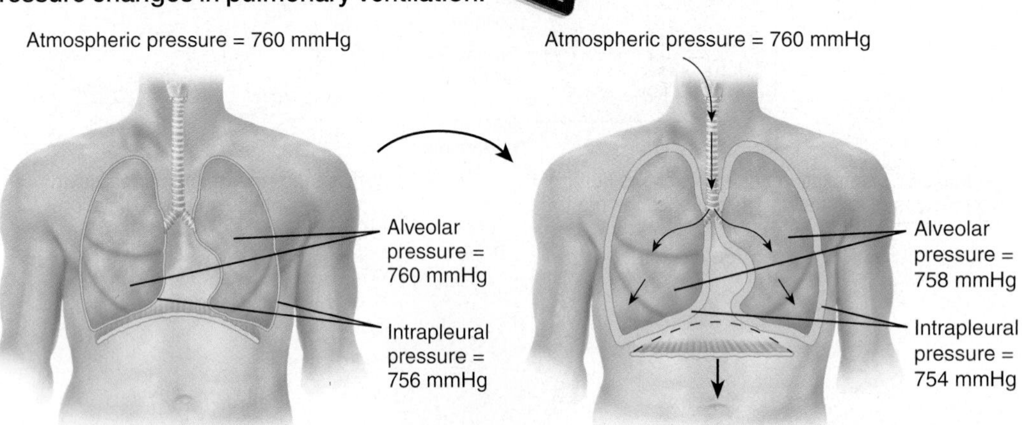


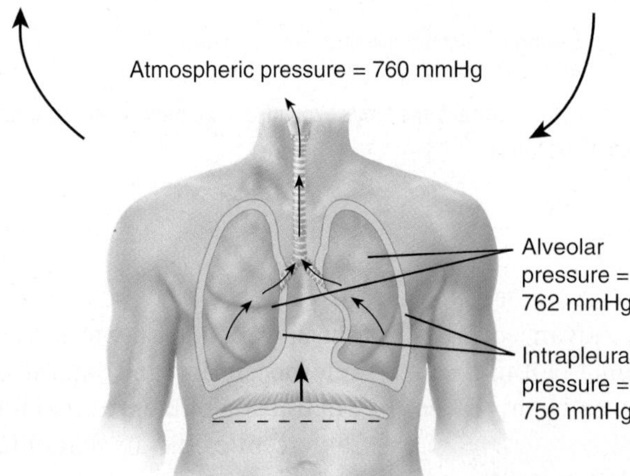


**Air moves into the lungs when alveolar pressure is less than atmospheric pressure, and out of the lungs when alveolar pressure is greater than atmospheric pressure.**

which elevate the sternum; the scalenes, which elevate the first two ribs; and the pectoralis minors, which elevate the third through fifth ribs. Because both normal quiet inhalation and inhalation during forced ventilation involve muscular contraction, the process of inhalation is said to be an *active process*.

Figure 22.14a summarizes the events of inhalation.

### Exhalation

Breathing out, called **exhalation (expiration)**, is also due to a pressure gradient, but in this case the gradient is in the opposite direction: The pressure in the lungs is greater than the pressure of the atmosphere. Normal exhalation during quiet breathing, unlike inhalation, is a *passive process* because no muscular contractions are involved. Instead, exhalation results from **elastic recoil** of the thoracic wall and lungs, both of which have a natural tendency to spring back after they have been stretched. Two inwardly directed forces contribute to elastic recoil: (1) the recoil of elastic fibers that were stretched during inhalation and (2) the inward pull of surface tension due to the film of alveolar fluid.

Exhalation starts when the inspiratory muscles relax. As the diaphragm relaxes, its dome moves superiorly owing to its elasticity. As the external intercostals relax, the ribs are depressed. These movements decrease the vertical, lateral, and anteroposterior diameters of the thoracic cavity, which decreases lung volume. In turn, the alveolar pressure increases, to about 762 mmHg. Air then flows from the area of higher pressure in the alveoli to the area of lower pressure in the atmosphere (Figure 22.13).

Exhalation becomes active only during forceful breathing, as occurs while playing a wind instrument or during exercise. During these times, muscles of exhalation—the abdominals and internal intercostals (Figure 22.12a)—contract, which increases pressure in the abdominal region and thorax, resulting in increased airflow out of the lungs. Contraction of the abdominal muscles moves the inferior ribs downward and compresses the abdominal viscera, thereby forcing the diaphragm superiorly. Contraction of the internal intercostals pulls the ribs inferiorly.

Figure 22.14b summarizes the events of exhalation.

## Other Factors Affecting Pulmonary Ventilation

As you have just learned, air pressure differences drive airflow during inhalation and exhalation. However, three other factors affect the rate of airflow and pulmonary ventilation: surface tension of the alveolar fluid, compliance of the lungs, and airway resistance.

FIGURE 22.14 **Summary of events of inhalation and exhalation.**

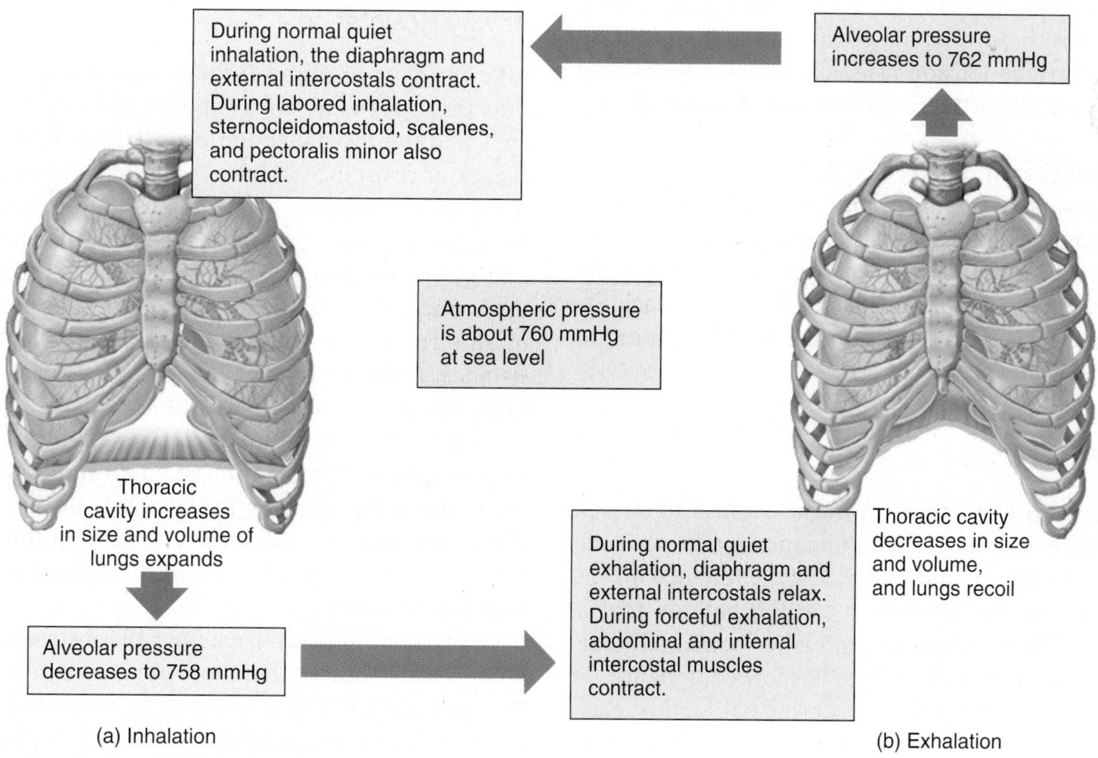
During normal quiet inhalation, the diaphragm and external intercostals contract. During labored inhalation, sternocleidomastoid, scalenes, and pectoralis minor also contract.
Alveolar pressure increases to 762 mmHg
Atmospheric pressure is about 760 mmHg at sea level
Thoracic cavity increases in size and volume of lungs expands
Alveolar pressure decreases to 758 mmHg
During normal quiet exhalation, diaphragm and external intercostals relax. During forceful exhalation, abdominal and internal intercostal muscles contract.
Thoracic cavity decreases in size and volume, and lungs recoil
(a) Inhalation
(b) Exhalation

**Inhalation and exhalation are caused by changes in alveolar pressure.**

## CLINICAL CONNECTION | *Respiratory Distress Syndrome*

**Respiratory distress syndrome (RDS)** is a breathing disorder of premature newborns in which the alveoli do not remain open due to a lack of surfactant. Recall that surfactant reduces surface tension and is necessary to prevent the collapse of alveoli during exhalation. The more premature the newborn, the greater the chance that RDS will develop. The condition is also more common in infants whose mothers have diabetes and in male infants, and occurs more often in European Americans than African Americans. Symptoms of RDS include labored and irregular breathing, flaring of the nostrils during inhalation, grunting during exhalation, and perhaps a blue skin color. Besides the symptoms, RDS is diagnosed on the basis of chest radiographs and a blood test. A newborn with mild RDS may require only supplemental oxygen administered through an oxygen hood or through a tube placed in the nose. In severe cases oxygen may be delivered by continuous positive airway pressure (CPAP) through tubes in the nostrils or a mask on the face. In such cases surfactant may be administered directly into the lungs.

### Surface Tension of Alveolar Fluid

As noted earlier, a thin layer of alveolar fluid coats the surface of alveoli and exerts a force known as **surface tension**. Surface tension arises at all air–water interfaces because the polar water molecules are more strongly attracted to each other than they are to gas molecules in the air. When liquid surrounds a sphere of air, as in an alveolus or a soap bubble, surface tension produces an inwardly directed force. Soap bubbles "burst" because they collapse inward due to surface tension. In the lungs, the inward force of surface tension causes the alveoli to assume the smallest possible diameter. During breathing, surface tension must be overcome to expand the lungs during each inhalation. Surface tension also accounts for two-thirds of lung elastic recoil, which decreases the size of alveoli during exhalation.

The surfactant present in alveolar fluid reduces its surface tension below the surface tension of pure water. A deficiency of surfactant in premature infants causes *respiratory distress syndrome*, in which the surface tension of alveolar fluid is greatly increased, so that many alveoli collapse at the end of each exhalation. Great effort is then needed at the next inhalation to reopen the collapsed alveoli.

### Compliance of the Lungs

**Compliance** refers to how much effort is required to stretch the lungs and thoracic wall. High compliance means that the lungs and thoracic wall expand easily; low compliance means that they resist expansion. By analogy, a thin balloon that is easy to inflate has high compliance, and a heavy and stiff balloon that takes a lot of effort to inflate has low compliance. In the lungs, compliance is related to two principal factors: elasticity and surface tension. The lungs normally have high compliance and expand easily because elastic fibers in lung tissue are easily stretched and surfactant in alveolar fluid reduces surface tension. Decreased compliance is a common feature in pulmonary conditions that scar lung tissue (for example, tuberculosis), cause lung tissue to become filled with fluid (pulmonary edema), produce a deficiency in surfactant, or impede lung expansion in any way (for example, paralysis of the intercostal muscles). Decreased lung compliance occurs in *emphysema* due to destruction of elastic fibers in alveolar walls (see the Chapter 22 WileyPLUS Clinical Connection entitled Chronic Obstructive Pulmonary Disease for a discussion of emphysema).

### Airway Resistance

Like the flow of blood through blood vessels, the rate of airflow through the airways depends on both the pressure difference and the resistance to flow: Airflow decreases when there is greater resistance in the airways and increases when there is decreased resistance. The walls of the airways, especially the bronchioles, offer some resistance to the normal flow of air into and out of the lungs. As the lungs expand during inhalation, the bronchioles enlarge because their walls are pulled outward in all directions. Larger-diameter airways have decreased resistance, allowing greater flow of air. Airway resistance then increases during exhalation as the diameter of bronchioles decreases. Airway diameter is also regulated by the degree of contraction or relaxation of smooth muscle in the walls of the airways. Signals from the sympathetic division of the autonomic nervous system cause relaxation of this smooth muscle, which results in bronchodilation, decreased resistance, and increasing airflow.

Any condition that narrows or obstructs the airways increases resistance, so that more pressure is required to maintain the same airflow. The distinguishing characteristic of asthma or chronic obstructive pulmonary disease (emphysema or chronic bronchitis) is increased airway resistance due to obstruction or collapse of airways.

## Modified Respiratory Movements

Respirations provide humans with methods for expressing emotions such as laughing, sighing, and sobbing. Moreover, respiratory air can be used to expel foreign matter from the lower air passages through actions such as sneezing and coughing. Respiratory movements are also modified and controlled during talking and singing. Some of the modified respiratory movements that express emotion or clear the airways are listed in Table 22.1. All of these movements are reflexes, but some of them also can be initiated voluntarily.

**TABLE 22.1**

Modified Respiratory Movements

| MOVEMENT | DESCRIPTION |
|---|---|
| **Coughing** | A long-drawn and deep inhalation followed by a complete closure of the glottis, which results in a strong exhalation that suddenly pushes the glottis open and sends a blast of air through the upper respiratory passages; stimulus may be a foreign body lodged in the larynx, trachea, or epiglottis |
| **Sneezing** | Spasmodic contraction of muscles of exhalation that forcefully expels air through the nose and mouth; stimulus may be an irritation of the nasal mucosa |
| **Sighing** | A long-drawn and deep inhalation immediately followed by a shorter but forceful exhalation |
| **Yawning** | A deep inhalation through the widely opened mouth producing an exaggerated depression of the mandible; it may be stimulated by drowsiness, or someone else's yawning, but the precise cause is unknown |
| **Sobbing** | A series of convulsive inhalations followed by a single prolonged exhalation; glottis closes earlier than normal after each inhalation so only a little air enters the lungs with each inhalation |
| **Crying** | An inhalation followed by many short convulsive exhalations, during which the glottis remains open and the vocal folds vibrate; accompanied by characteristic facial expressions and tears |
| **Laughing** | The same basic movements as crying, but the rhythm of the movements and the facial expressions usually differ from those of crying; laughing and crying are sometimes indistinguishable |
| **Hiccupping** | Spasmodic contraction of the diaphragm followed by a spasmodic closure of the glottis, which produces a sharp sound on inhalation; stimulus is usually irritation of the sensory nerve endings of the gastrointestinal tract |
| **Valsalva (val-SAL-va) maneuver** | Forced exhalation against a closed glottis as may occur during periods of straining while defecating |
| **Pressurizing the middle ear** | The nose and mouth are held closed and air from the lungs is forced through the pharyngotympanic tube into the middle ear; employed by those snorkeling or scuba diving during descent to equalize the pressure of the middle ear with that of the external environment |

### ✓ CHECKPOINT

**13.** If the volume of the cylinder in Figure 22.11 decreased from 1 liter to 1/4 liter, how would the pressure change?

**14.** What is the main muscle that powers your breathing as you sit and study your anatomy and physiology text?

**15.** Would contraction of the external intercostals increase or decrease alveolar pressure? Would it cause air to enter or leave the lungs? Would contraction of the internal intercostals increase or decrease alveolar pressure? Would it cause air to enter or leave the lungs?

**16.** Which happens first during inhalation: (a) air enters the lungs or (b) the lungs expand?

**17.** Which forces produce the elastic recoil of the thoracic wall and lungs?

**18.** Would airflow into alveoli be greater with higher or lower compliance of the lungs? Larger or smaller bronchioles?

## RETURN TO Cari's Story

By the end of the week Cari knew from experience that the cold had turned into a sinus infection. First her head had begun to ache, with pressure over her eyes and cheekbones. Her mucus had changed in color from clear to greenish, and she felt feverish. She was having difficulty hearing, her chest felt heavy, and her eyes burned and watered.

"Cari, how long have you had this cold?" the nurse asked as she listened to Cari's lungs.

"I've had the cold for a couple of weeks now, but I've had the sinus pressure and green mucus for about a week. The worst part is that I can't catch my breath," Cari replied. As she finished speaking, Cari again felt an overwhelming urge to cough, and couldn't stop.

"Are you still smoking, Cari?" The nurse moved the bell of her stethoscope over Cari's back, noticing Cari's decreased breath sounds and ronchi (wheezing, musical sounds).

"Yeah, I know I should quit; it's been so hard, though. I guess I use the stress of my job as an excuse."

"Let's get your temperature, Cari. It's a bit high—101.2." The nurse checked Cari's blood oxygen saturation as she spoke, using an oxygen saturation monitor.

With each breath, Cari struggled to get air into her lungs. Her neck muscle stood out as she raised her chest and shoulders in an effort to breathe.

"Cari, I'm going to give you some oxygen and get the doctor. I'll be right back."

The nurse found the doctor in the hall. "What have you got, Nancy?" the doctor asked.

"Dr. Sommers, I think we have a case of pneumonia. She's a 47-year-old smoker, with a history of chronic bronchitis and multiple upper respiratory tract infections. Had a cold that turned into a sinus infection. But her lung sounds are really abnormal, especially at the left base. $O_2$ saturation is 90 percent, productive cough, and she's febrile, with a temp over 101. I started her on some oxygen; she's rather uncomfortable."

The doctor looked over Cari's chart. "Sounds like we might need to admit her; let's get a chest X-ray."

A. *How could an infection in Cari's nasal passages and pharynx spread into her sinuses?*
B. *What is the cough reflex? Describe the process that Cari's respiratory system is using to clear her lungs by coughing.*
C. *Which structures found in the terminal bronchioles and alveoli normally would protect Cari's lungs from infectious pathogens and particulate matter?*
D. *How would the resistance of Cari's airways be affected by excess mucus and fluid in her lung?*
E. *How would Cari's lung compliance (the effort required to expand the lungs) be altered as her alveoli fill with fluid due to pneumonia?*

## 22.4 Lung volumes and capacities are measured to determine the respiratory status of an individual.

While at rest, a healthy adult averages 12 breaths a minute, with each inhalation and exhalation moving about 500 mL of air into and out of the lungs. The volume of one breath is called the **tidal volume**. The **minute ventilation (MV)**—the total volume of air inhaled and exhaled each minute—is equal to breathing rate multiplied by tidal volume:

$$MV = 12 \text{ breaths/min} \times 500 \text{ mL/breath} = 6 \text{ liters/min}$$

A lower-than-normal minute ventilation usually is a sign of pulmonary malfunction. The apparatus commonly used to measure the volume of air exchanged during breathing and to measure the respiratory rate is a **spirometer** (spī-ROM-e-ter; *spiro-* = breathe; *-meter* = measuring device). The record produced by a spirometer is called a **spirogram**. Inhalation is recorded as an upward deflection, and exhalation is recorded as a downward deflection (Figure 22.15).

In a typical adult, about 70 percent of the tidal volume (350 mL) actually reaches the respiratory zone of the respiratory system—the respiratory bronchioles, alveolar ducts, alveolar sacs, and alveoli—and participates in external respiration. The other 30 percent (150 mL) remains in the conducting airways of the nose, pharynx, larynx, trachea, bronchi, bronchioles, and terminal bronchioles. Collectively, the conducting airways containing air that does not undergo gas exchange are known as the **anatomic dead space**.

Lung volume varies considerably from one person to another and in the same person at different times. In general, these volumes are larger in males, taller individuals, and younger

FIGURE 22.15 **Spirogram of lung volumes and capacities.** The average values for a healthy adult male and female are indicated, with the values for a female in parentheses. Note that the spirogram is read from right (start of record) to left (end of record).

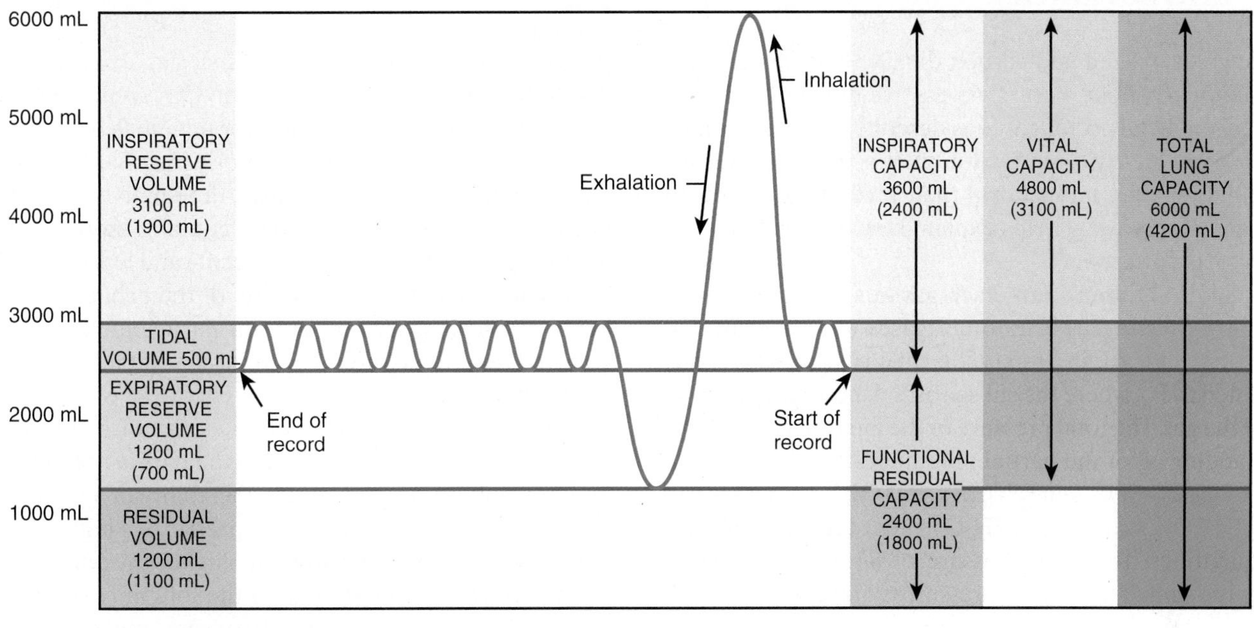


**Lung capacities are combinations of various lung volumes.**

adults, and smaller in females, shorter individuals, and the elderly. Various disorders also may be diagnosed by comparison of actual and predicted normal values for gender, height, and age. The values given here are averages for young adults.

By taking a very deep breath, you can inhale a good deal more than the 500 mL of tidal volume. This additional inhaled air, called the **inspiratory reserve volume**, is about 3100 mL in an average adult male and 1900 mL in an average adult female (Figure 22.15). Even more air can be inhaled if inhalation follows forced exhalation. If you inhale normally and then exhale as forcibly as possible, you should be able to push out considerably more air in addition to the 500 mL of tidal volume. The extra 1200 mL in males and 700 mL in females is called the **expiratory reserve volume**.

Even after the expiratory reserve volume is exhaled, considerable air remains in the lungs because the subatmospheric intrapleural pressure keeps the alveoli slightly inflated, and some air also remains in the noncollapsible airways. This volume, called the **residual volume** (re-ZID-ū-al), amounts to about 1200 mL in males and 1100 mL in females.

Lung capacities are combinations of specific lung volumes (Figure 22.15). **Inspiratory capacity** is the sum of tidal volume and inspiratory reserve volume (500 mL + 3100 mL = 3600 mL in males and 500 mL + 1900 mL = 2400 mL in females). **Functional residual capacity** is the sum of residual volume and expiratory reserve volume (1200 mL + 1200 mL = 2400 mL in males and 1100 mL + 700 mL = 1800 mL in females). **Vital capacity**, the maximum amount of air that can be exhaled after maximal inhalation, is the sum of inspiratory reserve volume, tidal volume, and expiratory reserve volume (4800 mL in males and 3100 mL in females). Finally, **total lung capacity** is the sum of vital capacity and residual volume (4800 mL + 1200 mL = 6000 mL in males and 3100 mL + 1100 mL = 4200 mL in females).

## ✓ CHECKPOINT

**19.** What is a spirometer?

**20.** What is the anatomic dead space?

**21.** What is the difference between tidal volume, inspiratory reserve volume, expiratory reserve volume, and residual volume?

## 22.5 Oxygen and carbon dioxide diffusion is based on partial pressure gradients and solubility.

The exchange of oxygen and carbon dioxide between alveolar air and pulmonary blood occurs via passive diffusion, which is governed by the behavior of gases as described by two gas laws, Dalton's law and Henry's law. Dalton's law is important for understanding how gases move down their pressure gradients by diffusion, and Henry's law helps explain how the solubility of a gas relates to its diffusion.

According to **Dalton's law**, each gas in a mixture of gases exerts its own pressure as if no other gases were present. The pressure of a specific gas in a mixture is called its **partial pressure** and is denoted as $P_X$, where the subscript X denotes the chemical formula of the gas. The total pressure of the mixture is calculated simply by adding all of the partial pressures. Atmospheric air is a mixture of gases—nitrogen ($N_2$), oxygen ($O_2$), argon (Ar), carbon dioxide ($CO_2$), water vapor ($H_2O$), plus other gases present in small quantities. The total pressure of air, the *atmospheric pressure*, is the sum of the partial pressures of all of these gases:

$$
\begin{aligned}
&P_{N_2}\,(597.4\text{ mmHg}) + P_{O_2}\,(158.8\text{ mmHg}) + P_{Ar}\,(0.7\text{ mmHg})\\
&\quad + P_{H_2O}\,(2.3\text{ mmHg}) + P_{CO_2}\,(0.3\text{ mmHg})\\
&\quad + P_{\text{other gases}}\,(0.5\text{ mmHg})\\
&\quad = \text{atmospheric pressure }(760\text{ mmHg})
\end{aligned}
$$

The partial pressure of each gas is proportional to its concentration in the mixture and is determined by multiplying the percentage of the individual gas by the total pressure of the air mixture. For example, atmospheric air is 20.9 percent oxygen ($O_2$) and 0.04 percent carbon dioxide. Therefore, inhaled air has a $P_{O_2}$ of 20.9 percent × 760 mmHg or 159 mmHg and a $P_{CO_2}$ of 0.04 percent × 760 mmHg or 0.3 mmHg.

Partial pressures determine the movement of $O_2$ and $CO_2$ between the atmosphere and lungs, between the lungs and blood, and between the blood and body cells. Each gas diffuses across a permeable membrane from the area where its partial pressure is greater to the area where its partial pressure is less. The greater the difference in partial pressure, the faster is the rate of diffusion.

Compared with inhaled air, alveolar air has less $O_2$ (13.6 percent versus 20.9 percent) and more $CO_2$ (5.2 percent versus 0.04 percent) for two reasons. First, gas exchange across the respiratory membrane increases the $CO_2$ content and decreases the $O_2$ content of alveolar air. Second, inhaled air becomes humidified as it passes along the moist mucosal linings. As water vapor content of the air increases, the relative percentage of $O_2$ decreases. In contrast, exhaled air contains more $O_2$ than alveolar air (16 percent versus 13.6 percent) and less $CO_2$ (4.5 percent versus 5.2 percent) because some of the exhaled air was in the anatomic dead space and did not participate in gas exchange. Exhaled air is a mixture of alveolar air and inhaled air that was in the anatomic dead space.

**Henry's law** states that the quantity of a gas that will dissolve in a liquid is proportional to the partial pressure of the gas and its solubility. In body fluids, the ability of a gas to stay in solution is greater when its partial pressure is higher and when it has a high solubility in water. The higher the partial pressure of a gas over a liquid and the higher the solubility, the more gas will stay in solution. In comparison to oxygen, much more $CO_2$ is dissolved in blood plasma because the solubility of $CO_2$ is 24 times greater than that of $O_2$. Even though the air we breathe contains mostly $N_2$, this gas has no known effect on bodily functions, and at sea level pressure very little of it dissolves in blood plasma because its solubility is very low.

An everyday experience gives a demonstration of Henry's law. You have probably noticed that a soft drink makes a hissing sound when the top of the container is removed, and bubbles rise to the surface for some time afterward. The gas dissolved in carbonated beverages is $CO_2$. Because the soft drink is bottled or canned under high pressure and capped, the $CO_2$ remains dissolved as long as the container is unopened. Once you remove the cap, the pressure decreases and the gas begins to bubble out of solution.

✓ **CHECKPOINT**

**22.** Explain the principles described by Dalton's law and Henry's law.

## 22.6 Respiration occurs between alveoli and pulmonary capillaries and between systemic capillaries and tissue cells.

**External respiration** is the diffusion of $O_2$ from air in the alveoli of the lungs to blood in pulmonary capillaries, and the diffusion of $CO_2$ in the opposite direction (Figure 22.16a). External respiration in the lungs converts **deoxygenated blood** (depleted of some $O_2$) coming from the right side of the heart into **oxygenated blood** (saturated with $O_2$) that returns to the left side of the heart (see Figure 20.15). As blood flows through the pulmonary capillaries, it picks up $O_2$ from alveolar air and unloads $CO_2$ into alveolar air. Although this process is commonly called an "exchange" of gases, each gas diffuses *independently* from the area where its partial pressure is higher to the area where its partial pressure is lower.

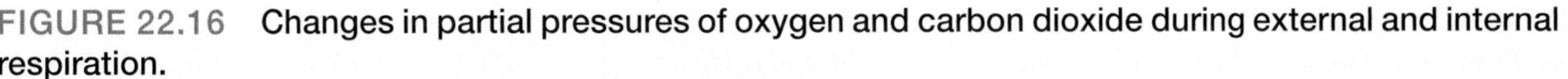

FIGURE 22.16 **Changes in partial pressures of oxygen and carbon dioxide during external and internal respiration.**

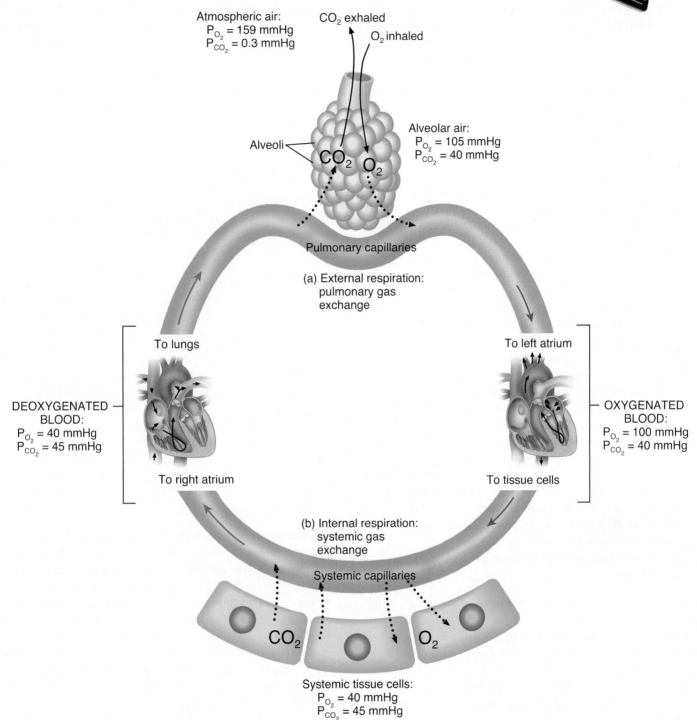



**Gases diffuse from areas of higher partial pressure to areas of lower partial pressure.**

As Figure 22.16a shows, $O_2$ diffuses from alveolar air, where its partial pressure is 105 mmHg, into the blood in pulmonary capillaries, where $P_{O_2}$ is only 40 mmHg in a resting person. If you have been exercising, the $P_{O_2}$ will be even lower because contracting muscle fibers are using more $O_2$. Diffusion continues until the $P_{O_2}$ of pulmonary capillary blood increases to match the $P_{O_2}$ of alveolar air, 105 mmHg. Blood leaving pulmonary capillaries near alveoli mixes with a small volume of blood that has flowed through conducting portions of the respiratory system, where gas exchange does not occur. Thus, the $P_{O_2}$ of blood in the pulmonary veins *leaving the lungs* is about 100 mmHg, slightly less than the $P_{O_2}$ in pulmonary capillaries.

While $O_2$ is diffusing from alveolar air into deoxygenated blood, $CO_2$ is diffusing in the opposite direction. The $P_{CO_2}$ of deoxygenated blood is 45 mmHg in a resting person, and the $P_{CO_2}$ of alveolar air is 40 mmHg. Because of this difference in $P_{CO_2}$, carbon dioxide diffuses from deoxygenated blood into the alveoli until the $P_{CO_2}$ of the blood decreases to 40 mmHg. Exhalation keeps alveolar $P_{CO_2}$ at 40 mmHg. Oxygenated blood returning to the left side of the heart in the pulmonary veins thus has a $P_{CO_2}$ of 40 mmHg.

The left ventricle pumps oxygenated blood into the aorta and through the systemic arteries to systemic capillaries. The exchange of $O_2$ and $CO_2$ between systemic capillaries and tissue cells is called **internal respiration** (Figure 22.16b). As $O_2$ leaves the bloodstream, oxygenated blood is converted into deoxygenated blood. Unlike external respiration, which occurs only in the lungs, internal respiration occurs in tissues throughout the body.

The $P_{O_2}$ of blood pumped into systemic capillaries is higher (100 mmHg) than the $P_{O_2}$ in tissue cells (about 40 mmHg at rest) because the cells constantly use $O_2$ to produce ATP. Due to this pressure difference, oxygen diffuses out of the capillaries into tissue cells and blood $P_{O_2}$ drops to 40 mmHg by the time the blood exits systemic capillaries.

While $O_2$ diffuses from the systemic capillaries into tissue cells, $CO_2$ diffuses in the opposite direction. Because tissue cells are constantly producing $CO_2$, the $P_{CO_2}$ of cells (45 mmHg at rest) is higher than that of systemic capillary blood (40 mmHg). As a result, $CO_2$ diffuses from tissue cells through interstitial fluid into systemic capillaries until the $P_{CO_2}$ in the blood increases to 45 mmHg. The deoxygenated blood then returns to the heart and is pumped to the lungs for another cycle of external respiration.

In a person at rest, tissue cells (on average) need only 25 percent of the available $O_2$ in oxygenated blood; despite its name, deoxygenated blood retains 75 percent of its $O_2$ content. During exercise, more $O_2$ diffuses from the blood into metabolically active cells, such as contracting skeletal muscle fibers. Active cells use more $O_2$ for ATP production, causing the $O_2$ content of deoxygenated blood to drop below 75 percent.

The *rate* of pulmonary and systemic gas exchange depends on several factors.

- ***Partial pressure difference of the gases.*** Alveolar $P_{O_2}$ must be higher than blood $P_{O_2}$ for oxygen to diffuse from alveolar air into the blood. The rate of diffusion is faster when the difference between $P_{O_2}$ in alveolar air and pulmonary capillary blood is larger; diffusion is slower when the difference is smaller. The differences between $P_{O_2}$ and $P_{CO_2}$ in alveolar air versus pulmonary blood increase during exercise. The larger partial pressure differences accelerate the rates of gas diffusion. The partial pressures of $O_2$ and $CO_2$ in alveolar air also depend on the rate of airflow into and out of the lungs. With increasing altitude, the total atmospheric pressure decreases, as does the partial pressure of $O_2$—from 159 mmHg at sea level, to 110 mmHg at 10,000 ft, to 73 mmHg at 20,000 ft. Although $O_2$ still is 20.9 percent of the total, the $P_{O_2}$ of inhaled air decreases with increasing altitude. Alveolar $P_{O_2}$ decreases correspondingly, and $O_2$ diffuses into the blood more slowly. The common signs and symptoms of *high altitude sickness*—shortness of breath, headache, fatigue, insomnia, nausea, and dizziness—are due to a lower level of oxygen in the blood.
- ***Surface area available for gas exchange.*** As you learned earlier in this chapter, the surface area of the alveoli is huge (about 70 $m^2$ or 750 $ft^2$). In addition, many capillaries surround each alveolus, so many that as much as 900 mL of blood is able to participate in gas exchange at any instant. Any pulmonary disorder that decreases the functional surface area of the respiratory membranes decreases the rate of external respiration. In emphysema, for example, alveolar walls disintegrate, so surface area is smaller than normal and pulmonary gas exchange is slowed.
- ***Diffusion distance.*** The respiratory membrane is very thin, so diffusion occurs quickly. Also, the capillaries are so narrow that the red blood cells must pass through them in single file, which minimizes the diffusion distance from an alveolar air space to hemoglobin inside red blood cells. Buildup of interstitial fluid between alveoli, as occurs in pulmonary edema, slows the rate of gas exchange because it increases diffusion distance (see the Chapter 22 WileyPLUS Clinical Connection entitled Pulmonary Edema).
- ***Molecular weight and solubility of the gases.*** Because $O_2$ has a lower molecular weight than $CO_2$, $O_2$ could be expected to diffuse across the respiratory membrane more quickly. However, the solubility of $CO_2$ in the fluid portions of the respiratory membrane is about 24 times greater than that of $O_2$. Taking both of these factors into account, net outward $CO_2$ diffusion occurs 20 times more rapidly than net inward $O_2$ diffusion. Consequently, when diffusion is slower than normal, for example, in emphysema or pulmonary edema, $O_2$ insufficiency (*hypoxia*) typically occurs before there is significant retention of $CO_2$ (*hypercapnia*).

### ✓ CHECKPOINT

**23.** What causes oxygen to enter pulmonary capillaries from alveolar air and to enter tissue cells from systemic capillaries?

**24.** Which factors affect the rates of diffusion of oxygen and carbon dioxide?

**25.** How does the partial pressure of atmospheric oxygen change as altitude changes?

## 22.7 Oxygen is primarily transported attached to hemoglobin, while carbon dioxide is transported in three different ways.

The blood transports gases between the lungs and body tissues (Figure 22.17). When $O_2$ and $CO_2$ enter the blood, certain chemical reactions occur that aid in gas transport and gas exchange.

### Oxygen Transport

Oxygen does not dissolve easily in water, so only about 1.5 percent of the inhaled $O_2$ is in blood is dissolved in blood plasma,

FIGURE 22.17 Transport of oxygen and carbon dioxide in the blood.

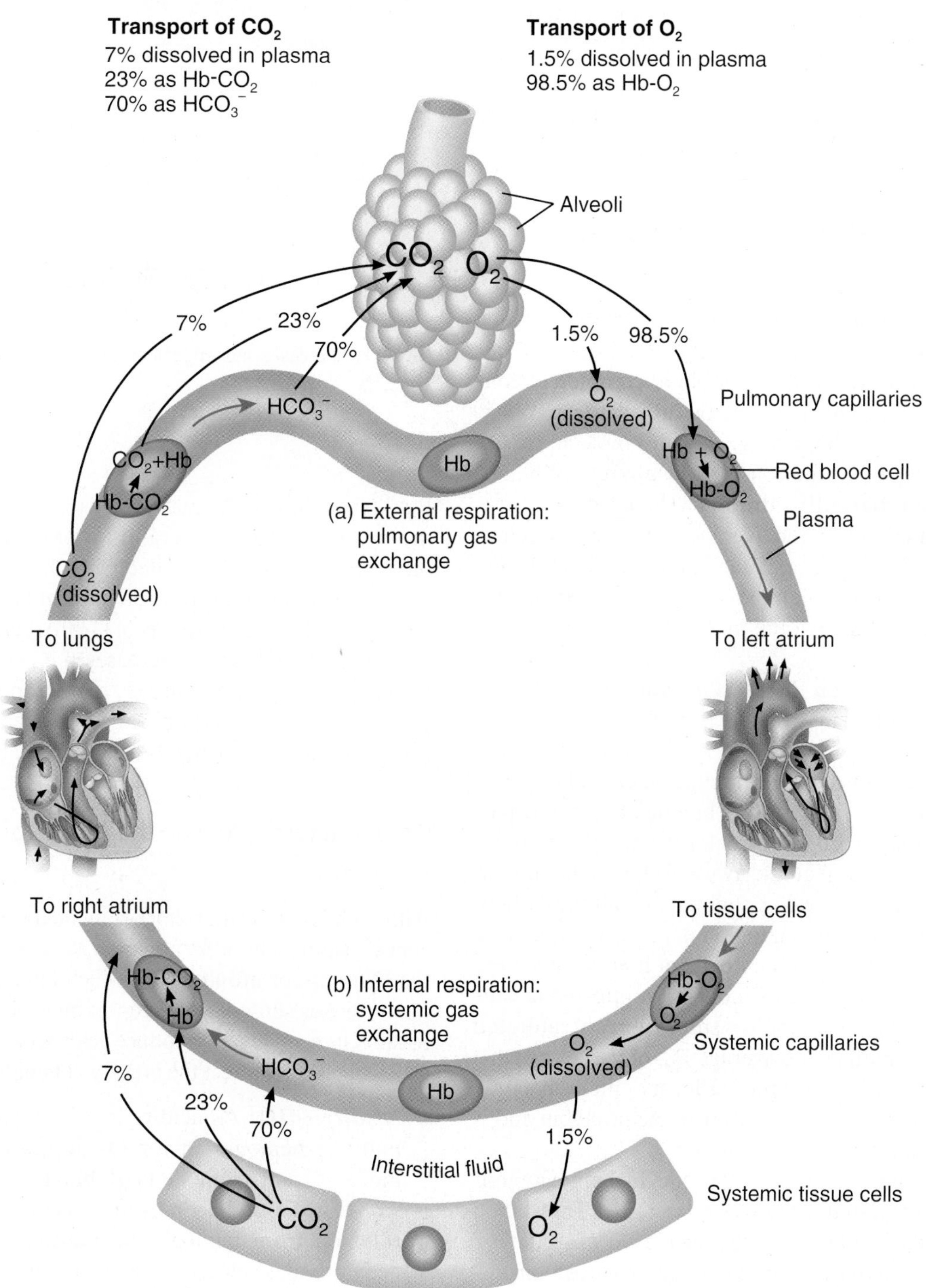


Most $O_2$ is transported by hemoglobin as oxyhemoglobin (Hb-$O_2$) within red blood cells; most $CO_2$ is transported in blood plasma as bicarbonate ions ($HCO_3^-$).

which is mostly water. The remaining 98.5 pcent of blood $O_2$ is bound to hemoglobin in red blood cells (Figure 22.17).

The heme portion of hemoglobin contains four atoms of iron, each capable of binding to a molecule of $O_2$ (see Figure 18.4b, c). Oxygen and hemoglobin (Hb) bind in an easily reversible reaction to form **oxyhemoglobin (Hb-$O_2$)** (ok′-sē-HĒ-mō-glō-bin):

$$\underset{\text{Hemoglobin}}{\text{Hb}} + \underset{\text{Oxygen}}{O_2} \underset{\text{Dissociation of } O_2}{\overset{\text{Binding of } O_2}{\rightleftharpoons}} \underset{\text{Oxyhemoglobin}}{\text{Hb-}O_2}$$

The 98.5 percent of the $O_2$ that is bound to hemoglobin is trapped inside RBCs, so only the dissolved $O_2$ (1.5 percent) can diffuse out of tissue capillaries into tissue cells. Thus, it is important to understand the factors that promote the binding and separation of $O_2$ with hemoglobin.

### Relationship between Hemoglobin and Oxygen Partial Pressure

The most important factor that determines how much $O_2$ binds to hemoglobin is the $P_{O_2}$; the higher the $P_{O_2}$, the more $O_2$ combines with hemoglobin. When hemoglobin is completely converted to Hb-$O_2$, the molecule is said to be **fully saturated**; that is, every available iron atom has combined with a molecule of $O_2$. When the hemoglobin molecule consists of a mixture of hemoglobin and Hb-$O_2$, it is **partially saturated**. The **percent saturation of hemoglobin** expresses the average saturation of hemoglobin with oxygen. For instance, if each hemoglobin molecule has bound two $O_2$ molecules, then the hemoglobin is 50 percent saturated because each hemoglobin molecule can bind a maximum of four $O_2$.

The relationship between the percent saturation of hemoglobin and $P_{O_2}$ is illustrated in the oxygen–hemoglobin dissociation (separation) curve in Figure 22.18. Note that when the $P_{O_2}$ is high, hemoglobin binds with large amounts of $O_2$ and is almost 100 percent saturated. When $P_{O_2}$ is low, hemoglobin is only partially saturated. In other words, the greater the $P_{O_2}$, the more $O_2$ will bind to hemoglobin, until all the available hemoglobin molecules are saturated. Therefore, in pulmonary capillaries, where $P_{O_2}$ is high, a lot of $O_2$ binds to hemoglobin. In tissue capillaries, where the $P_{O_2}$ is lower, hemoglobin does not hold as much $O_2$, and the dissolved $O_2$ is unloaded via diffusion into tissue cells (Figure 22.17). Note that hemoglobin is still 75 percent saturated with $O_2$ at a $P_{O_2}$ of 40 mmHg, the average $P_{O_2}$ of tissue cells in a person at rest. This is the basis for the earlier statement that only 25 percent of the available $O_2$ unloads from hemoglobin and is used by tissue cells under resting conditions.

When the $P_{O_2}$ is between 60 and 100 mmHg, hemoglobin is 90 percent or more saturated with $O_2$ (Figure 22.18). Thus, blood picks up a nearly full load of $O_2$ from the lungs even when the $P_{O_2}$ of alveolar air is as low as 60 mmHg. The oxygen–hemoglobin dissociation curve explains why people can still perform well at high altitudes or when they have certain cardiac and pulmonary diseases, even though $P_{O_2}$ may drop as low as 60 mmHg. Note also in the curve that at a considerably lower $P_{O_2}$ of 40 mmHg, hemoglobin is still 75 percent saturated with $O_2$. However, oxygen saturation of hemoglobin drops to 35 percent at 20 mmHg. Between 40 and 20 mmHg, large amounts of $O_2$ are released from hemoglobin in response to only small decreases in $P_{O_2}$. During exercise, the $P_{O_2}$ in actively contracting muscles may drop well below 40 mmHg. Then, a large percentage of the $O_2$ is released from hemoglobin, providing more $O_2$ to metabolically active tissues.

**FIGURE 22.18 Oxygen–hemoglobin dissociation curve showing the relationship between hemoglobin saturation and $P_{O_2}$ at normal body temperature.**

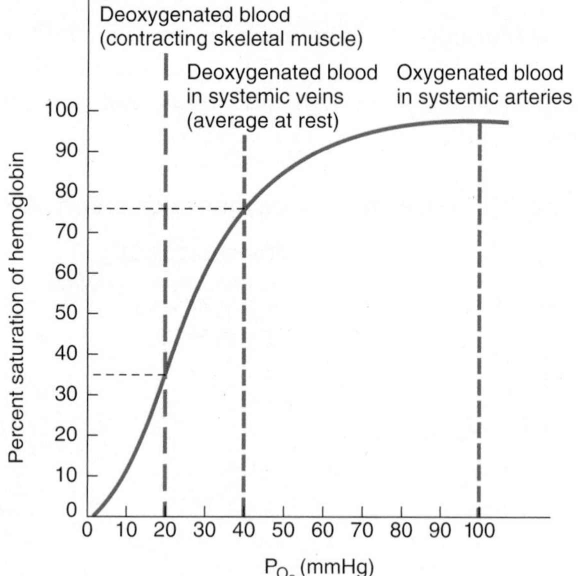


**As $P_{O_2}$ increases, more $O_2$ combines with hemoglobin.**

### Other Factors Affecting the Affinity of Hemoglobin for Oxygen

Although $P_{O_2}$ is the most important factor that determines the percent $O_2$ saturation of hemoglobin, several other factors influence the tightness, or **affinity**, with which hemoglobin binds $O_2$. Each factor makes sense if you keep in mind that metabolically active tissue cells need $O_2$ and produce acids, $CO_2$, and heat as wastes. The following factors affect the affinity of hemoglobin for $O_2$:

- ***Acidity (pH).*** As acidity increases (pH decreases), the affinity of hemoglobin for $O_2$ decreases, and $O_2$ dissociates more readily from hemoglobin (Figure 22.19a). In other words, as blood becomes more acidic, oxygen is more easily released from hemoglobin. During exercise, metabolically active tissues release lactic acid and carbonic acid. As blood becomes more acidic, $O_2$ is driven off hemoglobin, making more $O_2$ available to tissue cells that are the most active. This

FIGURE 22.19 **Oxygen–hemoglobin dissociation curves.** These curves show the relationship of (a) pH and (b) $P_{CO_2}$ to hemoglobin saturation at normal body temperature. As pH increases or $P_{CO_2}$ decreases, $O_2$ combines more tightly with hemoglobin, so that less is available to tissues. The broken lines emphasize these relationships.

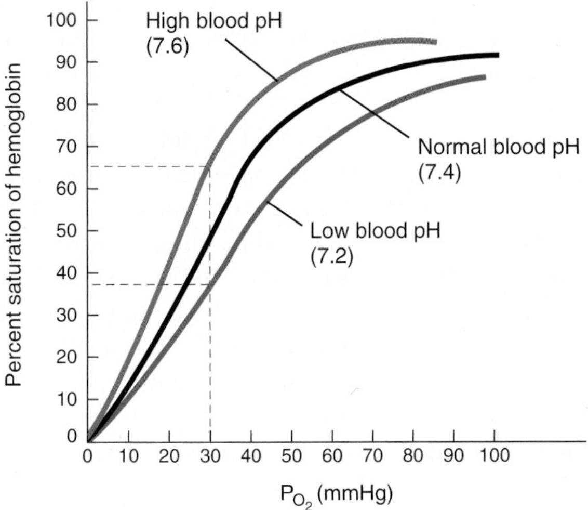


(a) Effect of pH on affinity of hemoglobin for oxygen

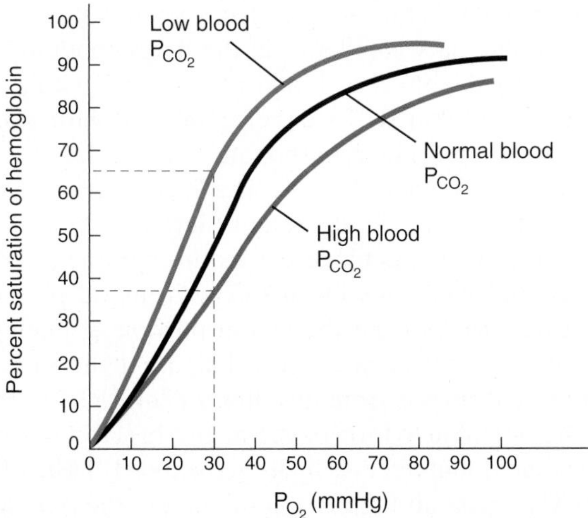


(b) Effect of $P_{CO_2}$ on affinity of hemoglobin for oxygen

**As pH decreases or $P_{CO_2}$ increases, the affinity of hemoglobin for $O_2$ declines, so less $O_2$ combines with hemoglobin and more is available to tissues.**

phenomenon, called the *Bohr effect*, is an example of how body activities are adjusted to meet moment-to-moment cellular needs.

- ***Partial pressure of carbon dioxide.*** $CO_2$ also can bind to hemoglobin, and the effect on $O_2$ binding is similar to that of $H^+$. As $P_{CO_2}$ rises, hemoglobin releases $O_2$ more readily (Figure 22.19b). $P_{CO_2}$ and pH are related factors because low blood pH (acidity) results from high $P_{CO_2}$. As $CO_2$ enters the blood from tissue cells, much of it is temporarily converted to carbonic acid ($H_2CO_3$), a reaction catalyzed by an enzyme in red blood cells called *carbonic anhydrase (CA)*:

$$\underset{\text{Carbon dioxide}}{CO_2} + \underset{\text{Water}}{H_2O} \overset{CA}{\rightleftharpoons} \underset{\text{Carbonic acid}}{H_2CO_3} \rightleftharpoons \underset{\text{Bicarbonate ion}}{HCO_3^-} + \underset{\text{Hydrogen ion}}{H^+}$$

As carbonic acid forms in red blood cells, it dissociates into bicarbonate ions and hydrogen ions. As the $H^+$ concentration increases, blood pH decreases. As blood flows through active tissues that are producing more $CO_2$, such as muscle tissue during exercise, an increased $P_{CO_2}$ produces a more acidic local environment, which helps release $O_2$ from hemoglobin. During exercise, lactic acid—a by-product of anaerobic metabolism within muscles—also decreases blood pH.

- ***Temperature.*** Within limits, as temperature increases, so does the amount of $O_2$ released from hemoglobin (Figure 22.20). Heat is a by-product of the metabolic reactions of all cells, and the heat released by contracting muscle fibers tends to raise body temperature. Active tissues liberate more heat, which promotes release of $O_2$ from oxyhemoglobin. Fever produces a similar result. In contrast, during hypothermia (lowered body temperature) cellular metabolism slows, the need for $O_2$ is reduced, and more $O_2$ remains bound to hemoglobin.
- ***BPG.*** **2,3-bisphosphoglycerate (BPG)** (bī′-fos-fō-GLIS-e-rāt) is formed in red blood cells when they break down glucose to produce ATP. The greater the level of BPG in red blood cells, the more $O_2$ is unloaded from hemoglobin. Some hormones increase the formation of BPG, as does living at higher altitudes.

FIGURE 22.20 **Oxygen–hemoglobin dissociation curves showing the effect of temperature changes.**

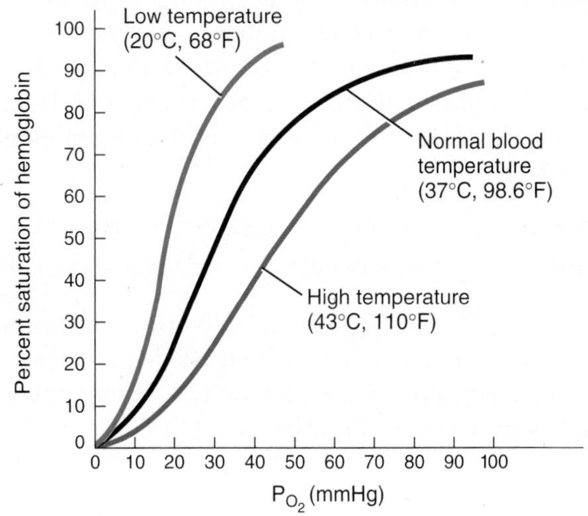


**As temperature increases, the affinity of hemoglobin for $O_2$ decreases.**

## CLINICAL CONNECTION | *Carbon Monoxide Poisoning*

Carbon monoxide (CO) is a colorless and odorless gas found in exhaust fumes from automobiles, gas furnaces, and space heaters and in tobacco smoke. It is a by-product of the combustion of carbon-containing materials such as coal, gas, and wood. CO binds to the heme group of hemoglobin, just as $O_2$ does, except that the binding of carbon monoxide to hemoglobin is over 200 times as strong as the binding of $O_2$ to hemoglobin. Thus, at a concentration as small as 0.1 percent ($P_{CO}$ = 0.5 mmHg), CO will combine with half the available hemoglobin molecules and reduce the oxygen-carrying capacity of the blood by 50 percent. Elevated blood levels of CO cause **carbon monoxide poisoning**, which can cause the lips and oral mucosa to appear bright, cherry-red (the color of hemoglobin with carbon monoxide bound to it). Without prompt treatment, carbon monoxide poisoning is fatal. It is possible to rescue a victim of CO poisoning by administering pure oxygen, which speeds up the separation of carbon monoxide from hemoglobin.

## Carbon Dioxide Transport

Carbon dioxide is transported in the blood in three main forms (Figure 22.17):

- ***Dissolved $CO_2$.*** The smallest percentage—about 7 percent—is dissolved in blood plasma. Upon reaching the lungs, it diffuses into alveolar air and is exhaled.
- ***Bound to hemoglobin.*** About 23 percent of the $CO_2$ is transported bound to hemoglobin inside red blood cells. The main $CO_2$ binding sites are the amino acids in the globin portions of the hemoglobin molecule. Hemoglobin that has bound $CO_2$ is termed **carbaminohemoglobin (Hb-$CO_2$)**:

$$\underset{\text{Hemoglobin}}{\text{Hb}} + \underset{\text{Carbon dioxide}}{CO_2} \rightleftharpoons \underset{\text{Carbaminohemoglobin}}{\text{Hb-}CO_2}$$

In tissue capillaries $P_{CO_2}$ is relatively high, which promotes formation of carbaminohemoglobin. But in pulmonary capillaries, $P_{CO_2}$ is relatively low, and the $CO_2$ readily splits apart from globin and enters the alveoli by diffusion.

- ***Bicarbonate ions.*** The greatest percentage of $CO_2$—about 70 percent—is transported in blood plasma as **bicarbonate ions ($HCO_3^-$)** (bī′-KAR-bō-nāt). As $CO_2$ diffuses into systemic capillaries and enters red blood cells, it reacts with water in the presence of the enzyme carbonic anhydrase (CA) to form carbonic acid, which dissociates into $H^+$ and $HCO_3^-$:

$$\underset{\text{Carbon dioxide}}{CO_2} + \underset{\text{Water}}{H_2O} \overset{CA}{\rightleftharpoons} \underset{\text{Carbonic acid}}{H_2CO_3} \rightleftharpoons \underset{\text{Hydrogen ion}}{H^+} + \underset{\text{Bicarbonate ion}}{HCO_3^-}$$

Thus, as blood picks up $CO_2$, $HCO_3^-$ forms and moves out into the blood plasma, down its concentration gradient. In exchange, chloride ions ($Cl^-$) move from plasma into the RBCs. This exchange of negative ions, which maintains the electrical balance between blood plasma and RBC cytosol, is known as the **chloride shift** (see Figure 22.21b). The net effect of these reactions is that $CO_2$ is removed from tissue cells and transported in blood plasma as $HCO_3^-$. As blood passes through pulmonary capillaries in the lungs, all of these reactions reverse and $CO_2$ is exhaled.

## Summary of Gas Exchange and Transport

As summarized in Figure 22.21a, deoxygenated blood returning to the pulmonary capillaries in the lungs contains carbon dioxide ($CO_2$) dissolved in blood plasma and within RBCs. $CO_2$ is carried within RBCs either combined with hemoglobin or incorporated into bicarbonate ions ($HCO_3^-$) forming carbaminohemoglobin (Hb-$CO_2$). The RBCs have also picked up $H^+$, some of which binds to and therefore is buffered by hemoglobin (Hb-H). As blood passes through the pulmonary capillaries, molecules of $CO_2$ dissolved in blood plasma and $CO_2$ that dissociates from the globin portion of hemoglobin diffuse into alveolar air and are exhaled. At the same time, inhaled $O_2$ is diffusing from alveolar air into RBCs where it binds to hemoglobin to form oxyhemoglobin (Hb-$O_2$). Binding of $O_2$ to hemoglobin releases hydrogen ions ($H^+$). Bicarbonate ions ($HCO_3^-$) pass into the RBC and bind to the released $H^+$, forming carbonic acid ($H_2CO_3$). The $H_2CO_3$ dissociates into water ($H_2O$) and $CO_2$, and the $CO_2$ diffuses from blood into alveolar air where it is exhaled. To maintain electrical balance, a chloride ion ($Cl^-$) exits the RBC for each $HCO_3^-$ that enters (a process called the *reverse chloride shift*). In summary, oxygenated blood leaving the lungs has increased $O_2$ content and decreased amounts of $CO_2$ and $H^+$.

At tissue cells, the chemical reactions reverse as cells use $O_2$ and produce $CO_2$ (Figure 22.21b). As oxygenated blood flows through systemic capillaries, $CO_2$ diffuses out of tissue cells that produce it and enters red blood cells where some of it binds to

FIGURE 22.21 **Summary of chemical reactions that occur during gas transport and exchange.** (a) In pulmonary capillaries, red blood cells unload $CO_2$ and pick up $O_2$ from alveolar air, and $O_2$ binds to hemoglobin. (b) In systemic capillaries, red blood cells release $O_2$ and pick up $CO_2$ from tissue cells.

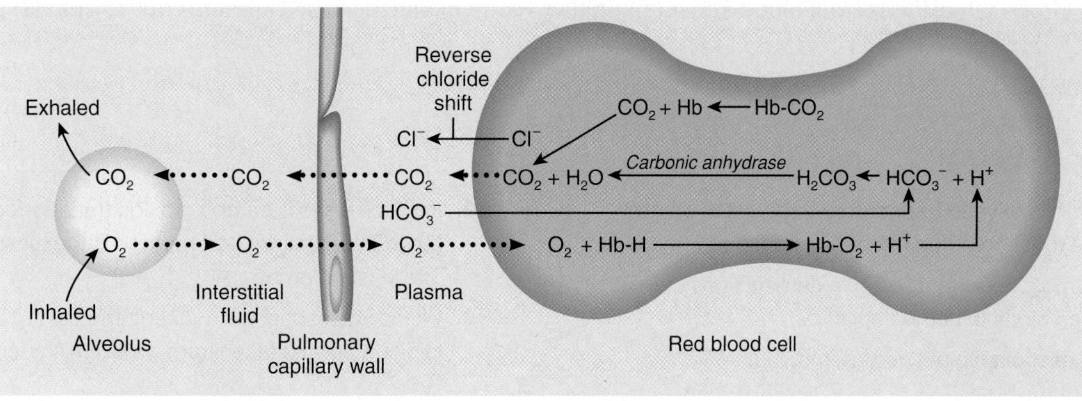


(a) Exchange of $O_2$ and $CO_2$ in pulmonary capillaries (external respiration)

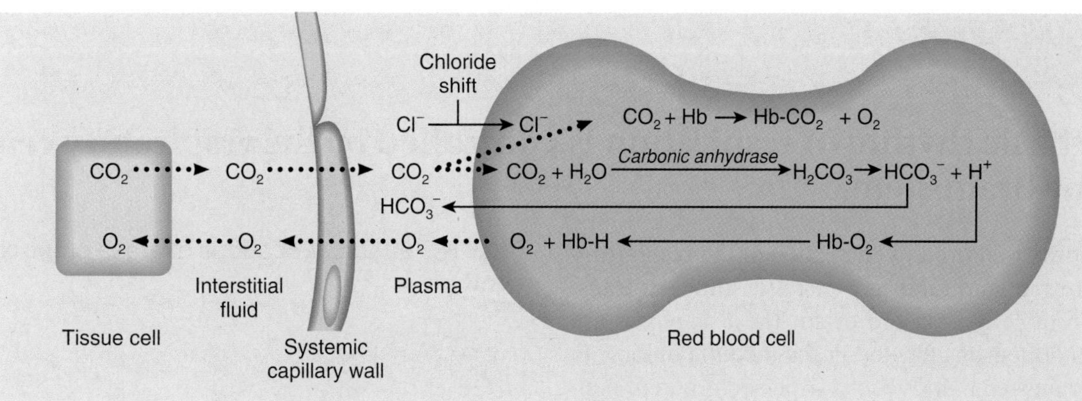


(b) Exchange of $O_2$ and $CO_2$ in systemic capillaries (internal respiration)

**Hemoglobin inside red blood cells transports $O_2$ and $CO_2$.**

## CLINICAL CONNECTION | *Effects of Smoking on the Respiratory System*

Smoking may cause a person to become easily "winded" during even moderate exercise because several factors decrease respiratory efficiency in smokers:

- Nicotine constricts terminal bronchioles, which decreases airflow into and out of the lungs.
- Carbon monoxide in smoke binds to hemoglobin and reduces its oxygen-carrying capability.
- Irritants in smoke cause increased mucus secretion by the mucosa of the bronchial tree and swelling of the mucosal lining, both of which impede airflow into and out of the lungs.
- Irritants in smoke also inhibit the movement of cilia and destroy cilia in the lining of the respiratory system. Thus, excess mucus and foreign debris are not easily removed, which further adds to the difficulty in breathing. The irritants can also convert the normal respiratory epithelium into stratified squamous epithelium, which lacks cilia and goblet cells.
- With time, smoking leads to destruction of elastic fibers in the lungs and is the prime cause of emphysema (see the Chapter 22 WileyPLUS Clinical Connection entitled Chronic Obstructive Pulmonary Disease for a discussion of emphysema). These changes cause collapse of small bronchioles and trapping of air in alveoli at the end of exhalation. The result is less efficient gas exchange.

hemoglobin, forming Hb-$CO_2$. This reaction causes $O_2$ to dissociate from Hb-$O_2$. Other molecules of $CO_2$ combine with $H_2O$ to produce $HCO_3^-$ and $H^+$. As Hb binds to and buffers $H^+$, the Hb releases $O_2$ (Bohr effect). $O_2$ diffuses from blood into interstitial fluid and then into tissue cells. To maintain electrical balance, a $Cl^-$ enters the RBC for each $HCO_3^-$ that exits (*chloride shift*). In summary, deoxygenated blood leaving tissues has decreased $O_2$ content and increased amounts of $CO_2$ and $H^+$.

**✓ CHECKPOINT**

**26.** In which ways is oxygen transported through the blood? Which is the most common method of transport?

**27.** What is the most important factor in determining how much oxygen binds to hemoglobin?

**28.** What is the relationship between hemoglobin and $P_{O_2}$?

**29.** In which ways is carbon dioxide transported through the blood? Which transport method is the most common? The least common?

**30.** Would you expect the concentration of $HCO_3^-$ to be higher in blood plasma taken from a systemic artery or a systemic vein?

# 22.8 The basic rhythm of respiration is controlled by the respiratory center in the brain stem.

The size of the thorax is altered by the action of the respiratory muscles, which contract and relax as a result of impulses transmitted to them from centers in the brain. These impulses are sent from clusters of neurons located in the medulla oblongata and pons of the brain stem. This widely dispersed group of neurons, collectively called the **respiratory center**, can be divided into three areas on the basis of their functions: the medullary rhythmicity area, the pneumotaxic area, and the apneustic area (Figure 22.22).

FIGURE 22.22 **Locations of areas of the respiratory center.**

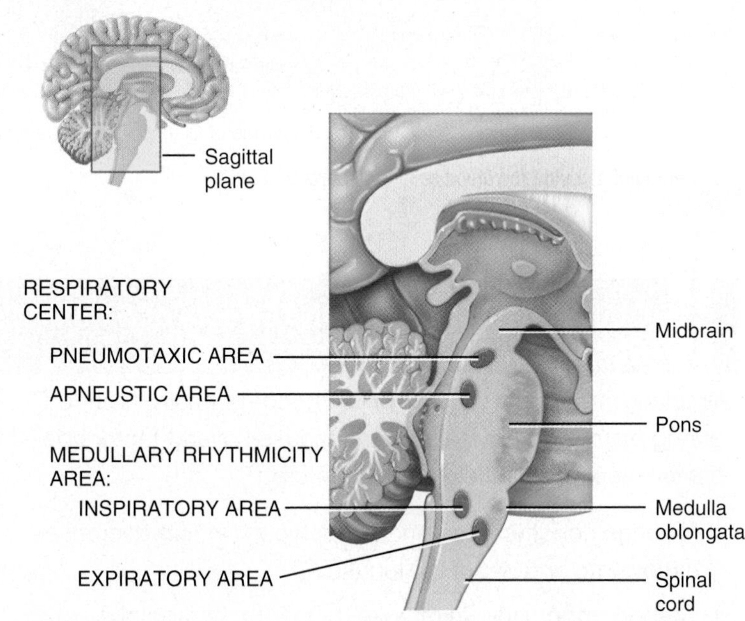


**The respiratory center includes the medullary rhythmicity area in the medulla oblongata plus the pneumotaxic and apneustic areas in the pons.**

## Medullary Rhythmicity Area

The **medullary rhythmicity area** (rith-MIS-i-tē) in the medulla oblongata controls the basic rhythm of respiration. There are inspiratory and expiratory areas within the medullary rhythmicity area. Figure 22.23 shows the relationships of the inspiratory and expiratory areas during normal quiet breathing and forceful breathing.

During quiet breathing, inhalation lasts for about 2 seconds and exhalation lasts for about 3 seconds. Impulses generated in the **inspiratory area** establish the basic rhythm of breathing. While the inspiratory area is active, it generates impulses for about 2 seconds (Figure 22.23a). The impulses propagate to the diaphragm via the phrenic nerves and to the external

FIGURE 22.23 **Activities of the medullary rhythmicity area.** Roles of the medullary rhythmicity area include controlling (a) the basic rhythm of respiration and (b) forceful breathing.

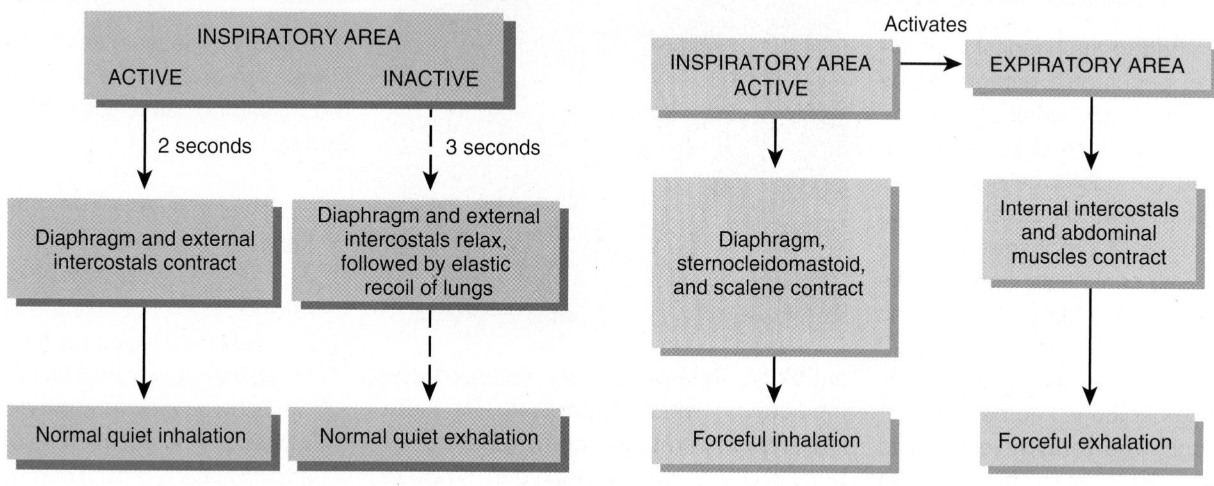


**During normal, quiet breathing, the expiratory area is inactive; during forceful breathing, the inspiratory area activates the expiratory area.**

intercostals via intercostal nerves. When the impulses reach the diaphragm and external intercostals, the muscles contract and inhalation occurs. At the end of 2 seconds, the inspiratory area becomes inactive and impulses cease. With no impulses arriving, the diaphragm and external intercostals relax for about 3 seconds, allowing passive elastic recoil of the lungs and thoracic wall. Then, the cycle repeats. Traumatic injury to both phrenic nerves causes paralysis of the diaphragm and respiratory arrest.

The neurons of the **expiratory area** remain inactive during quiet breathing. However, during forceful breathing, impulses from the inspiratory area activate the expiratory area (Figure 22.23b). Impulses from the expiratory area cause contraction of the internal intercostal and abdominal muscles, which decreases the size of the thoracic cavity and causes forceful exhalation.

## Pneumotaxic Area

Although the medullary rhythmicity area controls the basic rhythm of respiration, other sites in the brain stem help coordinate the transition between inhalation and exhalation. One of these sites is the **pneumotaxic area** (noo-mō-TAK-sik; *pneumo-* = air or breath; *-taxic* = arrangement) in the upper pons (Figure 22.22), which transmits inhibitory impulses to the inspiratory area. The major effect of these impulses is to help turn off the inspiratory area before the lungs become too full of air. In other words, the impulses shorten the duration of inhalation. When the pneumotaxic area is more active, breathing rate is more rapid.

## Apneustic Area

Another part of the brain stem that coordinates the transition between inhalation and exhalation is the **apneustic area** (ap-NOO-stik) in the lower pons (Figure 22.22). This area sends stimulatory impulses to the inspiratory area that activate it and prolong inhalation. The result is a long, deep inhalation. When the pneumotaxic area is active, it overrides signals from the apneustic area.

### ✓ CHECKPOINT

**31.** Which nerves convey impulses from the respiratory center to the diaphragm?

**32.** How does the medullary rhythmicity area regulate respiration?

**33.** How are the apneustic and pneumotaxic areas related to the control of respiration?

## RETURN TO Cari's Story

Cari was admitted to the hospital immediately. She had pneumonia brought on by an impaired pulmonary defense system. Cigarette smoke had paralyzed her cilia and stimulated the goblet cells lining her respiratory passages to secrete excess amounts of mucus. Macrophages and neutrophils had been continually drawn to the lung to combat the damaging effects of the smoke but had instead become trapped in the thick mucus secretions. Cari's only way of clearing her lungs had become her smoker's cough. The X-ray revealed fluid within the alveolar spaces, and a Gram stain of her *sputum* (mucus coughed up from the lung) had revealed pneumococcal pneumonia, the most common form. The doctor started Cari on intravenous antibiotics, continued her oxygen supplementation, and ran further tests.

Ken walked into her room. "How are you feeling?"

"I feel pretty much the same. The oxygen seems to help, but I feel like I'm panting like a dog, and my chest hurts," she replied.

Ken talked to the doctor. "Why isn't she getting better?"

The doctor sighed and replied, "Ken, we're doing everything we can. Cari's pneumonia wasn't responding to the initial antibiotic therapy so we changed her medication, which appears to be helping. Her temperature is coming down. Her body is still struggling for oxygen, so we're putting her on a ventilator to help get oxygen to her tissues and also address her blood alkalosis. The pH of her blood was rising because she was hyperventilating. The mechanical ventilator will help control the acute change in blood pH until her kidneys can compensate for the respiratory alkalosis."

Cari was lucky—she finally responded to the antibiotics. She was discharged after being in the hospital for nearly two weeks. Her chest still hurt, and she still had some difficulty breathing. The doctor warned her that if she didn't quit smoking, she was at risk for another serious infection. Pneumonia is related to impaired lung functioning. Chronic obstructive pulmonary disease, which has a direct correlation to long-term smoking, is the fourth leading cause of death in the United States.

F. *How would fluid in Cari's lungs affect her total lung capacity?*

G. *How does the elevation of Cari's respiratory rate alter her minute ventilation?*

H. *Normal blood oxygen saturation levels are greater than 94 percent; Cari's blood oxygen saturation level was 90 percent at the time of her exam and an initial arterial blood gas analysis done when she was admitted to the hospital revealed her arterial $P_{O_2}$ was 54 mmHg. How do these clinical findings relate to the internal respiration in Cari's body?*

I. *Which of the symptoms Cari has described are due to lack of oxygen and reduced oxygen exchange at her tissues?*

J. *As Cari's $P_{CO_2}$ rose, how was the oxygen-carrying capacity of hemoglobin affected?*

K. *How would you have expected Cari's decreased $P_{CO_2}$ and alkaline blood pH to have affected her breathing?*

L. *How would administration of oxygen enhance Cari's central drive to breathe?*

# 22.9 Respiration may be modified by cortical influences, chemical stimuli, proprioceptor input, and the inflation reflex.

At rest, about 200 mL of $O_2$ are used each minute by body cells. During strenuous exercise, however, $O_2$ use typically increases 15- to 20-fold in normal healthy adults, and as much as 30-fold in elite endurance-trained athletes. Several mechanisms help match respiratory effort to metabolic demand. Although the basic rhythm of respiration is set and coordinated by the inspiratory area, the rhythm can be modified in response to inputs from other brain regions, receptors in the peripheral nervous system, and other factors.

## Cortical Influences on Respiration

Because the cerebral cortex has connections with the respiratory center, we can voluntarily alter our pattern of breathing. We can even refuse to breathe at all for a short time. Voluntary control is protective because it enables us to prevent water or irritating gases from entering the lungs. The ability to not breathe, however, is limited by the buildup of $CO_2$ and $H^+$ in the body. When $P_{CO_2}$ and $H^+$ concentrations increase to a certain level, the inspiratory area is strongly stimulated, impulses are sent along the phrenic and intercostal nerves to inspiratory muscles, and breathing resumes, whether the person wants it to or not. It is impossible for small children to kill themselves by voluntarily holding their breath, even though many have tried in order to get their way. If the breath is held long enough to cause fainting, breathing resumes when consciousness is lost. Impulses from the hypothalamus and limbic system also stimulate the respiratory center, allowing emotional stimuli to alter respirations as, for example, in laughing and crying.

## Chemoreceptor Regulation of Respiration

Certain chemical stimuli modulate how quickly and how deeply we breathe. The respiratory system functions to maintain proper levels of $CO_2$ and $O_2$ and is very responsive to changes in the levels of these gases in body fluids. Sensory neurons that are responsive to chemicals are called **chemoreceptors** (kē′-mō-rē-SEP-tors). Chemoreceptors monitor levels of $CO_2$, $H^+$, and $O_2$ and provide input to the respiratory center (Figure 22.24). **Central chemoreceptors** are located in or near the medulla oblongata in the *central* nervous system. They respond to changes in $H^+$ concentration or $P_{CO_2}$ in cerebrospinal fluid. **Peripheral chemoreceptors** are part of the *peripheral* nervous system and are sensitive to changes in $P_{O_2}$, $H^+$, and $P_{CO_2}$ in the blood. Peripheral chemoreceptors are located in the **aortic bodies**, clusters of chemoreceptors located in the wall of the arch of the aorta, and in the **carotid bodies**, which are oval nodules in the wall of the left and right common carotid arteries, where they divide into the internal and external carotid arteries. Axons of sensory neurons from the aortic bodies are part of the vagus (X) nerves, and those from the carotid bodies are part of the right and left glossopharyngeal (IX) nerves.

Because $CO_2$ is lipid-soluble, it easily diffuses into cells where, in the presence of carbonic anhydrase, it combines with water ($H_2O$) to form carbonic acid ($H_2CO_3$). Carbonic acid quickly breaks down into $H^+$ and $HCO_3^-$. Thus, an increase in $CO_2$ in the blood causes an increase in $H^+$ inside cells, and a decrease in blood $CO_2$ level causes a decrease in $H^+$ inside cells.

Normally, the $P_{CO_2}$ in arterial blood is 40 mmHg. If even a slight increase in $P_{CO_2}$ occurs—a condition called **hypercapnia** (hī′-per-KAP-nē-a)—the central chemoreceptors are stimulated and respond vigorously to the resulting increase in $H^+$ level. The peripheral chemoreceptors also are stimulated by both the high $P_{CO_2}$ and the rise in $H^+$. In addition, the peripheral chemoreceptors (but not the central chemoreceptors) respond to a deficiency of $O_2$. When $P_{O_2}$ in arterial blood falls from a normal level of 100 mmHg but is still above 50 mmHg, the peripheral chemoreceptors are stimulated. Severe deficiency of $O_2$ depresses activity of the central chemoreceptors and inspiratory area, which then do not respond well to any inputs and send fewer impulses to the muscles of inhalation. As the breathing rate decreases or breathing ceases altogether, $P_{O_2}$ falls lower and lower, establishing a positive feedback cycle with a possibly fatal result.

The chemoreceptors participate in a negative feedback system that regulates the levels of $CO_2$, $O_2$, and $H^+$ in the blood

**FIGURE 22.24 Locations of peripheral chemoreceptors.**

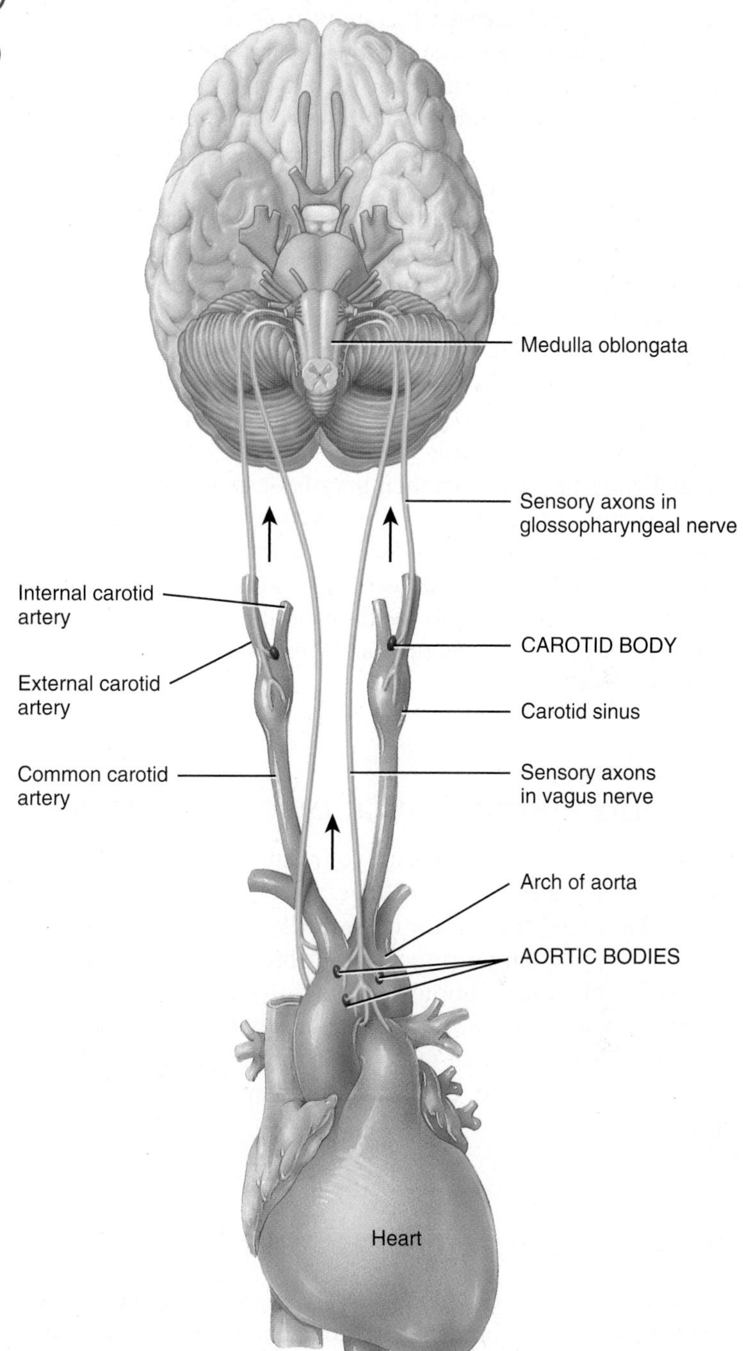


**Chemoreceptors are sensory neurons that respond to changes in the levels of certain chemicals in the body.**

(Figure 22.25). As a result of increased $P_{CO_2}$, decreased pH (increased $H^+$), or decreased $P_{O_2}$, input from the central and peripheral chemoreceptors causes the inspiratory area to become highly active, and the rate and depth of breathing increase. Rapid and deep breathing, called **hyperventilation**, allows the inhalation of more $O_2$ and exhalation of more $CO_2$ until $P_{CO_2}$ and $H^+$ are lowered to normal.

If arterial $P_{CO_2}$ is lower than 40 mmHg—a condition called **hypocapnia**—the central and peripheral chemoreceptors are not stimulated, and stimulatory impulses are not sent to the inspiratory area. As a result, the inspiratory area sets its own moderate pace until $CO_2$ accumulates and the $P_{CO_2}$ rises to 40 mmHg. The inspiratory center is more strongly stimulated when $P_{CO_2}$ is rising above normal than when $P_{O_2}$ is falling below normal. As a result, people who hyperventilate voluntarily and cause hypocapnia can hold their breath for an unusually long period. Swimmers were once encouraged to hyperventilate just before diving in to compete. However, this practice is risky because the $O_2$ level may fall dangerously low and cause fainting before the $P_{CO_2}$ rises high enough to stimulate inhalation. If you faint on land you may suffer bumps and bruises, but if you faint in the water you could drown.

## Proprioceptor Stimulation of Respiration

As soon as you start exercising, your rate and depth of breathing increase, even before changes in $P_{O_2}$, $P_{CO_2}$, or $H^+$ level occur. The main stimulus for these quick changes in respiratory effort is input from proprioceptors, which monitor movement of joints and muscles. Impulses from the proprioceptors stimulate the inspiratory area of the medulla oblongata. At the same time, motor neurons that originate in the primary motor cortex (precentral gyrus) also feed excitatory impulses into the inspiratory area.

## The Inflation Reflex

Similar to those in the blood vessels, stretch receptors are located in the walls of bronchi and bronchioles. When these receptors become stretched during overinflation of the lungs, impulses are sent to the inspiratory and apneustic areas. In response, the inspiratory area is inhibited directly, and the apneustic area is inhibited from activating the inspiratory area. As a result, exhalation begins. As air leaves the lungs during exhalation, the lungs deflate and the stretch receptors are no longer stimulated. Thus, the inspiratory and apneustic areas are no longer inhibited, and a new inhalation begins. This reflex, referred to as the **inflation (Hering–Breuer) reflex** (HER-ing BROY-er), is mainly a protective mechanism for preventing excessive inflation of the lungs rather than a key component in the normal regulation of respiration.

**FIGURE 22.25** **Regulation of breathing in response to changes in blood $P_{CO_2}$, $P_{O_2}$, and pH ($H^+$ concentration) via negative feedback.**

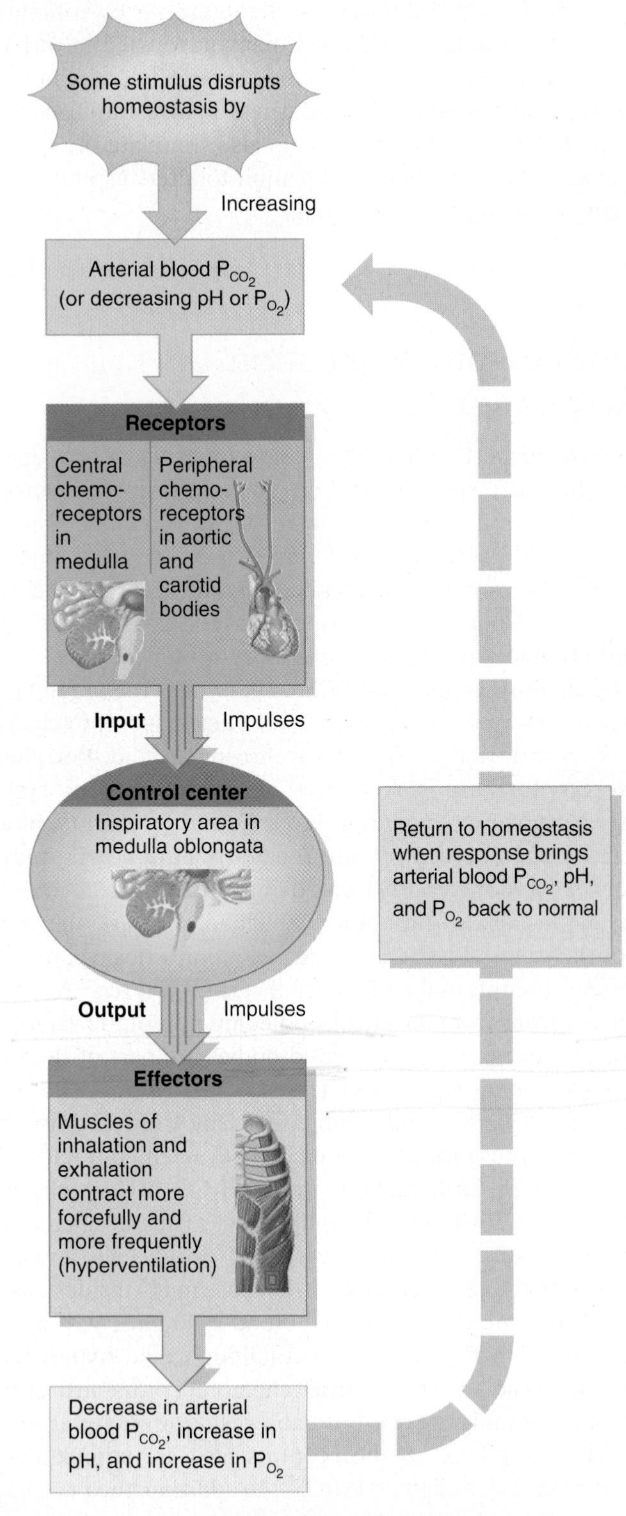

Some stimulus disrupts homeostasis by
Increasing
Arterial blood $P_{CO_2}$ (or decreasing pH or $P_{O_2}$)
Receptors
Central chemoreceptors in medulla
Peripheral chemoreceptors in aortic and carotid bodies
Input
Impulses
Control center
Inspiratory area in medulla oblongata
Output
Impulses
Effectors
Muscles of inhalation and exhalation contract more forcefully and more frequently (hyperventilation)
Decrease in arterial blood $P_{CO_2}$, increase in pH, and increase in $P_{O_2}$
Return to homeostasis when response brings arterial blood $P_{CO_2}$, pH, and $P_{O_2}$ back to normal

An increase in arterial blood $P_{CO_2}$ stimulates the inspiratory area.

## Other Influences on Respiration

Other factors that contribute to regulation of respiration include the following:

- ***Limbic system stimulation.*** Anticipation of activity or emotional anxiety may stimulate the limbic system, which then sends excitatory input to the inspiratory area, increasing the rate and depth of ventilation.
- ***Temperature.*** An increase in body temperature, as occurs during a fever or vigorous muscular exercise, increases the rate of respiration. A decrease in body temperature decreases respiratory rate. A sudden cold stimulus (such as plunging into cold water) causes temporary apnea (AP-nē-a; *a-* = without; *-pnea* = breath), an absence of breathing.
- ***Pain.*** A sudden, severe pain brings about brief apnea, but a prolonged somatic pain increases respiratory rate. Visceral pain may slow the rate of respiration.
- ***Stretching the anal sphincter muscle.*** This action increases the respiratory rate and is sometimes used to stimulate respiration in a newborn baby or a person who has stopped breathing.
- ***Irritation of airways.*** Physical or chemical irritation of the pharynx or larynx brings about an immediate cessation of breathing followed by coughing or sneezing.
- ***Blood pressure.*** The carotid and aortic baroreceptors that detect changes in blood pressure have a small effect on respiration. A sudden rise in blood pressure decreases the rate of respiration, and a drop in blood pressure increases the respiratory rate.

Table 22.2 summarizes the stimuli that affect the rate and depth of ventilation.

**TABLE 22.2**

**Stimuli That Affect Ventilation Rate and Depth**

| STIMULI THAT INCREASE VENTILATION RATE AND DEPTH | STIMULI THAT DECREASE VENTILATION RATE AND DEPTH |
|---|---|
| Voluntary hyperventilation controlled by the cerebral cortex and anticipation of activity by stimulation of the limbic system | Voluntary hypoventilation controlled by the cerebral cortex |
| Increase in arterial blood $P_{CO_2}$ above 40 mmHg (causes an increase in $H^+$) detected by peripheral and central chemoreceptors | Decrease in arterial blood $P_{CO_2}$ below 40 mmHg (causes a decrease in $H^+$) detected by peripheral and central chemoreceptors |
| Decrease in arterial blood $P_{O_2}$ from 100 mmHg to 50 mmHg | Decrease in arterial blood $P_{O_2}$ below 50 mmHg |
| Increased activity of proprioceptors | Decreased activity of proprioceptors |
| Increase in body temperature | Decrease in body temperature decreases the rate of respiration, and a sudden cold stimulus causes apnea |
| Prolonged pain | Severe pain causes apnea |
| Decrease in blood pressure | Increase in blood pressure |
| Stretching the anal sphincter | Irritation of pharynx or larynx by touch or chemicals causes brief apnea followed by coughing or sneezing |

### ✓ CHECKPOINT

**34.** Where are central chemoreceptors located?

**35.** Which chemicals stimulate peripheral chemoreceptors?

**36.** What is the normal arterial blood $P_{CO_2}$?

**37.** What happens to the rate and depth of breathing when the $P_{CO_2}$ becomes elevated?

**38.** How do proprioceptors, inflation reflex, temperature changes, pain, and irritation of the airways each modify respiration?

# 22.10 Acid–base balance is maintained by controlling the $H^+$ concentration of body fluids.

Various ions play different roles that help maintain homeostasis. A major homeostatic challenge is keeping the hydrogen ion ($H^+$) concentration, or pH, of body fluids at an appropriate level. This task—the maintenance of acid–base balance—is of critical importance to normal cellular function. For example, the three-dimensional shape of all body proteins, which enables them to perform specific functions, is very sensitive to pH changes. When the diet contains a large amount of protein, as is typical in North America, cellular metabolism produces more acids than bases, which tends to acidify the blood. Before proceeding with this section of the chapter, you may wish to review the discussion of acids, bases, and pH in Concept 2.4.

In a healthy person, several mechanisms help maintain the pH of systemic arterial blood between 7.35 and 7.45. Because metabolic reactions often produce a huge excess of $H^+$, the level of $H^+$ in body fluids would quickly rise to a lethal level if there were not a way to dispose of $H^+$. Maintenance of $H^+$ concentration within a narrow range is essential to our survival. The removal of $H^+$ from body fluids and its subsequent elimination from the body depend on three major mechanisms: buffer systems, exhalation of carbon dioxide, and excretion of $H^+$ in the urine. Excretion of $H^+$ in the urine will be discussed in Chapter 24.

## The Actions of Buffer Systems

Most buffer systems in the body consist of a weak acid and the salt of that acid, which functions as a weak base. Buffers act quickly to temporarily bind excess $H^+$, removing the highly reactive $H^+$ from solution. Buffers thus raise pH of body fluids but do not remove $H^+$ from the body. Buffers prevent rapid, drastic changes in the pH of body fluids by converting *strong* acids and bases into *weak* acids and weak bases within fractions of a second. Strong acids lower pH more than weak acids because strong acids release $H^+$ more readily and thus contribute more free hydrogen ions. Similarly, strong bases raise pH more than weak ones. Important buffer systems related to the respiratory system include the protein buffer system and the carbonic acid–bicarbonate buffer system.

### Protein Buffer System

The **protein buffer system** is the most abundant buffer in intracellular fluid and blood plasma. For example, the protein hemoglobin is an especially good buffer within red blood cells, and albumin is the main protein buffer in blood plasma. Proteins are composed of amino acids, organic molecules that contain at least one *carboxyl group* ($-COOH$) and at least one *amino group* ($-NH_2$); these groups are the functional components of the protein buffer system. The free carboxyl group at one end of a protein acts like an acid by releasing $H^+$ when pH rises ($H^+$ concentration decreases); it dissociates as follows (with R representing the rest of the protein molecule):

$$NH_2-\underset{H}{\overset{R}{\underset{|}{\overset{|}{C}}}}-COOH \longrightarrow NH_2-\underset{H}{\overset{R}{\underset{|}{\overset{|}{C}}}}-COO^- + H^+$$

The $H^+$ is then able to react with any excess hydroxide ions ($OH^-$) in the solution to form water. In other words, as the pH of a solution shifts into an alkaline (basic) range, $H^+$ released from protein buffers remove excess $OH^-$ to return the solution to a healthy pH:

$$\underset{\text{Hydrogen ion}}{H^+} + \underset{\text{Hydroxide ion}}{OH^-} \longrightarrow \underset{\text{Water}}{H_2O}$$

The free amino group at the other end of a protein can act as a base by combining with $H^+$ when pH falls ($H^+$ concentration increases). In other words, as the pH of a solution becomes acidic, protein buffers remove the excess $H^+$:

$$NH_2-\underset{H}{\overset{R}{\underset{|}{\overset{|}{C}}}}-COOH + H^+ \longrightarrow {}^+NH_3-\underset{H}{\overset{R}{\underset{|}{\overset{|}{C}}}}-COOH$$

So proteins can buffer both acids and bases. In addition to the terminal carboxyl and amino groups, side chains that can buffer $H^+$ are present on 7 of the 20 amino acids.

As we have already noted, the protein hemoglobin is an important buffer of $H^+$ in red blood cells (see **Figure 22.21**). As blood flows through the systemic capillaries, carbon dioxide ($CO_2$) passes from tissue cells into red blood cells, where it combines with water ($H_2O$) to form carbonic acid ($H_2CO_3$):

$$\underset{\substack{\text{Carbon dioxide}\\ \text{(entering RBCs)}}}{CO_2} + \underset{\text{Water}}{H_2O} \longrightarrow \underset{\text{Carbonic acid}}{H_2CO_3}$$

Once formed, $H_2CO_3$ dissociates into $HCO_3^-$ and $H^+$:

$$\underset{\text{Carbonic acid}}{H_2CO_3} \longrightarrow \underset{\text{Bicarbonate ion}}{HCO_3^-} + \underset{\text{Hydrogen ion}}{H^+}$$

At the same time that $CO_2$ is entering red blood cells, oxyhemoglobin (Hb-$O_2$) is giving up its oxygen to tissue cells. Hemoglobin picks up most of the $H^+$ as deoxyhemoglobin (Hb-H):

$$\underset{\substack{\text{Oxyhemoglobin}\\ \text{(in RBCs)}}}{\text{Hb-}O_2} + \underset{\substack{\text{Hydrogen ion}\\ \text{(from carbonic acid)}}}{H^+} \longrightarrow \underset{\text{Deoxyhemoglobin}}{\text{Hb-H}} + \underset{\text{Oxygen}}{O_2}$$

To summarize the buffering action of hemoglobin, $H^+$ ions that are released from $H_2CO_3$ are removed from the blood by hemoglobin to maintain blood pH in a healthy range.

### Carbonic Acid–Bicarbonate Buffer System

The **carbonic acid–bicarbonate buffer system** is based on *carbonic acid* ($H_2CO_3$), which can act as a weak acid, and *bicarbonate ion* ($HCO_3^-$), which can act as a weak base. $HCO_3^-$ is a significant anion in both intracellular and extracellular fluids. The $CO_2$ that is constantly being released from tissue cells during cellular respiration produces $H_2CO_3$ (see Figure 22.21b). When a body solution shifts into the acidic range, the $HCO_3^-$ can function as a weak base and remove the excess $H^+$ from solution:

$$\underset{\text{Hydrogen ion}}{H^+} + \underset{\text{Bicarbonate ion (weak base)}}{HCO_3^-} \longrightarrow \underset{\text{Carbonic acid}}{H_2CO_3}$$

Then, $H_2CO_3$ dissociates into water and carbon dioxide, and the $CO_2$ is exhaled from the lungs. Conversely, when a body solution starts to become too alkaline, the $H_2CO_3$ can function as a weak acid and provide needed $H^+$:

$$\underset{\text{Carbonic acid (weak base)}}{H_2CO_3} + \underset{\text{Bicarbonate ion}}{HCO_3^-} \longrightarrow \underset{\text{Hydrogen ion}}{H^+}$$

The released $H^+$ removes excess $OH^-$, allowing the alkaline solution to shift back into a healthy range. Because $CO_2$ and $H_2O$ combine to form $H_2CO_3$, this buffer system cannot protect against pH changes due to respiratory problems in which there is an excess or shortage of $CO_2$.

## Exhalation of Carbon Dioxide

The simple act of breathing also plays an important role in maintaining the pH of body fluids. An increase in the $CO_2$ concentration in body fluids increases $H^+$ concentration and thus lowers the pH (makes body fluids more acidic). Conversely, a decrease in the $CO_2$ concentration of body fluids raises the pH (makes body fluids more alkaline). This chemical interaction is illustrated by the following reversible reactions:

$$\underset{\text{Carbon dioxide}}{CO_2} + \underset{\text{Water}}{H_2O} \rightleftharpoons \underset{\text{Carbonic acid}}{H_2CO_3} \rightleftharpoons \underset{\text{Hydrogen ion}}{H^+} + \underset{\text{Bicarbonate ion}}{HCO_3^-}$$

Changes in the rate and depth of breathing can alter the pH of body fluids within a couple of minutes. With increased ventilation, more $CO_2$ is exhaled, reducing the level of carbonic acid in blood, which reduces blood $H^+$ level and raises the blood pH. If ventilation is slower than normal, less carbon dioxide is exhaled. As $CO_2$ levels rise, blood $H^+$ level increases and the blood pH falls. These examples show the powerful effect of alterations in breathing on the pH of body fluids.

The pH of body fluids and the rate and depth of breathing interact via a negative feedback loop (Figure 22.26). When the

**FIGURE 22.26 Negative feedback regulation of blood pH by the respiratory system.**

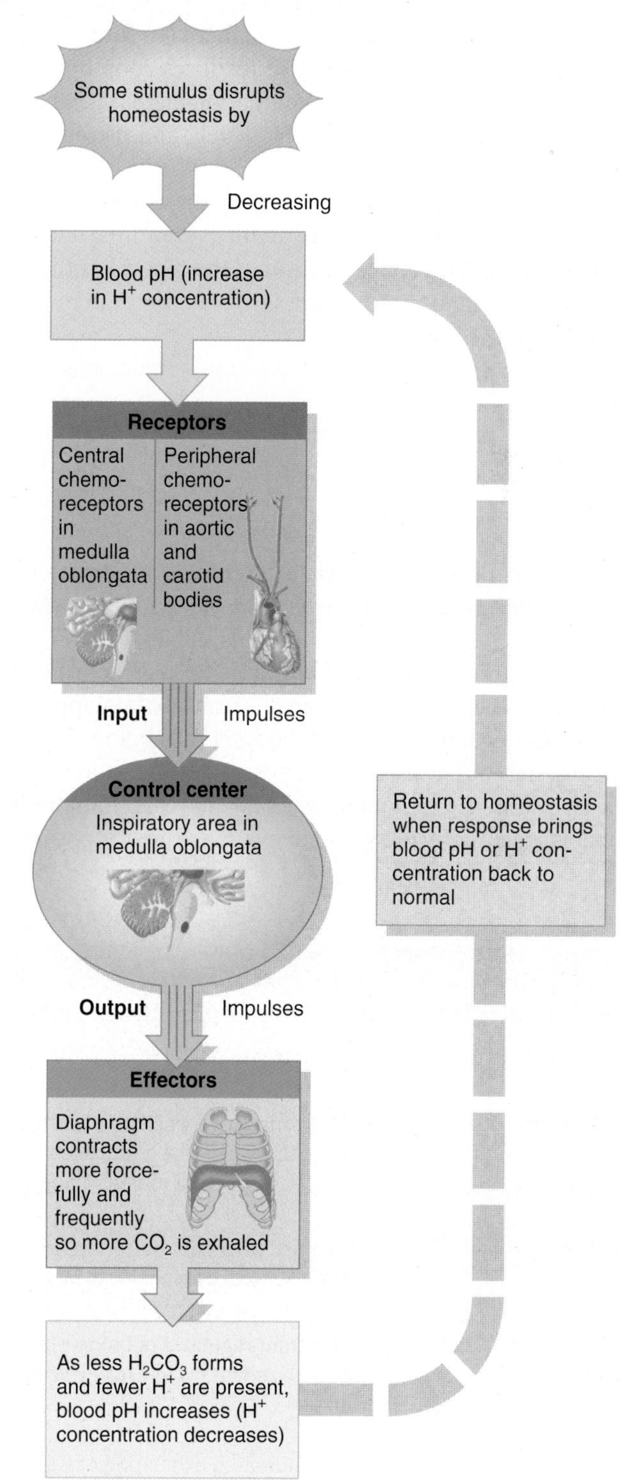

Some stimulus disrupts homeostasis by
Decreasing
Blood pH (increase in $H^+$ concentration)
Receptors
Central chemo-receptors in medulla oblongata
Peripheral chemo-receptors in aortic and carotid bodies
Input
Impulses
Control center
Inspiratory area in medulla oblongata
Output
Impulses
Effectors
Diaphragm contracts more force-fully and frequently so more $CO_2$ is exhaled
As less $H_2CO_3$ forms and fewer $H^+$ are present, blood pH increases ($H^+$ concentration decreases)
Return to homeostasis when response brings blood pH or $H^+$ con-centration back to normal

Exhalation of carbon dioxide lowers the $H^+$ concentration of blood.

blood acidity increases, the decrease in pH (increase in concentration of $H^+$) is detected by central chemoreceptors in the medulla oblongata and peripheral chemoreceptors in the aortic and carotid bodies, both of which stimulate the inspiratory area in the medulla oblongata. As a result, the diaphragm and other respiratory muscles contract more forcefully and frequently, so more $CO_2$ is exhaled. As less $H_2CO_3$ forms and fewer $H^+$ are present, blood pH increases. When the response brings blood pH ($H^+$ concentration) back to normal, there is a return to acid–base homeostasis. The same negative feedback loop operates if the blood level of $CO_2$ increases. Ventilation increases, which removes more $CO_2$, reducing blood $H^+$ levels, and increasing the blood's pH.

By contrast, if the pH of the blood increases, the respiratory center is inhibited and the rate and depth of breathing decreases. A decrease in the $CO_2$ concentration of the blood has the same effect. When breathing decreases, $CO_2$ accumulates in the blood and its $H^+$ concentration increases.

Table 22.3 summarizes the respiratory mechanisms that maintain a pH of body fluids.

## Acid–Base Imbalances

The normal pH range of systemic arterial blood is between 7.35 and 7.45. **Acidosis** is a condition in which blood pH is below 7.35; **alkalosis** is a condition in which blood pH is higher than 7.45.

The major physiological effect of acidosis is depression of the central nervous system through depression of synaptic transmission. If the systemic arterial blood pH falls below 7, depression of the nervous system is so severe that the individual becomes disoriented, then comatose, and may die. Patients with severe acidosis usually die while in a coma. A major physiological effect of alkalosis, by contrast, is overexcitability in both the central nervous system and peripheral nerves. Neurons conduct impulses repetitively, even when not prompted by normal stimuli; the results are nervousness, muscle spasms, and even convulsions and death.

A change in blood pH that leads to acidosis or alkalosis may be countered by **respiratory compensation**, a physiological response to an acid–base imbalance that acts to normalize arterial blood pH. If a person has altered blood pH due to metabolic causes, hyperventilation or hypoventilation can help bring blood pH back toward the normal range; respiratory compensation occurs within minutes and reaches its maximum within hours.

Both respiratory acidosis and respiratory alkalosis are disorders resulting from changes in the partial pressure of $CO_2$ ($P_{CO_2}$) in systemic arterial blood (normal range is 35–45 mmHg).

### Respiratory Acidosis

The hallmark of **respiratory acidosis** is an abnormally high $P_{CO_2}$ in systemic arterial blood—above 45 mmHg. Inadequate exhalation of $CO_2$ causes the blood pH to drop. Any condition that decreases the movement of $CO_2$ from the blood to the alveoli of the lungs to the atmosphere causes a buildup of $CO_2$, and therefore increased $H_2CO_3$, and $H^+$. Such conditions include emphysema, pulmonary edema, injury to the respiratory center of the medulla oblongata, airway obstruction, or disorders of the muscles involved in breathing. The goal in treatment of respiratory acidosis is to increase the exhalation of $CO_2$, as, for instance, by providing ventilation therapy. In addition, intravenous administration of $HCO_3^-$ may be helpful.

### Respiratory Alkalosis

In **respiratory alkalosis**, systemic arterial blood $P_{CO_2}$ falls below 35 mmHg. The cause of the drop in $P_{CO_2}$ and the resulting increase in pH is hyperventilation, which occurs in conditions that stimulate the inspiratory area in the brain stem. Such conditions include oxygen deficiency due to high altitude or pulmonary disease, cerebrovascular accident (stroke), or severe anxiety. Treatment of respiratory alkalosis is aimed at increasing the level of $CO_2$ in the body. One simple treatment is to have the person inhale and exhale into a paper bag for a short period; as a result, the person inhales air containing a higher-than-normal concentration of $CO_2$.

**TABLE 22.3**

**Respiratory Mechanisms That Maintain pH of Body Fluids**

| MECHANISM | COMMENTS |
|---|---|
| **Buffer Systems** | Most consist of a weak acid and the salt of that acid, which functions as a weak base; prevent drastic changes in body fluid pH |
| **Proteins** | The most abundant buffers in body cells and blood; hemoglobin inside red blood cells is a good buffer |
| **Carbonic acid–bicarbonate** | Important regulator of blood pH; most abundant buffers in extracellular fluid |
| **Exhalation of $CO_2$** | With increased exhalation of $CO_2$, pH rises (fewer $H^+$); with decreased exhalation of $CO_2$, pH falls (more $H^+$) |

✓ **CHECKPOINT**

39. How do proteins and bicarbonate ions help maintain the pH of body fluids?
40. What are acidosis and alkalosis? What are the major physiological effects of each condition?

## Cari's Story

## EPILOGUE AND DISCUSSION

Cari, a 47-year-old female with a 20-year history of pack-a-day smoking and chronic bronchitis, presents at the clinic complaining of a cold and potential sinus infection. She describes general muscle aches, fatigue, and sore throat, and has a temperature of 101.2 and a history of several weeks of flulike symptoms. Upon examination, it is found that her respiratory rate is elevated and she has a low oxygen saturation level. Her lung sounds are abnormal. A follow-up chest X-ray reveals fluid infiltrate within the lungs. A Gram stain reveals the presence of bacteria, resulting in a diagnosis of pneumococcal pneumonia. Cari is lucky—she makes a full recovery.

Like many smokers, Cari has a hard decision to make: Continue smoking and risk further infections, loss of respiratory function, emphysema, cancer, and chronic obstructive lung disease, or quit. Now that you have learned about the profound effects of smoking on the delicate tissue of the lungs and about the importance of the respiratory system to homeostasis, answer the following questions.

M. *Which anatomical structures in Cari's respiratory system were initially involved?*

N. *Why was Cari plagued with a chronic smoker's cough?*

O. *Which damaging effects of tobacco smoke led to Cari's impaired respiratory defense mechanisms?*

P. *How did the pneumonia affect Cari's lung function?*

# Concept and Resource Summary

| Concept | Resources  |
|---|---|
| **Introduction** | |
| **1.** Metabolic reactions need a continual supply of $O_2$, and generate $CO_2$ waste that must be removed rapidly. The **respiratory system** acts with the cardiovascular system to bring $O_2$ into the body and removes $CO_2$ from the body. | |
| **Concept 22.1** Inhaled air travels in the upper respiratory system through the nasal cavities and then through the pharynx. 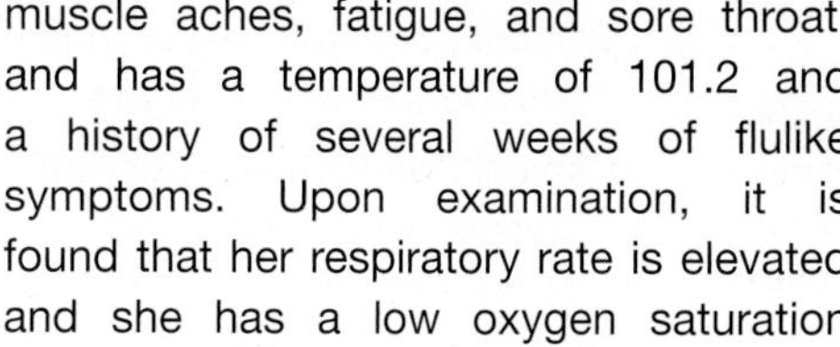 | |
| **1.** The respiratory system includes the nose, pharynx, larynx, trachea, bronchi, and lungs. The respiratory system includes a **conducting zone**, passageways that filter, warm, and moisten air, and conduct it to the lungs; and a **respiratory zone** within the lungs where gas exchange occurs. | Anatomy Overview—Nasal Cavity and Pharynx<br>Anatomy Overview—Respiratory Epithelium |
| **2.** The visible **external nose** is supported by bone and cartilage and is lined with a mucous membrane. Openings to the exterior are the **external nares**. The **nasal cavity** connects to the pharynx through the **internal nares** and is divided by the **nasal septum**. The nose warms, moistens, and filters incoming air; detects olfactory stimuli; and contributes to voice resonance. | Clinical Connection—Rhinoplasty |
| **3.** The **paranasal sinuses** are cavities in skull bones that open into the nasal cavity. They produce mucus and contribute to voice resonance. | Clinical Connection—Tonsillectomy |
| **4.** The pharynx is a muscular tube lined by a mucous membrane. The **nasopharynx** is a passageway for air from the nasal cavity. The **oropharynx** and **laryngopharynx** serve as common passageways for air, food, and drink. | |

## Concept

**Concept 22.2** Inhaled air travels in the lower respiratory system from the larynx to alveoli.

1. Structurally, the **lower respiratory system** consists of the larynx, trachea, bronchi, and lungs.
2. The **larynx** is a passageway connecting the laryngopharynx with the trachea. It includes the **thyroid cartilage**, the **cricoid cartilage** connecting the larynx and trachea, the **arytenoid cartilages**, and the epiglottis. The **epiglottis** moves to cover the larynx and airways during swallowing, which routes food and liquids into the esophagus. The **glottis** consists of the vocal folds and the space between them.
3. The **ventricular folds** are one of two pairs of mucous membrane folds of the larynx. When brought together, they hold the breath but do not produce sound. The other pair, the **vocal folds**, produces sound as they vibrate. Taut vocal folds produce high-pitched sounds, and relaxed ones produce low-pitched sounds.
4. The **trachea** is an air passageway extending from the larynx to the primary bronchi. Its walls contain C-shaped cartilage rings that give support to prevent the tracheal wall from collapsing.
5. The trachea divides into the **right primary bronchus** and the **left primary bronchus**, which enter the right and left lung, respectively. The bronchial tree consists of the **trachea**, **primary bronchi**, **secondary bronchi**, **tertiary bronchi**, **bronchioles**, and **terminal bronchioles**.
6. The **lungs** are paired organs separated from each other by the mediastinum. **Visceral pleura** cover each lung, and the **parietal pleura** line the thoracic cavity wall. Within the **pleural cavity** between the parietal and visceral pleura, pleural fluid reduces friction as the pleura move during breathing. Each lung is surrounded by a separate pleural cavity.
7. The right lung has three lobes separated by two fissures; the left lung has two lobes separated by one fissure and a depression, the **cardiac notch**. Secondary bronchi give rise to branches called tertiary bronchi, which supply segments of lung tissue called **bronchopulmonary segments**. Each bronchopulmonary segment consists of **lobules**, which contain lymphatic vessels, arterioles, venules, and terminal bronchioles, **respiratory bronchioles**, and **alveolar ducts**.
8. An **alveolar sac** consists of two or more cup-shaped alveoli sharing a common opening to an alveolar duct. Walls of alveoli consist of **type I alveolar cells** involved in gas exchange, simple squamous epithelial cells that form a continuous lining; **type II alveolar cells** that secrete alveolar fluid with **surfactant** to inhibit alveolar collapse; and **alveolar macrophages** to remove dust and other debris. Gas exchange occurs across the **respiratory membrane**.

## Resources

Anatomy Overview—Respiratory Organs
Anatomy Overview—Respiratory Tissues
Figure 22.3—Larynx
Figure 22.7—Relationship of the Pleural Membranes to the Lungs
Figure 22.10—Structural Components of an Alveolus
Clinical Connection—Laryngitis and Cancer of the Larynx
Clinical Connection—Coryza, Seasonal Influenza, and H1N1 Influenza
Clinical Connection—Pulmonary Edema
Clinical Connection—Asbestos-related Diseases
Clinical Connection—Asthma
Clinical Connection—Tuberculosis
Exercise—Paint the Lung
Exercise—Concentrate on the Respiratory Structures
Concepts and Connections—Functional Anatomy of the Respiratory System

---

**Concept 22.3** Inhalation and exhalation result from pressure changes caused by muscle contraction and relaxation.

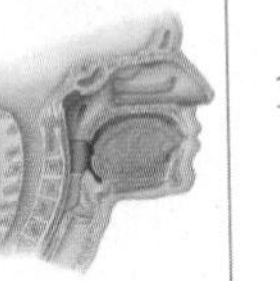

1. **Pulmonary ventilation**, or **breathing**, consists of inhalation and exhalation of air between the atmosphere and alveoli. **External respiration** is the exchange of gases across the respiratory membrane. **Internal respiration** is gas exchange between blood and tissue cells.
2. **Inhalation** occurs when **alveolar pressure** falls below atmospheric pressure. Contraction of the diaphragm and external intercostals increases the size of the lungs. Expansion of the lungs decreases alveolar pressure so that air moves down a pressure gradient from the atmosphere into the lungs.
3. During forceful inhalation, accessory muscles of inhalation (sternocleidomastoids, scalenes, and pectoralis minors) are used to allow greater airflow into the lungs than occurs during quiet inhalation.
4. **Exhalation** occurs when alveolar pressure is higher than atmospheric pressure. Relaxation of the diaphragm and external intercostals results in elastic recoil of the chest wall and lungs, which decreases the size of the lungs. As lung volume decreases, alveolar pressure increases, so that air moves down a pressure gradient from the lungs to the atmosphere.

Animation—Gas Exchange Introduction
Animation—Pulmonary Ventilation
Animation—Regulation of Ventilation: Basic Rhythms
Figure 22.13—Pressure Changes in Pulmonary Ventilation
Exercise—The Airway
Exercise—Build an Airway
Clinical Connection—Severe Acute Respiratory Syndrome

## Concept

5. Forceful exhalation involves contraction of the internal intercostals and abdominal muscles to force additional air out of the lungs.
6. The **surface tension** exerted by alveolar fluid produces an inward force on alveoli and accounts for much of the elastic recoil of exhalation. Surfactant in alveolar fluid reduces its surface tension.
7. **Compliance** is the amount of effort required to stretch the lungs and thoracic wall.
8. The walls of the airways offer some resistance to breathing. Resistance in air passageways decreases airflow.
9. Modified respiratory movements, such as coughing, sneezing, sighing, yawning, sobbing, crying, laughing, and hiccupping, are used to express emotions and to clear the airways.

### Concept 22.4 Lung volumes and capacities are measured to determine the respiratory status of an individual.

1. The volume of one breath is **tidal volume**. The total volume of air inhaled and exhaled each minute is the **minute ventilation**.
2. **Anatomic dead space** of the conducting airways contains air that does not undergo gas exchange.
3. Forceful inspiration can take in more air beyond the tidal volume; this additional air is the **inspiratory reserve volume**. The **expiratory reserve volume** is the amount of air beyond tidal volume that can be forced out of the lungs.
4. After the expiratory reserve volume is exhaled, air remaining in the lungs is the **residual volume**.
5. **Inspiratory capacity** is the tidal volume plus inspiratory reserve volume. **Functional residual capacity** is the residual volume plus expiratory reserve volume. **Vital capacity** is the sum of the inspiratory reserve volume, tidal volume, and expiratory reserve volume. **Total lung capacity** is the vital capacity plus residual volume.

### Concept 22.5 Oxygen and carbon dioxide diffusion is based on partial pressure gradients and solubility.

1. The **partial pressure** of a gas is the pressure exerted by that gas in a mixture of gases.
2. According to **Dalton's law**, each gas in a mixture of gases exerts its own pressure as if all of the other gases were not present. Each gas diffuses across a membrane by moving down its partial pressure gradient.
3. **Henry's law** states that the quantity of a gas that will dissolve in a liquid is proportional to the partial pressure of the gas and its solubility. In body fluids, a gas has a greater ability to stay in solution when it has a higher partial pressure and is more soluble in water.

### Concept 22.6 Respiration occurs between alveoli and pulmonary capillaries and between systemic capillaries and tissue cells.

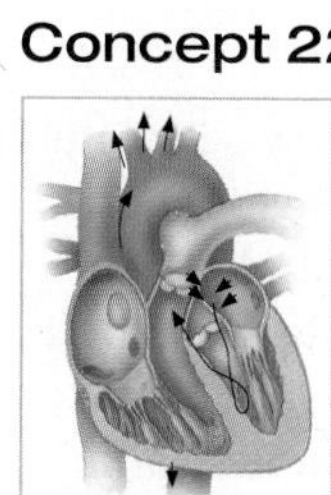

1. **External respiration** is the exchange of gases between alveoli and pulmonary blood capillaries. This exchange converts **deoxygenated blood** into **oxygenated blood**.
2. **Internal respiration** is the exchange of gases between systemic blood capillaries and tissue cells, converting oxygenated blood into deoxygenated blood.
3. External and internal respiration depend on partial pressure differences, the surface area for gas exchange, the diffusion distance across the membrane, and the molecular weight and solubility of the gases.

## Resources

Clinical Connection—Pneumothorax and Hemothorax

Clinical Connection—Cystic Fibrosis

Clinical Connection—Chronic Obstructive Pulmonary Disease

Animation—Regulation of Ventilation—Basic Rhythms

Animation—Gas Exchange

Exercise—Gas Exchange Match-up

Clinical Connection—Hyperbaric Oxygenation

Animation—Gas Exchange—Internal and External Respiration

Figure 22.16—Changes in Partial Pressures of Oxygen and Carbon Dioxide during External and Internal Respiration

Concepts and Connections—Gas Exchange

Clinical Connection—Pneumonia

## Concept

### Concept 22.7 Oxygen is primarily transported attached to hemoglobin, while carbon dioxide is transported in three different ways.

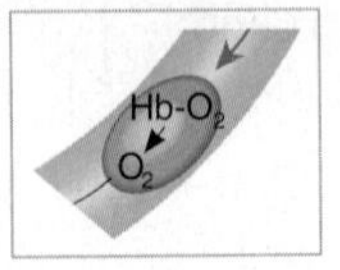


1. Most of the $O_2$ transported in the blood is bound to hemoglobin in the blood cells, and a small amount is dissolved in blood plasma. Hemoglobin has a heme portion that is capable of binding reversibly to a molecule of $O_2$, forming **oxyhemoglobin**. The amount of $O_2$ that binds to hemoglobin is primarily influenced by the partial pressure of oxygen; the higher the partial pressure of oxygen, the more $O_2$ combines with hemoglobin.
2. $O_2$ dissociates from hemoglobin more readily in an acidic environment, as the partial pressure of carbon dioxide rises in a tissue, and as temperature increases.
3. Deoxygenated blood returning to the lungs contains $CO_2$ dissolved in blood plasma, $CO_2$ combined with globin of hemoglobin as **carbaminohemoglobin**, and $CO_2$ incorporated into bicarbonate ions.
4. In the lungs, $CO_2$ dissociates from the globin portion of hemoglobin and diffuses into alveolar air. At the same time, $O_2$ diffuses from alveolar air and binds to the heme portion of hemoglobin, forming oxyhemoglobin.

### Concept 22.8 The basic rhythm of respiration is controlled by the respiratory center in the brain stem.

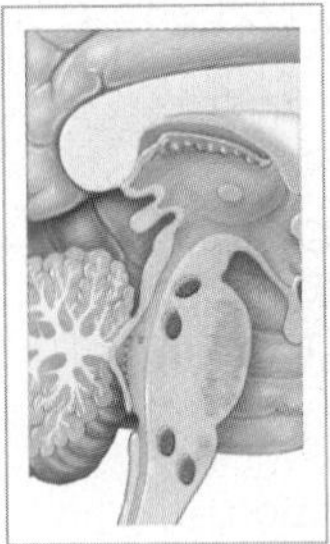

1. The **respiratory center** transmits impulses to the respiratory muscles that alter the size of the thorax.
2. The **medullary rhythmicity area** in the medulla oblongata has an **inspiratory area** that establishes the basic rhythm of breathing by stimulating contraction of the diaphragm and external intercostals. During forceful breathing, the **expiratory area** stimulates the internal intercostals and abdominal muscles to contract.
3. The **pneumotaxic area** in the pons inhibits the inspiratory area before the lungs become too full of air.
4. The **apneustic area** in the pons stimulates the inspiratory area to prolong inhalation.

### Concept 22.9 Respiration may be modified by cortical influences, chemical stimuli, proprioceptor input, and the inflation reflex.

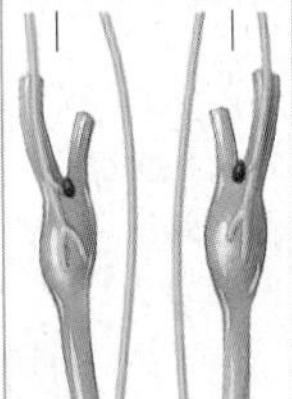

1. When metabolic demand for $O_2$ increases, several mechanisms help adjust the basic rhythm of breathing.
2. The cerebral cortex has limited influence on the respiratory center to allow voluntary alteration of the pattern of breathing.
3. **Chemoreceptors** respond to changes in $P_{O_2}$, $H^+$, and $P_{CO_2}$ in blood by stimulating the inspiratory area to alter breathing rate and depth.
4. During exercise, proprioceptors monitoring joint and muscle movement stimulate the inspiratory area to increase breathing rate and depth.
5. Stretch receptors detect stretch in the walls of bronchi and bronchioles. Overinflation of the lungs stimulates these receptors to inhibit the inspiratory and apneustic areas, resulting in exhalation.
6. Other factors affecting respiration include stimulation of the limbic system, body temperature, pain, stretching of the anal sphincter, irritation of airways, and changes in blood pressure.

## Resources

**Concept 22.7**

Animation—Gas Transport
Figure 22.21—Summary of Chemical Reactions That Occur During Gas Transport and Exchange
Exercise—Concentrate on Respiration
Exercise—Oxygen Transport Tryout
Exercise—Carbon Dioxide Transport Tryout
Concepts and Connections—Carbon Dioxide Transport
Concepts and Connections—Oxygen Transport
Clinical Connection—Hypoxia
Clinical Connection—Sudden Infant Death Syndrome
Clinical Connection—Lung Cancer

**Concept 22.8**

Anatomy Overview—Respiratory Control Center
Animation—Regulation of Ventilation

**Concept 22.9**

Anatomy Overview—Structures That Control Respiration
Animation—Control of Ventilation Rate and Blood Chemistry
Exercise—Respiration and pH Reflex
Concepts and Connections—Ventilation
Concepts and Connections—Respiratory Rate

Concept | Resources

## Concept 22.10 Acid–base balance is maintained by controlling the $H^+$ concentration of body fluids.

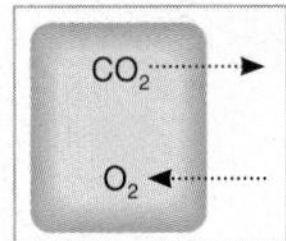


1. The acid–base balance of the body is maintained by controlling the $H^+$ concentration of body fluids. Removal of $H^+$ from the body depends on buffer systems, exhalation of carbon dioxide, and excretion of $H^+$ in the urine.
2. Buffers act to temporarily bind to, and remove, excess $H^+$ from solution to prevent rapid, drastic shifts in pH.
3. The **protein buffer system** utilizes blood proteins such as hemoglobin and albumin to buffer both acids and bases.
4. The **carbonic acid–bicarbonate buffer system** uses bicarbonate ions and carbonic acid to adjust blood pH.
5. The rate and depth of breathing contribute to the amount of $CO_2$ exhaled from the body and an alteration of the pH of body fluids. An increase in exhalation of carbon dioxide increases blood pH; a decrease in exhalation of $CO_2$ decreases blood pH.
6. Normal pH range of systemic arterial blood is 7.35–7.45. **Acidosis** is a systemic arterial blood pH below 7.35; its principal effect is depression of the central nervous system (CNS). **Alkalosis** is a systemic arterial blood pH above 7.45; its principal effect is overexcitability of the CNS.
7. **Respiratory acidosis** is abnormally high blood $P_{CO_2}$ due to inadequate exhalation of $CO_2$. **Respiratory alkalosis** is abnormally low blood $P_{CO_2}$ due to hyperventilation.

Animation—Acid-Base Imbalances
Animation—Role of the Respiratory System in pH Regulation

## Understanding the Concepts

1. What functions do the respiratory and cardiovascular systems share?
2. Compare the functions of the conducting portion and respiratory portion of the respiratory system.
3. How does the larynx function in respiration and voice production?
4. Why are the right and left lungs slightly different in size and shape?
5. Describe the path of $O_2$ from the air space within the alveolus to the capillary blood.
6. What are the basic differences among pulmonary ventilation, external respiration, and internal respiration?
7. Do the lungs expand or contract as the thoracic cavity expands? Explain your answer.
8. Compare what happens during quiet and forceful inhalation and exhalation.
9. Would increased surfactant assist inhalation or exhalation? Both? Neither? Explain your answer.
10. If you breathe in as deeply as possible and then exhale as much air as you can, which lung capacities have you demonstrated?
11. Why is the composition of inhaled air and alveolar air different?
12. What are the diffusion paths of oxygen and carbon dioxide during external and internal respiration?
13. Why can hemoglobin unload more oxygen as blood flows through capillaries of metabolically active tissues, such as skeletal muscle during exercise, than is unloaded at rest?
14. Which area of the respiratory center contains neurons that are active and then inactive in a repeating cycle?
15. Why is it impossible to voluntarily hold your breath indefinitely?
16. If you hold your breath for 30 seconds, what is likely to happen to your blood pH?

CHAPTER

# 23

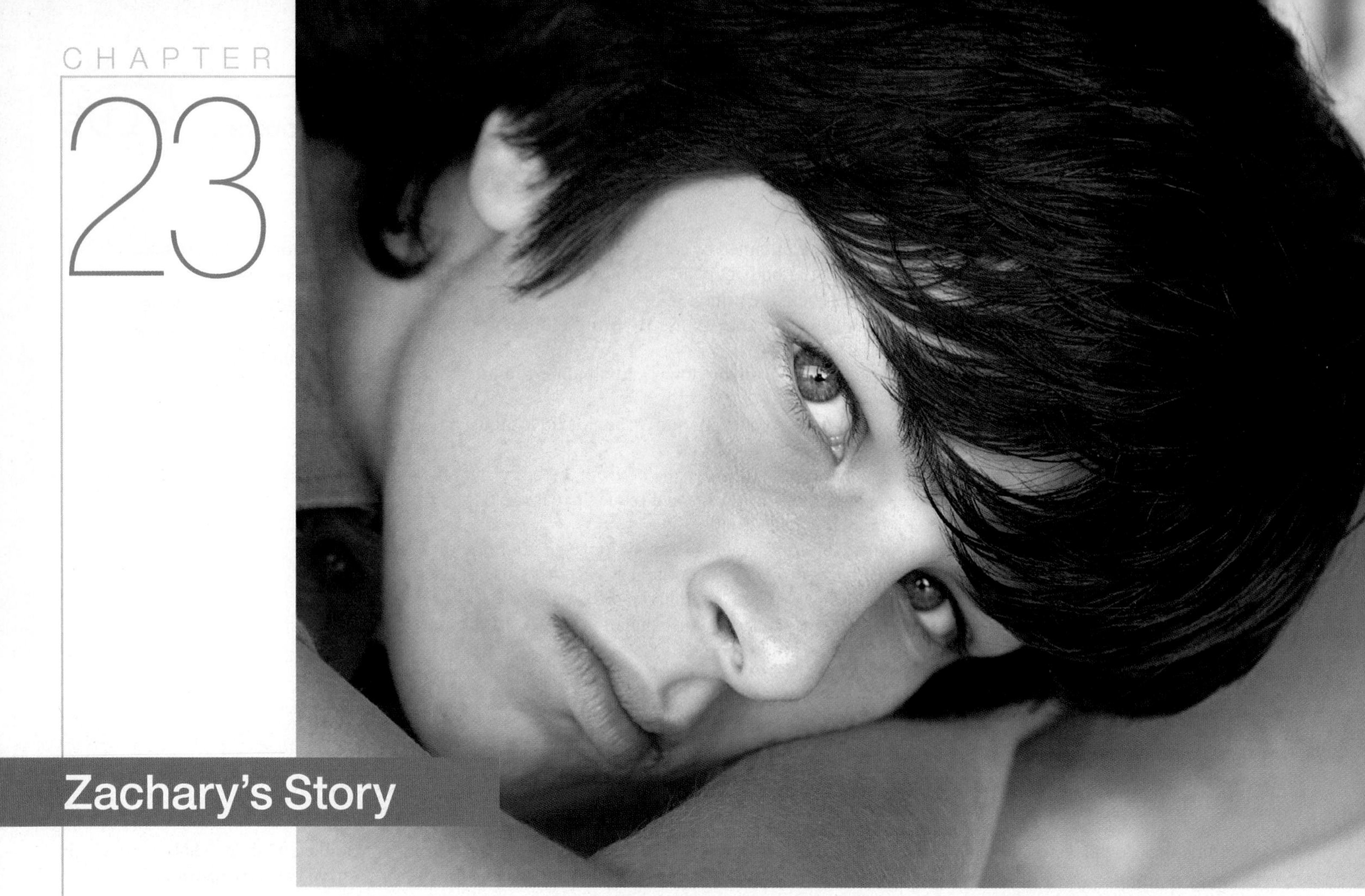

## Zachary's Story

"Mrs. Lewis, please calm down," Dr. Lee pleads.

"I will not calm down!" Sandra cries in frustration. "We've been in and out of this horrible hospital twelve times in the last two years. We've seen eight different doctors, and none of you knows what's wrong with my son!" If this was all happening to her, Sandra might have a little more patience left, but when it comes to her children, all bets are off. "Zachary's been in pain for two whole years! He's in pain right now and none of you are doing anything about it! It's not just ulcers, he doesn't have celiac disease, he's not lactose intolerant, and don't you dare try to tell me that it's growing pains again!" Somewhere in the back of her mind Sandra knew that screeching at Dr. Lee was not helping, but after watching her son slowly wasting away to nothing and vomiting blood, not for the first time, she had just about reached her limit.

"Mrs. Lewis, please sit down! I understand your frustrations, but we have some new scans back and I'd like to discuss them with you, if you would just calm down a bit." Dr. Lee squared his shoulders and looked her straight in the eye, hoping it would help calm Sandra enough to listen. Reluctantly she sat down, hoping that something concrete had finally been discovered. "Mrs. Lewis, Zachary has a very inflamed appendix that we need to take out immediately. Now wait, don't interrupt," Dr. Lee quickly interjects, as Sandra opens her mouth to respond. "I know that appendicitis does not explain all of the things that are happening to your son. I believe it is something else entirely, which we can confirm when we go in to take out the appendix."

Still skeptical, Sandra asks the question that frightens her the most. "What do you think he has?" Dr. Lee smiles gently and firmly states, "I believe that Zachary has Crohn's disease."

# The Digestive System

## INTRODUCTION

The food we eat contains a variety of nutrients—molecules needed for building new body tissues and repairing damaged tissues. Food is also vital to life because it is our only source of the energy that drives the chemical reactions occurring in every cell of our bodies. Most of the food we eat consists of molecules that are too large to be used by body cells. Therefore, foods must be broken down into molecules that are small enough to enter body cells, a process known as digestion. The passage of these smaller molecules through cells lining the stomach and intestines into the blood and lymph is termed absorption. The organs involved in the breakdown and absorption of food—collectively called the **digestive system**—are the focus of this chapter. The digestive system is a continuous tubular system that extends from the mouth to the anus. It provides extensive surface area in contact with the external environment, and its close association with the cardiovascular system is essential for processing the food that we eat.

## CONCEPTS

**23.1** The GI tract is a continuous multilayered tube extending from the mouth to the anus.

**23.2** The mouth lubricates and begins digestion of food, and maneuvers it to the pharynx for swallowing.

**23.3** Swallowing consists of voluntary oral, involuntary pharyngeal, and involuntary esophageal stages.

**23.4** The stomach mechanically breaks down the bolus and mixes it with gastric secretions.

**23.5** The pancreas secretes pancreatic juice, the liver secretes bile, and the gallbladder stores and concentrates bile.

**23.6** In the small intestine, chyme mixes with digestive juices from the small intestine, pancreas, and liver.

**23.7** In the large intestine, the final secretion and absorption of nutrients occur as chyme moves toward the rectum.

**23.8** Digestive activities occur in three overlapping phases: cephalic, gastric, and intestinal.

**23.9** Metabolism includes the catabolism and anabolism of molecules.

**23.10** Food molecules supply energy for life processes and serve as building blocks for complex molecules.

# 23.1 The GI tract is a continuous multilayered tube extending from the mouth to the anus.

## Overview of the Digestive System

Two groups of organs compose the digestive system (Figure 23.1): the gastrointestinal tract and the accessory digestive organs. The **gastrointestinal (GI) tract** is a continuous tube that extends from the mouth to the anus through the thoracic and abdominopelvic cavities. Organs of the GI tract include the mouth, most of the pharynx, esophagus, stomach, small intestine, and large intestine. The length of the adult GI tract is about 5–7 meters (16.5–23 ft). The **accessory digestive organs** include the teeth, tongue, salivary glands, liver, gallbladder, and pancreas. Teeth aid in the physical breakdown of food, and the tongue assists in chewing and swallowing. The other accessory digestive organs, however, never come into direct contact with food. They produce or store secretions that flow into the GI tract through ducts and aid in the chemical breakdown of food.

The GI tract contains food from the time it is eaten until it is digested and absorbed or eliminated. Muscular contractions in the wall of the GI tract physically break down food and propel it from the esophagus to the anus. The contractions also help dissolve food by mixing it with fluids secreted into the GI tract while enzymes break down the food chemically.

Overall, the digestive system performs six basic processes:

- ***Ingestion.*** This process involves taking foods and liquids into the mouth (eating).

FIGURE 23.1 **Organs of the digestive system.**

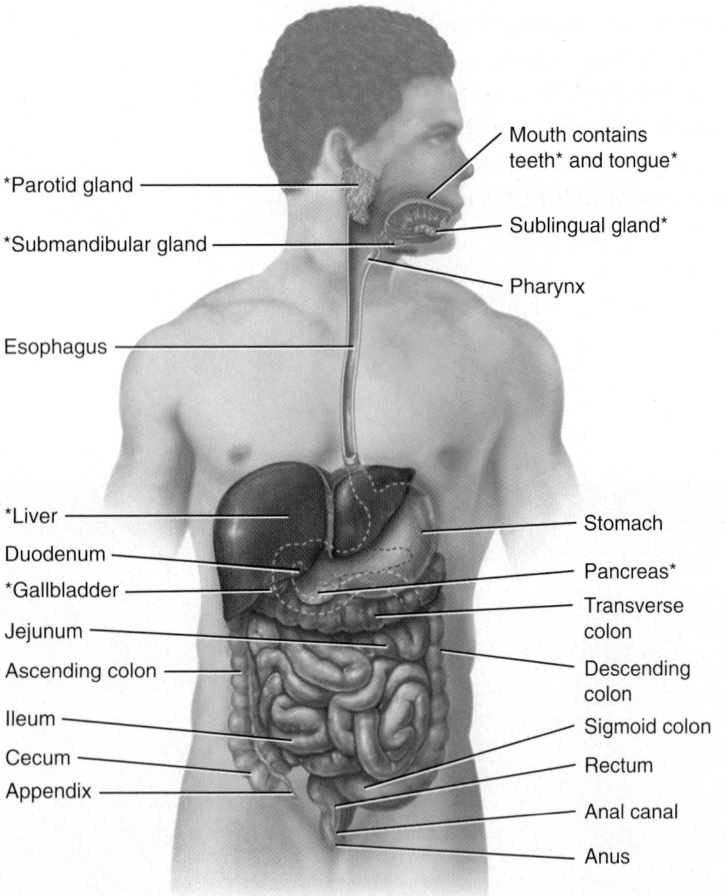


(a) Right lateral view of head and neck and anterior view of trunk

**FUNCTIONS OF THE DIGESTIVE SYSTEM**

1. Ingests food (takes food into the mouth).
2. Secretes water, acid, buffers, and enzymes into the lumen of the GI tract.
3. Mixes and propels food through the GI tract.
4. Digests food (breaks it down mechanically and chemically).
5. Absorbs digested products from the GI tract into the blood and lymph.
6. Eliminates feces from the GI tract.

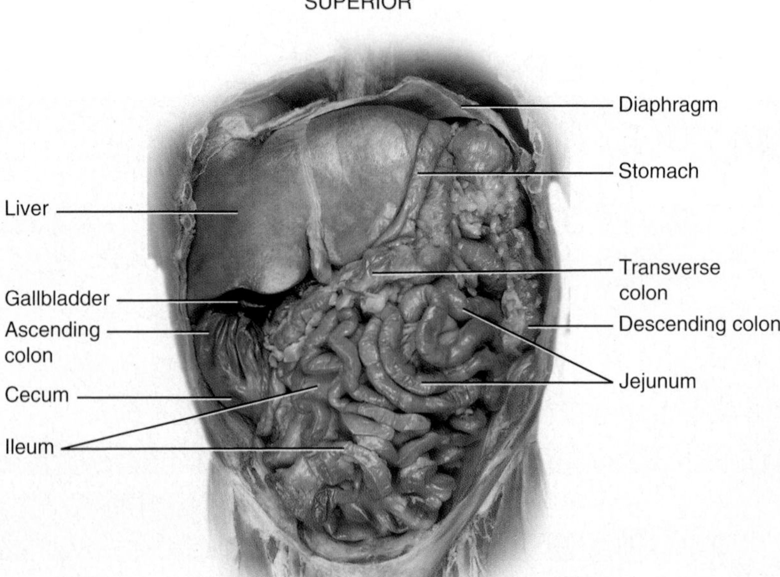


(b) Anterior view

**Organs of the gastrointestinal tract are the mouth, pharynx, esophagus, stomach, small intestine, and large intestine. Accessory digestive organs include the teeth, tongue, salivary glands, liver, gallbladder, and pancreas and are indicated with an asterisk (*).**

- ***Secretion.*** Each day, cells within the walls of the GI tract and accessory digestive organs secrete a total of about 7 liters of water, acid, buffers, and enzymes into the lumen (interior space) of the tract.
- ***Mixing and propulsion.*** Alternating contractions and relaxations of smooth muscle in the walls of the GI tract mix food and secretions and propel them toward the anus. This capability of the GI tract to mix and move material along its length is called **motility** (mō-TIL-i-tē).
- ***Digestion.*** Mechanical and chemical processes break down ingested food into small molecules. In **mechanical digestion**, the teeth cut and grind food before it is swallowed, and then smooth muscles of the stomach and small intestine churn the food to further assist the process. As a result, food molecules become dissolved and thoroughly mixed with digestive enzymes. In **chemical digestion** the large carbohydrate, lipid, protein, and nucleic acid molecules in food are split into smaller molecules. Digestive enzymes produced by the salivary glands, tongue, stomach, pancreas, and small intestine catalyze these catabolic reactions. A few substances in food can be absorbed without chemical digestion. These include vitamins, ions, cholesterol, and water.
- ***Absorption.*** The entrance of ingested and secreted fluids, ions, and the products of digestion into the epithelial cells lining the lumen of the GI tract is called **absorption** (ab-SŌRP-shun). The absorbed substances pass into blood or lymph and circulate to cells throughout the body.
- ***Defecation.*** Wastes, indigestible substances, bacteria, cells sloughed from the lining of the GI tract, and digested materials that were not absorbed in their journey through the digestive tract leave the body through the anus in a process called **defecation** (def′-e-KĀ-shun). The eliminated material is termed **feces** (FĒ-sēz).

## Layers of the GI Tract

The wall of the GI tract from the lower esophagus to the anal canal has the same basic, four-layered arrangement of tissues. The four layers of the tract, from deep to superficial, are the mucosa, submucosa, muscularis, and serosa (Figure 23.2).

FIGURE 23.2 **Layers of the gastrointestinal tract.** Variations of this basic plan may be seen in the esophagus (Figure 23.9), stomach (Figure 23.12), small intestine (Figure 23.19), and large intestine (Figure 23.24).

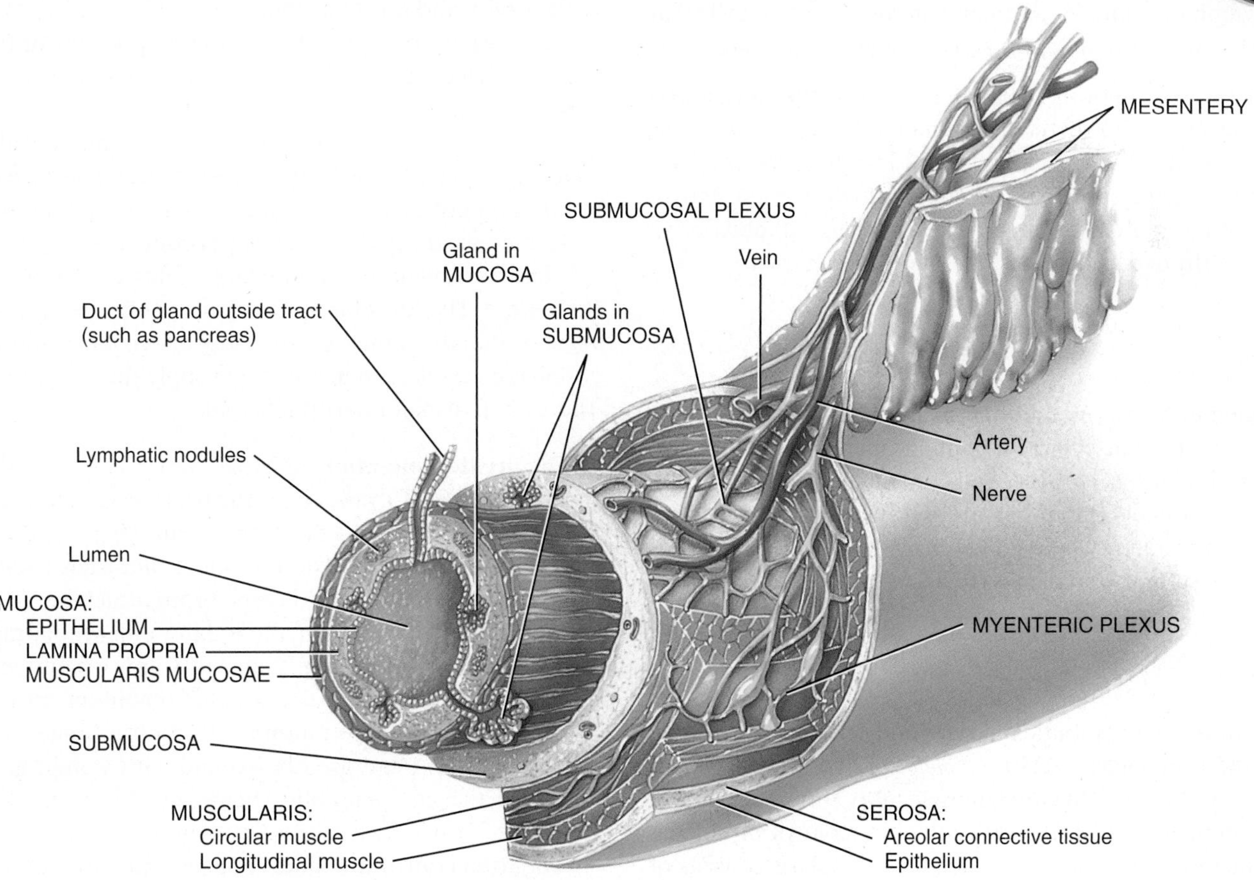


**The four layers of the GI tract, from deep to superficial, are the mucosa, submucosa, muscularis, and serosa.**

### Mucosa

The **mucosa**, or inner lining of the GI tract, is a mucous membrane. It is composed of a layer of epithelium in direct contact with the contents of the GI tract lumen that absorbs the digested components, a layer of areolar connective tissue called the lamina propria, and a thin layer of smooth muscle (muscularis mucosae).

- The **epithelium** in the mouth, pharynx, esophagus, and anal canal is mainly stratified squamous epithelium that serves a protective function. Simple columnar epithelium, which functions in secretion and absorption, lines the stomach and intestines. The rate of renewal of GI tract epithelial cells is rapid: Every 5 to 7 days they slough off and are replaced by new cells. Located among the epithelial cells are exocrine cells that secrete mucus and fluid into the lumen of the tract, and several types of endocrine cells, collectively called **enteroendocrine cells** (en′-ter-ō-EN-do-krin), that secrete hormones.
- The **lamina propria** (*lamina* = thin, flat plate; *propria* = one's own) is areolar connective tissue containing many blood and lymphatic vessels, which are the routes by which nutrients absorbed into the GI tract reach the other tissues of the body. This layer supports the epithelium and binds it to the muscularis mucosae (discussed next). The lamina propria also contains **mucosa-associated lymphatic tissue (MALT)**, prominent lymphatic nodules containing immune system cells that protect against entry of pathogens through the GI tract.
- A thin layer of smooth muscle fibers called the **muscularis mucosae** (myū-KŌ-sē) creates small folds in the mucous membrane of the stomach and small intestine; the folds increase the surface area for digestion and absorption. Movements of the muscularis mucosae ensure that all absorptive cells are fully exposed to the contents of the GI tract.

### Submucosa

The **submucosa** consists of areolar connective tissue that binds the mucosa to the muscularis. It contains many blood and lymphatic vessels that receive absorbed food molecules. In addition, the submucosa may contain glands and lymphatic tissue. Also located in the submucosa is an extensive network of neurons known as the submucosal plexus (to be described shortly).

### Muscularis

The **muscularis** of the mouth, pharynx, and superior and middle parts of the esophagus contains *skeletal muscle* that produces voluntary swallowing. Skeletal muscle also forms the external anal sphincter, which permits voluntary control of defecation. Throughout the rest of the tract, the muscularis consists of *smooth muscle* that is generally found in two sheets: an inner sheet of circular fibers and an outer sheet of longitudinal fibers. Involuntary contractions of the smooth muscle help break down food, mix it with digestive secretions, and propel it along the tract. Between the layers of the muscularis is a second plexus of neurons—the myenteric plexus (to be described shortly).

### Serosa

Those portions of the GI tract that are suspended in the abdominopelvic cavity have a superficial layer called the **serosa**. As its name implies, the serosa is a serous membrane composed of areolar connective tissue and simple squamous epithelium. The serosa secretes a slippery, watery fluid that allows the tract to glide easily against other organs. Inferior to the diaphragm, the serosa is also called the visceral peritoneum because it forms a portion of the peritoneum, which we examine next.

## Peritoneum

The **peritoneum** (per′-i-tō-NĒ-um; *peri-* = around) is the largest serous membrane of the body; it consists of a layer of simple squamous epithelium with an underlying supporting layer of connective tissue. The peritoneum is divided into the **parietal peritoneum**, which lines the wall of the abdominopelvic cavity, and the **visceral peritoneum**, which covers some of the organs in the cavity and serves as their serosa (Figure 23.3a). The slim space between the parietal and visceral portions of the peritoneum, which contains lubricating serous fluid, is called the **peritoneal cavity**.

Some organs lie on the posterior abdominal wall and are covered by peritoneum only on their anterior surfaces; they are not in the peritoneal cavity. Such organs, including the kidneys and pancreas, are said to be **retroperitoneal** (*retro-* = behind).

The peritoneum contains large folds that weave between the viscera. The folds bind the organs to one another and to the walls of the abdominal cavity. They also contain blood vessels, lymphatic vessels, and nerves that supply the abdominal organs. There are five major peritoneal folds:

- The **greater omentum** (ō-MEN-tum = fat skin), the largest peritoneal fold, drapes over the transverse colon and coils of the small intestine like a "fatty apron" (Figure 23.3a, d). The greater omentum is a double sheet that folds back on itself, giving it a total of four layers. From attachments along the stomach and duodenum, the greater omentum drapes down over the small intestine, then turns upward and attaches to the transverse colon. The greater omentum normally contains a considerable amount of adipose tissue. Its adipose tissue content can greatly expand with weight gain, giving rise to the characteristic "beer belly" seen in some overweight individuals. The many lymph nodes of the greater omentum contribute macrophages and antibody-producing plasma cells that help combat and contain infections of the GI tract.

- The **falciform ligament** (FAL-si-form; *falc-* = sickle-shaped) attaches the liver to the anterior abdominal wall and diaphragm (Figure 23.3b). The liver is the only digestive organ that is attached to the anterior abdominal wall.
- The **lesser omentum** arises as an anterior fold in the serosa of the stomach and duodenum, and it suspends the stomach and duodenum from the liver (Figure 23.3a, c).
- A fan-shaped fold of the peritoneum, called the **mesentery** (MEZ-en-ter′-ē; *mes-* = middle), binds the small intestine to the posterior abdominal wall (Figure 23.3a, d). It extends from the posterior abdominal wall to wrap around the small intestine and then returns to its origin, forming a double-layered structure.
- Two separate folds of peritoneum, the **mesocolon** (mez′-ō-KŌ-lon), bind the large intestine to the posterior abdominal wall (Figure 23.3a). Together, the mesentery and mesocolon hold the intestines loosely in place, allowing movement as muscular contractions mix and move the contents of the lumen along the GI tract.

## CLINICAL CONNECTION | *Peritonitis*

A common cause of **peritonitis** (per′-i-tō-NĪ-tis), an acute inflammation of the peritoneum, is contamination of the peritoneum by infectious microbes, which can result from accidental or surgical wounds in the abdominal wall, or from perforation or rupture of abdominal organs. If, for example, bacteria gain access to the peritoneal cavity through an intestinal perforation or rupture of the appendix, they can produce an acute, life-threatening form of peritonitis. A less serious (but still painful) form of peritonitis can result from the rubbing together of inflamed peritoneal surfaces. The increased risk of peritonitis is of particular concern to those who rely on peritoneal dialysis, a procedure in which the peritoneum is used to filter the blood when the kidneys do not function properly (see Clinical Connection entitled Dialysis in Concept 24.8).

FIGURE 23.3 **Relationship of the peritoneal folds to one another and to organs of the digestive system.** The size of the peritoneal cavity in (a) is exaggerated for emphasis.

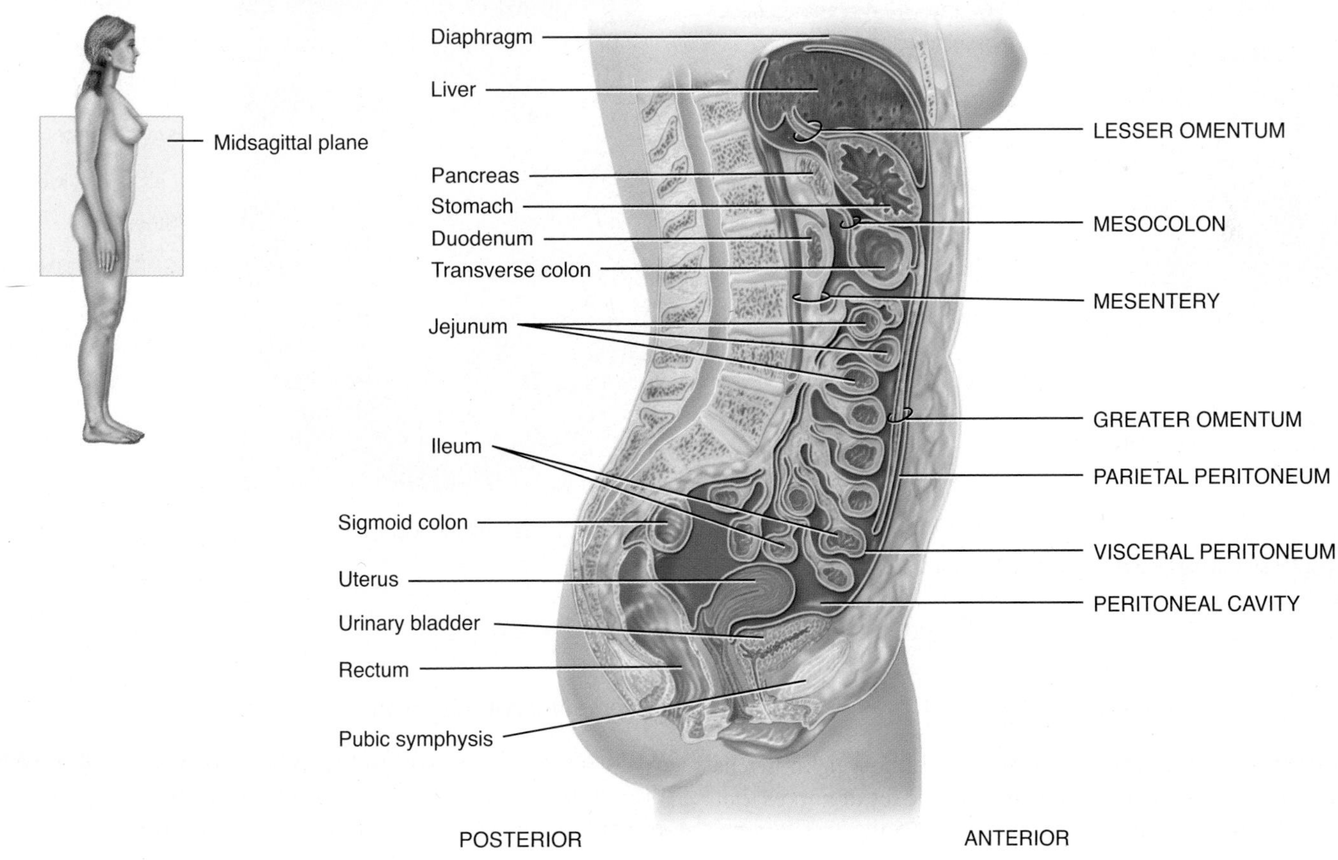


(a) Midsagittal section showing peritoneal folds

(continues)

FIGURE 23.3 (continued)

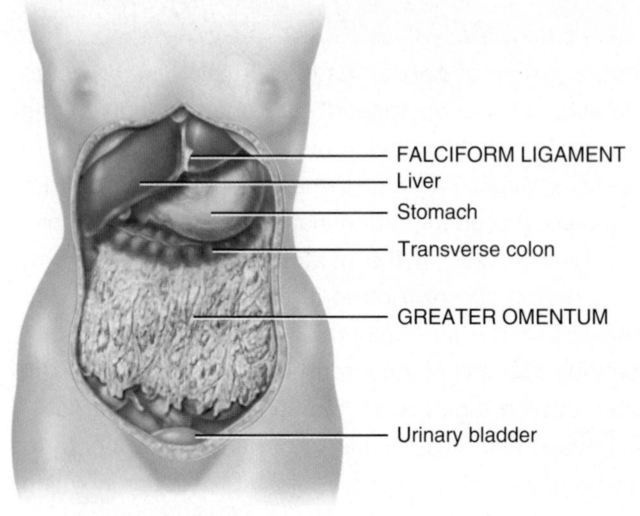


(b) Anterior view

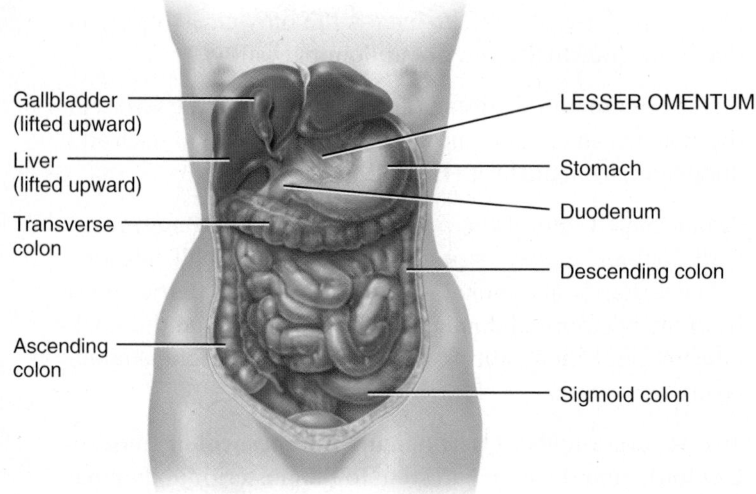


(c) Lesser omentum, anterior view (liver and gallbladder lifted)

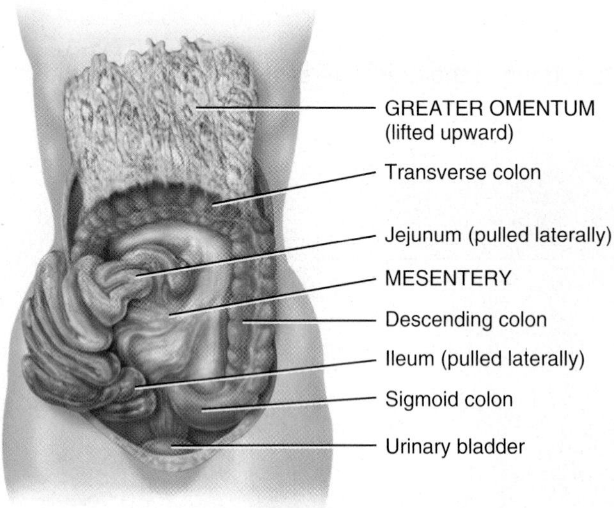


(d) Anterior view (greater omentum lifted and small intestine reflected to right side)

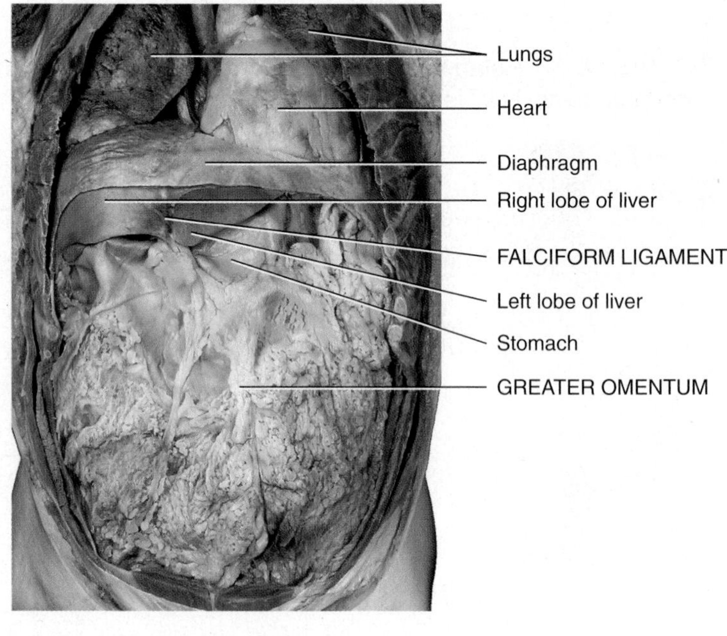


(e) Anterior view

**The peritoneum is the largest serous membrane in the body.**

# Neural Innervation of the GI Tract

The GI tract is regulated by an intrinsic set of nerves known as the enteric nervous system and by an extrinsic set of nerves that are part of the autonomic nervous system (ANS).

## Enteric Nervous System

We first introduced you to the **enteric nervous system (ENS)**, the "brain of the gut," in Concept 12.2. It consists of about 100 million neurons that extend from the esophagus to the anus.

Enteric neurons are arranged into two networks: the myenteric plexus and submucosal plexus (see Figure 23.2). The **myenteric plexus** (*myo-* = muscle) is located between the longitudinal and circular smooth muscle layers of the muscularis. The **submucosal plexus** is found within the submucosa. Enteric plexuses consist of sensory neurons, interneurons, and motor neurons (Figure 23.4). Because the motor neurons of the myenteric plexus supply the smooth muscle layers of the muscularis, this plexus mostly controls GI tract motility (movement), particularly the frequency and strength of contraction of the muscularis. The motor neurons of the submucosal plexus supply the secretory cells of the mucosal epithelium, controlling the secretions of the organs of the GI tract. The enteric interneurons interconnect the neurons of the myenteric and submucosal plexuses. Some enteric sensory neurons are *chemoreceptors*, receptors that are activated by the presence of certain chemicals in food located in the lumen of a GI tract organ. Other sensory neurons are *stretch receptors* that are activated when food distends (stretches) the wall of a GI tract organ.

FIGURE 23.4 **Organization of the enteric nervous system.**

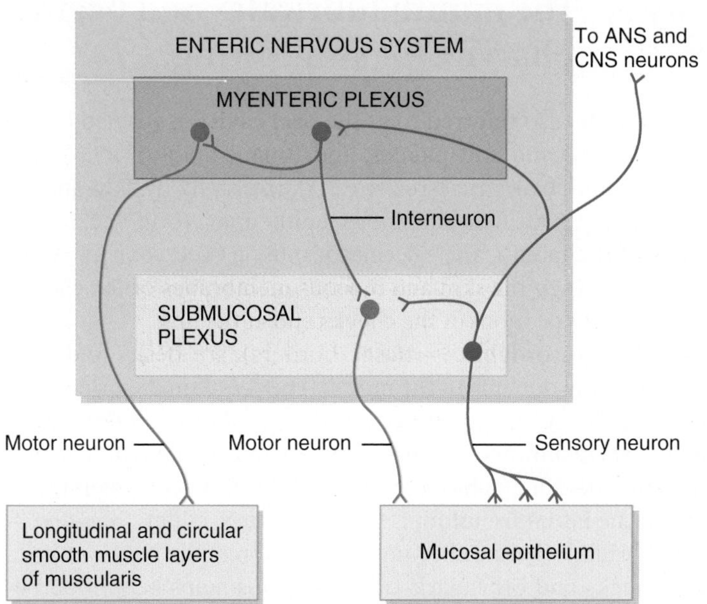


**The enteric nervous system consists of neurons arranged into the myenteric and submucosal plexuses.**

## Autonomic Nervous System

Although the enteric nervous system can function independently, it is subject to regulation by the autonomic nervous system. Parasympathetic nerves arise from the vagus (X) nerves and the sacral spinal cord; they either synapse with enteric neurons or directly innervate smooth muscle and glands within the wall of the GI tract (see Figure 14.24). Sympathetic nerves that supply the GI tract arise from the thoracic and lumbar regions of the spinal cord and synapse with enteric neurons of the myenteric plexus and the submucosal plexus. In general, parasympathetic stimulation increases GI secretion and motility by increasing the activity of enteric neurons, while sympathetic stimulation decreases GI secretion and motility by inhibiting the activity of enteric neurons. Emotions such as anger, fear, and anxiety may slow digestion as the central nervous system utilizes sympathetic nerves to inhibit GI tract glands and smooth muscle.

## Gastrointestinal Reflex Pathways

Many neurons of the ENS are components of *gastrointestinal reflex pathways* that regulate GI secretion and motility in response to stimuli in the lumen of the GI tract. Enteric sensory receptors (such as chemoreceptors and stretch receptors) transmit input to interneurons in the ENS, ANS, or central nervous system (CNS) about the nature of the contents and the degree of distension (stretching) of the GI tract. These interneurons process the incoming sensory information and subsequently activate or inhibit outgoing enteric motor neurons. The enteric motor neurons then activate or inhibit GI glands and smooth muscle, altering GI secretion and motility.

### ✓ CHECKPOINT

1. Give the name and function of each of the four layers of the GI tract.
2. Where along the GI tract is the muscularis composed of skeletal muscle? Is control of this skeletal muscle voluntary or involuntary?
3. What are the attachment sites of the mesentery, mesocolon, falciform ligament, lesser omentum, and greater omentum?
4. What are the functions of the myenteric and submucosal plexuses?

## 23.2 The mouth lubricates and begins digestion of food, and maneuvers it to the pharynx for swallowing.

The **mouth**, also referred to as the **oral cavity**, is formed by the cheeks, hard and soft palates, and tongue (Figure 23.5). The **cheeks**, muscular structures covered externally by skin and internally by stratified squamous epithelium, form the lateral walls of the mouth. The buccinator muscles and connective tissue lie between the skin and mucous membranes of the cheeks. The anterior portions of the cheeks end at the lips.

The **lips**, or **labia** (= fleshy borders), are fleshy folds surrounding the opening of the mouth. They contain the orbicularis oris muscle and are covered externally by skin and internally by a mucous membrane. The inner surface of each lip is attached to its corresponding gum by a midline fold of mucous membrane called the **labial frenulum** (LĀ-bē-al FREN-ū-lum; *frenulum* = small bridle). During chewing, contraction of the buccinators in the cheeks and orbicularis oris in the lips helps keep food between the upper and lower teeth. These muscles also assist in speech.

The **vestibule** (= entrance to a canal) of the mouth is a space bounded externally by the cheeks and lips and internally by the gums and teeth. The **oral cavity proper** is a space that extends from the gums and teeth to the **fauces** (FAW- sēs = passages), the opening between the mouth and the pharynx (throat).

The **palate** forms the roof of the mouth and separates the oral and nasal cavities. This important structure makes it possible to chew and breathe at the same time. The **hard palate**, the anterior, bony portion of the roof of the mouth, is formed by the maxillae and palatine bones. The **soft palate** is an arch-shaped muscular partition that forms the posterior portion of the roof of the mouth.

Hanging from the free border of the soft palate is a finger-like structure called the **uvula** (Ū-vū-la = little grape). During swallowing, the soft palate and uvula are drawn superiorly to close off the nasopharynx, which prevents swallowed foods and liquids from entering the nasal cavity. Lateral to the base of the uvula are two muscular folds that run down the lateral sides of the soft palate: Anteriorly, the **palatoglossal arch** (pal-a-tō-GLOS-al) extends to the base of the tongue; posteriorly, the **palatopharyngeal arch** (PAL-a-tō-fa-rin′-jē-al) extends to the

FIGURE 23.5 **Structures of the mouth (oral cavity).**

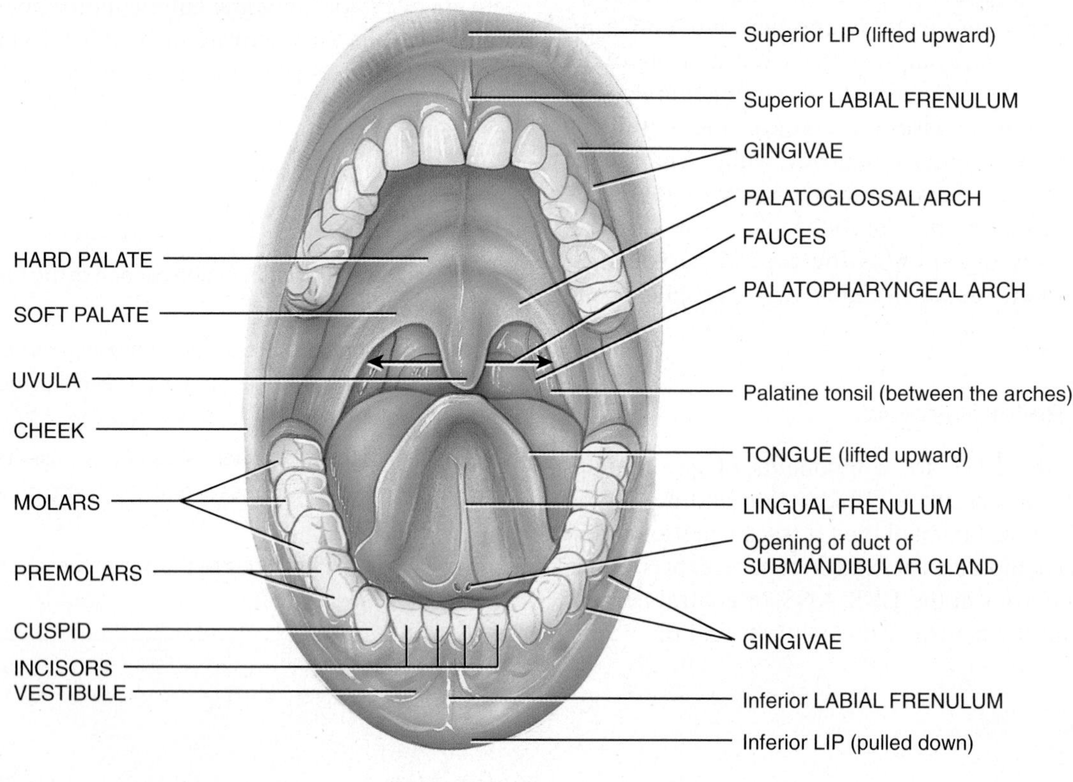


**The mouth is formed by the cheeks, hard and soft palates, and tongue.**

side of the pharynx. The palatine tonsils are situated between the arches, and the lingual tonsils are situated at the base of the tongue. As noted above, at the posterior border of the soft palate the mouth opens into the pharynx through the fauces (Figure 23.5).

## Tongue

The **tongue** is composed of skeletal muscle covered with mucous membrane. Together with its associated muscles, it forms the floor of the mouth (Figure 23.5). The tongue is divided into symmetrical lateral halves by a median septum that extends its entire length, and it is attached inferiorly to the hyoid bone, temporal bone, and mandible. Each half of the tongue consists of an identical complement of extrinsic and intrinsic muscles.

The **extrinsic muscles** of the tongue originate outside the tongue (attach to bones in the area) and insert into connective tissues in the tongue (see Figure 11.6). The extrinsic muscles move the tongue from side to side and in and out to maneuver food for chewing, shape the food into a rounded mass, and force the food to the back of the mouth for swallowing. They also form the floor of the mouth and hold the tongue in position. The **intrinsic muscles** originate inside and insert into connective tissues within the tongue. They alter the shape and size of the tongue for speech and swallowing. The **lingual frenulum** (*lingua* = the tongue), a fold of mucous membrane in the midline of the undersurface of the tongue, is attached to the floor of the mouth and limits the movement of the tongue posteriorly (Figures 23.5 and 23.6).

The dorsum (upper surface) and lateral surfaces of the tongue are covered with **papillae** (pa-PIL-ē = nipple-shaped projections), projections of the lamina propria covered with stratified squamous epithelium (see Figure 16.4). Many papillae contain taste buds, the receptors for gustation (taste). Some papillae lack taste buds, but they contain receptors for touch and increase friction between the tongue and food, making it easier for the tongue to move food in the mouth. Papillae and taste buds are described in detail in Concept 16.2.

FIGURE 23.6 **The three major salivary glands—parotid, sublingual, and submandibular.**

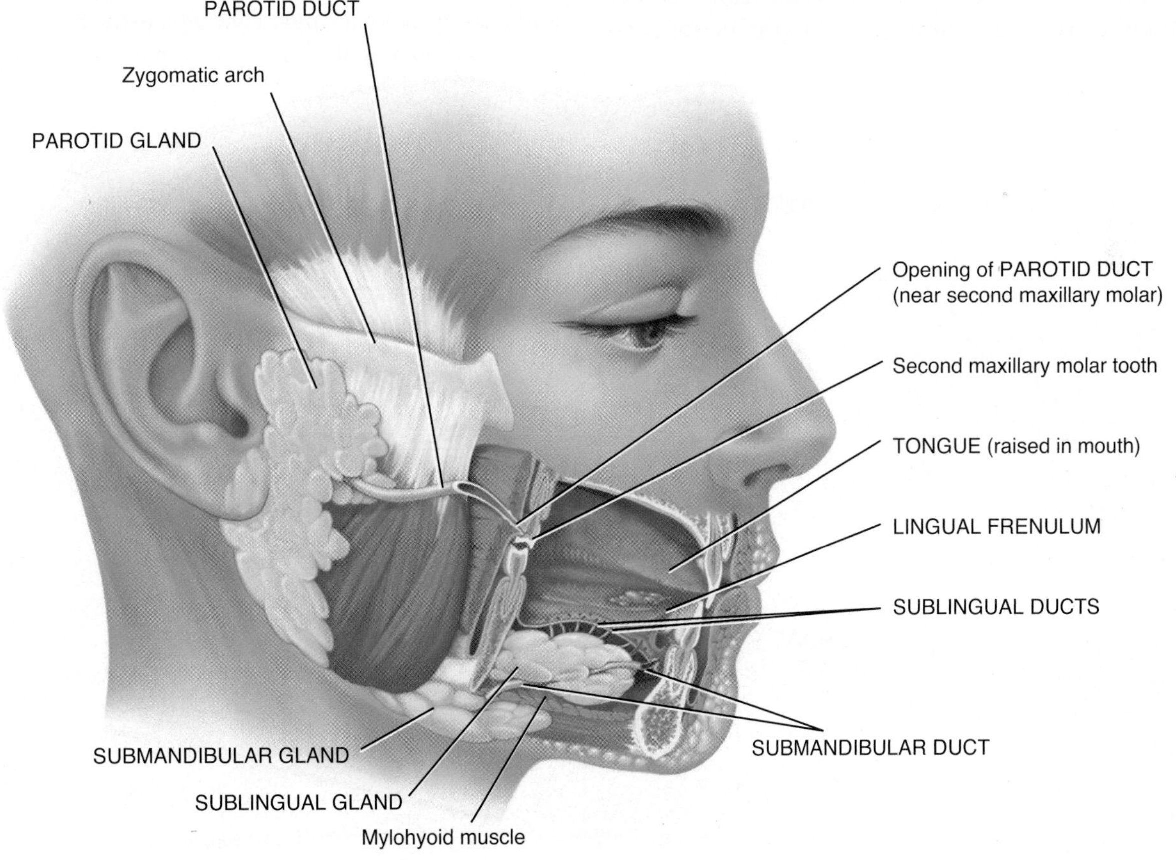


Location of salivary glands

**Saliva lubricates and dissolves foods and begins the chemical breakdown of carbohydrates and lipids.**

**Lingual glands** in the lamina propria of the tongue secrete both mucus and a watery serous fluid that contains the enzyme **lingual lipase** (LĪ-pās), which begins the digestion of triglycerides.

## Teeth

**Teeth** are accessory digestive organs located in sockets of the alveolar processes of the mandible and maxillae (Figure 23.7). The alveolar processes are covered by the **gingivae** (JIN-ji-vē), or gums, which extend slightly into each socket. The sockets are lined by the **periodontal ligament** (per′- ē-ō-DON-tal; *odont-* = tooth), which consists of dense fibrous connective tissue that anchors the teeth to the socket walls and acts as a shock absorber during chewing.

A typical **tooth** has three major external regions: the crown, root, and neck. The **crown** is the visible portion above the level of the gingiva. Embedded in the socket are one to three **roots**. The **neck** is the constricted junction of the crown and root near the gum line.

Internally, **dentin** forms the majority of the tooth. Dentin consists of a calcified connective tissue that gives the tooth its basic shape and rigidity. It is harder than osseous tissue in bone because of its higher content of calcium salts (70 percent of dry weight). The dentin of the root is covered by **cementum**, another bonelike substance, which attaches the root to the periodontal ligament.

The dentin of the crown is covered by **enamel**, which consists primarily of calcium phosphate and calcium carbonate. Enamel is also harder than osseous tissue because it has an even higher content of calcium salts (about 95 percent of dry weight) than dentin. In fact, enamel is the hardest substance in the body. It protects the tooth from the wear and tear of chewing and from acids that can easily dissolve dentin.

The dentin of a tooth encloses a space. The enlarged part of the space, the **pulp cavity**, lies within the crown and is filled with **pulp**, a connective tissue containing blood vessels, nerves, and lymphatic vessels. Narrow extensions of the pulp cavity, called **root canals**, run through the root of the tooth. Each root canal has an opening at its base, the **apical foramen**, through which blood vessels bring nourishment, lymphatic vessels offer protection, and nerves provide sensation.

Humans have two **dentitions**, or sets of teeth: deciduous and permanent. The first of these—the **deciduous (primary) teeth** (*deciduous* = falling out)—begin to erupt at about 6 months of age. Approximately two teeth appear each month thereafter, until all 20 are present (Figure 23.8a). The incisors, which are closest to the midline, are chisel-shaped and adapted for cutting into food. They are referred to as either **central** or **lateral incisors** based on their position. Next to the incisors, moving posteriorly, are the **cuspids (canines)**, which have a pointed surface, called a *cusp*, to tear and shred food. Incisors and cuspids have a single root. Posterior to the cuspids lie the **first** and **second molars**, which have four cusps. Maxillary

FIGURE 23.7 **A typical tooth and surrounding structures.**

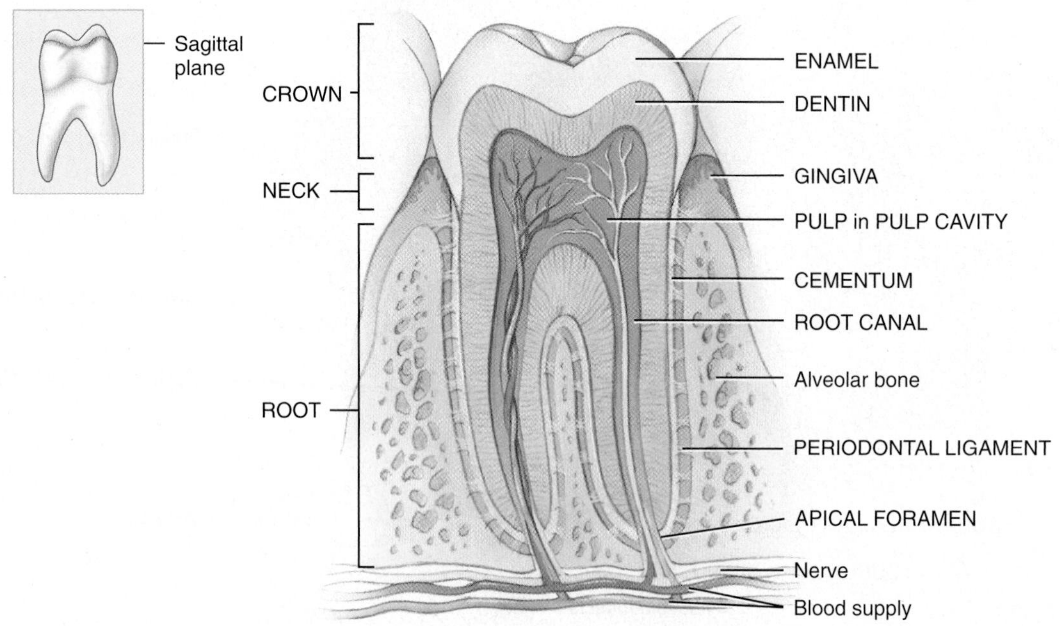


Sagittal section of a mandibular (lower) molar

**Teeth are anchored in sockets of the alveolar processes of the mandible and maxillae.**

FIGURE 23.8 **Dentitions and times of eruptions.** A designated letter (deciduous teeth) or number (permanent teeth) uniquely identifies each tooth. Times of eruptions are indicated in parentheses.

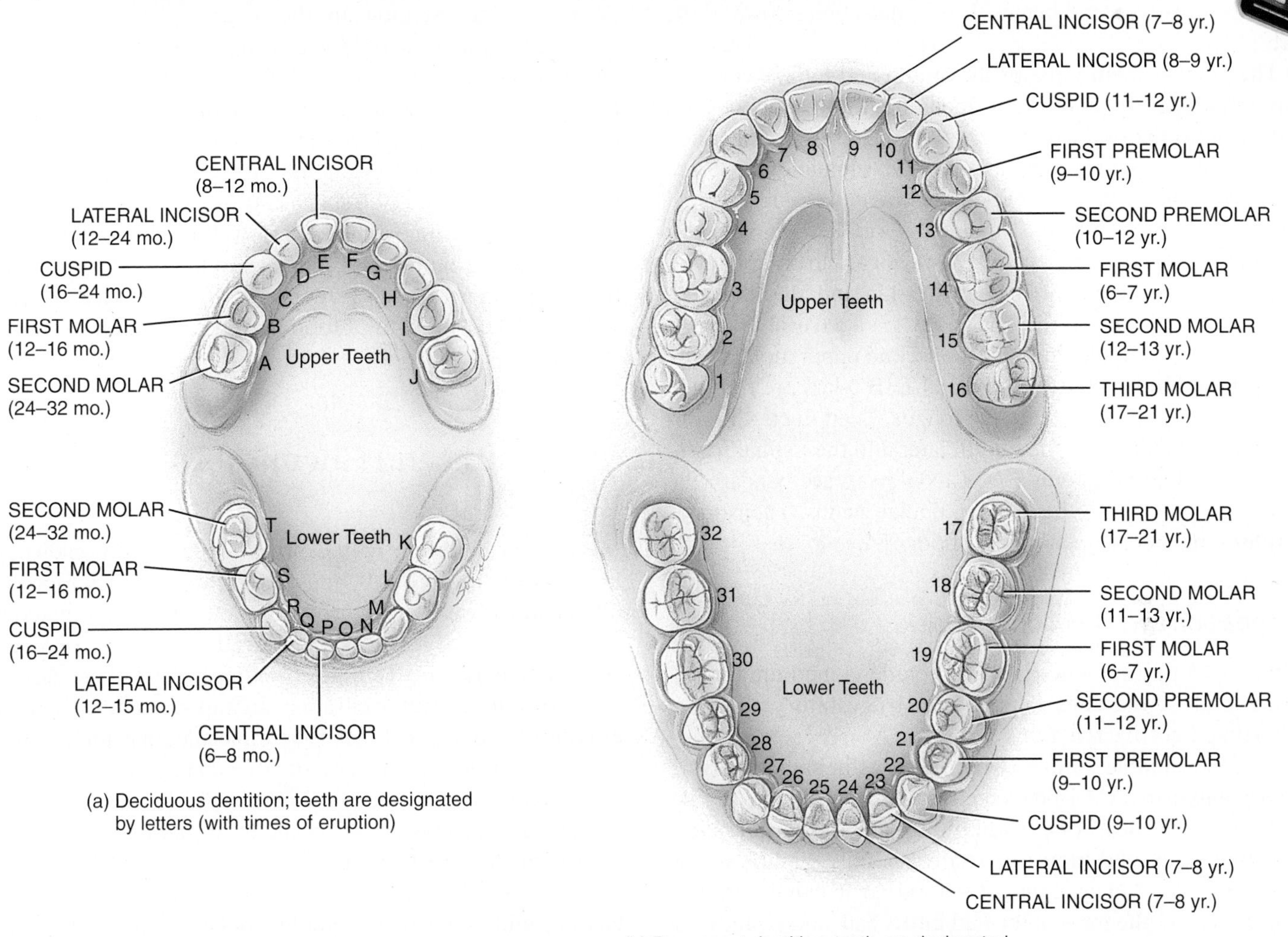


(a) Deciduous dentition; teeth are designated by letters (with times of eruption)

(b) Permanent dentition; teeth are designated by numbers (with times of eruption)

**There are 20 teeth in a complete deciduous set and 32 teeth in a complete permanent set.**

(upper) molars have three roots; mandibular (lower) molars have two roots. The molars crush and grind food to prepare it for swallowing.

All of the deciduous teeth are lost—generally between 6 and 12 years of age—and are replaced by the **permanent (secondary) teeth** (Figure 23.8b). The permanent dentition contains 32 teeth that erupt between age 6 and adulthood. The pattern resembles the deciduous dentition, with the following exceptions. The deciduous molars are replaced by the **first** and **second premolars (bicuspids)**, which have two cusps and one root (upper first premolars have two roots) and are used for crushing and grinding. The permanent molars, which erupt into the mouth posterior to the premolars, do not replace any deciduous teeth and erupt as the jaw grows to accommodate them—the **first molars** at age 6 (six-year molars), the **second molars** at age 12 (twelve-year molars), and the **third molars (wisdom teeth)** after age 17 or not at all.

Often the human jaw does not have enough room posterior to the second molars to accommodate the eruption of the third molars. In this case, the third molars remain embedded in the alveolar bone and are said to be "impacted." They often cause pressure and pain and must be removed surgically. In some people, third molars may be dwarfed in size or may not develop at all.

## Salivary Glands

A **salivary gland** (SAL-i-var-ē) is a gland that releases a secretion called saliva into the oral cavity. Ordinarily, just enough saliva is secreted to keep the mucous membranes of the mouth and pharynx moist and to cleanse the mouth and teeth. When food enters the mouth, however, secretion of saliva increases, and it lubricates, dissolves, and begins the chemical breakdown of the food.

The mucous membrane of the mouth and tongue contains many small salivary glands that open directly, or indirectly via short ducts, to the oral cavity. However, most saliva is secreted by three pairs of salivary glands—parotid, submandibular, and sublingual glands—that lie beyond the oral mucosa and secrete saliva through ducts into the mouth (Figure 23.6). The **parotid glands** (pa-ROT-id; *par-* = near; *to-* = ear) are located inferior and anterior to the ears, between the skin and the masseter. Each secretes saliva into the oral cavity via a **parotid duct** that opens into the vestibule opposite the second upper molar tooth. The **submandibular glands** (sub′-man-DIB-ū-lar) are found in the floor of the mouth beneath the tongue. Their ducts, the **submandibular ducts**, enter the mouth lateral to the lingual frenulum. The **sublingual glands** (sub-LING-gwal) are beneath the tongue and superior to the submandibular glands. Their ducts, the **sublingual ducts**, open into the floor of the mouth.

### Composition and Functions of Saliva

**Saliva** is 99.5 percent water, which provides a medium for dissolving foods so that they can be tasted by gustatory receptors and so that digestive reactions can begin. Also present in saliva are several valuable solutes. Chloride ions in saliva activate **salivary amylase** (AM-i-lās), an enzyme that starts the breakdown of starch. Bicarbonate and phosphate ions buffer acidic foods that enter the mouth, so saliva is only slightly acidic (pH 6.35–6.85). Mucus lubricates food so it can be moved around easily in the mouth, formed into a ball, and swallowed. Immunoglobulin A (IgA) antibodies (see Table 21.2) prevent attachment of microbes so they cannot penetrate the epithelium, and the enzyme lysozyme kills bacteria; however, these substances are not present in large enough quantities to eliminate all oral bacteria.

### Salivation

The secretion of saliva, or **salivation** (sal-i-VĀ-shun), is controlled by the autonomic nervous system. Normally, parasympathetic stimulation promotes continuous secretion of a moderate amount of saliva, which keeps the mucous membranes moist and lubricates the movements of the tongue and lips during speech. The saliva is then swallowed and helps moisten the esophagus. Eventually, most components of saliva are reabsorbed, which prevents fluid loss. Sympathetic stimulation dominates during stress, resulting in dryness of the mouth. If the body becomes dehydrated, the salivary glands stop secreting saliva to conserve water; the resulting dryness in the mouth contributes to the sensation of thirst. Drinking not only restores the homeostasis of body water but also moistens the mouth.

The feel and taste of food also are potent stimulators of salivary gland secretions. Chemicals in the food stimulate receptors in taste buds on the tongue, and impulses are conveyed from the taste buds to the brain stem. Returning parasympathetic impulses in the facial (VII) and glossopharyngeal (IX) nerves stimulate the secretion of saliva. Saliva continues to be secreted heavily for some time after food is swallowed; this flow of saliva washes out the mouth and dilutes and buffers the remnants of irritating chemicals such as that tasty (but hot!) salsa. The smell, sight, sound, or thought of food may also stimulate secretion of saliva.

## Mechanical and Chemical Digestion in the Mouth

Mechanical digestion in the mouth results from chewing, or **mastication** (mas′-ti-KĀ-shun = to chew), in which food is manipulated by the tongue, ground by the teeth, and mixed with saliva. As a result, the food is reduced to a soft, flexible, easily swallowed mass called a **bolus** (= lump). Food molecules begin to dissolve in the water in saliva, an important activity because enzymes can only react with food molecules in a liquid medium. Two such enzymes, salivary amylase and lingual lipase, contribute to chemical digestion in the mouth.

Dietary carbohydrates are monosaccharides (simple sugars such as glucose, fructose, or galactose), disaccharides (molecules composed of two simple sugars such as sucrose, table sugar, or lactose, milk sugar), or complex polysaccharides (molecules composed of many simple sugars such as starches). Most of the carbohydrates we eat are starches, but only monosaccharides can be absorbed into the bloodstream. Thus, ingested disaccharides and starches must be broken down into monosaccharides. The enzyme salivary amylase, which is secreted by the salivary glands, begins the breakdown of starches into smaller molecules, such as the disaccharide maltose. Even though food is usually swallowed too quickly for all starches to be broken down in the mouth, salivary amylase in the swallowed food continues to act on the starches for about another hour, at which time stomach acids inactivate it.

Saliva also contains lingual lipase, which is secreted by lingual glands in the tongue. This enzyme becomes activated in the acidic environment of the stomach and thus starts to work after food is swallowed. It breaks down dietary triglycerides (fats and oils) into fatty acids and diglycerides. A diglyceride consists of a glycerol molecule that is attached to two fatty acids.

Table 23.1 summarizes the digestive activities in the mouth.

**TABLE 23.1**

Digestive Activities in the Mouth

| STRUCTURE | ACTIVITY | RESULT |
|---|---|---|
| **Cheeks and lips** | Keep food between teeth | Foods uniformly chewed during mastication |
| **Salivary glands** | Secrete saliva | Lining of mouth and pharynx moistened and lubricated; saliva softens, moistens, and dissolves food and cleanses mouth and teeth; salivary amylase splits starch into smaller fragments (maltose, maltotriose, and α-dextrins) |
| **Tongue** | | |
| **Extrinsic tongue muscles** | Move tongue from side to side and in and out | Food maneuvered for mastication, shaped into bolus, and maneuvered for swallowing |
| **Intrinsic tongue muscles** | Alter shape of tongue | Swallowing and speech |
| **Taste buds** | Serve as receptors for gustation (taste) and presence of food in mouth | Secretion of saliva stimulated by nerve impulses from taste buds to brain stem to salivary glands |
| **Lingual glands** | Secrete lingual lipase | Triglycerides broken down into fatty acids and diglycerides |
| **Teeth** | Cut, tear, and pulverize food | Solid foods reduced to smaller particles for swallowing |

✓ **CHECKPOINT**

5. Which structures form the mouth?
6. What is the name of the cone-shaped process that hangs down from the roof of your mouth? What is its function?
7. How are the major salivary glands distinguished on the basis of location?
8. How is the secretion of saliva regulated?
9. Which functions do incisors, cuspids, premolars, and molars perform?
10. Define mastication, using the term *bolus.*

## 23.3 Swallowing consists of voluntary oral, involuntary pharyngeal, and involuntary esophageal stages.

### Pharynx

When food is first swallowed, it passes from the mouth into the **pharynx** (FAIR-inks), or throat, a funnel-shaped tube that extends from the internal nares to the esophagus posteriorly and to the larynx anteriorly (see Figure 22.2b). The pharynx is composed of skeletal muscle and lined by mucous membrane, and is divided into three parts: the nasopharynx, oropharynx, and laryngopharynx. The nasopharynx functions only in respiration, but both the oropharynx and laryngopharynx have digestive as well as respiratory functions. Swallowed food passes from the mouth into the oropharynx and laryngopharynx; muscular contractions of these areas help propel food into the esophagus and then into the stomach.

### Esophagus

The **esophagus** (e-SOF-a-gus = eating gullet) is a collapsible muscular tube, about 25 cm (10 in.) long, that lies posterior to the trachea. The esophagus begins at the inferior end of the laryngopharynx, descends through the mediastinum, extends through an opening in the diaphragm called the **esophageal hiatus** (e-sof-a-JĒ-al HĪ-ā-tus), and ends in the superior portion of the stomach (see Figure 23.1). Sometimes, part of the stomach protrudes above the diaphragm through the esophageal hiatus. This condition is termed a **hiatal hernia** (HER-nē-a).

### Histology of the Esophagus

The mucosa and submucosa of the esophagus consists of the same tissues as the rest of the GI tract (Figure 23.9). The stratified squamous epithelium of the mucosa in the lips, mouth, tongue, oropharynx, laryngopharynx, and esophagus affords considerable protection against abrasion and wear and tear from food particles that are chewed, mixed with secretions, and swallowed. The muscularis transitions from skeletal muscle in the superior portion of the esophagus to smooth muscle in the inferior portion. At each end of the esophagus, the muscularis becomes slightly more prominent and forms two sphincters—the **upper esophageal sphincter** (e-sof′-a-JĒ-al), which consists of skeletal muscle, and the **lower esophageal (cardiac) sphincter**, which consists of smooth muscle and is near the heart. The upper esophageal sphincter regulates the movement of food from the pharynx into the esophagus; the lower esophageal sphincter regulates the movement of food from the esophagus into the stomach. The superficial layer of the esophagus is known as the **adventitia** (ad-ven-TISH-a) rather than the serosa, because the connective tissue attaches to the surrounding structures of the mediastinum through which the esophagus passes.

FIGURE 23.9 **Histology of the esophagus.**

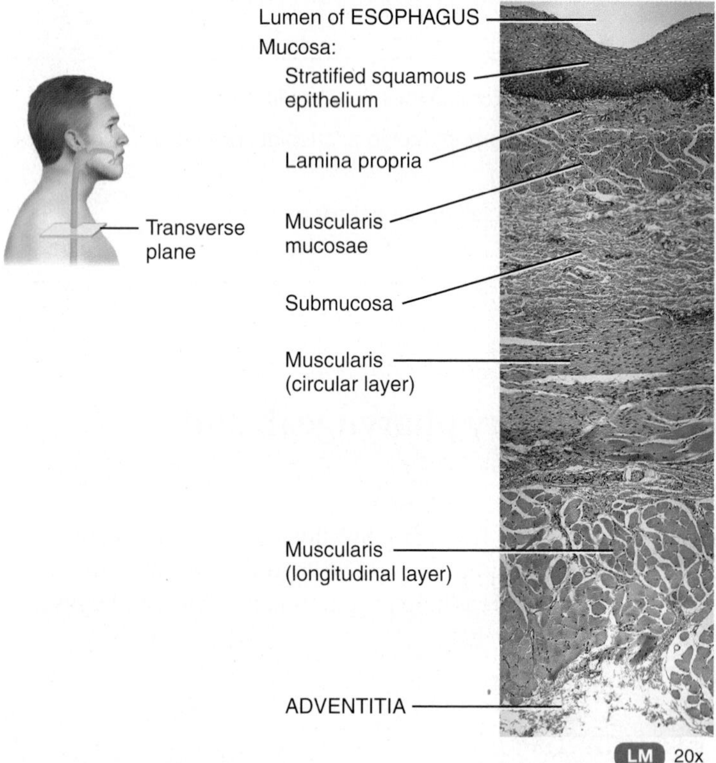


The esophagus secretes mucus and transports food to the stomach.

### Physiology of the Esophagus

The esophagus secretes mucus and transports food into the stomach. It does not produce digestive enzymes, and it does not carry out absorption.

## Deglutition

The movement of food from the mouth into the stomach is achieved by the act of swallowing, or **deglutition** (dē-gloo-TISH-un) (Figure 23.10). Deglutition is facilitated by the secretion of saliva and mucus and involves the mouth, pharynx, and esophagus. Swallowing occurs in three stages:

1. The **voluntary stage** of swallowing starts when the bolus is forced to the back of the mouth and into the oropharynx by the movement of the tongue upward and backward against the palate (Figure 23.10a).
2. The involuntary **pharyngeal stage** of swallowing passes the bolus through the oropharynx and then through the laryngopharynx into the esophagus (Figure 23.10b). The bolus stimulates receptors in the oropharynx, which send sensory impulses to the **deglutition center** in the brain stem (medulla oblongata and pons). The returning motor impulses cause the soft palate and uvula to move upward to close off the nasopharynx, which prevents swallowed foods and liquids from entering the nasal cavity. In addition, the epiglottis closes off the opening to the larynx, which prevents the bolus from entering the rest of the respiratory tract. Once the upper esophageal sphincter relaxes, the bolus moves into the esophagus.
3. The involuntary **esophageal stage** of swallowing passes the bolus through the esophagus into the stomach (Figure 23.10c). During this phase, **peristalsis** (per′-i-STAL-sis; *-stalsis* = constriction), a progression of coordinated contractions and relaxations of the circular and longitudinal

FIGURE 23.10 **Deglutition (swallowing).** During the pharyngeal stage of deglutition (b), the tongue rises against the palate, the nasopharynx is closed off, the larynx rises, the epiglottis seals off the larynx, and the bolus is passed into the esophagus. During the esophageal stage of deglutition (c), food moves through the esophagus into the stomach via peristalsis.

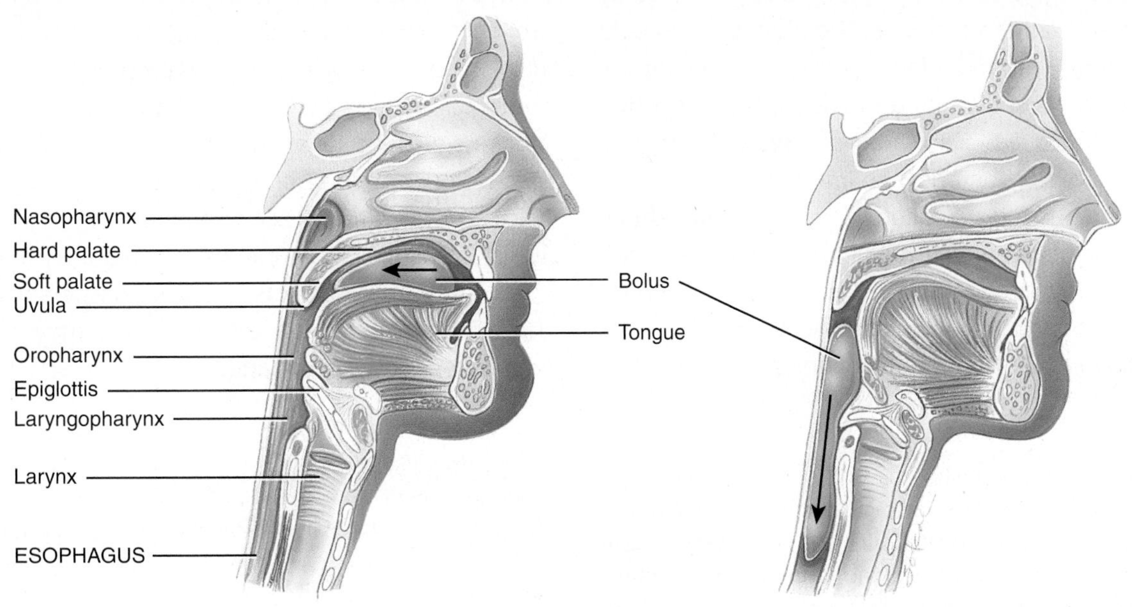


(a) Position of structures before swallowing

(b) During pharyngeal stage of swallowing

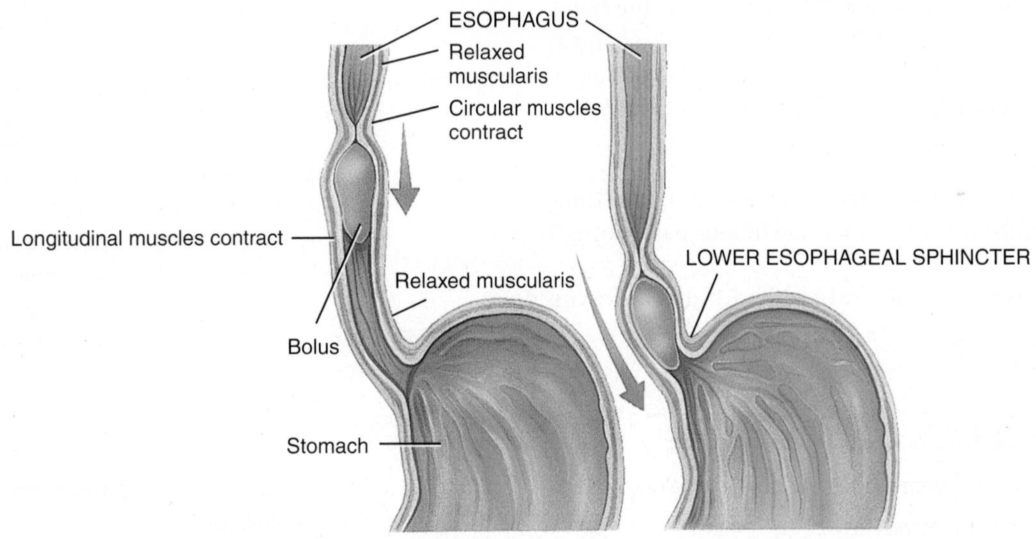


(c) Anterior view of frontal sections of peristalsis in esophagus

**Deglutition moves food from the mouth into the stomach.**

## CLINICAL CONNECTION | *Gastroesophageal Reflux Disease*

If the lower esophageal sphincter fails to close adequately after food has entered the stomach, the stomach contents can *reflux* (back up) into the inferior portion of the esophagus. This condition is known as **gastroesophageal reflux disease (GERD)** (gas′-trō-e-sof-a-JĒ-al). Hydrochloric acid (HCl) from the stomach contents can irritate the esophageal wall, resulting in a burning sensation that is called **heartburn** because it is experienced in a region very near the heart; however, it is unrelated to any cardiac problem. Drinking alcohol and smoking can cause the sphincter to relax, worsening the problem. GERD symptoms often can be controlled by avoiding foods that strongly stimulate stomach acid secretion (coffee, chocolate, tomatoes, fatty foods, orange juice, peppermint, spearmint, and onions). Other acid-reducing strategies include taking over-the-counter histamine-2 ($H_2$) blockers such as Tagamet HB® or Pepcid AC® 30 to 60 minutes before eating to block acid secretion, and neutralizing acid that has already been secreted with antacids such as Tums® or Maalox®. Symptoms are less likely to occur if food is eaten in smaller amounts and if the person does not lie down immediately after a meal. GERD may be associated with cancer of the esophagus.

layers of the muscularis, pushes the bolus onward. In the section of the esophagus just superior to the bolus, the circular muscle fibers contract, constricting the esophageal wall and squeezing the bolus toward the stomach. Meanwhile, longitudinal fibers inferior to the bolus also contract, which shortens this inferior section and pushes its walls outward so it can receive the bolus. The contractions are repeated in waves that push the food toward the stomach. As the bolus approaches the end of the esophagus, the lower esophageal sphincter relaxes and the bolus moves into the stomach. Along the way, mucus secreted by esophageal glands lubricates the bolus and reduces friction.

Passage of solid or semisolid food from the mouth to the stomach takes 4 to 8 seconds; very soft foods and liquids pass through in about 1 second.

Table 23.2 summarizes the digestive activities of the pharynx and esophagus.

**TABLE 23.2**

Digestive Activities in the Pharynx and Esophagus

| STRUCTURE | ACTIVITY | RESULT |
|---|---|---|
| **Pharynx** | Pharyngeal stage of deglutition | Moves bolus from oropharynx to laryngopharynx and into esophagus; closes air passageways |
| **Esophagus** | Relaxation of upper esophageal sphincter | Permits entry of bolus from laryngopharynx into esophagus |
| | Esophageal stage of deglutition (peristalsis) | Pushes bolus down esophagus |
| | Relaxation of lower esophageal sphincter | Permits entry of bolus into stomach |
| | Secretion of mucus | Lubricates esophagus for smooth passage of bolus |

### ✓ CHECKPOINT

11. To which two organ systems does the pharynx belong?
12. What does deglutition mean?
13. What are the functions of the upper and lower esophageal sphincters?
14. What occurs during the oral, pharyngeal, and esophageal stages of swallowing?
15. Is swallowing a voluntary action or an involuntary action?
16. Does peristalsis "push" or "pull" food along the GI tract?

## 23.4 The stomach mechanically breaks down the bolus and mixes it with gastric secretions.

The **stomach** (Figure 23.11) is a J-shaped enlargement of the GI tract in the abdomen directly inferior to the diaphragm (see Figure 23.1). The stomach connects the esophagus to the duodenum, the first part of the small intestine (Figure 23.11). Because a meal can be eaten much more quickly than the intestines can digest and absorb it, one of the functions of the stomach is to serve as a mixing chamber and holding reservoir. At appropriate intervals after food is ingested, the stomach forces a small

FIGURE 23.11 **External and internal anatomy of the stomach.**

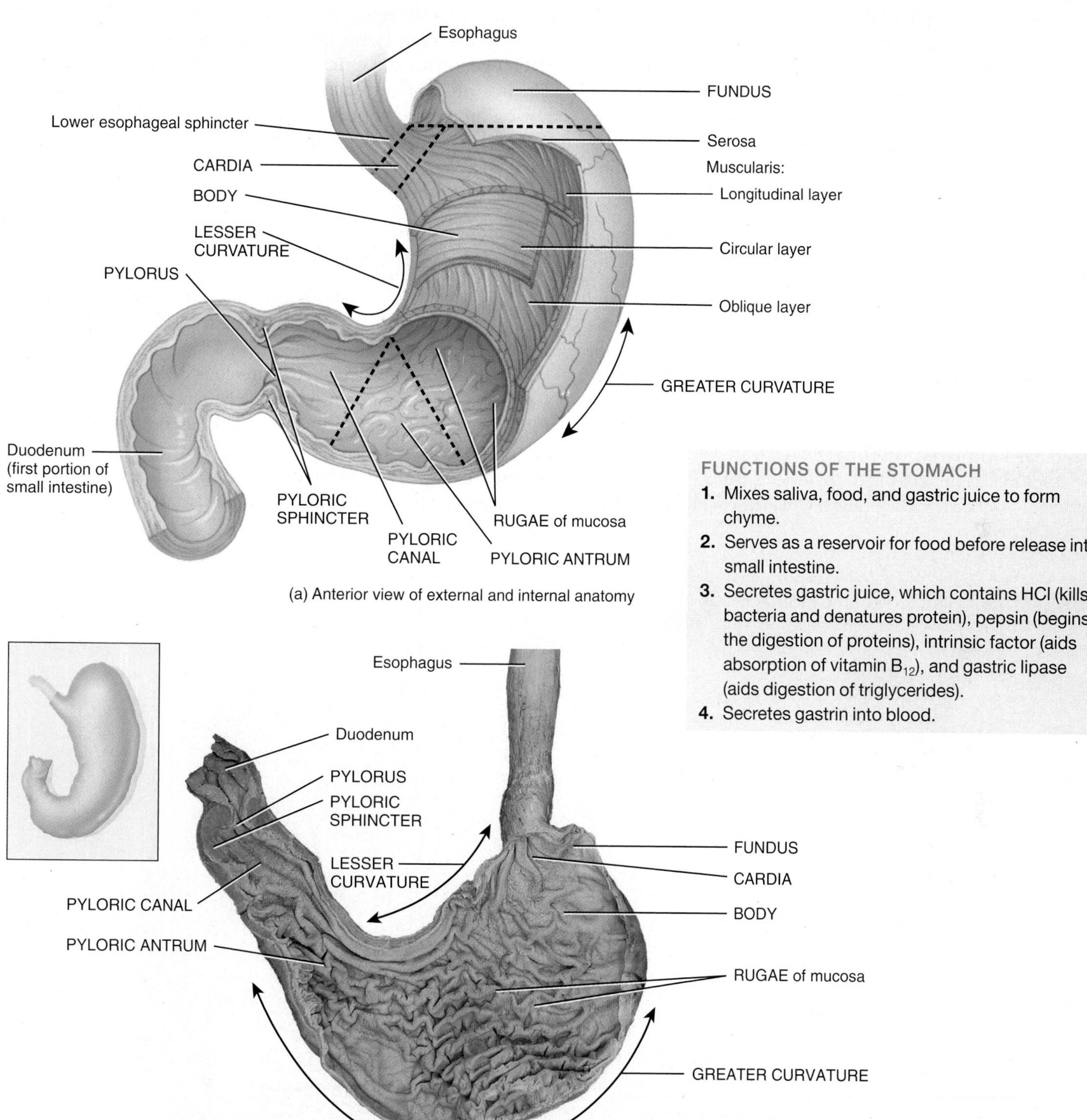


(a) Anterior view of external and internal anatomy

(b) Anterior view of internal anatomy

**FUNCTIONS OF THE STOMACH**

1. Mixes saliva, food, and gastric juice to form chyme.
2. Serves as a reservoir for food before release into small intestine.
3. Secretes gastric juice, which contains HCl (kills bacteria and denatures protein), pepsin (begins the digestion of proteins), intrinsic factor (aids absorption of vitamin $B_{12}$), and gastric lipase (aids digestion of triglycerides).
4. Secretes gastrin into blood.

**The four regions of the stomach are the cardia, fundus, body, and pylorus.**

quantity of material into the first portion of the small intestine. The position and size of the stomach vary continually. The diaphragm pushes it inferiorly with each inhalation and pulls it superiorly with each exhalation. Empty, it is about the size of a large sausage, but it is the most distensible part of the GI tract and can accommodate a large quantity of food. In the stomach, digestion of starch continues, digestion of proteins and triglycerides begins, the semisolid bolus is converted to a liquid, and certain substances are absorbed.

## Anatomy of the Stomach

The stomach has four main regions: the cardia, fundus, body, and pyloric part (Figure 23.11). The **cardia** (CAR-dē-a) surrounds the superior opening of the stomach. The rounded portion superior to and to the left of the cardia is the **fundus** (FUN-dus). Inferior to the fundus is the large central portion of the stomach, called the **body**. The region of the stomach that connects to the duodenum is the **pyloric part** (pī-LOR-ik; *pyl-* = gate; *-orus* = guard). The pyloric part consists of the **pyloric antrum** (AN-trum = cave), which connects to the body; the **pyloric canal**, which leads to the pylorus; and the **pylorus** (pī-LOR-us), which in turn connects to the duodenum, the first part of the small intestine. When the stomach is empty, the mucosa lies in large folds, called **rugae** (ROO-gē = wrinkles), that can be seen with the unaided eye. The pylorus communicates with the duodenum of the small intestine through a smooth muscle sphincter called the **pyloric sphincter**. The concave medial border of the stomach is called the **lesser curvature**, and the convex lateral border is called the **greater curvature**.

## Histology of the Stomach

The stomach wall is composed of the same four basic layers as the rest of the GI tract, with certain modifications (Figure 23.12). The surface of the mucosa is a layer of simple columnar epithelial cells called **surface mucous cells** (Figure 23.12b). Epithelial cells extend down into the lamina propria, where they form columns of secretory cells called **gastric glands**. Several gastric glands open into the bottom of narrow channels called **gastric pits**. Secretions from several gastric glands flow into each gastric pit and then into the lumen of the stomach.

**FIGURE 23.12 Histology of the stomach.**

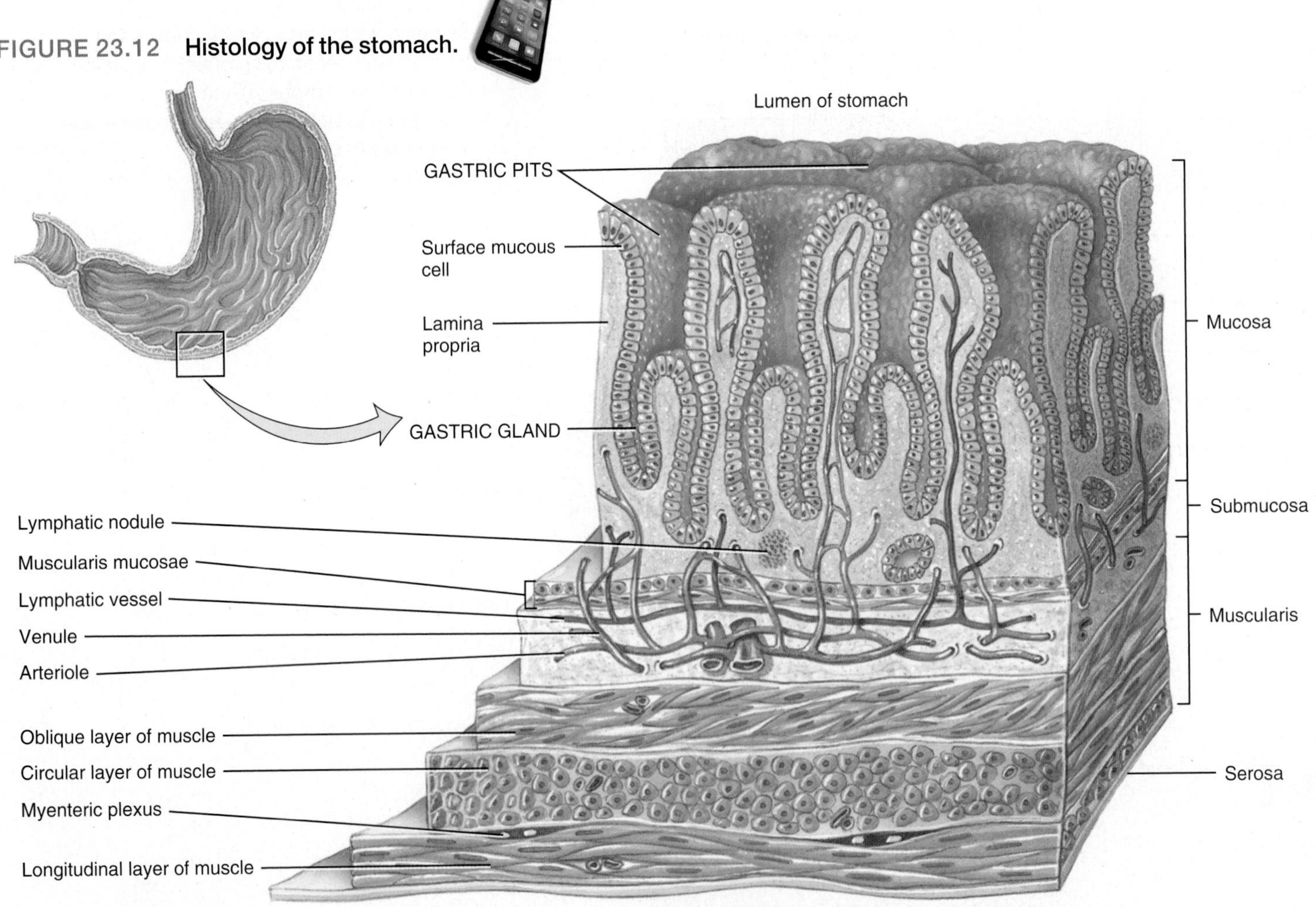


(a) Three-dimensional view of layers of stomach

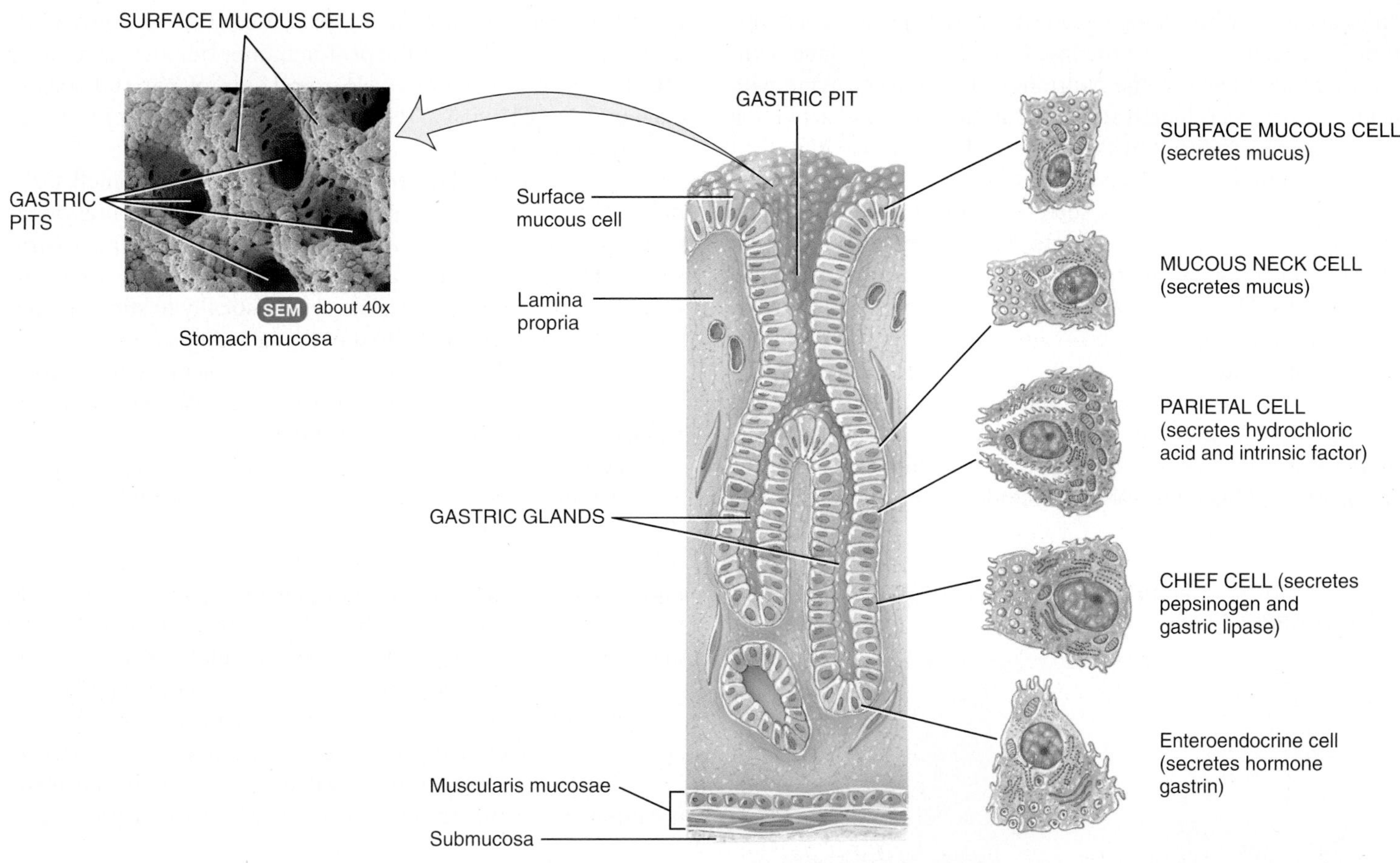


(b) Sectional view of stomach mucosa showing gastric glands and cell types

Gastric juice is the combined secretions of mucous cells, parietal cells, and chief cells.

The gastric glands contain *exocrine gland cells* that secrete 2000–3000 mL (roughly 2–3 qt) of **gastric juice** into the stomach lumen each day. Both surface mucous cells and **mucous neck cells** secrete mucus (Figure 23.12b). **Parietal cells** produce intrinsic factor (needed for absorption of vitamin $B_{12}$) and hydrochloric acid. **Chief cells** secrete pepsinogen and gastric lipase. In addition to the three types of exocrine gland cells, the gastric glands include enteroendocrine cells that secrete the hormone gastrin into the bloodstream. As we will see shortly, gastrin stimulates several aspects of gastric activity.

The muscularis of the stomach has three layers of smooth muscle: an outer longitudinal layer, a middle circular layer, and an inner oblique layer (Figure 23.12a). Although the muscularis of the esophagus, small intestine, and large intestine consists of only two layers, the additional oblique layer of the muscularis is needed to accommodate the mixing waves (described shortly) that are unique to the stomach.

The serosa covering the stomach is part of the visceral peritoneum. At the lesser curvature of the stomach, the visceral peritoneum extends upward to the liver as the lesser omentum. At the greater curvature of the stomach, the visceral peritoneum continues downward as the greater omentum and drapes over the intestines.

## Mechanical and Chemical Digestion in the Stomach

Several minutes after food enters the stomach, peristaltic movements called **mixing waves** pass over the stomach every 15 to 25 seconds. These waves macerate food, mix it with secretions of the gastric glands, and reduce it to a soupy liquid called **chyme** (KĪM = juice). When food initially enters the cardia, it is exposed to gentle rippling mixing waves. As digestion proceeds in the stomach, more vigorous mixing waves begin at the body of the stomach and intensify as they reach the pylorus. Most of the chyme that enters the pylorus is forced back into the body of the stomach, where mixing continues. These forward and backward movements of the gastric contents are responsible for most mixing in the stomach.

The pyloric sphincter normally remains almost, but not completely, closed. As food reaches the pylorus, each mixing wave periodically forces about 3 mL of chyme through the pyloric sphincter into the duodenum. The next wave pushes the chyme forward again and forces a little more into the duodenum.

Few mixing waves occur in the fundus, which primarily has a storage function. Foods may remain in the fundus for about an hour without becoming mixed with gastric juice. During this time, digestion by salivary amylase from the salivary glands continues. Soon, however, the churning action mixes chyme with acidic gastric juice, inactivating salivary amylase and activating lingual lipase produced by the tongue, which starts to digest triglycerides into fatty acids and diglycerides.

Although parietal cells secrete hydrogen ions ($H^+$) and chloride ions ($Cl^-$) separately into the stomach lumen, the net effect is secretion of hydrochloric acid (HCl). **Proton pumps** powered by $H^+/K^+$ ATPases actively transport $H^+$ into the stomach lumen in exchange for potassium ions ($K^+$) that are brought into the cell (Figure 23.13). At the same time, $Cl^-$ and $K^+$ diffuse out into the stomach lumen through $Cl^-$ and $K^+$ channels. Within the parietal cells, the enzyme *carbonic anhydrase* catalyzes the formation of carbonic acid ($H_2CO_3$) from water ($H_2O$) and carbon dioxide ($CO_2$). Carbonic acid dissociates into $H^+$ and bicarbonate ions ($HCO_3^-$), providing a ready source of $H^+$ for the proton pumps but also generating $HCO_3^-$. $HCO_3^-$ exits the basolateral side of parietal cells in exchange for $Cl^-$ via $Cl^-/HCO_3^-$ antiporters, and diffuses into nearby blood capillaries.

**FIGURE 23.13 Secretion of hydrochloric acid (HCl) by parietal cells in the stomach.**

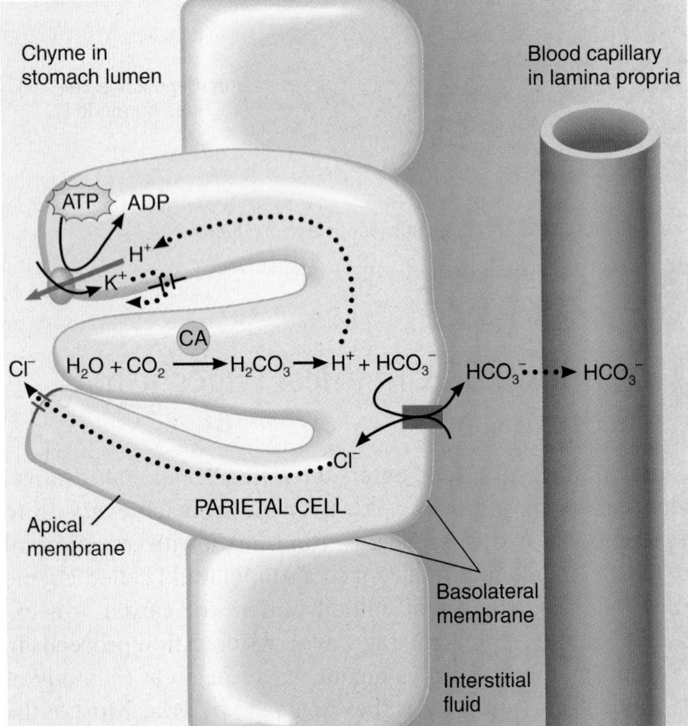


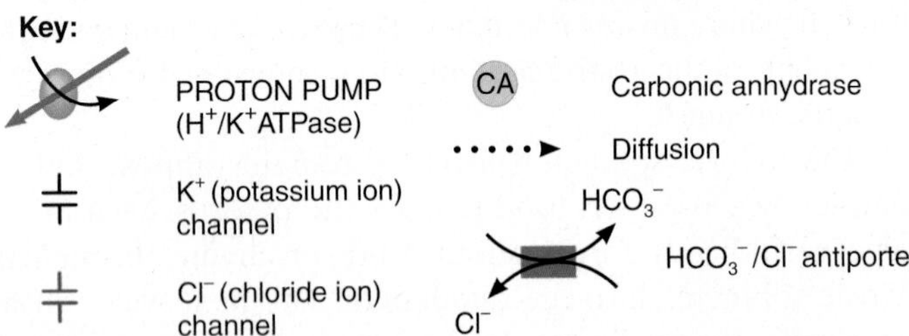


**$H^+$ is actively transported and $Cl^-$ diffuses into the stomach lumen; bicarbonate ions ($HCO_3^-$) diffuse into blood capillaries.**

Even before food enters the stomach, the sight, smell, taste, or thought of food initiates reflexes that stimulate parasympathetic neurons to release acetylcholine. Acetylcholine, gastrin secreted by gastric glands, and histamine released by mast cells in the nearby lamina propria act synergistically to stimulate parietal cells to secrete HCl into the stomach lumen (Figure 23.14). Receptors for all three substances are present in the plasma membrane of parietal cells. Gastrin also increases motility of the stomach and relaxes the pyloric sphincter.

The strongly acidic gastric juice kills many microbes in food. HCl partially denatures (unfolds) proteins in food and stimulates the secretion of hormones that promote the secretion of bile and pancreatic juice (described shortly). Enzymatic digestion of proteins also begins in the stomach by **pepsin**. Pepsin severs peptide bonds between amino acids, breaking down a protein chain of many amino acids into smaller peptide fragments. Pepsin is most effective in the very acidic environment of the stomach (pH 2).

What keeps pepsin from digesting the protein in the stomach wall along with the food in the stomach lumen? First, pepsin is secreted in an inactive form called *pepsinogen*; in this form, it cannot digest the proteins in the chief cells that produce it.

**FIGURE 23.14 Regulation of hydrochloric acid (HCl) secretion.**

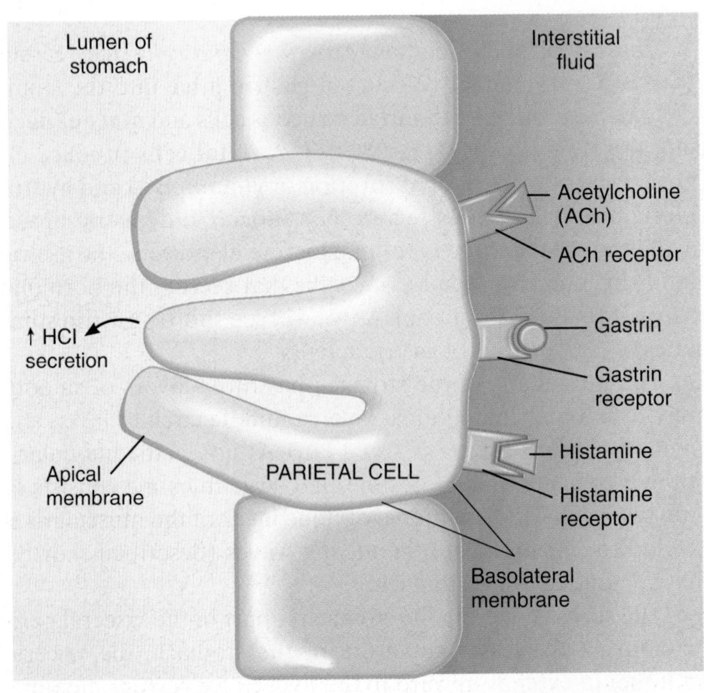


**HCl secretion by parietal cells is stimulated by acetylcholine, gastrin, and histamine.**

Pepsinogen is not converted into active pepsin until it comes in contact with hydrochloric acid in gastric juice within the stomach lumen. Second, the stomach epithelium is protected from gastric juices by a thick layer of alkaline mucus secreted by surface mucous cells and mucous neck cells.

Another enzyme of the stomach is **gastric lipase**, which splits triglycerides in fat molecules into fatty acids and monoglycerides. A monoglyceride consists of a glycerol molecule that is attached to one fatty acid molecule. Gastric lipase, which has a limited role in the stomach, operates best at a pH of 5–6. A more powerful enzyme than either lingual lipase or gastric lipase is pancreatic lipase, which is secreted by the pancreas into the small intestine.

Only a small amount of nutrients are absorbed in the stomach because its epithelial cells are impermeable to most materials. However, mucous cells of the stomach absorb some water, ions, and short-chain fatty acids, as well as certain drugs (especially aspirin) and alcohol.

Within 2 to 4 hours after eating a meal, the stomach has emptied its contents into the duodenum. Foods rich in carbohydrate spend the least time in the stomach; high-protein foods remain somewhat longer, and emptying is slowest after a fat-laden meal containing large amounts of triglycerides.

Table 23.3 summarizes the digestive activities of the stomach.

**TABLE 23.3**

Digestive Activities in the Stomach

| STRUCTURE | ACTIVITY | RESULT |
|---|---|---|
| **Mucosa** | | |
| **Surface mucous cells and mucous neck cells** | Secrete mucus | Forms protective barrier that prevents digestion of stomach wall |
| | Absorption | Small quantity of water, ions, short-chain fatty acids, and some drugs enter bloodstream |
| **Parietal cells** | Secrete intrinsic factor | Needed for absorption of vitamin $B_{12}$ (used in red blood cell formation, or erythropoiesis) |
| | Secrete hydrochloric acid | Kills microbes in food; denatures proteins; converts pepsinogen into pepsin |
| **Chief cells** | Secrete pepsinogen | Pepsin (activated form) breaks down proteins into peptides |
| | Secrete gastric lipase | Splits triglycerides into fatty acids and monoglycerides |
| **Enteroendocrine cells** | Secrete gastrin | Stimulates parietal cells to secrete HCl and chief cells to secrete pepsinogen; contracts lower esophageal sphincter, increases motility of stomach, and relaxes pyloric sphincter |
| **Muscularis** | Mixing waves (gentle peristaltic movements) | Churns and physically breaks down food and mixes it with gastric juice, forming chyme; forces chyme through pyloric sphincter |
| **Pyloric sphincter** | Opens to permit passage of chyme into duodenum | Regulates passage of chyme from stomach to duodenum; prevents backflow of chyme from duodenum to stomach |

## ✓ CHECKPOINT

**17.** Which stomach layer is in contact with swallowed food?

**18.** Which part of the stomach primarily serves as a food reservoir?

**19.** Which molecule is the source of the hydrogen ions that are secreted into gastric juice?

**20.** Which branch of the autonomic nervous system promotes digestion?

**21.** What is the role of pepsin? Why is it secreted in an inactive form?

**22.** Which substances are absorbed in the stomach?

## RETURN TO Zachary's Story

When Zach first started getting stomach aches, Sandra thought it was just something he had eaten, or maybe the flu. As the pain continued, she started to wonder if Zach might be faking it. Being a mother to a 12-year-old, she had to be skeptical when he didn't want to go to school because of a "tummy ache." She discarded this passing thought quickly, because it kept getting worse. When he used the bathroom it was almost exclusively diarrhea, it smelled terrible, and he said it was painful. When he stopped eating she really started to panic. As soon as her two older sons hit 12 years of age, it was almost impossible to stop them from eating. Zach was on the same path until he got sick, and then all of a sudden he was shrinking. She had thought that the first day he vomited blood was the worst day of her life, but now she really wishes it had been. They rushed Zach to the hospital and were immediately told it was "probably" an ulcer. Sandra asked the ER doctor if 12-year-olds can even get ulcers, and was told that it was increasingly more common for younger people to develop them. She knew in her bones that there was something else wrong. This event initiated two years of revolving door hospital visits—vomiting blood, blood in the toilet, passing out during school, high fevers, and not eating. Zachary went through so much. He would wake up at night crying from the pain, and Sandra tried to hide her own tears as she comforted him. It felt like she hadn't slept in thirty years.

A. *An ulcer starts by eroding the mucosa of the GI tract wall. What functions of digestion and/or reabsorption might be lost if this layer is no longer functional? What functions will be compromised if the ulcer eats through the submucosa and then the muscularis?*
B. *If Zach has a peptic ulcer affecting his stomach or duodenum, which components of the peritoneum will be affected (see the Chapter 23 WileyPLUS Clinical Connection entitled Peptic Ulcer Disease)?*
C. *How can Zach's stomach contribute to the formation of ulcers in other parts of the GI tract? Which cells directly participate in ulcer formation, and how do they contribute to the creation of lesions in the GI tract wall?*
D. *Why does Zach's GI tract need the substance that contributes to the formation of ulcers? How is this substance secreted by the cells within the gastric pits?*
E. *If Zach's only normal digestive enzymes come from his mouth, what substances will he be able to digest?*

# 23.5 The pancreas secretes pancreatic juice, the liver secretes bile, and the gallbladder stores and concentrates bile.

## Pancreas

From the stomach, chyme passes into the small intestine. Because chemical digestion in the small intestine depends on activities of the pancreas, liver, and gallbladder, we first consider the activities of these accessory digestive organs and their contributions to digestion in the small intestine.

### Anatomy of the Pancreas

The **pancreas** (*pan-* = all; *-creas* = flesh), a retroperitoneal gland that is about 12–15 cm (5–6 in.) long, lies posterior to the greater curvature of the stomach. The pancreas consists of a head, a body, and a tail and is usually connected to the duodenum by two ducts (Figure 23.15a). The **head** is the expanded portion of the organ near the curve of the duodenum; to the left of the head are the central **body** and the tapering **tail**.

Pancreatic juices are secreted by exocrine cells into small ducts that ultimately unite to form two larger ducts, the pancreatic duct and the accessory duct. These ducts in turn convey the secretions into the duodenum of the small intestine. The **pancreatic duct** is the larger of the two ducts. In most people, the pancreatic duct joins the common bile duct from the liver and gallbladder and enters the duodenum as a common duct called the **hepatopancreatic ampulla** (hep′-a-tō-pan-crē-A-tik), which opens about 10 cm (4 in.) inferior to the pyloric sphincter of the stomach. The passage of pancreatic juice and bile through the hepatopancreatic ampulla into the duodenum of the small intestine is regulated by the **sphincter of the hepatopancreatic ampulla**, a band of smooth muscle surrounding the ampulla. The other major duct of the pancreas, the **accessory duct**, empties into the duodenum about 2.5 cm (1 in.) superior to the hepatopancreatic ampulla.

### Histology of the Pancreas

The pancreas is made up of small clusters of glandular epithelial cells. About 99 percent of the clusters, called **acini** (AS-i-nē), constitute the *exocrine* portion of the organ (see Figure 17.17b, c).

**FIGURE 23.15 Relationship of the pancreas to the liver, gallbladder, and duodenum.** The inset (b) shows details of the common bile duct and pancreatic duct forming the hepatopancreatic ampulla that empties into the duodenum.



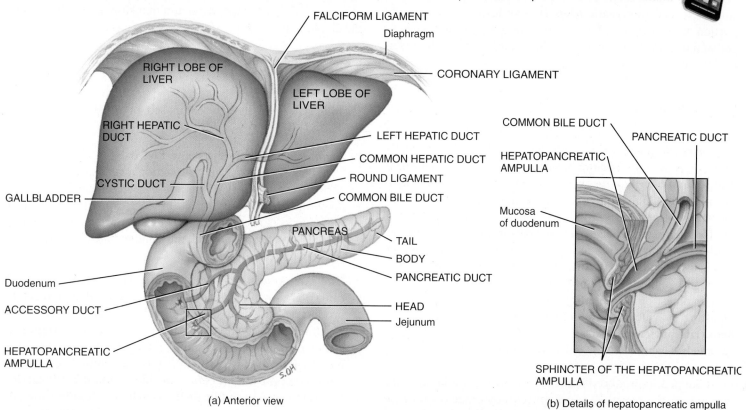

(a) Anterior view



(b) Details of hepatopancreatic ampulla

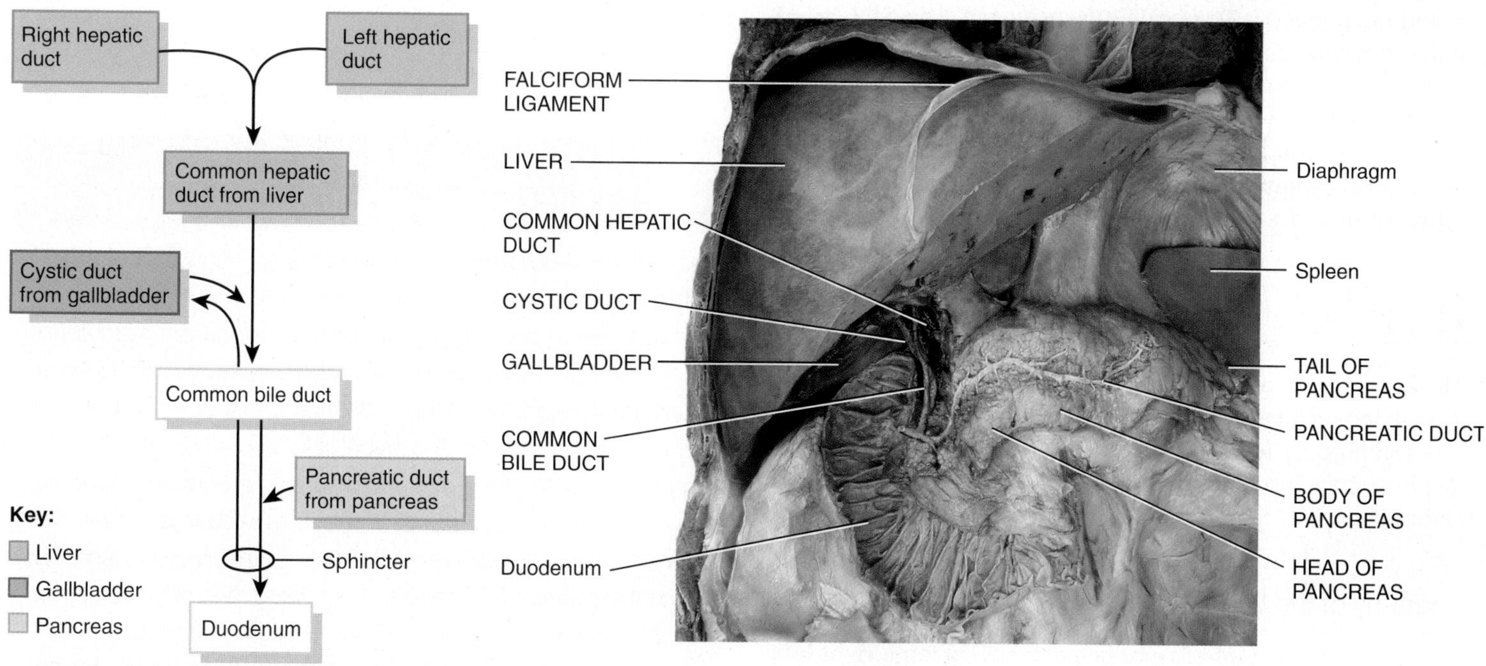


(c) Ducts carrying bile from liver and gallbladder and pancreatic juice from pancreas to duodenum

(d) Anterior view

**Pancreatic enzymes digest starches, proteins, triglycerides, and nucleic acids.**

The cells within acini secrete a mixture of fluid and digestive enzymes called **pancreatic juice**. The remaining 1 percent of the clusters, called **pancreatic islets** (Ī-lets), forms the *endocrine* portion of the pancreas. These cells secrete the hormones glucagon and insulin. The functions of these hormones are discussed in Concept 17.8.

### Pancreatic Juice

Each day the pancreas produces 1200–1500 mL (about 1.2–1.5 qt) of **pancreatic juice**, a clear, colorless liquid consisting mostly of water, some salts, sodium bicarbonate, and several enzymes. The **sodium bicarbonate** gives pancreatic juice a slightly alkaline pH (7.1–8.2) that buffers acidic gastric juice in chyme, stops the action of pepsin from the stomach, and creates the proper pH for the action of digestive enzymes in the small intestine. The enzymes in pancreatic juice include **pancreatic amylase**, which digests starches; **trypsin** (TRIP-sin), **chymotrypsin** (kī′-mō-TRIP-sin), and **carboxypeptidase** (kar-bok′-sē-PEP-ti-dās) for protein digestion; **pancreatic lipase**, the principal triglyceride-digesting enzyme; and **ribonuclease** (rī′-bō-NOO-klē-ās) and **deoxyribonuclease** (dē-oks-ē-rī′-bō-NOO-klē-ās), which digest ribonucleic acid (RNA) and deoxyribonucleic acid (DNA) into nucleotides.

Like pepsin produced in the stomach, the protein-digesting enzymes of the pancreas are produced in an inactive form so that they do not digest the cells of the pancreas. Trypsin is secreted in an inactive form called **trypsinogen** (trip-SIN-ō-jen). When trypsinogen reaches the lumen of the small intestine, it encounters an activating brush border enzyme called **enterokinase** (en′-ter-ō-KĪ-nās), which splits off part of the trypsinogen molecule to form trypsin. In turn, trypsin converts the inactive precursors **chymotrypsinogen** and **procarboxypeptidase** to their active forms, chymotrypsin and carboxypeptidase, respectively.

## Liver

The **liver** is the heaviest gland of the body, weighing about 1.4 kg (about 3 lb) in an average adult. Of all of the organs of the body, it is second only to the skin in size. The liver is inferior to the diaphragm and occupies most of the upper right quadrant of the abdominal cavity (see Figure 23.1).

### Anatomy of the Liver

The liver is divided into two principal lobes—a large **right lobe** and a smaller **left lobe**—by the **falciform ligament** (FAL-si-form), a fold of the mesentery (Figure 23.15a). An inferior *quadrate lobe* (kwa-DRĀT) and a posterior *caudate lobe* (KAW-dāt) are continuations of the left lobe. The falciform ligament extends from the undersurface of the diaphragm to the superior surface of the liver, helping to suspend the liver in the abdominal cavity. In the free border of the falciform ligament is the **round ligament**, a fibrous cord that extends from the liver to the umbilicus. The right and left **coronary ligaments** are narrow extensions of the parietal peritoneum that suspend the liver from the diaphragm.

### Histology of the Liver

Histologically, the liver is composed of several components (Figure 23.16):

- ***Hepatocytes*.** **Hepatocytes** (he-PAT-ō-cytes; *hepat-* = *liver; -cytes* = cell) are the major functional cells of the liver and perform a wide array of metabolic, secretory, and endocrine functions. Hepatocytes, which make up about 80 percent of the volume of the liver, are arranged in irregular, interconnected rows that are one cell thick and radiate out from a central vein (described shortly).
- ***Bile duct system*.** **Bile canaliculi** (kan-a-LIK-ū-li = small canals) are small ducts between hepatocytes that collect bile produced by the hepatocytes. From bile canaliculi, bile passes into **bile ducts**. The bile ducts merge into the larger **right** and **left hepatic ducts**, which unite and exit the liver as the **common hepatic duct** (Figure 23.15). The common hepatic duct joins the **cystic duct** (*cystic* = bladder) from the gallbladder to form the **common bile duct**. From here, bile enters the duodenum of the small intestine to participate in digestion.

### CLINICAL CONNECTION | *Jaundice*

**Jaundice** (JAWN-dis = yellowed) is a yellowish coloration of the sclerae (whites of the eyes), skin, and mucous membranes due to a buildup of a yellow compound called bilirubin. After bilirubin is formed from the breakdown of the heme pigment in aged red blood cells, it is transported to the liver, where it is processed and eventually excreted into bile. The three main categories of jaundice are (1) *prehepatic jaundice,* due to excess production of bilirubin; (2) *hepatic jaundice,* due to congenital liver disease, cirrhosis of the liver, or hepatitis; and (3) *extrahepatic jaundice,* due to blockage of bile drainage by gallstones or cancer of the bowel or the pancreas.

Because the liver of a newborn functions poorly for the first week or so, many babies experience a mild form of jaundice called *neonatal (physiological) jaundice* that disappears as the liver matures. Usually, it is treated by exposing the infant to blue light, which converts bilirubin into substances the kidneys can excrete.

FIGURE 23.16 **Histology of the liver.**

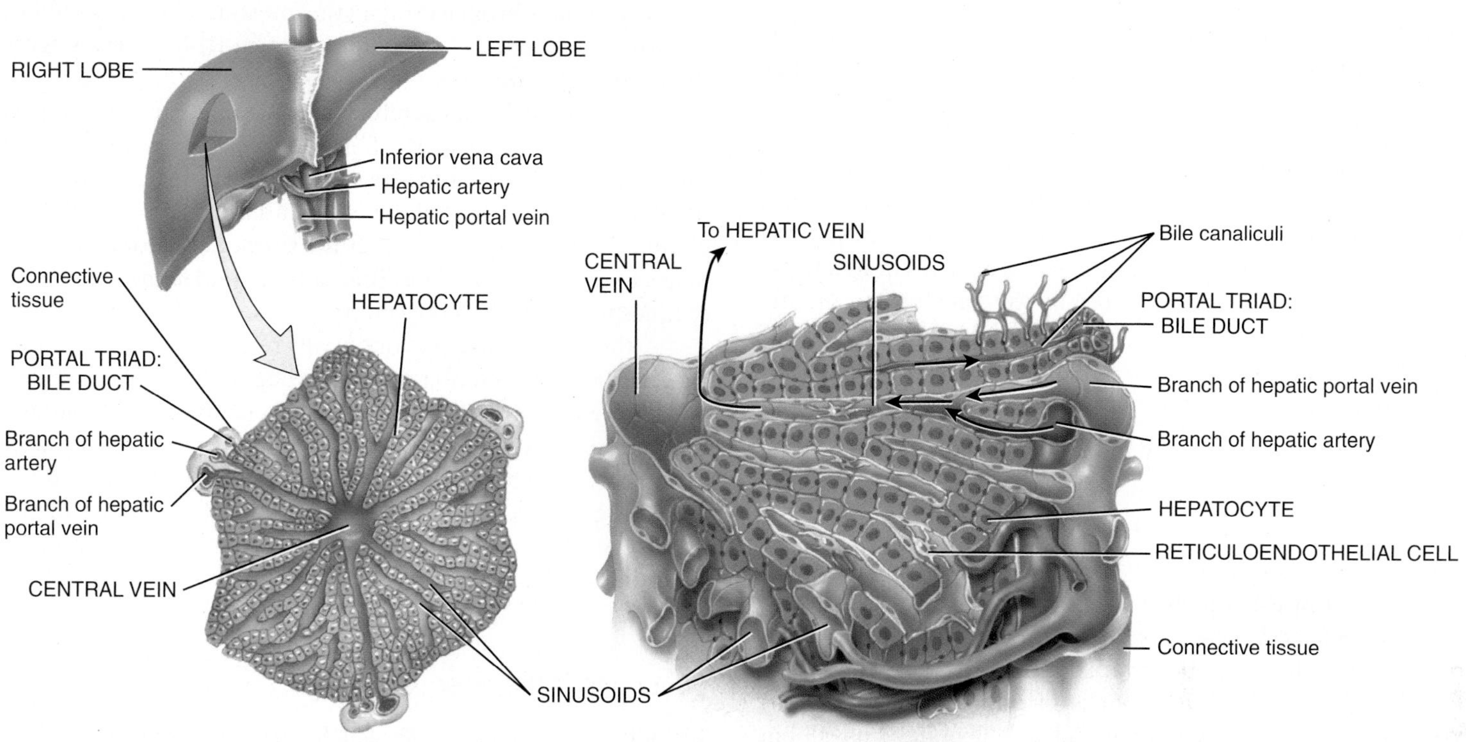


(a) Overview of histological components of liver

(b) Details of histological components of liver

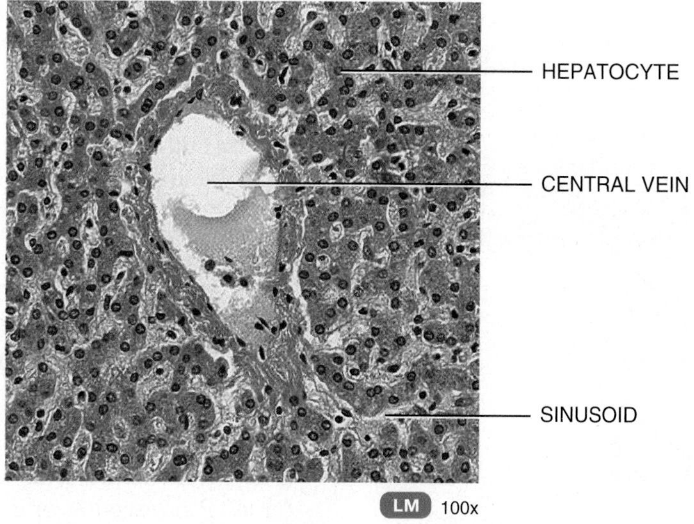


(c) Photomicrograph

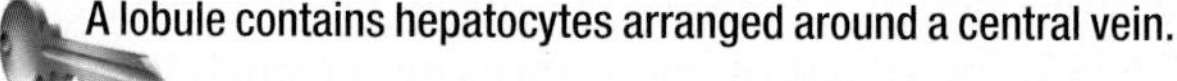

**A lobule contains hepatocytes arranged around a central vein.**

- ***Sinusoids.*** **Sinusoids** of the liver are highly permeable blood capillaries between rows of hepatocytes through which blood passes (see Concept 20.3). Also present in the sinusoids are phagocytes called **reticuloendothelial cells** (re-tik-ū-lō-en′-dō-THĒ-lē-al), also called hepatic macrophages, which destroy worn-out blood cells, bacteria, and other foreign matter in the venous blood draining from the GI tract. The sinusoids converge and deliver blood into a **central vein**. In contrast to blood, which flows toward a central vein, bile flows in the opposite direction toward bile ducts (Figure 23.16b). From the central veins, blood flows into the **hepatic veins**.
- ***Portal triads.*** Together, a bile duct, branch of the hepatic artery, and branch of the hepatic portal vein are referred to as a **portal triad** (*tri* = three).

## Blood Supply of the Liver

The liver receives blood from two sources (Figure 23.17). The hepatic artery supplies oxygenated blood from the abdominal aorta. The hepatic portal vein supplies deoxygenated blood from the GI tract containing newly absorbed nutrients, drugs, and possibly microbes and toxins (see Figure 20.28). Branches of both the hepatic artery and the hepatic portal vein carry blood into the sinusoids of the liver, where oxygen, most of the nutrients, and certain toxic substances are taken up by the hepatocytes. Products manufactured by the hepatocytes and nutrients needed by other body cells are secreted into the blood, which then drains into the central vein and eventually passes into a hepatic vein. Hepatic veins, in turn, drain into the inferior vena cava.

## Bile

Each day, hepatocytes secrete 800–1000 mL (about 1 qt) of **bile**, a yellow, brownish, or olive-green liquid. Bile consists mostly of water, bile acids, bile salts, cholesterol, a phospholipid called lecithin, bile pigments, and several ions.

The principal bile pigment is **bilirubin** (bil′-i-ROO-bin). The phagocytosis of aged red blood cells liberates iron, globin, and bilirubin (derived from heme) (see Figure 18.5). The iron and globin are recycled; the bilirubin is secreted into the bile and is eventually broken down in the intestine. One of bilirubin's breakdown products—*stercobilin* (ster-kō-BĪ-lin)—gives feces their normal brown color.

Bile is partially an excretory product and partially a digestive secretion. Bile salts, which are sodium salts and potassium salts, play a role in **emulsification** (ē-mul-si-fi-KĀ-shun), the breakdown of large lipid globules into a suspension of small lipid globules, which greatly increases the surface area for pancreatic lipase to digest triglycerides. Bile salts also aid in the absorption of lipids following their digestion.

Although hepatocytes continually release bile, they increase production and secretion when the hepatic portal blood contains more bile salts. Thus, as digestion and absorption continue in the small intestine, bile release increases. Between meals, after most absorption has occurred, bile secreted from the liver flows into the gallbladder for storage because the sphincter of the hepatopancreatic ampulla (Figure 23.15) closes off the entrance to the duodenum.

FIGURE 23.17 **Hepatic blood flow: sources, path through the liver, and return to the heart.**

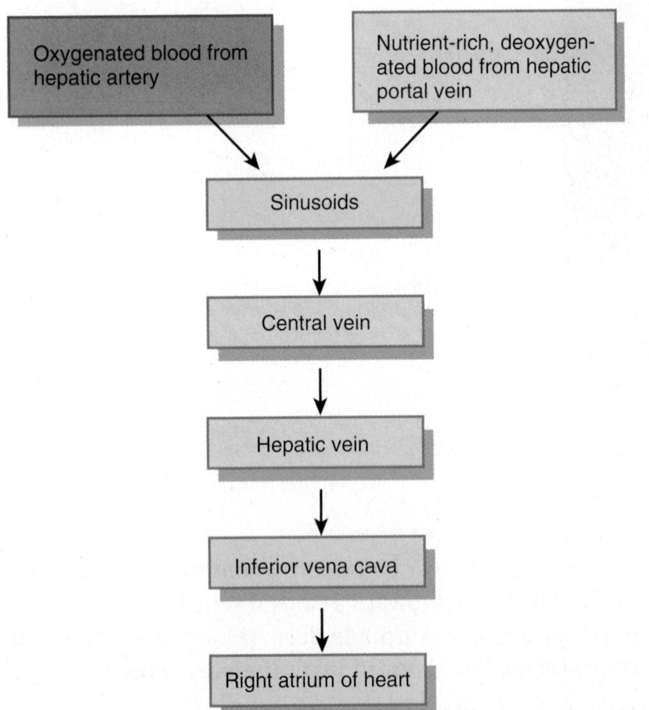


**The liver receives oxygenated blood via the hepatic artery and nutrient-rich deoxygenated blood via the hepatic portal vein.**

## Functions of the Liver

In addition to producing bile, which is needed for the emulsification and absorption of dietary fats in the small intestine, the liver performs many other vital functions:

- ***Carbohydrate metabolism.*** The liver is especially important in maintaining a normal blood glucose level. When blood glucose is low, the liver breaks down glycogen to glucose and releases glucose into the bloodstream. The liver can also convert certain amino acids, lactic acid, and other sugars into glucose. When blood glucose is high, as occurs just after eating a meal, the liver stores glucose as glycogen and triglycerides.
- ***Lipid metabolism.*** Hepatocytes store some triglycerides; break down fatty acids to generate ATP; synthesize **lipoproteins** (lip′-ō-PRŌ-tēns), which transport fatty acids, triglycerides, and cholesterol to and from body cells; synthesize cholesterol; and use cholesterol to make bile salts.
- ***Protein metabolism.*** Hepatocytes remove the amino group ($NH_2$) from amino acids so that the amino acids can be used for ATP production or converted to carbohydrates or fats. The resulting toxic ammonia is then converted into the much less toxic urea, which is excreted in urine. Hepatocytes also synthesize most plasma proteins, such as alpha and beta globulins, albumin, prothrombin, and fibrinogen.
- ***Processing of drugs and hormones.*** The liver can detoxify substances such as alcohol and excrete drugs such as penicillin, erythromycin, and sulfonamides into bile. It can also inactivate thyroid hormones and steroid hormones such as estrogens and aldosterone.
- ***Excretion of bilirubin.*** As previously noted, bilirubin derived from the heme of aged red blood cells is absorbed by the liver

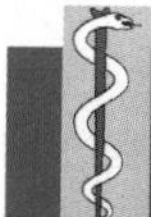

## CLINICAL CONNECTION | *Gallstones*

If bile contains insufficient bile salts or lecithin or an excessive amount of cholesterol, the cholesterol may crystallize to form **gallstones.** As they grow in size and number, gallstones may cause minimal, intermittent, or complete obstruction to the flow of bile from the gallbladder into the duodenum. Treatment consists of using gallstone-dissolving drugs, lithotripsy (shock-wave therapy), or surgery. For people with recurrent gallstones or for whom drugs or lithotripsy are not indicated, **cholecystectomy** (kō-le-sis-TEK-tō-mē)—the removal of the gallbladder and its contents—is necessary. More than half a million cholecystectomies are performed each year in the United States. To prevent side effects resulting from a loss of the gallbladder, patients should make lifestyle and dietary changes, including the following: (1) limit intake of saturated fat; (2) avoid consumption of alcoholic beverages; (3) eat smaller amounts of food during a meal and eat five to six smaller meals per day instead of two to three larger meals; and (4) take vitamin and mineral supplements.

from the blood and secreted into bile. Most of the bilirubin in bile is metabolized in the small intestine by bacteria and eliminated in feces.

- ***Storage.*** In addition to glycogen, the liver is a prime storage site for certain vitamins (A, $B_{12}$, D, E, and K) and minerals (iron and copper), which are released from the liver when needed elsewhere in the body.
- ***Phagocytosis.*** The reticuloendothelial cells of the liver phagocytize aged blood cells and some bacteria.
- ***Activation of vitamin D.*** The skin, liver, and kidneys participate in synthesizing the active form of vitamin D.

## Gallbladder

The **gallbladder** (*gall-* = bile) is a pear-shaped sac that is located in a depression of the posterior surface of the liver. It is 7–10 cm (3–4 in.) long and typically hangs from the anterior inferior margin of the liver (Figure 23.15a).

Bile entering the cystic duct is temporarily stored in the gallbladder. Contraction of the smooth muscle fibers of the gallbladder wall ejects the contents of the gallbladder back into the cystic duct. The functions of the gallbladder are to store and concentrate the bile produced by the liver (up to tenfold) until it is needed in the small intestine. In the concentration process, water and ions are absorbed by the gallbladder mucosa.

### ✓ CHECKPOINT

23. What are the structures through which pancreatic juice might travel on its way to the duodenum?
24. What are the functions of the components of pancreatic juice?
25. How are the protein-digesting enzymes in pancreatic juice activated?
26. Which type of fluid is found in the pancreatic duct? The common bile duct? The hepatopancreatic ampulla?
27. What is the function of bile?
28. The first few hours after a meal, how does the chemical composition of blood change as it flows through the liver's sinusoids?

# 23.6 In the small intestine, chyme mixes with digestive juices from the small intestine, pancreas, and liver.

The **small intestine** begins at the pyloric sphincter of the stomach, coils through the central and inferior part of the abdominal cavity, and eventually opens into the large intestine. This long, convoluted tube is about 3 m (10 ft) long in a living person and about 6.5 m (21 ft) long in a cadaver due to the loss of smooth muscle tone after death. Most digestion and absorption of nutrients occur in the small intestine. Because of this, its structure is specially adapted for these functions. Its length alone provides a large surface area for digestion and absorption, and that area is further increased by circular folds, villi, and microvilli.

## Anatomy of the Small Intestine

The small intestine is divided into three regions. The **duodenum** (doo′-ō-DĒ-num *or* doo-OD-e-num), the first part of the small

**FIGURE 23.18 Anatomy of the small intestine.** Regions of the small intestine are the duodenum, jejunum, and ileum.

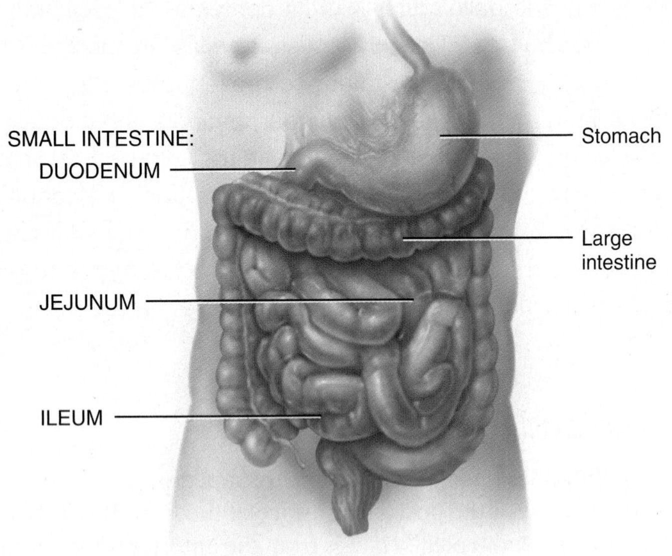


**FUNCTIONS OF THE SMALL INTESTINE**

1. Mixes chyme with digestive juices and brings food into contact with the mucosa for absorption; propels chyme through the small intestine.
2. Completes the digestion of carbohydrates, proteins, and lipids; begins and completes the digestion of nucleic acids.
3. Absorbs about 90 percent of nutrients and water that pass through the digestive system.

**Most digestion and absorption occur in the small intestine.**

intestine, is the shortest region and is retroperitoneal (Figure 23.18). It starts at the pyloric sphincter of the stomach and extends about 25 cm (10 in.) until it merges with the jejunum. *Duodenum* means "12"; it is so named because it is about as long as the width of 12 fingers. The next portion, the **jejunum** (je-JOO-num), is about 1 m (3 ft) long and extends to the ileum. *Jejunum* means "empty," which is how it is found at death. The final and longest region of the small intestine, the **ileum** (IL-ē-um = twisted), measures about 2 m (6 ft) and joins the large intestine at the *ileocecal sphincter* (see Figure 23.23).

## Histology of the Small Intestine

Even though the wall of the small intestine is composed of the same four layers that make up most of the GI tract, special modifications of both the mucosa and the submucosa facilitate the processes of digestion and absorption (Figure 23.19).

The epithelium of the mucosa consists of simple columnar epithelium containing several types of specialized cells (Figure 23.19c). **Absorptive cells** of the epithelium release enzymes that digest nutrients and absorb them from the small intestinal chyme. Also present are **goblet cells**, which secrete mucus. The small intestinal mucosa contains many deep crevices lined with glandular epithelium. Cells lining the crevices form the **intestinal glands**, which secrete intestinal juice (to be discussed shortly). Besides absorptive cells and goblet cells, the intestinal glands also contain Paneth cells and enteroendocrine cells. **Paneth cells** secrete lysozyme, a bactericidal enzyme, and are capable of phagocytosis. Paneth cells may have a role in regulating the microbial population in the intestines. Enteroendocrine cells found in the intestinal glands secrete the hormones secretin and cholecystokinin. The functions of these hormones will be described in Concept 23.8.

The lamina propria of the mucosa has an abundance of mucosa-associated lymphoid tissue (MALT) and **lymphatic**

**FIGURE 23.19 Histology of the small intestine.** Circular folds increase the surface area for digestion and absorption in the small intestine.

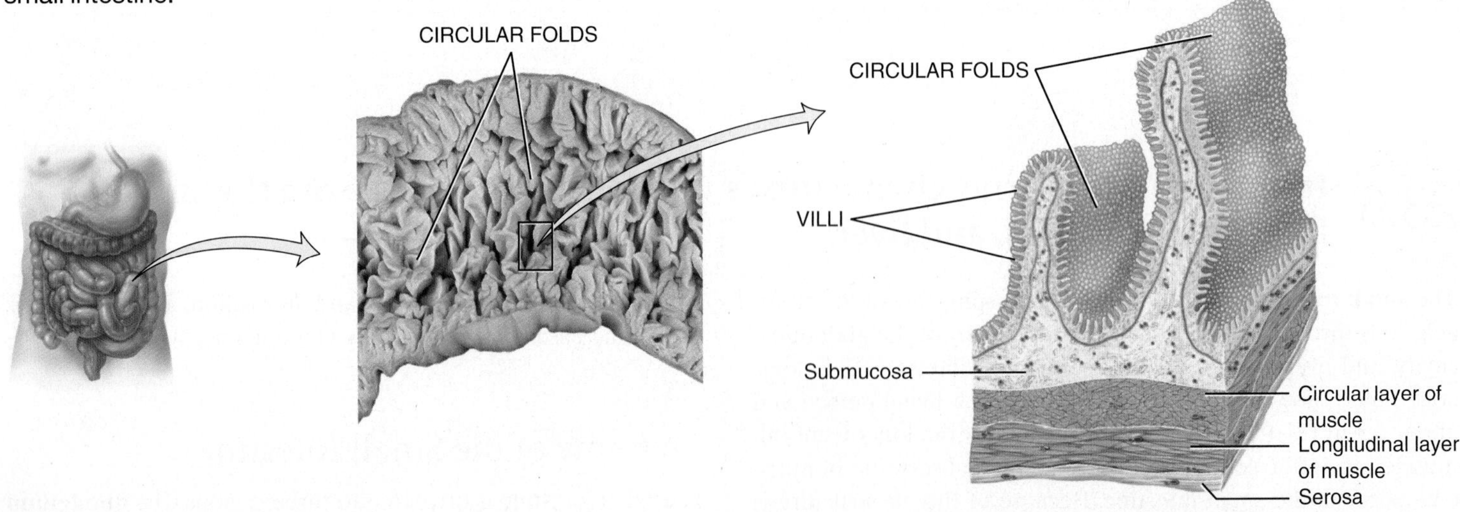


(a) Relationship of villi to circular folds

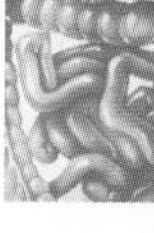

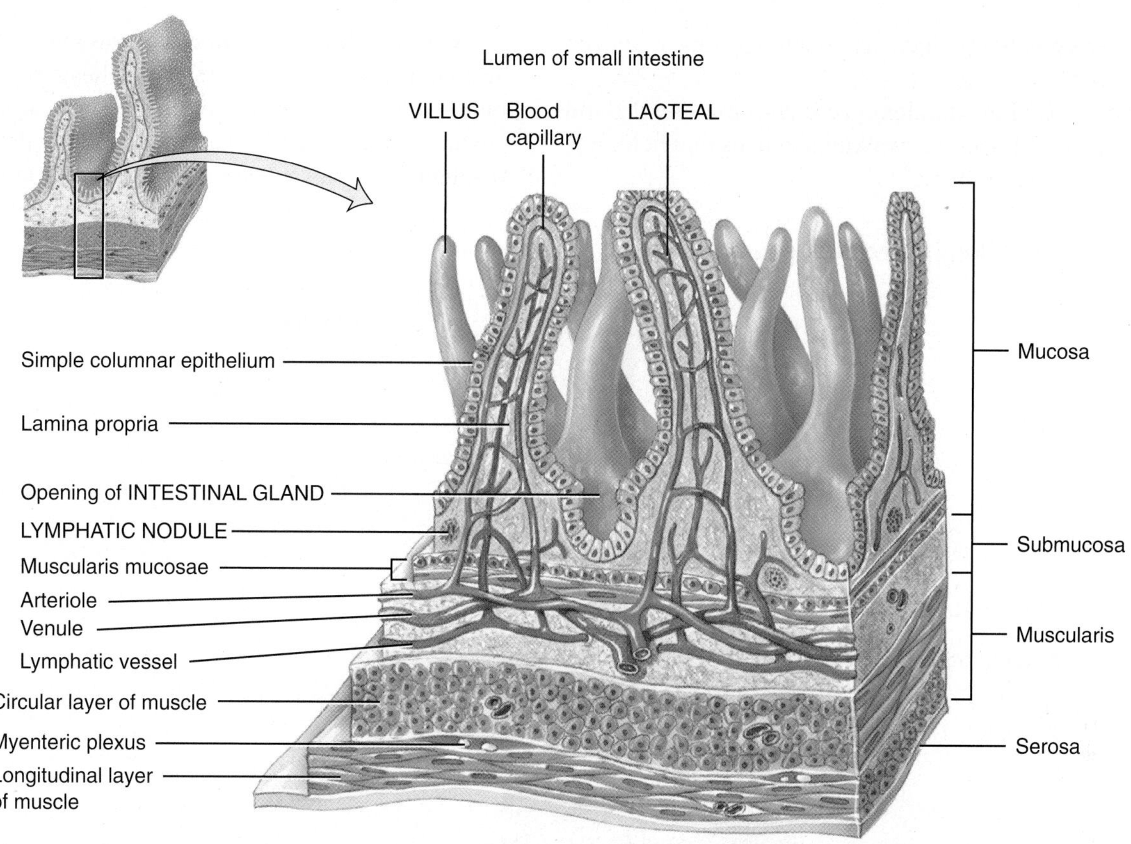


(b) Three-dimensional view of layers of small intestine showing villi

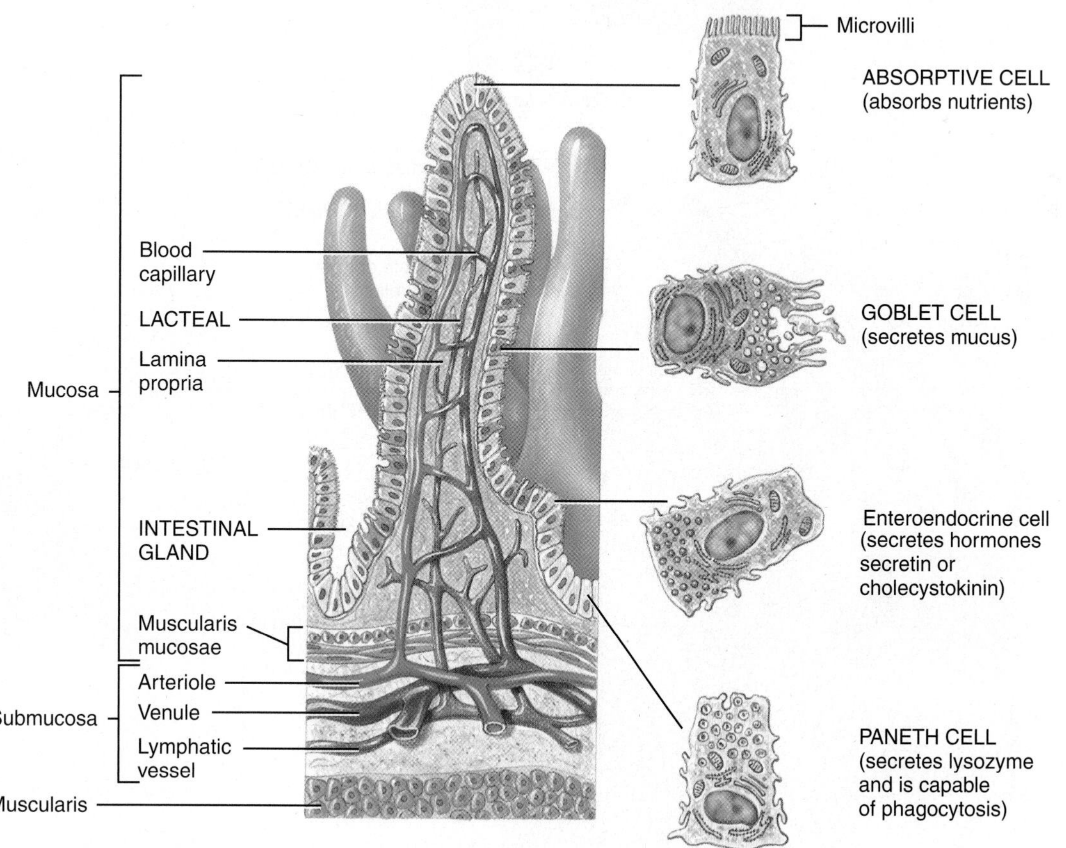


(c) Enlarged villus showing lacteal, capillaries, intestinal glands, and cell types

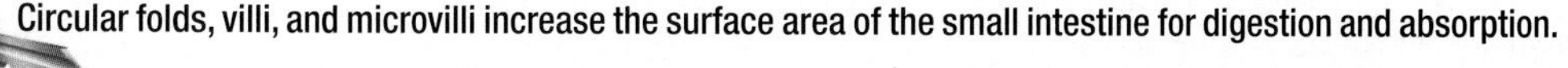

**Circular folds, villi, and microvilli increase the surface area of the small intestine for digestion and absorption.**

**nodules (Peyer's patches)**, which help defend against pathogens in food (Figure 23.20c).

The submucosa of the duodenum contains **duodenal glands** (Figure 23.20a), which secrete an alkaline mucus that helps neutralize gastric acid in the chyme.

**Circular folds (plicae circulares)** are folds of the mucosa and submucosa (Figure 23.19a). These permanent ridges enhance absorption by increasing surface area and causing the chyme to spiral, rather than move in a straight line, as it passes through the small intestine. Also present in the small intestine

FIGURE 23.20 **Histology of the duodenum and ileum.**

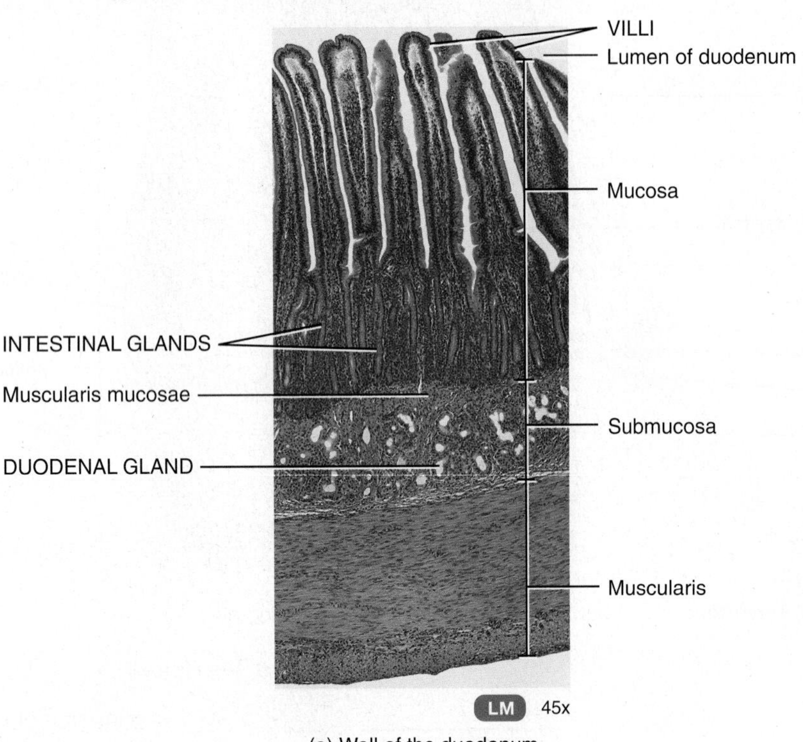


(a) Wall of the duodenum

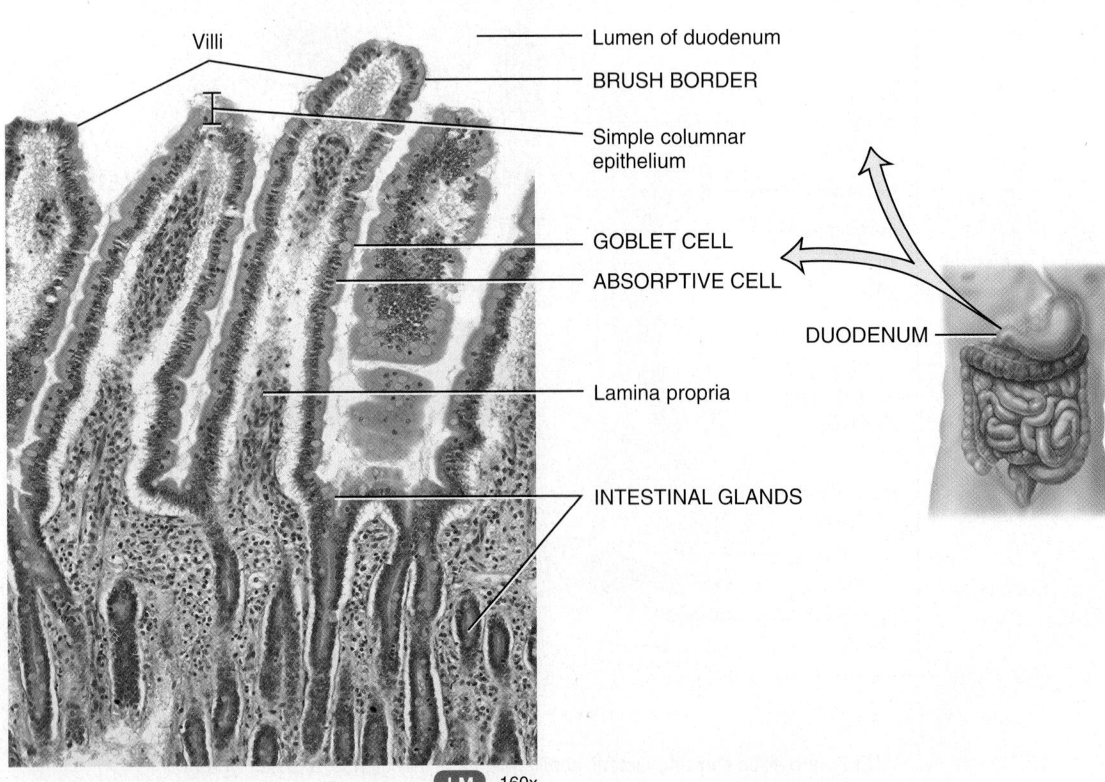


(b) Three villi from duodenum

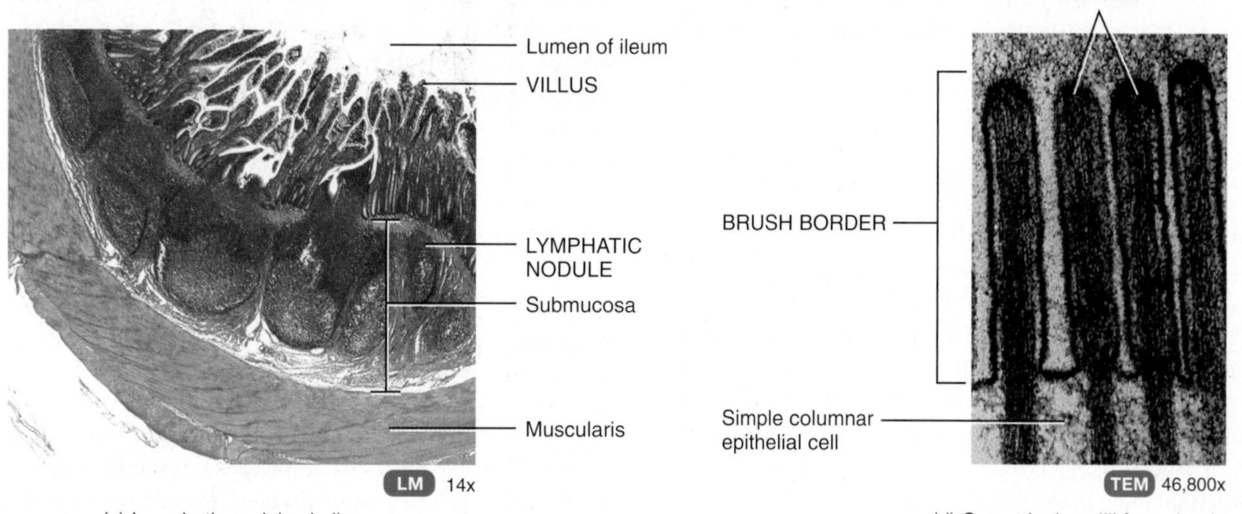


(c) Lymphatic nodules in ileum

(d) Several microvilli from duodenum

**Microvilli in the small intestine contain brush border enzymes that help digest nutrients.**

are **villi** (= tufts of hair; singular is *villus*), fingerlike projections of the mucosa that are 0.5–1 mm long. The large number of villi (20–40 per square millimeter) vastly increases the surface area of the epithelium available for absorption and digestion and gives the intestinal mucosa a velvety appearance. Each villus is covered by epithelium and has a core of lamina propria; embedded in the connective tissue of the lamina propria are an arteriole, a venule, a blood capillary network, and a **lacteal** (LAK-tē-al = milky), which is a lymphatic capillary. Nutrients absorbed by the epithelial cells covering the villus pass through the wall of a capillary or a lacteal to enter blood or lymph, respectively.

Besides circular folds and villi, the small intestine also has **microvilli** (mī-krō-VIL-ī; *micro-* = small), tiny projections of the apical (free) membrane of the absorptive cells. When viewed through a light microscope, the microvilli are too small to be seen individually; instead they form a fuzzy line, called the **brush border**, extending into the lumen of the small intestine (Figure 23.20b, d). There are an estimated 200 million microvilli per square millimeter of small intestine. Because the microvilli greatly increase the surface area of the plasma membrane, larger amounts of digested nutrients can diffuse into absorptive cells. The brush border also contains several brush border enzymes that have digestive functions (discussed shortly).

## Role of Intestinal Juice and Brush Border Enzymes

About 1–2 liters (1–2 qt) of **intestinal juice** are secreted each day. Intestinal juice contains water and mucus and is slightly alkaline (pH 7.6). The alkaline pH of intestinal juice is due to its high concentration of bicarbonate ions ($HCO_3^-$). Together, pancreatic and intestinal juices provide a liquid medium that aids the absorption of substances from chyme in the small intestine. The absorptive cells of the epithelium synthesize several digestive enzymes, called **brush border enzymes**, and insert them in the plasma membrane of the microvilli. Thus, some enzymatic digestion occurs at the surface of the absorptive cells that cover the villi, rather than in the lumen, as occurs in other parts of the GI tract. Among the brush border enzymes are enzymes that digest carbohydrates, proteins, and nucleotides.

## Mechanical Digestion in the Small Intestine

The two types of movements of the small intestine—segmentations and peristalsis—are governed mainly by the myenteric plexus. **Segmentations** are localized, mixing contractions that occur in portions of intestine distended by a large volume of chyme. Segmentations do not push the intestinal contents along the tract. Instead, they mix chyme with digestive juices and bring the particles of food into contact with the mucosa for absorption. A segmentation starts with the contractions of circular muscle fibers in a portion of the small intestine, an action that constricts the intestine into segments. Next, muscle fibers that encircle the middle of each segment also contract, dividing each segment again. Finally, the fibers that first contracted relax, and each small segment unites with an adjoining small segment so that large segments are formed again. As this sequence of events repeats, the chyme sloshes back and forth. Segmentations are similar to alternately squeezing the middle and then the ends of a capped tube of toothpaste.

After most of a meal has been absorbed, which lessens distension of the wall of the small intestine, segmentation stops and peristalsis begins. Each peristaltic wave begins in the lower portion of the stomach and slowly pushes chyme forward along the small intestine, with peristaltic migrations reaching the end of the ileum in 90 to 120 minutes. Then another wave of peristalsis begins in the stomach. Altogether, chyme remains in the small intestine for 3–5 hours.

## Chemical Digestion in the Small Intestine

In the mouth, salivary amylase converts starch (a polysaccharide) to disaccharides and short-chain glucose polymers called α-dextrins. In the stomach, pepsin converts proteins to peptides (small fragments of proteins), and lingual and gastric lipases convert some triglycerides into fatty acids, diglycerides, and monoglycerides. Thus, chyme entering the small intestine contains partially digested carbohydrates, proteins, and lipids. The completion of digestion is a collective effort of pancreatic juice, bile, and intestinal juice in the small intestine.

### Digestion of Carbohydrates

Even though the action of salivary amylase may continue in the stomach for a while, the acidic pH of the stomach destroys salivary amylase and ends its activity. Thus, only a few starches are broken down by the time chyme leaves the stomach. Those starches not already broken down are cleaved by pancreatic amylase, an enzyme in pancreatic juice that acts in the small intestine. Although pancreatic amylase acts on both glycogen and starches, it has no effect on another polysaccharide called cellulose, an indigestible plant fiber that is commonly referred to as "roughage" as it moves through the digestive system.

After amylase (either salivary or pancreatic) has split starch into smaller fragments, brush border enzymes act on the resulting glucose subunits, clipping off one glucose unit at a time. Ingested disaccharide molecules of sucrose, lactose, and maltose are not acted on until they reach the small intestine. Three brush border enzymes digest the disaccharides into monosaccharides. **Sucrase** breaks sucrose into a molecule of glucose and a molecule of fructose; **lactase** digests lactose into a molecule of glucose and a molecule of galactose; and **maltase** splits maltose into two molecules of glucose. Digestion of carbohydrates ends with the production of monosaccharides (glucose, fructose, and galactose), which the digestive system is able to absorb.

### Digestion of Proteins

Recall that protein digestion starts in the stomach, where proteins are fragmented into peptides by the action of **pepsin**. Enzymes in pancreatic juice—**trypsin**, **chymotrypsin**, and **carboxypeptidase**—continue to break down proteins into peptides, though their actions differ somewhat because each pancreatic juice enzyme splits peptide bonds between different amino acids. Trypsin and chymotrypsin cleave the peptide bond between a specific amino acid and its neighbor; carboxypeptidase splits off the amino acid at the carboxyl end of a peptide. Protein digestion is completed by **peptidases** in the brush border that break peptides into single amino acids, dipeptides, and tripeptides.

### Digestion of Lipids

The most abundant lipids in the diet are triglycerides, which consist of a molecule of glycerol bonded to three fatty acid molecules (see Figure 2.16). Enzymes that split triglycerides and phospholipids are called **lipases**. Recall that there are three types of lipases that can participate in lipid digestion: lingual lipase, gastric lipase, and pancreatic lipase. Although some lipid digestion occurs in the stomach through the action of lingual and gastric lipases, most occurs in the small intestine through the action of pancreatic lipase. Triglycerides are broken down by pancreatic lipase into fatty acids and monoglycerides. The liberated fatty acids can be either short-chain fatty acids (with fewer than 10–12 carbons) or long-chain fatty acids.

Before a large lipid globule containing triglycerides can be digested in the small intestine, it must first undergo **emulsification**—a process in which the large lipid globule is broken down into several small lipid globules. Recall that bile contains bile salts, the sodium salts and potassium salts of bile acids. Bile salts are **amphipathic** (am′-fē-PATH-ik), which means that each bile salt has a hydrophobic (nonpolar) region and a hydrophilic (polar) region. The hydrophobic regions of bile salts interact with the large lipid globule, while the hydrophilic regions of bile salts interact with the watery intestinal chyme. The amphipathic nature of bile salts allows them to break apart a large lipid globule into several small lipid globules, each about 1 μm in diameter. The small lipid globules formed from emulsification provide a large surface area that allows pancreatic lipase to function more effectively.

### Digestion of Nucleic Acids

Pancreatic juice contains two nucleases: ribonuclease, which digests RNA, and deoxyribonuclease, which digests DNA. The nucleotides that result from the action of the two nucleases are further digested by brush border enzymes called **nucleosidases** (noo′-klē-ō-SĪ-dās-ez) and **phosphatases** (FOS-fă-tās-ez) into pentoses, phosphates, and nitrogenous bases. These products are absorbed by active transport.

Table 23.4 summarizes the digestive enzymes and their functions for the digestive system.

**TABLE 23.4**

Digestive Enzymes

| ENZYME | SOURCE | SUBSTRATES | PRODUCTS |
|---|---|---|---|
| **SALIVA** | | | |
| **Salivary amylase** | Salivary glands | Starches (polysaccharides) | Maltose (disaccharide), maltotriose (trisaccharide), and $\alpha$-dextrins |
| **Lingual lipase** | Lingual glands in tongue | Triglycerides (fats and oils) and other lipids | Fatty acids and diglycerides |
| **GASTRIC JUICE** | | | |
| **Pepsin** (activated from pepsinogen by pepsin and hydrochloric acid) | Stomach chief cells | Proteins | Peptides |
| **Gastric lipase** | Stomach chief cells | Triglycerides (fats and oils) | Fatty acids and monoglycerides |
| **PANCREATIC JUICE** | | | |
| **Pancreatic amylase** | Pancreatic acinar cells | Starches (polysaccharides) | Maltose (disaccharide), maltotriose (trisaccharide), and $\alpha$-dextrins |
| **Trypsin** (activated from trypsinogen by enterokinase) | Pancreatic acinar cells | Proteins | Peptides |
| **Chymotrypsin** (activated from chymotrypsinogen by trypsin) | Pancreatic acinar cells | Proteins | Peptides |
| **Elastase** (activated from proelastase by trypsin) | Pancreatic acinar cells | Proteins | Peptides |
| **Carboxypeptidase** (activated from procarboxypeptidase by trypsin) | Pancreatic acinar cells | Amino acid at carboxyl end of peptides | Amino acids and peptides |
| **Pancreatic lipase** | Pancreatic acinar cells | Triglycerides (fats and oils) that have been emulsified by bile salts | Fatty acids and monoglycerides |
| **Nucleases** | | | |
| **Ribonuclease** | Pancreatic acinar cells | Ribonucleic acid | Nucleotides |
| **Deoxyribonuclease** | Pancreatic acinar cells | Deoxyribonucleic acid | Nucleotides |
| **BRUSH-BORDER ENZYMES** | | | |
| **$\alpha$-Dextrinase** | Small intestine | $\alpha$-Dextrins | Glucose |
| **Maltase** | Small intestine | Maltose | Glucose |
| **Sucrase** | Small intestine | Sucrose | Glucose and fructose |
| **Lactase** | Small intestine | Lactose | Glucose and galactose |
| **Enterokinase** | Small intestine | Trypsinogen | Trypsin |
| **Peptidases** | | | |
| **Aminopeptidase** | Small intestine | Amino acid at amino end of peptides | Amino acids and peptides |
| **Dipeptidase** | Small intestine | Dipeptides | Amino acids |
| **Nucleosidases and phosphatases** | Small intestine | Nucleotides | Nitrogenous bases, pentoses, and phosphates |

## Absorption in the Small Intestine

All of the chemical and mechanical phases of digestion from the mouth through the small intestine are directed toward changing food into forms that can pass through the absorptive cells of the epithelium and into the underlying blood and lymphatic vessels. These absorptive forms are monosaccharides from carbohydrates; single amino acids, dipeptides, and tripeptides from proteins; and fatty acids, glycerol, and monoglycerides from triglycerides. Passage of these digested nutrients from the GI tract into the blood or lymph constitutes absorption. About 90 percent of all absorption of nutrients occurs in the small intestine; the other 10 percent occurs in the stomach and large intestine. Any undigested or unabsorbed material left in the small intestine passes on to the large intestine. Absorption of materials occurs via diffusion, facilitated diffusion, osmosis, and active transport (see Concept 3.3).

### Absorption of Monosaccharides

All carbohydrates are absorbed as monosaccharides. The capacity of the small intestine to absorb monosaccharides is huge—an estimated 120 grams per hour. As a result, all dietary carbohydrates that are digested normally are absorbed, leaving only indigestible cellulose and fibers in the feces. Fructose, a monosaccharide found in fruits, is transported by *facilitated diffusion*; glucose and galactose are transported into epithelial cells by *secondary active transport* that is coupled to the active transport of $Na^+$ (Figure 23.21a). Monosaccharides then move across the basolateral surfaces of the epithelial cells by *facilitated diffusion* and enter the capillaries of the villi (see Figure 23.21a, b).

### Absorption of Amino Acids, Dipeptides, and Tripeptides

About half of the absorbed amino acids are present in food; the other half come from the body itself in the form of proteins in digestive juices and dead cells that slough off the mucosal surface! Amino acids, dipeptides, and tripeptides enter epithelial cells of the small intestine via active transport processes. Some amino acids enter epithelial cells of the villi via $Na^+$-dependent secondary active transport processes; other amino acids are actively transported by themselves (Figure 23.21a). Dipeptides and tripeptides are transported into epithelial cells by *secondary active transport* that is coupled to the active transport of $H^+$; the peptides then are digested into single amino acids inside the epithelial cells. Amino acids move out of the epithelial cells by *simple diffusion* and enter capillaries of the villi (Figure 23.21a, b). Both monosaccharides and amino acids are transported in the blood to the liver by way of the hepatic portal system. If not removed by hepatocytes, they enter the general circulation.

### Absorption of Lipids and Bile Salts

All dietary lipids are absorbed via *simple diffusion*. Following emulsification and digestion, triglycerides are broken down into monoglycerides and fatty acids, which can be either short-chain or long-chain fatty acids. Although small short-chain fatty acids (with fewer than 10-12 carbons) are hydrophobic, their very small size allows them to dissolve in the watery intestinal chyme, pass through the epithelial cells via simple diffusion, and follow the same route taken by monosaccharides and amino acids into a blood capillary of a villus (Figure 23.21a).

Large short-chain fatty acids, long-chain fatty acids, and monoglycerides are large hydrophobic molecules and, since they are not water-soluble, they have difficulty being suspended in the watery environment of the intestinal chyme. Besides their role in emulsification, bile salts also help to make these large short-chain fatty acids, long-chain. fatty acids, and monoglycerides more soluble. The bile salts in intestinal chyme surround them, forming tiny spheres called **micelles** (mī-SELZ = small morsels) (Figure 23.21a). Micelles are formed due to the amphipathic nature of bile salts: The hydrophobic regions of bile salts interact with the long-chain fatty acids and monoglycerides, and the hydrophilic regions of bile salts interact with the watery intestinal chyme. From micelles, these lipids diffuse into epithelial cells of the villi where they are recombined and packaged into **chylomicrons** (kī-lō-MĪ-krons), large spherical particles that are coated with proteins. Chylomicrons leave the epithelial cell via *exocytosis*. Because they are so large and bulky, chylomicrons cannot enter blood capillaries—the pores in the walls of blood capillaries are too small. Instead, chylomicrons enter lacteals, which have much larger pores than blood capillaries. From lacteals, chylomicrons are transported in lymph through lymphatic vessels to the thoracic duct and enter the blood at the left subclavian vein (see Figure 23.21b).

The hydrophilic protein coat that surrounds each chylomicron keeps the chylomicrons suspended in blood and prevents them from sticking to each other. As blood passes through capillaries in adipose tissue and the liver, chylomicrons are removed and their lipids are stored for future use. Within 10 minutes after absorption, about half of the chylomicrons have already been removed from the blood by the liver and adipose tissue by converting the triglycerides in chylomicrons into fatty acids and glycerol. Two or three hours after a meal, few chylomicrons remain in the blood.

There are several benefits to including some healthy fats in the diet. For example, fats delay gastric emptying and this helps

**FIGURE 23.21 Absorption of digested nutrients in the small intestine.** For simplicity, all digested foods are shown in the lumen of the small intestine, even though some nutrients are digested by brush border enzymes.

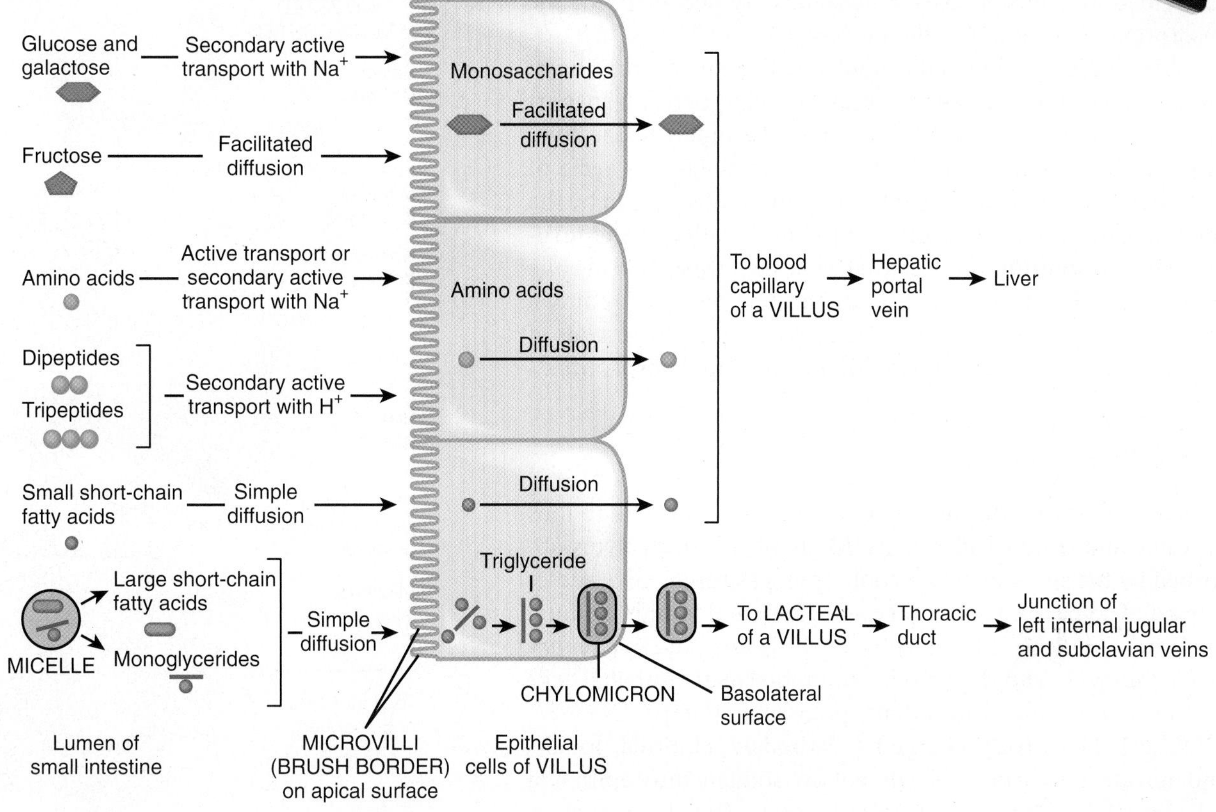


(a) Mechanisms for movement of nutrients through absorptive epithelial cells of villi

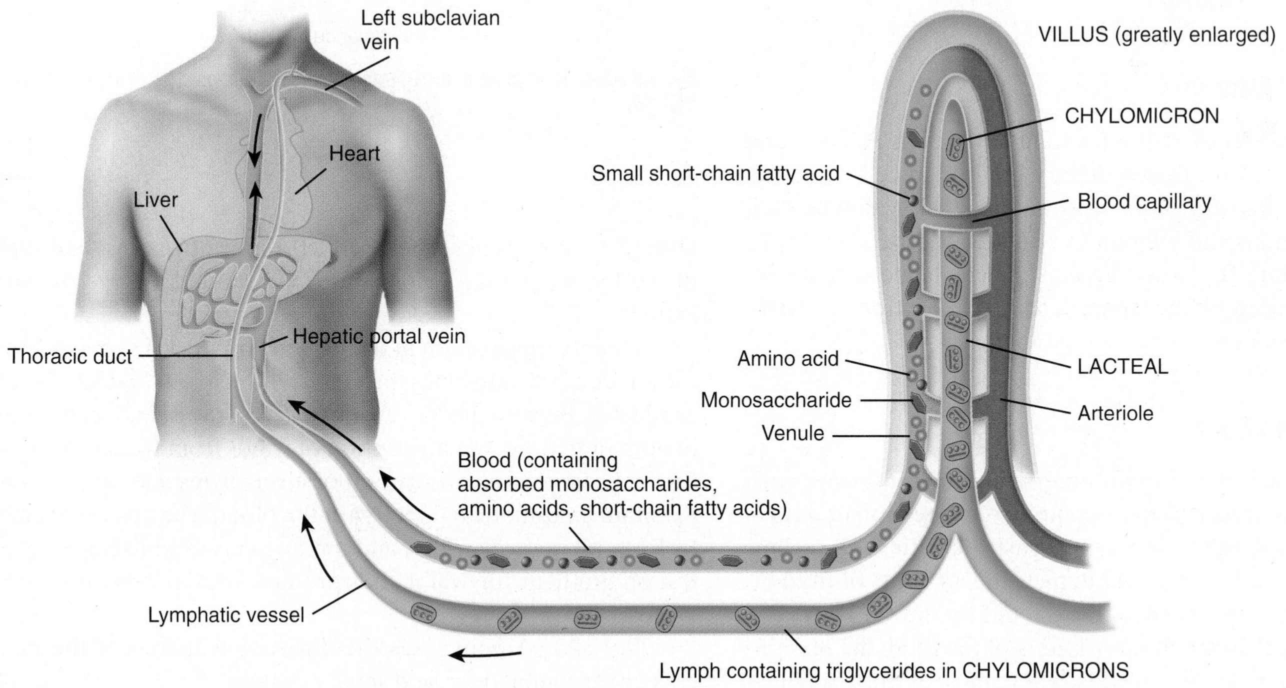


(b) Movement of absorbed nutrients into blood and lymph

**Long-chain fatty acids and monoglycerides are absorbed into lacteals; other products of digestion enter blood capillaries.**

a person to feel full. Fats also enhance the feeling of fullness by triggering the release of a hormone called cholecystokinin (described in Concept 23.8). Finally, fats are necessary for the absorption of fat-soluble vitamins (see Concept 23.10).

After participating in the emulsification and absorption of lipids, most of the bile salts are reabsorbed by active transport in the small intestine and returned by the blood to the liver through the hepatic portal system for recycling. This cycle of bile salt secretion by hepatocytes into bile, reabsorption by the small intestine, and resecretion into bile is called the **enterohepatic circulation** (en′-ter-ō-HEP-a-tik). Insufficient bile salts, due either to obstruction of the bile ducts or removal of the gallbladder, can result in the loss of up to 40 percent of dietary lipids in the feces due to diminished lipid absorption.

## Absorption of Electrolytes

Recall that electrolytes are compounds that separate into ions in water and conduct electricity. Many of the electrolytes absorbed by the small intestine come from gastrointestinal secretions, and some are part of ingested foods and liquids. Sodium ions move into epithelial cells by *diffusion* and *secondary active transport*, and then move out of intestinal epithelial cells by *active transport* via sodium–potassium pumps ($Na^+$–$K^+$ ATPases). Negatively charged bicarbonate, chloride, iodide, and nitrate ions can passively follow sodium into epithelial cells or can be actively transported. Other electrolytes such as calcium, iron, potassium, magnesium, and phosphate ions are absorbed by active transport.

## Absorption of Vitamins

As you have just learned, the fat-soluble vitamins A, D, E, and K are included with ingested dietary lipids in micelles and are absorbed by *simple diffusion*. Most water-soluble vitamins, such as most B vitamins and vitamin C, also are absorbed via simple diffusion. Vitamin $B_{12}$, however, must be combined with intrinsic factor produced by the stomach for its absorption via active transport in the ileum.

## Absorption of Water

The total volume of fluid that enters the small intestine each day—about 9.3 liters (9.8 qt)—comes from ingestion of liquids (about 2.3 liters) and from various gastrointestinal secretions (about 7.0 liters). Figure 23.22 depicts the amounts of fluid ingested, secreted, absorbed, and excreted by the GI tract. The small intestine absorbs about 8.3 liters of the fluid; the remainder passes into the large intestine, where most of the rest of it—about 0.9 liter—is absorbed. Only 0.1 liter of water is excreted in the feces each day. Most water is excreted via the urinary system.

**FIGURE 23.22 Daily volumes of fluid ingested, secreted, absorbed, and excreted from the GI tract.**

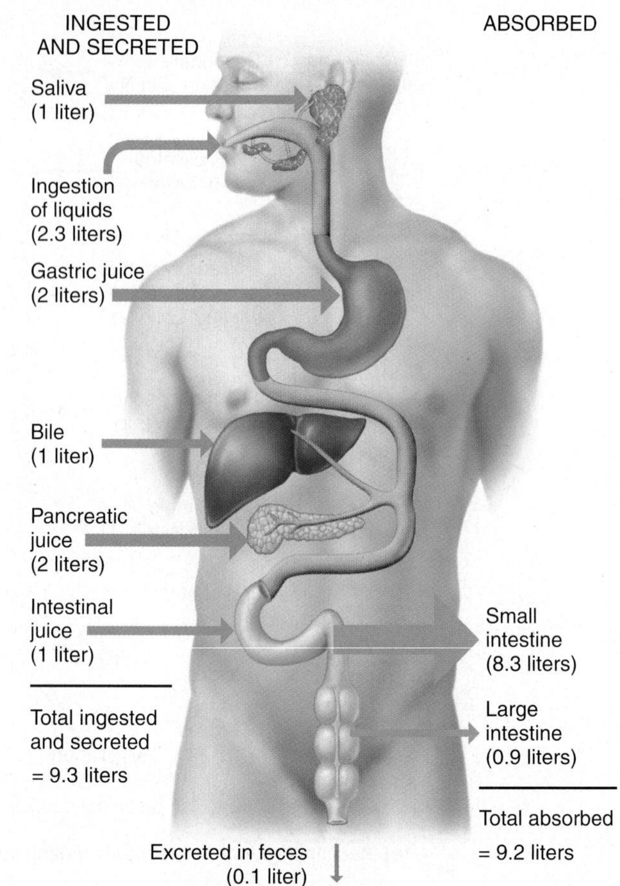


**All water absorption in the GI tract occurs via osmosis.**

All water absorption in the GI tract occurs via *osmosis* from the lumen of the intestines through epithelial cells and into blood capillaries. Because water can move across the intestinal mucosa in both directions, the absorption of water from the small intestine depends on the absorption of electrolytes and nutrients to maintain an osmotic balance with the blood. The absorbed electrolytes, monosaccharides, and amino acids establish a concentration gradient for water that promotes water absorption via osmosis.

Table 23.5 summarizes the digestive activities of the pancreas, liver, gallbladder, and small intestine.

**TABLE 23.5**

Digestive Activities in the Pancreas, Liver, Gallbladder, and Small Intestine

| STRUCTURE | ACTIVITY |
|---|---|
| **Pancreas** | Delivers pancreatic juice into duodenum via pancreatic duct (see Table 23.4 for pancreatic enzymes and their functions) |
| **Liver** | Produces bile (bile salts) necessary for emulsification and absorption of lipids |
| **Gallbladder** | Stores, concentrates, and delivers bile into duodenum via common bile duct |
| **Small intestine** | Major site of digestion and absorption of nutrients and water in gastrointestinal tract |
| **Mucosa/submucosa** | |
| **Intestinal glands** | Secrete intestinal juice to assist absorption |
| **Duodenal glands** | Secrete alkaline fluid to buffer stomach acids, and mucus for protection and lubrication |
| **Microvilli** | Microsopic, membrane-covered projections of epithelial cells that contain brush border enzymes (listed in Table 23.4) and that increase surface area for digestion and absorption |
| **Villi** | Fingerlike projections of mucosa that are sites of absorption of digested food and increase surface area for digestion and absorption |
| **Circular folds** | Folds of mucosa and submucosa that increase surface area for digestion and absorption |
| **Muscularis** | |
| **Segmentation** | Localized, alternating contractions of circular smooth muscle fibers of small intestine that mix chyme with digestive juices and bring food into contact with mucosa for absorption |
| **Peristalsis** | Waves of contraction and relaxation of circular and longitudinal smooth muscle fibers passing length of small intestine; moves chyme toward ileocecal sphincter |

✓ **CHECKPOINT**

**29.** What are the regions of the small intestine in order from stomach to large intestine? Which region is the longest?

**30.** What purpose do the circular folds of the small intestine serve?

**31.** What is the function of Paneth cells? Of the fluid secreted by duodenal glands?

**32.** What are the functions of segmentation and peristalsis in the small intestine?

**33.** Explain the functions of pancreatic amylase, brush border enzymes, pancreatic lipase, and deoxyribonuclease.

**34.** Why are bile salts needed for lipid digestion?

**35.** How are the end products of carbohydrate, protein, and lipid digestion absorbed?

**36.** Why are triglycerides circulated through the body in chylomicrons?

**37.** By which routes do absorbed nutrients reach the liver?

## 23.7 In the large intestine, the final secretion and absorption of nutrients occur as chyme moves toward the rectum.

The **large intestine** is the terminal portion of the GI tract. The overall functions of the large intestine are the completion of absorption, the production of certain vitamins, the formation of feces, and the expulsion of feces from the body.

### Anatomy of the Large Intestine

The large intestine is about 1.5 m (5 ft) long and extends from the ileum to the anus. Structurally, the four major regions of the large intestine are the cecum, colon, rectum, and anal canal (Figure 23.23a).

The opening from the ileum into the large intestine is guarded by a fold of mucous membrane called the **ileocecal sphincter** (il′-ē-ō-SĒ-kal), which controls passage of materials from the small intestine into the large intestine. Hanging inferior to the ileocecal sphincter is the **cecum**, a small pouch about 6 cm (2.4 in.) long. Attached to the cecum is a twisted, coiled tube, measuring about 8 cm (3 in.) in length, called the **vermiform appendix** or simply **appendix** (*vermiform* = worm-shaped; *appendix* = appendage).

The open end of the cecum merges with a long tube called the **colon** (= food passage), which is divided into ascending, transverse, descending, and sigmoid portions. Both the ascending and descending colon are retroperitoneal; the transverse and sigmoid colon are not. True to its name, the **ascending colon** ascends on the right side of the abdomen, reaches the inferior surface of the liver, and turns abruptly to the left to form the **right colic (hepatic) flexure**. The colon continues across the abdomen to the left side as the **transverse colon**. It curves beneath the inferior end of the spleen on the left side as the **left colic (splenic) flexure** and passes inferiorly to the level of the iliac crest as the **descending colon**. The **sigmoid colon** (*sigm-* = S-shaped) begins near the left iliac crest, projects medially to the midline, and terminates as the rectum at about the level of the third sacral vertebra.

The **rectum**, the last 20 cm (8 in.) of the GI tract, lies anterior to the sacrum and coccyx. The terminal 2–3 cm (1 in.) of the rectum is called the **anal canal** (Figure 23.23b). The mucous membrane of the anal canal is arranged in longitudinal folds

**FIGURE 23.23 Anatomy of the large intestine.**

**FUNCTIONS OF THE LARGE INTESTINE**

1. Drives the contents of the colon into the rectum.
2. Converts proteins to amino acids, breaks down amino acids, and produces some B vitamins and vitamin K.
3. Absorbs some water, ions, and vitamins.
4. Forms feces.
5. Discharges feces from the rectum.

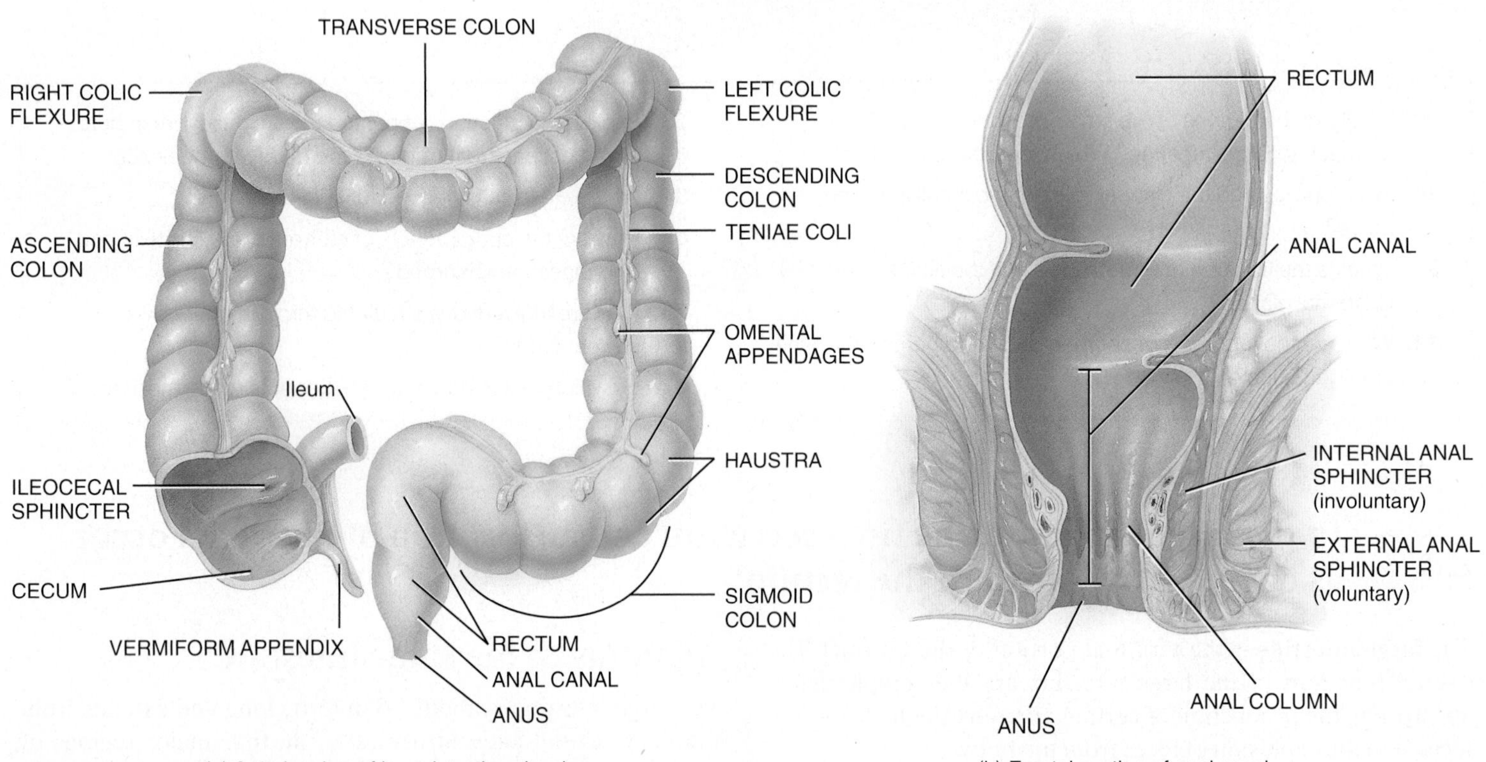


(a) Anterior view of large intestine showing major regions

(b) Frontal section of anal canal

**The regions of the large intestine are the cecum, colon, rectum, and anal canal.**

## CLINICAL CONNECTION | *Appendicitis*

Inflammation of the appendix, termed **appendicitis**, is preceded by obstruction of the lumen of the appendix by chyme, inflammation, a foreign body, a carcinoma of the cecum, stenosis, or kinking of the organ. It is characterized by high fever, elevated white blood cell count, and a neutrophil count higher than 75 percent. The infection that follows may result in edema and ischemia and may progress to gangrene and perforation within 24 hours. Typically, appendicitis begins with referred pain in the umbilical region of the abdomen, followed by anorexia (loss of appetite for food), nausea, and vomiting. After several hours the pain localizes in the right lower quadrant; is continuous, dull or severe; and is intensified by coughing, sneezing, or body movements. Early appendectomy (removal of the appendix) is recommended because it is safer to operate than to risk rupture, peritonitis, and gangrene. Although it required major abdominal surgery in the past, today appendectomies are usually performed laparoscopically.

called **anal columns**. The opening of the anal canal to the exterior, called the **anus**, is guarded by an **internal anal sphincter** of smooth muscle (involuntary) and an **external anal sphincter** of skeletal muscle (voluntary). Normally these sphincters keep the anus closed except during the elimination of feces.

## Histology of the Large Intestine

The wall of the large intestine contains the typical four layers found in the rest of the GI tract: mucosa, submucosa, muscularis, and serosa (Figure 23.24a). The epithelium of the mucosa contains mostly absorptive and goblet cells (Figure 23.24b, d). The absorptive cells function primarily in water absorption; the goblet cells secrete mucus that lubricates the passage of the colonic contents. Both absorptive and goblet cells are located in long, tubular intestinal glands that extend the full thickness of the mucosa. Solitary lymphatic nodules are also found in the lamina propria of the mucosa and may extend through the muscularis mucosae into the submucosa. Compared to the small intestine, the mucosa of the large intestine does not have as many structural adaptations that increase surface area. There are no

**FIGURE 23.24 Histology of the large intestine.**

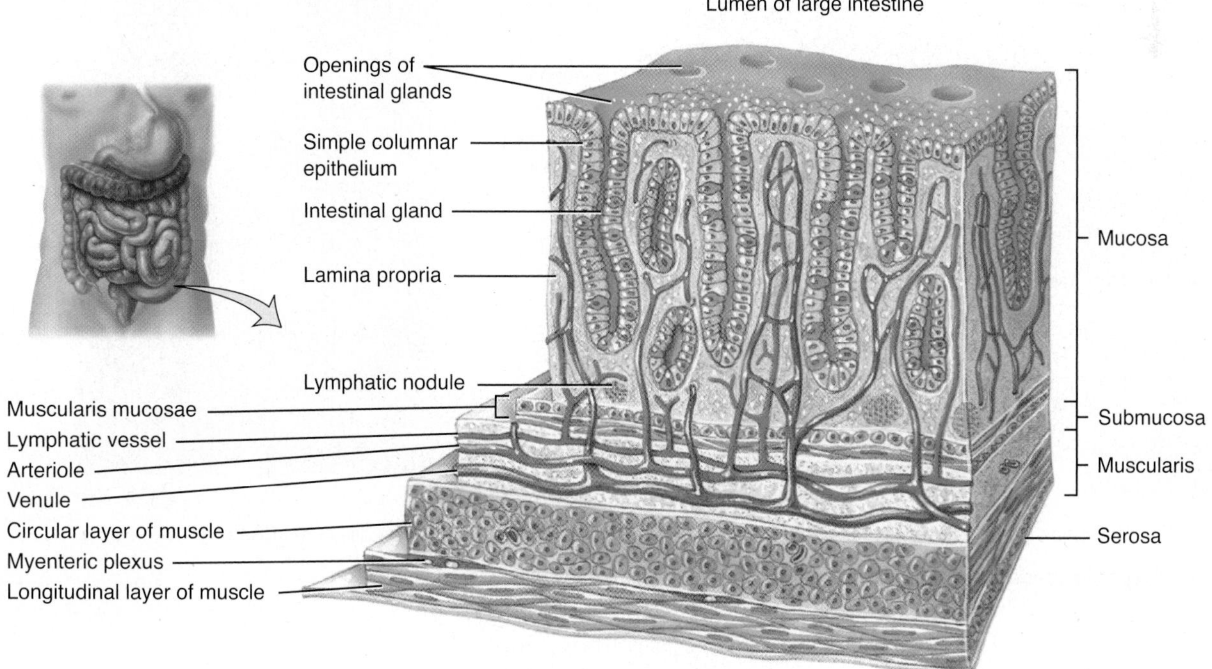


(a) Three-dimensional view of layers of large intestine

(continues)

**FIGURE 23.24 (continued)**

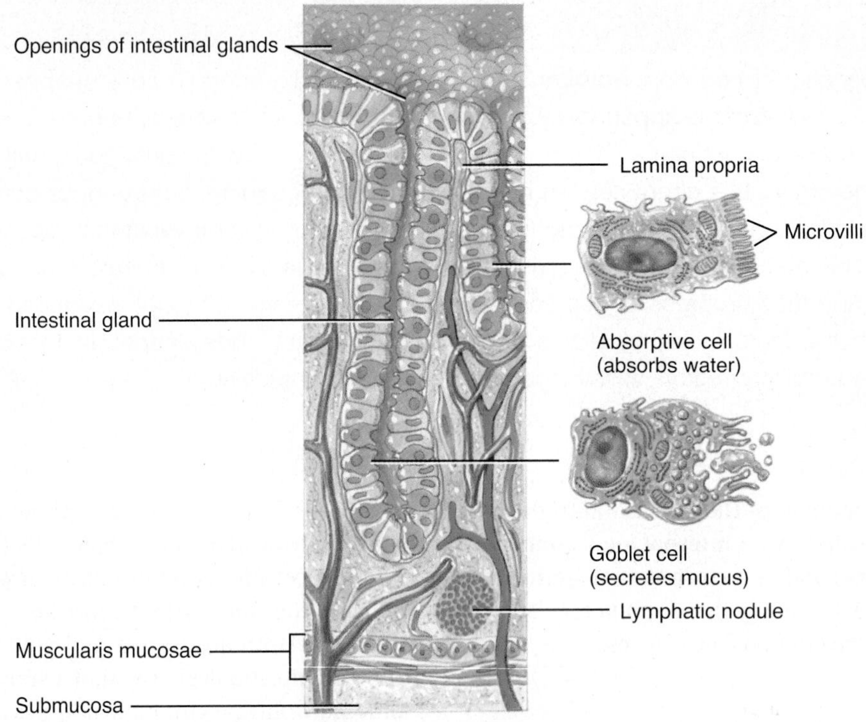


(b) Sectional view of intestinal glands and cell types

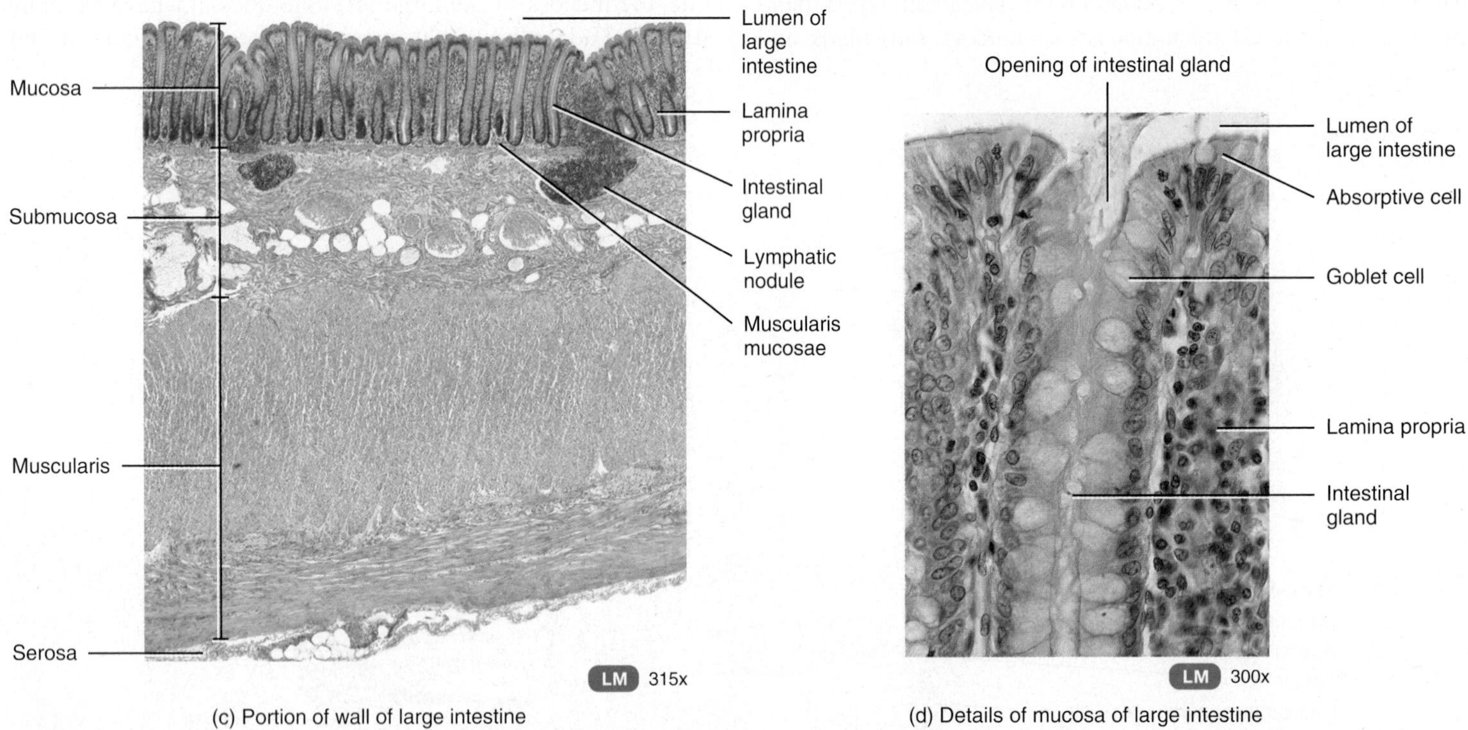


(c) Portion of wall of large intestine

(d) Details of mucosa of large intestine

**Intestinal glands containing absorptive cells and goblet cells extend the full thickness of the mucosa.**

circular folds or villi; however, microvilli are present on the epithelial cells. Consequently, much more absorption occurs in the small intestine than in the large intestine.

The muscularis consists of an external layer of longitudinal smooth muscle and an internal layer of circular smooth muscle. Unlike other parts of the GI tract, portions of the longitudinal muscles are thickened, forming three conspicuous bands called the **teniae coli** (TĒ-nē-ē KŌ-lī; *teniae* = flat bands) that run most of the length of the large intestine (Figure 23.23a). Tonic contractions of the bands gather the colon into a series of pouches called **haustra** (HAWS-tra = shaped like pouches; singular is *haustrum*), which give the colon a puckered appearance.

The serosa of the large intestine is part of the visceral peritoneum. Small pouches of visceral peritoneum filled with fat are attached to teniae coli and are called **omental appendages**.

## Mechanical Digestion in the Large Intestine

The passage of chyme from the ileum into the cecum is regulated by the action of the ileocecal sphincter. Normally, the sphincter remains partially closed so that the passage of chyme into the cecum occurs slowly. Immediately after a meal, a **gastroileal reflex** (gas′-trō-IL-ē-al) intensifies peristalsis in the ileum and forces any remaining chyme into the cecum. At the same time, the hormone gastrin relaxes the sphincter. Whenever the cecum is distended, the degree of contraction of the ileocecal sphincter intensifies.

Movements of the colon begin when substances pass the ileocecal sphincter. As food passes through the ileocecal sphincter, it fills the cecum and accumulates in the ascending colon. One movement characteristic of the large intestine is **haustral churning**. In this process, the haustra remain relaxed and become distended while they fill up. When the distension reaches a certain point, the walls contract and squeeze the contents into the next haustrum. Peristalsis also occurs, although at a slower rate than in more proximal portions of the GI tract. A final type of movement is **mass peristalsis**, a strong peristaltic wave that begins at about the middle of the transverse colon and quickly drives the contents of the colon into the rectum. Because food in the stomach initiates this **gastrocolic reflex** in the colon, mass peristalsis usually takes place three or four times a day, during or immediately after a meal.

## Chemical Digestion in the Large Intestine

The final stage of digestion occurs in the colon through the activity of bacteria that inhabit the lumen. Mucus is secreted by the glands of the large intestine, but no enzymes are secreted. Chyme is prepared for elimination by the action of bacteria, which ferment any remaining carbohydrates and release hydrogen, carbon dioxide, and methane gases. These gases contribute to flatus (gas) in the colon, termed *flatulence* when it is excessive. Bacteria also break down the remaining proteins to amino acids and decompose bilirubin to simpler pigments, including stercobilin, which gives feces their brown color. Bacterial products that are absorbed in the colon include several vitamins needed for normal metabolism, among them some B vitamins and vitamin K.

## Absorption and Feces Formation in the Large Intestine

Although 90 percent of all water absorption occurs in the small intestine, the large intestine absorbs enough to make it an

### CLINICAL CONNECTION | *Occult Blood*

The term **occult blood** refers to blood that is hidden; it is not detectable by the human eye. The main diagnostic value of occult blood testing is to screen for colorectal cancer. Two substances often examined for occult blood are feces and urine. Several types of products are available for at-home testing for hidden blood in feces. The tests are based on color changes when reagents are added to feces. The presence of occult blood in urine may be detected at home by using dip-and-read reagent strips.

important organ in maintaining the body's water balance. Of the 0.5–1.0 liters of water that enter the large intestine, all but about 100–200 mL is normally absorbed via osmosis. The large intestine also absorbs ions, including sodium and chloride, and some vitamins.

By the time chyme has remained in the large intestine 3–10 hours, it has become solid or semisolid because of water absorption and is now called **feces**. Chemically, feces consist of water, inorganic salts, sloughed-off epithelial cells from the mucosa of the GI tract, bacteria, products of bacterial decomposition, unabsorbed digested materials, and indigestible parts of food.

## The Defecation Reflex

Mass peristaltic movements push fecal material from the sigmoid colon into the rectum. The resulting distension of the rectal wall stimulates stretch receptors, which initiates a **defecation reflex** that empties the rectum. In response to distension of the rectal wall, receptors send sensory impulses to the sacral spinal cord. Motor impulses from the spinal cord travel along parasympathetic nerves back to the large intestine. The resulting contraction of the longitudinal rectal muscles shortens the rectum, thereby increasing the pressure within it. This pressure, along with voluntary contractions of the diaphragm and abdominal muscles, plus parasympathetic stimulation, opens the internal anal sphincter.

The external anal sphincter is voluntarily controlled. If it is voluntarily relaxed, the elimination of feces from the rectum occurs through the anus; if it is voluntarily constricted, defecation can be postponed. Voluntary contractions of the diaphragm and abdominal muscles aid defecation by increasing the pressure within the abdomen, which pushes the walls of the sigmoid colon and rectum inward. If defecation does not occur, the feces back up into the sigmoid colon until the next wave of mass peristalsis stimulates the stretch receptors, again creating the urge to defecate. In infants, the defecation reflex causes automatic emptying of the rectum because voluntary control of the external anal sphincter has not yet developed.

The amount of bowel movements that a person has over a given period of time depends on various factors such as diet, health, and stress. The normal range of bowel activity varies from two or three bowel movements per day to three or four bowel movements per week.

**Diarrhea** (dī-a-RĒ-a; *dia-* = through; *rrhea* = flow) is an increase in the frequency, volume, and fluid content of the feces caused by *increased* motility of and *decreased* absorption by the intestines. When chyme passes too quickly through the small intestine and feces pass too quickly through the large intestine, there is not enough time for absorption. Frequent diarrhea can result in dehydration and electrolyte imbalances. Excessive motility may be caused by lactose intolerance, stress, and microbes that irritate the gastrointestinal mucosa.

**Constipation** (kon-sti-PĀ-shun; *con-* = together; *stip-* = to press) refers to infrequent or difficult defecation caused by decreased motility of the intestines. When feces remain in the colon for prolonged periods, excessive water absorption occurs, and the feces become dry and hard. Constipation may be caused by poor habits (delaying defecation), insufficient fiber in the diet, inadequate fluid intake, lack of exercise, emotional stress, and certain drugs.

Table 23.6 summarizes the digestive activities in the large intestine, and Table 23.7 summarizes the functions of all digestive system organs.

**TABLE 23.6**

Digestive Activities in the Large Intestine

| STRUCTURE | ACTIVITY | FUNCTION(S) |
|---|---|---|
| **Lumen** | Bacterial activity | Breaks down undigested carbohydrates, proteins, and amino acids into products that can be expelled in feces or absorbed and detoxified by liver; synthesizes certain B vitamins and vitamin K |
| **Mucosa** | Secretes mucus | Lubricates colon; protects mucosa |
| | Absorption | Water absorption solidifies feces and contributes to body's water balance; solutes absorbed include ions and some vitamins |
| **Muscularis** | Haustral churning | Moves contents from haustrum to haustrum by muscular contractions |
| | Peristalsis | Moves contents along length of colon by contractions of circular and longitudinal muscles |
| | Mass peristalsis | Forces contents into sigmoid colon and rectum |
| | Defecation reflex | Eliminates feces by contractions in sigmoid colon and rectum |

**TABLE 23.7**

Organs of the Digestive System and Their Functions

| ORGAN | FUNCTIONS |
|---|---|
| **Mouth** | See other listings in this table for the functions of the tongue, salivary glands, and teeth, all of which are in the mouth; additionally, lips and cheeks keep food between the teeth during mastication, and buccal glands lining the mouth produce saliva |
| **Tongue** | Maneuvers food for mastication, shapes food into a bolus, maneuvers food for deglutition, detects taste and touch sensations, and initiates digestion of triglycerides |
| **Salivary glands** | Produce saliva, which softens, moistens, and dissolves foods; cleanses mouth and teeth; and initiates the digestion of starch |
| **Teeth** | Cut, tear, and pulverize food to reduce solids to smaller particles for swallowing |
| **Pharynx** | Receives a bolus from the oral cavity and passes it into the esophagus |
| **Esophagus** | Receives a bolus from the pharynx and moves it into the stomach; requires relaxation of the upper esophageal sphincter and secretion of mucus |
| **Stomach** | Mixing waves macerate food, mix it with secretions of gastric glands (gastric juice), and reduce food to chyme; gastric juice activates pepsin and kills many microbes in food; intrinsic factor aids absorption of vitamin $B_{12}$; stomach serves as a reservoir for food before release into small intestine |
| **Pancreas** | Pancreatic juice buffers acidic gastric juice in chyme (creating the proper pH for digestion in the small intestine), stops the action of pepsin from the stomach, and contains enzymes that digest carbohydrates, proteins, triglycerides, and nucleic acids |
| **Liver** | Produces bile, which is needed for the emulsification and absorption of lipids in the small intestine |
| **Gallbladder** | Stores and concentrates bile and releases it into the small intestine |
| **Small intestine** | Segmentations mix chyme with digestive juices; peristalsis propels chyme toward the ileocecal sphincter; digestive secretions from the small intestine, pancreas, and liver complete digestion of carbohydrates, proteins, lipids, and nucleic acids; circular folds, villi, and microvilli increase surface area for absorption; site where about 90% of nutrients and water are absorbed |
| **Large intestine** | Haustral churning, peristalsis, and mass peristalsis drive contents of colon into rectum; bacteria produce some B vitamins and vitamin K; absorption of some water, ions, and vitamins; defecation |

✓ **CHECKPOINT**

**38.** What are the major regions of the large intestine?

**39.** What is the function of the goblet cells in the large intestine?

**40.** How does mechanical digestion occur in the large intestine?

**41.** Which primary activity occurs in the large intestine to change its contents into feces?

# 23.8 Digestive activities occur in three overlapping phases: cephalic, gastric, and intestinal.

Digestive activities occur in three overlapping phases: the cephalic phase, the gastric phase, and the intestinal phase.

## Cephalic Phase

During the **cephalic phase** of digestion, the smell, sight, thought, or initial taste of food activates neural centers in the cerebral cortex, hypothalamus, and brain stem. The brain stem then activates the facial (VII), glossopharyngeal (IX), and vagus (X) nerves. The facial and glossopharyngeal nerves stimulate the salivary glands to secrete saliva, while the vagus (X) nerve stimulates the gastric glands to secrete gastric juice. The purpose of the cephalic phase of digestion is to prepare the mouth and stomach for food that is about to be eaten.

## Gastric Phase

Once food reaches the stomach, the **gastric phase** of digestion begins. Neural and hormonal mechanisms regulate the gastric phase of digestion to promote gastric secretion and gastric motility.

### Neural Regulation

Food of any kind distends the stomach and stimulates stretch receptors in its walls. Chemoreceptors in the stomach monitor the pH of the stomach chyme. When the stomach walls are distended or pH increases because proteins have entered the stomach and buffered some of the stomach acid, the stretch receptors and chemoreceptors are activated, and a neural negative feedback loop is set in motion (Figure 23.25). From the stretch receptors and chemoreceptors, impulses propagate to the submucosal plexus, where they activate parasympathetic and enteric neurons. The resulting impulses cause waves of peristalsis and stimulate the flow of gastric juice from gastric glands. The peristaltic waves mix the food with gastric juice; when the waves become strong enough, a small quantity of chyme is passed into the duodenum. As chyme passes from the stomach to the duodenum, the pH of the stomach chyme decreases (becomes more acidic) and the distension of the stomach walls lessens, suppressing secretion of gastric juice.

### Hormonal Regulation

Gastric secretion during the gastric phase is also regulated by the hormone **gastrin**. Gastrin is released from gastric glands in response to several stimuli including distension of the stomach by chyme, partially digested proteins in chyme, or a high pH of chyme due to the presence of certain foods in the stomach. Once gastrin is released, it enters the bloodstream, makes a round-trip through the body, and finally reaches its target organs in the digestive system. Gastrin stimulates gastric glands to secrete large amounts of gastric juice. It also strengthens the contraction of the lower esophageal sphincter to prevent reflux of acid chyme into the esophagus, increases motility of the stomach, and relaxes the pyloric sphincter, which promotes gastric emptying. Gastrin secretion is inhibited when the pH of gastric juice drops below 2.0 and is stimulated when the pH rises. This negative feedback mechanism helps provide an optimal low pH for the functioning of pepsin, the killing of microbes, and the breakdown of proteins in the stomach.

## Intestinal Phase

The **intestinal phase** of digestion begins once food enters the small intestine. In contrast to reflexes initiated during the cephalic and gastric phases, which stimulate stomach secretory activity and motility, those occurring during the intestinal phase have inhibitory effects that slow the exit of chyme from the stomach. This prevents the duodenum from being overloaded with more chyme than it can handle. In addition, responses during the intestinal phase promote the continued digestion

**FIGURE 23.25 Neural negative feedback regulation of the pH of gastric juice and gastric motility during the gastric phase of digestion.**

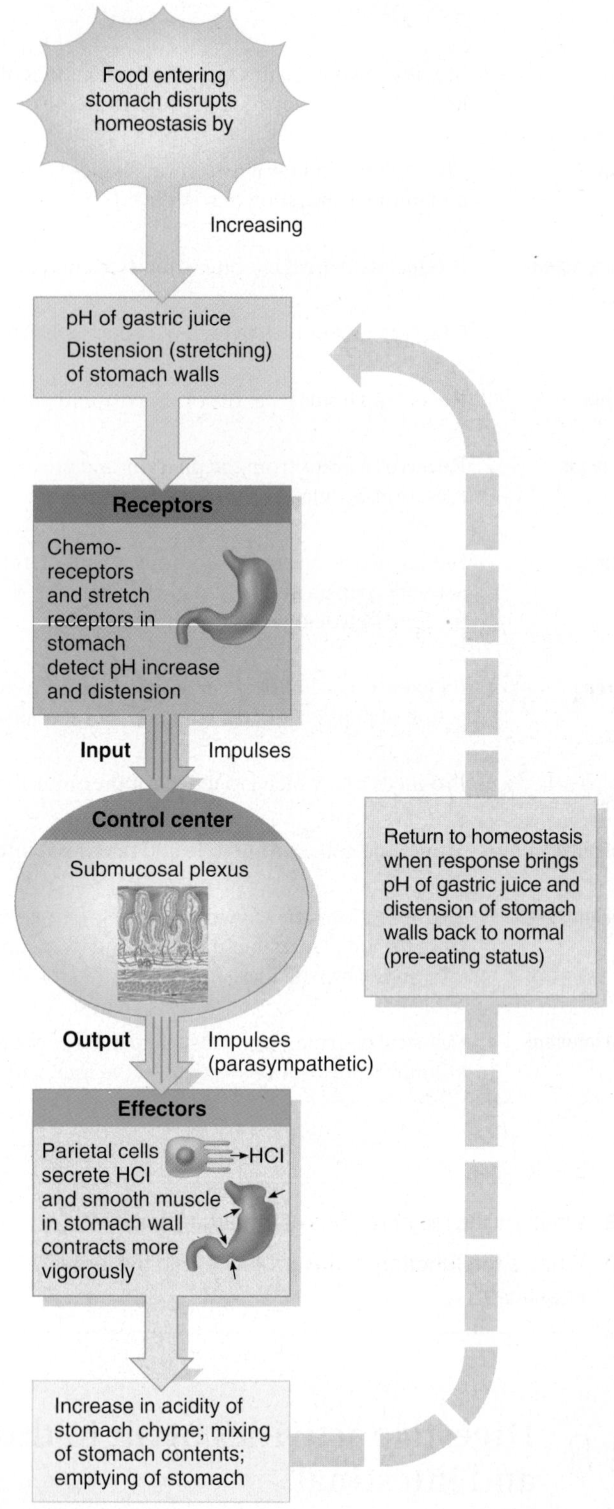


**Food entering the stomach stimulates secretion of gastric juice and causes vigorous waves of peristalsis.**

of foods that have reached the small intestine. These activities of the intestinal phase of digestion are regulated by neural and hormonal mechanisms.

### Neural Regulation

Distension of the duodenum by the presence of chyme causes stretch receptors in the duodenal wall to send impulses to the medulla oblongata, where they inhibit parasympathetic stimulation and stimulate sympathetic stimulation of the stomach. As a result, gastric motility is inhibited and there is an increase in the contraction of the pyloric sphincter, which decreases gastric emptying.

### Hormonal Regulation

The intestinal phase of digestion is mediated by two major hormones secreted by enteroendocrine cells of the small intestinal glands: secretin and cholecystokinin. Acidic chyme entering the duodenum stimulates the release of **secretin** from the small intestinal glands. Secretin stimulates the flow of pancreatic juice that is rich in bicarbonate ions to buffer the acidic chyme that enters the duodenum from the small intestine. Secretin also inhibits secretion of gastric juice. Overall, secretin decreases the acidity of chyme entering the duodenum by buffering acids in chyme that reaches the duodenum and slowing production of acid in the stomach.

**Cholecystokinin (CCK)** (kō-lē-sis′-tō-KĪN-in) is secreted in response to chyme containing amino acids from partially digested proteins and fatty acids from partially digested triglycerides. CCK stimulates secretion of pancreatic juice that is rich in digestive enzymes. It causes the gallbladder to contract, ejecting its stored bile into the cystic duct and through the common bile duct. In addition, CCK slows gastric emptying by promoting contraction of the pyloric sphincter and produces satiety (a feeling of fullness) by acting on the hypothalamus in the brain.

Table 23.8 summarizes the major hormones that control digestion.

**TABLE 23.8**

Major Hormones That Control Digestion

| HORMONE | STIMULUS AND SITE OF SECRETION | ACTIONS |
|---|---|---|
| **Gastrin** | Distension of stomach, partially digested proteins and caffeine in stomach, and high pH of stomach chyme stimulate gastrin secretion by enteroendocrine cells of stomach | ***Major effects:*** Promotes secretion of gastric juice, increases gastric motility, promotes growth of gastric mucosa<br>***Minor effects:*** Constricts lower esophageal sphincter, relaxes pyloric sphincter |
| **Secretin** | Acidic (high $H^+$ level) chyme that enters small intestine stimulates secretion of secretin by enteroendocrine cells of the duodenum | ***Major effects:*** Stimulates secretion of pancreatic juice and bile that are rich in $HCO_3^-$ (bicarbonate ions)<br>***Minor effects:*** Inhibits secretion of gastric juice, promotes normal growth and maintenance of pancreas, enhances effects of CCK |
| **Cholecystokinin (CCK)** | Partially digested proteins (amino acids), triglycerides, and fatty acids that enter small intestine stimulate secretion of CCK by enteroendocrine cells of small intestine | ***Major effects:*** Stimulates secretion of pancreatic juice rich in digestive enzymes, causes ejection of bile from gallbladder and opening of sphincter of the hepatopancreatic ampulla, induces satiety (feeling full to satisfaction)<br>***Minor effects:*** Inhibits gastric emptying, promotes normal growth and maintenance of pancreas, enhances effects of secretin |

✓ **CHECKPOINT**

**42.** What is the result and purpose of the cephalic phase of digestion?

**43.** Why does food initially cause the pH of the gastric juice to rise?

**44.** Why is stomach activity inhibited during the intestinal phase of digestion?

**45.** Explain the roles of cholecystokinin and secretin in the intestinal phase of digestion.

## RETURN TO Zachary's Story

"Mrs. Lewis, these are Zachary's scans. I ordered a CT when we suspected appendicitis and noticed that there was some inflammation of the bowel where the appendix is attached. I ordered an MRI of his abdominal cavity and as I suspected, there is some pretty heavy inflammation of large sections of his small intestine." Dr. Lee pauses after this last sentence, giving Sandra a chance to absorb what he is saying.

"Mrs. Lewis, inflammation aside, there is something else on these scans that concerns me." As Dr. Lee speaks these words, every irrational fear Sandra has ever experienced comes flooding in and she holds her breath. "Mrs. Lewis, may I continue?" asks Dr. Lee, clearly seeing her distress. After she nods, Dr. Lee continues. "There is inflammation surrounding an opening in the walls of the small intestine where digestive juices made in the liver and pancreas enter, and this inflammation may be causing a blockage of the opening. These juices contain enzymes that break down the food Zachary eats into a form his body can absorb. Without them, even though Zachary eats food, his body is breaking down because he is not getting enough nutrients."

"I need to sit down," Sandra murmurs, and Dr. Lee quickly pulls up a chair. Sandra takes a deep breath and asks Dr. Lee very frightening questions: "Will he be okay? Can you fix this?" Dr. Lee's face fills with compassion for Sandra. He crouches down to be at eye-level and places his hand on her shoulder. "Crohn's disease is something we call a chronic inflammatory disease. This word *chronic* means that Zachary will most likely struggle with these symptoms throughout his entire life. I don't want to sugarcoat things for you, Mrs. Lewis, so I need to tell you that malnutrition and malabsorption caused by the inflammation are not the only complications Zachary may face. You are already familiar with the fatigue, nausea, pain, fever, and weight loss, but you should also be prepared for the possibility of skin, liver, and eye problems, as well as the possibility of fistulas, which are inappropriate connections between two body structures that may have to be corrected surgically."

F. *What do you think the ultimate fate of Zach's pancreas would be if the hepatopancreatic ampulla continued to be blocked? What do you think would happen to the liver and then eventually to the rest of Zach's body?*

G. *What enzymes has Zachary's body been unable to use because of the blockage of the hepatopancreatic ampulla? What are the specific molecules these enzymes work on?*

H. *Selecting one of Zach's symptoms of fever, diarrhea, or weight loss, explain how inflammation of one section of the small intestine could lead to that symptom.*

## 23.9 Metabolism includes the catabolism and anabolism of molecules.

The food we eat is our only source of energy for running, walking, and even breathing. Many molecules needed to maintain cells and tissues can be made from simpler precursors by the body's metabolic reactions; others—the essential amino acids, essential fatty acids, vitamins, and minerals—must be obtained from our food. To review, carbohydrates, lipids, and proteins in food are digested by enzymes into monosaccharides, fatty acids, glycerol, monoglycerides, and amino acids that are absorbed by epithelial cells of the GI tract mucosa. Some minerals and many vitamins are part of enzyme systems that catalyze the breakdown and synthesis of carbohydrates, lipids, and proteins. Once absorbed by the GI, food molecules have three main fates:

- Most food molecules are used to *supply energy* for sustaining life processes, such as active transport, DNA replication, protein synthesis, muscle contraction, maintenance of body temperature, and mitosis.
- Some food molecules *serve as building blocks* for the synthesis of more complex structural or functional molecules, such as muscle proteins, hormones, and enzymes.
- Other food molecules are *stored for future use.* For example, glycogen is stored in liver cells, and triglycerides are stored in adipose cells.

### Metabolic Reactions

**Metabolism** (me-TAB-ō-lizm; *metabol-* = change) refers to all of the chemical reactions that occur in the body. There are two types of metabolism: catabolism and anabolism (Figure 23.26). Those chemical reactions that break down complex organic molecules into simpler ones are collectively known as **catabolism** (ka-TAB-ō-lizm; *cata-* = downward). Overall, catabolic

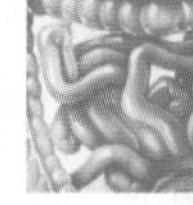

FIGURE 23.26 **Role of ATP in linking anabolic and catabolic reactions.** When complex molecules are split apart (catabolism, at left), some of the energy is transferred to form ATP and the rest is given off as heat. When simple molecules are combined to form complex molecules (anabolism, at right), ATP provides the energy for synthesis, and again some energy is given off as heat.

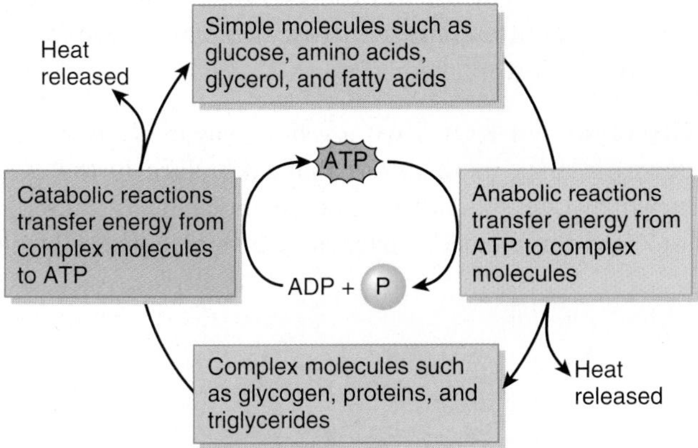


**The coupling of energy-releasing and energy-requiring reactions is achieved through ATP.**

(decomposition) reactions are *exergonic*; they produce more energy than they consume, releasing the chemical energy stored in organic molecules. Important sets of catabolic reactions occur in glycolysis, the Krebs cycle, and the electron transport chain, each of which will be discussed later in the chapter.

Chemical reactions that combine simple molecules to form the body's complex structural and functional components are collectively known as **anabolism** (a-NAB-ō-lizm; *ana-* = upward). Examples of anabolic reactions are the formation of peptide bonds between amino acids during protein synthesis, the building of fatty acids into phospholipids that form the plasma membrane bilayer, and the linkage of glucose molecules to form glycogen. Anabolic reactions are *endergonic*; they consume more energy than they produce.

**Enzymes** serve as catalysts to speed up metabolism. Some enzymes require the presence of an ion such as calcium, iron, or zinc. Other enzymes work together with **coenzymes**, which function as temporary carriers of atoms being removed from or added to a substrate during a reaction. Many coenzymes are derived from vitamins. Examples include the coenzyme $NAD^+$, derived from the B vitamin niacin, and the coenzyme *FAD*, derived from vitamin $B_2$ (riboflavin).

## Coupling of Catabolism and Anabolism by ATP

Metabolism is an energy-balancing act between catabolic (decomposition) reactions, and anabolic (synthesis) reactions. The molecule that participates most often in energy exchanges in living cells is **ATP (adenosine triphosphate)**, which couples energy-releasing catabolic reactions to energy-requiring anabolic reactions. The energy released from catabolic reactions is transferred to molecules of ATP and then used to power anabolic reactions.

The chemical reactions of living systems depend on the efficient transfer of manageable amounts of energy from one molecule to another. The molecule that most often performs this task is ATP, the "energy currency" of a living cell. Like money, ATP is readily available to "buy" cellular activities; it is spent and earned over and over. A typical cell has about a billion molecules of ATP, each of which typically lasts for less than a minute before being used. Thus, ATP is not a long-term storage form of currency, like gold in a vault, but rather convenient cash for moment-to-moment transactions.

Recall from Chapter 2 that a molecule of ATP consists of an adenine molecule, a ribose molecule, and three phosphate groups bonded to one another (see Figure 2.24). Figure 23.26 shows how ATP links anabolic and catabolic reactions. When the terminal phosphate group is split off ATP in anabolic reactions, adenosine diphosphate (ADP) and a phosphate group (symbolized as Ⓟ) are formed. Some of the energy released is used to drive anabolic reactions such as the formation of glycogen from glucose. In addition, energy from complex molecules is used in catabolic reactions to combine ADP and a phosphate group to resynthesize ATP:

$$ADP + Ⓟ + \text{energy} \rightarrow ATP$$

About 40 percent of the energy released in catabolism is used for cellular functions; the rest is converted to heat, some of which helps maintain normal body temperature. Excess heat is lost to the environment. Compared with machines, which typically convert only 10–20 percent of energy into work, the 40 percent efficiency of the body's metabolism is impressive. Still, the body has a continuous need to take in and process external sources of energy so that cells can synthesize enough ATP to sustain life.

# Carbohydrate Metabolism

During the digestion of carbohydrates, both polysaccharide and disaccharide are catabolized into monosaccharides—glucose, fructose, and galactose—which are absorbed in the small intestine. Shortly after their absorption into the blood, however, hepatocytes convert fructose and galactose to glucose. Thus, the story of carbohydrate metabolism is really the story of glucose metabolism.

Because glucose is the body's preferred source for synthesizing ATP, the fate of glucose absorbed from the diet depends on the needs of body cells, which include the following:

- ***ATP production.*** In body cells that require immediate energy, glucose is broken down to produce ATP. Glucose that is not needed for immediate ATP production can enter one of several other metabolic pathways.

- ***Amino acid synthesis.*** Cells throughout the body can use glucose to form several amino acids, which then can be incorporated into proteins.
- ***Glycogen synthesis.*** Hepatocytes and muscle fibers can combine glucose molecules to form glycogen.
- ***Triglyceride synthesis.*** When the glycogen storage areas are filled up, hepatocytes can transform the glucose to glycerol and fatty acids that can be used to synthesize triglycerides. Triglycerides then are deposited in adipose tissue, which has virtually unlimited storage capacity.

Before glucose can be used by body cells, it must first pass through the plasma membrane and enter the cytosol. Glucose absorption in the GI tract (and kidney tubules) is accomplished via secondary active transport. Glucose entry into most other body cells occurs via facilitated diffusion. Insulin increases the rate of facilitated diffusion of glucose into cells.

## Glucose Catabolism

The catabolism of glucose to produce ATP is known as cellular respiration. Four interconnecting sets of chemical reactions contribute to cellular respiration (Figure 23.27):

1. **Glycolysis** (glī-KOL-i-sis; *glyco-* = sugar; *-lysis* = breakdown) is a set of reactions that take place in the cytosol in which one six-carbon glucose molecule is catabolized into two three-carbon pyruvic acid molecules. Glycolysis

FIGURE 23.27 **Cellular respiration.**

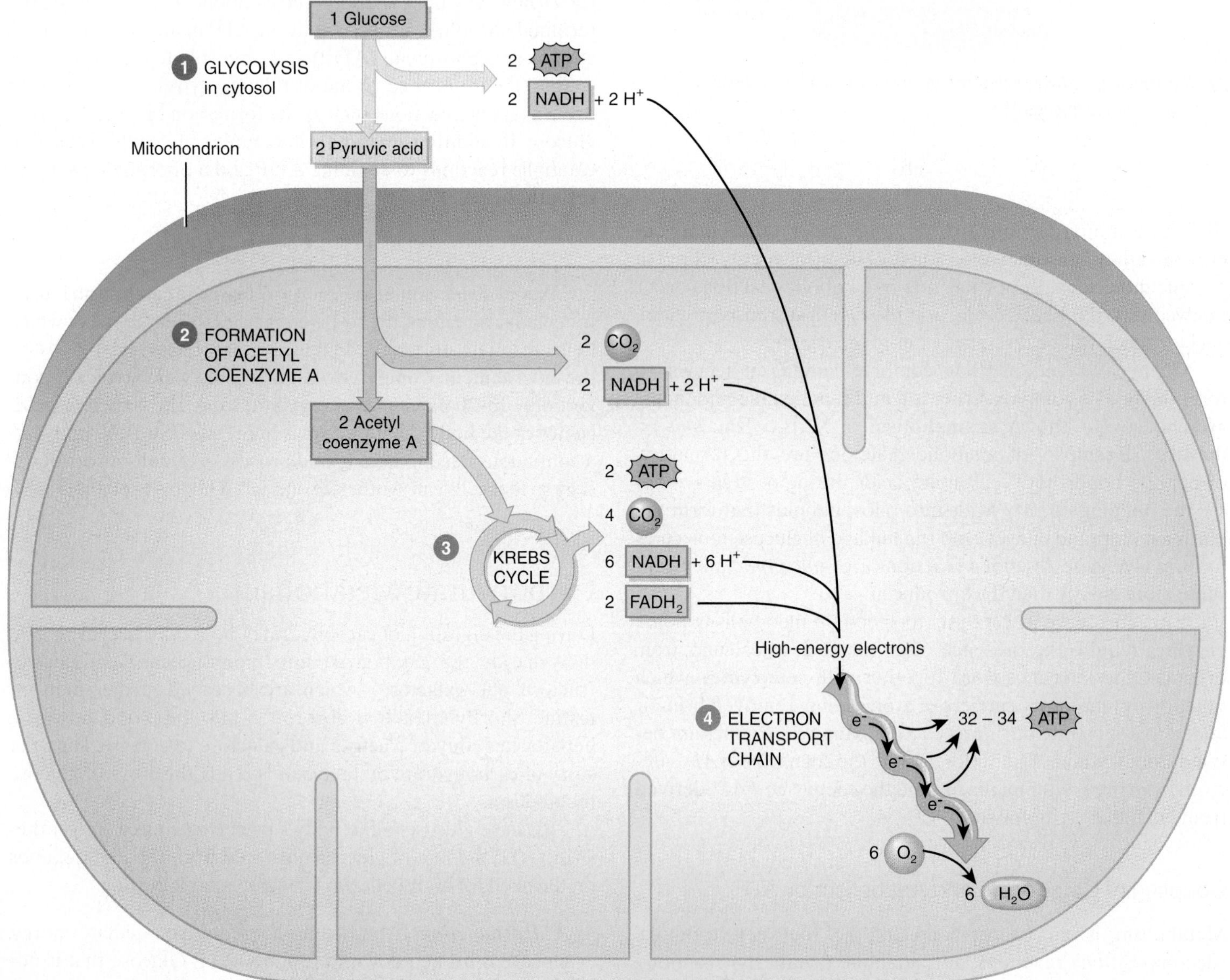


**The catabolism of glucose to produce ATP involves glycolysis, the formation of acetyl coenzyme A, the Krebs cycle, and the electron transport chain.**

produces two molecules of ATP and two energy-containing molecules of ($NADH + H^+$). Because glycolysis does not require oxygen, it is a way to produce ATP anaerobically (without oxygen) and is known as **anaerobic cellular respiration** (an-ar-Ō-bik; *an-* = not; *aer-* = air; *-bios* = life). If oxygen is available, however, most cells next convert pyruvic acid to acetyl coenzyme A.

2. **Formation of acetyl coenzyme A** is a transition step that prepares pyruvic acid for entrance into the Krebs cycle. First, pyruvic acid enters a mitochondrion and is converted to a two-carbon fragment by removing a molecule of carbon dioxide ($CO_2$). Molecules of $CO_2$ produced during glucose catabolism diffuse into the blood and are eventually exhaled. Then, the coenzyme $NAD^+$ removes a hydrogen molecule ($H_2$) from pyruvic acid, in the process becoming the energy-containing molecule ($NADH + H^+$). Finally, the remaining atom, called an acetyl group (a-SĒT-il), is attached to coenzyme A, forming acetyl coenzyme A.

3. **Krebs cycle reactions** transfer chemical energy from acetyl coenzyme A to two other coenzymes—$NAD^+$ and FAD—thereby forming ($NADH + H^+$) and $FADH_2$. Krebs cycle reactions also produce $CO_2$ and one ATP for each acetyl coenzyme A that enters the Krebs cycle. To harvest the energy in NADH and $FADH_2$, their high-energy electrons must first go through the electron transport chain.

4. **Electron transport chain** reactions use the energy in ($NADH + H^+$) and $FADH_2$ to synthesize ATP. As the coenzymes pass their high-energy electrons through a series of "electron carriers," ATP is synthesized. Finally, lower-energy electrons are passed to oxygen in a reaction that produces water. Because the Krebs cycle and the electron transport chain together require oxygen to produce ATP, they are collectively known as **aerobic cellular respiration**.

During cellular respiration, 36 or 38 ATPs can be generated from one molecule of glucose. The many reactions of cellular respiration can be summarized as follows:

$$\underset{\text{Glucose}}{C_6H_{12}O_6} + \underset{\text{Oxygen}}{6\,O_2} + 36 \text{ or } 38 \text{ ADPs} + 36 \text{ or } 38 \text{ Ⓟ} \longrightarrow \underset{\text{Carbon dioxide}}{6\,CO_2} + \underset{\text{Water}}{6\,H_2O} + 36 \text{ or } 38 \text{ ATPs}$$

Glycolysis, the Krebs cycle, and the electron transport chain provide all of the ATP for cellular activities. Because the Krebs cycle and electron transport chain are aerobic processes, cells cannot carry on their activities for long if oxygen is lacking.

## Glucose Anabolism

Although most of the glucose in the body is catabolized to generate ATP, glucose may be formed by several anabolic reactions. One is the synthesis of glycogen; another is the synthesis of new glucose molecules from some of the products of protein and lipid breakdown.

If glucose is not needed immediately for ATP production, it combines with many other glucose molecules to form **glycogen**, a polysaccharide that is the only stored form of carbohydrate in our bodies (Figure 23.28). The synthesis of glycogen is stimulated by the hormone insulin released from the pancreas. The body can store about 500 grams (about 1.1 lb) of glycogen, roughly 75 percent in skeletal muscle fibers and the rest in hepatocytes.

**FIGURE 23.28 Reactions of glucose anabolism: synthesis of glycogen, breakdown of glycogen, and synthesis of glucose from amino acids, lactic acid, or glycerol.**

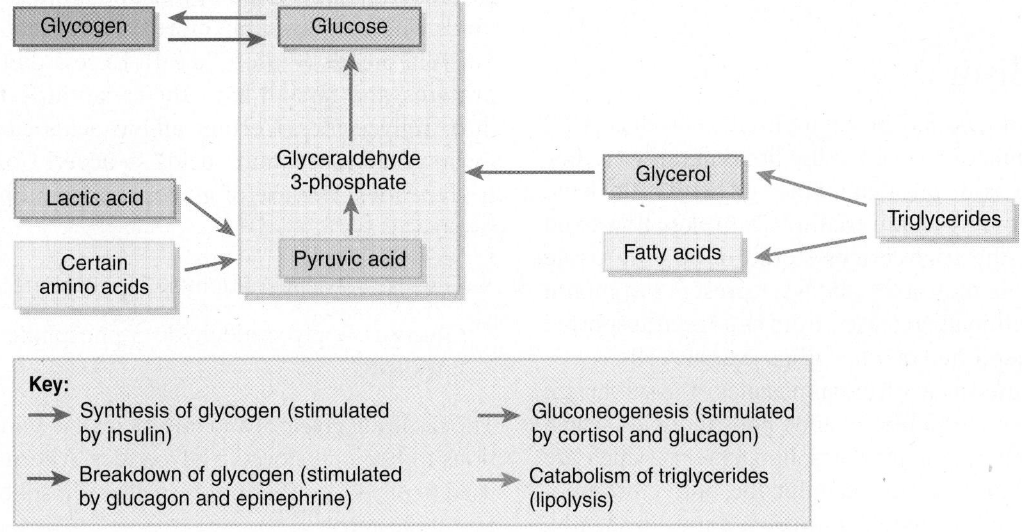


**The body stores glucose as glycogen in skeletal muscles and the liver.**

If blood glucose level falls below normal, glucagon is released from the pancreas and epinephrine is released from the adrenal medulla. These hormones stimulate the breakdown of glycogen into glucose molecules (Figure 23.28). When body activities require ATP, glycogen stored in hepatocytes is broken down into glucose and released into the blood to be transported to cells, where it will be catabolized by the processes of cellular respiration already described. Glycogen catabolism usually occurs between meals.

When your liver runs low on glycogen, it is time to eat. If you don't, your body starts catabolizing triglycerides and proteins. Actually, the body normally catabolizes some of its triglycerides and proteins, but large-scale triglyceride and protein catabolism does not happen unless you are starving, eating very few carbohydrates, or suffering from an endocrine disorder.

Lactic acid, certain amino acids, and the glycerol part of triglycerides can be converted in the liver to glucose (Figure 23.28). The series of reactions in which glucose is formed from these noncarbohydrate sources is called **gluconeogenesis** (gloo′-kō-nē′-ō-JEN-e-sis; *-neo-* = new). Glycerol may be converted into glyceraldehyde 3-phosphate, which may form pyruvic acid or may be used to synthesize glucose. About 60 percent of the amino acids in the body can be used for gluconeogenesis. These amino acids and lactic acid are converted to pyruvic acid, which may be synthesized into glyceraldehyde 3-phosphate, and then glucose. Gluconeogenesis releases glucose into the blood, thereby keeping blood glucose level normal during the hours between meals when glucose is not being absorbed. Gluconeogenesis occurs when the liver is stimulated by cortisol from the adrenal cortex, and by glucagon from the pancreas. In addition, cortisol stimulates the breakdown of proteins into amino acids, thus expanding the pool of amino acids available for gluconeogenesis. Thyroid hormones (thyroxine and triiodothyronine) also mobilize proteins and may mobilize triglycerides from adipose tissue, thereby making glycerol available for gluconeogenesis.

## Lipid Metabolism

Lipids, like carbohydrates, may be catabolized to produce ATP. If the body has no immediate need to use lipids in this way, they are stored in adipose tissue throughout the body and in the liver. Triglycerides in adipose tissue are continually broken down and resynthesized. Thus, the triglycerides stored in adipose tissue today are not the same molecules that were present last month because they are continually released from storage, transported in the blood, and redeposited in other adipose tissue cells.

A few lipids are used as structural molecules or to synthesize other substances. Some examples include phospholipids, which are constituents of plasma membranes; lipoproteins, which are used to transport cholesterol throughout the body; thromboplastin, which is needed for blood clotting; and myelin sheaths, which speed up impulse conduction. Two **essential fatty acids** that the body cannot synthesize are linoleic acid and linolenic acid. Dietary sources of these lipids include vegetable oils and leafy vegetables.

### Lipid Catabolism

Muscle, liver, and adipose cells routinely catabolize fatty acids from triglycerides to produce ATP. To do so, the triglycerides must first be split into glycerol and fatty acids, a process called **lipolysis** (li-POL-i-sis) (Figure 23.29). Epinephrine and norepinephrine enhance triglyceride breakdown into fatty acids and glycerol. These hormones are released when sympathetic tone increases, as occurs, for example, during exercise. Cortisol, thyroid hormones, and insulinlike growth factors also stimulate lipolysis. By contrast, insulin inhibits lipolysis.

The glycerol and fatty acids that result from lipolysis are catabolized via different pathways (Figure 23.29). Glycerol is converted by many cells of the body to glyceraldehyde 3-phosphate, one of the compounds also formed during the catabolism of glucose. If ATP supply in a cell is high, glyceraldehyde 3-phosphate is converted into glucose, an example of gluconeogenesis. If the ATP supply in a cell is low, glyceraldehyde 3-phosphate enters the catabolic pathway to pyruvic acid.

Fatty acid catabolism begins within mitochondria as enzymes remove two carbon atoms at a time from the fatty acid and attach them to molecules of coenzyme A, forming acetyl coenzyme A (acetyl CoA). Then acetyl CoA enters the Krebs cycle (Figure 23.29).

As part of normal fatty acid catabolism, hepatocytes convert some acetyl CoA molecules into substances known as **ketone bodies** (KĒ-tōn) (Figure 23.29). Ketone bodies freely diffuse out of hepatocytes and enter the bloodstream.

### Lipid Anabolism

Insulin stimulates hepatocytes and adipose cells to synthesize lipids when we consume more calories than we need to satisfy our ATP needs (Figure 23.29). Excess dietary carbohydrates, proteins, and fats all have the same fate—they are converted into triglycerides. Certain amino acids can undergo the following reactions: amino acids → acetyl CoA → fatty acids → triglycerides. The use of glucose to form lipids takes place via two pathways:

- Glucose → glyceraldehyde 3-phosphate → glycerol
- Glucose → glyceraldehyde 3-phosphate → acetyl CoA → fatty acids

The resulting glycerol and fatty acids can undergo anabolic reactions to become stored triglycerides. Alternatively, they can be used to produce other lipids such as phospholipids, lipoproteins, and cholesterol.

FIGURE 23.29 **Metabolism of lipids.** Lipolysis is the breakdown of triglycerides into glycerol and fatty acids. Glycerol may be converted to glyceraldehyde 3-phosphate, which can then be converted to glucose or enter the Krebs cycle. Fatty acid fragments enter the Krebs cycle as acetyl coenzyme A. Fatty acids also can be converted into ketone bodies.

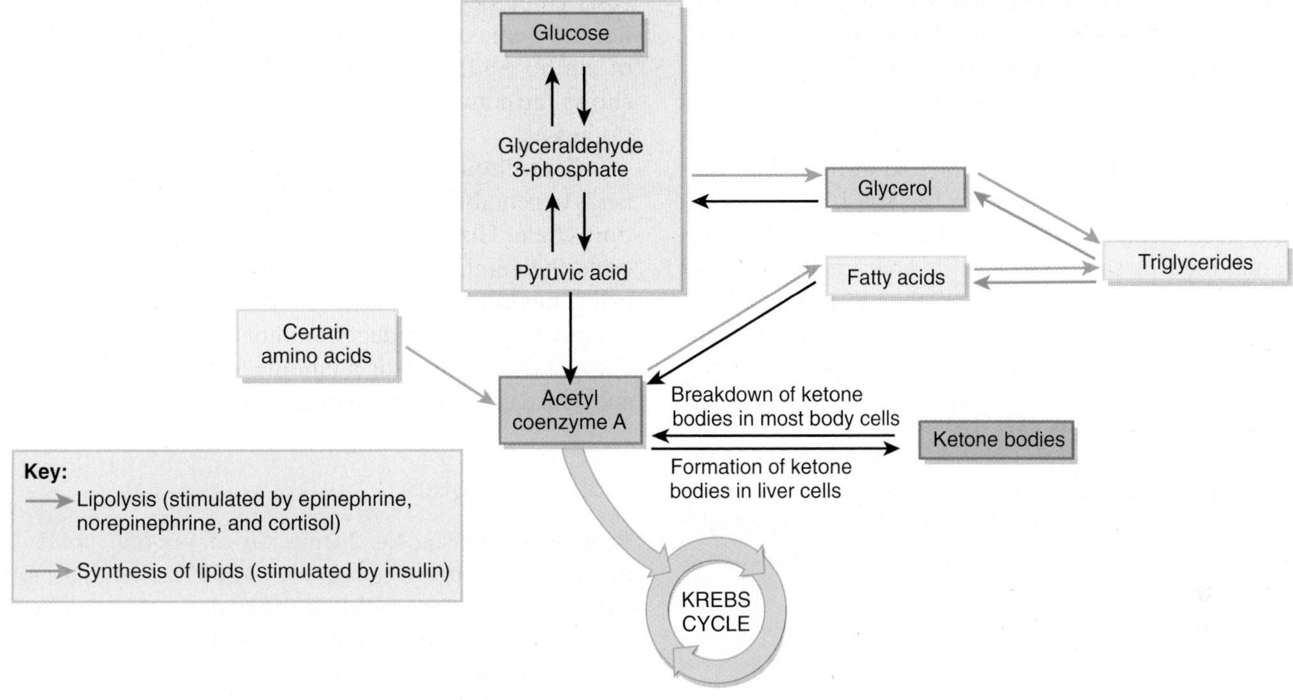


**Glycerol and fatty acids are catabolized in separate pathways.**

## Lipid Transport in Blood

Most lipids, such as triglycerides, are nonpolar and therefore very hydrophobic molecules. They do not dissolve in water. To be transported in watery blood, such molecules first must be made more water-soluble by combining them with proteins. The lipid and protein combinations thus formed are *lipoproteins*, spherical particles with an outer shell of proteins, phospholipids, and cholesterol molecules surrounding an inner core of triglycerides and other lipids. The proteins in the outer shell help the lipoprotein particles dissolve in body fluids. Lipoproteins are transport vehicles: They provide delivery and pickup services so that lipids can be available when cells need them or removed from circulation when they are not needed. Lipoproteins are categorized and named mainly according to their size and density, which varies with the ratio of lipids (which have a low density) to proteins (which have a high density). From largest and lightest to smallest and heaviest, the four major types of lipoproteins are chylomicrons, very low-density lipoproteins, low-density lipoproteins, and high-density lipoproteins.

- Chylomicrons form in absorptive epithelial cells of the small intestine and transport dietary lipids to adipose tissue for storage. As you learned in Concept 23.6, chylomicrons enter lacteals of intestinal villi and are carried by lymph into venous blood and then into the systemic circulation. They remain in the blood for only a few minutes. As chylomicrons circulate through the capillaries of adipose tissue, fatty acids from chylomicron triglycerides are taken up by adipocytes for synthesis and storage as triglycerides and by muscle cells for ATP production. Chylomicron remnants are removed from the blood by hepatocytes.
- **Very low-density lipoproteins (VLDLs)**, which form in hepatocytes, transport triglycerides to adipose tissue for storage and muscle cells for ATP production. As they deposit some of their triglycerides in adipose tissue, VLDLs are converted to LDLs.
- **Low-density lipoproteins (LDLs)** carry about 75 percent of the total cholesterol in blood and deliver it to cells throughout the body for use in repair of cell membranes and synthesis of steroid hormones and bile salts. LDLs bind to LDL receptors on the plasma membrane of body cells so that LDL can enter the cell by endocytosis. Within the cell, the LDL is broken down, and the cholesterol is released to serve the cell's needs.

Once a cell has sufficient cholesterol for its activities, a negative feedback system inhibits the cell's synthesis of new LDL receptors.

When present in excessive numbers, LDLs also deposit cholesterol in and around smooth muscle fibers in arteries, forming fatty plaques that increase the risk of coronary artery disease (see the Chapter 19 WileyPLUS Clinical Connection entitled Coronary Artery Disease). For this reason, the cholesterol in LDLs, called *LDL cholesterol*, is known as "bad" cholesterol. Because some people have too few LDL receptors, their body cells remove LDL from the blood less efficiently; as a result, their plasma LDL level is abnormally high, and they are more likely to develop fatty plaques. Eating a high-fat diet increases the production of VLDLs, which elevates the LDL level and increases the formation of fatty plaques.

- **High-density lipoproteins (HDLs)** remove excess cholesterol from body cells and the blood and transport it to the liver for elimination. Because HDLs prevent accumulation of cholesterol in the blood, a high HDL level is associated with decreased risk of coronary artery disease. For this reason, HDL cholesterol is known as "good" cholesterol.

## Protein Metabolism

During digestion, proteins are broken down into amino acids. Unlike carbohydrates and triglycerides, which are stored, proteins are not warehoused for future use. Instead, amino acids are either oxidized to produce ATP or used to synthesize new proteins for body growth and repair. Excess dietary amino acids are not excreted in the urine or feces but instead are converted into glucose or triglycerides.

The active transport of amino acids into body cells is stimulated by insulinlike growth factors and insulin. Almost immediately after digestion, amino acids are reassembled into proteins. Many proteins function as enzymes; others are involved in transportation (hemoglobin) or serve as antibodies, clotting chemicals (fibrinogen), hormones (insulin), or contractile elements in muscle fibers (actin and myosin). Several proteins serve as structural components of the body (collagen, elastin, and keratin).

### Protein Catabolism

A certain amount of protein catabolism occurs in the body each day, stimulated mainly by cortisol from the adrenal cortex. Proteins from worn-out cells (such as red blood cells) are broken down into amino acids. Some amino acids are converted into other amino acids, peptide bonds are reformed, and new proteins are synthesized as part of the recycling process. Hepatocytes convert some amino acids to fatty acids, ketone bodies, or glucose. Figure 23.28 shows the conversion of amino acids into glucose (gluconeogenesis). Figure 23.29 shows the conversion of amino acids into fatty acids or ketone bodies.

Cells throughout the body oxidize a small amount of amino acids to generate ATP via the Krebs cycle and the electron transport chain. However, before amino acids can enter the Krebs cycle, their amino group ($-NH_2$) must first be removed, a process called **deamination** (dē-am′-i-NĀ-shun). Deamination occurs in hepatocytes and produces ammonia ($NH_3$). Hepatocytes then convert the highly toxic ammonia to urea, a relatively harmless substance that is excreted in the urine.

### Protein Anabolism

Protein anabolism, the formation of peptide bonds between amino acids to produce new proteins, is carried out on the ribosomes of almost every cell in the body, as directed by the cells' DNA and RNA. Insulinlike growth factors, thyroid hormones, insulin, estrogens, and testosterone stimulate protein synthesis. Because proteins are a main component of most cell structures, adequate dietary protein is especially essential during the growth years, during pregnancy, and when tissue has been damaged by disease or injury. Once dietary intake of protein is adequate, eating more protein will not increase bone or muscle mass; only a regular program of forceful, weight-bearing muscular activity accomplishes that goal.

Of the 20 amino acids in the human body, 10 are **essential amino acids**: It is *essential* to include them in your diet because they cannot be synthesized in the body in adequate amounts. A **complete protein** contains sufficient amounts of all essential amino acids. Beef, fish, poultry, eggs, and milk are examples of foods that contain complete proteins. An **incomplete protein** does not contain all essential amino acids. Examples of incomplete proteins are leafy green vegetables, legumes (beans and peas), and grains. **Nonessential amino acids** can be synthesized by the body. They are formed by the transfer of an amino group from an amino acid to pyruvic acid or to an acid in the Krebs cycle. Once the appropriate essential and nonessential amino acids are present in cells, protein synthesis occurs rapidly.

Table 23.9 summarizes the processes occurring in catabolism and anabolism of carbohydrates, lipids, and proteins.

**TABLE 23.9**

Metabolism

| PROCESS | COMMENTS |
|---|---|
| **CARBOHYDRATES** | |
| **Glucose catabolism** | Complete oxidation of glucose (cellular respiration) is chief source of ATP in cells; consists of glycolysis, Krebs cycle, and electron transport chain; complete oxidation of 1 molecule of glucose yields maximum of 36 or 38 molecules of ATP |
| **Glycolysis** | Conversion of glucose into pyruvic acid results in production of some ATP; reactions do not require oxygen (anaerobic cellular respiration) |
| **Krebs cycle** | Cycle includes series of oxidation–reduction reactions in which coenzymes ($NAD^+$ and FAD) pick up hydrogen ions and hydride ions from oxidized organic acids; some ATP produced; $CO_2$ and $H_2O$ are by-products; reactions are aerobic |
| **Electron transport chain** | Third set of reactions in glucose catabolism: another series of oxidation–reduction reactions, in which electrons are passed from one carrier to next; most ATP produced; reactions require oxygen (aerobic cellular respiration) |
| **Glucose anabolism** | Some glucose is converted into glycogen (glycogenesis) for storage if not needed immediately for ATP production; glycogen can be reconverted to glucose (glycogenolysis); conversion of amino acids, glycerol, and lactic acid into glucose is called gluconeogenesis |
| **LIPIDS** | |
| **Triglyceride catabolism** | Triglycerides are broken down into glycerol and fatty acids; glycerol may be converted into glucose (gluconeogenesis) or catabolized via glycolysis; fatty acids are catabolized via beta oxidation into acetyl coenzyme A that can enter Krebs cycle for ATP production or be converted into ketone bodies (ketogenesis) |
| **Triglyceride anabolism** | Synthesis of triglycerides from glucose and fatty acids is called lipogenesis; triglycerides are stored in adipose tissue |
| **PROTEINS** | |
| **Protein catabolism** | Amino acids are oxidized via Krebs cycle after deamination; ammonia resulting from deamination is converted into urea in liver, passed into blood, and excreted in urine; amino acids may be converted into glucose (gluconeogenesis), fatty acids, or ketone bodies |
| **Protein anabolism** | Protein synthesis is directed by DNA and utilizes cells' RNA and ribosomes |

✓ **CHECKPOINT**

**46.** What is metabolism? What is the difference between anabolism and catabolism?

**47.** How does ATP link anabolism and catabolism?

**48.** Which components of cellular respiration produce ATP during the complete catabolism of a molecule of glucose? How many molecules of ATP are produced?

**49.** What happens during glycolysis? In the electron transport chain?

**50.** What is gluconeogenesis, and why is it important?

**51.** What are the functions of the proteins in lipoproteins? Which lipoprotein particles contain "good" and "bad" cholesterol, and what are the meanings of these terms?

# 23.10 Food molecules supply energy for life processes and serve as building blocks for complex molecules.

**Nutrients** are chemical substances in food that body cells use for growth, maintenance, and repair. The six main types of nutrients are water, carbohydrates, lipids, proteins, minerals, and vitamins. Water is the nutrient needed in the largest amount—about 2–3 liters per day. As the most abundant compound in the body, water provides the medium in which most metabolic reactions occur, and it also participates in some reactions (for example, hydrolysis reactions). The important roles of water in the body can be reviewed in Concept 2.4. Carbohydrates, lipids, and proteins provide the energy needed for metabolic reactions and serve as building blocks to make body structures. Some minerals and many vitamins are components of the enzyme systems that catalyze metabolic reactions. *Essential nutrients* are specific nutrient molecules that the body cannot make in sufficient quantity to meet its needs and thus must be obtained from the diet. Some amino acids, fatty acids, vitamins, and minerals are essential nutrients.

## Guidelines for Healthy Eating

On June 2, 2011, the United States Department of Agriculture (USDA) introduced a new icon called **MyPlate** based on revised guidelines for healthy eating. It replaces the USDA MyPyramid, which first appeared in 2005. As shown in Figure 23.30, the plate is divided into four different-sized colored sections: green (vegetables), red (fruits), orange (grains, preferably whole grains), purple (protein), and adjacent to the plate is a blue cup (dairy).

The Dietary Guidelines for Americans released in January 2011 are the basis for MyPlate. Among the guidelines are the following:

- Enjoy food but balance calories by eating less.
- Avoid oversized portions.
- Make half of your plate vegetables and fruits.
- Switch to fat-free or low-fat (1 percent) milk.
- Make at least half of your grains whole grains.
- Compare sodium in foods like soup, breads, and frozen meals and choose foods that have low sodium content.
- Drink water instead of sugary drinks.

The MyPlate food guide emphasizes the importance of proportionality, variety, and moderation in a healthy diet. Proportionality means eating more of some types of foods than others. The MyPlate icon shows how much of your plate should be filled with foods from various food groups. Note that the vegetables and fruits take up one half of the plate, while protein and grains take up the other half. Note also that vegetables and grains represent the largest portions. The blue cup (dairy) adjacent to the plate icon is a reminder that meals should include dairy foods. Even though the plate is divided into four proportionally sized sections, the daily servings do not have to be exactly proportional since each person has different nutritional needs based on factors such as age and general health. The sections are designed to be visual cues to help you make healthier eating choices.

Variety is important for a healthy diet because no one food or food group provides all of the nutrients and food components your body needs. A variety of foods should be selected from within each of the five food groups. Vegetable choices should include dark green vegetables such as broccoli, collard greens, and kale; red and orange vegetables such as carrots, sweet potatoes, and red peppers; starchy vegetables such as corn, green peas, and potatoes; other vegetables such as cabbage, asparagus, and artichokes; and beans such as lentils, lima beans, pinto beans,

FIGURE 23.30 **MyPlate.** The MyPlate icon shows what a balanced meal should look like.

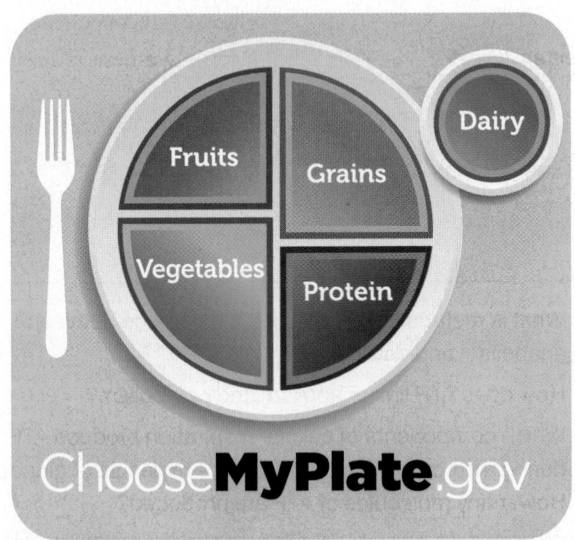


**The MyPlate icon illustrates the proportions of food recommended from each of five food groups.**

chickpeas, and black beans. Beans are good sources of the nutrients found in both vegetables and protein foods so they can be counted in either food group. Protein food choices are extremely varied, including meat, poultry, seafood, beans and peas, eggs, processed soy products, nuts, and seeds. Grains include whole grains such as whole-wheat bread, oatmeal, and brown rice as well as refined grains such as white bread, white rice, and white pasta. Fruits include fresh, canned, or dried fruit and 100 percent fruit juice. Dairy includes milk, cheese, yogurt, and pudding, as well as calcium-fortified soy products.

## Minerals

**Minerals** are inorganic elements that occur naturally in the Earth's crust. In the body they appear in combination with one another, in combination with organic compounds, or as ions in solution. Minerals constitute about 4 percent of the total body weight and are concentrated most heavily in the skeleton. Minerals with known functions in the body include calcium, phosphorus, potassium, sulfur, sodium, chloride, magnesium, iron, iodide, manganese, copper, cobalt, zinc, fluoride, selenium, and chromium. Other minerals—aluminum, boron, silicon, and molybdenum—are present but their functions are unclear. Typical diets supply adequate amounts of potassium, sodium, chloride, and magnesium. Some attention must be paid to eating foods that provide enough calcium, phosphorus, iron, and iodide. Excess amounts of most minerals are excreted in the urine and feces.

Calcium and phosphorus form part of the matrix of bones. Because minerals do not form long-chain compounds, they are otherwise poor building materials. A major role of minerals is to help regulate enzymatic reactions. Calcium, iron, magnesium, and manganese are part of some coenzymes. Magnesium also serves as a catalyst for the conversion of ADP to ATP. Minerals such as sodium and phosphorus work in buffer systems, which help control the pH of body fluids. Sodium also helps regulate the osmosis of water and, along with other ions, is involved in the generation of impulses.

## Vitamins

**Vitamins** are organic nutrients required in small amounts to maintain growth and normal metabolism. Unlike carbohydrates, lipids, or proteins, vitamins do not provide energy or serve as the body's building materials. Most vitamins with known functions are coenzymes.

Most vitamins cannot be synthesized by the body and must be ingested in food. Other vitamins, such as vitamin K, are produced by bacteria in the GI tract and then absorbed. The body can assemble some vitamins if the raw materials, called **provitamins**, are provided. For example, vitamin A is produced by the body from the provitamin beta-carotene, a chemical present in yellow vegetables such as carrots and in dark green vegetables such as spinach. No single food contains all of the required vitamins—one of the best reasons to eat a varied diet.

Vitamins are divided into two main groups: fat-soluble and water-soluble. The **fat-soluble vitamins**, vitamins A, D, E, and K, are absorbed along with other dietary lipids in the small intestine and packaged into chylomicrons. They cannot be absorbed in adequate quantity unless they are ingested with other lipids. Fat-soluble vitamins may be stored in cells, particularly hepatocytes. The **water-soluble vitamins**, including several B vitamins and vitamin C, are dissolved in body fluids. Excess quantities of these vitamins are not stored but instead are excreted in the urine.

Besides their other functions, three vitamins—C, E, and beta carotene (a provitamin)—are termed **antioxidant vitamins** because they inactivate oxygen free radicals. *Free radicals* are highly reactive ions or molecules that carry an unpaired electron in their outermost electron shell (see Figure 2.3). Free radicals damage cell membranes, DNA, and other cellular structures and contribute to the formation of artery-narrowing atherosclerotic plaques. Some free radicals arise naturally in the body, and others come from environmental hazards such as tobacco smoke and radiation. Antioxidant vitamins are thought to play a role in protecting against some kinds of cancer, reducing the buildup of atherosclerotic plaque, delaying some effects of aging, and decreasing the chance of cataract formation in the lenses of the eyes.

### ✓ CHECKPOINT

**52.** What does the MyPlate icon tell you about a healthy diet?

**53.** Briefly describe the functions of the following minerals: calcium, iron, magnesium, manganese, sodium, and phosphorus.

**54.** What is a vitamin? What is the general function of vitamins? How do we obtain vitamins?

**55.** How are fat-soluble and water-soluble vitamins different?

**56.** Which functions are served by antioxidant vitamins?

## Zachary's Story — EPILOGUE AND DISCUSSION

Thankfully, Zach did not need surgery this time. He was admitted into the hospital and had to stay for a little over a week to manage his bleeding and bring down the inflammation. His hepatopancreatic ampulla was partially blocked by inflammation and was causing malabsorption and prevention of efficient digestion, which ultimately led to severe abdominal discomfort and diarrhea. His diet and nutrition must be carefully monitored for the rest of his life, because as Dr. Lee explained, Crohn's disease is a chronic inflammatory disease affecting the bowels. Foods high in fiber and fats have to be limited or not ingested at all, and foods that cause an increase in gas production must be avoided as well. There is no known absolute cause of Crohn's disease, but it is widely considered an autoimmune disease. Zachary's immune system may be attacking the cells of his intestines, causing inflammation and his symptoms. Unfortunately for Zachary, there is no known cure for Crohn's disease. He may have to adhere to a regimen of immunosuppressant and anti-inflammatory drugs to manage his inflammatory reactions, and it is also possible that he may experience long periods of remission and live a mostly pain-free life. For now, Sandra is still vigilant in monitoring his energy levels and overall nutrition. Because Zach is much more comfortable now and sleeps through the night almost every night, so does Sandra.

I. *With Zach's new diet, which type of lipoprotein will decrease in circulation the most? Which type of lipoprotein do most people wish they could decrease, and why? Which one do most people wish they could increase?*

J. *In a healthy person, other molecules can be substituted into the pathways of glucose catabolism when the blood glucose level is low. What specific molecules will Zach now have problems making during times of low glucose and what is the normal source of these molecules?*

K. *If Zach is no longer ingesting foods high in lipid content, how will his body continue to supply itself with phospholipids, lipoproteins, and cholesterol? Without an adequate supply of lipids in the body, what process will most likely increase in his hepatocytes to ensure proper ATP production in times of low blood glucose?*

# Concept and Resource Summary

| Concept | Resources |
|---|---|

### Introduction

1. Food contains nutrients needed by the body for building and repair of tissues, and for sustaining metabolic reactions. The **digestive system** is responsible for digestion of food into smaller molecules capable of entering body cells, and for absorption of these smaller molecules through cells lining the stomach and intestines into the blood and lymph.

### Concept 23.1 The GI tract is a continuous multilayered tube extending from the mouth to the anus.

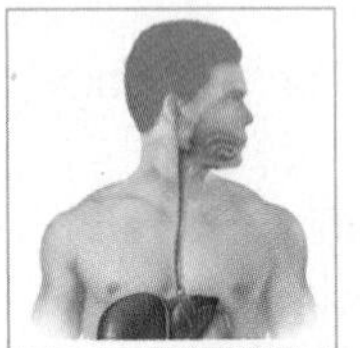

1. The digestive tract is comprised of the gastrointestinal tract and accessory digestive organs. The **gastrointestinal (GI) tract** is a continuous tube extending from the mouth to the anus. GI tract organs include the mouth, pharynx, esophagus, stomach, small intestine, and large intestine. The **accessory digestive organs** include the teeth, tongue, salivary glands, liver, gallbladder, and pancreas.
2. Digestion includes six basic processes: ingestion, secretion, mixing and propulsion, mechanical and chemical digestion, **absorption**, and **defecation**. **Mechanical digestion** breaks up and churns food, and **chemical digestion** enzymatically breaks down large nutrients into smaller molecules.
3. The basic arrangement of layers in most of the GI tract, from deep to superficial, is the **mucosa** (consisting of **epithelium**, **lamina propria**, and **muscularis mucosae**), **submucosa**, **muscularis**, and **serosa**.

Resources:

Anatomy Overview—GI Tract Histology

Animation—Neural Regulation of Mechanical Digestion

Figure 23.2—Layers of the Gastrointestinal Tract

**Concept** | **Resources**

4. The **peritoneum** is the largest serous membrane of the body. The **parietal peritoneum** lines the abdominopelvic cavity wall, and **visceral peritoneum** covers some of the cavity organs. The **peritoneal cavity** is a thin space containing serous fluid that is located between the parietal peritoneum and visceral peritoneum.
5. Folds of the peritoneum include the **greater omentum** draping over the transverse colon and small intestines, the **falciform ligament** attaching the liver to the abdominal wall and diaphragm, the **lesser omentum** suspending the stomach and duodenum from the liver, the **mesentery** binding the small intestine to the posterior abdominal wall, and the **mesocolon** binding the large intestine to the posterior abdominal wall.
6. The GI tract is regulated by an intrinsic set of nerves known as the **enteric nervous system (ENS)** and by an extrinsic set of nerves that is part of the autonomic nervous system (ANS). The ENS consists of neurons arranged into two plexuses: the **myenteric plexus** of the muscularis controls GI tract motility, and the **submucosal plexus** of the submucosa controls GI tract secretions.
7. Although the ENS can function independently, it is subject to regulation by the ANS. Parasympathetic stimulation increases GI tract secretion and motility by increasing the activity of ENS neurons. Sympathetic stimulation decreases GI tract secretion and motility by inhibiting ENS neurons.

## Concept 23.2 The mouth lubricates and begins digestion of food, and maneuvers it to the pharynx for swallowing.

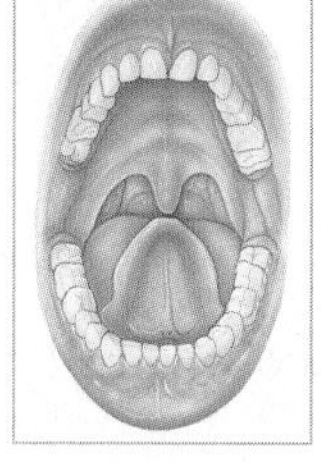

1. The **mouth** is formed by muscular cheeks, hard and soft palates, and tongue. The **lips** and **cheeks** help keep food between the teeth during chewing. The **vestibule** is the space bounded externally by the cheeks and lips and internally by the teeth and gums. The **oral cavity proper** extends from the vestibule to the **fauces**, the opening between the oral cavity and throat. The **hard palate** is the bony anterior roof of the mouth, and the **soft palate** is the muscular posterior roof of the mouth. The **uvula** hangs from the border of the soft palate and helps to close off the nasopharynx during swallowing.
2. The **tongue** forms the floor of the oral cavity. Muscles of the tongue maneuver food and assist with swallowing and speech. The upper surface and sides of the tongue are covered with **papillae**, some of which contain taste buds.
3. The three regions of a **tooth** are the visible **crown**, one to three **roots** embedded in sockets of the mandible and maxillae and surrounded by **gingivae**, and the constricted **neck** at the gum line. **Dentin** composes most of the tooth and encloses the **pulp cavity**. **Enamel**, the hardest substance in the body, covers the crown and protects the tooth from wear and tear and acids. **Cementum** and the **periodontal ligament** anchor the tooth in the socket.
4. There are two **dentitions**: the 20 **deciduous (primary) teeth** and 32 **permanent (secondary) teeth**. The permanent teeth include **incisors**, **cuspids**, **premolars**, and **molars**.
5. **Salivary glands** secrete saliva through ducts into the oral cavity. There are three pairs of major salivary glands: **parotid**, **submandibular**, and **sublingual glands**. **Saliva** moistens food, which helps form it into a ball for swallowing; dissolves foods so they can be tasted; and begins chemical digestion of carbohydrates.
6. **Salivation** is controlled by the autonomic nervous system.
7. Through **mastication**, food is mixed with saliva and shaped into a soft, flexible mass called a **bolus**. **Salivary amylase** begins the digestion of starches into disaccharides. **Lingual lipase** breaks down triglycerides into fatty acids and diglycerides.

Resources:

Anatomy Overview—Oral Cavity
Anatomy Overview—Salivary Glands
Animation—Mastication
Animation—Carbohydrate Digestion in the Mouth
Animation—Lipid Digestion in the Mouth
Animation—Chemical Digestion—Enzymes

Figure 23.8—Dentitions and Times of Eruptions

Clinical Connection—Mumps
Clinical Connection—Root Canal Therapy
Clinical Connection—Dental Caries
Clinical Connection—Periodontal Disease

## Concept 23.3 Swallowing consists of voluntary oral, involuntary pharyngeal, and involuntary esophageal stages.

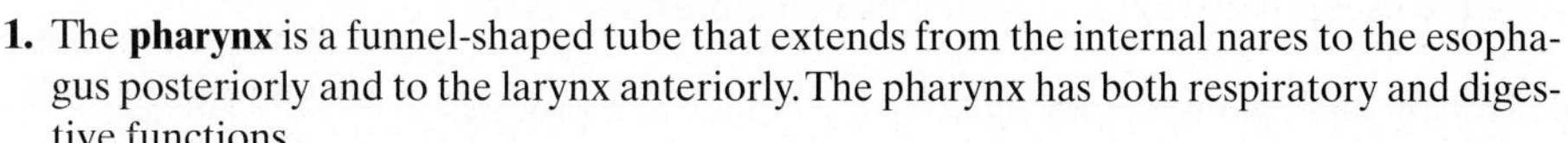

1. The **pharynx** is a funnel-shaped tube that extends from the internal nares to the esophagus posteriorly and to the larynx anteriorly. The pharynx has both respiratory and digestive functions.
2. The **esophagus** is a collapsible, muscular tube that connects the pharynx to the stomach. The **upper esophageal sphincter** regulates movement of food from the pharynx into the esophagus, and the **lower esophageal sphincter** regulates movement of food from the esophagus into the stomach. The esophagus is not involved in chemical digestion or absorption.

Resources:

Anatomy Overview—Pharynx and Esophagus
Anatomy Overview—Esophagus Histology
Animation—Deglutition

| Concept | Resources  |
|---|---|
| 3. **Deglutition** moves a bolus from the mouth to the stomach. The **voluntary stage** passes the bolus from the mouth into the oropharynx. The involuntary **pharyngeal stage** passes the bolus from the pharynx to the esophagus. During the involuntary **esophageal stage**, coordinated sequences of contraction and relaxation of the muscularis, called **peristalsis**, push food through the esophagus toward the stomach. | |

## Concept 23.4 The stomach mechanically breaks down the bolus and mixes it with gastric secretions.

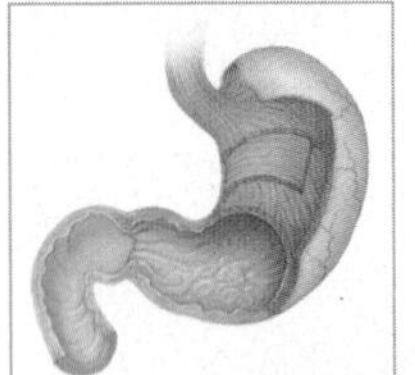

1. The stomach lies inferior to the diaphragm and connects the esophagus to the duodenum. The principal regions of the stomach are the **cardia**, **fundus**, **body**, and **pyloric part**. The pyloric part connects to the small intestine through the **pyloric sphincter**.
2. The stomach mucosa contains **gastric glands** that open into **gastric pits**. Gastric glands secrete **gastric juice** composed of mucus, hydrochloric acid, pepsinogen, gastric lipase, and intrinsic factor. The stomach muscularis has three alternately arranged layers of smooth muscle.
3. Mechanical digestion in the stomach consists of **mixing waves** that mix food with gastric juice and macerate it to a soupy **chyme**.
4. Hydrochloric acid in gastric juice kills microbes, partially denatures proteins, and stimulates hormone secretion, promoting flow of bile and pancreatic juice.
5. Chemical digestion consists mostly of the conversion of proteins into peptides by pepsin. **Chief cells** secrete pepsinogen, activated to **pepsin** by acidic pH, which begins protein digestion. **Gastric lipase** allows digestion of triglycerides into fatty acids and monoglycerides.
6. Absorption by the stomach is limited to water, certain ions, drugs, and alcohol.

**Resources:**
Anatomy Overview—The Stomach
Anatomy Overview—Stomach Histology
Animation—Stomach Peristalsis
Animation—Protein Digestion in the Stomach
Animation—Lipid Digestion in the Stomach
Animation—Chemical Digestion—Gastric Acid
Figure 23.12—Histology of the Stomach
Figure 23.13—Secretion of HCl by Parietal Cells in the Stomach
Clinical Connection—Pylorospasm and Pyloric Stenosis
Clinical Connection—Vomiting
Clinical Connection—Peptic Ulcer Disease

## Concept 23.5 The pancreas secretes pancreatic juice, the liver secretes bile, and the gallbladder stores and concentrates bile.

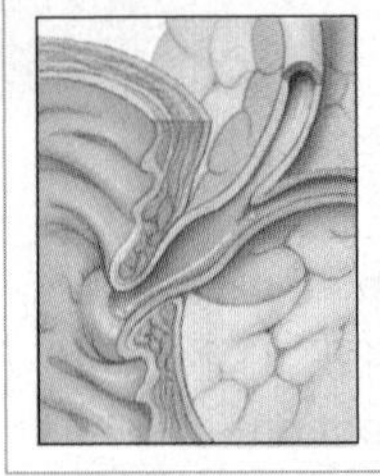

1. The **pancreas** consists of a **head**, a **body**, and a **tail** and secretes pancreatic juice through the **pancreatic duct** and **accessory duct** into the duodenum. Endocrine **pancreatic islets** secrete the hormones glucagon and insulin.
2. **Pancreatic juice** contains **sodium bicarbonate**, which makes it slightly alkaline to buffer the acidic gastric juice in chyme and inactivate pepsin. Pancreatic juice contains enzymes that digest starch (**pancreatic amylase**), proteins (**trypsin**, **chymotrypsin**, and **carboxypeptidase**), triglycerides (**pancreatic lipase**), and RNA and DNA (**ribonuclease** and **deoxyribonuclease**).
3. The liver is composed of a **right lobe** and **left lobe** that is continuous with the quadrate and caudate lobes. The lobes of the liver contain **hepatocytes** radiating out from a **central vein** and many **sinusoids** containing phagocytic **reticuloendothelial cells**.
4. Hepatocytes secrete bile into **bile canaliculi**, which drain into **bile ducts**. Bile travels in bile ducts to **right** and **left hepatic ducts**, then to the **common hepatic duct**, which joins the **cystic duct** of the gallbladder to form the **common bile duct**.
5. The liver receives oxygenated blood from the hepatic artery; nutrient-rich venous blood from the GI tract enters from the hepatic portal vein. The liver processes and detoxifies blood before it enters the venous circulation.
6. In the small intestine, **bile** breaks down large lipid globules into small lipid globules in a process called **emulsification**. Between meals, bile enters the gallbladder for storage.
7. Besides making bile, the liver functions in carbohydrate, lipid, and protein metabolism; processing of drugs and hormones; excretion of bilirubin; storage of vitamins and minerals; phagocytosis of old blood cells, and activation of vitamin D.
8. The **gallbladder** is a sac located on the posterior surface of the liver that stores and concentrates bile.

**Resources:**
Anatomy Overview—Pancreas
Anatomy Overview—Liver and Gallbladder
Anatomy Overview—Liver Histology
Anatomy Overview—Pancreas Histology
Animation—Carbohydrate Digestion—Pancreas
Animation—Protein Digestion—Pancreatic Juice
Animation—Lipid Digestion: Bile Salts and Pancreatic Lipase
Animation—Chemical Digestion—Bile
Figure 23.15—Relationship of the Pancreas to the Liver, Gallbladder, and Duodenum
Clinical Connection—Pancreatitis and Pancreatic Cancer
Clinical Connection—Hepatitis
Clinical Connection—Cirrhosis

## Concept

## Resources

### Concept 23.6 In the small intestine, chyme mixes with digestive juices from the small intestine, pancreas, and liver.

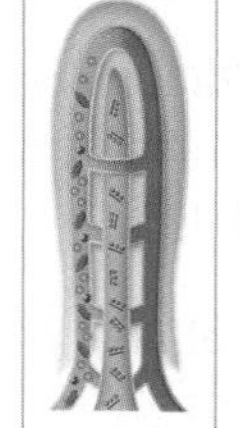

1. The **small intestine** is divided into the **duodenum**, the **jejunum**, and the **ileum** that joins the large intestine at the ilocecal sphincter. The small intestine wall has **absorptive cells** that absorb nutrients, **goblet cells** that secrete mucus, phagocytic **Paneth cells**, enteroendocrine cells that secrete secretin and cholecystokinin, **intestinal glands** that secrete intestinal juice, and **lymphatic nodules** that attack pathogens in food. **Circular folds** cause chyme to spiral and expose more of the mucosa to the nutrients. **Villi** and **microvilli** provide a large surface area for digestion and absorption, and the microvilli form a **brush border**.
2. **Intestinal juice** and pancreatic juice enhance absorption of substances from chyme. **Brush border enzymes** digest carbohydrates, proteins, and nucleotides.
3. **Segmentations** mix chyme with digestive juices and bring food into contact with the mucosa. Peristalsis moves chyme forward through the small intestine.
4. Digestion of carbohydrates begins by salivary amylase from the mouth and pancreatic amylase in pancreatic juice. The resulting glucose subunits are digested by brush border enzymes into monosaccharides.
5. Digestion of proteins begins by **pepsin** from the stomach and continues in the small intestine by the action of **trypsin**, **chymotrypsin**, and **carboxypeptidase** in pancreatic juice. Protein digestion is completed by brush border **peptidases**.
6. Some digestion of lipids occurs in the stomach by lingual lipase and gastric lipase. Once in the small intestine, larger lipid globules are **emulsified** into smaller lipid globules by the bile salts in bile. Pancreatic lipase attacks the smaller lipid globules, completing the digestion of triglycerides into fatty acids and a monoglyceride.
7. Pancreatic juice contains ribonuclease that digests RNA and deoxyribonuclease that digests DNA. Brush border **nucleosidases** and **phosphatases** complete nucleic acid breakdown.
8. Monosaccharides and amino acids, dipeptides, tripeptides, and short-chain fatty acids are absorbed into the small intestine epithelium then pass from the epithelium into blood capillaries.
9. Following lipid digestion, bile salts surround the fatty acids and monoglycerides, forming **micelles**. The fatty acids and monoglycerides diffuse out of micelles into the small intestine epithelium, where they recombine into **chylomicrons**. The chylomicrons pass from the epithelium into lacteals, where lymph transports them to the bloodstream.
10. The small intestine also absorbs electrolytes, vitamins, and water.

Anatomy Overview—Small Intestine
Anatomy Overview—Small Intestine Histology
Animation—Segmentation
Animation—Carbohydrate Digestion in the Small Intestine
Animation—Protein Digestion in the Small Intestine
Animation—Nucleic Acid Digestion in the Small Intestine
Animation—Carbohydrate Absorption in the Small Intestine
Animation—Protein Absorption in the Small Intestine
Animation—Nucleic Acid Absorption in the Small Intestine
Animation—Lipid Absorption in the Small Intestine
Figure 23.21—Absorption of Digested Nutrients in the Small Intestine
Clinical Connection—Lactose Intolerance
Clinical Connection—Absorption of Alcohol

### Concept 23.7 In the large intestine, the final secretion and absorption of nutrients occur as chyme moves toward the rectum.

1. The **large intestine** extends from the **ileocecal sphincter** to the anus. Its regions include the cecum, colon, rectum, and anal canal. The **cecum** is a small pouch hanging inferior to the ileocecal valve and has the **vermiform appendix** attached to it. The **colon** is divided into **ascending**, **transverse**, **descending**, and **sigmoid** portions. The terminal end of the **rectum** is the **anal canal**, with an exterior opening called the **anus**. Opening and closing of the anus is controlled by the involuntary **internal anal sphincter** and the voluntary **external anal sphincter**.
2. The mucosa has absorptive cells and goblet cells. The muscularis has smooth muscle bands called **teniae coli** that pucker its wall into **haustra**.
3. Food is moved through the large intestine by **haustral churning** and peristalsis. **Mass peristalsis** drives colon contents into the rectum.
4. Bacteria inhabiting the lumen break down remaining carbohydrates, proteins, and bilirubin; and produce vitamin K and some B vitamins.
5. The large intestine absorbs water, ions, and vitamins. It contributes to the final formation of **feces** and its elimination from the body.
6. When mass peristalsis pushes feces into the rectum, distension of the rectal wall initiates the **defecation reflex**. Parasympathetic stimulation opens the internal anal sphincter. Voluntary relaxation of the external anal sphincter allows feces to be expelled through the anus.

Anatomy Overview—The Large Intestine
Anatomy Overview—Large Intestine Histology
Animation—Mechanical Digestion in the Large Intestine
Exercise—Digestive Enzyme Activity
Exercise—Concentrate on Digestion
Exercise—Paint the Digestive Tract
Exercise—Digestion Reaction Race
Exercise—Role of Chemicals
Exercise—Match the Movement
Exercise—Nutrient Absorption Routes
Clinical Connection—Polyps in the Colon
Clinical Connection—Dietary Fiber
Clinical Connection—Diverticular Disease
Clinical Connection—Colorectal Cancer

## Concept

**Resources** WILEY PLUS

### Concept 23.8 Digestive activities occur in three overlapping phases: cephalic, gastric, and intestinal.

1. Digestive activities occur in three overlapping phases: cephalic phase, gastric phase, and intestinal phase.
2. During the **cephalic phase** of digestion, salivary glands secrete saliva and gastric glands secrete gastric juice in order to prepare the mouth and stomach for food that is about to be eaten.
3. The presence of food in the stomach causes the **gastric phase**, involving peristalsis and gastric juice secretion from gastric glands. The hormone **gastrin** stimulates gastric juice secretion.
4. During the **intestinal phase** of digestion, food is digested in the small intestine. In addition, **reflex** activity slows gastric motility in order to slow the exit of chyme from the stomach, which prevents the small intestine from being overloaded with more chyme than it can handle. **Secretin** and **cholecystokinin** stimulate the secretion of pancreatic juice into the small intestine.

Resources:
- Animation—Neural Regulation of Mechanical Digestion
- Animation—Enterogastric Reflex
- Animation—Hormonal Control of Digestive Activities
- Exercise—Digestive Hormone Activities
- Concepts and Connections—Digestive Hormones

### Concept 23.9 Metabolism includes the catabolism and anabolism of molecules.

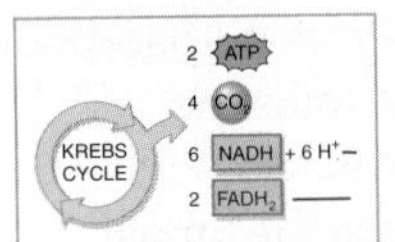


1. **Metabolism** refers to all of the chemical reactions of the body. **Catabolism** is the exergonic reactions that break down complex organic molecules into simple molecules. **Anabolism** is the endergonic reactions that synthesize complex molecules from simpler molecules. **Enzymes** are the catalysts for all metabolic reactions. Some enzymes require **coenzymes** to function properly.
2. **ATP** links anabolic and catabolic reactions. Energy to run anabolic reactions comes from energy that is released when ATP is separated into ADP and a phosphate group Ⓟ. Energy released from catabolic reactions allows bonding of ADP and a phosphate group to resynthesize ATP.
3. The glucose that results from carbohydrate digestion is oxidized by cells to provide ATP. Glucose not needed for immediate ATP production is converted to amino acids, glycogen, or triglycerides.
4. **Cellular respiration**, the complete oxidation of glucose to $CO_2$ and $H_2O$, involves glycolysis, the Krebs cycle, and the electron transport chain. **Glycolysis** breaks down glucose into two molecules of pyruvic acid. Each molecule of pyruvic acid that enters the Krebs cycle is first converted to **acetyl coenzyme A**. In the **Krebs cycle** acetyl coenzyme A is used to produce $CO_2$, (NADH + $H^+$), $FADH_2$, and ATP. In the **electron transport chain** the energy in (NADH + $H^+$) and $FADH_2$ is liberated and transferred to ATP. The electron transport chain produces ATP and $H_2O$.
5. The liver can form glucose from lactic acid, amino acids, and glycerol in a process called **gluconeogenesis** when stimulated by cortisol from the adrenal cortex and glucagon from the pancreas.
6. In **lipolysis**, triglycerides are split into glycerol and fatty acids, which then can be further catabolized by different pathways to produce ATP. Insulin stimulates liver cells and adipose cells to synthesize triglycerides when more calories are consumed than required by the body's ATP demands.
7. Lipoproteins transport lipids in the bloodstream. Types of lipoproteins include chylomicrons, which carry dietary lipids to adipose tissue; **very low-density lipoproteins**, which carry triglycerides from the liver to adipose tissue; **low-density lipoproteins**, which deliver cholesterol to body cells; and **high density lipoproteins**, which remove excess cholesterol from body cells and transport it to the liver for elimination.
8. Proteins are not stored in the body. Amino acids are either oxidized to produce ATP or used to synthesize new proteins for growth and repair. Excess dietary amino acids are converted into glucose or triglycerides. Protein synthesis is stimulated by insulinlike growth factors, thyroid hormones, insulin, estrogens, and testosterone. **Essential amino acids** must be present in the diet; **nonessential amino acids** can be synthesized in the body. A **complete protein** contains all of the essential amino acids; an **incomplete protein** does not contain all of the essential amino acids.

Resources:
- Animation—Introduction to Metabolism
- Animation—Carbohydrate Metabolism
- Animation—Lipid Metabolism
- Animation—Protein Metabolism
- Exercise—Predict ATP Production
- Exercise—Glucose Catabolism Sequence
- Exercise—Glucose Catabolism Substrates and Products
- Concepts and Connections—Glucose Catabolism

### Concept 23.10 Food molecules supply energy for life processes and serve as building blocks for complex molecules.

1. Body cells use **nutrients** in food for growth, maintenance, and repair. Nutrients include water, carbohydrates, lipids, proteins, minerals, and vitamins.
2. The **MyPlate** guide recommends that vegetables and fruits take up one half of each meal, protein and grains take up the other half, and dairy is included in every meal.

Resources:
- Anatomy Overview—Role of Nutrients

| Concept | Resources |
|---|---|
| 3. **Minerals** important to the body include calcium, phosphorus, potassium, sulfur, sodium, chloride, magnesium, iron, iodide, manganese, copper, cobalt, zinc, fluoride, selenium, and chromium. Many minerals play key roles as components of coenzymes, are components of buffers that help control body pH, and are needed for impulses. | |
| 4. **Vitamins** maintain growth and normal metabolism. Most function as coenzymes rather than nutrients. Most vitamins must be obtained from the diet. Some vitamins can be assembled by the body if **provitamins** are provided in the diet. **Fat-soluble vitamins** must be ingested with lipids for proper absorption. **Water-soluble vitamins** are dissolved in blood fluids. **Antioxidant vitamins** inactivate oxygen free radicals that damage cell structures. | |

## Understanding the Concepts

1. What are the similarities between mechanical and chemical digestion? How do they differ?
2. What are the functions of the muscles and papillae of the tongue? Of saliva?
3. How does the mouth participate in chemical digestion?
4. What is the role of the esophagus in digestion?
5. What is the importance of surface mucous cells, mucous neck cells, chief cells, parietal cells, and enteroendocrine cells in the stomach?
6. What are the functions of gastrin, hydrochloric acid, gastric lipase, and lingual lipase in the stomach?
7. How do the functions of pancreatic acini differ from those of the pancreatic islets?
8. Once the liver has formed bile, how is bile transported to the gallbladder for storage?
9. What is the pathway of blood flow into, through, and out of the liver?
10. In which ways are the mucosa and submucosa of the small intestine adapted for digestion and absorption?
11. What is the functional significance of the blood capillary network and lacteal in the center of each villus of the small intestine?
12. What is the difference between digestion and absorption? Which organ of the GI tract absorbs the most nutrients?
13. Describe the absorption of electrolytes, vitamins, and water by the small intestine.
14. Referring to Figure 23.22, which two organs of the digestive system secrete the most fluid? Which organ of the digestive system absorbs the most fluid?
15. How does the defecation reflex occur?
16. Describe the role of gastrin in the gastric phase of digestion.
17. What are the major events in the catabolism of lipids to produce ATP?
18. What is a nutrient? What are the general functions of water, carbohydrates, lipids, proteins, and minerals?

CHAPTER

# 24

## Sam's Story

I will be right down, Cindy says and hangs up her office phone. She does not bother to finish what she is doing, because when the ER calls her she knows the case will be serious. Cindy has been a social worker in a large county hospital for six years now and has seen a lot of things, but she consciously walks into every new case with an optimistic attitude. The elevator opens and she sees the ER doctor who called her.

"What do you have for me?" Cindy asks.

"We have an eleven-year-old girl who has a bladder infection and possibly a kidney infection, which most likely started from a lower urethral urinary tract infection. We're almost positive that her infection has progressed to the level it's at right now because of severe neglect."

Cindy's optimism falters slightly. "What's her name, and how bad is it right now?"

"Her name is Samantha, and she's having a lot of pain in her urogenital, abdominopelvic, and lower back areas. She's thrown up once and she's shivering. She has a fever of 103.4°F. We've taken urine and blood samples, and her urine came out very bloody." As she approaches the bed pointed out by the doctor, Cindy reminds herself that it is her job to create a relationship based on trust and ensure that Samantha feels safe. She pulls the curtain back slowly and sees Samantha flinch.

"Hi, Samantha. Can I call you Sam? My name is Cindy. I don't really have a nickname but you can make one up if you like. I'm a social worker here at the hospital." Samantha does not respond right away, her face clearly expressing some of the pain she is feeling.

"Where's my Mommy and Daddy?" Samantha barely whispers, without making eye contact with Cindy.

Resisting the impulse to give her what appears to be a much-needed hug, Cindy knows Samantha is not ready for a stranger to touch her so she squats down to be at eye level and replies, "I don't know, but I will find out for you, okay? I want you to know that it's my job to make sure that you're safe and I will do everything I can to help you."

# The Urinary System

## INTRODUCTION

As body cells carry out their metabolic functions, they consume oxygen and nutrients and produce substances such as carbon dioxide and nitrogenous wastes that have no useful functions and need to be eliminated from the body. While the respiratory system rids the body of carbon dioxide, the urinary system disposes of most other unneeded substances. The **urinary system** consists of two kidneys, two ureters, one urinary bladder, and one urethra (see Figure 24.1). After the kidneys filter blood plasma, they return most of the water and needed solutes to the bloodstream while selectively eliminating unneeded substances. The remaining water and solutes constitute **urine**, which passes through the ureters and is stored in the urinary bladder until it is excreted from the body through the urethra. Besides forming urine, the urinary system contributes to homeostasis by altering the composition, pH, volume, and pressure of blood, and by producing hormones.

## CONCEPTS

**24.1** The kidneys regulate the composition of the blood, produce hormones, and excrete wastes.

**24.2** As urine forms, it travels through the renal medulla, calyces, and renal pelvis.

**24.3** Each nephron consists of a renal corpuscle and a renal tubule.

**24.4** Urine is formed by glomerular filtration, tubular reabsorption, and tubular secretion.

**24.5** Water and solutes are forced through the filtration membrane during glomerular filtration.

**24.6** Tubular reabsorption reclaims needed substances from the filtrate, while tubular secretion discharges unneeded substances.

**24.7** Five hormones regulate tubular reabsorption and tubular secretion.

**24.8** The kidneys regulate the rate of water loss in urine.

**24.9** The kidneys help maintain the overall fluid and acid–base balance of the body.

**24.10** The ureters transport urine from the renal pelvis to the urinary bladder, where it is stored until micturition.

## 24.1 The kidneys regulate the composition of the blood, produce hormones, and excrete wastes.

The kidneys do the major work of the urinary system. The other parts of the system (ureters, urinary bladder, urethra) are mainly passageways and storage areas (Figure 24.1).

Functions of the kidneys include the following:

- ***Regulation of blood ionic composition.*** The kidneys help regulate the blood levels of several ions, most importantly sodium ions ($Na^+$), potassium ions ($K^+$), calcium ions ($Ca^{2+}$), chloride ions ($Cl^-$), and phosphate ions ($HPO_4^{2-}$).
- ***Regulation of blood pH.*** The kidneys excrete a variable amount of hydrogen ions ($H^+$) into the urine and conserve bicarbonate ions ($HCO_3^-$), which are an important buffer of $H^+$ in the blood. Both of these activities help regulate blood pH.
- ***Regulation of blood volume.*** The kidneys adjust blood volume by conserving or eliminating water in the urine. An increase in blood volume increases blood pressure; a decrease in blood volume decreases blood pressure.
- ***Regulation of blood pressure.*** The kidneys also help regulate blood pressure by secreting the enzyme renin, which activates the renin–angiotensin–aldosterone pathway (see Concept 24.7 and Figure 17.15). Increased renin causes an increase in blood pressure.
- ***Production of hormones.*** The kidneys produce two hormones. *Calcitriol*, the active form of vitamin D, helps regulate calcium homeostasis (see Figure 17.13), and *erythropoietin* stimulates the production of red blood cells (see Figure 18.5).
- ***Regulation of blood glucose level.*** Like the liver, the kidneys can use the amino acid glutamine in *gluconeogenesis*, the synthesis of new glucose molecules. They can then release glucose into the blood to help maintain a normal blood glucose level.
- ***Excretion of wastes and foreign substances.*** By forming urine, the kidneys help excrete **wastes**—substances that have no useful function in the body. Some wastes excreted in urine result from metabolic reactions in the body. These include ammonia and urea from the breakdown of amino acids; bilirubin from the breakdown of hemoglobin; creatinine from the breakdown of creatine phosphate in muscle fibers; and uric acid from the breakdown of nucleic acids. Other wastes excreted in urine are foreign substances from the diet, such as drugs and environmental toxins.

**✓ CHECKPOINT**

1. Which organs constitute the urinary system?
2. How do the kidneys alter blood chemistry?

FIGURE 24.1 **Organs of the urinary system in a female.**

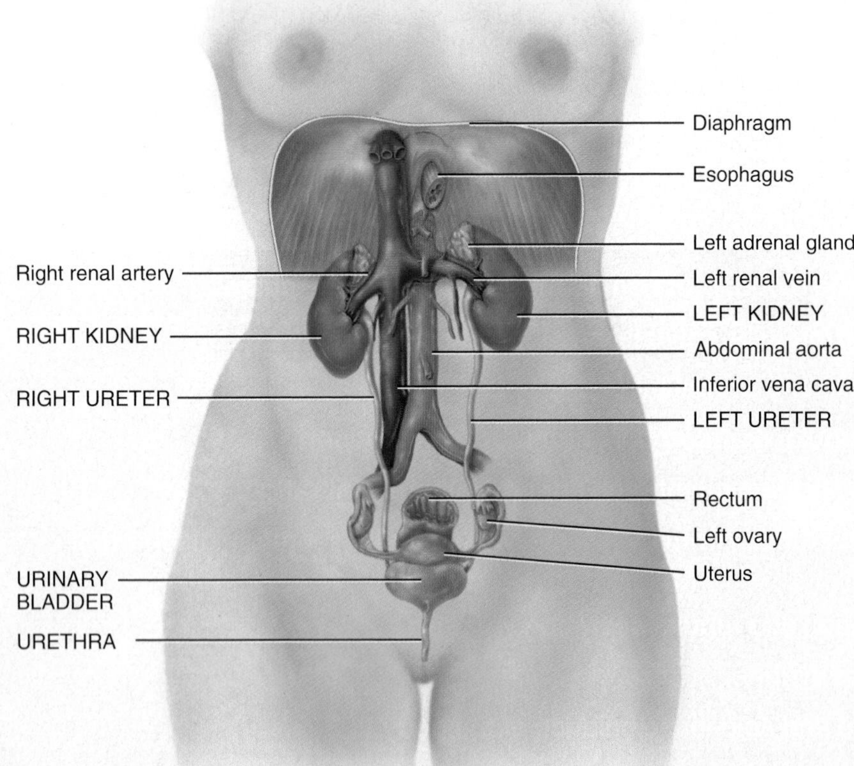


**FUNCTIONS OF THE URINARY SYSTEM**

1. Regulates blood volume and composition.
2. Helps regulate blood pressure.
3. Synthesizes glucose.
4. Secretes erythropoietin.
5. Participates in vitamin D synthesis.
6. Excretes wastes in the urine.
7. Stores urine in urinary bladder.
8. Discharges urine from the body through the urethra.

**Urine formed by the kidneys passes first into the ureters, then to the urinary bladder for storage, and finally through the urethra for elimination from the body.**

# 24.2 As urine forms, it travels through the renal medulla, calyces, and renal pelvis.

The paired **kidneys** are reddish, kidney-bean-shaped organs located just above the waist between the peritoneum and the posterior wall of the abdomen. Because they are positioned posterior to the peritoneum of the abdominal cavity, they are said to be **retroperitoneal** (re-trō-per-i-tō-NĒ-al; *retro-* = behind) organs (Figure 24.2). The kidneys are located between the levels of the twelfth thoracic and third lumbar vertebrae, a position where they are partially protected by the lower thoracic cage. The right kidney is slightly lower than the left (see Figure 24.1) because the liver occupies considerable space on the right side superior to the kidney.

FIGURE 24.2 **Position and coverings of the kidneys.**

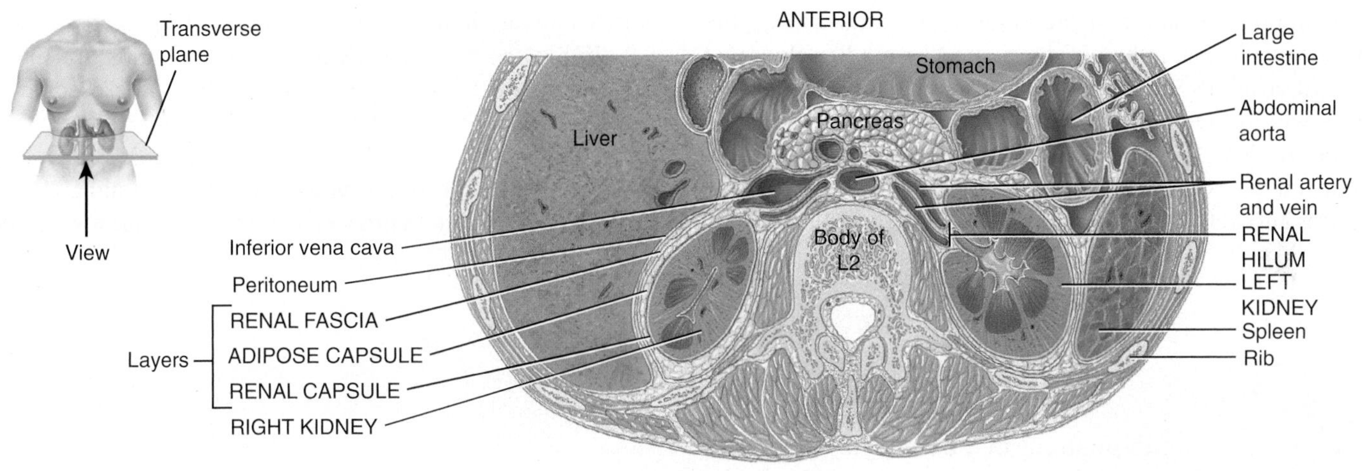


(a) Inferior view of transverse section of abdomen (L2)

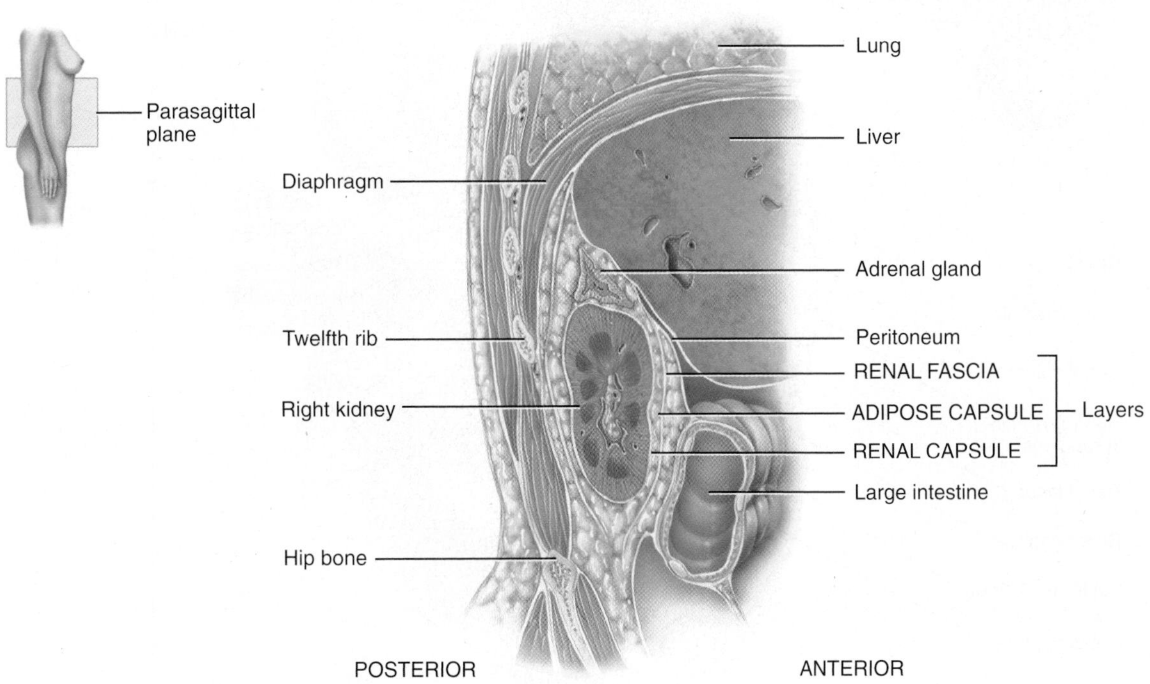


(b) Sagittal section through right kidney

The kidneys are surrounded by a renal capsule, adipose capsule, and renal fascia.

## External Anatomy of the Kidneys

A typical adult kidney is 10–12 cm (4–5 in.) long, 5–7 cm (2–3 in.) wide, and 3 cm (1 in.) thick and has a mass of 135–150 g (4.5–5 oz). The concave medial border of each kidney faces the vertebral column (see Figure 24.1). Near the center of the concave border is an indentation called the **renal hilum** (RĒ-nal HĪ-lum; *renal* = kidney) (Figure 24.3), through which the ureter leaves the kidney, and blood vessels, lymphatic vessels, and nerves enter and exit.

Three layers of tissue surround each kidney (Figure 24.2). The deep layer, the **renal capsule**, is a smooth, transparent sheet of dense irregular connective tissue that is continuous with the outer coat of the ureter. It serves as a barrier against trauma and helps maintain the shape of the kidney. The middle layer, the **adipose capsule**, is a mass of fatty tissue surrounding the renal capsule. It cushions the kidney from trauma and holds it firmly in place within the abdominal cavity. The superficial layer, the **renal fascia** (FASH-ē-a), is another thin layer of dense irregular connective tissue that anchors the kidney to the surrounding structures and to the abdominal wall.

## Internal Anatomy of the Kidneys

A frontal section through the kidney reveals two distinct regions: a superficial, light red area called the **renal cortex** (*cortex* = rind or bark) and a deep, darker reddish-brown region called the **renal medulla** (*medulla* = inner portion) (Figure 24.3a). The renal medulla consists of several cone-shaped **renal pyramids**. The base (wider end) of each pyramid faces the renal cortex, and its apex (narrower end), called a **renal papilla**, points toward the renal hilum. The renal cortex is the smooth-textured area extending from the renal capsule to the bases of the renal pyramids and into the spaces between them. Those portions of the renal cortex that extend between renal pyramids are called **renal columns**. A **renal lobe** consists of a renal pyramid, its overlying area of renal cortex, and one-half of each adjacent renal column.

Within the renal cortex and renal pyramids are the functional units of the kidney—about 1 million microscopic structures called **nephrons** (NEF-rons; *nephros* = kidney). Urine formed by the nephrons drains into large **papillary ducts**, which extend through the renal papillae of the pyramids. The papillary ducts drain into cuplike structures called **minor** and **major**

FIGURE 24.3 **Internal anatomy of the kidneys.**

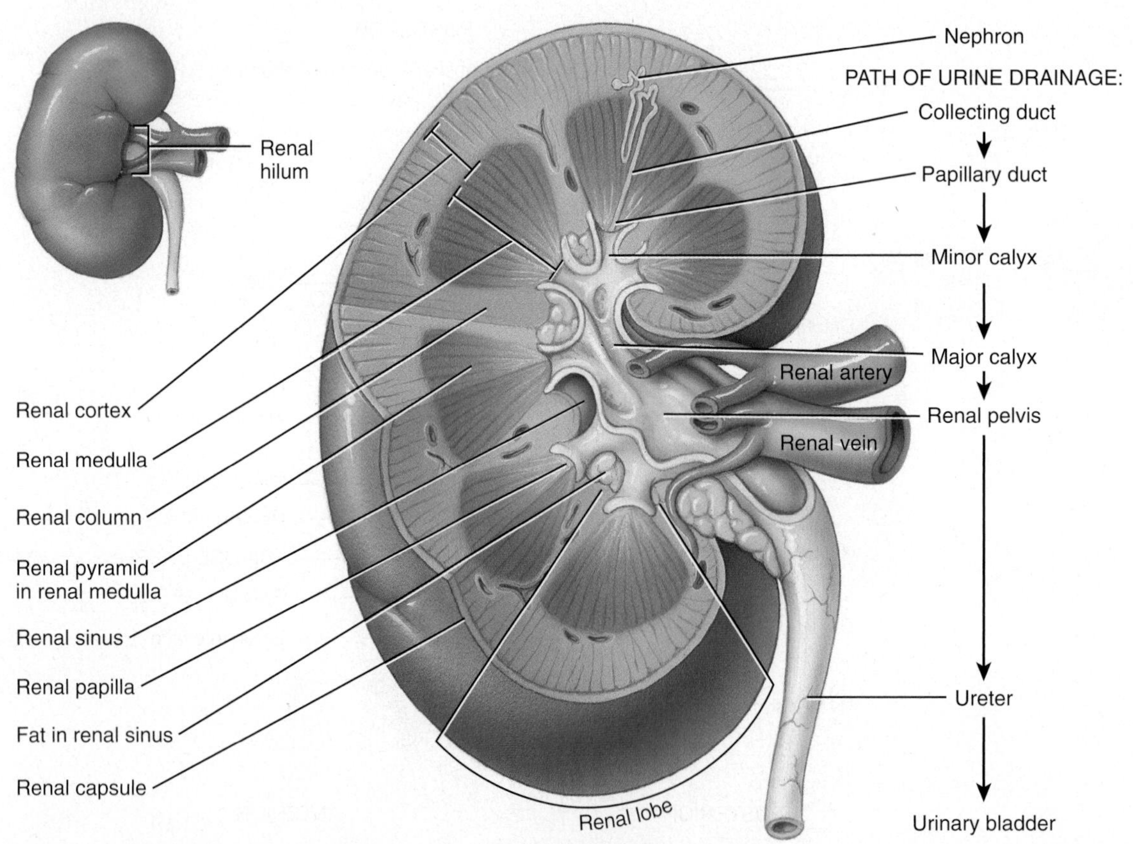


(a) Frontal section of right kidney

**calyces** (KĀ-li-sēz = cups; singular is *calyx,* pronounced KĀ-liks). Each kidney has 8 to 18 minor calyces and 2 or 3 major calyces. A minor calyx receives urine from the papillary ducts of one renal papilla and delivers it to a major calyx. From the major calyces, urine drains into a single large cavity called the **renal pelvis** (*pelv-* = basin) and then out through the ureter to the urinary bladder.

The hilum expands into a cavity within the kidney called the **renal sinus**, which contains part of the renal pelvis, the calyces, and branches of the renal blood vessels and nerves. Adipose tissue helps stabilize the position of these structures in the renal sinus.

## Blood Supply of the Kidneys

Because the kidneys remove wastes from the blood and regulate its volume and ionic composition, it is not surprising that they are abundantly supplied with blood vessels. Although the kidneys constitute less than 0.5 percent of total body mass, they receive 20–25 percent of the resting cardiac output via the right and left **renal arteries** (Figure 24.4). In adults, blood flow through both kidneys is about 1200 mL per minute.

Within the kidney, the renal artery divides into several **segmental arteries** (seg-MEN-tal)**,** which supply different segments (areas) of the kidney. Each segmental artery gives off several branches that pass through the renal columns between the renal pyramids as the **interlobar arteries** (in′-ter-LŌ-bar). At the bases of the renal pyramids, the interlobar arteries arch between the renal medulla and cortex; here they are known as the **arcuate arteries** (AR-kū-āt = shaped like a bow). Divisions of the arcuate arteries divide into small **cortical radiate arteries** (also called *interlobular arteries*) that extend into the renal cortex and give off branches called **afferent arterioles** (AF-er-ent; *af-* = toward; *-ferrent* = to carry).

Each nephron receives one afferent arteriole, which divides into a tangled, ball-shaped capillary network called the **glomerulus** (glō-MER-ū-lus = little ball; plural is *glomeruli*). The glomerular capillaries then reunite to form an **efferent arteriole** (EF-er-ent; *ef-* = out) that carries blood out of the glomerulus. Glomerular capillaries are unique among capillaries in the body because they are positioned between two arterioles, rather than between an arteriole and a venule. The glomeruli are considered part of both the cardiovascular system and the urinary system because they are capillary networks that also play an important role in urine formation.

The efferent arterioles divide to form the **peritubular capillaries** (per-i-TOOB-ū-lar; *peri-* = around), which surround tubular parts of the nephron in the renal cortex. Extending from some efferent arterioles are long loop-shaped capillaries called **vasa recta** (VĀ-sa REK-ta; *vasa* = vessels; *recta* = straight) that supply tubular portions of the nephron in the renal medulla (see Figure 24.5c).

The peritubular capillaries eventually drain into the **cortical radiate veins** (also called *interlobular veins*), which also receive blood from the vasa recta. Then the blood drains through the **arcuate veins** to the **interlobar veins** running between the renal pyramids. Blood leaves the kidney through the **renal vein** that exits at the renal hilum and carries venous blood to the inferior vena cava. Figure 24.4b summarizes the path of blood flow through the kidneys.

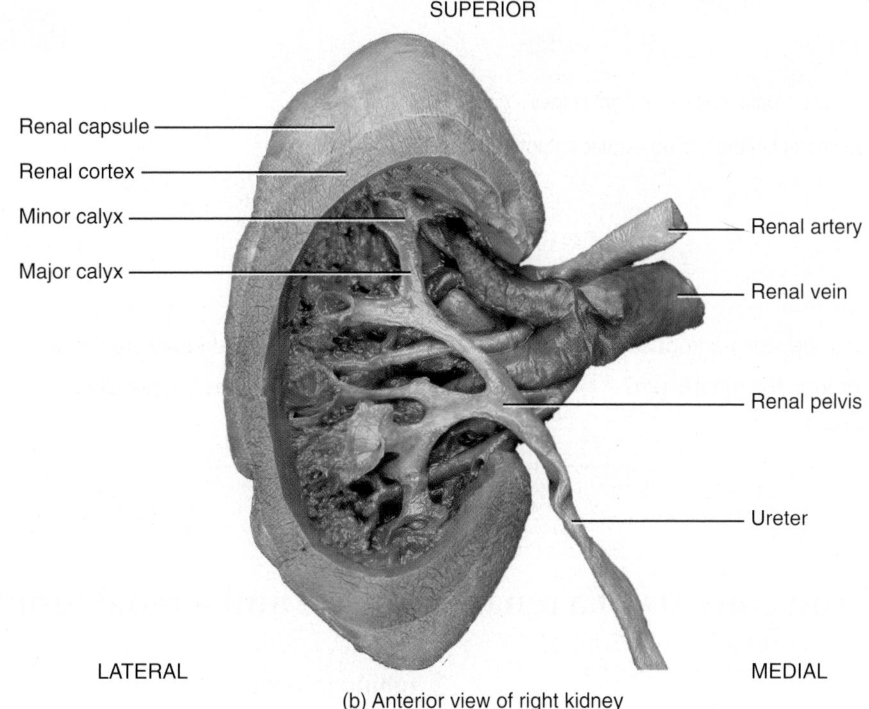


(b) Anterior view of right kidney

The two main regions of the kidney are the renal cortex and the renal medulla.

FIGURE 24.4 **Blood supply of the kidneys.**

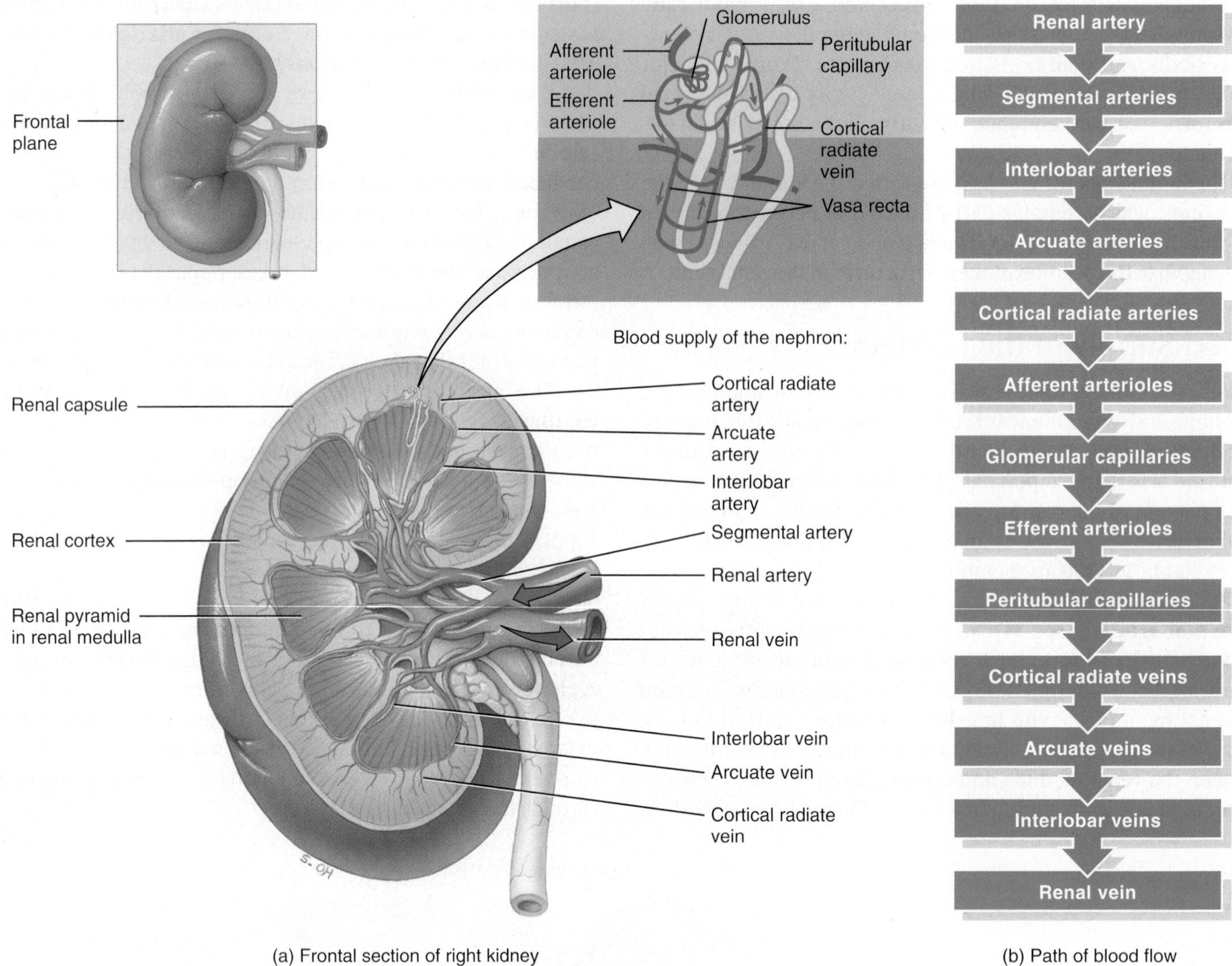


(a) Frontal section of right kidney

(b) Path of blood flow

**The renal arteries deliver 20–25 percent of the resting cardiac output to the kidneys.**

✓ **CHECKPOINT**

3. Why are the kidneys said to be retroperitoneal?
4. Which structures pass through the renal hilum?
5. How is the kidney protected?
6. Where in the kidney are the renal pyramids located?
7. Which arteries supply blood to the cortical radiate arteries?

# 24.3 Each nephron consists of a renal corpuscle and a renal tubule.

## Parts of a Nephron

Nephrons are the functional units of the kidneys. Each nephron (Figure 24.5) consists of two parts: a **renal corpuscle** (KOR-pus-sul = tiny body), where blood plasma is filtered, and a **renal tubule** into which the filtered fluid passes. Closely associated with a nephron is its blood supply. The two components of a renal corpuscle are the **glomerulus** (glō-MER-ū-lus) (capillary network) and the **glomerular capsule** (or *Bowman's capsule*), a double-walled cup of epithelial cells that surrounds the glomerular capillaries. Blood plasma is filtered into the glomerular capsule, and then the filtered fluid passes into the renal tubule, which has three main sections. In the order that fluid passes through them, the renal tubule consists of a (1) **proximal convoluted tubule** (kon′-vō-LOOT-ed), (2) **nephron loop** (or *loop of Henle*), and (3) **distal convoluted tubule**. *Proximal* denotes the part of the tubule attached to the glomerular capsule, and *distal* denotes the part that is farther away. *Convoluted* means that the tubule is tightly coiled rather than straight.

The renal corpuscle and both convoluted tubules lie within the renal cortex. The nephron loop, which connects the proximal and distal convoluted tubules, extends into the renal medulla. The first part of the nephron loop begins in the renal cortex and extends downward into the renal medulla, where it is called the **descending limb of the nephron loop** (Figure 24.5). It then makes that hairpin turn and returns to the renal cortex as the **ascending limb of the nephron loop**.

About 85 percent of the nephrons are **cortical nephrons** (KOR-ti-kul), which have *short* nephron loops that lie mainly in the cortex and penetrate only into the outer region of the renal medulla (Figure 24.5b). The short nephron loops receive their blood supply from peritubular capillaries. The other 15 percent of the nephrons are **juxtamedullary nephrons**

FIGURE 24.5 **The structure of nephrons and associated blood vessels.** (a) Nephron structure. (b) A cortical nephron. (c) A juxtamedullary nephron.

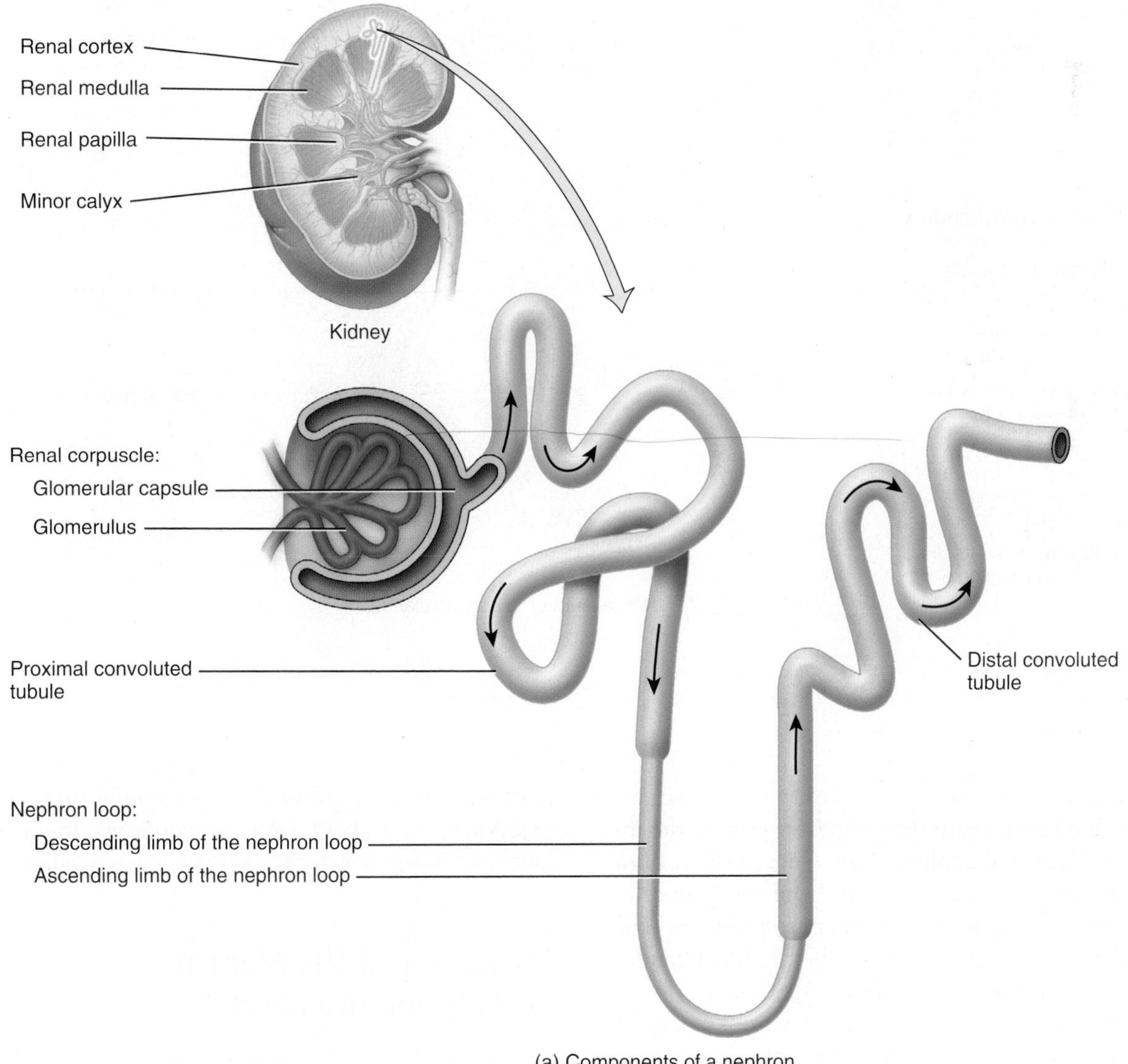


(a) Components of a nephron

(continues)

FIGURE 24.5 (continued)

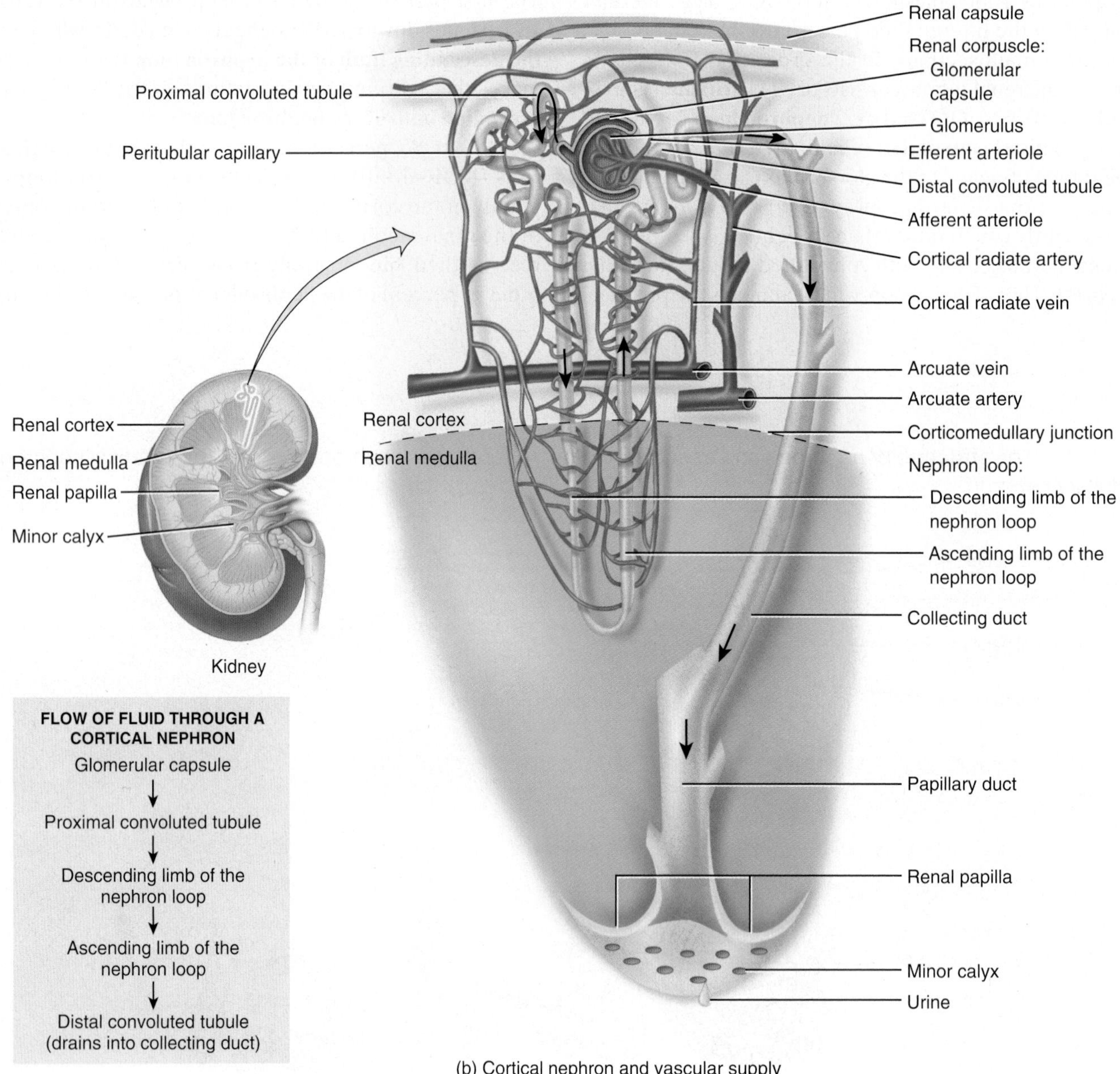


(b) Cortical nephron and vascular supply

(juks′-ta-MED-ū-lar′-e; *juxta-* = near to), which have *long* nephron loops that extend into the deepest region of the medulla (Figure 24.5c). Long nephron loops receive their blood supply from peritubular capillaries and from the vasa recta. In addition, the ascending limb of the nephron loop of juxtamedullary nephrons consists of two portions: a **thin ascending limb** followed by a **thick ascending limb** (Figure 24.5c).

The distal convoluted tubules of several nephrons empty into a single **collecting duct**. Collecting ducts then drain into several hundred large **papillary ducts** (PAP-i-lar′-ē), which drain into the minor calyces. The collecting ducts and papillary ducts extend from the renal cortex through a renal pyramid in the renal medulla. One kidney has about 1 million nephrons, but a much smaller number of collecting ducts, and even fewer papillary ducts.

## Histology of the Nephron and Collecting Duct

A single layer of epithelial cells forms the entire wall of the glomerular capsule, renal tubule, and ducts. However, each part has distinctive histological features that reflect its particular functions.

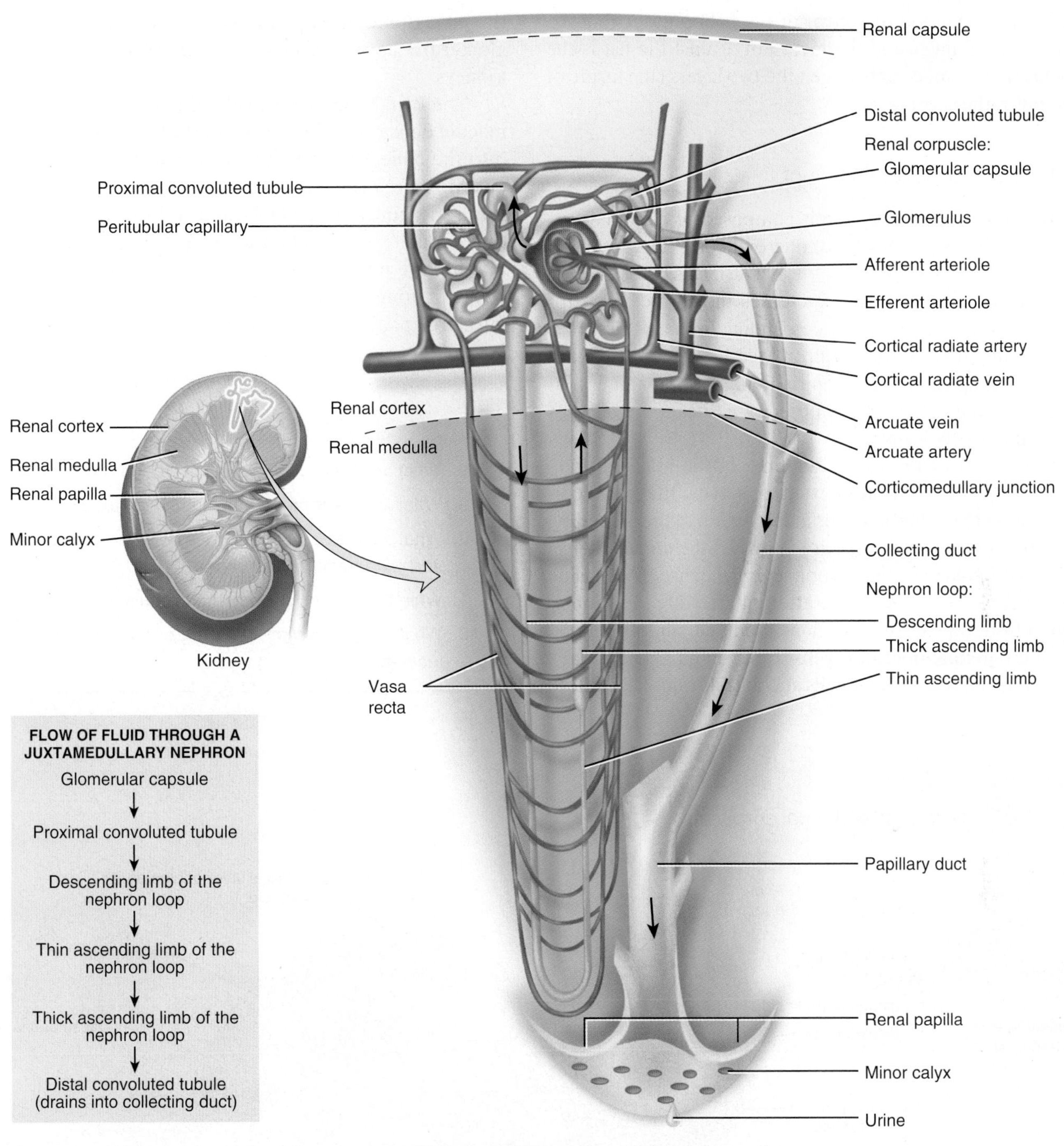


(c) Juxtamedullary nephron and vascular supply

**Nephrons are the functional units of the kidneys.**

We will discuss them in the order that fluid flows through them: glomerular capsule, renal tubule, and collecting duct.

## Glomerular Capsule

The glomerular capsule consists of outer and inner layers. The *parietal layer* forms the outer wall of the glomerular capsule and consists of simple squamous epithelium. The *visceral layer* forms the inner wall of the capsule and consists of modified squamous cells called **podocytes** (PŌ-dō-cīts; *podo-* = foot; *-cytes* = cells). The many footlike projections of these cells (*pedicels*) wrap tightly around the endothelial cells of the glomerulus. Fluid filtered from the glomerular capillaries enters the **capsular space**, the space between the two layers of the glomerular capsule. Think of the glomerulus as a fist punched into a limp balloon (the glomerular capsule) until the fist is covered by two layers of balloon; the layer of the balloon touching the fist is the visceral layer and the layer not against the fist is the parietal layer. The space in between the two layers (the inside of the balloon) is the capsular space.

## Renal Tubule and Collecting Duct

The proximal convoluted tubule is composed of cuboidal epithelial cells with a brush border of microvilli on their apical surface (surface facing the lumen) (Figure 24.6). This brush border, like that of the small intestine, increases the surface area for reabsorption and secretion. The descending limb and first part of the ascending limb of the nephron loop have very thin walls composed of simple squamous epithelium. The remaining portion of the ascending limb is composed of a thicker cuboidal or even columnar epithelium.

The final part of the ascending limb of the nephron loop makes contact with the afferent arteriole serving that renal corpuscle. Because the columnar cells in this region of the ascending limb are crowded together, they are known as the **macula densa** (MAK-ū-la DEN-sa; *macula* = spot; *densa* = dense). Alongside the macula densa, the wall of the afferent arteriole (and sometimes the efferent arteriole) contains modified smooth muscle fibers called **juxtaglomerular cells** (juks′-ta-glō-MER-ū-lar). Together, the juxtaglomerular cells and macula densa constitute the **juxtaglomerular apparatus**. As you will see later, the juxtaglomerular apparatus helps regulate blood pressure within the kidneys.

The distal convoluted tubule (DCT) begins just past the macula densa cells. In the last part of the DCT and continuing into the collecting duct, two different types of cells are present. Most are **principal cells**, which have receptors for both antidiuretic hormone (ADH) and aldosterone, two hormones that regulate their functions (see Concept 24.7). A smaller number are **intercalated cells** (in-TER-ka-lā-ted), which play a role in homeostasis of blood pH. The collecting ducts drain into large papillary ducts, which are lined by simple columnar epithelium.

### ✓ CHECKPOINT

8. What are the two main parts of a nephron?
9. What are the components of the renal tubule?
10. What are the basic differences between cortical nephrons and juxtamedullary nephrons?
11. Which structures enclose the capsular space?
12. What is the juxtaglomerular apparatus, and where is it located?

FIGURE 24.6 Histology of a renal corpuscle.

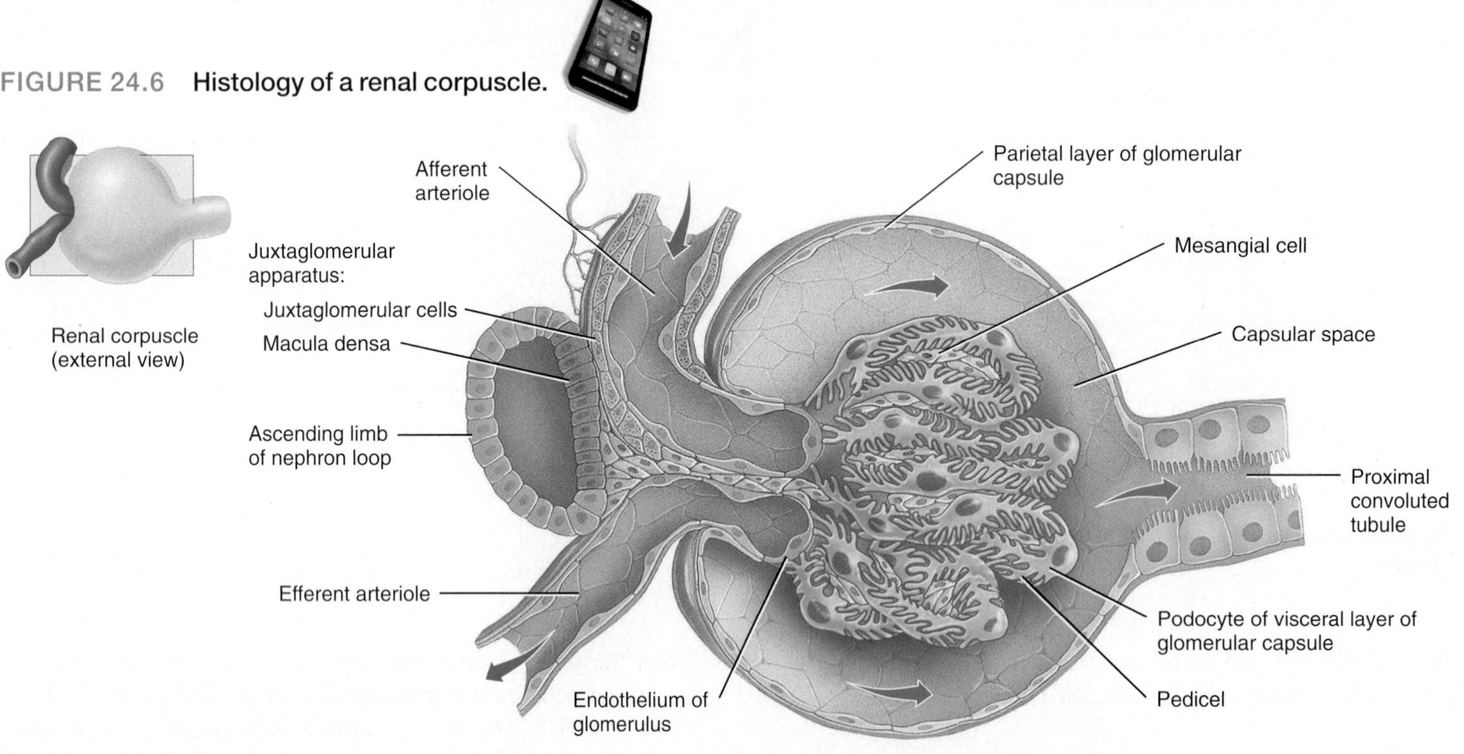


A renal corpuscle consists of a glomerular capsule and a glomerulus.

# 24.4 Urine is formed by glomerular filtration, tubular reabsorption, and tubular secretion.

The formation of urine involves three basic processes in the nephrons and collecting ducts (Figure 24.7):

1. ***Glomerular filtration.*** Filtration uses pressure to force substances across capillary walls. In the first step of urine production, glomerular filtration uses blood pressure to force water and most solutes in blood plasma across the wall of glomerular capillaries and into the capsular space.
2. ***Tubular reabsorption.*** As filtered fluid flows along the renal tubule and the collecting duct, tubule cells reabsorb about 99 percent of the filtered water and many useful solutes. The water and solutes return to the blood flowing through the peritubular capillaries and vasa recta. Note that tubular reabsorption *returns* substances to the bloodstream. In contrast, the term *absorption* means entry of *new* substances into the body, as occurs in the gastrointestinal tract.
3. ***Tubular secretion.*** As filtered fluid flows through the renal tubule and collecting duct, the renal tubule and collecting duct cells secrete other materials, such as wastes, drugs, and excess ions, into the fluid. Notice that tubular secretion *removes* a substance from the blood.

Solutes and the fluid that drain into the renal pelvis constitute urine and are excreted. The rate of urinary excretion of any solute is equal to its rate of glomerular filtration, plus its rate of secretion, minus its rate of reabsorption.

By filtering, reabsorbing, and secreting, nephrons help maintain homeostasis of the blood's volume and composition. The situation is somewhat analogous to a recycling center: Garbage trucks dump refuse into an input hopper that separates smaller items from larger items. Smaller refuse passes through the input hopper onto a conveyor belt (glomerular filtration of plasma). As the conveyor belt carries the garbage along, workers remove useful items, such as aluminum cans and glass containers (tubular reabsorption). Other workers place additional garbage left at the center onto the conveyor belt (tubular secretion). All garbage that remains at the end of the conveyor belt falls into a truck for transport to the landfill (excretion of wastes in urine).

### ✓ CHECKPOINT

**13.** If the rate a drug such as penicillin is excreted from the body in urine is greater than the rate at which it is filtered at the glomerulus, how else is it getting into the urine?

**14.** When cells of the renal tubules secrete the drug penicillin, is the drug being added to or removed from the bloodstream?

FIGURE 24.7 **Relationship of a nephron's structure to its three basic functions: glomerular filtration, tubular reabsorption, and tubular secretion.**

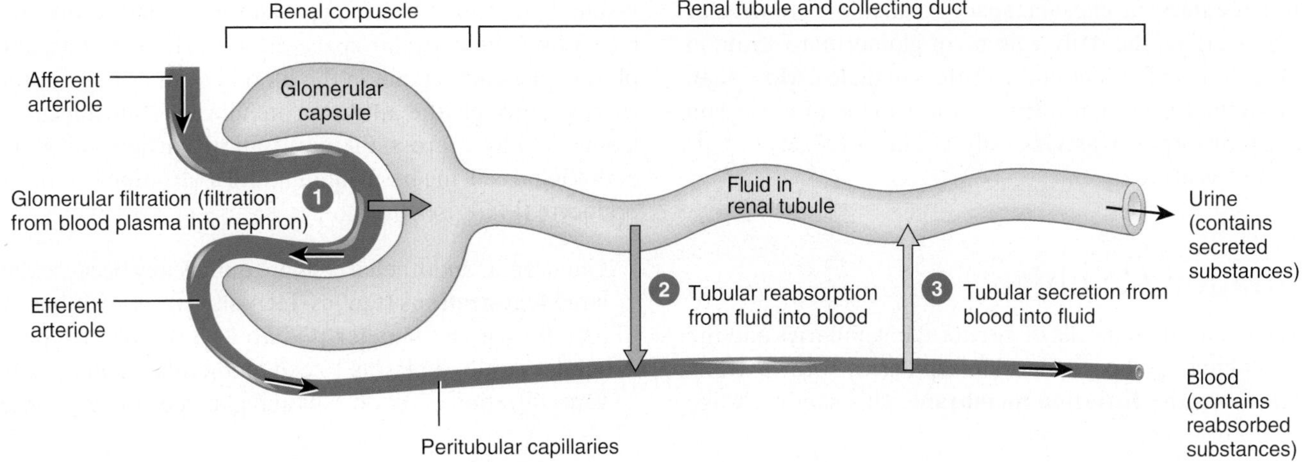


**Glomerular filtration occurs in the renal corpuscle; tubular reabsorption and tubular secretion occur along the renal tubule and collecting duct.**

## RETURN TO Sam's Story

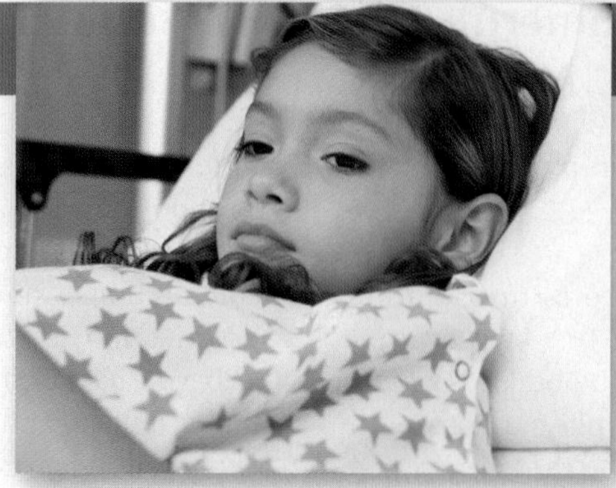

Several things have happened in the few hours since Samantha was first brought into the ER. The most positive changes are that Samantha is able to make eye contact and that "Sam," as Cindy is able to call her now, is constantly holding Cindy's hand. However, Sam's physical condition has taken a turn for the worse. Her initial blood work and urinalysis have come back and it is confirmed that Sam has a kidney infection in her left kidney. The doctors are also concerned that Sam may be malnourished, anemic, and have other "imbalances" stemming from kidney dysfunction. Sam has had a catheter inserted to monitor her urine production, and an intravenous line installed to keep her blood pressure up and help her stay hydrated. She is in less pain and is mostly uncomfortable, but Cindy waits with her until she is able to fall asleep. Cindy knows that she needs to go and speak with Sam's parents and gingerly releases Sam's grip on her hand. Along with the team of doctors on Sam's case, she will need to determine whether a call to the police will be necessary. The doctors have ordered more tests and a few diagnostic scans to fully assess Sam's condition.

On Sam's chart Cindy sees a request for an imaging technique that she's unfamiliar with and asks the nearest doctor to explain it to her. "Dr Jacobs, what is an 'intravenous pyelogram'?"

Dr Jacobs replies, "A substance will be injected into Sam's blood that will be 'cleaned' out by her kidneys. Once the kidneys have taken this substance from the blood it will make the physical structures of the urinary system visible in an x-ray. It will allow us to be able to visualize any blockages, malformations, or abnormal growths."

Cindy thanks Dr. Jacobs and heads to the waiting room to speak to Sam's parents.

A. *The doctor has told Cindy that not only does Sam have a kidney infection, but she might possibly have other "imbalances." Based on her symptoms, what imbalances does Sam already seem to have?*

B. *Sam will have a substance injected into her blood that will expose the physical aspects of her urinary system in an x-ray. This substance will move through her kidneys and be excreted as would normal urine. Please describe the pathway this substance would take traveling through the kidneys in the blood.*

C. *Once the substance referred to in question B has reached the functional unit of the kidney, what process must happen in order for the substance to appear in the urine? What structures will it pass in order to be excreted out of the kidneys?*

## 24.5 Water and solutes are forced through the filtration membrane during glomerular filtration.

The fluid that enters the capsular space is called the **glomerular filtrate**. On average, the daily volume of glomerular filtrate in adults is 150 liters in females and 180 liters in males. More than 99 percent of the glomerular filtrate returns to the bloodstream via tubular reabsorption, leaving only 1–2 liters (about 1–2 qt) to be excreted as urine.

### The Filtration Membrane

Together, the endothelial cells of glomerular capillaries and the podocytes encircling the glomerular capillaries form a leaky barrier known as the **filtration membrane**. This sandwich-like assembly permits filtration of water and small solutes from the blood into the capsular space. Blood cells, platelets, and most plasma proteins remain in the blood because they are too large to pass through the filtration membrane. Substances filtered from the blood cross three filtration barriers—a glomerular endothelial cell, the basal lamina, and a filtration slit formed by a podocyte (Figure 24.8):

- Glomerular endothelial cells are quite leaky because they have large **fenestrations** (fen′-es-TRĀ-shuns) (pores) that measure 0.07–0.1 μm in diameter (Figure 24.8a). This size permits all solutes in blood plasma to exit glomerular capillaries but prevents filtration of blood cells and platelets. Located among the

**FIGURE 24.8 The filtration membrane.** The size of the endothelial fenestrations and filtration slits in (a) have been exaggerated for emphasis.

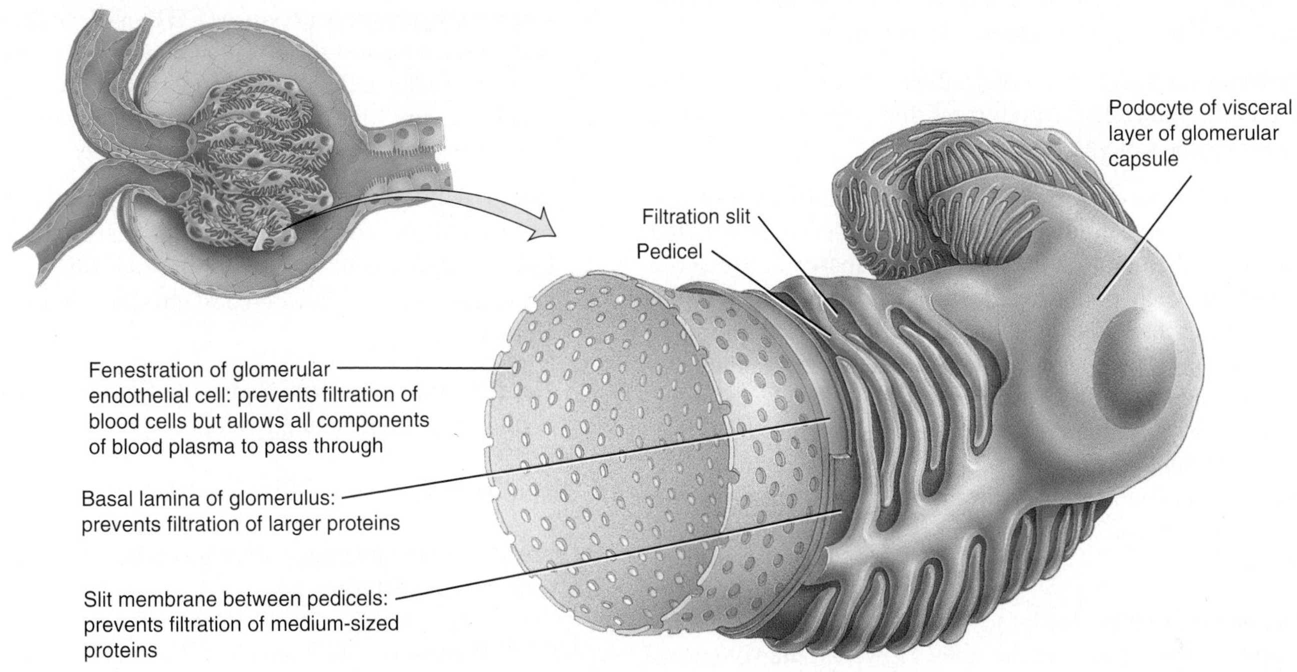


(a) Diagram of details of filtration membrane

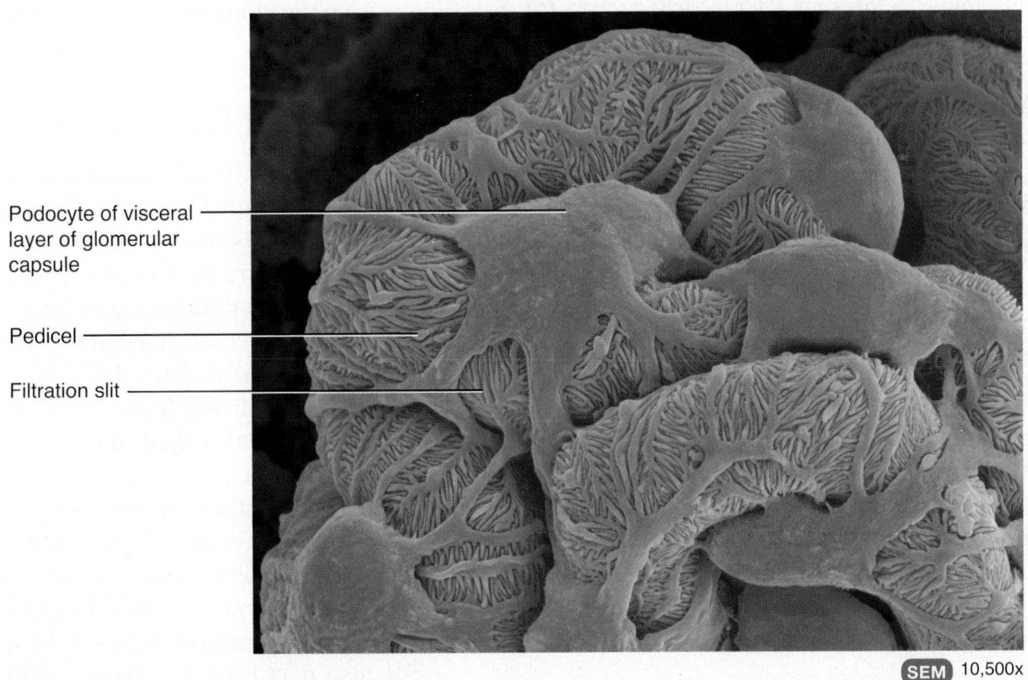


(b) Scanning electron micrograph of filtration membrane

**During glomerular filtration, water and solutes pass from blood plasma into the capsular space.**

glomerular capillaries and in the cleft between afferent and efferent arterioles are **mesangial cells** (mes-AN-jē-al; *mes-* = in the middle; *-angi-* = blood vessel) (see Figure 24.6). These contractile cells help regulate glomerular filtration.

- **The basal lamina** is a layer of acellular glycoprotein between the endothelium and the podocytes that prevents filtration of larger plasma proteins.
- Extending from each podocyte are thousands of foot-like processes termed **pedicels** (PED-i-sels = little feet) that wrap around glomerular capillaries. The spaces between the pedicels of each podocyte are the **filtration slits**. A thin membrane, the *slit membrane,* extends across each filtration slit; it permits the passage of molecules with a diameter smaller than 0.006–0.007 μm, including water, glucose, vitamins, amino acids, very small plasma proteins, ammonia, urea, and ions. Less than 1 percent of albumin, the most plentiful plasma protein, passes the slit membrane because, with a diameter of 0.007 μm, albumin is slightly too big to get through.

The process of filtration is the same in glomerular capillaries as in capillaries elsewhere in the body. However, the volume of fluid filtered by the renal corpuscle is much larger than in other capillaries of the body for three reasons:

1. Glomerular capillaries present a large surface area for filtration because they are long and extensive. The mesangial cells regulate how much of this surface area is available for filtration. When mesangial cells are relaxed, surface area is maximal, and glomerular filtration is very high. Contraction of mesangial cells reduces the available surface area, and glomerular filtration decreases.
2. The filtration membrane is thin and porous. Despite having several layers, the thickness of the filtration membrane is only 0.1 μm. Glomerular capillaries also are about 50 times leakier than capillaries in most other tissues, mainly because of their large fenestrations.
3. Glomerular capillary blood pressure is high. Because the efferent arteriole is smaller in diameter than the afferent arteriole, resistance to the outflow of blood from the glomerulus is high. As a result, blood pressure in glomerular capillaries is considerably higher than in capillaries elsewhere in the body.

## Net Filtration Pressure

Glomerular filtration depends on three main pressures. One pressure *promotes* filtration and two other pressures *oppose* filtration (Figure 24.9).

1. **Glomerular blood hydrostatic pressure (GBHP)** is the blood pressure in the glomerulus. It promotes filtration by forcing water and solutes in blood plasma through the filtration membrane.
2. **Capsular hydrostatic pressure (CHP)** is the hydrostatic pressure exerted against the filtration membrane by fluid already in the capsular space. CHP opposes filtration by serving as a "back pressure" that pushes water and solutes from the filtrate back into the plasma.
3. **Blood colloid osmotic pressure (BCOP)** is due to the presence of proteins in blood plasma such as albumin, globulins, and fibrinogen. The BCOP opposes filtration by drawing water from the filtrate back into the plasma within the glomerulus.

**Net filtration pressure (NFP)**, the total pressure that promotes filtration, is determined as follows:

$$\text{Net filtration pressure (NFP)} = \text{GBHP} - (\text{CHP} + \text{BCOP})$$

By substituting the average values given in Figure 24.9, normal NFP may be calculated:

$$\text{NFP} = 55 \text{ mm Hg} - (15 \text{ mm Hg} + 30 \text{ mm Hg}) = 10 \text{ mm Hg}$$

Thus, under normal conditions, a net filtration pressure of only 10 mm Hg causes blood plasma (without plasma proteins) to filter from the glomerulus into the capsular space.

## Glomerular Filtration Rate

The amount of filtrate formed each minute in all of the renal corpuscles of both kidneys is the **glomerular filtration rate (GFR)**. In adults, the GFR averages 125 mL/min in males and 105 mL/min in females. Homeostasis of body fluids requires that the kidneys maintain a relatively constant GFR. If the GFR is too high, needed substances may pass so quickly through the renal tubules that some are not reabsorbed and are lost in the urine. If the GFR is too low, nearly all the filtrate may be reabsorbed and some waste products may not be excreted adequately.

GFR is directly related to the pressures that influence net filtration pressure; any change in net filtration pressure will affect GFR. Severe blood loss, for example, reduces arterial blood pressure and decreases the glomerular blood hydrostatic pressure. As glomerular blood hydrostatic pressure drops, GFR also decreases. Glomerular filtration ceases altogether if glomerular blood hydrostatic pressure drops to 45 mm Hg because, as shown above, the opposing pressures add up to 45 mm Hg.

Glomerular filtration rate is regulated in two main ways: (1) by adjusting blood flow into and out of the glomerulus and (2) by altering the glomerular capillary surface area available for filtration. GFR increases when blood flow into the glomerular

**FIGURE 24.9 The pressures that drive glomerular filtration.** Taken together, these pressures determine net filtration pressure (NFP).

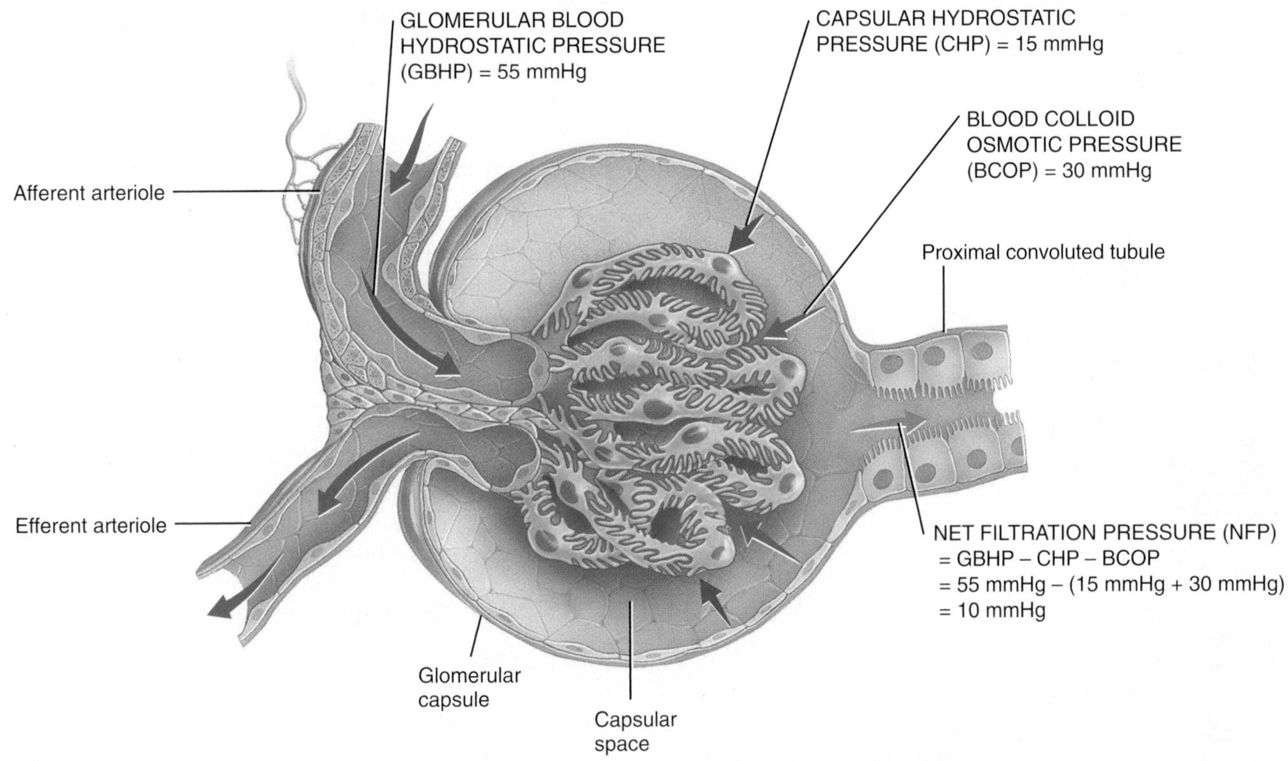


**Glomerular blood hydrostatic pressure promotes filtration; capsular hydrostatic pressure and blood colloid osmotic pressure oppose filtration.**

capillaries increases. Coordinated control of the diameter of both afferent and efferent arterioles regulates glomerular blood flow. Constriction of the afferent arteriole decreases blood flow into the glomerulus; dilation of the afferent arteriole increases it. Three mechanisms control GFR: renal autoregulation, neural regulation, and hormonal regulation.

## Renal Autoregulation of GFR

The kidneys themselves help maintain a constant renal blood flow and GFR despite normal, everyday changes in blood pressure, like those that occur during exercise. This capability is called **renal autoregulation** (aw′-tō-reg′-ū-LĀ-shun) and consists of two mechanisms—the myogenic mechanism and tubuloglomerular feedback. Working together, they can maintain nearly constant GFR over a wide range of systemic blood pressures.

The **myogenic mechanism** (mī-ō-JEN-ik; *myo-* = muscle; *-genic* = producing) occurs when stretching triggers contraction of smooth muscle cells in the walls of afferent arterioles. As arterial blood pressure rises, the walls of the afferent arterioles are stretched. In response, smooth muscle fibers in the wall of the afferent arteriole contract, which narrows the arteriole's lumen. As a result, renal blood flow decreases, thus reducing GFR to its previous level. Conversely, when arterial blood pressure drops, the walls of the afferent arterioles are stretched less. As the smooth muscle cells in the wall relax, the afferent arterioles dilate, renal blood flow increases, and GFR increases.

The second contributor to renal autoregulation, **tubuloglomerular feedback** (too′-bū-lō-glō-MER-ū-lar), is so named because part of the renal tubules—the macula densa of the juxtaglomerular apparatus—provides feedback to the glomerulus (Figure 24.10). When GFR is above normal due to elevated systemic blood pressure, filtered fluid flows more rapidly along the renal tubules. As a result, the proximal convoluted tubule and nephron loop have less time to reabsorb $Na^+$, $Cl^-$, and water. Macula densa cells are thought to detect the increased delivery of $Na^+$, $Cl^-$, and water in the filtered fluid and respond by inhibiting nitric oxide (NO) release from cells in the juxtaglomerular apparatus. When NO is released, it stimulates afferent arterioles to dilate; when NO release is inhibited, afferent arterioles constrict, resulting in decreased blood flow into the

FIGURE 24.10 **Tubuloglomerular feedback.**

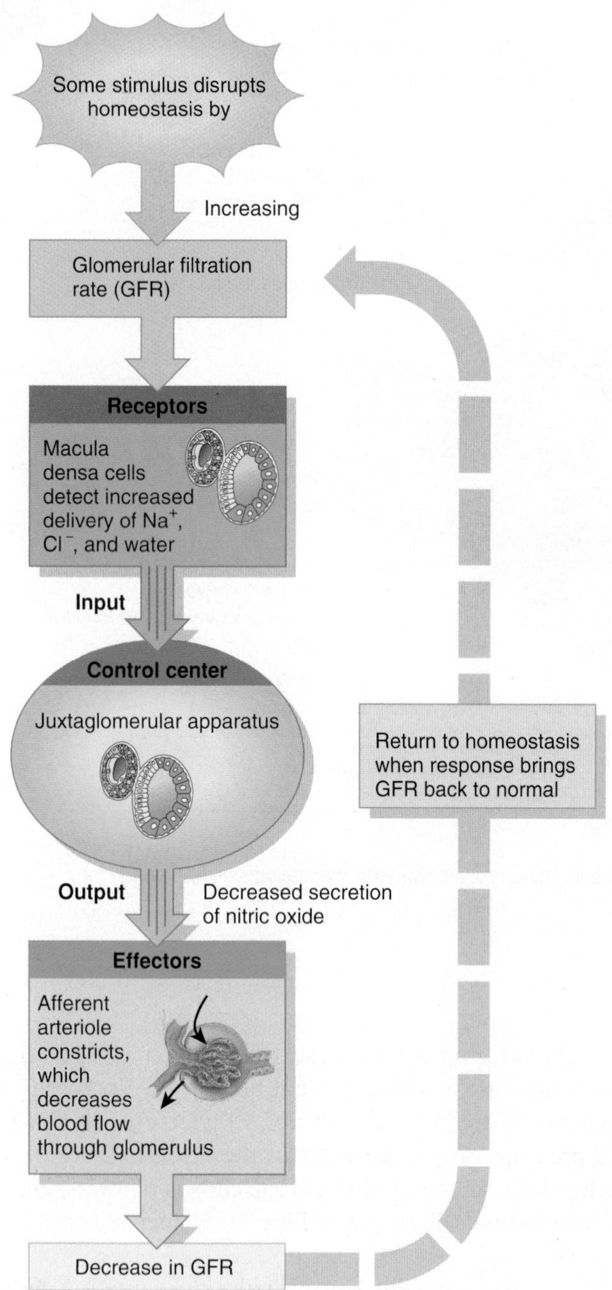


**Macula densa cells of the juxtaglomerular apparatus provide negative feedback regulation of glomerular filtration rate.**

glomerular capillaries, and decreased GFR. On the other hand, as blood pressure falls (resulting in decreased GFR), filtered fluid flows more slowly along the renal tubules (allowing time for reabsorption), and the release of NO from the juxtaglomerular apparatus is no longer inhibited by the macula densa cells. As the level of NO increases, afferent arterioles dilate, increasing blood flow into the glomerulus, and increasing GFR. Macula densa cells also inhibit release of renin from juxtaglomerular cells when increased $Na^+$ and $Cl^-$ is detected in filtered fluid. As you will see in Concept 24.7, this links renal autoregulation of GFR with hormonal regulation of tubular reabsorption and secretion.

### Neural Regulation of GFR

Like most blood vessels of the body, those of the kidneys are supplied by sympathetic neurons of the autonomic nervous system (ANS) that release norepinephrine. Norepinephrine produces vasoconstriction of the afferent arterioles. At rest, sympathetic stimulation is low, afferent arterioles are relatively dilated, and renal autoregulation of GFR prevails. With greater sympathetic stimulation, as occurs during exercise or hemorrhage, afferent arterioles constrict. As a result, blood flow into glomerular capillaries decreases, glomerular filtration decreases, and GFR drops. This lowering of renal blood flow has two consequences:

1. It reduces urine output, which helps conserve blood volume.
2. It permits greater blood flow to other body tissues.

### Hormonal Regulation of GFR

Two hormones contribute to regulation of GFR. **Angiotensin II** (an′-jē-ō-TEN-sin) is a potent vasoconstrictor that narrows both afferent and efferent arterioles and reduces renal blood flow, thereby decreasing GFR. As blood volume increases, the atria of the heart secrete **atrial natriuretic peptide** (na′-trē-ū-RET-ik), which relaxes the glomerular mesangial cells, and increases the capillary surface area available for filtration. Glomerular filtration rate rises as the surface area increases.

Table 24.1 summarizes the regulation of glomerular filtration rate.

### ✓ CHECKPOINT

**15.** What is the major chemical difference between blood plasma and glomerular filtrate?

**16.** Why is there much greater filtration through glomerular capillaries than through capillaries elsewhere in the body?

**17.** How does blood pressure promote filtration of blood in the kidneys?

**18.** How do the myogenic mechanism and tubuloglomerular feedback regulate the glomerular filtration rate? Why are these processes called autoregulation?

**19.** How is the glomerular filtration rate regulated by the autonomic nervous system and by hormones?

**TABLE 24.1**

Regulation of Glomerular Filtration Rate (GFR)

| TYPE OF REGULATION | MAJOR STIMULUS | MECHANISM AND SITE OF ACTION | EFFECT ON GFR |
|---|---|---|---|
| **RENAL AUTOREGULATION** | | | |
| **Myogenic mechanism** | Increased stretching of smooth muscle fibers in afferent arteriole walls due to increased blood pressure | Stretched smooth muscle fibers contract, thereby narrowing the lumen of the afferent arterioles | Decrease |
| **Tubuloglomerular feedback** | Rapid delivery of $Na^+$ and $Cl^-$ to the macula densa due to high systemic blood pressure | Decreased release of nitric oxide (NO) by the juxtaglomerular apparatus causes constriction of afferent arterioles | Decrease |
| **Neural regulation** | Increase in level of activity of renal sympathetic nerves releases norepinephrine | Constriction of afferent arterioles | Decrease |
| **HORMONAL REGULATION** | | | |
| **Angiotensin II** | Decreased blood volume or blood pressure stimulates production of angiotension II | Constriction of both afferent and efferent arterioles | Decrease |
| **Atrial natriuretic peptide (ANP)** | Stretching of the atria of the heart stimulates secretion of ANP | Relaxation of mesangial cells in glomerulus increases capillary surface area available for filtration | Increase |

## CLINICAL CONNECTION | *Renal Failure*

**Renal failure** is a decrease or cessation of glomerular filtration. In **acute renal failure (ARF)**, the kidneys abruptly stop working entirely (or almost entirely). The main feature of ARF is the suppression of urine flow, usually characterized either by *oliguria* (ol-i-GŪ-rē-a), daily urine output between 50 mL and 250 mL, or by *anuria* (an-Ū-rē-a), daily urine output less than 50 mL. Causes include low blood volume (for example, due to hemorrhage), decreased cardiac output, damaged renal tubules, kidney stones, the dyes used to visualize blood vessels in angiograms, nonsteroidal anti-inflammatory drugs, and some antibiotic drugs. It is also common in people who suffer a devastating illness or overwhelming traumatic injury; in such cases it may be related to a more general organ failure known as *multiple organ dysfunction syndrome (MODS).*

Renal failure causes a multitude of problems. There is edema due to salt and water retention and metabolic acidosis due to an inability of the kidneys to excrete acidic substances. In the blood, urea builds up due to impaired renal excretion of metabolic waste products and potassium level rises, which can lead to cardiac arrest. Often, there is anemia because the kidneys no longer produce enough erythropoietin for adequate red blood cell production. Because the kidneys are no longer able to convert vitamin D to calcitriol, which is needed for adequate calcium absorption from the small intestine, osteomalacia also may occur.

**Chronic renal failure (CRF)** refers to a progressive and usually irreversible decline in glomerular filtration rate (GFR). CRF may result from chronic glomerulonephritis, pyelonephritis, polycystic kidney disease, or traumatic loss of kidney tissue. CRF develops in three stages. In the first stage, *diminished renal reserve,* nephrons are destroyed until about 75 percent of the functioning nephrons are lost. At this stage, a person may have no signs or symptoms because the remaining nephrons enlarge and take over the function of those that have been lost. Once 75 percent of the nephrons are lost, the person enters the second stage, called *renal insufficiency,* characterized by a decrease in GFR and increased blood levels of nitrogen-containing wastes and creatinine. Also, the kidneys cannot effectively concentrate or dilute the urine. The final stage, called *end-stage renal failure,* occurs when about 90 percent of the nephrons have been lost. At this stage, GFR diminishes to 10–15 percent of normal, oliguria is present, and blood levels of nitrogen-containing wastes and creatinine increase further. People with end-stage renal failure need dialysis therapy (see Concept 24.8) and are possible candidates for a **kidney transplant** operation (see WileyPLUS Chapter 24 Clinical Connection).

# 24.6 Tubular reabsorption reclaims needed substances from the filtrate, while tubular secretion discharges unneeded substances.

## Principles of Tubular Reabsorption and Secretion

The volume of fluid entering the proximal convoluted tubules in just half an hour is greater than the total volume of plasma in the blood because the glomerular filtration rate is normally so high. Obviously, some of this fluid must be returned to the bloodstream. Tubular reabsorption—the return of most of the filtered water and many of the filtered solutes to the bloodstream—is the second basic function of the nephron and collecting duct. Normally, about 99 percent of the filtered water is reabsorbed. Each day our kidneys produce about 90 two-liter soda bottles of filtrate, yet we excrete less than a single one of those bottles as urine. Tubular reabsorption not only salvages needed components of plasma, but it also keeps us from being in the restroom all day!

Epithelial cells all along the renal tubule and duct carry out reabsorption. Solutes that are reabsorbed include glucose, amino acids, urea, and ions such as sodium, potassium, calcium, chloride, bicarbonate, and phosphate. Once fluid passes through the proximal convoluted tubule, cells located in distal portions of the renal tubule fine-tune the reabsorption processes to maintain homeostatic balances of water and selected ions. Most small proteins and peptides that pass through the filtration membrane are also reabsorbed. To appreciate the magnitude of tubular reabsorption, look at Table 24.2 and compare the amounts of substances that are typically filtered, reabsorbed, and excreted in urine.

The third basic function of nephrons and collecting ducts is tubular secretion, the return of materials from the blood and tubule cells into the glomerular filtrate. Secreted substances include hydrogen, potassium, and ammonium ions, creatinine, and certain drugs such as penicillin. Tubular secretion has two important outcomes: (1) The secretion of hydrogen ions helps control blood pH; (2) The secretion of other substances helps eliminate them from the body.

As a result of tubular secretion, certain substances pass from blood into urine and may be detected by a urinalysis (see WileyPLUS Chapter 24 Clinical Connection). It is especially important to test athletes for the presence of performance-enhancing drugs such as anabolic steroids, plasma expanders, erythropoietin, hCG, hGH, and amphetamines. Urine tests can also be used to detect the presence of alcohol or illegal drugs such as marijuana, cocaine, and heroin.

### Reabsorption Routes

A substance being reabsorbed from the fluid in the renal tubule lumen can take one of two routes before entering a peritubular capillary: It can move *between* adjacent tubule cells or *through* an individual tubule cell (Figure 24.11). Along the renal tubule, cell junctions join neighboring cells to one another, much like the plastic rings that hold a six-pack of soda cans together. The **apical membrane** (the tops of the soda cans) contacts the tubular fluid, and the

**TABLE 24.2**

Substances Filtered, Reabsorbed, and Excreted in Urine

| SUBSTANCE | FILTERED* (ENTERS GLOMERULAR CAPSULE PER DAY) | REABSORBED (RETURNED TO BLOOD PER DAY) | URINE (EXCRETED PER DAY) |
|---|---|---|---|
| **Water** | 180 liters | 178–179 liters | 1–2 liters |
| **Proteins** | 2.0 g | 1.9 g | 0.1 g |
| **Sodium ions ($Na^+$)** | 579 g | 575 g | 4 g |
| **Chloride ions ($Cl^-$)** | 640 g | 633.7 g | 6.3 g |
| **Bicarbonate ions ($HCO_3^-$)** | 275 g | 274.97 g | 0.03 g |
| **Glucose** | 162 g | 162 g | 0 g |
| **Urea** | 54 g | 24 g | 30 g[†] |
| **Potassium ions ($K^+$)** | 29.6 g | 29.6 g | 2.0 g[‡] |
| **Uric acid** | 8.5 g | 7.7 g | 0.8 g |
| **Creatinine** | 1.6 g | 0 g | 1.6 g |

*Assuming GFR is 180 liters per day.

[†]In addition to being filtered and reabsorbed, urea is secreted.

[‡]After virtually all filtered $K^+$ is reabsorbed in the convoluted tubules and nephron loop, a variable amount of $K^+$ is secreted by principal cells in the collecting duct.

FIGURE 24.11 **Reabsorption routes.** (a) Paracellular reabsorption. (b) Transcellular reabsorption.

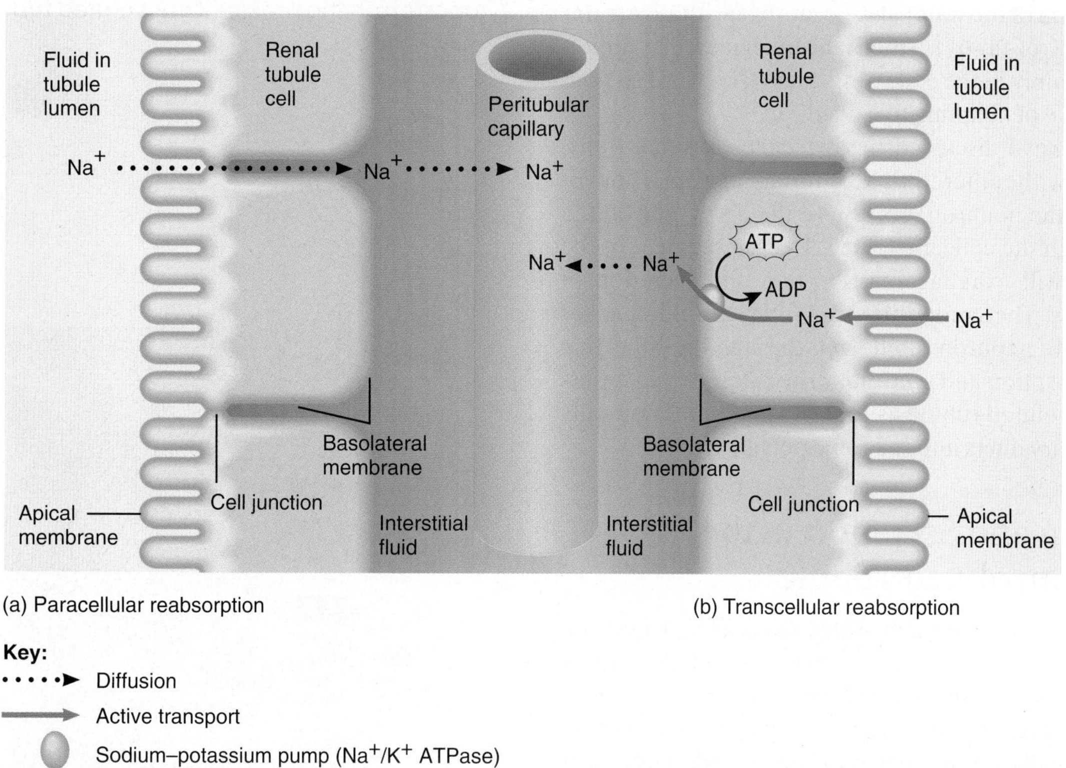


**In paracellular reabsorption, water and solutes in tubular fluid return to the bloodstream by moving between tubule cells; in transcellular reabsorption, water and solutes in tubular fluid return to the bloodstream by passing through a tubule cell.**

**basolateral membrane** (the bottoms and sides of the soda cans) contacts interstitial fluid at the base and sides of the tubule cell.

Even though tubule cells are connected by cell junctions, the cell junctions can be "leaky" and permit some fluid to pass *between* the cells into peritubular capillaries by a passive process known as **paracellular reabsorption** (par′-a-SEL-ū-lar; *para-* = beside). In **transcellular reabsorption** (trans′-SEL-ū-lar; *trans-* = across), a substance passes from the fluid in the tubular lumen through the apical membrane of a tubule cell, across the cytosol, and out into interstitial fluid through the basolateral membrane.

## Transport Mechanisms

When tubule cells transport solutes out of or into tubular fluid, they move specific substances in one direction only. Not surprisingly, various carrier proteins are present in the apical and basolateral membranes to assist transport movement. Reabsorption of $Na^+$ by the renal tubules is especially important because of the large number of sodium ions that pass through the filtration membrane. Cells lining the renal tubules, like other cells throughout the body, have a low concentration of $Na^+$ in their cytosol due to the activity of sodium–potassium pumps ($Na^+/K^+$ ATPases). These pumps are located in the basolateral membranes and eject $Na^+$ from the renal tubule cells into interstitial fluid (Figure 24.11). The absence of sodium–potassium pumps in the apical membrane ensures that reabsorption of $Na^+$ is a one-way process.

As noted in Chapter 3, transport of materials across membranes may be either active or passive. Recall that in **primary active transport** the energy derived from hydrolysis of ATP is used to "pump" a substance across a membrane; the sodium–potassium pump is one such pump. In **secondary active transport** the energy stored in an ion's electrochemical gradient, rather than hydrolysis of ATP, drives a substance across a membrane. Secondary active transport couples the movement of one substance "downhill" along its electrochemical gradient to the movement of a second substance "uphill" against its electrochemical gradient. *Symporters* are membrane proteins that move two or more transported substances in the *same* direction across a membrane. *Antiporters* move two or more transported substances in *opposite* directions across a membrane.

All water reabsorption occurs via osmosis. Recall from Chapter 3 that osmosis passively moves water from an area of lower solute concentration to an area of higher solute concentration. Therefore, solute reabsorption drives water reabsorption. The movement of solutes into peritubular capillaries *decreases* the solute concentration of the tubular fluid but *increases* the solute concentration in the peritubular capillaries. As a result, water moves by osmosis into peritubular capillaries. Water reabsorbed with solutes in tubular fluid is termed **obligatory water reabsorption** (ob-LIG-a-tor′-ē) because the water is "obliged" to follow the solutes when they are reabsorbed. This type of water reabsorption occurs in the proximal convoluted tubule and the descending limb of the nephron loop because these segments of the nephron are always permeable to water. Reabsorption of the water also occurs by **facultative water reabsorption** (FAK-ul-tā′-tiv), a mechanism regulated by antidiuretic hormone mainly in the collecting ducts. The word *facultative* means "capable of adapting to a need."

Now that we have discussed the principles of renal transport, we can follow the filtered fluid from the proximal convoluted tubule, into the nephron loop, on to the distal convoluted tubule, and through the collecting ducts. In each segment of the renal tubule, we will examine reabsorption and secretion of specific substances. The composition of filtered fluid changes as it flows along the nephron tubule and through the collecting duct due to reabsorption and secretion. The filtered fluid enters the proximal convoluted tubule as *tubular fluid* and eventually drains from papillary ducts into the renal pelvis as *urine*.

## Reabsorption and Secretion in the Proximal Convoluted Tubule

The largest amount of solute and water reabsorption from filtered fluid occurs in the proximal convoluted tubules, where most absorptive processes involve sodium ions. $Na^+$ transport in the proximal convoluted tubules occurs via symport and antiport mechanisms. Normally, filtered glucose, amino acids, lactic acid, water-soluble vitamins, and other nutrients are not lost in the urine. Rather, they are reabsorbed in the proximal convoluted tubules by **$Na^+$ symporters** located in the apical membrane. Figure 24.12 depicts the operation of one such $Na^+$ symporter, the **$Na^+$–glucose symporter**. Two $Na^+$ and a molecule of glucose attach to the symporter protein, which carries them from the tubular fluid into the tubule cell in the proximal convoluted tubules. The glucose molecules exit the basolateral membrane via facilitated diffusion, and then the glucose and sodium ions diffuse into peritubular capillaries. Other $Na^+$ symporters in the proximal convoluted tubules reclaim additional filtered solutes in a similar way.

In another secondary active transport process, the **$Na^+/H^+$ antiporters** carry filtered $Na^+$ into a proximal convoluted tubule cell while $H^+$ is moved from the cell into the tubule lumen (Figure 24.13a). As a result of the exchange, $Na^+$ is reabsorbed into peritubular blood and $H^+$ is secreted into tubular fluid. Proximal convoluted tubule cells produce the $H^+$ needed to keep the antiporters running in the following way: (1) Carbon dioxide ($CO_2$) diffuses from peritubular blood or tubular fluid, or is produced by metabolic reactions within the cells; (2) the enzyme *carbonic anhydrase* (CA) (an-HĪ-drās) catalyzes the reaction of $CO_2$ with water ($H_2O$) to form carbonic acid ($H_2CO_3$), which then dissociates into $H^+$ and bicarbonate ions ($HCO_3^-$):

$$CO_2 + H_2O \xrightarrow{\text{Carbonic anhydrase}} H_2CO_3 \longrightarrow H^+ + HCO_3^-$$

Most of the bicarbonate ions in filtered fluid are reabsorbed in the proximal convoluted tubules, thereby safeguarding the body's supply of an important buffer (Figure 24.13b). After $H^+$ is secreted into the lumen of the proximal convoluted tubule, it reacts with filtered $HCO_3^-$ to form $H_2CO_3$, which readily dissociates into $CO_2$ and $H_2O$:

$$H^+ + HCO_3^- \longrightarrow H_2CO_3 \xrightarrow{\text{Carbonic anhydrase}} CO_2 + H_2O$$

Carbon dioxide then diffuses into the tubule cells and rejoins $H_2O$ to form $H_2CO_3$, which this time dissociates again into $H^+$ and $HCO_3^-$:

$$CO_2 + H_2O \xrightarrow{\text{Carbonic anhydrase}} H_2CO_3 \longrightarrow H^+ + HCO_3^-$$

The bicarbonate ion exits the basolateral membrane via facilitated diffusion and diffuses into peritubular blood with $Na^+$. Thus, for every $H^+$ secreted into the tubular fluid of the proximal convoluted tubule, one $HCO_3^-$ and one $Na^+$ are reabsorbed.

**FIGURE 24.12 Reabsorption of glucose by $Na^+$–glucose symporters in the proximal convoluted tubule.**

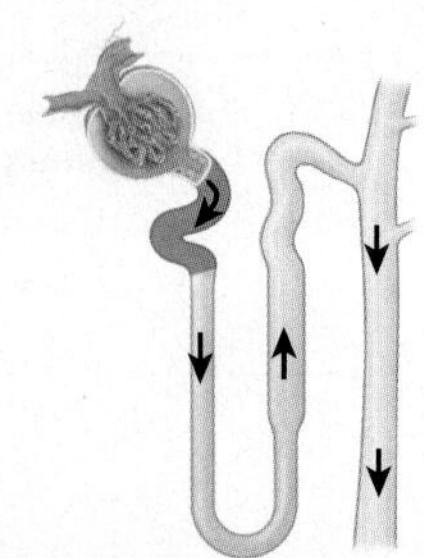

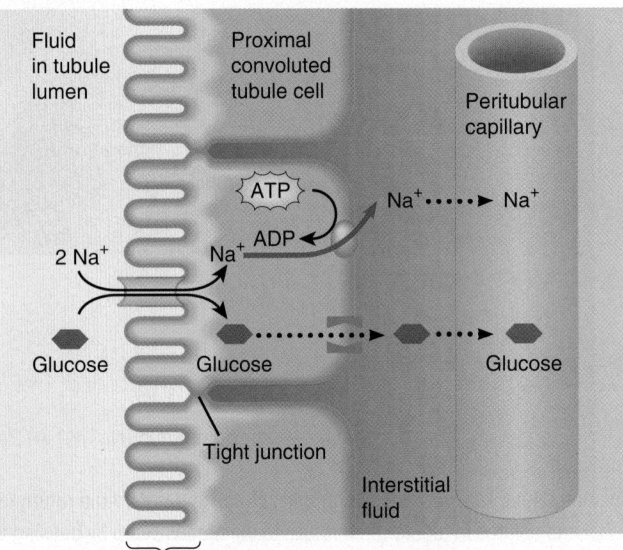


Key:
- $Na^+$–glucose symporter
- Glucose facilitated diffusion transporter
- Diffusion
- Sodium–potassium pump

**Normally, all filtered glucose is reabsorbed in the proximal convoluted tubule.**

FIGURE 24.13 **Actions of $Na^+$–$H^+$ antiporters in the proximal convoluted tubule.** (a) Reabsorption of sodium ions ($Na^+$) and secretion of hydrogen ions ($H^+$) via secondary active transport through the apical membrane; (b) reabsorption of bicarbonate ions ($HCO_3^-$) via facilitated diffusion through the basolateral membrane.

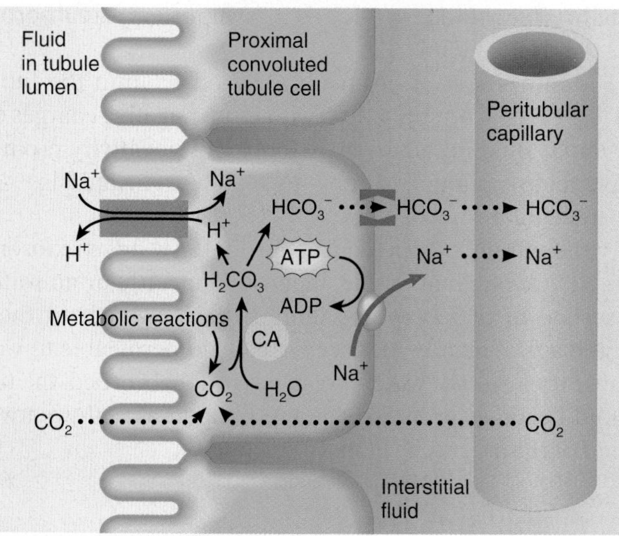


(a) $Na^+$ reabsorption and $H^+$ secretion

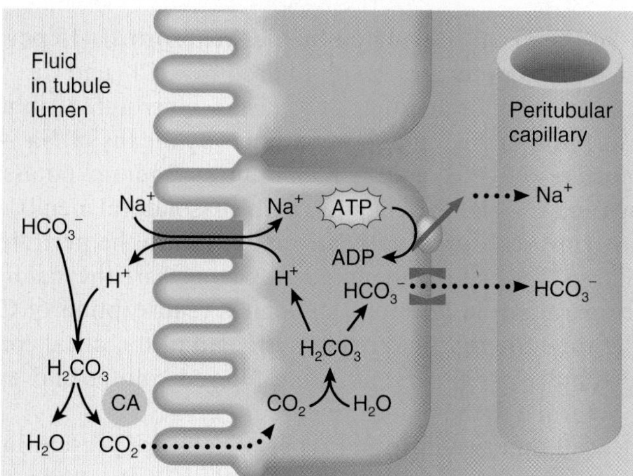


(b) $HCO_3^-$ reabsorption

**Key:**

$Na^+$–$H^+$ antiporter

$HCO_3^-$ facilitated diffusion transporter

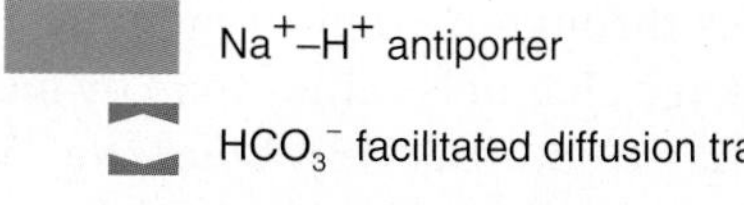


**$Na^+$–$H^+$ antiporters promote transcellular reabsorption of $Na^+$ and secretion of $H^+$.**

Solute reabsorption in proximal convoluted tubules promotes osmosis of water. Each reabsorbed solute increases the osmotic pressure, first inside the tubule cell, then in interstitial fluid, and finally in the blood. Water thus moves rapidly from the tubular fluid into the peritubular capillaries, via both the paracellular and transcellular routes, and restores osmotic balance (Figure 24.14). In other words, reabsorption of solutes creates an osmotic gradient that promotes the reabsorption of water via osmosis.

As water leaves the tubular fluid, the concentrations of solutes remaining in the tubule lumen increases. Increasing electrochemical gradients for $Cl^-$, $K^+$, $Ca^{2+}$, $Mg^{2+}$, and urea promote their diffusion into peritubular capillaries, via both the paracellular and transcellular routes. Diffusion of negatively charged $Cl^-$ into interstitial fluid makes the interstitial fluid electrically more negative than the tubular fluid. This negativity promotes passive reabsorption of cations (positively charged ions), such as $K^+$, $Ca^{2+}$, and $Mg^{2+}$.

Ammonia ($NH_3$) is a poisonous waste product that is produced when amino groups are removed from amino acids, a reaction that occurs mainly in the liver. Hepatocytes (liver cells) convert most of this ammonia to urea, a less toxic compound. Most excretion of these nitrogen-containing waste products occurs via the urine. Urea and ammonia in blood are filtered at the glomerulus and secreted by proximal convoluted tubule cells into the tubular fluid.

## Reabsorption in the Nephron Loop

Glucose, amino acids, and other nutrients, and about 65 percent of the filtered water are reabsorbed as filtered fluid moves through the proximal convoluted tubules. As a result, the chemical composition

FIGURE 24.14 **Passive reabsorption of $Cl^-$, $K^+$, $Ca^{2+}$, $Mg^{2+}$, urea, and water in the proximal convoluted tubule.**

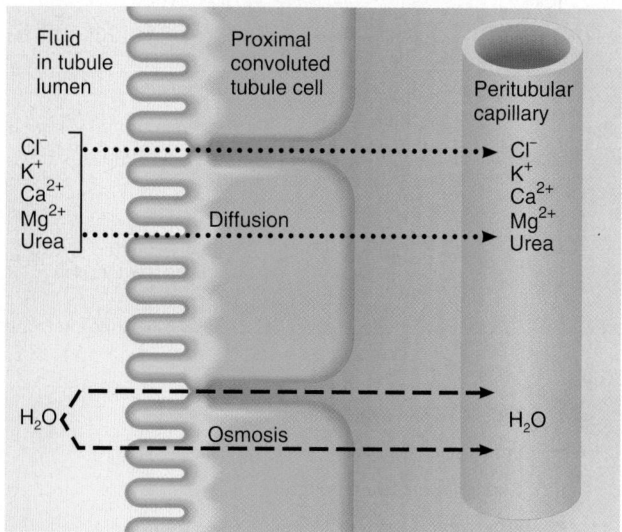


**Electrochemical gradients promote passive reabsorption of solutes via both paracellular and transcellular routes.**

of the tubular fluid entering the next part of the nephron, the nephron loop, is quite different from that of glomerular filtrate.

Of filtered substances, the nephron loop reabsorbs about 15 percent of the water along with a variable amount of $Na^+$, $K^+$, $Cl^-$, $HCO_3^-$, $Ca^{2+}$, and $Mg^{2+}$. Here, for the first time, reabsorption of water via osmosis is *not* automatically coupled to reabsorption of solutes because part of the nephron loop is relatively impermeable to water.

The apical membranes of cells in the thick ascending limb of the nephron loop have **$Na^+$–$K^+$–$2Cl^-$ symporters** that simultaneously reclaim one $Na^+$, one $K^+$, and two $Cl^-$ from the tubular fluid (Figure 24.15). $Na^+$ is actively transported across the basolateral membrane into interstitial fluid, and then diffuses into the vasa recta. $Cl^-$ moves across the basolateral membrane through leakage channels (plasma membrane channels that randomly open and close), and then diffuses from interstitial fluid into the vasa recta. Because many $K^+$ leakage channels are present in the apical membrane, most $K^+$ brought in by the symporters moves down its concentration gradient back into the tubular fluid. Thus, the main effect of the $Na^+$–$K^+$–$2Cl^-$ symporters is reabsorption of $Na^+$ and $Cl^-$.

The movement of positively charged $K^+$ into the tubular fluid leaves the interstitial fluid with more negative charges than the tubular fluid in the nephron loop. This negativity promotes passive reabsorption of cations—$Na^+$, $K^+$, $Ca^{2+}$, and $Mg^{2+}$—via the paracellular route.

Although about 15 percent of the filtered water is reabsorbed in the *descending* limb of the nephron loop, little or no water is reabsorbed in the *ascending* limb. In this segment of the tubule, the apical membranes are virtually impermeable to water. Because ions but not water molecules are reabsorbed, the tubular fluid becomes progressively more dilute as it flows toward the end of the ascending limb.

FIGURE 24.15 $Na^+$–$K^+$–$2Cl^-$ symporter in the thick ascending limb of the nephron loop.

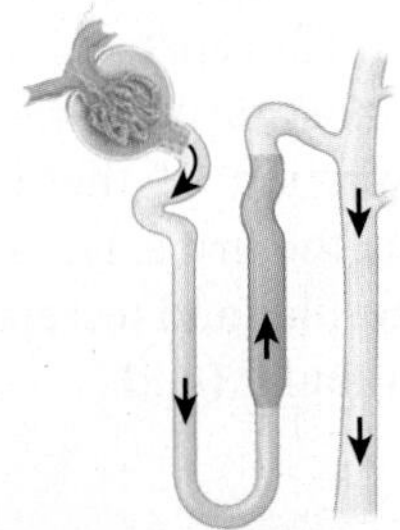

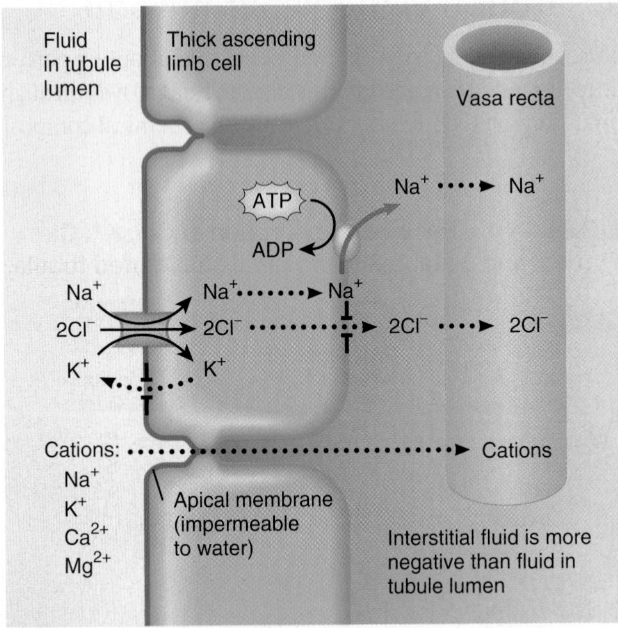


Cells in the thick ascending limb have symporters that simultaneously reabsorb one $Na^+$, one $K^+$, and two $Cl^-$.

## Reabsorption and Secretion in the Distal Convoluted Tubule and Collecting Duct

Cells of the distal convoluted tubules reabsorb 10–15 percent of the filtered water along with filtered $Na^+$, $Cl^-$, and $Ca^{2+}$. As filtered fluid flows along the distal convoluted tubules, reabsorption of $Na^+$ and $Cl^-$ continues by means of **$Na^+$–$Cl^-$ symporters** in the apical membranes. Sodium–potassium pumps and $Cl^-$ leakage channels in the basolateral membranes then permit reabsorption of $Na^+$ and $Cl^-$ into the peritubular capillaries. The distal convoluted tubules also are the major site where parathyroid hormone stimulates reabsorption of $Ca^{2+}$. By the time filtered fluid reaches the end of the distal convoluted tubule, 90–95 percent of the filtered solutes and water have returned to the bloodstream.

Recall that two different types of cells are present in the last part of the distal convoluted tubule and throughout the collecting duct: **principal cells** and **intercalated cells**. Principal cells reabsorb $Na^+$ and secrete $K^+$; intercalated cells reabsorb $K^+$ and $HCO_3^-$ and secrete $H^+$. In contrast to earlier segments of the nephron, $Na^+$ passes through the apical membrane of principal cells via $Na^+$ leakage channels rather than by means of symporters or antiporters (Figure 24.16). Once $Na^+$ is in the principal cells, sodium–potassium pumps actively transport it across the basolateral membranes. Then $Na^+$ passively diffuses into the peritubular capillaries from the interstitial fluid.

Normally, most filtered $K^+$ is returned to the bloodstream by reabsorption in the proximal convoluted tubule and nephron loop. To adjust for varying dietary intakes of potassium and to maintain a stable level of $K^+$ in body fluids, principal cells secrete a variable amount of $K^+$ into the tubule fluid

FIGURE 24.16 **Reabsorption of $Na^+$ and secretion of $K^+$ by principal cells in the last part of the distal convoluted tubule and in the collecting duct.**

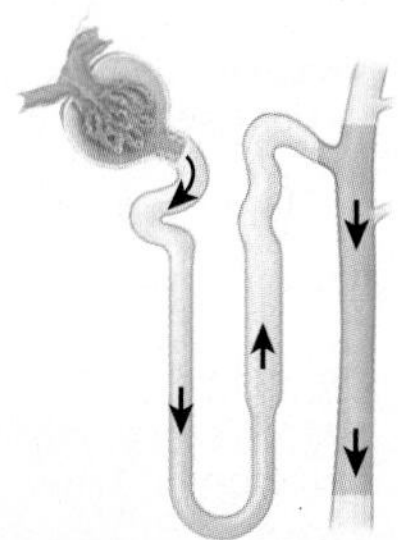

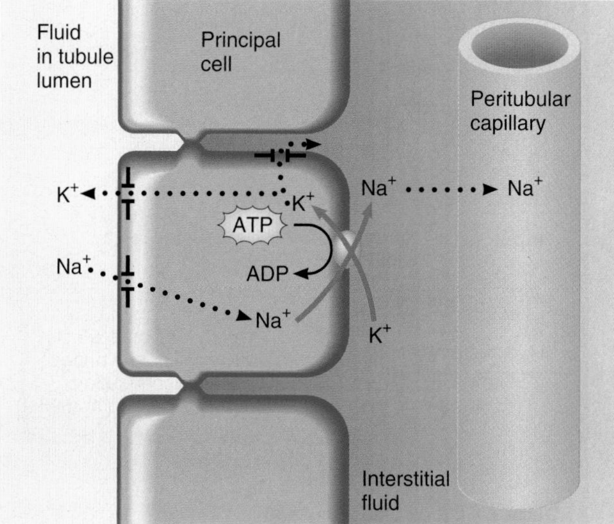

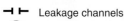

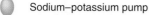

Key:
Diffusion
Leakage channels
Sodium–potassium pump

**In the apical membrane of principal cells, $Na^+$ leakage channels allow entry of $Na^+$ while $K^+$ leakage channels allow exit of $K^+$ into the tubular fluid.**

(Figure 24.16). Because the sodium–potassium pumps continually bring $K^+$ into principal cells, their concentration of $K^+$ remains high. $K^+$ diffuses down its concentration gradient through $K^+$ leakage channels in both the apical and basolateral membranes. Some $K^+$ is returned to the interstitial fluid but most $K^+$ diffuses from a high $K^+$ concentration in the principal cells into the tubular fluid, where the $K^+$ concentration is very low. This secretion mechanism is the main source of $K^+$ excreted in the urine. By contrast, intercalated cells salvage $K^+$ by reabsorbing it from tubular fluid while secreting $H^+$ into the tubular fluid.

✓ CHECKPOINT

20. Describe the two ways in which substances can be absorbed across the tubule cells.
21. On which tubule cell membrane are the sodium–potassium pumps located? What is the significance of sodium–potassium pumps not being located on the other membrane?
22. What is obligatory water reabsorption, and in which parts of the nephron does it occur?
23. What are two mechanisms in the proximal convoluted tubule for reabsorption of $Na^+$? Which other solutes are reabsorbed or secreted with $Na^+$ in each mechanism?
24. What is the mechanism for reabsorption of $Na^+$ in the nephron loop? Which other solutes are reabsorbed or secreted with $Na^+$ in this mechanism?
25. Why does the tubular fluid become more dilute as it travels through the nephron loop? In which part of the nephron does the most reabsorption of water occur? Which part of the nephron is unable to reabsorb water?

# 24.7 Five hormones regulate tubular reabsorption and tubular secretion.

Five hormones influence tubular reabsorption of $Na^+$, $Cl^-$, $Ca^{2+}$, and water as well as tubular secretion of $K^+$. These hormones are angiotensin II, aldosterone, antidiuretic hormone, atrial natriuretic peptide, and parathyroid hormone.

## Renin–Angiotensin–Aldosterone System

When blood volume and blood pressure decrease, the walls of the afferent arterioles are stretched less, which stimulates the juxtaglomerular cells to secrete the enzyme **renin** (RĒ-nin) into the blood. In addition, as systemic blood pressure falls, sympathetic innervation *directly* stimulates juxtaglomerular cells to release renin. Renin converts angiotensinogen, which is synthesized in the liver, into angiotensin I. *Angiotensin-converting enzyme* then converts angiotensin I to **angiotensin II**, the active form of the hormone.

Angiotensin II affects renal physiology in three main ways:

1. It decreases glomerular filtration rate by causing vasoconstriction of the afferent arterioles.
2. It enhances reabsorption of $Na^+$, $Cl^-$, and water in the proximal convoluted tubule by stimulating the activity of $Na^+/H^+$ antiporters.
3. It stimulates the adrenal cortex to release **aldosterone** (al-DOS-ter-ōn), a hormone that in turn stimulates principal cells in the collecting ducts to reabsorb more $Na^+$ and $Cl^-$ and secrete more $K^+$. As more $Na^+$ and $Cl^-$ are reabsorbed, then more water is also reabsorbed by osmosis. As water levels in the blood rise, blood volume and blood pressure increase.

## Antidiuretic Hormone

**Antidiuretic hormone (ADH)** is released by the posterior pituitary. It regulates facultative water reabsorption by increasing the water permeability of principal cells in the last part of the distal convoluted tubule and the collecting duct. In the absence of ADH, the apical membranes of principal cells have a very low permeability to water. ADH stimulates insertion of water channel proteins called **aquaporins** into their apical membranes. As a result, the water permeability of the principal cell's apical membrane increases, and water molecules move more rapidly from the tubular fluid into the cells. Because the basolateral membranes are always permeable to water, water molecules then move rapidly into the blood. The kidneys can produce a small volume of very concentrated urine each day when ADH concentration is maximal, such as during severe dehydration. When ADH level is low, the aquaporins are removed from the apical membrane by endocytosis, and the kidneys produce a large volume of dilute urine.

ADH regulates facultative water reabsorption through a negative feedback system (Figure 24.17). When the concentration of water in the blood decreases by as little as 1 percent, **osmoreceptors** (receptors that sense osmotic pressure) in the hypothalamus stimulate the posterior pituitary to secrete more ADH into the blood, and the principal cells become more permeable to water. As facultative water reabsorption increases, the water concentration in the blood increases to normal. A second powerful stimulus for ADH secretion is a decrease in blood volume, as occurs in hemorrhaging or severe dehydration. In the pathological absence of ADH activity, a condition known as *diabetes insipidus,* a person may excrete up to 20 liters of very dilute urine daily.

**FIGURE 24.17 Negative feedback regulation of water reabsorption by antidiuretic hormone (ADH).**

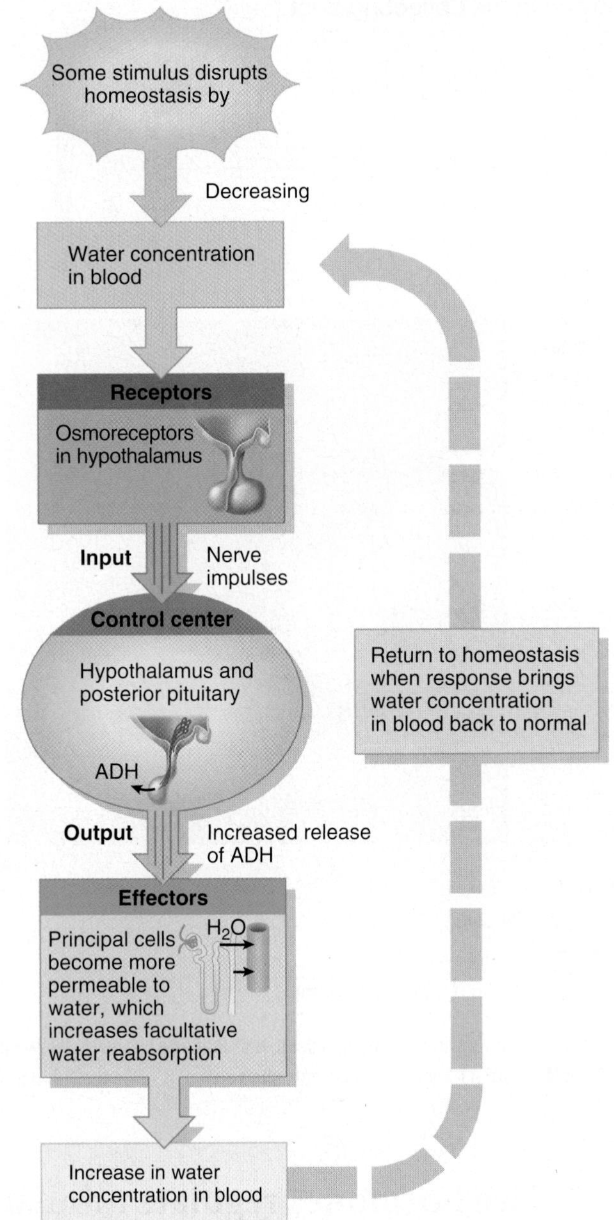


**ADH stimulates the kidneys to return more water to the blood.**

## Atrial Natriuretic Peptide

A large increase in blood volume promotes release of **atrial natriuretic peptide** (nā′-trē-ū-RET-ik) from the heart. Atrial natriuretic peptide inhibits reabsorption of $Na^+$ and water in the proximal convoluted tubule and collecting duct. Atrial natriuretic peptide also suppresses the secretion of aldosterone and ADH. These effects increase the excretion of $Na^+$ and water in urine, which increases urine output and decreases blood volume and blood pressure.

## Parathyroid Hormone

Although the hormones mentioned there far involve regulation of water loss in urine, the kidney tubular also respond to a hormone that regulates ionic composition. For example, a lower-than-normal level of $Ca^{2+}$ in the blood stimulates the parathyroid glands to release **parathyroid hormone**. Parathyroid hormone in turn stimulates cells in the distal convoluted tubules to reabsorb more $Ca^{2+}$ into the blood. Parathyroid hormone also inhibits $HPO_4^{2-}$ (phosphate) reabsorption in proximal convoluted tubules, thereby promoting phosphate excretion.

Table 24.3 summarizes hormonal regulation of tubular reabsorption and tubular secretion.

### ✓ CHECKPOINT

**26.** How is the juxtaglomerular apparatus involved in blood pressure regulation by the kidneys?

**27.** How do angiotensin II and aldosterone regulate tubular reabsorption and secretion?

**28.** How does antidiuretic hormone stimulate water reabsorption by principal cells? Is this process obligatory or facultative water reabsorption?

**29.** Which hormones stimulate water reabsorption, and which hormone inhibits it?

TABLE 24.3

Hormonal Regulation of Tubular Reabsorption and Tubular Secretion

| HORMONE | MAJOR STIMULI THAT TRIGGER RELEASE | MECHANISM AND SITE OF ACTION | EFFECTS |
|---|---|---|---|
| **Angiotensin II** | Low blood volume or low blood pressure stimulates renin-induced production of angiotensin II | Stimulates activity of $Na^+/H^+$ antiporters in proximal tubule cells | Increases reabsorption of $Na^+$, other solutes, and water, which increases blood volume and blood pressure |
| **Aldosterone** | Increased angiotensin II level and increased level of plasma $K^+$ promote release of aldosterone by adrenal cortex | Enhances activity of sodium–potassium pumps in basolateral membrane and $Na^+$ channels in apical membrane of principal cells in collecting duct | Increases secretion of $K^+$ and reabsorption of $Na^+$, $Cl^-$; increases reabsorption of water, which increases blood volume and blood pressure |
| **Antidiuretic hormone (ADH) or vasopressin** | Increased osmotic pressure of extracellular fluid or decreased blood volume promotes release of ADH from posterior pituitary gland | Stimulates insertion of water-channel proteins (aquaporin-2) into apical membranes of principal cells in distal convoluted tubule and collecting duct | Increases facultative reabsorption of water, which decreases osmotic pressure of body fluids and increases blood volume and pressure |
| **Atrial natriuretic peptide (ANP)** | Stretching of atria of heart stimulates ANP secretion | Suppresses reabsorption of $Na^+$ and water in proximal convoluted tubule and collecting duct; inhibits secretion of aldosterone and ADH | Increases excretion of $Na^+$ in urine (natriuresis); increases urine output (diuresis) and thus decreases blood volume and blood pressure |
| **Parathyroid hormone (PTH)** | Decreased level of plasma $Ca^{2+}$ promotes release of PTH from parathyroid glands | Stimulates opening of $Ca^{2+}$ channels in apical membranes of distal convoluted tubule cells | Increases reabsorption of $Ca^{2+}$ |

# 24.8 The kidneys regulate the rate of water loss in urine.

Even though your fluid intake can be highly variable, the total volume of fluid in your body normally remains stable. Homeostasis of body fluid volume depends in large part on the ability of the kidneys to regulate the rate of water loss in urine. Normally functioning kidneys produce a large volume of dilute urine (light yellow) when fluid intake is high, and a small volume of concentrated urine (dark yellow) when fluid intake is low or fluid loss is large. As you learned in the last section, ADH controls whether dilute urine or concentrated urine is formed. When the ADH level is low, urine is very dilute. However, a high level of ADH produces a concentrated urine by stimulating the reabsorption of water into blood.

## Formation of Dilute Urine

Glomerular filtrate has the same ratio of water and solute particles as blood plasma but the water and solute concentrations in filtered fluid are adjusted dramatically as the fluid flows through the nephron and collecting duct. When *dilute* urine is being formed (Figure 24.18), the following conditions occur along the path of tubular fluid:

1. ***Proximal convoluted tubule.*** As solutes are reabsorbed from the filtered fluid in the proximal convoluted tubule, water follows the solutes by osmosis. Because this obligatory water reabsorption maintains osmotic balance with the filtered fluid, the tubular fluid remaining in the proximal convoluted tubule is still isotonic to blood plasma.
2. ***Descending limb of the nephron loop.*** The osmotic gradient of the interstitial fluid in the renal medulla becomes progressively greater as the nephron loop descends in the renal pyramid. (The source of this osmotic gradient is explained shortly.) As tubular fluid flows down the descending limb of the nephron loop, more and more water is reabsorbed by osmosis, which progressively concentrates the fluid remaining in the lumen.
3. ***Ascending limb of the nephron loop.*** In the thick ascending limb of the nephron loop, symporters actively reabsorb $Na^+$, $K^+$, and $Cl^-$ from the tubular fluid (see Figure 24.15). The ions pass from the tubule cells into interstitial fluid, and then diffuse into the blood inside the vasa recta. The water permeability of the thick ascending limb is quite low, so

FIGURE 24.18 **Formation of dilute urine.** Heavy green lines in the ascending limb of the nephron loop and in the distal convoluted tubule indicate impermeability to water; heavy blue lines indicate the last part of the distal convoluted tubule and the collecting duct, which are impermeable to water when ADH level is low; light blue areas around the nephron represent interstitial fluid. Numbers indicate solute concentration in milliosmoles per liter (mOsm/liter).

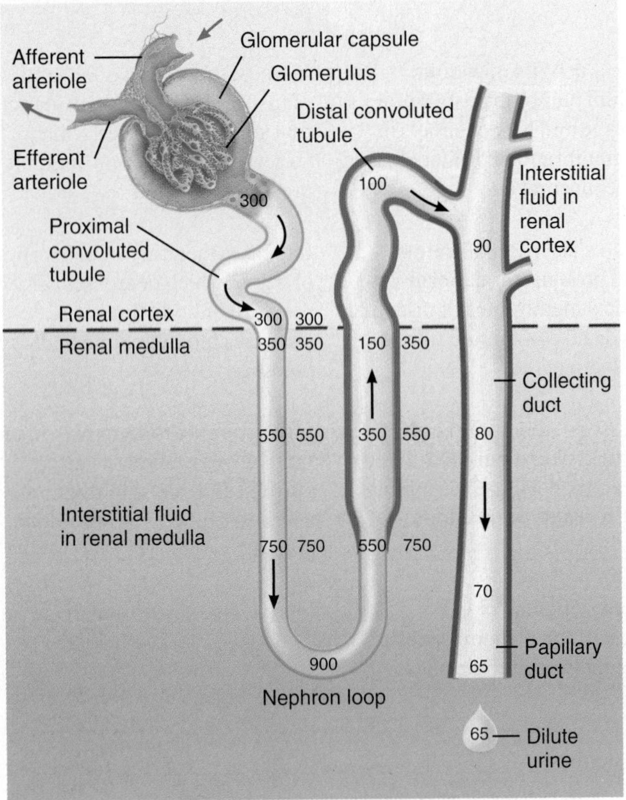


**When ADH level is low, urine is dilute and is less concentrated than blood plasma.**

water cannot follow the reabsorbed solutes by osmosis. As ions and other solutes—but not water molecules—are leaving the tubular fluid, the fluid becomes progressively *more* dilute as it flows up the ascending limb. The tubular fluid entering the distal convoluted tubule is thus more dilute than plasma.

4. ***Distal convoluted tubule and collecting ducts.*** Most of the distal convoluted tubule is not very permeable to water. As the tubular fluid flows through the distal convoluted tubule, additional solutes but only a few water molecules are reabsorbed; hence the fluid becomes more dilute. The water permeability of the late distal convoluted tubules and collecting ducts are regulated by ADH; when the ADH level is low, they are impermeable to water. As the tubular fluid flows onward, it becomes progressively more dilute as additional solutes (but not water molecules) are reabsorbed. By the time the tubular fluid drains into the minor calyx, it can be four times more dilute than blood plasma.

## Formation of Concentrated Urine

When water intake is low or water loss is high (such as during heavy sweating), the kidneys must conserve water while still eliminating wastes and excess ions. Under the influence of ADH, the kidneys produce a small volume of highly concentrated urine. Urine can be as much as four times more concentrated than blood plasma.

The ability of ADH to cause excretion of concentrated urine depends on the presence of an **osmotic gradient** of solutes in the interstitial fluid of the renal medulla. The blue arrow at the left of Figure 24.19 shows that the solute concentration of the interstitial fluid in the kidney increases from about 300 mOsm/liter (milliosmoles per liter) in the renal cortex to about 1200 mOsm/liter deep in the renal medulla. The three major solutes that contribute to this high osmotic gradient are $Na^+$, $Cl^-$, and urea. Two main factors contribute to building and maintaining this osmotic gradient: (1) differences in permeability and reabsorption in different sections of the nephron loops and collecting duct, and (2) the countercurrent flow of fluid in neighboring descending and ascending limbs of the nephron loop and vasa recta.

*Countercurrent flow* refers to the flow of fluid in opposite directions. This occurs when fluid flowing in one tube runs counter (opposite) to fluid flowing in a nearby parallel tube. Note in Figure 24.19a that the descending limb of the nephron loop carries tubular fluid from the renal cortex deep into the medulla, and the ascending limb carries it in the opposite direction. Countercurrent flow through the descending and ascending limbs of the nephron loop establishes a progressively increasing osmotic gradient in the interstitial fluid of the renal medulla. Using this osmotic gradient, the kidneys produce concentrated urine in the following way (Figure 24.19):

1. ***Symporters in thick ascending limb cells of the nephron loop create a buildup of $Na^+$ and $Cl^-$ in the renal medulla.*** In the thick ascending limb of the nephron loop, the $Na^+$–$K^+$–$2Cl^-$ symporters reabsorb $Na^+$ and $Cl^-$ from the tubular fluid (Figure 24.19a). Water is not reabsorbed in this segment, however, because the cells are impermeable to water. As a result, there is a buildup of $Na^+$ and $Cl^-$ ions in the interstitial fluid of the medulla.

2. ***Countercurrent flow through the descending and ascending limbs of the nephron loop establishes an osmotic gradient in the renal medulla.*** As $Na^+$ and $Cl^-$ are constantly reabsorbed by the thick ascending limb, these ions become increasingly concentrated in the interstitial fluid of the medulla, which forms an osmotic gradient (see blue arrow at left) that increases from the outer to deeper renal medulla. The descending limb of the nephron loop is very permeable to water but impermeable to most solutes. Due to the osmotic gradient established by $Na^+$ and $Cl^-$, water is drawn out of the descending limb by osmosis, making the tubular fluid increasingly concentrated as it approaches the hairpin turn of the loop. As the tubular fluid continues on

FIGURE 24.19 **Formation of concentrated urine.** The green line indicates the presence of $Na^+$–$K^+$–$2C^-$ symporters that simultaneously reabsorb these ions into the interstitial fluid of the renal medulla; this portion of the nephron is also relatively impermeable to water and urea. All solute concentrations are in milliosmoles per liter (mOsm/liter).

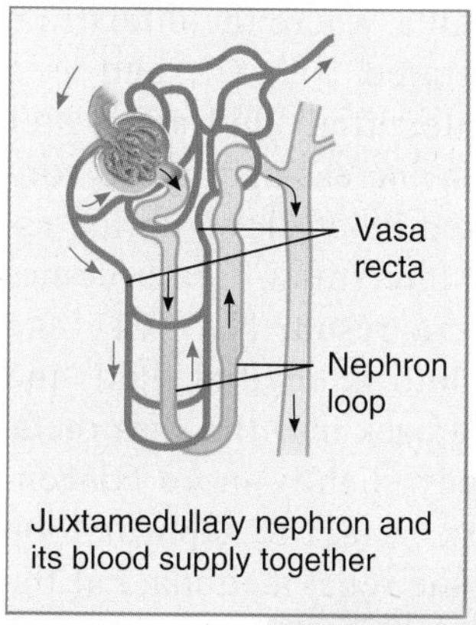

through the ascending limb, its symporters reabsorb $Na^+$ and $Cl^-$ from the tubular fluid, causing the tubular fluid to become progressively more dilute as it flows toward the junction of the medulla and cortex. To summarize, tubular fluid becomes progressively more concentrated as it flows down the descending limb and progressively more dilute as it moves up the ascending limb.

3. ***Cells in the collecting ducts reabsorb more water under influence of ADH.*** When ADH increases the water permeability of the principal cells of the collecting duct, water quickly moves by osmosis out of tubular fluid, into the interstitial fluid, and then into the vasa recta. As water leaves, the tubular fluid left behind in the collecting duct becomes increasingly concentrated. Because the collecting

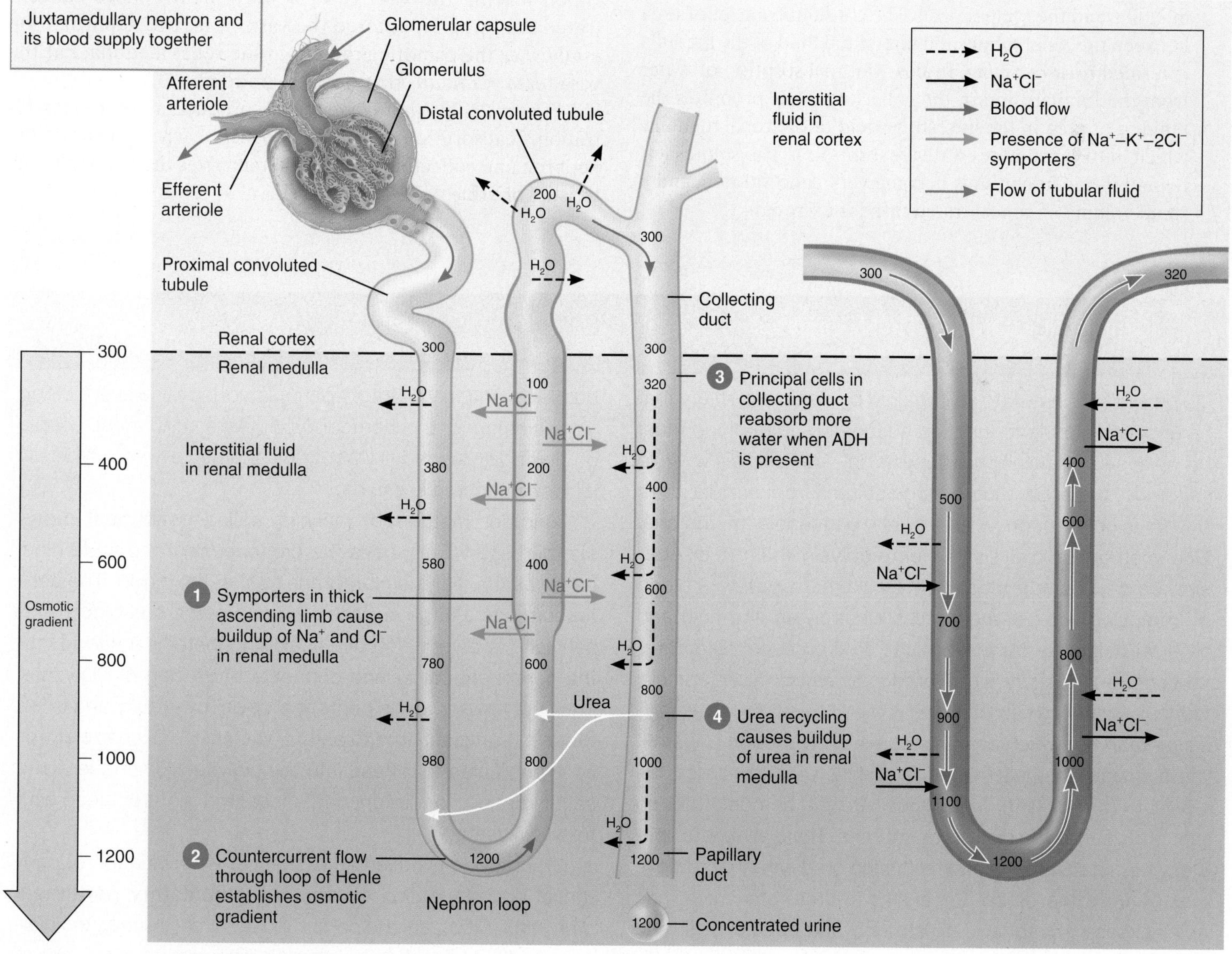

(a) Reabsorption of $Na^+$, $Cl^-$, and water in juxtamedullary nephron

(b) Recycling of salts and urea in vasa recta

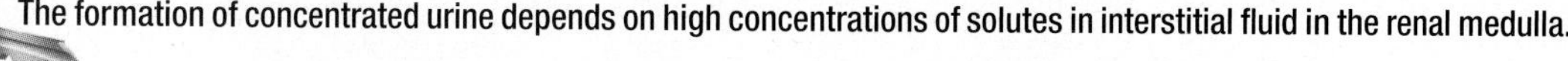

**The formation of concentrated urine depends on high concentrations of solutes in interstitial fluid in the renal medulla.**

duct passes through the high osmotic gradient of the renal medulla, more water can be reabsorbed.

4 ***Urea recycling causes a buildup of urea in the renal medulla.*** Collecting ducts deep in the renal medulla are permeable to urea, allowing it to diffuse from the tubular fluid into the interstitial fluid of the medulla. As urea accumulates in the interstitial fluid, some of it diffuses into the tubular fluid in the descending and thin ascending limbs of the nephron loops, which also are permeable to urea (Figure 24.19a). However, while tubular fluid flows through the thick ascending limb, distal convoluted tubule, and cortical portion of collecting duct, urea remains in the lumen because cells in these segments are impermeable to urea. As water reabsorption continues via osmosis in the presence of ADH, the concentration of urea in the tubular fluid *further increases*. More urea diffuses into the interstitial fluid of the renal medulla, and the cycle repeats. The constant transfer of urea between the renal tubule and interstitial fluid of the medulla is termed *urea recycling*. In this way, reabsorption of water from the tubular fluid of the collecting ducts promotes the buildup of urea in the interstitial fluid of the renal medulla, which in turn promotes water reabsorption. The solutes left behind in the lumen thus become very concentrated, and a small volume of concentrated urine is excreted.

Countercurrent flow also allows solutes and water to passively exchange between the blood of the vasa recta and interstitial fluid of the renal medulla. Note in Figure 24.19b that just as tubular fluid flows in opposite directions in the nephron loop, blood flows in opposite directions in parallel ascending and descending limbs of the vasa recta. Blood entering the vasa recta is fairly dilute. As it flows down the descending limb into the renal medulla, where the interstitial fluid becomes increasingly concentrated, $Na^+$, $Cl^-$, and urea diffuse into the blood from the interstitial fluid and water flows out of the blood, resulting in increasingly more concentrated blood. As the concentrated blood flows up the ascending loop of the vasa recta, the interstitial fluid becomes increasingly less concentrated. As a result, $Na^+$, $Cl^-$, and urea diffuse from the blood back into interstitial fluid, and water diffuses from interstitial fluid back into the vasa recta. Blood leaving the vasa recta is only slightly more concentrated than when it entered the vasa recta. The nephron loop *establishes* the osmotic gradient in the renal medulla, but the vasa recta *maintains* that osmotic gradient.

Figure 24.20 summarizes urine production; correlates filtration, reabsorption, and secretion with each segment of the nephron and collecting duct; and summarizes the major characteristics of urine.

## CLINICAL CONNECTION | *Dialysis*

If a person's kidneys are so impaired by disease or injury that he or she is unable to function adequately, then blood must be cleansed artificially by **dialysis** (dī-AL-i-sis; *dialyo* = to separate), the separation of large solutes from smaller ones by diffusion through a selectively permeable membrane. One method of dialysis is **hemodialysis** (hē′-mō-dī-AL-i-sis; *hemo-* = blood), which directly filters the patient's blood by removing wastes and excess electrolytes and fluid and then returning the cleansed blood to the patient. Blood removed from the body is delivered to a *hemodialyzer* (artificial kidney). Inside the hemodialyzer, blood flows through a *dialysis membrane,* which contains pores large enough to permit the diffusion of small solutes. A special solution, called the *dialysate* (dī-AL-i-sāt), is pumped into the hemodialyzer so that it surrounds the dialysis membrane. The dialysate is specially formulated to maintain diffusion gradients that remove wastes from the blood (for example, urea, creatinine, uric acid, excess phosphate, potassium, and sulfate ions) and add needed substances (for example, glucose and bicarbonate ions) to it. The cleansed blood is passed through an air embolus detector to remove air and then returned to the body. An anticoagulant (heparin) is added to prevent blood from clotting in the hemodialyzer. As a rule, most people on hemodialysis require about 6–12 hours a week, typically divided into three sessions.

Another method of dialysis, called **peritoneal dialysis** (per′-i-tō-NĒ-al)**,** uses the peritoneum of the abdominal cavity as the dialysis membrane to filter the blood. The peritoneum has a large surface area and numerous blood vessels, and is a very effective filter. A catheter is inserted into the peritoneal cavity and connected to a bag of dialysate. The fluid flows into the peritoneal cavity by gravity and is left there for sufficient time to permit wastes and excess electrolytes and fluids to diffuse into the dialysate. Then the dialysate is drained out into a bag, discarded, and replaced with fresh dialysate.

Each cycle is called an *exchange.* One variation of peritoneal dialysis, called **continuous ambulatory peritoneal dialysis (CAPD),** can be performed at home. Usually, the dialysate is drained and replenished four times a day and once at night during sleep. Between exchanges the person can move about freely with the dialysate in the peritoneal cavity.

FIGURE 24.20 **Summary of filtration, reassorption, and secretion in the nephron and collecting duct.**

RENAL CORPUSCLE

**Glomerular filtration rate:** 105–125 mL/min of fluid that is isotonic to blood

**Filtered substances:** water and all solutes present in blood (except proteins) including ions, glucose, amino acids, creatinine, uric acid

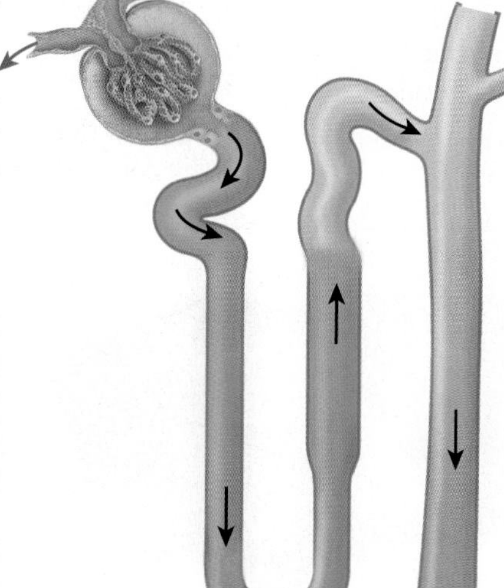

NEPHRON LOOP

**Reabsorption** (into blood) of:

| | |
|---|---|
| Water | 15% (osmosis in descending limb) |
| $Na^+$ | 20–30% (symporters in ascending limb) |
| $K^+$ | 20–30% (symporters in ascending limb) |
| $Cl^-$ | 35% (symporters in ascending limb) |
| $HCO_3^-$ | 10–20% (facilitated diffusion) |
| $Ca^{2+}$, $Mg^{2+}$ | variable (diffusion) |

**Secretion** (into urine) of:

| | |
|---|---|
| Urea | variable (recycling from collecting duct) |

At end of nephron loop, tubular fluid is hypotonic (100–150 mOsm/liter).

PROXIMAL CONVOLUTED TUBULE

**Reabsorption** (into blood) of:

| | |
|---|---|
| Water | 65% (osmosis) |
| $Na^+$ | 65% (sodium–potassium pumps, symporters, antiporters) |
| $K^+$ | 65% (diffusion) |
| Glucose | 100% (symporters and facilitated diffusion) |
| Amino acids | 100% (symporters and facilitated diffusion) |
| $Cl^-$ | 50% (diffusion) |
| $HCO_3^-$ | 80–90% (facilitated diffusion) |
| Urea | 50% (diffusion) |
| $Ca^{2+}$, $Mg^{2+}$ | variable (diffusion) |

**Secretion** (into urine) of:

| | |
|---|---|
| $H^+$ | variable (antiporters) |
| $NH_4^+$ | variable, increases in acidosis (antiporters) |
| Urea | variable (diffusion) |
| Creatinine | small amount |

At end of PCT, tubular fluid is still isotonic to blood (300 mOsm/liter).

Urine

NORMAL CHARACTERISTICS OF URINE

| | |
|---|---|
| Volume | one to two liters per 24 hours; considerable variation in normal volume |
| Color | yellow or amber color; color is darker in concentrated urine |
| Turbidity | transparent in freshly voided urine; microbes, pus, epithelial cells, or crystals may cause cloudiness |
| Odor | aromatic when fresh; ammonia-like after standing because of breakdown of urea to ammonia by bacteria |
| pH | normal range is 4.6–8.0; high protein diets produce an acidic urine; vegetarian diets produce an alkaline urine |
| Specific gravity | normal range is 1.001–1.035; low specific gravity represents dilute urine, higher values represent a concentrated urine |

DISTAL CONVOLUTED TUBULE AND COLLECTING DUCT

**Reabsorption** (into blood) of:

| | |
|---|---|
| Water | 20% (osmosis, water channels stimulated by ADH) |
| $Na^+$ | 5–15% (symporters, sodium–potassium pumps, sodium channels stimulated by aldosterone) |
| $Cl^-$ | 5% (symporters) |
| $HCO_3^-$ | variable, depends on $H^+$ secretion (antiporters) |
| Urea | variable (recycling of nephron loop) |
| $Ca^{2+}$ | variable (stimulated by parathyroid hormone) |

**Secretion** (into urine) of:

| | |
|---|---|
| $K^+$ | variable, adjusts to dietary intake (leakage channels) |
| $H^+$ | variable, adjusts to maintain acid–base homeostasis ($H^+$ pumps) |

Tubular fluid leaving the collecting duct is dilute when ADH level is low and concentrated when ADH level is high.

**Filtration occurs in the renal corpuscle; reabsorption occurs all along the renal tubule and collecting ducts.**

## ✓ CHECKPOINT

**30.** What force draws water out of the nephron loop and into the interstitial fluid of the renal medulla?

**31.** Which portions of the renal tubule and collecting duct reabsorb more solutes than water to produce dilute urine?

**32.** Which solutes are the main contributors to the high osmotic gradient of interstitial fluid in the renal medulla?

**33.** Where in the renal tubule and/or collecting duct is water reabsorbed to produce concentrated urine? Which hormone controls the production of concentrated urine?

**34.** How does urea recycling contribute to the production of concentrated urine?

## RETURN TO Sam's Story

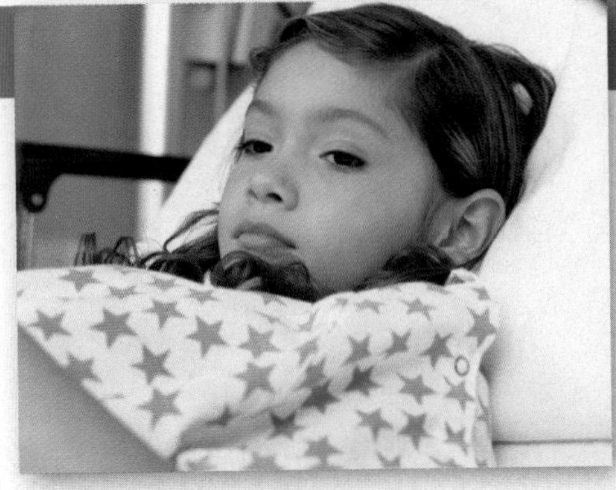

Sam's parents have a very common story: recently divorced, both working multiple jobs just to make ends meet, each assuming that the other was taking care of Sam. When Cindy enters the waiting room, Sam's parents are in the middle of a screaming match.

"Excuse me! Are you Samantha's parents?" Cindy interjects a little too loudly.

"Who are you?" snaps Sam's mother.

"My name is Cindy. I am the social worker assigned to your daughter's case," Cindy replies.

Sam's father looks livid. "Why do we need a social worker? We didn't do anything wrong. At least, I didn't do anything wrong." With that the parents launch into another verbal sparring match.

"Stop!" Cindy nearly screams. "Let's not talk about who is to blame right now. Samantha is very sick and if you would just stop yelling, I can update you on her condition." Sam's parents sit down dejectedly, reality finally hitting them. Regretting her earlier tone, Cindy says calmly, "Look, anyone can see that you both care a lot. We all just need to focus on your daughter, because right now, her problems are bigger than yours."

Sam's parents stay quiet long enough for Cindy to go through the developments in her case. After Cindy finishes, Sam's mother, with tears streaming down her face, says, "I just don't understand why they have to take her urine when she's having such a hard time making any." Cindy thinks and says, "It's because there are abnormal things in her urine that shouldn't be there. What these things are and how much of them will tell the doctors how bad her condition is. I'll go get a doctor who can explain the details more clearly. I'll be right back," Cindy says as she leaves the room. For Sam's sake, it's important to get all the facts right.

D. *How does the nephron select what gets filtered and what stays behind in the blood plasma? Why might Sam's filtration system not be functioning properly? What might the doctors find in her urine to support your hypothesis?*

E. *If Sam's glomerular filtration rate (GFR) is found to be lower than normal, what component of her net filtration pressure (NFP) is most likely the culprit? What do you think may be causing the decreased GFR?*

F. *How would electrolyte imbalances and obligatory water loss explain Sam's decreased urine output?*

G. *Of the five hormones that regulate tubular reabsorption and tubular secretion, which ones do you expect to be in high concentration, and which ones to be in low concentration in Sam's case?*

H. *What hormone is needed if Sam's body is to produce concentrated urine and combat her dehydration?*

# 24.9 The kidneys help maintain the overall fluid and acid–base balance of the body.

## Fluid Balance

One important function of the kidneys is to help maintain fluid balance in the body. The water and dissolved solutes in the body constitute the **body fluids**. Regulatory mechanisms involving the kidneys and other organs normally maintain homeostasis of the body fluids to keep body organs functioning normally.

About two-thirds of body fluid is **intracellular fluid** or **cytosol**, the fluid within cells. The other third, called **extracellular fluid**, is outside cells and includes all other body fluids. Extracellular fluid includes interstitial fluid between tissue cells, synovial fluid in joints, cerebrospinal fluid in the nervous system, aqueous humor and vitreous body in the eyes, endolymph and perilymph in the ears, plasma in blood, lymph in lymphatic vessels, and pleural, pericardial, and peritoneal fluids between serous membranes.

The body is in **fluid balance** when the required amounts of water and solutes are present and are correctly distributed throughout the body. **Water** is by far the largest single component of the body, making up 45–75 percent of total body mass, depending on age and gender. Most solutes in body fluids are **electrolytes**, inorganic compounds that break apart into ions when dissolved in water. Because osmosis is the primary means of water movement between intracellular and interstitial fluids, water moves toward the fluid where solutes (mostly electrolytes) are more concentrated. Because intake of water and electrolytes varies, the ability of the kidneys to excrete excess water in dilute urine, or to excrete excess electrolytes in concentrated urine, is of utmost importance in the maintenance of fluid balance.

## Regulation of Body Water Gain

The body can gain water by ingesting liquids and moist foods and by metabolic synthesis of water. Body water gain is regulated mainly by how much fluid you drink. When water loss is greater than water gain, **dehydration**—a decrease in volume and an increase in osmotic pressure of body fluids—stimulates thirst (Figure 24.21). As blood volume decreases, blood pressure drops, which stimulates the kidneys to release renin. Renin promotes the formation of angiotensin II. Increased angiotensin II in the blood, along with stimulation of osmoreceptors by the increased osmotic pressure, activate an area in the hypothalamus known as the **thirst center**. As a result, the sensation of thirst increases, which usually leads to increased fluid intake and restoration of normal fluid volume.

FIGURE 24.21 **Pathways through which dehydration stimulates thirst.**

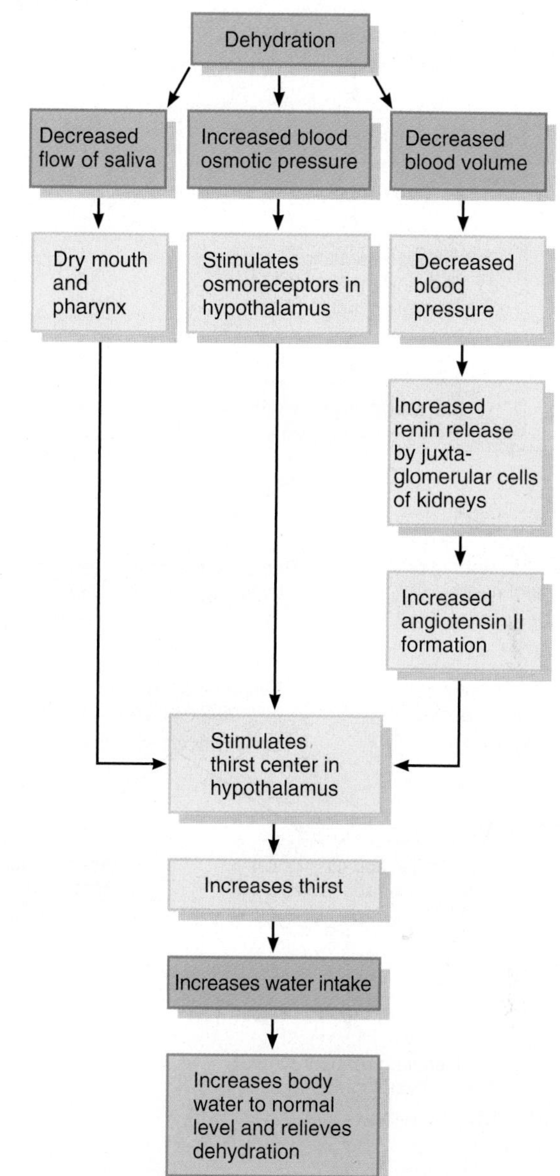


**Dehydration occurs when water loss is greater than water gain.**

## Regulation of Water and Solute Loss

Normally, body fluid volume remains constant because water loss equals water gain. Water is lost in urine, perspiration, feces, as water vapor exhaled from the lungs, and as menstrual flow in women of reproductive age. Elimination of *excess* body water or solutes occurs mainly by control of their loss in urine.

The two primary solutes in extracellular fluid (and in urine) are sodium ions ($Na^+$) and chloride ions ($Cl^-$). The extent of *urinary salt (NaCl) loss* is the main influence on body fluid *volume* because water "follows solutes" in osmosis. Because the amount of NaCl in our daily diet is highly variable, urinary excretion of $Na^+$ and $Cl^-$ must also vary to maintain homeostasis.

Figure 24.22 depicts the sequence of changes that occur after a salty meal. The increased intake of NaCl produces an increase in blood levels of $Na^+$ and $Cl^-$. As a result, the osmotic pressure of interstitial fluid increases, which moves water from intracellular fluid into interstitial fluid, and then into plasma. As water enters plasma, blood volume increases.

An increase in blood volume, as might occur after you finish a supersized drink, stretches the atria of the heart and promotes release of atrial natriuretic peptide. Atrial natriuretic peptide increases *natriuresis*, the urinary loss of $Na^+$. As $Na^+$ is excreted, followed by $Cl^-$ and water, blood volume decreases. An increase in blood volume also slows release of renin from kidney juxtaglomerular cells. When renin level drops, less angiotensin II is formed. Decline in angiotensin II increases glomerular filtration rate and reduces $Na^+$, $Cl^-$, and water reabsorption in the kidney tubules. In addition, less angiotensin II leads to less aldosterone, which decreases $Na^+$ and $Cl^-$ reabsorption in the collecting ducts. As more $Na^+$ and $Cl^-$ ions are excreted in the urine, more water is lost in urine, which decreases blood volume. By contrast, when someone becomes dehydrated, higher levels of angiotensin II and aldosterone promote urinary reabsorption of $Na^+$ and $Cl^-$, and water by osmosis with the solutes, and thereby conserve the volume of body fluids.

The major hormone that regulates water loss is antidiuretic hormone (ADH). An increase in the osmotic pressure of body fluids not only stimulates the thirst mechanism, as previously discussed, but also stimulates release of ADH (see Figure 24.17). ADH promotes water reabsorption by the collecting ducts. As a result, water is reabsorbed from the tubular fluid and a small

FIGURE 24.22 **Hormonal regulation of renal $Na^+$ and $Cl^-$ reabsorption.**

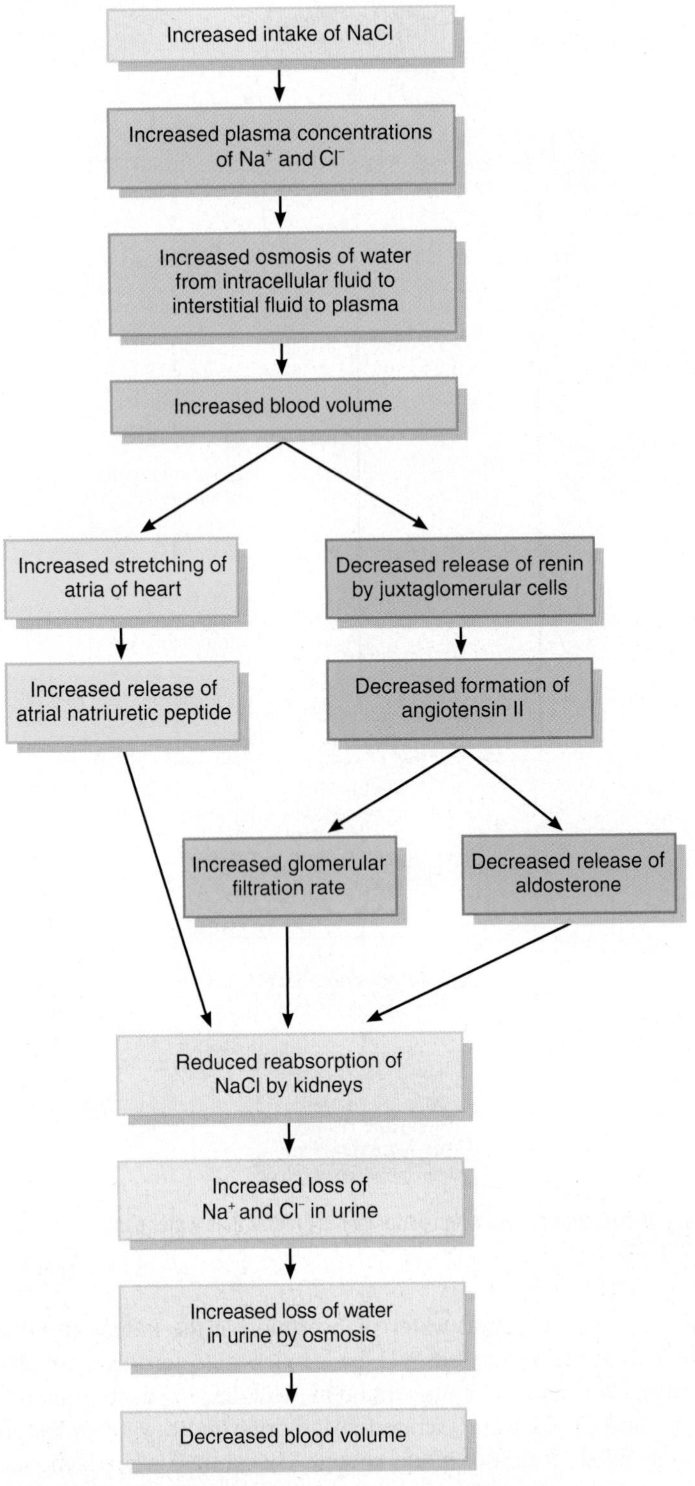


**The three main hormones that regulate renal $Na^+$ and $Cl^+$ reabsorption (and thus the amount lost in the urine) are angiotensin II, aldosterone, and atrial natriuretic peptide.**

volume of concentrated urine is produced. By contrast, intake of water decreases the osmotic pressure of body fluids. Within minutes, ADH secretion shuts down, collecting ducts become less permeable to water, and more water is lost in the urine.

### Movement of Water into and out of Cells

Normally, intracellular and interstitial fluids have the same osmotic pressure, so cells neither shrink nor swell. When the body loses more water than it gains, osmotic pressure of interstitial fluid increases, which draws water out of cells, causing them to shrink. Normally at these times, ADH is released by the posterior pituitary and water is salvaged by the kidneys, allowing cells to shrink only slightly before quickly returning to normal size. By contrast, a decrease in the osmotic pressure of interstitial fluid, as may occur after drinking a large volume of water, draws water into cells, making them swell. As low osmotic pressure inhibits secretion of ADH, the kidneys excrete excess water in the urine, which raises the osmotic pressure of body fluids to normal. As a result, cells swell only slightly, and only for a brief time. But when a person steadily consumes water faster than the kidneys can excrete it, the result may be **water intoxication**, a state in which excessive body water causes cells to swell dangerously. If the body water and $Na^+$ lost during blood loss or excessive sweating, vomiting, or diarrhea is replaced by drinking plain water, then body fluids become more dilute. This dilution can cause the $Na^+$ concentration of interstitial fluid to fall below the normal range. The resulting decrease in the osmotic pressure of interstitial fluid moves water into cells, and can quickly lead to convulsions, coma, and possibly death. To prevent this dire sequence of events in cases of severe electrolyte and water loss, solutions given for intravenous or oral rehydration therapy include a small amount of table salt (NaCl).

## Acid–Base Balance

A major challenge for the kidneys is keeping the $H^+$ level (pH) of body fluids in the appropriate range. This task—the maintenance of acid–base balance—is of critical importance to normal cellular function. For example, the three-dimensional shapes of all body proteins, which enable them to perform specific functions, are very sensitive to pH changes. In a healthy person, the pH of systemic arterial blood remains between 7.35 and 7.45. Because metabolic reactions often produce a huge excess of $H^+$, the lack of any mechanism for the disposal of $H^+$ would cause $H^+$ in body fluids to rise quickly to a lethal level. Maintaining $H^+$ concentration within a narrow range is thus essential to survival. The removal of $H^+$ from body fluids and its subsequent elimination from the body depend on three major mechanisms: buffer systems, excretion of $H^+$ in the urine, and exhalation of carbon dioxide (discussed in Chapter 22). The removal of $H^+$ from body fluids and its subsequent elimination from the body is accomplished by buffer systems, by the kidneys through excretion of $H^+$ in the urine, and through the cellular exchange of $K^+$ with $H^+$. As was discussed in Chapter 22, the respiratory system also helps maintain acid–base balance by increasing the rate and depth of breathing to exhale additional carbon dioxide. As blood carbon dioxide levels drop, so does the blood level of carbonic acid, which raises the blood pH by reducing blood $H^+$ level.

## The Actions of Buffer Systems

Buffers prevent rapid, drastic changes in the pH of body fluids by converting strong acids and bases into weak acids and weak bases within fractions of a second. Strong acids lower pH more than weak acids because strong acids release $H^+$ more readily and thus contribute more free hydrogen ions. Similarly, strong bases raise pH more than weak ones. **Buffers** act quickly to temporarily bind $H^+$, removing the highly reactive, excess $H^+$ from solution.

An important buffer system related to the urinary system is the **phosphate buffer system**. Phosphate ions are common in intracellular and extracellular fluids. The components of the phosphate buffer system are *dihydrogen phosphate* ($H_2PO_4^-$) and *monohydrogen phosphate* ($HPO_4^{2-}$). If there is an excess of $H^+$, the monohydrogen phosphate ion acts as a weak base and removes the excess $H^+$ as follows:

$$\underset{\text{Hydrogen ion}}{H^+} + \underset{\text{Monohydrogen phosphate}}{HPO_4^{2-}} \longrightarrow \underset{\text{Dihydrogen phosphate}}{H_2PO_4^-}$$

Conversely, if body fluids are becoming too basic, the dihydrogen phosphate ion acts as a weak acid and dissociates to release hydrogen ions:

$$\underset{\text{Dihydrogen phosphate}}{H_2PO_4^-} \longrightarrow \underset{\text{Hydrogen ion}}{H^+} + \underset{\text{Monohydrogen phosphate}}{HPO_4^{2-}}$$

Because the concentration of phosphates is highest in intracellular fluid, the phosphate buffer system is an important regulator of pH within the cytosol. It also acts to a smaller degree in extracellular fluids, and buffers acids in urine. $H_2PO_4^-$ is formed when excess $H^+$ in the kidney tubule fluid combines with $HPO_4^{2-}$ (Figure 24.23a). The $H^+$ that becomes part of the $H_2PO_4^-$ passes into the urine, helping maintain blood pH by excreting $H^+$ in the urine.

## Kidney Excretion of $H^+$

The only way to eliminate most acids that form in the body is to excrete $H^+$ in the urine. Given the magnitude of the contributions of the kidneys to acid–base balance, it's not surprising that renal failure can quickly cause death.

Cells in the proximal convoluted tubules secrete $H^+$ as they reabsorb $Na^+$ (see Figure 24.13). Even more important for regulation of pH of body fluids, however, are the intercalated cells of the collecting ducts. The *apical* membranes of some intercalated cells include **proton pumps ($H^+$ ATPases)** that secrete $H^+$ into the tubular fluid (Figure 24.23). The bicarbonate ion ($HCO_3^-$), the important buffer produced by dissociation of carbonic acid ($H_2CO_3$) inside intercalated cells, crosses the basolateral membrane by means of **$Cl^-/HCO_3^-$ antiporters** and then diffuses into peritubular capillaries (Figure 24.23b).

**FIGURE 24.23 Secretion of $H^+$ by intercalated cells in the collecting duct.**

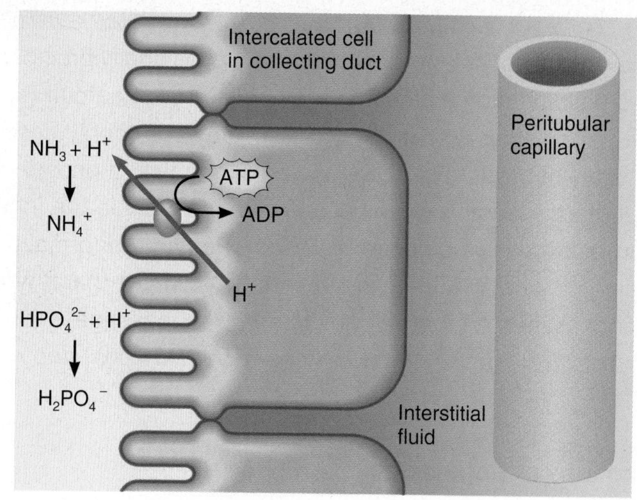


(a) Buffering of $H^+$ in urine

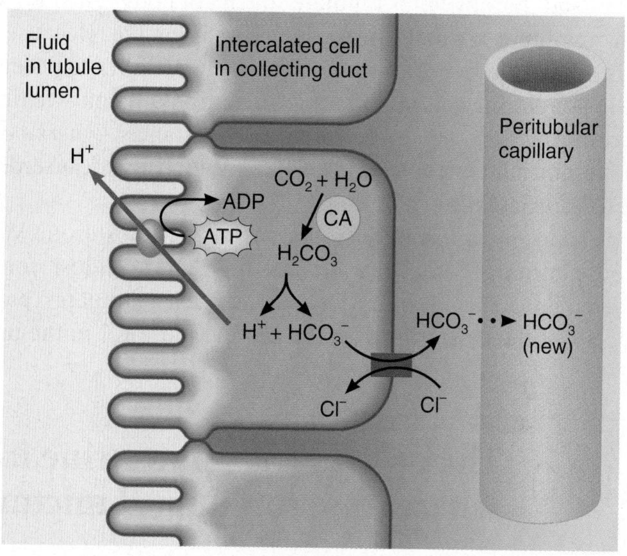


(b) Secretion of $H^+$

**Key:**

- Proton pump ($H^+$ ATPase) in apical membrane
- $Cl^-/HCO_3^-$ antiporter in basolateral membrane
- Diffusion

**Urine can be up to 1000 times more acidic than blood due to the operation of the proton pumps in the collecting ducts of the kidneys.**

Some hydrogen ions secreted into the tubular fluid of the collecting duct are buffered, not by $HCO_3^-$, but by two other buffers (Figure 24.23b). As noted above, $H^+$ combines with $HPO_4^{2-}$ (monohydrogen phosphate ion) to form $H_2PO_4^-$ (dihydrogen phosphate ion); $H^+$ also combines with $NH_3$ (ammonia) to form $NH_4^+$ (ammonium ion).

## CLINICAL CONNECTION | *Renal Calculi*

The crystals of salts present in urine occasionally precipitate and solidify into insoluble stones called **renal calculi** (KAL-kū-lī = pebbles) or *kidney stones*. They commonly contain crystals of calcium oxalate, uric acid, or calcium phosphate. Conditions leading to calculus formation include the ingestion of excessive calcium, low water intake, abnormally alkaline or acidic urine, and overactivity of the parathyroid glands. When a stone lodges in a narrow passage, such as a ureter, the pain can be intense. *Shock-wave lithotripsy* (LITH-ō-trip′-sē; *litho-* = stone) is a procedure that uses high-energy shock waves to disintegrate kidney stones and offers an alternative to surgical removal. Once the kidney stone is located using x-rays, a device called a *lithotripter* delivers brief, high-intensity sound waves through a water- or gel-filled cushion placed under the back. Over a period of 30 to 60 minutes, 1000 or more shock waves pulverize the stone, creating fragments that are small enough to wash out in the urine.

### Functions of Potassium

Potassium ions ($K^+$) are the most abundant cations in intracellular fluid. When $K^+$ moves into or out of cells, it often is exchanged for $H^+$ and thereby helps regulate the pH of body fluids. $K^+$ also helps maintain normal intracellular fluid volume. Because $K^+$ plays a key role in establishing the resting membrane potential and in generating action potentials in neurons and muscle fibers, abnormal $K^+$ levels can be lethal. For instance, *hyperkalemia* (above-normal concentration of $K^+$ in blood) can cause death due to ventricular fibrillation.

$K^+$ concentration is controlled mainly by aldosterone. When blood plasma $K^+$ concentration is high, more aldosterone is secreted into the blood. Aldosterone then stimulates principal cells of the collecting ducts to secrete more $K^+$ in the urine. Conversely, when blood plasma $K^+$ concentration is low, aldosterone secretion decreases and less $K^+$ is excreted in urine.

### ✓ CHECKPOINT

**35.** What are body fluids? What are the percentage distributions of body fluids? What is the primary influence on body fluid volume?

**36.** How do atrial natriuretic peptide, angiotensin II, aldosterone, and antidiuretic hormone regulate the volume and osmotic pressure of body fluids?

**37.** How do phosphate ions help maintain the pH of body fluids?

**38.** How do intercalated cells maintain the pH of body fluids?

# 24.10 The ureters transport urine from the renal pelvis to the urinary bladder, where it is stored until micturition.

Urine drains from collecting ducts into papillary ducts, minor calyces, major calyces, and finally the renal pelvis (see Figure 24.3). From the renal pelvis, urine enters the ureters and then flows into the urinary bladder. Urine is discharged from the body through the single urethra (see Figure 24.1).

## Ureters

Each of the two **ureters** (Ū-rē-ters) is a 25–30 cm (10–12 in.) long, thick-walled, narrow tube that transports urine to the urinary bladder. Peristaltic contractions of the muscular walls of the ureters push urine toward the urinary bladder, but hydrostatic pressure and gravity also contribute. Like the kidneys, the ureters are retroperitoneal (see Concept 24.2). At the base of the urinary bladder, the ureters curve medially and obliquely penetrate the posterior wall of the urinary bladder (Figure 24.24).

Even though there is no anatomical valve at the opening of each ureter into the urinary bladder, a physiological one is quite effective. As the urinary bladder fills with urine, pressure within it compresses the openings into the ureters and prevents the backflow of urine. When this physiological valve is not operating properly, it is possible for microbes to travel up the ureters from the urinary bladder to infect one or both kidneys.

The mucosa of the ureters is a mucous membrane composed of **transitional epithelium** (see Table 4.1). Transitional epithelium is able to stretch—a marked advantage for any organ that must accommodate a variable volume of fluid. Mucus secreted by the goblet cells of the mucosa prevents the cells from coming in contact with urine, which can vary drastically in solute concentration and pH from the cytosol of cells forming the wall of the ureters. The muscularis of the ureters is composed of longitudinal and circular layers of smooth muscle fibers that function in peristalsis. The superficial adventitia, a layer of areolar connective tissue, anchors the ureters to surrounding connective tissue.

FIGURE 24.24 **Ureters, urinary bladder, and urethra in a female.**

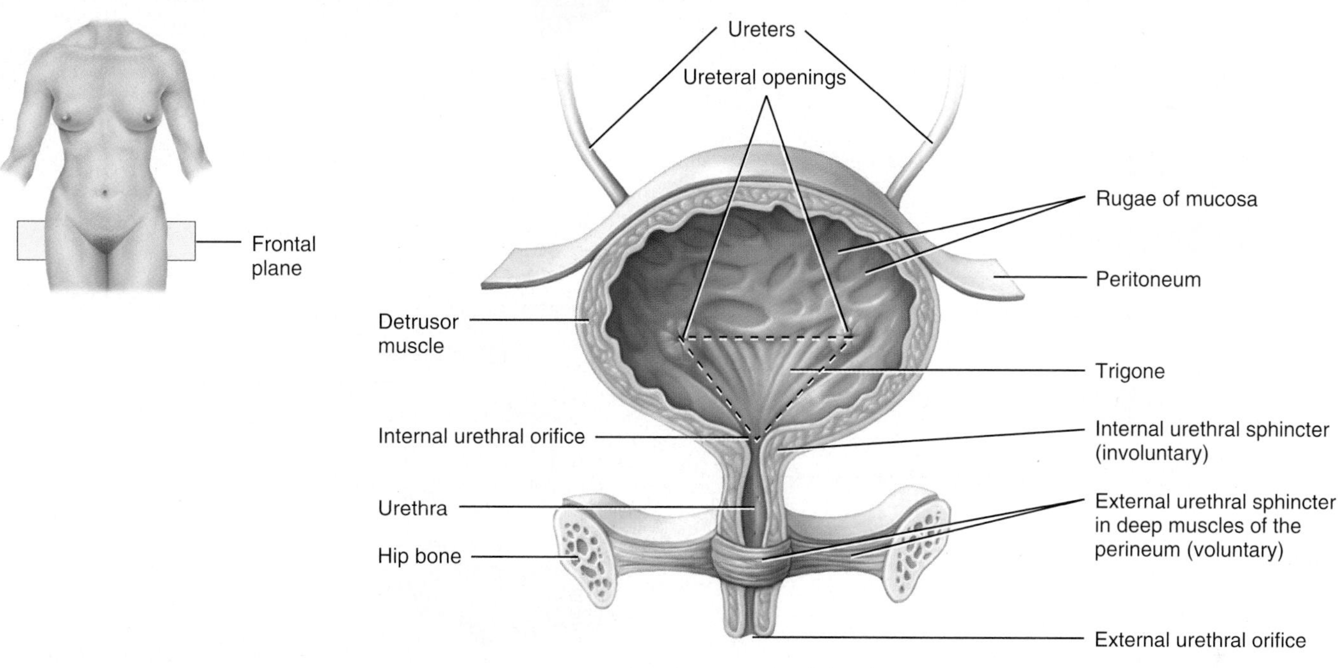


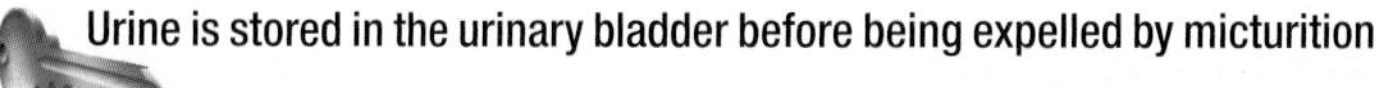
**Urine is stored in the urinary bladder before being expelled by micturition.**

# Urinary Bladder

The **urinary bladder** is a hollow, distensible muscular organ situated in the pelvic cavity posterior to the pubic symphysis. In males, it is directly anterior to the rectum; in females, it is anterior to the vagina and inferior to the uterus (see Figure 24.25). When empty, the urinary bladder collapses like a deflated balloon. As urine volume increases, it becomes pear-shaped and rises into the abdominal cavity. The urinary bladder is smaller in females because the uterus occupies the space just superior to the urinary bladder.

In the floor of the urinary bladder is a small triangular area called the **trigone** (TRĪ-gōn = triangle). The two posterior corners of the trigone contain the two ureteral openings, while the anterior corner contains the **internal urethral orifice** (OR-i-fis), the opening into the urethra. Because its mucosa is firmly bound to the muscularis, the trigone has a smooth surface, allowing it to funnel urine into the urethra as the muscularis contracts.

The mucosa of the urinary bladder is composed of transitional epithelium, allowing the bladder wall to stretch as it fills with urine. **Rugae** (folds in the mucosa) permit expansion of the urinary bladder. The muscularis of the bladder wall consists of three layers of smooth muscle and is also called the **detrusor muscle** (de-TROO-ser = to push down). Around the opening to the urethra the muscularis forms an **internal urethral sphincter**; being smooth muscle, its opening and closing is involuntary. Inferior to the urethral internal sphincter is the **external urethral sphincter**, which is composed of skeletal muscle and is under voluntary control. (See Figures 11.11 and 11.12.)

## The Micturition Reflex

Discharge of urine from the urinary bladder is called **micturition** (mik′-too-RISH-un; *mictur-* = urinate), also known as *urination or voiding*. Micturition occurs via a combination of involuntary and voluntary muscle contractions. As the volume of urine in the urinary bladder increases, stretch receptors in its wall transmit nerve impulses into the spinal cord. These impulses propagate to the sacral region of the spinal cord and trigger a spinal reflex called the **micturition reflex**. In this reflex arc, parasympathetic impulses from the spinal cord stimulate *contraction* of the detrusor muscle and *relaxation* of the internal urethral sphincter muscle. Simultaneously, the spinal cord inhibits somatic motor neurons, causing relaxation of skeletal muscle in the external urethral sphincter. As the urinary bladder wall contracts and both sphincters relax, urination takes place. Although emptying of the urinary bladder is a reflex, in early childhood we learn to initiate it and stop it voluntarily. Through learned control of the external urethral sphincter muscle and muscles of the pelvic floor, the cerebral cortex can initiate micturition or delay its occurrence for a limited period.

# Urethra

The **urethra** (ū-RĒ-thra) is a small tube leading from the internal urethral orifice in the floor of the urinary bladder to the exterior of the body. In both males and females, the urethra is the passageway for discharging urine from the body. In males, it also discharges semen (fluid that contains sperm) (Figure 24.25).

**FIGURE 24.25** **Comparison between female and male urethras.**

Sagittal
plane
Uterus
Urinary bladder
Pubic symphysis
Rectum
Vagina
Urethra
External urethral
orifice

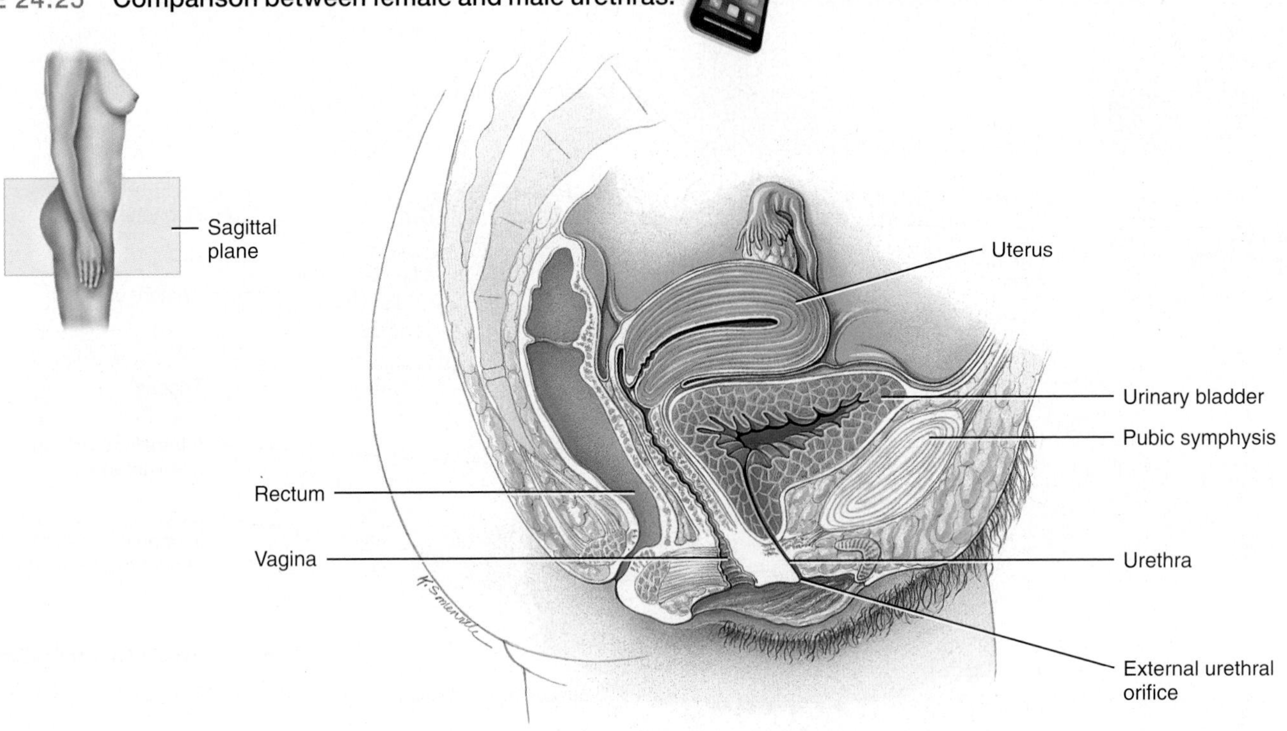

(a) Sagittal section of a female urethra

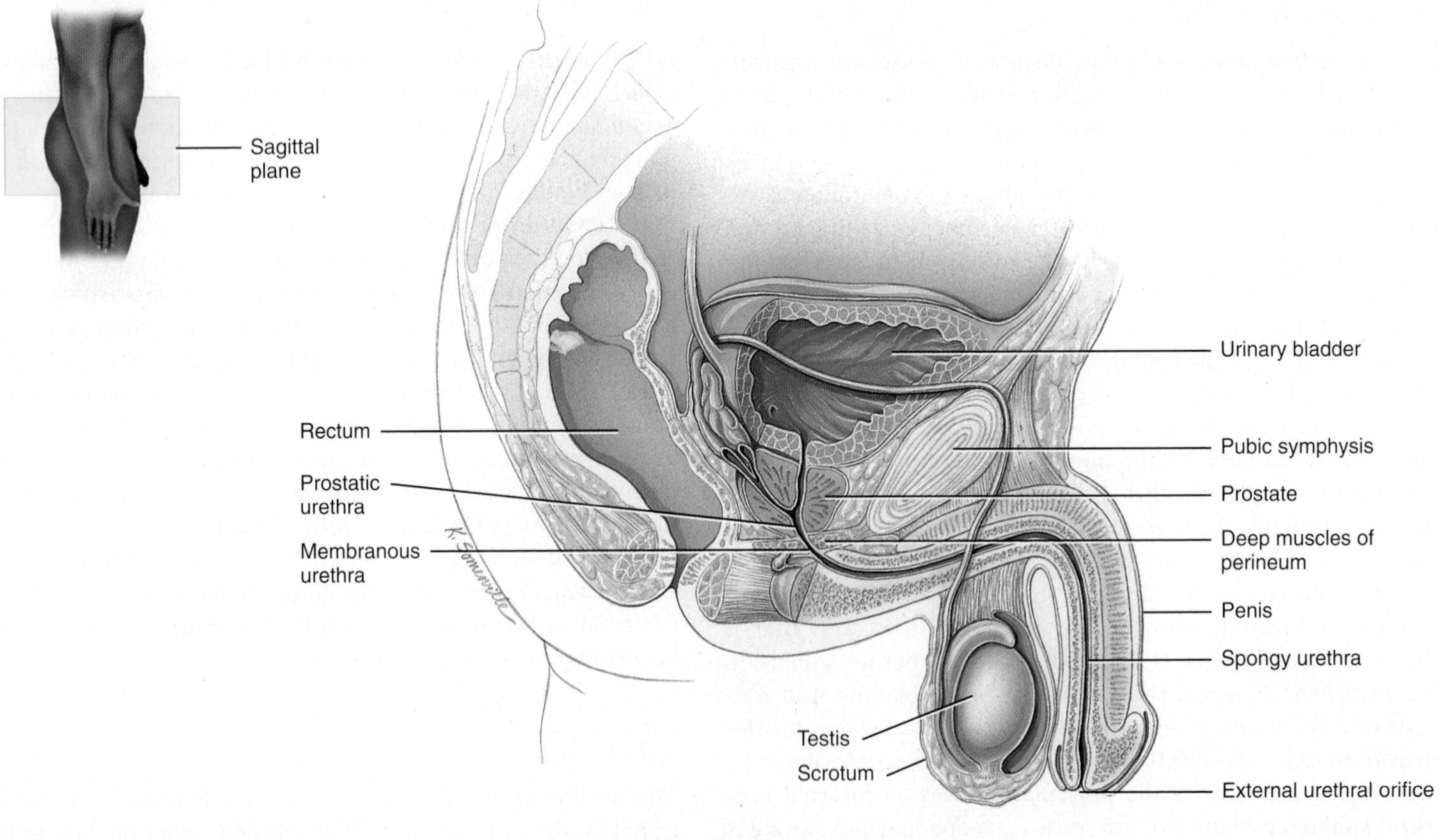

Sagittal
plane
Urinary bladder
Rectum
Pubic symphysis
Prostatic
urethra
Prostate
Deep muscles of
perineum
Membranous
urethra
Penis
Spongy urethra
Testis
Scrotum
External urethral orifice

(b) Sagittal section of a male urethra

**The female urethra is about 4 cm (1.5 in.) in length; the male urethra is about 20 cm (8 in.) in length.**

## CLINICAL CONNECTION | *Urinary Tract Infections*

The term **urinary tract infection (UTI)** is used to describe either an infection of a part of the urinary system or the presence of large numbers of microbes in urine. UTIs are more common in females due to the shorter length of the urethra. Symptoms include painful or burning urination, urgent and frequent urination, low back pain, and bedwetting. UTIs include *urethritis* (ū-rē-THRĪ-tis), inflammation of the urethra; *cystitis* (sis-TĪ-tis), inflammation of the urinary bladder; and *pyelonephritis* (pī-e-lō-ne-FRĪ-tis), inflammation of the kidneys. If pyelonephritis becomes chronic, scar tissue can form in the kidneys and severely impair their function. Drinking cranberry juice can prevent the attachment of *E. coli* bacteria to the lining of the urinary bladder so that they are more readily flushed away during urination.

In females, the urethra lies directly posterior to the pubic symphysis and has a length of 4 cm (1.5 in.) (Figure 24.25a). The opening of the urethra to the exterior, the **external urethral orifice**, is located between the clitoris and the vaginal opening (see Figure 25.11a).

In males, the urethra also extends from the internal urethral orifice to the exterior, but its length and passage through the body are considerably different than in females (Figure 24.25b). The male urethra is about 20 cm (8 in.) long and passes through the prostate, through the deep muscles of the perineum, and finally through the penis. The male urethra is subdivided into three anatomical regions: (1) The **prostatic urethra** passes through the prostate; (2) the **membranous urethra** passes through the deep muscles of the perineum; (3) the **spongy urethra** passes through the penis and opens to the exterior as the external urethral orifice. Several glands and other structures associated with reproduction (described in detail in Chapter 25) deliver their contents into the male urethra.

Near the urinary bladder, the mucosa contains transitional epithelium that is continuous with that of the urinary bladder; near the external urethral orifice, the epithelium is nonkeratinized stratified squamous epithelium. Between these areas, the mucosa contains primarily pseudostratified columnar epithelium. The muscularis is composed of circularly arranged smooth muscle fibers that help form the internal urethral sphincter of the urinary bladder.

### ✓ CHECKPOINT

**39.** Which forces help propel urine from the renal pelvis to the urinary bladder?

**40.** What is the functional significance of transitional epithelium?

**41.** Which urethral sphincter is under conscious control?

**42.** How do the location and length of the urethra compare in males and females?

**43.** What are the three subdivisions of the male urethra?

## Sam's Story

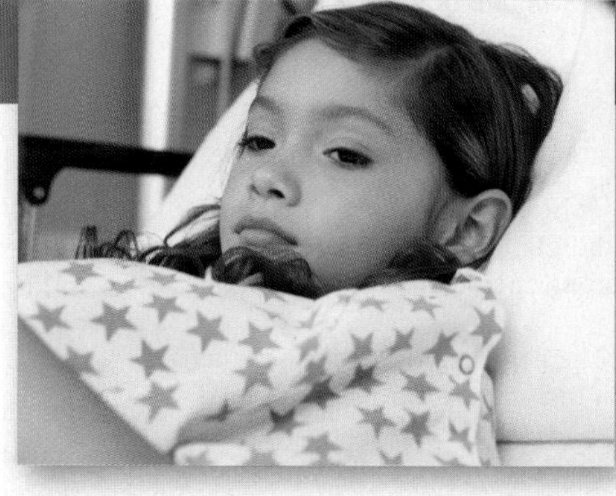

## EPILOGUE AND DISCUSSION

Over the past several hours, Cindy has had a long talk with Sam's parents, and all three of them are waiting in Cindy's office for the test results.

Dr. Jacobs knocks quietly on the door and enters. "We have most of the results back. Sam's urinalysis and blood work confirm that she is malnourished, anemic, and severely dehydrated. Her x-rays show that she has microfractures along her long bones and what looks to be some buildup nearly obstructing her left ureter near the renal pelvis. Antibiotics have been added to her IV drip to combat the infection. I don't think that she will lose her kidney, but I am worried about some short-term issues she may have while she regains full renal function. Cindy, may I see you outside?"

Once the door closes behind them, Cindy blurts out, "I don't think that Sam needs to be removed from her parents' custody just yet."

The doctor smiles and says, "Sam has to stay in the hospital for a while longer so we can manage her antibiotics and get her nutrition back to normal, but I agree with you."

Cindy breathes a sigh of relief. She thanks the doctor and goes back to Sam's room, determined to be there when she wakes up.

I. *Sam's kidney infection most likely started as a bacterial infection located near the urethral opening. How did it make its way into her left kidney?*

J. *If Sam were a boy, why might this situation never have happened?*

K. *If Sam's dehydration is not alleviated, what fluid balance issue might arise?*

L. *How might Sam's kidney infection result in a drop in the pH of her blood?*

# Concept and Resource Summary

| Concept | Resources |
|---|---|

## Introduction

1. The organs of the **urinary system** are the kidneys, ureters, urinary bladder, and urethra.
2. After the kidneys filter blood and return most water and many solutes to the bloodstream, the remaining water and solutes constitute **urine**.

## Concept 24.1 The kidneys regulate the composition of the blood, produce hormones, and excrete wastes.

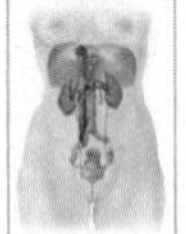

1. The kidneys regulate blood ionic composition, pH, volume, pressure, and glucose level.
2. The kidneys produce hormones (*calcitriol* and *erythropoietin*).
3. The kidneys excrete **wastes** and foreign substances such as drugs.

Resources:
- Anatomy Overview–Kidney Overview

## Concept 24.2 As urine forms, it travels through the renal medulla, calyces, and renal pelvis.

1. The paired **kidneys** are **retroperitoneal** organs attached to the posterior abdominal wall.
2. The **renal hilum** is an indentation where the ureter exits, and where blood vessels, lymphatic vessels, and nerves enter and exit. From superficial to deep, three layers of tissue surround the kidney: **renal fascia**, **adipose capsule**, and **renal capsule**.
3. Internally, the kidneys include a superficial **renal cortex** and a deep **renal medulla** containing **renal pyramids** and **renal columns** extending between the renal pyramids.
4. Urine produced by **nephrons** drains into **papillary ducts, minor calyces, major calyces**, and then into the **renal pelvis**.
5. Blood flows into the kidney through the **renal artery**, from which blood flows sequentially through **segmental arteries**, **interlobar arteries**, **arcuate arteries**, **cortical radiate arteries**, and **afferent arterioles**. Each afferent arteriole enters a nephron and supplies a **glomerulus**, where blood is filtered. From the glomerulus, blood flows through **efferent arterioles**, **peritubular capillaries** (some of which form long loop-shaped capillaries called the **vasa recta**), **cortical radiate veins**, **arcuate veins**, **interlobar veins**, and finally empties into a **renal vein**, where blood flows out of the kidney.

Resources:
- Clinical Connection–Kidney Transplant

## Concept 24.3 Each nephron consists of a renal corpuscle and a renal tubule.

1. The nephron is the functional unit of the kidneys. A nephron consists of a **renal corpuscle** (consisting of the **glomerulus** surrounded by the **glomerular capsule**); and a **renal tubule** (consisting of the **proximal convoluted tubule**, **nephron loop**, and **distal convoluted tubule**).
2. The renal corpuscle and convoluted tubules lie in the renal cortex. The nephron loop begins in the renal cortex, extends into the medulla as the **descending limb**, then turns and returns to the renal cortex as the **ascending limb**.
3. **Cortical nephrons** have short nephron loops that dip only into the superficial region of the renal medulla. **Juxtamedullary nephrons** have long nephron loops that extend deep into the renal medulla, almost to the renal papilla.
4. Distal convoluted tubules empty into **collecting ducts**, which drain urine into papillary ducts.
5. A layer of epithelial cells forms the wall of the glomerular capsule, renal tubule, and ducts. The glomerular capsule consists of **podocytes** that wrap around endothelial cells of the glomerulus, forming the inner wall of the glomerular capsule, and simple squamous epithelium that forms the outer wall of the glomerular capsule. Fluid filtered from the glomerulus enters the **capsular space** between the two layers of the glomerular capsule.
6. The **juxtaglomerular apparatus** consists of the **juxtaglomerular cells** of an afferent arteriole and the **macula densa** of the ascending limb of the nephron loop.

Resources:
- Anatomy Overview–Types of Nephrons
- Anatomy Overview–Fluid Flow through a Nephron
- Figure 24.6–Histology of a Renal Corpuscle
- Exercise–Paint the Nephron
- Exercise–Assemble the Urinary Tract

Clinical Connection–Polycystic Kidney Disease

**Concept** | **Resources**

## Concept 24.4 Urine is formed by glomerular filtration, tubular reabsorption, and tubular secretion.

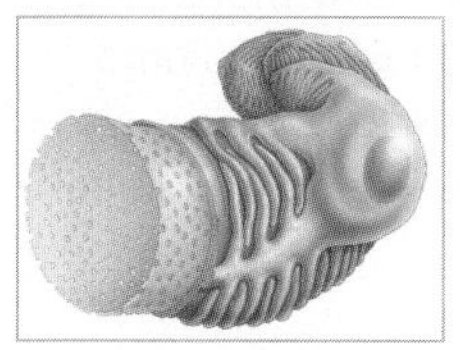

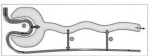

1. During urine formation, nephrons and collecting ducts perform **glomerular filtration**, **tubular reabsorption**, and **tubular secretion**. These processes maintain homeostasis of the volume and composition of the blood.
2. The rate of urinary excretion of any solute is equal to its rate of glomerular filtration, plus its rate of secretion, minus its rate of reabsorption.

- Anatomy Overview–Functional Anatomy of a Nephron

## Concept 24.5 Water and solutes are forced through the filtration membrane during glomerular filtration.

1. Fluid that enters the capsular space is **glomerular filtrate**.
2. Substances that cross the **filtration membrane** must pass through **fenestrations** of the glomerular endothelium, the **basal lamina**, and **filtration slits** between **pedicels** of podocytes. Water and small solutes are capable of passing from the blood into the capsular space; blood cells, platelets, and most plasma proteins are too large to pass through the filtration membrane.
3. **Glomerular blood hydrostatic pressure** in glomerular blood promotes filtration. Filtration is opposed by **capsular hydrostatic pressure**, caused by fluid in the capsular space, and **blood colloid osmotic pressure** from plasma proteins in glomerular blood. The overall relationship of these three pressures, the **net filtration pressure**, favors glomerular filtration.
4. The **glomerular filtration rate** is the amount of filtrate formed in both kidneys per minute. The glomerular filtration rate can be regulated by adjusting blood flow into and out of the glomerulus, and by altering the glomerular capillary surface. Three mechanisms maintain a constant glomerular filtration rate: renal autoregulation, neural regulation, and hormonal regulation.
5. The kidneys maintain a constant renal blood flow, despite daily blood pressure fluctuations, through **renal autoregulation**, in which blood flow to glomerular capillaries is adjusted by smooth muscle cells in the afferent arteriole wall and adjustments in nitric oxide release by the juxtaglomerular apparatus.
6. Neural regulation from sympathetic stimulation by the autonomic nervous system produces vasoconstriction of afferent arterioles, which leads to decreased urine output and greater blood flow to other body tissues.
7. Hormonal regulation of glomerular filtration involves the release of **angiotensin II**, which reduces renal blood flow, decreasing the glomerular filtration rate. **Atrial natriuretic peptide** increases the glomerular filtration rate by increasing capillary surface area.

- Anatomy Overview–Renal Corpuscle
- Animation–Renal Filtration
- Figure 24.9–The Pressures that Drive Glomerular Filtration
- Exercise–Filtration Finale

Clinical Connection–Loss of Plasma Proteins in Urine Causes Edema

Clinical Connection–Glomerular Diseases

## Concept 24.6 Tubular reabsorption reclaims needed substances from the filtrate, while tubular secretion discharges unneeded substances.

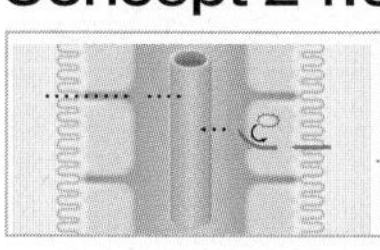

1. Tubular reabsorption is a selective process that reclaims materials from tubular fluid and returns them to the bloodstream. Reabsorbed substances include water, glucose, amino acids, urea, and ions. Some substances not needed by the body are removed from the blood and discharged into the urine via tubular secretion. Included are some ions, urea, creatinine, and certain drugs.
2. Tubular fluid can enter a peritubular capillary either by **paracellular reabsorption** between tubule cells or by **transcellular reabsorption** through tubule cells.
3. About 90% of water reabsorption is **obligatory water reabsorption**; it occurs via osmosis, together with reabsorption of solutes, and is not hormonally regulated. The remaining 10% is **facultative water reabsorption**, which varies according to body needs and is regulated by antidiuretic hormone.
4. In the proximal convoluted tubule (PCT), most reabsorption involves $Na^+$. **$Na^+$ symporters** in the PCT reabsorb glucose, amino acids, lactic acid, and water-soluble vitamins. **$Na^+$–$H^+$ antiporters** reabsorb $Na^+$ into the blood while secreting $H^+$ into tubular fluid. Also in the PCT, $Cl^-$, $K^+$, $Ca^{2+}$, $Mg^{2+}$, and urea are reabsorbed by diffusion. Urea and ammonia in blood are filtered at the glomerulus and secreted by cells of the PCT into tubular fluid.
5. The nephron loop reabsorbs about 15% of the water; 25% of the $Na^+$ and $K^+$; 35% of the $Cl^-$; 15% of the $HCO_3^-$: and a variable amount of the $Ca^{2+}$ and $Mg^{2+}$. Water can leave the descending limb of the

- Anatomy Overview–Overview of Fluids
- Animation–Renal Reabsorption and Secretion
- Exercise–Concentrate on Solvent and Solutes
- Exercise–Pick the Urinary Process
- Exercise–Concentrate on the Urinary System
- Exercise–The Renal Trip
- Concepts and Connections–The Functions of Ions

Clinical Connection–Glucosuria

| Concept | Resources  |
|---|---|
| nephron loop to be reabsorbed but cannot leave the ascending limb of the nephron loop. This results in more dilute tubular fluid as it reaches the end of the ascending limb.<br>**6.** The distal convoluted tubule reabsorbs $Na^+$ and $Cl^-$ via $Na^+$–$Cl^-$ symporters. The distal convoluted tubule is the major site of reabsorption of $Ca^{2+}$, which is stimulated by parathyroid hormone.<br>**7.** In the collecting duct, principal cells reabsorb $Na^+$ and secrete $K^+$; intercalated cells reabsorb $K^+$ and $HCO_3^-$ and secrete $H^+$. | |

## Concept 24.7 Five hormones regulate tubular reabsorption and tubular secretion.

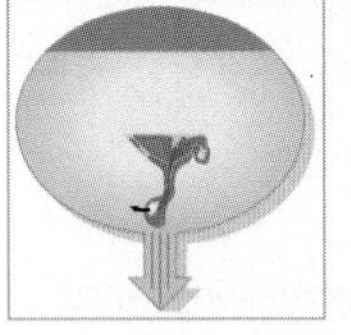

1. Decreases in blood volume and blood pressure cause juxtaglomerular cells to release **renin**, which induces production of angiotensin II. **Angiotensin II** decreases glomerular filtration rate by vasoconstriction of afferent arterioles, enhances reabsorption of $Na^+$, $Cl^-$, and water, and stimulates the adrenal cortex to release **aldosterone**, which in turn stimulates reabsorption of $Na^+$ and $Cl^-$ and secretion of $K^+$. The reabsorption of $Na^+$ and $Cl^-$ results in greater reabsorption of water.
2. **Antidiuretic hormone** increases reabsorption of water. A decrease in the concentration of water in the blood and a decrease in blood volume stimulate release of antidiuretic hormone from the posterior pituitary.
3. Increased blood volume promotes release of **atrial natriuretic peptide** from the heart, which inhibits reabsorption of $Na^+$ and water. Atrial natriuretic peptide also suppresses aldosterone and antidiuretic hormone secretion, leading to increased excretion of $Na^+$ in urine and increased urine output.
4. Decreased blood levels of $Ca^{2+}$ promotes release of **parathyroid hormone** by the parathyroid glands. Parathyroid hormone stimulates $Ca^{2+}$ reabsorption and $HPO_4^{2-}$ excretion.

Resources:
- Animation–Hormonal Control of Blood Volume and Pressure
- Exercise–Renal Regulation

## Concept 24.8 The kidneys regulate the rate of water loss in urine.

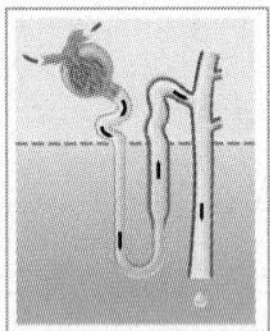

1. Antidiuretic hormone (ADH) controls whether the kidneys form dilute urine or concentrated urine.
2. In the absence of ADH, the kidneys produce dilute urine; renal tubules absorb more solutes than water. Formation of dilute urine involves the following: The descending limb of the nephron loop reabsorbs more water, resulting in more concentrated tubular fluid. The ascending limb of the nephron loop reabsorbs more solutes but not water, resulting in progressively dilute tubular fluid. The distal convoluted tubule absorbs solutes but very little water, again further diluting tubular fluid. Without ADH, collecting ducts are not permeable to water yet solutes continue to be absorbed, making the tubular fluid increasingly dilute as it drains into the renal pelvis.
3. In the presence of ADH, the kidneys produce concentrated urine; large amounts of water are reabsorbed from the tubular fluid into interstitial fluid, increasing solute concentration of the urine. Formation of concentrated urine involves the following: the ascending limb of the nephron loop establishes an **osmotic gradient** of solutes in the interstitial fluid of the renal medulla; with ADH present, collecting ducts reabsorb more water and urea; urea recycling causes a buildup of urea in the renal medulla, which promotes further reabsorption of water.
4. The **countercurrent mechanism** establishes an osmotic gradient in the interstitial fluid of the renal medulla that enables production of concentrated urine when ADH is present.

Resources:
- Animation–Water Homeostasis
- Figure 24.19–Mechanism of Urine Concentration in Juxtamedullary Nephrons

Clinical Connection–Diuretics

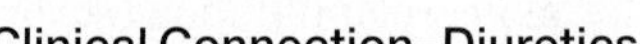

## Concept 24.9 The kidneys help maintain the overall fluid and acid–base balance of the body.

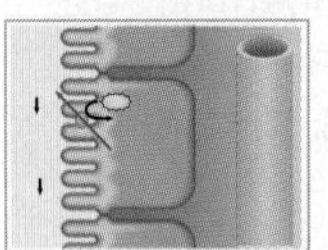

1. **Body fluids** are composed of water and dissolved solutes, primarily **electrolytes**. About two-thirds of body fluid is **intracellular fluid** within cells; one-third of body fluid is **extracellular fluid** outside of cells. **Fluid balance** means that the required amounts of water and solutes are present and are correctly proportioned throughout the body.
2. Water is the largest single constituent in the body. Water is gained from ingested liquids and foods and by metabolic synthesis of water. Elimination of excess body water or solutes occurs mainly on their excretion in urine. $Na^+$ and $Cl^-$ are the two main solutes in extracellular fluid and urine. The extent of their loss in urine is the main determinant of body fluid volume because water follows solutes during osmosis.
3. Angiotensin II and aldosterone reduce urinary loss of $Na^+$ and $Cl^-$, and thereby increase the volume of body fluids. Atrial natriuretic peptide increases urinary loss of $Na^+$ and $Cl^-$ and water, which decreases blood volume.

Resources:
- Anatomy Overview–Overview of Fluids
- Animation–Water and Fluid Flow
- Animation–Regulation of pH
- Figure 24.23–Secretion of $H^+$ by Intercalated Cells in the Collecting Duct

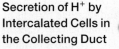

## Concept

## Resources

4. An increase in the osmotic pressure of body fluids stimulates ADH secretion, decreasing urinary loss of water. An increase in osmotic pressure of interstitial fluid draws water out of cells, and they shrink slightly. A decrease in osmotic pressure of interstitial fluid causes cells to swell.
5. The kidneys play an important role in maintaining appropriate $H^+$ levels (pH) of body fluids. Homeostasis of pH is maintained by buffer systems, excretion of $H^+$ in the urine, and exhalation of carbon dioxide.
6. A **buffer** binds to $H^+$ and removes excess $H^+$ from solution, preventing rapid, drastic changes in the pH of a body fluid. The **phosphate buffer system** is composed of the weak acid *dihydrogen phosphate*, ($H_2PO_4^-$), which removes excess $OH^-$ from solution, and the weak base *monohydrogen phosphate* ($HPO_4^{2-}$), which removes excess $H^+$ from solution.
7. Cells in the proximal convoluted tubule, distal convoluted tubule, and collecting ducts of the kidneys are able to secrete $H^+$ into the tubular fluid, thereby increasing urinary loss of $H^+$. The kidneys also produce the bicarbonate ion buffer ($HCO_3^-$), which diffuses into peritubular capillaries to neutralize excess $H^+$ throughout the body.
8. **Potassium ions** ($K^+$) are the most abundant cations in intracellular fluid. They play a key role in the resting membrane potential and action potential of neurons and muscle fibers; help maintain intracellular fluid volume; and contribute to regulation of pH. $K^+$ level is controlled by aldosterone.

- Exercise–Phix the pH
- Concepts and Connections–Mechanisms of pH Balance

Clinical Connection–Urinalysis

### Concept 24.10 The ureters transport urine from the renal pelvis to the urinary bladder, where it is stored until micturition.

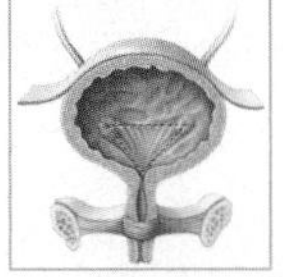

1. Two retroperitoneal tubes called **ureters** transport urine from the renal pelvis of the kidneys to the urinary bladder, primarily by peristalsis. The **mucosa** is composed of **transitional epithelium** that stretches to accommodate varying volumes of urine and goblet cells that secrete protective mucus.
2. The **urinary bladder** is a distensible muscular organ located in the pelvic cavity posterior to the pubic symphysis. Its function is to store urine before micturition. The **trigone** in the bladder floor contains the two ureteral openings and the **internal urethral orifice**. The **mucosa** of the bladder wall contains **transitional epithelium** and rugae to allow expansion. The **muscularis** has smooth muscle called the **detrusor muscle**.
3. **Micturition** is the discharge of urine from the urinary bladder. As urine begins to fill the urinary bladder, stretch receptors trigger the **micturition reflex**. The micturition reflex discharges urine from the urinary bladder through parasympathetic impulses that cause contraction of the detrusor muscle and relaxation of the involuntary, smooth muscle **internal urethral sphincter**. The reflex also inhibits somatic motor neurons, leading to relaxation of the voluntary, skeletal muscle **external urethral sphincter** to allow urination.
4. The **urethra** is a tube leading from the bladder to the exterior. In both sexes, the urethra functions to discharge urine from the body; in males, it also discharges semen. The male urethra is subdivided into three regions: the prostatic urethra, the membranous urethra, and the spongy urethra.

- Anatomy Overview–Ureters, Urinary Bladder, and Urethra
- Figure 24.25–Comparison between Female and Male Urethras

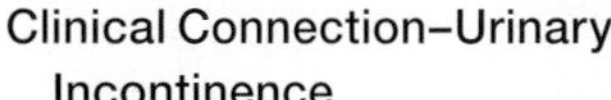

Clinical Connection–Urinary Incontinence

Clinical Connection–Urinary Retention

## Understanding the Concepts

1. Describe the path of blood from its entrance into the kidney to its exit from the kidney, and of urine from the nephron to the urinary bladder.
2. A water molecule has just entered the proximal convoluted tubule of a cortical nephron. Which parts of the nephron will it travel through in order to reach the renal pelvis in a drop of urine?
3. How are the kidneys involved in the regulation of blood pressure?
4. A patient is retaining electrolytes, metabolic wastes, and water in her body; which of the three basic processes of urine formation are most likely involved in her condition?
5. A tumor is pressing on and obstructing the right ureter. What effect might this have on capsular hydrostatic pressure and net filtration pressure in the right kidney? Would the left kidney also be affected?
6. How does filtered glucose enter and leave a proximal convoluted tubule cell?
7. How does reabsorption of $Cl^-$ promote reabsorption of nutritionally important cations?
8. How are the waste products urea and ammonia removed from the blood? In which part of the nephron does this occur?
9. What is the mechanism for reabsorption of $Na^+$ in the distal convoluted tubule and collecting duct?
10. Would the level of antidiuretic hormone in the blood be higher or lower than normal in a person who has just completed a 5-km run without drinking any water? Explain your answer.
11. How do symporters in the ascending limb of the nephron loop and principal cells in the collecting duct contribute to the formation of concentrated urine?
12. What is the countercurrent mechanism? Why is it important?
13. What is micturition? How does the micturition reflex occur?
14. What would be the effects of a drug that blocks the activity of carbonic anhydrase?

CHAPTER 25

# Ryan and Megan's Story

"Ryan Wilson, the doctor will see you now. Please come this way." Ryan sets down his magazine and follows the nurse into an examination room.

"You say that you're here today because of a penile discharge, is that correct?" Ryan flushes, looks down at the floor, and nods his head. She takes the college junior's vitals and asks him to have a seat on the examination table.

"Hi, I'm Doctor Kincaid. What seems to be the problem today?" He is a large man with a warm smile and shakes Randy's hand calmly as he takes a seat.

"Well . . . I've had a milky substance coming out of my penis for the past few days. At first I thought it was just left over from, well, . . ." Dr. Kincaid interrupts him by asking if Ryan has been having pain when he urinates. "Yeah, actually, that's been kind of uncomfortable for the past few weeks. My girlfriend, Megan, has us on this weird macrobiotic diet and I thought that might be the reason."

Dr. Kincaid gently shakes his head and says, "I don't think diet is the cause of your problems. Tell me, Ryan, are you and Megan sexually active?"

Ryan admits that he's had sexual intercourse with Megan for the past few months they have been dating, and other sexual encounters during the six months before that. Dr. Kincaid suspects that Ryan has a sexually transmitted disease. He tells Ryan that he'd like to inspect the affected tissues and collect a sample for examination.

# The Reproductive Systems and Development

## INTRODUCTION

The male and female reproductive systems work together to produce offspring. Males and females have anatomically distinct reproductive organs that are adapted for producing gametes, facilitating fertilization, and, in females, sustaining the growth of the embryo and fetus. The **gonads**—testes in males and ovaries in females—produce germ cells called **gametes** (GAM-ēts = spouses). Various **ducts** then store and transport the gametes, and **accessory sex glands** produce substances that protect the gametes and facilitate their movement. Finally, **supporting structures**, such as the penis in males and the uterus in females, assist the delivery of gametes, and the uterus is also the site for the growth of the embryo and fetus during pregnancy.

Following its deposit in the vagina, a male gamete (sperm cell) may penetrate the female gamete (secondary oocyte) and pregnancy can occur. **Pregnancy** is a sequence of events that begins with fertilization, proceeds to implantation, embryonic development, and fetal development, and ideally ends with birth about 38 weeks later, or 40 weeks after the last menstruation.

In addition to producing gametes, the gonads secrete sex hormones that influence prenatal development and are responsible for the development and maintenance of reproductive structures, secondary sex characteristics, and sexual function.

## CONCEPTS

**25.1** The scrotum supports and regulates the temperature of the testes, which produce spermatozoa.
**25.2** Sperm travel through the epididymis, ductus deferens, ejaculatory ducts, and urethra.
**25.3** After a secondary oocyte is discharged from an ovary, it may undergo fertilization and implantation in the uterus.
**25.4** The vagina is a passageway for childbirth; the mammary glands secrete milk.
**25.5** The female reproductive cycle includes the ovarian and uterine cycles.
**25.6** The zygote divides into a morula and then a blastocyst that implants in the endometrium of the uterus.
**25.7** Major tissues and organs develop during embryonic development and grow and differentiate during fetal development.
**25.8** During pregnancy the uterus expands, displacing and compressing maternal organs.
**25.9** Labor includes dilation of the cervix and expulsion of the fetus and placenta.
**25.10** Lactation is influenced by prolactin, estrogens, progesterone, and oxytocin.

# 25.1 The scrotum supports and regulates the temperature of the testes, which produce spermatozoa.

The organs of the male reproductive system include the testes, a system of ducts (including the epididymis, ductus deferens, ejaculatory ducts, and urethra), accessory sex glands (seminal vesicles, prostate, and bulbourethral glands), and several supporting structures, including the scrotum and the penis (Figure 25.1). The testes (male gonads) produce sperm and secrete hormones. The duct system transports and stores sperm, assists in their maturation, and conveys them to the exterior. Semen contains sperm plus the secretions provided by the accessory sex glands. The penis delivers sperm into the female reproductive tract, and the scrotum supports the testes.

## Scrotum

The **scrotum** (SKRŌ-tum = bag), the supporting structure for the testes, consists of loose skin and underlying hypodermis that hangs from the root (attached portion) of the penis (Figure 25.1a). Externally, the scrotum looks like a single pouch of skin separated into lateral portions by a median ridge called the **raphe** (RĀ-fē = seam). Internally, the **scrotal septum** divides the scrotum into two sacs, each containing a single testis (Figure 25.2). The septum is made up of hypodermis and smooth muscle tissue called the **dartos muscle** (DAR-tōs = skinned). Associated with

FIGURE 25.1 **Male organs of reproduction and surrounding structures.**

FUNCTIONS OF THE MALE REPRODUCTIVE SYSTEM
1. Produces sperm and the male sex hormone testosterone.
2. Transports, stores, and assists in maturation of sperm.
3. Secretes seminal fluid, which mixes with sperm to form semen.
4. Discharges semen from the penis.

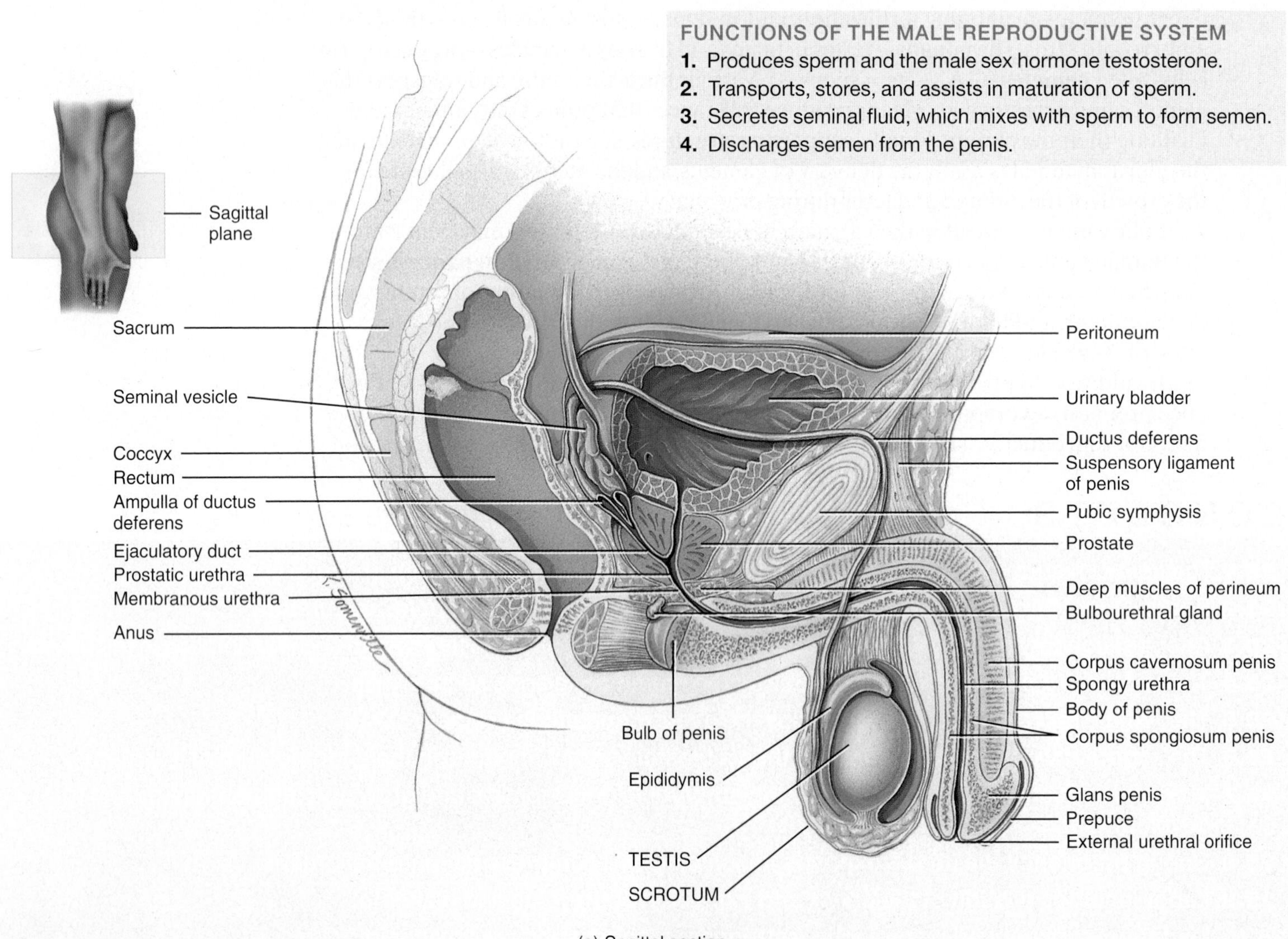


(a) Sagittal section

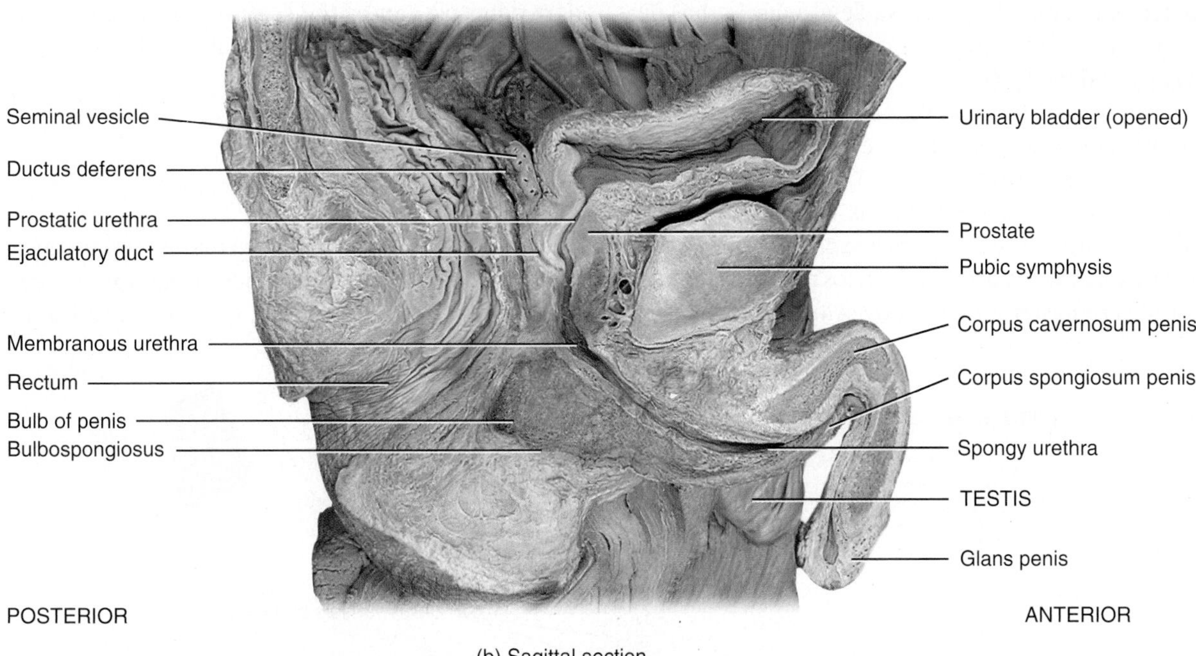


(b) Sagittal section

**Reproductive organs are adapted for producing new individuals and passing on genetic material from one generation to the next.**

**FIGURE 25.2** **The scrotum, the supporting structure for the testes.**

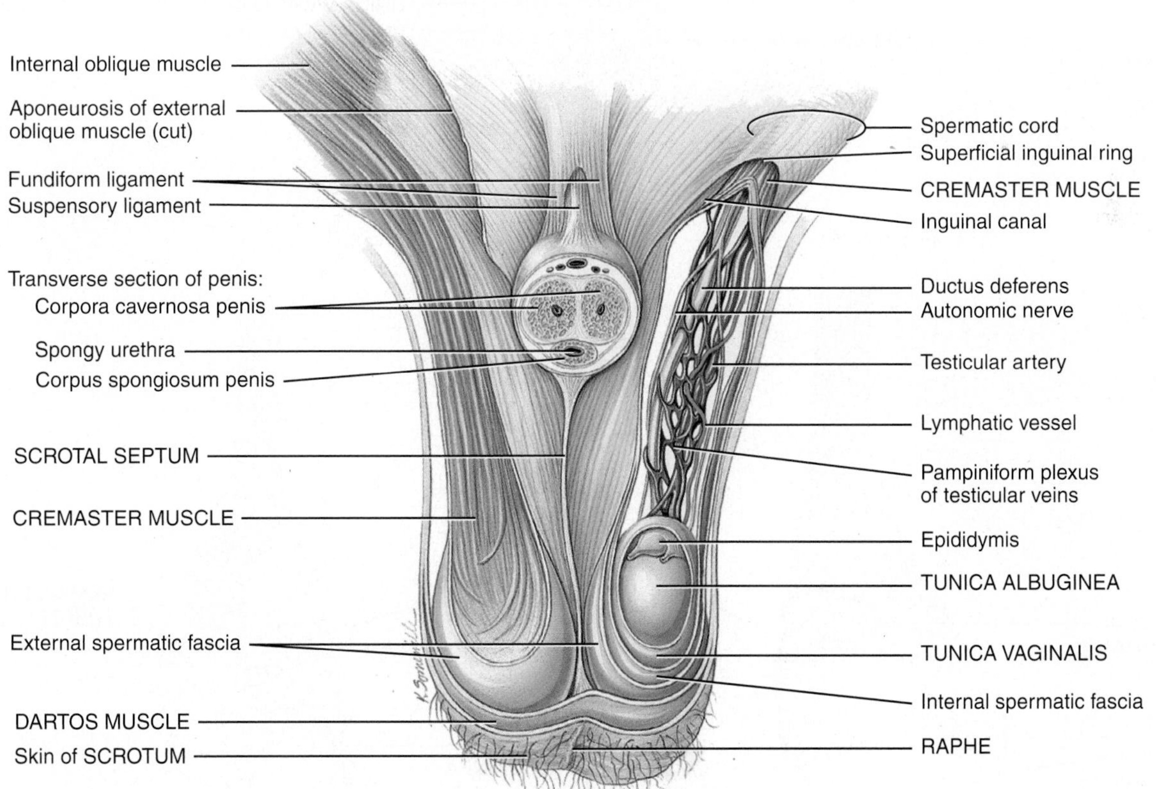


Anterior view of scrotum and testes and transverse section of penis

**The scrotum consists of loose skin and an underlying hypodermis, and supports the testes.**

each testis in the scrotum is the **cremaster muscle** (krē-MAS-ter = suspender), a series of small bands of skeletal muscle that descends as an extension of the internal oblique through the spermatic cord to surround the testes.

The location of the scrotum and the contraction of its muscle fibers regulate the temperature of the testes. Normal sperm production requires a temperature about 2–3°C below core body temperature. This lowered temperature is maintained within the scrotum because it is outside the pelvic cavity. In response to cold temperatures, the cremaster and dartos muscles contract. Contraction of the cremaster muscles moves the testes closer to the body, where they can absorb body heat. Contraction of the dartos muscle causes the scrotum to become tight (wrinkled in appearance), which reduces heat loss. Exposure to warmth reverses these actions.

## Testes

The **testes** (TES-tēz; singular is *testis*), or **testicles**, are paired oval glands in the scrotum measuring about 5 cm (2 in.) long and 2.5 cm (1 in.) in diameter (Figure 25.3). The testes develop

FIGURE 25.3 **Internal and external anatomy of a testis.**

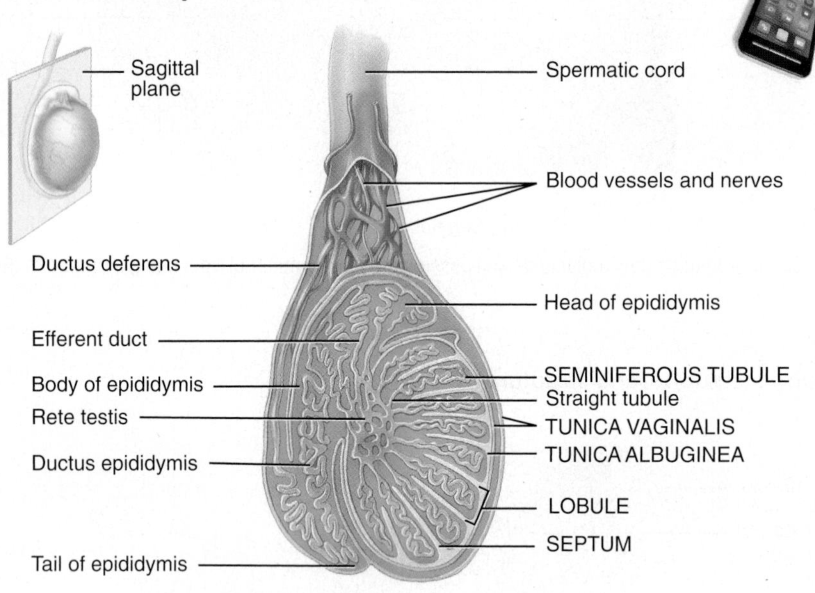


(a) Sagittal section of testis showing seminiferous tubules

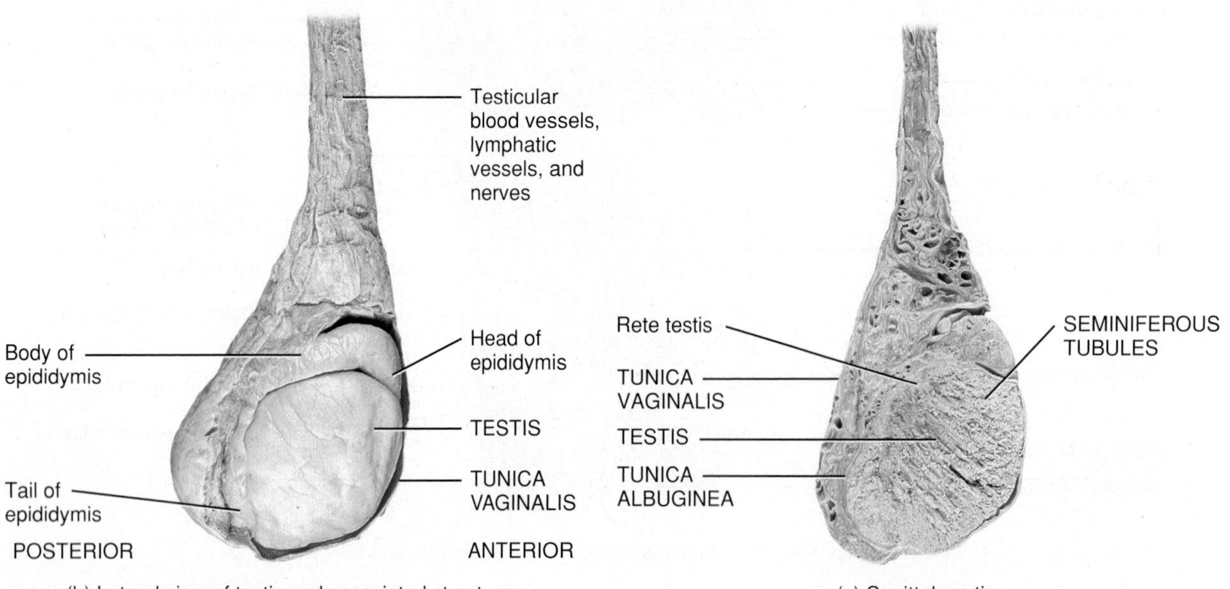


(b) Lateral view of testis and associated structures

(c) Sagittal section

**The testes are the male gonads, which produce haploid sperm.**

near the kidneys, in the posterior portion of the abdomen, and they usually begin their descent into the scrotum through the inguinal canals (passageways in the anterior abdominal wall; see Figure 25.2) during the seventh month of fetal development.

A serous membrane called the **tunica vaginalis** (*tunica* = sheath) partially covers the testes. Internal to the tunica vaginalis is a dense white fibrous capsule composed of dense irregular connective tissue, the **tunica albuginea** (al′-bū-JIN-ē-a; *albu-* = white); it extends inward, forming septa that divide the testis into internal compartments called **lobules**. Each of the 200–300 lobules contains one to three tightly coiled **seminiferous tubules** (*semin-* = seed; *-fer-* = to carry), where sperm are produced. The process by which the seminiferous tubules of the testes produce sperm is called **spermatogenesis** (sper′-ma-tō-JEN-e-sis; *-genesis* = beginning process or production).

The seminiferous tubules contain two types of cells: **spermatogenic cells**, the sperm-forming cells, and Sertoli cells, which have several functions in supporting spermatogenesis (Figure 25.4). Stem cells called **spermatogonia** (sper′-ma-tō-GŌ-nē-a; *-gonia* = offspring; singular is *spermatogonium*) are spermatogenic cells that remain dormant during childhood and actively begin producing sperm at puberty. Toward the lumen of the seminiferous tubule are layers of progressively more mature spermatogenic cells. In order of advancing maturity, these

FIGURE 25.4 **Microscopic anatomy of the seminiferous tubules and stages of sperm production (spermatogenesis).** Arrows in (b) indicate the progression of spermatogenic cells from least mature to most mature. The (*n*) and (2*n*) refer to haploid and diploid numbers of chromosomes, respectively.

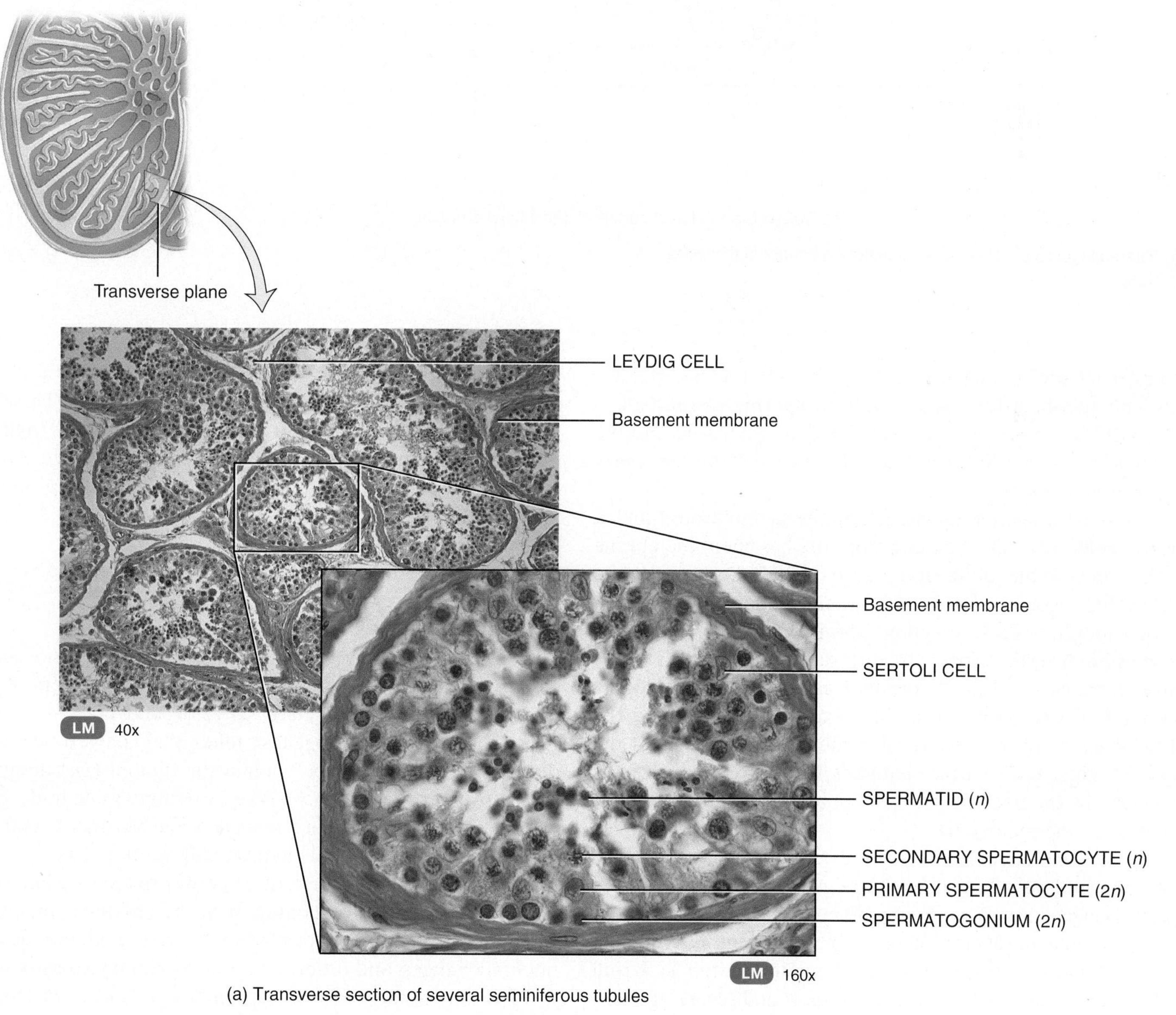


(a) Transverse section of several seminiferous tubules

(continues)

FIGURE 25.4 (continued)

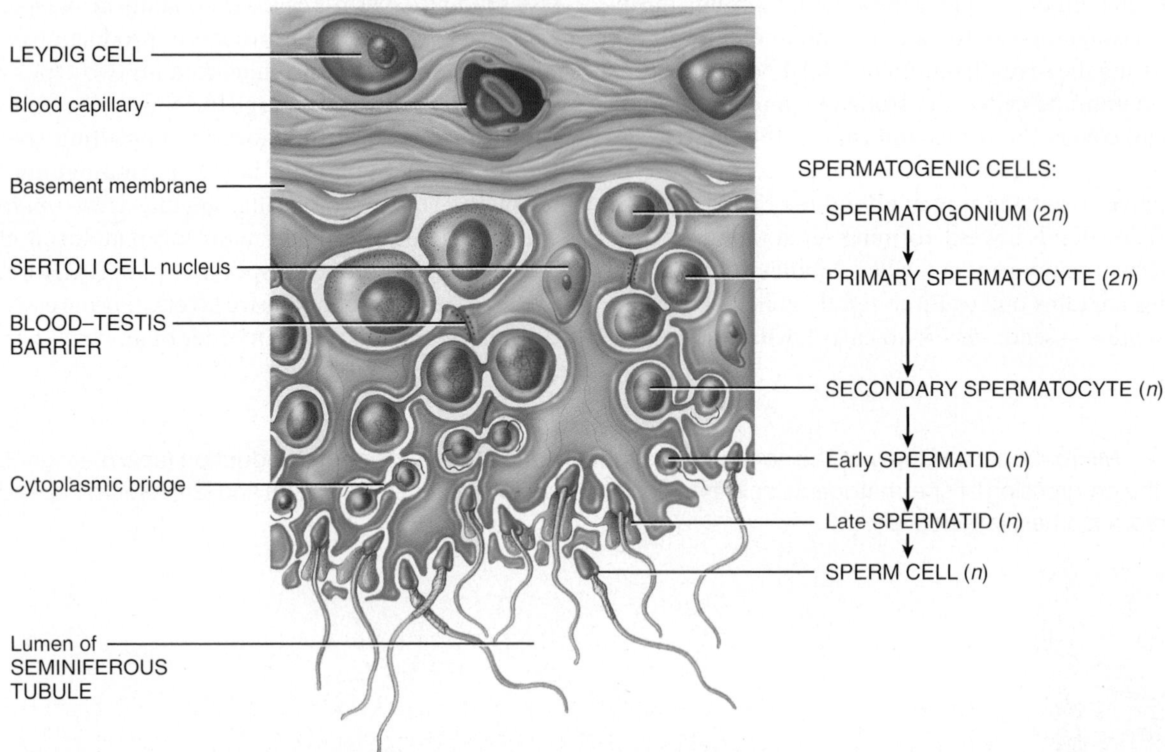


(b) Transverse section of portion of seminiferous tubule

**Spermatogenesis occurs in the seminiferous tubules of the testes.**

are primary spermatocytes, secondary spermatocytes, spermatids, and sperm. After a **sperm cell**, or **spermatozoon** (sper′-ma-tō-ZŌ-on; *-zoon* = life), has formed, it is released into the lumen of the seminiferous tubule. (The plural terms are *sperm* and *spermatozoa*.)

Among the spermatogenic cells in the seminiferous tubules, **Sertoli cells** (ser-TŌ-lē) extend from the basement membrane to the lumen of the tubule. Internal to the basement membrane and spermatogonia, tight junctions join neighboring Sertoli cells to one another. These junctions form an obstruction known as the **blood–testis barrier** because substances must first pass through the Sertoli cells before they can reach the developing sperm. By isolating the developing gametes from the blood, the blood–testis barrier prevents an immune response against the spermatogenic cell's surface antigens, which are recognized as "foreign" by the immune system. The blood–testis barrier does not include spermatogonia.

Sertoli cells support and protect developing spermatogenic cells in several ways. They nourish spermatocytes, spermatids, and sperm; phagocytize excess spermatid cytoplasm as development proceeds; and control movements of spermatogenic cells and the release of sperm into the lumen of the seminiferous tubule. Sertoli cells also produce fluid for sperm transport and secrete the hormone inhibin to decrease the rate of spermatogenesis.

In the spaces between adjacent seminiferous tubules are clusters of cells called **Leydig cells** (Figure 25.4). These cells secrete testosterone, the most prevalent androgen (male sex hormone). An **androgen** is a hormone that promotes the development of masculine characteristics. Testosterone also promotes a man's *libido* (sexual desire).

## Spermatogenesis

Before you read this section, please review the topics of somatic cell division and reproductive cell division in Concept 3.7. Pay particular attention to Figures 3.31, 3.32, and 3.33.

In humans, spermatogenesis takes 65–75 days. It begins with the spermatogonia, which contain the diploid (2*n*) number of chromosomes (Figure 25.5). When spermatogonia undergo mitosis, some spermatogonia remain near the basement membrane of the seminiferous tubule in an undifferentiated state to serve as a reservoir of cells for future sperm production. The rest of the spermatogonia lose contact with the basement membrane, squeeze through the blood–testis barrier, undergo developmental changes, and differentiate into **primary spermatocytes** (SPER-ma-tō-sīts′). Primary spermatocytes, like spermatogonia, are diploid (2*n*); that is, they have 46 chromosomes.

FIGURE 25.5 **Events in spermatogenesis.** Diploid cells (2*n*) have 46 chromosomes; haploid cells (*n*) have 23 chromosomes.

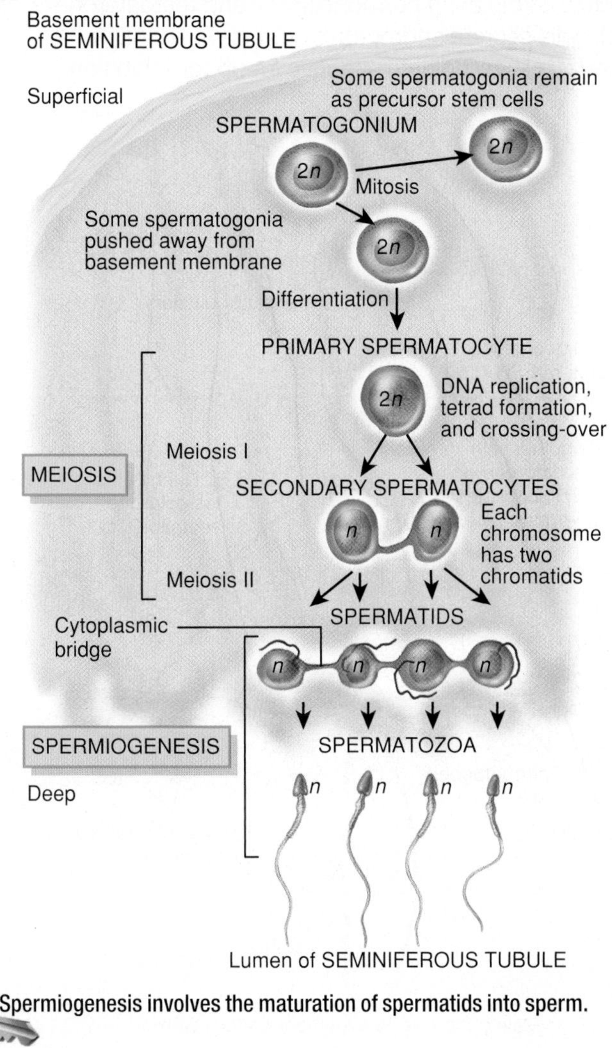


**Spermiogenesis involves the maturation of spermatids into sperm.**

Shortly after it forms, each primary spermatocyte replicates its DNA and then meiosis begins (Figure 25.5). In meiosis I, homologous pairs of chromosomes line up at the metaphase plate, and crossing-over occurs to rearrange genes between chromatids. Then, the meiotic spindle pulls one (duplicated) chromosome of each pair to an opposite pole of the dividing cell. The two cells formed by meiosis I are called **secondary spermatocytes**. Each secondary spermatocyte has 23 chromosomes, the haploid number (*n*). However, after meiosis I, each chromosome within a secondary spermatocyte is made up of two chromatids (two copies of the DNA) still attached by a centromere.

In meiosis II, the chromosomes line up in single file along the metaphase plate, and the two chromatids of each chromosome separate. The four haploid cells resulting from meiosis II are called **spermatids**. Therefore, a single primary spermatocyte with 46 chromosomes undergoes two rounds of cell division (in meiosis I and meiosis II) to produce four spermatids, each with 23 chromosomes.

A unique process occurs during spermatogenesis. As spermatogenic cells proliferate, they fail to complete cytoplasmic separation (cytokinesis). The four daughter cells remain in contact via *cytoplasmic bridges* through their entire development (Figures 25.4b and 25.5). This pattern of development most likely accounts for the synchronized production of sperm in any given area of a seminiferous tubule. It may also have survival value in that half of the sperm contain an X chromosome and half contain a Y chromosome. The larger X chromosome may carry genes needed for spermatogenesis that are lacking on the smaller Y chromosome.

The final stage of spermatogenesis, **spermiogenesis** (sper′-mē-ō-JEN-e-sis), is the development of haploid spermatids into sperm. No cell division occurs in spermiogenesis; each spherical haploid spermatid becomes an elongated, slender **sperm cell** (Figure 25.6). An acrosome (described shortly) forms atop the condensed, elongated nucleus, a flagellum develops, and mitochondria multiply. Sertoli cells dispose of the excess cytoplasm that sloughs off. Finally, sperm are released from their connections to Sertoli cells and enter the lumen of the seminiferous tubule. Fluid secreted by Sertoli cells pushes sperm along their way, toward the ducts of the testes. At this point, sperm are not yet able to swim.

FIGURE 25.6 **Parts of a sperm cell.**

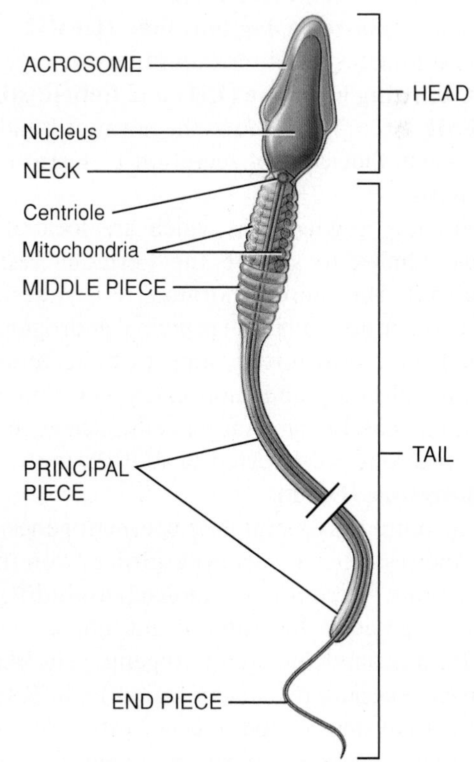


**About 300 million sperm mature each day.**

## Sperm

Each day about 300 million sperm complete the process of spermatogenesis. A sperm cell contains several structures that are highly adapted for reaching and penetrating a secondary oocyte (Figure 25.6) (see Concept 25.3). The major parts of a sperm cell are the head and the tail. The flattened, pointed **head** of the sperm cell contains a nucleus with 23 highly condensed chromosomes. Covering the anterior two-thirds of the nucleus is the **acrosome** (AK-rō-sōm; *acro-* = atop; *-some* = body), a caplike vesicle filled with enzymes that help a sperm cell penetrate a secondary oocyte to bring about fertilization. The **tail** (flagellum) of a sperm cell is subdivided into four parts: neck, middle piece, principal piece, and end piece. The **neck** is the constricted region just behind the head that contains centrioles. The centrioles form the microtubules that comprise the remainder of the tail. The **middle piece** contains mitochondria arranged in a spiral, which provide the energy (ATP) for locomotion of sperm to the secondary oocyte and for sperm metabolism. The **principal piece** is the longest portion of the tail, and the **end piece** is the terminal, tapering portion of the tail. Once ejaculated, most sperm do not survive more than 48 hours within the female reproductive tract.

# Hormonal Regulation of Male Reproductive Function

At the onset of puberty, the hypothalamus increases its secretion of **gonadotropin-releasing hormone (GnRH)**. This hormone in turn stimulates the anterior pituitary to increase its secretion of **luteinizing hormone (LH)** and **follicle-stimulating hormone (FSH)**. Figure 25.7 shows the hormones and negative feedback systems that control secretion of testosterone and spermatogenesis.

LH stimulates Leydig cells, which are located between seminiferous tubules, to secrete the hormone **testosterone** (tes-TOS-te-rōn). This steroid hormone is synthesized from cholesterol in the testes and is the principal androgen. Through negative feedback, testosterone suppresses secretion of LH by the anterior pituitary and suppresses secretion of GnRH by the hypothalamus. In some target cells, such as those in the prostate, testosterone is converted to another androgen called **dihydrotestosterone (DHT)**.

FSH acts indirectly to stimulate spermatogenesis (Figure 25.7). FSH and testosterone act synergistically on the Sertoli cells to stimulate secretion of **androgen-binding protein (ABP)** into the lumen of the seminiferous tubules and into the interstitial fluid around the spermatogenic cells. ABP binds to testosterone, keeping its concentration high. Testosterone stimulates the final steps of spermatogenesis in the seminiferous tubules. Once the degree of spermatogenesis required for male reproductive functions has been achieved, Sertoli cells release **inhibin**, a hormone named for its role in inhibiting

FIGURE 25.7 **Hormonal control of spermatogenesis and actions of testosterone and dihydrotestosterone (DHT).** In response to stimulation by follicle-stimulating hormone (FSH) and testosterone, Sertoli cells secrete androgen-binding protein (ABP). Dashed red lines indicate negative feedback inhibition.

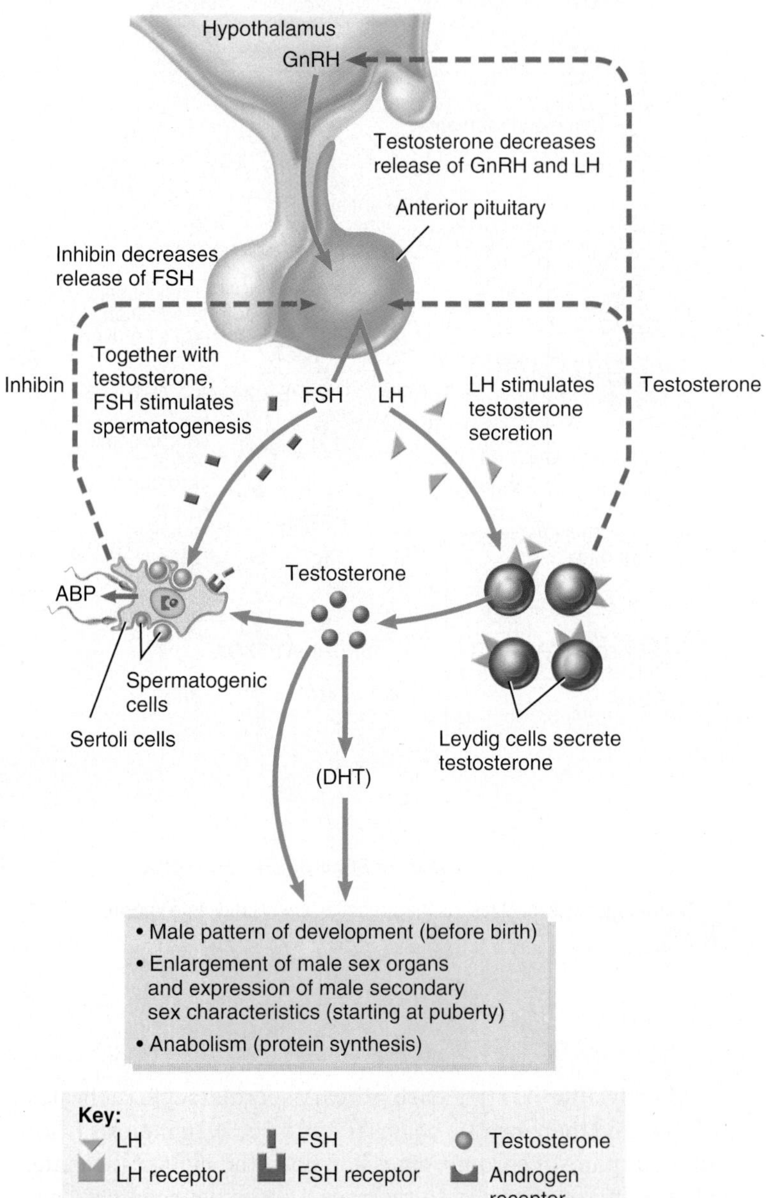


**Release of FSH is stimulated by gonadotropin-releasing hormone (GnRH) and inhibited by inhibin; release of luteinizing hormone (LH) is stimulated by GnRH and inhibited by testosterone.**

FSH secretion by the anterior pituitary (Figure 25.7). If spermatogenesis is proceeding too slowly, less inhibin is released, which permits more FSH secretion and an increased rate of spermatogenesis.

Testosterone and dihydrotestosterone both bind to the same androgen receptors within the nuclei of target cells. The hormone–receptor complex turns some genes on and turns others off. Because of these changes, the androgens produce several effects:

- ***Prenatal development***. Before birth, testosterone stimulates the male pattern of development of reproductive system ducts and the descent of the testes. Dihydrotestosterone stimulates development of the male **external genitals** (penis and scrotum). Testosterone also is converted in the brain to estrogens (feminizing hormones), which may play a role in the development of certain regions of the brain in males.
- ***Development of male sexual characteristics***. At puberty, testosterone and dihydrotestosterone bring about development and enlargement of the male sex organs and muscular and skeletal growth that results in masculine secondary sexual characteristics. **Secondary sex characteristics** are traits that distinguish males and females but do not have a direct role in reproduction. In males these include wide shoulders and narrow hips, facial and chest hair (within hereditary limits) and more hair on other parts of the body, thickening of the skin, increased sebaceous (oil) gland secretion, and enlargement of the larynx and consequent deepening of the voice.
- ***Development of sexual function***. Androgens contribute to male sexual behavior and spermatogenesis and to sex drive (libido) in both males and females. Recall that the adrenal cortex is the main source of androgens in females.
- ***Stimulation of anabolism***. Androgens are anabolic hormones; that is, they stimulate protein synthesis. This effect is obvious in the heavier muscle and bone mass of most men as compared to women.

A negative feedback system regulates testosterone production (Figure 25.8). When testosterone concentration in the blood increases to a certain level, it inhibits the release of GnRH by the hypothalamus. As a result, there is less GnRH traveling from the hypothalamus to the anterior pituitary. The anterior pituitary then releases less LH. With less stimulation by LH, the Leydig cells in the testes secrete less testosterone, and there is a return to homeostasis. If the testosterone concentration in the blood falls too low, however, GnRH is again released by the hypothalamus and stimulates secretion of LH by the anterior pituitary. LH, in turn, stimulates testosterone production by the testes.

FIGURE 25.8 **Negative feedback control of blood level of testosterone.**

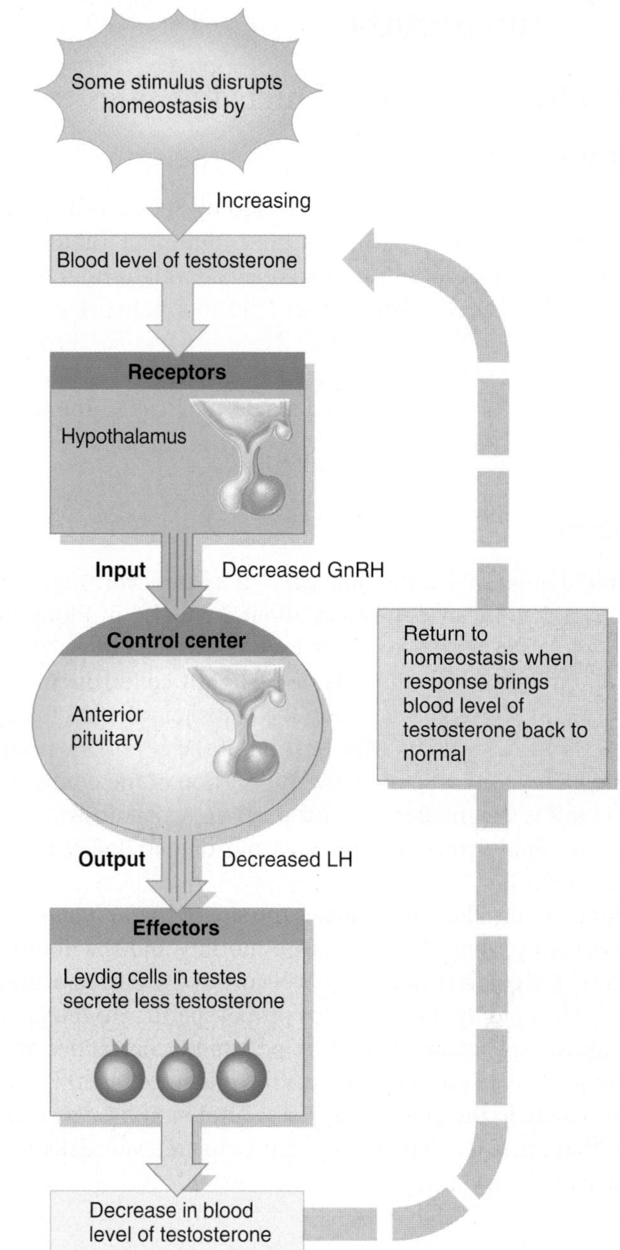


**As testosterone levels rise, decreased gonadotropin-releasing hormone (GnRH) results in decreased luteinizing hormone (LH), inhibiting testosterone secretion.**

## ✓ CHECKPOINT

1. Name the male reproductive structures and state their general functions.
2. What tissue layers cover and protect the testes? Where in the testes are sperm cells produced?
3. What are the functions of Sertoli cells and Leydig cells?
4. What is "reduced" during meiosis I?
5. What are the functions of the sperm cell acrosome and middle piece?
6. What is the source of luteinizing hormone and follicle-stimulating hormone?
7. Which cells of the male reproductive system secrete testosterone? What are the functions of testosterone?

# 25.2 Sperm travel through the epididymis, ductus deferens, ejaculatory ducts, and urethra.

## Reproductive System Ducts in Males

### Ducts of the Testis

Pressure generated by the fluid secreted by Sertoli cells pushes sperm and fluid along the lumen of seminiferous tubules and then into a series of very short ducts called **straight tubules** (see Figure 25.3a). The straight tubules lead to a network of ducts in the testis called the **rete testis** (RĒ-tē = network). From the rete testis, sperm move into a series of coiled **efferent ducts** in the epididymis that empty into a single tube called the ductus epididymis.

### Epididymis

The **epididymis** (ep′-i-DID-i-mis; *epi-* = above or over; *-didymis* = testis) is a comma-shaped organ about 4 cm (1.5 in.) long that lies along the posterior border of each testis (see Figure 25.3a). Each epididymis consists mostly of the tightly coiled **ductus epididymis**. The efferent ducts from the testis join the ductus epididymis at the larger, superior portion of the epididymis called the **head**. The **body** is the narrow midportion of the epididymis, and the **tail** is the smaller, inferior portion. At its distal end, the tail of the epididymis continues as the ductus deferens (discussed shortly).

Functionally, the epididymis is the site of *sperm maturation*, the process by which sperm acquire motility and the ability to penetrate a secondary oocyte. This occurs over a period of about 14 days. The epididymis also helps propel sperm into the ductus deferens during sexual arousal by peristaltic contraction of its smooth muscle. In addition, the epididymis stores sperm, which remain viable in the epididymis for several months. Any stored sperm that are not ejaculated by that time are eventually reabsorbed by phagocytes.

### Ductus Deferens

Within the tail of the epididymis, the ductus epididymis becomes less convoluted, and its diameter increases. Beyond this point, the duct is known as the **ductus deferens** or *vas deferens* (see Figure 25.3a). The ductus deferens, which is about 45 cm (18 in.) long, ascends along the posterior border of the epididymis through the spermatic cord and then enters the pelvic cavity. There it loops over the ureter and passes over the side and down the posterior surface of the urinary bladder (see Figure 25.1a). The dilated terminal portion of the ductus deferens is the **ampulla** (am-PUL-la = little jar; Figure 25.9).

Functionally, the ductus deferens conveys sperm during sexual arousal from the epididymis toward the urethra by peristaltic contractions of its muscularis. Like the epididymis, the ductus deferens can store sperm for several months. Sperm that are stored longer are eventually phagocytized.

### Spermatic Cord

The **spermatic cord** is a supporting structure of the male reproductive system that ascends out of the scrotum (see Figure 25.2). It contains the ductus deferens as it travels through the scrotum, the testicular artery, veins that drain the testes and carry testosterone into circulation, autonomic nerves, lymphatic vessels, and the cremaster muscle. The spermatic cord passes through the **inguinal canal** (IN-gwi-nal = groin), an oblique passageway in the anterior abdominal wall that ends at the *superficial inguinal ring*, an opening in the aponeurosis of the external oblique muscle.

### Ejaculatory Ducts

Each **ejaculatory duct** (e-JAK-ū-la-tō′-re; *ejacul-* = to expel) is about 2 cm (1 in.) long and is formed by the union of the duct from the seminal vesicle and the ampulla of the ductus deferens (Figure 25.9). The ejaculatory ducts form just superior to the prostate and pass inferiorly and anteriorly through the prostate. They terminate in the prostatic urethra, where they eject sperm and seminal vesicle secretions just before the release of semen from the urethra to the exterior.

### Urethra

In males, the **urethra** is the shared terminal duct of the reproductive and urinary systems, serving as a passageway for both semen and urine. About 20 cm (8 in.) long, it passes through the prostate and the penis, and is subdivided into three parts (see Figure 25.1). The **prostatic urethra** passes through the prostate. As this duct continues inferiorly, it passes through the deep muscles of the perineum, where it is known as the **membranous urethra**. As the duct passes through the corpus spongiosum of the penis, it is known as the **spongy urethra** or *penile urethra*. The spongy urethra ends at the **external urethral orifice**. The histology of the male urethra may be reviewed in Concept 24.10.

## Accessory Sex Glands

The ducts of the male reproductive system store and transport sperm, but the **accessory sex glands** secrete most of the liquid portion of semen that protects sperm and facilitates their movement. The accessory sex glands include the seminal vesicles, the prostate, and the bulbourethral glands.

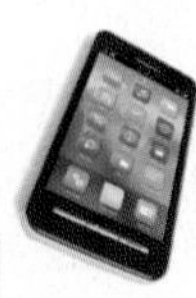

FIGURE 25.9 **Locations of several accessory reproductive organs in males.** The prostate, urethra, and penis have been sectioned to show internal details.

**FUNCTIONS OF MALE ACCESSORY SEX GLAND SECRETIONS**

1. Neutralize acid in the male urethra and the female reproductive tract.
2. Provide fructose for ATP production by sperm.
3. Contribute to sperm motility and viability.
4. Help semen coagulate after ejaculation and subsequently break down the clot.
5. Lubricate the lining of the urethra and the tip of the penis during sexual intercourse.

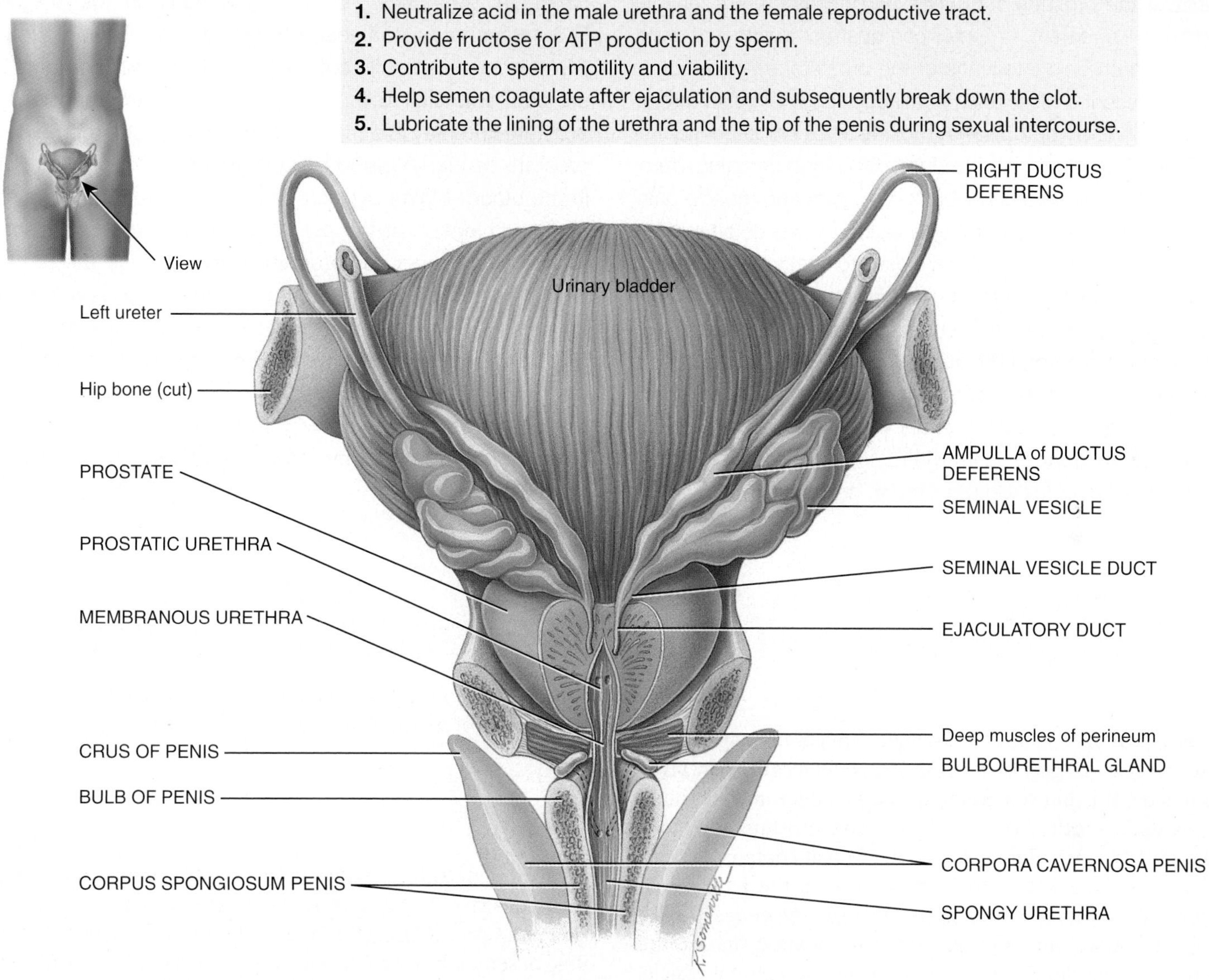

Posterior view of male accessory organs of reproduction

**The male urethra has three subdivisions: the prostatic, membranous, and spongy urethra.**

## Seminal Vesicles

The paired **seminal vesicles** (VES-i-kuls) are convoluted pouchlike structures, about 5 cm (2 in.) in length, lying posterior to the base of the urinary bladder and anterior to the rectum (Figure 25.9). Through the seminal vesicle ducts they secrete an alkaline, viscous fluid that contains fructose (a monosaccharide sugar), prostaglandins, and clotting proteins that are different from those in blood. The alkaline nature of the seminal fluid helps to neutralize the acidic environment of the male urethra and female reproductive tract that otherwise would inactivate and kill sperm. The fructose is used for ATP production by sperm. Prostaglandins contribute to sperm motility and viability and may stimulate smooth muscle contractions within the female reproductive tract. The clotting proteins help semen coagulate after ejaculation. Fluid secreted by the seminal vesicles normally constitutes about 60 percent of the volume of semen.

## CLINICAL CONNECTION | *Prostate Disorders*

Because the prostate surrounds part of the urethra, any infection, enlargement, or tumor can obstruct the flow of urine. Acute and chronic infections of the prostate are common in postpubescent males, often in association with inflammation of the urethra. Symptoms may include fever, chills, urinary frequency, frequent urination at night, difficulty in urinating, burning or painful urination, low back pain, joint and muscle pain, blood in the urine, or painful ejaculation. However, often there are no symptoms. Antibiotics are used to treat most cases that result from a bacterial infection. In **acute prostatitis**, the prostate becomes swollen and tender. **Chronic prostatitis** is one of the most common chronic infections in men of the middle and later years. On examination, the prostate feels enlarged, soft, and very tender, and its surface outline is irregular.

**Prostate cancer** is the leading cause of death from cancer in men in the United States, having surpassed lung cancer in 1991. Each year it is diagnosed in almost 200,000 U.S. men and causes nearly 40,000 deaths. The amount of PSA (prostate-specific antigen), which is produced only by prostate epithelial cells, increases with enlargement of the prostate and may indicate infection, benign enlargement, or prostate cancer. A blood test can measure the level of PSA in the blood. Males over the age of 40 should have an annual examination of the prostate gland. In a **digital rectal exam**, a physician palpates the gland through the rectum with the fingers (digits). Many physicians also recommend an annual PSA test for males over age 50. Treatment for prostate cancer may involve surgery, cryotherapy, radiation, hormonal therapy, and chemotherapy. Because many prostate cancers grow very slowly, some urologists recommend “watchful waiting” before treating small tumors in men over age 70.

### Prostate

The **prostate** (PROS-tāt) is a single, doughnut-shaped gland about the size of a golf ball measuring about 4 cm (1.6 in.) from side to side. It is inferior to the urinary bladder and surrounds the prostatic urethra (Figure 25.9). The prostate secretes a milky, slightly acidic fluid (pH about 6.5) that contains (1) *citric acid*, used by sperm for ATP production via the Krebs cycle; (2) *proteolytic enzymes*, such as *prostate-specific antigen* (*PSA*), pepsinogen, lysozyme, amylase, and hyaluronidase, that eventually break down the clotting proteins from the seminal vesicles; and (3) seminalplasmin, an antibiotic that may help decrease the number of naturally occurring bacteria in semen and in the lower female reproductive tract. Secretions of the prostate enter the prostatic urethra through many prostatic ducts. Prostatic secretions make up about 25 percent of the volume of semen and contribute to sperm motility and viability.

### Bulbourethral Glands

The paired **bulbourethral glands** (bul′-bō-ū-RĒ-thral), or *Cowper's glands*, are about the size of peas. They are located inferior to the prostate on either side of the membranous urethra, and their ducts open into the spongy urethra (Figure 25.9). During sexual arousal, the bulbourethral glands secrete an alkaline fluid into the urethra that protects the passing sperm by neutralizing acids from urine in the urethra. They also secrete mucus that lubricates the end of the penis and the lining of the urethra, decreasing the number of sperm damaged during ejaculation.

## Semen

**Semen** (= seed) is a mixture of sperm and **seminal fluid**, a liquid that consists of the secretions of the seminiferous tubules, seminal vesicles, prostate, and bulbourethral glands. The volume of semen in a typical ejaculate is 2.5–5 milliliters (mL), with 50–150 million sperm per milliliter. A very large number of sperm is required for successful fertilization because only a tiny fraction ever reaches the secondary oocyte. When the number of sperm falls below 20 million per milliliter of semen, the male is likely to be infertile.

Despite the slight acidity of prostatic fluid, semen has a slightly alkaline pH of 7.2–7.7 due to the higher pH and larger volume of fluid from the seminal vesicles. The prostatic secretion gives semen a milky appearance, and fluids from the seminal vesicles and bulbourethral glands give it a sticky consistency. Seminal fluid provides sperm with a transportation medium, nutrients, and protection from the hostile acidic environment of the male's urethra and the female's vagina.

Once ejaculated, liquid semen coagulates within 5 minutes due to the presence of clotting proteins from the seminal vesicles. The functional role of semen coagulation is not known. After about 10 to 20 minutes, semen reliquefies because proteolytic enzymes produced by the prostate break down the clot. Abnormal or delayed liquefaction of clotted semen may cause complete or partial immobilization of sperm, thereby inhibiting their movement through the cervix of the uterus.

## Penis

The **penis** contains the urethra and is a passageway for the ejaculation of semen and the excretion of urine (Figure 25.10). It is cylindrical in shape and consists of a body, glans penis, and root.

FIGURE 25.10 **Internal structure of the penis.** The inset in (b) shows details of the skin and fasciae.

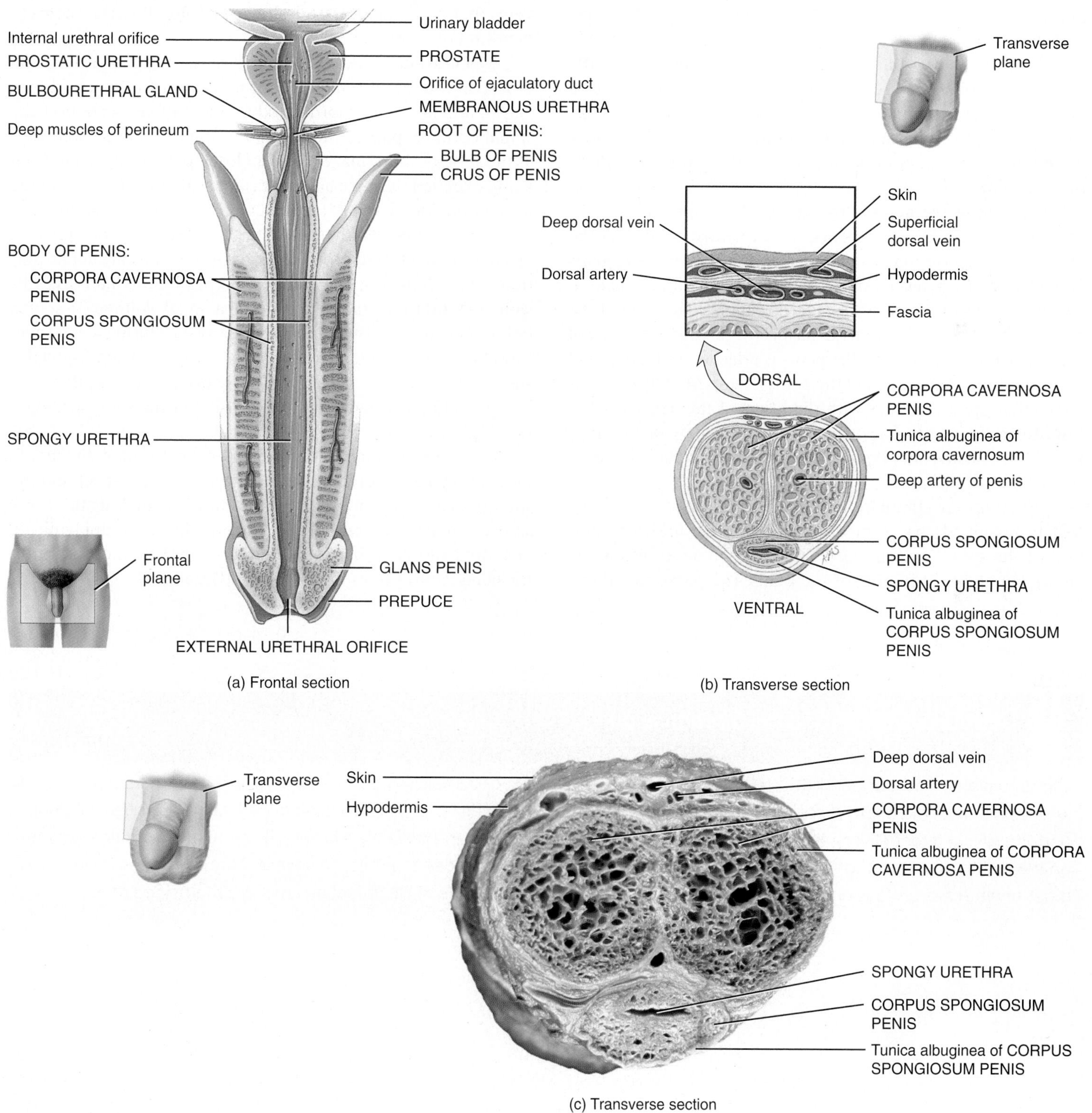


(a) Frontal section

(b) Transverse section

(c) Transverse section

**The penis contains the urethra, a pathway for semen and urine.**

The **body of the penis** is composed of three cylindrical masses of tissue, each surrounded by the tunica albuginea (Figure 25.10b, c). The two dorsolateral masses are called the **corpora cavernosa penis** (*corpora* = main bodies; *cavernosa* = hollow). The smaller midventral mass, the **corpus spongiosum penis**, contains the spongy urethra and keeps it open during ejaculation. Skin and hypodermis enclose all three masses, which consist of erectile tissue. *Erectile tissue* is composed of numerous blood sinuses (vascular spaces) surrounded by smooth muscle and elastic connective tissue.

The distal end of the corpus spongiosum penis is a slightly enlarged, acorn-shaped region called the **glans penis**. The distal urethra enlarges within the glans penis and forms a terminal slitlike opening, the **external urethral orifice**. Covering the glans in an uncircumcised penis is the loosely fitting **prepuce** (PRĒ-poos), or *foreskin*.

The **root of the penis** is the attached, proximal portion. It consists of the **bulb of the penis**, the expanded portion of the base of the corpus spongiosum penis, and the **crura of the penis** (singular is *crus* = resembling a leg), the two separated and tapered portions of the corpora cavernosa penis. The bulb of the penis is attached to the inferior surface of the deep muscles of the perineum. Each crus of the penis is attached to the ischium of the hip bone. The weight of the penis is supported by two ligaments that are continuous with the fascia of the penis: (1) The **fundiform ligament** arises from the inferior part of the linea alba; (2) the **suspensory ligament** arises from the pubic symphysis (see Figure 25.2).

Upon sexual stimulation (visual, tactile, auditory, olfactory, or imagined), parasympathetic fibers from the sacral portion of the spinal cord initiate and maintain an **erection**, the enlargement and stiffening of the penis. The parasympathetic fibers cause smooth muscle in the walls of arterioles supplying erectile tissue to relax, which allows these blood vessels to dilate. This in turn causes large amounts of blood to enter the erectile tissue of the penis. Parasympathetic fibers also cause the smooth muscle within the erectile tissue to relax, resulting in widening of the blood sinuses. The combination of increased blood flow and widening of the blood sinuses results in an erection. Expansion of the blood sinuses also compresses the veins that drain the penis; the slowing of blood outflow helps to maintain the erection.

**Ejaculation** (ē-jak-ū-LĀ-shun; *ejectus-* = to throw out), the powerful release of semen from the urethra to the exterior, is a sympathetic reflex coordinated by the lumbar portion of the spinal cord. As part of the reflex, the smooth muscle sphincter at the base of the urinary bladder closes, preventing urine from being expelled during ejaculation, and semen from entering the urinary bladder. Even before ejaculation occurs, peristaltic contractions in the epididymis, ductus deferens, seminal vesicles, ejaculatory ducts, and prostate propel semen into the penile portion of the urethra (spongy urethra). Typically, this leads to **emission** (ē-MISH-un), the discharge of a small volume of semen before ejaculation. Emission may also occur during sleep (nocturnal emission). The musculature of the penis (bulbospongiosus, ischiocavernosus, and superficial transverse perineal), which is supplied by the pudendal nerve, also contracts at ejaculation (see Figure 11.12).

Once sexual stimulation of the penis has ended, the arterioles supplying the erectile tissue of the penis constrict and the smooth muscle within erectile tissue contracts, making the blood sinuses smaller. This relieves pressure on the veins supplying the penis and allows the blood to drain through them. Consequently, the penis returns to its flaccid (relaxed) state.

## CLINICAL CONNECTION | *Circumcision*

**Circumcision** (= to cut around) is a surgical procedure in which part of or the entire prepuce is removed. It is usually performed just after delivery, three to four days after birth, or on the eighth day as part of a Jewish religious rite. Although most health-care professionals find no medical justification for circumcision, some feel that it has benefits, such as a lower risk of urinary tract infections, protection against penile cancer, and possibly a lower risk for sexually transmitted diseases. Indeed, studies in several African villages have found lower rates of HIV infection among circumcised men.

## ✓ CHECKPOINT

8. What are the functions of the epididymis, ductus deferens, and ejaculatory duct?
9. Name the structures within the spermatic cord. How does the spermatic cord travel through the abdominal wall?
10. What are the locations of the three subdivisions of the male urethra?
11. What are the locations and functions of the seminal vesicles, prostate, and bulbourethral glands?
12. Which accessory sex gland contributes the majority of the seminal fluid?
13. Which tissue masses form the erectile tissue in the penis, and how do they become rigid during sexual arousal?
14. Explain the physiological processes involved in ejaculation.

## RETURN TO Ryan and Megan's Story

Ryan walks into Dr. Kincaid's office for his next appointment and sits down in the chair facing the doctor's desk.

"Ryan, you have tested positive for gonorrhea, a sexually-transmitted bacterium. It's a fairly common sexually transmitted disease or STD. That's what has been causing your discharges and pain during urination." Ryan seems surprised, and says that he and Megan have been careful during sex. When Dr. Kincaid asks him to elaborate, Ryan explains how he would withdraw before ejaculating. "I heard that sperm do not live long outside the body."

"Ryan, withdrawal might help reduce the chance of pregnancy, somewhat unreliably, but it doesn't protect at all against the transmission of STDs." Ryan looks embarrassed and asks about treatment.

"I'm going to prescribe some antibiotics for you. They have been particularly effective and should clear up your signs and symptoms fairly quickly. Has Megan had any problems that might indicate she is infected?" Dr. Kincaid asks.

Ryan thinks for a moment and said that Megan has complained about pain in her belly, but nothing else.

"Well, that is important. Many STD infections in women can manifest themselves as abdominal pain. She needs to come in for an examination. You should also warn any other sexual partners you have had that they should get checked out for gonorrhea," advises the doctor.

A. *Ryan mentioned a semen discharge. What four glands contribute to semen?*

B. *One reliable birth control method that Ryan is unlikely to select at this time in his life is a vasectomy. What is a vasectomy, and what happens to the man's sperm after this procedure?*

C. *Why do sperm have at most 48 hours to live after ejaculation?*

## 25.3 After a secondary oocyte is discharged from an ovary, it may undergo fertilization and implantation in the uterus.

The organs of the female reproductive system include the ovaries, the uterine tubes, the uterus, the vagina, and external organs, which are collectively called the vulva. The mammary glands are considered part of both the integumentary system and the female reproductive system.

### Ovaries

The **ovaries** (= egg receptacles), which are the female gonads, are paired glands that resemble almonds in size and shape

**FIGURE 25.11** **Female organs of reproduction and surrounding structures.**

**FUNCTIONS OF THE FEMALE REPRODUCTIVE SYSTEM**

1. Produces secondary oocytes and hormones, including progesterone and estrogens (female sex hormones), inhibin, and relaxin.
2. Provides sites of fertilization of a secondary oocyte, implantation of a fertilized ovum, development of the fetus during pregnancy, and labor.
3. Receives the penis during sexual intercourse.
4. Synthesizes, secretes, and ejects milk for nourishment of the newborn.

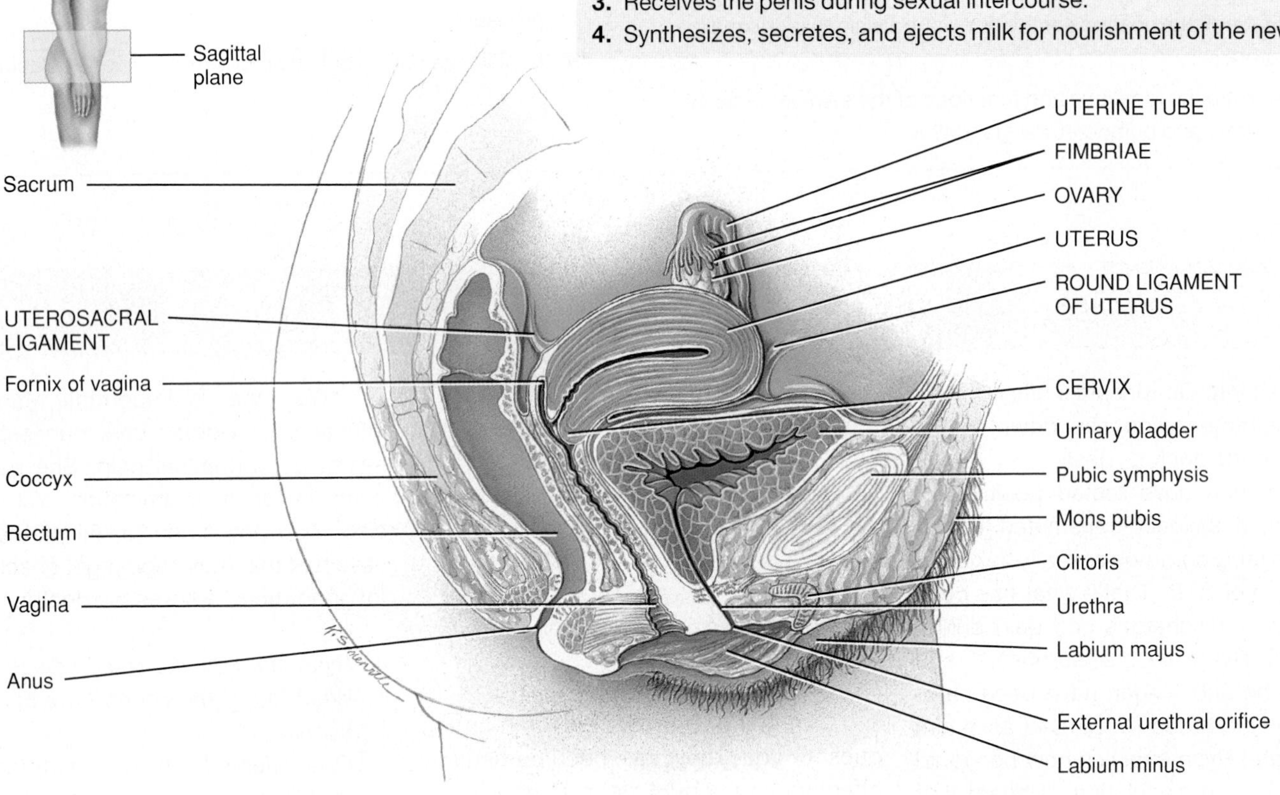


(a) Sagittal section

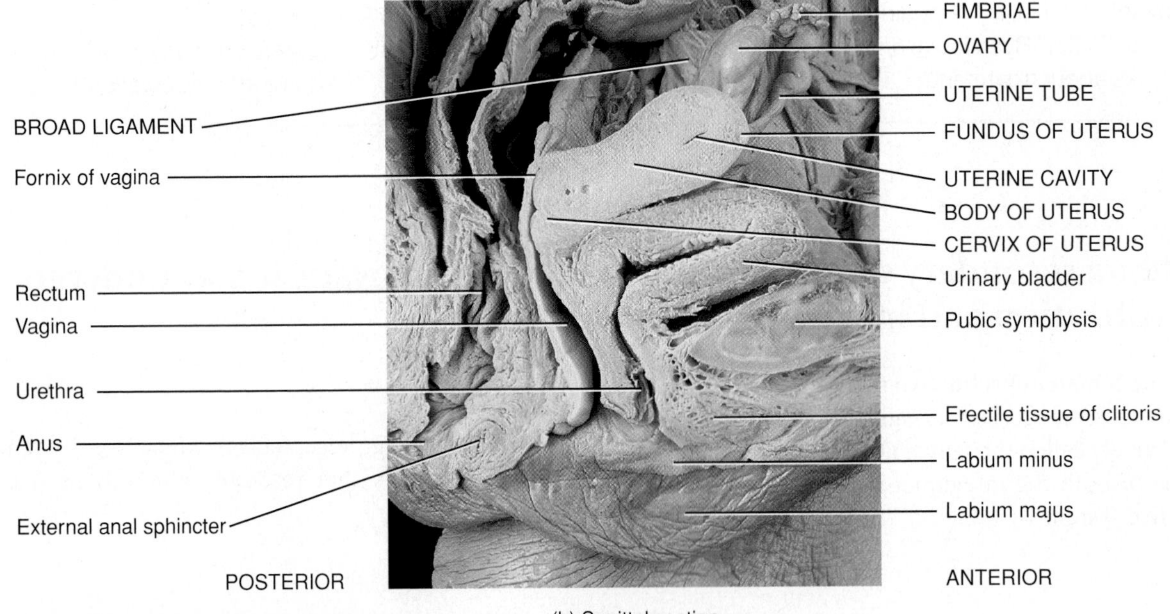


(b) Sagittal section

**The organs of reproduction in females include the ovaries, uterine tubes, uterus, vagina, vulva, and mammary glands.**

(Figure 25.11); they are homologous to the testes. (Here *homologous* means that two organs have the same embryonic origin.) The ovaries produce several hormones and gametes, secondary oocytes that develop into mature ova (eggs).

A series of ligaments holds the ovaries in position (Figure 25.12). The **broad ligament** of the uterus encloses the ovaries and attaches them to the pelvic wall, where the broad ligament continues as the parietal peritoneum. A component of the broad ligament, a double-layered fold of peritoneum called the **mesovarium**, attaches to and further stabilizes the ovaries. The **ovarian ligament** anchors the ovaries to the uterus, and the **suspensory ligament** attaches the ovaries laterally to the pelvic wall.

## Histology of the Ovary

Each ovary includes the following parts (Figure 25.13):

- The **germinal epithelium** (*germen* = sprout or bud) is a layer of simple epithelium (cuboidal or squamous) that covers the surface of the ovary.
- The *tunica albuginea* is a capsule of dense connective tissue located immediately deep to the germinal epithelium (see Figure 25.14e).
- The **ovarian cortex** is a region just deep to the tunica albuginea. It consists of ovarian follicles (described shortly) surrounded by dense connective tissue.
- **Ovarian follicles** (*folliculus* = little bag) in the ovarian cortex consist of **oocytes** in various stages of development, plus the cells surrounding them. When the surrounding cells form a single layer, they are called **follicular cells**; later in development, when they form several layers, they are referred to as **granulosa cells**. The surrounding cells nourish the developing oocyte and begin to secrete estrogen as the follicle grows larger.
- A **mature follicle** or *graafian follicle* is a large, fluid-filled ovarian follicle that is ready to rupture and expel its secondary oocyte, a process known as ovulation.
- A **corpus luteum** (= yellow body) contains the remnants of a mature follicle after ovulation. The corpus luteum produces progesterone, estrogens, relaxin, and inhibin until it degenerates into fibrous scar tissue called the **corpus albicans** (= white body).
- The **ovarian medulla** is deep to the ovarian cortex and contains blood vessels, lymphatic vessels, and nerves.

FIGURE 25.12 **Relative positions of the ovaries, the uterus, and the ligaments that support them.**

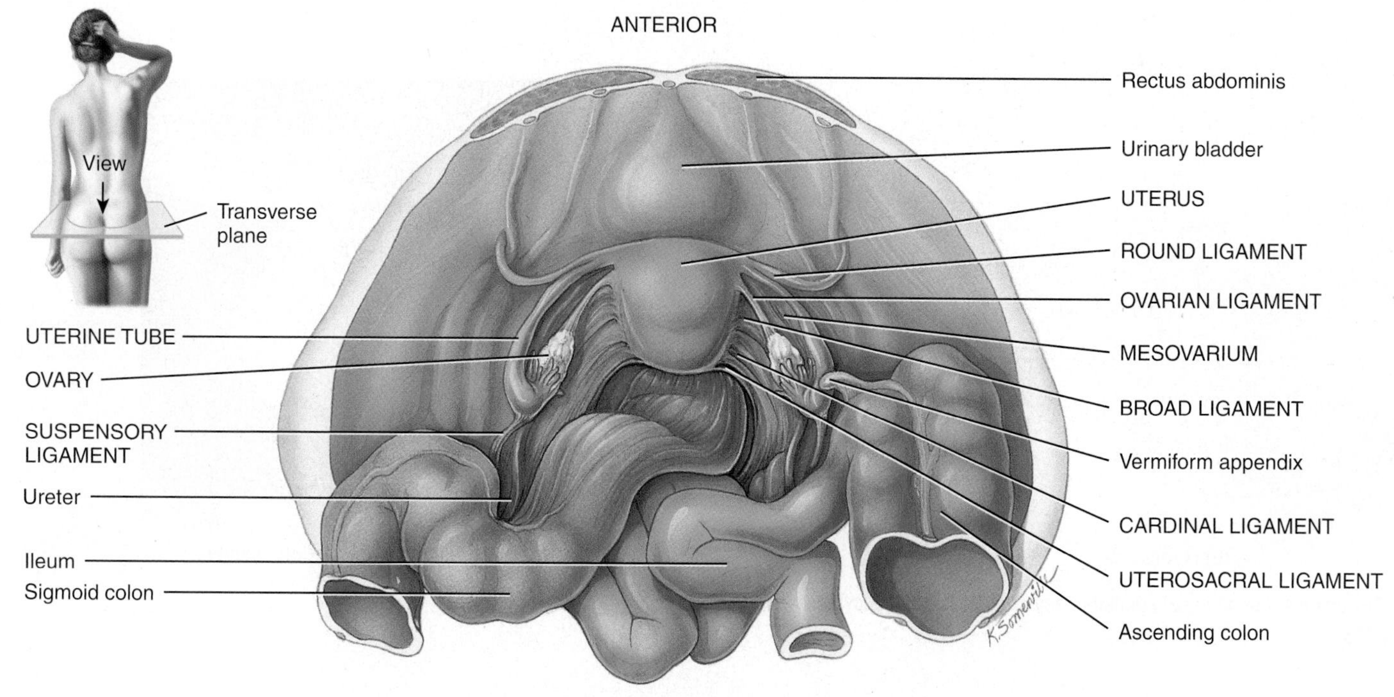


**Ligaments holding the ovaries in position are the mesovarium, the ovarian ligament, and the suspensory ligament.**

**FIGURE 25.13 Histology of the ovary.** The arrows in (a) indicate the sequence of developmental stages that occur as part of the maturation of an ovum during the ovarian cycle.

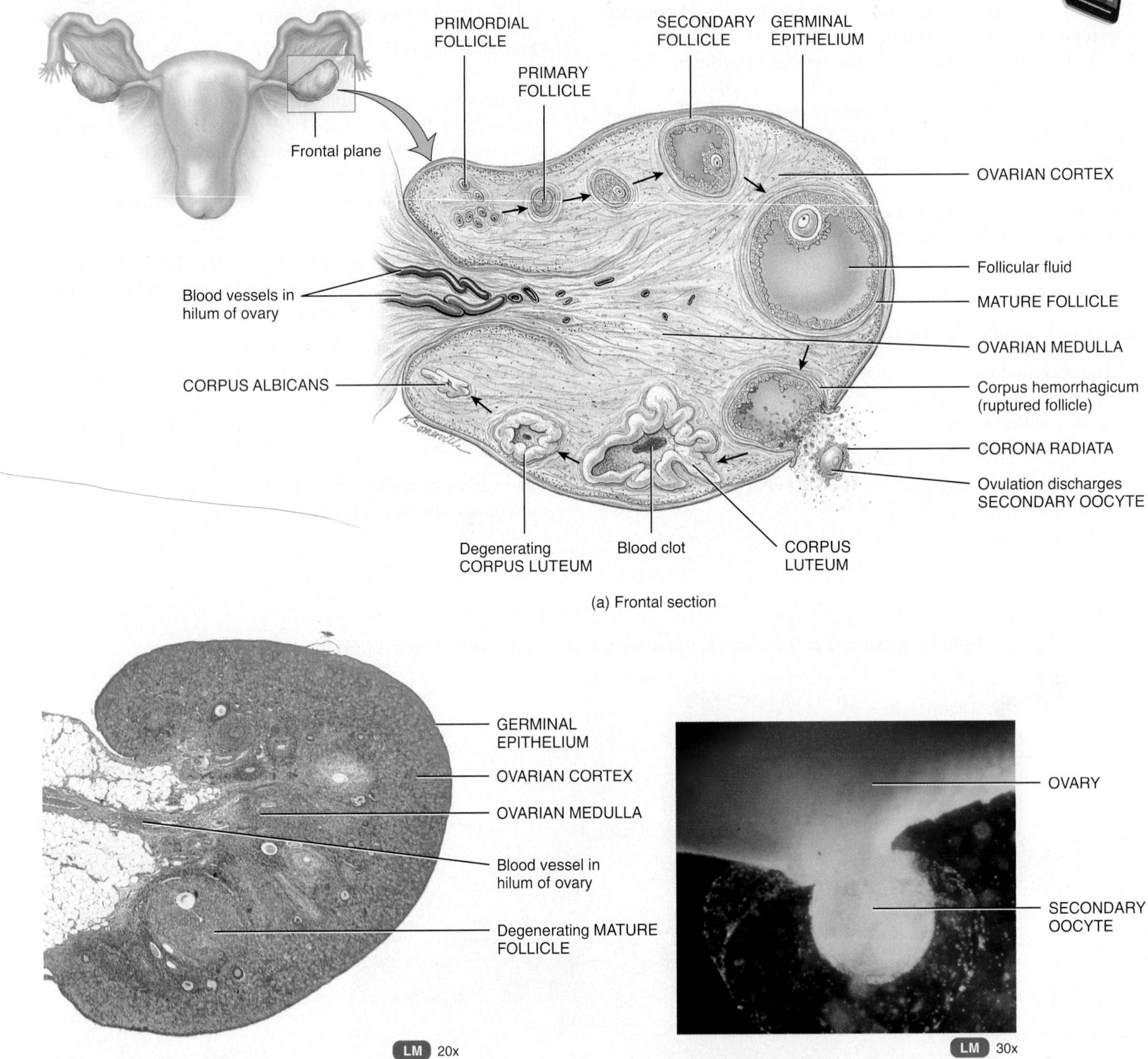


(a) Frontal section

(b) Hemisection

(c) Ovulation of secondary oocyte

**The ovaries are the female gonads; they produce haploid oocytes.**

## Oogenesis and Follicular Development

The formation of gametes in the ovaries is termed **oogenesis** (ō′-ō-JEN-e-sis; *oo-* = egg). Once males reach puberty, spermatogenesis begins and they can produce new sperm throughout life. In contrast to spermatogenesis, which begins in males at puberty, oogenesis begins in females before they are even born. Oogenesis occurs in essentially the same manner as spermatogenesis; meiosis (see Concept 3.7) takes place and the resulting germ cells undergo maturation.

During early fetal development, primordial (primitive) germ cells differentiate within the ovaries into diploid ($2n$) stem cells called **oogonia** (ō′-ō-GŌ-nē-a; singular is *oogonium*). Oogonia are diploid ($2n$) stem cells that divide mitotically to produce millions of oogonia. Even before birth, most oogonia degenerate and are reabsorbed. A few, however, develop into **primary oocytes** (Ō-ō-sits) that enter prophase of meiosis I during fetal development but do not complete that phase until after puberty. During this arrested stage of development, each primary oocyte is surrounded by a single layer of flat follicular cells, and the entire structure is called a **primordial follicle** (Figure 25.14a). At birth, 200,000 to 2,000,000 primary oocytes remain in each ovary.

FIGURE 25.14 Ovarian follicles.

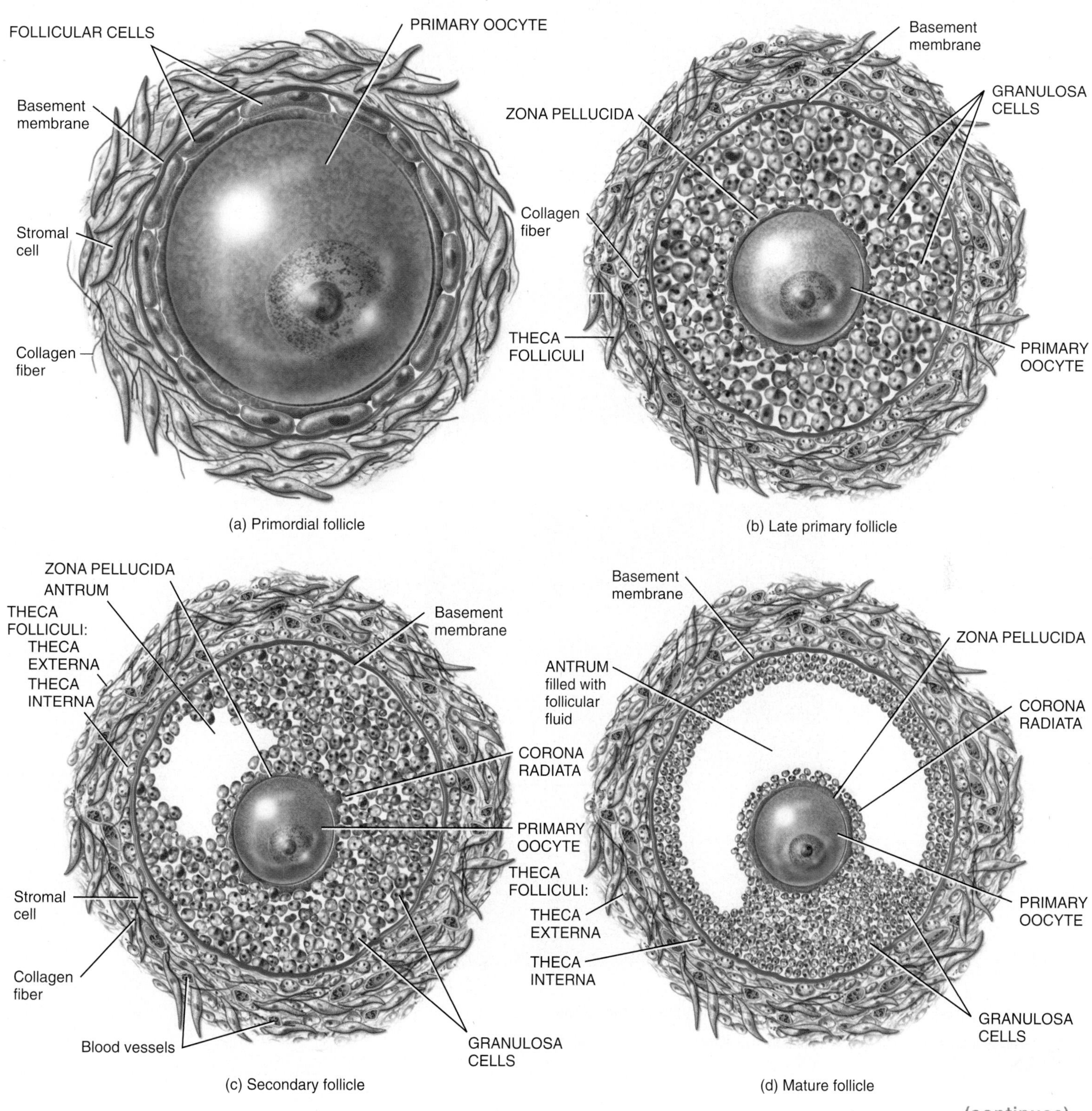


(a) Primordial follicle

(b) Late primary follicle

(c) Secondary follicle

(d) Mature follicle

(continues)

**FIGURE 25.14 (continued)**

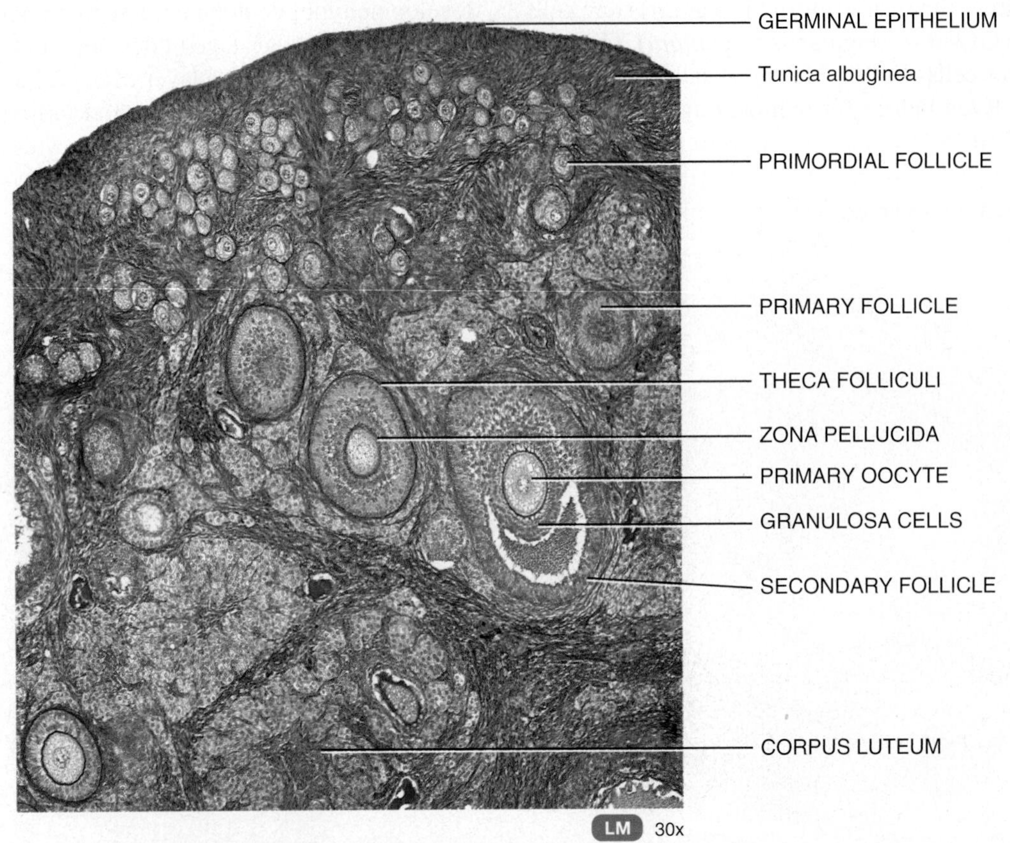


(e) Ovarian cortex

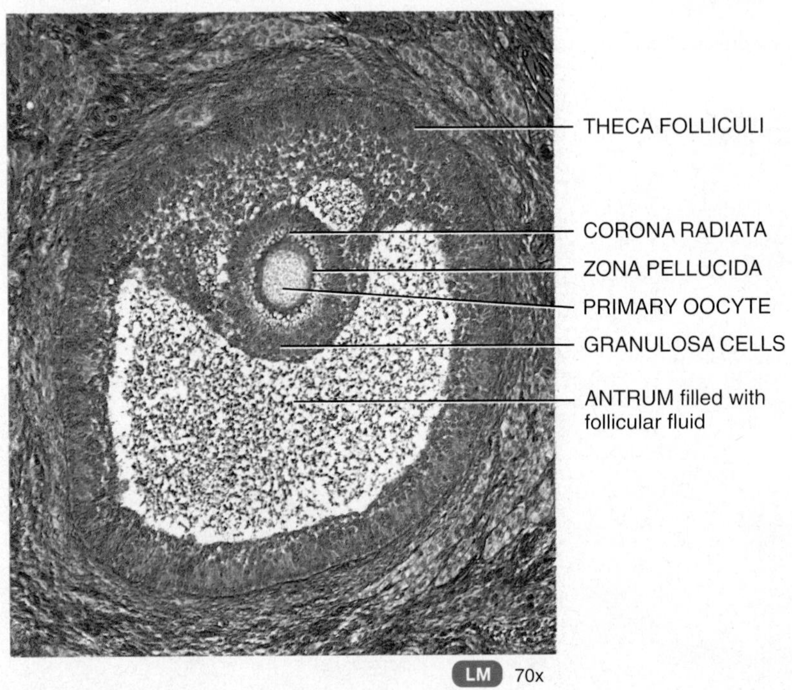


(f) Secondary follicle

**As an ovarian follicle enlarges, follicular fluid accumulates in a cavity called the antrum.**

Of these, about 40,000 remain at puberty, and around 400 will mature and ovulate during a woman's reproductive lifetime. The remainder of the primary oocytes degenerate.

Each month after puberty until menopause, FSH and LH secreted by the anterior pituitary further stimulate the development of several primordial follicles, although only one will typically reach the maturity needed for ovulation. A few primordial follicles start to grow, developing into **primary follicles** (Figure 25.14b). Each primary follicle consists of a primary oocyte that is surrounded by several layers of follicular cells called **granulosa cells**. As the primary follicle grows, it forms a clear glycoprotein layer, called the **zona pellucida** (ZŌ-na = zone; pe-LOO-si-da = allowing passage of light), between the primary oocyte and the granulosa cells. In addition, ovarian cortex cells called *stromal cells* surround the follicle, forming an organized layer called the **theca folliculi** (THĒ-ka fo-LIK-ū-li).

With continuing maturation, a primary follicle develops into a **secondary follicle** (Figure 25.14c). In a secondary follicle, the theca differentiates into the **theca interna**, a *vascularized* (containing many blood vessels) internal layer of cuboidal cells that secrete estrogens, and the **theca externa**, an outer layer of stromal cells and collagen fibers. The granulosa cells begin to secrete follicular fluid, which builds up in a cavity called the **antrum** in the center of the secondary follicle. In addition, the innermost layer of granulosa cells becomes firmly attached to the zona pellucida and is now called the **corona radiata** (kō-RŌ-na = crown; rā-dē-A-ta = radiation).

The secondary follicle eventually becomes larger, turning into a mature follicle (Figure 25.14d) that is capable of ejecting its oocyte, a process known as ovulation. Just before ovulation, the diploid primary oocyte completes meiosis I, producing two haploid ($n$) cells of unequal size—each with 23 chromosomes (Figure 25.15). The smaller cell produced by meiosis I, called the *first polar body*, is essentially a packet of discarded nuclear material. The larger cell, known as the secondary oocyte, receives most of the cytoplasm. Once a *secondary oocyte* is formed, it begins meiosis II but then stops in metaphase. The mature follicle soon ruptures and releases its secondary oocyte during ovulation (see Figure 25.13).

**FIGURE 25.15 Oogenesis.** Diploid cells (**2*n***) have 46 chromosomes; haploid cells ($n$) have 23 chromosomes.

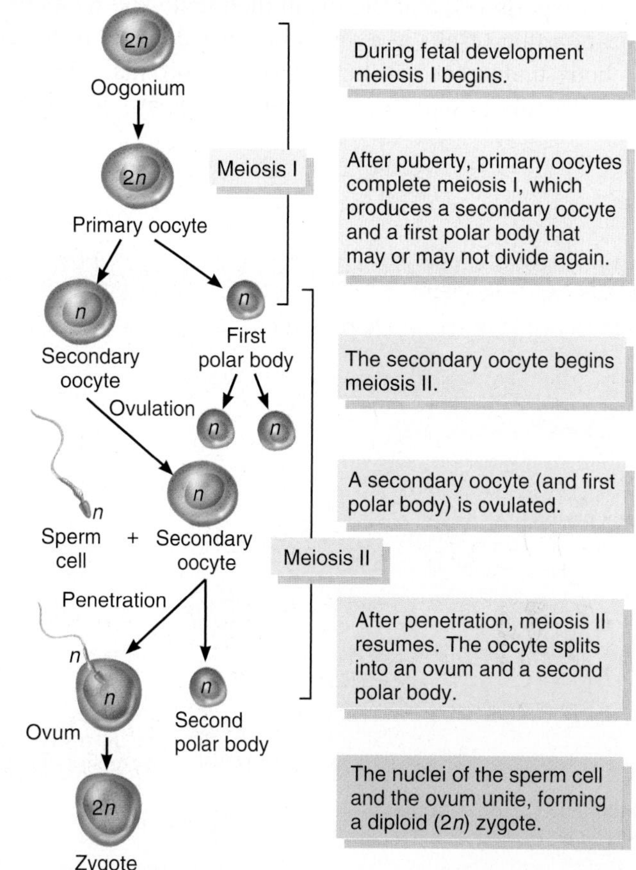
2n
Oogonium
During fetal development meiosis I begins.
2n
Meiosis I
Primary oocyte
After puberty, primary oocytes complete meiosis I, which produces a secondary oocyte and a first polar body that may or may not divide again.
n
Secondary oocyte
n
First polar body
The secondary oocyte begins meiosis II.
Ovulation
n
n
n
n
Sperm cell
+
Secondary oocyte
A secondary oocyte (and first polar body) is ovulated.
Meiosis II
Penetration
n
n
Ovum
n
Second polar body
After penetration, meiosis II resumes. The oocyte splits into an ovum and a second polar body.
2n
Zygote
The nuclei of the sperm cell and the ovum unite, forming a diploid (2n) zygote.

**In a secondary oocyte, meiosis II is completed only if fertilization occurs.**

At ovulation, the secondary oocyte is expelled into the pelvic cavity together with the first polar body and corona radiata. Normally these cells are swept into the uterine tube. If fertilization does not occur, the cells degenerate. However, if sperm are present in the uterine tube and one penetrates the secondary oocyte, meiosis II resumes. The secondary oocyte splits into two

## CLINICAL CONNECTION | *Ovarian Cysts*

An **ovarian cyst** is a fluid-filled sac in or on an ovary. Such cysts are relatively common, are usually noncancerous, and frequently disappear on their own. Cancerous cysts are more likely to occur in women over 40. Ovarian cysts may cause pain, pressure, a dull ache, or fullness in the abdomen; pain during sexual intercourse; delayed, painful, or irregular menstrual periods; abrupt onset of sharp pain in the lower abdomen; and/or vaginal bleeding. Most ovarian cysts require no treatment, but larger ones (more than 2 in.) may be removed surgically.

haploid cells, again of unequal size. The larger cell is the **ovum**, or mature egg; the smaller one is the **second polar body**. The nuclei of the sperm cell and the ovum then unite, the moment of fertilization that results in a diploid fertilized ovum. If the first polar body undergoes another division to produce two polar bodies, then the primary oocyte ultimately gives rise to three haploid polar bodies, which all degenerate, and a single haploid ovum. Thus, one primary oocyte gives rise to a single gamete (an ovum). By contrast, recall that in males one primary spermatocyte produces four gametes (sperm).

Table 25.1 summarizes the events of oogenesis and follicular development.

**TABLE 25.1**

Oogenesis and Follicular Development

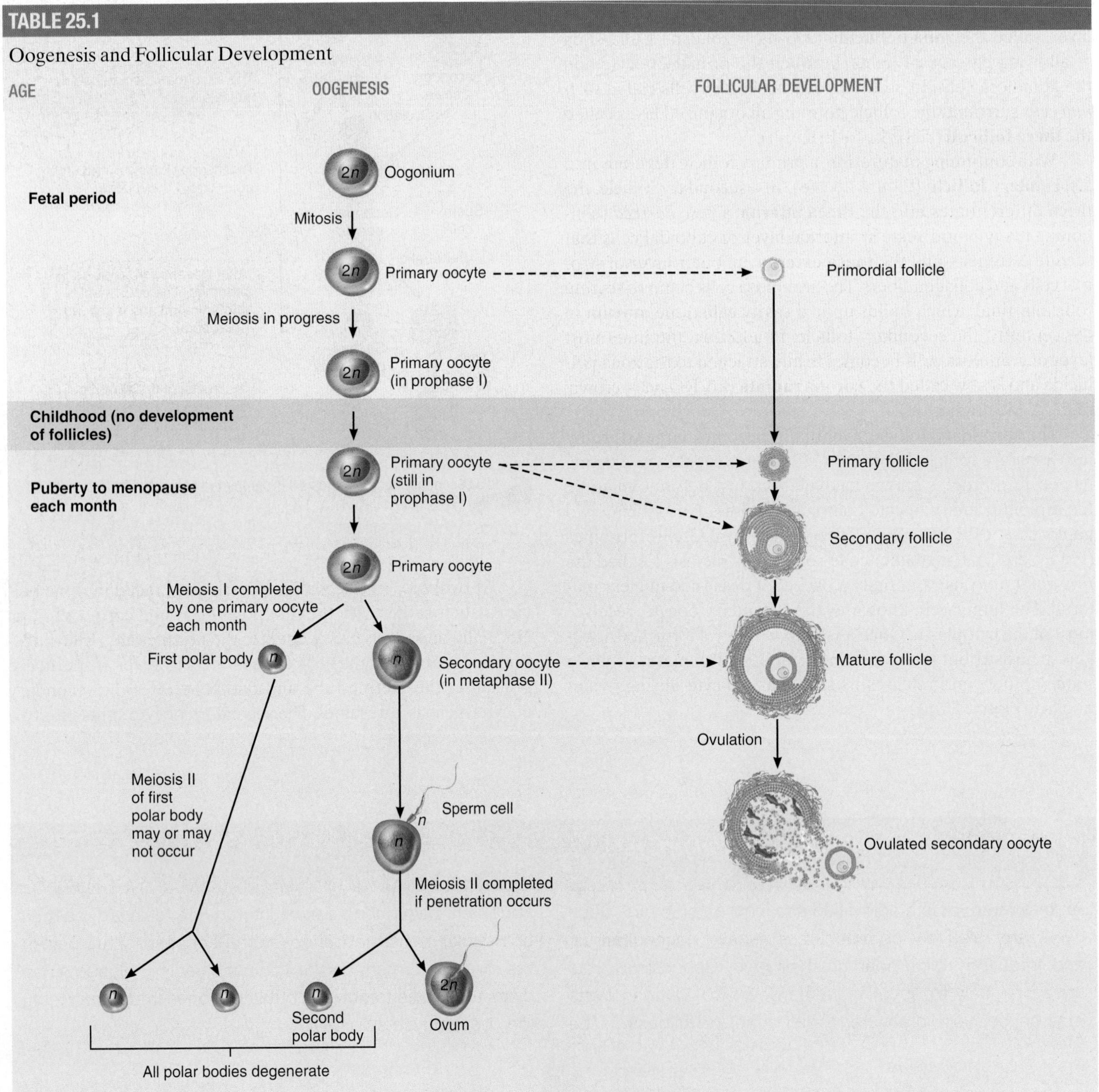

## Uterine Tubes

Two **uterine (fallopian) tubes**, or **oviducts**, extend laterally from the uterus (Figure 25.16). The tubes, which measure about 10 cm (4 in.) long, lie between the folds of the broad ligaments of the uterus. They provide a route for sperm to reach an ovum and transport secondary oocytes and fertilized ova from the ovaries to the uterus. The funnel-shaped portion of each tube, called the **infundibulum**, is close to the ovary but is open to the pelvic cavity. It ends in a fringe of fingerlike projections called **fimbriae** (FIM-brē-ē = fringe), one of which is attached to the lateral end of the ovary. From the infundibulum, the uterine tube extends medially and attaches to the upper, outer corners of the uterus. The **ampulla** (am-PUL-la) of the uterine tube is the widest, longest portion, making up about the lateral two-thirds of its length. The **isthmus** (IS-mus) of the uterine tube is the short, narrow, thick-walled portion that joins the uterus.

FIGURE 25.16 **Relationship of the uterine tubes to the ovaries, uterus, and associated structures.** In the left side of the drawing, the uterine tube and uterus have been sectioned to show internal structures.

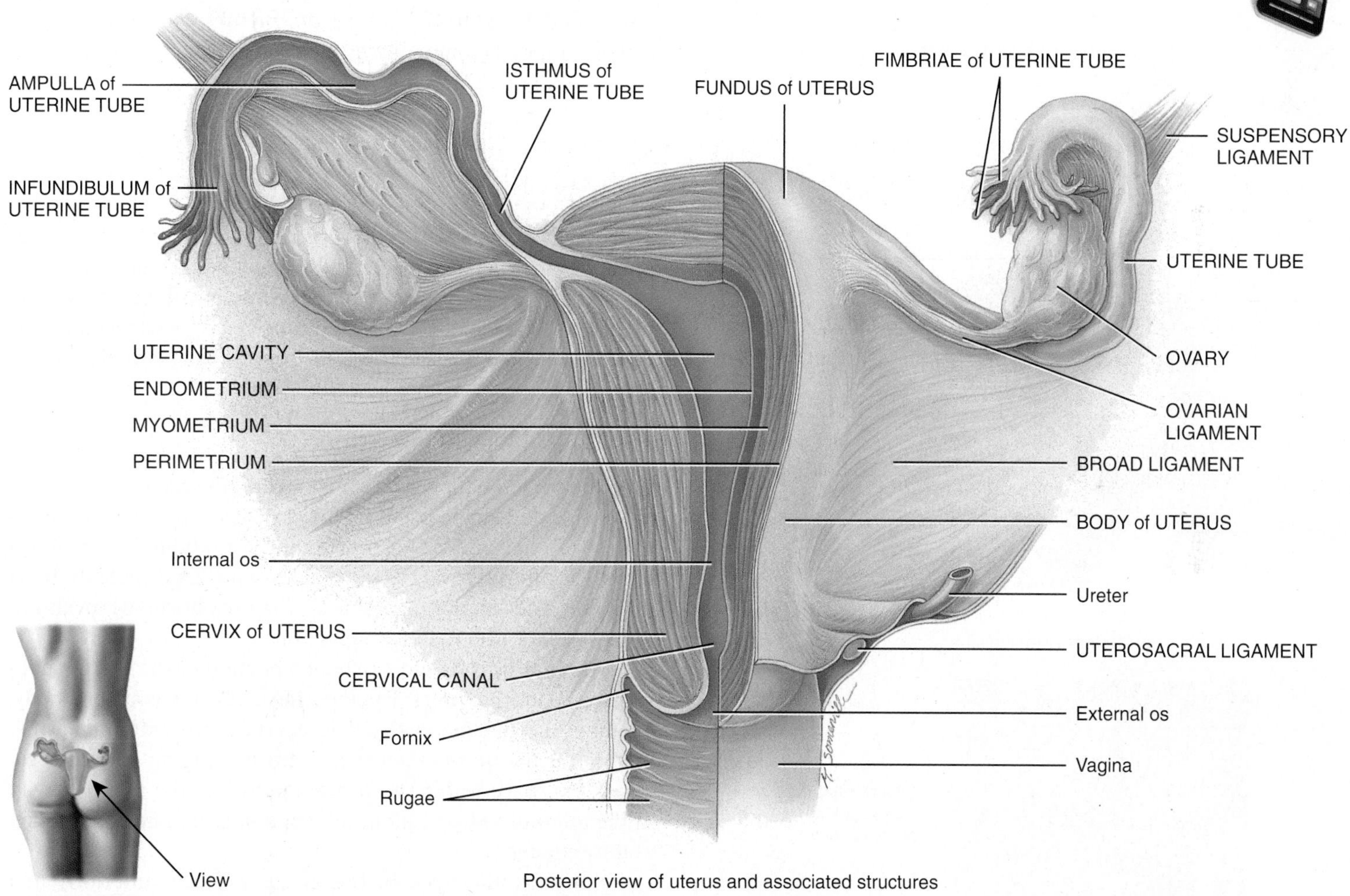


Posterior view of uterus and associated structures

**The uterus is the site of menstruation, implantation of a fertilized ovum, development of the fetus, and labor.**

The mucosal lining of the uterine tube consists of epithelium and lamina propria (areolar connective tissue) (Figure 25.17a). The epithelium contains ciliated simple columnar cells, which function as a "ciliary conveyor belt" to help move a fertilized ovum (or secondary oocyte) within the tube toward the uterus (Figure 25.17b). The corona radiata of the ovulated oocyte serves as a rough surface around the secondary oocyte; it provides friction to facilitate movement of the oocyte by the ciliated columnar cells of the uterine tube. The epithelium also contains nonciliated *peg cells*, which have microvilli and secrete a fluid that provides nutrition for the ovum. The muscularis is composed of an inner, circular layer of smooth muscle and an outer layer of longitudinal smooth muscle tissue. Peristaltic contractions of the muscularis and the ciliary action of the mucosa help move the oocyte, or fertilized ovum, toward the uterus.

Local currents produced by movements of the fimbriae, which surround the ovary during ovulation, sweep the ovulated secondary oocyte from the pelvic cavity into the uterine tube. A sperm cell usually encounters and penetrates a secondary oocyte in the ampulla of the uterine tube, although penetration in the pelvic cavity is not uncommon. *Penetration* can occur at any time up to about 24 hours after ovulation. Some hours after penetration, *fertilization* occurs as the nuclear materials of the *haploid* ovum and sperm unite to form a *diploid* fertilized ovum—now called a **zygote** (ZĪ-got; *zygon* = yolk). The zygote begins cell divisions while moving through the uterine tube toward the uterus, where it arrives 6 to 7 days after ovulation.

**FIGURE 25.17 Histology of the uterine tube.**

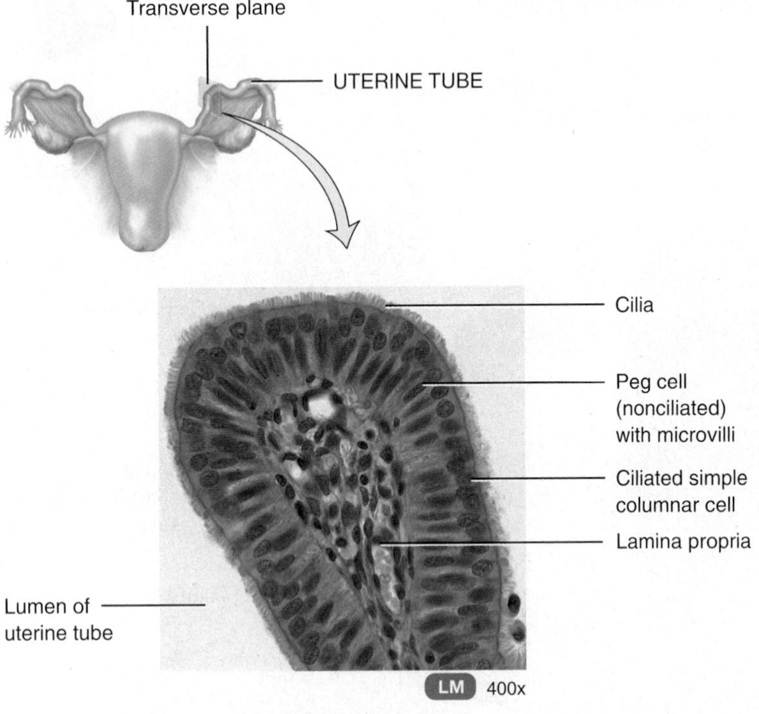


(a) Details of epithelium

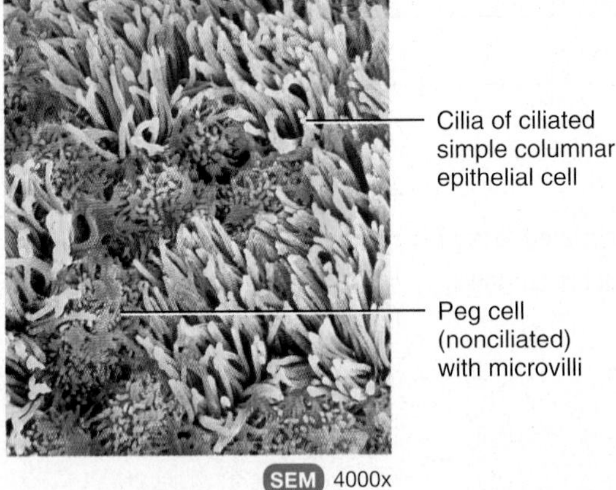


(b) Details of epithelium in surface view

**Peristaltic contractions of the muscularis and ciliary action of the mucosa of the uterine tube help move the oocyte or fertilized ovum toward the uterus.**

# Uterus

The **uterus** (womb) serves as part of the pathway for sperm deposited in the vagina to reach the uterine tubes. It is also the site of implantation of a fertilized ovum, development of the fetus during pregnancy, and labor. During reproductive cycles when implantation does not occur, the uterus is the source of menstrual flow.

## Anatomy of the Uterus

Situated between the urinary bladder and the rectum, the uterus is the size and shape of an inverted pear (Figure 25.16). In females who have never been pregnant, it is about 7.5 cm (3 in.) long. The uterus is larger in females who have recently been pregnant, and smaller (atrophied) when sex hormone levels are low, as occurs after menopause.

The **fundus** is the dome-shaped portion of the uterus superior to the uterine tubes. The tapering central portion, the **body**, encloses the **uterine cavity**. The **cervix** is the inferior narrow portion of the uterus that opens into the vagina. The **cervical canal**, the space within the cervix, opens into the uterine cavity at the *internal os* (*os* = mouthlike opening) and into the vagina at the *external os*.

Normally, the body of the uterus projects anteriorly and superiorly over the urinary bladder in a position called **anteflexion**. The cervix projects inferiorly and posteriorly and enters the anterior wall of the vagina at nearly a right angle (see Figure 25.11). Several ligaments maintain the anteflexed position of the uterus (see Figure 25.12). The paired broad ligaments attach the uterus to either side of the pelvic cavity. The paired **uterosacral ligaments** lie on either side of the rectum and connect the uterus to the sacrum. The **cardinal ligaments** extend from the cervix and vagina to the lateral pelvic wall. The **round ligaments** extend between the layers of the broad ligament from the uterus just inferior to the uterine tubes to the labia majora described in Concept 25.4. Although the ligaments normally maintain the anteflexed position of the uterus, they also allow the uterine body enough movement that the uterus may shift into a less than desirable position. A posterior tilting of the uterus, called **retroflexion** (*retro-* = backward or behind), is a harmless variation of the normal position of the uterus. There is often no cause for the condition, but it may occur after childbirth.

## Histology of the Uterus

Histologically, the uterus consists of three layers of tissue: the perimetrium, myometrium, and endometrium (Figure 25.18). The outer layer—the **perimetrium** (*peri-* = around; *-metrium* = uterus) or serosa—is part of the visceral peritoneum. Laterally, it becomes the broad ligament. The middle layer of the uterus, the **myometrium** (*myo-* = muscle), consists of three layers of smooth muscle tissue and forms the bulk of the uterine wall. During labor and childbirth, coordinated contractions of the myometrium help expel the fetus from the uterus.

The lumen of the uterus is lined with the **endometrium** (*endo-* = within), a highly vascularized mucosa composed of simple columnar epithelium with a thick underlying lamina propria of areolar connective tissue. **Endometrial glands** open onto the surface of the endometrium and extend almost to the myometrium. The endometrium is divided into two layers. The **stratum functionalis** (*functional layer*) lines the uterine cavity and

FIGURE 25.18 **Histology of the uterus.**

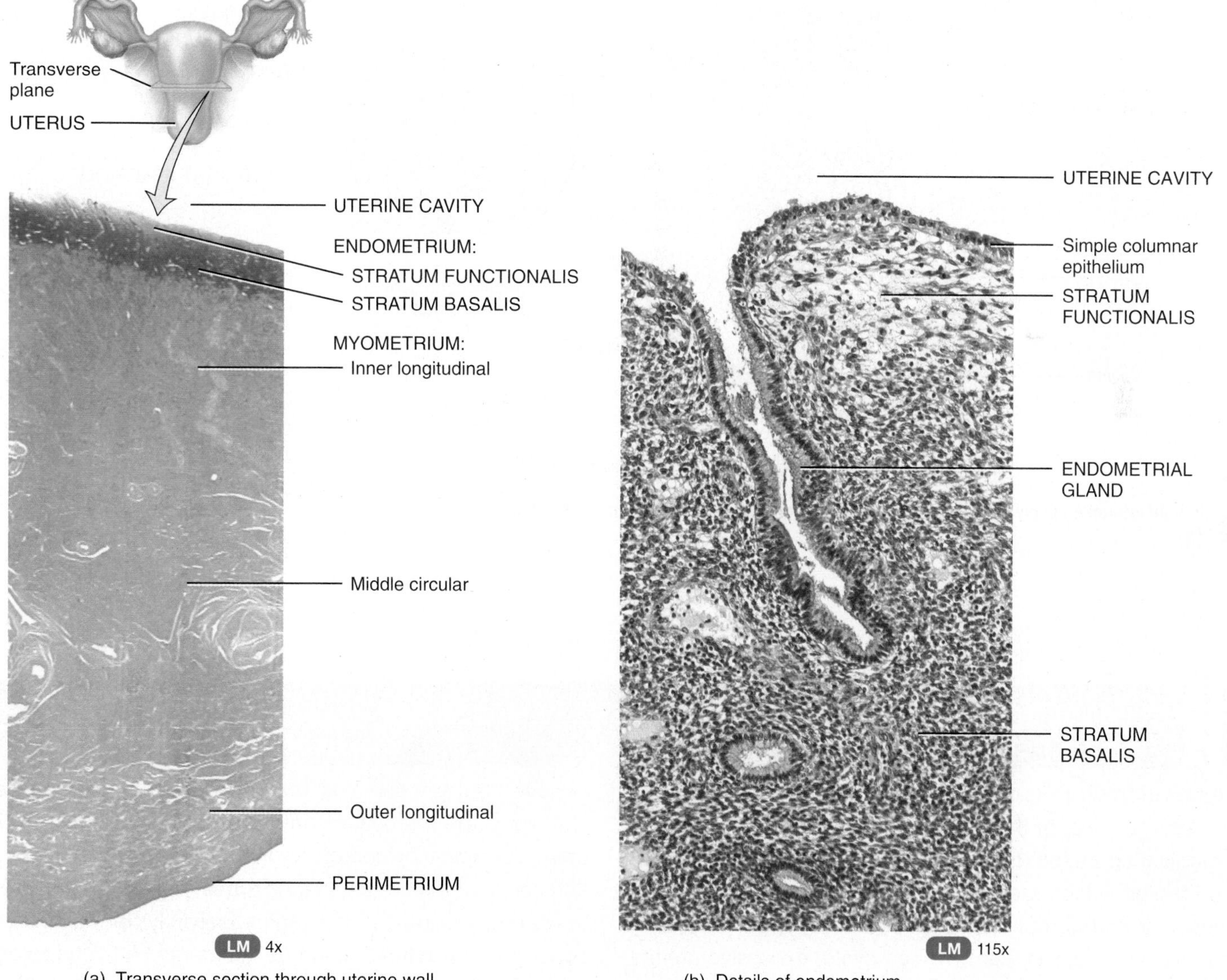


(a) Transverse section through uterine wall

(b) Details of endometrium

**The three layers of the uterus from superficial to deep are the perimetrium, myometrium, and endometrium.**

sloughs off during menstruation. The deeper layer, the **stratum basalis** (*basal layer*), is permanent and gives rise to a new stratum functionalis after each menstruation.

Branches of the internal iliac artery called **uterine arteries** (Figure 25.19) supply blood to the uterus. Uterine arteries give off branches called **arcuate arteries** (= shaped like a bow) that are arranged in a circular fashion within the myometrium. Arcuate arteries branch into **radial arteries** that penetrate deeply into the myometrium. Just before the branches enter the endometrium, they divide into arterioles. **Straight arterioles** supply the stratum basalis with the materials needed to regenerate the stratum functionalis. **Spiral arterioles** supply the stratum functionalis and play

FIGURE 25.19 **Blood supply of the uterus.** The inset shows histological details of the blood vessels of the endometrium.

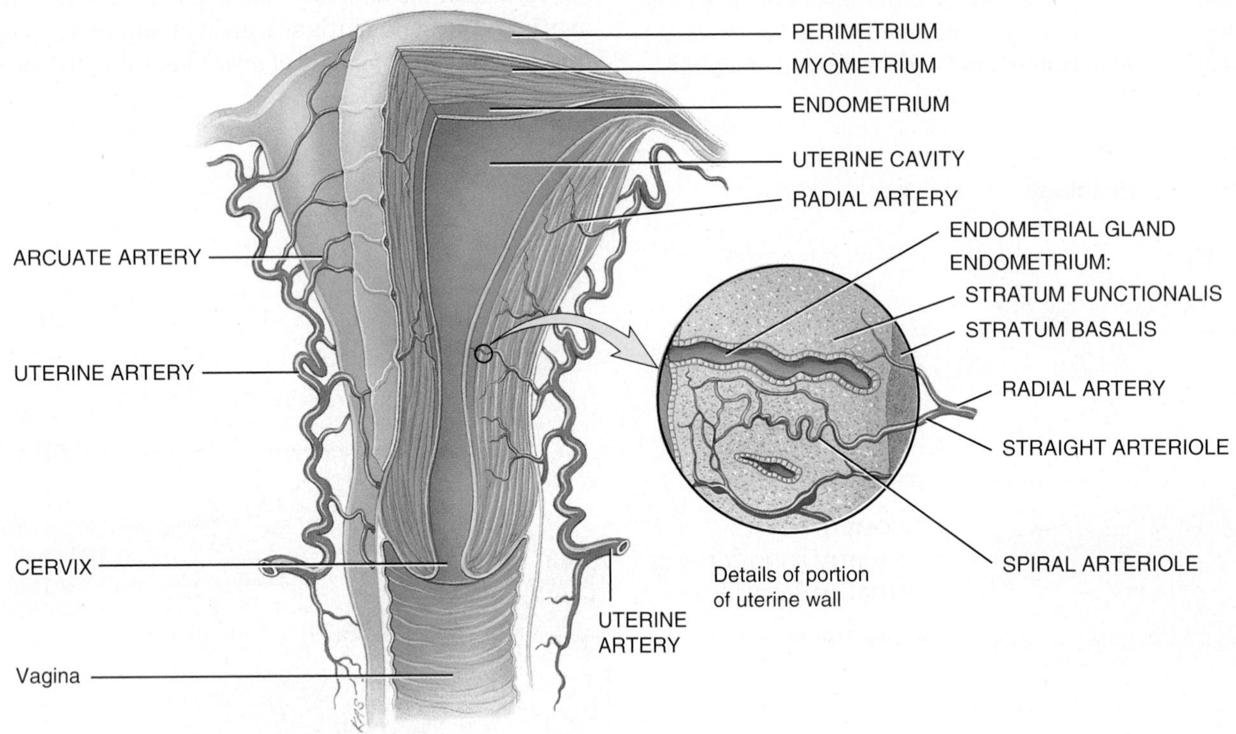

**Straight arterioles supply the materials needed for regeneration of the stratum functionalis.**

## CLINICAL CONNECTION | *Hysterectomy*

**Hysterectomy** (hiss-te-REK-tō-mē; *hyster-* = uterus), the surgical removal of the uterus, is the most common gynecological operation. It may be indicated in conditions such as fibroids, which are noncancerous tumors composed of muscular and fibrous tissue, endometriosis, pelvic inflammatory disease, recurrent ovarian cysts, excessive uterine bleeding, and cancer of the cervix, uterus, or ovaries. In a *partial (subtotal) hysterectomy*, the body of the uterus is removed but the cervix is left in place. *A complete hysterectomy* is the removal of both the body and cervix of the uterus. A *radical hysterectomy* includes removal of the body and cervix of the uterus, uterine tubes, possibly the ovaries, the superior portion of the vagina, pelvic lymph nodes, and supporting structures, such as ligaments. A hysterectomy can be performed either through an incision in the abdominal wall or through the vagina.

an active role in menstruation. Blood leaving the uterus is drained by the **uterine veins** into the internal iliac veins. The extensive blood supply of the uterus is essential to support regrowth of a new stratum functionalis after menstruation, implantation of a fertilized ovum, and development of the placenta.

### Cervical Mucus

Secretory cells of the cervical canal mucosa secrete **cervical mucus**. During their reproductive years, females secrete 20–60 mL of cervical mucus per day. Cervical mucus is more hospitable to sperm at or near the time of ovulation because it is then less viscous and more alkaline (pH 8.5). At other times, a more viscous mucus forms a cervical plug that physically impedes sperm penetration. Cervical mucus supplements the energy needs of sperm, and protects sperm from phagocytes and the hostile environment of the vagina and uterus. Cervical mucus may also play a role in *capacitation*—a series of changes that sperm undergo in the female reproductive tract before they are able to penetrate a secondary oocyte.

### ✓ CHECKPOINT

15. What are the functions of the ovaries? How are they held in position in the pelvic cavity?
16. Which structures in the ovaries serve an endocrine function, and which hormones do they secrete?
17. How does the age of a primary oocyte in a female compare with the age of a primary spermatocyte in a male?
18. How are primordial, primary, secondary, and mature follicles structurally different? From which follicle does ovulation occur?
19. Where are the uterine tubes located? What are the functions of the cells that line the uterine tubes?
20. What are the principal parts of the uterus? Where are they located in relationship to one another?
21. How is the uterus held in its normal position in the pelvic cavity?
22. Which structural features of the endometrium and myometrium contribute to their functions?

## 25.4 The vagina is a passageway for childbirth; the mammary glands secrete milk.

### Vagina

The **vagina** (= sheath) is a tubular 10 cm (4 in.) long canal lined with mucous membrane that extends from the exterior of the body to the uterine cervix (see Figures 25.11 and 25.16). It is the receptacle for the penis during sexual intercourse, the outlet for menstrual flow, and the passageway for childbirth. The vagina is situated between the urinary bladder and the rectum. A recess called the **fornix** (= arch or vault) surrounds the vaginal attachment to the cervix. When properly inserted, a contraceptive diaphragm rests on the fornix, where it is held in place as it covers the cervix.

The mucosa of the vagina is continuous with that of the uterus. It consists of stratified squamous epithelium that lies in a series of transverse folds called **rugae** (ROO-gē). The mucosa contains large stores of glycogen, the decomposition of which produces organic acids. The resulting acidic environment retards microbial growth, but it also is harmful to sperm. Alkaline components of semen, mainly from the seminal vesicles, raise the pH of fluid in the vagina and increase viability of the sperm. The muscularis (Figure 25.20a) consists of two layers of smooth muscle tissue that can stretch considerably to accommodate the penis during sexual intercourse and a child during birth. The adventitia, the superficial layer of the vagina, anchors the vagina to adjacent organs such as the urethra and urinary bladder anteriorly and the rectum and anal canal posteriorly.

A thin fold of mucous membrane, called the **hymen** (= membrane), forms a border around and partially closes the inferior end of the vaginal opening to the exterior, the **vaginal orifice** (see Figure 25.20b). After its rupture, usually during the first sexual intercourse, only remnants of the hymen remain.

## Vulva

The term **vulva** (VUL-va = to wrap around), or **pudendum** (pū-DEN-dum), refers to the *external genitals* of the female (Figure 25.20b). The following components make up the vulva:

- Anterior to the vaginal and urethral openings is the **mons pubis** (MONZ PŪ-bis; *mons* = mountain), an elevation of adipose tissue covered by skin and coarse pubic hair that cushions the pubic symphysis.
- From the mons pubis, two longitudinal folds of skin, the **labia majora** (LĀ-bē-a ma-JŌ-ra; *labia* = lips; *majora* = larger), extend inferiorly and posteriorly to protectively enclose medial components of the vulva. The singular term is *labium majus*. The labia majora are covered by pubic hair and contain an abundance of adipose tissue, sebaceous (oil) glands, and apocrine sudoriferous (sweat) glands. They are homologous (develop from common embryonic tissue) to the scrotum in males.
- Medial to the labia majora are two smaller protective folds of skin, the **labia minora** (mī-NŌ-ra; *minora* = smaller). The singular term is *labium minus*. Unlike the labia majora, the labia minora are devoid of pubic hair and fat and have few sudoriferous glands, but they do contain many sebaceous glands. The labia minora are homologous to the spongy urethra.
- The **clitoris** (KLI-to-ris) is a small cylindrical mass of erectile tissue and numerous nerves and blood vessels. The clitoris is located at the anterior junction of the labia minora. A layer of skin called the *prepuce of the clitoris* is formed at the point where the labia minora unite and cover the body of the clitoris. The exposed portion of the clitoris is the *glans*

**FIGURE 25.20** The vagina and vulva.

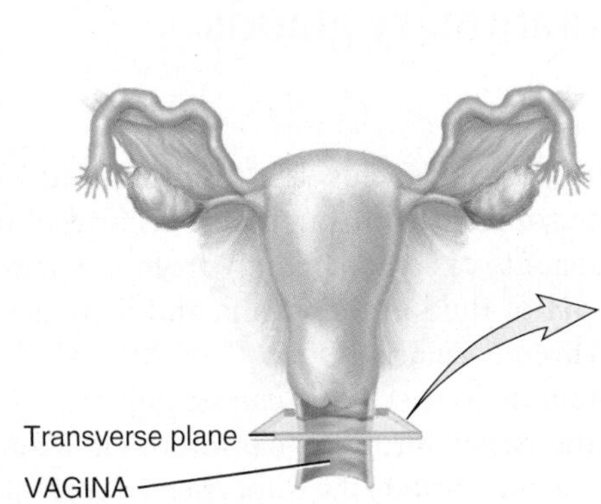


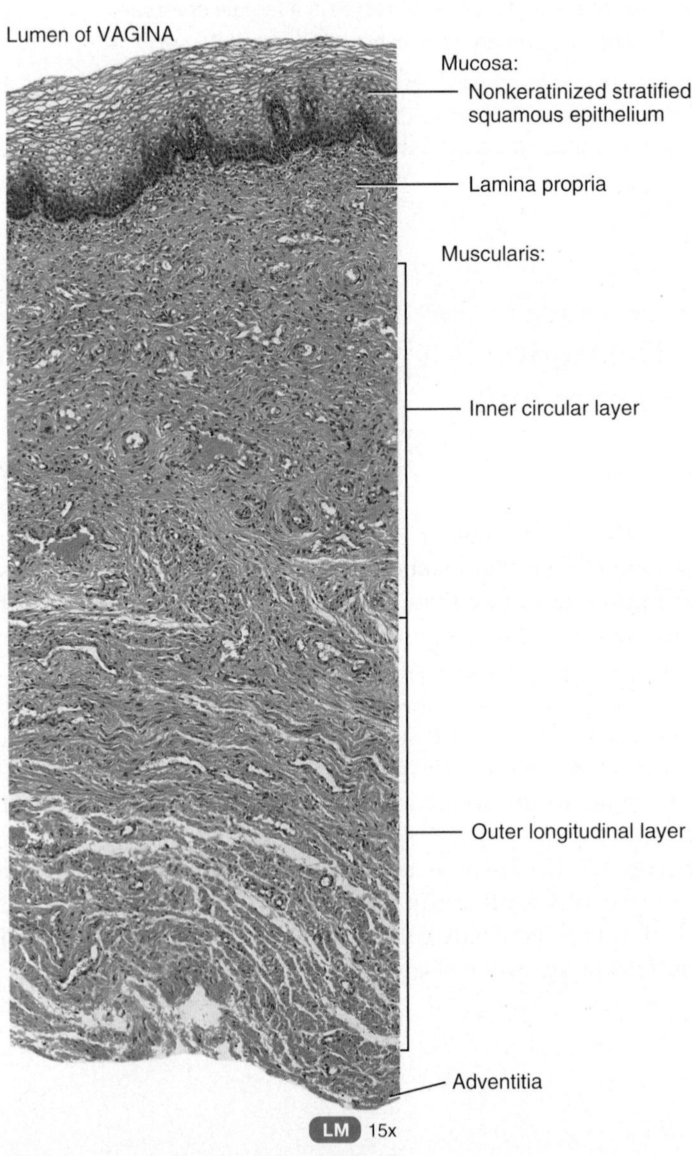


(a) Transverse section through vaginal wall

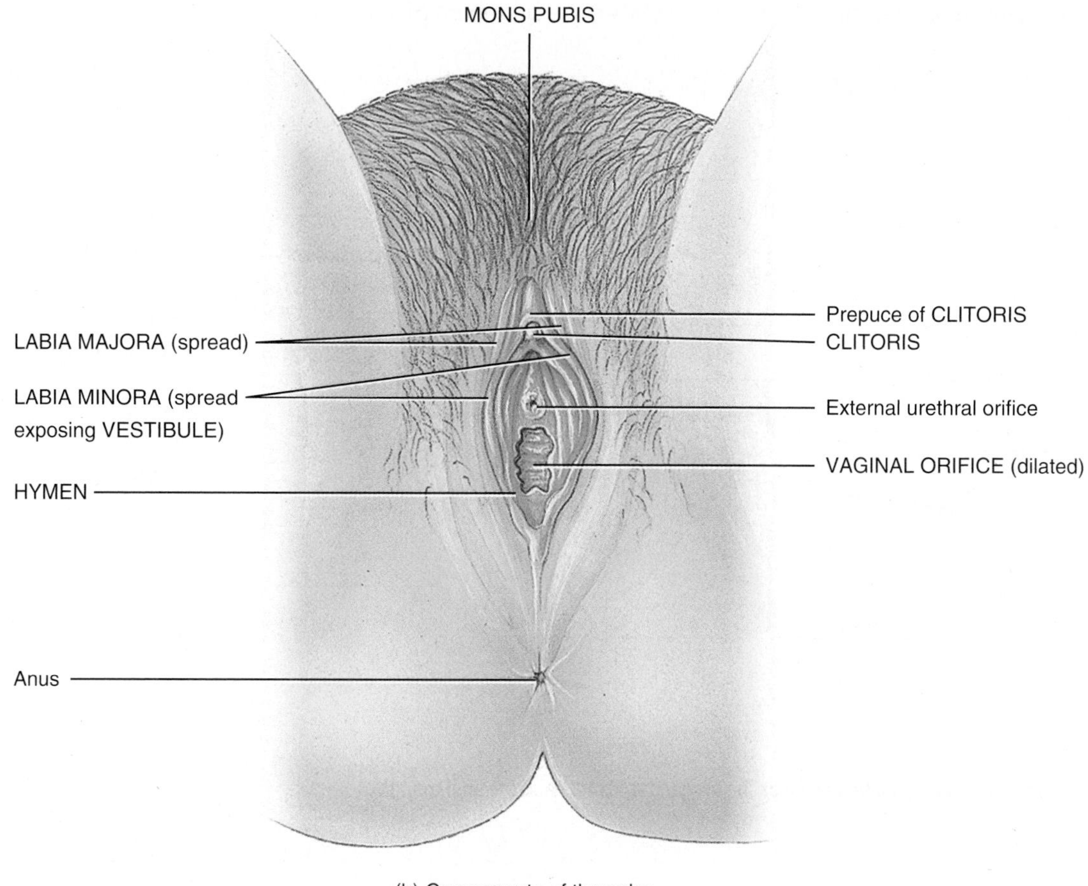


(b) Components of the vulva

**The vulva refers to the external genitals of the female.**

*clitoris*. The clitoris is homologous to the glans penis in males. Like the male structure, the clitoris is capable of enlargement upon tactile or other stimulation and has a role in sexual excitement in the female.

- The region between the labia minora is the **vestibule**. Within the vestibule are the vaginal orifice, the external urethral orifice, and the openings of the ducts of several glands. The vaginal orifice, the opening of the vagina to the exterior, occupies the greater portion of the vestibule and is bordered by the hymen. Anterior to the vaginal orifice and posterior to the clitoris is the *external urethral orifice*, the opening of the urethra to the exterior. On either side of the external urethral orifice are the openings of the ducts of the **paraurethral glands**. These mucus-secreting glands are embedded in the wall of the urethra. The paraurethral glands are homologous to the prostate. On either side of the vaginal orifice itself are

## CLINICAL CONNECTION | *Episiotomy*

During childbirth, the emerging fetus stretches the perineal region. To prevent excessive stretching and even tearing of this region, a physician sometimes performs an **episiotomy** (e-piz′-ē-OT-ō-mē; *episi-* = vulva or pubic region; *-otomy* = incision), a perineal cut made with surgical scissors. The cut may be made along the midline or at an angle of approximately 45 degrees to the midline. In effect, a straight, more easily sutured cut is substituted for the jagged tear that would otherwise be caused by passage of the fetus. The incision is closed in layers with sutures that are absorbed within a few weeks, so that the busy new mom does not have to worry about making time to have them removed.

FIGURE 25.21 **Perineum of a female.** (Figure 11.12 shows the perineum of a male.)

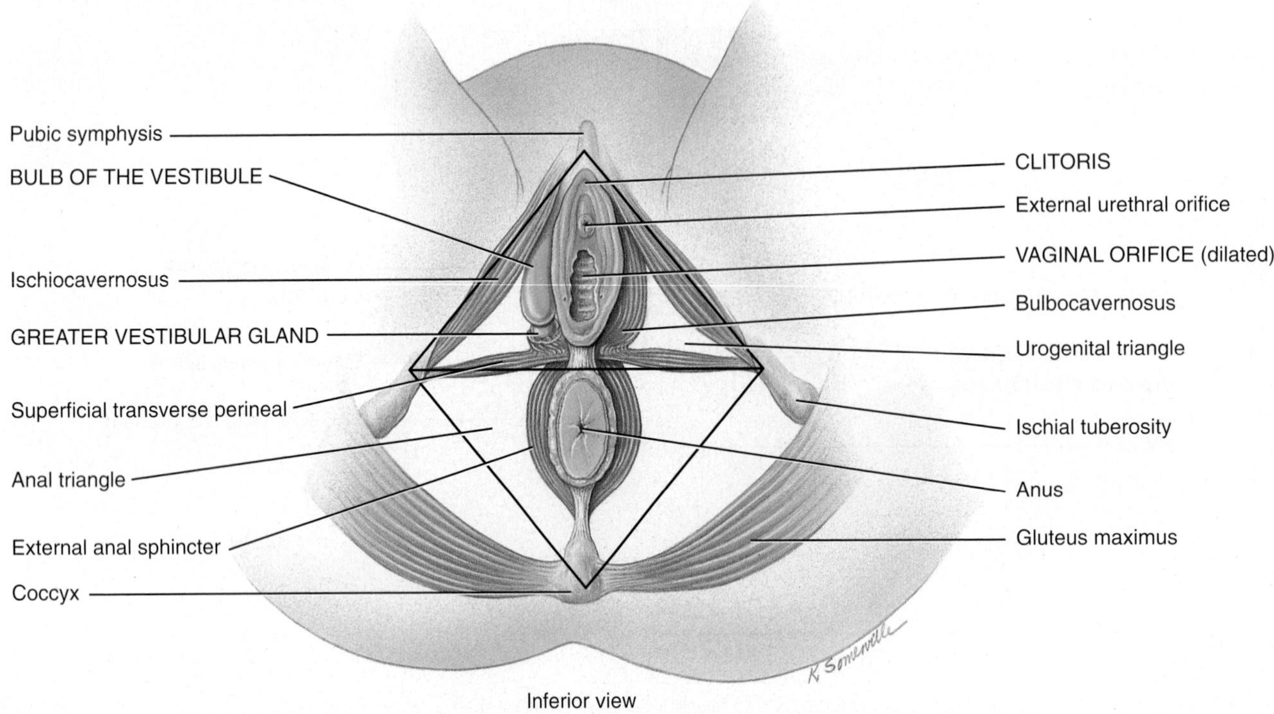


**The perineum is a diamond-shaped area that includes the urogenital triangle and the anal triangle.**

the **greater vestibular glands** (Figure 25.21), which produce a small quantity of lubricating mucus during sexual arousal and intercourse. The greater vestibular glands are homologous to the bulbourethral glands in males.

- The **bulb of the vestibule** (Figure 25.21) consists of two elongated masses of erectile tissue just deep to the labia on either side of the vaginal orifice. The bulb of the vestibule becomes engorged with blood during sexual arousal, narrowing the vaginal orifice and placing pressure on the penis during intercourse. The bulb of the vestibule is homologous to the corpus spongiosum and bulb of the penis in males.

Table 25.2 summarizes the homologous structures of the female and male reproductive systems.

## Perineum

The **perineum** (per′-i-NĒ-um) is the diamond-shaped area medial to the thighs and buttocks of both males and females (Figure 25.21). It contains the external genitals and anus. The perineum is bounded anteriorly by the pubic symphysis, laterally by the ischial tuberosities, and posteriorly by the coccyx. A transverse line drawn between the ischial tuberosities divides the perineum into an anterior *urogenital triangle* (ū′-rō-JEN-i-tal) that contains the external genitalia and a posterior *anal triangle* that contains the anus.

**TABLE 25.2**

Homologous Structures of the Female and Male Reproductive Systems

| FEMALE STRUCTURE | MALE STRUCTURE |
|---|---|
| Ovaries | Testes |
| Ovum | Sperm cell |
| Labia majora | Scrotum |
| Labia minora | Spongy urethra |
| Bulb of vestibule | Corpus spongiosum penis and bulb of penis |
| Clitoris | Glans penis |
| Paraurethral glands | Prostate |
| Greater vestibular glands | Bulbourethral glands |

## Mammary Glands

Each **breast** is a hemispheric projection of variable size anterior to the pectoralis major and serratus anterior and attached to them by a layer of fascia (Figure 25.22).

Each breast has one pigmented projection, the **nipple**. The circular pigmented area of skin surrounding the nipple is called the **areola** (a-RĒ-ō-la = small space); it appears rough because it contains modified sebaceous (oil) glands. Strands of connective tissue called the **suspensory ligaments of the breast** run between the skin and fascia and support the breast. These ligaments become looser with age or with the excessive strain that can occur in long-term jogging or high-impact aerobics. Wearing a supportive bra can slow this process and help maintain the strength of the suspensory ligaments.

Within each breast is a **mammary gland**, a modified sudoriferous gland that produces milk. A mammary gland consists of 15 to 20 lobes, or compartments, separated by a variable amount of adipose tissue. In each lobe are several smaller compartments called **lobules**, composed of grapelike clusters of milk-secreting glands termed **alveoli** (= small cavities) embedded in connective tissue. When milk is being produced, it passes from the alveoli into a series of **secondary tubules** and then into the **mammary ducts**. Near the nipple, the mammary ducts expand to form sinuses called *lactiferous sinuses* (*lact-* = milk), where some milk may be stored before draining into a lactiferous duct. Each **lactiferous duct** carries milk from one of the lobes to the exterior through an opening at the nipple. The functions of the mammary glands are the production and ejection of milk; these functions, called lactation, are associated with pregnancy and childbirth. Lactation is discussed in depth in Concept 25.10.

### ✓ CHECKPOINT

23. What are the functions of the vagina?
24. What is the vulva? What are the structures and functions of each part of the vulva?
25. Which structures in males are homologous to the ovaries, the clitoris, the greater vestibular glands, and the bulb of the vestibule?
26. Which surface structures are anterior to the vaginal opening? Lateral to it?
27. How are breasts supported?
28. Through which structures does milk pass from its production site to its exit from the breast?

**FIGURE 25.22** Mammary glands.

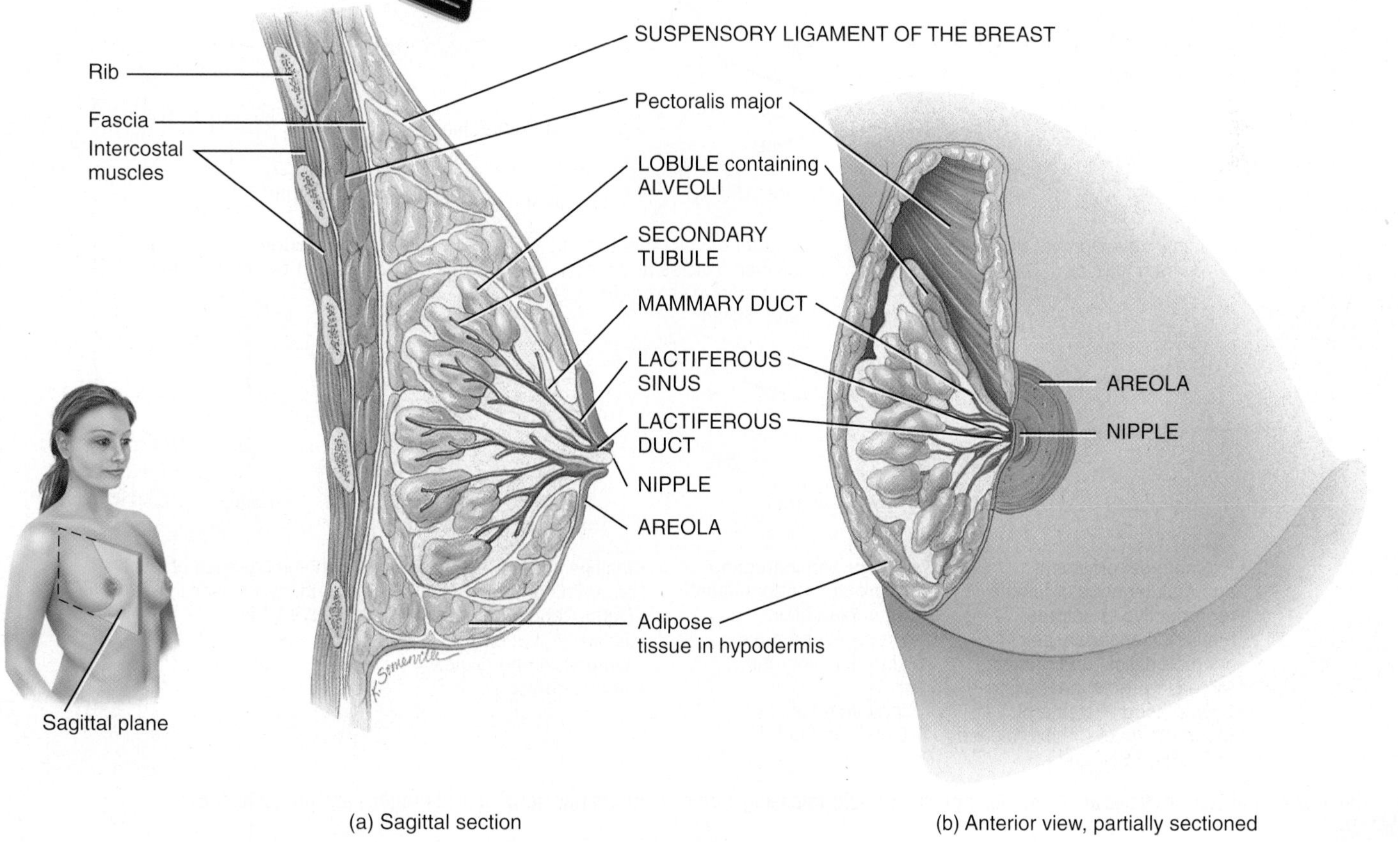

Rib
Fascia
Intercostal muscles
SUSPENSORY LIGAMENT OF THE BREAST
Pectoralis major
LOBULE containing ALVEOLI
SECONDARY TUBULE
MAMMARY DUCT
LACTIFEROUS SINUS
LACTIFEROUS DUCT
NIPPLE
AREOLA
Adipose tissue in hypodermis
Sagittal plane
AREOLA
NIPPLE

(a) Sagittal section

(b) Anterior view, partially sectioned

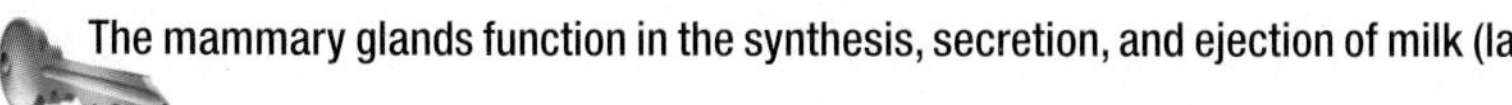

The mammary glands function in the synthesis, secretion, and ejection of milk (lactation).

# 25.5 The female reproductive cycle includes the ovarian and uterine cycles.

During their reproductive years, nonpregnant females normally exhibit cyclical changes in the ovaries and uterus. Each cycle takes about a month and involves both oogenesis and preparation of the uterus to receive a fertilized ovum. Hormones secreted by the hypothalamus, anterior pituitary, and ovaries control the main events. The **ovarian cycle** is a series of events in the ovaries that occur during and after the maturation of an oocyte. Steroid hormones released by the ovaries control the **uterine (menstrual) cycle**, a concurrent series of changes in the endometrium of the uterus to prepare it for the arrival of a fertilized ovum, which will develop there until birth. If fertilization does not occur, ovarian hormones wane; this causes the stratum functionalis of the endometrium to slough off. The general term **female reproductive cycle** encompasses the ovarian and uterine cycles, the hormonal changes that regulate them, and the related cyclical changes in the breasts and cervix.

## Hormonal Regulation of the Female Reproductive Cycle

*Gonadotropin-releasing hormone* (*GnRH*) secreted by the hypothalamus controls the ovarian and uterine cycles (Figure 25.23). GnRH stimulates the release of *follicle-stimulating hormone* (*FSH*) and *luteinizing hormone* (*LH*) from the anterior pituitary. FSH initiates growth of ovarian follicles, while LH stimulates their further development. In addition, both FSH and LH stimulate the ovarian follicles to secrete estrogens. LH stimulates the theca folliculi cells to produce androgens, and FSH causes the granulosa cells to convert the androgens into estrogens. At midcycle, LH triggers ovulation and then promotes formation of the corpus luteum, the reason for the name luteinizing hormone. Stimulated by LH, the corpus luteum produces and secretes estrogens, progesterone, relaxin, and inhibin.

FIGURE 25.23 **Secretion and physiological effects of estrogens, progesterone, relaxin, and inhibin in the female reproductive cycle.** Dashed red lines indicate negative feedback inhibition.

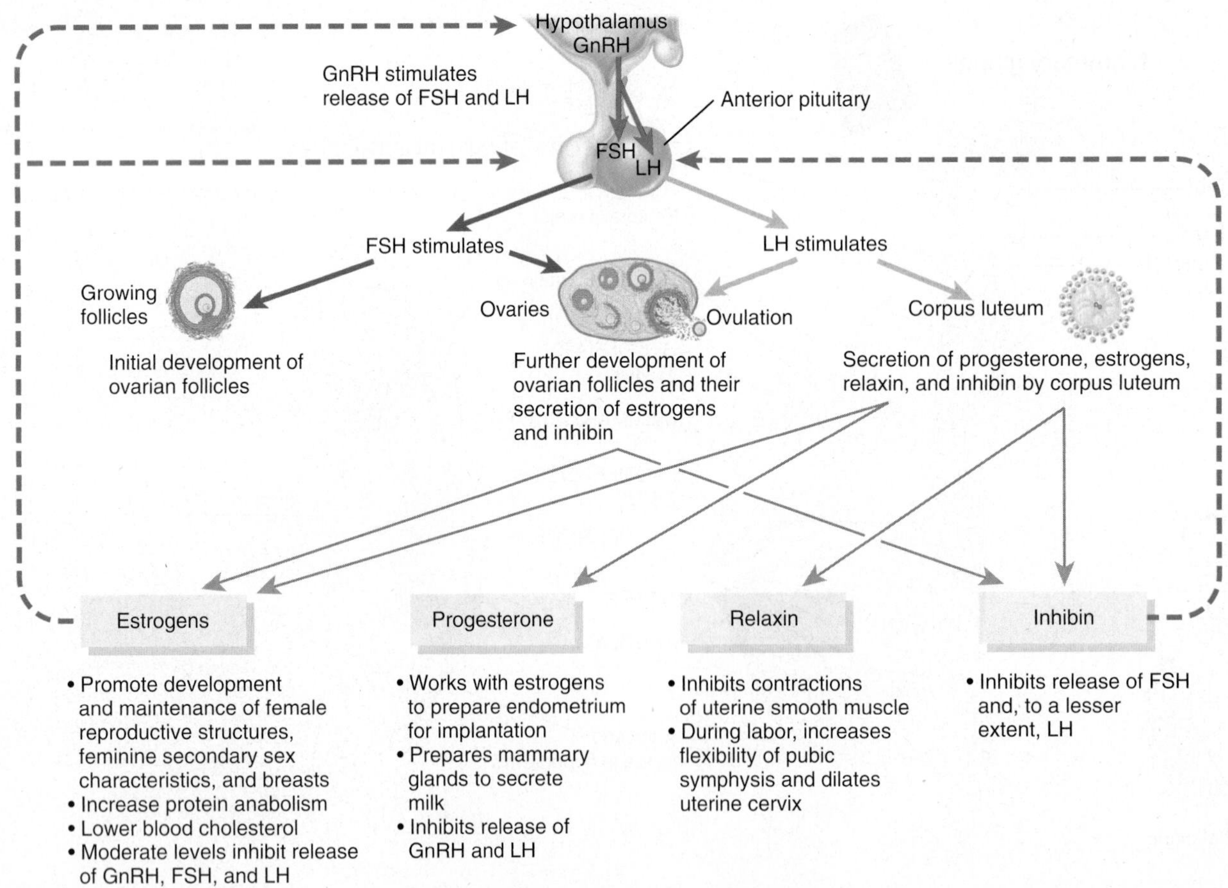


**The ovarian and uterine cycles are controlled by gonadotropin-releasing hormone and ovarian hormones (estrogens and progesterone).**

At least six different estrogens have been isolated from the plasma of human females, but only three are present in significant quantities: *estradiol*, *estrone*, and *estriol*. In a nonpregnant woman, the most abundant estrogen is estradiol, which is synthesized from cholesterol in the ovaries. **Estrogens** are secreted by ovarian follicles and have several important functions (Figure 25.23):

- Promote the development and maintenance of female reproductive structures, secondary sex characteristics, and the breasts. The secondary sex characteristics include distribution of adipose tissue in the breasts, abdomen, mons pubis, and hips; voice pitch; a broad pelvis; and pattern of hair growth on the head and body.
- Work synergistically with human growth hormone to increase protein anabolism, including the building of strong bones.
- Lower blood cholesterol level, which is probably the reason that women under age 50 have a much lower risk of coronary artery disease than do men of comparable age.

**Progesterone** is secreted by the corpus luteum. It cooperates with estrogens to prepare and maintain the endometrium for implantation of a fertilized ovum and to prepare the mammary glands for milk secretion.

The small quantity of **relaxin** produced by the corpus luteum during each monthly cycle relaxes the uterus by inhibiting contractions of the myometrium. Presumably, implantation of a fertilized ovum occurs more readily in a "quiet," relaxed uterus. During pregnancy, the placenta produces much more relaxin, and it continues to relax uterine smooth muscle. At the end of pregnancy, relaxin also increases the flexibility of the pubic symphysis and may help dilate the uterine cervix, both of which ease delivery of the baby.

Three hormones provide feedback for the female reproductive cycle. Moderate blood levels of estrogens inhibit both the release of GnRH by the hypothalamus and secretion of LH and FSH by the anterior pituitary. High levels of progesterone inhibit secretion of GnRH and LH. In addition, *inhibin* (secreted by growing follicles and by the corpus luteum) inhibits secretion of FSH and, to a lesser extent, LH.

# Phases of the Female Reproductive Cycle

The duration of the female reproductive cycle typically ranges from 24 to 35 days. For this discussion, we assume a duration of 28 days and divide it into four phases: the menstrual phase, the preovulatory phase, ovulation, and the postovulatory phase (Figure 25.24). Because they occur at the same time, the events of the ovarian cycle and uterine cycle are discussed together.

## Menstrual Phase

The **menstrual phase** (MEN-stroo-al), also called **menstruation** (men′-stroo-Ā-shun), lasts for roughly the first five days of the cycle. (By convention, the first day of menstruation is day 1 of a new cycle.)

**Ovarian events** Under the influence of FSH, several primordial follicles develop into primary follicles and then into secondary follicles. This developmental process may take several months to occur. Therefore, a follicle that begins to develop at the beginning of a particular menstrual cycle may not reach maturity and ovulate until several menstrual cycles later.

**Uterine events** Menstrual flow from the uterus consists of 50–150 mL of blood, tissue fluid, mucus, and epithelial cells shed from the endometrium. This discharge occurs because the declining levels of progesterone and estrogens stimulate release of prostaglandins that cause the uterine spiral arterioles to constrict. As a result, the cells they supply become oxygen-deprived and start to die. Eventually, the entire stratum functionalis sloughs off. At this time the endometrium is very thin, about 2–5 mm, because only the stratum basalis remains. The menstrual flow passes from the uterine cavity through the cervix and vagina to the exterior.

## Preovulatory Phase

The **preovulatory phase** is the time between the end of menstruation and ovulation. On average, it lasts from days 6 to 13 in a 28-day cycle.

**Ovarian events** Some of the secondary follicles in the ovaries begin to secrete estrogens and inhibin. By about day 6, a single secondary follicle in one of the two ovaries has outgrown all the others to become the **dominant follicle**. Estrogens and inhibin secreted by the dominant follicle decrease the secretion of FSH, which causes other, less well-developed follicles to stop growing and undergo *atresia* (degeneration). Dizygotic (fraternal) twins or triplets result when two or three secondary follicles become codominant and later are ovulated and fertilized at about the same time.

Normally, the one dominant secondary follicle becomes the mature follicle, which continues to enlarge until it is more than 20 mm in diameter and ready for ovulation (see Figure 25.13). This follicle forms a blisterlike bulge on the surface of the ovary due to the swelling antrum. During the final maturation process, the mature follicle continues to increase its production of estrogens (Figure 25.24).

**Uterine events** Estrogens liberated into the blood by growing ovarian follicles stimulate the repair of the endometrium; cells of the stratum basalis undergo mitosis and produce a new stratum functionalis. As the endometrium thickens, endometrial glands develop, and arterioles coil and lengthen as they penetrate the stratum functionalis. The thickness of the endometrium approximately doubles, to about 4–10 mm.

## Ovulation

**Ovulation**, the rupture of the mature follicle and the release of the secondary oocyte into the pelvic cavity, usually occurs on

**FIGURE 25.24 The female reproductive cycle.** The length of the female reproductive cycle typically is 24 to 36 days. (a) Events in the ovarian and uterine cycles and the release of anterior pituitary hormones are correlated with the cycle's phases. In the cycle shown, fertilization and implantation have not occurred. (b) Relative concentrations of anterior pituitary hormones (follicle-stimulating hormone and luteinizing hormone) and ovarian hormones (estrogens and progesterone) during the phases of a cycle.

Hypothalamus
GnRH
Anterior pituitary
FSH
LH
Ovarian cycle
Primordial follicles
Primary follicles
Secondary follicle
Mature follicle
Ovulation
Corpus hemorrhagicum
Corpus luteum
Corpus albicans
Progesterone and estrogens
Estrogens
Uterine cycle
Stratum functionalis
Stratum basalis
Menstruation
Menstruation
Days 1 2 3 4 5 6 7 8 9 10 11 12 13 14 15 16 17 18 19 20 21 22 23 24 25 26 27 28 1 2
Menstrual phase
Preovulatory phase
Ovulation
Postovulatory phase

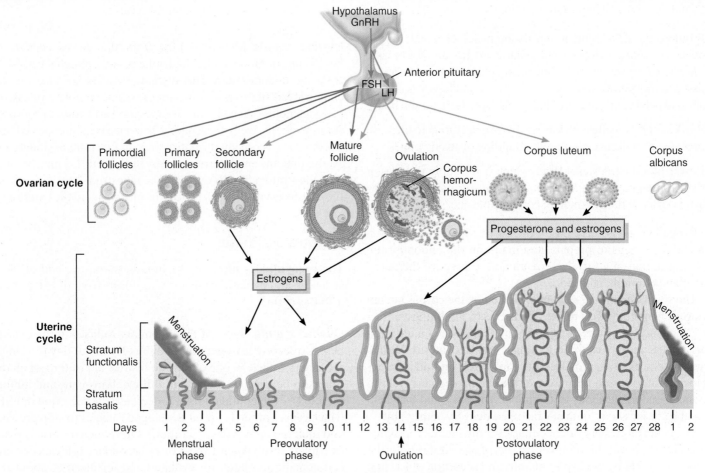

(a) Hormonal regulation of changes in the ovary and uterus

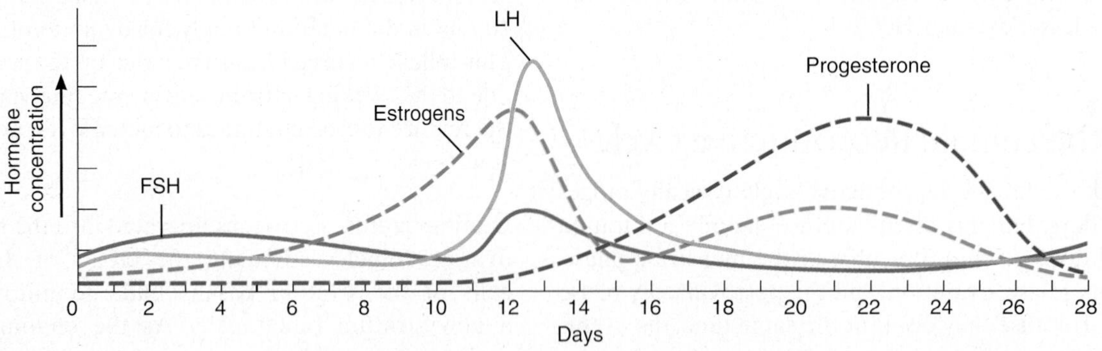


(b) Changes in concentration of anterior pituitary and ovarian hormones

**Estrogens are secreted by follicles before ovulation; after ovulation, both progesterone and estrogens are secreted by the corpus luteum.**

day 14 in a 28-day cycle. During ovulation, the secondary oocyte remains surrounded by its zona pellucida and corona radiata.

The *high levels of estrogens* during the last part of the preovulatory phase exert a *positive* feedback effect on the cells that secrete LH and GnRH and cause ovulation, as follows (Figure 25.25):

1. A high blood level of estrogens stimulates the hypothalamus to release more GnRH. It also directly stimulates the anterior pituitary to secrete LH.
2. GnRH stimulates the anterior pituitary to secrete FSH and even more LH.
3. The resulting surge of LH causes rupture of the mature follicle and expulsion of a secondary oocyte. The ovulated oocyte and its corona radiata are usually swept into the uterine tube.

FIGURE 25.25 **Positive feedback control of ovulation.** High levels of estrogens exert a positive feedback effect (green arrows) on the hypothalamus and anterior pituitary, thereby increasing secretion of gonadotropin-releasing hormone and luteinizing hormone.

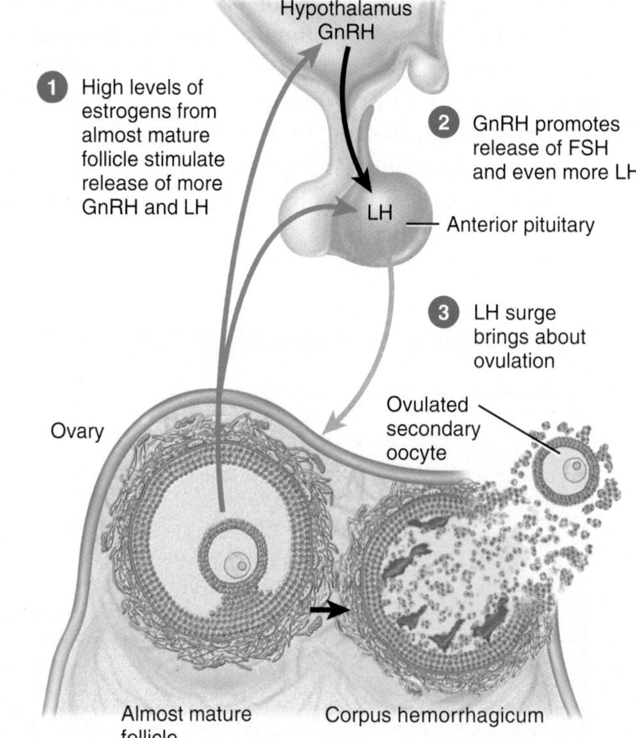
Hypothalamus
GnRH
1 High levels of estrogens from almost mature follicle stimulate release of more GnRH and LH
2 GnRH promotes release of FSH and even more LH
LH
Anterior pituitary
3 LH surge brings about ovulation
Ovulated secondary oocyte
Ovary
Almost mature follicle
Corpus hemorrhagicum

At midcycle, a surge of luteinizing hormone triggers ovulation.

## Postovulatory Phase

The **postovulatory phase** of the female reproductive cycle is the time between ovulation and onset of the next menstruation. It lasts for 14 days in a 28-day cycle, from day 15 to day 28 (Figure 25.24).

Ovarian events After ovulation, the mature follicle collapses. Once a blood clot forms from minor bleeding of the ruptured follicle, the follicle becomes the **corpus hemorrhagicum** (hem′-o-RAJ-i-kum; *hemo-* = blood; *-rrhagic-* = bursting forth) (see Figure 25.13). Stimulated by LH, the corpus hemorrhagicum transforms into the corpus luteum, which secretes progesterone, estrogen, relaxin, and inhibin.

Later ovarian events depend on whether the oocyte is fertilized. If the oocyte *is not fertilized*, the corpus luteum has a life span of only two weeks. Then, it degenerates into a corpus albicans, which does not secrete ovarian hormones (see Figure 25.13). As the levels of progesterone, estrogens, and inhibin quickly decrease, release of GnRH, FSH, and LH rises due to loss of negative feedback suppression by the ovarian hormones. Earlier in the cycle, high blood levels of the ovarian hormones progesterone, estrogens, and inhibin had suppressed secretion of GnRH by the hypothalamus and FSH and LH by the pituitary

### CLINICAL CONNECTION | *Ectopic Pregnancy*

**Ectopic pregnancy** (ek-TOP-ik; *ec-* = out of; *-topic* = place) is the development of an embryo or fetus outside the uterine cavity. An ectopic pregnancy usually occurs when movement of the fertilized ovum through the uterine tube is impaired by scarring due to a prior tubal infection, when there is decreased movement of the uterine tube smooth muscle, or when there is abnormal tubal anatomy. Although the most common site of ectopic pregnancy is the uterine tube, ectopic pregnancies may also occur in the ovary, abdominal cavity, or uterine cervix. Women who smoke are twice as likely to have an ectopic pregnancy because nicotine in cigarette smoke paralyzes the cilia in the lining of the uterine tube (as it does those in the respiratory airways). Scars from pelvic inflammatory disease, previous uterine tube surgery, and previous ectopic pregnancy may also hinder movement of the fertilized ovum.

The signs and symptoms of ectopic pregnancy include one or two missed menstrual cycles followed by bleeding and acute abdominal and pelvic pain. Unless removed, the developing embryo can rupture the uterine tube, often resulting in death of the mother. Treatment options include surgery or the use of a cancer drug called methotrexate, which causes embryonic cells to stop dividing and eventually disappear.

gland. Following ovulation, blood levels of progesterone, estrogens, and inhibin quickly drop, which allows increased release of GnRH, FSH, and LH. With elevated FSH and LH, follicular growth resumes and a new ovarian cycle begins.

If the secondary oocyte is *fertilized* and begins to divide, the corpus luteum persists past its normal two-week life span. It is "rescued" from degeneration by **human chorionic gonadotropin** (hCG) (kō-rē-ON-ik gō′-nad-o-TRŌ-pin). This hormone is produced by the chorion of the embryo beginning about eight days after fertilization. Like LH, hCG stimulates the secretory activity of the corpus luteum. The presence of hCG in maternal blood or urine is an indicator of pregnancy and is the hormone detected by home pregnancy tests.

**Uterine events** Progesterone and estrogens produced by the corpus luteum promote growth of the endometrial glands, and blood vessel formation and thickening of the endometrium to 12–18 mm (0.48–0.72 in.). These preparatory changes peak about one week after ovulation, at the time a fertilized ovum might arrive in the uterus. If fertilization does not occur, the levels of progesterone and estrogens decline due to degeneration of the corpus luteum. Withdrawal of progesterone and estrogens causes menstruation.

Figure 25.26 summarizes the hormonal interactions and cyclical changes in the ovaries and uterus during the embtyo and uterine cycles.

FIGURE 25.26 **Hormonal interactions in the ovarian and uterine cycles.**

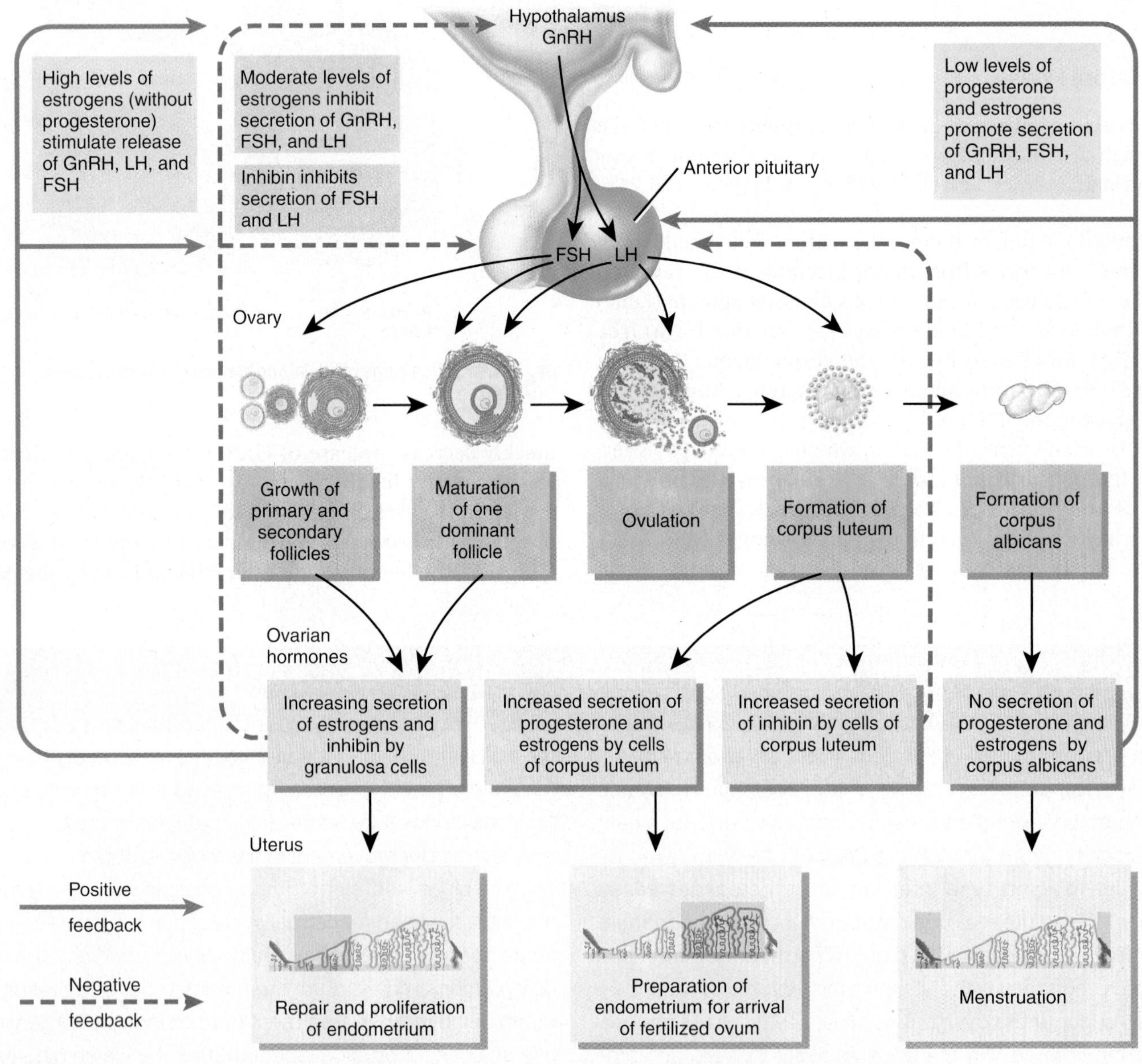


**Hormones from the anterior pituitary regulate ovarian function, and hormones from the ovaries regulate changes in the endometrium.**

## ✓ CHECKPOINT

**29.** Of the several estrogens, which one exerts the major effect?

**30.** What functions do estrogen and progesterone have outside the uterus? What are the functions of relaxin?

**31.** Describe the function of each of the following hormones in the uterine and ovarian cycles: GnRH, FSH, LH, estrogens, progesterone, and inhibin.

**32.** Why does only one follicle usually become a mature follicle?

**33.** Which hormones are responsible for endometrial growth, for ovulation, for growth of the corpus luteum, and for the surge of LH at midcycle?

**34.** Is stimulation of secretion of GnRH by declining levels of estrogens and progesterone a positive or negative feedback effect? Explain your answer.

**35.** What causes follicles to start growing again in a new ovarian cycle?

## RETURN TO **Ryan and Megan's Story**

After agonizing over how to tell her, Ryan informs Megan about his diagnosis, and she promptly makes an appointment with the clinic and comes in for an examination. In Dr. Kincaid's office, Megan admits that she has had abdominal cramps for the past week.

"I thought it was from my new diet. I took a nutrition class last semester and decided to try a macrobiotic diet. Did that have anything with my developing this infection?" Dr. Kincaid repeats what he told Ryan and asks, "Have you had any other signs or symptoms?"

Megan thinks about it for a few moments then replies. "Well, my periods have been a little weird over the past two months. The consistency of the discharge has been different, and I've noticed discharge at times other than during my period. I thought that was just from the diet, or maybe from my last Pap smear, which I had about three months ago . . . but I guess that's not right either?" Dr. Kincaid confirms that diet isn't the likely cause and explains that they need to examine her and test the vaginal secretions.

After the examination, Dr. Kincaid informs Megan that she does have gonorrhea. She mentions that she has felt extra tired, assuming that it was due to a very busy semester, but now wondered if that might be connected.

"Definitely. The infection generally presents with fatigue and cramping. I can prescribe an antibiotic to clear up your infection, but you'll have to be more careful in the future. Megan, if you have had any sexual partners other than Ryan, they should be warned that they might have gonorrhea," replies Dr. Kincaid.

Megan seems relieved momentarily, but then something suddenly occurs to her. "Doctor, could this affect my chances of having a baby in the future? I don't know if Ryan's the one, but someday I'd like to have children."

Dr. Kincaid explains that gonorrhea can, in some cases, infect the uterine tubes. "If that happens, scarring can occur which could complicate successful fertilization and pregnancy. Once you complete the course of antibiotics, we can see if there is any permanent damage."

D. *Megan is hoping to become pregnant someday. When that day comes, in which part of her body would fertilization of her secondary oocyte likely occur?*

E. *How could damage to Megan's uterine tubes complicate fertilization or pregnancy?*

F. *The Pap smear that Megan mentions is a routine examination of the female reproductive tract through biopsy of the vagina and cervix (see the Clinical Connection entitled Papanicolaou Test found in Concept 4.3). Why is the cervix an important location to test?*

# 25.6 The zygote divides into a morula and then a blastocyst that implants in the endometrium of the uterus.

The **gestation period** (jes-TĀ-shun) is the time of development from fertilization to birth and includes both embryological and fetal development. Although our discussion will focus on weekly development, we usually think of the gestation period divided into three **trimesters**, each three calendar months long.

1. The *first trimester* is the most critical stage of development during which the rudiments of all the major organ systems appear. Throughout this stage the developing organism is the most vulnerable to the effects of drugs, radiation, and microbes.
2. The *second trimester* is characterized by the nearly complete development of organ systems. By the end of this stage, the fetus assumes distinctively human features.
3. The *third trimester* involves rapid fetal growth. Early in this stage, most of the organ systems become fully functional.

## First Week of Development

The first embryonic week of development is characterized by several significant events, including fertilization, cleavage of the zygote, blastocyst formation, and implantation.

### Fertilization

Once sperm and a secondary oocyte have developed through meiosis and matured, and sperm have been deposited in the vagina, fertilization can occur. During **fertilization** (fer′-ti-li-ZĀ-shun; *fertil-* = fruitful), the genetic material from a *haploid* sperm cell and a *haploid secondary* oocyte merges into a single *diploid* nucleus. Of the approximate 200 million sperm introduced into the vagina, fewer than 2 million (1 percent) reach the cervix of the uterus and only about 200 reach the secondary oocyte. Fertilization normally occurs in the uterine tube within 12 to 24 hours after ovulation. Sperm can remain viable for about 48 hours after deposition in the vagina, although a secondary oocyte is viable for only about 24 hours after ovulation. Thus, pregnancy is *most likely* to occur if intercourse takes place during a three-day "window"—from two days before ovulation to one day after ovulation.

Sperm swim from the vagina into the cervical canal by the whiplike movements of their tails. The passage of sperm through the rest of the uterus and then into the uterine tube results mainly from contractions of the walls of these organs. Prostaglandins in semen are believed to stimulate uterine motility at the time of intercourse and to aid in the movement of sperm through the uterus and into the uterine tube. Sperm that reach the vicinity of the oocyte within minutes after ejaculation *are not capable* of penetrating it until about seven hours later. During this time in the female reproductive tract, mostly in the uterine tube, sperm undergo **capacitation** (ka-pas′-i-TĀ-shun; *capacit-* = capable of), a series of changes in sperm that make them capable of penetrating the oocyte. During capacitation the sperm cell's tail begins to beat even more vigorously and secretions of the female reproductive tract weaken the plasma membrane around the head of the sperm cell.

For fertilization to occur, a sperm cell first must penetrate the corona radiata surrounding the oocyte as well as the *zona pellucida*, the glycoprotein layer between the corona radiata and the oocyte's plasma membrane (Figure 25.27a). The helmet-like *acrosome* that covers the head of a sperm contains several enzymes (see Figure 25.6). Acrosomal enzymes and strong tail

**FIGURE 25.27 Selected structures and events in fertilization.**

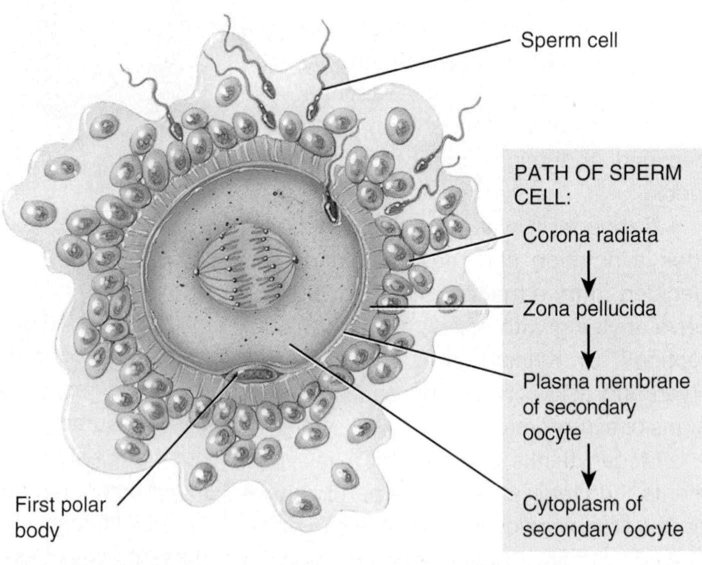


(a) Sperm cell penetrating secondary oocyte

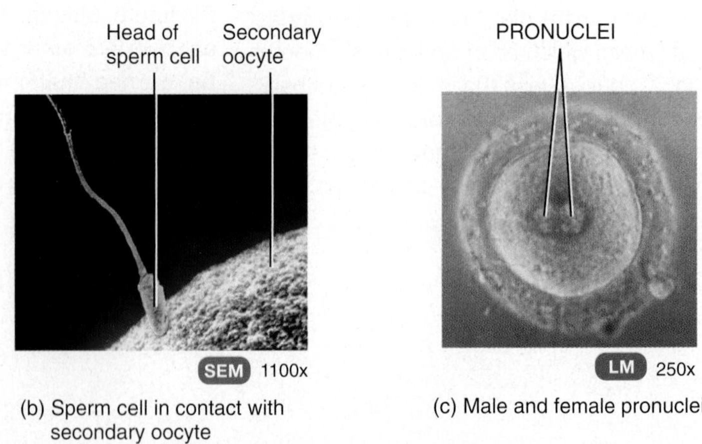


(b) Sperm cell in contact with secondary oocyte

(c) Male and female pronuclei

**During fertilization, genetic material from a sperm cell and a secondary oocyte merge to form a single diploid nucleus.**

movements help sperm penetrate the corona radiata and come in contact with the zona pellucida. One of the glycoproteins in the zona pellucida binds to the sperm cell's head, triggering the **acrosomal reaction**, the release of the contents of the acrosome. The acrosomal enzymes digest a path through the zona pellucida as the lashing tail pushes the sperm cell onward. Although many sperm bind to the zona pellucida and release their acrosomal enzymes, only the first sperm cell to penetrate the entire zona pellucida and reach the oocyte's plasma membrane fuses with the oocyte.

The fusion of a sperm cell with a secondary oocyte sets in motion events that block **polyspermy**, penetration of the oocyte by more than one sperm cell. Within a few seconds, the cell membrane of the oocyte depolarizes, which acts as a *fast block to polyspermy*—a depolarized oocyte cannot fuse with another sperm cell. Depolarization also triggers exocytosis from the oocyte of molecules that harden the zona pellucida, preventing additional sperm from binding to or penetrating it, events called the *slow block to polyspermy*.

Once a sperm cell penetrates a secondary oocyte, the oocyte completes meiosis II, dividing into an ovum and a second polar body that disintegrates (see Figure 25.15). The nucleus in the head of the sperm cell develops into the **male pronucleus** (prō-NOO-klē-us; plural is *pronuclei*), and the nucleus of the ovum develops into the **female pronucleus** (Figure 25.27c). Fertilization occurs as the male and female pronuclei fuse, producing a single diploid nucleus that contains 23 chromosomes from each pronucleus. Thus, the fusion of the haploid ($n$) pronuclei restores the diploid number ($2n$) of 46 chromosomes. The fertilized ovum now is called a zygote.

**Dizygotic (fraternal) twins** are produced from the independent release of two secondary oocytes and the subsequent fertilization of each by different sperm. Dizygotic twins are the same age and are in the uterus at the same time, but genetically they are as dissimilar as any other siblings. Dizygotic twins may or may not be the same sex. Because **monozygotic (identical) twins** develop from a single fertilized ovum, they contain exactly the same genetic material and are always the same sex. Monozygotic twins arise from separation of the developing cells into two embryos, which in 99 percent of the cases occurs before eight days have passed. Separations that occur later than eight days are likely to produce **conjoined twins**, a situation in which the twins are physically joined together and share some body structures.

## Early Embryonic Development

From fertilization through the eighth week of development the developing human is called an **embryo** (EM-brē-ō; *em-* = into; *-bryo* = grow). The embryo begins its development as the single-celled zygote.

After fertilization, a series of rapid mitotic cell divisions of the zygote called **cleavage** (KLĒV-ij) takes place (Figure 25.28). The first division of the zygote begins about 24 hours after fertilization and is completed about 6 hours later. Each succeeding division takes slightly less time and produces progressively smaller cells called **blastomeres** (BLAS-tō-merz; *blasto-* = germ; *-meres* = parts). By the second day after fertilization, the second cleavage is completed and there are four blastomeres (Figure 25.28b). By the end of the third day, there are 16 blastomeres. Successive cleavages eventually produce a solid sphere of blastomeres called the **morula** (MOR-ū-la; *morula* = mulberry). The morula is still surrounded by the zona pellucida and is about the same size as the original zygote (Figure 25.28c).

By the end of the fourth day, the number of cells in the morula increases as it continues to move through the uterine tube toward the uterine cavity. When the morula enters the uterine cavity on day 4 or 5, a glycogen-rich secretion from the glands of the uterine

**FIGURE 25.28 Cleavage and the formation of the morula and blastocyst.**

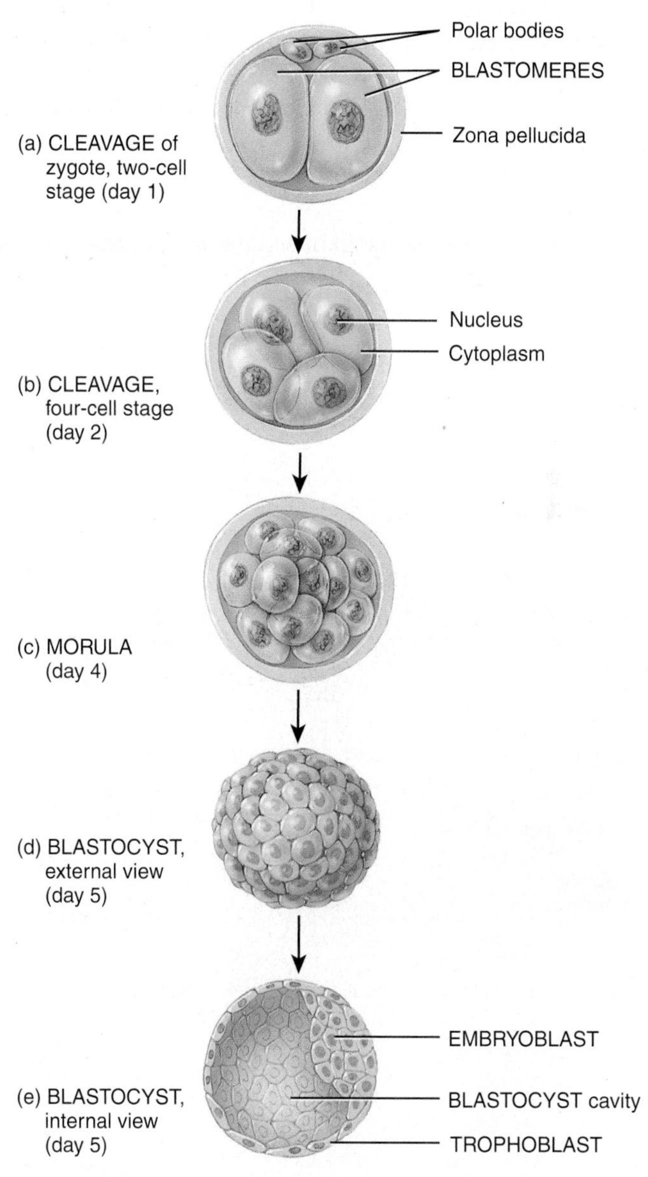


**Cleavage refers to the early, rapid mitotic divisions of a zygote.**

endometrium passes into the uterine cavity and enters the morula through the zona pellucida. The morula obtains nourishment from this endometrial fluid and from nutrients stored in the cytoplasm of its blastomeres. At the 32-cell stage, the fluid accumulates internally in a large fluid-filled cavity (Figure 25.28e). Once the blastocyst cavity is formed, the developing hollow sphere is called the **blastocyst** (BLAS-tō-sist; *blasto-* = germ or sprout; *-cyst* = bag). Though it now has hundreds of cells, the blastocyst is still about the same size as the original zygote.

During the formation of the blastocyst, two distinct cell populations arise: the embryoblast and trophoblast (Figure 25.28e). The **embryoblast** is located internally and eventually develops into the embryo. The **trophoblast** (TRŌF-ō-blast; *tropho-* = develop or nourish) is the outer superficial layer of cells that forms the sphere-like wall of the blastocyst. It will ultimately develop into the chorion, which later becomes the *fetal* part of the placenta, the site of exchange of nutrients and wastes between the mother and fetus. On about the fifth day after fertilization, the blastocyst "hatches" from the zona pellucida by digesting a hole in it with an enzyme, and then squeezing through the hole. This shedding of the zona pellucida is necessary to permit the next step, implantation.

## Implantation

The blastocyst remains free within the uterine cavity for about two days before it attaches to the uterine wall. About six days after fertilization, the blastocyst loosely attaches to the endometrium to begin **implantation**, attachment of the blastocyst to the endometrium (Figure 25.29). As the blastocyst implants, usually in either the fundus or body of the uterus, it orients with the embryoblast closest to the endometrium (Figure 25.29b). About seven days after fertilization, the blastocyst attaches to the endometrium more firmly. At the point of attachment, the endometrium quickly enlarges, its endometrial glands enlarge, and additional blood vessels develop. The blastocyst eventually secretes enzymes, burrows into the endometrium, and becomes surrounded by it.

The endometrium surrounding the implanted blastocyst is known as the **decidua** (dē-SID-ū-a = falling off). The decidua separates from the endometrium after the fetus is delivered, much as the stratum basalis does in normal menstruation. The decidua provides large amounts of glycogen and lipids for the developing embryo and fetus and later becomes the *maternal* part of the placenta.

The major events associated with the first week of development are summarized in Figure 25.30.

**FIGURE 25.29** **Relationship of a blastocyst to the endometrium of the uterus at the time of implantation.**

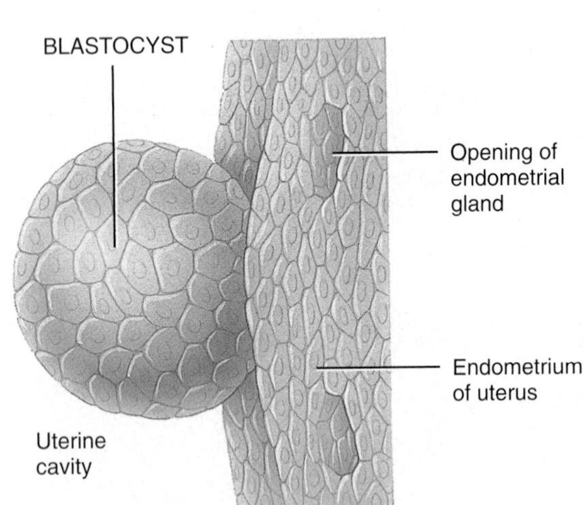


(a) External view of blastocyst, about six days after fertilization

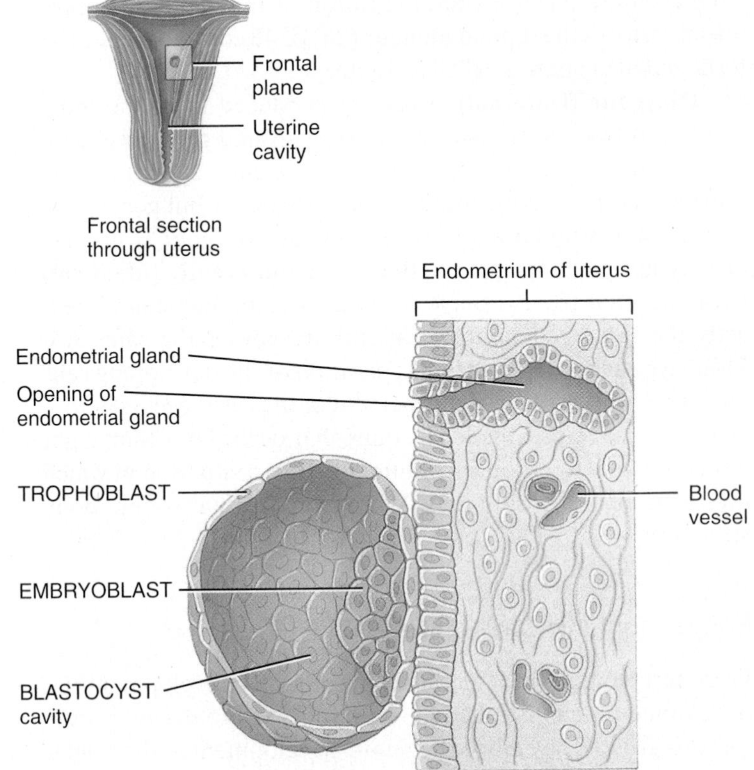


(b) Frontal section through endometrium of uterus and blastocyst, about six days after fertilization

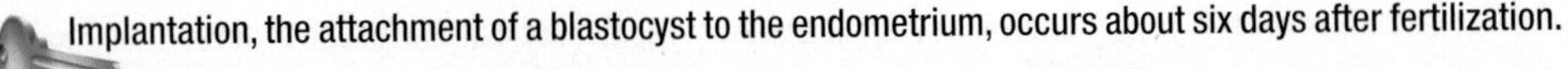

**Implantation, the attachment of a blastocyst to the endometrium, occurs about six days after fertilization.**

FIGURE 25.30 **Summary of events associated with the first week of development.**

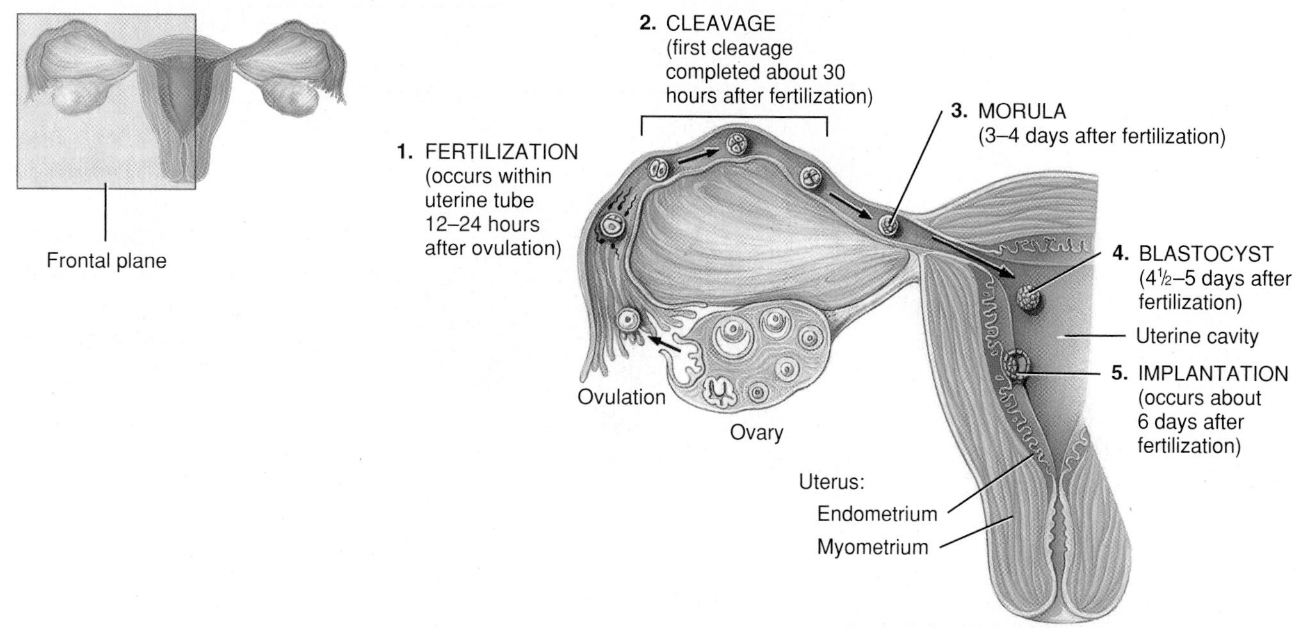


**After ovulation, a secondary oocyte and its corona radiata move from the pelvic cavity into the uterine tube, where fertilization usually occurs.**

## ✓ CHECKPOINT

**36.** Which mechanisms move sperm from the vagina to the uterine tube?

**37.** How does a sperm cell penetrate the covering of the secondary oocyte?

**38.** How is polyspermy prevented?

**39.** What is a morula and how is it formed?

**40.** Describe the layers of a blastocyst and their eventual fates.

**41.** In which phase of the uterine cycle does implantation occur? Where does it usually occur? How is the blastocyst oriented at implantation?

**42.** What are the components of the decidua? Which part of the decidua helps form the maternal part of the placenta?

# 25.7 Major tissues and organs develop during embryonic development and grow and differentiate during fetal development.

## Second Week of Development

### Development of the Trophoblast

About eight days after fertilization, the trophoblast develops into two layers in the region of contact between the blastocyst and endometrium. The cells of the outer layer, the **syncytiotrophoblast** (sin-sīt′-ē-ō-TRŌF-ō-blast), contain no distinct cell boundaries (Figure 25.31a). The layer closest to the embryoblast is the **cytotrophoblast** (sī-tō-TRŌF-ō-blast) and is composed of cells with distinct boundaries. The two layers of trophoblast become part of the chorion as they undergo further growth (see Figure 25.32a, b). During implantation, the syncytiotrophoblast secretes enzymes that digest and liquefy the endometrial cells, allowing the blastocyst to penetrate the uterine lining. Eventually, the blastocyst becomes buried in the endometrium and inner one-third of the myometrium. Another secretion of the trophoblast is *human chorionic gonadotropin* (*hCG*), which rescues the corpus luteum from degeneration, allowing it to continue secreting progesterone and estrogens. Progesterone and estrogens, in turn, maintain the stability of the uterine lining, preventing menstruation and sustaining the pregnancy. Peak secretion of hCG occurs about the ninth week of pregnancy, at which time the placenta is fully developed and assumes the function of producing progesterone and estrogens directly.

**FIGURE 25.31 Principal events of the second week of development.**

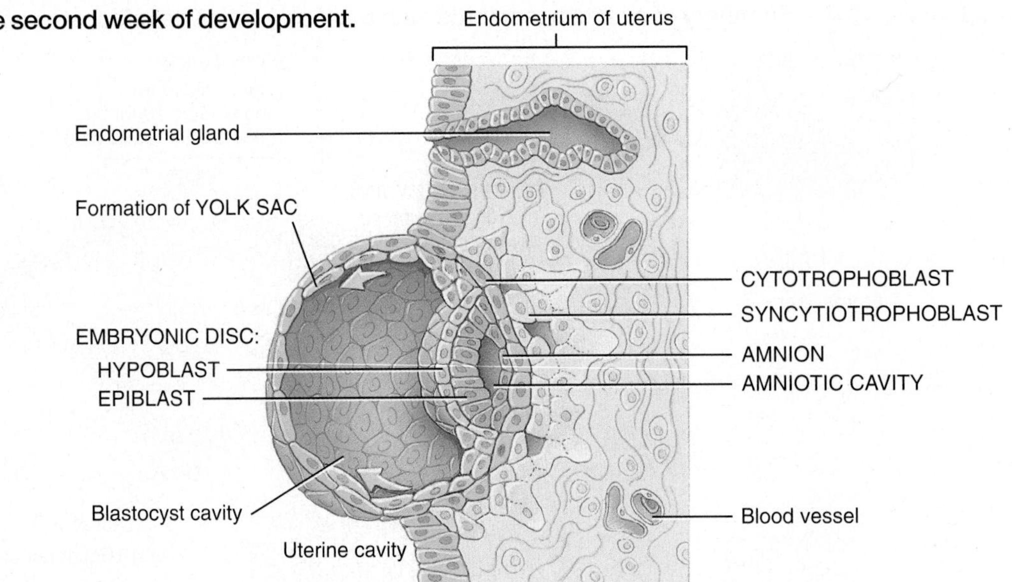


(a) Frontal section through endometrium of uterus showing blastocyst, about 8 days after fertilization

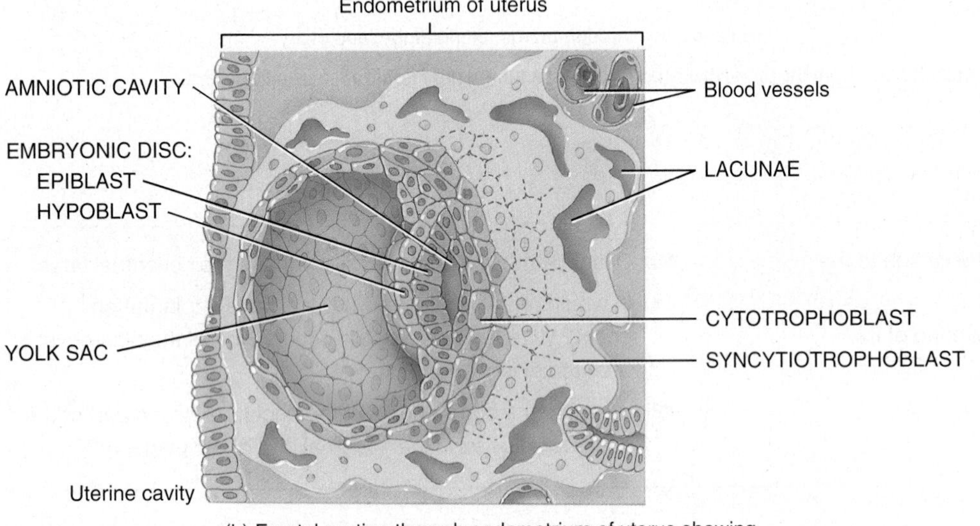


(b) Frontal section through endometrium of uterus showing blastocyst, about 9 days after fertilization

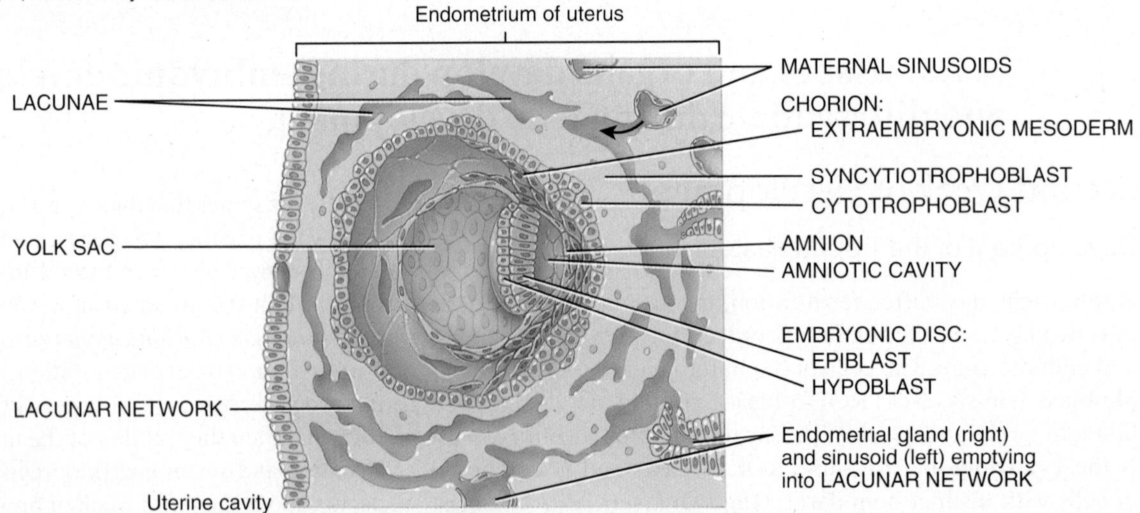


(c) Frontal section through endometrium of uterus showing blastocyst, about 12 days after fertilization

**About eight days after fertilization, the trophoblast develops into a syncytiotrophoblast and a cytotrophoblast; the embryoblast develops into an embryonic disc.**

### Development of the Embryonic Disc

Like those of the trophoblast, cells of the embryoblast also differentiate into two layers around eight days after fertilization: a **hypoblast** and **epiblast** (Figure 25.31a). Cells of the hypoblast and epiblast together form a flat disc referred to as the **embryonic disc**. Soon, a small cavity appears within the epiblast and eventually enlarges to form the **amniotic cavity** (am′-nē-OT-ik; *amnio-* = lamb).

### Development of the Amnion

As the amniotic cavity enlarges, a single cell layer of the epiblast forms a domelike roof above the epiblast cells called the **amnion** (AM-nē-on) (Figure 25.31a). Thus, the amnion forms the roof of the amniotic cavity, and the epiblast forms the floor. Initially, the amnion overlies only the embryonic disc. However, as the embryo develops, the amnion eventually surrounds the entire embryo (see Figure 25.33), creating the amniotic cavity that becomes filled with **amniotic fluid**. Most amniotic fluid is initially derived from maternal blood. Later, the fetus contributes to the fluid by excreting urine into the amniotic cavity. Amniotic fluid serves as a shock absorber for the fetus and helps regulate fetal body temperature, prevents the fetus from drying out, and prevents adhesions between the skin of the fetus and surrounding tissues. The amnion usually ruptures just before birth; it and its fluid constitute the "bag of waters." Embryonic cells are normally sloughed off into amniotic fluid. They can be examined in a procedure called *amniocentesis*, which involves withdrawing some of the amniotic fluid that bathes the developing fetus and analyzing the fetal cells and dissolved substances (see the Chapter 25 WileyPLUS Clinical Connection entitled Amniocentesis).

### Development of the Yolk Sac

Also on the eighth day after fertilization, cells at the edge of the hypoblast migrate and cover the inner surface of the blastocyst wall (Figure 25.31a). The migrating cells form a thin membrane that, together with the hypoblast, becomes the wall of the **yolk sac**, the former blastocyst cavity (Figure 25.31b). As a result, the embryonic disc is now positioned between the amniotic cavity and yolk sac.

In most animals the yolk sac is a major source of nutrients for the embryo, but since human embryos receive their nutrients from the endometrium, our yolk sac is relatively empty, small, and decreases in size as development progresses (see Figure 25.33a inset). Nevertheless, the yolk sac has several important functions in humans. It supplies nutrients to the embryo during the second and third weeks of development, is the source of embryonic blood cells, contains the first (primordial) germ cells that will migrate into the developing gonads and eventually form gametes, and forms part of the gastrointestinal (GI) tract. Finally, the yolk sac functions as a shock absorber and helps prevent the embryo from drying out.

### Development of Maternal Sinusoids

On the ninth day after fertilization, the blastocyst becomes completely embedded in the endometrium. As the syncytiotrophoblast expands, small spaces called **lacunae** (la-KOO-nē = little lakes) develop within it (Figure 25.31b). By the 12th day of development, the lacunae fuse to form larger, interconnecting spaces called **lacunar networks** (Figure 25.31c). Endometrial capillaries around the developing embryo become dilated and are referred to as **maternal sinusoids**. As the syncytiotrophoblast invades the uterine wall, some of the maternal sinusoids and endometrial glands erode. The resulting flow of maternal blood and endometrial gland secretions into the lacunar networks serves as both a rich source of materials for embryonic nutrition and a disposal site for the embryo's wastes.

### Development of the Chorion

Also about the 12th day after fertilization, the **extraembryonic mesoderm** develops from the yolk sac and forms a connective tissue layer around the amnion and yolk sac (Figure 25.31c). The extraembryonic mesoderm, cytotrophoblast, and syncytiotrophoblast form the **chorion** (KOR-ē-on = membrane) (Figure 25.31c). The chorion surrounds the embryo and, later, the fetus (see Figure 25.33). Eventually the chorion becomes the principal embryonic part of the placenta, the structure for exchange of materials between mother and fetus. The chorion also protects the embryo and fetus from the immune responses of the mother in two ways: (1) It secretes proteins that block antibody production by the mother; and (2) it promotes the production of T lymphocytes that suppress the normal immune response in the uterus. Finally, the chorion produces hCG, an important hormone of pregnancy (see Figure 25.35).

By the end of the second week of development, the embryonic disc becomes connected to the trophoblast by a band of extraembryonic mesoderm called the **connecting stalk** (see Figure 25.32). The connecting stalk is the future umbilical cord.

## Third Week of Development

The third embryonic week begins a six-week period of rapid development and differentiation. During the third week, the three primary germ layers (ectoderm, mesoderm, and endoderm) are established and lay the groundwork for organ development in weeks 4 through 8.

### Development of the Primary Germ Layers

The first major event of the third week of development, **gastrulation** (gas′- troo-LĀ-shun), occurs about 15 days after fertilization. In this process, a third layer of cells forms between the epiblast and hypoblast of the embryonic disc. These three layers, now called the endoderm, mesoderm, and, ectoderm, are referred to collectively as the primary germ layers. The **primary germ layers** are the major embryonic tissues from which the various tissues and organs of the body develop.

**TABLE 25.3**

Structures Produced by the Three Primary Germ Layers

| ENDODERM | MESODERM | ECTODERM |
|---|---|---|
| Epithelial lining of gastrointestinal tract (except oral cavity and anal canal) and epithelium of its glands<br>Epithelial lining of urinary bladder, gallbladder, and liver<br>Epithelial lining of pharynx, auditory tubes, tonsils, tympanic (middle ear) cavity, larynx, trachea, bronchi, and lungs<br>Epithelium of thyroid gland, parathyroid glands, pancreas, and thymus<br>Epithelial lining of prostate and bulbourethral glands, vagina, vestibule, urethra, and associated glands such as greater vestibular and lesser vestibular glands<br>Gametes (sperm and oocytes) | All skeletal and cardiac muscle tissue and most smooth muscle tissue<br>Cartilage, bone, and other connective tissues<br>Blood, red bone marrow, and lymphatic tissue<br>Blood vessels and lymphatic vessels<br>Dermis of skin<br>Fibrous tunic and vascular tunic of eye<br>Mesothelium of thoracic, abdominal, and pelvic cavities<br>Kidneys and ureters<br>Adrenal cortex<br>Gonads and genital ducts (except germ cells)<br>Dura mater | All nervous tissue<br>Epidermis of skin<br>Hair follicles, arrector pili muscles, nails, epithelium of skin glands (sebaceous and sudoriferous), and mammary glands<br>Lens, cornea, and internal eye muscles<br>Internal and external ear<br>Epithelium of sense organs<br>Epithelium of oral cavity, nasal cavity, paranasal sinuses, salivary glands, and anal canal<br>Epithelium of pineal gland, pituitary gland, and adrenal medullae<br>Melanocytes (pigment cells)<br>Almost all skeletal and connective tissue components of head<br>Arachnoid mater and pia mater |

As the embryo develops, the **endoderm** (*endo-* = inside; *-derm* = skin) ultimately becomes the epithelial lining of the gastrointestinal tract, respiratory tract, and several other organs. The **mesoderm** (*meso-* = middle) gives rise to muscles, bones, and other connective tissues, and the peritoneum. The **ectoderm** (*ecto-* = outside) develops into the epidermis of the skin and the nervous system. Table 25.3 provides further details about the fates of the primary germ layers.

## Development of the Cardiovascular System

At the beginning of the third week, blood vessels begin to form in the extraembryonic mesoderm of the yolk sac, connecting stalk, and chorion. This early development is necessary because the nutrients in the yolk sac and ovum are insufficient for adequate nutrition of the rapidly developing embryo. Cells of the extraembryonic mesoderm form an extensive system of blood vessels throughout the embryo. Other mesodermal cells develop into pluripotent stem cells that form the formed elements of the blood. The heart forms from mesoderm on days 18 and 19. Initially, a single primitive heart tube develops. By the end of the third week, the primitive heart tube bends on itself, becomes S-shaped, and begins to beat. It then joins blood vessels in other parts of the embryo, connecting stalk, chorion, and yolk sac to form a primitive cardiovascular system that circulates blood.

## Development of the Chorionic Villi

By the end of the second week of development, **chorionic villi** (ko-rē-ON-ik VIL-ī) begin to develop. These fingerlike projections of the chorion project into the endometrial wall of the uterus (Figure 25.32a). By the end of the third week, embryonic

FIGURE 25.32 Development of chorionic villi.

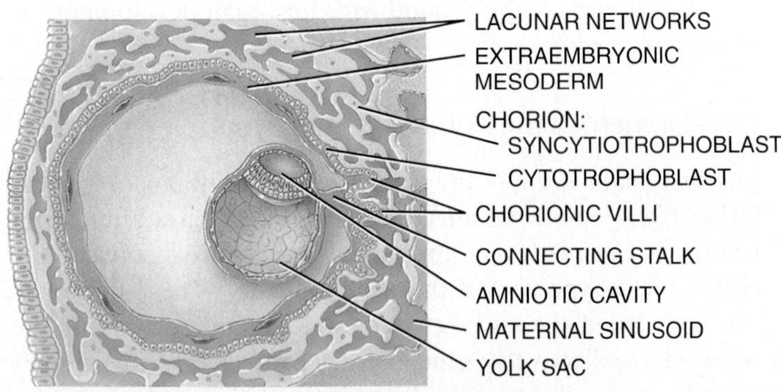


(a) Frontal section through uterus showing blastocyst, about 13 days after fertilization

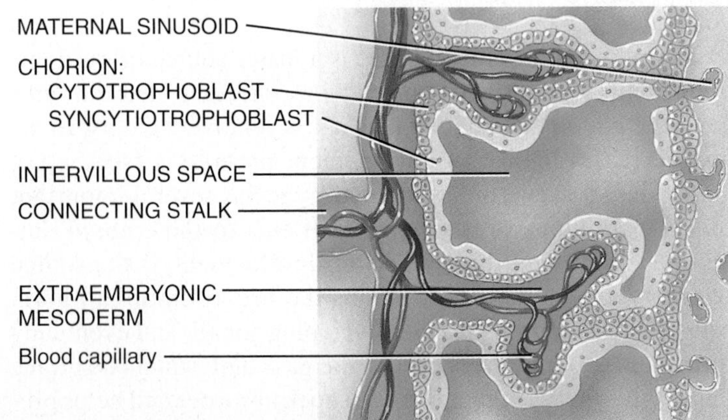


(b) Details of two chorionic villi, about 21 days after fertilization

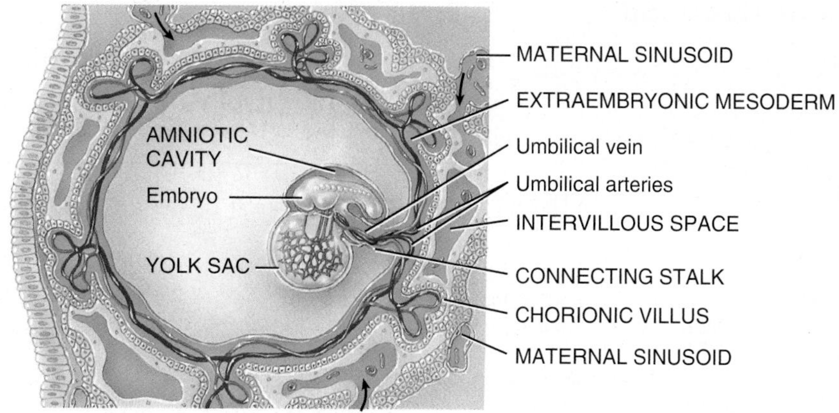


(c) Frontal section through uterus showing an embryo and its vascular supply, about 21 days after fertilization

**Blood vessels in chorionic villi connect to the embryonic heart via the umbilical arteries and umbilical vein.**

blood capillaries develop in the chorionic villi (Figure 25.32b). Blood vessels from the chorionic villi connect to the embryonic heart by way of the umbilical arteries and umbilical vein within the connecting stalk (Figure 25.32c). The fetal blood capillaries within the chorionic villi project into lacunae, which unite to form the **intervillous spaces** between chorionic villi that bathe the chorionic villi with maternal blood. As a result, maternal and fetal blood vessels are in close proximity. Note, however, that maternal and fetal blood vessels do not join, and the blood they carry does not normally mix. Instead, oxygen and nutrients in the mother's blood diffuse from the intervillous spaces into the capillaries of the villi. Waste products, such as carbon dioxide, diffuse in the opposite direction into the mother's blood.

### Development of the Placenta and Umbilical Cord

The **placenta** (pla-SEN-ta = flat cake) is the site of exchange of nutrients and wastes between the mother and fetus.

During the third week, the placenta begins its unique development from two separate individuals, the mother and the fetus. By the beginning of the twelfth week, the placenta has two distinct parts: (1) the *fetal portion* formed by the chorionic villi of the chorion and (2) the *maternal portion* formed by the decidua basalis of the endometrium (Figure 25.33a). When fully developed, the placenta is shaped like a pancake (Figure 25.33b).

Functionally, the placenta allows oxygen and nutrients to diffuse from maternal blood into fetal blood while carbon dioxide and wastes diffuse from fetal blood into maternal blood. The placenta also is a protective barrier because most microorganisms cannot pass through it. However, certain viruses, such as those that cause AIDS, German measles, chickenpox, measles, encephalitis, and poliomyelitis, can cross the placenta. Many drugs, alcohol, and some substances that can cause birth defects also pass freely. The placenta stores nutrients such as carbohydrates, proteins, calcium, and iron, which are released into fetal circulation as required. The placenta also produces hormones needed to sustain the pregnancy (see Figure 25.35).

The actual connection between the placenta and embryo, and later the fetus, is through the **umbilical cord** (um-BIL-i-kul = navel), which develops from the connecting stalk and is usually about 50–60 cm (20–24 in.) long. The umbilical cord contains two umbilical arteries that carry deoxygenated embryonic blood to the placenta and into maternal blood, and one umbilical vein that carries oxygen and nutrients acquired from the mother's intervillous spaces into the fetus (Figure 25.33b). A layer of amnion surrounds the entire umbilical cord and gives it a shiny appearance. In some cases, the umbilical vein is used to transfuse blood into a fetus or to introduce drugs for various medical treatments.

In about 1 in 200 newborns, only one of the two umbilical arteries is present in the umbilical cord. It may be due to failure of the artery to develop or degeneration of the vessel early in development. Nearly 20 percent of infants with this condition develop cardiovascular defects.

After the birth of the baby, the placenta detaches from the uterus and is therefore termed the **afterbirth**. At this time, the umbilical cord is tied off and then severed. The small portion (about an inch) of the cord that remains attached to the infant begins to wither and falls off, usually within 12 to 15 days after birth. The area where the cord was attached becomes covered by a thin layer of skin, and scar tissue forms. The scar is the *umbilicus* (navel).

Pharmaceutical companies use human placentas as a source of hormones, drugs, and blood; portions of placentas are also used for burn coverage. The placental and umbilical cord veins can also be used in blood vessel grafts, and cord blood can be frozen to provide a future source of pluripotent stem cells, for example, to repopulate red bone marrow following radiotherapy for cancer.

## Fourth through Eighth Weeks of Development

The fourth through eighth weeks of development are very significant in embryonic development because all major organs

FIGURE 25.33 **Placenta and umbilical cord.**

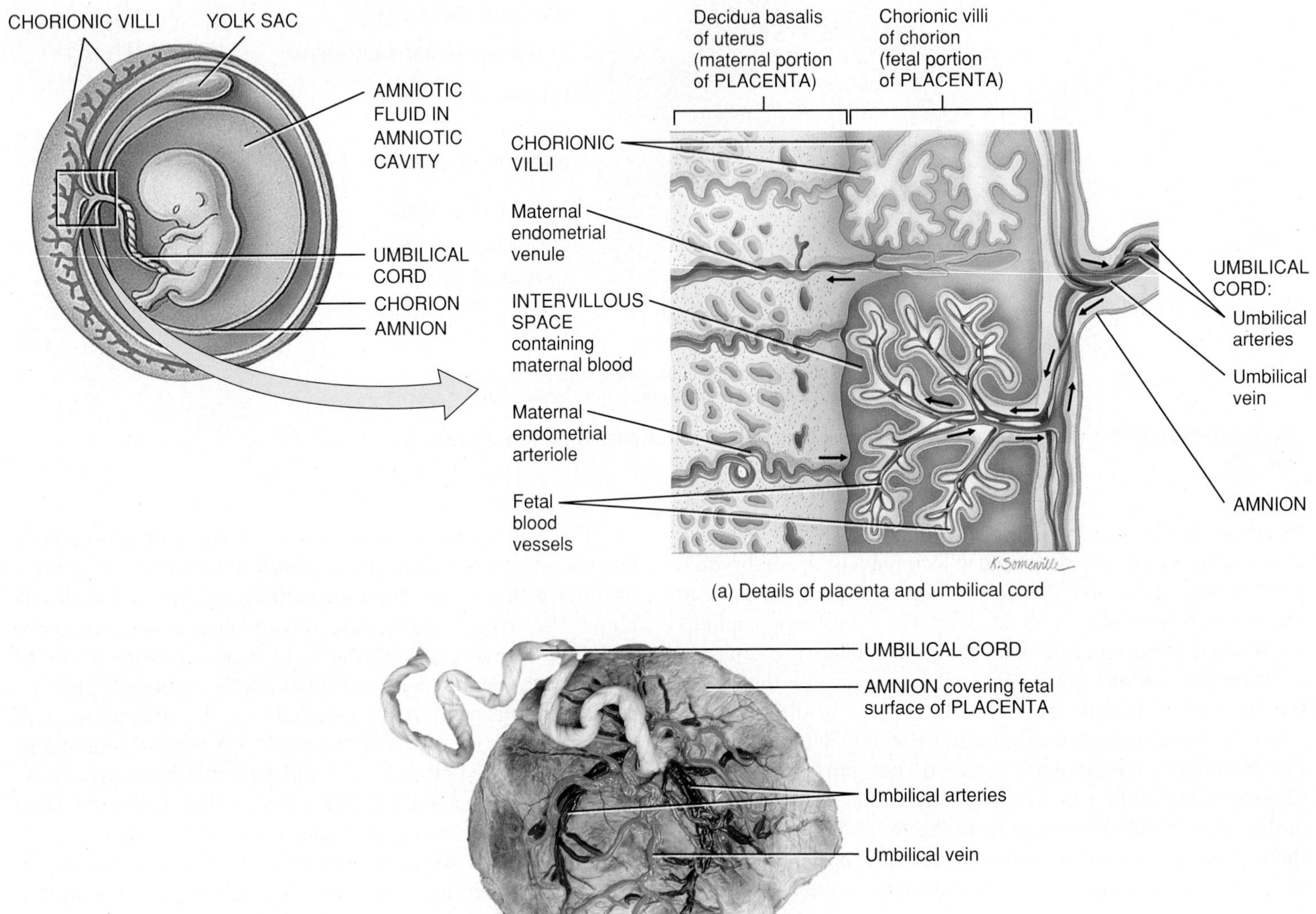


(a) Details of placenta and umbilical cord

(b) Fetal surface of placenta

**The placenta is formed by the chorionic villi of the embryo and the decidua basalis of the endometrium of the mother.**

appear during this time. By the end of the eighth week, all of the major body systems have begun to develop, although their functions for the most part are minimal.

During the fourth week after fertilization, the embryo undergoes very dramatic changes in shape and size, nearly tripling in size. It is essentially converted from a flat, two-dimensional embryonic disc to a three-dimensional cylinder consisting of endoderm in the center (gut), ectoderm on the outside (epidermis), and mesoderm in between. Due to the different rates of growth of various parts of the embryo, the embryo curves into a C-shape. At the beginning of the fourth week, the first sign of the developing ears (otic placodes) and eyes (lens placodes) can be distinguished. By the middle of the fourth week, the upper limbs begin their development as outgrowths of ectoderm-covered mesoderm called upper limb buds, and the lower limb buds develop by the end of the fourth week. During the fourth week the primitive GI tract also forms. At the end of the fourth week the embryo has a distinctive tail.

During the fifth week of development, there is very rapid development of the brain, so growth of the head is considerable. By the end of the sixth week, the head grows even larger relative to the trunk, and the limbs show substantial development. In addition, the neck and trunk begin to straighten, and the heart is now four-chambered. By the seventh week, the various regions of the limbs become distinct and the beginnings of fingers appear. At the start of the eighth week, the fingers are short and webbed, the tail is shorter but still visible, the eyes are open, and the auricles of the ears are visible. By the end of the eighth week, all regions of limbs are apparent; the fingers are distinct and no longer webbed. Also, the eyelids come together and may fuse, the tail disappears, and the external genitals begin to differentiate. The embryo now has clearly human characteristics.

## Ninth through Thirty-Eighth Week of Development

The developing human is called a **fetus** (FĒ-tus = offspring) from week 9 until birth. During this time, embryonic tissues and organs grow and differentiate. Very few new structures appear, but the rate of body growth of the fetus is remarkable, especially during the second half of intrauterine life. For example, during the last two and one-half months of intrauterine life, half of the full-term weight is added. At the beginning of the ninth week, the head is half the length of the body. At birth, the head size is only one quarter the length of the body. During the same period, the limbs also increase in size from one-eighth to one-half the fetal length. The fetus is also less vulnerable to the damaging effects of drugs, radiation, and microbes than it was as an embryo.

A summary of the major events during embryonic and fetal development is illustrated in Figure 25.34 and presented in Table 25.4.

FIGURE 25.34 **Developmental events of the embryo and fetus.** The embryos and fetuses are not shown at their actual sizes.

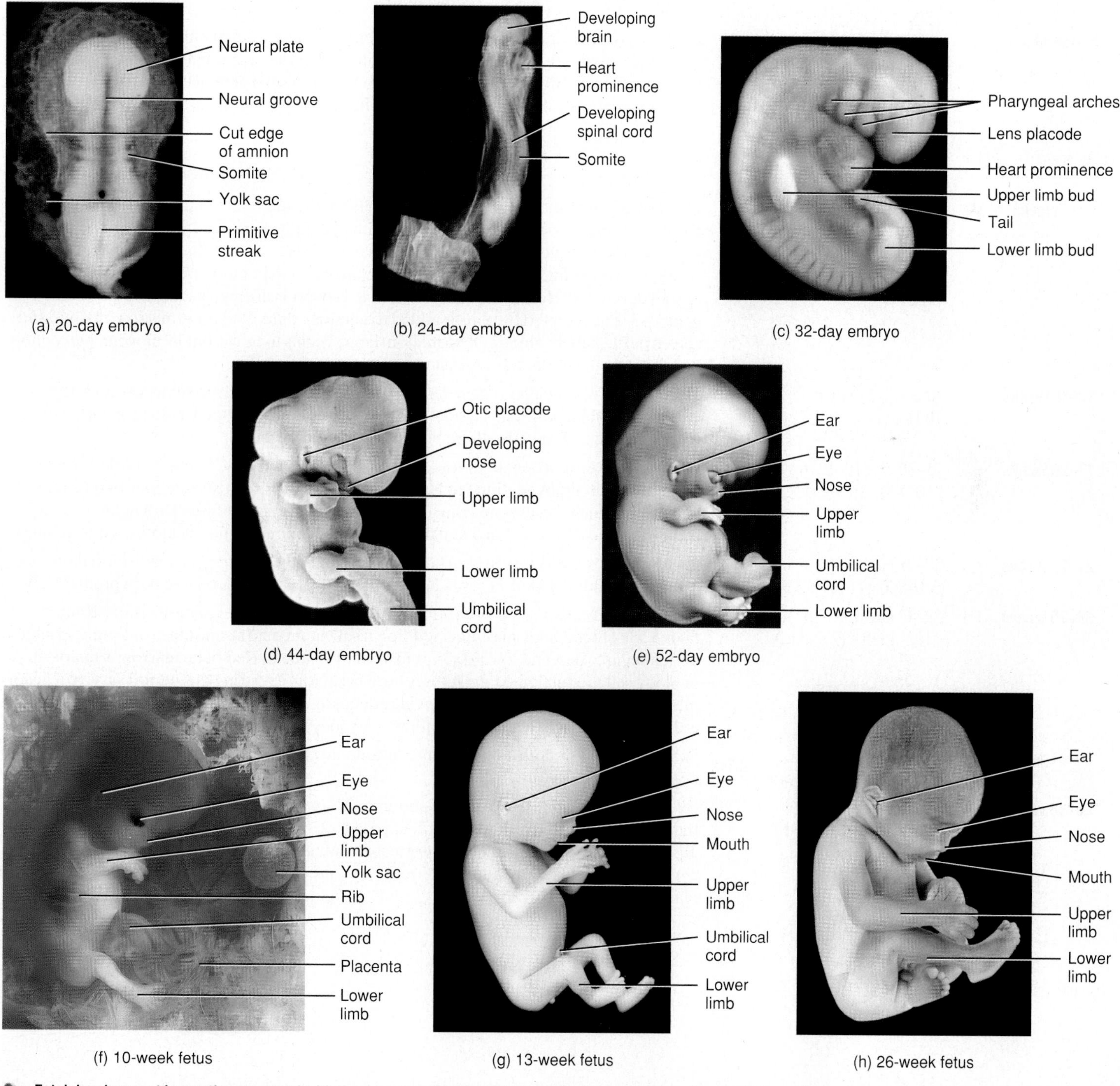


**Fetal development is mostly concerned with the growth and differentiation of tissues and organs formed in the embryo.**

**TABLE 25.4**

Changes during Embryonic and Fetal Development

| TIME | APPROXIMATE SIZE AND WEIGHT | REPRESENTATIVE CHANGES |
|---|---|---|
| **EMBRYONIC PERIOD** | | |
| **1–4 weeks** | 0.6 cm (3/16 in.) | Primary germ layers and notochord develop. Neurulation occurs. Primary brain vesicles, somites, and intraembryonic coelom develop. Blood vessel formation begins and blood forms in yolk sac, allantois, and chorion. Heart forms and begins to beat. Chorionic villi develop and placental formation begins. The embryo folds. The primitive gut, pharyngeal arches, and limb buds develop. Eyes and ears begin to develop, tail forms, and body systems begin to form. |
| **5–8 weeks** | 3 cm (1.25 in.)<br>1 g (1/30 oz) | Limbs become distinct and digits appear. Heart becomes four-chambered. Eyes are far apart and eyelids are fused. Nose develops and is flat. Face is more humanlike. Bone formation begins. Blood cells start to form in liver. External genitals begin to differentiate. Tail disappears. Major blood vessels form. Many internal organs continue to develop. |
| **FETAL PERIOD** | | |
| **9–12 weeks** | 7.5 cm (3 in.)<br>30 g (1 oz) | Head constitutes about half the length of fetal body, and fetal length nearly doubles. Brain continues to enlarge. Face is broad, with eyes fully developed, closed, and widely separated. Nose develops a bridge. External ears develop and are low set. Bone formation continues. Upper limbs almost reach final relative length but lower limbs are not quite as well developed. Heartbeat can be detected. Gender is distinguishable from external genitals. Urine secreted by fetus is added to amniotic fluid. Red bone marrow, thymus, and spleen participate in blood cell formation. Fetus begins to move, but its movements cannot be felt yet by the mother. Body systems continue to develop. |
| **13–16 weeks** | 18 cm (6.5–7 in.)<br>100 g (4 oz) | Head is relatively smaller than rest of body. Eyes move medially to final positions, and ears move to final positions on sides of head. Lower limbs lengthen. Fetus appears even more humanlike. Rapid development of body systems occurs. |
| **17–20 weeks** | 25–30 cm (10–12 in.)<br>200–450 g (0.5–1 lb) | Head is more proportionate to rest of body. Eyebrows and head hair are visible. Growth slows but lower limbs continue to lengthen. Vernix caseosa (fatty secretions of oil glands and dead epithelial cells) and lanugo (delicate fetal hair) cover fetus. Brown fat forms and is the site of heat production. Fetal movements are commonly felt by mother (quickening). |
| **21–25 weeks** | 27–35 cm (11–14 in.)<br>550–800 g (1.25–1.5 lb) | Head becomes even more proportionate to rest of body. Weight gain is substantial, and skin is pink and wrinkled. Fetuses 24 weeks and older usually survive if born prematurely. |
| **26–29 weeks** | 32–42 cm (13–17 in.)<br>1100–1350 g (2.5–3 lb) | Head and body are more proportionate and eyes are open. Toenails are visible. Body fat is 3.5% of total body mass and additional subcutaneous fat smoothes out some wrinkles. Testes begin to descend toward scrotum at 28 to 32 weeks. Red bone marrow is major site of blood cell production. Many fetuses born prematurely during this period survive if given intensive care because lungs can provide adequate ventilation and central nervous system is developed enough to control breathing and body temperature. |
| **30–34 weeks** | 41–45 cm (16.5–18 in.)<br>2000–2300 g (4.5–5 lb) | Skin is pink and smooth. Fetus assumes upside-down position. Body fat is 8% of total body mass. |
| **35–38 weeks** | 50 cm (20 in.)<br>3200–3400 g (7–7.5 lb) | By 38 weeks, circumference of fetal abdomen is greater than that of head. Skin is usually bluish-pink, and growth slows as birth approaches. Body fat is 16% of total body mass. Testes are usually in scrotum in full-term male infants. Even after birth, an infant is not completely developed; an additional year is required, especially for complete development of nervous system. |

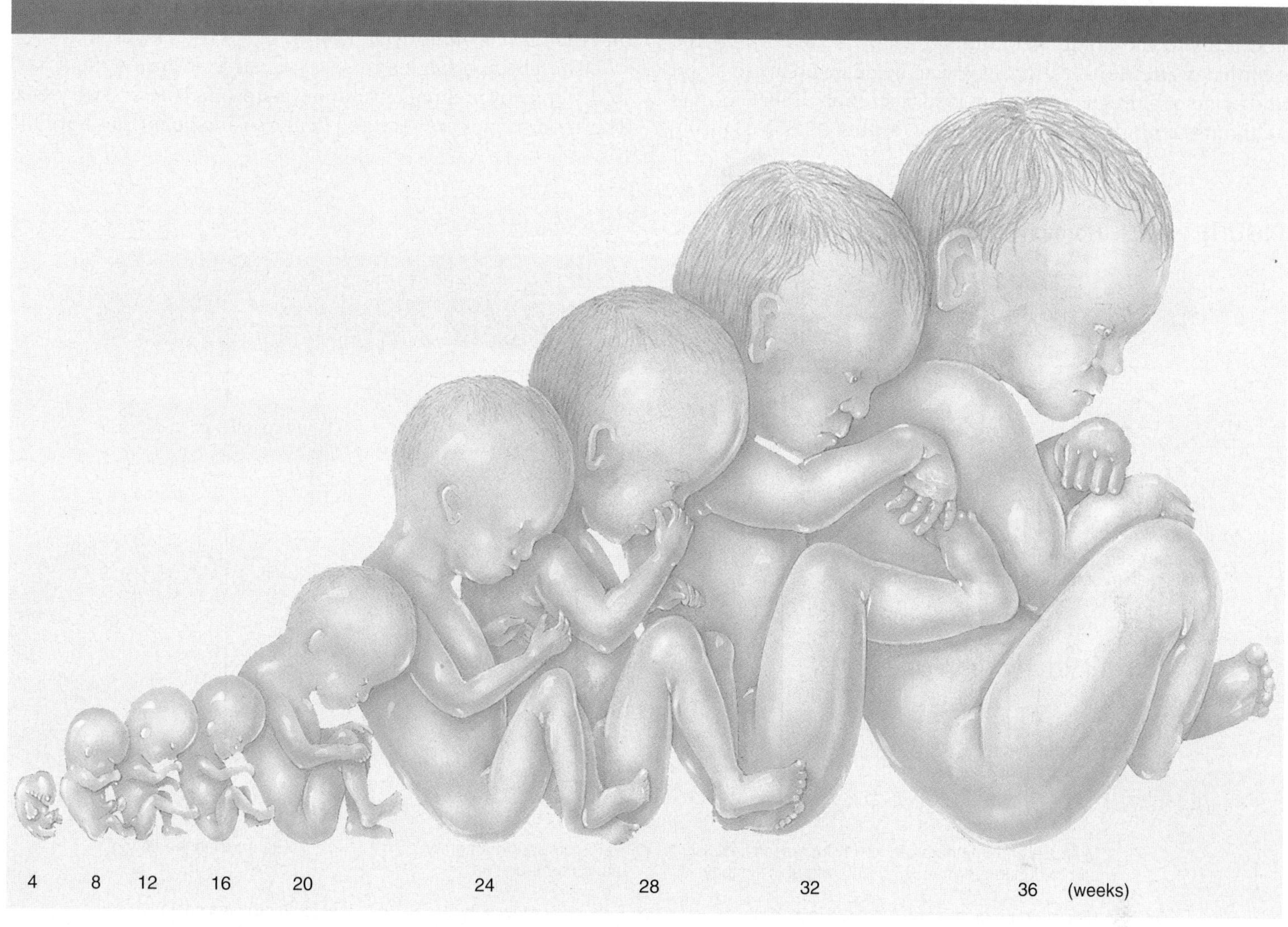


✓ **CHECKPOINT**

**43.** How does the embryo burrow into the endometrium?

**44.** How is the embryonic disc formed? How is it connected to the trophoblast?

**45.** How does the placenta develop and what is its importance?

**46.** Describe the components of the umbilical cord.

**47.** What is the afterbirth?

# 25.8 During pregnancy the uterus expands, displacing and compressing maternal organs.

## Hormones of Pregnancy

During the first three to four months of pregnancy, the corpus luteum in the ovary continues to secrete progesterone and estrogens, which maintain the lining of the uterus during pregnancy and prepare the mammary glands to secrete milk. The amounts secreted by the corpus luteum, however, are only slightly more than those produced after ovulation in a normal menstrual cycle. From the third month through the remainder of the pregnancy, the placenta itself provides the needed high levels of progesterone and estrogens. As noted previously, the chorion of the placenta secretes hCG into the blood. In turn, hCG stimulates the corpus luteum to continue production of progesterone and estrogens—an activity required to prevent menstruation and for the continued attachment of the embryo and fetus to the lining of the uterus (Figure 25.35a). By the eighth day after fertilization, hCG can be detected in the blood and urine of a pregnant woman. Peak secretion of hCG occurs at about the ninth week of pregnancy (Figure 25.35b). During the fourth and fifth months, the hCG level decreases sharply and then levels off until childbirth.

The chorion begins to secrete estrogens after the first 3 or 4 weeks of pregnancy and progesterone by the sixth week. These hormones are secreted in increasing quantities until the

**FIGURE 25.35 Hormones during pregnancy.**

(a) Sources and functions of hormones



(b) Blood levels of hormones during pregnancy

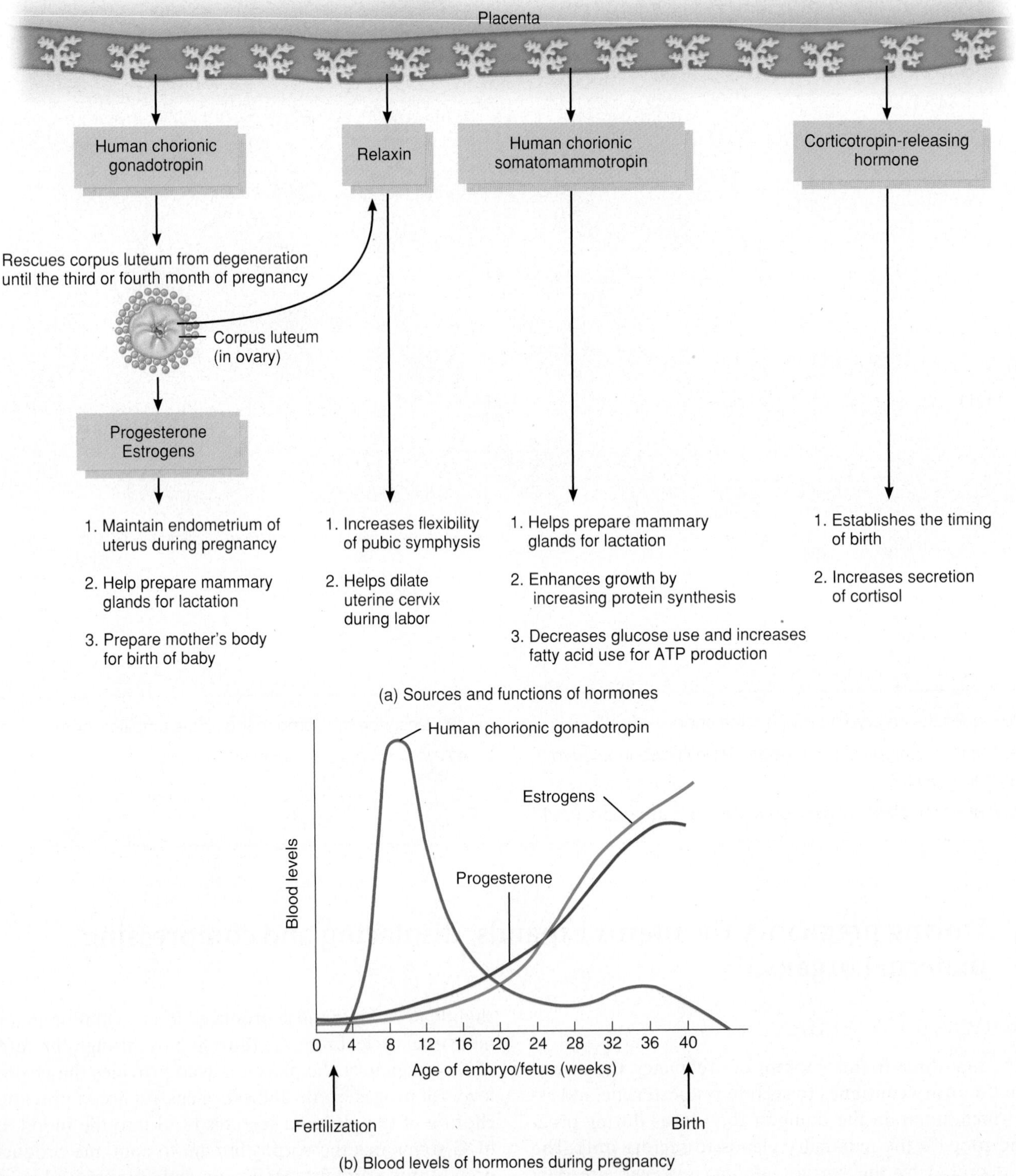

**Progesterone and estrogens are produced by the corpus luteum during the first 3–4 months of pregnancy, after which time the placenta produces them.**

time of birth (Figure 25.35b). By the fourth month, when the placenta is fully established, the secretion of hCG is greatly reduced, and the secretions of the corpus luteum are no longer essential. A high level of estrogens prepares the myometrium for labor, and progesterone ensures that the uterine myometrium is relaxed and that the cervix is tightly closed. After delivery, estrogens and progesterone in the blood decrease to normal levels.

Relaxin, a hormone produced first by the corpus luteum of the ovary and later by the placenta, increases the flexibility of the pubic symphysis and pelvic ligaments and helps dilate the uterine cervix during labor. All of these actions ease delivery of the baby.

A third hormone produced by the chorion of the placenta is **human chorionic somatomammotropin (hCS)**. The rate of secretion of hCS increases in proportion to the size of the placenta, reaching maximum levels after 32 weeks and remaining relatively constant after that. It is thought to help prepare the mammary glands for lactation, enhance maternal growth by increasing protein synthesis, and regulate certain aspects of metabolism in both mother and fetus. For example, hCS decreases the use of glucose by the mother and promotes the release of fatty acids from her adipose tissue, making more glucose available to the fetus.

The hormone most recently found to be produced by the placenta is **corticotropin-releasing hormone (CRH)**, which in nonpregnant people is secreted only by the hypothalamus. CRH is now thought to be part of the "clock" that establishes the timing of birth. Secretion of CRH by the placenta begins at about 12 weeks and increases enormously toward the end of pregnancy. Women who have higher levels of CRH earlier in pregnancy are more likely to deliver prematurely; those who have low levels are more likely to deliver after their due date. CRH from the placenta has a second important effect: It increases secretion of cortisol, which is needed for maturation of the fetal lungs and the production of surfactant to keep alveoli (air sacs) open (see Concept 22.2).

## Changes during Pregnancy

Near the end of the third month of pregnancy, the uterus occupies most of the pelvic cavity. As the fetus continues to grow, the uterus extends higher and higher into the abdominal cavity. Toward the end of a full-term pregnancy, the uterus fills nearly the entire abdominal cavity, reaching nearly to the xiphoid process of the sternum (Figure 25.36). It pushes the maternal intestines, liver, and stomach superiorly, elevates the diaphragm, and widens the thoracic cavity. In the pelvic cavity, compression of the ureters and urinary bladder occurs.

Pregnancy-induced physiological changes also occur, including weight gain due to the fetus, amniotic fluid, the placenta, uterine enlargement, and increased total body water; increased storage of proteins, triglycerides, and minerals; marked breast enlargement in preparation for lactation; and lower back pain due to lordosis (hollow back).

Several changes occur in the maternal cardiovascular system. Stroke volume and cardiac output rise due to increased maternal blood flow to the placenta and increased metabolism. Heart rate and blood volume increase, mostly during the second half of pregnancy. These increases are necessary to meet the additional demands of the fetus for nutrients and oxygen. Compression of the inferior vena cava decreases venous return, which leads to edema in the lower limbs and may produce varicose veins. Compression of the renal artery can lead to renal hypertension.

Respiratory function is also altered during pregnancy to meet the added oxygen demands of the fetus. Tidal volume (the volume of air in one breath) and minute ventilation (the volume of air inhaled and exhaled each minute) can increase, and airway resistance in the bronchial tree can decline. But residual volume (the volume of air remaining in the lungs after exhalation) can decline as the expanding uterus pushes on the diaphragm, with resulting dyspnea (*disp*-NĒ-a) (difficult breathing).

The digestive system also undergoes changes. Pregnant women experience an increase in appetite due to the added nutritional demands of the fetus. Pressure on the stomach may force the stomach contents superiorly into the esophagus, resulting in heartburn. A general decrease in GI tract motility can cause constipation, delay gastric emptying time, and produce nausea and vomiting.

Pressure on the urinary bladder by the enlarging uterus can produce urinary symptoms, such as increased frequency and urgency of urination, and stress incontinence. Greater urine ouput from the kidneys allows faster elimination of the extra wastes produced by the fetus.

Changes in the skin during pregnancy are more apparent in some women than in others. Some women experience increased pigmentation around the eyes and cheekbones in a masklike pattern, in the areolae of the breasts, and in the linea alba of the lower abdomen. **Striae** (stretch marks) over the abdomen can occur as the uterus enlarges, and hair loss increases.

Changes in the reproductive system include edema and increased vascularity of the vulva and increased pliability and vascularity of the vagina. The uterus increases from its nonpregnant mass of 60–80 g to 900–1200 g at term.

**FIGURE 25.36** **Normal fetal location and position at the end of a full-term pregnancy.**

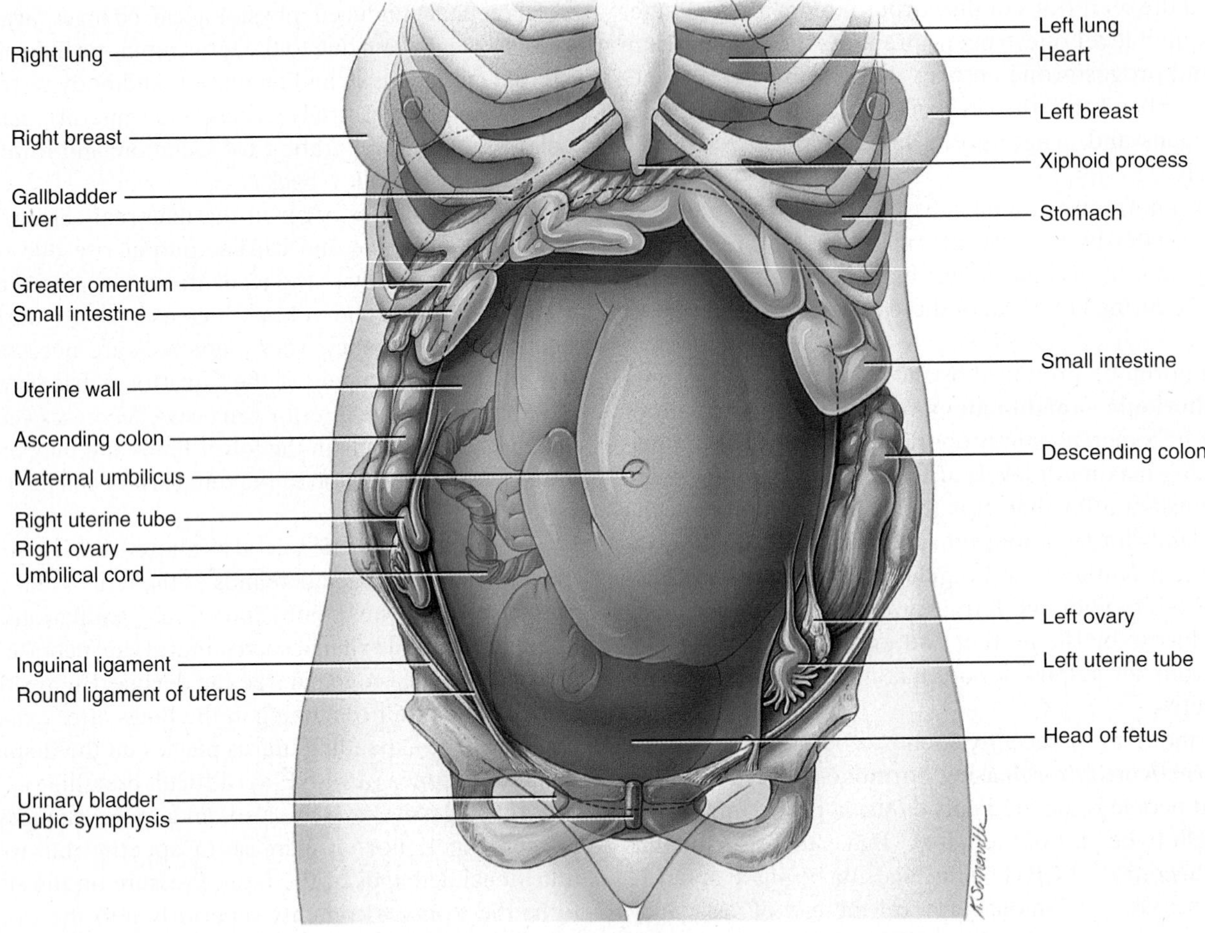


Anterior view of position of organs at end of full-term pregnancy

**The gestation period is the time interval (about 38 weeks) from fertilization to birth.**

**✓ CHECKPOINT**

**48.** What is the source of progesterone and estrogens during early pregnancy? During late pregnancy?

**49.** What is the source and function of human chorionic gonadotropin? Of human chorionic somatomammotropin?

**50.** Which hormone helps ease delivery of the baby? How?

**51.** Which hormone helps control the timing of birth?

# 25.9 Labor includes dilation of the cervix and expulsion of the fetus and placenta.

## Labor and Delivery

**Labor** is the process by which the fetus is expelled from the uterus through the vagina, and is also referred to as **parturition** (par′-too-RISH-un; *parturit-* = childbirth) or, more simply, "giving birth."

The onset of labor is determined by complex interactions of several placental and fetal hormones. Because progesterone inhibits uterine contractions, labor cannot take place until its effects are diminished. Toward the end of gestation, the levels of estrogens in the mother's blood rise sharply, producing changes that overcome the inhibiting effects of progesterone. High levels of estrogens increase the number of oxytocin receptors on uterine muscle fibers, and cause uterine muscle fibers to form gap junctions with one another. **Oxytocin** released by the posterior pituitary stimulates uterine contractions, and relaxin from the placenta assists by increasing the flexibility of the pubic symphysis and helping dilate the uterine cervix. Estrogen also stimulates the placenta to release prostaglandins, which induce production of enzymes that digest collagen fibers in the cervix, causing the cervix to soften.

Control of labor contractions occurs via a positive feedback cycle (see Figure 1.4). Contractions of the uterine myometrium force the baby's head or body into the cervix, stretching the cervical wall. Stretch receptors in the cervix send impulses to the hypothalamus, causing it to release oxytocin through the posterior pituitary. Oxytocin is then carried in the blood to the uterus, where it stimulates the myometrium to contract more forcefully. As the contractions intensify, the baby's body stretches the cervix even more. The resulting impulses from the cervix stimulate the secretion of yet more oxytocin. With birth of the infant, the positive feedback cycle is broken because cervical distension suddenly lessens once the baby has passed through the birth canal.

Uterine contractions occur in waves (quite similar to the peristaltic waves of the gastrointestinal tract) that start at the top of the uterus and move downward, eventually expelling the fetus. **True labor** begins when uterine contractions occur at regular intervals, usually producing pain. As the interval between contractions shortens, the contractions intensify. Another symptom of true labor in some women is localization of pain in the back that is intensified by walking. The most reliable indicator of true labor is dilation of the cervix and the "show," a discharge of blood-containing mucus into the cervical canal. In **false labor**, pain is felt in the abdomen at irregular intervals, but it does not intensify and walking does not alter it significantly; there is no "show" and no cervical dilation.

True labor can be divided into three stages (Figure 25.37):

1. ***Dilation stage***. The time from the onset of labor to the complete dilation of the cervix is the **dilation stage**. This stage, which typically lasts 6–12 hours, features regular contractions of the uterus, usually a rupturing of the amniotic sac, and complete dilation (to 10 cm) of the cervix. If the amniotic sac does not rupture spontaneously, it is ruptured intentionally.

2. ***Expulsion stage***. The time (10 minutes to several hours) from complete cervical dilation to delivery of the baby is the **expulsion stage**.

3. ***Placental stage***. The time (5–30 minutes or more) after delivery until the placenta or afterbirth is expelled by powerful uterine contractions is the **placental stage**. These contractions also constrict blood vessels that were torn during delivery, reducing the likelihood of hemorrhage.

As a rule, labor lasts longer with first babies, typically about 14 hours. For women who have previously given birth, the average duration of labor is about 8 hours—although the time varies enormously among births. Because the fetus may be squeezed through the birth canal (cervix and vagina) for up to several hours, the fetus is stressed during childbirth: The fetal head is compressed, and the fetus undergoes some degree of intermittent hypoxia due to compression of the umbilical cord and the placenta during uterine contractions. In response to this stress, the fetal adrenal medullae secrete very high levels of epinephrine and norepinephrine, the "fight-or-flight" hormones. Much of the protection against the stresses of parturition, as well as preparation of the infant for

FIGURE 25.37 **Stages of true labor.**

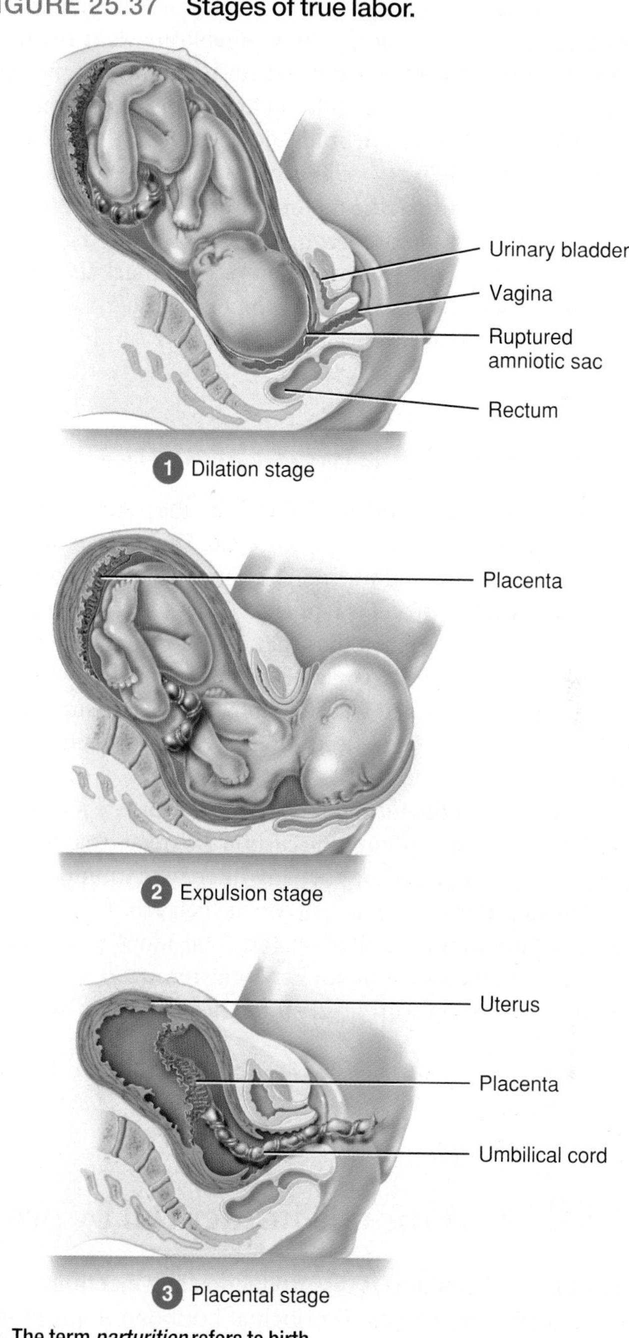


**The term *parturition* refers to birth.**

surviving extrauterine life, is provided by these hormones. Among other functions, epinephrine and norepinephrine clear the lungs and alter their physiology in readiness for breathing air, mobilize readily usable nutrients for cellular metabolism, and promote an increased flow of blood to the brain and heart.

Following the delivery of the baby and placenta is a six-week period during which the maternal reproductive organs and physiology return to the prepregnancy state. Through a process of tissue catabolism, the uterus undergoes a remarkable reduction in size, especially in lactating women. The cervix loses its elasticity and regains its prepregnancy firmness.

## Adjustments of the Infant at Birth

During pregnancy, the embryo (and later the fetus) is totally dependent on the mother for its existence. The mother supplies the fetus with oxygen and nutrients, eliminates its carbon dioxide and other wastes, protects it against shocks and temperature changes, and provides antibodies that confer protection against certain harmful microbes. At birth, the newborn's body systems must make adjustments to become much more self-supporting. The most dramatic changes occur in the respiratory and cardiovascular systems.

The reason that the fetus depends entirely on the mother for obtaining oxygen and eliminating carbon dioxide is that the fetal lungs are either collapsed or partially filled with amniotic fluid. After delivery, the baby's supply of oxygen from the mother ceases, and any amniotic fluid in the fetal lungs is absorbed. Because carbon dioxide is no longer being removed, it builds up in the blood. A rising $CO_2$ level stimulates the medulla oblongata, causing the respiratory muscles to contract, and the baby to draw his or her first breath. Because the first inspiration is unusually deep, as the lungs contain no air, the baby also exhales vigorously and naturally cries. A full-term baby may breathe 45 times a minute for the first two weeks after birth. Breathing rate gradually declines until it approaches a normal rate of 12 breaths per minute.

After the baby's first inspiration, several changes occur to the fetal heart and vessels to divert deoxygenated blood to the lungs for the first time. After the umbilical cord is tied off and severed and blood no longer flows through the umbilical arteries, they fill with connective tissue, and their distal portions become the medial umbilical ligaments. The umbilical vein then becomes the round ligament of the liver. In the fetus, the umbilical vein carries blood directly to the inferior vena cava, allowing blood from the placenta to bypass the fetal liver. When the umbilical cord is severed, venous blood from the viscera of the fetus is diverted into the hepatic portal vein to the liver and then via the hepatic vein to the inferior vena cava. At birth, an infant's pulse may range from 120 to 160 beats per minute and may go as high as 180 upon excitation.

✓ **CHECKPOINT**

**52.** Which hormonal changes induce labor?

**53.** Which hormone controls labor contractions? Why is control of labor considered a positive feedback loop?

**54.** What is the difference between false labor and true labor?

# 25.10 Lactation is influenced by prolactin, estrogens, progesterone, and oxytocin.

**Lactation** (lak′-TĀ-shun) is the production and ejection of milk from the mammary glands. A principal hormone in promoting milk production is **prolactin**, which is secreted from the anterior pituitary gland. Even though prolactin levels increase as pregnancy progresses, no milk secretion occurs because progesterone inhibits the effects of prolactin. After delivery, the levels of estrogens and progesterone in the mother's blood decrease, and the inhibition is removed. The principal stimulus in maintaining prolactin secretion during lactation is the sucking action of the infant. Suckling stimulates stretch receptors in the nipples, which stimulate the hypothalamus, resulting, in turn, in the release of more prolactin from the anterior pituitary.

Oxytocin causes release of milk into the mammary ducts via the **milk ejection reflex** (Figure 25.38). Milk formed by the glandular cells of the breasts is stored until the baby begins active suckling. Stimulation of touch receptors in the nipple initiates sensory impulses that are relayed to the hypothalamus. In response, secretion of oxytocin from the posterior pituitary increases. The bloodstream carries oxytocin to the mammary glands where it stimulates contraction of **myoepithelial cells**, smooth-muscle-like cells that surround the alveoli. The resulting compression moves the milk from the alveoli into the mammary ducts, where it can be suckled. This process is termed **milk ejection (letdown)**. Even though the actual ejection of milk does not occur until 30–60 seconds after *nursing* (feeding an infant at the breast) begins, some milk stored in lactiferous sinuses near the nipple is initially available. Stimuli other than suckling, such as hearing a baby's cry or touching the mother's genitals, also can trigger oxytocin release and milk ejection. The suckling stimulation that produces the release of oxytocin also increases secretion of prolactin, which maintains lactation.

FIGURE 25.38 **The milk ejection reflex. Milk ejection is controlled by a positive feedback cycle.**

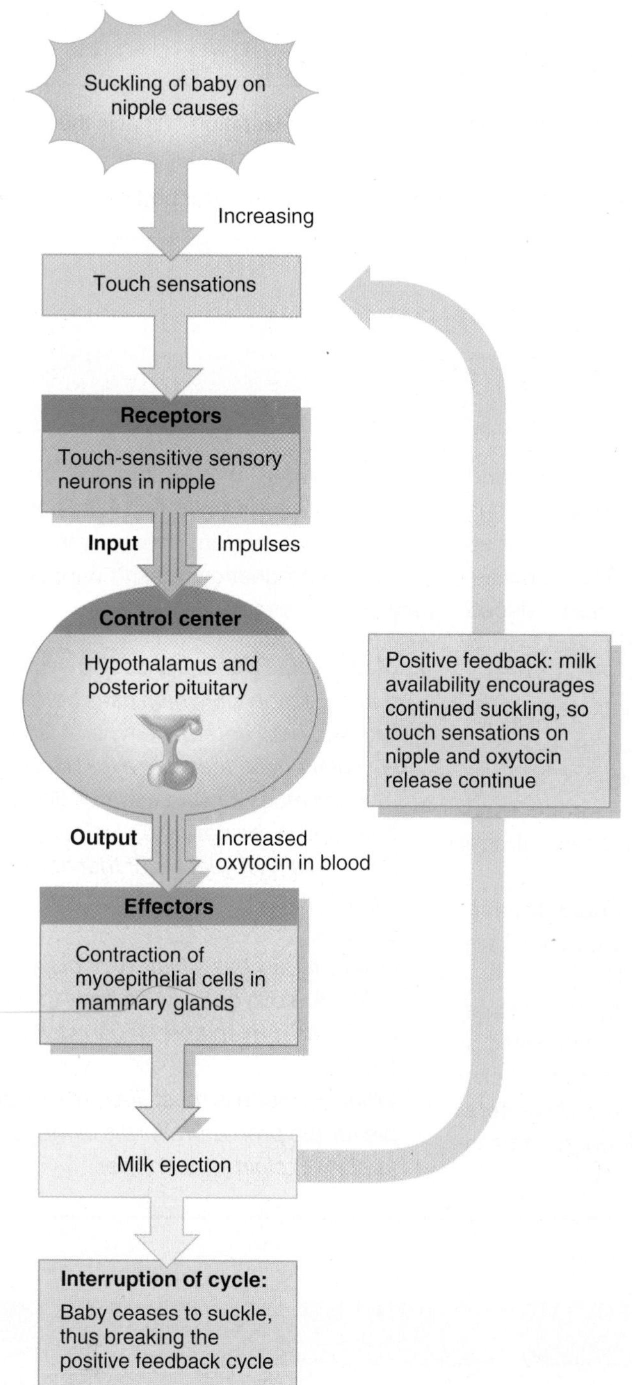

Suckling of baby on nipple causes
Increasing
Touch sensations
Receptors
Touch-sensitive sensory neurons in nipple
Input
Impulses
Control center
Hypothalamus and posterior pituitary
Positive feedback: milk availability encourages continued suckling, so touch sensations on nipple and oxytocin release continue
Output
Increased oxytocin in blood
Effectors
Contraction of myoepithelial cells in mammary glands
Milk ejection
Interruption of cycle:
Baby ceases to suckle, thus breaking the positive feedback cycle

**Oxytocin stimulates contraction of myoepithelial cells in the breasts, which causes milk ejection.**

During late pregnancy and the first few days after birth, the mammary glands secrete a cloudy fluid called **colostrum**. Although it is not as nutritious as milk—it contains less lactose and virtually no fat—colostrum serves adequately until the appearance of true milk on about the fourth day. Colostrum and maternal milk contain important antibodies that protect the infant during the first few months of life.

Following birth of the infant, the prolactin level starts to return to the nonpregnant level. However, each time the mother nurses the infant, impulses from the nipples to the hypothalamus result in a tenfold increase in prolactin secretion by the anterior pituitary that lasts about an hour. Prolactin acts on the mammary glands to provide milk for the next nursing period. If this surge of prolactin is blocked by injury or disease, or if nursing is discontinued, the mammary glands lose their ability to secrete milk in only a few days. Even though milk secretion normally decreases considerably within 7–9 months after birth, it can continue for several years if nursing continues.

A primary benefit of **breast milk** is nutritional: Human milk is a sterile solution that contains amounts of fatty acids, lactose, amino acids, minerals, vitamins, and water that are ideal for the baby's digestion, brain development, and growth. Breast milk also benefits infants by providing the following:

- ***Beneficial cells.*** Several types of white blood cells are present in breast milk. *Neutrophils* and *macrophages* serve as phagocytes, ingesting microbes in the baby's gastrointestinal tract. Macrophages also produce lysozyme and other immune system components. *Plasma cells* produce antibodies against specific microbes, and *T lymphocytes* kill microbes directly or help mobilize other immune defenses.
- ***Beneficial molecules.*** Breast milk also contains an abundance of beneficial molecules. *Maternal IgA antibodies* in breast milk bind to microbes in the baby's gastrointestinal tract and prevent their migration into other body tissues. Because a mother produces antibodies to whatever disease-causing microbes are present in her environment, her breast milk affords protection against the specific infectious agents to which her baby is also exposed. Additionally, two *milk proteins* bind to nutrients such as vitamin $B_{12}$ and iron that many bacteria need to grow and survive. Some *fatty acids* can kill certain viruses by disrupting their membranes, and *lysozyme* kills bacteria by disrupting their cell walls. Finally, *interferons* enhance the antimicrobial activity of immune cells.
- ***Decreased incidence of diseases later in life.*** Breast milk provides children with a slight reduction in risk of lymphoma, heart disease, allergies, respiratory and gastrointestinal infections, ear infections, diarrhea, diabetes mellitus, and meningitis.
- ***Miscellaneous benefits.*** Breast milk supports optimal infant growth, enhances intellectual and neurological development, and fosters mother–infant relations by establishing early and prolonged contact between them. Compared to cow's milk, the fats and iron in breast milk are more easily absorbed, the proteins in breast milk are more readily metabolized, and the lower sodium content of breast milk is better suited to an infant's needs. Premature infants benefit even more from nursing because the milk produced by mothers of

premature infants seems to be specially adapted to the infant's needs; it has a higher protein content than the milk of mothers of full-term infants. Finally, a baby is less likely to have an allergic reaction to its mother's milk than to milk from another source.

Years before oxytocin was discovered, it was common practice in midwifery to let a first-born twin nurse at the mother's breast to speed the birth of the second child. Now we know why this practice is helpful—it stimulates the release of oxytocin. Even after a single birth, nursing promotes expulsion of the placenta (afterbirth) and helps the uterus return to its normal size.

✓ **CHECKPOINT**

**55.** Which hormones contribute to lactation? What is the function of each?

**56.** What are the advantages of nursing over bottle-feeding?

## Ryan and Megan's Story

## EPILOGUE AND DISCUSSION

Ryan and Megan have completed their antibiotic regimens and all signs and symptoms of their STD infections have subsided. A more in-depth examination has shown that there is no scar formation in Megan's uterine tubes.

The two are still together. Doctor Kincaid informed Ryan and Megan they could resume their sexual activities and the couple has become more knowledgeable and diligent about practicing safe sex. Megan had previously tried birth control pills, but found that they affected her mood and energy level too much. Instead, Megan and Ryan are using condoms and spermicidal sponges. They have purchased an ovulation detection kit, which will help them to improve their chances of not getting pregnant until they are ready.

As Megan and Ryan grow closer, they share stories about their families. Ryan tells Megan that fraternal twins have occurred quite often in his family tree. Megan confides that, although she is excited about the prospect of one day being a mother, she is concerned about the changes to her body during pregnancy.

"These books say that I could develop stretch marks and hair loss, two things I am definitely not looking forward to."

Ryan reassures her, saying "I heard that doesn't always happen, and that, even if they do, they're reversible." Megan admits that her biggest fear is having a premature birth, which had occurred enough times in her family to be a concern.

They agree that neither of them is ready to have a family just yet, and that what's most important right now is making sure that a pregnancy doesn't happen until they are both ready.

G. *Megan and Ryan purchase an ovulation detection kit, which uses hormone levels to help determine when Megan is ovulating, so they can avoid sex until they are ready to start a family. What hormone is the detection kit looking for? Explain why levels of that hormone would indicate that ovulation was occurring.*

H. *Ryan's family has a history of dizygotic twins. Are dizygotic twins likely to be produced if Ryan and Megan start a family?*

I. *What hormone is most likely responsible for the premature births in Megan's family? Explain your answer.*

# Concept and Resource Summary

| Concept | Resources  |
|---|---|
| **Introduction**<br>1. **Gonads** produce **gametes** and sex hormones, **ducts** store and transport gametes, and **accessory sex glands** produce materials that protect and help move gametes. **Supporting structures** assist the delivery of gametes, and the uterus is the site of the embryo and fetus during pregnancy.<br>2. **Pregnancy** begins with fertilization and is a time when the fertilized ovum undergoes implantation, embryonic and fetal development, and birth. | |

## Concept | Resources

### Concept 25.1 The scrotum supports and regulates the temperature of the testes, which produce spermatozoa.

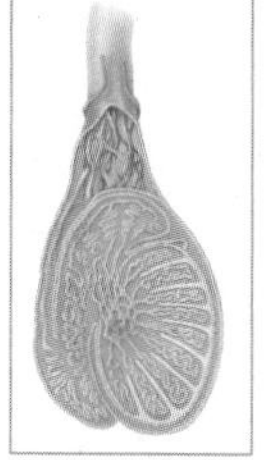

1. Male reproductive organs include the testes, a system of ducts (epididymis, ductus deferens, ejaculatory ducts, and urethra), accessory glands (seminal vesicles, prostate, and bulbourethral glands), and supporting structures (scrotum and penis).
2. The **scrotum** is a sac that hangs from the root of the penis and consists of loose skin and underlying hypodermis; it supports the testes. The temperature of the testes is regulated by contraction of the **cremaster muscle** and **dartos muscle**, which either elevates them and brings them closer to the warmth of the pelvic cavity or relaxes and moves them farther from the pelvic cavity for cooling.
3. The **testes** are paired glands in the scrotum containing **seminiferous tubules**, which produce sperm during **spermatogenesis**. The seminiferous tubules contain the **spermatogenic cells** that form sperm; **Sertoli cells**, which nourish spermatogenic cells and secrete inhibin; and **Leydig cells**, which produce the male sex hormone testosterone.
4. **Spermatogonia** begin producing sperm at puberty. Some spermatogonia develop into diploid **primary spermatocytes** that sequence through spermatogenesis (meiosis I, meiosis II, and spermiogenesis) to form haploid **sperm cells**. During **spermiogenesis**, developing sperm form an acrosome and flagellum, shed excess cytoplasm, increase mitochondria, and become elongated sperm cells that are released into the lumen of the seminiferous tubules. Sertoli cells secrete fluid that pushes sperm toward the ducts of the testes.
5. Mature sperm have a pointed **head**, a caplike **acrosome** filled with enzymes for penetrating the secondary oocyte, and a **tail** with a **middle piece** containing mitochondria that provide energy for movement toward the secondary oocyte.
6. At puberty, the hypothalamus secretes **gonadotropin-releasing hormone**, which stimulates anterior pituitary secretion of **luteinizing hormone** and **follicle-stimulating hormone**. Luteinizing hormone stimulates Leydig cells to secrete **testosterone**. Follicle-stimulating hormone and testosterone stimulate Sertoli cells to secrete **androgen-binding protein**, which binds to testosterone and keeps its concentration high in the seminiferous tubule.
7. Sertoli cells secrete **inhibin**, which inhibits follicle-stimulating hormone release to help regulate the rate of spermatogenesis.
8. Testosterone stimulates development of male reproductive structures and **secondary sex characteristics**, development of male sexual behavior, male and female libido, bone growth, protein anabolism, and sperm maturation.

Resources:

Anatomy Overview—Male Reproductive Anatomy
Anatomy Overview—Histology of the Testes
Anatomy Overview—Testicular Hormones
Animation—Spermatogenesis
Animation—Hormonal Control of Male Reproduction
Figure 25.3—Internal and External Anatomy of a Testis
Figure 25.7—Hormonal Control of Spermatogenesis and Actions of Testosterone and Dihydrotestosterone (DHT)
Exercise—Match the Male Hormones
Exercise—Spermatogenesis Selections
Concepts and Connections—Regulation of Male Reproduction
Clinical Connection—Testicular Cancer
Clinical Connection—Cryptorchidism

### Concept 25.2 Sperm travel through the epididymis, ductus deferens, ejaculatory ducts, and urethra.

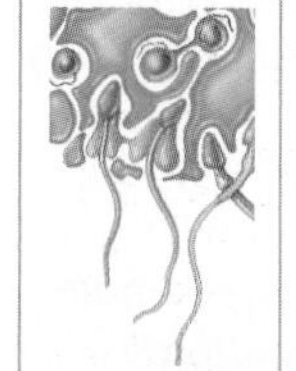

1. Fluid from Sertoli cells pushes sperm through the testis in the **straight tubules**, through the **rete testis**, then through the epididymis in the **efferent ducts**.
2. The **epididymis** lies along the posterior border of each testis and contains the tightly coiled **ductus epididymis**, the site of sperm maturation and storage. The ductus epididymis propels sperm into the ductus deferens.
3. The **ductus deferens** stores sperm and propels them to the urethra for ejaculation.
4. The **spermatic cord** passes the ductus deferens, the testicular artery and veins, autonomic nerves, lymphatic vessels, and cremaster muscle through the **inguinal canal**.
5. The ducts of the ductus deferens and seminal vesicle unite to form the **ejaculatory duct**, the passageway for ejection of sperm and secretions of the seminal vesicles into the first portion of the urethra, the prostatic urethra.
6. The male **urethra** serves as a passage for both sperm and urine. The urethra extends from the urinary bladder to the tip of the penis and is subdivided into the **prostatic**, **membranous**, and **spongy urethra**.
7. Paired **seminal vesicles** lie posterior to the base of the urinary bladder. Their secretion is an alkaline, viscous fluid that contains fructose (used by sperm for ATP production). Seminal fluid constitutes about 60% of the volume of semen and contributes to sperm viability.
8. The **prostate** lies inferior to the urinary bladder and surrounds the prostatic urethra. Its secretion constitutes about 25% of the volume of semen and contributes to sperm motility and viability.
9. Paired **bulbourethral glands** lie inferior to the prostate on either side of the membranous urethra. They secrete mucus for lubrication and an alkaline fluid that neutralizes acids from urine in the urethra.

Resources:

Anatomy Overview—Male Reproductive Anatomy
Figure 25.9—Locations of Several Accessory Reproductive Organs in Males
Clinical Connection—Priapism
Clinical Connection—Vasectomy
Clinical Connection—Erectile Dysfunction
Clinical Connection—Premature Ejaculation

**Concept** | **Resources** 

10. **Semen** is a mixture of sperm and **seminal fluid**, which consists of secretions from the seminiferous tubules, seminal vesicles, prostate, and bulbourethral glands. Semen provides the fluid in which sperm are transported, supplies nutrients for sperm, and neutralizes the acidity of the male urethra and the vagina.
11. The **penis** contains the urethra and is a passageway for ejaculation of semen and excretion of urine. The **root of the penis** attaches to the deep muscles of the perineum. The **body of the penis** is composed of three masses of erectile tissue: two **corpora cavernosa penis** and a **corpus spongiosum penis** that contains the spongy urethra and keeps it open during ejaculation. The **glans penis** is the expanded distal tip.
12. Engorgement of the penile blood sinuses under the influence of sexual excitation is called **erection.** **Ejaculation** is the expulsion of semen from the urethra to the exterior of the body.

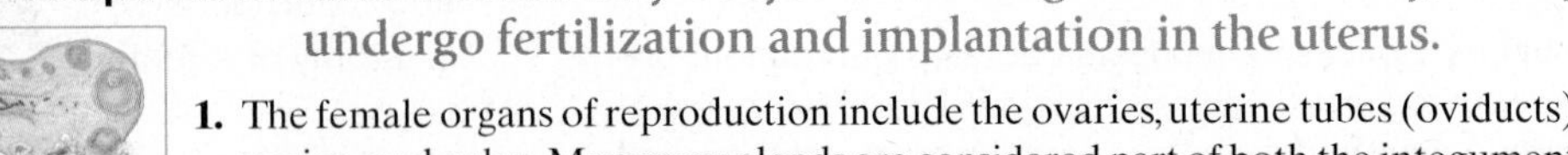

## Concept 25.3 After a secondary oocyte is discharged from an ovary, it may undergo fertilization and implantation in the uterus.

1. The female organs of reproduction include the ovaries, uterine tubes (oviducts), uterus, vagina, and vulva. Mammary glands are considered part of both the integumentary and reproductive systems.
2. The **ovaries** are located on each side of the uterus and held in position by the **broad**, **ovarian**, and **suspensory ligaments**. Ovaries secrete hormones and produce secondary oocytes.
3. The **ovarian cortex** contains **ovarian follicles** with oocytes in different stages of development. A **mature follicle** is ready to expel its secondary oocyte during **ovulation**. A **corpus luteum**, the remains of a follicle after ovulation, degenerates into the **corpus albicans**.
4. **Oogenesis**, the formation of female gametes, begins in the ovaries. The oogenesis sequence includes meiosis I and meiosis II, which stops in metaphase. The secondary oocyte is released from the ovary during ovulation and is usually swept into the uterine tube. If a sperm cell penetrates the oocyte, meiosis II resumes, and a haploid **ovum** is produced. The sperm and ovum nuclei unite during fertilization to become a diploid **zygote**.
5. The **uterine tubes** extend laterally from the uterus and are the normal sites of fertilization. Their distal end is an open, funnel-shaped **infundibulum** with fingerlike **fimbriae** that sweep the oocyte into the uterine tube. Ciliated cells and peristaltic contractions help move a secondary oocyte or zygote toward the uterus.
6. The **uterus** functions in menstruation, implantation of a fertilized ovum, fetal development, and labor. It also is part of the pathway for sperm to reach the uterine tubes to fertilize a secondary oocyte. The uterus has a **fundus**, **body**, and **cervix** that opens into the vagina. It is suspended by the broad, **uterosacral**, **cardinal**, and **round ligaments**. The uterus wall is composed of an outer **perimetrium**; a middle **myometrium** consisting of three layers of smooth muscle important during labor; and the inner **endometrium** that lines the lumen of the uterus. There is an extensive blood supply to support regrowth of sloughed-off lining after menstruation, implantation of the fertilized ovum, and development of the placenta.

**Resources:**

Anatomy Overview—Female Reproductive Anatomy
Anatomy Overview—Histology of the Female Reproductive System
Animation—Oogenesis
Figure 25.13—Histology of the Ovary
Figure 25.16—Relationship of the Uterine Tubes to the Ovaries, Uterus, and Associated Structures
Exercise—Organize Oogenesis
Clinical Connection—Uterine Prolapse
Clinical Connection—Endometriosis
Clinical Connection—Cervical Cancer
Clinical Connection—Ovarian Cancer

## Concept 25.4 The vagina is a passageway for childbirth; the mammary glands secrete milk.

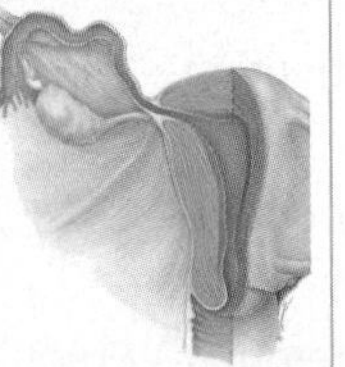

1. The **vagina** extends from the uterine cervix to the exterior of the body. It is a receptacle for the penis during sexual intercourse, an outlet for menstrual flow, and a passageway for childbirth.
2. The external genitals of the female are termed the **vulva** and include the **mons pubis**, **labia majora**, **labia minora**, **clitoris**, **vestibule**, **vaginal orifice**, urethral orifice, and **bulb of the vestibule**. Also part of the vulva are the **paraurethral glands**, which secrete mucus into the urethra; and the **greater vestibular glands**, which secrete lubricating mucus into the vestibule to facilitate intercourse.
3. The **perineum** is the diamond-shaped area medial to the thighs and buttocks containing the external genitals and anus.
4. Each **breast** lies anterior to the pectoralis major and serratus anterior. Within each breast is a **mammary gland** that functions in lactation, and the production and ejection of milk.

**Resources:**

Anatomy Overview—Female Reproductive Anatomy
Anatomy Overview—Histology of the Mammary Glands
Figure 25.22—Mammary Glands
Clinical Connection—Vulvovaginal Candidiasis
Clinical Connection—Breast Cancer
Clinical Connection—Breast Augmentation and Reduction
Clinical Connection—Fibrocystic Disease of the Breasts

| Concept | Resources |
|---|---|

## Concept 25.5 The female reproductive cycle includes the ovarian and uterine cycles.

1. The **ovarian cycle** occurs in the ovaries during and after maturation of an oocyte. The **uterine cycle** is a series of concurrent changes in the endometrium of the uterus to prepare it to receive the fertilized ovum and support ovum development. The **female reproductive cycle** includes both the ovarian and uterine cycles.
2. The female reproductive cycle is controlled by several hormones. Gonadotropin-releasing hormone from the hypothalamus stimulates the anterior pituitary to release follicle-stimulating hormone and luteinizing hormone, which stimulate development of follicles and secretion of estrogens by the follicles. Luteinizing hormone also stimulates ovulation, formation of the corpus luteum, and secretion of estrogens, progesterone, relaxin, and inhibin from the corpus luteum.
3. **Estrogens** stimulate the development and maintenance of female reproductive structures, secondary sex characteristics, and protein anabolism, and decrease blood cholesterol levels. **Progesterone** works with estrogens to prepare the endometrium for implantation and the mammary glands for milk secretion. **Relaxin** relaxes the myometrium at the time of possible implantation and, at the end of pregnancy, increases the flexibility of the pubic symphysis to facilitate delivery. Inhibin inhibits secretion of follicle-stimulating hormone and luteinizing hormone.
4. The female reproductive cycle ranges from 24 to 35 days and has four phases: the menstrual phase, preovulatory phase, ovulation, and postovulatory phase.
5. During the **menstrual phase**: (1) In the ovaries, primordial follicles develop into primary, then secondary follicles; (2) in the uterus, the stratum functionalis of the endometrium is shed through menstrual flow.
6. The **preovulatory phase** occurs between **menstruation** and ovulation. In the ovaries, one follicle becomes **dominant** while the others degenerate. The dominant follicle becomes the mature follicle and secretes estrogens and inhibin, which decreases FSH secretion and stops the other follicles from growing. In the uterus, estrogens stimulate growth of a new stratum functionalis.
7. **Ovulation**, the rupture of the mature follicle and release of the secondary oocyte into the pelvic cavity, is induced by high levels of estrogens.
8. The **postovulatory phase** occurs between ovulation and menstruation. In the ovaries, LH stimulates development of the corpus luteum from the ruptured follicle and stimulates the corpus luteum to secrete progesterone, estrogen, relaxin, and inhibin. In the uterus, the endometrium thickens in readiness for implantation.
9. If fertilization and implantation do not occur: (1) In the ovaries, the corpus luteum degenerates, and the levels of progesterone and estrogens decrease; (2) in the uterus, the resulting low levels of progesterone and estrogens result in menstruation followed by the initiation of another reproductive cycle.
10. If fertilization and implantation do occur: (1) In the ovaries, **human chorionic gonadotropin** secreted by the embryo maintains the corpus luteum so it continues to be a source of progesterone and estrogens; (2) in the uterus, endometrial development continues under the influence of progesterone and estrogen in preparation for implantation of the embryo.

**Resources**

Anatomy Overview—Ovarian Hormones
Anatomy Overview—Anterior Pituitary Reproductive Hormones
Animation—Oogenesis
Animation—Hormonal Control of Female Reproduction
Figure 25.24—The Female Reproductive Cycle
Exercise—Assemble the Cycle
Exercise—Match the Female Hormones
Concepts and Connections—Regulation of Female Reproduction
Clinical Connection—Premenstrual Syndrome and Premenstrual Dysphoric Disorder
Clinical Connection—Female Athlete Triad: Disordered Eating, Amenorrhea, and Premature Osteoporosis
Clinical Connection—Sexually Transmitted Diseases

## Concept 25.6 The zygote divides into a morula and then a blastocyst that implants in the endometrium of the uterus.

1. The **gestation period** is from fertilization to birth and is divided into three **trimesters**.
2. **Fertilization**, the union of the genetic material from a haploid sperm cell and a haploid secondary oocyte into a single diploid nucleus, typically occurs within the uterine tube 12 to 24 hours after ovulation.
3. Sperm use flagella to swim from the vagina through the cervical canal to the uterine tube, where sperm undergo **capacitation** to assist penetration of the oocyte. After penetrating the corona radiata, the zona pellucida triggers the **acrosomal reaction**, the release of acrosomal enzymes that digest a path through the zona pellucida. Normally, only one sperm cell penetrates a secondary oocyte because of the fast and slow blocks to **polyspermy**.
4. When the sperm cell enters the secondary oocyte, it triggers completion of meiosis II. The pronuclei of the sperm cell and secondary oocyte unite. The resulting cell is a zygote.
5. If two secondary oocytes are released and fertilized by sperm, **dizygotic twins** are produced. **Monozygotic twins** develop from a single fertilized ovum that separates into two embryos. If the developing cells do not separate within eight days, **conjoined twins** will likely result.

**Resources**

Animation—Fertilization
Clinical Connection—Stem Cell Research and Therapeutic Cloning
Clinical Connection—Infertility
Clinical Connection—Birth Control Methods
Clinical Connection—Abortion

**Concept** | **Resources** 

6. The developing human is an **embryo** from fertilization to the eighth week of development. Early rapid cell division of a zygote is called **cleavage**, and the cells produced by cleavage are called **blastomeres**. The solid sphere of blastomeres is a **morula**. The morula develops into a **blastocyst**, a hollow ball of cells differentiated into an **embryoblast** that develops into the embryo and a **trophoblast** that becomes part of the placenta.
7. During **implantation**, the blastocyst attaches to the endometrium, secretes enzymes that degrade the endometrium, and burrows into the endometrium. After implantation, the endometrium surrounding the implanted blastocyst is known as the **decidua**.

## Concept 25.7 Major tissues and organs develop during embryonic development and grow and differentiate during fetal development.

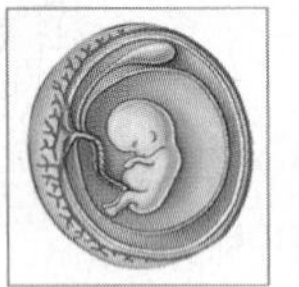

1. The trophoblast develops into the **syncytiotrophoblast** and **cytotrophoblast**, both of which become part of the chorion. The syncytiotrophoblast secretes enzymes that implant the embryo deeper in the endometrium, and secretes human chorionic gonadotropin to maintain the corpus luteum and prevent menstruation.
2. The embryoblast differentiates into the **embryonic disc** composed of two layers, the **hypoblast** and **epiblast**. The **amniotic cavity** develops within the epiblast.
3. The thin membrane called the **amnion** eventually surrounds the entire embryo, creating the amniotic cavity filled with amniotic fluid, derived as a filtrate of maternal blood, and later, fetal urine. **Amniotic fluid** serves as a shock absorber, prevents the fetus from drying out or forming adhesions with surrounding tissues, and helps maintain fetal body temperature.
4. The **yolk sac** develops from hypoblast and transfers nutrients to the embryo, forms blood cells, produces germ cells that will eventually form gametes, and forms part of the GI tract.
5. Once the blastocyst is completely implanted, endometrial capillaries around the embryo dilate and become **maternal sinusoids**, where embryonic nutrients are obtained and wastes eliminated.
6. The **extraembryonic mesoderm**, cytotrophoblast, and syncytiotrophoblast form the **chorion**, which surrounds the embryo and eventually becomes the embryonic part of the placenta. The chorion protects the fetus from immune responses of the mother and produces hCG.
7. During **gastrulation** the embryonic disc develops into the **endoderm**, **mesoderm**, and **ectoderm**, the **primary germ layers** that develop into the various body tissues and organs.
8. Mesoderm develops into embryonic blood vessels, formed elements, and the heart. By the end of the third week, the embryonic heart beats and circulates blood.
9. **Chorionic villi**, projections of the chorion, connect to the embryonic heart, allowing the exchange of nutrients and wastes between maternal and fetal blood.
10. The **placenta** is the site of exchange of nutrients and wastes between the mother and fetus. It has a fetal portion formed by chorionic villi and a maternal portion formed by the endometrium. The placenta supplies oxygen and removes carbon dioxide from fetal blood, functions as a protective barrier, stores nutrients, and produces several hormones to maintain pregnancy. The **umbilical cord** develops from the connecting stalk and connects the placenta and the embryo/fetus. After childbirth, the placenta detaches from the uterus and is called the **afterbirth**.
11. All major organs develop in the embryo between the fourth and eighth weeks.
12. The developing human is a **fetus** from week nine until birth. While a fetus, embryonic tissues and organs grow and differentiate.

Anatomy Overview—Developmental Stages
Animation—Embryonic and Fetal Development
Clinical Connection—Placenta Previa
Clinical Connection—Teratogens
Clinical Connection—Down Syndrome
Clinical Connection—Congenital Defects
Clinical Connection—Prenatal Diagnostic Tests

## Concept 25.8 During pregnancy the uterus expands, displacing and compressing maternal organs.

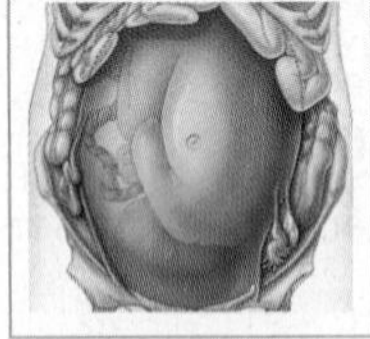

1. Progesterone and estrogens maintain the uterine lining and the preparation of mammary glands for milk secretion. They are secreted by the corpus luteum, then the chorion, and later, the placenta. High estrogen levels prepare the myometrium for labor, while progesterone ensures the myometrium is relaxed and the cervix tightly closed.
2. The corpus luteum, and later the placenta, produces relaxin, which increases the flexibility of the pubic symphysis and pelvic ligaments, and helps dilate the cervix for delivery. The chorion produces **human chorionic somatomammotropin**, which prepares the mammary glands for lactation, increases protein synthesis for maternal growth, and influences maternal and fetal metabolism. **Corticotropin-releasing hormone** from the placenta probably functions in the timing of the birth.

Animation—Hormonal Regulation of Pregnancy
Figure 25.35—Hormones during Pregnancy
Clinical Connection—Pregnancy-induced Hypertension
Clinical Connection—Early Pregnancy Tests

| Concept | Resources |
|---|---|

3. As the uterus occupies more of the pelvic cavity, maternal abdominal organs get pushed superiorly against the diaphragm while the ureters and urinary bladder get compressed. During pregnancy there is weight gain, breast enlargement, and lower back pain resulting from lordosis.
4. To meet the fetal demands for nutrients and oxygen, cardiac output, heart rate, and blood volume increase, as do respiratory rate and volume. Several other anatomical and physiological changes also occur in the pregnant woman.

**Concept 25.9** Labor includes dilation of the cervix and expulsion of the fetus and placenta.

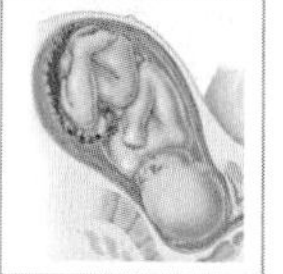

1. **Labor** is the process by which the fetus is expelled from the uterus through the vagina, and depends on complex interactions of several hormones. Rising levels of estrogens near the end of pregnancy stimulate release of prostaglandins by the placenta, which cause softening of the cervix, and estrogens increase the number of oxytocin receptors in uterine muscle fiber. **Oxytocin** stimulates uterine contractions via a positive feedback cycle. Relaxin increases flexibility of the pubic symphysis and helps dilate the cervix.
2. **True labor** involves uterine contractions at regular intervals and dilation of the cervix, expulsion of the fetus, and delivery of the placenta. **False labor** involves pain at irregular intervals that does not intensify.
3. Following delivery, maternal reproductive organs return to the prepregnancy state, including reduction in size of the uterus.
4. The fetus depends on the mother's body for gas and waste exchange, nutrients, and protection. Following birth, an infant's respiratory and cardiovascular systems undergo changes to enable them to become self-supporting.

Resources:
Anatomy Overview—Hypothalamic Reproductive Hormones
Animation—Regulation of Labor and Birth
Clinical Connection—Dystocia and Cesarean Section
Clinical Connection—Premature Infants

**Concept 25.10** Lactation is influenced by prolactin, estrogens, progesterone, and oxytocin.

1. **Lactation** refers to the production and ejection of milk by the mammary glands.
2. Milk production is influenced by **prolactin**, estrogens, and progesterone. Milk ejection is stimulated by oxytocin.
3. A few of the many benefits of **breast milk** include ideal nutrition for the infant, protection from disease, and decreased likelihood of developing allergies.

Resources:
Anatomy Overview—Hypothalamic Reproductive Hormones
Animation—Regulation of Lactation
Exercise—Pregnancy, Birth, and Lactation

## Understanding the Concepts

1. Describe the principal events of spermatogenesis. Which spermatogenic cells in a seminiferous tubule are least mature? Most mature?
2. What are the roles of luteinizing hormone, follicle-stimulating hormone, testosterone, and inhibin in the male reproductive system? How is secretion of these hormones controlled?
3. What is the course of sperm through the ducts of the male reproductive system from the seminiferous tubules through the urethra?
4. Describe the principal events of oogenesis. How do the results of oogenesis and spermatogenesis differ?
5. What is the functional significance of the stratum basalis of the endometrium?
6. Why is an abundant blood supply important to the uterus?
7. How does the histology of the vagina contribute to its function?
8. Describe the influence of rising levels of estrogens on the secretion of GnRH, LH, and FSH.
9. What are the major events of each of the three phases of the uterine cycle? How are they correlated with the events of the ovarian cycle?
10. How does the corpus luteum develop? Why does it last more than two weeks when fertilization occurs?
11. Why is fertilization most likely to occur within a three-day period of the female reproductive cycle?
12. What is capacitation?
13. What are the functions of the trophoblast?
14. Why are maternal sinusoids and the development of chorionic villi important to the embryo?
15. Describe the formation of the amnion, yolk sac, and chorion. What are their functions?
16. Explain the structural and functional changes that occur in the mother during pregnancy.
17. Describe the dilation, expulsion, and placental stages of labor.
18. What is a function of oxytocin other than its involvement in lactation?

# Appendix A
## Measurements

**U.S. Customary System**

| Parameter | Unit | Relation to Other U.S. Units | SI (Metric) Equivalent |
|---|---|---|---|
| **Length** | | | |
| | inch | 1/12 foot | 2.54 centimeters |
| | foot | 12 inches | 0.305 meter |
| | yard | 36 inches | 0.9144 meter |
| | mile | 5,280 feet | 1.609 kilometers |
| **Mass** | | | |
| | grain | 1/1000 pound | 64.799 milligrams |
| | dram | 1/16 ounce | 1.772 grams |
| | ounce | 16 drams | 28.350 grams |
| | pound | 16 ounces | 453.6 grams |
| | ton | 2,000 pounds | 907.18 kilograms |
| **Volume (Liquid)** | | | |
| | ounce | 1/16 pint | 29.574 milliliters |
| | pint | 16 ounces | 0.473 liter |
| | quart | 2 pints | 0.946 liter |
| | gallon | 4 quarts | 3.785 liters |
| **Volume (Dry)** | | | |
| | pint | 1/2 quart | 0.551 liter |
| | quart | 2 pints | 1.101 liters |
| | peck | 8 quarts | 8.810 liters |
| | bushel | 4 pecks | 35.239 liters |

**International System (SI)**

**Base Units**

| Unit | Quantity | Symbol |
|---|---|---|
| **meter** | length | m |
| **kilogram** | mass | kg |
| **second** | time | s |
| **liter** | volume | L |
| **mole** | amount of matter | mol |

**Prefixes**

| Prefix | Multiplier | Symbol |
|---|---|---|
| tera- | $10^{12}$ = 1,000,000,000,000 | T |
| giga- | $10^{9}$ = 1,000,000,000 | G |
| mega- | $10^{6}$ = 1,000,000 | M |
| kilo- | $10^{3}$ = 1,000 | k |
| hecto- | $10^{2}$ = 100 | h |
| deca- | $10^{1}$ = 10 | da |
| deci- | $10^{-1}$ = 0.1 | d |
| centi- | $10^{-2}$ = 0.01 | c |
| milli- | $10^{-3}$ = 0.001 | m |
| micro- | $10^{-6}$ = 0.000001 | $\mu$ |
| nano- | $10^{-9}$ = 0.000000001 | n |
| pico- | $10^{-12}$ = 0.000000000001 | p |

## Temperature Conversion

**Fahrenheit (F) To Celsius (C)**

°C = (°F − 32) ÷ 1.8

**Celsius (C) To Fahrenheit (F)**

°F = (°C × 1.8) + 32

## U.S. to SI (Metric) Conversion

| When You Know | Multiply By | To Find |
|---|---|---|
| **inches** | 2.54 | centimeters |
| **feet** | 30.48 | centimeters |
| **yards** | 0.91 | meters |
| **miles** | 1.61 | kilometers |
| **ounces** | 28.35 | grams |
| **pounds** | 0.45 | kilograms |
| **tons** | 0.91 | metric tons |
| **fluid ounces** | 29.57 | milliliters |
| **pints** | 0.47 | liters |
| **quarts** | 0.95 | liters |
| **gallons** | 3.79 | liters |

## SI (Metric) to U.S. Conversion

| When You Know | Multiply By | To Find |
|---|---|---|
| **millimeters** | 0.04 | inches |
| **centimeters** | 0.39 | inches |
| **meters** | 3.28 | feet |
| **kilometers** | 0.62 | miles |
| **liters** | 1.06 | quarts |
| **cubic meters** | 35.32 | cubic feet |
| **grams** | 0.035 | ounces |
| **kilograms** | 2.21 | pounds |

# Appendix B
## Periodic Table

The periodic table lists the known **chemical elements**, the basic units of matter. The elements in the table are arranged left to right in rows in order of their **atomic number**, the number of protons in the nucleus. Each horizontal row, numbered from 1 to 7, is a **period**. All elements in a given period have the same number of electron shells as their period number. For example, an atom of hydrogen or helium each has one electron shell, while an atom of potassium or calcium each has four electron shells. The elements in each column, or **group**, share chemical properties. For example, the elements in column IA are very chemically reactive, whereas the elements in column VIIIA have full electron shells and thus are chemically inert.

Scientists now recognize 117 different elements; 92 occur naturally on Earth, and the rest are produced from the natural elements using particle accelerators or nuclear reactors. Elements are designated by **chemical symbols**, which are the first one or two letters of the element's name in English, Latin, or another language.

Twenty-six of the 92 naturally occurring elements normally are present in your body. Of these, just four elements—oxygen (O), carbon (C), hydrogen (H), and nitrogen (N) (coded blue)—constitute about 96% of the body's mass. Eight others—calcium (Ca), phosphorus (P), potassium (K), sulfur (S), sodium (Na), chlorine (Cl), magnesium (Mg), and iron (Fe) (coded pink)—contribute 3.8% of the body's mass. An additional 14 elements, called **trace elements** because they are present in tiny amounts, account for the remaining 0.2% of the body's mass. The trace elements are aluminum, boron, chromium, cobalt, copper, fluorine, iodine, manganese, molybdenum, selenium, silicon, tin, vanadium, and zinc (coded yellow). Table 2.1 on page 30 provides information about the main chemical elements in the body.

23 ← Atomic number
V ← Chemical symbol
50.942 ← Atomic mass (weight)

Percentage of body mass
- 96% (4 elements)
- 3.8% (8 elements)
- 0.2% (14 elements)

| | IA | IIA | IIIB | IVB | VB | VIB | VIIB | VIIIB | VIIIB | VIIIB | IB | IIB | IIIA | IVA | VA | VIA | VIIA | VIIIA |
|---|---|---|---|---|---|---|---|---|---|---|---|---|---|---|---|---|---|---|
| 1 | 1 Hydrogen **H** 1.0079 | | | | | | | | | | | | | | | | | 2 Helium **He** 4.003 |
| 2 | 3 Lithium **Li** 6.941 | 4 Beryllium **Be** 9.012 | | | | | | | | | | | 5 Boron **B** 10.811 | 6 Carbon **C** 12.011 | 7 Nitrogen **N** 14.007 | 8 Oxygen **O** 15.999 | 9 Fluorine **F** 18.998 | 10 Neon **Ne** 20.180 |
| 3 | 11 Sodium **Na** 22.989 | 12 Magnesium **Mg** 24.305 | | | | | | | | | | | 13 Aluminum **Al** 26.9815 | 14 Silicon **Si** 28.086 | 15 Phosphorus **P** 30.974 | 16 Sulfur **S** 32.066 | 17 Chlorine **Cl** 35.453 | 18 Argon **Ar** 39.948 |
| 4 | 19 Potassium **K** 39.098 | 20 Calcium **Ca** 40.08 | 21 Scandium **Sc** 44.956 | 22 Titanium **Ti** 47.87 | 23 Vanadium **V** 50.942 | 24 Chromium **Cr** 51.996 | 25 Manganese **Mn** 54.938 | 26 Iron **Fe** 55.845 | 27 Cobalt **Co** 58.933 | 28 Nickel **Ni** 58.69 | 29 Copper **Cu** 63.546 | 30 Zinc **Zn** 65.41 | 31 Gallium **Ga** 69.723 | 32 Germanium **Ge** 72.59 | 33 Arsenic **As** 74.992 | 34 Selenium **Se** 78.96 | 35 Bromine **Br** 79.904 | 36 Krypton **Kr** 83.80 |
| 5 | 37 Rubidium **Rb** 85.468 | 38 Strontium **Sr** 87.62 | 39 Yttrium **Y** 88.905 | 40 Zirconium **Zr** 91.22 | 41 Niobium **Nb** 92.906 | 42 Molybdenum **Mo** 95.94 | 43 Technetium **Tc** (98) | 44 Ruthenium **Ru** 101.07 | 45 Rhodium **Rh** 102.905 | 46 Palladium **Pd** 106.42 | 47 Silver **Ag** 107.868 | 48 Cadmium **Cd** 112.40 | 49 Indium **In** 114.82 | 50 Tin **Sn** 118.69 | 51 Antimony **Sb** 121.75 | 52 Tellurium **Te** 127.60 | 53 Iodine **I** 126.904 | 54 Xenon **Xe** 131.30 |
| 6 | 55 Cesium **Cs** 132.905 | 56 Barium **Ba** 137.33 | | 72 Hafnium **Hf** 178.49 | 73 Tantalum **Ta** 180.948 | 74 Tungsten **W** 183.85 | 75 Rhenium **Re** 186.2 | 76 Osmium **Os** 190.2 | 77 Iridium **Ir** 192.22 | 78 Platinum **Pt** 195.08 | 79 Gold **Au** 196.967 | 80 Mercury **Hg** 200.59 | 81 Thallium **Tl** 204.38 | 82 Lead **Pb** 207.19 | 83 Bismuth **Bi** 208.980 | 84 Polonium **Po** (209) | 85 Astatine **At** (210) | 86 Radon **Rn** (222) |
| 7 | 87 Francium **Fr** (223) | 88 Radium **Ra** (226) | | 104 Rutherfordium **Rf** (261) | 105 Dubnium **Db** (262) | 106 Seaborgium **Sg** (266) | 107 Bohrium **Bh** (264) | 108 Hassium **Hs** (269) | 109 Meitnerium **Mt** (268) | 110 Darmstadtium **Ds** (271) | 111 Roentgenium **Rg** (272) | 112 Copernicium **Cn** (277) | 113 **Uut** (?) | 114 **Uuq** (285) | 115 **Uup** (?) | 116 **Uuh** (289) | 117 **Uus** (?) | 118 **Uuo** (294) |

57–71, Lanthanides

| 57 Lanthanum **La** 138.91 | 58 Cerium **Ce** 140.12 | 59 Praseodymium **Pr** 140.907 | 60 Neodymium **Nd** 144.24 | 61 Promethium **Pm** 144.913 | 62 Samarium **Sm** 150.35 | 63 Europium **Eu** 151.96 | 64 Gadolinium **Gd** 157.25 | 65 Terbium **Tb** 158.925 | 66 Dysprosium **Dy** 162.50 | 67 Holmium **Ho** 164.930 | 68 Erbium **Er** 167.26 | 69 Thulium **Tm** 168.934 | 70 Ytterbium **Yb** 173.04 | 71 Lutetium **Lu** 174.97 |
|---|---|---|---|---|---|---|---|---|---|---|---|---|---|---|

89–103, Actinides

| 89 Actinium **Ac** (227) | 90 Thorium **Th** 232.038 | 91 Protactinium **Pa** (231) | 92 Uranium **U** 238.03 | 93 Neptunium **Np** (237) | 94 Plutonium **Pu** 244.064 | 95 Americium **Am** (243) | 96 Curium **Cm** (247) | 97 Berkelium **Bk** (247) | 98 Californium **Cf** 242.058 | 99 Einsteinium **Es** (254) | 100 Fermium **Fm** 257.095 | 101 Mendelevium **Md** 258.10 | 102 Nobelium **No** 259.10 | 103 Lawrencium **Lr** 260.105 |
|---|---|---|---|---|---|---|---|---|---|---|---|---|---|---|

# Appendix C
## Normal Values for Selected Blood Tests

The system of international (SI) units (Système Internationale d'Unités) is used in most countries and in many medical and scientific journals. Clinical laboratories in the United States, by contrast, usually report values for blood and urine tests in conventional units. The laboratory values in this Appendix give conventional units first, followed by SI equivalents in parentheses. Values listed for various blood tests should be viewed as reference values rather than absolute "normal" values for all well people. Values may vary due to age, gender, diet, and environment of the subject or the equipment, methods, and standards of the lab performing the measurement.

**KEY TO SYMBOLS**

| | |
|---|---|
| g = gram | mL = milliliter |
| mg = milligram = $10^{-3}$ gram | μL = microliter |
| μg = microgram = $10^{-6}$ gram | mEq/L = milliequivalents per liter |
| U = units | mmol/L = millimoles per liter |
| L = liter | μmol/L = micromoles per liter |
| dL = deciliter | > = greater than; < = less than |

Blood Tests

| TEST (SPECIMEN) | U.S. REFERENCE VALUES (SI UNITS) | VALUES INCREASE IN | VALUES DECREASE IN |
|---|---|---|---|
| **Aminotransferases** (serum) | | | |
| **Alanine aminotransferase (ALT)** | 0–35 U/L (same) | Liver disease or liver damage due to toxic drugs | |
| **Aspartate aminotransferase (AST)** | 0–35 U/L (same) | Myocardial infarction, liver disease, trauma to skeletal muscles, severe burns | Beriberi, uncontrolled diabetes mellitus with acidosis, pregnancy |
| **Ammonia** (plasma) | 20–120 μg/dL (12–55 μmol/L) | Liver disease, heart failure, emphysema, pneumonia, hemolytic disease of newborn | Hypertension |
| **Bilirubin** (serum) | Conjugated: <0.5 mg/dL (<5.0 μmol/L)<br>Unconjugated: 0.2–1.0 mg/dL (18–20 μmol/L)<br>Newborn: 1.0–12.0 mg/dL (<200 μmol/L) | Conjugated bilirubin: liver dysfunction or gallstones<br>Unconjugated bilirubin: excessive hemolysis of red blood cells | |
| **Blood urea nitrogen (BUN)** (serum) | 8–26 mg/dL (2.9–9.3 mmol/L) | Kidney disease, urinary tract obstruction, shock, diabetes, burns, dehydration, myocardial infarction | Liver failure, malnutrition, overhydration, pregnancy |
| **Carbon dioxide content** (bicarbonate + dissolved $CO_2$) (whole blood) | Arterial: 19–24 mEq/L (19–24 mmol/L)<br>Venous: 22–26 mEq/L (22–26 mmol/L) | Severe diarrhea, severe vomiting, starvation, emphysema, aldosteronism | Renal failure, diabetic ketoacidosis, shock |
| **Cholesterol, total** (plasma)<br>**HDL cholesterol** (plasma)<br>**LDL cholesterol** (plasma) | <200 mg/dL (<5.2 mmol/L) is desirable<br>>40 mg/dL (>1.0 mmol/L) is desirable<br><130 mg/dL (<3.2 mmol/L) is desirable | Hypercholesterolemia, uncontrolled diabetes mellitus, hypothyroidism, hypertension, atherosclerosis, nephrosis | Liver disease, hyperthyroidism, fat malabsorption, pernicious or hemolytic anemia, severe infections |

| | | | |
|---|---|---|---|
| **Creatine** (serum) | Males: 0.15–0.5 mg/dL (10–40 μmol/L)<br>Females: 0.35–0.9 mg/dL (30–70 μmol/L) | Muscular dystrophy, damage to muscle tissue, electric shock, chronic alcoholism. | |
| **Creatine kinase (CK),** also known as **creatine phosphokinase (CPK)** (serum) | 0–130 U/L (same) | Myocardial infarction, progressive muscular dystrophy, hypothyroidism, pulmonary edema. | |
| **Creatinine** (serum) | 0.5–1.2 mg/dL<br>(45–105 μmol/L) | Impaired renal function, urinary tract obstruction, giantism, acromegaly. | Decreased muscle mass, as occurs in muscular dystrophy or myasthenia gravis. |
| **Gamma-glutamyl transferase (GGT)** (serum) | 0–30 U/L (same) | Bile duct obstruction, cirrhosis, alcoholism, metastatic liver cancer, congestive heart failure. | |
| **Glucose** (plasma) | 70–110 mg/dL (3.9–6.1 mmol/L) | Diabetes mellitus, acute stress, hyperthyroidism, chronic liver disease, Cushing's disease. | Addison's disease, hypothyroidism, hyperinsulinism. |
| **Hemoglobin** (whole blood) | Males: 14–18 g/100 mL (140–180 g/L)<br>Females: 12–16 g/100 mL (120–160 g/L)<br>Newborns: 14–20 g/100 mL<br>(140–200 g/L) | Polycythemia, congestive heart failure, chronic obstructive pulmonary disease, living at high altitude. | Anemia, severe hemorrhage, cancer, hemolysis, Hodgkin's disease, nutritional deficiency of vitamin $B_{12}$, systemic lupus erythematosus, kidney disease. |
| **Iron, total** (serum) | Males: 80–180 μg/dL (14–32 μmol/L)<br>Females: 60–160 μg/dL (11–29 μmol/L) | Liver disease, hemolytic anemia, iron poisoning. | Iron-deficiency anemia, chronic blood loss, pregnancy (late), chronic heavy menstruation. |
| **Lactic dehydrogenase (LDH)** (serum) | 71–207 U/L (same) | Myocardial infarction, liver disease, skeletal muscle necrosis, extensive cancer. | |
| **Lipids** (serum)<br>**Total**<br>**Triglycerides** | <br>400–850 mg/dL (4.0–8.5 g/L)<br>10–190 mg/dL (0.1–1.9 g/L) | Hyperlipidemia, diabetes mellitus. | Fat malabsorption, hypothyroidism. |
| **Platelet (thrombocyte) count** (whole blood) | 150,000–400,000/μL | Cancer, trauma, leukemia, cirrhosis. | Anemias, allergic conditions, hemorrhage. |
| **Protein** (serum)<br>**Total**<br>**Albumin**<br>**Globulin** | <br>6–8 g/dL (60–80 g/L)<br>4–6 g/dL (40–60 g/L)<br>2.3–3.5 g/dL (23–35 g/L) | Dehydration, shock, chronic infections. | Liver disease, poor protein intake, hemorrhage, diarrhea, malabsorption, chronic renal failure, severe burns. |
| **Red blood cell (erythrocyte) count** (whole blood) | Males: 4.5–6.5 million/μL<br>Females: 3.9–5.6 million/μL | Polycythemia, dehydration, living at high altitude. | Hemorrhage, hemolysis, anemias, cancer, overhydration. |
| **Uric acid (urate)** (serum) | 2.0–7.0 mg/dL<br>(120–420) μmol/L | Impaired renal function, gout, metastatic cancer, shock, starvation. | |
| **White blood cell (leukocyte) count, total** (whole blood) | 5,000–10,000/μL (See Table 18.5 for relative percentages of different types of WBCs.) | Acute infections, trauma, malignant diseases, cardiovascular diseases. (See also Table 18.4.) | Diabetes mellitus, anemia. (See also Table 18.4.) |

# Appendix D
## Normal Values for Selected Urine Tests

Urine Tests

| TEST (SPECIMEN) | U.S. REFERENCE VALUES (SI UNITS) | CLINICAL IMPLICATIONS |
|---|---|---|
| **Amylase** (2 hour) | 35–260 Somogyi units/hr (6.5–48.1 units/hr) | Values increase in inflammation of the pancreas (pancreatitis) or salivary glands, obstruction of the pancreatic duct, and perforated peptic ulcer. |
| **Bilirubin*** (random) | Negative | Values increase in liver disease and obstructive biliary disease. |
| **Blood*** (random) | Negative | Values increase in renal disease, extensive burns, transfusion reactions, and hemolytic anemia. |
| **Calcium ($Ca^{2+}$)** (random) | 10 mg/dL (2.5 mmol/liter); up to 300 mg/24 hr (7.5 mmol/24 hr) | Amount depends on dietary intake; values increase in hyperparathyroidism, metastatic malignancies, and primary cancer of breasts and lungs; values decrease in hypoparathyroidism and vitamin D deficiency. |
| **Casts** (24 hour) | | |
| **Epithelial** | Occasional | Values increase in nephrosis and heavy metal poisoning. |
| **Granular** | Occasional | Values increase in nephritis and pyelonephritis. |
| **Hyaline** | Occasional | Values increase in kidney infections. |
| **Red blood cell** | Occasional | Values increase in glomerular membrane damage and fever. |
| **White blood cell** | Occasional | Values increase in pyelonephritis, kidney stones, and cystitis. |
| **Chloride ($Cl^-$)** (24 hour) | 140–250 mEq/24 hr (140–250 mmol/24 hr) | Amount depends on dietary salt intake; values increase in Addison's disease, dehydration, and starvation; values decrease in pyloric obstruction, diarrhea, and emphysema. |
| **Color** (random) | Yellow, straw, amber | Varies with many disease states, hydration, and diet. |
| **Creatinine** (24 hour) | Males: 1.0–2.0 g/24 hr (9–18 mmol/24 hr)<br>Females: 0.8–1.8 g/24 hr (7–16 mmol/24 hr) | Values increase in infections; values decrease in muscular atrophy, anemia, and kidney diseases. |
| **Glucose*** | Negative | Values increase in diabetes mellitus, brain injury, and myocardial infarction. |
| **Hydroxycorticosteroids** (17-hydroxysteroids) (24 hour) | Males: 5–15 mg/24 hr (13–41 μmol/24 hr)<br>Females: 2–13 mg/24 hr (5–36 μmol/24 hr) | Values increase in Cushing's syndrome, burns, and infections; values decrease in Addison's disease. |

| | | |
|---|---|---|
| **Ketone bodies*** (random) | Negative | Values increase in diabetic acidosis, fever, anorexia, fasting, and starvation. |
| **17-ketosteroids** (24 hour) | Males: 8–25 mg/24 hr (28–87 μmol/24 hr)<br>Females: 5–15 mg/24 hr (17–53 μmol/24 hr) | Values decrease in surgery, burns, infections, adrenogenital syndrome, and Cushing's syndrome. |
| **Odor** (random) | Aromatic | Becomes acetonelike in diabetic ketosis. |
| **Osmolality** (24 hour) | 500–1400 mOsm/kg water<br>(500–1400 mmol/kg water) | Values increase in cirrhosis, congestive heart failure (CHF), and high-protein diets; values decrease in aldosteronism, diabetes insipidus, and hypokalemia. |
| **pH*** (random) | 4.6–8.0 | Values increase in urinary tract infections and severe alkalosis; values decrease in acidosis, emphysema, starvation, and dehydration. |
| **Phenylpyruvic acid** (random) | Negative | Values increase in phenylketonuria (PKU). |
| **Potassium ($K^+$)** (24 hour) | 40–80 mEq/24 hr (40–80 mmol/24 hr) | Values increase in chronic renal failure, dehydration, starvation, and Cushing's syndrome; values decrease in diarrhea, malabsorption syndrome, and adrenal cortical insufficiency. |
| **Protein*** (albumin) (random) | Negative | Values increase in nephritis, fever, severe anemias, trauma, and hyperthyroidism. |
| **Sodium ($Na^+$)** (24 hour) | 75–200 mg/24 hr (75–200 mmol/24 hr) | Amount depends on dietary salt intake; values increase in dehydration, starvation, and diabetic acidosis; values decrease in diarrhea, acute renal failure, emphysema, and Cushing's syndrome. |
| **Specific gravity*** (random) | 1.001–1.035 (same) | Values increase in diabetes mellitus and excessive water loss; values decrease in absence of antidiuretic hormone (ADH) and severe renal damage. |
| **Urea** (random) | 25–35 g/24 hr (420–580 mmol/24 hr) | Values increase in response to increased protein intake; values decrease in impaired renal function. |
| **Uric acid** (24 hour) | 0.4–1.0 g/24 hr (1.5–4.0 mmol/24 hr) | Values increase in gout, leukemia, and liver disease; values decrease in kidney disease. |
| **Urobilinogen*** (2 hour) | 0.3–1.0 Ehrlich units (1.7–6.0 μmol/24 hr) | Values increase in anemias, hepatitis A (infectious), biliary disease, and cirrhosis; values decrease in cholelithiasis and renal insufficiency. |
| **Volume, total** (24 hour) | 1000–2000 mL/24 hr (1.0–2.0 liters/24 hr) | Varies with many factors. |

*Test often performed using a **dipstick**, a plastic strip impregnated with chemicals that is dipped into a urine specimen to detect particular substances. Certain colors indicate the presence or absence of a substance and sometimes give a rough estimate of the amount(s) present.

# Appendix E
## Answers to Checkpoint Questions

### Chapter 1

**1.** (a) molecules, (b) chemical level **2.** chemical, cellular, tissues **3.** organ **4.** organs: composed of different kinds of tissues which join together; tissues: groups of cells and the materials around them which work together to perform a particular function **5.** movement of blood, food through the GI tract, skeletal muscles **6.** responsiveness **7.** intracellular fluid: inside cells; extracellular fluid: outside cells; interstitial fluid: between cells of tissues **8.** cerebrospinal fluid **9.** nervous system, endocrine system **10.** effector **11.** nervous system, endocrine system **12.** (a) both occur in response to a stimulus to regulate a controlled condition; (b) negative feedback reverses the stimulus; positive feedback enhances the stimulus (c) negative feedback systems **13.** if unstopped, the mechanism can produce life-threatening conditions in the body **14.** subject stands erect facing the observer, head is level, eyes face forward, feet are flat on the floor and directed forward, upper limbs are at the sides with palms turned forward **15.** front of the elbow **16.** (a) posterior; (b) medial; (c) distal; (d) superficial **17.** (a) proximal and distal; (b) medial and lateral; (c) medial and lateral; (d) superior and inferior **18.** planes are imaginary flat surfaces that pass through various parts of the body; a section is a cut through a structure along a plane **19.** thoracic and abdominopelvic **20.** urinary bladder, P; stomach, A; heart, T; small intestine, A; lungs, T; thymus, T; liver, A **21.** heart, esophagus, trachea, thymus **22.** (a) pericardial; (b) pleural **23.** large intestine **24.** reduce friction **25.** between the abdominopelvic wall and the peritoneum **26.** liver, epigastric; colon, umbilical; bladder, hypogastric; spleen, left lumbar **27.** the right or left of the person you are facing

### Chapter 2

**1.** electrons **2.** (a) proton, positive; neutron, zero; electron, negative (b) the neutral atom has a zero charge because the negatively charged electrons and positively charged protons equal each other **3.** (a) atomic number, 6; mass number, 12; atomic mass, 12.011; (b) atomic number is the number of protons in an atom; the mass number is the number of protons plus the number of neutrons in an atom; the atomic mass is the average mass of all of the naturally occurring isotopes of the element **4.** (a) isotopes are atoms of an element that have different numbers of neutrons, and therefore different mass numbers (b) a free radical is an electrically charged atom or group of atoms with an unpaired electron in the outermost shell **5.** the outermost electron shell around an atom; the number of electrons in the valence shell will determine whether or not another atom will form a chemical bond with that atom **6.** ionic bond involves transfer of electrons from one atom to another atom; covalent bond is sharing of electrons between two atoms **7.** a slightly positive charge of a hydrogen atom forms a weak bond with a negative charge of a neighboring atom **8.** (a) reactants are the starting substances in a reaction; products are the ending substances; (b) metabolism **9.** the reaction shown releases more energy than it absorbs **10.** they lower the amount of energy needed to start the reaction **11.** (a) anabolism: a chemical reaction that forms a new and larger molecule; catabolism: a chemical reaction that splits larger molecules into smaller chemical components (b) anabolism reactions **12.** reactions in which products can revert back to the original reactants **13.** inorganic compounds lack carbon and are held together by ionic or covalent bonds; organic compounds always contain carbon and always have covalent bonds **14.** acts as a solvent, acts as a lubricant; actively participates in some decomposition and synthesis reactions; can release or absorb a relatively large amount of heat with only a slight change in its own temperature **15.** solvent is a liquid or gas in which some other substance (a solute) is dissolved; solute is the dissolved substance in a solution **16.** hydrophilic molecules are easily dissolved in water because they are charged or contain polar covalent bonds; hydrophobic molecules are not very water-soluble because they contain mainly nonpolar covalent bonds **17.** breakdown of large molecules into smaller molecules by the addition of water molecules **18.** 6.82 **19.** They convert strong acids or bases into weak acids or bases, thereby maintaining the pH of fluids inside and outside body cells. **20.** it can form bonds with thousands of other carbon atoms to produce large molecules, and it has four electrons in its valence shell, so can bond covalently with a variety of atoms **21.** monomers are small, building-block molecules; polymers are macromolecules made up of monomers bonded together **22.** primary source of chemical energy for generating ATP **23.** monosaccharides contain from three to seven carbon atoms; a disaccharide is a combination of two monosaccharide molecules; polysaccharides contain tens or hundreds of monosaccharides joined together **24.** liver cells and skeletal muscle cells **25.** they have few polar covalent bonds and are hydrophobic **26.** (a) saturated fats contain only single covalent bonds between fatty acid -carbon atoms; monounsaturated fats contain fatty acids with one double covalent bond between two fatty acid carbon atoms; polyunsaturated fats contain more than one double covalent bond between fatty acid carbon atoms; (b) saturated fats: meats, non-skim dairy products, a few plant products; monounsaturated fats: olive oil, peanut oil, canola oil, most nuts, and avocadoes; polyunsaturated fats: corn oil, safflower oil, sunflower oil, soybean oil, and fatty fish **27.** (a) a triglyceride consists of a single glycerol backbone with three fatty acid molecules attached to it; a triglyceride has a glycerol backbone with two fatty acid molecules and a phosphate group attached to it (b) The polar phosphate group and charged group form the hydrophilic head and the nonpolar fatty acids form the hydrophobic tails. **28.** triglycerides are the body's most highly concentrated form of chemical energy; phospholipids make up much of the membrane that surrounds each cell; steroids are required for cell membrane structure (cholesterol), regulating sexual functions (estrogen and progesterone), maintaining normal blood sugar levels (cortisol), digesting and absorbing lipids (bile salts), and bone growth (vitamin D) **29.** proteins contain carbon, hydrogen, oxygen and nitrogen and are much more complex than carbohydrates or lipids **30.** (a) the covalent bond joining two amino acids together; (b) 2 **31.** peptideis a chain of four to nine amino acids; polypeptide is a chain of 10 to 2000 or more amino acids; protein is a polypeptide that contains as few as 50 or as many as 2000 amino acids **32.** specificity; efficiency; control **33.** they lower the activation energy of a chemical reaction by decreasing the randomness of the collisions between molecules and bring the substrates together in the proper orientations so that the reaction can occur **34.** DNA forms the inherited genetic material inside each human cell; RNA relays instructions from the genes to guide each cell's synthesis of proteins from amino acids **35.** a nitrogenous base, a five-carbon sugar, and a phosphate group **36.** muscle contractions, movement of chromosomes during cell division, movement of structures within cells, transport of substances across cell membranes, and synthesis of larger molecules from smaller ones **37.** three phosphate groups attached to adenosine

## Chapter 3

**1.** cytoplasm is all the cell components between the plasma membrane and nucleus while cytosol is the liquid portion of the cytoplasm **2.** the hydrophobic ends of the lipids face inward and the hydrophilic ends face outward **3.** the carbohydrate portions of glycolipids and glycoproteins which enable cells to recognize one another **4.** uncharged, non-polar molecules and some small uncharged, polar molecules can pass through the lipid bilayer; ions and large, uncharged polar molecules **5.** selective permeability allows some substances to pass readily across the membrane and prohibits others from crossing **6.** differences in chemical concentration and electrical charges between the cytosol and extracellular fluid sides of the plasma membrane **7.** energy input from the cell **8.** steepness of the concentration gradient, temperature, mass of the diffusing substances, surface area, diffusion distance; **9.** a force proportional to the concentration of the solute particles that cannot cross the plasma membrane **10.** the key difference is the source of energy used to drive the process: energy obtained from hydrolysis of ATP powers primary active transport, and energy stored in an ionic concentration gradient powers secondary active transport **11.** symporters move substances in the same direction; antiporters move -substances in opposite directions **12.** an active process by which vesicles are used to successively move a substance into, through, and out of a cell **13.** (a) microfilaments; (b) microtubules **14.** cilia move substances across the cell surface; a flagellum moves the entire cell **15.** they are formed in the nucleolus and assembled in the cytoplasm **16.** rough ER has ribosomes attached to the membrane and is involved in the synthesis of proteins; smooth ER lacks ribosomes and is involved in the synthesis of lipids **17.** extracellular fluid, plasma membrane, other cytoplasmic organelles such as lysosomes **18.** process by which lysosomes digest an entire cell **19.** ATP is produced by the aerobic phase of cellular respiration **20.** via nuclear pores **21.** in chromosomes **22.** double-stranded DNA, linker DNA, and histone proteins **23.** transcription is the copying of a region of DNA into a complementary copy called mRNA; translation is the production of a protein based on the mRNA copy **24.** mRNA enters the cytoplasm to direct the synthesis of a specific protein; rRNA joins with ribosomal proteins to make ribosomes; tRNA binds to an amino acid and holds it in place until it is incorporated into a protein during translation **25.** P site holds the first tRNA with its amino acid; A site holds the next tRNA with another amino acid **26.** in somatic cell division, a cell undergoes mitosis and cytokinesis to produce two identical cells; reproductive cell division is a mechanism that produces gametes which have half the chromosome number of the somatic cells **27.** $G_1$ phase: cell growth, organelle duplication, beginning of centriole replication; S phase: DNA replication; $G_2$ phase: synthesis of enzymes and other proteins, completion of centrosome replication **28.** prophase: chromatin fibers condense into paired chromatids, nucleolus and nuclear enveolope disappear, centrosomes move to opposite poles of the cell; metaphase: centromeres of chromatid pairs line up at the metaphase plate; anaphase: centromeres split, identical sets of chromosomes move to opposite poles of the cell; telophase: nuclear envelopes and nucleoli reappear, chromosomes resume chromatin form, mitotic spindle disappears **29.** in late anaphase **30.** in anaphase I of meiosis chromosome pairs are separated; in mitotic anaphase and anaphase of meiosis II, the chromatids of the individual chromosomes are separated **31.** to reduce the chromosome number by one-half

## Chapter 4

**1.** groups of cells and the materials surrounding them that work together to perform a particular function **2.** epithelial tissue, connective tissue, muscular tissue, nervous tissue **3.** they form tight seals between cells, anchor cells to one another, provide channels for ions and molecules to pass from one cell to another, allow communication between cells **4.** apical surface **5.** (a) attach to and support the overlying epithelial tissues, anchors it to underlying connective tissue, form a surface along which epithelial cells migrate during growth or wound healing, restrict passage of molecules between epithelium and connective tissue; (b) basal lamina containing glycoproteins and reticular lamina containing collagen proteins **6.** (a) simple, pseudostratified, or stratified epithelium (b) squamous, cuboidal, or columnar **7.** (a) endothelium is found lining the heart chambers and the lumen of blood vessels and lymphatic vessels; mesothelium is the epithelial layer of serous membranes; (b) both are composed of simple squamous epithelium **8.** goblet cells secrete mucus; microvilli increase surface area; cilia sweep substances across epithelial apical surfaces **9.** to help protect the skin and underlying tissues from heat, microbes, and chemicals **10.** it binds together, supports, and strengthens other body tissues; protects and insulates internal organs; compartmentalizes structures such as skeletal muscles; is the major transport system within the body; is the major site of stored energy; and is the main source of immune responses **11.** these are cells that retain the capacity for cell division and secrete the extracellular matrix **12.** (a) adipocytes (b) white blood cells (c) blast cells **13.** (a) collagen fibers; (b) reticular fibers **14.** (a) fibrocartilage; (b) elastic cartilage; (c) hyaline cartilage **15.** since epithelium is avascular, it depends on the blood vessels in connective tissue for oxygen, nutrients, and waste disposal **16.** a flat sheet of pliable tissue that covers or lines a part of the body and consists of a combination of an epithelial layer and an underlying connective tissue layer **17.** mucous membranes line body cavities that open directly to the exterior and function as a barrier that microbes and other pathogens have difficulty penetrating; serous membranes line body cavities that do not open directly to the exterior and cover the organs that lie within the cavity and functions to allow organs to glide easily over one another or to slide against the walls of cavities; cutaneous membranes cover the surface of the body and regulate body temperature, protect underlying tissues, provide cutaneous sensations, excrete body wastes, and synthesize vitamin D **18.** (a) prevents the cavities from drying out, traps particles in the respiratory passageways, and lubricates food as it moves through the GI tract (b) mucous membranes (c) goblet cells and other epithelial cells **19.** (a) visceral layer of the peritoneum; (b) line body cavities that do not open directly to the exterior and cover the organs that lie within the cavity to allow -organs to glide easily over one another or to slide against the walls of the cavities **20.** (a) lining the cavities of freely moving joints; (b) lubricates and nourishes the cartilage covering the bones at movable joints, contains macrophages that remove microbes and debris from the joint cavity, line cushioning sacs called bursae and tendon sheaths to ease the movement of muscle tendons **21.** produce movements, maintain posture, generate heat, provide protection, pump blood throughout the body **22.** (a) skeletal muscle has striations and long fibers; (b) cardiac muscle has branched fibers and intercalated discs; (c) smooth muscle has spindle-shaped fibers and lacks striations **23.** (a) smooth muscle; (b) cardiac muscle; (c) skeletal muscle **24.** strengthen cardiac muscle tissue, hold the fibers together during contractions, and provide a route for quick conduction of electrical impulses from one cardiac muscle cell to another **25.** (a) dendrite (b) axon **26.** neurons generate and conduct nerves impulses; neuroglia support, nourish, and protect the neurons **27.** epithelial, connective, muscle, nervous tissue **28.** osseous tissue is vascular and cartilage is avascular **29.** parenchyma cells, which constitute the functioning part of a tissue or organ, can regenerate; stroma, the supporting connective tissue, will form scar tissue when injured **30.** stromal repair results in scar tissue; parenchymal repair results in tissue regeneration **31.** provides a supporting framework through which epithelial cells migrate to fill the wounded area and secretes a fluid that kills bacteria

## Chapter 5

**1.** skin, hair, skin glands, nails **2.** areolar and adipose tissues **3.** keratinized stratified squamous epithelium **4.** (a) they release a lipid-rich water-repellent secretion that fills the spaces between cells of the stratum granulosum, stratum lucidum, and stratum corneum; (b) contributes to skin color and absorbs damaging UV light **5.** stratum basale **6.** stratum basale: a single row of cuboidal or columnar keratinocytes just above the dermis; stratum spinosum: layers of polyhedral keratinocytes that appear to have spines when viewed on a microscope slide; stratum granulosum: presence of membrane-enclosed lamellar granules within the keratinocytes; stratum lucidum: clear layer found only in thick skin; stratum corneum: multiple layers of dead, flat keratinocytes **7.** papillary region contains dermal papillae and thin collagen and fine elastic fibers; reticular region contains thick collagen fibers, some elastic fibers, fibroblasts, adipose cells, and a variety of wandering cells **8.** provide the skin with strength, extensibility, and elasticity **9.** the cut is only in the epidermal region **10.** they are areas of accumulated melanin **11.** melanin: gives the skin color ranging from pale yellow to tan to black; carotene: gives the skin color ranging from yellow to red to tan to black; hemoglobin: gives skin color ranging from pink to red **12.** protection, decrease in heat loss, sensing light touch **13.** medulla, cortex, cuticle **14.** contains blood vessels that nourish the growing hair follicle **15.** autonomic stimulation of the arrector pili muscle causing contraction and pulling the hair shafts into an upright position **16.** sebum: coats the surface of hairs and helps keep them from drying and becoming brittle, prevents excessive evaporation of water from the skin, keeps the skin soft and pliable, inhibits growth of certain bacteria; eccrine sweat: helps regulate body temperature through evaporation; cerumen: provides a sticky barrier that impedes the entrance of foreign bodies into the external auditory canal **17.** help grasp and manipulate small objects, provide protection against trauma to the ends of the fingers and toes, allow us to scratch various parts of the body **18.** thick skin has more stratum corneum layers **19.** by liberating sweat at the skin surface and by adjusting the flow of blood in the dermis **20.** keratinocyte activity, release of lipids by lamellar granules, retention of moisture and bactericidal action of sebum, the acidic pH of sweat, vasoconstriction **21.** touch, pressure, vibration, tickling, heat, cold, pain **22.** stratum basale **23.** it prepares the wound for repair by helping eliminate microbes, foreign material, and dying tissue and increases the permeability and diameter of local blood vessels **24.** (a) scar tissue formation; (b) epidermal tissue is mitotic tissue; its stem cells can divide and replace the lost tissue

## Chapter 6

**1.** calcium and phosphorus **2.** (a) short bone; (b) long bone **3.** flat bone **4.** (a) sesamoid bone **5.** diaphysis is the long cylindrical main portion of the bone; epiphyses are at the proximal and distal ends of the bone; metaphyses are between the diaphysis and each of the epiphyses **6.** epiphyseal plate: located in the metaphysis, composed of hyaline cartilage, is the growth plate; articular cartilage: covers the part of the epiphysis where the bone forms an articulation with another bone, composed of hyaline cartilage, reduces friction and absorbs shock at freely movable joints; periosteum: surrounds the bone surface wherever it is not covered by articular cartilage, is a sheath of connective tissue containing osteoblasts, functions to protect the bone, assists in fracture repair, nourishes bone tissue, and serves as an attachment point for ligaments and tendons **7.** water, collagen fibers, and crystalized mineral salts called hydroxyapatite (mainly, calcium phosphate and calcium hydroxide) **8.** (a) osteogenic cells; (b) osteoblasts **9.** osteoclasts **10.** osteoblasts form extracellular matrix ; osteocytes maintain daily metabolism; osteoclasts: break down extracellular matrix **11.** Compact bone tissue is arranged into osteons packed closely together while spongy bone tissue contains trabeculae arranged in a lattice-like network with spaces filled with red marrow. Compact bone tissue found beneath the periosteum of all bones and makes up the bulk of the diaphyses of long bones. Spongy bone tissue is found in the epiphyses of long bones and in the interior of short, flat, and irregular bones. Compact bone tissue protects and supports while spongy bone tissue supports and protects red bone marrow.**12.** hemopoietic (red) bone marrow **13.** through numerous perforating canals **14.** near the center of the diaphysis via the nutrient foramen **14.** via nutrient veins and periosteal veins in the diaphysis, metaphyseal veins in the metaphysis, and epiphyseal veins in the epiphysis **15.** (a) development of the ossification center, calcification, formation of trabeculae, and development of the periosteum; (b) development of the cartilage model, growth of the cartilage model, development of the primary ossification center; (c) development of the medullary cavity, development of the secondary ossification centers, formation of articular cartilage and the epiphyseal plate **16.** primary ossification center forms in the diaphysis, secondary ossification centers form in the epiphysis **17.** through the action of osteoclasts **18.** (a) endochondrial ossification; (b) intramembraneous ossification **19.** new chondrocytes are formed on the epiphyseal side of the plate, while old chondrocytes on the diaphyseal side of the plate are replaced by osseous tissue **20.** the epiphyseal cartilage cells stop dividing and bone replaces the cartilage **21.** osteoblasts beneath the periosteum add new extracellular matrix to the surface of the bone **22.** osteoclasts in the endosteum breaking down osseous tissue lining the medullary cavity **23.** (a) the ongoing replacement of old osseous tissue by new osseous tissue (b) osteoblasts produce new extracellular matrix and osteoclasts breakdown old extracellular matrix **24.** it renews osseous tissue before deterioration sets in, it heals injured bone, it redistributes bone along lines of mechanical stress **25.** calcium, phosphorus, fluoride, magnesium iron, manganese, vitamins A, C, $B_{12}$, D, and K **26.** human growth hormone stimulates the production and release of insulinlike growth factors, which stimulate bone growth **27.** bone growth and remodeling, muscle and nerve cell function, and blood clotting

## Chapter 7

**1.** axial skeleton: skull, vertebral column; appendicular skeleton: clavicle, shoulder girdle, humerus, pelvic girdle, femur **2.** allow passage of blood vessels and nerves or help form joints **3.** form joints or attachment points for connective tissue **4.** cranial bones: enclose and protect the brain, provide large areas for muscle attachment, protect and support the special sense organs; facial bones: form the face, protect and support the special sense organs **5.** passageway for blood vessels, nerves, or ligaments **6.** frontal, parietal, temporal, occipital, sphenoid, ethmoid, inferior nasal conchae, vomer **7.** (a) frontal, sphenoid, temporal, occipital; (b) sphenoid, parietal, occipital, mandible, zygomatic **8.** (a) temporal process of the zygomatic bone and zygomatic process of the temporal bone; (b) mandibular fossa of the temporal bone and the condylar process of the mandible **9.** (a) occipital condyles; (b) external occipital protuberance **10.** ethmoid, frontal, parietal, temporal and occipital bones **11.** (a) internal auditory meatus allows impulses for hearing and balance to be carried to the brain; external auditory meatus directs sound waves into the ear (b) temporal bone **12.** (a) increase the vascular and mucous membrane surface area in the nasal cavities, which aids in the sense of smell and warms, moistens, and filters inhaled air before it passes into the lungs; (b) superior and middle nasal concha **13.** nasal bones **14.** (a) maxillae and palatine bones; (b) hard palate **15.** Lacrimal bones **16.** alveolar processes **17.** vomer **18.** mandible **19.** condylar process of the mandible articulates with the mandibular fossa of the temporal bone **20.** (a) vomer, septal cartilage, perpendicular plate of the ethmoid bone; (b) hyaline cartilage **21.** (a) frontal, ethmoid, sphenoid, and maxillary bones; (b) produce mucus, light the skull, serve as resonating chambers for sound **22.** (a) parietal and temporal bones; (b) both parietal bones and the occipital bone; (c) frontal bone and both parietal bones; (d) the two parietal bones **23.** anterolateral fontanels **24.** 18 to 24 months after birth **25.** it supports the tongue and provides attachment sites for some tongue muscles and for muscles of the pharynx and larynx **26.** encloses and protects the spinal cord, supports the head, serves as a point of attachment for the ribs, pelvic, girdle, and muscles of the back **27.** (a) cervical (7), thoracic (12), lumbar (5), sacrum (1), coccyx (1) **28.** (a) increase its strength, help maintain balance in the upright position, absorb shocks during walking, and help protect the vertebrae from fracture; (b) thoracic and sacral **29.** form strong joints, permit various movements of the vertebral column, absorb vertical shock **30.** vertebral body **31.** pedicles, laminae **32.** vertebral foramen **33.** (a) two transverse processes, a spinous process; (b) two superior articular processes and two inferior articular processes; (c) facets or demifacets **34.** (a) cervical; (b) allow passage of the vertebral artery to supply blood to the brain **35.** cervical: slender and often bifid processes; thoracic: long and fairly thick processes with most projecting inferiorly; lumbar: short, blunt, and project posteriorly rather than inferiorly **36.** atlas (first) and axis (second) **37.** (a) occipital bone and atlas; (b) axis and atlas **38.** (a) thoracic vertebrae; (b) the body and transverse processes **39.** (a) sacral canal and sacral hiatus; (b) auricular surface and sacral tuberosity **40.** manubrium, body, xiphoid process **41.** (a) clavicles, costal cartilage of first and second ribs, and body of the sternum (b) costal cartilages of the second through tenth ribs, manubrium, xiphoid process **42.** provide attachment for some abdominal muscles **43.** costal cartilage **44.** (a) by its head and the tubercle; (b) intercostal space; (c) it provides surgical access to the lungs or other structures in the thoracic cavity

## Chapter 8

**1.** (a) sternal end and manubrium of the sternum (b) acromial end and acromion of the scapula **2.** glenoid cavity and head of the humerus **3.** supraspinous fossa and infraspinous fossa **4.** 30 bones **5.** greater tubercle **6.** olecranon **7.** (a) head of radius to capitulum of humerus, trochlear notch of ulna to trochlea of humerus; (b) coronoid fossa of the humerus receives coronoid process of ulna; (c) olecranon fossa of humerus receives olecranon of ulna **8.** pisiform and hamate (on ulnar side) and scaphoid and trapezium (on radial side) **9.** (a) metacarpals; (b) distal ends of metacarpals **10.** (a) accepts the bones of the lower limbs and connects them to the axial skeleton; (b) provides a strong, stable support for the vertebral column and protects the pelvic viscera **11.** hip bone: fused ilium, ischium, and pubis; pelvic girdle: the two hip bones; bony pelvis: complete ring composed of the hip bones, pubic symphysis, and sacrum **12.** (a) anterior superior iliac spine, anterior inferior iliac spine, posterior superior iliac spine, posterior inferior iliac spine; (b) serve as points of attachment for the muscles of the trunk, hips, and thighs **13.** (a) auricular surface of the ilium; (b) acetabulum formed by the ilium, ischium, and pubis **14.** it is the route taken by the baby's head as it descends through the pelvis **15.** because of the female's adaptations to the requirements of pregnancy and childbirth **16.** more space in the true pelvis especially in the pelvic inlet and pelvic outlet **17.** femur **18.** (a) greater and lesser trochanters, (b) greater trochanter **19.** (a) sesamoid bone; (b) it is enclosed within the tendon of anterior thigh muscles **20.** (a) talus; (b) calcaneus

## Chapter 9

**1.** based on either structure or function **2.** (a) cartilaginous (b) synarthrosis (c) diarthrosis **3.** (a) skull; (b) articulations between the roots of the teeth and the alveolar sockets in the maxillae and mandible **4.** A synchondrosis is composed of hyaline cartilage but a symphysis is composed of hyaline cartilage and fibrocartilage **5.** (a) synchondroses; (b) synostosis **6.** articular cartilage **7.** reducing friction by lubrication of the joint, absorbing shocks, supplying oxygen and nutrients to and removing carbon dioxide and metabolic wastes from the chondrocytes within articular cartilage, and supplying phagocytic cells that remove microbes and debris which result from normal wear and tear in the joint **8.** diarthroses **9.** ligaments and articular menisci **10.** (a) they resemble joint capsules because their walls consist of connective tissue lined by a synovial membrane and they are also filled with synovial fluid; they are different from joint capsules because they are located outside joints; (b) cushion the movement of body parts over one another **11.** (a) extension; (b) circumduction; (c) lateral rotation; (d) adduction; (e) elevation **12.** Adduction has "add" in it's name and it adds a bone toward the midline.**13.** abduction and adduction **14.** opposition **15.** because the movement between the atlas and axis is allowed by the atlantoaxial joint which is a pivot joint **16.** condyloid joints are found in the wrists and saddle joints are found in the fingers **17.** one axis: hinge, pivot; two axes: condyloid, saddle; three axes: ball-and-socket; no axes: planar **18.** (a) coracohumeral ligaments; (b) glenohumeral ligaments **19.** rotator cuff muscles **20.** (a) ulnar collateral ligament; (b) radial collateral ligament **21.** flexion and extension **22.** (a) iliofemoral ligament; (b) pubofemoral ligament **23.** (a) patellar ligament; (b) fibular collateral ligament

## Chapter 10

**1.** (a) skeletal; (b) cardiac and smooth **2.** smooth muscle is spindle shaped and does not have striations **3.** producing body movements, stabilizing body positions, storing and moving substances within the body, generating heat **4.** electrical excitability, contractility, extensibility, elasticity **5.** electrical excitability **6.** supports and surrounds muscles and other organs, allows free movement of muscles, carries nerves and blood and lymphatic vessels, and fills spaces **7.** A tendon is a cord of dense connective tissue, an aponeurosis is a broad, flat sheet of dense connective tissue **8.** T tubules: propagate action potential into the muscle fiber; myoglobin: binds oxygen molecules that diffuse into muscle fibers from interstitial fluid and releases oxygen when mitochondria need it for ATP production; sarcoplasmic reticulum: stores calcium ions **9.** thick filament, myofibril, muscle fiber **10.** (a) I band contains thin filaments; A band contains thick and thin filaments; H zone contains thick filaments (b) thick filament and titin (c) thin filament and titin **11.** (a) myosin (b) actin, tropomyosin, and troponin **12.** motor end plate **13.** Step 1- calcium ions enter synaptic end bulb which leads to acetylcholine release into synaptic cleft; Step 2-ACh binds to ACh receptors causing their activation; Step 3- inflow of sodium ions across the sarcolemma cause the generation of a muscle action potential **14.** it is broken down by acetylcholinesterase **15.** (a) it decreases; (b) it increases **16.** other myosin heads on the myosin filament are still attached **17.** Z discs pull on neighboring sarcomeres, whole muscle fiber shortens, entire muscle shortens, which tightens tendons, moving the associated bone **18.** calcium ion release channels in the SR close and calcium ion active transport pumps use ATP to restore low levels of calcium ions in the sarcoplasm, the troponin–tropomyosin complex slides back into position where it blocks the myosin-binding sites on the actin molecules, then the muscle relaxes **19.** provide energy for myosin head movement, detachment of the myosin head from actin, and for operation of the calcium active transport pump **20.** one motor neuron and all the skeletal muscle fibers it stimulates **21.** (a) the muscle action potential sweeps over the sarcolemma, calcium ions are released from the sarcoplasmic reticulum and bind to troponin which allows the myosin heads to start binding to actin; (b) calcium ions are transported back into the sarcoplasmic reticulum, the level of calcium ions decreases in the sarcoplasm, myosin-binding sites are covered by tropomyosin, myosin heads detach from actin, power strokes cease, the muscle fiber relaxes **22.** (a) unfused tetanus: a sustained but wavering contraction; fused tetanus: a sustained contraction in which individual twitches cannot be detected;(b) fused tetanus **23.** both kinds of tetanus result from the release of additional calcium ions from the sarcoplasmic reticulum; due to the calcium ion increase in the sarcoplasm, fused tetanus contractions are more forceful **24.** produces smooth movements rather than a series of jerky movements **25.** ATP **26.** (a) energy transfer from creatine phosphate, anaerobic cellular respiration, and aerobic cellular respiration; (b) aerobic cellular respiration **27.** the inability of a muscle to contract forcefully after prolonged activity **28.** Because of how quickly ATPase in myosin heads split (hydrolyze) ATP **29.** (a) skeletal and cardiac; (b) skeletal and cardiac; (c) smooth; (d) smooth **30.** (a) connect cardiac muscle fibers through desmosomes and gap junctions; (b) cardiac muscle **31.** skeletal: sarcoplasmic reticulum; cardiac and smooth: sarcoplasmic reticulum and interstitial fluid **32.** (a) visceral (single-unit), (b) multiunit **33.** speed of onset is slower and duration of contraction is longer in smooth muscle compared to skeletal muscle

## Chapter 11

**1.** the origin of the muscle attaches to the bone that will remain immovable while the contraction or shortening of the muscle belly pulls on the movable bone at the muscle insertion **2.** fulcrum: knee joint; load: weight of the lower leg **3.** large range of motion **4.** direction of fibers: rectus abdominis, shape: trapezoid, action: depressor labii inferioris, size: gluteus maximus, origin and insertion: sternocleidomastoid, location: temporalis, number of tendons of origin: biceps brachii **5.** they insert into skin or other muscles **6.** (a) buccinator; (b) orbicularis oculi; (c) orbicularis oris; (d) platysma **7.** (a) lateral rectus; (b) medial rectus **8.** superior rectus **9.** aids in mastication, taste perception, deglutition, and speech **10.** stabilize the hyoid bone **11.** (a) elevate the hyoid, floor of the oral cavity and tongue during swallowing; (b) depress the hyoid bone and move the thyroid cartilage of the larynx during swallowing and speech **12.** external oblique: inferiorly and medially; internal oblique: extend at right angles to those of external oblique; transverses abdominis: transversely around the abdominal wall; rectus abdominis: down the entire length of the anterior abdominal wall **13.** rectus abdominis **14.** aorta, esophagus, inferior vena cava **15.** pubis (anteriorly), coccyx (posteriorly), one lateral wall of the pelvis to the other **16.** (a) bulbospongiosus, deep transverse perineal and external urethral sphincter; (b) bulbospongiosus and ischiocavernosus **17.** from the pubic symphysis anteriorly, to the coccyx posteriorly, to the -ischial tuberosities laterally **18.** urogenital triangle: external genitals; anal triangle: anus **19.** flex: deltoid and coracobrachialis, extend: pectoralis major, latissimus dorsi, deltoid, teres major and teres minor, adduct: pectoralis major, latissimus dorsi, infraspinatus, teres major, teres minor and coracobrachialis **20.** subscapularis, supraspinatus, infraspinatus, teres minor **21.** (a) biceps brachii, brachialis, brachioradialis; (b) triceps brachii, anconeus **22.** supinate: supinator, biceps brachii, brachoradialis; pronate: pronator teres, pronator quadratus, brachioradialis **23.** flexor retinaculum: secures the tendons of the flexors of the fingers and wrist; extensor retinaculum: stabilizes tendons of the extensors of the wrist and fingers **24.** (a) flexor pollicis brevis; (b) abductor pollicis brevis **25.** flex: abductor digiti minimi, flexor digiti minimi brevis, lumbricals: -adduct: palmar interossei **26.** iliocostalis, longissimus, and spinalis **27.** laterally flex head: splenius capitis, splenius cervicis, anterior scalene, middle scalene, posterior scalene; laterally flex vertebral column: iliocostalis cervicis, iliocostalis thoracis, iliocostalis lumborum, longissimus capitis, longissimus cervicis, longissimus thoracis, multifidus, intertransversarii, anterior scalene, middle scalene, posterior scalene **28.** upper limb muscles are characterized by versatility of movement; lower limb muscles are larger and more powerful to provide stability, locomotion, and maintenance of posture, they also often cross two joints and act equally on both **29.** quadriceps femoris (rectus femoris, vastus lateralis, vastus medialis, -vastus intermedius) **30.** iliopsoas: psoas major and iliacus; quadriceps femoris: rectus femoris, vastus lateralis, vastus medialis, vastus intermedius; hamstrings: biceps femoris, semitendinosus, semimembranosus **31.** medially: tendons of semitendinosus and semimembranosus; laterally: tendons of biceps femoris **32.** dorsiflexion: tibialis anterior, extensor hallucis longus, extensor digitorum longus, fibularis tertius; plantar flexion: fibularis longus, fibularis brevis, gastrocnemius, soleus, plantaris, tibialis posterior, flexor digitorum longus, flexor hallucis longus **33.** (a) abductor hallucis, flexor digitorum brevis, abductor digiti minimi, quadratus plantae, lumbricals, flexor hallucis brevis, adductor hallucis, flexor digiti minimus brevis and dorsal interossei; (b) extensor digitorum brevis, lumbricals and plantar interossei **34.** muscles of the hand are specialized for precise and intricate movements; muscles of the foot move the toes and contribute to the longitudinal arch of the foot

## Chapter 12

**1.** sensory function: your eyes see the clock; integrative function: processes this visual information that it is lunch time and the decision to pull your lunch bag out; motor function: stimulating muscles to reach for and open lunch bag. **2.** eyes and legs **3.** CNS: brain, spinal cord; PNS: cranial nerves and their branches, spinal nerves and their branches, ganglia, enteric plexuses, sensory receptors **4.** SNS: sensory neurons conducting information from somatic and special senses receptors to CNS, and motor neurons conducting impulses from CNS to skeletal muscles; ANS: sensory neurons conducting information from autonomic sensory receptors to CNS, and motor neurons conducting impulses from CNS to smooth muscle, cardiac muscle, and glands; ENS: enteric sensory neurons conducting information from enteric receptors, and enteric motor neurons conducting information to GI tract smooth muscle and secretory cells **5.** (a) cell body: biosynthetic center and receptive region; (b) dendrites: receiving or input processing; (c)) axon: propagates nerves impulses toward another neuron, muscle fiber, or gland **6.** synapse **7.** dendrites: input region that receives information; cell body: integration region that analyzes information; axon: output region that transmits information toward another cell **8.** multipolar neurons usually have several dendrites and one axon; bipolar neurons have one main dendrite and one axon; unipolar neurons have one process that is a fused dendrite and another process which is an axon **9.** (a) sensory; (b) motor **10.** (a) microglia; (b) CNS **11.** (a) astrocytes; (b) CNS **12.** ependymal cells **13.** the sheath electrically insulates the axon of a neuron and increases the speed of nerve impulse conduction **14.** (a) Schwann cells, oligodendrocytes; (b) Schwann cells: PNS; oligodendrocytes: CNS **15.** myelin sheath: multilayered lipid and protein covering axons of most mammalian neurons and produced by neuroglia; neurolemma: outer, nucleated cytoplasmic layer of the Schwann cell enclosing the myelin sheath **16.** white matter consists primarily of myelinated axons and possibly some unmyelinated axons, whereas gray matter consists of neuron cell bodies, dendrites, unmyelinated axons, axon terminals, and neuroglia **17.** the plasma membrane has a resting membrane potential and specific ion channels **18.** (a) voltage change, a ligand that binds to a receptor, mechanical stimulation; (b) ions specific to that channel move across the plasma membrane, along their electrochemical gradient **19.** the gates randomly alternate between open and closed positions **20.** maintaining small buildup of negative ions on the cytosol side of plasma membrane and an equal buildup of positive ions on extracellular fluid side of plasma membrane **21.** it increases **22.** they spread along the plasma membrane for a short distance and then die out **23.** dendrites and cell bodies **24.** ligand-gated and mechanically gated **25.** graded potentials can generate an action potential **26.** (a) when depolarization reaches threshold, the voltage-gated channels open, and an action potential that is always the same size (amplitude) occurs; (b) no **27.** (a) $Na^+$ and $K^+$; (b) $Na^+$ channels; (c) $K^+$ channels **28.** (a) the period of time after an action potential begins during which an excitable cell cannot generate another action potential; (b) absolute refractory period occurs from the time the $Na^+$ channel activation gates open to when the $Na^+$ channel inactivation gates close and is a time when a second action potential cannot be initiated; the relative refractory period is the period of time during which a second action potential can be initiated but only by a larger-than-normal stimulus **29.** because in salutatory conduction, ions flow from node to node along the axon, moving at a faster pace than they would if the ions had to flow through their voltage-gated channels in each adjacent segment of the membrane **30.** because of the greater surface area in large diameter axons **31.** by the frequency of impulses **32.** graded potentials arise mainly in dendrites and cell bodies via ligand-gated or mechanically gated ion channels and are localized responses, whereas action potentials arise in the axon via voltage-gated channels for $Na^+$ and $K^+$ and permit communication over a long distance **33.** in electrical synapses, action potentials are conducted directly between adjacent cells through gap junctions giving a faster communication and synchronization, whereas in chemical synapses, a chemical bridges the synapse between neurons or a neuron and an effector **34.** (a) a nerve impulse arrives at the synaptic end bulb of a presynaptic neuron; (b) the nerve impulse opens voltage-gated $Ca^{2+}$ channels in the synaptic end bulbs; (c) $Ca^{2+}$ ions then flow into the end bulb and trigger exocytosis of some synaptic vesicles **35.** because the synaptic vesicles, which hold neurotransmitter, can only be activated to release neurotransmitter by $Ca^{2+}$ **36.** an excitatory postsynaptic potential will depolarize the cell membrane and possibly produce an action potential, whereas an inhibitory postsynaptic potential will hyperpolarize the membrane and prevent an action potential from occurring **37.** Nissal bodies break up into granular masses during chromatolysis, degeneration of distal portions of axon and myelin sheath during Wallerian degeneration, associated Schwann cells form regeneration tube to guide growth of new axon **38.** form regeneration tube and new myelin sheath **39** in the PNS, regeneration is possible but in the CNS, CNS myelin and scar tissue formed by astrocytes prevents neural repair

## Chapter 13

**1.** (a) cerebrum; (b) medulla oblongata of the brain stem **2.** (a) between the spinal cord and the vertebra; (b) dura mater, arachnoid mater, pia mater **3.** epidural: above the dura; subdural: between the dura and the arachnoid; subarachnoid: between the arachnoid and the pia mater **4.** subarachnoid **5.** internal jugular veins **6.** serves as a shock-absorbing medium and protects from blows (mechanical protection); provides an optimal chemical environment for accurate neuronal signaling (chemical protection) **7.** (a) brain stem (medulla oblongata); (b) cerebellum **8.** choroid plexuses produce CSF and are located in the walls of the ventricles **9.** through arachnoid villi into dural sinuses **10.** gray matter **11.** brain folds: gyri; shallow grooves: sulci; deep grooves: fissures **12.** (a) longitudinal fissure (b) central sulcus (separates frontal lobe from parietal lobe); lateral cerebral sulcus (separates frontal lobe from temporal lobe); parieto-occipital sulcus (separates parietal lobe from occipital lobe) **13.** association fibers (between gyri in same hemisphere); commissural fibers (between gyri in opposite hemispheres); projection fibers (between cerebrum and lower brain areas) **14.** refers to the functional asymmetry of the brain which means that the two hemispheres specialize in performing certain unique functions **15.** Broca's area is in only one hemisphere, usually the left hemisphere **16.** intermediate mass **17.** (a) hypothalamus; (b) serve as relay stations for reflexes related to the sense of smell **18.** control of the ANS; production of hormones; regulation of emotional and behavioral patterns, regulation of eating and drinking, control of body temperature, regulation of circadian rhythms and states of concentration **19.** produces melatonin which is thought to promote sleepiness and appears to contribute to the setting of the body's biological clock **20.** the midbrain is inferior to the diencephalon and superior to the pons; the pons is inferior to the midbrain and superior to the medulla oblongata; the medulla oblongata is inferior to the pons and superior to the foramen magnum **21.** (a) midbrain; (b) they contain axons of motor neurons that conduct impulses from motor areas in the cerebral cortex to the spinal cord, medulla, and pons **22.** pons and medulla oblongata **23.** (a) it is the crossing over of axons in the right pyramid to the left side and axons in the left pyramid to the right side; (b) it explains why each side of the brain controls movements on the opposite side of the body **24.** helps maintain consciousness and filter incoming sensory messages **25.** it has a highly folded surface**26.** anterior and posterior lobes govern subconscious aspects of skeletal muscle movements, and the flocculonodular lobe contributes to equilibrium and -balance **27.** encircles upper portion of brain stem and corpus callosum **28.** (a) cervical enlargement (b) lumbar enlargement **29.** conus medullaris: tapered end of the spinal cord; filum terminale: an extension of the pia mater that extends inferiorly and anchors the spinal cord to the coccyx; cauda equina: roots of the nerves that arise from the inferior part of the spinal cord and do not leave the vertebral column at the same level as they exit from the cord **30.** (a) sensory axons; (b) cell bodies of sensory neurons; (c) motor axons **31.** (a) gray commissure; (b) anterior (ventral) white commissure **32.** (a) anterior gray horns: cells bodies of somatic motor neurons; posterior gray horns: somatic and autonomic sensory neuron cell bodies; lateral gray horns: cell bodies of autonomic motor neurons; (b) horns are composed of gray matter; columns are composed of white matter **33.** in the columns **34.** (a) posterior columns; (b) spinothalamic tracts **35.** (a) direct; (b) indirect

## Chapter 14

**1.** a bundle of hundreds to thousands of axons plus associated connective -tissue and blood vessels that lie outside the brain and spinal cord **2.** Nerves are covered by epineurium, perineurium, endoneurium **3.** (a) names designate a nerve's distribution or function and numbers indicate the order, from anterior to posterior, in which the nerves arise from the brain; (b) 12 pairs **4.** mixed nerves have both motor and sensory components, whereas sensory nerves have sensory components only **5.** somatic motor axons innervate skeletal muscles; autonomic motor axons innervate glands, smooth muscle, and cardiac muscle **6.** pons **7.** jugular foramen **8.** according to the region and level of the vertebral column from which they emerge **9.** the posterior and anterior roots connect spinal nerves to the spinal cord gray matter **10.** they are formed from both sensory axons and motor axons **11.** originate from the spinal nerve proper; posterior rami serve the skin and muscles of the back, anterior rami serve the upper and lower limbs and the skin of the lateral and anterior trunk **12.** cervical plexus supplies head, neck, and superior shoulders and chest; brachial plexus supplies shoulders and upper limbs; lumbar plexus supplies anterior and lateral abdominal wall, external genitals, and part of lower limbs; sacral plexus supplies buttocks, perineum, and lower limbs; coccygeal plexus supplies coccygeal region **13.** intercostal muscles between ribs, thorax, and abdominal wall **14.** cranial reflexes involve integration in the brain stem and spinal reflexes involve integration in the spinal cord **15.** somatic reflexes involve contraction of skeletal muscles; autonomic reflexes involve responses of smooth muscle, cardiac muscle, and glands **16.** sensory receptor, sensory neuron, integrating center, motor neuron, effector **17.** monosynaptic: a reflex pathway having only one synapse in the CNS; polysynaptic: a reflex pathway involving more than two types of neurons and more than one CNS synapse; reciprocal innervation: a reflex arc that occurs when the stretched muscle contracts during a stretch reflex and antagonistic muscles that oppose the contraction simultaneously relax; intersegmental: a type of reflex where nerve impulses from one sensory neuron ascend and -descend in the spinal cord and activate interneurons in several segments of the spinal cord **18.** input: autonomic sensory neurons in visceral organs and blood vessels; output: autonomic motor neurons to smooth muscle, cardiac muscle, and glands **19.** (a) skeletal muscles; (b) cardiac muscles, smooth muscles, glands **20.** it consists of two motor neurons in series, one following the other **21.** organs receive impulses from both sympathetic and parasympathetic motor neurons **22.** preganglionic: synapses with the postganglionic neuron; postganglionic: targets the effector **23.** sympathetic trunk ganglia: lie in a vertical row on either side of the vertebral column; prevertebral ganglia: lie anterior to the vertebral column; terminal ganglia: located close to or actually within the wall of a visceral organ **24.** (a) sympathetic trunk ganglia and prevertebral ganglia (b) terminal ganglia **25.** (a) parasympathetic (b) because the axons of parasympathetic preganglionic axons extend from the CNS to a terminal ganglion near or in the associated organ **26.** to control motility and secretory activities of the gastrointestinal tract **27.** cholinergic neurons release the neurotransmitter acetylcholine while adrenergic neurons release norepinephrine **28.** cholinergic receptors bind with acetylcholine and adrenergic receptors bind both norepinephrine and epinephrine **29.** (a) cholinergic; (b) because acetylcholine is quickly inactivated by the enzyme acetylcholinesterase **30.** hypothalamus **31.** (a) pupils of the eyes dilate; heart rate, force of heart contraction, blood pressure increase; airways dilate; blood vessels that supply the kidneys and GI tract constrict; blood vessels that supply organs involved in exercise or fighting off danger dilate; liver increases breakdown of glycogen to glucose and adipose tissue increases breakdown of triglycerides to fatty acids and glycerol; liver -release of glucose increases blood glucose levels; processes not essential for meeting the emergency are inhibited; (b) sympathetic **32.** salivation, lacrimation, urination, digestion, and defecation **33.** iris of eye: dilates; lungs: increase respiration; urinary bladder: relaxation of wall and contraction of sphincter; stomach: decreased motility; intestines: decreased motility; heart: increased heart rate and force of contractions; arterioles of the abdominal viscera: constrict, reducing blood flow to the organs; arterioles of skeletal muscles: dilate to increase blood flow **34.** gastric, intestinal and salivary glands: sympathetic division decreases secretion while parasympathetic division increases secretion; heart: sympathetic division increases heart rate while parasympathetic division decreases heart rate; stomach and intestines: sympathetic division decreases motility while parasympathetic division increases motility **35.** blood pressure, digestion, defecation, and urination **36.** by axons of neurons that form tracts from the hypothalamus to sympathetic and to parasympathetic and sympathetic nuclei in the brain stem and spinal cord through relays in the reticular formation

## Chapter 15

**1.** each unique type of sensation such as touch, pain, vision, or hearing **2.** general senses are somatic senses and visceral senses, whereas special senses are smell, taste, vision, hearing, and equilibrium **3.** any given receptor responds to only one type of stimulus **4.** (a) stimulation of the sensory receptor; (b) transduction of the stimulus; (c) generation of nerve impulses; (d) integration of sensory input **5.** conversion of energy from a stimulus into a graded potential **6.** free nerve endings are bare dendrites; encapsulated nerve endings have dendrites enclosed within a connective tissue capsule; specialized separate cells **7.** the generator potential or receptor potential decreases in amplitude during a maintained, constant stimulus resulting in the sensation fading or disappearing even though the stimulus persists **8.** rapidly adapting receptors adapt very quickly and are specialized for signaling changes in a stimulus; slowly adapting receptors adapt slowly and continue to trigger nerve impulses as long as the stimulus persists **9.** adapt slowly: tactile discs, Ruffini corpuscles, nociceptors; adapt rapidly: corpuscles of touch, hair root plexuses, Pacinian corpuscles, cold and warm receptors, and joint kinesthetic receptors **10.** (a) pain serves a protective function by signaling the presence of noxious, -tissue-damaging conditions; (b) by intense thermal, mechanical, or chemical stimuli **11.** superficial somatic pain arises from stimulation of receptors in the skin; deep somatic pain arises from stimulation of receptors in skeletal muscles, joints, tendons, and fascia; visceral pain arises from stimulation of nocicceptors in visceral organs **12.** muscle spindles: muscle length; tendon organs: muscle tension **13.** monitoring joint position and movement **14.** touch, pressure, vibration and proprioception **15.** (a) type of sensations conveyed, (b) destination of first-order neuron, 3) area of crossing over **16.** type of sensory information: proprioceptive; usefulness: posture, balance, and coordination of skilled movements **17.** lips, face, tongue, and hand **18.** muscles in the hand, lips, tongue, and vocal cords **19.** because those regions involve skilled, complex or delicate movements **20.** provide output from the CNS to skeletal muscles **21.** (a) local circuit neurons help coordinate rhythmic activity in specific muscle groups; (b) upper motor neurons execute voluntary movements and help maintain muscle tone, posture, and balance; (c) basal nuclei neurons help initiate and terminate movements, suppress unwanted movements, and establish a normal level of muscle tone; (d) cerebellar neurons coordinate body movements and help maintain normal posture and balance **22.** direct motor pathways provide input to lower motor neurons via axons coming directly from the cerebral cortex; indirect motor pathways provide input to lower motor neurons from motor centers in the brain stem **23.** RAS stimulation increases activity of the cerebral cortex to maintain wakefulness; inhibition of RAS neurons results in sleep **24.** (a) memory is the process by which learned information is stored and retrieved; (b) immediate memory to recall ongoing experiences for a few seconds, short-term memory to recall information for seconds to minutes; long-term memory to recall information for days to years

## Chapter 16

**1.** to provide physical support, nourishment, and electrical insulation for the olfactory receptor cells, and to help detoxify chemicals that come in contact with the olfactory epithelium **2.** they produce new olfactory receptor cells **3.** cell division **4.** to moisten the surface of the olfactory epithelium and dissolve odorants **5.** about one minute **6.** sweet (sugar), sour (sour pickle), salty (potato chips), bitter (coffee), umami (meat) **7.** (a) circumvallate: form an inverted V-shaped row at the back of the tongue; (b) fungiform: scattered over the entire surface of the tongue; (c) foliate: located in small trenches on the lateral margins of the tongue; (d) filiform: over the entire surface of the tongue **8.** as with the nasal epithelium, they provide physical support, nourishment, and electrical insulation for the olfactory receptor cells **9.** chemoreceptors **10.** binding of tastant generates receptor potential, neurotransmitter molecule released from the gustatory receptor cell, neurotransmitter triggers action potentials in first-order neurons **11.** (a) levator palpebre superioris: raises upper eyelid; (b) orbicularis oculi: closes eyelid **12.** Meibomian glands secrete a fluid that keeps the eyelids from adhering to each other **13.** bulbar conjunctiva **14.** a watery solution containing salts, mucus, and lysozyme and functions to protect, clean, lubricate, and moisten the eyeball **15.** the lacrimal glands produce excessive lacrimal fluid that may spill over the edges of the eyelids and even fill the nasal cavity with fluid **16.** fibrous tunic: cornea and sclera; vascular tunic: choroids, ciliary body, iris **17.** it absorbs stray light rays which prevents reflection and scattering of light within the eyeball **18.** central fovea **19.** (a) nourishes the lens and cornea; (b) holds the retina flush against the choroid **20.** (a) iris; (b) lens **21.** Blood to ciliary processes, through posterior chamber, through pupil, through anterior chamber, through scleral venous sinus to blood **22.** (a) bending of light rays as they pass through substances of differing densities (b) cornea and lens **23.** accommodation **24.** (a) as ciliary muscle relaxes, ciliary processes moves away from lens, tension on zonular fibers increases, lens stretches into flatter shape (b) as ciliary muscle contracts, ciliary process moves toward lens, tension on zonular fibers decreases, lens recoils into more spherical shape **25.** (a) the medial movement of the two eyeballs so that both are directed toward the object being viewed; (b) it allows for binocular vision **26.** transduction of light energy into a receptor potential occurs in the outer segment of both rods and cones; both contain photopigments; both synapse with bipolar neurons **27.** (a) isomerization; (b) regeneration **28.** in light more and more photopigment is bleached, whereas in darkness more and more photopigment regenerates **29.** the glutamate is an inhibitory neurotransmitter that causes hyperpolarization in the bipolar cells **30.** axons of the ganglion cells **31.** nasal half **32.** sound waves could not be directed into the external auditory canal **33.** (a) tympanic membrane; (b) oval window and round window **34.** (a) malleus, incus, anvil; (b) malleus: tympanic membrane and incus; incus: malleus and stapes; stapes: incus and oval window **35.** semicircular canals, vestibule, cochlea **36.** spiral organ **37.** (a) vestibulocochlear; (b) vestibular; (c) cochlear **38.** (a) high frequency; (b) high intensity **39.** from auricles through external auditory canal, tympanic membrane, auditory ossicles, oval window where fluid pressure waves are generated in scala vestibuli, which vibrates vestibular membrane, creating pressure waves inside cochlear duct **40.** (a) it vibrates more gently; (b) it vibrates more slowly **41.** they are absorbed by the round window **42.** cochlea to cochlear branch of vestibulocochlear nerve, medulla oblongata, pons, midbrain, thalamus, and finally to primary auditory area in temporal lobe of the cerebral cortex **43.** static equilibrium refers to the maintenance of the position of the body relative to the force of gravity and linear acceleration and deceleration, whereas dynamic equilibrium is the maintenance of body position in response to rotational acceleration and deceleration **44.** they both involve hair cells as the sensory receptors; receptors synapse on different branches of the vestibulocochlear nerve **45.** add weight to the otolithic membrane, amplifying the pull of gravity during movements **46.** the ducts lie at right angles to one another and allow them to respond to virtually any rotation movement of the head **47.** semicircular ducts sense dynamic equilibrium; utricle and saccule sense static equilibrium **48.** (a) medulla oblongata, pons, and cerebellum; (b) the medulla oblongata and pons control eye, head, and neck movements, and muscle tone while the cerebellum adjusts skeletal muscle movements.

## Chapter 17

**1.** because it is released by a motor neuron and not an endocrine gland **2.** (a) nervous system; (b) endocrine system **3.** sweat gland is an exocrine gland that releases its secretion to skin surface via a duct, while an adrenal gland is an endocrine gland that releases its hormones into surrounding interstitial fluid **4.** because the thyroid does not have specific receptors for that hormone **5.** lipid-soluble (steroid hormones and thyroid hormones); water-soluble (amines, peptides and proteins, and eicosanoids) **6.** (a) within the target cell; (b) on a receptor on the target cell's plasma membrane **7.** because cAMP rather than the hormone brings about changes within the cell **8.** hormone concentration, abundance of target cell's hormone receptors, -influence exerted by other hormones **9.** signals from the nervous system, chemical changes in the blood, other hormones **10.** they are the direct route taken by hypothalamic hormones to act immediately on anterior pituitary cells, before the hormones are diluted or destroyed in the general circulation **11.** FSH, LH, TSH, ACTH **12.** hGH **13.** Hyperglycemia because high levels of hGH would raise blood glucose **14.** (a) hypothalamic neurosecretory cells; (b) they are packed into vesicles and transported down the axons to the axon terminals in the posterior pituitary until exocytosis is triggered, then the hormones are released and diffuse into the capillaries of the posterior pituitary **15.** both are direct routes for hormone release that do not involve the general circulation **16.** stretching of the cervix and suckling **17.** (a) lower osmotic pressure; (b) decrease level of ADH **18.** (a) follicular cells; (b) parafollicular cells; (c) $T_3$ and $T_4$ **19.** synthesis: iodide is brought into follicles where oxidated then binds to tyrosine component of thyroglobulin to ultimately form $T_3$ and $T_4$; storage: in follicle; blood transport: bound to transport protein TGB **20.** low blood levels of $T_3$ and $T_4$ or low metabolic rate stimulates TRH release, which stimulates TSH release, which then stimulates synthesis and release of $T_3$ and $T_4$ **21.** low blood levels of $T_3$ and $T_4$ would stimulate increase in levels of TRH, which would stimulate increased levels of TSH; however, removal of the thyroid gland would prevent production of $T_3$ and $T_4$ **22.** thyroid hormone stimulates the growth and development of nervous tissue **23.** (a) osteoclasts (b) inhibit bone breakdown and accelerate uptake of calcium and phosphates into bone extracellular matrix **24.** PTH acts on bones and kidneys and calcitriol acts on the GI tract **25.** blood calcium levels **26.** similarity: both increase blood calcium ion levels: differences: PTH increases number and activity of osteoclasts, slows the rate of calcium ion loss from blood into urine, and stimulates the kidneys to produce calcitriol, whereas calcitriol acts on the GI tract to increase the rate of calcium ion absorption from food into the blood **27.** (a) stimulates vasoconstriction of arterioles and secretion of aldosterone (b) arterioles and adrenal cortex **28.** (a) dehydration, low blood sodium or hemorrhage; (b) increase in blood pressure; (c) negative **29.** hypothalamus produces CRH which targets the anterior pituitary to produce ACTH which targets the adrenal cortex to produce glucocorticoids **30.** (a) provide additional energy sources for ATP production (b) additional ATP helps the body combat a range of stresses **31.** norepinephrine **32.** glucocorticoids, mineralocorticoids, adrenal cortex androgens, and adrenal medulla norepinephrine and epinephrine **33.** it produces hormones (endocrine function) and digestive enzymes (exocrine function) **34.** they flow into the GI tract by a series of ducts **35.** effect of high levels of insulin on the body **36.** (a) parasympathetic; (b) insulin **37.** (a) sympathetic; (b) glucagons **38.** adrenal cortex and testes **39.** adrenal cortex **40.** anterior pituitary **41.** Melatonin is thought to help reset the body's biological clock **42.** The sympathetic innervation stimulates the pineal gland to secrete melatonin in a rhythmic pattern **43.** They stimulate maturation of T cells, a type of white blood cell

## Chapter 18

**1.** oxygen, carbon dioxide, nutrients, waste products, hormones **2.** pH, body temperature, water content of cells **3.** red blood cells **4.** white blood cells and platelets **5.** albumin helps maintain proper blood osmotic pressure; globulins are involved in immune responses or transport of other molecules; fibrinogen is essential in blood clotting **6.** platelets **7.** red bone marrow **8.** they are capable of developing into other types of cells **9.** no nucleus, thus more room; no mitochondria, thus generate ATP anaerobically; biconcave shape allows greater surface area for diffusion; contain hemoglobin for better transport of oxygen **10.** four **11.** (a) the iron ion at the center of the heme ring; (b) amino acids in the globin part of hemoglobin **12.** it binds to and carries iron in the bloodstream. **13.** (a) globin is broken down into amino acids for reuse by the body, (b) iron is removed from the heme portion and attached to transferrin, then carried to red bone marrow to be used in hemoglobin synthesis, (c) the non-iron portion is converted to biliverdin, then to bilirubin, which is transported to the liver, then secreted by hepatocytes into bile which eventually enters the large intestine and is converted into urobilinogen, some of which is absorbed back into the blood and eventually excreted in urine, most is eliminated in feces **14.** reticulocytes and erythrocytes **15.** it might increase because more red cells would be needed due to the -reduced oxygen content of air at high altitudes **16.** (a) type AB blood has A and B antigens; (b) type O blood has neither A nor B antigens **17.** in the blood plasma **18.** (a) type A blood has anti-B antibodies; (b) type AB blood has neither anti-A nor anti-B antibodies; (c) type O blood has both anti-A and anti-B antibodies **19.** A, B, AB, or O **20.** hemolysis is the rupture of red blood cells; it occurs if an Rh$^-$ person -receives Rh1 blood and begins to make anti-Rh antibodies, and then receives a second transfusion of Rh1 blood **21.** (a) lymphocytes and monocytes; (b) because their granules are not visible under a light microscope **22.** lymphocytes continually recirculate from blood to interstitial fluid to lymphatic fluid and back to blood **23.** Emigration allows WBC's to leave the bloodstream; chemotaxis attracts phagocytes to pathogen infection and inflamed tissue; phagocytosis destroys pathogens and disposes of dead matter **24.** lysozyme, hydrogen peroxide and hypochlorite are capable of digesting the pathogen while defensins "poke" holes in the pathogen's membrane causing loss of cellular contents **25.** a count of each of the five types of WBCs **26.** help stop blood loss by forming a temporary plug and promote blood clot formation **27.** vascular spasm: occurs when a blood vessel is damaged and the smooth muscle in its wall contracts immediately; platelet plug formation: platelets adhere to the damaged blood vessel (platelet adhesion), become activated, and begin to release chemicals (platelet release reaction) making other platelets sticky, gathering of platelets (platelet aggregation) forming the platelet plug **28.** formation of prothrombinase **29.** thrombin activates more platelets, increasing their release of platelet phospholipids, and directly accelerates the formation of prothrombinase, which accelerates the production of more thrombin; this cycle continues, causing the fibrin clot to grow **30.** it is required for several stages of the blood clotting pathways **31.** as the clot retracts, it pulls the edges of the damaged blood vessel closer together.

## Chapter 19

**1.** it rests on the diaphragm, near the midline of the thoracic cavity, with about two-thirds of the mass lying to the left of the body's midline **2.** reduces friction between the membranes as the heart moves **3.** fibrous pericardium, serous pericardium, epicardium, myocardium, endocardium **4.** slightly increases the capacity of the atria so that they can hold a greater volume of blood **5.** superior atria and inferior ventricles **6.** right and left ventricles **7.** (a) interatrial septum; (b) interventricular septum **8.** the blood flowing from areas of higher blood pressure to areas of lower blood pressure **9.** back-flowing blood in the ventricles force the semilunar valves to close **10.** (a) the right atrium receives blood from the vena cava and coronary sinus and the left atria receives blood from the pulmonary veins (b) the right ventricle sends blood into the pulmonary arteries and the left ventricle sends blood into the aorta **11.** pulmonary veins and the aorta carry oxygenated blood; superior and inferior vena cava, coronary sinus, and pulmonary arteries carry deoxygenated blood **12.** ascending aorta **13.** (a) circumflex branch; (b) great cardiac vein; (c) small cardiac vein **14.** specialized cardiac muscle fibers that are self-excitable and repeatedly and rhythmically generate action potentials that trigger heart contractions; they act as a pacemaker and form the conduction system throughout the heart **15.** atrioventricular node **16.** sinoatrial node, atrioventricular node, atrioventricular bundle, right and left bundle branches, Purkinje fibers **17.** sinoatrial node **18.** ATP, fatty acids, glucose, lactic acid, amino acids, ketone bodies **19.** a recording of the electrical changes that accompany the heartbeat **20.** P wave: atrial depolarization; QRS complex: onset of ventricular depolarization; T wave: ventricular repolarization **21.** useful in diagnosing abnormal cardiac rhythms and conduction patterns, determining whether the heart is damaged or enlarged, determining the cause of chest pain, and in following the course of recovery from a heart attack **22.** blood turbulence from the AV valves closing after ventricular systole begins (lubb); blood turbulence associated with the SL valves closing at the end of ventricular systole (dupp) **23.** (a) systole; (b) diastole **24.** diastole **25.** (a) end-diastolic volume; (b) end-systolic volume **26.** volume of blood ejected each minute from the left ventricle into the aorta; equals the stroke volume (mL/beat) times heart rate (beats/min) **27.** (a) volume of blood ejected by the ventricle during each contraction; (b) preload, contractility and afterload **28.** cardiovascular center in the medulla oblongata **29.** (a) proprioceptors, chemoreceptors baroreceptors; (b) proprioceptors: physical activity; chemoreceptors: chemical changes in the blood; baroreceptors: changes in blood pressure **30.** they can increase the heart's pumping effectiveness (heart rate and contractility)

## Chapter 20

**1.** elastic fibers allow the walls of the arteries to stretch and recoil in response to changes in blood pressure; smooth muscle regulates the diameter of the lumen **2.** elastic arteries: more elastic fibers and less smooth muscle in their tunica media; muscular arteries: more smooth muscle and fewer elastic fibers **3.** (a) elastic arteries help propel blood onward while the ventricles are relaxing; (b) muscular arteries adjust the rate of blood flow and distribute blood to organs; (c) arterioles regulate the flow of blood into capillaries **4.** an anastomosis is the union of two or more arteries supplying the same body region; collateral circulation is the alternate circulatory route of blood flow to a body part through an anastomosis **5.** exchange is mainly by diffusion across capillary walls through either intercellular clefts, fenestrations or the plasma membranes of the endothelial cells, but for large, lipid insoluble molecules transcytosis via pinocytic vesicles **6.** blood flow through the capillary bed can be controlled and regulated by precapillary sphincters and can by-pass the capillary bed via the thoroughfare channel **7.** continuous capillaries only have intercellular clefts this restricts movement across the walls; fenestrated capillaries contain intercellular clefts and fenestrations that allow greater exchange of materials; sinusoids have large fenestrations and intercellular clefts and an incomplete basement membrane, which allows the greatest amount of capillary exchange **8.** autoregulation: the ability of a tissue to automatically adjust its blood flow to match its metabolic demands **9.** diffusion and transcytosis **10.** they enter lymphatic capillaries and are eventually returned by the lymphatic system to the bloodstream **11.** high blood pressure "pushes" fluid out of the capillaries, whereas high blood colloid osmotic pressure "pulls" fluid back into the capillaries **12.** the femoral artery has the thicker wall; the femoral vein has the wider lumen **13.** arteries: thick walls with smooth muscle and elastic fibers, conduct blood from the heart to the body; capillaries: walls are single-celled with only an endothelium and basement membrane, permit exchange of nutrients and wastes between blood and interstitial fluid; veins: thin walls with smooth muscle and elastic fibers with valves, return blood to the heart **14.** venous return is due to the pressure generated by contraction of the left ventricle, by the skeletal muscle pump (contraction of skeletal muscles in the limbs), by the respiratory pump alternating compression and decompression of veins during inhalation and exhalation **15.** blood volume in venules and veins is about 64% of 5 liters, or 3.2 liters; blood volume in capillaries is about 7% of 5 liters, or 350 mL **16.** (a) systole (contraction); (b) diastole (relaxation) **17.** opposition to blood flow due to friction between blood and the walls of the blood vessels; factors: size of the lumen, blood viscosity, total blood vessel length **18.** vasodilation and vasoconstriction of arterioles are the main regulators of systemic vascular resistance **19.** input: the brain, proprioceptors, baroreceptors, and chemoreceptors; output: parasympathetic output along vagus nerves, sympathetic and parasympathetic neurons of the ANS **20.** cardiac muscle in the heart and smooth muscle in blood vessel walls **21.** by decreasing parasympathetic stimulation of the heart, increasing sympathetic stimulation of the heart, stimulating adrenal medullae to increase secretion of epinephrine and norepinephrine and increasing vasoconstriction of vascular smooth muscle **22.** to inform the cardiovascular center about changes in blood levels of $O_2$, $CO_2$, and $H^+$ **23.** altering cardiac output, changing systemic vascular resistance, or adjusting the total blood volume **24.** (a) tachycardia is a rapid heart beat or pulse; (b) bradycardia is a slow heartbeat or pulse **25.** a cuff is placed on a bared arm; the cuff is inflated until the brachial artery is compressed and blood flow stops; the cuff is deflated slowly while the technician listens at the artery with a stethoscope; when the cuff is deflated enough to allow the artery to open, the spurt of blood passing through the artery is heard, and the pressure is read as the systolic blood pressure; as the cuff is deflated further, when the sounds become faint, the pressure is measured as the diastolic blood pressure **26.** 142 is systolic, 95 is diastolic **27.** (a) pulmonary arteries; (b) pulmonary veins **28.** blood in pulmonary circulation only has to be pumped to the lungs, but blood in the systemic circulation has to be pumped to the rest of the body **29.** carries oxygenated blood from the heart to all parts of the body and returns deoxygenated blood to the heart **30.** ascending aorta, arch of the aorta, thoracic aorta, abdominal aorta **31.** ascending aorta: the heart; arch of the aorta: the head, neck and arms; thoracic aorta: structures that form the thoracic cavity or are contained within the thoracic cavity; abdominal aorta: inferior surface of diaphragm, structures contained within the abdominal and pelvic cavity and the legs **32.** (a) brachiocephalic trunk, left common carotid artery, left subclavian artery; (b) brachiocephalic trunk–head, neck, arm thoracic wall; left common carotid artery–head and neck; left subclavian artery–brain, spinal cord, neck, shoulder, thoracic muscle wall and scapular muscles **33.** diaphragm **34.** paired visceral branches: suprarenal, renal, gonadal arteries; unpaired visceral branches: celiac trunk, superior mesenteric and inferior mesenteric arteries; paired visceral branches supply the kidneys, adrenal glands, ureters, male and female reproductive organs; unpaired visceral branches supply the stomach and esophagus, pancreas, liver, gallbladder, small intestine, large intestine **35.** (a) paired parietal branch: inferior phrenic arteries and lumbar arteries; unpaired parietal branch: median sacral artery (b) paired parietal branch supplies diaphragm and adrenal glands; unpaired parietal branch supplies the sacrum, coccyx, sacral spinal nerves and piriformis muscle **36.** internal iliac arteries and external iliac arteries; internal iliac arteries supply the pelvis, buttocks, external genitals, thigh; external iliac arteries supply the muscles of the anterior abdominal wall, the cremaster muscle in males, round ligament of the uterus in females, lower limbs **37.** the superior vena cava drains regions above the diaphragm, and the inferior vena cava drains regions below the diaphragm **38.** great cardiac, middle cardiac, small cardiac and anterior cardiac vein **39.** internal and external jugular and vertebral veins **40.** internal jugular: brain, bones of cranium, muscles and tissues of face and neck; external jugular: scalp and skin of face and neck, oral cavity and pharynx; vertebral veins: cervical vertebrae, cervical spinal cord and meninges and some neck muscles **41.** cephalic: lateral aspect of upper limbs; basilic: medial aspect of upper limbs; median antebrachial: palms and forearms; radial: lateral aspects of the forearms; ulnar: medial aspect of the forearms **42.** median cubital vein **43.** the median cubital vein **44.** Junctions of subclavian and internal jugular veins **45.** may serve as a bypass for the inferior vena cava **46.** (a) hemiazygos vein and accessory hemiazygos vein; (b) azygos vein **47.** inferior vena cava **48.** lumbar: both sides of the posterior abdominal wall, vertebral canal, spinal cord, and meninges; gonadal: testes in the male and ovaries in the female; renal: kidneys; suprarenal: adrenal glands; inferior phrenic: diaphragm; hepatic: liver **49.** (a) great saphenous vein; (b) anterior tibial veins; (c) popliteal vein **50.** prolonged administration of intravenous fluids, coronary bypass grafting **51.** (a) hepatic portal vein; (b) common hepatic artery **52.** hepatic veins

## Chapter 21

**1.** specialized form of reticular connective tissue that contains large numbers of lymphocytes **2.** interstitial fluid that has entered lymphatic vessels **3.** have thinner walls and more valves **4.** the ends of the endothelial cells overlap, so that when pressure is greater in the interstitial fluid than in lymph, the cells separate slightly and interstitial fluid enters the capillary but when pressure is greater within the lymph vessel then the endothelial cells adhere more closely together, which prevents lymph from leaving **5.** (a) thoracic duct—left side of the head, neck, and chest, the left upper limb, and the entire body inferior to the ribs; (b) right lymphatic duct—upper right side of the body; (c) thoracic duct: the left internal jugular and left subclavian veins; right lymphatic duct: at the junction of the right internal jugular and right subclavian veins **6.** receives lymph from the left and right lumbar trunks and intestinal trunk and drains into the thoracic duct **7.** blood capillaries to interstitial spaces to lymphatic capillaries to lymphatic vessels to lymph nodes to lymphatic vessels to lymphatic ducts to junction of the internal jugular and subclavian veins **8.** mmaturation of T cells **9.** atrophies with age (gets smaller) **10.** afferent lymphatic vessels to subcapsular sinus to trabecular sinuses to medullary sinuses to efferent lymphatic vessels **11.** they are trapped by reticular fibers within the sinuses, and macrophages destroy some by phagocytosis while lymphocytes destroy others by immune responses **12.** spleen: immune responses, destruction of pathogens, removal of worn-out or damaged blood cells and platelets, platelet storage, and blood cell production; tonsils: participate in immune responses against inhaled or ingested foreign substances **13.** non-specific defense against infection by means of physical and chemical barriers, antimicrobial proteins, phagocytes, inflammation and fever **14.** physical factors: closely packed, keratinized epithelial skin cells, shedding of epithelial skin cells, mucus, hairs, cilia, and washing surfaces with tears, saliva, urine; chemical factors: sebum, lysozyme, gastric juice, vaginal secretions **15.** antimicrobial proteins, natural killer cells, phagocytes, inflammation, fever **16.** interferons **17.** (a) lymphocytes; (b) neutrophils and macrophages (enlarged monocytes) **18.** chemotaxis, adherence, ingestion, digestion, killing **19.** they are broken down by lysosomal enzymes within the phagolysome **20.** in the liver **21.** phagocytosis, cytolysis, inflammation **22.** by enzymes splitting the inactive form into active fragments **23.** it is able to carry out adaptive immune responses when properly stimulated; B and T lymphocytes **24.** B cells develop in red bone marrow; T cells develop in the bone marrow and then in the thymus **25.** cell-mediated immunity: cytotoxic T cells directly attack invading antigens; antibody-mediated immunity: B cells transform into plasma cells which synthesize and secrete specific antibodies that inactivate specific antigens **26.** B cells that have the proper receptor can bind, recognize and respond to an antigen that is in interstitial fluid, lymph or blood plasma but T cells can only recognize and respond to antigen fragments that have been processed and presented to them by an antigen presenting cell (APC) **27.** help T cells recognize foreign antigens **28.** viral proteins, toxins from intracellular bacteria, and abnormal proteins synthesized by a cancerous cell **29.** first signal: binding of an antigen–MHC complex by a TCR with CD4 or CD8 proteins; second signal: costimulation by cytokines **30.** helper T cells: secrete cytokines; cytoxic T cells: destroy intruders; memory T cells: recognize the original invading antigen and initiate a very swift reaction against it **31.** tumor cells, tissue transplant cells **32.** cause target cells to undergo apoptosis or cytolysis, or destroy the cell's DNA **33.** helper T cells **34.** (1) antigen taken into B cell and combined with MHC protein to form an antigen–MHC complex on the B cell, (2) helper T cells deliver the costimulation needed for B cell division and differentiation, (3) B cell differentiates into plasma cells that secrete antibodies **35.** two **36.** (a) IgG: protects against bacteria and viruses by enhancing phagocytosis, neutralizing toxins, and triggering the complement system; (b) IgA: provides localized protection on mucous membranes against bacteria and viruses; (c) IgM: activates complement and causes agglutination and lysis of microbes; (d) IgD: involved in activation of B cells; (e) IgE: involved in allergic and hypersensitivity reactions, protection against parasitic worms **37.** neutralizing the antigen, immobilizing bacteria, agglutinating and precipitating the antigen, activating complement, enhancing phagocytosis of the antigen **38.** they can proliferate and differentiate into helper T cells, plasma cells, or cytotoxic T cells very quickly, thus providing a quicker response to a second exposure to an antigen **39.** There are more antibodies, they're produced more quickly, and have a higher affinity for the antigen and so are more successful in eliminating it **40.** IgG

## Chapter 22

**1.** nostrils, vestibule, nasal cavity (past nasal conchae), internal naris, nasopharynx **2.** (a) warming, moistening, and filtering incoming air; (b) detecting olfactory stimuli; (c) modify speech vibrations as they pass through the large, hollow resonating chambers **3.** external and internal nares **4.** produce mucus, serve as resonating chambers for sound, lighten the weight of the skull **5.** filtered by hairs in the nose and mucus secreted by goblet cells; warmed by blood in the capillaries; moistened by mucus **6.** superior border: nasal cavity; inferior border: larynx **7.** by closing over the glottis when swallowing **8.** produce sounds during speaking and singing **9.** (a) it accommodates slight expansion of the esophagus into the trachea during swallowing; (b) to maintain a patent (open) airway **10.** primary bronchi, secondary (lobar) bronchi, tertiary (segmental) bronchi, bronchioles, terminal bronchioles **11.** parietal pleura: lines the wall of the thoracic cavity; visceral pleura: -covers the lungs themselves; pleural cavity: potential space between the parietal and visceral pleura **12.** terminal bronchioles to respiratory bronchioles to alveolar ducts to alveoli **13.** it would increase **14.** diaphragm **15.** (a) decrease; (b) enter; (c) increase; (d) leave **16.** (b) lungs expand **17.** the recoil of elastic fibers that were stretched during inhalation and the inward pull of surface tension due to the film of alveolar fluid **18.** (a) higher compliance; (b) larger bronchioles **19.** apparatus commonly used to measure the volume of air exchanged during breathing and measure the respiratory rate **20.** conducting airways containing air that do not undergo gas exchange **21.** tidal volume: amount of air inhaled or exhaled with each quiet breath; inspiratory reserve: amount of air that can be forcefully inhaled above tidal; expiratory reserve: amount of air that can be forcefully exhaled above tidal; residual volume: amount of air remaining in the lungs after a forced exhalation **22.** Dalton's law: the total pressure exerted by a mixture of gases is the sum of the pressures exerted independently by each gas in the mixture; Henry's law: the quantity of a gas that will dissolve in a liquid is proportional to the partial pressure of the gas and its solubility **23.** it diffuses from an area where partial pressure is highest to an area where partial pressure is lower, thus alveoli to capillaries and then capillaries to tissue cells **24.** partial pressure difference of the gases, surface area available for gas -exchange, molecular weight and solubility of the gases **25.** partial pressure of $O_2$ decreases with increasing altitude **26.** (a) as oxyhemoglobin and dissolved in plasma; (b) attached to hemoglobin (oxyhemoglobin) **27.** the partial pressure of oxygen **28.** the higher the partial pressure of $O_2$, the more $O_2$ combines with hemoglobin **29.** (a) dissolved in blood plasma, attached to hemoglobin, and as bicarbonate ions; (b) as bicarbonate ions; (c) as dissolved $CO_2$ **30.** systemic vein **31.** phrenic **32.** it controls the basic rhythm of respiration **33.** apneustic area: sends stimulatory impulses to the inspiratory area that activate it and prolong inhalation; pneumotaxic area: transmits inhibitory impulses to the inspiratory area to help turn off the inspiratory area before the lungs become too full of air **34.** in or near the medulla oblongata in the central nervous system **35.** $P_{O_2}$, $H^+$ and $P_{CO_2}$ **36.** 40 mmHg **37.** rate and depth of breathing increase **38.** proprioceptors: rate and depth of breathing increase; inflation reflex: exhalation begins; temperature: increase in temperature results in increase in respiratory rate; pain: sudden, severe pain brings about brief apnea, but a prolonged somatic pain increases respiratory rate; irritation of the airways: an immediate cessation of breathing followed by coughing or sneezing **39.** they serve as buffers for $H^+$ **40.** (a) acidosis is a condition in which blood pH is below 7.35, and alkalosis is a condition in which blood pH is higher than 7.45; (b) acidosis: depression of the CNS through depression of synaptic transmission; alkalosis: overexcitability in both the CNS and peripheral nerves resulting in nervousness, muscle spasms, and even convulsions and death

## Chapter 23

**1.** (a) mucosa: inner surface for secretion of digestive juices and absorption of digestive end products; (b) submucosa: site of glands, blood vessels, lymph vessels, and nerve plexuses; (c) muscularis: circular and longitudinal muscles for propulsion and mixing; (d) serosa: serous membrane to reduce friction with movement **2.** (a) mouth, pharynx, superior and middle parts of the esophagus and external anal sphincter; (b) voluntary **3.** mesentery: binds large intestine to posterior abdominal wall; falciform ligament: attaches liver to anterior abdominal wall and diaphragm; lesser omentum: suspends stomach and duodenum from the liver; greater omentum: from stomach and duodenum to transverse colon **4.** myenteric plexus: innervates smooth muscle of muscularis; submucosal plexus: innervates secretory cells of mucosal epithelium to control the secretions of the organs of the GI tract **5.** cheeks, hard and soft palate, tongue **6.** (a) uvula; (b) closes off nasopharynx during swallowing **7.** parotid glands: located inferior and anterior to the ears, between the skin and masseter muscle; submandibular glands: located beneath the base of the tongue in the posterior part of the floor of the mouth; sublingual glands: superior to submandibular glands **8.** via ANS: parasympathetic fibers increase salivation, sympathetic fibers decrease salivation; feel and taste of food; smell, sight, sound, or thought of food **9.** incisors: grasping, tearing of food; cuspids: tear and shred food; premolars: crushing and grinding; molars: crush and grind food in preparation for swallowing **10.** chewing **11.** respiratory and digestive **12.** swallowing **13.** upper esophageal sphincter: regulates movement of food from the pharynx into the esophagus; lower esophageal sphincter: regulates movement of food from the esophagus to the stomach **14.** oral: voluntary passage of bolus into the oropharynx; pharyngeal: involuntary passage of bolus through pharynx into the esophagus; esophageal: -involuntary passage of bolus through the esophagus into the stomach **15.** both: oral phase is voluntary; pharyngeal and esophageal phases are -involuntary **16.** push **17.** mucosa **18.** fundus **19.** carbonic acid **20.** parasympathetic division **21.** (a) break peptide bonds between amino acids; (b) in the inactive form pepsinogen, it cannot digest the proteins in the chief cells that produce it **22.** water, ions, short-chain fatty acids, certain drugs **23.** exocrine cells to small ducts to large ducts to pancreatic duct and accessory duct; pancreatic duct joins common bile duct to form hepatopancreatic ampulla to sphincter of Oddi to duodenum; accessory ducts directly enter duodenum **24.** digest carbohydrates, proteins, lipids, and nucleic acids and buffer acidity (bicarbonate ions) **25.** the brush border enzyme enterokinase activates inactive trypsinogen to the active protein trypsin; trypsin activates the other pancreatic proteases **26.** (a) enzymes and bicarbonate ions; (b) bile; (c) pancreatic juices and bile **27.** energy released from catabolic reactions is transferred to molecules of ATP and then used to power anabolic reactions **28.** $O_2$, most nutrients, and certain toxic substances are taken up by hepatocytes **29.** (a) duodenum, jejunum, ileum; (b) ileum **30.** enhance absorption by increasing surface area and causing chyme to spiral as it moves through the small intestine **31.** (a) regulate microbial population in the intestines via lysozyme and phagocytosis; (b) help neutralize gastric acid in chyme **32.** segmentation: mix chyme with the digestive juices and bring food particles into contact with the mucosa for absorption; peristalsis: push chyme forward down the small intestine **33.** pancreatic amylase: chemically break down glycogen and starches; brush border enzymes: chemically break down disaccharides into monosaccharides; pancreatic lipases: break triglycerides into fatty acids and monoglycerides; deoxyribonuclease: digest DNA **34.** to emulsify fats, i.e., break a large fat globule into smaller lipid globules **35.** carbohydrate end products: by facilitated diffusion or secondary active transport into blood; protein end products: by active transport processes into blood; lipids: by simple diffusion into lacteals (lymph system) **36.** the hydrophilic protein coat that surrounds each chylomicron keeps the chylomicrons suspended in blood and prevents them from sticking to each other **37.** via the hepatic portal system (all absorbed nutrients except triglycerides) or via the lymphatic vessels to the blood vascular system and then to the liver (triglycerides) **38.** cecum, ascending colon, transverse colon, descending colon, sigmoid colon, rectum, anal canal, anus **39.** secrete mucus that lubricates the passage of the colonic contents **40.** through haustral churning, peristalsis, and mass peristalsis **41.** water absorption **42.** result: activation of neural centers in the cerebral cortex, hypothalamus, and brain stem; purpose: to prepare the mouth and stomach for food that is about to be eaten **43.** it buffers the stomach acid **44.** it prevents the duodenum from being overloaded with more chyme that it can handle **45.** cholecystokinin stimulates secretion of pancreatic juice rich in digestive enzymes, causes ejection of bile from the gallbladder, causes opening of the sphincter of the hepatopancreatic ampulla, induces satiety, and inhibits gastric emptying; secretin stimulates secretion of pancreatic juice and bile that are rich in bicarbonate ions and inhibits secretion of gastric juice **46.** (a) refers to all the chemical reactions of the body; (b) anabolism: chemical reactions that combine simple substances into more complex molecules; catabolism: chemical reactions that break down complex organic molecules into simpler ones **47.** when complex molecules and polymers are split apart, some of the energy is transferred to form ATP and the rest is given off as heat; when simple molecules and monomers are combined to form complex molecules, ATP provides the energy for synthesis, and again some energy is given off as heat **48.** (a) glycolysis, Krebs cycle, electron transport chain; (b) 36 or 38 **49.** (a) glucose is catabolized into 2 pyruvic acid and results in the production of two ATP molecules and 2 NADH + $H^+$ molecules; (b) 32–34 ATP molecules are formed from the energy contained in the NADH + $H^+$ and $FADH_2$ molecules **50.** glucose is formed from noncarbohydrate sources which helps maintains normal blood glucose levels between meals **51.** (a) to transport lipids in blood by making them temporarily water-soluble; (b) "good" cholesterol: HDL; "bad" cholesterol: VLDL and LDL; (c) HDL: high-density lipoprotein (higher percentage of protein, lower percentage of lipid); LDL: low-density lipoprotein (lower percentage of protein, higher percentage of lipid); VLDL: very low-density lipoprotein (lowest percentage of protein and highest percentage of lipid) **52.** each meal should be half fruits and vegetables with a bigger serving of vegetables, half grains and proteins with a bigger serving of grains, and include dairy **53.** Calcium, iron, magnesium, manganese are important in enzymatic reactions; sodium and phosphorus are important components in buffer systems; sodium is important in water regulation and neural function **54.** (a) organic nutrients required in small amounts to maintain growth and normal metabolism; (b) function as coenzymes; (c) ingested in food except for vitamin K and B complex vitamins which are synthesized by bacteria in the large intestine **55.** fat-soluble: are absorbed along with other dietary lipids in the small intestine and packaged into chylomicrons; water-soluble: are dissolved in body fluids **56.** inactivate oxygen free radicals, possibly protect against some cancers, reduce build-up of plaque and delaying some effects of aging

## Chapter 24

**1.** kidneys, ureters, urinary bladder, urethra **2.** by regulating blood ionic composition, pH, volume, pressure, and glucose level **3.** their position in the body is posterior to the peritoneum of the abdominal cavity **4.** ureter, renal artery, renal vein, lymphatic vessels, nerves **5.** deep renal capsule, middle adipose capsule, superficial renal fascia **6.** medulla **7.** arcuate arteries **8.** renal corpuscle (consisting of glomerulus and glomerular capsule) and renal tubules (proximal convoluted tubule, loop of Henle, distal convoluted tubule) **9.** proximal convoluted tubule, nephron loop, and distal convoluted tubule **10.** cortical nephrons: most numerous type, renal corpuscles in outer portion of renal cortex, short loops of Henle lying mainly in cortex and penetrate only into outer region of medulla, loop blood supply is from peritubular capillaries; juxtamedullay nephrons: least numerous, corpuscles lie deep in cortex close to medulla, long loop of Henle extending into deepest region of medulla, loop blood supply from vasa recta **11.** parietal and visceral layers of glomerular capsule **12.** structure: macula densa cells in the wall of the distal convoluted tubule and juxtaglomerular cells (modified smooth muscle cells) in the afferent arteriole wall; location: where the distal convoluted tubule makes contact with the afferent arteriole **13.** by secretion **14.** removed from the bloodstream **15.** blood plasma: contains blood cells, platelets, most plasma proteins and nutrients; glomerular filtrate: contains nutrients but no formed elements and very few, if any, of the plasma proteins **16.** glomerular capillaries present a large surface area for filtration, are about 50 times leakier than capillaries in most other tissues due to their large fenestrations, and glomerular capillary blood pressure is high **17.** by forcing water and solutes in blood plasma through the filtration membrane **18.** (a) by vasodilation of afferent arteriole which increases GRF; (b) kidneys are adjusting their own resistance to blood flow **19.** ANS: increased activity of sympathetic nerves results in release of norepinephrine which causes constriction of afferent arterioles which increases systemic blood pressure which increases GFR; hormones: angiotensin reduces GFR and atrial natriuretic peptide increases GFR **20.** paracellular reabsorption (fluids leaking between the cells passively) and transcellular reabsorption (substance passes from the fluid in the tubular lumen through the apical membrane of a tubule cell, across the cytosol, and out into interstitial fluid through the basolateral membrane) **21.** (a) basolateral surface; (b) it ensures that reabsorption of $Na^+$ is a one-way process **22.** water reabsorbed with solutes in tubular fluid which occurs in the proximal convoluted tubule and the descending limb of the loop of Henle **23.** (a) $Na^+$ symporters and $Na^+/H^+$ antiporters; (b) $Na^+$ symporters: virtually all nutrients such as glucose and amino acids; $Na^+/H^+$ antiporters: bicarbonate ions **24.** (a) $Na^+$–$K^+$–$2Cl^-$ symporters; (b) $K^+$, $Cl^-$, $Ca^{2+}$, and $Mg^{2+}$ **25.** (a) because little or no water is reabsorbed in the ascending limb of the nephron loop and ions are reabsorbed in the ascending limb, the tubular fluid becomes progressively more dilute as it flows toward the end of the ascending limb; (b) proximal convoluted tubule and descending limb of the nephron loop; (c) ascending limb of the nephron loop **26.** When blood pressure entering the renal tubule drops, juxtaglomerular cells secrete renin, which converts angiotensinogen to angiotensin I; angiotensin I becomes angiotensin II which stimulates the adrenal cortex to release aldosterone. Aldosterone increases $Na^+$ and $Cl^-$ reabsorption. As water follows back into the blood, blood pressure increases to a normal level **27.** angiotensin II stimulates the adrenal cortex to release aldosterone, which stimulates the principal cells in the collecting ducts to reabsorb more $Na^+$ and $Cl^-$ and secrete more $K^+$ with the reabsorption of $Na^+$ and $Cl^-$ **28.** (a) ADH increases the water permeability of principal cells in the last part of the distal convoluted tubule and throughout the collecting duct; (b) facultative **29.** stimulate: angiotensin II, aldosterone, ADH; inhibit: atrial natriuretic peptide **30.** osmotic pressure **31.** ascending limb of nephron loop and entire collecting duct **32.** $Na^+$, $K^+$, and urea **33.** (a) long nephron loops and collecting duct; (b)ADH **34.** the constant transfer of urea between segments of the renal tubule and the interstitial fluid of the medulla promotes water reabsorption with the solutes left behind in the lumen becoming very concentrated resulting in a small volume of concentrated urine **35.** (a) water and dissolved solutes in the body; (b) ICF (66%), ECF (33%); of ECF, interstitial fluid (80%) and plasma (20%); (c) water **36.** ANP: decreases blood volume and raises blood osmotic pressure; angiotensin II: increases blood volume and lowers blood osmotic pressure; aldosterone: decreases blood volume and raises blood osmotic pressure; ADH: decreases blood volume and raises blood osmotic pressure **37.** dihydrogen phosphate ion acts as a weak acid and is capable of buffering strong bases; monohydrogen phosphate ion acts as a weak base and is capable of buffering $H^+$ released by a strong acid **38.** by excreting excess $H^+$ when pH of body fluids is too low and by excreting excess $HCO_3^-$ when pH is too high **39.** peristaltic contractions of the muscular walls of the ureters **40.** it allows the organ lumen to stretch **41.** external **42.** males: extends from the internal urethral orifice and is elongated; females: lies directly posterior to the pubic symphysis and is shortened **43.** prostatic, membranous, spongy

## Chapter 25

**1.** (a) testes: produce sperm; (b) epididymis: sperm maturation and storage; (c) ductus deferens: sperm storage; (d) ejaculatory duct: sperm release; (e) urethra: sperm release channel; accessory sex glands (seminal vesicles, prostate, urethra): contribute to semen; (f) scrotum: house testes and epididymis, keep sperm at acceptable temperature; (g) penis: male copulatory organ **2.** (a) tunica vaginalis, tunica albuginea; (b) seminiferous tubules **3.** Sertoli cells: form blood–testis barrier, support and protect developing spermatogenic cells; produce fluid for sperm transport and secrete the hormone inhibin; Leydig cells: secrete testosterone **4.** chromosome number **5.** acrosome: a caplike vesicle filled with enzymes that help a sperm penetrate a secondary oocyte to bring about fertilization; middle piece: contains mitochondria arranged in a spiral, which provide the energy for locomotion of sperm to the secondary oocyte and for sperm metabolism **6.** anterior pituitary gland **7.** (a) Leydig cells; (b) development of the male fetus, male sexual characteristics, sexual function and anabolism **8.** epididymis: site of sperm maturation; ductus deferens: stores sperm and conveys sperm from the epididymis toward the urethra; ejaculatory duct: eject sperm and seminal vesicle secretions **9.** (a) ductus deferens, testicular artery, veins that drain the testes, autonomic nerves, lymphatic vessels, cremaster muscle; (b) passes through the inguinal canal **10.** prostatic urethra: passes through the prostate; membranous urethra: passes through the deep muscles of the perineum; spongy (penile) urethra: passes through the corpus spongiosum of the penis **11.** (a) seminal vesicles: lie posterior to the base of the urinary bladder and anterior to the rectum; functions to secrete an alkaline, viscous fluid that contains fructose, prostaglandins, and clotting proteins; (b) prostate: located inferior to the urinary bladder and surrounds the prostatic urethra; functions to secrete a milky, slightly acidic fluid containing citric acid, proteolytic enzymes, and seminalplasmin; (c) bulbourethral glands: located inferior to the prostate on either side of the membranous urethra and their ducts open into the spongy urethra; function to secrete an alkaline fluid into the urethra and secrete mucus that lubricates the end of the penis and the lining of the urethra **12.** seminal vesicles **13.** (a) corpus spongiosum and corpus cavernosa; (b) under parasympathetic stimulation, erectile tissue arterioles dilate becoming engorged with blood and smooth muscle within the erectile tissue relaxes resulting in widening of the blood sinuses, with the combination of increased blood flow and widening of the blood sinuses resulting in an erection **14.** in a sympathetic reflex, smooth muscle sphincter at the base of the urinary bladder closes, peristaltic waves occur in the ducts of the male, and semen is propelled into and through the male urethra **15.** (a) produce gametes and hormones; (b) by the broad ligament, mesovarium, ovarian ligament, and suspensory ligament **16.** follicles which produce estrogens and corpus luteum which produces -estrogens and progesterone **17.** primary oocytes are present before birth, whereas primary spermatocytes develop during puberty **18.** (a) primordial follicle: a single layer of follicular cells surround the -primary oocyte; primary follicle: primary oocyte surround by several layers of follicular cells called granulosa cells; secondary follicle: granulosa cells secrete follicular fluid, which builds up in the antrium, and the innermost layer of granulosa cells becomes firmly attached to the zona pellucida and is called the corona radiate; mature follicle: forms as secondary follicle enlarges and undergoes meiosis; (b) mature follicle **19.** (a) location: extend laterally from the uterus and lie between the folds of the broad ligaments of the uterus; function: provide a route for sperm to reach an ovum and transport secondary oocytes and fertilized ova from the ovaries to the uterus; (b) ciliated simple columnar cells function to help move a fertilized ovum along the tube toward the uterus; nonciliated cells with microvilli secrete a fluid that provides nutrition for the ovum **20.** (a) fundus, body, cervix; (b) fundus: superior to the uterine tubes; body: central portion; cervix: inferior narrow portion that opens into the vagina **21.** broad ligaments attach the uterus to either side of the pelvic cavity; uterosacral ligaments lie on either side of the rectum and connect the uterus to the sacrum, cardinal ligaments extend from the cervix and vagina to the lateral pelvic wall, round ligaments extend between the layers of the broad ligament from the uterus just inferior to the uterine tubes to the labia majora of the external genitalia **22.** endometrium: has two layers, the stratum basalis, which is permanent and gives rise to a new stratum functionalis after each menstruation, and the stratum functionalis, which lines the uterine cavity and sloughs off during menstruation; myometrium: consists of three layers of smooth muscle fibers that function in labor and childbirth **23.** receptacle for the penis during sexual intercourse, outlet for menstrual flow, passageway for childbirth **24.** (a) the external genitals of the female; (b) mons pubis: cushions the pubic symphysis; labia majora: protectively enclose medial components of the vulva; labia minora: contain sebaceous glands; clitoris: is capable of enlargement upon tactile stimulation and has a role in sexual excitement in the female; vestibule: contains the hymen, vaginal orifice, external urethral orifice, openings of the ducts of several glands; bulb of the vestibule: consists of two elongated masses of erectile tissue which become engorged with blood during sexual arousal resulting in narrowing the vaginal orifice and placing pressure on the penis during intercourse **25.** ovaries: testes; clitoris: glans penis; greater vestibular glands: bulbourethral glands; bulb of the vestibule: corpus spongiosum penis and bulb of penis **26.** (a) external urethral orifice, clitoris, mons pubis; (b) labia majora and labia minora **27.** by strands of connective tissue called the suspensory ligaments of the breast **28.** alveoli to secondary tubules to mammary ducts to lactiferous sinuses to lactiferous duct to exterior **29.** estradiol **30.** (a) estrogen: promote the development and maintenance of female reproductive structures, secondary sex characteristics, and breasts; increase protein anabolism; lower blood cholesterol level; (b) progesterone: prepare the mammary glands for milk secretion; (c) relax the uterus by inhibiting muscle contractions, increase the flexibility of the pubic symphysis, help dilate the uterine cervix **31.** (a) GnRH: stimulates release of FSH and LH; (b) FSH: stimulates follicle (and ovum) development; (c) LH: helps complete development of follicle and stimulates ovulation and corpus luteum formation; (d) estrogens: promote -uterine lining development and development of mature follicle: (e) progesterone: promotes development of uterine lining and maintains lining if pregnancy occurs; (f) inhibin: inhibits FSH and LH secretion **32.** a follicle that begins to develop at the beginning of a particular menstrual cycle may not reach maturity and ovulate until several menstrual cycles later since the developmental process may take several months to occur **33.** endometrial growth: estrogens; ovulation: estrogens, LH, GnRH; corpus luteum: LH; LH midcycle surge: estrogens **34.** negative **35.** decreased levels of progesterone, estrogens, and inhibin result in the -release of GnRH, FSH, and LH which results in follicles starting to grow again **36.** sperm flagella movement and contractions of the walls of the uterus and uterine tube **37.** sperm receptors in the zona pellucida bind to specific membrane proteins in the sperm head, triggering the release of sperm acrosomal enzymes that digest a path through the zona pellucida as the lashing sperm tail pushes the sperm cell onward **38.** fast block to polyspermy: oocyte cell membrane depolarizes on fertilization; slow block to polyspermy: intracellular calcium ions are released that stimulate exocytosis of chemicals which inactivate sperm receptors and harden the entire zona pellucida **39.** a solid sphere of cells formed by the mitotic divisions of the zygote **40.** inner cell mass forms the embryoblast which forms the embryo; trophoblast forms the fetal portion of the placenta **41.** (a) postovulatory phase; (b) in the fundus or body of the uterus; (c) with the embryoblast toward the endometrium **42.** (a) decidua basalis is the portion of the endometrium between the embryo and the stratum basalis; decidua capsularis is the portion of the endometrium located between the embryo and the uterine cavity; decidua parietalis is the remaining modified endometrium that lines the noninvolved areas of the rest of the uterus; (b) decidua basalis **43.** syncytiotrophoblast cells secrete enzymes that enable the blastocyst to penetrate the uterine lining by digesting and liquefying the endometrial cells **44.** (a) differentiation of the embryoblast cells into hypoblast and epiblast that form a flat disc; (b) by a band of extraembryonic mesoderm called the connection stalk **45.** (a) from the fetal portion formed by the chorionic villi of the chorion and the maternal portion formed by the decidua basalis of the endometrium; (b) it allows oxygen and nutrients to diffuse from maternal blood into fetal blood while carbon dioxide and wastes diffuse from fetal blood into maternal blood, is a protective barrier against most microorganisms, stores nutrients which are released into fetal circulation as required, and produces hormones needed to sustain the pregnancy **46.** two umbilical arteries, one umbilical vein, and supporting mucous connective tissue **47.** placenta detached from the uterus **48.** (a) corpus luteum; (b) placenta **49.** (a) source: chorion; function: stimulate corpus luteum to continue production of progesterone and estrogens; (b) source: chorion; function: help prepare the mammary glands for lactation, enhance maternal growth by increasing protein synthesis, regulate certain aspects of metabolism in both mother and fetus **50.** (a) relaxin; (b) increasing the flexibility of pelvic ligaments, helping -dilate the uterine cervix during labor **51.** corticotropin-releasing hormone **52.** rise in estrogen levels, release of oxytocin accompanied by increase in oxytocin receptors, release of prostaglandins, relaxin increase **53.** (a) oxytocin; (b) the actual birth process works to increase the levels of oxytocin, with oxytocin levels falling only after the birth of the child **54.** false labor is pain felt in the abdomen at irregular intervals, does not intensify, and is not accompanied by cervical dilation; true labor begins when uterine contractions occur at regular intervals, usually producing pain with back pain intensified by walking, and dilation of the cervix occurring accompanied by a discharge of blood-containing mucus into the cervical canal **55.** (a) prolactin and oxytocin; (b) prolactin: milk synthesis; oxytocin: milk ejection **56.** presence of beneficial cells for the infant, presence of beneficial molecules in breast milk; decreased incidence of diseases in later life, support of optimal infant growth, enhancement of intellectual and neurological development, as well as other benefits

# Glossary

## Pronunciation Key

**1.** The most strongly accented syllable appears in capital letters, for example, bilateral (bī-LAT-er-al) and diagnosis (dī-ag-NŌ-sis).

**2.** If there is a secondary accent, it is noted by a prime (′), for example, constitution (kon′-sti-TOO-shun) and physiology (fiz′-ē-OL-ō-jē). Any additional secondary accents are also noted by a prime, for example, decarboxylation (dē′-kar-bok′-si-LĀ-shun).

**3.** Vowels marked by a line above the letter are pronounced with the long sound, as in the following common words:

*ā* as in *māk*
*ē* as in *bē*
*ī* as in *īvy*
*ō* as in *pōle*
*ū* as in *cūte*

**4.** Vowels not marked by a line above the letter are pronounced with the short sound, as in the following words:

*a* as in *above* or *at*
*e* as in *bet*
*i* as in *sip*
*o* as in *not*
*u* as in *bud*

**5.** Other vowel sounds are indicated as follows:

*oy* as in *oil*
*oo* as in *root*

**6.** Consonant sounds are pronounced as in the following words:

*b* as in *bat*
*ch* as in *chair*
*d* as in *dog*
*f* as in *father*
*g* as in *get*
*h* as in *hat*
*j* as in *jump*
*k* as in *can*
*ks* as in *tax*
*kw* as in *quit*
*l* as in *let*
*m* as in *mother*
*n* as in *no*
*p* as in *pick*
*r* as in *rib*
*s* as in *so*
*t* as in *tea*
*v* as in *very*
*w* as in *welcome*
*z* as in *zero*
*zh* as in *lesion*

---

## A

**Abdomen** (ab-DŌ-men *or* AB-dō-men) The area between the diaphragm and pelvis.

**Abdominal** (ab-DŌM-i-nal) **cavity** Superior portion of the abdominopelvic cavity that contains the stomach, spleen, liver, gallbladder, pancreas, kidneys, small intestine, and part of the large intestine.

**Abdominal thrust maneuver** A first-aid procedure for choking. Employs a quick, upward thrust against the diaphragm that forces air out of the lungs with sufficient force to eject any lodged material.

**Abdominopelvic** (ab-dom′-i-nō-PEL-vik) **cavity** Inferior to the diaphragm and subdivided into a superior abdominal cavity and an inferior pelvic cavity.

**Abduction** (ab-DUK-shun) Movement away from the midline of the body.

**Abortion** (a-BOR-shun) The premature loss **(spontaneous)** or removal **(induced)** of the embryo or nonviable fetus; miscarriage due to a failure in the normal process of developing or maturing.

**Abscess** (AB-ses) A localized collection of pus and liquefied tissue in a cavity.

**Absorption** (ab-SORP-shun) Intake of fluids or other substances by cells of the skin or mucous membranes; the passage of digested foods from the gastrointestinal tract into blood or lymph.

**Accessory duct** A duct of the pancreas that empties into the duodenum about 2.5 cm (1 in.) superior to the hepatopancreatic ampulla. Also called the **duct of Santorini** (san′-tō-RĒ-nē).

**Acetabulum** (as′-e-TAB-ū-lum) The rounded cavity on the external surface of the hip bone that receives the head of the femur.

**Acetylcholine** (as′-ē-til-KŌ-lēn) **(ACh)** A neurotransmitter liberated by many peripheral nervous system neurons and some central nervous system neurons. It is excitatory at neuromuscular junctions but inhibitory at some other synapses (for example, it slows heart rate).

**Achalasia** (ak′-a-LĀ-zē-a) A condition, caused by malfunction of the myenteric plexus, in which the lower esophageal sphincter fails to relax normally as food approaches. A whole meal may become lodged in the esophagus and enter the stomach very slowly. Distension of the esophagus results in chest pain that is often confused with pain originating from the heart.

**Achilles tendon** *See* **Calcaneal tendon.**

**Acini** (AS-i-nē) Groups of cells in the pancreas that secrete digestive enzymes. *Singular* is **acinus** (AS-i-nus).

**Acoustic** (a-KOOS-tik) Pertaining to sound or the sense of hearing.

**Acquired immunodeficiency syndrome (AIDS)** A fatal disease caused by the human immunodeficiency virus (HIV). Characterized by a positive HIV-antibody test, low helper T cell count, and certain indicator diseases (for example, Kaposi's sarcoma, pneumocystis carinii pneumonia, tuberculosis, fungal diseases). Other symptoms include fever or night sweats, coughing, sore throat, fatigue, body aches, weight loss, and enlarged lymph nodes.

**Acrosome** (AK-rō-sōm) A lysosomelike organelle in the head of a sperm cell containing enzymes that facilitate the penetration of a sperm cell into a secondary oocyte.

**Actin** (AK-tin) A contractile protein that is part of thin filaments in muscle fibers.

**Action potential (AP)** An electrical signal that propagates along the membrane of a neuron or muscle fiber (cell); a rapid change in membrane potential that involves a depolarization followed by a repolarization. Also called a **nerve action potential** or **nerve impulse** as it relates to a neuron, and a **muscle action potential** as it relates to a muscle fiber.

**Activation** (ak′-ti-VĀ-shun) **energy** The minimum amount of energy required for a chemical reaction to occur.

**Active transport** The movement of substances across cell membranes against a concentration gradient, requiring the expenditure of cellular energy (ATP).

**Acute** (a-KŪT) Having rapid onset, severe symptoms, and a short course; not chronic.

**Adaptation** (ad′-ap-TĀ-shun) The adjustment of the pupil of the eye to changes in light intensity. The property by which a sensory neuron relays a decreased frequency of action potentials from a receptor, even though the strength of the stimulus remains constant; the decrease in perception of a sensation over time while the stimulus is still present.

**Adduction** (ad-DUK-shun) Movement toward the midline of the body.

**Adenoids** (AD-e-noyds) The pharyngeal tonsils.

**Adenosine triphosphate** (a-DEN-ō-sēn trī-FOS-fāt) **(ATP)** The main energy currency in living cells; used to transfer the chemical energy needed for metabolic reactions. ATP consists of the purine base *adenine* and the five-carbon sugar *ribose*, to which are added, in linear array, three *phosphate* groups.

**Adhesion** (ad-HĒ-zhun) Abnormal joining of parts to each other.

**Adipocyte** (AD-i-pō-sīt) Fat cell, derived from a fibroblast. Also called **fat cell** *or* **adipose cell.**

**Adipose** (AD-i-pōz) **tissue** Tissue composed of adipocytes specialized for triglyceride storage and present in the form of soft pads between various organs for support, protection, and insulation.

**Adrenal cortex** (a-DRĒ-nal KOR-teks) The outer portion of an adrenal gland, divided into three zones; the zona glomerulosa secretes mineralocorticoids, the zona fasciculata secretes glucocorticoids, and the zona reticularis secretes androgens.

**Adrenal glands** Two glands located superior to each kidney. Also called the **suprarenal** (soo′-pra-RĒ-nal) **glands.**

**Adrenal medulla** (me-DUL-a) The inner part of an adrenal gland, consisting of cells that secrete epinephrine, norepinephrine, and a small amount of dopamine in response to stimulation by sympathetic preganglionic neurons.

**Adrenergic** (ad′-ren-ER-jik) **neuron** A neuron that releases epinephrine (adrenaline) or norepinephrine (noradrenaline) as its neurotransmitter.

**Adrenocorticotropic** (ad-rē′-nō-kor-ti-kō-TRŌP-ik) **hormone (ACTH)** A hormone produced by the anterior pituitary that influences the production and secretion of certain hormones of the adrenal cortex. Also called **corticotropin** (kor′-ti-kō-TRŌ-pin).

**Adventitia** (ad-ven-TISH-a) The outermost covering of a structure or organ.

**Aerobic** (air-Ō-bik) Requiring molecular oxygen.

**Aerobic** (ār-Ō-bik) **cellular respiration** The production of ATP (36 molecules) from the complete oxidation of pyruvic acid in mitochondria. Carbon dioxide, water, and heat are also produced.

**Afferent arteriole** (AF-er-ent ar-TĒ-rē-ōl) A blood vessel of a kidney that divides into the capillary network called a glomerulus; there is one afferent arteriole for each glomerulus.

**Agglutination** (a-gloo-ti-NĀ-shun) Clumping of microorganisms or blood cells, typically due to an antigen–antibody reaction.

**Aggregated lymphatic follicles** Clusters of lymph nodules that are most numerous in the ileum. Also called **Peyer's** (PĪ-erz) **patches.**

**Albinism** (AL-bin-izm) Abnormal, nonpathological, partial, or total absence of pigment in skin, hair, and eyes.

**Aldosterone** (al-DOS-ter-ōn) A mineralocorticoid produced by the adrenal cortex that promotes sodium and water reabsorption by the kidneys and potassium excretion in urine.

**All-or-none principle** If a stimulus depolarizes a neuron to threshold, the neuron fires at its maximum voltage (all); if threshold is not reached, the neuron does not fire at all (none). Given above threshold, stronger stimuli do not produce stronger action potentials.

**Allantois** (a-LAN-tō-is) A small, vascularized outpouching of the yolk sac that serves as an early site for blood formation and development of the urinary bladder.

**Alleles** (a-LĒLZ) Alternate forms of a single gene that control the same inherited trait (such as type A blood) and are located at the same position on homologous chromosomes.

**Allergen** (AL-er-jen) An antigen that evokes a hypersensitivity reaction.

**Alopecia** (al′-ō-PĒ-shē-a) The partial or complete lack of hair as a result of factors such as genetics, aging, endocrine disorders, chemotherapy, and skin diseases.

**Alpha** (AL-fa) **cell** A type of cell in the pancreatic islets (islets of Langerhans) that secretes the hormone glucagon. Also termed an **A cell.**

**Alpha (α) receptor** A type of receptor for norepinephrine and epinephrine; present on visceral effectors innervated by sympathetic postganglionic neurons.

**Alveolar duct** Branch of a respiratory bronchiole around which alveoli and alveolar sacs are arranged.

**Alveolar macrophage** (MAK-rō-fāj) Highly phagocytic cell found in the alveolar walls of the lungs. Also called a **dust cell.**

**Alveolar sac** A cluster of alveoli that share a common opening.

**Alveolus** (al-VĒ-ō-lus) A small hollow or cavity; an air sac in the lungs; milk-secreting portion of a mammary gland. *Plural* is **alveoli** (al-VĒ-ol-ī).

**Alzheimer** (ALTZ-hī-mer) **disease (AD)** Disabling neurological disorder characterized by dysfunction and death of specific cerebral neurons, resulting in widespread intellectual impairment, personality changes, and fluctuations in alertness.

**Amenorrhea** (ā-men-ō-RĒ-a) Absence of menstruation.

**Amnesia** (am-NĒ-zē-a) A lack or loss of memory.

**Amnion** (AM-nē-on) A thin, protective fetal membrane that develops from the epiblast; holds the fetus suspended in amniotic fluid. Also called the "**bag of waters.**"

**Amniotic** (am′-nē-OT-ik) **fluid** Fluid in the amniotic cavity, the space between the developing embryo (or fetus) and amnion; the fluid is initially produced as a filtrate from maternal blood and later includes fetal urine. It functions as a shock absorber, helps regulate fetal body temperature, and helps prevent desiccation.

**Amphiarthrosis** (am′-fē-ar-THRŌ-sis) A slightly movable joint, in which the articulating bony surfaces are separated by fibrous connective tissue or fibrocartilage to which both are attached; types are syndesmosis and symphysis.

**Ampulla** (am-PUL-la) A saclike dilation of a canal or duct.

**Ampulla of Vater** *See* **Hepatopancreatic ampulla.**

**Anabolism** (a-NAB-ō-lizm) Synthetic, energy-requiring reactions whereby small molecules are built up into larger ones.

**Anaerobic** (an-ar-Ō-bik) Not requiring oxygen.

**Anal** (Ā-nal) **canal** The last 2 or 3 cm (1 in.) of the rectum; opens to the exterior through the anus.

**Anal column** A longitudinal fold in the mucous membrane of the anal canal that contains a network of arteries and veins.

**Anal triangle** The subdivision of the female or male perineum that contains the anus.

**Analgesia** (an-al-JĒ-zē-a) Pain relief; absence of the sensation of pain.

**Anaphase** (AN-a-fāz) The third stage of mitosis in which the chromatids that have separated at the centromeres move to opposite poles of the cell.

**Anaphylaxis** (an′-a-fi-LAK-sis) A hypersensitivity (allergic) reaction in which IgE antibodies attach to mast cells and basophils, causing them to produce mediators of anaphylaxis (histamine, leukotrienes, kinins, and prostaglandins) that bring about increased blood permeability, increased smooth muscle contraction, and increased mucus production. Examples are hay fever, hives, and anaphylactic shock.

**Anastomosis** (a-nas′-tō-MŌ-sis) An end-to-end union or joining of blood vessels, lymphatic vessels, or nerves.

**Anatomic dead space** Spaces of the nose, pharynx, larynx, trachea, bronchi, and bronchioles totaling about 150 mL of the 500 mL in a quiet breath (tidal volume); air in the anatomic dead space does not reach the alveoli to participate in gas exchange.

**Anatomical** (an′-a-TOM-i-kal) **position** A position of the body universally used in anatomical descriptions in which the body is erect, the head is level, the eyes face forward, the upper limbs are at the sides, the palms face forward, and the feet are flat on the floor.

**Anatomy** (a-NAT-ō-mē) The structure or study of the structure of the body and the relationship of its parts to each other.

**Androgens** (AN-drō-jenz) Masculinizing sex hormones produced by the testes in males and the adrenal cortex in both sexes; responsible for libido (sexual desire); the two main androgens are testosterone and dihydrotestosterone.

**Anemia** (a-NĒ-mē-a) Condition of the blood in which the number of functional red blood cells or their hemoglobin content is below normal.

**Anesthesia** (an′-es-THĒ-zē-a) A total or partial loss of feeling or sensation; may be general or local.

**Aneurysm** (AN-ū-rizm) A saclike enlargement of a blood vessel caused by a weakening of its wall.

**Angina pectoris** (an-JĪ-na *or* AN-ji-na PEK-tō-ris) A pain in the chest related to reduced coronary circulation due to coronary artery disease (CAD) or spasms of vascular smooth muscle in coronary arteries.

**Angiogenesis** (an′-jē-ō-JEN-e-sis) The formation of blood vessels in the extraembryonic mesoderm of the yolk sac, connecting stalk, and chorion at the beginning of the third week of development.

**Ankylosis** (ang′-ki-LŌ-sis) Severe or complete loss of movement at a joint as the result of a disease process.

**Antagonist** (an-TAG-ō-nist) A muscle that has an action opposite that of the prime mover (agonist) and yields to the movement of the prime mover.

**Antagonistic** (an-tag-ō-NIST-ik) **effect** A hormonal interaction in which the effect of one hormone on a target cell is opposed by another hormone. For example, calcitonin (CT) lowers blood calcium level, whereas parathyroid hormone (PTH) raises it.

**Anterior** (an-TĒR-ē-or) Nearer to or at the front of the body. Equivalent to ventral in bipeds.

**Anterior pituitary** Anterior lobe of the pituitary gland. Also called the **adenohypophysis** (ad′-e-nō-hī-POF-i-sis).

**Anterior root** The structure composed of axons of motor (efferent) neurons that emerges from the anterior aspect of the spinal cord and extends laterally to join a posterior root, forming a spinal nerve. Also called a **ventral root.**

**Anterolateral** (an′-ter-ō-LAT-er-al) **pathway** Sensory pathway that conveys information related to pain, temperature, tickle, and itch. Also called **spinothalamic pathway.**

**Antibody** (AN-ti-bod′-ē) **(Ab)** A protein produced by plasma cells in response to a specific antigen; the antibody combines with that antigen to neutralize, inhibit, or destroy it. Also called an **immunoglobulin** (im-ū-nō-GLOB-ū-lin) or **Ig.**

**Anticoagulant** (an-tī-cō-AG-ū-lant) A substance that can delay, suppress, or prevent the clotting of blood.

**Antidiuretic** (an′-ti-dī-ū-RET-ik) Substance that inhibits urine formation.

**Antidiuretic hormone (ADH)** Hormone produced by neurosecretory cells in the paraventricular and supraoptic nuclei of the hypothalamus that stimulates water reabsorption from kidney tubule cells into the blood and vasoconstriction of arterioles. Also called **vasopressin** (vāz-ō-PRES-in).

**Antigen** (AN-ti-jen) **(Ag)** A substance that has immunogenicity (the ability to provoke an immune response) and reactivity (the ability to react with the antibodies or cells that result from the immune response); contraction of *anti*body *gen*erator. Also termed a **complete antigen.**

**Antigen-presenting cell (APC)** Special class of migratory cell that processes and presents antigens to T cells during an immune response; APCs include macrophages, B cells, and dendritic cells, which are present in the skin, mucous membranes, and lymph nodes.

**Antioxidant** A substance that inactivates oxygen-derived free radicals. Examples are selenium, zinc, beta carotene, and vitamins C and E.

**Antrum** (AN-trum) Any nearly closed cavity or chamber, especially one within a bone, such as a sinus.

**Anuria** (an-Ū-rē-a) Absence of urine formation or daily urine output of less than 50 mL.

**Anus** (Ā-nus) The distal end and outlet of the rectum.

**Aorta** (ā-OR-ta) The main systemic trunk of the arterial system of the body that emerges from the left ventricle.

**Aortic** (ā-OR-tik) **body** Cluster of chemoreceptors on or near the arch of the aorta that respond to changes in blood levels of oxygen, carbon dioxide, and hydrogen ions ($H^+$).

**Aortic reflex** A reflex that helps maintain normal systemic blood pressure; initiated by baroreceptors in the wall of the ascending aorta and arch of the aorta. Nerve impulses from aortic baroreceptors reach the cardiovascular center via sensory axons of the vagus (X) nerves.

**Apex** (Ā-peks) The pointed end of a conical structure, such as the apex of the heart.

**Aphasia** (a-FĀ-zē-a) Loss of ability to express oneself properly through speech or loss of verbal comprehension.

**Apnea** (AP-nē-a) Temporary cessation of breathing.

**Apneustic** (ap-NOO-stik) **area** A part of the respiratory center in the pons that sends stimulatory nerve impulses to the inspiratory area that activate and prolong inhalation and inhibit exhalation.

**Apocrine** (AP-ō-krin) **gland** A type of gland in which the secretory products gather at the free end of the secreting cell and are pinched off, along with some of the cytoplasm, to become the secretion, as in mammary glands.

**Aponeurosis** (ap′-ō-noo-RŌ-sis) A sheetlike tendon joining one muscle with another or with bone.

**Apoptosis** (ap′-ōp-TŌ-sis *or* ap-ō-TŌ-sis) Programmed cell death; a normal type of cell death that removes unneeded cells during embryological development, regulates the number of cells in tissues, and eliminates many potentially dangerous cells such as cancer cells. During apoptosis, the DNA fragments, the nucleus condenses, mitochondria cease to function, and the cytoplasm shrinks, but the plasma membrane remains intact. Phagocytes engulf and digest the apoptotic cells, and an inflammatory response does not occur.

**Appositional** (a-pō-ZISH-o-nal) **growth** Growth due to surface deposition of material, as in the growth in diameter of cartilage and bone. Also called **exogenous** (eks-OJ-e-nus) **growth.**

**Aqueous humor** (A-kwē-us HŪ-mor) The watery fluid, similar in composition to cerebrospinal fluid, that fills the anterior cavity of the eye.

**Arachnoid** (a-RAK-noyd) **mater** The middle of the three meninges (coverings) of the brain and spinal cord. Also termed the **arachnoid.**

**Arachnoid villus** (VIL-us) Berrylike tuft of the arachnoid mater that protrudes into the superior sagittal sinus and through which cerebrospinal fluid is reabsorbed into the bloodstream.

**Arbor vitae** (AR-bor VĪ-tē) The white matter tracts of the cerebellum, which have a treelike appearance when seen in midsagittal section.

**Arch of the aorta** The most superior portion of the aorta, lying between the ascending and descending segments of the aorta.

**Areola** (a-RĒ-ō-la) Any tiny space in a tissue. The pigmented ring around the nipple of the breast.

**Arm** The part of the upper limb from the shoulder to the elbow.

**Arousal** (a-ROW-zal) Awakening from sleep, a response due to stimulation of the reticular activating system (RAS).

**Arrector pili** (a-REK-tor PĪ-lē) Smooth muscles attached to hairs; contraction pulls the hairs into a vertical position, resulting in "goose bumps."

**Arrhythmia** (a-RITH-mē-a) An irregular heart rhythm. Also called a **dysrhythmia.**

**Arteriole** (ar-TĒ-rē-ōl) A small, almost microscopic, artery that delivers blood to a capillary.

**Arteriosclerosis** (ar-tē-rē-ō-skle-RŌ-sis) Group of diseases characterized by thickening of the walls of arteries and loss of elasticity.

**Artery** (AR-ter-ē) A blood vessel that carries blood away from the heart.

**Arthritis** (ar-THRĪ-tis) Inflammation of a joint.

**Arthrology** (ar-THROL-ō-jē) The study or description of joints.

**Arthroplasty** (AR-thrō-plas′-tē) Surgical replacement of joints, for example, the hip and knee joints.

**Arthroscopy** (ar-THROS-kō-pē) A procedure for examining the interior of a joint, usually the knee, by inserting an arthroscope into a small incision; used to determine extent of damage, remove torn cartilage, repair cruciate ligaments, and obtain samples for analysis.

**Arthrosis** (ar-THRŌ-sis) A joint or articulation.

**Articular** (ar-TIK-ū-lar) **capsule** Sleevelike structure around a synovial joint composed of a fibrous capsule and a synovial membrane.

**Articular cartilage** (KAR-ti-lij) Hyaline cartilage attached to articular bone surfaces.

**Articular disc** Fibrocartilage pad between articular surfaces of bones of some synovial joints. Also called a **meniscus** (men-IS-kus).

**Articulation** (ar-tik-ū-LĀ-shun) A joint; a point of contact between bones, cartilage and bones, or teeth and bones.

**Arytenoid** (ar′-i-TĒ-noyd) **cartilages** A pair of small, pyramidal cartilages of the larynx that attach to the vocal folds and intrinsic pharyngeal muscles and can move the vocal folds.

**Ascending colon** (KŌ-lon) The part of the large intestine that passes superiorly from the cecum to the inferior border of the liver, where it bends at the right colic (hepatic) flexure to become the transverse colon.

**Ascites** (a-SĪ-tēz) Abnormal accumulation of serous fluid in the peritoneal cavity.

**Association areas** Large cortical regions on the lateral surfaces of the occipital, parietal, and temporal lobes and on the frontal lobes anterior to the motor areas connected by many motor and sensory axons to other parts of the cortex. The association areas are concerned with motor patterns, memory, concepts of word-hearing and word-seeing, reasoning, will, judgment, and personality traits.

**Asthma** (AZ-ma) Usually allergic reaction characterized by smooth muscle spasms in bronchi resulting in wheezing and difficult breathing. Also called **bronchial asthma.**

**Astigmatism** (a-STIG-ma-tizm) An irregularity of the lens or cornea of the eye causing the image to be out of focus and producing faulty vision.

**Astrocyte** (AS-trō-sīt) A neuroglial cell having a star shape that participates in brain development and the metabolism of neurotransmitters, helps form the blood–brain barrier, helps maintain the proper balance of $K^+$ for generation of nerve impulses, and provides a link between neurons and blood vessels.

**Ataxia** (a-TAK-sē-a) A lack of muscular coordination, lack of precision.

**Atherosclerosis** (ath-er-ō-skle-RŌ-sis) A progressive disease characterized by the formation in the walls of large and medium-sized arteries of lesions called atherosclerotic plaques.

**Atherosclerotic** (ath′-er-ō-skle-RO-tik) **plaque** (PLAK) A lesion that results from accumulated cholesterol and smooth muscle fibers (cells) of the tunica media of an artery; may become obstructive.

**Atom** Unit of matter that makes up a chemical element; consists of a nucleus (containing positively charged protons and uncharged neutrons) and negatively charged electrons that orbit the nucleus.

**Atresia** (a-TRĒ-zē-a) Degeneration and reabsorption of an ovarian follicle before it fully matures and ruptures; abnormal closure of a passage, or absence of a normal body opening.

**Atrial fibrillation** (Ā-trē-al fib-ri-LĀ-shun) **(AF)** Asynchronous contraction of cardiac muscle fibers in the atria that results in the cessation of atrial pumping.

**Atrial natriuretic** (nā′-trē-ū-RET-ik) **peptide (ANP)** Peptide hormone, produced by the atria of the heart in response to stretching, that inhibits aldosterone production and thus lowers blood pressure; causes natriuresis, increased urinary excretion of sodium.

**Atrioventricular (AV)** (ā′-trē-ō-ven-TRIK-ū-lar) **bundle** The part of the conduction system of the heart that begins at the atrioventricular (AV) node, passes through the cardiac skeleton separating the atria and the ventricles, then extends a short distance down the interventricular septum before splitting into right and left bundle branches. Also called the **bundle of His** (HISS).

**Atrioventricular (AV) node** The part of the conduction system of the heart made up of a compact mass of conducting cells located in the septum between the two atria.

**Atrioventricular (AV) valve** A heart valve made up of membranous flaps or cusps that allows blood to flow in one direction only, from an atrium into a ventricle.

**Atrium** (Ā-trē-um) A superior chamber of the heart.

**Atrophy** (AT-rō-fē) Wasting away or decrease in size of a part, due to a failure, abnormality of nutrition, or lack of use.

**Auditory ossicle** (AW-di-tō-rē OS-si-kul) One of the three small bones of the middle ear called the malleus, incus, and stapes.

**Auditory tube** The tube that connects the middle ear with the nose and nasopharynx region of the throat. Also called the **eustachian** (ū-STĀ-kē-an *or* ū-STĀ-shun) **tube** or **pharyngotympanic tube.**

**Auscultation** (aws-kul-TĀ-shun) Examination by listening to sounds in the body.

**Autoimmunity** An immunological response against a person's own tissues.

**Autolysis** (aw-TOL-i-sis) Self-destruction of cells by their own lysosomal digestive enzymes after death or in a pathological process.

**Autonomic ganglion** (aw′-tō-NOM-ik GANG-lē-on) A cluster of cell bodies of sympathetic or parasympathetic neurons located outside the central nervous system.

**Autonomic nervous system (ANS)** Visceral sensory (afferent) and visceral motor (efferent) neurons. Autonomic motor neurons, both sympathetic and parasympathetic, conduct nerve impulses from the central nervous system to smooth muscle, cardiac muscle, and glands. So named because this part of the nervous system was thought to be self-governing or spontaneous.

**Autonomic plexus** (PLEK-sus) A network of sympathetic and parasympathetic axons; examples are the cardiac, celiac, and pelvic plexuses, which are located in the thorax, abdomen, and pelvis, respectively.

**Autophagy** (aw-TOF-a-jē) Process by which worn-out organelles are digested within lysosomes.

**Autopsy** (AW-top-sē) The examination of the body after death.

**Autorhythmic** (aw′-tō-RITH-mik) **cells** Cardiac or smooth muscle fibers that are self-excitable (generate impulses without an external stimulus); act as the heart's pacemaker and conduct the pacing impulse through the conduction system of the heart; self-excitable neurons in the central nervous system, as in the inspiratory area of the brain stem.

**Autosome** (AW-tō-sōm) Any chromosome other than the X and Y chromosomes (sex chromosomes).

**Axilla** (ak-SIL-a) The small hollow beneath the arm where it joins the body at the shoulders. Also called the **armpit.**

**Axon** (AK-son) The usually single, long process of a nerve cell that propagates a nerve impulse toward the axon terminals.

**Axon terminal** Terminal branch of an axon where synaptic vesicles undergo exocytosis to release neurotransmitter molecules. Also called **telodendria** (tel′-o-DEN-drea).

## B

**B cell** A lymphocyte that can develop into a clone of antibody-producing plasma cells or memory cells when properly stimulated by a specific antigen.

**Babinski** (ba-BIN-skē) **sign** Extension of the great toe, with or without fanning of the other toes, in response to stimulation of the outer margin of the sole; normal up to 18 months of age and indicative of damage to descending motor pathways such as the corticospinal tracts after that.

**Back** The posterior part of the body; the dorsum.

**Ball-and-socket joint** A synovial joint in which the rounded surface of one bone moves within a cup-shaped depression or socket of another bone, as in the shoulder or hip joint. Also called a **spheroid** (SFĒ-royd) **joint.**

**Baroreceptor** (bar′-ō-re-SEP-tor) Neuron capable of responding to changes in blood or air or fluid pressure. Also called a **pressoreceptor** or **stretch receptor.**

**Basal nuclei** Paired clusters of gray matter deep in each cerebral hemisphere including the globus pallidus, putamen, and caudate nucleus. Together, the caudate nucleus and putamen are known as the corpus striatum. Nearby structures that are functionally linked to the basal nuclei are the substantia nigra of the midbrain and the subthalamic nuclei of the diencephalon.

**Basement membrane** Thin, extracellular layer between epithelium and connective tissue consisting of a basal lamina and a reticular lamina.

**Basilar** (BĀS-i-lar) **membrane** A membrane in the cochlea of the internal ear that separates the cochlear duct from the scala tympani and on which the spiral organ (organ of Corti) rests.

**Basophil** (BĀ-sō-fil) A type of white blood cell characterized by a pale nucleus and large granules that stain blue-purple with basic dyes.

**Belly** The abdomen. The gaster or prominent, fleshy part of a skeletal muscle.

**Beta** (BĀ-ta) **cell** A type of cell in the pancreatic islets (islets of Langerhans) in the pancreas that secretes the hormone insulin. Also called a **B cell.**

**Beta ($\beta$) receptor** A type of adrenergic receptor for epinephrine and norepinephrine; found on visceral effectors innervated by sympathetic postganglionic neurons.

**Bicuspid** (bī-KUS-pid) **valve** Atrioventricular (AV) valve on the left side of the heart. Also called the **mitral valve.**

**Bilateral** (bī-LAT-er-al) Pertaining to two sides of the body.

**Bile** (BĪL) A secretion of the liver consisting of water, bile salts, bile pigments, cholesterol, lecithin, and several ions; it emulsifies lipids prior to their digestion.

**Bilirubin** (bil-ē-ROO-bin) An orange pigment that is one of the end products of hemoglobin breakdown in the hepatocytes and is excreted as a waste material in bile.

**Biopsy** (BĪ-op-sē) The removal of a sample of living tissue to help diagnose a disorder, for example, cancer.

**Blastocyst** (BLAS-tō-sist) In the development of an embryo, a hollow ball of cells that consists of a blastocele (the internal cavity), trophoblast (outer cells), and embryoblast (inner cell mass).

**Blastocyst** (BLAS-tō-sist) **cavity** The fluid-filled cavity within the blastocyst. Also called the **blastocele.**

**Blastomere** (BLAS-tō-mēr) One of the cells resulting from the cleavage of a fertilized ovum.

**Blastula** (BLAS-tyū-la) An early stage in the development of a zygote.

**Blind spot** Area in the retina at the end of the optic (II) nerve in which there are no photoreceptors. Also called **optic disc.**

**Blood** The fluid that circulates through the heart, arteries, capillaries, and veins and that constitutes the chief means of transport within the body.

**Blood–brain barrier (BBB)** A barrier consisting of specialized brain capillaries and astrocytes that prevents the passage of materials from the blood to the cerebrospinal fluid and brain.

**Blood clot** A gel that consists of the formed elements of blood trapped in a network of insoluble protein fibers.

**Blood island** Isolated mass of mesoderm derived from angioblasts and from which blood vessels develop.

**Blood pressure (BP)** Force exerted by blood against the walls of blood vessels due to contraction of the heart and influenced by the elasticity of the vessel walls; clinically, a measure of the pressure in arteries during ventricular systole and ventricular diastole.

**Blood reservoir** (REZ-er-vwar) Systemic veins and venules that contain large amounts of blood that can be moved quickly to parts of the body requiring the blood.

**Blood–testis barrier (BTB)** A barrier formed by Sertoli cells that prevents an immune response against antigens produced by spermatogenic cells by isolating the cells from the blood.

**Body cavity** A space within the body that contains various internal organs.

**Bolus** (BŌ-lus) A soft, rounded mass, usually food, that is swallowed.

**Bone remodeling** Replacement of old bone by new bone tissue.

**Bony labyrinth** (LAB-i-rinth) A series of cavities within the petrous portion of the temporal bone forming the vestibule, cochlea, and semicircular canals of the inner ear.

**Bowman's capsule** *See* **Glomerular capsule.**

**Brachial plexus** (BRĀ-kē-al PLEK-sus) A network of nerve axons of the ventral rami of spinal nerves C5, C6, C7, C8, and T1. The nerves that emerge from the brachial plexus supply the upper limb.

**Bradycardia** (brād′-i-KAR-dē-a) A slow resting heart or pulse rate (under 50 beats per minute).

**Brain** The part of the central nervous system contained within the cranial cavity.

**Brain stem** The portion of the brain immediately superior to the spinal cord, made up of the medulla oblongata, pons, and midbrain.

**Brain waves** Electrical signals that can be recorded from the skin of the head due to electrical activity of brain neurons.

**Broad ligament** A double fold of parietal peritoneum attaching the uterus to the side of the pelvic cavity.

**Broca's** (BRŌ-kaz) **speech area** Motor area of the brain in the frontal lobe that translates thoughts into speech. Also called the **motor speech area.**

**Bronchi** (BRON-kī) Branches of the respiratory passageway including primary bronchi (the two divisions of the trachea), secondary or lobar bronchi (divisions of the primary bronchi that are distributed to the lobes of the lung), and tertiary or segmental bronchi (divisions of the secondary bronchi that are distributed to bronchopulmonary segments of the lung). *Singular* is **bronchus.**

**Bronchial** (BRON-kē-al) **tree** The trachea, bronchi, and their branching structures up to and including the terminal bronchioles.

**Bronchiole** (BRONG-kē-ōl) Branch of a tertiary bronchus further dividing into terminal bronchioles (distributed to lobules of the lung), which divide into respiratory bronchioles (distributed to alveolar sacs).

**Bronchitis** (brong-KĪ-tis) Inflammation of the mucous membrane of the bronchial tree; characterized by hypertrophy and hyperplasia of seromucous glands and goblet cells that line the bronchi, which results in a productive cough.

**Bronchopulmonary** (brong′-kō-PUL-mō-ner-ē) **segment** One of the smaller divisions of a lobe of a lung supplied by its own branches of a bronchus.

**Brunner's gland** *See* **Duodenal gland.**

**Buccal** (BUK-al) Pertaining to the cheek or mouth.

**Buffer system** A weak acid and the salt of that acid (that functions as a weak base). Buffers prevent drastic changes in pH by converting strong acids and bases to weak acids and bases.

**Bulb of penis** Expanded portion of the base of the corpus spongiosum penis.

**Bulbourethral** (bul′-bō-ū-RĒ-thral) **gland** One of a pair of glands located inferior to the prostate on either side of the urethra that secretes an alkaline fluid into the cavernous urethra. Also called a **Cowper's** (KOW-perz) **gland.**

**Bulimia** (boo-LIM-ē-a *or* boo-LĒ-mē-a) A disorder characterized by overeating at least twice a week followed by purging by self-induced vomiting, strict dieting or fasting, vigorous exercise, or use of laxatives or diuretics. Also called **binge–purge syndrome.**

**Bulk-phase endocytosis** A process by which most body cells can ingest membrane-surrounded droplets of interstitial fluid.

**Bundle branch** One of the two branches of the atrioventricular (AV) bundle made up of specialized muscle fibers (cells) that transmit electrical impulses to the ventricles.

**Bundle of His** *See* **Atrioventricular (AV) bundle.**

**Burn** Tissue damage caused by excessive heat, electricity, radioactivity, or corrosive chemicals that denature (break down) proteins in the skin.

**Bursa** (BUR-sa) A sac or pouch of synovial fluid located at friction points, especially about joints.

**Bursitis** (bur-SĪ-tis) Inflammation of a bursa.

**Buttocks** (BUT-oks) The two fleshy masses on the posterior aspect of the inferior trunk, formed by the gluteal muscles.

## C

**Calcaneal** (kal-KĀ-nē-al) **tendon** The tendon of the soleus, gastrocnemius, and plantaris muscles at the back of the heel. Also called the **Achilles** (a-KIL-ēz) **tendon.**

**Calcification** (kal′-si-fi-KĀ-shun) Deposition of mineral salts, primarily hydroxyapatite, in a framework formed by collagen fibers in which the tissue hardens. Also called **mineralization** (min′-e-ral-i-ZĀ-shun).

**Calcitonin** (kal-si-TŌ-nin) **(CT)** A hormone produced by the parafollicular cells of the thyroid gland that can lower the amount of blood calcium and phosphates by inhibiting bone resorption (breakdown of bone extracellular matrix) and by accelerating uptake of calcium and phosphates into bone matrix.

**Calculus** (KAL-kū-lus) A stone, or insoluble mass of crystallized salts or other material, formed within the body, as in the gallbladder, kidney, or urinary bladder.

**Callus** (KAL-lus) A growth of new bone tissue in and around a fractured area, ultimately replaced by mature bone. An acquired, localized thickening.

**Calyx** (KĀL-iks) Any cuplike division of the kidney pelvis. *Plural* is **calyces** (KĀ-li-sēz).

**Canal** (ka-NAL) A narrow tube, channel, or passageway.

**Canaliculus** (kan′-a-LIK-ū-lus) A small channel or canal, as in bones, where they connect lacunae. *Plural* is **canaliculi** (kan′-a-LIK-ū-lī).

**Canal of Schlemm** *See* **Scleral venous sinus.**

**Cancer** A group of diseases characterized by uncontrolled or abnormal cell division.

**Capacitation** (ka-pas′-i-TĀ-shun) The functional changes that sperm undergo in the female reproductive tract that allow them to fertilize a secondary oocyte.

**Capillary** (KAP-i-lar′-ē) A microscopic blood vessel located between an arteriole and venule through which materials are exchanged between blood and interstitial fluid.

**Carbohydrate** Organic compound consisting of carbon, hydrogen, and oxygen; the ratio of hydrogen to oxygen atoms is usually 2:1. Examples include sugars, glycogen, starches, and glucose.

**Carcinogen** (kar-SIN-ō-jen) A chemical substance or radiation that causes cancer.

**Cardiac** (KAR-dē-ak) **arrest** Cessation of an effective heartbeat in which the heart is completely stopped or in ventricular fibrillation.

**Cardiac cycle** A complete heartbeat consisting of systole (contraction) and diastole (relaxation) of both atria plus systole and diastole of both ventricles.

**Cardiac muscle** Striated muscle fibers (cells) that form the wall of the heart; stimulated by an intrinsic conduction system and autonomic motor neurons.

**Cardiac notch** An angular notch in the anterior border of the left lung into which part of the heart fits.

**Cardiac output (CO)** The volume of blood ejected from the left ventricle (or the right ventricle) into the aorta (or pulmonary trunk) each minute.

**Cardinal ligament** A ligament of the uterus, extending laterally from the cervix and vagina as a continuation of the broad ligament.

**Cardiogenic area** (kar-dē-ō-JEN-ik) A group of mesodermal cells in the head end of an embryo that gives rise to the heart.

**Cardiology** (kar-dē-OL-ō-jē) The study of the heart and diseases associated with it.

**Cardiovascular** (kar-dē-ō-VAS-kū-lar) **(CV) center** Groups of neurons scattered within the medulla oblongata that regulate heart rate, force of contraction, and blood vessel diameter.

**Cardiovascular physiology** Study of the functions of the heart and blood vessels.

**Cardiovascular system** System that consists of blood, the heart, and blood vessels.

**Carotene** (KAR-ō-tēn) Antioxidant precursor of vitamin A, which is needed for synthesis of photopigments; yellow-orange pigment present in the stratum corneum of the epidermis. Accounts for the yellowish coloration of skin. Also termed **beta carotene.**

**Carotid** (ka-ROT-id) **body** Cluster of chemoreceptors on or near the carotid sinus that respond to changes in blood levels of oxygen, carbon dioxide, and hydrogen ions.

**Carotid sinus** A dilated region of the internal carotid artery just superior to where it branches from the common carotid artery; it contains baro-receptors that monitor blood pressure.

**Carpal bones** The eight bones of the wrist. Also called **carpals.**

**Carpus** (KAR-pus) A collective term for the eight bones of the wrist.

**Cartilage** (KAR-ti-lij) A type of connective tissue consisting of chondrocytes in lacunae embedded in a dense network of collagen and elastic fibers and an extracellular matrix of chondroitin sulfate.

**Cartilaginous** (kar′-ti-LAJ-i-nus) **joint** A joint without a synovial (joint) cavity where the articulating bones are held tightly together by cartilage, allowing little or no movement.

**Catabolism** (ka-TAB-ō-lizm) Chemical reactions that break down complex organic compounds into simple ones, with the net release of energy.

**Cataract** (KAT-a-rakt) Loss of transparency of the lens of the eye or its capsule or both.

**Cauda equina** (KAW-da ē-KWĪ-na) A tail-like array of roots of spinal nerves at the inferior end of the spinal cord.

**Caudal** (KAW-dal) Pertaining to any tail-like structure; inferior in position.

**Cecum** (SĒ-kum) A blind pouch at the proximal end of the large intestine that attaches to the ileum.

**Celiac plexus** (SĒ-lē-ak PLEK-sus) A large mass of autonomic ganglia and axons located at the level of the superior part of the first lumbar vertebra. Also called the **solar plexus.**

**Cell** The basic structural and functional unit of all organisms; the smallest structure capable of performing all of the activities vital to life.

**Cell biology** The study of cellular structure and function. Also called **cytology.**

**Cell cycle** Growth and division of a single cell into two identical cells; consists of interphase and cell division.

**Cell division** Process by which a cell reproduces itself that consists of a nuclear division (mitosis) and a cytoplasmic division (cytokinesis); types include somatic and reproductive cell division.

**Cell junction** Point of contact between plasma membranes of tissue cells.

**Cellular respiration** The oxidation of glucose to produce ATP that involves glycolysis, acetyl coenzyme A formation, the Krebs cycle, and the electron transport chain.

**Cementum** (se-MEN-tum) Calcified tissue covering the root of a tooth.

**Central canal** A microscopic tube running the length of the spinal cord in the gray commissure. A circular channel running longitudinally in the center of an osteon (haversian system) of mature compact bone, containing blood and lymphatic vessels and nerves. Also called a **haversian** (ha-VER-shun) **canal.**

**Central fovea** (FŌ-vē-a) A depression in the center of the macula lutea of the retina, containing cones only and lacking blood vessels; the area of highest visual acuity (sharpness of vision).

**Central nervous system (CNS)** That portion of the nervous system that consists of the brain and spinal cord.

**Centrioles** (SEN-trē-ōlz) Paired, cylindrical structures of a centrosome, each consisting of a ring of microtubules and arranged at right angles to each other.

**Centromere** (SEN-trō-mēr) The constricted portion of a chromosome where the two chromatids are joined; serves as the point of attachment for the microtubules that pull chromatids during anaphase of cell division.

**Centrosome** (SEN-trō-sōm) A dense network of small protein fibers near the nucleus of a cell, containing a pair of centrioles and pericentriolar material.

**Cephalic** (se-FAL-ik) Pertaining to the head; superior in position.

**Cerebellar peduncle** (ser-e-BEL-ar pe-DUNG-kul) A bundle of nerve axons connecting the cerebellum with the brain stem.

**Cerebellum** (ser′-e-BEL-um) The part of the brain lying posterior to the medulla oblongata and pons; governs balance and coordinates skilled movements.

**Cerebral aqueduct** (SER-ē-bral AK-we-dukt) A channel through the midbrain connecting the third and fourth ventricles and containing cerebrospinal fluid. Also termed the **aqueduct of the midbrain.**

**Cerebral arterial circle** A ring of arteries forming an anastomosis at the base of the brain between the internal carotid and basilar arteries and arteries supplying the cerebral cortex. Also called the **circle of Willis.**

**Cerebral cortex** The surface of the cerebral hemispheres, 2–4 mm thick, consisting of gray matter; arranged in six layers of neuronal cell bodies in most areas.

**Cerebral peduncle** (pe-DUNG-kul or PĒ-dung-kul) One of a pair of nerve axon bundles located on the anterior surface of the midbrain, conducting nerve impulses between the pons and the cerebral hemispheres.

**Cerebrospinal** (se-rē′-brō-SPĪ-nal) **fluid (CSF)** A fluid produced by ependymal cells that cover choroid plexuses in the ventricles of the brain; the fluid circulates in the ventricles, the central canal, and the subarachnoid space around the brain and spinal cord.

**Cerebrovascular** (se- rē′-brō-VAS-kū-lar) **accident (CVA)** Destruction of brain tissue (infarction) resulting from obstruction or rupture of blood vessels that supply the brain. Also called a **stroke** or **brain attack.**

**Cerebrum** (se-RĒ-brum *or* SER-e-brum) The two hemispheres of the forebrain (derived from the telencephalon), making up the largest part of the brain.

**Cerumen** (se-ROO-men) Waxlike secretion produced by ceruminous glands in the external auditory meatus (ear canal). Also termed **earwax.**

**Ceruminous** (se-RŪ-mi-nus) **gland** A modified sudoriferous (sweat) gland in the external auditory meatus that secretes cerumen (ear wax).

**Cervical ganglion** (SER-vi-kul GANG-glē-on) A cluster of cell bodies of postganglionic sympathetic neurons located in the neck, near the vertebral column.

**Cervical plexus** (PLEK-sus) A network formed by nerve axons from the ventral rami of the first four cervical nerves and receiving gray rami communicantes from the superior cervical ganglion.

**Cervix** (SER-viks) Neck; any constricted portion of an organ, such as the inferior cylindrical part of the uterus.

**Chemical reaction** The formation of new chemical bonds or the breaking of old chemical bonds between atoms.

**Chemistry** (KEM-is-trē) The science of the structure and interactions of matter.

**Chemoreceptor** (kē′-mō-rē-SEP-tor) Sensory receptor that detects the presence of a specific chemical.

**Chiasm** (KĪ-azm) A crossing; especially the crossing of axons in the optic (II) nerve.

**Chief cell** The secreting cell of a gastric gland that produces pepsinogen, the precursor of the enzyme pepsin, and the enzyme gastric lipase. Also called a **zymogenic** (zī′-mō-JEN-ik) **cell.** Cell in the parathyroid glands that secretes parathyroid hormone (PTH). Also called a **principal cell.**

**Cholecystectomy** (kō′-lē-sis-TEK-tō-mē) Surgical removal of the gallbladder.

**Cholecystitis** (kō′-lē-sis-TĪ-tis) Inflammation of the gallbladder.

**Cholesterol** (kō-LES-te-rol) Classified as a lipid, the most abundant steroid in animal tissues; located in cell membranes and used for the synthesis of steroid hormones and bile salts.

**Cholinergic** (kō′-lin-ER-jik) **neuron** A neuron that liberates acetylcholine as its neurotransmitter.

**Chondrocyte** (KON-drō-sīt) Cell of mature cartilage.

**Chondroitin** (kon-DROY-tin) **sulfate** An amorphous extracellular matrix material found outside connective tissue cells.

**Chordae tendineae** (KOR-dē TEN-di-nē-ē) Tendonlike, fibrous cords that connect atrioventricular valves of the heart with papillary muscles.

**Chorion** (KŌ-rē-on) The most superficial fetal membrane that becomes the principal embryonic portion of the placenta; serves a protective and nutritive function.

**Chorionic villi** (kō-rē-ON-ik VIL-li) Fingerlike projections of the chorion that grow into the decidua basalis of the endometrium and contain fetal blood vessels.

**Chorionic villi sampling (CVS)** The removal of a sample of chorionic villus tissue by means of a catheter to analyze the tissue for prenatal genetic defects.

**Choroid** (KŌ-royd) One of the vascular coats of the eyeball.

**Choroid plexus** (PLEK-sus) A network of capillaries located in the roof of each of the four ventricles of the brain; ependymal cells around choroid plexuses produce cerebrospinal fluid.

**Chromaffin** (KRŌ-maf-in) **cell** Cell that has an affinity for chrome salts, due in part to the presence of the precursors of the neurotransmitter epinephrine; found, among other places, in the adrenal medulla.

**Chromatid** (KRŌ-ma-tid) One of a pair of identical connected nucleoprotein strands that are joined at the centromere and separate during cell division, each becoming a chromosome of one of the two daughter cells.

**Chromatin** (KRŌ-ma-tin) The threadlike mass of genetic material, consisting of DNA and histone proteins, that is present in the nucleus of a nondividing or interphase cell.

**Chromatolysis** (krō′-ma-TOL-i-sis) The breakdown of Nissl bodies into finely granular masses in the cell body of a neuron whose axon has been damaged.

**Chromosome** (KRŌ-mō-sōm) One of the small, threadlike structures in the nucleus of a cell, normally 46 in a human diploid cell, that bears the genetic material; composed of DNA and proteins (histones) that form a delicate chromatin thread during interphase; becomes packaged into compact rodlike structures that are visible under the light microscope during cell division.

**Chronic** (KRON-ik) Long term or frequently recurring; applied to a disease that is not acute.

**Chronic obstructive pulmonary disease (COPD)** A disease, such as bronchitis or emphysema, in which there is some degree of obstruction of airways and consequent increase in airway resistance.

**Chyle** (KĪL) The milky-appearing fluid found in the lacteals of the small intestine after absorption of lipids in food.

**Chyme** (KĪM) The semifluid mixture of partly digested food and digestive secretions found in the stomach and small intestine during digestion of a meal.

**Ciliary** (SIL-ē-ar′-ē) **body** One of the three parts of the vascular tunic of the eyeball, the others being the choroid and the iris; includes the ciliary muscle and the ciliary processes.

**Ciliary ganglion** (GANG-glē-on) A very small parasympathetic ganglion whose preganglionic axons come from the oculomotor (III) nerve and whose postganglionic axons carry nerve impulses to the ciliary muscle and the sphincter muscle of the iris.

**Cilium** (SIL-ē-um) A hair or hairlike process projecting from a cell that may be used to move the entire cell or to move substances along the surface of the cell. *Plural* is **cilia.**

**Circadian** (ser-KĀ-dē-an) **rhythm** The pattern of biological activity on a 24-hour cycle, such as the sleep–wake cycle.

**Circle of Willis** *See* **Cerebral arterial circle.**

**Circular folds** Permanent, deep, transverse folds in the mucosa and submucosa of the small intestine that increase the surface area for absorption. Also called **plicae circulares** (PLĪ-kē SER-kū-lar-ēs).

**Circulation time** The time required for a drop of blood to pass from the right atrium, through pulmonary circulation, back to the left atrium, through systemic circulation down to the foot, and back again to the right atrium.

**Circumduction** (ser-kum-DUK-shun) A movement at a synovial joint in which the distal end of a bone moves in a circle while the proximal end remains relatively stable.

**Cirrhosis** (si-RŌ-sis) A liver disorder in which the parenchymal cells are destroyed and replaced by connective tissue.

**Cisterna chyli** (sis-TER-na KĪ-lē) The origin of the thoracic duct.

**Cleavage** (KLĒV-ij) The rapid mitotic divisions following the fertilization of a secondary oocyte, resulting in an increased number of progressively smaller cells, called blastomeres.

**Clitoris** (KLI-to-ris) An erectile organ of the female, located at the anterior junction of the labia minora, that is homologous to the male penis.

**Clone** (KLŌN) A population of identical cells.

**Coarctation** (kō′-ark-TĀ-shun) **of the aorta** A congenital heart defect in which a segment of the aorta is too narrow. As a result, the flow of oxygenated blood to the body is reduced, the left ventricle is forced to pump harder, and high blood pressure develops.

**Coccyx** (KOK-siks) The fused bones at the inferior end of the vertebral column.

**Cochlea** (KOK-lē-a) A winding, cone-shaped tube forming a portion of the inner ear and containing the spiral organ (organ of Corti).

**Cochlear duct** The membranous cochlea consisting of a spirally arranged tube enclosed in the bony cochlea and lying along its outer wall. Also called the **scala media** (SCA-la MĒ-dē-a).

**Collagen** (KOL-a-jen) A protein that is the main organic constituent of connective tissue.

**Collateral circulation** The alternate route taken by blood through an anastomosis.

**Colliculus** (ko-LIK-ū-lus) A small elevation.

**Colon** The portion of the large intestine consisting of ascending, transverse, descending, and sigmoid portions.

**Colony-stimulating factor (CSF)** One of a group of molecules that stimulates development of white blood cells. Examples are macrophage CSF and granulocyte CSF.

**Colostrum** (kō-LOS-trum) A thin, cloudy fluid secreted by the mammary glands a few days prior to or after delivery before true milk is produced.

**Column** (KOL-um) Group of white matter tracts in the spinal cord.

**Common bile duct** A tube formed by the union of the common hepatic duct and the cystic duct that empties bile into the duodenum at the hepatopancreatic ampulla (ampulla of Vater).

**Compact (dense) bone tissue** Bone tissue that contains few spaces between osteons (haversian systems); forms the external portion of all bones and the bulk of the diaphysis (shaft) of long bones; is found immediately deep to the periosteum and external to spongy bone.

**Concha** (KON-ka) A scroll-like bone found in the skull. *Plural* is **conchae** (KON-kē).

**Concussion** (kon-KUSH-un) Traumatic injury to the brain that produces no visible bruising but may result in abrupt, temporary loss of consciousness.

**Conduction system** A group of autorhythmic cardiac muscle fibers that generates and distributes electrical impulses to stimulate coordinated contraction of the heart chambers; includes the sinoatrial (SA) node, the atrioventricular (AV) node, the atrioventricular (AV) bundle, the right and left bundle branches, and the Purkinje fibers.

**Condyloid** (KON-di-loyd) **joint** A synovial joint structured so that an oval-shaped condyle of one bone fits into an elliptical cavity of another bone, permitting side-to-side and back-and-forth movements, such as the joint at the wrist between the radius and carpals. Also called an **ellipsoidal** (ē-lip-SOYD-al) **joint.**

**Cone** (KŌN) The type of photoreceptor in the retina that is specialized for highly acute color vision in bright light.

**Congenital** (kon-JEN-i-tal) Present at the time of birth.

**Conjunctiva** (kon′-junk-TĪ-va) The delicate membrane covering the eyeball and lining the eyes.

**Connective tissue** One of the most abundant of the four basic tissue types in the body, performing the functions of binding and supporting; consists of relatively few cells in a generous extracellular matrix (the ground substance and fibers between the cells).

**Consciousness** (KON-shus-nes) A state of wakefulness in which an individual is fully alert, aware, and oriented, partly as a result of feedback between the cerebral cortex and reticular activating system.

**Continuous conduction** (kon-DUK-shun) Propagation of an action potential (nerve impulse) in a step-by-step depolarization of each adjacent area of an axon membrane.

**Contraception** (kon′-tra-SEP-shun) The prevention of fertilization or impregnation without destroying fertility.

**Contractility** (kon′-trak-TIL-i-tē) The ability of cells or parts of cells to actively generate force to undergo shortening for movements. Muscle fibers (cells) exhibit a high degree of contractility.

**Contralateral** (KON-tra-lat-er-al) On the opposite side; affecting the opposite side of the body.

**Conus medullaris** (KŌ-nus med-ū-LAR-is) The tapered portion of the spinal cord inferior to the lumbar enlargement.

**Convergence** (con-VER-jens) A synaptic arrangement in which the synaptic end bulbs of several presynaptic neurons terminate on one postsynaptic neuron. The medial movement of the two eyeballs so that both are directed toward a near object being viewed in order to produce a single image.

**Cornea** (KOR-nē-a) The nonvascular, transparent fibrous coat through which the iris of the eye can be seen.

**Corona** (kō-RŌ-na) Margin of the glans penis.

**Corona radiata** The innermost layer of granulosa cells that is firmly attached to the zona pellucida around a secondary oocyte.

**Coronary artery disease (CAD)** A condition such as atherosclerosis that causes narrowing of coronary arteries so that blood flow to the heart is reduced. The result is **coronary heart disease (CHD),** in which the heart muscle receives inadequate blood flow due to an interruption of its blood supply.

**Coronary circulation** The pathway followed by the blood from the ascending aorta through the blood vessels supplying the heart and returning to the right atrium. Also called **cardiac circulation.**

**Coronary sinus** (SĪ-nus) A wide venous channel on the posterior surface of the heart that collects the blood from the coronary circulation and returns it to the right atrium.

**Corpus albicans** (KOR-pus AL-bi-kanz) A white fibrous patch in the ovary that forms after the corpus luteum regresses.

**Corpus callosum** (kal-LŌ-sum) The great commissure of the brain between the cerebral hemispheres.

**Corpuscle of touch** *See* **Meissner corpuscle.**

**Corpus luteum** (LOO-tē-um) A yellowish body in the ovary formed when a follicle has discharged its secondary oocyte; secretes estrogens, progesterone, relaxin, and inhibin.

**Corpus striatum** (strī-Ā-tum) An area in the interior of each cerebral hemisphere composed of the caudate and putamen of the basal ganglia and white matter of the internal capsule, arranged in a striated manner.

**Cortex** (KOR-teks) An outer layer of an organ. The convoluted layer of gray matter covering each cerebral hemisphere.

**Costal** (KOS-tal) Pertaining to a rib.

**Cowper's gland** *See* **Bulbourethral gland.**

**Cramp** A spasmodic, usually painful contraction of a muscle.

**Cranial** (KRĀ-nē-al) **cavity** A body cavity formed by the cranial bones and containing the brain.

**Cranial nerve** One of 12 pairs of nerves that leave the brain; pass through foramina in the skull; and supply sensory and motor neurons to the head, neck, part of the trunk, and viscera of the thorax and abdomen. Each is designated by a Roman numeral and a name.

**Craniosacral** (krā-nē-ō-SĀK-ral) **outflow** The axons of parasympathetic preganglionic neurons, which have their cell bodies located in nuclei in the brain stem and in the lateral gray matter of the sacral portion of the spinal cord.

**Cranium** (KRĀ-nē-um) The skeleton of the skull that protects the brain and the organs of sight, hearing, and balance; includes the frontal, parietal, temporal, occipital, sphenoid, and ethmoid bones.

**Crista** (KRIS-ta) A crest or ridged structure. A small elevation in the ampulla of each semicircular duct that contains receptors for dynamic equilibrium. *Plural* is **cristae.**

**Crossing-over** The exchange of a portion of one chromatid with another during meiosis. It permits an exchange of genes among chromatids and is one factor that results in genetic variation of progeny.

**Crus** (KRUS) of **penis** Separated, tapered portion of the corpora cavernosa penis. *Plural* is **crura** (KROO-ra).

**Crypt of Lieberkühn** *See* **Intestinal gland.**

**Cryptorchidism** (krip-TOR-ki-dizm) The condition of undescended testes.

**Cuneate** (KŪ-nē-āt) **nucleus** A group of neurons in the inferior part of the medulla oblongata in which axons of the cuneate fasciculus terminate.

**Cupula** (KU-pū-la) A mass of gelatinous material covering the hair cells of a crista; a sensory receptor in the ampulla of a semicircular canal stimulated when the head moves.

**Cushing's syndrome** Condition caused by a hypersecretion of glucocorticoids characterized by spindly legs, "moon face," "buffalo hump," pendulous abdomen, flushed facial skin, poor wound healing, hyperglycemia, osteoporosis, hypertension, and increased susceptibility to disease.

**Cutaneous** (kū-TĀ-nē-us) Pertaining to the skin.

**Cyanosis** (sī-a-NŌ-sis) A blue or dark purple discoloration, most easily seen in nail beds and mucous membranes, that results from an increased concentration of deoxygenated (reduced) hemoglobin (more than 5 g/dL).

**Cyst** (SIST) A sac with a distinct connective tissue wall, containing a fluid or other material.

**Cystic** (SIS-tik) **duct** The duct that carries bile from the gallbladder to the common bile duct.

**Cystitis** (sis-TĪ-tis) Inflammation of the urinary bladder.

**Cytokinesis** (sī′-tō-ki-NĒ-sis) Distribution of the cytoplasm into two separate cells during cell division; coordinated with nuclear division (mitosis).

**Cytolysis** (sī-TOL-i-sis) The rupture of living cells in which the contents leak out.

**Cytoplasm** (SĪ-tō-plasm) Cytosol plus all organelles except the nucleus.

**Cytoskeleton** Complex internal structure of cytoplasm consisting of microfilaments, microtubules, and intermediate filaments.

**Cytosol** (SĪ-tō-sol) Semifluid portion of cytoplasm in which organelles and inclusions are suspended and solutes are dissolved. Also called **intracellular fluid.**

## D

**Dartos** (DAR-tōs) The contractile smooth muscular tissue deep to the skin of the scrotum.

**Decidua** (dē-SID-ū-a) That portion of the endometrium of the uterus (all but the deepest layer) that is modified during pregnancy and shed after childbirth.

**Deciduous** (dē-SID-ū-us) Falling off or being shed seasonally or at a particular stage of development. In the body, referring to the first set of teeth.

**Decussation** (dē′-ku-SĀ-shun) A crossing-over to the opposite (contralateral) side; an example is the crossing of 90% of the axons in the large motor tracts to opposite sides in the medullary pyramids.

**Decussation** (dē′-ku-SĀ-shun) of pyramids The crossing of most axons (90%) in the left pyramid of the medulla to the right side and the crossing of most axons (90%) in the right pyramid to the left side.

**Deep** Away from the surface of the body or an organ.

**Deep abdominal inguinal** (IN-gwi-nal) **ring** A slitlike opening in the aponeurosis of the transversus abdominis muscle that represents the origin of the inguinal canal.

**Deep vein thrombosis (DVT)** The presence of a thrombus in a vein, usually a deep vein of the lower limbs.

**Defecation** (def-e-KĀ-shun) The discharge of feces from the rectum.

**Deglutition** (dē-gloo-TISH-un) The act of swallowing.

**Dehydration** (dē-hī-DRĀ-shun) Excessive loss of water from the body or its parts.

**Delta cell** A cell in the pancreatic islets (islets of Langerhans) that secretes somatostatin. Also termed a **D cell.**

**Demineralization** (dē-min′-er-al-i-ZĀ-shun) Loss of calcium and phosphorus from bones.

**Dendrite** (DEN-drīt) A neuronal process that carries electrical signals, usually graded potentials, toward the cell body.

**Dendritic** (den-DRIT-ik) **cell** One type of antigen-presenting cell with long branchlike projections that commonly is present in mucosal linings such as the vagina, in the skin (Langerhans cells in the epidermis), and in lymph nodes (follicular dendritic cells).

**Dental caries** (KA-rēz) Gradual demineralization of the enamel and dentin of a tooth that may invade the pulp and alveolar bone. Also called **tooth decay.**

**Denticulate** (den-TIK-ū-lāt) Finely toothed or serrated; characterized by a series of small, pointed projections.

**Dentin** (DEN-tin) The bony tissues of a tooth enclosing the pulp cavity.

**Dentition** (den-TI-shun) The eruption of teeth. The number, shape, and arrangement of teeth.

**Deoxyribonucleic** (dē-ok′-sē-rī-bō-nū-KLĒ-ik) **acid (DNA)** A nucleic acid constructed of nucleotides consisting of one of four bases (adenine, cytosine, guanine, or thymine), deoxyribose, and a phosphate group; encoded in the nucleotides is genetic information.

**Depression** (de-PRESH-un) Movement in which a part of the body moves inferiorly.

**Dermal papilla** (pa-PIL-a) Fingerlike projection of the papillary region of the dermis that may contain blood capillaries or corpuscles of touch (Meissner corpuscles).

**Dermatology** (der′-ma-TOL-ō-jē) The medical specialty dealing with diseases of the skin.

**Dermatome** (DER-ma-tōm) The cutaneous area developed from one embryonic spinal cord segment and receiving most of its sensory innervation from one spinal nerve. An instrument for incising the skin or cutting thin transplants of skin.

**Dermis** (DER-mis) A layer of dense irregular connective tissue lying deep to the epidermis.

**Descending colon** (KŌ-lon) The part of the large intestine descending from the left colic (splenic) flexure to the level of the left iliac crest.

**Detrusor** (de-TROO-ser) muscle Smooth muscle that forms the wall of the urinary bladder.

**Developmental biology** The study of development from the fertilized egg to the adult form.

**Deviated nasal septum** A nasal septum that does not run along the midline of the nasal cavity. It deviates (bends) to one side.

**Diabetes mellitus** (dī-a-BĒ-tēz MEL-i-tus) An endocrine disorder caused by an inability to produce or use insulin. It is characterized by the three "polys": polyuria (excessive urine production), polydipsia (excessive thirst), and polyphagia (excess eating).

**Diagnosis** (dī′-ag-NŌ-sis) Distinguishing one disease from another or determining the nature of a disease from signs and symptoms by inspection, palpation, laboratory tests, and other means.

**Dialysis** (dī-AL-i-sis) The removal of waste products from blood by diffusion through a selectively permeable membrane.

**Diaphragm** (DĪ-a-fram) Any partition that separates one area from another, especially the dome-shaped skeletal muscle between the thoracic and abdominal cavities. Also a dome-shaped device that is placed over the cervix, usually with a spermicide, to prevent conception.

**Diaphysis** (dī-AF-i-sis) The shaft of a long bone.

**Diarrhea** (dī-a-RĒ-a) Frequent defecation of liquid feces caused by increased motility of the intestines.

**Diarthrosis** (dī-ar-THRŌ-sis) A freely movable joint; types are gliding, hinge, pivot, condyloid, saddle, and ball-and-socket.

**Diastole** (dī-AS-tō-lē) In the cardiac cycle, the phase of relaxation or dilation of the heart muscle, especially of the ventricles.

**Diastolic** (dī-as-TOL-ik) **blood pressure (DBP)** The force exerted by blood on arterial walls during ventricular relaxation; the lowest blood pressure measured in the large arteries, normally about 70 mmHg in a young adult.

**Diencephalon** (DĪ-en-sef′-a-lon) A part of the brain consisting of the thalamus, hypothalamus, and epithalamus.

**Differentiation** (dif′-er-en-shē-Ā-shun) Development of a cell from an unspecialized to a specialized one.

**Diffusion** (di-FŪ-zhun) A passive process in which there is a net or greater movement of molecules or ions from a region of high concentration to a region of low concentration until equilibrium is reached.

**Digestion** (dī-JES-chun) The mechanical and chemical breakdown of food to simple molecules that can be absorbed and used by body cells.

**Digestive system** A system that consists of the gastrointestinal tract (mouth, pharynx, esophagus, stomach, small intestine, and large intestine) and accessory digestive organs (teeth, tongue, salivary glands, liver, gallbladder, and pancreas). Its function is to break down foods into small molecules that can be used by body cells.

**Dilate** (DĪ-lāt) To expand or swell.

**Diploid** (DIP-loyd) **cell** Having the number of chromosomes characteristically found in the somatic cells of an organism; having two haploid sets of chromosomes, one each from the mother and father. Symbolized $2n$.

**Direct motor pathways** Collections of upper motor neurons with cell bodies in the motor cortex that project axons into the spinal cord, where they synapse with lower motor neurons or interneurons in the anterior horns. Also called the **pyramidal pathways.**

**Disease** Any change from a state of health. Any illness characterized by a recognizable set of signs and symptoms.

**Dislocation** (dis′-lō-KĀ-shun) Displacement of a bone from a joint with tearing of ligaments, tendons, and articular capsules. Also called **luxation** (luks-Ā-shun).

**Dissect** (di-SEKT) To separate tissues and parts of a cadaver or an organ for anatomical study.

**Distal** (DIS-tal) Farther from the attachment of a limb to the trunk; farther from the point of origin or attachment.

**Diuretic** (dī-ū-RET-ik) A chemical that increases urine volume by decreasing reabsorption of water, usually by inhibiting sodium reabsorption.

**Divergence** (dī-VER-jens) A synaptic arrangement in which the synaptic end bulbs of one presynaptic neuron terminate on several postsynaptic neurons.

**Diverticulum** (dī′-ver-TIK-ū-lum) A sac or pouch in the wall of a canal or organ, especially in the colon.

**Dorsal ramus** (RĀ-mus) A branch of a spinal nerve containing motor and sensory axons supplying the muscles, skin, and bones of the posterior part of the head, neck, and trunk.

**Dorsiflexion** (dor-si-FLEK-shun) Bending the foot in the direction of the dorsum (upper surface).

**Down-regulation** Phenomenon in which there is a decrease in the number of receptors in response to an excess of a hormone or neurotransmitter.

**Dual innervation** The concept by which most organs of the body receive impulses from sympathetic and parasympathetic neurons.

**Duct of Santorini** *See* **Accessory duct.**

**Duct of Wirsung** *See* **Pancreatic duct.**

**Ductus arteriosus** (DUK-tus ar-tē-rē-O-sus) A small vessel connecting the pulmonary trunk with the aorta; found only in the fetus.

**Ductus** (vas) deferens (DEF-er-ens) The duct that carries sperm from the epididymis to the ejaculatory duct. Also called the **seminal duct.**

**Ductus epididymis** (ep′-i-DID-i-mis) A tightly coiled tube inside the epididymis, distinguished into a head, body, and tail, in which sperm undergo maturation.

**Ductus venosus** (ve-NŌ-sus) A small vessel in the fetus that helps the circulation bypass the liver.

**Duodenal** (doo-ō-DĒ-nal) **gland** Gland in the submucosa of the duodenum that secretes an alkaline mucus to protect the lining of the small intestine from the action of enzymes and to help neutralize the acid in chyme. Also called a **Brunner's** (BRUN-erz) **gland.**

**Duodenal papilla** (pa-PIL-a) An elevation on the duodenal mucosa that receives the hepatopancreatic ampulla (ampulla of Vater).

**Duodenum** (doo′-ō-DĒ-num *or* doo-OD-e-num) The first 25 cm (10 in.) of the small intestine, which connects the stomach and the ileum.

**Dura mater** (DOO-ra MĀ-ter) The outermost of the three meninges (coverings) of the brain and spinal cord.

**Dynamic equilibrium** (ē-kwi-LIB-rē-um) The maintenance of body position, mainly the head, in response to sudden movements such as rotation.

**Dysmenorrhea** (dis′-men-ō-RĒ-a) Painful menstruation.

**Dysplasia** (dis-PLĀ-zē-a) Change in the size, shape, and organization of cells due to chronic irritation or inflammation; may either revert to normal if stress is removed or progress to neoplasia.

**Dyspnea** (DISP-nē-a) Shortness of breath; painful or labored breathing.

## E

**Ectoderm** The primary germ layer that gives rise to the nervous system and the epidermis of skin and its derivatives.

**Ectopic** (ek-TOP-ik) Out of the normal location, as in ectopic pregnancy.

**Edema** (e-DĒ-ma) An abnormal accumulation of interstitial fluid.

**Effector** (e-FEK-tor) An organ of the body, either a muscle or a gland, that is innervated by somatic or autonomic motor neurons.

**Efferent arteriole** (EF-er-ent ar-TĒ-rē-ōl) A vessel of the renal vascular system that carries blood from a glomerulus to a peritubular capillary.

**Efferent** (EF-er-ent) **ducts** A series of coiled tubes that transport sperm from the rete testis to the epididymis.

**Ejaculation** (ē-jak-ū-LĀ-shun) The reflex ejection or expulsion of semen from the penis.

**Ejaculatory** (ē-JAK-ū-la-tō-rē) **duct** A tube that transports sperm from the ductus (vas) deferens to the prostatic urethra.

**Elasticity** (e-las-TIS-i-tē) The ability of tissue to return to its original shape after contraction or extension.

**Electrocardiogram** (e-lek′-trō-KAR-dē-ō-gram) **(ECG** or **EKG)** A recording of the electrical changes that accompany the cardiac cycle that can be detected at the surface of the body; may be resting, stress, or ambulatory.

**Elevation** (el-e-VĀ-shun) Movement in which a part of the body moves superiorly.

**Embolus** (EM-bō-lus) A blood clot, bubble of air or fat from broken bones, mass of bacteria, or other debris or foreign material transported by the blood.

**Embryo** (EM-brē-ō) The young of any organism in an early stage of development; in humans, the developing organism from fertilization to the end of the eighth week of development.

**Embryoblast** (EM-brē-ō-blast) A region of cells of a blastocyst that differentiates into the three primary germ layers—ectoderm, mesoderm, and endoderm—from which all tissues and organs develop; also called an **inner cell mass.**

**Embryology** (em′-brē-OL-ō-jē) The study of development from the fertilized egg to the end of the eighth week of development.

**Emesis** (EM-e-sis) Vomiting.

**Emigration** (em′-i-GRĀ-shun) Process whereby white blood cells (WBCs) leave the bloodstream by rolling along the endothelium, sticking to it, and squeezing between the endothelial cells. Adhesion molecules help WBCs stick to the endothelium. Also known as **migration** or **extravasation.**

**Emission** (ē-MISH-un) Propulsion of sperm into the urethra due to peristaltic contractions of the ducts of the testes, epididymides, and ductus (vas) deferens as a result of sympathetic stimulation.

**Emphysema** (em-fi-SĒ-ma) A lung disorder in which alveolar walls disintegrate, producing abnormally large air spaces and loss of elasticity in the lungs; typically caused by exposure to cigarette smoke.

**Emulsification** (e-mul-si-fi-KĀ-shun) The dispersion of large lipid globules into smaller, uniformly distributed particles in the presence of bile.

**Enamel** (e-NAM-el) The hard, white substance covering the crown of a tooth.

**Endocardium** (en-dō-KAR-dē-um) The layer of the heart wall, composed of endothelium and smooth muscle, that lines the inside of the heart and covers the valves and tendons that hold the valves open.

**Endochondral** (en′-dō-KON-dral) **ossification** Bone formation within hyaline cartilage that develops from mesenchyme.

**Endocrine** (EN-dō-krin) **gland** A gland that secretes hormones into interstitial fluid and then the blood; a ductless gland.

**Endocrine system** (EN-dō-krin) All endocrine glands and hormone-secreting cells.

**Endocrinology** (en′-dō-kri-NOL-ō-jē) The science concerned with the structure and functions of endocrine glands and the diagnosis and treatment of disorders of the endocrine system.

**Endocytosis** (en′-dō-sī-TŌ-sis) The uptake into a cell of large molecules and particles in which a segment of plasma membrane surrounds the substance, encloses it, and brings it in; includes phagocytosis, pinocytosis, and receptor-mediated endocytosis.

**Endoderm** (EN-dō-derm) A primary germ layer of the developing embryo; gives rise to the gastrointestinal tract, urinary bladder, urethra, and respiratory tract.

**Endodontics** (en′-dō-DON-tiks) The branch of dentistry concerned with the prevention, diagnosis, and treatment of diseases that affect the pulp, root, periodontal ligament, and alveolar bone.

**Endolymph** (EN-dō-limf′) The fluid within the membranous labyrinth of the internal ear.

**Endometriosis** (en′-dō-me′-trē-Ō-sis) The growth of endometrial tissue outside the uterus.

**Endometrium** (en′-dō-MĒ-trē-um) The mucous membrane lining the uterus.

**Endomysium** (en′-dō-MĪZ-ē-um) Invagination of the perimysium separating each individual muscle fiber (cell).

**Endoneurium** (en′-dō-NOO-rē-um) Connective tissue wrapping around individual nerve axons.

**Endoplasmic reticulum** (en′-dō-PLAS-mik re-TIK-ū-lum) **(ER)** A network of channels running through the cytoplasm of a cell that serves in intracellular transportation, support, storage, synthesis, and packaging of molecules. Portions of ER where ribosomes are attached to the outer surface are called **rough ER;** portions that have no ribosomes are called **smooth ER.**

**End organ of Ruffini** *See* **Type II cutaneous mechanoreceptor.**

**Endosteum** (end-OS-tē-um) The membrane that lines the medullary (marrow) cavity of bones, consisting of osteogenic cells and scattered osteoclasts.

**Endothelium** (en′-dō-THĒ-lē-um) The layer of simple squamous epithelium that lines the cavities of the heart, blood vessels, and lymphatic vessels.

**Enteric** (en-TER-ik) **nervous system** A portion of the autonomic nervous system within the wall of the gastrointestinal tract, pancreas, and gallbladder. Its sensory neurons monitor tension in the intestinal wall and assess the composition of intestinal contents; its motor neurons exert control over the motility and secretions of the gastrointestinal tract.

**Enteroendocrine** (en-ter-ō-EN-dō-krin) **cell** A cell of the mucosa of the gastrointestinal tract that secretes a hormone that governs function of the GI tract; hormones secreted include gastrin, cholecystokinin, glucose-dependent insulinotropic peptide (GIP), and secretin.

**Enzyme** (EN-zīm) A substance that accelerates chemical reactions; an organic catalyst, usually a protein.

**Eosinophil** (ē-ō-SIN-ō-fil) A type of white blood cell characterized by granules that stain red or pink with acid dyes.

**Ependymal** (ep-EN-de-mal) **cells** Neuroglial cells that cover choroid plexuses and produce cerebrospinal fluid (CSF); they also line the ventricles of the brain and probably assist in the circulation of CSF.

**Epicardium** (ep′-i-KAR-dē-um) The thin outer layer of the heart wall, composed of serous tissue and mesothelium. Also called the **visceral pericardium.**

**Epidemiology** (ep′-i-dē-mē-OL-ō-jē) Study of the occurrence and transmission of diseases and disorders in human populations.

**Epidermis** (ep′-i-DERM-is) The superficial, thinner layer of skin, composed of keratinized stratified squamous epithelium.

**Epididymis** (ep′-i-DID-i-mis) A comma-shaped organ that lies along the posterior border of the testis and contains the ductus epididymis, in which sperm undergo maturation. *Plural* is **epididymides** (ep′-i-di-DIM-i-dēz).

**Epidural** (ep′-i-DOO-ral) space A space between the spinal dura mater and the vertebral canal, containing areolar connective tissue and a plexus of veins.

**Epiglottis** (ep′-i-GLOT-is) A large, leaf-shaped piece of cartilage lying on top of the larynx, attached to the thyroid cartilage; its unattached portion is free to move up and down to cover the glottis (vocal folds and rima glottidis) during swallowing.

**Epimysium** (ep-i-MĪZ-ē-um) Fibrous connective tissue around muscles.

**Epinephrine** (ep-ē-NEF-rin) Hormone secreted by the adrenal medulla that produces actions similar to those that result from sympathetic stimulation. Also called **adrenaline** (a-DREN-a-lin).

**Epineurium** (ep′-i-NOO-rē-um) The superficial connective tissue covering around an entire nerve.

**Epiphyseal** (ep′-i-FIZ-ē-al) **line** The remnant of the epiphyseal plate in the metaphysis of a long bone.

**Epiphyseal plate** The hyaline cartilage plate in the metaphysis of a long bone; site of lengthwise growth of long bones.

**Epiphysis** (e-PIF-i-sis) The end of a long bone, usually larger in diameter than the shaft (diaphysis).

**Epiphysis cerebri** (se-RĒ-brē) Pineal gland.

**Episiotomy** (e-piz′-ē-OT-ō-mē) A cut made with surgical scissors to avoid tearing of the perineum at the end of the second stage of labor.

**Epistaxis** (ep′-i-STAK-sis) Loss of blood from the nose due to trauma, infection, allergy, neoplasm, and bleeding disorders. Also called **nosebleed.**

**Epithalamus** (ep′-i-THAL-a-mus) Part of the diencephalon superior and posterior to the thalamus, comprising the pineal gland and associated structures.

**Epithelial** (ep-i-THĒ-lē-al) **tissue** The tissue that forms the innermost and outermost surfaces of body structures and forms glands.

**Eponychium** (ep′-o-NIK-ē-um) Narrow band of stratum corneum at the proximal border of a nail that extends from the margin of the nail wall. Also called the **cuticle**.

**Equilibrium** (ē-kwi-LIB-rē-um) The state of being balanced.

**Erectile dysfunction (ED)** Failure to maintain an erection long enough for sexual intercourse. Previously known as **impotence** (IM-pō-tens).

**Erection** (ē-REK-shun) The enlarged and stiff state of the penis or clitoris resulting from the engorgement of the spongy erectile tissue with blood.

**Eructation** (e-ruk′-TĀ-shun) The forceful expulsion of gas from the stomach. Also called **belching.**

**Erythema** (er-e-THĒ-ma) Skin redness usually caused by dilation of the capillaries.

**Erythrocyte** (e-RITH-rō-sīt) A mature red blood cell.

**Erythropoietin** (e-rith′-rō-POY-e-tin) **(EPO)** A hormone released by the juxtaglomerular cells of the kidneys that stimulates red blood cell production.

**Esophagus** (e-SOF-a-gus) The hollow muscular tube that connects the pharynx and the stomach.

**Estrogens** (ES-tro-jenz) Feminizing sex hormones produced by the ovaries; govern development of oocytes, maintenance of female reproductive structures, and appearance of secondary sex characteristics; also affect fluid and electrolyte balance, and protein anabolism. Examples are $\beta$-estradiol, estrone, and estriol.

**Eupnea** (ŪP-nē-a) Normal quiet breathing.

**Eustachian tube** *See* **Auditory tube.**

**Eversion** (ē-VER-zhun) The movement of the sole laterally at the ankle joint or of an atrioventricular valve into an atrium during ventricular contraction.

**Excitability** (ek-sīt′-a-BIL-i-tē) The ability of muscle fibers to receive and respond to stimuli; the ability of neurons to respond to stimuli and generate nerve impulses.

**Excretion** (eks-KRĒ-shun) The process of eliminating waste products from the body; also the products excreted.

**Exercise physiology** Study of the changes in cell and organ function due to muscular activity.

**Exhalation** (eks-ha-LĀ-shun) Breathing out; expelling air from the lungs into the atmosphere. Also called **expiration.**

**Exocrine** (EK-sō-krin) **gland** A gland that secretes its products into ducts that carry the secretions into body cavities, into the lumen of an organ, or to the outer surface of the body.

**Exocytosis** (ek-sō-sī-TŌ-sis) A process in which membrane-enclosed secretory vesicles form inside the cell, fuse with the plasma membrane, and release their contents into the interstitial fluid; achieves secretion of materials from a cell.

**Extensibility** (ek-sten′-si-BIL-i-tē) The ability of muscle tissue to stretch when it is pulled.

**Extension** (eks-TEN-shun) An increase in the angle between two bones; restoring a body part to its anatomical position after flexion.

**External** Located on or near the surface.

**External auditory** (AW-di-tōr-ē) **canal** or **meatus** (mē-Ā-tus) A curved tube in the temporal bone that leads to the middle ear.

**External ear** The **outer ear,** consisting of the pinna, external auditory canal, and tympanic membrane (eardrum).

**External nares** (NĀ-rez) The openings into the nasal cavity on the exterior of the body. Also called the **nostrils.**

**External respiration** The exchange of respiratory gases between the lungs and blood. Also called **pulmonary respiration** or **pulmonary gas exchange.**

**Exteroceptor** (EKS-ter-ō-sep′-tor) A sensory receptor adapted for the reception of stimuli from outside the body.

**Extracellular fluid (ECF)** Fluid outside body cells, such as interstitial fluid and plasma.

**Extracellular matrix** (MĀ-triks) The ground substance and fibers between cells in a connective tissue.

**Eyebrow** The hairy ridge superior to the eye.

## F

**F cell** A cell in the pancreatic islets (islets of Langerhans) that secretes pancreatic polypeptide.

**Face** The anterior aspect of the head.

**Falciform ligament** (FAL-si-form LIG-a-ment) A sheet of parietal peritoneum between the two principal lobes of the liver. The ligamentum teres, or remnant of the umbilical vein, lies within its fold.

**Fallopian tube** *See* **Uterine tube.**

**Falx cerebelli** (FALKS′ ser-e-BEL-lī) A small triangular process of the dura mater attached to the occipital bone in the posterior cranial fossa and projecting inward between the two cerebellar hemispheres.

**Falx cerebri** (FALKS SER-e-brē) A fold of the dura mater extending deep into the longitudinal fissure between the two cerebral hemispheres.

**Fascia** (FASH-ē-a) Large connective tissue sheet that wraps around groups of muscles.

**Fascicle** (FAS-i-kul) A small bundle or cluster, especially of nerve or muscle fibers (cells). Also called a **fasciculus** (fa-SIK-ū-lus). *Plural* is **fasciculi** (fa-SIK-yoo-lī).

**Fasciculation** (fa-sik-ū-LĀ-shun) Abnormal, spontaneous twitch of all skeletal muscle fibers in one motor unit that is visible at the skin surface; not associated with movement of the affected muscle; present in progressive diseases of motor neurons, for example, poliomyelitis.

**Fat** A triglyceride that is a solid at room temperature.

**Fatty acid** A simple lipid that consists of a carboxyl group and a hydrocarbon chain; used to synthesize triglyceride and phospholipids.

**Fauces** (FAW-sēs) The opening from the mouth into the pharynx.

**Feces** (FĒ-sēz) Material discharged from the rectum and made up of bacteria, excretions, and food residue. Also called **stool**.

**Feedback system (loop)** A cycle of events in which the status of a body condition is monitored, evaluated, changed, remonitored, reevaluated, and so on.

**Female reproductive cycle** General term for the ovarian and uterine cycles, the hormonal changes that accompany them, and cyclic changes in the breasts and cervix; includes changes in the endometrium of a nonpregnant female that prepares the lining of the uterus to receive a fertilized ovum. Less correctly termed the **menstrual cycle.**

**Fertilization** (fer-til-i-ZĀ-shun) Penetration of a secondary oocyte by a sperm cell, meiotic division of a secondary oocyte to form an ovum, and subsequent union of the nuclei of the gametes.

**Fetal circulation** The cardiovascular system of the fetus, including the placenta and special blood vessels involved in the exchange of materials between fetus and mother.

**Fetus** (FĒ-tus) In humans, the developing organism *in utero* from the beginning of the third month to birth.

**Fever** An elevation in body temperature above the normal temperature of 37°C (98.6°F) due to a resetting of the hypothalamic thermostat.

**Fibroblast** (FĪ-brō-blast) A large, flat cell that secretes most of the extracellular matrix of areolar and dense connective tissues.

**Fibrosis** The process by which fibroblasts synthesize collagen fibers and other extracellular matrix materials that aggregate to form scar tissue.

**Fibrous** (FĪ-brus) **joint** A joint that allows little or no movement, such as a suture or a syndesmosis.

**Fibrous tunic** (TOO-nik) The superficial coat of the eyeball, made up of the posterior sclera and the anterior cornea.

**Fight-or-flight response** The effects produced on stimulation of the sympathetic division of the autonomic nervous system.

**Filiform papilla** (FIL-i-form pa-PIL-a) One of the conical projections that are distributed in parallel rows over the anterior two-thirds of the tongue and lack taste buds.

**Filtration** (fil-TRĀ-shun) The flow of a liquid through a filter (or membrane that acts like a filter) due to a hydrostatic pressure; occurs in capillaries due to blood pressure.

**Filum terminale** (FĪ-lum ter-mi-NAL-ē) Nonnervous fibrous tissue of the spinal cord that extends inferiorly from the conus medullaris to the coccyx.

**Fimbriae** (FIM-brē-ē) Fingerlike structures, especially the lateral ends of the uterine (fallopian) tubes.

**Fissure** (FISH-ur) A groove, fold, or slit that may be normal or abnormal.

**Fixator** A muscle that stabilizes the origin of the prime mover so that the prime mover can act more efficiently.

**Fixed macrophage** (MAK-rō-fāj) Stationary phagocytic cell found in the liver, lungs, brain, spleen, lymph nodes, subcutaneous tissue, and red bone marrow. Also called a **tissue macrophage** or **histiocyte** (HIS-tē-ō-sīt).

**Flaccid** (FLAK-sid) Relaxed, flabby, or soft; lacking muscle tone.

**Flagellum** (fla-JEL-um) A hairlike, motile process on the extremity of a bacterium, protozoan, or sperm cell. *Plural* is **flagella** (fla-JEL-a).

**Flatus** (FLĀ-tus) Gas in the stomach or intestines; commonly used to denote expulsion of gas through the anus.

**Flexion** (FLEK-shun) Movement in which there is a decrease in the angle between two bones.

**Follicle** (FOL-i-kul) A small secretory sac or cavity; the group of cells that contains a developing oocyte in the ovaries.

**Follicle-stimulating hormone (FSH)** Hormone secreted by the anterior pituitary; it initiates development of ova and stimulates the ovaries to secrete estrogens in females, and initiates sperm production in males.

**Fontanel** (fon-ta-NEL) A mesenchyme-filled space where bone formation is not yet complete, especially between the cranial bones of an infant's skull.

**Foot** The terminal part of the lower limb, from the ankle to the toes.

**Foramen** (fō-RĀ-men) A passage or opening; a communication between two cavities of an organ, or a hole in a bone for passage of vessels or nerves. *Plural* is **foramina** (fō-RAM-i-na).

**Foramen ovale** (fō-RĀ-men ō-VAL-ē) An opening in the fetal heart in the septum between the right and left atria. A hole in the greater wing of the sphenoid bone that transmits the mandibular branch of the trigeminal (V) nerve.

**Forearm** (FOR-arm) The part of the upper limb between the elbow and the wrist.

**Fornix** (FOR-niks) An arch or fold; a tract in the brain made up of association fibers, connecting the hippocampus with the mammillary bodies; a recess around the cervix of the uterus where it protrudes into the vagina.

**Fossa** (FOS-a) A furrow or shallow depression.

**Fourth ventricle** (VEN-tri-kul) A cavity filled with cerebrospinal fluid within the brain lying between the cerebellum and the medulla oblongata and pons.

**Fracture** (FRAK-choor) Any break in a bone.

**Free radical** An atom or group of atoms with an unpaired electron in the outermost shell. It is unstable, highly reactive, and destroys nearby molecules.

**Frontal plane** A plane at a right angle to a midsagittal plane that divides the body or organs into anterior and posterior portions. Also called a **coronal** (kō-RŌ-nal) **plane.**

**Fundus** (FUN-dus) The part of a hollow organ farthest from the opening.

**Fungiform papilla** (FUN-ji-form pa-PIL-a) A mushroomlike elevation on the upper surface of the tongue appearing as a red dot; most contain taste buds.

**Furuncle** (FŪ-rung-kul) A boil; painful nodule caused by bacterial infection and inflammation of a hair follicle or sebaceous (oil) gland.

## G

**Gallbladder** A small pouch, located inferior to the liver, that stores bile and empties by means of the cystic duct.

**Gallstone** A solid mass, usually containing cholesterol, in the gallbladder or a bile-containing duct; formed anywhere between bile canaliculi in the liver and the hepatopancreatic ampulla (ampulla of Vater), where bile enters the duodenum. Also called a **biliary calculus.**

**Gamete** (GAM-ēt) A male or female reproductive cell; a sperm cell or secondary oocyte.

**Ganglion** (GANG-glē-on) Usually, a group of neuronal cell bodies lying outside the central nervous system (CNS). *Plural* is **ganglia** (GANG-glē-a).

**Gastric** (GAS-trik) **glands** Glands in the mucosa of the stomach composed of cells that empty their secretions into narrow channels called gastric pits. Types of cells are chief cells (secrete pepsinogen), parietal cells (secrete hydrochloric acid and intrinsic factor), surface mucous and mucous neck cells (secrete mucus), and G cells (secrete gastrin).

**Gastroenterology** (gas′-trō-en′-ter-OL-ō-jē) The medical specialty that deals with the structure, function, diagnosis, and treatment of diseases of the stomach and intestines.

**Gastrointestinal** (gas-trō-in-TES-ti-nal) **(GI) tract** A continuous tube running through the ventral body cavity extending from the mouth to the anus. Also called the **alimentary** (al′-i-MEN-tar-ē) **canal.**

**Gastrulation** (gas-troo-LĀ-shun) The migration of groups of cells from the epiblast that transform a bilaminar embryonic disc into a trilaminar embryonic disc with three primary germ layers; transformation of the blastula into the gastrula.

**Gene** (JĒN) Biological unit of heredity; a segment of DNA located in a definite position on a particular chromosome; a sequence of DNA that codes for a particular mRNA, rRNA, or tRNA.

**Genetic engineering** The manufacture and manipulation of genetic material.

**Genetics** The study of genes and heredity.

**Genome** (JĒ-nōm) The complete set of genes of an organism.

**Genotype** (JĒ-nō-tīp) The genetic makeup of an individual; the combination of alleles present at one or more chromosomal locations, as distinguished from the appearance, or phenotype, that results from those alleles.

**Geriatrics** (jer′-ē-AT-riks) The branch of medicine devoted to the medical problems and care of elderly persons.

**Gestation** (jes-TĀ-shun) The period of development from fertilization to birth.

**Gingivae** (jin-JI-vē) Gums. They cover the alveolar processes of the mandible and maxilla and extend slightly into each socket.

**Gland** Specialized epithelial cell or cells that secrete substances; may be exocrine or endocrine.

**Glans penis** (glanz PĒ-nis) The slightly enlarged region at the distal end of the penis.

**Glaucoma** (glaw-KŌ-ma) An eye disorder in which there is increased intraocular pressure due to an excess of aqueous humor.

**Glomerular** (glō-MER-ū-lar) **capsule** A double-walled globe at the proximal end of a nephron that encloses the glomerular capillaries. Also called **Bowman's** (BŌ-manz) **capsule.**

**Glomerular filtrate** (glō-MER-ū-lar FIL-trāt) The fluid produced when blood is filtered by the filtration membrane in the glomeruli of the kidneys.

**Glomerular filtration** The first step in urine formation in which substances in blood pass through the filtration membrane and the filtrate enters the proximal convoluted tubule of a nephron.

**Glomerular filtration rate (GFR)** The amount of filtrate formed in all renal corpuscles per minute. It averages 125 mL/min in males and 105 mL/min in females.

**Glomerulus** (glō-MER-ū-lus) A rounded mass of nerves or blood vessels, especially the microscopic tuft of capillaries that is surrounded by the glomerular (Bowman's) capsule of each kidney tubule. *Plural* is **glomeruli** (glō-MER-ū-li).

**Glottis** (GLOT-is) The vocal folds (true vocal cords) in the larynx plus the space between them (rima glottidis).

**Glucagon** (GLOO-ka-gon) A hormone produced by the alpha cells of the pancreatic islets (islets of Langerhans) that increases blood glucose level.

**Glucocorticoids** (gloo′-kō-KOR-ti-koyds) Hormones secreted by the cortex of the adrenal gland, especially cortisol, that influence glucose metabolism.

**Glucose** (GLOO-kōs) A hexose (six-carbon sugar), $C_6H_{12}O_6$, that is a major energy source for the production of ATP by body cells.

**Glucosuria** (gloo′-kō-SOO-rē-a) The presence of glucose in the urine; may be temporary or pathological. Also called **glycosuria.**

**Glycogen** (GLĪ-kō-jen) A highly branched polymer of glucose containing thousands of subunits; functions as a compact store of glucose molecules in liver and muscle fibers (cells).

**Goblet cell** A goblet-shaped unicellular gland that secretes mucus; present in epithelium of the airways and intestines.

**Goiter** (GOY-ter) An enlarged thyroid gland.

**Golgi** (GOL-jē) **complex** An organelle in the cytoplasm of cells consisting of four to six flattened sacs (cisternae), stacked on one another, with expanded areas at their ends; functions in processing, sorting, packaging, and delivering proteins and lipids to the plasma membrane, lysosomes, and secretory vesicles.

**Golgi tendon organ** *See* **Tendon organ.**

**Gomphosis** (gom-FŌ-sis) A fibrous joint in which a cone-shaped peg fits into a socket.

**Gonad** (GŌ-nad) A gland that produces gametes and hormones; the ovary in the female and the testis in the male.

**Gonadotropic hormone** Anterior pituitary hormone that affects the gonads.

**Gout** (GOWT) Hereditary condition associated with excessive uric acid in the blood; the acid crystallizes and deposits in joints, kidneys, and soft tissue.

**Graafian follicle** *See* **Mature follicle.**

**Gracile** (GRAS-īl) **nucleus** A group of nerve cells in the inferior part of the medulla oblongata in which axons of the gracile fasciculus terminate.

**Gray commissure** (KOM-mi-shur) A narrow strip of gray matter connecting the two lateral gray masses within the spinal cord.

**Gray matter** Areas in the central nervous system and ganglia containing neuronal cell bodies, dendrites, unmyelinated axons, axon terminals, and neuroglia; Nissl bodies impart a gray color and there is little or no myelin in gray matter.

**Gray ramus communicans** (RĀ-mus kō-MŪ-ni-kans) A short nerve containing axons of sympathetic postganglionic neurons; the cell bodies of the neurons are in a sympathetic chain ganglion, and the unmyelinated axons extend via the gray ramus to a spinal nerve and then to the periphery to supply smooth muscle in blood vessels, arrector pili muscles, and sweat glands. *Plural* is **rami communicantes** (RĀ-mē kō-mū-ni-KAN-tēz).

**Greater omentum** (ō-MEN-tum) A large fold in the serosa of the stomach that hangs down like an apron anterior to the intestines.

**Greater vestibular** (ves-TIB-ū-lar) **glands** A pair of glands on either side of the vaginal orifice that open by a duct into the space between the hymen and the labia minora. Also called **Bartholin's** (BAR-to-linz) **glands**.

**Groin** (GROYN) The depression between the thigh and the trunk; the inguinal region.

**Gross anatomy** The branch of anatomy that deals with structures that can be studied without using a microscope. Also called **macroscopic anatomy.**

**Growth** An increase in size due to an increase in (1) the number of cells, (2) the size of existing cells as internal components increase in size, or (3) the size of intercellular substances.

**Gustation** (gus-TĀ-shun). The sense of taste.

**Gustatory** (GUS-ta-tō′-rē) Pertaining to taste.

**Gynecology** (gī′-ne-KOL-ō-jē) The branch of medicine dealing with the study and treatment of disorders of the female reproductive system.

**Gynecomastia** (gīn′-e-kō-MAS-tē-a) Excessive growth (benign) of the male mammary glands due to secretion of estrogens by an adrenal gland tumor (feminizing adenoma).

**Gyrus** (JI-rus) One of the folds of the cerebral cortex of the brain. *Plural* is **gyri** (JĪ-rī). Also called a **convolution.**

## H

**Hair** A threadlike structure produced by hair follicles that develops in the dermis. Also called a **pilus** (PĪ-lus).

**Hair follicle** (FOL-li-kul) Structure composed of epithelium and surrounding the root of a hair from which hair develops.

**Hair root plexus** (PLEK-sus) A network of dendrites arranged around the root of a hair as free or naked nerve endings that are stimulated when a hair shaft is moved.

**Hand** The terminal portion of an upper limb, including the carpals, metacarpals, and phalanges.

**Haploid** (HAP-loyd) **cell** Having half the number of chromosomes characteristically found in the somatic cells of an organism; characteristic of mature gametes. Symbolized *n*.

**Hard palate** (PAL-at) The anterior portion of the roof of the mouth, formed by the maxillae and palatine bones and lined by mucous membrane.

**Haustra** (HAWS-tra) A series of pouches that characterize the colon; caused by tonic contractions of the teniae coli. *Singular* is **haustrum.**

**Haversian canal** *See* **Central canal.**

**Haversian system** *See* **Osteon.**

**Head** The superior part of a human, cephalic to the neck. The superior or proximal part of a structure.

**Hearing** The ability to perceive sound.

**Heart** A hollow muscular organ lying slightly to the left of the midline of the chest that pumps the blood through the cardiovascular system.

**Heart block** An arrhythmia (dysrhythmia) of the heart in which the atria and ventricles contract independently because of a blocking of electrical impulses through the heart at some point in the conduction system.

**Heart murmur** (MER-mer) An abnormal sound that consists of a flow noise that is heard before, between, or after the normal heart sounds, or that may mask normal heart sounds.

**Hemangioblast** (hē-MAN-jē-ō-blast) A precursor mesodermal cell that develops into blood and blood vessels.

**Hematocrit** (he-MAT-ō-krit) **(Hct)** The percentage of blood made up of red blood cells. Usually measured by centrifuging a blood sample in a graduated tube and then reading the volume of red blood cells and dividing it by the total volume of blood in the sample.

**Hematology** (hēm-a-TOL-ō-jē) The study of blood.

**Hematoma** (hē′-ma-TŌ-ma) A tumor or swelling filled with blood.

**Hemiplegia** (hem-i-PLĒ-jē-a) Paralysis of the upper limb, trunk, and lower limb on one side of the body.

**Hemodialysis** (hē-mō-dī-AL-i-sis) Direct filtration of blood by removing wastes and excess electrolytes and fluid and then returning the cleansed blood.

**Hemodynamics** (hē-mō-dī-NAM-iks) The forces involved in circulating blood throughout the body.

**Hemoglobin** (hē′-mō-GLŌ-bin) **(Hb)** A substance in red blood cells consisting of the protein globin and the iron-containing red pigment heme that transports most of the oxygen and some carbon dioxide in blood.

**Hemolysis** (hē-MOL-i-sis) The escape of hemoglobin from the interior of a red blood cell into the surrounding medium; results from disruption of the cell membrane by toxins or drugs, freezing or thawing, or hypotonic solutions.

**Hemolytic disease of the newborn** (HDN) A hemolytic anemia of a newborn child that results from the destruction of the infant's erythrocytes (red blood cells) by antibodies produced by the mother; usually the antibodies are due to an Rh blood type incompatibility. Also called **erythroblastosis fetalis** (e-rith′-rō-blas-TŌ-sis fe-TAL-is).

**Hemophilia** (hē′-mō-FIL-ē-a) A hereditary blood disorder in which there is a deficient production of certain factors involved in blood clotting, resulting in excessive bleeding into joints, deep tissues, and elsewhere.

**Hemopoiesis** (hēm-ō-poy-Ē-sis) Blood cell production, which occurs in red bone marrow after birth. Also called **hematopoiesis** (hem′-a-tō-poy-Ē-sis).

**Hemorrhage** (HEM-o-rij) Bleeding; the escape of blood from blood vessels, especially when the loss is profuse.

**Hemorrhoids** (HEM-ō-royds) Dilated or varicosed blood vessels (usually veins) in the anal region. Also called **piles**.

**Hepatic** (he-PAT-ik) Refers to the liver.

**Hepatic duct** A duct that receives bile from the bile capillaries. Small hepatic ducts merge to form the larger right and left hepatic ducts that unite to leave the liver as the common hepatic duct.

**Hepatic portal circulation** The flow of blood from the gastrointestinal organs to the liver before returning to the heart.

**Hepatocyte** (he-PAT-ō-cyte) A liver cell.

**Hepatopancreatic** (hep′-a-tō-pan′-krē-A-tik) **ampulla** A small, raised area in the duodenum where the combined common bile duct and main pancreatic duct empty into the duodenum. Also called the **ampulla of Vater** (FAH-ter).

**Hernia** (HER-nē-a) The protrusion or projection of an organ or part of an organ through a membrane or cavity wall, usually the abdominal cavity.

**Herniated** (HER-nē-ā′-ted) **disc** A rupture of an intervertebral disc so that the nucleus pulposus protrudes into the vertebral cavity. Also called a **slipped disc.**

**Hiatus** (hī-Ā-tus) An opening; a foramen.

**Hilum** (HĪ-lum) An area, depression, or pit where blood vessels and nerves enter or leave an organ. Also called a **hilus.**

**Hinge joint** A synovial joint in which a convex surface of one bone fits into a concave surface of another bone, such as the elbow, knee, ankle, and interphalangeal joints. Also called a **ginglymus** (JIN-gli-mus) **joint.**

**Hirsutism** (HER-soo-tizm) An excessive growth of hair in females and children, with a distribution similar to that in adult males, due to the conversion of vellus hairs into large terminal hairs in response to higher-than-normal levels of androgens.

**Histamine** (HISS-ta-mēn) Substance found in many cells, especially mast cells, basophils, and platelets, that is released when the cells are injured; results in vasodilation, increased permeability of blood vessels, and constriction of bronchioles.

**Histology** (his′-TOL-ō-jē) Microscopic study of the structure of tissues.

**Holocrine** (HŌ-lō-krin) **gland** A type of gland in which entire secretory cells, along with their accumulated secretions, make up the secretory product of the gland, as in the sebaceous (oil) glands.

**Homeostasis** (hō′-mē-ō-STĀ-sis) The condition in which the body's internal environment remains relatively constant within physiological limits.

**Homologous** (hō-MOL-ō-gus) **chromosomes** Two chromosomes that belong to a pair. Also called **homologs.**

**Hormone** (HOR-mōn) A secretion of endocrine cells that alters the physiological activity of target cells of the body.

**Horn** An area of gray matter (anterior, lateral, or posterior) in the spinal cord.

**Human chorionic gonadotropin** (kō-rē-ON-ik gō-nad-ō-TRŌ-pin) **(hCG)** A hormone produced by the developing placenta that maintains the corpus luteum.

**Human chorionic somatomammotropin** (sō-mat-ō-mam-ō-TRŌ-pin) **(hCS)** Hormone produced by the chorion of the placenta that stimulates breast tissue for lactation, enhances body growth, and regulates metabolism. Also called **human placental lactogen (hPL).**

**Human growth hormone (hGH)** Hormone secreted by the anterior pituitary that stimulates growth of body tissues, especially skeletal and muscular tissues. Also known as **somatotropin** (sō′-ma-tō-TRŌ-pin) and **somatotropic hormone (STH).**

**Hyaluronic** (hī′-a-loo-RON-ik) **acid** A viscous, amorphous extracellular material that binds cells together, lubricates joints, and maintains the shape of the eyeballs.

**Hymen** (HĪ-men) A thin fold of vascularized mucous membrane at the vaginal orifice.

**Hyperextension** (hī′-per-ek-STEN-shun) Continuation of extension beyond the anatomical position, as in bending the head backward.

**Hyperplasia** (hī-per-PLĀ-zē-a) An abnormal increase in the number of normal cells in a tissue or organ, increasing its size.

**Hypersecretion** (hī′-per-se-KRĒ-shun) Overactivity of glands resulting in excessive secretion.

**Hypersensitivity** (hī′-per-sen-si-TI-vi-tē) Overreaction to an allergen that results in pathological changes in tissues. Also called **allergy.**

**Hypertension** (hī′-per-TEN-shun) High blood pressure.

**Hyperthermia** (hī′-per-THERM-ē-a) An elevated body temperature.

**Hypertonia** (hī′-per-TŌ-nē-a) Increased muscle tone that is expressed as spasticity or rigidity.

**Hypertonic** (hī′-per-TON-ik) Solution that causes cells to shrink due to loss of water by osmosis.

**Hypertrophy** (hī-PER-trō-fē) An excessive enlargement or overgrowth of tissue without cell division.

**Hyperventilation** (hī′-per-ven-ti-LĀ-shun) A rate of inhalation and exhalation higher than that required to maintain a normal partial pressure of carbon dioxide in the blood.

**Hyponychium** (hī′-pō-NIK-ē-um) Free edge of the fingernail.

**Hypophyseal fossa** (hī′-pō-FIZ-ē-al FOS-a) A depression on the superior surface of the sphenoid bone that houses the pituitary gland.

**Hypophyseal** (hī′-pō-FIZ-ē-al) **pouch** An outgrowth of ectoderm from the roof of the mouth from which the anterior pituitary develops. Also called **Rathke's pouch.**

**Hypophysis** (hī-POF-i-sis) Pituitary gland.

**Hyposecretion** (hī′-pō-se-KRĒ-shun) Underactivity of glands resulting in diminished secretion.

**Hypothalamohypophyseal** (hī′-pō-thal′-a-mō-hī-pō-FIZ-ē-al) **tract** A bundle of axons containing secretory vesicles filled with oxytocin or antidiuretic hormone that extend from the hypothalamus to the posterior pituitary.

**Hypothalamus** (hī′-pō-THAL-a-mus) A portion of the diencephalon, lying beneath the thalamus and forming the floor and part of the wall of the third ventricle.

**Hypothermia** (hī′-pō-THER-mē-a) Lowering of body temperature below 35°C (95°F); in surgical procedures, it refers to deliberate cooling of the body to slow down metabolism and reduce oxygen needs of tissues.

**Hypotonia** (hī′-pō-TŌ-nē-a) Decreased or lost muscle tone in which muscles appear flaccid.

**Hypotonic** (hī′-pō-TON-ik) Solution that causes cells to swell and perhaps rupture due to gain of water by osmosis.

**Hypoventilation** (hī-pō-ven-ti-LĀ-shun) A rate of inhalation and exhalation lower than that required to maintain a normal partial pressure of carbon dioxide in plasma.

**Hypoxia** (hī-POKS-ē-a) Lack of adequate oxygen at the tissue level.

**Hysterectomy** (hiss-te-REK-tō-mē) The surgical removal of the uterus.

## I

**Ileocecal** (il-ē-ō-SĒ-kal) **sphincter** A fold of mucous membrane that guards the opening from the ileum into the large intestine. Also called the **ileocecal valve.**

**Ileum** (IL-ē-um) The terminal part of the small intestine.

**Immunity** (i-MŪ-ni-tē) The state of being resistant to injury, particularly by poisons, foreign proteins, and invading pathogens.

**Immunoglobulin** (im-ū-nō-GLOB-ū-lin) **(Ig)** An antibody synthesized by plasma cells derived from B lymphocytes in response to the introduction of an antigen. Immunoglobulins are divided into five kinds (IgG, IgM, IgA, IgD, IgE).

**Immunology** (im′-ū-NOL-ō-jē) The study of the responses of the body when challenged by antigens.

**Imperforate** (im′-PER-fō-rāt) Abnormally closed.

**Implantation** (im′-plan-TĀ-shun) The insertion of a tissue or a part into the body. The attachment of the blastocyst to the stratum basalis of the endometrium about 6 days after fertilization.

**Incontinence** (in-KON-ti-nens) Inability to retain urine, semen, or feces through loss of sphincter control.

**Indirect motor pathways** Motor tracts that convey information from the brain down the spinal cord for automatic movements, coordination of body movements with visual stimuli, skeletal muscle tone and posture, and balance. Also known as **extrapyramidal pathways.**

**Induction** (in-DUK-shun) The process by which one tissue (inducting tissue) stimulates the development of an adjacent unspecialized tissue (responding tissue) into a specialized one.

**Infarction** (in-FARK-shun) A localized area of necrotic tissue, produced by inadequate oxygenation of the tissue.

**Infection** (in-FEK-shun) Invasion and multiplication of microorganisms in body tissues, which may be inapparent or characterized by cellular injury.

**Inferior** (in-FĒR-ē-or) Away from the head or toward the lower part of a structure. Also called **caudal** (KAW-dal).

**Inferior vena cava** (VĒ-na KĀ-va) **(IVC)** Large vein that collects blood from parts of the body inferior to the heart and returns it to the right atrium.

**Infertility** Inability to conceive or to cause conception. Also called **sterility** in males.

**Inflammation** (in′-fla-MĀ-shun) Localized, protective response to tissue injury designed to destroy, dilute, or wall off the infecting agent or injured tissue; characterized by redness, pain, heat, swelling, and sometimes loss of function.

**Infundibulum** (in-fun-DIB-ū-lum) The stalk-like structure that attaches the pituitary gland to the hypothalamus of the brain. The funnel-shaped, open, distal end of the uterine (fallopian) tube.

**Ingestion** (in-JES-chun) The taking in of food, liquids, or drugs, by mouth.

**Inguinal** (IN-gwin-al) Pertaining to the groin.

**Inguinal canal** An oblique passageway in the anterior abdominal wall just superior and parallel to the medial half of the inguinal ligament that transmits the spermatic cord and ilioinguinal nerve in the male and round ligament of the uterus and ilioinguinal nerve in the female.

**Inhalation** (in-ha-LĀ-shun) The act of drawing air into the lungs. Also termed **inspiration.**

**Inheritance** The acquisition of body traits by transmission of genetic information from parents to offspring.

**Inhibin** A hormone secreted by the gonads that inhibits release of follicle-stimulating hormone (FSH) by the anterior pituitary.

**Inhibiting hormone** Hormone secreted by the hypothalamus that can suppress secretion of hormones by the anterior pituitary.

**Insertion** (in-SER-shun) The attachment of a muscle tendon to a movable bone or the end opposite the origin.

**Insula** (IN-soo-la) A triangular area of the cerebral cortex that lies deep within the lateral cerebral fissure, under the parietal, frontal, and temporal lobes.

**Insulin** (IN-soo-lin) A hormone produced by the beta cells of a pancreatic islet (islet of Langerhans) that decreases the blood glucose level.

**Integrins** (IN-te-grinz) A family of transmembrane glycoproteins in plasma membranes that function in cell adhesion; they are present in hemidesmosomes, which anchor cells to a basement membrane, and they mediate adhesion of neutrophils to endothelial cells during emigration.

**Integumentary** (in-teg-ū-MEN-tar-ē) Relating to the skin.

**Integumentary** (in-teg-ū-MEN-tar-ē) **system** A system composed of organs such as the skin, hair, oil and sweat glands, nails, and sensory receptors.

**Intercalated** (in-TER-ka-lāt-ed) **disc** An irregular transverse thickening of sarcolemma that contains desmosomes, which hold cardiac muscle fibers (cells) together, and gap junctions, which aid in conduction of muscle action potentials from one fiber to the next.

**Intercostal** (in′-ter-KOS-tal) **nerve** A nerve supplying a muscle located between the ribs. Also called **thoracic nerve.**

**Intermediate** (in′-ter-MĒ-dē-at) Between two structures, one of which is medial and one of which is lateral.

**Intermediate filament** Protein filament, ranging from 8 to 12 nm in diameter, that may provide structural reinforcement, hold organelles in place, and give shape to a cell.

**Internal** Away from the surface of the body.

**Internal capsule** A large tract of projection fibers lateral to the thalamus that is the major connection between the cerebral cortex and the brain stem and spinal cord; contains axons of sensory neurons carrying auditory, visual, and somatic sensory signals to the cerebral cortex plus axons of motor neurons descending from the cerebral cortex to the thalamus, subthalamus, brain stem, and spinal cord.

**Internal ear** The inner ear or labyrinth, lying inside the temporal bone, containing the organs of hearing and balance.

**Internal nares** (NĀ-rez) The two openings posterior to the nasal cavities opening into the nasopharynx. Also called the **choanae** (kō-Ā-nē).

**Internal respiration** The exchange of respiratory gases between blood and body cells. Also called **tissue respiration** or **systemic gas exchange.**

**Interneurons** (in′-ter-NOO-ronz) Neurons whose axons extend only for a short distance and contact nearby neurons in the brain, spinal cord, or a ganglion; they comprise the vast majority of neurons in the body. Also called **association neurons.**

**Interoceptor** (IN-ter-ō-sep′-tor) Sensory receptor located in blood vessels and viscera that provides information about the body's internal environment. Also called **visceroceptor.**

**Interphase** (IN-ter-fāz) The period of the cell cycle between cell divisions, consisting of the $G_0$ phase; $G_1$ (gap or growth) phase, when the cell is engaged in growth, metabolism, and production of substances required for division; S (synthesis) phase, during which chromosomes are replicated; and $G_2$ phase.

**Interstitial cell of Leydig** *See* **Interstitial endocrinocyte.**

**Interstitial** (in′-ter-STISH-al) **endocrinocyte** A cell that is located in the connective tissue between seminiferous tubules in a mature testis that secretes testosterone. Also called an **interstitial cell of Leydig** (LĪ-dig).

**Interstitial** (in′-ter-STISH-al) **fluid** The portion of extracellular fluid that fills the microscopic spaces between the cells of tissues; the internal environment of the body. Also called **intercellular** or **tissue fluid.**

**Interstitial growth** Growth from within, as in the growth of cartilage. Also called **endogenous** (en-DOJ-e-nus) **growth.**

**Interventricular** (in′-ter-ven-TRIK-ū-lar) **foramen** A narrow, oval opening through which the lateral ventricles of the brain communicate with the third ventricle. Also called the **foramen of Monro.**

**Intervertebral** (in′-ter-VER-te-bral) **disc** A pad of fibrocartilage located between the bodies of two vertebrae.

**Intestinal gland** A gland that opens onto the surface of the intestinal mucosa and secretes digestive enzymes. Also called a **crypt of Lieberkühn** (LĒ-ber-kūn).

**Intracellular** (in′-tra-SEL-yū-lar) **fluid (ICF)** Fluid located within cells.

**Intrafusal** (in′-tra-FŪ-sal) **fibers** Three to ten specialized muscle fibers (cells), partially enclosed in a spindle-shaped connective tissue capsule, that make up a muscle spindle.

**Intramembranous** (in′-tra-MEM-bra-nus) **ossification** Bone formation within mesenchyme arranged in sheetlike layers that resemble membranes.

**Intramuscular injection** An injection that penetrates the skin and subcutaneous layer to enter a skeletal muscle. Common sites are the deltoid, gluteus medius, and vastus lateralis muscles.

**Intraocular** (in′-tra-OK-ū-lar) **pressure (IOP)** Pressure in the eyeball, produced mainly by aqueous humor.

**Intrinsic factor (IF)** A glycoprotein, synthesized and secreted by the parietal cells of the gastric mucosa, that facilitates vitamin $B_{12}$ absorption in the small intestine.

**Invagination** (in-vaj′-i-NĀ-shun) The pushing of the wall of a cavity into the cavity itself.

**Inversion** (in-VER-zhun) The movement of the sole medially at the ankle joint.

**In vitro** (VĒ-trō) Literally, in glass; outside the living body and in an artificial environment such as a laboratory test tube.

**Ipsilateral** (ip-si-LAT-er-al) On the same side, affecting the same side of the body.

**Iris** The colored portion of the vascular tunic of the eyeball seen through the cornea that contains circular and radial smooth muscle; the hole in the center of the iris is the pupil.

**Irritable bowel syndrome (IBS)** Disease of the entire gastrointestinal tract in which a person reacts to stress by developing symptoms (such as cramping and abdominal pain) associated with alternating patterns of diarrhea and constipation. Excessive amounts of mucus may appear in feces, and other symptoms include flatulence, nausea, and loss of appetite. Also known as **irritable colon** or **spastic colitis.**

**Ischemia** (is-KĒ-mē-a) A lack of sufficient blood to a body part due to obstruction or constriction of a blood vessel.

**Islet of Langerhans** *See* **Pancreatic islet.**

**Isotonic** (ī′-sō-TON-ik) Having equal tension or tone. A solution having the same concentration of impermeable solutes as cytosol.

**Isthmus** (IS-mus) A narrow strip of tissue or narrow passage connecting two larger parts.

## J

**Jaundice** (JON-dis) A condition characterized by yellowness of the skin, the white of the eyes, mucous membranes, and body fluids because of a buildup of bilirubin.

**Jejunum** (je-JOO-num) The middle part of the small intestine.

**Joint** A point of contact between two bones, between bone and cartilage, or between bone and teeth. Also called an **articulation** or **arthrosis.**

**Joint kinesthetic** (kin′-es-THET-ik) **receptor** A proprioceptive receptor located in a joint, stimulated by joint movement.

**Juxtaglomerular** (juks-ta-glō-MER-ū-lar) **apparatus (JGA)** Consists of the macula densa (cells of the distal convoluted tubule adjacent to the afferent and efferent arteriole) and juxtaglomerular cells (modified cells of the afferent and sometimes efferent arteriole); secretes renin when blood pressure starts to fall.

## K

**Keratin** (KER-a-tin) An insoluble protein found in the hair, nails, and other keratinized tissues of the epidermis.

**Keratinocyte** (ker-a-TIN-ō-sīt) The most numerous of the epidermal cells; produces keratin.

**Kidney** (KID-nē) One of the paired reddish organs located in the lumbar region that regulates the composition, volume, and pressure of blood and produces urine.

**Kidney stone** A solid mass, usually consisting of calcium oxalate, uric acid, or calcium phosphate crystals, that may form in any portion of the urinary tract. Also called a **renal calculus.**

**Kinesiology** (ki-nē-sē′-OL-ō-jē) The study of the movement of body parts.

**Kinesthesia** (kin′-es-THĒ-zē-a) The perception of the extent and direction of movement of body parts; this sense is possible due to nerve impulses generated by proprioceptors.

**Kinetochore** (ki-NET-ō-kor) Protein complex attached to the outside of a centromere to which kinetochore microtubules attach.

**Kupffer's cell** *See* **Stellate reticuloendothelial cell.**

**Kyphosis** (kī-FŌ-sis) An exaggeration of the thoracic curve of the vertebral column, resulting in a "round-shouldered" appearance. Also called **hunchback.**

## L

**Labial frenulum** (LĀ-bē-al FREN-ū-lum) A medial fold of mucous membrane between the inner surface of the lip and the gums.

**Labia majora** (LĀ-bē-a ma-JŌ-ra) Two longitudinal folds of skin extending downward and backward from the mons pubis of the female.

**Labia minora** (min-OR-a) Two small folds of mucous membrane lying medial to the labia majora of the female.

**Labium** (LĀ-bē-um) A lip. A liplike structure. *Plural* is **labia** (LA-bē-a).

**Labor** The process of giving birth in which a fetus is expelled from the uterus through the vagina.

**Labyrinth** (LAB-i-rinth) Intricate communicating passageway, especially in the internal ear.

**Lacrimal canal** A duct, one on each eyelid, beginning at the punctum at the medial margin of an eyelid and conveying tears medially into the nasolacrimal sac.

**Lacrimal gland** Secretory cells, located at the superior anterolateral portion of each orbit, that secrete tears into excretory ducts that open onto the surface of the conjunctiva.

**Lacrimal sac** The superior expanded portion of the nasolacrimal duct that receives the tears from a lacrimal canal.

**Lactation** (lak-TĀ-shun) The secretion and ejection of milk by the mammary glands.

**Lacteal** (LAK-tē-al) One of many lymphatic vessels in villi of the intestines that absorb triglycerides and other lipids from digested food.

**Lacuna** (la-KOO-na) A small, hollow space, such as that found in bones in which the osteocytes lie. *Plural* is **lacunae** (la-KOO-nē).

**Lambdoid** (LAM-doyd) **suture** The joint in the skull between the parietal bones and the occipital bone; sometimes contains sutural (Wormian) bones.

**Lamellae** (la-MEL-ē) Concentric rings of hard, calcified extracellular matrix found in compact bone.

**Lamellated corpuscle** *See* **Pacinian corpuscle.**

**Lamina** (LAM-i-na) A thin, flat layer or membrane, as the flattened part of either side of the arch of a vertebra. *Plural* is **laminae** (LAM-i-nē).

**Lamina propria** (PRŌ-prē-a) The connective tissue layer of a mucosa.

**Langerhans** (LANG-er-hans) **cell** Epidermal dendritic cell that functions as an antigen-presenting cell (APC) during an immune response.

**Lanugo** (la-NOO-gō) Fine downy hairs that cover the fetus.

**Large intestine** The portion of the gastrointestinal tract extending from the ileum of the small intestine to the anus, divided structurally into the cecum, colon, rectum, and anal canal.

**Laryngopharynx** (la-rin′-gō-FAIR-inks) The inferior portion of the pharynx, extending downward from the level of the hyoid bone that divides posteriorly into the esophagus and anteriorly into the larynx. Also called the **hypopharynx.**

**Larynx** (LAIR-inks) The **voice box,** a short passageway that connects the pharynx with the trachea.

**Lateral** (LAT-er-al) Farther from the midline of the body or a structure.

**Lateral ventricle** (VEN-tri-kul) A cavity within a cerebral hemisphere that communicates with the lateral ventricle in the other cerebral hemisphere and with the third ventricle by way of the interventricular foramen.

**Leg** The part of the lower limb between the knee and the ankle.

**Lens** A transparent organ constructed of proteins (crystallins) lying posterior to the pupil and iris of the eyeball and anterior to the vitreous body.

**Lesion** (LĒ-zhun) Any localized, abnormal change in a body tissue.

**Lesser omentum** (ō-MEN-tum) A fold of the peritoneum that extends from the liver to the lesser curvature of the stomach and the first part of the duodenum.

**Lesser vestibular** (ves-TIB-ū-lar) **gland** One of the paired mucus-secreting glands with ducts that open on either side of the urethral orifice in the vestibule of the female.

**Leukemia** (loo-KĒ-mē-a) A malignant disease of the blood-forming tissues characterized by either uncontrolled production and accumulation of immature leukocytes in which many cells fail to reach maturity (acute) or an accumulation of mature leukocytes in the blood because they do not die at the end of their normal life span (chronic).

**Leukocyte** (LOO-kō-sīt) A white blood cell.

**Leydig** (LĪ-dig) **cell** A type of cell that secretes testosterone; located in the connective tissue between seminiferous tubules in a mature testis. Also known as **interstitial cell of Leydig** or **interstitial endocrinocyte.**

**Ligament** (LIG-a-ment) Dense regular connective tissue that attaches bone to bone.

**Ligand** (LĪ-gand) A chemical substance that binds to a specific receptor.

**Limbic system** A part of the forebrain, sometimes termed the visceral brain, concerned with various aspects of emotion and behavior; includes the limbic lobe, dentate gyrus, amygdala, septal nuclei, mammillary bodies, anterior thalamic nucleus, olfactory bulbs, and bundles of myelinated axons.

**Lingual frenulum** (LIN-gwal FREN-ū-lum) A fold of mucous membrane that connects the tongue to the floor of the mouth.

**Lipase** An enzyme that splits fatty acids from triglycerides and phospholipids.

**Lipid** (LIP-id) An organic compound composed of carbon, hydrogen, and oxygen that is usually insoluble in water, but soluble in alcohol, ether, and chloroform; examples include triglycerides (fats and oils), phospholipids, steroids, and eicosanoids.

**Lipid bilayer** Arrangement of phospholipid, glycolipid, and cholesterol molecules in two parallel sheets in which the hydrophilic "heads" face outward and the hydrophobic "tails" face inward; found in cellular membranes.

**Lipoprotein** (lip′-ō-PRŌ-tēn) One of several types of particles containing lipids (cholesterol and triglycerides) and proteins that make it water-soluble for transport in the blood; high levels of **low-density lipoproteins (LDLs)** are associated with increased risk of atherosclerosis, whereas high levels of **high-density lipoproteins (HDLs)** are associated with decreased risk of atherosclerosis.

**Liver** Large organ under the diaphragm that occupies most of the right hypochondriac region and part of the epigastric region. Functionally, it produces bile and synthesizes most plasma proteins; interconverts nutrients; detoxifies substances; stores glycogen, iron, and vitamins; carries on phagocytosis of worn-out blood cells and bacteria; and helps synthesize the active form of vitamin D.

**Long-term potentiation** (po-ten′-shē-Ā-shun) **(LTP)** Prolonged, enhanced synaptic transmission that occurs at certain synapses within the hippocampus of the brain; believed to underlie some aspects of memory.

**Lordosis** (lor-DŌ-sis) An exaggeration of the lumbar curve of the vertebral column. Also called **hollow back.**

**Lower limb** The appendage attached at the pelvic (hip) girdle, consisting of the thigh, knee, leg, ankle, foot, and toes. Also called the **lower extremity** or **lower appendage.**

**Lumbar** (LUM-bar) Region of the back and side between the ribs and pelvis; loin.

**Lumbar plexus** (PLEK-sus) A network formed by the anterior (ventral) branches of spinal nerves L1 through L4.

**Lumen** (LOO-men) The space within an artery, vein, intestine, renal tubule, or other tubular structure.

**Lungs** Main organs of respiration that lie on either side of the heart in the thoracic cavity.

**Lunula** (LOO-noo-la) The moon-shaped white area at the base of a nail.

**Luteinizing** (LOO-tē-in′-īz-ing) **hormone (LH)** A hormone secreted by the anterior pituitary that stimulates ovulation, stimulates progesterone secretion by the corpus luteum, and readies the mammary glands for milk secretion in females; stimulates testosterone secretion by the testes in males.

**Lymph** (LIMF) Fluid confined in lymphatic vessels and flowing through the lymphatic system until it is returned to the blood.

**Lymph node** An oval or bean-shaped structure located along lymphatic vessels.

**Lymphatic** (lim-FAT-ik) **capillary** Closed-ended microscopic lymphatic vessel that begins in spaces between cells and converges with other lymphatic capillaries to form lymphatic vessels.

**Lymphatic system** (lim-FAT-ik) A system consisting of a fluid called lymph, vessels called lymphatics that transport lymph, a number of organs containing lymphatic tissue (lymphocytes within a filtering tissue), and red bone marrow.

**Lymphatic tissue** A specialized form of reticular tissue that contains large numbers of lymphocytes.

**Lymphatic vessel** A large vessel that collects lymph from lymphatic capillaries and converges with other lymphatic vessels to form the thoracic and right lymphatic ducts.

**Lymphocyte** (LIM-fō-sīt) A type of white blood cell that helps carry out cell-mediated and antibody-mediated immune responses; found in blood and in lymphatic tissues.

**Lysosome** (LĪ-sō-sōm) An organelle in the cytoplasm of a cell, enclosed by a single membrane and containing powerful digestive enzymes.

**Lysozyme** (LĪ-sō-zīm) A bactericidal enzyme found in tears, saliva, and perspiration.

## M

**Macrophage** (MAK-rō-fāj) Phagocytic cell derived from a monocyte; may be fixed or wandering.

**Macula** (MAK-ū-la) A discolored spot or a colored area. A small, thickened region on the wall of the utricle and saccule that contains receptors for static equilibrium.

**Macula lutea** (LOO-tē-a) The yellow spot in the center of the retina.

**Major histocompatibility (MHC) antigens** Surface proteins on white blood cells and other nucleated cells that are unique for each person (except for identical siblings); used to type tissues and help prevent rejection of transplanted tissues. Also known as **human leukocyte antigens (HLA).**

**Malignant** (ma-LIG-nant) Referring to diseases that tend to become worse and cause death, especially the invasion and spreading of cancer.

**Mammary** (MAM-ar-ē) **gland** Modified sudoriferous (sweat) gland of the female that produces milk for the nourishment of the young.

**Mammillary** (MAM-i-ler-ē) **bodies** Two small rounded bodies on the inferior aspect of the hypothalamus that are involved in reflexes related to the sense of smell.

**Marrow** (MAR-ō) Soft, spongelike material in the cavities of bone. *Red bone marrow* produces blood cells; *yellow bone marrow* contains adipose tissue that stores triglycerides.

**Mast cell** A cell found in areolar connective tissue that releases histamine, a dilator of small blood vessels, during inflammation.

**Mastication** (mas′-ti-KĀ-shun) Chewing.

**Mature follicle** A large, fluid-filled follicle containing a secondary oocyte and surrounding granulosa cells that secrete estrogens. Also called a **graafian** (GRAF-ē-an) **follicle.**

**Meatus** (mē-Ā-tus) A passage or opening, especially the external portion of a canal.

**Mechanoreceptor** (me-KAN-ō-rē-sep-tor) Sensory receptor that detects mechanical deformation of the receptor itself or adjacent cells; stimuli so detected include those related to touch, pressure, vibration, proprioception, hearing, equilibrium, and blood pressure.

**Medial** (MĒ-dē-al) Nearer the midline of the body or a structure.

**Medial lemniscus** (lem-NIS-kus) A white matter tract that originates in the gracile and cuneate nuclei of the medulla oblongata and extends to the thalamus on the same side; sensory axons in this tract conduct nerve impulses for the sensations of proprioception, fine touch, vibration, hearing, and equilibrium.

**Median aperture** (AP-er-choor) One of the three openings in the roof of the fourth ventricle through which cerebrospinal fluid enters the subarachnoid space of the brain and cord. Also called the **foramen of Magendie.**

**Median plane** A vertical plane dividing the body into right and left halves. Situated in the middle.

**Mediastinum** (mē′-dē-as-TĪ-num) The anatomical region on the thoracic cavity between the pleurae of the lungs that extends from the sternum to the vertebral column and from the first rib to the diaphragm.

**Medulla** (me-DOOL-la) An inner layer of an organ, such as the medulla of the kidneys.

**Medulla oblongata** (me-DOOL-la ob′-long-GA-ta) The most inferior part of the brain stem. Also termed the **medulla.**

**Medullary** (MED-ū-lar′-ē) cavity The space within the diaphysis of a bone that contains yellow bone marrow. Also called the **marrow cavity.**

**Medullary rhythmicity** (rith-MIS-i-tē) **area** The neurons of the respiratory center in the medulla oblongata that control the basic rhythm of respiration.

**Meibomian gland** *See* **Tarsal gland.**

**Meiosis** (mī-Ō-sis) A type of cell division that occurs during production of gametes, involving two successive nuclear divisions that result in cells with the haploid ($n$) number of chromosomes.

**Meissner** (MĪS-ner) **corpuscle** A sensory receptor for touch; found in dermal papillae, especially in the palms and soles. Also called a **corpuscle of touch.**

**Melanin** (MEL-a-nin) A dark black, brown, or yellow pigment found in some parts of the body such as the skin, hair, and pigmented layer of the retina.

**Melanocyte** (MEL-a-nō-sīt′) A pigmented cell, located between or beneath cells of the deepest layer of the epidermis, that synthesizes melanin.

**Melanocyte-stimulating hormone (MSH)** A hormone secreted by the anterior pituitary that stimulates the dispersion of melanin granules in melanocytes in amphibians; continued administration produces darkening of skin in humans.

**Melatonin** (mel-a-TŌN-in) A hormone secreted by the pineal gland that helps set the timing of the body's biological clock.

**Membrane** A thin, flexible sheet of tissue composed of an epithelial layer and an underlying connective tissue layer, as in an epithelial membrane, or of areolar connective tissue only, as in a synovial membrane.

**Membranous labyrinth** (mem-BRA-nus LAB-i-rinth) The part of the labyrinth of the internal ear that is located inside the bony labyrinth and separated from it by the perilymph; made up of the semicircular ducts, the saccule and utricle, and the cochlear duct.

**Memory** The ability to recall thoughts; commonly classifed as short-term (activated) and long-term.

**Menarche** (me-NAR-kē) The first menses (menstrual flow) and beginning of ovarian and uterine cycles.

**Meninges** (me-NIN-jēz) Three membranes covering the brain and spinal cord, called the dura mater, arachnoid mater, and pia mater. *Singular* is **meninx** (MEN-inks).

**Menopause** (MEN-ō-pawz) The termination of the menstrual cycles.

**Menstruation** (men′-stroo-Ā-shun) Periodic discharge of blood, tissue fluid, mucus, and epithelial cells that usually lasts for 5 days; caused by a sudden reduction in estrogens and progesterone. Also **called the menstrual phase** or **menses.**

**Merkel** (MER-kel) **cell** *See* **Tactile cell.**

**Merkel disc** *See* **Tactile disc.**

**Merocrine** (MER-ō-krin) **gland** Gland made up of secretory cells that remain intact throughout the process of formation and discharge of the secretory product, as in the salivary and pancreatic glands.

**Mesenchyme** (MEZ-en-kīm) An embryonic connective tissue from which all other connective tissues arise.

**Mesentery** (MEZ-en-ter′-ē) A fold of peritoneum attaching the small intestine to the posterior abdominal wall.

**Mesocolon** (mez′-ō-KŌ-lon) A fold of peritoneum attaching the colon to the posterior abdominal wall.

**Mesoderm** The middle primary germ layer that gives rise to connective tissues, blood and blood vessels, and muscles.

**Mesothelium** (mez′-ō-THĒ-lē-um) The layer of simple squamous epithelium that lines serous membranes.

**Mesovarium** (mez′-ō-VAR-ē-um) A short fold of peritoneum that attaches an ovary to the broad ligament of the uterus.

**Metabolism** (me-TAB-ō-lizm) All of the biochemical reactions that occur within an organism, including the synthetic (anabolic) reactions and decomposition (catabolic) reactions.

**Metacarpus** (met′-a-KAR-pus) A collective term for the five bones that make up the palm.

**Metaphase** (MET-a-fāz) The second stage of mitosis, in which chromatid pairs line up on the metaphase plate of the cell.

**Metaphysis** (me-TAF-i-sis) Region of a long bone between the diaphysis and epiphysis that contains the epiphyseal plate in a growing bone.

**Metarteriole** (met′-ar-TĒ-rē-ōl) A blood vessel that emerges from an arteriole, traverses a capillary network, and empties into a venule.

**Metastasis** (me-TAS-ta-sis) The spread of cancer to surrounding tissues (local) or to other body sites (distant).

**Metatarsus** (met′-a-TAR-sus) A collective term for the five bones located in the foot between the tarsals and the phalanges.

**Microfilament** (mī-krō-FIL-a-ment) Rodlike protein filament about 6 nm in diameter; constitutes contractile units in muscle fibers (cells) and provides support, shape, and movement in nonmuscle cells.

**Microglia** (mī-KROG-lē-a) Neuroglial cells that carry on phagocytosis.

**Microtubule** (mī-krō-TOO-būl) Cylindrical protein filament, from 18 to 30 nm in diameter, consisting of the protein tubulin; provides support, structure, and transportation.

**Microvilli** (mī-krō-VIL-ī) Microscopic, fingerlike projections of the plasma membranes of cells that increase surface area for absorption, especially in the small intestine and proximal convoluted tubules of the kidneys.

**Micturition** (mik′-choo-RISH-un) The act of expelling urine from the urinary bladder. Also called **urination** (ū-ri-NĀ-shun).

**Midbrain** The part of the brain between the pons and the diencephalon. Also called the **mesencephalon** (mes′-en-SEF-a-lon).

**Middle ear** A small, epithelial-lined cavity hollowed out of the temporal bone, separated from the external ear by the eardrum and from the internal ear by a thin bony partition containing the oval and round windows; extending across the middle ear are the three auditory ossicles. Also called the **tympanic** (tim-PAN-ik) **cavity.**

**Midline** An imaginary vertical line that divides the body into equal left and right sides.

**Midsagittal plane** A vertical plane through the midline of the body that divides the body or organs into *equal* right and left sides. Also called a **median plane.**

**Mineralocorticoids** (min′-er-al-ō-KOR-ti-koyds) A group of hormones of the adrenal cortex that help regulate sodium and potassium balance.

**Mitochondrion** (mī-tō-KON-drē-on) A double-membraned organelle that plays a central role in the production of ATP; known as the "powerhouse" of the cell. *Plural* is **mitochondria.**

**Mitosis** (mī-TŌ-sis) The orderly division of the nucleus of a cell that ensures that each new nucleus has the same number and kind of chromosomes as the original nucleus. The process includes the replication of chromosomes and the distribution of the two sets of chromosomes into two separate and equal nuclei.

**Mitotic spindle** Collective term for a football-shaped assembly of microtubules (nonkinetochore, kinetochore, and aster) that is responsible for the movement of chromosomes during cell division.

**Modality** (mō-DAL-i-tē) Any of the specific sensory entities, such as vision, smell, taste, or touch.

**Modiolus** (mō-DĪ-ō′-lus) The central pillar or column of the cochlea.

**Molecule** (mol′-e-KŪL) A substance composed of two or more atoms chemically combined.

**Monocyte** (MON-ō-sīt′) The largest type of white blood cell, characterized by agranular cytoplasm.

**Monounsaturated fat** A fatty acid that contains one double covalent bond between its carbon atoms; it is not completely saturated with hydrogen atoms. Plentiful in triglycerides of olive and peanut oils.

**Mons pubis** (MONZ PŪ-bis) The rounded, fatty prominence over the pubic symphysis, covered by coarse pubic hair.

**Morula** (MOR-ū-la) A solid sphere of cells produced by successive cleavages of a fertilized ovum about 4 days after fertilization.

**Motor area** The region of the cerebral cortex that governs muscular movement, particularly the precentral gyrus of the frontal lobe.

**Motor end plate** Region of the sarcolemma of a muscle fiber (cell) that includes acetylcholine (ACh) receptors, which bind ACh released by synaptic end bulbs of somatic motor neurons.

**Motor neurons** (NOO-ronz) Neurons that conduct impulses from the brain toward the spinal cord or out of the brain and spinal cord into cranial or spinal nerves to effectors that may be either muscles or glands. Also called **efferent neurons.**

**Motor unit** A motor neuron together with the muscle fibers (cells) it stimulates.

**Mucosa associated lymphatic tissue (MALT)** Lymphatic nodules scattered throughout the lamina propria (connective tissue) of mucous membranes lining the gastrointestinal tract, respiratory airways, urinary tract, and reproductive tract.

**Mucous** (MŪ-kus) **cell** A unicellular gland that secretes mucus. Two types are mucous neck cells and surface mucous cells in the stomach.

**Mucous membrane** A membrane that lines a body cavity that opens to the exterior. Also called the **mucosa** (mū-KŌ-sa).

**Mucus** The thick fluid secretion of goblet cells, mucous cells, mucous glands, and mucous membranes.

**Muscarinic** (mus′-ka-RIN-ik) **receptor** Receptor for the neurotransmitter acetylcholine found on all effectors innervated by parasympathetic postganglionic axons and on sweat glands innervated by cholinergic sympathetic postganglionic axons; so named because muscarine activates these receptors but does not activate nicotinic receptors for acetylcholine.

**Muscle** An organ composed of one of three types of muscle tissue (skeletal, cardiac, or smooth), specialized for contraction to produce voluntary or involuntary movement of parts of the body.

**Muscle action potential** A stimulating impulse that propagates along the sarcolemma and transverse tubules; in skeletal muscle, it is generated by acetylcholine, which increases the permeability of the sarcolemma to cations, especially sodium ions ($Na^+$).

**Muscle fatigue** (fa-TĒG) Inability of a muscle to maintain its strength of contraction or tension; may be related to insufficient oxygen, depletion of glycogen, and/or lactic acid buildup.

**Muscle spindle** An encapsulated proprioceptor in a skeletal muscle, consisting of specialized intrafusal muscle fibers and nerve endings; stimulated by changes in length or tension of muscle fibers.

**Muscle strain** Tearing of skeletal muscle fibers or tendon. Also called a **muscle pull** or **muscle tear.**

**Muscle tone** A sustained, partial contraction of portions of a skeletal or smooth muscle in response to activation of stretch receptors or a baseline level of action potentials in the innervating motor neurons.

**Muscular dystrophies** (DIS-trō-fēz) Inherited muscle-destroying diseases, characterized by degeneration of muscle fibers (cells), which causes progressive atrophy of the skeletal muscle.

**Muscularis** (MUS-kū-la′-ris) A muscular layer (coat or tunic) of an organ.

**Muscularis mucosae** (mū-KŌ-sē) A thin layer of smooth muscle fibers that underlie the lamina propria of the mucosa of the gastrointestinal tract.

**Muscular system** Usually refers to the approximately 100 voluntary muscles of the body that are composed of skeletal muscle tissue.

**Muscular tissue** A tissue specialized to produce motion in response to muscle action potentials by its qualities of contractility, extensibility, elasticity, and excitability; types include skeletal, cardiac, and smooth.

**Musculoskeletal** (mus′-kyū-lō-SKEL-e-tal) **system** An integrated body system consisting of bones, joints, and muscles.

**Mutation** (mū-TĀ-shun) Any change in the sequence of bases in a DNA molecule resulting in a permanent alteration in some inheritable trait.

**Myasthenia** (mī-as-THĒ-nē-a) **gravis** Weakness and fatigue of skeletal muscles caused by antibodies directed against acetylcholine receptors.

**Myelin** (MĪ-e-lin) **sheath** Multilayered lipid and protein covering, formed by Schwann cells and oligodendrocytes, around axons of many peripheral and central nervous system neurons.

**Myenteric** (mī-en-TER-ik) **plexus** A network of autonomic axons and postganglionic cell bodies located in the muscularis of the gastrointestinal tract. Also called the **plexus of Auerbach** (OW-er-bak).

**Myocardial infarction** (mī′-ō-KAR-dē-al in-FARK-shun) **(MI)** Gross necrosis of myocardial tissue due to interrupted blood supply. Also called a **heart attack.**

**Myocardium** (mī′-ō-KAR-dē-um) The middle layer of the heart wall, made up of cardiac muscle tissue, lying between the epicardium and the endocardium and constituting the bulk of the heart.

**Myofibril** (mī-ō-FĪ-bril) A threadlike structure, extending longitudinally through a muscle fiber (cell), consisting mainly of thick filaments (myosin) and thin filaments (actin, troponin, and tropomyosin).

**Myoglobin** (mī-ō-GLŌB-in) The oxygen-binding, iron-containing protein present in the sarcoplasm of muscle fibers (cells); contributes the red color to muscle.

**Myogram** (MĪ-ō-gram) The record or tracing produced by a myograph, an apparatus that measures and records the force of muscular contractions.

**Myology** (mī-OL-ō-jē) The study of muscles.

**Myometrium** (mī′-ō-MĒ-trē-um) The smooth muscle layer of the uterus.

**Myopathy** (mī-OP-a-thē) Any abnormal condition or disease of muscle tissue.

**Myopia** (mī-Ō-pē-a) Defect in vision in which objects can be seen distinctly only when very close to the eyes; nearsightedness.

**Myosin** (MĪ-ō-sin) The contractile protein that makes up the thick filaments of muscle fibers.

**Myotome** (MĪ-ō-tōm) A group of muscles innervated by the motor neurons of a single spinal segment. In an embryo, the portion of a somite that develops into some skeletal muscles.

## N

**Nail** A hard plate, composed largely of keratin, that develops from the epidermis of the skin to form a protective covering on the dorsal surface of the distal phalanges of the fingers and toes.

**Nail matrix** (MĀ-triks) The part of the nail beneath the body and root from which the nail is produced.

**Nasal** (NĀ-zal) **cavity** A mucosa-lined cavity on either side of the nasal septum that opens onto the face at the external nares and into the nasopharynx at the internal nares.

**Nasal septum** (SEP-tum) A vertical partition composed of bone (perpendicular plate of ethmoid and vomer) and cartilage, covered with a mucous membrane, separating the nasal cavity into left and right sides.

**Nasolacrimal** (nā′-zō-LAK-ri-mal) **duct** A canal that transports the lacrimal secretion (tears) from the nasolacrimal sac into the nose.

**Nasopharynx** (nā′-zō-FAR-inks) The superior portion of the pharynx, lying posterior to the nose and extending inferiorly to the soft palate.

**Neck** The part of the body connecting the head and the trunk. A constricted portion of an organ, such as the neck of the femur or uterus.

**Necrosis** (ne-KRŌ-sis) A pathological type of cell death that results from disease, injury, or lack of blood supply in which many adjacent cells swell, burst, and spill their contents into the interstitial fluid, triggering an inflammatory response.

**Negative feedback system** A feedback cycle that reverses a change in a controlled condition.

**Neonatal** (nē-ō-NĀ-tal) Pertaining to the first 4 weeks after birth.

**Neoplasm** (NĒ-ō-plazm) A new growth that may be benign or malignant.

**Nephron** (NEF-ron) The functional unit of the kidney.

**Nerve** A cordlike bundle of neuronal axons and/or dendrites and associated connective tissue coursing together outside the central nervous system.

**Nerve fiber** General term for any process (axon or dendrite) projecting from the cell body of a neuron.

**Nerve impulse** A wave of depolarization and repolarization that self-propagates along the plasma membrane of a neuron; also called a **nerve action potential.**

**Nervous system** A network of billions of neurons and even more neuroglia that is organized into two main divisions, central nervous system (brain and spinal cord) and peripheral nervous system (nerves, ganglia, enteric plexuses, and sensory receptors outside the central nervous system).

**Nervous tissue** Tissue containing neurons that initiate and conduct nerve impulses to coordinate homeostasis, and neuroglia that provide support and nourishment to neurons.

**Neuralgia** (noo-RAL-jē-a) Attacks of pain along the entire course or branch of a peripheral sensory nerve.

**Neural plate** A thickening of ectoderm, induced by the notochord, that forms early in the third week of development and represents the beginning of the development of the nervous system.

**Neural tube defect (NTD)** A developmental abnormality in which the neural tube does not close properly. Examples are spina bifida and anencephaly.

**Neuritis** (noo-RĪ-tis) Inflammation of one or more nerves.

**Neurofibral node** *See* **Node of Ranvier.**

**Neuroglia** (noo-RŌG-lē-a) Cells of the nervous system that perform various supportive functions. The neuroglia of the central nervous system are the astrocytes, oligodendrocytes, microglia, and ependymal cells; neuroglia of the peripheral nervous system include Schwann cells and satellite cells. Also called **glial** (GLĒ-al) **cells.**

**Neurohypophyseal** (noo′-rō-hī′-pō-FIZ-ē-al) **bud** An outgrowth of ectoderm located on the floor of the hypothalamus that gives rise to the posterior pituitary.

**Neurolemma** (noo-rō-LEM-ma) The peripheral, nucleated cytoplasmic layer of the Schwann cell. Also called **sheath of Schwann** (SCHWON).

**Neurology** (noo-ROL-ō-jē) The study of the normal functioning and disorders of the nervous system.

**Neuromuscular** (noo-rō-MUS-kū-lar) **junction (NMJ)** A synapse between the axon terminals of a motor neuron and the sarcolemma of a muscle fiber (cell).

**Neuron** (NOO-ron) A nerve cell, consisting of a cell body, dendrites, and an axon.

**Neurophysiology** (NOOR-ō-fiz-ē-ol′-ō-jē) Study of the functional properties of nerves.

**Neurosecretory** (noo-rō-SĒK-re-tō-rē) **cell** A neuron that secretes a hypothalamic releasing hormone or inhibiting hormone into blood capillaries of the hypothalmus; a neuron that secretes oxytocin or antidiuretic hormone into blood capillaries of the posterior pituitary.

**Neurotransmitter** (noo′-rō-trans′-MIT-er) One of a variety of molecules within axon terminals that are released into the synaptic cleft in response to a nerve impulse and that change the membrane potential of the postsynaptic neuron.

**Neurulation** (noor-oo-LĀ-shun) The process by which the neural plate, neural folds, and neural tube develop.

**Neutrophil** (NOO-trō-fil) A type of white blood cell characterized by granules that stain pale lilac with a combination of acidic and basic dyes.

**Nicotinic** (nik′-ō-TIN-ik) **receptor** Receptor for the neurotransmitter acetylcholine found on both sympathetic and parasympathetic postganglionic neurons and on skeletal muscle in the motor end plate; so named because nicotine activates these receptors but does not activate muscarinic receptors for acetylcholine.

**Nipple** A pigmented, wrinkled projection on the surface of the breast that is the location of the openings of the lactiferous ducts for milk release.

**Nociceptor** (nō′-sē-SEP-tor) A free (naked) nerve ending that detects painful stimuli.

**Node of Ranvier** (RON-vē-ā ) A space along a myelinated axon between the individual Schwann cells that form the myelin sheath and the neurolemma. Also called a **neurofibral node.**

**Norepinephrine** (nor′-ep-ē-NEF-rin) **(NE)** A hormone secreted by the adrenal medulla that produces actions similar to those that result from sympathetic stimulation. Also called **noradrenaline** (nor-a-DREN-a-lin).

**Notochord** (NŌ-tō-cord) A flexible rod of mesodermal tissue that lies where the future vertebral column will develop and plays a role in induction.

**Nucleic** (noo-KLĒ-ik) **acid** An organic compound that is a long polymer of nucleotides, with each nucleotide containing a pentose sugar, a phosphate group, and one of four possible nitrogenous bases (adenine, cytosine, guanine, and thymine or uracil).

**Nucleolus** (noo′-KLĒ-ō-lus) Spherical body within a cell nucleus composed of protein, DNA, and RNA that is the site of the assembly of small and large ribosomal subunits. *Plural* is **nucleoli.**

**Nucleosome** (NOO-klē-ō-sōm) Structural subunit of a chromosome consisting of histones and DNA.

**Nucleus** (NOO-klē-us) A spherical or oval organelle of a cell that contains the hereditary factors of the cell, called genes. A cluster of unmyelinated nerve cell bodies in the central nervous system. The central part of an atom made up of protons and neutrons.

**Nucleus pulposus** (pul-PŌ-sus) A soft, pulpy, highly elastic substance in the center of an intervertebral disc; a remnant of the notochord.

**Nutrient** A chemical substance in food that provides energy, forms new body components, or assists in various body functions.

## O

**Obesity** (ō-BĒS-i-tē) Body weight more than 20% above a desirable standard due to excessive accumulation of fat.

**Oblique** (ō-BLĒK) **plane** A plane that passes through the body or an organ at an angle between the transverse plane and either the midsagittal, parasagittal, or frontal plane.

**Obstetrics** (ob-STET-riks) The specialized branch of medicine that deals with pregnancy, labor, and the period of time immediately after delivery (about 6 weeks).

**Olfaction** (ōl-FAK-shun) The sense of smell.

**Olfactory** (ōl-FAK-tō-rē) Pertaining to smell.

**Olfactory bulb** A mass of gray matter containing cell bodies of neurons that form synapses with neurons of the olfactory (I) nerve, lying inferior to the frontal lobe of the cerebrum on either side of the crista galli of the ethmoid bone.

**Olfactory receptor** A bipolar neuron with its cell body lying between supporting cells located in the mucous membrane lining the superior portion of each nasal cavity; transduces odors into neural signals.

**Olfactory tract** A bundle of axons that extends from the olfactory bulb posteriorly to olfactory regions of the cerebral cortex.

**Oligodendrocyte** (OL-i-gō-den′-drō-sīt) A neuroglial cell that supports neurons and produces a myelin sheath around axons of neurons of the central nervous system.

**Oliguria** (ol′-i-GŪ-rē-a) Daily urinary output usually less than 250 mL.

**Olive** A prominent oval mass on each lateral surface of the superior part of the medulla oblongata.

**Oncogene** (ON-kō-jēn) Cancer-causing gene; it derives from a normal gene, termed a proto-oncogene, that encodes proteins involved in cell growth or cell regulation but has the ability to transform a normal cell into a cancerous cell when it is mutated or inappropriately activated. One example is *p53*.

**Oncology** (on-KOL-ō-jē) The study of tumors.

**Oogenesis** (ō-ō-JEN-e-sis) Formation and development of female gametes (oocytes).

**Oophorectomy** (ō′-of-ō-REK-tō-me) Surgical removal of the ovaries.

**Ophthalmic** (of-THAL-mik) Pertaining to the eye.

**Ophthalmologist** (of′-thal-MOL-ō-jist) A physician who specializes in the diagnosis and treatment of eye disorders using drugs, surgery, and corrective lenses.

**Ophthalmology** (of-thal-MOL-ō-jē) The study of the structure, function, and diseases of the eye.

**Optic** (OP-tik) Refers to the eye, vision, or properties of light.

**Optic chiasm** (kī-AZM) A crossing point of the two branches of the optic (II) nerve, anterior to the pituitary gland. Also called **optic chiasma.**

**Optic disc** A small area of the retina containing openings through which the axons of the ganglion cells emerge as the optic (II) nerve. Also called the **blind spot.**

**Optic tract** A bundle of axons that carry nerve impulses from the retina of the eye between the optic chiasm and the thalamus.

**Ora serrata** (Ō-ra ser-RĀ-ta) The irregular margin of the retina lying internal and slightly posterior to the junction of the choroid and ciliary body.

**Orbit** (OR-bit) The bony, pyramidal-shaped cavity of the skull that holds the eyeball.

**Organ** A structure composed of two or more different kinds of tissues with a specific function and usually a recognizable shape.

**Organelle** (or-ga-NEL) A permanent structure within a cell with characteristic morphology that is specialized to serve a specific function in cellular activities.

**Organism** (OR-ga-nizm) A total living form; one individual.

**Organogenesis** (or′-ga-nō-JEN-e-sis) The formation of body organs and systems. By the end of the eighth week of development, all major body systems have begun to develop.

**Orifice** (OR-i-fis) Any aperture or opening.

**Origin** (OR-i-jin) The attachment of a muscle tendon to a stationary bone or the end opposite the insertion.

**Oropharynx** (or′-ō-FAR-inks) The intermediate portion of the pharynx, lying posterior to the mouth and extending from the soft palate to the hyoid bone.

**Orthopedics** (or′-thō-PĒ-diks) The branch of medicine that deals with the preservation and restoration of the skeletal system, articulations, and associated structures.

**Osmoreceptor** (oz′-mō-rē-CEP-tor) Receptor in the hypothalamus that is sensitive to changes in blood osmolarity and, in response to high osmolarity (low water concentration), stimulates synthesis and release of antidiuretic hormone (ADH).

**Osmosis** (oz-MŌ-sis) The net movement of water molecules through a selectively permeable membrane from an area of higher water concentration to an area of lower water concentration until equilibrium is reached.

**Osseous** (OS-ē-us) Bony.

**Ossicle** (OS-si-kul) One of the small bones of the middle ear (malleus, incus, stapes).

**Ossification** (os′-i-fi-KĀ-shun) Formation of bone. Also called **osteogenesis.**

**Ossification** (os′-i-fi-KĀ-shun) **center** An area in the cartilage model of a future bone where the cartilage cells hypertrophy, secrete enzymes that calcify their extracellular matrix, and die, and the area they occupied is invaded by osteoblasts that then lay down bone.

**Osteoblast** (OS-tē-ō-blast′) Cell formed from an osteogenic cell that participates in bone formation by secreting some organic components and inorganic salts.

**Osteoclast** (OS-tē-ō-klast′) A large, multinuclear cell that resorbs (destroys) bone matrix.

**Osteocyte** (OS-tē-ō-sīt′) A mature bone cell that maintains the daily activities of bone tissue.

**Osteogenic** (os′-tē-ō-JEN-ik) **cell** Stem cell derived from mesenchyme that has mitotic potential and the ability to differentiate into an osteoblast.

**Osteogenic layer** The inner layer of the periosteum that contains cells responsible for forming new bone during growth and repair.

**Osteology** (os-tē-OL-ō-jē) The study of bones.

**Osteon** (OS-tē-on) The basic unit of structure in adult compact bone, consisting of a central (haversian) canal with its concentrically arranged lamellae, lacunae, osteocytes, and canaliculi. Also called a **haversian** (ha-VER-shun) **system.**

**Osteoporosis** (os′-tē-ō-pō-RŌ-sis) Age-related disorder characterized by decreased bone mass and increased susceptibility to fractures, often as a result of decreased levels of estrogens.

**Otic** (Ō-tik) Pertaining to the ear.

**Otolith** (Ō-tō-lith) A particle of calcium carbonate embedded in the otolithic membrane that functions in maintaining static equilibrium.

**Otolithic** (ō-tō-LITH-ik) **membrane** Thick, gelatinous, glycoprotein layer located directly over hair cells of the macula in the saccule and utricle of the internal ear.

**Otorhinolaryngology** (ō-tō-rī′-nō-lar-in-GOL-ō-jē) The branch of medicine that deals with the diagnosis and treatment of diseases of the ears, nose, and throat.

**Oval window** A small, membrane-covered opening between the middle ear and inner ear into which the footplate of the stapes fits.

**Ovarian** (ō-VAR-ē-an) **cycle** A monthly series of events in the ovary associated with the maturation of a secondary oocyte.

**Ovarian follicle** (FOL-i-kul) A general name for oocytes (immature ova) in any stage of development, along with their surrounding epithelial cells.

**Ovarian ligament** (LIG-a-ment) A rounded cord of connective tissue that attaches the ovary to the uterus.

**Ovary** (Ō-var-ē) Female gonad that produces oocytes and the estrogen, progesterone, inhibin, and relaxin hormones.

**Ovulation** (ov′-ū-LĀ-shun) The rupture of a mature ovarian (graafian) follicle with discharge of a secondary oocyte into the pelvic cavity.

**Ovum** (Ō-vum) The female reproductive or germ cell; an egg cell; arises through completion of meiosis in a secondary oocyte after penetration by a sperm.

**Oxyhemoglobin** (ok′-sē-HĒ-mō-glō-bin) **($Hb–O_2$)** Hemoglobin combined with oxygen.

**Oxytocin** (ok′-sē-TŌ-sin) **(OT)** A hormone secreted by neurosecretory cells in the paraventricular and supraoptic nuclei of the hypothalamus that stimulates contraction of smooth muscle in the pregnant uterus and myoepithelial cells around the ducts of mammary glands.

## P

**P wave** The deflection wave of an electrocardiogram that signifies atrial depolarization.

**Pacinian corpuscle** (pa-SIN-ē-an) Oval-shaped pressure receptor located in the dermis or subcutaneous tissue and consisting of concentric layers of a connective tissue wrapped around the dendrites of a sensory neuron. Also called a **lamellated corpuscle.**

**Palate** (PAL-at) The horizontal structure separating the oral and the nasal cavities; the roof of the mouth.

**Palpate** (PAL-pāt) To examine by touch; to feel.

**Pancreas** (PAN-krē-as) A soft, oblong organ lying along the greater curvature of the stomach and connected by a duct to the duodenum. It is both an exocrine gland (secreting pancreatic juice) and an endocrine gland (secreting insulin, glucagon, somatostatin, and pancreatic polypeptide).

**Pancreatic** (pan′-krē-AT-ik) **duct** A single large tube that unites with the common bile duct from the liver and gallbladder and drains pancreatic juice into the duodenum at the hepatopancreatic ampulla (ampulla of Vater). Also called the **duct of Wirsung** (VĒR-sung).

**Pancreatic islet** (Ī-let) Cluster of endocrine gland cells in the pancreas that secretes insulin, glucagon, somatostatin, and pancreatic polypeptide. Also called an **islet of Langerhans** (LAHNG-er-hanz).

**Papanicolaou** (pa-pa-NI-kō-lō) **test** A cytological staining test for the detection and diagnosis of premalignant and malignant conditions of the female genital tract. Cells scraped from the epithelium of the cervix of the uterus are examined microscopically. Also called a **Pap test** or **Pap smear.**

**Papilla** (pa-PIL-a) A small nipple-shaped projection or elevation.

**Paralysis** (pa-RAL-a-sis) Loss or impairment of motor function due to a lesion of nervous or muscular origin.

**Paranasal sinus** (par′-a-NĀ-zal SĪ-nus) A mucus-lined air cavity in a skull bone that communicates with the nasal cavity. Paranasal sinuses are located in the frontal, maxillary, ethmoid, and sphenoid bones.

**Paraplegia** (par-a-PLĒ-jē-a) Paralysis of both lower limbs.

**Parasagittal plane** (par-a-SAJ-i-tal) A vertical plane that does not pass through the midline and that divides the body or organs into *unequal* left and right portions.

**Parasympathetic** (par′-a-sim-pa-THET-ik) **division** One of the two subdivisions of the autonomic nervous system, having cell bodies of preganglionic neurons in nuclei in the brain stem and in the lateral gray horn of the sacral portion of the spinal cord; primarily concerned with activities that conserve and restore body energy.

**Parathyroid** (par′-a-THĪ-royd) **gland** One of usually four small endocrine glands embedded in the posterior surfaces of the lateral lobes of the thyroid gland.

**Parathyroid hormone (PTH)** A hormone secreted by the chief (principal) cells of the parathyroid glands that increases blood calcium level and decreases blood phosphate level. Also called **parathormone.**

**Paraurethral** (par′-a-ū-RĒ-thral) **gland** Gland embedded in the wall of the urethra whose duct opens on either side of the urethral orifice and secretes mucus. Also called **Skene's** (SKĒNZ) **gland.**

**Parenchyma** (pa-RENG-ki-ma) The functional parts of any organ, as opposed to tissue that forms its stroma or framework.

**Parietal** (pa-RĪ-e-tal) Pertaining to or forming the outer wall of a body cavity.

**Parietal cell** A type of secretory cell in gastric glands that produces hydrochloric acid and intrinsic factor. Also called an **oxyntic cell.**

**Parietal pleura** (PLOO-ra) The outer layer of the serous pleural membrane that encloses and protects the lungs; the layer that is attached to the wall of the pleural cavity.

**Parkinson disease (PD)** Progressive degeneration of the basal nuclei and substantia nigra of the cerebrum resulting in decreased production of dopamine (DA) that leads to tremor, slowing of voluntary movements, and muscle weakness.

**Parotid** (pa-ROT-id) **gland** One of the paired salivary glands located inferior and anterior to the ears and connected to the oral cavity via a duct (Stensen's) that opens into the inside of the cheek opposite the maxillary (upper) second molar tooth.

**Pars intermedia** A small avascular zone between the anterior and posterior pituitary glands.

**Parturition** (par-toor-ISH-un) Act of giving birth to young; childbirth, delivery.

**Patent** (PĀ-tent) **ductus arteriosus (PDA)** A congenital heart defect in which the ductus arteriosus remains open. As a result, aortic blood flows into the lower-pressure pulmonary trunk, increasing pulmonary trunk pressure and overworking both ventricles.

**Pathogen** (PATH-ō-jen) A disease-producing microbe.

**Pathological** (path′-ō-LOJ-i-kal) **anatomy** The study of structural changes caused by disease.

**Pathophysiology** (PATH-ō-fez-ē-ol-ō-jē) Study of functional changes associated with disease and aging.

**Pectinate** (PEK-ti-nāt) **muscles** Projecting muscle bundles of the anterior atrial walls and the lining of the auricles.

**Pectoral** (PEK-tō-ral) Pertaining to the chest or breast.

**Pedicel** (PED-i-sel) Footlike structure, as on podocytes of a glomerulus.

**Pelvic** (PEL-vik) **cavity** Inferior portion of the abdominopelvic cavity that contains the urinary bladder, sigmoid colon, rectum, and internal female and male reproductive structures.

**Pelvic splanchnic** (PEL-vik SPLANGK-nik) **nerves** Consist of preganglionic parasympathetic axons from the levels of S2, S3, and S4 that supply the urinary bladder, reproductive organs, and the descending and sigmoid colon and rectum.

**Pelvis** The basinlike structure formed by the two hip bones, the sacrum, and the coccyx. The expanded, proximal portion of the ureter, lying within the kidney and into which the major calyces open.

**Penis** (PĒ-nis) The organ of urination and copulation in males; used to deposit semen into the female vagina.

**Pepsin** Protein-digesting enzyme secreted by chief cells of the stomach in the inactive form pepsinogen, which is converted to active pepsin by hydrochloric acid.

**Peptic ulcer** An ulcer that develops in areas of the gastrointestinal tract exposed to hydrochloric acid; classified as a gastric ulcer if in the lesser curvature of the stomach and as a duodenal ulcer if in the first part of the duodenum.

**Percussion** (pur-KUSH-un) The act of striking (percussing) an underlying part of the body with short, sharp taps as an aid in diagnosing the part by the quality of the sound produced.

**Perforating canal** A minute passageway by means of which blood vessels and nerves from the periosteum penetrate into compact bone. Also called **Volkmann's** (FŌLK-mans) **canal.**

**Pericardial** (per′-i-KAR-dē-al) **cavity** Small potential space between the visceral and parietal layers of the serous pericardium that contains pericardial fluid.

**Pericardium** (per′-i-KAR-dē-um) A loose-fitting membrane that encloses the heart, consisting of a superficial fibrous layer and a deep serous layer.

**Perichondrium** (per′-i-KON-drē-um) A covering of dense irregular connective tissue that surrounds the surface of most cartilage.

**Perilymph** (PER-i-limf) The fluid contained between the bony and membranous labyrinths of the inner ear.

**Perimetrium** (per′-i-MĒ-trē-um) The serosa of the uterus.

**Perimysium** (per-i-MĪZ-ē-um) Invagination of the epimysium that divides muscles into bundles.

**Perineum** (per′-i-NĒ-um) The pelvic floor; the space between the anus and the scrotum in the male and between the anus and the vulva in the female.

**Perineurium** (per′-i-NOO-rē-um) Connective tissue wrapping around fascicles in a nerve.

**Periodontal** (per′-ē-ō-DON-tal) **disease** A collective term for conditions characterized by degeneration of gingivae, alveolar bone, periodontal ligament, and cementum.

**Periodontal ligament** The periosteum lining the alveoli (sockets) for the teeth in the alveolar processes of the mandible and maxillae.

**Periosteum** (per′-ē-OS-tē-um) The covering of a bone that consists of connective tissue, osteogenic cells, and osteoblasts; is essential for bone growth, repair, and nutrition.

**Peripheral** (pe-RIF-er-al) Located on the outer part or a surface of the body.

**Peripheral nervous system (PNS)** The part of the nervous system that lies outside the central nervous system, consisting of nerves and ganglia.

**Peristalsis** (per′-i-STAL-sis) Successive muscular contractions along the wall of a hollow muscular structure.

**Peritoneum** (per′-i-tō-NĒ-um) The largest serous membrane of the body that lines the abdominal cavity and covers the viscera within it.

**Peritonitis** (per′-i-tō-NĪ-tis) Inflammation of the peritoneum.

**Peroxisome** (pe-ROKS-i-sōm) Organelle similar in structure to a lysosome that contains enzymes that use molecular oxygen to oxidize various organic compounds; such reactions produce hydrogen peroxide; abundant in liver cells.

**Perspiration** Sweat; produced by sudoriferous (sweat) glands and containing water, salts, urea, uric acid, amino acids, ammonia, sugar, lactic acid, and ascorbic acid. Helps maintain body temperature and eliminate wastes.

**Peyer's patches** *See* **Aggregated lymphatic follicles.**

**pH** A measure of the concentration of hydrogen ions (H1) in a solution. The **pH scale** extends from 0 to 14, with a value of 7 expressing neutrality, values lower than 7 expressing increasing acidity, and values higher than 7 expressing increasing alkalinity.

**Phagocytosis** (fag′-ō-sī-TŌ-sis) The process by which phagocytes ingest and destroy microbes, cell debris, and other foreign matter.

**Phalanx** (FĀ-lanks) The bone of a finger or toe. *Plural* is **phalanges** (fa-LAN-jēz).

**Pharmacology** (far′-ma-KOL-ō-jē) The science of the effects and uses of drugs in the treatment of disease.

**Pharynx** (FAR-inks) The throat; a tube that starts at the internal nares and runs partway down the neck, where it opens into the esophagus posteriorly and the larynx anteriorly.

**Phenotype** (FĒ-nō-tīp) The observable expression of genotype; physical characteristics of an organism determined by genetic makeup and influenced by interaction between genes and internal and external environmental factors.

**Phlebitis** (fle-BĪ-tis) Inflammation of a vein, usually in a lower limb.

**Photopigment** A substance that can absorb light and undergo structural changes that can lead to the development of a receptor potential. An example is rhodopsin. In the eye, also called **visual pigment.**

**Photoreceptor** Receptor that detects light shining on the retina of the eye.

**Physiology** (fiz′-ē-OL-ō-jē) Science that deals with the functions of an organism or its parts.

**Pia mater** (PĪ-a MĀ-ter *or* PĒ-a MA-ter) The innermost of the three meninges (coverings) of the brain and spinal cord.

**Pineal** (PĪN-ē-al) **gland** A cone-shaped gland located in the roof of the third ventricle that secretes melatonin. Also called the **epiphysis cerebri** (ē-PIF-i-sis se-RĒ-brē).

**Pinealocyte** (pin-ē-AL-ō-sīt) Secretory cell of the pineal gland that releases melatonin.

**Pinna** (PIN-na) The projecting part of the external ear composed of elastic cartilage and covered by skin and shaped like the flared end of a trumpet. Also called the **auricle** (OR-i-kul).

**Pituicyte** (pi-TOO-i-sīt) Supporting cell of the posterior pituitary.

**Pituitary** (pi-TOO-i-tār-ē) **gland** A small endocrine gland occupying the hypophyseal fossa of the sphenoid bone and attached to the hypothalamus by the infundibulum. Also called the **hypophysis** (hī-POF-i-sis).

**Pivot joint** A synovial joint in which a rounded, pointed, or conical surface of one bone articulates with a ring formed partly by another bone and partly by a ligament, as in the joint between the atlas and axis and between the proximal ends of the radius and ulna. Also called a **trochoid** (TRŌ-koyd) **joint.**

**Placenta** (pla-SEN-ta) The special structure through which the exchange of materials between fetal and maternal circulations occurs. Also called the **afterbirth.**

**Plane joint** A synovial joint having articulating surfaces that are usually flat, permitting only side-to-side and back-and-forth movements, as between carpal bones, tarsal bones, and the scapula and clavicle. Also called an **arthrodial** (ar-THRŌ-dē-al) **joint.**

**Plantar flexion** (PLAN-tar FLEK-shun) Bending the foot in the direction of the plantar surface (sole).

**Plaque** (PLAK) A layer of dense proteins on the inside of a plasma membrane in adherens junctions and desmosomes. A mass of bacterial cells, dextran (polysaccharide), and other debris that adheres to teeth (dental plaque). *See also* **Atherosclerotic plaque.**

**Plasma** (PLAZ-ma) The extracellular fluid found in blood vessels; blood minus the formed elements.

**Plasma cell** Cell that develops from a B cell (lymphocyte) and produces antibodies.

**Plasma (cell) membrane** Outer, limiting membrane that separates the cell's internal parts from extracellular fluid or the external environment.

**Platelet** (PLĀT-let) A fragment of cytoplasm enclosed in a cell membrane and lacking a nucleus; found in the circulating blood; plays a role in hemostasis. Also called a **thrombocyte** (THROM-bō-sīt).

**Platelet plug** Aggregation of platelets (thrombocytes) at a site where a blood vessel is damaged that helps stop or slow blood loss.

**Pleura** (PLOO-ra) The serous membrane that covers the lungs and lines the walls of the chest and the diaphragm.

**Pleural cavity** Small potential space between the visceral and parietal pleurae.

**Plexus** (PLEK-sus) A network of nerves, veins, or lymphatic vessels.

**Plexus of Auerbach** *See* **Myenteric plexus.**

**Plexus of Meissner** *See* **Submucosal plexus.**

**Pluripotent** (ploo-RI-pō-tent) **stem cell** Immature stem cell in red bone marrow that gives rise to precursors of all of the different mature blood cells.

**Pneumotaxic** (noo-mō-TAK-sik) **area** A part of the respiratory center in the pons that continually sends inhibitory nerve impulses to the inspiratory area, limiting inhalation and facilitating exhalation.

**Polycythemia** (pol′-ē-sī-THĒ-mē-a) Disorder characterized by an above-normal hematocrit (above 55%) in which hypertension, thrombosis, and hemorrhage can occur.

**Polyunsaturated fat** A fatty acid that contains more than one double covalent bond between its carbon atoms; abundant in triglycerides of corn oil, safflower oil, and cottonseed oil.

**Polyuria** (pol′-ē-Ū-rē-a) An excessive production of urine.

**Pons** (PONZ) The part of the brain stem that forms a "bridge" between the medulla oblongata and the midbrain, anterior to the cerebellum.

**Portal system** The circulation of blood from one capillary network into another through a vein.

**Positive feedback system** A feedback cycle that strengthens or reinforces a change in a controlled condition.

**Postcentral gyrus** Gyrus of cerebral cortex located immediately posterior to the central sulcus; contains the primary somatosensory area.

**Posterior** (pos-TĒR-ē-or) Nearer to or at the back of the body. Equivalent to **dorsal** in bipeds.

**Posterior column–medial lemniscus pathway** Sensory pathway that carries information related to proprioception, fine touch, two-point discrimination, pressure, and vibration. First-order neurons project from the spinal cord to the ipsilateral medulla in the posterior columns (gracile fasciculus and cuneate fasciculus). Second-order neurons project from the medulla to the contralateral thalamus in the medial lemniscus. Third-order neurons project from the thalamus to the somatosensory cortex (postcentral gyrus) on the same side.

**Posterior pituitary** Posterior lobe of the pituitary gland. Also called the **neurohypophysis** (noo-rō-hī-POF-i-sis).

**Posterior root** The structure composed of sensory axons lying between a spinal nerve and the dorsolateral aspect of the spinal cord. Also called the **dorsal (sensory) root.**

**Posterior root ganglion** (GANG-glē-on) A group of cell bodies of sensory neurons and their supporting cells located along the posterior root of a spinal nerve. Also called a dorsal **(sensory) root ganglion.**

**Postganglionic neuron** (pōst′-gang-lē-ON-ik NOO-ron) The second autonomic motor neuron in an autonomic pathway, having its cell body and dendrites located in an autonomic ganglion and its unmyelinated axon ending at cardiac muscle, smooth muscle, or a gland.

**Postsynaptic** (pōst-sin-AP-tik) **neuron** The nerve cell that is activated by the release of a neurotransmitter from another neuron and carries nerve impulses away from the synapse.

**Pouch of Douglas** *See* **Rectouterine pouch.**

**Precapillary sphincter** (SFINGK-ter) The distal-most muscle fiber (cell) at the metarteriol—capillary junction that regulates blood flow into capillaries.

**Precentral gyrus** Gyrus of cerebral cortex located immediately anterior to the central sulcus; contains the primary motor area.

**Preganglionic** (pre′-gang-lē-ON-ik) **neuron** The first autonomic motor neuron in an autonomic pathway, with its cell body and dendrites in the brain or spinal cord and its myelinated axon ending at an autonomic ganglion, where it synapses with a postganglionic neuron.

**Pregnancy** Sequence of events that normally includes fertilization, implantation, embryonic growth, and fetal growth, and terminates in birth.

**Premenstrual syndrome (PMS)** Severe physical and emotional stress ocurring late in the post-ovulatory phase of the menstrual cycle and sometimes overlapping with menstruation.

**Prepuce** (PRĒ-poos) The loose-fitting skin covering the glans of the penis and clitoris. Also called the **foreskin.**

**Presbyopia** (prez-bē-Ō-pē-a) A loss of elasticity of the lens of the eye due to advancing age with resulting inability to focus clearly on near objects.

**Presynaptic** (prē-sin-AP-tik) **neuron** A neuron that propagates nerve impulses toward a synapse.

**Prevertebral ganglion** (prē-VER-te-bral GANG-glē-on) A cluster of cell bodies of postganglionic sympathetic neurons anterior to the spinal column and close to large abdominal arteries. Also called a **collateral ganglion.**

**Primary germ layer** One of three layers of embryonic tissue, called ectoderm, mesoderm, and endoderm, that give rise to all tissues and organs of the body.

**Primary motor area** A region of the cerebral cortex in the precentral gyrus of the frontal lobe of the cerebrum that controls specific muscles or groups of muscles.

**Primary somatosensory area** A region of the cerebral cortex posterior to the central sulcus in the postcentral gyrus of the parietal lobe of the cerebrum that localizes exactly the points of the body where somatic sensations originate.

**Prime mover** The muscle directly responsible for producing a desired motion. Also called an **agonist** (AG-ō-nist).

**Primitive gut** Embryonic structure formed from the dorsal part of the yolk sac that gives rise to most of the gastrointestinal tract.

**Primordial** (prī-MŌR-dē-al) Existing first; especially primordial egg cells in the ovary.

**Principal cell** Cell type in the distal convoluted tubules and collecting ducts of the kidneys that is stimulated by aldosterone and antidiuretic hormone.

**Proctology** (prok-TOL-ō-jē) The branch of medicine concerned with the rectum and its disorders.

**Progeny** (PROJ-e-nē) Offspring or descendants.

**Progesterone** (prō-JES-te-rōn) A female sex hormone produced by the ovaries that helps prepare the endometrium of the uterus for implantation of a fertilized ovum and the mammary glands for milk secretion.

**Prognosis** (prog-NŌ-sis) A forecast of the probable results of a disorder; the outlook for recovery.

**Prolactin** (prō-LAK-tin) **(PRL)** A hormone secreted by the anterior pituitary that initiates and maintains milk secretion by the mammary glands.

**Prolapse** (PRŌ-laps) A dropping or falling down of an organ, especially the uterus or rectum.

**Proliferation** (prō-lif′-er-Ā-shun) Rapid and repeated reproduction of new parts, especially cells.

**Pronation** (prō-NĀ-shun) A movement of the forearm in which the palm is turned posteriorly.

**Prophase** (PRŌ-fāz) The first stage of mitosis during which chromatid pairs are formed and aggregate around the metaphase plate of the cell.

**Proprioception** (prō-prē-ō-SEP-shun) The perception of the position of body parts, especially the limbs, independent of vision; this sense is possible due to nerve impulses generated by proprioceptors.

**Proprioceptor** (PRŌ-prē-ō-sep′-tor) A receptor located in muscles, tendons, joints, or the internal ear (muscle spindles, tendon organs, joint kinesthetic receptors, and hair cells of the vestibular apparatus) that provides information about body position and movements.

**Prostaglandin** (pros′-ta-GLAN-din) **(PG)** A membrane-associated lipid; released in small quantities and acts as a local hormone.

**Prostate** (PROS-tāt) A doughnut-shaped gland inferior to the urinary bladder that surrounds the superior portion of the male urethra and secretes a slightly acidic solution that contributes to sperm motility and viability.

**Proteasome** (PRŌ-tē-a-sōm) Tiny cellular organelle in cytosol and nucleus containing proteases that destroy unneeded, damaged, or faulty proteins.

**Protein** An organic compound consisting of carbon, hydrogen, oxygen, nitrogen, and sometimes sulfur and phosphorus; synthesized on ribosomes and made up of amino acids linked by peptide bonds.

**Prothrombin** (prō-THROM-bin) An inactive blood-clotting factor synthesized by the liver, released into the blood, and converted to active thrombin in the process of blood clotting by the activated enzyme prothrombinase.

**Proto-oncogene** (prō′-tō-ON-kō-jēn) Gene responsible for some aspect of normal growth and development; it may transform into an oncogene, a gene capable of causing cancer.

**Protraction** (prō-TRAK-shun) The movement of the mandible or shoulder girdle forward on a plane parallel with the ground.

**Proximal** (PROK-si-mal) Nearer the attachment of a limb to the trunk; nearer to the point of origin or attachment.

**Pseudopod** (SOO-dō-pod) Temporary protrusion of the leading edge of a migrating cell; cellular projection that surrounds a particle undergoing phagocytosis.

**Pterygopalatine ganglion** (ter′-i-gō-PAL-a-tīn GANG-glē-on) A cluster of cell bodies of parasympathetic postganglionic neurons ending at the lacrimal and nasal glands.

**Ptosis** (TŌ-sis) Drooping, as of the eyelid or the kidney.

**Puberty** (PŪ-ber-tē) The time of life during which the secondary sex characteristics begin to appear and the capability for sexual reproduction is possible; usually occurs between the ages of 10 and 17.

**Pubic symphysis** A slightly movable cartilaginous joint between the anterior surfaces of the hip bones.

**Puerperium** (pū-er-PER-ē-um) The period immediately after childbirth, usually 4–6 weeks.

**Pulmonary** (PUL-mo-ner′-ē) Concerning or affected by the lungs.

**Pulmonary circulation** The flow of deoxygenated blood from the right ventricle to the lungs and the return of oxygenated blood from the lungs to the left atrium.

**Pulmonary edema** (e-DĒ-ma) An abnormal accumulation of interstitial fluid in the tissue spaces and alveoli of the lungs due to increased pulmonary capillary permeability or increased pulmonary capillary pressure.

**Pulmonary embolism** (EM-bō-lizm) **(PE)** The presence of a blood clot or a foreign substance in a pulmonary arterial blood vessel that obstructs circulation to lung tissue.

**Pulmonary ventilation** The inflow (inhalation) and outflow (exhalation) of air between the atmosphere and the lungs. Also called **breathing.**

**Pulp cavity** A cavity within the crown and neck of a tooth, which is filled with pulp, a connective tissue containing blood vessels, nerves, and lymphatic vessels.

**Pulse** (PULS) The rhythmic expansion and elastic recoil of a systemic artery after each contraction of the left ventricle.

**Pupil** The hole in the center of the iris, the area through which light enters the posterior cavity of the eyeball.

**Purkinje** (pur-KIN-jē) **fiber** Muscle fiber (cell) in the ventricular tissue of the heart specialized for conducting an action potential to the myocardium; part of the conduction system of the heart.

**Pus** The liquid product of inflammation containing leukocytes or their remains and debris of dead cells.

**Pyloric** (pī-LOR-ik) **sphincter** A thickened ring of smooth muscle through which the pylorus of the stomach communicates with the duodenum. Also called the **pyloric valve.**

**Pyorrhea** (pī-ō-RĒ-a) A discharge or flow of pus, especially in the alveoli (sockets) and the tissues of the gums.

**Pyramid** (PIR-a-mid) A pointed or cone-shaped structure. One of two roughly triangular structures on the anterior aspect of the medulla oblongata composed of the largest motor tracts that run from the cerebral cortex to the spinal cord. A triangular structure in the renal medulla.

**Pyramidal** (pi-RAM-i-dal) **tracts (pathways)** *See* **Direct motor pathways.**

## Q

**QRS** complex The deflection waves of an electrocardiogram that represent onset of ventricular depolarization.

**Quadrant** (KWOD-rant) One of four parts.

**Quadriplegia** (kwod′-ri-PLĒ-jē-a) Paralysis of four limbs: two upper and two lower.

## R

**Radiographic** (rā′-dē-ō-GRAF-ik) **anatomy** Diagnostic branch of anatomy that includes the use of x-rays.

**Rami communicantes** (RĀ-mē kō-mū-ni-KAN-tēz) Branches of a spinal nerve. *Singular* is **ramus communicans** (RĀ-mus kō-MŪ-ni-kans).

**Rathke's pouch** *See* **Hypophyseal pouch.**

**Receptor** A specialized cell or a distal portion of a neuron that responds to a specific sensory modality, such as touch, pressure, cold, light, or sound, and converts it to an electrical signal (generator or receptor potential). A specific molecule or cluster of molecules that recognizes and binds a particular ligand.

**Receptor-mediated endocytosis** A highly selective process whereby cells take up specific ligands, which usually are large molecules or particles, by enveloping them within a sac of plasma membrane. Ligands are eventually broken down by enzymes in lysosomes.

**Recombinant DNA** Synthetic DNA, formed by joining a fragment of DNA from one source to a portion of DNA from another.

**Rectouterine** (rek-tō-Ū-ter-in) **pouch** A pocket formed by the parietal peritoneum as it moves posteriorly from the surface of the uterus and is reflected onto the rectum; the most inferior point in the pelvic cavity. Also called the **pouch of Douglas.**

**Rectum** (REK-tum) The last 20 cm (8 in.) of the gastrointestinal tract, from the sigmoid colon to the anus.

**Recumbent** (re-KUM-bent) Lying down.

**Red bone marrow** A highly vascularized connective tissue located in microscopic spaces between trabeculae of spongy bone tissue.

**Red nucleus** A cluster of cell bodies in the midbrain, occupying a large part of the tectum from which axons extend into the rubroreticular and rubrospinal tracts.

**Red pulp** That portion of the spleen that consists of venous sinuses filled with blood and thin plates of splenic tissue called splenic (Billroth's) cords.

**Referred pain** Pain that is felt at a site remote from the place of origin.

**Reflex** Fast response to a change (stimulus) in the internal or external environment that attempts to restore homeostasis.

**Reflex arc** The most basic conduction pathway through the nervous system, connecting a receptor and an effector and consisting of a receptor, a sensory neuron, an integrating center in the central nervous system, a motor neuron, and an effector. Also called **reflex circuit.**

**Regional anatomy** The division of anatomy dealing with a specific region of the body, such as the head, neck, chest, or abdomen.

**Regurgitation** (rē-gur′-ji-TĀ-shun) Return of solids or fluids to the mouth from the stomach; backward flow of blood through incompletely closed heart valves.

**Relaxin (RLX)** A female hormone produced by the ovaries and placenta that increases flexibility of the pubic symphysis and helps dilate the uterine cervix to ease delivery of a baby.

**Releasing hormone** Hormone secreted by the hypothalamus that can stimulate secretion of hormones of the anterior pituitary.

**Renal** (RĒ-nal) Pertaining to the kidneys.

**Renal corpuscle** (KOR-pus-l) A glomerular (Bowman's) capsule and its enclosed glomerulus.

**Renal pelvis** A cavity in the center of the kidney formed by the expanded, proximal portion of the ureter, lying within the kidney, and into which the major calyces open.

**Renal physiology** Study of the functions of the kidneys.

**Renal pyramid** A triangular structure in the renal medulla containing the straight segments of renal tubules and the vasa recta.

**Reproduction** (rē-prō-DUK-shun) The formation of new cells for growth, repair, or replacement; the production of a new individual.

**Reproductive cell division** Type of cell division in which gametes (sperm and oocytes) are produced; consists of meiosis and cytokinesis.

**Respiration** (res-pi-RĀ-shun) Overall exchange of gases between the atmosphere, blood, and body cells consisting of pulmonary ventilation, external respiration, and internal respiration.

**Respiratory center** Neurons in the pons and medulla oblongata of the brain stem that regulate the rate and depth of pulmonary ventilation.

**Respiratory** (RES-pir-a-tōr-ē) **physiology** Study of the functions of the air passageways and lungs.

**Respiratory** (RES-pi-ra-tōr-ē) **system** System composed of the nose, pharynx, larynx, trachea, bronchi, and lungs that obtains oxygen for body cells and eliminates carbon dioxide from them.

**Retention** (rē-TEN-shun) A failure to void urine due to obstruction, nervous contraction of the urethra, or absence of sensation of desire to urinate.

**Rete** (RĒ-tē) **testis** The network of ducts in the testes.

**Reticular** (re-TIK-ū-lar) **activating system (RAS)** A portion of the reticular formation that has many ascending connections with the cerebral cortex; when this area of the brain stem is active, nerve impulses pass to the thalamus and widespread areas of the cerebral cortex, resulting in generalized alertness or arousal from sleep.

**Reticular formation** A network of small groups of neuronal cell bodies scattered among bundles of axons (mixed gray and white matter) beginning in the medulla oblongata and extending superiorly through the central part of the brain stem.

**Reticulocyte** (re-TIK-ū-lō-sīt) An immature red blood cell.

**Reticulum** (re-TIK-ū-lum) A network.

**Retina** (RET-i-na) The deep coat of the posterior portion of the eyeball consisting of nervous tissue (where the process of vision begins) and a pigmented layer of epithelial cells that contact the choroid.

**Retinaculum** (ret-i-NAK-ū-lum) A thickening of fascia that holds structures in place, for example, the superior and inferior retinacula of the ankle.

**Retraction** (rē-TRAK-shun) The movement of a protracted part of the body posteriorly on a plane parallel to the ground, as in pulling the lower jaw back in line with the upper jaw.

**Retroperitoneal** (re′-trō-per-i-tō-NĒ-al) External to the peritoneal lining of the abdominal cavity.

**Rh factor** An inherited antigen on the surface of red blood cells in $Rh^+$ individuals; not present in $Rh^-$ individuals.

**Rhinology** (rī-NOL-ō-jē) The study of the nose and its disorders.

**Ribonucleic** (rī-bō-noo-KLĒ-ik) **acid (RNA)** A single-stranded nucleic acid made up of nucleotides, each consisting of a nitrogenous base (adenine, cytosine, guanine, or uracil), ribose, and a phosphate group; major types are messenger RNA (mRNA), transfer RNA (tRNA), and ribosomal RNA (rRNA), each of which has a specific role during protein synthesis.

**Ribosome** (RĪ-bō-sōm) A cellular structure in the cytoplasm of cells, composed of a small subunit and a large subunit that contain ribosomal RNA and ribosomal proteins; the site of protein synthesis.

**Right lymphatic** (lim-FAT-ik) **duct** A vessel of the lymphatic system that drains lymph from the upper right side of the body and empties it into the right subclavian vein.

**Rigidity** (ri-JID-i-tē) Hypertonia characterized by increased muscle tone, but reflexes are not affected.

**Rigor mortis** State of partial contraction of muscles after death due to lack of ATP; myosin heads (cross-bridges) remain attached to actin, thus preventing relaxation.

**Rod** One of two types of photoreceptors in the retina of the eye; specialized for vision in dim light.

**Root canal** A narrow extension of the pulp cavity lying within the root of a tooth.

**Root of penis** Attached portion of penis that consists of the bulb and crura.

**Rotation** (rō-TĀ-shun) Moving a bone around its own axis, with no other movement.

**Rotator cuff** Refers to the tendons of four deep shoulder muscles (subscapularis, supraspinatus, infraspinatus, and teres minor) that form a complete circle around the shoulder; they strengthen and stabilize the shoulder joint.

**Round ligament** (LIG-a-ment) A band of fibrous connective tissue enclosed between the folds of the broad ligament of the uterus, emerging from the uterus just inferior to the uterine tube, extending laterally along the pelvic wall and through the deep inguinal ring to end in the labia majora.

**Round window** A small opening between the middle and internal ear, directly inferior to the oval window, covered by the secondary tympanic membrane.

**Ruffini corpuscle** A sensory receptor embedded deeply in the dermis and deeper tissues that detects the stretching of the skin.

**Rugae** (ROO-gē) Large folds in the mucosa of an empty hollow organ, such as the stomach or vagina.

## S

**Saccule** (SAK-ūl) The inferior and smaller of the two chambers in the membranous labyrinth inside the vestibule of the internal ear containing a receptor organ for static equilibrium.

**Sacral plexus** (SĀ-kral PLEK-sus) A network formed by the ventral branches of spinal nerves L4 through S3.

**Sacral promontory** (PROM-on-tor′-ē) The superior surface of the body of the first sacral vertebra that projects anteriorly into the pelvic cavity; a line from the sacral promontory to the superior border of the pubic symphysis divides the abdominal and pelvic cavities.

**Saddle joint** A synovial joint in which the articular surface of one bone is saddle-shaped and the articular surface of the other bone is shaped like the legs of the rider sitting in the saddle, as in the joint between the trapezium and the metacarpal of the thumb.

**Sagittal** (SAJ-i-tal) **plane** A plane that divides the body or organs into left and right portions. Such a plane may be **midsagittal (median),** in which the divisions are equal, or **parasagittal,** in which the divisions are unequal.

**Saliva** (sa-LĪ-va) A clear, alkaline, somewhat viscous secretion produced mostly by the three pairs of salivary glands; contains various salts, mucin, lysozyme, salivary amylase, and lingual lipase (produced by glands in the tongue).

**Salivary amylase** (SAL-i-ver-ē AM-i-lās) An enzyme in saliva that initiates the chemical breakdown of starch.

**Salivary gland** One of three pairs of glands that lie external to the mouth and pour their secretory product (saliva) into ducts that empty into the oral cavity; the parotid, submandibular, and sublingual glands.

**Sarcolemma** (sar′-kō-LEM-ma) The cell membrane of a muscle fiber (cell), especially of a skeletal muscle fiber.

**Sarcomere** (SAR-kō-mēr) A contractile unit in a striated muscle fiber (cell) extending from one Z disc to the next Z disc.

**Sarcoplasm** (SAR-kō-plazm) The cytoplasm of a muscle fiber (cell).

**Sarcoplasmic reticulum** (sar′-kō-PLAZ-mik re-TIK-ū-lum) **(SR)** A network of saccules and tubes surrounding myofibrils of a muscle fiber (cell), comparable to endoplasmic reticulum; functions to reabsorb calcium ions during relaxation and to release them to cause contraction.

**Satellite** (SAT-i-līt) **cell** Flat neuroglial cells that surround cell bodies of peripheral nervous system ganglia to provide structural support and regulate the exchange of material between a neuronal cell body and interstitial fluid.

**Saturated fat** A fatty acid that contains only single bonds (no double bonds) between its carbon atoms; all carbon atoms are bonded to the maximum number of hydrogen atoms; prevalent in triglycerides of animal products such as meat, milk, milk products, and eggs.

**Scala tympani** (SKA-la TIM-pan-ē) The inferior spiral-shaped channel of the bony cochlea, filled with perilymph.

**Scala vestibuli** (ves-TIB-ū-lē) The superior spiral-shaped channel of the bony cochlea, filled with perilymph.

**Schwann** (SCHVON or SCHWON) **cell** A neuroglial cell of the peripheral nervous system that forms the myelin sheath and neurolemma around a nerve axon by wrapping around the axon in a jellyroll fashion.

**Sciatica** (sī-AT-i-ka) Inflammation and pain along the sciatic nerve; felt along the posterior aspect of the thigh extending down the inside of the leg.

**Sclera** (SKLE-ra) The white coat of fibrous tissue that forms the superficial protective covering over the eyeball except in the most anterior portion; the posterior portion of the fibrous tunic.

**Scleral venous sinus** A circular venous sinus located at the junction of the sclera and the cornea through which aqueous humor drains from the anterior chamber of the eyeball into the blood. Also called the **canal of Schlemm** (SHLEM).

**Sclerosis** (skle-RŌ-sis) A hardening with loss of elasticity of tissues.

**Scoliosis** (skō-lē-Ō-sis) An abnormal lateral curvature from the normal vertical line of the backbone.

**Scrotum** (SKRŌ-tum) A skin-covered pouch that contains the testes and their accessory structures.

**Sebaceous** (se-BĀ-shus) **gland** An exocrine gland in the dermis of the skin, almost always associated with a hair follicle, that secretes sebum. Also called an **oil gland.**

**Sebum** (SĒ-bum) Secretion of sebaceous (oil) glands.

**Secondary sex characteristic** A characteristic of the male or female body that develops at puberty under the influence of sex hormones but is not directly involved in sexual reproduction; examples are distribution of body hair, voice pitch, body shape, and muscle development.

**Secretion** (se-KRĒ-shun) Production and release from a cell or a gland of a physiologically active substance.

**Selective permeability** (per′-mē-a-BIL-i-tē) The property of a membrane by which it permits the passage of certain substances but restricts the passage of others.

**Semen** (SĒ-men) A fluid discharged at ejaculation by a male that consists of a mixture of sperm and the secretions of the seminiferous tubules, seminal vesicles, prostate, and bulbourethral (Cowper's) glands.

**Semicircular canals** Three bony channels (anterior, posterior, lateral), filled with perilymph, in which lie the membranous semicircular canals filled with endolymph. They contain receptors for equilibrium.

**Semicircular ducts** The membranous semicircular canals filled with endolymph and floating in the perilymph of the bony semicircular canals; they contain cristae that are concerned with dynamic equilibrium.

**Semilunar** (sem′-ē-LOO-nar) **(SL) valve** A valve between the aorta or the pulmonary trunk and a ventricle of the heart.

**Seminal vesicle** (SEM-i-nal VES-i-kul) One of a pair of convoluted, pouchlike structures, lying posterior and inferior to the urinary bladder and anterior to the rectum, that secrete a component of semen into the ejaculatory ducts. Also termed **seminal gland.**

**Seminiferous tubule** (sem′-i-NI-fer-us TOO-būl) A tightly coiled duct, located in the testis, where sperm are produced.

**Sensation** A state of awareness of external or internal conditions of the body.

**Sensory area** A region of the cerebral cortex concerned with the interpretation of sensory impulses.

**Sensory neurons** (NOO-ronz) Neurons that carry sensory information from cranial and spinal nerves into the brain and spinal cord or from a lower to a higher level in the spinal cord and brain. Also called **afferent neurons.**

**Septal defect** An opening in the atrial septum (atrial septal defect) because the foramen ovale fails to close, or the ventricular septum (ventricular septal defect) due to incomplete development of the ventricular septum.

**Septum** (SEP-tum) A wall dividing two cavities.

**Serous** (SĒR-us) **membrane** A membrane that lines a body cavity that does not open to the exterior. The external layer of an organ formed by a serous membrane. The membrane that lines the pleural, pericardial, and peritoneal cavities. Also called a **serosa** (se-RŌ-sa).

**Sertoli** (ser-TŌ-lē) **cell** A supporting cell in the seminiferous tubules that secretes fluid for supplying nutrients to sperm and the hormone inhibin, removes excess cytoplasm from spermatogenic cells, and mediates the effects of FSH and testosterone on spermatogenesis. Also called a **sustentacular** (sus′-ten-TAK-ū-lar) **cell.**

**Serum** Blood plasma minus its clotting proteins.

**Sesamoid** (SES-a-moyd) **bones** Small bones usually found in tendons.

**Sex chromosomes** The twenty-third pair of chromosomes, designated X and Y, which determines the genetic sex of an individual; in males, the pair is XY; in females, XX.

**Sexual intercourse** The insertion of the erect penis of a male into the vagina of a female. Also called **coitus** (KŌ-i-tus).

**Sheath of Schwann** *See* **Neurolemma.**

**Shock** Failure of the cardiovascular system to deliver adequate amounts of oxygen and nutrients to meet the metabolic needs of the body due to inadequate cardiac output. It is characterized by hypotension; clammy, cool, and pale skin; sweating; reduced urine formation; altered mental state; acidosis; tachycardia; weak, rapid pulse; and thirst. Types include hypovolemic, cardiogenic, vascular, and obstructive.

**Shoulder joint** A synovial joint where the humerus articulates with the scapula.

**Sigmoid colon** (SIG-moyd KŌ-lon) The S-shaped part of the large intestine that begins at the level of the left iliac crest, projects medially, and terminates at the rectum at about the level of the third sacral vertebra.

**Sign** Any objective evidence of disease that can be observed or measured, such as a lesion, swelling, or fever.

**Sinoatrial** (si-nō-Ā-trē-al) (SA) node A small mass of cardiac muscle fibers (cells) located in the right atrium inferior to the opening of the superior vena cava that spontaneously depolarize and generate a cardiac action potential about 100 times per minute. Also called the natural **pacemaker.**

**Sinus** (SĪ-nus) A hollow in a bone (paranasal sinus) or other tissue; a channel for blood (vascular sinus); any cavity having a narrow opening.

**Sinusoid** (SĪ-nū-soyd) A large, thin-walled, and leaky type of capillary, having large intercellular clefts that may allow proteins and blood cells to pass from a tissue into the bloodstream; present in the liver, spleen, anterior pituitary, parathyroid glands, and red bone marrow.

**Skeletal muscle** An organ specialized for contraction, composed of striated muscle fibers (cells), supported by connective tissue, attached to a bone by a tendon or an aponeurosis, and stimulated by somatic motor neurons.

**Skeletal system** Framework of bones and their associated cartilages, ligaments, and tendons.

**Skene's gland** *See* **Paraurethral gland.**

**Skin** The external covering of the body that consists of a superficial, thinner epidermis (epithelial tissue) and a deep, thicker dermis (connective tissue) that is anchored to the subcutaneous layer. Also called **cutaneous membrane.**

**Skin graft** The transfer of a patch of healthy skin taken from a donor site to cover a wound.

**Skull** The skeleton of the head consisting of the cranial and facial bones.

**Sleep** A state of partial unconsciousness from which a person can be aroused; associated with a low level of activity in the reticular activating system.

**Sliding filament mechanism** A model that describes muscle contraction in which thin filaments slide past thick ones so that the filaments overlap, causing shortening of a sarcomere, and thus shortening of muscle fibers and alternately shortening of the entire muscle.

**Small intestine** A long tube of the gastrointestinal tract that begins at the pyloric sphincter of the stomach, coils through the central and inferior part of the abdominal cavity, and ends at the large intestine; divided into three segments: duodenum, jejunum, and ileum.

**Smooth muscle** A tissue specialized for contraction, composed of smooth muscle fibers (cells), located in the walls of hollow internal organs, and innervated by autonomic motor neurons.

**Sodium–potassium ATPase** An active transport pump located in the plasma membrane that transports sodium ions out of the cell and potassium ions into the cell at the expense of cellular ATP. It functions to keep the ionic concentrations of these ions at physiological levels. Also called the **sodium–potassium pump.**

**Soft palate** (PAL-at) The posterior portion of the roof of the mouth, extending from the palatine bones to the uvula. It is a muscular partition lined with mucous membrane.

**Somatic** (sō-MAT-ik) **cell division** Type of cell division in which a single starting cell duplicates itself to produce two identical cells; consists of mitosis and cytokinesis.

**Somatic motor pathway** Pathway that carries information from the cerebral cortex, basal nuclei, and cerebellum that stimulates contraction of skeletal muscles.

**Somatic nervous system (SNS)** The portion of the peripheral nervous system consisting of somatic sensory (afferent) neurons and somatic motor (efferent) neurons.

**Somatic sensory pathway** Pathway that carries information from somatic sensory receptor to the primary somatosensory area in the cerebral cortex and cerebellum.

**Somite** (SŌ-mīt) Block of mesodermal cells in a developing embryo that is distinguished into a myotome (which forms most of the skeletal muscles), dermatome (which forms connective tissues), and sclerotome (which forms the vertebrae).

**Spasm** (SPAZM) A sudden, involuntary contraction of large groups of muscles.

**Spasticity** (spas-TIS-i-tē) Hypertonia characterized by increased muscle tone, increased tendon reflexes, and pathological reflexes (Babinski sign).

**Spermatic** (sper-MAT-ik) **cord** A supporting structure of the male reproductive system, extending from a testis to the deep inguinal ring, that includes the ductus (vas) deferens, arteries, veins, lymphatic vessels, nerves, cremaster muscle, and connective tissue.

**Spermatogenesis** (sper′-ma-tō-JEN-e-sis) The formation and development of sperm in the seminiferous tubules of the testes.

**Sperm cell** A mature male gamete. Also termed spermatozoon (sper′-ma-tō-ZŌ-on).

**Spermiogenesis** (sper′-mē-ō-JEN-e-sis) The maturation of spermatids into sperm.

**Sphincter** (SFINGK-ter) A circular muscle that constricts an opening.

**Sphincter of Oddi** *See* **Sphincter of the hepatopancreatic ampulla.**

**Sphincter of the hepatopancreatic ampulla** A circular muscle at the opening of the common bile and main pancreatic ducts in the duodenum. Also called the **sphincter of Oddi** (OD-ē).

**Spinal** (SPĪ-nal) **cord** A mass of nerve tissue located in the vertebral canal from which 31 pairs of spinal nerves originate.

**Spinal nerve** One of the 31 pairs of nerves that originate on the spinal cord from posterior and anterior roots.

**Spinal shock** A period from several days to several weeks following transection of the spinal cord that is characterized by the abolition of all reflex activity.

**Spinothalamic** (spī-nō-tha-LAM-ik) **tract** Sensory (ascending) tract that conveys information up the spinal cord to the thalamus for sensations of pain, temperature, itch, and tickle.

**Spinous** (SPĪ-nus) **process** A sharp or thornlike process or projection. Also called a **spine.** A sharp ridge running diagonally across the posterior surface of the scapula.

**Spiral organ** The organ of hearing, consisting of supporting cells and hair cells that rest on the basilar membrane and extend into the endolymph of the cochlear duct. Also called the **organ of Corti** (KOR-tē).

**Splanchnic** (SPLANK-nik) Pertaining to the viscera.

**Spleen** (SPLĒN) Large mass of lymphatic tissue between the fundus of the stomach and the diaphragm that functions in formation of blood cells during early fetal development, phagocytosis of ruptured blood cells, and proliferation of B cells during immune responses.

**Spongy (cancellous) bone tissue** Bone tissue that consists of an irregular latticework of thin plates of bone called trabeculae; spaces between trabeculae of some bones are filled with red bone marrow; found inside short, flat, and irregular bones and in the epiphyses (ends) of long bones.

**Sprain** Forcible wrenching or twisting of a joint with partial rupture or other injury to its attachments without dislocation.

**Squamous** (SKWĀ-mus) Flat or scalelike.

**Starvation** (star-VĀ-shun) The loss of energy stores in the form of glycogen, triglycerides, and proteins due to inadequate intake of nutrients or inability to digest, absorb, or metabolize ingested nutrients.

**Static equilibrium** (ē-kwi-LIB-rē-um) The maintenance of posture in response to changes in the orientation of the body, mainly the head, relative to the ground.

**Stellate reticuloendothelial** (STEL-āt re-tik′-ū-lō-en′-dō-THĒ-lē-al) **cell** Phagocytic cell bordering a sinusoid of the liver. Also called a **Kupffer** (KOOP-fer) **cell.**

**Stem cell** An unspecialized cell that has the ability to divide for indefinite periods and give rise to a specialized cell.

**Stenosis** (sten-Ō-sis) An abnormal narrowing or constriction of a duct or opening.

**Stereocilia** (ste′-rē-ō-SIL-ē-a) Groups of extremely long, slender, nonmotile microvilli projecting from epithelial cells lining the epididymis.

**Sterile** (STE-ril) Free from any living microorganisms. Unable to conceive or produce offspring.

**Sterilization** (ster′-i-li-ZĀ-shun) Elimination of all living microorganisms. Any procedure that renders an individual incapable of reproduction (for example, castration, vasectomy, hysterectomy, or oophorectomy).

**Stimulus** Any stress that changes a controlled condition; any change in the internal or external environment that excites a sensory receptor, a neuron, or a muscle fiber.

**Stomach** The J-shaped enlargement of the gastrointestinal tract directly inferior to the diaphragm in the epigastric, umbilical, and left hypochondriac regions of the abdomen, between the esophagus and small intestine.

**Straight tubule** (TOO-būl) A duct in a testis leading from a convoluted seminiferous tubule to the rete testis.

**Stratum** (STRĀ-tum) A layer.

**Stratum basalis** (ba-SAL-is) The layer of the endometrium next to the myometrium that is maintained during menstruation and gestation and produces a new stratum functionalis following menstruation or parturition.

**Stratum functionalis** (funk′-shun-AL-is) The layer of the endometrium next to the uterine cavity that is shed during menstruation and that forms the maternal portion of the placenta during gestation.

**Stretch receptor** Receptor in the walls of blood vessels, airways, or organs that monitors the amount of stretching. Also termed **baroreceptor.**

**Striae** (STRĪ-ē) Internal scarring due to overstretching of the skin in which collagen fibers and blood vessels in the dermis are damaged. Also called **stretch marks.**

**Stroma** (STRŌ-ma) The tissue that forms the ground substance, foundation, or framework of an organ, as opposed to its functional parts (parenchyma).

**Subarachnoid** (sub′-a-RAK-noyd) **space** A space between the arachnoid mater and the pia mater that surrounds the brain and spinal cord and through which cerebrospinal fluid circulates.

**Subcutaneous** (sub′-kū-TĀ-nē-us) Beneath the skin. Also called **hypodermic** (hī-pō-DER-mik).

**Subcutaneous (subQ) layer** A continuous sheet of areolar connective tissue and adipose tissue between the dermis of the skin and the deep fascia of the muscles. Also called the **hypodermis.**

**Subdural** (sub-DOO-ral) **space** A space between the dura mater and the arachnoid mater of the brain and spinal cord that contains a small amount of fluid.

**Sublingual** (sub-LING-gwal) **gland** One of a pair of salivary glands situated in the floor of the mouth deep to the mucous membrane and to the side of the lingual frenulum, with a duct (Rivinus') that opens into the floor of the mouth.

**Submandibular** (sub′-man-DIB-ū-lar) **gland** One of a pair of salivary glands found inferior to the base of the tongue deep to the mucous membrane in the posterior part of the floor of the mouth, posterior to the sublingual glands, with a duct (Wharton's) situated to the side of the lingual frenulum. Also called the **submaxillary** (sub′-MAK-si-ler-ē) **gland.**

**Submucosa** (sub-mū-KŌ-sa) A layer of connective tissue located deep to a mucous membrane, as in the gastrointestinal tract or the urinary bladder; the submucosa connects the mucosa to the muscularis layer.

**Submucosal plexus** A network of autonomic nerve fibers located in the superficial part of the submucous layer of the small intestine. Also called the **plexus of Meissner** (MĪZ-ner).

**Substrate** A molecule on which an enzyme acts.

**Subthalamus** (sub-THAL-a-mus) Part of the diencephalon inferior to the thalamus; the substantia nigra and red nucleus extend from the midbrain into the subthalamus.

**Sudoriferous** (soo′-dor-IF-er-us) **gland** An apocrine or eccrine exocrine gland in the dermis or subcutaneous layer that produces perspiration. Also called a **sweat gland.**

**Sulcus** (SUL-kus) A groove or depression between parts, especially between the convolutions of the brain. *Plural* is **sulci** (SUL-sī).

**Superficial** (soo′-per-FISH-al) Located on or near the surface of the body or an organ. Also called **external.**

**Superficial subcutaneous inguinal** (IN-gwi-nal) **ring** A triangular opening in the aponeurosis of the external oblique muscle that represents the termination of the inguinal canal.

**Superior** (soo-PĒR-ē-or) Toward the head or upper part of a structure. Also called **cephalic** or **cranial.**

**Superior vena cava** (VĒ-na KĀ-va) **(SVC)** Large vein that collects blood from parts of the body superior to the heart and returns it to the right atrium.

**Supination** (soo-pi-NĀ-shun) A movement of the forearm in which the palm is turned anteriorly.

**Surface anatomy** The study of the structures that can be identified from the outside of the body.

**Surfactant** (sur-FAK-tant) Complex mixture of phospholipids and lipoproteins, produced by type II alveolar (septal) cells in the lungs, that decreases surface tension.

**Suspensory ligament** (sus-PEN-so-rē LIG-a-ment) A fold of peritoneum extending laterally from the surface of the ovary to the pelvic wall.

**Sustentacular cell** *See* **Sertoli cell.**

**Sutural** (SOO-chur-al) **bone** A small bone located within a suture between certain cranial bones. Also called **Wormian** (WER-mē-an) **bone.**

**Suture** (SOO-chur) An immovable fibrous joint that joins skull bones.

**Sympathetic** (sim′-pa-THET-ik) **division** One of the two subdivisions of the autonomic nervous system, having cell bodies of preganglionic neurons in the lateral gray columns of the thoracic segment and the first two or three lumbar segments of the spinal cord; primarily concerned with processes involving the expenditure of energy.

**Sympathetic trunk ganglion** (GANG-glē-on) A cluster of cell bodies of sympathetic postganglionic neurons lateral to the vertebral column, close to the body of a vertebra. These ganglia extend inferiorly through the neck, thorax, and abdomen to the coccyx on both sides of the vertebral column and are connected to one another to form a chain on each side of the vertebral column. Also called **sympathetic chain ganglia, vertebral chain ganglia,** or **paravertebral ganglia.**

**Symphysis** (SIM-fi-sis) A line of union. A slightly movable cartilaginous joint such as the pubic symphysis.

**Symptom** (SIMP-tum) A subjective change in body function not apparent to an observer, such as pain or nausea, that indicates the presence of a disease or disorder of the body.

**Synapse** (SIN-aps) The functional junction between two neurons or between a neuron and an effector, such as a muscle or gland; may be electrical or chemical.

**Synapsis** (sin-AP-sis) The pairing of homologous chromosomes during prophase I of meiosis.

**Synaptic** (sin-AP-tik) **cleft** The narrow gap at a chemical synapse that separates the axon terminal of one neuron from another neuron or muscle fiber (cell) and across which a neurotransmitter diffuses to affect the postsynaptic cell.

**Synaptic end bulb** Expanded distal end of an axon terminal that contains synaptic vesicles. Also called a **synaptic knob.**

**Synaptic vesicle** Membrane-enclosed sac in a synaptic end bulb that stores neurotransmitters.

**Synarthrosis** (sin′-ar-THRŌ-sis) An immovable joint such as a suture, gomphosis, or synchondrosis.

**Synchondrosis** (sin′-kon-DRŌ-sis) A cartilaginous joint in which the connecting material is hyaline cartilage.

**Syndesmosis** (sin′-dez-MŌ-sis) A slightly movable joint in which articulating bones are united by fibrous connective tissue.

**Synergist** (SIN-er-jist) A muscle that assists the prime mover by reducing undesired action or unnecessary movement.

**Synergistic** (syn-er-JIS-tik) **effect** A hormonal interaction in which the effects of two or more hormones acting together is greater or more extensive than the sum of each hormone acting alone.

**Synostosis** (sin′-os-TŌ-sis) A joint in which the dense fibrous connective tissue that unites bones at a suture has been replaced by bone, resulting in a complete fusion across the suture line.

**Synovial** (si-NŌ-vē-al) **cavity** The space between the articulating bones of a synovial joint, filled with synovial fluid. Also called a **joint cavity.**

**Synovial fluid** Secretion of synovial membranes that lubricates joints and nourishes articular cartilage.

**Synovial joint** A fully movable or diarthrotic joint in which a synovial (joint) cavity is present between the two articulating bones.

**Synovial membrane** The deeper of the two layers of the articular capsule of a synovial joint, composed of areolar connective tissue that secretes synovial fluid into the synovial (joint) cavity.

**System** An association of organs that have a common function.

**Systemic** (sis-TEM-ik) Affecting the whole body; generalized.

**Systemic anatomy** The anatomical study of particular systems of the body, such as the skeletal, muscular, nervous, cardiovascular, or urinary systems.

**Systemic circulation** The routes through which oxygenated blood flows from the left ventricle through the aorta to all the organs of the body and deoxygenated blood returns to the right atrium.

**Systole** (SIS-tō-lē) In the cardiac cycle, the phase of contraction of the heart muscle, especially of the ventricles.

**Systolic** (sis-TOL-ik) **blood pressure (SBP)** The force exerted by blood on arterial walls during ventricular contraction; the highest pressure measured in the large arteries, about 110 mmHg under normal conditions for a young adult.

## T

**T cell** A lymphocyte that becomes immunocompetent in the thymus and can differentiate into a helper T cell or a cytotoxic T cell, both of which function in cell-mediated immunity.

**T wave** The deflection wave of an electrocardiogram that represents ventricular repolarization.

**Tachycardia** (tak′-i-KAR-dē-a) An abnormally rapid resting heartbeat or pulse rate (over 100 beats per minute).

**Tactile** (TAK-tīl) Pertaining to the sense of touch.

**Tactile cell** Type of cell in the epidermis of hairless skin that makes contact with a tactile disc, which functions in touch. Also called **Merkel** (MER-kel) **cell.**

**Tactile disc** Soucer-shaped free nerve endings that make contact with tactile cells in the epidermis and function as touch receptors. Also called **Merkel disc.**

**Target cell** A cell whose activity is affected by a particular hormone.

**Tarsal bones** The seven bones of the ankle. Also called **tarsals.**

**Tarsal gland** Sebaceous (oil) gland that opens on the edge of each eyelid. Also called a **Meibomian** (mī-BŌ-mē-an) **gland.**

**Tarsal plate** A thin, elongated sheet of connective tissue, one in each eyelid, giving the eyelid form and support. The aponeurosis of the levator palpebrae superioris is attached to the tarsal plate of the superior eyelid.

**Tarsus** (TAR-sus) A collective term for the seven bones of the ankle.

**Tectorial** (tek-TŌ-rē-al) **membrane** A gelatinous membrane projecting over and in contact with the hair cells of the spiral organ (organ of Corti) in the cochlear duct.

**Teeth** (TĒTH) Accessory structures of digestion, composed of calcified connective tissue and embedded in bony sockets of the mandible and maxilla, that cut, shred, crush, and grind food. Also called **dentes** (DEN-tēz).

**Telophase** (TEL-ō-fāz) The final stage of mitosis.

**Tendon** (TEN-don) A white fibrous cord of dense regular connective tissue that attaches muscle to bone.

**Tendon organ** A proprioceptive receptor, sensitive to changes in muscle tension and force of contraction, found chiefly near the junctions of tendons and muscles. Also called a **Golgi** (GOL-jē) **tendon organ.**

**Tendon reflex** A polysynaptic, ipsilateral reflex that protects tendons and their associated muscles from damage that might be brought about by excessive tension. The receptors involved are called tendon organs (Golgi tendon organs).

**Teniae coli** (TĒ-nē-ē KŌ-lī) The three flat bands of thickened, longitudinal smooth muscle running the length of the large intestine, except in the rectum. *Singular* is **tenia coli.**

**Tentorium cerebelli** (ten-TŌ-rē-um ser′-e-BEL-ī) A transverse shelf of dura mater that forms a partition between the occipital lobe of the cerebral hemispheres and the cerebellum and that covers the cerebellum.

**Teratogen** (TER-a-tō-jen) Any agent or factor that causes physical defects in a developing embryo.

**Terminal ganglion** (TER-min-al GANG-glē-on) A cluster of cell bodies of parasympathetic postganglionic neurons either lying very close to the visceral effectors or located within the walls of the visceral effectors supplied by the postganglionic neurons. Also called **intramural ganglion.**

**Testis** (TES-tis) Male gonad that produces sperm and the hormones testosterone and inhibin. Also called a **testicle.**

**Testosterone** (tes-TOS-te-rōn) A male sex hormone (androgen) secreted by interstitial (Leydig) cells of a mature testis; needed for development of sperm; together with a second androgen termed **dihydrotestosterone** (dī-hī-drō-tes-TOS-ter-ōn) **(DHT),** controls the growth and development of male reproductive organs, secondary sex characteristics, and body growth.

**Tetralogy of Fallot** (tet-RAL-ō-jē of fal-Ō) A combination of four congenital heart defects: (1) constricted pulmonary semilunar valve, (2) interventricular septal opening, (3) emergence of the aorta from both ventricles instead of from the left only, and (4) enlarged right ventricle.

**Thalamus** (THAL-a-mus) A large, oval structure located bilaterally on either side of the third ventricle, consisting of two masses of gray matter organized into nuclei; main relay center for sensory impulses ascending to the cerebral cortex.

**Thermoreceptor** (THER-mō-rē-sep-tor) Sensory receptor that detects changes in temperature.

**Thermoregulation** Homeostatic regulation of body temperature through sweating and adjustment of blood flow in the dermis.

**Thigh** The portion of the lower limb between the hip and the knee.

**Third ventricle** (VEN-tri-kul) A slitlike cavity between the right and left halves of the thalamus and between the lateral ventricles of the brain.

**Thoracic** (thor-AS-ik) **cavity** Cavity superior to the diaphragm that contains two pleural cavities, the mediastinum, and the pericardial cavity.

**Thoracic duct** A lymphatic vessel that begins as a dilation called the cisterna chyli, receives lymph from the left side of the head, neck, and chest, left arm, and the entire body below the ribs, and empties into the junction between the internal jugular and left subclavian veins. Also called the **left lymphatic** (lim-FAT-ik) **duct.**

**Thoracolumbar** (thōr′-a-kō-LUM-bar) **outflow** The axons of sympathetic preganglionic neurons, which have their cell bodies in the lateral gray columns of the thoracic segments and first two or three lumbar segments of the spinal cord.

**Thorax** (THŌ-raks) The chest.

**Thrombosis** (THROM-BŌ-sis) The formation of a clot in an unbroken blood vessel, usually a vein.

**Thrombus** (THROM-bus) A stationary clot formed in an unbroken blood vessel, usually a vein.

**Thymus** (THĪ-mus) A bilobed organ, located in the superior mediastinum posterior to the sternum and between the lungs, in which T cells develop immunocompetence.

**Thyroid cartilage** (THĪ-royd KAR-ti-lij) The largest single cartilage of the larynx, consisting of two fused plates that form the anterior wall of the larynx. Also called **Adam's apple.**

**Thyroid follicle** (FOL-i-kul) Spherical sac that forms the parenchyma of the thyroid gland and consists of follicular cells that produce thyroxine ($T_4$) and triiodothyronine ($T_3$).

**Thyroid gland** An endocrine gland with right and left lateral lobes on either side of the trachea connected by an isthmus; located anterior to the trachea just inferior to the cricoid cartilage; secretes thyroxine ($T_4$), triiodothyronine ($T_3$), and calcitonin.

**Thyroid-stimulating hormone (TSH)** A hormone secreted by the anterior pituitary that stimulates the synthesis and secretion of thyroxine ($T_4$) and triiodothyronine ($T_3$). Also called **thyrotropin** (THĪ-rō-TRŌ-pin).

**Thyroxine** (thī-ROK-sēn) ($T_4$) A hormone secreted by the thyroid gland that regulates metabolism, growth and development, and the activity of the nervous system. Also called **tetraiodothyronine** (tet-ra-ī-ō-dō-THĪ-rō-nēn).

**Tic** Spasmodic, involuntary twitching of muscles that are normally under voluntary control.

**Tissue** A group of similar cells and their intercellular substance joined together to perform a specific function.

**Tissue rejection** Phenomenon by which the body recognizes the protein (HLA antigens) in transplanted tissues or organs as foreign and produces antibodies against them.

**Tongue** A large skeletal muscle covered by a mucous membrane located on the floor of the oral cavity.

**Tonsil** (TON-sil) An aggregation of large lymphatic nodules embedded in the mucous membrane of the throat.

**Topical** (TOP-i-kal) Applied to the surface rather than ingested or injected.

**Torn cartilage** A tearing of an articular disc (meniscus) in the knee.

**Trabecula** (tra-BEK-ū-la) Irregular latticework of thin plates of spongy bone tissue. Fibrous cord of connective tissue serving as supporting fiber by forming a septum extending into an organ from its wall or capsule. *Plural* is **trabeculae** (tra-BEK-ū-lē).

**Trabeculae carneae** (KAR-nē-ē) Ridges and folds of the myocardium in the ventricles.

**Trachea** (TRĀ-kē-a) Tubular air passageway extending from the larynx to the fifth thoracic vertebra. Also called the **windpipe.**

**Tract** A bundle of nerve axons in the central nervous system.

**Transplantation** (tranz-plan-TĀ-shun) The transfer of living cells, tissues, or organs from a donor to a recipient or from one part of the body to another in order to restore a lost function.

**Transverse colon** (trans-VERS KŌ-lon) The portion of the large intestine extending across the abdomen from the right colic (hepatic) flexure to the left colic (splenic) flexure.

**Transverse fissure** (FISH-er) The deep cleft that separates the cerebrum from the cerebellum.

**Transverse plane** A plane that divides the body or organs into superior and inferior portions. Also called a **cross-sectional** or **horizontal plane.**

**Transverse tubules** (TOO-būls) **(T tubules)** Small, cylindrical invaginations of the sarcolemma of striated muscle fibers (cells) that conduct muscle action potentials toward the center of the muscle fiber.

**Tremor** (TREM-or) Rhythmic, involuntary, purposeless contraction of opposing muscle groups.

**Triad** (TRĪ-ad) A complex of three units in a muscle fiber composed of a transverse tubule and the sarcoplasmic reticulum terminal cisterns on both sides of it.

**Tricuspid** (trī-KUS-pid) **valve** Atrioventricular (AV) valve on the right side of the heart.

**Triglyceride** (trī-GLI-ser-īd) A lipid formed from one molecule of glycerol and three molecules of fatty acids that may be either solid (fats) or liquid (oils) at room temperature; the body's most highly concentrated source of chemical potential energy. Found mainly within adipocytes. Also called a **neutral fat** or a **triacylglycerol.**

**Trigone** (TRĪ-gōn) A triangular region at the base of the urinary bladder.

**Triiodothyronine** (trī-ī-ō-dō-THĪ-rō-nēn) **($T_3$)** A hormone produced by the thyroid gland that regulates metabolism, growth and development, and the activity of the nervous system.

**Trophoblast** (TRŌF-ō-blast) The superficial covering of cells of the blastocyst.

**Tropic** (TRŌ-pik) **hormone** A hormone whose target is another endocrine gland.

**Trunk** The part of the body to which the upper and lower limbs are attached.

**Tubal ligation** (lī-GĀ-shun) A sterilization procedure in which the uterine (fallopian) tubes are tied and cut.

**Tubular reabsorption** The movement of filtrate from renal tubules back into blood in response to the body's specific needs.

**Tubular secretion** The movement of substances in blood into renal tubular fluid in response to the body's specific needs.

**Tumor-suppressor gene** A gene coding for a protein that normally inhibits cell division; loss or alteration of a tumor suppressor gene called *p53* is the most common genetic change in a wide variety of cancer cells.

**Tunica albuginea** (TOO-ni-ka al′-bū-JIN-ē-a) A dense white fibrous capsule covering a testis or deep to the surface of an ovary.

**Tunica externa** (eks-TER-na) The superficial coat of an artery or vein, composed mostly of elastic and collagen fibers. Also called the **adventitia.**

**Tunica interna** (in-TER-na) The deep coat of an artery or vein, consisting of a lining of endothelium, basement membrane, and internal elastic lamina. Also called the **tunica intima** (IN-ti-ma).

**Tunica media** (MĒ-dē-a) The intermediate coat of an artery or vein, composed of smooth muscle and elastic fibers.

**Tympanic antrum** (tim-PAN-ik AN-trum) An air space in the middle ear that leads into the mastoid air cells or sinus.

**Tympanic** (tim-PAN-ik) **membrane** A thin, semitransparent partition of fibrous connective tissue between the external auditory meatus and the middle ear. Also called the **eardrum.**

## U

**Umbilical** (um-BIL-i-kul) **cord** The long, ropelike structure containing the umbilical arteries and vein that connect the fetus to the placenta.

**Umbilicus** (um-BIL-i-kus *or* um-bi-LĪ-kus) A small scar on the abdomen that marks the former attachment of the umbilical cord to the fetus. Also called the **navel**.

**Upper limb** The appendage attached at the shoulder girdle, consisting of the arm, forearm, wrist, hand, and fingers. Also called **upper extremity** or **upper appendage.**

**Uremia** (ū-RĒ-mē-a) Accumulation of toxic levels of urea and other nitrogenous waste products in the blood, usually resulting from severe kidney malfunction.

**Ureter** (Ū-rē-ter) One of two tubes that connect the kidney with the urinary bladder.

**Urethra** (ū-RĒ-thra) The duct from the urinary bladder to the exterior of the body that conveys urine in females and urine and semen in males.

**Urinalysis** (ū-ri-NAL-i-sis) An analysis of the volume and physical, chemical, and microscopic properties of urine.

**Urinary** (Ū-ri-ner-ē) **bladder** A hollow, muscular organ situated in the pelvic cavity posterior to the pubic symphysis; receives urine via two ureters and stores urine until it is excreted through the urethra.

**Urinary system** A system that consists of the kidneys, ureters, urinary bladder, and urethra. The system regulates the ionic composition, pH, volume, pressure, and osmolarity of blood.

**Urine** The fluid produced by the kidneys that contains wastes and excess materials; excreted from the body through the urethra.

**Urogenital** (ū′-rō-JEN-i-tal) **triangle** The region of the pelvic floor inferior to the pubic symphysis, bounded by the pubic symphysis and the ischial tuberosities, and containing the external genitalia.

**Urology** (ū-ROL-ō-jē) The specialized branch of medicine that deals with the structure, function, and diseases of the male and female urinary systems and the male reproductive system.

**Uterine cycle** A series of changes in the endometrium of a nonpregnant female that prepares the lining of the uterus to receive a fertilized ovum. Also called **menstrual cycle.**

**Uterine** (Ū-ter-in) **tube** Duct that transports ova from the ovary to the uterus. Also called the **fallopian** (fal-LŌ-pē-an) **tube** or **oviduct.**

**Uterosacral ligament** (ū-ter-ō-SĀ-kral LIG-a-ment) A fibrous band of tissue extending from the cervix of the uterus laterally to the sacrum.

**Uterus** (Ū-te-rus) The hollow, muscular organ in females that is the site of menstruation, implantation, development of the fetus, and labor. Also called the **womb.**

**Utricle** (Ū-tri-kul) The larger of the two divisions of the membranous labyrinth located inside the vestibule of the inner ear, containing a receptor organ for static equilibrium.

**Uvea** (Ū-vē-a) The three structures that together make up the vascular tunic of the eye.

**Uvula** (Ū-vū-la) A soft, fleshy mass, especially the U-shaped pendant part, descending from the soft palate.

## V

**Vagina** (va-JĪ-na) A muscular, tubular organ that leads from the uterus to the vestibule, situated between the urinary bladder and the rectum of the female.

**Vallate papilla** (VAL-āt pa-PIL-a) One of the circular projections that is arranged in an inverted V-shaped row at the back of the tongue; the largest of the elevations on the upper surface of the tongue containing taste buds. Also called **circumvallate papilla.**

**Varicocele** (VAR-i-kō-sēl) A twisted vein; especially, the accumulation of blood in the veins of the spermatic cord.

**Varicose** (VAR-i-kōs) Pertaining to an unnatural swelling, as in the case of a varicose vein.

**Vas** A vessel or duct.

**Vasa recta** (VĀ-sa REK-ta) Extensions of the efferent arteriole of a juxtamedullary nephron that run alongside the nephron loop (loop of Henle) in the medullary region of the kidney.

**Vasa vasorum** (va-SŌ-rum) Blood vessels that supply nutrients to the larger arteries and veins.

**Vascular** (VAS-kū-lar) Pertaining to or containing many blood vessels.

**Vascular (venous) sinus** A vein with a thin endothelial wall that lacks a tunica media and externa and is supported by surrounding tissue.

**Vascular spasm** Contraction of the smooth muscle in the wall of a damaged blood vessel to prevent blood loss.

**Vascular tunic** (TOO-nik) The middle layer of the eyeball, composed of the choroid, ciliary body, and iris. Also called the uvea (Ū-vē-a).

**Vasectomy** (va-SEK-tō-mē) A means of sterilization of males in which a portion of each ductus (vas) deferens is removed.

**Vasoconstriction** (vāz-ō-kon-STRIK-shun) A decrease in the size of the lumen of a blood vessel caused by contraction of the smooth muscle in the wall of the vessel.

**Vasodilation** (vāz′-ō-dī-LĀ-shun) An increase in the size of the lumen of a blood vessel caused by relaxation of the smooth muscle in the wall of the vessel.

**Vein** A blood vessel that conveys blood from tissues back to the heart.

**Vena cava** (VĒ-na KĀ-va) One of two large veins that open into the right atrium, returning to the heart all of the deoxygenated blood from the systemic circulation except from the coronary circulation.

**Ventral** (VEN-tral) Pertaining to the anterior or front side of the body; opposite of dorsal.

**Ventral ramus** (RĀ-mus) The anterior branch of a spinal nerve, containing sensory and motor fibers to the muscles and skin of the anterior surface of the head, neck, trunk, and the limbs.

**Ventricle** (VEN-tri-kul) A cavity in the brain filled with cerebrospinal fluid. An inferior chamber of the heart.

**Ventricular fibrillation** (ven-TRIK-ū-lar fib-ri-LĀ-shun) (**VF** or **V-fib**) Asynchronous ventricular contractions; unless reversed by defibrillation, results in heart failure.

**Venule** (VEN-ūl) A small vein that collects blood from capillaries and delivers it to a vein.

**Vermiform appendix** (VER-mi-form a-PEN-diks) A twisted, coiled tube attached to the cecum.

**Vermis** (VER-mis) The central constricted area of the cerebellum that separates the two cerebellar hemispheres.

**Vertebrae** (VER-te-brē) Bones that make up the vertebral column.

**Vertebral** (VER-te-bral) **canal** A cavity within the vertebral column formed by the vertebral foramina of all the vertebrae and containing the spinal cord. Also called the **spinal canal.**

**Vertebral column** The 26 vertebrae of an adult and 33 vertebrae of a child; encloses and protects the spinal cord and serves as a point of attachment for the ribs and back muscles. Also called the **backbone, spine,** or **spinal column.**

**Vesicle** (VES-i-kul) A small bladder or sac containing liquid.

**Vesicouterine** (ves′-ik-ō-Ū-ter-in) **pouch** A shallow pouch formed by the reflection of the peritoneum from the anterior surface of the uterus, at the junction of the cervix and the body, to the posterior surface of the urinary bladder.

**Vestibular** (ves-TIB-ū-lar) **apparatus** Collective term for the organs of equilibrium, which include the saccule, utricle, and semicircular ducts.

**Vestibular membrane** The membrane that separates the cochlear duct from the scala vestibuli.

**Vestibule** (VES-ti-būl) A small space or cavity at the beginning of a canal, especially the inner ear, larynx, mouth, nose, and vagina.

**Villus** (VIL-lus) A projection of the intestinal mucosal cells containing connective tissue, blood vessels, and a lymphatic vessel; functions in the absorption of the end products of digestion. *Plural* is **villi** (VIL-ī).

**Viscera** (VIS-er-a) The organs inside the ventral body cavity. *Singular* is **viscus** (VIS-kus).

**Visceral** (VIS-er-al) Pertaining to the organs or to the covering of an organ.

**Visceral effectors** (e-FEK-torz) Organs of the ventral body cavity that respond to neural stimulation, including cardiac muscle, smooth muscle, and glands.

**Vision** The act of seeing.

**Vitamin** An organic molecule necessary in trace amounts that acts as a catalyst in normal metabolic processes in the body.

**Vitreous** (VIT-rē-us) **body** A soft, jellylike substance that fills the vitreous chamber of the eyeball, lying between the lens and the retina.

**Vocal folds** Pair of mucous membrane folds below the ventricular folds that function in voice production. Also called **true vocal cords.**

**Volkmann's canal** *See* **Perforating canal.**

**Vulva** (VUL-va) Collective designation for the external genitalia of the female. Also called the pudendum (poo-DEN-dum).

## W

**Wallerian** (wal-LE-rē-an) **degeneration** Degeneration of the portion of the axon and myelin sheath of a neuron distal to the site of injury.

**Wandering macrophage** (MAK-rō-fāj) Phagocytic cell that develops from a monocyte, leaves the blood, and migrates to infected tissues.

**White matter** Aggregations or bundles of myelinated and unmyelinated axons located in the brain and spinal cord.

**White pulp** The regions of the spleen composed of lymphatic tissue, mostly B lymphocytes.

**White ramus communicans** (RĀ-mus kō-MŪ-ni-kans) The portion of a preganglionic sympathetic axon that branches from the anterior ramus of a spinal nerve to enter the nearest sympathetic trunk ganglion.

**Wormian bone** *See* **Sutural bone.**

## X

**Xiphoid** (ZĪ-foyd) Sword-shaped.

**Xiphoid** (ZĪ-foyd) **process** The inferior portion of the sternum **(breastbone).**

## Y

**Yolk sac** An extraembryonic membrane composed of the exocoelomic membrane and hypoblast. It transfers nutrients to the embryo, is a source of blood cells, contains primordial germ cells that migrate into the gonads to form primitive germ cells, forms part of the gut, and helps prevent desiccation of the embryo.

## Z

**Zona fasciculata** (ZŌ-na fa-sik′-ū-LA-ta) The middle zone of the adrenal cortex consisting of cells arranged in long, straight cords that secrete glucocorticoid hormones, mainly cortisol.

**Zona glomerulosa** (glo-mer′-ū-LŌ-sa) The outer zone of the adrenal cortex, directly under the connective tissue covering, consisting of cells arranged in arched loops or round balls that secrete mineralocorticoid hormones, mainly aldosterone.

**Zona pellucida** (pe-LOO-si-da) Clear glycoprotein layer between a secondary oocyte and the surrounding granulosa cells of the corona radiata.

**Zona reticularis** (ret-ik′-ū-LAR-is) The inner zone of the adrenal cortex, consisting of cords of branching cells that secrete sex hormones, chiefly androgens.

**Zygote** (ZĪ-gōt) The single cell resulting from the union of male and female gametes; the fertilized ovum.

# Credits

## ILLUSTRATIONS

**Chapter 1** Figure 1.1, 1.10–1.12: Kevin Somerville/Imagineering. 1.2–1.4, 1.8: Imagineering. 1.5: Molly Borman. 1.6: Kevin Somerville. 1.7, 1.9, Table 1.1: DNA Illustrations

**Chapter 2** Figure 2.1–2.24: Imagineering.

**Chapter 3** Figure 3.1, 3.2, 3.15, 3.16. 3.18, 3.19, 3.21–3.23: Tomo Narashima. 3.3, 3.5–3.14, 3.17, 3.20, 3.24–3.33: Imagineering.

**Chapter 4** Figure 4.1, 4.4, 4.5, 4.8: Kevin Somerville/Imagineering. 4.2: Hilda Muinos. 4.3, 4.6, Tables 4.1–4.9: Imagineering.

**Chapter 5** Figure 5.1, 5.3–5.5: Kevin Somerville. 5.2: Imagineering.

**Chapter 6** Figure 6.1, 6.2, 6.5, 6.7: John Gibb. 6.3, 6.8: Imagineering. 6.4, 6.6: Kevin Somerville.

**Chapter 7** Figure 7.1–7.11, 7.13, 7.14bc–7.22: John Gibb. 7.12: John Gibb/Imagineering. 7.14a: Imagineering.

**Chapter 8** Figure 8.1–8.16: John Gibb.

**Chapter 9** Figure 9.1–9.3, 9.10–9.14: John Gibb.

**Chapter 10** Figure 10.1, 10.2a-c, 10.16: Kevin Somerville. 10.2d, 10.3, 10.5, 10.7–10.13, 10.15, 10.17, Table 10.2: Imagineering. 10.6: Kevin Somerville/Imagineering

**Chapter 11** Figure 11.1–11.19a–d, 11.20a–d, 11.21: John Gibb. 11.19e, 11.20e: Imagineering. Table 11.1: Kevin Somerville.

**Chapter 12** Figure 12.1, 12.3, 12.5, 12.21: Kevin Somerville/Imagineering. 12.6, 12.7: Kevin Somerville. 12.2, 12.4, 12.8–12.20, 12.22–12.24, Table 12.2: Imagineering.

**Chapter 13** Figure 13.1–13.5, 13.7, 13.9a, 13.11, 13.12, 13.14–13.17: Kevin Somerville. 13.6, 13.9b, 13.10, 13.18, 13.19: Kevin Somerville/Imagineering.

**Chapter 14** Figure 14.1, 14.2, 14.13, 14.14, 14.26: Kevin Somerville. 14.3, 14.4, 14.23, 14.24: Imagineering. 14.5–14.12: Richard Coombs/Imagineering. 14.15, 14.16a, 14.17a, 14.18: Imagineering/Steve Oh. 14.16b, 14.17b: John Gibb. 14.19, 14.22, 14.25, 14.27: Kevin Somerville/Imagineering. 14.20, 14.21: Leonard Dank/Imagineering.

**Chapter 15** Figure 15.1, 15.3, 15.4, 15.6, 15.8: Imagineering. 15.2: Kevin Somerville. 15.5, 15.7, 15.9: Kevin Somerville/Imagineering.

**Chapter 16** Figure 16.1, 16.8, 16.11, 16.18–16.22: Tomo Narashima. 16.2, 16.9, 16.10, 16.12–16.17, 16.23, 16.26, Tables 16.1–16.3: Imagineering. 16.3: Tomo Narashima/Imagineering. 16.4: Molly Borman. 16.5: Molly Borman/Imagineering. 16.7: Sharon Ellis. 16.24, 16.25: Tomo Narashima/Sharon Ellis.

**Chapter 17** Figure 17.1: Kevin Somerville. 17.2, 17.3, 17.5, 17.6, 17.8, 17.10, 17.11, 17.13, 17.15, 17.16, Tables 17.4–17.10: Imagineering. 17.4, 17.7, 17.9, 17.18: Lynn O'Kelley. 17.12, 17.14, 17.17: Lynn O'Kelley/Imagineering.

**Chapter 18** Figure 18.1, 18.3–18.7, 18.10, 18.11, 18.13, Table 18.5: Imagineering.

**Chapter 19** Figure 19.1–19.5, 19.7: Kevin Somerville. 19.6b, 19.8, 19.11, 19.13: Kevin Somerville/Imagineering. 19.9, 19.10, 19.12: Imagineering.

**Chapter 20** Figure 20.1, 20.4, 20.7, 20.14, 20.16a, 20.20b–d, 20.22, 20.29: Kevin Somerville. 20.2, 20.3, 20.5, 20.6, 20.8–20.10, 20.12, 20.13, 20.23a, 20.28, Table 20.3: Imagineering. 20.11, 20.15, 20.17: Kevin Somerville/Imagineering. 20.16b, 20.23b, 20.25b, 20.27b: John Gibb. 20.18, 20.19, 20.20a, 20.21, 20.24, 20.25a, 20.26, 20.27a: John Gibb/Imagineering.

**Chapter 21** Figure 21.1, 21.6: Kevin Somerville. 21.2, 21.4, 21.8–21.18, Table 21.2: Imagineering. 21.3: John Gibb. 21.5, 21.7: Steve Oh.

**Chapter 22** Figure 22.1, 22.2a, 22.9, 22.10: Kevin Somerville. 22.2b, 22.3, 22.4, 22.6, 22.8: Molly Borman. 22.11, 22.13, 22.14: Imagineering. 22.12: John Gibb/Imagineering.

**Chapter 23** Figure 23.1, 23.2, 23.12, 23.16, 23.18, 23.19, 23.24: Kevin Somerville. 23.3, 23.4, 23.13, 23.14, 23.17, 23.21, 23.22, 23.25–23.29: Imagineering. 23.5, 23.7, 23.8, 23.10: Nadine Sokol. 23.6: DNA Illustrations. 23.11, 23.15: Steve Oh. 23.23: Molly Borman.

**Chapter 24** Figure 24.1, 24.2, 24.6, 24.8, 24.9, 24.24, 24.25: Kevin Somerville. 24.3, 24.4, 24.21: Steve Oh/Imagineering. 24.5, 24.10–24.23: Imagineering. Orientation Diagrams and Focus on Homeostasis icons: Imagineering.

**Chapter 25** Figure 25.1–25.4, 25.6, 25.9–25.14, 25.16, 25.19–25.22, 25.27–25.33, 25.36, 25.37, Table 25.4: Kevin Somerville. 25.5, 25.7, 25.8, 25.15, 25.23–25.26, 25.35, 25.38, Table 25.1: Imagineering.

## PHOTOGRAPHS

Repeat smartphone icon photo Oleksiy Mark/iStockphoto

**Chapter 1** Opener: Masterfile. Figure 1.1: Kenneth Eward/Rubberball Productions/Photo Researchers, Inc. Figure 1.1: Rubberball Productions/Getty Images. Figure 1.8a: Dissection Shawn Miller; Photograph

Mark Nielsen. Figure 1.8b: Dissection Shawn Miller; Photograph Mark Nielsen. Figure 1.8c: Dissection Shawn Miller; Photograph Mark Nielsen. Figure 1.12a: Andy Washnik

**Chapter 2** Opener: blackred/Getty Images, Inc.

**Chapter 3** Opener: Thinkstock/Getty Images, Inc. Figure 3.4: Andy Washnik. Figure 3.8b: David Phillips/Photo Researchers, Inc. Figure 3.12b,c: Omikron/Photo Researchers, Inc. Figure 3.14a: Albert Tousson/Phototake 3.14b: Albert Tousson/Phototake. Figure 3.14c: Alexey Khodjakov/Photo Researchers, Inc. Figure 3.15c: Don W. Fawcett/Photo Researchers, Inc. Figure 3.16b: P. Motta/Photo Researchers, Inc. Figure 3.16c: Dennis Kunkel Microscopy, Inc./Phototake. Figure 3.18b: D. W. Fawcett/Photo Researchers, Inc. Figure 3.19b: Biophoto Associates/Photo Researchers, Inc. Figure 3.21b: Dr. Gopal Murti/Phototake. Figure 3.22b: Don W. Fawcett/Photo Researchers, Inc. Figure 3.23c: D. W. Fawcett/Photo Researchers, Inc. Figure 3.24b: Andrew Syred/Photo Researchers, Inc. Figure 3.31: Courtesy Michael Ross, University of Florida.

**Chapter 4** Opener: Masterfile. Table 4.1a: Mark Nielsen. Table 4.1b: Mark Nielsen. Table 4.1c: Mark Nielsen. Table 4.1d: Mark Nielsen. Table 4.1e: Mark Nielsen. Table 4.1f: Mark Nielsen. Table 4.1g: Mark Nielsen. Table 4.2a: Mark Nielsen. Table 4.2b: Mark Nielsen. Table 4.3a: Mark Nielsen. Table 4.3b: Mark Nielsen. Table 4.3c: Mark Nielsen. Table 4.4a: Mark Nielsen. Table 4.4b: Mark Nielsen. Table 4.4c: Mark Nielsen. Table 4.5a: Mark Nielsen. Table 4.5b: Mark Nielsen. Table 4.5c: Mark Nielsen. Table 4.6: Mark Nielsen. Table 4.7: Mark Nielsen. Figure 4.7a: Mark Nielsen. Figure 4.7b: Mark Nielsen. Table 4.8a: Courtesy Michael Ross, University of Florida. Table 4.8b: Mark Nielsen. Table 4.8c: Mark Nielsen. Table 4.9: Mark Nielsen.

**Chapter 5** Opener: Image Source/Getty Images. Figure 5.1b: Courtesy Michael Ross, University of Florida. Figure 5.1c: David Becker/Photo Researchers, Inc. Figure 5.1d: Courtesy Andrew J. Kuntzman. Figure 5.2b: Mark Nielsen. Figure 5.3b: VVG/Science Photo Library/Photo Researchers, Inc.

**Chapter 6** Opener: ©2009 Levy Carneiro Jr./Getty Images, Inc. Figure 6.1: Mark Nielsen. Figure 6.2b: Mark Nielsen. Figure 6.3: (left) CNRI/Photo Researchers, Inc. Figure 6.3: (center) Dennis Kunkel Microscopy, Inc./Phototake. Figure 6.3: (right) Ed Reschke/Peter Arnold, Inc. Figure 6.7: Scott Camazine/Photo Researchers, Inc. Figure 6.8a: The Bergman Collection. Figure 6.8b: Mark Nielsen.

**Chapter 7** Opener: Masterfile. Figure 7.18b: Mark Nielsen. Figure 7.18c: Mark Nielsen. Figure 7.19b: Mark Nielsen. Figure 7.19c: Mark Nielsen. Figure 7.21b: Mark Nielsen.

**Chapter 8** Opener: Masterfile.

**Chapter 9** Opener: Michael Newman/PhotoEdit. Figure 9.3b: Mark Nielsen. Figure 9.4: Mark Nielsen. Figure 9.5: John Wilson White. Figure 9.6: John Wilson White. Figure 9.7: John Wilson White. Figure 9.8: John Wilson White. Figure 9.9a-h: John Wilson White. Figure 9.9i: Andy Washnik.

**Chapter 10** Opener: Stockbyte/Getty Images, Inc. Figure 10.4: Courtesy D. E. Kelley. Figure 10.6d: Don Fawcett/Photo Researchers, Inc. Figure 10.7: Courtesy Hiroyouki Sasaki, Yale E. Goldman and Clara Franzini-Armstrong. Figure 10.14: John Wiley & Sons. Table 10.1: Biophoto Associates/Photo Researchers, Inc.

**Chapter 11** Opener: Masterfile.

**Chapter 12** Opener: Uppercut Images/Getty Images, Inc. Figure 12.3b: Mark Nielsen. Figure 12.7c: Dennis Kunkel/Phototake. Figure 12.7d: Martin Rotker/Phototake. Figure 12.21: Eye of Science/Photo Researchers, Inc.

**Chapter 13** Opener: digitalskillet/Getty Images, Inc. Figure 13.1c: Dissection Shawn Miller, Photograph Mark Nielsen. Figure 13.4b: Dissection Shawn Miller, Photograph Mark Nielsen. Figure 13.4c: Dissection Shawn Miller, Photograph Mark Nielsen. Figure 13.6a: Dissection Shawn Miller, Photograph Mark Nielsen. Figure 13.8: N. Gluhbegovic and T. H. Williams, The Human Brain: A Photographic Guide, Harper and Row, Publishers, Inc. Hagerstown, MD, 1980. Figure 13.13: Dissection Shawn Miller, Photograph Mark Nielsen. Figure 13.18b: Courtesy Michael Ross, University of Florida. Table 13.2: Dissection Shawn Miller, Photograph Mark Nielsen.

**Chapter 14** Opener: Digital Vision/Getty Images. Figure 14.1: Mark Nielsen.

**Chapter 15** Opener: selimaksan/iStockphoto.

**Chapter 16** Opener: Comstock Images/PictureQuest. Figure 16.1c: Courtesy Michael Ross, University of Florida. Figure 16.4d: Mark Nielsen. Figure 16.6: Geirge Diebold/Getty Images, Inc. Figure 16.1: Mark Nielsen. Figure 16.17: N. Gluhbegovic and T. H. Williams, The Human Brain: A Photographic Guide, Harper and Row, Publishers, Inc., Hagerstown, MD, 1980.

**Chapter 17** Opener: Denkou Images/Getty Images, Inc. Figure 17.9b: Mark Nielsen. Figure 17.9c: Dissection Shawn Miller, Photograph Mark Nielsen. Figure 17.9d: Dissection Shawn Miller, Photograph Mark Nielsen. Figure 17.12b: Mark Nielsen. Figure 17.12d: Dissection Shawn Miller, Photograph Mark Nielsen. Figure 17.14b: Dissection Shawn Miller, Photograph Mark Nielsen. Figure 17.14d: Mark Nielsen. Figure 17.17c: Mark Nielsen. Figure 17.17d: Mark Nielsen.

**Chapter 18** Opener: Digital Vision/Getty Images. Figure 18.2: Juergen Berger/Photo Researchers, Inc. Figure 18.2b: Mark Nielsen. Figure 18.8: Jean Claude Revy/Phototake. Figure 18.9a: Courtesy Michael Ross, University of Florida. Figure 18.9b: Courtesy Michael Ross, University of Florida. Figure 18.9c: Courtesy Michael Ross, University of Florida. Figure 18.9d: Courtesy Michael Ross, University of Florida. Figure 18.9e: Courtesy Michael Ross, University of Florida. Figure 18.12a: Dennis Kunkel Microscopy, Inc./Phototake. Figure 18.12b: Dennis Kunkel Microscopy, Inc./Phototake.

**Chapter 19** Opener: Masterfile. Figure 19.3b: Dissection Shawn Miller, Photograph Mark Nielsen. Figure 19.4b: Mark Nielsen. Figure 19.4c: Mark Nielsen. Figure 19.5c: Dissection Shawn Miller, Photograph Mark Nielsen. Figure 19.5f: Dissection Shawn Miller, Photograph Mark Nielsen. Figure 19.5g: Dissection Shawn Miller, Photograph Mark Nielsen. Figure 19.7c: Dissection Shawn Miller, Photograph Mark Nielsen.

**Chapter 20** Opener: SW Productions/Photodisc Green/PhotoDisc, Inc./Getty Images. Figure 20.1d: Dennis Strete. Figure 20.1e: Courtesy Michael Ross, University of Florida. Figure 20.6: Dissection Shawn Miller, Photograph Mark Nielsen. Figure 20.18e: Dissection Shawn Miller, Photograph Mark Nielsen. Figure 20.20e: Dissection Shawn Miller, Photograph Mark Nielsen. Figure 20.24e: Mark Nielsen.

**Chapter 21** Opener: Corbis/Jupiter Images. Figure 21.6b: Francis Leroy, Biocosmos/Photo Researchers, Inc. Figure 21.6c: Dissection Shawn Miller, Photograph Mark Nielsen. Figure 21.7: Mark Nielsen. Figure 21.8: Science Photo Library/Photo Researchers, Inc.

**Chapter 22** Opener: Emma Lee/Life File/PhotoDisc, Inc./Getty Images. Figure 22.1b: Dissection Shawn Miller, Photograph Mark Nielsen. Figure 22.4: CNRI/Photo Researchers, Inc. Figure 22.5: Dissection Shawn Miller, Photograph Mark Nielsen. Figure 22.7: Dissection Shawn Miller, Photograph Mark Nielsen. Figure 22.9: Biophoto Associates/Photo Researchers, Inc. Figure 22.10c: Biophoto Associates/Photo Researchers, Inc.

**Chapter 23** Opener: Masterfile. Figure 23.1b: Dissection Shawn Miller, Photograph Mark Nielsen. Figure 23.3e: Dissection Shawn Miller, Photograph Mark Nielsen. Figure 23.9: Mark Nielsen. Figure 23.11b: Dissection Shawn Miller, Photograph Mark Nielsen. Figure 23.12: Steve Gschmeissner/Photo Researchers, Inc. Figure 23.15d: Dissection Shawn Miller, Photograph Mark Nielsen.

**Chapter 24** Opener: Masterfile. Figure 24.3b: Dissection Shawn Miller, Photograph Mark Nielsen. Figure 24.8b: Dennis Kunkel Microscopy, Inc./Phototake.

**Chapter 25** Opener: Masterfile. Figure 25.1b: Dissection Shawn Miller, Photograph Mark Nielsen. Figure 25.3b: Dissection Shawn Miller, Photograph Mark Nielsen. Figure 25.3c: Dissection Shawn Miller, Photograph Mark Nielsen. Figure 25.4a: Mark Nielsen. Figure 25.4a: (inset) Mark Nielsen. Figure 25.10c: Dissection Shawn Miller, Photograph Mark Nielsen. Figure 25.11b: Dissection Shawn Miller, Photograph Mark Nielsen. Figure 25.13b: Courtesy Michael Ross, University of Florida. Figure 25.13c: Claude Edelmann/Photo Researchers, Inc. Figure 25.14e: Mark Nielsen. Figure 25.14f: Mark Nielsen. Figure 25.17a: Mark Nielsen. Figure 25.17b: P. Motta/Photo Researchers, Inc. Figure 25.18a: Courtesy Michael Ross, University of Florida. Figure 25.18b: Courtesy Michael Ross, University of Florida. Figure 25.20b: Courtesy Michael Ross, University of Florida. Figure 25.27b: Don W. Fawcett/Photo Researchers, Inc. Figure 25.27c: Myriam Wharman/Phototake. Figure 25.33b: Preparation and photography Mark Nielsen. Figure 25.34a: Photo provided courtesy of Kohei Shiota, Congenital Anomaly Research Center, Kyoto University, Graduate School of Medicine. Figure 25.34b: Courtesy National Museum of Health and Medicine, Armed Forces Institute of Pathology. Figure 25.34c: Courtesy National Museum of Health and Medicine, Armed Forces Institute of Pathology. Figure 25.34d: Courtesy National Museum of Health and Medicine, Armed Forces Institute of Pathology. Figure 25.34e: Courtesy National Museum of Health and Medicine, Armed Forces Institute of Pathology. Figure 25.34f: Photo by Lennart Nilsson/Scanpix Sweden AB. Figure 25.34g: Photo provided courtesy of Kohei Shiota, Congenital Anomaly Research Center, Kyoto University, Graduate School of Medicine. Figure 25.34h: Photo provided courtesy of Kohei Shiota, Congenital Anomaly Research Center, Kyoto University, Graduate School of Medicine.

# Index

**Note:** Page numbers in boldface type indicate a major discussion. A *t* following a page number indicates a table, and an *f* following a page number indicates a figure.

## B

## Q



## R

## S

T

W



X



Y



Z